Pronouncing Organism Names (*continued*)

C. fetus fē-tus
C. jejuni jē-jū'nē
Candida albicans kan'did-ä al'bi-kanz
C. milleri mil'ler-ē
Caulobacter crescentus kō-lō-bak'tėr kre-sen'tus
Cephalosporium acremonium sef-ä-lō-spô'rē-um ac-re-mō'nē-um
Chlamydia trachomatis kla-mi'dē-a trä-kō'mä-tis
Chlamydophila pneumoniae kla-mi'dof-i-la nü-mō'nē-ī
C. psittaci sit'a-sē
Chlamydomonas klam-i-dō-mō'äs
Chromobacter violaceum krō-mō-bak'-tėr vī-ō-lā'sē-um
Citrobacter freundii sit'rō-bak'tėr frun'dē-ē
C. intermedium in-tėr-mē'dē-um
Claviceps purpurea kla'vi-seps pür-pü-rē'ä
Clostridium acetobutylicum klôs-tri'dē-um ā-sē-tō-bū-til'i-kum
C. botulinum bot-ū-lī'num
C. difficile dif'fi-sil-ē
C. perfringens pėr-frin'jens
C. tetani te'tän-ē
Coccidioides immitis kok-sid-ē-oi'dēz im'mi-tis
C. posadasii pō-sä-da'sē-ē
Corynebacterium diphtheriae kôr'ē-nē-bak-ti-rē-um dif-thi'rē-ī
Coxiella burnetii käks'ē-el-lä bėr-ne'tē-ē
Cryptococcus neoformans krip'tō-kok-kus nē-ō-fôr'manz
Cryptosporidium coccidi krip'tō-spô-ri-dē-um kok'sid-ē
C. hominis hō'mi-nis
C. parvum pär'vum
Cyclospora cayetanensis sī'klō-spô-rä kī'ē-tan-en-sis

Deinococcus radiodurans dī'nō-kok-kus rā-dē-ō-dür'anz
Desulfovibrio dē'sul-fō-vib-rē-ō
Dicrocoelium dendriticum dī-krō-sē'lē-um den-dri'ti-kum

Echinococcus granulosus ē-kīn-ō-kok'kus gra-nū-lō'sis
Ehrlichia chaffeensis ėr'lik-ē-ä chäf-fen'sis
E. phagocytophila fā-gō-cī-to'fī-lä
Emmonsiella capsulata em'mon-sē-el-lä cap-sül-ä'tä
Entamoeba histolytica en-tä-mē'bä his-tō-li'ti-kä
Enterobacter aerogenes en-te-ro-bak'ter a-ra'jen-ēz
E. cloacae klō-ā'ki
Enterobius vermicularis en-te-rō'bē-us ver-mi-kū-lar'is
Enterococcus faecalis en-tė-rō-kok'kus fē-kā'lis
E. faecium fē'sē-um
Epidermophyton ep-ē-der-mō-fī'ton
Erysipelothrix rhusiopathiae ār-ē-sip'e-lō-thriks rü'sī-ō-pa-thē
Escherichia coli esh-ėr-ē'kē-ä kō'lī (or kō'lē)
Euglena ū-glē'nä

Fasciola hepatica fä-sē-ō'lä he-pat'ik-ä
Fasciolopsis buski fä-sē-ō-lop'sis bus'kī
Francisella tularensis fran'sis-el-lä tü'lä-ren-sis

Gambierdiscus toxicus gam'bē-ėr-dis-kus toks'i-kus
Gardnerella intestinalis gärd-nė-rel'lä in-tes-ti-nal'is
G. vaginalis va-jin-al'is
Geobacillus stearothermophilus jē-ō-bä-sil'lus ste-är-ō-thėr-mä'fil-us
Giardia lamblia jē-är'dē-ä lam'lē-ä
Gluconobacter glü'kon-ō-bak-tėr
Gonyaulax catanella gon-ē-ō'laks kat-ä-nel'lä
Gymnodinitum jim-nō-din'i-tum

Haemophilus aegyptius hē-mä'fil-us ē-jip'tē-us
H. ducrcyi dü-krä'ē
H. influenzae in-flü-en'zī
H. vaginalis va-jin-al'is
Hartmannella vermiformis hart-mä-nel'lä vêr-mi-fôr'mis
Helicobacter pylori hē'lik-ō-bak-tėr pī'lō-rē
Histoplasma capsulatum his-tō-plaz'mä kap-su-lä'tum

Klebsiella pneumoniae kleb-sē-el'lä nü-mō'ne-ī

Lactobacillus acidophilus lak-tō-bä-sil'lus a-sid-o'fil-us
L. bulgaricus bul-gā'ri-kus
L. caseii kā'sē-ē
L. plantarum plan-tär'um
L. sanfranciscensis san-fran-si-sen'-sis
Lactococcus lactis lak-tō-kok'kus lak'tis
Lagenidium giganteum la-je-ni'dē-um jī-gan'tē-üm
Legionella pneumophila lē-jä-nel'lä nü-mō'fi-lä
Leishmania donovani lish'mä-nē-ä don'ō-vän-ē
L. tropica trop'i-kä
Leptospira interrogans lep-tō-spī'rä in-tėr'rō-ganz

(continued on inside back cover)

John T. Mattison

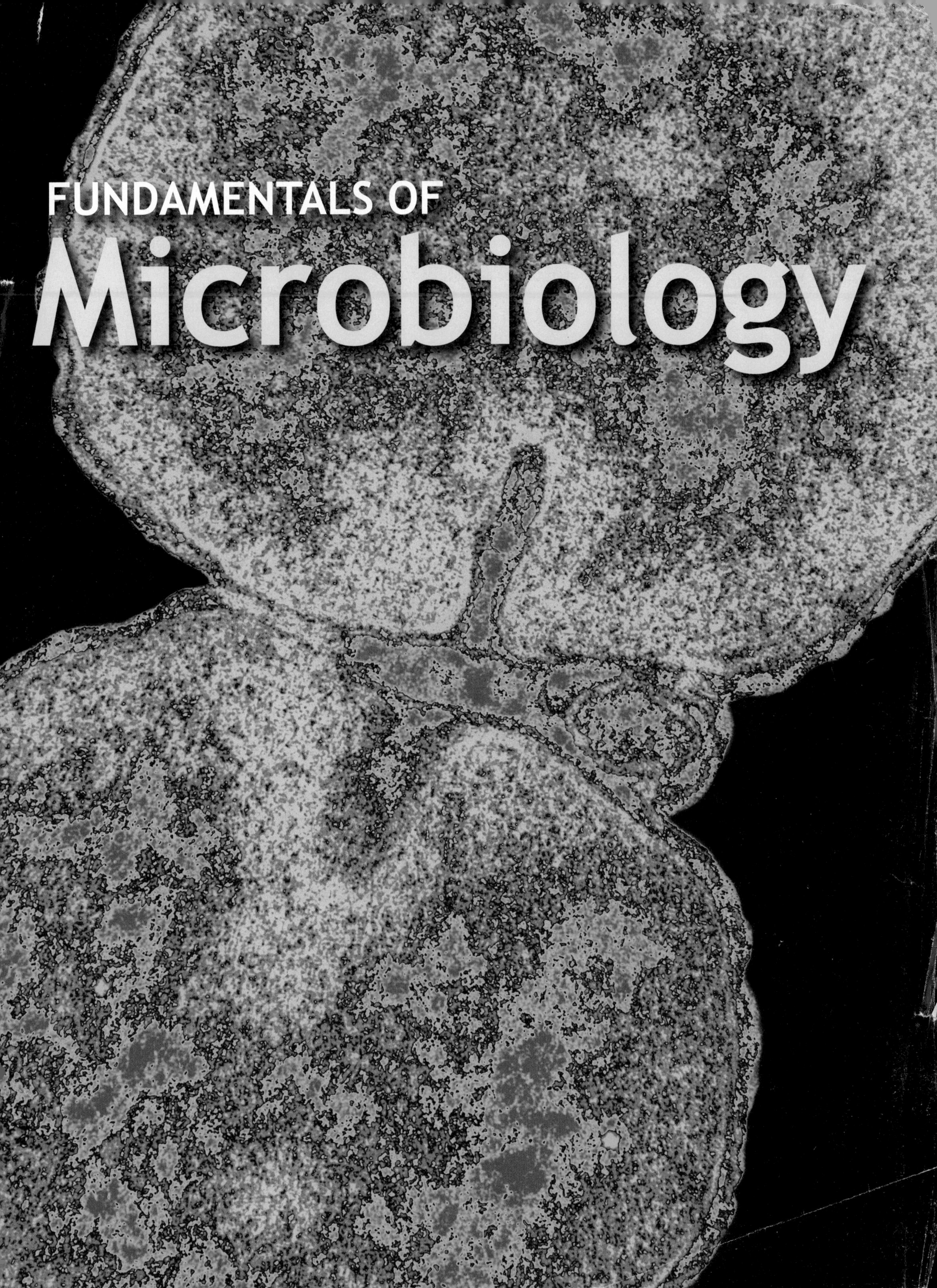
FUNDAMENTALS OF
Microbiology

Jones & Bartlett Learning Titles in Biological Science

AIDS: Science and Society, Seventh Edition
Hung Fan, Ross F. Conner, & Luis P. Villarreal

AIDS: The Biological Basis, Fifth Edition
Benjamin S. Weeks & I. Edward Alcamo

Alcamo's Fundamentals of Microbiology, Body Systems Edition, Second Edition
Jeffrey C. Pommerville

Alcamo's Fundamentals of Microbiology, Ninth Edition
Jeffrey C. Pommerville

Alcamo's Microbes and Society, Third Edition
Benjamin S. Weeks

Biochemistry
Raymond S. Ochs

Bioethics: An Introduction to the History, Methods, and Practice, Third Edition
Nancy S. Jecker, Albert R. Jonsen, & Robert A. Pearlman

Bioimaging: Current Concepts in Light and Electron Microscopy
Douglas E. Chandler & Robert W. Roberson

Biomedical Graduate School: A Planning Guide to the Admissions Process
David J. McKean & Ted R. Johnson

Biomedical Informatics: A Data User's Guide
Jules J. Berman

Botany: An Introduction to Plant Biology, Fifth Edition
James D. Mauseth

Botany: A Lab Manual
Stacy Pfluger

Case Studies for Understanding the Human Body, Second Edition
Stanton Braude, Deena Goran, & Alexander Miceli

Electron Microscopy, Second Edition
John J. Bozzola & Lonnie D. Russell

Encounters in Microbiology, Volume 1, Second Edition
Jeffrey C. Pommerville

Encounters in Microbiology, Volume 2
Jeffrey C. Pommerville

Encounters in Virology
Teri Shors

Essential Genetics: A Genomics Perspective, Sixth Edition
Daniel L. Hartl

Essentials of Molecular Biology, Fourth Edition
George M. Malacinski

Evolution: Principles and Processes
Brian K. Hall

Exploring Bioinformatics: A Project-Based Approach
Caroline St. Clair & Jonathan E. Visick

Exploring the Way Life Works: The Science of Biology
Mahlon Hoagland, Bert Dodson, & Judy Hauck

Genetics: Analysis of Genes and Genomes, Eighth Edition
Daniel L. Hartl & Maryellen Ruvolo

Genetics of Populations, Fourth Edition
Philip W. Hedrick

Guide to Infectious Diseases by Body System, Second Edition
Jeffrey C. Pommerville

Human Biology, Seventh Edition
Daniel D. Chiras

Human Biology Laboratory Manual
Charles Welsh

Human Body Systems: Structure, Function, and Environment, Second Edition
Daniel D. Chiras

Human Embryonic Stem Cells, Second Edition
Ann A. Kiessling & Scott C. Anderson

Laboratory Investigations in Molecular Biology
Steven A. Williams, Barton E. Slatko, & John R. McCarrey

Lewin's CELLS, Second Edition
Lynne Cassimeris, Vishwanath R. Lingappa, & George Plopper

Lewin's Essential GENES, Third Edition
Jocelyn E. Krebs, Elliott S. Goldstein, & Stephen T. Kilpatrick

Lewin's GENES XI
Jocelyn E. Krebs, Elliott S. Goldstein, & Stephen T. Kilpatrick

Microbial Genetics, Second Edition
Stanley R. Maloy, John E. Cronan, Jr., & David Freifelder

The Microbial Challenge, Third Edition
Robert I. Krasner & Teri Shors

Molecular Biology: Genes to Proteins, Fourth Edition
Burton E. Tropp

Neoplasms: Principles of Development and Diversity
Jules J. Berman

Precancer: The Beginning and the End of Cancer
Jules J. Berman

Principles of Cell Biology
George Plopper

Principles of Modern Microbiology
Mark Wheelis

Principles of Molecular Biology
Burton E. Tropp

Science and Society: Scientific Thought and Education for the 21st Century
Peter Daempfle

Strickberger's Evolution, Fifth Edition
Brian K. Hall

Symbolic Systems Biology: Theory and Methods
M. Sriram Iyengar

20th Century Microbe Hunters
Robert I. Krasner

Understanding Viruses, Second Edition
Teri Shors

FUNDAMENTALS OF

Microbiology

Tenth Edition

Jeffrey C. Pommerville
Glendale Community College

World Headquarters
Jones & Bartlett Learning
5 Wall Street
Burlington, MA 01803
978-443-5000
info@jblearning.com
www.jblearning.com

Jones & Bartlett Learning books and products are available through most bookstores and online booksellers. To contact Jones & Bartlett Learning directly, call 800-832-0034, fax 978-443-8000, or visit our website, www.jblearning.com.

Substantial discounts on bulk quantities of Jones & Bartlett Learning publications are available to corporations, professional associations, and other qualified organizations. For details and specific discount information, contact the special sales department at Jones & Bartlett Learning via the above contact information or send an email to specialsales@jblearning.com.

Production Credits
Chief Executive Officer: Ty Field
President: James Homer
SVP, Editor-in-Chief: Michael Johnson
SVP, Chief Marketing Officer: Alison M. Pendergast
Executive Publisher: Kevin Sullivan
Senior Acquisitions Editor: Erin O'Connor
Editorial Assistant: Rachel Isaacs
Editorial Assistant: Michelle Bradbury
Production Editor: Leah Corrigan
Senior Marketing Manager: Andrea DeFronzo
V.P., Manufacturing and Inventory Control: Therese Connell
Composition: Circle Graphics, Inc.
Cover Design: Scott Moden
Rights & Photo Research Associate: Lauren Miller
Cover Image: © Dr. David Phillips/Visuals Unlimited, Inc./Corbis
Printing and Binding: Courier Companies
Cover Printing: John Pow Company
Image on pages v, vi, xviii, xxii, xxiii, xxv, and xxxi courtesy of Dr. Fred Murphy/CDC.

To order this product, use ISBN: 978-1-4496-8861-5

Library of Congress Cataloging-in-Publication Data

Pommerville, Jeffrey C.
Fundamentals of microbiology / Jeffrey Pommerville.—10th ed.
p. ; cm.
Rev. ed. of: Alcamo's fundamentals of microbiology / Jeffrey C. Pommerville. c2011.
Includes index.
ISBN 978-1-4496-4796-4 (alk. paper)
I. Pommerville, Jeffrey C. Alcamo's fundamentals of microbiology. II. Title.
[DNLM: 1. Microbiological Phenomena. 2. Microbiology. QW 4]
616.9'041—dc23

6048 2012021741

Printed in the United States of America
17 16 15 14 13 10 9 8 7 6 5 4 3 2

Brief Contents

PART 1 FOUNDATIONS OF MICROBIOLOGY 1

Chapter 1 Microbiology: Then and Now 3

Chapter 2 The Chemical Building Blocks of Life 35

Chapter 3 Concepts and Tools for Studying Microorganisms 64

Chapter 4 Structure of Bacterial and Archaeal Cells 99

Chapter 5 Microbial Growth and Nutrition 133

Chapter 6 Microbial Metabolism 162

Chapter 7 Control of Microorganisms: Physical and Chemical Methods 195

PART 2 THE GENETICS OF MICROORGANISMS 229

Chapter 8 Microbial Genetics 231

Chapter 9 Gene Transfer, Genetic Engineering, and Genomics 266

PART 3 BACTERIAL DISEASES OF HUMANS 304

Chapter 10 Airborne Bacterial Diseases 307

Chapter 11 Foodborne and Waterborne Bacterial Diseases 343

Chapter 12 Soilborne and Arthropodborne Bacterial Diseases 380

Chapter 13 Sexually Transmitted and Contact Transmitted Bacterial Diseases 406

PART 4 VIRUSES AND EUKARYOTIC MICROORGANISMS 449

Chapter 14 The Viruses and Virus-Like Agents 451

Chapter 15 Viral Infections of the Respiratory Tract and Skin 490

Chapter 16 Viral Infections of the Blood, Lymphatic, Gastrointestinal, and Nervous Systems 522

Chapter 17 Eukaryotic Microorganisms: The Fungi 552

Chapter 18 Eukaryotic Microorganisms: The Parasites 586

PART 5 INTERACTIONS AND IMPACT OF MICROORGANISMS WITH HUMANS 627

Chapter 19 Infection and Disease 629

Chapter 20 Resistance and the Immune System: Innate Immunity 668

Chapter 21 Resistance and the Immune System: Adaptive Immunity 695

Chapter 22 Immunity and Serology 725

Chapter 23 Immune Disorders and AIDS 761

Chapter 24 Antimicrobial Drugs 804

PART 6 ENVIRONMENTAL AND APPLIED MICROBIOLOGY 847
Available Online with access code

Chapter 25 Microbiology of Foods 849

Chapter 26 Environmental Microbiology 877

Chapter 27 Industrial Microbiology and Biotechnology 901

Appendix A Metric Measurement A-1

Appendix B Temperature Conversion Chart A-1

Glossary G-1

Index I-1

Contents

Preface xviii
Acknowledgments xxii
About the Author xxiii
To the Student—Study Smart xxv
A Tribute to I. Edward Alcamo xxxi

PART 1 FOUNDATIONS OF MICROBIOLOGY 1

Chapter 1 Microbiology: Then and Now 3

1.1 The Discovery of Microbes Leads to Questioning Their Origins 6
Microscopy—Discovery of the Very Small 6
Do Animalcules Arise Spontaneously? 7

Courtesy of Jean Roy/CDC.

1.2 Disease Transmission Can Be Interrupted 9
Vaccination Prevents Infectious Disease 9
Disease Transmission Does Not Result from a Miasma 13
The Stage Is Set 14

1.3 The Classical Golden Age of Microbiology Reveals the Germ 16
Louis Pasteur Proposes That Germs Cause Infectious Disease 16
Pasteur's Work Stimulates Disease Control and Reinforces Disease Causation 17
Robert Koch Formalizes Standards to Equate Germs with Infectious Disease 17
Competition Fuels the Study of Infectious Disease 18

1.4 With the Discovery of Other Microbes, the Microbial World Expands 20
Other Global Pioneers Contribute to New Disciplines in Microbiology 20
The Microbial World Can Be Catalogued into Five Major Groups 22

1.5 A Second Golden Age of Microbiology Involves the Birth of Molecular Biology and Chemotherapy 25
Molecular Biology Relies on Microorganisms As Model Systems 25
Two Types of Cellular Organization Are Realized 25
Antibiotics Are Used to Cure Infectious Disease 25

1.6 The Third Golden Age of Microbiology Is Now 28
Microbiology Continues to Face Many Challenges 28
Microbial Ecology and Evolution Are Helping to Drive the New Golden Age 30
Chapter Review 31

Chapter 2 The Chemical Building Blocks of Life 35

2.1 Organisms Are Composed of Atoms 37
Atoms Are Composed of Charged and Uncharged Particles 37

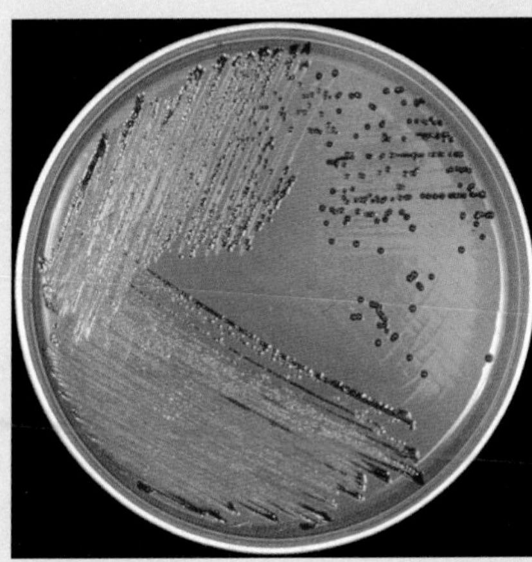

Courtesy of CDC.

Atoms Can Vary in the Number of Neutrons or Electrons 38
Electron Placement Determines Chemical Reactivity 38

2.2 Chemical Bonds Form Between Reactive Atoms 39
Ionic Bonds Form Between Oppositely Charged Ions 39
Covalent Bonds Share Electrons 40
Hydrogen Bonds Form Between Polar Groups or Molecules 42
Chemical Reactions Change Bonding Partners 42

2.3 All Living Organisms Depend on Water 43
Water Has Several Unique Properties 44
Acids and Bases Affect a Solution's pH 45
Cell Chemistry Is Sensitive to pH Changes 45

2.4 Living Organisms Are Composed of Four Types of Organic Compounds 47
Functional Groups Define Molecular Behavior 47
Carbohydrates Consist of Sugars and Sugar Polymers 47
Lipids Are Water-Insoluble Compounds 50
Nucleic Acids Are Large, Information-Containing Polymers 51
Proteins Are the Workhorse Polymers in Cells 53
Chapter Review 60

Chapter 3 Concepts and Tools for Studying Microorganisms 64

3.1 The Prokaryotes Are Not Simple, Primitive Organisms 65
Prokaryotic Cells Exhibit Some Remarkable and Widespread Behaviors 66
Prokaryotic and Eukaryotic Cells Share Similarities in Organizational Patterns 70
Prokaryotic and Eukaryotic Cells Also Have Structural Distinctions 72

3.2 Classifying Microorganisms Reveals Relationships Between Organisms 73
Classification Attempts to Catalog Organisms 73
Kingdoms and Domains: Making Sense of Taxonomic Relationships 77
Nomenclature Gives Scientific Names to Organisms 78
Classification Uses a Hierarchical System 79
Many Methods Are Available to Identify and Classify Microorganisms 80

3.3 Microscopy Is Used to Visualize the Structure of Cells 83
Many Microbial Agents Are in the Micrometer Size Range 84
Light Microscopy Is Used to Observe Most Microorganisms 85
Staining Techniques Provide Contrast 87
Other Light Microscopy Optics Can Also Enhance Contrast 91
Electron Microscopy Provides Detailed Images of Cells, Cell Parts, and Viruses 93
Chapter Review 95

Chapter 4 Structure of Bacterial and Archaeal Cells 99

4.1 There is Tremendous Diversity Among the Bacteria and Archaea 100
The Domain Bacteria Contains Some of the Most Studied Microbial Organisms 100
The Domain Archaea Contains Organisms with Diverse Physiologies 104

4.2 Prokaryotes Can Be Distinguished by Their Cell Shape and Arrangements 106
Variations in Cell Shape and Cell Arrangement Exist 106

4.3 An Overview to Bacterial and Archaeal Cell Structure 108
Cell Structure Organizes Cell Function 108

4.4 External Cell Structures Interact with the Environment 110
Pili Are Primarily Used for Attachment 110
Flagella Provide Motility 110
The Glycocalyx Serves Several Functions 115

4.5 Most Bacterial and Archaeal Cells Have a Cell Envelope 116
The Bacterial Cell Wall Is a Tough and Protective External Shell 116
The Archaeal Cell Wall Also Provides Mechanical Strength 118
The Cell Membrane Represents a Selectively Permeable Barrier 120

Archaeal Membranes Are Structurally Unique 121

4.6 The Cell Cytoplasm Is Packed with Internal Structures 122

The Nucleoid Represents a Subcompartment Containing the Chromosome 122

Plasmids Are Found in Many Bacterial and Archaeal Cells 123

Other Subcompartments Exist in the Cell Cytoplasm 123

Cytoskeletal Proteins Regulate Cell Division and Help Determine Cell Shape 125

Chapter Review 130

Chapter 5 Microbial Growth and Nutrition 133

5.1 Microbial Reproduction Is Part of the Cell Cycle 135

Binary Fission Is Part of the Cell Cycle 135

Bacterial and Archaeal Cells Can Grow Exponentially 135

5.2 Microbial Growth Progresses Through Distinct Phases 139

A Bacterial Growth Curve Illustrates the Dynamics of Growth 139

Bacterial Cells Can Exist In Metabolically Inactive States 140

Optimal Microbial Growth Is Dependent on Several Physical Factors 142

5.3 Culture Media Are Used to Grow Microbes and Measure Their Growth 151

Culture Media Are of Two Basic Types 151

Culture Media Can Be Modified to Select for or Differentiate Between Microbial Species 151

Population Measurements Are Made Using Pure Cultures 155

Population Growth Can Be Measured in Several Ways 155

Chapter Review 159

Chapter 6 Microbial Metabolism 162

6.1 Enzymes and Energy Drive Cellular Metabolism 163

Enzymes Catalyze All Chemical Reactions in Cells 164

Enzymes Act Through Enzyme-Substrate Complexes 165

Enzymes Often Team Up in Metabolic Pathways 166

Enzyme Activity Can Be Inhibited 167

Energy in the Form of ATP Is Required for Metabolism 168

6.2 Glucose Catabolism Generates Cellular Energy 170

Glucose Contains Stored Energy That Can Be Extracted 171

Glycolysis Is the First Stage of Energy Extraction 171

The Citric Acid Cycle Extracts More Energy from Pyruvate 173

Oxidative Phosphorylation Is the Process by Which Most ATP Molecules Form 175

6.3 There Are Other Pathways to ATP Production 180

Other Nutrients Represent Potential Energy Sources 180

Anaerobic Respiration Produces ATP Using Other Final Electron Acceptors 182

Fermentation Produces ATP Using an Organic Final Electron Acceptor 183

6.4 Photosynthesis Converts Light Energy to Chemical Energy 185

Photosynthesis Is a Process to Acquire Chemical Energy 185

6.5 Microbes Exhibit Metabolic Diversity 188

Autotrophs and Heterotrophs Get Their Energy and Carbon in Different Ways 188

Chapter Review 191

Chapter 7 Control of Microorganisms: Physical and Chemical Methods 195

7.1 Microbial Growth Can Be Controlled in Several Ways 197

Sterilization and Sanitization Are Key to Good Public Health 197

7.2 There Are Various Physical Methods to Control Microbial Growth 198

Heat Is the Most Common Physical Control Methods 198

Dry Heat Has Useful Applications 198

Moist Heat Is More Versatile Than Dry Heat 198

Filtration Traps Microorganisms 204

Ultraviolet Light Can Be Used to Control Microbial Growth 206

Other Types of Radiation Also Can Sterilize Materials 206

Preservation Methods Retard Spoilage by Microorganisms in Foods 208

7.3 Chemical Control Usually Involves Disinfection 209

Chemical Control Methods Are Dependent on the Object to Be Treated 209

Chemical Agents Are Important to Laboratory and Hospital Safety 212

Antiseptics and Disinfectants Can Be Evaluated for Effectiveness 213

7.4 A Variety of Chemical Methods Can Control Microbial Growth 214

Halogens Oxidize Proteins 217

Phenol and Phenolic Compounds Denature Proteins 218

Heavy Metals Interfere with Microbial Metabolism 218

Alcohols Denature Proteins and Disrupt Membranes 219

Soaps and Detergents Act as Surface-Active Agents 219

Peroxides Damage Cellular Components 219

Some Chemical Agents Combine with Nucleic Acids and/or Cell Proteins 221

Chapter Review 225

PART 2 THE GENETICS OF MICROORGANISMS 229

Chapter 8 Microbial Genetics 231

8.1 The Hereditary Molecule in All Organisms Is DNA 233

Bacterial and Archaeal DNA Is Organized Within the Nucleoid 233

DNA Within a Nucleoid Is Highly Compacted 233

Many Microbial Cells Also Contain Plasmids 237

8.2 DNA Replication Occurs Before a Cell Divides 238

DNA Replication Occurs in Three Stages 238

DNA Polymerase Only Reads in the 3′ to 5′ Direction 240

8.3 Gene Expression Produces RNA and Protein for Cell Function 241

Transcription Copies Genetic Information into Complementary RNA 242

The Genetic Code Consists of Three-Letter Words 243

Translation Is the Process of Making the Polypeptide 245

Antibiotics Interfere with Gene Expression 246

Gene Expression Can Be Controlled in Several Ways 246

Transcription and Translation Are Localized 251

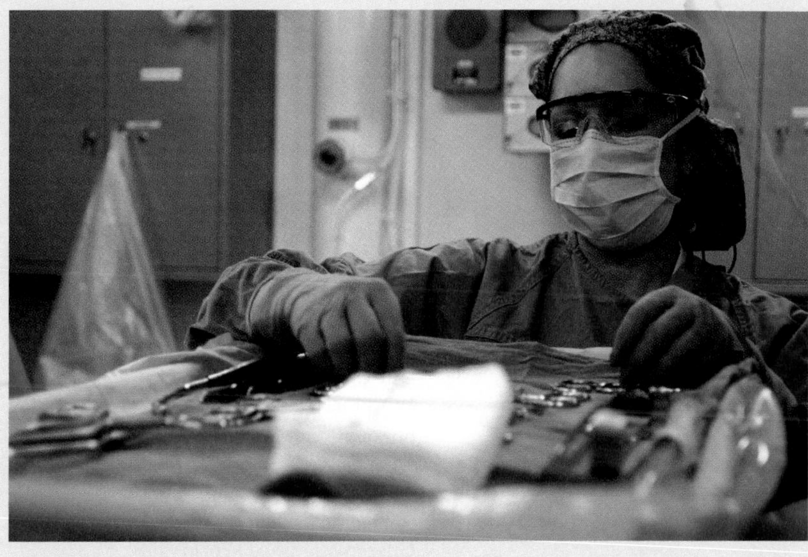

Courtesy of Journalist 2nd Class Shane Tuc/U.S. Navy.

8.4 Mutations Are Permanent Changes in a Cell's DNA 252

Mutations Are the Result of Heritable Changes in a Genome 252

Point Mutations Can Affect Protein Structure and Function 254

Repair Mechanisms Attempt to Correct Mistakes or Damage in the DNA 255

Transposable Genetic Elements Can Also Cause Mutations 256

8.5 Techniques Exist for Identifying Mutants 260

Plating Techniques Select for Specific Mutants or Characteristics 260

The Ames Test Can Identify Potential Mutagens 261

Chapter Review 263

Chapter 9 Gene Transfer, Genetic Engineering, and Genomics 266

9.1 Bacterial Cells Can Recombine Genes in Several Ways 268

Genetic Information Can Be Transferred Vertically and Horizontally 268

Transformation Is the Uptake and Expression of DNA in a Recipient Cell 268

Conjugation Involves Cell-to-Cell Contact 271

Conjugation Also Can Transfer Chromosomal DNA 274

Transduction Involves Viruses as Agents for Horizontal Transfer of DNA 274

9.2 Genetic Engineering Involves the Deliberate Transfer of Genes Between Organisms 279

Genetic Engineering Was Born from Genetic Recombination 279

Biotechnology Has Spawned Many Commercial and Practical Products 281
DNA Probes Can Identify a Cloned Gene or DNA Segment 287

9.3 Microbial Genomics Studies Genes at the Single Cell to Community Levels 291
Many Microbial Genomes Have Been Sequenced 291
Segments of the Human Genome Have "Microbial Ancestors" 293
Microbial Genomics Will Advance Our Understanding of the Microbial World 294
Comparative Genomics Brings a New Perspective to Defining Infectious Diseases and Studying Evolution 296
Metagenomics Is Identifying the Previously Unseen Microbial World 298
Chapter Review 301

PART 3 BACTERIAL DISEASES OF HUMANS 304

Chapter 10 Airborne Bacterial Diseases 307

10.1 The Respiratory System Possesses an Indigenous Microbiota 309
Upper Respiratory Tract Defenses Limit Microbe Colonization of the Lower Respiratory Tract 309

10.2 Several Bacterial Diseases Affect the Upper Respiratory Tract 311
Pharyngitis Is an Inflammation of the Throat 311
Diphtheria Is a Life-Threatening Illness 312
The Epiglottis Is Subject to Infection, Especially in Children 314
The Nose Is the Most Commonly Infected Region of the Upper Respiratory Tract 314
Ear Infections Are Common Illnesses in Early Childhood 315

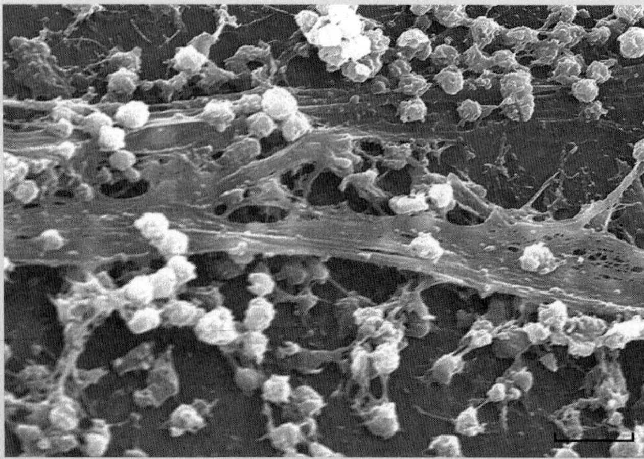

Courtesy of Rodney M. Donlan, Ph.D. and Janice Carr/CDC.

Acute Bacterial Meningitis Is a Rapidly Developing Inflammation 316
A Few Bacterial Species Cause Neonatal Meningitis 319

10.3 Many Bacterial Diseases of the Lower Respiratory Tract Can Be Life Threatening 320
Pertussis (Whooping Cough) Is Highly Contagious 320
Tuberculosis Remains a Major Cause of Death Worldwide 321
Infectious Bronchitis Is an Inflammation of the Bronchi 328
Pneumonia Can Be Caused by Several Bacterial Species 328
Community-Acquired Pneumonia Is Also Caused by Intracellular Pathogens 336
Inhalational Anthrax Is an Occupational Hazard 337
Chapter Review 340

Chapter 11 Foodborne and Waterborne Bacterial Diseases 343

11.1 The Digestive System Has an Extensive Indigenous Microbiota 345
The Digestive System Is Composed of Two Separate Categories of Organs 345
Our Understanding of the Human Oral and Gastrointestinal Microbiome Is Rapidly Improving 346

11.2 Bacterial Diseases of the Oral Cavity Can Affect One's Overall Health 350
Dental Caries Causes Pain and Tooth Loss in Affected Individuals 350
Periodontal Disease Can Arise from Bacteria in Dental Plaque 352

11.3 Bacterial Diseases of the GI Tract Are Usually Spread Through Food and Water 355
GI Tract Diseases May Arise from Intoxications or Infections 355
There Are Several Ways Foods or Water Become Contaminated 355

11.4 Some Bacterial Diseases Are the Result of Foodborne Intoxications 357
Food Poisoning Can Be the Result of Enterotoxins 357

11.5 GI Infections Can Be Caused by Several Bacterial Pathogens 360
Bacterial Gastroenteritis Often Produces an Inflammatory Condition 360
Several Bacterial Species Can Cause an Invasive Gastroenteritis 365
Gastric Ulcer Disease Can Be Spread Person to Person 371
Chapter Review 377

Chapter 12 Soilborne and Arthropodborne Bacterial Diseases 380

12.1 Several Soilborne Bacterial Diseases Develop from Endospores 382
Anthrax Is an Enzootic Disease 382
Tetanus Causes Hyperactive Muscle Contractions 383
Gas Gangrene Causes Massive Tissue Damage 384
Leptospirosis Is an Emerging Zoonotic Disease 386

12.2 Bacterial Diseases Can Be Transmitted by Arthropods 388
Plague Can Be a Highly Fatal Disease 388
Tularemia Has More Than One Disease Presentation 390
Lyme Disease and Relapsing Fever Are Transmitted by Spirochetes 393

12.3 Rickettsial and Ehrlichial Diseases Are Arthropodborne 396
Rickettsial Infections Are Transmitted by Arthropods 396
Other Tickborne Zoonoses Are Emerging Diseases in the United States 399
Chapter Review 402

Chapter 13 Sexually Transmitted and Contact Transmitted Bacterial Diseases 406

13.1 Portions of the Female and Male Reproductive Systems Contain an Indigenous Microbiota 408
The Male and Female Reproductive Systems Consist of Primary and Accessory Sex Organs 408
The Female Reproductive System Is Prone to More Infections than the Male Reproductive System 408
Common Vaginal Infections Come from Indigenous Microbiota 409

13.2 Many Sexually Transmitted Diseases Are Caused by Bacteria 410
Chlamydial Urethritis Is the Most Frequently Reported STD 411
Gonorrhea Can Be an Infection in Any Sexually Active Person 414
Syphilis Is a Chronic, Infectious Disease 416
Other Sexually Transmitted Diseases Also Exist 418

13.3 Urinary Infections Are the Second Most Common Body Infection 420
The Urinary System Removes Waste Products from the Blood and Helps Maintain Homeostasis 420
Part of the Urinary Tract Harbors an Indigenous Microbiota 420
Urinary Tract Infections Occur Primarily in the Urethra and Bladder 421

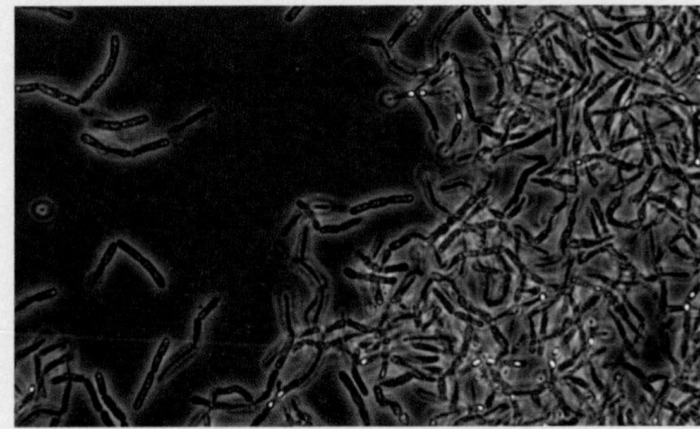
Courtesy of Larry Stauffer, Oregon State Public Health Laboratory/CDC.

13.4 Contact Diseases Can Be Caused by Indigenous Bacterial Species 424
The Skin Protects Underlying Tissues from Microbial Colonization 424
The Skin Harbors an Indigenous Microbiota 426
Acne Is the Most Common Skin Condition in the Developed World 429
Indigenous Microbiota Can Form Biofilms 431

13.5 Contact Diseases Can Also Be Caused by Exogenous Bacterial Species 432
Staphylococcal Contact Diseases Have Several Manifestations 432
Streptococcal Diseases Can Be Mild to Severe 434
Other Wounds Also Can Cause Skin Infections 437
Leprosy (Hansen Disease) Is a Chronic, Systemic Infection 438

13.6 Several Contact Diseases Affect the Eye 441
Some Bacterial Eye Infections Can Cause Blindness 441
Chapter Review 445

PART 4 VIRUSES AND EUKARYOTIC MICROORGANISMS 449

Chapter 14 The Viruses and Virus-Like Agents 451

14.1 Filterable Infectious Agents Cause Disease 453
Many Scientists Contributed to the Early Understanding of Viruses 453

14.2 Viruses Have a Simple Structural Organization 454
Viruses Are Tiny Infectious Agents 454
Viruses Are Grouped by Their Shape 456
Viruses Have a Host Range and Tissue Specificity 458

14.3 Viruses Can Be Classified by Their Genome 460
A Taxonomic Scheme for All Viruses Has Yet to Be Universally Adopted 460

14.4 Viral Replication Follows a Set of Common Steps 462
The Replication of Bacteriophages Can Follow One of Two Cycles 463
Animal Virus Replication Often Results in a Productive Infection 465
Some Animal Viruses Produce a Latent Infection 469

14.5 Viruses and Their Infections Can Be Detected in Various Ways 473
Detection of Viruses Often Is Critical to Disease Identification 473

14.6 Some Viruses Are Associated with Human Tumors and Cancers 476
Cancer Is an Uncontrolled Growth and Spread of Cells 476
Viruses Are Associated with About 20% of Human Tumors 476
Oncogenic Viruses Transform Infected Cells 477

14.7 Emerging Viruses Arise from Genetic Recombination and Mutation 481
Emerging Viruses Usually Arise Through Natural Phenomena 481

14.8 Virus-Like Agents Include Viroids and Prions 482
Viroids Are Infectious RNA Particles 482
Prions Are Infectious Proteins 484
Chapter Review 486

Chapter 15 Viral Infections of the Respiratory Tract and Skin 490

15.1 Viruses Account for Most Upper Respiratory Tract Infections 492
Rhinovirus Infections Are the Chief Cause of the Common Cold 492
Adenovirus Infections Also Produce Symptoms Typical of a Common Cold 493

15.2 Viral Infections of the Lower Respiratory Tract Can Be Severe 496
Influenza Is a Highly Communicable Acute Respiratory Infection 496
Some Paramyxovirus Infections Can Cause Serious Respiratory Disease 500
Other Respiratory Viruses Also Produce Pneumonia 502

15.3 Herpesviruses Cause Several Human Skin Diseases 505
Human Herpes Simplex Infections Are Widespread and Often Recurrent 505
Chickenpox Is No Longer as Prevalent a Disease in the United States 508
Other Herpesviruses Also Cause Human Disease 509

15.4 Several Other Viral Diseases Affect the Skin 511
A Few Viruses Cause Typical Childhood Illnesses 511
Some Human Papillomavirus Infections Cause Warts 514
Poxvirus Infections Have Had Great Medical Impacts on Populations 516
Chapter Review 519

Chapter 16 Viral Infections of the Blood, Lymphatic, Gastrointestinal, and Nervous Systems 522

16.1 Viral Infections Can Affect the Blood and the Lymphatic System 524
Two Herpesviruses Cause Blood Diseases 524
Several Hepatitis Viruses Are Bloodborne 526

16.2 Some Viral Diseases Cause Hemorrhagic Fevers 529
Flaviviruses Can Cause a Terrifying and Severe Illness 529
Members of the Filoviridae Produce Severe Hemorrhagic Lesions 533
Members of the Arenaviridae Are Associated with Chronic Infections in Rodents 533

16.3 Viral Infections of the Gastrointestinal Tract Are Major Global Health Problems 535
Hepatitis Viruses A and E Are Transmitted by the Gastrointestinal Tract 535
Several Unrelated Viruses Can Cause Viral Gastroenteritis 536

16.4 Viral Diseases of the Nervous System Can Be Deadly 539
The Rabies Virus Is of Great Medical Importance Worldwide 539
The Polio Virus May Be the Next Infectious Disease Eradicated 542
Arboviruses Can Cause a Type of Primary Encephalitis 544
Chapter Review 548

Chapter 17 Eukaryotic Microorganisms: The Fungi 552

17.1 The Kingdom Fungi Includes the Molds and Yeasts 554
Fungi Share a Combination of Characteristics 554
Fungal Growth Is Influenced by Several Factors 555
Reproduction in Fungi Involves Spore Formation 557

17.2 Fungi Have Evolved into a Variety of Forms 560
Fungi Can Be Classified into Several Major Groups 560
Yeasts Represent a Term for Any Single-Celled Stage of a Fungus 567

17.3 Some Fungi Cause Intoxications 570
Some Fungi Can Be Poisonous or Even Deadly When Consumed 570
Some Mushrooms Produce Mycotoxins 570

17.4 Some Fungi Can Invade the Skin 572
Dermatophytosis Is an Infection of the Skin, Hair, and Nails 572
Candidiasis Often Is a Mild, Superficial Infection 573
Sporotrichosis Is an Occupational Hazard 574

17.5 Many Fungal Pathogens Cause Lower Respiratory Tract Diseases 575
Cryptococcosis Usually Occurs in Immunocompromised Individuals 575
Histoplasmosis Can Produce a Systemic Disease 576
Blastomycosis Usually Is Acquired Via the Respiratory Route 577
Coccidioidomycosis Can Become a Potentially Lethal Infection 577
Pneumocystis Pneumonia Can Cause a Lethal Pneumonia 577
Other Fungi Also Cause Mycoses 578
Chapter Review 582

Chapter 18 Eukaryotic Microorganisms: The Parasites 586

18.1 Protists Exhibit Great Structural and Functional Diversity 588
Most Protists Are Unicellular and Nutritionally Diverse 588
The Protists Encompass a Variety of Parasitic Lifestyles 591

18.2 Protistan Parasites Attack the Skin, and the Digestive and Urinary Tracts 595
Leishmania Can Cause a Cutaneous or Visceral Infection 595
Several Protistan Parasites Cause Diseases of the Digestive System 596
A Protistan Parasite Also Infects the Urinary Tract 599

18.3 Many Protistan Diseases of the Blood and Nervous System Can Be Life Threatening 601
Plasmodium Can Be a Deadly Blood Parasite 601
The *Trypanosoma* Parasites Can Cause Life-Threatening Systemic Diseases 603
Babesia Is an Apicomplexan Parasite 605
Toxoplasma Causes a Relatively Common Blood Infection 605
Naegleria Can Infect the Central Nervous System 608

18.4 Parasitic Helminths Cause Substantial Morbidity Worldwide 610
There Are Two Groups of Parasitic Helminths 611
Several Trematodes Can Cause Human Illness 612
Tapeworms Survive in the Human Intestines 612
Humans Are Hosts to at Least 50 Roundworm Species 614
Roundworms Also Infect the Lymphatic System 619
Chapter Review 621

PART 5 INTERACTIONS AND IMPACT OF MICROORGANISMS WITH HUMANS 627

Chapter 19 Infection and Disease 629

19.1 The Host and Microbe Form an Intimate Relationship in Health and Disease 631
The Human Body Maintains a Symbiosis with Its Microbiota 631
The Human Microbiome Begins at Birth 633
Pathogens Differ in Their Ability to Cause Disease 633
Several Events Must Occur for Disease to Develop in the Host 636

19.2 Establishment of Infection and Disease Involves Host and Pathogen 637
Diseases Progress Through a Series of Stages 637
Pathogen Entry into the Host Depends on Cell Adhesion and the Infectious Dose 641
Breaching the Host Barriers Can Establish Infection and Disease 643
Successful Invasiveness Requires Pathogens to Have Virulence Factors 643
Pathogens Must Be Able to Leave the Host to Spread Disease 647

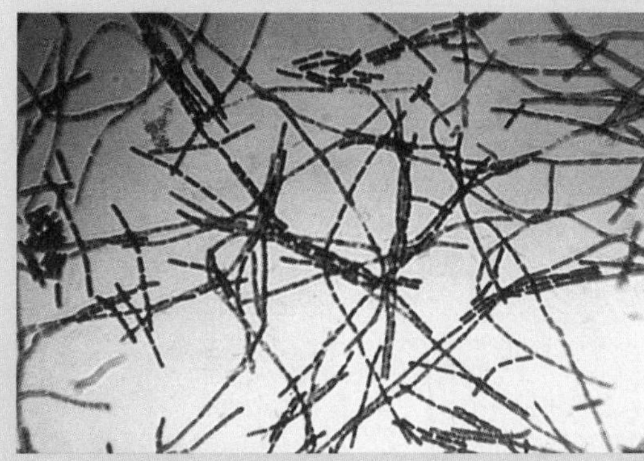

Courtesy of CDC.

Courtesy of Prof. Frank Hadley Collins, Director, Center for Global Health and Infectious Diseases, University of Notre Dame/CDC.

19.3 Infectious Disease Epidemiology Includes Frequency and Spread of Disease 648

Epidemiologists Often Have to Identify the Reservoir of an Infectious Disease 648

Epidemiologists Have Several Terms that Apply to the Infectious Disease Process 649

Infectious Diseases Can Be Transmitted in Several Ways 649

Diseases Also Are Described by How They Occur Within a Population 651

Nosocomial Infections Are Serious Health Threats Within the Healthcare System 651

Infectious Diseases Continue to Challenge Public Health Organizations 660

Chapter Review 665

Chapter 20 Resistance and the Immune System: Innate Immunity 668

20.1 The Immune System Is a Network of Cells and Molecules to Defend Against Foreign Substances 670

Blood Cells Form an Important Defense for Innate and Adaptive Immunity 670

The Lymphatic System Is Composed of Cells and Tissues Essential to Immune Function 672

© nazira_g/ShutterStock, Inc.

Innate and Adaptive Immunity Compose a Fully Functional Human Immune System 672

20.2 Surface Barriers Are Part of Innate Immunity 675

Host Defensive Barriers Limit Pathogen Entry 675

20.3 Coordinated Cellular Defenses Respond to Pathogen Invasion 678

Innate Immunity Depends on Receptor Recognition of Common Pathogen-Associated Molecules 678

Phagocytosis Is a Nonspecific Defense Mechanism to Clear Microbes from Infected Tissues 678

Inflammation Plays an Important Role in Fighting Infection 681

Moderate Fever Benefits Host Defenses 683

Natural Killer Cells Recognize and Kill Abnormal Cells 684

20.4 Effector Molecules Damage Invading Pathogens 685

Complement Marks Pathogens for Destruction 686

Interferon Puts Cells in an Antiviral State 687

Chapter Review 692

Chapter 21 Resistance and the Immune System: Adaptive Immunity 695

21.1 The Adaptive Immune Response Targets the Specific Invading Pathogen 697

The Ability to Eliminate Pathogens Requires a Multifaceted Approach 697

Adaptive Immunity Generates Two Complementary Responses to Most Pathogens 698

Lymphoid Progenitors Differentiate into Several Types of Lymphocytes 700

Clonal Selection Activates the Appropriate B and T Cells 700

21.2 Humoral Immunity Is an Antibody Response to Pathogens in Body Fluids 702

Antibodies Share a Common Structure 702

There Are Five Immunoglobulin Classes 704

Antibody Responses to Pathogens Are of Two Types 705

Antibody Diversity Is a Result of Gene Rearrangements 705

Antibody Interactions Mediate the Disposal of Antigens (Pathogens) 709

21.3 Cell-Mediated Immunity Detects and Eliminates Intracellular Pathogens 712

Cell-Mediated Immunity Relies on T Lymphocyte Receptors and Recognition 712

Naive T Cells Mature into Effector T Cells 714
Activated Cytotoxic T Cells Destroy Virus-Infected Cells 714
Some Antigens Are T-Cell Independent 719
Chapter Review 722

Chapter 22 Immunity and Serology 725

22.1 Immunity to Disease Can Be Generated Naturally or Artificially 727
Adaptive Immunity Can Result by Actively Producing Antibodies to an Antigen 727
Whole Agent Vaccines Contain Weakened or Inactivated Antigens 728
Newer Vaccines Contain Only Subunits or Fragments of Antigens 730
Some Vaccines Are Specifically Recommended for Adults 732
Adaptive Immunity Also Can Result by Passively Receiving Antibodies to an Antigen 734
Herd Immunity Results from Effective Vaccination Programs 737
Do Vaccines Have Dangerous Side Effects? 739

22.2 Serological Reactions Can Be Used to Diagnose Disease 743
Serological Reactions Have Certain Characteristics 743
Neutralization Involves Antigen-Antibody Reactions 744
Precipitation Requires the Formation of a Lattice between Soluble Antigen and Antibody 744
Agglutination Involves the Clumping of Antigens 746
Complement Fixation Can Detect Antibodies to a Variety of Pathogens 746
Labeling Methods Are Used to Detect Antigen-Antibody Binding 747

22.3 Monoclonal Antibodies Are Used for Immunotherapy 754
Monoclonal Antibodies Are Becoming a "Magic Bullet" in Biomedicine 754
Chapter Review 757

Chapter 23 Immune Disorders and AIDS 761

23.1 Type I Hypersensitivity Represents a Familiar Allergic Response 763
Type I Hypersensitivity Is Induced by Allergens 763
Type I Hypersensitivities Can Be Localized or Systemic 766
Allergic Reactions Also Are Responsible for Triggering Many Cases of Asthma 768
Why Do Only Some People Have Allergies? 770
Therapies Sometimes Can Control Allergies 772

23.2 Other Types of Hypersensitivity Represent Immediate or Delayed Reactions 774
Type II Hypersensitivity Involves Cytotoxic Reactions 774
Type III Hypersensitivity Involves an Immune Complex Reaction 776
Type IV Hypersensitivity Is Mediated by Antigen-Specific T Cells 778

23.3 Autoimmune Disorders and Transplantation Are Immune Responses to "Self" 781
An Autoimmune Disorder Is a Failure to Distinguish Self from Nonself 782
Transplantation of Tissues or Organs Is an Important Medical Therapy 784
Immunosuppressive Agents Prevent Allograft Rejection 786

23.4 Immunodeficiency Disorders Can Be Inherited or Acquired 787
Immunodeficiencies Can Involve Any Aspect of the Immune System 787
The Human Immunodeficiency Virus (HIV) Is Responsible for HIV Infection and AIDS 789
Chapter Review 800

Chapter 24 Antimicrobial Drugs 804

24.1 Antimicrobial Agents Are Chemical Substances Used to Treat Infectious Disease 806
The History of Chemotherapy Originated with Paul Ehrlich 806
Fleming's Observation of the Penicillin Effect Ushered in the Era of Antibiotics 807
Antimicrobial Agents Have a Number of Important Properties 808
Antibiotics Are More Than Agents of Natural Biological Warfare 809

24.2 Synthetic Antibacterial Agents Primarily Inhibit DNA Synthesis and Cell Wall Formation 811
Sulfonamides Target Specific Metabolic Reactions 811
Other Synthetic Antimicrobials Are Commonly Prescribed 812

24.3 Beta-Lactam Antibiotics Inhibit Bacterial Cell Wall Synthesis 812
Penicillin Has Remained the Most Widely Used Antibiotic 812
Other Beta-Lactam Antibiotics Also Inhibit Cell Wall Synthesis 814

24.4 Other Bacterially Produced Antibiotics Inhibit Some Aspect of Metabolism 815
Vancomycin Also Inhibits Cell Wall Synthesis 815

Polypeptide Antibiotics Affect the Cell Envelope 815
Many Antibiotics Affect Protein Synthesis 816
Some Antibiotics Inhibit Nucleic Acid Synthesis 818

24.5 Other Antimicrobial Drugs Target Viruses, Fungi, and Parasites 819
Antiviral Drugs Interfere with the Viral Replication Cycle 819
Several Classes of Antifungal Drugs Cause Membrane Damage 821
The Goal of Antiprotistan Agents Is to Eradicate the Parasite 823
Antihelminthic Agents Target Nondividing Helminths 824

24.6 Antibiotic Drug Resistance Is a Growing Challenge 825
There Are Several Antibiotic Susceptibility Assays 825
Antibiotic Resistance Can Develop and Spread in Several Ways 827
Antibiotic Resistance Is of Vital Concern in the Medical Community 829
New Approaches to Antibiotic Therapy Are Needed 834
Chapter Review 842

PART 6 ENVIRONMENTAL AND APPLIED MICROBIOLOGY 847
Available Online with access code

Chapter 25 Microbiology of Foods 849

25.1 Food Spoilage Is Generally a Result of Microbial Contamination and Growth 851
Food Spoilage Comes from Many Microbial Sources 851
Several Conditions Can Determine if Spoilage Will Occur 851
The Microorganisms Responsible for Spoilage Produce Specific Products 853
Meat and Seafood Can Become Contaminated in Several Ways 853
Poultry and Eggs Can Spoil Quickly 854
Breads and Bakery Products Can Support Bacterial and Fungal Growth 855
Some Grains Are Susceptible to Spoilage 855
Milk and Dairy Products Sometimes Spoil 855

25.2 Food Preservation Inhibits Foodborne Pathogens and Spoilage Microorganisms 858
Heat Denatures Proteins 859
Low Temperatures Slow Microbial Growth 861
Drying and Osmotic Pressure Help Preserve Foods 862
Chemical Preservatives Inhibit Microbial Growth 864
Foods May Be Irradiated for Pest and Pathogen Elimination 864
Foodborne Disease Can Result from an Infection or Intoxication 865
HACCP Systems Attempt to Identify Potential Contamination Points 868

25.3 Many Foods Are the Product of Microbial Metabolism 869
Many Foods Are Fermented Products 869
Many Milk Products Are the Result of Fermentation 871
Chapter Review 873

Chapter 26 Environmental Microbiology 877

26.1 Water Pollution Includes Biological Changes Harmful to Water Quality 879
Unpolluted and Polluted Water Contain Different Microbial Populations 879
There Are Three Types of Water Pollution 881
Diseases Can Be Transmitted by Water 882

26.2 Proper Treatment of Water and Sewage Ensures Safe Drinking Water 884
Water Purification Is a Three-Step Process 884
Sewage Treatment Can Be a Multistep Process 885
Biofilms Are Prevalent in the Environment 888
The Bacteriological Analysis of Water Tests for Indicator Organisms 889

26.3 Microbes Are Indispensable for Recycling Major Chemical Elements 891
The Carbon Cycle Is Influenced by Microorganisms 892
The Sulfur Cycle Recycles Sulfate Molecules 894
The Nitrogen Cycle Is Dependent on Microorganisms 894
Chapter Review 898

Chapter 27 Industrial Microbiology and Biotechnology 901

27.1 Microorganisms Are Used to Produce Many Industrial Products 903
Microorganisms Produce Many Useful Organic Compounds 903
Microorganisms Also Produce Important Enzymes and Other Products 905

27.2 Alcoholic Beverages Are Products of Fermentation 907

Beer Is Produced by the Fermentation of Malted Barley 907

Wine Is Produced by the Fermentation of Fruit or Plant Extracts 908

Distilled Spirits Contain More Alcohol than Beer or Wine 911

27.3 Microorganisms Also Produce Many Other Valuable Commercial Products 912

Many Antibiotics Are the Result of Industrial Production 912

Some Microbial Products Can Be Used to Control Insects 913

Fungal Organisms Also Are Being Commercially Developed 915

Bioremediation Helps Clean Up Pollution Naturally 916

Industrial Genetic Engineering Continues to Make Advances 919

Chapter Review 922

Appendix A Metric Measurement A-1

Appendix B Temperature Conversion Chart A-1

Glossary G-1

Index I-1

Preface

The Importance of a Biology (Microbiology) Education

Introductory college courses in the biological sciences (and most other sciences, too) are often seen as courses where students *are told about science* and then *asked to remember "the facts."* Today's world of instant news facts and Internet access provides a mass of readily accessible information about science, some good and some so bad it is counter to logic and evidence. To interpret and use this information correctly and intelligently, you, the student, must know more than a pile of memorized facts. Bruce Alberts, the past president of the U.S. National Academy of Sciences and an acknowledged expert in science education policy, the American Association for the Advancement of Science and the National Science Foundation report *Vision and Change in Undergraduate Biology Education: A Call to Action*, the American Society for Microbiology's *Curriculum Guidelines for Undergraduate Microbiology*, and many other organizations have identified four goals for any biology and microbiology course. The overarching aim of these goals is to prepare students to:

- Know, use, and interpret scientific explanations of the natural world.
- Produce and evaluate scientific evidence.
- Understand the nature and development of scientific knowledge.
- Participate productively in scientific practices and discourse.

These goals require the development of proper critical thinking skills; that is, actively and skillfully applying, analyzing, evaluating, and/or synthesizing information clearly and rationally. A student with critical thinking skills is capable then of achieving the four goals listed above. Importantly, the mastery of such skills is essential in today's job market and that certainly includes the fields of nursing and allied health, pharmacy, medicine, and other health-related fields, all related in part to microbiology.

Because there is a substantial amount of information you will need to learn and understand in microbiology, attaining these skills requires a multifaceted approach. First and foremost, your microbiology instructor is the one who can guide you with "knowing the specifics" and mentor you as you develop these critical thinking skills. In addition, a microbiology textbook should not only support and expand your new knowledge about the microbial world, but also provide examples, evaluations, and assessments to gauge your progress. To facilitate this understanding and coordinate it with class material, this textbook contains a "critical learning design" format (described below) to make reading easier, studying more efficient, and active learning straightforward. Most importantly, the design allows you to better evaluate your learning and provides you with the tools needed to probe your understanding—that is, chapter learning aids and assessment drills to evaluate your progress. The "learning design" format accelerates your success in microbiology—and your career.

I am excited that you are using and reading this new, tenth edition of *Fundamentals of Microbiology*. I hope the textbook is very useful in your studies and also enjoyable to read. Always take time to read many of the sidebars (MicroFocus boxes) whether they are assigned or not. They will help in your overall microbiology experience and the realization that microorganisms do rule the world!

AUDIENCE

Fundamentals of Microbiology, Tenth Edition is written for introductory microbiology courses having an emphasis in the health sciences. It is geared toward students in health and allied health science curricula such as nursing, dental hygiene, medical assistance, sanitary science, and medical laboratory technology. It also will be an asset to students studying food science, agriculture, environmental science, and health administration. In addition, the text provides a firm foundation for advanced programs in biological sciences, as well as medicine, pharmacy, dentistry, and other health professions.

ORGANIZATION

Fundamentals of Microbiology, Tenth Edition is divided into several areas of concentration. These areas reflect the *Curriculum Guidelines for Undergraduate Microbiology* as recommended by the American Society for Microbiology and provide a framework for the unity and diversity of microbiology. Therefore, among the overarching principles explored (evolution is woven throughout the textbook) are:

Cell Structure and Function

- The evolutionary relatedness of organisms is best reflected in phylogenetic trees.
- While microscopic eukaryotes (fungi and protists) carry out many of the same basic processes as prokaryotes, there are fundamental structural differences within and between them.
- Prokaryotes have specialized structures that often confer critical capabilities and can be targets for antibiotics and immune system defenses.

Metabolic Pathways

- The interactions of microorganisms among themselves and with their environment are determined by their metabolic abilities and diversity.
- Cells, organelles (e.g., mitochondria and chloroplasts), and all major metabolic pathways evolved from early prokaryotic cells.
- The growth of microorganisms can be controlled by physical and chemical means.

Information Flow and Genetics

- Genetic variation can impact microbial functions and the regulation of gene expression is influenced by external and internal molecular cues and/or signals.
- Mutations and horizontal gene transfer, with the immense variety of microenvironments, have provided the means through evolution for an amazing diversity of microorganisms.
- Microbial genomes can be manipulated to alter cell function such that humans can use and harness microorganisms and their products.

Microbial Systems

- The synthesis of viral genetic material and proteins is dependent on interactions with host cells.
- Microorganisms and viruses interact with both human and nonhuman hosts in beneficial, neutral, and detrimental ways, the latter often responsible for infection and disease.
- The replication cycles of viruses differ among virus families and are determined by their unique structures and genomes.

Impact of Microorganisms

- Microbes are essential for life as we know it and the processes that support life.
- Microorganisms provide essential models that give us fundamental knowledge about life processes.
- Human impact on the environment influences the evolution of microorganisms as seen through emerging diseases and the selection of antibiotic resistance.

WHAT'S NEW AND IMPORTANT

When you read this text, you get a global perspective on microbiology and infectious disease that is found in no other similar textbook. Therefore, this edition provides detailed updates to microbial structure and function, microbial genomics, and, of course, disease information and statistics. In addition, this new edition has expanded coverage of:

- **The human microbiome**, the vast assemblage of microbes that we all possess on us and in us. This truly fascinating topic and research area will, in the near future, dramatically change the way we treat physiological and metabolic disorders in the human body and how we can improve our general health and wellbeing.
- **Infectious diseases** that continue to plague many parts of the human population. This includes AIDS, influenza, tuberculosis, and

malaria, sometimes referred to as "the big four" when considering the global mortality numbers.
- **Evolution**, which is rooted into much of the material in many of the chapters.

The "Learning Design" Concept mentioned above has been expanded to provide more tools for developing critical thinking skills. The new features include:

- **Chapter Challenge.** Each activity is designed to test some part of the four goals listed above for science understanding. The challenges help you to think deeper about the chapter material and provide you with the critical thinking skills needed for your future career in nursing or the health sciences. Each challenge includes a critical-thinking question related to the chapter material.
- **Investigating the Microbial World.** Although most of you are not going to become practicing microbiologists, the ability to use, interpret, and evaluate scientific explanations and evidence of the natural world will be important in your careers. Therefore, almost every chapter relates an actual experiment (on an interesting topic) that requires you to apply the process of science and use quantitative reasoning.
- **Figure Art.** Many figures with multipart microbiological processes or figures of organism life cycles have been redesigned in a more readily understandable and visual from.

This new tenth edition of *Fundamentals of Microbiology* retains all of the reading and learning activities designed to encourage student interaction and assessment. These design elements form an integrated study and learning package (learning tools) for student understanding and assessment.

BEING SKEPTICAL

One of the types of essay boxes in this book is titled **Being Skeptical**. A good scientist is a skeptic and skepticism is an important part of science. Skepticism, unlike cynicism, is not unwillingness to accept a claim or observation. Skepticism simply says "Prove it!" Science applies scientific reasoning as the method for proof. Thus, a scientist, such as a microbiologist, must see the evidence and it must be compelling before the observation or statement is provisionally accepted. The claim is still open to further examination and experimentation.

NAVIGATE FUNDAMENTALS OF MICROBIOLOGY

Navigate Fundamentals of Microbiology transforms how students learn and instructors teach by bringing together authoritative and interactive content aligned to course objectives, with student practice activities and assessments, adaptive study planning and remediation, and learning analytics reporting tools. **Navigate Fundamentals of Microbiology** empowers faculty and students with easy-to-use web-based curriculum solutions that optimize student success, identify retention risks, and improve completion rates. Using best practices in instructional design, **Navigate Fundamentals of Microbiology** uniquely focuses on developing students' cognitive intelligence and practical skills needed for success in the 21st century workforce.

Navigate Fundamentals of Microbiology also includes an interactive eBook, interactive course lessons, practice activities, and robust course management tools. It is a fully hosted and supported online learning solution delivered in the Moodle® Course Delivery System. **Navigate Fundamentals of Microbiology** component modules can also be configured for locally hosted learning management systems such as Blackboard, Desire2Learn, and others. Please contact your Jones & Bartlett sales representative for more information.

ADDITIONAL RESOURCES

Jones & Bartlett Learning also offers an array of traditional ancillaries to assist instructors and students in teaching and mastering the concepts in this text. Additional information and review copies of any of the following items are available through your Jones & Bartlett sales representative or by going to http://www.jblearning.com/science/.

For the Student

The website we developed exclusively for the tenth edition of this text, http://microbiology

.jbpub.com/10e, offers a variety of resources to enhance understanding of microbiology. The site contains chapter summaries, an interactive glossary for key term review, interactive flashcards, crossword puzzles, short study quizzes, and more. Part 6 of this book, "Environmental and Applied Microbiology," is also available for download through the student companion site. Access to this site is free with every new print copy of the text, and is also available for purchase separately.

The *Student Study Guide* to accompany this textbook contains important information to help you study, take effective class notes, prepare properly for exams, and even to manage your time effectively. The latter is the single most common reason for poor performance in college courses. The *Student Study Guide* also contains over 3,000 practice exercises and study questions of various types to help you learn and retain the information in the text.

Alcamo's Laboratory Fundamentals of Microbiology, Ninth Edition, is a series of over 30 multipart laboratory exercises providing basic training in the handling of microorganisms and reinforcing ideas and concepts described in the textbook.

Guide to Infectious Diseases by Body Systems is an excellent tool for learning about microbial diseases. Each of the fifteen body systems units presents a brief introduction to the anatomical system and the bacterial, viral, fungal, or parasitic organism infecting the system.

An anthology called *Encounters in Microbiology* (*Volume I, Second Edition*, and *Volume II*) brings together "Vital Signs" articles from *Discover* magazine in which health professionals use their knowledge of microbiology in their medical cases.

For the Instructor

Compatible with Windows® and Macintosh® platforms, the Instructor's Media CD provides instructors with the following traditional ancillaries:

- The *PowerPoint® Image Bank* provides the illustrations, photographs, and tables (to which Jones & Bartlett Learning holds the copyright or has permission to reproduce digitally) inserted into PowerPoint slides. You can quickly and easily copy individual images or tables into your existing lecture presentations.
- The *PowerPoint Lecture Outline* presentation package provides lecture notes and images for each chapter of *Fundamentals of Microbiology*. Instructors with the Microsoft PowerPoint software can customize the outlines, art, and order of presentation.

The following materials are also available online, at http://www.jblearning.com/catalog/9781449688615/.

- The *Instructor's Manual*, provided as a text file, includes chapter summaries and complete chapter lecture outlines and answers to all the end-of-chapter assessments.
- The *Test Bank* is available as straight text files.

Acknowledgments

Putting together a new edition of a textbook always requires the input of a whole team and so I want to once again thank everyone at Jones & Bartlett Learning who helped put together this new edition of *Fundamentals of Microbiology*. As publisher, Cathleen Sether was always there to answer questions and provide guidance. Erin O'Connor and Rachel Isaacs provided valuable feedback. At the production end, Leah Corrigan kept me on a timetable and managed the process expertly; Scott Moden put together the design format; Lauren Miller found the photos needed to illustrate the pages; Deborah Patton organized the Index; Linda DeBruyn did the copyediting while Jan Cocker proofread the textbook pages.

Throughout all my years of teaching at universities and colleges, I would not be able to be the instructor I am without the great students I have had the fortune to have in my many classes. They keep me on my toes in the classroom, require me to always be prepared, and let me know when a topic or concept was not conveyed in as clear and understandable way as it could be. Their suggestions and evaluations have encouraged me to continually assess my instruction, and make it the best it can be. I salute all my former students — and I hope those of you who read this text will let me know what works and what still needs improvement to make your learning effective, enjoyable, and most of all successful.

Jeff Pommerville

About the Author

Courtesy of Jeffrey Pommerville.

Today I am a microbiologist, researcher, and science educator. My plans did not start with that intent. While in high school in Santa Barbara, California, I wanted to play professional baseball, study the stars, and own a '66 Corvette. None of these desires would come true—my batting average was miserable (but I was a good defensive fielder), I hated the astronomy correspondence course I took, and I never bought that Corvette.

I found an interest in biology at Santa Barbara City College. After squeaking through college calculus, I transferred to the University of California at Santa Barbara (UCSB) where I received a B.S. in biology and stayed on to pursue a Ph.D. degree in the lab of Ian Ross studying cell communication and sexual pheromones in a water mold. After receiving my doctorate in cell and organismal biology, my graduation was written up in the local newspaper as a native son who was a fungal sex biologist—an image that was not lost on my three older brothers!

While in graduate school at UCSB, I rescued a secretary in distress from being licked to death by a German shepherd. Within a year, we were married (the secretary and I). When I finished my doctoral thesis, I spent several years as a postdoctoral fellow at the University of Georgia. Worried that I was involved in too many research projects, a faculty member told me something I will never forget. He said, "Jeff, it's when you can't think of a project or what to do that you need to worry." Well, I have never had to worry!

I then moved on to Texas A&M University, where I spent 8 years in teaching and research—and telling Aggie jokes. Toward the end of this time, after publishing over 30 peer-reviewed papers in national and international research journals, I realized I had a real interest in teaching and education. Leaving the sex biologist nomen behind, I headed farther west to Arizona to join the biology faculty at Glendale Community College, where I continue to teach introductory biology and microbiology.

I have been lucky to be part of several educational research projects and have been honored, with two of my colleagues, with a Team Innovation of the Year Award by the League of Innovation in the Community Colleges. In 2000, I became project director and lead principal investigator for a National Science Foundation grant to improve student outcomes in science through changes in curriculum and pedagogy. I had a fascinating three years coordinating more than 60 science faculty members (who at times were harder to manage than students) in designing and field testing 18 interdisciplinary science units. This culminated with me being honored in 2003 with the Gustav Ohaus Award (College Division) for Innovations in Science Teaching from the National Science Teachers Association.

I am the Perspectives Editor for *Microbiology Education* (now the *Journal of Microbiology and Biology Education*), the education research journal of the American Society for Microbiology (ASM) and in 2004 was co-chair for the ASM Conference for Undergraduate Educators. From 2006 to 2007, I was the chair of Undergraduate Education Division of ASM. In 2006, I was selected as one of four outstanding instructors at Glendale Community College. The culmination of my teaching career came in 2008 when I was nationally recognized by being awarded the Carski Foundation Distinguished Undergraduate Teaching Award for distinguished teaching of microbiology to undergraduate students and encouraging them to subsequent achievement.

I mention all this not to impress but to show how the road of life sometimes offers opportunities in unexpected and unplanned ways. The key though is keeping your "hands on the wheel and your eyes on the prize;" then unlimited opportunities will come your way. With the untimely passing of my friend and professional colleague Ed Alcamo, also the Carski recipient in 2000, I was privileged in 2003 to be offered the opportunity to take over the authorship of *Fundamentals of Microbiology*. It is an undertaking I continue to relish as I (along with the wonderful folks at Jones & Bartlett Learning) try to evolve a new breed of microbiology textbook reflecting the pedagogy change occurring in science classrooms today. And, hey, who knows—maybe that '66 Corvette could be in my garage yet.

DEDICATION

This is the fourth edition of *Fundamentals of Microbiology* that I have authored. Over these 10+ years, I have spent countless months in revisions and updating to the text, often unintentionally neglecting my wife. Therefore, I am proud to dedicate this landmark tenth edition of the book to my wife, Yvonne. Without her continual support and encouragement, these editions of the text would have been almost insurmountable. Yvonne, you are my best friend, my soul mate, my lover—my wife. We are but one.

To the Student—Study Smart

Your success in microbiology and any college or university course will depend on your ability to study effectively and efficiently. Therefore, this textbook was designed with you, the student, in mind. The text's organization will help you improve your learning and understanding and, ultimately, your grades. The learning design concept described in the Preface and illustrated below reflects this organization. Study it carefully, and, if you adopt the flow of study shown, you should be a big step ahead in your preparation and understanding of microbiology—and for that matter any subject you are taking.

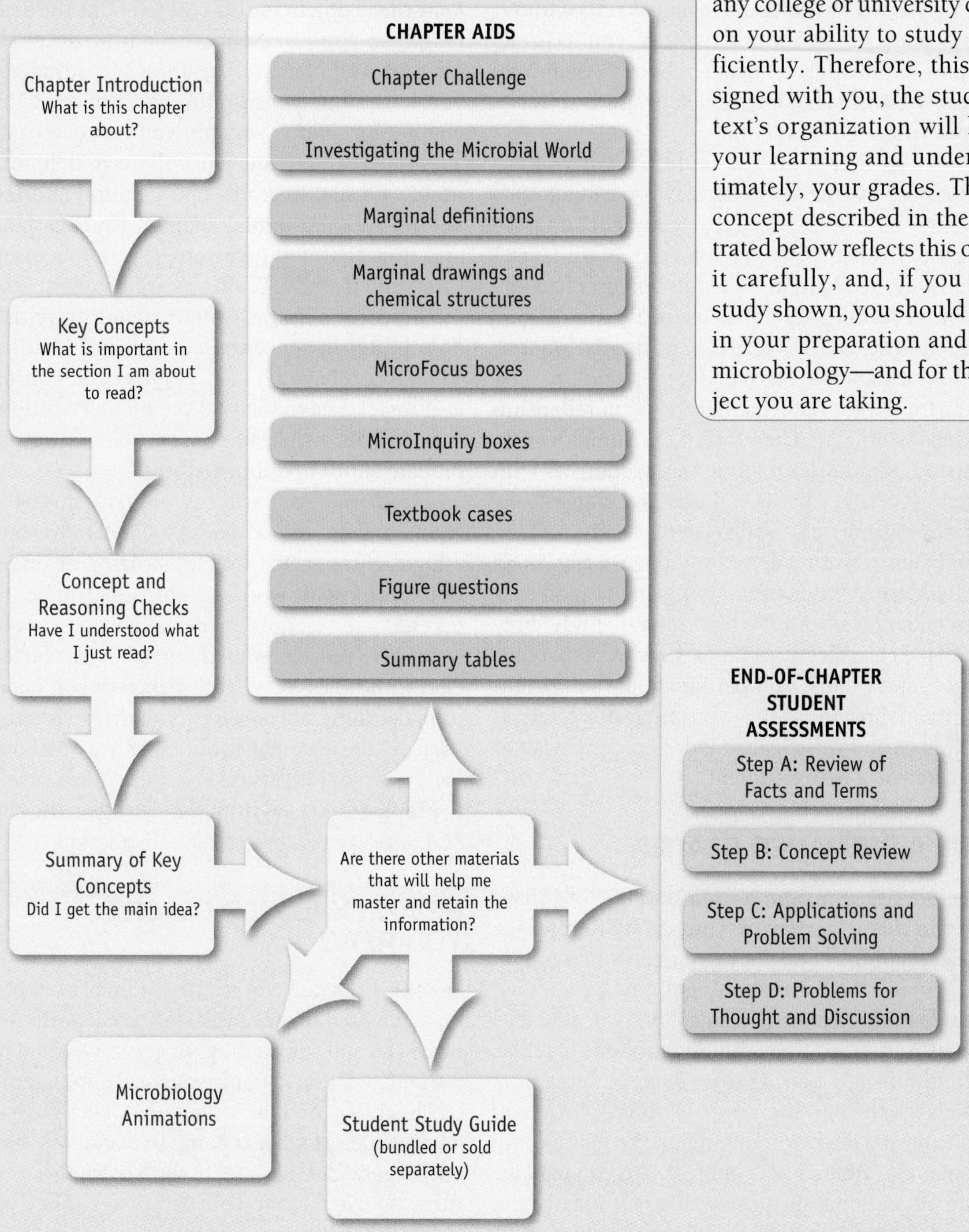

Chapter Introduction
What is this chapter about?

Key Concepts
What is important in the section I am about to read?

Concept and Reasoning Checks
Have I understood what I just read?

Summary of Key Concepts
Did I get the main idea?

Are there other materials that will help me master and retain the information?

CHAPTER AIDS

- Chapter Challenge
- Investigating the Microbial World
- Marginal definitions
- Marginal drawings and chemical structures
- MicroFocus boxes
- MicroInquiry boxes
- Textbook cases
- Figure questions
- Summary tables

END-OF-CHAPTER STUDENT ASSESSMENTS

- Step A: Review of Facts and Terms
- Step B: Concept Review
- Step C: Applications and Problem Solving
- Step D: Problems for Thought and Discussion

Microbiology Animations

Student Study Guide (bundled or sold separately)

When I was an undergraduate student, I hardly ever read the "To the Student" section (if indeed one existed) in my textbooks because the section rarely contained any information of importance. This one does, so please read on.

In college, I was a mediocre student until my junior year. Why? Mainly because I did not know how to study properly, and, important here, I did not know how to read a textbook effectively. My textbooks were filled with underlined sentences (highlighters hadn't been invented yet!) without any plan on how I would use this "emphasized" information. In fact, most textbooks *assume* you know how to read a textbook properly. I didn't and you might not, either.

Reading a textbook is difficult if you are not properly prepared. So that you can take advantage of what I learned as a student and have learned from instructing thousands of students, I have worked hard to make this text user friendly with a reading style that is not threatening or complicated. Still, there is a substantial amount of information to learn and understand, so having the appropriate reading and comprehension skills is critical. Therefore, I encourage you to spend 30 minutes reading this section, as I am going to give you several tips and suggestions for acquiring those skills. Let me show you how to be an active reader. Note: the *Student Study Guide* also contains similar information on how to take notes from the text, how to study, how to take class (lecture) notes, how to prepare for and take exams, and perhaps most important for you, how to manage your time effectively. It all is part of this "learning design," my wish to make you a better student.

BE A PREPARED READER

Before you jump into reading a section of a chapter in this text, prepare yourself by finding the place and time and having the tools for study.

Place. Where are you right now as you read these lines? Are you in a quiet library or at home? If at home, are there any distractions, such as loud music, a blaring television, or screaming kids? Is the lighting adequate to read? Are you sitting at a desk or lounging on the living room sofa? Get where I am going? When you read for an educational purpose—that is, to learn and understand something—you need to maximize the environment for reading. Yes, it should be comfortable but not to the point that you will doze off.

Time. All of us have different times during the day when we perform some skill, be it exercising or reading, the best. The last thing you want to do is read when you are tired or simply not "in tune" for the job that needs to be done. You cannot learn and understand the information if you fall asleep or lack a positive attitude. I have kept the chapters in this text to about the same length so you can estimate the time necessary for each and plan your reading accordingly. If you have done your preliminary survey of the chapter or chapter section, you can determine about how much time you will need. If 40 minutes is needed to read—and comprehend (see below)—a section of a chapter, find the place and time that will give you 40 minutes of uninterrupted study. Brain research suggests that most people's brains cannot spend more than 45 minutes in concentrated, technical reading. Therefore, I have avoided lengthy presentations and instead have focused on smaller sections, each with its own heading. These should accommodate shorter reading periods.

Reading Tools. Lastly, as you read this, what study tools do you have at your side? Do you have a highlighter or pen for emphasizing or underlining important words or phrases? Notice, the text has wide margins, which allow you to make notes or to indicate something that needs further clarification. Do you have a pencil or pen handy to make these notes? Or, if you do not want to "deface" the text, make your notes in a notebook. Lastly, some students find having a ruler is useful to prevent your eyes from wandering on the page and to read each line without distraction.

BE AN EXPLORER BEFORE YOU READ

When you sit down to read a section of a chapter, do some preliminary exploring. Look at the section head and subheadings to get an idea of what is discussed. Preview any diagrams, photographs, tables, graphs, or other visuals used. They give you a better idea of what is going to occur. We have used a good deal of space in the text for these features, so use them to your advantage. They will help you learn the written information and comprehend its meaning. Do not try to understand all the visuals, but try to generate a mental "big picture" of what is to come. Familiarize yourself with any symbols or technical jargon that might be used in the visuals.

The end of each chapter contains a **Summary of Key Concepts** for that chapter. It is a good idea to read the summary before delving into the chapter. That way you will have a framework for the chapter before filling in the nitty-gritty information.

BE A DETECTIVE AS YOU READ

Reading a section of a textbook is not the same as reading a novel. With a textbook, you need to uncover the important information (the terms and concepts) from the forest of words on the page. So, the first thing to do is read the complete paragraph. When you have determined the main ideas, highlight or underline them. However, I have seen students highlighting the entire paragraph in yellow, including every *a*, *the*, and *and*. This is an example of highlighting before knowing what is important. So, I have helped you out somewhat. Important terms and concepts are in **bold face** followed by the definition (or the definition might be in the margin). So only highlight or underline with a pen essential ideas and key phrases—not complete sentences, if possible. By the way, the important microbiological terms and major concepts also are in the **Glossary** at the back of the text.

What if a paragraph or section has no boldfaced words? How do you find what is important here? From an English course, you may know that often the most important information is mentioned first in the paragraph. If it is followed by one or more examples, then you can backtrack and know what was important in the paragraph. In addition, I have added section "speed bumps" (called **Concept and Reasoning Checks**) to let you test your learning and understanding before getting too far ahead in the material. These checks also are clues to what was important in the section you just read.

BE A REPETITIOUS STUDENT

Brain research has shown that each individual can only hold so much information in short-term memory. If you try to hold more, then something else needs to be removed—sort of like a full computer disk. So that you do not lose any of this important information, you need to transfer it to long-term memory—to the hard drive if you will. In reading and studying, this means retaining the term or concept; so, write it out in your notebook *using your own words*. Memorizing a term does not mean you have learned the term or understood the concept. By actively writing it out in your own words, you are forced to think and actively interact with the information. This repetition reinforces your learning.

BE A PATIENT STUDENT

In textbooks, you cannot read at the speed that you read your e-mail or a magazine story. There are unfamiliar details to be learned and understood—and this requires being a patient, slower reader. Actually, if you are not a fast reader to begin with, as I am, it may be an advantage in your learning process. Identifying the important information from a textbook chapter requires you to *slow down* your reading speed. Speed-reading is of no value here.

KNOW THE WHAT, WHY, AND HOW

Have you ever read something only to say, "I have no idea what I read!" As I've already mentioned, reading a microbiology text is not the same as reading *Sports Illustrated* or *People* magazine. In these entertainment magazines, you read passively for leisure or perhaps amusement. In *Fundamentals of Microbiology, Tenth Edition*, you must read actively for learning and understanding—that is, for *comprehension*. This can quickly lead to boredom unless you engage your brain as you read—that is, be an active reader. Do this by knowing the *what, why*, and *how* of your reading.

- *What* is the general topic or idea being discussed? This often is easy to determine because the section heading might tell you. If not, then it will appear in the first sentence or beginning part of the paragraph.
- *Why* is this information important? If I have done my job, the text section will tell you why it is important or the examples provided

will drive the importance home. These surrounding clues further explain why the main idea was important.

- *How* do I "mine" the information presented? This was discussed under being a detective.

A MARKED UP READING EXAMPLE

So let's put words into action. Below is a passage from the text. I have marked up the passage as if I were a student reading it for the first time. It uses many of the hints and suggestions I have provided. Remember, it is important to read the passage slowly, and concentrate on the main idea (concept) and the special terms that apply.

HAVE A DEBRIEFING STRATEGY

After reading the material, be ready to debrief. Verbally summarize what you have learned. This will start moving the short-term information into the long-term memory storage—that is, *retention*. Any notes you made concerning confusing material should be discussed as soon as possible with your instructor. For microbiology, allow time to draw out diagrams. Again, repetition makes for easier learning and better retention.

In many professions, such as sports or the theater, the name of the game is practice, practice, practice. The hints and suggestions I have given you form a skill that requires practice to

Many Foodborne and Waterborne Diseases Have a Bacterial Cause

KEY CONCEPT

- Bacterial gastrointestinal diseases may arise from intoxications or infections.

We categorize the foodborne and waterborne bacterial diseases as either intoxications or infections. **Intoxications** are illnesses in which bacterial toxins are ingested in food or water. Examples are the toxins causing botulism, staphylococcal food poisoning, and clostridial food poisoning. By contrast, **infections** refer to illnesses in which live bacterial pathogens in food and water are ingested and subsequently grow in the body. Salmonellosis, shigellosis, and cholera are examples. Toxins may be produced, but they are the result of infection.

Etiology: The study of the causes (origins) of disease.

Determining the etiology of a bacterial disease depends on several factors.

Incubation period. If an individual ingests and swallows a contaminated food or beverage, there is a delay, called the **incubation period**, before the symptoms appear. This period can range from hours to days, depending on the bacterial species and on the infectious dose. During the incubation period, the toxins or microbes pass through the stomach into the intestine where they may directly affect gastrointestinal function or be absorbed into the bloodstream.

Infectious dose: The number of organisms consumed to give rise to symptoms of an illness.

Clinical symptoms. The symptoms produced by an intoxication or infection depend perfect and use efficiently. Be patient, things will not happen overnight; perseverance and willingness though will pay off with practice. You might also check with your college or university academic (or learning) resource center. These folks will have more ways to help you to read a textbook better and to study well overall.

CONCEPT MAPS

In science as well as in other subjects you take at the college or university, there often are concepts that appear abstract or simply so complex that they are difficult to understand. A **concept map** is one tool to help you enhance your abilities to think and learn. Critical reasoning and the ability to make connections between complex, nonlinear information are essential to your studies and career.

Concept maps are a learning tool designed to represent complex or abstract information visually. Neurobiologists and psychologists tell us that the brain's primary function is to take incoming information and interpret it in a meaningful or practical way. They also have found that the brain has an easier time making sense of information when it is presented in a visual format. Importantly, concept maps not only present the information in "visual sentences" but also take paragraphs of material and present it in an "at-a-glance" format. Therefore, you can use concept maps to

- Communicate and organize complex ideas in a meaningful way
- Aid your learning by seeing connections within or between concepts and knowledge
- Assess your understanding or diagnose misunderstanding

There are many different types of concept maps. The two most used in this textbook are the *process map* or *flow chart* and the *hierarchical map*. The hierarchical map starts with a general concept (the most inclusive word or phrase) at the top of the map and descends downward using more specific, less general words or terms. In several chapters in this textbook process or hierarchical maps are drawn—and you have the opportunity to construct your own hierarchical maps as well.

Concept mapping is the strategy used to produce a concept map. So, let's see how one makes a hierarchical map.

How to Construct a Concept Map

1. Print the central idea (concept or question to be mapped) in a box at the top center of a blank, unlined piece of paper. Use uppercase letters to identify the central idea.
2. Once the concept has been selected, identify the key terms (words or short phrases) that apply to or stem from the concept. Often these may be given to you as a list. If you have read a section of a text, you can extract the terms from that material, as the words are usually boldfaced or italicized.
3. Now, from this list, try to create a hierarchy for the terms you have identified; that is, list them from the most general, most inclusive to the least general, most specific. This ranking may only be approximate and subject to change as you begin mapping.
4. Construct a preliminary concept map. This can be done by writing all of the terms on Post-its®, which can be moved around easily on a large piece of paper. This is necessary as one begins to struggle with the process of building a good hierarchical organization.
5. The concept map connects terms associated with a concept in the following way:
   - The relationship between the concept and the first term(s), and between terms, is connected by an arrow pointing in the direction of the relationship (usually downward or horizontal if connecting related terms).
   - Each arrow should have a label, a very short phrase that explains the relationship with the next term. In the end, each link with a label reads like a sentence.
6. Once you have your map completed, redraw it in a more permanent form. Box in all terms that were on the sticky notes. Remember there may be more than one way to draw a good concept map, and don't be scared off if at first you have some problems mapping; mapping will become more apparent to you after you have practiced this technique a few times using the opportunities given to you in the early chapters of the textbook.

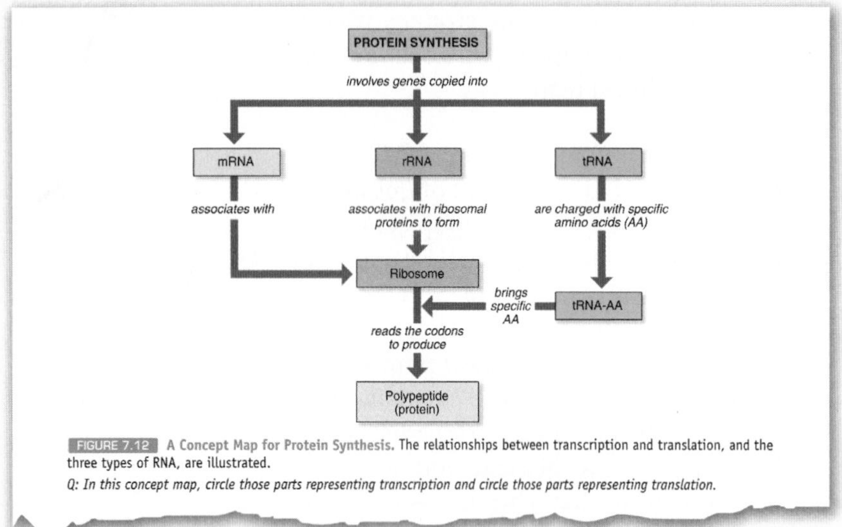

FIGURE 7.12 **A Concept Map for Protein Synthesis.** The relationships between transcription and translation, and the three types of RNA, are illustrated.

Q: In this concept map, circle those parts representing transcription and circle those parts representing translation.

7. Now look at the map and see if it answers the following. Does it:
   - Define clearly the central idea by positioning it in the center of the page?
   - Place all the terms in a logical hierarchy and indicate clearly the relative importance of each term?
   - Allow you to figure out the relationships among the key ideas more easily?
   - Permit you to see all the information visually on one page?
   - Allow you to visualize complex relationships more easily?
   - Make recall and review more efficient?

Example

After reading the section on "Protein Synthesis," a student makes a list of the terms used and maps the concept. Using the steps outlined above, the student produces the following hierarchical map. Does it satisfy all the questions asked in (7)?

Practical Uses for Mapping

- **Summarizing textbook readings.** Use mapping to summarize a chapter section or a whole chapter in a textbook. This purpose for mapping is used many times in this text.
- **Summarizing lectures.** Although producing a concept map during the classroom period may not be the best use of the time, making a concept map or maps from the material after class will help you remember the important points and encourage high-level, critical reasoning, which is so important in university and college studies.
- **Reviewing for an exam.** Having concept maps made ahead of time can be a very useful and productive way to study for an exam, particularly if the emphasis of the course is on understanding and applying abstract, theoretical material, rather than on simply reproducing memorized information.
- **Working on an essay.** Mapping also is a powerful tool to use during the early stages of writing a course essay or term paper. Making a concept map before you write the first rough draft can help you see and ensure you have the important points and information you will want to make.

SEND ME A NOTE

In closing, I would like to invite you to write me and let me know what is good about this textbook so I can build on it and what may need improvement so I can revise it. Also, I would be pleased to hear about any news of microbiology in your community, and I'd be happy to help you locate any information not covered in the text.

I wish you great success in your microbiology course. Welcome! Let's now plunge into the wonderful and sometimes awesome world of microorganisms.

—Dr. P

Email: jeffrey.pommerville@gccaz.edu
Web site: http://web.gcca3.edu/~jpommerv/

A Tribute to I. Edward Alcamo

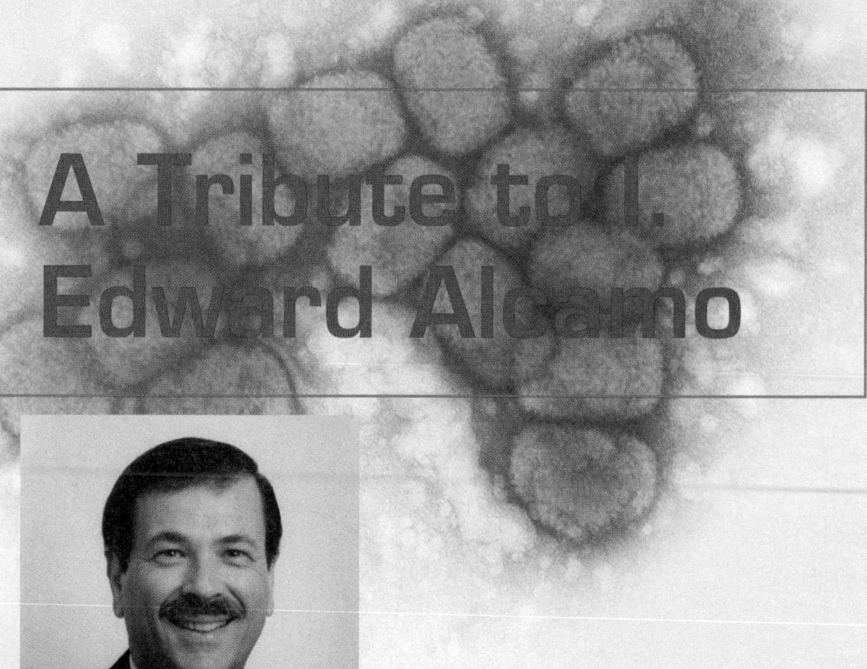

Courtesy of I. Edward Alcamo.

Dr. Ignazio Edward Alcamo was a long-time Professor of Microbiology at the State University of New York at Farmingdale and the author of numerous textbooks, lab kits, and educational materials. He was the 2000 recipient of the Carski Foundation Distinguished Undergraduate Teaching Award, the highest honor bestowed upon microbiology educators by the American Society for Microbiology.

Ed Alcamo was educated at Iona College and St. John's University and held a deep belief in the partnership between research scientists and allied health educators. He sought to teach the scientific basis of microbiology in an accessible manner as well as to inspire students with a sense of topical relevance. Michael Vinciguerra, Provost at the SUNY Farmingdale wrote, "In 1970, when I joined the faculty as a chemistry professor, Ed's reputation as an excellent biology educator was already well known."

A prolific author, Dr. Alcamo produced a broad array of publications including several learning guides and textbooks—*Fundamentals of Microbiology*, now in its 10th edition, and the recently published *Microbes and Society, 3rd Edition*. He also prepared the *Encarta* encyclopedia entry entitled "Procaryotes," as well as *The Microbiology Coloring Book*, and *Schaum's Outline of Microbiology*. His other books published within the past several years include *AIDS: The Biological Basis, DNA Technology: The Awesome Skill, The Biology Coloring Workbook*, and *Anatomy and Physiology the Easy Way*. In December 2002, after a six-month illness, Dr. Alcamo died of acute myeloid leukemia.

Dr. Alcamo's teaching career was dedicated to the proposition that emphasizing quality in education is central to turning back the tide of fear and uncertainty and enabling doctors to find cures for disease. In the early 1980s, when the early cases of an unknown acquired immunodeficiency syndrome were turning into a mysterious and intractable epidemic, Dr. Alcamo told this to his class:

> *One afternoon, about 350 years ago, in the countryside near London, a clergyman happened to meet Plague.*
>
> *"Where are you going?" asked the clergyman.*
>
> *"To London," responded Plague, "to kill a thousand."*
>
> *They chatted for another few moments, and each went his separate way.*
>
> *Some time later, they chanced to meet again. The clergyman said, "I see you decided to show no mercy in London. I heard that 10,000 died there."*
>
> *"Ah, yes," Plague replied, "but I only killed a thousand. Fear killed the rest."*

Cells of *Vibrio cholerae*, transmitted to humans in contaminated water and food, are the cause of cholera. Courtesy of CDC.

PART 1

Foundations of Microbiology

CHAPTER 1 Microbiology: Then and Now

CHAPTER 2 The Chemical Building Blocks of Life

CHAPTER 3 Concepts and Tools for Studying Microorganisms

CHAPTER 4 Structure of Bacterial and Archaeal Cells

CHAPTER 5 Microbial Growth and Nutrition

CHAPTER 6 Microbial Metabolism

CHAPTER 7 Control of Microorganisms: Physical and Chemical Methods

In 1676, a century before the Declaration of Independence, a Dutch merchant named Antony van Leeuwenhoek sent a noteworthy letter to the Royal Society of London. Writing in the vernacular of his home in the United Netherlands, Leeuwenhoek described how he used a simple microscope to observe vast populations of minute, living creatures. His reports opened a chapter of science that would evolve into the study of microscopic organisms and the discipline of microbiology. During the next three centuries scientists would discover how profoundly these organisms influence the quality of our lives and the environment around us.

We begin our study of the microorganisms by exploring the grassroot developments that led to the establishment of microbiology as a science. These developments are surveyed in Chapter 1, where we focus on some of the individuals who stood at the forefront of discovery. Today we are in the midst of a third Golden Age of microbiology and our understanding of microorganisms continues to grow even as you read this book.

Chapter 2 reviews basic chemistry, inasmuch as microbial growth, metabolism, and control are grounded in the molecules and macromolecules these organisms contain and in the biological processes they undergo. Chapter 3 sets down some basic microbiological concepts and describes one of the major tools for studying microorganisms. We will concentrate on the bacterial organisms in Chapter 4, where we survey their structural frameworks. In Chapter 5, we build on these frameworks by examining microbial growth patterns and nutritional requirements. Chapter 6 describes the metabolism of microbial cells, including those chemical reactions that produce and use energy. Part 1 concludes by considering the physical and chemical methods used to control microbial growth and metabolism (Chapter 7).

Much as the alphabet applies to word development, in each succeeding chapter we will formulate words into sentences and sentences into ideas as we construct an understanding of microorganisms and concentrate on their importance to public health and human welfare.

MICROBIOLOGY PATHWAYS

Being a Scientist

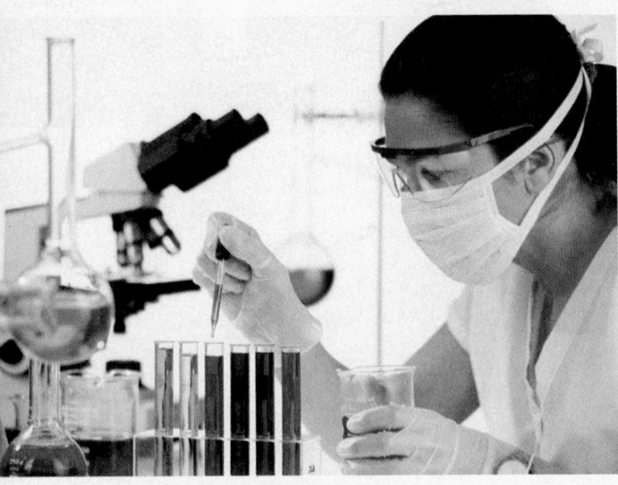

© Comstock/Thinkstock.

Science may not seem like the most glamorous profession. So, as you read many of the chapters in this text, you might wonder why many scientists have the good fortune to make key discoveries. At times, it might seem like it is the luck of the draw, but actually many scientists have a set of characteristics that put them on the trail to success.

Robert S. Root-Bernstein, a physiology professor at Michigan State University, points out that many prominent scientists like to goof around, play games, and surround themselves with a type of chaos aimed at revealing the unexpected. Their labs may appear to be in disorder, but they know exactly where every tube or bottle belongs. Scientists also identify intimately with the organisms or creatures they study (it is said that Louis Pasteur actually dreamed about microorganisms), and this identification brings on an intuition—a "feeling for the organism." In addition, there is the ability to recognize patterns that might bring a breakthrough. (Pasteur had studied art as a teenager and, therefore, he had an appreciation of patterns.)

The geneticist and Nobel laureate Barbara McClintock once remarked, *"I was just so interested in what I was doing I could hardly wait to get up in the morning and get at it. One of my friends, a geneticist, said I was a child, because only children can't wait to get up in the morning to get at what they want to do."* Clearly, another characteristic of a scientist is having a child-like curiosity for the unknown.

Another Nobel laureate and immunologist, Peter Medawar, once said *"Scientists are people of very dissimilar temperaments doing different things in very different ways. Among scientists are collectors, classifiers, and compulsive tidiers-up; many are detectives by temperament and many are explorers; some are artists and others artisans. There are poet-scientists and philosopher-scientists and even a few mystics."* In other words, scientists come from all walks of life.

For this author, I too have found science to be an extraordinary opportunity to discover and understand something never before known. Science is fun, yet challenging—and at times arduous, tedious, and frustrating. As with most of us, we will not make the headlines for a breakthrough discovery or find a cure for a disease. However, as scientists we all hope our hard work and achievements will contribute to a better understanding of a biological (or microbiological) phenomenon and will push back the frontiers of knowledge and have a positive impact on society.

Like any profession, being a scientist is not for everyone. Besides having a bachelor's degree in biology or microbiology, you should be well read in the sciences and capable of working as part of an interdisciplinary team. Of course, you should have good quantitative and communication skills, have an inquisitive mind, and be goal oriented. If all this sounds interesting, then maybe you fit the mold of a scientist. Why not consider pursuing a career in microbiology? Some possibilities are described in other Microbiology Pathways included in this book, but you should also visit with your instructor. Simply stop by the student union, buy two cups of coffee, and you are on your way.

1

Microbiology: Then and Now

CHAPTER PREVIEW

1.1 The Discovery of Microbes Leads to Questioning Their Origins
Investigating the Microbial World 1: Can Life Arise Spontaneously?
MicroInquiry 1: Experimentation and Scientific Inquiry

1.2 Disease Transmission Can Be Interrupted

1.3 The Classical Golden Age of Microbiology Reveals the Germ

1.4 With the Discovery of Other Microbes, the Microbial World Expands

1.5 A Second Golden Age of Microbiology Involves the Birth of Molecular Biology and Chemotherapy

1.6 The Third Golden Age of Microbiology Is Now

Microorganisms account for most of the biomass on the planet and are an essential foundation on which the global ecosystem rests. They play an absolutely essential role in the survival of the human race.
—Carl Woese (Professor of Microbiology, University of Illinois at Urbana-Champaign)

Space. The final frontier! Really? *The* final frontier? There are an estimated 350 billion large galaxies and more than 10^{23} stars in the visible universe. However, the invisible microbial universe consists of more than 10^{30} **microorganisms** (or **microbes** for short) scattered among an estimated 2 to 3 billion species. They may be microscopic in size but they are magnificent in their evolutionary diversity and astounding in their sheer numbers. Existing in such diversity and numbers in the oceans, the land masses, and the atmosphere means they must possess some amazing powers that contribute to the very survival of other organisms on planet Earth. So, could understanding these microscopic organisms on Earth be as important to us and all earthly creatures as studying stars and galaxies in space? Let's uncover a few examples of what a "day in the life of a microorganism" is like.

A Day in the Life of a Microorganism

The oceans and seas cover 70% of planet Earth and are swarming with microbes—some 3×10^{29}—and helping regulate life on Earth. Floating near the surface are photosynthetic groups that are part of the marine food chain on which all fish and ocean mammals depend. In addition, they provide up to 50% of the oxygen gas we breathe and other organisms use to stay alive (FIGURE 1.1A). Other diverse marine and freshwater microbes are the engines that drive nutrient and mineral recycling needed to provide the building blocks to sustain all life. One particular microbe, called *Pelagibacter ubique*, accounts for 20% (2.4×10^{28} cells) of marine microbes—and 50% of the microbes in the surface waters of temperate oceans in the summer. What is its daily routine?

Image courtesy of Dr. Fred Murphy/CDC.

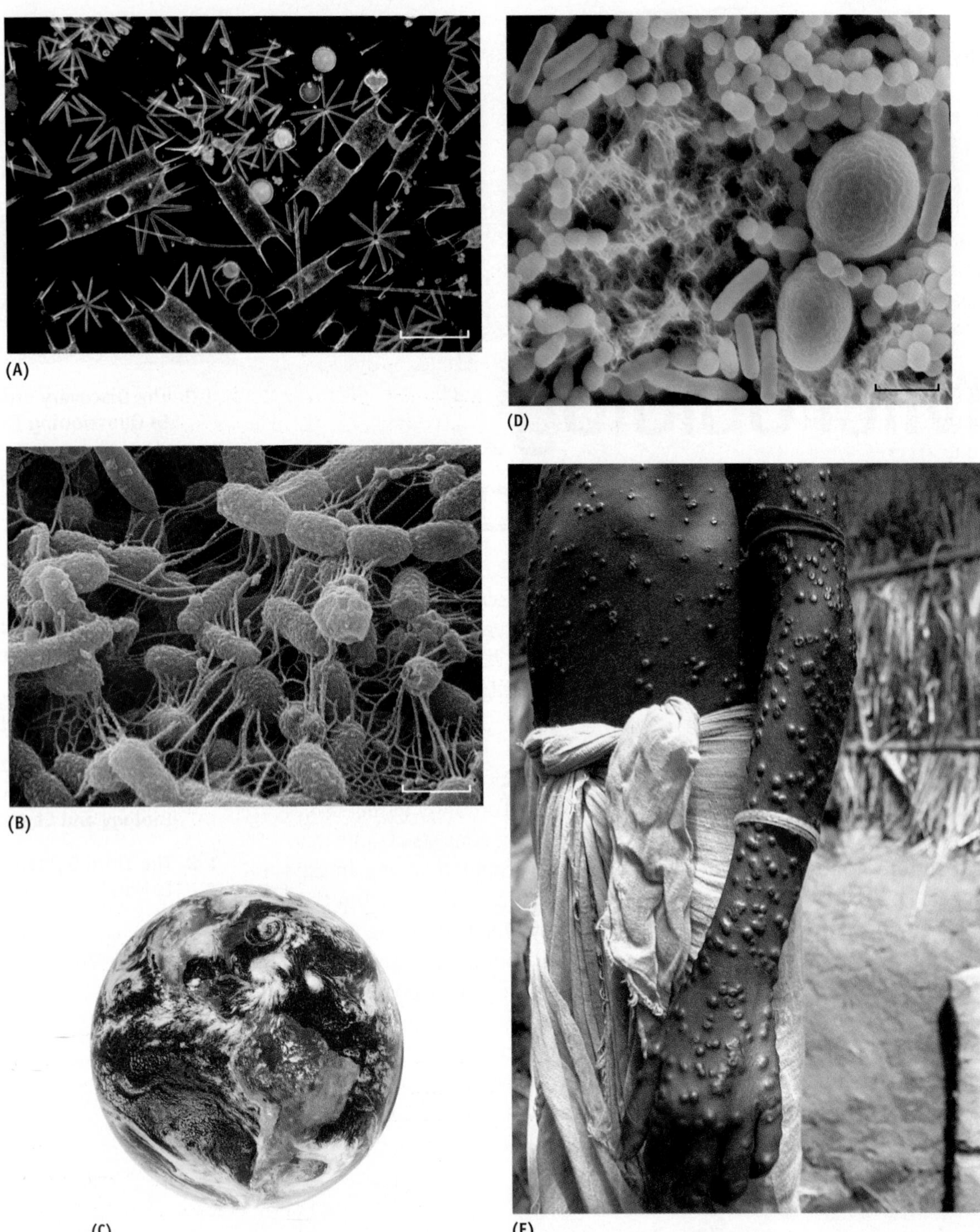

FIGURE 1.1 **Daily Life in the Microbial World.** Microbes play many roles. For example, (**A**) photosynthetic microbes inhabit the upper sunlit layer of almost all oceans and bodies of fresh water where they produce food molecules that sustain the aquatic food web and generate oxygen gas. (Bar = 5 µm.) © D.P. Wilson/FLPA/Photo Researchers, Inc. (**B**) In the soil, microbes degrade dead plants and animals, form beneficial partnerships with plants, and recycle carbon, nitrogen, and sulfur. (Bar = 5 µm.) © David Scharf/Photo Researchers, Inc. (**C**) Besides their involvement in the formation of rain drops and snowflakes, microbes in the atmosphere are important for water vapor to condense into clouds that help cool the Earth. © Loskutnikov/Shutterstock, Inc. (**D**) Large numbers of microbes can be found on and in the body where most play beneficial roles for our health. (Bar = 5 µm.) © Visuals Unlimited/Corbis. (**E**) A few microbes play disease roles and have affected world health. This 1974 photo of a Bengali boy shows the effects of smallpox, which was responsible for 300–500 million deaths during the 20th century. Smallpox has since been eliminated globally through vaccination during the Smallpox Eradication Campaign (1966–1979). Courtesy of Jean Roy/CDC. »» What would happen to life on Earth if each of the examples above (A–D) was devoid of microbes?

Some scientists, such as Steven Giovannoni who discovered the organism, believe it is responsible for up to 10% of all nutrient recycling on the planet, affecting the cycling of carbon and even influencing climate change.

What we do know about the daily lives of marine microbes today represents only a small fraction of what the marine microbial workforce actually does on a daily basis. Craig Venter and his team of scientists have discovered more than 148 new microbial species in just the waters of the Sargasso Sea and from them have isolated 1.2 million new gene sequences. What are the daily occupations of these and other mysterious microbes that are still being identified and catalogued? Undoubtedly, many play useful and probably intimate roles with important global consequences we have yet to discover.

Their partners on dry land are no less impressive in their daily activities and can be found in every imaginable place, from the tops of the highest mountains to the deepest caves. In fact, a single spoonful of garden soil may have up to 10^9 microbes (FIGURE 1.1B). This diverse soil workforce is responsible for such daily activities as: recycling carbon and other nutrients, decomposing animal and plant matter, purifying water, detoxifying harmful substances, recycling wastes, returning carbon dioxide to the atmosphere, providing nitrogen in a useful form for plants, and providing us with a source of antibiotics to fight their disease-causing brethren.

Not to be forgotten are the microbes found in the atmosphere (FIGURE 1.1C). Every cubic meter of air holds up to 10^8 microorganisms. Some of these are commuters, traveling on the wind from one location to another. But most have important daily functions to perform. They are integral to the formation of water vapor (clouds) and help form raindrops and snowflakes. As cloud dwellers, they can affect the chemical composition of the atmosphere, influence weather cycles, alter the composition of rain and snow, and ultimately be a factor in the atmospheric processes we encounter every day.

So you say, "Okay, they are diverse and are found in typical habitats on Earth. I am impressed." But realize these are only the "usual habitats" that are most familiar to us. In fact, a diverse microbial workforce is being catalogued anywhere there is an energy source. Drill down more than 1,000 meters into the Earth's crust on land or below the ocean floor and you will find microbes that are sealed off from the usual habitats. These "intraterrestrials" (microbes living in sediment and rock) are another diverse workforce that, even at these deep depths, are involved with the daily recycling of minerals and stabilizing the health of our planet.

In all, the global workforce of microbes is still being catalogued and their numbers and daily activities keep growing. Thus, as the opening quote also asserts, it is safe to say that the global community of microbes makes up more than half of Earth's **biomass**. Microbial evolution has influenced all life on Earth and has outpaced that of the more familiar plants and animals. We, all life, and our planet are dependent on the daily lives of microbes!

■ Biomass: **The total weight of living organisms within a defined environment.**

Some microbial members of the global workforce have been recruited by industry and have been purposely put into many of the foods we eat (e.g., yogurt, cheeses) or have employed their talents to produce numerous other foods (e.g., sauerkraut, bread, chocolate) or medicines (antibiotics). Much closer to home, some 10^{11} microorganisms inhabit our bodies; these seeming intimate strangers have established themselves since birth on our skin and in our digestive tract (FIGURE 1.1D). Fortunately, the majority of these endogenous microbes, collectively called the human **microbiome**, spend each day helping us resist disease, regulating our digestion, maintaining a strong immune system—and even influencing our risk of obesity, asthma, and allergies. To be human and healthy, we must share our daily lives with this homegrown microbial workforce.

Finally, when most of us hear the word microbe or "bacterium" or "virus," we think infection or disease. Although such disease-causing agents, called **pathogens**, are rare, some—causing diseases like plague, malaria, and smallpox (FIGURE 1.1E)—throughout history have swept through cities and villages, devastated populations, killed great leaders and commoners alike, and, as a result, have transformed politics, economies, and public health worldwide.

A major focus of this introductory chapter is to give you an introspective "first look" at microbiology—

then and now. We will see how microbes were first discovered and how those that cause infectious disease preoccupied the minds and efforts of so many. Along the way, we will see how curiosity and scientific inquiry stimulated the quest to understand the microbial world just as the science of microbiology does today. To begin our story, we reach back to the 1600s, where we encounter some very inquisitive individuals.

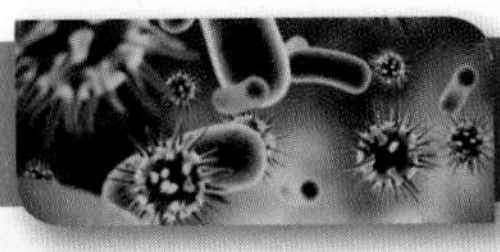

Chapter Challenge

How was this diverse global workforce of microorganisms and pathogens revealed, and what challenges do such organisms pose for microbiology today? Let's investigate!

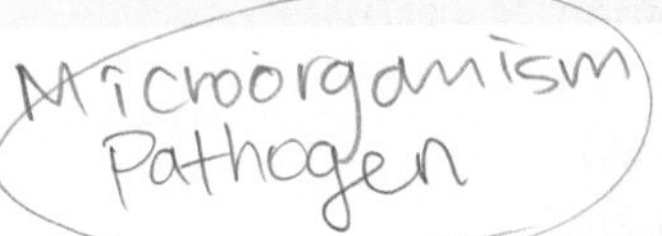

KEY CONCEPT 1.1 The Discovery of Microbes Leads to Questioning Their Origins

As the 17th century arrived, an observational revolution was about to begin: Dutch spectacle maker Zacharias Janssen was one of several individuals who discovered that if two convex lenses were put together, small objects could be magnified. Many individuals in Holland, England, and Italy further developed this combination of lenses that in 1625 would go by the term *microscopio* or "microscope." This new invention would be the forerunner of the modern-day microscope.

■ **Convex:** Referring to a surface that curves outward.

Microscopy—Discovery of the Very Small

Robert Hooke, an English natural philosopher (the term "scientist" was not coined until 1833), was one of the most inventive and ingenious minds in the history of science. As the Curator of Experiments for the Royal Society of London, Hooke took advantage of the magnification abilities of the early compound microscope and made detailed studies of many living objects. One of the most important observations is contained in his *Micrographia,* published in 1665, where he describes and draws the structure of cork. Seeing a "great many little boxes," he called these spaces *cella* (= rooms) and from that observation today we have the word "cell."

Micrographia represents one of the most important books in science history because it awakened the learned and general population of Europe to the world of the very small, revolutionized the art of scientific investigation, showed that the microscope was an important tool for unlocking the secrets of nature, and, notably, opened the door to a completely new world: the world of the cell.

At this same time, across the North Sea in Delft, Holland, Antony van Leeuwenhoek, a successful tradesman and dry goods dealer, was using hand lenses to inspect the quality of his cloth. As such, and without any scientific training, Leeuwenhoek became skilled at grinding single pieces of glass into fine magnifying lenses. Placing such a lens between two metal plates riveted together, Leeuwenhoek's "simple microscope" could greatly out magnify Hooke's microscope (FIGURE 1.2A, B).

Beginning in 1673 and lasting until his death in 1723, Leeuwenhoek communicated his microscope observations through letters to England's Royal Society. In 1674, one letter described a sample of cloudy surface water from a marshy lake. Placing the sample before his lens, he described hundreds of what he thought were tiny, living animals, which he called **animalcules**. His curiosity aroused, Leeuwenhoek soon located even smaller animalcules in such materials as rainwater, scrapings from his teeth, and even his own feces. In fact, among the 165 letters sent to the Royal Society, he outlined structural details of yeast cells, described thread-like fungi and microscopic algae and protozoa, and importantly was the first to describe

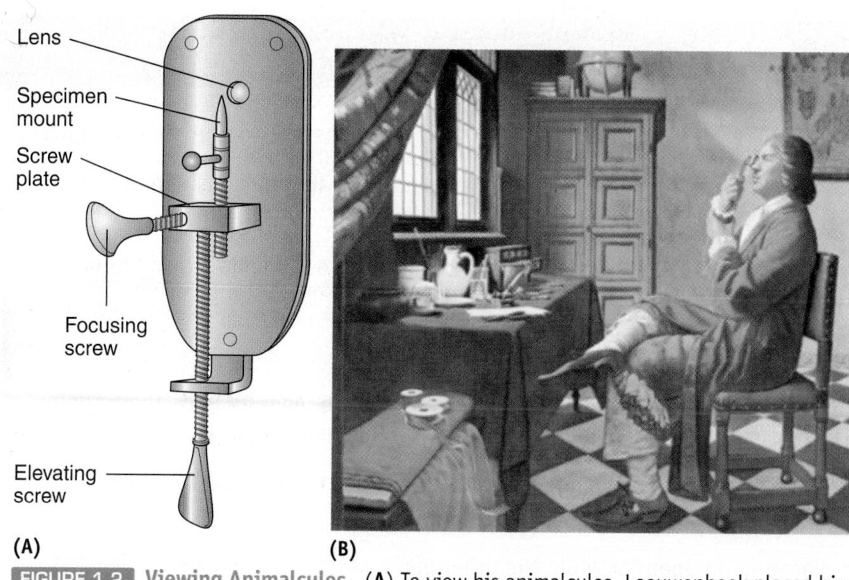

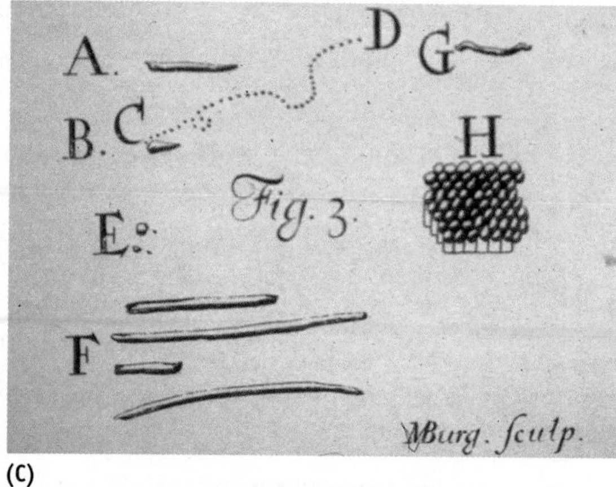

FIGURE 1.2 **Viewing Animalcules.** (**A**) To view his animalcules, Leeuwenhoek placed his sample on the tip of the specimen mount that was attached to a screw plate. An elevating screw moved the specimen up and down while the focusing screw pushed against the metal plate, moving the specimen toward and away from the lens. (**B**) Holding the microscope up to the bright light, Leeuwenhoek then looked through the lens to view his subject. Collection of the University of Michigan Health System, Gift of Pfizer, Inc. (UMHS.15). (**C**) From such observations, Leeuwenhoek drew the animalcules (bacteria in this drawing) he saw. © Royal Society, London. »» Why would Leeuwenhoek believe his living (and often moving) creatures were tiny animals?

and illustrate what we know today as the smallest living microbes, the bacteria (FIGURE 1.2C).

The process of "observation" is an important skill for all scientists, including microbiologists, and Hooke and Leeuwenhoek are excellent models of individuals with sound observation skills—a requirement that remains a cornerstone of all science inquiry today. Unfortunately, Leeuwenhoek invited no one to work with him, nor did he show anyone how he ground his lenses. Thus, naturalists at the time found it difficult to repeat and verify his observations, which also are key components of scientific inquiry. Still, Leeuwenhoek's observations on the presence and diversity of his animalcules opened yet a second door to another entirely new world: the world of the microbe.

Do Animalcules Arise Spontaneously?

In the early 1600s, most naturalists were "vitalists," individuals who thought life depended on a mysterious and pervasive "vital force" in the air. This force provided the basis for the doctrine of **spontaneous generation**, which suggested that some forms of life could arise from nonliving, decaying matter. Others also embraced the idea, for they too witnessed toads that appeared from mud, snakes coming from the marrow of a decaying human spine, and rats arising from garbage wrapped in rags.

Resolving the reality of such bizarre beliefs would require a new form of investigation—"experimentation"—and a new generation of experimental naturalists arose.

Among the first was the Italian naturalist Francesco Redi who, in 1668, performed one of history's first biological experiments, which was designed to test the belief that worm-like maggots (fly larvae) could arise from rotting meat (Investigating the Microbial World 1).

Although Redi's experiments verified that spontaneous generation could not produce larger living creatures, what about the mysterious and minute animalcules that appeared to straddle the boundary between the nonliving and living world? Could they arise spontaneously?

In 1745, a British clergyman and naturalist, John Needham, proposed that the spontaneous generation of animalcules, first seen by Leeuwenhoek some 70 years earlier, resulted from a vital force that reorganized decaying matter. To test this idea, Needham heated several flasks of animal broth and sealed the flasks with corks after they cooled. After several days, Needham proclaimed that the *"gravy swarm'd with life, with microscopical animals of most dimensions."* He was convinced that putrefaction could generate the vital force needed for spontaneous generation.

■ **Broth:** A watery solution containing nutrients that support the growth of microbes (if present).

Investigating the Microbial World 1

Can Life Arise Spontaneously?

For centuries, many people, learned and not, believed that some forms of life could arise spontaneously from non-living, decaying matter. For example:

■ **OBSERVATION:** In the 17th century many people believed that fly maggots (larvae) arose spontaneously from rotting meat. Francesco Redi set out to find the answer.

■ **QUESTION:** ***Do fly maggots arise from rotting meat?***

■ **HYPOTHESIS:** Redi proposed that fly maggots arise from hatched eggs laid in decaying meat by flies. If so, then preventing flies from laying eggs in the rotting meat should result in no maggots being generated.

■ **EXPERIMENTAL DESIGN:** Redi obtained similar pieces of rotting meat and jars in which the meat would be placed.

■ **EXPERIMENT:** One piece of meat was placed in an open jar while the other piece was placed in a similar jar that was then covered with a piece of gauze to keep out any flying insects, including flies. The meat in each jar was allowed to rot.

■ **RESULTS:**

See figure.

■ **CONCLUSIONS:**

QUESTION 1: ***Was Redi's hypothesis validated? Explain using the figure.***

QUESTION 2: ***What is the control in this experiment and why was it important to have a control?***

QUESTION 3: ***Why was it important that Redi used gauze to cover the one jar? That is, why not seal it completely?*** **Hint: Remember what people believed about the 'vital force.'**

Answers can be found on the Student Companion Website in **Appendix D.**

Adapted from: Redi, F. (1688) as reprinted by Open Court Publishing Company, Chicago (1909). Icon image © Tischenko Irina/ShutterStock, Inc.

Experiments often can be subject to varying interpretations. As such, the Italian cleric and naturalist Lazzaro Spallanzani challenged Needham's conclusions and suggested that the animalcules came from the air and would therefore grow in the broth of the cooled flasks. So, in 1765, he repeated Needham's experiments but with the few changes. He left some flasks with broth open to the air, others were stoppered loosely with corks, and the remaining flasks were sealed. All were then boiled. After 2 days, the open flasks were swarming with animalcules, but the loosely stoppered ones had many fewer—and the sealed ones contained no animalcules. Spallanzani concluded that *"the number of animalcula developed is proportional to the communication with the external air."*

Needham and others countered that Spallanzani's experiments had destroyed the vital force because sealing the flasks prevented entry of this force necessary for the spontaneous generation of animalcules.

The controversy over spontaneous generation of animalcules continued into the mid-1800s and only deepened when Rudolf Virchow, a German pathologist, put forward, without direct evidence, the idea of **biogenesis**, which said that life only arises from life. To solve the debate, a new experimental strategy would be needed.

Louis Pasteur, a French chemist and scientist, took up the challenge in 1861 and, through an elegant series of experiments that were a variation of the methods of Needham and Spallanzani, discredited the idea. MicroInquiry 1 outlines the process of scientific inquiry and Pasteur's experiments.

Although Pasteur's experiments generated considerable debate for several years, his exacting and carefully designed experiments marked the end of the belief in spontaneous generation and validated the idea of biogenesis.

However, today there is another form of "spontaneous generation"—this time occurring in the laboratory (MicroFocus 1.1).

CONCEPT AND REASONING CHECKS 1

a. If you were alive in Leeuwenhoek's time, how would you explain the origin for the animalcules he saw with his simple microscope?

b. Evaluate the role of experimentation as an important skill to the eventual rejection of spontaneous generation as an origin for animalcules.

KEY CONCEPT 1.2 Disease Transmission Can Be Interrupted

In the 13th century, people knew diseases could be transmitted between individuals, so quarantines were used to combat disease spread.

■ Quarantine: Enforced isolation of people or animals with a highly communicable disease.

By the mid-1700s, the prevalent belief among naturalists and laypersons alike was that disease resulted from an altered chemical quality of the atmosphere or from tiny poisonous particles of decomposed matter in the air, an entity called **miasma** (the word malaria comes from *mala aria,* meaning "bad air"). To protect oneself from the black plague, for example, plague doctors in Europe often wore an elaborate costume they thought would protect them from the plague miasma (FIGURE 1.3).

Vaccination Prevents Infectious Disease

In the 1700s, smallpox was prevalent throughout Europe. In England, smallpox epidemics were so severe that one third of the children died before the age of three and many victims who recovered often were blinded and left pockmarked.

However, since the 14th century, the Chinese had practiced **variolation**, which involved blowing a ground smallpox powder into the individual's nose. By the 18th century, Europeans were inoculating dried smallpox scabs under the skin of the arm. Although some individuals did get smallpox, most contracted only a mild form of the disease and, upon recovery, were resistant to future smallpox infections.

As an English country surgeon, Edward Jenner learned that milkmaids who occasionally contracted cowpox would subsequently be protected from deadly smallpox. Jenner hypothesized that intentionally giving cowpox to people should protect them against smallpox. So, in 1796, he took a cowpox lesion from a milkmaid's hand and scratched it into the skin of a young boy's arm. The boy soon developed a slight fever, but recovered. Six weeks later Jenner infected the boy with smallpox pus. Within days, the boy developed a reaction at the skin site but failed to show any sign of smallpox.

■ Cowpox: A localized skin infection coming from contact with an infected animal (e.g., cow).

In 1798, Jenner repeated his experiments with others, verifying his therapeutic technique

MICROINQUIRY 1

Experimentation and Scientific Inquiry

Science certainly is a body of knowledge as you can see from the thickness of this textbook! However, science also is a process—a way of learning. Often we accept and integrate into our understanding new information because it appears consistent with what we believe is true. But, are we confident our beliefs are always in line with what is actually true? To test or challenge current beliefs, scientists must present logical arguments supported by well-designed and carefully executed experiments.

The Components of Scientific Inquiry

There are many ways of finding out the answer to a problem. In science, **scientific inquiry**—or what has been called the "scientific method"—is the way problems are investigated. Let's understand how scientific inquiry works by following the logic of the experiments Louis Pasteur published in 1861 to refute the idea of spontaneous generation.

When studying a problem, the inquiry process usually begins with **observations.** For spontaneous generation, Pasteur's earlier observations suggested that organisms do not appear from nonliving matter (see text discussion of the early observations supporting spontaneous generation).

Next comes the **question**, which can be asked in many ways but usually as a "what," "why," or "how" question. For example, "What accounts for the generation of microorganisms in the animal broth?"

From the question, various hypotheses are proposed that might answer the question. A **hypothesis** is a provisional but testable explanation for an observed phenomenon. In almost any scientific question, several hypotheses can be proposed to account for the same observation. However, previous work or observations usually bias which hypothesis looks most promising, and scientists then put their "pet hypothesis" to the test first.

Pasteur's previous work suggested that the purported examples of life arising spontaneously in the broths were simply cases of airborne microorganisms in dust landing on a suitable substance and then multiplying in such profusion that they could be seen as a cloudy liquid.

Pasteur's Experiments

Pasteur set up a series of **experiments** to test the hypothesis that "Life only arises from other life" (see facing page).

Experiment 1A and **1B**: Pasteur sterilized animal broths in glass flasks by heating. He then either left the neck open to the air (**A**) or sealed the glass neck (**B**). Organisms only appeared (turned the broth cloudy) in the open flask.

Experiment 2A and **2B**: Pasteur sterilized a meat broth in swan-necked flasks (**A**), so named because their S-shaped necks resembled a swan's neck. No organisms appeared, even after many days. However, if the neck was snapped off or the broth tipped to come in contact with the neck (**B**), organisms (cloudy broth) soon appeared.

Analysis of Pasteur's Experiments

Let's analyze the experiments. Pasteur had a preconceived notion of the truth and designed experiments to test his hypothesis. In his experiments, only one **variable** (an adjustable condition) changed. In experiment 1, the flask was open or sealed; in experiment 2, the neck was left intact or exposed to the unsterile air. Pasteur kept all other factors the same; that is, the broth was the same in each experiment; it was heated the same length of time; and similar flasks were used. Thus, the experiments had rigorous **controls** (the comparative condition). For example, in experiment 1, the control was the flask left open. Such controls are pivotal when explaining an experimental result. Pasteur's finding that no life appeared in the sealed flask (experiment 2A) is interesting, but tells us very little by itself. Its significance comes by comparing this to the broken neck (or tipped flask) where life quickly appeared.

Also note that the idea of spontaneous generation could not be dismissed by just one experiment (see "His critics" on facing page). Pasteur's experiments required the accumulation of many experiments, all of which supported his hypothesis.

Hypothesis and Theory

When does a hypothesis become a theory? The answer is that there is no set time or amount of evidence that specifies the change from hypothesis to theory. A **theory** is defined as a hypothesis that has been tested and shown to be correct every time by many separate investigators. So, at some point, sufficient evidence exists to say a hypothesis is now a theory. However, theories are not written in stone. They are open to further experimentation and so can be refuted.

As a side note, today a theory often is used incorrectly in everyday speech and in the news media. In these cases, a theory is equated incorrectly with a hunch or belief—whether or not there is evidence to support it. In science, a theory is a general set of principles supported by large amounts of experimental evidence.

Discussion Point

Based on Pasteur's experiments, could one still argue that spontaneous generation could occur? Explain. Also see end of chapter question 45.

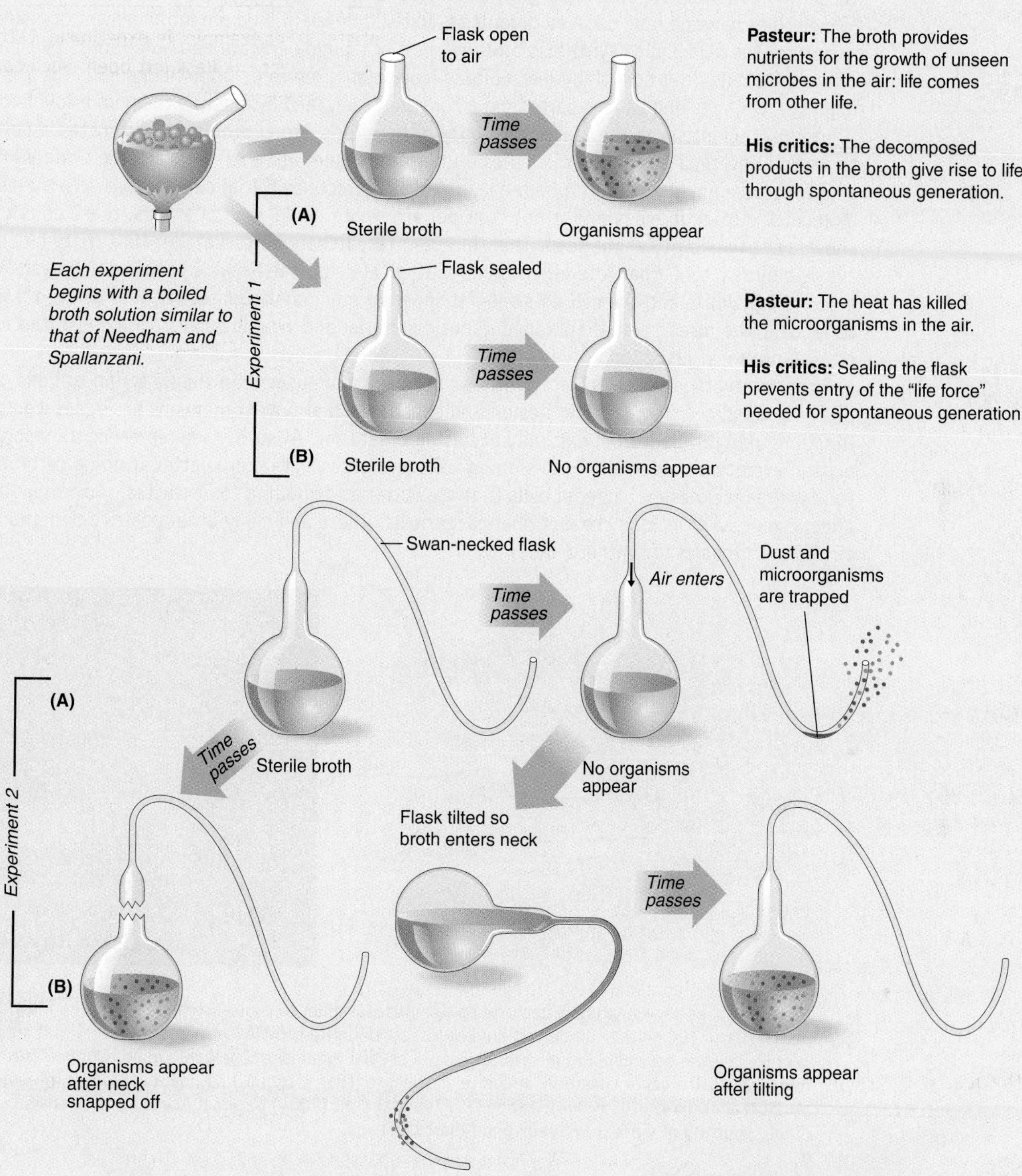

Pasteur and the Spontaneous Generation Controversy. (**1A**) When a flask of sterilized broth is left open to the air, organisms appear. (**1B**) When a flask of broth is boiled and sealed, no organisms appear. (**2A**) If broth sterilized in a swan-necked flask is left open to the air, the curvature of the neck traps dust particles and microorganisms, preventing them from reaching the broth. (**2B**) If the neck is snapped off to allow in air or the flask is tipped so broth enters the neck, organisms come in contact with the broth and grow.

Pasteur: No life will appear in the flask because microorganisms cannot reach the broth.

His critics: If the "life force" has free access to the flask, life will appear, given enough time.

Many days later the intact flask is still free of any life. Pasteur has refuted the doctrine of spontaneous generation.

MICROFOCUS 1.1: Biotechnology

Generating Life—Today

Those who believed in spontaneous generation proposed that animalcules arose from the rearrangement of molecules released from decayed organisms. Today, a different kind of rearrangement of molecules is occurring. The field, called **synthetic biology**, aims to rebuild or create new "life forms" (such as viruses or bacterial cells) from scratch by recombining molecules taken from different species. It is like fashioning a new car by taking various parts from a Ford and Chevy, and assembling them on a Toyota chassis.

In 2002, scientists at the State University of New York, Stony Brook, reconstructed a poliovirus by assembling separate poliovirus genes and proteins (see figure A). A year later, Craig Venter and his group assembled a bacteriophage—a virus that infects bacterial cells—from "off-the-shelf" biomolecules. Although many might not consider viruses to be "living" microbes, these constructions showed the feasibility of the idea. Then in 2004, researchers at Rockefeller University created small "vesicle bioreactors" that resembled crude biological cells (see figure B). The vesicle walls were made of egg white and the cell contents, stripped of any genetic material, were derived from a bacterial cell. The researchers then added genetic material and viral enzymes, which resulted in the cell making proteins, just as in a live cell.

Importantly, these steps toward synthetic life have more uses than simply trying to build something like a bacterial cell from scratch. Design and construction of novel organisms or viruses would have functions very different from naturally occurring organisms. As such, they represent the opportunity to expand evolution's repertoire by designing cells or organisms that are better at doing certain jobs. Can we, for example, design bacterial cells that are better at degrading toxic wastes, providing alternative energy sources, or making cheaper pharmaceuticals? These and many other positive benefits are envisioned as outcomes of synthetic biology.

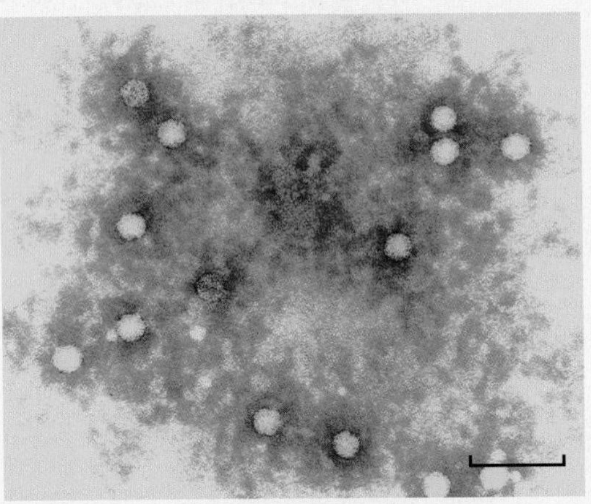

(A)

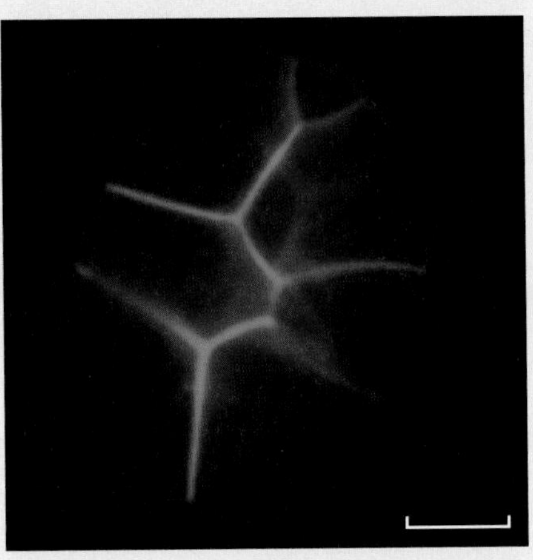

(B)

(**A**) This image shows naturally occurring polioviruses, similar to those assembled from the individual parts. (Bar = 100 nm.) © Dr. Dennis Kunkel/Visuals Unlimited. (**B**) A "vesicle bioreactor" that simulates a crude cell was assembled from various parts of several organisms. The green glow is from a protein produced by the genetic material added to the vesicle. (Bar = 10 µm.) Reproduced from V. Noireaux and A. Libchaber, *PNAS*, 101 (2004), 17669–17674; Copyright (2004) National Academy of Science, U.S.A. Photo courtesy of Vincent Noireaux and Albert Libchabe.

of **vaccination** (*vacca* = "cow"). Prominent physicians soon confirmed his findings, and within a few years, Jenner's method of vaccination spread through Europe and abroad. President Thomas Jefferson wrote to Jenner, *"You have erased from the calendar of human afflictions one of its greatest. Yours is the comfortable reflection that mankind can never forget that you have lived."* However, Jefferson was some 284 years premature in his pronouncement. It would not be until 1980 that the World Health Organization (WHO) would certify that smallpox had been eradicated globally through a massive vaccination effort carried out between 1966 and 1979.

In retrospect, it is remarkable that without any knowledge of microbes or disease causation, Jenner accomplished what he did. Again, hallmarks of a scientist—keen observational skills and insight—led to a therapeutic intervention against disease.

Disease Transmission Does Not Result from a Miasma

As the 19th century unfolded, more scientists relied on keen observations and experimentation as a way of understanding and explaining disease transmission.

Epidemiology, as applied to infectious diseases, is another example of scientific inquiry—in this case to identify the source, cause, and mode of transmission of disease. The first such epidemiological studies, carried out by Ignaz Semmelweis and John Snow, were instrumental in suggesting how diseases were transmitted—and how simple measures could interrupt transmission.

Ignaz Semmelweis was a Hungarian obstetrician who, after accepting a position in a large Vienna maternity hospital, was shocked by the numbers of women who were dying of puerperal fever (a type of blood poisoning also called childbed fever) following labor. His investigations revealed that the disease was 20 times more prevalent and deadly in the ward handled by medical students than in the adjacent ward run by midwifery students and in a similar maternity hospital in Dublin, Ireland (FIGURE 1.4). The comparative studies suggested to Semmelweis that disease transmission must involve his medical students and that the source

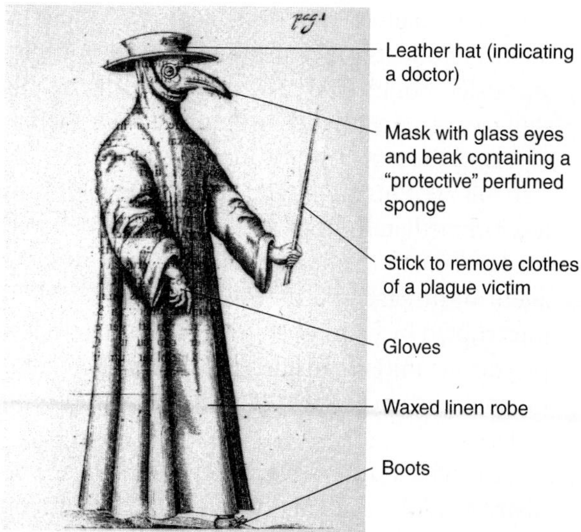

FIGURE 1.3 **Dressed for Protection.** This dress was thought to protect a plague doctor from the air (miasma) that caused the plague. © National Library of Medicine. »» How would each item of dress offer protection?

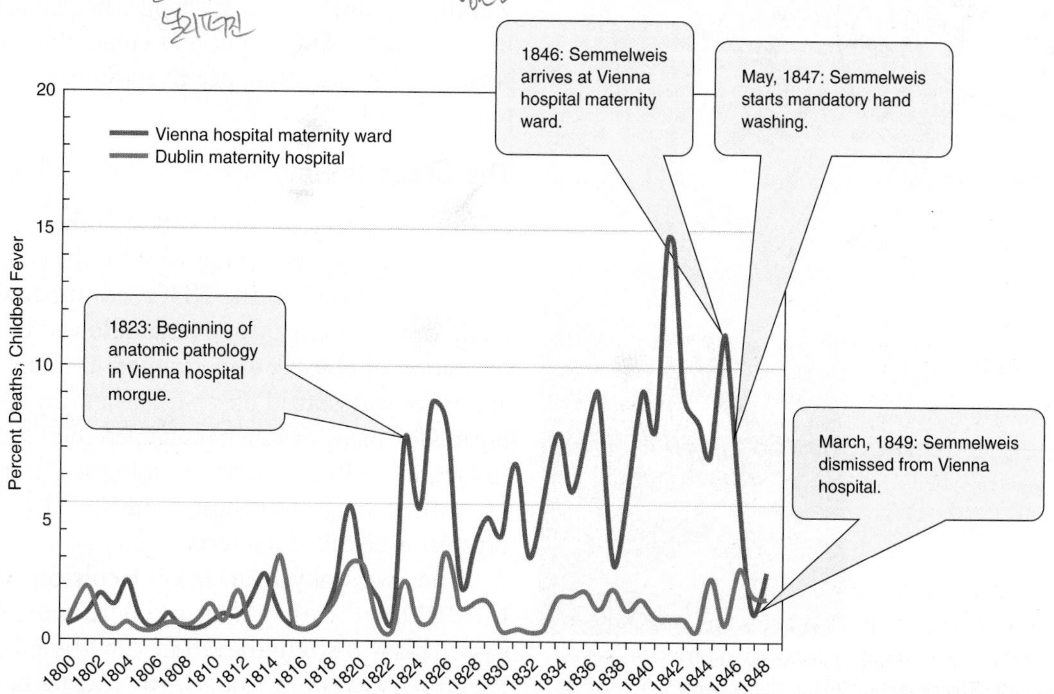

FIGURE 1.4 **Yearly Mortality for Childbed Fever 1800–1849.** Semmelweis collected data on deaths of birth-giving mothers in the Vienna maternity hospital (red line) and compared the deaths to those in a large Dublin maternity hospital (blue line) where doctors and students focused on obstetrics and did not concern themselves with anatomic pathology. Data from Semmelweis (1861). »» Why do Semmelweis' observations on the prevalence of childbed fever not support a miasma as the cause of and transmission for the disease?

of contagion must be from cadavers on which the medical students previously had been performing autopsies. So, in 1847, Semmelweis directed his staff to wash their hands in chlorine water before entering the maternity ward. Begrudgingly, his staff followed his orders and deaths from childbed fever immediately dropped. Semmelweis believed "cadaver matter" on his doctors' hands was the agent of disease and its transmission could be interrupted by hand washing. Unfortunately, few physicians initially heeded Semmelweis' recommendations and he was relieved of his position in 1849.

In 1854, a cholera epidemic hit London's Soho district. With residents dying, English surgeon John Snow set out to discover the source and reason for cholera's spread. He carried out one of the first thorough epidemiological studies by interviewing sick and healthy Londoners and plotting the location of each cholera case on a district map (FIGURE 1.5). The results indicated most cholera cases were linked to a sewage-contaminated street pump from which many local residents obtained their drinking water.

Hypothesizing that the pump was the source of the cholera, Snow instituted the first known example of a public health measure to interrupt disease transmission—he requested the parish Board of Guardians to remove the street pump handle! Cholera cases dropped and again disease spread was broken by a simple procedure.

Snow went on to propose that cholera was not inhaled as a miasma but rather was waterborne. In fact, he asserted that "organized particles" caused cholera—another hypothesis that proved to be correct even though the causative agent would not be identified for another 29 years.

It is important to realize that although the miasma premise was incorrect, the fact that disease was associated with bad air and filth led to new hygiene measures, such as cleaning streets, laying new sewer lines in cities, and improving working conditions. These changes helped usher in the Sanitary Movement and create the infrastructure for the public health systems we have today (MicroFocus 1.2).

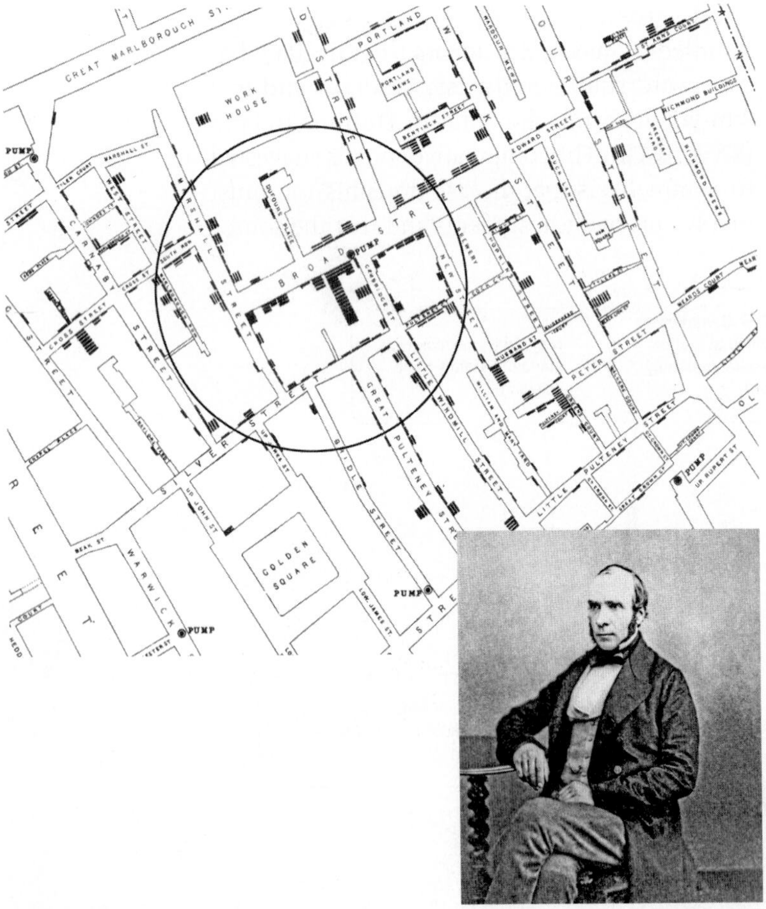

FIGURE 1.5 **Blocking Disease Transmission.** John Snow (inset) produced a map plotting all the cholera cases (black rectangles) in the London Soho district and observed a cluster near the Broad Street pump (circle). Courtesy of Frerichs, R. R. John Snow website: http://www.ph.ucla.edu/epi/snow.html, 2006. Inset © National Library of Medicine. »» Why would removing the pump handle stop the spread of cholera?

The Stage Is Set

During the early years of the 1800s, other events occurred that helped set the stage for the coming "germ revolution." In the 1830s, advances were made in microscope optics that allowed better resolution of objects. This resulted in improved and more widespread observations of tiny living organisms, many of which resembled short sticks. In fact, in 1838 the German biologist Christian Ehrenberg suggested these "rod-like" looking organisms be called **bacteria** (*bakter* = "rod").

The Swiss physician Jacob Henle reported in 1840 that living organisms could cause disease. This was strengthened in 1854 by Filippo Pacini's discovery of rod-shaped cholera bacteria in stool samples from cholera patients. Still, scientists debated whether bacterial organisms could cause disease because such living organ-

MICROFOCUS 1.2: Public Health

Epidemiology Today

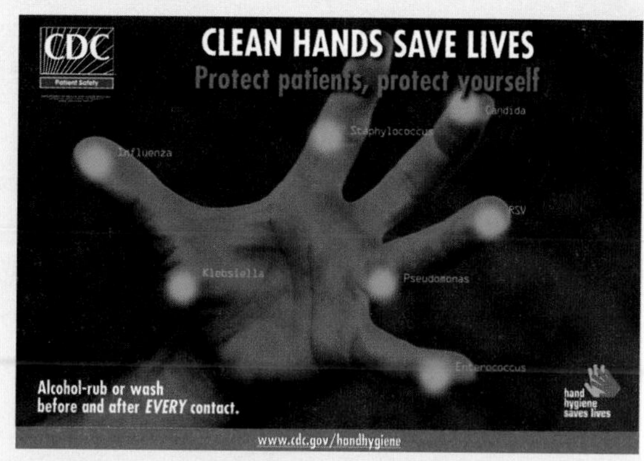

Courtesy of the Public Health Foundation/CDC.

In October 2010, the Haitian Ministry of Public Health and Population reported the first cholera epidemic in Haiti in over a century. This outbreak was quickly confirmed by the US Centers for Disease Control and Prevention (CDC). In the footsteps of Semmelweis and Snow, epidemiologists needed to quickly locate the source of the outbreak. A French and Haitian team identified sewage-contaminated river water downstream from a United Nations military camp (set up to help victims from the January earthquake) and found that soldiers from Nepal probably imported cholera accidentally. Unfortunately, once established, eradication of the cholera-causing microbe is difficult. Thus, by January 2012, more than 440,000 cases, over 234,000 hospitalizations, and some 7,000 deaths had been reported.

Today, even though we have a good grasp of disease transmission, sanitation, and public health, access to clean water remains as elusive in some parts of the developing world as it was in Snow's time. In addition, almost 160 years after Semmelweis' suggestions, a lack of hand washing by hospital staff, even in developed nations, remains a major mechanism for disease transmission (see figure). The simple process of washing one's hands still could reduce substantially disease transmission among the public and in hospitals.

Two of the most important epidemiological organizations today are the CDC in Atlanta, Georgia, and, on a global level, the World Health Organization (WHO) in Geneva, Switzerland. Both employ numerous epidemiologists, popularly called "disease detectives," who, like Snow (but with more expertise), systematically gather information about disease outbreaks in an effort to discover how the disease agent is introduced, how it is spread in a community or population, and how the spread can be stopped. With an ever-present danger of emerging disease, epidemiology remains a critical tool in the fight against infectious disease.

isms sometimes were found in healthy people. Therefore, how could these bacterial cells possibly cause disease?

To understand clearly the nature of infectious disease, a new concept of disease had to emerge. In doing so, it would be necessary to demonstrate that a specific bacterial organism was associated with a specific infectious disease. This would require some very insightful work, guided by Louis Pasteur in France and Robert Koch in Germany.

CONCEPT AND REASONING CHECKS 2

a. Evaluate the role of variolation and vaccination as ways to interrupt disease transmission.

b. Contrast the observations and studies of Semmelweis and Snow toward providing a better understanding of disease transmission.

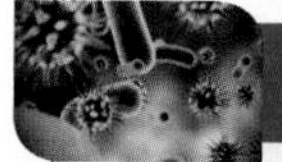

Chapter Challenge A

We are beginning to see how microbes and pathogens were revealed through the observations and studies of natural philosophers, a vaccine pioneer, and the fathers of epidemiology.

QUESTION A:

Identify the three diseases that were thought to be caused by a miasma. What types of studies and actions suggested they were not the product of a miasma?

Answers can be found on the Student Companion Website in **Appendix D.**

KEY CONCEPT 1.3 The Classical Golden Age of Microbiology Reveals the Germ

Beginning around 1854, the association of microbes in the disease process blossomed and continued until the advent of World War I. Over these 60 years, the foundations were laid for the maturing process that has led to the modern science of microbiology. We refer to this period as the first, or classical, Golden Age of microbiology.

Louis Pasteur Proposes That Germs Cause Infectious Disease

Trained as a chemist, Louis Pasteur was among the first scientists who believed that problems in science could be solved in the laboratory with the results having practical applications (FIGURE 1.6A).

Always one to tackle big problems, Pasteur soon set out to understand the chemical process of fermentation. The prevailing theory held that fermentation was strictly a chemical reaction with the air. However, Pasteur's microscope observations consistently revealed large numbers of tiny yeast cells in fermented juice that were overlooked by other scientists. When he mixed yeast in a sugar-water solution in the absence of air, the yeast grew and converted the sugar to alcohol. Yeast, therefore, must be one of the living "ferments" responsible for the fermentation process.

■ Fermentation: A splitting of sugar molecules into simpler products, including alcohol, acid, and gas (CO_2).

Pasteur also demonstrated that wines, beers, and vinegar each contained different and specific types of microorganisms. For example, in studying a local problem of wine souring, he observed that only soured wines contained populations of bacterial cells (FIGURE 1.6B). These cells must have contaminated a batch of yeast and produced the acids that caused the souring.

Pasteur recommended a practical solution for the "wine disease" problem: heat the wine gently to kill the bacterial cells but not to affect the quality of the wine. His controlled heating technique, known as **pasteurization**, soon was applied to other products, such that today pasteurization is used to kill pathogens and retard spoilage in milk and other beverages.

Pasteur's experiments demonstrated that yeast and bacterial cells are tiny, living factories in which important chemical changes take place. Therefore, if microorganisms represented agents of change, perhaps human infections could be caused by other microorganisms in the air—what he called **germs**. Thus, from his fermentation studies and his experiments refuting spontaneous generation, Pasteur formulated the **germ theory of disease**, which holds that some microorganisms are responsible for infectious disease.

(A)

(B)

FIGURE 1.6 **Louis Pasteur and Fermentation Bacteria.** (**A**) Louis Pasteur as a 46-year-old professor of chemistry at the University of Paris. © National Library of Medicine. (**B**) The following is part of a description of the living bacterial cells he observed. *"A most beautiful object: vibrios all in motion, advancing or undulating. They have grown considerably in bulk and length since the 11th; many of them are joined together in long sinuous chains . . ."* Pasteur concluded these bacterial cells can live without air or free oxygen; in fact, *"the presence of gaseous oxygen operates prejudicially against the movements and activity of those vibrios."* »» Why would such cells exist in a wine bottle that had soured?

Pasteur's Work Stimulates Disease Control and Reinforces Disease Causation

Pasteur had reasoned that if germs were acquired from the environment, their spread could be controlled and the chain of disease transmission broken.

Joseph Lister was Professor of Surgery at Glasgow Royal Infirmary in Scotland, where more than half his amputation patients died—not from the surgery—but rather from postoperative infections. Hearing of Pasteur's germ theory, Lister hypothesized that these surgical infections resulted from germs in the air. Knowing that carbolic acid had been effective on sewage control, in 1865 he used a carbolic acid spray in surgery and on surgical wounds (FIGURE 1.7). The result was spectacular—the wounds healed without infection. His technique would soon not only revolutionize medicine and the practice of surgery, but also lead to the practice of **antisepsis**, the use of chemical methods for disinfection of external living surfaces, such as the skin. So, germs can come from the environment and they can be controlled.

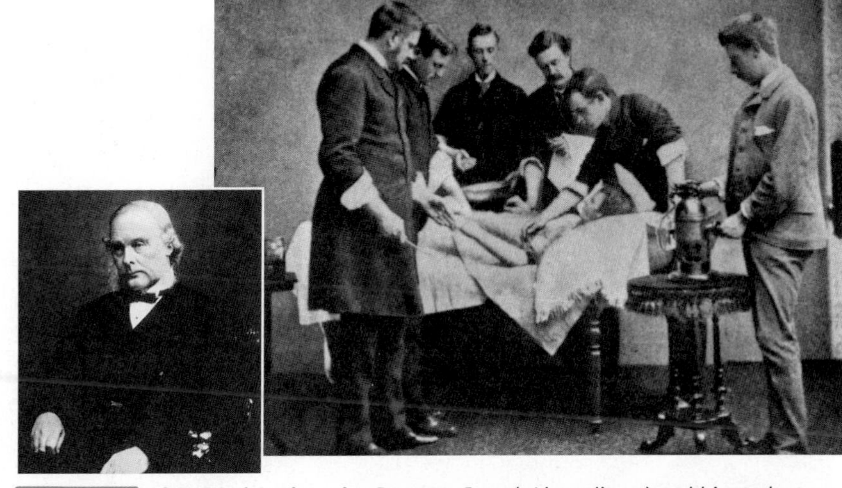

FIGURE 1.7 **Lister and Antisepsis.** By 1870, Joseph Lister (inset) and his students were using a carbolic acid spray in surgery and on surgical wounds to prevent postoperative infections. © Mary Evans Picture Library/Alamy Images. Inset © National Library of Medicine. »» Hypothesize how carbolic acid prevented surgical infections.

In an effort to familiarize himself with biological problems, Pasteur had the opportunity to study pébrine, a disease of silkworms. After several setbacks, he finally identified a new type of germ, unlike the bacterial cells and yeast he had observed with his microscope. These tiny globules, called "corpuscular parasites" were the infectious agent in silkworms and on the mulberry leaves fed to the worms. By separating the healthy silkworms from the diseased silkworms and their food, he managed to quell the spread of disease. The identification of the pathogen was crucial to supporting the germ theory and Pasteur would never again doubt the ability of germs to cause infectious disease. Now infectious disease would be his only interest.

In 1865, cholera engulfed Paris, killing 200 people a day. Pasteur tried to capture the responsible pathogen by filtering the hospital air and trapping the bacterial cells in cotton. Unfortunately, Pasteur could not grow or separate one bacterial species apart from the others because his broth cultures allowed the organisms to mix freely. Although Pasteur demonstrated that bacterial inoculations made animals ill, he could not pinpoint an exact cause.

To completely validate the germ theory, what was missing was the ability to isolate a specific germ from a diseased individual and demonstrate the isolated germ caused the same disease.

Robert Koch Formalizes Standards to Equate Germs with Infectious Disease

Robert Koch (FIGURE 1.8A) was a German country doctor who was well aware of anthrax, a deadly disease that periodically ravaged cattle and sheep, and could cause disease in humans.

In 1875, Koch injected mice with the blood from such diseased sheep and cattle. He then performed meticulous autopsies and noted the same symptoms in the mice that had appeared in the sheep and cattle. Next, he isolated from the blood a few rod-shaped bacterial cells and, with his microscope, watched for hours as the bacterial cells multiplied, formed tangled threads, and finally reverted to spores. He then took several spores on a sliver of wood and injected them into healthy mice. The symptoms of anthrax soon appeared and when Koch autopsied the animals, he found their blood swarming with the same bacterial cells. He reisolated the cells in fresh aqueous humor. The cycle was now complete. Here was the first evidence that a specific germ was the causative agent of anthrax.

Growing bacterial cells was not very convenient. Then, in 1880, Koch observed a slice of potato on which small masses of bacterial cells, which he termed **colonies**, were growing and multiplying. So, Koch tried adding gelatin to his broth to prepare a similar solid culture surface. He then inoculated bacterial cells on the surface and set the dish aside to incubate. Within 24 hours,

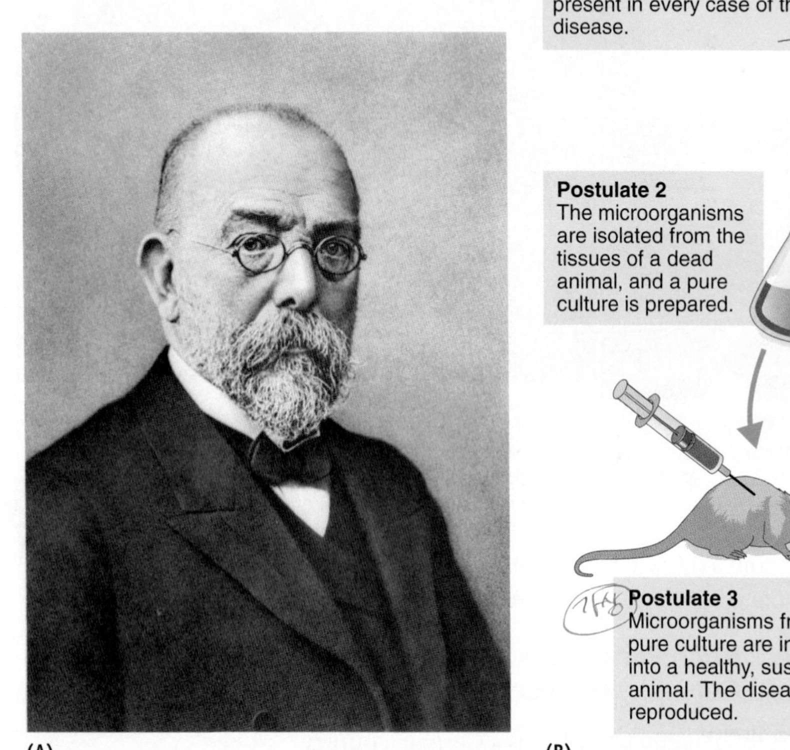

(A)

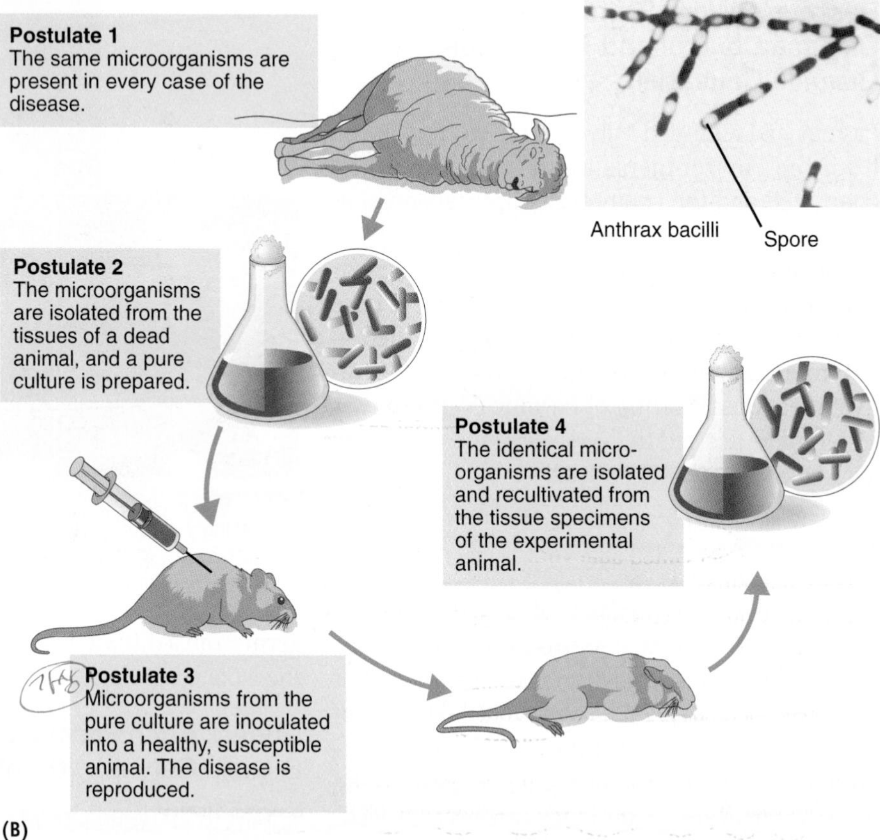

(B)

FIGURE 1.8 **A Demonstration of Koch's Postulates.** Robert Koch (**A**) developed what became known as Koch's postulates (**B**) that were used to relate a single microorganism to a single disease. The insert (in the upper right) is a photo of the rod-shaped anthrax bacterial cells. Many rods are swollen with spores (white ovals). (A) © National Library of Medicine. (Inset) Courtesy of CDC. »» What is the relationship between postulate 2 and postulate 4?

■ Agar: A complex polysaccharide derived from marine algae.

visible colonies would be growing on the surface, each colony representing a **pure culture** containing only one bacterial type. By 1884, agar replaced gelatin as the preferred solidifying agent (MicroFocus 1.3).

Applying the same procedure he had used for his anthrax studies, in 1884 Koch proclaimed he had found a rod-shaped bacterial cell as the cause of tuberculosis. When Koch presented his work, scientists were astonished. Here was the verification of the germ theory that had eluded Pasteur. Koch's procedures became known as **Koch's postulates** and were quickly adopted as the formalized standards for implicating a specific germ to a specific disease (FIGURE 1.8B). This work culminated in his being awarded the Nobel Prize in Physiology or Medicine in 1905.

Competition Fuels the Study of Infectious Disease

Research studies conducted in a laboratory were becoming the normal method of work. Whereas Koch's lab focused on the methods to isolate, grow, and identify specific pathogens of disease, such as those responsible for anthrax, tuberculosis, and cholera, Pasteur's lab was more concerned with preventing disease through vaccination.

In 1881, Pasteur and his coworker Charles Chamberland developed a vaccine for chicken cholera. When the weakened bacterial cells were inoculated into chickens and later followed by a dose of lethal pathogen, the animals did not develop cholera. Pasteur also developed a vaccine for anthrax and, in a public demonstration, found he could protect sheep against this disease as well (FIGURE 1.9).

In 1885 Pasteur and his coworker Émile Roux tested a rabies vaccine with success in dogs—all immunized animals survived a rabies exposure. Then, a 9-year-old boy, who had been bitten and mauled by a rabid dog was brought to Pasteur. A doctor gave the boy the untested (in humans) rabies vaccine (MicroFocus 1.4). The treatment lasted 10 days and the boy recovered and remained

MICROFOCUS 1.3: History

Jams, Jellies, and Microorganisms

One of the major developments in microbiology was Robert Koch's use of a solid culture surface on which bacterial colonies would grow. He accomplished this by solidifying beef broth with gelatin. When inoculated onto the surface of the nutritious medium, bacterial cells grew vigorously at room temperature and produced discrete, visible colonies.

On occasion, however, Koch was dismayed to find that the gelatin turned to liquid. It appeared that certain bacterial species were producing a chemical substance to digest the gelatin. Moreover, gelatin liquefied at the warm incubator temperatures commonly used to cultivate certain bacterial species.

Walther Hesse, an associate of Koch's, mentioned the problem to his wife and laboratory assistant, Fanny Hesse. She had a possible solution. For years, she had been using a seaweed-derived powder called agar (pronounced ah'gar) to solidify her jams and jellies. Agar was valuable because it mixed easily with most liquids and once gelled, it did not liquefy, even at the warm incubator temperatures.

Fanny Hesse. © National Library of Medicine.

In 1880, Hesse was sufficiently impressed to recommend agar to Koch. Soon Koch was using it routinely to grow bacterial species, and in 1884 he first mentioned agar in his paper on the isolation of the bacterial organism responsible for tuberculosis. It is noteworthy that Fanny Hesse may have been among the first Americans (she was originally from New Jersey) to make a significant contribution to microbiology.

Another key development, the common petri dish (plate), also was invented about this time (1887) by Julius Petri, one of Koch's former assistants.

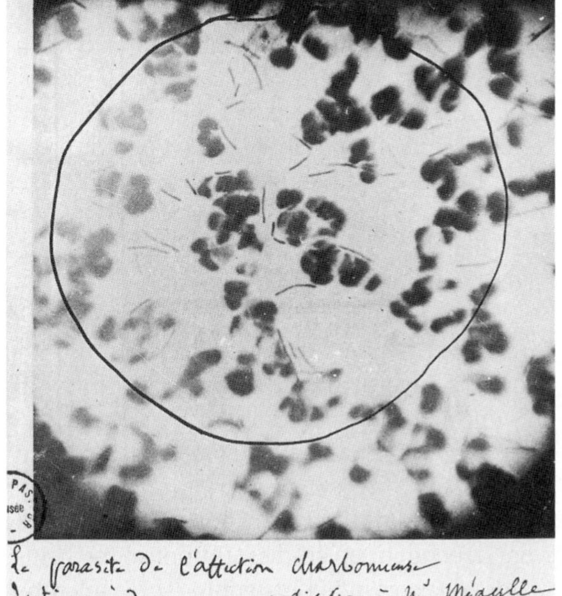

FIGURE 1.9 The Anthrax Bacterial Cells. A photomicrograph of the anthrax bacterial cells taken by Louis Pasteur in 1885. Pasteur circled the bacilli in tissue and annotated the photograph, "the parasite of Charbonneuse." ("Charbonneuse" is the French equivalent of anthrax.) © Institut Pasteur, Paris.
»» Identify the anthrax bacterial cells in this image.

healthy. The rabies vaccine was a triumph because it fulfilled Pasteur's dream of applying the principles of science to practical problems. Such successes helped establish the Pasteur Institute in Paris, one of the world's foremost scientific establishments.

In summary, the germ theory set a new course for studying and treating infectious disease. The studies carried out by Pasteur and Koch made the discipline of **bacteriology**, the study of bacterial organisms, a well-respected field of study. In fact, a new generation of international scientists, including several from the Pasteur and Koch labs, stepped in to expand the work on infectious disease (TABLE 1.1).

CONCEPT AND REASONING CHECKS 3

a. How did Pasteur's studies of wine fermentation and souring suggest to him that germs may cause disease?
b. Assess Lister's antisepsis procedures and Pasteur's studies of pébrine to supporting the germ theory.
c. Why was the pure culture crucial to Koch's validation of the germ theory?
d. What were the major discoveries made in Pasteur's lab and Koch's lab during the Golden Age of microbiology.

MICROFOCUS 1.4: History
The Private Pasteur

The notebooks of Louis Pasteur had been an enduring mystery of science ever since the scientist himself requested his family not to show them to anyone. But in 1964, Pasteur's last surviving grandson donated the notebooks to the National Library in Paris, and after soul-searching for a decade, the directors made them available to a select group of scholars. Among the group was Gerald Geison of Princeton University. What Geison found stripped away part of the veneration conferred on Pasteur and showed another side to his work.

In 1881, Pasteur conducted a trial of his new anthrax vaccine by inoculating half a flock of animals with the vaccine, then exposing the entire flock to the disease. When the vaccinated half survived, Pasteur was showered with accolades. However, Pasteur's notebooks, according to Geison, reveal that he had prepared the vaccine not by his own method, but by a competitor's.

Pasteur also apparently sidestepped established protocols when he inoculated two boys with a rabies vaccine before it was tested on animals. Fortunately, the two boys survived, possibly because they were not actually infected or because the vaccine was, indeed, safe and effective. Nevertheless, the untested treatment should not have been used, says Geison. His book, *The Private Science of Louis Pasteur* (Princeton University Press, 1995) places the scientist in a more realistic light and shows that today's pressures to succeed in research are little different than they were more than a century ago.

KEY CONCEPT 1.4 With the Discovery of Other Microbes, the Microbial World Expands

Although the list of identified microbes was growing, the agents responsible for diseases such as measles, mumps, smallpox, and yellow fever continued to elude identification.

Other Global Pioneers Contribute to New Disciplines in Microbiology

In 1892, a Russian scientist, Dimitri Ivanowsky, used a ceramic filter developed by Pasteur's group to trap what he thought were bacterial cells responsible for tobacco mosaic disease, which produces mottled and stunted tobacco leaves. Surprisingly, Ivanowsky discovered that when he applied the liquid that passed through the filter to healthy tobacco plants, the leaves became mottled and stunted. Ivanowsky, not understanding the significance of this, simply assumed bacterial cells somehow had slipped through the filter.

Unaware of Ivanowsky's work, Martinus Beijerinck, a Dutch investigator, did similar experiments in 1898 but suggested tobacco mosaic disease was a "contagious, living liquid" that acted like a poison or virus (*virus* = "poison"). Also in 1898, the causative agent for an animal disease—hoof-and-mouth disease—was found to be another filterable liquid, and in 1901 American Walter Reed concluded that the agent responsible for yellow fever in humans also was a virus. With these discoveries, the discipline of **virology**, the study of viruses, was launched.

While scientists like Pasteur and Koch were investigating the bacterial contribution to the infectious disease process, others were identifying other types of disease-causing microbes. That fungi could cause plant diseases was known since 1767 and such diseases were studied extensively by Anton de Bary in the 1860s. As already mentioned in this chapter, Pasteur identified the role of fungal yeasts (first seen by Leeuwenhoek) with fermentation. Importantly, the recognition that some fungi were linked to human skin diseases was proposed as early as 1841 when a Hungarian physician, David Gruby, discovered a fungus associated with human scalp infections.

The realization that infectious disease could be caused by yet another group of microbes, the protozoa (again first seen by Leeuwenhoek), was another major milestone in understanding infectious disease. In fact, Pasteur's "corpuscular parasites" of pébrine were protozoa. Other advances in the study of these types of microbes were dependent on studies in tropical medicine. Major advances in understanding these microbes included Charles Laveran's discovery (1880) that the protozoan parasite causing malaria could be found in human blood and David Bruce's studies (1903) that another protozoan parasite was the agent of human sleeping sickness. These and many other investigations with the fungi and protozoa

TABLE 1.1 **Other International Scientists and Their Accomplishments During the Classical Golden Age of Microbiology**

Investigator (Year)	Country	Accomplishment
Otto Obermeier (1868)	Germany	Observed bacterial cells in relapsing fever patients
Ferdinand Cohn (1872)	Germany	Established bacteriology as a science; produced the first bacterial taxonomy scheme
Gerhard Hansen (1873)	Norway	Observed bacterial cells in leprosy patients
Albert Neisser (1879)	Germany	Discovered the bacterium that causes gonorrhea
*Charles Laveran (1880)	France	Discovered that malaria is caused by a protozoan
Hans Christian Gram (1884)	Denmark	Introduced staining system to identify bacterial cells
Pasteur Lab		
Elie Metchnikoff (1884)	Ukraine	Described phagocytosis
Émile Roux and Alexandre Yersin (1888)	France	Identified the diphtheria toxin
Koch Lab		
Friedrich Löeffler (1883)	Germany	Isolated the diphtheria bacillus
Georg Gaffky (1884)	Germany	Cultivated the typhoid bacillus
*Paul Ehrlich (1885)	Germany	Suggested some dyes might control bacterial infections
Shibasaburo Kitasato (1889)	Japan	Isolated the tetanus bacillus
Emil von Behring (1890)	Germany	Developed the diphtheria antitoxin
Theodore Escherich (1885)	Germany	Described the bacterium responsible for infant diarrhea
Daniel E. Salmon (1886)	United States	Developed the first heat-killed vaccine
Richard Pfeiffer (1892)	Germany	Identified a bacterial cause of meningitis
William Welch and George Nuttall (1892)	United States	Isolated the gas gangrene bacillus
Theobald Smith and F. Kilbourne (1893)	United States	Proved that ticks transmit Texas cattle fever
S. Kitasato and A. Yersin (1894)	Japan France	Independently discovered the bacterium causing plague
Emile van Ermengem (1896)	Belgium	Identified the bacterium causing botulism
*Ronald Ross (1898)	Great Britain	Showed mosquitoes transmit malaria to birds
Kiyoshi Shiga (1898)	Japan	Isolated a cause of bacterial dysentery
Walter Reed (1901)	United States	Studied mosquito transmission of yellow fever
David Bruce (1903)	Great Britain	Proved that tsetse flies transmit sleeping sickness
Fritz Schaudinn and Erich Hoffman (1903)	Germany	Discovered the bacterium responsible for syphilis
*Jules Bordet and Octave Gengou (1906)	France	Cultivated the pertussis bacillus
Albert Calmette and Camille Guérin (1906)	France	Developed immunization process for tuberculosis
Howard Ricketts (1906)	United States	Proved that ticks transmit Rocky Mountain spotted fever
Charles Nicolle (1909)	France	Proved that lice transmit typhus fever
George McCoy and Charles Chapin (1911)	United States	Discovered the bacterial cause of tularemia

*Nobel Prize winners in Physiology or Medicine.

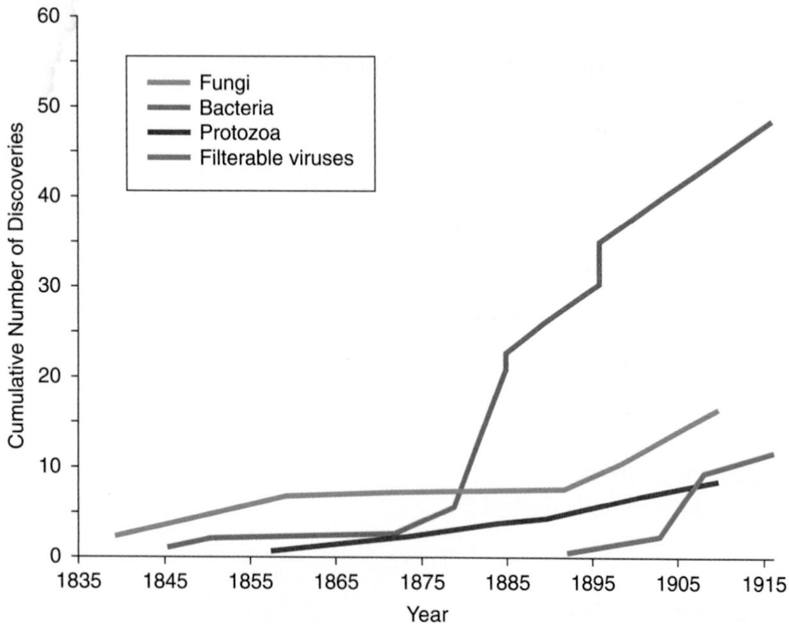

FIGURE 1.10 **The Discovery of Microbes 1840–1915.** The discovery of microbes, especially bacterial pathogens, rose rapidly during the first Golden Age of microbiology. Reproduced from Flint et al, *Principles of Virology*, 3rd edition, ASM Press, Washington DC (2009). »» What key event (a) around 1880 triggered the rise in identified bacterial organisms and (b) in the late 1890s started the rise in identified viruses?

served to increase the pace of discovery in the microbial world (FIGURE 1.10).

Realize that not all microbes are pathogens; in fact relatively few are. So, during the first Golden Age of microbiology, other scientists and microbiologists devoted their studies to the environmental and ecological importance of nonpathogenic microbes. The Russian soil scientist Sergei Winogradsky, a student of de Bary's, discovered bacterial organisms that metabolized sulfur and developed the concept of **nitrogen fixation**, a process whereby bacterial cells convert inert atmospheric nitrogen gas (N_2) into biologically useable ammonia (NH_3). Beijerinck, besides working on tobacco mosaic disease, was the first to obtain pure cultures of microorganisms from soil and water by enriching the growth conditions. Together with Winogradsky, they developed many of the laboratory methods essential to the study of microbial ecology, while revealing the physiological significance of soil microbes and discovering the essential roles such microorganisms play in the recycling of matter on a global scale. As the founders of microbial ecology, we must acknowledge Winogradsky and Beijerinck for providing much of the foundation for what we know today about many of the so-called microbial workforce microbes mentioned in the chapter opener.

So, by the end of the first Golden Age of microbiology, the world of the microorganism had greatly broadened beyond the organism to applied fields, such as **immunology**, the study of the body's response to viruses and microorganisms, and microbial ecology. What started as a curious observation of animalcules by Leeuwenhoek had been resolved within 200 years into an increasingly diverse menagerie of microorganisms. And today, many microbiologists are still searching for, finding, and trying to understand the roles of microorganisms in the environment as well as in health and disease. In fact, with less than 2% of all microorganisms on Earth having been identified and many fewer cultured, there is still a lot to be discovered and studied in the microbial world!

The Microbial World Can Be Catalogued into Five Major Groups

Besides bacteriology and virology, other disciplines in microbiology also were developing. This included: **mycology**, the study of fungi; **protozoology**, the study of protozoa; and **phycology**, the study of the algae (FIGURE 1.11). Let's briefly survey what we know about these major groups.

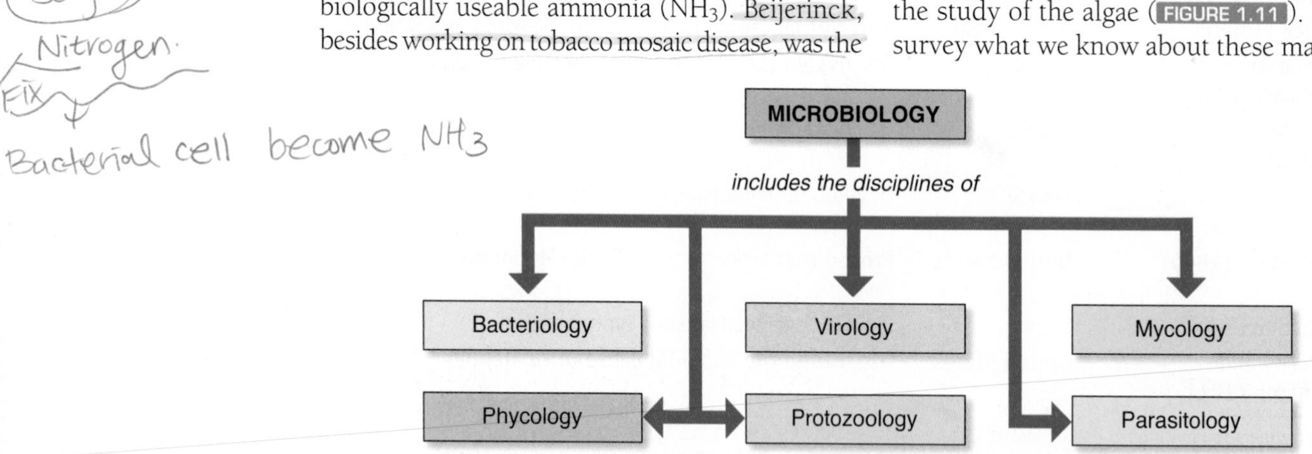

FIGURE 1.11 **Microbiology Disciplines.** This simple concept map shows the relationship between microbiology and members that make up the various disciplines. Parasitology is the study of animal parasites, which traditionally includes the parasitic protozoa and the animal parasites (worms).

Bacteria and Archaea. It is estimated that there may be more than 10 million bacterial species. Most are very small, single-celled (unicellular) organisms (although some form filaments, and most associate in a bacterial mass called a "biofilm") that have a rigid cell wall. The cells may be spherical, spiral, or rod-shaped (FIGURE 1.12A), and they lack the cell nucleus and most of the typical membrane-enclosed cellular compartments typical of other microbes and multicellular organisms. Many get their food from the environment, although some make their own food through photosynthesis (FIGURE 1.12B).

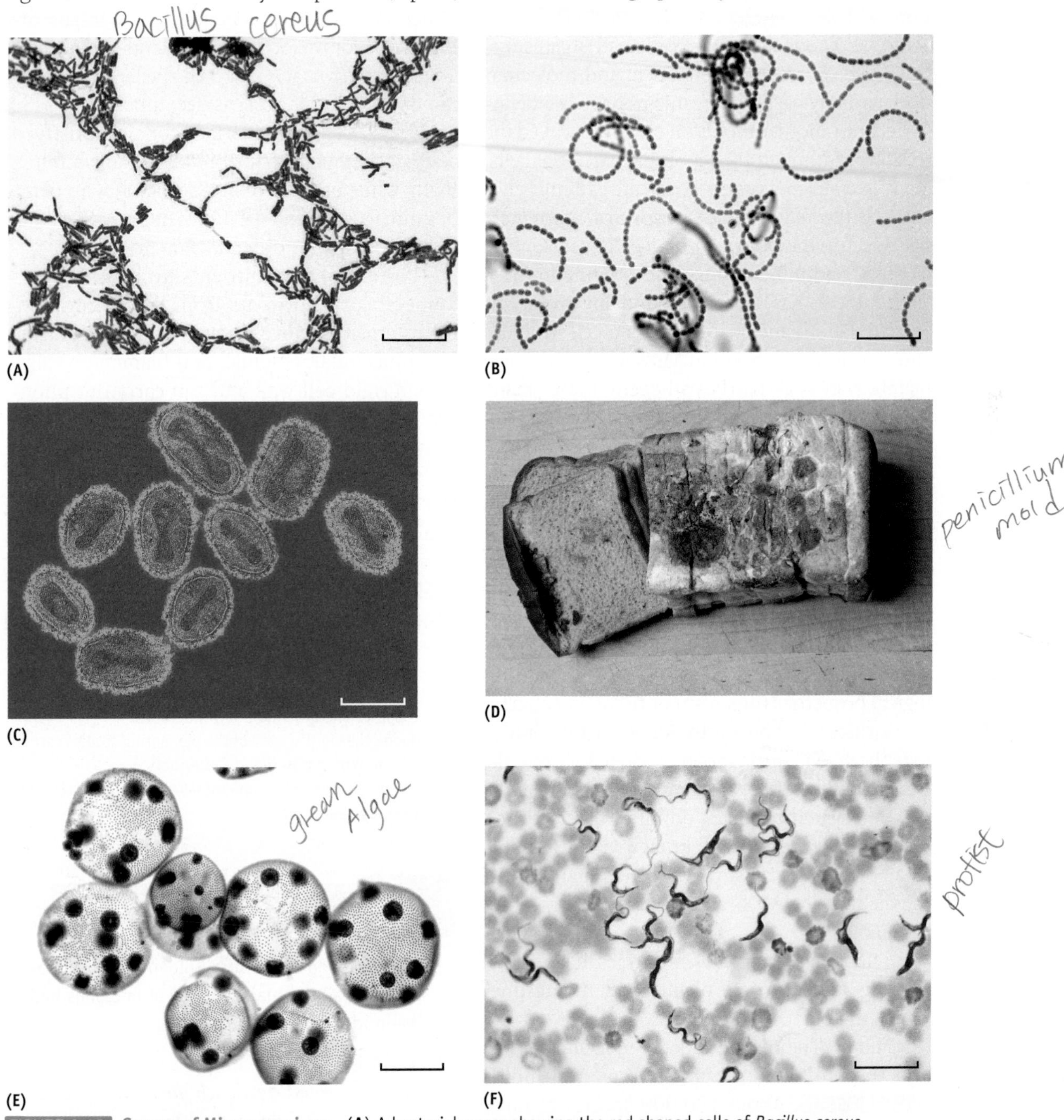

FIGURE 1.12 **Groups of Microorganisms.** (**A**) A bacterial smear showing the rod shaped cells of *Bacillus cereus* (stained purple), a normal inhabitant of the soil. (Bar = 10 μm.) Courtesy of Jeffrey Pommerville. (**B**) Filamentous strands of *Anabaena*, a cyanobacterium that carries out photosynthesis. (Bar = 100 μm.) Courtesy of Jeffrey Pommerville. (**C**) Smallpox viruses. (Bar = 100 nm.) © Dr. Hans Gelderblom/Visuals Unlimited/Getty. (**D**) A typical blue-gray *Penicillium* mold growing on a loaf of bread. © Jones & Bartlett Learning. Photographed by Kimberly Potvin. (**E**) The colonial green alga, *Volvox*. (Bar = 300 μm.) Courtesy of Jeffrey Pommerville. (**F**) The ribbon-like cells of the protist *Trypanosoma*, the causative agent of African sleeping sickness. (Bar = 10 μm.) Courtesy of Jeffrey Pommerville. »» Within these groups, why don't organisms like *Anabaena* and *Volvox* cause disease?

Bacterial cells are found in most all environments, making up a large percentage of the Earth's microbial workforce.

Besides the disease-causing members, some are responsible for food spoilage while others are useful in the food industry. Many bacterial members, along with several fungi, are **decomposers**, organisms that recycle nutrients from dead organisms.

Based on recent biochemical and molecular studies, many bacterial organisms have been reassigned into another evolutionary group, called the **Archaea**. Although they look like bacterial cells, many grow in environments that are extremely hot (such as the Yellowstone hot springs), extremely salty (such as the Dead Sea), or of extremely low pH (such as acid mine drainage). Surviving in these environments has brought about many evolutionary adaptations and changes to their cell structure and chemical composition. As such, no archaeal members are known to be pathogens. In fact, many normally grow in soils and water, and are an integral part of animal digestive tracts.

Viruses. Although not correctly labeled as microorganisms, currently there are more than 3,600 known types of viruses. Viruses are not cellular and cannot be grown in pure culture. They have a core of nucleic acid (DNA or RNA) surrounded by a protein coat. Among the features used to identify viruses are morphology (size, shape), genetic material (RNA, DNA), and biological properties (organism or tissue infected).

Viruses infect organisms for one reason only—to replicate. Viruses in the air or water, for example, cannot replicate because they need the metabolic machinery and chemical building blocks found inside living cells. Of the known viruses, only a small percentage causes disease in humans. Polio, the flu, measles, AIDS, and smallpox are examples (FIGURE 1.12C).

The other groups of microbes have a cell nucleus and a variety of internal, membrane-bound cellular compartments.

Fungi. The fungi include the unicellular yeasts and the multicellular mushrooms and molds (FIGURE 1.12D). About 100,000 species of fungi have been described; however, there may be as many as 1.5 million species in nature.

Other than the yeasts, molds tend to grow as filaments with unique rigid cell walls. Most fungi grow best in warm, moist places and secrete digestive enzymes that break down nutrients into smaller bits that can be absorbed easily across a rigid cell wall. Fungi thus live in their own food supply. If that "food supply" is a human, diseases such as ringworm or vaginal yeast infections may result.

For the pharmaceutical industry, some fungi are sources for useful products, such as antibiotics. Others are used in the food industry to impart distinctive flavors in foods such as Roquefort cheeses. Together with many bacterial species, numerous molds play a major role as decomposers.

Protists. The protists consist mostly of single-celled algae and protozoa. Some are free-living while others live in association with plants or animals. Movement, if present, is achieved by flagella or cilia, or by a crawling movement.

Protists obtain nutrients in different ways. Some absorb nutrients from the surrounding environment or ingest smaller microorganisms. The unicellular, colonial, and filamentous algae have a rigid cell wall and can carry out photosynthesis (FIGURE 1.12E). The aquatic protists also provide energy and organic compounds for the lower trophic levels of the food web. Some protists (the protozoa) are capable of causing diseases in animals, including humans; these include malaria, several types of diarrhea, and sleeping sickness (FIGURE 1.12F).

CONCEPT AND REASONING CHECKS 4

- **a.** Describe how viruses were discovered as disease-causing agents.
- **b.** What significant discoveries added the fungi and protozoa to the growing list of microbes?
- **c.** Judge the significance of the pioneering studies carried out by Winogradsky and Beijerinck.
- **d.** Identify which of the microbial groups were originally seen by Leeuwenhoek.

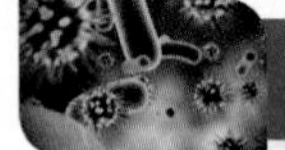

Chapter Challenge B

With the end of the first Golden Age of microbiology, all the major groups of microbes had been identified.

QUESTION B:

Briefly recount how each group of microbes in the diverse global workforce was revealed and identify the characteristics that separate them into different groups.

Answers can be found on the Student Companion Website in **Appendix D.**

KEY CONCEPT 1.5 A Second Golden Age of Microbiology Involves the Birth of Molecular Biology and Chemotherapy

The 1940s brought the birth of molecular genetics to biology. Many biologists focused on understanding the genetics of organisms, including the nature of the genetic material and its regulation.

Molecular Biology Relies on Microorganisms as Model Systems

In 1943, the Italian-born microbiologist Salvador Luria and the German physicist Max Delbrück carried out a series of experiments with bacterial cells and viruses that marked the second Golden Age of microbiology. They used a common gut-inhabiting bacterial organism, *Escherichia coli*, to address a basic question regarding evolutionary biology: Do mutations occur spontaneously or does the environment induce them? Luria and Dulbrück showed that bacterial cells could develop spontaneous mutations that generate resistance to viral infection. Such uses of microbial model systems showed to other researchers that microorganisms could be used to study general principles of biology.

■ Mutations: Permanent alterations in DNA base sequences.

So, biologists were quick to jump on the "microbial bandwagon." For example:

- Americans George Beadle and Edward Tatum (1941) use the fungus *Neurospora* to demonstrate that "one gene codes for one enzyme";
- Oswald Avery, Colin MacLeod, and Maclyn McCarty (1944) use the bacterial organism *Streptococcus pneumoniae* to suggest that DNA is the genetic material in cells;
- Alfred Hershey and Martha Chase (1952) use a virus that infects bacterial cells to assert that DNA is the substance of the genetic material;
- Francis Crick (1958) uses *E. coli* and a virus to show how the DNA genetic code works to make individual proteins.

Again, microbes were at the forefront in answering fundamental questions that applied to all of biology. In addition, during this second Golden Age, other studies illuminated the basic structure of microbes and led to one of the greatest breakthroughs (antibiotics) that would revolutionize medicine's ability to treat and eliminate infectious disease.

Two Types of Cellular Organization Are Realized

The small size of bacterial cells hindered scientists' abilities to confirm whether these cells were similar to other cellular organisms in organization. In the 1940s and 1950s, a new type of microscope—the electron microscope—was being developed that could magnify objects and cells thousands of times better than typical light microscopes. With the electron microscope, for the first time bacterial cells were seen as being cellular like all other microbes, plants, and animals. However, studies showed that they were organized in a structurally different way from other organisms.

It was known that animal and plant cells contained a cell nucleus that houses the genetic instructions in the form of chromosomes and was separated physically from other cell structures by a membrane envelope (FIGURE 1.13A). This type of cellular organization is called **eukaryotic** (*eu* = "true"; *karyon* = "nucleus"). Microscope observations of the protists and fungi had revealed that these organisms also have a eukaryotic organization. Thus, not only are all plants and animals eukaryotes, so are the microorganisms that comprise the fungi and protists.

Studies with the electron microscope revealed that bacterial (and archaeal) cells had few of the membranous compartments typical of eukaryotic cells. They lacked a cell nucleus, indicating the bacterial chromosome (DNA) was not surrounded by a membrane envelope (FIGURE 1.13B). Therefore, members of the Bacteria and Archaea have a **prokaryotic** (*pro* = "before") type of cellular organization and represent prokaryotes. Importantly, there are also many differences between bacterial and archaeal cells, accounting for their split into separate microbial groups. By the way, because viruses lack a cellular organization, they are neither prokaryotes nor eukaryotes.

Antibiotics Are Used to Cure Infectious Disease

In 1910, another coworker of Koch's, Paul Ehrlich, synthesized the first "magic bullet"—a chemical

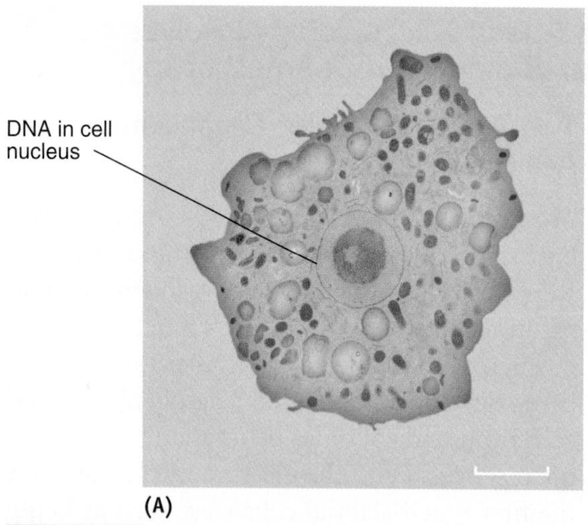

(A)

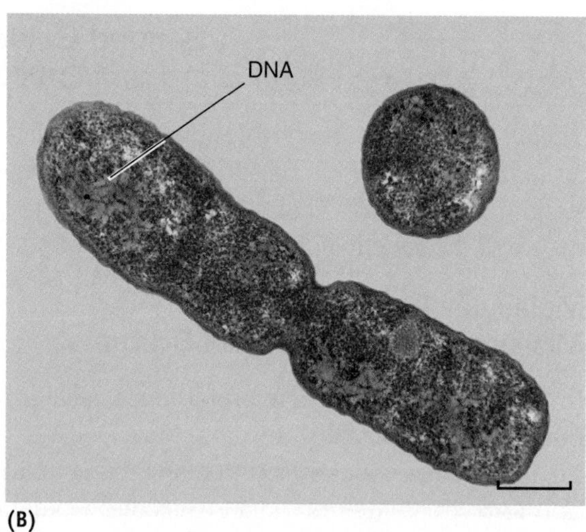

(B)

FIGURE 1.13 **False Color Images of Eukaryotic and Prokaryotic Cells.** (**A**) An electron microscope image of a protozoan cell. All eukaryotes, including the protists and fungi, have their DNA enclosed in a cell nucleus. (Bar = 3 μm.) © London School of Hygiene & Tropical Medicine/Photo Researchers, Inc. (**B**) An electron microscope image of a dividing *Escherichia coli* cell. (Bar = 0.5 μm.) © Dennis Kunkel Microscopy, Inc./Visuals Unlimited/Corbis. »» How is the DNA in the eukaryotic cell structurally different from the DNA in the bacterial cell?

that could kill pathogens without damaging the surrounding tissue. Called salvarsan, Ehrlich showed that this arsenic-containing compound cured syphilis, a sexually transmitted disease. Antibacterial **chemotherapy**, the use of antimicrobial chemicals to kill microbes, was born.

In 1928, Alexander Fleming, a Scottish scientist, discovered a mold growing in one of his bacterial cultures (FIGURE 1.14A, B). His curiosity aroused, Fleming observed that the mold, a species of *Penicillium*, killed the bacterial colonies that were near the mold. He named the antimicrobial substance penicillin and soon discovered penicillin would kill other bacterial pathogens causing diphtheria, scarlet fever, and gonorrhea. In 1940, biochemists Howard Florey and Ernst Chain further showed the antimicrobial potential of the natural drug and developed a way to mass produce penicillin (MicroFocus 1.5).

Additional magic bullets also were being discovered. The German chemist Gerhard Domagk discovered a synthetic chemical dye, called prontosil, which was effective in treating *Streptococcus* infections. Examination of soil bacteria led Selman Waksman to the discovery of actinomycin and streptomycin, the latter being the first effective agent against tuberculosis. He coined the term **antibiotic** to refer to those antimicrobial substances naturally produced by mold and bacterial species that inhibit growth or kill other microorganisms.

The push to market effective antibiotics was stimulated by a need to treat potentially deadly infections in casualties of World War II (FIGURE 1.14C). By the 1950s, penicillin and several additional antibiotics were established treatments in medical practice. In fact, the growing arsenal of antibiotics convinced many that the age of infectious disease was waning. By the mid-1960s, many believed all major infections would soon disappear due to antibiotic chemotherapy.

Partly due to the perceived benefits of antibiotics, interest in microbes was waning by the end of the 1960s as the knowledge gained from bacterial studies was being applied to eukaryotic organisms, especially animals. What was ignored was the mounting evidence that bacterial species were becoming resistant to antibiotics.

Still, antibiotics represent one of the greatest breakthroughs in medicine and have saved millions of lives since their introduction.

CONCEPT AND REASONING CHECKS 5

a. What roles did microorganisms and viruses play in understanding general principles of biology?

b. Distinguish between prokaryotic and eukaryotic cells.

c. Contrast Ehrlich's salvarsan and Domagk's prontosil from those drugs developed by Fleming, Florey and Chain, and Waksman.

(A)

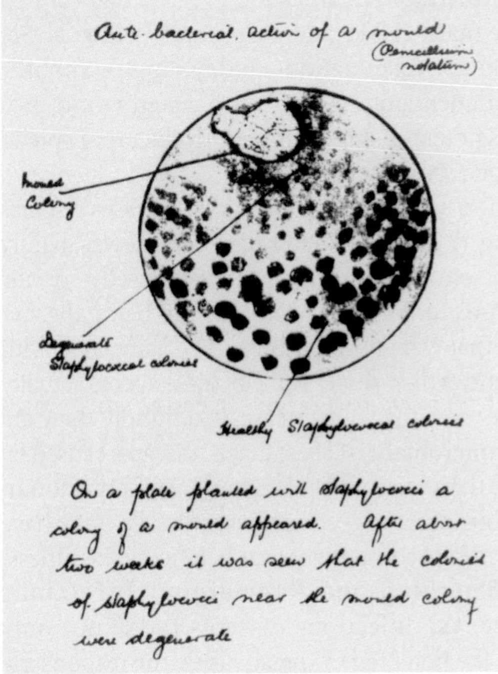

(B)

(C)

FIGURE 1.14 **Fleming and Penicillin.** (**A**) A painting by Dean Fausett of Fleming in his laboratory. © Science Source, photo by Dean Fausett/Photo Researchers, Inc. (**B**) Fleming's notes on the inhibition of bacterial growth by the fungus *Penicillium*. © St. Mary's Hospital Medical School/Science Photo Library. (**C**) A World War II poster touting the benefits of penicillin and illustrating the great enthusiasm in the United States for treating infectious diseases in war casualties. Used with permission from Pfizer, Inc. »» In (B), why didn't penicillin kill the bacterial colonies farther away from the fungus?

MICROFOCUS 1.5: History

Hiding a Treasure

Their timing could not have been worse. Howard Florey, Ernst Chain, Norman Heatley, and others of the team had rediscovered penicillin, purified it, and proved it useful in infected patients. But it was 1939, and German bombs were falling on London. This was a dangerous time to be doing research into new drugs and medicines. What would they do if there was a German invasion of England? If the enemy were to learn the secret of penicillin, the team would have to destroy all their work. So, how could they preserve the vital fungus yet keep it from falling into enemy hands?

Heatley made a suggestion. Each team member would rub the mold on the inside lining of his coat. The *Penicillium* mold spores would cling to the rough coat surface where the spores could survive for years (if necessary) in a dormant form. If an invasion did occur, hopefully at least one team member would make it to safety along with his "moldy coat." Then, in a safe country, the spores would be used to start new cultures and the research could continue. Of course, a German invasion of England did not occur, but the plan was an ingenious way to hide the treasured organism.

The whole penicillin story is well told in *The Mold in Dr. Flory's Coat* by Eric Lax (Henry Holt Publishers, 2004).

KEY CONCEPT 1.6 The Third Golden Age of Microbiology Is Now

Microbiology finds itself on the world stage again, in part from the biotechnology advances made in the latter part of the 20th century. Industry frequently uses the natural and genetically engineered abilities of microbial agents to carry out biological processes for industrial/commercial/medical applications. It has revolutionized the way microorganisms are genetically manipulated to act as tiny factories producing human proteins, such as insulin, or new synthetic vaccines, such as the hepatitis B vaccine. In the latest Golden Age, microbiology again is making important contributions to the life sciences and humanity.

However, the third Golden Age of microbiology also faces several challenges, many of which still concern infectious diseases that today are responsible for 26% of all deaths globally (FIGURE 1.15).

Microbiology Continues to Face Many Challenges

The resurgence of infectious disease has brought the subject back into the mainstream of epidemiology. Even in the United States, more than 100,000 people die each year from bacterial infections, making them the fourth leading cause of death. In fact, on a global scale, infectious diseases are spreading geographically faster than at any time in history.

A New Infectious Disease Paradigm. Experts estimate that more than 4 billion people traveled by air in 2011, making an outbreak or epidemic in one part of the world only a few airline hours away from becoming a potentially dangerous threat in another region of the globe. It is a sobering thought to realize that since 2002, the World Health Organization (WHO) has verified more than 1,100 epidemic events worldwide. So, unlike past generations, today's highly mobile, interdependent, and interconnected world provides potential opportunities for the rapid spread of infectious diseases.

Today, our view of infectious diseases also has changed. In Pasteur and Koch's time, it was mainly a problem of finding the agent that caused a specific disease. Today, new pathogens are being discovered that were never known to be associated with infectious disease and some of these agents actually cause more than one disease. In addition, there are **polymicrobial diseases**; that is, diseases caused by more than one infectious agent. Even some noninfectious diseases, such as heart disease, may have a microbial component that heightens the illness.

Emerging and Reemerging Infectious Diseases. Infectious diseases today not only have the potential to spread faster, the responsible pathogens are subject to evolutionary pressures. Since the 1970s, new diseases have been identified at the unprecedented rate of one or more per year. There are now nearly 40 infectious diseases that were unknown a generation ago.

Emerging infectious diseases are those that have come from somewhere else (i.e., animals) and have recently surfaced in the human population. Among the more newsworthy have been AIDS, severe acute respiratory syndrome, Lyme disease, mad cow disease, and most recently swine flu. There is no cure for any of these and undoubtedly there are more ready to emerge. **Reemerging infectious diseases** are ones that have existed in the past but are now showing a resurgence in resistant forms or a spread in geographic range. Among the more prominent reemerging diseases are drug-resistant tuberculosis, cholera, dengue fever, and, for the first time in the Western Hemisphere, West Nile

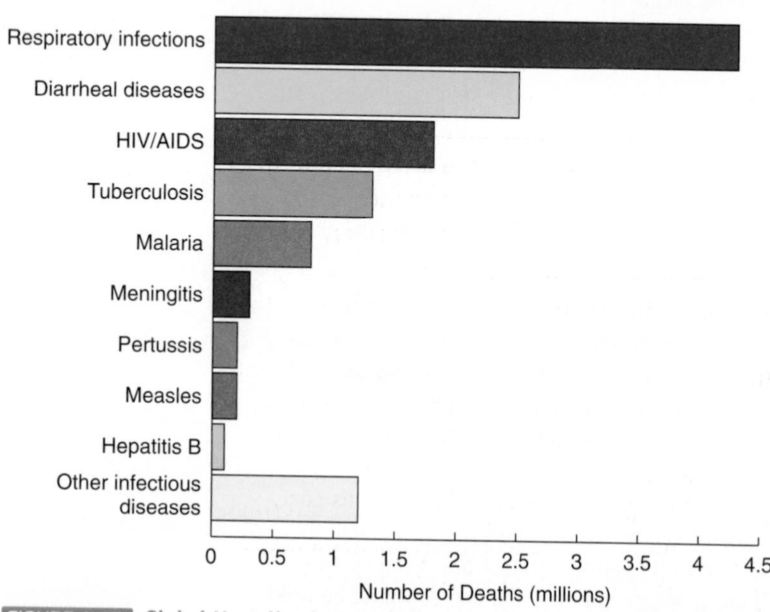

FIGURE 1.15 **Global Mortality from Infectious Diseases—2011.** On a global scale, of the 59 million annual deaths globally, some 15 million are caused by infectious diseases. Noninfectious causes include chronic diseases, injuries, nutritional deficiencies, and maternal and perinatal conditions. *Source:* World Health Statistics 2008: World Health Organization. Data from: World Health Statistics 2011: World Health Organization. »» Identify some of the "other infectious diseases."

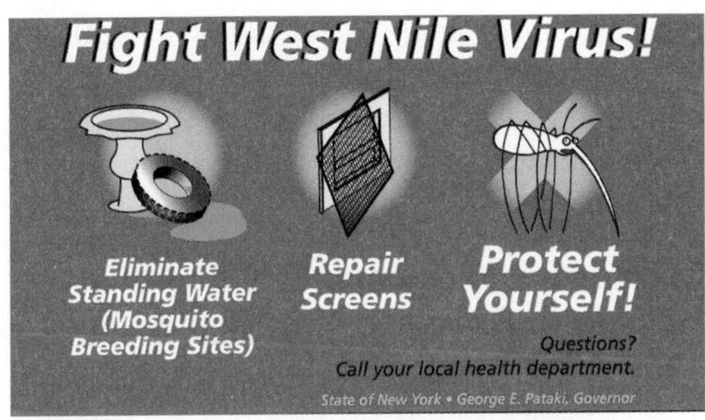

(A)

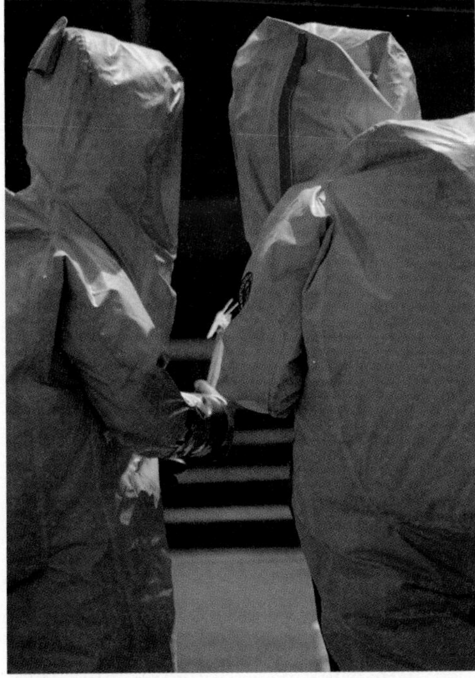

(B)

FIGURE 1.16 **Emerging Disease Threats: Natural and Intentional.** **(A)** There have been and will continue to be natural disease outbreaks. West Nile virus (WNV) is just one of several agents responsible for emerging or reemerging diseases. Methods have been designed that individuals can use to protect themselves from mosquitoes that spread the WNV. Reproduced with permission of the New York State Department of Health. **(B)** Combating the threat of bioterrorism often requires special equipment and protection because many agents seen as possible bioweapons could be spread through the air. © Photodisc. »» How do these suits compare to those to protect from a miasma (Figure 1.3)?

virus disease (FIGURE 1.16A). Therefore, emerging and reemerging diseases will remain as perpetual challenges to public health and microbiology.

Increased Antibiotic Resistance. Another challenge concerns our increasing inability to fight infectious disease because most pathogens are now resistant to one or more antibiotics and such antibiotic resistance is developing faster than new antibiotics are being discovered. Ever since it was recognized that pathogens could mutate into "superbugs," a crusade has been waged to restrain the inappropriate use of these drugs by doctors and to educate patients not to demand them in uncalled-for situations.

■ Superbug: A microbe resistant to many antimicrobial drugs.

The challenge facing microbiologists and drug companies is to find new and effective antibiotics to which pathogens will not quickly develop resistance before the current arsenal is completely useless. Unfortunately, the growing threat of antibiotic resistance has been accompanied by a decline in new drug discovery and an increase in the time to develop a drug from discovery to market. Thus, antibiotic resistance has become a major health threat and a significant challenge for microbiology today. If actions are not taken to contain and reverse resistance, the world could be faced with previously treatable diseases that have again become untreatable, as in the days before antibiotics were developed.

Bioterrorism. Perhaps it is the potential misuse of microbiology that has brought microbiology to the attention of the life science community and the public. **Bioterrorism** involves the intentional or threatened use of biological agents to cause fear in or actually inflict death or disease upon a large population. Most of the recognized biological agents are microorganisms, viruses, or microbial toxins that are bringing diseases like anthrax, smallpox, and plague back into the human psyche (FIGURE 1.16B). To minimize the use of these agents to inflict mass casualties, the challenge to the scientific community and microbiologists is to improve the ways that bioterror agents are detected, discover effective measures to protect the public, and develop new and effective treatments for individuals or whole populations. If there is anything good to come out of such challenges, it is that we will be better prepared for potential natural emerging infectious disease outbreaks, which initially might be difficult to tell apart from a bioterrorist attack.

Climate Change and Infectious Disease. A very controversial issue is how **climate change** (including warming temperatures and altering rainfall patterns) may affect the frequency and distribution of infectious diseases around the world. Many scientists believe that as temperatures rise in various regions of the world, mosquitoes that can transmit diseases like malaria and dengue fever will broaden their range, especially into more temperate climates such as North America. Warming ocean

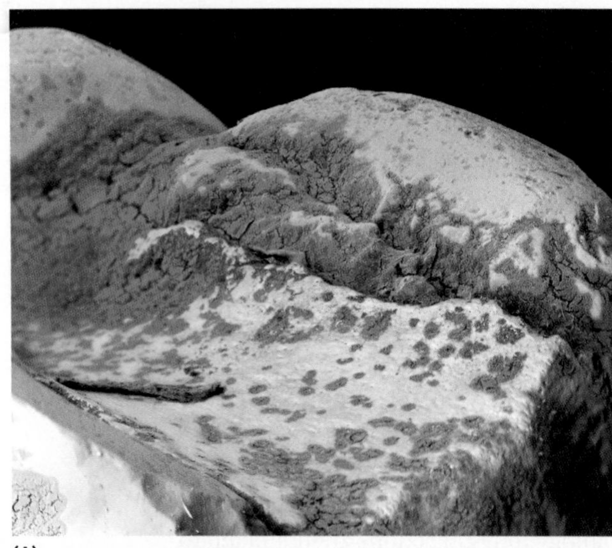

(A)

(B)

FIGURE 1.17 **Microbial Ecology—Biofilms and Bioremediation.** (**A**) Plaque (the false-colored brown crust in the photo) typically found on tooth surfaces is an example of a biofilm. © Eye of Science/Photo Researchers, Inc. (**B**) Microbes can be used to clean up toxic spills. A shoreline coated with oil from an oil spill can be sprayed with microorganisms that, along with other measures, help degrade oil. Courtesy of the Exxon Valdez Oil Spill Trustee Council/NOAA. »» How do these examples of microbes fit into the concept of the global microbial workforce?

■ Bioremediation: The use of microorganisms to remove or decontaminate toxic materials in the environment.

waters may also provide ideal environments for cholera. As such, microbiologists, epidemiologists, climate scientists, and many others are studying new strategies to limit potential pandemics before they can get started. In so doing, we need to better understand the dynamics of climate change and how such changes might affect the behaviors of potentially emerging and reemerging diseases that could affect the health of humans, livestock, plants, and wildlife. Climate change coupled with rapid air travel and the expanding evolution of antibiotic resistance, for example, could work synergistically to cause major epidemics, or even pandemics, if we don't find ways to constantly survey the microbial landscape and decrease the burden of infectious disease.

Microbial Ecology and Evolution Are Helping to Drive the New Golden Age

Since the time of the first Golden Age of microbiology, microbiologists have wanted to know how a microbe interacts, survives, and thrives in the environment. Today, microbiology is less concerned with a specific microbe and more concerned with the relationships among microorganisms and with their environment.

Microbial Ecology. Traditional methods of microbial ecology require organisms from an environment be isolated and cultivated in the laboratory so that they can be characterized and identified. However, up to 99% of microorganisms do not grow well in the lab (if at all) and therefore could not be studied. Today, many microbiologists, armed with genetic, molecular, and biotechnological tools, can study and characterize these uncultured microbes. These investigators have found more microbial diversity in a sample of seawater than in all the diverse microbes that have been cultured!

Today we are learning that most microbes do not act as individual entities; rather, in nature they survive in complex, often polymicrobial communities called a **biofilm** (FIGURE 1.17A). For example, if you or someone you know has had a middle ear infection, the cause was a bacterial biofilm. Microbes in biofilms act very differently than individual cells and can be difficult to treat when biofilms cause infectious disease.

The discovered versatility of many bacterial and archaeal species is being applied to problems that have the potential to benefit the planet. Bioremediation is one example where the understanding of microbial ecology has produced a useful outcome (FIGURE 1.17B). Other microbes also are playing increasingly important roles in the health of the planet. As the highly respected naturalist E. O. Wilson has stated: "*If I could do it all over again, and relive my vision in the twenty-first century, I would be a microbial ecologist.*"

Microbial Evolution. It was Charles Darwin—another of the scientists who combined observation with a "prepared mind"—who in 1859 first described the principles of evolution, which today is the unifying force in modern biology, tying together such distinct fields as genetics, ecology, medicine—and yes, microbiology.

Like all life, microorganisms evolve. In fact, they were evolving for some 2 billion years before the first truly eukaryotic cells appeared on the planet. Because most microbes have relatively short generation times, they represent experimental (model) systems in which evolutionary processes can be observed and tested and, in so doing, help us better understand the origin of all microorganisms—and larger organisms as well. Although challenging to study, it is possible today to "replay history" by following the accumulation of unpredictable, chance events that lead to evolutionary novelty. In addition, when considering the challenges facing microbiology today, current research on microbial evolution is putting together new approaches for the treatment of infectious disease, to improve agricultural productivity, to monitor and assess climate change, and even to produce clean fuels and energy.

Today, thanks to the availability of sequenced genomes for groups of related and unrelated microbes, and new analytical approaches, microbiologists and microbial ecologists are studying the processes that drive evolutionary diversification and constructing a family tree that more clearly illustrates evolutionary relationships among all organisms. Such developments are giving us a better appreciation for the roles microbes have played and are continuing to play in Earth's evolution. I wonder what Darwin would make of the microbial world if he were alive today?

CONCEPT AND REASONING CHECKS 6

a. Give some examples of how microbial ecology and evolution are helping drive the new Golden Age of microbiology.

Chapter Challenge C

Many of the microbes that first occupied the minds and work of the early microbial pioneers like Pasteur and Koch still challenge microbiologists today.

QUESTION C:

Describe the natural and intentional disease threats challenging microbiology today, explaining why they are still so prominent even with all the advances in medical science and microbiology.

Answers can be found on the Student Companion Website in **Appendix D**.

In conclusion, microbiology (from then until now) has gone from observing the first microbes (Leeuwenhoek) to identifying and studying individual microorganisms (Pasteur and Koch) to sequencing all species in a sample of seawater. Yet, over these 300+ years, microbiologists have only discovered perhaps 1% of all microbial species. Microbiology from then until now has come a long way, but has a much longer way yet to go.

SUMMARY OF KEY CONCEPTS

1.1 The Discovery of Microbes Leads to Questioning Their Origins

1. The observations with the microscope made by Hooke and especially Leeuwenhoek, who reported the existence of **animalcules** (microorganisms), sparked interest in an unknown world of microscopic life. (Fig. 1.2)
2. The controversy over **spontaneous generation** initiated the need for accurate scientific experimentation, which then provided the means to refute the concept.

1.2 Disease Transmission Can Be Interrupted

3. Edward Jenner determined that disease (smallpox) could be prevented through **vaccination** with a similar but milder disease-causing agent.
4. Semmelweis and Snow believed that infectious disease could be caused by particles transmitted from the environment and that the transmission could be interrupted. (Figs. 1.4, 1.5)

1.3 The Classical Golden Age of Microbiology Reveals the Germ

5. Pasteur's fermentation experiments indicated that microorganisms could induce chemical changes. He proposed the **germ theory of disease**, which stated that human disease could be due to chemical changes brought about by microorganisms in the body.
6. Lister's use of **antisepsis** techniques and Pasteur's studies of pébrine supported the germ theory and showed how diseases can be controlled.

7. Koch's work with anthrax allowed him to formalize the methods (**Koch's postulates**) for relating a specific microorganism to a specific disease. These postulates were only valid after he discovered how to make **pure cultures** of bacterial species. (Fig. 1.8)
8. Laboratory science arose as Pasteur and Koch hunted down the microorganisms of infectious disease. Pasteur's lab studied the mechanisms for infection and developed vaccines for chicken cholera, animal anthrax, and human rabies. Koch's lab focused on isolation, cultivation, and identification of pathogens such as those responsible for cholera and tuberculosis.

1.4 With the Discovery of Other Microbes, the Microbial World Expands

9. Ivanowsky and Beijerinck provided the first evidence for viruses as infectious agents.
10. Winogradsky and Beijerinck were the first to recognize the beneficial roles played by microorganisms found in the environment.
11. Microbes include the "bacteria" (Bacteria and Archaea), viruses, fungi, and protists. (Fig. 1.11)

1.5 A Second Golden Age of Microbiology Involves the Birth of Molecular Biology and Chemotherapy

12. Many of the advances toward understanding molecular biology and general principles in biology were based on experiments using microbial model systems.
13. With the advent of the electron microscope, microbiologists realized that there were two basic types of cellular organization: **eukaryotic** and **prokaryotic**. (Fig. 1.13)
14. Following from the initial work by Ehrlich, **antibiotics** were developed as "magic bullets" to cure many infectious diseases.

1.6 The Third Golden Age of Microbiology Is Now

15. In the 21st century, fighting infectious disease, identifying **emerging** and **reemerging infectious diseases**, combating increasing antibiotic resistance, countering the **bioterrorism** threat, and addressing the potential spread of infectious diseases due to **climate change** are challenges facing microbiology, healthcare systems, and society. (Fig. 1.15)
16. **Microbial ecology** is providing new clues to the roles of microorganisms in the environment. The understanding of **microbial evolution** through the use of genomic technologies has expanded our understanding of microorganism relationships.

CHAPTER SELF-TEST

For **STEPS A–D**, answers to even-numbered questions and problems can be found in **Appendix C** on the Student Companion Website at **http://microbiology.jbpub.com/10e**. In addition, the site features eLearning, an online review area that provides quizzes and other tools to help you study for your class. You can also follow useful links for in-depth information, read more MicroFocus stories, or just find out the latest microbiology news.

STEP A: REVIEW OF FACTS AND TERMS

Multiple Choice

Read each question carefully, then select the **one** answer that best fits the question or statement.

1. Who was the first person to see bacterial cells with the microscope?
   A. Pasteur
   B. Koch
   C. Leeuwenhoek - 1674
   D. Hooke
2. What process was studied by Redi and Spallanzani?
   A. Spontaneous generation
   B. Fermentation
   C. Variolation
   D. Antisepsis
3. The process of _____ involved the inoculation of dried smallpox scabs under the skin.
   A. vaccination
   B. antisepsis
   C. variolation
   D. immunization
4. What is the name for the field of study established by Semmelweis and Snow in the mid 1800s?
   A. Immunology
   B. Bacteriology
   C. Virology
   D. Epidemiology
5. The process of controlled heating, called _____, was used to keep wine from spoiling.
   A. curdling
   B. fermentation
   C. pasteurization
   D. variolation
6. What surgical practice was established by Lister?
   A. Antisepsis
   B. Chemotherapy
   C. Variolation
   D. Sterilization

7. Which one of the following statements is NOT part of Koch's postulates?
   A. The microorganism must be isolated from a dead animal and pure cultured.
   B. The microorganism and disease can be identified from a mixed culture.
   C. The pure cultured organism is inoculated into a healthy, susceptible animal.
   D. The same microorganism must be present in every case of the disease.
8. Match the lab with the correct set of identified diseases.
   A. Pasteur: tetanus and tuberculosis
   B. Koch: anthrax and rabies
   C. Koch: cholera and tuberculosis
   D. Pasteur: diphtheria and typhoid fever
9. What group of microbial agents would eventually be identified from the work of Ivanowsky and Beijerinck?
   A. Viruses
   B. Fungi
   C. Protists
   D. Bacteria
10. What microbiological field was established by Winogradsky and Beijerinck?
   A. Virology
   B. Microbial ecology
   C. Bacteriology
   D. Mycology
11. What group of microorganisms has a variety of internal cell compartments and where some members act as decomposers?
   A. Bacteria
   B. Viruses
   C. Archaea
   D. Fungi
12. Which one of the following organisms was NOT a model organism related to the birth of molecular genetics?
   A. *Streptococcus*
   B. *Penicillium*
   C. *Escherichia*
   D. *Neurospora*
13. Which group of microbial agents is eukaryotic?
   A. Bacteria
   B. Viruses
   C. Archaea
   D. Algae
14. The term antibiotic was coined by _____ to refer to antimicrobial substances naturally derived from _____.
   A. Waksman; bacteria and fungi
   B. Domagk; other living organisms
   C. Fleming; fungi and bacteria
   D. Ehrlich; bacteria
15. Which one of the following is NOT considered an emerging infectious disease?
   A. Polio
   B. Severe acquired respiratory syndrome
   C. Lyme disease
   D. AIDS
16. A _____ is a mixture of _____ that form as a complex community.
   A. genome; genes
   B. biofilm; microbes
   C. biofilm; chemicals
   D. miasma; microbes

True-False

Each of the following statements is true (T) or false (F). If the statement is false, substitute a word or phrase for the underlined word or phrase to make the statement true.

17. _____ Leeuwenhoek believed that animalcules arose spontaneously from decaying matter.
18. _____ Pasteur proposed that "wine disease" was a souring of wine caused by yeast cells.
19. _____ Separate bacterial colonies can be observed in a broth culture.
20. _____ Semmelweis proposed that cholera was a waterborne disease.
21. _____ Some bacterial cells can convert nitrogen gas (N_2) into ammonia (NH_3).
22. _____ Fungi are eukaryotic microorganisms, some of which are decomposers.
23. _____ Koch proposed the germ theory.
24. _____ Variolation involved inoculating individuals with smallpox scabs.
25. _____ Mycology is the scientific study of viruses.

STEP B: CONCEPT REVIEW

26. Describe the concept of **spontaneous generation** and distinguish between the experiments that supported and refuted the belief. (**Key Concept 1**)
27. Assess the importance of the work carried out by Semmelweis and by Snow that went against the **miasma** view of disease and established the field of **epidemiology.** (**Key Concept 2**)
28. Compare Jenner's work on smallpox and Pasteur's studies on anthrax and rabies to the concept of preventing disease through **vaccination.** (**Key Concepts 2 and 3**)
29. Discuss Pasteur's early studies and Lister's surgical work suggesting that **germs** cause disease. (**Key Concept 3**)
30. Judge the importance of (a) the **germ theory of disease** and (b) **Koch's postulates** to the identification of microbes as agents of infectious disease. (**Key Concept 3**)
31. Provide evidence to support the statement: "Not all microbes cause disease; many play important roles in the environment." (**Key Concept 4**)
32. Construct a concept map for **Microbial Agents**, using the following terms. (**Key Concept 4**)

Algae	Fungi	Protists
Archaea	Microorganisms	Protozoa
Bacteria	Nucleated cells	Viruses
Decomposers	Pathogens (germs)	

33. Distinguish between the "new generation" of scientists in the second Golden Age of microbiology that set the stage for the antibiotic revolution. (**Key Concept 5**)
34. Assess the importance of **microbial ecology** and **microbial evolution** to the current Golden Age of microbiology. (**Key Concept 6**)

STEP C: APPLICATIONS AND PROBLEM SOLVING

35. As a microbiologist in the 1940s, you are interested in discovering new antibiotics that will kill bacterial pathogens. You have been given a liquid sample of a chemical substance to test in order to determine if it kills bacterial cells. Drawing on the culture techniques of Robert Koch, design an experiment that would allow you to determine the killing properties of the sample substance.

36. As a microbial ecologist, you discover a new species of microbe. How could you determine if it has a prokaryotic or eukaryotic cell structure? Suppose it has a eukaryotic structure. What information would be needed to determine if it is a member of the protista or fungi?

37. One of the foundations of scientific inquiry is proper experimental design involving the use of controls. What is the role of a control in an experiment? For each of the experiments described in the section on spontaneous generation, identify the control(s) and explain how the interpretation of the experimental results would change without such controls.

38. You isolate and pure culture a bacterial organism from ill humans that you believe causes the disease. However, you cannot find a susceptible animal for testing that contracts the disease. What would you conclude from these observations?

STEP D: QUESTIONS FOR THOUGHT AND DISCUSSION

39. Many people are fond of pinpointing events that alter the course of history. In your mind, which single event described in this chapter had the greatest influence on (a) the development of microbiology and (b) medicine?

40. Louis Pasteur once stated: "In the field of observation, chance favors only the prepared mind." How does this quote apply to the work done by (a) Semmelweis, (b) Snow, and (c) Fleming?

41. One reason for the rapid advance in knowledge concerning molecular biology during the second Golden Age of microbiology was because many researchers used microorganisms as model systems. Why would bacterial cells be more advantageous to use for research than, say, rats or guinea pigs?

42. When you tell a friend that you are taking microbiology this semester, she asks, *"Exactly what is microbiology?"* How do you answer her?

43. As microbiologists continue to explore the microbial universe, it is becoming more apparent that microbes are "invisible emperors" that rule the world. Now that you have completed Chapter 1, provide examples to support the statement: Microbes Rule!

44. Who would you select as the "first microbiologist?" (a) Leeuwenhoek, (b) Hooke, or (c) Pasteur and Koch. Support your decision.

45. Felix Pouchet was a French biologist and science writer who believed in spontaneous generation. As such, he was often in debate with Pasteur because he was not convinced that Pasteur's experiments refuted the idea of spontaneous generation. As proof, Pouchet set up a series of swan-necked flasks identical to those used by Pasteur to refute spontaneous generation. He then filled the flasks with a broth made from hay, boiled the flasks for one hour, and allowed them to cool. Everything was identical to Pasteur's experiments except Pasteur used a sugar and yeast extract broth and only boiled the flasks for a few minutes.

In all cases, Pouchet saw growth of microorganisms in all his flasks, even with boiling for one hour. Propose a solution for the contradictory results of Pasteur and Pouchet, knowing that (a) what both scientists saw was valid and correct, and (b) spontaneous generation does not occur.

If you remain stumped, check out the short paper entitled: John Tyndall and the Spontaneous Generation Debate (**Microbiology Today, November 2005-http:www.sgm.ac.uk/pubs/micro_today/pdf/110501.pdf**).

CHAPTER PREVIEW

2.1 Organisms Are Composed of Atoms

2.2 Chemical Bonds Form Between Reactive Atoms

2.3 All Living Organisms Depend on Water

2.4 Living Organisms Are Composed of Four Types of Organic Compounds

TEXTBOOK CASE 2: An Outbreak of *Salmonella* Food Poisoning

Investigating The Microbial World 2: Chemical Evolution

MICROINQUIRY 2: Is Protein or DNA the Genetic Material?

The Chemical Building Blocks of Life

The significant chemicals in living tissue are rickety and unstable, which is exactly what is needed for life.
—Isaac Asimov (1920–1992)

The origin of life is one of the great unsolved problems of science. We are fairly certain that microbial life established itself on Earth about 3.5 billion years ago, although no one can definitely say how or where life originated. In fact, there are many hypotheses. About all that is known is that a common primitive life form gave rise to bacterial and archaeal cells. But even this is not certain. Assuming that life did arise here on Earth, did such life arise but once or was there opportunity for "life" arising more than once—and in a different chemical form?

Paul Davies is an award-winning physicist and director of BEYOND: Center for Fundamental Concepts in Science at Arizona State University. One of the "big questions" he and others are pursuing is whether "alien" life may be hiding right in front of our noses. His hypothesis is that perhaps life formed several times on planet Earth and still exists here today in a so-called "shadow biosphere." To pursue this controversial idea, scientists have begun searching high and low (literally in the air and deep in the crust of the earth) for evidence of "alien" life-forms—specifically microorganisms that, as the result of a "second genesis," would differ chemically from all known microbial life because they arose independently. Davies believes an alternate and distinctive chemistry would most likely make alien life different. Microbes and all known life essentially use the same chemical building blocks and work with an almost identical genetic code. Perhaps any undiscovered life forms would look like prokaryotic cells, but their biochemistry might single them out as "alien."

■ Biosphere: That part of the earth—including the air, soil, and water—where life occurs.

Image courtesy of Dr. Fred Murphy/CDC.

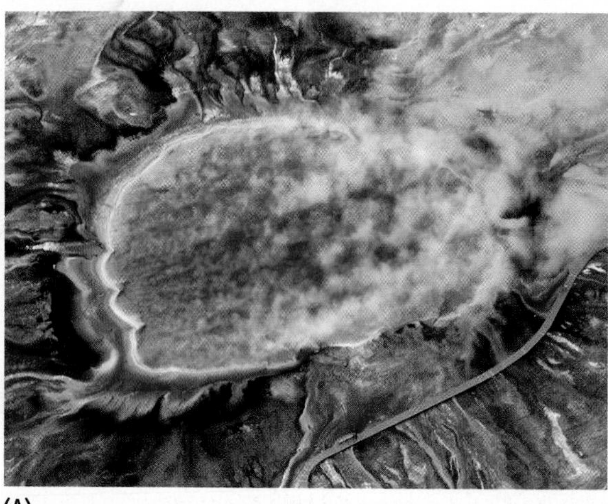
(A)

(B)

FIGURE 2.1 **Cellular Chemistry Allows Microbes to Colonize and Survive in Earth's Extreme Environments.** **(A)** Yellowstone's Grand Prismatic Spring. The gentle flow of heated water spreads out in terraces, with green- and red-pigmented bacterial mats thriving in the warm, shallow water toward the edge. Courtesy of Jim Peaco/Yellowstone National Park. **(B)** Microbial communities can thrive in extremely alkaline and salty waters, such as Mono Lake in California. © Geir Olav Lyngfjell/ShutterStock, Inc. »» What other extreme environments can you identify where microbes might survive?

Microbiologists have identified only about 1% of the organisms forming the microbial universe here on Earth. Because we know so little about the diversity of these microbes, it is possible there is or are other alien forms dispersed or hidden among the undiscovered. Importantly, that is what **science** is all about—the field of study that based on evidence tries to describe and comprehend the nature of the universe in whole or part, wherever that might lead.

Microorganisms are found in most, if not all, habitats on Earth. Some microbes can survive high temperatures of a hot spring, the acidic runoff from a mine, or alkaline, hypersaline conditions of some lakes (FIGURE 2.1). In these cases, as with all habitats where microbes exist, their survival depends on **cellular chemistry**, the chemical reactions between atoms and molecules that provide for the unique metabolism found in microbial cells.

Therefore, studying the basic principles of chemistry enables microbial ecologists, biochemists, and scientists in many other disciplines to understand the daily lives of the microbial workforce. Take another look at Figure 2.1. How does microbial chemistry in a hot spring differ from that in an alkaline lake? Or, in medicine, how can understanding the biochemistry of a pathogen lead to the development of better antibiotics and vaccines to cure or prevent infectious disease?

This chapter serves as a primer or review of the fundamental concepts of chemistry that form a foundation for the chapters ahead. We will identify the elements making up all known substances and show how these elements combine to form the major groups of organic compounds found in all "known" forms of life.

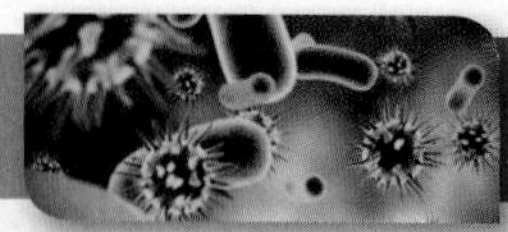

Chapter Challenge

As we will see in this chapter, we have a relatively good understanding of the chemical building blocks needed for life—as we know it. But if some unrecognized form of microbial life exists on Earth, it must be composed of some unique chemical building blocks to build large substances. How would we recognize them? What are the potential markers by which such "shadow life" could be detected? Let's investigate!

© qcontrol/ShutterStock, Inc.

KEY CONCEPT 2.1 Organisms Are Composed of Atoms

As far as scientists know, all matter in the physical universe—be it a rock, a tree, or a microbe—is built of substances called chemical elements. **Chemical elements** are the most basic forms of matter and they cannot be broken down into simpler substances by ordinary chemical means. Ninety-two naturally occurring elements have been discovered, while additional elements have been made in the laboratory or nuclear reactor.

Only about 25 of the 92 naturally occurring elements are essential to the survival of living organisms. Many of these are major elements needed in relatively large amounts (TABLE 2.1). Note that just six of these elements—carbon, hydrogen, nitrogen, oxygen, phosphorus, and sulfur—make up about 98% of the weight in both human and bacterial cells. (The acronym CHNOPS is helpful in remembering these six important elements.) Another five elements make up most of the remaining 2%.

In addition, there are a number of elements needed in much smaller amounts. These so-called "trace elements" vary from organism to organism, but often include such elements as manganese (Mn), iron (Fe), copper (Cu), and zinc (Zn).

Atoms Are Composed of Charged and Uncharged Particles

An **atom** is the smallest unit of an element having the properties of that element; that is, if you split an atom like carbon into simpler parts, it no longer has the properties of carbon. Simply stated, carbon consists of carbon atoms, oxygen of oxygen atoms, and so forth.

An atom consists of a positively charged core, the **atomic nucleus** (FIGURE 2.2A). The atomic nucleus makes up most of the atom's mass and

■ Matter:
Anything that occupies space and has mass.

■ Mass:
The quantity of matter in a sample.

TABLE 2.1 Some of the Major Elements of Bacterial and Human Cells

		Percent by Mass			
Element	Symbol	Bacterial Cell	Human Cell	Atomic Number	Mass Number
Oxygen	O	72	65	8	16
Carbon	C	12	18	6	12
Hydrogen	H	10	10	1	1
Nitrogen	N	3	3	7	14
Phosphorus	P	0.6	1	15	31
Sulfur	S	0.3	0.3	16	32
Sodium	Na	1.0	0.2	11	23
Magnesium	Mg	0.5	0.1	12	24
Calcium	Ca	0.5	1.5	20	40
Potassium	K	1.0	0.4	19	39
Chlorine	Cl	0.05	0.15	17	35

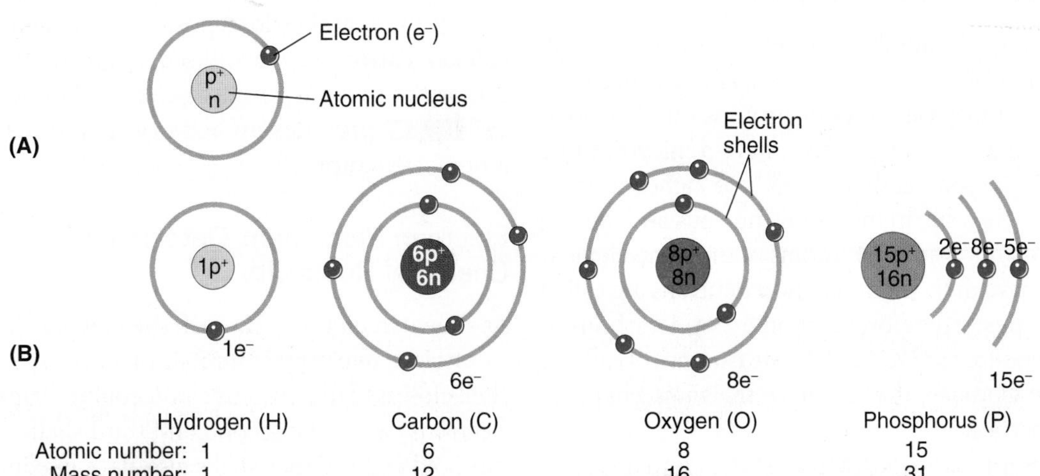

FIGURE 2.2 **Atomic Structure and the Electron Configurations for Four Biologically Important Elements.** (**A**) The atom is composed of protons and neutrons in the atomic nucleus, and electrons that are found in electron shells. (**B**) The atomic structure of four biologically essential elements illustrates that the number of protons equals the number of electrons (though not necessarily equal to the number of neutrons). »» Knowing the mass number for an element, what does that tell you about the mass of an electron?

contains two kinds of tightly packed particles called **protons** and **neutrons**. Although these particles have about the same mass, protons bear a positive electrical charge (value = +1), while neutrons have no charge.

The number of protons in an atom defines each element. For example, carbon atoms always have six protons. If there are seven protons, it is no longer carbon but rather the element nitrogen. The number of protons also represents the **atomic number** of the atom. As shown in Table 2.1, carbon with six protons has an atomic number of 6. The **mass number** is the total number of protons and neutrons in the nucleus. Because carbon atoms have six protons and usually six neutrons, the mass number of carbon would be 12.

Surrounding the atomic nucleus is a negatively charged cloud of fast-moving particles called **electrons** (value = −1). In any uncharged atom, the number of electrons is equal to the number of protons; that is, an atom has no net electrical charge. Because electrons move so fast, it is impossible at any moment to predict where a particular electron might be located. However, we can identify the spaces within the atom where electrons are usually found. These spaces are called **electron shells**, each shell representing a different energy level. FIGURE 2.2B provides a simple diagram of the structures, atomic numbers, and mass numbers of four atoms essential to life.

Atoms Can Vary in the Number of Neutrons or Electrons

Although the number of protons is the same for all atoms in an element, the number of neutrons in an element may vary, altering its mass number. Most carbon atoms, for example, have a mass number of 12, but some carbon atoms have eight neutrons, rather than six, in the atomic nucleus and, hence, a mass number of 14. Atoms of the same element that have different numbers of neutrons are called **isotopes**. Therefore, carbon-12 and carbon-14 (symbolized as ^{12}C and ^{14}C) are isotopes of carbon. Such isotopes, though, have the same chemical properties.

Some isotopes are unstable and give off energy in the form of radiation. Such **radioisotopes** are useful in research and medicine. ^{14}C can be incorporated into an organic substance and isotopes of other elements can be used as radioactive **tracers** to follow the fate of a substance, as MicroInquiry 2 demonstrates (at the end of this chapter).

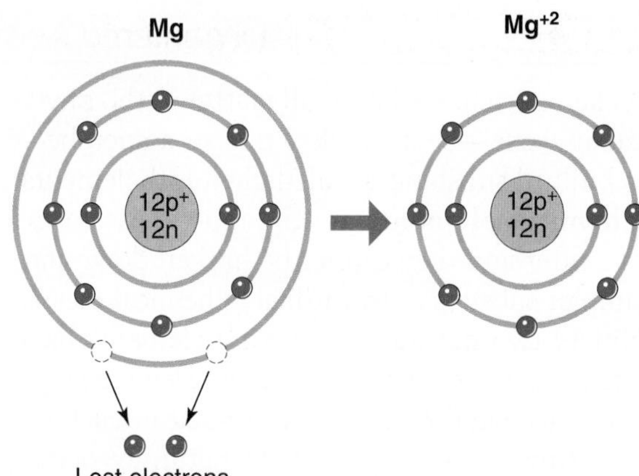

FIGURE 2.3 **Formation of an Ion.** Ions can be formed by the loss or gain of one or more electrons. Here, a magnesium (Mg) atom has lost two electrons to become a magnesium ion (Mg^{+2}). »» For Mg, what does the superscript denote?

Atoms are uncharged when they contain equal numbers of electrons and protons. Should an atom acquire an electrostatic charge, it is called an **ion** (FIGURE 2.3). The addition of one or more electrons to an atom means there is/are more negatively charged electrons than positively charged protons. Such a negatively charged ion is called an **anion**. By contrast, the loss of one or more electrons leaves the atom with extra protons and yields a positively charged ion, called a **cation**. As we will see, ion formation is important to some forms of chemical bonding. FIGURE 2.4 provides an activity to summarize atomic structure.

Electron Placement Determines Chemical Reactivity

As shown in Figure 2.2B, each shell (energy level) can hold a maximum number of electrons. The shell closest to the nucleus can accommodate two electrons, while the second and third shells each can hold eight. Other shells also have maximum numbers but usually no more than 18 are present in those outer shells. Because the 25 essential elements are of lower mass number, only the first

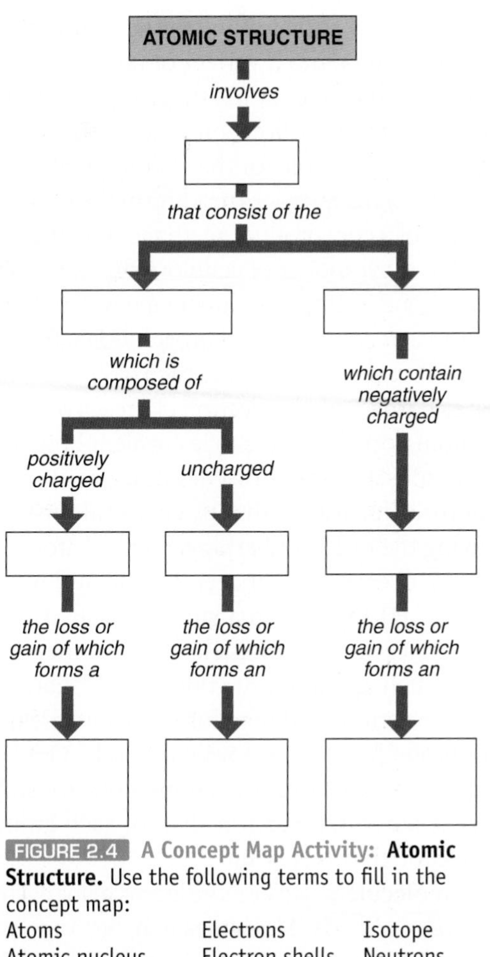

FIGURE 2.4 **A Concept Map Activity: Atomic Structure.** Use the following terms to fill in the concept map:

Atoms	Electrons	Isotope
Atomic nucleus	Electron shells	Neutrons
Different element	Ion	Protons

few shells are of significance to life. Inner shells are filled first and, if there are not enough electrons to completely fill the shell, the outermost shell is left incompletely filled.

Atoms with an unfilled outer electron shell are unstable and tend to be chemically reactive. They can become stable by interacting with another unstable atom. A carbon atom, with six electrons, has two electrons in its first shell and only four in the second (see Figure 2.2B). For this reason, carbon is extremely reactive in "finding" four more electrons and, as we will see, forms innumerable combinations with other elements. Therefore, only atoms with unfilled outer shells will participate in a chemical reaction.

The shells of a few elements normally are filled completely. Each of these elements, called an "inert gas," is chemically stable and unreactive. Helium (atomic number 2) and neon (atomic number 10) are examples, as each has its outermost electron shell filled.

CONCEPT AND REASONING CHECKS 1

a. Explain how the atomic number differs from the mass number.

b. Describe how an isotope differs from an ion.

c. Looking at Figure 2.2B, do these atoms have filled outer electron shells? Explain.

KEY CONCEPT 2.2 Chemical Bonds Form Between Reactive Atoms

Isaac Asimov's chapter-opening quote that chemicals are "rickety and unstable" applies to how atoms interact. When the electron shells of two unstable atoms come close, the electron shells overlap, an energy exchange takes place, and each of the participating atoms assumes a more stable electron configuration. When two or more atoms are linked together, the force holding them is called a **chemical bond**. Chemical bonds are the result of these rickety, unstable atoms filling their outer electron shells.

The rearrangement of atoms through chemical bonding can occur in one of two major ways: atoms, as ions, can interact electrostatically; or, each reactive atom can share electrons with one or more other reactive atoms. In both cases, the result is atoms having full electron shells.

Ionic Bonds Form Between Oppositely Charged Ions

In the formation of an **ionic bond**, one atom gives up its outermost electrons to another. The reaction between sodium and chlorine is illustrative of how these atoms become ions and then form ionic bonds (FIGURE 2.5). Both atoms now have their outermost shell filled. Because opposite electrical charges attract each other, the chloride ions and sodium ions come together to form stable sodium chloride (NaCl).

Salts are typically formed through ionic bonding. Besides sodium and chloride, important salts are formed from other ions, including calcium (Ca^{+2}), potassium (K^{+2}), magnesium (Mg^{+2}), and iron (Fe^{+2} or Fe^{+3}). Although ionic bonds are relatively weak, they play important roles in protein

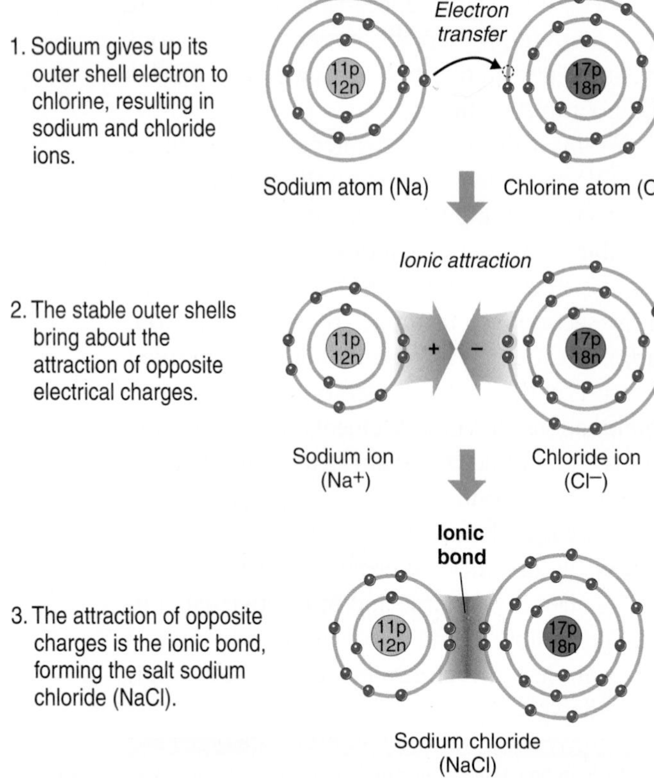

FIGURE 2.5 **Ion Formation and Ionic Bonding.** The transfer of an electron from sodium to chlorine generates oppositely charged ions that are attracted to one another by forming an ionic bond. »» Explain why the sodium ion has a net positive charge and the chloride ion has a net negative charge.

structure and the reactions between antigens and antibodies in the immune response.

When two or more different elements interact with one another to achieve stability, they form a **compound**. Each compound, like each element, has a definite formula and set of properties that distinguish it from its components. For example, sodium (Na) is an explosive metal and chlorine (Cl) is a poisonous gas, but the compound they form is crystals of edible table salt (NaCl).

Covalent Bonds Share Electrons

Reactive atoms also can achieve stability by sharing electrons between the atoms, the sharing producing a **covalent bond**. Such strong bonds are very important in biology because the CHNOPS elements of life usually enter into covalent bonds with themselves and one another.

Covalent bonding occurs frequently in carbon because this element has four electrons in its unfilled outer shell. The carbon atom is not strong enough to acquire four additional electrons, but it is sufficiently strong to retain the four it has. It therefore enters into a variety of covalent bonds with other reactive atoms or groups of atoms. The vast array of carbon compounds that can be formed is responsible for the chemistry of life.

Many of the microbes residing in the ruminant stomach of a cow produce methane or natural gas (CH_4) as a byproduct of cellulose digestion. This gas is a good example to illustrate covalent bonding between carbon and hydrogen (FIGURE 2.6A). A reactive carbon atom shares each of its four outer shell electrons with the electron of a reactive hydrogen atom, forming four single covalent bonds.

Scientists often draw chemical structures as **structural formulas**; that is, chemical diagrams showing the order and arrangement of atoms. In Figure 2.6, each line between carbon and hydrogen (C—H) represents a single covalent bond between a pair of shared electrons. Other molecules, such as carbon dioxide (CO_2), share two pairs of electrons and therefore two lines are used to indicate the "double covalent bond": O=C=O. However, in all cases, the atoms now are stable because the outer electron shell of each atom is filled through this sharing.

A **molecule** is two or more atoms held together by covalent bonds. Molecules may be composed of only one kind of atom, as in oxygen gas (O_2), or they may consist of different kinds of atoms in substances such as water (H_2O), carbon dioxide (CO_2), and the simple sugar glucose ($C_6H_{12}O_6$). As shown by these examples, the kinds and amounts of atoms (the subscript) in a molecule is called the **molecular formula**. (Note that the presence of one atom is represented without the subscript "1.")

The simplest derivatives of carbon are the **hydrocarbons**, molecules consisting solely of carbon and hydrogen. Methane is the most fundamental hydrocarbon. Other hydrocarbons consist of chains of carbon atoms and, in some cases, the chains may be closed to form a ring (FIGURE 2.6B). Many form the basis for crude oil (MicroFocus 2.1).

When atoms bond together to form a molecule, they establish a geometric relationship determined largely by the electron configuration. Notice in the hydrocarbons drawn in Figure 2.6B that the covalent bonds are distributed equally around each carbon atom. Each of these examples of the "equal sharing" of electron pairs represents a **nonpolar molecule**—there are no electrical charges (poles) and the bonds are called "nonpolar covalent bonds."

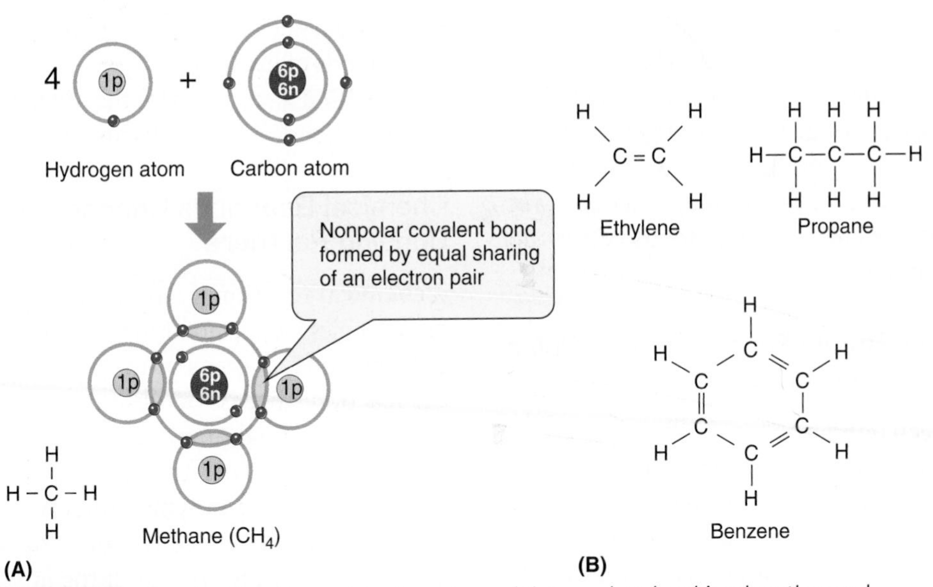

FIGURE 2.6 **Chemical Bonding and Hydrocarbons.** (**A**) A covalent bond involves the equal sharing of electron pairs between atoms, the example shown here being the simple organic compound methane. (**B**) The molecular formulas for a few relatively simple hydrocarbons.
»» Supply the molecular formula for each of the structural formulas shown in (B).

MICROFOCUS 2.1: Environmental Microbiology

Microbes to the Rescue!

On April 20, 2010, the Deepwater Horizon drilling rig that was working on a well for the British Petroleum oil company blew up in the Gulf of Mexico. Four days later, it was discovered that the wellhead was damaged and was leaking oil and methane gas into the Gulf. For 3 months, oil spilled into the Gulf and contaminated nearby shores and wetlands (see figure), making it the largest accidental oil spill in history. According to federal government estimates, some 5 million barrels, or 780 million liters, were spilled and it was not until September that the well was declared sealed. Where did all those hydrocarbons go?

Petroleum or crude oil consists of a complex mixture of hydrocarbons of various sizes and liquid, gaseous, and solid forms. An oil well, such as the one in the Gulf, produces primarily crude oil, with some natural gas (primarily methane) dissolved in it. For decades, scientists have tried to use bioremediation, the breakdown (biodegradation) of contaminating compounds using microorganisms, as a natural method for cleaning up some of the environment's worst chemical hazards, including oil spills.

Sunlight is reflected off the Deepwater Horizon oil spill in the Gulf of Mexico on May 24, 2010. The image was taken by NASA's Terra satellite. Courtesy of NASA/GSFC, MODIS Rapid Response.

But what about all the hydrocarbons and methane gas that formed the deep-water dispersed oil plume from the ruptured Deepwater Horizon well? Could indigenous bacterial species act as natural bioremediation agents? It appears they did! The entry of oil profoundly altered the microbial community by significantly stimulating deep-sea, oil-hungry microbes to consume much of the oil and methane gas. It remains to be determined just how efficient these oil denizens were in mopping up the hydrocarbons, but it is clear they were intimately involved in the clean up. Still, microbes cannot eliminate or digest all the hydrocarbons present in oil and it may be several years before all the oil is completely gone.

Not all molecules are nonpolar. Indeed, one of the most important molecules to life, water, is a **polar molecule**—it has electrically charged poles (FIGURE 2.7A). Oxygen has a stronger "pull" on the shared electrons and thus has a slight negative charge. The hydrogen atoms are then left with a slight positive charge. The water molecule therefore consists of "polar covalent bonds."

Hydrogen Bonds Form Between Polar Groups or Molecules

A **hydrogen bond** involves the attraction of a partially positive hydrogen atom that is covalently bonded to one polar molecule toward another polar molecule having either a partially negative oxygen atom (H^+–O^-) or nitrogen atom (H^+–N^-). Although hydrogen bonds are much weaker than covalent bonds, hydrogen bonds provide the "glue" to hold water molecules together (FIGURE 2.7B). These bonds also are important to the structure of proteins and nucleic acids, two of the major organic compounds of living cells.

TABLE 2.2 summarizes the different types of chemical bonds we have discussed.

Chemical Reactions Change Bonding Partners

A **chemical reaction** is a process in which atoms or molecules interact to form new bonds. Different combinations of atoms or molecules result from the reaction; that is, bonding partners change. However, the total number of interacting atoms remains constant. For chemical reactions, an arrow is used to indicate in which direction the reaction will proceed. By convention, the atoms or molecules drawn to the left of the arrow are the **reactants** and those to the right are the **products** of the reaction.

In biology, many chemical reactions are based on the assembly of larger compounds or the break-

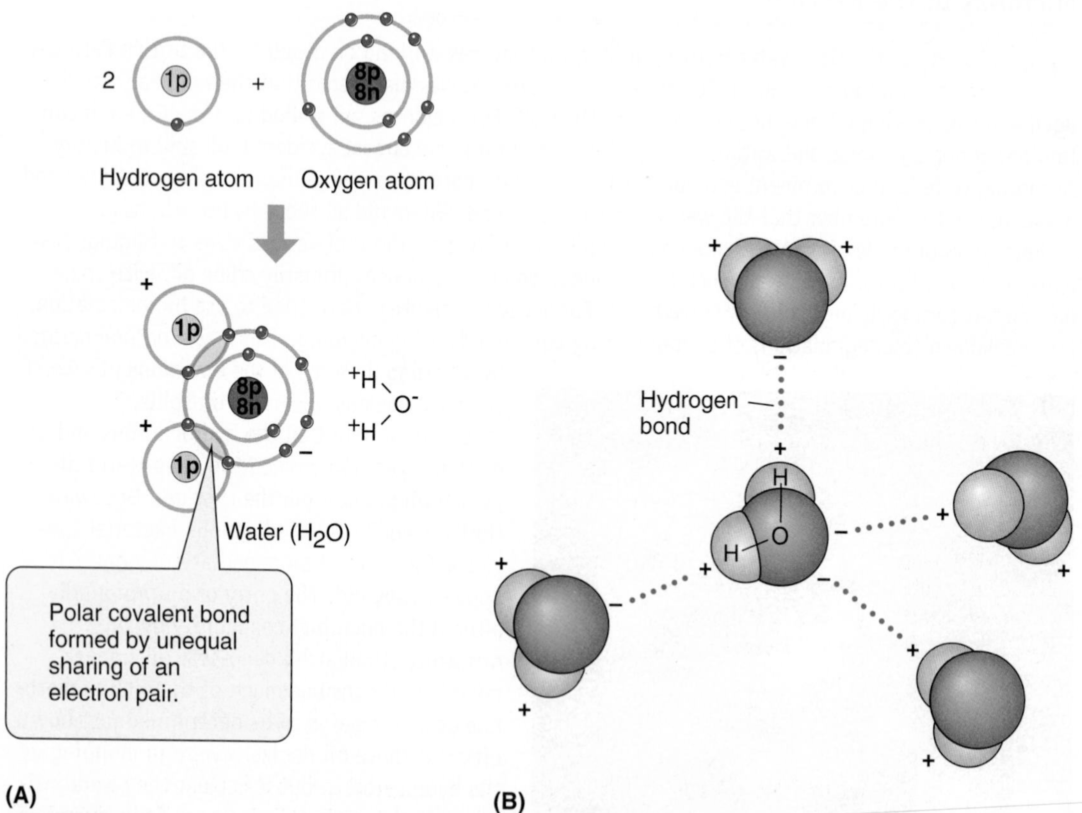

FIGURE 2.7 **Chemical Bonding and Water.** (**A**) Due to the unequal sharing of electron pairs between oxygen and hydrogen, a polar molecule results. (**B**) Due to the charged regions of each polar water molecule, hydrogen bonds form between the hydrogen of one water molecule and the oxygen of another water molecule. »» Why is there attraction between a hydrogen and oxygen atom?

TABLE

2.2 Three Types of Chemical Bonds in Living Organisms

Type	Chemical Basis	Strength	Example
Ionic	Attraction between oppositely charged ions	Weak	Sodium chloride; salts
Covalent	Sharing of electron pairs between atoms	Strong	Glucose
Hydrogen	Attraction of a hydrogen nucleus (a proton) to negatively charged oxygen or nitrogen atoms in the same or neighboring molecules	Weak	Water

ing apart of larger compounds into smaller ones. In a "synthesis" reaction, smaller reactants are put together into larger products. If water is involved as a product, often it is called a **dehydration synthesis (condensation) reaction**:

$$\underset{\text{Glucose}}{C_6H_{12}O_6} + \underset{\text{Glucose}}{C_6H_{12}O_6} \rightarrow \underset{\text{Maltose}}{C_{12}H_{22}O_{11}} + \underset{\text{Water}}{H_2O}$$

The reverse is a "decomposition" reaction, where a larger reactant is broken into smaller products. Often in biology, water is one of the reactants used to break a molecule, so it is referred to as a **hydrolysis reaction** (*hydro* = "water"; *lysis* = "break"):

$$\underset{\text{Maltose}}{C_{12}H_{22}O_{11}} + \underset{\text{Water}}{H_2O} \rightarrow \underset{\text{Glucose}}{C_6H_{12}O_6} + \underset{\text{Glucose}}{C_6H_{12}O_6}$$

Most importantly, the new products formed have the same number and types of atoms that were present in the reactants. In forming new products, chemical reactions only involve a change in the bonding partners. No atoms have been gained or lost from any of these reactions.

CONCEPT AND REASONING CHECKS 2

a. Construct a diagram to show how the salt calcium chloride ($CaCl_2$) is formed.

b. Why do atoms share electron pairs?

c. Construct a diagram to show how hydrogen bonding occurs in a molecule of liquid ammonia (NH_3).

d. In the dehydration synthesis and hydrolysis reactions drawn above, what are the reactants and products in each reaction?

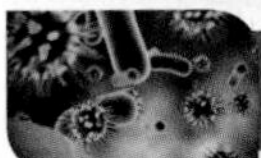

Chapter Challenge A

At this point it should be clear that all life as we know it is carbon based. With four electrons to share, carbon can form many bonds with like or different elements. Some scientists have proposed that the element silicon (Si) could replace carbon. After all, silicon is the second most abundant element in the Earth's crust (oxygen is first and carbon is fifteenth) and one characteristic feature of diatoms (a type of alga) is the presence of a cell wall containing silica (hydrated silicon dioxide). A few scientists have even suggested that the first life on Earth (microbial, we assume) might have been silicon based.

QUESTION A:

Knowing that silicon (Si) has an atomic number of 14 and a mass number of 28, how similar is silicon to carbon that some individuals would propose it as a potential substitute for carbon in "shadow life?" Draw the structural formula for silicon that has four hydrogen atoms attached (called silane). How similar is it to methane? As a note: An original **Star Trek** ***episode ("Devil in the Dark") concerns a silicon-based "life-form," called the "Horta."***

Answers can be found on the Student Companion Website in **Appendix D.**

KEY CONCEPT 2.3 All Living Organisms Depend on Water

All organisms are composed primarily of water. Human cells, as well as bacterial and archaeal cells, are about 70% water by weight. No organism can survive and grow without water, and many organisms, including microbes, live in water.

Water Has Several Unique Properties

■ Solvent:
The liquid doing the dissolving to form a solution.

■ Solute:
The substance dissolved in water.

Liquid water is the medium in which all cellular chemical reactions occur. Being polar, water molecules are attracted to other polar molecules and act as the universal solvent in cells. Take for example what happens when you put a solute like salt in water. The salt is **hydrophilic** because it easily dissolves into separate sodium and chloride ions as water molecules break the weak ionic bonds and surround each ion in a sphere of water molecules (FIGURE 2.8). An **aqueous solution**, which consists of solutes in water, is essential for chemical reactions to occur (MicroFocus 2.2). Molecules that do not dissolve in water are **hydrophobic**.

Water molecules also are reactants in many chemical reactions. The example of the hydrolysis reaction shown on the previous page involved water in splitting maltose into two molecules of glucose.

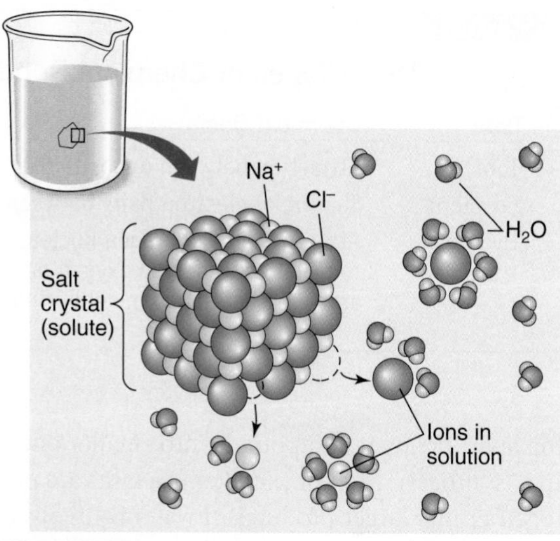

FIGURE 2.8 **Solutes Dissolve in Water.** Water molecules surround Na^+ and Cl^- ions, facilitating their dissolving into solution. »» When dissolving, why do the H's of water surround Cl^- while the O's surround the Na^+?

As you have learned, the polar nature of water molecules leads to hydrogen bonding. By forming a large number of hydrogen bonds between water molecules, it takes a large amount of heat energy to increase the temperature of water. Likewise, a large amount of heat must be lost before water decreases temperature. So, by being about 70% water, cells are bathed in a solvent that maintains a more consistent temperature even when the environmental temperatures change.

MICROFOCUS 2.2: Tools

The Relationship Between Mass Number and Molecular Weight

■ Dalton:
The unit used to measure the weight of atomic particles or molecules; equivalent to an atomic mass unit used in chemistry (one twelfth the weight of an atom of ^{12}C).

Often scientists need to make solutions that have a specific concentration of solutes. To make these solutions, one needs to know how much a particular molecule or solute weighs, which is referred to as the **molecular weight.** The calculation of the molecular weight simply consists of adding together the mass number of all the individual atoms in a substance, such as water, carbon dioxide, salt, or glucose. Thus, the molecular weight of a water molecule is 18 daltons, while the molecular weight of a glucose molecule is 180 daltons. Other molecules can reach astonishing proportions—antibodies of the immune system may have a molecular weight of 150,000 daltons and the bacterial toxin causing botulism is over 900,000 daltons.

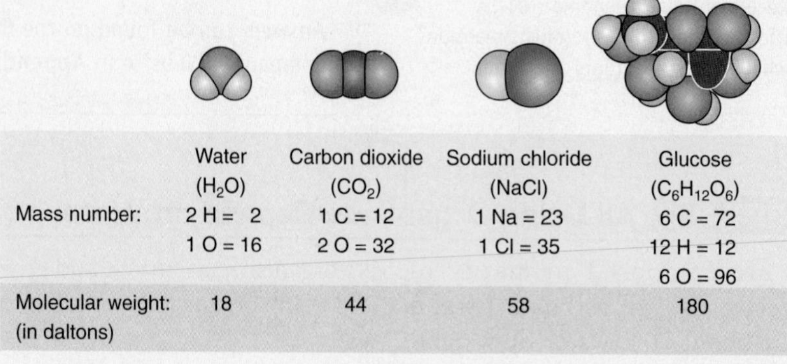

	Water (H_2O)	Carbon dioxide (CO_2)	Sodium chloride (NaCl)	Glucose ($C_6H_{12}O_6$)
Mass number:	2 H = 2 1 O = 16	1 C = 12 2 O = 32	1 Na = 23 1 Cl = 35	6 C = 72 12 H = 12 6 O = 96
Molecular weight: (in daltons)	18	44	58	180

Acids and Bases Affect a Solution's pH

In an aqueous solution, most of the water molecules remain intact. However, some can dissociate spontaneously into hydrogen ions (H^+; protons) and hydroxide ions (OH^-) only to rapidly recombine (the double arrow indicates a reversible reaction):

$$H_2O \leftrightarrow H^+ + OH^-$$

Besides water, other compounds in cells can release H^+ when they dissolve in water. For our purposes, an **acid** (such as hydrochloric acid; HCl) is a chemical substance that donates H^+ to a solution, increasing the hydrogen ion concentration [H^+]:

$$HCl \leftrightarrow H^+ + Cl^-$$

By contrast, a **base** (such as sodium hydroxide; NaOH) is a substance that combines with H^+ in solution:

$$NaOH \rightarrow Na^+ + OH^-$$
$$OH^- + H^+ \rightarrow H_2O$$

To indicate the concentration of H^+ in a solution, the Danish chemist Søren P. L. Sørensen introduced the symbol **pH** (power of hydrogen ions) and the **pH scale**. This numerical scale extends from 0 (extremely acidic; high H^+) to 14 (extremely basic or alkaline; low H^+) and is based on actual calculations of the number of hydrogen ions present when a substance mixes with water. A substance with a pH of 7, such as pure water, is said to be neutral; solutions that gain H^+ are said to be "acidic" and have a pH lower than 7; solutions that lose H^+ are "basic" (or alkaline) and have a pH greater than 7.

The pH scale is logarithmic; that is, every time the pH changes by one unit, the [H^+] changes 10 times. For example, lemon juice (pH 2) and black coffee (pH 5) differ a thousandfold (10^3) in [H^+]. FIGURE 2.9 summarizes the pH values of several common substances. With regard to organisms, fungi prefer a slightly acidic environment compared to the more neutral environment preferred by most microbes—although there are some spectacular exceptions (MicroFocus 2.3).

Cell Chemistry Is Sensitive to pH Changes

As microorganisms—and all organisms—take up or ingest nutrients and undergo metabolism, chemical reactions occur that use up or produce H^+. It is important for all organisms to balance the acids and bases in their cells because chemical reactions and organic compounds are very sensitive to pH shifts. Proteins are especially vulnerable, as we will soon see. If the internal cellular pH is not maintained, these proteins may be destroyed. Likewise, when most microbes grow in a microbiological nutrient medium, the waste products produced may lower the pH of the medium, which could kill the organisms.

To prevent pH shifts, cells and the growth medium contain **buffers**, which are substances that maintain a specific pH. The buffer does not necessarily maintain a neutral pH, but rather whatever pH is required for that environment.

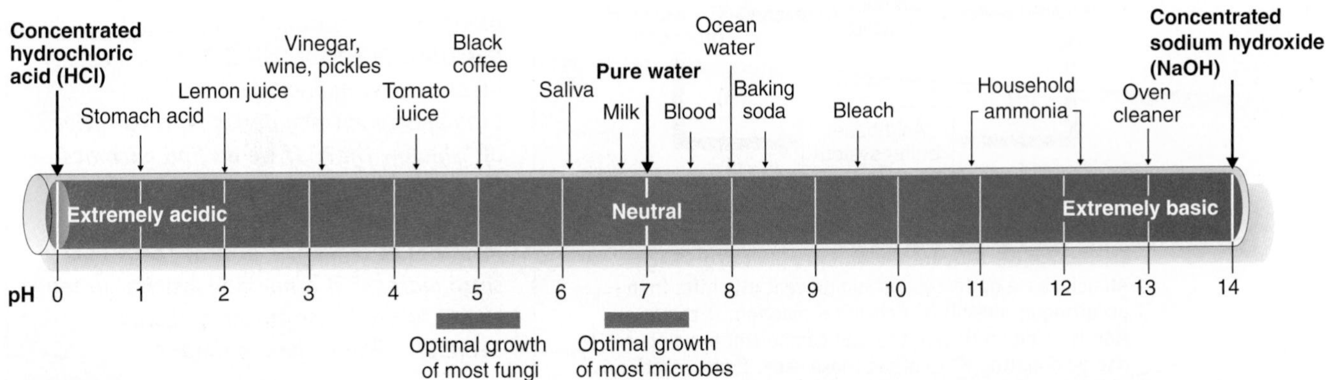

FIGURE 2.9 **A Sample of pH Values for Some Common Substances.** Most fungi prefer a slightly acidic pH for growth compared to most microbes. »» On the pH scale, notice that many of the beverages we drink (e.g., wine, tomato juice, coffee) are fairly acidic. However, we would never normally drink equally alkaline solutions (e.g., commercial bleach, ammonia). Propose an explanation for these observations.

MICROFOCUS 2.3: Environmental Microbiology

Just South of Chicago

All you need is a map, some pH paper, and a few collection vials. When in Chicago, use your map to find the Lake Calumet region just southeast of Chicago. When you arrive, pull out your pH paper and sample some of the groundwater in the region near the Calumet River. You will be shocked to discover the pH is greater than 12—almost as alkaline as oven cleaner! In fact, this might be one of the most extreme pH environments on Earth.

How did the water get this alkaline and could anything possibly live in the groundwater?

The groundwater in the area near Lake Calumet became strongly alkaline as a result of the steel slag that was dumped into the area for more than 100 years. Used to fill the wetlands and lakes, water and air chemically react with the slag to produce lime [calcium hydroxide, $Ca(OH)_2$]. It is estimated that 10 trillion cubic feet of slag and the resulting lime has pushed the pH to such a high value.

Now use your collection vials to collect some samples of the water. Back in the lab you will be surprised to find that there are bacterial communities present in the water. Hydrogeologists who have collected such samples have discovered some bacterial species that until then had only been found in Greenland and deep gold mines of South Africa. Other identified species appear to use the hydrogen resulting from the corrosion of the iron for energy.

How did these bacterial organisms get there? The hydrogeologists propose that the bacterial species have always been there and have simply adapted to the environment over the last 100 years as the slag accumulated. Otherwise, the microbes must have been imported in some way.

So, once again, provide a specific environment and they will come (or evolve)—the microbes that is.

Most biological buffers consist of a weak acid and a weak base (FIGURE 2.10). If an excessive number of H^+ are produced (potential pH drop), the base can absorb them. Alternatively, if there is a decrease in H^+ (potential pH increase), the weak acid can dissociate, replacing the lost H^+.

CONCEPT AND REASONING CHECKS 3

- **a.** Explain the difference between a solution, a solvent, and a solute.
- **b.** What are the properties of acids and bases?
- **c.** If the pH does drop in a cell, what does that tell you about the buffer system?

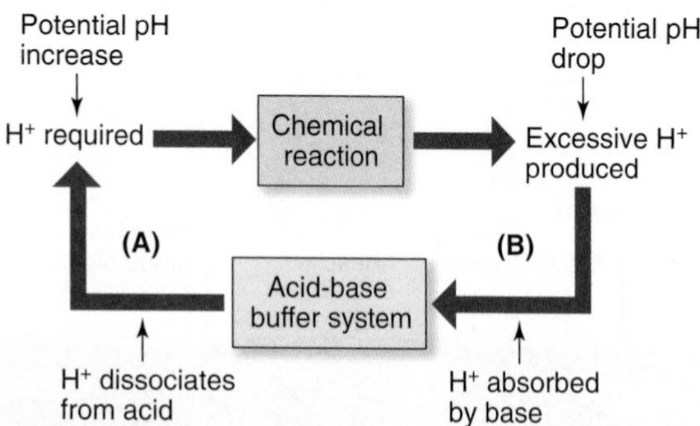

FIGURE 2.10 A Hypothetical Example for pH Shifts. An acid/base buffer system can prevent pH shifts from occurring as a result of a chemical reaction. If the reaction is using up H^+ (**A**), the acid component prevents a pH rise by donating H^+ to offset those used. If the reaction is producing excess H^+ (**B**), the base can prevent a pH drop by "absorbing" them. »» Propose what would happen if a chemical reaction continued to release excessive H^+ for a prolonged period of time.

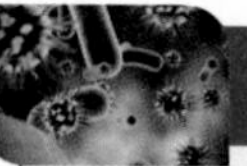

Chapter Challenge B

In trying to find "markers" to detect "shadow life," we run into the problem of microbial diversity. Microbes occupy so many diverse and often extreme habitats, many are called "extremophiles." It is hard to know what "shadow life" really means in the context of life on Earth.

QUESTION B:

True microbial life as we know it can be found, for example, at extreme temperatures (including substantially above the boiling point of water), extreme pressures (in excess of 1,000× atmospheric pressure), and extreme pHs (both exceptionally acidic and, as described in MicroFocus 2.3, tremendously alkaline environments). Are these examples of extremophiles representative of "shadow life?" If we do find microbes that use silicon, is that "alien" or just another example of the flexibility of microbes to make the best of their surroundings? If a microbe lost its carbon source, could it use silicon if it was available? What's your opinion?

Answers can be found on the Student Companion Website in **Appendix D**.

KEY CONCEPT 2.4 Living Organisms Are Composed of Four Types of Organic Compounds

As mentioned in the last section, a typical prokaryotic cell is about 70% water. If all the water is evaporated, the predominant "dry weight" remaining consists of **organic compounds**, which are those molecules related to or having a carbon basis: the carbohydrates, lipids, proteins, and nucleic acids (FIGURE 2.11). Except for the lipids, each class represents a **polymer** (*poly* = many; *mer* = part) built from a very large number of building blocks called **monomers** (*mono* = one).

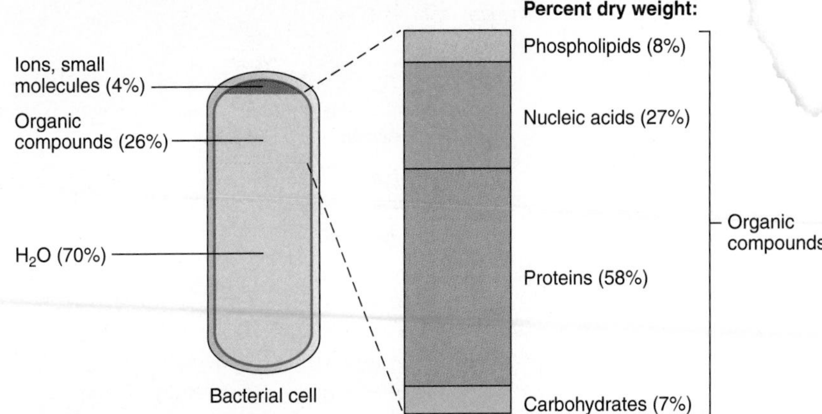

FIGURE 2.11 **Organic Compounds in Bacterial Cells.** Organic compounds are abundant in cells. The approximate composition of these compounds in a bacterial cell is similar to the percentages found in other microbes. »» Propose a reason why proteins make up almost 60% of the dry weight of a bacterial cell.

Functional Groups Define Molecular Behavior

Before we look at the major classes of organic compounds, we need to address one question. The monomers building carbohydrates, nucleic acids, and proteins are essentially stable, unreactive molecules because their outer shells are filled through covalent bonding. Why then should these molecules take part in chemical reactions to build polymers?

The answer is that these monomers are not completely stable. Projecting from the carbon skeletons or other atoms on these biological molecules are groups of atoms called functional groups. **Functional groups** represent points where further chemical reactions can occur if facilitated by a specific enzyme. The reactions will not happen spontaneously.

There is a small number of functional groups but their differences and placement on compounds makes possible a large variety of chemical reactions. The important functional groups in living organisms are identified in TABLE 2.3.

Functional groups on monomers can interact to form larger molecules or polymers through dehydration synthesis reactions. In addition, functional groups can be critical for the decomposition of larger polymers into monomers through hydrolysis reactions.

If you are unsure about the role of these functional groups, do not worry. We will see how specific functional groups interact through dehydration synthesis reactions as we now visit each of the four classes of organic compounds.

Carbohydrates Consist of Sugars and Sugar Polymers

Carbohydrates are organic compounds composed of carbon, hydrogen, and oxygen atoms that build sugars and starches. In simple sugars, like glucose ($C_6H_{12}O_6$), the ratio of hydrogen to oxygen is 2 to 1, the same as in water. The carbohydrates therefore are considered "hydrated carbon." However, the atoms are not present as water molecules bound to carbon but rather carbon covalently bonded to hydrogen and hydroxyl groups (H–C–OH).

■ Enzyme: A protein that facilitates a specific chemical reaction.

TABLE 2.3 **Common Functional Groups on Organic Compounds**

Functional Group	Shorthand	Structural Formula
Hydroxyl	—OH	—O—H
Carboxyl	—COOH	O=C(—)—OH
Carbonyl	—CO—	O=C(—)—
Amino	$—NH_2$	—N(H)—H
Sulfhydryl	—SH	—S—H
Phosphate	$—H_2PO_4$	—O—P(OH)(OH)=O

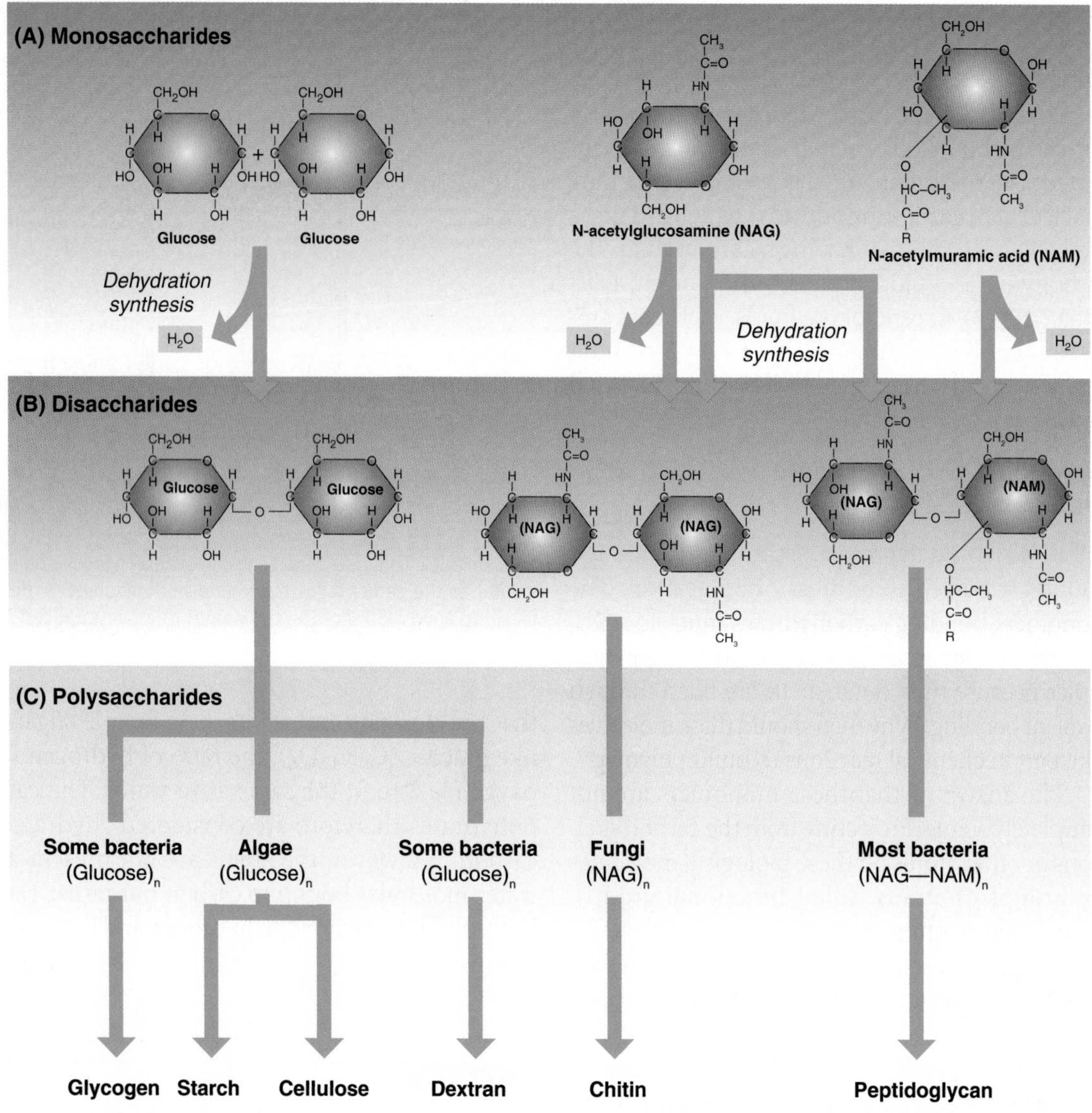

Note: The polysaccharides of glucose [(Glucose)$_n$] vary in the carbon bonding between glucose monomers and branching of the polymer chains.

FIGURE 2.12 **Carbohydrate Monomers Are Built into Polymers.** (**A**) There are many monosaccharides used by organisms that can be combined into disaccharides (**B**) or assembled into long polymers called polysaccharides (**C**).
»» What type of chemical reaction is required to link glucose into a long polymer such as cellulose?

Carbohydrates function as major fuel sources in cells. They also function as structural molecules in cell walls and nucleic acids. Often the carbohydrates are termed saccharides (*sacchar* = "sugar") and are divided into three groups.

Monosaccharides and **disaccharides** are simple sugars; many represent the monomers to build more complex polymers. Glucose, a six-carbon sugar, is one of the most widely encountered monosaccharides (FIGURE 2.12A). Glucose serves as the basic supply for cellular energy. Estimates vary, but many scientists estimate half the world's carbon exists as glucose. Such sugars are synthesized from water and carbon dioxide through the process of **photosynthesis**. Algae and cyanobacteria are microorganisms that have the chemical machinery for this process.

Disaccharides (*di* = two) are composed of two monosaccharides held together by a covalent bond. Sucrose (table sugar) is an example. It is constructed from a glucose and fructose molecule through a dehydration synthesis reaction. Sucrose is a starting point in wine fermentations. Maltose, another disaccharide, is composed of two glucose monomers (FIGURE 2.12B). This disaccharide occurs in cereal grains, such as barley, and is fermented by yeasts for energy. An important byproduct of the fermentation is the formation of alcohol in beer. Lactose, a third common disaccharide, is composed of the monosaccharides glucose and galactose. Lactose is known as "milk sugar" because it is the principal sugar in milk. Under controlled industrial conditions, microorganisms digest the lactose for energy; in the process, they produce the acid in yogurt, sour cream, and other sour dairy products.

In the microbial world, the real significance of monosaccharides is as building blocks for polysaccharides (FIGURE 2.12C). **Polysaccharides** (*poly* = many) are complex carbohydrates formed by joining together hundreds of thousands of similar monomers. Covalent bonds resulting from the reactions link the units together.

Glucose units can be joined together to form starch and glycogen, two common storage polysaccharides in algal and some bacteria cells, where they function as a stored energy source. Cellulose, a structural polysaccharide, is a component of the cell walls of many algae while chitin, built from chains of another glucose derivative, N-acetylglucosamine (NAG), forms the cell walls of fungi. Some bacterial cells also produce dextran that enables the cells to attach to surfaces (MicroFocus 2.4). In most bacterial cells, the cell wall is composed of carbohydrate and protein. The

MICROFOCUS 2.4: Public Health

Sugars, Acid, and Dental Cavities

Each of us at some time probably has feared the dentist's drill after a cavity has been detected. Dental cavities (caries) usually result from eating too much sugar or sweets. These sugars contribute to cavity formation only in an indirect way. The real culprits are oral bacteria.

Many species of microorganisms normally inhabit the mouth (see figure). Some of the bacterial species, along with saliva and food debris, form a gummy layer called "dental plaque," a type of biofilm. If not removed, plaque accumulates on the grooved chewing surfaces of back molars and at the gum line.

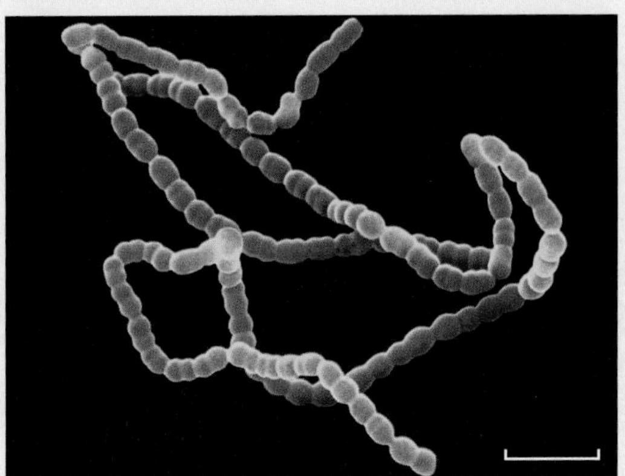

Cells of *Streptococcus mutans,* one of the major agents in dental plaque that produces cavity-causing acid. (Bar = 10 μm.) © Dr. David Phillips/Visuals Unlimited/ Getty.

Plaque starts to accumulate within 20 minutes after eating. As the bacterial cells multiply, they digest the sucrose (table sugar) in sweets for energy. The metabolism of sucrose has two consequences. Some bacterial cells produce dextran, an adhesive polysaccharide that increases the thickness of plaque. They also produce lactic acid as a byproduct of sugar metabolism. Being trapped under the plaque, the acid is not neutralized by the saliva. When the pH drops to 5.5 or lower, the hydrogen ions start to dissolve or demineralize the dental enamel. Over time, a depression or cavity forms. When the soft dental tissues underneath the enamel are reached, toothache pain results from the exposure of the sensitive nerve endings in the soft tissues.

Good oral hygiene, including flossing, brushing, and regular professional dental cleaning, can keep plaque to a minimum. At home, watch what you eat. Consuming sugary foods with a meal or for dessert is less likely to cause cavities because the increased saliva produced while eating helps wash food debris off the tooth surface and neutralize any acids produced. However, snacking on sugary foods that are sticky, like caramel, toffee, dried fruit, or candies, allows the food debris to cling to teeth for a longer time, causing the formation of more plaque and providing a continuous acid attack on your teeth. No wonder cavities are one of the most prevalent infectious diseases, second only to the common cold.

carbohydrate building block is a disaccharide of N-acetylglucosamine and N-acetylmuramic acid (NAM) linked in long chains.

Lipids Are Water-Insoluble Compounds

The **lipids** are a broad group of nonpolar organic compounds that are **hydrophobic**; they do not dissolve in water. Like carbohydrates, lipids are composed of carbon, hydrogen, and oxygen, but the proportion of oxygen is much lower. Lipids serve many microorganisms, but not bacterial species, as important stored energy sources.

Lipids consist of a three-carbon glycerol molecule and up to three long-chain fatty acids (triglyceride) (FIGURE 2.13A). Each fatty acid is a long nonpolar hydrocarbon chain containing between 16 and 18 carbon atoms. Bonding of each fatty acid to the glycerol molecule occurs by a dehydration synthesis reaction between the hydroxyl and carboxyl functional groups.

A fatty acid is considered to be **saturated** if it contains the maximum number of hydrogen atoms extending from the carbon backbone; that is, no double covalent bonds between carbon atoms. A fatty acid is **unsaturated** if it contains less than the maximum hydrogen atoms; that is, there is one or more double covalent bonds between a few carbon atoms.

Another type of lipid found in cell membranes is the **phospholipids**, which have only two fatty acid tails attached to glycerol (FIGURE 2.13B). In place of the third fatty acid there is a phosphate group, representing a functional group that is polar and can actively interact with other polar mole-

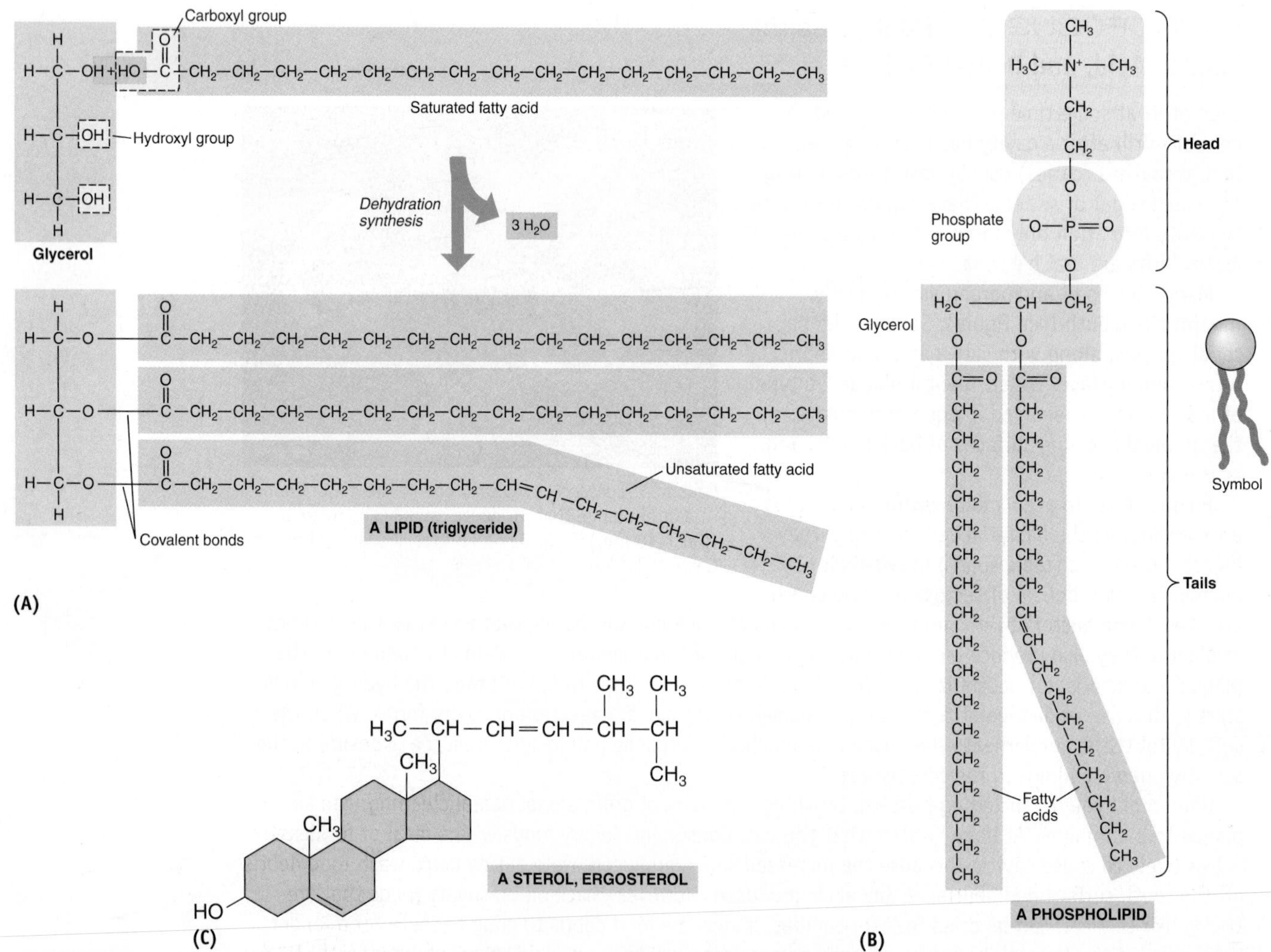

FIGURE 2.13 **Lipid and Lipid-Related Compounds.** (**A**) A lipid, such as a triglyceride, consists of glycerol and fatty acids. (**B**) A phospholipid consists of glycerol attached to two fatty acids and a phosphate head group. Inset: The symbol for the structure of a phospholipid. (**C**) The sterol ergosterol. »» Why are all these compounds considered "lipids"?

Textbook CASE 2

An Outbreak of *Salmonella* Food Poisoning

1. During the morning of October 17, a restaurant employee prepared a Caesar salad dressing, cracking fresh eggs into a large bowl containing olive oil.
2. Anchovies, garlic, and warm water then were mixed into the eggs and oil.
3. The warm water raised the temperature of the mixture slightly before the dressing was placed in the refrigerator.
4. Later that day, the Caesar dressing was placed at the salad bar in a cooled compartment having a temperature of about 16°C. The dressing remained at the salad bar until the restaurant closed, a period of 8 to 10 hours. During that time, many patrons helped themselves to the Caesar salad.
5. Within 3 days, 15 restaurant patrons experienced gastrointestinal illness. Symptoms included diarrhea, fever, abdominal cramps, nausea, and chills. Thirteen sought medical care, and eight (all elderly over 65 years of age) required intravenous rehydration.
6. From the stool samples of all 13 patrons who sought medical attention, bacterial colonies were cultured (see figure) and laboratory tests identified *Salmonella enterica* serotype Enteritidis as the causative agent. All patrons recovered within 7 days.
7. *S. enterica* serotype Enteritidis produces a lipopolysaccharide toxin that causes the symptoms experienced by all the affected patrons.

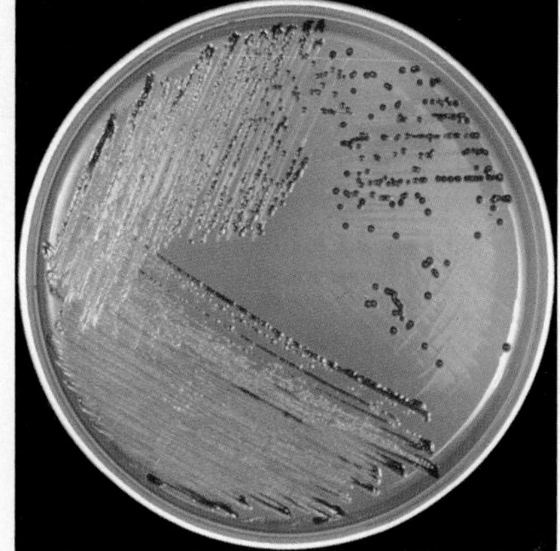

A culture plate of *Salmonella*. Courtesy of CDC.

Questions:

(Answers can be found on the Student Companion Website in **Appendix D.**)

A. What might have been the origin of the bacterial contamination?

B. What conditions would have encouraged bacterial growth?

C. How could the outbreak have been prevented?

D. What types of organic compounds form the lipopolysaccharide toxin?

E. Why did so many of the elderly patrons develop a serious illness?

For additional information see www.cdc.gov/nczved/divisions/dfbmd/diseases/salmonella_enteritidis.

cules. Some bacterial toxins are a combination of polysaccharide and lipid (Textbook Case 2).

Other types of lipids include the waxes and sterols. Waxes are composed of long chains of fatty acids and form part of the cell wall in *Mycobacterium tuberculosis*, the bacterium causing tuberculosis. **Sterols** are very different from lipids and are included with lipids solely because they too are hydrophobic molecules (FIGURE 2.13C). Sterols, such as ergosterol, are composed of several rings of carbon atoms with side chains. They stabilize membranes of protists and fungi. Other sterols are found in the cell membrane of the bacterium *Mycoplasma*.

Nucleic Acids Are Large, Information-Containing Polymers

The **nucleic acids** are organic compounds composed of carbon, hydrogen, oxygen, nitrogen, and phosphorus atoms. Two types function in all living organisms: **deoxyribonucleic acid (DNA)** and **ribonucleic acid (RNA)**.

Both DNA and RNA are composed of repeating monomers called **nucleotides** (FIGURE 2.14A). Each nucleotide has three components: a sugar molecule, a phosphate group, and a nucleobase. The sugar in DNA is deoxyribose, while in RNA it is ribose. The **nucleobases** are nitrogen-containing molecules. In DNA, the purine nucleobases are adenine (A) and guanine (G), while the pyrimidine nucleobases are cytosine (C) and thymine (T). In RNA, adenine, guanine, and cytosine also are present, but uracil (U) is found instead of thymine.

Nucleotides are covalently joined through dehydration synthesis reactions between the sugar of one nucleotide and the phosphate of the adja-

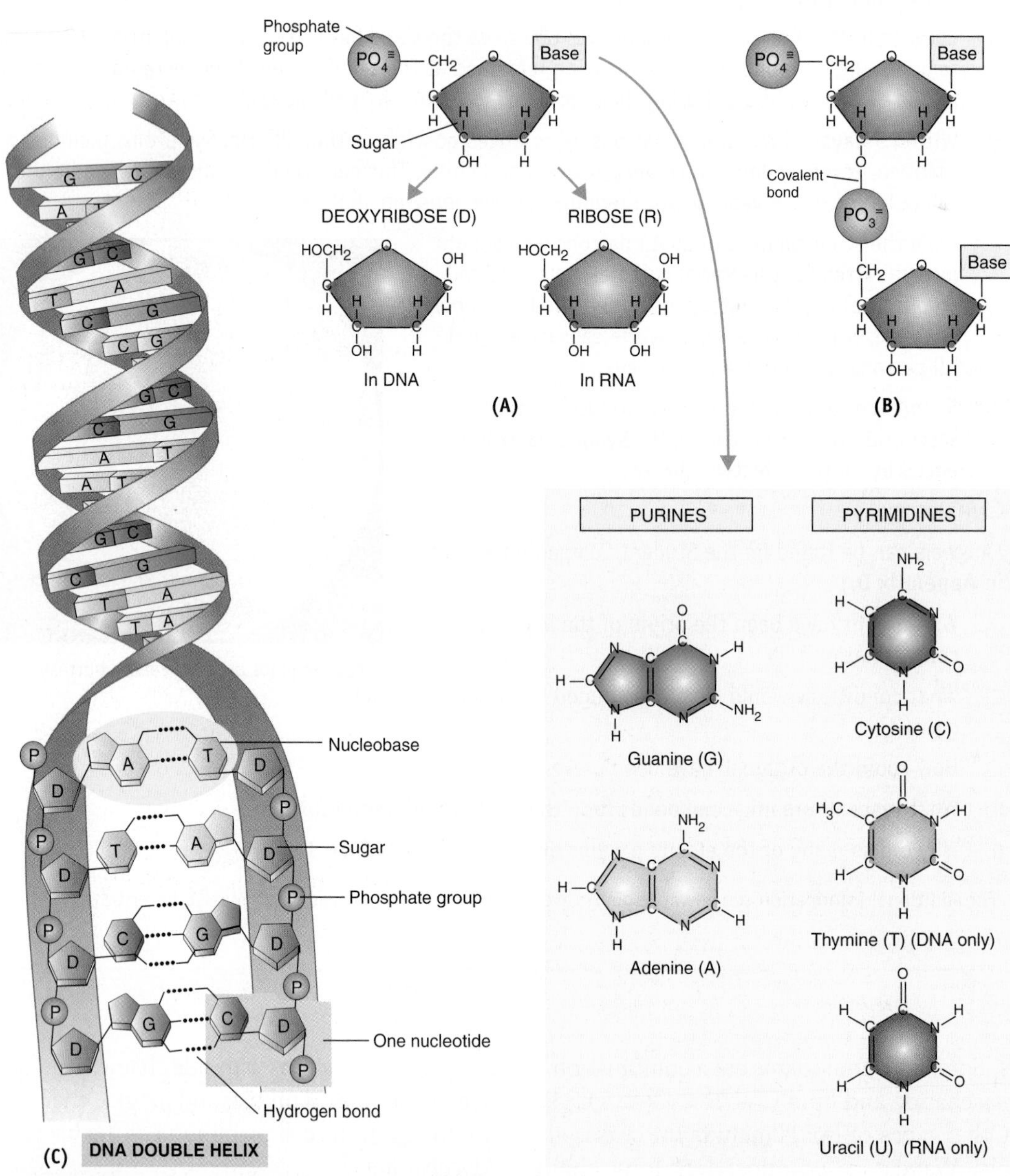

FIGURE 2.14 **The Molecular Structures of Nucleotide Components and the Construction of DNA.** **(A)** The sugars in nucleotides are ribose and deoxyribose, which are identical except for one additional oxygen atom in RNA. The nucleobases include adenine and guanine, which are large purine molecules, and thymine, cytosine, and uracil, which are smaller pyrimidine molecules. Note the similarities in the structures of these bases and the differences in the side groups. **(B)** Nucleotides are bonded together by dehydration synthesis reactions. **(C)** The two polynucleotides of DNA are held together by hydrogen bonds between adenine (A) and thymine (T) or guanine (G) and cytosine (C) to form a double helix. »» If a segment of one strand of DNA has the bases TTAGGCACG, what would be the sequence of bases in the complementary strand?

cent nucleotide to eventually form a **polynucleotide** (FIGURE 2.14B).

DNA. In 1953, James Watson, Francis Crick, Rosalind Franklin, and Maurice Wilkins published papers describing how a complete DNA molecule consists of two polynucleotide strands opposed to each other in a ladder-like arrangement (FIGURE 2.14C). Guanine and cytosine line up opposite one another, and thymine and adenine oppose each other in the two strands. The complementary base pairs in the double-stranded DNA molecule are held together by hydrogen bonds. The double strand then twists to form a spiral arrangement called the **DNA double helix**.

DNA is the genetic material in all living organisms. This genetic information exists in discrete units called **genes**, which are sequences of nucleotides that encode information to regulate and synthesize proteins. In bacterial and archaeal cells, these genes are usually found on a circular chromosome, while in most eukaryotic microbes, the genes are located on several linear chromosomes.

■ Chromosome: A DNA molecule containing the hereditary information in the form of genes.

RNA. Besides having uracil as a base and ribose as the sugar, RNA molecules in cells are single-stranded polynucleotides. Biologists once viewed RNAs as the intermediaries, involved in carrying gene information or as structural molecules needed to construct proteins. This certainly is a major role for RNA but not the only role.

In viruses such as the influenza and measles viruses, RNA is the genetic information, not DNA. Other RNA molecules play key roles in regulating gene activity, while several small RNAs control various cellular processes in microbial cells.

The nucleic acids cannot be altered without injuring the organism or killing it. Ultraviolet (UV) light damages DNA, and thus it can be used to control microbes on an environmental surface. Chemicals, such as formaldehyde, alter the nucleic acids of viruses and can be used in the preparation of vaccines. Certain antibiotics interfere with DNA or RNA function and thereby kill bacteria.

It also is important to point out that nucleotides have other roles in cells besides being part of DNA or RNA. Adenine nucleotides with three attached phosphate groups form **adenosine triphosphate (ATP)**, which is the cellular energy currency in all cells. Other nucleotides can be part of the structure of some enzymes, while independent, modified nucleotides called cyclic adenosine monophosphate (cAMP) act as chemical signals in many microbes.

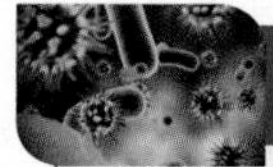

Chapter Challenge C

An article published online November 2010 in *Science* made the news—NASA scientists announced they had found a bacterial organism that they believed used arsenic in place of "traditional" phosphate. The scientists reported that an apparently unusual (alien?) bacterial organism, called GFAJ-1, had been discovered in California's Mono Lake that had replaced phosphorus in its DNA backbone with arsenic, if no phosphate was available. If true, it would redefine the chemistry of life as discussed in this chapter. As of this writing, scientists have not been able to verify the finding of the NASA scientists and many believe arsenic cannot replace phosphate. Be that as it may, again here is the question: If GFAJ-1 can use arsenic in place of phosphate in DNA, does that represent a marker for "shadow life?" Or, once again is this just another example of the power of the microbe to eke out an existence no matter what extreme conditions are present?

QUESTION C:

The atomic number for arsenic is 33 and the mass number is 75. Draw a segment of a DNA double helix (use Figure 2.14C as a template) to show how arsenic would replace phosphate. What two other phosphate-containing molecules or structures discussed in this chapter might also be replaced with arsenic?

Answers can be found on the Student Companion Website in **Appendix D**.

Proteins Are the Workhorse Polymers in Cells

Proteins are the most abundant organic compounds in microorganisms and all living organisms, making up about 58% of a bacterial cell's dry weight. The high percentage of protein indicates their essential and diverse roles. Many proteins function as structural components of cells and cell walls, and as transport agents in membranes. A large number of proteins serve as enzymes. Proteins are composed of carbon, hydrogen, oxygen, nitrogen, and, usually, sulfur atoms.

MICROFOCUS 2.5: Environmental Microbiology
Triple Play—Bacteria to Plants to Humans

About 80% of the atmosphere is nitrogen gas (N_2). Nitrogen gas, as you have discovered, contains a triple covalent bond that is very hard to break. Yet, one of the essential elements in nucleic acids and proteins in all organisms is nitrogen. So, how can the gaseous form of nitrogen be converted into a form that can be used to make essential biological compounds?

The most important biological process to break the triple covalent bond in N_2 is accomplished by a few bacterial species commonly found in root nodules of pea and bean plants (legumes) or in the soil close to the plant roots. These bacterial organisms contain an enzyme, called nitrogenase, which converts N_2 into ammonia that then can be further converted by microbial action into forms used by legumes and other plants. The process is called **nitrogen fixation:**

$$N_2 \rightarrow NH_3 \text{ (ammonia)}$$

In contact with water, the gaseous ammonia is converted to ammonium ions (NH_4^+), which serve as a source of nitrogen for nucleic acid and amino acid synthesis by bacterial cells and plants. We then get our nitrogen for amino acids and nucleobases from eating plants or through exchange reactions of carbohydrate metabolism that convert sugars into nucleotides or amino acids.

The important point to remember in all this chemistry is that the initial fixation of nitrogen is dependent on bacterial chemistry. In fact, without nitrogen fixation, life as we know it would not exist.

Proteins are polymers built from nitrogen-containing monomers called **amino acids** (MICROFOCUS 2.5). At the center of each amino acid is a carbon atom attached to two functional groups: an amino group ($-NH_2$) and a carboxyl group ($-COOH$) (FIGURE 2.15A). Also attached to the carbon is a side chain, called the **R group**. Each of the 20 amino acids differs only by the atoms composing the R group. These side chains, many being functional groups, are essential in determining the final shape, and therefore function, of the protein.

In protein formation, two amino acids (sometimes called peptides) are joined together by a covalent bond when the amino group of one amino acid is linked to the carboxyl of another amino acid through a dehydration synthesis reaction (Figure 2.15A). Repeating the reaction hundreds of times produces a long chain of amino acids called a **polypeptide** and the covalent bond between amino acids therefore is called a **peptide bond**. However, the sequence of amino acids is critical because a single amino acid improperly positioned may change the three-dimensional shape and function of the protein. Because proteins have tremendously diverse roles, they come in many sizes and shapes. The final shape depends on several factors associated with the amino acids.

The sequence of amino acids in the polypeptide represents the **primary structure** (FIGURE 2.15B). Each protein that has a different function will have a different primary structure. However, the sequence of amino acids alone is not sufficient to confer function.

Many polypeptides have regions folded into a corkscrew shape or **alpha helix**. These regions represent part of the protein's **secondary structure** (FIGURE 2.15C). Hydrogen bonds between amino groups (–NH) and carbonyl groups (–CO) on nearby amino acids maintain this structure. A secondary structure also may form when the hydrogen bonds cause portions of the polypeptide chain to zigzag in a flat plane, forming a **pleated sheet**. Other regions may not interact and remain in a random coil.

Many polypeptides also have a **tertiary structure** (FIGURE 2.15D). Such a three-dimensional (3-D) shape of a polypeptide is folded back on itself much like a spiral telephone cord. Ionic and hydrogen bonds between R groups on amino acids in proximity to each other help form and maintain the polypeptide in its tertiary structure. In addition, covalent bonds, called **disulfide bridges**, between sulfhydryl (SH) groups in R groups are important in stabilizing tertiary structure.

The ionic and hydrogen bonds helping hold a protein in its 3-D shape are relatively weak associations. As such, these interactions in a protein are influenced by environmental conditions. When

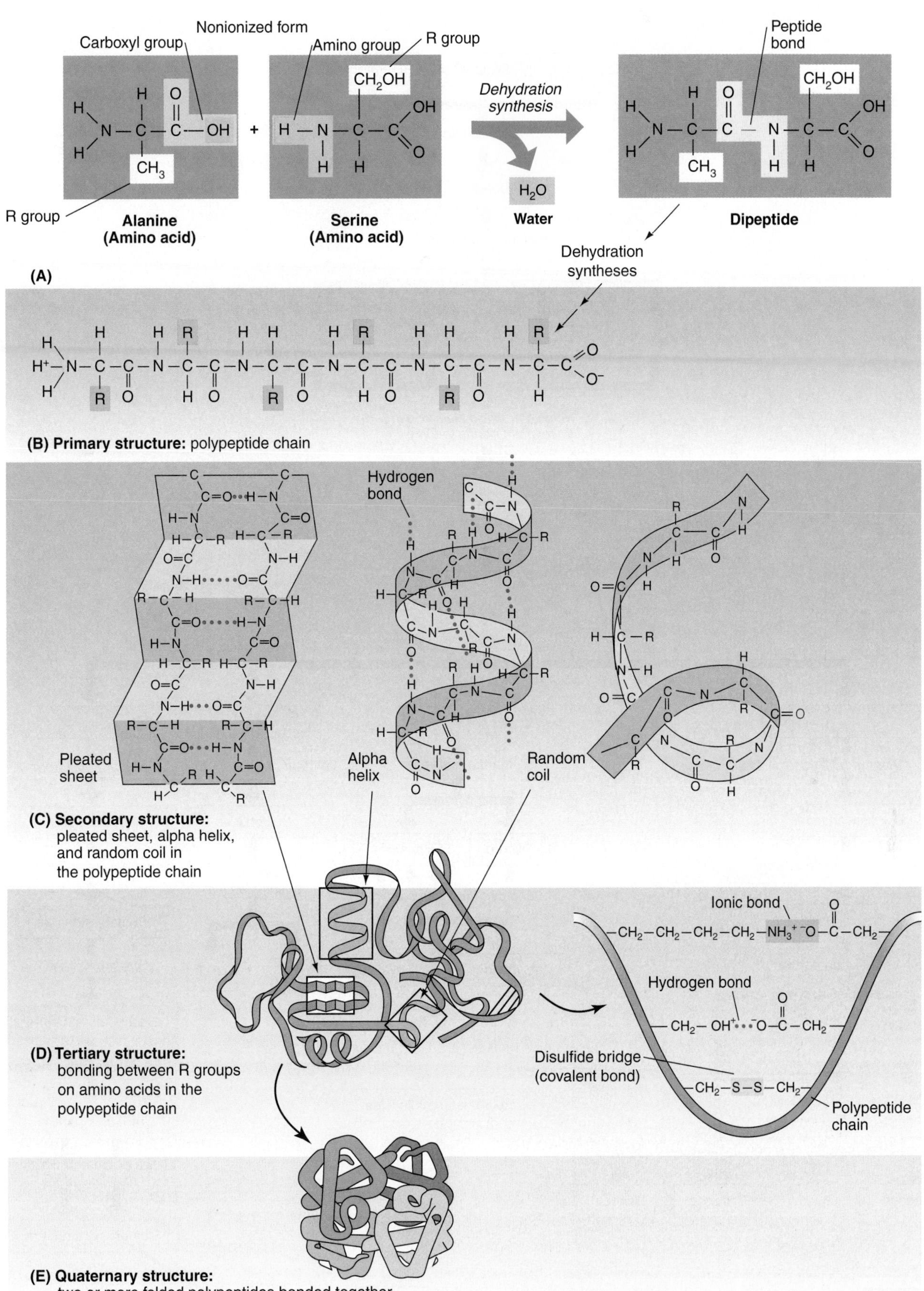

FIGURE 2.15 **Amino Acids and Their Assembly into Polypeptides.** (**A**) Amino acids are linked together by dehydration synthesis reactions. (**B**) As they get longer, the sequence of amino acids forms the primary structure, which takes on a secondary structure (**C**). (**D**) The whole polypeptide folds into a tertiary structure through bonding between R groups. Some proteins consist of more than one polypeptide, forming a quaternary structure (**E**). »» Explain why each and every protein must have three or four levels of folding.

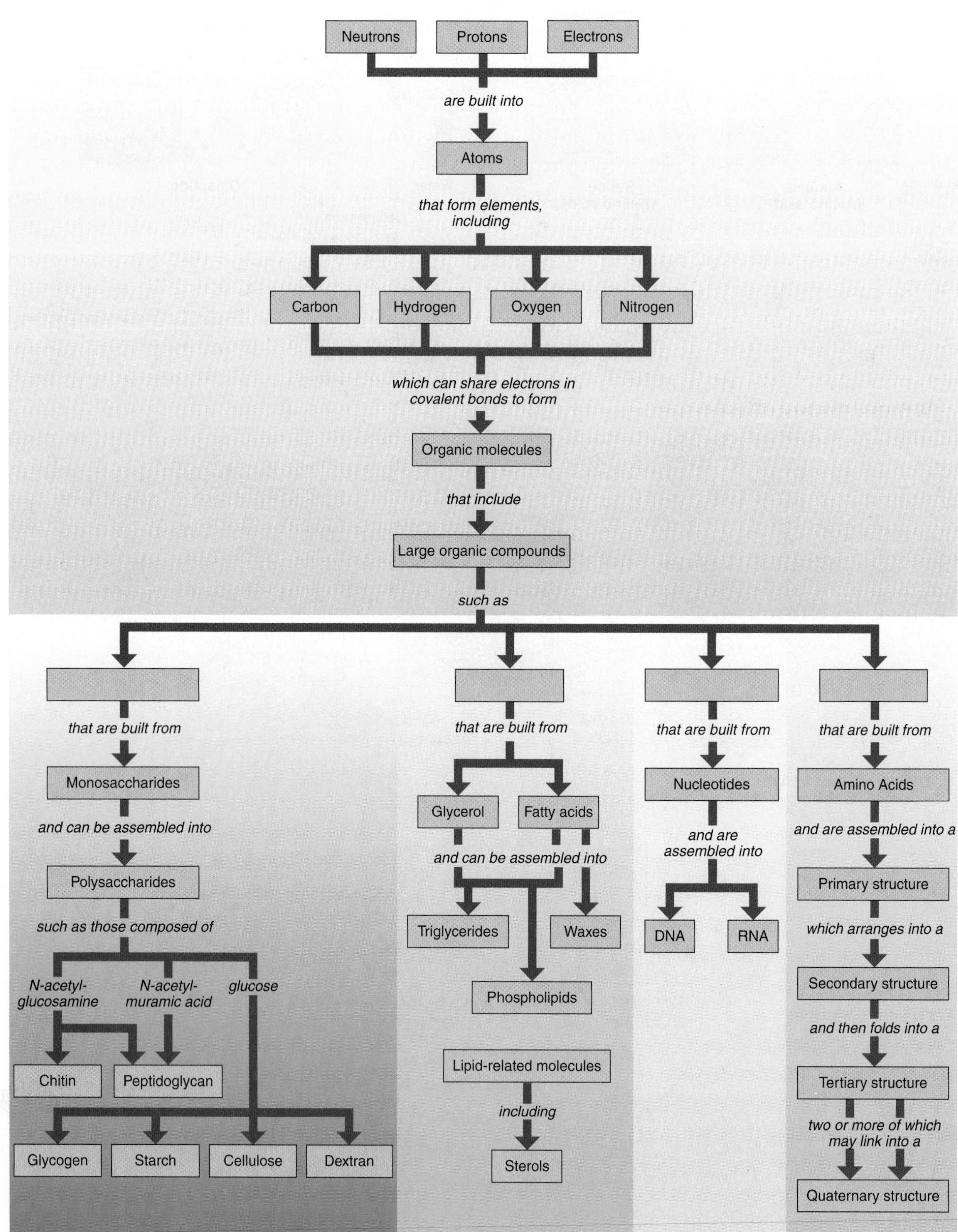

FIGURE 2.16 A Concept Map Summarizing Atoms, Elements, and Organic Compounds. »» Finish the concept map by filling in the four empty rectangles with the correct type of organic compound.

subjected to heat, pH changes, or certain chemicals, these bonds may break, causing the polypeptide to unfold and lose its biological activity. This loss of 3-D shape is referred to as **denaturation**. For example, the white of a boiled egg is denatured egg protein (albumin) and cottage cheese is denatured milk protein. Should enzymes be denatured in living cells, the important chemical reactions they facilitate will be interrupted and death of the organism may result. Viruses also can be destroyed by denaturing the proteins found on and in the viruses.

Now you should understand the importance of buffers in cells; by preventing pH shifts, they prevent protein denaturation and maintain protein function.

Many proteins are single polypeptides. However, other proteins contain two or more polypeptides to form the complete and functional protein; this is called the **quaternary structure** (FIGURE 2.15E). Each polypeptide chain is folded into its tertiary structure and the unique association between separate polypeptides produces the quaternary structure. The same types of chemical bonds are involved as in tertiary structure.

The four major classes of organic compounds are summarized in FIGURE 2.16. Investigating the Microbial World 2 looks at the origins of the monomers discussed in this chapter, while MicroInquiry 2 uses the radioactive attributes of two chemical elements to discover whether protein or DNA is the genetic material.

CONCEPT AND REASONING CHECKS 4

a. What is common to all functional groups except the carbonyl group?
b. Explain how stable monosaccharides are assembled into energy storage and structural polymers. Give examples.
c. Why are lipids not considered polymers in the sense that polysaccharides are?
d. How does the structure of DNA differ from that of RNA?
e. Why does a denatured protein no longer have biological activity?

Investigating the Microbial World 2

Chemical Evolution

Life today is full of organic monomers that assemble into larger organic compounds that are characteristic of all life, including microorganisms. Knowing this, in the 1920s, Aleksandr Oparin, a Russian biochemist, and J. B. S. Haldane, a British geneticist, wondered whether these small monomers could have nonliving (abiotic) origins that were then used to build the first cell, a precursor to the prokaryotic cell.

■ **OBSERVATION:** Because the primitive atmosphere of Earth contained hydrogen and other substances that readily provide electrons (an atmosphere lacking oxygen gas), Oparin and Haldane suggested independently that with an appropriate supply of energy, such as lightning or ultraviolet light, a variety of organic monomers might be formed. In 1953, this suggestion for chemical evolution was put to the test by Stanley Miller, an American chemist/biologist, and Harold Urey, a physical chemist.

■ **QUESTION:** ***Can organic monomers be formed in an atmosphere mimicking the conditions on the primitive Earth?***

■ **HYPOTHESIS:** Organic monomers, specifically amino acids, can be formed using the prebiotic atmospheric conditions of early Earth. If so, then a combination of the prebiotic atmospheric gases, along with an energy source to drive the reactions, should give rise to a variety of amino acids.

■ **EXPERIMENTAL DESIGN:** The apparatus illustrated on the next page depicts the design of a closed environment to mimic the prebiotic conditions believed to represent Earth's primitive atmosphere [methane (CH_4), ammonia (NH_3), hydrogen (H_2), and water (H_2O)].

■ **EXPERIMENT:** The water in the closed apparatus was boiled and the water vapor allowed to mix with the three gases and the electrical discharge. The reaction was run continuously for one week. Then, the contents of the water in the apparatus were analyzed for the presence of amino acids.

■ **RESULTS:** During the run, the water in the apparatus turned pink and then deep red and turbid (cloudy). The red color was presumably due to the presence of organic molecules. The figure shows the results for amino acid detection along with the pattern for known amino acids.

(continued on next page)

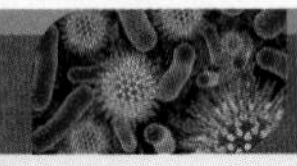

Investigating the Microbial World 2 (continued)

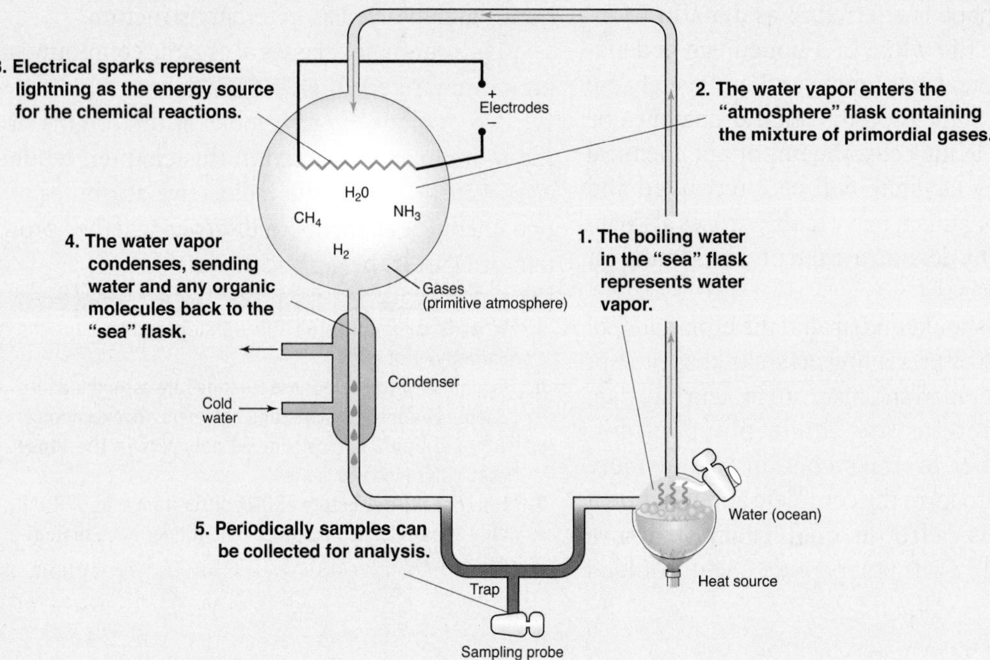

The Miller-Urey apparatus.

Identification of known amino acids

Lysine
Argenine and Histimine
Glycine
Alamine
Valine
Serine
Froline
Lencine and
Threonine
Isofencine
Cystine
Glutamine
Phenylalanine
Asparagine
Tryptophan

Results of experiment

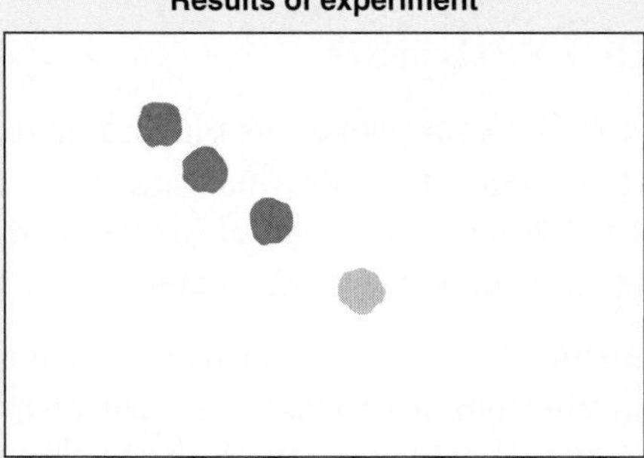

■ **CONCLUSION:**

QUESTION 1: *Was the hypothesis validated?*

QUESTION 2: *What amino acids were detected?*

QUESTION 3: *How do you know that the amino acids detected were not produced by living bacterial cells in the apparatus?*

Answers can be found on the Student Companion Website in **Appendix D.**

Note: Science often is a process of repetition and refinement. The validity of the Miller-Urey experiment was later questioned when evidence suggested that earth's primitive atmosphere was actually composed of different gases, ones released from volcanic eruptions [CO_2, nitrogen (N_2), hydrogen sulfide (H_2S), and sulfur dioxide (SO_2)]. However, modern experiments using these gases, along with those used in the original Miller–Urey experiment, have produced up to 22 amino acids, all five nucleobases in DNA and RNA, and the sugar ribose. So, whatever the gases, chemical evolution and abiotic synthesis of organic molecules appear to be validated as at least one source for organic monomers.

Adapted from: Miller, S. L. (1953). *Science* **117** (3046): 528–529; Johnson, A. P. *et al.* (2008) *Science* **322** (5900): 404. Icon image © Tischenko Irina/ShutterStock, Inc.

MICROINQUIRY 2

Is Protein or DNA the Genetic Material?

In the early 1950s, there were scientists who still debated whether protein or DNA was the genetic material in cells. To settle the controversy, in 1952 Alfred Hershey and Martha Chase carried out a series of experiments to trace the fates of protein and DNA, and in so doing hopefully settle the debate.

It was known that some viruses that infect bacterial cells were composed of DNA and protein, and that the virus genetic material needed to enter the bacterial cells to direct the production of more viruses. Because the viruses left a viral coat on the surface, what actually entered the cells—protein or DNA? Whichever did must be the genetic material.

Several biologically important elements have isotopes that are radioactive. The table to the right lists a few such elements. Hershey and Chase decided to radioactively label the viruses such that the protein and DNA could be identified by their unique radioactive profiles.

Some Radioactive Isotopes

Element	Common Form	Radioactive Form
Hydrogen	^{1}H	^{3}H (tritium)
Carbon	^{12}C	^{14}C
Phosphorus	^{31}P	^{32}P
Sulfur	^{32}S	^{35}S

2a. Which of the radioactive elements would only label protein?

2b. Which of the radioactive elements would only label DNA?

2c. Could ^{3}H or ^{14}C have been used? Explain.

The Hershey and Chase experiment is outlined in the figure below. From the two experiments, they could then measure the radioactivity in the pellet (bacterial cells) and the fluid (virus coats) and determine which radioactive isotope was associated with the bacterial cells.

2d. If protein is the genetic material, which isotope should be associated with the pellet?

2e. If DNA is the genetic material, which isotope should be associated with the pellet?

2f. So, did the experiments carried out by Hershey and Chase support or refute the hypothesis that DNA was the genetic material?

Answers can be found on the Student Companion Website in **Appendix D.**

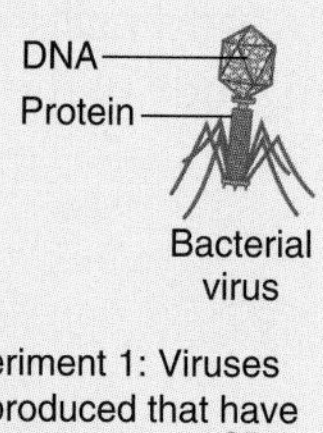

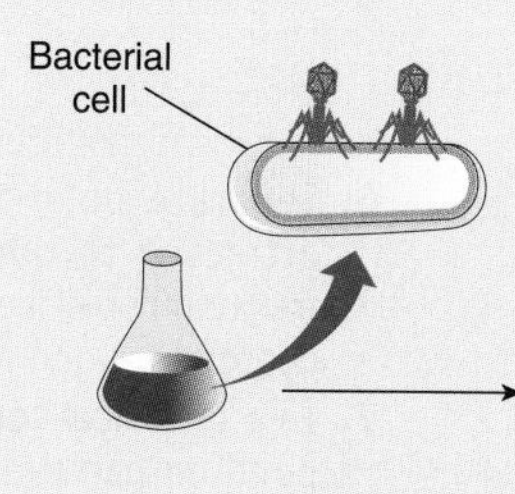

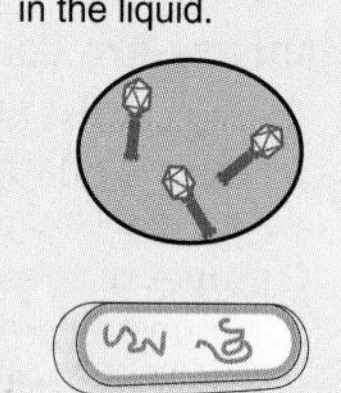

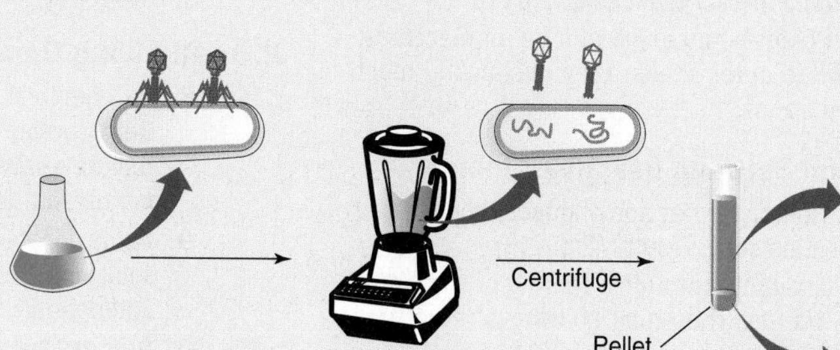

Is Protein or DNA the Genetic Material? The Hershey-Chase experiment.

Chapter Challenge D

In this chapter challenge we have been investigating if life arose on Earth more than once and, if so, how would we detect it. If there truly is a molecule capable of making life different chemically, it might be in the amino acids used to make proteins. Most all known life uses the same 20 amino acids to synthesize proteins. However, other amino acids exist and some, like isovaline and pseudoleucine, have been identified in meteorites. Therefore, is it possible that a marker for "shadow life" microbes could be unusual amino acids in their proteins?

QUESTION D:

The Miller-Urey experiment described in the Investigating the Microbial World may have produced more than the 20 traditional amino acids. If one or more of the rare amino acids were formed in the primordial Earth, and incorporated into protein, would the existing organisms be candidates for "shadow life?"

Scientists have found some bacterial organisms like Salmonella can become antibiotic resistant if they synthesize and use a rare form of the amino acid lysine when making proteins. Would you consider this a marker for "shadow life?"

Some extremophile members of the prokaryotic world that make methane gas possess an unusual amino acid called pyrrolysine. Would you consider this a marker for "shadow life?"

Answers can be found on the Student Companion Website in **Appendix D.**

■ **In conclusion**, this chapter described atoms and elements, and their interactions through chemical bonding to construct compounds and molecules. The information discussed forms the underlying foundation for many of the topics we will examine in microbiology, as growth, metabolism, and genetics are based on cellular chemical reactions. In fact, many microbial pathogens that infect and damage human cells and tissues do so by means of special metabolic reactions they carry out or by the metabolism they inhibit in the infected cells and tissues. Realize the time invested now to understand or refresh your memory about chemistry will make subsequent information easier and prepare you for a rewarding learning experience as you continue your study of the microbial world.

SUMMARY OF KEY CONCEPTS

2.1 Organisms Are Composed of Atoms

1. **Atoms** consist of an **atomic nucleus** (with **neutrons** and positively charged **protons**) surrounded by a cloud of negatively charged **electrons.** (Fig. 2.2)
2. **Isotopes** of an element have different numbers of neutrons. Some unstable ones, called **radioisotopes**, are useful in research and medicine. If an atom gains or loses electrons, it becomes an electrically charged **ion.** Many ions are important in microbial metabolism. (Fig. 2.3)
3. Each electron shell holds a maximum number of electrons. If atoms have unfilled outer shells, they chemically react with other reactive atoms.

2.2 Chemical Bonds Form Between Reactive Atoms

4. **Ionic bonds** result from the attraction of oppositely charged ions. Compounds called **salts** result. (Fig. 2.5)
5. Most atoms achieve stability through a sharing of electrons, forming **covalent bonds.** The equal sharing of electrons produces **nonpolar molecules** (no electrical charge). Atomic interactions between hydrogen and oxygen (or nitrogen) produce unequal sharing of electrons, which generates **polar molecules** (have electrical charges). (Figs. 2.6, 2.7)
6. Separate polar molecules, like water, are electrically attracted to one another and form **hydrogen bonds,** involving positively charged hydrogen atoms and negatively charged oxygen atoms. (Fig. 2.7)
7. In a **chemical reaction**, the atoms in the **reactant** change bonding partners in forming one or more **products.** Two common chemical reactions in cells are **dehydration synthesis reactions** and **hydrolysis reactions.** In these reactions, the number of atoms is the same in the reactants and products.

2.3 All Living Organisms Depend on Water

8. All chemical reactions in organisms occur in liquid water. Being polar, water has unique properties. These include its role as a **solvent**, as a chemical reactant, and as a factor to maintain a fairly constant temperature. (Fig. 2.8)
9. **Acids** donate hydrogen ions (H^+) while **bases** acquire H^+ from a solution. The **pH scale** indicates the number of H^+ in a solution and denotes the relative acidity of a solution. (Fig. 2.9)
10. **Buffers** are a mixture of a weak acid and a weak base that maintain acid/base balance in cells. Excess H^+ can be absorbed by the base and too few H^+ can be provided by the acid. (Fig. 2.10)

2.4 Living Organisms Are Composed of Four Types of Organic Compounds

11. The building of large **organic compounds** depends on the **functional groups** found on the building blocks called **monomers**. Functional groups on monomers interact through dehydration synthesis reactions to form a covalent bond between monomers. (Table 2.3)
12. **Carbohydrates** include **monosaccharides** such as glucose, which can be linked into **polysaccharides** that represent energy and structural molecules. (Fig. 2.12)
13. **Lipids** serve as energy sources, but also have a major role as **phospholipids** in cell membranes. Other lipids include the **sterols**. (Fig. 2.13)
14. The genetic instructions for living organisms are composed of two types of **nucleic acids: deoxyribonucleic acid (DNA)**, which stores and encodes the hereditary information; and **ribonucleic acid (RNA)**, which transmits the information to make proteins, controls genes, and helps regulate genetic activity. (Fig. 2.14)
15. **Proteins** are chains of **amino acids** connected by **peptide bonds**. Proteins are used as enzymes and as structural components of cells. **Primary, secondary**, and **tertiary structures** form the functional shape of many proteins, which can unfold by **denaturation**. Many proteins are the result of two or more polypeptides bonding together (**quaternary structure**). (Fig. 2.15)

CHAPTER SELF-TEST

For **STEPS A–D**, answers to even-numbered questions and problems can be found in **Appendix C** on the Student Companion Website at **http://microbiology.jbpub.com/10e**. In addition, the site features eLearning, an online review area that provides quizzes and other tools to help you study for your class. You can also follow useful links for in-depth information, read more MicroFocus stories, or just find out the latest microbiology news.

STEP A: REVIEW OF FACTS AND TERMS

Multiple Choice

Read each question carefully, then select the ***one*** answer that best fits the question or statement.

1. These positively charged particles are found in the atomic nucleus.
   A. Protons
   B. Electrons
   C. Protons and neutrons
   D. Neutrons
2. Atoms of the same element that have different numbers of neutrons are called ______.
   A. isotopes
   B. ions
   C. isomers
   D. inert elements
3. If an element has two electrons in the first shell and seven in the second shell, the element is said to be what?
   A. Stable
   B. Unreactive
   C. Unstable
   D. Inert
4. For ______ bonding, one or more electrons are transferred between atoms.
   A. hydrogen
   B. ionic
   C. peptide
   D. covalent
5. The covalent bonding of atoms forms a/an ______.
   A. molecule
   B. ion
   C. element
   D. isomer
6. The ______ bond is a weak bond that can exist between poles of adjacent molecules.
   A. hydrogen
   B. ionic
   C. polar covalent
   D. nonpolar covalent
7. In what type of chemical reaction are the products of water removed during the formation of covalent bonds?
   A. Hydrolysis
   B. Ionization
   C. Dehydration synthesis
   D. Decomposition
8. A ______ dissolves in water.
   A. solvent
   B. hydrophobic molecule
   C. solute
   D. nonpolar molecule
9. The pH scale relates the measure of ______ of a chemical substance.
   A. ionization
   B. denaturation
   C. acidity
   D. buffering
10. Which one of the following statements about buffers is FALSE?
    A. They work inside cells.
    B. They consist of a weak acid and weak base.
    C. They prevent pH shifts.
    D. They enhance chemical reactions.
11. A functional group designated—COOH is known as a/an ______.
    A. carboxyl
    B. carbonyl
    C. amino
    D. hydroxyl
12. Which one of the following is NOT a polysaccharide?
    A. Chitin
    B. Glycogen
    C. Cellulose
    D. Lipid
13. How do the lipids differ from the other organic compounds?
    A. They are the largest organic compounds.
    B. They are nonpolar compounds.
    C. They have no biological role.
    D. They are not used for energy storage.

14. Both DNA and RNA are composed of ______.
A. polynucleotides
B. genes
C. polysaccharides
D. polypeptides

15. The ______ structure of a protein is the sequence of amino acids.
A. primary
B. secondary
C. tertiary
D. quaternary

STEP B: CONCEPT REVIEW

16. Assess the importance of the three types of atomic particles to atomic structure and reactivity. (**Key Concept 1**)

17. Construct a concept map for **chemical bonds** using the following terms: (**Key Concept 2**)

Hydrocarbon	Methane	Polar covalent bonds
Hydrogen bonds	NaCl	Salts
Ionic bonds	Nonpolar covalent bonds	Water

18. Support the statement: "Life as we know it could not exist without water." (**Key Concept 3**)

19. Predict the chemical reactivity potential for a molecule having five functional groups versus one having only one functional group. Explain your prediction. (**Key Concept 4**)

20. Construct a table providing the name of each type of organic compound and its function or functions. (**Key Concept 4**)

STEP C: APPLICATIONS AND PROBLEM SOLVING

21. You want to grow a bacterial species that is acid-loving; that is, it grows best in very acid environments. Would you want to grow it in a culture that has a pH of 2.0, 6.8, or 11.5? Explain.

22. You are given two beakers of a broth growth medium. However, only one of the beakers of broth is buffered. How could you determine which beaker contains the buffered broth solution? Hint: You are provided with a bottle of concentrated HCl and pH papers that indicate a solution's pH.

23. The microbial community in a termite's gut contains the enzyme cellulase. How does this benefit the termite and the termite's microbial community?

24. Use the following list to identify the structures (in 25 i–v) drawn below.
A. Amino acid
B. Monosaccharide
C. Nucleotide
D. Lipid
E. Disaccharide
F. Polysaccharide
G. Sterol

25. Identify any and all functional groups on each structure (i–v).

(i)

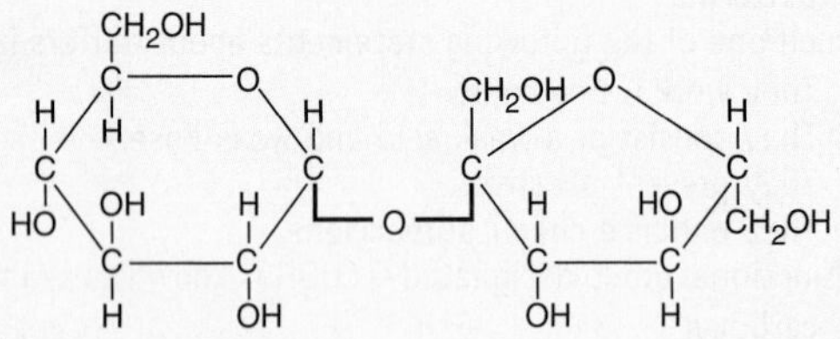

(ii)

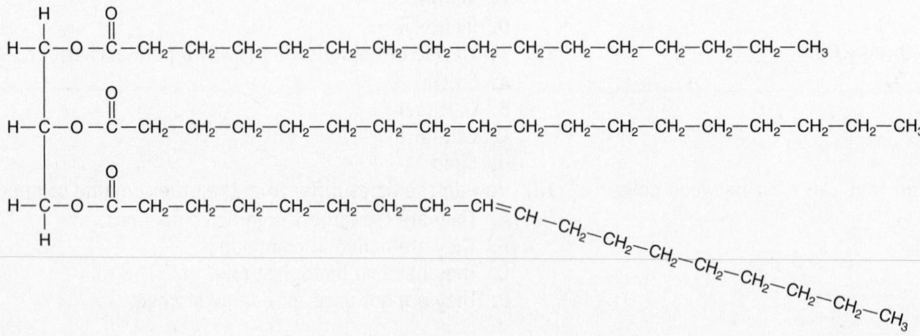

$H-C-O-C(=O)-CH_2-CH_2-CH_2-CH_2-CH_2-CH_2-CH_2-CH_2-CH_2-CH_2-CH_2-CH_2-CH_2-CH_2-CH_3$

$H-C-O-C(=O)-CH_2-CH_2-CH_2-CH_2-CH_2-CH_2-CH_2-CH_2-CH_2-CH_2-CH_2-CH_2-CH_2-CH_2-CH_2-CH_2-CH_3$

$H-C-O-C(=O)-CH_2-CH_2-CH_2-CH_2-CH_2-CH_2-CH_2-CH=CH-CH_2-CH_2-CH_2-CH_2-CH_2-CH_2-CH_2-CH_3$

(iii)

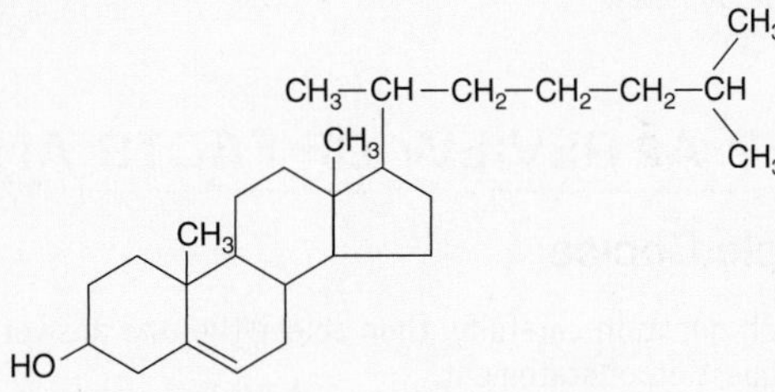

(iv)

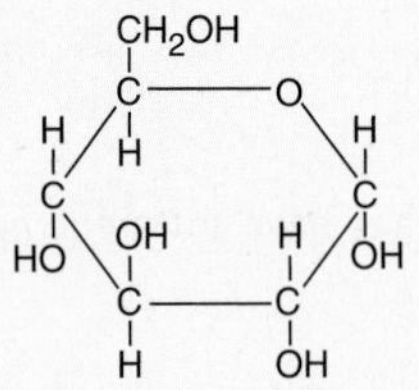

(v)

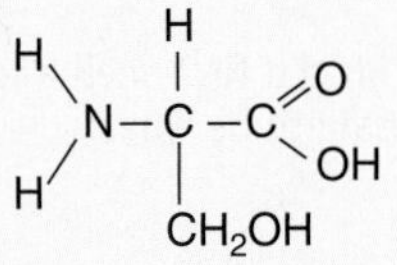

STEP D: QUESTIONS FOR THOUGHT AND DISCUSSION

26. Propose a reason why organic molecules tend to be so large.
27. Bacterial cells do not grow on bars of soap even though the soap is wet and covered with bacterial organisms after one has washed. Explain this observation.
28. Suppose you had the choice of destroying one class of organic compounds in bacterial cells to prevent their spread. Which class would you choose? Why?
29. Milk production typically has the bacterium *Lactobacillus* added to the milk before it is delivered to market. This organism produces lactic acid. (a) Why would this organism be added to the milk and (b) why was it chosen?
30. The toxin associated with the foodborne disease botulism is a protein. To avoid botulism, home canners are advised to heat preserved foods to boiling for at least 12 minutes. How does the heat help?
31. Justify Isaac Asimov's quote, "The significant chemicals in living tissue are rickety and unstable, which is exactly what is needed for life," to the atoms, molecules, and organic compounds described in this chapter.

CHAPTER PREVIEW

3.1 The Prokaryotes Are Not Simple, Primitive Organisms

3.2 Classifying Microorganisms Reveals Relationships Between Organisms

MicroInquiry 3: The Evolution of Eukaryotic Cells

3.3 Microscopy Is Used to Visualize the Structure of Cells

Textbook Case 3: Bacterial Meningitis and a Misleading Gram Stain

Concepts and Tools for Studying Microorganisms

We think we have life down; we think we understand all the conditions of its existence; and then along comes an upstart bacterium, live or fossilized, to tweak our theories or teach us something new.
—Jennifer Ackerman in *Chance in the House of Fate* (2001)

The oceans of the world are a teeming but invisible forest of microorganisms and viruses. For example, one liter of seawater contains more than 25,000 different bacterial species although most are in low abundance.

A substantial portion of these marine microbes represent the **phytoplankton** (*phyto* = "plant"; *plankto* = "wandering"), which are floating communities of cyanobacteria and eukaryotic algae. Besides forming the foundation for the marine food web, the phytoplankton account for 50% of the photosynthesis on Earth and, in so doing, supply about half the oxygen gas we breathe and other organisms breathe or use.

While sampling ocean water, scientists from MIT's Woods Hole Oceanographic Institution discovered that many of their samples contained one especially abundant organism, a tiny marine cyanobacterium they named *Prochlorococcus*. Inhabiting tropical and subtropical oceans, a typical sample often contained more than 200,000 (2×10^5) cells in one drop of seawater (FIGURE 3.1).

Studies with *Prochlorococcus* suggest the organism is responsible for almost 50% of the photosynthesis in the open oceans. This makes *Prochlorococcus* the smallest and most abundant marine photosynthetic organism yet discovered.

Image courtesy of Dr. Fred Murphy/CDC.

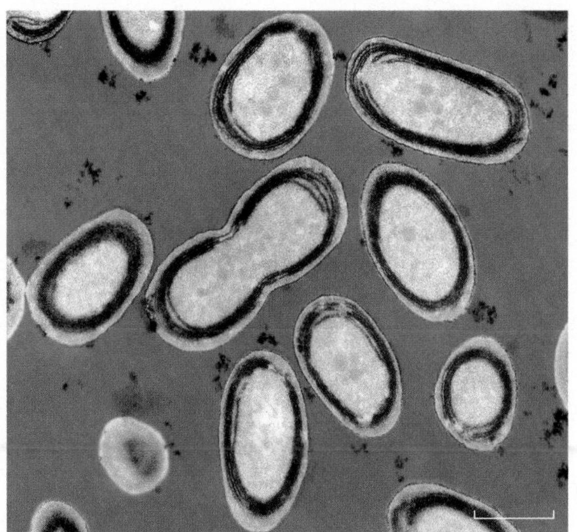

FIGURE 3.1 *Prochlorococcus.* This false-color transmission electron microscope image of a section through several *Prochlorococcus* cells reveals bands of membranes (green) that contain the chlorophyll used in photosynthesis. (Bar = 0.5 µm.) © Claire Ting/Photo Researchers, Inc.
»» Do these cells look structurally complex or simple?

The success of *Prochlorococcus* is due, in part, to the presence of different ecotypes inhabiting different ocean depths. For example, the high sunlight ecotype occurs in the surface waters while the low-light type is found below 50 meters. This latter ecotype compensates for the decreased light by increasing the amount of cellular chlorophyll that can capture the available light.

In terms of nitrogen sources, the high-light ecotype only uses ammonium ions (NH_4^+). At increasing depth, NH_4^+ is less abundant so the low-light ecotype compensates by using a wider variety of nitrogen sources.

These and other attributes of *Prochlorococcus* illustrate how microbes survive and adapt to environmental change. As part of the global microbial workforce, they are important to the functioning of the biosphere and play key roles in maintaining the "health" of the planet.

■ Biosphere: That part of the earth—including the air, soil, and water—where life occurs.

Once again, we encounter an interdisciplinary group of scientists studying how microorganisms influence life on this planet. Microbial ecologists study how the phytoplankton communities help in the natural recycling and use of chemical elements such as nitrogen. Evolutionary microbiologists look at these microorganisms to learn more about their taxonomic relationships, while microscopists, biochemists, and geneticists study how *Prochlorococcus* cells compensate for a changing environment of sunlight and nutrients.

This chapter focuses on many of the aspects described above. We examine how microbes maintain a stable internal state and how they can exist in "multicellular," complex communities. Throughout the chapter we are concerned with the relationships between microorganisms and the many attributes they share. Along the way, we explore the methods used to name and catalog microorganisms and become familiar with some basic tools and techniques used to observe the microbial world.

■ Ecotype: A subgroup of a species that has special characteristics to survive in its ecological surroundings.

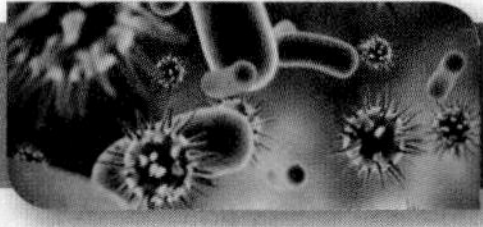

Chapter Challenge

Look again at Figure 3.1. Bacterial cells often are described and referred to as "simple" cells, often thought of as not much more than bags of enzymes and chemicals—even if they have transformed planet Earth! Our challenge, as we proceed through this chapter, is to decide whether this perception of prokaryotic simplicity is valid. Who are we calling simple? Let's find out!

© qcontrol/ShutterStock, Inc.

KEY CONCEPT 3.1 The Prokaryotes Are Not Simple, Primitive Organisms

Despite their microscopic size, prokaryotes share a common set of characteristics, or emerging properties, with all living organisms. These include:

- DNA as the hereditary material.
- Complex biochemical patterns of growth and energy conversions.
- Complex responses to stimuli.
- Reproduction to produce offspring.
- Adaptation from one generation to the next.
- Interactions with other organisms and the environment.

The last property is the focus in the first part of this chapter.

Prokaryotic Cells Exhibit Some Remarkable and Widespread Behaviors

Historically, when one looks at bacterial cells even with an electron microscope, often there is little to see (FIGURE 3.2A). The lack of visible "cell structure," representing the cell's physical appearance or its components, and the "pattern of organization," referring to the configuration of those structures and their relationships to one another, do give the impression of simpler cells.

But what has been overlooked is the "cellular process," the activities all cells carry out for the continued survival of the cell (and organism). At this level, the complexity is just as intricate as in any eukaryotic cell. So, in reality, prokaryotic cells carry out many of the same cellular processes as eukaryotic cells. Let's look at a few cases illustrating some remarkable cellular processes.

Homeostasis. All organisms continually battle their external environment, where factors such as temperature, sunlight, or toxic chemicals can have serious consequences. Organisms strive to maintain a stable internal state by making appropriate metabolic or structural adjustments. This ability to adjust yet maintain a relatively steady internal state is called **homeostasis** (*homeo* = "similar"; *stasis* = "state"). Two examples illustrate the concept (FIGURE 3.2B).

The low-light *Prochlorococcus* ecotype mentioned in the chapter introduction lives at depths below 50 meters. At these depths, transmitted sunlight decreases and any one nitrogen source is less accessible. The ecotype compensates for the light reduction and nitrogen limitation by (1) increasing the amount of cellular **chlorophyll** to capture light and (2) using a wider variety of available nitrogen sources. These adjustments in structure and pattern organization maintain a steady internal state.

For our second example, suppose a patient is given an antibiotic to combat a bacterial infection. In response, the infecting pathogen compensates for the change by breaking the structure of the antibiotic. The adjustment, antibiotic resistance, maintains homeostasis in the bacterial cell.

In both these examples, the internal environment is maintained despite a changing external environment. However, homeostasis involves more than a simple response to current conditions. Compensation can be planned for ahead of time.

Believe it or not, bacterial cells can "think," not in the cognitive way you and I think, but rather they can anticipate and prepare for future conditions to ensure that homeostasis is maintained. Remember how Pavlov had conditioned his dogs to anticipate a meal by ringing a bell? The dogs, anticipating a meal, would start salivating before the meal arrived. Well, microbiologists have shown that bacterial organisms demonstrate a similar type of associative learning. Here is just one example. When lab cultures of *Escherichia coli* cells are repeatedly shifted from 25°C and 20% oxygen to 37°C and 0% oxygen, the bacterial cells over a few weeks "learn" to anticipate the coming oxygen drop. As the temperature increased, the bacterial cells altered their metabolism in preparation for the coming oxygen change. In other words, the cells had associated the coming oxygen change with the change in temperature. So, one would predict that in the natural environment, microbes can also respond to environmental cues that precede coming events. Bacterial cells can plan ahead and ensure homeostasis—not something that one would find in a simple, primitive organism!

■ Associative learning: A type of learning by which an experience is learned through association with a separate, pre-occurring experience.

Biofilms and Cell Communication. The last emerging property of life listed on the previous page stated that cells must interact with one another and with the environment. This is certainly true in multicellular animals and plants—and it is true of most prokaryotic organisms as well. Prokaryotic cells possess a sophisticated chemical language involved with coordinating behavior.

For example, until recently the prevailing idea was that bacterial cells were self-contained, independent, noncommunicative units. In their natural environments however, we now know that few bacterial organisms live such a solitary life. In fact, it has been estimated that up to 99% of bacterial organisms live in highly organized communities called **biofilms**; that is, in a "multicellular state" where survival requires "listening" and "talking" through chemical communication with neighboring cells.

As a biofilm forms, the cells become embedded in an extracellular matrix of proteins, sugars, and other substances produced and secreted by the bacterial cells themselves (FIGURE 3.2C). This matrix holds the biofilm together and helps it spread on nonliving or living surfaces, such as metals, plastics, soil particles, medical indwelling devices, or human tissue. The mature, fully functioning biofilm is like a living tissue with a primitive circulatory system made of water channels to

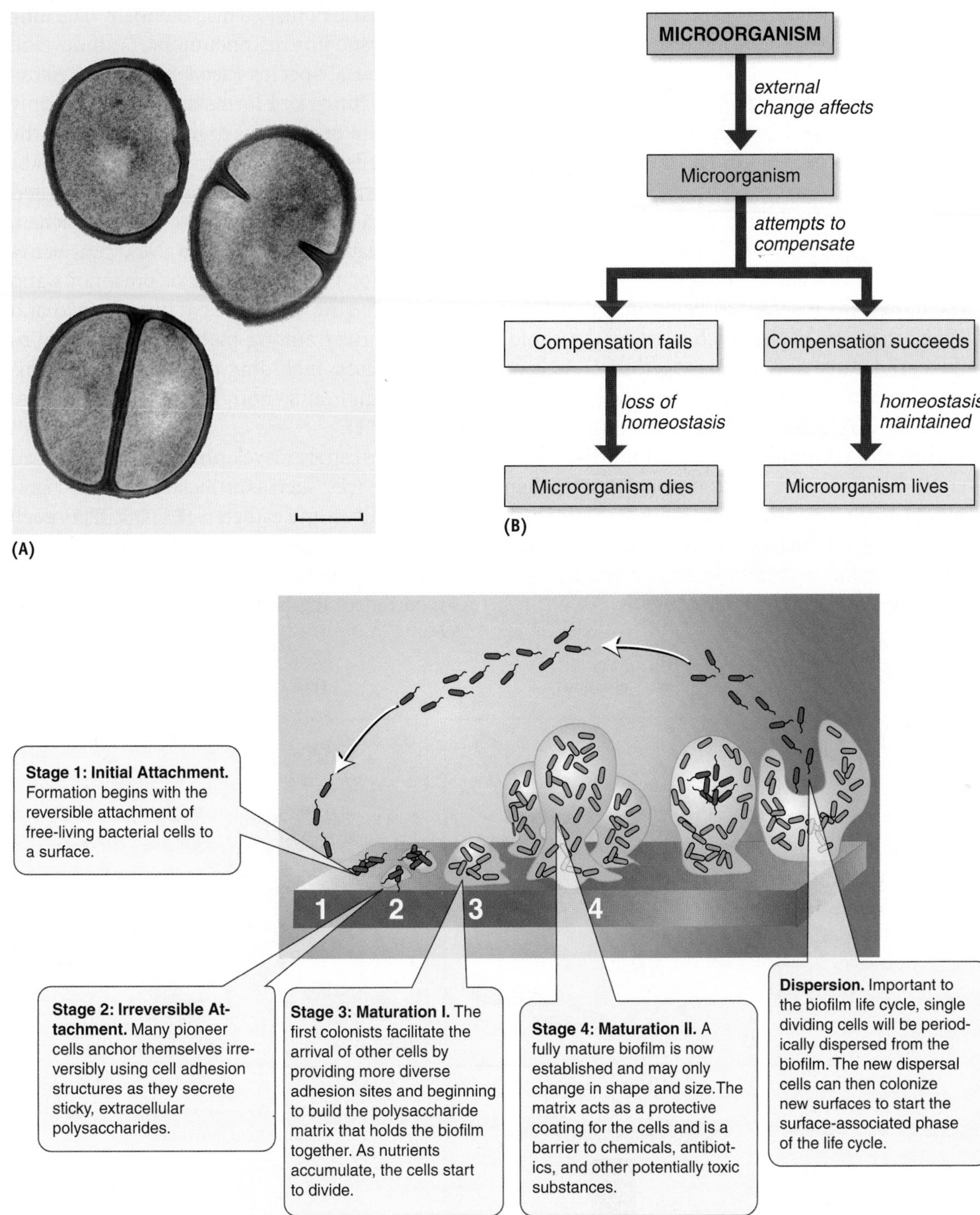

FIGURE 3.2 **Simpler, Unicellular Organisms?** (**A**) This false-color electron microscope image of *Staphylococcus aureus* gives the impression of simplicity in structure. (Bar = 0.5 μm.) © SPL/Photo Researchers, Inc. (**B**) A concept map illustrating how bacterial organisms, like all microorganisms, have to compensate for environmental changes. Survival depends on such homeostatic abilities. (**C**) The biofilm life cycle is an example of intercellular cooperation in the development of a multicellular structure. Modified from David G. Davies, Binghamton University, Binghamton, NY. »» Using the concept map in (B), explain how *Prochlorococcus* survives low-light conditions in its environment.

■ Cystic fibrosis: A hereditary disease that starts in childhood and results in secretion and accumulation of abnormally thick mucus in the lungs.

bring in nutrients and eliminate wastes. Therefore, a biofilm is a complex, metabolically cooperative community made up of peacefully coexisting species that share resources and respond to environmental changes.

It is during colonization that the cells "speak to each other" and cooperate through chemical communication. This process, called **quorum sensing (QS)**, involves the ability of bacterial cells to sense their numbers by producing and responding to extracellular chemicals. The greater the number of signaling molecules a cell detects, the more cells it senses as being in the vicinity. When these molecules reach a critical threshold, the community of cells acts together. New gene expression occurs triggering a specific behavioral response (FIGURE 3.3). Thus, in biofilms the cells within the community are profoundly different in behavior and function from those of their independent, free-living cousins. For example, in nonpathogenic bacterial species, QS can regulate such behaviors as bioluminescence and nutrient foraging (MicroFocus 3.1).

Biofilms and QS also are associated with infections. One example involves individuals suffering from cystic fibrosis. This potentially fatal lung disease (cystic fibrosis pneumonia) can develop if the bacterial species *Pseudomonas aeruginosa* infects the lungs and forms a biofilm. Not only is the biofilm impervious to most antibiotics, the bacterial cells in the community are keeping tabs of their numbers through signaling molecules and QS. When a threshold number of cells is reached, the group launches into action. New gene activity codes for the production of poisonous and tissue damaging molecules. This scenario and social behavior among pathogens is typical of other diseases, including middle ear infections (otitis media), and tooth decay (dental caries) (FIGURE 3.4A).

Biofilms can also develop on improperly cleaned medical devices, such as artificial joints, mechanical heart valves, and catheters (FIGURE 3.4B), such that when implanted into the body, the result is a slow developing but persistent biofilm infection. As mentioned, the polysaccharide matrix acts as

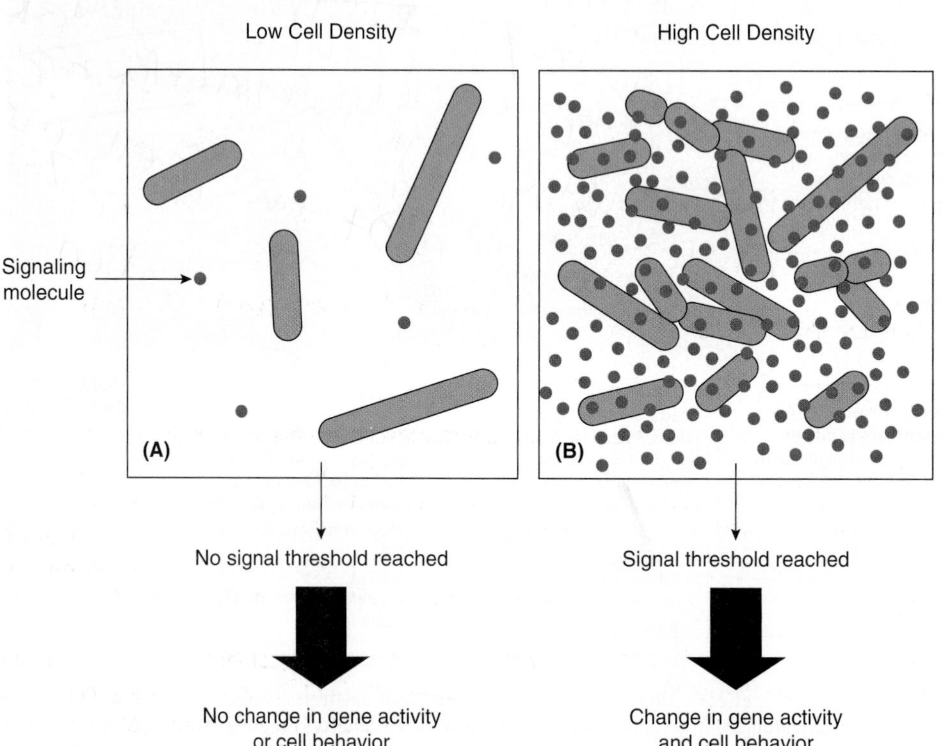

FIGURE 3.3 **Cell Communication and Quorum Sensing.** (**A**) At low cell densities few signal molecules accumulate. (**B**) As the cell density increases (quorum sensing), signaling molecules reach a threshold concentration that, when detected by the cells, leads to a change in cell behavior through new gene activity. »» Why must there be a critical concentration of signal before the cells respond?

MICROFOCUS 3.1: Environmental Microbiology

Social Networking—Bacterial Style

As the chapter opener stated, the microbial world is truly immense and we are continually surprised by what we find. Take quorum sensing for example. The discovery that bacterial cells can carry out social networking with each other changed our general perception of bacterial species as single, simple organisms. Here are two remarkable examples that helped change our perception.

Aliivibrio fischeri

The Hawaiian bobtail squid is a nocturnal species. It hides in the sand during the day and comes out to hunt at night in shallow marine waters. On moonlit nights, the swimming squid would appear as a dark silhouette easily detected by its predators from below. Therefore, the squid needs some type of camouflage so it can blend in with the moonlit background.

What the squid does is collect cells of *Aliivibrio* (formerly *Vibrio*) *fischeri*, a light-emitting, marine bacterial species found in the world's oceans at very low concentrations. At these low densities, the cells do not emit any light. However, by capturing and confining these cells in the squid's light organ, called the photophore, the squid allows the *A. fischeri* cells to reproduce and increase tremendously in density. At this high concentration (more than 1,000 cells/ml), which is reached at night, the *A. fischeri* cells start chemically "chatting" with one another (quorum sensing) and produce a signaling molecule that triggers the synthesis of the bacterial enzyme luciferase. This enzyme oxidizes bacterial luciferin and in the process gives off energy in the form of cold light (bioluminescence) similar to the light produced by fireflies (see figure). In other words, the squid's photophore generates light and by directing the "bacterial glow" from its lower body surface, the squid eliminates any silhouette and, on moonlit nights, camouflages itself from any predators.

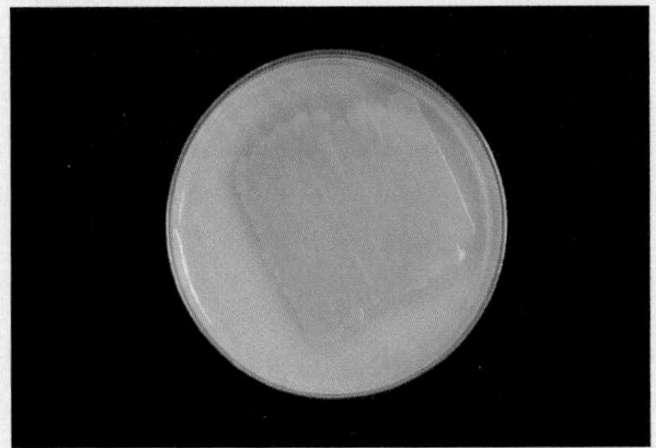

Photographs of *Aliivibrio fischeri* growing in a culture plate (left) and triggered to bioluminesce (right).

Myxobacteria

The myxobacteria are a bacterial group that predominantly lives in the soil. Individual cells are always evaluating, through quorum sensing, both their own nutritional status and that of their community. On sensing food (bacterial, yeast, or algal cells), the myxobacterial cells, which move by a gliding movement, aggregate and travel in "swarms" (also known as "wolf packs") to attack the food source. This form of quorum sensing coordinates feeding behavior and triggers the production of extracellular enzymes from the "multicellular" swarm to digest the prey. Like a lone wolf, a single cell could not effectively carry out this behavior.

Under nutrient starvation, a different behavior occurs as a result of quorum sensing—the cells aggregate into spore-producing structures called "fruiting bodies." During this developmental program, approximately 100,000 cells coordinately construct the macroscopic fruiting body. In *Myxococcus xanthus*, the myxobacterial cells first respond by triggering a quorum-sensing signaling molecule that helps the cells assess starvation and induce the first stage of aggregation. Later, another signaling molecule helps to coordinate fruit body development, as many myxobacterial cells die in forming the stalk while the remaining viable cells differentiate into environmentally resistant and metabolically quiescent myxospores. Again, social networking, through quorum sensing, is the key to an appropriate behavioral response.

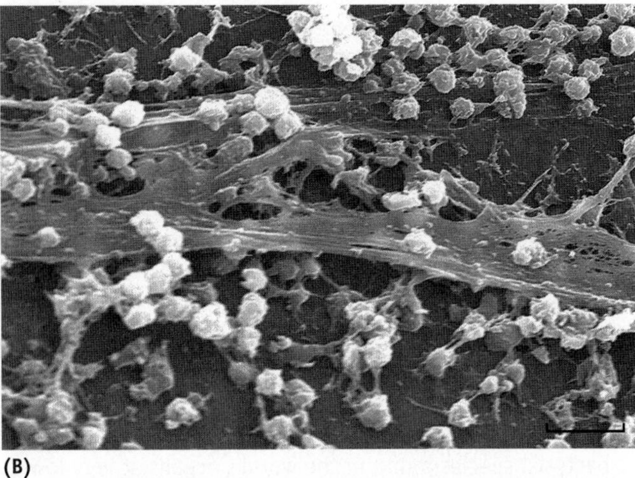

(A) (B)

FIGURE 3.4 **Biofilms in Disease.** (**A**) A false-color electron microscope image of a tooth surface showing the plaque biofilm (purple) containing bacteria cells. The red cells are red blood cells. (Bar = 60 µm.) © Mona Lisa Production/Photo Researchers, Inc. (**B**) An electron microscope image of *Staphylococcus aureus* contamination on a catheter. The fibrous-looking substance is part of the biofilm. (Bar = 3 µm.) Courtesy of Rodney M. Donlan, Ph.D. and Janice Carr/CDC. »» What is the best way to minimize such biofilms on the teeth?

a protective coating for the embedded cells and impedes penetration by antibiotics and other antimicrobial substances. As a result, the infection can be extremely hard to eradicate.

Not all the "bacterial chatter" is limited to the biofilm community. Bacterial cells also can intercept chemical signals from other species of microbes, or, in the case of a human infection, even from cells of the immune system. In the latter case, such "listening" can trigger bacterial behavior that disrupts the immune response.

On the other hand, not all biofilms are harmful as some can have positive effects on the environment. For example, sewage treatment plants use biofilms to remove contaminants from water. Likewise, **bioremediation** uses microorganisms to help remove or clean up chemically contaminated environments, such as oil spills or toxic waste sites. Perchlorate (ClO_4^-) is a soluble anion that is a component in rocket fuels, fireworks, explosives, and airbag manufacture. It is toxic to humans and is highly persistent in drinking water, especially in the western United States. Natural subterranean biofilms are being genetically modified so the cells contain the genes needed to degrade perchlorate from groundwater.

The important concept here is that bacterial cells undergo chemical "cross-talk" through QS and multicellular communities, leading to some form of community behavior when their numbers reach a threshold. Bacterial cells, like their counterparts in the plant and animal world, have sophisticated and complex communication networks to coordinate their behavior.

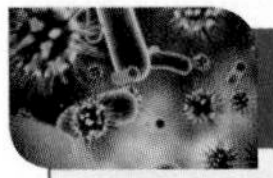

Chapter Challenge A

Up to this point we have learned that many prokaryotic cells can maintain in internal homeostasis and chemically communicate with neighboring cells, often forming a multicellular biofilm.

QUESTION A:

Based on what you have read up to this point, would you consider prokaryotic cells to be simple, primitive life forms? Explain.

Answers can be found on the Student Companion Website in **Appendix D.**

Prokaryotic and Eukaryotic Cells Share Similarities in Organizational Patterns

In the 1830s, Matthias Schleiden and Theodor Schwann developed part of the **cell theory** by demonstrating all plants and animals are composed of one or more cells, making the cell the fundamental unit of life. (Note: about 20 years later, Rudolph Virchow added that all cells arise from pre-existing cells.) Although the concept of a microorganism was just in its infancy at the time, the theory suggests that there are certain organizational patterns common to all cells, be they prokaryotic or eukaryotic.

Genetic Organization. All organisms have a similar genetic organization whereby the hereditary material (DNA) is communicated or expressed using an almost universal genetic code. The organizational pattern for the DNA is in the form of one or more **chromosomes**.

Structurally, eukaryotic cells have multiple, linear chromosomes enclosed by the membrane envelope of the cell nucleus. Most prokaryotic cells have a single, circular DNA molecule without an enclosing membrane (FIGURE 3.5), although one bacterial group does appear to have its DNA enclosed by a membrane envelope.

Compartmentation. All cells have an organizational pattern separating the internal compartments from the surrounding environment but allowing for the exchange of solutes and wastes. The pattern is for all cells to be surrounded by a **cell membrane** (known as the **plasma membrane** in eukaryotes), where the phospholipids form the impermeable boundary to solutes while membrane proteins are the gates through which the exchange of solutes and wastes occurs, and across which chemical signals are communicated. No exceptions here.

Metabolic Organization. The process of metabolism is a consequence of compartmentation. By being enclosed by a membrane, all cells have an internal environment in which chemical reactions occur and ATP, the cellular energy source, is generated. This space, called the **cytoplasm**, represents everything surrounded by the membrane and, in eukaryotic cells, exterior to the cell nucleus. If the cell structures are removed from the cytoplasm, what remains is the **cytosol**, which consists of water, salts, ions, and small organic molecules.

■ Metabolism: All the chemical reactions occurring in an organism or cell.

Protein Synthesis. All organisms must make proteins, which are the workhorses of cells and organisms. The structure common in all cells is the **ribosome**, an RNA-protein machine that manufactures proteins based on the genetic instructions it receives from the DNA. Although the pattern for protein synthesis is identical, prokaryotic ribosomes are slightly smaller than their counterparts in eukaryotic cells.

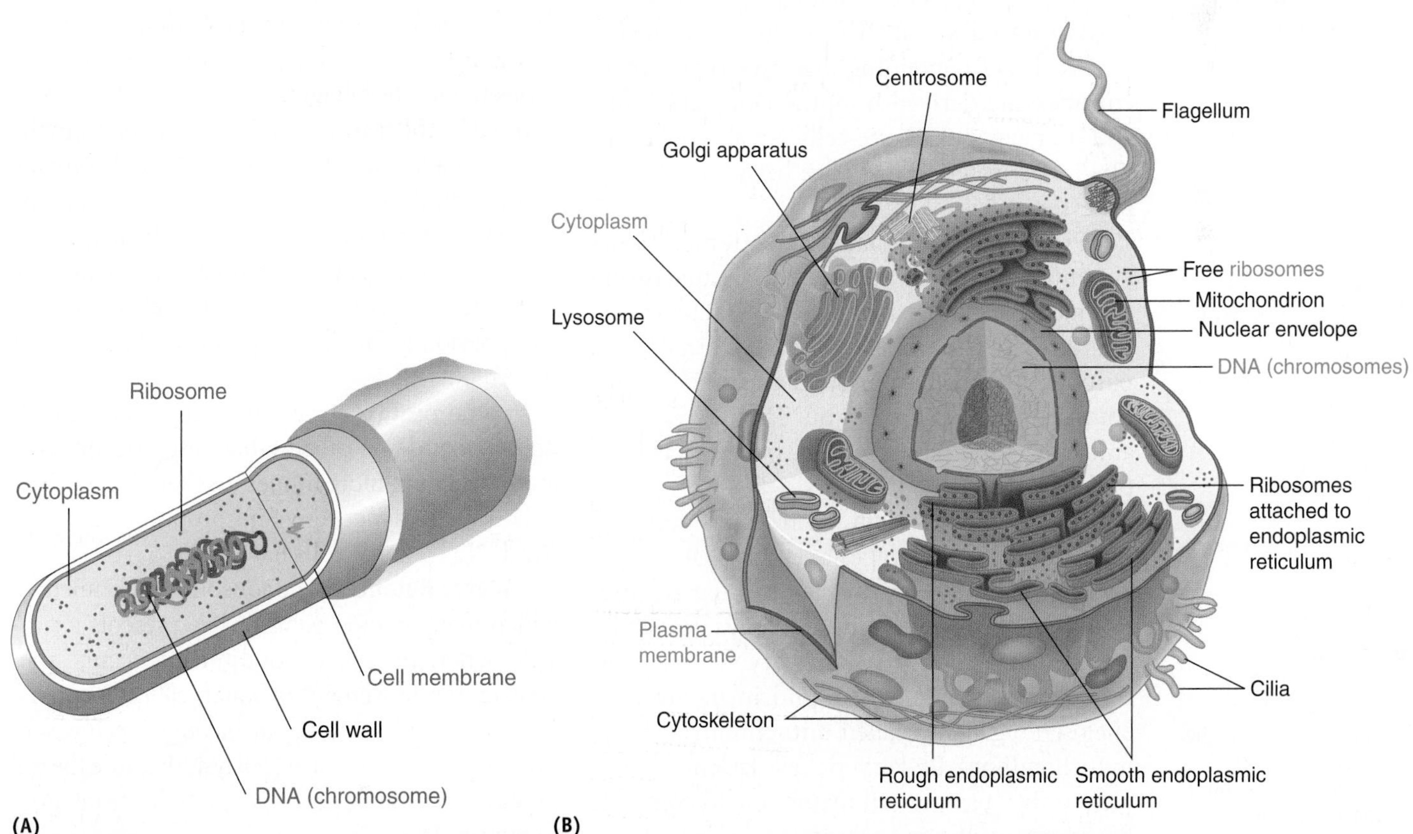

FIGURE 3.5 **A Comparison of a Bacterial and Eukaryotic Cell.** (**A**) A typical bacterial cell has relatively few visual compartments. (**B**) A protistan cell is a typical eukaryotic cell. Note the variety of cellular subcompartments, many of which are discussed in the text. Universal structures are indicated in red. »» List the ways you could microscopically distinguish a eukaryotic microbial cell from a bacterial cell.

■ Vesicle: A membrane-enclosed sphere involved with secretion, transport, and storage.

■ Diffusion: The movement of a substance from where it is in a higher concentration to where it is in a lower concentration.

Prokaryotic and Eukaryotic Cells Also Have Structural Distinctions

In the cytoplasm, eukaryotic microbes have a variety of structurally discrete, often membrane-enclosed, subcellular compartments called **organelles** to carry out specialized functions (Figure 3.5). Bacterial cells also have subcellular compartments—they just are not readily visible or membrane-enclosed.

Protein/Lipid Transport. Eukaryotic microbes have a series of membrane-enclosed organelles that compose the cell's **endomembrane system**, which is dedicated to transporting protein and lipid cargo through and out of the cell. This system includes the **endoplasmic reticulum (ER)**, which consists of flat membranes to which ribosomes are attached (rough ER) and tube-like membranes without ribosomes (smooth ER). These portions of the ER are involved in protein and lipid synthesis and transport, respectively.

The **Golgi apparatus** is a group of independent stacks of flattened membranes and vesicles where the proteins and lipids coming from the ER are processed, sorted, and packaged for transport. **Lysosomes**, somewhat circular, membrane-enclosed sacs containing digestive (hydrolytic) enzymes, are derived from the Golgi apparatus and, in many eukaryotic cells, break down captured food materials that have been brought into the cytoplasm.

Prokaryotic cells lack an endomembrane system, yet they are capable of manufacturing and modifying molecules just as their eukaryotic relatives do. In fact, many have internal compartments. Some have protein-enclosed "microcompartments" that carry out specific reactions and protect the cell from toxic byproducts from the reaction.

Energy Metabolism. Cells and organisms carry out one or two types of energy transformations. Through a process called **cellular respiration**, all cells convert chemical energy into cellular energy for cellular work. In eukaryotic microbes, this occurs in the cytosol and in membrane-enclosed organelles called **mitochondria** (sing., mitochondrion). Prokaryotic cells lack mitochondria so they use the cell membrane to complete the energy converting process.

A second energy transformation, **photosynthesis**, involves the conversion of light energy into chemical energy. In algal protists, photosynthesis occurs in membrane-bound **chloroplasts**. Some bacterial groups, such as the cyanobacteria we have mentioned, also carry out almost identical energy transformations. Again, the cell membrane or elaborations of the membrane represent the chemical workbench for the process.

Cell Structure and Transport. The eukaryotic **cytoskeleton** is organized into an interconnected system of cytoplasmic fibers, threads, and interwoven molecules that give structure to the cell and assist in the transport of materials throughout the cell. The main components of this internal cytoskeleton are microtubules that originate from the **centrosome** and microfilaments, each assembled from different protein subunits. Prokaryotic cells also have a cytoskeleton made of proteins related to those that construct microtubules and microfilaments and function to aid in determining cell shape in some prokaryotic cells and positioning structures in other types of bacterial cells.

Cell Motility. Many microbial organisms live in watery or damp environments and use the process of cell motility to move from one place to another. Some protists have long, thin protein projections called **flagella** (sing., flagellum) that, covered by the plasma membrane, extend from the cell. By beating back and forth, the flagella provide a mechanical force for motility. Many prokaryotic cells also exhibit motility; however, the flagella are structurally different and without a cell membrane covering. The pattern of motility also is different, providing a rotational propeller-like force for movement.

Some protists have other membrane-enveloped appendages called **cilia** (sing., cilium) that are shorter and more numerous than flagella. The cilia wave in synchrony and propel the cell forward. No bacterial cells have cilia.

Water Balance. The aqueous environment in which many microorganisms live presents a situation where the process of diffusion occurs, specifically the movement of water, called **osmosis**, into the cell. Continuing unabated, the cell would eventually swell and burst (cell lysis) because the cell or plasma membrane does not provide the integrity to prevent lysis.

Most prokaryotic and some eukaryotic cells (fungi, algae) contain a **cell wall** exterior to the cell

TABLE

3.1 Comparison of Prokaryotic and Eukaryotic Cell Structure

	Cell Structure or Compartment	
Characteristic	**Prokaryotic**	**Eukaryotic**
Genetic organization	Circular DNA chromosome	Linear DNA chromosomes
Cell compartmentation	Cell membrane	Plasma membrane
Metabolic organization	Cytoplasm	Cytoplasm
Protein synthesis	Ribosomes	Ribosomes
Protein/lipid transport	Cytoplasm	Endomembrane system
Energy metabolism	Cell membrane	Mitochondria and chloroplasts
Cell structure and transport	Thin protein filaments in cytoplasm	Protein tubules and filaments in cytoplasm
Cell motility	Bacterial flagella	Eukaryotic flagella or cilia
Water balance	Cell wall	Cell wall (algae, fungi)

or plasma membrane. Although the structure and organization of the wall differs between groups, all cell walls provide support for the cells, give them shape, and help them resist cell lysis.

A summary of the prokaryotic and eukaryotic processes and their associated structures is presented in (TABLE 3.1). MicroInquiry 3 examines a scenario for the evolution of the eukaryotic cell.

CONCEPT AND REASONING CHECKS 1

a. How is quorum sensing similar to the human social medium network Twitter?
b. List the universal structures and process that all cells possess as living organisms.
c. Explain how variation in cell structure between prokaryotic and eukaryotic cells can be compatible with a similarity in cellular processes between these cells.

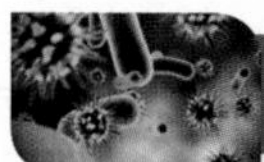

Chapter Challenge B

We have now covered the basic comparison between prokaryotic and eukaryotic cell structure and major metabolic processes (cell respiration and photosynthesis).

QUESTION B:

From what we have covered in terms of prokaryotic "social behavior" and now cell structure, how has your opinion changed regarding the so-called simple cells? Provide evidence to back up your opinion.

Answers can be found on the Student Companion Website in **Appendix D**.

KEY CONCEPT 3.2 Classifying Microorganisms Reveals Relationships Between Organisms

If you open any catalog, items are separated by types, styles, or uses. For example, in a fashion catalog, watches are separated from shoes and, within the shoes, men's, women's, and children's styles are separated from one another. Even the brands of shoes or their use (e.g., dress, casual, athletic) may be separated.

With such an immense diversity of organisms on planet Earth, the human drive to catalog these organisms has not been very different from cataloging watches and shoes; both are based on shared characteristics. In this section, we shall explore the principles on which microorganisms are classified and cataloged.

Classification Attempts to Catalog Organisms

In the 18th century, Carolus Linnaeus, a Swedish scientist, began identifying living organisms according to similarities in form (resemblances) and placing organisms in one of two "kingdoms"—Vegetalia and Animalia (FIGURE 3.6). This system was well accepted until the mid-1860s when a German naturalist, philosopher, and physician, Ernst Haeckel,

MICROINQUIRY 3

The Evolution of Eukaryotic Cells

Biologists and geologists have speculated for decades about the chemical evolution that led to the origins of the first prokaryotic cells on Earth. Whatever the origin, the first ancestral prokaryotes arose about 3.8 billion years ago and remained the sole inhabitants for some 1.5 billion years.

Scientists also have proposed various scenarios to account for the origins of the first eukaryotic cells. The oldest known fossils thought to be eukaryotic are about 2 billion years old.

A key concern here is figuring out how different membrane compartments arose to evolve into the organelles that are found in the eukaryotic cells today. Debate on this long intractable problem continues, so here we present some of the ideas that have fueled such discussions.

At some point around 2 billion years ago, the increasing number of metabolic reactions occurring in presumably larger bacterial and archaeal cells started to interfere with one another. As cells increased in size, the increasing volume of cell cytoplasm outpaced the ability of the cell surface (membrane) to be an effective "workbench" for servicing the metabolic needs of the whole cell. Complexity would necessitate more extensive workbench surface through compartmentation.

The Endomembrane System May Have Evolved Through Invagination

Similar to today's bacterial and archaeal cells, the cell membrane of an ancestral cell may have had specialized regions involved in protein synthesis, lipid synthesis, and nutrient hydrolysis. If the invagination of these regions occurred, the result could have been the internalization of these processes as independent internal membrane systems. For example, the membranes of the endoplasmic reticulum may have originated by multiple invagination events of the cell membrane (**Figure A1**).

Biologists have suggested that the elaboration of the evolving ER surrounded the nuclear region and DNA, creating the nuclear envelope. Surrounded and protected by a double membrane, greater genetic complexity could occur as the primitive eukaryotic cell continued to evolve in size and function. Other internalized membranes could give rise to the Golgi apparatus.

Chloroplasts and Mitochondria Arose from a Symbiotic Union with Engulfed Bacterial Cells

Mitochondria and chloroplasts are not part of the extensive endomembrane system. Therefore, these energy-converting organelles probably originated in a different way.

The structure of modern-day chloroplasts and mitochondria is very similar to a bacterial cell. In fact, mitochondria, chloroplasts, and bacterial cells share a large number of similarities (see table). In addition, there are bacterial cells alive today that carry out cellular respiration similarly to mitochondria and other bacterial cells (the cyanobacteria) that can carry out photosynthesis similarly to chloroplasts.

These similar functional patterns, along with other chemical and molecular similarities, suggested to the late Lynn Margulis at the University of Massachusetts, Amherst, that present-day chloroplasts and mitochondria represent modern representatives of what were once, many eons ago, free-living bacterial cells. Margulis, therefore, proposed the "endosymbiont model" for the origin of mitochondria and chloroplasts. The hypothesis suggests, in part, that billions of years ago a free-living bacterial cell that carried out cellular respiration was "swallowed" (engulfed) by a primitive eukaryotic cell (see Figure 3.7). The bacterial partner then lived within (*endo*) the eukaryotic cell in a mutually beneficial association (*symbiosis*) (**Figure A2**).

Likewise, a photosynthetic bacterial cell, perhaps a primitive cyanobacterium, was engulfed and evolved into the chloroplasts present in plants and algae today (**Figure A3**). The theory also would explain why both organelles have two membranes. One was the cell membrane of the engulfed bacterial cell and the other was the plasma membrane resulting from the engulfment process. By engulfing these bacterial cells and not destroying them, the evolving eukaryotic cell gained energy-conversion abilities, while the symbiotic bacterial cells gained a protected home.

If the first ancestral cell appeared about 3.8 billion years ago and the first single-celled eukaryote about 2 billion years ago, then it took some 1.5 billion years of evolution for the events described above to occur. With the appearance of the first eukaryotic cells, a variety of single-celled forms evolved, many of which were the very ancient ancestors of the single-celled eukaryotic organisms that exist today.

Obviously, laboratory studies can only hypothesize at mechanisms to explain how cells evolved and can only suggest—not prove—what might have happened billions of years ago. Short of inventing a time machine, we may never know the exact details for the origin of eukaryotic cells and organelles.

Discussion Point

Determine which endosymbiotic event must have come first: the engulfment of the bacterial progenitor of the chloroplast or the engulfment of the bacterial progenitor of the mitochondrion.

Answers (and comments) can be found on the Student Companion Website in **Appendix D**.

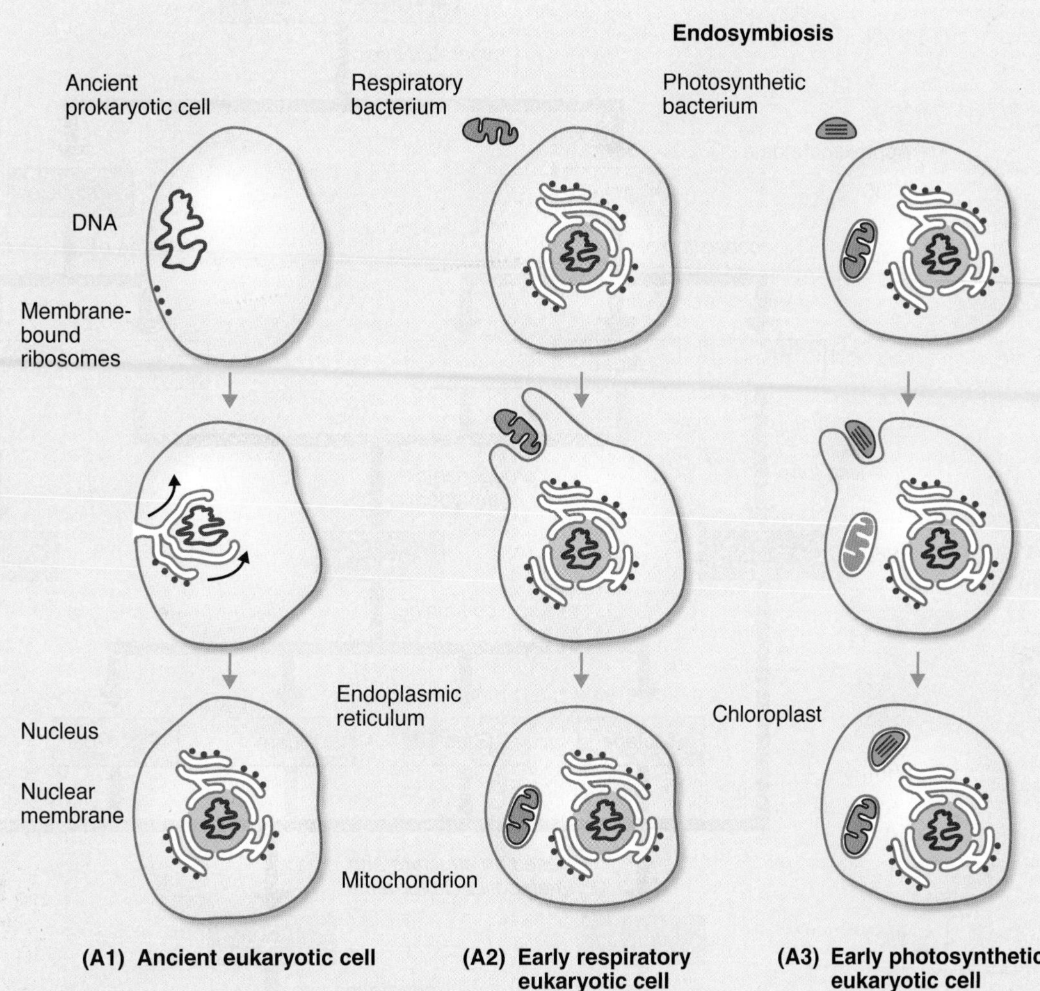

FIGURE A **Possible Origins of Eukaryotic Cell Compartments.** (**A1**) Invagination of the cell membrane from an ancient eukaryotic cell may have led to the development of the cell nucleus as well as to the membranes of the endomembrane system, including the endoplasmic reticulum. (**A2**) The mitochondrion may have resulted from the uptake and survival of a bacterial cell that carried out cellular respiration. (**A3**) A similar process, involving a bacterial cell that carried out photosynthesis, could have accounted for the origin of the chloroplast.

TABLE

Similarities Between Mitochondria, Chloroplasts, Bacteria, and Microbial Eukaryotes

Characteristic	Mitochondria	Chloroplasts	Bacteria	Microbial Eukaryotes
Average size	1–5 μm	1–5 μm	1–5 μm	10–20 μm
Nuclear envelope present	No	No	No	Yes
DNA molecule shape	Circular	Circular	Circular	Linear
Ribosomes	Yes; bacterial-like	Yes; bacterial-like	Yes	Yes; eukaryotic-like
Protein synthesis	Make some of their proteins	Make some of their proteins	Make all of their proteins	Make all of their proteins
Reproduction	Binary fission	Binary fission	Binary fission	Mitosis and cytokinesis

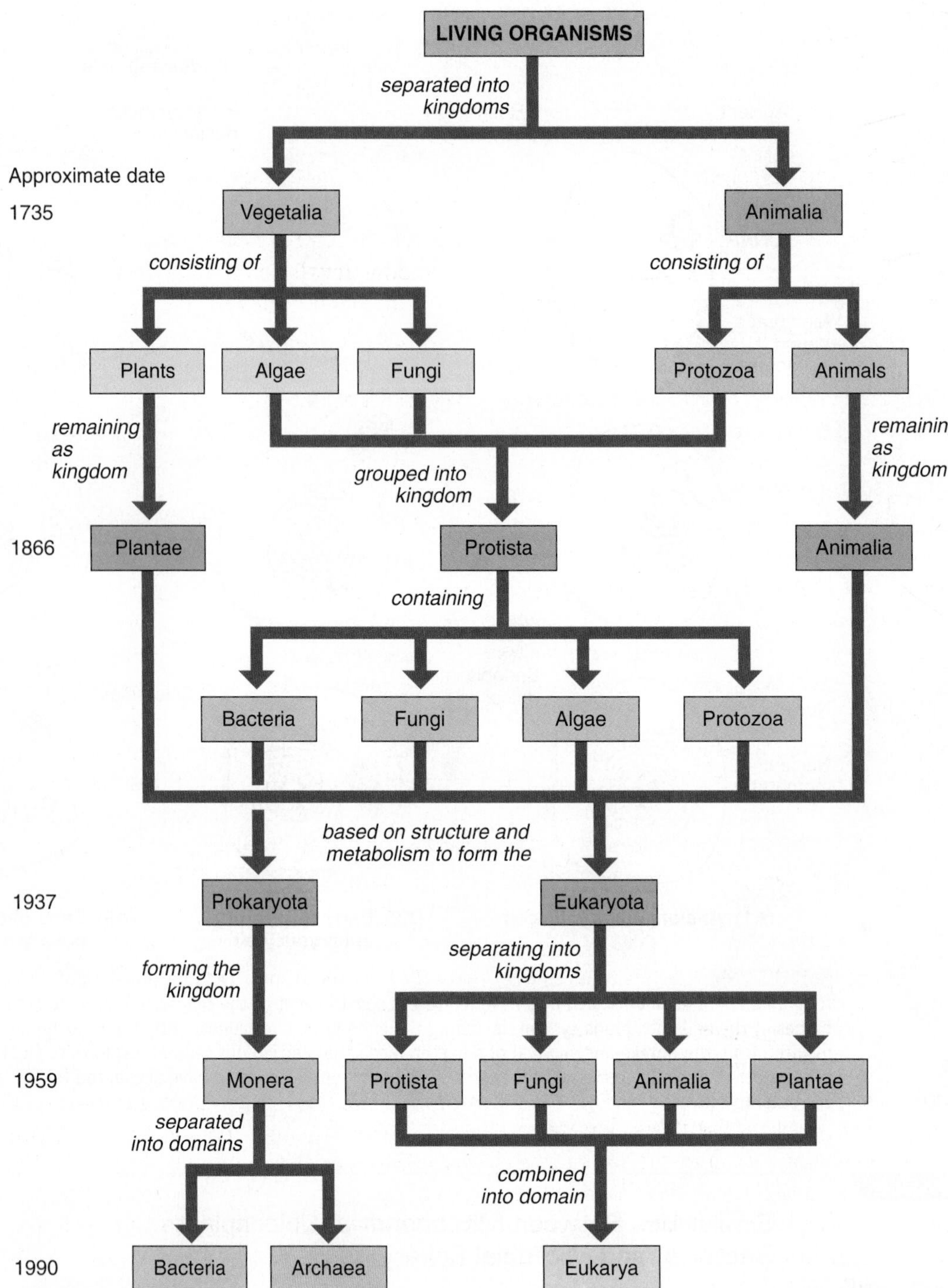

FIGURE 3.6 **A Concept Map Illustrating the Development of Classification for Living Organisms.** Since about 1866, new observations and techniques have been used to reclassify and reorganize living organisms. Modified from Schaechter, Ingraham, and Neidhardt. *Microbe*. ASM Press, 2006, Washington, D.C. »» Of the plants, algae, fungi, bacteria, protists, and animals, which are in each of the three domains?

identified a fundamental problem in the two-kingdom system. The unicellular (microscopic) organisms being identified by Haeckel, Pasteur, Koch, and their associates did not conform to the two-kingdom system of multicellular organisms. Haeckel constructed a third kingdom, the Protista, in which all the known unicellular organisms were placed. The bacterial organisms, which he called "moneres," were near the bottom of the tree, closest to the root of the tree.

With improvements in the design of light microscopes, more observations were made of bacterial and protist organisms. In 1937, a French biologist, Edouard Chatton, proposed that there was a fundamental dichotomy among the Protista. He saw bacterial cells as having distinctive properties (not articulated in his writings) in "the prokaryotic nature of their cells" and should be separated from all other protists "which have eukaryotic cells." With the development of the electron microscope in the 1950s, it became apparent that protists had a membrane-enclosed nucleus and were identified, along with the plants and animals, as being eukaryotes while other protists (the bacterial organisms) lacked this structure and were considered to be prokaryotes. Thus, in 1956, Herbert Copland suggested bacterial organisms be placed in a fourth kingdom, the Monera.

But there was still one more problem with the kingdom Protista. Robert H. Whittaker, a botanist at the University of California, saw the fungi as yet another kingdom of organisms. The fungi are the only eukaryotic group that must externally digest their food prior to absorption and, as such, live in the food source. For this and other reasons, Whittaker in 1959 refined the four-kingdom system into five kingdoms, identifying the kingdom Fungi as a separate, multicellular, eukaryotic group distinguished by an absorptive mode of nutrition.

The five kingdom system rested safely for about 15 years. In the late 1970s, Carl Woese, an evolutionary biologist at the University of Illinois, began a molecular analysis of living organisms based on comparisons of nucleotide sequences of genes coding for the small subunit ribosomal RNA (rRNA) found in all organisms. These analyses revealed yet another dichotomy, this time within the kingdom Monera. By 1990, it was clear that the kingdom Monera contained two fundamentally unrelated groups, what Woese initially called the Bacteria and Archaebacteria. These two groups were as different from each other as they were different from the eukaryotes.

Kingdoms and Domains: Making Sense of Taxonomic Relationships

What many of these scientists are or were doing is **systematics**; that is, studying the diversity of life and its evolutionary relationships. Systematic biologists—systematists for short—identify, name, and classify organisms (**taxonomy**), and organize their observations within a framework that shows taxonomic relationships.

Often it is difficult to make sense of taxonomic relationships because new information that is more detailed keeps being discovered about organisms. This then motivates taxonomists to figure out how the new information fits into the known classification schemes—or how the schemes need to be modified to fit the new information. This is no clearer than the most recent taxonomic revolution that, as the opening quote states, has come along to "tweak our theories or teach us something new."

Carl Woese, along with George Fox and coworkers at the University of Illinois, Urbana-Champaign, proposed a new classification scheme with a new most inclusive taxon, the **domain**. The new scheme initially came from work that compared the DNA nucleotide base sequences for the RNA in ribosomes, those protein manufacturing machines needed by all cells. Woese and Fox's results were especially relevant when comparing those sequences from a group of bacterial organisms formerly called the archaebacteria (*archae* = "ancient"). Many of these bacterial forms are known for their ability to live under extremely harsh environments. Woese discovered that the nucleotide sequences in these archaebacteria were different from those in other prokaryotes and in eukaryotes. After finding other differences, including cell wall composition, membrane lipids, and sensitivity to certain antibiotics, the evidence pointed to there being three taxonomic lines to the "tree of life."

■ Taxon (pl., taxa): A subdivision used to classify organisms.

One goal of systematics, and the main one of interest here, is to reconstruct the **phylogeny** (*phylo* = "tribe"; *geny* = "production"), the evolutionary history of a species or group of species. Systematists illustrate phylogenies with **phylogenetic trees**, which identify inferred relationships among species. In Woese's phylogeny, one branch of the phylogenetic tree includes the former archaebacteria and is called the domain **Archaea** (FIGURE 3.7). The second encompasses all the remaining prokaryotes and is called the domain **Bacteria**. The third domain, the **Eukarya**, includes three multicellular kingdoms (Plants, Fungi, and Animals and a diverse set of single-celled eukaryotes that make up many groups of protists). Because, like all theories, they are revised as scientists gather new data, the tree of life continues to change as our knowledge of diversity increases.

Several important events are shown in the phylogenetic tree. First, the Archaea branch off

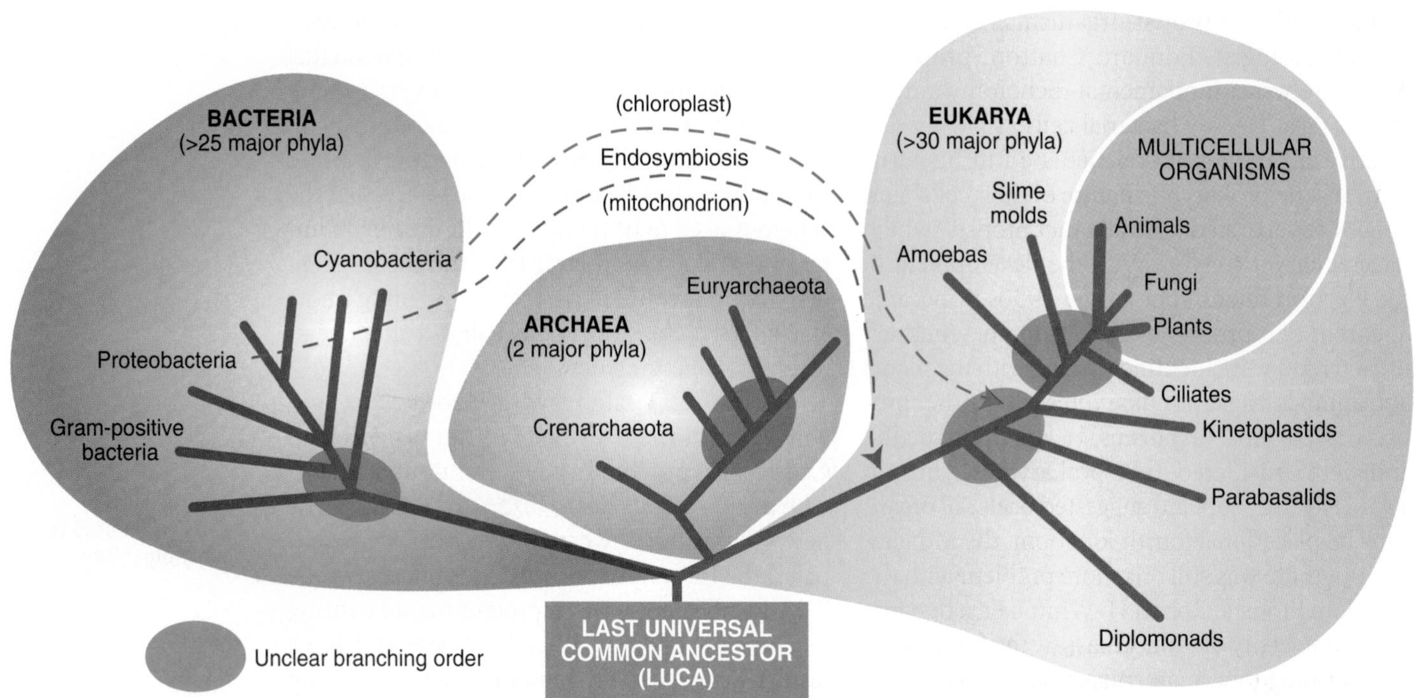

FIGURE 3.7 **Three Domains Form the "Tree of Life."** Fundamental differences in genetic endowments are the basis for the three domains of all organisms on Earth. Some 3.8 billion years ago, a universal ancestor arose from which all modern day organisms descended. All organisms in the yellow portion of the Eukarya represent protists. »» What portions of the phylogenetic tree represent microorganisms?

the lineage leading to the Eukarya, meaning members of this domain share a more recent ancestor with the Archaea than they do with the Bacteria. Second, the mitochondria of the Eukarya (and chloroplasts of plants and algae in the Eukarya) evolved, through the process of **endosymbiosis**, from different, once free-living, bacterial species (see MicroInquiry 3). Third, because, like all theories, the tree of life is revised as scientists gather new data concerning life's diversity, there are many tentative branches on the tree; that is, the divergence of many lineages remains uncertain.

The last aspect on the tree concerns the origin of all three domains. According to one scenario, once life got going, there was a single life form, a very ancient prokaryote called a "**progenote**." Results from molecular studies done by many scientists suggest that this "last universal common ancestor" or LUCA had a cell wall, a simple but leaky cell membrane, a highly active metabolism to break down and extract energy from a variety of food sources, and perhaps an internal compartment. It appears LUCA did not have the ability to synthesize or use DNA. Instead, it may have used RNA to store information and control metabolism. Most interesting, LUCA was probably so error-prone in making proteins that such cells must have "shared" their RNA and good proteins with other such cells. Thus, LUCA functioned in Earth's oceans as a giant mega-organism, passing useful molecules back and forth between cells, made easier by having leaky membranes. Then, at some point about 3.8 billion years ago, some cells evolved ways to go it alone and, as time passed, two lineages of life evolved, one of which in another 500 million years would further branch into the Archaea and Eukarya.

Nomenclature Gives Scientific Names to Organisms

Another goal of systematics is the naming of organisms and their placement in a classification. In his *Systema Naturae,* Linnaeus popularized a two-word (binomial) scheme of nomenclature, the two words usually derived from Latin or Greek stems. Each organism's name consists of the **genus** to which the organism belongs and a **specific epithet**, a descriptor that further describes the genus name. Together these two words make up the **species** name. For example, the common bacterium *Escherichia coli* resides in the gut of all humans (*Homo sapiens*) (MicroFocus 3.2).

MICROFOCUS 3.2: Tools

Naming Names

As you read this book, you have and will come across many scientific names for microbes, where a species name is a combination of the genus and specific epithet. Not only are many of these names tongue twisting to pronounce (many are listed with their pronunciation inside the front and back covers), but how in the world did the organisms get those names? Here are a few examples.

Genera Named After Individuals

Escherichia coli: named after Theodore Escherich who isolated the bacterial cells from infant feces in 1885. Being in feces, it commonly is found in the colon.

Neisseria gonorrhoeae: named after Albert Neisser who discovered the bacterial organism in 1879. As the specific epithet points out, the disease it causes is gonorrhea.

Genera Named for a Microbe's Shape

Vibrio cholerae: vibrio means "comma-shaped," which describes the shape of the bacterial cells that cause cholera.

Staphylococcus epidermidis: staphylo means "cluster" and *coccus* means "spheres." So, these bacterial cells form clusters of spheres that are found on the skin surface (epidermis).

Genera Named After an Attribute of the Microbe

Saccharomyces cerevisiae: in 1837, Theodor Schwann observed yeast cells and called them *Saccharomyces* (*saccharo* = "sugar"; *myce* = "fungus") because the yeast converted grape juice (sugar) into alcohol; *cerevisiae* (from *cerevisia* = "beer") refers to the use of yeast since ancient times to make beer.

Myxococcus xanthus: myxo means "slime," so these are slime-producing spheres that grow as yellow (*xantho* = "yellow") colonies on agar.

Thiomargarita namibiensis: see MicroFocus 3.5.

Notice in these examples that when a species name is written, only the first letter of the genus name is capitalized, while the specific epithet is not. In addition, both words are printed in italics or underlined. After the first time a species name has been spelled out, biologists usually abbreviate the genus name using only its initial genus letter or some accepted substitution, together with the full specific epithet; that is, *E. coli* or *H. sapiens*. A cautionary note: often in magazines and newspapers, proper nomenclature is not followed, so our gut bacterium would be written as Escherichia coli.

Classification Uses a Hierarchical System

Linnaeus' cataloging of plants and animals used shared and common characteristics. Such similar organisms that could interbreed were related as a species, which formed the least inclusive level of the hierarchical system. Part of Linnaeus' innovation was the grouping of species into higher taxa that also were based on shared, but more inclusive, similarities.

Today several similar species are grouped together into a genus (pl., genera). A collection of similar genera makes up a **family** and families with similar characteristics make up an **order**. Different orders may be placed together in a **class** and classes are assembled together into a **phylum** (pl., phyla). All phyla would be placed together in a kingdom and/or domain, the most inclusive level of classification. TABLE 3.2 outlines the taxonomic hierarchy for three organisms.

Many prokaryotic organisms may belong to a rank below the species level to indicate a special characteristic exists within a subgroup of the species. Such ranks have practical usefulness in helping to identify an organism. For example, two biotypes of the cholera bacterium, *Vibrio cholerae,* are known: *Vibrio cholerae* classic and *Vibrio cholerae* El Tor. Other designations of ranks include subspecies, serotype, strain, morphotype, and variety.

■ Biotype: A population or group of individuals having the same genetic constitution (genotype).

David Hendricks Bergey devised one of the first systems of classification for the bacterial species in 1923. Today, the proper taxonomic classification for the Bacteria and Archaea can be found in

TABLE

3.2 Taxonomic Classification of Humans, Brewer's Yeast, and a Common Bacterium

	Humans	Brewer's Yeast	*Escherichia coli*
Domain	Eukarya	Eukarya	Bacteria
Kingdom	Animalia	Fungi	
Phylum	Chordata	Ascomycota	Proteobacteria
Class	Mammalia	Saccharomycotina	Gammaproteobacteria
Order	Primates	Saccharomycetales	Enterobacteriales
Family	Hominidae	Saccharomycetaceae	Enterobacteriaceae
Genus	*Homo*	*Saccharomyces*	*Escherichia*
Species	*H. sapiens*	*S. cerevisiae*	*E. coli*

the second edition of *Bergey's Manual of Systematic Bacteriology*. The tremendous changes that have taken place in taxonomy is evident by the addition of more than 2,200 new species and 390 new genera to just the first volume of the five volumes of the second edition.

Many Methods Are Available to Identify and Classify Microorganisms

There are several traditional and more modern criteria that microbiologists can use to identify and classify microorganisms. For example, a hospital's clinical microbiology lab (CML) is involved in the diagnosis and treatment of infectious diseases. The bacteriology section will use physical and biochemical methods (metabolic tests) for the identification of a bacterial pathogen. In fact, *Bergey's Manual of Determinative Bacteriology*, now in its ninth edition, is the primary source for making routine medical identifications of bacterial pathogens. On the other hand, taxonomists rely more on biochemical, molecular, and nucleic acid sequencing technologies (the CML is becoming more reliant on these methods as well).

Let's briefly review some of the more determinative methods and a few molecular methods available.

Physical Characteristics. These include differential staining reactions to help determine the organism's shape (morphology), and the size and arrangement of cells. Other characteristics can include: oxygen, pH, and growth temperature requirements; spore-forming ability; and motility. Unfortunately, there are many prokaryotes that have the same physical characteristics, so other distinguishing features are usually needed.

Biochemical Tests. As microbiologists better understood bacterial physiology, they discovered there were certain metabolic properties that were present only in certain groups.

Today, a large number of biochemical tests exist and often a specific test can be used to eliminate certain groups from the identification process. Among the more common tests are: fermentation of carbohydrates; the use of a specific substrate; and the production of specific byproducts or waste products. But, as with the physical characteristics, often several biochemical tests are needed to differentiate between species.

The vast number of tests and analyses available for bacterial cells can make it difficult to know which are relevant for pathogen identification purposes. One widely used technique in many disciplines is the **dichotomous key**. There are various forms of dichotomous keys, but one very useful construction is a flow chart where a series of positive or negative test procedures are listed down the page. Based on the dichotomous nature of the test (always a positive or negative result), the flow chart immediately leads to the next test result. The result is the identification of a specific organism. A simplified example is shown in MicroFocus 3.3.

These identification tests are important clinically, as they can be part of the arsenal available to the CML that is trying to identify a pathogen. Today, many of these tests use rapid, miniaturized identification methods (MicroFocus 3.4) or automated systems (FIGURE 3.8).

Serological Tests. The immunology section within the CML often diagnoses infectious agents, detecting antigens or antibodies in clinical

MICROFOCUS 3.3: Tools
Dichotomous Key Flow Chart

A medical version of a taxonomic key (in the form of a dichotomous flow chart) can be used to identify very similar bacterial species based on physical and biochemical characteristics.

In this simplified scenario, an unknown bacterium has been cultured and several tests run. The test results are shown in the box at the top. Using the test results and the flow chart, identify the bacterial species that has been cultured.

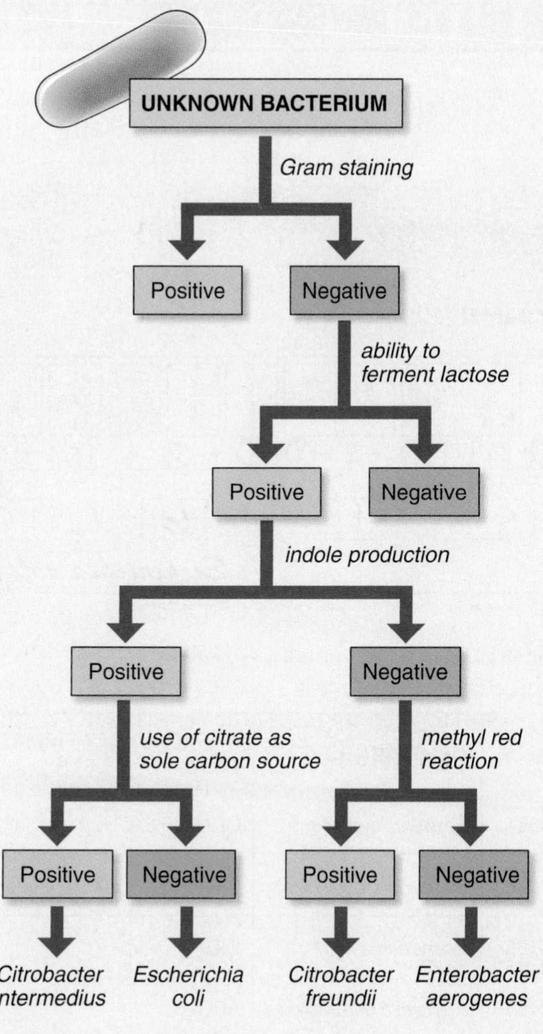

■ Enteric: Referring to the intestines.

MICROFOCUS 3.4: Tools

Rapid Identification of Enteric Bacteria

In recent years, a number of miniaturized systems have been made available to microbiologists for the rapid identification of enteric bacteria. One such system is the Enterotube™ II, a self-contained, sterile, compartmentalized plastic tube containing 12 different media and an enclosed inoculating wire. This system permits the inoculation of all media and the performance of 15 standard biochemical tests using a single bacterial colony. The media in the tube indicate by color change whether the organism can carry out the metabolic reaction. After 24 hours of incubation, the *positive* tests are circled and all the circled numbers in each boxed section are added to yield a 5-digit ID for the organism being tested. This 5-digit number is looked up in a reference book or computer software to determine the identity of the species.

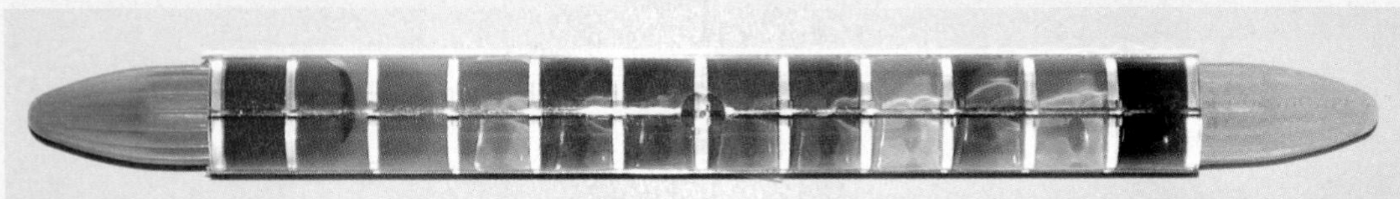

(A) An uninoculated tube.

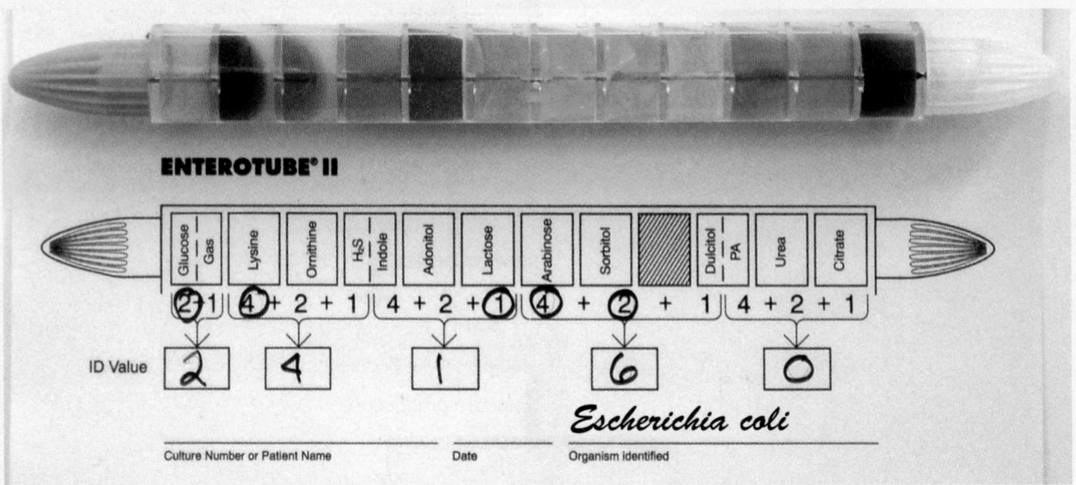

(B) An inoculated tube incubated for 24 hours. Courtesy of Jeffrey Pommerville.

VALUE	IDENTIFICATION	ATYPICAL TESTS	CONFIRMATORY TESTS
24061	*Serratia liquifaciens*	ORN–	
24070	*Escherichia coli (AD)*	IND–	Pungent odor
24160	*Escherichia coli*	IND–	
24163	*Klebsiella pneumoniae*	ADO–	

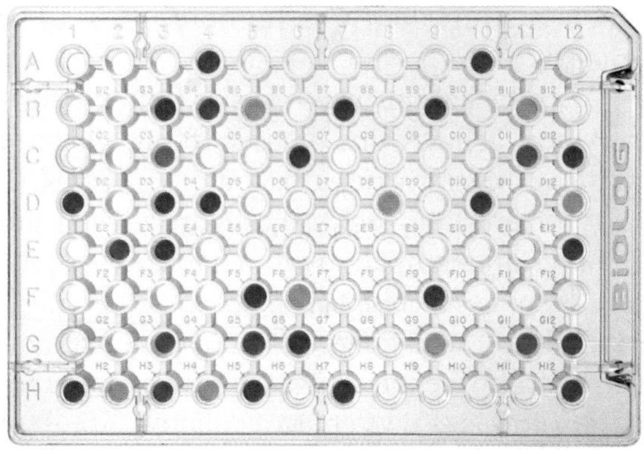

FIGURE 3.8 **A Biolog MicroPlate®.** The Biolog system is capable of identifying hundreds of bacterial species by assessing the organism's ability to use any of 95 different substrates in a 96-well microtiter plate. The use of any substrate results in a reduction of the dye in that well, producing purple color development. The intensity of the purple coloration indicates the degree of substrate usage and is read by a computer-linked automated microtiter reader. The first well (upper left) is a negative control with no substrate. Courtesy of Biolog, Inc. »» Of the methods described in this section, which is/are most likely to be used in this more automated system? Explain.

specimens. Microorganisms are antigenic, meaning they are capable of triggering the production of antibodies. Solutions of such collected antibodies, called "antisera," are commercially available for many medically important pathogens. For example, mixing a *Salmonella* antiserum with *Salmonella* cells will cause the cells to clump together or agglutinate. Therefore if a foodborne illness occurs, the antiserum may be useful in identifying if *Salmonella* is the pathogen. More information about serological testing will be presented in a later chapter.

■ Antibody: A protein produced by the immune system in response to a specific chemical configuration (antigen).

Nucleic Acid Analysis. In 1984, the editors of *Bergey's Manual of Systematic Bacteriology* noted that there is no "official" classification of bacterial species and that the closest approximation to an official classification is the one most widely accepted by the community of microbiologists. The editors stated that a comprehensive classification might one day be possible. Today, the fields of molecular genetics and genomics have advanced the analysis and sequencing of nucleic acids. This has given rise to a new era of molecular taxonomy.

Molecular taxonomy is based on the universal presence of ribosomes in all living organisms. In particular, it is the RNAs in the ribosome, called ribosomal RNA (rRNA), which are of most interest and the primary basis of Woese's construction of the three domains of life. Many taxonomists today believe the genes for rRNA are one of the most accurate measures for precise bacterial classification. Other techniques, including the polymerase chain reaction and nucleic acid hybridization, will be mentioned in later chapters.

CONCEPT AND REASONING CHECKS 2

a. What four events changed the cataloging of microorganisms by shared characteristics?

b. With regard to the three domains, it has been said that Woese "lifted a whole submerged continent out of the ocean." What is the "submerged continent" and why is the term "lifted" used?

c. Which one of the following is a correctly written scientific name for the bacterium that causes anthrax? (i) *bacillus Anthracis*, (ii) *Bacillus Anthracis*, or (iii) *Bacillus anthracis*.

d. How would you describe an order in the taxonomic classification?

e. Why are so many tests often needed to identify a specific bacterial species?

Chapter Challenge C

Many different types of tools and techniques have been used to classify the prokaryotes and eukaryotes in one of the three domains in the current "tree of life."

QUESTION C:

Based on the previous two chapter challenges, what do all the similarities, differences, and "exceptions to the rule" with regard to prokaryotes and eukaryotes tell you about the nature of early life on Earth? How does LUCA factor into your judgment?

Answers can be found on the Student Companion Website in **Appendix D.**

KEY CONCEPT 3.3 Microscopy Is Used to Visualize the Structure of Cells

In the previous sections, microscopy was the tool that provided the images to contrast the structure of prokaryotic and eukaryotic cells. Before we examine the modern instruments used to "see" and describe microorganisms and viruses, we need to be familiar with the units of measurement.

Many Microbial Agents Are in the Micrometer Size Range

One morphological characteristic used to study microorganisms and viruses is their size. Because they are so small, a convenient system of measurement is used that is the scientific standard around the world. The measurement system is the metric system, where the standard unit of length is the meter and is a little longer than a yard (see **Appendix A**). To measure microorganisms, we need to use units that are a fraction of a meter. In microbiology, the common unit for measuring length is the **micrometer (μm)**, which is equivalent to a millionth (10–6) of a meter.

Microbial agents range in size from the relatively large, almost visible protists (100 μm) down to the incredibly tiny viruses (0.02 μm) (FIGURE 3.9). Most bacterial and archaeal cells are about 1 μm to 5 μm in length, although notable exceptions have been discovered recently (MicroFocus 3.5). Because most viruses are a fraction of one micrometer, their size is expressed in nanometers. A **nanometer (nm)** is equivalent to a billionth (10–9) of a meter; that is, 1/1,000 of a μm. Using nanometers, the size of the poliovirus, among the smaller viruses, measures 20 nm (0.02 μm) in diameter.

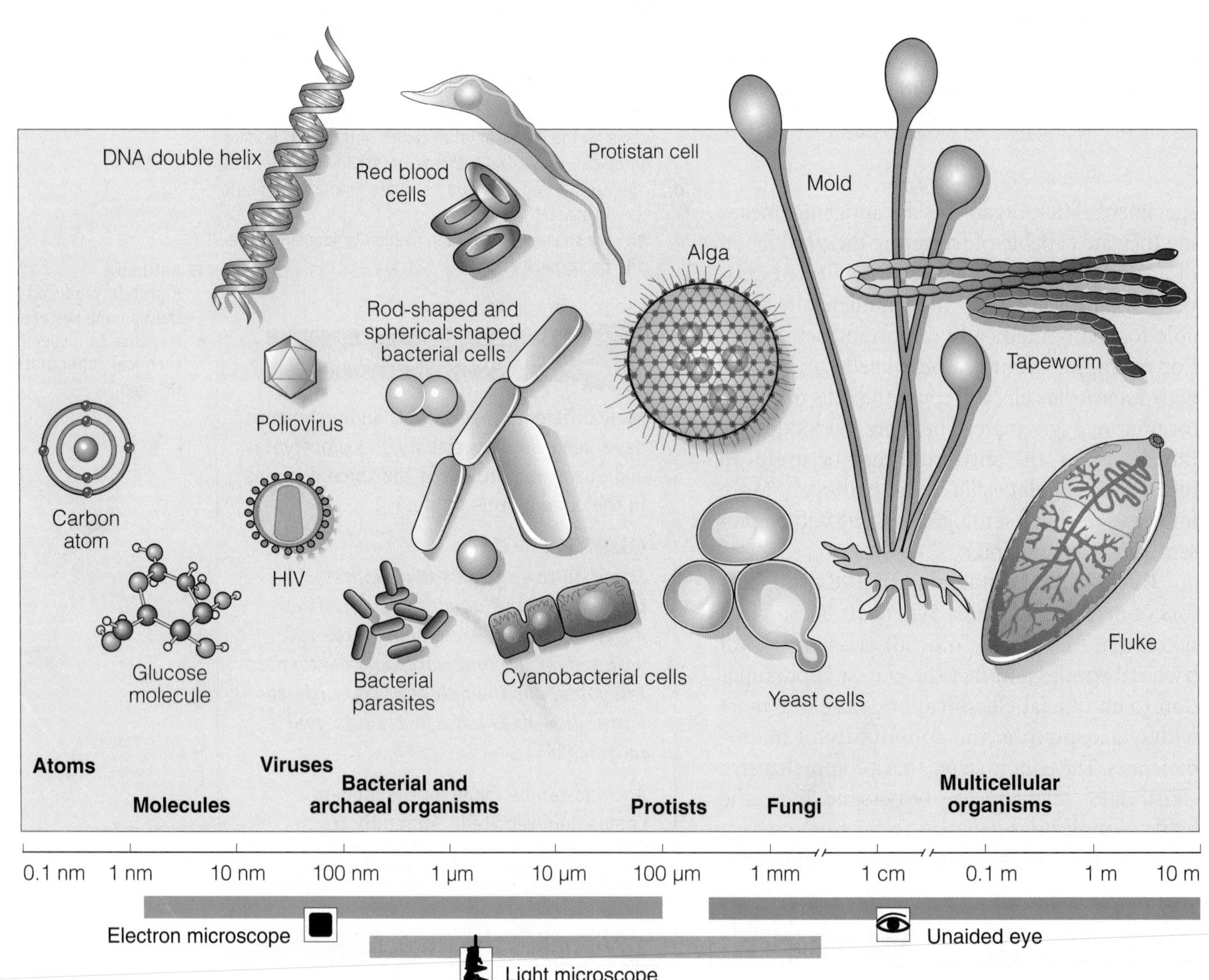

FIGURE 3.9 **Size Comparisons Among Various Atoms, Molecules, Viruses, and Microorganisms (not drawn to scale).** Although tapeworms and flukes usually are macroscopic, the diseases these parasites cause are often studied by microbiologists. »» Referring to MicroFocus 3.5, where would you place an ARMAN cell and a *Thiomargarita* cell?

MICROFOCUS 3.5: Environmental Microbiology

Size Extremes

Copper mines can be acidic caldrons. Take, for example, the Richmond Mine at Iron Mountain near Redding, California. It is the source of the most acidic water naturally found on Earth. And believe it or not, there are acidophilic (acid-loving) prokaryotes that form a pink, floating biofilm several millimeters thick on the surface of the hot, toxic water that has a pH of 0.8. If that isn't amazing enough, in 2006 University of California scientists identified one of these microbes, a member of the domain Archaea, as the smallest living organism yet discovered. Called ARMAN (Archaeal Richmond Mine Acidophilic Nanoorganism), it is only 0.2 to 0.4 μm in diameter (about the size of a large virus). By comparison, *Escherichia coli* is three times this diameter and up to 100 times the cell volume. ARMAN cells are free-living, they contain few ribosomes, and possess a relatively small number of genes.

At the other extreme, while on an expedition off the coast of Namibia (western coast of southern Africa) in 1997, scientists from the Max Planck Institute for Marine Microbiology in Bremen, Germany, found a bacterial monster in sediment samples from the sea floor. These chains of spherical cells (see figure) were 100 μm to 300 μm in diameter—but some as large as 750 μm—about the diameter of the period in this sentence. Their volume is about 3 million times greater than that of *E. coli.* The cells, shining white with enclosed sulfur granules, looked like a string of pearls. Thus, the bacterial species was named *Thiomargarita namibiensis* (meaning "sulfur pearl of Namibia"). Another closely related strain was discovered in the Gulf of Mexico in 2005.

Yes, the vast majority of microorganisms are of typical size, but exceptions have been found in some exotic places.

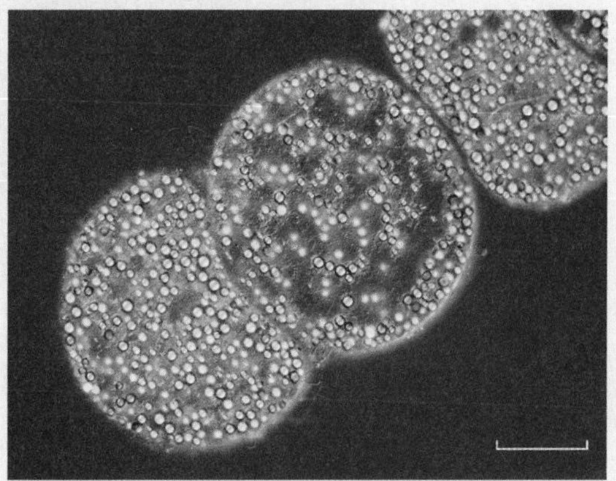

A phase microscopy image showing a chain of *Thiomargarita namibiensis* cells. (Bar = 100 μm.) Courtesy of Heide Schulz-Vogt, Max Planck Institute of Marine Microbiology, Germany.

Light Microscopy Is Used to Observe Most Microorganisms

The basic microscope system used in the microbiology laboratory is the **light microscope**, in which visible light passes directly through the lenses and specimen (FIGURE 3.10A). Such an optical configuration is called **bright-field microscopy**. Visible light is projected through a condenser lens, which focuses the light into a sharp cone (FIGURE 3.10B). The light then passes through the opening in the stage. When hitting the glass slide, the light is reflected or refracted as it goes through the specimen. Then, the light enters the objective lens to form a magnified intermediate image inverted from that of the specimen. This intermediate image becomes the object magnified by the ocular lens (eyepiece) and seen by the observer.

Two important factors to consider with the microscope are magnification and resolution.

Magnification. The increase in the apparent size of the specimen being observed is called **magnification**. A light microscope usually has at least three objective lenses: the low-power, high-power, and oil-immersion lenses. In general, these lenses magnify an object 10, 40, and 100 times, respectively. (Magnification is represented by the multiplication sign, ×.) The ocular lens then magnifies the intermediate image produced by the objective lens by 10×. Therefore, the total magnification achieved is 100×, 400×, and 1,000×, respectively.

■ Total magnification: The magnification of the ocular multiplied by the magnification of the objective lens being used.

Resolution. For a magnified object to be seen distinctly, the lens system must have good **resolution**; that is, it must transmit light without variation and allow closely spaced objects to be clearly distinguished as two objects. For example, a car seen in the distance at night may appear to have a single headlight because at that distance the unaided eye lacks sufficient resolution. However,

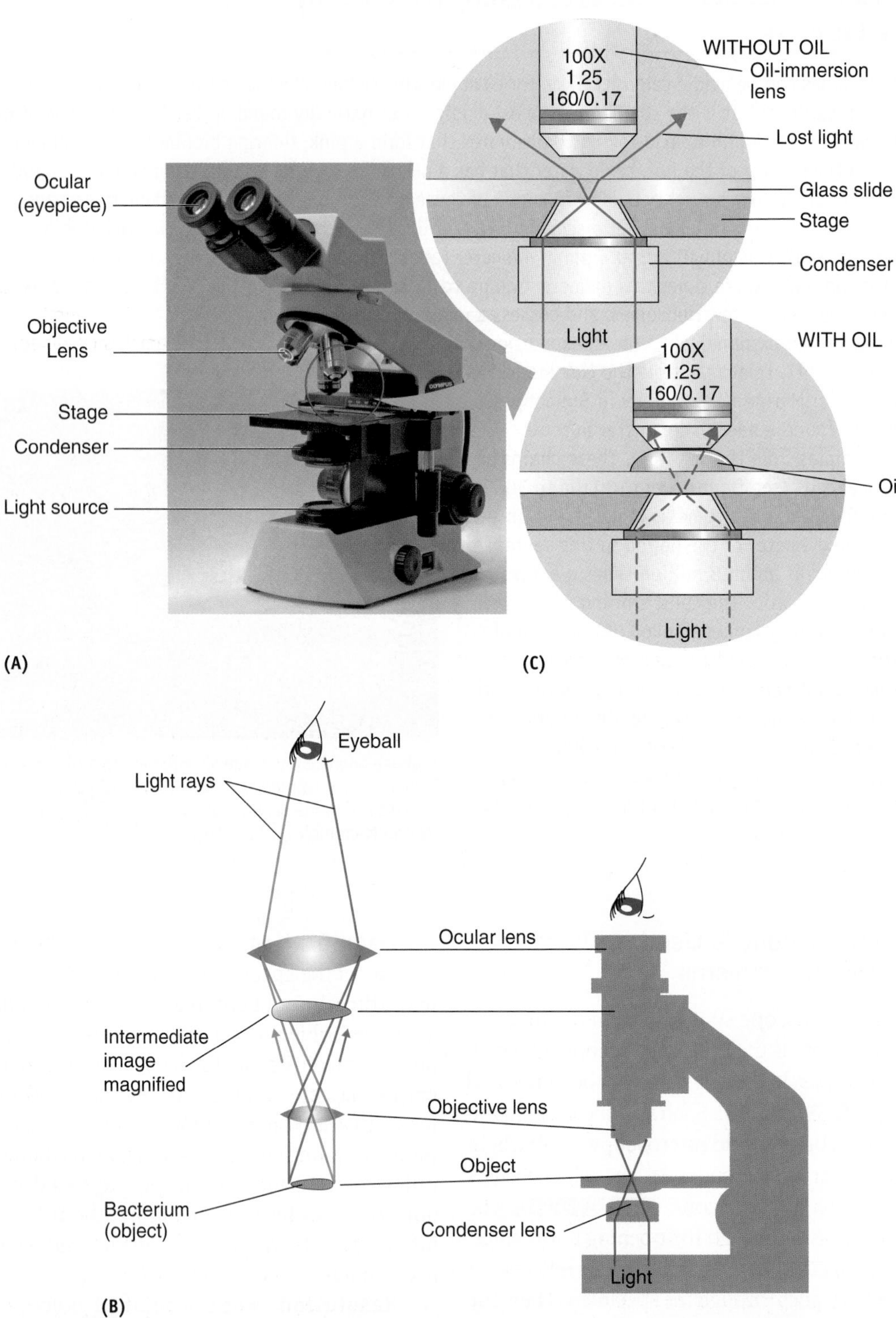

FIGURE 3.10 **The Light Microscope.** (**A**) The light microscope is used in many instructional and clinical laboratories. Note the important features of the microscope that contribute to the visualization of the object. Courtesy of Jeffrey Pommerville. (**B**) Image formation in the light microscope requires light to pass through the objective lens, forming an intermediate image. This image serves as an object for the ocular lens, which further magnifies the image and forms the final image the eye perceives. (**C**) When using the oil immersion lens (100×), oil must be placed between and continuous with the slide and objective lens. »» Why must oil be used with the 100× oil-immersion lens?

MICROFOCUS 3.6: Tools

Calculating Resolving Power

The resolution or resolving power (RP) of a lens system is important in microscopy because it indicates the size of the smallest object that can be seen clearly. The resolving power varies for each objective lens and is calculated using the following formula:

$$RP = \frac{\lambda}{2 \times NA}$$

In this formula, the Greek letter λ (lambda) represents the wavelength of light; for white light, it averages about 550 nm. The symbol NA stands for the numerical aperture of the lens and refers to the size of the cone of light that enters the objective lens after passing through the specimen. This number generally is printed on the side of the objective lens (see Figure 3.10C). For an oil-immersion objective with an NA of 1.25, the resolving power may be calculated as follows:

$$RP = \frac{550\,\text{nm}}{2 \times 1.25} = \frac{550}{2.5} = 220\text{ nm or }0.22\text{ µm}$$

Because the resolution limit for this lens system is 220 nm, any object smaller than 220 nm could not be seen as a clear, distinct object. An object larger than 220 nm would be resolved.

using binoculars, the two headlights can be seen clearly because the binoculars have higher resolving ability than the eye.

However, when switching from the low-power (10×) or high-power (40×) lens to the oil-immersion lens (100×), one quickly finds that the image has become fuzzy. The object lacks resolution, and the resolving power of the lens system appears to be poor. The poor resolution results from the refraction of light.

Both low-power and high-power objectives are wide enough to capture sufficient light for viewing. The oil-immersion objective, on the other hand, is so narrow that most light bends away and would miss the objective lens FIGURE 3.10C. The **index of refraction** (or refractive index) is a measure of the light-bending ability of a substance. Immersion oil has an index of refraction of 1.5, which is almost identical to that of glass. Therefore, by immersing the 100× lens in oil, the light does not bend away from the lens as it passes from the glass slide and the specimen.

The oil thus provides a homogeneous pathway for light from the slide to the objective, and the resolution of the object increases. With the oil-immersion lens, the highest resolution possible with the light microscope is attained, which is near 0.2 µm (200 nm) (MicroFocus 3.6). Still, at this resolution little internal detail can be seen in the cell cytoplasm.

Staining Techniques Provide Contrast

Microbiologists commonly stain bacterial cells before viewing them with bright-field microscopy because the cytoplasm usually lacks color, making it hard to see the colorless cells on the bright microscope background, called the **field**. Several staining techniques have been developed to provide **contrast**. However, the stain preparation kills the cells.

Before staining the bacterial cells they are first smeared on a glass slide and the slide air-dried. Next, the slide is passed briefly through a flame in a process called **heat fixation**, which bonds the cells to the slide, kills any cells still alive, and increases stain absorption.

Simple Stains. In the **simple stain technique**, the smear is flooded with a **basic** (cationic) **dye** such as methylene blue (FIGURE 3.11A). Because cationic dyes have a positive charge, the dye is attracted to the cytoplasm and cell wall, which primarily have negative charges. By contrasting the blue cells against the bright background, the staining procedure allows the observer to measure cell size and determine **cell morphology**; that is, the shape and arrangement of cells with respect to one another.

Negative Stains. The **negative stain technique** works in the opposite manner (FIGURE 3.11B). Bacterial cells are mixed on a slide with an **acidic** (anionic) **dye** such as nigrosin (a black stain) or India ink (a black drawing ink). The mixture is

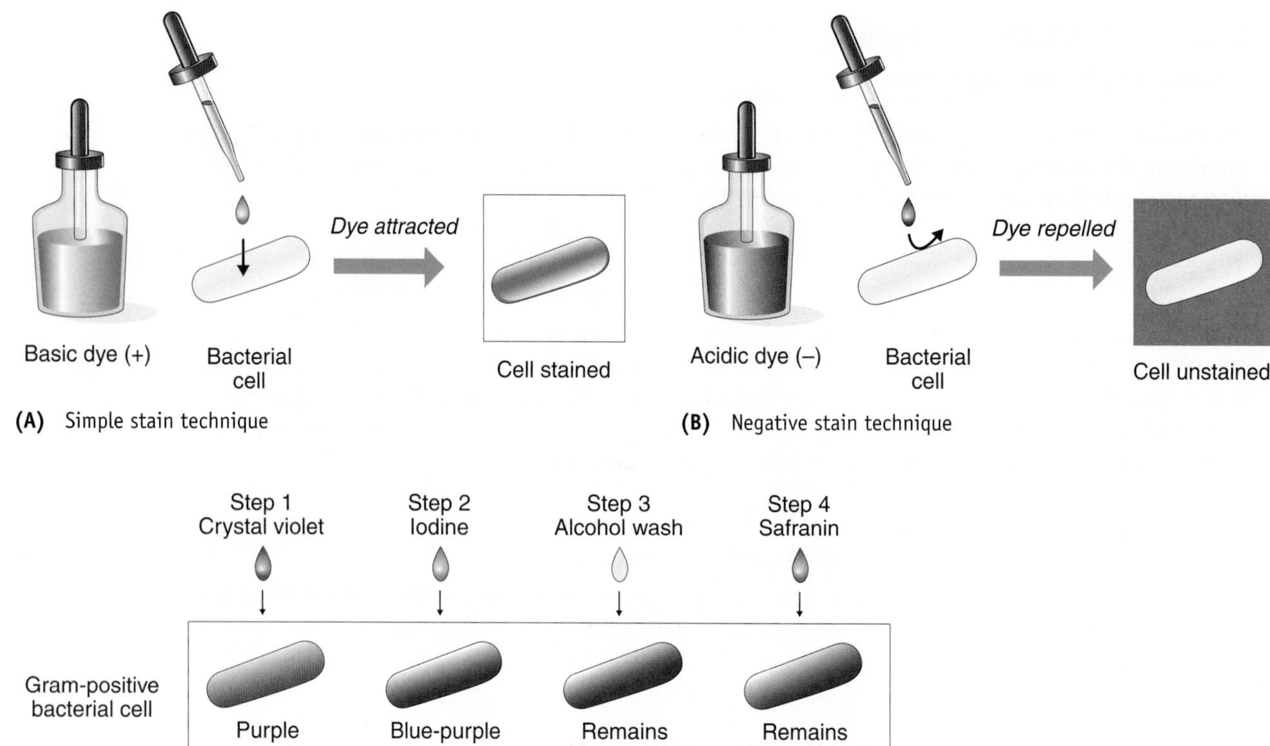

FIGURE 3.11 **Some Important Staining Reactions in Microbiology.** (**A**) In the simple stain technique, the cells of the smear are stained and contrasted against a light background whereas in a negative stain (**B**) the cells remain unstained and contrasted against a dark background. In the Gram stain (**C**), cells are stained either purple (Gram positive) or red-orange (Gram negative). »» What common cell characteristics can be determined by all three staining techniques?

then spread across the surface of the slide and allowed to air-dry. Because the anionic dye carries a negative charge, it is repelled from the cells and the observer sees clear or white cells on a stained black or gray background. Because this technique avoids chemical reactions and heat fixation, the cells appear less shriveled than in a simple stain and better resemble their natural condition.

Website Animation: Gram Staining

Differential Stains. Unlike simple staining, a **differential staining procedure** allows the observer to differentiate (separate) bacterial cells visually into two groups based on staining differences. The **Gram stain technique**, named for Hans Christian Gram, the Danish physician who first perfected the technique in 1884, uses air-dried and heat-fixed smears.

As depicted in FIGURE 3.11C, a smear is (1) stained with crystal violet (a purple, basic dye), rinsed, and then (2) a special Gram's iodine solution is added. All bacterial cells would appear blue-purple if the procedure was stopped and the sample viewed with the light microscope. Next, the smear is (3) rinsed with a decolorizer, such as 95% alcohol or an alcohol-acetone mixture. Observed at this point, certain bacterial cells may lose their color and become transparent. These are the **gram-negative** bacterial cells. Others retain the crystal violet and represent the **gram-positive** bacterial cells. The last step (4) uses safranin (a red, basic dye) to counterstain the gram-negative organisms; that is, give them an orange-red color. So, at the technique's conclusion, gram-positive cells are blue-purple while gram-negative cells are orange-red. Similar to simple staining, gram staining also allows the observer to determine size and cell morphology. FIGURE 3.12A–C shows examples of the staining outcomes.

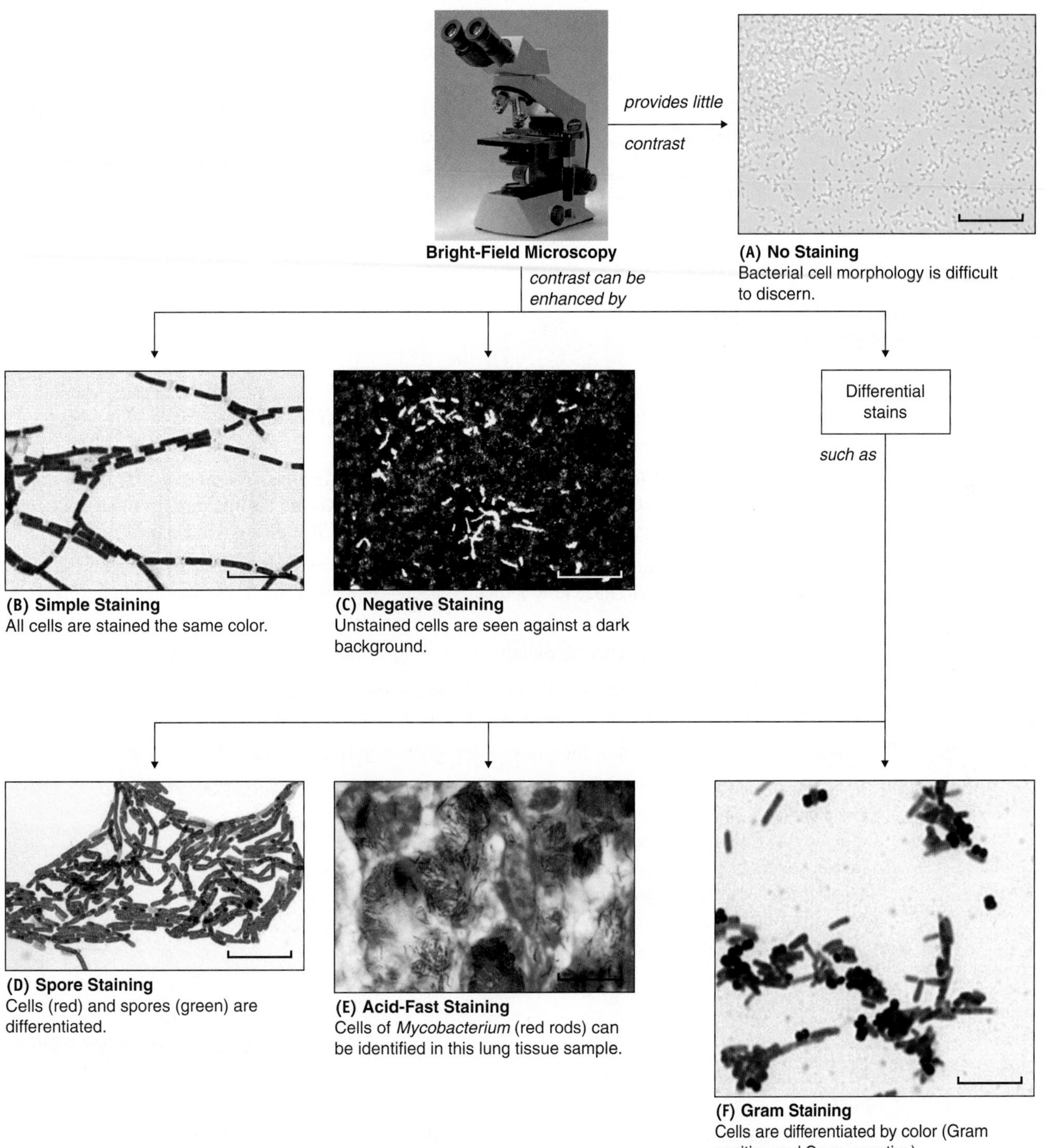

FIGURE 3.12 Observing Stained Cells with Bright-Field Microscopy. Several bacterial staining techniques provide the contrast needed when observing the stained cells by bright-field microscopy. (**A–F**, Bar = 10 μm.) (Light microscope and **A–C**) Courtesy of Jeffrey Pommerville (**D**) © Dr. John D. Cunningham/Visuals Unlimited, Inc. (**E**) Courtesy of James Anderson, Leica Biosystems, Wetzlar, Germany. (**F**) Courtesy of Larry Stauffer, Oregon State Public Health Laboratory/CDC. »» What advantages does a differential stain have over a simple stain?

■ Toxin: A chemical substance that is poisonous.

Knowing whether a bacterial cell is gram positive or gram negative is important for microbiologists and clinical technicians in the CML who use the results from the Gram stain technique to classify it in *Bergey's Manual* or aid in the identification of an unknown disease pathogen (TEXTBOOK CASE 3).

Gram-positive and gram-negative bacterial cells also differ in their susceptibility to chemical substances such as antibiotics (gram-positive cells are more susceptible to penicillin, gram-negative cells to tetracycline), so Gram staining can be important for treatment decisions. Also, gram-positive and gram-negative bacterial species can produce different types of toxins.

Two other common differential staining procedures should be mentioned. The **acid-fast technique** is used to identify members of the genus *Mycobacterium*, one species of which causes tuberculosis. These bacterial cells are normally difficult

Textbook CASE 3

Bacterial Meningitis and a Misleading Gram Stain

1 A woman comes to the hospital emergency room complaining of severe headache, nausea, vomiting, and pain in her legs. On examination, cerebral spinal fluid (CSF) was observed leaking from a previous central nervous system (CNS) surgical site.

2 The patient indicates that 6 weeks and 8 weeks ago she had undergone CNS surgery after complaining of migraine headaches and sinusitis. Both surgeries involved a spinal tap. In the clinical microbiology lab (CML), analysis of cultures prepared from the CSF indicated no bacterial growth.

3 The patient was taken to surgery where a large amount of CSF was removed from underneath the old incision site. The pinkish, hazy fluid indicated bacterial meningitis, so among the laboratory tests ordered was a Gram stain.

4 The patient was placed on antibiotic therapy, consisting of vancomycin and cefotaxime.

5 CML findings from the gram-stained CSF smear showed a few gram-positive, spherical bacterial cells that often appeared in pairs. The results suggested a *Streptococcus pneumoniae* infection.

6 However, upon reexamination of the smear, a few gram-negative spheres were observed.

7 When transferred to a blood agar plate, growth occurred and a prepared smear showed many gram-negative spheres (see figure). Further research indicated that several genera of gram-negative bacteria, including *Acinetobacter,* can appear gram-positive due to under-decolorization during the alcohol wash step.

8 Although complicated by the under-decolorization outcome, the final diagnosis was bacterial meningitis due to *Acinetobacter baumanii.*

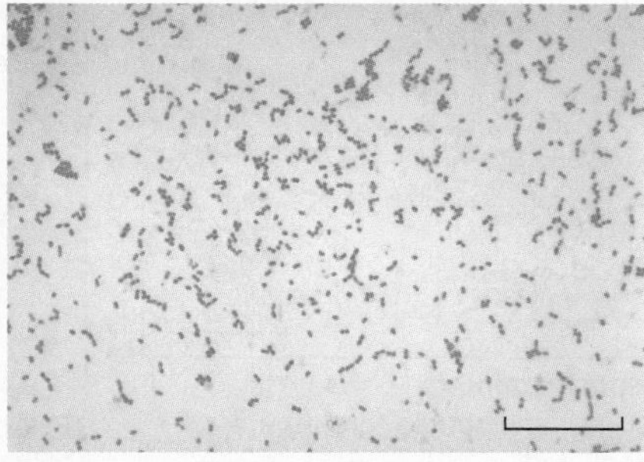
A gram-stained preparation from the blood agar plate. (Bar = 10 μm.) Courtesy of Dr. W. A. Clark/CDC.

Questions:

(Answers can be found on the Student Companion Website in **Appendix D.**)

A. From the gram-stained CSF smear, what color were the gram-positive bacterial spheres?

B. After reexamination of the CSF smear, assess the reliability of the gram-stained smear.

C. What reagent is used for the decolorization step in the Gram stain?

Adapted from: Harrington, B. J. and Plenzler, M., 2004. Misleading gram stain findings on a smear from a cerebrospinal fluid specimen. *Lab. Med.* 35(8): 475–478.

For additional information see www.cdc.gov/HAI/organisms/acinetobacter.html.

to stain with the Gram stain because the cells have very waxy walls that repel the dyes. However, the cells will stain red when subjected to the acid-fast procedure (FIGURE 3.12D).

Some gram-positive bacterial species produce endospores that can be easily differentiated from the bacterial cells by performing the **spore stain technique** (FIGURE 3.12E).

Other Light Microscopy Optics Can Also Enhance Contrast

A light microscope can be outfitted with other optical systems to improve contrast of microorganisms. Three systems commonly employed are mentioned here.

Phase-contrast microscopy uses a special condenser and objective lenses. The condenser lens splits the light beam and throws the light rays slightly out of phase. The separated beams of light then pass through and around the specimen, and small differences in the refractive index within the specimen show up as different degrees of brightness and contrast. With phase-contrast microscopy, microbiologists can see organisms alive and unstained when suspended in water (FIGURE 3.13A).

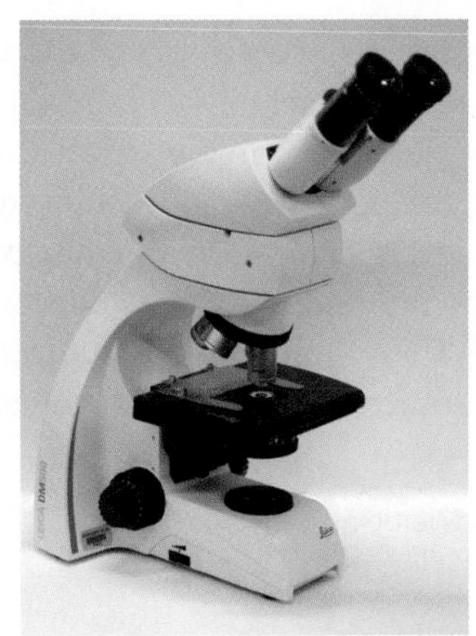

Other Types of Light Microscopy

contrast can be enhanced by

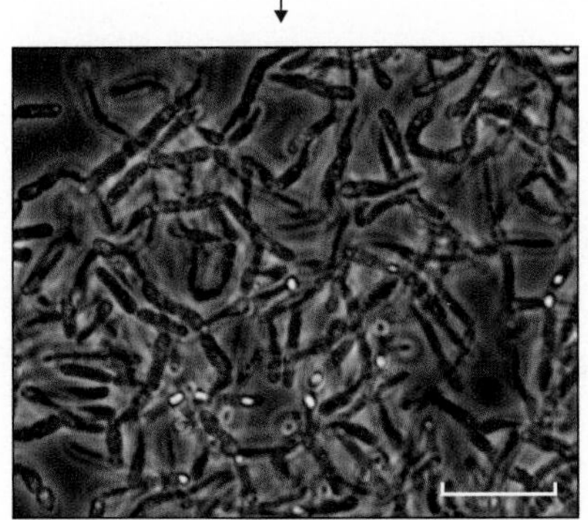

(A) Phase Contrast
Cells (dark) are contrasted from the lighter dots (spores).

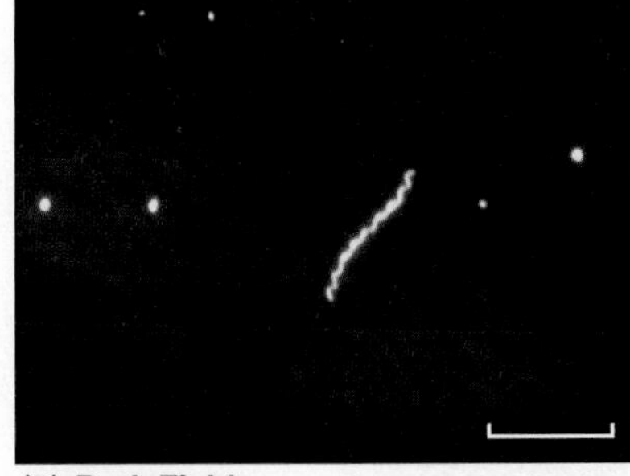

(B) Dark Field
Unstained cells are seen against a dark background.

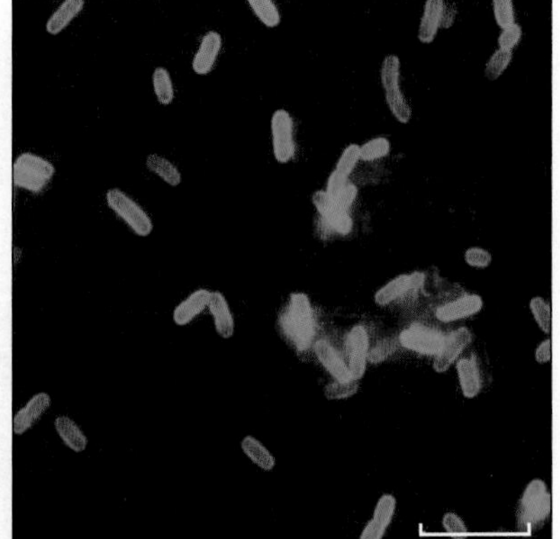

(C) Flourescence
Cells "glow" due to the presence of a flourescent antibody that binds to the cells.

FIGURE 3.13 **Observing Cells with Other Types of Light Microscopy.** The three images show some common techniques for contrasting bacterial cells. These methods require special optical configurations added onto the light microscope. (**A–C**, Bar = 10 μm.) (**A and C**) Courtesy of Larry Stauffer, Oregon State Public Health Laboratory/CDC. (**B**) Courtesy of Schwartz/CDC. Microscope image courtesy of Jeffrey Pommerville. »» Are the bacterial cells killed in preparing the sample for each of these three types of microscopy? Explain.

Dark-field microscopy also uses a special condenser lens mounted under the stage. The condenser scatters the light and causes it to hit the specimen from the side. Only light bouncing off the specimen and into the objective lens makes the specimen visible, as the surrounding area appears dark because it lacks background light.

In the CML, dark-field microscopy helps in the diagnosis of diseases caused by organisms near the limit of resolution of the light microscope. For example, the spiral bacterium *Treponema pallidum*, the cause of syphilis, has a diameter of only about 0.15 μm. Therefore, this bacterial species may be observed in scrapings taken from a lesion of a person who has the disease and observed with dark-field microscopy (FIGURE 3.13B).

Fluorescence microscopy is a major asset to clinical and research laboratories. The technique has been applied to the identification of many microorganisms and is a mainstay of modern microbial ecology and especially clinical microbiology.

For fluorescence microscopy, fluorescent dyes, such as fluorescein are used. When illuminated with an ultraviolet (UV) light source attached to the microscope, the energy in UV light excites electrons in fluorescein, causing them to move to higher energy levels. However, the electrons quickly drop back to their original energy levels and give off the excess energy as visible light; in the case of fluorescein, a greenish yellow glow is observed in the microscope. Other dyes produce other colors.

An important application of fluorescence microscopy in the CML is the **fluorescent antibody technique** used to identify an unknown organism in a clinical specimen (FIGURE 3.14). In one variation of this procedure, fluorescein is chemically attached to antibodies, the protein molecules produced by the body's immune system. These "tagged" antibodies are mixed with a sample of the unknown organism. If the antibodies are specific for that organism, they will bind to it and coat the cells with the dye(FIGURE 3.13C). When subjected to UV light, the cells, such as those that cause plague, will fluoresce when viewed with the microscope (FIGURE 3.13D). If the organisms fail to fluoresce, the antibodies were not specific to that organism and the test result is negative.

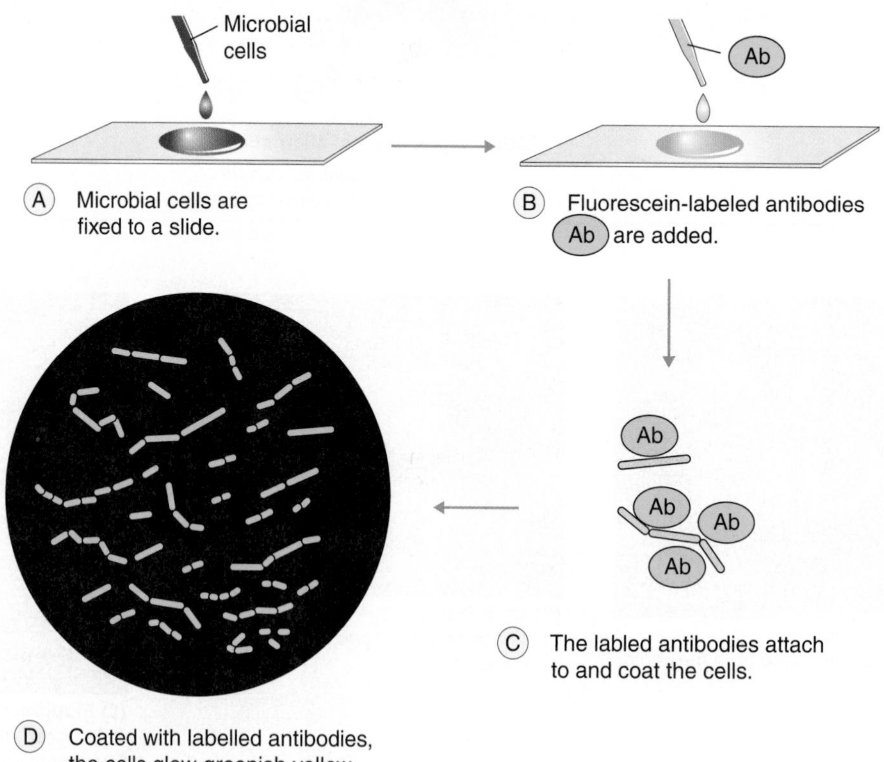

FIGURE 3.14 **The Fluorescent Antibody Technique.** Bacterial cells in a sample can be identified by mixing them with a fluorescent antibody that, when bound to the cells, generates a visible greenish-yellow "glow" (fluorescence) when observed with ultraviolet light on the microscope.
»» What would be seen in the microscope if the antibody did not bind to the bacterial cells?

In summary, specimens for viewing in the light microscope can be prepared through staining or using alternative optical configurations that provide sufficient contrast to study cells. Since the 1990s, a number of innovations and technologies have been introduced that produce better resolution and contrast with the light microscope, especially suitable for studying eukaryotic cell structure. In addition, some advances now permit using the light microscope to measure chemical substances, like proteins, inside cells—in some cases almost down to the level of a single molecule.

Amazing resolution also can be gained with the electron microscope, the last topic for this chapter.

Electron Microscopy Provides Detailed Images of Cells, Cell Parts, and Viruses

The **electron microscope** grew out of an engineering design made in 1932 by the German physicist Ernst Ruska (winner of the 1986 Nobel Prize in Physics). Ruska showed that electrons will flow in a sealed tube if a vacuum is maintained to prevent electron scattering. Magnets, rather than glass lenses, pinpoint the flow onto an object, where the electrons are absorbed, deflected, or transmitted depending on the density of structures within the object (FIGURE 3.15). When projected onto a screen underneath, the electrons form a final image that outlines the structures. Such images recorded on electron-sensitive film are called "electron micrographs."

The power of electron microscopy is the extraordinarily short wavelength of the beam of electrons. Measured at 0.005 nm (compared to 550 nm for visible light), the short wavelength dramatically increases the resolution of the system and makes possible the visualization of viruses and detailed cellular structures, often called the **ultrastructure** of cells. The practical limit of resolution of biological samples with the electron microscope is about 2 nm, which is 100× better than the resolution of the light microscope. The drawback of the electron microscope is that the method needed to prepare a specimen kills the cells or organisms.

Two types of electron microscopes are commonly in use. The **transmission electron microscope (TEM)** is used to view and record detailed structures within cells (FIGURE 3.16A). Ultrathin sections of the prepared specimen must be cut because the electron beam can penetrate matter only a very short distance. After embedding the specimen in a suitable plastic mounting medium or freezing it, the specimen is cut into sections with a diamond knife. In this manner, a single microbial cell can be sliced, like a loaf of bread, into hundreds of thin (100 nm thick) sections.

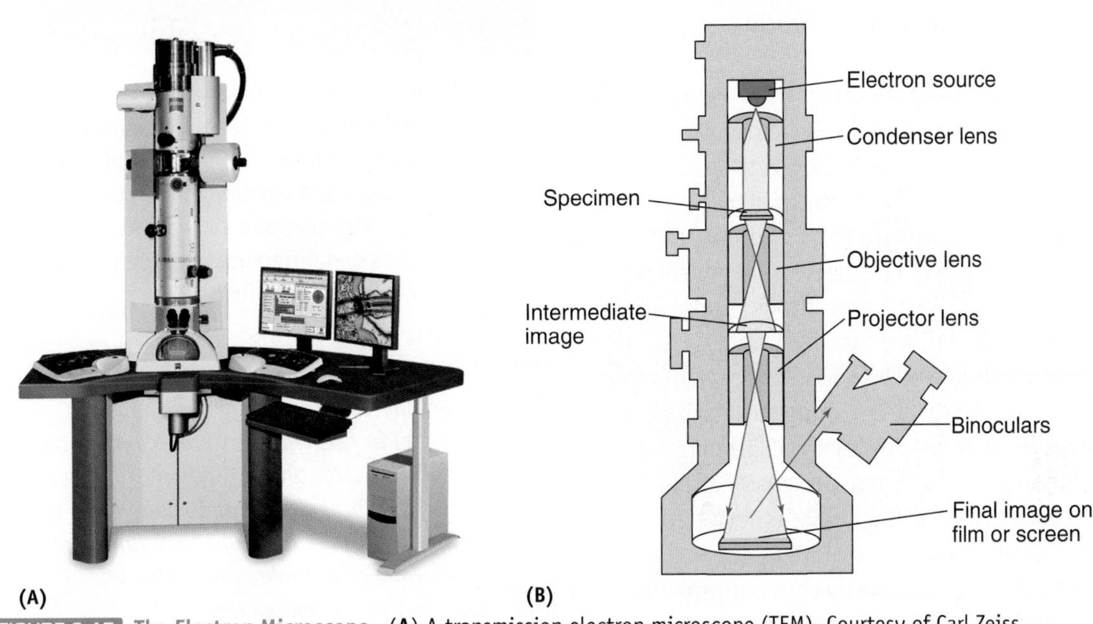

FIGURE 3.15 **The Electron Microscope.** (**A**) A transmission electron microscope (TEM). Courtesy of Carl Zeiss MicroImaging, LLC. (**B**) A schematic of the vacuum tube. A beam of electrons is emitted from the electron source and electromagnets function as lenses to focus the beam on the specimen. The image is magnified by objective and projector lenses. The final image is projected on a screen, television monitor, or electron-sensitive film. »» How does the path of the image for the transmission electron microscope compare with that of the light microscope (Figure 3.10B)?

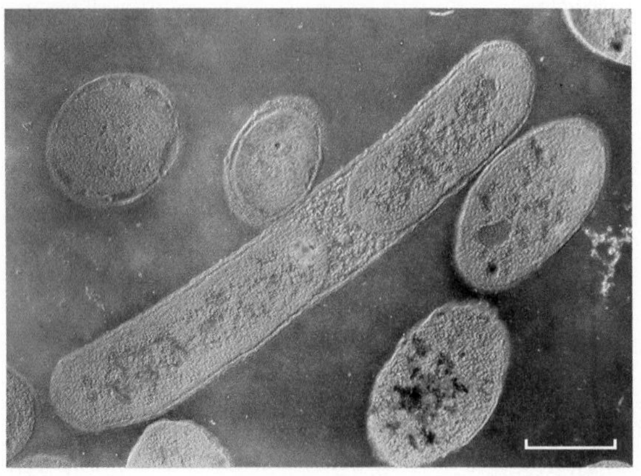

(A)

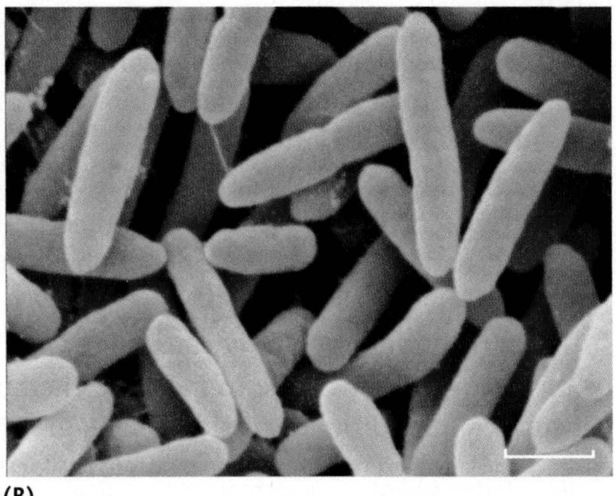

(B)

FIGURE 3.16 **Transmission and Scanning Electron Microscopy Compared.** The bacterium *Pseudomonas aeruginosa* (false-color images) as seen with two types of electron microscopy. (**A**) A view of sectioned cells seen with the transmission electron microscope. (Bar = 1.0 µm.) © CNRI/Photo Researchers, Inc. (**B**) A view of whole cells seen with the scanning electron microscope. (Bar = 2.0 µm.) © SciMAT/Photo Researchers, Inc. »» What types of information can be gathered from each of these electron micrographs?

Several of the sections are placed on a small grid and stained with heavy metals such as lead and osmium to provide contrast. The microscopist then inserts the grid into the vacuum tube of the microscope and focuses a 100,000-volt electron beam on one portion of a section at a time. As some electrons pass through while others are blocked by the specimen, an image forms on the screen below the tube or can be recorded on film. The electron micrograph (maximum resolving power is about 200,000×) may be enlarged to achieve a final magnification approaching 2 million×.

The **scanning electron microscope (SEM)** was developed in the late 1960s to enable researchers to see the surfaces of objects in the natural state and without sectioning. The specimen is placed in the vacuum tube and covered with a thin coat of gold. The electron beam then scans across the specimen and knocks loose showers of electrons that are captured by a detector. An image builds line by line, as in a television receiver. Electrons that strike a sloping surface yield fewer electrons, thereby producing a darker contrasting spot and a sense of three dimensions. The resolution of the conventional SEM is about 10 nm and magnifications with the SEM are limited to about 20,000×. However, the instrument provides vivid and undistorted views of an organism's surface (FIGURE 3.16B).

The various types of light and electron microscopy are compared in TABLE 3.3.

CONCEPT AND REASONING CHECKS 3

a. If a bacterial cell is 0.75 µm in length, what is its length in nanometers?

b. Why is resolution just as important as magnification?

c. Explain the difference between a cationic dye and an anionic dye in terms of cell staining.

d. Which techniques for improving contrast do not result in killing the cells?

e. What type of electron microscope would be used to examine (i) the surface structures on a *Paramecium* cell and (ii) the organelles in an algal cell?

Chapter Challenge D

This chapter challenge has had you question the widely held view that prokaryotic cells are simple and primitive. Decades of studies with the light and electron microscopes have provided extensive and rich detail into the similarities and differences between prokaryotic and eukaryotic cells.

QUESTION D:

As your final challenge, summarize how taxonomy and microscopy (two major concepts in this chapter) have contributed to both a better understanding of microbial cells (both prokaryotic and eukaryotic) and to a better realization of the organization and behavior of prokaryotic cells, as exemplified by the bacterial cell.

Answers can be found on the Student Companion Website in **Appendix D**.

TABLE

3.3 Comparison of Various Types of Microscopy

Type of Microscopy	Special Feature	Appearance of Object	Magnification Range	Uses
Light				
Bright-field	Visible light illuminates object	Stained microorganisms on clear background	100×–1,000×	Morphology and size of killed microorganisms (except viruses)
Phase-contrast	Special condenser throws light rays "out of phase"	Unstained microorganisms with contrasted structures	100×–1,000×	Internal structures of live, unstained eukaryotic microorganisms
Dark-field	Special condenser scatters light	Unstained microorganisms on dark background	100×–1,000×	Live, unstained microorganisms; motility of live cells
Fluorescence	UV light illuminates fluorescent-coated objects	Fluorescing microorganisms on dark background	100×–1,000×	Identification of microorganisms coated with fluorescent-tagged antibodies
Electron				
Transmission	Short-wavelength electron beam penetrates sections	Alternating light and dark areas contrasting internal cell structures	100×–200,000×	Ultrathin slices of nonliving microorganisms, internal components, and viruses
Scanning	Short-wavelength electron beam knocks loose electron showers	Microbial surfaces	10×–20,000×	Surfaces and textures on nonliving microorganisms, cell components, and viruses

■ **In conclusion**, this chapter emphasized the fact that prokaryotic organisms, although they may look structurally simple, are not so when it comes to organismal processes and behaviors. In fact, with recent advances made to the light and electron microscopes, we now are finding more structure to prokaryotic cells than ever imagined. You will discover this for yourself as you continue to learn about these organisms.

■ SUMMARY OF KEY CONCEPTS

3.1 Prokaryotes Are Not Simple, Primitive Organisms

1. All living organisms share the common emergent properties of life, attempt to maintain a stable internal state called **homeostasis**, and interact through a multicellular association (a **biofilm**) involving chemical communication and cooperation between cells (**quorum sensing**). (Fig. 3.3)
2. Bacterial and eukaryotic cells share certain organizational patterns, including genetic organization, compartmentation, metabolic organization, and protein synthesis. (Fig. 3.5)
3. Although bacterial and eukaryotic cells carry out many similar processes, eukaryotic cells often contain a variety of subcellular compartments (**organelles**) to accomplish the processes. (Table 3.1)

3.2 Classifying Microorganisms Reveals Relationships Between Organisms

4. Many systems of classification have been devised to catalog organisms based on shared characteristics. (Fig. 3.6)
5. Based on several molecular and biochemical differences, Woese proposed **three domains of life** that include the Bacteria and Archaea. The protists, fungi, plants, and animals are placed in the domain Eukarya. (Fig. 3.7)
6. Part of an organism's binomial name is the **genus** name; the remaining part is the **specific epithet** that describes the genus name. Thus, a **species** name consists of the genus and specific epithet.
7. Organisms are properly classified using a standardized hierarchical system from species (the least inclusive) to domain (the most inclusive). (Table 3.2)
8. *Bergey's Manual* is the standard reference to identify and classify bacterial species. Criteria have included traditional characteristics, but modern molecular methods have led to a reconstruction of evolutionary events and organism relationships.

3.3 Microscopy Is Used to Visualize the Structure of Cells

9. Another criterion of a microorganism is its size, a characteristic that varies among members of different groups. The **micrometer** (**μm**) is used to measure the dimensions of bacterial, protist, and fungal cells. The **nanometer** (**nm**) is commonly used to express viral sizes. (Fig. 3.9)

10. The instrument most widely used to observe microorganisms is the **light microscope**. Light passes through several lens systems that magnify and resolve the object being observed. Although **magnification** is important, **resolution** is key. The light microscope can magnify up to 1,000× and resolve objects as small as 0.2 μm.
11. For bacterial cells, staining generally precedes observation. The **simple, negative, Gram, acid-fast**, and other staining techniques can be used to impart contrast and determine structural or physiological properties. (Figs. 3.11, 3.12)
12. Light microscopes employing **phase-contrast, dark-field**, and **fluorescence** optics have specialized uses in microbiology to contrast cells without staining. (Fig. 3.13)
13. To increase resolving power and achieve extremely high magnification, the electron microscope employs a beam of electrons to magnify and resolve specimens. To observe internal details (**ultrastructure**), the **transmission electron microscope** is most often used; to study whole objects or surfaces, the **scanning electron microscope** is employed. (Fig. 3.16)

CHAPTER SELF-TEST

For **STEPS A–D**, answers to even-numbered questions and problems can be found in **Appendix C** on the Student Companion Website at **http://microbiology.jbpub.com/10E**. In addition, the site features eLearning, an online review area that provides quizzes and other tools to help you study for your class. You can also follow useful links for in-depth information, read more MicroFocus stories, or just find out the latest microbiology news.

STEP A: REVIEW OF FACTS AND TERMS

Multiple Choice

Read each question carefully, then select the ***one*** answer that best fits the question or statement.

1. What is the term that describes the ability of organisms to maintain a stable internal state?
   A. Metabolism
   B. Homeostasis
   C. Biosphere
   D. Ecotype
2. Which one of the following is NOT an organizational pattern common to all organisms?
   A. Genetic organization
   B. Protein synthesis
   C. Compartmentation
   D. Endomembrane system
3. Which one of the following is NOT found in bacterial cells?
   A. Ribosomes
   B. DNA
   C. Mitochondria
   D. Cytoplasm
4. Who is considered to be the father of modern taxonomy?
   A. Woese
   B. Whittaker
   C. Haeckel
   D. Linnaeus
5. ______ was first used to catalog organisms into one of three domains.
   A. Photosynthesis
   B. Ribosomal RNA genes
   C. Nuclear DNA genes
   D. Cell respiration
6. Several classes of organisms would be classified into one ______.
   A. order
   B. genus
   C. phylum
   D. family
7. An important automated method used in the rapid identification of a pathogen is ______.
   A. rRNA gene sequencing
   B. polymerase chain reaction
   C. molecular taxonomy
   D. biochemical tests
8. Most bacterial cells are measured using what metric system of length?
   A. Millimeters (mm)
   B. Micrometers (μm)
   C. Nanometers (nm)
   D. Centimeters (cm)
9. Before bacterial cells are simple stained and observed with the light microscope, they must be ______.
   A. smeared on a slide
   B. heat fixed
   C. air dried
   D. All the above (**A–C**) are correct.
10. If you wanted to study the surface of a bacterial cell, you would use a ______.
   A. transmission electron microscope
   B. light microscope with phase-contrast optics
   C. scanning electron microscope
   D. light microscope with dark-field optics

Matching

Match the statement on the left to the term on the right by placing the letter of the term in the available space.

Statement

11. ______ The domain containing organisms whose cells have no cell nucleus or mitochondria in the cytoplasm.
12. ______ How microbes monitor the presence and number of neighboring cells.
13. ______ Type of electron microscope for which cell sectioning is not required.
14. ______ The structure that carries out protein synthesis in all cells.
15. ______ The organelle, absent in bacteria, that carries out the conversion of chemical energy to cellular energy in eukaryotes.
16. ______ Domain in which fungi and protists are classified.
17. ______ Staining technique that differentiates bacterial cells into two groups.
18. ______ Category into which two or more genera are grouped.
19. ______ The staining technique employing a single cationic dye.
20. ______ Type of microscopy using UV light to excite a dye.

Term

A. Bacteria
B. Chloroplast
C. Cyanobacteria
D. Dark-field
E. Eukarya
F. Family
G. Fluorescence
H. Fungi
I. Gram
J. Homeostasis
K. Mitochondrion
L. Negative
M. Phase-contrast
N. Quorum sensing
O. Ribosome
P. Scanning
Q. Simple
R. Transmission

Label Identification

21. Identify the cell structures (a–p) indicated in drawings (A) and (B) below. What is a function for each structure?

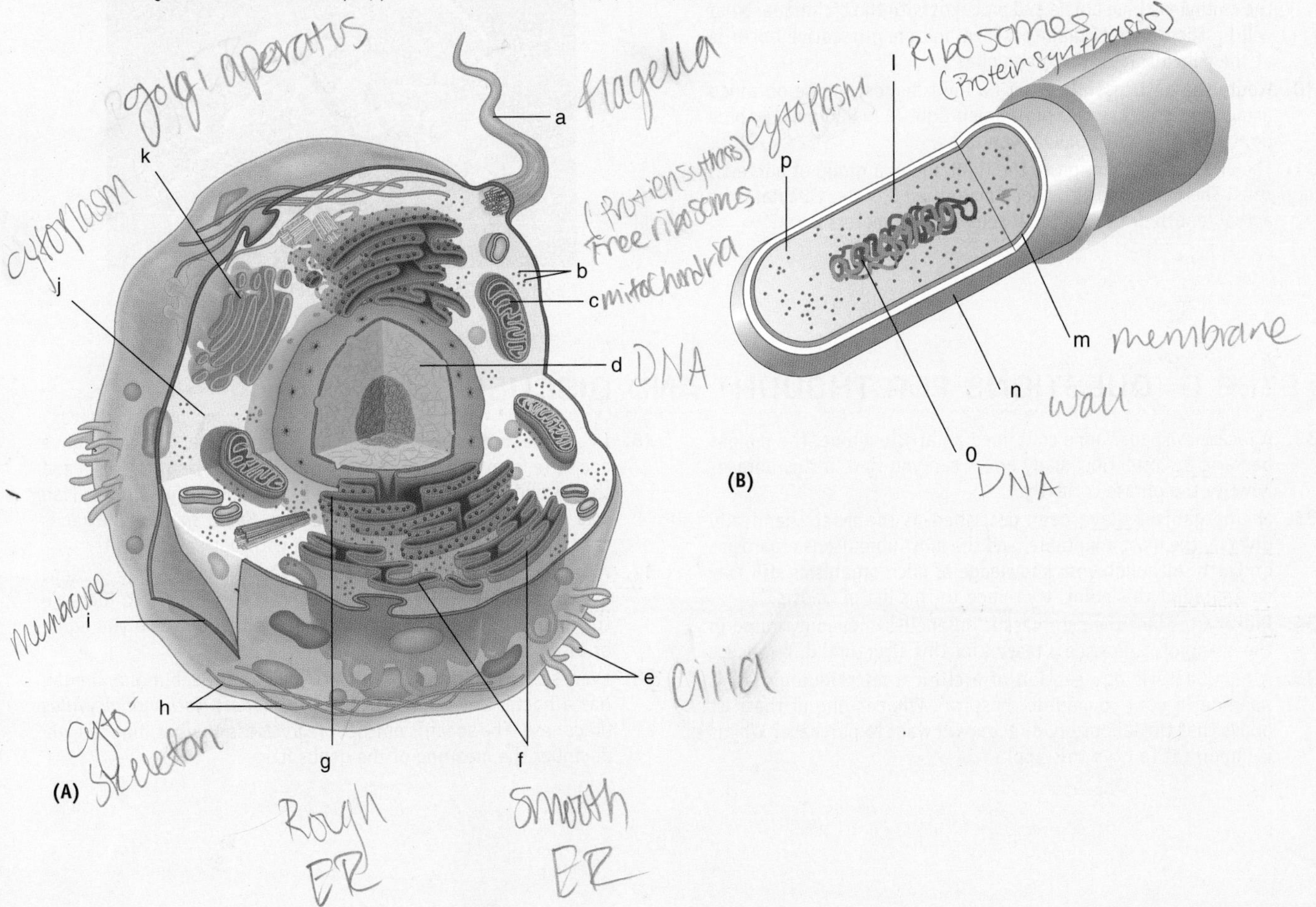

STEP B: CONCEPT REVIEW

22. Assess the importance of **homeostasis** and **quorum sensing** to cell (organismal) survival and behavior, and contrast bacteria as **unicellular** and **multicellular** organisms. (**Key Concept 1**)
23. Construct a concept map for **Living Organisms** using the following terms (terms can be used more than once). (**Key Concept 1**)

Bacterial cells	Golgi apparatus
Cell membrane	Lysosomes
Chloroplasts	Microcompartments
Cytoplasm	Mitochondria
Cytoskeleton	Nucleus
Cytosol	RER
DNA region	Ribosomes
Eukaryotic cells	SER
Flagella	

24. Explain the assignment of organisms to one of the three domains in the "tree of life." (**Key Concept 2**)
25. Write scientific names of organisms using the **binomial system.** (**Key Concept 2**)
26. Assess the importance of **magnification** and **resolution** to microscopy. (**Key Concept 3**)
27. Construct a concept map for **staining techniques** using the following terms only once. (**Key Concept 3**)

Acid-fast technique	Differential stain procedure
Acidic dye	Gram negative
Basic dye	Gram positive
Blue-purple cells	Gram stain technique
Cell arrangement	*Mycobacterium*
Cell shape	Negative stain technique
Cell size	Orange-red cells
Contrast	Simple stain technique

28. Compare the uses of the **transmission** and **scanning electron microscopes.** (**Key Concept 3**)

STEP C: APPLICATIONS AND PROBLEMS

29. A student is performing the Gram stain technique on a mixed culture of gram-positive and gram-negative bacterial cells. In reaching for the counterstain in step 4, he inadvertently takes the methylene blue bottle and proceeds with the technique. What will be the colors of gram-positive and gram-negative bacteria at the conclusion of the technique?
30. Would the best resolution with a light microscope be obtained using red light ($\lambda = 680$ nm), green light ($\lambda = 520$ nm), or blue light ($\lambda = 500$ nm)? Explain your answer.
31. The electron micrograph to the right shows a group of bacterial cells. The micrograph has been magnified 5,000×. Calculate the actual length of the bacterial cells in micrometers (μm)?

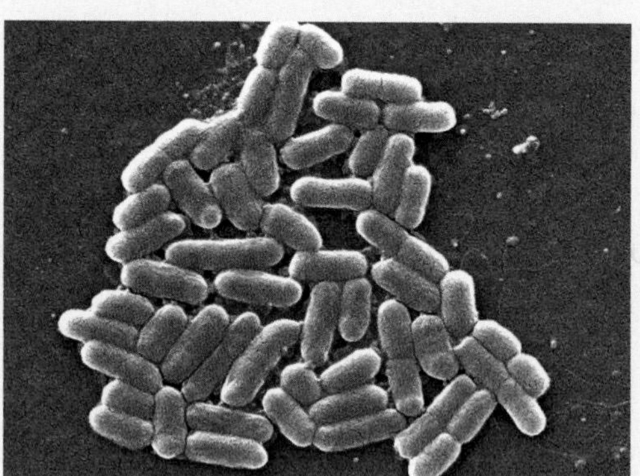

Courtesy of Janice Hanley Carr/CDC.

STEP D: QUESTIONS FOR THOUGHT AND DISCUSSION

32. A local newspaper once contained an article about "the famous bacteria E. Coli." How many errors can you find in this phrase? Rewrite the phrase correctly.
33. Microorganisms have been described as the most chemically diverse, the most adaptable, and the most ubiquitous organisms on Earth. Although your knowledge of microorganisms still may be limited at this point, try to add to this list of "mosts."
34. Prokaryotes lack the cytoplasmic organelles commonly found in the eukaryotes. Provide a reason for this structural difference.
35. A new bacteriology section of a clinical microbiology lab is opening in your community hospital. What is one of the first books that the laboratory director will want to purchase? Why is it important to have this book?
36. In a respected science journal, an author wrote, "Linnaeus gave each life form two Latin names, the first denoting its genus and the second its species." A few lines later, the author wrote, "Man was given his own genus and species *Homo sapiens.*" What is conceptually and technically wrong with both statements?
37. A student of general biology observes a microbiology student using immersion oil and asks why the oil is used. "To increase the magnification of the microscope" is the reply. Do you agree or disagree? Why?
38. Every state has an official animal, flower, or tree, but only Oregon has a bacterial species named in its honor: *Methanohalophilus oregonese*. The specific epithet *oregonese* is obvious, but can you decipher the meaning of the genus name?

Structure of Bacterial and Archaeal Cells

CHAPTER PREVIEW

4.1 There Is Tremendous Diversity Among the Bacteria and Archaea

4.2 Prokaryotes Can Be Distinguished by Their Cell Shape and Arrangements

4.3 An Overview to Bacterial and Archaeal Cell Structure

4.4 External Cell Structures Interact with the Environment

Investigating the Microbial World 4: The Role of Pili

Textbook Case 4: An Outbreak of *Enterobacter cloacae* Associated with a Biofilm

4.5 Most Bacterial and Archaeal Cells Have a Cell Envelope

4.6 The Cell Cytoplasm Is Packed with Internal Structures

MicroInquiry 4: The Prokaryote/Eukaryote Model

Our planet has always been in the "Age of Bacteria," ever since the first fossils—bacteria of course—were entombed in rocks more than 3 billion years ago. On any possible, reasonable criterion, bacteria are—and always have been—the dominant forms of life on Earth.
—Paleontologist Stephen J. Gould (1941–2002)

"**Double, double toil and trouble**; *Fire burn, and cauldron bubble*" is the refrain repeated several times by the chanting witches in Shakespeare's *Macbeth* (Act IV, Scene 1). This image of a hot, boiling cauldron actually describes the environment in which many bacterial, and especially archaeal, species happily grow! For example, some species can be isolated from hot springs or the hot, acidic mud pits of volcanic vents (FIGURE 4.1).

When the eminent evolutionary biologist and geologist Stephen J. Gould wrote the opening quote of this chapter, he, as well as most microbiologists at the time, had no idea that embedded in these "bacteria" was another whole domain of organisms. Thanks to the pioneering studies of Carl Woese and his colleagues, it now is quite evident there are two distinctly different groups of "prokaryotes"—the Bacteria and the Archaea. Many of the organisms Woese and others studied are organisms that would live a happy life in a witch's cauldron because they can grow at high temperatures, produce methane gas, or survive in extremely acidic and hot environments—a real cauldron! Termed **extremophiles**, these members of the domains Bacteria and Archaea have a unique genetic makeup and have adapted to extreme environmental conditions.

In fact, Gould's "first fossils" may have been archaeal species. Many microbiologists believe the ancestors of today's archaeal species might represent a type of organism that first inhabited planet Earth when it was a young, hot place. These unique characteristics led Woese to propose

Image courtesy of Dr. Fred Murphy/CDC.

(A)

(B)

FIGURE 4.1 **Life at the Edge.** Bacterial and archaeal extremophiles have been isolated from the edges of natural cauldrons, including (**A**) the Grand Prismatic Spring in Yellowstone National Park, Wyoming, where the water of the hot spring is over 70°C, or (**B**) the mud pools surrounding sulfurous steam vents of the Solfatara Crater in Pozzuoli, Italy, where the mud has a very low pH and a temperature above 90°C. (**A**) © Aleix Ventayol Farrés/ShutterStock, Inc. (**B**) © Corbis/age fotostock. »» How do extremophiles survive under these extreme conditions?

these organisms be lumped together and called the Archaebacteria (*archae* = "ancient").

Since then, the domain name has been changed to Archaea because (1) not all members are extremophiles or related to these possible ancient ancestors and (2) they are not Bacteria—they are Archaea. Some might also debate using the term prokaryotes when referring to both domains, as organisms in the two domains are as different from each other as they are from the Eukarya.

In this chapter, we examine briefly some of the organisms in the domains Bacteria and Archaea. However, because so many pathogens of humans are in the domain Bacteria, we emphasize structure within this domain. As we see in this chapter, a study of the structural features of bacterial cells provides a window to their activities and illustrates how the Bacteria relate to other living organisms.

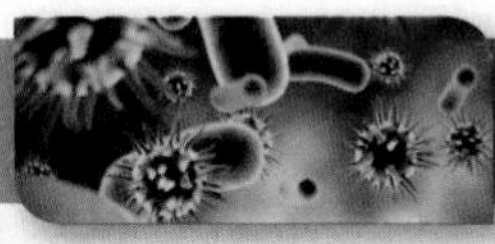
© qcontrol/ShutterStock, Inc.

Chapter Challenge

As microbes have been studied in more detail using special microscopy optics and have had their genes sequenced, it is clear that there are unique as well as shared characteristics between species in the domains Bacteria, Archaea, and Eukarya. Once you finish studying this chapter on the structure of cells in the Bacteria and Archaea domains, you will be better prepared to wisely respond to this chapter challenge: "Studying the diversity of life only accentuates life's unity." Let's search for the evidence!

KEY CONCEPT 4.1 There Is Tremendous Diversity Among the Bacteria and Archaea

Currently, there are some 7,000 known bacterial and archaeal species and a suspected 10 million species. In this section, we will highlight a few phyla and groups using the phylogenetic tree in FIGURE 4.2.

The Domain Bacteria Contains Some of the Most Studied Microbial Organisms

There are about 25 assigned phyla of Bacteria that have been identified from culturing or nucleotide sequencing. It should come as no shock to you by now to read that the vast majority of these phyla play a positive role in nature (MicroFocus 4.1). Of course, we know from personal experience that some bacterial organisms are harmful—many human pathogens are members of the domain Bacteria. Certain species multiply within the human body, where they disrupt tissues or produce toxins that result in disease.

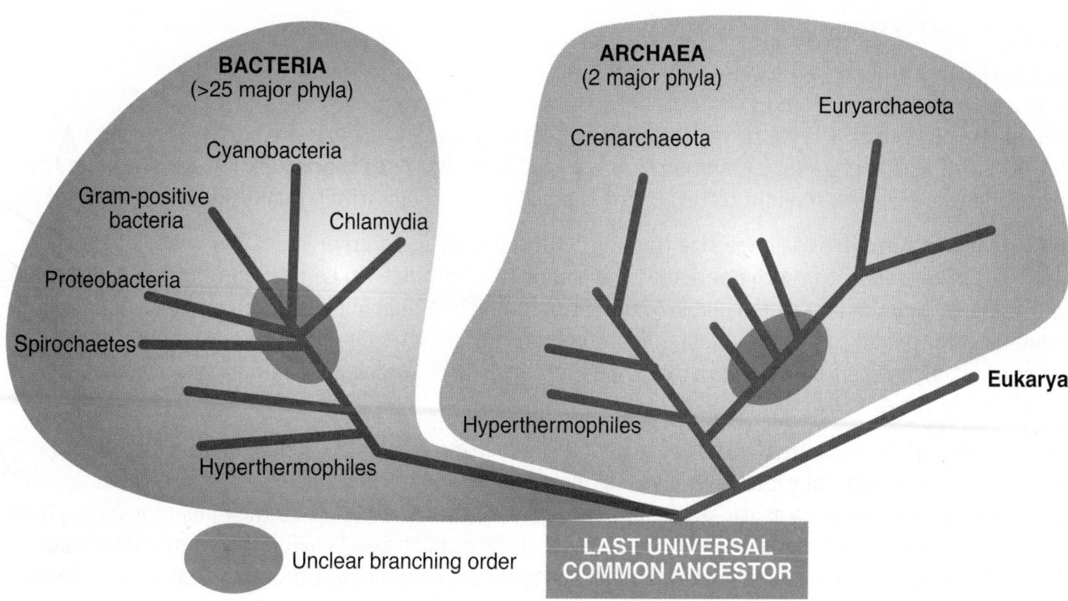

FIGURE 4.2 **The Phylogenetic Tree of Bacteria and Archaea.** The tree shows several of the bacterial and archaeal lineages discussed in this chapter. »» What is common to the branch base of both the Bacteria and Archaea?

The Bacteria have adapted to the diverse environments on Earth, inhabiting the air, soil, and water, and they exist in enormous numbers on the surfaces of virtually all plants and animals. They can be isolated from Arctic ice, thermal hot springs, the fringes of space, and the tissues of animals. Bacterial species, along with their archaeal relatives, have so completely colonized every part of the Earth that their mass is estimated to outweigh the mass of all plants and animals combined. Let's look briefly at some of the major phyla and other groups.

Proteobacteria. The **Proteobacteria** (*proteo* = "first") contains the largest and most diverse group of species. The phylum, spread among five classes, includes many familiar gram-negative genera, such as *Escherichia* (FIGURE 4.3A) and some of the most recognized human pathogens, including species of *Shigella, Salmonella, Neisseria* (responsible for gonorrhea), *Yersinia* (responsible for plague), and *Vibrio* (responsible for cholera). It is likely that the mitochondria of the Eukarya evolved through **endosymbiosis** from a free-living ancestor of the Proteobacteria.

The group also includes the **rickettsiae** (sing., rickettsia), which are called **obligate, intracellular parasites** because they can only reproduce once inside a host cell. These tiny bacterial cells are transmitted among humans primarily by arthropods, and are cultivated only in living tissues such as chick embryos. Different species cause a number of important diseases, including Rocky Mountain spotted fever and typhus fever.

The gram-positive bacteria rival the Proteobacteria in diversity and are divided into two phyla.

Firmicutes. The **Firmicutes** (*firm* = "strong"; *cuti* = "skin") consists of many species that are gram-positive. As we will see in this chapter, they share a similar thick "skin," which refers to their cell wall structure. Genera include *Bacillus* and *Clostridium,* specific species of which are responsible for anthrax and botulism, respectively. Species within the genera *Staphylococcus* (FIGURE 4.3B) and *Streptococcus* are responsible for several mild to life-threatening human illnesses.

Also within the Firmicutes is the genus *Mycoplasma,* which lacks a cell wall but is otherwise phylogenetically related to the gram-positive bacterial species (FIGURE 4.3C). Among the smallest free-living bacterial organisms, one species causes a form of pneumonia while another mycoplasmal illness represents a sexually-transmitted disease.

Actinobacteria. Another phylum of gram-positive species is the **Actinobacteria**. Often called the "actinomycetes," these soil organisms form a system of branched filaments that somewhat resemble the growth form of fungi (FIGURE 4.3D). The genus *Streptomyces* is the source for important antibiotics. Another medically important genus is *Mycobacterium,* one species of which is responsible for tuberculosis.

■ Host:
An organism on or in which a pathogen infects.

■ Arthropod:
An animal having jointed appendages and segmented body (e.g., ticks, lice, fleas, mosquitoes).

MICROFOCUS 4.1: Bacteria in Eight Easy Lessons[1]

Mélanie Hamon, an assistante de recherché at the Institut Pasteur in Paris, says that when she introduces herself as a bacteriologist, she often is asked, "Just what does that mean?" To help explain her discipline, she gives us, in eight letters, what she calls "some demystifying facts about bacteria."

Basic principles: Their average size is 1/25,000th of an inch. In other words, hundreds of thousands of bacteria fit into the period at the end of this sentence. In comparison, human cells are 10 to 100 times larger with a more complex inner structure. While human cells have copious amounts of membrane-contained subcompartments, bacteria more closely resemble pocketless sacs. Despite their simplicity, they are self-contained living beings, unlike viruses, which depend on a host cell to carry out their life cycle.

Astonishing: Bacteria are the root of the evolutionary tree of life, the source of all living organisms. Quite successful evolutionarily speaking, they are ubiquitously distributed in soil, water, and extreme environments such as ice, acidic hot springs or radioactive waste. In the human body, bacteria account for 10% of dry weight, populating mucosal surfaces of the oral cavity, gastrointestinal tract, urogenital tract and surface of the skin. In fact, bacteria are so numerous on earth that scientists estimate their biomass to far surpass that of the rest of all life combined.

Crucial: It is a little known fact that most bacteria in our bodies are harmless and even essential for our survival. Inoffensive skin settlers form a protective barrier against any troublesome invader while approximately 1,000 species of gut colonizers work for our benefit, synthesizing vitamins, breaking down complex nutrients and contributing to gut immunity. Unfortunately for babies (and parents!), we are born with a sterile gut and "colic" our way through bacterial colonization.

Tools: Besides the profitable relationship they maintain with us, bacteria have many other practical and exploitable properties, most notably, perhaps, in the production of cream, yogurt and cheese. Less widely known are their industrial applications as antibiotic factories, insecticides, sewage processors, oil spill degraders, and so forth.

Evil: Unfortunately, not all bacteria are "good," and those that cause disease give them all an often undeserved and unpleasant reputation. If we consider the multitude of mechanisms these "bad" bacteria—pathogens—use to assail their host, it is no wonder that they get a lot of bad press. Indeed, millions of years of coevolution have shaped bacteria into organisms that "know" and "predict" their hosts' responses. Therefore, not only do bacterial toxins know their target, which is never missed, but bacteria can predict their host's immune response and often avoid it.

Resistant: Even more worrisome than their effectiveness at targeting their host is their faculty to withstand antibiotic therapy. For close to 50 years, antibiotics have revolutionized public health in their ability to treat bacterial infections. Unfortunately, overuse and misuse of antibiotics have led to the alarming fact of resistance, which promises to be disastrous for the treatment of such diseases.

Ingenious: The appearance of antibiotic-resistant bacteria is a reflection of how adaptable they are. Thanks to their large populations they are able to mutate their genetic makeup, or even exchange it, to find the appropriate combination that will provide them with resistance. Additionally, bacteria are able to form "biofilms," which are cellular aggregates covered in slime that allow them to tolerate antimicrobial applications that normally eradicate free-floating individual cells.

Along tradition: Although "little animalcules" were first observed in the 17th century, it was not until the 1850s that Louis Pasteur fathered modern microbiology. From this point forward, research on bacteria has developed into the flourishing field it is today. For many years to come, researchers will continue to delve into this intricate world, trying to understand how the good ones can help and how to protect ourselves from the bad ones. It is a great honor to be part of this tradition, working in the very place where it was born.

---

[1]Republished with permission of the author, the Institut Pasteur, and the Pasteur Foundation. The original article appeared in *Pasteur Perspectives* Issue 20 (Spring 2007), the newsletter of the Pasteur Foundation, which may be found at www.pasteurfoundation.org © Pasteur Foundation.

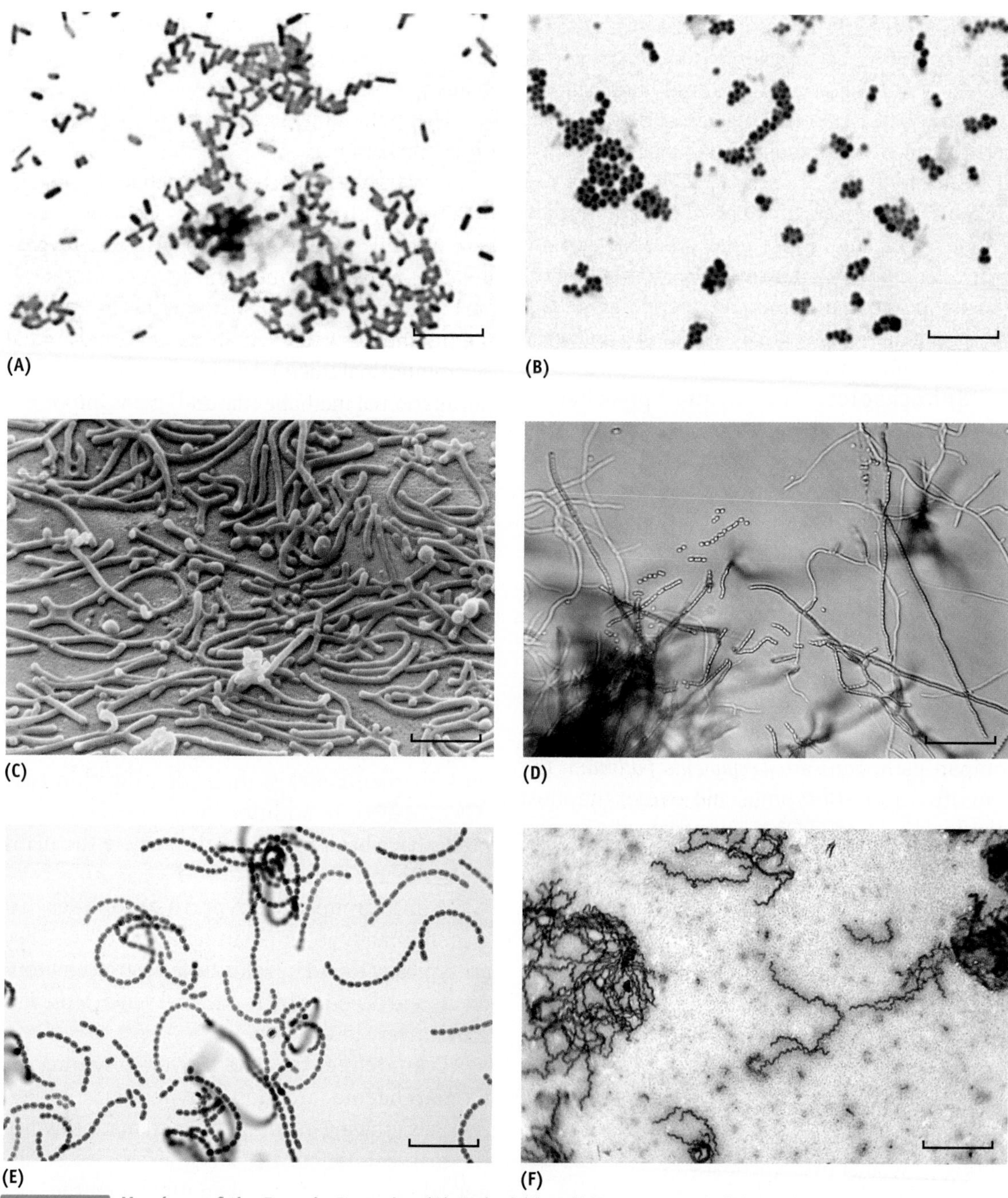

FIGURE 4.3 **Members of the Domain Bacteria.** (**A**) *Escherichia coli* (Bar = 10 µm.), (**B**) *Staphylococcus aureus* (Bar = 10 µm.), (**C**) *Mycoplasma* species (Bar = 2 µm.), (**D**) *Streptomyces* species (Bar = 20 µm.), (**E**) *Anabaena* species (Bar = 100 µm.), and (**F**) *Treponema pallidum* (Bar = 10 µm.). All images are light micrographs except (**C**), a false-color scanning electron micrograph. (**A, B, E, F**) Courtesy of Jeffrey Pommerville. (**C**) © Don W. Fawcett/Photo Researchers, Inc. (**D**) Courtesy of Dr. David Berd/CDC. »» What is the Gram staining result for *E. coli* and *S. aureus*?

Cyanobacteria. The members of cyanobacteria are phylogenetically related to the gram-positive species and can exist as unicellular, filamentous, or colonial forms (FIGURE 4.3E). Once known as blue-green algae because of their pigmentation, pigments also may be black, yellow, green, or red. The periodic redness of the Red Sea, for example, is due to blooms of those cyanobacterial species that contain large amounts of red pigment.

■ Bloom: A sudden increase in the number of cells of an organism in an environment.

The phylum **Cyanobacteria** is unique among bacterial groups because its members carry out photosynthesis similar to unicellular algae using the light-trapping pigment chlorophyll. Their evolution on Earth was responsible for the "oxygen revolution" that transformed life on the young planet some 2 billion years ago. In addition, chloroplasts probably evolved through endosymbiosis from a free-living cyanobacterial ancestor.

Chlamydiae. Roughly half the size of the rickettsiae, members of the phylum **Chlamydiae** also are obligate, intracellular parasites and are cultivated only within living cells. Most species in the phylum are pathogens and one species causes the gonorrhea-like sexually transmitted disease (STD) chlamydia.

Spirochaetes. The phylum **Spirochaetes** contains more than 340 gram-negative species that possess a unique cell body that coils into a long helix and moves in a corkscrew pattern. The ecological niches for the spirochetes is diverse: from free-living species found in mud and sediments, to symbiotic species present in the digestive tracts of insects, to the pathogens found in the urogenital tracts of vertebrates. Many spirochetes are found in the human oral cavity; in fact, some of the first animalcules seen by Leeuwenhoek were probably spirochetes from his teeth scrapings. Among the human pathogens are *Treponema pallidum,* the causative agent of syphilis and one of the most common STDs (FIGURE 4.3F) and specific species of *Borrelia,* which are transmitted by ticks or lice and are responsible for Lyme disease and relapsing fever.

Other Phyla. There are many other phyla within the domain Bacteria. Several lineages branch off near the root of the domain. The common link between these organisms is that they are **hyperthermophiles**; they grow at high temperatures (70°C–85°C). Examples include *Aquifex* and *Thermotoga,* which typically are found in earthly cauldrons such as hot springs and other hydrothermal sites.

The Domain Archaea Contains Organisms with Diverse Physiologies

Classification within the domain Archaea has been more difficult than within the domain Bacteria, in large part because they have not been studied as long as their bacterial counterparts.

Archaeal organisms are found throughout the biosphere. Many genera are extremophiles, growing best at environmental extremes, such as very high temperatures, high salt concentrations, or extremes of pH. However, many more species exist in very cold environments. There also are archaeal genera that thrive under more modest conditions but there are no known species that cause disease in any plants or animals.

The archaeal genera can be placed into one of two phyla.

Euryarchaeota. The **Euryarchaeota** contain organisms with varying physiologies, many being extremophiles. Some groups, such as the **methanogens** (*methano* = "methane"; *gen* = "produce") are killed by oxygen gas and therefore are found in marine and freshwater environments (and animal gastrointestinal tracts) devoid of oxygen gas. The production of methane (natural) gas is important in their energy metabolism (FIGURE 4.4A). In fact, these archaeal species release more than 2 billion tons of methane gas into the atmosphere every year. About a third comes from the archaeal species living in the stomach (rumen) of cows.

Another group is the **extreme halophiles** (*halo* = "salt"; *phil* = "loving"). They are distinct from the methanogens in that they require oxygen gas for energy metabolism and need high concentrations of salt (up to 30% NaCl) to grow and reproduce. The fact that they often contain pink pigments makes their identification easy (FIGURE 4.4B). In addition, some extreme halophiles have been found in lakes where the pH is greater than 11.

A third group is the **hyperthermophiles** that grow optimally at temperatures above 80°C. They are typically found in volcanic terrestrial environments and deep-sea hydrothermal vents. Most also grow at very low pHs.

Crenarchaeota. The second phylum, the **Crenarchaeota**, are mostly hyperthermophiles, typically growing in hot springs and marine hydrothermal vents (FIGURE 4.4C). Other species are dispersed in open oceans, often inhabiting the cold ocean waters (–3°C) of the deep sea environments and polar seas.

(TABLE 4.1) summarizes some of the characteristics that are shared or are unique among the three domains.

CONCEPT AND REASONING CHECKS 1

a. What three unique events occurred within the Proteobacteria and Cyanobacteria that contributed to the evolution of the Eukarya and the oxygen-rich atmosphere of Earth?

b. Compared to the more moderate environments in which some archaeal species grow, why have others adapted to such extreme environments?

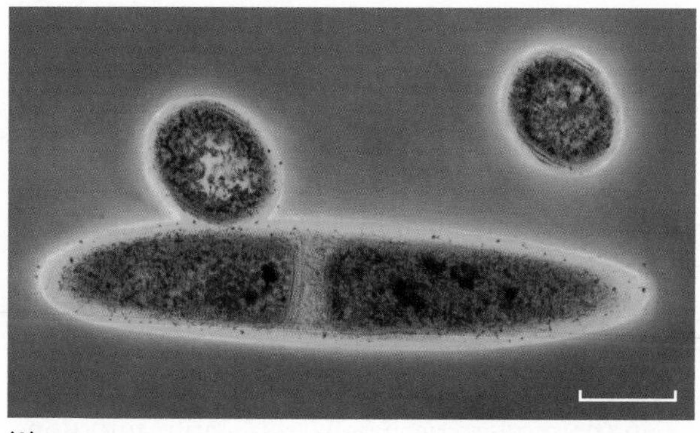

(A)

(B)

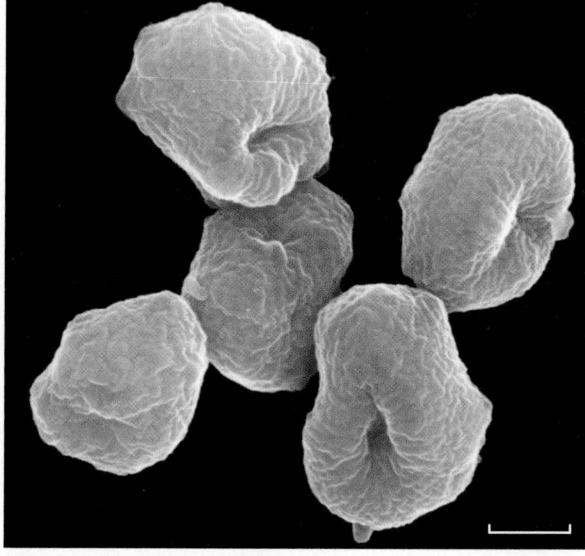

(C)

FIGURE 4.4 **Members of the Domain Archaea.** (**A**) A false-color transmission electron micrograph of the methanogen *Methanospirillum hungatei.* (Bar = 0.5 μm.) © Dr. Kari Lounatmaa/Photo Researchers, Inc. (**B**) An aerial view above Redwood City, California, of the salt ponds whose color is due to high concentrations of extreme halophiles. © Aerial Archives/Alamy Images. (**C**) A false-color scanning electron micrograph of *Sulfolobus,* a hyperthermophile that grows in waters as hot as 90°C. (Bar = 0.5 μm.) © Eye of Science/Photo Researchers, Inc. »» What advantage is afforded these species that grow in such extreme environments?

TABLE

4.1 Some Major Differences Between Bacteria, Archaea, and Eukarya

Characteristic	Bacteria	Archaea	Eukarya
Cell nucleus	No	No	Yes
Chromosome form	Single, circular	Single, circular	Multiple, linear
Histone proteins present	No	Yes	Yes
Peptidoglycan cell wall	Yes	No	No
Membrane lipids	Ester-linked	Ether-linked	Ester-linked
Ribosome sedimentation value	70S	70S	80S
Ribosome sensitivity to diphtheria toxin	No	Yes	Yes
First amino acid in a protein	Formylmethionine	Methionine	Methionine
Chlorophyll-based photosynthesis	Yes (cyanobacteria)	No	Yes (algae)
Growth above 80°C	Yes	Yes	No
Growth above 100°C	No	Yes	No
Pathogens	Yes	No	Yes

KEY CONCEPT 4.2 Prokaryotes Can Be Distinguished by Their Cell Shape and Arrangements

Bacterial and archaeal cells come in a variety of shapes that, in almost all members of the two domains, is determined by the cell wall and underlying cytoskeletal proteins. However, when viewing stained cells with the light microscope most, including the clinically significant ones, appear in one of three basic shapes: the rod, the sphere, or the spiral.

Variations in Cell Shape and Cell Arrangement Exist

A prokaryotic cell with a rod shape is called a **bacillus** (pl., bacilli) and depending on the species may be as long as 20 µm or as short as 0.5 µm. Certain bacilli are slender, such as those of *Salmonella typhi* that cause typhoid fever. Others, such as the agent of anthrax (*Bacillus anthracis*), are rectangular with squared ends; still others, such as the diphtheria bacilli (*Corynebacterium diphtheriae*), are club shaped. Most rods occur singly, in pairs called **diplobacillus**, or arranged into a long chain called **streptobacillus** (*strepto* = "chains") (FIGURE 4.5A). Realize there are two ways to use the word "bacillus": to denote a rod-shaped bacterial cell, and as a genus name (*Bacillus*).

A spherically shaped bacterial cell is known as a **coccus** (pl., cocci; *kokkos* = "berry") and tends to be quite small, being only 0.5 µm to 1.0 µm in diameter. Although they are usually round, they also may be oval, elongated, or indented on one side.

Many bacterial species that are cocci stay together after division and take on cellular arrangements characteristic of the species (FIGURE 4.5B). Cocci remaining in a pair after reproducing represent a **diplococcus**. The organism that causes gonorrhea, *Neisseria gonorrhoeae,* and one type of bacterial meningitis (*N. meningitidis*) are diplococci. Cocci that remain in a chain are called **streptococcus**. Certain species of streptococci are involved in strep throat (*Streptococcus pyogenes*) and tooth decay (*S. mutans*). Another arrangement of cocci is the **tetrad**, consisting of four spheres forming a square. A cube-like packet of eight cocci is called a **sarcina** (*sarcina* = "bundle"). *Micrococcus luteus,* a common inhabitant of the skin, is one example. Other cocci may divide randomly and form an irregular grape-like cluster of cells called a **staphylococcus** (*staphylo* = "cluster"). A well-known example, *Staphylococcus aureus,* is often a cause of food poisoning, toxic shock syndrome, and several skin infections. The latter are known in the modern vernacular as "staph" infections. Notice again that the words "streptococcus" and "staphylococcus" can be used to describe cell shape and arrangement, or a bacterial genus (*Streptococcus* and *Staphylococcus*).

The third common morphology of bacterial cells is the **spiral**, which can take one of three forms (FIGURE 4.5C). The **vibrio** is a curved rod that resembles a comma. The cholera-causing organism *Vibrio cholerae* is typical. Another spiral form called **spirillum** (pl., spirilla) has a helical shape with a thick, rigid cell wall. The spiral-shaped form known as **spirochete** has a thin, flexible cell wall. The organism causing syphilis, *Treponema pallidum,* typifies a spirochete. Spiral-shaped bacterial cells can be from 1 µm to 100 µm in length.

In addition to the bacillus, coccus, and spiral shapes, other morphologies exist. Some bacterial species have appendaged bacterial cells while others consist of branching filaments; and some archaeal species have square and star shapes.

CONCEPT AND REASONING CHECKS 2

a. What morphology does a bacterial cell have that is said to be coccobacillus?

Chapter Challenge A

So far you have learned more about the diverse groups in the domains Bacteria and Archaea, and have become familiar with their dissimilar cell shapes.

QUESTION A:

Based on what you have read up to this point, how has this diversity of life highlighted its unity?

Answers can be found on the Student Companion Website in **Appendix D.**

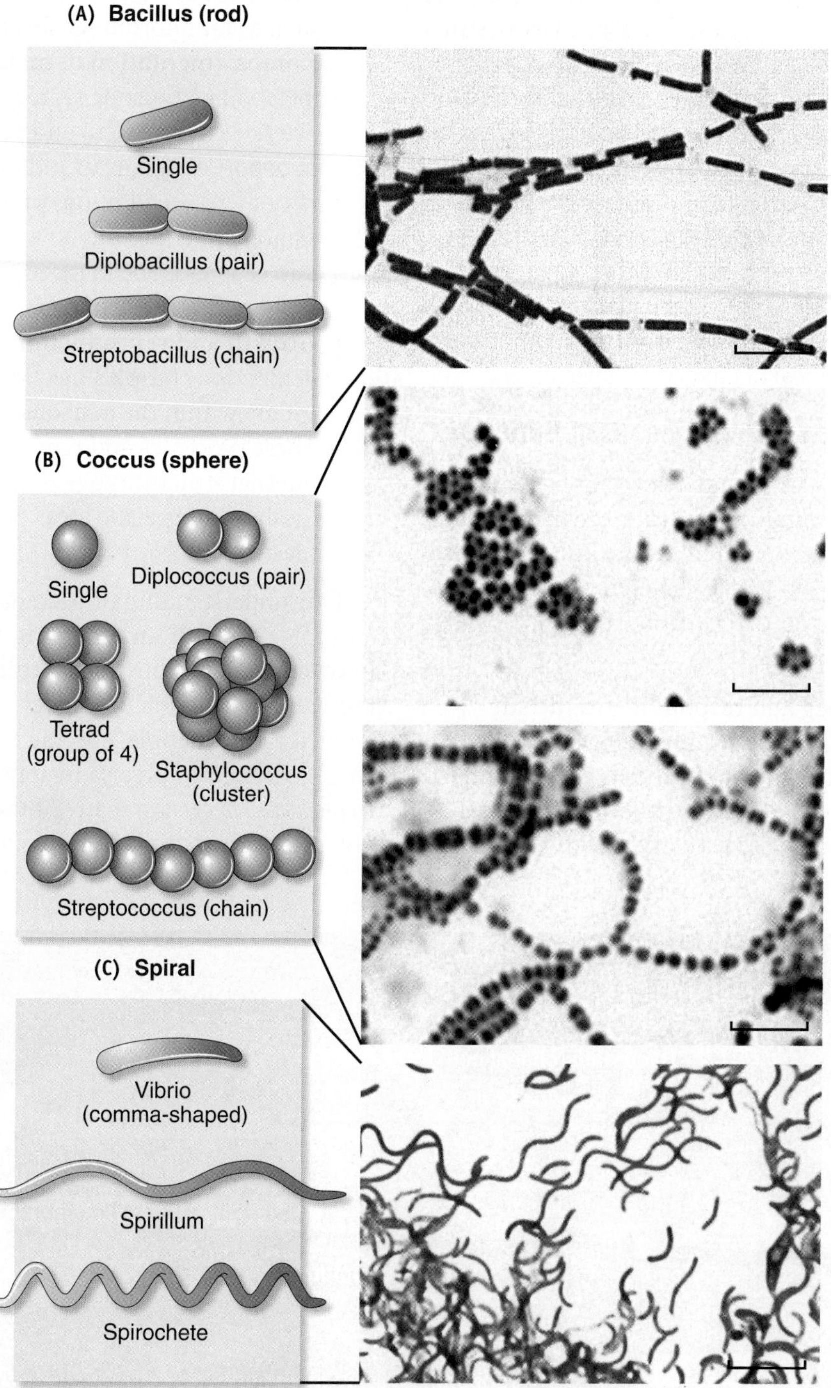

FIGURE 4.5 **Variation in Shape and Cell Arrangements.** Many bacterial and archaeal cells have a bacillus (**A**) or coccus (**B**) shape. Most spiral shaped-cells (**C**) are not organized into a specific arrangement (All bars = 10 µm.). Courtesy of Jeffrey Pommerville.

»» In Panel C, identify the vibrio and the spirillum forms in the photograph.

KEY CONCEPT 4.3 An Overview to Bacterial and Archaeal Cell Structure

In another chapter we discovered that bacterial and archaeal cells appear to have little visible structure when observed with a light microscope. This, along with their small size, gave the impression they are homogeneous, static structures with an organization very different from eukaryotic cells.

However, bacterial and archaeal species still have all the complex processes typical of eukaryotic cells. It is simply a matter that, in most cases, the structure and sometimes pattern to accomplish these processes is different from the membranous organelles typical of eukaryotic species.

Cell Structure Organizes Cell Function

Recent advances in understanding bacterial and archaeal cell biology indicate these organisms exhibit a highly ordered intracellular organization. This organization is centered on three specific processes that need to be carried out (FIGURE 4.6). These are:

- **Sensing and responding to the surrounding environment.** Because most bacterial and archaeal cells are surrounded by a cell wall, some pattern of "external structures" is necessary to sense their environment and respond to it or other cells such as for quorum sensing and motility.
- **Compartmentation of metabolism.** Cell metabolism must be segregated from the exterior environment and yet be able to transport materials to and from that environment. In addition, protection from osmotic pressure due to water movement into cells must be in place. The "cell envelope" fulfills those roles.
- **Growth and reproduction.** Cell survival demands a complex metabolism that occurs within the aqueous "cytoplasm." These processes and reproduction exist as internal structures or subcompartments localized to specific areas within the cytoplasm.

Our understanding of bacterial and archaeal cell biology is still an emerging field of study. However, there is more to cell structure than previously thought—smallness does not equate with simplicity. For example, specific cellular proteins can be localized to precise regions of the cell. In *Streptococcus pyogenes*, many of the proteins that confer its pathogenic nature in causing diseases

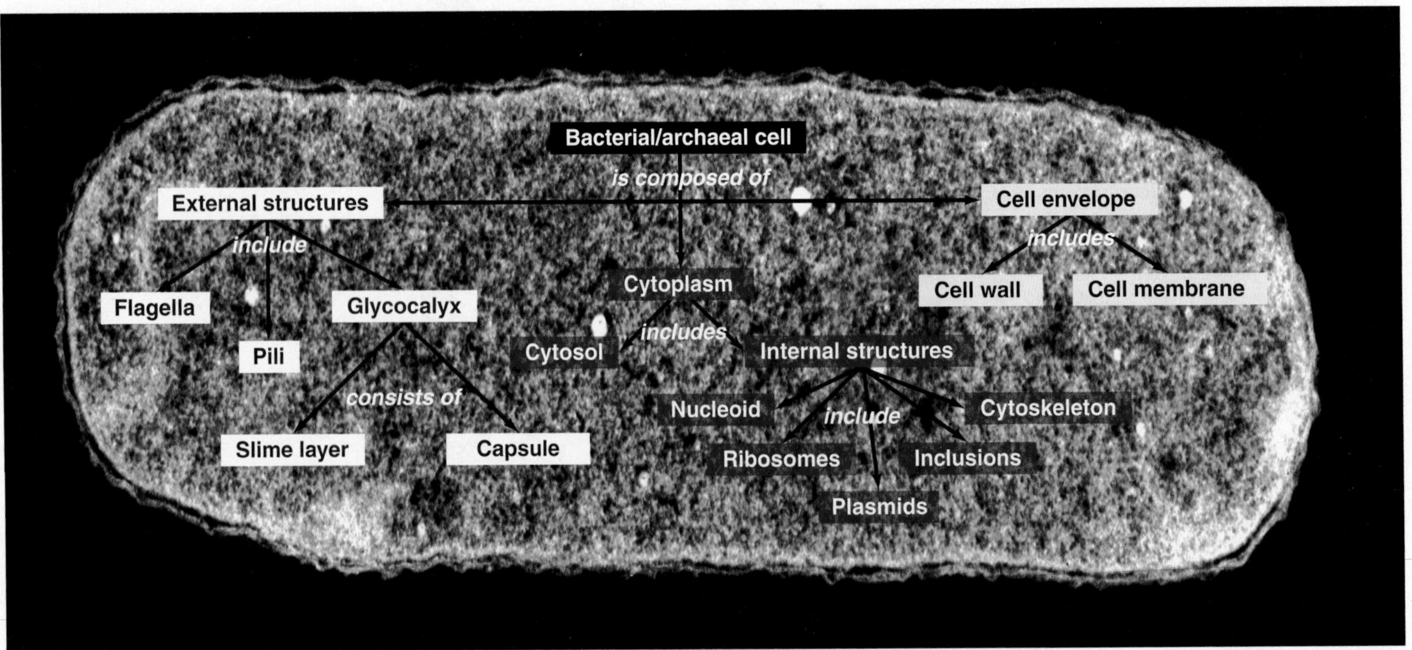

FIGURE 4.6 **A Concept Map for Studying Bacterial and Archaeal Cell Structure.** Not all cells have all the structures shown here. © George Chapman/Visuals Unlimited. »» Why can't we see in the TEM image of a bacterial cell all the structures outlined in the concept map?

like strep throat are secreted from a specific area of the surface. *Yersinia pestis,* which is the agent responsible for plague, contains a specialized secretion apparatus through which toxic proteins are released. This apparatus only exists on the bacterial surface that is in contact with the target host cells.

The cell biology studies also may have important significance to clinical microbiology and the fight against infectious disease. As more is discovered about these cells and how they truly differ from eukaryotic cells, the better equipped we will be to develop new antimicrobial agents that will target the subcellular organization of pathogens. In an era when we have fewer effective antibiotics to fight infections, understanding of cell structure and function may be very important.

On the following pages, we examine some of the common structures found in an idealized bacterial cell, as no single species always contains all the structures (FIGURE 4.7). Our journey starts by examining the structures on or protruding from the surface of the cell. Then, we examine the cell envelope before plunging into the cell cytoplasm. Note: images of cells give the impression of a static, motionless structure. Realize such images are but a "snapshot" of a highly active living cell.

CONCEPT AND REASONING CHECKS 3

a. What is gained by bacterial and archaeal cells being organized into three general sets of structures—external, envelope, and cytoplasmic?

Chapter Challenge B

Prokaryotic cell structure is organized to efficiently carry out those functions required by the organism; that is, sensing and responding to the surrounding environment, compartmentalizing metabolism, and carrying out growth and reproduction.

QUESTION B:

Do you believe these three major functions are different from those in the eukaryotic microbial cell? Explain, keeping in mind the challenge of diversity emphasizing unity.

Answers can be found on the Student Companion Website in **Appendix D.**

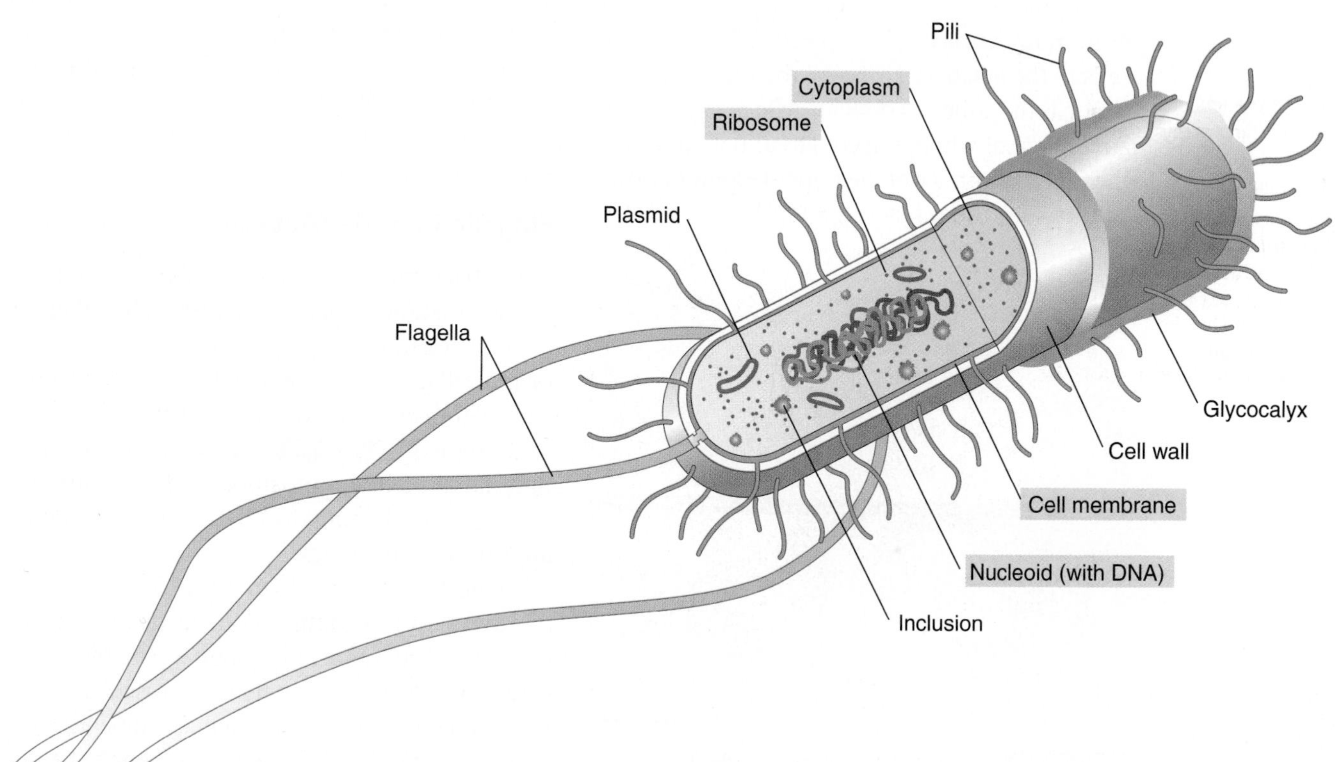

FIGURE 4.7 **Bacterial Cell Structure.** The structural features of a composite, "idealized" bacterial cell. Structures highlighted in blue are found in all bacterial and archaeal species. »» Which structures represent (a) external structures, (b) the cell envelope, and (c) cytoplasmic structures?

KEY CONCEPT 4.4 External Cell Structures Interact with the Environment

Bacterial and archaeal cells need to respond to and monitor their external environment. This is made difficult by having a cell wall that "blindfolds" the cell. Many cells have solved this sensing problem by possessing structures that extend from the cell surface into the environment.

Pili Are Primarily Used for Attachment

Numerous short, thin fibers, called **pili** (sing., pilus; *pilus* = "hair"), protrude from the surface of most gram-negative bacteria (FIGURE 4.8). It should be noted that microbiologists often use the term "pili" interchangeably with "fimbriae" (sing., fimbria; *fimbria* = "fiber"). The rigid fibers, composed of a protein called pilin, act as a scaffolding onto which specific adhesive molecules, called **adhesins**, are attached at the tip. Therefore, the primary function of most pili is to attach the cells to environmental surfaces or, in the case of human pathogens, to appropriate host cells and tissues. This requires that the pili on different bacterial species have specialized adhesins to recognize the appropriate cell. For example, the pili adhesins on *Neisseria gonorrhoeae* cells specifically anchor the cells to the mucosal surface of the urogenital tract whereas the adhesins on *Bordetella pertussis* (causative agent of whooping cough) adhere to cells of the mucosal surface of the upper respiratory tract. In this way, the pili act as a virulence factor by facilitating colonization and biofilm formation, and possibly leading to disease development. Without the chemical mooring line lashing the bacterial cells to host cells, it is less likely the cells could infect host tissue. Investigating the Microbial World 4 looks at pili as a virulence factor.

- Mucosal: Referring to the mucous membranes lining many body cavities exposed to the environment.
- Virulence factor: A pathogen-produced molecule or structure that allows the cell to invade or evade the immune system and possibly cause disease.

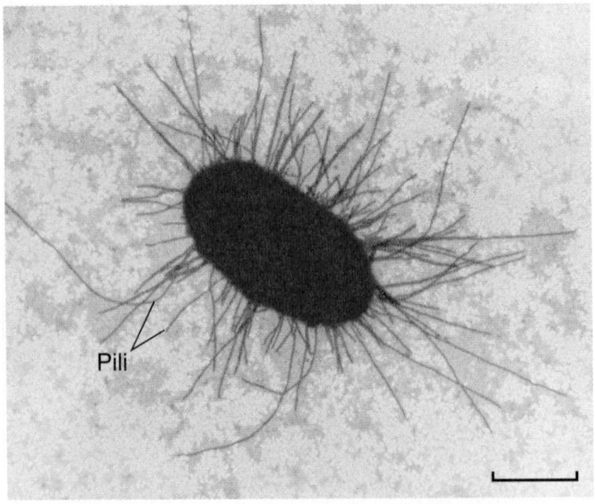

Pili

FIGURE 4.8 **Bacterial Pili.** False-color transmission electron micrograph of an *Escherichia coli* cell (blue) with many pili (green). (Bar = 0.5 µm.) © Dennis Kunkel Microscopy, Inc./Visuals Unlimited/Corbis. »» What function do pili play?

Until recently, attachment pili were thought to be specific to only certain species of gram-negative bacteria. However, extremely thin pili are present on at least some gram-positive bacteria, such as *Streptococcus*. These pili play a very similar role to the pili on gram-negative cells.

Besides these attachment pili, termed "type I pili," some bacterial species produce flexible **conjugation pili** that establish contact between appropriate cells, facilitating the transfer of genetic material from donor to recipient through a process called "conjugation." Conjugation pili are longer than attachment pili and only one or a few are produced on a cell. Pili in some cases function in cell movement. Several bacterial species possess "type IV pili" which, by pilus extension, attachment, and retraction, pull the bacterial cell toward the site of attachment. This movement, called either "twitching motility" or "gliding motility" depending on the species involved, helps cells move through drier environments, over cell tissues, or within biofilms.

Flagella Provide Motility

Many bacterial and archaeal cells are motile by using remarkable "nanomachines" called **flagella** (sing., flagellum). Depending on the species, one or more flagella may be attached to one or both ends of the cell, or at positions distributed over the cell surface (FIGURE 4.9A). In the domain Archaea, flagellar protein composition and structure differs significantly from that of the Bacteria; motility appears similar though.

Flagella range in length from 10 µm to 20 µm and are many times longer than the diameter of the cell. Because they are only about 20 nm thick, they cannot be seen with the light microscope unless stained. However, their existence can be inferred by using dark-field microscopy to watch the live cells dart about.

In the domain Bacteria, each flagellum is composed of a helical filament, hook, and basal body

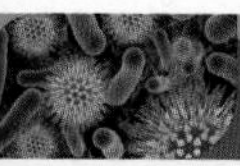

Investigating the Microbial World 4

The Role of Pili

The cell surface of *Escherichia coli* (and many other bacterial species) is covered with tiny hair-like appendages called pili that function to attach the cells to human host cells—in this case, epithelial cells of the gut.

■ **OBSERVATION:** There are *E. coli* strains that normally exist in our gut but cause us no ill as long as they stay there. However, there are other strains that, if ingested from a contaminated food source, can cause illness and disease (such as diarrhea). So, what a group of researchers at Stanford University wanted to investigate is whether the pili on an enteropathogenic (intestinal disease-causing) strain of *E. coli* are necessary to initiate the disease process.

■ **QUESTION:** ***Is* E. coli *attachment by pili to the intestinal epithelium required for the onset of gastrointestinal disease (diarrhea)?***

■ **HYPOTHESIS:** Pili-mediated adhesion of *E. coli* cells is necessary for the infection process leading to disease. If correct, then mutants of *E. coli* that lack pili should not cause diarrhea while mutants with an excessive number of pili should cause a more intense illness.

■ **EXPERIMENTAL DESIGN:** First, what was needed was a marker for infection. With the strain of *E. coli* used, this was easy—a good case of diarrhea would develop (as measured by the volume of liquid stools produced). Next, the Stanford researchers needed experimental subjects who would not complain if they developed a case of diarrhea. So, a tolerant group of healthy Stanford student volunteers were enrolled in the study—and paid $300 for their "contribution" to science! The volunteers were randomized into separate groups, each group receiving a slightly different form or dose of *E. coli*. Neither the volunteers nor the Stanford researchers knew who was drinking which liquid mixture; it was a so-called double-blind study. A number of nurses and doctors were on hand to help the volunteers through their ordeal.

■ **EXPERIMENT 1:** Each of three groups drank three doses of a fruit-flavored cocktail containing a mutant strain (muT) from the wild type having few pili, another mutant strain (muA) from the wild type also having few pili, or the diarrhea-causing strain (wild-type; wt) with normal numbers of pili. Each group was further divided into three subgroups based on the dose they would receive (5×10^8, 2.5×10^9, or 2×10^{10} *E. coli* cells). Over 48 hours, the cumulative volume of liquid stool collected from each volunteer was recorded.

■ **EXPERIMENT 2:** Two additional groups drank three doses of a fruit-flavored cocktail containing a mutant strain (muF) from the wild type having excessive pili that causes the cells to aggregate together in culture. One group received doses of 2.5×10^9 cells while the other group received doses of 2×10^{10} cells. Again, the volume of liquid stool collected from each volunteer over a 48 hour period was recorded.

■ **RESULTS:** See figure. In the figure, each vertical bar represents one volunteer and the cumulative volume of liquid stool. The short bars below the horizontal X axis represent volunteers who produced no liquid stools in the 48 hour period of the experiments.

■ **CONCLUSIONS:**

QUESTION 1: ***Was the hypothesis validated? Explain using the figure and identify the control in experiment 1.***

QUESTION 2: ***In experiment 1, explain why the volunteers drinking the cocktail with muT and muA did not suffer severe diarrhea? Were there any anomalies within these groups?***

QUESTION 3: ***From experiment 2, propose a reason why the volunteers drinking the cocktail with muF did not suffer severe diarrhea? Were there any anomalies within these groups?***

Answers can be found on the Student Companion Website in **Appendix D**.

Adapted from: Bieber, D. *et al.* (1998). *Science* **280** (5372): 2114–2118. Icon image © Tischenko Irina/ShutterStock, Inc.

(continued on next page)

Investigating the Microbial World 4 (continued)

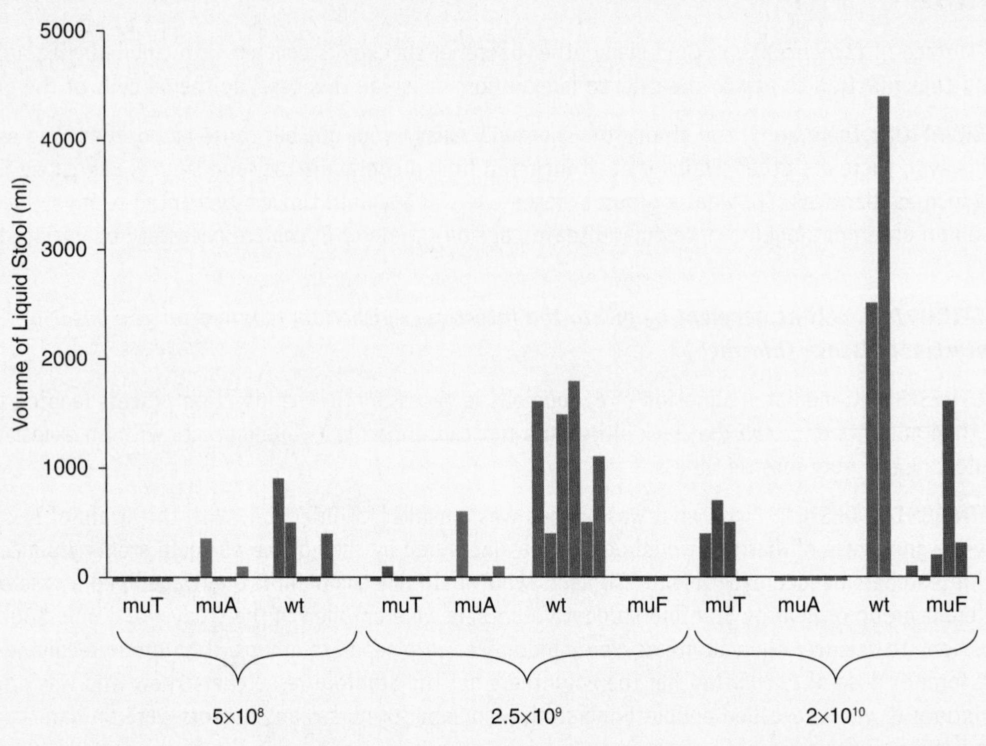

Adapted from: Bieber, D. et al. (1998). *Science* 280 (5372): 2114–2118.

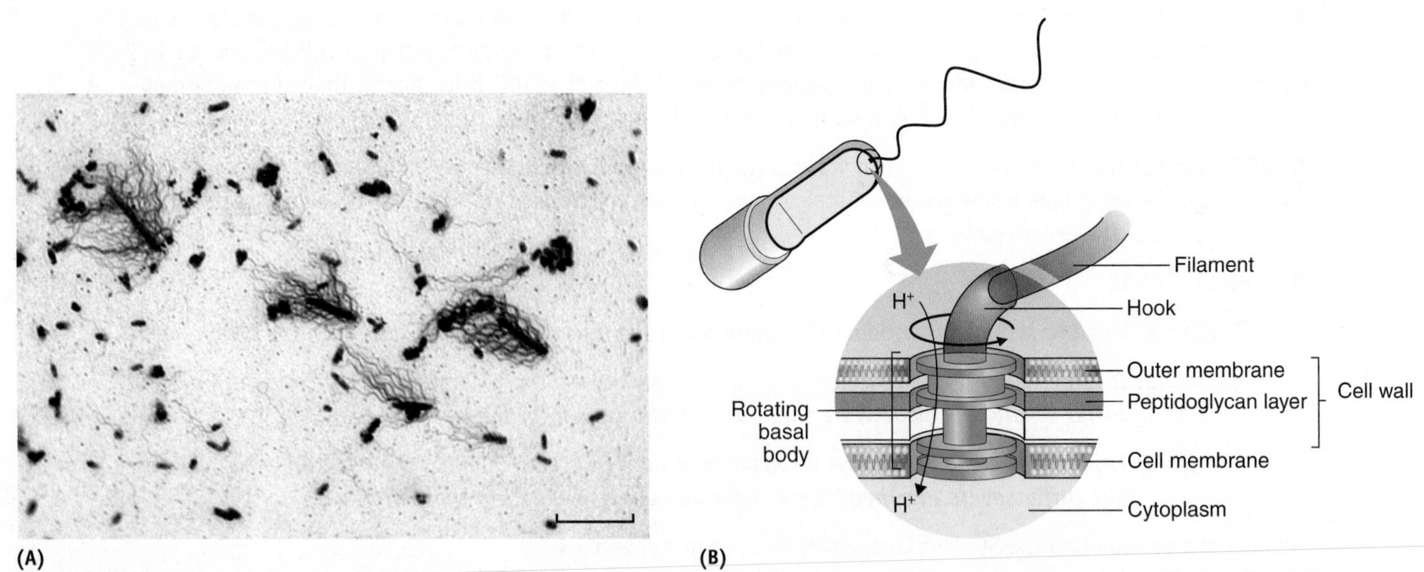

FIGURE 4.9 **Bacterial Flagella.** (**A**) A light micrograph of stained *Proteus vulgaris* showing numerous flagella extending from the cell surface. (Bar = 10 µm.) Note that the length of a flagellum is many times the width of the cell. Courtesy of Jeffrey Pommerville. (**B**) The flagellum on a gram-negative bacterial cell is attached to the cell wall and membrane by two pairs of protein rings in the basal body. »» Why is the flagellum referred to as a "nanomachine"?

(FIGURE 4.9B). The hollow filament is composed of long, rigid strands of protein while the hook attaches the filament to a basal body (motor) anchored in the cell envelope.

The basal body is an assembly of more than 20 different proteins that form a central rod and set of enclosing rings. Gram-positive cells have a pair of rings embedded in the cell membrane and one ring in the cell wall, while gram-negative cells have a pair of rings embedded in the cell membrane and another pair in the cell wall.

The basal body represents a powerful biological motor or rotary engine that generates a propeller-type rotation of the rigid filament. The energy for rotation comes from the diffusion of protons (hydrogen ions; H^+) into the cell through proteins associated with the basal body. This energy is sufficient to produce up to 1,500 rpm by the filament, driving the cell forward.

What advantage is gained by cells having flagella? In nature, there are many chemical nutrients in the environment that cells need to survive. Cells will search out such **attractants** by using their flagella to move up the concentration gradient; that is, toward the attractant. The process is called **chemotaxis**.

Being so small, the cells sense their chemical surroundings using a temporal sensing system. In the absence of a gradient, the flagella all rotate as a bundle counterclockwise and the cell moves straight ahead in short bursts called "runs" (FIGURE 4.10A). These runs can last a few seconds and the cells can move up to 10 body lengths per second (the fastest human can run about 5–6 body lengths per second). A reversal of flagellar rotation (clockwise rotation) causes the cell to "tumble" randomly for a second as the flagella become unbundled and uncoordinated. Then, the motor again reverses direction and another run occurs in a new direction.

If an attractant gradient is present, cell behavior changes and the cells moving up the gradient now undergo longer periods when the motor turns counterclockwise (lengthened runs) and shorter periods when it turns clockwise (shortened tumbles) (FIGURE 4.10B). The combined result is a net movement toward the attractant; that is, up the concentration gradient.

Similar types of motile behavior are seen in photosynthetic organisms moving toward light (phototaxis) or other cells moving toward oxygen gas (aerotaxis). MicroFocus 4.2 investigates how flagella may have evolved.

One additional type of flagellar organization is found in the spirochetes, the group of gram-negative, coiled bacterial species described earlier in this chapter. The cells are motile by flagella that extend from one or both poles of the cell

■ Temporal sensing: One that compares the chemical environment and concentration from one moment to the next.

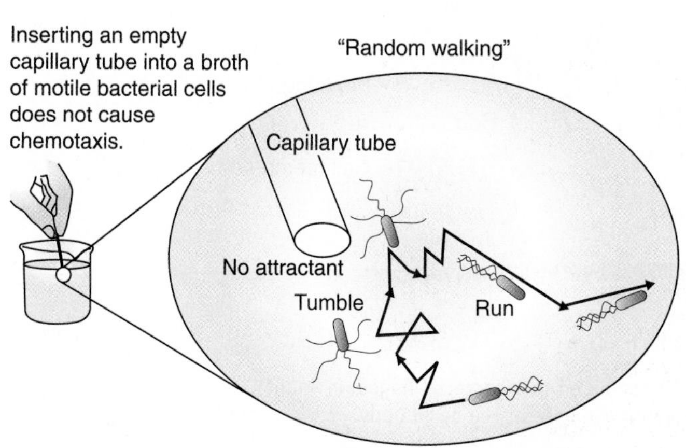

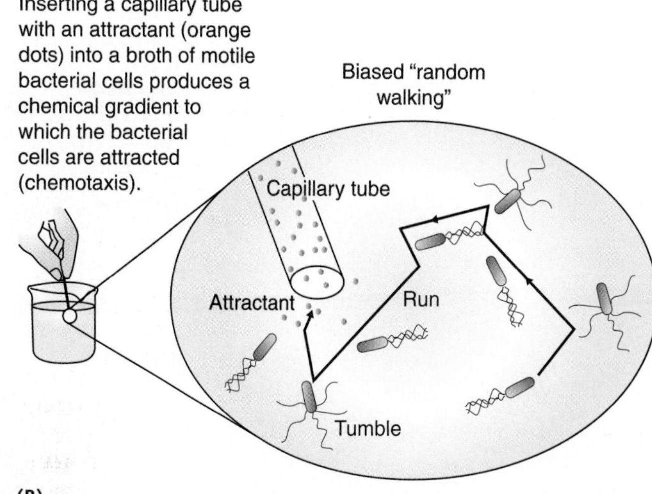

FIGURE 4.10 **Chemotaxis.** Chemotaxis represents a behavioral response to chemicals. (**A**) In a "random walk," rotation of the flagellum counterclockwise causes the bacterial cell to "run," while rotation of the flagellum clockwise causes the bacterial cell to "tumble," as shown. (**B**) During chemotaxis (biased "random walk") to an attractant, such as sugar, cellular behavior leads to longer runs and fewer tumbles, which will result in biased movement toward the attractant. »» Predict the behavior of a bacterial cell if it sensed a repellant; that is a potential harmful or lethal chemical.

MICROFOCUS 4.2: Evolution

The Origin of the Bacterial Flagellum

Flagella are an assembly of protein parts forming a rotary engine that, like an outboard motor, propel the cell forward through its moist environment. Recent work has shown how such a nanomachine may have evolved.

Several bacterial species, including *Yersinia pestis,* the agent of bubonic plague, contain structures to inject toxins into an appropriate eukaryotic host cell. These bacterial cells have a hollow tube or needle to accomplish this process, just as the bacterial flagellum and filament are hollow (see diagram below). In addition, many of the flagellar proteins are similar to some of the injection proteins. In 2004, investigations discovered that *Y. pestis* cells actually contain all the genes needed for a flagellum—but the cells have lost the ability to use these genes for that purpose. *Y. pestis* is nonmotile and it appears that the cells use a subset of the flagellar proteins to build the injection device.

One scenario then is that an ancient cell evolved a structure that was the progenitor of the injection and flagellar systems. In fact, many of the proteins in the basal body of flagellar and injection systems are similar to proteins involved in proton (hydrogen ion; H^+) transport. Therefore, a proton transport system may have evolved into the injection device and, through diversification events, evolved into the motility structure present on many bacterial cells today.

The fascinating result of these investigations and proposals is it demonstrates that structures can evolve from other structures with a different function. It is not necessary that evolution "design" a structure from scratch but rather it can modify existing structures for other functions.

Individuals have proposed that the complexity of structures like the bacterial flagellum are just too complex to arise gradually through a step-by-step process. However, the investigations being conducted illustrate that a step-by-step evolution of a specific structure is not required. Rather, there can be cooperation, where one structure is modified to have other functions. The bacterial flagellum almost certainly falls into that category.

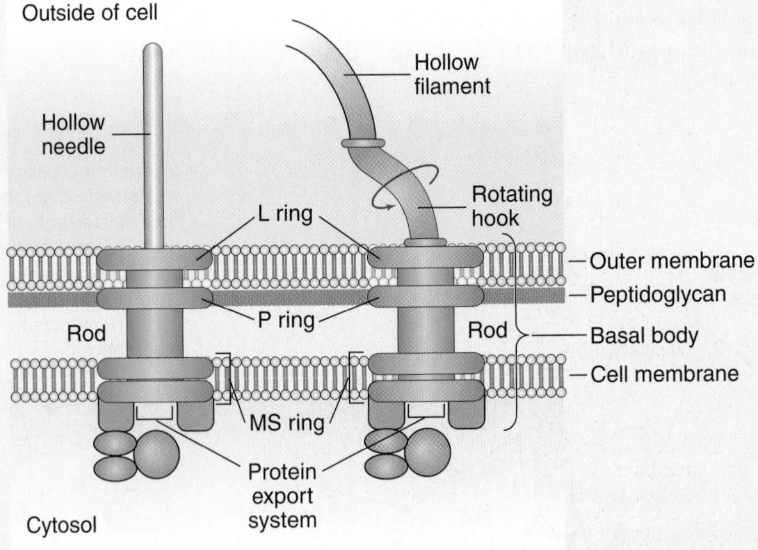

A bacterial injection device (left) compared to a bacterial flagellum (right). Both have a protein export system in the base of the basal body.

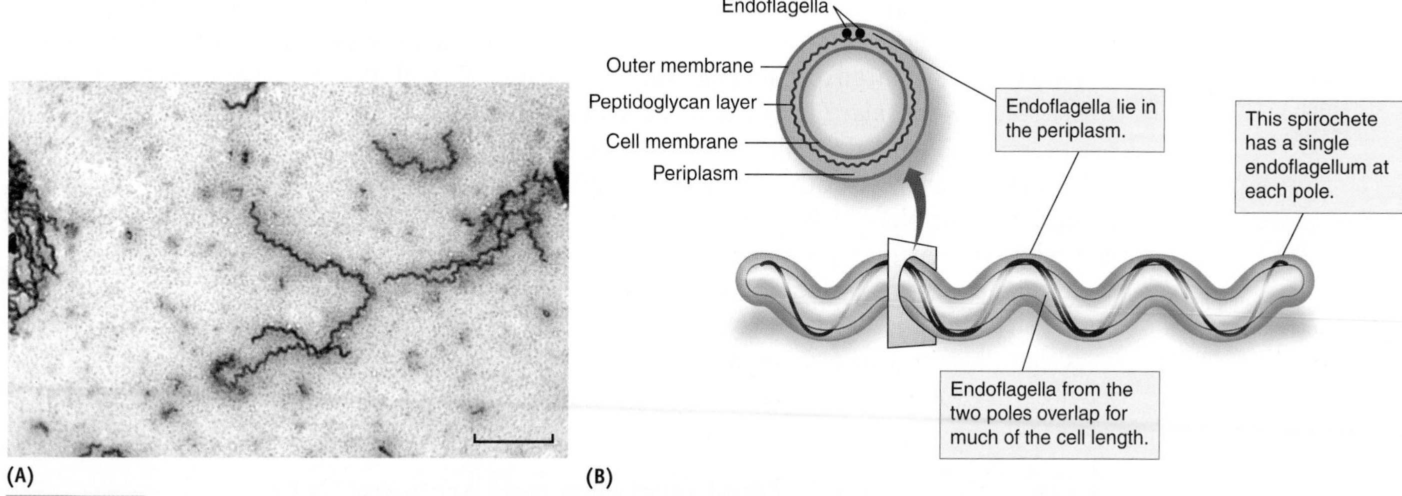

FIGURE 4.11 **The Spirochete Endoflagella.** **(A)** A light micrograph of *Treponema pallidum* shows the corkscrew-shaped spirochete cell. (Bar = 10 μm.) Courtesy of Jeffrey Pommerville. **(B)** Diagram showing the positioning of endoflagella in a spirochete. »» How are endoflagella different from true bacterial flagella?

but fold back along the cell body (FIGURE 4.11). Such **endoflagella** lie in an area called the "periplasm" (see below). Motility results from the torsion generated on the cell by the normal rotation of the flagella. The resulting motility is less regular and more jerky than with flagellar motility.

The Glycocalyx Serves Several Functions

Many bacterial species secrete an adhering layer of polysaccharides, or polysaccharides and small proteins, called the **glycocalyx** (*glyco* = "sweet"; *calyx* = "coat"). The layer can be thick and covalently bound to the cell, in which case it is known as a **capsule**. A thinner, loosely attached layer is referred to as a **slime layer**. Colonies containing cells with a glycocalyx appear moist and glistening. The actual capsule can be seen by light microscopy when observing cells in a negative stain preparation or by transmission electron microscopy (FIGURE 4.12).

The glycocalyx serves as a buffer between the cell and the external environment. Because of its high water content, the glycocalyx can protect cells from desiccation. Another major role of the glycocalyx is to allow the cells to stick to surfaces. The glycocalyx of *V. cholerae*, for example, permits the cells to attach to the intestinal wall of the host.

Website Animation: Bacterial Cell Motility and Chemotaxis

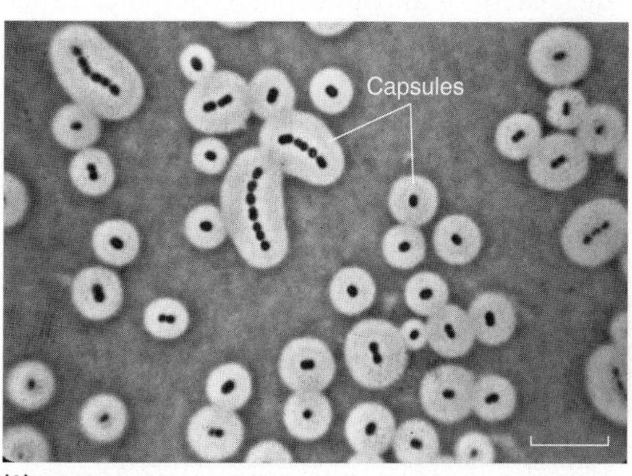

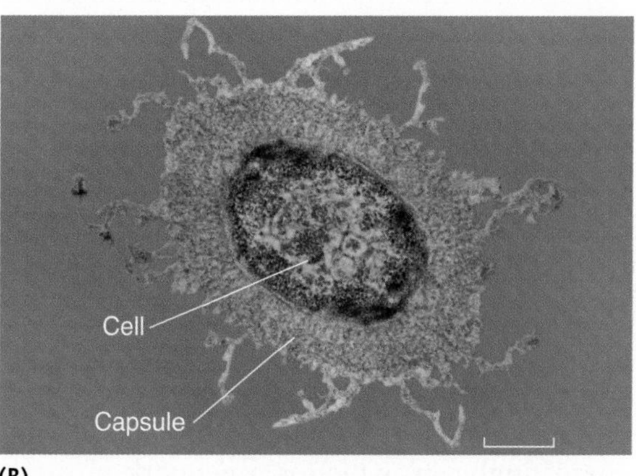

FIGURE 4.12 **The Bacterial Glycocalyx.** **(A)** Demonstration of the presence of a capsule in an *Acinetobacter* species by negative staining and observed by phase-contrast microscopy. (Bar = 10 μm.) Courtesy of Elliot Juni, Department of Microbiology and Immunology, The University of Michigan. **(B)** A false-color transmission electron micrograph of *Escherichia coli*. The cell is surrounded by a thick capsule (pink). (Bar = 0.5 μm.) © George Musil/Visuals Unlimited. »» How does the capsule provide protection for the bacterial cell?

■ Encapsulated: Referring to a cell having a capsule.

■ Phagocytosis: A process whereby certain white blood cells (phagocytes) engulf foreign matter and often destroy microorganisms.

The glycocalyx of pathogens therefore represents another virulence factor.

Other encapsulated pathogens, such as *Streptococcus pneumoniae* (a principal cause of bacterial pneumonia) and *Bacillus anthracis*, evade the immune system because they cannot be easily engulfed by white blood cells during phagocytosis. Scientists believe the repulsion between bacterial cell and phagocyte is due to strong negative charges on the capsule and phagocyte surface.

A slime layer usually contains a mass of tangled fibers of a polysaccharide called **dextran**. The fibers attach the bacterial cell to tissue surfaces. A case in point is *Streptococcus mutans*, an important cause of tooth decay. This species forms dental plaque, which represents a type of biofilm on the tooth surface. Textbook Case 4 details a medical consequence of a biofilm.

CONCEPT AND REASONING CHECKS 4

a. What is the primary function for pili? What other roles can they have?
b. Explain how flagella move a cell during a "run."
c. Under what circumstances might it be advantageous to a bacterial cell to have a capsule rather than a slime layer?

KEY CONCEPT 4.5 Most Bacterial and Archaeal Cells Have a Cell Envelope

The **cell envelope** is a complex structure that forms the two "wrappers"—the **cell wall** and the **cell membrane**—surrounding the cell cytoplasm. The cell wall is relatively porous to the movement of substances whereas the cell membrane regulates transport of nutrients and metabolic products.

The Bacterial Cell Wall Is a Tough and Protective External Shell

■ Hypertonic: A solution with more dissolved material (solutes) than the surrounding solution.

■ Autolytic enzyme: An enzyme that breaks bonds in the peptidoglycan, thereby causing lysis of the cell (e.g., lysozyme).

The fact that most bacterial and archaeal cells have a semi-rigid cell wall suggests the critical role this structure must play. By covering the entire cell surface, the cell wall acts as an exoskeleton to protect the cell from injury and damage. It helps, along with the cytoskeleton, to maintain the shape of the cell and reinforce the cell envelope against the high intracellular water (osmotic) pressure pushing against the cell membrane. Most microbes live in an environment where there are more dissolved materials inside the cell than outside. This hypertonic condition in the cell means water diffuses inward, accounting for the increased osmotic pressure. Without a cell wall, the cell would rupture or undergo **osmotic lysis** (FIGURE 4.13). It is similar to blowing so much air into a balloon that the air pressure bursts the balloon.

The bacterial cell wall differs markedly from the walls of archaeal cells and cells of eukaryotic microorganisms (algae and fungi) in containing **peptidoglycan**, which is a network of disaccharide chains (glycan strands) cross-linked by short, elastic peptides (FIGURE 4.14A). Each disaccharide unit in this very large molecule is composed of two alternating monosaccharides, *N*-acetylglucosamine (NAG) and *N*-acetylmuramic acid (NAM). The carbohydrate backbone can occur in multiple layers connected by side chains of four amino acids and peptide cross-bridges.

There is more to a bacterial cell wall than just peptidoglycan, so several forms of cell wall architecture exist.

Gram-Positive Walls. Most gram-positive bacterial cells have a very thick, rigid peptidoglycan cell wall (FIGURE 4.14B). The abundance and thickness (25 nm) of this material may be one reason why they retain the crystal violet in the Gram stain procedure. The multiple layers of glycan strands are cross-linked to one another both in the same layer as well as between layers forming a three-dimensional mesh.

The gram-positive cell wall also contains a sugar-alcohol and phosphate polymer called **teichoic acid**. Teichoic acids, which are bound to the glycan chains or to the cell membrane, are essential for cell viability—if the genes for teichoic acid synthesis are deleted, cell death occurs. Still, the function of the teichoic acids remains unclear. They may protect the cell envelope by establishing a surface charge on the cell wall, controlling the activity of autolytic enzymes acting on the peptidoglycan, and/or acting as a virulence factor that triggers fever and inflammation.

The actinomycetes are phylogenetically related to the gram-positive bacteria. However, these bacterial cells have evolved another type of wall architecture to protect the cell membrane from rupture. In organisms like *Mycobacterium*, the cell wall is composed of a waxy lipid called

Textbook CASE 4

An Outbreak of *Enterobacter cloacae* Associated with a Biofilm

Hemodialysis is a treatment for people with severe chronic kidney disease (kidney failure). The treatment filters the patient's blood to remove wastes and excess water. Before a patient begins hemodialysis, an access site is created on the lower part of one arm. Similar to an intravenous (IV) site, a tiny tube runs from the arm to the dialysis machine. The patient's blood is pumped through the dialysis machine, passed through a filter or artificial kidney called a dialyzer, and the cleaned blood returned to the patient's body at the access site. The complete process can take 3 to 4 hours.

1 During September 1995, a patient at an ambulatory hemodialysis center in Montreal, Canada received treatment on a hemodialysis machine to help relieve the effects of kidney disease. The treatment was performed without incident.

2 The next day, a second patient received treatment on the same hemodialysis machine. His treatment also went normally, and he returned to his usual activities after the session was completed.

3 In the following days, both patients experienced bloodstream infections (BSIs). They had high fever, muscular aches and pains, sore throat, and impaired blood circulation. Because the symptoms were severe, the patients were hospitalized. The clinical microbiology lab reported that both patients had infections of *Enterobacter cloacae,* a gram-negative rod.

4 In the following months, an epidemiological investigation reviewed other hemodialysis patients at that center. In all, seven additional adult patients were identified who had used the same hemodialysis machine. They discovered all seven had similar BSIs.

5 Inspection of the hemodialysis machine used by these nine patients indicated the presence of biofilms containing *Enterobacter cloacae,* which was identical to those samples taken from the patients' bloodstreams (see figure).

6 Further study indicated that the dialysis machine was contaminated with *E. cloacae,* specifically where fluid flows.

7 It was discovered that hospital personnel were disinfecting the machines correctly. The problem was that the valves in the drain line were malfunctioning, allowing a backflow of contaminated material.

8 Health officials began a hospital education program to ensure that further outbreaks of infection would be minimized.

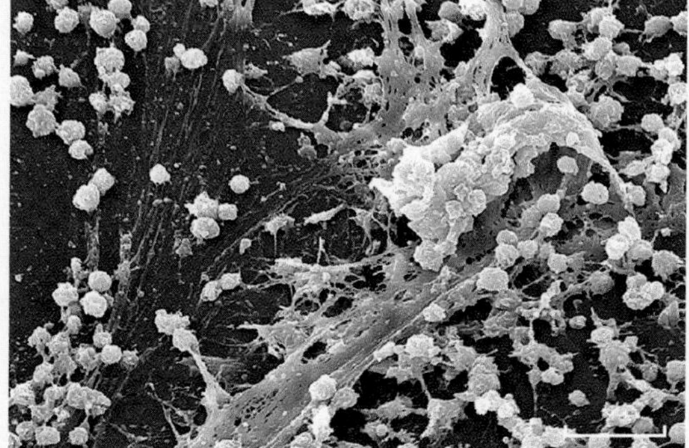

Similar to the description in this textbook case, biofilms consisting of *Staphylococcus* cells can contaminate hemodialysis machines. (Bar = 5 μm.) Courtesy of Dr. Rodney M. Donlan and Janice Carr/CDC.

Questions:

(Answers can be found on the Student Companion Website in **Appendix D.**)

A. Suggest how the hemodialysis machine originally became contaminated.

B. Why weren't the other five cases of BSI correlated with the hemodialysis machine until the epidemiological investigation was begun?

C. How could future outbreaks of infection be prevented?

For additional information see www.cdc.gov/mmwr/preview/mmwrhtml/00051244.htm.

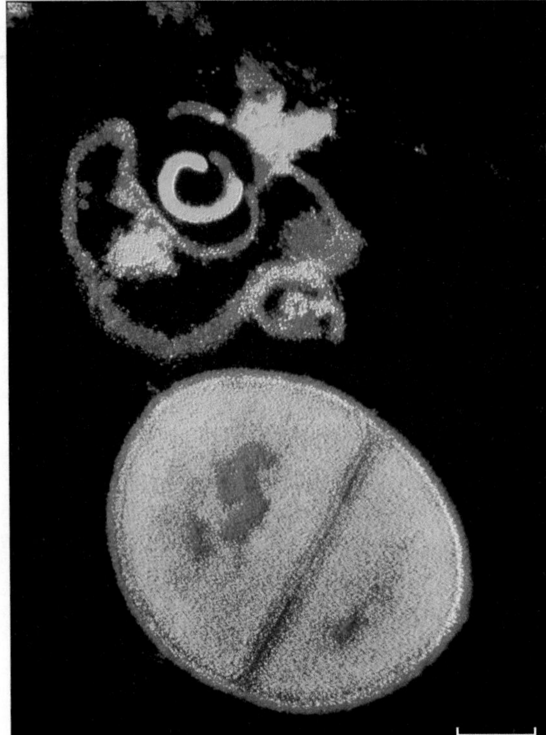

FIGURE 4.13 **Cell Rupture (Lysis).** A false-color electron micrograph showing the lysis of a *Staphylococcus aureus* cell. The addition of the antibiotic penicillin interferes with the construction of the peptidoglycan in new cells, and they quickly burst (top cell). (Bar = 0.25 µm.) © CNRI/Photo Researchers, Inc. »» Where is the concentration of dissolved substances (solutes) higher, inside the cell or outside? Explain how this leads to cell lysis.

■ Endotoxin: A poison that can activate inflammatory responses, leading to high fever, shock, and organ failure.

■ Hydrophilic: Pertaining to molecules or parts of molecules that are soluble in water.

■ Hydrophobic: Pertaining to molecules or parts of molecules that are not soluble in water.

mycolic acid that is arranged in two layers that are covalently attached to the underlying peptidoglycan. Such a hydrophobic layer is impervious to the Gram stains, so stain identification of *M. tuberculosis* is carried out using the acid-fast stain procedure.

Gram-Negative Walls. The cell wall of gram-negative bacterial cells is structurally quite different from that of the gram-positive wall (FIGURE 4.14C). The peptidoglycan strands compose just a single layer or two making the cell more susceptible to lysis. This is one reason why it loses the crystal violet dye during the Gram stain procedure. Also, there is no teichoic acid present.

The unique feature of the gram-negative cell wall is the presence of an **outer membrane**, which is separated by a gap, called the **periplasm**, from the cell membrane. This gel-like compartment contains digestive enzymes and transport proteins to speed entry of nutrients into the cell. The peptidoglycan layer is located in the periplasm and attached to lipoproteins in the outer membrane.

The inner half of the outer membrane contains phospholipids similar to the cell membrane. However, the outer half is composed primarily of **lipopolysaccharide (LPS)**, which consists of polysaccharide attached to a unique anchoring lipid molecule known as **lipid A**. The so-called O polysaccharide is used to identify variants of a species (e.g., strain O157:H7 of *E. coli*). On cell death, lipid A is released and represents an endotoxin that can be toxic if ingested.

The outer membrane also contains unique proteins called **porins**. These proteins form pores in the outer membrane through which small, hydrophilic molecules (sugars, amino acids, some ions) diffuse passively into the periplasm. The outer membrane also contains substrate-specific active transporters for some needed compounds, such as vitamins. However, other larger, hydrophobic molecules cannot pass, partly accounting for the resistance of gram-negative cells to many antimicrobial agents, dyes, disinfectants, and lysozyme.

Before leaving the bacterial cell walls, a brief mention should be made of bacterial species that lack a cell wall. The mycoplasmas are a wall-less genus that is again phylogenetically related to the gram-positive bacteria. Taxonomists believe that the mycoplasmas once had a cell wall but lost it because of their parasitic relationship with their host. To help protect the cell membrane from rupture, the mycoplasmas are unusual in containing sterols in the cell membrane.

TABLE 4.2 summarizes the major differences between the three major types of bacterial cell walls.

The Archaeal Cell Wall Also Provides Mechanical Strength

Archaeal species vary in the type of wall they possess. None have the peptidoglycan typical of the Bacteria. Some species have a **pseudopeptidoglycan** where the NAM is replaced by *N*-acetyltalosamine uronic acid (NAT). Other archaeal cells have walls made of polysaccharide, protein, or both.

A common component of the cell wall among most archaeal species is a surface layer called the **S-layer**, which is also quite common in many

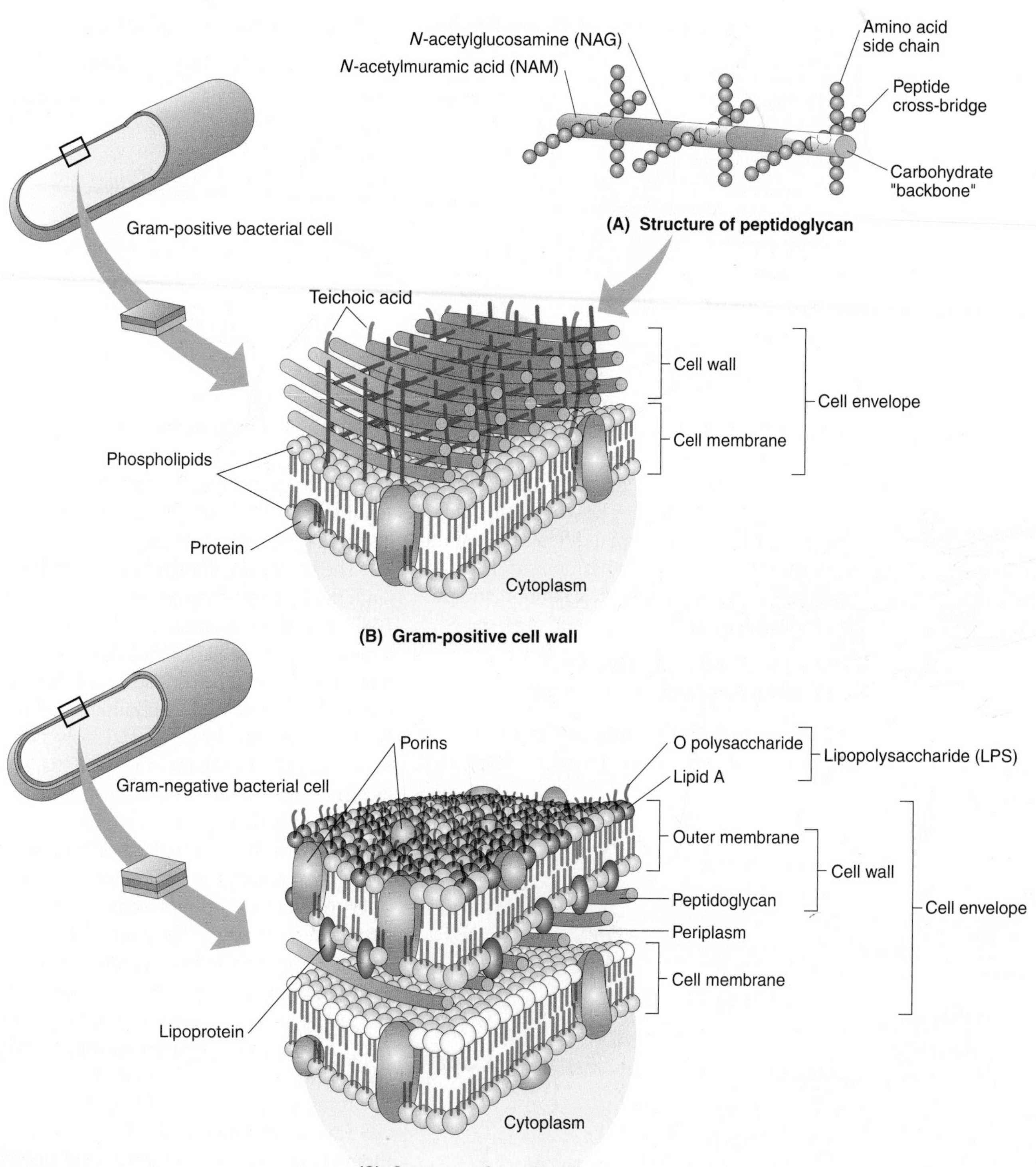

FIGURE 4.14 **A Comparison of the Cell Walls of Gram-Positive and Gram-Negative Bacterial Cells.** (**A**) The structure of peptidoglycan is shown as units of NAG and NAM joined laterally by amino acid cross-bridges and vertically by side chains of four amino acids. (**B**) The cell wall of a gram-positive bacterial cell is composed of multiple peptidoglycan layers combined with teichoic acid molecules. (**C**) In the gram-negative cell wall, the peptidoglycan layer is much thinner, and there is no teichoic acid. Moreover, an outer membrane overlies the peptidoglycan layer such that both comprise the cell wall. Note the structure of the outer membrane in this figure. It contains porin proteins and the outer half is unique in containing lipopolysaccharide.
»» Simply based on cell wall structure, assess the potential of gram-positive and gram-negative cells as pathogens.

TABLE

4.2 A Comparison of Gram-Positive and Gram-Negative Cell Walls

Characteristic	Gram Positive	Gram Positive Acid-fast	Gram Negative
Peptidoglycan	Yes, thick layer	Yes	Yes, thin layer
Teichoic acids	Yes	No[1]	No
Outer membrane	No	Yes	Yes
Mycolic acid	No	Yes	No
Lipopolysaccharides (LPS)	No	No	Yes
Porin proteins	No	Yes	Yes
Periplasm	No	Yes	Yes

[1]Have a different type of glycopolymer.

groups of the Bacteria. It consists of hexagonal patterns of protein or glycoprotein that self-assemble into a crystalline lattice 5 nm to 25 nm thick.

Although the walls may be structurally different and the molecules form a different structural pattern, the function is the same as in bacterial species—to provide mechanical support and prevent osmotic lysis.

The Cell Membrane Represents a Selectively Permeable Barrier

A **cell** (or **plasma**) **membrane** is a universal structure that separates external from internal (cytoplasmic) environments, preventing soluble materials from simply diffusing into and out of the cell. One exception is water, which due to its small size and overall lack of charge can diffuse slowly across the membrane.

The bacterial cell membrane, which is about 7 nm thick, is 25% phospholipid (by weight) and 75% protein. In illustrations, the cell membrane appears very rigid (FIGURE 4.15). In reality, it is quite fluid, having the consistency of olive oil. This means the mosaic of phospholipids and proteins are not cemented in place, but rather they can move laterally in the membrane. This dynamic model of membrane structure therefore is called the **fluid mosaic model**.

The phospholipid molecules, typical of most biological membranes, are arranged in two parallel layers (a bilayer) and represent the barrier function of the membrane. The phospholipids contain a charged phosphate head group attached to two hydrophobic fatty acid chains. The fatty acid "tails" are the portion that forms the permeability barrier. In contrast, the hydrophilic head groups are exposed to the aqueous external or cytoplasmic environments.

Several antimicrobial substances act on the membrane bilayer. The antibiotic polymyxin pokes holes in the bilayer, while some detergents and alcohols dissolve the bilayer. Such action allows the cytoplasmic contents to leak out of bacterial cells, resulting in death through cell lysis.

A diverse population of membrane proteins populates the phospholipid bilayer. These mem-

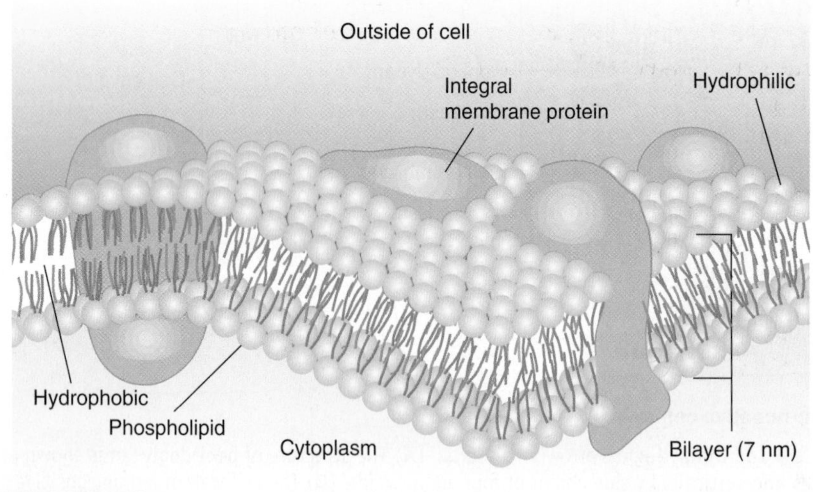

FIGURE 4.15 **The Structure of the Bacterial Cell Membrane.** The cell membrane of a bacterial cell consists of a phospholipid bilayer in which are embedded integral membrane proteins. Other proteins and ions may be associated with the integral proteins or the phospholipid heads. »» Why is the cell membrane referred to as a fluid mosaic structure?

brane proteins often have stretches of hydrophobic amino acids that interact with the hydrophobic fatty acid chains in the membrane. These proteins span the width of the bilayer and are referred to as "integral membrane proteins." Other proteins, called "peripheral membrane proteins," are associated with the polar heads of the bilayer or exposed parts of integral membrane proteins.

Bacterial and archaeal cell membrane proteins carry out numerous important functions, some of which are absent from the eukaryotic plasma membrane. This includes enzymes needed for cell wall synthesis and energy metabolism. As mentioned, bacterial and archaeal cells lack mitochondria and part of that organelle's function is carried out by the cell membrane. Other membrane proteins help anchor the DNA to the membrane during replication or act as receptors of chemical information, sensing changes in environment conditions and triggering appropriate behavioral responses as demonstrated through quorum sensing.

Perhaps the largest group of integral membrane proteins is involved as transporters of charged solutes, such as amino acids, simple sugars, and ions across the lipid bilayer. The transport proteins are highly specific though, only transporting a single molecular type or a very similar class of molecules. Therefore, there are many different transport proteins to regulate the diverse molecular traffic that must flow into or out of a cell.

The transport process can be passive or active. In **facilitated diffusion**, integral membrane proteins facilitate the movement of materials down their concentration gradient; that is, from an area of higher concentration to one of lower concentration (FIGURE 4.16). By acting as a conduit for diffusion or as a transporter through the hydrophobic bilayer, hydrophilic solutes can enter or leave without the need for cellular energy.

Unlike facilitated diffusion, **active transport** allows different concentrations of solutes to be established outside or inside of the cell against the concentration gradient. These membrane proteins act as "pumps" and, as such, demand an energy input from the cell. Cellular processes such as cell energy production and flagella rotation also depend on active transport.

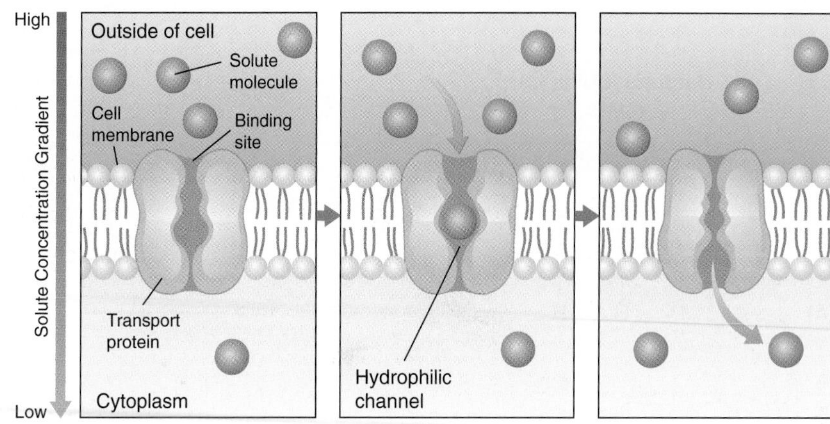

FIGURE 4.16 **Facilitated Transport Through a Membrane Protein.** Many transport proteins facilitate the diffusion of nutrients across the lipid bilayer. The transport protein forms a hydrophilic channel through which a specific solute can diffuse. »» Why would a solute move through a membrane protein rather than simply across the lipid bilayer?

Archaeal Membranes Are Structurally Unique

Besides the differences in gene sequences for ribosomal RNA in the domain Archaea, another major difference used to separate the archaeal organisms into their own domain is the chemical nature of the cell membrane.

The manner in which the hydrophobic lipid tails are attached to the glycerol is different in the Archaea. The tails are bound to the glycerol by "ether linkages" rather than the "ester linkages" found in the domains Bacteria and Eukarya (FIGURE 4.17A).

Also, typical fatty acid tails are absent from the membranes; instead, repeating five-carbon units are linked end-to-end to form lipid tails longer than the fatty acid tails. The result can be a lipid bilayer or **monolayer** (FIGURE 4.17B). A monolayer provides an advantage to the hyperthermophiles by preventing a peeling in two of the membrane, which would occur with a typical bilayer structure under high temperature conditions.

CONCEPT AND REASONING CHECKS 5

a. Penicillin and lysozyme primarily affect peptidoglycan synthesis in gram-positive bacterial cells. Why are these agents less effective against gram-negative bacterial cells?

b. Distinguish between peptidoglycan and pseudopeptidoglycan cell walls.

c. Justify the necessity for phospholipids and proteins in the cell membrane.

d. What is unique about archaeal membrane structure?

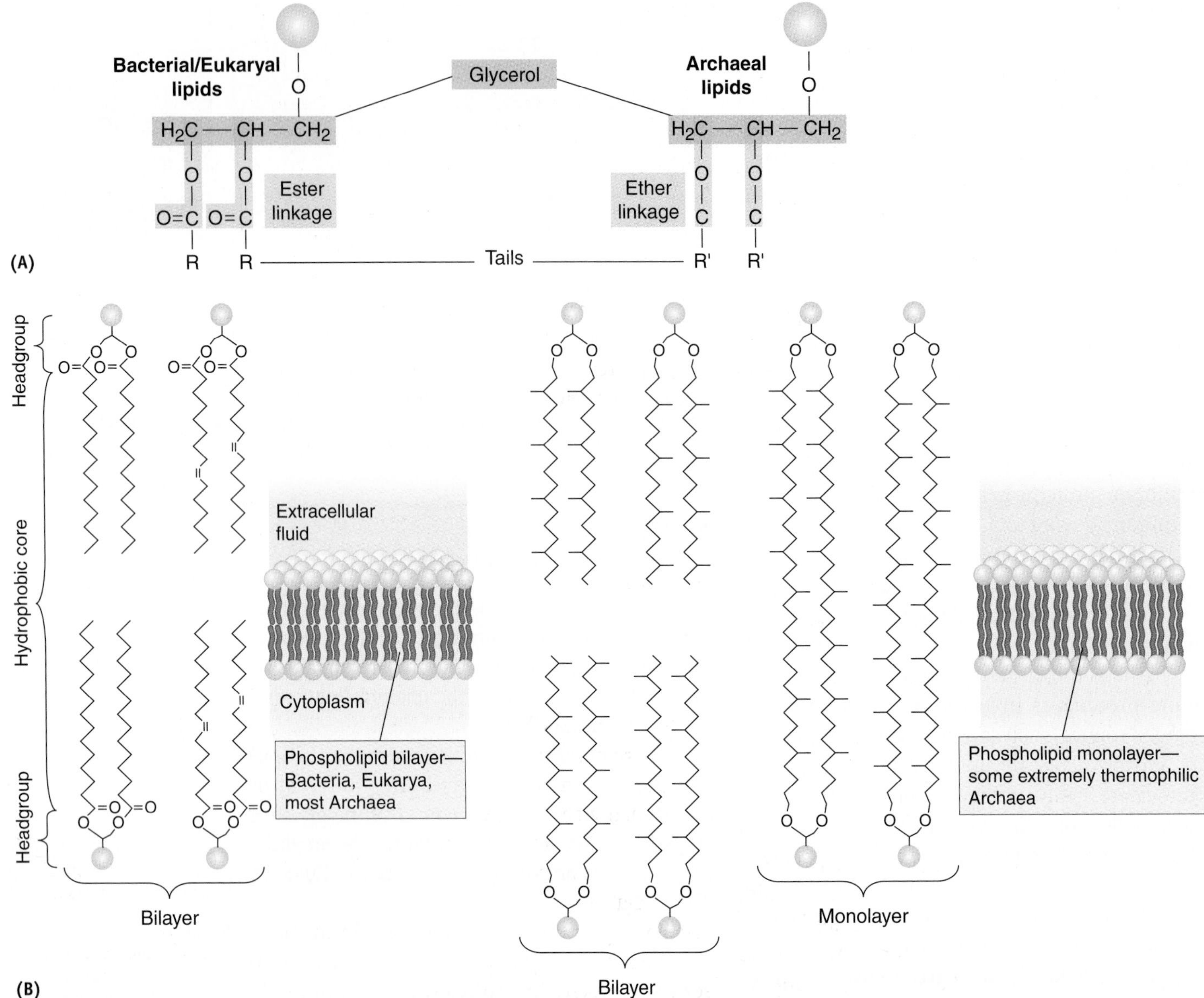

FIGURE 4.17 **Structure of Cell Membranes.** (**A**) Bacterial and eukaryotic cell membranes involve an ester linkage joining the glycerol to the fatty acid tails (R) while archaeal membranes have an ether linkage to the isoprenoid tails (R′). (**B**) Bacterial and eukaryal membranes form a bilayer while archaeal membranes may be a bilayer or a monolayer. »» What identifies an ester linkage from an ether linkage?

KEY CONCEPT 4.6 The Cell Cytoplasm Is Packed with Internal Structures

The cell membrane encloses the **cytoplasm**, which is the compartment within which most growth and metabolism occurs. The cytoplasm consists of the **cytosol**, a semifluid mass of proteins, amino acids, sugars, nucleotides, salts, vitamins, and ions—all dissolved in water—and several bacterial structures, subcompartments, and filaments, each with a specific function.

The Nucleoid Represents a Subcompartment Containing the Chromosome

The chromosome region in bacterial and archaeal cells appears as a diffuse mass termed the **nucleoid** (FIGURE 4.18). The nucleoid is not surrounded by a membrane envelope; rather, it represents a

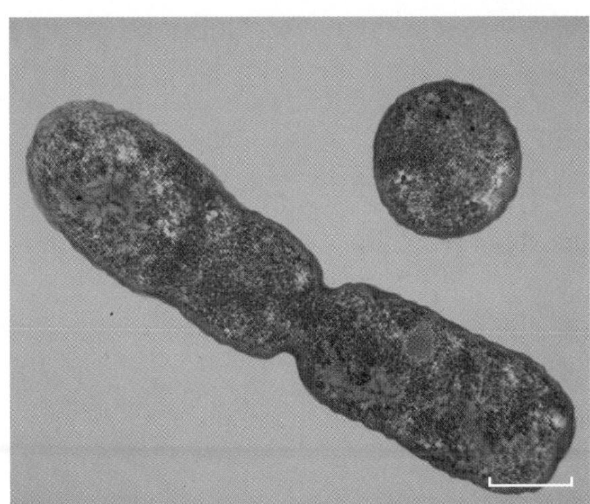

FIGURE 4.18 **The Bacterial Nucleoid.** In this false-color transmission electron micrograph of *Escherichia coli,* nucleoids (orange) occupy a large area in this dividing bacterial cell. Both longitudinal (center cells) and cross sections (cell upper right) of *E. coli* are visible. (Bar = 0.5 μm.) © Dennis Kunkel Microscopy, Inc./Visuals Unlimited/Corbis.
»» How does a nucleoid differ from the eukaryotic cell nucleus described in the previous chapter?

subcompartment in the cytoplasm where the DNA aggregates and ribosomes are absent. Usually there is a single chromosome per cell and, with few exceptions, exists as a closed loop of DNA and protein.

The DNA contains the essential hereditary information for cell growth, metabolism, and reproduction. Because most cells only have one chromosome, the cells are genetically haploid. Unlike eukaryotic microorganisms and other eukaryotes, the nucleoid and chromosome do not undergo mitosis and having but the one set of genetic information cannot undergo meiosis.

The complete set of genes in an organism or virus is called the **genome**. For example, the genome of *E. coli,* a typical bacterial species in the mid-size range, contains about 4,300 genes. In all cases, these genes determine what proteins and enzymes the cell can make; that is, what metabolic reactions and activities can be carried out. For *E. coli,* this equates to some 2,000 different proteins.

Plasmids Are Found in Many Bacterial and Archaeal Cells

Besides a nucleoid, many bacterial and archaeal cells also contain smaller molecules of DNA called **plasmids**. About a tenth the size of the chromosome, these stable, extrachromosomal DNA molecules exist as closed loops containing 5 to 100 genes. There can be one or more plasmids in a cell and these may contain similar or different genes. Plasmids replicate independently of the chromosome and can be transferred between cells during recombination. They also represent important vectors in industrial technologies that use genetic engineering.

■ Vector:
A genetic element capable of incorporating and transferring genetic information.

Although plasmids may not be essential for cellular growth, they provide a level of genetic flexibility. For example, some plasmids possess genes for disease-causing toxins and many carry genes for chemical or antibiotic resistance. For this latter reason, these genetic elements often are called **R plasmids** (R for resistance).

Other Subcompartments Exist in the Cell Cytoplasm

For a long time the cytoplasm was looked at as a bag enclosing the genetic machinery and biochemical reactions. As more studies were carried out with the electron microscope and with biochemical techniques, it became evident that the cytoplasm contained "more than meets the eye."

Ribosomes. One of the universal structures in all cells is the **ribosome**. There are thousands of these nearly spherical particles in the cell cytoplasm, which gives it a granular appearance when viewed with the electron microscope (FIGURE 4.19A). In addition, there also are ribosomes loosely associated with the cell membrane. Their relative size is measured by how fast they settle when spun in a centrifuge. Measured in Svedberg units (S), bacterial and archaeal ribosomes represent 70S particles.

■ Haploid:
Having a single set of genetic information.

■ Centrifuge:
An instrument that spins particles suspended in liquid at high speed.

The ribosomes are built from RNA and protein and are composed of a small subunit (30S) and a large subunit (50S) (FIGURE 4.19B). For proteins to be synthesized, the two subunits come together to form a 70S functional ribosome. Some antibiotics, such as streptomycin and tetracycline, prevent bacterial and archaeal ribosomes from carrying out protein synthesis.

The free ribosomes make soluble proteins that are used in the cell, while the membrane-associated ones produce proteins for the cell envelope and for secretion.

Microcompartments. Recently, some bacterial species have been found to contain **microcompartments**. The microcompartments appear to be unique to the Bacteria and consist of a polyprotein shell 100–200 nm in diameter (FIGURE 4.20).

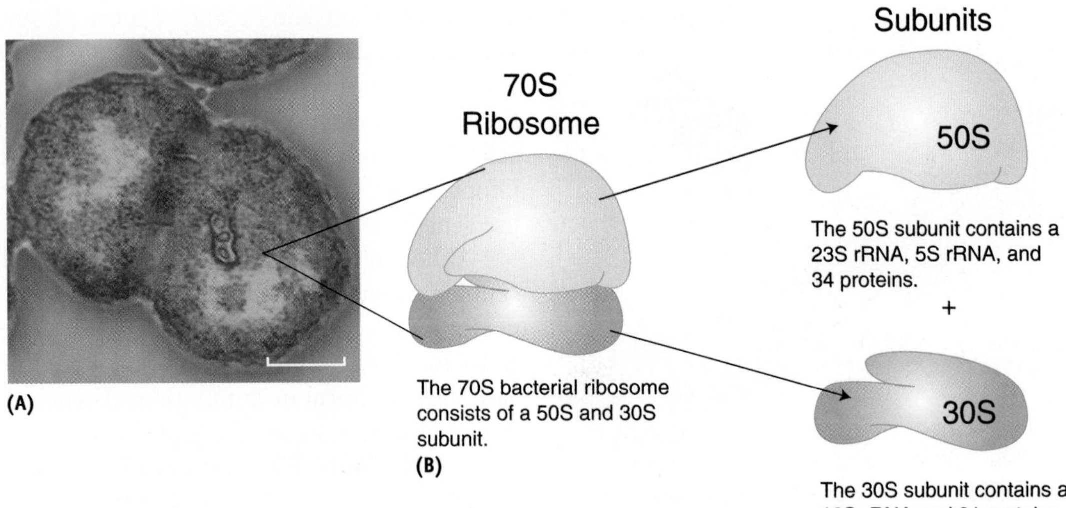

FIGURE 4.19 **The Bacterial Ribosome.** (**A**) A false-color transmission electron micrograph of *Neisseria gonorrhoeae,* showing the nucleoid (yellow) and ribosomes (blue). (Bar = 1 µm.) © Science Source/Photo Researchers, Inc. (**B**) The functional 70S ribosome is assembled from a small (30S) and large (50S) subunit. »» How many rRNA molecules and proteins construct a 70S ribosome?

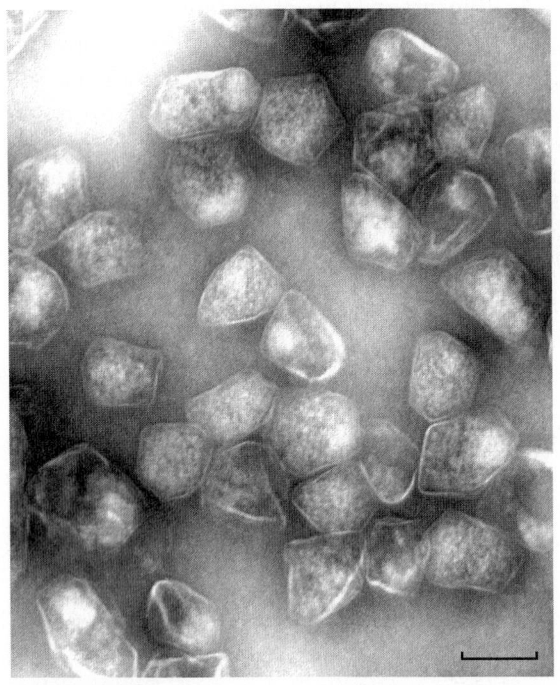

FIGURE 4.20 **Microcompartments.** Purified bacterial microcompartments from *Salmonella enterica* are composed of a complex protein shell that encases metabolic enzymes. (Bar = 100 nm.) Reprinted with permission from the American Society for Microbiology (Microbe, January, 2006, p. 20-24). Photo courtesy of Dr. Thomas A. Bobik, Department of Biochemistry, Biophysics and Molecular, Iowa State University. »» How do these bacterial microcompartments differ structurally from a eukaryotic organelle?

In the cyanobacteria, microcompartments called "carboxysomes" function to enhance carbon dioxide fixation and in some non-photosynthetic species, microcompartments limit diffusion of volatile or toxic metabolic products. In general, microcompartments represent localized areas where enzymes can more directly interact with their substrates or potentially sequester harmful reaction products.

Inclusions. Cytoplasmic structures, called **inclusions**, can be found in the cytoplasm. Many of these bodies store nutrients or the monomers for cellular structures. For example, some inclusions function as nutrient reserves. They consist of aggregates or granules of polysaccharides (glycogen), globules of elemental sulfur, or lipid. Other inclusions can serve as important identification characters for bacterial pathogens. One example is the diphtheria bacilli that contain **metachromatic granules**, also called **volutin granules**, which are deposits of polyphosphate (long chains of inorganic phosphate) along with calcium and other ions. These granules stain with dyes such as methylene blue. Similar organelles called "acidocalcisomes" are found in the protists.

Some aquatic and marine forms achieve buoyancy using **gas vacuoles**, cytoplasmic compartments built from a water-tight protein shell. These vacuoles are filled with a gas to decrease the density of the cell, which generates and regulates their buoyancy.

The **magnetosome**, another type of inclusion or subcompartment, is described in MicroFocus 4.3.

MICROFOCUS 4.3: Environmental Microbiology

Magnetic Navigation

To get from place to place, humans often require the assistance of maps, GPS systems, or gas station attendants. In the microbial world, life is generally more simple, and traveling is no exception.

In the early 1980s, Richard P. Blakemore and his colleagues at the University of New Hampshire observed mud-dwelling bacterial cells gathering at the north end of water droplets. On further study, they discovered each cell had a chain of aligned magnetic particles acting like a compass directing the organism's movements (magnetotaxis). Additional interdisciplinary investigations by microbiologists and physicists have shown the magnetotactic bacteria contain a linear array of 15–20 membrane-bound vesicles along the cell's long axis (see figure). Each vesicle, called a magnetosome, is an invagination of the cell membrane and contains the protein machinery to nucleate and grow a magnetic crystal of magnetite (Fe_3O_4) or greigite (Fe_3S_4). The chain of magnetosomes is organized by filaments that are similar to eukaryotic actin. By running parallel to the magnetosome chain, the filaments organize the vesicles into a chain. As each vesicle accumulates the magnetite or greigite crystal to form a magnetosome, magnetostatic interactions between vesicles stabilizes the linear aggregation. As an aside, magnetite is also found in migrating animals such as butterflies, homing pigeons, and dolphins.

To date, all magnetotactic bacterial cells are motile, gram-negative cells common in aquatic and marine habitats, including sediments where oxygen is absent. This last observation is particularly noteworthy because it explains why these organisms have magnetosomes—they act as a compass needle to orient the cell.

Recent studies have shown that magnetotactic bacteria prefer low concentrations of oxygen. So, a current theory is that both magnetotaxis and aerotaxis work together to allow cells to "find" the optimal oxygen concentration, permitting the bacteria cells to reach a sort of biological nirvana and settle in for a life of environmental bliss.

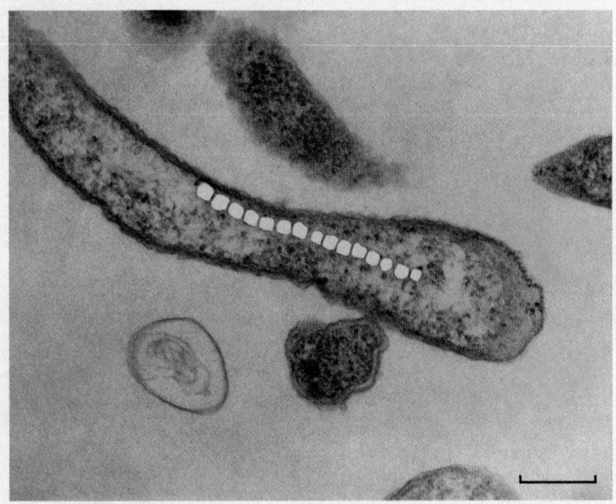

Bacterial magnetosomes (yellow) are seen in this false-color transmission electron micrograph of a magnetotactic marine spirillum. (Bar = 1 µm.) © Dennis Kunkel Microscopy, Inc./Visuals Unlimited/Corbis.

These inclusions, found in some aquatic bacterial species, are invaginations of the cell membrane, which are coordinated and positioned by cytoskeletal filaments similar to eukaryotic microfilaments.

Cytoskeletal Proteins Regulate Cell Division and Help Determine Cell Shape

Until recently, the dogma was that bacterial and archaeal cells lacked a cytoskeleton, which is a common feature in eukaryotic cells. However, it is now clear that cytoskeletal proteins that resemble those in the eukaryotic cytoskeleton are present in bacterial and archaeal cells.

The first protein discovered was a **homolog** of the eukaryotic protein tubulin, which forms filaments that assemble into microtubules. The homolog forms filaments similar to those in microtubules but the filaments do not assemble into microtubules. These tubulin-like proteins have been found in all bacterial and archaeal cells examined and appear to function in the regulation of cell division. During this process, the protein localizes around the neck of the dividing cell where it recruits other proteins needed for the deposition of a new cell wall between the dividing cells (MICROFOCUS 4.4).

Protein homologs remarkably similar in three-dimensional structure to eukaryotic microfilaments assemble into filaments that help determine cell shape in *E. coli* and *Bacillus subtilis*. These homologs have been found in most nonspherical cells where they form a helical network beneath the cell membrane to guide the proteins involved

■ **Homolog:** An entity with similar attributes due to shared ancestry.

MICROFOCUS 4.4: **Public Health**

The Wall-less Cytoskeleton

Sometimes the lack of something can speak loudly. Take for example the mycoplasmas such as *Mycoplasma pneumoniae* that causes primary atypical pneumonia (walking pneumonia). This species, as well as other *Mycoplasma* species, lack a cell wall. How then can they maintain a defined cell shape (see figure)?

Transmission electron microscopy has revealed that mycoplasmal cells contain a very complex cytoskeleton and further investigations indicate the cytoskeletal proteins are very different from the typical cytoskeletal homologs found in other groups of the Firmicutes. For example, *Spiroplasma citri,* which causes infections in other animals, has a fibril protein cytoskeleton that is laid down as a helical ribbon. This fibril protein has not been found in any other organisms. Because the cells are spiral shaped, the ribbon probably is laid down in such a way to determine cell shape, suggesting that shape does not have to be totally dependent on a cell wall.

In *Mycoplasma genitalium,* which is closely related to *M. pneumoniae* and causes human urethral infections, a eukaryotic-like tubulin homolog has been identified, but none of the other proteins have been identified that it recruits at the division neck for cell division. Surprising? Not really. Important? Immensely! Because mycoplasmas do not have a cell wall, why would they require those proteins that lay down a peptidoglycan cross wall between cells? So the lack of something (wall-forming proteins) tells us what those proteins must do in their gram-positive relatives that do have walls.

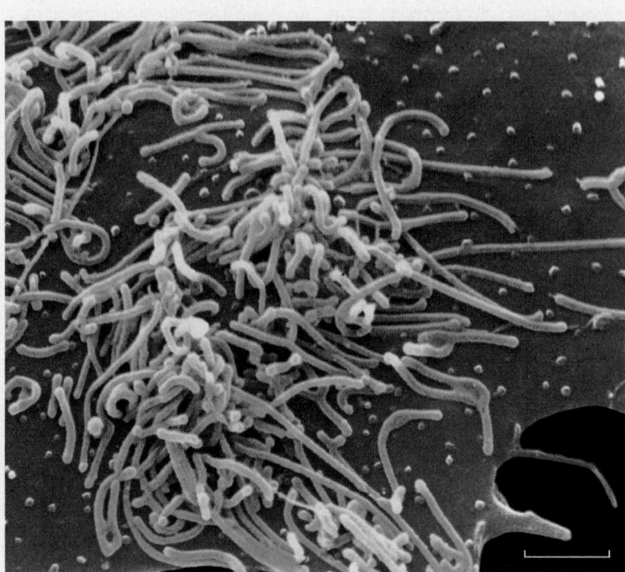

False-color scanning electron micrograph of *Mycoplasma pneumoniae* cells. (Bar = 2.5 µm.)

in cell wall formation (FIGURE 4.21A). The homologs also are involved with chromosome segregation during cell division and magnetosome formation.

Intermediate filaments (IF), another component of the eukaryotic cytoskeleton in some metazoans, have a homolog as well. The protein, called crescentin, helps determine the characteristic crescent shape of *Caulobacter crescentus* cells. In older cells that become filamentous, crescentin maintains the helical shape of the cells by aligning with the inner cell curvature beneath the cytoplasmic membrane (FIGURE 4.21B).

■ Metazoan: A member of the vertebrates, nematodes, and mollusks.

Even though the evolutionary relationships are quite distant between bacterial/archaeal and eukaryotic cytoskeletal proteins based on protein sequence data, the similarity of their three-dimensional structure and function is strong evidence supporting homologous cytoskeletons. (TABLE 4.3) summarizes the structural features of bacterial and archaeal cells.

CONCEPT AND REASONING CHECKS 6

- **a.** Why do we say that the bacterial chromosome contains the "essential hereditary information"?
- **b.** What properties distinguish the bacterial chromosome from a plasmid?
- **c.** Provide the roles for the subcompartments found in bacterial cells.
- **d.** Evaluate the relationship between the eukaryotic cytoskeleton and the cytoskeletal protein homologs in bacterial and archaeal cells.

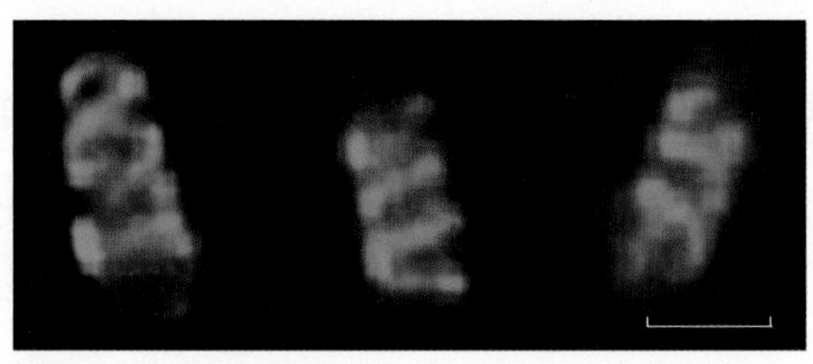

(A)

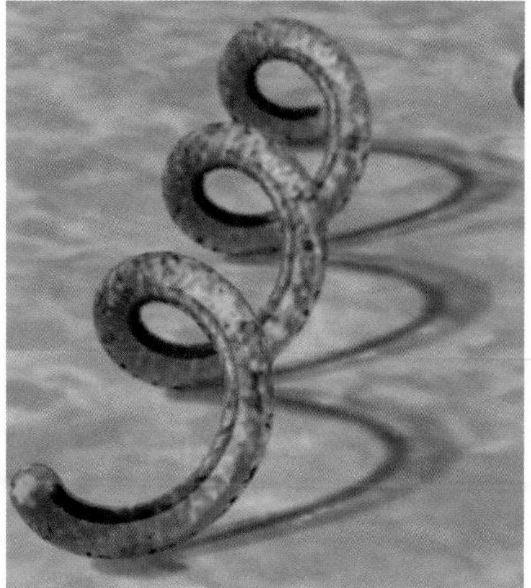

(B)

FIGURE 4.21 **A Bacterial Cytoskeleton.** Bacterial cells have proteins similar to those that form the eukaryotic cytoskeleton. (**A**) Microfilament-like proteins form helical filaments that curve around the edges of these cells of *Bacillus subtilis*. (Bar = 1.5 μm.) Courtesy of Rut Carballido-López, University of Oxford. (**B**) Three-dimensional model of a helical *Caulobacter crescentus* cell (green) with a helical cytoskeletal filament of crescentin (pink). Reprinted from *Cell*, vol 115, Jacobs-Wagner, C., cover, copyright 2003, with permission from Elsevier. Photo courtesy of Christine Jacobs-Wagner, Yale University. »» What would be the shape of these cells without the cytoskeletal proteins?

TABLE

4.3 A Summary of the Structural Features of Bacterial and Archaeal Cells

Structure	Chemical Composition	Function	Comment
External Structures			
Pilus	Protein	Attachment to surfaces Genetic transfer	Found primarily in gram-negative bacteria
Flagellum	Protein	Motility	Present in many rods and spirilla; few cocci; vary in number and placement
Glycocalyx	Polysaccharides and small proteins	Buffer to environment Attachment to surfaces	Capsule and slime layer Contributes to disease development Found in plaque bacteria and biofilms
Cell Envelope			
Cell wall		Cell protection Shape determination Cell lysis prevention	
Bacterial	Gram positives: thick peptidoglycan and teichoic acid Gram negatives: little peptidoglycan and an outer membrane		Site of activity of penicillin and lysozyme Gram-negative bacteria release endotoxins
Archaeal	Pseudopeptidoglycan Protein		S-layer
Cell membrane			
Bacterial	Protein and phospholipid	Cell boundary	Lipid bilayer
Archaeal		Transport into/out of cell Site of enzymatic reactions	Lipid bilayer/monolayer

(continued on next page)

TABLE

4.3 A Summary of the Structural Features of Bacterial and Archaeal Cells (continued)

Structure	Chemical Composition	Function	Comment
Internal Structures			
Nucleoid	DNA	Site of essential genes	Exists as single, closed loop chromosome
Plasmid	DNA	Site of nonessential genes	R plasmids
Ribosome	RNA and protein	Protein synthesis	Inhibited by certain antibiotics
Microcompartment	Various metabolic enzymes	Carbon dioxide fixation Retention of volatile or toxic metabolites	Enzymes are enclosed in a protein shell
Inclusion	Glycogen, sulfur, lipid	Nutrient storage	Used as nutrients during starvation periods
Metachromatic granule	Polyphosphate	Storage of polyphosphate and calcium ions	Found in diphtheria bacilli
Gas vacuole	Protein shell	Buoyancy	Helps cells remain buoyant
Magnetosome	Magnetite/greigite	Cell orientation	Helps locate preferred habitat
Cytoskeleton	Proteins	Cell division, chromosomal segregation, cell shape	Functionally similar to eukaryotic cytoskeletal proteins

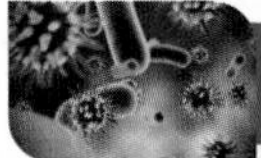

Chapter Challenge C

Congratulations! You have made it to the end of the chapter on bacterial and archaeal cell structure. In these last three sections you discovered the exterior structures, studied the organization of the cell envelope, and surveyed the cytoplasmic structures.

QUESTION C:

The chapter challenge was: "Studying the diversity of life only accentuates life's unity." Summarize your opinion as to the validity of this statement, using cell structure as your reference.

Answers can be found on the Student Companion Website in **Appendix D**.

■ **In conclusion**, one of the take-home lessons from the discussions of cell structure and function explored in this chapter is the ability of bacterial and archaeal organisms to carry out the "complex" metabolic and biochemical processes typically associated with eukaryotic cells—usually without the need for elaborate membrane-enclosed subcompartments.

However, there is now clear evidence for an intricate and sophisticated subcellular architecture in the domains Bacteria and Archaea. But what about other major cellular processes such as making proteins? This requires two processes, that of transcription and translation. In eukaryotic cells, these processes are spatially separated into the cell nucleus (transcription) and the cytoplasm (translation).

In bacterial and archaeal cells, there also can be spatial separation between transcription and translation (FIGURE 4.22). The RNA polymerase molecules needed for transcription are localized to a region separate from the ribosomes and other proteins that perform translation. In fact, studies indicate that transcription is localized to the periphery of the nucleoid. So, even without a

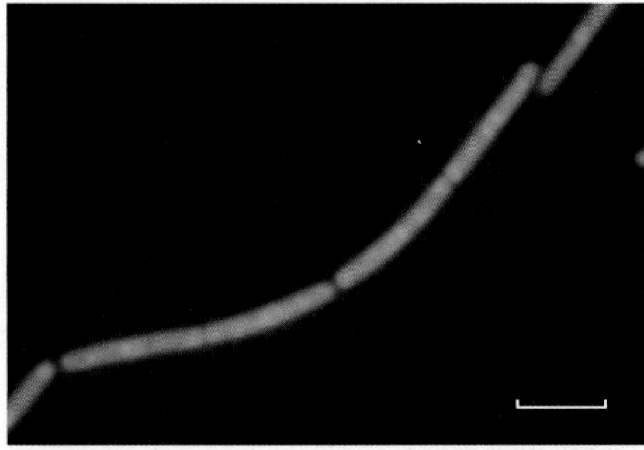

FIGURE 4.22 **Spatial Separation of Transcription and Translation.** In these cells of *Bacillus subtilis,* fluorescence microscopy was used to identify RNA polymerase (transcription) using a red fluorescent protein and ribosomes (translation) using a green fluorescent protein. Separate subcompartments are evident. (Bar = 3 μm.) Courtesy of Dr. Peter Lewis; School of Environmental and Life Sciences, University of Newcastle. »» What does the spatial separation indicate concerning compartmentation?

MICROINQUIRY 4

The Prokaryote/Eukaryote Model

"It is now clear that among organisms there are two different organizational patterns of cells, which Chatton . . . called, with singular prescience, the eukaryotic and prokaryotic type. The distinctive property of bacteria and blue-green algae is the prokaryotic nature of their cells. It is on this basis that they can be clearly segregated from all other protists (namely, other algae, protozoa, and fungi), which have eukaryotic cells."

Stanier and van Niel (1962)
—*The concept of a bacterium*

The idea of a tree of life extends back centuries and originates not with scientific thinking, but rather with folklore and culture, and often focused on immortality or fertility (see figure).

The development of the three domains in tree of life, on the other hand, represents the evolutionary relationships between species. Its development has made a profound change in biology. Instead of two kinds of organisms—prokaryotes and eukaryotes—there are three: Bacteria, Archaea, and Eukarya.

In 2006, Norm Pace, a molecular biologist turned evolutionist at the University of Colorado, Boulder, suggested that the massive data bank of gene sequences identified since 1995 shows just how different archaeal organisms are from bacterial organisms and, in some ways, the archaeal ones are more similar to eukaryotic organisms. Therefore, Pace says, "we need to reassess our understanding of the course of evolution at the most fundamental level." Among items needing reassessment is the prokaryotic/eukaryotic paradigm—the tradition (folklore) if you will—that if an organism is not a eukaryote, it must be a prokaryote.

Glass mosaic of tree of life on a wall of the 16th century Sim Wat Xiang Thong Luang Prabang UNESCO World Heritage Site, Laos. © Carlos Arguelles/ShutterStock, Inc.

The quote at the top of the page refers to Edouard Chatton who coined the terms "prokaryotic" and "eukaryotic." Interestingly, neither he, nor Stanier and van Niel, ever really made mention as to what a prokaryotic cell is. Yet if you look in any introductory biology textbook, prokaryote is defined as a group of organisms that lack a cell nucleus. According to Pace, "the prokaryote/eukaryote model for biological diversity and evolution is invalid." How can all organisms without a cell nucleus be called "prokaryotic," especially because the eukaryotic cell nucleus appears to be descended from as ancient a line of cells as the Archaea? Yes, the concept of a nuclear membrane (or not) is important, but no more important than other cellular properties. And the problem is that the word "prokaryote" is so engrained in the culture of biology and in the scientific mind of biologists—and students—that inappropriate inferences about organisms are made using this term. Pace does not buy the argument that the term "prokaryote" can be used to identify organisms that are not eukaryotes because the Bacteria are very different from the Archaea and, therefore, should not be put under the umbrella of "prokaryote."

Pace believes that saying prokaryotes lack a cell nucleus is a scientifically invalid description; although open to debate, he says no one can define what a prokaryote is—only what it is not (e.g., no nucleus, no mitochondria, no chloroplasts, no endomembrane system, etc.). Therefore, lumping the Bacteria and Archaea conceptually dismisses the fundamental and important differences between these two kinds of organisms and reinforces an incorrect understanding of biological organization and evolution.

Pace believes it is time to delete the term prokaryote as a term for bacterial and archaeal organisms. Because it has long been used by all biology texts, including this one (although in many cases "bacteria" and "archaea" have replaced the term prokaryote), Pace says he realizes "it is hard to stop using the word prokaryote."

Discussion Point

There is no doubt that bacterial and archaeal organisms are very different entities. So, if "prokaryote" is to be deleted from the biological vocabulary, what can we call the Bacteria and Archaea? Can you think of some positive characters that would define both bacterial and archaeal cells? If so, then how about inventing a common noun and adjective for both? Or do we simply speak of the bacteria, archaea, and eukaryotes separately?

Answers (and comments) can be found on the Student Companion Website in **Appendix D.**

nuclear membrane, these cells can separate the process involved in making cellular proteins, in a manner similar to eukaryotic cells.

Traditionally, a prokaryote was an organism without a cell nucleus; that is, without a membrane surrounding the DNA or chromosome. But is that a fair way to describe all bacterial and archaeal organisms? In this chapter, you have learned that there are some differences between bacterial and archaeal cells, yet do we lump them together just because they both lack a cell nucleus? Some scientists say no—the terms "prokaryote" and "prokaryotic" are not appropriate for these two domains of life.

So, to finish this chapter, take a look at MicroInquiry 4, which discusses the prokaryotic/eukaryotic model for living organisms.

SUMMARY OF KEY CONCEPTS

4.1 There Is Tremendous Diversity Among the Bacteria and Archaea

1. The **phylogenetic tree** of life contains many bacterial phyla and groups, including the **Proteobacteria, Gram-positive bacteria, Cyanobacteria, Chlamydiae**, and **Spirochaetes.** (Fig. 4.2)
2. Many organisms in the domain Archaea live in extreme environments. The **Euryarchaeota** (**methanogens, extreme halophiles, hyperthermophiles**, and the **thermoacidophiles**) and the **Crenarchaeota** are the two phyla.

4.2 Prokaryotes Can Be Distinguished by Their Cell Shape and Arrangements

3. **Bacilli** have a cylindrical shape and can remain as single cells or be arranged into **diplobacilli** or chains (**streptobacilli**). **Cocci** are spherical and form a variety of arrangements, including the **diplococcus, streptococcus**, and **staphylococcus**. The spiral-shaped bacteria can be curved rods (**vibrios**) or spirals (**spirochetes** and **spirilla**). Spirals generally appear as single cells. (Fig. 4.5)

4.3 An Overview to Bacterial and Archaeal Cell Structure

4. Cell organization is centered on three specific processes: sensing and responding to environmental changes, compartmentalizing metabolism, and growing and reproducing. (Figs. 4.6, 4.7)

4.4 External Cell Structures Interact with the Environment

5. **Pili** are short hair-like appendages found on many gram-negative bacteria to facilitate attachment to a surface. **Conjugation pili** are used for genetic transfer of DNA.
6. One or more **flagella**, found on many rods and spirals, provide for cell motility. Each flagellum consists of a **basal body** attached to the flagellar filament. In nature, flagella propel bacterial cells toward nutrient sources (**chemotaxis**). Spirochetes have **endoflagella**. (Figs. 4.10, 4.11)
7. The **glycocalyx** is a sticky layer of polysaccharides that protects the cell against desiccation, attaches it to surfaces, and helps evade immune cell attack. The glycocalyx can be thick and tightly bound to the cell (**capsule**) or thinner and loosely bound (**slime layer**).

4.5 Most Bacterial and Archaeal Cells Have a Cell Envelope

8. The **cell wall** provides structure and protects against cell lysis. Gram-positive bacteria have a thick wall of **peptidoglycan** strengthened with **teichoic acids**. Gram-negative cells have a single layer of peptidoglycan and an **outer membrane** containing **lipopolysaccharide** and **porin proteins.** (Fig. 4.14)
9. Archaeal cell walls lack peptidoglycan but may have either a **pseudopeptidoglycan** or **S-layer.**
10. The **cell membrane** represents a permeability barrier and the site of transfer for nutrients and metabolites into and out of the cell. The cell membrane reflects the **fluid mosaic model** for membrane structure in that the lipids are fluid and the proteins are a mosaic that can move laterally in the bilayer. (Fig. 4.15)
11. The archaeal cell membrane links lipids through an ether linkage and the lipid tails may be bonded together into a single **monolayer.** (Fig. 4.17)

4.6 The Cell Cytoplasm Is Packed with Internal Structures

12. The DNA (**bacterial chromosome**), located in the **nucleoid**, is the essential genetic information and represents most of the organism's **genome.**
13. Bacterial and archaeal cells may contain one or more **plasmids**, circular pieces of nonessential DNA that replicate independently of the chromosome.
14. **Ribosomes** carry out protein synthesis, **microcompartments** sequester species-specific processes, while **inclusions** store nutrients or structural building blocks.
15. The **cytoskeleton**, containing protein homologs to the cytoskeletal proteins in eukaryotic cells, helps determine cell shape, regulates cell division, and controls chromosomal segregation during cell division.

CHAPTER SELF-TEST

For **STEPS A–D**, answers to even-numbered questions and problems can be found in **Appendix C** on the Student Companion Website at **http://microbiology.jbpub.com/10e**. In addition, the site features eLearning, an online review area that provides quizzes and other tools to help you study for your class. You can also follow useful links for in-depth information, read more MicroFocus stories, or just find out the latest microbiology news.

STEP A: REVIEW OF FACTS AND TERMS

Multiple Choice

Read each question carefully, then select the ***one*** answer that best fits the question or statement.

1. Which one of the following is NOT a genus within the gram-positive bacteria?
   A. *Staphylococcus* ✓
   B. Methanogens
   C. *Mycoplasma*
   D. *Bacillus* and *Clostridium* ✓
2. The domain Archaea includes all the following groups *except* the _______.
   A. mycoplasmas
   B. extreme halophiles
   C. Crenarchaeota
   D. Euryarchaeota
3. Spherical bacterial cells in chains would be referred to as a ______ arrangement.
   A. vibrio
   B. streptococcus
   C. staphylococcus
   D. tetrad

4. Intracellular organization in bacterial and archaeal species is centered around ______.
   A. compartmentation of metabolism
   B. growth and reproduction
   C. sensing and responding to environment
   D. All the above (A–C) are correct.
5. Which one of the following statements does NOT apply to pili?
   A. Pili are made of protein.
   B. Pili allow for attachment to surfaces.
   C. Pili facilitate nutrient transport.
   D. Pili contain adhesins.
6. Flagella are ______.
   A. made of carbohydrate and lipid
   B. found on all bacterial cells
   C. shorter than pili
   D. important for chemotaxis
7. Capsules are similar to pili because both ______.
   A. contain DNA
   B. are made of protein
   C. contain dextran fibers
   D. permit attachment to surfaces
8. Gram-negative bacterial cells would stain _____ with the Gram stain and have _____ in the wall.
   A. orange-red; teichoic acid
   B. orange-red; lipopolysaccharide
   C. purple; peptidoglycan
   D. purple; teichoic acid
9. The cell membrane of archaeal hyperthermophiles contains ______.
   A. a monolayer
   B. sterols
   C. ester linkages
   D. All the above (A–C) are correct.
10. The movement of glucose into a cell occurs by ______.
   A. facilitated diffusion
   B. active transport
   C. simple diffusion
   D. phospholipid exchange
11. When comparing bacterial and archaeal cell membranes, only archaeal cell membranes ______.
   A. have three layers of phospholipids
   B. have a phospholipid bilayer
   C. are fluid
   D. have ether linkages
12. Which one of the following statements about the nucleoid is NOT true?
   A. It contains a DNA chromosome.
   B. It represents a nonmembranous subcompartment.
   C. It represents an area devoid of ribosomes.
   D. It contains nonessential genetic information.
13. Plasmids ______.
   A. replicate with the bacterial chromosome
   B. contain essential growth information
   C. may contain antibiotic resistance genes
   D. are as large as the bacterial chromosome
14. Which one of the following is NOT a structure or subcompartment found in bacterial cells?
   A. Microcompartments
   B. Volutin
   C. Ribosomes
   D. Mitochondria
15. The bacterial cytoskeleton ______.
   A. transports vesicles
   B. helps determine cell shape
   C. is organized identical to its eukaryotic counterpart
   D. centers the nucleoid

Label Identification

16. Identify and label the structure on the accompanying bacterial cell from each of the following descriptions. Some separate descriptions may apply to the same structure.

Descriptions

a. Is involved with nutrient storage.
b. An essential structure for chemotaxis, aerotaxis, or phototaxis.
c. Contains nonessential genetic information that provides genetic variability.
d. The structure that synthesizes proteins.
e. The protein structures used for attachment to surfaces.
f. Contains essential genes for metabolism and growth.
g. Prevents cell desiccation.
h. A 70S particle.
i. Contains peptidoglycan.
j. Regulates the passage of substances into and out of the cell.
k. Extrachromosomal loops of DNA.
l. Represents a capsule or slime layer.
m. The semifluid mass of proteins, amino acids, sugars, salts, and ions dissolved in water.

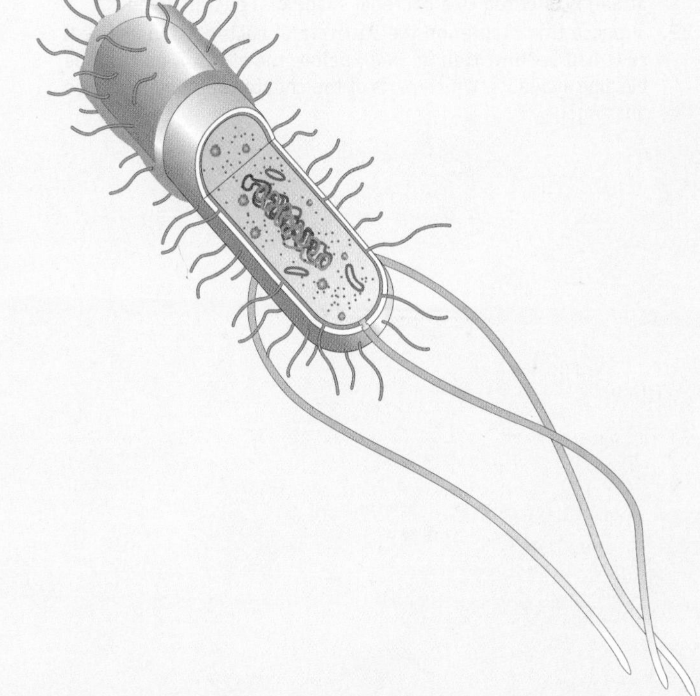

STEP B: CONCEPT REVIEW

17. Construct a concept map for the **domain Bacteria** using the following terms. (**Key Concept 1**)

Actinobacteria	Hyperthermophiles
Bacillus	*Mycoplasma*
Blooms	Proteobacteria
Chlamydiae	Rickettsiae
Cyanobacteria	Spirochaetes
Escherichia	*Staphylococcus*
Firmicutes	*Streptomyces*
Gram-negative species	*Treponema*
Gram-positive species	

18. Compare the various shapes and arrangements of bacterial and archaeal cells. (**Key Concept 2**)
19. Summarize how the processes of sensing and responding to the environment, compartmentation of metabolism, and growth and metabolism are linked to cell structure. (**Key Concept 3**)
20. Assess the role of the three external cell structures to cell function and survival. (**Key Concept 4**)
21. Construct a concept map for the **Cell Envelope** using the following terms. (**Key Concept 5**)

Active transport	Membrane proteins
Cell membrane	NAG
Cell wall	NAM
Endotoxin	Outer membrane
Facilitated transport	Peptidoglycan
Fluid-mosaic model	Periplasm
Gram-negative wall	Phospholipids
Gram-positive wall	Polysaccharide
Lipid A	Porin proteins
Lipopolysaccharide (LPS)	Teichoic acid

22. Justify the need for a **cell membrane** surrounding all bacterial and archaeal cells. (**Key Concept 5**)
23. Describe the function of the nucleoid and plasmids to cell metabolism and cell survival. (**Key Concept 6**)
24. Explain the roles for subcellular compartments in the bacterial cell. (**Key Concept 6**)

STEP C: APPLICATIONS AND PROBLEMS

25. A bacterial species has been isolated from a patient and identified as a gram-positive rod. Knowing that it is a human pathogen, what structures would it most likely have? Explain your reasons for each choice.
26. Another patient has a blood infection caused by a gram-negative bacterial species. Why might it be dangerous to prescribe an antibiotic to treat the infection?
27. In the research lab, the gene for the cytoskeletal protein similar to eukaryotic tubulin is transferred into the DNA chromosome of a coccus-shaped bacterium. When this cell undergoes cell division, predict what shape the daughter cells will exhibit. Explain your answer.

STEP D: QUESTIONS FOR THOUGHT AND DISCUSSION

28. In reading a story about a bacterial species that causes a human disease, the word "bacillus" is used. How would you know if the article is referring to a bacterial shape or a bacterial genus?
29. Suppose this chapter on the structure of bacterial and archaeal cells had been written in 1940, before the electron microscope became available. Which parts of the chapter would probably be missing?
30. Why has it taken so long for microbiologists to discover microcompartments and a cytoskeleton in bacterial and archaeal cells?
31. Apply the current understanding of the bacteria/eukaryote paradigm to the following statement: "Studying the diversity of life only accentuates life's unity."

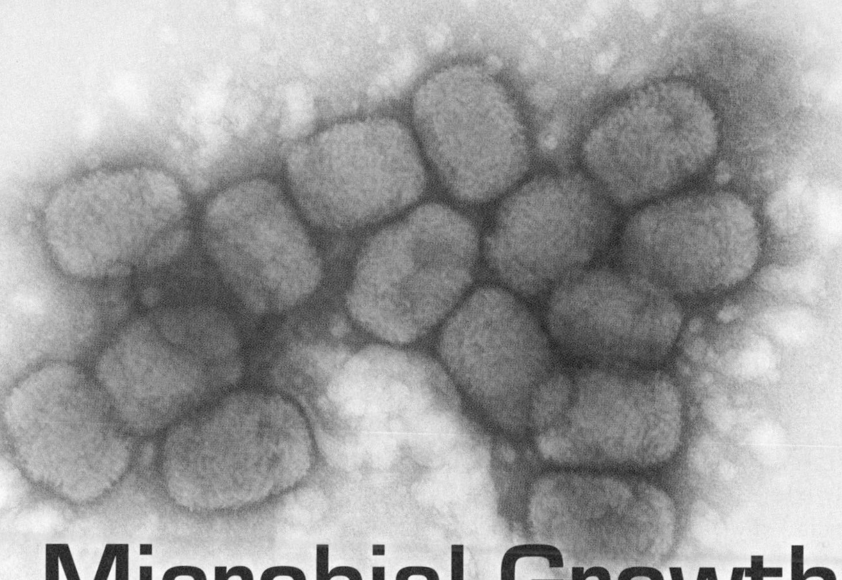

5

CHAPTER PREVIEW

5.1 Microbial Reproduction Is Part of the Cell Cycle

5.2 Microbial Growth Progresses Through Distinct Phases

TEXTBOOK CASE 5: An Outbreak of Food Poisoning Caused by *Campylobacter jejuni*

5.3 Culture Media Are Used to Grow Microbes and Measure Their Growth

MICROINQUIRY 5: Identification of a Bacterial Species

Investigating the Microbial World 5: The Great Plate Count Anomaly

Microbial Growth and Nutrition

But who shall dwell in these worlds if they be inhabited? . . . Are we or they Lords of the World? . . .
—Johannes Kepler (quoted in *The Anatomy of Melancholy*)

Books have been written about it; movies have been made; even a radio play in 1938 about it frightened thousands of Americans. What is it? Martian life. In 1877 the Italian astronomer, Giovanni Schiaparelli, saw lines on Mars, which he and others assumed were canals built by intelligent beings. It wasn't until well into the 20th century that this notion was disproved. Still, when we gaze at the red planet, we wonder: Did life ever exist there?

We are not the only ones wondering. Astronomers, geologists, and many other scientists have asked the same question. Today microbiologists have joined their scientific colleagues, wondering if microbial life once existed on the Red Planet or, for that matter, elsewhere in our Solar System. In 1996, NASA scientists reported finding what looked like fossils of microbes inside a meteorite thought to have come from Mars. Although most now believe these "fossils" are not microbial, it only fueled the debate.

Could microbes, as we know them here on Earth, survive on Mars where the temperatures are far below 0°C, the atmosphere contains little oxygen gas, and the surface is bombarded with ultraviolet radiation? Researchers, using a device to simulate the Martian environment, placed in it microbes known to survive extremely cold environments here on Earth. Their results indicated that members of the Archaea, specifically the methanogens, could grow in the cold, low oxygen atmosphere, especially if they were buried just under the soil surface.

So, microbiologists have joined the search for extraterrestrial life. This seems a valid pursuit because the **extremophiles** found here on Earth survive, and even require, living in extreme environments (TABLE 5.1)—some not so different from Mars (FIGURE 5.1). If life (as we know it) did or does exist on Mars, it almost certainly was or is bacterial/archaeal.

Image courtesy of Dr. Fred Murphy/CDC.

TABLE

5.1 Some Microbial Record Holders

Hottest environment (Juan de Fuca ridge)—121°C: Strain 121 (Archaea)
Coldest environment (Antarctica)—15°C: Cryptoendoliths (Bacteria and lichens)
Highest radiation survival—5MRad, or 5000× what kills humans: *Deinococcus radiodurans* (Bacteria)
Deepest—3.2 km underground: Many bacterial and archaeal species
Most acid environment (Iron Mountain, CA)—pH 0.0 (most life is at least a factor of 100,000 less acidic): *Ferroplasma acidarmanus* (Archaea)
Most alkaline environment (Lake Calumet, IL)—pH 12.8 (most life is at least a factor of 1000 less basic): Proteobacteria (Bacteria)
Longest in space (NASA satellite)-6 years: *Bacillus subtilis* (Bacteria)
High pressure environment (Mariana Trench)—1200 times atmospheric pressure: *Moritella, Shewanella*, and others (Bacteria)
Saltiest environment (Eastern Mediterranean basin)—47% salt (15 times human blood saltiness): Several bacterial and archaeal species

FIGURE 5.1 **The Martian Surface?** This barren-looking landscape is not Mars but the Atacama Desert in Chile. It looks similar to photos taken by the Mars rovers *Spirit* and *Opportunity*. © Photodisc/age fotostock. »» Does this area look like a habitable place for life, even microbial life?

In 2004, NASA sent two spacecrafts to Mars to look for indirect signs of past life. Scientists here on Earth monitored instruments on the Mars rovers, *Spirit* and *Opportunity*, designed to search for signs suggesting water once existed on the planet. Some findings suggest there are areas where salty seas once washed over the plains of Mars, creating a life-friendly environment.

Opportunity found evidence for ancient shores on what once was a sea. Scientists reported in 2008 that a more recent spacecraft, the *Phoenix Mars Lander*, detected water ice near the Martian soil surface.

Now another exploration of the Red Planet is underway, as the Mars rover *Curiosity* is searching to see "who shall dwell in these worlds if they be inhabited . . ."

Whether microorganisms are here on Earth in a moderate or extreme environment, or on Mars, there are certain physical and chemical requirements they must possess to survive, reproduce, and grow. In this chapter, we explore the process of microbial cell reproduction, examine the physical and chemical conditions required for growth, and discover the ways that microbial growth can be measured.

As we have been emphasizing in this text, the domains of organisms may have different structures and patterns, but they carry out many of the same processes.

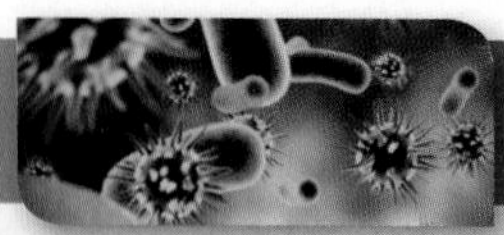

Chapter Challenge

The United States has sent several spacecraft to Mars since the first Viking landers in 1976. Recently, an international team of scientists carried out studies suggesting terrestrial microbes could hitch a ride to Mars on such a craft—and even survive the journey. The team believes most spacecraft that have touched down on Mars were not thoroughly sterilized, so they could have carried living microbes from Earth. NASA scientists have assumed Mars' thin atmosphere, which allows intense ultraviolet (UV) radiation to reach the planet's surface—triple Earth's intensity—would kill any life inadvertently carried on the spacecraft. However, in laboratory tests, the scientists found that some spore-forming bacterial species could survive UV bombardment that was equivalent to that on Mars—if the spores were buried just a few millimeters in the soil. Could such an earthly extremophile survive on Mars or there actually be Martian microbes? Are there conditions here on Earth that might give us an idea? Let's examine what it takes for a bacterial species to survive both "normal" and "extreme" earthly environments.

KEY CONCEPT 5.1 Microbial Reproduction Is Part of the Cell Cycle

Growth in the microbial world usually refers to an increase in the numbers of individuals; that is, an increase in the population size with each cell carrying the identical genetic instructions of the parent cell. **Asexual reproduction** is the process to maintain this genetic constancy. In eukaryotic microbes, an elaborate interaction of microtubules and proteins with pairs of chromosomes in the cell nucleus allows for the precise events of mitosis and cytokinesis. Bacterial and archaeal cells divide without the microtubular involvement but still possess similar protein filaments as part of their cytoskeleton.

Binary Fission Is Part of the Cell Cycle

The series of events involving growth, DNA replication, and cell division is called the **cell cycle**. For a bacterial cell like *E. coli*, the cycle can be divided into three periods (FIGURE 5.2A).

B Period. Before a bacterial cell actually divides into two cells, it goes through a phase of metabolic "growth," called the **B period**, where it increases in cell mass and cell size. During this period, the cell is preparing for chromosome replication, so in most species the chromosome remains a single, circular DNA molecule.

C Period. Once the cell has increased in size, DNA replication occurs. During this **C period**, the bacterial chromosome is copied, which will ensure each daughter cell will have one complete set of genetic information when cell division is completed. Chromosome segregation involves the cell's cytoskeleton but, unlike eukaryotic cells, lacks the mitotic spindle to segregate replicated chromosomes. DNA replication will be examined in a later chapter.

D Period. As DNA replication ends, the cell undergoes **binary fission**. During this **D period**, a partition or septum forms at midcell between segregated chromosomes (FIGURE 5.2B). The splitting of the cell in half is coordinated by cytoskeletal proteins organized into a fission ring apparatus, which ensures that two nearly identical daughter cells are formed. Depending on the growth conditions, the septum material may dissolve at a slow rate allowing pairs, chains, or clusters of cells to form that represent characteristic arrangements for many bacterial species. With other species, like *E. coli*, the cells completely separate from one another after the septum is complete. Each of the attached or free cells then enters another B period.

Reproduction by binary fission seems to confer immortality because there is never a moment at which the first bacterial cell has died. Each mother cell undergoes binary fission to become the two young daughter cells. However, the perception of immortality has been challenged by experiments suggesting bacterial cells do age (MicroFocus 5.1).

Bacterial and Archaeal Cells Can Grow Exponentially

The interval of time between successive binary fissions of a cell or population of cells is known as the **generation time** (or doubling time). Under optimal conditions, some species have a very fast generation time; for others, it is much slower (TABLE 5.2).

Website Animation: Binary Fission

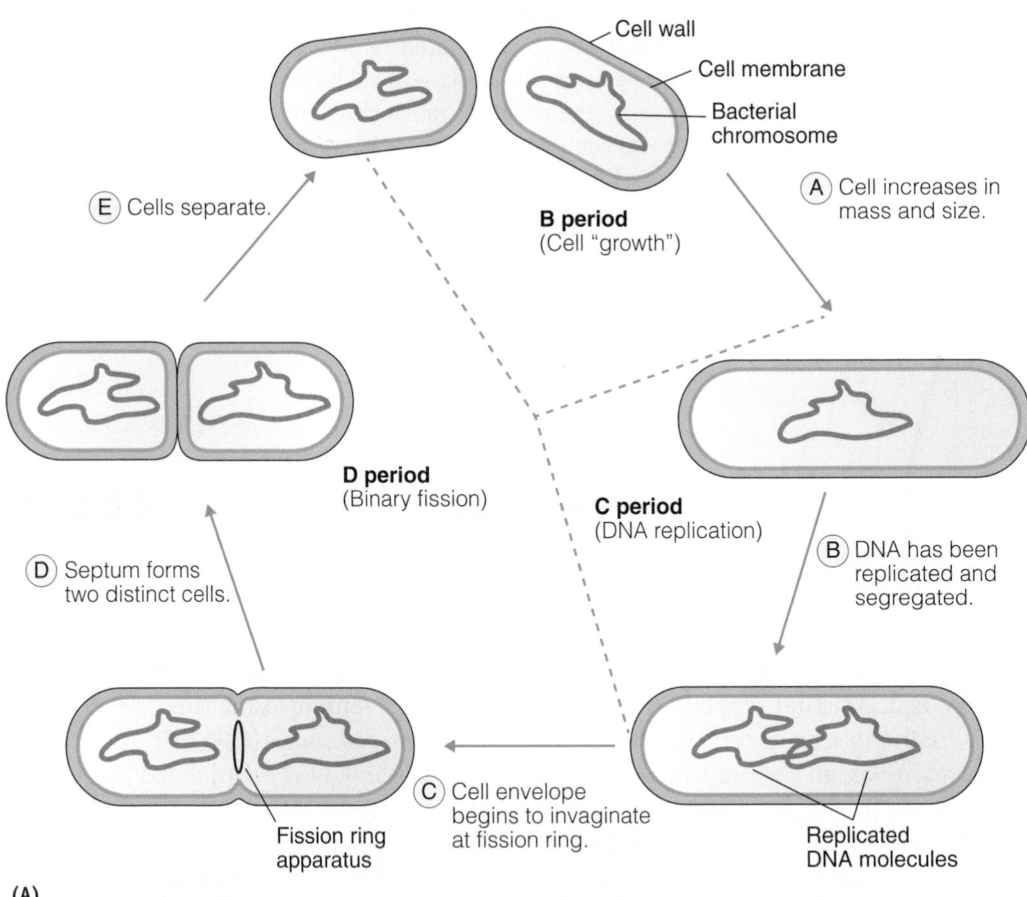

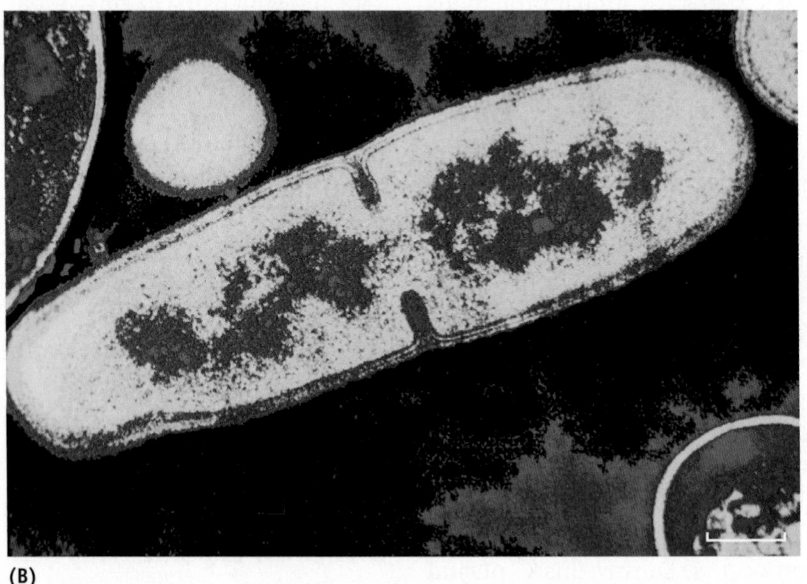

FIGURE 5.2 **The Bacterial Cell Cycle.** (**A**) The bacterial cell cycle can be separated into three stages or periods involved with cell growth (B period), chromosome replication (C period), and binary fission (D period). (**B**) A false-color transmission electron micrograph of a cell of *Bacillus licheniformis* undergoing binary fission. The inward growth of the cell envelope is evident at midcell between the segregated chromosomes. (Bar = 0.25 μm.) © Lee Simon/Photo Researchers, Inc. »» How would binary fission differ for a prokaryotic organism having cells arranged in chains and another that forms single cells?

MICROFOCUS 5.1: Evolution/Environmental Microbiology

A Microbe's Life

It used to be thought that bacterial cells do not age—they are immortal. This might seem obvious considering a mother cell divides at midcell by binary fission to become the two genetically equal daughter cells. However, recent research suggests that although the DNA may be identical, after several generations of binary fission, the population consists of cells of different ages and the oldest ones have the longest generation time.

Eric Stewart and his collaborators at INSERM, the National Institute of Health and Medical Research in Paris, filmed *Escherichia coli* cells as they divided into daughter cells on a specially designed microscope slide. A record of every daughter cell (total of 35,000 individual cells) was recorded for nine generations over a period of 6 hours. Then, a custom-designed computer system analyzed the micrographs.

The group's results suggest that the cells divide asymmetrically, so daughter cells are not morphologically or physiologically symmetrical. Each contains cellular poles of different ages.

When a mother cell divides, each daughter cell inherits one end or pole of the mother cell. The region where the cells split develops into the other pole (see figure). For example, in the first division, the mother cell splits with a new wall (red). When the daughter cells grow in size they contain an old pole (brown) and a new one (red). When each of these cells divides, two have the oldest pole (brown) and youngest pole (green) while the two other cells have a younger pole (red) and a youngest pole (green). So after just two divisions, there are two populations of daughter cells: two have oldest and youngest poles while two have younger poles and youngest poles. According to Stewart's group, the two cells with the oldest/youngest poles grew 2.2 percent slower than the cells with younger/youngest poles.

As more and more binary fissions occur, the difference in age between daughter cells will continue to increase. The bottom line is that cells inheriting older and older poles experience longer generation times, reduced rates of offspring formed, and increased risk of dying compared to cells with younger, newer poles. This loss of fitness is called "senescence." Note: Stewart's group could not follow any cells to actual death because the cell populations eventually had so many cells, even their computer program could not keep them all independently recorded.

Exactly why the older cells senesce is not understood. One idea is that the older cell retains damaged proteins not found in the younger "renewed" daughter cell. So, it appears that although a microbe's life is limited, the lineage of younger cells in each generation maintains immortality.

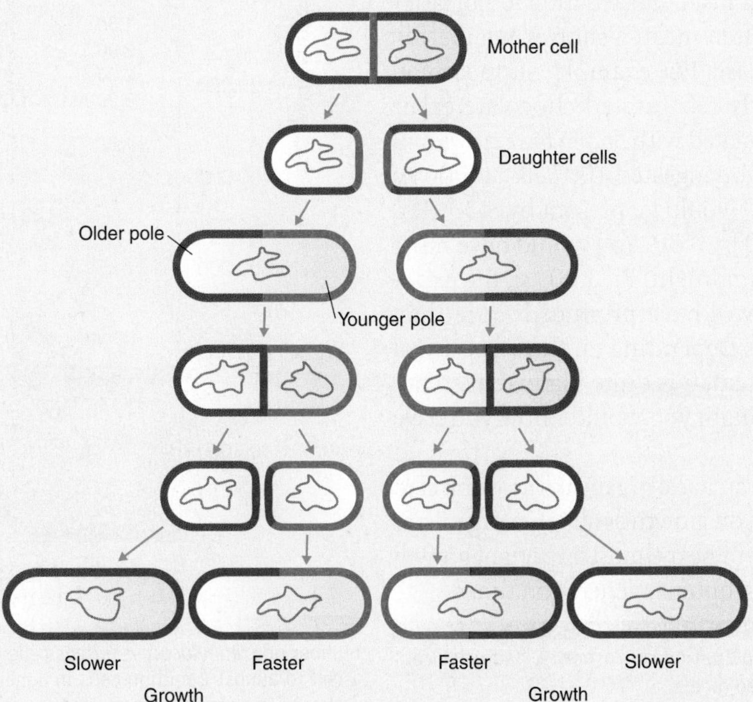

Two successive binary fissions produce daughter cells with various aged poles (brown = oldest; red = younger; green = youngest).

TABLE

5.2 Examples of Generation Times—Common Bacterial Species Growing Under Optimal Conditions[1]

Species	Growth Medium	Generation Time (minutes)
Escherichia coli	Synthetic	17
Bacillus megaterium	Synthetic	25
Staphylococcus aureus	Complex	27–30
Streptococcus lactis	Milk	26
Streptococcus lactis	Complex	48
Lactobacillus acidophilus	Milk	66–87
Mycobacterium tuberculosis	Synthetic	792–932 (= 12–15.5 hours)
Treponema pallidum	Rabbit testes	1,980 (= 33 hours)

[1]Table modified from microblog.me.uk/138

One enterprising mathematician calculated that if *E. coli* binary fissions were to continue at their optimal generation time (15 minutes) for 36 hours, the bacterial cells would cover the surface of the Earth! Thankfully, this will not occur because of the limitation of nutrients and the loss of ideal physical factors required for growth. The majority of the bacterial cells would starve to death or die in their own waste.

The generation time is useful in determining the amount of time that passes before disease symptoms appear in an infected individual; faster division times often mean a shorter **incubation period** for a disease. For example, suppose you eat an improperly refrigerated chocolate eclair that was contaminated with *Staphylococcus aureus* (FIGURE 5.3). If you ingested 100 cells at 8:00 PM this evening, 200 would be present by 8:30, 400 by 9:00, and 800 by 9:30. You would have more than 25,000 by midnight. By 3:00 AM, the exponential growth will have produced more than 1.6 million cells. Depending on the response of the immune system, it is quite likely that sometime during the night you would know you have food poisoning.

■ Incubation period: The time from entry of a pathogen into the body until the first symptoms appear.

Bacterial and archaeal organisms are subject to the same controls on growth as all other organisms on Earth. Let's examine the most important growth factors conferring optimal generation times.

CONCEPT AND REASONING CHECKS 1

a. Propose an explanation as to how a bacterial cell "knows" when to divide.

b. If it takes *E. coli* 7 hours of exponential growth to reach some 2 million cells, how long would it take *T. pallidum* to reach that same number under optimal conditions?

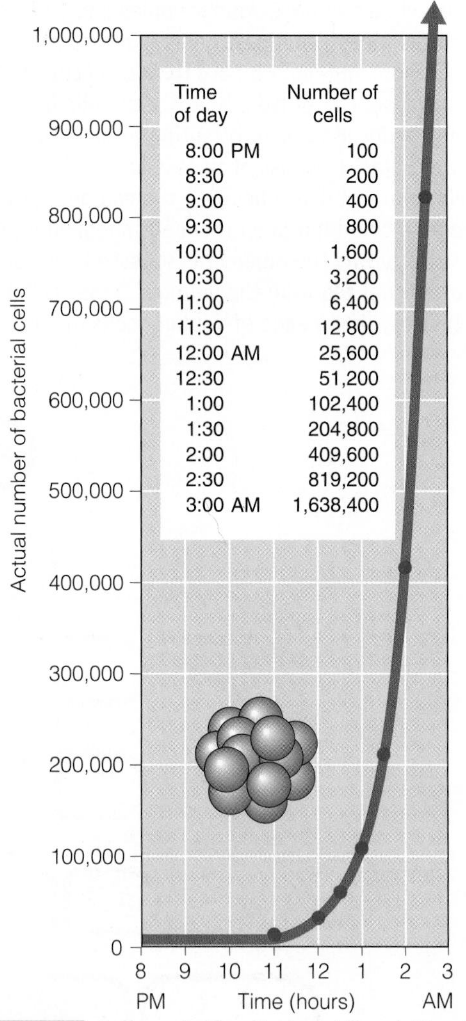

FIGURE 5.3 **A Skyrocketing Bacterial Population.** The number of *Staphylococcus aureus* cells progresses from 1 cell to almost 2 million cells in a mere 7 hours. The J-shaped growth curve gets steeper and steeper as the hours pass. Only a depletion of food, buildup of waste, or some other limitation will halt the progress of the curve. »» What is the generation time for *S. aureus* in this figure?

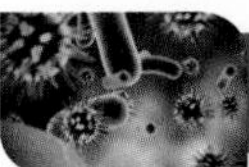

Chapter Challenge A

We have already seen that bacterial and archaeal organisms can survive in some hostile environments—as defined by us "temperate" organisms. And yet these extremophiles not only survive in these environments, they actually can thrive there. That means they need to go through binary fission as part of their cell cycle to maintain their population numbers.

QUESTION A:
Propose how these extremophiles can carry out DNA replication and reproduce in an environment that would otherwise kill a temperate microbe? With that, what are the chances that an earthly extremophile would have been on a spacecraft to Mars? And if it did, would it survive?

Answers can be found on the Student Companion Website in **Appendix D**.

KEY CONCEPT 5.2 Microbial Growth Progresses Through Distinct Phases

In the previous section, we discovered how fast some bacterial cells can grow under ideal circumstances. Let's look at the dynamics of bacterial growth in a little more detail.

A Bacterial Growth Curve Illustrates the Dynamics of Growth

A typical **bacterial growth curve** for a population illustrates the events occurring over time (FIGURE 5.4). If a sample of bacterial cells is transferred to a tube of fresh broth, four distinct phases of growth occur: the lag phase; the logarithmic phase; the stationary phase; and the decline phase.

The Lag Phase. The first portion of the growth curve during which time the bacterial cells are adapting to their new environment and compensating for changes in nutritional conditions is called the **lag phase**. In the broth, some cells may actually die from the shock of transfer or the inability to adapt to the new environment. The actual length of the lag phase (**B period** of the cell cycle) depends on the metabolic activity of the microbial population. They must grow in size, take up nutrients, and replicate their DNA (**C period**)—all in preparation for binary fission (**D period**).

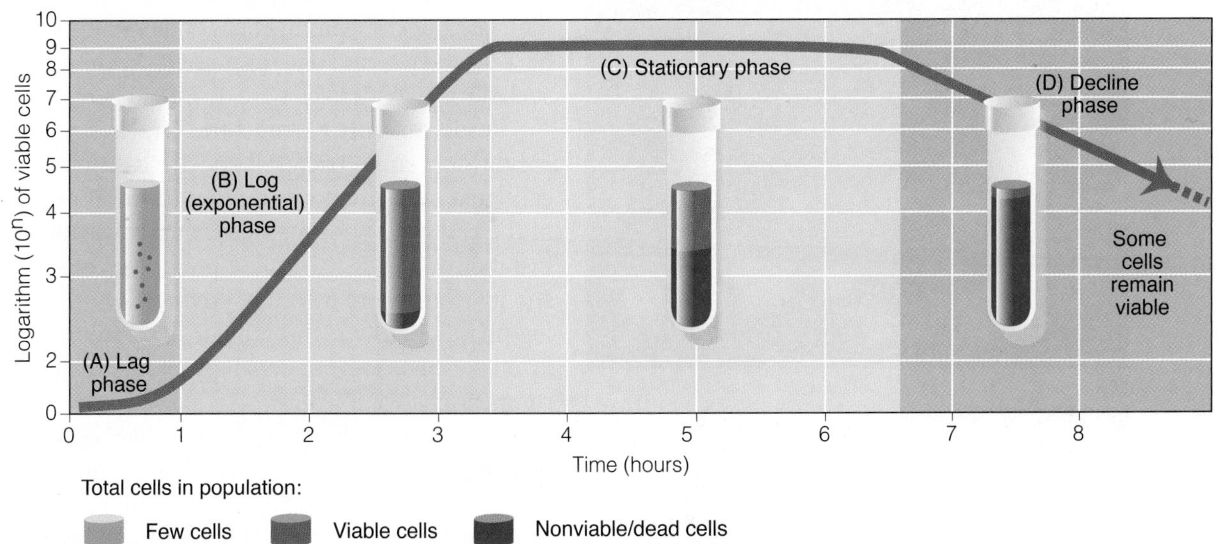

FIGURE 5.4 **The Growth Curve for a Bacterial Population.** (**A**) During the lag phase, the population numbers remain stable as bacterial cells prepare for division. (**B**) During the logarithmic (exponential) phase, the numbers double with each cell cycle. Environmental factors later lead to stationary phase (**C**), which involves a stabilizing population size. (**D**) The decline phase is the period during which cell death becomes substantial. »» Why would antibiotics work best to kill or inhibit cells in the log phase?

The Log Phase. The population now enters an active stage of growth called the **logarithmic (log) phase (exponential)**. In the log phase, all cells are undergoing binary fission and the generation time is dependent on the species and environmental conditions present. The cells exhibit balanced growth because all aspects of metabolism and physiology remain constant. As each generation time passes, the number of cells doubles and the graph rises accordingly on a logarithmic scale.

In a broth tube, the medium becomes cloudy (turbid) due to increasing cell numbers. If plated on solid growth medium, bacterial growth will be so vigorous that visible colonies appear within 24 to 48 hours and each colony may consist of millions of cells (FIGURE 5.5). Vulnerability to antibiotics is also highest at this active stage of growth because many antibiotics affect metabolic processes like protein synthesis in dividing cells.

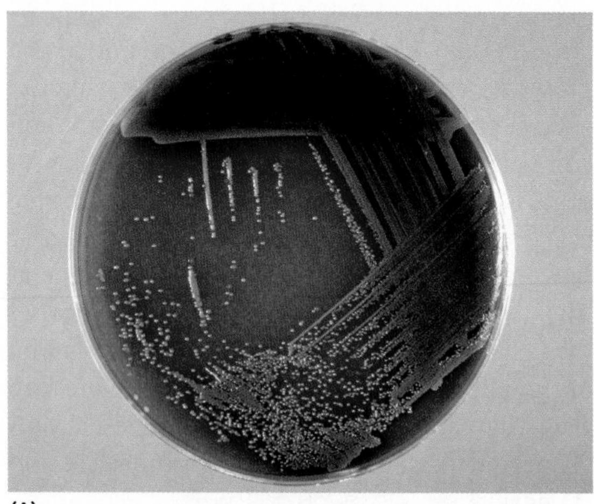

(A)

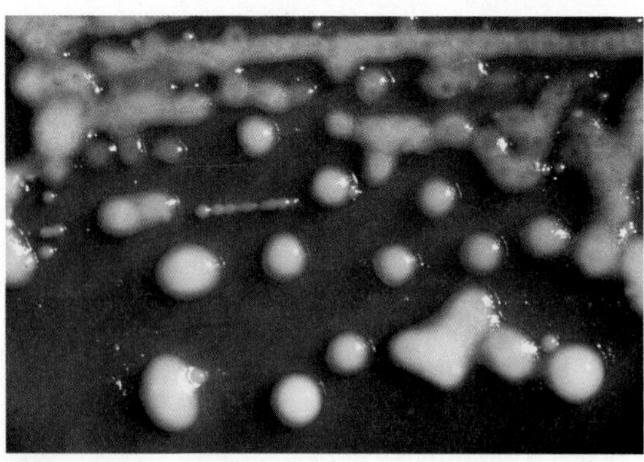

(B)

FIGURE 5.5 **Two Views of Bacterial Colonies.** (**A**) Bacterial colonies cultured on blood agar in a culture dish. Blood agar is a mixture of nutrient agar and blood cells. It is often used for growing bacterial colonies. Courtesy of Dr. J. J. Farmer/CDC. (**B**) Close-up of typhoid bacterial (*Salmonella enterica* serotype Typhi) colonies being cultured on a growth medium. Courtesy of CDC. »» How did each colony in (A) or (B) start?

The Stationary Phase. After some hours in a broth tube, available nutrients become scarce and waste products accumulate. Factors such as oxygen also may be in short supply. This limitation of nutrients and buildup of waste materials leads to a decline in the growth rate. The vigor of the population changes and the cells enter a plateau, called the **stationary phase**. Now the number of viable cells equals the number of nonviable cells.

The Decline Phase. If nutrients in the external environment remain limited or the quantities become exceedingly low, the population enters a **decline phase** (or death phase). Here the number of dying cells far exceeds the number of viable cells. For many species, the history of the population ends with the death of the last cell.

For some bacterial species, the population as a whole can escape death by entering a state of dormancy.

Bacterial Cells Can Exist In Metabolically Inactive States

Environmental conditions vary tremendously and often they might not be favorable to active population growth. Such unfavorable conditions include the presence of toxic chemicals, such as antibiotics, and nutrient limitation (potential starvation). Therefore, microbes must constantly monitor or sense their surroundings to ensure the conditions are capable of supporting continued growth. For bacterial species, if the environment is not favorable, the cells respond by entering a dormant (nondividing) state. When environmental conditions improve, the cells "revive" and once again start dividing. Two such dormancy strategies are described here.

Persister Cell Formation. Under stressful conditions, some bacterial species can produce cells that stop dividing (remain in the B period of the cell cycle) but maintain a very low rate of metabolism. These, so-called **persister cells**, are spontaneously and naturally produced during the log phase of growth and are not the result of unfavorable environmental change. Should the environment change for the worse, the persister cells survive. For example, such "altruistic-like" behavior benefits pathogens like *M. tuberculosis*. If antibiotics are used to treat the infection, the persister cells can "hide out" in the body as a latent infection because the persister cells would not be affected by antibiotics, which target only highly active, log phase

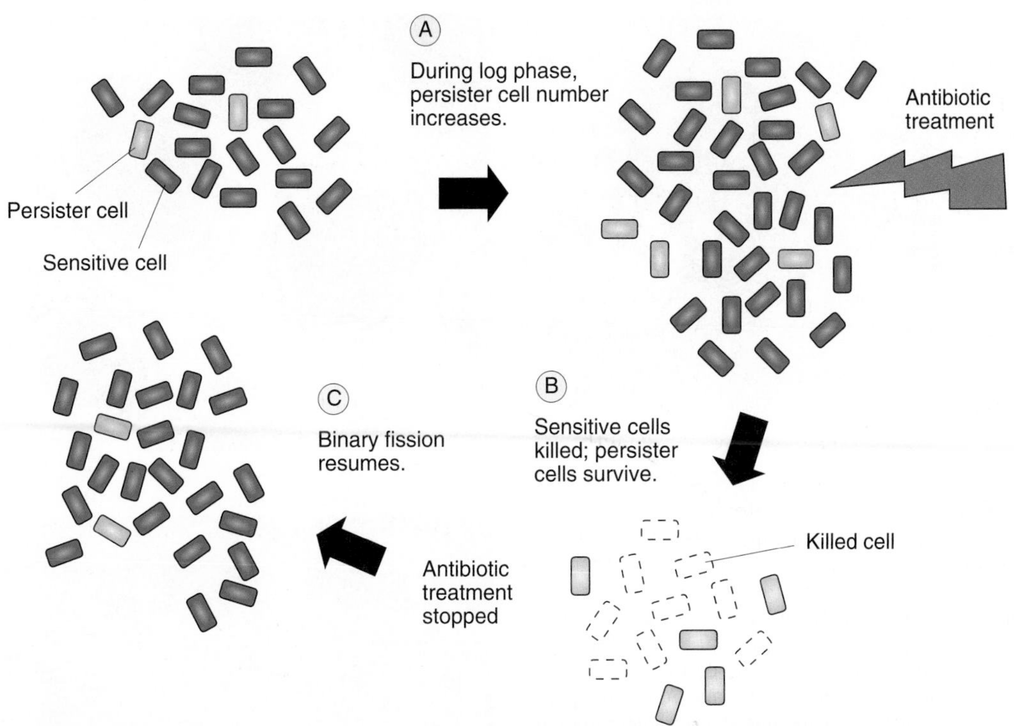

FIGURE 5.6 **Persister Cells.** The presence of persister cells within a bacterial cell population allows the species to survive environmental changes, in this case the presence of an antibiotic. »» What other way could a susceptible bacterial cell population survive in the presence of an antibiotic?

cells (FIGURE 5.6). Although most log phase cells are killed, the persisters survive, often remaining viable for decades in this dormant state. Should the antibiotics be withdrawn at some later date, the persister cells become active again and repopulate the infection.

Endospore Formation. A few gram-positive genera, especially *Bacillus* and *Clostridium,* have a different dormancy scenario that plays out when the cells experience nutrient depletion. Species, such as *Bacillus anthracis* (causative agent of anthrax) and *Clostridium botulinum* (the causative agent of botulism) enter the stationary phase, begin spore formation or **sporulation** that produces dormant structures called **endospores** (FIGURE 5.7).

Unlike persister cells, endospores are the result of a focused morphological differentiation process specifically tied to nutrient limitation (starvation) in the environment and is initiated by **quorum sensing** within the "starving" population.

Spore formation begins when the bacterial chromosome replicates and binary fission is characterized by an asymmetric cell division (FIGURE 5.8). The smaller cell, the prespore, will become the mature endospore, while the larger mother cell will commit itself to maturation of the endospore before rupturing.

Depending on the exact asymmetry of cell division, the single endospore may develop at the end of the mother cell, near the end, or at the center of the cell (the position is useful for species identification purposes).

The highly dehydrated free spore contains cytoplasm, highly compacted DNA, and a large amount of **dipicolinic acid**, a unique organic substance that, linked with calcium ions, helps stabilize the proteins and DNA. Thick layers of peptidoglycan form the cortex and some 70 proteins form the layers of the spore coat to protect the cytoplasmic contents. It should be stressed that sporulation is not a reproductive process. Rather, the endospore represents a "resting stage" produced in response to environmental "hard times."

Endospores are probably the most resistant living structures known. Desiccation has little effect on the spore. By containing little water, endospores also are heat resistant and undergo very few chemical reactions. These properties make them difficult to eliminate from contaminated medical materials and food products. For example, endospores can remain viable in boiling water (100°C) for 2 hours. When placed in 70% ethyl alcohol, endospores have survived for 20 years. Humans can barely withstand 500 rems of radiation, but endospores

■ Rem (Roentgen Equivalent in Man): A measure of radiation dose related to biological effect.

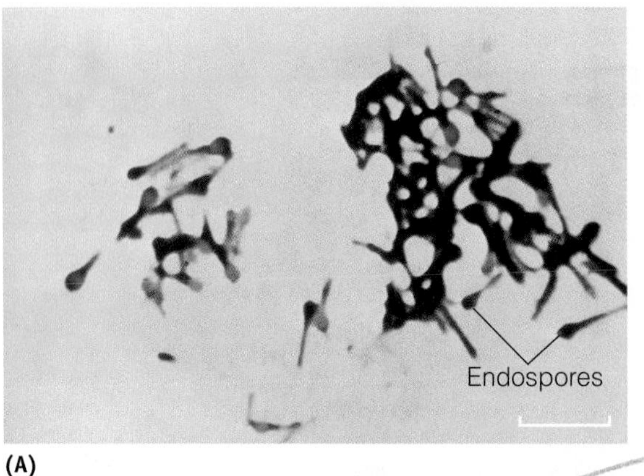

(A)

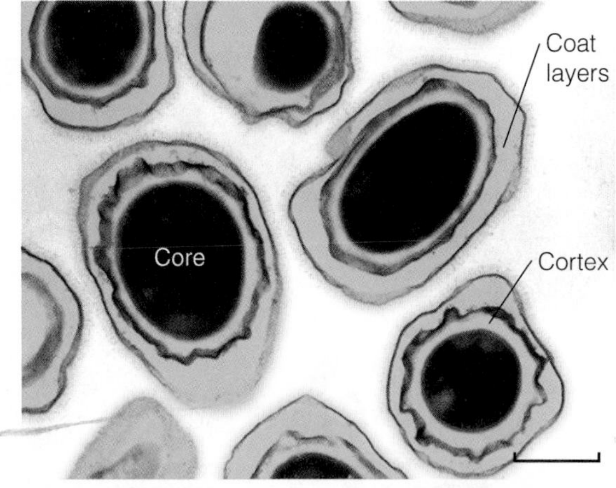

(B)

(C)

FIGURE 5.7 **Three Different Views of Bacterial Endospores.** (**A**) A light microscope image of *Clostridium* cells showing terminal spore formation. Note the characteristic drumstick appearance of the spores. (Bar = 5.0 μm.) Courtesy of CDC. (**B**) A false-color transmission electron micrograph of *Bacillus anthracis* spores. The visible spore structures include the core, cortex, and coat layers. (Bar = 0.5 μm.) © Scott Camazine/Alamy Images. (**C**) A scanning electron microscope view of a germinating spore (arrow). Note that the spore coat divides equatorially along the long axis, and as it separates, the vegetative cell emerges. (Bar = 2.0 μm.) Courtesy of Janice Carr/CDC. »» If an endospore is resistant to so many environmental conditions, how does a spore "know" conditions are favorable for germination?

can survive one million rems. In this dormant condition, endospores can "survive" for extremely long periods of time (MicroFocus 5.2).

When the environment is favorable for cell growth, the spore coat and cortex break down and each endospore rapidly germinates and grows out as a vegetative cell (FIGURE 5.7C).

A few serious diseases in humans are caused by spore formers. The most newsworthy has been *B. anthracis*, the agent of the 2001 anthrax bioterror attack through the mail. This potentially deadly disease, originally studied by Koch and Pasteur, develops when inhaled spores germinate in the lower respiratory tract and the resulting vegetative cells secrete two deadly toxins. Botulism, gas gangrene, and tetanus are diseases caused by different species of *Clostridium*. Clostridial endospores often are found in soil, as well as in human and animal intestines. However, the environment must be free of oxygen for the spores to germinate to vegetative cells. Dead tissue in a wound provides such an environment for the development of tetanus and gas gangrene, and a vacuum-sealed can of food is suitable for the development of botulism.

Killing endospores can be a tough task. Heating them for many hours under high pressure will do the trick. If they contaminate machinery, such as they did in mail sorting equipment in the 2001 anthrax attacks, there are potent but highly dangerous chemical methods to kill the spores.

Optimal Microbial Growth Is Dependent on Several Physical Factors

Now that we have examined the bacterial cell cycle and dormancy, let's examine the essential physical and chemical factors influencing cell growth.

Temperature. Temperature is one of the most important factors governing growth. Every microbial species has an optimal growth temperature and an approximate 30°C operating range, from minimum to maximum, over which the cells will grow albeit with a slower generation time (FIGURE 5.9). In general, most microbes can be assigned to one of three groups—psychrophiles, mesophiles, or thermophiles—based on their optimal growth temperature as well as their minimal and maximal growth temperatures.

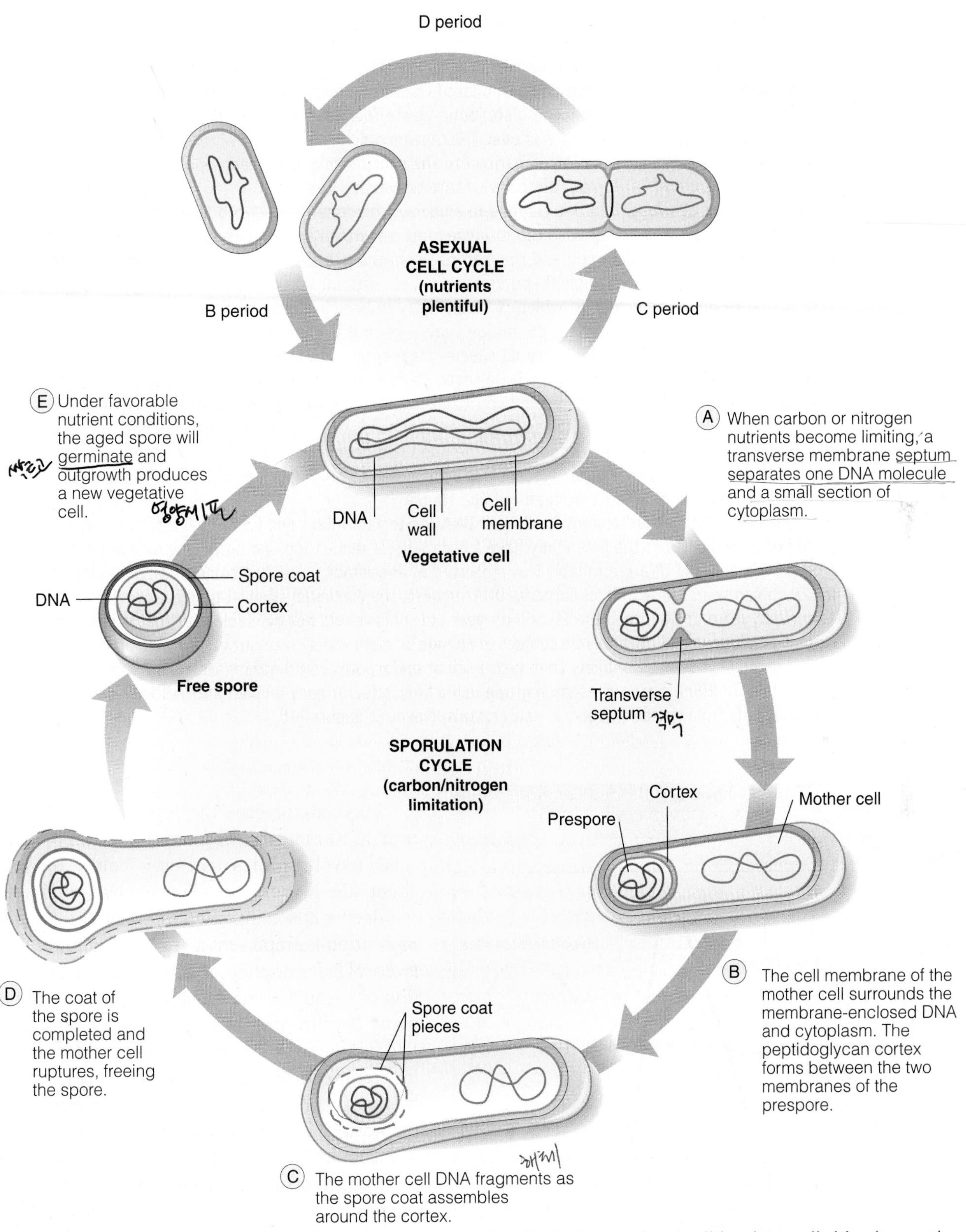

FIGURE 5.8 **The Formation of a Bacterial Spore by *Bacillus subtilis*.** **(A–E)** When nutrient conditions become limiting (e.g., carbon, nitrogen), endospore formers, such as *B. subtilis*, enter the sporulation cycle shown here. »» Hypothesize how a vegetative cell "knows" nutrient conditions are limiting.

MICROFOCUS 5.2: Being Skeptical

Germination of 25 Million-Year-Old Endospores?

Endospores have been discovered in various archaeological sites and environments around the world. Living spores have been recovered and germinated from the intestines of Egyptian mummies several thousand years old. In 1983, archaeologists found viable *Thermoactinomyces* spores in sediment lining Minnesota's Elk Lake. The sediment was over 7,500 years old.

All these reports though pale in comparison to the controversial discovery reported in 1995 by researcher Raul Cano of California Polytechnic State University, San Luis Obispo. Cano found bacterial spores in the gut of a fossilized bee trapped in amber—a hardened resin—produced from a tree in the Dominican Republic. The amber with the fossilized bee was dated as being about 25 million years old. When the amber was cracked open and the material from the gut of the bee extracted and placed in nutrient medium, the equally ancient spores germinated. With microscopy, the cells from a colony were very similar to *Bacillus sphaericus,* which is found today in bees in the Dominican Republic. Is it possible for an endospore to survive for 25 million years—even if it is encased in amber?

Critics were quick to claim the bacterial species may represent a modern-day species that contaminated the amber sample being examined. However, Professor Cano had carried out appropriate and rigorous decontamination procedures and sterilized the amber sample before cracking it open. He also carried out all the procedures in a class II laminar flow hood, which prevents outside contamination from entering the working area. In addition, the hood had never been used for any other bacterial extraction processes. Several other precautions were added to eliminate any chance that the spores were modern-day contaminants from an outside source.

The major question that remains is whether DNA can remain intact and functional after so long a period of dormancy. Does the DNA really have a capability of replication and producing new vegetative growth? Granted, the DNA presumably was protected in a resistant spore, but could DNA remain intact for 25 million years? Research on bacterial DNA suggests the maximum survival time is about 400,000 to 1.5 million years. If true, then the 25 million-year-old spores could not be viable. But that is based on current predictions and they may be subject to change as more research is carried out with ancient DNA.

The verdict? It seemed unlikely that such ancient endospores could germinate after 25 million years. Then, in 2000, another research group using Cano's techniques revived 250-million-year-old bacterial cells from spores trapped in salt crystals. Maybe it is possible.

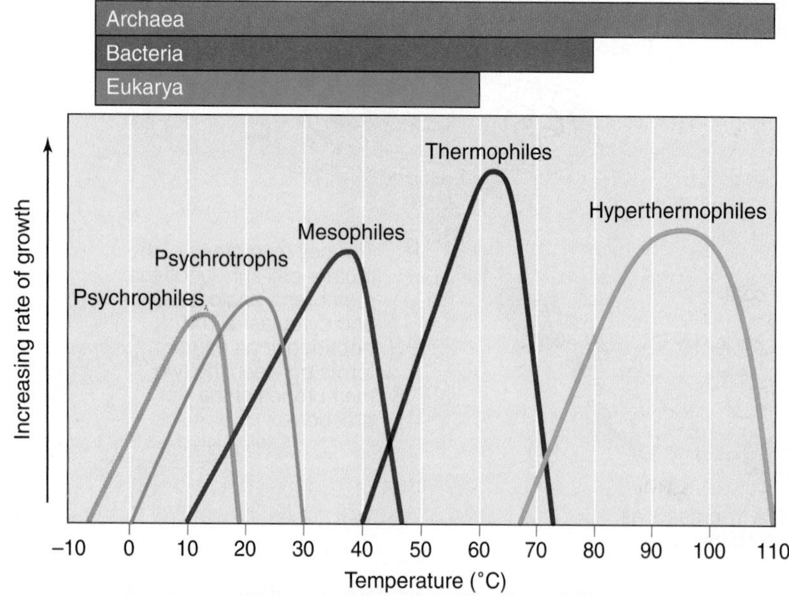

FIGURE 5.9 Growth Rates for Different Microorganisms in Response to Temperature. Temperature optima and ranges define the growth rates for Bacteria, Archaea, and Eukarya. Notice that the growth rates decline quite rapidly to either side of the optimal growth temperature. »» Propose what adaptations are needed for microbes to survive at the psychrophilic or thermophilic extremes.

Microbes that have their optimal growth rates near 15°C but can still grow at 0°C to 20°C are called **psychrophiles** (*psychro* = "cold"). Because about 70% of the Earth is covered by oceans having deep water temperatures below 5°C, many psychrophiles represent a group of bacterial and archaeal extremophiles that make up a large portion of the global microbial community. In fact, many psychrophiles can grow as fast at 4°C as *E. coli* does at 37°C. On the other hand, at these low temperatures, psychrophiles could not be human pathogens because they cannot grow at the warmer 37°C body temperature.

Another group of "cold-loving" microbes are the **psychrotrophs** or **psychrotolerant** microorganisms. These species have a higher optimal growth temperature (25°C) as well as a higher minimal and maximal growth temperature. Psychrotrophs can be found in water and soil in temperate regions of the world but are perhaps most commonly encountered on spoiled refrigerated

Textbook CASE 5

An Outbreak of Food Poisoning Caused by *Campylobacter jejuni*

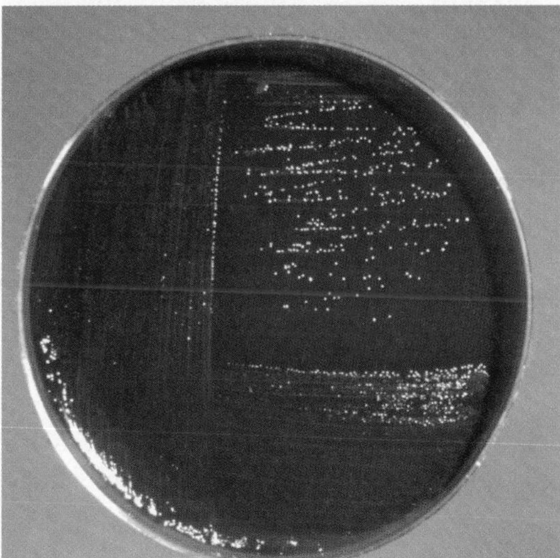

Campylobacter jejuni agar culture. Courtesy of Sheila Mitchell/CDC.

1. On August 15, a cook began his day by cutting up raw chickens to be roasted for dinner.
2. He also cut up lettuce, tomatoes, cucumbers, and other salad ingredients on the same countertop. The countertop surface where he worked was unusually small.
3. For lunch that day, the cook prepared sandwiches on the same countertop. Most were garnished with lettuce.
4. Restaurant patrons enjoyed sandwiches for lunch and roasted chicken for dinner. Many patrons also had a portion of salad with their meal.
5. During the next 3 days, 14 people experienced stomach cramps, nausea, and vomiting.
6. Public health officials learned that all the affected patrons had eaten salad with lunch or dinner. *Campylobacter jejuni*, a bacterial pathogen of the intestines, was isolated from their stools (see figure).

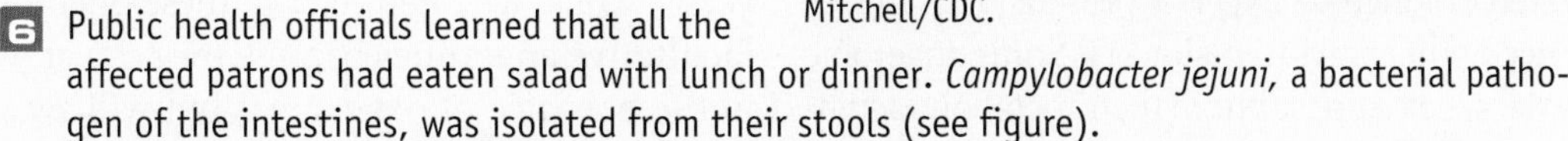

7. On inspection, microbiologists concluded that the chicken was probably contaminated with *C. jejuni*. However, the microbiologists suspected that the cooked chicken was not the cause of the illness. Rather, *C. jejuni* from the raw chicken was the source.

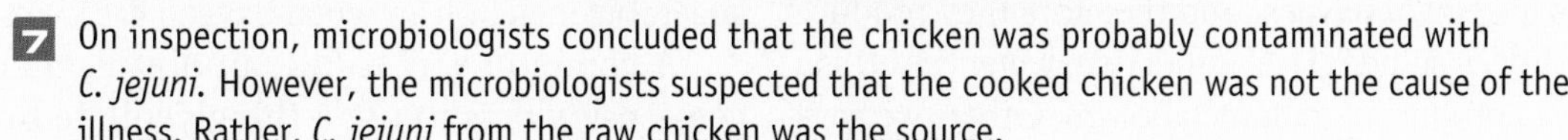

Questions:

(Answers can be found on the Student Companion Website in **Appendix D.**)

A. Why would the cooked chicken not be the source for the illness?

B. Why was the raw chicken identified as the source?

C. How, in fact, did the patrons become ill?

For additional information see www.cdc.gov/mmwr/preview/mmwrhtml/00051427.htm.

foods (4°C). Some bacterial and archaeal species are psychrotolerant as are several microbial eukaryotic species, including fungi (molds). When such contaminated foods are consumed without heating, the toxins may cause food poisoning. One example is *Campylobacter*, the most frequently identified cause of infective diarrhea (Textbook Case 5).

At the opposite extreme are the **thermophiles** (*thermo* = "heat") that multiply best at temperatures around 60°C but still multiply from 40°C to 70°C. Thermophiles are present in compost heaps and hot springs, and can be contaminants in dairy products because they survive pasteurization temperatures. However, thermophiles pose little threat to human health because they do not grow well at the cooler temperature of the body.

There also are many archaeal species that grow optimally at temperatures that exceed 80°C and may have optima near 95°C. These **hyperthermophiles** have been isolated from seawater brought up from hot-water vents along rifts on the floor of the Pacific Ocean. Because the high pressure keeps the water from boiling, some archaeal species can grow at an astonishing 121°C (Table 5.1).

Most of the best-characterized microbial species are **mesophiles** (*meso* = "middle"), which thrive at the middle temperature range of 10°C to 45°C. This includes the pathogens that grow in warm-blooded animals, including humans, as well as those species found in aquatic and soil environments in temperate and tropical regions of the world. *E. coli* is a typical mesophile.

Oxygen. The growth of many microbes depends on a plentiful supply of oxygen gas, and in this respect, such **obligate aerobes** must use the gas as a final electron acceptor to make cellular energy (ATP). Other species, such as *Treponema pallidum*, the agent of syphilis, are termed **microaerophiles** because they survive in environments where the concentration of oxygen is relatively low. In the body, certain microaerophiles cause disease of the oral cavity, urinary tract, and gastrointestinal tract. Conditions can be established in the laboratory to study these microbes (FIGURE 5.10A).

The **anaerobes**, by contrast, are microbes that do not or cannot use oxygen. Some are **aerotolerant**, meaning they are insensitive to oxygen. Many bacterial and archaeal species, as well as a few fungal and protistan species, are **obligate anaerobes**, which are inhibited or killed if oxygen is present. This means they need other ways to make ATP. Some anaerobic bacterial species use sulfur in their metabolic activities instead of oxygen, and therefore they produce hydrogen sulfide (H_2S) rather than water (H_2O) as a waste product of their metabolism. Others we have already encountered, such as the ruminant archaeal organisms that produce methane as the byproduct of the energy conversions. In fact, life originated on Earth in an anaerobic environment consisting of methane and other gases (MicroFocus 5.3).

Some anaerobic bacterial species cause disease in humans. For example, the *Clostridium* species that cause tetanus and gas gangrene multiply in the dead, anaerobic tissue of a wound and produce toxins causing tissue damage. Another species of *Clostridium* multiplies in the oxygen-free environment of a vacuum-sealed can of food, where it produces the lethal toxin of botulism.

Among the most widely used methods to establish anaerobic conditions in the laboratory is the GasPak system, in which hydrogen reacts with oxygen in the presence of a catalyst to form water, thereby creating an oxygen-free atmosphere (FIGURE 5.10B, C).

Many microbes are neither strictly aerobic nor anaerobic but rather can grow in either the presence or a reduced concentration of oxygen. This group includes *E. coli,* many staphylococci and streptococci, members of the genus *Bacillus,* and the fungal yeasts. Thus, we often refer to these organisms as **facultative anaerobes** because they can grow best in the presence of oxygen gas but will switch to anaerobic metabolism when oxygen gas is absent.

A common way to test an organism's oxygen sensitivity is to use a **thioglycollate broth**, which binds free oxygen so that only fresh oxygen entering at the top of the tube would be available (FIGURE 5.11).

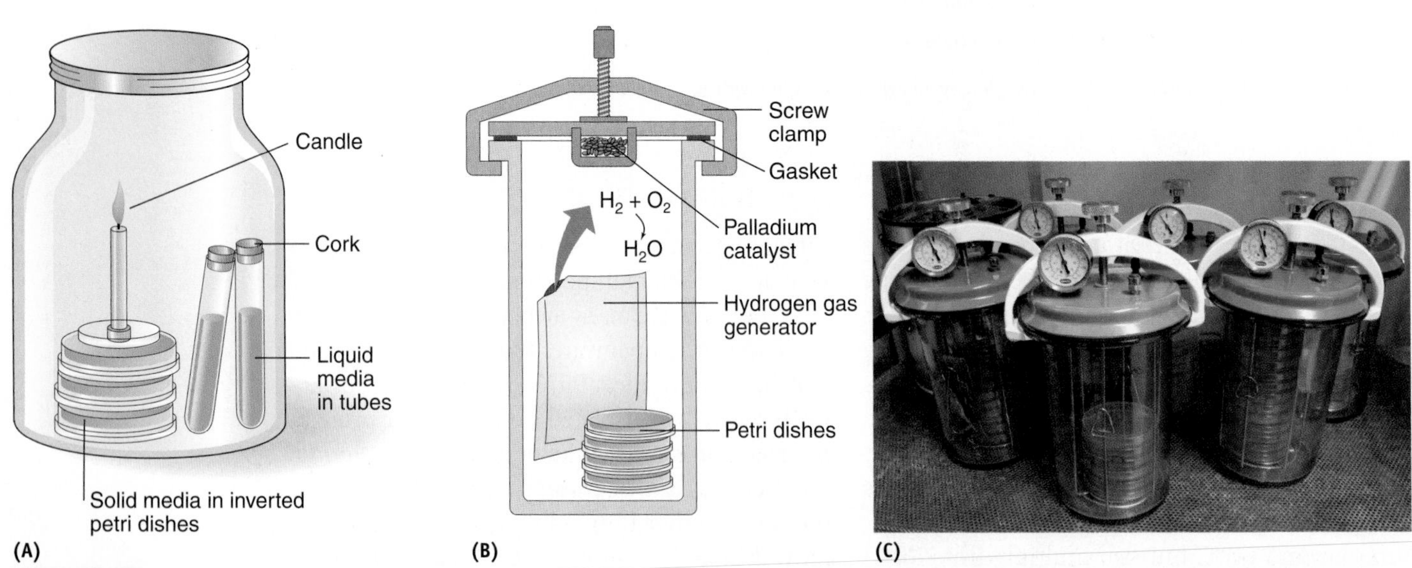

FIGURE 5.10 **Bacterial Cultivation in Different Gas Environments.** Two types of cultivation methods are shown for bacterial species that grow poorly in an oxygen-rich environment. **(A)** A candle jar, in which microaerophilic bacterial species grow in an atmosphere where the oxygen is reduced by the burning candle. **(B, C)** An anaerobic jar, in which hydrogen is released from a generator and then combines with oxygen through a palladium catalyst to form water and create an anaerobic environment. **(C)** © Scott Coutts/Alamy Images. »» In which jar would a facultative anaerobe grow?

MICROFOCUS 5.3: Evolution

"It's Not Toxic to Us!"

It's hard to think of oxygen as a poisonous gas considering how many organisms need it to survive. Yet billions of years ago, oxygen was extremely toxic. One whiff by an organism and a cascade of highly destructive oxidation reactions was set into motion. Death followed quickly.

Difficult to believe? Not if you realize that the ancient Bacteria and Archaea relied on anaerobic chemistry for their energy needs. The atmosphere was full of methane and other gases that they could use to generate energy. But no oxygen. And it was that way for hundreds of millions of years.

Then, some 2.5 billion years ago, along came the cyanobacteria with their ability to perform photosynthesis. Chlorophyll and chlorophyll-like pigments evolved, and the bacterial cells could now trap radiant energy from the sun and convert it to chemical energy in carbohydrates. But there was a downside: oxygen was a waste product of the process—and it was deadly because the oxygen radicals (O_2^-, OH•) produced could disrupt cellular metabolism by "tearing away" electrons from other molecules.

Unable to cope with the toxic conditions, enormous numbers of microbial species became extinct. Others "escaped" to oxygen-free environments, e.g., lake sediments, deep ocean depths that are still in existence today. The cyanobacteria survived because they evolved the enzymes to safely tuck away oxygen atoms in a nontoxic form—that form was water.

One of the survivors of these first communities were gigantic, shallow-water colonies called "stromatolites." In fact, these rock-like looking structures are still found in a few places on Earth, such as Shark Bay off the western coast of Australia (see figure). These structures formed from ocean sediments and calcium carbonate that became trapped in the microbial community as a biofilm, building a rock-like fortress. The top few inches in the crown of a stromatolite contain the oxygen-evolving, photosynthetic cyanobacteria, while below these species are other bacterial species that can tolerate oxygen and sunlight. Buried beneath these organisms are other bacterial species that survive the anaerobic, dark niche of the stromatolite where neither oxygen nor sunlight can reach. A couple of billion years would pass before one particularly well-known species of oxygen-breathing creature evolved: *Homo sapiens*.

Stromatolites, Shark Bay, Western Australia. © Jon Nightingale/ShutterStock, Inc.

Finally, there are bacterial species said to be **capnophilic** (*capno* = "smoke"); they require an atmosphere low in oxygen but rich in carbon dioxide gas. Members of the genera *Neisseria* and *Streptococcus* are capnophiles.

pH. The cytoplasm of most microorganisms has a pH near 7.0. This means that the majority of species are **neutrophiles**, growing optimally at neutral pH and having a growth pH range that covers three pH units (1,000-fold change in H^+ concentration). However, some bacterial species, such as *Vibrio cholerae*, can tolerate acidic conditions as low as pH 2.0 and alkaline conditions as high as pH 9.5.

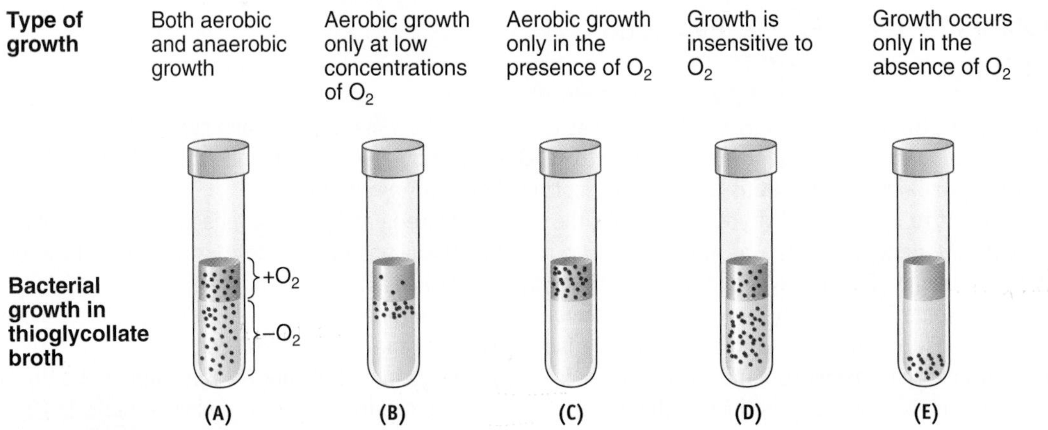

Type of growth	Both aerobic and anaerobic growth	Aerobic growth only at low concentrations of O_2	Aerobic growth only in the presence of O_2	Growth is insensitive to O_2	Growth occurs only in the absence of O_2
Bacterial growth in thioglycollate broth	(A) $+O_2$ / $-O_2$	(B)	(C)	(D)	(E)

FIGURE 5.11 The Effect of Oxygen on Microbial Growth. Each tube contains a thioglycollate broth into which was inoculated a different bacterial species. »» Identify the O_2 requirement in each thioglycollate tube based on the growth density [example: (A) represents facultative anaerobe species].

Acid-tolerant bacteria called **acidophiles** grow best at pHs below 5 and are valuable in the food and dairy industries. For example, certain species of *Lactobacillus* and *Streptococcus* produce the acid that converts milk to buttermilk and cream to sour cream. These species pose no threat to good health even when consumed in large amounts. The "active cultures" in a cup of yogurt are actually acidophilic bacterial species. **Extreme acidophiles**, preferring pHs of 1–2, are found among the Archaea. At the opposite extreme are the **alkaliphiles** that grow best at pHs above 9.

The majority of known bacterial species, however, do not grow well under acidic conditions. Thus, the acidic environment of the stomach helps deter disease, while providing a natural barrier to the organs beyond. In addition, you may have noted certain acidic foods such as lemons, oranges, and other citrus fruits as well as tomatoes and many vegetables are hardly ever contaminated by bacterial growth. However, such damaged produce may be subject to fungal growth because many fungi grow well at a pH of 5 or lower.

Hydrostatic and Osmotic Pressure. Further environmental factors can influence the growth of microbial cells. Psychrophiles in deep ocean waters and sediments are under extremely high hydrostatic pressure. In some deep marine trenches the hydrostatic pressure is tremendous—as high as 16,000 pounds per square inch (psi). Some extremophiles may be the only organisms adapted to the pressure. Such **barophiles** in fact will die quite quickly at normal atmospheric pressures (14.7 psi).

■ Hydrostatic pressure: The pressure exerted by the weight of water.

Other microbes have adapted to saline or highly saline environments such as the Great Salt Lake in Utah, the Dead Sea, and in evaporation ponds. These **halophiles** (*halo* = "salt") are characterized by their need for hypersaline conditions for growth. They include species, such as *Vibrio cholerae,* that grow optimally at 2% to 5% NaCl and other genera that grow optimally at 5% to 20% NaCl. **Extreme halophiles**, like *Halobacterium salinarium* and the eukaryotic green alga *Dunaliella salina,* grow optimally at 20% to 30% NaCl. In contrast, species like *E. coli* are **nonhalophiles** because they grow optimally at less than 2% NaCl and genera, such as *Staphylococcus*, are **halotolerant** because they can grow in slightly saline as well as nonsaline environments.

The hypersaline conditions can be deadly to microbial cells because water will be lost to the hypertonic external medium. To prevent the loss of cellular water, halophiles accumulate high cytoplasmic concentrations of solutes, such as sugars and amino acids, to balance out the osmotic difference. A major exception is for organisms like *H. salinarium,* which accumulate cytoplasmic KCl equal to the external concentration of NaCl.

The fact that microbial life can survive almost anywhere is illustrated in MicroFocus 5.4. FIGURE 5.12 can be used to summarize the physical factors influencing microbial growth.

CONCEPT AND REASONING CHECKS 2

a. In a broth tube, describe the status of the bacterial cell population in each phase of the growth curve.
b. Explain how the trigger to persister cell survival is different from the trigger to endospore formation.
c. Identify what would be extremophile-type conditions for each of the physical factors described in this section.

MICROFOCUS 5.4: Environmental Microbiology

Drilling for Microbes

There are some 140 known subglacial, freshwater lakes beneath the Antarctic ice sheet. Some of these lakes have been sealed off from the outside world for at least 15 to 20 million years. The largest of these lakes so far discovered is Lake Vostok that lies at the depth of more than 3 kilometers below Vostok Station, a Russian research outpost some 1,300 kilometers southeast of the South Pole in the central part of the continent. Many scientists believe that Lake Vostok has been isolated from the atmosphere for 15 to 25 million years.

Lake Vostok was discovered by seismic soundings and radar surveys in the 1960s and 1970s. It is estimated to be 260 kilometers long and 48 kilometers across at its widest point, similar in area to Lake Ontario. The water, thought to be near −4°C, is kept from freezing by the tremendous ice pressure and because the solid block of ice above it acts like a blanket, keeping in heat generated by geothermal energy from beneath the lake. Most interesting from a microbiological perspective is that the presence of liquid water means there is the possibility of microbial life existing within the ancient waters. In fact, Lake Vostok may contain unknown species of microbes (extremophiles, of course) because previous ice samples from meters above the lake were found to contain bacterial organisms (see figure).

But how do you discover if such life is there? The answer is to drill down through the ice block in such a fashion that the drilling apparatus (a five-inch bore hole) and surrounding materials do not contaminate the lake. Thus, such an undertaking is a tremendous challenge in the coldest spot in the world, where the winter surface temperature has been recorded as low as −89°C.

For more than two decades and 57 Antarctic expeditions, Russian scientists have been carefully drilling down through the ice toward the lake. And then on Sunday, February 5, 2012, just a day before the expedition would end its season, it was announced they had hit the lake! However, when the drill contacted the lake, the lake water, being under such great pressure from the ice above, shot up the bore hole and froze. At this writing, the scientists don't know what the frozen water contains because the scientists had to quickly leave the research station to avoid the coming harsh winter. They plan to return in the next Antarctic summer of 2013 to collect and analyze the samples.

If microbes exist in samples from this pristine body of water, microbiologists believe the microbial community could provide a unique insight into how life on Earth evolved in the distant past. In addition, NASA scientists hope the analysis of microbial survival under these extreme conditions might suggest whether life could exist under similar conditions thought to be found under the ice crust on Mars, Jupiter's moon Europa, and Saturn's moon Enceladus. "In the simplest sense, it can transform the way we think about life," NASA's chief scientist Waleed Abdalati said. Robin Bell, a glaciologist with the Lamont-Doherty Earth Observatory of Columbia University said, "It's like exploring another planet, except this one is ours."

Perhaps, by the time you read this MicroFocus, the Russians will have collected those frozen samples and we will have an answer to whether microbes are present in the lake. I bet they do, because, as you now know, microbes seem to always find a way to survive and thrive in seemingly extreme places.

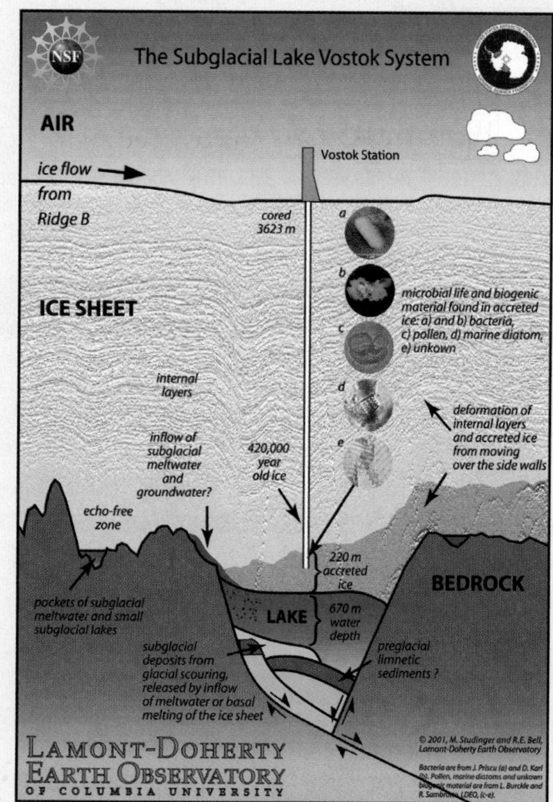

Courtesy of Michael Studinger/Lamont-Doherty Earth Observatory.

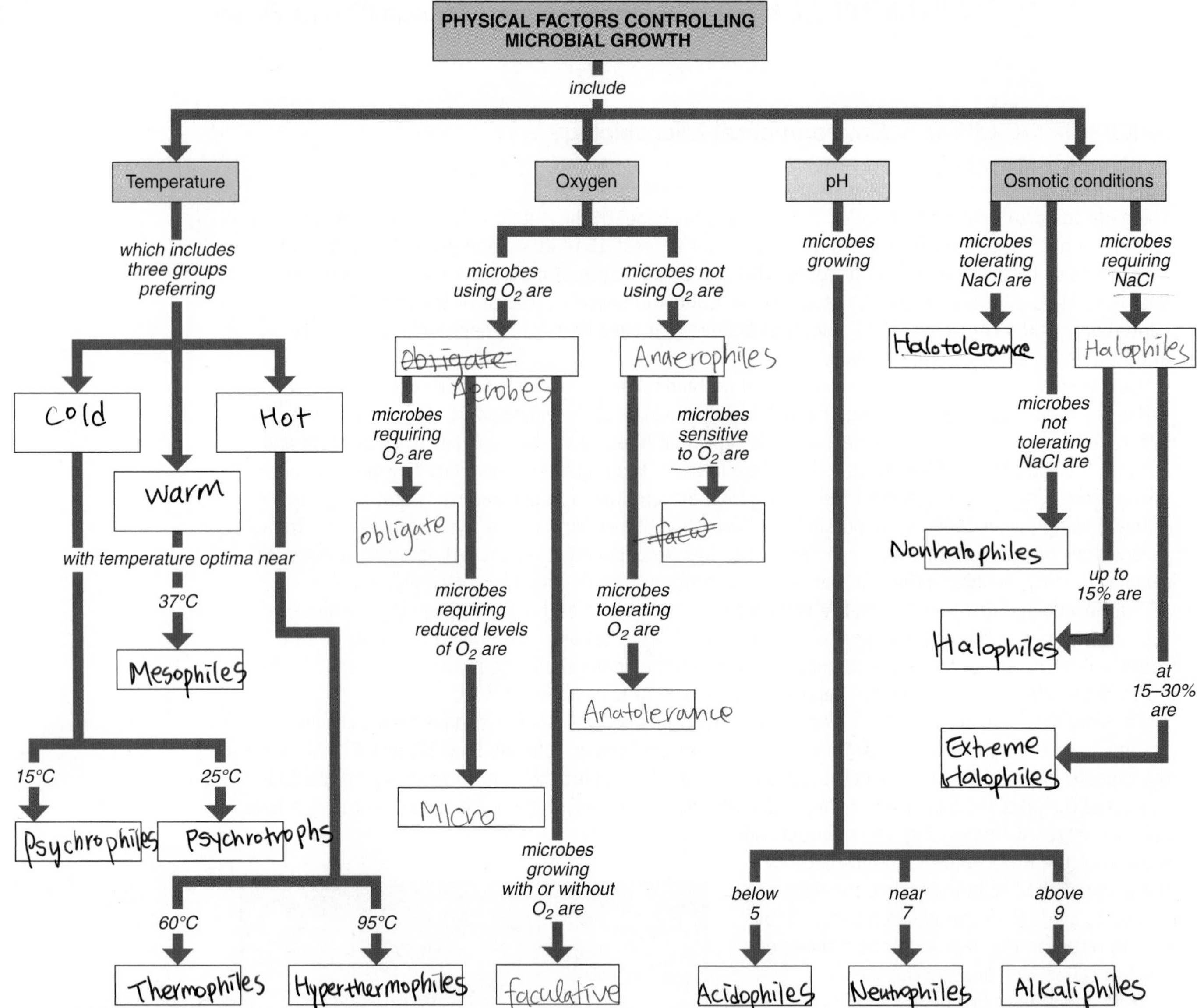

FIGURE 5.12 **Types of Microbes Based on Physical Factors.** This incomplete concept map shows the four major physical factors controlling microbial growth. »» Fill in each of the empty boxes with the correct term for that type of microbe.

Chapter Challenge B

You have now studied many of the major physical factors (temperature, oxygen, pH, and salt) that influence microbial growth and reproduction. Here on Earth, there are many microbes that can survive these physical extremes (see Table 5.1).

QUESTION B:

Could a microbe survive the physical extremes on Mars? What kind of earthly microbe might this Martian microbe be like? **Here are some useful facts about Mars.**

Surface temperature: Estimated to be from a warm 27°C (81°F) to –143°C (–225°F) at the winter polar caps.

Atmosphere: 95% carbon dioxide, but also present are nitrogen (2.7%), argon (1.6%), and oxygen (0.13%; 21% on Earth) gases; mean surface pressure is much lower than Earth's. UV radiation is 3× that on Earth (note: there are earthly bacterial species that can survive 5,000× the dose that would kill human cells).

Soil: Existence of water ice confirmed; soil pH = about 7.7; several chemicals found that could serve as nutrients for life forms, including magnesium, potassium, and chloride.

Salt: Dark, finger-like features could be the flow of salty water perhaps equivalent in salinity to Earth's oceans.

Answers can be found on the Student Companion Website in **Appendix D**.

KEY CONCEPT 5.3 Culture Media Are Used to Grow Microbes and Measure Their Growth

Microorganisms can be grown (cultured) in a liquid medium (*pl.* media) or on a solid medium. Liquid media, called **broths**, are contained in tubes and consist of the growth nutrients dissolved in water. After sterilization and specimen inoculation, the broth will eventually become **turbid** or cloudy as the microbial population grows in the tube. Solid media consist of the liquid media to which has been added a solidifying agent called **agar**. Agar, a polysaccharide derived from red seaweed, contains no nutrients and, like gelatin, melts when heated and then solidifies when cooled. After sterilization of the nutrient agar medium, it is poured into a Petri dish or tube where it will solidify on cooling. After specimen inoculation, the specimen will grow as colonies (bacterial organisms and yeasts) or as filaments (molds).

Culture Media Are of Two Basic Types

Since the time of Pasteur and Koch, microbiologists have been growing bacterial and other microbial species in artificial media; that is, in nutrients designed to mimic the natural environment. Today, different culture media recipes are used in the clinical microbiology lab (CML) and research lab because species vary in their nutritional requirements and no single medium will grow all microorganisms.

For the isolation and identification of most microorganisms, two types of culture media are commonly used. A chemically undefined medium, called a **complex medium**, contains nutrients in which the exact components or their quantity is not completely known. Such media typically contain animal or plant digests (e.g., beef extract, soybean extract) or yeast extracts of an undefined nature (TABLE 5.3A). Complex media are commonly used in the teaching laboratory because most of the organisms used will grow in these nutrient rich conditions.

The second type of medium is a **chemically defined medium**. In this medium, the precise chemical composition and amount of all components are known (TABLE 5.3B). Chemically defined media are used when trying to determine an organism's specific growth requirements.

Culture Media Can Be Modified to Select for or Differentiate Between Microbial Species

In the CML, the basic ingredients of growth media can be modified in one of three ways to provide fast and critical information about a pathogen causing an infection or disease (TABLE 5.4).

A **selective medium** contains ingredients to inhibit the growth of certain microbes in a mixture while allowing (selecting for) the growth of others. The basic growth medium may contain extra salt (NaCl) or a dye to inhibit the growth of intolerant or sensitive organisms but permit the growth of those species or pathogens one wants to isolate.

Another modification to a basic growth medium is the addition of one or more compounds that allow one to differentiate between very similar species based on specific biochemical or physiological properties growing in or on a culture medium. This **differential medium** contains specific chemicals to indicate which

TABLE

5.3 Composition of a Complex and a Synthetic Growth Medium

Ingredient	Nutrient Supplied	Amount
A. Complex Agar Medium		
Peptone	Amino acids, peptides	5.0 g
Beef extract	Vitamins, minerals, other nutrients	3.0 g
Sodium chloride (NaCl)	Sodium and chloride ions	8.0 g
Agar		15.0 g
Water		1.0 liter
B. Synthetic Broth Medium		
Glucose	Simple sugar	5.0 g
Ammonium phosphate ($(NH_4)_2HPO_4$)	Nitrogen, phosphate	1.0 g
Sodium chloride (NaCl)	Sodium and chloride ions	5.0 g
Magnesium sulfate ($MgSO_4 \cdot 7H_2O$)	Magnesium ions, sulphur	0.2 g
Potassium phosphate (K_2HPO_4)	Potassium ions, phosphate	1.0 g
Water		1.0 liter

TABLE

5.4 A Comparison of Special Culture Media

Name	Components	Uses	Examples
Selective medium	Growth stimulants Growth inhibitors	Selecting certain microbes out of a mixture	Mannitol salt agar for staphylococci
Differential medium	Dyes Growth stimulants Growth inhibitors	Distinguishing different microbes in a mixture	MacConkey agar for gram-negative bacteria
Enriched medium	Growth stimulants	Cultivating fastidious microbes	Blood agar for streptococci; chocolate agar for *Neisseria* species

■ Fastidious: Having special nutritional requirements.

species possess and which lack a particular biochemical process. Such indicators make it easy to distinguish visually colonies of one organism from colonies of other similar organisms on the same culture plate. Look at FIGURE 5.13 and see if you can determine which medium was used in each example. MicroInquiry 5 presents another example using these two approaches to identify and differentiate between similar bacterial species.

Although many microorganisms grow well in nutrient broth and nutrient agar, certain so-called fastidious organisms may require an **enriched medium** containing extra vitamins or amino acids to promote growth. For example, the causative agent of gonorrhea, *Neisseria gonorrhoeae*, requires the addition of powdered hemoglobin. MicroFocus 5.5 details a particularly difficult culture identification.

Many bacterial and archaeal species are impossible to cultivate in any artificial culture medium yet devised. In fact, less than 1% of the species in natural water and soil samples can be cultured.

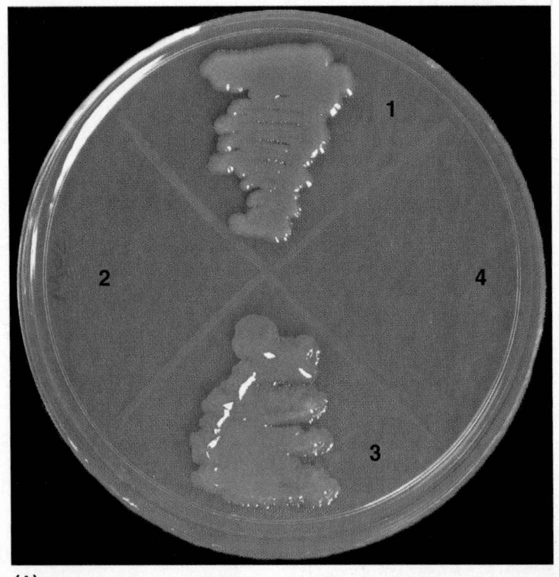

(A)

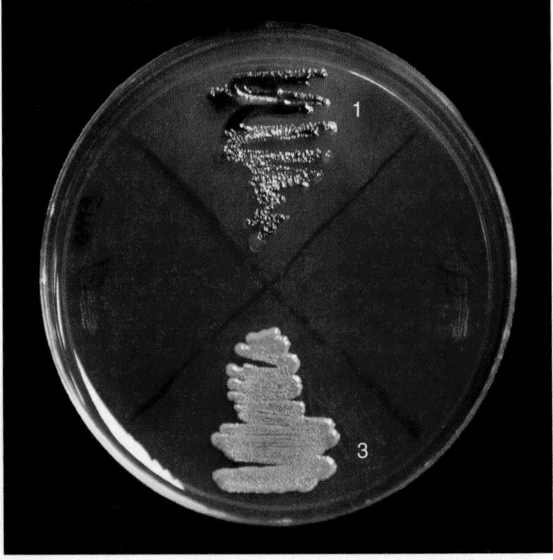

(B)

FIGURE 5.13 **Special Media Formulations.** (**A**) Four different bacterial species (1–4), two gram positive and two gram negative, were streaked onto separate sections of a MacConkey agar plate and allowed to incubate for 48 hours. MacConkey agar only supports the growth of gram-negative species. (**B**) Because the two gram-negative species cannot be visually distinguished from one another, they were then streaked into an eosin methylene blue agar (EMB) plate and incubated for 48 hours. EMB allows one to distinguish between human enteric bacteria, where *Escherichia coli* (1) produces a green metallic sheen while species like *Enterobacter aerogenes* (3) produce a pink color. Courtesy of Jeffrey Pommerville. »» Which special medium is selective and which is differential? Explain your reasoning.

MICROINQUIRY 5

Identification of a Bacterial Species

It often is necessary to identify a bacterial species or be able to tell the difference between similar-looking species in a mixture. In microbial ecology, it might be necessary to isolate certain naturally growing species from others in a mixture. In the clinical and public health setting, microbes might be pathogens associated with disease or poor sanitation. In addition, some may be resistant to standard antibiotics normally used to treat an infection. In all these cases, identification can be accomplished by modifying the composition of a complex or synthetic growth medium. Let's go through two scenarios.

- Suppose you are an undergraduate student in a marine microbiology course. On a field trip, you collect some seawater samples and, now back in the lab, you want to grow only photosynthetic marine microbes.

  How would you select for photosynthetic microbes? First, you know the photosynthetic organisms manufacture their own food, so their energy source will be sunlight and not the organic compounds typically found in culture media (see Table 5.3). So, you would need to use a synthetic medium but leave out the glucose. Also, knowing the salts typically found in ocean waters, you would want to add them to the medium. You would then inoculate a sample of the collected material into a broth tube, place the tube in the light, and incubate for one week at a temperature typical of where the organisms were collected.

5a. What would you expect to find in the broth tube after one week's incubation?

What you have used in this scenario is a selective medium; that is, one that will encourage the growth of photosynthetic microbes (light and sea salts) and suppress the growth of nonphotosynthetic microorganisms (no carbon = no energy source). So, only marine photosynthetic microbes should be present.

- As an infection disease officer in a local hospital, you routinely swab critical care areas to determine if there are any antibiotic resistant bacteria present. You are especially concerned about methicillin-resistant *Staphylococcus aureus* (MRSA) as it frequently can cause disease outbreaks in a hospital setting. One swab you put in a broth tube showed turbidity after 48 hours.

5b. Knowing that *Staphylococcus* species are halotolerant, how could you devise an agar medium to visually determine if any of the growth is due to *Staphylococcus aureus*?

Again, a selective medium would be used. It would be prepared by adding 7.5% salt to a complex agar medium. A sample from the broth tube would be streaked on the plate and incubated at 37°C for 48 hours.

5c. What would you expect to find on the agar plate after 48 hours?

Your selective medium contained 10 discrete colonies. You do a Gram stain and discover that all the colonies contain clusters of purple spheres; they are gram-positive. However, there are other species of *Staphylococcus* that do not cause disease. One is *S. epidermidis,* a common skin bacterium. A Gram stain therefore is of no use to differentiate *S. aureus* from *S. epidermidis.*

5d. Knowing that only *S. aureus* will produce acid in the presence of the sugar mannitol, how could you design a differential broth medium to determine if any of the colonies are *S. aureus*? (Hint: phenol red is a pH indicator that is red at neutral pH and yellow at acid pH).

You can identify each bacterial species by taking a complex broth medium, such as nutrient broth, and adding salt and mannitol (mannitol salt broth) and phenol red. Next, you inoculate a sample of each colony into a separate tube. You inoculate the 10 tubes and incubate them for 48 hours at 37°C.

5e. The broth tubes are shown below. What do the results signify? Which tubes contain which species of *Staphylococcus*?

This method is an example of a differential medium because it allowed you to visually differentiate or distinguish between two very similar bacterial species. Knowing which colonies on the original selective medium plate are *S. aureus,* you need to determine which, if any, are resistant to the antibiotic methicillin.

5f. How could you design an agar medium to identify any MRSA colonies?

5g. If the plates are devoid of growth, what can you conclude?

Again, you have used a selective medium; the addition of methicillin will permit the growth of any MRSA bacteria and suppress the growth of staphylococci sensitive to methicillin.

The answers to 5e can be found on the Student Companion Website in **Appendix D.**

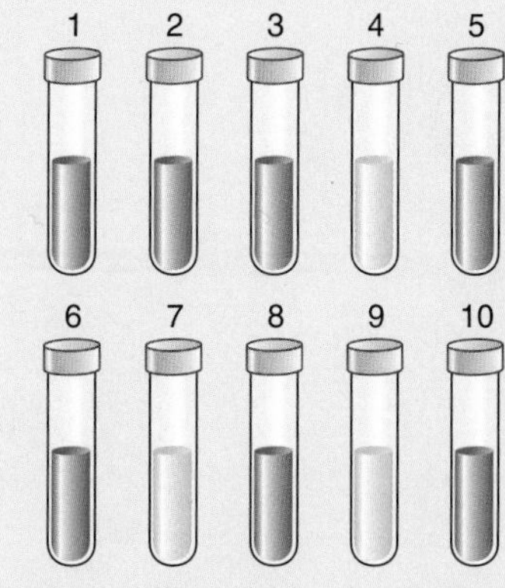

Results from differential broth tubes.

MICROFOCUS 5.5: Public Health

"Enriching" Koch's Postulates

On July 21–23, 1976, some 5,000 Legionnaires attended the Bicentennial Convention of the American Legion in Philadelphia, PA. About 600 of the Legionnaires stayed at the Bellevue Stratford Hotel. As the meeting was ending, several Legionnaires who stayed at the hotel complained of flu-like symptoms. Four days after the convention ended, an Air Force veteran who had stayed at the hotel died. He would be the first of 34 Legionnaires over several weeks to succumb to a lethal pneumonia, which became known as Legionnaires' disease.

As with any new disease, epidemiological studies look for the source of the disease. The Centers for Disease Control and Prevention (CDC) had an easy time tracing the source back to the Bellevue Stratford Hotel. Epidemiological studies also try to identify the causative agent. Using Koch's postulates, CDC staff collected tissues from lung biopsies and sputum samples. However, no microbes could be detected on slides of stained material. By December 1976, they were no closer to identifying the infectious agent.

How can you verify Koch's postulates if you have no infectious agent? It was almost like being back in the times of Pasteur and Koch. Why was this bacterial species so difficult to culture on bacteriological media? Perhaps it was a virus.

After trying 17 different culture media formulations, the infectious agent was finally cultured. It turns out it was a bacterial species, but one with fastidious growth requirements. The initial agar medium contained a beef infusion, amino acids, and starch. When this medium was enriched with 1% hemoglobin and 1% isovitalex, small, barely visible colonies were seen after five days of incubation at 37°C. Investigators then realized the hemoglobin was supplying iron to the bacterium and the isovitalex was a source of the amino acid cysteine. Using these two chemicals in pure form, along with charcoal to absorb bacterial waste, a pH of 6.9, and an atmosphere of 2.5% CO_2, bacterial growth was significantly enhanced (see figure). From a microscope examination of these cultures, a gram-negative rod was confirmed and the organism was appropriately named *Legionella pneumophila*.

With an enriched medium to pure culture the organism, susceptible animals (guinea pigs) could be injected as required by Koch's postulates. *L. pneumophila* then was recovered from infected guinea pigs, verifying the organism as the causative agent of Legionnaires' disease.

Today, we know *L. pneumophila* is found in many aquatic environments, both natural and artificial. At the Bellevue Stratford Hotel, epidemiological studies indicated guests were exposed to *L. pneumophila* as a fine aerosol emanating from the air-conditioning system. Through some type of leak, the organism gained access to the system from the water cooling towers.

Koch's postulates are still useful—it's just hard sometimes to satisfy the postulates without an isolated pathogen.

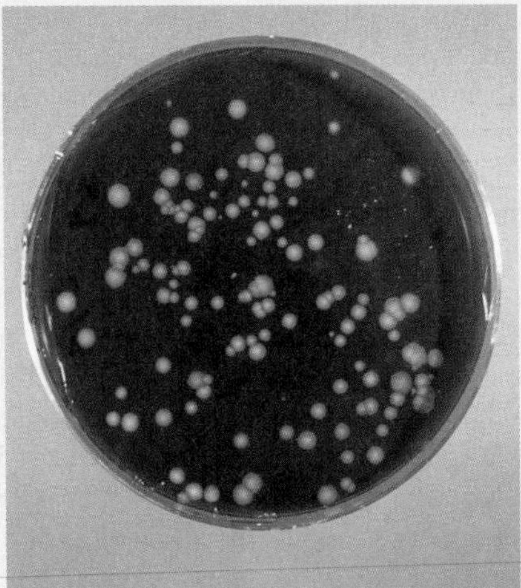

Colonies of *L. pneumophila* on an enriched medium. Courtesy of Dr. Jim Feeley/CDC.

So, it is impossible to estimate accurately microbial diversity in an environment based solely on culturability. Such "uncultured" organisms are said to be in a **viable but noncultured (VBNC)** state. Procedures for identifying VBNC organisms include direct microscopic examination and, most commonly, amplification of diagnostic gene sequences or 16S rRNA gene sequences.

Why do these organisms remain uncultured? Microbiologists believe that part of the reason may be due to their presence in a "foreign" environment. These species have adapted to their own familiar and specific environment; an artificial medium is not their typical home. Therefore, these species go into a type of dormancy state and do not divide; that is, they are viable, but cannot be cultured (Investigating the Microbial World 5). Studies on VBNC bacterial and archaeal species present a vast and as yet unexplored field, which is important not only for detection of human pathogens, but also to reveal the tremendous diversity in the microbial world.

Population Measurements Are Made Using Pure Cultures

Microorganisms rarely occur in nature as a single species. Rather, they are mixed with other species, in a so-called "mixed culture" most often as a **biofilm**. Therefore, to study a species, microbiologists and laboratory technologists must use a **pure culture**—that is, a population consisting of only one species.

If one has a mixed broth culture of bacterial species, how can the organisms be isolated as pure colonies? Two established methods are available. The first is the **pour-plate method**. Here, diluted samples of the mixed culture in molten nutrient agar tubes are each poured into a sterile Petri dish and allowed to harden. During a 24 hour to 48 hour incubation, the cells divide to form discrete colonies on and in the agar (FIGURE 5.14).

A second, more commonly used technique, called the **streak-plate method**, uses a single plate of nutrient agar (FIGURE 5.15A–D). An inoculum from a mixed culture is removed with a sterile loop using aseptic technique, and a series of streaks is made on the surface of one area of the plate. The loop is flamed, touched to the first area, and a second series is made in a second area. Similarly, streaks are made in the third and fourth areas, thereby spreading out the individual cells so they grow as separated colonies. After a 24 hour to 48 hour incubation, discrete colonies will be present on the plate (FIGURE 5.15E).

In both methods, each colony is a pure culture because the colony is derived from an original single cell that underwent numerous binary fissions. The researcher, medical technologist, or student can select samples of the colonies for further testing and subculturing.

■ Subculturing: The process of transferring bacterial cells from one culture medium to another.

■ Aseptic technique: The practice of transferring microorganisms to a sterile culture medium without introducing other contaminating organisms.

Population Growth Can Be Measured in Several Ways

Microbial growth in a culture medium can be measured by direct and indirect methods.

Direct Counting Methods. There are a number of ways to directly measure cell numbers. Scientists may wish to perform a **direct microscopic count** using a known volume of the liquid sample that has been placed on a specially designed counting chamber (FIGURE 5.16). However, this procedure will count both live and dead cells.

In the **most probable number test**, microbial samples diluted 10× and 100× are added to a set of lactose broth tubes and the presence or absence of gas formed in fermentation gives a rough statistical estimation of the cell number; that is, the most probable number. This technique has been used for measuring water quality.

In the **standard plate count procedure**, a sample of a broth culture is placed in a sterile Petri dish and melted nutrient agar is added (pour-plate method) (FIGURE 5.17). The assumption is that each cell will undergo multiple rounds of cell division to produce separate colonies on the plate. Because two or more cells could clump together on a plate and grow as a single colony, the standard plate count is expressed as the number of **colony-forming units (CFUs)**. After incubation, the number of CFUs will be used to estimate the number of viable cells originally plated.

Indirect Measurement Methods. Indirect methods include measuring the dry weight of the cell population, which gives an indication of the cell mass. Oxygen uptake and CO_2 production in metabolism also can be measured as an indication of metabolic activity and therefore cell number.

Another indirect method uses a spectrophotometer to measure the cloudiness, or **turbidity**,

Investigating the Microbial World 5

The Great Plate Count Anomaly

About 99% of bacterial species from the environment cannot be grown using known culture media.

■ **OBSERVATION:** Take a sediment sample from an environmental water source, mix it with saline solution (or water), and wait for the sediment to settle. Now take a drop from the liquid and place it on a nutrient agar culture dish (or any type of complex medium). Place another drop on a slide and add stain. On the stained slide, you will undoubtedly be able to count hundreds of cells and find dozens of different bacterial morphologies. On the plate, if you are lucky, maybe one or two colonies will appear in a few days. The majority of the cells will not grow even though they are inundated in nutrients. This is the so-called "great plate count anomaly." Standard laboratory culture techniques fail to support the growth of these viable but noncultured (VBNC) species that reside in environmental sediments.

■ **QUESTION:** ***Why won't 99% of the bacterial species grow in laboratory culture media?***

■ **HYPOTHESIS:** VBNCs need some essential "nutrient" from their neighbor species in the natural environment. If so, then growing the VBNCs in their natural environment will supply the needed nutrient and the uncultured should grow.

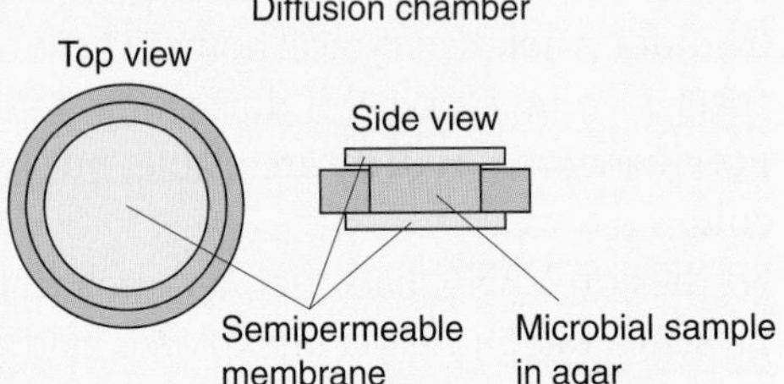

■ **EXPERIMENTAL DESIGN:** A diffusion apparatus is designed that sandwiches a microbial sediment sample in agar between two semipermeable membranes that allow for the free diffusion of "nutrients" and waste products through the chamber (see figure).

■ **EXPERIMENT:** Sediment from a freshwater pond sample is mixed with agar and placed within the diffusion chamber. The chamber is then placed back in its natural environment. Another pond sediment sample from the same environment is cultured on standard agar culture medium in a culture dish. Both the diffusion chamber culture and the lab culture are left undisturbed for 4 weeks. After the incubation period, the colonies isolated on the culture plate and on the agar in the diffusion chamber (but not the biofilm growing on the semipermeable membrane) are identified. Phylum identification is carried out by sequencing each isolate's 16S ribosomal RNA gene.

■ **RESULTS:** See table.

■ **CONCLUSIONS:**

QUESTION ***1: Was the hypothesis supported? Explain using the table.***

QUESTION ***2: Are the majority of the isolates representative of gram-positive or gram-negative organisms? Explain.***

QUESTION ***3: What might have been the result if the biofilm organisms growing on the membranes were included in the analysis?***

TABLE

Phylogenetic Strains Obtained by Culture Dish and Diffusion-Chamber Methods

	Number of strains isolated by			
Phylum	Culture dish only	Diffusion chamber only	Both methods	Total number of isolated strains
Alphaproteobacteria	20	36	6	62
Betaproteobacteria	3	63	3	69
Gammaproteobacteria	2	4	1	7
Deltaproteobacteria		1		1
Bacteroidetes	5	6	1	12
Spirochaetes		4		4
Firmicutes	5	1		6
Actinobacteria	5	1		6
Total	40	116	11	167

■ **FURTHER QUESTION:** What is it that the previously uncultured colonies in the diffusion chamber recognize? New evidence suggests that one "nutrient" is iron, which is needed for ATP generation and other biochemical processes. To grow, the uncultured must "capture" the iron (presumably from their neighbors) in the bound form that is otherwise unavailable in a culture dish "environment."

Answers can be found on the Student Companion Website in **Appendix D.**

Adapted from: Bollman, A., Lewis, K., and Epstein, S. S. (2007) *Appl. Environ. Microbiol.* **73**(20): 6386–6390.

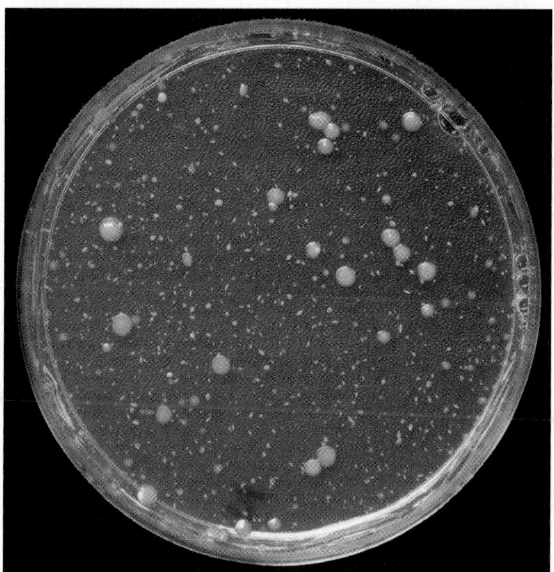

FIGURE 5.14 **A Pour Plate.** The dispersed bacterial cells grow as individual, discrete colonies. Courtesy of Jeffrey Pommerville. »» By looking at this plate, how would you know the original broth culture was a mixture of bacterial species?

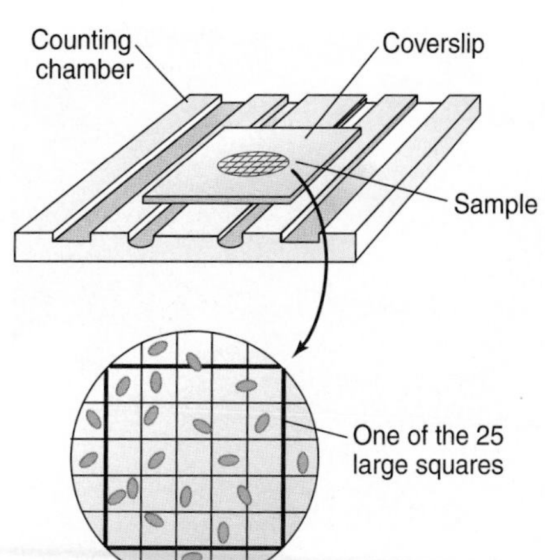

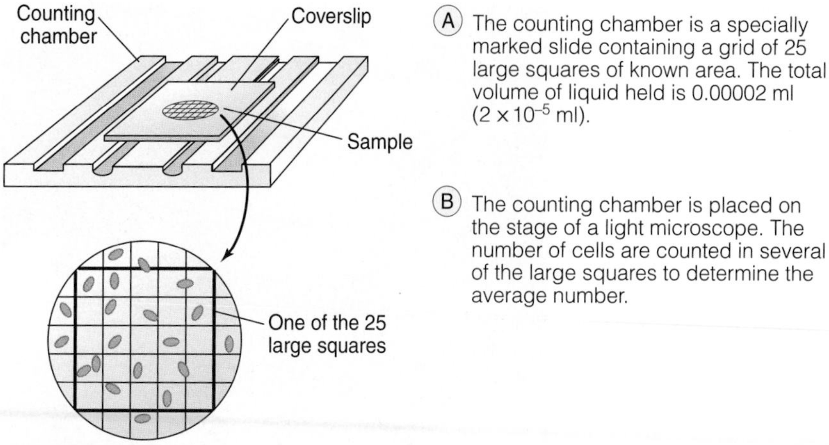

Counting chamber
Coverslip
Sample
One of the 25 large squares
(A) The counting chamber is a specially marked slide containing a grid of 25 large squares of known area. The total volume of liquid held is 0.00002 ml (2×10^{-5} ml).
(B) The counting chamber is placed on the stage of a light microscope. The number of cells are counted in several of the large squares to determine the average number.

FIGURE 5.16 **Direct Microscopic Count.** This procedure can be used to estimate the total number of live and dead cells in a culture sample. »» Suppose the average number of cells per large square was 14. Calculate the number of cells in a 10 ml sample.

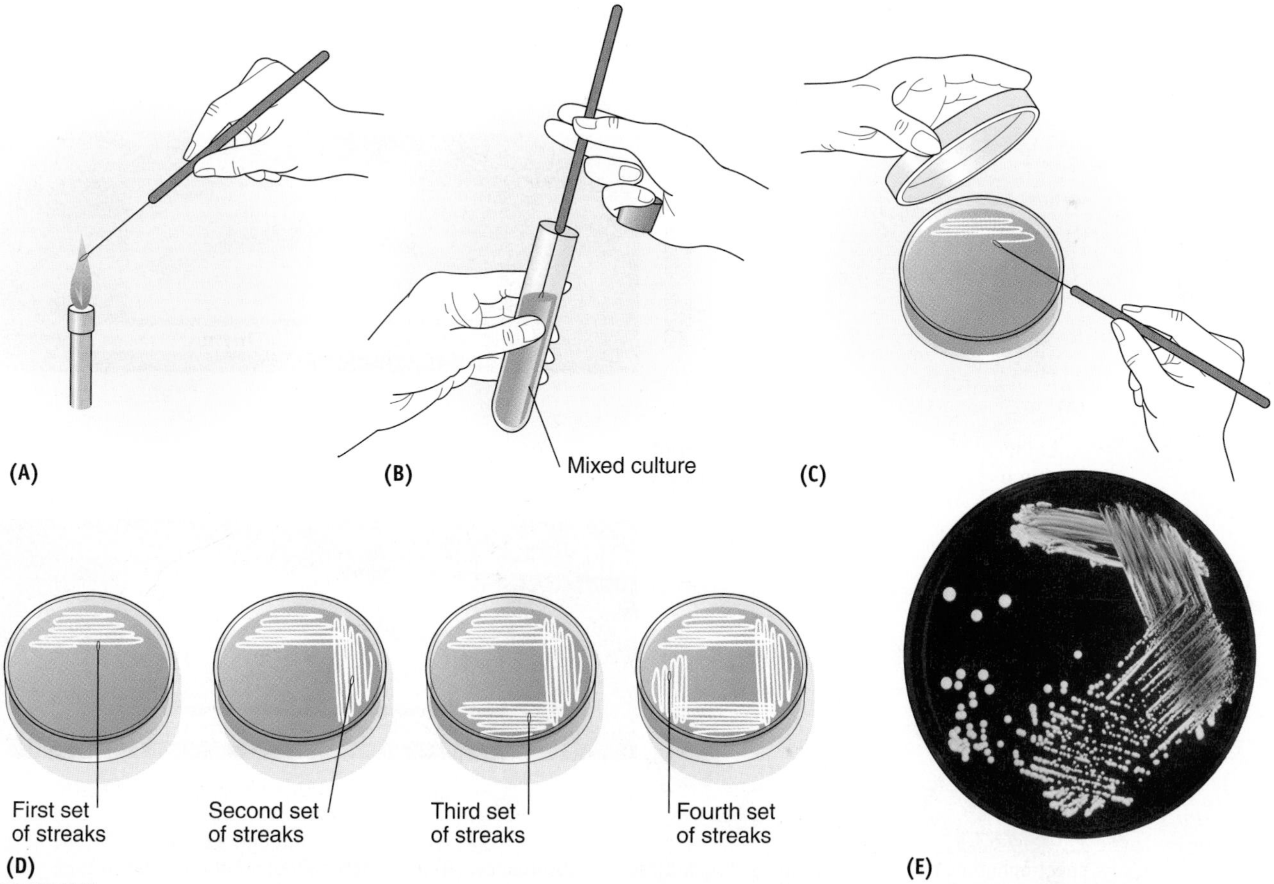

(A)
(B)
Mixed culture
(C)
First set of streaks
Second set of streaks
Third set of streaks
Fourth set of streaks
(D)
(E)

FIGURE 5.15 **The Streak-Plate Method.** (**A**) A loop is sterilized, (**B**) a sample of cells is obtained from a mixed culture, and (**C**) streaked near one edge of the agar medium. (**D**) Successive streaks are performed, and the plate is incubated. (**E**) Well-isolated and defined colonies illustrate a successful isolation. Courtesy of James Gathany/CDC. »» Justify the need to streak a mixed sample over four areas on a culture plate.

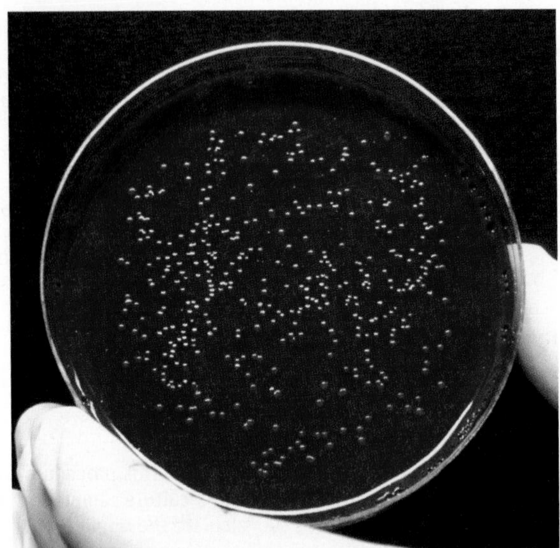

FIGURE 5.17 **The Standard Plate Count.** Individual bacterial colonies have grown on this blood agar plate. Each colony represents a colony-forming unit (CFU). © R.A. Longuehaye/Photo Researchers, Inc. »» If a 0.1 ml sample of a 10^4 dilution contained 250 colonies, how many bacterial cells were in 10 ml of the original broth culture?

of a broth culture. This instrument detects the amount of light scattered by a suspension of cells placed in the spectrophotometer such that the amount of light scatter (optical density, OD) is a function of the cell number; that is, the more cells present, the more light is scattered or absorbed, resulting in a higher absorbance reading on the spectrophotometer (FIGURE 5.18). A standard curve can be generated to serve as a measure of cell numbers. However, because more than 10 million cells are needed to make a reading on the spectrophotometer, turbidity is not a useful way to study the growth of small populations of bacterial cells.

CONCEPT AND REASONING CHECKS 3

a. Compare and contrast complex and chemically undefined media.

b. Explain how microbiologists have figured that some 99% of the microbial world has not been "seen."

c. Why might it be more difficult to isolate a colony from a pour-plate than from a streak-plate culture?

d. Distinguish between direct and indirect methods to measure population growth.

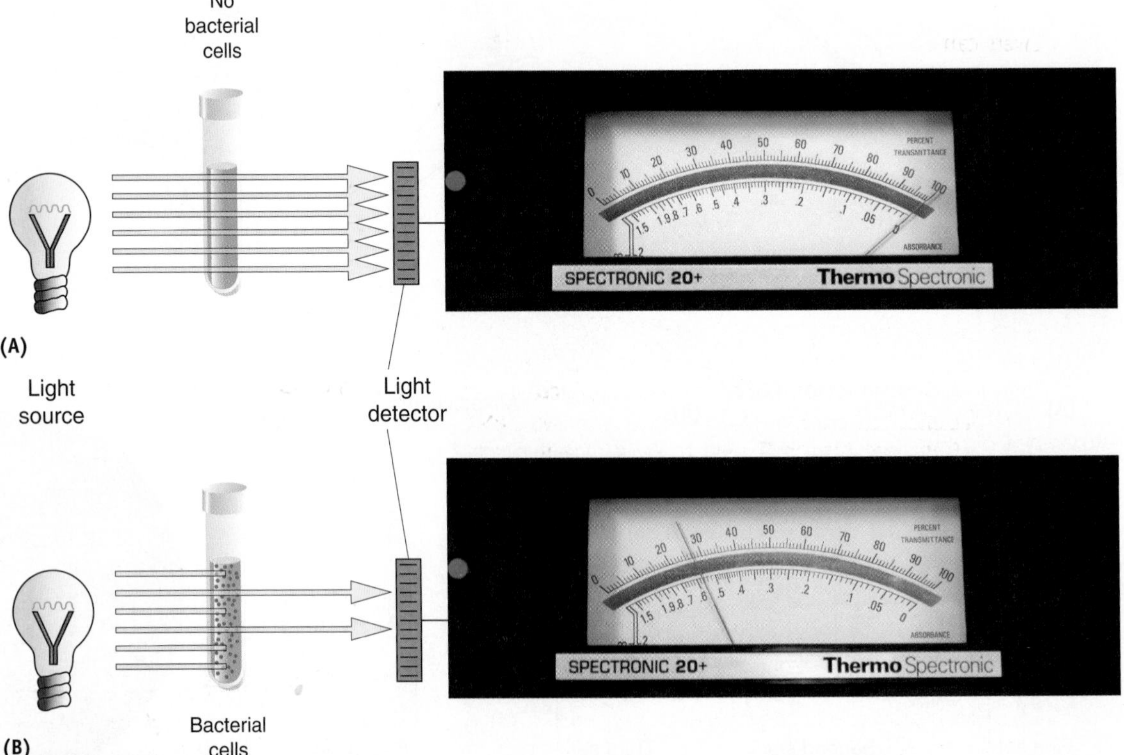

FIGURE 5.18 **Using Turbidity to Measure Population Growth.** (**A**) As light passes through a sterile broth tube in the spectrophotometer, the instrument is standardized at 0 absorbance. (**B**) As a bacterial population in a broth tube grows, the cells will scatter more of the light, which on the spectrophotometer is detected as an increase in absorbance. Courtesy of Jeffrey Pommerville. »» Why do turbidity measurements represent an indirect method to measure population growth?

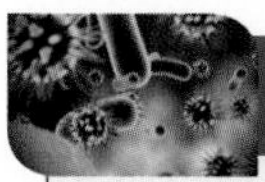

Chapter Challenge C

On November 13, 1971, *Mariner 9* went into orbit around the Red Planet. Before its arrival, we thought Mars was a dead world. Then *Mariner 9* sent back pictures of a landscape where it appeared water once flowed. But the craft found no signs of biological material in the Martian soil. As more spacecraft have gone to Mars, the evidence for life on the Red Planet has waxed and waned—yet some are hopeful that microbial life may exist near the surface or deeper underground.

QUESTION C:

As a result, several missions to Mars are being planned by American, European, Russian, and Chinese space agencies. As an exomicrobiologist (one who looks and searches for microbial life beyond Earth), what types of experiments would you design (based on this chapter) to see if microbial life does exist on Mars? The spacecraft will not return to Earth, so the experiments need to be completed on the Red Planet.

Answers can be found on the Student Companion Website in **Appendix D**.

In conclusion, we examined the major physical factors and nutrient media formulations that affect the rate at which microbial populations grow. In particular, the generation time for a bacterial species is dependent on the physical factors (temperature, oxygen need, pH, and osmotic pressure) and nutrients available in the environment, be it the soil, a laboratory culture dish, or a human host. In addition, many pathogenic bacteria are fastidious at least in a laboratory setting, so that if they are taken from different parts of the body, the clinical microbiology lab may have to use different growth media for isolation and identification. This brings up an interesting problem when considering the diagnosis and treatment of an infectious disease. Such pathogens, including *Treponema pallidum,* have to be identified from actual clinical samples using direct microscopic observation. So, one has to wonder how many human diseases may go undiagnosed today because they are VBNC? In fact, microbiologists have argued whether members of the Archaea can cause human disease. None are currently known to cause disease, yet we know from microscopic observation that archaeal organisms are present in the human body. Could some of them be VBNC pathogens?

SUMMARY OF KEY CONCEPTS

5.1 Microbial Reproduction Is Part of the Cell Cycle

1. The bacterial **cell cycle** involves "metabolic" growth, DNA replication, and **binary fission** to produce genetically identical daughter cells. (Fig. 5.2A)
2. Binary fissions occur at intervals called the **generation time**, which may be as short as 20 minutes. (Fig. 5.3)

5.2 Microbial Growth Progresses Through Distinct Phases

3. The dynamics of the **bacterial growth curve** show how a microbial population grows logarithmically, reaches a certain peak and levels off, and then may decline. (Fig. 5.4)
4. Dormancy is a response to potential or actual environmental change. (Figs. 5.6, 5.8)
5. Temperature, oxygen, pH, and hydrostatic/osmotic pressure are physical factors that influence microbial growth. Away from the optimal condition, growth slows within a set range. (Figs. 5.9, 5.11)

5.3 Culture Media Are Used to Grow Microbes and Measure Their Growth

6. **Complex** and **chemically-defined media** contain the nutrients for microbial growth.
7. Growth media can be modified to select for a desired microbial species, to differentiate between two similar species, or to enrich for species requiring special nutrients. (Fig. 5.13)
8. **Pure cultures** can be produced from a mixed culture by the **pour-plate method** or the **streak-plate method.** In both cases, discrete colonies can be identified that represent only one microbial species. (Figs. 5.14, 5.15)
9. Microbial growth can be measured by **direct microscopic count**, the **most probable number test**, and the **standard plate count** procedure. Indirect methods include dry weight, oxygen uptake, CO_2 production, and **turbidity** measurements. (Fig. 5.18)

CHAPTER SELF-TEST

For **STEPS A–D**, answers to even-numbered questions and problems can be found in **Appendix C** on the Student Companion Website at **http://microbiology.jbpub.com/10e**. In addition, the site features eLearning, an online review area that provides quizzes and other tools to help you study for your class. You can also follow useful links for in-depth information, read more MicroFocus stories, or just find out the latest microbiology news.

STEP A: REVIEW OF FACTS AND TERMS

Multiple Choice

Read each question carefully, then select the ***one*** answer that best fits the question or statement.

1. Which one of the following statements does NOT apply to bacterial reproduction?
   A. A fission ring apparatus is present.
   B. Septum formation occurs.
   C. A spindle apparatus is used.
   D. Symmetrical cell division occurs.
2. If a bacterial cell in a broth tube has a generation time of 40 minutes, how many cells will there be after 6 hours of optimal growth?
   A. 18
   B. 64
   C. 128
   D. 512
3. A bacterial species generation time would be determined during the _____ phase.
   A. decline
   B. lag
   C. log
   D. stationary
4. Which one of the following is NOT an example of dormancy?
   A. Log phase
   B. Endospore formation
   C. VBNCs
   D. Persister cells
5. A microbe that is a microaerophilic mesophile would grow optimally at _____ and _____.
   A. high O_2; 30°C
   B. low O_2; 20°C
   C. no O_2; 30°C
   D. low O_2; 37°C
6. If the carbon source in a growth medium is beef extract, the medium must be an example of a/an _____ medium.
   A. complex
   B. chemically-defined
   C. enriched
   D. differential
7. A _____ medium would involve the addition of the antibiotic methicillin to identify methicillin-resistant bacteria.
   A. differential
   B. selective
   C. thioglycollate
   D. VBNC
8. Which one of the following is NOT part of the streak-plate method?
   A. Making four sets of streaks on a plate
   B. Diluting a mixed culture in molten agar
   C. Using a mixed culture
   D. Using a sterilized loop
9. Direct methods to measure bacterial growth would include all the following except _____.
   A. total bacterial count
   B. direct microscopic count
   C. turbidity measurements
   D. most probable number

True-False

Each of the following statements is true (T) or false (F). If the statement is false, substitute a word or phrase for the underlined word or phrase to make the statement true.

10. _____ Endospores are produced by some gram-negative bacterial species.
11. _____ Obligate aerobes use oxygen gas as a final electron acceptor in energy production.
12. _____ The most common growth medium used in the teaching laboratory is a complex medium.
13. _____ The majority of bacterial and archaeal organisms can be cultured in growth media.
14. _____ In attempting to culture a fastidious bacterial pathogen, a differential medium would be used.
15. _____ Acidophiles grow best at pHs greater than 9.
16. _____ Mesophiles have their optimal growth near 37°C.
17. _____ Bacterial and archaeal cells lack a mitotic spindle to separate chromosomes.
18. _____ The fastest doubling time would be found in the lag phase of a bacterial growth curve.
19. _____ If *E. coli* cells are placed in distilled water, they will burst.
20. _____ Halophiles would dominate in marine environments.

STEP B: CONCEPT REVIEW

21. Describe the three phases of a bacterial **cell cycle**. (**Key Concept 1**)
22. Compare the events of each phase of a **bacterial growth curve**. (**Key Concept 2**)
23. Explain the importance of bacterial **dormancy**. (**Key Concept 2**)
24. Identify the four major physical factors governing microbial growth and describe how microorganisms have adapted to these physical environments. (**Key Concept 2**)
25. Explain how **selective** and **differential media** are each constructed. (**Key Concept 3**)
26. Explain the procedures used in the **pour-plate** and **streak-plate methods**. (**Key Concept 3**)
27. Construct a concept map for **Growth Measurements** using the following terms. (**Key Concept 3**)

CO_2 production	Indirect measurement methods
Cell mass	Metabolic activity
Colony-forming unit	Most probable number test
Direct-counting method	Oxygen uptake
Direct microscopic count	Standard plate count
Dry weight	Turbidity

STEP C: APPLICATIONS AND PROBLEMS

28. Use the log phase growth curves (1, 2, or 3) below to answer each of the following questions (a–c).

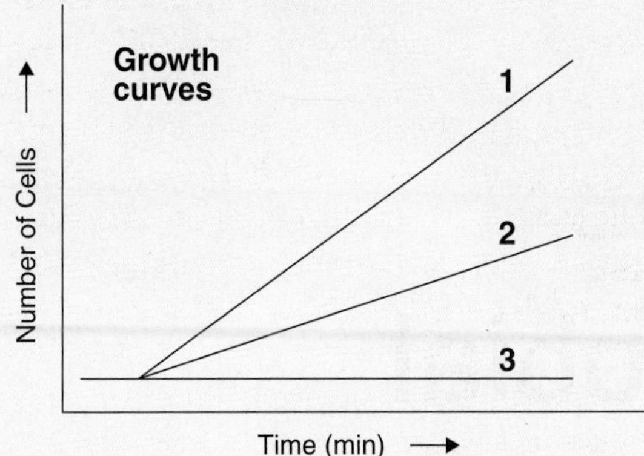

_____a. Which curve (1, 2, or 3) best represents the growth curve for a mesophile incubated at 60°C?

_____b. Which curve (1, 2, or 3) best represents a nonhalophile growing in 5% salt?

_____c. Which curve (1, 2, or 3) best represents an acidophile growing at pH 4?

29. Consumers are advised to avoid stuffing a turkey the night before cooking, even though the turkey is refrigerated. A homemaker questions this advice and points out that the bacterial species of human disease grow mainly at warm temperatures, not in the refrigerator. What explanation might you offer to counter this argument?

30. Public health officials found that the water in a Midwestern town was contaminated with sewage bacteria. The officials suggested that homeowners boil their water for a couple of minutes before drinking it. (a) Would this treatment sterilize the water? Why? (b) Is it important that the water be sterile? Explain.

STEP D: QUESTIONS FOR THOUGHT AND DISCUSSION

31. To prevent decay by bacterial species and to display the mummified remains of ancient peoples, museum officials place the mummies in glass cases where oxygen has been replaced with nitrogen gas. Why do you think nitrogen is used?

32. Extremophiles are of interest to industrial corporations, who see these organisms as important sources of enzymes that function at temperatures of 100°C and pH levels of 10 (the enzymes have been dubbed "extremozymes"). What practical uses can you foresee for these enzymes?

33. During the filming of the movie *Titanic,* researchers discovered at least 20 different bacterial and archaeal species literally consuming the ship, especially a rather large piece of the midsection. What type of environmental conditions are these bacterial and archaeal species subjected to at the wreck's depth of 12,600 feet?

34. Every year news media report cases of skin and lung infections in people sitting in hot tubs. How might such infections occur in hot tubs?

6

CHAPTER PREVIEW

6.1 Enzymes and Energy Drive Cellular Metabolism

6.2 Glucose Catabolism Generates Cellular Energy

MicroInquiry 6: The Machine that Makes ATP

6.3 There Are Other Pathways to ATP Production

6.4 Photosynthesis Converts Light Energy to Chemical Energy

Investigating the Microbial World 6: Microbes Provide the Answer

6.5 Microbes Exhibit Metabolic Diversity

Image courtesy of Dr. Fred Murphy/CDC.

Microbial Metabolism

Life is like a fire; it begins in smoke and ends in ashes.
—Ancient Arab proverb connecting energy to life

Charlie Swaart had been a social drinker for years. A few beers or drinks with his pals, but no lasting alcoholic consequences. Then, in 1945, he began a nightmare that would make medical history.

One October day, while stationed in Tokyo after World War II, Swaart suddenly became drunk for no apparent reason. He had not had any alcohol for days, but suddenly he felt like he had been partying all night. After sleeping it off, he would be fine the next day.

Unfortunately, this "behavior" returned time and time again. For years after his return to the United States, the episodes continued—bouts of drunkenness and monumental hangovers without drinking so much as a beer! Doctors were puzzled as they could detect alcohol on his breath or in his blood. Was this some type of internal metabolism gone haywire? Was it the result of a bacterial infection? It didn't seem likely. They warned him though not to drink any additional alcohol for fear of damaging his liver. Swaart followed their advice to the letter; still, he experienced periods of drunkenness.

Twenty years passed before Swaart, known as the "drinkless drunk," learned of a similar case in Japan. A Japanese businessman had endured years of social and professional disgrace before doctors discovered a yeast-like fungus in his intestine. Studying this eukaryotic microbe showed that the fungal cells were fermenting carbohydrates to alcohol right there in his intestine. The fungus was identified as *Candida albicans* (FIGURE 6.1). Now having *C. albicans* in one's intestine is not uncommon; but, finding fermenting *C. albicans* was historic. With this knowledge, Swaart approached his doctor. Sure enough, lab tests showed massive colonies of *C. albicans* in Swaart's intestine too. The sugar in a cup of coffee or any carbohydrate in pasta, cake, or candy could bring on drunkenness.

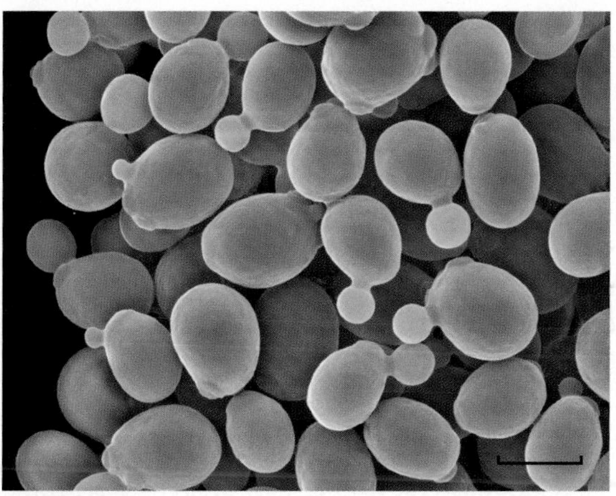

FIGURE 6.1 *Candida albicans* **Cells.** This false-color scanning electron micrograph of *C. albicans* shows daughter cells (yellow) budding from the mother cells. (Bar = 10 µm.) © Dennis Kunkel Microscopy, Inc./Visuals Unlimited/Corbis. »» To what domain of organisms does *C. albicans* belong?

Swaart then learned that an antibiotic had worked to kill the yeast cells in the Japanese man's gut. However, to cure his illness, Swaart had to travel back to Japan to get the effective antibiotic.

Researchers believed the atomic blasts of Hiroshima and Nagasaki in 1945 may have caused a normal *C. albicans* to mutate to a fermenting form, which somehow found its way into Swaart's digestive system. One can only wonder if there are many other individuals who have become living fermentation vats for the fungus. For Charlie Swaart, though, the nightmare was finally over.

The process of fermentation described here was in a eukaryotic microbe. However, many other types of fermentation processes also occur in microorganisms and they are but one aspect of the broad topic of microbial metabolism.

In this chapter, we examine some of the major metabolic pathways carried out by microorganisms. Much of the chapter discusses the catabolic reactions involved in energy conversions forming adenosine triphosphate (ATP). Because the chapter emphasizes the role of carbohydrates in the energy conversions, it might be worthwhile to review the material on the biomacromolecules (carbohydrates, lipids, and proteins).

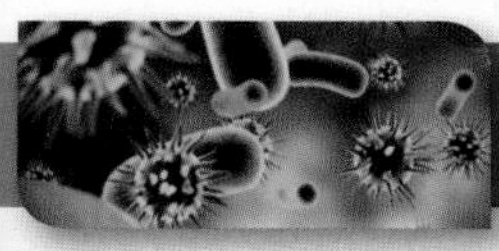

Chapter Challenge

What do a termite's gut, a cow's rumen, and the warm, waterlogged soil in a rice paddy have in common? They all produce substantial amounts of methane (CH_4) gas. In Africa, Australia, and South America, there are large areas inhabited by mound-building termites whose mounds can be up to 9 meters tall. The methane gas produced from the termites in these mounds contributes some 5% of the atmospheric methane. Methane produced from ruminant livestock, like dairy cattle, goats, and sheep, is currently estimated to contribute 16% of the atmospheric methane. However, rice production currently accounts for approximately 20% of global methane emissions. So, methane emissions from these three "natural" sources account for more than 40% of the methane gas being released every day into the atmosphere. Where is this methane gas coming from? You have probably already guessed—microbes! Still how do termite guts, cow rumens, and waterlogged rice patty soils produce the methane gas? Let's find out!

© qcontrol/ShutterStock, Inc.

KEY CONCEPT 6.1 Enzymes and Energy Drive Cellular Metabolism

Metabolism refers to all the biochemical reactions taking place in an organism and, as such, they manage the material and energy resources of the cell. Some of these reactions are called **anabolism**, which is the building of larger organic compounds (**polymers**) from simpler building blocks (**monomers**). For example, photosynthesis is an anabolic process because the sugar glucose is formed from carbon dioxide and water. Building reactions consume energy in forming the bonds between the monomers, so, from an energy perspective, anabolic reactions are said to be **endergonic** (*end* = "inner"; *ergon* = "work") **reactions**. The opposite biochemical reactions are called **catabolism**, which breaks down (hydrolyzes) polymers into simpler molecules. A major catabolic pathway in cells is

TABLE

6.1 A Comparison of Two Key Aspects of Cellular Metabolism

Anabolism	Catabolism
Synthesis of larger molecules	Breakdown of large molecules
Products are large molecules	Products are small molecules
Photosynthesis	Glycolysis, citric acid cycle
Mediated by enzymes	Mediated by enzymes
Energy generally is required (endergonic)	Energy generally is released (exergonic)

cellular respiration where sugars like glucose are broken down into carbon dioxide and water. Such catabolic reactions from an energy perspective release energy and are referred to as **exergonic** (*ex* = "outside of") **reactions.** TABLE 6.1 compares anabolism and catabolism, which are reactions often taking place simultaneously in cells and organisms. Realize metabolism also includes numerous conversion reactions that transform one molecule into another without any type of catabolic or anabolic event taking place.

Enzymes Catalyze All Chemical Reactions in Cells

Microbial growth depends on metabolic processes that occur in the cell cytosol, on the cell (plasma) membrane, in the periplasm (gram-negative bacterial cells), in eukaryotic organelles, and outside the cell. To carry out these reactions, cells need a large variety of enzymes. Therefore, we begin our study of metabolism with a discussion of these essential proteins, which have been known only since the early 1900s (MICROFOCUS 6.1).

Enzymes are proteins (or in a few instances RNA molecules) that increase the probability of chemical reactions while themselves remaining unchanged. They accomplish in fractions of a second what otherwise might take hours, days, or longer to happen spontaneously under normal biological conditions. For example, even though organic molecules like amino acids have functional groups, it is highly unlikely they would randomly bump into one another in the precise way needed for a chemical reaction (dehydration synthesis) to occur and for a new peptide bond to be formed.

MICROFOCUS 6.1: History
"Hans, Du Wirst Das Nicht Glauben!"

Louis Pasteur's discovery of the role yeast cells play in fermentation heralded the beginnings of microbiology because it showed that tiny organisms could bring about important chemical changes. However, it also opened debate on how yeasts accomplished fermentation. Lively controversies ensued among scientists; some suggesting sugars from grape juice entered yeast cells to be fermented; others believing fermentation occurred outside the cells. The question would not be resolved until a fortunate accident happened in the late 1890s.

In 1897, two German chemists, Eduard and Hans Buchner, were preparing yeast as a nutritional supplement for medicinal purposes. They ground yeast cells with sand and collected the cell-free "juice." To preserve the juice, they added a large quantity of sugar (as was commonly done at that time) and set the mixture aside. Several days later Eduard noticed an unusual alcoholic aroma coming from the mixture. Excitedly, he called to his brother, "Hans, Du wirst das nicht glauben!" ("Hans, you won't believe this!") One taste confirmed their suspicion: The sugar had fermented to alcohol.

The discovery by the Buchner brothers was momentous because it demonstrated that a chemical substance inside yeast cells brings about fermentation, and that fermentation can occur without living cells. The chemical substance came to be known as an "enzyme," meaning "in yeast."

In 1905, the English chemist Arthur Haden expanded the Buchner study by showing that "enzyme" is really a multitude of chemical compounds and should better be termed "enzymes." Thus, he added to the belief that fermentation is a chemical process. Soon, many chemists became biochemists, and biochemistry gradually emerged as a new scientific discipline.

Thus, the reaction rate would be very slow were it not for the activity of enzymes.

Enzymes have several common characteristics.

1. Enzymes are reusable. Once a chemical reaction has occurred, the enzyme is released to participate in another identical reaction. In fact, the same enzyme can catalyze the same type of reaction 100 to 1 million times each second.

2. Enzymes are highly specific. An enzyme that functions in one type of chemical reaction usually will not participate in another type of reaction. That means there must be thousands of different enzymes to catalyze the thousands of different chemical reactions of metabolism occurring in a microbial cell.

3. Enzymes have an active site. Each enzyme has a special pocket or cleft called an **active site**, which has a specific three-dimensional shape complementary to a reactant (called a **substrate**). The active site positions the substrate such that it is highly likely a chemical reaction will occur to form one or more **products**.

4. Enzymes are required in minute amounts. Because an enzyme can be used thousands of times to catalyze the same reaction, only minute amounts of a particular enzyme are needed to ensure that a fast and efficient metabolic effect occurs.

Many enzymes can be identified by their names, which often end in "-ase." For example, "sucrase" is the enzyme that breaks down sucrose and "ribonuclease" digests ribonucleic acid. In terms of anabolic metabolism, "polymerases" link together nucleotides and "transferases" link together the NAG and NAM units to build the bacterial cell wall peptidoglycan.

Enzymes Act Through Enzyme-Substrate Complexes

Website Animation: Mechanism of Enzyme Action

Enzymes function by aligning substrate molecules in such a way that a reaction is highly favorable. In the hydrolysis reaction shown in FIGURE 6.2, the three-dimensional shape of the enzyme's active site recognizes and holds the substrate in an **enzyme-substrate complex**. While in the complex,

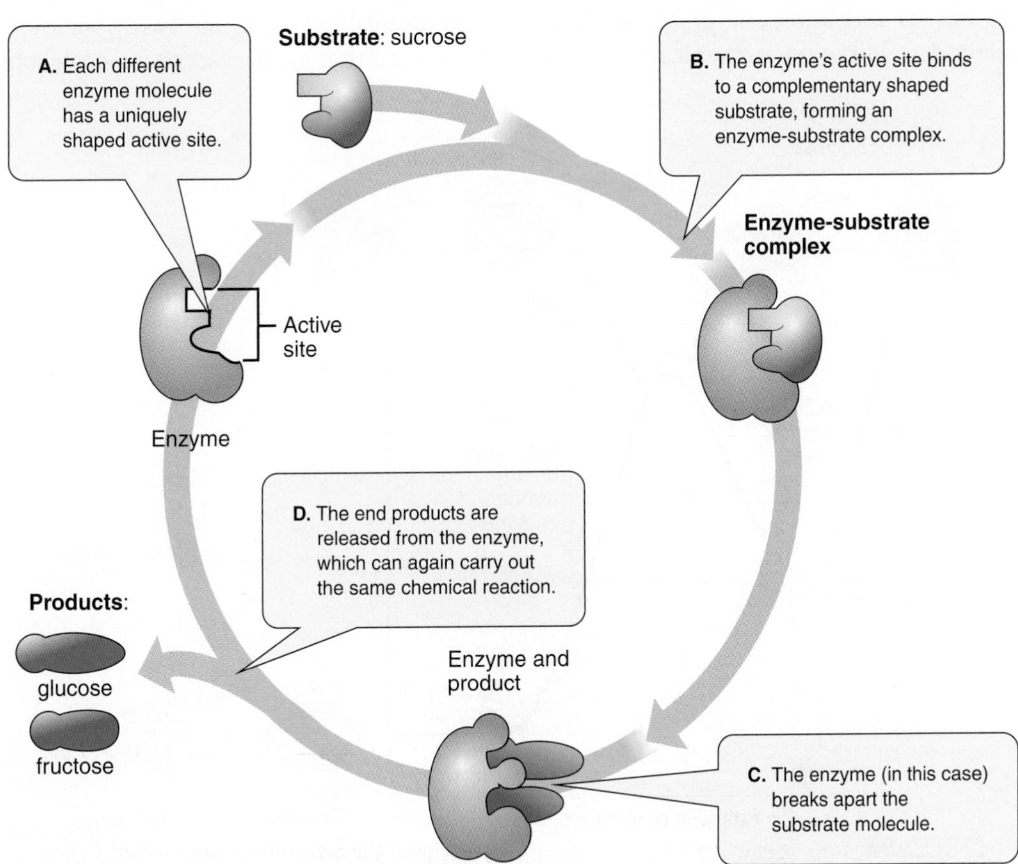

FIGURE 6.2 **The Mechanism of Enzyme Action.** Although this example shows an enzyme hydrolyzing a substrate (sucrose), enzymes also catalyze dehydration reactions, which, in this case, would combine glucose and fructose into sucrose. »» How do enzymes recognize specific substrates?

chemical bonds in the substrate are stretched or weakened by the enzyme, causing the bond to break. In a synthesis reaction, by contrast, the electron shells of the substrates in the enzyme-substrate complex are forced to overlap in the spot where the chemical bond will form.

Thus, in a hydrolysis or dehydration reaction, recognition of the substrate(s) is a precisely controlled, nonrandom event.

Looking at sucrose again, the bonds holding glucose and fructose together will not break spontaneously. The reason is the bond between the monosaccharides is stable and there is a substantial energy barrier preventing such a reaction (FIGURE 6.3A). The job of sucrase is to bind the substrate and lower the energy barrier so that it is much more likely that the reaction will occur. In other words, the bond holding glucose to fructose needs to be destabilized (i.e., stretched, weakened) by the enzyme (FIGURE 6.3B). This energy barrier is called the **activation energy**. Enzymes, then, play a key role in metabolism because they provide an alternate reaction pathway of less resistance; that is, with a lower activation energy barrier. They assist in the destabilization of chemical bonds and the formation of new ones by separating or joining atoms in a carefully orchestrated fashion.

Some enzymes are made up entirely of protein. An example is **lysozyme**, the enzyme in human tears and saliva that hydrolyzes the bond between NAG and NAM in the cell walls of gram-positive bacterial cells. Other enzymes, however, may contain small, nonprotein substances, called **cofactors**, that participate in the catalytic reaction. Inorganic cofactors are metal ions, such as magnesium (Mg^{2+}), iron (Fe^{2+}), and zinc (Zn^{2+}). When the nonprotein cofactor is a small organic molecule, it is referred to as a **coenzyme**, most of which are derived from vitamins. Examples of two important coenzymes are **nicotinamide adenine dinucleotide (NAD^+)** and **flavin adenine dinucleotide (FAD)**. These coenzymes play a significant role as electron carriers in metabolism, and we shall encounter them in our ensuing study of microbial metabolism.

Enzymes Often Team Up in Metabolic Pathways

There are many examples, such as the sucrose example, where an enzymatic reaction is a single substrate

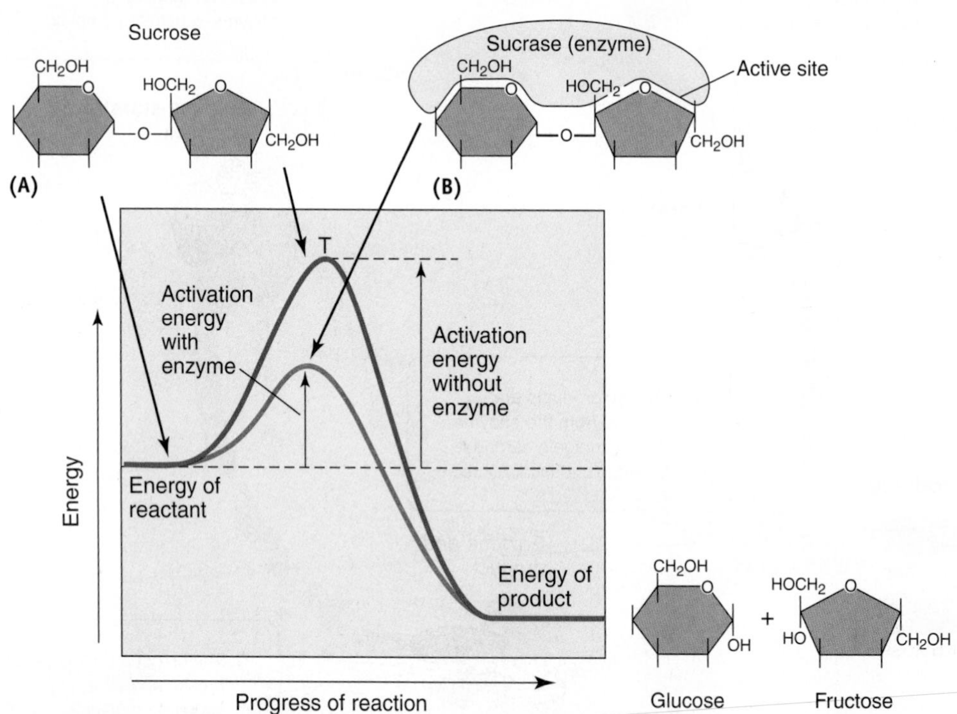

FIGURE 6.3 **Enzymes and Activation Energy.** Enzymes lower the activation energy barrier required for chemical reactions of metabolism. (**A**) The hydrolysis of sucrose is unlikely because of the high activation energy barrier. (**B**) When sucrase is present, the enzyme effectively lowers the activation energy barrier at the transition state (**T**), making the hydrolysis reaction highly favorable. »» How does the enzyme lower the activation energy of a stable substrate?

to product reaction. However, cells more often use metabolic pathways. A **metabolic pathway** is a sequence of chemical reactions, each reaction catalyzed by a different enzyme, in which the product (output) of one reaction serves as a substrate (input) for the next reaction (FIGURE 6.4A). The pathway starts with the initial substrate and finishes with the final end product. The products of "in-between" stages are referred to as "intermediates."

Metabolic pathways can be anabolic, where larger molecules are synthesized from smaller monomers. In contrast, other pathways are catabolic because they break larger molecules into smaller ones. Such pathways may be linear, branched, or cyclic. We will see many of these pathways in the microbial metabolism sections ahead.

Enzyme Activity Can Be Inhibited

Both environmental factors by themselves and in concert with metabolic pathways can inhibit enzyme activity permanently or temporarily.

Enzyme Inhibition by Environmental Factors. In another chapter, we spent some time investigating the physical factors affecting microbial growth. For example, temperature affects an enzyme's reaction rate, which slows down the more the temperature deviates from the optimal. Because most enzymes are proteins, they are sensitive to changes in temperature; indeed, high temperature can denature a protein, perhaps bringing metabolism to a sudden halt.

We also discussed pH and how it affects microbial growth. Again, an increase or decrease in protons (H^+) will interfere with an enzyme's reaction rate, extreme change leading to not only enzyme denaturation, but to enzyme and metabolic inhibition.

In addition, chemicals applied "environmentally" may inhibit enzyme action. Alcohols and phenol inactivate enzymes and precipitate proteins, making these chemical agents effective antiseptics or disinfectants. Other natural chemicals interfere with enzyme action (e.g., penicillin) or with a cell's ability to carry out a critical enzyme reaction (e.g., sulfa drugs), making these agents effective antibiotics.

Enzyme Inhibition Through Pathway Modulation. The same chemical reaction does not occur in a cell all the time, even if the substrate is present. Rather, cells regulate the enzymes so that they are present or active only at the appropriate time during metabolism.

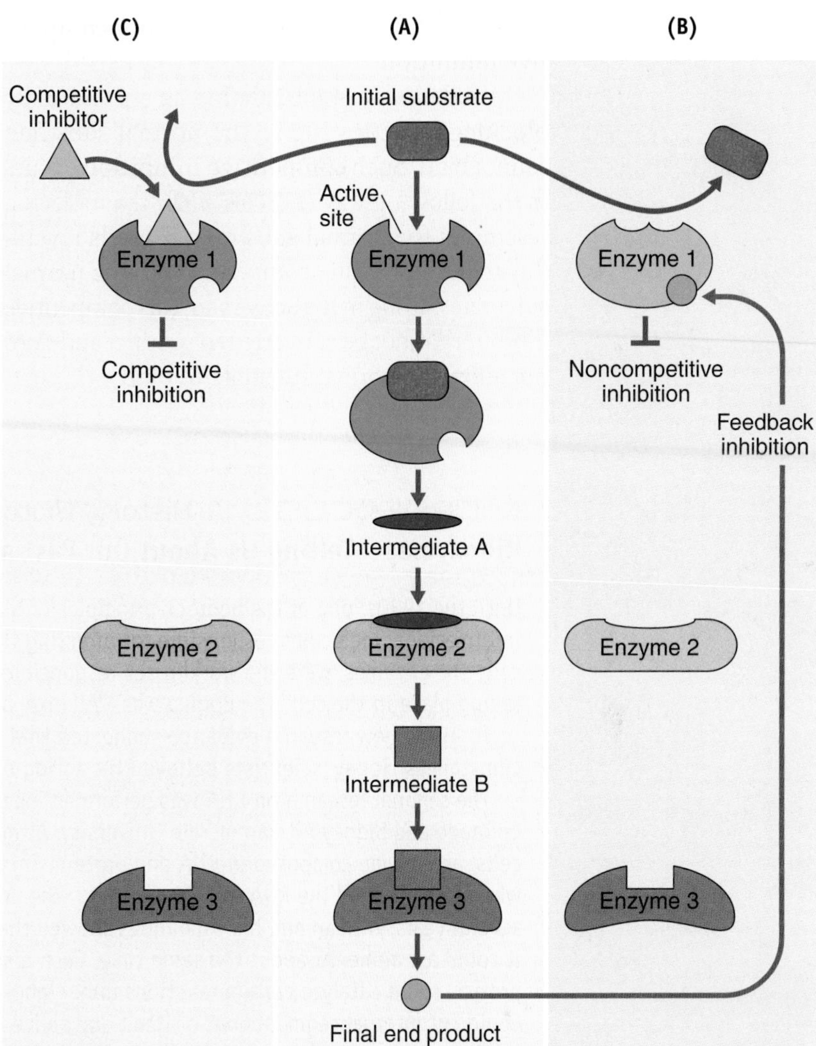

FIGURE 6.4 **Metabolic Pathways and Enzyme Inhibition.** (**A**) In a metabolic pathway, a series of enzymes transforms an initial substrate into a final end product. (**B**) If excess final end product accumulates, it "feeds back" on the first enzyme in the pathway and inhibits the enzyme by binding at another site on the enzyme. (**C**) In competitive inhibition, a substrate that resembles the normal substrate competes with the substrate for the enzyme's active site. Competitive inhibition would reduce the productivity of the metabolic pathway by slowing down or stopping the pathway. In both (**B**) and (**C**), the whole pathway can become temporarily inoperative. »» Hypothesize why most substrates cannot be converted into a final end product in one enzymatic step.

One of the most common ways of modulating enzyme activity is for the final end product of a metabolic pathway to inhibit an enzyme in that pathway (FIGURE 6.4B). If the first enzyme in the pathway is inhibited, then no more product is available as input for the rest of the pathway. Such **feedback inhibition** is typical of many metabolic pathways in cells. In general, when the final end product or any molecule binds to a nonactive site on the enzyme, the shape of the active site changes and can no longer bind substrate. This type of modulation is referred to as **noncompetitive inhibition**.

Another way of modulating an enzyme is by blocking its active site so the normal substrate cannot bind. Such **competitive inhibition** occurs in the following way (FIGURE 6.4C). If a molecule resembles the normal substrate, it binds reversibly to the active site, competing with the normal substrate. Sitting in the active site, this competitive inhibitor cannot be converted to product and does not allow the normal substrate to bind.

The dogma in biology used to say that all enzymes were proteins. Although this statement usually is true, there are cases where ribonucleic acid (RNA) can have catalytic effects (MICROFOCUS 6.2).

Energy in the Form of ATP Is Required for Metabolism

In many metabolic reactions, energy is needed, along with enzymes, for the reactions to occur. The cellular "energy currency" is a compound called

MICROFOCUS 6.2: History/Biotechnology

Ribozymes—Telling Us About Our Past and Helping with Our Future

Until the 1980s, one of the bedrock principles of biology held that nucleic acids (DNA and RNA) were the informational molecules responsible for directing the metabolic reactions in the cell. Proteins, specifically the enzymes, were the workhorses responsible for catalyzing the thousands of chemical reactions taking place in the cell. The dogma was, "All enzymes are proteins."

In 1981, new research evidence suggested that RNA molecules could act as catalysts in certain circumstances. Today, scientists believe RNA acting by itself can trigger specific chemical reactions.

The seminal research on RNA was performed independently by Thomas R. Cech of the University of Colorado and Sidney Altman of Yale University. Altman had found an unusual enzyme in some bacterial cells, an enzyme composed of RNA and protein. Initially, he thought the RNA was a contaminant, but when he separated the RNA from the protein, the protein by itself could not function as an enzyme. After several years, Altman and his colleagues showed that RNA was the enzyme's key component because it could act alone. At about the same time, Cech discovered that RNA molecules from *Tetrahymena*, a protist, could catalyze certain reactions under laboratory conditions. He showed that a molecule of RNA could cut internal segments out of itself and splice together the remaining segments.

Many biologists responded to the findings of Cech and Altman with disbelief. The implication of the research was that proteins and nucleic acids are not necessarily interdependent, as had been assumed. The research also opened the possibility that RNA could have evolved on Earth without protein. In fact, a number of scientists have proposed the hypothesis that life may have started in a primeval "RNA world." This world would have been swarming with self-catalyzing forms of RNA having the ability to reproduce and carry genetic information. In essence, there arose a whole new way of imagining how life might have begun. The Nobel Prize committee was equally impressed. In 1989, it awarded the Nobel Prize in Chemistry to Cech and Altman.

By 1990, these self-reproducing molecules of RNA had a name—**ribozymes**. They share many similarities with their protein counterparts, including the presence of binding pockets that, like active sites on enzymes, recognize specific molecular shapes. Biochemists at Massachusetts General Hospital showed that one type of ribozyme could join together separate short nucleotide segments. The research was a step toward designing a completely self-copying RNA molecule.

Today, the understanding of catalytic ribozymes goes beyond the research laboratory. Several companies are using new molecular techniques to construct new catalytic ribozymes in what is termed "directed evolution." Development of these ribozymes may have uses in clinical diagnostics and as therapeutic agents. For example, in diagnostic applications, ribozymes are being developed to identify potential new drugs. Other companies are using ribozymes as biosensors to detect viral contaminants in blood. These catalytic molecules also may be useful in fighting infectious diseases by inactivating RNA molecules in viruses or other pathogens.

So, ribozymes have much to offer in understanding our very distant past as well as providing for a healthier future.

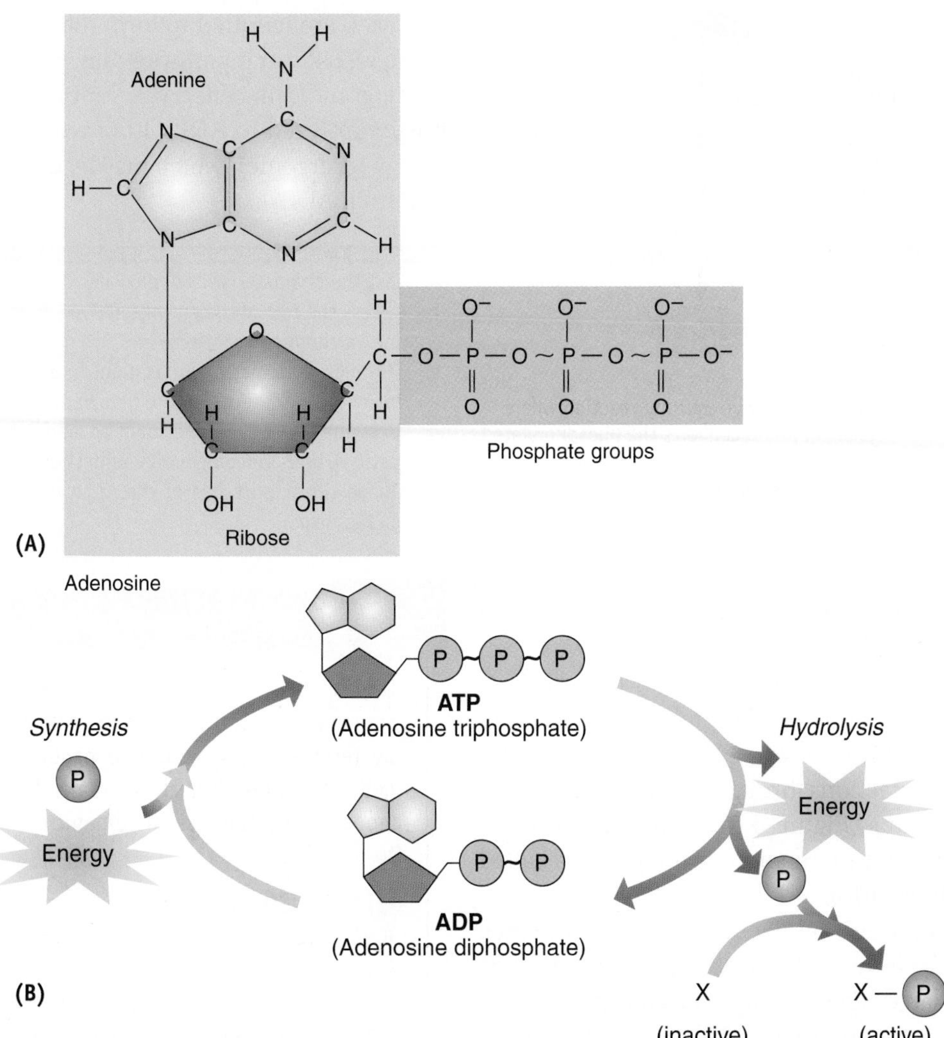

FIGURE 6.5 **Adenosine Triphosphate and the ATP/ADP Cycle.** Adenosine triphosphate (ATP) is a key immediate energy source for microbial cells and all other living organisms. (**A**) The ATP molecule is composed of adenine and ribose bonded to one another and to three phosphate groups. (**B**) When the ATP molecule breaks down, it releases a phosphate group and energy, and becomes adenosine diphosphate (ADP). The freed phosphate can activate another chemical reaction through phosphorylation. For the synthesis of ATP, energy and a phosphate group must be supplied to an ADP molecule. »» What genetic molecule closely resembles ATP?

adenosine triphosphate (ATP) (FIGURE 6.5A). In bacterial and archaeal cells, most of the ATP is formed on the cell membrane, while in eukaryotes the reactions occur primarily in the mitochondria. An ATP molecule acts like a portable battery. It provides the needed energy for activities such as binary fission, flagellar motion, active transport, and spore formation. On a more chemical level, it fuels protein synthesis and carbohydrate breakdown. It is safe to say that a major share of microbial functions depends on a continual supply of ATP. Should the supply be cut off, the cell dies very quickly, as ATP cannot be stored.

ATP molecules are relatively unstable. In Figure 6.5A, notice that the three phosphate groups all have negative charges on an oxygen atom. Like charges repel, so the phosphate groups in ATP, being tightly packed together, are very unstable. Breaking the so-called "high-energy bond" holding the last phosphate group on the molecule produces a more stable **adenosine diphosphate (ADP)** molecule and a free phosphate group (FIGURE 6.5B). ATP hydrolysis is analogous to a spring compacted in a box. Open the box (hydrolyze the phosphate group) and you have a more stable spring (a more stable ADP molecule). The

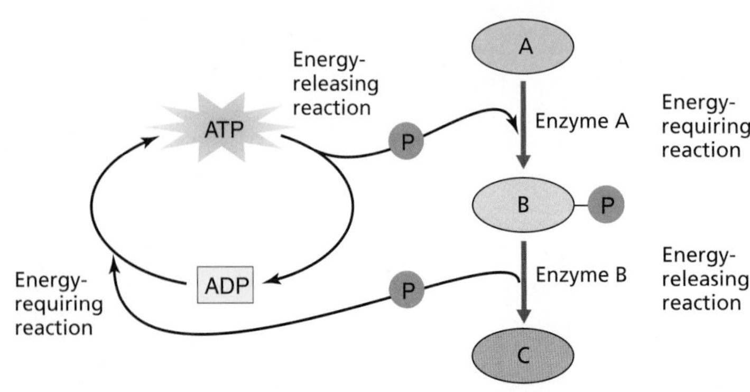

FIGURE 6.6 **A Metabolic Pathway Coupled to the ATP/ADP Cycle.** In this metabolic pathway, enzyme A catalyzes an energy-requiring reaction where the energy comes from ATP hydrolysis. Enzyme B converts the phosphorylated substrate to the final end product C. Being an energy-releasing reaction, the free phosphate can be coupled to the reformation of ATP. »» What are the terms for energy-requiring and energy-releasing reactions?

release of the spring (the freeing of a phosphate group) provides the means by which work can be done. Thus, the hydrolysis of the unstable phosphate groups in ATP molecules to a more stable condition is what drives other energy-requiring reactions through the transfer of phosphate groups (Figure 6.5B). The addition of a phosphate group to another molecule is called **phosphorylation**.

Because ATP molecules are unstable, they cannot be stored. Therefore, microbial cells synthesize large organic compounds like glycogen or lipids for energy storage. As needed, the chemical energy in these molecules can be released in catabolic reactions and used to reform ATP from ADP and phosphate (Figure 6.5B). This **ATP/ADP cycle** occurs continuously in cells. It has been estimated that a typical bacterial cell must reform about 3 million ATP molecules per second from ADP and phosphate to supply its energy needs.

It might be a good idea to review what we have covered to this point, which is summarized in FIGURE 6.6. Enzymes regulate metabolic reactions by binding an appropriate substrate at its active site. Often a series of metabolic steps (metabolic pathway) are required to form the final product. Some steps in the pathway may require energy (endergonic); this energy is supplied by ATP as it is hydrolyzed to ADP. Other reactions release energy (exergonic), which may be used to reform ATP from ADP.

CONCEPT AND REASONING CHECKS 1

a. List the characteristics of enzymes.
b. Assess the role of the enzyme active site in "stimulating" a chemical reaction.
c. If a metabolic pathway has eight intermediates, how many different enzymes are involved? Explain.
d. Describe how an enzyme could be modulated by competitive and noncompetitive inhibition.
e. Judge the importance of the ATP cycle to microbial metabolism.

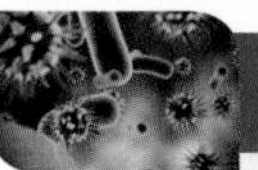

Chapter Challenge A

Our chapter challenge is to discover the source and reason for methane gas production by termites, cows, and rice paddy fields. To get started, termites and cows need to take in "food" and catabolize the organic compounds to generate ATP.

QUESTION A:

Being vegetarians, what do termites typically eat and cows in the field ingest? Now, what is the basic structural polysaccharide that all these "plant cell wall products" contain? The problem is that termites and cows cannot digest these polysaccharides. Their cells of the digestive system lack the ability to synthesize the enzymes to carry out the catabolic reactions—yet the termites and cows survive off these substances by breaking them down. So, if they are to get energy out of these polysaccharides, what in their digestive tracts must be helping out?

Answers can be found on the Student Companion Website in **Appendix D**.

KEY CONCEPT 6.2 Glucose Catabolism Generates Cellular Energy

Since the early part of the twentieth century, the chemistry of glucose catabolism has been the subject of intense investigation by biochemists because glucose is a key source of energy for ATP production. Moreover, the process of glucose catabolism is very similar in all organisms, making this "metabolic interlock" one feature that unites all life.

Glucose Contains Stored Energy That Can Be Extracted

A mole of glucose (180 g) contains about 686,000 calories of energy. This fact can be demonstrated in the laboratory by setting fire to a mole of glucose and measuring the energy released. In a cell, however, not all the energy is set free from glucose, nor can the cell trap all that is released. The process accounts for the transfer of about 40% of the glucose energy to ATP energy; that is, chemical energy to cellular energy.

Virtually all cells make ATP by harvesting energy from exergonic metabolic pathways, such as the hydrolysis of food molecules. Such a process is called **cellular respiration**. If cells consume oxygen in making ATP, the process is called **aerobic respiration**. In other instances, cells can still carry out cellular respiration without using oxygen, in which case it is called **anaerobic respiration**. In these instances, another inorganic molecule will replace oxygen.

A form of "anaerobic metabolism," different from anaerobic respiration, is **fermentation**, which we will examine later in this chapter.

The catabolism of glucose or another molecule does not take place in one chemical reaction, nor do ATP molecules form all at once. Rather, the energy in glucose is extracted and transferred slowly to ATP through metabolic pathways (FIGURE 6.7). It is similar to the proverb quoted at the beginning of this chapter, "*Life is like a fire; it begins in smoke and ends in ashes.*" The catabolism of glucose starts with a little energy being converted to ATP (the "smoke"), which builds to a point where large amounts of energy are converted to ATP (the "fire"), and the original glucose molecule has been depleted of its useful energy (the "ashes").

To begin our study of cellular respiration, we shall follow the process of aerobic respiration as it occurs in obligate aerobes. The process is represented in the following summary form:

$$\underset{\text{Glucose}}{C_6H_{12}O_6} + \underset{\text{Oxygen}}{6\ O_2} + 32\ \text{ADP} + 32\ \text{P}$$

$$\downarrow$$

$$\underset{\text{Carbon dioxide}}{6\ CO_2} + \underset{\text{Water}}{6\ H_2O} + 32\ \text{ATP}$$

The events summarized in the overall reaction are conveniently divided into three stages: glycolysis, the citric acid cycle, and oxidative phosphorylation. Let's examine each of these in sequence. To simplify our discussion of glucose catabolism, we shall follow the fate of one glucose molecule.

- **Mole:** The molecular weight of a substance expressed in grams.
- **Calorie:** A unit of energy defined as the amount of heat required to raise one gram of water 1°C.

Glycolysis Is the First Stage of Energy Extraction

The splitting of glucose, called **glycolysis** (*glyco* = "sweet"), occurs in the cytosol of all microorganisms and involves a metabolic pathway that converts an initial 6-carbon substrate, **glucose**, into two 3-carbon molecules (final end products) called **pyruvate**. Between glucose and pyruvate, there are eight intermediates formed, each catalyzed by a specific enzyme. FIGURE 6.8 illustrates the process. For easy referral, numbers in circles identify each

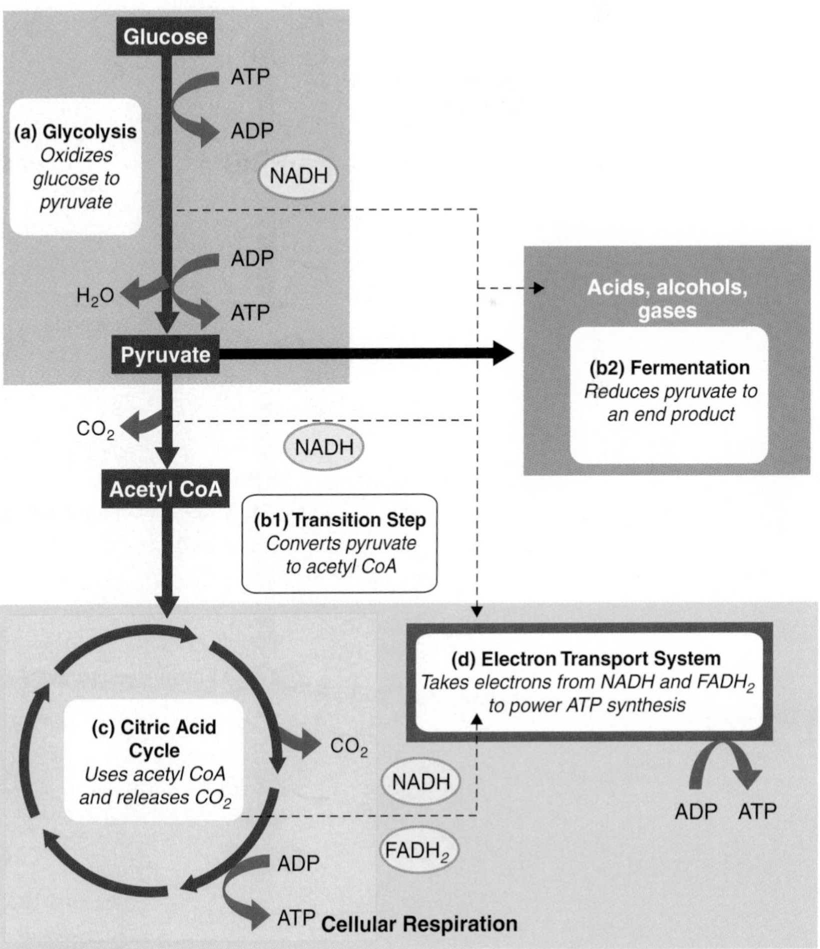

FIGURE 6.7 **A Metabolic Map of Aerobic and Anaerobic Pathways for ATP Production.** The production of ATP by microorganisms can be achieved through glycolysis (**a**) following a cellular respiration pathway (**b1, c, d**) or fermentation pathway (**b2**). »» Using this map, show how *"Life is like a fire; it begins in smoke and ends in ashes."*

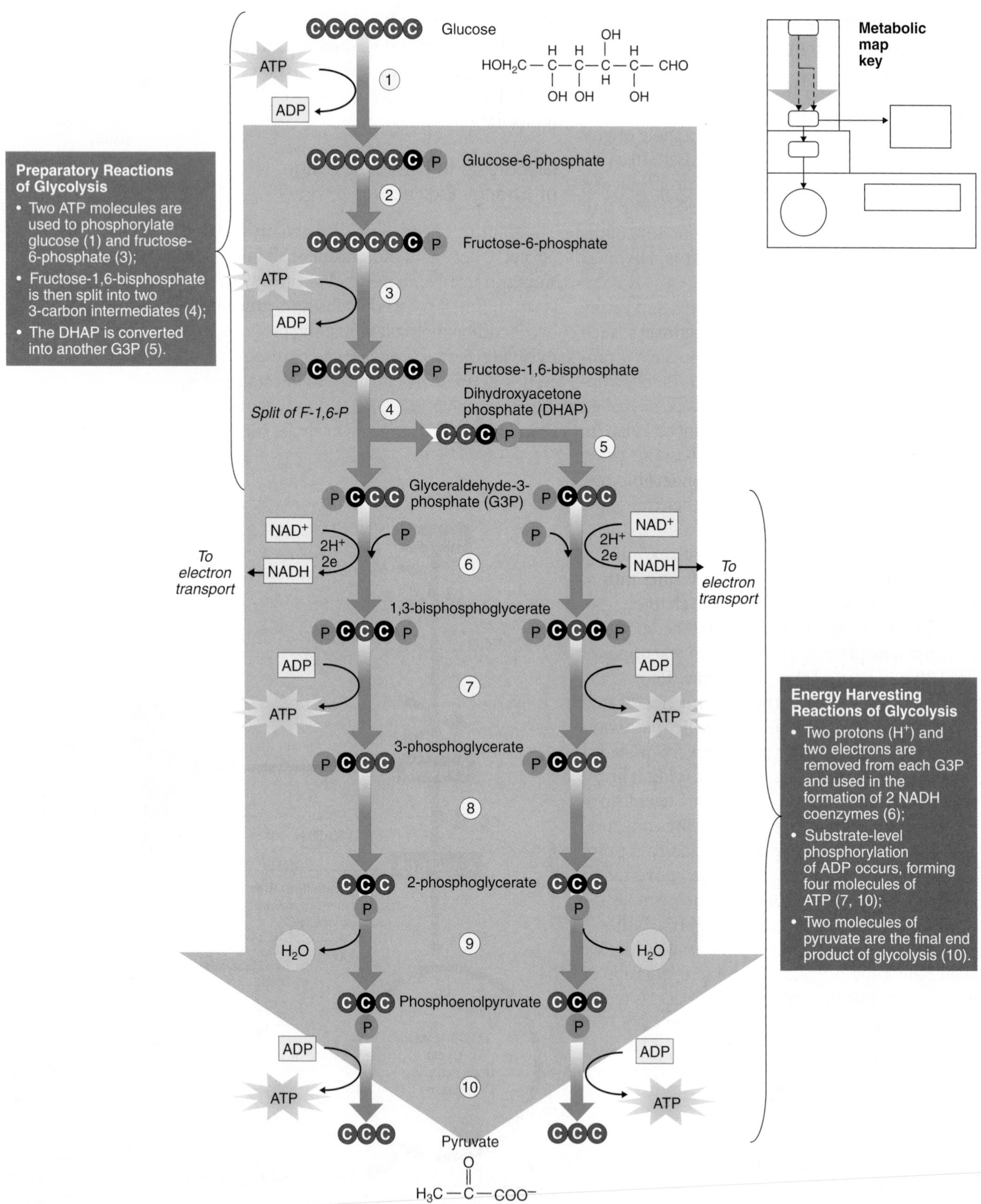

FIGURE 6.8 **The Reactions of Glycolysis.** Glycolysis is a metabolic pathway that converts glucose, a 6-carbon sugar, into two 3-carbon pyruvate products. In the process, two NADH coenzymes and a net gain of two ATP molecules occur. Carbon atoms are represented by circles. The dark circles represent carbon atoms bonded to phosphate groups. »» How many substrate-level phosphorylation events occur when one glucose molecule is broken into two pyruvate molecules?

reaction, and it would be helpful to refer to the figure as the discussion proceeds.

The first part of glycolysis (the preparatory reactions) is endergonic; one molecule of ATP is hydrolyzed (consumed) in reaction (**1**) and a second in reaction (**3**). In both cases, the phosphate group from ATP attaches to the product. Thus, reaction (**1**) produces glucose-6-phosphate, and reaction (**3**) yields fructose-1,6-bisphosphate (bis means "two separate"; that is, two separate phosphate molecules).

After the splitting of fructose-1,6-bisphosphate into two 3-carbon molecules, each passes through an additional series of conversions (the energy harvesting reactions) that ultimately form pyruvate. During reactions (**7**) and (**10**), ATP is generated. In both exergonic steps, enough energy is released to synthesize an ATP molecule from ADP and phosphate, resulting in a total of four ATP molecules. Because these ATP molecules were the result of the transfer of a phosphate from a substrate to ADP, we say these ATP molecules were the result of **substrate-level phosphorylation**. Considering two ATP molecules were consumed in reactions (**1**) and (**3**), the net gain from glycolysis is two molecules of ATP.

Before we proceed, take note of reaction (**6**). This enzymatic reaction releases two high-energy electrons and two protons (H^+), which are picked up by the coenzyme NAD^+, reducing each to NADH. This and similar events will have great significance shortly as an additional source to generate ATP.

The Citric Acid Cycle Extracts More Energy from Pyruvate

The **citric acid cycle** (also called the **Krebs cycle** in honor of Hans Krebs and colleagues who worked out the pathway) is a series of chemical reactions that are referred to as a cycle because the end product formed is used as one substrate to initiate the pathway. All of the reactions are catalyzed by enzymes, and all take place along the cell membrane of bacterial and archaeal cells. In eukaryotic microbes, including the protists and fungi, the cycle occurs in the mitochondria.

The citric acid cycle is somewhat like a constantly turning wheel. Each time the wheel comes back to the starting point, something must be added to spin it for another rotation. That something is the pyruvate molecule derived from glycolysis. FIGURE 6.9 shows the citric acid cycle. The reactions are identified by capital letters in circles to guide us through the cycle.

Before pyruvate molecules enter the cycle, they undergo oxidation, indicated in reaction (**A**). An enzyme removes a carbon atom from each of the two pyruvate molecules and releases the carbons as two carbon dioxide molecules ($2CO_2$). The remaining two carbon atoms of pyruvate are combined with **coenzyme A (CoA)** to form **acetyl-CoA**. Equally important, the lost electrons from pyruvate, along with two protons are transferred to NAD^+ to form NADH.

■ Oxidation: The process of removing electron pairs from a substance.

The two remaining carbons from pyruvate are now ready to enter the citric acid cycle. In reaction (**B**), each acetyl-CoA unites with a 4-carbon oxaloacetate to form citrate, a 6-carbon molecule. (Citrate, or citric acid, may be familiar to you as a component of soft drinks.) Citrate undergoes a series of reactions (**C**, **D**), forming a 4-carbon succinate. The two lost carbons are released as CO_2. Succinate then undergoes a series of modifications (reactions **E**, **F**, and **G**) reforming oxaloacetate. The cycle is now complete, and oxaloacetate is ready to unite with another molecule of acetyl-CoA.

Several features of the cycle merit closer scrutiny. First, we shall follow the carbon. Pyruvate, with three carbon atoms, emerges from glycolysis, but after one turn of the cycle, three molecules of CO_2 are produced. There are two molecules of pyruvate from glycolysis, so when the second molecule enters the cycle, three more CO_2 molecules are generated. Remember that we began with a 6-carbon glucose molecule; six CO_2 molecules now have been produced. This fulfills part of the equation for aerobic respiration:

■ Reducing (reduction): Referring to the process whereby a substance gains electron pairs.

$$\underline{C}_6H_{12}O_6 + 6\ O_2 + 32\ ADP + 32\ P$$
$$\downarrow$$
$$\underline{6}\ \underline{C}O_2 + 6\ H_2O + 32\ ATP$$

The second feature of the cycle is reaction (**D**). Here a molecule of ATP forms. Because we have two pyruvate molecules entering the cycle (per molecule of glucose), a second ATP molecule will form when the second pyruvate passes through the cycle.

Last, and perhaps of most importance, are reactions **C**, **D**, **E**, and **G**. Reactions **C**, **D**, and **G**,

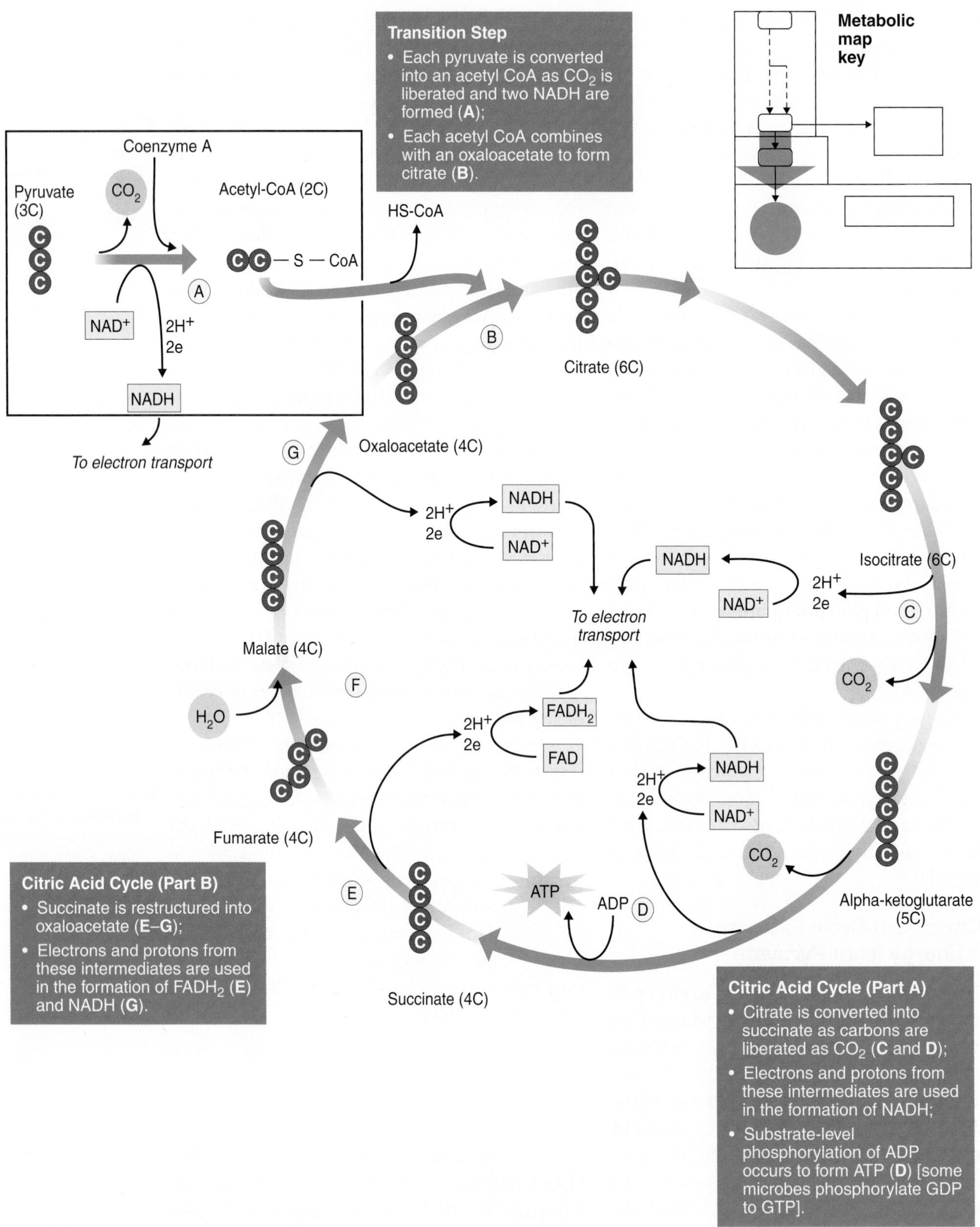

FIGURE 6.9 **The Reactions of the Citric Acid Cycle.** Pyruvate from glycolysis combines with coenzyme A to form acetyl-CoA (transition step). This molecule then joins with oxaloacetate to form citrate (citric acid cycle—Part A). Each turn of the cycle releases CO_2, produces ATP, and forms NADH and $FADH_2$ coenzymes as oxaloacetate is replaced (citric acid cycle—Part B). »» Why are so many reactions required to extract energy out of pyruvate?

like reaction 6 in glycolysis and the transition step (**A**), are associated with NAD^+ and again are reduced to NADH. Two NADH molecules are produced in each step for a total of six. In addition, reaction **E** accomplishes much of the same result except it is associated with another coenzyme, FAD. It too receives two electrons and two protons from the reaction, being reduced to $FADH_2$. For the two pyruvate molecules starting the process, two $FADH_2$ molecules are formed.

In summary, the central metabolic pathways of glycolysis and the citric acid cycle have extracted as much energy as possible from glucose and pyruvate (FIGURE 6.10). This has amounted to a small gain of ATP molecules formed from one glucose molecule. However, the 10 NADH and 2 $FADH_2$ molecules formed are most significant. Let's see how.

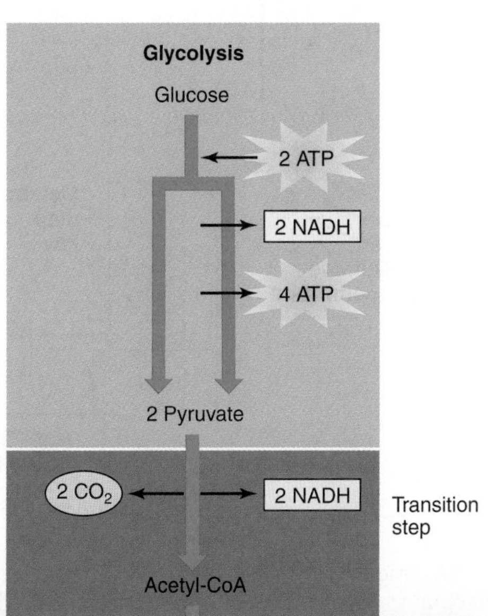
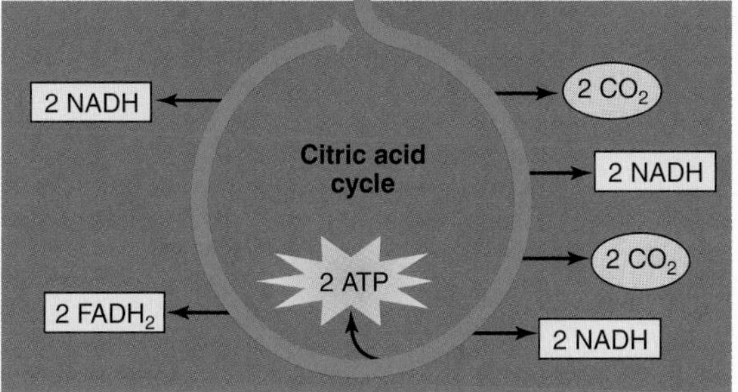

FIGURE 6.10 Summary of Glycolysis and the Citric Acid Cycle. Glycolysis and the citric acid cycle are the central metabolic pathways to extract chemical energy from glucose to generate cellular energy (ATP). »» Every time one glucose molecule is broken down to CO_2 and water, how many ATP, NADH, and $FADH_2$ molecules are gained?

Oxidative Phosphorylation Is the Process by Which Most ATP Molecules Form

Oxidative phosphorylation refers to a sequence of reactions in which two events happen: Pairs of electrons are passed from one chemical substance to another (electron transport), and the energy released during their passage is used to combine phosphate with ADP to form ATP (ATP synthesis). The adjective "oxidative" is derived from the term **oxidation**, which, as defined earlier, refers to the loss of electron pairs from molecules. Its counterpart is **reduction**, which refers to a gain of electron pairs by molecules. "Phosphorylation," as we already have seen, implies adding a phosphate to another molecule. So, in oxidative phosphorylation, the loss and transport of electrons will enable ADP to be phosphorylated to ATP. Oxidative phosphorylation takes place at the cell membrane in bacterial and archaeal cells, and in the mitochondrial inner membrane of eukaryotic microbes.

Oxidative phosphorylation is responsible for producing 28 molecules of ATP per glucose. The overall sequence involves the NAD^+ and FAD coenzymes that underwent reduction to NADH and $FADH_2$ during glycolysis and the citric acid cycle; remember, they gained two electrons in those metabolic pathways. In oxidative phosphorylation, the coenzymes will be reoxidized by transferring those two electrons to a series of electron carriers (FIGURE 6.11). These carriers, called **cytochromes** (*cyto* = "cell"; *chrome* = "color"), are a set of proteins containing iron cofactors that accept and release electron pairs. Together, the cytochrome complexes (I–IV) form an **electron transport chain**. The last link in the chain is oxygen gas.

In the oxidative phosphorylation process shown in Figure 6.11, the two electrons with each NADH and $FADH_2$ are passed to cytochrome complex I in the chain (**A**). The reoxidized coenzymes, NAD^+ and FAD, return to the cytosol (**B**) to be used again in glycolysis or the citric acid cycle. Like walking along stepping stones, each electron pair is passed from one cytochrome complex (I–IV) to the next down the chain (**C**) until the electron pair is finally transferred from complex IV to the final electron acceptor, oxygen. Oxygen also acquires protons from the cytosol (**D**) and forms the final end product, water. Oxygen's role is of great significance because if oxygen was not present, there would be no way for cytochromes to "unload" their

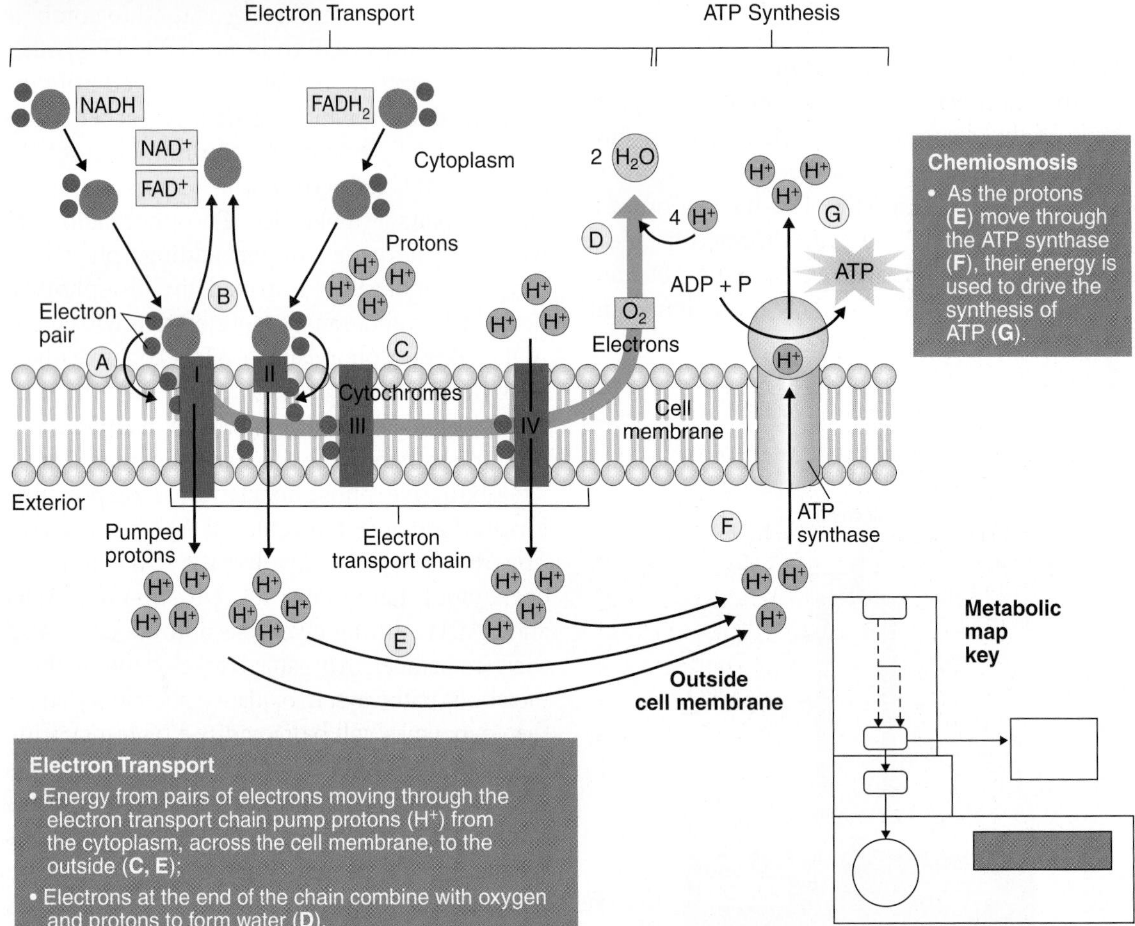

FIGURE 6.11 **Oxidative Phosphorylation in Bacterial Cells.** (**A**) Originating in glycolysis and the citric acid cycle, coenzymes NADH and $FADH_2$ transport electron pairs to the electron transport chain in the cell membrane, which fuels the transport of protons (H^+) across the cell membrane. Protons then reenter the cytosol through a protein channel in the ATP synthase enzyme. ADP molecules join with phosphates as protons move through the channel, producing ATP.
»» What would happen to the oxidative phosphorylation process if this cell were deprived of oxygen?

electrons and the entire system would soon back up like a jammed conveyer belt and come to a halt. Oxygen's role also is reflected in the equation for aerobic respiration:

$$C_6\underline{H_{12}}O_6 + \underline{6}\ \underline{O_2} + 32\ ADP + 32\ P$$
$$\downarrow$$
$$6\ CO_2 + \underline{6}\ \underline{H_2O} + 32\ ATP$$

So, what is the importance of the electron transport chain to ATP synthesis? The actual mechanism for ATP synthesis comes from the pumping of protons by a process called **chemiosmosis** (*osmos* = "push"). First proposed by Nobel Prize winner Peter Mitchell, chemiosmosis uses the power of proton movement across a membrane to conserve energy for ATP synthesis.

What happens in chemiosmosis also is shown in Figure 6.11. As the electrons pass between cytochrome complexes, the electrons gradually lose energy. The energy, however, is not lost in the sense that it is gone forever. Instead, the energy is used at three transition points to "pump" (push) protons (H^+) across the membrane from the cytosol to the area outside of the cell membrane (**E**). Soon a large number of protons have built up outside the membrane, and because they cannot easily reenter the cell, they represent a large concentration of potential energy (much like a boulder at the top of a hill). The protons are positively charged, so there also is a buildup of charges outside the membrane.

Suddenly, a series of channels opens and the proton flow reverses (**F**). Each "channel" is contained within a large, membrane-spanning enzyme complex called **ATP synthase**, which has binding sites for ADP and phosphate. As the protons rush through the channel, they release their energy, and the energy is used to synthesize ATP molecules from ADP and phosphate ions (**G**), as MicroInquiry 6 explains. Based on current experimental data, 2.5 molecules of ATP can be synthesized for each pair of electrons originating from NADH; 1.5 molecules of ATP are produced for each pair of electrons from $FADH_2$ because the coenzyme interacts further down the chain. MicroFocus 6.3 highlights a novel way of using the bacterial respiratory process to generate electricity.

If the cell membrane is damaged so chemiosmosis cannot take place, the synthesis of ATP ceases and the organism rapidly dies. This is one reason why damage to the bacterial cell membrane, such as with antibiotics or detergent disinfectants, is so harmful.

The ATP yield from aerobic respiration is summarized in FIGURE 6.12. The reactions can generate

MICROINQUIRY 6

The Machine That Makes ATP

Every day, an adult human weighing 160 pounds uses up about 80 pounds of ATP (about half of his or her weight). The ATP is changed to its two breakdown products, ADP and phosphate, and enormous amounts of energy are made available to do metabolic work.

However, the body's weight does not go down, nor does it change perceptibly because the cells are constantly regenerating ATP from the breakdown products. Discovering how this is accomplished and how the recycling works were the seminal achievements of 1997's Nobel Prize winners in Chemistry.

The Chemiosmotic Basis for ATP Synthesis

One of the three 1997 winners was Paul D. Boyer at the University of California at Los Angeles. Boyer's work expanded the pioneering work of Peter Mitchell, who developed the concept of chemiosmosis. Chemiosmosis proposes that electron transport between cytochrome complexes provides the energy to "pump" protons (H^+) across the membrane; in the case of bacterial and archaeal cells, this is from the cytosol to the environment. As explained in the text, this proton gradient provides the force or potential to drive the protons back into the cell through an enzyme called ATP synthase. This flow of rapidly streaming H^+ brings together ADP and phosphate to form ATP.

How Does Proton Flow Cause ATP Synthesis?

The groundbreaking research as to how the ATP synthase works came from studies with *Escherichia coli* cells. Today, we know that the ATP synthase consists of two polypeptide complexes (see **Figure A**). The headpiece (F_1) faces into the cytosol and consists of nine polypeptides of five different types (α, β, γ, ε, and δ) and represents the catalytic complex for converting ADP + P to ATP. The basal unit (F_0) is embedded in the cell membrane and consists of 15 polypeptides of three different types (a, b, and c). The basal unit contains the proton-transporting channel through the membrane. So, an ATP synthase consists of 24 polypeptides—a veritable nanomachine.

Boyer took the three complexes and hypothesized how they could manufacture ATP. His ideas plus newer findings have been merged into the current model (see **Figure B**):

1. The flow of protons through the basal unit c proteins causes the basal unit to spin (somewhat similar to the turning of a waterwheel).
2. The γ and ε polypeptides in the headpiece also spin and, as they spin, δ makes contact with each of the β subunits.
3. Each β subunit changes shape, and like an enzyme's active site, allows the subunit to bind an ADP + P and catalyze the production of an ATP.
4. When a β subunit returns to its original shape, it releases the ATP.

Because there are three β subunits in the headpiece, three ATP molecules are produced each time the basal unit and the γ and ε polypeptides make one complete rotation.

Discussion Point

Ribosomes, flagella, and ATP synthase all represent "nanomachines" to carry out specific functions in cells. Discuss the concept of a bacterial cell as being an assemblage of nanomachines.

Answers (and comments) can be found on the Student Companion Website in **Appendix D**.

(continued on next page)

MICROINQUIRY 6

The Machine That Makes ATP (continued)

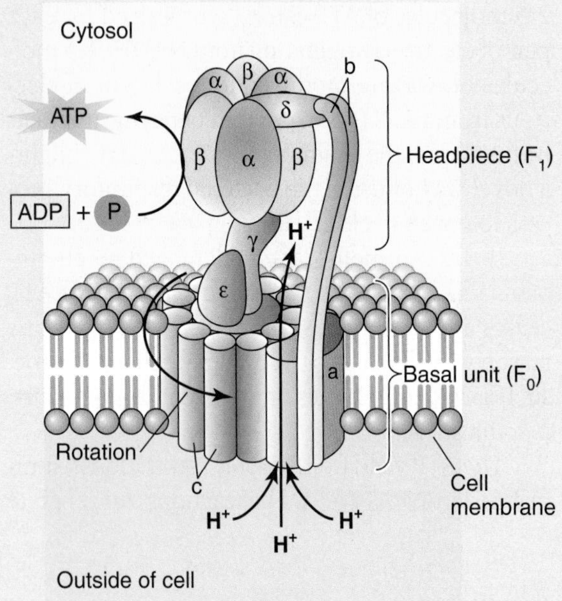

FIGURE A A bacterial ATP synthase enzyme consists of 24 polypeptides in two complexes, the basal unit embedded in the cell membrane and the headpiece that projects into the cytosol.

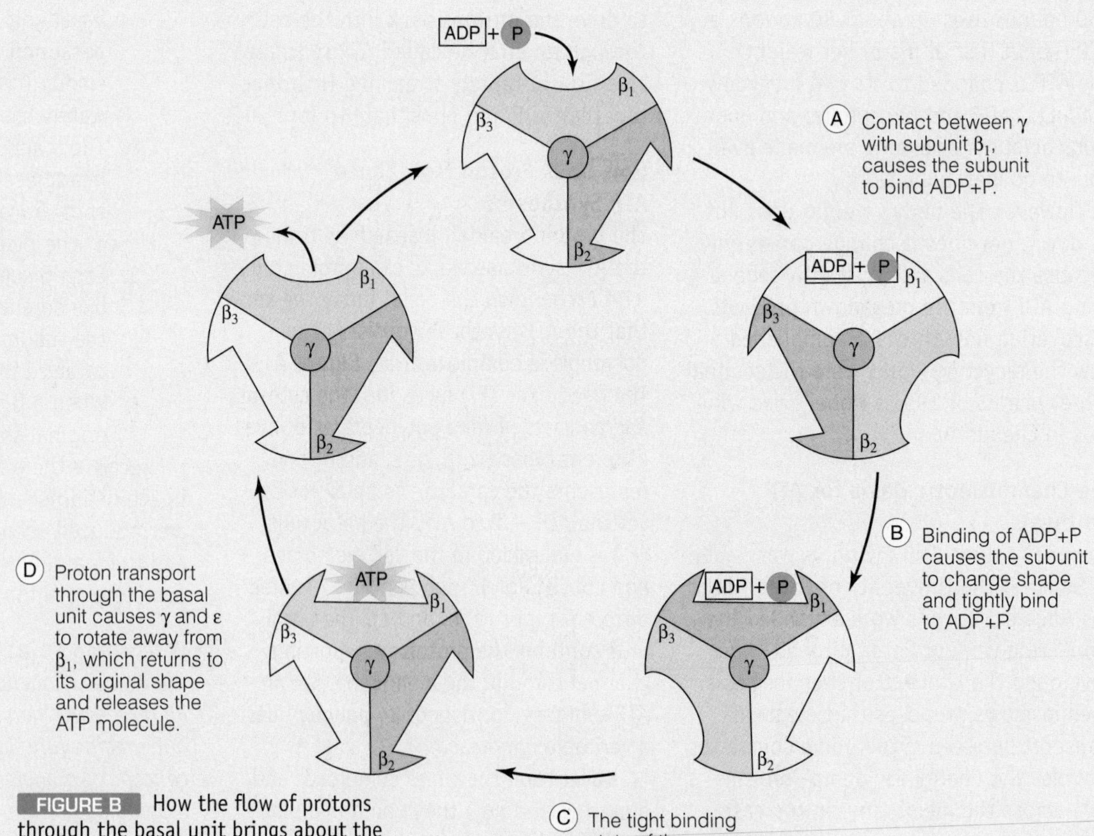

FIGURE B How the flow of protons through the basal unit brings about the ATP synthesis is shown for one of the three β subunits in the headpiece.

MICROFOCUS 6.3: Biotechnology

Bacteria Not Included

How many toys (child or adult) or electronic devices do you purchase each year where batteries are needed to run the device? And often batteries are not included. Today, a new type of battery is being developed—one that converts sugar not into ATP but rather into electricity. The battery is one packed with bacterial cells.

Realize that cellular respiration involves the generation of minute electrical currents. During cellular respiration, electrons are transferred to cofactors like NAD^+ and then passed along a chain of cytochrome complexes during oxidative phosphorylation. Swades Chaudhuri and Derek Lovely of the University of Massachusetts at Amherst have taken this idea and applied it to developing a new type of fuel cell or battery.

The scientists mixed the bacterial species *Rhodoferax ferrireducens*, which they found in aquifer sediments in Virginia, with a variety of common sugars. When placed in a chamber with a graphite electrode, *R. ferrireducens* metabolized the sugar, stripped off the electrons, and transferred them directly to the electrode. The result: an electrical current was produced. In addition, the bacterial cells continued to grow, so a stable current could be produced with high efficiency.

Although it is still a long way from producing a reliable, long-lasting bacterial battery, the researchers believe much of the agricultural or industrial waste produced today could be the "sugar" used in making these bacterial batteries. So, as Sarah Graham reported for *ScientificAmerican.com*, "Perhaps one day electronics will be sold with the caveat 'bacteria not included.' "

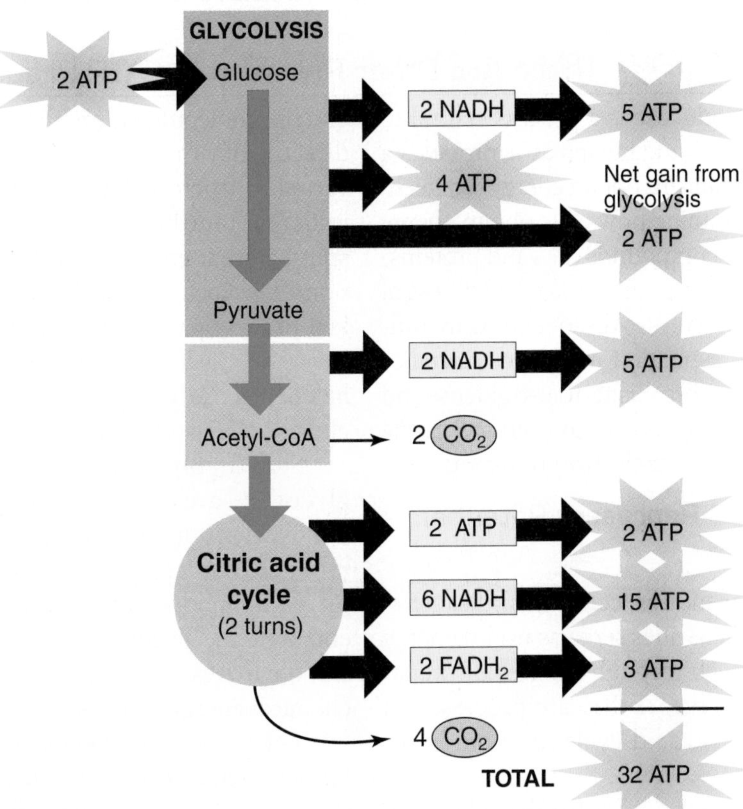

FIGURE 6.12 **The ATP Yield from Aerobic Respiration.** In a microbial cell, 32 molecules of ATP can result from the metabolism of a molecule of glucose. Each NADH molecule accounts for the formation of 2.5 molecules of ATP; each molecule of $FADH_2$ accounts for 1.5 ATP molecules. »» From this diagram, what is the single most important reactant for ATP synthesis?

up to 32 ATP molecules from one glucose molecule. It also completes the equation for aerobic respiration:

$$C_6H_{12}O_6 + 6\ O_2 + \underline{32\ ADP} + \underline{32\ P}$$
$$\downarrow$$
$$6\ CO_2 + 6\ H_2O + \underline{32\ ATP}$$

CONCEPT AND REASONING CHECKS 2

a. Write the summary equation for cellular respiration and indicate where each of the substrate atoms ends up in the products.
b. At the end of glycolysis, where is the energy that was originally in glucose?
c. Identify the initial substrates and final end products of the citric acid cycle.
d. Why are NADH and $FADH_2$ so critical to cell energy metabolism?

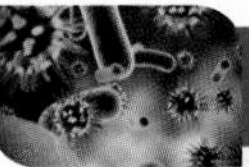

Chapter Challenge **B**

In our Chapter Challenge, hopefully you have figured out which structural polysaccharide of plant cell walls was the common denominator in the metabolism of termites and cows—and that the polysaccharide is being broken down by microbes that reside in the termite gut and cow rumen.

QUESTION B:
What is the building block, or monomer, that builds this structural polysaccharide?

And if this monomer could be produced through a catabolic process, and then used for aerobic respiration, what are the final products formed besides ATP? Is one of them methane gas? Explain. So, do the guts of termites and the rumens of cows carry out aerobic respiration? Explain. I believe we need to keep looking.

Answers can be found on the Student Companion Website in **Appendix D**.

KEY CONCEPT 6.3 There Are Other Pathways to ATP Production

The catabolism of glucose is a process central to the metabolism of all organisms as it provides a glimpse of how organisms obtain energy for life. In this section, we examine how cells obtain energy from other organic compounds (fats and proteins) by directing those compounds into the process of cellular respiration. We also discover how modifications to cellular respiration and glucose metabolism allow anaerobic organisms to use glucose and generate ATP without having oxygen gas as the final electron acceptor in electron transport.

Other Nutrients Represent Potential Energy Sources

A wide variety of monosaccharides, disaccharides, and polysaccharides serve as useful energy sources. All must go through a series of preparatory conversions before they are processed in glycolysis, the citric acid cycle, and oxidative phosphorylation.

In preparation for entry into the scheme of metabolism, different carbohydrates use different pathways (FIGURE 6.13). Sucrose, for example, is first digested by the enzyme sucrase into its constituent molecules, glucose and fructose. The glucose molecule enters the glycolysis pathway directly, but the fructose molecule first is converted to fructose-1-phosphate. The latter then undergoes further conversions and a molecular split before it enters the scheme as DHAP [reaction (**4**)]. Lactose, another disaccharide, is broken in two by the enzyme lactase to glucose and galactose. Galactose undergoes a series of changes before it is ready to enter glycolysis in the form of glucose-6-phosphate [reaction (**2**)].

Stored polysaccharides, such as starch and glycogen, are metabolized by enzymes that remove one glucose unit at a time and convert it to glucose-1-phosphate. An enzyme converts this compound to glucose-6-phosphate, ready for entry into the glycolysis pathway. The point is that carbohydrates other than glucose also are used as chemical energy sources.

The economy of metabolism is demonstrated further when we consider protein and fat catabolism (Figure 6.13). Fats are extremely valuable energy sources because their chemical bonds contain enormous amounts of chemical energy. Although proteins are generally not considered energy sources, cells use them for energy when carbohydrates and

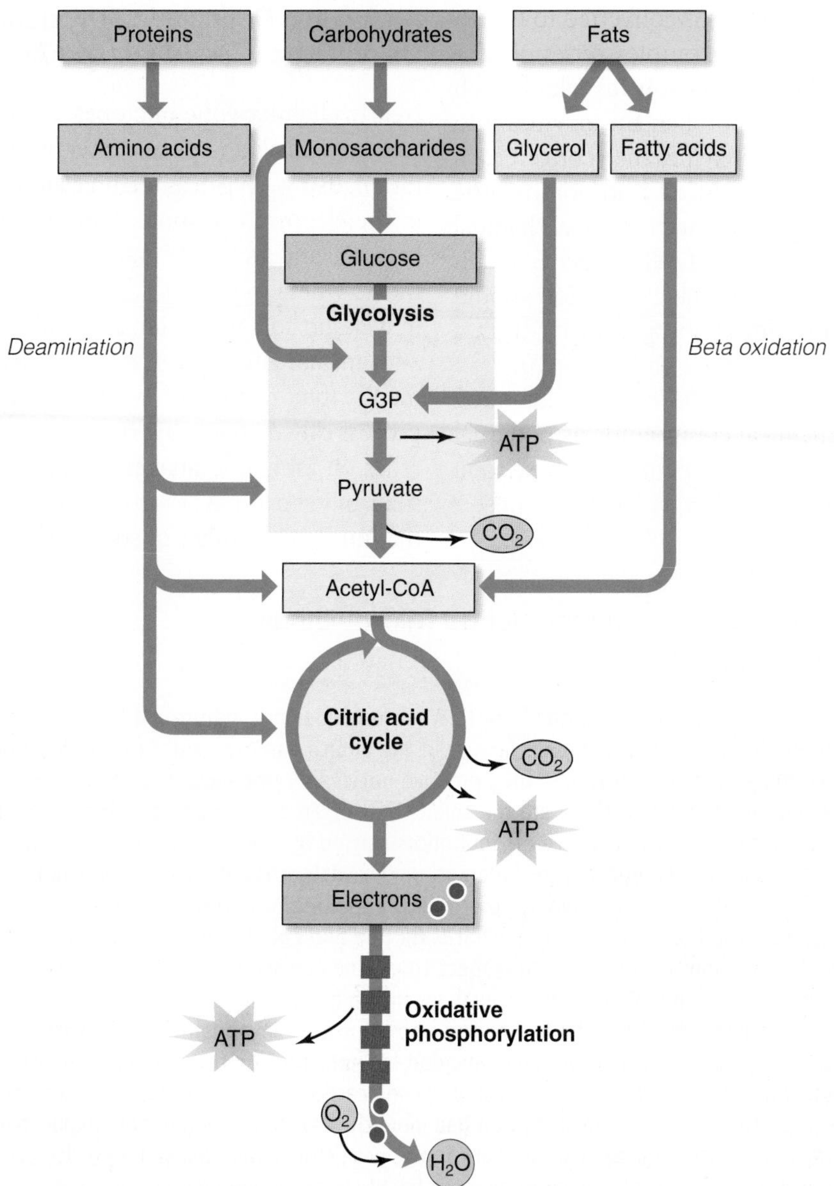

FIGURE 6.13 **Carbohydrate, Protein, and Fat Metabolism.** Besides glucose, other carbohydrates as well as proteins and fats can be sources of energy by providing electrons and protons for cellular respiration. The intermediates enter the pathway at various points. »» Why would more ATP be produced from the products of one fatty acid entering the pathway at acetyl-CoA than from one amino acid entering at acetyl-CoA?

fats are in short supply. Both fats and proteins are broken down through glucose catabolism as well as through other pathways. Basically, the proteins and fats undergo a series of enzyme-catalyzed conversions and form components normally occurring in carbohydrate metabolism. These components then continue along the metabolic pathways as if they originated from carbohydrates.

Proteins are broken down to amino acids. Enzymes then convert many amino acids to pathway components by removing the amino group and substituting a carbonyl group. This process is called **deamination**. For example, alanine is converted to pyruvate and aspartic acid is converted to oxaloacetate. For certain amino acids, the process is more complex, but the result is the same: The amino acids become pathway intermediates of cellular respiration.

Fats consist of three fatty acids bonded to a glycerol molecule. To be useful for energy purposes, the fatty acids are separated from the glycerol by the enzyme lipase. Once this has taken place, the glycerol portion is converted to DHAP. For fatty acids, there is a complex series of conversions called **beta oxidation**, in which each long-chain fatty acid is broken by enzymes into 2-carbon units. Other enzymes then convert each unit to a molecule of acetyl-CoA ready for the citric acid cycle. We previously noted that for each initial 6C glucose, 24 molecules of ATP are generated in the citric acid cycle. A quick calculation should illustrate the substantial energy output from a 16-carbon fatty acid (eight 2-carbon units).

In most habitats, microbes can use a large and diverse set of chemical compounds as potential energy sources. When competing for these food resources, some bacterial organisms can "raise a stink" (MicroFocus 6.4).

Anaerobic Respiration Produces ATP Using Other Final Electron Acceptors

Nearly all eukaryotic microbes, as well as multicellular animals and plants, carry out aerobic respiration, using oxygen as the final electron acceptor in the electron transport chain. However, many bacterial and archaeal organisms exist in environments where oxygen is scarce or absent, such as in wetland soil and water, and within the human and animal digestive tracts. In these environments, the organisms have evolved a respiratory process called **anaerobic respiration** that usually relies on inorganic final electron acceptors other than oxygen for ATP production. Considering the immense number of species that live in such

MICROFOCUS 6.4: Environmental Microbiology
Microbes "Raise a Stink"

We are all familiar with body odor and bad breath. When one does not maintain a level of cleanliness or hygiene, bacterial species on the skin surface or in the mouth can grow out of proportion and, as they metabolize compounds like proteins, they produce noticeably unpleasant, smelly odors.

On a more environmental level, there are the smells that often come from decaying or rotting foods. As decomposers, these microbes also give off foul odors caused by the presence of bacterial species colonizing the dead carrion. Competing in the environment with other animal scavengers for food, do these bacterially produced odors have a useful role in repelling or deterring animal species from consuming important food resources? This is what Mark Hay and collaborators at Georgia Institute of Technology wanted to know, especially with respect to marine ecosystems. Their hypothesis: Decaying food resources become repugnant to larger animal species like crabs or fish.

To test their hypothesis, the research team baited crab traps near Savannah, Georgia, with menhaden, a typical bait-fish for crabs. Some traps contained microbe-laden menhaden that had been allowed to rot for one or two days, while other traps contained freshly thawed carrion having relatively few microbes. When the traps were inspected, those with fresh carrion had more than twice the number of animals per trap than did the traps with microbe-laden carrion. Lab studies with stone crabs showed they also avoided the microbe-laden, rotting food, but readily consumed the freshly thawed menhaden.

To examine the role of bacterial organisms in the avoidance behavior by stone crabs (see figure), some menhaden was allowed to rot in water without the antibiotic chloramphenicol while other samples contained the antibiotic in the water to prevent or inhibit microbial growth. Again, the study observed that the crabs readily ate the antibiotic-incubated menhaden but avoided the menhaden without antibiotic; the bacterial organisms were in some way responsible for the aversion.

Finally, the researchers used organic extracts prepared from the microbe-laden carrion and mixed these chemical substances with freshly thawed menhaden. Again, the crabs were repelled. Exactly what chemical compounds were responsible for the behavior were not evident from the study. In summary, it appears that bacterial species not only act as decomposers and pathogens in the environment, but also can compete very successfully with relatively large animal consumers for mutually attractive food sources.

A stone crab. Courtesy of Catherine Billick, www.flickr.com/photos/catbcorner/.

anaerobic environments, anaerobic respiration is extremely important ecologically.

The human enteric, facultative species, for example, uses nitrate (NO_3^-) with which electrons combine to form nitrite (NO_2^-) or another nitrogen product. The obligate anaerobe *Desulfovibrio* uses sulfate (SO_4^{2-}) for anaerobic respiration. The sulfate combines with the electrons from the cytochrome chain and is reduced to hydrogen sulfide (H_2S). This gas gives a rotten egg smell to the environment (as in a tightly compacted landfill). A final example is exhibited by the archaeal methanogens. These obligate anaerobes use carbonate (CO_3) or CO_2 as a final electron acceptor and, with hydrogen nuclei, form large amounts of methane gas (CH_4).

In anaerobic respiration, the amount of ATP produced is less than in aerobic respiration. There are several reasons for this. First, only a portion of the citric acid cycle functions in anaerobic respiration, so fewer reduced coenzymes are available to the electron transport chain. Also, not all of the cytochrome complexes function during anaerobic respiration, so the ATP yield will be less.

Fermentation Produces ATP Using an Organic Final Electron Acceptor

In environments that are anoxic and without the alternative final electron acceptors needed by anaerobes, much of the organic material will be catabolized through fermentation. **Fermentation**, probably the most ancient form of energy metabolism, is the enzymatic process for producing ATP by substrate-level phosphorylation, using endogenous organic compounds as both electron donors and acceptors—exogenous final electron acceptors (O_2, NO_3^-, SO_4^{2-}, CO_3) are absent.

The chemical process of fermentation makes a few ATP molecules in the absence of cellular respiration. However, the citric acid cycle and oxidative phosphorylation are shut down, so the products of glycolysis (pyruvate) are shuttled through a pathway that produces other final end products. In these pathways, pyruvate is the intermediary accepting the electrons.

In all cases, no matter what the end product, fermentation ensures a constant supply of NAD^+ for glycolysis and the production of two ATP molecules per glucose (FIGURE 6.14A). For example, in the fermentation of glucose by *Streptococcus lactis*, the conversion of pyruvate to lactic acid is a way to reform NAD^+ coenzymes so glycolysis can still make two ATP molecules for every glucose molecule consumed (FIGURE 6.14B).

The diversity of fermentation chemistry extends to some eukaryotic microbes as well. In yeasts like *Saccharomyces*, when pyruvate is converted to ethyl alcohol (ethanol), NAD^+ is reformed.

The energy benefits to fermentative organisms are far less than in cellular respiration. In fermentation, each glucose passing through glycolysis yields two ATP molecules and the production of fermentation end products. This is in sharp contrast to the 32 molecules from cellular respiration. It is clear that cellular respiration is the better choice for energy conservation, but under anoxic conditions, there may be little alternative if life for *S. lactis*, *Saccharomyces*, or any fermentative microorganism is to continue.

Although fermentation end products are waste products to the microbes producing them, industries see many of these products in a very different light (Figure 6.14B).

In a dairy plant, the process of **lactic acid fermentation** by *S. lactis* is carefully controlled so the acid will curdle fresh milk to make buttermilk or yogurt. The liquor industry uses the ethyl alcohol produced in **alcoholic fermentation** to make alcoholic beverages such as beer and wine. Fermentation of carbohydrates to alcohol by *C. albicans* also may take place in the human body, such as that of Charlie Swaart's, as described in the opener of this chapter.

As indicated in Figure 6.14B, other industries also make use of microbial fermentations. For example, Swiss cheese develops its flavor from the propionic acid produced during fermentation and gets its holes from trapped carbon dioxide gas resulting from fermentation.

The chemical industries also have harnessed the power of fermentation in the production of acetone, butanol, and other industrial solvents. Thus, fermentation is useful not only to microorganisms, but also to consumers who enjoy its products and industry that benefits from the products.

The ability of microbes to carry out different fermentation reactions that produce different end products can be very useful in species identification. Therefore, specific tests have been developed to detect particular end products. For example, with enteric species the "methyl red test" maintains a red colored solution if a

■ Anoxic:
Without oxygen gas (O_2).

■ Enteric:
Referring to the intestines.

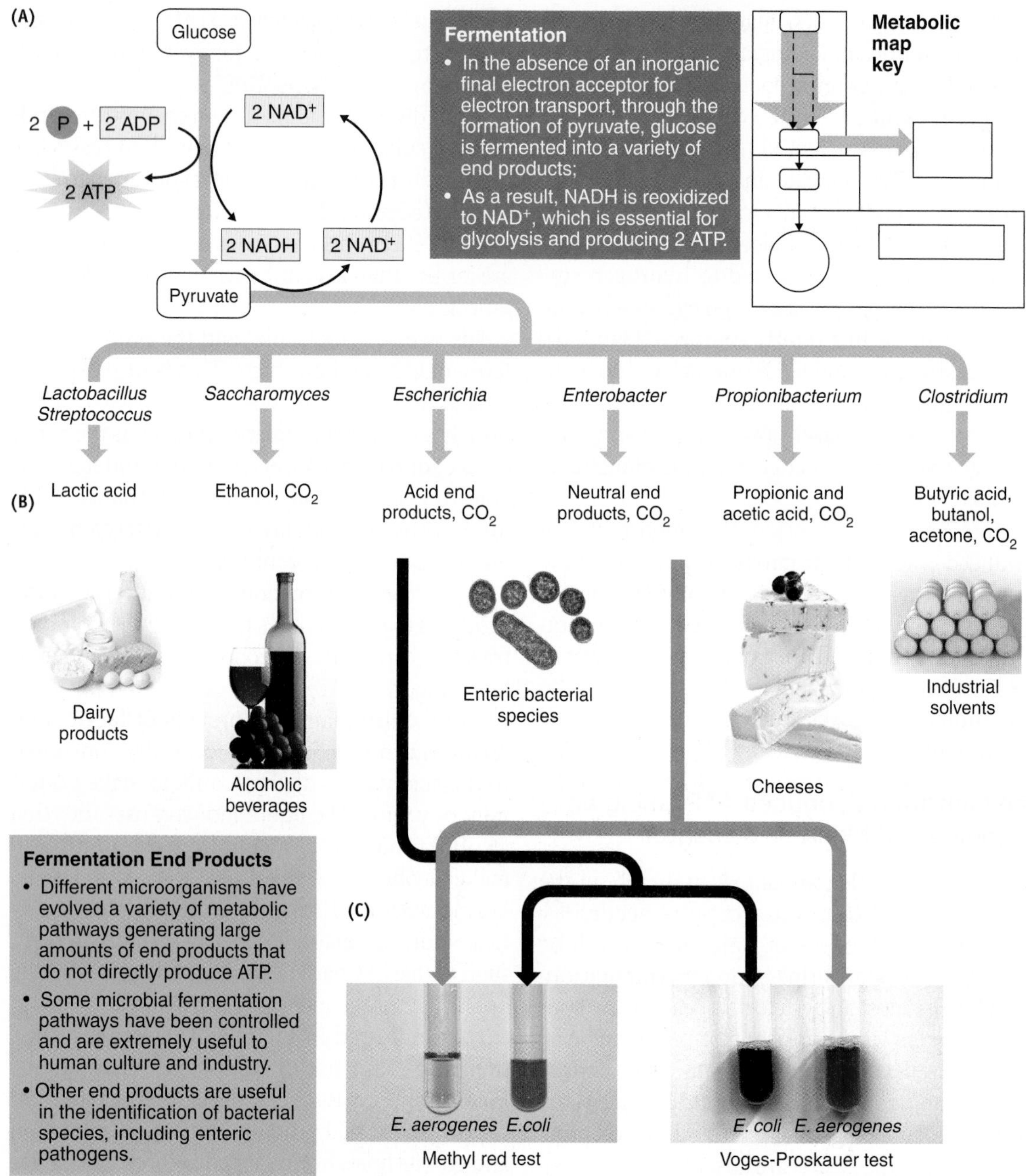

FIGURE 6.14 **Microbial Fermentation.** (**A**) Fermentation is an anaerobic process that reoxidizes NADH to NAD^+ by converting organic materials into fermentation end products (**B**). Fermentation end products also provide a way to help identify bacterial species (**C**). (**B**) © niderlander/ShutterStock, Inc.; © drKaczmar/ShutterStock, Inc.; © SPL/Photo Researchers, Inc.; © Elena Elisseeva/ShutterStock, Inc.; © Esteban De Armas/ShutterStock, Inc. (**C**) Courtesy of Jeffrey Pommerville. »» How are lactic acid and alcohol fermentation identical in purpose?

species can ferment glucose to acid end products, while the "Voges-Proskauer test" produces a brownish-red colored solution if a species forms neutral end products from the acids produced through glucose fermentation (FIGURE 6.14C). These and other physiological and biochemical tests are often essential to the clinical lab microbiologist to identify a potential pathogen—and most likely for you in your microbiology lab class to identify an unknown bacterial species.

CONCEPT AND REASONING CHECKS 3

a. Describe how lipids and proteins are prepared for entry into the cellular respiration pathway.

b. Why would obligate anaerobes tend to grow slower than obligate aerobes?

c. Justify the need for some microbes to produce fermentation end products.

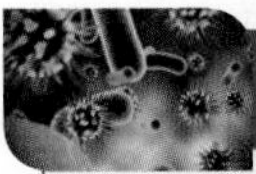

Chapter Challenge C

So we have decided that aerobic respiration is not the source for the methane that is produced by the termite gut and cow rumen microbes. Yet, the structural polysaccharide we have been referring to still is broken down into glucose and is used as a substrate for ATP generation.

QUESTION C:

In what other catabolic pathway can glucose be used to generate ATP? If you have figured it out, what is the nature of the environment in the termite's gut and the cow's rumen? And by association, what must the waterlogged rice paddy field be like?

Answers can be found on the Student Companion Website in **Appendix D.**

KEY CONCEPT 6.4 Photosynthesis Converts Light Energy to Chemical Energy

Although the anabolism or synthesis of carbohydrates takes place through various mechanisms in microorganisms, the unifying feature is the requirement for energy.

Photosynthesis Is a Process to Acquire Chemical Energy

Photosynthesis is a process by which light energy is converted to chemical energy that is then stored as carbohydrate or other organic compounds. In the cyanobacteria, the process takes place in special **thylakoid membranes**, which contain chlorophyll or chlorophyll-like pigments (FIGURE 6.15A). Among eukaryotes, photosynthesis occurs in the chloroplasts of such organisms as diatoms, dinoflagellates, and green algae (FIGURE 6.15B). In all cases, these microbes carry out **oxygenic photosynthesis**; that is, where oxygen gas (O_2) is a byproduct of the process. The Investigating the Microbial World 6 looks at an experiment that used microbes to make an important contribution toward understanding the photosynthetic process.

The phases of photosynthesis in the cyanobacteria are shown in FIGURE 6.16, where the sequence of stages is labeled by number.

The Energy-Fixing Reactions. In the first part of photosynthesis, light energy is converted into or "fixed" as chemical energy in the form of ATP and NADPH. Thus, the process is referred to as the **energy-fixing reactions**, or "light-dependent" reactions, because light energy is required for the process.

Like the reactions of cellular respiration, the energy-fixing reactions of photosynthesis are dependent on electrons and protons. The source of

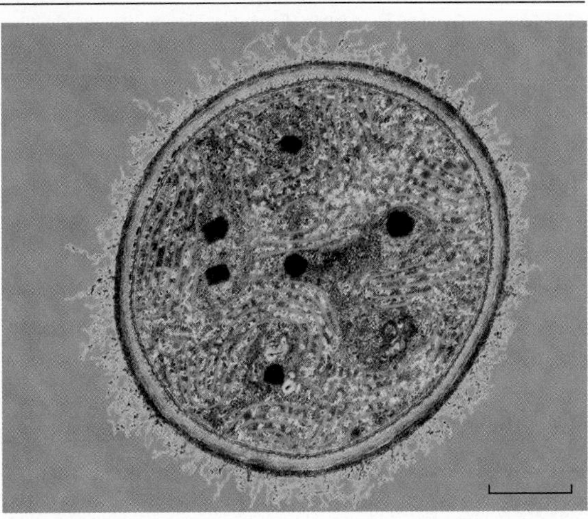

(A)

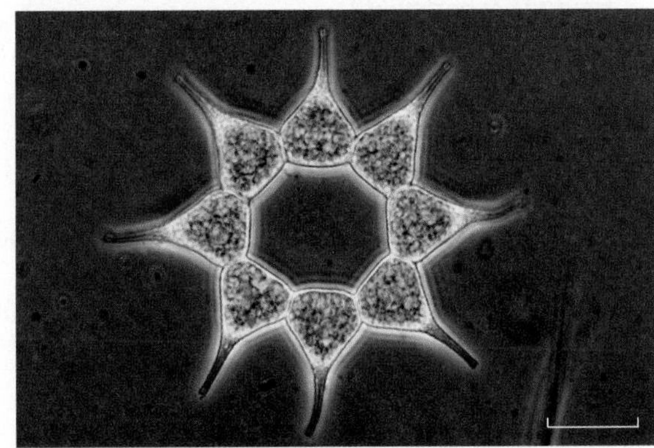

(B)

FIGURE 6.15 **Photosynthetic Microbes.** (**A**) False-color transmission electron micrograph of a cyanobacterium displaying the membranes (green) along which photosynthetic pigments are located. Carboxysomes are the polyhedral, black bodies. (Bar = 2 μm.) © Dennis Kunkel Microscopy, Inc./Visuals Unlimited/Corbis. (**B**) A light micrograph of the colonial green alga *Pediastrum*. (Bar = 5 μm). © blickwinkel/Alamy. »» What membranes in algae are analagous to the cyanobacterial membranes?

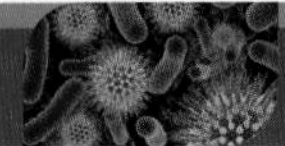

Investigating the Microbial World 6

Microbes Provide the Answer

It was known by the late 1700s that plants and algae produced oxygen gas during photosynthesis. In the late 1800s, Theodor W. Engelmann, a German physiologist, botanist, and microbiologist, confirmed that oxygen production was dependent on chlorophyll that is found in chloroplasts of plant and algal cells. He further observed bacterial aerotaxis, the movement of oxygen-sensing cells toward higher concentrations of oxygen gas.

- **OBSERVATION:** In 1881, Engelmann observed that when white light was shining on the green alga *Spirogyra,* the motile bacterial cells moved toward the chloroplasts in a filament of the green alga. Knowing that the bacterial cells exhibited aerotaxis and that white light is composed of a spectrum of colors (red, orange, yellow, green, blue, indigo, and violet), he wondered:
- **QUESTION:** ***Which of these colors (wavelengths) in white light stimulated photosynthesis?***
- **HYPOTHESIS:** Photosynthesis is more efficient with certain colors within the spectrum of white light. If so, then the aerotactic bacterial cells can be used to identify which colors (wavelengths) are more efficient (as measured by the production of oxygen gas).
- **EXPERIMENTAL DESIGN:** In 1882, Engelmann had a specially designed microscope equipped with a prism that would break white light into its spectrum of colors on a microscope slide.
- **EXPERIMENT:** A filament of the green alga *Cladophora* was placed on a slide on the special microscope, which exposed different segments of the algal filament to different colors (wavelengths) of visible light. Motile cells of the aerotactic bacterial species were then added to the slide and the algal filament was observed to see the effect the different colors (wavelengths) of light had on the accumulation of the aerotactic bacterial cells.
- **RESULTS:** See figure.
- **CONCLUSION:**

  QUESTION 1: ***What could Engelmann conclude from his experiment?***

  QUESTION 2: ***Did the results support his hypothesis? Explain.***

  QUESTION 3: ***What would have been the response of the aerotactic bacteria if only green light was used to illuminate the algal filament?***

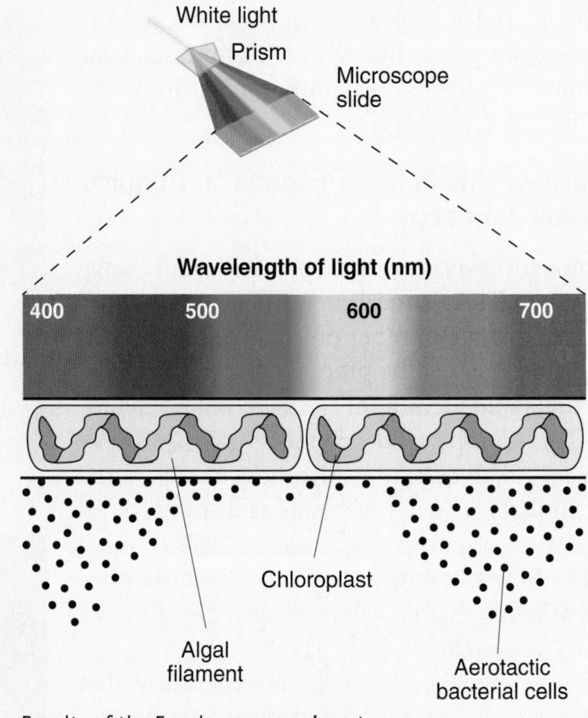

Results of the Engelmann experiment.

Answers can be found on the Student Companion Website in **Appendix D.**

Adapted from: Engelmann, T. W. (1882). *Bot. Zeit.* **40:** 419–426. Icon image © Tischenko Irina/ShutterStock, Inc.

these atomic particles is water. In the cyanobacteria and algae, the splitting of water not only produces the needed atomic particles, it releases oxygen as a byproduct:

$$2\,H_2O \rightarrow 4\,H^+ + 4e^- + O_2$$

The ability of the ancestors of modern-day cyanobacteria to generate oxygen gas profoundly changed the atmosphere of Earth some 3.8 billion years ago, leading to the evolution of aerobic organisms carrying out aerobic respiration.

Light energy is absorbed by the green pigment **chlorophyll *a***, a magnesium-containing, lipid-soluble compound (Figure 6.16A). Chlorophylls and accessory pigments make up light-receiving complexes called **photosystems**.

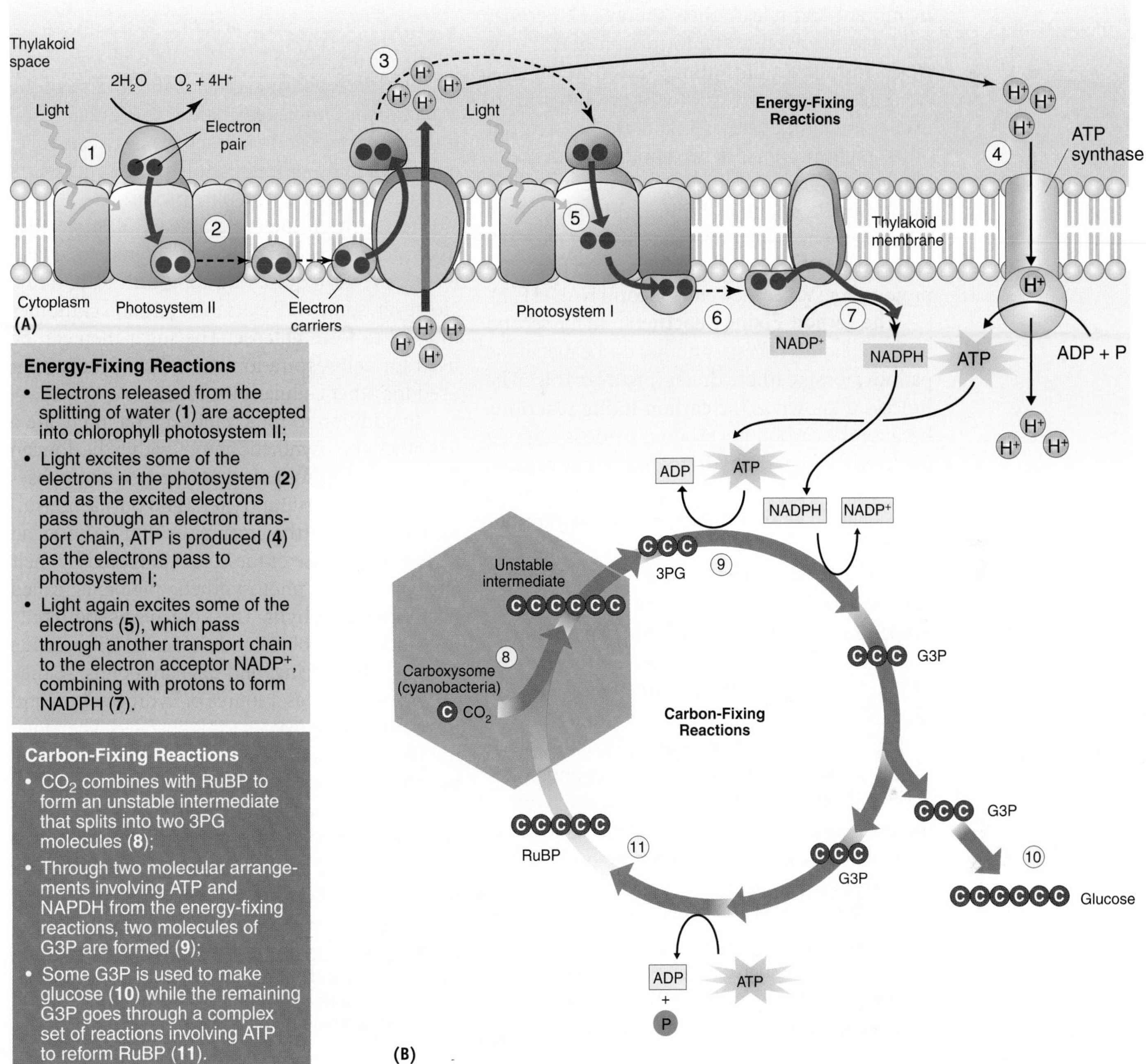

FIGURE 6.16 **Photosynthesis in Cyanobacteria and Algae.** (**A**) The energy-fixing reactions generate ATP and NADPH. (**B**) The carbon-fixing reactions unite carbon dioxide with ribulose bisphosphate (RuBP) to form two molecules of 3-phosphoglycerate (3PG) then glyceraldehyde-3-phosphate (G3P). ATP and NADPH from the energy-fixing reactions are used in the latter conversion. Condensations of two 3-carbon G3P molecules yields glucose, and the remainder are used to form RuBP to continue the process. »» How are the carbon-fixing reactions of photosynthesis dependent on the energy-fixing reactions?

The light excites pigment molecules in photosystem II, resulting in the loss of electrons (**1**). These electrons are replaced from the splitting of water.

Excited electrons are immediately accepted by the first of a series of electron carriers (**2**). The electrons are passed along the membrane carriers and eventually the electrons are taken up by other chlorophyll pigments that form photosystem I (**5**).

As the electrons move between cytochromes, energy is made available for proton (H^+) pumping across the thylakoid membrane of the cyanobacterium (**3**), followed by chemiosmosis. As described for oxidative phosphorylation, ATP is formed when protons pass back across the membrane and release their energy (**4**). Because light was involved in the formation of ATP, this process is called **photophosporylation**.

The electrons in photosystem I again are excited by light energy (**5**) and are boosted out of the pigment molecules to another set of membrane carriers (**6**), and finally to a coenzyme called nicotinamide adenine dinucleotide phosphate ($NADP^+$). The coenzyme functions much like NAD^+ in that $NADP^+$ receives pairs of electrons and protons from water molecules to form NADPH (**7**).

The Carbon-Fixing Reactions. In the second stage of photosynthesis, another cyclic metabolic pathway forms carbohydrates (Figure 6.16B). The process is known as the **carbon-fixing reactions** because the carbon in carbon dioxide is trapped (or fixed) into carbohydrates and other organic compounds. It also is called the "Calvin-Benson cycle," named after the biologists who worked out the sequence of reactions.

The first part of this metabolic pathway in cyanobacterial cells is associated with **carboxysomes**, a type of microcompartment found in many bacterial cells. The carboxysomes contain the enzyme ribulose-1,5-bisphosphate carboxylase/oxidase (RuBisCO), which catalyzes the bonding of CO_2 with a 5-carbon sugar ribulose-1,5-bisphosphate (RuBP) to form two molecules of 3-phosphoglycerate (3PG) (**8**). The 3PG then is transported out of the carboxysomes and into the cytoplasm where the remaining carbon-fixing reactions occur. In the next step, the products of the energy-fixing reactions, ATP and NADPH, drive the conversion of 3PG to glyceraldehyde-3-phosphate (G3P)(**9**). Two molecules of G3P then condense with each other to form a molecule of glucose (**10**). Thus, the overall formula for photosynthesis may be expressed as:

$$6\ CO_2 + 6\ H_2O + ATP \xrightarrow{\text{light}} C_6H_{12}O_6 + 6\ O_2 + ADP + P$$

Notice that this reaction is the reverse of the equation for aerobic respiration. The fundamental difference is that aerobic respiration is a catabolic, energy-yielding process, while photosynthesis is an anabolic, energy-trapping process.

To finish off the cycle, most G3P molecules undergo a complex series of enzyme-catalyzed reactions that require ATP to reform RuBP (**11**). However, some G3P exits the cycle and combines in pairs to form glucose. The sugar then can be used for cell respiration, stored as glycogen, or used for other cellular purposes.

In addition to the Cyanobacteria, there are a few other photosynthetic groups within the domain Bacteria. These groups include the green sulfur bacteria, the purple sulfur bacteria and purple nonsulfur bacteria, and the green nonsulfur bacteria, all so named because of the colors imparted by their chlorophyll-like photosynthetic pigments, called **bacteriochlorophylls**. These organisms commonly live under anaerobic conditions in environments such as sulfur springs and stagnant ponds, so they do not use water as a source of hydrogen ions and electrons, and no oxygen gas is liberated. Therefore, the photosynthetic process is called **anoxygenic photosynthesis**. Instead of water, these organisms use hydrogen gas (H_2) or hydrogen sulfide gas (H_2S) as a source of electrons and hydrogen ions. Lastly, the green sulfur bacteria have only photosystem I from which NADPH is generated, while the other groups mentioned above have only photosystem II.

CONCEPT AND REASONING CHECKS 4

a. Compare and contrast the processes of glucose catabolism (aerobic respiration) with glucose anabolism (photosynthesis).

KEY CONCEPT 6.5 Microbes Exhibit Metabolic Diversity

In this text, we examined the processes and patterns that all organisms, including microbes, use to generate cellular energy. Provided there is a carbon source and some form of energy that can be harvested to extract cellular energy, organisms can survive and thrive. The unique aspect is that in the microbes, and especially the domains Bacteria and Archaea, the sources can be a diverse group of organic or inorganic substances. This we refer to as their nutritional pattern; that is, their source of carbon and energy.

Autotrophs and Heterotrophs Get Their Energy and Carbon in Different Ways

Two different patterns exist for satisfying an organism's metabolic needs. These patterns are called **autotrophy** and **heterotrophy**. They are primarily based on the preferred source of carbon and energy (FIGURE 6.17).

Autotrophs. Organisms that synthesize their own foods from simple carbon sources such as

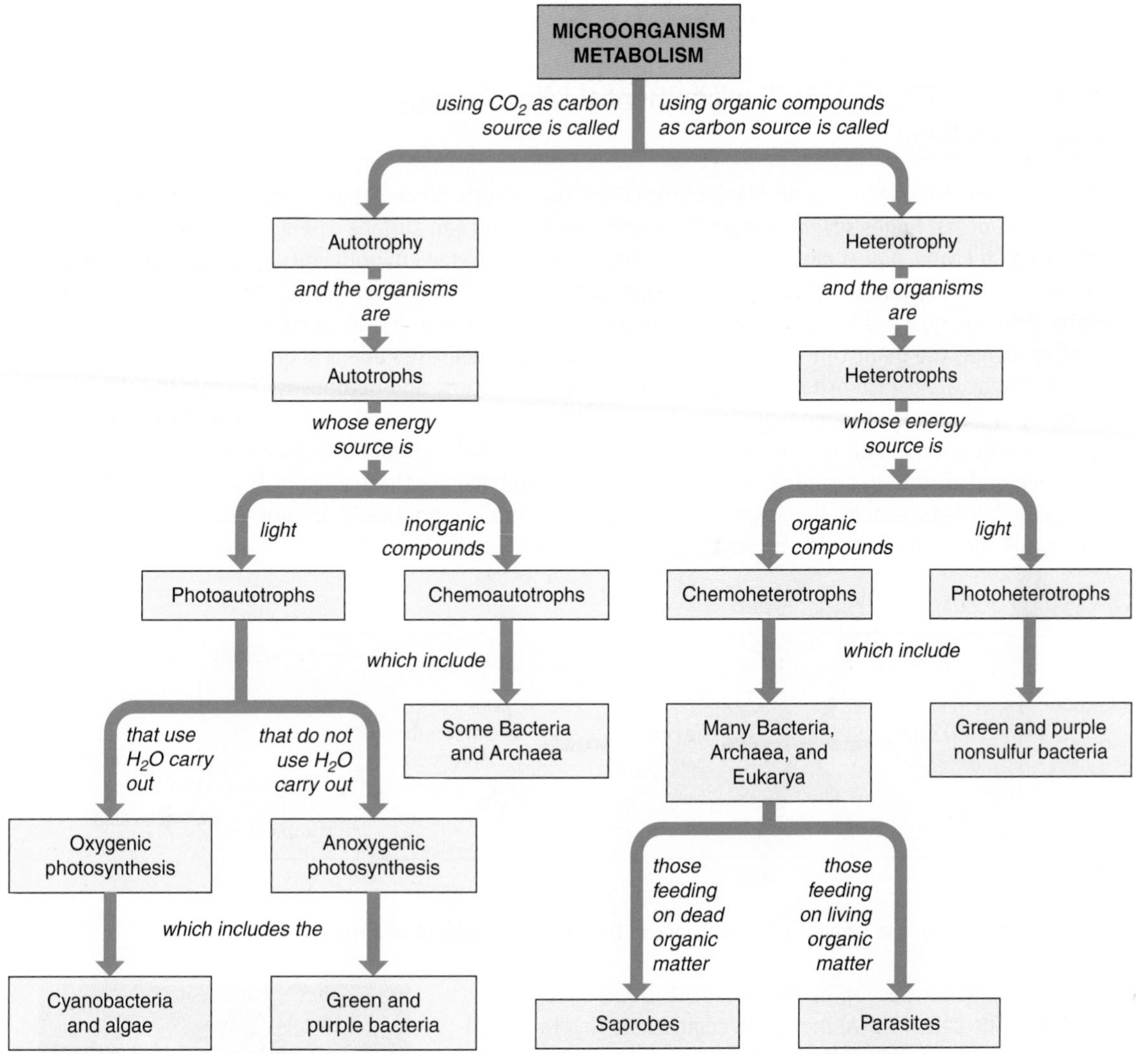

FIGURE 6.17 **Microbial Metabolic Diversity.** This concept map summarizes microbial metabolism based on carbon sources (CO_2 or chemical compounds) and energy sources (light or inorganic/organic compounds). »» Why do the Bacteria and Archaea show such diversity of metabolic types?

CO_2 are referred to as **autotrophs** (*auto* = "self"; *troph* = "nourish"). Those that use light as the energy source, such as the cyanobacteria, are **photoautotrophs**. Microorganisms, including the cyanobacteria and algae, can carry out photosynthesis using water and producing oxygen gas (oxygenic photosynthesis), while other bacterial species carry out anoxygenic photosynthesis as described above.

Another group of autotrophs do not use light as an energy source. Instead, they use inorganic compounds and are referred to as **chemoautotrophs**. For example, species of *Nitrosomonas* convert ammonium ions (NH_4^+) into nitrite ions (NO_2^-) under aerobic conditions, thereby obtaining ATP. The genus *Nitrobacter* then converts the nitrite ions into nitrate ions (NO_3^-), also as an ATP-generating mechanism. In addition to providing energy to both bacterial species, these reactions have great significance in the environment as a critical part of the nitrogen cycle. By preserving nitrogen in the soil in the form of nitrate or ammonia, it can be used by green plants to form amino acids. Another example of chemoautotrophy involving a symbiosis between animal and bacterial cells is described in MicroFocus 6.5.

■ Nitrogen cycle: A biogeochemical process that cycles nitrogen gas into nitrogenous compounds and back again.

Heterotrophs. Many microorganisms are **heterotrophs** (*hetero* = "other"). Such heterotrophic organisms obtain their energy and carbon in one of two ways. The **photoheterotrophs** use light as their energy source and preformed organic compounds such as fatty acids and alcohols as sources of carbon. Photoheterotrophs include certain green nonsulfur and purple nonsulfur bacteria.

MICROFOCUS 6.5: Environmental Microbiology

Life in the Deep

Deep-sea hydrothermal vents, or black smokers, are truly exotic places, representing volcanically active mid-ocean ridges several thousand meters below the ocean surface where the Earth's crust has cracked as tectonic plates move apart. The vents in these extreme environments spew forth extremely hot water that can be as high as 300°C, compared to the chilling 2°C for the surrounding deep ocean water. And you guessed it—microbes are found here often in symbiotic association with one of the vent residents, the giant tube worm *Riftia pachyptila*. These deep-sea aliens share the cooler parts of the vent ecosystem with other vent creatures like crabs, lobsters, and octopuses, and live in colonies made up of hundreds of individuals (see photo). *Riftia* lacks a mouth and a gut, yet somehow is able to grow to more than 2 meters in length. To accomplish this growth, they wave their bright red plumes, exposing and absorbing chemicals from the vent fluids and making these chemicals available to the symbiotic bacterial species living within the tube worm. The bacterial cells, in turn, use these chemicals to provide the tube worm with sustenance.

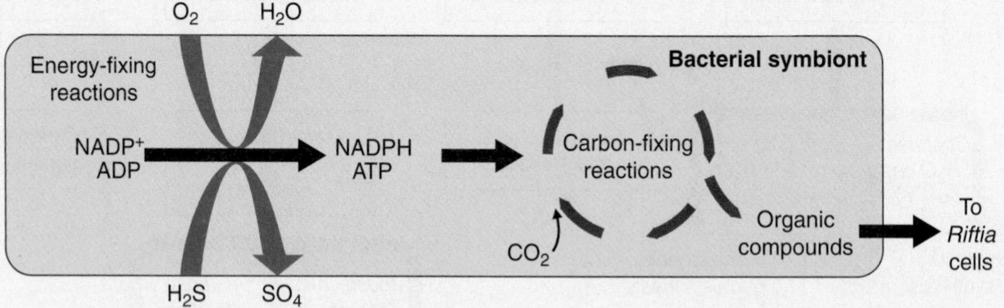

The energy- and carbon-fixing reactions in the bacterial symbiont of *Riftia*.

These communities represent chemoautotrophic proteobacteria. The cells receive inorganic compounds [hydrogen sulfide (H_2S) and O_2] from *Riftia's* circulatory system, the chemicals then serving as the energy source, along with carbon dioxide gas as the carbon source, for growth (see drawing).

Specifically, the worms remove hydrogen sulfide and oxygen from the vent seawater and deliver it to the bacterial cells that have been gathered together in an organ called the trophosome to population densities surpassing 10 billion cells per gram of worm tissue. The bacterial symbionts oxidize the sulfide and use some of the released energy to fix CO_2 via the carbon-fixing reactions into organic compounds, some of which are translocated back to the tube worm's tissues, supporting its growth. Life in the deep is truly remarkable—and microbes are key members of the deep sea community.

Tube worms on a hydrothermal vent. Courtesy of OAR/National Undersea Research Program (NURP)/NOAA.

The **chemoheterotrophs** use preformed organic compounds for both their energy and carbon sources. Glucose would be one example. Those chemoheterotrophic microorganisms that feed exclusively on dead organic matter are commonly called **saprobes**. In contrast, chemoheterotrophs that feed on living organic matter, such as human tissues, are commonly known as **parasites**. The term **pathogen** is used if the parasite causes disease in its host organism. We will certainly see many examples of this in upcoming chapters.

CONCEPT AND REASONING CHECKS 5

a. Why are there no autotrophic parasites or pathogens?

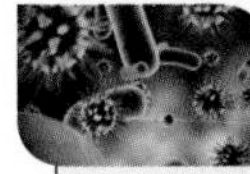

Chapter Challenge D

Hopefully you have figured out the chapter challenge. The reason all these animal and soil "environments" were producing methane gas was because the environments are highly anaerobic. The breakdown of cellulose (the polysaccharide) to glucose requires the anaerobes to use a different final electron acceptor to generate ATP through anaerobic respiration. For some of these microbes, especially the archaeal members, carbonate is the final electron acceptor, forming ATP and methane gas as a final end product.

QUESTION D:

As a last step, use Figure 6.17 to trace the microorganism metabolism that these anaerobic microbes possess. Where do you end up?

Answers can be found on the Student Companion Website in **Appendix D.**

■ **In conclusion**, we saw that cellular respiration is an efficient energy metabolism process. Cell respiration couples the oxidation of organic compounds (such as sugars and lipids) through the central metabolic pathways (glycolysis and the citric acid cycle) to the reduction of a separate final electron acceptor (usually oxygen or another inorganic acceptor) in oxidative phosphorylation. Thus, cell respiration represents a substantial source of ATP generation compared to the substrate-level phosphorylation characteristic of fermentation. Having said this, realize that nearly all organisms (microbial and multicellular) have pretty much the same sets of chemical reactions to carry out cell respiration. The substrates and enzymes of glycolysis and the citric acid cycle are practically the same whether we are discussing *E. coli* or a human liver cell. Even the chemical reactions by which organic compounds are assembled from their monomers are very similar in all organisms. Furthermore, oxygenic photosynthesis also is the same in cyanobacteria, algae, and green plants. We can say that the processes of energy metabolism are almost universal. The difference? Structural organization. The diverse members of the Bacteria and Archaea lack mitochondria and chloroplasts, yet they carry out these metabolic processes almost identically to that found in the diverse members of the Eukarya. These common processes illustrate once again that bacterial and archaeal species are not simple, primitive organisms and, perhaps more importantly, present powerful evidence that the evolution of all life on the "tree of life" has occurred from a single, universal common ancestor.

■ SUMMARY OF KEY CONCEPTS

6.1 Enzymes and Energy Drive Cellular Metabolism

1. The two major themes of microbial metabolism are **catabolism** (the breakdown of organic molecules) and **anabolism** (the synthesis of organic molecules).
2. Microorganisms and all living organisms use **enzymes**, protein molecules that speed up a chemical change, to control cellular reactions.
3. Enzymes bind to **substrates** at their **active site** (enzyme-substrate complex) where functional groups are destabilized, lowering the **activation energy**. (Figs. 6.2, 6.3)
4. Many metabolic processes occur in **metabolic pathways**, where a sequence of chemical reactions is catalyzed by different enzymes. (Fig. 6.4)
5. Enzymes can be inhibited by physical agents, which can denature enzymes and inhibit their action in cells. Enzymes are modulated through **feedback inhibition**. The final end product often inhibits the first enzyme in the pathway.
6. Many metabolic reactions in cells require energy in the form of **adenosine triphosphate** (**ATP**). The breaking of the terminal phosphate produces enough energy to supply an **endergonic reaction**, and often involves the addition of the phosphate to another molecule (**phosphorylation**). (Fig. 6.5)

6.2 Glucose Catabolism Generates Cellular Energy

7. **Cellular respiration** is a series of metabolic pathways in which chemical energy is converted to cellular energy (ATP). It may require oxygen gas (**aerobic respiration**) or another inorganic final electron acceptor (**anaerobic respiration**). (Fig. 6.7)
8. **Glycolysis**, the catabolism of glucose to pyruvate, extracts some energy from which two ATP and 2 NADH molecules result. (Fig. 6.8)
9. The catabolism of pyruvate into carbon dioxide and water in the **citric acid cycle** extracts more energy as ATP, NADH, and $FADH_2$. Carbon dioxide gas is released. (Fig. 6.9)
10. The process of **oxidative phosphorylation** involves the oxidation of NADH and $FADH_2$, the transport of freed electrons along a cytochrome chain, the pumping of protons across the cell membrane, and the synthesis of ATP from a reversed flow of protons. These last two steps are referred to as **chemiosmosis.** (Fig. 6.11)

6.3 There Are Other Pathways to ATP Production

11. Other carbohydrates, such as sucrose, lactose, and polysaccharides, represent energy sources that can be metabolized through cellular respiration. Besides carbohydrates, proteins and fats can be metabolized through the cellular respiratory pathways to produce ATP. (Fig. 6.13)
12. The **anaerobic respiration** of glucose uses different final electron acceptors in oxidative phosphorylation. Glycolysis and parts of the citric acid cycle still function, and ATP synthesis occurs.
13. In **fermentation**, the catabolism of glucose can continue without a functional citric acid cycle or oxidative phosphorylation process. To maintain a steady supply of NAD^+ for glycolysis and ATP synthesis, pyruvate is redirected into other pathways that reoxidize NADH to NAD^+. End products include lactic acid or ethanol. Only the two ATP molecules of glycolysis are synthesized in fermentation from each molecule of glucose. (Fig. 6.14)

6.4 Photosynthesis Converts Light Energy to Chemical Energy

14. The anabolism of carbohydrates can occur by **photosynthesis**, the process whereby light energy is used to synthesize ATP, and the latter is then used to fix atmospheric carbon dioxide into carbohydrate molecules. (Fig. 6.16)

6.5 Microbes Exhibit Metabolic Diversity

15. **Autotrophs** synthesize their own food from carbon dioxide and light energy (**photoautotrophs**) or carbon dioxide and inorganic compounds (**chemoautotrophs**). **Heterotrophs** obtain their carbon from organic compounds and energy from light (**photoheterotrophs**) or from organic compounds (**chemoheterotrophs**). (Fig. 6.17)

CHAPTER SELF-TEST

For **STEPS A–D**, answers to even-numbered questions and problems can be found in **Appendix C** on the Student Companion Website at **http://microbiology.jbpub.com/10e**. In addition, the site features eLearning, an online review area that provides quizzes and other tools to help you study for your class. You can also follow useful links for in-depth information, read more MicroFocus stories, or just find out the latest microbiology news.

STEP A: REVIEW OF FACTS AND TERMS

Multiple Choice

Read each question carefully, then select the ***one*** answer that best fits the question or statement.

1. Enzymes are _____.
   A. inorganic compounds
   B. destroyed in a reaction
   C. proteins
   D. vitamins
2. Enzymes combine with a _____ at the _____ site to lower the activation energy.
   A. substrate; active
   B. product; noncompetitive
   C. product; active
   D. coenzyme; active
3. Which one of the following is NOT a metabolic pathway?
   A. Citric acid cycle
   B. The carbon-fixing reactions
   C. Glycolysis
   D. Sucrose → glucose + fructose
4. If an enzyme's active site becomes deformed, _____ inhibition was likely responsible.
   A. metabolic
   B. competitive
   C. noncompetitive
   D. cellular
5. Which one of the following is NOT part of an ATP molecule?
   A. Phosphate groups
   B. Cofactor
   C. Ribose
   D. Adenine
6. The use of oxygen gas (O_2) in an exergonic pathway generating ATP is called _____.
   A. anaerobic respiration
   B. photosynthesis
   C. aerobic respiration
   D. fermentation
7. Which one of the following is NOT produced during glycolysis?
   A. ATP
   B. NADH
   C. Pyruvate
   D. Glucose
8. All the following are produced during the citric acid cycle *except* _____.
   A. CO_2
   B. O_2
   C. ATP
   D. NADH
9. The electron transport chain is directly involved with _____.
   A. ATP synthesis
   B. CO_2 production
   C. H^+ pumping
   D. generating oxygen gas

10. Which one of the following macromolecules would NOT normally be used for microbial energy metabolism?
    A. DNA
    B. Proteins
    C. Carbohydrates
    D. Fats
11. Anaerobic respiration does NOT ______.
    A. use an electron transport system
    B. use oxygen gas (O_2)
    C. occur in bacterial cells
    D. generate ATP molecules
12. In fermentation, the conversion of pyruvate into a final end product is critical for the production of ______.
    A. CO_2
    B. glucose
    C. NAD^+
    D. O_2
13. Which one of the following is the correct sequence for the flow of electrons in the energy-fixing reactions of photosynthesis?
    A. Water—photosystem I—photosystem II—NADPH
    B. Photosystem I—NADPH—water—photosystem II
    C. Water—photosystem II—photosystem I—NADPH
    D. NADPH—photosystem II—photosystem I—water
14. Microorganisms that use organic compounds as energy and carbon sources are ______.
    A. chemoheterotrophs
    B. chemoautotrophs
    C. photoautotrophs
    D. photoheterotrophs

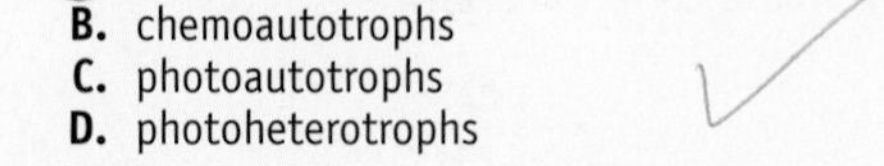

Identification

15. Identify on the metabolic map key where each reactant is used and where each product is produced in the aerobic cellular respiration summary equation.

$$C_6H_{12}O_6 + 6\ O_2 + 32\ ADP + 32\ P \rightarrow 6\ CO_2 + 6\ H_2O + 32\ ATP$$

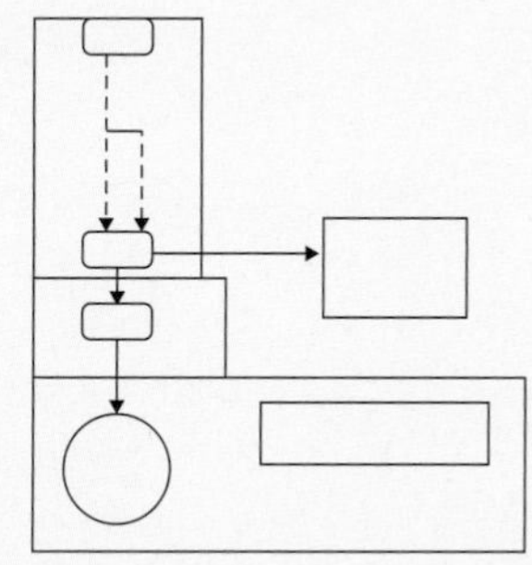

Term Selection

For each choice, circle the word or term that best completes each of the following statements.

16. The sum total of all an organism's biochemical reactions is known as (catabolism, metabolism); it includes all the (synthesis, digestion) reactions called anabolism and all the breakdown reactions known as (inactivation, catabolism).
17. Enzymes are a group of (carbohydrate, protein) molecules that generally (slow down, speed up) a chemical reaction by converting the (substrate, active site) to end products.
18. The aerobic respiration of glucose begins with the process of (oxidative phosphorylation, glycolysis) and requires that (amino acids, energy) be supplied by (ATP, NADH) molecules.
19. The process of (fermentation, the citric acid cycle) takes place in the absence of (oxygen, carbon dioxide) and begins with a molecule of (glucose, protein) and ends with molecules of (amino acids, an organic end product).
20. In oxidative phosphorylation, pairs of (protons, electrons) are passed among a series of (chromosomes, cytochromes) with the result that (oxygen, energy) is released for (NAD^+, ATP) synthesis.
21. In the citric acid cycle, (glucose, pyruvate) undergoes a series of changes and releases its (carbon, nitrogen) as (carbon dioxide, nitrous oxide) and its electrons to (NAD^+, ATP).
22. For use as energy compounds, proteins are first digested to (uric, amino) acids, which then lose their (carboxyl, amino) groups in the process of (fermentation, deamination) and become intermediates of cellular respiration.
23. Ribulose 1,5-bisphosphate bonds with (carbon monoxide, carbon dioxide) molecules during (fermentation, photosynthesis), a process that ultimately results in molecules of (pyruvate, glucose).
24. Chemoautotrophs use energy from (light, inorganic compounds) to synthesize (carbohydrates, oxygen gas) and are typified by species of (*Staphylococcus,* some Bacteria and Archaea).
25. Fats are broken down to (fatty acids, coenzymes), which are converted through (beta oxidation, deamination) reactions to (glucose, two-carbon units) and eventually enter (cellular respiration, photosynthesis).

STEP B: CONCEPT REVIEW

26. State the role of an **enzyme-substrate complex** to regulating metabolism. (**Key Concept 1**)
27. Judge the importance of **metabolic pathways** in microbial cells. (**Key Concept 1**)
28. Assess the role of **ATP** and the **ATP/ADP cycle** in cell metabolism. (**Key Concept 1**)
29. Explain the importance of **glucose** to energy metabolism. (**Key Concept 2**)
30. Summarize the important steps of **glycolysis.** (**Key Concept 2**)
31. Identify the importance of the **citric acid cycle** to **cellular respiration.** (**Key Concept 2**)
32. Construct an **electron transport pathway**, indicating the important steps in the production of ATP. (**Key Concept 2**)
33. Identify what other compounds, besides glucose, can be used to supply chemical energy for ATP production. (**Key Concept 3**)
34. Compare and contrast **aerobic** and **anaerobic respiration.** (**Key Concept 3**)
35. Summarize the steps in **fermentation** and identify the reason why pyruvate is converted into a final end product. (**Key Concept 3**)
36. Summarize the importance of (a) the **energy-fixing reactions** and (b) the **carbon-fixing reactions** of **photosynthesis.** (**Key Concept 4**)
37. Distinguish between the energy and carbon sources for the four nutritional classes of microorganisms. (**Key Concept 5**)

STEP C: APPLICATIONS AND PROBLEMS

38. You have two flasks with broth media. One contains a species of cyanobacteria. The other flask contains *E. coli*. Both flasks are sealed and incubated under optimal growth conditions for 2 days. Assuming the cell volume and metabolic rate of the bacterial cells is identical in each flask, why would the carbon dioxide concentration be higher in the *E. coli* flask than in the cyanobacteria flask after the 2-day incubation?

39. A stagnant pond usually has a putrid odor because hydrogen sulfide has accumulated in the water. A microbiologist recommends that tons of green sulfur bacteria be added to remove the smell. What chemical process does the microbiologist have in mind? Do you think it will work?

40. Citrase is the enzyme that converts citrate to α-ketoglutarate in the citric acid cycle. A chemical company has isolated a mutant microorganism that cannot produce this enzyme and proposes to use the microorganism to manufacture a particular product. What do you suppose the product is? How might this product be useful?

STEP D: QUESTIONS FOR THOUGHT AND DISCUSSION

41. A student goes on a college field trip and misses the microbiology exam covering microbial metabolism. Having made prior arrangements with the instructor for a make-up exam, he finds one question on the exam: "Discuss the interrelationships between anabolism and catabolism." How might you have answered this question?

42. If ATP is such an important energy source for microbes, why do you think it is not added routinely to the growth medium for these organisms?

43. One of the most important steps in the evolution of life on Earth was the appearance of certain organisms in which photosynthesis takes place. Why was this critical?

44. A population of a *Bacillus* species is growing in a soil sample. Suppose glycolysis came to a halt in these bacterial cells. Would this mean that the citric acid cycle would also stop? Why?

45. While you are taking microbiology, a friend is enrolled in a general biology course. You both are studying cell energy metabolism. Your friend looks puzzled when you tell him that the citric acid cycle and electron transport occur in the bacterial cytoplasm and cell membrane. Your friend insists these processes occur in the mitochondrion. Who is correct and why?

Control of Microorganisms: Physical and Chemical Methods

7

CHAPTER PREVIEW

7.1 Microbial Growth Can Be Controlled in Several Ways

7.2 There Are Various Physical Methods to Control Microbial Growth

MicroInquiry 7: Exploring Heat as an Effective Control Mechanism

Textbook Case 7: An Outbreak of Endophthalmitis in Thailand

7.3 Chemical Control Usually Involves Disinfection

7.4 A Variety of Chemical Methods Can Control Microbial Growth

Investigating the Microbial World 7: The Antibacterial Effects of Garlic

Advances in our understanding of hygiene, sanitation, and pathology that followed the development of the germ theory have done more to extend life expectancy and change the nature of society than any other medical innovation.

—Dr. Harry Burns, Chief Medical Officer for Scotland

For personal hygiene, washing our hands, taking regular showers or baths, brushing our teeth with fluoride toothpaste, and using an underarm deodorant are common practices we use to control microorganisms on our bodies. In our homes, we try to keep microbes in check by cooking and refrigerating foods, cleaning our kitchen counters and bathrooms with disinfectant chemicals, and washing our clothes with detergents.

In our attempt to be hygiene-minded consumers, sometimes we have become excessively "germophobic." The news media regularly report about this scientific study or that survey identifying places in our homes (e.g., toilets, kitchen drains) or environment (e.g., public bathrooms, drinking-water fountains) that represent infectious dangers. Consumer groups distribute pamphlets on "microbial awareness" and numerous companies have responded by manufacturing dozens of household antimicrobial products—some useful, but many unnecessary (FIGURE 7.1A).

Our desire to protect ourselves from microbes also stems from events beyond our doorstep. The news media again report about dangerous disease outbreaks, many of which result from a lack of sanitary controls or a

Image courtesy of Dr. Fred Murphy/CDC.

(A) (B)

FIGURE 7.1 **Controlling Microorganisms.** (**A**) Household cleaning products are diverse and formulated for every cleaning need to maintain a sanitary condition. © Jones & Bartlett Learning. Photographed by Kimberly Potvin. (**B**) This African shantytown has slum houses, open sewage, and littered walkways. It is not surprising that in these unsanitary conditions diseases such as cholera and typhoid are common. © Flat Earth/FotoSearch.

»» In these examples, does controlling microorganisms mean sterilization or simply reducing the number of microbes to a safe level? Explain.

lack of vigilance to maintain those controls. In our communities, we expect our drinking water to be clean. That goes for our hospitals as well. Nowhere is this more important than in the operating rooms and surgical wards. Here, hospital personnel must maintain scrupulous levels of cleanliness and have surgical instruments that are sterile. Yet, **nosocomial** (hospital-acquired) **infections** do occur when hygiene barriers are breached.

Microbial control also is a global endeavor. Government and health agencies in many developing nations often lack the means (financial, medical, social) to maintain proper sanitary conditions. Cholera, for example, tends to be associated with poverty-stricken areas where overcrowding and inadequate sanitation practices generate contaminated water supplies (FIGURE 7.1B).

We do need to be hygiene conscious. If the procedures and methods to control pathogens fail or are not monitored properly, serious threats to health and well-being may occur. Yet, the successful control of microorganisms usually requires only simple methods and procedures. As the opening quote reminds us, proper hygiene and sanitation have extended our lives and afforded us the opportunity to contribute more fully to society.

Now that we have a good understanding of microbial growth and metabolism, we can apply that knowledge to controlling the growth and spread of microorganisms. Our study begins by outlining some general principles and terminology and then identifies physical methods commonly used today. We also explore chemical methods for microbial control and discuss the spectrum of antiseptics and disinfectants. Whether the methods are physical or chemical, they are integral in public health practices to ensure continued good health and protection from infectious disease.

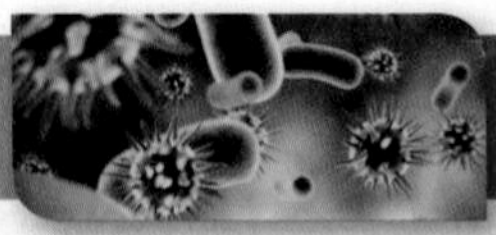

Chapter Challenge

© qcontrol/ShutterStock, Inc.

Most households have stocked up with various and sundry types of household cleaning agents. In addition, every day we almost unconsciously use physical methods and chemical agents to ensure we have a microbially safe environment both in terms of the foods we eat and the home we live in. So, just how many everyday "activities" are involved in controlling potential pathogens in our homes or daily work? Let's identify the ways as we consider the physical methods and chemical agents in this chapter.

KEY CONCEPT 7.1 Microbial Growth Can Be Controlled in Several Ways

The effective control of disease-causing microbes is important not only in hospitals and healthcare facilities, but also in the food industry, in restaurants, and in our own homes. It requires an understanding of the physical and chemical methods used to limit microbial growth and/or microbial transmission.

First, let's establish some basic vocabulary generally used by the public and by health officials when talking about microbial control in the environment.

Sterilization and Sanitization Are Key to Good Public Health

Microbiologically speaking, **sterilization** involves the destruction or removal of all living microbes, spores, and viruses on an object or in an area. For example, in a surgical operation, the surgeon uses "sterile" instruments previously treated in some way to kill any microbes present on them (FIGURE 7.2).

Everyday experiences bring us in contact with sterile materials. An unopened can of corn or peas is sterile inside. During the canning process, companies use special sterilization procedures to kill all the microbes on the vegetables and in the tin can. Agents that kill microbes are **microbicidal** (*-cide* = "kill") or more simply called "germicides." If the agent specifically kills bacterial cells, it is **bactericidal**; if it kills fungi, it is **fungicidal**. Many physical methods and some chemical agents like disinfectants are capable of eliminating microbes on nonliving materials. However, once exposed to the air and surroundings, sterile objects will again become contaminated with microbes in the air or the surrounding area.

More often, in our daily experiences we are likely to encounter materials where microbial populations have been reduced or where their growth has been inhibited. **Sanitization** involves those procedures reducing the numbers of pathogenic microbes to a level considered safe by public health standards. It does not sterilize and, given enough time, pathogens will again grow on the sanitized object. The toilet bowl that has been "sanitized for your protection" today contains few, if any, pathogens. Tomorrow it may again be an area with increased numbers of bacterial species. Many chemical agents, such as the disinfectants and antiseptics, are **microbiostatic** (*-static* = "remain in place"); they reduce microbial numbers or inhibit their growth. Again, specific agents can be **bacteriostatic** or **fungistatic**.

Sanitary measures, both physical and chemical, to control pathogens are very important in areas frequented by the public. City and state sanitation agencies monitor drinking water quality and the preparation of food in restaurants to ensure pathogen elimination. Public health depends on good sanitary practices at home and in the workplace.

CONCEPT AND REASONING CHECK 1

a. Explain the difference between maintaining sterility and maintaining sanitary conditions.

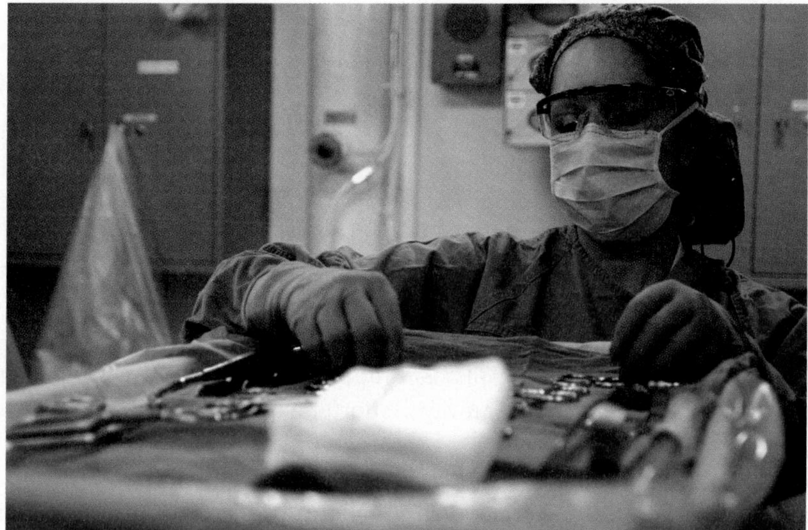

FIGURE 7.2 **Sterile Surgical Instruments.** Here a surgical nurse is unwrapping sterile surgical instruments. Courtesy of Journalist 2nd Class Shane Tuc/U.S. Navy. »» How would you sterilize these instruments?

KEY CONCEPT 7.2 There Are Various Physical Methods to Control Microbial Growth

With this background, let's now examine some specific physical methods that kill microorganisms or inhibit their growth. In hospitals, healthcare facilities, and laboratories, these methods generally include heat, filtration, radiation, osmotic pressure, and low temperatures.

■ Foot-and-mouth disease: A highly contagious viral disease affecting cattle, sheep, and pigs, in which the animal develops ulcers in the mouth and near the hooves.

Heat Is the Most Common Physical Control Method

The killing effect of elevated temperature on microorganisms has long been known. Heat is fast, reliable, and relatively inexpensive. Above the growth range temperature for a microbe, enzymes and other proteins as well as nucleic acids are denatured. Heat also drives off water, and because all organisms depend on water, this loss may be fatal.

The effectiveness of heat for sterilization is a function of time and temperature. For example, bacilli of *Mycobacterium tuberculosis* are destroyed in 30 minutes at 58°C, but in only 2 minutes at 65°C, and in a few seconds at 72°C. Each microbial species has a **thermal death time**, the time necessary for killing the population at a specified temperature. Each species also has a **thermal death point**, the lowest temperature that will kill all the microbes in 10 minutes. These measurements are particularly important in the food industry, where heat is used for preservation. MicroInquiry 7 examines heat and microbial killing.

Dry Heat Has Useful Applications

The form in which dry heat is used to sterilize depends on the nature of the substance to be treated.

Incineration. Using a direct flame can incinerate (burn to ashes) microbes very rapidly. For example, the flame of the Bunsen burner (1600°C) in the microbiology lab is employed for a few seconds to sterilize the bacteriological loop before removing a sample from a culture tube (FIGURE 7.3). Flaming the lip of the tube also destroys organisms that happen to contact the lip, while burning away lint and dust.

Today, disposable hospital gowns and certain plastic apparatus are examples of materials that may be incinerated. In past centuries, the bodies of disease victims were burned to prevent spread of the plague. It still is common practice to incinerate the carcasses of cattle that have died of anthrax and to put the contaminated field to the torch because anthrax spores cannot be destroyed adequately by other means. The 2001 outbreak of foot-and-mouth disease in British cattle required the mass incineration of thousands of cattle as a means to stop the spread of the disease (FIGURE 7.4).

Hot Air Sterilization. The hot-air oven uses radiating dry heat for sterilization. This type of energy does not penetrate materials easily, and therefore, long periods of exposure to high temperatures are necessary. For example, at a temperature of 160°C, a period of 2 hours is required for the destruction of bacterial spores. The hot-air method is also useful for sterilizing dry powders and water-free oily substances, as well as for many types of glassware, such as pipettes, flasks, and syringes. Hot air sterilization does not corrode sharp instruments as steam often does, nor does it erode the ground glass surfaces of nondisposable syringes.

The effect of dry heat on microorganisms is equivalent to that of baking. The heat changes microbial proteins by oxidation reactions and creates an arid internal environment, thereby burning microorganisms slowly. It is essential that organic matter such as oil or grease films be removed from the materials, because such substances insulate against heat penetration. Moreover, the time required for heat to reach sterilizing temperatures varies according to the material. This factor must be considered in determining the total exposure time.

Moist Heat Is More Versatile Than Dry Heat

There are several ways that moist heat is used to control microbes and sterilize materials.

Boiling Water. Boiling water and steam are examples of moist heat that penetrates materials much more rapidly than dry heat because water molecules conduct heat better than air. Therefore, moist heat can be used at a lower temperature and shorter exposure time than for dry heat.

Moist heat kills the vegetative forms of most microorganisms by denaturing their proteins. **Denaturation** is a change in the chemical or physical property of a protein. It includes structural alterations due to destruction of the chemical bonds holding proteins in a three-dimensional form. The

MICROINQUIRY 7

Exploring Heat as an Effective Control Method

Is that can of unopened peas in your pantry sterile? Yes, because companies, like General Mills (manufacturer of Green Giant® products) and the food industry in general, have established appropriate procedures for sterilizing commercial foods.

Sterilization depends on several factors. Identifying the type(s) of microbes in a product can determine whether the heating process will sterilize or eliminate only potential disease-causing species. In many foods, the microbes usually are not in water but rather in the food material. Microbes in powders or dry materials will require a different length of time to sterilize the product than microbes in organic matter.

Environmental conditions also influence the sterilization time. Microbes in acidic or alkaline materials decrease sterilization times while microbes in fats and oils, which slow heat penetration, increase sterilization times. It must be remembered that sterilization times are not precise values. However, by knowing these factors, heating the product to temperatures above the maximal range for microbial growth will kill microbes rapidly and effectively. Let's explore the factors of time and temperature.

Microbial death occurs in an exponential fashion. Look at the table to the right. The table records the death of a microbial population by heating. Notice that the cells die at a constant rate. In this generalized example, each minute 90% of the cells die (10% survive). Therefore, if you know the initial number of microorganisms, you can predict the thermal death time (TDT), which is defined as the length of time, at a specified temperature, required to kill a population of microorganisms.

The food industry depends on knowing a microorganism's heat sensitivity when planning the canning or packaging of many foods as excessive heat can affect the texture and flavor of the food product. One way the industry determines this sensitivity is by using standard curves that take into account the factors mentioned above. The graph drawn below represents three such curves, each representing a different bacterial species treated at the same temperature (60°C) in the same food material (curve B represents the plotted data from the table).

7a. If you had to sterilize this food product that initially contained 10^6 bacteria, how long would it take for each bacterial species? (Hint: 10° = 1)

Another value of importance is the **decimal reduction time** (**DRT**) or **D value**, which is the time required at a specific temperature to kill 90% of the viable organisms. These are the values typically used in the canning industry. D values are usually identified by the temperature used for killing. In the graph to the right, the temperature was 60°C, so D is written as D60. Look at the graph again.

7b. Calculate the D60 values for the three bacterial populations depicted in curves A, B, and C.

On another day in the canning plant, you need to sterilize a food product. However, the only information you have is a D70 = 12 minutes for the microorganism in the food product. Assuming the D value is for the volume you have to sterilize and there are 10^8 bacteria in the food product:

7c. At what temperature will you treat the food product?

7d. How long will it take to sterilize the product?

Answers can be found on the Student Companion Website in **Appendix D.**

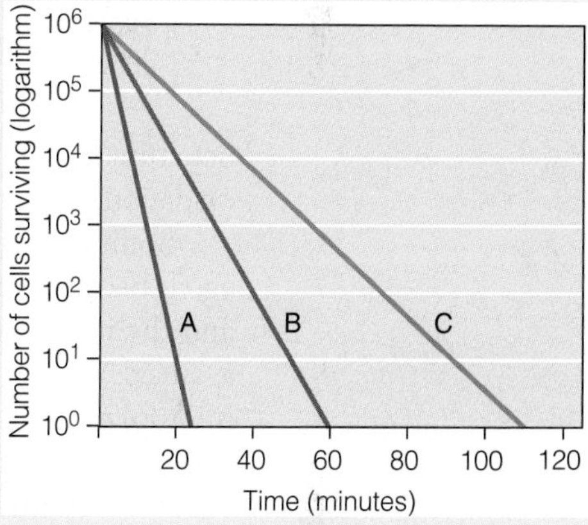

Standard curves for death of three microbial species (A, B, C).

TABLE

Microbial Death Rate

Time (minutes)	Number of Cells Surviving	% Killed
0	1,000,000	—
10	100,000	90
20	10,000	90
30	1,000	90
40	100	90
50	10	90
60	1	90

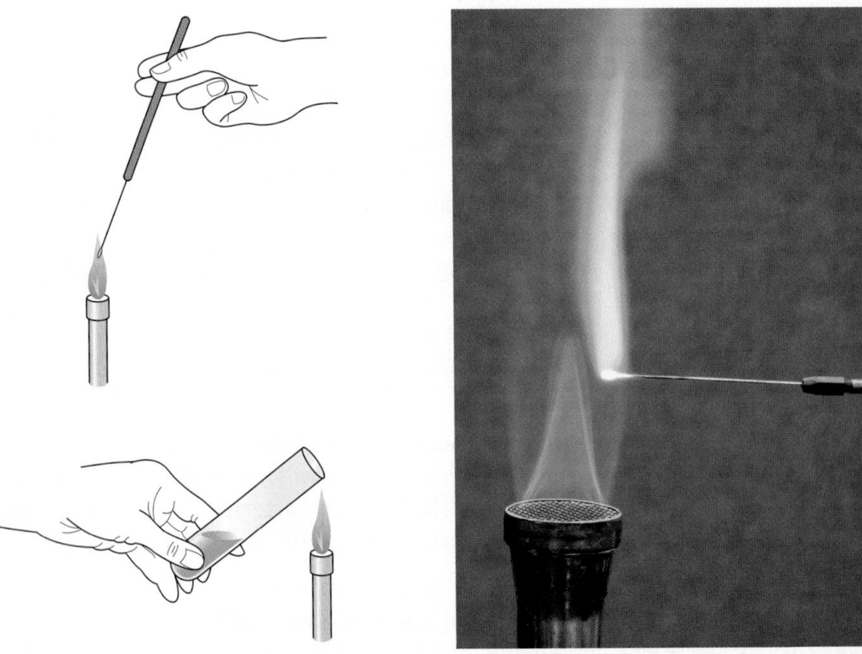

FIGURE 7.3 **Use of the Direct Flame as a Sterilizing Agent.** A few seconds in the flame of a laboratory Bunsen or Fisher burner is usually sufficient to effect sterilization of a culture tube lip or inoculation loop. Courtesy of Jeffrey Pommerville. »» Why is it necessary to flame a culture tube lip?

denaturation of proteins requires less energy than oxidation, and, therefore, less heat need be applied.

Boiling water is not considered a sterilizing agent because the destruction of bacterial spores and the inactivation of viruses cannot always be assured. Under ordinary circumstances, with microorganisms at concentrations of less than 1 million per milliliter, most species of microorganisms can be killed within 10 minutes. Indeed, the process may require only a few seconds. However, fungal spores, protistan cysts, and large concentrations of hepatitis A viruses require a 30 minute exposure. Bacterial spores often require 2 hours or more (FIGURE 7.5).

FIGURE 7.4 **Incineration Is an Extreme Form of Dry Heat.** Nearly four million hoofed animals infected with or exposed to foot-and-mouth disease in England in 2001 were incinerated and buried. © Simon Fraser/Photo Researchers, Inc. »» How is incineration similar to flame sterilization?

If boiling water must be used to destroy microorganisms, then materials must be thoroughly cleaned to remove traces of organic matter, such as blood or feces. The minimum exposure period should be 30 minutes, except at high altitudes, where it should be increased to compensate for the lower boiling point of water. All materials should be well covered. Baking soda may be added at a 2% concentration to increase the efficiency of the process.

Pressurized Steam. Moist heat in the form of pressurized steam is regarded as the most dependable method for sterilization, including the destruction of bacterial spores. This method is incorporated into a device called the **autoclave** (FIGURE 7.6). Working similar to a home pressure cooker, when the pressure of the steam increases, the temperature of the gas also increases proportionally. Importantly, the sterilizing agent is the heat, not the pressure. This principle is used to reduce cooking time in the home pressure cooker and to reduce sterilizing time in the autoclave.

Autoclaves contain a sterilizing chamber into which articles are placed. As steam flows into the sterilizing chamber, a special valve increases the pressure to 15 pounds/square inch (psi) and the temperature rises to 121.5°C. The time for destruction of the most resistant bacterial species is about 15 minutes (see Figure 7.5). For denser objects or larger volumes, more than 30 minutes of exposure may be required.

The autoclave is used in hospitals to sterilize blankets, bedding, utensils, instruments, intravenous solutions, and a broad variety of other heat-resistant objects. In the laboratory, it is used to sterilize glassware and metalware, growth media, and other solutions.

The autoclave has certain limitations. For example, some plasticware melts in the high heat and sharp instruments often become dull. Moreover, many chemicals break down during the sterilization process and oily substances cannot be treated because they do not mix with water.

The conditions must be carefully controlled to assure sterilization has been accomplished. Autoclaves have temperature and pressure gauges visible from the outside and most models can produce a paper record of the temperature, time, and pressure. To gauge the success of sterilization, materials usually are autoclaved with autoclave tape, which turns color if the object inside the material has been autoclaved correctly (TEXTBOOK CASE 7).

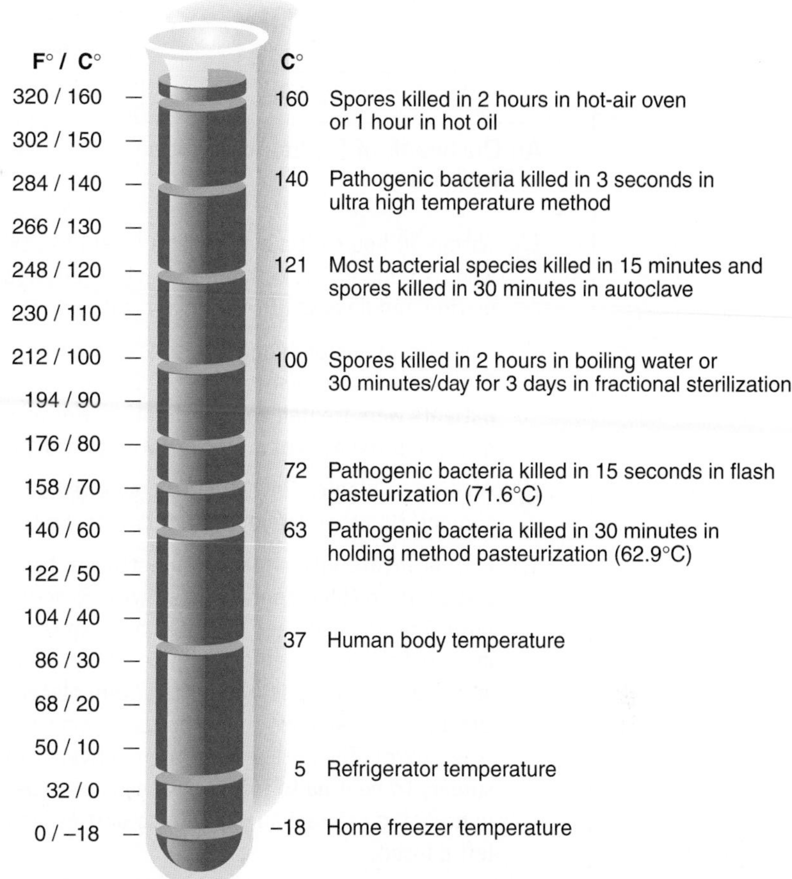

FIGURE 7.5 **Temperature and the Physical Control of Microorganisms.** Notice that materials containing bacterial endospores require longer exposure times and higher temperatures for killing. »» Pure water boils and freezes at what temperatures on the Celsius scale?

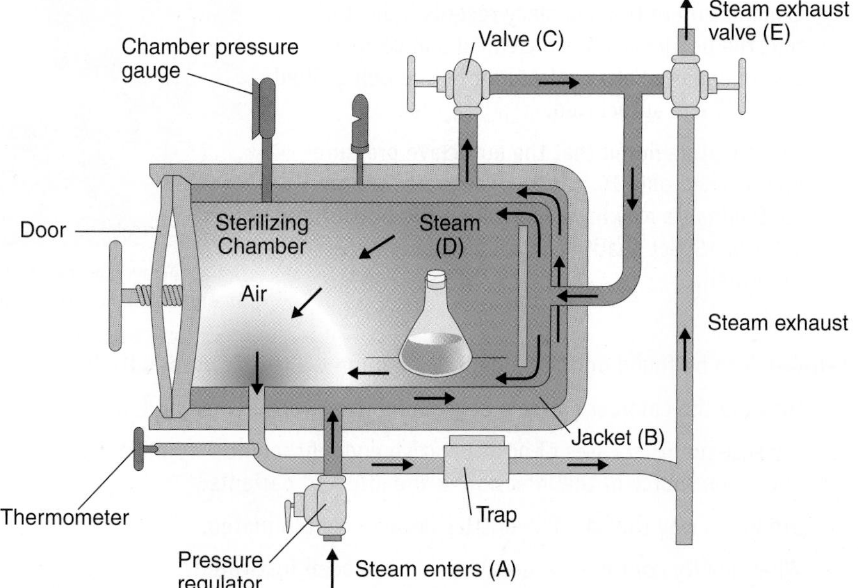

FIGURE 7.6 **Operating an Autoclave.** Steam enters through the port (A) and passes into the jacket (B). After the air has been exhausted through the vent, a valve (C) opens to admit pressurized steam (D), which circulates among and through the materials, thereby sterilizing them. At the conclusion of the cycle, steam is exhausted through the steam exhaust valve (E). »» Is it the steam or the pressure that kills microorganisms in an autoclave? Explain.

Textbook CASE 7

An Outbreak of Endophthalmitis in Thailand

1 Within 30 hours after surgery at a local hospital in Tak Province, Thailand, three cataract patients who had undergone cataract extraction developed postoperative endophthalmitis, an eye inflammation and disease of the ocular cavity and the adjacent structures.

2 The clinical microbiology lab isolated a pathogenic strain of the gram-negative bacterial species *Pseudomonas aeruginosa* from the intraocular fluid taken from two of the patients. The infected patients were treated with antibiotics. However, due to antibiotic resistance, antibiotic therapy failed and the eyes of all three patients had to be eviscerated.

3 An epidemiological investigation reviewed all hospital records of all patients who had cataract surgery. No other cataract patients with postoperative endophthalmitis were identified.

4 Further investigation identified bottles of basal salt solution (BSS) contaminated with *P. aeruginosa*. The BSS had been prepared by hospital pharmacists for use in the hospital operating rooms. To sterilize the solutions, the bottles were placed in the autoclave and left to run on its automatic cycle (see figure). The bottles were then delivered to surgery to be used to irrigate the eyes of patients undergoing cataract surgery. Some bottles were left unused.

5 Health investigators tested the unused bottles of BSS as well as the tubes attached to the now-empty bottles. They found the identical strain of *P. aeruginosa*.

6 Examination of the pharmacy records indicated that the bottles had been cleaned and correctly placed under an ultraviolet light before being filled with BSS and autoclaved.

7 Investigators noted that the autoclave pressure had reached only 10 to 12 psi, suggesting that contaminants may have survived in the bottles and solution. Strict quality control measures were instituted.

An autoclave can be used to sterilize liquid materials. © Huntstock, Inc/Alamy.

Questions:

(Answers can be found on the Student Companion Website in **Appendix D.**)

A. How did the cataract patients become infected with *P. aeruginosa?*

B. Because no other cases of postoperative endophthalmitis were identified, what does that say about the source of the infection in the affected patients?

C. Propose a way that the BSS bottles became contaminated.

D. What quality control measures should have been instituted?

For additional information see www.cdc.gov/mmwr/preview/mmwrhtml/00042645.htm.

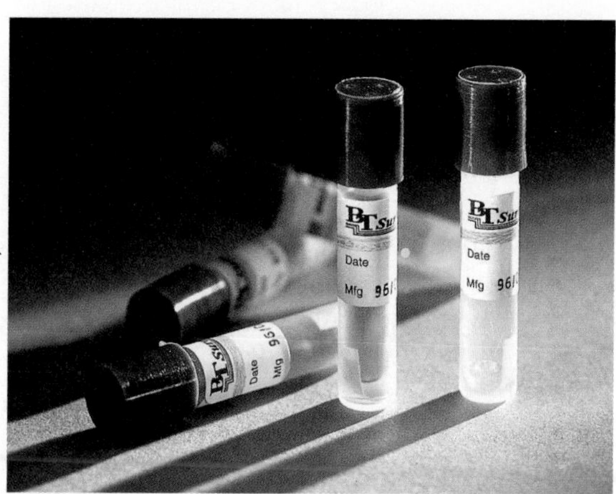

FIGURE 7.7 **Monitoring the Effectiveness of Autoclave Sterilization.** Vials containing spores of *Geobacillus stearothermophilus* are suspended in a growth medium containing a pH indicator. Following autoclaving, the vial is incubated at 35°C. If the autoclave run was effective, all spores will be killed and the solution will remain purple. If the spores were not killed (an ineffective sterilization process), they will germinate and grow. The acid the cells produce will turn the pH indicator yellow. Courtesy of Thermo Scientific. »» What do you know about this bacterial species from its scientific name?

Biological indicators also can be used (FIGURE 7.7). A nutrient vial containing spores of *Geobacillus stearothermophilus* can be included with the objects treated. At the conclusion of the cycle, the vial is incubated at 37°C. If the sterilization process has been unsuccessful, the spores will germinate and their metabolism will change the color of a pH indicator in the growth medium.

Another type of autoclave, called the **prevacuum autoclave**, has been developed for sterilization procedures. This machine draws air out of the sterilizing chamber at the beginning of the cycle. Saturated steam then is used at a temperature of 132°C to 134°C at a pressure of 28 psi to 30 psi. The time for sterilization is now reduced to as little as 4 minutes. A vacuum pump operates at the end of the cycle to remove the steam and dry the load. The major advantages of the prevacuum autoclave are the minimal exposure time for sterilization and the reduced time to complete the cycle.

Sterilization Without Pressurized Steam. In the years before the development of the autoclave, liquids and other objects were sterilized by exposure to free-flowing steam at 100°C for 30 minutes on each of three successive days, with incubation periods at room temperature between intermittent steaming. The method is called **intermittent sterilization**. It was also called **tyndallization** after its developer, John Tyndall.

Sterilization by the intermittent method is achieved as follows. During the first day's exposure, steam kills virtually all organisms except bacterial spores. During overnight incubation, the spores germinate and the viable cells multiply only to be killed on the second day's 100°C exposure. Again, the material is cooled and any remaining spores may germinate, the resulting cells only to be killed on the third day.

Intermittent sterilization has assumed renewed importance in modern microbiology with the development of high-technology instrumentation and new chemical substances. Often, these materials cannot be sterilized at autoclave temperatures, or by long periods of boiling or baking, or with chemicals.

Pasteurization. The final example of moist heat involves the process of **pasteurization**, which reduces the bacterial population in liquids, such as milk, and destroys organisms that may cause spoilage and human disease (FIGURE 7.8). Spores are not affected by pasteurization.

One method for milk pasteurization, called the **holding** (or **batch**) **method**, involves heating at 63°C for 30 minutes. For decades, pasteurization has been aimed at destroying *M. tuberculosis*, long considered the most heat-resistant bacterial species. More recently, attention has shifted to destruction of *Coxiella burnetii*, the agent of Q fever, because this bacterial organism has a higher resistance to heat. Because both organisms are eliminated by pasteurization, dairy

FIGURE 7.8 **The Pasteurization of Milk.** Milk is pasteurized by passing the liquid through a heat exchanger. The flow rate and temperature are monitored carefully. Following heating, the liquid is rapidly cooled. © Apply Pictures/Alamy Images. »» Why is the liquid rapidly cooled?

microbiologists assume other pathogenic bacteria also are destroyed. Pasteurization also is used to eliminate any *Salmonella* and *Escherichia coli* that could contaminate fruit juices.

Two other methods are the **flash pasteurization method** where the liquid is held at 71.6°C for 15 seconds and the **ultra high temperature (UHT) method** using 140°C for 3 seconds. UHT is the only method that sterilizes the liquid if done under aseptic conditions (MicroFocus 7.1).

Although heat is a valuable physical agent for controlling microorganisms, sometimes it is impractical to use. For example, no one would suggest removing the microbial population from a tabletop by using a Bunsen burner, nor can heat-sensitive solutions be subjected to an autoclave. In instances such as these and numerous others, a heat-free method must be used.

Filtration Traps Microorganisms

Filters came into prominent use in microbiology as interest in viruses grew during the 1890s. Previous to that time, filters were used to trap airborne organisms and sterilize growth media. Among the early pioneers of filter technology was Charles Chamberland, an associate of Pasteur. His porcelain filter was important to early virus research. Another pioneer was Julius Petri (inventor of the Petri dish), who developed a sand filter to separate bacterial cells from the air.

Filtration is a mechanical method used to remove microorganisms suspended in liquids or gases. Several types of filters are used in the microbiology laboratory.

The most common is the **membrane filter**, which consists of a pad of nitrocellulose acetate or polycarbonate mounted in a holding device

MICROFOCUS 7.1: Being Skeptical
Milk Stays Fresh Longer If It's Organic

If you ever go shopping for milk at the local market, or especially one specializing in "natural and organic foods," you might notice that the "Best by" date expires much sooner on a carton of regular milk than on one that is organic milk. In fact, the date on the organic milk may be some 3 weeks longer than on the regular milk, which typically is 5 to 7 days from the store delivery date. So, being organic, does that ensure a longer shelf life?

A carton of organic milk. Courtesy of Jeffrey Pommerville.

The fact that the milk is organic has nothing to do with its longer shelf life. Labeling the milk as "organic" only means that the cows on the dairy farm were not given antibiotics or hormones like bovine growth hormone (BGH), which stimulates a cow's milk production (see figure). The reason it has a longer shelf life is due to the pasteurization process. Organic milk is subjected to the ultra high temperature (UHT) process (ultra-pasteurized) where the milk is heated to 140°C for 3 seconds. This kills all the microorganisms that may be in the liquid—it is sterile. Most regular milk today is subjected to the flash pasteurization method where the milk is at about 72°C for 15 seconds. This "high temperature, short duration" process does not kill all microbes that may be in the milk; only the pathogens have been eliminated. Because there are bacterial species that are psychrotolerant, the milk can spoil if left on the refrigerated shelf too long.

Regular milk also could be subjected to UHT. However, it usually is not because it has to travel only a short distance to market; organic products, on the other hand, are not often produced throughout the country, so they have further to travel to reach the consumer. Therefore, UHT preserves the product longer. Although not found commonly within the United States, room-temperature Parmalat milk is a product of UHT and can be found commonly in Europe, Mexico, and other parts of the world.

The verdict? "Organic" is not defined as "longer shelf life." If shelf life is important to you, simply look for products treated by UHT. Also of note: UHT does burn some of the sugars in the milk, so the milk may taste slightly "caramelized," something some people find less tasty.

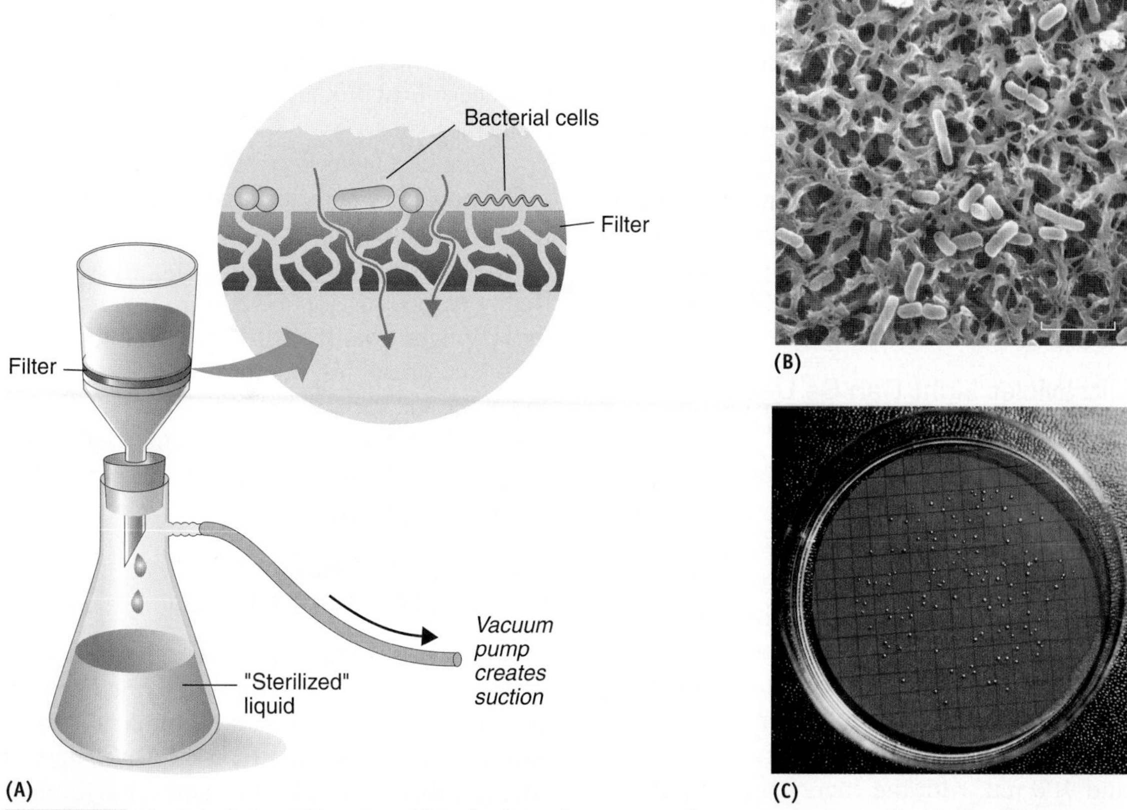

FIGURE 7.9 **The Principle of Filtration.** Filtration is used to remove microorganisms from a liquid. The effectiveness of the filter is proportional to the size of its pores. (**A**) Bacteria-laden liquid is poured into a filter, and a vacuum pump helps pull the liquid through and into the flask below. The bacterial cells are larger than the pores of the filter, and they become trapped. (**B**) A scanning electron micrograph of *Escherichia coli* cells trapped in the pores of a 0.45-μm membrane filter. (Bar = 5 μm.) Courtesy of Pall Corporation. (**C**) *E. coli* colonies growing on a membrane filter. © Kathy Talaro/Visuals Unlimited/Corbis. »» Why would most viruses not be trapped on a filter with 0.45 μm pores?

(FIGURE 7.9A). As fluid passes through the filter, organisms are trapped on the membrane because they are too large to pass through the pores. The solution dripping through the filter into the receiving container is decontaminated or, in some cases, sterilized. Membrane filters are used to purify such heat-sensitive liquids as beverages, some growth media, toxoids, many pharmaceuticals, and blood solutions.

The membrane filter is particularly valuable because bacterial cells trapped on the filter multiply and form colonies when the filter pad is placed on an agar culture plate. Microbiologists then can count the colonies to determine the number of bacterial cells originally present (FIGURE 7.9B, C).

Air also can be purified to remove microorganisms. The filter generally used is a **high-efficiency particulate air (HEPA) filter**, which consists of a mat of randomly arranged fibers that trap particles, microorganisms, and spores. As part of a **biological safety cabinet** (FIGURE 7.10),

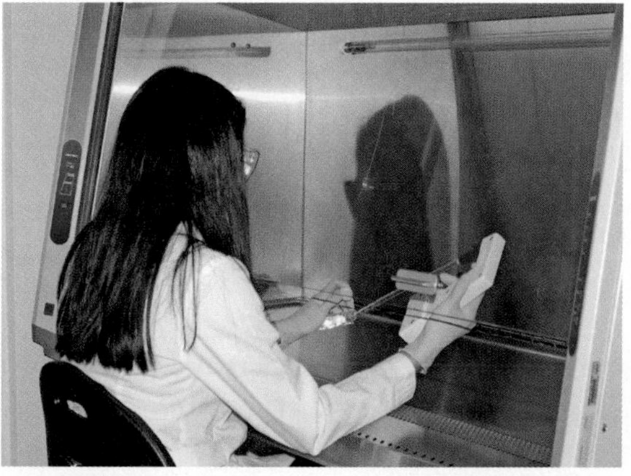

FIGURE 7.10 **A Biological Safety Cabinet.** The cabinet shown has a metal grid at the top that covers a HEPA filter through which air enters the cabinet. As the filtered air, free of contaminants and microbes, moves into and across the workspace, it exits out the bottom front and rear. A UV light is also positioned at the top rear to decontaminate the metal surfaces maintaining a contaminant-free workspace when the cabinet is not in use. Courtesy of Jeffrey Pommerville. »» Why is air moved out of the cabinet rather than into the cabinet?

HEPA filters prevent over 99% of all particles, including microorganisms and spores with a diameter larger than 0.3 μm from escaping as aerosols into the laboratory air. The air entering or exiting surgical units and specialized treatment facilities, such as burn units, also are HEPA filtered to exclude microorganisms. In some hospital wards, such as for respiratory diseases and in certain pharmaceutical filling rooms, the air is recirculated through HEPA filters to ensure air purity.

Ultraviolet Light Can Be Used to Control Microbial Growth

Visible light is a type of radiant energy detected by the light-sensitive cells of the eye. The wavelength of this energy is between 400 and 750 nm. Other types of radiation have wavelengths longer or shorter than that of visible light, and therefore cannot be detected by the human eye.

One type of radiant energy, **ultraviolet (UV) light**, is useful for controlling microorganisms. Ultraviolet light has a wavelength between 100 and 400 nm, with the energy at about 265 nm most destructive to bacterial cells (FIGURE 7.11). When microorganisms are subjected to UV light (often a "germicidal lamp"), cellular DNA absorbs the energy, and adjacent thymine molecules (in the same strand) link together, kinking the double helix and disrupting DNA replication and transcription. The damaged organism can no longer produce critical proteins or reproduce, and eventually dies.

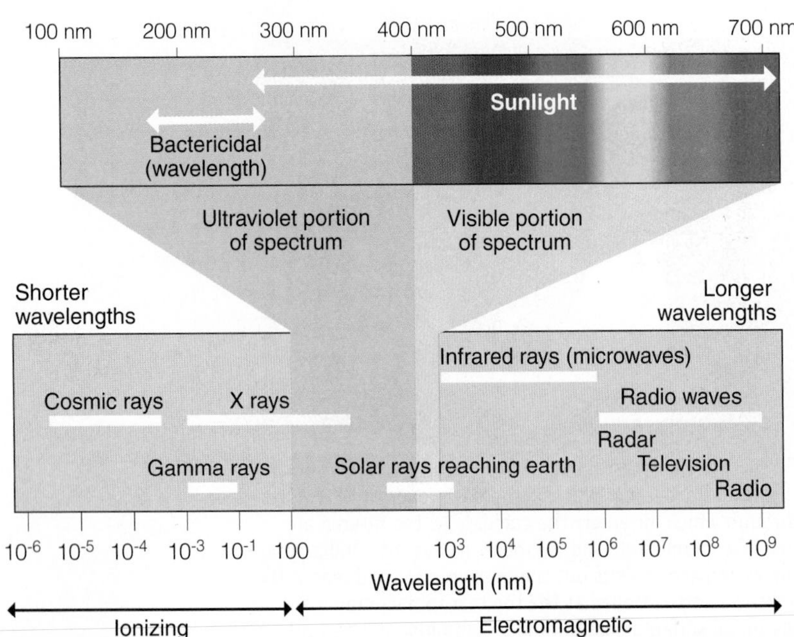

FIGURE 7.11 The Ionizing and Electromagnetic Spectrum of Energies. The complete spectrum is presented at the bottom of the chart, and the ultraviolet and visible sections are expanded at the top. »» How does UV light kill bacterial cells?

Ultraviolet light effectively reduces the microbial population where direct exposure takes place. It is used to limit airborne or surface contamination in a hospital room, morgue, pharmacy, toilet facility, or food service operation. It is noteworthy that UV light from the sun may be an important factor in controlling microorganisms in the air and at the soil surface, but it may not be effective against all bacterial spores. Ultraviolet light does not penetrate liquids or solids, and it can lead to human skin cancer.

Other Types of Radiation Also Can Sterilize Materials

Looking again at Figure 7.11, there are two other forms of radiation useful for destroying microorganisms. These are **X rays** and **gamma rays**. Both have wavelengths shorter than the wavelength of UV light. As X rays and gamma rays pass through microbial molecules, they force electrons out of their shells, thereby creating ions. For this reason, the radiations are called **ionizing radiations**. The ions quickly combine, mainly with cellular water, and the free radicals generated affect cell metabolism and physiology, and often cause mutations. Ionizing radiations currently are used to sterilize such heat-sensitive pharmaceuticals as vitamins, hormones, vaccines, and antibiotics as well as certain plastics and suture materials.

Ionizing radiations also have been approved for controlling microorganisms, and for preserving foods, as noted in MicroFocus 7.2. The approval has generated some controversy, fueled by activists concerned with the safety of factory workers and consumers. First used in 1921 to inactivate *Trichinella spiralis*, the agent of trichinellosis, irradiation now is used as a food preservation method in more than 40 countries for over 100 food items, including potatoes, onions, cereals, flour, fresh fruit, and poultry (FIGURE 7.12A). The U.S. Food and Drug Administration (FDA) approved cobalt-60 and cesium-137 irradiation to preserve or extend the shelf life of several foods. This includes irradiating poultry and red meats such as beef, lamb, and pork. In 2008, the FDA approved the irradiation

MICROFOCUS 7.2: Public Health

"No, the Food Does Not Glow!"

In the United States, there are more than 76 million cases of foodborne disease accounting for more than 300,000 hospitalizations and 5,000 deaths each year. One major source of foodborne disease is agricultural produce contaminated with intestinal pathogens. Another source is from improperly cooked or handled meats, or poultry harboring human intestinal pathogens, such as *Escherichia coli* 0157:H7, *Campylobacter, Listeria,* and *Salmonella.*

Irradiation has the potential to greatly limit such illnesses. The Centers for Disease Control and Prevention (CDC) have estimated that if just 50% of the meat and poultry sold in the United States was irradiated, there would be 900,000 fewer cases of foodborne illness and 350 fewer deaths each year.

Irradiated strawberries displaying the "radura" symbol, meaning they have been treated by irradiation. © Jones & Bartlett Learning. Photographed by Kimberly Potvin.

Yet today manufacturers continue to wrestle with the concept of food irradiation as they constantly confront a leery public, some of who still have visions of Hiroshima and Nagasaki. In the United States, just 10% of the herbs and spices are irradiated and only 0.002% of fruits, vegetables, meats, and poultry are irradiated.

Food irradiation is entirely different from atomic radiation. The irradiation comes from gamma rays produced during the natural decay of cobalt-60 or cesium-137. The most common method involves electron beams (e-beams) not unlike those used in an electron microscope. None of these types of radiations produce radioactivity—the irradiated food does not glow (see figure).

Low doses of irradiation are used for disinfestations and extending the shelf life of packaged foods. As mentioned in the chapter narrative, a pasteurizing dose is used on meats, poultry, and other foods. Such levels do not eliminate all microbes in the food, but, similar to pasteurization, help to reduce the dangers of pathogen-contaminated or cross-contaminated meats and poultry.

During the irradiation, the gamma rays or electrons penetrate the food, and, just as in cooking, cause molecular changes in any contaminating microorganisms, which ultimately leads to their death.

Irradiation of foods also has its limitations. The irradiation dose will not kill bacterial endospores, inactivate viruses, or neutralize toxins. Therefore, irradiated food still must be treated in a sanitary fashion. Nutritional losses are similar to those occurring in cooking and/or freezing. Otherwise, there are virtually no known changes in the food, and there is no residue.

(A)

(B)

FIGURE 7.12 Food Irradiation. (**A**) The FDA has approved irradiation as a preservation method for some foods, including many fruits and vegetables, as well as poultry and red meats. © Jones & Bartlett Learning. Photographed by Kimberly Potvin. (**B**) Many otherwise perishable foods eaten by NASA astronauts are prepared by irradiation. Courtesy of US Army Natick Soldier Center. »» Does irradiation sterilize the treated product? Explain.

of fresh and bagged spinach, and iceberg lettuce, to reduce potential foodborne illness. Irradiation has been used to prepare many meals for the U.S. military and the American astronauts (FIGURE 7.12B). What is called a **pasteurizing dose** is used on meats, poultry, and other foods. Such levels are not intended to eliminate all microbes in the food, but, like pasteurization, to eliminate the pathogens. The foods are not necessarily sterile.

Another form of energy, the microwave, has a wavelength longer than that of ultraviolet light and visible light. In a microwave oven, microwaves are absorbed by water molecules, which are set into high-speed motion, and the heat of friction from these excited molecules is transferred to foods. In fact, the microwave can be an excellent way to sterilize your kitchen sponge. Just heat the damp sponge in the microwave oven for 2 minutes.

Preservation Methods Retard Spoilage by Microorganisms in Foods

Over the course of many centuries, various physical methods have evolved for controlling microorganisms in food. Though valuable for preventing the spread of infectious agents, these procedures are used mainly to retard spoilage and prolong the shelf life of foods, rather than for sterilization. Irradiation is an example of a preservation method.

Drying (dessication) is useful in the preservation of various meats, fish, cereals, and other foods. Because water is necessary for life, it follows that where there is no water, there is virtually no life. Many nonperishable foods (such as cereals, rice, and sugar) in the kitchen pantry represent such shelf-stable products.

Preservation by salting is based upon the principle of osmotic pressure. When food is salted (usually sodium chloride), water diffuses out of microorganisms toward the higher salt concentration and lower water concentration in the surrounding environment. This flow of water, called **osmosis**, leaves the microorganisms dehydrated, and they die. The same phenomenon occurs in highly sugared foods such as syrups, jams, and jellies. However, fungal contamination (molds and yeasts) and growth at the surface will eventually occur because the microbes can tolerate low water and high sugar concentrations.

■ Osmotic pressure: The pressure applied to a solution to stop the inward diffusion (osmosis) of a solvent through a semipermeable membrane.

Cold temperatures found in the refrigerator and freezer retard spoilage by lowering the metabolic rate of bacterial cells and molds and thereby substantially reducing their rate of growth. Spoilage is not totally eliminated in cold foods, however, and many psychrotrophs remain alive, even at freezer temperatures (0°C). These organisms multiply rapidly when food thaws, which is why prompt cooking to the same degree of doneness as fresh food is recommended.

Note in these examples that there are significant differences between killing microorganisms, holding them in check, and reducing their numbers. The preservation methods are described as bacteriostatic because they prevent the further multiplication of foodborne pathogens such as *Salmonella* and *Clostridium*. TABLE 7.1 and FIGURE 7.13 summarize the physical agents used for controlling microorganisms.

CONCEPT AND REASONING CHECKS 2

a. How does the thermal death time differ from the thermal death point?
b. Explain how dry heat can be used to "eliminate" microorganisms.
c. Summarize the ways that moist heat controls or sterilizes materials or beverages.
d. Determine the uses for filtration in a healthcare setting.
e. Identify some uses for UV light as a physical control method.
f. Identify some uses for X rays and gamma rays as a physical control method.
g. Explain how salting foods acts as a preservation method.

Chapter Challenge A

So, far we have examined the physical methods used to control microbes. In most cases, we are primarily concerned about controlling and eliminating any potential pathogens.

QUESTION A:

Thinking about the physical methods, which of the following do you use at home or in the workplace (or in products you have purchased, such as food, etc.), and how are the methods used?

a. Dry and moist heat:
b. Filtration:
c. Ultraviolet light:
d. Spoilage preservation:

Answers can be found on the Student Companion Website in **Appendix D.**

TABLE

7.1 A Summary of Physical Agents Used to Control Microorganisms

Physical Method	Conditions	Instrument	Examples of Uses	Comment
Incineration	A few seconds	Flame	Laboratory instruments	Object must be disposable or heat-resistant
Hot air	160°C for 2 hr	Oven	Glassware Powders Oily substances	Not useful for fluid materials
Boiling water	100°C for 10 min 100°C for 2 hr+	— —	Wide variety of objects Spores	Total immersion and precleaning necessary
Pressurized steam	121°C for 15 min at 15 psi	Autoclave	Instruments Surgical materials Solutions and media	Broad application in microbiology
Intermittent sterilization	30 min/day for 3 successive days	Sterilizer	Materials not sterilized by other methods	Long process Sterilization not assured
Pasteurization	Holding method Flash method UHT method	Pasteurizer	Dairy products Beverages	Sterilization achieved with UHT under aseptic conditions
Filtration	Entrapment in pores	Membrane filter HEPA filter	Fluids Air	Many adaptations
Ultraviolet light	265 nm energy	Generator	Surface and air sterilization	Not useful in fluids
X rays Gamma rays	Short wavelength energy	Generator	Heat-sensitive materials	Extending food shelf life
Dehydration	Osmotic conditions	—	Salted and sugared foods	Food preservation
Refrigeration/Freezing	5°C/–10°C	Refrigerator/Freezer	Numerous foods	Spoilage/Food preservation

KEY CONCEPT 7.3 Chemical Control Usually Involves Disinfection

Sanitation and disinfection methods are not unique to the modern era. The Bible refers often to cleanliness and prescribes certain dietary laws to prevent consumption of what was believed to be contaminated food. Egyptians used resins and aromatics for embalming, and ancient peoples burned sulfur for deodorizing and sanitary purposes. Arabian physicians first suggested using mercury to treat syphilis. Over the centuries, spices were used as preservatives as well as masks for foul odors, making Marco Polo's trips to Asia for new spices a necessity as well as an adventure. And fans of Western movies probably have noted that American cowboys practiced disinfection by pouring whiskey onto gunshot wounds.

Chemical Control Methods Are Dependent on the Object to Be Treated

As early as 1830, the United States Pharmacopoeia listed tincture of iodine as a valuable antiseptic, and soldiers in the Civil War used it in plentiful amounts. Joseph Lister in the 1860s established the principles of aseptic surgery using carbolic acid (phenol) for treating wounds.

As we have discussed, the physical agents for controlling microorganisms generally are intended to achieve sterilization. Chemical agents, by contrast, rarely achieve sterilization. Instead, they are expected only to destroy the pathogenic organisms on or in an object or area. The process of eliminating or reducing pathogens (except endospores) is called **disinfection** and the object is said to be "disinfected." If the object treated is nonliving, such as a tabletop, the chemical agent used is called a **disinfectant**. However, if the object treated is living, such as human skin tissue, the chemical agent used is an **antiseptic** (FIGURE 7.14). It is important to note that even though a particular chemical may be used as a disinfectant as well as an antiseptic (e.g., iodine), the precise formulations are very

■ Tincture: A substance dissolved in alcohol.

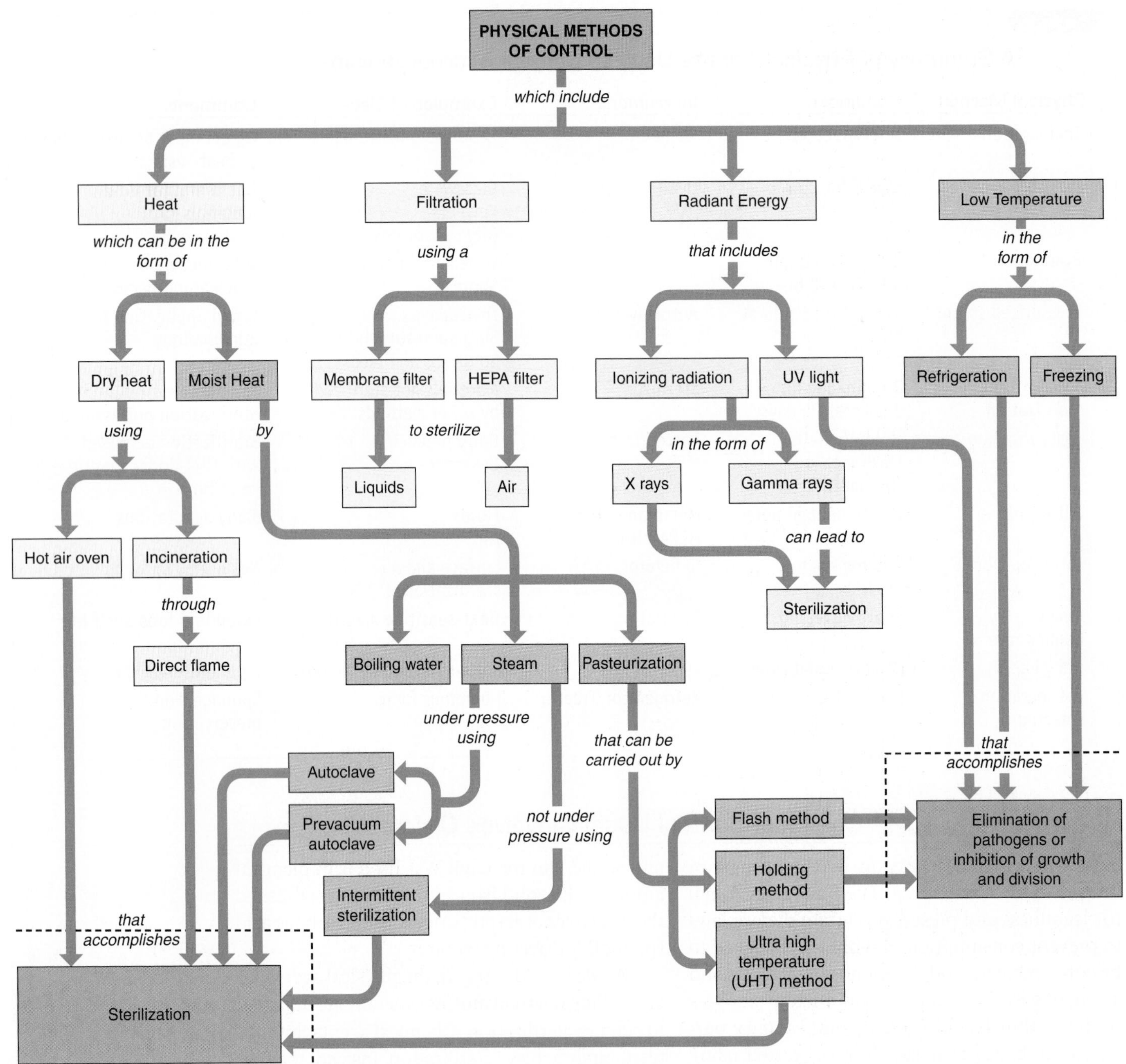

FIGURE 7.13 **A Concept Map Summarizing the Physical Methods of Microbial Control.** Note that some methods sterilize while others tend to inhibit growth and division. »» What is common to most of the sterilization methods?

different so that as an antiseptic it is not toxic or damaging to human tissue.

Antiseptics and disinfectants are usually microbicidal; they inactivate the major enzymes of an organism and interfere with its metabolism. A chemical also may be microbiostatic, disrupting minor chemical reactions and slowing the metabolism, which results in a longer time between cell divisions. Although a subtle difference sometimes exists between the two chemical agents, the terms indicate effectiveness in a particular situation.

The word **sepsis** (*seps* = "putrid") refers to a condition in which microbes or their toxins are present in tissues or the blood; thus, we have **septicemia**, meaning "microbial growth in the blood,"

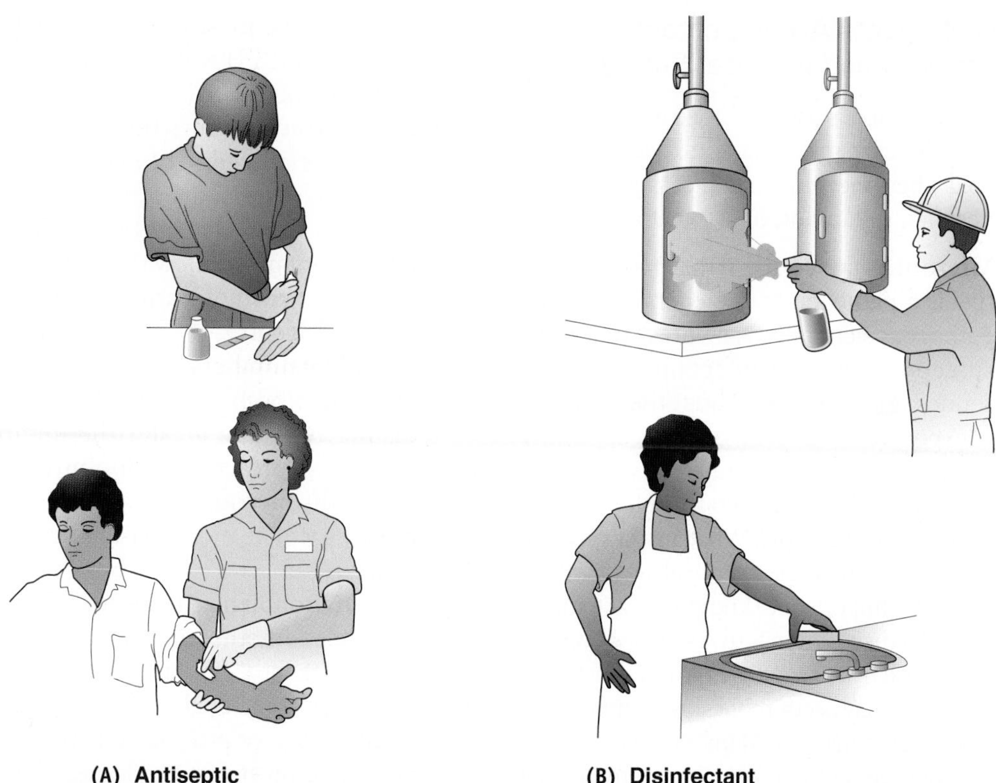

FIGURE 7.14 **Sample Uses of Antiseptics and Disinfectants.** (**A**) Antiseptics are used on body tissues, such as on a wound or before piercing the skin to take blood. (**B**) A disinfectant is used on inanimate objects, such as equipment used in an industrial process or tabletops and sinks. »» Why aren't disinfectants normally used as antiseptics?

and antiseptic, "against infection." It also is the origin of the term **asepsis**, meaning "free of disease-causing microbes."

Other expressions are associated with chemical control. As mentioned in the introduction to this chapter, sanitization of a location or object reduces the microbial population to a safe level as determined by public health standards. For example, in dairy and food-processing plants, the equipment usually is sanitized through the process of sanitization. Commercial establishments, such as restaurants, depend on disinfectants to maintain a sanitary kitchen and work establishment. To **degerm** an object is to mechanically remove organisms from its surface. Washing with soap and water, or using alcohol to prepare the skin surface for an injection, degerms the skin surface but has little effect on microorganisms deep in the skin pores.

FIGURE 7.15 illustrates the extent of "clean-up" necessary in an operation room following an operation on an infected patient.

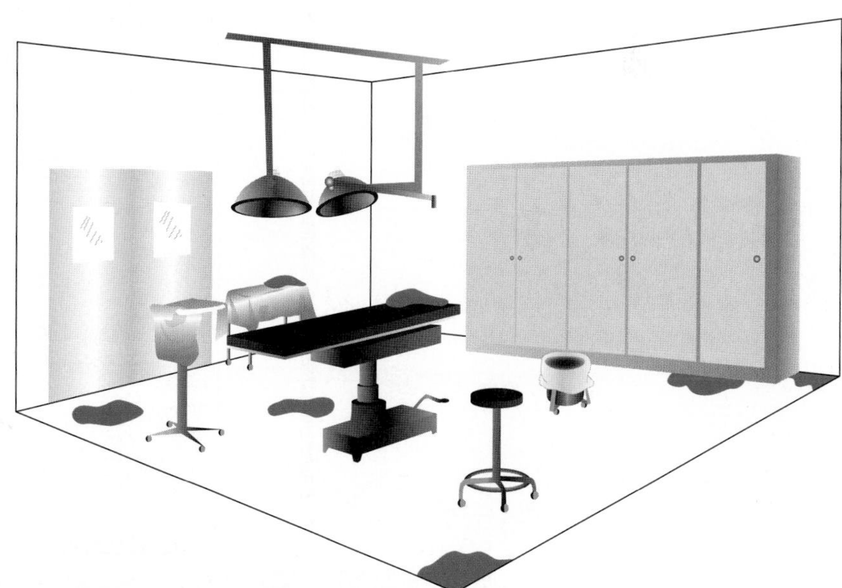

FIGURE 7.15 **A Pathogen-Contaminated Operating Room.** Following a "dirty operation" (for an infectious disease), the operating room (OR) may have become broadly contaminated with the pathogen. The purple areas indicate where the pathogen was cultured from a swab sample. »» Will the OR need to be sterilized, disinfected, or sanitized before the next operation? Explain.

Chemical Agents Are Important to Laboratory and Hospital Safety

Because disinfection is usually aimed at controlling pathogens, there are several factors that should be considered when selecting a chemical agent for use in the hospital, restaurant, or at home.

- **Microbial susceptibility**. How sensitive is a pathogen to the chemical agent? Pathogens generally vary greatly in their susceptibility to disinfectants and antiseptics (FIGURE 7.16). Thus, the duration of exposure will depend on the accessibility of the agent to the organism as well as the actual microbe(s) being targeted.
- **Temperature and pH**. What are the physical conditions where the chemical agent will be used? It is important to know at what temperature the disinfection is to take place because a chemical reaction occurring at warmer temperatures may occur more slowly or not at all at low temperatures. Many chemical agents also have an optimal pH. For example, one agent may be most effective at neutral or slightly acidic pH while another works best at an alkaline pH.
- **Concentration**. Does the chemical agent need to be diluted to be effective? Usually, the higher the concentration, the more effective it is as a chemical agent. Therefore, it is important to follow the manufacturer's directions when preparing dilutions of the chemical agent.
- **Microbial numbers**. How contaminated is the surface or instruments that need to be disinfected? The greater the number of organisms present (the **bioburden**), the longer disinfection will take. Microbes in biofilms will be more resistant to chemical attack.
- **The environment**. Are the microorganisms embedded in other organic matter? Pathogens in healthcare facilities may be associated with organic materials, such as blood, feces, bodily fluids, or pus such that the object to be treated must first be cleaned to remove

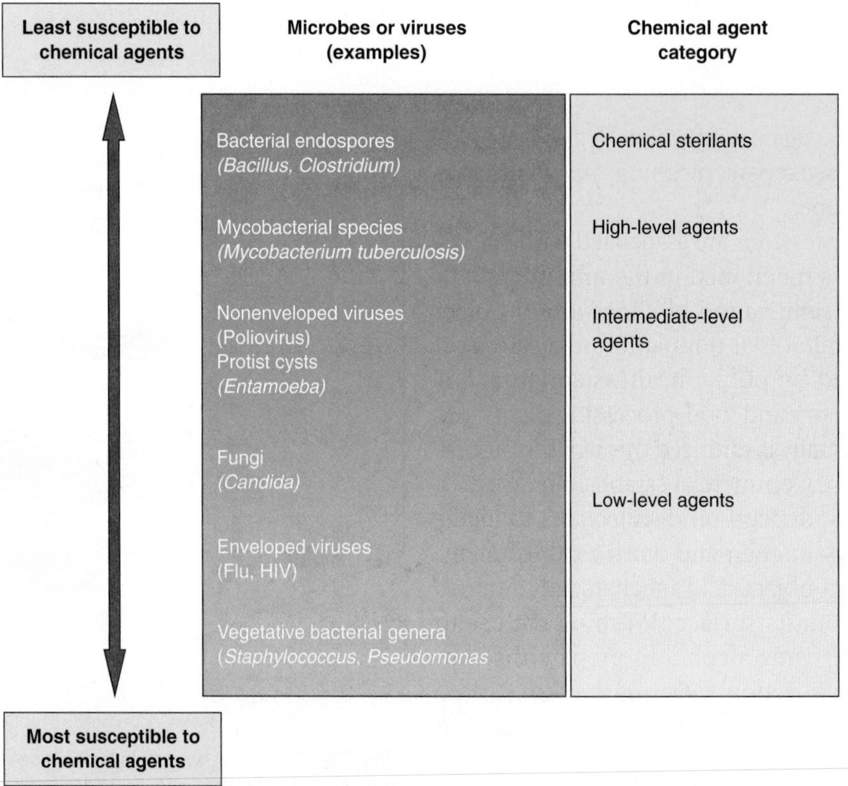

FIGURE 7.16 **Microbial Susceptibility to Chemical Agents.** Microbes and viruses vary in their susceptibility to chemical agents. »» Which chemical agent category (or categories) would not destroy (kill) endospores? Polioviruses?

organic matter so the chemical agent has access to the targeted microbes. Therefore, a chemical applied to a smooth and hard laboratory bench (disinfectant) is considerably different from one applied to the porous skin surface or to a wound (antiseptic).

- **Endospore formers**. Are the targeted organisms endospore formers? Endospores are resistant to many of the chemical agents used to inhibit microbial growth, so endospore formers can evade destruction by these agents (see Figure 7.16). This can be especially important when considered in the context of the hospital environment because removal of bacterial spores often requires more vigorous treatment than the removal of vegetative cells.

To be useful as a disinfectant or antiseptic, the chemical agent must also have a number other properties.

- It can kill or slow the growth of a wide variety of microorganisms while being nontoxic to animals or humans.
- It should be soluble in water or alcohol, easy to apply, and have a substantial shelf life.
- It is useful in diluted form and acts in a relatively short time.
- It is nonstaining and noncorrosive.
- It is odorless, easy to obtain, and relatively inexpensive.

Lastly, it is important to know how a chemical agent kills or inhibits the growth of microbes or inactivates viruses. Some agents (e.g., alcohols, phenolic compounds) target and disrupt cell membranes. Others (e.g., halogens, heavy metals, ethylene oxide) destroy (denature) enzymes and structural proteins.

Antiseptics and Disinfectants Can Be Evaluated for Effectiveness

Currently, there are more than 8,000 disinfectants and antiseptics for hospital use and thousands more for general use. Evaluating these chemical agents is a tedious process because of the broad diversity of conditions under which they are used.

One measure of effectiveness for chemical agents is the **phenol coefficient** (**PC**). This is a number indicating the disinfecting ability of an antiseptic or disinfectant in comparison to phenol under identical conditions, using *Staphylococcus aureus* or *Salmonella enterica* subtype Typhi as test organism. (TABLE 7.2). A PC higher than 1 indicates the chemical is more effective than phenol; a number less than 1 indicates it is less effective than phenol. For example, chemical A may have a PC of 78.5, while chemical B has a PC of 0.28. These numbers can then be used relative to each other.

The test has many drawbacks, especially because it is performed in the laboratory rather than in a real-life situation. Nor does it take into account many of the factors cited above, such as tissue toxicity, activity in the presence of organic matter, or temperature variations.

A more direct way of determining the value of a chemical agent is to take swab samples from the affected area before and after the application of a chemical agent. Another method, the **use dilution test**, uses standardized cultures of a bacterial

TABLE

7.2 Phenol Coefficients of Some Common Antiseptics and Disinfectants[a]

Chemical Agent	*Staphylococcus aureus*	*Salmonella enterica* subtype Typhi
Phenol	1	1
Chloramine	133	100
Tincture of iodine	6.3	5.8
Lysol	3.5	1.9
Mercurochrome	5.3	2.7
Ethyl alcohol	0.04	0.04
Formalin	0.3	0.7
Cetylpyridinium chloride	337	228

[a]All PC values were determined at 37°C.

species on small stainless steel cylinders. The dried cylinders are exposed to the chemical agent at various dilutions. In both these methods, the swabs and cylinders are then placed in broth growth medium to determine the presence or absence of growth. These methods of standardization have value under certain circumstances. However, it is conceivable that a universal test may never be developed, in view of the huge variety of chemical agents available and the numerous conditions under which they are used.

In the teaching laboratory, the **disk diffusion method** is a simple way to examine the effect of different chemical agents against a specific bacterial species. Small absorbent paper disks are each dipped in a separate solution of the chemical agents to be tested. Each damp disk is then placed on an agar plate that has been inoculated with the bacterial species in question. Incubation of the plate for 24 hours to 48 hours allows the chemical to diffuse out from the disk and into the agar, and possibly affect bacterial growth. Therefore, inhibition of growth will be seen as a clear halo, called a **zone of inhibition**, around a disk (FIGURE 7.17). Resistant bacteria will show no zone of inhibition. Realize, zone diameters on the plate are not comparable because the chemicals may be used at different concentrations and the molecular size of the chemical agent might affect its diffusion through the agar (larger molecules will diffuse from the disk more slowly, producing smaller zone diameters while smaller molecules will move faster and produce larger zone diameters). Still, this method can be used to determine if a particular chemical agent affects bacterial growth.

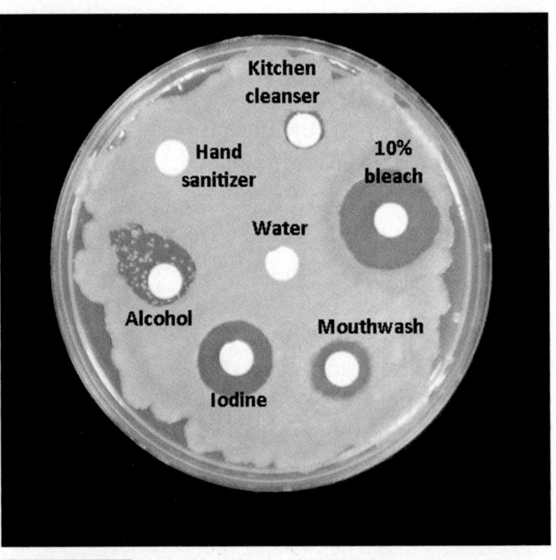

FIGURE 7.17 **The Disk Diffusion Method.** Paper disks containing different chemical agents can be compared by their zone diameters. Courtesy of Jeffrey Pommerville. »» Hypothesize what the small bacterial colonies represent in the clear zone around the paper disk containing alcohol.

CONCEPT AND REASONING CHECKS 3

a. Distinguish between (i) an antiseptic and a disinfectant and (ii) disinfection and sanitization.

b. Summarize the properties important in the selection of a disinfectant or antiseptic.

c. Assess the need to know a chemical agent's effectiveness.

Chapter Challenge B

You may use a variety of disinfectants in your home or workplace that are designed to eliminate or control potential pathogens. Use may depend on the object being treated.

QUESTION B:

Look at several disinfectants or antiseptics you have in your home or workplace. Do these products give you any information as to (a) microbial susceptibility (what is the product designed to eliminate or control) and (b) how the product should be used and stored?

Answers can be found on the Student Companion Website in **Appendix D.**

KEY CONCEPT 7.4 A Variety of Chemical Methods Can Control Microbial Growth

The chemical agents currently in use for controlling microorganisms range from very simple substances, such as household bleach, to very complex compounds, typified by cationic detergents. Many of these agents have been used for generations, while others represent the latest modern chemical products. In this section, we shall survey several groups of chemical agents and indicate how they are best applied in the control of microorganisms. MicroFocus 7.3 identifies some common but surprising antiseptics and Investigating the Microbial World 7 tests the antimicrobial properties of garlic.

MICROFOCUS 7.3: Being Skeptical

Antiseptics in Your Pantry?

Today, we live in an age when alternative and herbal medicine claims are always in the news, and these reports have generated a whole industry of health products that often make unbelievable claims. With regard to "natural products," are there some that have genuine medicinal and antiseptic properties?

Courtesy of Jeffrey Pommerville.

Cinnamon

Professor Daniel Y. C. Fung, Professor of Food Science and Food Microbiology at Kansas State University in Manhattan, Kansas, believes cinnamon might be an antiseptic that can control pathogens, at least in fruit beverages. Fung's group added cinnamon to commercially pasteurized apple juice. They then added typical foodborne pathogens (*Salmonella typhimurium, Yersinia enterocolitica,* and *Staphylococcus aureus*) and viruses. After one week of monitoring the juice at refrigerated and room temperatures, the investigators discovered the pathogens were killed more readily in the cinnamon blend than in the cinnamon-free juice. In addition, more bacterial organisms and viruses were killed in the juice at room temperature than when refrigerated.

Honey

For nearly three decades, Professor Peter Molan, associate professor of biochemistry and director of the Waikato Honey Research Unit at the University of Waikato, New Zealand, has been studying the medicinal properties of and uses for honey. Its acidity, between 3.2 and 4.5, is low enough to inhibit many pathogens. Its low water content (15% to 21% by weight) means that it osmotically ties up free water and "drains water" from wounds, helping to deprive pathogens of an ideal environment. In addition, in 2009–2010 other researchers discovered two proteins in honey, one interfering with bacterial cell wall synthesis and the other interacting with the bacterial cell membrane.

But beware! Not all honey is alike. The antibacterial properties of honey depend on the kind of nectar, or plant pollen, that bees use to make honey. At least manuka honey from New Zealand and honeydew from central Europe are thought to contain useful levels of antiseptic potency. Professor Molan is convinced that "honey belongs in the medicine cabinet as well as the pantry." In 2011, the United States Food and Drug Administration (FDA) approved wound dressings containing manuka honey.

Licorice Root

Dried licorice root has been used in traditional Chinese medicine for centuries. In 2012, scientists reported they had identified two substances in licorice root that can kill the two most common bacterial species causing tooth decay and one species responsible for gum disease. But don't run out and start eating lots of licorice candy because the licorice root extract originally in candies has been replaced by anise oil, which has a similar flavor but no antimicrobial properties.

Wasabi

The green, pungent, Japanese horseradish called wasabi may be more than a spicy condiment for sushi. Professor Hedeki Masuda, director of the Material Research and Development Laboratories at Ogawa & Co. Ltd., in Tokyo, Japan, and his colleagues have found that natural chemicals in wasabi, called isothiocyanates, inhibit the growth of *Streptococcus mutans*—one of the bacterial species causing tooth decay. At this point, these are only test-tube laboratory studies and the results will need to be proven in clinical trials.

The verdict? There appear to be products having genuine antimicrobial properties—and there are many more than can be described here.

Investigating the Microbial World 7

The Antibacterial Effects of Garlic

In 1858, Louis Pasteur examined the effects of garlic as an antiseptic. During World War II, when penicillin and sulfa drugs were in short supply, garlic was used as an antiseptic to disinfect open wounds and prevent gangrene.

- **OBSERVATION:** New research studies suggest that organosulfur compounds from garlic (see figure), some of which give garlic its characteristic taste and smell, block key enzymes that bacterial cells need to invade and damage host cells.

- **QUESTION:** ***Do these organosulfur compounds (OSCs) have a bactericidal effect on growth and survival of foodborne pathogens?***

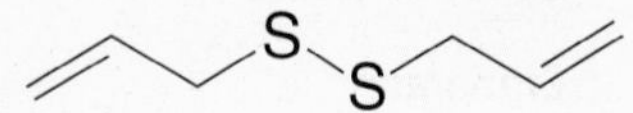

- **HYPOTHESIS:** Garlic-derived OSCs have a bactericidal effect on foodborne pathogens growing in culture. If so, then treating cultures of a foodborne pathogen with garlic-derived OSCs will adversely affect bacterial growth.

- **EXPERIMENTAL DESIGN:** The OSCs were extracted from garlic (procedure not given here) and produced as a concentrate in sterile saline water or a nutrient broth medium. The foodborne pathogen *Campylobacter jejuni,* one of the most common human foodborne pathogens, was prepared as a suspension culture in saline water.

- **EXPERIMENT 1:** Different concentrations of OSCs (0, 25, and 50 μl/ml) in saline water suspensions were added to *C. jejuni* cultures to give a final cell concentration of 10^5 cells/ml. Tubes were incubated at 4°, 22°, and 35°C for 24 hours.

- **EXPERIMENT 2:** Different concentrations of OSCs (0, 25, 50, and 100 μl/ml) in nutrient broth medium were added to *C. jejuni* cultures to give a final cell concentration of 10^5 cells/ml. Tubes were incubated at 4°, 22°, and 35°C for 24 hours.

After incubation, samples from both experiments were serially diluted and plated on a *Campylobacter* culture medium for 48 hours at 42°C. The number of viable cells, as colony-forming units (CFUs), was counted and total viable cells/ml estimated.

- **RESULTS:** See table.

- **CONCLUSION:**

QUESTION 1: ***Do the OSCs have an affect on the C. jejuni viability? Explain using the table.***

QUESTION 2: ***How did the concentration of OSCs and incubation temperature affect bacterial cell viability?***

QUESTION 3: ***Thinking about food preservation, what was the point of using different incubation temperatures?***

QUESTION 4: ***Being that C. jejuni is a strict mesophile, what is the significance of the growth at 35°C in broth (Experiment 2; 38,900,000) but not in saline (Experiment 1; 0) at 24h?***

TABLE

Effects of Garlic Organosulfur Compounds (OSC) on Viability of *Campylobacter jejuni* at Different Temperatures and OSC Concentrations

Temperature (°C)	OSC Concentration (μl/ml)	Experiment 1 Number of Viable Cells/ml		Experiment 2 Number of Viable Cells/ml	
		0h	24h	0h	24h
4	0	89,100	17,800	95,500	30,200
	25	74,100	500	112,200	5,600
	50	70,800	100	93,300	800
	100	ND	ND	125,900	500
22	0	104,700	40,700	83,200	8,700
	25	107,200	0	131,800	400
	50	117,500	0	95,500	0
	100	ND	ND	123,000	0
35	0	97,700	0	79,400	38,900,000
	25	104,700	0	77,600	0
	50	102,300	0	97,700	0
	100	ND	ND	131,800	0

Answers can be found on the Student Companion Website in **Appendix D.**

Adapter from: Lu, X. *et al.* (2011). *Appl. Environ. Microbiol.* **77**(15): 5257–5269. Icon image © Tischenko Irina/ShutterStock, Inc.

Halogens Oxidize Proteins

The **halogens** are a group of highly reactive elements. Two halogens, chlorine and iodine, represent intermediate-level chemical agents. In microorganisms, halogens are oxidizing agents or protein synthesis inhibitors that combine with and inactivate certain cytoplasmic proteins, such as enzymes. Killing almost always occurs within 30 minutes after application.

Chlorine (Cl_2) is effective against a broad variety of organisms, including most bacterial and fungal species, and many viruses. At high concentrations, it can be sporicidal.

Chlorine is available in a gaseous form and as both organic and inorganic compounds (FIGURE 7.18). It is widely used in municipal water supplies and swimming pools, where it keeps microbial populations at low levels. Chlorine combines readily with numerous ions in water; therefore, enough chlorine must be added to ensure a residue remains for antimicrobial activity. In municipal water, the residue of chlorine is usually about 0.2 to 1.0 parts per million (ppm) of free chlorine. One ppm is equivalent to 0.0001 percent, an extremely small amount.

Chlorine also is available as sodium hypochlorite (Clorox® bleach) or calcium hypochlorite. The latter, also known as chlorinated lime, was used by Semmelweis in his studies (see text on the history of microbiology). Hypochlorite compounds cause cellular proteins to denature, destroying their function. Hypochlorites also are useful in very dilute solutions for sanitizing commercial and factory equipment.

$Na^+ [OCl]^-$

Sodium hypochlorite

$Ca^{2+} [OCl]^-_2$

Calcium hypochlorite

The chloramines, such as chloramine-T, are organic compounds used as bactericides and for the disinfection of drinking water.

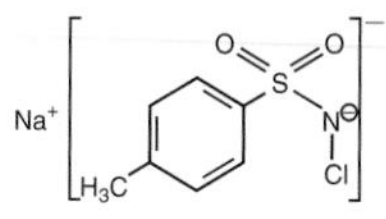

Chloramine-T

Iodine (I_2) is slightly larger than the chlorine atom and is more reactive and more germicidal. A tincture of iodine, a commonly used antiseptic for wounds, consists of 2% iodine. Iodine damages microbes and many endospores by reacting with enzymes and with proteins in the cell membrane and cell wall. **Iodophors** are iodine linked to a solubilizing agent, such as a detergent. These water-soluble complexes release iodine over a long period of time and have the added advantage of not irritating the skin. Some examples of iodophors are Wescodyne®, used in preoperative skin preparations; and Betadine®, which is iodine combined with a nondetergent carrier

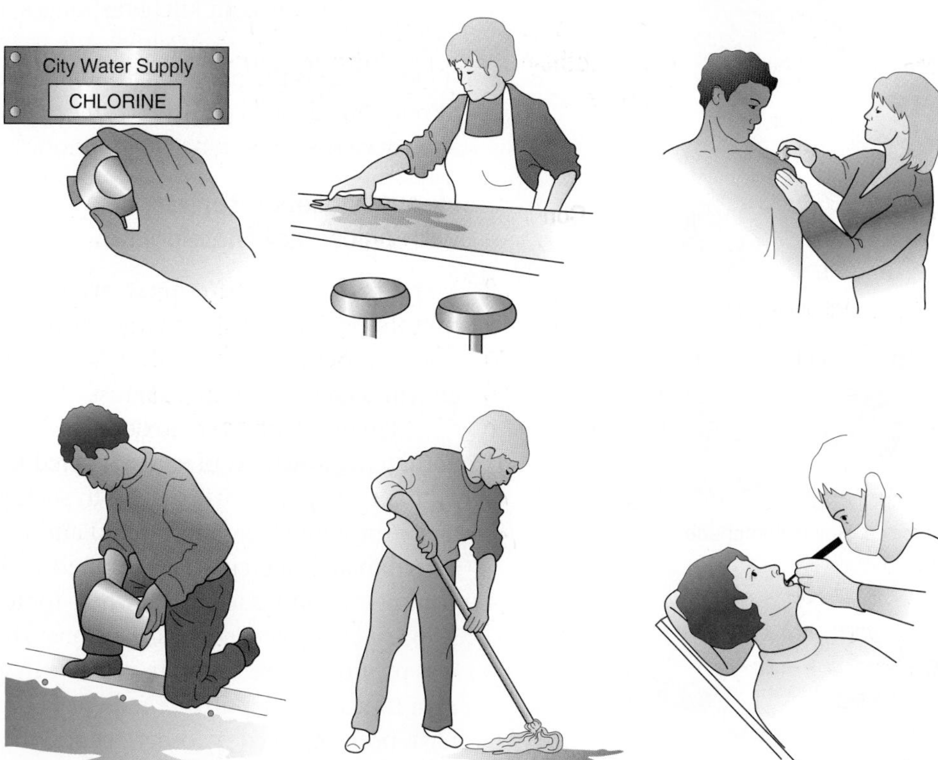

FIGURE 7.18 **Some Practical Applications of Disinfection with Chlorine Compounds.** Different chlorine compounds have been used as both disinfectants and antiseptics. »» In each of the above illustrations, is the chemical agent being used as a disinfectant or an antiseptic?

called povidone, which stabilizes the iodine and releases it slowly.

Phenol and Phenolic Compounds Denature Proteins

Phenol (carbolic acid) and phenolic compounds are low-level to intermediate-level chemical agents (FIGURE 7.19). Phenol, which has played a key role in disinfection practices since Joseph Lister first used it in the 1860s, remains the standard against which other antiseptics and disinfectants are evaluated using the phenol coefficient test. It is active against many bacterial and fungal species, and some viruses. Phenol and its derivatives act by denaturing proteins, especially in the cell membrane.

Chlorhexidine

Phenol is expensive, has a pungent odor, and is caustic to the skin; therefore, the usefulness of phenol as an antiseptic has diminished. However, phenol derivatives have greater germicidal activity and lower toxicity than the parent compound. Therefore, combinations of two phenol molecules, called **bisphenols**, have become prominent in modern disinfection and antisepsis. Orthophenylphenol, for example, is used in Lysol. Another bisphenol, hexachlorophene, was used extensively during the 1950s and 1960s in toothpastes, under-arm deodorants, and bath soaps. One product, pHisoHex, combined hexachlorophene with a pH-balanced detergent cream. However, studies indicated that excessive amounts of the bisphenol can be absorbed through the skin and cause neurological damage, so hexachlorophene has been removed from over-the-counter products.

An important bisphenol relative is **chlorhexidine**. This compound is used as a surgical hand scrub and superficial skin wound cleanser. A 4% chlorhexidine solution in isopropyl alcohol is commercially available as Hibiclens®. Another bisphenol in widespread use is **triclosan**, a broad-spectrum antimicrobial agent that destroys bacterial cells by disrupting cell membranes (and possibly cell walls) by blocking the synthesis of lipids. Triclosan is fairly mild and nontoxic, and it is effective against pathogenic bacteria (but only partially against viruses and fungi). The chemical is included in such products as antibacterial soaps, lotions, mouthwashes, toothpastes, toys, food trays, underwear, kitchen sponges, utensils, and cutting boards. A negative side to extensive triclosan use is the possibility of bacterial species developing resistance to the chemical, just as they have developed resistance to antibiotics.

(A) Phenol

(B) Orthophenylphenol, Hexachlorophene, Triclosan

FIGURE 7.19 Phenol and Some Derivatives. (**A**) The chemical structure of phenol and (**B**) some important derivatives. »» Why are most phenolic compounds only used as disinfectants?

Heavy Metals Interfere with Microbial Metabolism

Mercury, silver, and copper are called **heavy metals** because of their large atomic weights. They are low-level chemical agents that inactivate enzymes and structural proteins.

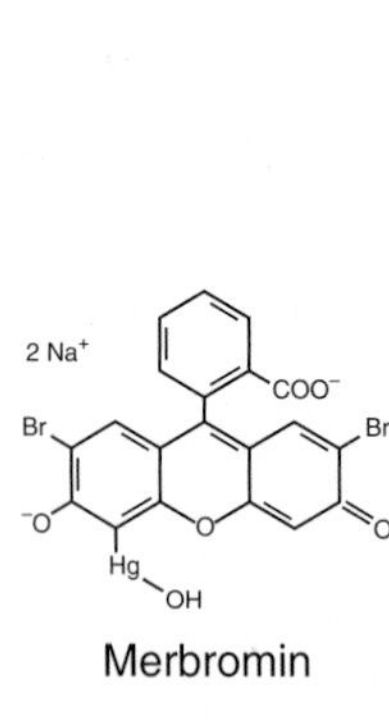

Thimerosal

Mercury (Hg) is very toxic to the host and the antimicrobial activity of mercury is reduced when other organic matter is present. In such products as merbromin (Mercurochrome) and thimerosal (Merthiolate), mercury is combined with carrier compounds and is less toxic when applied to the skin, especially after surgical incisions. Thimerosal was previously used as a preservative in vaccines. As a topical antiseptic, the mercury compounds have been replaced by other agents, such as the iodophors.

Copper (Cu) is a potent inhibitor of algae. As copper sulfate ($CuSO_4$), it is incorporated into algicides and is used in swimming pools and municipal water supplies. Copper ions also are very toxic to bacterial cells. Hospital fixtures containing copper were found to harbor 70% to 90% fewer microbes than did non-copper fixtures and reduced the risk of acquiring an infection by 40%.

Silver (Ag) in the form of 1% silver nitrate ($AgNO_3$) had been placed in the eyes of newborns to protect against infection by *Neisseria gonorrhoeae*, which could be contracted by a newborn during passage through an infected mother's birth canal. Today, antibiotic drops are used.

Alcohols Denature Proteins and Disrupt Membranes

Alcohols are intermediate-level chemical agents that affect most bacterial and fungal species, and viruses. They have no effect on spores. Alcohols denature proteins and dissolve lipids, an action leading to cell membrane disintegration. Ethyl alcohol also is a strong dehydrating agent.

Because alcohols react readily with organic matter, medical instruments and thermometers must be thoroughly cleaned before exposure. Alcohols are most effective when diluted in water because water prevents rapid evaporation. Thus, 70% alcohol is more effective than 95% alcohol in controlling microbes.

Ethyl alcohol (ethanol) is the active ingredient in many popular hand sanitizers and is used as a degerming agent to treat skin before a venipuncture or injection. It mechanically removes bacterial cells from the skin (degerms) and dissolves lipids. Isopropyl alcohol, or rubbing alcohol, also has high bactericidal activity.

Soaps and Detergents Act as Surface-Active Agents

Soaps are chemical compounds of fatty acids combined with potassium or sodium hydroxide. The pH of the compounds is usually about 8.0, and some microbial destruction is due to the alkaline conditions they establish on the skin. However, the major activity of soaps is as degerming agents for the mechanical removal of microorganisms from the skin surface. Soaps, therefore, are surface-active agents called **surfactants**; that is, they emulsify and solubilize particles clinging to a surface and reduce the surface tension. Soaps also remove skin oils, further reducing the surface tension and increasing the cleaning action. MicroFocus 7.4 discusses the antibacterial soaps.

Detergents are synthetic chemicals acting as strong surfactants. Because they are actively attracted to the phosphate groups of cellular membranes, they also alter the membranes and cause cell lysis. The most useful detergents to control microorganisms and many viruses are cationic (positively charged) derivatives of ammonium chloride. In these detergents, four organic groups replace the four hydrogens, and at least one is a long hydrocarbon chain (FIGURE 7.20). Such compounds often are called **quaternary ammonium compounds** or, simply, **quats**. They represent low-level chemical agents.

Quats have rather long, complex names, such as benzalkonium chloride in Zephiran® and cetylpyridinium chloride in Cēpacol®. They are used as sanitizing agents for industrial equipment and food utensils, as skin antiseptics, as disinfectants in mouthwashes and contact lens cleaners, and as disinfectants for use on hospital walls and floors. Their use as disinfectants for food-preparation surfaces can help reduce contamination incidents.

H H
H—C—C—OH
H H
Ethanol

Peroxides Damage Cellular Components

Peroxides are high-level chemical agents that contain oxygen-oxygen single bonds. **Hydrogen peroxide** (H_2O_2), a common household antiseptic, has been used as a rinse for wounds, scrapes, and abrasions. However, H_2O_2 applied to such areas foams and effervesces, as catalase in the tissue breaks down hydrogen peroxide to oxygen and water. Therefore, it is not recommended as an antiseptic for open wounds. However, the furious bubbling loosens dirt, debris, and dead tissue, and the oxygen gas is effective against anaerobic bacterial species. Hydrogen peroxide decomposition also results in a reactive form of oxygen—the superoxide radical—which is highly toxic to microorganisms and viruses.

New forms of H_2O_2 are more stable than traditional forms, do not decompose spontaneously, and therefore can be used topically. Such inanimate materials as soft contact lenses, utensils, heat-sensitive plastics, and food-processing equipment can be disinfected within 30 minutes.

Benzoyl peroxide is an active ingredient in teeth whitening products and at low concentrations (2.5%) is used to treat acne.

H—O—O—H
Hydrogen peroxide

■ Venipuncture: The piercing of a vein to take blood, to feed somebody intravenously, or to administer a drug.

H OH H
H—C—C—C—H
H H H
Isopropyl alcohol

Benzoyl peroxide

MICROFOCUS 7.4: Public Health/Being Skeptical

Are Antibacterial Soaps Worth the Money?

All of us want to be as clean as possible. In fact, hand washing is one of the best ways to protect oneself and prevent the spread of disease-causing microbes. To that end, numerous consumer product companies have provided us with many different types of antimicrobial cleaning and hygiene items. Perhaps the most pervasive are the antibacterial soaps, which usually contain about 0.2% triclosan.

Washing hands with soap and water is a key to preventing disease transmission. © Jones & Bartlett Learning. Photographed by Kimberly Potvin.

It is estimated that 75% of liquid and 30% of bar soaps on the market today are of the antibacterial type. The question though is: Are these products any better than regular soaps? The short answer is—no.

Numerous studies have shown these antibacterial soaps do little against foodborne pathogens such as *Salmonella* and *Escherichia coli*. In addition, they do nothing to reduce the chances of picking up and harboring infectious microbes.

A 2005 study gathered together over 200 families with children. Each family was given cleaning and hygiene supplies—soaps, detergents, and household cleaners—to use for one year. Half of the families (controls) received regular products without added antibacterial chemicals, while the other half used products with the antibacterial chemicals.

When the families were surveyed after one year, those using the antibacterial products were just as likely to get sick, as identified by such symptoms as coughs, fevers, sore throats, vomiting, and diarrhea.

You may say that this is not surprising, as many of these symptoms are the result of a viral infection—and the antibacterial products are not effective on viruses. However, further analysis of the families indicated there were just as many bacterial infections in the antibacterial group as there were in the control group.

A 2007 study that reviewed 27 reports on the effectiveness of triclosan-containing antibacterial soaps concluded that antibacterial soaps were no more effective than regular soap and water.

Ammonium ion

$[H-N^{+}(H)(H)-H]^{+}$

Benzalkonium chloride (Zephiran): $[C_6H_5-CH_2-N^{+}(CH_3)_2-C_{18}H_{37}]\ Cl^{-}$

Cetylpyridinium chloride (Ceepryn): $[C_5H_5N^{+}-C_{16}H_{33}]\ Cl^{-}$

FIGURE 7.20 **Cationic Detergents.** The chemical structures of some important quaternary ammonium compounds (quats) used in disinfection and antisepsis. »» How has the basic ammonium ion been modified to generate these two quats?

Some Chemical Agents Combine with Nucleic Acids and/or Cell Proteins

The chemical agents we discussed in previous sections are used for disinfection. However, a few other chemical agents can be used as chemical sterilants.

Aldehydes. Aldehydes are highly active agents that react with amino and hydroxyl groups of nucleic acids and proteins. The resulting cross linking inactivates the proteins and nucleic acids in microbes, viruses, and spores.

Formaldehyde is a gas at high temperatures and a solid at room temperature. As a 37% solution it is called **formalin**. For over a century, formalin was used in embalming fluid for anatomical specimens (though rarely used anymore) and by morticians for disinfecting purposes. In microbiology, formalin has been used for inactivating viruses in certain vaccines and producing toxoids from toxins.

Formalin can be used to disinfect surgical instruments, isolation rooms, and dialysis equipment. However, it leaves a residue, and instruments must be rinsed before use. Many allergic individuals develop a contact dermatitis to this compound.

Glutaraldehyde is a small, organic molecule that as a 2% solution destroys bacterial and fungal cells within 10 minutes and spores in 10 hours. It does not damage delicate objects, so it can be used to disinfect or sterilize optical equipment, such as the fiber-optic endoscopes used for arthroscopic surgery. Glutaraldehyde gives off irritating fumes, however, and instruments must be rinsed thoroughly in sterile water.

Sterilizing Gases. The development of plastics for use in microbiology required a suitable method for sterilizing these heat-sensitive materials. In the 1950s, research scientists discovered the antimicrobial abilities of ethylene oxide, which essentially made the plastic Petri dish and disposable plastic syringe possible.

Ethylene oxide is a small molecule with excellent penetration capacity, and, by destroying proteins and cross linking DNA, it is microbicidal as well as sporicidal. However, it can induce human tumors and is highly explosive. Its explosiveness is reduced by mixing it with Freon gas or carbon dioxide gas, but its toxicity remains a problem for those who work with it. Ethylene oxide chambers, called "gas autoclaves," have become the chemical counterparts of heat- and pressure-based autoclaves for sterilization procedures.

Ethylene oxide is used to sterilize paper, leather, wood, metal, and rubber products as well as plastics. In medicine, it is used to sterilize catheters, artificial heart valves, heart-lung machine components, and optical equipment. The National Aeronautics and Space Administration (NASA) uses the gas for sterilization of interplanetary space capsules.

Chlorine dioxide has properties very similar to chloride gas and sodium hypochlorite but, unlike ethylene oxide, it produces nontoxic byproducts and is not tumor-causing. Chlorine dioxide can be used as a gas or liquid. In a gaseous form, with proper containment and humidity, a 15-hour fumigation can be used to sanitize air ducts, food and meat processing plants, and hospital areas. It was the gas used to decontaminate the 2001 anthrax-contaminated mail and office buildings (MicroFocus 7.5).

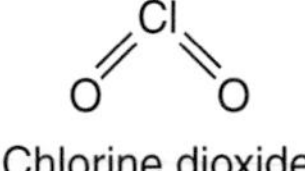

Chlorine dioxide

O
H C H
Formalin

O O
H H
Glutaraldehyde

TABLE 7.3 and FIGURE 7.21 summarize the chemical agents used in controlling microorganisms.

CONCEPT AND REASONING CHECKS 4

a. Compare the uses for chlorine and iodine chemical agents.
b. Explain why bisphenols are preferred as disinfectants and antiseptics.
c. Evaluate the use of heavy metals as antiseptics and disinfectants.
d. Why is 70% ethanol preferable to 95% ethanol as an antiseptic?
e. How do soaps differ from quats as chemical agents of control?
f. Judge the advantages and disadvantages of using hydrogen peroxide as an antiseptic.
g. Summarize the uses for aldehydes, ethylene oxide, and chloride dioxide for sterilization.

Chapter Challenge C

In this last section of the chapter, you have studied several groups of chemical agents that accomplish disinfection (in the form of disinfectants or antiseptics).

QUESTION C:

For the products you examined in part B of the chapter challenge, as well as other personal products like a mouthwash, toothpaste, or deodorant, look at the label and especially pay attention to the "active ingredients" or "contains" section. What types of chemical agents are included in each of these products?

Answers can be found on the Student Companion Website in **Appendix D.**

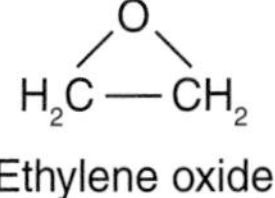

Ethylene oxide

MICROFOCUS 7.5: Tools/Bioterrorism

Decontamination of Anthrax-Contaminated Mail and Office Buildings

This chapter has examined the chemical procedures and methods used to control the numbers of microorganisms on inanimate and living objects. These control measures usually involve a level of sanitation, although some procedures may sterilize. Some examples were given for their use in the home and workplace. However, what about real instances where large-scale and extensive decontamination has to be carried out?

(A) Anthrax Letter Sent to Senator Tom Daschle. © Reuters/Department of Justice/Handout/Landov.

(B) Mail sorting machines. Courtesy of the US Census Bureau, Public Information Office.

In October 2001, the United States experienced a bioterrorist attack. The perpetrator(s) used anthrax spores as the bioterror agent. Four anthrax-contaminated letters were sent through the mail on the same day, addressed to NBC newscaster Tom Brokaw, the *New York Post,* and to two United States Senators, Senator Patrick Leahy and Senate Majority Leader Tom Daschle (see Figure A). The Centers for Disease Control and Prevention (CDC) confirmed that anthrax spores from at least the Daschle letter contaminated the Hart Senate Office Building and several post office sorting facilities in Trenton, New Jersey (from where the letters were mailed), and Washington, D.C., areas. This resulted in the closing of the Senate building and the postal sorting facilities. With the Senate building and postal sorting facilities closed, the CDC, the Environmental Protection Agency (EPA), other governmental agencies, and commercial companies had to devise and implement a strategy to decontaminate these buildings and the mail sorting machines (see Figure B).

As mentioned in this chapter, most sanitation procedures do not require a high technology solution. In fact, all of the methods actually used for this situation are described in this chapter.

The Hart Senate Office Building and the post office sorting facilities were contaminated with *Bacillus anthracis* endospores. These are large, multiroom facilities with many pieces of furniture and instruments, including computers, copy machines, and mail sorting machines. To decontaminate these buildings, chemical disinfectants such as bleach solutions could have been used. However, spores may have gotten into the office machinery and sorting machines. Liquids would not work here.

Therefore, a gas was needed that could permeate the air ducts as well as all the office machinery and sorting machines. The gas chosen was chlorine dioxide. Essentially, the buildings were sealed as if they were going to be fumigated for termites. The gas was pumped in, and after a time that was believed to be sufficient to kill any anthrax spores, the gas was evacuated. Swabs were taken from the buildings and plated on nutrient media. If any spores were still alive, they would germinate on the plates and the vegetative cells would grow into visible colonies. Such results would require retreatment of the facility.

To protect the mail from similar attacks in the future, a system was devised using ultraviolet (UV) light to kill any spores that might be found in a piece of mail moving through the sorting machines.

It took months, and even years, for some of the postal facilities to be declared safe and free of anthrax spores. Still, simple physical and chemical methods worked to decontaminate the buildings and equipment.

TABLE

7.3 Summary of Chemical Agents Used to Control Microorganisms

Chemical Agent	Mechanism of Activity	Applications	Limitations	Antimicrobial Spectrum
Chlorine (Chlorine gas, sodium hypochlorite, chloramines)	Protein inactivation	Water treatment Skin antisepsis Equipment spraying Food processing	Inactivated by organic matter Objectionable taste, odor	Microbicidal Viricidal Sporocidal
Iodine (Tincture of iodine, iodophors)	Reacts with enzymes and structural proteins	Skin antisepsis Medical/lab disinfection	Inactivated by organic matter Objectionable taste, odor	Microbicidal Viricidal Sporocidal
Phenol and derivatives (Hexachlorophene, chlorhexidine, triclosan)	Denatures proteins Disrupts cell membranes	Surface disinfection Skin antisepsis with detergent	Toxic to tissues Disagreeable odor	Microbicidal Viricidal
Mercury (Mercuric chloride, merthiolate, merbromin)	Inactivates enzymes and structural proteins	Skin antiseptics Disinfectants	Inactivated by organic matter Toxic to tissues Slow acting	Microbiostatic
Copper (Copper sulfate)	Combines with proteins	Algicide in swimming pools Municipal water supplies	Inactivated by organic matter	Microbiostatic Fungistatic
Silver (Silver nitrate)	Binds proteins	Antiseptic in eyes of newborns	Skin irritation	Microbiostatic
Alcohol (70% ethyl, 75% isopropyl)	Denatures proteins Dissolves lipids Dehydrating agent	Instrument disinfectant Skin antiseptic	Precleaning necessary Skin irritation	Microbicidal Viricidal
Cationic detergents (Quaternary ammonium compounds)	Dissolve lipids in cell membranes	Instrument disinfection Skin antisepsis	Neutralized by soap	Microbicidal Viricidal
Peroxides (Hydrogen peroxide, benzoyl peroxide)	Formation of superoxide radicals	Skin antisepsis Surface disinfection	Limited use	Microbicidal Viricidal Sporocidal
Formaldehyde (Formalin)	Reacts with functional groups in proteins and nucleic acids	Embalming Vaccine production Gaseous sterilant	Poor penetration Allergenic Toxic to tissues Neutralized by organic matter	Microbicidal Viricidal Sporocidal
Glutaraldehyde (Glutaraldehyde)	Reacts with functional groups in proteins and nucleic acids	Sterilization of surgical supplies	Unstable Toxic to skin Respiratory irritation	Microbicidal Viricidal Sporocidal
Ethylene oxide (Ethylene oxide gas)	Reacts with functional groups in proteins and nucleic acids	Sterilization of instruments, equipment, heat-sensitive objects	Explosive Toxic to skin Requires constant humidity	Microbicidal Viricidal Sporocidal
Chlorine dioxide (Chlorine dioxide gas)	Reacts with functional groups in proteins and nucleic acids	Sanitizes equipment, rooms, buildings	Burns skin and eyes on contact	Microbicidal Viricidal Sporocidal

■ **In conclusion**, we have examined the various physical and chemical methods developed to control microbial growth. The control methods described are based on the knowledge about and understanding of the biochemistry, cell organization and structure, growth, and metabolic activities of microbes. One of the most important historical discoveries in the field of medicine was Semmelweis' insistence on hand washing to inhibit the spread of pathogens. And today, hand washing remains the single most important chemical method used in hospitals, clinics, restaurants, public restrooms, and at home. Much of the effectiveness of hand washing is simply taking the time (15 to 20 seconds) to wash with soap and warm water (see MicroFocus 7.4 on the use of antibacterial soaps). This simple chemical method is perhaps the most effective way to prevent the transmission of pathogens. So, as the Centers for Disease Control and Prevention proclaim: *Remember to Wash—Clean Hands Save Lives!*

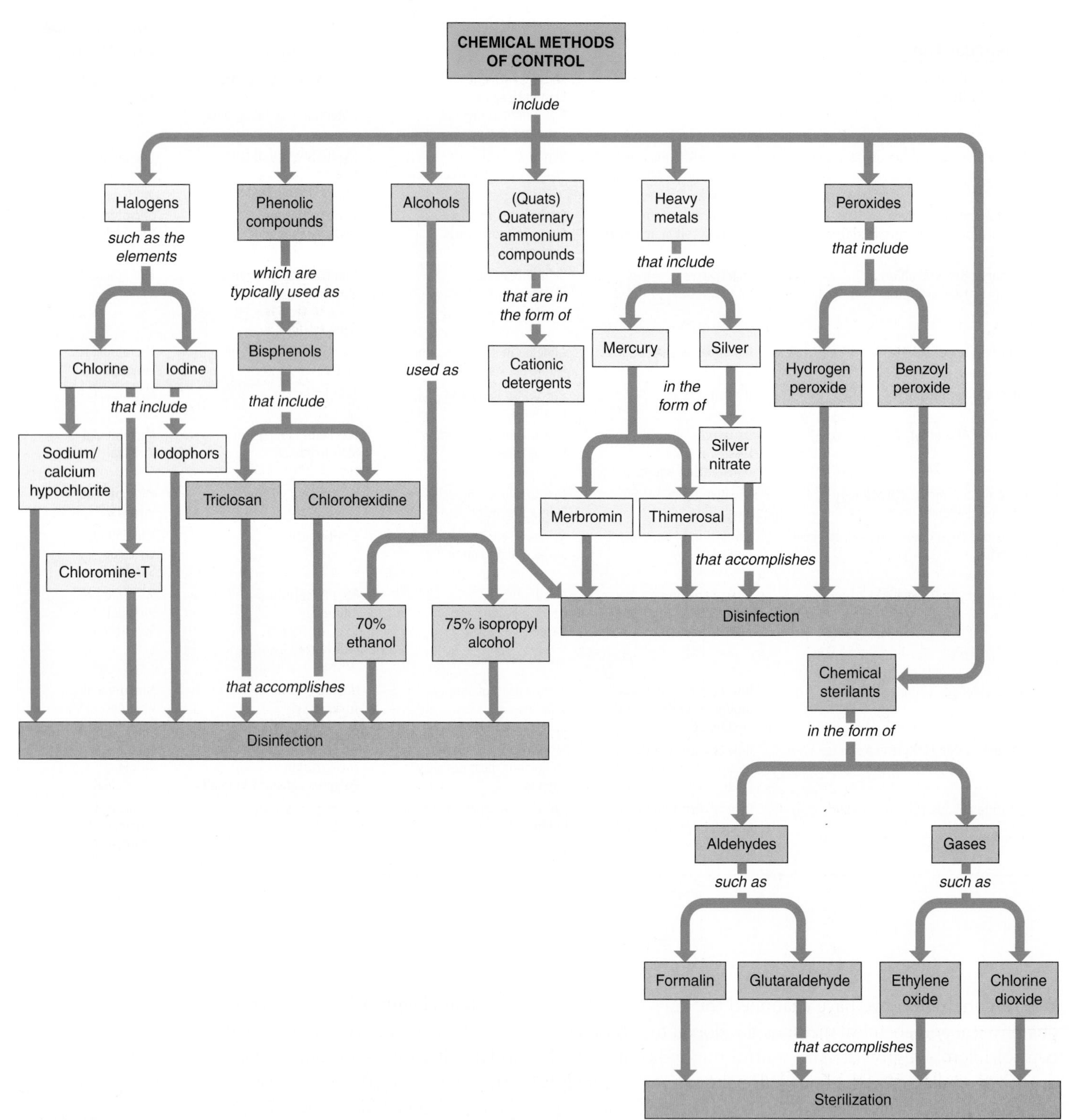

FIGURE 7.21 **A Concept Map Summarizing the Chemical Methods of Microbial Control.** The chemical methods predominantly disinfect, although a few can sterilize. »» If you wanted to sanitize a kitchen counter, which chemicals might be selected?

SUMMARY OF KEY CONCEPTS

7.1 Microbial Growth Can Be Controlled in Several Ways

1. The physical methods for controlling microorganisms are generally intended to achieve **sanitization**, which involves methods to reduce the numbers of, or inhibit the growth of, microbes.

7.2 There Are Various Physical Methods to Control Microbial Growth

2. Heat is a common control method. Used in the food industry, **thermal death point** and **thermal death time** are used to determine how long it takes to kill a population of microbial cells.
3. **Incineration** using a direct flame achieves sterilization in a few seconds. (Fig. 7.3)
4. Moist heat has several applications: The exposure to boiling water at 100°C for 2 hours can result in sterilization, but spore destruction cannot always be assured; the **autoclave** sterilizes materials in about 15 minutes, while a **prevacuum sterilizer** shortens this time using higher temperatures and pressures. **Intermittent sterilization** sterilizes through successive exposures to steam on 3 days; **pasteurization** reduces the microbial population in a liquid but is not intended to be a sterilization method. (Fig. 7.6)
5. **Filtration** uses various materials to trap microorganisms within or on a filter. **Membrane filters** are the most common. Air can be filtered using a **high-efficiency particulate air (HEPA) filter.** (Figs. 7.9, 7.10)
6. **Ultraviolet (UV) light** is an effective way of killing microorganisms on a dry surface and in a confined air space. (Fig. 7.11)
7. **X rays** and **gamma rays** are two forms of **ionizing radiation** used to sterilize heat-sensitive objects. Irradiation also is used in the food industry to control microorganisms on perishable foods. (Fig. 7.12)
8. For food preservation, drying, salting, and low temperatures can be used to control microorganisms.

7.3 Chemical Control Usually Involves Disinfection

9. Chemical agents are effectively used to control the growth of microorganisms. A chemical agent used on a living object is an **antiseptic**; one used on a nonliving object is a **disinfectant.** (Fig. 7.14)
10. Both antiseptics and disinfectants are selected according to certain criteria, including an ability to kill microorganisms or interfere with their metabolism. (Fig. 7.16)
11. The **phenol coefficient test** can be used to evaluate antiseptics and disinfectants. **Use dilution tests** are more practical for everyday applications of antiseptics and disinfectants. The **disk diffusion method** is also used to determine bacterial susceptibility to various chemical agents. (Fig. 7.17)

7.4 A Variety of Chemical Methods Can Control Microbial Growth

12. **Halogens** (chlorine and iodine) are useful for water disinfection, wound antisepsis, and various forms of sanitation.
13. **Phenol** derivatives, such as hexachlorophene, are valuable skin antiseptics and active ingredients in presurgical scrubs. (Fig. 7.19)
14. **Heavy metals** (silver and copper) are useful as antiseptics and disinfectants, respectively.
15. **Alcohol** (70% ethyl alcohol) is an effective skin antiseptic.
16. Soaps and detergents are effective degerming agents. **Quats** are more effective as a disinfectant than as an antiseptic. (Fig. 7.20)
17. **Hydrogen peroxide** acts by releasing oxygen to cause an effervescing cleansing action. It is better as a disinfectant than an antiseptic.
18. **Formaldehyde** and **glutaraldehyde** are sterilants that alter the biochemistry of microorganisms. **Ethylene oxide** gas under controlled conditions is an effective sterilant for plasticware. **Chlorine dioxide** gas can be used to sanitize air ducts, food and meat processing plants, and hospital areas.

CHAPTER SELF-TEST

For **STEPS A–D**, answers to even-numbered questions and problems can be found in **Appendix C** on the Student Companion Website **http://microbiology.jbpub.com/10e**. In addition, the site features eLearning, an online review area that provides quizzes and other tools to help you study for your class. You can also follow useful links for in-depth information, read more MicroFocus stories, or just find out the latest microbiology news.

STEP A: REVIEW OF FACTS AND TERMS

Multiple-Choice

Read each question carefully, then select the ***one*** answer that best fits the question or statement.

1. All the following terms apply to microbial killing *except* _____.
   - **A.** sterilization
   - **B.** microbicidal
   - **C.** bactericidal
   - **D.** fungistatic
2. The thermal death time is _____.
   - **A.** the time to kill a microbial population at a given temperature
   - **B.** the time to kill a microbial population in boiling water
   - **C.** the temperature to kill all pathogens
   - **D.** the minimal temperature needed to kill a microbial population
3. Which one of the following statements is NOT true of dry heat?
   - **A.** The energy does not penetrate materials easily.
   - **B.** Long periods of exposure are needed.
   - **C.** The heat oxidizes microbial proteins.
   - **D.** The heat corrodes sharp instruments.
4. An autoclave normally sterilizes material by heating the material to _____°C for _____ minutes at _____ psi.
   - **A.** 100; 10; 30
   - **B.** 121.5; 15; 15
   - **C.** 100; 15; 0
   - **D.** 110; 30; 5

5. Air filtration typically uses a _____ filter.
   A. HEPA
   B. membrane
   C. sand
   D. diatomaceous earth
6. For bactericidal activity, _____ has/have the ability to cause thymine dimer formation.
   A. X rays
   B. ultraviolet (UV) light
   C. gamma rays
   D. microwaves
7. The elimination of pathogens in foods by irradiation is called _____.
   A. the D value
   B. the pasteurizing dose
   C. incineration
   D. sterilization
8. Preservation methods such as salting result in the _____ microbial cells.
   A. loss of salt from
   B. gain of water into
   C. loss of water from
   D. lysis of
9. Which one of the following statements does NOT apply to antiseptics?
   A. They are used on living objects.
   B. They usually are microbicidal.
   C. They should be useful as dilute solutions.
   D. They can sanitize objects.
10. All the following are chemical parameters considered when selecting an antiseptic or disinfectant *except* _____.
    A. dehydration
    B. temperature
    C. stability
    D. pH
11. If a chemical has a phenol coefficient (PC) of 63, it means the chemical _____.
    A. is better than one with a PC of 22
    B. will kill 63% of bacteria
    C. kills microbes at 63°C
    D. will kill all bacteria in 63 minutes
12. Which one of the following is NOT a halogen?
    A. Iodine
    B. Mercury
    C. Clorox bleach
    D. Chlorine
13. Phenolics include chemical agents _____.
    A. such as the iodophores
    B. derived from carbolic acid
    C. used as tinctures
    D. such as formaldehyde
14. Heavy metals, such as _____ work by _____.
    A. mercury; disrupting membranes
    B. copper; producing toxins
    C. iodine; denaturing proteins
    D. silver; inactivating proteins
15. Alcohols are _____.
    A. surfactants
    B. heavy metals
    C. denaturing agents
    D. detergents
16. All the following statements apply to quats *except* _____.
    A. they react with cell membranes
    B. they are positively charged molecules
    C. they are types of soaps
    D. they can be used as disinfectants
17. Hydrogen peroxide _____.
    A. is an effective sterilant
    B. cross-links proteins and nucleic acids
    C. can emulsify and solubilize pathogens
    D. is not recommended for use on open wounds
18. Ethylene oxide can be used to _____.
    A. kill bacterial spores
    B. clean wounds
    C. sanitize work surfaces
    D. treat water supplies

Fill-in: Physical Methods

Use the following syllables to form the term that answers the clue pertaining to physical methods of control. The number of letters in the term is indicated by the dashes, and the number of syllables in the term is shown by the number in parentheses. Each syllable is used only once.
A AU BER BRANE CLAVE CU DE DER DRY HOLD ING ING LET LO MEM NA O POW SIS TION TO TRA TU TUR UL VI

19. Instrument for sterilization (3) _ _ _ _ _ _ _ _ _ _
20. Type of filter (2) _ _ _ _ _ _ _ _
21. Sterilized in an oven (2) _ _ _ _ _ _
22. Occurs with moist heat (5) _ _ _ _ _ _ _ _ _ _ _ _ _
23. Preserves meat, fish (2) _ _ _ _ _ _
24. Method of pasteurization (2) _ _ _ _ _ _ _
25. "Light" for air sterilization (5) _ _ _ _ _ _ _ _ _ _ _ _
26. Disease prevented by pasteurization (5) _ _ _ _ _ _ _ _ _ _ _ _ _

Matching: Chemical Agents

Chemical agents are a broad and diverse group, as this chapter has demonstrated. To test your knowledge over the chemical methods of control, match the chemical agent on the right to the statement on the left by placing the correct letter in the available space. A letter may be used once, more than once, or not at all.

Statement

____ **27.** The halogen in bleach
____ **28.** Sterilizes heat-sensitive materials
____ **29.** A 70% concentration is recommended
____ **30.** Quaternary compounds, or quats
____ **31.** Often used as a tincture
____ **32.** Rinse for wounds and scrapes
____ **33.** Example of a heavy metal
____ **34.** Triclosan is a derivative
____ **35.** Used by Joseph Lister
____ **36.** Used for sterilizing plastic culture dishes
____ **37.** Broken down by catalase

Chemical agent

A. Cationic detergent
B. Chlorine
C. Ethyl alcohol
D. Ethylene oxide
E. Formaldehyde
F. Glutaraldehyde
G. Hydrogen peroxide
H. Iodine
I. Phenol
J. Silver
K. Soap

STEP B: CONCEPT REVIEW

38. Distinguish between **sterilization** and **sanitization**. (**Key Concept 1**)
39. Contrast **thermal death time** and **thermal death point**. (**Key Concept 2**)
40. Discuss the four ways that moist heat can be used to control microbial growth. (**Key Concept 2**)
41. Summarize the **filtration** methods used to sterilize a liquid and decontaminate air. (**Key Concept 2**)
42. Summarize how **ultraviolet (UV) light** works to control microbial growth and explain how **X rays** and **gamma rays** are used as physical control agents. (**Key Concept 2**)
43. Compare and contrast an **antiseptic** and a **disinfectant**. (**Key Concept 3**)
44. List the desirable chemical properties of antiseptics and disinfectants. (**Key Concept 3**)
45. Describe how the effectiveness of a chemical agent can be measured. (**Key Concept 3**)
46. Evaluate the usefulness of **halogens** and phenolic derivatives as disinfectants and/or antiseptics. (**Key Concept 4**)
47. Summarize the uses for **heavy metals** in the chemical control of microorganisms. (**Key Concept 4**)
48. Justify why **alcohol** is not a method for skin sterilization. (**Key Concept 4**)
49. Distinguish between soaps, detergents, and **quats**. (**Key Concept 4**)
50. Estimate the value of **hydrogen peroxide** as a bacteriostatic agent. (**Key Concept 4**)
51. Identify the uses of aldehydes and gases as chemical sterilants. (**Key Concept 4**)

STEP C: APPLICATIONS AND PROBLEMS

52. When the local drinking water is believed to be contaminated, area residents are advised to boil their water before drinking. Often, however, they are not told how long to boil it. As a student of microbiology, what is your recommendation?
53. You need to sterilize a liquid. What methods could you devise using only the materials found in the average household?
54. Suppose you were in charge of a clinical microbiology laboratory where instruments are routinely disinfected and equipment is sanitized. A salesperson from a disinfectant company stops in to spur your interest in a new chemical agent. What questions might you ask the salesperson about the product?
55. A portable room humidifier can incubate and disseminate infectious microorganisms. If a friend asked for your recommendations on disinfecting the humidifier, what would you suggest?
56. A student has finished his work in the laboratory and is preparing to leave. He remembers the instructor's precautions to wash his hands and disinfect the lab bench before leaving. However, he cannot remember whether to wash first then disinfect, or to disinfect then wash. What advice would you give?
57. As a hospital infection officer, one of your duties is to ensure that the examination rooms are disinfected after patient treatment and prior to receiving new patients. Having hired new staff to specifically carry out the examination room "cleaning," what will you tell the new employees with regard to examination room disinfection?

STEP D: QUESTIONS FOR THOUGHT AND DISCUSSION

58. Instead of saying that food has been irradiated, manufacturers indicate that it has been "cold pasteurized." Why do you believe they must use this deception?

59. The label on the container of a product in the dairy case proudly proclaims, "This dairy product is sterilized for your protection." However, a statement in small letters below reads: "Use within 30 days of purchase." Should this statement arouse your suspicion about the sterility of the product? Why?

60. In view of all the sterilization methods we have discussed in this chapter, why do you think none has been widely adapted to the sterilization of milk?

61. A liquid that has been sterilized may be considered pasteurized, but one that has been pasteurized may not be considered sterilized. Why not?

62. Before taking a blood sample from the finger, the blood bank technician commonly rubs the skin with a pad soaked in alcohol. Many people think that this procedure sterilizes the skin. Are they correct? Why?

63. Suppose you had just removed the thermometer from the mouth of your sick child and confirmed your suspicion of fever. Before checking the temperature of your other child, how would you treat the thermometer to disinfect it?

64. The water in your home aquarium always seems to resemble pea soup, but your friend's is crystal clear. Not wanting to appear stupid, you avoid asking him his secret. But one day, in a moment of desperation, you break down and ask, whereupon he knowledgeably points to a few pennies among the gravel. What is the secret of the pennies?

Escherichia coli cells, such as these gram-negative stained cells, are a model organism for genetic studies. Courtesy of CDC.

The Genetics of Microorganisms

PART 2

CHAPTER 8 Microbial Genetics

CHAPTER 9 Gene Transfer, Genetic Engineering, and Genomics

The microorganisms that make up a substantial portion of the Earth's biomass have been evolving for billions of years. They exist in virtually every environment on planet Earth; many survive—and even thrive—in extremes of heat, cold, radiation, pressure, salt, and acidity. This diversity and its range of environmental conditions show that microbes long ago "solved" many of the challenges of adaptation in these environments. In fact, whether we examine cell structure and function, growth and nutrition, microbial metabolism, or virtually any microbial characteristic we might wish to consider, including antibiotic resistance, all are the result of inherited information. This information is stored as deoxyribonucleic acid (DNA) and is passed on from generation to generation. The study is called microbial genetics.

The contributions of microbial genetics have been numerous, diverse, and far reaching. Model microbial systems, such as *Escherichia coli*, have established many of the principles of molecular biology. In addition, many of the molecular techniques in the geneticist's toolbox (e.g., polymerases, restriction enzymes, cloning vectors) are derived from genetic studies of microbes. Despite this legacy and the huge clinical significance of microbial genetics, the potential of the field is only just beginning to be tapped. This is, in part, why we are in the third Golden Age of microbiology!

In the next two chapters, we will examine the microbial genome in detail. Chapter 8 is devoted to the basics of microbial genetics. We will see how microbial DNA is replicated and how this information codes for and directs protein synthesis. We will also examine the mechanisms of genome regulation and how it is affected by mutation. Then, Chapter 9 introduces us to the fields of genetic engineering, biotechnology, and microbial genomics. We will discover how DNA information can be transferred laterally from one microbe to another. We will examine the techniques of genetic engineering, the applications of biotechnology, and finish with a discussion of microbial genomics; that is, the study of genes and their function. Genomic analyses of microorganisms have broad significance not only for the science of microbiology, but also in our daily lives.

MICROBIOLOGY PATHWAYS

Biotechnology

© Jon Feingersh Photogr/Blend Images/age fotostock.

During the 1980s, the editors of *Time Magazine* referred to DNA technology as "the most awesome skill acquired by man since the splitting of the atom." Indeed, the work with DNA, begun in the 1950s and continuing today, has opened vistas previously unimagined. Scientists can now remove bits of DNA from organisms, snip and rearrange the genes, and insert them into different species, where the genes will express themselves. Practical results of these experiments have led to the mass production of hormones, blood-clotting factors, and other pharmaceutical products. They also have given us diagnostic methods based on DNA fingerprinting; advances in gene therapy; a revolution in agricultural research; barnyard animals producing human hemoglobin; and a colossal project that has mapped the entire human genome.

Industrial microbiology or microbial biotechnology (the terms "industrial microbiology" and "biotechnology" are often one and the same) applies scientific and engineering principles to the processing of materials by microorganisms and viruses, or plant and animal cells, to create useful products or processes. Because biotechnology essentially uses the basic ingredients of life to make new products, it is both a cutting-edge technology and an applied science.

If you would like to be part of what analysts predict will be one of the most important applied sciences of this century, then microbiology is the place to start. You would be well advised to take a course in biochemistry as well as one in genetics. Courses in physiology and cell biology are also helpful. Employers will be looking for individuals with good laboratory skills, so be sure to take as many lab courses as you can.

You may enter the biotechnology field with an associate's, bachelor's, master's, or doctoral degree. This is because there are so many levels at which individuals are hired. Most professional levels of employment require a college degree (BS) in biology, microbiology, or biotechnology with minors in one or more of the complementary sciences. Persons who have project responsibilities often have one or more advanced degrees (MS and/or PhD) in biology, microbiology, or some other allied field such as molecular biology, biochemistry, biotechnology, chemical engineering, or genetics.

An employer also will be looking for work experience, which you can obtain by assisting a senior scientist, doing an internship, or working summers in a biotech firm (usually for slave wages). The campus research lab is another good place to obtain work experience. It also might be a good idea to sharpen your writing skills, because you will be preparing numerous reports.

As Chapter 8 explains, the novel and imaginative research that established biotechnology was founded in microbiology, and it continues to call on microbiology for its continuing growth.

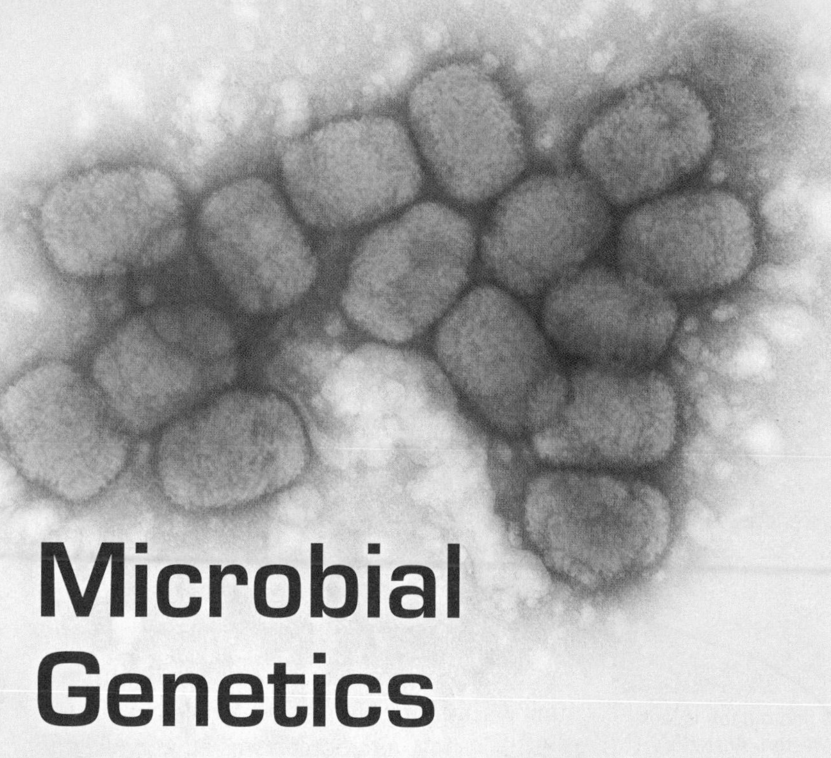

Microbial Genetics

8

CHAPTER PREVIEW

8.1 The Hereditary Molecule in All Organisms Is DNA

Investigating the Microbial World 8: Survival in a "Wet" Desert

8.2 DNA Replication Occurs Before a Cell Divides

8.3 Gene Expression Produces RNA and Protein for Cell Function

MicroInquiry 8: The Operon Theory and the Control of Protein Synthesis

8.4 Mutations Are Permanent Changes in a Cell's DNA

Textbook Case 8: *Klebsiella pneumoniae* in a Hospitalized Patient

8.5 Techniques Exist for Identifying Mutants

Image courtesy of Dr. Fred Murphy/CDC.

We wish to suggest a structure for the salt of deoxyribose nucleic acid (DNA). This structure has novel features which are of considerable biological interest.

—In the first 1953 paper by Watson and Crick describing the structure of DNA

In our fast-paced world, we often measure time in minutes and seconds, so our minds find it difficult to imagine the colossal 4.5 billion years that the Earth has been in existence. It may help, however, to think of Earth's history as a single year.

In the month of January, Earth was a hot, volcanic, lifeless ball of rock bombarded by material left over from the formation of the solar system. As the earth cooled during February, water vapor condensed into oceans and seas, providing conditions more amenable for the origin of life. Around March, something akin to the Bacteria or Archaea first appeared (FIGURE 8.1). As they evolved, they thrived and diversified in environments without oxygen gas and, by mid-June (2.6 billion years ago), in environments with oxygen gas. Members of the Bacteria and Archaea were the only organisms on Earth until August, when single-celled eukaryotes, such as the algae, emerged. These organisms flourished and represent ancestors of present-day species. About October, multicellular eukaryotes arose, whose descendants would evolve into diverse plants, fungi, and animals. Not until December did the first of the plants, fungi, and animals move out of the sea onto the land. The dinosaurs were in existence from December 19 to December 25, and by December 27, the Earth bore a resemblance to modern Earth. Finally, on December 31, close to midnight, humans appeared.

We take this trek through geologic time to help us appreciate why microorganisms have prospered genetically and in evolutionary terms. They have been successful primarily because they have been around the longest and have adapted well. In fact, the Bacteria and Archaea domains

FIGURE 8.1 **Fossil Microbes.** This photograph is looking down on an ancient sea floor in western Australia's Pilbara region. The wavy markings and cone-shaped formations may be evidence for a microbial reef made of cyanobacteria that existed 3.4 billion years ago. Courtesy of Abigail Allwood, Geologist at NASA Jet Propulsion Laboratory. »» What would it mean for the environment once oxygen-producing cyanobacteria predominated?

have been on Earth about 3.8 billion years (versus about 200,000 years for humans), as FIGURE 8.2 shows. During this time, gene changes have been occurring regularly and nature has used the microbes to test its newest genetic traits. Evolution has eliminated the detrimental traits (together with the organisms unlucky enough to have them), while the beneficial traits have thrived and have been passed on to the next generation—and on to the present day.

Present day microorganisms enjoy the fruits of genetic change. Because of their ability to reshape their genetic makeup, they can thrive in the varied environments on Earth, whether it is the ice of the Arctic, the boiling hot volcanic vents of the ocean depths, or in the human body. No other group of organisms has this diverse capacity.

Finally, consider a bacterium's multiplication rate—a new generation every half hour in some cases—and it is easy to see how a useful genetic change (such as drug resistance) can be propagated quickly.

Any one of these factors—time on Earth, sheer numbers, multiplication rate—would be sufficient to explain how microorganisms have evolved to their current form. However, when taken together, the factors help us appreciate why they have done very well in the evolutionary lottery—very well, indeed.

In this chapter, we examine mutation, one of the two processes that have brought ancient microorganisms to the myriad forms we observe on Earth today. However, to understand the material in these topics, we first must look at DNA replication and how the information in DNA is processed and regulated in making proteins—something alluded to in the opening quote by the remarkable discovery of DNA structure made in 1953.

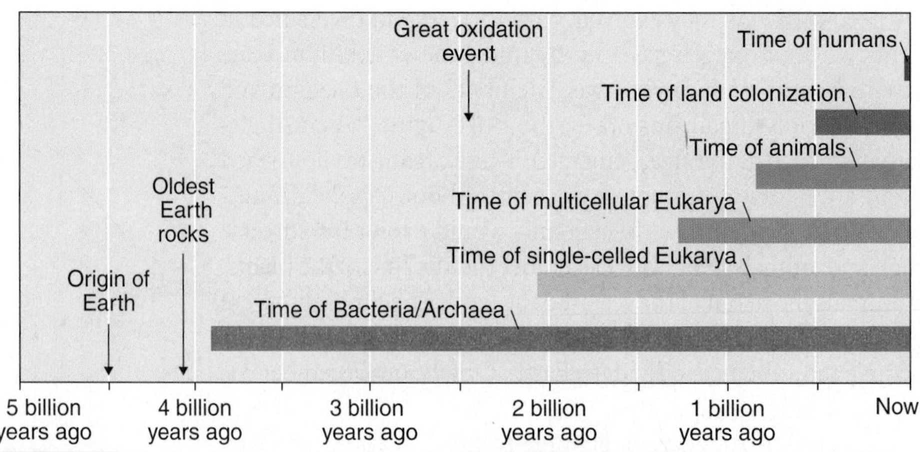

FIGURE 8.2 **The Appearance of Life on Earth.** This time line shows the relative amounts of time that various groups of organisms have existed on Earth. The Bacteria and Archaea have been in existence for a notably longer period than any other group, particularly humans. They have adapted well to Earth simply because they have had ample time and numbers. »» Why did it take so long for eukaryotes to appear on Earth?

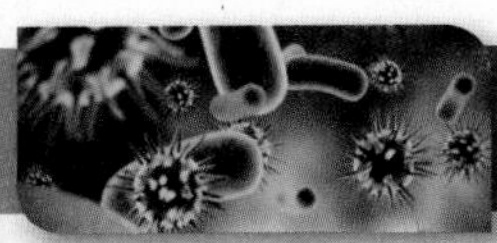

Chapter Challenge

As we are about to see, there is great divergence in the number of genes that species can have. Some free-living prokaryotes have half the genes found in other free-living species like *Escherichia coli* while others have twice as many genes. Interestingly, most obligate symbionts/parasites (meaning they must live within a host cell to survive) have even fewer genes than the free-living lineages. Even within the same genus, some species may be free living while others exist as parasites. How do these microbes manage to survive, quite well as a matter of fact, with substantially fewer genes? Let's see if we can uncover the answer.

KEY CONCEPT 8.1 The Hereditary Molecule in All Organisms Is DNA

Genetics is the science of heredity and covers the structure, replication, and transmission of that genetic information between organisms. This hereditary information is **deoxyribonucleic acid (DNA)**.

In 1953, James Watson and Francis Crick worked out DNA's double helix structure based, in part, on the X-ray studies of Rosalind Franklin (MicroFocus 8.1). This discovery has formed the foundation for all we know today about DNA replication, protein synthesis, and gene control.

Recall that DNA is a macromolecule composed of repeating monomers called **nucleotides**. Each nucleotide is composed of three components: a five-carbon sugar molecule (deoxyribose), a phosphate group, and a nitrogen-containing nucleobase [adenine (A), guanine (G), cytosine (C), or thymine (T)]. Nucleotides are covalently joined through dehydration synthesis reactions between the sugar of one nucleotide and the phosphate of the adjacent nucleotide to eventually form a **polynucleotide**. A DNA molecule consists of two polynucleotide strands opposed to each other in a ladder-like arrangement called the DNA **double helix**. These two polynucleotides are hydrogen bonded to one another via the nucleobases: G and C pair together as do A and T on opposing polynucleotides. Therefore, knowing the sequence of bases in one strand allows the identification of the nucleobases in the other strand. Thus, the two strands are "complementary."

A unit of heredity (except for some viruses) is a DNA segment called a **gene**, which codes for a functional product, be it ribonucleic acid (RNA) or a single polypeptide. The complete set of genetic information for an organism or virus is called the **genome**.

Bacterial and Archaeal DNA Is Organized Within the Nucleoid

Most of the genetic information in the Bacteria, Archaea, and Eukarya is contained within **chromosomes**, the cell's intracellular source of genetic information. In the Bacteria and Archaea, this is usually a single, circular molecule of DNA, although a few species may have a single genome contained in two or more chromosomes.

The chromosome exists as thread-like fibers associated with some protein and is localized in the cytosol within a space called the **nucleoid**. Remember that one of the unique features defining the nucleoid area is the absence of a surrounding membrane envelope typical of the cell nucleus in eukaryotic cells.

The circular chromosome of *Escherichia coli* probably has been studied more thoroughly than that of any other microbe. The genome has about 4,300 genes coding for proteins and RNA needed for growth and metabolic activities. By contrast, many other bacterial and archaeal genomes are much smaller, especially those of obligate symbionts/parasites (FIGURE 8.3). Such "genome reduction" in a free-living bacterial genus is examined in Investigating the Microbial World 8. At the other extreme are the genomes of the eukaryotes, which, along with other characteristics listed in TABLE 8.1, can be much larger in size. Note in this table some of the archaeal/eukaryote similarities.

■ Genome: The complete set of genes in an organism or virus.

DNA Within a Nucleoid Is Highly Compacted

In most bacterial cells, including *E. coli*, the DNA occupies about one-third of the total volume of the cell, and when extended its full length, it is about

MICROFOCUS 8.1: History

The Tortoise and the Hare

Photo of Rosalind Franklin. © Vittorio Luzzati/ Photo Researchers, Inc.

We all remember the children's fable of the tortoise and the hare. The moral of the story was those who plod along slowly and methodically (the tortoise) will win the race over those who are speedy and impetuous (the hare). The race to discover the structure of DNA is a story of collaboration and competition—a science "tortoise and the hare."

Rosalind Franklin (the tortoise) was 31 when she arrived at King's College in London in 1951 to work in J. T. Randell's lab. Having received a Ph.D. in physical chemistry from Cambridge University, she moved to Paris where she learned the art of X-ray crystallography. At King's College, Franklin was part of Maurice Wilkins's group and she was assigned the job of using X-ray crystallography to work out the structure of DNA fibers. Her training and constant pursuit of excellence allowed her to produce excellent, high-resolution X-ray images of DNA.

Meanwhile, at the Cavendish Laboratory in Cambridge, James Watson (the hare) was working with Francis Crick on the structure of DNA. Watson, who was in a rush for honor and greatness that could be gained by figuring out the structure of DNA, had a brash "bull in a china shop" attitude. This was in sharp contrast to Franklin's philosophy where you don't make conclusions until all of the experimental facts have been analyzed. Therefore, until she had all the facts, Franklin was reluctant to share her data with Wilkins—or anyone else.

Feeling left out, Wilkins was more than willing to help Watson and Crick. Because Watson thought Franklin was "incompetent in interpreting X-ray photographs" and he was better able to use the data, Wilkins shared with Watson an X-ray image and report that Franklin had filed. From these materials, it was clear that DNA was a helical molecule. It also seems clear Franklin knew this as well but, perhaps being a physical chemist, she did not grasp its importance because she was concerned with getting all the facts first and making sure they were absolutely correct. But, looking through the report that Wilkins shared, the proverbial "light bulb" went on when Crick saw what Franklin had missed; that the two DNA strands were antiparallel. This knowledge, together with Watson's ability to work out the base pairing, led Watson and Crick to their "leap of imagination" and the structure of DNA.

In her book entitled, *Rosalind Franklin: The Dark Lady of DNA* (HarperCollins, 2002), author Brenda Maddox suggests it is uncertain if Franklin could have made that leap as it was not in her character to jump beyond the data in hand. In this case, the leap of intuition won out over the methodical data collecting in research—the hare beat the tortoise this time. However, it cannot be denied that Franklin's data provided an important key from which Watson and Crick made the historical discovery.

In 1962, Watson, Crick, and Wilkins received the Nobel Prize in Physiology or Medicine for their work on the structure of DNA. Should Franklin have been included? The Nobel Prize committee does not make awards posthumously and Franklin had died 4 years earlier from ovarian cancer. So, if she had lived, did Rosalind Franklin deserve to be included in the award?

1.5 millimeters (mm) long. This is approximately 500 times the length of the bacterial cell. So, how can a 1.5-mm-long circular chromosome fit into a 1.0 to 2.0 μm *E. coli* cell?

The answer is **supercoiling**, a twisting and tight packing caused by a number of abundant **nucleoid-associated proteins (NAPs)**. Thus, the DNA double helix twists on itself like a wound-up rubber band. The NAPs assist the coils in folding further into loops of 10,000 bases, each forming a **supercoiled domain** (FIGURE 8.4A) and there are about 400 such domains in an *E. coli* chromosome,

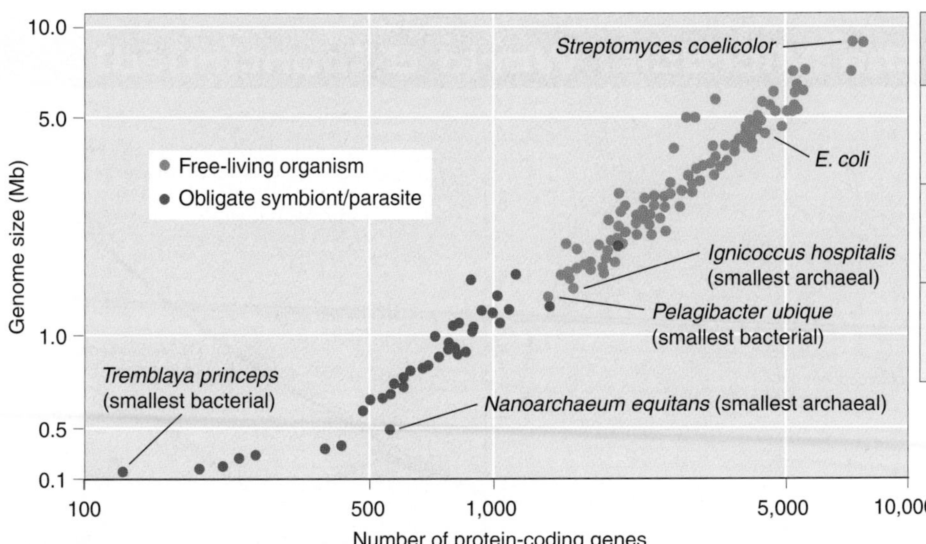

Organism	Genome size (Mb)	Protein-coding genes
Mycobacterium tuberculosis (free-living)	4.4	3,900
Staphylococcus aureus (free-living)	2.8	2,700
Chlamydia trachomatis (obligate/parasite)	1.0	936

FIGURE 8.3 **Genome Size Among the Bacteria and Archaea.** This graph illustrates the relationship between genome size (Mb = millions of base pairs) and number of protein-coding genes for some organisms currently sequenced. Note the extremely small size of *Tremblaya princeps.* »» Plot the location for each of the three organisms listed in the table.

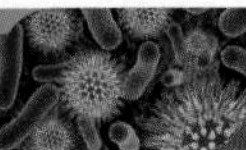

Investigating the Microbial World 8

Survival in a "Wet" Desert

Evolution of organisms is often described as a progression toward increasing complexity. If you look back at Figure 8.3, in general, increased genome size is equated with increased physiological and/or biochemical complexity. However, as the figure indicates, many of the obligate prokaryotic symbionts/parasites have undergone "reductive evolution" where they have lost genes (have smaller genomes) because their symbiotic partner/host organism can supply the needed metabolites formerly supplied by the organism's own (now lost) genes. But what about free-living microbes?

■ **OBSERVATION:** *Prochlorococcus* is a cyanobacterial genus that is the most abundant photosynthetic marine microbe. From a complexity standpoint, it should have substantially more genes than it actually has, especially considering that the organism has to survive in marine surface waters that usually are low in nutrients—a so-called "wet desert." Yet in *Prochlorococcus,* evolution has resulted in a loss of genes (genome reduction). For example, when exposed to sunlight, the surface ocean water generates toxic hydrogen peroxide (HOOH). Because of genome reduction, *Prochlorococcus* has lost the gene to make catalase, the enzyme that can degrade HOOH; that is, genome reduction has made the organism vulnerable to HOOH, which should cause a loss of cell integrity and photosynthetic capacity. Yet the organism survives amazing well!

■ **QUESTION:** ***How can a HOOH-defenseless organism like* Prochlorococcus *thrive in a marine environment where it is continually faced with such a potentially destructive chemical?***

■ **HYPOTHESIS:** *Prochlorococcus* has evolved such that it is dependent on other marine community members ("helper" cells) to degrade HOOH in the common surrounding environment. If so, then growing *Prochlorococcus* in pure culture in the presence of HOOH will kill *Prochlorococcus* but adding a "helper" organism that can degrade HOOH will protect *Prochlorococcus.*

■ **EXPERIMENTAL DESIGN:** A strain of *Prochlorococcus* sensitive to HOOH was cultured. The "helper" organism, a strain of the heterotroph *Alteromonas* that produces catalase to degrade HOOH, was also cultured in the lab. HOOH concentrations used were similar to what would naturally be produced in a typical marine environment. Growth of *Prochlorococcus* cells was monitored by counting the number of cells/ml in the experimental culture.

■ **EXPERIMENT 1: Does coculturing the "helper" organism with *Prochlorococcus* protect *Prochlorococcus* from HOOH damage?** Two cultures containing HOOH were established: (1) *Prochlorococcus* only and (2) "helper" organism and *Prochlorococcus.* Cell growth was measured over 16 days.

(continued on next page)

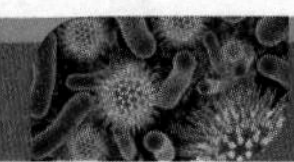

Investigating the Microbial World 8 (continued)

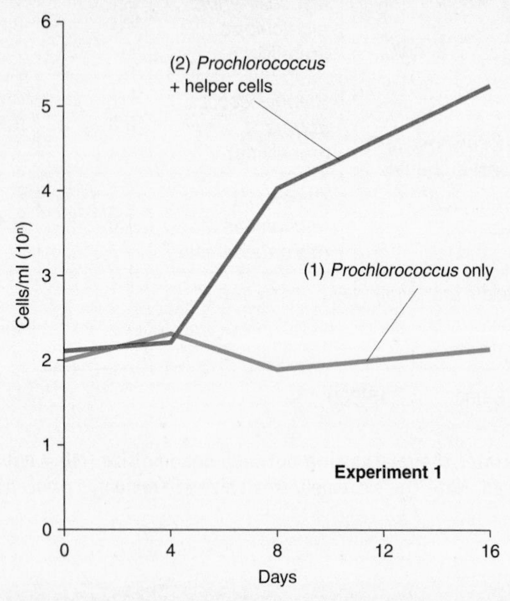

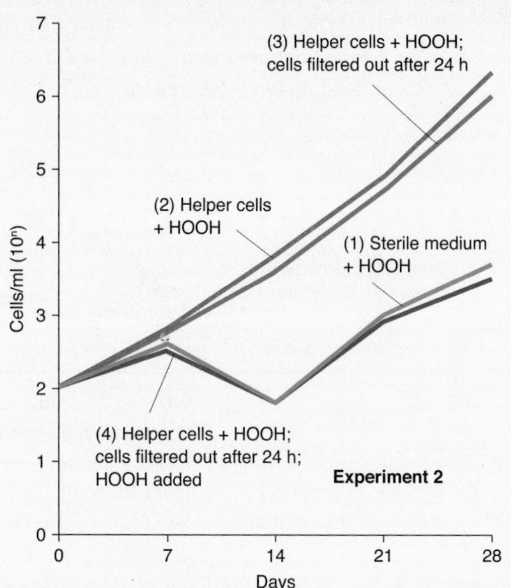

■ **EXPERIMENT 2: Is removal of HOOH from the growth medium necessary and sufficient to allow *Prochlorococcus* growth?** Four cultures were established. Prior to inoculation of *Prochlorococcus:* (1) the sterile growth medium was supplemented with HOOH; (2) the "helper" organism was added to the medium with HOOH; (3) the "helper" organism was filtered out after 24 h; and (4) HOOH was added back to the filtered medium of culture 3. Cell growth was measured over 28 days.

■ **EXPERIMENT 3: If growth was inhibited in experiments 1 and 2, were the *Prochlorococcus* cells actually killed or were they simply not dividing?** Three cultures were established. (1) *Prochlorococcus* culture without added HOOH; (2) *Prochlorococcus* culture supplemented with HOOH; and (3) *Prochlorococcus* and the "helper" organism supplemented with HOOH. Viability was monitored over 7 days using a fluorescent green dye that is taken up into dead cells only.

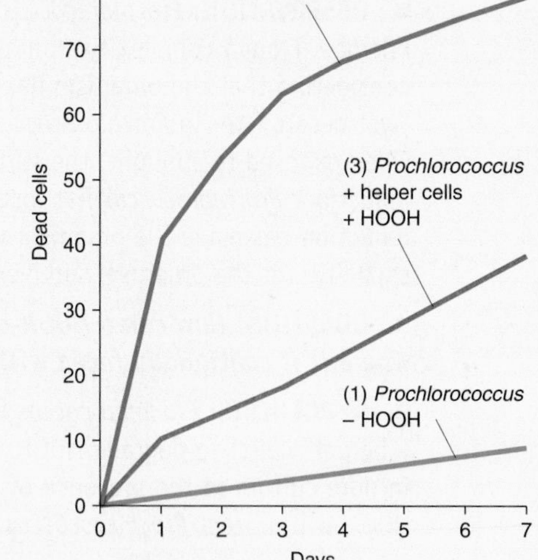

■ **RESULTS:** See figures.

■ **CONCLUSION:** From experiment 1, it can be seen that the presence of the "helper" organism did "protect" the *Prochlorococcus* cells from cell damage by HOOH.

QUESTION 1: ***From experiment 2, are the "helper" cells physically required to protect* Prochlorococcus *from HOOH damage? Explain what the "helper" cells are doing.***

QUESTION 2: ***From experiment 3, does HOOH actually kill the unprotected* Prochlorococcus *cells? Explain.***

QUESTION 3: ***Protection from HOOH damage is an energy demanding process. What advantage has* Prochlorococcus *gained from genome reduction (e.g., loss of the gene to destroy HOOH)? Remember, the marine environment in which* Prochlorococcus *exists is oligotrophic, meaning available nutrients are limited and in low concentrations. It's a "wet" desert.***

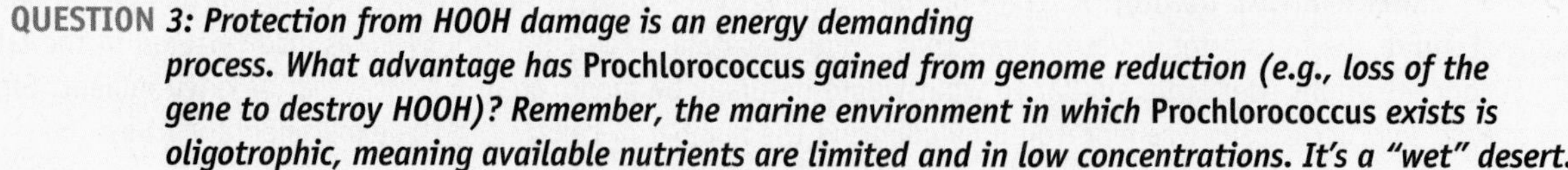

Answers can be found on the Student Companion Website in **Appendix D**.

Adapted from: Morris, J. J. *et al.* (2011). *PLoS ONE* **6**(2): e16805. doi:10.1371/journal.pone.0016805. Icon image © Tischenko Irina/ShutterStock, Inc.

giving the molecule an overall "flower" structure called the **looped domain structure**. The high level of compaction is evident when the cell envelope is broken, releasing the DNA in a looped form (FIGURE 8.4B). It is only when the supercoiling is relaxed through enzyme action that processes like gene expression and DNA replication can occur.

TABLE

8.1 Characteristics of Bacteria, Archaeal, and Eukaryotic Chromosomes

Characteristic	Bacteria	Archaea	Eukarya
Organization	Nucleoid	Nucleoid	Nucleus
Chromosome morphology	Usually circular	Usually circular	Linear
Ploidy	Usually haploid	Usually haploid	Haploid, diploid, polyploid
Genome size (Mb)[1]	0.16–10	0.5–6	10–100,000
Average protein-coding genes	Few thousand	Few thousand	Tens of thousands
Presence of histone proteins	Histone-like	Yes	Yes
Presence of introns	No	Present in rRNA and tRNA genes	Yes
Replication	Just prior to binary fission	Just prior to binary fission	Few hours before mitosis
Replication origins	Single	Multiple	Multiple

[1]Mb = Millions of base pairs.

Many Microbial Cells Also Contain Plasmids

Many bacterial, archaeal, and fungal cells contain **plasmids**, which are stable extrachromosomal DNA elements that do not carry genetic information essential for normal structure, growth, and metabolism. This means a plasmid could be removed from a cell without affecting its viability, assuming the cell is in a nutrient-rich environment free of toxic materials.

Most plasmids are circular and supercoiled, and they are easily transferred between cells. Plasmids exist and replicate as independent genetic elements in the cytosol where they typically contain about 2% of the total genetic information of the cell. Exceptions include some plasmids that can be quite large because they can integrate into a chromosome and excise from it many additional chromosomal genes, some of which may be essential for cell growth.

In most cases though, plasmids are not essential to the normal survival of the cell but they can confer selective advantages and provide genetic flexibility for those organisms possessing plasmids. For example, some bacterial plasmids, called **F plasmids**, allow for the transfer of genetic material from donor to recipient through a recombination process. Other plasmids, called

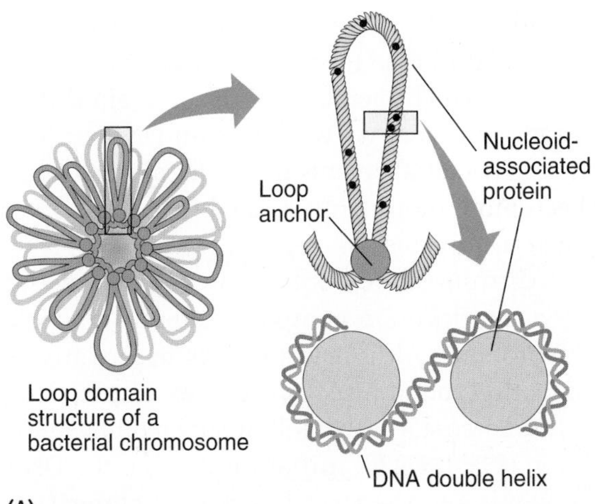

(A)

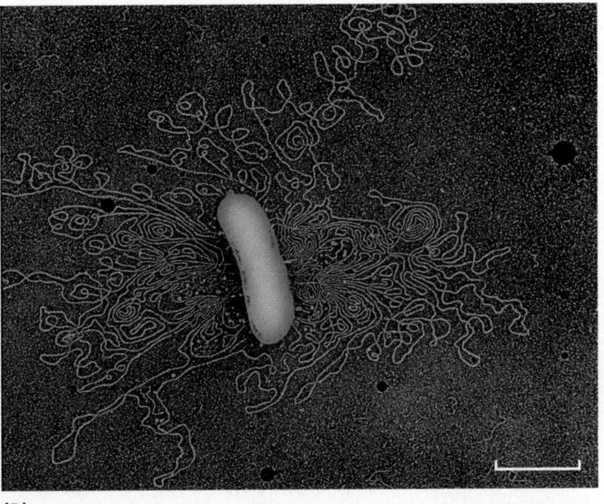
(B)

FIGURE 8.4 **DNA Packing.** (**A**) The loop domain structure of the chromosome as seen head-on. The loops in DNA help account for the compacting of a large amount of DNA in a relatively small cell. (**B**) An electron micrograph of an *E. coli* cell immediately after cell lysis. The uncoiled DNA fiber exists in loops attached to the disrupted cell envelope. (Bar = 1 µm.) (**B**) © G. Murti/Photo Researchers, Inc. »» How does plasmid structure compare to that of the chromosome in a bacterial cell?

R plasmids ("resistance" factors), carry genes for antibiotic resistance. Others contain genes for resistance to potentially toxic heavy metals (e.g., silver, mercury).

Another category of plasmids provide offensive abilities. For example, species of *Streptomyces* carry plasmids for the production of antibiotics while plasmids in other bacteria contain genes for the production of **bacteriocins**, a group of proteins that inhibit or kill other bacterial species.

Finally, there are plasmids containing genes coding for protein toxins affecting human cells and disease processes. The genes encoding the toxins responsible for anthrax are carried on a plasmid contained in *Bacillus anthracis*. We shall have much to say about these extrachromosomal units when we discuss recombination and genetic engineering in the next chapter.

CONCEPT AND REASONING CHECKS 1

a. Describe the basic structure of a bacterial chromosome.
b. Justify the necessity for DNA supercoiling and looped domains.
c. What does it mean to say plasmids carry nonessential genetic information?

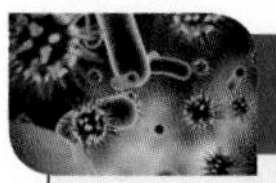

Chapter Challenge A

In our chapter challenge, we are examining why obligate parasites have such small genomes. Looking at Figure 8.3, there are a number of such obligate bacterial and archaeal species with extremely small genomes and very few genes. For example, *Nanoarchaeum equitans*, the only known archaeal hyperthermophile exhibiting parasitic life style, has only some 586 protein-coding genes. Then, there is *Tremblaya princeps*, a bacterial symbiont in aphids, which has only 110 protein-coding genes. It is the smallest genome yet discovered.

QUESTION A:

Propose why these obligate symbionts/parasites have such small genomes and thus few genes? Clue: If free-living species need substantially more genes, how do the obligate types get by with so many fewer genes?

Answers can be found on the Student Companion Website in **Appendix D.**

KEY CONCEPT 8.2 DNA Replication Occurs Before a Cell Divides

Website Animation: Bacterial DNA Replication

Watson and Crick's 1953 paper on the structure of DNA provided a glimpse of how DNA might be copied. They concluded, "*It has not escaped our notice that the specific pairing we have postulated immediately suggests a possible copying mechanism for the genetic material.*" In fact, the copying process for the genetic material, called **DNA replication**, occurs with such precision that the two daughter cells resulting from binary fission are genetically identical.

DNA Replication Occurs in Three Stages

The archaeal organisms appear to be an interesting mosaic of bacterial, eukaryotic, and unique features. This applies to the chromosome (see Table 8.1) and to DNA replication as well. Most archaeal proteins involved in DNA replication are more similar in sequence to those found in eukaryotic cells than to analogous replication proteins in bacterial cells. The archaeal DNA replication apparatus also contains features not found in other organisms, which is probably a result of the broad range of environmental conditions in which members of this domain thrive. That being said, we will focus on DNA replication in bacterial organisms like *E. coli*, which has been more thoroughly studied than most other bacterial organisms.

Chromosome replication is carried out by a multiprotein complex made up of some 15 proteins. Although it occurs in a relatively smooth, synchronized process, we can separate it into three stages that follow supercoiling relaxation (FIGURE 8.5): **initiation** is when the DNA unwinds and the strands separate; **elongation** involves when enzymes synthesize a new polynucleotide strand of DNA for each of the two old template (parental) strands; and **termination** occurs, when each of the two DNA helices

DNA Replication: Initiation

- Replication of the circular bacterial chromosome begins at a fixed region on the DNA called ***oriC***.
- At *oriC*, copies of initiation proteins bind to the *oriC* DNA sequences.
- The other proteins needed for replication will then add to this so-called "**replication factory.**"

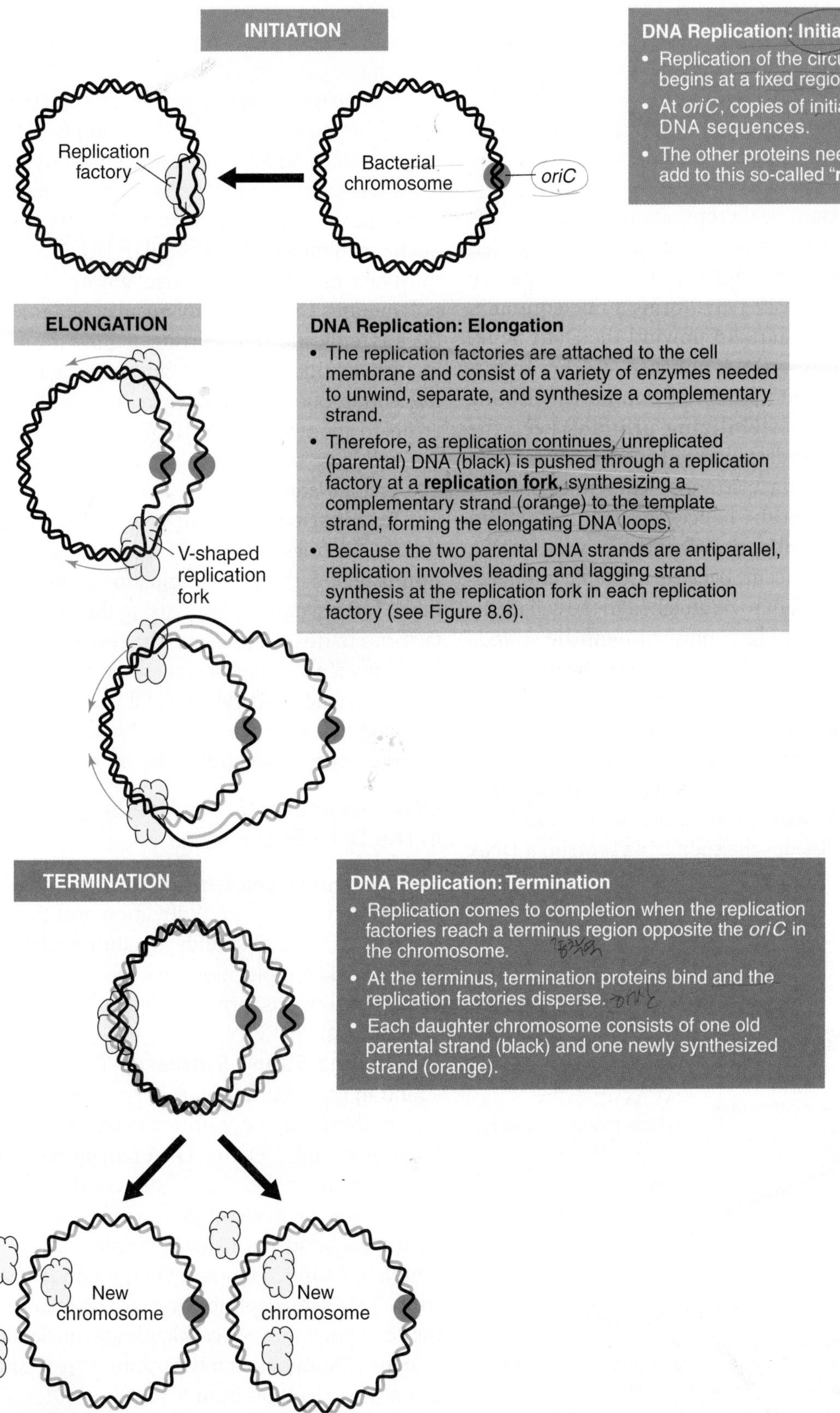

FIGURE 8.5 **Replication of the Circular Chromosome of *Escherichia coli*.** DNA replication involves the addition of complementary bases to the parental (template) strand within replication factories that are attached to the cell membrane. »» Why is DNA replication considered to be semiconservative?

separate from one another. This combination of a new and old strand is called **semiconservative replication** because each old strand of the replicated DNA is conserved in each new chromosome and one strand is newly synthesized. Let's look at each stage in more detail using Figure 8.5 to guide us.

■ Mutation: A permanent alteration in a DNA base sequence.

Initiation. DNA replication in bacterial cells starts at a fixed region on the chromosome called the **replication origin** (*oriC*). This sequence of about 250 base pairs forms a nucleoprotein complex that starts to unwind the DNA helix. Subsequently, a **DNA helicase** attaches to further unwind and unzip the two polynucleotide strands, while **stabilizing proteins** keep the template strands separated for the replication of complementary strands. Therefore, each of the two unwinding helices is contained within a so-called **replication factory** in which DNA synthesis will occur. Because the replication factories are thought to be attached to the cell membrane, the yet to be replicated template strands are threaded through a V-shaped **replication fork** in each factory.

Elongation. Synthesis of DNA in each factory then occurs on each old strand, which represents a template for the synthesis of a new complementary strand. Many proteins are involved in DNA synthesis. Besides the stabilizing protein, a **DNA polymerase** moves along each strand, catalyzing the insertion of new complementary nucleotides to each template strand.

In *E. coli*, DNA synthesis takes about 40 minutes, which means at each replication fork DNA polymerase is adding new complementary bases at the rate of about 1,000 per second! At this pace, errors occur where an incorrect base is added. Such potential mutations could be lethal, so there must be a mechanism to correct any errors. DNA polymerases can detect any mismatched nucleotides as replication is occurring, remove the incorrect nucleotide in the pair, and add the correct nucleotide. Such proofreading reduces replication errors to about 1 in every 10 billion bases added. We will have more to say about mutations and DNA repair later in this chapter.

Termination. In about 40 minutes, the replication factories meet 180° from *oriC*. At the **terminus** region, there are termination proteins that block further replication, causing the replication factories to dissociate. Then, the two intertwined DNA molecules (chromosomes) are separated by other enzymes and cytoskeletal proteins, guaranteeing that each daughter cell will inherit one complete chromosome after binary fission.

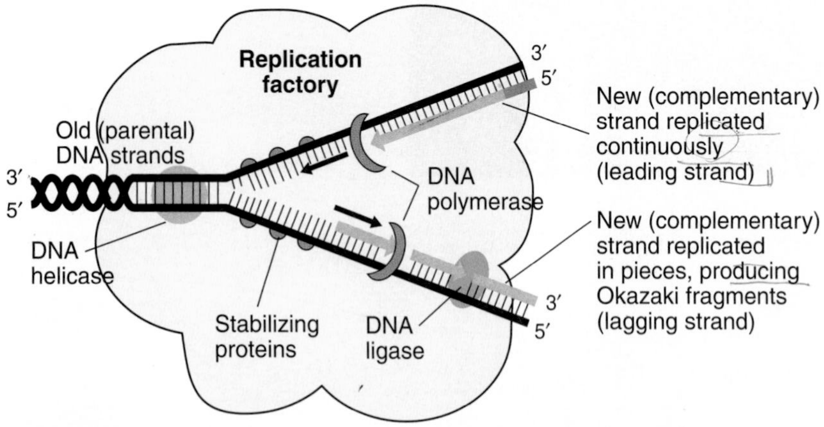

FIGURE 8.6 **A Replication Factory.** This diagram outlines the general events occurring at one replication fork, showing both the leading and lagging strand synthesis. The discontinuous synthesis on the lagging strand results from the DNA polymerase moving away from the replication fork, resulting in the formation of short DNA fragments, called Okazaki fragments, which are eventually joined by a DNA ligase. »» Why is the DNA polymerase on the lagging strand moving away from the replication fork?

DNA Polymerase Only Reads in the 3′ to 5′ Direction

Because DNA polymerase can "read" the template DNA only in the 3′ to 5′ direction and the two parental (template) strands are antiparallel, this means at each replication fork the complementary DNA strand is formed in opposite directions (FIGURE 8.6).

Leading Strand Synthesis. One parental strand in each replication factory is the template for synthesizing a continuous complementary **leading strand**. Here the DNA polymerase reads the template in the 3′ to 5′ direction, bringing in triphosphate nucleotides (A, T, G, and C) that hydrogen bond with their complement in the template strand. The high-energy bonds in the triphosphate nucleotides provide the energy for the DNA polymerase to covalently join nucleotides into the continuous strand, forming an elongating chain of nucleotides from 5′ to 3′.

Lagging Strand Synthesis. The other template strand in each fork of a replication factory must be read "backwards"; that is, as the DNA polymerase moves away from the replication fork, a discontinuous process of starts and stops occurs, with the new strand always lagging behind the leading strand. This piecemeal strand, therefore, is called the **lagging strand**. These segments, which are about 1,000 to 2,000 nucleobases long, are called **Okazaki fragments**, after Reiji Okazaki, who discovered them in 1968. As these new polynucleotide segments are produced, the gaps between segments are eventually joined into a complete and elongating single strand with the help of an enzyme called **DNA ligase**.

CONCEPT AND REASONING CHECKS 2

a. Describe the role for replication factories in DNA synthesis.

b. Why are there leading and lagging strands in each replication fork?

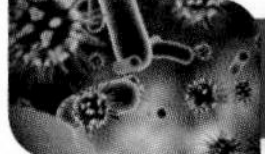

Chapter Challenge B

No matter the number of protein-coding genes found in an obligate symbiont/parasite, the genome contains DNA in the form of a chromosome, and the organism still needs to reproduce within the host.

QUESTION B:

When considering the obligate symbionts/parasites, would the genes required for the multiprotein complex found within the replication factories be part of the parasite's genome? Explain. What would be the minimal number of genes required for DNA replication?

Answers can be found on the Student Companion Website in **Appendix D**.

KEY CONCEPT 8.3 Gene Expression Produces RNA and Protein for Cell Function

The discovery of the structure of DNA also provided a glimpse into understanding how a cell uses its genes to make proteins. The process, called **gene expression**, requires not only DNA, but also ribonucleic acid (RNA). TABLE 8.2 summarizes the characteristics of DNA and RNA. A review of their structure is recommended so the discussion to follow can be fully comprehended.

One of the central truths in genetics states that the genetic information in DNA first is expressed as RNA by a process called **transcription**. One type of RNA then functions as a messenger by carrying the genetic code to areas of the cytosol where the ribosomes are located. There, amino acids are joined together by ribosomes into a precise sequence to form the protein (polypeptide). This

■ Genetic code: The sequence of bases in the DNA or codons in the RNA that specify a specific polypeptide.

TABLE

8.2 A Comparison of DNA and RNA

DNA (Deoxyribonucleic Acid)	RNA (Ribonucleic Acid)
In Bacteria and Archaea, found in the nucleoid and plasmids; in Eukarya, found in the nucleus, mitochondria, and chloroplasts	In all organisms, found in the cytosol and in ribosomes; in Eukarya, also found in the nucleolus
Always associated with chromosome (genes); each chromosome has a fixed amount of DNA	Found mainly in combinations with proteins in ribosomes (ribosomal RNA), as messenger RNA, and as transfer RNA
Contains a 5-carbon sugar called deoxyribose	Contains a 5-carbon sugar called ribose
Contains bases adenine, guanine, cytosine, and thymine	Contains bases adenine, guanine, cytosine, and uracil
Contains phosphorus (in phosphate groups) that connects deoxyribose sugars with one another	Contains phosphorus (in phosphate groups) that connects ribose sugars with one another
Functions as the molecule of inheritance	Functions in protein synthesis and gene regulation
Large double-stranded molecule	Small single-stranded molecule

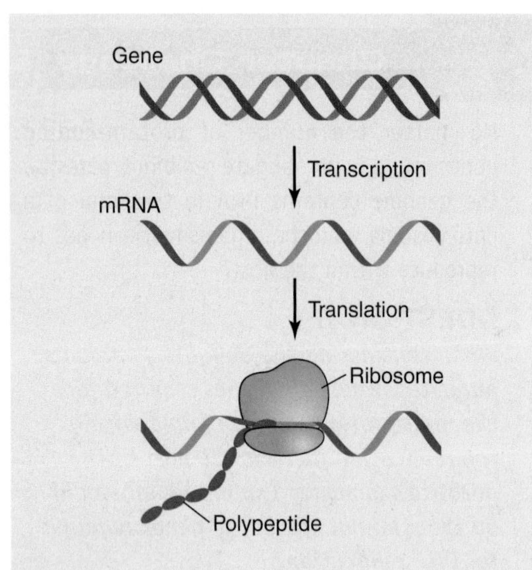

FIGURE 8.7 **The Central Dogma.** The flow of genetic information proceeds from DNA to RNA to protein (polypeptide). »» Where in a bacterial cell and in a eukaryotic cell do transcription and translation occur?

sequencing process, called **translation**, reflects the genetic information in the DNA. This **central "dogma"** (*dogma* = "opinion") of biology is shown in FIGURE 8.7.

Transcription Copies Genetic Information into Complementary RNA

Transcription is the first, and perhaps most regulated, step in gene expression.

RNA polymerase is a large enzyme that carries out transcription (FIGURE 8.8). In the Bacteria and Archaea, there is but one RNA polymerase with the archaeal enzyme being more similar to the eukaryotic polymerases. Like DNA polymerases, the RNA polymerase "reads" the DNA template strand in the 3′ to 5′ direction. However, unlike DNA replication, only one of the two DNA strands within a gene is transcribed.

Transcription begins when RNA polymerase recognizes the DNA template strand in a gene by a sequence of bases called the **promoter** located on the template strand. The polymerase binds to the promoter (initiation), unwinds the helix, and separates the two strands within the gene. As the enzyme moves along the DNA template strand (elongation), complementary pairing at the active site brings RNA triphosphate nucleotides to the template strand—guanine (G) and cytosine (C) pair with one another and thymine (T) in the DNA template pairs with adenine (A) in the RNA. However, an adenine base on the DNA template pairs with a uracil (U) base in the RNA because RNA nucleotides contain no thymine bases.

Like initiation, termination of transcription occurs at a specific base sequence, called the **terminator**, on the DNA template strand. The RNA transcription product released represents a complementary image to the base sequence in the DNA template strand.

Transcription produces three types of RNA that are needed for gene expression.

Messenger RNA (mRNA). This RNA carries the genetic information or "blueprint" that ribosomes "read" to manufacture a polypeptide. Each mRNA transcribed from a different gene carries a different message; that is, a different sequence of nucleotides coding for a different polypeptide. The message is encoded in a series of three-base combinations called **codons** that are found along the length of the mRNA. Each codon specifies an individual amino acid to be slotted into position during translation.

Ribosomal RNA (rRNA). Three rRNAs are transcribed from specific regions of the DNA. Together with more than 50 proteins, these RNAs serve a structural role as the framework of the ribosomes, which are the sites at which amino acids assemble into proteins. They also serve a functional role in the translation process.

Transfer RNA (tRNA). The conventional drawing for a tRNA is in a shape roughly like a cloverleaf (FIGURE 8.9). One region consists of three bases, which functions as an **anticodon**; that is, a sequence that complementary binds to an mRNA codon. The tRNAs have a structural role in delivering amino acids to the ribosome for assembly into proteins. Each tRNA has a specific amino acid attached to it through an enzymatic reaction involving ATP. For example, the amino acid alanine binds only to the tRNA specialized to transport alanine; glycine is transported by a different tRNA.

There is one important difference between microbial RNAs. In most bacterial and some archaeal cells, all of the bases in a gene are transcribed and used to specify a particular protein or RNA (FIGURE 8.10A). However, in many archaeal and eukaryotic cells certain portions of the RNA coding sequence of a gene are not part of the final RNA and are removed from the RNA before the

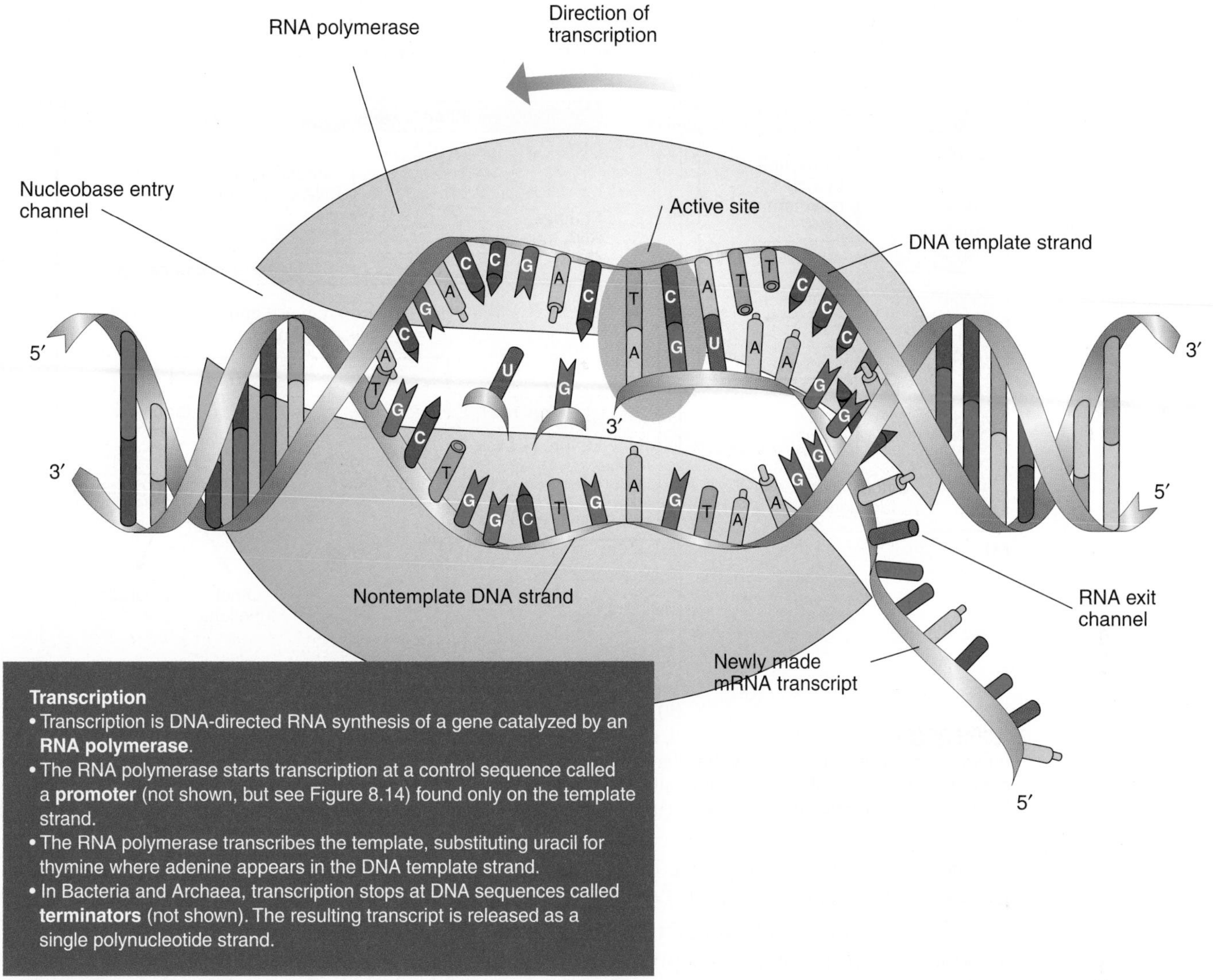

Transcription

- Transcription is DNA-directed RNA synthesis of a gene catalyzed by an **RNA polymerase**.
- The RNA polymerase starts transcription at a control sequence called a **promoter** (not shown, but see Figure 8.14) found only on the template strand.
- The RNA polymerase transcribes the template, substituting uracil for thymine where adenine appears in the DNA template strand.
- In Bacteria and Archaea, transcription stops at DNA sequences called **terminators** (not shown). The resulting transcript is released as a single polynucleotide strand.

FIGURE 8.8 **The Transcription Process During Elongation.** For transcription to occur, a gene must be unwound and the base pairs separated. The enzyme RNA polymerase does this as it moves along the template strand of the DNA and adds complementary RNA nucleotides. Note that the other DNA strand of the gene is not transcribed. »» Justify the need for a promoter sequence in a gene.

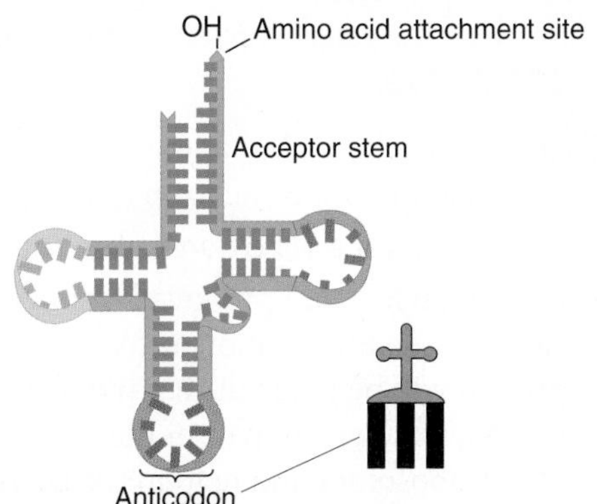

FIGURE 8.9 **Structure of a Transfer RNA (tRNA).** The traditional "cloverleaf" configuration for tRNA. The appropriate amino acid attaches to the end of the acceptor arm. Insert: the symbol for a tRNA. »» What part of the tRNA is critical for the complementary binding to a codon in a mRNA?

molecule can function (FIGURE 8.10B). These intervening DNA segments removed after transcription are called **introns**, while the remaining, amino acid-coding segments are called **exons**.

While the exons have the "standard" information coding for a polypeptide (or protein), the introns also may have roles in gene regulation, or in metabolic control through association with other RNAs or proteins.

The Genetic Code Consists of Three-Letter Words

By now you should have the idea that the information to specify the amino acid sequence for a polypeptide is encoded in the DNA template strand

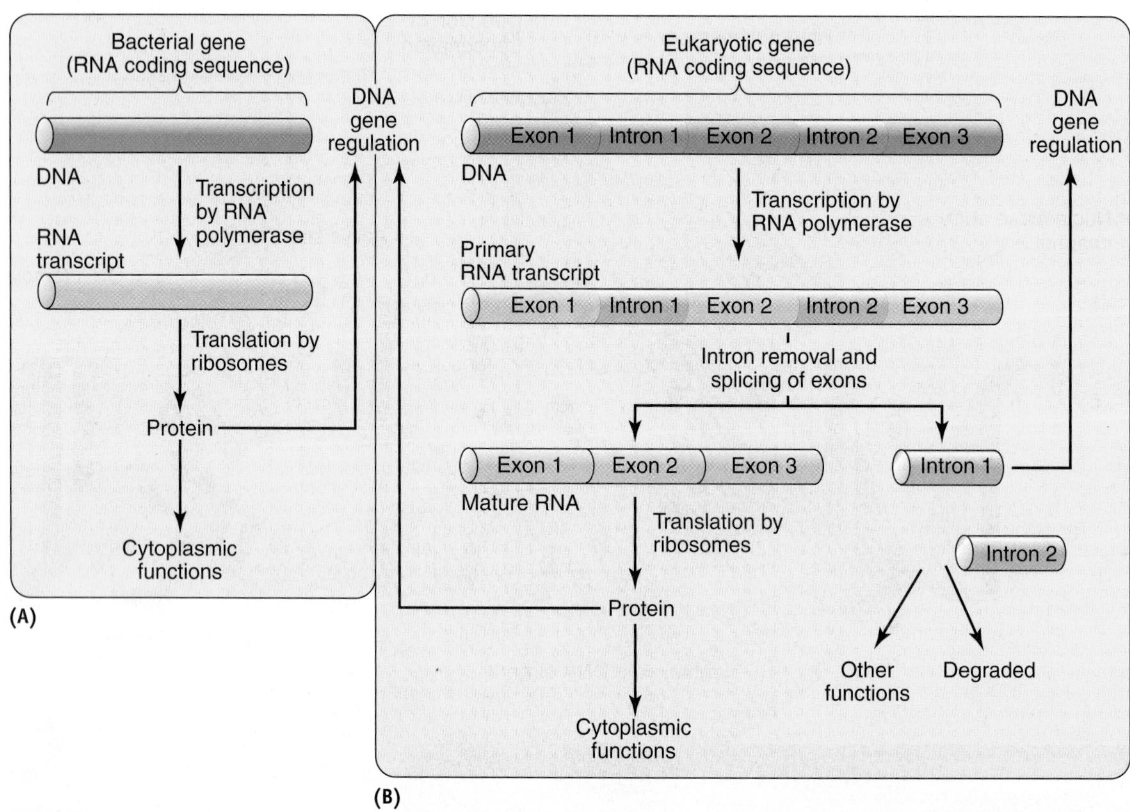

FIGURE 8.10 **RNA Processing and Gene Activity.** (**A**) In bacterial cells, gene DNA is almost entirely protein-coding information that is transcribed and translated into a protein having structural or functional roles in the cytoplasm, or gene regulation roles. (**B**) In eukaryotic cells, some of the intron RNA may be degraded, used to regulate gene function, or be associated with other RNAs or proteins in the cytoplasm. »» How do the roles of exonic RNA and intronic RNA differ?

of a gene. This specific sequence of nucleotide bases is called the **genetic code** and each sequence is made up of "three-letter words" that are called a codon. To synthesize a polypeptide, the DNA gene sequence must first be transcribed into RNA codons, as was depicted in Figure 8.8.

One of the startling discoveries of biochemistry is that the genetic code in most cases contains more than one codon for each amino acid. Because there are four nucleobases, mathematics tells us 64 possible combinations can be made of the four bases, using three at a time. But there are only 20 amino acids for which a code must be supplied. How do scientists account for the remaining 44 codons?

It is now known that 61 of the 64 codons are **sense codons** that specify an amino acid, and most of those amino acids have multiple codons (as shown in TABLE 8.3). For example, GCU, GCC, GCA, and GCG all code for the amino acid alanine (ala). This lack of a one-to-one relationship between codon and amino acid generates **redundancy**. In Table 8.3, notice one of the 64 codons, AUG, represents the **start codon** that sets the "reading frame" for making a protein. This codon usually specifies the amino acid methionine (met). Three additional codons, which do not code for an amino acid (UGA, UAG, UAA), are called **stop codons** because they terminate the addition of amino acids to a growing polypeptide chain.

Before we proceed to the last stage, translation, let's summarize the gene expression process to this point (FIGURE 8.11).

1. Each gene of the DNA contains information to manufacture a specific form of RNA.
2. The information can be transcribed into:
   - mRNAs, which are produced from genes carrying the information as to what protein will be made during translation;
   - rRNAs, which form part of the structure of the ribosomes and help in the translation of the mRNA; and
   - tRNAs, each of which carries a specific amino acid needed for the translation process.
3. With the tRNAs and mRNAs present in the cytosol, they can combine within ribosomes to

manufacture specific cellular proteins. Recall that what is actually made from the ribosome is a polypeptide. This polypeptide may represent the functional protein (tertiary structure) or first combine with one or more other polypeptides to form the functional protein (quaternary structure).

Translation Is the Process of Making the Polypeptide

In the process of translation, the language of the genetic code (nucleotides) is translated into the language of proteins (amino acids). As with DNA replication and RNA transcription, translation occurs in three steps: initiation, elongation, and termination.

Chain Initiation. Translation begins with the association of a small ribosomal subunit with an initiator tRNA at the AUG start codon (FIGURE 8.12A). Then, the large ribosomal subunit is added to form the functional ribosome with three tRNA binding sites, called A, P, and E. In the Bacteria, the first amino acid is formylmethionine (fMet) while in the Archaea and Eukarya, it is methionine (Met). Once formed, a second tRNA can complementary bind at the A site and a ribozyme transfers the fMet to the amino acid on the second tRNA.

Chain Elongation. With the second tRNA attached, the first tRNA is released from the E site (FIGURE 8.12B). Moving right one codon, the ribosome exposes the next codon (GCC), and the appropriate tRNA with the amino acid alanine (Ala) attached. Again, a ribozyme transfers the dipeptide fMet–Ser to alanine. The tRNA that carried serine exited the ribosome and the process of chain elongation continues as the ribosome moves to expose the next codon.

Chain Termination/Release. The process of adding tRNAs and transferring the elongating polypeptide to the entering amino acid/tRNA at the A site continues until the ribosome reaches a stop codon (UGA in this example). There is no tRNA to recognize any of these stop codons (FIGURE 8.12C). Rather, proteins called **termination factors** bind where the tRNA would normally attach. This triggers the release of the polypeptide and a disassembly of the ribosome subunits, which can be reassembled for translation of another mRNA.

TABLE

8.3 The Genetic Code Decoder

The genetic code embedded in an mRNA is decoded by knowing which codon specifies which amino acid. On the far left column, find the first letter of the codon; then find the second letter from the top row; finally read up or down from the right-most column to find the third letter. The three-letter abbreviations for the amino acids are given. Note: In the Bacteria, AUG codes for formylmethionine (fMet) when starting a polypeptide.

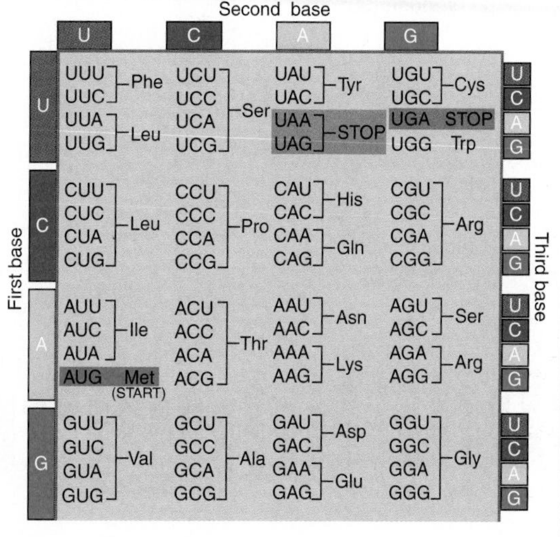

First base	Second base U	Second base C	Second base A	Second base G	Third base
U	UUU Phe	UCU Ser	UAU Tyr	UGU Cys	U
U	UUC Phe	UCC Ser	UAC Tyr	UGC Cys	C
U	UUA Leu	UCA Ser	UAA STOP	UGA STOP	A
U	UUG Leu	UCG Ser	UAG STOP	UGG Trp	G
C	CUU Leu	CCU Pro	CAU His	CGU Arg	U
C	CUC Leu	CCC Pro	CAC His	CGC Arg	C
C	CUA Leu	CCA Pro	CAA Gln	CGA Arg	A
C	CUG Leu	CCG Pro	CAG Gln	CGG Arg	G
A	AUU Ile	ACU Thr	AAU Asn	AGU Ser	U
A	AUC Ile	ACC Thr	AAC Asn	AGC Ser	C
A	AUA Ile	ACA Thr	AAA Lys	AGA Arg	A
A	AUG Met (START)	ACG Thr	AAG Lys	AGG Arg	G
G	GUU Val	GCU Ala	GAU Asp	GGU Gly	U
G	GUC Val	GCC Ala	GAC Asp	GGC Gly	C
G	GUA Val	GCA Ala	GAA Glu	GGA Gly	A
G	GUG Val	GCG Ala	GAG Glu	GGG Gly	G

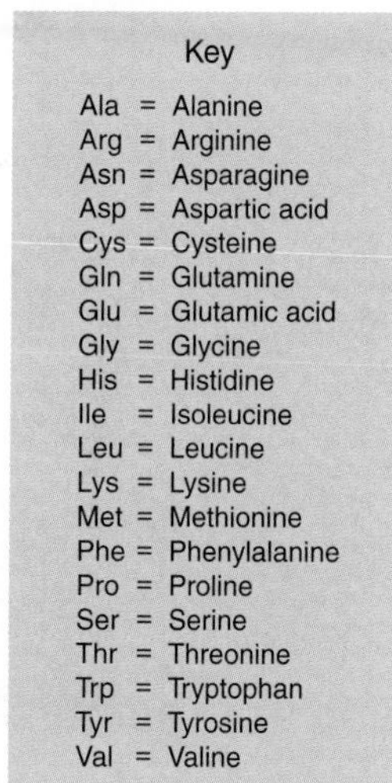

Key

Abbreviation	Amino acid
Ala	Alanine
Arg	Arginine
Asn	Asparagine
Asp	Aspartic acid
Cys	Cysteine
Gln	Glutamine
Glu	Glutamic acid
Gly	Glycine
His	Histidine
Ile	Isoleucine
Leu	Leucine
Lys	Lysine
Met	Methionine
Phe	Phenylalanine
Pro	Proline
Ser	Serine
Thr	Threonine
Trp	Tryptophan
Tyr	Tyrosine
Val	Valine

During synthesis, the polypeptide already may start to twist into its secondary and tertiary structure. For many polypeptides, groups of cytoplasmic proteins called **chaperones** ensure the folding process occurs correctly.

- Formylmethionine: The presence of a formyl group (H-CO—) attached to methionine.
- Ribozyme: An RNA molecule capable of carrying out a chemical reaction.

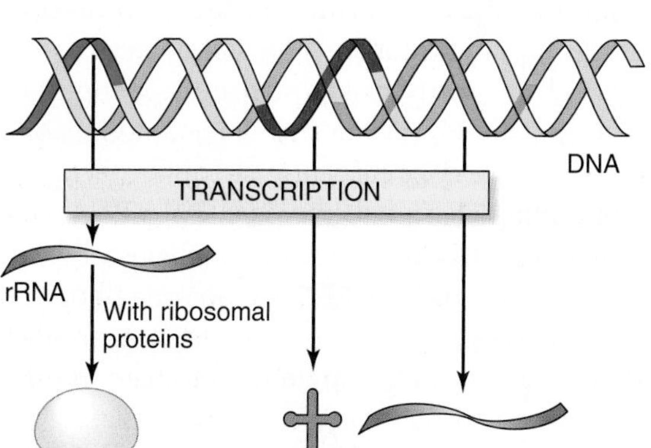

FIGURE 8.11 **The Transcription of the Three Types of RNA.** Genes in the DNA contain the information to produce three types of RNA needed for translation: rRNA, tRNA, and mRNA. »» What is each type of RNA used for in a microbial cell?

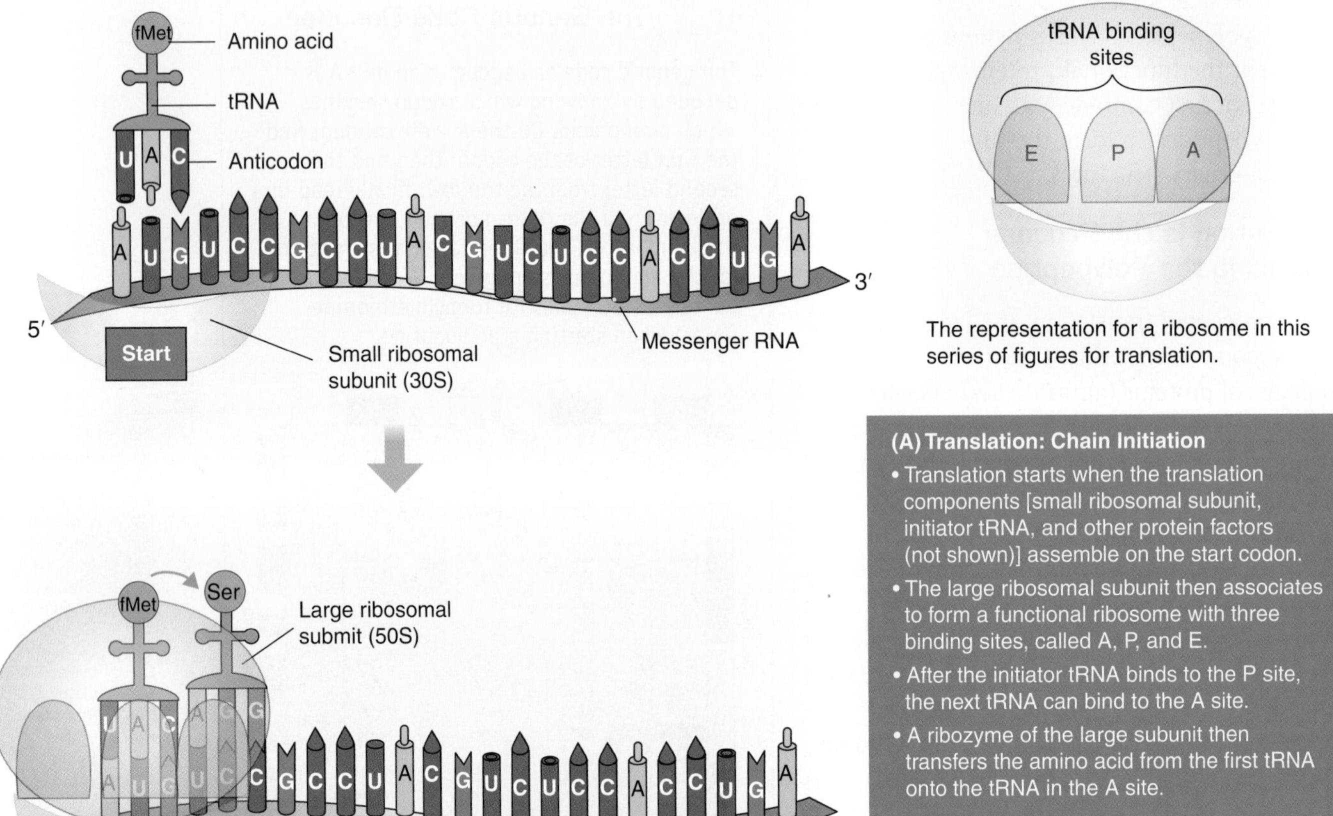

(A)

FIGURE 8.12 **Protein Synthesis in a Bacterial Cell.** The steps of (**A**) chain initiation, (**B**) chain elongation, and (**C**) chain termination are outlined. »» How do the A, P, and E sites in the ribosome differ?

MicroFocus 8.2 describes how the understanding of protein synthesis has been used to block "harmful" proteins from being made.

Cells typically make thousands of copies of each protein. Producing such large amounts of a protein can be done efficiently and quite quickly by transcribing large numbers of identical mRNA molecules that can be translated simultaneously by several ribosomes (FIGURE 8.13). Once one ribosome has moved far enough along the mRNA, another small subunit can "jump on" and initiate translation. Such a string of ribosomes all translating the same mRNA at the same time is called a **polysome**.

Antibiotics Interfere with Gene Expression

Many antibiotics affect gene expression in bacterial cells and therefore are clinically useful in treating human infections and disease. A few antibiotics interfere with transcription. For example, rifampin binds to the RNA polymerase so that transcription cannot be initiated.

A very large number of antibiotics inhibit translation by binding to the bacterial 30S or 50S ribosomal subunit. For example, tetracycline prevents chain initiation by binding to the 30S subunit, while drugs like chloramphenicol and erythromycin inhibit chain elongation by binding to the 50S subunit. We will have much more to learn about antibiotics in another chapter.

Gene Expression Can Be Controlled in Several Ways

Because transcription is the first step leading to protein manufacture in cells, one way to control what proteins and enzymes are present in bacterial

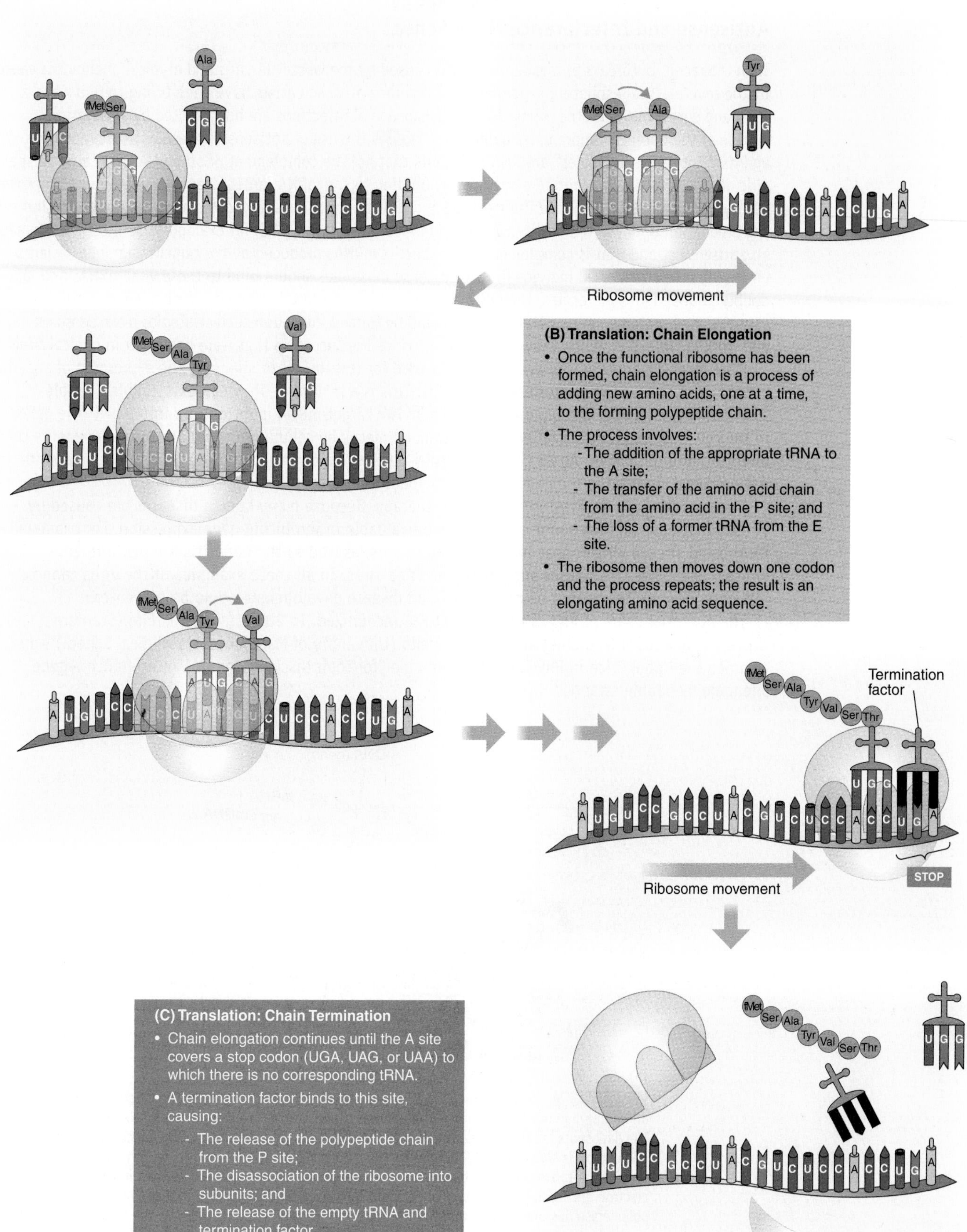

FIGURE 8.12 Continued.

MICROFOCUS 8.2: Biotechnology
Antisense and Interference Make Sense

With the recent outbreaks of fatal encephalitis caused by the West Nile virus and atypical pneumonia caused by the severe acute respiratory syndrome (SARS) coronavirus, scientists have been trying to find ways to treat and cure these and other viral diseases because viral infections are not affected by antibiotics.

One of the potential approaches being considered is the use of antisense molecules as therapeutic agents. "Antisense molecules" are RNA fragments that are the complement of an mRNA that carries a specific genetic message for protein synthesis. By binding to the mRNA, antisense molecules should block the ability of ribosomes to translate the message and thus the antisense molecules have the ability to shut off the production of unwanted or disease-causing proteins. To treat AIDS, for example, scientists could create an antisense strand that is complementary to specific mRNAs produced by the human immunodeficiency virus (HIV). In an infected individual, the antisense molecules should bind to these viral mRNAs and, as double-stranded RNA molecules, the mRNAs could not be translated by the cell's ribosomes. Without these essential viral proteins, no new HIV particles could be formed. Although such strategies make sense on paper and work in lab experiments, they have yet to produce the successes that were hoped for in clinical trials.

More recently, another way has been discovered for turning off or silencing the expression of specific genes. This is called "RNA interference" (RNAi). This is a technique in which extracellular, double-stranded (ds) RNA that is complementary to a known target mRNA is introduced into a cell. The dsRNA in the cell is chopped into smaller, single-stranded pieces by cellular enzymes and these fragments then bind to the target mRNA. Again these new dsRNA pieces are degraded and the protein or polypeptide is not produced. Indirectly, the gene for that polypeptide has been silenced.

One potential use for RNAi is for antiviral therapy. Because many human diseases are caused by viruses that have an RNA genome, RNAi may be valuable in inhibiting gene expression. For example, RNAi could silence viruses that induce human tumors, as well as the hepatitis A virus, influenza viruses, and other RNA viruses such as the measles virus. In all these examples, if the virus cannot replicate, new viruses cannot be produced—and disease development would be prevented.

The potential value of RNAi has recently been recognized. In 2006, Andrew Z. Fire (Stanford University School of Medicine) and Craig C. Mello (University of Massachusetts Medical School) were awarded the Nobel Prize in Physiology or Medicine "for their discovery of RNA interference—gene silencing by double-stranded RNA."

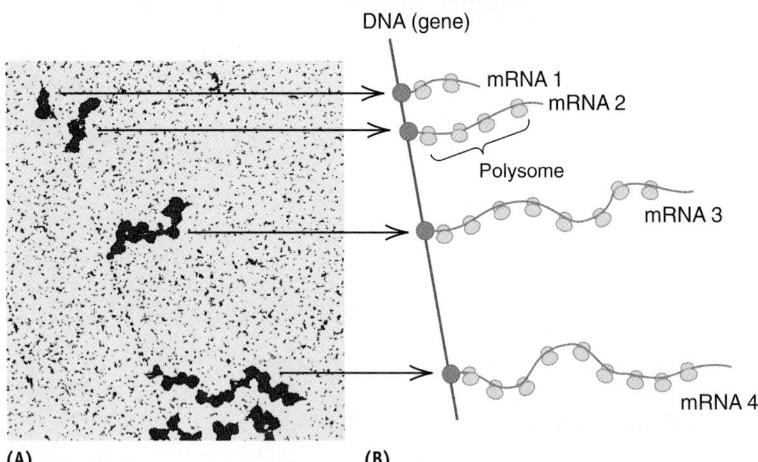

FIGURE 8.13 **Coupled Transcription and Translation in *E. coli*.** (**A**) The electron micrograph shows transcription of a gene in *E. coli* and translation of the mRNA. The dark spots are ribosomes, which coat the mRNA. An interpretation of the electron micrograph is shown in (**B**). Each mRNA has ribosomes attached along its length. The large red dots are the RNA polymerase molecules; they are too small to be seen in the electron micrograph. The length of each mRNA is equal to the distance that each RNA polymerase has progressed from the transcription-initiation site. For clarity, the polypeptides elongating from the ribosomes are not shown. Reproduced from O. L. Miller, B. A. Hamkalo, and C. A. Thomas, *Science* 169 (1977): 392. Reprinted with permission from AAAS. »» From the interpretation of the micrograph, (a) how many times has this gene been transcribed and (b) how many identical polypeptides are being translated?

and archaeal cells is to regulate the mechanisms that induce ("turn on") or repress ("turn off") transcription of a gene or set of genes.

In 1961, two Pasteur Institute scientists, Françoise Jacob and Jacques Monod, proposed such a mechanism for controlling protein synthesis. They suggested segments of bacterial DNA are organized into "transcriptional units" called **operons** (FIGURE 8.14). Their pioneering research along with more recent studies indicates that each operon consists of a cluster of **structural genes** providing genetic codes for proteins often having metabolically related functions. In this way, the cells can coregulate genes needed in the same functional or metabolic process. Adjacent to the structural genes is the **operator**, which is a sequence of bases controlling the expression (transcription) of the structural genes. Next to the operator is a **promoter**, which represents another sequence of bases to which the RNA polymerase binds to initiate transcription. Also important, but not part of the operon, is a distant **regulatory gene** that codes for a **repressor protein**.

In the operon model, the repressor protein binds to the operator, which prevents the RNA polymerase from moving down the operon and thus cannot transcribe the structural genes. This is called **negative control** because the repressor protein inhibits or "turns off" gene transcription within the operon. When the repressor in some way is prevented from binding to the operator, the RNA polymerase has clear sailing and transcribes the structural genes, which then are translated into the final polypeptides.

MicroInquiry 8 presents two contrasting examples of how an operon works to induce or repress gene transcription.

Website Animation: The Operon and Negative Control

(A) An operon consists of one or more structural genes, an operator, and a promoter. A regulatory gene is found at a site separate from the operon.

(B) The regulatory gene encodes a repressor protein that binds to the operator, thereby preventing the transcription of the structural genes by the RNA polymerase.

(C) If the repressor protein is blocked somehow from binding to the operator, the RNA polymerase is free to transcribe the structural genes.

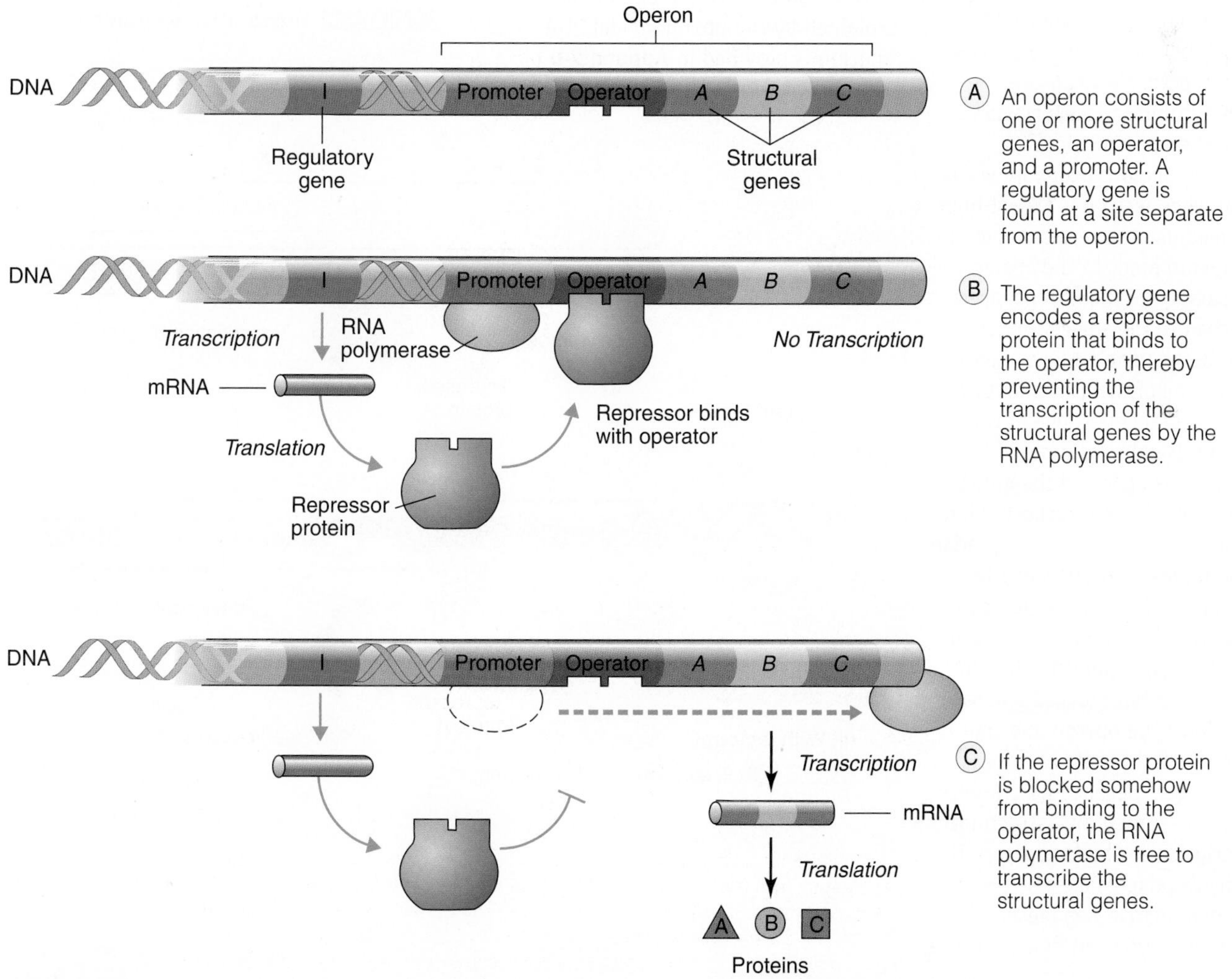

FIGURE 8.14 **The Operon and Negative Control.** An operon consists of a group of structural genes that are under the control of a single operator. Negative control exists if the operator prevents the RNA polymerase from transcribing the structural genes. »» How does the operator prevent structural gene transcription in negative control?

MICROINQUIRY 8

The Operon Theory and the Control of Protein Synthesis

The best way to visualize and understand the operon model for control of protein synthesis is by working through a couple of examples.

The Lactose *(lac)* Operon

Here is a piece of experimental data. The disaccharide lactose represents a potential energy source for *E. coli* cells if it can be broken into its monomers of glucose and galactose. One of the enzymes involved in the metabolism of lactose is β-galactosidase. If *E. coli* cells are grown in the absence of lactose, β-galactosidase activity cannot be detected as shown in the graph (**Figure A**).

However, when lactose is added to the nutrient broth, very quickly enzyme activity is detected. How can this change from inhibition to expression be explained in the operon model?

Based on the operon theory, we would propose that when lactose is absent from the growth medium, the repressor protein for the *lac* operon binds to the operator and blocks passage of the RNA polymerase that is attached to the adjacent promoter (**Figure Bi**). Being unable to move past the operator, the polymerase cannot transcribe the structural genes, one of which (*lacZ*) codes for β-galactosidase.

When lactose is added to the growth medium, lactose will be transported into the bacterial cell, where the disaccharide binds to the repressor protein and inactivates it (**Figure Bii**). With the repressor protein inactive, it no longer can recognize and bind to the operator. The RNA polymerase now is not blocked and can translocate down the operon and transcribe the structural genes. Lactose is called an "inducer" because its presence has induced, or "turned on," structural gene transcription in the *lac* operon. It explains why β-galactosidase activity increases when lactose was present.

Now let's see if you can figure out this scenario.

Tryptophan *(trp)* Operon

E. coli cells have a cluster of structural genes that code for five enzymes in the metabolic pathway for the synthesis of the amino acid tryptophan (trp). Therefore, if *E. coli* cells are grown in a broth culture lacking trp, they continue to grow normally by synthesizing their own tryptophan, as shown in the graph (**Figure A**).

However, as the graph shows, when trp is added to the growth medium, new enzyme synthesis is repressed or "turned off" and cells use the trp supplied in the growth medium.

How can enzyme repression be explained by the operon model? The solution is provided in **Appendix D** on the Student Companion Website.

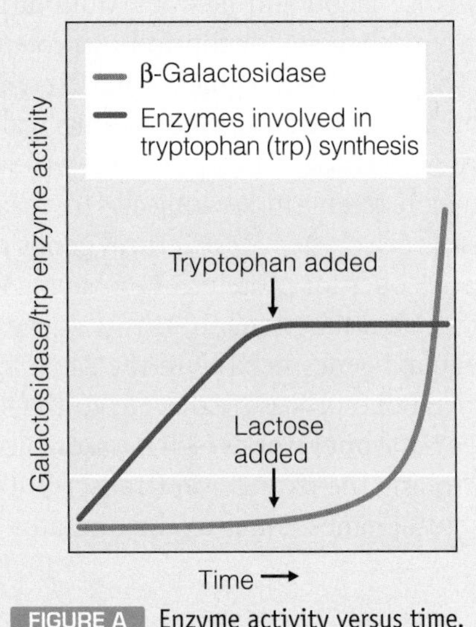

FIGURE A Enzyme activity versus time.

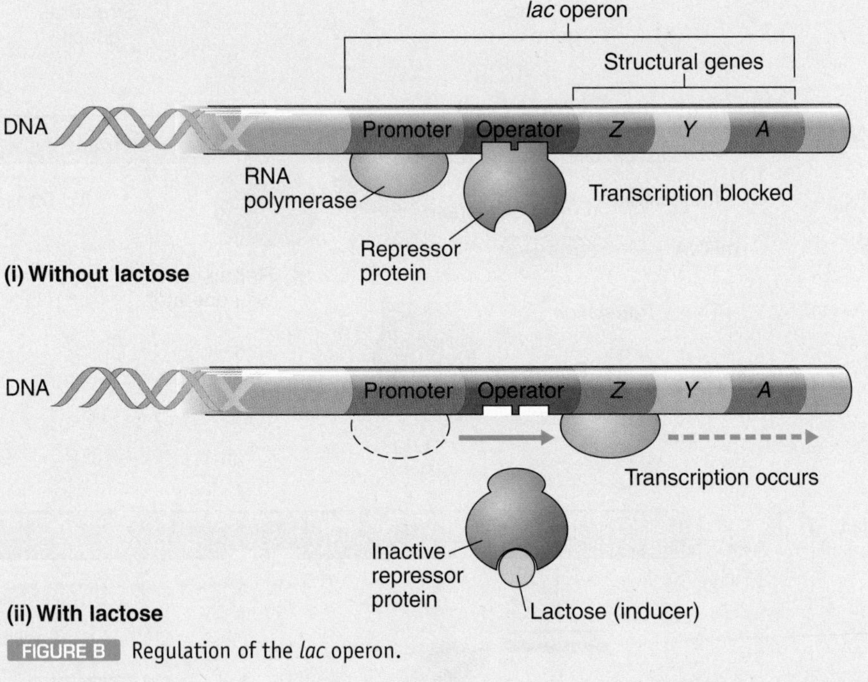

FIGURE B Regulation of the *lac* operon.

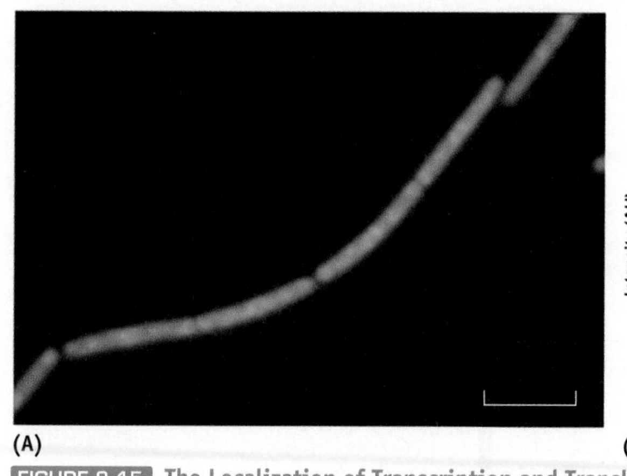

(A)

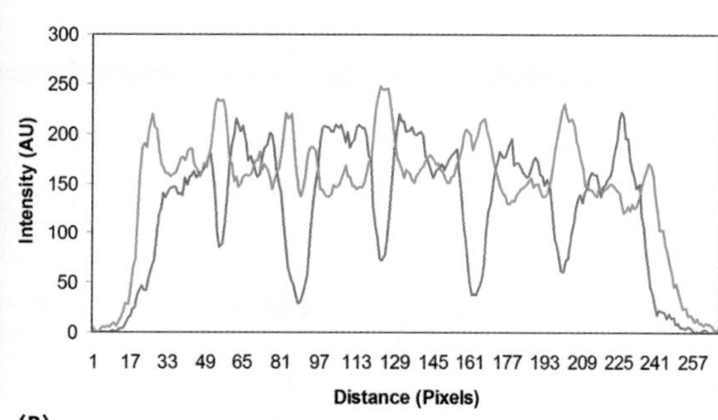

(B)

FIGURE 8.15 The Localization of Transcription and Translation in *Bacillus subtilis* Cells. (**A**) In these dividing *B. subtilis* cells, ribosomal subunits have been labeled with a green fluorescent protein and RNA polymerase subunits with a label that fluoresces red. The RNA polymerase (transcription) is found mainly in the nucleoid core while the ribosomes (translation) are concentrated at the poles of the cell. (Bar = 3 μm.) (**B**) This linescan through the cells confirms the interpretation that where polymerase RNA polymerase fluorescence is high (red line), ribosomal fluorescence is low (green line) and vice versa. Courtesy of Dr. Peter Lewis; School of Environmental and Life Sciences, University of Newcastle. »» How does fluorescence microscopy aid in the identification of spatially separated compartments?

Transcription and Translation Are Localized

The nucleoid is an amorphous area containing the cell's chromosome. Although it lacks a nuclear envelope, nucleoid and cytoplasmic activities can be segregated much as they are in eukaryotic cells.

In *Bacillus subtilis,* for example, RNA polymerases would be expected to be concentrated within the nucleoid where transcription occurs. In fact, RNA polymerase is often localized to specific regions of the nucleoid, somewhat similar to their localization in eukaryotic nuclei that are carrying out transcription. This is in contrast to the localization of ribosomes, which, at least in *B. subtilis,* are absent in the nucleoid region and primarily concentrated at the cell poles (FIGURE 8.15). Such observations with *B. subtilis* suggest that the ability to segregate transcription and translation without the need for a nuclear envelope is analogous to that of the Eukarya.

How the gene products, mRNAs, and the proteins, are targeted to specific locations in a bacterial cell depends on localization sequences (sort of like zip codes) within the mRNA or protein. In *E. coli,* synthesized proteins diffuse away from the ribosomes and can be captured by regional structures that recognize the localization sequence. In addition, mRNAs themselves may contain localization sequences and be translated only at the cellular site where they are "captured."

Once again we see that the cytoplasm of a bacterial cell is intricate in design and dynamic in its actions. FIGURE 8.16 summarizes the protein synthesis process.

CONCEPT AND REASONING CHECKS 3

a. How is the coding sequence of a bacterial gene processed differently from that in a eukaryotic gene?

b. What is meant by the genetic code being redundant? Give two examples.

c. Explain why the ribosome can be portrayed as a "cellular translator."

d. Propose a hypothesis to explain why so many antibiotics specifically affect translation.

e. In the *lac* operon, why is transcription referred to as negative control?

f. How is compartmentation in bacterial and eukaryotic cells similar in regard to transcription and translation?

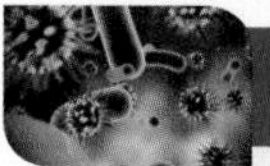

Chapter Challenge C

In this section of the chapter, you studied gene expression in bacterial species. This involved both transcription by an RNA polymerase and translation by ribosomes. Again, both the transcription and translation processes require many proteins to regulate and carry out the expression of a gene.

QUESTION C:

When considering the obligate symbionts/parasites, would the genes required for transcription and translation be part of the parasite's genome? Explain. Remember, a bacterial ribosome is built from four rRNAs and some 55 proteins.

Answers can be found on the Student Companion Website in **Appendix D**.

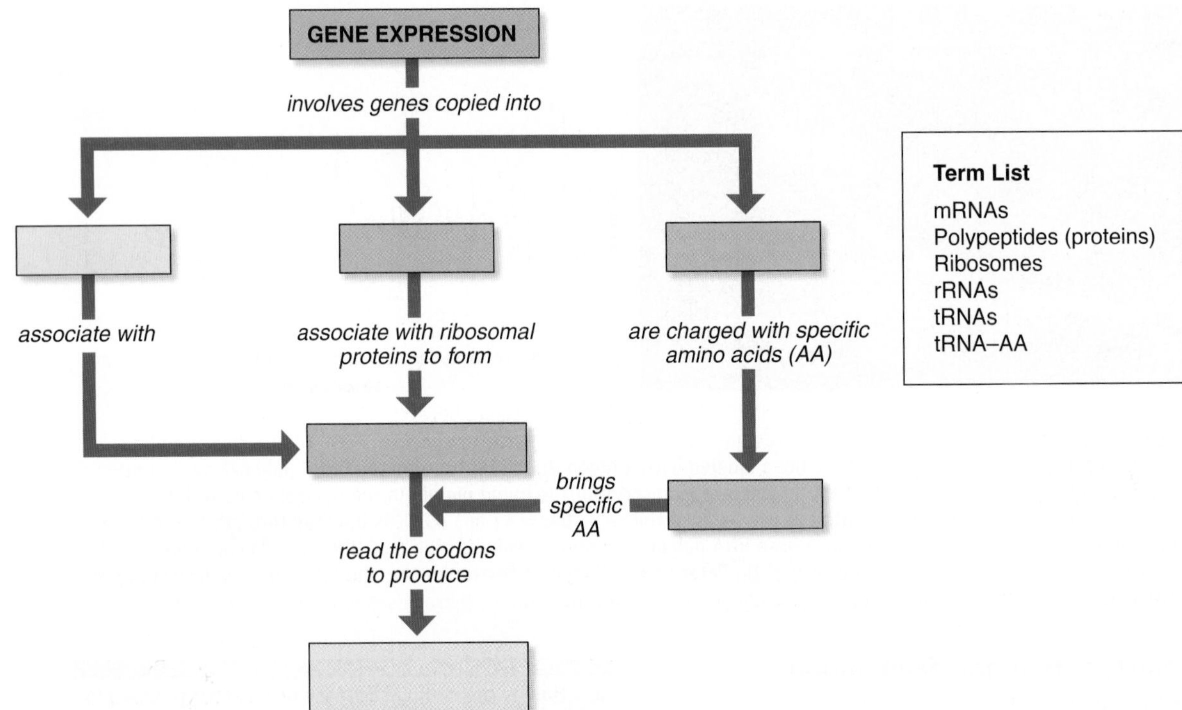

FIGURE 8.16 **A Concept Map for Gene Expression.** Gene expression is a combination of the transcription and translation processes. »» Use the terms in the box to fill in this concept map.

KEY CONCEPT 8.4 Mutations Are Permanent Changes in a Cell's DNA

■ Wild type: The common or native form of a gene or organism.

■ Niche: The functioning of a species in relation to other species and its physical environment.

The information in a chromosome may be altered through a permanent change in the DNA, which is called a **mutation**. In most cases, a mutation involves a change in or disruption to a base sequence in the DNA molecule. The result is the production of a miscoded mRNA, and ultimately a change in one or more amino acids found in the polypeptide during translation. Because proteins govern numerous cellular activities, mutations may alter some aspect of these activities—for better or worse (MicroFocus 8.3).

Mutations Are the Result of Heritable Changes in a Genome

Spontaneous mutations are heritable, random changes to the base sequence in the DNA that result from natural phenomena. These changes could be from everyday radiation penetrating the atmosphere or errors made and not corrected by DNA polymerase during replication. It has been estimated that one such mutation may occur for every 10^6 to 10^{10} divisions of a microbial cell.

A mutant cell arising from a spontaneous mutation usually is masked by the normal wild type cells in the population. However, should the environment favor the mutant, it may multiply and emerge as the predominant form. For example, for many decades doctors used penicillin to treat gonorrhea. Then, in 1976, a penicillin-resistant strain of *Neisseria gonorrhoeae* emerged. Many investigators believe the resistant bacterial strain had been present for perhaps centuries, but only now with heavy use of penicillin could it arise and fill the niche once held by the penicillin-sensitive forms.

Most of our understanding of mutations has come from experiments involving **induced mutations** that are produced by chemical or physical agents called **mutagens**.

Physical Mutagens. Ultraviolet (UV) light is a physical mutagen whose energy induces adjacent thymine (or cytosine) bases in the DNA to covalently link together forming dimers (FIGURE 8.17). If these dimers occur in a protein-coding gene, the

MICROFOCUS 8.3: Evolution

Evolution of an Infectious Disease

Could the Black Death of the 14th century and the 25 million Europeans that succumbed to plague have been the result of a few genetic changes to a bacterial cell? Could the entire course of Western civilization have been affected by these changes?

Possibly so, maintain researchers from the federal Rocky Mountain Laboratory in Montana. In 1996, a research group led by Joseph Hinnebusch reported that three genes missing in the plague bacillus *Yersinia pestis* are present in a related species *(Y. pseudotuberculosis)* that causes mild food poisoning. Thus, it is possible that the entire story of plague's pathogenicity revolves around a small number of gene changes.

Bubonic, septicemic, and pneumonic plague are caused by *Y. pestis,* a rod-shaped bacterium transmitted by the rat flea. In an infected flea, the bacterial cells eventually amass in its foregut and obstruct its gastrointestinal tract. Soon the flea is starving, and it starts biting victims (humans and rodents) uncontrollably and feeding on their blood. During the bite, the flea can regurgitate some 24,000 plague bacilli into the bloodstream of the unfortunate victim.

At least three genes are important in the evolution of plague. The nonpathogenic *Y. pseudotuberculosis* bacilli have these genes, which encourage the bacilli to remain harmlessly in the midgut of the flea. Pathogenic plague bacilli, by contrast, do not have the genes. Free of their control, the bacterial cells migrate from the midgut to the foregut, allowing the bacilli to be passed on to the victim in a flea bite.

In 2002, Hinnebusch and colleagues published evidence that another gene, carried on a plasmid, codes for an enzyme that is required for the initial survival of *Y. pestis* bacilli in the flea midgut. By acquiring this gene from another unrelated organism, *Y. pestis* made a crucial jump in its host range. It now could survive in fleas and became adapted to relying on its blood-feeding host for transmission. So, a few genetic changes may have been a key force leading to the evolution and emergence of plague. This is just another example of the flexibility that many microbes have to repackage themselves constantly into new and, sometimes, more dangerous agents of infectious disease.

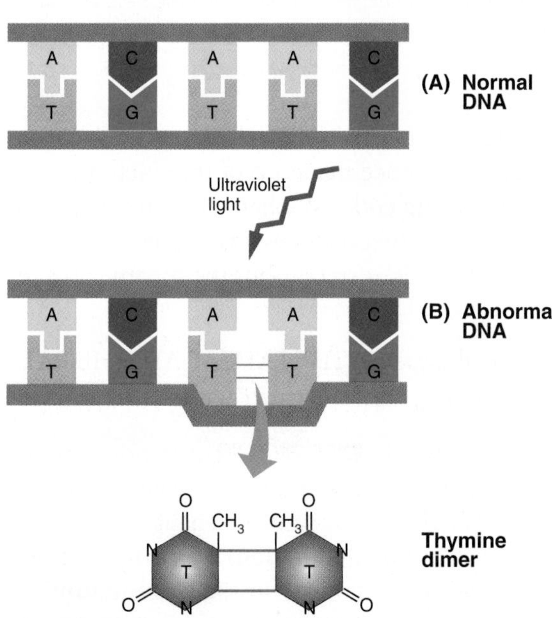

FIGURE 8.17 **Ultraviolet Light and DNA.** (**A**) When cells are irradiated with ultraviolet (UV) light either naturally or through experiment, the radiations may affect the cell's DNA. (**B**) UV light can cause adjacent thymine molecules to pair within the DNA strand to form a thymine dimer. »» How might a thymine dimer block the movement of RNA polymerase?

RNA polymerase cannot insert the correct bases (A–A) in mRNA molecules where the dimers are located. In addition, ionizing radiations, such as gamma and X rays, can cause double-strand breaks in the DNA.

Chemical Mutagens. Nitrous acid is an example of a chemical mutagen that converts DNA's adenine bases to hypoxanthine bases (FIGURE 8.18A). Adenine would normally base pair with thymine, but the presence of hypoxanthine causes a base pairing with cytosine during replication. Later, should replication occur from the gene with the cytosine mutation, the mRNA will contain a guanine rather than an adenine.

Mutations also are induced by **base analogs**, such as 5-bromouracil (5-BU), which bears a close chemical resemblance to thymine (FIGURE 8.18B). If 5-BU was incorporated into a gene, the base analog pairs with guanine, rather than adenine, when the gene is replicated or transcribed.

Other base analogs resemble other DNA bases and are useful as antiviral agents in the treatment of diseases caused by DNA viruses, such as the herpesviruses. Acyclovir, for example, is a base

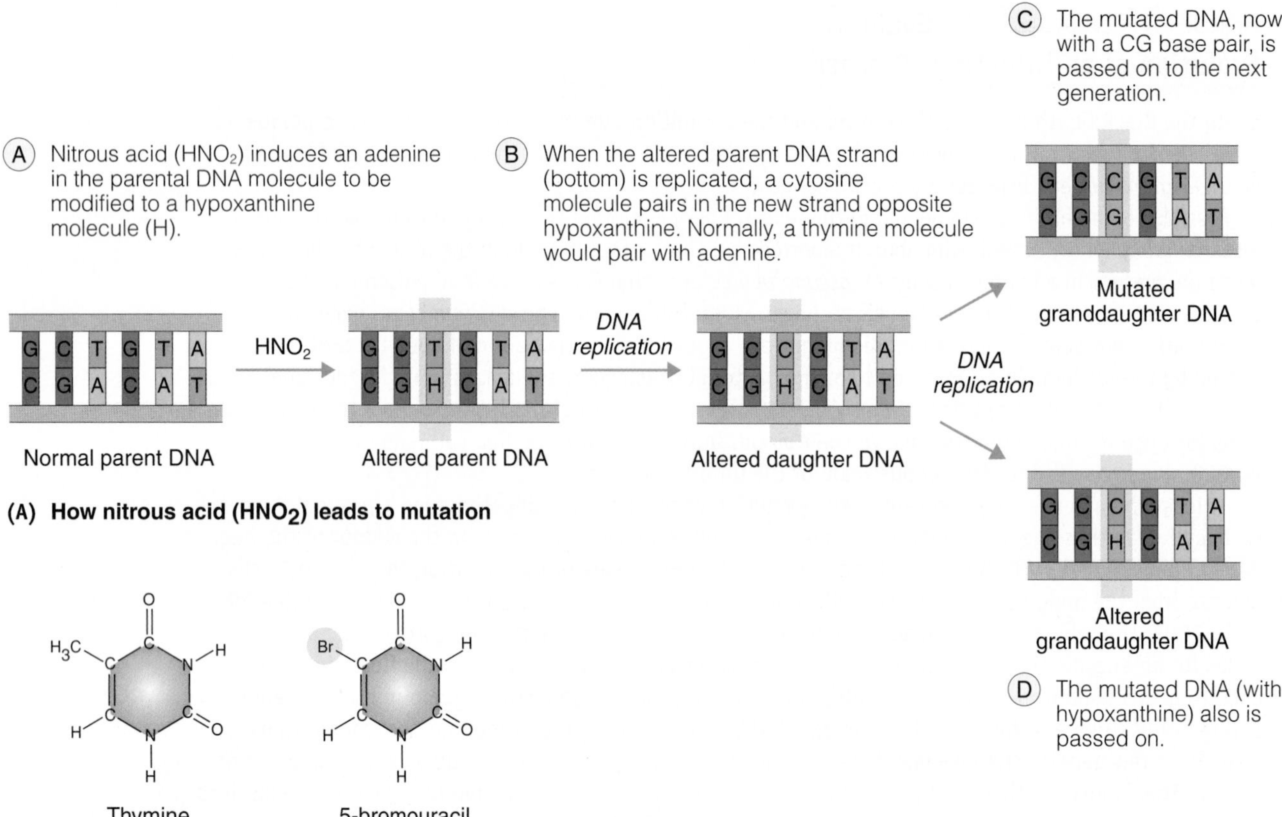

(A) How nitrous acid (HNO_2) leads to mutation

(B) A nitrogenous base and its mutation-causing base analog

FIGURE 8.18 The Effect of Chemical Mutagens. (**A**) Nitrous acid induced an adenine to hypoxanthine change. After replication of the hypoxanthine-containing strand, the granddaughter DNA has a mutated C–G base pair. (**B**) Base analogs induce mutations by substituting for nitrogenous bases in the synthesis of DNA. Note the similarity in chemical structure between thymine and the base analog 5-bromouracil. »» How does nitrous acid differ from 5-bromouracil in inducing mutations?

analog that can substitute for guanine during viral replication. The presence of acyclovir blocks viral replication, so new virus particles cannot be produced. As such, treatment with acyclovir can be effective in decreasing the frequency and severity of fever blisters (cold sores).

Point Mutations Can Affect Protein Structure and Function

Regardless of the cause of the mutation, one of the most common results is a **point mutation**, which usually affects just one point (base pair) in a gene. Such mutations may be a change to or substitution of a different base pair, or a deletion or addition of a base pair.

Base-Pair Substitutions. If a point mutation causes a base-pair substitution, then the transcription of that gene will have one incorrect base in the mRNA sequence of codons. Perhaps one way to see the effects of such changes is using an English sentence made up of three-letter words (representing codons) where one letter has been changed. As three-letter words, the letter substitution still reads correctly, but the sentence makes less sense.

Normal sequence: THE FAT CAT ATE THE RAT

Substitution: THE FAT BAT ATE THE RAT

As shown in FIGURE 8.19A, depending on the placement of the substituted base, when the mRNA is translated this may cause no change due to code redundancy (**silent mutation**), lead to the insertion of one wrong amino acid (**missense mutation**), or generate a stop codon (**nonsense mutation**), prematurely terminating the polypeptide.

Base-Pair Deletion or Insertion. Point mutations also can cause the loss or addition of a base in a gene, resulting in an inappropriate num-

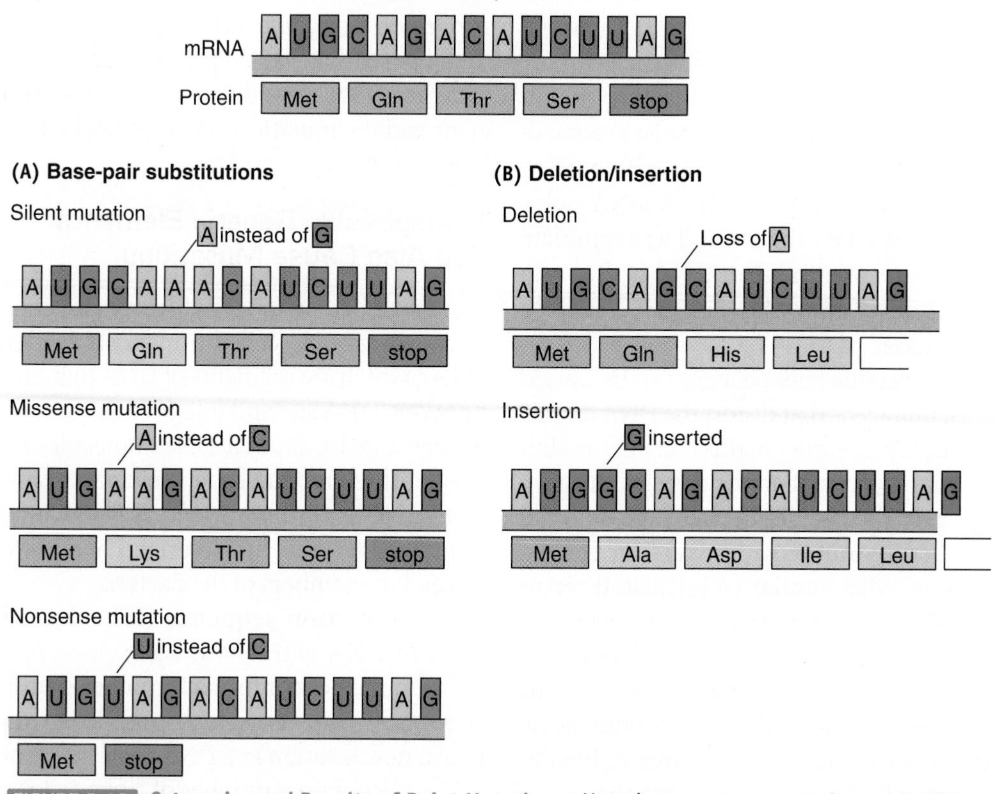

FIGURE 8.19 **Categories and Results of Point Mutations.** Mutations are permanent changes in DNA, but they are represented here as they are reflected in mRNA and its protein product. (**A**) Base-pair substitutions can produce silent, missense, or nonsense mutations. (**B**) Deletions or insertions shift the reading frame of the ribosome. »» Determine the normal sequence of bases in the template strand of the gene and the base change that gave rise to each of the "mutated" mRNAs.

ber of bases. Again, using our English sentence, we can see how a deletion or insertion of one letter affects the reading frame of the three-letter word sentence.

Normal sequence: THE FAT CAT ATE THE RAT

Deletion: THE F_TC ATA TET HER AT

Insertion: THE FAT ACA TAT ETH ERA T

As you can see, the "sentence mutations" are nonsense when reading the sentence as three-letter words. The same is true in a cell. Ribosomes always read three letters (one codon) at one time, generating potentially extensive mistakes in the amino acid sequence (FIGURE 8.19B). Thus, like our English sentence, the deletion or addition of a base will cause a "reading frameshift" because the ribosome always reads the genetic code in groups of three bases. Therefore, loss or addition of a base shifts the reading of the code by one base. The result is serious sequence errors in the amino acids, which will probably produce an abnormal protein (nonsense) unable to carry out its role in metabolism.

Repair Mechanisms Attempt to Correct Mistakes or Damage in the DNA

The fact that DNA is double stranded is not a fluke. By being double stranded, one polynucleotide strand can act as a template to repair mismatches between strands or distortions within one strand. Bacterial cells, such as *E. coli*, have two repair systems to mend these forms of damage.

One repair mechanism is called **mismatch repair**. Realize that during the life of a microbial cell (indeed, of every cell), cellular DNA endures thousands of potentially damaging events resulting from DNA replication errors. Even though DNA polymerase is very accurate in proofreading during replication, errors (mismatched nucleotides) are sometimes missed. Mismatch repair can be used to detect and repair these errors. First, a special mismatch correction enzyme scans the DNA for any missed mismatched pairs. Finding such a pair, another enzyme cuts out (excises) a small segment of the polynucleotide strand containing the mismatched nucleotide. A DNA polymerase then uses the other strand as a template to replace the excised nucleotide segment with the correct set of bases, which is sealed in place by a DNA ligase.

Other nucleotide base changes can be caused by physical mutagens that distort the DNA double helix. We discussed earlier in this chapter the ability of UV light to cause thymine dimer formation. Such distortions in the double helix are detected and corrected by **nucleotide excision repair**, which is somewhat similar to mismatch repair (FIGURE 8.20). First, nucleases excise the thymine dimer along with adjacent nucleotides in that strand. Then, DNA polymerase replaces the missing nucleotides with the correct ones, again using the other strand as the template. Finally, DNA ligase seals the new strand into the rest of the polynucleotide.

Thymine dimer distorts DNA strand.

(A) The damaged strand and adjacent nucleotides are cut and removed by a nuclease enzyme.

(B) The complementary (undamaged) strand serves as a template to repair the damaged strand.

DNA polymerase

(C) The repaired strand is sealed to the polynucleotide by DNA ligase.

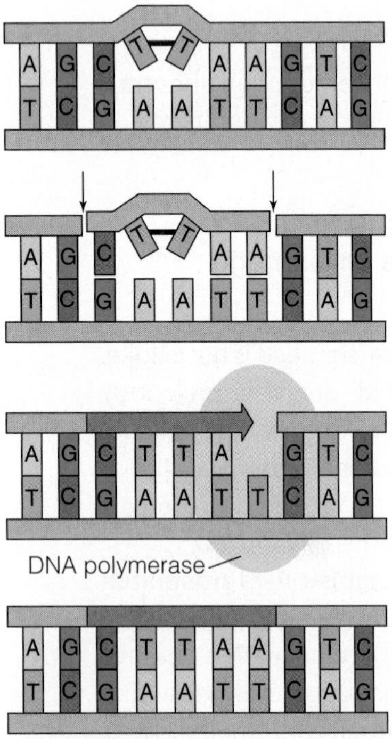

FIGURE 8.20 Nucleotide Excision Repair. Thymine dimer distortion triggers nuclease repair enzymes that excise the damaged DNA and permit resynthesis of the correct nucleotides. »» How might excision nucleases recognize an error in a DNA fragment?

An impressive example of DNA repair is seen in *Deinococcus radiodurans* (MICROFOCUS 8.4). Still, realize that DNA repair is seldom 100% perfect. Sometimes "shoddy repairs" fail to correct an error and the mutation becomes "locked in"—for better or worse.

Transposable Genetic Elements Can Also Cause Mutations

Mutations of a different nature may be caused by fragments of DNA called **transposable elements (TEs)**, which are segments of DNA that can move from one DNA site and integrate into another site in the same or another DNA molecule. For this reason, these mobile genetic elements have been referred to as "jumping genes." The two types of transposable elements described here have been studied in members of the Bacteria.

An **insertion sequence (IS)** is a small segment of DNA with about 1,000 base pairs. An IS has no genetic information other than for the enzyme **transposase**, which is needed to move the IS to a new location in a DNA molecule. The ends of the IS contain a number of "inverted repeats"; that is, identical, inverted base pair sequences (FIGURE 8.21A).

A second type of TE, called a **transposon**, is larger than an IS and carries additional genes for various functions, such as antibiotic resistance. Like IS, the transposon has inverted repeats at each end of the element (FIGURE 8.21B).

The actual movement (transposition) of TEs often involves a simple "cut and paste" process. The TE is cut out of the DNA by the transposase and then, using the inverted repeats, it is inserted into a new site; nothing takes the place where the TE originated. However, other transposons are replicative; that is, the transposon remains at the original site and a replicated copy of that transposon then jumps to a new target DNA site.

The movement of TEs can have great significance. First, when such a random transposition occurs, it could interrupt the coding sequence in an essential gene such that protein synthesis produces a nonfunctional protein or, more likely due to the disruption, no protein at all. Thus, TEs can be a prime force behind spontaneous mutations.

TEs can move from chromosome to chromosome, plasmid to plasmid, plasmid to chromosome, or chromosome to plasmid. Although

MICROFOCUS 8.4: Evolution

Shattered Chromosomes

It has been called "Conan the Bacterium"[1] and has been listed in *The Guinness Book of World Records* as "the world's toughest bacterium." The organism: *Deinococcus radiodurans* (*deino* = "terrible"; *coccus* = "sphere"; *radio* = "ray"; *dura* = "hard"), referring to its resistance to gamma ray radiation, a spherical bacterial species arranged in tetrads (see figure), is easily cultured and does not cause any known disease.

D. radiodurans can survive up to 5,000 gray (Gy; formerly called rad = 0.01 Gy) of ionizing radiation, which should cause several hundred double-stranded breaks—one of the hardest types of DNA damage to repair—in the organism's DNA. For comparison, 10 Gy can kill a human and 100 Gy would kill *Escherichia coli*.

Analysis of *D. radiodurans*' genome indicates that it consists of two circular chromosomes, one of 2.6 Mb and the other of 0.41 Mb, containing about 3,200 genes. During stationary phase of its growth curve, each bacterial cell contains four copies of this genome; when in log phase, each bacterium contains 8 to 10 copies of the genome.

A false-color scanning electron micrograph of *Deinococcus radiodurans*. (Bar = 2 μm) © Dennis Kunkel Microscopy, Inc./Phototake/Alamy Images.

So, how does *D. radiodurans* survive a dose of radiation that produces "shattered chromosomes" consisting of hundreds of short DNA fragments? The key appears to be the presence of multiple copies of its genome and a rapid, novel two-step DNA repair mechanism that can repair double-stranded breaks in its chromosomes within just a few hours. First, repair requires at least two genome copies broken at different positions. Therefore, *D. radiodurans* undergoes massive and rapid DNA synthesis to produce a mosaic of new and old fragments with single-stranded ends that can then reconnect accurately into larger chromosomal segments. In the second step, a protein, called RecA, efficiently joins the double-strand breaks into functional chromosomes. It is a remarkably efficient process that occurs in an equally incredible short period of time.

A genetically engineered *D. radiodurans* has been used for bioremediation to consume and digest solvents and heavy metals, especially in highly radioactive sites. In addition, bacterial genes from *E. coli* have been introduced into *D. radiodurans* so it can detoxify ionic mercury and toluene, chemicals often included in the radioactive waste from nuclear weapons manufacture.

A few other bacterial genera, including *Chroococcidiopsis* (phylum Cyanobacteria) and *Rubrobacter* (phylum Actinobacteria), and the archaeal species *Thermococcus gammatolerans,* also are gamma-radiation resistant. However, with its added genes for bioremediation, *D. radiodurans* rightly retains its title of "the world's toughest bacterium."

---

[1]Huyghe, P. 1998. Conan the bacterium. *The Sciences* 38:4, 16–19.

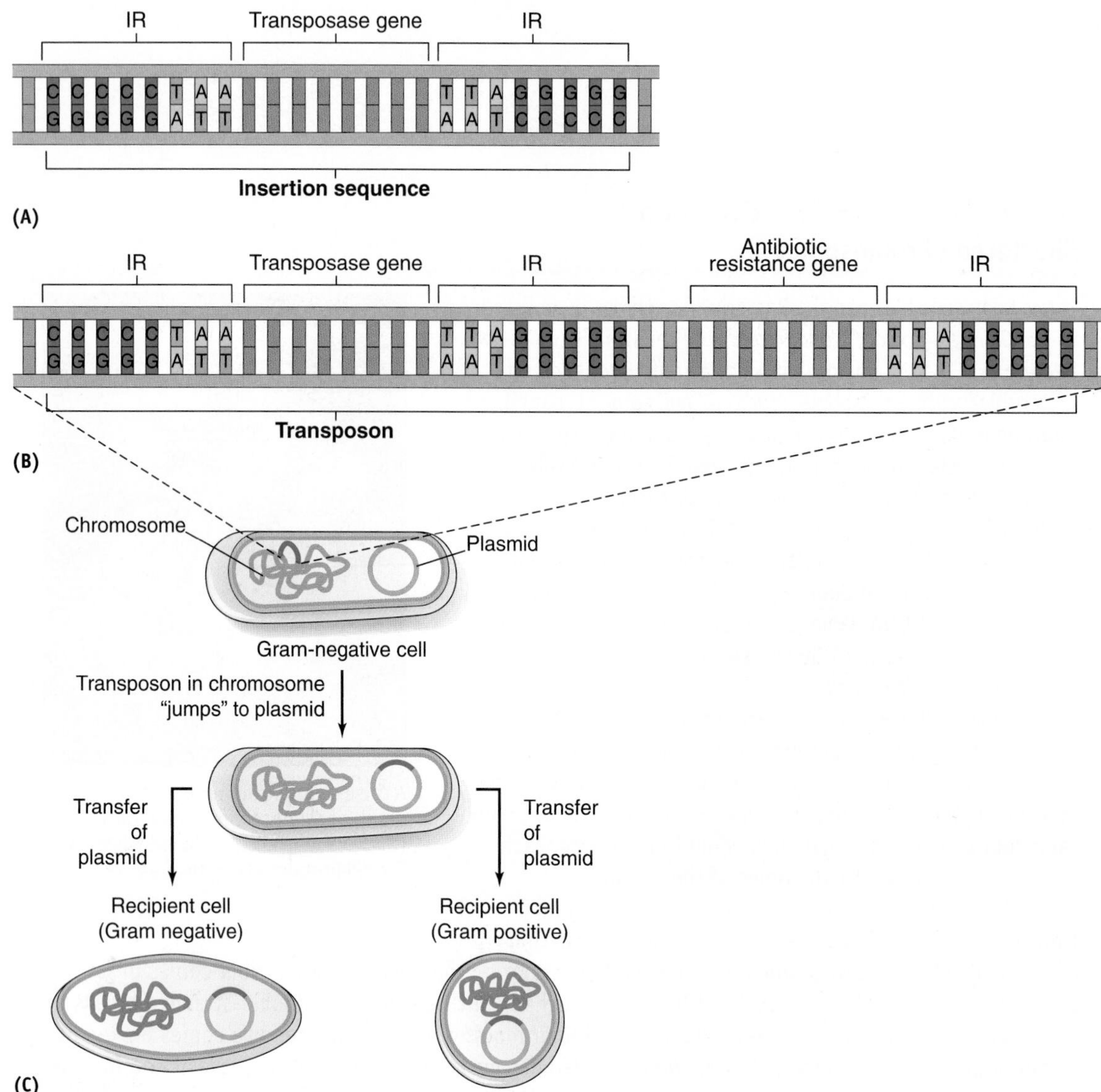

FIGURE 8.21 **Transposable Elements.** (**A**) An insertion element consists only of the transposase gene and the adjacent inverted repeats (IRs). (**B**) A transposon contains additional genes, such as antibiotic resistance. (**C**) Transposons can "jump" to another DNA molecule, such as a plasmid, which then can be transferred to another cell conferring new genetic capabilities (e.g., antibiotic resistance) on the recipient. »» What is required for the transposon to "jump" to another DNA molecule?

these events are rare, they are of particular significance when transposons carrying genes for antibiotic resistance are transferred to similar or a different bacterial species (FIGURE 8.21C). For example, if a chromosome carries a transposon with an antibiotic resistance gene, that transposon could jump to a plasmid in that cell. If that plasmid is then transferred to another bacterial cell, as plasmids are known to do, the transposon will move along with it, thus spreading the gene for antibiotic resistance to the recipient cell. In fact, the movement of transposons among plasmids is a major mechanism for the spread of antibiotic resistance among bacterial species (TEXTBOOK CASE 8).

CONCEPT AND REASONING CHECKS 4

- **a.** How do chemical mutagens interfere with DNA replication or gene expression?
- **b.** Justify the statement: "A base-pair deletion (or insertion) potentially is more dangerous to an organism's viability than a base-pair substitution."
- **c.** Explain why cells need at least two repair mechanisms (mismatch and excision).
- **d.** What harm or benefit is conferred by insertion sequences and transposons?

Textbook CASE 8

Klebsiella pneumoniae in a Hospitalized Patient

1 In March 2008, a 91-year-old man with dialysis-dependent end-stage renal disease, congestive heart failure, anemia, and peptic ulcer disease was admitted to a hospital in Tel Aviv, Israel. Following amputation of the left leg below the knee due to an infected heel wound, the patient developed sepsis. The patient was treated with a variety of antibiotics, including ertapenem, a carbapenem drug.

2 During treatment, an acute inflammation of the gallbladder developed, so the patient had a surgical incision made in his gallbladder to help drain the infection.

3 During his hospital stay, the patient was screened for carbapenem-resistant Enterobacteriacae (CRE) as part of the hospital's routine infection control program aimed at limiting the spread of these often antibiotic-resistant organisms. As such, CRE is an important health challenge in healthcare settings.

4 Two rectal swabs were collected one week apart for analysis by the hospital's clinical microbiology lab. The first swab specimen was negative for CRE by culture. However, the second swab specimen grew colonies (see figure). Microscopy identified small, gram-stained rods, which on further analysis were shown to be facultatively anaerobic and resistant to carbapenem, as well as all cephalosporin and monobactam antibiotics. Identification was made as a carbapenem-resistant *Klebsiella pneumoniae* (CRKP) strain.

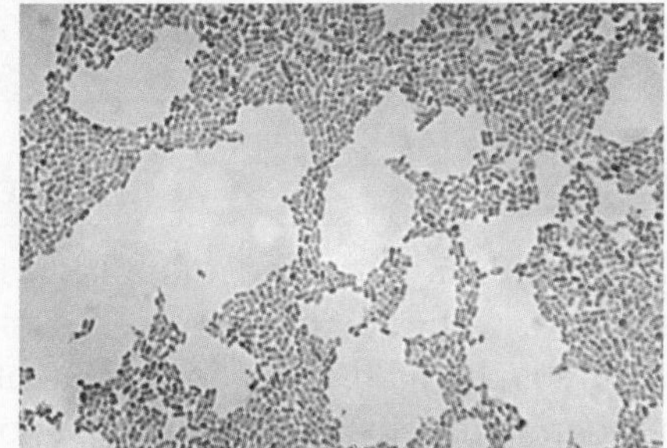

A gram-stained preparation of *Klebsiella pneumoniae*. © Eye of Science/Photo Researchers, Inc.

5 Alarm was raised as CRKP is associated with increased mortality, particularly in patients with prolonged hospitalization. An aggressive infection control strategy was instituted.

6 Two weeks later, a carbapenem-resistant *E. coli* strain was recovered from the patient's gallbladder drainage. Further infection control strategies were instituted in the hospital.

Questions:

(Answers can be found on the Student Companion Website in **Appendix D.**)

A. From the figure, what is the Gram stain reaction for the small rods?

B. From the material covered in this chapter, propose a mechanism for the subsequent emergence of the carbapenem-resistant *E. coli* strain.

C. What infection control measures should the hospital institute?

D. Why was the presence of a carbapenem-resistant *E. coli* strain of even greater concern as a healthcare threat?

For additional information see www.cdc.gov/eid/content/16/6/1014.htm.

Chapter Challenge D

The purpose of this challenge was to try to figure out what genes obligate symbionts/parasites must contain when they remain dependent on living in a host. As you have seen, there is a great divergence in genome size and gene number among the Bacteria and Archaea (as Figure 8.3 illustrated). For the obligate symbionts/parasites and all free-living species, there is a set of core genes that is absolutely necessary to support any lifeform. This would include the genes for DNA replication, transcription, and translation. In addition, the genes for ribosomal subunits also would be part of the core. In addition, genes necessary for cell membrane biogenesis would also be included because the membrane separates the symbiont/parasite from the cytoplasm of the host. On the other hand, genes for many aspects of metabolism may have been lost over time. In fact, it appears that the tiny genomes of symbionts/parasites are a result of outsourcing. The genes have been transferred to the host's chromosome or the genes have simply been lost because these organisms can rely on many of the host's genes to supply essential cellular products. Therefore, their own genes are not needed and probably have been lost over evolutionary time. Perhaps the mitochondria and chloroplasts, once free-living bacterial cells that evolved into organelles through endosymbiosis, are the ultimate examples where symbiosis has resulted in the loss of many genes or transfer of the genes to the host.

QUESTION D:

So, to summarize, put together a list of core genes that must be part of a universal minimal genome. Scientists do not yet know what any such universal minimal genome size is (if such a genome exists). Some have estimated it could be as small as 40 genes. Mitochondria, for example, have some 40 to 80 genes.

Answers can be found on the Student Companion Website in **Appendix D.**

KEY CONCEPT 8.5 Techniques Exist for Identifying Mutants

Any organism carrying a mutation is called a **mutant**, while the normal strain isolated from nature is the **wild type**. Some mutants are easy to identify because the **phenotype**, or physical appearance, of the organism or the colony has changed from the wild type. For example, some bacterial colonies appear red because they produce a red pigment. Treat the colonies with a mutagen and, after plating on nutrient agar, mutants form colorless colonies. However, not all mutants can be identified solely by their "looks."

Plating Techniques Select for Specific Mutants or Characteristics

Selection is a very useful technique to identify and isolate a single mutant from among thousands of possible cells or colonies. Let's look at two selection techniques that make this search possible.

First, in both techniques, the chemical composition of the transfer plate is key to visual identification of the colonies being hunted. The use of a replica plating device makes the identification possible. The device consists of a sterile velveteen cloth or filter paper mounted on a solid support. When an agar plate (master plate) with bacterial colonies is gently pressed against the surface of the velveteen, some cells from each colony stick to the velveteen. If another agar plate then is pressed against this velveteen cloth, some cells will be transferred (replicated) in the same pattern as on the master plate.

Now, suppose you want to find a nutritional mutant unable to grow without the amino acid histidine. This mutant (written his^-) has lost the ability that the wild type strain (his^+) has to make its own histidine. Such a mutant having a nutritional requirement for growth is called an **auxotroph** (*auxo* = "grow"; *troph* = "nourishment"), while the wild type is a **prototroph** (*proto* = "original"). Phenotypically, there is no difference between the two strains when they grow on a complete medium with histidine. However, you can visually identify the auxotroph using a **negative selection** plating technique (FIGURE 8.22). Any colonies missing on the minimal medium plate (lacking histidine) must be his^-.

As another example, suppose you want to see if there are any bacterial organisms in a hospital ward that are resistant to the antibiotic carbapenem

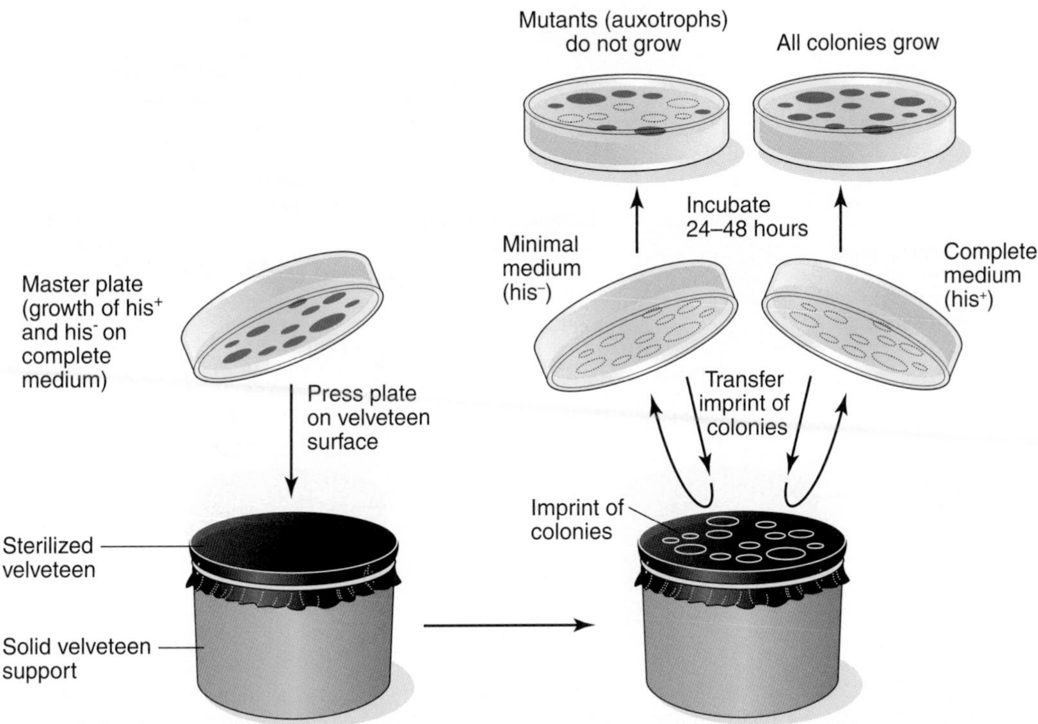

FIGURE 8.22 **Negative Selection Identifies Auxotrophs.** Negative selection plating techniques can be used to detect nutritional mutants (auxotrophs) that fail to grow when replica plated on minimal medium (in this example, a growth medium lacking histidine). Comparison to replica plating on complete medium visually identifies the auxotrophic mutants. »» In this example, which colonies transferred to the complete medium represent the auxotrophs (his⁻)?

(see Textbook Case 8). Again, phenotypically there is no difference between those strains sensitive to ertapenem and those resistant to the antibiotic. However, a **positive selection** plating technique permits visual identification of such carbapenem resistant mutants because they will grow on a complete medium containing ertapenem (FIGURE 8.23).

The Ames Test Can Identify Potential Mutagens

Some years ago, scientists observed that about 90% of human **carcinogens**—agents causing tumors in humans—also induce mutations in bacterial cells. Working on this premise, Bruce Ames of the University of California developed a procedure to help identify potential human carcinogens by determining whether the agent can mutate bacterial auxotrophs. The procedure, called the **Ames test**, is a widely used, relatively inexpensive, rapid, and accurate screening test. For the Ames test, an auxotrophic, histidine-requiring strain (his⁻) of *Salmonella enterica* serotype Typhimurium is used. If inoculated onto a plate of nutrient medium lacking histidine, no colonies will appear because in this auxotrophic strain the gene inducing histidine synthesis is mutated and hence not active.

In preparation for the Ames test, the potential carcinogen is mixed with a liver enzyme preparation. The reason for doing this is because often chemicals only become tumor causing and mutagenic in humans after they have been modified by liver enzymes.

To perform the Ames test, the his⁻ strain is inoculated onto an agar plate lacking histidine (FIGURE 8.24). A well is cut in the middle of the agar, and the potential liver-modified carcinogen is added to the well (or a filter paper disk with the chemical is placed on the agar surface). The chemical diffuses into the agar. The plate is incubated for 24 hours to 48 hours. If bacterial colonies appear, one may conclude the agent mutated the bacterial his⁻ gene back to the wild type (his⁺); that is, **revertants** were generated that could again encode the enzyme needed for histidine synthesis. Because the agent is a mutagen, it is therefore a possible carcinogen in humans. If bacterial colonies fail to appear, one assumes that no mutation took place. However, it is possible the mutation did occur,

■ Screening test: A process for detecting mutants by examining numerous colonies.

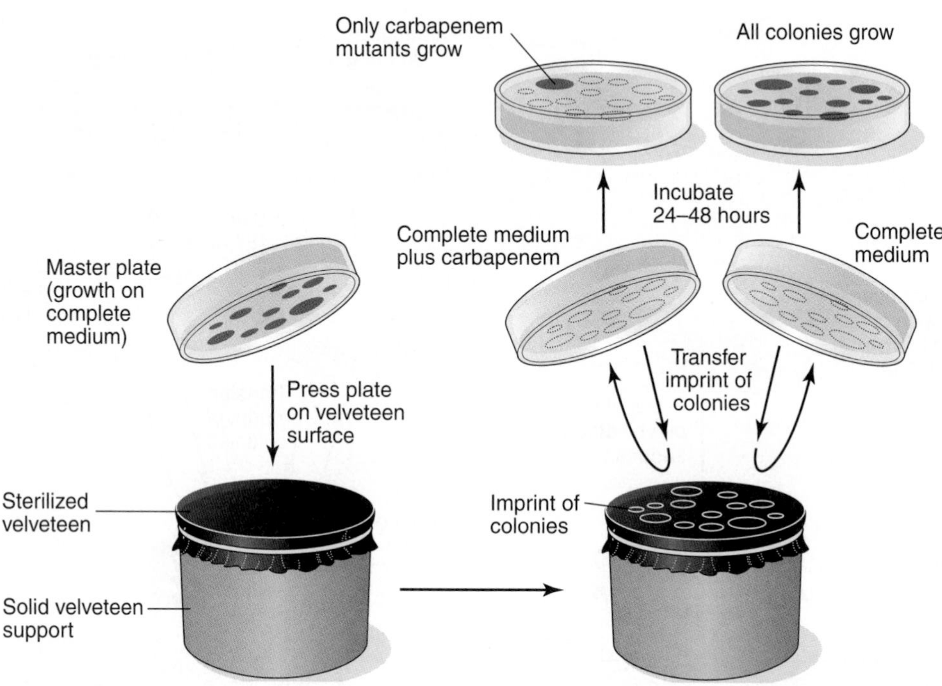

FIGURE 8.23 **Positive Selection of Mutants.** Positive selection plating techniques can be used to identify antibiotic resistant mutants. »» What do the "vacant spots" on the complete medium plus carbapenem represent?

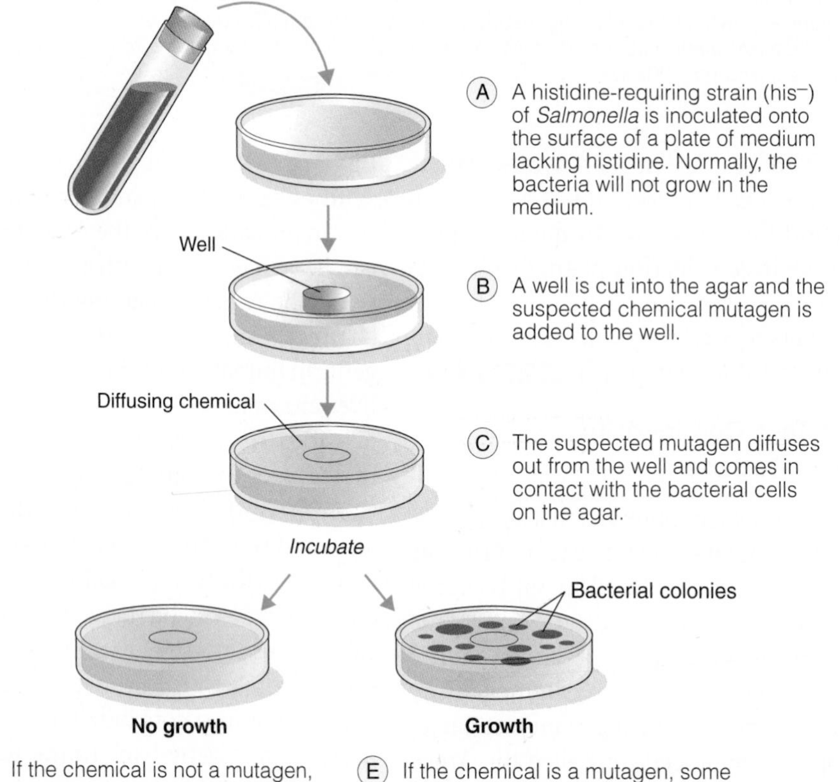

FIGURE 8.24 **Using the Ames Test.** The Ames test is a screening technique to identify mutants that reverted back to the wild type because of the presence of a mutagen. »» As a control, a plate with *Salmonella* (his−) but without mutagen is incubated. After 24–48 hours, a few colonies are seen on the plate. Explain this observation.

but was repaired by a DNA repair enzyme. This possibility has been overcome by using bacterial strains known to be inefficient at repairing errors.

CONCEPT AND REASONING CHECKS 5

a. How does negative selection differ from positive selection?

b. Just because a potential carcinogen generates revertants in *Salmonella,* does that always mean it is tumor causing in humans? Explain.

In conclusion, this chapter has discussed the basics of microbial genetics with an emphasis on the bacterial organisms. One of the important take-home lessons emphasizes, once again, the similarity between the process occurring in the Bacteria, Archaea, and Eukarya domains—this time at the genetic level.

The generation of mutations was a common way that members of all three domains of life could affect genetic change, providing the mutation was a beneficial one. But is this the only way to accomplish genetic variation? In the eukaryotic microbes, plants, and animals, most organisms can go through sexual reproduction, so gene recombination between parents is another mechanism for acquiring new traits or phenotypes. However, remember the prokaryotes do not undergo sexual reproduction, so genetic recombination between "parents" is not an option. Knowing the abilities of the prokaryotes, I bet you can guess that they must have another way to affect genetic change besides mutations. And boy do they—three ways to be exact for sharing genes with variations within these mechanisms. These recombination mechanisms are explored in the chapter on Gene Transfer, Genetic Engineering, and Genomics.

SUMMARY OF KEY CONCEPTS

8.1 The Hereditary Molecule in All Organisms Is DNA

1. The bacterial and archaeal **chromosomes** are circular, haploid structures in the cell's **nucleoid**.
2. The DNA of a microorganism's chromosome is **supercoiled** and folded into a series of **looped domains**. Each loop consists of 10,000 bases. (Fig. 8.4)
3. Many microbial cells may have one or more **plasmids**. These small closed loops of DNA carry information that can confer selective advantages (e.g., antibiotic resistance, protein toxins) to the cells.

8.2 DNA Replication Occurs Before a Cell Divides

4. DNA replicates by a **semiconservative** mechanism where each strand of the original DNA molecule acts as a template to synthesize a new strand. DNA replication starts with initiator proteins binding at the **replication origin**, forming two **replication "factories." DNA polymerase** moves along the strands inserting the correct DNA nucleotide to complementary bind with the template strand. (Fig. 8.5)
5. At each **replication fork**, one of the two strands is synthesized in a continuous fashion while the other strand is formed in a discontinuous fashion, forming **Okazaki fragments**, which are joined by a **DNA ligase**. (Fig. 8.6)

8.3 Gene Expression Produces RNA and Protein for Cell Function

6. **Transcription** occurs when the **RNA polymerase** binds to a **promoter** sequence on the DNA template strand. Various forms of RNA, including **mRNA, tRNA**, and **rRNA**, are transcribed from the DNA and are important in the translation process. (Fig. 8.8)
7. The **genetic code**, a series of three nucleotides to specify a polypeptide, is redundant because often more than one **codon** can specify an amino acid. (Table 8.3)
8. **Translation** occurs on ribosomes, which bring together mRNA and tRNAs. The ribosome "reads" the mRNA codons and inserts the correct tRNA to match codon and **anticodon**. A **polyribosome** is a string of ribosomes all translating the same mRNA. (Figs. 8.12, 8.13)
9. Many antibiotics interfere with protein synthesis by binding to RNA polymerase, or to either the small or large ribosomal subunit.
10. Different control factors influence protein synthesis. The best understood is the bacterial **operon** model where binding of a **repressor protein** to the operon represses transcription. (Fig. 8.14)
11. Research studies suggest that at least in some bacterial species, transcription and translation are spatially separated.

8.4 Mutations Are Permanent Changes in a Cell's DNA

12. **Mutations** are permanent changes in the cellular DNA. This can occur spontaneously in nature resulting from a replication error or the effects of natural radiation. In the laboratory, physical and chemical mutagens can induce mutations. (Figs. 8.17, 8.18)
13. Base pairs in the DNA can change in one of two ways. A base-pair substitution does not change the reading frame of an mRNA but can result in a **silent, missense**, or **nonsense mutation**. A **point mutation** also can occur from the loss or gain of a base pair. Such mutations change the reading frame and often lead to loss of protein function. (Fig. 8.19)
14. Replication errors or other damage done to the DNA often can be repaired. **Mismatch repair** replaces an incorrectly matched base pair with the correct pair. **Excision repair** removes a section of damaged (distorted) DNA and replaces it with the correctly paired bases. (Fig. 8.20)
15. **Transposable genetic elements** exist in many microbial cells. **Insertion sequences** (**ISs**) only carry information to move the sequences and insert them into another location in the DNA. **Transposons** are similar to IS except they contain additional genes, such as antibiotic resistance. (Fig. 8.21)

8.5 Techniques Exist for Identifying Mutants

16. Auxotrophic mutants can be identified by **negative selection** plating techniques. **Positive selection** can be used to identify mutants having certain attributes, such as antibiotic resistance. (Fig. 8.22)

17. The **Ames test** is a method of using an auxotrophic bacterial species to identify mutagens that may be carcinogens in humans. The test is based on the ability of a potential mutagen to revert an auxotrophic mutant to its prototrophic form. (Fig. 8.24)

CHAPTER SELF-TEST

For **STEPS A–D**, answers to even-numbered questions and problems can be found in **Appendix C** on the Student Companion Website at **http://microbiology.jbpub.com/10e**. In addition, the site features eLearning, an online review area that provides quizzes and other tools to help you study for your class. You can also follow useful links for in-depth information, read more MicroFocus stories, or just find out the latest microbiology news.

STEP A: REVIEW OF FACTS AND TERMS

Multiple Choice

Read each question carefully, then select the ***one*** answer that best fits the question or statement.

1. Which one of the following statements is NOT true of the bacterial chromosome?
   A. It is located in the nucleoid.
   B. It usually is a single, circular molecule.
   C. Some genes are dominant to others.
   D. It usually is haploid.
2. DNA compaction involves _____.
   A. a twisting and packing of the DNA
   B. supercoiling
   C. the formation of looped domains
   D. All the above (**A–C**) are correct.
3. Plasmids are _____.
   A. another name for transposons
   B. accessory genetic information
   C. domains within a chromosome
   D. daughter chromosomes
4. The enzyme _____ adds complementary bases to the DNA template strand during replication.
   A. ligase
   B. helicase
   C. DNA polymerase III
   D. RNA polymerase
5. At a chromosome replication fork, the lagging strand consists of _____ that are joined by _____.
   A. RNA sequences; DNA ligase
   B. Okazaki fragments; RNA polymerase
   C. RNA sequences; ribosomes
   D. Okazaki fragments; DNA ligase
6. In a eukaryotic microbe, those sections of a primary RNA transcript that will NOT be translated are called _____.
   A. introns
   B. anticodons
   C. "jumping genes"
   D. exons
7. Which one of the following codons would terminate translation?
   A. AUG
   B. UUU
   C. UAA
   D. UGG
8. The translation of a mRNA by multiple ribosomes is called _____ formation.
   A. Okazaki
   B. polysome
   C. plasmid
   D. transposon
9. If an antibiotic binds to a 50S subunit, what cellular process will be inhibited?
   A. DNA replication
   B. Intron excision
   C. Translation
   D. Transcription
10. Which one of the following is NOT part of an operon?
    A. Regulatory gene
    B. Operator
    C. Promoter
    D. Structural genes
11. Being compartmentalized, bacterial RNA polymerases are localized in the _____ and ribosomes are found _____.
    A. nucleoid; at the nucleoid periphery
    B. cytosol; in the cytosol
    C. cytosol; at the cell poles
    D. nucleoid; in the nucleoid
12. Spontaneous mutations could arise from _____.
    A. DNA replication errors
    B. atmospheric radiation
    C. addition of insertion sequences
    D. All the above (**A–C**) are correct.
13. Which one of the following could NOT cause a change in the mRNA "reading frame"?
    A. Insertion sequence
    B. Base-pair substitution
    C. Base addition
    D. Base deletion
14. Nucleotide excision repair would correct DNA damage caused by _____.
    A. antibiotics
    B. UV light
    C. transcription
    D. a DNA replication error
15. Transposons _____.
    A. were first discovered by Watson and Crick
    B. are smaller than insertion sequences
    C. are examples of plasmids
    D. may have information for antibiotic resistance
16. Nutritional mutants are referred to as _____.
    A. prototrophs
    B. wild type
    C. revertants
    D. auxotrophs
17. The Ames test is used to _____.
    A. identify potential human carcinogens
    B. discover auxotrophic mutants
    C. find pathogenic bacterial species
    D. identify antibiotic resistant mutants

STEP B: CONCEPT REVIEW

18. Describe the contents of the bacterial **nucleoid** and summarize the packing of the DNA and **chromosome** within the nucleoid. (**Key Concept 1**)
19. Assess the role of **plasmids** in microbial cells. (**Key Concept 1**)
20. Identify and explain the events of the three phases of **DNA replication.** (**Key Concept 2**)
21. Describe the role of **RNA polymerase** in the **transcription** process. (**Key Concept 3**)
22. Label the sequences composing a bacterial **operon** and compare the functions of each sequence to the transcription process. (**Key Concept 3**)
23. Construct a concept map for **translation** using the following terms. (**Key Concept 3**)

Amino acid	Ribosome
Chain elongation	Sense codons
Chain initiation	Small subunit
Chain termination	Start codon
Large subunit	Stop codon
mRNA	Termination factors
Polypeptide	tRNAs

24. Compare and contrast **spontaneous** and **induced mutations**, and differentiate between physical and chemical **mutagens.** (**Key Concept 4**)
25. Explain how **mismatch repair** works and describe how cells use **excision repair** to correct UV-induced mutations. (**Key Concept 4**)
26. Compare and contrast **negative** and **positive selection** plating techniques for identifying mutants. (**Key Concept 5**)
27. Evaluate the use of the **Ames test** to identify chemicals that are potential **carcinogens** in humans. (**Key Concept 5**)

STEP C: APPLICATIONS AND PROBLEMS

Answer the following questions that pertain to (1) transcription and translation, and (2) mutations. Use the genetic code (Table 8.3).

28. The following base sequence is a complete polynucleotide made in a bacterial cell.

AUGGCGAUAGUUAAACCCGGAGGGUGA

With this sequence, answer the following questions.

A. Provide the sequence of nucleotide bases found in the inactive DNA strand of the gene.

B. How many codons will be transcribed in the mRNA made from the template DNA strand?

C. How many amino acids are coded by the mRNA made and what are the specific amino acids?

D. Why isn't the number of codons in the template DNA the same as the number of amino acids in the polypeptide?

29. Use the base sequence to answer the following questions about mutations.

TACACGATGGTTTTGAAGTTACGTATT

A. Is the sequence above a single strand of DNA or RNA? Why?

B. Using the sequence above, show the translation result if a mutation results in a C replacing the T at base 12 from the left end of the sequence. Is this an example of a silent, missense, or nonsense mutation?

C. Using the sequence above, show the translation result if a mutation results in an A inserted between the T (base 12) and the T (base 13) from the left end of the sequence. Is this an example of a silent, missense, or nonsense mutation?

30. You are interested in identifying mutants of *E. coli;* specifically mutations occurring in the promoter and operator regions of the *lac* operon such that their respective molecules cannot bind to the region. You have agar plates containing lactose or glucose as the energy source. How can plating the potential mutants on these media help you identify (a) promoter region mutants and (b) operator region mutants?
31. A chemical is tested with the Ames test to see if the chemical is mutagenic and therefore possibly a tumor-causing chemical in humans. On the test plate containing the chemical, no his^+ colonies are seen near the central well. However, many colonies are growing some distance from the well. If these colonies truly represent his^+ colonies, why are there no colonies closer to the central well?
32. Bioremediation is a process that uses bacteria to degrade environmental pollutants. You want to use one of these organisms to clean up a toxic waste site (biochemical refinery) that contains benzene in the soil around the refinery. Benzene (molecular formula $= C_6H_6$) is a major contaminant found around many of these chemical refineries. You have a culture of bacterial cells growing on a culture plate that were derived from a soil sample from the refinery area. You also have a supply of benzene. Explain how you could visually identify chemoheterotrophic bacterial colonies on agar that use benzene as their sole carbon and energy source for metabolism.
33. Suppose you now have such benzene colonies growing on agar. However, it also has been discovered that material containing radioactive phosphorus (^{32}P) is in the soil around the refinery and this radioactive material can kill bacterial organisms. Because you want to identify colonies that might be sensitive to ^{32}P, you obtain a sample of the material containing ^{32}P. Explain how you could visually determine if any of your colonies are sensitive to ^{32}P.

STEP D: QUESTIONS FOR THOUGHT AND DISCUSSION

34. The author of a general biology textbook writes in reference to the development of antibiotic resistance, "The speed at which bacteria reproduce ensures that sooner or later a mutant bacterium will appear that is able to resist the poison." How might this mutant bacterial cell appear? Do you agree with the statement? Does this bode ill for the future use of antibiotics?
35. Many viruses have double-stranded DNA as their genetic information while many others have single-stranded RNA as the genetic material. Which group of viruses do you believe is more likely to efficiently repair its genetic material? Explain.
36. Some scientists suggest that mutation is the single most important event in evolution. Do you agree? Why or why not?

CHAPTER PREVIEW

9.1 Bacterial Cells Can Recombine Genes in Several Ways

Investigating the Microbial World 9: Spontaneous Generation?

Textbook Case 9: Vancomycin-Resistant *Staphylococcus aureus*

9.2 Genetic Engineering Involves the Deliberate Transfer of Genes Between Organisms

MicroInquiry 9: Molecular Cloning of a Human Gene into Bacterial Cells

9.3 Microbial Genomics Studies Genes at the Single Cell to Community Levels

Image courtesy of Dr. Fred Murphy/CDC.

Gene Transfer, Genetic Engineering, and Genomics

Genetic engineering is the most powerful and awesome skill acquired by man since the splitting of the atom.
—The editors of *Time* magazine describing the potential for genetic engineering

Medicinal microbe's genome sequenced. How often have we read in the newspaper or heard from the news media about an organism's genes (genome) being sequenced or its DNA being mapped? It must be significant, right? After all, it made the news! But what is the underlying significance of such sequencing? Here is a good example.

In late spring of 2003, a group of British scientists announced they had mapped the genome of a very important bacterial species, *Streptomyces coelicolor*. This is a gram-positive organism commonly found in the soil. The mapping project began in 1997 and took 6 years to complete partly because the organism's genome is one of the largest ever sequenced. It has 8.6 million base pairs and some 7,825 genes. On completion of the sequencing, one of the scientists on the project said, "*It is a fabulous resource for scientists.*" Why?

Well, here is where microbial genomics shows its power. *S. coelicolor* belongs to the phylum Actinobacteria, which is responsible for producing over 65% of the naturally known antibiotics used today. This includes tetracycline and erythromycin. By analyzing the genome of *S. coelicolor* and other *Streptomyces* species, additional metabolic pathways may be discovered for the production of other, yet unidentified and perhaps novel antibiotics. In fact, the researchers identified 18 gene clusters they suspect are involved with the production of antibiotics. If correct, knowing the genome and its organization might allow scientists to transform the organism into an "antibiotic factory" and add to the dwindling armada of usable antibiotics to which pathogens are not yet resistant (FIGURE 9.1). Using genetic engineering techniques, they could rearrange gene clusters and perhaps produce even more useful and potent antibiotics than naturally possible.

One example illustrating the need for newer antibiotics concerns the recent rise in antimicrobial resistance among healthcare-associated pathogens. According to the National Nosocomial Infections Surveillance System, a number of hospital or healthcare pathogens, such as *Staphylococcus aureus, Pseudomonas aeruginosa, Acinetobacter baumannii,* and *Clostridium difficile,* are showing increased resistance to many of the established and newest antibiotics (TABLE 9.1). As a consequence, doctors are finding it harder to treat bacterial infections caused by these **multidrug-resistant** (**MDR**) species. So, new and unique antibiotics might be useful for treating these drug resistant infections.

TABLE

9.1 Healthcare-Associated Infections and Deaths—United States, 2008*

Number of healthcare infections	1.7 million
Number of healthcare deaths	99,000
Number of healthcare multidrug-resistant (MDR) infections	272,000

*Data from: Kallen, A. J. et al. (2010). *Journal of the American Medical Association,* **304** (6): 641–648.

FIGURE 9.1 ***Streptomyces coelicolor* Colonies.** In this photograph of *S. coelicolor,* colonies growing on agar are secreting droplets of liquid containing antibiotics. Courtesy of John Innes Centre www.jic.ac.uk. »» What is the advantage to the bacterial cells to secrete a chemical with antibiotic properties?

But there is more. *S. coelicolor* is a close relative of the tuberculosis, leprosy, and diphtheria bacilli. By comparing genomes, scientists hope to learn why *S. coelicolor* is not pathogenic, while the other three are pathogens. What is different about their genomes might be important in understanding infectious nature of their relatives and perhaps even designing new antibiotics through genetic engineering to attack these pathogens.

Genetic engineering involves the manipulating of genes in organisms or between organisms in order to introduce new characteristics into the recipient to either produce a useful product or to actually generate **genetically modified organisms** (**GMOs**). **Genomics** is the study of an organism's genome; its study has the potential of offering new therapeutic methods for the treatment of several human diseases. So, this chapter addresses more than simply research techniques; importantly, we discuss how their applications can have far-reaching consequences for all of us.

Before we can explore these topics, we need to understand the natural mechanism by which bacterial cells transfer genetic information from one to another. Its understanding provides a unique perspective on microbial evolution, ecology, and molecular biology while providing insights for the techniques of genetic engineering and the field of microbial genomics.

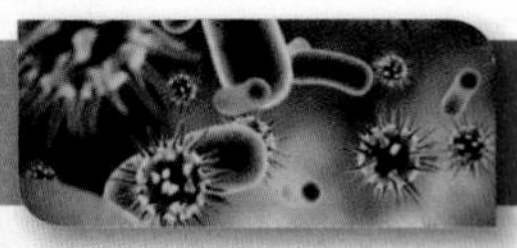

Chapter Challenge

The Human Genome Project (HGP) was an international scientific investigation that began in 1990. It had as its primary goal to determine the sequence of nucleobases that compose human DNA and then to understand the genetic makeup of the human species. A working draft of the sequence was announced in 2000 and published in 2001. In the end, it was discovered that the human genome consists of some 23,000 genes. However, the work on interpretation of the genome sequence data is still being analyzed and studied. It is anticipated that detailed knowledge of the human genome will provide new medicines to help fight human diseases and produce a more healthy population. Some advances have been made in medicine toward producing therapeutics and diagnostics related to human health and toward better understanding the origins of the human genome. In all cases, microbes are playing an important role. Are there microbes in our genome? Let's find out.

KEY CONCEPT 9.1 Bacterial Cells Can Recombine Genes in Several Ways

Traditionally, when one thinks about the inheritance of genetic information, one envisions genes passed from parent to offspring. However, imagine being able to directly transfer genes between members of your own family, or between you and one of your classmates. These mechanisms of remixing genes, called **genetic recombination**, occur in many bacterial and archaeal species and provide tremendous genetic and evolutionary flexibility. We will focus on bacterial species because they have been most extensively studied.

Genetic Information Can Be Transferred Vertically and Horizontally

Mutations are one of the ways by which the genetic material in a cell can be permanently altered. Because the permanent change occurred in the parent cell, all future generations derived by binary fission from the parent also will have the mutation. This form of asexual genetic transfer is referred to as **vertical gene transfer** (FIGURE 9.2A).

The Bacteria lack sexual reproduction as a mechanism for genetic diversity. However, they still possess ways by which genetic recombination and most evolutionary change can arise. This is through the process of **horizontal gene transfer** (**HGT**), a type of genetic recombination that involves the lateral intercellular transfer of DNA from a donor cell to a recipient cell (FIGURE 9.2B). If, for example, the recipient cell receives a piece of DNA for antibiotic resistance from the donor, the new DNA pairs with a complementary region of recipient DNA and replaces it. In this case, genetic recombination has made an antibiotic sensitive cell into a resistant one. In fact, the increasing resistance to antibiotics by MDR pathogens is one example of the prevalence of HGT.

Website Animation: Transformation

HGT can occur between members of the same or related species, or between species phylogenetically distant. In any case, three distinctive mechanisms mediate the horizontal transfer of DNA between bacterial cells: transformation, conjugation, and transduction. All three processes involve a similar four steps. The donor DNA must be:

1. Readied for transfer.
2. Transferred to the recipient cell.
3. Taken up successfully by the recipient cell.
4. Recombined in a stable state in the recipient.

Let's now look at each of the three recombination mechanisms involving HGT.

Transformation Is the Uptake and Expression of DNA in a Recipient Cell

Read through this chapter's Investigating the Microbial World 9. It presents one of the first glimpses into DNA as being the genetic material in cells and that such information can be transferred between cells or organisms.

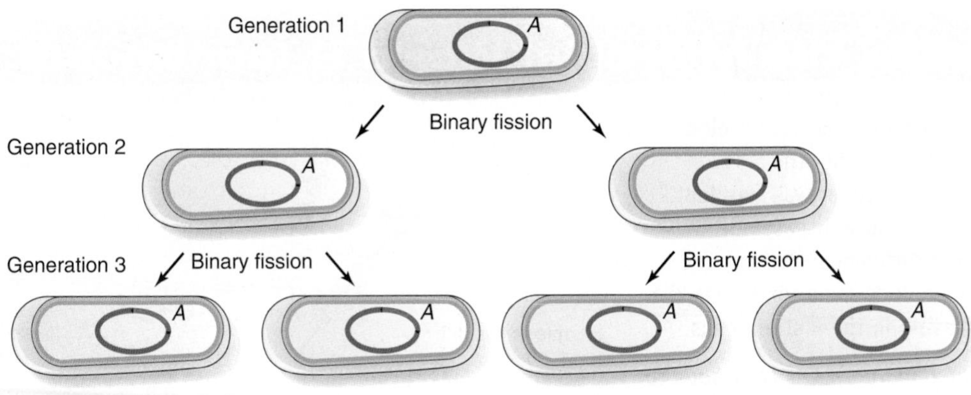

(A) Vertical gene transfer

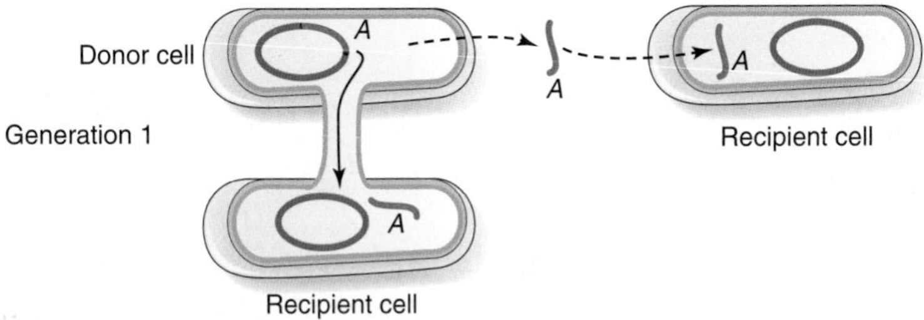

(B) Horizontal gene transfer

FIGURE 9.2 **Gene Transfer Mechanisms.** Genes can be transferred between cells in two ways. **(A)** In vertical transfer, a cell undergoes binary fission and the daughter cells of the next generation contain the identical genes found in the parent. **(B)** In horizontal transfer, genes are transferred by various mechanisms to other individual cells. »» Determine which transfer mechanism (vertical or horizontal) provides the potential for the most genetic diversity.

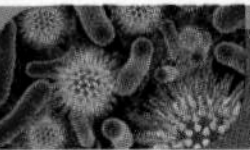

Investigating the Microbial World 9

Spontaneous Generation?

In the 1920s, Frederick Griffith, an English physician, was studying the pathogen *Streptococcus pneumoniae,* or pneumococcus, in the hope of developing a vaccine against the devastating pneumonia that the bacterial species produced in humans.

■ **OBSERVATION:** Griffith discovered that when he first isolated the pneumococcus from the lungs of mice with pneumonia, the bacterial colonies that grew on the agar plates had a glistening, mucus-looking appearance and the diplococcus pairs had a smooth capsule (the S strain). When he transferred these colonies repeatedly from one agar plate to another, however, mutant colonies would appear that were much smaller, drier-looking, and the diplococci were rough-looking in appearance because they lacked a capsule (the R strain). When Griffith injected mice with the S strain they contracted pneumonia, and S strain cells could be reisolated from the infected mice (see figure part A).

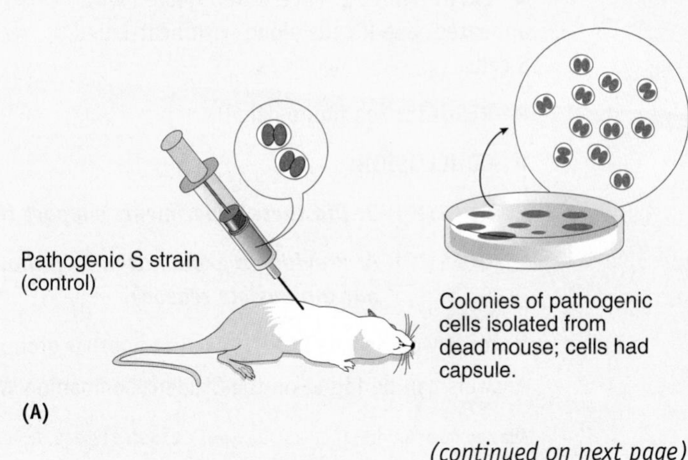

(A)

(continued on next page)

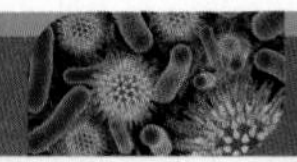

Investigating the Microbial World 9 (continued)

Mice infected with the R strain did not develop the disease (see figure part B). So, the S strain was capable of causing disease, or was "virulent," while the R strain did not cause disease and was "avirulent." What is the difference? Perhaps the S strain cells actively produce some toxin that kills the mice, which is missing in the R strain cells.

■ **QUESTION:** ***Do the pneumococcus cells have to be alive to cause disease?***

■ **HYPOTHESIS:** Cells have to be alive to cause disease. If so, then heat-killing S strain cells will not produce pneumonia after injection into mice.

■ **EXPERIMENT 1:** Griffith prepared cultures of the S strain cells and heated the material for 1 hour. He then injected the heat-treated cells into mice.

■ **RESULTS:** See figure part C.

■ **CONCLUSION:**

QUESTION ***1: What do these results suggest about the S strain cells?***

QUESTION ***2: Griffith had noticed that in some of his human patients he could isolate more than one form of the pneumococcus from sputum samples. What happens if a mixed culture of pneumococcus strains is injected into mice?***

■ **HYPOTHESIS:** Injecting both avirulent heat-killed S strain cells along with living, unheated R strain cells will produce no disease. If so, then the mice should survive and not contract pneumonia.

■ **EXPERIMENT 2:** Mice were injected with unheated, live R cells along with heat-killed, S cells.

■ **RESULTS:** See figure part D.

■ **CONCLUSION:**

QUESTION ***3: Did these experiments support the hypothesis? Explain.***

QUESTION ***4: Provide an explanation as to how living S cells could be produced (spontaneous generation is not a plausible reason).***

Harmless R strain (control)

Colonies of harmless cells from healthy mouse; cells had no capsule.

(B)

Heat-killed S strain (control)

No colonies isolated from healthy mouse.

(C)

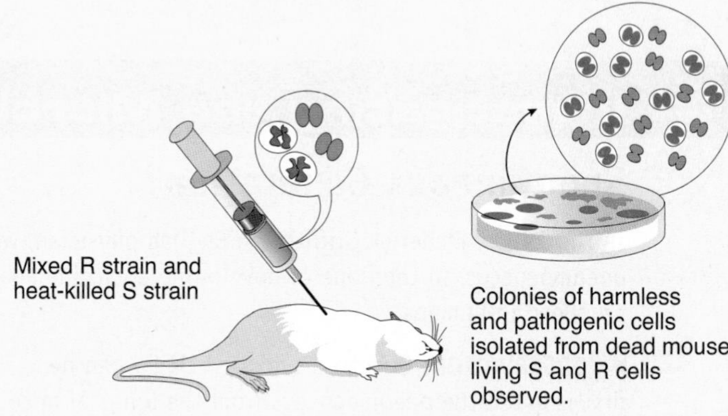

(D)

Adapted from: Griffith, F. (1928). *J. Hygiene* 27 (2): 113–159.

Note: It wouldn't be until 1944 that another group of researchers would identify the transforming agent.

Answers can be found on the Student Companion Website in **Appendix D**.

Adapted from: Griffith, F. (1928). *J. Hygiene* **27** (2): 113–159.

The experiments you studied illustrate one form of HGT called **transformation**, which involves the uptake of a small piece of DNA from the surrounding environment and the expression of that genetic information in a recipient cell.

Microbiologists regard transformation as an important genetic recombination method even though it is an inefficient, random process that takes place in less than 1% of a cell population.

The ability of some bacterial genera, like *Bacillus*, *Streptococcus*, and *Staphylococcus* to undergo transformation depends on their **competence**, which refers to the ability of a recipient cell to take up extracellular DNA from the environment. In such cases, transformation starts when a DNA fragment, composed of about 10 to 20 genes, binds to proteins at the cell pole (FIGURE 9.3). Internalization of the DNA is an ATP-dependent process and one strand of the DNA is degraded as the other strand is taken into the cell. Such single-stranded DNA is then recombined with a homologous region of the recipient's chromosome.

In Griffith's pneumococci experiments, the live R strain cells acquired the genes for capsule formation from the dead S strains, which allowed the organism to avoid body defenses and thus cause disease. Microbiologists also have demonstrated that when mildly pathogenic bacterial strains take up DNA from other mildly pathogenic strains, there is a cumulative effect, and the recipient often becomes more virulent. Observations such as these may help explain why highly pathogenic bacterial strains appear from time to time and how transformation can contribute to the dispersal of antibiotic-resistance genes.

Extracellular DNA can be quite prevalent. MicroFocus 9.1 identifies a massive source of such "dead" DNA.

Conjugation Involves Cell-to-Cell Contact

In the recombination process called **conjugation**, two bacterial cells come together and the donor cell directly transfers plasmid DNA to the recipient cell. This process was first observed in 1946 by Joshua Lederberg and Edward Tatum in a series of experiments with *Escherichia coli*. Lederberg and Tatum mixed two different strains of *E. coli* and found genetic traits could be transferred if cell contact occurred.

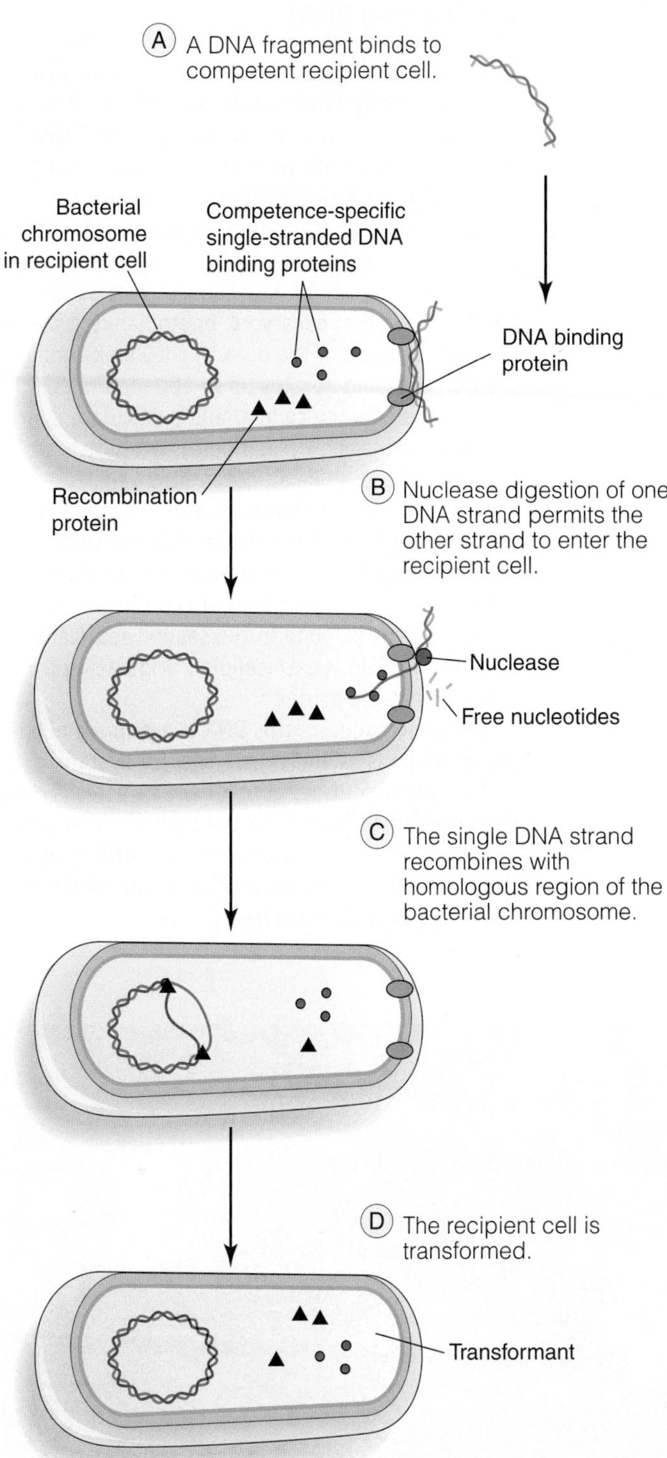

FIGURE 9.3 **Transformation in a Gram-Positive Cell.** Transformation (A–D) is the process in which DNA from the environment binds to a competent recipient cell, passes into the recipient, and incorporates into the recipient's chromosome. »» Using this diagram, illustrate how Griffith's experiment with mixed heat-killed S strain and live R stain pneumococci produced live S strain encapsulated cells.

MICROFOCUS 9.1: Environmental Microbiology

It's Snowing DNA!

Throughout this text, we have observed on several occasions the massive populations of microbes that live and thrive in the world's oceans. These marine environments are home to a huge variety of Bacteria and Archaea. So what happens to all the genetic material in these organisms when they die? Where does the DNA go?

A decade's worth of data suggest that as cyanobacteria and other eukaryotic phytoplankton die in the surface layers of the oceans, they sink through the water column at rates up to 100 meters per day. This free-falling cellular and particulate debris is called marine snow (see figure). Often it is so thick that it looks like an ocean blizzard!

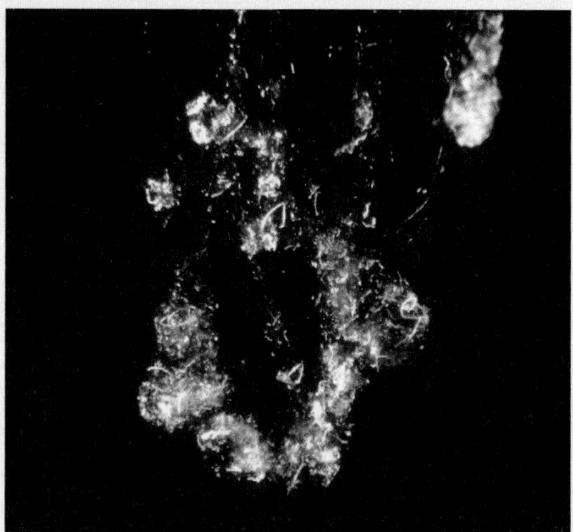

Marine snow represents microbial cellular debris. Courtesy of Chris Gotschalk, University of California, Santa Barbara.

In 2005, Roberto Danovaro and his colleague Antonio Dell'Anno of the Marine Science Institute of the University of Ancona in Italy published a paper suggesting about 65% of the DNA in the world's oceans is found in the seabed sediments; 90% of that DNA is extracellular, originating from organisms in the marine snow.

Danovaro suggests this DNA is a primary nutritional source for carbon, nitrogen, and phosphorus for seabed organisms and is essential for sustaining the microbial communities on and in the sediments.

But what about horizontal gene transfer? Could this DNA represent a source of genetic diversity through transformation for the bacterial and archaeal populations on the ocean bottom? Danovaro hypothesizes that most of the snowy DNA is of eukaryotic origin. Still, with all the marine cyanobacteria, why couldn't they be making a substantial contribution? And if so, is transformation a significant phenomenon on the sea beds?

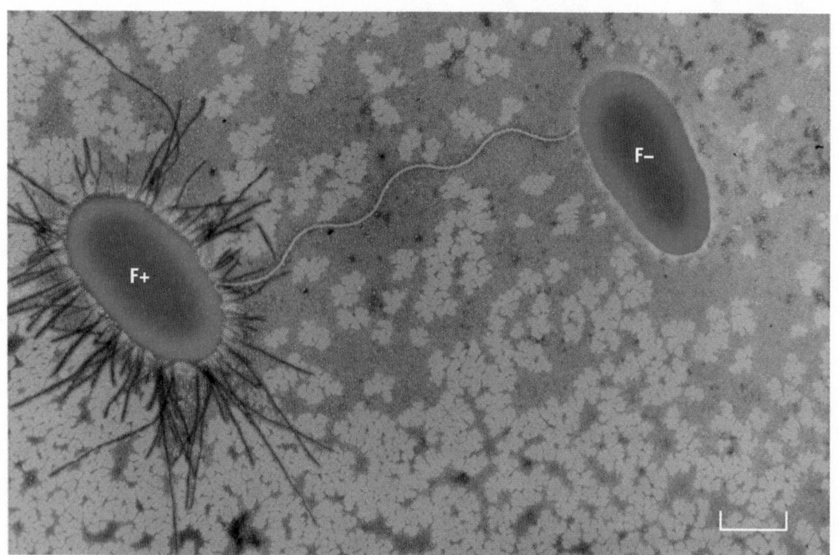

FIGURE 9.4 **Bacterial Conjugation in *E. coli*.** In this false-color transmission electron micrograph, the donor (F^+ cell) (left) has produced a conjugation pilus that has contacted the recipient (F^- cell). Contraction of the pilus will bring donor and recipient close together. (Bar = 0.5 μm.) © Dennis Kunkel Microscopy, Inc./Visuals Unlimited/Corbis. »» What are the other structures projecting from the F^+ cell?

In gram-negative species, like *E. coli*, the process of conjugation requires a special conjugation apparatus called the **conjugation pilus** (FIGURE 9.4). For cell-to-cell contact, the donor cell, designated F^+, produces the conjugation pilus that contacts the recipient cell, known as an F^- cell. The pilus shortens to bring the two cells close together and a conjugation bridge connects the two cytoplasms. The donor cell is called F^+ because it contains an **F factor**, which is a transmissible plasmid containing about 100 genes, most of which are associated with plasmid DNA replication and production of the conjugation pilus. The F^- cell lacks the plasmid.

Following conjugation bridge formation (FIGURE 9.5A), the F factor DNA replicates by a **rolling-circle mechanism**; one strand of plasmid DNA remains in a closed loop, while an enzyme nicks the other strand at a point called the **origin of transfer (*oriT*)**. This single-stranded DNA then

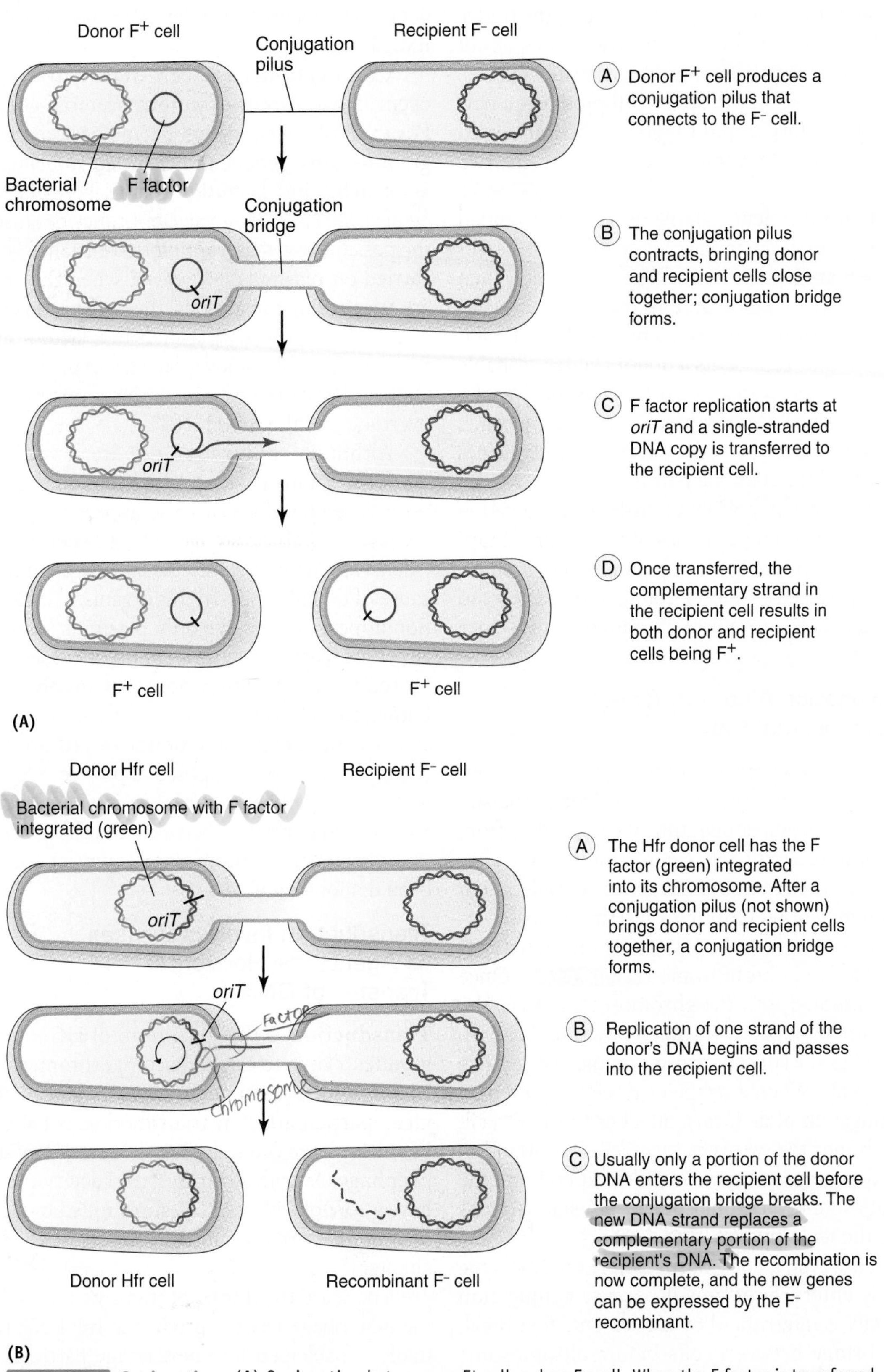

FIGURE 9.5 Conjugation. (**A**) Conjugation between an F+ cell and an F− cell. When the F factor is transferred from a donor (F+) cell to a recipient (F−) cell, the F− cell becomes an F+ cell as a result of the presence of an F factor. (**B**) Conjugation between an Hfr and an F− cell allows for the transfer of some chromosomal and plasmid DNA from donor to recipient cell. »» Propose a hypothesis to explain why only a single-stranded DNA molecule is transferred across the conjugation pilus.

"rolls off" the loop and passes through the bridge to the recipient cell; the transfer takes about 5 minutes. As the horizontal transfer occurs, DNA synthesis in the donor cell produces a new complementary strand to replace the transferred strand. Once DNA transfer is complete, the two cells separate.

In the recipient cell, the new single-stranded DNA serves as a template for synthesis of a complementary polynucleotide strand, which then circularizes to reform an F factor. This completes the conversion of the recipient from F^- to F^+ and this cell now represents a donor cell (F^+) capable of conjugating with another F^- recipient. Transfer of the F factor does not involve the chromosome; therefore, the recipient does not acquire new genes other than those on the F factor.

The efficiency of DNA transfer by conjugation shows that conjugative plasmids can spread rapidly, converting a whole population into plasmid-containing cells. Indeed, conjugation appears to be the major mechanism for antibiotic resistance transfer.

Conjugation Also Can Transfer Chromosomal DNA

Bacterial species also can undergo a type of conjugation that accounts for the horizontal transfer of some chromosomal and plasmid DNA from donor to recipient cell. Cells exhibiting the ability to donate chromosomal genes are called **high frequency of recombination** (**Hfr**) strains.

In Hfr strains, the F factor is integrated into the bacterial chromosome (FIGURE 9.5B). Once incorporated into the chromosomal DNA, the F factor no longer controls its own replication. However, the Hfr cell triggers conjugation just like an F^+ cell. When a recipient F^- cell is present, a conjugation pilus forms, attaches to the F^- cell, and brings the two cells together. One strand of the donor chromosome is nicked at *oriT* and the single-stranded chromosomal DNA starts to pass into the recipient cell.

However, a copy of the whole chromosome rarely enters the recipient because conjugation usually is interrupted by movements that break the bridge between cells before chromosome transfer is complete. The new DNA replaces a similar piece of the recipient's chromosome. Thus, the F^- cell usually remains F^-, although it now has a few recombined donor chromosomal genes. The recipient is referred to as a **recombinant F^- cell**.

Conjugation has been demonstrated to occur between cells of various bacterial genera. For example, conjugation occurs between such gram-negative genera as *Escherichia* and *Shigella, Salmonella* and *Serratia,* and *Escherichia* and *Salmonella.* HGT has great significance because of the possible transfer of antibiotic-resistance genes carried on plasmids. Moreover, when the genes are attached to transposons, the transposons may "jump" from ordinary plasmids to F factors, after which transfer by conjugation may occur. TEXTBOOK CASE 9 describes one case with serious medical overtones resulting from HGT.

Although conjugation pili are found only on some gram-negative bacteria, some gram-positive bacterial species also appear capable of conjugation. Microbiologists have experimented extensively with *Streptococcus mutans,* a common cause of dental caries. In this organism, conjugation appears to involve only plasmids, particularly those carrying genes for antibiotic resistance. Moreover, conjugation does not involve pili. Rather, the recipient cell apparently secretes substances encouraging the donor cell to produce "clumping factors" composed of protein. The factors bring together (clump) the donor and recipient cell, and pores form between the cells to permit plasmid transfer. Chromosomal transfer has not been demonstrated.

Transduction Involves Viruses as Agents for Horizontal Transfer of DNA

Transduction is the third form of HGT and it requires a virus to carry a bacterial chromosomal DNA fragment from donor to recipient cell. The virus participating in transduction is called a **bacteriophage** (literally "bacteria eater") or simply **phage**. As with all viruses, the bacteriophages have a core of DNA or RNA surrounded by a coat of protein (viruses will be covered in another chapter).

There are two forms of transduction. When the new phage DNA is produced, the DNA normally is packaged into new phage particles. In **generalized transduction** a random fragment of bacterial DNA may accidentally be packaged into a phage rather than phage DNA (FIGURE 9.6A). Such phages are fully formed though and are

Textbook CASE 9

Vancomycin-Resistant *Staphylococcus aureus*

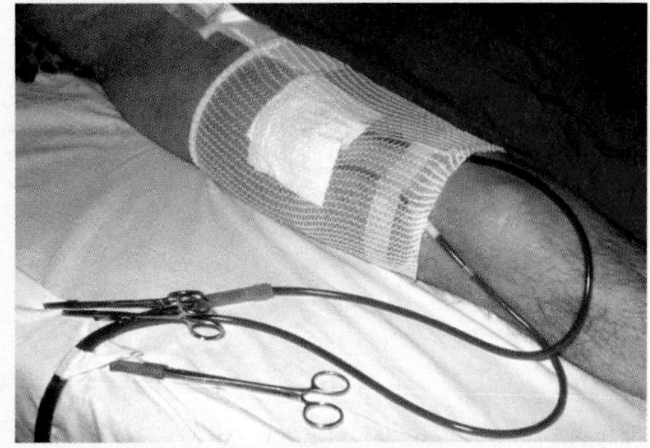

An arteriovenous graft for hemodialysis. © Medical-on-Line/Alamy Images.

1 In June 2002, a 40-year-old Michigan resident with diabetes, peripheral vascular disease, and chronic renal failure developed a suspected catheter exit site infection. The clinical microbiology lab used a swab from the catheter exit site to isolate vancomycin-resistant *Staphylococcus aureus* (VRSA).

2 In April 2001, the patient had been treated for chronic foot ulcerations with multiple courses of antimicrobial therapy, some of which included the antibiotic vancomycin. In April 2002, the patient underwent amputation of a gangrenous toe and subsequently developed a methicillin-resistant *S. aureus* (MRSA) blood infection that resulted when hemodialysis was performed. In the form of hemodialysis used, an artificial vessel was used to join the artery and vein (called an arteriovenous graft; see figure), which often is subject to infection. The infection was treated with the antibiotics vancomycin and rifampin, and the infected graft was removed.

3 With the identification of VRSA from the swab, cultures from the exit site and catheter tip were made and subsequently grew *S. aureus* resistant to oxacillin and vancomycin.

4 A week after the patient's catheter was removed, the exit site appeared healed; however, the patient's chronic foot ulcer appeared infected. VRSA, vancomycin-resistant *Enterococcus faecalis* (VRE), and *Klebsiella oxytoca* also were recovered from a culture of the ulcer. Swab cultures of the patient's healed catheter exit site and anterior nares did not grow VRSA. The ulcer was cleaned of dead and contaminated tissue. The patient was urged to maintain aggressive wound care and, as an outpatient, put on systemic antimicrobial therapy with the sulfa drug trimethoprim/sulfamethoxazole (Bactrim).

5 The VRSA isolate recovered from the catheter exit site was identified initially at a local hospital clinical microbiology laboratory and was confirmed by the Michigan Department of Community Health and the Centers for Disease Control and Prevention (CDC).

6 Further molecular analysis indicated the VRSA isolate contained the *vanA* vancomycin resistance gene typically found in enterococci.

7 Epidemiologic and laboratory investigations were undertaken to assess the risk for transmission of VRSA to other patients, healthcare workers, and close family and other contacts. No VRSA transmission was identified.

8 Infection-control practices in the local dialysis center were assessed; all healthcare workers followed standard precautions consistent with CDC guidelines.

Questions:

(Answers can be found on the Student Companion Website in **Appendix D.**)

A. This report describes the first documented case of infection caused by vancomycin-resistant *S. aureus* (VRSA) in a patient in the United States. Why was this patient so susceptible to infection with *S. aureus?*

B. Because vancomycin resistance determinants had not previously been identified in clinical isolates of *S. aureus* in the United States, how did the *vanA* gene get "transferred" into *S. aureus* in this patient?

C. Why was a culture swab from the patient's anterior nares tested for VRSA?

D. Why was the patient put on a sulfa drug?

E. Besides standard precautions, what other procedures should be in place to prevent transmission of antimicrobial resistant microorganisms in healthcare settings?

For additional information see www.cdc.gov/mmwr/preview/mmwrhtml/mm5126a1.htm.

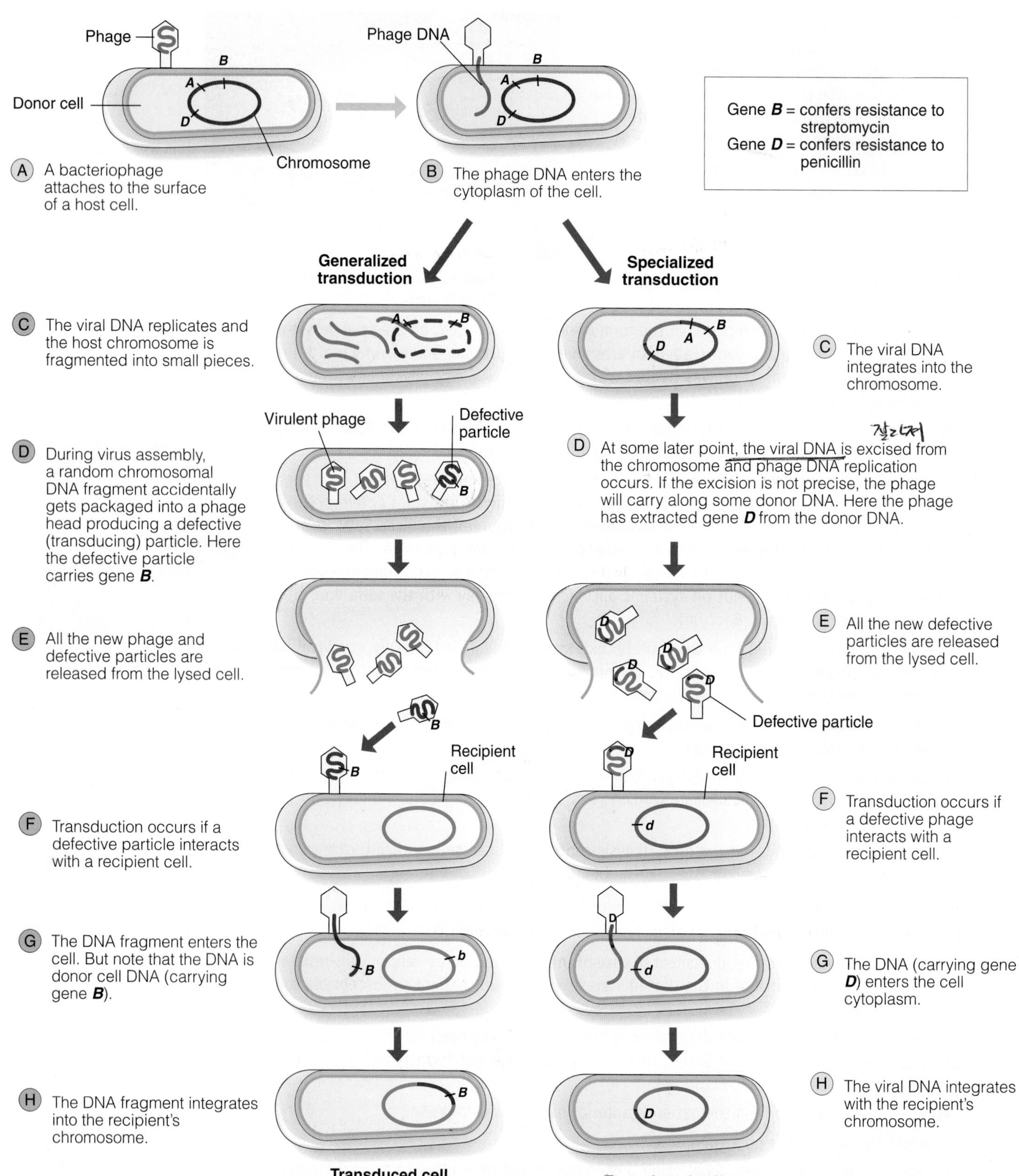

FIGURE 9.6 Generalized and Specialized Transduction. Phage can transfer bacterial genes by generalized transduction (can involve any bacterial gene) or specialized transduction (can involve genes from a specialized region). »» Using genes ***B*** and ***D*** for antibiotic resistance, which form of transduction, generalized or specialized, would likely spread antibiotic resistance faster? Explain.

MICROFOCUS 9.2: Environmental Microbiology
Gene Swapping in the World's Oceans

Many of us are familiar with the accounts of microorganisms in and around us, but we are less familiar with the massive numbers of microbes in the world's oceans. For example, microbial ecologists estimate there are an estimated 10^{29} Bacteria and Archaea in the world's oceans. At the Axial Seamount, a Pacific deep-sea volcano, researchers have discovered some 3,000 archaeal species and more than 37,000 different bacterial species. Also, there are some 10^{30} viruses called bacteriophages in the oceans that infect these oceanic microbes.

Courtesy of Jeffrey Pommerville.

In the infection process, sometimes bacteriophages by mistake carry pieces of the bacterial chromosome (rather than viral DNA) from the infected cell to another recipient cell. In the recipient cell, the new DNA fragment can be swapped for an existing part of the recipient's chromosome. It is a fairly rare event, occurring only once in every 100 million (10^8) virus infections. That doesn't seem very significant until you now consider the number of bacteriophages and susceptible bacteria existing in the oceans. Working with these numbers and the potential number of virus infections, scientists suggest that if only one in every 100 million infections brings a fragment of DNA to a recipient cell, there are about 10 million billion (that's 10,000,000,000,000,000 or 10^{16}) such gene transfers per second in the world's oceans. That is about 10^{21} infections per day!

We do not understand what all this recombination means. What we can conclude is there's an awful lot of gene swapping going on!

capable of infecting another cell. However, they are called "defective particles" because they carry no phage genes.

When a defective phage attaches to a new (recipient) cell, it injects the donor DNA into the cell. Once in the recipient, the genes can recombine with a section of the recipient's DNA and replace the section in a fashion similar to conjugation. The recipient has now been transduced (changed) using genes from the donor cell. MicroFocus 9.2 provides a spectacular example of generalized transduction.

Specialized transduction, unlike generalized transduction, results in the transfer of only specific genes (FIGURE 9.6B). Again, these are defective phages because they are missing a few phage genes needed for replication. Such a transducing phage can infect another cell and transfer its genes, which can integrate into the recipient chromosome. As before, the recipient cell acquired genes from the original donor cell and the recipient is now considered transduced.

Specialized transduction can be of medical significance. For example, the diphtheria bacillus, *Corynebacterium diphtheriae*, produces a toxin that is encoded on the defective phage DNA genes transferred through transduction. Other toxins including staphylococcal enterotoxins in food poisoning, clostridial toxins in some forms of botulism, and streptococcal toxins in scarlet fever, also are introduced into the avirulent strain by transduction.

FIGURE 9.7 compares the three forms of genetic recombination through horizontal gene transfer.

CONCEPT AND REASONING CHECKS 1

- **a.** How does HGT differ from vertical gene transfer?
- **b.** Why is competence key to the transformation process?
- **c.** Identify the genes that must be transferred to an F^- cell to convert it to F^+.
- **d.** Unlike conjugation between F^+ and F^-, where the recipient F^- becomes F^+, why doesn't an Hfr and F^- conjugation result in the F^- becoming an Hfr?
- **e.** What is the major difference between generalized and specialized transduction?

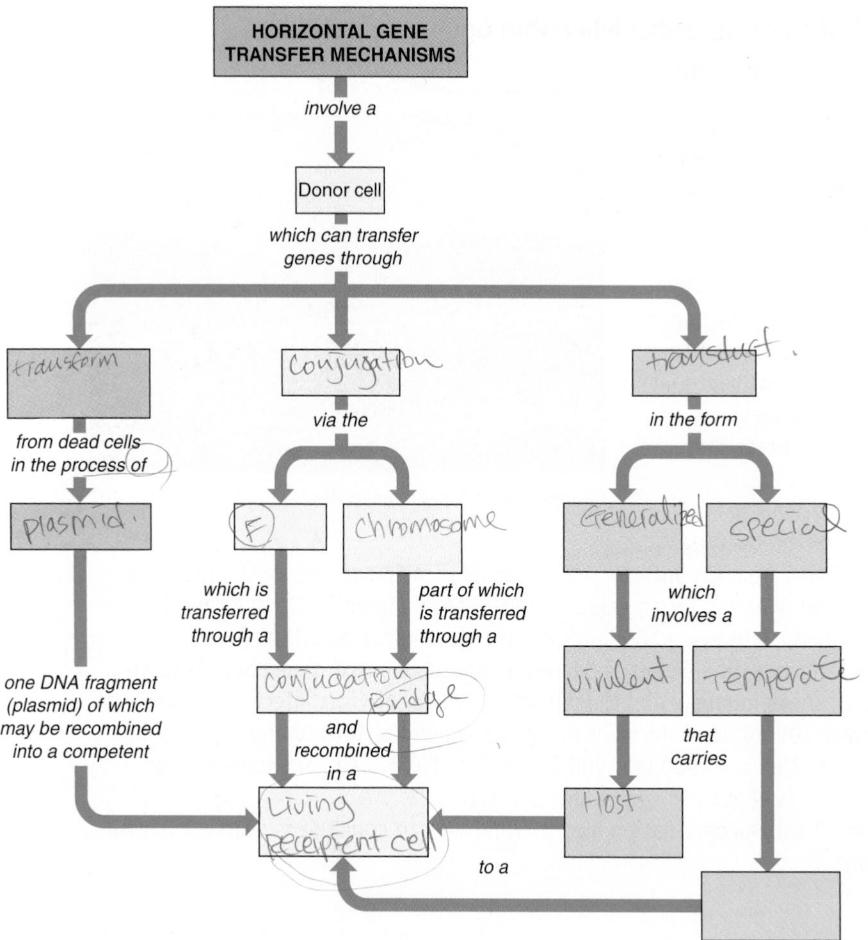

Term List

Defective phage
Bacterial chromosome
Conjugation
Conjugation bridge
DNA fragments
F factor
Generalized transduction
Living recipient cell
Random DNA fragment
Specialized transduction
Specific genes
Transduction
Transformation

FIGURE 9.7 **A Summary of Genetic Recombination Through Horizontal Gene Transfer.** This concept map summarizes the three mechanisms of transfer of genes from donor to recipient cells. »» Fill in the boxes with the correct term provided. Terms can be used more than once.

Chapter Challenge A

When the HGP draft sequence was published in 2001, much excitement was generated by a passage in the middle of the long manuscript.

"An interesting category is a set of 223 proteins that have significant similarity to proteins from bacteria, but no comparable similarity to proteins from yeast, worm, fly, and mustard weed, or indeed from any other (nonvertebrate) eukaryote. These sequences should not represent bacterial contamination in the draft human sequence . . . A more detailed computational analysis indicated that at least 113 of these [human] genes are widespread among bacteria, but, among eukaryotes, appear to be present only in vertebrates . . . A more [interesting] explanation is that these genes entered the vertebrate (or prevertebrate) lineage by horizontal transfer from bacteria." So, the HGP scientists had identified more than a hundred genes that appeared to have come into our genome from bacteria.

QUESTION A:

If HGT did occur (and the idea is very controversial), it apparently was not through some intermediary organism. As the quote mentioned, no similar sequences could be found in yeasts, worms, flies, plants, or in any nonvertebrate eukaryote. Having studied the three types of HGT (and transposon jumping) among bacterial species, propose which mechanism might have been most likely to transfer donor bacterial genes to the chromosomes in a recipient vertebrate cell.

Answers can be found on the Student Companion Website in **Appendix D.**

KEY CONCEPT 9.2 Genetic Engineering Involves the Deliberate Transfer of Genes Between Organisms

Prior to the 1970s, bacterial species having special or unique metabolic properties were detected through mutant analysis or by screening for cells with certain metabolic talents (MicroFocus 9.3). Experiments in genetic recombination entered a new era in the late 1970s, when it became possible to insert genes into bacterial DNA and thereby establish a genetically identical population that would produce proteins from the inserted genes. **Genetic engineering**, the use of microbial genetics, including the isolation, manipulation, and control of gene expression, was born.

Genetic Engineering Was Born from Genetic Recombination

The field of genetic engineering surfaced in the early 1970s when the techniques became available to manipulate DNA. In 1971, Paul Berg and his coworkers at Stanford University opened the circular DNA molecule from simian virus-40 (SV40) and

MICROFOCUS 9.3: History/Biotechnology
Clostridium acetobutylicum and the Jewish State

In 1999, scientists completed sequencing the genome of *Clostridium acetobutylicum,* a nonpathogenic bacterial species. Because some other species of *Clostridium* are major pathogens (one produces the food toxin that causes botulism, and another is responsible for tetanus), the scientists hope their sequencing work will yield insights into what enables some species to become pathogens while others remain harmless. However, *C. acetobutylicum's* ability to convert starch into the organic solvents is what has a prominent place in history—and represents the beginning of the modern age of biotechnology.

In 1900, an outstanding chemist named Chaim Weizmann, a Russian-born Jew, completed his doctorate at the University of Geneva in Switzerland and in 1904, he began working in the laboratory of Professor William Perkin in Manchester, England, where he attempted to use microbial fermentation to produce industrially useful substances. He discovered that *C. acetobutylicum* converted starch to a mixture of ethanol, acetone, and butanol, the latter an important ingredient in rubber manufacture. As World War I broke out in 1914, the favored propellant for rifle bullets and artillery projectiles was a material called cordite. To produce it, a mixture of cellulose nitrate and nitroglycerine was combined into a paste using acetone and petroleum jelly. Before 1914, acetone was obtained through the destructive distillation of wood. However, the supply was inadequate for wartime needs, and by 1915, there was a serious shell shortage, mainly due to the lack of acetone for making cordite.

After his inquiries to serve the British government were not returned, a friend of Weizmann's went to Lloyd George, the Minister of Munitions, and was told about Weizmann's work and how he could synthesize acetone in a new way. The conversation resulted in a London meeting between Weizmann, Lloyd George, and Winston Churchill. After explaining the capabilities of *C. acetobutylicum,* Weizmann became director of the British admiralty laboratories where he instituted the full-scale production of acetone from corn. Additional distilleries soon were added in Canada and India. The shell shortage ended.

After the war ended, now British Prime Minister Lloyd George wished to honor Weizmann for his contributions to the war effort. Weizmann declined any honors but asked for support of a Jewish homeland in Palestine. Discussions with Foreign Minister Earl Balfour led to the Balfour Declaration of 1917, which committed Britain to help establish the Jewish homeland. Weizmann went on to make significant contributions to science—he suggested that other organisms be examined for their ability to produce industrial products and is considered the father of industrial fermentation. Weizmann also laid the foundations for what would become the Weizmann Institute of Science, one of Israel's leading scientific research centers. His political career also moved upward—he was elected the first President of Israel in 1949. Chaim Weizmann died in 1952.

spliced it into a bacterial chromosome. In doing so, they constructed the first **recombinant DNA molecule**—a DNA molecule containing DNA segments spliced together from two or more organisms. This human-manipulated genetic recombination process was extremely tedious though.

The process became easier after Herbert Boyer and his group at the University of California isolated a **restriction endonuclease** enzyme that recognizes and cuts specific short stretches of nucleotides. Importantly, the enzyme leaves the DNA with single-stranded extensions, called "sticky ends" that easily attached to complementary ends protruding from another fragment of DNA.

Today, there is a vast array of such restriction enzymes from the Bacteria and Archaea, each recognizing a specific nucleotide sequence (TABLE 9.2). Enzyme designations are derived from the species from which they were isolated. For example, the restriction enzyme *Eco*RI stands for ***E****scherichia* ***co****li* **R**estriction enzyme **I**.

Each enzyme cuts both strands of the DNA because the sequences recognized are all palindromes. Thus, restriction enzymes are "molecular scissors" used by genetic engineers to open a bacterial chromosome or plasmid at specific locations and into which a DNA segment from another organism can be inserted. To seal the recombinant DNA segments, **DNA ligase** was used.

■ Palindrome: A series of letters reading the same left to right and right to left.

Putting this all together, one can take a plasmid from a bacterial species such as *E. coli* and open it with a restriction enzyme (FIGURE 9.8). Then, one can insert a segment of foreign DNA into the plasmid and seal the segment using DNA ligase. Mimicking natural genetic recombination (transformation), the recombinant DNA molecule is inserted into fresh *E. coli* cells. It thus became possible to easily manipulate genes from a wide variety of species and splice them together.

The best way to understand how genetic engineering operates is to follow an actual gene cloning procedure.

Today, over 18 million people in the United States have diabetes, a group of diseases resulting from abnormally high blood glucose levels. In the case of insulin-dependent diabetes (also called juvenile or type I diabetes), diabetics must receive regular injections of insulin to control their blood glucose level. Before 1982, diabetics received purified insulin extracted from the pancreas of cattle and pigs, or even cadavers. However, this can pose a problem because animal insulin is not human insulin and could trigger allergic reactions. In addition, the animal insulin could contain unknown disease-causing viruses that had infected the animal. The solution was to produce human insulin using genetic engineering techniques.

MicroInquiry 9 describes one such method—by cloning the human gene for insulin into bacterial cells. This involves:

1. Isolating the piece of DNA containing the human insulin gene and precisely cutting the gene out;
2. Splicing the insulin gene into a bacterial plasmid (**cloning vector**);

TABLE

9.2 Examples of Restriction Endonuclease Recognition Sequences

Organism	Restriction Enzyme	Recognition Sequence*
Escherichia coli	*Eco*RI	G ↓ AATTC CTTAA ↑ G
Streptomyces albus	*Sal*I	G ↓ TCGAC CAGCT ↑ G
Haemophilus influenzae	*Hind*III	A ↓ AGCTT TTCGA ↑ A
Bacillus amyloliquefaciens	*Bam*HI	G ↓ GATC^{m}C CC^{m}TAG ↑ G
Providencia stuartii	*Pst*I	CTGCA ↓ G G ↑ ACGTC

*Arrows indicate where the restriction enzyme cuts the two strands of the recognition sequence; C^{m} = methylcytosine.

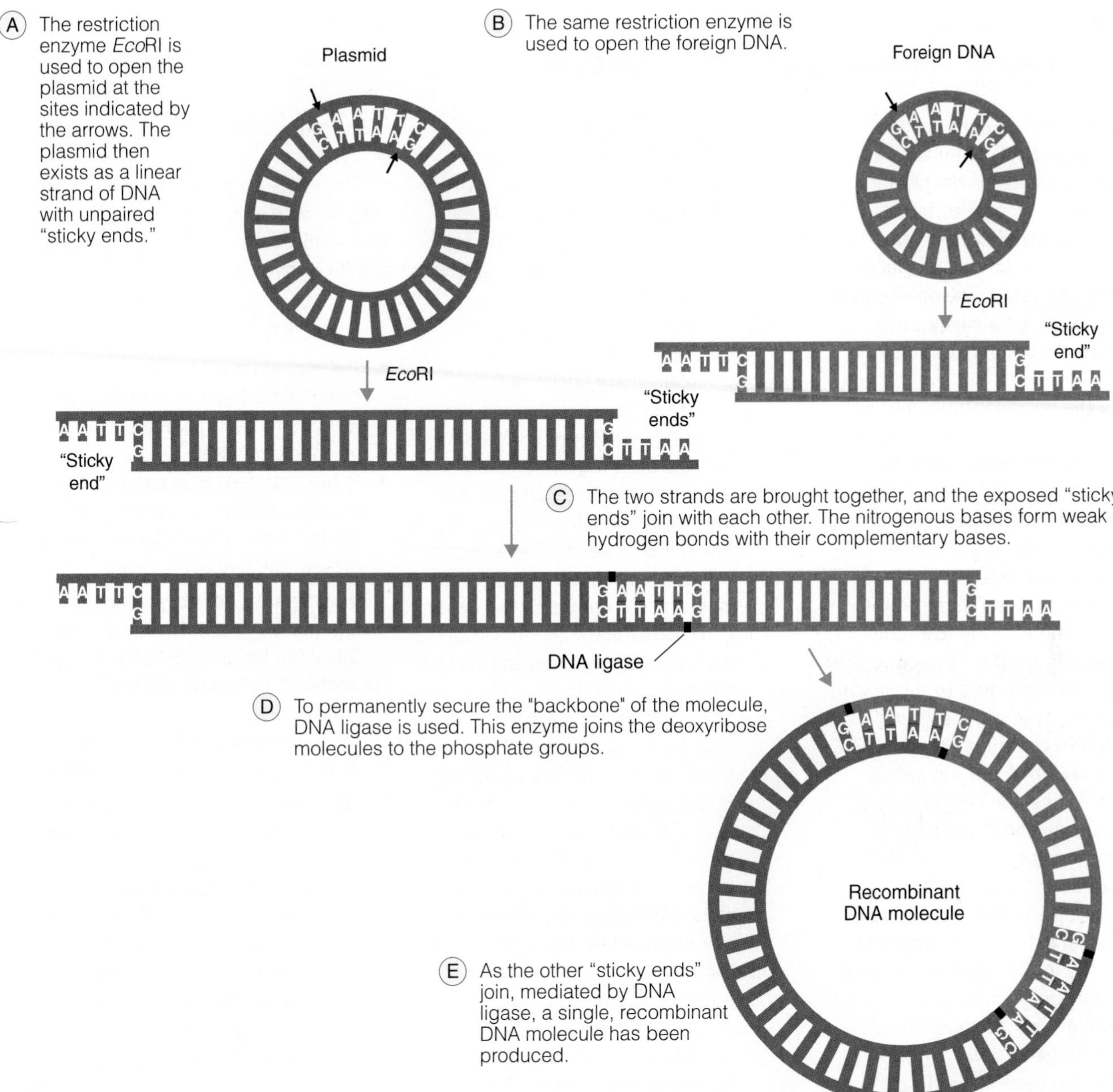

FIGURE 9.8 **Construction of a Recombinant DNA Molecule.** In this construction, two unrelated plasmids (loops of DNA) are united to form a single plasmid representing a recombinant DNA molecule. »» Why is the product of genetic engineering called a recombinant DNA molecule?

3. Placing the recombinant plasmid in bacterial cells to form clones;
4. Screening for recombinant plasmids;
5. Identifying and isolating clones carrying the insulin gene.

Biotechnology Has Spawned Many Commercial and Practical Products

Eli Lilly marketed the first synthetic human insulin, called Humulin, in 1982. Since then, other genetically engineered insulin products have been developed. Such commercial successes of biotechnology were a sign of things to come. **Biotechnology** then can be defined as the field of applying the techniques of genetic engineering to manipulate living organisms (usually microorganisms) to manufacture products important to human health and the environment.

Besides insulin, many other proteins of important pharmaceutical value to humans have been

MICROINQUIRY 9

Molecular Cloning of a Human Gene into Bacterial Cells

Genetic engineering has been used to produce pharmaceuticals of human benefit. One example concerns the need for insulin injections in people suffering from diabetes (an inability to produce the protein insulin to regulate blood glucose level). Prior to the 1980s, the only source for insulin was through a complicated and expensive extraction procedure from cattle or pig pancreases. But, what if you could isolate the human insulin gene and, through transformation, place it in bacterial cells? These cells would act as factories churning out large amounts of the pure protein that diabetics could inject.

To do molecular cloning of a gene, besides the bacterial cells, we need three ingredients: a cloning vector, the human gene of interest, and restriction enzymes. Plasmids are the "cloning vector," a genetic element used to introduce the gene of interest into the bacterial cells. Human DNA containing the insulin gene must be obtained. We will not go into detail as to how the insulin gene can be "found" from among 25,000 human genes. Suffice it to say that there are standard procedures to isolate known genes. Restriction enzymes will cut open the plasmids and cut the gene fragments that contain the insulin gene, generating complementary sticky ends.

The following description represents one procedure to genetically engineer the human insulin gene into cells of *Escherichia coli*.

1. Plasmids often carry genes, such as antibiotic resistance. We are going to use the cloning vector shown in **Figure A** because it contains a gene for resistance to ampicillin (amp^R) and the *lacZ* gene that encodes the enzyme β-galactosidase (B-gal) that splits lactose into glucose and galactose. This will be important for identification of clones that have been transformed. In addition, this vector has a single restriction sequence for the restriction enzyme *Sal*I (see Table 9.2). Importantly, this cut site is within the *lacZ* gene. Also, the plasmid will replicate independently in *E. coli* cells and can be placed in the cells by transformation.
2. The vector and human DNA are cut with *Sal*I to produce complementary sticky ends on both the opened vector and the insulin gene (**Figure B**). Vector and the insulin gene then are mixed together in a solution with DNA ligase, which will covalently link the sticky ends. Some vectors will be recombinant plasmids; that is, plasmids containing the insulin gene. Other plasmids will close back up without incorporating the gene.
3. The plasmids are placed in *E. coli* cells by transformation. The plasmids replicate independently in the bacterial cells, but as the bacterial cells multiply, so do the plasmids. By allowing the plasmids to replicate, we have cloned the plasmids, including any that contain the insulin gene.
4. Because we do not know which plasmids contain the insulin gene, we need to screen the clones to identify what colonies contain the recombinant plasmids. This is why we selected a plasmid with amp^R and *lacZ*.

Because the *Sal*I cut site is within *lacZ*, any recombinant plasmids will have a defective *lacZ* gene and the bacterial cells carrying those plasmids will not be able to produce B-gal. Plasmids without the insulin gene have an intact *lacZ* gene and will have B-gal activity. Thus, using the positive selection technique described in the text on microbial genetics and a selective medium as described in the text on microbial growth and nutrition, we can identify which bacterial cells contain the recombinant plasmids.

Therefore, we plate all our bacterial cells onto an agar plate containing ampicillin and a substrate called X-gal that B-gal can hydrolyze.

$$\text{X-gal} \xrightarrow{\text{B-gal}} \text{X} + \text{galactose}$$

- The X product is blue in color, so any colonies having an intact *lacZ* gene will hydrolyze X-gal and appear blue on the agar plate;
- If the *lacZ* gene is inactive due to the presence of the insulin gene, then no product will be formed and those colonies on the plate will appear white.

Question 9a. Identify what colonies on the plate contain the insulin gene.

Question 9b. Explain why the bacteria without a plasmid did not grow on the plate.

The answers can be found on the Student Companion Website in **Appendix D**.

5. These colonies with the insulin gene can now be isolated and grown in larger batches of liquid medium. These batch cultures then are inoculated into large "production vats," called bioreactors, in which the cells grow to massive numbers while secreting large quantities of insulin into the liquid.

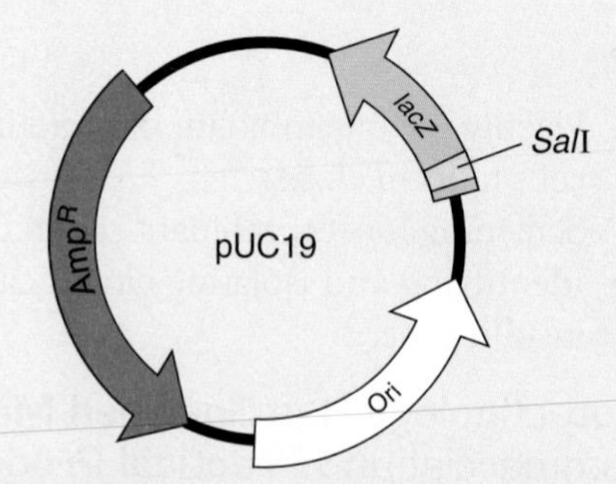

FIGURE A The Cloning Vector, pUC19.

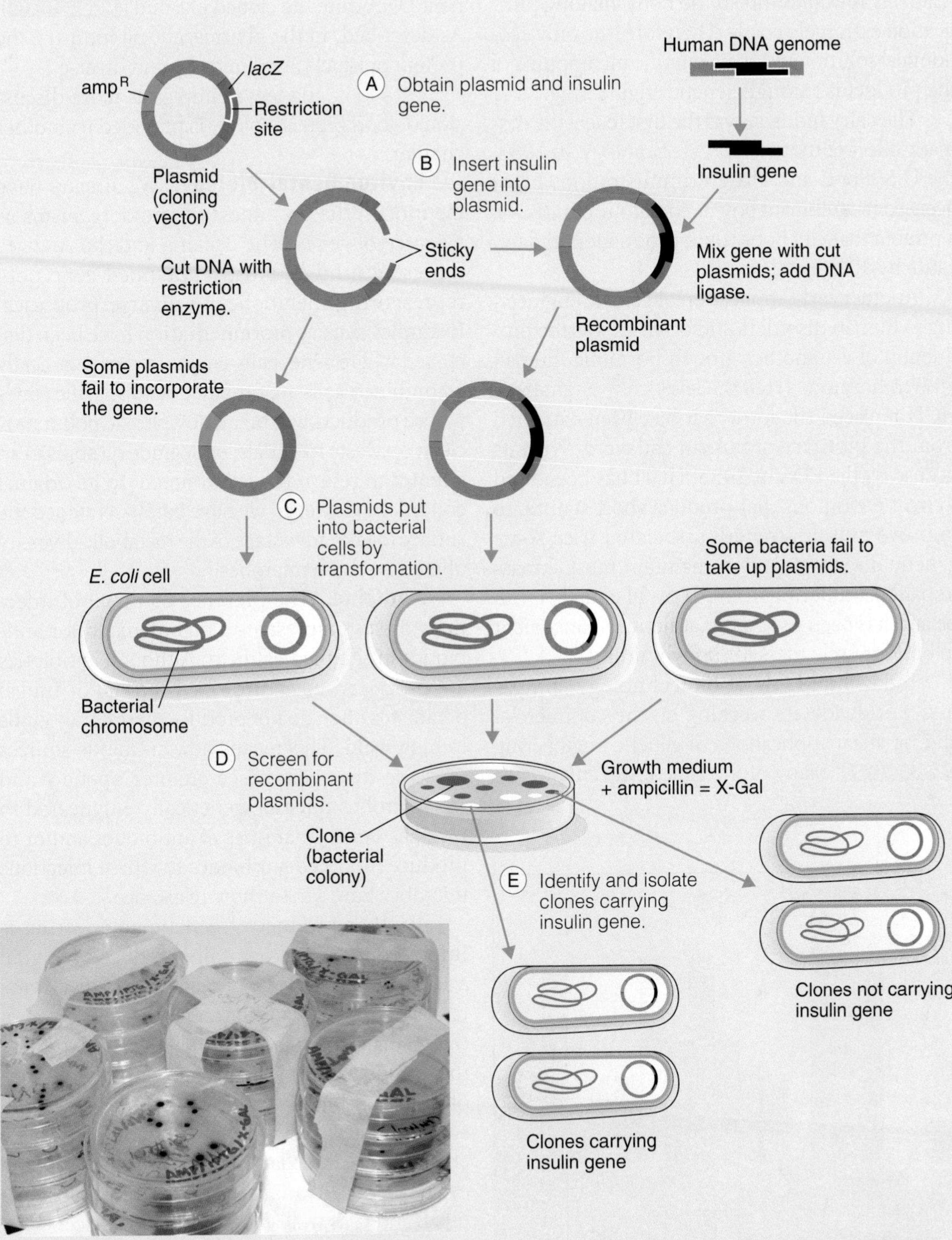

FIGURE B **The Sequence of Steps to Engineer the Insulin Gene into *Escherichia coli* Cells.** © Martin Shields/ Alamy Images.

produced by genetically engineered microorganisms. Many of these proteins are produced in relatively low amounts in the body, making purification extremely costly. Therefore, the only economical solution to obtain significant amounts of the product is through genetic engineering.

The dairy industry was the first to feel the dramatic effect of the new DNA technology. In 1982, the U.S. Food and Drug Administration (FDA) licensed recombinant bovine somatotropin (rBST), a protein that can boost milk production in dairy cattle by 40%.

Another early application of genetic engineering to human disorders and diseases was the production of yet another growth hormone, human growth hormone (HGH). Genetically engineered HGH replaced the form that had been extracted from the pituitary of human cadavers. With its license by the FDA in 1985, HGH has been used to treat conditions that produce short stature, to improve muscle strength associated with some genetic disorders, and to maintain muscle mass in patients suffering from AIDS. Of course, it can be and has been used as an athletic enhancement to build muscle mass for bodybuilding.

Today, hundreds of biotechnology companies worldwide are working on the commercial and practical applications of genetic engineering (FIGURE 9.9). Many of the genetically engineered products are either proteins expressed by the recombinant DNA in the bacterial cells or the recombinant DNA from the cloned plasmid (FIGURE 9.10). As described, in the pharmaceutical industry, the protein products are numerous and diverse.

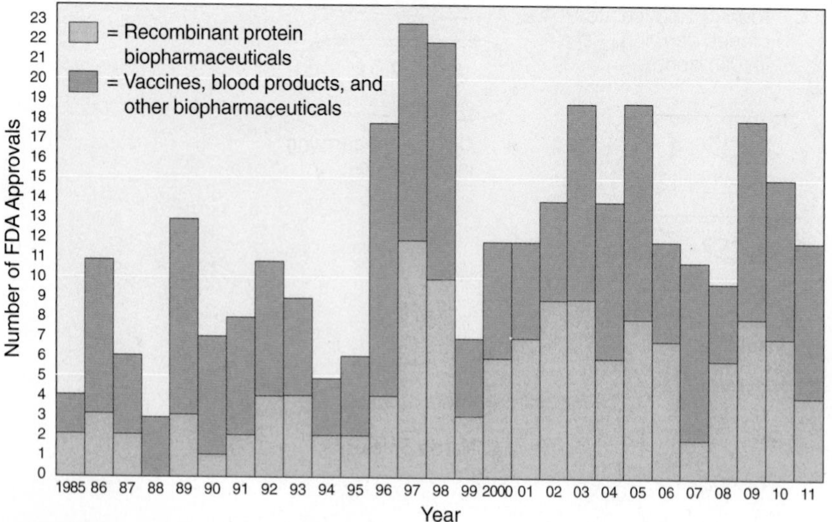

FIGURE 9.9 **FDA Approvals of New Pharmaceutical Products, 1985–2011.** This graph shows the number of approvals of new biopharmaceutical products since 1985. Data redrawn from: Biotechnology Information Institute (http://www.biopharma.com/approvals_2011.html). »» Can you come up with a reason for the drop in recombinant protein biopharmaceuticals in 2007?

Let's look at a few examples as more discussion of food biotechnology is provided in another chapter.

Environmental Biology. We already have mentioned the usefulness of microorganisms as a source of genes. The Bacteria and the Archaea represent a huge, mostly untapped gene pool representing metabolically diverse processes. Examples such as **bioremediation** have been discussed where genetically engineered or genetically recombined cells are provided with specific genes whose products will break down toxic pollutants, clean up waste materials, or degrade oil spills in an attempt to return the environment to its original condition. However, we have barely scratched the surface to take advantage of the metabolic diversity offered by these microbes.

Medicine. The presence or threat of infectious disease represents a high demand for antibiotics in the medical field. Although antibiotics are produced in nature, the bacterial or fungal organisms often do not produce these compounds in high yield. This means new antibiotic sources must be discovered (see chapter opener) and the microbes must be genetically engineered to produce larger quantities of antibiotics and/or to produce modified antibiotics to which infectious microbes have yet to show resistance.

Another product of genetic engineering is **interferon**, a set of naturally produced antiviral agents produced by the human body to block viral replication. As with insulin and HGH, the body produces small amounts of these chemicals, so prior to the introduction of genetic engineering thousands of units of human blood were needed to obtain sufficient interferon to treat a patient. With genetic engineering, much larger amounts of pure protein can be produced to aid patients suffering from hepatitis B and C as well as some forms of cancer.

Vaccine production is now safer as a result of genetic engineering. By making a vaccine that only contains a part of the whole microbial agent, or isolating a gene that will stimulate the immune system to generate protective immunity, makes the vaccine much safer because the patient is not exposed to the active virus or microbe that can

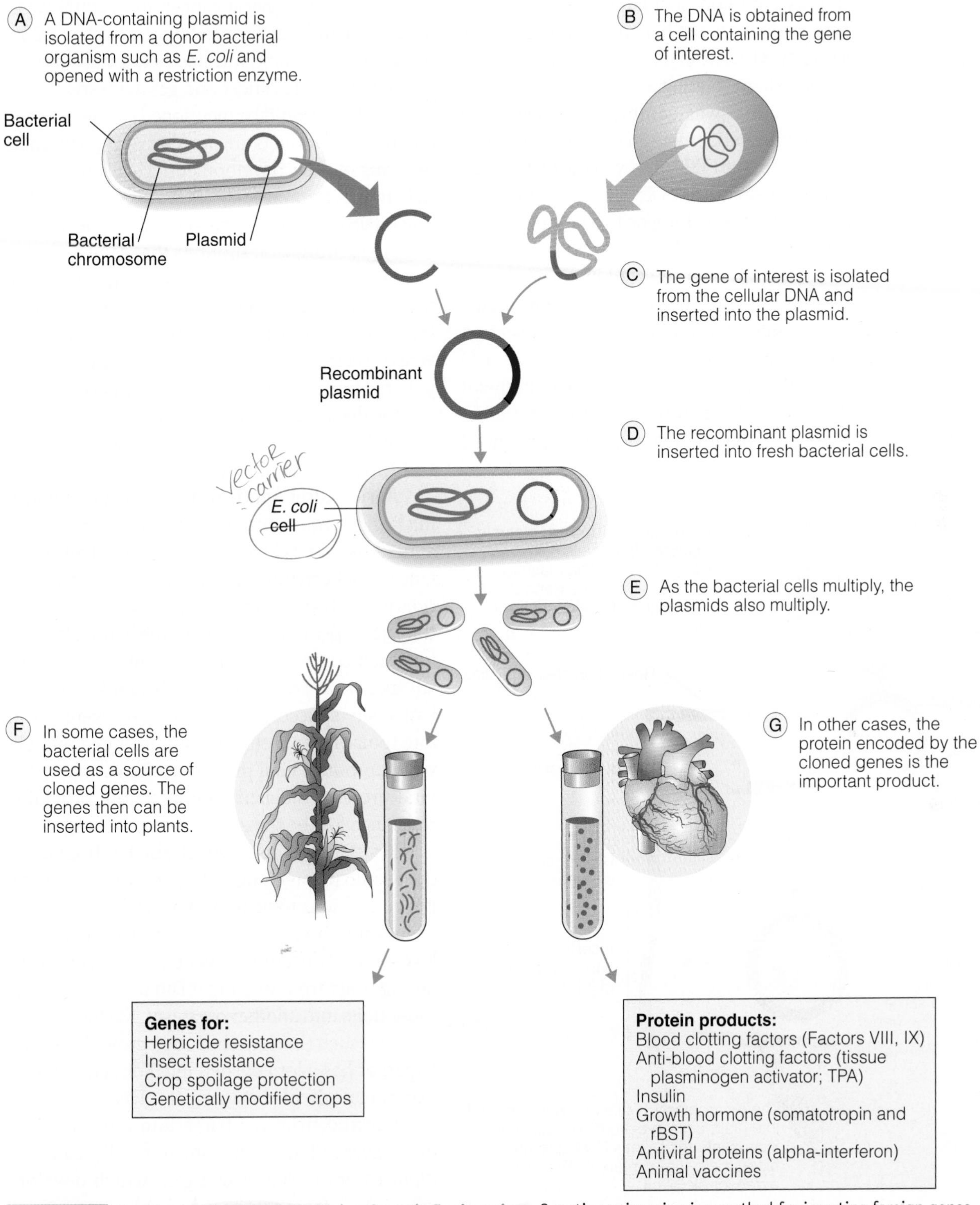

FIGURE 9.10 Developing New Products Using Genetic Engineering. Genetic engineering is a method for inserting foreign genes into bacterial cells and obtaining chemically useful products. »» How do bacterial cells that have been genetically modified to encode gene products (F) differ from those modified cells encoding protein products (G)?

cause the disease. For example, hepatitis B, a serious viral infection spread by contact with infected blood, kills 2 million people globally each year. The first vaccine against hepatitis B was made by extracting the virus from infected blood of chronic carriers and then isolating a viral surface protein. Such a procedure was complex and posed the risk of possible contamination by other infectious agents. In 1986, the FDA approved the first recombinant hepatitis B vaccine (Recombivax HB®) that was made by inserting the gene for the viral surface protein into common baker's yeast, *Saccharomyces cerevisiae*. The protein produced by the yeast was identical to the natural viral protein and made for a much safer vaccine for humans. A second recombinant-type hepatitis B vaccine (Engerix-B®) was licensed in 1989. A more recent recombinant vaccine, Gardasil®, also uses recombinant DNA technology by having a viral protein produced in yeast cells. The vaccine generates protective immunity against many of the papilloma viruses responsible for cervical cancer and genital warts in women and was licensed by the FDA in 2006. In 2009, the vaccine was licensed for use in men to prevent penile cancer and genital warts.

■ Transgenic: Referring to an organism containing a stable gene from another organism.

Lastly, it should be mentioned that genetically engineered products are not always a "no brainer" in terms of their development. This is no clearer than in the attempts to develop a vaccine for AIDS. Since 1987, scientists and genetic engineers have tried to identify viral subunits that can be used to develop an AIDS vaccine. However, it is not so much that genetic engineering can't be done as it is that the virus just seems to find ways to circumvent a vaccine and the immunity developing in the patient. Still, scientists hope a safe and effective genetically engineered vaccine can be developed. We will have much more to say about vaccines and AIDS in other chapters.

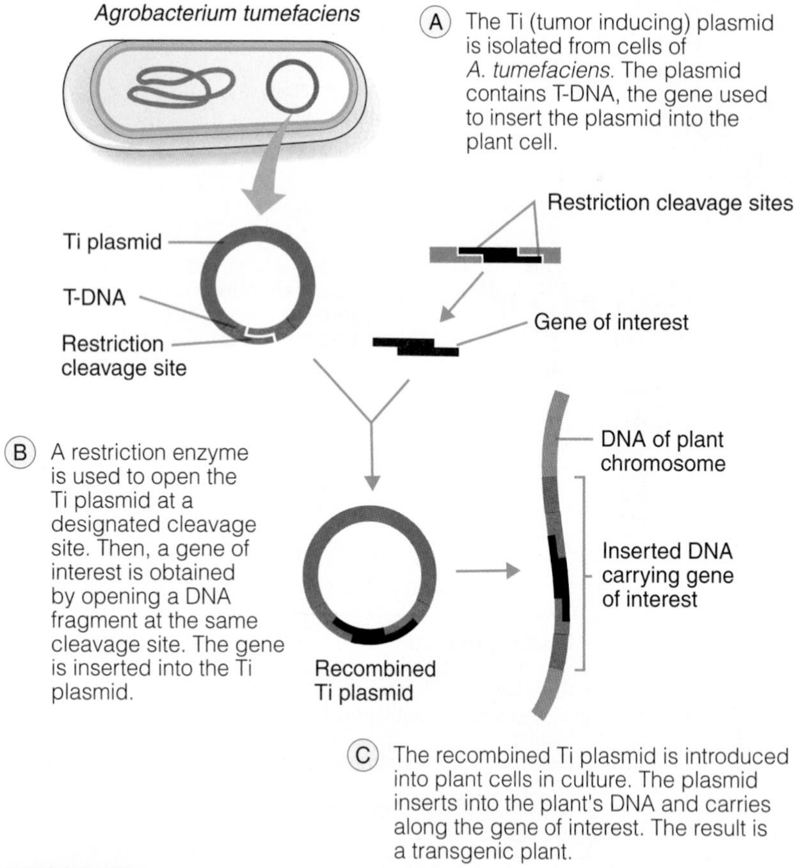

Agricultural Applications. Genetic engineering has extended into many realms of science. In agriculture, for example, genes for herbicide resistance have been transplanted from cloned bacterial cells into tobacco plants, demonstrating that these transgenic plants better tolerate the herbicides used for weed control. For tomato growers, a notable advance was made when researchers at Washington University spliced genes from a pathogenic virus into tomato plant cells and demonstrated the cells would produce viral proteins at their surface. The viral proteins blocked viral infection, providing resistance for the transgenic tomato plants.

Resistance to insect attack also has been introduced into plants using a plasmid carrying a bacterial gene that is toxic to beetle and fly larvae.

As much as 60% of the food we eat today has some connection to genetic engineering. By taking traits from one organism and introducing those traits into another organism, the food can be changed such that it tastes better, grows faster and larger, or has a longer shelf life. For gene transfer experiments in plants, the vector DNA often used is a plasmid from the bacterium *Agrobacterium tumefaciens*. This organism normally causes a plant tumor called crown gall, which develops when DNA from the bacterial cells inserts itself into the plant cell's chromosomes (FIGURE 9.11). Genetic engineering techniques remove the tumor-inducing (Ti) gene from the plasmid and then splice the desired gene into the plasmid and allow

FIGURE 9.11 **The Ti Plasmid as a Vector in Plant Genetic Engineering.** *Agrobacterium tumefaciens* induces tumors in plants and causes a disease called crown gall. The catalyst for infection is a tumor-inducing (Ti) plasmid. This plasmid, without the tumor-causing gene, is used to carry a gene of interest into plant cells. »» Why must the tumor-inducing gene in the Ti plasmid be removed before the plasmid is used in genetic engineering procedures?

the bacterial cells to infect the plant. The Ti system works well with broadleaf plants such as tomato, potato, soybeans, and cotton.

DNA Probes Can Identify a Cloned Gene or DNA Segment

The genes of an organism contain the essential information responsible for its behavior and characteristics. Bacterial and viral pathogens, for example, contain specific sequences of nucleotides that give the pathogen the ability to infect and cause disease. Because these nucleotide sequences are distinctive and often unique, if detectable, they can be used as a diagnostic tool to identify a pathogen.

In the medical laboratory, diagnosticians are optimistic about the use of **DNA probes**, single-stranded DNA molecules that recognize and bind to a distinctive and unique nucleotide sequence of a pathogen. To use a DNA probe effectively, it is valuable to increase the amount of DNA to be searched. This can be done through the **polymerase chain reaction** (**PCR**); the technique is outlined in MicroFocus 9.4.

The DNA probe binds (hybridizes) to its complementary nucleotide sequence from the pathogen, much like strips of Velcro stick together. To make a probe, scientists first identify the segment (or gene) in the pathogen that will be the target of a probe. Using this segment, they construct the single-stranded DNA probe (FIGURE 9.12). More than 100 DNA probes have been developed for the detection of pathogens.

One example of where DNA probes and PCR have been useful is in the detection of the human immunodeficiency virus (HIV). T lymphocytes, in which HIV replicates, are obtained from the patient and disrupted to obtain the cellular DNA. The DNA then is amplified by PCR and the DNA probe is added. The probe is a segment of DNA that complements the DNA in the virus synthesized from the genome of HIV. If the person is infected with HIV, the probe will locate the viral DNA, bind to it, and emit radioactivity. An accumulation of radioactivity constitutes a positive test.

A DNA probe also is available for detecting human papilloma virus (HPV). The test uses a DNA probe to detect viral DNA in a sample of tissue obtained from a woman's cervix. Because certain forms of HPV have been linked to cervical tumors, the test has won acceptance as an important preventive technique, and it has been licensed by the FDA. It is commercially available as the ViraPap test. Clearly, the use of DNA probes represents a reliable and rapid method for detecting and diagnosing many human infectious diseases (MicroFocus 9.5).

A similar technique can be used to conduct water-quality tests based on the detection of coliform bacteria such as *E. coli*. Traditionally, *E. coli* had to be cultivated in the laboratory and identified biochemically. With DNA probe technology, a sample of water can be filtered, and the bacterial cells trapped on the filter can be broken open to release their DNA for PCR and DNA probe analysis. Not only is the process time saving (many days by the older method, but only a few hours with the newer method), it is also extremely sensitive: a single *E. coli* cell can be detected in a 100-mL sample of water.

Another useful tool in biotechnology is the **DNA microarray**, a small slide surface on which genes or DNA segments are attached and arranged spatially in a known pattern that can be used to assess gene expression in microorganisms. The technique is described in MicroFocus 9.6.

CONCEPT AND REASONING CHECKS 2

a. List the steps required to form a recombinant DNA molecule.
b. Give several examples of how genetic engineering has benefited the fields of environmental biology, medicine, and agriculture.
c. Why are DNA probes such useful tools in the detection of disease-causing agents?

Chapter Challenge B

In 2008, *Discover* magazine published an article by Shara Yurkiewicz and Susannah F. Locke entitled *10 Ways Genetically Engineered Microbes Could Help Humanity* [http://discovermagazine.com/2008/sep/06-10-ways-genetically-engineered-microbes-could-help]. In this chapter, we have seen several examples of how genetically engineered microbes have had definite effects in trying to circumvent disease conditions arising from the human genome.

QUESTION B:

Provide your own list of five ways genetically engineered microbes have helped to alleviate disease conditions caused by faulty genes in the human genome.

Answers can be found on the Student Companion Website in **Appendix D.**

MICROFOCUS 9.4: Tools

The Polymerase Chain Reaction

The polymerase chain reaction (PCR) is a technique that takes a segment of DNA and replicates it millions of times in just a few hours. The technique was developed in 1984 by Kary Mullis working for the Cetus Corporation, a biotechnology company in Emeryville, California; in 1993, he shared the Nobel Prize in Chemistry for his discovery.

The PCR process is a repeating three-step process (see figure). Target DNA is mixed with DNA polymerase (the enzyme that synthesizes DNA), short strands of primer DNA, and DNA nucleotides. The mixture is then alternately heated and cooled during which time the double-stranded DNA unravels, is duplicated, and then reforms the double helix.

The process is repeated over and over again in a highly automated PCR machine, which is the biochemist's equivalent of an office copier. Each cycle takes about 5 minutes, and each new DNA segment serves as a template for producing many additional identical copies, which in turn serve as templates for producing more identical copies.

PCR is now a common and often essential tool in medical and biological research. Applications for PCR are numerous and include its use for DNA cloning, organismal DNA phylogeny studies, and (as described in the text) diagnosis of infectious diseases.

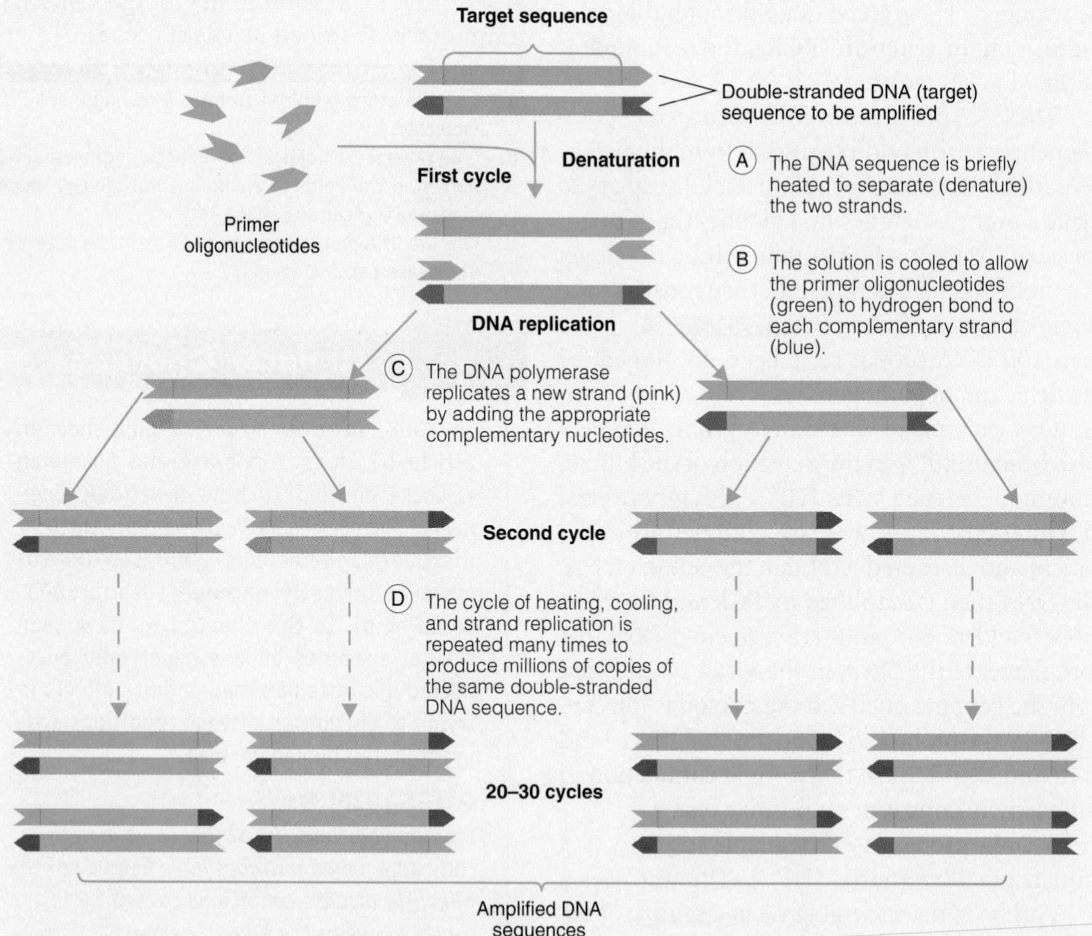

The polymerase chain reaction produces billions of copies of a DNA sequence that are identical to the starting, target sequence.

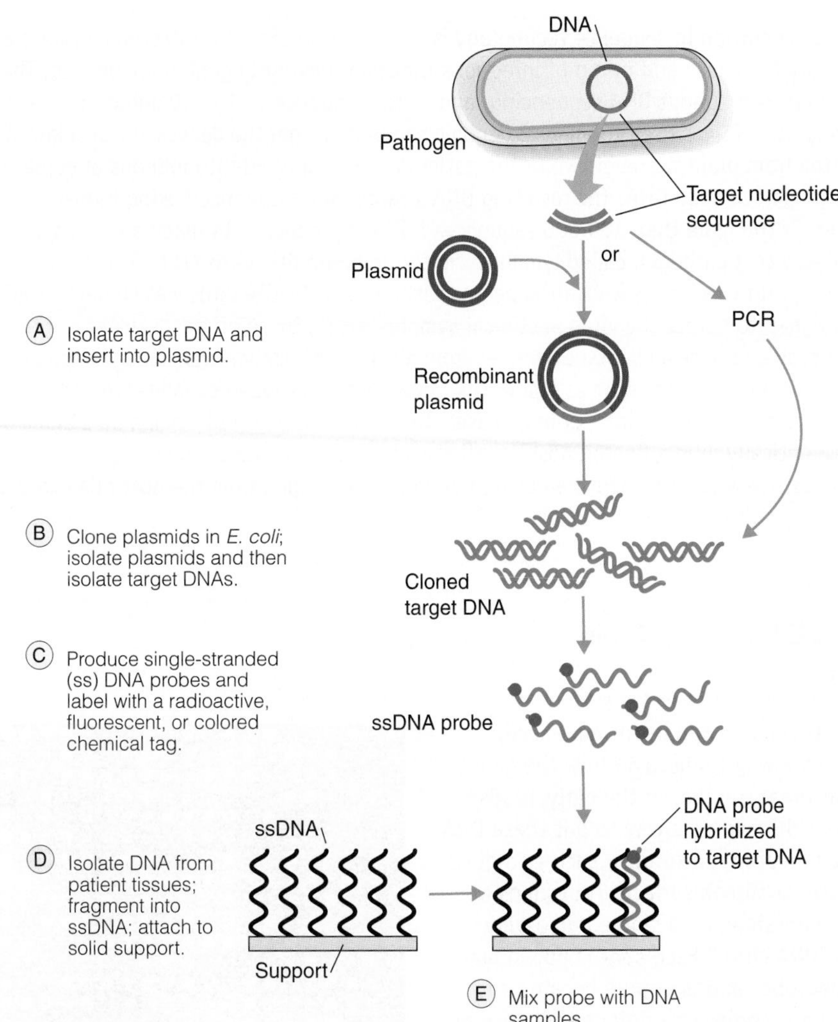

FIGURE 9.12 **DNA Probes.** Construction of a DNA probe and its use in disease detection and diagnosis. »» Why must the DNA probes be single stranded?

MICROFOCUS 9.5: Tools

Discovering Emerging Pathogens

In April 2007, three Australian transplant patients received organs from a single 57-year-old donor, who had died of a stroke and was not thought to have had an infectious disease. Two women, aged 44 and 63, each received one of the donor's kidneys while a 64-year-old woman received the donor's liver. Each of the women soon developed fever and encephalitis, a swelling of the brain. Within 6 weeks of the operation, all three were dead. Health officials at the time were unable to identify the cause of the patients' deaths. In fact, up to 40% of central nervous system infectious diseases and 30% to 60% of respiratory illnesses cannot be traced back to a specific pathogen.

(continued on next page)

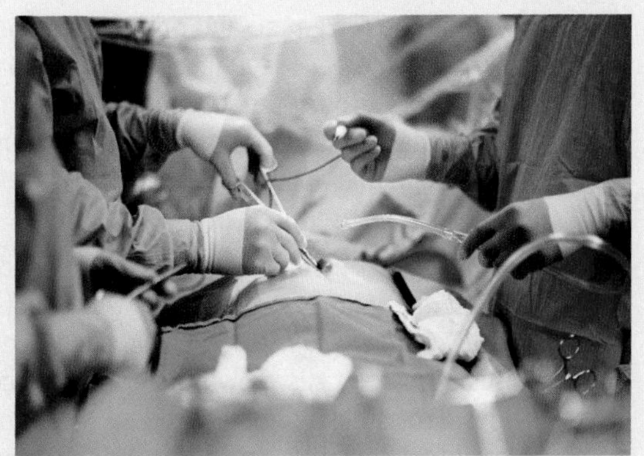

A surgical team performing a kidney transplant. © Photos.com.

Now the next generation in sequence technology is making it possible for infectious disease epidemiologists to identify the cause and origin of infections that had previously gone undiagnosed. The technique is called high-throughput DNA sequencing, and it can sequence up to 100 million nucleotide bases of DNA per 7-hour run. It was used to analyze genome sequences from the deceased transplant patients.

RNA, extracted from brain tissue of two of the patients, was amplified into millions of copies of the corresponding single-stranded DNA. The resulting DNA strands were sequenced using high-throughput DNA sequencing. Of the more than 100,000 sequences initially produced, 14 matched viral sequences that were closely related to a pathogen called lymphocytic choriomeningitis virus (LCMV). This virus usually causes only a minor flu-like illness in healthy people. Once the LCMV-like virus was characterized, DNA probes were designed to detect the virus in clinical samples. Using these probes, evidence of the LCMV-like virus was discovered in several tissue samples from all three Australian transplant recipients.

The new sequencing method is seen as a powerful new tool for pathogen surveillance and for diagnosing mysterious illnesses and emerging infectious diseases. In 2008, a 70-year-old woman died and a 57-year-old man became critically ill in a Boston hospital after each received a kidney from a donor infected with the LCMV-like virus. The virus was identified using the probes developed from the Australian cases.

MICROFOCUS 9.6: Tools

Microarrays

Once a microbial genome had been sequenced, scientists needed a way to discover how the genes interact in that organism. So, in the early 1990s, biotechnologists discovered a way to put these DNA segments on a miniaturized surface, such as a glass slide, or a plastic or silicon surface. Thousands of spots could be contained on a single microarray, often called a "DNA chip." Each spot is put in place by a mechanical robot and is unique because the spot contains many copies of a unique sequence of single-stranded nucleotides produced through the polymerase chain reaction (PCR) from the DNA of the sequenced organism. Such an array may contain DNA from one species or many species.

DNA microarray showing locations of differently labeled DNA molecules. © Alfred Pasieka/Photo Researchers, Inc.

As the probe, scientists use single-stranded, fluorescently labeled DNA that is complementary to a target DNA sequence. Often two different samples of probes are used, each labeled with a different colored fluorescent dye (often red and green). When exposed to (washed over) the microarray, a probe will bind to any complementary sequences (the target) and the spot will fluoresce the color (red or green) of the bound probe. If both probes bind, they will fluoresce the intermediate color, yellow; or if more of one probe binds than another, the spot may be more orange or light green (see figure). The results usually are scanned and analyzed by computer.

Microarrays can be used in many ways.

- **Gene expression:** Microarrays can be developed and used to study gene expression. For example, a microbiologist might want to study the effects of oxygen on gene expression in the facultative anaerobe *Escherichia coli.* Using a microarray containing the *E. coli* genome, the microbiologist could isolate *E. coli* cells grown under aerobic conditions and another sample under anaerobic conditions. From these samples, the mRNAs (from active genes under the two conditions) would be isolated and copied (reverse transcribed) into single-stranded DNA. The DNA probes from the aerobic condition may be labeled with the red dye while the anaerobic condition may be labeled with the green dye. The two probes allow the scientist to study the same cell (genes) under two different conditions to see which genes are active with or without oxygen gas.

- **Organism detection:** Because of the constant threat posed by emerging infectious diseases and the limitations of existing approaches available to identify new pathogens, microarrays offer a rapid and accurate method for viral discovery. DNA microarray-based platforms have been developed to detect a wide range of known viruses as well as novel members of some viral families. For example, during the outbreak of severe acute respiratory syndrome (SARS) in March 2003, a viral isolate cultivated from a SARS patient was made into a DNA probe and, on the microarray, produced a spot representing the SARS virus.

  A microarray consisting of human DNA can be probed with a DNA sequence from an unknown pathogen to see if that person has been infected. For example, early infection with the malarial parasite *Plasmodium falciparum* can be identified by isolating DNA from the tissues of a suspected patient, fragmenting the DNA into single-stranded DNA segments, and attaching these to the solid support. This microarray is then washed with a fluorescently labeled *P. falciparum* probe. Any fluorescent spots detected on the microarray indicate a match to *P. falciparum* DNA, confirming the patient is infected.
- **Phylogenetic relationships:** The extent of microbial diversity in an environment can be assessed by producing a microarray, called a "Phylochip," containing oligonucleotides (short sequences of nucleotides) that are complementary to 16S rRNA gene sequence probes of different bacterial species. Any lit spot on the microarray is an indicator that that species is part of the microbial community in that environment.

KEY CONCEPT 9.3 Microbial Genomics Studies Genes at the Single Cell to Community Levels

In April 2003, exactly 50 years to the month after Watson and Crick announced the structure of DNA, a publicly financed, $3 billion international consortium of biologists, industrial scientists, computer experts, engineers, and ethicists completed perhaps the most ambitious project to date in the history of biology. The Human Genome Project, as it was called, had succeeded in mapping the **human genome**—that is, the 3 billion nucleobases (equivalent to 750 megabytes of data) in a human cell were identified and strung together (sequenced) in the correct order. The completion of the project represents a scientific milestone with unimaginable potential health benefits.

Many Microbial Genomes Have Been Sequenced

If the human genome was represented by a rope 2 inches in diameter, it would be 32,000 miles long. The genome of a bacterial species like *E. coli* at this scale would be only 1,600 miles long or about 1/20 the length of human DNA. So, being substantially smaller, microbial genomes have been easier and much faster to sequence.

In May 1995, the first complete genome of a free-living organism was sequenced: the 1.8 million base pairs (1.8 Mb) in the genome of the bacterial species *Haemophilus influenzae* (MicroFocus 9.7). In a few short months, the genome for a second organism, *Mycoplasma genitalium* was reported. This reproductive tract pathogen has one of the smallest known bacterial genomes, consisting of only 580,000 base pairs and 485 protein-coding genes. *E. coli* by comparison has about 4,300 protein-coding genes. As more genomes were sequenced, Craig Venter and his team then at The Institute for Genomic Research (TIGR) in Maryland wanted to know: What is the minimal number of genes an organism actually needs to grow and reproduce? They answered this question by disrupting the *M. genitalium* genes one-by-one to discover what genes were not essential. The analysis suggested that some 382 of the 485 protein-coding genes are essential (under laboratory conditions). So, scientists have found the genetic "essence of life"; that is, the minimal genetic information needed to operate a cell—bacterial at least.

MICROFOCUS 9.7: Biotechnology

Putting Humpty Back Together Again

Humpty Dumpty sat on a wall.
Humpty Dumpty had a great fall.
All the king's horses,
And all the king's men,
Couldn't put Humpty together again.

We all remember this nursery rhyme, but it can be an analogy for the efforts required in sequencing the genome of an organism. To sequence a whole genome, you have to take thousands of small DNA fragments and, after sequencing them, try to "put the fragments together again." The good news in genomics is that you can "put Humpty together again."

Small fragments must be used when sequencing a whole genome because current methods will not work with the tremendously long stretches of DNA, even those shorter ones found in bacterial genomes. Therefore, one strategy is to break the genome into small fragments. These fragments then are sequenced using extremely fast sequencing machines and the fragments reassembled into the full genome. This technique is called the "whole-genome shotgun method."

The "shotgun" strategy first was used in 1995 by Craig Venter, Hamilton Smith, Claire Fraser, and their colleagues to sequence the genome of *Haemophilus influenzae* and *Mycoplasma genitalium*. To sequence these bacterial genomes, segments of the DNA were cut into 1,600 to 2,000 base pair segments. The segments then were partly sequenced at both ends, using automated sequencing machines. These base-pair sequences—with their many overlaps—became the sequence information that was entered into the computer. Using innovative computer software, the thousands of DNA fragments generated were compared, clustered, and matched for assembling the genome of each organism.

Once assembled, the genes could be located, compared to known genes, and a detailed map developed (see figure). Sequencing of each genome took about a year but demonstrated that "the king's horses" (supercomputers and shotgun sequencing) and "the king's men" (the large group of collaborators) could "put Humpty together again"—and with speed and accuracy. Note: Since 1995, great strides have been made in sequencing technology. If *H. influenzae* were to be sequenced today, it would take about 5 days, rather than an entire year.

A linear map of the *Mycoplasma genitalium* genome. The horizontal arrows identify protein-coding genes. The direction of the arrow indicates the direction of transcription. Reproduced from: Claire M. Fraser et al., *Science* 270 (1995): 397-404. Reprinted with permission from AAAS.

By the end of 2011, some 2,000 genomes (primarily microbial) had been completely sequenced (FIGURE 9.13). Although 60% of these sequencing projects represent clinically important human pathogens, sequences by themselves do not tell us much. So, what can these sequences tell us and what practical use might be derived from this information?

Segments of the Human Genome Have "Microbial Ancestors"

With the sequencing of the human genome, one interesting development was to compare the human nucleotide gene sequences to known bacterial and viral sequences. Are there any similarities in the genes each contains?

Some comparisons indicate as many as 200 of our 23,000 genes are essentially identical to those found in members of the Bacteria; 25% or some 6,000 genes are found in yeast. However, these human genes were not acquired directly from these species, but rather were genes picked up by early ancestors of humans. So important were these genes, they have been preserved and passed along from organism to organism throughout evolution; they are life's oldest genes. For example, several researchers suggest some genes coding for brain signaling chemicals and for communication between cells did not evolve gradually in human ancestors; rather, these ancestors acquired the genes directly from bacterial organisms. This provocative claim remains highly controversial though and more work is needed to better analyze this possibility.

Another discovery coming from the decade-long **Encyclopedia of DNA Elements (ENCODE) project** is that 80% of the human genome is of functional importance. Other studies suggest that almost 10% of our genome (and substantial parts of other animal genomes) is composed of self-replicating fragments of viral DNA called **human endogenous retroviruses (HERVs)**. Retroviruses are RNA viruses that can reverse transcribe themselves into double-stranded DNA and then insert into a human chromosome. The HERV infections presumably occurred in germ cells thousands or perhaps millions of years ago, integrated their viral genes in the human genome, and have been transmitted vertically to future generations ever

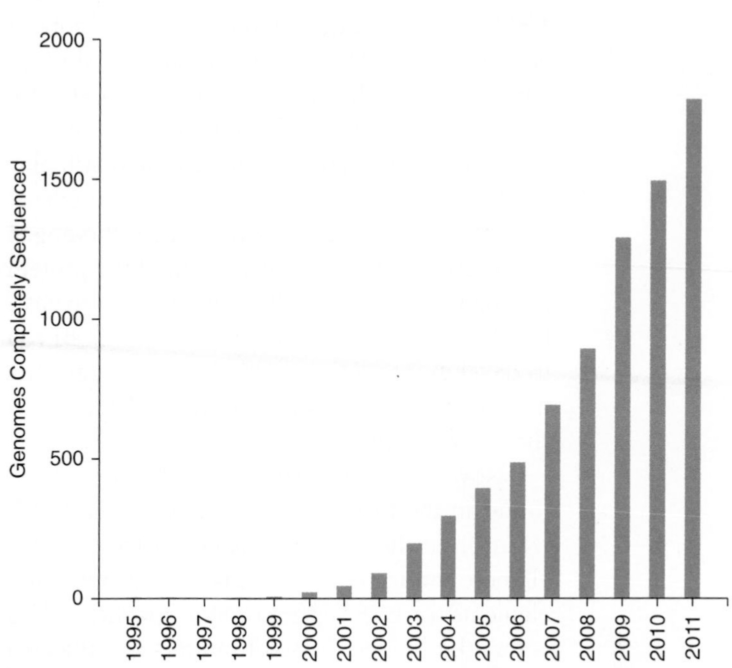

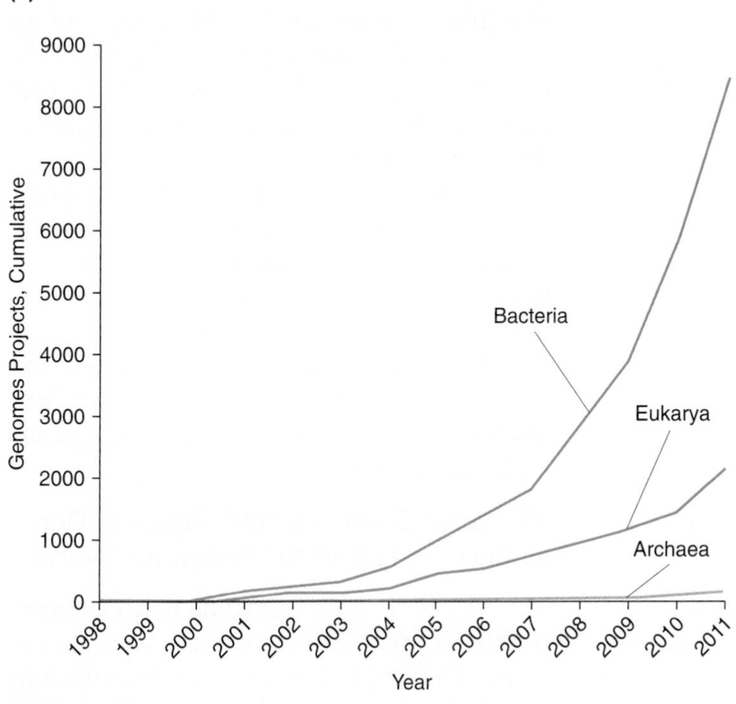

FIGURE 9.13 **Genomics Activities.** **(A)** The total number of genomes completely sequenced will easily surpass 2,000 in 2012. **(B)** The total number of genome sequencing projects is increasing rapidly, especially for the bacterial organisms. Almost half are bacterial species in the phylum Proteobacteria. Modified from: Genomes Online Database. »» Why have substantially fewer genomes been sequenced from the domain Eukarya and Archaea?

since—and these are the remnants we see in our DNA today. The HERVs, of which there are some 100,000 copies scattered in our genome, consist of more than 30 families of retroviruses. Most no longer have an effect in the body to cause disease but undoubtedly have affected human evolution.

Some of these HERV sequences may still play a critical role in our bodies today. For example, at least six genes interact in the normal functioning of the placenta. One HERV gene codes for a protein that allows cells in the outer layer of the placenta to fuse together, a needed event for a fertilized egg to develop into a fetus. Other HERV fragments may play roles in cancer, multiple sclerosis, and rheumatoid arthritis.

Now even more virus "signatures" have been found in the human genome that are very different from the HERVs. The collective term for all these endogenous viruses is "endogenous viral elements" or EVEs. One scenario for these EVEs is that the viruses crossed from rats and infected evolving primates. In primates, the viral genes at some point managed to infect the genomes of egg and sperm cells, allowing the viral genes to be passed through vertical transmission. Then, over time, the viral genes became mutated and disabled, and could no longer produce new viruses.

So, viral genomes might be a legacy of long distant infections and epidemics, and genomics analysis makes it clear that our genome is a hybrid of vertebrate and viral genetic information. Further analysis of our genome has identified other viruses that have integrated into our DNA. Certainly microbial genomics will have much more to tell us about our distant past and evolution as a species.

Microbial Genomics Will Advance Our Understanding of the Microbial World

Microorganisms have existed on Earth for about 3.8 billion years, although we have known about them for little more than 300 years. Over this long period of evolution, they have become established in almost every environment on Earth, make up a significant percentage of Earth's biomass, and, although they are the smallest organisms on the planet, they influence—if not control—some of the largest events. Yet, with few exceptions, we do not know a great deal about any of these microbes, and we have been able to culture and study in the laboratory only about 1% of all microorganism species.

However, our limited knowledge is changing. With the advent of **microbial genomics**, the discipline of sequencing, analyzing, and comparing microbial genomes, we have begun the third Golden Age of microbiology, a time when remarkable scientific discoveries will be made toward understanding the workings and interactions of the microbial world. Some potential consequences from the understanding of microbial genomes are outlined below.

Safer Food Production. Because microorganisms play important roles in our foods both as contamination and spoilage agents, understanding how they get into the food product and how they produce dangerous foodborne toxins, will help produce safer foods. However, a major limitation with traditional food safety surveillance is that food-contaminating or spoiling microbes can only be identified after a long time-consuming approach that requires a number of days for colony counting of surviving microorganisms or bacterial toxin production using agar culture media. In addition, separate tests need to be run for each potential foodborne pathogen. This whole process can be greatly speeded up with microbial genomic technologies, enabling a quick and reliable prediction of the safety of our foods.

Microbial genomics can provide for a quick identification of microorganisms present in a (raw) food product. For example, until recently, there was little research into why *Campylobacter jejuni*, a bacterial pathogen that is a common bacterial cause of food poisoning, is so virulent. Microbial genomics studies have now shown that *C. jejuni* has over 1,700 genes. So, which of these genes are important to the organism when it faces different environmental challenges, such as contaminating raw chicken, residing in fecal matter, or surviving in water? There are a variety of toxins it produces, as well as adherence and invasive factors needed for infection. So knowing the sequences for these genes means one could detect their presence in a food sample. As explained earlier, microarrays would be one way to rapidly detect the presence of these disease-enhancing genes not only from *C. jejuni*, but from all potential foodborne pathogens. Microarrays would provide a very rapid response to potential food contamination and enable the removal of the product or strengthen food-chain control strategies before the pathogen could cause illness to consumers.

Overall, genomics of food microbes generates valuable knowledge that can be used to protect our agriculture produce and meats, and also could be applied as a tool to trace potential food contamination between farm and table, a real problem today for fruits and vegetables coming from both within and outside the United States.

Identification of Uncultured Microorganisms. Because the vast majority of bacterial and archaeal species have not been cultured, genomics offers a way to identify these organisms. Craig Venter is just one of many scientists studying the gene sequences of these **viable but noncultured (VBNC)** organisms. Venter and others are attempting to sequence and identify the collective genomes of all bacterial and archaeal species in a microbial community. This ability to identify uncultured organisms allows microbiologists to better understand how microbial communities function and how the organisms interact with one another (see below).

Genomic information also is being used to discover if there are uncultured microbial representatives that could be used as alternative energy sources to solve critical environmental problems, including climate change and the development of renewable energy sources such as hydrogen and methane.

Microbial Forensics. The advent of the anthrax bioterrorism events of 2001 and the continued threat of bioterrorism has led many researchers to look for ways to more efficiently and more rapidly detect the presence of such bioweapons. Many of the diseases caused by these potential biological agents cause no symptoms for at least several days after infection, and when symptoms appear, initially they are flu-like. Therefore, it can be difficult to differentiate between a natural outbreak and the intentional release of a potentially deadly pathogen.

Such concerns have lead to a relatively new and emerging area in microbiology called **microbial forensics**, the discipline that attempts to recognize patterns in a disease outbreak, identify the responsible pathogen, control the pathogen's spread, and discover the source of the pathogenic agent. Investigative tools, like gene sequencing, DNA probes, microarrays, and PCR often are sufficient if such a disease outbreak is a natural one. However, if the "outbreak" is the result of a purposeful release—a bioterrorism attack—then tracking down the source of the microbe (and perpetrator) is critical. For example, the anthrax letter attacks of 2001 generated panic among the public and showed the need to establish "attribution" (who is responsible for the crime) for fear that another such attack might occur. The 2009 swine flu pandemic at one point was proposed by some to be the result of an accidental release of the virus from a research lab doing vaccine experimentation. Microbial forensic investigations did not support this claim.

The science behind microbial forensics includes classical microbiology, genetic engineering, microbial genomics, and phylogenetics. Forensic microbiological investigations are essentially the same as any other forensic investigation as they involve a crime scene(s) investigation, evidence collection, chain of custody for the collection, handling and preservation of evidence, interpretation of results and—unique to the scientist—court presentation. Importantly, microbial forensic data must be undeniable and must hold up to the scrutiny of judges and juries in a court of law in order to bring a perpetrator(s) to justice and to deter future attacks.

Microbial forensics also is involved in other types of medical and hygiene cases. Unique pathogen genetic sequences can be used to track infections in hospitals. Did a patient have the infection when admitted or was it acquired in the hospital? Cases of medical negligence, where a hospital's inadequate or improper hygiene can lead to a patient contracting a postsurgical or hospital-acquired infection (and perhaps die), could be settled through forensics analysis.

Outbreaks of foodborne disease have brought lawsuits against companies alleging negligence in sanitary practices. And the potential for intentional contamination through foodborne terrorism is certainly possible.

In all these cases, tracing the infecting microbe to the company or person(s) of origin is critical. This places the fields of genetic engineering, biotechnology, and microbial genomics at the forefront of this emerging field.

Metabolic Engineering. As described above, many of the products of biotechnology are derived from bacterial cells that have been turned into tiny factories to manufacture the desired protein or vaccine. However, if more than one gene is needed to produce the molecule (perhaps a whole metabolic pathway is required) then another technique of biotechnology can be used. Called **metabolic engineering**, the process involves altering a set of genes and metabolic pathways in a cell to specifically produce the identified product. Often this means engineering the metabolic pathway from a hard to grow organism into one more amenable to normal growth conditions. With the coming shortage of fossil fuels, researchers in the field see the cellular production of gasoline and other fuels as a target for metabolic engineering. Other potentially valuable pharmaceuticals that are too complex and expensive to chemically synthesize might be more reasonably produced with metabolically engineered microbes.

Comparative Genomics Brings a New Perspective to Defining Infectious Diseases and Studying Evolution

Sequencing the DNA bases of a microorganism (or any other organism) has and still does provide important information concerning the number of bases and genes comprising the organism. However, such sequences provide little understanding of how these genes work together to run the metabolism of an organism. One needs to understand how the organism uses its genome to form a functioning unit. Sequencing is only the first part of a deeper understanding.

Having sequenced a microbial genome, the next step is to discover the functions for the genes. Sequences need to be analyzed (called **genome annotation**) to identify the location of the genes and the function of their RNA or protein products. For example, in most of the microbial genomes sequenced to date, nearly 50% of the identified genes encoding proteins have not yet been connected with a cellular function. About 30% of these proteins are unique to each species. The challenging discipline of **functional genomics** attempts to discover what these proteins do, and how those genes interact with others and the environment to maintain microbial growth and reproduction.

■ Pathogenicity: The ability of a pathogen to cause disease.

One of the most important areas beyond DNA sequencing is the field of **comparative genomics**, which compares the DNA sequence from one microbe with the DNA sequence of another similar or dissimilar organism. Comparing sequences of similar genes indicates how genomes have evolved over time and provides clues to the relationships between microbes on the phylogenetic tree of life.

Comparisons indicate, for example, that some strains of a bacterial species contain **genomic islands**, clusters of up to 25 genes that are absent from other strains of the same species. Many of these islands can be identified as having come from an altogether different species, suggesting some form of HGT, such as conjugation, occurred in the past. It is believed the nonpathogenic bacterial species *Thermotoga maritima* has acquired about 25% of its genome from HGT. In addition, sequence analysis indicates its genome is a mixture of bacterial and archaeal genes and suggests *T. maritima* evolved before the split of the Bacteria and Archaea domains.

One of the most interesting aspects of comparative genomics relates to infectious disease. By comparing the genomes of pathogenic and nonpathogenic bacterial species, or between pathogens with different host ranges, microbiologists are learning a lot about pathogen evolution. Here are a few examples.

There are three bacterial species of *Bordetella*. *B. pertussis* causes whooping cough in humans, *B. parapertussis* causes whooping cough in infants, but also infects sheep, and *B. bronchiseptica* produces respiratory infections in other animals (FIGURE 9.14A). Comparative genomic analysis of these three species reveals that *B. pertussis* and *B. parapertussis* are missing large segments of DNA (1,719 genes), which are present in *B. bronchiseptica*. This analysis suggests that (1) *B. pertussis* and *B. parapertussis* evolved from a *B. bronchiseptica*-like ancestor; and (2) the adaptation of *B. pertussis* and *B. parapertussis* to their more restrictive hosts is due to the loss of the 1,719 genes. In fact, only *B. bronchiseptica* is capable of surviving outside the host. So, in this genome comparison between similar species, survival of *B. pertussis* and *B. parapertussis* requires they infect organisms supplying them with the materials they no longer can make; that is, pathogenicity has evolved from the loss of gene function.

At the opposite extreme is the evolution of pathogenicity through the acquisition of new genes (FIGURE 9.14B). *Corynebacterium diphtheriae* is the causative agent for diphtheria. Genome analysis indicates that an ancestor species acquired through HGT 13 genetic regions, each representing a genomic island. These islands are called **pathogenicity islands** because they encode many of the pathogenic characteristics of the bacterial species (e.g., pili formation and iron uptake).

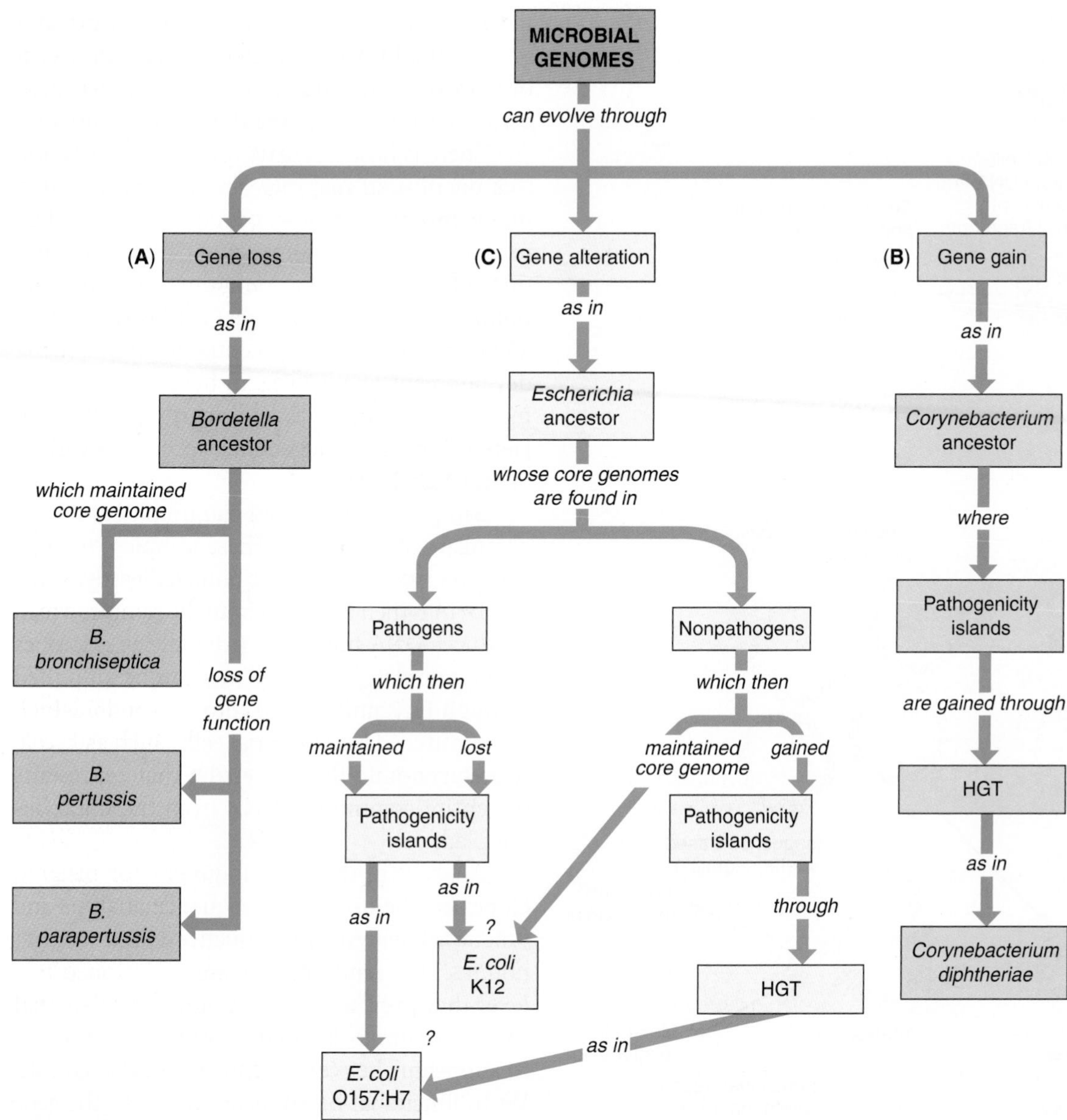

FIGURE 9.14 Comparative Genomics Suggests How Microbial Genomes Can Evolve. This concept map summarizes the evolution of three microbial genomes through loss or gain of gene function. »» What advantage is there to losing or gaining genes, or whole sets of genes (pathogenicity islands)?

A strain of *E. coli*, called O157:H7, has been a dangerous threat to human health worldwide, causing severe gastrointestinal ailments. One of the most recent outbreaks involved the contamination of bagged spinach. Some 200 Americans became ill and at least two died. When the genome of *E. coli* 0157:H7 was compared to the nucleotide sequence of a nonpathogenic strain (K12), another example for the presence of pathogenicity islands was discovered (FIGURE 9.14C). Both strains have a large genome and have evolved from a common ancestor. Both have genomic islands acquired through HGT. However, the genomic islands in *E. coli* O157:H7 code for the known pathogenicity genes (e.g., pili and toxins) and therefore represent pathogenicity islands. The genomic islands in the nonpathogenic strain lack these pathogenicity genes. What is not clear is if the pathogenicity islands were acquired only by the O157:H7 strain or the nonpathogenic strain lost the pathogenicity islands.

These few examples represent examples of the power of comparative genomics to resolve differences between species and shed light on the evolution of bacterial pathogens.

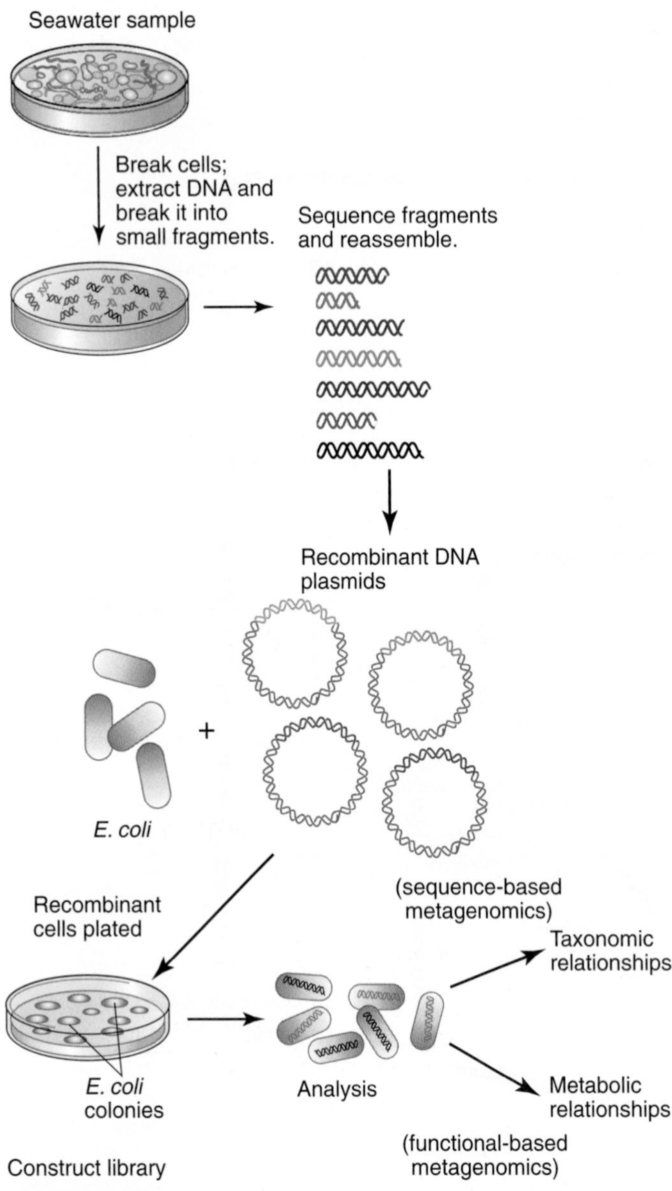

FIGURE 9.15 **The Process of Metagenomics.** Metagenomics allows the simultaneous sequencing of a whole community of microorganisms without growing each species in culture. Fragments can be either sequenced or subjected to functional analysis. »» What type of information is supplied by sequence-based metagenomics versus functional-based metagenomics?

Metagenomics Is Identifying the Previously Unseen Microbial World

Existing within, on, and around every living organism, and in most all environments on Earth, are microorganisms. Yet, as some 99% of the bacterial and archaeal species found within us and in the environment will not grow on traditional growth media; we referred to these organisms as viable, but noncultured (VBNC). That means most organisms found in the soil, in oceans, and even in the human body have never been seen or named—and certainly not studied. Yet these organisms represent a wealth of genetic variety.

There is now a genetic process for analyzing this uncultured majority. The process is called **metagenomics** (*meta* = "beyond"), and it refers to identifying the genome sequences within mixtures of organisms in a community (the **metagenome**); that is, "beyond" what has been cultured. Metagenomics has the potential to stimulate the development of advances in fields as diverse as medicine, agriculture, and energy production. Here is how the metagenomics process is carried out (FIGURE 9.15).

Samples of the desired community of microorganisms are collected. These samples could be from soil, water, or even the human digestive tract. The DNA is then extracted from the sample, which produces DNA fragments from all the microbes in the samples. The fragments can be amplified through PCR and cloned into plasmids, which are introduced into bacterial cells, such as *E. coli*. A metagenomic library results that represents the entire community DNA from the microbes sampled.

Analysis of the DNA fragments or plasmid clones can be used to do sequence analysis and functional analysis. In "sequence-based metagenomics," the random fragments are cloned to a level that produces vast amounts of DNA that can, for example, be used to compare taxonomic relationships between organisms in the sample. With "functional-based metagenomics," the gene products from the cloned plasmids in the bacterial cells can be used to compare metabolic relationships within the community or to search for new enzymes, vitamins, antibiotics, or other potential chemicals of therapeutic or industrial use.

As Venter's explorations and those of others have shown, metagenomics already has opened our eyes to the diversity of microbes in ocean environments. The process also makes it possible to harness the power of microbial communities to help solve some of the most complex medical, environmental, agricultural, and economic challenges in today's world.

Medicine. Understanding how the microbial communities that inhabit our bodies affect human health could lead to new strategies for diagnosing, treating, and preventing diseases.

Ecology and the Environment. Exploring how microbial communities in soil and the oceans affect the atmosphere and environmental conditions could help us understand, predict, and address climate change.

Energy. Harnessing the power of microbial communities might result in sustainable and eco-friendly bioenergy sources, as exemplified by the explorations of Craig Venter and others.

Bioremediation. Adding to the arsenal of microorganism-based environmental tools can help in monitoring environmental damage and cleaning up oil spills, groundwater, sewage, nuclear waste, and other hazards.

Biotechnology. Taking advantage of the functions of microbial communities might lead to the development of newer and safer food and health products.

Agriculture. Understanding the roles of beneficial microorganisms living in, on, and around domesticated plants and animals could enable detection of diseases in crops and livestock, and aid in the development of improved farming practices.

Biodefense. The addition of metagenomics tools to microbial forensics will help to monitor pathogens, create more effective vaccines and therapeutics against bioterror agents, and reconstruct attacks that involve microorganisms.

The ability to manipulate genomes is increasing at an ever rapid pace and this is no clearer than in the creation of nonnatural life (MicroFocus 9.8).

CONCEPTS AND REASONING CHECKS 3

- **a.** Why have so many bacterial species been sequenced?
- **b.** Why do we say the human genome is a blending of vertebrate and viral genomes?
- **c.** How is microbial genomics contributing to a better world?
- **d.** Describe how comparative genomics helps explain some aspects of pathogenicity.
- **e.** What does metagenomics offer that previous sequencing techniques have not been able to provide?

MICROFOCUS 9.8: Biotechnology/Evolution
A Bacterial Imposter

Invasion of the Body Snatchers is a classic sci-fi film of the 1950s (remade in 1978 and 1994). In the original, the town's doctor discovers that many of his patients are being invaded and replaced by emotionless alien imposters and he must try to combat and quell the deadly, indestructible threat.

In 2010, a bacterial imposter was announced—but of a purposeful kind. Researchers at the J. Craig Venter Institute (JCVI) in Rockville, Maryland and San Diego, California reported they were able to carry out genome transplantation and in the process transform one bacterial species into another species. In their work, the researchers built the entire genome of *Mycoplasma mycoides* from scratch; that is, they fed the genetic sequence into a DNA synthesizer and produced fragments of the complete *M. mycoides* genome. Then, with the help of yeast cells, the fragments were assembled into the complete genome.

The Venter team then transferred the synthetic genome into a genome-less *Mycoplasma capricolum* cell. The result was that the transplanted genome then dictated the behavior of the recipient *M. capricolum* cell—it had the phenotypic characteristics of *M. mycoides* and was able to reproduce.

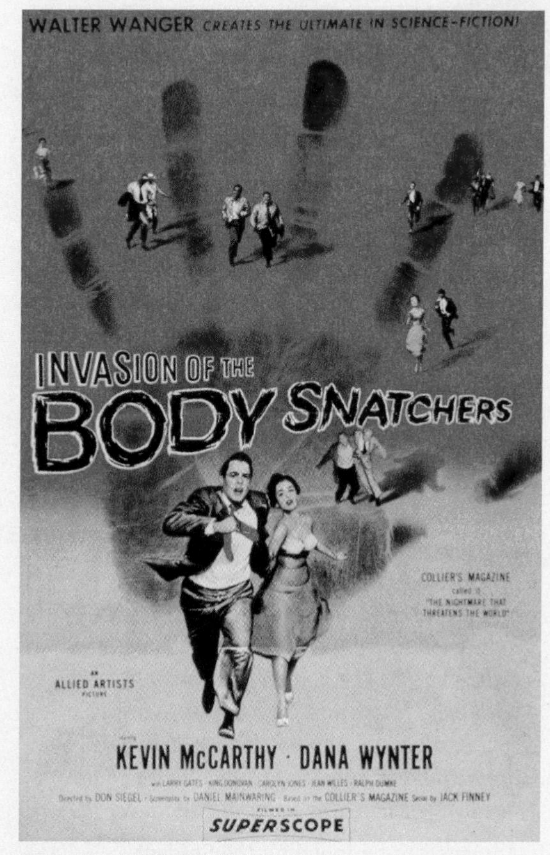

JCVI refers to this as the world's first "synthetic life form." However, it is not really a "synthetic life form" because the recipient cell was a normal *M. capricolum* cell and the genome was a normal (but synthetic) genome of *M. mycoides*. So, it is like erasing a computer's old operating system and reprogramming it with a new operating system. Importantly, the JCVI folks have not created life à la spontaneous generation and have not made a completely new organism. Perhaps it is better to say that they have made the world's first "nonnatural life form."

So, what is to be gained from this type of "genetically engineered transplantation?" The JCVI scientists' goal was not to create some deadly, indestructible group of alien imposters as in *Invasion of the Body Snatchers*, but to endow a synthetic genome with genes that in a recipient cell will allow those cells to carry out useful functions. According to Venter, in the lab one could start with a naturally occurring microbe and strip it down to its minimal genome, producing what Venter calls *Mycoplasma laboratorium*. With such an organism, the researchers could then add nonnative genes to build an organism with unique biological properties and commercial utility. For example, genes could be added to the cells to produce ethanol, hydrogen, or another form of renewable biofuels. According to Venter, such "artificial life" could also go a long way toward helping resolve the climate crisis by building organisms that could soak up carbon dioxide from the atmosphere. Certainly there are numerous other potential uses that could be programmed into *M. laboratorium*—hopefully for the good of humankind.

Which brings up a more related *Invasion* scenario. Beyond the amazing genetic engineering and biotechnological breakthroughs that might come in the near future from genome transplantation, some critics are concerned that such nonnatural organisms could be turned into biological weapons by inserting genes that make the cells the ultimate doomsday pathogen. As this area of synthetic biology continues to advance, certainly there are bioethical issues to consider—and Venter would be the first to say such discourse should continue.

Chapter Challenge C

From your reading in this section, you should appreciate that the human genome is a hybrid of human and viral genes. Throw in a few hundred bacterial genes, and it becomes hard to fathom exactly what a human genome is. To many, the idea that the human genome could have evolved from viruses and other microbial genes seems somehow uncomforting.

QUESTION C:

So what is your impression, feeling, or response to the following statement, which is based in the science you have learned from this text? In the human individual: (1) there are 10× more microbial cells in your body than human cells; (2) the millions of mitochondria in all body cells have evolved from once free-living bacterial cells; and (3) the human genome is littered with gene fragments from ancient viruses. And in all cases, these intimate strangers are helping us survive as humans. We could not live without this inner microbial world.

Answers can be found on the Student Companion Website in **Appendix D.**

■ **In conclusion**, genetic engineering and the various molecular techniques it has spun off through biotechnology have revolutionized the field of microbiology—and stimulated the third Golden Age of microbiology. Examining and comparing the gene sequences of cultured and uncultured microbes (in isolation or from whole communities) have provided microbiologists with a better understanding of the processes microbes carry out and their pathogenic capabilities. With regard to taxonomic relationships, the information gleaned from genome sequencing and comparative genomics has turned the "tree of life" into a somewhat "tangled bush" because of the ability of Bacteria and Archaea to undergo horizontal gene transfer in an almost promiscuous way—genes or whole genomic islands can be shared between dissimilar species. In addition, since the first microbe was sequenced in 1995, the majority of sequenced genomes have been biased, for obvious reasons, toward microorganisms of medical and economic interest. Therefore, today we do not have a set of sequenced genomes representing a balanced and broad spectrum of species in the Bacteria and Archaea from which to generate a consensus "tree of life."

SUMMARY OF KEY CONCEPTS

9.1 Bacterial Cells Can Recombine Genes in Several Ways

1. Recombination implies a horizontal transfer of DNA fragments between bacterial cells and an acquisition of genes by the recipient cell. All three forms of recombination are characterized by the recombination of new genes to a recipient cell by **horizontal gene transfer**. (Fig. 9.2)
2. In **transformation**, a competent recipient cell takes up "naked" DNA from the local environment. The new DNA displaces a segment of equivalent DNA in the recipient cell, and new genetic characteristics may be expressed. (Fig. 9.3)
3. In one form of **conjugation**, a live donor (F^+) cell transfers an **F factor** (plasmid) to a recipient cell (F^-), which then becomes F^+. (Fig. 9.5A)
4. In another form of conjugation, **Hfr** strains contribute a portion of the donor's chromosomal genes to the recipient cell. (Fig. 9.5B)
5. **Transduction** involves a virus entering a bacterial cell and later replicating within it. In **generalized transduction**, a bacterial DNA fragment is mistakenly incorporated into an assembling phage. In **specialized transduction**, the replicating virus takes a segment of chromosomal DNA with it. In both forms, the phage transports the DNA to a new recipient (transduced) cell. (Fig. 9.6)

9.2 Genetic Engineering Involves the Deliberate Transfer of Genes Between Organisms

6. **Genetic engineering** is an outgrowth of studies in bacterial genetic recombination. The ability to construct **recombinant DNA molecules** was based on the ability of **restriction endonucleases** to form **sticky ends** on DNA fragments. (Fig. 9.8)
7. **Plasmids** can be isolated from a bacterial cell, spliced with foreign genes, then inserted into fresh bacterial cells where the foreign genes are expressed as protein. The cells become biochemical factories for the synthesis of such proteins as insulin, interferon, and human growth hormone. (Figs. 9.10, 9.11)
8. **DNA probes** can be used to detect pathogens. (Fig. 9.12)

9.3 Microbial Genomics Studies Genes at the Single Cell to Community Levels

9. Since 1995, increasingly more microbial genomes have been sequenced; that is, the linear sequence of bases has been identified. (Fig. 9.13)
10. A comparison of bacterial genomes with the human genome has shown there may be some 200 genes in common between these organisms. Comparisons between microbial genomes indicate almost 50% of the identified genes have yet to be associated with a protein or function in the cell.
11. With the understanding of the relationships between sequenced microbial DNA molecules comes the potential for safer food production, the identification of uncultured microorganisms, a cleaner environment, and improved monitoring of pathogens through **microbial forensics.**
12. Sequencing is only the first step in understanding the behaviors and capabilities of microorganisms. **Functional genomics** attempts to determine the functions of the sequenced genes and how those genes interact with one another and with the environment. **Comparative genomics** compares the similarities and differences between microbial genome sequences. Such information provides an understanding of the evolutionary past and how pathogens might have arose through the gain or loss of **pathogenicity islands.**
13. **Metagenomics** is providing new insights into the function of diverse genomes in microbial communities. (Fig. 9.15)

CHAPTER SELF-TEST

For **STEPS A–D**, answers to even-numbered questions and problems can be found in **Appendix C** on the Student Companion Website at **http://microbiology.jbpub.com/10e**. In addition, the site features eLearning, an online review area that provides quizzes and other tools to help you study for your class. You can also follow useful links for in-depth information, read more MicroFocus stories, or just find out the latest microbiology news.

STEP A: REVIEW OF FACTS AND TERMS

Multiple Choice

Read each question carefully, then select the ***one*** answer that best fits the question or statement.

1. Which one of the following is NOT an example of genetic recombination?
   A. Conjugation
   B. Binary fission
   C. Transduction
   D. Transformation
2. Transformation refers to ______.
   A. using a virus to transfer DNA fragments
   B. DNA fragments transferred between live donor and recipient cells
   C. the formation of an F^- recombinant cell
   D. the transfer of "naked" DNA
3. An F^- cell is unable to initiate conjugation because it lacks ______.
   A. double-stranded DNA
   B. a prophage
   C. an F factor
   D. DNA polymerase
4. An Hfr cell ______.
   A. has a free F factor in the cytoplasm
   B. has a chromosomally integrated F factor
   C. contains a prophage for conjugation
   D. cannot conjugate with a F^- recombinant
5. A _____ is NOT associated with specialized transduction.
   A. virulent phage
   B. lysogenic cycle
   C. prophage
   D. recipient cell

6. Which complementary sequence would NOT be recognized by a restriction endonuclease?
   A. GAATTC
      CTTAAG
   B. AAGCTT
      TTCGAA
   C. GTCGAC
      CAGCTG
   D. AATTCC
      TTAAGG
7. A _____ seals sticky ends of recombinant DNA segments.
   A. DNA ligase
   B. restriction endonuclease
   C. protease
   D. RNA polymerase
8. _____ are single-stranded DNA molecules that can recognize and bind to a distinctive nucleotide sequence of a pathogen.
   A. Prophages
   B. Plasmids
   C. Cloning vectors
   D. DNA probes
9. The first completely sequenced genome from a free-living organism was from _____.
   A. humans
   B. *E. coli*
   C. *Haemophilus*
   D. *Bordetella*
10. What percentage of the human genome is identical to the yeast genome?
   A. 5%
   B. 10%
   C. 25%
   D. 50%
11. A metagenome refers to _____.
   A. a large genome in an organism
   B. the collective genomes of many organisms
   C. the genome of a metazoan
   D. two identical genomes in different species
12. Genomic islands are _____.
   A. gene sequences not part of the chromosomal genes
   B. adjacent gene sequences unique to one or a few strains in a species
   C. acquired by HGT
   D. Both **B** and **C** are correct.
13. Craig Venter's sampling of ocean microorganisms is an example of _____.
   A. microarrays
   B. horizontal gene transfer
   C. microbial forensics
   D. metagenomics

Fill-in

Use the following syllables to compose the term that answers each of the clues below. The number of letters in each term is indicated by the blank lines, and the number of syllables is shown by the number in parentheses. Each syllable is used only once.
ASE BAC CLE COC COM CON CUS DO DROME EN FITH GA GASE GE GRIF HOR I I IN JU LENT LI MIDS MO NOME NU O PAL PE PHAGE PLAS PNEU TAL TENCE TER TION U VIR ZON

14. Closed loops of DNA (2) _ _ _ _ _ _ _
15. Restriction recognition sequence (3) _ _ _ _ _ _ _ _ _ _
16. Transforming property (3) _ _ _ _ _ _ _ _ _ _
17. Transduction virus (5) _ _ _ _ _ _ _ _ _ _ _ _ _
18. Recombinant DNA enzyme (5) _ _ _ _ _ _ _ _ _ _ _ _
19. Transformed bacterium (4) _ _ _ _ _ _ _ _ _ _ _ _
20. DNA linking enzyme (2) _ _ _ _ _ _
21. Discovered transformation (2) _ _ _ _ _ _ _ _
22. Type of HGT (4) _ _ _ _ _ _ _ _ _ _ _ _
23. Phage that causes lysis (3) _ _ _ _ _ _ _ _ _
24. Complete set of genes (2) _ _ _ _ _ _
25. Type of gene transfer (4) _ _ _ _ _ _ _ _ _ _

STEP B: CONCEPT REVIEW

26. Contrast **vertical** and **horizontal gene transfer** mechanisms. (**Key Concept 1**)
27. Describe and assess the role of **transformation** as a **genetic recombination** mechanism. (**Key Concept 1**)
28. Explain how an **F factor** is transferred during **conjugation**. (**Key Concept 1**)
29. Distinguish between an **Hfr strain** and a **F$^-$ recombinant** cell. (**Key Concept 1**)
30. Summarize the steps involved in (a) **generalized** and (b) **specialized transduction**. (**Key Concept 1**)
31. Differentiate between **genetic engineering** and **biotechnology**. (**Key Concept 2**)
32. Identify the role of **plasmids** and **restriction endonucleases** in the genetic engineering process. (**Key Concept 2**)
33. Explain how **DNA probes** are used to (1) identify pathogens and (2) conduct water-quality tests. (**Key Concept 2**)
34. Assess the importance of **microbial genomics** in understanding the human genome. (**Key Concept 3**)
35. Summarize how microbial genomics can contribute to food safety, microorganism identification, and **microbial forensics**. (**Key Concept 3**)
36. Justify the need for **comparative genomics** as related to pathogen evolution. (**Key Concept 3**)
37. Identify how information from **metagenomics** can contribute to medicine, energy production, agriculture, and biodefense. (**Key Concept 3**)

STEP C: APPLICATIONS AND PROBLEMS

38. In 1976, an outbreak of pulmonary infections among participants at an American Legion convention in Philadelphia led to the identification of a new disease, Legionnaires' disease. The bacterial organism, *Legionella pneumophila*, responsible for the disease had never before been known to be pathogenic. From your knowledge of bacterial genetics, can you postulate how it might have acquired the ability to cause disease?

39. You are going to do a genetic engineering experiment, but the labels have fallen off the bottles containing the restriction endonucleases. One loose label says *Eco*RI and the other says *Pvu*I. How could you use the plasmid shown in MicroInquiry 9 to determine which bottle contains the *Pvu*I restriction enzyme?

40. As a research member of a genomics company, you are asked to take the lead on sequencing the genome of *Legionella pneumophila* (see Question 38). (a) Why is your company interested in sequencing this bacterial species, and (b) what possible applications are possible from knowing its DNA sequence?

STEP D: QUESTIONS FOR THOUGHT AND DISCUSSION

41. Which of the recombination processes (transformation, conjugation, or transduction) would most likely occur in the natural environment? What factors would encourage or discourage your choice from taking place?

42. Many bacterial cells can take up DNA via the transformation process. From an evolutionary perspective, what might have been the original advantage for the cells taking up naked DNA fragments from the extracellular environment?

43. The world is continually plagued by a broad series of influenza viruses that differ genetically from one another. For example, we have heard of Hong Kong flu, avian flu, and now swine flu. How might the process of transduction help explain this variability?

44. It is not uncommon for students of microbiology to confuse the terms reproduction and recombination. How do the terms differ?

45. While studying for the microbiology exam covering the material in this chapter, a friend and biology major asks you why genomics, and especially microbial genomics, was emphasized. How would you answer this question?

46. In 2011, researchers reported that they had discovered for the first time that bacterial cells can acquire human genes. They found that more than 10% of a *Neisseria gonorrhoeae* population, the species responsible for gonorrhea, contained a human DNA fragment. This fragment was not found in other species of *Neisseria*. How might a bacterial cell have acquired this human DNA fragment?

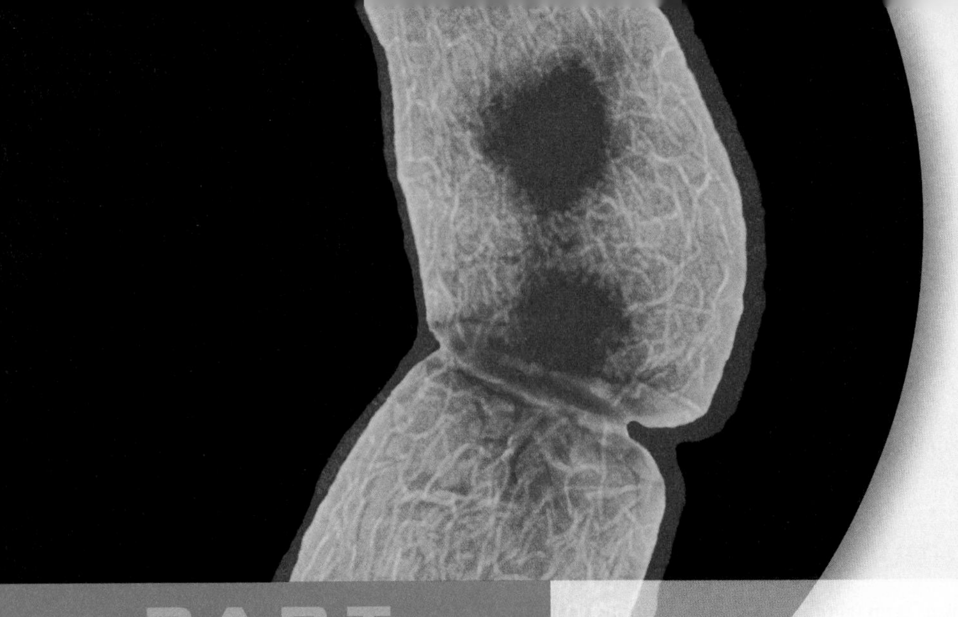

False-color transmission electron microscope image of *Mycobacterium tuberculosis,* which has infected one-third of the human population.

PART 3 Bacterial Diseases of Humans

CHAPTER 10 Airborne Bacterial Diseases

CHAPTER 11 Foodborne and Waterborne Bacterial Diseases

CHAPTER 12 Soilborne and Arthropodborne Bacterial Diseases

CHAPTER 13 Sexually Transmitted and Contact Transmitted Bacterial Diseases

Throughout history, bacterial diseases have posed a formidable challenge to humans and often swept through populations virtually unchecked. In the eighteenth century, the first European visitors to the South Pacific found the islanders robust, happy, and well adapted to their environment. But the explorers introduced syphilis, tuberculosis, and pertussis (whooping cough) to a susceptible population. Soon these diseases spread like wildfire. For example, the Hawaiian population was about 300,000 when Captain Cook landed in 1778; by 1860, disease had reduced the population to fewer than 37,000.

With equally devastating results, the Great Plague came to Europe from Asia, and cholera spread westward from India. Together with tuberculosis, diphtheria, and dysentery, these bacterial diseases ravaged European populations for centuries and insidiously wove themselves into the pattern of life. Infant mortality was particularly shocking: England's Queen Anne, who reigned in the early 1700s, lost 16 of her 17 babies to disease; and until the mid-1800s, only half the children born in the United States reached their fifth year.

Today, humans can cope better with bacterial diseases. Though credit often is given to antimicrobial drugs, the major health gains have resulted from understanding disease and the body's resistance mechanisms, coupled with modern sanitary methods to prevent microorganisms from reaching their targets. Immunization also has played a key role in preventing disease. Indeed, very few people in our society die of the bacterial diseases that once accounted for the majority of all deaths.

In Part 3 of this text, we study the bacterial diseases of humans. The diseases have been grouped according to their major mode of transmission. Airborne diseases are discussed in Chapter 10; foodborne and waterborne diseases in Chapter 11; soilborne and arthropodborne diseases in Chapter 12; and sexually transmitted and contact transmitted diseases in Chapter 13. Many of these diseases are of historical interest and are currently under control. However, the human body is continually confronted with newly emerging or resurgent infectious diseases. In this regard, disease has not changed; only the pattern of disease has changed.

MICROBIOLOGY PATHWAYS

Clinical Microbiology

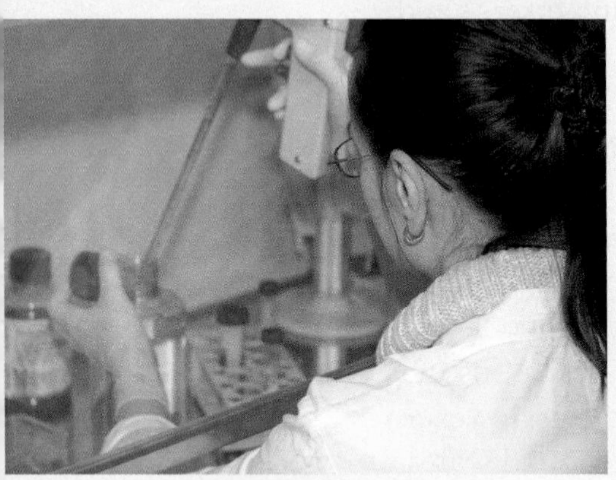

© Viktor Pryymachuk/ShutterStock, Inc.

One of the most famous books of the twentieth century was *Microbe Hunters* by Paul de Kruif. First published in the 1920s, de Kruif's book describes the joys and frustrations of Pasteur, Koch, Ehrlich, von Behring, and many of the original microbe hunters. The exploits of these scientists make for fascinating reading and help us understand how the concepts of microbiology were formulated. I would urge you to leaf through the book at your leisure.

Microbe hunters did not come to an end with Pasteur, Koch, and their contemporaries, nor did the stories of microbe hunters end with the publication of de Kruif's book. Approximately 25% of all deaths worldwide and 68% of all deaths in children under 5 years of age are due to infectious agents. Today, clinical microbiology is concerned with the microbiology of infectious diseases, and the men and women working in hospital, public, and private laboratories are today's diseases detectives. These individuals search for the pathogens of disease. Many travel to far corners of the world studying organisms, and many more remain close to home, identifying the pathogens in samples sent by physicians, identifying their interactions with the immune system, and working out the diagnosis and epidemiology of these diseases.

In fact, a well-developed knowledge of clinical microbiology is critical for the physician and medical staff who are faced with the concepts of disease and antimicrobial therapy. Microbiologists even work in dental clinical labs, since many bacterial species are involved in tooth decay and periodontal disease. Microbiology is one of the few courses where much of the fundamentals of microbiology are used regularly. This includes the clinical aspects of infectious diseases: manifestations (signs and symptoms), diagnosis, treatment, and prevention.

A career in clinical microbiology requires at least a bachelor's degree in a health-related field and preferably a master of science in infection control. In fact, about 70% of infection control practioners are registered nurses and 25% are medical technologists. With such a degree, jobs include supervisory positions in medical centers or private reference laboratories, infection control positions in state and public health labs, marketing and sales in the pharmaceutical and biotechnology industries, teaching at community colleges or technical colleges, or research in university and medical schools, as well as in government or industry (pharmaceutical and biotechnology) settings.

The microbe hunters have not changed materially in the past 100 years. The objectives of the search may be different, but the fundamental principles of the detective work remain the same. The clinical microbiologist is today's version of the great masters of a bygone era.

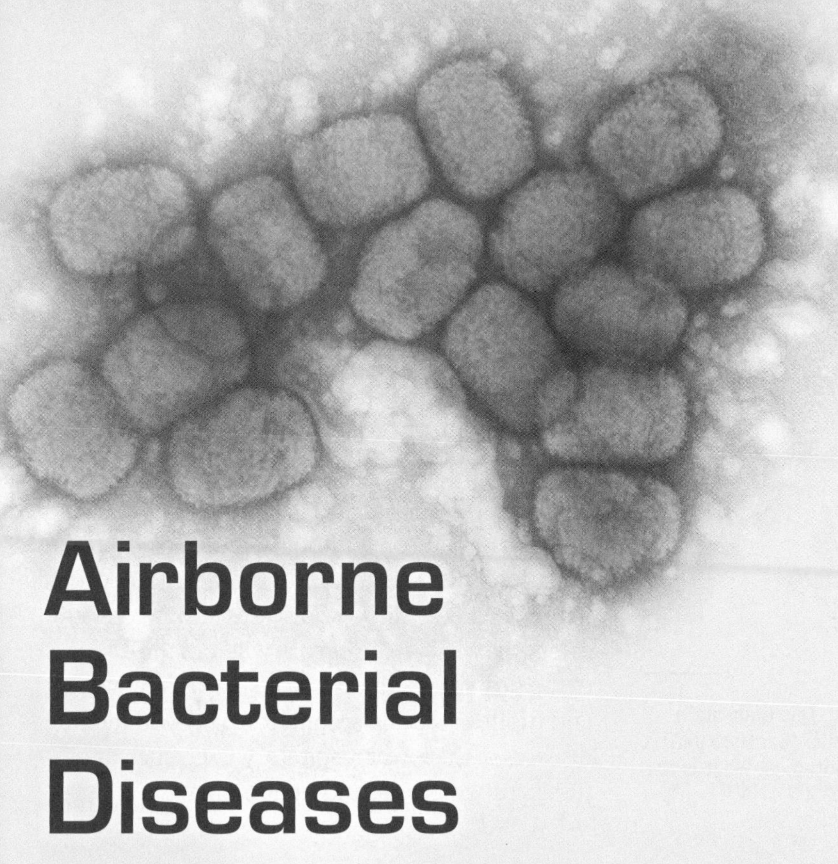

Airborne Bacterial Diseases

10

CHAPTER PREVIEW

10.1 **The Respiratory System Possesses an Indigenous Microbiota**

10.2 **Several Bacterial Diseases Affect the Upper Respiratory Tract**

10.3 **Many Bacterial Diseases of the Lower Respiratory Tract Can Be Life Threatening**

Investigating the Microbial World 10: Cleanliness Is Next to Disinfectedness

Textbook Case 10: Legionnaires' Disease Outbreak—Bogalusa, Louisiana

MicroInquiry 10: Infectious Disease Identification

Pertussis is the only vaccine-preventable childhood illness that has continued to rise since the 1980s with an increasing proportion of cases in adolescents and adults.
—Centers for Disease Control and Prevention

In March 2010, the California Department of Public Health (CDPH) began receiving an abnormally large number of reports detailing patients with symptoms of runny nose, low-grade fever, and a mild, occasional cough that persisted for 7 to 14 days. For many, the symptoms led to a series of rapid coughs followed by a high-pitched "whoop." It appeared that an outbreak of pertussis, commonly called whooping cough, was occurring in California. Caused by *Bordetella pertussis,* this highly communicable bacterial disease was spreading throughout the state.

In the early parts of the 20th century, one of the most common childhood diseases and causes of death in the United States was pertussis. Before the introduction of a pertussis vaccine in 1940, *B. pertussis* was responsible for infection and disease in 150 out of every 100,000 people. By 1980, the **incidence**, or frequency with which the disease occurs, had dropped to one in every 100,000 individuals. The vaccine had almost eliminated the pathogen.

Pertussis outbreaks can be difficult to identify, so usually culture confirmation that *B. pertussis* is circulating in the outbreak or potential epidemic is necessary (FIGURE 10.1). By the end of June, the CDPH said there had been a 418% increase in cases reported from the same period in 2009. Five infants had died and it was obvious that a major epidemic of pertussis was in full swing in California. Meanwhile, other states, including Michigan and Ohio, were also reporting historically high numbers of pertussis cases. Medical experts urged parents to vaccinate their young children, who are the most vulnerable to infection,

Image courtesy of Dr. Fred Murphy/CDC.

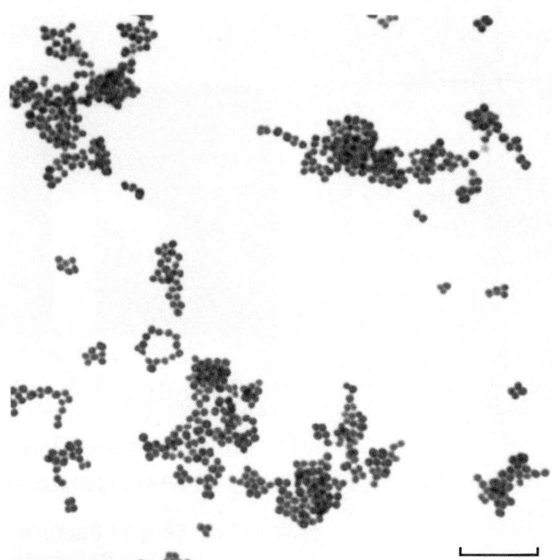

FIGURE 10.1 ***Bordetella pertussis.*** This Gram stain shows chains of small *B. pertussis* cells. (Bar = 20 μm.) © Dr. George J. Wilder/Visuals Unlimited. »» What is the Gram reaction of these stained cells?

and recommended those who care for or are in contact with children be immunized to prevent further spread.

Still, by the end of 2010, the CDPH had reported 9,273 cases of pertussis (including ten infant deaths), making this the most cases reported in 65 years and the highest incidence in 52 years (FIGURE 10.2). Disease activity levels in 2011 in California declined but were still above the norm.

Pertussis is but one among a group of bacterial infectious diseases affecting the respiratory system. We will divide these diseases into two general categories. The first category will include diseases of the upper respiratory tract, such as strep throat and diphtheria. The second category will include diseases of the lower respiratory tract, such as pertussis, tuberculosis, and pneumonia.

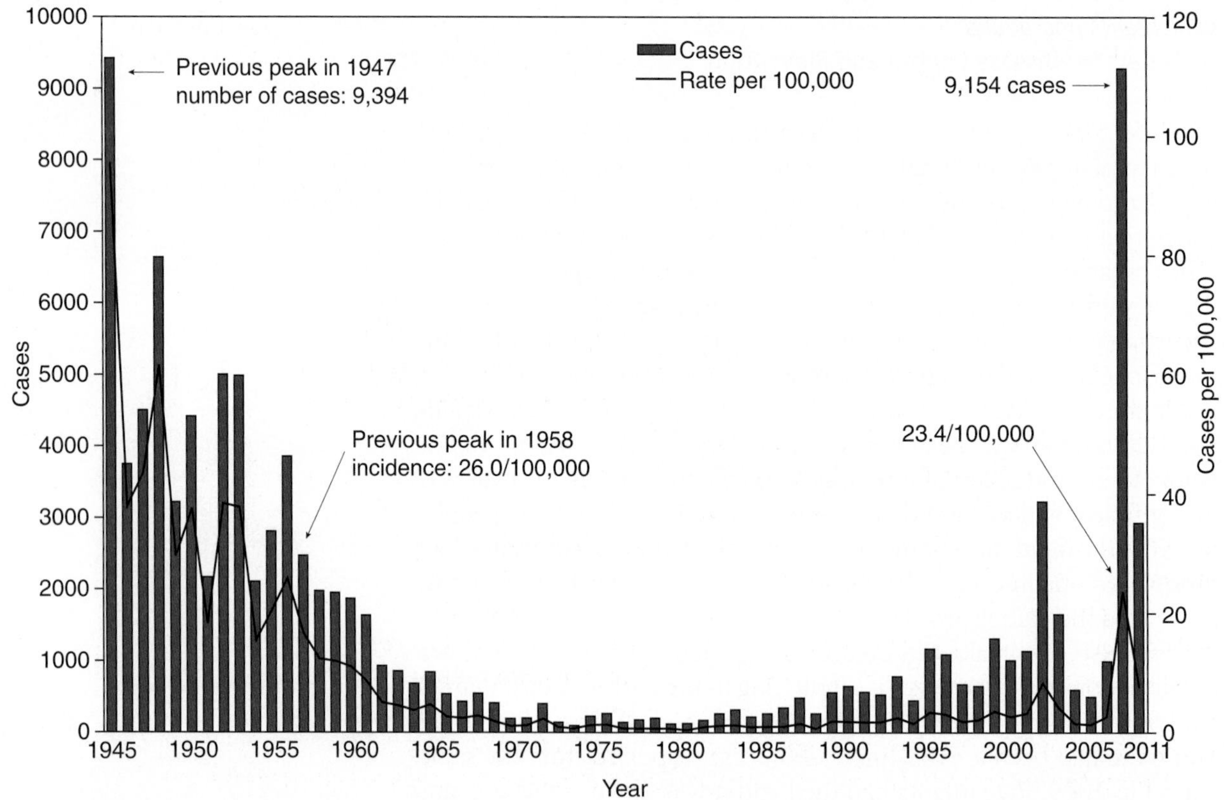

FIGURE 10.2 Reported Cases of Pertussis—California, 1945–2011. In 2011, the California Department of Public Health reported the largest number of pertussis cases since 1947 and the highest incidence rate since 1958. Modified from California Department of Public Health, Immunization branch. »» To what does the incidence rate refer?

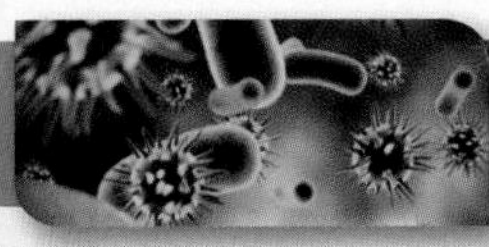

Chapter Challenge

Streptococcus is a gram-positive, nonmotile, encapsulated bacterial genus in the phylum Firmicutes. Members of the genus are very heterogeneous and widely distributed in nature. The cells have always been thought to be very sensitive to the environment and thus require close personal contact for transmission. The most common species are *S. pyogenes,* which 5% to 15% of normal healthy individuals carry as part of their normal microbiota. The other species, *S. pneumoniae* (solely a human organism), is carried in up to 40% of children. Fewer adults carry the organism. To confirm the presence of *S. pyogenes,* throat swabs are used while sputum or blood samples plated on blood agar can be used to identify *S. pneumoniae.* Both species can be severe pathogens, so as we progress through this chapter on airborne bacterial diseases, let's discover how many diseases are attributable to these two species.

KEY CONCEPT 10.1 The Respiratory System Possesses an Indigenous Microbiota

The **respiratory system** is composed of a conducting portion that brings oxygen to the lungs and a respiratory portion that exchanges oxygen and carbon-dioxide gasses with the bloodstream. Because air typically contains microbes and viruses carried on dust and droplet nuclei, it should not be surprising that the respiratory system is the most common portal of entry for many infectious agents.

Upper Respiratory Tract Defenses Limit Microbe Colonization of the Lower Respiratory Tract

The respiratory system is divided into the upper respiratory tract and the lower respiratory tract (FIGURE 10.3). The **upper respiratory tract (URT)** is composed of the nose, sinus cavities, pharynx (throat), and larynx, while the **lower respiratory tract (LRT)** is composed of the trachea, bronchi, and lungs. The lungs contain the alveoli where gas exchange occurs. The average adult inhales and exhales approximately 10,000 liters of air per day. Given that the inspired air contains microbes and microbe-laden particulate matter that could potentially bring microbes that cause infection, the respiratory system has evolved effective defense mechanisms to minimize colonization and infection.

During breathing, the URT and bronchi play a critical role in defending against and filtering out foreign material, such as bacterial cells, viruses, and the dust particles that might carry these microbes. In particular, the airway epithelium lining the URT surfaces (FIGURE 10.4A) is involved in a defensive process called **mucociliary clearance**. Mucus, consisting of charged glycoproteins called **mucins**, is secreted from the goblet cells in the airway epithelium. By overlying the epithelial cells, the mucus traps microbes and particulate matter, which is then moved by ciliated epithelial cells toward the pharynx where it is either swallowed or expectorated (FIGURE 10.5). Mucociliary clearance is supplemented by the presence and activity of several antimicrobial substances, including interferon, lactoferrin, and several human defensins. Cells of the airway epithelium also release several cytokines that regulate both innate and acquired immunity by recruiting immune cells in the defense against pathogens (FIGURE 10.4B).

■ Cytokine:
A small, short-lived protein that is released by one cell to regulate the function of another cell.

Sneezing and coughing are additional mechanical methods to eliminate microbes trapped in the mucus of the respiratory tract.

In the LRT, the epithelial cells lining the alveolar and respiratory bronchioles are not ciliated. However, the region is covered by alveolar fluid, which contains a number of antimicrobial components, including immunoglobulins. Should larger numbers of microbes enter the alveoli, the alveolar macrophages recruit neutrophils from the pulmonary capillaries to help clear the invaders.

The constant exposure to the environment means that many different microorganisms can form part of the commensal microbiota of the URT.

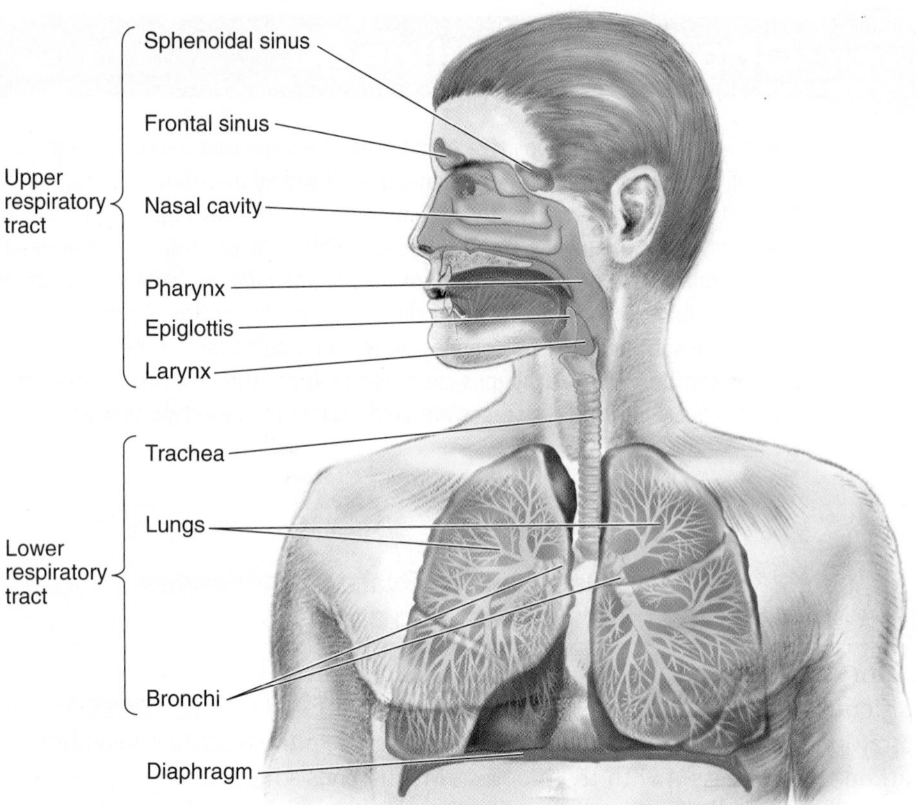

FIGURE 10.3 **Respiratory System Anatomy.** The major parts of the respiratory system are organized into the upper and lower respiratory tracts. »» Which part of the respiratory system would be the most susceptible to colonization and infection?

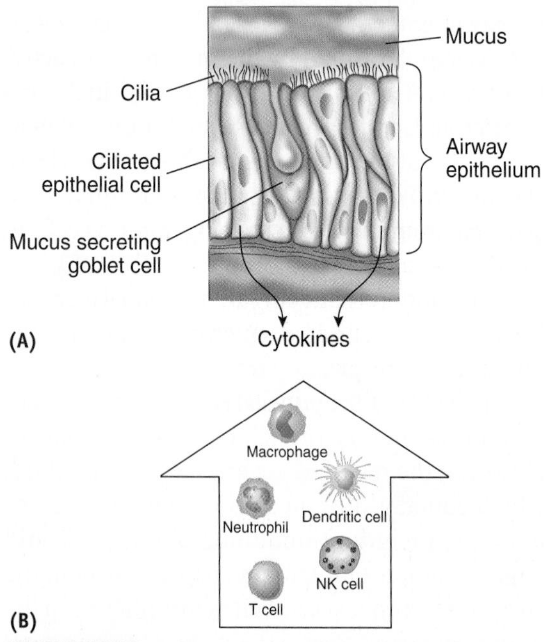

FIGURE 10.4 **Defenses of the Airway Epithelium.** (**A**) Epithelial cells provide a physical, chemical, and cellular barrier against potential pathogens. (**B**) In addition, immune cells limit pathogen spread. »» Identify the specific physical, chemical, and cellular components of the airway defenses.

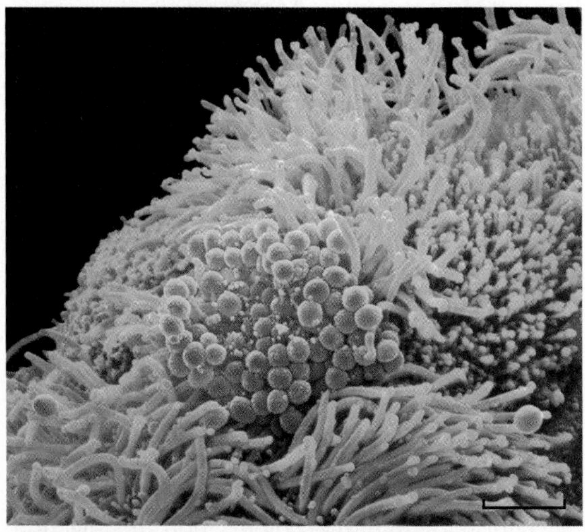

FIGURE 10.5 ***Staphylococcus* in the Ciliated Epithelium.** False-color scanning electron micrograph of a *Staphylococcus* colony (yellow) on the epithelial cells of the trachea. The cilia (blue hair-like projections) keep the trachea free of dust and other irritants. (Bar = 4 μm.) © Juergen Berger/Photo Researchers, Inc. »» How does the ciliated epithelium work to eliminate these bacterial cells?

This indigenous community of microbes will vary depending on the specific part of the tract because of the "environmental differences" (see Figure 10.3). For example, colonization of the nostrils primarily involves the members of the phylum Actinobacteria, such as the corynebacteria, and phylum Firmicutes. Although some 30% of people carry *Staphylococcus aureus* (phylum Firmicutes) in the nose, keeping it under control may depend on the competitiveness afforded by the Actinobacteria.

In the nasopharynx (the upper portion of the pharynx), the mucosal surface is mainly colonized by streptococci, and species of *Neisseria* and *Haemophilus*. In the oropharynx (the middle portion of the pharynx), members of the phyla Firmicutes, Proteobacteria, and Bacteroidetes prevail. As a site that can carry important bacterial pathogens, it is not surprising that genera such as *Haemophilus* and *Neisseria* (both phylum Proteobacteria) as well as α-hemolytic and nonhemolytic streptococci (phylum Firmicutes) are common inhabitants.

Until recently it was thought that the LRT and the lungs of healthy individuals were sterile. However, by using metagenomics to analyze the microbiome of the lungs of healthy individuals and those of patients with genetic conditions, like cystic fibrosis (CF), or chronic obstructive pulmonary disorder (COPD), this belief appears mistaken.

The analysis of expectorated sputum from healthy subjects and those with CF indicates that the lungs of both healthy individuals and CF patients contain a core of 19 bacterial virus genomes, the majority of which were unknown to science. However, the virus profiles between healthy volunteers appear more diverse and dependent on the exterior environment than those in CF patients who, because of the disorder, retain the viruses for a longer period of time. A few eukaryotic viruses were also detected.

In another study that investigated the lung microbiome in healthy nonsmokers and smokers, a significant core community of resident microbes was again identified (FIGURE 10.6). This community was dominated by three bacterial phyla (and a few dominant genera): Proteobacteria (*Pseudomonas, Haemophilus*), Firmicutes (*Streptococcus, Veillonella*), and Bacteroidetes (*Prevotella, Porphyromonas*). However, at the species level, diversity was very limited in the LRT. Interestingly, in volunteers with advanced COPD, the bacterial community composition varied between different sites, even in the same lung.

So, the human respiratory microbiome is slowly becoming better understood and the major players identified.

■ Cystic fibrosis: A life-threatening disorder caused by a defective gene that leads to difficulty breathing and to lung infections.

■ COPD: A progressive and potentially fatal lung disease caused predominantly by smoking.

CONCEPT AND REASONING CHECKS 1

a. Why are microorganisms primarily found only in the upper respiratory tract?

Chapter Challenge A

Streptococcus can be found as part of the normal microbiota in adults.

QUESTION A:

Where in the human respiratory system are the streptococci most likely to be found?

Answers can be found on the Student Companion Website in **Appendix D**.

KEY CONCEPT 10.2 Several Bacterial Diseases Affect the Upper Respiratory Tract

The bacterial diseases of the URT are usually mild but can be more serious if a pathogen in the respiratory tract spreads to the blood, and from there to other sensitive internal organs.

Pharyngitis Is an Inflammation of the Throat

A sore throat, known medically as **pharyngitis**, is an inflammation of the pharynx and sometimes the tonsils (tonsilitis). The inflammation usually is a symptom of a viral infection, such as the common cold or the flu. However, sometimes **group A streptococci (GAS)**, in particular, *Streptococcus pyogenes*, a facultative, gram-positive coccus, is responsible for a potentially more dangerous form of pharyngitis, **streptococcal pharyngitis**, popularly known as **strep throat**.

Streptococcus pyogenes

The *S. pyogenes* cells are highly transmissible and reach the URT within respiratory droplets expelled by infected persons during coughing

■ Respiratory droplet: A relatively large mucus particle that travels less than one meter.

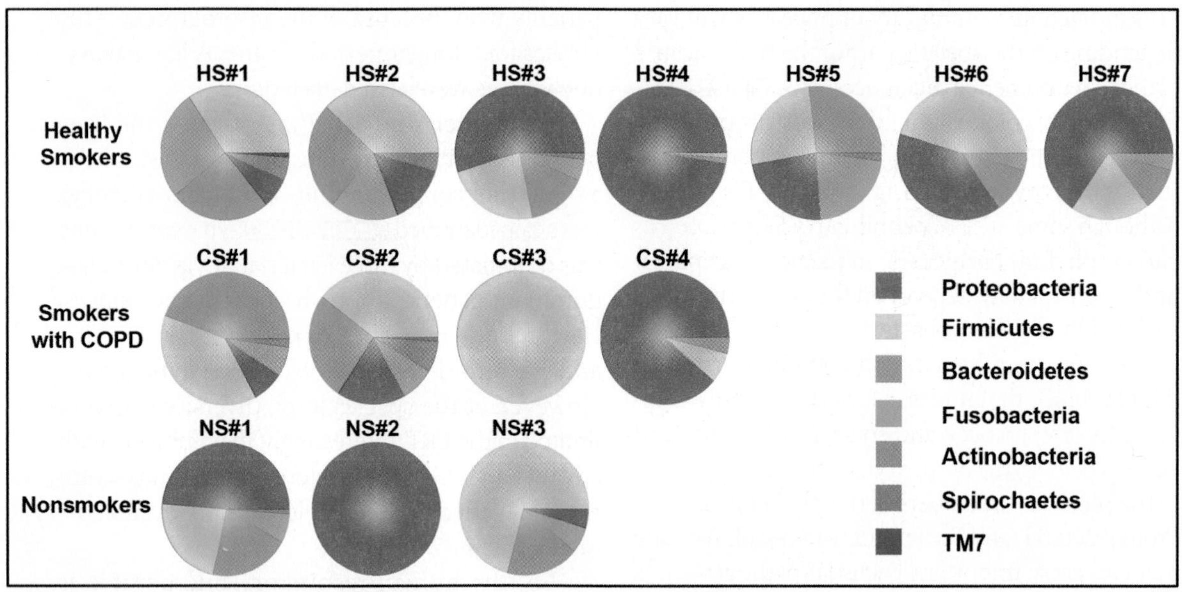

FIGURE 10.6 **Bacterial Phyla Present in the Lungs of Volunteers.** The Proteobacteria, Firmicutes, and Bacteroidetes phyla dominate but show a varied distribution between individual volunteers. TM7 is a major bacterial lineage that has not been cultured in the lab. Reproduced from Erb-Downward, et al. (2011). *PLoS ONE* 6(2): e16384. »» From these phylum-level microbiome analyses, does it appear that COPD patients have a unique bacterial community profile? Explain.

and sneezing. If the cells grow and secrete toxins, inflammation of the oropharynx and tonsils can occur. Besides a sore throat, patients may develop a fever, headache, swollen lymph nodes and tonsils, and a beefy red appearance to pharyngeal tissues owing to tissue damage. More than a million Americans, primarily children, suffer strep throat annually. An antibiotic, such as amoxicillin or ampicillin, is often prescribed to lessen the duration and severity of the inflammation and to prevent possible complications. Hand hygiene is the best prevention.

■ Sequela:
A pathological condition resulting from a disease.

Scarlet fever is a disease arising in about 10% of children with streptococcal pharyngitis or a streptococcal skin infection. Some strains of *S. pyogenes* carry genes coding for so-called **erythrogenic** (*erythro* = "red") exotoxins that cause a pink-red skin rash on the neck, chest, and soft-skin areas of the arms (FIGURE 10.7A). The rash, which usually occurs in children under 15 years of age, results from blood leaking through the walls of capillaries damaged by the toxins. Other symptoms include a sore throat, fever, and a "strawberry-like" inflamed tongue (FIGURE 10.7B). Normally, an individual experiences only one case of scarlet fever in a lifetime because recovery generates immunity.

■ Exotoxin:
A harmful protein released from living bacterial cells into the environment.

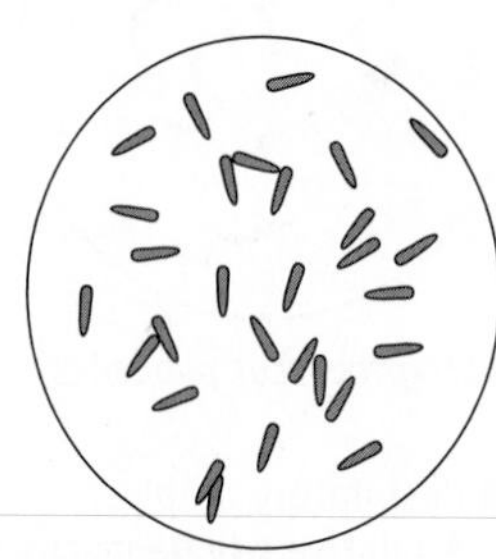

Corynebacterium diphtheriae

Individuals with scarlet fever usually get better within 2 weeks without treatment (MicroFocus 10.1). Treatment with antibiotics, such as penicillin or clarithromycin, can shorten the duration of symptoms.

A serious complication resulting from a lack of treatment is **rheumatic fever**, which is most common in young school-age children. This postinfective sequela is an inflammation in response to the throat infection and primarily affects the joints and heart. It is characterized by fever and joint pain. The most significant long-range effect is permanent scarring and distortion of the heart valves, a condition called **rheumatic heart disease**. The damage arises from a response of the body's antibodies to streptococcal M proteins (found on the surface of the bacterial cells) recognizing similar proteins on heart muscle. Rheumatic fever cases have been declining in the United States due to antibiotic treatment. However, in developing nations, rheumatic fever remains a serious problem.

Streptococcal infections of the LRT are described later in this chapter.

Diphtheria Is a Life-Threatening Illness

Causative Agent and Epidemiology. Diphtheria is an infection of the URT that is caused by *Corynebacterium diphtheriae*, an aerobic, club-shaped, gram-positive rod (*coryne* = "club") that forms a characteristic picket-fence-like arrangement of cells called a "palisade arrangement." Due

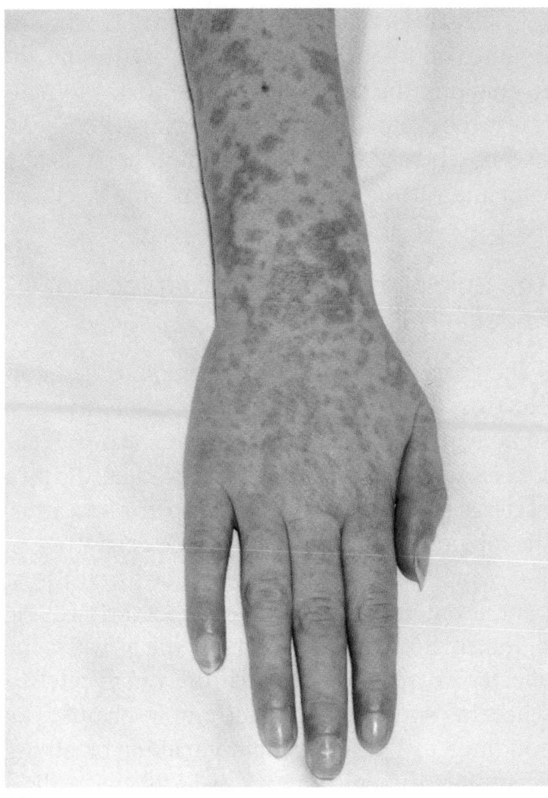

(A)

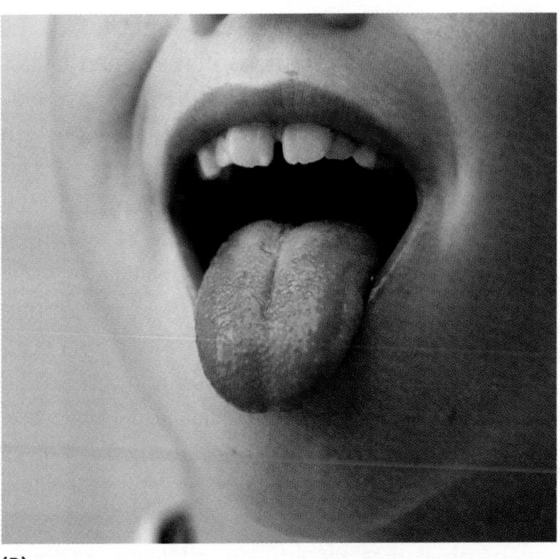

(B)

FIGURE 10.7 Scarlet Fever. Among the early symptoms of scarlet fever are (**A**) a pink-red skin rash and (**B**) a bright red tongue with a "strawberry" appearance. (**A**) © Medical-on-Line/Alamy Images. (**B**) © imagebroker/Alamy Images.
»» What causes the skin rash seen with scarlet fever? What other symptoms are typical of scarlet fever?

MICROFOCUS 10.1: Public Health

A Wakeup Call

In life, sometimes the best way to move past a road block is to simply face it head on—or attack the defensive team by running right at them. Often in the microbial world, bacterial species may have the same idea—but with a twist.

A blood clot, showing red blood cells trapped in a fibrin mesh. © Volker Steger/Photo Researchers, Inc.

One of the most common human host responses to a group A streptococci (GAS) infection is to try to contain the infection by forming a blood clot around the infected area (see figure)—in other words, the host attempts to entomb the bacterial cells in the clot. Unfortunately, one of the abilities GAS possess is to simply break through the defensive wall set up to contain the infection. How do they do this?

Normally a blood clot stays as a clot because the protein plasmin that would dissolve the clot remains in an inactive form called plasminogen. What *Streptococcus pyogenes* does is to "wake up" or activate plasminogen.

Trapped within a clot, the *S. pyogenes* cells secrete an enzyme called streptokinase. Streptokinase then catalyzes the conversion of inactive plasminogen into active plasmin—in other words, it wakes up the protein. As plasmin, the protein triggers a series of reactions leading to the dissolving of the clot. Now the bacterial cells can escape and perhaps cause a serious infection elsewhere in the body.

to immunization starting in early childhood, the number of cases of diphtheria in the United States is essentially zero (the last confirmed case was in 2003). However, the disease remains a health problem in many regions of the world and booster doses are required. The World Health Organization (WHO) estimates there are about 4,000 annual cases of diphtheria worldwide.

■ Toxoid: An exotoxin that has been rendered harmless but retains the ability to stimulate an antibody response.

Clinical Presentation. Diphtheria is acquired by inhaling respiratory droplets or by direct contact with the skin from an infected person. Initial symptoms include a sore throat and low-grade fever. On the surface of the mucus membrane of the throat or mouth, the bacterial cells secrete a potent exotoxin that inhibits the translation process by ribosomes, resulting in cell death. A prominent feature is the accumulation of dead tissue, mucus, white blood cells, and fibrous material, called a **pseudomembrane** ("pseudo" because it does not fit the definition of a true membrane) on the tonsils or pharynx (FIGURE 10.8). Mild cases fade after a week while more severe cases can persist for 2 to 6 weeks.

Complications can arise if the thickened pseudomembrane results in respiratory blockage. If the exotoxin spreads to the bloodstream, heart and peripheral nerve destruction can lead to cardiac arrhythmia and coma. Left untreated, 5% to 10% of respiratory cases result in death.

■ Arrhythmia: An irregular heart beat.

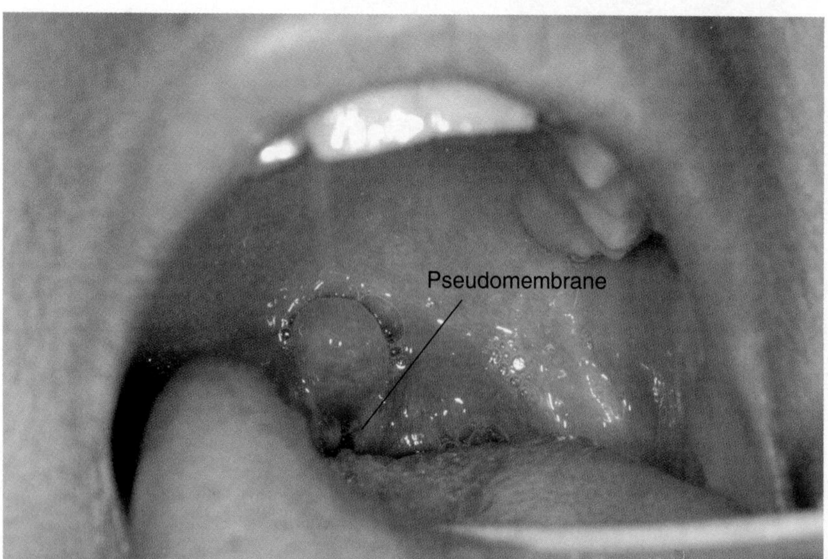

FIGURE 10.8 **Diphtheria Pseudomembrane.** Diphtheria is an upper respiratory infection of mucous membranes caused by a toxigenic strain of *Corynebacterium diphtheriae*. The infection is characterized by the formation of a pseudomembrane on the tonsils or pharynx. © Medical-on-Line/Alamy Images. »» What is the pseudomembrane composed of?

Treatment and Prevention. Treatment requires antibiotics (penicillin or erythromycin) to eradicate the pathogen and antitoxins to neutralize the exotoxins. Protection against diphtheria may be rendered by injection of diphtheria toxoid, which is part of the diphtheria-tetanus-acellular pertussis (DTaP) vaccine.

The Epiglottis Is Subject to Infection, Especially in Children

A life-threatening but rare condition, called **epiglottitis**, involves infection of the epiglottis—the small flap of cartilage covering the trachea. If the area around the epiglottis becomes infected with a bacterial pathogen, such as *Haemophilus influenzae*, the inflammation can spread to the epiglottis.

Symptoms of epiglottitis include severe throat pain, fever, and a muffled voice. As the swelling of the epiglottis starts to narrow the airways, the affected person exhibits **stridor**, a high pitched wheezing sound when breathing in or out. The condition can progress rapidly making breathing even more difficult. In rare cases where medical intervention does not occur, complete airway blocking can result in death. The infection is most common and most dangerous in children because they have a smaller airway than adults. If, or once, the individual is breathing freely, intravenous antibiotics are given. Immunization with the Hib vaccine is the most effective way to prevent epiglottitis in children younger than age 5, which has made epiglottitis a rare infection in the United States.

The Nose Is the Most Commonly Infected Region of the Upper Respiratory Tract

Because of its prominent position in the URT, the nose is a major portal of entry for infectious organisms and viruses. In fact, the microbiota in the nose can be a sign of potential illness. For example, healthy individuals contain primarily bacterial species belonging to the Actinobacteria (68%) and Firmicutes (27%) while hospitalized patients exhibit the opposite—Firmicutes (71%) and Actinobacteria (27%).

Sinusitis is inflammation in any of the sinuses, the air-filled hollow cavities around the nose and nasal passages (FIGURE 10.9). The condition nearly always begins with a viral infection of the nasal passages (**rhinitis**). About 10 to 15 million people each year develop a so-called "sinus infection."

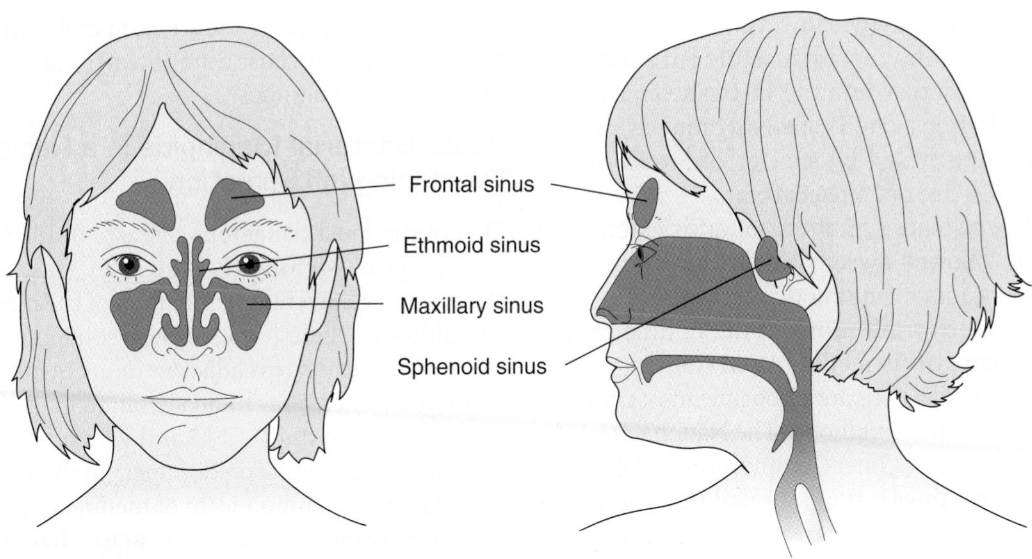

FIGURE 10.9 **The Sinuses.** The sinuses are hollow cavities within the facial bones. »» Which sinuses are most frequently "felt" by a person with a sinus infection?

Acute sinusitis may be caused by a variety of indigenous microbiota of the URT. The condition often develops from a blockage at the openings to the sinuses, resulting from a common cold infection of the URT. Trapped fluid becomes a nutrient medium for bacterial growth, increases the pressure in the sinuses, and causes pain, tenderness, and swelling over the affected sinuses. Treatment of acute sinusitis is aimed at improving sinus drainage and curing the infection. Nasal sprays can be used for a short time, and antibiotics, such as amoxicillin or trimethoprim-sulfamethoxazole, can be prescribed for a bacterial infection.

If untreated, acute sinusitis may develop into **chronic sinusitis** that can last for 8 to 12 weeks. The symptoms of chronic sinusitis are more subtle and pain occurs less often. The most common symptoms are nasal obstruction, nasal congestion, and postnasal drip. The treatment is the same as with acute sinusitis, except antibiotic use, if bacterial, may be for a longer period of time. Treating cold symptoms immediately and using decongestants also may help prevent the development of a chronic condition.

Ear Infections Are Common Illnesses in Early Childhood

As part of the URT, the ears, nose, and throat are located near each other and, as such, allow infections to spread from one to the other. The ear, which is the organ of hearing and balance, consists of the outer, middle, and inner ear (FIGURE 10.10). The Eustachian tube vents the middle ear to the nasopharynx, which explains why URT infections (such as the common cold) often result in infections to the middle ear.

A bacterial infection of the external auditory canal sends 2.4 million Americans to the doctor every year. Such an inflammation, referred to as **acute otitis externa** (*oti* = "ear"), can affect the entire ear canal or just one small area, as in a boil or pimple. Normally, the ceruminous glands in the ear canal produce cerumen (earwax) that has

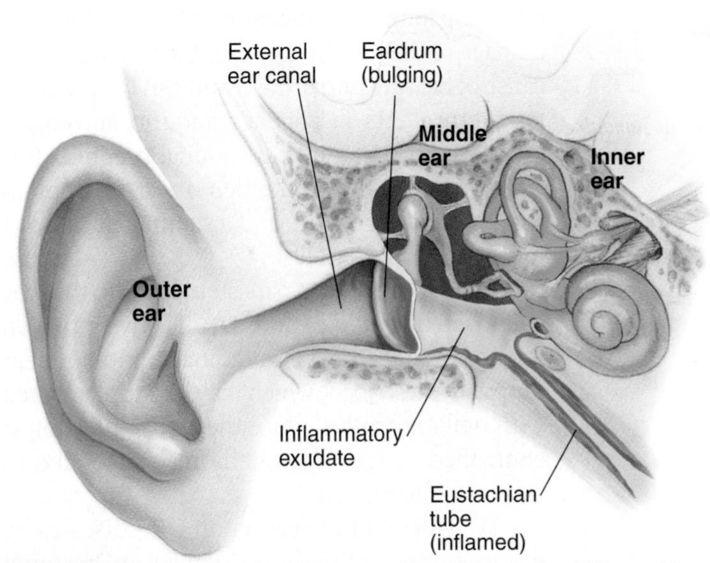

FIGURE 10.10 **Ear Anatomy.** The ear consists of external, middle, and inner structures. »» In a middle ear infection, how do the infecting microbes reach the middle ear?

■ Meninges: The membranes surrounding and protecting the brain and spinal chord.

antibacterial activity. However, outer ear infections commonly occur in children especially after extended swimming in freshwater pools. Such infections, often called **swimmer's ear**, are most often caused by species of *Streptococcus, Staphylococcus*, or *Pseudomonas.*

The primary symptoms of otitis externa are itching followed by ear pain. Treatment involves the application of antibiotic ear drops.

Short-term infections of the middle ear are called **acute otitis media** (**AOM**) (*media* = "middle"). Such infections are among the most common illnesses of early childhood. The National Institute on Deafness and Other Communication Disorders reports that three out of four children contract at least one middle ear infection by 3 years of age. Bacterial species typically responsible for middle ear infections include *S. pneumoniae* and *H. influenzae.*

Middle ear infections usually start with a common cold or other respiratory viral infection of the URT. Inflammation of the Eustachian tube allows bacterial cells to infect the sterile environment of the middle ear. Fluid buildup then provides an environment for bacterial growth, which results in the middle ear becoming inflamed. This is followed by ear pain with a red, bulging eardrum. Children with AOM may develop a fever, produce a fluid that drains from the ears, or have headaches.

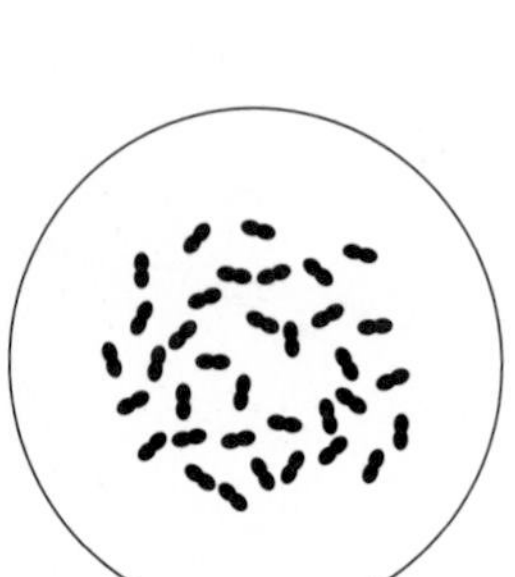

Neisseria meningitidis

The American Academy of Pediatrics (AAP) and the American Academy of Family Physicians (AAFP) report that 80% of children with AOM get better without antibiotics.

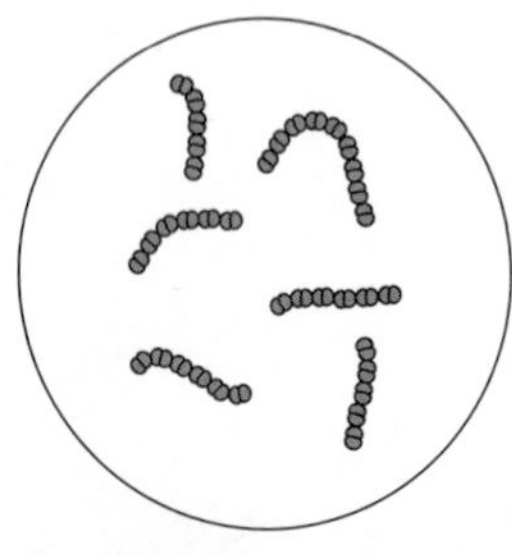

Streptococcus pneumoniae

Chronic otitis media (COM) is a condition involving long-term infection, inflammation, and damage to the middle ear. COM is a major global cause of hearing impairment and can have serious long-term effects on language, auditory and cognitive development, and educational progress; it is a major public health problem in many populations around the world, and a significant cause of morbidity and mortality.

COM stems from a persistent **biofilm** that has colonized the middle ear tissue. If active infection is present in the form of ear drainage, antibiotic ear drops are prescribed, which may be supplemented with oral antibiotics. Once the active infection is controlled, surgery is usually recommended to clear the obstruction.

We will end this section on URT diseases by discussing one of the most dangerous bacterial diseases spread through the air—meningitis. An infection can spread to the central nervous system (CNS) where, without a resident microbiota, the bacterial cells may cause a life-threatening inflammation of the meninges.

Acute Bacterial Meningitis Is a Rapidly Developing Inflammation

One of the most dangerous diseases of the CNS is meningitis. **Meningitis** is an acute, subacute, chronic, or recurrent inflammation of the meninges, although there usually is also some involvement of the brain areas adjacent to the meninges. In addition, there may be an alteration in the flow of cerebrospinal fluid (CSF) and with a bacterial infection, neutrophils predominate in the CSF.

The most common form of meningitis, causing some 80% of all cases, is **acute bacterial meningitis** (**ABM**), a rapidly developing infection affecting individuals over 3 months of age.

Causative Agents and Epidemiology. The species most commonly associated with ABM are *Neisseria meningitidis, Streptococcus pneumoniae, Haemophilus influenzae*, and *Listeria monocytogenes.*

Perhaps the most dangerous form of ABM is **meningococcal meningitis** caused by *N. meningitidis* (FIGURE 10.11), a small, encapsulated, aerobic, gram-negative diplococcus that attaches to the nasopharyngeal mucosa by pili. There are more than 14

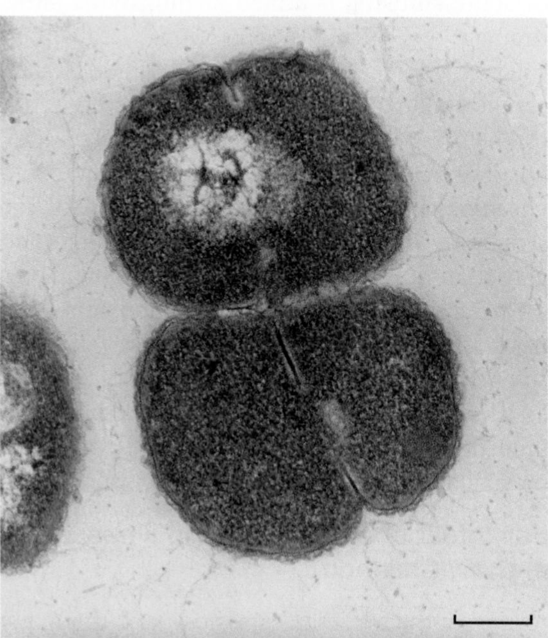

FIGURE 10.11 ***Neisseria meningitidis.*** A false-color transmission electron micrograph of *N. meningitidis* cells. The cytoplasm (blue) and nucleoid (pink) are shown. (Bar = 0.5 µm.) © CAMR/A.B. DOWSETT/Photo Researchers, Inc. »» Living *N. meningitidis* cells have a capsule. What roles do the capsules play in the disease process?

serogroups based on the chemical nature of the capsule with serogroups A, B, C, Y, and W-135 responsible for the majority of infections. Although the Centers for Disease Control and Prevention (CDC) reported fewer than 850 cases of meningococcal meningitis in Americans in 2010, globally meningococcal meningitis occurs in epidemic proportions. According to the WHO, meningococcal meningitis is responsible for more than 700,000 cases each year in Africa, and more than 90% of the meningitis epidemics are caused by *N. meningitidis* group A, which primarily attacks infants, children, and young adults. These cases occur during the dry season from November to June in the 22 countries forming sub-Saharan Africa's so-called "meningitis belt" (FIGURE 10.12).

The *N. meningitidis* cells do not survive long in the environment, so they can be found in the noses or the back of the throats of 10% to 25% of individuals worldwide. This means the organisms will be transmitted by respiratory droplets from person to person. Therefore, meningococcal meningitis is prevalent where people are in close proximity for long periods of time, such as grade-school classrooms, military camps, college dorms, and prisons. Crowding in highly populated African countries of the meningitis belt facilitates its spread in epidemic proportions.

By being one of the major causes of bacterial pneumonia (more than 16,500 cases reported by CDC in 2010), the gram-positive diplococci of *Streptococcus pneumoniae* can spread and develop into **pneumococcal meningitis**. The inflammation usually occurs as a community-acquired meningitis affecting infants, and middle-aged and elderly adults. In addition, patients with a diseased or absent spleen, sickle cell disease, chronic alcoholism, and patients with recent skull fractures or head injury are at high risk. Pneumoccocal meningitis is responsible for about 30% of ABM cases and has a high mortality rate (20% to 30%).

Haemophilus influenzae type b (Hib), an encapsulated, gram-negative coccobacillus, has six serogroups based on the capsular material. Hib once was the most prevalent bacterial species causing meningitis (***Haemophilus* meningitis**) in American

■ Serogroup: A group of bacterial strains containing a structure capable of generating a similar antibody response.

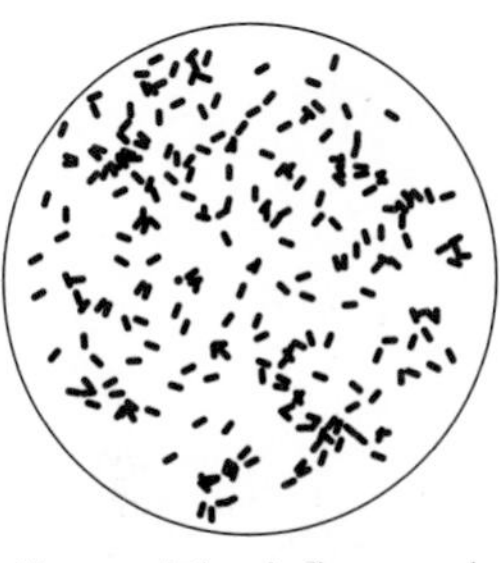

Haemophilus influenzae b

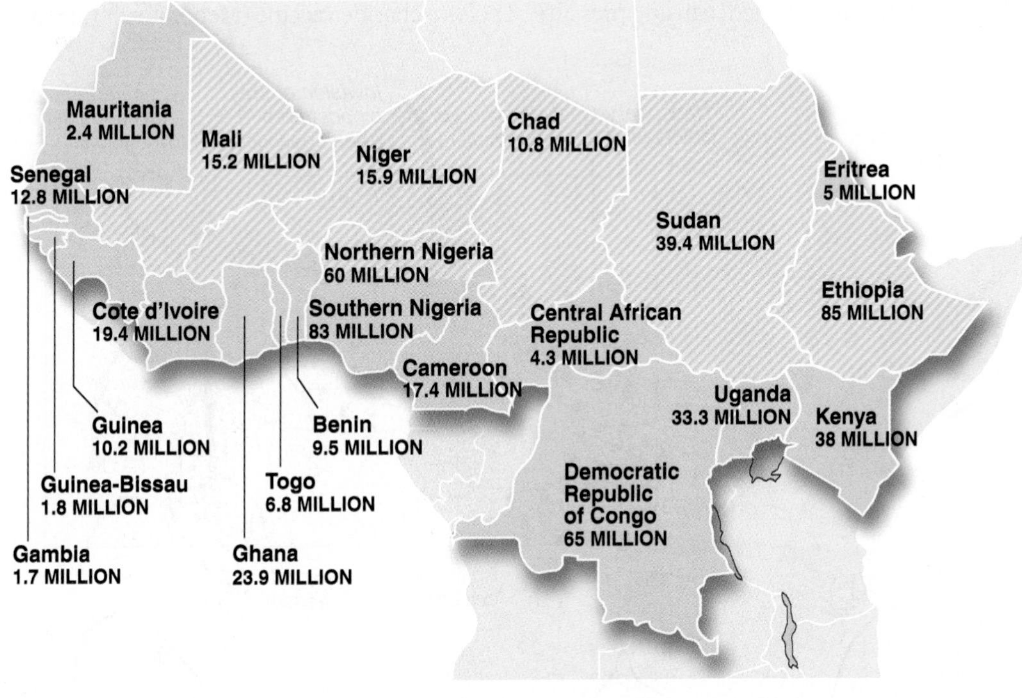

FIGURE 10.12 **The African Meningitis Belt.** This map of sub-Saharan Africa shows the so-called "meningitis belt" where deadly epidemics of meningitis are most prevalent. From Program for Appropriate Technology in Health (PATH)/David Simpson. »» Why is a new meningitis vaccine, called MenAfriVac™, being initially used to immunize 1 to 29-year olds in Mali, Burkina Faso, and Niger?

children under 5 years of age. In 1986, about 18,000 cases of *Haemophilus* meningitis occurred in the United States. At that time, however, a vaccine was licensed and the epidemic peaked. As of 1993, the vaccine was combined with the DTaP vaccine for distribution to children as Tetramune®, and in 2010, the number of reported cases among children under 5 years of age in the United States was down to 23. However, there are other non-b type strains that continue to cause substantial disease among infants and adults over 65 years of age.

Clinical Presentation. ABM begins with a localized infection. For example, it could start as an upper respiratory tract infection caused by *N. meningitidis, S. pneumoniae,* or *H. influenzae;* lobar pneumonia due to *S. pneumoniae;* or otitis media caused by *H. influenzae*. However, the organism may then invade the epithelium of the nasopharynx and spread to the blood (FIGURE 10.13). In the case of *N. meningitidis,* the blood infection is called **meningococcemia**. A rash also appears on the skin, beginning as bright red patches, which progress to blue-black spots.

Once in the blood, all three pathogens are capable of crossing the blood-brain barrier. The meninges then become inflamed, causing pressure on the spinal cord and brain such that the individual experiences a fever and stiff neck. Symptoms rapidly evolve into a pounding headache, nausea and vomiting, and often sensitivity to bright light. Left untreated, infection through the meninges, brain, and spinal cord can be very rapid, resulting in death within hours. Surviving ABM can result in lasting disabilities, such as deafness, blindness, and paralysis.

Treatment and Prevention. ABM, especially meningococcal meningitis, represents a medical emergency. Early diagnosis and treatment are crucial to prevent disabilities or death. A principal criterion for diagnosis is the observation and/or cultivation of the *N. meningitidis* cells in samples of CSF obtained by a spinal tap. However, the seriousness of the disease usually demands treatment before the results of diagnostic procedures are known. Treatment with antibiotics, such as penicillin, cefotaxime, or ceftriaxone, usually is recommended, often in large intravenous doses.

Although no single vaccine provides immunity to all forms of ABM, vaccinations have reduced reported cases by more than 31%. A meningococcal polysaccharide vaccine (Menomune®) and a conju-

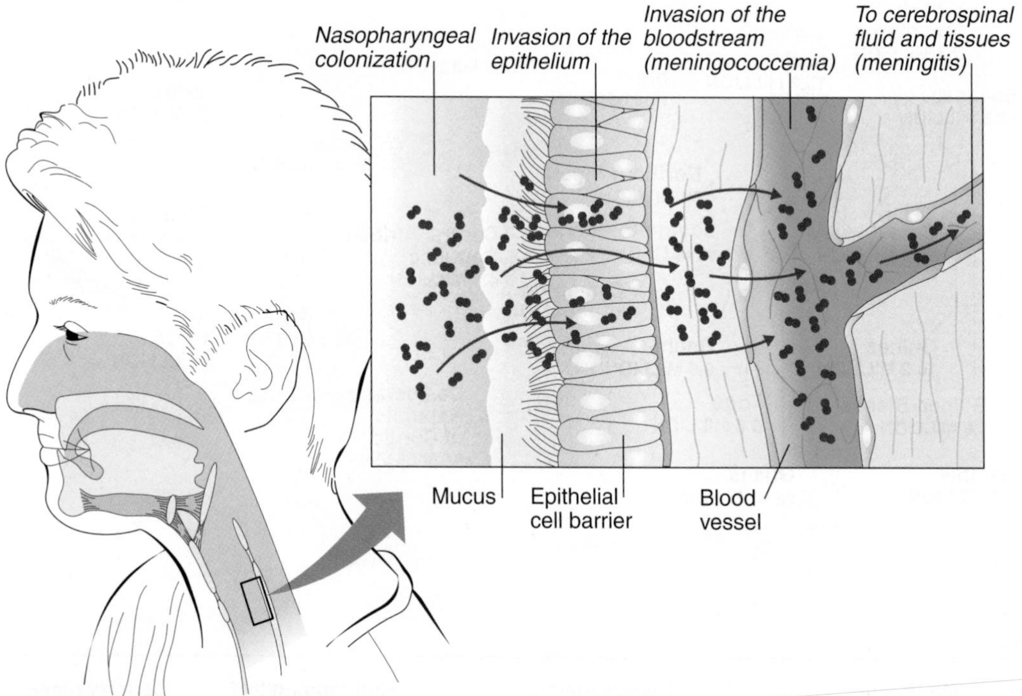

FIGURE 10.13 **Pathogenic Steps Leading to Meningitis.** The bacterial species capable of causing meningitis (*N. meningitidis, S. pneumoniae,* and *H. influenzae*) can colonize the nasopharynx, and then invade the epithelium causing respiratory distress. They then pass into the bloodstream. Finally, they disseminate to tissues near the spinal cord, causing inflammation and meningitis. »» What bacterial virulence factors would facilitate (a) attachment to the nasopharynx and (b) survival in the bloodstream?

gate vaccine (Menactra®) to serogroups A, C, Y, and W-135 are available. In December 2010, a meningococcal vaccine (MenAfriVac®), especially designed for African meningococcal meningitis (serogroup A), was introduced and appears capable of eliminating meningitis throughout the meningitis belt (see Figure 10.12). There currently are two types of pneumococcal vaccines: a pneumococcal conjugate vaccine (PCV13) for children under 2 years of age and a pneumococcal polysaccharide vaccine (PPSV) for all adults over 65 years of age. There is also a Hib vaccine for children under 5 years of age.

A Few Bacterial Species Cause Neonatal Meningitis

Neonatal meningitis is a rare but dangerous disease in both full-term and especially low-birth-weight neonates. The most common cause of neonatal meningitis is *Streptococcus agalactiae*, a group B streptococcus (GBS); about 50% of cases are due to this bacterial species that normally colonizes up to 30% of female genital tracts. Another 20% of cases are due to the K1 strain of *Escherichia coli*, and an additional 20% are caused by *Listeria monocytogenes*. A growing number of babies who suffer from neonatal meningitis are premature and their prognosis is poor, such that even with aggressive antibiotic treatment brain damage may result.

The pathogens are transmitted from the mother (who experiences no disease) via the uterus or, more typically, during passage through the birth canal. Transmission leads to a blood infection and neonatal sepsis, and if vulnerable tissues in the CNS are infected, neonatal meningitis develops. Ceftriaxone is usually administered intravenously along with ampicillin. Early-onset neonatal GBS infection can be prevented by screening mothers to identify high-risk pregnancies and administering penicillin during delivery.

The airborne bacterial diseases of the URT are summarized in TABLE 10.1.

TABLE

10.1 A Summary of the Major Bacterial URT Diseases

Disease	Causative Agent	Signs and Symptoms	Transmission	Treatment	Prevention
Streptococcal pharyngitis	*Streptococcus pyogenes*	Sore throat, fever, headache, swollen lymph nodes and tonsils	Respiratory droplets	Amoxicillin Ampicillin	Practicing good hand hygiene
Scarlet fever	*Streptococcus pyogenes*	Pink-red rash on neck, chest, arms Strawberry-like tongue	Respiratory droplets	Penicillin Clarithromycin	Practicing good hygiene
Diphtheria	*Corynebacterium diphtheriae*	Pseudomembrane	Respiratory droplets	Penicillin Erythromycin	Vaccinating with DTaP
Epiglottitis	*Haemophilus influenzae*	Severe throat pain, fever, muffled voice	Respiratory droplets	Intravenous antibiotics	Vaccinating with Hib
Sinusitis	Indigenous microbiota	Pain, tenderness, and swelling	Respiratory droplets	Nasal sprays Antibiotics	Minimizing contact with individuals with colds
Otitis externa	*Streptococcus, Staphylococcus, Pseudomonas* species	Itching and ear pain	Contaminated water	Lifestyle modifications Topical and oral medications	Keeping ears dry
Acute otitis media	*Streptococcus pneumoniae* *Haemophilus influenzae*	Ear pain Red, bulging eardrum	Airborne Direct contact	Wait and see Antibiotics	Limiting time in childcare
Acute bacterial meningitis	*Neisseria meningitidis* *Streptococcus pneumoniae* *Haemophilus influenzae* type b	Fever, stiff neck, severe headache, vomiting and nausea, sensitivity to light	Respiratory droplets from prolonged contact	Antibiotics	Vaccination

CONCEPT AND REASONING CHECKS 2

- **a.** What makes *S. pyogenes* such a potentially dangerous pathogen in the upper respiratory tract?
- **b.** In 17th century Spain, diphtheria was called "el garatillo" = the strangler. Why was it given this name?
- **c.** How does an *H. influenzae* infection affect the epiglottis and potentially lead to a life-threatening condition?
- **d.** How do acute and chronic sinusitis differ?
- **e.** In most cases with otherwise healthy children, why is the use of antibiotics not recommended for acute otitis media?
- **f.** What species are most commonly associated with neonatal meningitis?
- **g.** What is the common pathogenesis for meningococcal, pneumococcal, and *Haemophilus* meningitis?

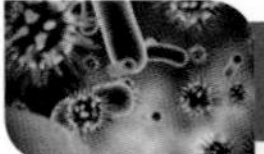

Chapter Challenge B

Both *S. pyogenes* and *S. pneumoniae* can cause infections and disease in the URT.

QUESTION B:

With which respiratory diseases of the URT are each of the Streptococcus species associated?

Answers can be found on the Student Companion Website in **Appendix D.**

KEY CONCEPT 10.3 Many Bacterial Diseases of the Lower Respiratory Tract Can Be Life Threatening

In the lower respiratory tract (LRT), a number of bacterial diseases affect the lung tissues. As injury occurs, fluid builds up in the lung cavity, and the space for obtaining oxygen and eliminating carbon dioxide is reduced. This is the basis for a possibly fatal pneumonia.

Pertussis (Whooping Cough) Is Highly Contagious

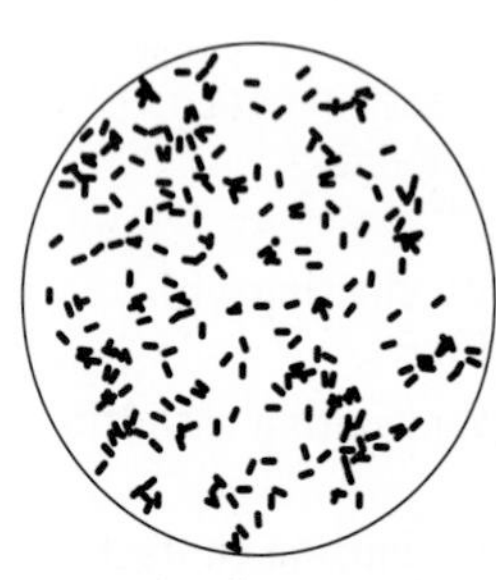

Bordetella pertussis

Causative Agent and Epidemiology. Pertussis (*per* = "through"; *tussi* = "cough"), also known as **whooping cough**, is caused by *Bordetella pertussis*, a small, aerobic, gram-negative rod that is strictly a human pathogen.

Pertussis is one of the most dangerous and highly contagious bacterial diseases. While most cases historically have occurred in children under 5 years of age, severe disease and death most often occur in infants only weeks to a few months old (see chapter opener). Since the early 2000s, the incidence of disease has been on the increase. In 2009, globally there were 16 million cases reported and 195,000 children died. In the United States, although the incidence of pertussis had declined substantially after the introduction of the first whole agent vaccine in 1949, there still were more than 27,000 cases reported, about 4,300 cases in infants, in 2010 (FIGURE 10.14). That means the majority of cases were in adolescents and adults and thus they are the major source of transmission of *B. pertussis* to unvaccinated infants.

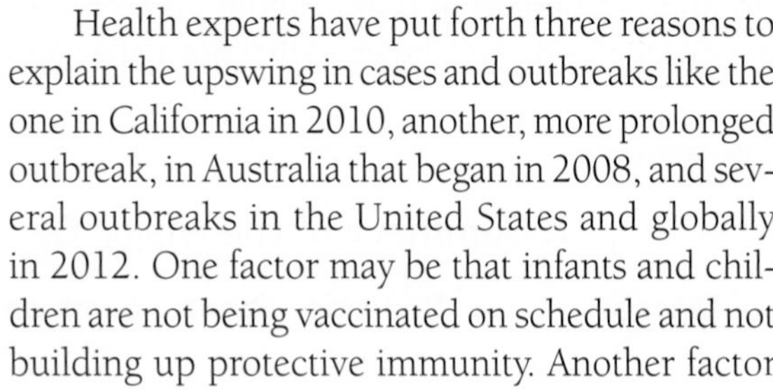

Health experts have put forth three reasons to explain the upswing in cases and outbreaks like the one in California in 2010, another, more prolonged outbreak, in Australia that began in 2008, and several outbreaks in the United States and globally in 2012. One factor may be that infants and children are not being vaccinated on schedule and not building up protective immunity. Another factor

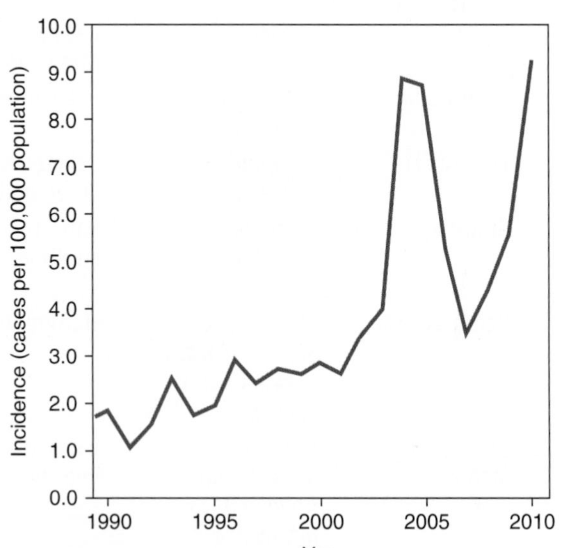

FIGURE 10.14 **Pertussis Cases—United States, 1990–2010.** The incidence of cases has increased abruptly beginning about 2002. Reproduced from: *Morbidity and Mortality Weekly Report*, 58(53);1-100. »» Propose a reason for the abrupt increase in pertussis cases considering that the first acellular DTaP vaccine was introduced in 1997.

explaining the increase in cases in adolescents and adults is waning immunity. It used to be thought that vaccination would provide life-long immunity. It now appears that immunity declines in vaccinated children between ages 8 and 12. Third, an Australian team of scientists have reported that the current outbreak strain of *B. pertussis* is a new one that is more resistant to vaccine-generated immunity. This suggests that new or modified vaccines may need to be developed.

Clinical Presentation. The bacilli are spread by respiratory droplets and by having pili the cells can adhere to and aggregate on the cilia of epithelial cells in the mouth and throat (FIGURE 10.15). Exotoxin production then paralyzes the ciliated cells and impairs mucus movement, potentially causing pneumonia. Following an average 9 to 10 day incubation period, typical pertussis cases occur in three stages. The initial **(catarrhal)** stage lasts 1 to 2 weeks and is marked by general malaise, low-grade fever, and increasingly severe cough. It is during this phase that the individual is most contagious and carries a high bacterial load.

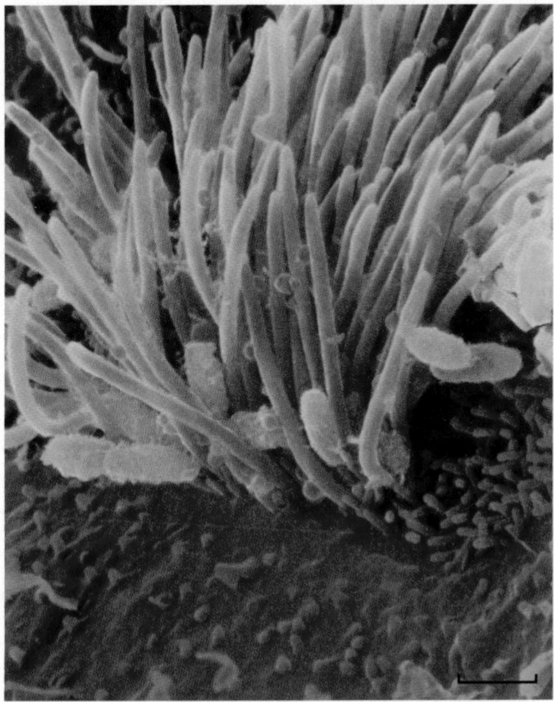

FIGURE 10.15 ***Bordetella pertussis* Associated with the Ciliated Epithelium.** False-color scanning electron micrograph of human tracheal epithelium. The pertussis cells (yellow) cause a dramatic loss of cilia function, which is essential to protect the respiratory tract from dust and particles. (Bar = 1 µm.) © NIBSC/Photo Researchers, Inc.
»» Propose a hypothesis to explain why cilia loss would lead to fits of coughing.

During the second **(paroxysmal)** stage, which lasts 2 to 4 weeks, disintegrating cells and mucus accumulate in the airways and cause labored breathing. Children experience multiple **paroxysms**, which consist of rapid-fire coughs all in one exhalation, followed by a forced inhalation over a partially closed glottis. This intake of breath results in a high-pitched "whoop" (hence, the name whooping cough). Some 10 to 15 paroxysms may occur daily, and exhaustion usually follows each. Adolescents and adults may not experience these characteristic symptoms. The third (**convalescent**) stage involves sporadic coughing that gradually decreases over several weeks, even after the pathogen has vanished. (Doctors call it the "100-day cough.")

Treatment and Prevention. Treatment is generally successful when erythromycin is administered during the catarrhal stage. However, antibiotic treatment only reduces the duration and severity of the illness.

The relatively low incidence of pertussis in developed nations stems partly from use of a pertussis vaccine. The older vaccine (diphtheria-pertussis-tetanus, or DPT) contained merthiolate (thimerosal)-killed *B. pertussis* cells and was considered risky because about 1 in every 300,000 vaccinated individuals suffered adverse reactions such as high fevers and seizures. Now, public health officials recommended the newer acellular pertussis (aP) vaccine prepared from *B. pertussis* chemical extracts. Combined with diphtheria and tetanus toxoids, the triple vaccine has the acronym DTaP; commercially, it is known as Tripedia®. In 2005, the U.S. Food and Drug Administration (FDA) licensed a reduced dose acellular pertussis vaccine in combination with tetanus and diphtheria (Tdap), which is recommended as a booster dose for children at age 11 or 12 and as an every 10-year booster for those over 20 years of age to provide protection against pertussis and its transmission to susceptible infants.

■ Merthiolate: A mercury derivative compound formerly used in vaccines as a disinfectant and preservative.

Tuberculosis Remains a Major Cause of Death Worldwide

Tuberculosis is a disease that has been with us for thousands of years yet continues to evolve and resist our best drugs.

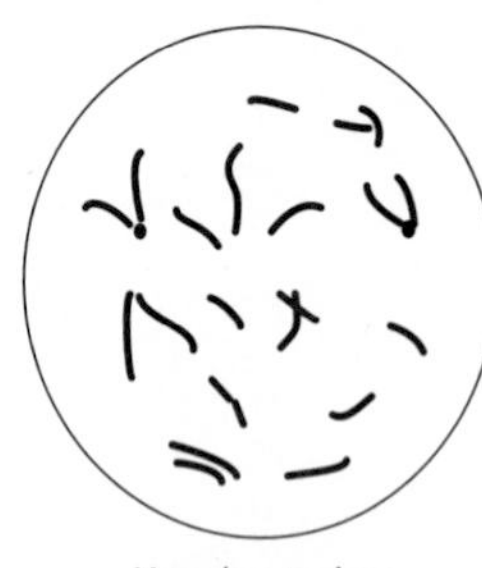
Mycobacterium tuberculosis

Causative Agent and Epidemiology. **Tuberculosis** (TB) is caused by *Mycobacterium tuberculosis*, the "tubercle" bacillus first isolated by Robert Koch in 1882. It is a small, aerobic, nonmotile rod whose cell wall forms a waxy cell surface that greatly enhances resistance to drying, chemical disinfectants, and many antibiotics. TB is an ancient disease that has been identified in Egyptian mummies more than 4,400 years old. Over the centuries, TB has continued to be a "slate wiper" in the human population. During the first half of the 20th century, TB was called "consumption" or "white plague" because the disease wasted away the body and made the patient appear pale.

The somewhat good news today is that the number of TB cases appears to be declining. For 2010, the CDC reported an all-time low of about 11,000 cases in the United States. The WHO reported that globally there has been a 2% drop in the number of TB cases since 2005 (8.8 million from 9 million), a decline in TB deaths to the lowest level since 2003 (1.4 million from 1.8 million), and a 40% drop in the death rate since 1990 (FIGURE 10.16). Still, some 2 billion people worldwide—about 30% of the world's population—are infected with the latent stage of the disease.

Tuberculosis is primarily an airborne disease and, as such, the bacilli are transmitted from person to person in small, aerosolized droplets when a person with active pulmonary disease sneezes, coughs, spits, or even sings. The infectious dose is quite small, perhaps only one to three bacilli. However, individuals with prolonged, frequent, or intense contact with a diseased individual are at most risk of becoming infected, with an estimated 30% infection rate. Thus, people who live in overcrowded, urban ghettoes often contract TB. Malnutrition and a generally poor quality of life also contribute to the establishment of disease.

Clinical Presentation. Unlike many other infectious diseases where an individual becomes ill after several days or a week, the incubation period for TB is much longer. In addition, the illness has two separate stages: an infection stage and a disease stage (FIGURE 10.17). If a person has a pulmonary infection (85% of infections are respiratory), the bacterial cells enter the alveoli where pathogen

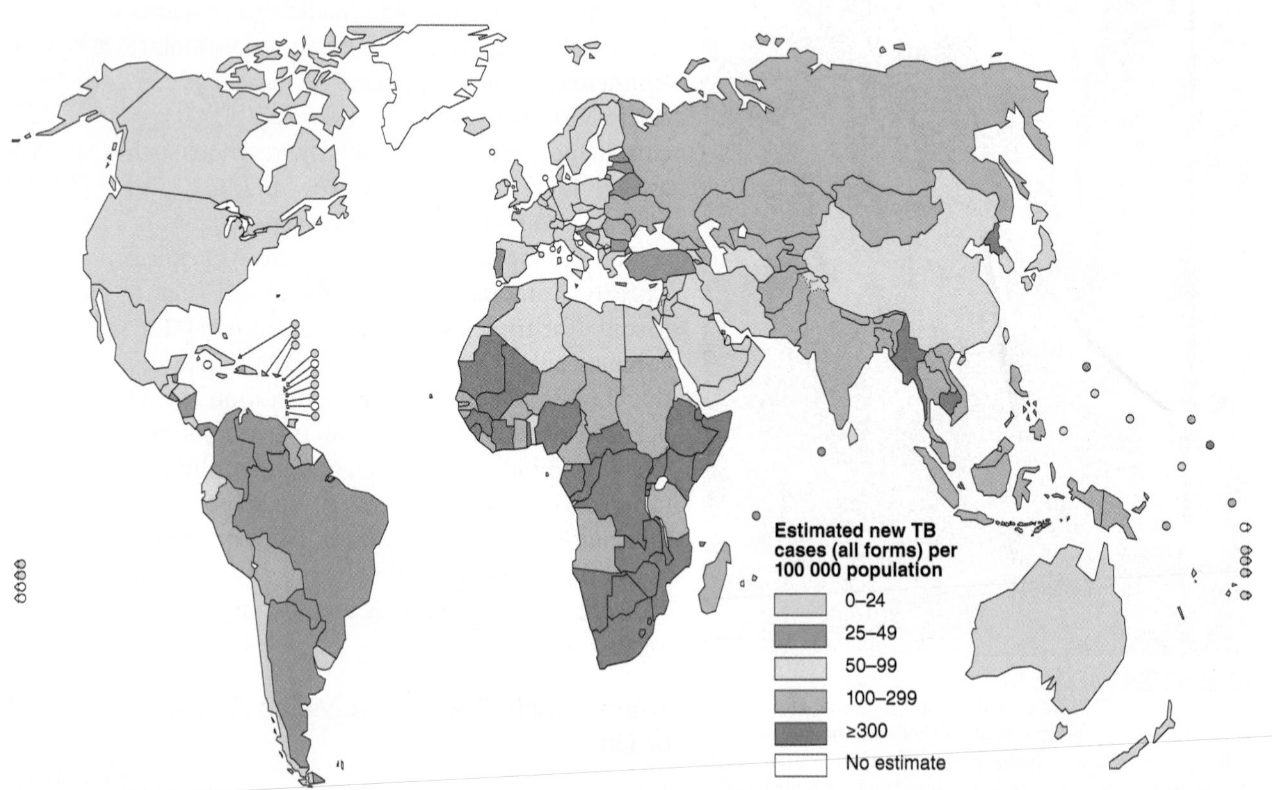

FIGURE 10.16 **Estimated Global Incidence of Tuberculosis, 2008.** About 85% of the new cases of TB occurred in Southeast Asia (34%), Africa (30%), and the Western Pacific (21%). Source: Global Tuburculosis Control, 2011, WHO. »» Where was the highest burden of new TB cases?

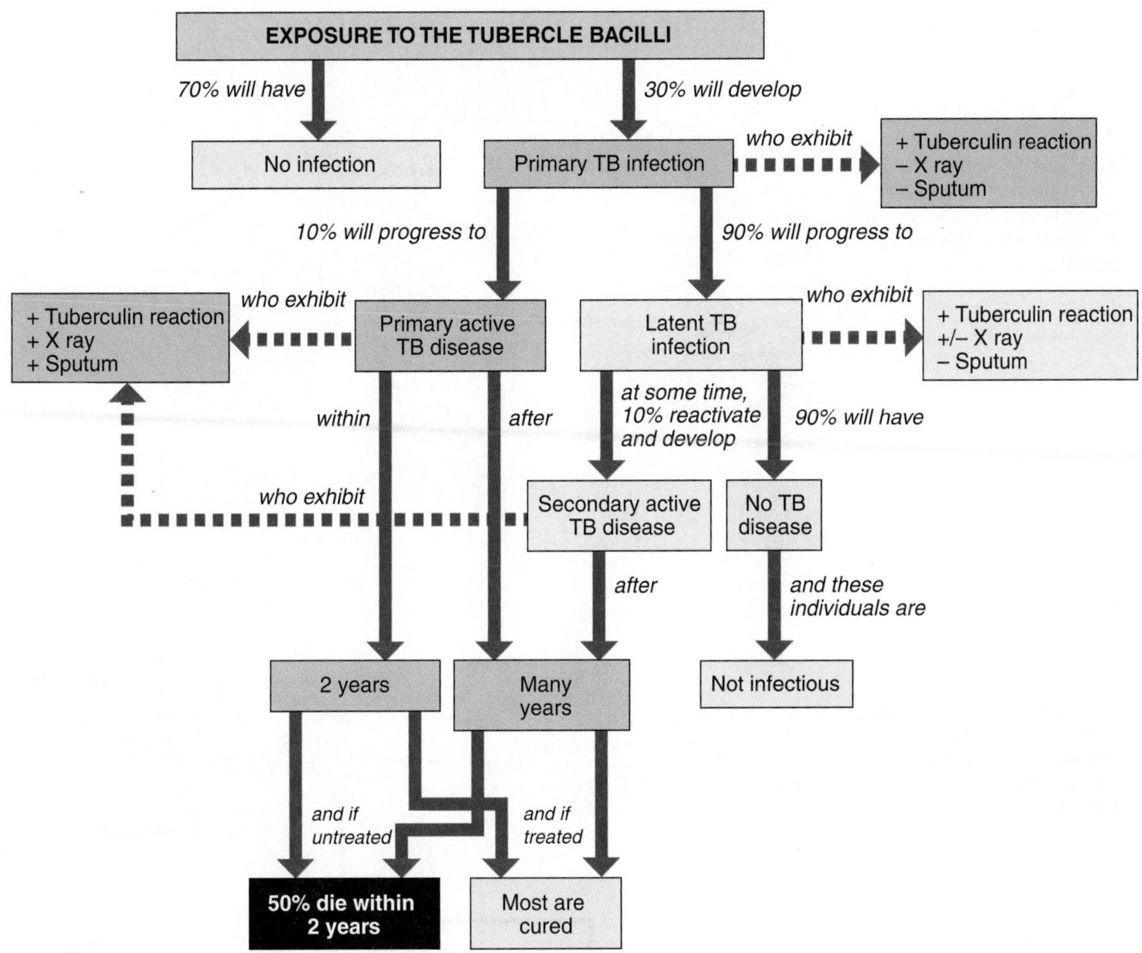

FIGURE 10.17 **A Concept Map for Tuberculosis.** The stages of tuberculosis infection and disease are shown. ("+" or "–" represents positive or negative test results.) »» How does TB infection differ from active TB disease?

interactions occur (FIGURE 10.18). This individual is now said to have a **primary TB infection**. If tested, the person would have a positive tuberculin reaction, but a chest X ray often is negative and a sputum test would be negative (see Disease Detection, below).

In the alveoli, macrophages respond to the infection by ingesting the bacilli. Unfortunately, the bacilli are not killed in the macrophages and as more macrophages arrive, they too phagocytize bacilli but are incapable of destroying them. An inflammatory condition ensues. After about 4 weeks, the cell-mediated immune response localizes the infection, forming a central area of large, multicellular giant cells. Recruited lymphocytes and fibroblasts surround the mass in the lung, forming a type of granuloma called a **tubercle** (hence the name tuberculosis). In 90% of primary TB infections, the infection becomes arrested and the individual usually has no clinical signs of infection. This dormant form of TB is referred to as a **latent TB infection** and is carried by 2 billion people worldwide. Of these, 90% will never develop active disease and will not be infectious.

Up to 10% of individuals who have a primary or latent TB infection will develop the second stage of the illness: a clinical disease. Primary TB infections can develop into **primary active TB disease** in 1 to 2 years. The disease usually becomes extrapulmonary and is disseminated through the body. Due to immune system dysfunction, latent TB infection can undergo reactivation developing into **secondary active TB disease**. Individuals will become ill within 3 months; experience chronic cough, chest pain, and high fever; and continue to expel sputum that accumulates in the LRT. (Often the sputum is rust colored, indicating that blood has entered the lung cavity.)

In active TB disease, the immune defenses cannot keep the tubercle bacilli in check. Many

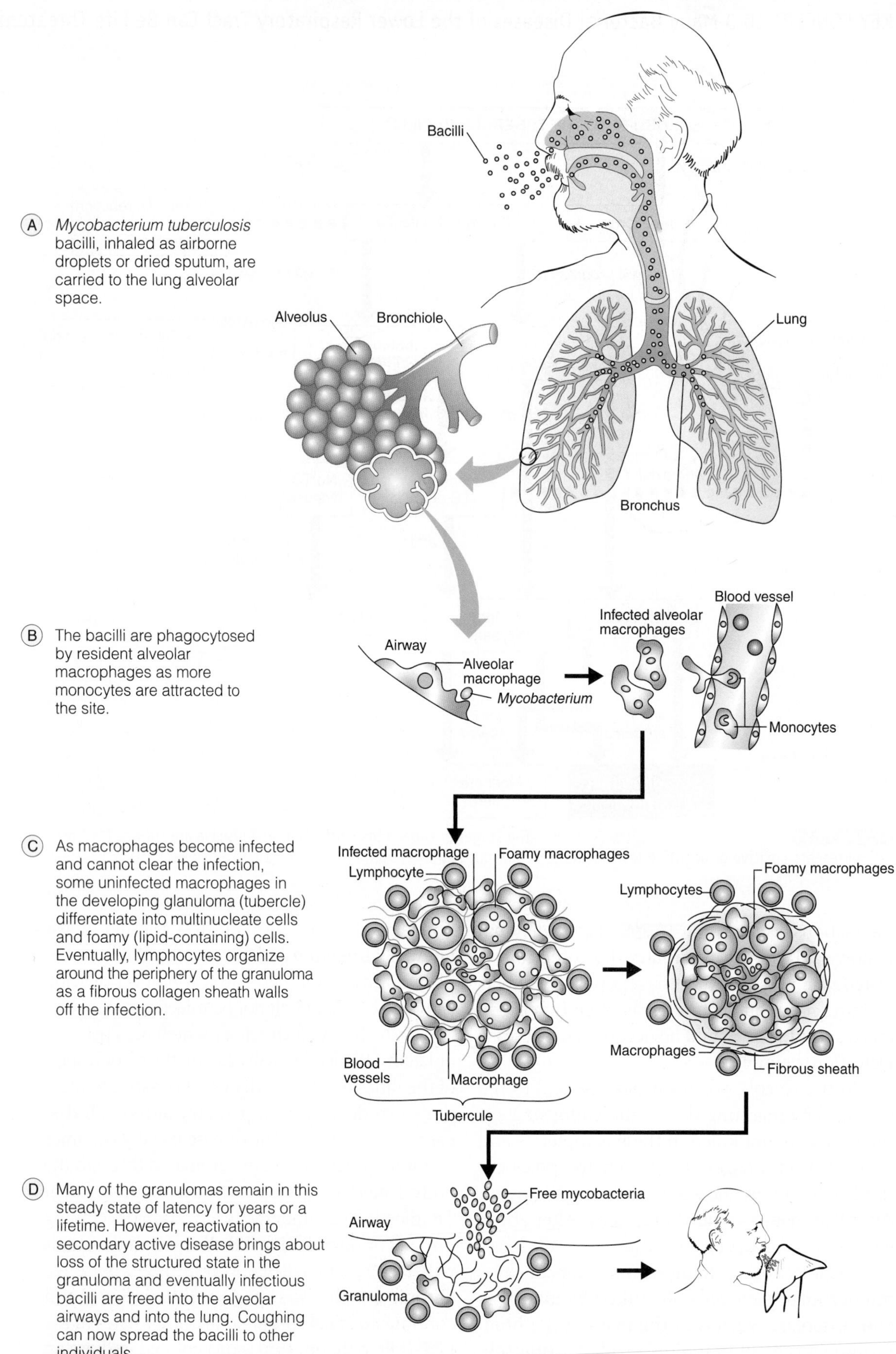

FIGURE 10.18 **The Progress of Tuberculosis.** Following invasion of the alveoli, the tubercle bacilli are taken up by macrophages and the immune system attempts to "wall off" the bacilli. »» What is the immune system attempting to do by forming tubercles?

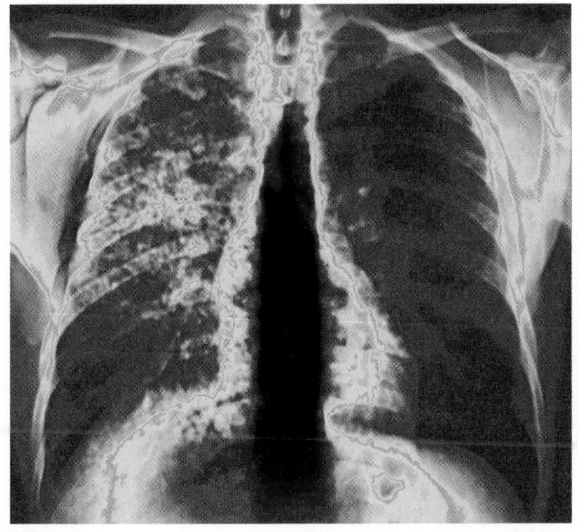

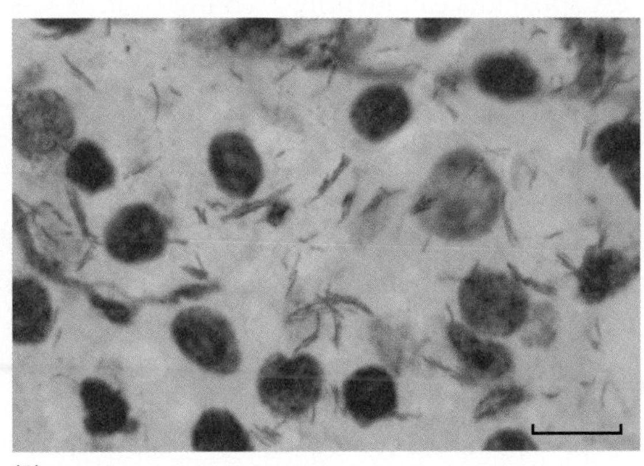

(A) (B)

FIGURE 10.19 **Pulmonary Tuberculosis.** **(A)** This false color X ray shows the extensive fibrosis (fuzzy yellow color in the lung cavity) typical in patients with advanced, active tuberculosis. © James Cavallini/Photo Researchers, Inc. **(B)** This light microscope image shows *M. tuberculosis* cells stained with the acid-fast procedure. In sputum samples, the bacterial cells often exhibit growth in thick strings. (Bar = 10 μm.) © Biophoto Associates/Photo Researchers, Inc. »» Why isn't the Gram stain used to identify *M. tuberculosis*?

of the infected macrophages die, releasing bacilli and producing a caseous (cheese-like) center in the tubercle. Live bacterial cells rupture from the tubercles, and spread and multiply throughout the LRT. These individuals will now have a positive tuberculin reaction, chest X ray, and sputum test; these individuals can transmit the disease to others (FIGURE 10.19).

Because the bacilli in individuals with primary active TB disease are not killed, tubercle erosion can allow the bacterial cells to spread through the blood and lymph to other organs such as the liver, kidney, meninges, and bone. If active tubercles develop throughout the body, the disease is called **miliary (disseminated) tuberculosis** (*milium* = "seed"; in reference to the tiny lesions resembling the millet seeds in bird food). Tubercle bacilli produce no known toxins, but growth is so unrelenting that the respiratory and other body tissues are literally consumed, a factor that gave tuberculosis its alternate name of "consumption." However, miliary TB is not infectious.

Disease Detection. Early detection of tuberculosis is aided by the tuberculin reaction, a delayed hypersensitivity that begins with the application of a purified protein derivative (PPD) of *M. tuberculosis* to the skin. One method of application, called the **Mantoux test**, uses an injection of PPD intradermally into the forearm. Depending on the patient's risk of exposure, the skin becomes thick, and a raised, red welt, termed an **induration**, of a defined diameter develops (FIGURE 10.20). For an individual never before exposed to *M. tuberculosis*, an induration greater than 15 mm is interpreted as a positive test. However, a positive test does not necessarily reflect the presence of active TB disease, but may indicate a recent immunization, previous tuberculin test, or past exposure to *M. tuberculosis*. It suggests a

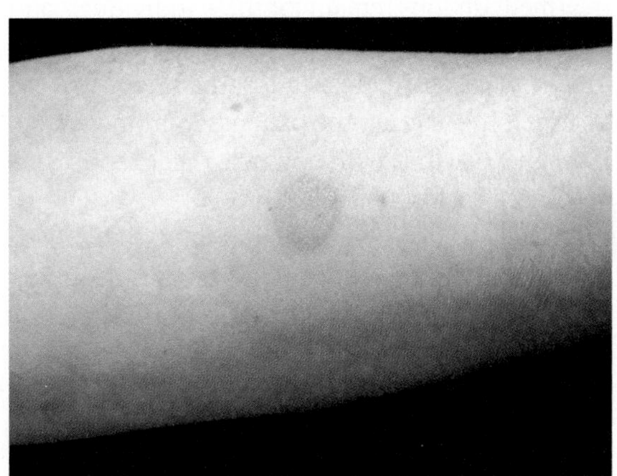

FIGURE 10.20 **Tuberculin Skin Test for Tuberculosis.** This is an example of a positive reaction to the Mantoux skin test. An induration of less than 15 mm is considered negative. © Phototake, Inc./Alamy Images. »» What is an induration?

need for further tests, such as a chest X ray or acid-fast staining of a sputum sample. In 2010, a new TB test system was endorsed by WHO. The test apparatus can identify a TB infection in less than 2 hours versus the 2 to 3 months required to verify a positive sputum test.

Treatment. Tuberculosis is an extremely stubborn disease especially with the development of antibiotic resistance. TB has been traditionally treated with such first-line drugs as isoniazid and rifampin. Ethambutol, pyrazinamide, and streptomycin also are used to help delay the emergence of resistant strains. Still, the reported cases of **multidrug-resistant tuberculosis (MDR-TB)** continue to increase. In 2010, 3.4% of TB cases worldwide were multidrug resistant and among patients who had previously been treated for TB, 19.8% of cases were multidrug resistant. MDR-TB has necessitated a switch to a group of second-line drugs, including fluoroquinolones and kanamycin. For MDR-TB cases, antimicrobial drug therapy is intensive and must be extended over a period of 6 to 9 months or more, partly because the organism multiplies at a very slow rate (its generation time is about 18 hours). Early relief, complacency, and forgetfulness often cause the patient to stop taking the medication, and the disease flares anew. MicroFocus 10.2 describes the WHO treatment strategy to maintain patient compliance.

An increasing number of MDR-TB cases have now evolved into **extensively drug-resistant tuberculosis (XDR-TB)**, meaning almost all drugs used to treat TB, including isoniazid, rifampin, fluoroquinolones, and kanamycin are useless. Few treatment options remain for these individuals and a successful outcome depends upon the extent of *M. tuberculosis* drug resistance, the severity of the disease, and whether the patient's immune system is weakened. In the 38 countries that report surveillance data for XDR-TB, almost 10% of MDR-TB cases have evolved into XDR-TB. In the United States, the number of MDR-TB cases is dropping and only 1.1% of new TB cases reported in 2010 were multidrug resistant.

If this is not bad enough, in 2007, two cases of so-called "totally drug-resistant TB" (TDR-TB) or "extremely drug-resistant TB" (XXDR-TB) were reported in Italy that had resistance to all first-line and second-line anti-TB drugs tested. In 2009, a report identified 15 patients in Iran who were resistant to all anti-TB drugs tested. In 2012, 12 patients in India were reported to have TDR-TB (three have died). Importantly, in 2012 the WHO decided not to recommend the creation of a new TDR-TB category because drug susceptibility testing is technically challenging and the reported cases of XXDR-TB or TDR-TB so far reported may actually be XDR-TB patients who have resistance to additional second-line anti-TB drugs.

A weakened immune system is especially worrisome because TB is a particularly insidious problem to those who have AIDS. In these coinfected patients, the T lymphocytes that normally mount a response to *M. tuberculosis* are being destroyed by HIV, and the patient cannot respond to the bacterial infection. Unlike most other TB patients, those with HIV usually develop miliary tuberculosis in the lymph nodes, bones, liver, and numerous other organs. Ironically, AIDS patients often test negative for the tuberculin skin test because without T lymphocytes, they cannot produce the telltale red welt signaling exposure. Today, TB is the leading cause of death in HIV-infected patients and is the causative agent in 13% of AIDS deaths worldwide.

Prevention. Vaccination against TB may be rendered by intradermal injections of an attenuated strain of *Mycobacterium bovis,* a species that causes tuberculosis in cows as well as humans. The attenuated strain is called **Bacille Calmette-Guérin (BCG)**, after Albert Calmette and Camille Guérin, the two French investigators who developed it in the 1920s. Though the vaccine is used in parts of the world where the disease causes significant mortality and morbidity, health officials in the United States generally do not recommend the BCG vaccine because it has limited effectiveness for preventing TB in adults and produces occasional side effects. Twelve new antimicrobial compounds, consisting of subunits, molecules of DNA, and attenuated strains of mycobacteria, are currently being developed.

Other *Mycobacterium* Species. Several other species of *Mycobacterium* deserve a brief mention. *M. chelonae,* another pathogenic species frequently found in soil and water, can cause lung diseases, wound infections, arthritis, and skin abscesses. *M. haemophilum* is a slow-growing pathogen often found in immunocompromised individuals including those with AIDS. Cutaneous ulcerating lesions and respiratory symptoms are typical in the patients. *M. kansasii* causes infections that are indistinguish-

MICROFOCUS 10.2: Public Health

Tragic Endings but Hopeful Futures

Tuberculosis (TB) is a contagious disease caused by *Mycobacterium tuberculosis*. In fact, over 33% of the world's population currently is infected with the tubercle bacillus and TB is killing more people every year; in 2008 someone was dying of TB every 15 seconds. Among the reasons for the rise in TB cases is that infected individuals are not completing the full course of antibiotics once they start to feel better. Not only does this behavior fail to cure the disease, it also helps generate multidrug-resistant TB (MDR-TB).

© Scott Camazine/Alamy Images.

To address these issues, the World Health Organization (WHO) has developed a TB treatment program called DOTS (Direct Observation Treatment System) to detect and cure TB. Once a patient with infectious TB is identified by a sputum smear, their treatment follows the DOTS strategy. A physician, community worker, or trained volunteer observes and records the patient swallowing four basic medications over a 6- to 8-month period. A sputum smear is repeated after 2 months to check progress and again at the end of treatment.

DOTS appears to be very effective when it is used. Take for example the following case. In the early 1990s, a powerful strain of drug-resistant TB emerged in New York City. Affecting hundreds of people in hospitals and prisons, the outbreak killed 80% of the infected patients. The city responded with DOTS, which seemed to work because the outbreak subsided.

However, in 1997 it was back, this time in South Carolina. A New York patient (not on DOTS) moved to South Carolina. His TB had lingered and he infected three family members in his new community. Soon, another six members of the community were sick with TB. However, these six individuals had not had contact with the family; in fact, they did not even know the family. Investigators from the Centers for Disease Control and Prevention (CDC) were called in to investigate. They learned that one infected family member had been in the hospital, where he was examined with a bronchoscope (a lighted tube extended into the air passageways). Unfortunately, the bronchoscope was not disinfected properly after the examination and was used to examine these other individuals.

Many stories have happy endings, but this is not one of them. Of the six patients infected with the drug-resistant strain of *M. tuberculosis,* two died from TB and three died from other causes while battling TB. Only one recovered.

Thus, one sees the importance and effectiveness of DOTS. The WHO reports that in 2009 DOTS produced an 87% cure rate; that means 2.3 million people were cured. DOTS also prevents the development of MDR-TB by making sure TB patients take the full course of treatment.

Since DOTS was introduced in 1995, more than 10 million infectious patients have been treated successfully. In China, there has been a 96% cure rate and in Peru a 91% cure rate for new cases of TB. Such successes offer a hopeful future.

able from tuberculosis and, in the United States, is most commonly found in the central states.

The group known as *M. avium* complex (MAC) consists of two species that are difficult to tell apart: *M. avium* and *M. intracellulare*. MAC is primarily a pulmonary pathogen that affects individuals who are immunocompromised, such as AIDS patients and individuals on immunosuppressive chemotherapy. Interestingly, cervical lymphadenitis (an inflammation of the lymph nodes in the neck) is the most common nontuberculous mycobacterial infection in immunocompetent children under 5 years of age, while the condition occurs rarely in immunocompetent adults.

For all species mentioned here, there is no evidence for spread between individuals; rather, infection comes from contacting soil, or ingesting food or water contaminated with the organism.

Infectious Bronchitis Is an Inflammation of the Bronchi

Infectious bronchitis occurs most often during the winter and can be caused by bacteria following a URT viral infection, such as the common cold. *Mycoplasma pneumoniae* and *Chlamydophila pneumoniae* often cause bacterial bronchitis in young adults, while *Streptococcus pneumoniae* and *Haemophilus influenzae* are the primary agents among middle-aged and older individuals. Influenza viruses also can cause a form of viral bronchitis.

Infectious bronchitis generally begins with the symptoms of a common cold: runny nose, sore throat, chills, general malaise, and perhaps a slight fever. The onset of a dry cough usually signals the beginning of **acute bronchitis**, a condition that occurs when the inner walls lining the main airways of the lungs become infected and inflamed. Inflammation increases the production of mucus, which then narrows the air passages (FIGURE 10.21). Most cases of acute bronchitis disappear within a few days without any adverse effects, although a cough can linger for several weeks. If the condition persists for more than 3 months, it is referred to as **chronic bronchitis**. The changing of the clear or white mucus (phlegm) to a yellow or green color usually indicates a bacterial infection.

Antibiotics may be prescribed for bacterial bronchitis. Because many cases result from influenza, getting a yearly flu vaccination may reduce the risk of bronchitis. Other preventative measures include good hygiene, including hand washing, to reduce the chance of transmission.

Pneumonia Can Be Caused by Several Bacterial Species

The term **pneumonia** refers to microbial disease of the bronchial tubes and lungs. A wide variety of bacterial species may cause pneumonia. In the United States, there are 5 to 10 million pneumonia infections each year that result in 1 million hospitalizations and 40,000 to 50,000 deaths. For discussion purposes, we can divide pneumonia into

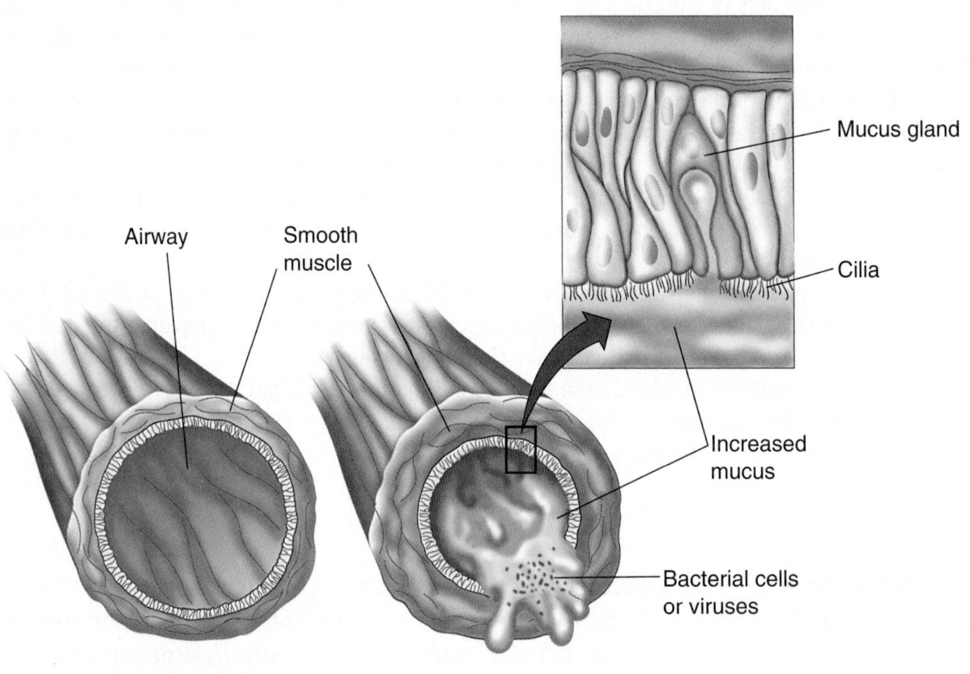

FIGURE 10.21 **Bronchus Inflammation.** Infectious bronchitis develops as the bronchial wall narrows due to inflammation and swelling. Increased mucus production, which contains bacterial cells or viruses, also narrows the airway. Adapted from: Merck, "Acute Bronchitis: Lung and Airway Disorders," *Merck Manual Home Edition,* December 09, 2008, http://www.merck.com/mmhe/sec04/ch041/ch041a.html. »» How would such pathogens in the mucus be transmitted to another person?

healthcare-acquired and community-acquired infections.

Healthcare-Acquired Pneumonia (HCAP). Pneumonia is the second most common healthcare-acquired infection after urinary tract infections, accounting for 250,000 to 300,000 cases per year in the United States. HCAP is defined as an inflammation of one or both lungs that develops at least 48 hours after administration to a hospital or other healthcare facility. In the United States, 86% of healthcare pneumonias are associated with mechanical ventilation (e.g., respirator or ventilator) to help patients breathe. In these cases, the intubation process itself can be the source of infection. That is why it is important that healthcare workers wash their hands and follow standard precautions so as not to transmit the pathogens from one patient to another.

One of the most common causes of HCAP results from an infection by *Staphylococcus aureus*, a facultatively anaerobic, gram-positive coccus. **Staphylococcal pneumonia** most commonly occurs in infants, young children, and immunocompromised patients. If bacterial cells infect the lungs, a severe, necrotizing pneumonia may occur exacerbated by a bacterial toxin that can itself cause pneumonia and kill healthy tissue. Symptoms include a short period of fever followed by rapid onset of respiratory distress, which may include rapid breathing and bluish skin. *S. aureus* is most often spread to others by contaminated hands. Persons who are immunocompromised or who have invasive medical devices are particularly vulnerable to infection.

With increased antibiotic resistance, infections can be difficult to treat and can progress to life-threatening infections because there are fewer effective antibiotics available for treatment.

Klebsiella pneumoniae is a gram-negative rod with a prominent capsule. The bacillus is acquired by respiratory droplets, and often it occurs naturally in the URT of humans. ***Klebsiella* pneumonia** may be a primary infection or a secondary infection in alcoholics or people with impaired pulmonary function. As a primary lobar pneumonia, it is characterized by sudden onset and gelatinous reddish-brown sputum. The bacterial cells grow over the lung surface and rapidly destroy the tissue, often causing death. In its secondary form, *K. pneumoniae* occurs in already ill individuals and is a hospital-acquired disease spread by such routes as clothing, intravenous solutions, foods, and the hands of healthcare workers. In 2010, scientists reported the discovery in India of a new *K. pneumoniae* gene, and another that has been present in the United States for many years, that can be horizontally transferred to many other bacterial species making them resistant to most all available antibiotics.

Pseudomonas aeruginosa, a gram-negative, aerobic rod, is one of the most dangerous opportunistic pathogens because of its ability to cause severe or fatal nosocomial infections, especially in immunocompromised patients. ***Pseudomonas* pneumonia** is another common cause of HCAP, accounting for more than 10% of all cases. In addition, it is often resistant to commonly used antibiotics. The *P. aeruginosa* cells can be transmitted via aspiration from contaminated ventilator tubing or other healthcare devices, such as bronchoscopes, mechanical ventilators, and nebulizer equipment. Symptoms in immunocompromised patients include breathing problems, productive cough, fever, chills, and bluish skin. The treatment of this condition may include a combination of drugs like carbenicillin and gentamicin.

Acinetobacter species are gram-negative, aerobic rods (similar in appearance to *Haemophilus influenzae* on Gram stain) that frequently colonize the respiratory tract. Thus, these species, called the *Acinetobacter calcoaceticus-A. baumannii* complex (Abc) account for 80% of clinical infections and are becoming a major cause of HCAP, especially in intensive care units. ***Acinetobacter* pneumonia** occurs in outbreaks and, like with *P. aeruginosa*, is usually associated with contaminated respiratory-support equipment or fluids. Likewise, the symptoms are similar. The Abc is becoming increasingly resistant to antibiotics, presenting a significant challenge in treating these infections. Relatively few antibiotics are effective and physicians often have to rely on older agents, such as polymyxins, for treatment. Because the Abc is relatively susceptible to disinfectants and antiseptics, aggressive cleaning of the hospital environment and respiratory equipment is required.

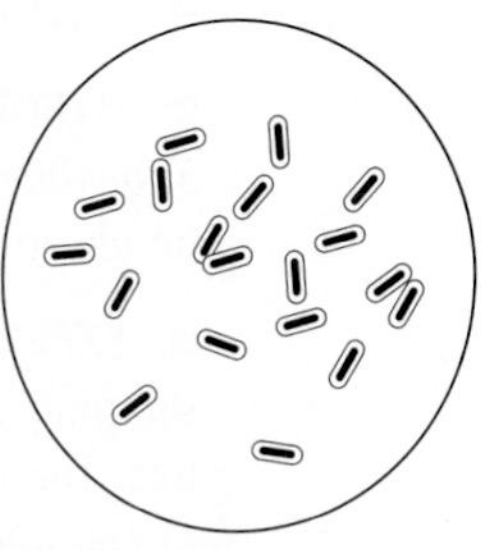
Klebsiella pneumoniae

Community-Acquired Pneumonia (CAP). In the United States, between 350,000 and 620,000 cases of CAP occur each year in the elderly. Over 80% of CAP cases are caused by *Streptococcus pneumoniae*, a gram-positive, encapsulated chain of diplococci (FIGURE 10.22). **Pneumococcal pneumonia**, being community acquired, exists in all age groups, although the mortality rate is highest among infants, the elderly, and those with underlying medical conditions.

S. pneumoniae can be acquired by aerosolized droplets or as a part of the normal microbiota in the URT of many individuals. However, Investigating the Microbial World 10 presents another potential form of transmission. Because mucociliary clearance is at work and the natural resistance of the body is high, pneumococcal pneumonia usually does not develop until the defenses are compromised. Malnutrition, smoking, viral infections, and treatment with immune-suppressing drugs most often predispose one to *S. pneumoniae* infections.

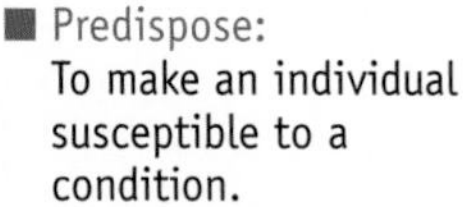

■ Predispose: To make an individual susceptible to a condition.

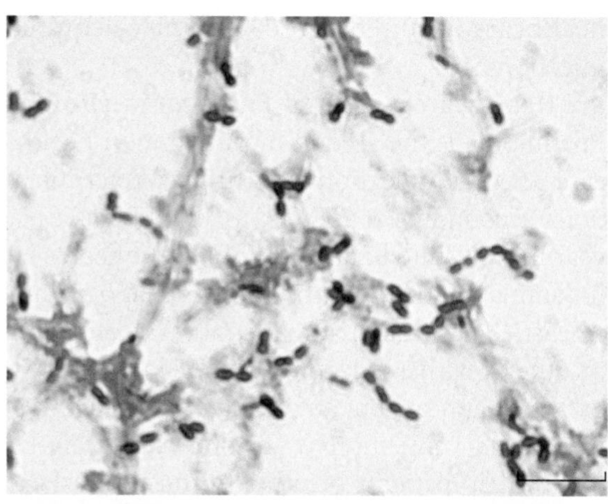

FIGURE 10.22 ***Streptococcus pneumoniae.*** Light microscope image of gram-stained *S. pneumoniae* cells (dark spheres), the cause of pneumococcal pneumonia. (Bar = 10 µm.) Courtesy of Dr. Mike Miller/CDC »» How would you describe the arrangement of *S. pneumoniae* cells?

Investigating the Microbial World 10

Cleanliness Is Next to Disinfectedness

A common inhabitant of the human nasopharynx is the gram-positive, encapsulated bacterium *Streptococcus pneumoniae*. As discussed in this chapter, its location in the upper respiratory tract makes it one of the prime candidates for meningitis, pneumonia, and middle ear infections should it switch from a harmless existence to a pathogenic one.

- **OBSERVATIONS:** As described in this chapter and by the Centers for Disease Control and Prevention (CDC), *S. pneumoniae* is transmitted "directly from person to person through close contact via respiratory droplets" during coughing and sneezing. Yet other respiratory pathogens are also spread by nonliving objects or materials, such as environmental surfaces.

- **QUESTION:** Can **S. pneumoniae** also survive for a prolonged time on a dry, desiccated environmental surface?

- **HYPOTHESIS:** Dry, desiccated environmental surfaces are not an alternate source for the spread of *S. pneumoniae*. If so, then after a "prolonged period" on a dry surface, the bacterial cells will die (no colonies will be observed when plated on blood agar), and consequently will no longer be infective.

- **EXPERIMENTAL DESIGN:** The desiccation protocol involved growing *S. pneumoniae* on blood agar overnight, scraping the bacterial cells off the plate, and spreading a thin, even layer of cells (10^8) onto polystyrene petri dish lids. The number of viable cells per lid was determined by plating dilutions on blood agar and counting the number of colonies growing.

- **EXPERIMENT 1:** Cell preparations were spread on lids and incubated at ambient temperature and humidity for 1 hour to 28 days. A control preparation (0 hour) was spread on a lid and immediately diluted and plated on blood agar.

- **EXPERIMENT 2:** Because survivability could be a result of starvation on polystyrene, cell preparations for desiccation were spread on lids and incubated at ambient temperature and humidity. Cell preparations for starvation were spread on agar plates containing only phosphate-buffered saline (PBS). All lids and plates were incubated for 6, 24, or 48 hours.

- **EXPERIMENT 3:** Because *S. pneumoniae* has a capsule, survivability could be directly related to having this type of glycocalyx. Two identical cell preparations were set up and incubated for 4 to 168 hours. One preparation used the encapsulated strain and the other used an nonencapsulated strain.

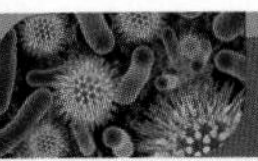

Investigating the Microbial World 10 (continued)

- **EXPERIMENT 4:** If *S. pneumoniae* cells survive desiccation, they may not be infectious. Samples of 48-hour nondesiccated and desiccated cells in PBS were inoculated intranasally into mice. A control group was inoculated intranasally with only PBS. Three days postinoculation, the nasal lavage fluid was plated on blood agar plates containing the antibiotic gentamicin to which *S. pneumoniae* is naturally resistant. Colonies on agar were then counted.
- **RESULTS:**

EXPERIMENT 1: See figure.

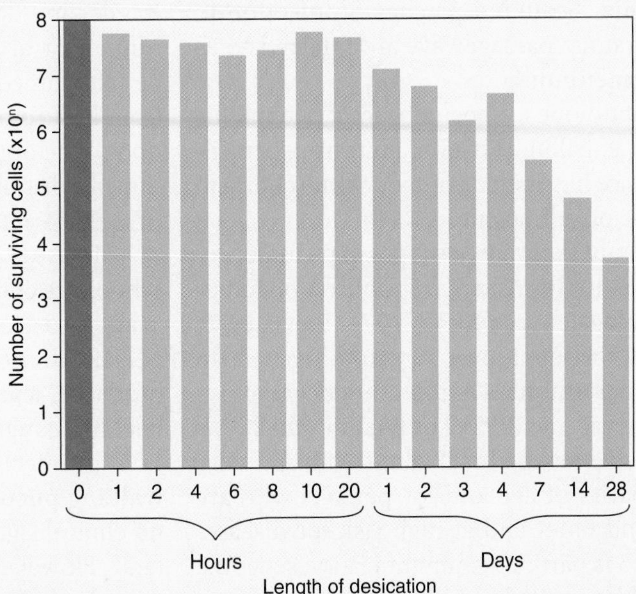

FIGURE *S. pneumoniae* survival after desiccation. Bacteria were rehydrated and plated after 0 hours to 28 days of desiccation to determine viability. Data for the medians are shown. 0 hour = control lid (=10^8 cells). Adapted from: Walsh, R.L. and Camilli, A. (2011). *mBio* 2(3): e00092-11.doi:10.1128/mBio.00092-11.

EXPERIMENT 2: Analysis of the desiccated and starved samples indicated that there were fewer colonies on the agar plated with starved samples than on the agar plated with desiccated samples.

EXPERIMENT 3: No significant difference was observed in bacterial viability for the nonencapsulated strain as compared to the encapsulated strain when colonies were counted on agar.

EXPERIMENT 4: No detectable *S. pneumoniae* colonies on agar could be detected from the control mice. However, 80% of mice receiving the nondesiccated sample and 75% of mice receiving the desiccated sample produced colonies on the agar plates.

- **CONCLUSIONS:** The medical view has been that *S. pneumoniae* cells must be transmitted through direct contact of airborne respiratory secretions.

QUESTION ***1: Was the hypothesis supported concerning desiccation and infectivity? Explain.***

QUESTION ***2: What can you conclude from experiment 2 regarding the influence of starvation on survivability through desiccation?***

QUESTION ***3: How important is the* S. pneumoniae *capsule in providing protection against desiccation?***

QUESTION ***4: Based on the results from this work, what are your recommendations concerning the transmissibility of* S. pneumoniae *from nonliving surfaces such as handkerchiefs, utensils, and hospital surfaces?***

Answers can be found on the Student Companion Website in **Appendix D.**

Adapted from: Walsh, R. L. and Camilli, A. (2011). *mBio* **2**(3): e00092-11.doi:10.1128/mBio.00092-11.

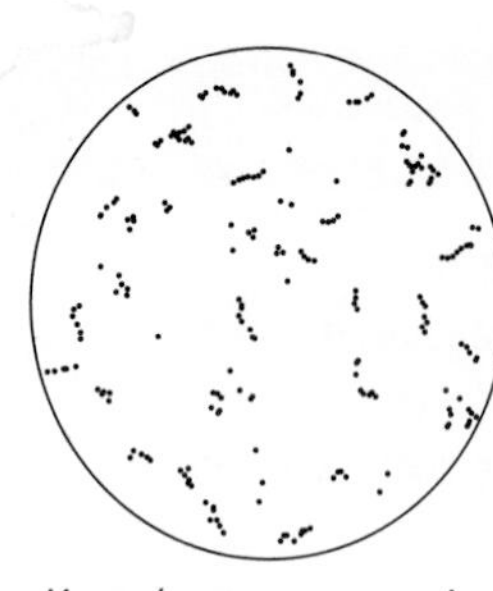
Mycoplasma pneumoniae

Patients with pneumococcal pneumonia experience high fever, sharp chest pains, difficulty breathing, and rust-colored sputum. The color results from blood seeping into the alveolar sacs of the lung as bacterial cells multiply and cause the tissues to deteriorate. The involvement of an entire lobe of the lung is called **lobar pneumonia**. If both left and right lungs are involved, the condition is called **double pneumonia**. Scattered patches of infection in the respiratory passageways are referred to as **bronchopneumonia**.

The antibiotic for pneumococcal pneumonia has been penicillin. However, increasing penicillin resistance has shifted antibiotic drug choice to cefotaxime or ceftriaxone.

Currently, there are two types of pneumococcal vaccines for the American public. A pneumococcal conjugate vaccine (PCV13) or Prevnar 13® is available for people over 50 years of age to help prevent pneumonia. A pneumococcal polysaccharide vaccine (PPSV) or Pneumovax 23® is also available for use in all adults who are older than 65 years of age and for persons who are 2 years and older and at high risk for disease. Prevnar 13 is part of the immunization schedule for all infants and should be given to children under 5 years of age who missed their shots or started the series late.

Every year some 1.6 million children worldwide die from pneumonia (MicroFocus 10.3). Pneumococcal pneumonia kills 800,000 children—lives that could be saved with an effective vaccine. Now such vaccines are becoming available. The Global Alliance for Vaccines and Immunization (GAVI Alliance) estimates that 700,000 children's lives can be saved by 2015 and 7 million by 2030 through the introduction of Prevnar 13 and another vaccine called Synflorix® to the world's poorest countries where 98% of the pneumonia deaths occur.

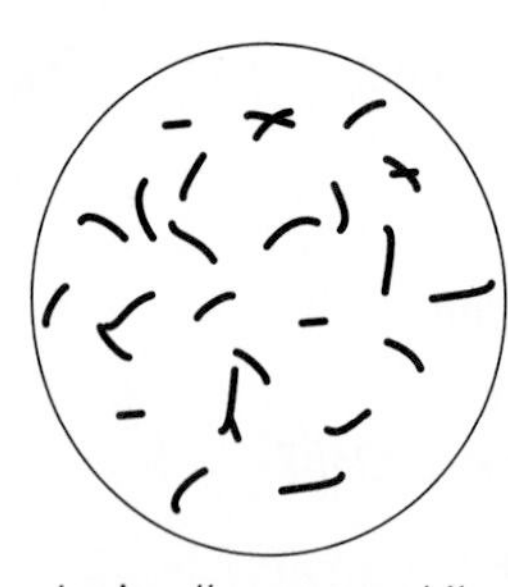
Legionella pneumophila

Some 10% of CAP cases, especially in the elderly and compromised individuals, are caused by inhaling respiratory droplets containing unencapsulated *H. influenzae* strains. These gram-negative coccobacilli can colonize the nasopharynx and cause otitis media and sinusitis. Spread to the LRT can cause pneumonia. Infections are treated with trimethoprim-sulfamethoxazole.

■ Insidious:
Refers to a disease that progresses gradually.

There is a more insidious form of CAP that is not caused by the typical pathogens described above. Several bacterial species can cause this form of **"atypical" pneumonia**.

Primary atypical pneumonia—"primary" because it occurs in previously healthy individuals—is caused by *Mycoplasma pneumoniae*. This community-acquired disease causes about 20% of CAP cases.

The cells of *M. pneumoniae* are among the smallest bacterial cells (FIGURE 10.23A), are very fragile, and do not survive for long outside the human or animal host. Therefore, they must be maintained in nature by passage in droplets from host to host. Diagnosis is assisted by isolation of the organism on blood agar and observation of a distinctive "fried egg" colony appearance (FIGURE 10.23B).

Most *M. pneumoniae* patients have been school-age children and teens, although in recent years it has accounted for up to 15% of CAP in persons over 65 years of age. Symptoms include headache, fever, fatigue, and a characteristic dry, hacking cough. Pneumonia develops in 5% to 10% of infected individuals. Often it is called **walking pneumonia** (even though the term has no clinical significance) and the disease is rarely fatal. However, epidemics are common where crowded conditions exist, such as in college dormitories, military bases, and urban ghettoes. Erythromycin and azithromycin are commonly used as treatments.

MicroFocus 10.4 describes an interesting application for another bacterial species known to cause respiratory infections.

Another form of CAP first surfaced in July 1976, when Pennsylvania's chapter of the American Legion held its annual convention in Philadelphia. Toward the end of the convention, 140 conventioneers and 72 other people in or near the convention hotel became ill with headaches, fever, coughing, and pneumonia. Thirty-four individuals died of the disease or its complications. In January 1977, investigators from the CDC isolated the infecting bacterial species, called *Legionella pneumophila,* from the lung tissue of one of the patients. This form of CAP became known as **Legionnaires' disease** (Textbook Case 10).

The *L. pneumophila* cells are aerobic, gram-negative rods that exist where water collects, such as cooling towers, industrial air-conditioning systems, and stagnant pools, and some protists

MICROFOCUS 10.3: Public Health

The Killer of Children

Global Health Magazine recently reported the following: "Chitra Kumal knows the pain of losing a child. When her daughter, Sunita, was 15 months old, she developed a respiratory infection that quickly progressed into pneumonia. With no health facilities in her Nepalese village, Kumal depended on the advice and treatment of a traditional healer or shaman. After just 3 days of fever, fast breathing, and chest indrawing, her only daughter died."

Similar stories are reported everyday around the world. According to the World Health Organization (WHO), pneumonia kills 2 million children under 5 years of age each year—more than AIDS, malaria, and measles combined—accounting for nearly one in five child deaths globally (see figure). However, this number may be an underestimate as nearly half of all pneumonia cases occur in malarious parts of the world where pneumonia often is misdiagnosed as malaria.

The WHO estimates that more than 150 million episodes of pneumonia occur every year among children under 5 in developing countries, accounting for more than 95% of all new cases worldwide, and between 11 and 20 million of these episodes require hospitalization. The highest incidence of pneumonia cases among children under 5 occurs in South Asia and sub-Saharan Africa, accounting for more than half the total number of pneumonia episodes worldwide.

Preventing and treating childhood pneumonia obviously is critical to reducing childhood mortality. However, only about one in four caregivers knows the two key symptoms of pneumonia: fast breathing and difficult breathing (indrawing). Estimates suggest that if antibiotics were universally available and given to children with pneumonia, around 600,000 lives could be saved each year. But this represents only about 25% of the annual cases. Clearly, other control measures are needed.

At the beginning of the 20th century, pneumonia accounted for 19% of childhood deaths in the United States, a statistic remarkably similar to the rate in developing countries today. Control in the United States was achieved largely without antibiotics and vaccines. Therefore, other control measures and strategies are needed on a global scale.

Key prevention measures include promoting balanced nutrition (including breastfeeding, vitamin A supplementation, and zinc intake), reducing environmental air pollution, and increasing immunization rates with vaccines, such as those against *Haemophilus influenza* type b (Hib) and *Streptococcus pneumoniae* (pneumococcus). However, only about 50% of pneumonia cases in Africa and Asia are caused by these two organisms, so other vaccines need to be developed against other bacterial species (and viruses) that cause pneumonia. And of course—hand washing, like in all areas of infectious disease, can play an important role in reducing the incidence of pneumonia.

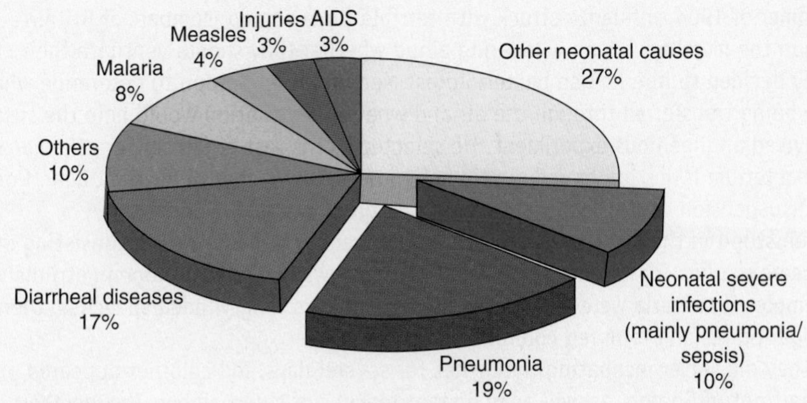

Pneumonia accounts for almost 20% of global childhood mortality.

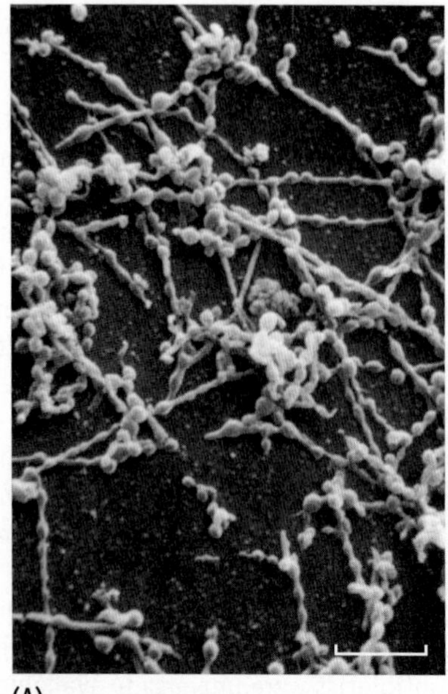

(A)

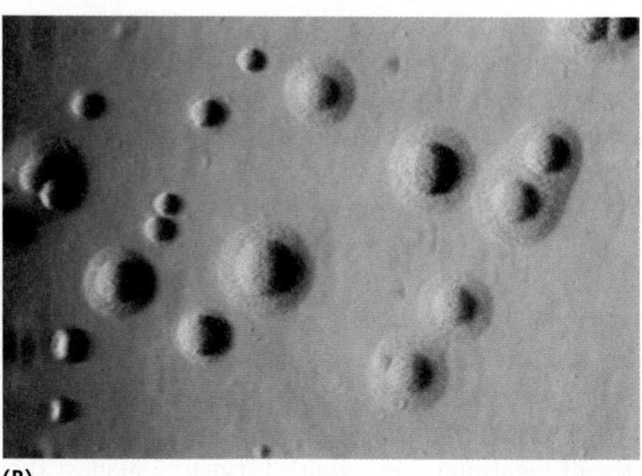

(B)

FIGURE 10.23 *Mycoplasma pneumoniae.* Two views of *M. pneumoniae,* the agent of primary atypical pneumonia. (**A**) A scanning electron micrograph, demonstrating the pleomorphic shape exhibited by mycoplasmas. (Bar = 2 μm.) © Michael Gabridge/Visuals Unlimited. (**B**) Colony morphology on agar shows the typical "fried egg" appearance. © ASM/Science Source/Photo Researchers, Inc. »» What structural feature is missing from the mycoplasma cells that allows for their pleomorphic shape?

MICROFOCUS 10.4: History

"Keep It Short, Please!"

Defining, developing, and proving the germ theory of disease was one of the great triumphs of scientists in the late 1800s. Applying the theory to practical problems was another matter, however, because people were reluctant to change their ways. It would take some rather persuasive evidence to move them.

In the summer of 1904, influenza struck with terrible force among members of Britain's House of Commons. Soon the members began wondering aloud whether they should ventilate their crowded chamber. They decided to hire British bacteriologist Mervyn Henry Gordon to determine whether "germs" were being transferred through the air and whether ventilation would help the situation.

Gordon devised an ingenious experiment. He selected as his test organism *Serratia marcescens* because the bacterium forms bright-red visible colonies in Petri dishes of nutrient agar. Gordon prepared a liquid suspension of the bacterial cells and gargled with it!

Gordon then stood in the Debating Chamber and delivered a 2-hour oration consisting of selections from Shakespeare's *Julius Caesar* and *Henry V.* His audience was hundreds of open Petri dishes. The theory was simple: If bacteria were transferred during Gordon's long-winded speeches, then they would land on the agar plates and form red colonies.

And land they did. After incubating the dishes for several days, red colonies appeared on plates placed right in front of Gordon, as well as in distant reaches of the chamber. The members were impressed. They proposed a more constant flow of fresh air to the chamber, as well as shorter speeches. No one was about to object to either solution, especially the latter.

Textbook CASE 10

Legionnaires' Disease Outbreak—Bogalusa, Louisiana

1 On October 31, 1989, two physicians in Bogalusa, Louisiana reported an outbreak of more than 50 cases of acute pneumonia to the state department of health. Most cases occurred within a 3-week interval in mid- to late October; six persons had died. All cases had occurred in older adults and 76% of the cases were female. Lab analysis confirmed 33 cases were caused by *Legionella pneumophila,* the agent of Legionnaires' disease.

2 When investigators from the Centers for Disease Control and Prevention (CDC) arrived, it was imperative to determine quickly the source and mode of transmission of *L. pneumophila.*

3 Most cases were among residents of Bogalusa. A total of 28 patients and 56 controls were interviewed. Patients and controls were asked about exposures to cooling towers and nearby buildings.

4 Of the 28 patients, three reported visiting hospital B with a cooling tower; of the 56 controls, seven reported visiting hospital B. Similarly, 7 of the patients and 12 of the controls reported having visited the nearby post office.

5 Microbiological analysis was unable to confirm any contamination in the hospital B cooling tower. In addition, visiting the post office was eliminated as a possible source.

6 Further interviews identified two other potential sources, butcher shop A and grocery store B. Butcher shop A was visited by 12 patients and 19 controls, while 25 patients and 28 controls had visited grocery store B.

7 A detailed microbiological investigation of grocery store B was undertaken. Several days later, the CDC investigators had the answer. *L. pneumophila* had been isolated from an ultrasonic misting machine close to where shoppers selected produce in the vegetable section (see figure). The machine's aerosol action had sprayed bacterial cells into the air to which shoppers were exposed. No cases were reported among employees.

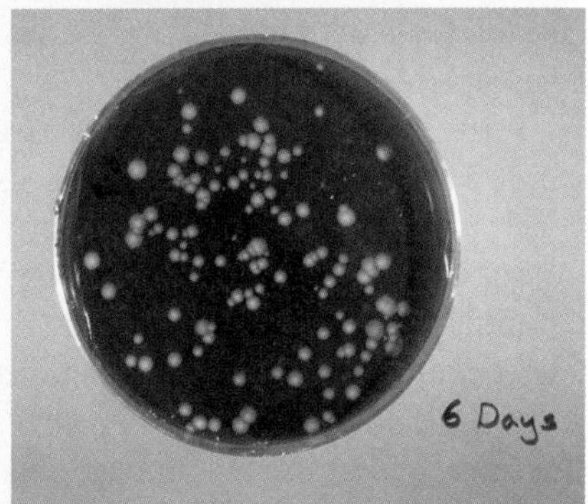

Colonies of *L. pneumophila* growing on an enriched agar medium. Courtesy of CDC.

Questions:

(Answers can be found on the Student Companion Website in **Appendix D.**)

A. Why was it important to know that most cases were in the city of Bogalusa? Who are the controls?

B. Why was a cooling tower first suspected as the source of the outbreak?

C. Why was the post office and butcher shop A eliminated as possible sources?

D. Who needs to know about these findings? How would you go about reporting the findings?

E. Explain the significance of (i) 76% of the cases being female and (ii) no cases occurring among grocery store employees.

For additional information see: www.cdc.gov/mmwr/preview/mmwrhtml/00001563.htm.

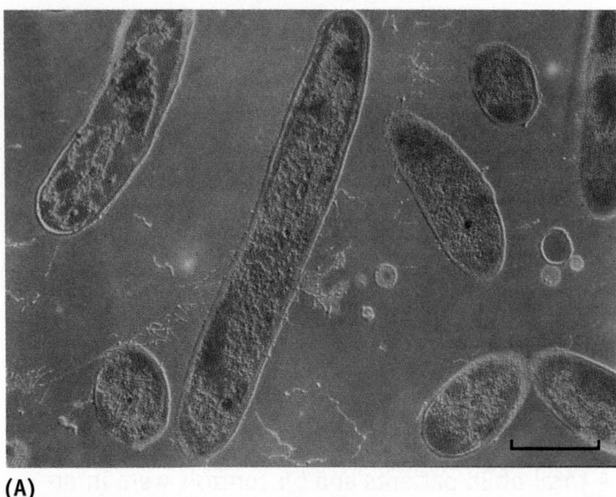

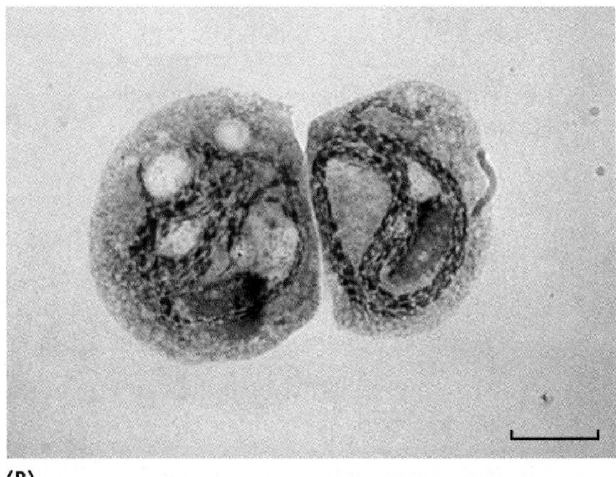

(A) (B)

FIGURE 10.24 *Legionella pneumophila.* Two views of *L. pneumophila,* the agent of Legionnaires' disease. (**A**) A false-color transmission electron micrograph of *L. pneumophila* cells. (Bar = 1 µm.) © Phototake/Alamy Images. (**B**) A false-color transmission electron micrograph of the protist *Tetrahymena pyriformis* infected with chains of *L. pneumophila* cells (dark red chains). (Bar = 10 µm.) Courtesy of Don Howard/CDC. »» How does infecting a protist benefit the bacterial species when in its natural environment?

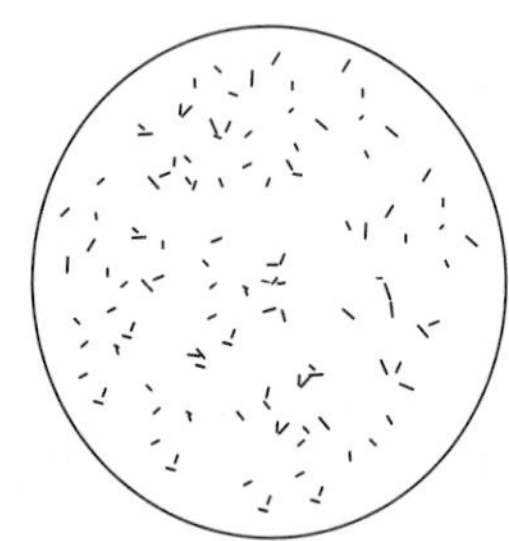

Coxiella burnetii

are its reservoir (FIGURE 10.24). Older adults and individuals with weak immune systems are most susceptible to infection.

In 2010, the CDC reported more than 3,300 cases of Legionnaires' disease. This was a 217% increase in the number of reported cases since 2000 and may reflect a rise in the elderly population, individuals with a high risk for infection, and increased detection and reporting.

Legionnaires' disease usually causes a severe atypical pneumonia characterized by headache, fever, a dry cough with little sputum, and some diarrhea and vomiting. In addition, chest X rays show a characteristic pattern of lung involvement, and necrotizing pneumonia is the most dangerous effect of the disease. Erythromycin is effective for treatment. Prevention requires that water sources be kept chlorinated.

After *L. pneumophila* was isolated in early 1977, microbiologists realized the organism was responsible for another milder infection called **Pontiac fever**. This is an influenza-like illness that lasts 2 to 5 days but does not cause pneumonia. Symptoms disappear without treatment. The term "legionellosis" encompasses both Legionnaires' disease and Pontiac fever.

Community-Acquired Pneumonia Is Also Caused by Intracellular Pathogens

Pneumonia also can be caused by some of the smallest bacterial organisms, the chlamydiae. Most are obligate, intracellular parasites, meaning they only grow inside host cells.

Q fever (the "Q" is derived from "query," originally referring to the unknown cause of the disease) is caused by *Coxiella burnetii,* which is not a strict intracellular pathogen (FIGURE 10.25). This form of atypical pneumonia is prevalent worldwide among dairy cows, sheep, and goats and human outbreaks may occur wherever infected animals are raised, housed, or transported. In 2009, more than 2,300 cases of Q fever, including six deaths, were reported in the Netherlands where intense goat farming occurs. In 2010, the CDC reported 131 confirmed cases of Q fever.

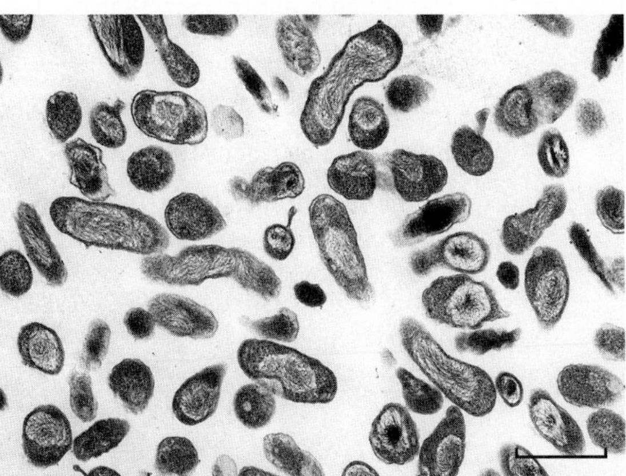

FIGURE 10.25 *Coxiella burnetii.* An electron micrograph of *C. burnetii,* the agent of Q fever. Note the oval-shaped rods of the organism. (Bar = 1 µm.) Courtesy of Rocky Mountain Laboratories, NIAID, NIH »» What does it mean to say *C. burnetii* is an intracellular pathogen?

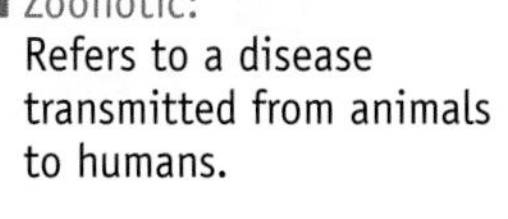

Being a zoonotic disease, transmission to humans occurs primarily by inhaling the organisms in dust particles or handling infected animals. In addition, humans may acquire the disease by consuming raw milk or cheese infected with *C. burnetii* or milk that has been improperly pasteurized. Although most cases are asymptomatic, some patients experience a bronchopneumonia characterized by severe headache, high fever, dry cough, and occasionally, lesions on the lung surface. The mortality rate is low, and treatment with doxycycline is effective for chronic infections. A vaccine is available for workers in high-risk occupations.

Psittacosis is another zoonotic disease caused by *Chlamydophila* (formerly *Chlamydia*) *psittaci*. These obligate, intracellular pathogens are transmitted to humans by infected parrots, parakeets, canaries, and other members of the psittacine family of birds (*psittakos* = "parrot"). The disease also occurs in pigeons, chickens, turkeys, and seagulls, and some microbiologists prefer to call it **ornithosis** (*ornith* = "bird") to reflect the more widespread occurrence in bird species. There were only four cases reported to the CDC in 2010.

Humans acquire *C. psittaci* by inhaling airborne dust or dried droppings from contaminated bird feces. Sometimes the disease is transmitted by a bite from a bird or via the respiratory droplets from another human. The symptoms of psittacosis resemble those of primary atypical pneumonia. Fever is accompanied by headaches, dry cough, and scattered patches of lung infection. Doxycycline is commonly used in therapy.

Chlamydial pneumonia is another type of atypical pneumonia and is caused by *Chlamydophila pneumoniae*. The organism is transmitted human-to-human by respiratory droplets and causes a mild CAP, principally in young adults and college students. The disease is clinically similar to psittacosis and is characterized by fever, headache, and nonproductive cough. Treatment with doxycycline or erythromycin hastens recovery from the infection.

Inhalational Anthrax Is an Occupational Hazard

Anthrax is caused by *Bacillus anthracis*, a spore-forming, aerobic, gram-positive rod. The disease is primarily found in large, domestic herbivores, such as cattle, sheep, and goats. Animals ingest the *B. anthracis* spores from the soil during grazing, and soon they are overwhelmed with vegetative bacterial cells as their organs fill with bloody black fluid. About 80% of untreated animals die.

Humans acquire **inhalational anthrax** from contaminated animal products or dust. For example, workers who tan hides, shear sheep, or process wool may inhale the spores and contract a pulmonary infection as a form of pneumonia called **woolsorters' disease**. It initially resembles a common cold (fever, chills, cough, chest pain, headache, and malaise). After several days, the symptoms may progress to severe breathing problems and shock. Inhalation anthrax is usually fatal without early treatment with penicillin or ciprofloxacin.

Anthrax also is considered a biological weapon for bioterrorism and biological warfare. The seriousness of using biological agents as a means for bioterrorism was underscored in October 2001 when *B. anthracis* spores were distributed intentionally as a powder through the United States mail. In all, 22 cases of anthrax (11 inhalation and 11 cutaneous) were identified. In the end, five individuals with inhalational anthrax died while the six other individuals with inhalation anthrax and all the individuals with cutaneous anthrax recovered. Had it not been for antibiotic therapy, many more might have been stricken.

The airborne bacterial diseases of the LRT are summarized in TABLE 10.2. MicroInquiry 10 presents several case studies concerning some of the bacterial diseases discussed in this chapter.

CONCEPT AND REASONING CHECKS 3

a. Why have reported cases of pertussis been increasing in the United States?
b. Explain how a primary or latent tuberculosis infection is different from primary or secondary tuberculosis disease.
c. Why does acute bronchitis often produce symptoms of breathlessness and wheezing?
d. What are the common characteristics to all forms of healthcare-acquired pneumonia?
e. Hypothesize why some pneumonia-causing species are community acquired and others are healthcare acquired?
f. Summarize (1) the unique characteristics and (2) the mode of transmission of the chlamydiae.
g. What cellular factor makes *B. anthracis* a dangerous pathogen and bioterror agent?

Bacillus anthracis

TABLE

10.2 A Summary of the Major Bacterial LRT Diseases

Disease	Causative Agent	Signs and Symptoms	Transmission	Treatment	Prevention
Bacterial					
Pertussis	*Bordetella pertussis*	Malaise, low-grade fever, severe cough Multiple paroxysms	Respiratory droplets	Erythromycin	Vaccinating with DTaP, Tdap
Tuberculosis	*Mycobacterium tuberculosis*	Active TB: Cough, weight loss, fatigue, fever, night sweats, chills, breathing pain	Respiratory droplets	Combination therapy with antibiotics	Preventing exposure to active TB patients BCG vaccine
Infectious bronchitis	*Mycoplasma pneumoniae* *Chlamydophila pneumoniae* *Streptococcus pneumoniae* *Haemophilus influenzae*	Runny nose, sore throat, chills, general malaise, slight fever, and dry cough	Respiratory droplets	Antibiotics	Annual flu vaccination Good hygiene
Healthcare-acquired pneumonia	*Streptococcus pneumoniae* *Staphylococcus aureus* *Klebsiella pneumoniae* *Pseudomonas aeruginosa*	Chills, high fever, sweating, shortness of breath, chest pain, cough with thick, greenish or yellow sputum	Respiratory droplets	Antibiotics	Practicing good hand hygiene
Community-acquired pneumonia	*Streptococcus pneumoniae*	High fever, sharp chest pains, difficulty breathing, rust-colored sputum	Respiratory droplets	Penicillin Cefotaxime	Vaccinating Hand hygiene
"Atypical" CAP	*Mycoplasma pneumoniae* *Legionella pneumophila*	Headache, fever, fatigue, dry hacking cough	Respiratory droplets Via water systems, whirlpool spas, air conditioning systems	Antibiotics	Extreme cleaning and disinfecting of water systems, pools, and spas
Q fever	*Coxiella burnetii*	Headache, fever, dry cough	Dust particles Contact with infected animals	Doxycycline	Vaccine for high risk occupations
Psittacosis	*Chlamydophila psittaci*	Headache, fever, dry cough	Contact with infected psittacine birds	Doxycycline	Keeping susceptible birds away from the infecting agent
Chlamydial pneumonia	*Chlamydophila pneumoniae*	Headache, fever, dry cough	Respiratory droplets	Doxycycline Erythromycin	Practicing good hygiene
Inhalational anthrax	*Bacillus anthracis*	Fever, chills, cough, chest pain, headache, and malaise Severe breathing and shock can develop	Airborne endospores	Penicillin Ciprofloxacin	Avoiding contact with infected livestock and animal products

MICROINQUIRY 10

Infectious Disease Identification

Below are several descriptions of infectious diseases based on material presented in this chapter. Read the case history and then answer the questions posed. Answers can be found on the Student Companion Website in **Appendix D**.

Case 1

The patient, a 33-year-old male, arrives at a local health clinic complaining that he has felt "out of sorts," has a fever, and has lost over 10% of his body weight in the last month. He also has a cough that produced rust-colored sputum. The patient is referred to a local hospital where a chest X ray and sputum sample are taken. Upon further questioning, the patient admits to having tested HIV-positive about 1 year ago. A tuberculin test also is ordered. Additional questioning of the patient reveals he had been living with two roommates for 2 years. Before that, he had lived for 8 years with another roommate who had tested positive for tuberculosis about 6 months before the onset of the patient's symptoms. The sputum samples are negative for the two roommates, but both have a positive tuberculin test result. Both test negative for HIV.

10.1a. Why was a chest X ray ordered?

10.1b. Why was a sputum sample taken?

10.1c. What should a positive tuberculin skin test look like?

10.1d. What does such a test result indicate?

10.1e. Based on the symptoms and laboratory results, from what infectious disease does the patient suffer? What is the agent?

10.1f. How did the patient contract the disease?

10.1g. Why is the infectious agent more virulent in HIV-infected patients?

Following diagnosis, the patient was placed on isoniazid (INH) for 12 months.

10.1h. Most treatment procedures call for a 6- to 8-month program. Why was the patient placed on INH for an extended period?

Case 2

The parents bring their 2-year-old daughter to the hospital emergency room. She appears to have an upper respiratory infection that her parents think started about 1 week previously. They say that their daughter had lost her appetite and appeared especially sleepy about 4 days ago. She complained of a sore throat. Examination indicates that she has a moderate fever but no chest congestion. Throat and blood cultures are taken and their daughter is put on penicillin.

On returning to the hospital 3 days later, her throat culture shows gram-positive rods. The blood culture is negative. Her parents remark that this morning their daughter started breathing harder. Examination of her pharynx indicates the presence of a leathery membrane. On questioning the parents, it is discovered that the child has had no immunizations. The child is admitted to the hospital and treatment immediately begun.

10.2a. What infectious agent does the child have?

10.2b. The medical staff is concerned about the seriousness of the disease. Why does the presence of gram-positive rods in the throat cause such concern?

10.2c. How could this disease have been prevented?

10.2d. What is the prescribed treatment protocol?

Case 3

A 63-year-old retired steel worker who is a heavy smoker and alcoholic comes to the emergency room complaining of having a fever and shortness of breath for the last 2 days. This morning he has developed a cough with rust-colored sputum. A chest X ray is taken and shows involvement in the left lower lobe of the lungs. A sputum sample and blood sample are taken for Gram staining, and the patient is checked into the hospital where he is placed on penicillin. Two days later the patient is feeling much improved. His physician tells him that both sputum and blood cultures indicate the presence of gram-positive diplococci. The patient is released from the hospital and recovers completely after finishing antibiotic therapy.

10.3a. What organism is responsible for the patient's infection?

10.3b. Why is this patient at high risk for becoming infected with the bacterial organism?

10.3c. How could the patient likely have prevented contracting the disease?

10.3d. What bacterial factors are responsible for virulence?

10.3e. If plated on blood agar, what type of hemolytic reaction should be seen?

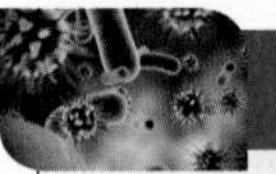

Chapter Challenge C

You should now realize that there are several infections and diseases that can be caused by the streptococci.

QUESTION C:

Finish off your list by identifying the diseases of the LRT that are caused by* S. pneumoniae*? What age groups seems to be the most affected and why?

Answers can be found on the Student Companion Website in **Appendix D.**

In conclusion, we have seen that the respiratory system can be a major portal of entry and infection site for bacterial pathogens. With the number of pathogens that exist (including the viruses and fungi), it is surprising that our respiratory system is not subjected to an even greater number of infections. The ability to limit the number of infections reflects on the strong immune defenses present in healthy individuals. Should those defenses not be fully developed, as in infants and children, or have weakened, such as in immunocompromised individuals or older adults, then the infectious agents, especially opportunistic pathogens, may gain the upper hand. This is especially true for infections of the LRT, as evidenced by pertussis, pneumonia, and tuberculosis.

SUMMARY OF KEY CONCEPTS

10.1 The Respiratory System Possesses an Indigenous Microbiota

- The **lower respiratory tract** (LRT) is virtually free of microbes and particulate matter due to the mechanical **(mucociliary clearance)** and chemical defenses (lysozyme, lactoferrin, other antimicrobial peptides, IgA and IgG antibodies, and human defensins) of the URT.
- In the LRT, lysozyme, IgG, and alveolar macrophages help eliminate any pathogens.
- Among the **microbiota** of the **upper respiratory tract** (URT) are opportunist species that can cause serious illnesses in individuals with a weakened immune system. Microbes inhabiting the respiratory tract are easily disseminated in respiratory secretions during breathing, talking, coughing, sneezing, spitting, and kissing.

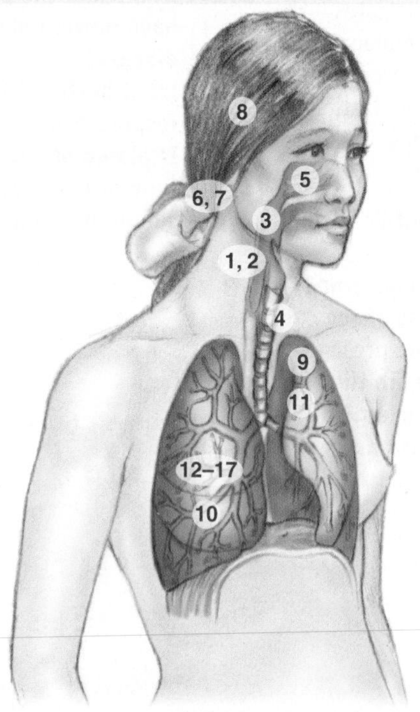

10.2 Several Bacterial Diseases Affect the URT

- Streptococcal pharyngitis
  1. *Streptococcus pyogenes*
- Scarlet fever
  2. *Streptococcus pyogenes*
- Diphtheria
  3. *Corynebacterium diphtheriae*
- Epiglottitis
  4. *Haemophilus influenzae*
- Sinusitis
  5. Various bacterial species
- Otitis externa
  6. *Streptococcus, Staphylococcus, Pseudomonas* species
- Otitis media
  7. *Streptococcus pneumoniae, Haemophilus influenzae*
- Acute meningitis
  8. *Neisseria meningitidis, Streptococcus pneumoniae, Haemophilus influenzae* type b

10.3 Many Bacterial Diseases of the LRT Can Be Life Threatening

- Pertussis
  9. *Bordetella pertussis*
- Tuberculosis
  10. *Mycobacterium tuberculosis*
- Infectious bronchitis
  11. *Mycoplasma pneumoniae, Chlamydophila pneumoniae, Streptococcus pneumoniae, Haemophilus influenzae*

- Healthcare-acquired and community-acquired pneumonia
  12. *Streptococcus pneumoniae, Haemophilus influenzae, Staphylococcus aureus, Pseudomonas aeruginosa, Klebsiella pneumoniae*
- "Atypical" pneumonia
  13. *Mycoplasma pneumoniae, Legionella pneumophila*
- Q fever
  14. *Coxiella burnetii*
- Psittacosis
  15. *Chlamydophila psittaci*
- Chlamydial pneumonia
  16. *Chlamydophila pneumoniae*
- Inhalational anthrax
  17. *Bacillus anthracis*

CHAPTER SELF-TEST

For **STEPS A–D**, answers to even-numbered questions and problems can be found in **Appendix C** on the Student Companion Website at **http://microbiology.jbpub.com/10e**. In addition, the site features eLearning, an online review area that provides quizzes and other tools to help you study for your class. You can also follow useful links for in-depth information, read more MicroFocus stories, or just find out the latest microbiology news.

STEP A: REVIEW OF FACTS AND TERMS

Multiple Choice

Read each question carefully, then select the **one** answer that best fits the question or statement.

1. Which one of the following is NOT part of the lower respiratory system?
   A. Lungs
   B. Pharynx
   C. Bronchi
   D. Trachea
2. Which one of the following is a complication of streptococcal pharyngitis?
   A. Rheumatic fever
   B. Pseudomembrane blockage
   C. Strawberry tongue
   D. Chest, back, and leg pain
3. A prominent feature of diphtheria is ________.
   A. meningitis
   B. a pseudomembrane
   C. pneumonia
   D. rheumatic fever
4. Severe throat pain, fever, muffled voice, and stridor are symptoms of ________.
   A. sinusitis
   B. epiglottitis
   C. bronchitis
   D. diphtheria
5. Which one of the following illnesses is characterized by yellow or green pus discharged from the nose?
   A. Pertussis
   B. Diphtheria
   C. Bronchitis
   D. Acute sinusitis
6. Swimmer's ear is a common name for a ________ infection of the ________ ear.
   A. bacterial; outer
   B. viral; outer
   C. bacterial; middle
   D. viral; inner
7. Acute meningitis ________.
   A. is an LRT infection
   B. is a disease affecting the membranes of the heart
   C. can be caused be *Corynebacterium diphtheriae*
   D. often starts as a nasopharynx infection
8. A catarrhal and paroxysmal stage is typical of which one of the following bacterial diseases?
   A. Tuberculosis
   B. Pneumonia
   C. Pertussis
   D. Q fever
9. Acid-fast staining is typically used to stain which bacterial genus?
   A. *Haemophilus*
   B. *Streptococcus*
   C. *Klebsiella*
   D. *Mycobacterium*
10. A person whose inner walls lining the main airways of the lungs become inflamed and who develops a dry cough for a few days, probably has ________.
    A. acute bronchitis
    B. epiglottitis
    C. pneumonia
    D. chronic bronchitis
11. Which one of the following is a gram-positive bacterial species commonly causing hospital-acquired pneumonia?
    A. *Haemophilus influenzae*
    B. *Klebsiella pneumoniae*
    C. *Staphylococcus aureus*
    D. *Chlamydophila pneumoniae*
12. Humans can acquire which one of the following diseases from the droppings of infected birds?
    A. Q fever
    B. Legionellosis
    C. Inhalational anthrax
    D. Psittacosis

Fill-in

Answer each of the following by filling in the blank with the correct word or phrase.

13. Scarlet fever is caused by a species of ________________.
14. ________________ is caused by a species of *Mycobacterium.*
15. The bacterial species ________________ causes psittacosis and pneumonia.
16. ________________ is caused by transmission in droplets of airborne water.
17. ________________ is a gram-negative rod that causes whooping cough.
18. The development of active tuberculosis throughout the body is called ________________ tuberculosis.
19. ________________ is a disease of parrots, parakeets, and canaries as well as humans.
20. Another name of *Streptococcus pneumoniae* is ________________.
21. A bacterial form of ________________ may be caused by *Neisseria* or *Haemophilus.*
22. *Mycoplasma* species have no ________________, which often gives them a pleomorphic shape.
23. *Coxiella burnetti* is the causative agent of ________________.
24. The genus ________________ consists of acid-fast rods.
25. The agent of pneumococcal pneumonia is a gram-____________ organism.

STEP B: CONCEPT REVIEW

26. Explain (a) how the URT maintains sterility in the LRT and (b) what portions of the URT are normally colonized by **indigenous microbiota.** (**Key Concept 1**)
27. Summarize the clinical aspects of **strep throat** and the complications arising from **streptococcal pharyngitis.** (**Key Concept 2**)
28. Name the bacterial species responsible for, and describe the clinical aspects of treatment and prevention of, **diphtheria.** (**Key Concept 2**)
29. Discuss why **epiglottitis** can be a life-threatening inflammation. (**Key Concept 2**)
30. Distinguish between **acute** and **chronic sinusitis.** (**Key Concept 2**)
31. Recognize the symptoms of outer and middle ear infections. (**Key Concept 2**)
32. Distinguish between the bacterial species responsible for **acute meningitis.** (**Key Concept 2**)
33. Justify why **pertussis** is viewed as one of the more dangerous contagious diseases. (**Key Concept 3**)
34. Summarize (a) the clinical aspects of *Mycobacterium tuberculosis* as an infection and disease, and (b) the problems concerning antibiotic resistance. (**Key Concept 3**)
35. Distinguish between **acute** and **chronic bronchitis.** (**Key Concept 3**)
36. Distinguish between the bacterial species responsible for **healthcare-acquired** and **community-acquired pneumonia.** (**Key Concept 3**)
37. Summarize the mode of transmission and types of pneumonia caused by intracellular parasites. (**Key Concept 3**)

STEP C: APPLICATIONS AND PROBLEMS

38. A patient is admitted to the hospital with high fever and a respiratory infection. Pneumococci and streptococci were eliminated as causes. Penicillin was ineffective. The most unusual sign was a continually dropping count of red blood cells. Can you make the final diagnosis?
39. One of the major world health stories of 1995 was the outbreak of diphtheria in the New Independent and Baltic States of the former Soviet Union. If you were in charge of this international public health emergency, what would be your plan to help quell the spread of *Corynebacterium diphtheriae?*
40. The CDC reports that an estimated 40,000 people in the United States die annually from pneumococcal pneumonia. Despite this high figure, only 30% of older adults who could benefit from the pneumococcal vaccine are vaccinated (compared to over 50% who receive an influenza vaccine yearly). As an epidemiologist in charge of bringing the pneumonia vaccine to a greater percentage of older Americans, what would you do to convince older adults to be vaccinated?

STEP D: QUESTIONS FOR THOUGHT AND DISCUSSION

41. Between 1986 and 1996 *Haemophilus* meningitis was virtually eliminated in the United States. Indeed, at the beginning of the period there were 18,000 cases annually, but in 1996, only 254 cases were reported. What factors probably contributed to the decline of the disease?
42. A bacterial virus is responsible for the ability of the diphtheria bacillus to produce the toxin that leads to disease. Do you believe that having the virus is advantageous to the infecting bacillus? Why or why not?
43. A children's hospital in Salt Lake City reported a dramatic increase in the number of rheumatic fever cases. Doctors were alerted to start monitoring sore throats more carefully. Why do you suppose this prevention method was recommended?

Foodborne and Waterborne Bacterial Diseases

11

CHAPTER PREVIEW

11.1 The Digestive System Has an Extensive Indigenous Microbiota

Investigating the Microbial World 11: An Apple a Day Keeps the Doctor Away

11.2 Bacterial Diseases of the Oral Cavity Can Affect One's Overall Health

11.3 Bacterial Diseases of the GI Tract Are Usually Spread Through Food and Water

11.4 Some Bacterial Diseases Are the Result of Foodborne Intoxications

11.5 GI Infections Can Be Caused by Several Bacterial Pathogens

Textbook Case 11: Outbreak of *E. coli* 0157:H7 Infection

MicroInquiry 11: The Bad and Good of a *Helicobacter pylori* Infection

"People are alive and well in the morning and then dead by the afternoon."
—Dr. David Sack, Johns Hopkins University epidemiologist, speaking on the Haitian cholera outbreak

On October 20, 2010, an outbreak of cholera was confirmed in Haiti. This report was quite a surprise because cholera had not been reported in Haiti for decades and, since no cholera cases were reported after the January 2010 earthquake, health authorities deemed cholera as an unlikely disease to develop. Yet within a few days after the report of the outbreak, hundreds of people became ill with stomachache, vomiting, and diarrhea so profuse that victims started dying from dehydration (see quote above).

Through his pioneering epidemiological studies, John Snow, a London physician, pinpointed a water pump on Broad Street as the source of transmission for the 1854 epidemic in London. Snow believed the only way to stop the spread was to remove the pump handle so no one could get water from the pump. On September 8, 1854, city officials took Snow's advice and had the pump handle removed. The action was successful, supporting Snow's belief that cholera was a waterborne, contagious disease. Few at the time believed Snow's theory for a waterborne disease, and it would not be until 1883 that the cholera bacterium, *Vibrio cholerae*, was isolated.

The year 2004 marked the 150th anniversary of John Snow's landmark epidemiological studies. Unfortunately, cholera still exists today and Haiti is only the most recent epidemic to make the news (FIGURE 11.1). In 2010–2011, one of the worst ever outbreaks of cholera occurred in west and central Africa. More than 105,000 cases were reported and nearly 3,000 died.

Image courtesy of Dr. Fred Murphy/CDC.

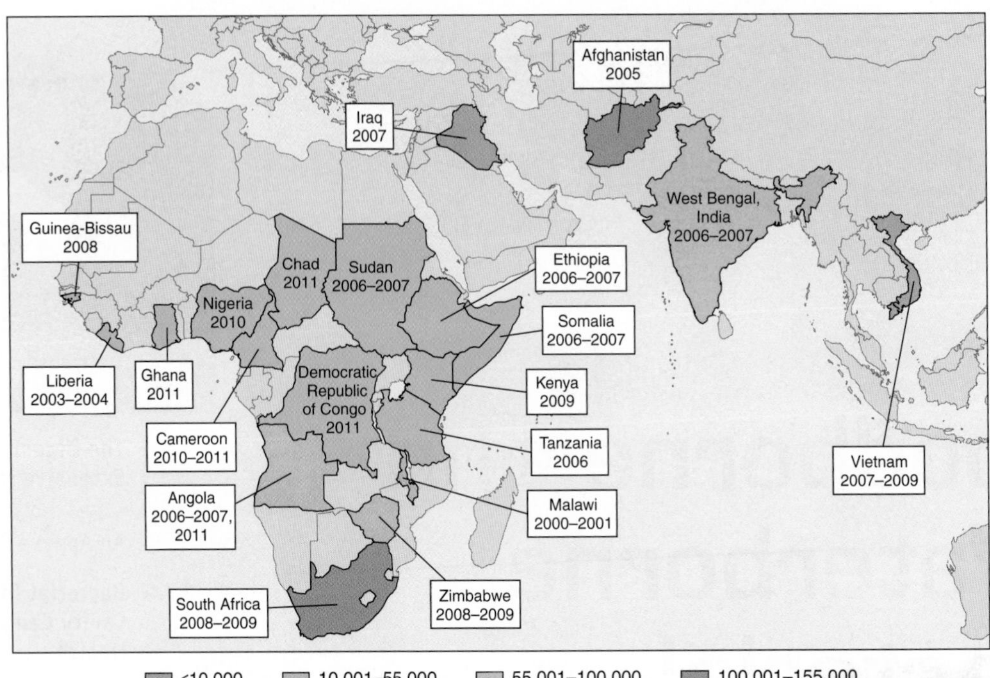

FIGURE 11.1 **Recent Cholera Outbreaks—Africa and Asia, 2000–2011.** There have been numerous outbreaks of cholera, especially in Africa. Adapted from Waldor, M.K., Hotez, P.J., and Clemens, J.D. (2010) *N Engl J Med* **363** (24):2279-2282. »» Medical experts are concerned that the rainy season will continue to spread the disease. Why would the rainy season spread cholera?

For a cholera outbreak to occur anywhere, be it in London in 1854 or Haiti in 2010, two conditions must exist: (1) cholera must be present in the local population and (2) poor hygiene or some type of unsanitary practice leads to contamination of water by *V. cholerae* cells. This water is then used by groups of unsuspecting people, which allows for large-scale exposure and further contamination of food and water. In central Africa, cholera is endemic, so seasonal factors, such as the rainy season with flooding will contaminate additional water sources. In addition, poor hygiene and population movements typical in Africa are contributing to this epidemic.

In May, 2011, French researchers suggested that the importation of *V. cholerae* into Haiti may be linked to the arrival of United Nations peacekeepers from Nepal. What we do know for sure is the hospitalized cases and deaths continued into the fall of 2012 and health experts fear the epidemic may continue for many more months. As of November 2012, there have been more than 580,000 cases and almost 7,700 deaths reported.

Cholera is just one of many illnesses affecting the digestive system. In this chapter, we will examine the intoxications and infections caused by bacterial species contaminating food and water.

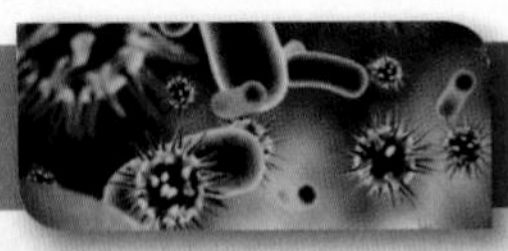

Chapter Challenge

Foodborne illnesses are a major cause of infectious disease in the United States. Every year there are an estimated 48 million foodborne infections, leading to 128,000 hospitalizations and 3,000 deaths. Yet many of us don't think about food safety until we or someone we know gets sick from eating contaminated food or drinking contaminated water. Because everyone is at risk for food poisoning, we need to think about ways we can try to reduce the risk and prevent illnesses of the digestive system. As we describe many of the more common foodborne and waterborne infections, let's try to think of ways to prevent such infections from the farm to the table.

KEY CONCEPT 11.1 The Digestive System Has an Extensive Indigenous Microbiota

When we eat a meal or drink fluids, our bodies take in the nutrients necessary for survival. During the ingestive and digestive processes, any pathogens that happen to be in the food or liquid we take in can upset the digestive system and lead to some form of intoxication or disease. Thus, the digestive system represents another major portal of entry for pathogens. In fact, acute infections of the digestive system are among the most frequent of all illnesses, exceeded only by some respiratory tract infections such as the common cold.

The Digestive System Is Composed of Two Separate Categories of Organs

The **digestive system** includes the organs that ingest the food, transport the food, digest the food into smaller usable components, absorb the necessary nutrients into the bloodstream, and expel the waste products from the body (FIGURE 11.2). The **gastrointestinal (GI) tract**, also called the digestive tract or alimentary (*aliment* = "nourishment") canal, contains the digestive organs (oral cavity, pharynx, esophagus, stomach, small intestine, and large intestine) and forms a continual tube from mouth to anus.

The **accessory digestive organs** that are outgrowths from and are connected to the GI tract include the salivary glands, liver, gallbladder, and pancreas.

Digestive system defenses are essential for preventing colonization and potential infections and disease. In the oral cavity, chewing food generates substantial mechanical and hydrodynamic forces that, along with salivary flow, dislodges many microbes from epidermal and tooth surfaces. The presence in saliva of mucins and various other proteins, including the antimicrobial proteins lysozyme, lactoferrin, and defensins, destroy microbes, while secretory IgA antibody binds to microbes keeping them in suspension. Swallowing delivers

■ Mucin:
A glycoprotein secreted on the surface of the gut (mucosa).

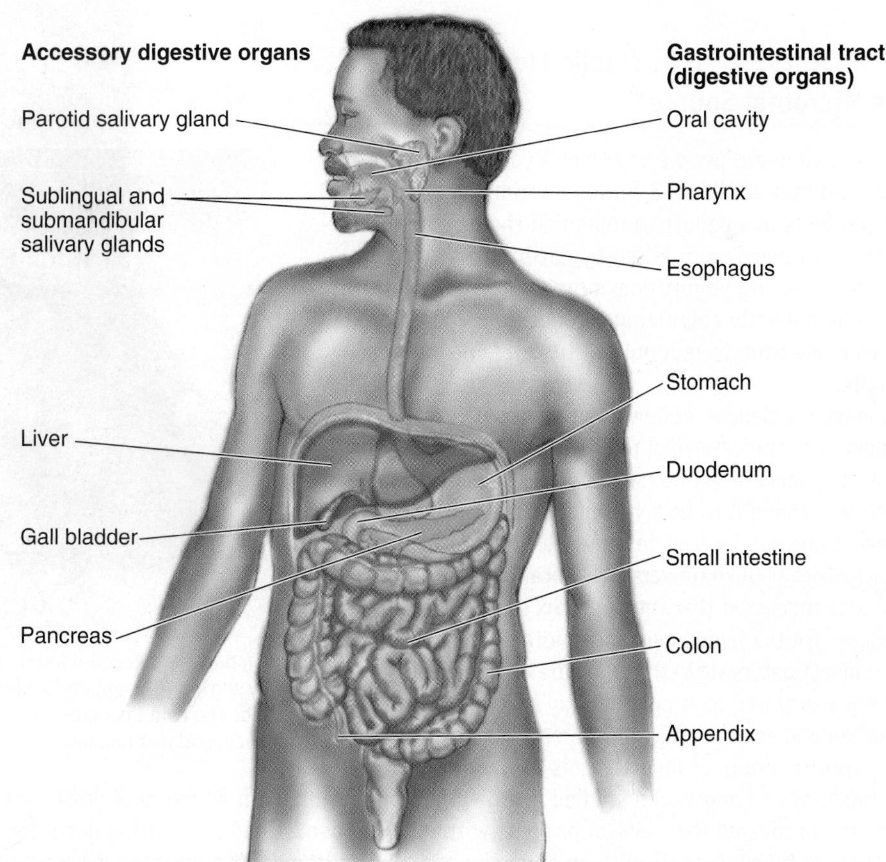

FIGURE 11.2 **The Human Digestive System.** The digestive system consists of the gastrointestinal (GI) tract and accessory digestive organs. »» Why are organs such as the teeth, liver, and pancreas considered accessory digestive organs?

the dislodged microbes to the stomach, where the low pH of the stomach fluid kills most microbes.

In the intestines, the surfaces are coated with mucus that traps bacterial cells. Along with secretory IgA antibody found on most intestinal surfaces, these chemical defenses usually prevent colonization and invasion of the underlying epithelium.

Like all mucosal surfaces in the body, the intestinal epithelium is a shedding surface. If bacterial cells colonize epithelial cells, the cells will be removed when the epithelial cells are shed. The small and large intestines also contain bile and proteolytic enzymes capable of killing a variety of microbes.

■ Bile:
A mixture of cholesterol, phospholipids, acids, and immunoglobulins that are stored in the gallbladder.

Finally, the mucosa of the small intestine, primarily in the adult ileum, contains large collections of lymphatic nodules called **Peyer's patches**, which represent part of the **mucosa-associated lymphatic tissue** (MALT). As food enters the GI tract, any antigenic material is transported and presented to the T and B cells in the lymphatic nodules, resulting in antibody secretion. Another MALT organ, the **appendix**, also may be important to maintaining an indigenous gut microbiota (MicroFocus 11.1).

The accessory digestive organs are sterile and normally free of microorganisms, similar to all other internal organs in the body.

Our Understanding of the Human Oral and Gastrointestinal Microbiome Is Rapidly Improving

Until recently, the indigenous microbiota inhabiting the human body, the so-called **human microbiome**, and its role were poorly understood. We knew that there is an enormous population of commensal microbiota in the GI tract, especially in the large intestine, and that these communities provide protection through microbial antagonism. However, through recent studies initiated by the **Human Gut Microbiome Initiative** (HGMI), we are rapidly learning the many beneficial roles the gastrointestinal microbiome play in our ability to digest and process foods, trigger immune reactions, and even affect brain development. In addition, **dysbiosis**, changes to the composition

MICROFOCUS 11.1: Public Health
The Gut's Microbial Source

You probably have heard people say that if you have a case of diarrhea or even take antibiotics for a bacterial infection, you should eat products like yogurt to replenish the gut microbiota lost through diarrhea or killed by antibiotics. Although this cannot hurt, eating yogurt does not supply nearly all the microorganisms normally colonizing the gut. So, where do all the microbes come from to repopulate the gut? The answer—your appendix!

The appendix is a slender, hollow, blind-ended pouch that projects from the posterior-medial region of the cecum, near its junction with the small intestine (see figure). For a long time, the appendix was thought to be a vestigial or useless organ that may have had a purpose far back in our evolutionary past. But in 2007, investigators at Duke University Medical Center and Arizona State University suggested that the appendix serves as an internal "safe house" for the indigenous microbiota normally living in the gut. The investigators suggested that the bacterial cells in the appendix form a very well developed biofilm that can survive a bout of diarrhea and emerge afterward to repopulate the gut.

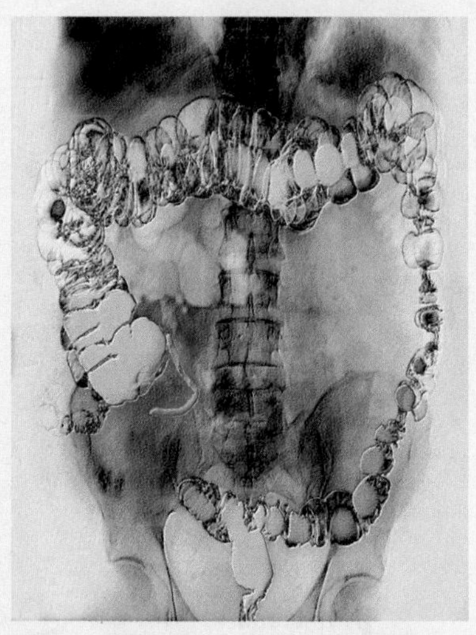

A computer-enhanced barium enema X ray showing the appendix highlighted in red. The colon is blue-gray. © Scott Camazine/Alamy Images.

In 2011, another group of investigators at Winthrop University Hospital on Long Island studied a large group of patients with a history of gut infections. The researchers discovered that 45% of patients without an appendix had reoccurring gut infections, while only 18% of those patients with an appendix had a reoccurrence. It appears that the appendix can supply "good bacteria" to reestablish a normal gut microbiota.

of the microbiota, is helping us understand how the gut microbiome governs our susceptibility to other diseases. In 2012, the HGMI finished sequencing 600 human indigenous microbial genomes, completing a collection of more than 1,000 microbial genomes. The results will allow us to better understand the interrelationships between the human gut and its microbial inhabitants and, importantly, provide new ways to diagnose and treat disease.

Already, the metagenomic analysis of the adult human intestine microbiome suggests that there are trillions of bacterial cells, outnumbering the cells of the human body by at least 10-fold. FIGURE 11.3A lists some of the more prominent bacterial genera in the oral cavity and the gastrointestinal tract that represent the 500 to 1,000 different bacterial species present. Little is known about the selective pressures shaping the community's species. Certainly differences in the anatomy, physiology, organization, and location in each part of the GI tract generate unique "environmental" conditions for potential microbial colonizers and, therefore, each location has a distinctive indigenous microbiota (FIGURE 11.3B).

The oral cavity, for example, has some unique features (such as teeth) that render it very different from the GI tract that consists of soft tissues; in addition, the intestinal regions are anaerobic. Many species in the mouth are members of the phyla Bacteroidetes, Firmicutes, Actinobacteria, and Proteobacteria, while the adult GI tract, especially the large intestine, is just as diverse but is dominated by members of the Bacteroidetes and Firmicutes. Interestingly, the community members and their numbers can vary depending on one's diet (Investigating the Microbial World 11).

Some scientists believe the human gut microbiome may be categorized into one of three distinctive **enterotypes** based on the relative abundance of three bacterial genera found in the gut. High levels of *Bacteroides* = enterotype 1; *Prevotella* = enterotype 2; and *Ruminococcus* = enterotype 3. These combinations of microbes are like blood types in that they appear to be independent of gender, age, nationality, and body weight. Just how distinct the

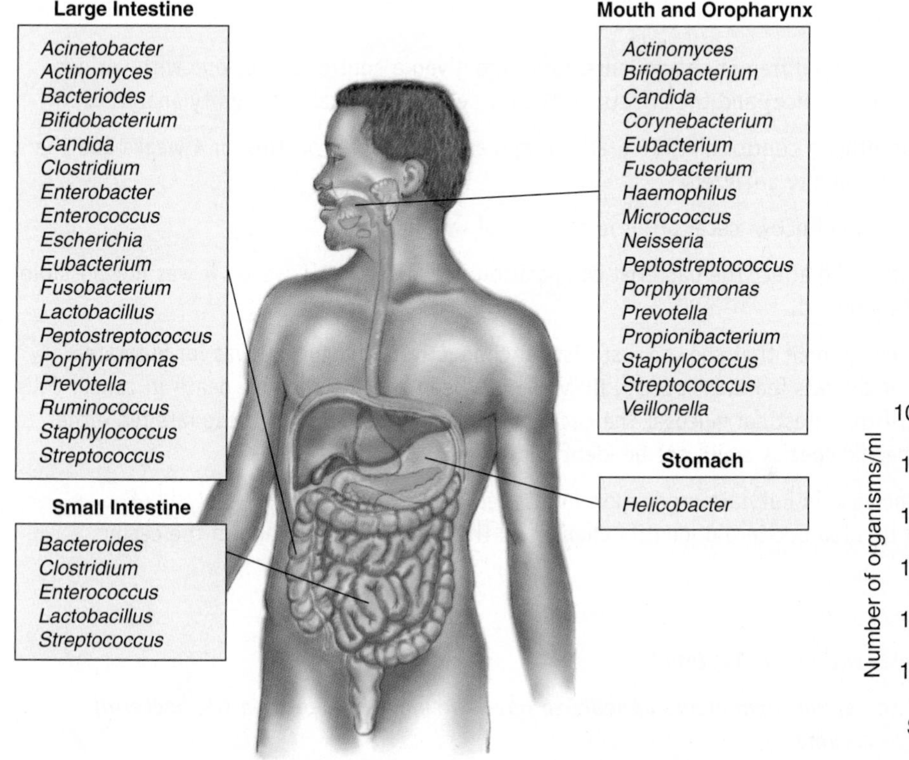

(A)

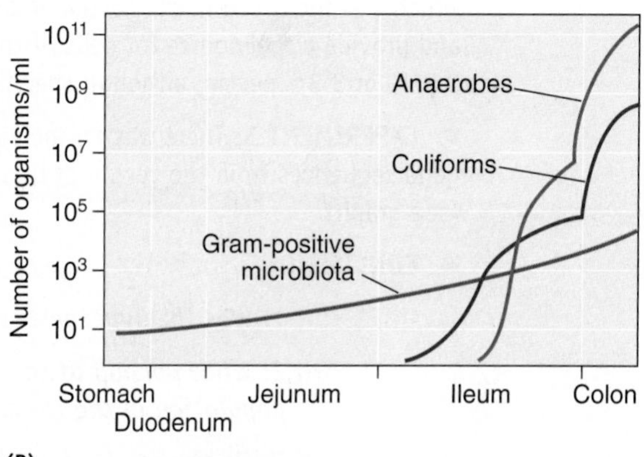

(B)

FIGURE 11.3 **Microbiota of the Digestive System.** (**A**) A variety of bacterial genera (and the fungus *Candida*) comprise the predominant indigenous microbiota of the digestive system. (**B**) The number of microbes found in the stomach, small intestine (duodenum, jejunum, and ileum), and large intestine (colon) increases from stomach toward colon. »» Why do the number of anaerobes and coliforms (facultative anaerobes) increase in the ileum and colon?

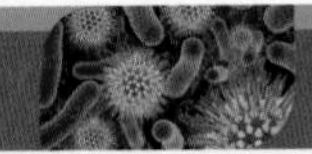

Investigating the Microbial World 11

An Apple a Day Keeps the Doctor Away

An apple a day keeps the doctor away. Many of us remember this saying. Its origin is a mystery. Some believe it might have come from an ancient Roman proverb because the Romans thought apples could cure some illnesses. Another origin, according to the Phrase Finder (www.phrases.org.uk/), suggests the phrase first appeared in print in *Notes and Queries,* a Welsh magazine that in February 1866 advocated, "Eat an apple on going to bed, And you'll keep the doctor from earning his bread." In any case, the idea was that apples are good for your health. Is there any truth to the maxim?

■ **OBSERVATIONS:** We continually hear about the benefits of fruits and vegetables in our diet. Many supposedly contain useful amounts of antioxidants, fiber, and other substances that help us stay healthy by eating healthy. In addition, reports have been published suggesting that vegetables and fruits affect the bacterial populations in the gut in a positive way. The mechanisms are largely unknown, although increases in members of the gram-positive Firmicutes and decreases in numbers of the gram-negative Bacteroidetes have been correlated with positive health gains.

■ **QUESTION:** ***If eating apples or apple products are so good for us, what is it about apples that provides a health advantage?***

■ **HYPOTHESIS:** Eating apples and apple products alters the composition of the intestinal microbiota. If so, then feeding rats apples or apple products should alter their gut microbiota.

■ **EXPERIMENTAL DESIGN:** Rats were used as test animals and the changes in the bacterial microbiota in the rat cecum (an extension of the terminal portion of the small intestine and home to anaerobic bacterial species) after apple feeding (compared to a control diet) were monitored by ribosomal RNA (rRNA) gene sequencing to identify the microbial members.

■ **EXPERIMENT 1:** Rats were fed either a control diet or 10 grams of apples a day for 14 weeks and the effects on the rat cecum microbial community analyzed.

■ **EXPERIMENT 2:** Because apples contain different components, rats were given a control diet or one with various apple products (puree, pomace, pectin, or juice) and the effects on the rat cecum microbial community analyzed.

■ **EXPERIMENT 3:** Rats were given either a control diet, 10 grams of apples a day, or 7% pectin, for 4 weeks and the effects on the rat cecum microbial community analyzed.

■ **RESULTS:** Apple consumption had no effect in cecal pH or relative cecal weight.

■ **EXPERIMENT 1:** Apple consumption did affect the bacterial composition in the cecum. However, it was not possible to identify specific bacterial species affected.

■ **EXPERIMENT 2:** The only apple component that showed a statistical change from the control was for butyrate, which was increased in the cecum of the rats fed 3.3% pectin. Butyrate has been shown to induce death in cancer cells and provide a fuel source for cells of the intestinal mucosa. The cecal population was altered in those rats fed either 0.33% or 3.3% pectin, although specific species could not be identified.

■ **EXPERIMENT 3:** There was an increase in butyrate production in rats fed apples or 7% pectin. Analysis of gene sequences from the cecum of rats fed pectin did identify changes in the bacterial population in the cecum (see graph).

■ **CONCLUSIONS:**

QUESTION ***1: Was the hypothesis validated? Explain.***

QUESTION ***2: What portion of the apple components appears to have the most influence on the bacterial population in the rat cecum?***

QUESTION ***3: Do the results from experiment 3 (see graph) tend to support or refute the idea that apples (or apple pectin) have a health promoting effect? Explain.***

QUESTION ***4: How does the fact that*** **Clostridium** ***species are known butyrate producers fit into the overall picture of apples and good health?***

Investigating the Microbial World 11 (continued)

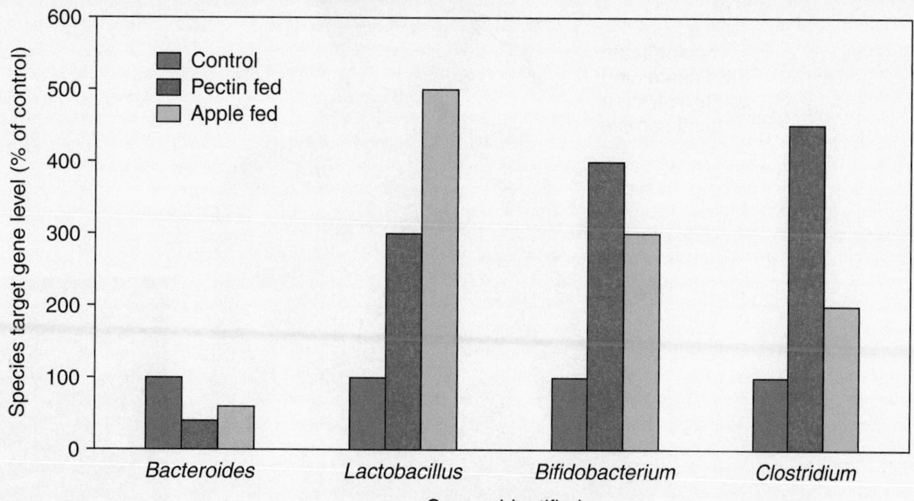

FIGURE Relative amount of target genes comparing the control group (blue), pectin-fed (red), and apple-fed (green) rats. Adapted from: Licht, T.R. et al. (2010). *BMC Microbiol.* 10:13.

Note: Rats metabolize apple components differently from humans, so such studies on apple or apple component consumption in rats needs to be interpreted with caution.

Answers can be found on the Student Companion Website in **Appendix D**.

Adapted from: Licht, T. R. *et al.* (2010). *BMC Microbiol.* **10**:13. Icon image © Tischenko Irina/ShutterStock, Inc.

three enterotypes are remains to be elucidated. Still, scientists hope to eventually tailor diets and drug medications to an individual's enterotype. Here is just one example of the potential.

Currently, *Clostridium difficile* infections, discussed later in this chapter, are on the rise and can be life threatening. Though the bacterial cells can be held in check by our normal indigenous gut microbiota, problems arise when antibiotics are used to treat the infection. By eliminating much of the indigenous microbiota, the antibiotic resistant *C. difficile* cells survive and flourish. Today, **fecal transplants** from a close relative are being used to cure individuals with severe, long-term *C. difficile* infections (FIGURE 11.4). The bacterial community from a stool sample is introduced through an enema or nasogastric tube (ick!) to reestablish the normal microbiota in the patient. Amazingly, more than 90% of patients are cured through fecal transplants that reestablish one's enterotype and rapidly displace the pathogenic intruder. Such transplants are also showing great promise for conditions like irritable bowel syndrome and Crohn's disease.

Certainly, even with our indigenous gut microbiota, we are still subject to infectious diseases. In our survey of diseases affecting the digestive system, we will start with the oral cavity.

CONCEPT AND REASONING CHECKS 1

a. Prepare a list of the chemical, mechanical, and cellular defense mechanisms found in the GI tract.

b. Why would the diversity of indigenous microbiota be highest in the oral cavity and in the large intestine?

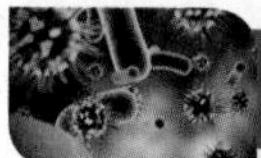

Chapter Challenge A

When we ingest food, at times it might be accidently contaminated with a foodborne pathogen. Most of us have had at least one case of food poisoning in our lives that has resulted from such an "exposure" that may have occurred somewhere between the farm and the table.

QUESTION A:

How is our digestive system set up to try to prevent entry and colonization by these pathogens?

Answers can be found on the Student Companion Website in **Appendix D.**

■ Irritable bowel syndrome: A bowel disorder that involves recurrent pain with constipation or diarrhea or alternating attacks of both.

■ Crohn's disease: A chronic inflammation of the gastrointestinal tract that is linked to an immune system weakness.

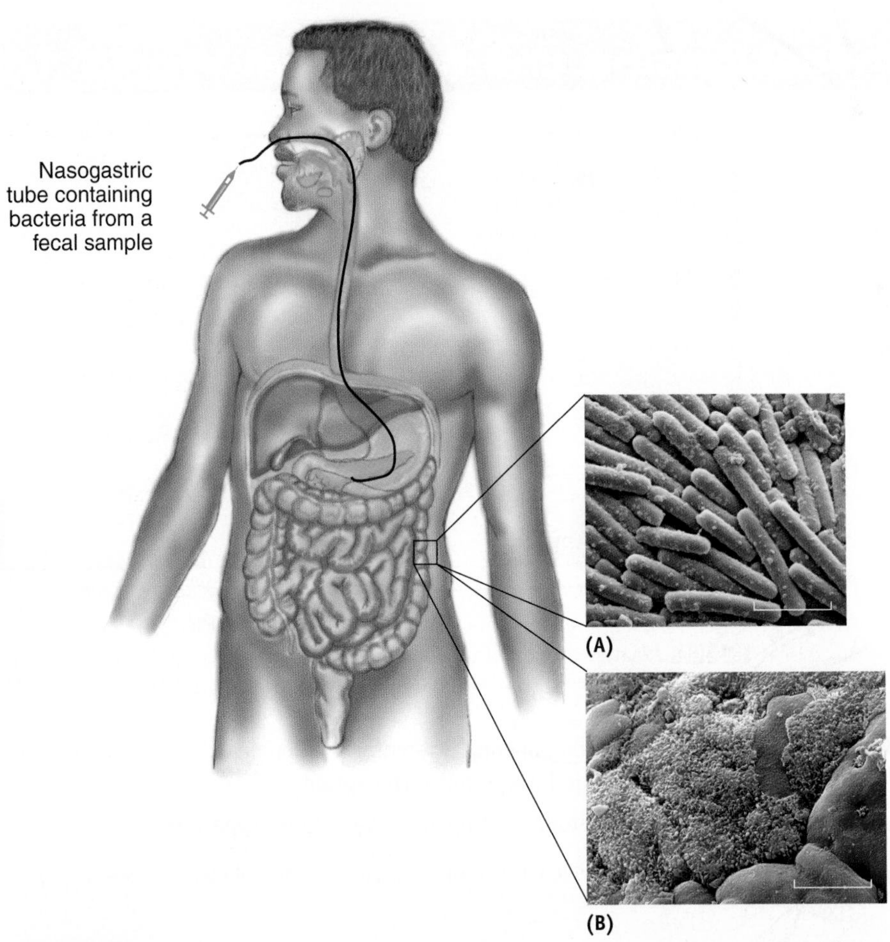

FIGURE 11.4 **Fecal Transplant.** For a person suffering from a severe *Clostridium difficile* infection (**A**), a fecal transplant could be carried out. The bacterial community from a donor stool sample is introduced through a nasogastric tube to the stomach. There, the bacterial cells pass into and recolonize the small intestine, reestablishing the normal microbiota. (**B**). (A) (Bar = 5 µm.); (B) (Bar = 20 µm.) (**A**) © Paul Gunning/Photo Researchers, Inc. (**B**) © Stephanie Schuller/Photo Researchers, Inc. »» How could the transplanted bacterial cells eliminate the infection?

KEY CONCEPT 11.2 Bacterial Diseases of the Oral Cavity Can Affect One's Overall Health

■ Sequela: A pathological condition resulting from a prior disease.

Dental caries, periodontal disease, and the sequelae of these two diseases constitute the majority of oral and dental infections. They, in turn, can have more systemic effects.

Dental Caries Causes Pain and Tooth Loss in Affected Individuals

The **oral cavity** or mouth represents a diverse microbial ecosystem, with complex interrelationships among the members of the resident population of microorganisms and the moist oral environment. The cavity has various ecological niches, each with a different physical property and nutrient supply dictating the number and type of microorganisms that can survive.

Some 50 to 100 billion bacterial cells, representing between 100 and 500 species, inhabit the oral cavity, many adhering to and colonizing the tooth surface, called the **dental pellicle** (FIGURE 11.5A). If ingested food material is not cleaned off the teeth through regular dental hygiene, bacterial cells on the pellicle will start metabolizing the carbohydrates into acids. It may only be time before tooth decay, or **dental caries** (*cario* = "rottenness"), begins (FIGURE 11.5B).

Dental caries affects more than 20% of American children aged 2 to 4, 50% of those aged 6 to 8, and nearly 60% of those aged 15. The disease develops if three factors are present: a caries-susceptible tooth with a buildup of plaque; dietary carbohy-

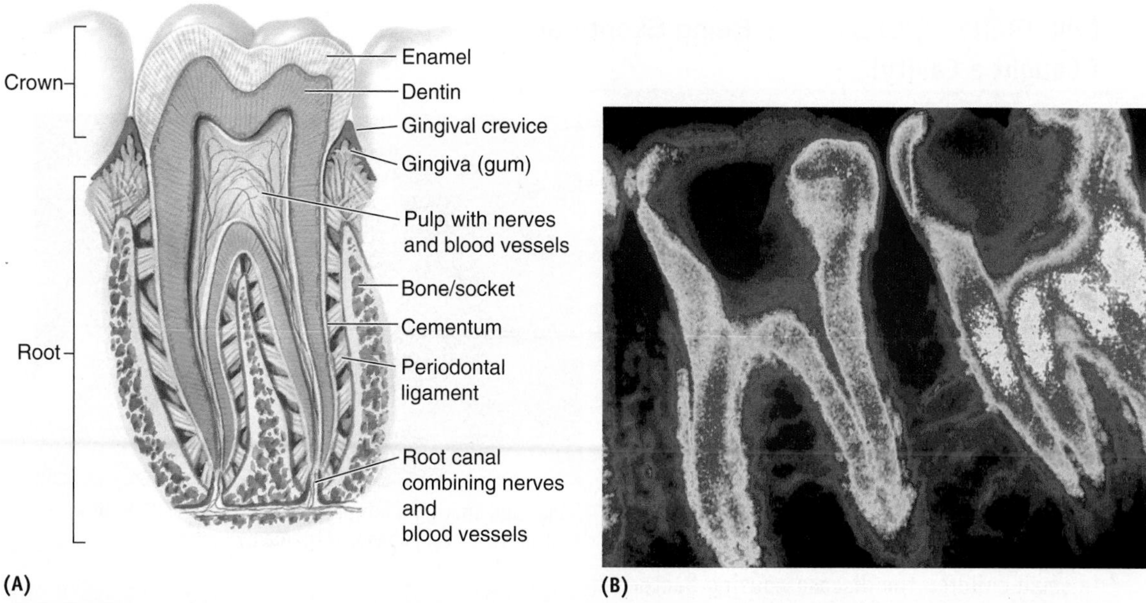

FIGURE 11.5 **The Anatomy of a Tooth.** (**A**) Different diseases can affect each part of the human molar shown in cross section. (**B**) In this false-color dental X ray, the caries appear as black spots on the molars. © BSIP/Photo Researchers, Inc. »» What parts of the tooth surface would be most susceptible to colorization and infection?

drate, usually in the form of sucrose (sugar); and acidogenic (acid-producing) bacterial species (FIGURE 11.6). The gram-positive streptococci, *Streptococcus mutans* and *S. sobrinus*, are the main acid producers, although a mixed species community including other streptococci and lactobacilli also is involved. In fact, some of these bacterial species may be contagious (MicroFocus 11.2).

Once colonization starts, a succession of bacterial species interact and form **dental plaque** (FIGURE 11.7A). Plaque is essentially a **biofilm**, a deposit of dense gelatinous material consisting of salivary proteins, trapped food debris, and an enormous mass of bacterial cells and their products. The plaque is slightly rough and is more noticeable on the molars, especially along the gumline. As the plaque thickens, it becomes dominated by anaerobic species. It has been suggested that mature dental plaque contains more than 300 bacterial species (FIGURE 11.7B).

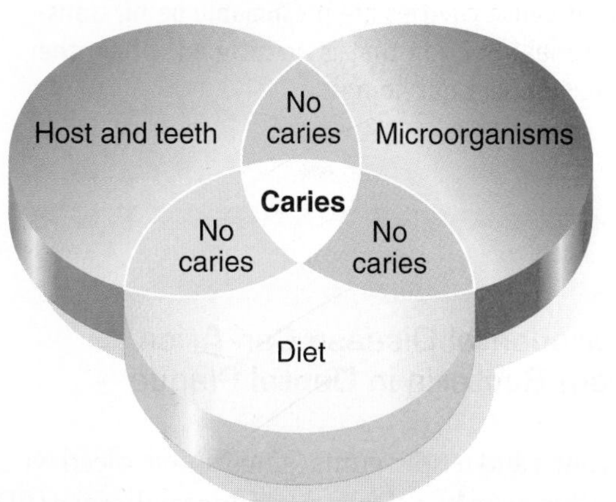

FIGURE 11.6 **Dental Caries.** Overlapping circles depicting the interrelationships of the three factors that lead to caries activity. »» What role does *S. mutans* play in the development of dental caries?

Streptococcus mutans

The acids in plaque attack minerals in the tooth's hard, outer surface, called the **enamel** (Figure 11.5A). This demineralization eventually leads to a cavity, which is a tiny hole or lesion in the enamel. Such cavities most often occur on the enamel-coated crown or the cementum of the exposed root surface of the molars or premolars. As the decay progresses into the pulp of the tooth, which contains nerves and blood vessels, severe toothache pain and sensitivity occur. The immune system may respond to the bacterial invaders by sending white blood cells to the blood vessels to fight the infection. This may result in a tooth abscess.

Treatment of dental caries depends on the progress and severity of the disease. Most of us probably have one or more fillings, which have been the main treatment option for decay. Lacking treatment, extensive decay may require either a

MICROFOCUS 11.2: Being Skeptical

I Caught a Cavity!

First, we all know that many infectious diseases are communicable—they can be passed person to person. Just look at the number of people who get colds and the flu every year. Second, cavities and tooth decay (dental caries) rank near or at the top among the world's most common health problems. In fact, the single most common infectious childhood disease in the United States is severe tooth decay and some see this as an epidemic problem. But are dental cavities contagious? Can you catch a cavity?

Sugar and sharing a lollipop. Saliva transfer? © Beijing Eastphoto stockimages Co.Ltd/Alamy.

Dental caries is caused primarily by *Streptococcus mutans*. Research indicates that in small children the disease-causing bacterium is usually absent. So where do the bacterial cells come from? The proposal is that children pick up *S. mutans* from an adult—usually a family member, friend, or other caretaker. Importantly, in children there appears to be a "window of infectivity" between 6 to 36 months of age, although all children maintain some level of susceptibility. That means *S. mutans* can be passed vertically from mother to child by kissing, sharing food utensils, or through any other activity involving an exchange of saliva from one mouth to another. In fact, dental studies report that many young mothers have not seen a dentist in 5 to 7 years before giving birth, meaning they carry a large "infectious dose" of *S. mutans* in their mouths that potentially can be transmitted to their children, often before the first teeth appear.

Transmission also occurs horizontally between children. For example, playmates at day care centers may pass saliva between one another from sharing toys or eating utensils that have been placed in the mouth (see figure). And maybe it can also be spread horizontally between adults. Dr. Margaret Mitchell, a Chicago cosmetic dentist reports, "In one instance, a patient in her 40s who had never had a cavity suddenly developed two cavities and was starting to get some gum disease," she said. Why at this age? The woman was now dating a man who hadn't seen a dentist in 18 years and had gum disease.

The verdict? Realize that dental caries involves more than simply having *S. mutans* present. As the chapter discussion describes, a person's dietary habits, fluoride use, salivary flow and composition, and oral hygiene are contributing factors. So, cavities are not being transferred. Rather, the *S. mutans* bacterial cells and perhaps other oral bacterial species that cause cavities are presumably being transmitted. So, dental decay is just another example of a bacterial infection that is no different from other infections—if the environment is right for infection, it may impact our bodies.

crown to replace the treated cavity, a root canal if the decay has reached the pulp, or tooth extraction.

Prevention means good oral and dental hygiene to keep the biofilm at a minimum between dental visits. Frequent snacking and drinking sugary beverages place the teeth under constant bacterial acid attack, so limiting such foods and drinks is important.

Periodontal Disease Can Arise from Bacteria in Dental Plaque

Swollen and tender gums (gingiva) that bleed when brushing can be a sign of **periodontal disease (PD)** that in some form affects about 80% of American adults. One of the most common forms of early-stage PD is **gingivitis**, which develops when bacterial

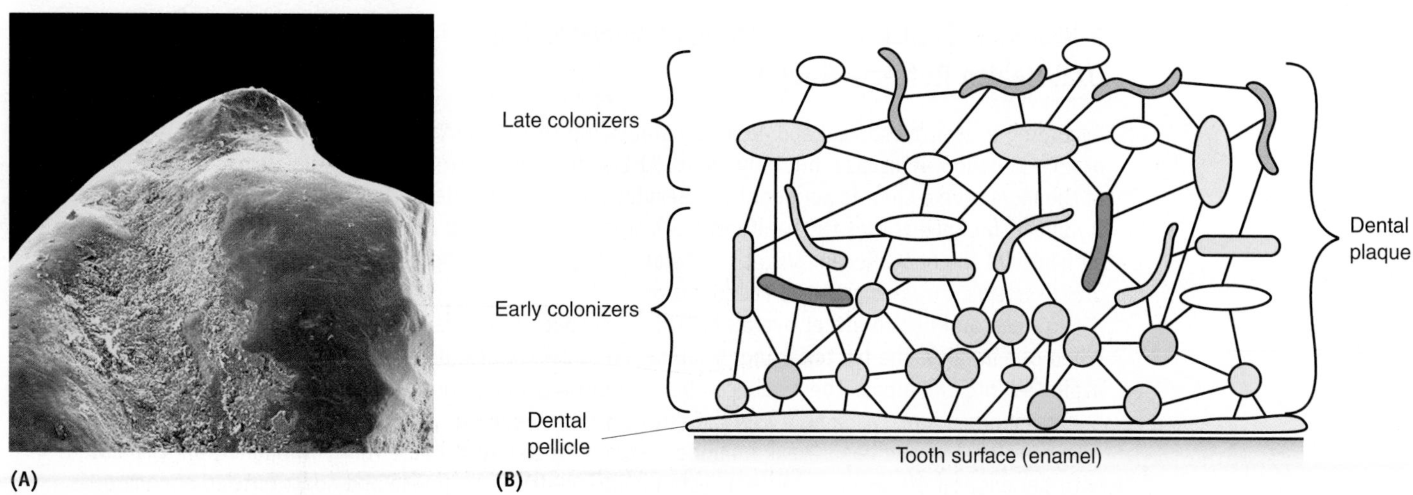

FIGURE 11.7 **Dental Plaque.** (**A**) This false-color scanning electron micrograph of a tooth surface shows dental plaque (yellow) coating the gray enamel of the tooth. © Dr. Tony Brain/Photo Researchers, Inc. (**B**) In this model of plaque formation, an initial set of colonizing bacterial species establish a biofilm with the dental pellicle. A more developed biofilm representing dental plaque occurs with the addition of late colonizers. Data from: Kolenbrander *et al.* (2010). *Nature Rev Microbiol* 8:471–480. »» Hypothesize what would happen if the early colonizers were absent from the oral cavity and the tooth surface?

cells in the plaque multiply and build up between the teeth and gums (FIGURE 11.8A). Gram-negative rods, such as *Fusobacterium, Porphyromonas, Prevotella,* and *Peptostreptococcus* species, secrete toxins and enzymes, including collagenase and hyaluronidase that directly or indirectly injure the periodontal tissues. The resulting immune inflammatory response attracts immune white blood cells to the damaged connective tissue, resulting in irritation and bleeding. Because there is no direct bacterial invasion of the gingival tissue, gingivitis is both treatable and preventable through professional cleaning to remove the plaque. As the most common cause of gingivitis is poor oral hygiene, daily brushing and flossing, along with regular professional cleanings, can greatly reduce the risk of developing more serious complications.

However, if left untreated, gingivitis can progress to **periodontitis**, a serious disease of the soft tissue and bone supporting the teeth; bone resorption occurs and the periodontal ligament may be lost (FIGURE 11.8B). This can result in loosening of the tooth and tooth loss. Left unchecked, long-term periodontitis can lead to even more serious problems, including higher blood sugar levels,

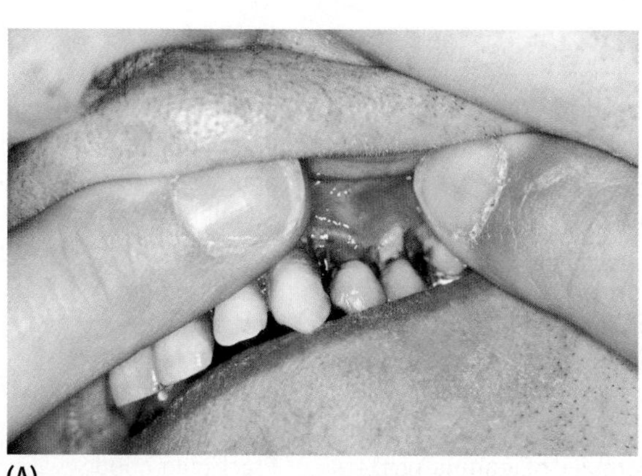
(A)

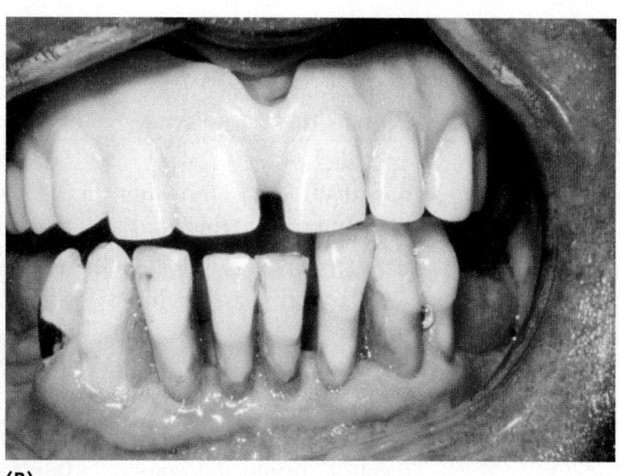
(B)

FIGURE 11.8 **Periodontal Disease.** (**A**) Gingivitis is an inflammation of the gums (gingiva) around the teeth, which is evident in this patient. © Medical-on-Line/Alamy Images. (**B**) Periodontitis refers to conditions when gum disease affects the structures supporting the teeth. In this example, gum recession and bone loss has occurred, which can lead to loosening of the teeth and even tooth loss. © SPL/Photo Researchers, Inc. »» What are the best ways to prevent periodontal diseases?

MICROFOCUS 11.3: Clinical Microbiology

Oral Health Reflects Body Health

According to several medical and dentistry groups, up to 75% of Americans have some form of gingivitis or mild periodontal disease, but only about 60% realize they have a problem and almost 30% show signs of the more severe chronic periodontitis. Besides the simple problems that can occur with these diseases, it appears fairly clear that the health of one's teeth and gums can have a significant impact on the overall health of the body. Recent scientific literature suggests a strong relationship between oral disease and other systemic diseases and medical conditions.

How can oral health affect overall health? First, bacterial cells on the gums can enter the saliva. When one inhales, the bacterial-laden saliva can stick to water droplets within the inhaled air and be aspirated into the lungs. The result can be a potentially dangerous pulmonary infection or pneumonia.

Dental procedures, as simple as a cut during dental cleaning, can provide an entry for bacterial cells into the blood. If one has a weak immune system or a damaged heart valve, inflammation associated with periodontal disease can cause a systemic inflammatory response and contribute to endocarditis and heart disease, diabetes, orthopedic implant failure, and kidney disease. Research also suggests that chronic periodontitis can predispose one to heart disease, clogged arteries, and stroke, all resulting from the entry of oral bacterial cells into the circulatory system.

Therefore, it is important to protect one's overall good health by maintaining good oral hygiene every day. In fact, dentists and oral hygienists are becoming more candid with their patients regarding oral health and the role it can play in the overall health of each patient. The American Dental Hygienists' Association says that "dental hygienists are preventive oral health professionals, licensed in dental hygiene, who specialize in the prevention and treatment of oral diseases in order to protect total health."

■ Fusiform: Refers to cells that are spindle shaped.

and an increased risk of heart attack and stroke, as bacterial cells now have access to the blood (MicroFocus 11.3). Periodontitis may even affect a pregnant woman's fetus. Studies have shown that pregnant women with periodontitis are more likely to give birth to premature babies than are women with healthy gums.

TABLE 11.1 summarizes the bacterial diseases of the oral cavity.

CONCEPT AND REASONING CHECKS 2

a. From the bacterial perspective, how does good oral hygiene help prevent dental caries?

b. What effects do bacterial disease of the oral cavity have on human overall health?

TABLE 11.1 A Summary of Bacterial Diseases of the Oral Cavity

Disease	Causative Agent	Signs and Symptoms	Transmission	Treatment	Prevention
Dental Caries	*Streptococcus mutans* *Streptococcus sobrinus* Other species	Toothache Sensitivity and pain when drinking and eating	Due to normal indigenous microbiota	Fluoride treatment Fillings Extraction	Practicing good oral hygiene Regular dental examinations
Gingivitis	*Bacteroides,* *Fusobacterium,* *Porphyromonas,* *Prevotella,* *Peptostreptococcus*	Swollen, soft, and red gums Bleeding gums	Due to normal indigenous microbiota	Cleaning of teeth to remove plaque	Practicing good oral hygiene Regular dental examinations
Periodontitis	*Actinobacillus* *Porphyromonas* *Bacteroides* *Treponema*	Swollen, bright red gums that are tender when touched Gums pulled away from teeth New spaces between teeth	Due to normal indigenous microbiota	Cleaning pockets of bacteria	Practicing good oral hygiene Regular dental examinations

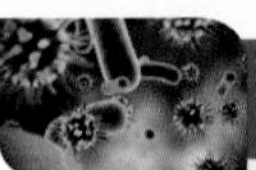

Chapter Challenge B

Although the foods we eat are not usually contaminated with a pathogen, we still can develop a disease if we do not maintain good oral hygiene after eating.

QUESTION B:

How is it possible that we can develop dental caries or periodontal disease even though the foods we eat are "clean" microbiologically speaking? Where do the bacterial species causing these diseases of the oral cavity come from?

Answers can be found on the Student Companion Website in **Appendix D.**

KEY CONCEPT 11.3 Bacterial Diseases of the GI Tract Are Usually Spread Through Food and Water

GI tract diseases may be mild or life-threatening for millions of people in the United States and around the world. The Centers for Disease Control and Prevention (CDC) estimates 48 million people in the United States suffer foodborne illnesses each year, accounting for 128,000 hospitalizations and more than 3,000 deaths—primarily infants, the elderly, and immunocompromised individuals. According to the World Health Organization (WHO), waterborne diseases account for an estimated 1.7 million deaths worldwide each year. Most of these deaths are from diarrheal diseases, especially among children in developing nations.

GI Tract Diseases May Arise from Intoxications or Infections

Most illnesses of the GI tract represent some form of **gastroenteritis**, an inflammation of the stomach and the intestines, usually with vomiting and diarrhea. Such inflammations can arise from either intoxications or infections. **Intoxications** are illnesses in which bacterial toxins are ingested in food or water. Examples are the toxins causing botulism, staphylococcal food poisoning, and clostridial food poisoning. By contrast, **infections** refer to illnesses in which live bacterial pathogens in food and water are ingested and subsequently grow in the body. Salmonellosis, shigellosis, and cholera are examples. Toxins may be produced, but they are the result of infection.

Determining the etiology of a bacterial disease depends on several factors.

Incubation Period. If an individual ingests and swallows a contaminated food or beverage, there is a delay, called the **incubation period**, before the symptoms appear. This period can range from hours to days, depending on the bacterial species. During the incubation period, the toxins or microbes pass through the stomach into the intestine where they may directly affect gastrointestinal function or be absorbed into the bloodstream.

Clinical Symptoms. The symptoms produced by an intoxication or infection depend on the specific toxin or microbe, and the number of toxins or cells ingested (toxic or infectious dose). Although the intoxications and infections may have different symptoms, nausea, abdominal cramps, vomiting, and diarrhea often are common. Because these symptoms are so universal, it can be difficult to identify the microbe causing a disease unless the disease is part of a recognized outbreak, or laboratory tests are done to identify the causative agent.

■ Toxic or infectious dose: The number of toxins or organisms consumed to give rise to symptoms of an illness.

Duration of Illness. Intoxications and infections can have very different lengths of time during which the symptoms persist. Some may be very abrupt and disappear quite quickly, while others may linger for a longer period of time.

Demographics. Certain individuals within a population may be more prone to infections or the effects of a toxin. Often infants, the elderly, and immunocompromised patients may be more vulnerable. Also, populations living in unsanitary and overcrowded conditions where public health measures are lacking will be more likely to become ill from contaminated food or water.

■ Immunocompromised: Referring to a person unable to develop a normal immune response, usually as a result of another disease or condition.

■ Etiology: The study of the causes (origins) of disease.

There Are Several Ways Foods or Water Become Contaminated

We live in a microbial world and there are many opportunities for food to become contaminated (FIGURE 11.9A). Many foodborne microbes are present in healthy animals (usually in their

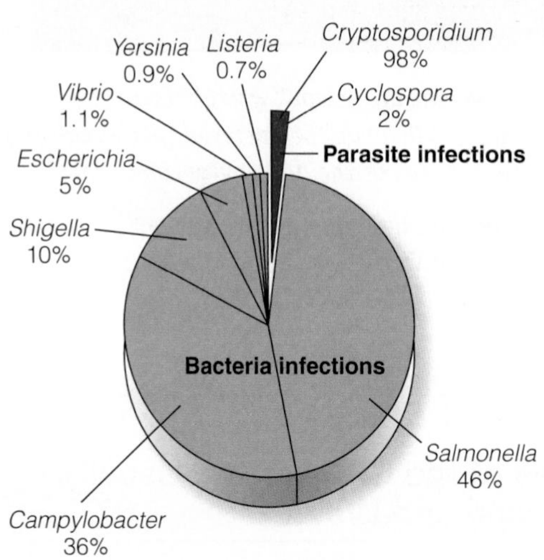

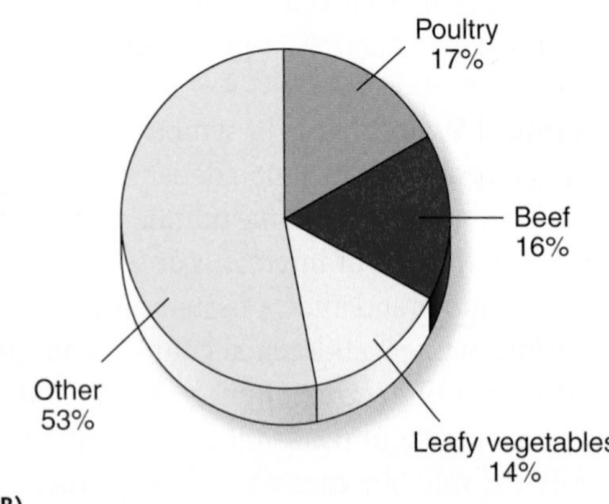

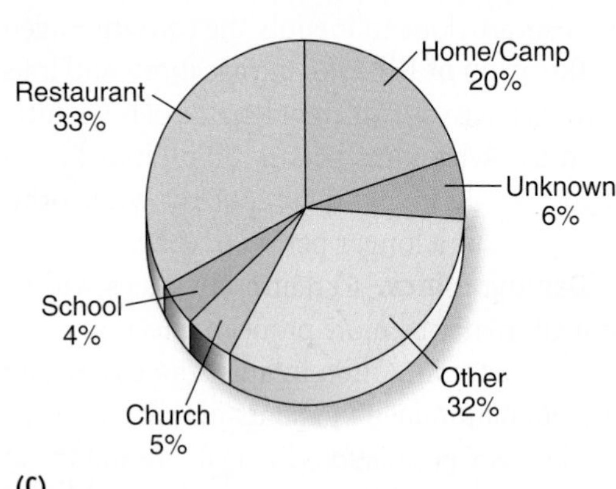

FIGURE 11.9 **Foodborne Illness Surveillance.** (**A**) Among the more than 19,000 lab-identified cases of foodborne illnesses in 2010, bacterial causes predominated. Data from CDC. (**B**) Among the foodborne outbreaks, the commodities most associated with illness were poultry, beef, and leafy vegetables. Data from CDC. (**C**) The most common source of illness was restaurants. Data from CDC. »» Looking at (B), what commodities might be represented by the "other" segment?

intestines) raised for food. The carcasses of cattle and poultry can become contaminated during slaughter if they are exposed to small amounts of intestinal contents. Fresh fruits and vegetables can be contaminated if they are washed or irrigated with water contaminated with animal manure or human sewage (FIGURE 11.9B).

Other foodborne microbes can be introduced through the fecal-oral route; from infected humans who handle the food, or by cross-contamination from some other raw agricultural product (FIGURE 11.9C). For example, *Shigella* species can contaminate foods from the unwashed hands of food handlers who are infected. In the home kitchen, microbes can be transferred from one food to another food by using the same knife, cutting board, or other utensil to prepare both without washing the surface or utensil between uses. A food fully cooked can become recontaminated if it touches raw foods or drippings containing pathogens.

Water can become contaminated in several ways. A common example is when an ill child or adult swimmer with diarrhea has an "accident" in a public pool. If the pool is not sufficiently chlorinated, such recreational water diseases can be passed to other individuals if they swallow the feces-contaminated water. Disease pathogens also can be spread by surface or groundwater contaminated with untreated or poorly treated sewage. In either case, individuals can become sick.

Some bacterial pathogens cause foodborne or waterborne bacterial diseases only when they are in large numbers. With warm, moist conditions and plenty of nutrients, lightly contaminated food left out overnight can be highly contaminated by the next day. *Escherichia coli* cells, for instance, dividing every 30 minutes can produce 17 million new cells in 12 hours. If the food had been refrigerated promptly, the bacterial cells would not multiply at all.

Several bacterial species are commonly found on many raw foods. Therefore, illness can be prevented by (1) controlling the initial number of organisms present, (2) preventing the small number from growing, (3) destroying the pathogens by proper cooking, and (4) avoiding recontamination.

CONCEPT AND REASONING CHECKS 3

a. What types of etiologic information may help in identifying the cause of an intoxication or infection?

b. List the ways food and water can become contaminated with pathogens.

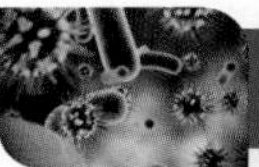

Chapter Challenge C

In this section of the chapter, the point was made that foods can become contaminated somewhere along the production line from farm to table. Some parts of the production we have little if any control over.

QUESTION C:

Present ways that foods can be protected from pathogen contamination on the farm, during manufacturing, at a restaurant, the grocery store, and at home. We will deal with more specific examples of prevention in the rest of the chapter.

Answers can be found on the Student Companion Website in **Appendix D.**

KEY CONCEPT 11.4 Some Bacterial Diseases Are the Result of Foodborne Intoxications

A few bacterial species secrete preformed microbial toxins that, when present in food, result in **food poisoning**, which is a **noninflammatory gastroenteritis**. These foodborne intoxications involve a brief incubation period and quick resolution. Botulism, caused by *Clostridium botulinum*, also is an example of food poisoning. However, the toxins affect the nervous system in several ways.

Food Poisoning Can Be the Result of Enterotoxins

Staphylococcal Food Poisoning. Ingestion of an exotoxin produced by *Staphylococcus aureus*, a facultatively anaerobic, gram-positive sphere can cause **staphylococcal food poisoning** (FIGURE 11.10A). Today, staphylococcal food

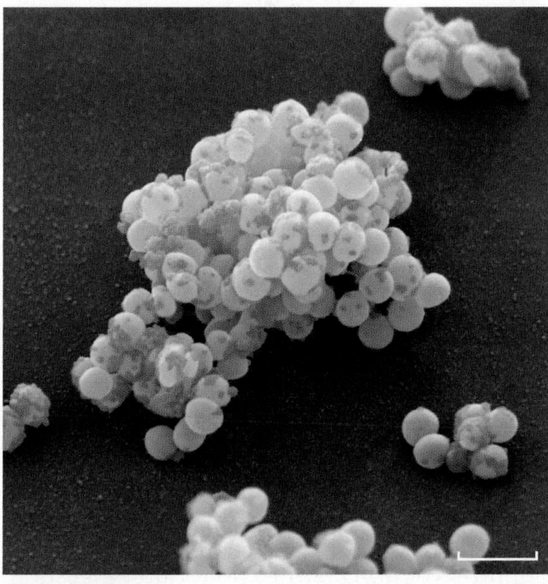

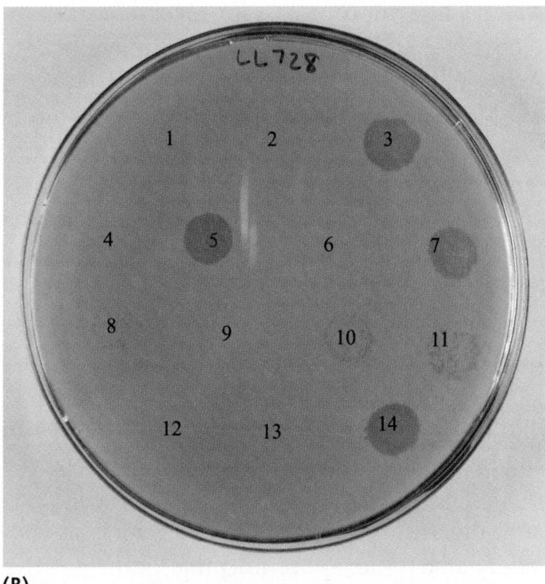

(A) (B)

FIGURE 11.10 *Staphylococcus aureus.* **(A)** A false-color scanning electron micrograph of *S. aureus* illustrating the typical grape-like cluster of cocci. (Bar = 1 μm.) © Eye of Science/Photo Researchers, Inc. **(B)** Typing of *S. aureus* strains with bacterial viruses. The plate of nutrient medium was seeded with the unknown strain of *Staphylococcus,* and numbered bacterial viruses were then placed into different areas. The clear areas indicate which of the viruses interacted specifically with the bacteria and killed them, leaving a clear circle. In this case, the strain of *S. aureus* is one that interacts with viruses 3, 5, 7, and 14. Courtesy of Julio Martin, University of Florida. »» Why is it important to type *S. aureus* strains?

poisoning ranks first among reported cases of foodborne intoxications. Alarmingly, in 2012 7% of sampled raw pork products contained the methicillin-resistant strain (MRSA).

■ Reservoir: The natural host or habitat of a pathogen.

A key reservoir of *S. aureus* in humans is the nose. Thus, an errant sneeze by a food handler may be the source of staphylococcal contamination. Studies indicate, however, that the most common mode of transmission is from boils or abscesses on the skin that shed staphylococci into the food product where the cocci soon begin to secrete the toxin.

When investigators locate staphylococci, they can identify the organisms by growth on mannitol salt agar, Gram staining, and testing with bacterial viruses to learn the strain involved, as FIGURE 11.10B illustrates.

The incubation period is a brief 1 to 6 hours, so the individual usually can think back and pinpoint the source, which often is a protein-rich food, such as meats or fish. Contaminated dairy products, cream-filled pastries, or salads such as egg salad also can be a cause.

Because the symptoms are restricted to the intestinal tract, the toxin is called an **enterotoxin** (*entero* = "intestine"). Patients experience abdominal cramps, nausea, vomiting, prostration, and diarrhea as the toxin encourages the release of water. The symptoms last for several hours, and recovery is usually rapid and complete in 1 to 2 days.

Staphylococci grow over a broad temperature range of 8°C to 45°C, and because refrigerator temperatures are generally set at about 5°C, refrigeration is not an absolute safeguard against growth in contaminated foods. Also, the staphylococcal enterotoxin is among the most heat resistant of all exotoxins. Heating at 100°C for 30 minutes will not denature the protein toxin.

Clostridial Food Poisoning. Since its recognition in the 1960s, **clostridial food poisoning** has risen to prominence as a common type of food poisoning in the United States. The causative organism, *Clostridium perfringens,* is an obligately anaerobic, spore-forming, gram-positive rod (FIGURE 11.11). Most commonly, it contaminates protein-rich foods such as meat, poultry, and fish. If the endospores survive the cooking process, they germinate to vegetative cells, which produce an enterotoxin. Consumption of the toxin leads to illness.

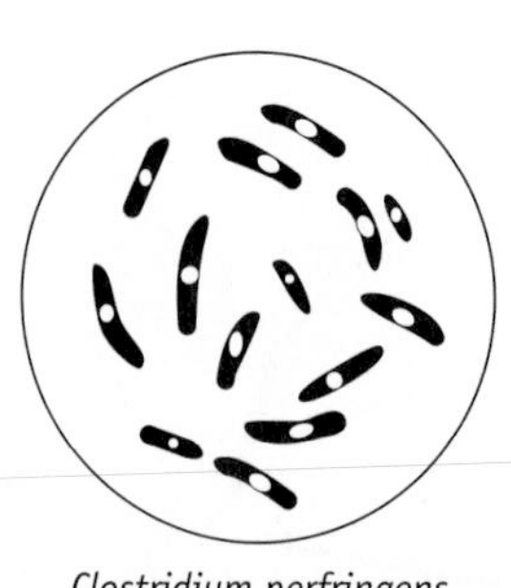

Clostridium perfringens

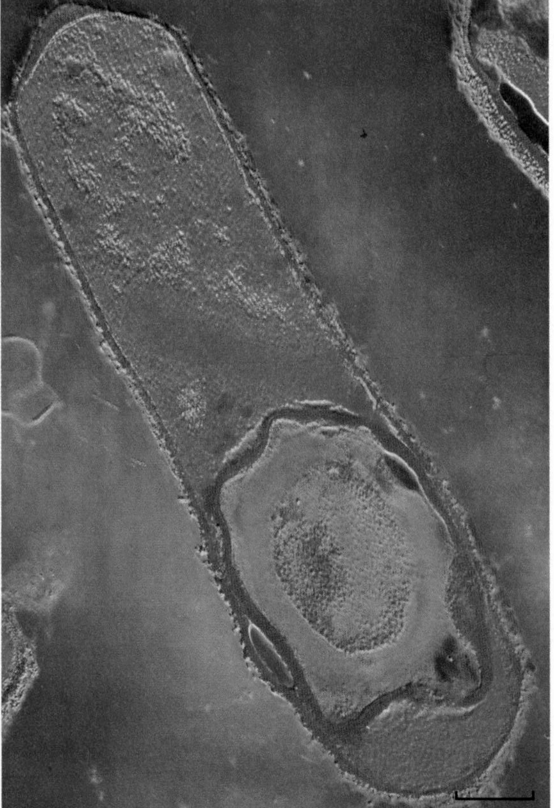

FIGURE 11.11 *Clostridium perfringens.* False-color transmission electron micrograph of a *C. perfringens* cell. It produces a heat-resistant enterotoxin that can contaminate protein-rich foods. (Bar = 0.5 μm.) © CNRI/Photo Researchers, Inc. »» What is the "green" structure in the bottom portion of the cell?

The incubation period for clostridial food poisoning is 8 to 24 hours, a factor that distinguishes it from staphylococcal food poisoning. Moderate to severe abdominal cramping and watery diarrhea are common symptoms. Recovery is rapid, often within 1 to 2 days, and antibiotic therapy is generally unnecessary.

***Bacillus cereus* Food Poisoning.** The spore-forming, gram-positive rod, *Bacillus cereus* causes food poisoning by producing one of two toxins. If the organism contaminates meats, poultry, or vegetables, it produces an enterotoxin in the small intestine that causes a watery diarrhea but little vomiting 6 to 15 hours after ingestion. If, on the other hand, it contaminates cooked grains, like rice, it produces a heat-stable toxin that causes substantial vomiting 30 minutes to 5 hours after ingestion. Most affected individuals recover after 2 days. Fluid replacement is standard treatment.

Botulism. Of all the foodborne intoxications in humans, none is more dangerous than **botulism**. It is a rare but serious illness caused by *Clostridium botulinum,* a spore-forming, obligately anaerobic, gram-positive bacillus. The endospores exist in the intestines of humans as well as fish, birds, and barnyard animals. They reach the soil in manure, organic fertilizers, and sewage, and often, they cling to harvested products. When spores enter the anaerobic environment of cans or jars, they germinate to vegetative bacilli, and the bacilli produce the exotoxin.

The bacterial cells themselves are of little consequence, but the botulinum neurotoxin is so powerful that it is thought to be the deadliest substance known to science. Scientists have identified seven types of *C. botulinum,* depending on the variant of toxin produced. Types A, B, and E cause most human disease, with type E associated with most cases of foodborne transmission. The annual number of cases in the United States is low (112 cases in 2010 and accounts for about 8% of all botulism cases).

The symptoms of foodborne botulism usually develop within 18 to 36 hours after ingesting the toxin-contaminated food. Being a preformed neurotoxin, patients suffer neurologic manifestations, including blurred vision, slurred speech, difficulty swallowing and chewing, and labored breathing. The limbs lose their tone and become flabby, a condition called **flaccid paralysis**. These symptoms result from the toxin's effects on the nervous system. The neurotoxin affects the nerve synapse and inhibits the release of the neurotransmitter acetylcholine. Without acetylcholine, nerve impulses cannot pass across the synapse into the muscles, and the muscles do not contract. Failure of the diaphragm and rib muscles to function leads to respiratory paralysis and death within a day or two.

Because botulism is an intoxication, antibiotics are of no value as a treatment against the toxin. Instead, if treated early, large doses of specific antibodies called **antitoxins** can be administered to neutralize the unbound toxins. Therefore, knowing which type of botulinum toxin is causing the disease is important because antitoxin therapy must be type specific. Life-support systems such as mechanical ventilators also are used. For example, in September 2006, four cases of foodborne botulism (three in the United States and one in Canada) were reported in individuals who consumed improperly refrigerated carrot juice. The lives of these patients depended on botulinum antitoxin and assistance of hospital ventilators. Complete recovery can take up to 1 year.

Botulism can be avoided by heating foods before eating them, because the toxin is destroyed on exposure to temperatures of 90°C for 10 minutes. However, experience shows that most outbreaks are related to home-canned foods having a low acid content, such as asparagus, green beans, beets, and corn, and from foods eaten cold. Other foods linked to botulism include mushrooms, olives, salami, and sausage. In fact, the word botulism is derived from the Latin *botulus,* for "sausage."

■ Neurotoxin: A substance that damages, destroys, or impairs the functioning of nerve tissue.

Although foodborne botulism is very dangerous, other forms of botulism exist. **Wound botulism** is caused by toxins produced in the anaerobic tissue of a wound infected with *C. botulinum.* Penicillin is an effective treatment.

Infant botulism is the most common form of botulism in the United States, accounting for over 70% of the total annual cases reported. Infant botulism results from the ingestion of soil or food contaminated with *C. botulinum* endospores. Thus, parents should not give their infant honey, which is the most common food triggering infant botulism. Spores that are in about 10% of honey germinate and grow in the colon where the botulinum cells release toxin. This form of botulism typically affects infants 3 to 24 months old because they have not established the normal balance of bowel microbes. Because the toxin produces lethargy and poor muscle tone, infant botulism often is referred to as floppy baby syndrome. Hospitalization along with mechanically assisted ventilation may be necessary, with antitoxin treatment reducing recovery time.

■ Syndrome: A collection of signs and symptoms.

In recent years, one of the botulinum toxins has been put to practical use. In minute doses, **Botox®** or **Dysport®** (botulism toxin type A) can relieve temporarily a number of movement disorders (so-called dystonias) caused by involuntary sustained muscle contractions. For example, the toxin is used to treat strabismus, or misalignment of the eyes, commonly known as crossed-eye; it also is used against blepharospasm, or involuntarily clenched eyelids. The toxin may be valuable in relieving stuttering, uncontrolled blinking, and musician's cramp. The toxin also has been

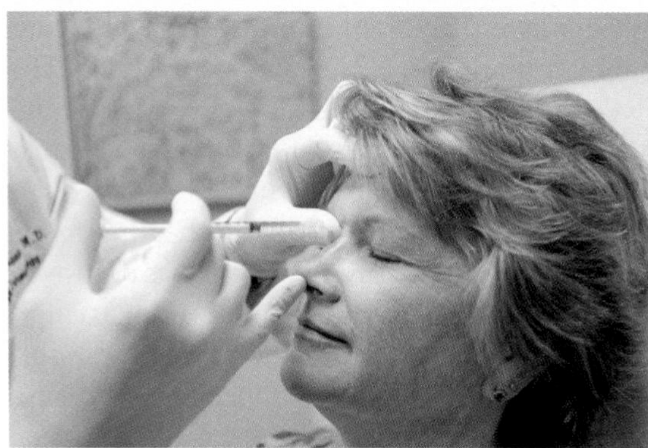

FIGURE 11.12 **Cosmetic Injection of Botulinum Toxin.** The botulinum toxin in controlled doses can be used to temporarily minimize wrinkles. © Michael N. Paras/age fotostock. »» How does the toxin work to "remove wrinkles"?

approved for use in the temporary relief of facial wrinkles and frown lines (FIGURE 11.12). In 2004, it was approved for temporary relief of hyperhidrosis (excessive body sweating) and in 2010 for chronic migraine headaches.

CONCEPT AND REASONING CHECKS 4

a. What characteristics are shared by all three foodborne intoxications?

b. Why are tetanospasmin and botulism toxin considered to be the most powerful toxins known to science?

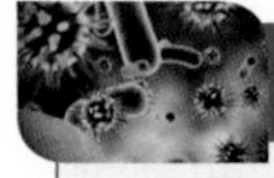

Chapter Challenge D

Foodborne intoxications involve foods that inadvertently become contaminated with microbes that produce dangerous intestinal toxins.

QUESTION D:

In our farm to table scenario, what are the most common ways that foods become contaminated with bacterial toxins or with pathogens that produce such toxins?

Answers can be found on the Student Companion Website in **Appendix D**.

KEY CONCEPT 11.5 GI Infections Can Be Caused by Several Bacterial Pathogens

Bacterial GI infections usually have a longer incubation period than intoxications because bacterial cells must first establish themselves in the body after ingestion of the contaminated food or water.

Bacterial GI infections can be one of two types. **Inflammatory gastroenteritis** is characterized by diarrhea and/or vomiting, and usually a fever, but there is no blood in the stool. **Invasive gastroenteritis** involves bacterial invasion beyond the intestinal lumen. Signs and symptoms may include fever, diarrhea or vomiting, and dysentery (the passage of blood and mucus in the feces). We will examine several infections characteristic of each type of gastroenteritis.

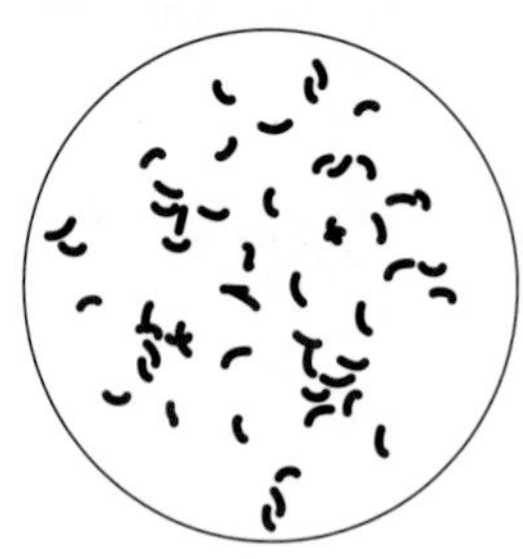

Vibrio cholerae

■ Electrolyte: An ion in cells, blood, or other organic material.

Bacterial Gastroenteritis Often Produces an Inflammatory Condition

The most familiar illnesses causing a bacterial inflammatory gastroenteritis are cholera, caused by *Vibrio cholerae,* and *Escherichia coli* gastrointestinal infections, such as traveler's diarrhea.

Cholera. No diarrheal disease can compare with the extensive diarrhea associated with **cholera** (see chapter opener). The WHO estimates there are more than 100,000 cases and over 1,900 deaths annually. The organism, *Vibrio cholerae,* was first isolated by Robert Koch in 1883.

In the past 200 years, seven pandemics have been documented, and they have been responsible for taking the lives of millions of people. The current seventh pandemic, caused by *V. cholerae* biotype El Tor, began in 1961 in Indonesia and now involves about 35 countries (MicroFocus 11.4).

The cells *V. cholerae* are motile, aerobic, gram-negative, curved rods (FIGURE 11.13). They enter the intestinal tract in contaminated food, such as raw oysters, or in water. Because *V. cholerae* is extremely susceptible to stomach acid, most infections are asymptomatic (TABLE 11.2). However, if high numbers are ingested, enough survive to colonize the small intestines. There the noninvasive cells move along the intestinal epithelium, secreting an enterotoxin (**cholera toxin**) that stimulates the unrelenting loss of fluid and electrolytes.

The majority of cases are mild and are characterized by watery diarrhea that lasts a few days. In more severe cases, extreme dehydration may occur because a patient may lose up to 1 liter of fluid every

MICROFOCUS 11.4: Biotechnology/Genomics

The Seventh Wave

Cholera has been afflicting the human species for centuries, causing untold deaths and suffering. Thanks to the work of John Snow and Robert Koch in the last half of the 19th century, the mechanism for cholera's spread (in water) and the discovery of the infectious agent (*Vibrio cholerae*) became known. Still today, *V. cholerae* produces some 3 to 5 million reported illnesses every year and causes 100,000 to 120,000 deaths. Over recorded history, there have been seven cholera pandemics with the current outbreaks in Haiti and Africa being just the latest in the seventh and continuing pandemic.

The current seventh pandemic is caused by the El Tor strain of *V. cholerae* serogroup 01. Because it has spread globally, it is of interest to study the origin of the strain as it can shed light on how a disease spreads around the world and the steps that such a pathogen must go through to be the serious pathogen that it is today. Past studies have suggested that the El Tor strain arose in Indonesia around 1961.

In 2011, a group of international scientists set out to use genome analysis to trace back the source of the seventh cholera pandemic. The *V. cholerae* genome from 154 cholera patients from around the world over the last 40 years were sequenced. The researchers were looking for single DNA changes in the genome that could be used to map the transmission routes of *V. cholerae* and thus work back to the origin of the El Tor strain and the seventh pandemic.

Their analysis identified the Bay of Bengal (orange dot on map) as the source for the ancestor of the El Tor strain, which originated in Indonesia around 1950. Perhaps more interestingly, they show that cholera from the Bay of Bengal has spread repeatedly to different parts of the world in independent, overlapping waves that comprise the seventh pandemic (see figure). These "seventh waves" infected people in South Asia and Indonesia, the Middle East, parts of Europe, Africa, and South America—and now the Caribbean (Haiti).

The international group was also able to determine through gene sequencing that the current El Tor strain first gained resistance to antibiotics about 1982, which other investigators have shown was due to horizontal gene transfer (transduction) from a bacterial virus. The abrupt gain in resistance then triggered another wave of global transmission from the Bay of Bengal, which continues to be the source for each new seventh wave.

What is the best explanation for the transmission of *V. cholerae*? Almost certainly it is through human activity, including human infections being carried by jet travel.

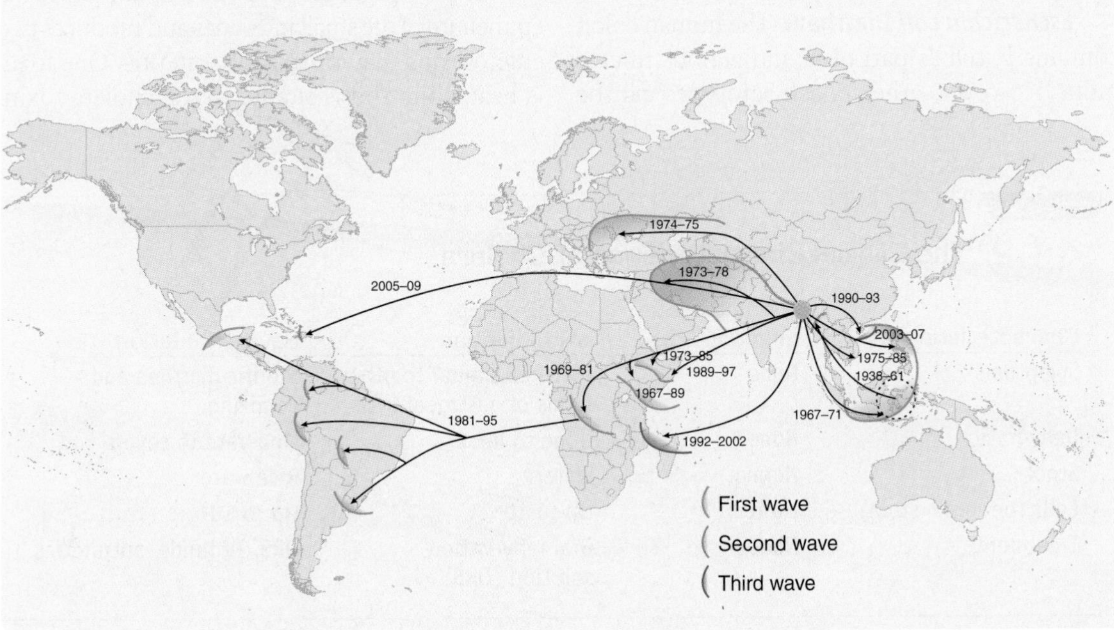

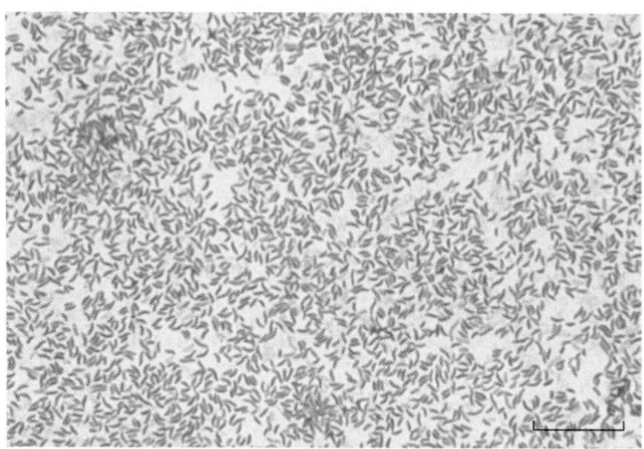

FIGURE 11.13 *Vibrio cholerae.* This light microscope image shows the stained *V. cholerae* cells. (Bar = 1 μm.) © Michael Abbey/Photo Researchers, Inc. »» What is the shape of these cells?

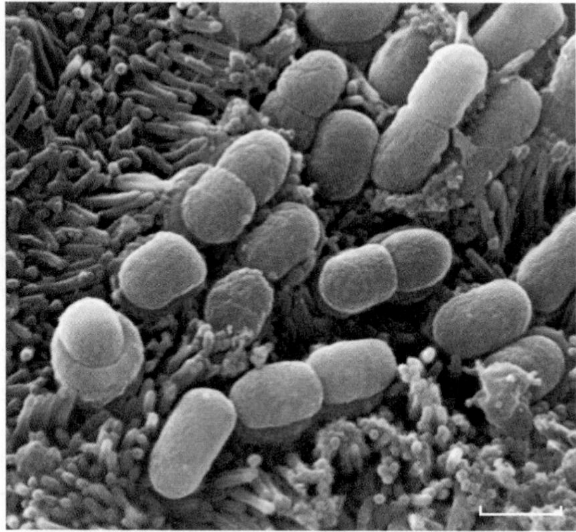

FIGURE 11.14 *Escherichia coli.* False-color scanning electron micrograph of *E. coli* cells inhabiting the surface of the intestines. When exogenous strains of *E. coli* enter the intestines through contaminated food or water, infections such as gastroenteritis may occur. (Bar = 2 μm.) © Stephanie Schuller/Photo Researchers, Inc. »» Why doesn't the normal intestinal *E. coli* cause gastroenteritis?

hour for several hours. The fluid, referred to as **rice-water stools**, is colorless and watery, reflecting the conversion of the intestinal contents to a thin liquid likened to water from boiled rice. Despite continuous thirst, sufferers cannot hold fluids. The blood thickens, urine production ceases, and the sluggish blood flow to the brain leads to shock and coma. If untreated, an individual can die with 24 hours of the onset of symptoms.

For severe cases, antibiotics such as tetracycline may be used to kill the bacterial cells, but the key treatment is restoration of the body's water and electrolyte balance. This entails **oral rehydration therapy**, involving a solution of electrolytes and glucose designed to restore the normal balances in the body. For the most severe cases, intravenous injections of the solution are required.

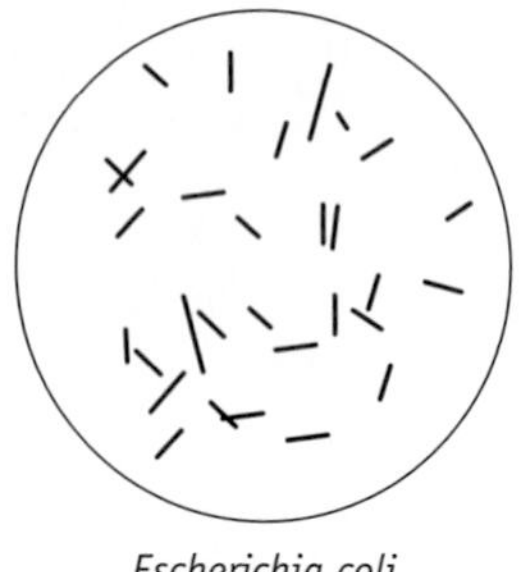

Escherichia coli

***Escherichia coli* Diarrhea.** The human colon contains *E. coli* as part of its indigenous microbiota. However, other *E. coli* serotypes can be pathogenic and are one of the major causes of infantile diarrhea.

E. coli is a facultatively anaerobic, gram-negative rod (FIGURE 11.14). Transmission follows the fecal-oral route where contaminated food or water represents the vehicle for transmission. The pathogenic serotypes may induce watery diarrhea in several ways.

Enterotoxigenic *E. coli* (ETEC) adheres to the epithelium of the small intestines and produces two enterotoxins that cause gastroenteritis. One toxin is heat labile and is similar to the cholera toxin.

TABLE

11.2 The Clinical Characteristics of Cholera

Characteristic	Asymptomatic Infection	Mild Infection	Severe Infection
Symptoms	None	Diarrhea similar to other forms of gastroenteritis	Prolific diarrhea and vomiting
Dehydration	None	None to mild	Moderate to severe
Stool	Normal	Watery	Rice water
Cells (per gram stool)	Up to 10^5	Up to 10^8	Up to 10^9
Treatment	None	Oral rehydration solution (ORS)	ORS, IV fluids, antibiotics

The other toxin is heat stable. The illness, typically referred to as **traveler's diarrhea**, affects 20% to 50% of travelers within 2 weeks of traveling to a tropical and subtropical destination where ETEC occurs. (A number of bacterial species, viruses, and protists may cause traveler's diarrhea, but recent studies point to *E. coli* as the principal agent.) The volume of fluid lost is usually low, but vomiting, cramps, nausea, and a low-grade fever may occur. The diarrhea lasts 3 to 7 days. The possibility of traveler's diarrhea may be reduced by careful hygiene and attention to the food and water consumed during visits to endemic countries.

Enteropathogenic *E. coli* (EPEC) causes a potentially fatal form of diarrhea in infants, especially where sanitation conditions are poor. Transmission is by the fecal-oral route and the *E. coli* cells cause a watery diarrhea that results from mucosal deformation caused by the binding of *E. coli* cells to the mucosa. Fluid replacement is critical to prevent dehydration.

An enterohemorrhagic serotype of *E. coli* causes an invasive gastroenteritis and is described later in this chapter.

***Clostridium difficile* Infections.** One of the most widespread and potentially serious illnesses is caused by *Clostridium difficile,* often simply called *C. diff,* an anaerobic, spore-forming, gram-positive rod. Since about 2004, ***C. difficile* infections (CDIs)** have reached epidemic proportions around the world but especially in the United States, Canada, and Europe. Published reports suggest that there are more than 330,000 cases of CDIs each year in the United States that take the lives of up to 30,000 Americans.

Those most at risk are older adults and people under medical care who have been taking certain antibiotics for a prolonged time for another illness or condition. Thus, about 75% of CDIs develop outside the hospital or healthcare setting. As a result of antibiotic therapy, the intestinal microbiota is damaged, allowing resident or ingested *C. difficile* to hypercolonize regions of the gastrointestinal tract. The bacterial cells are shed in feces, so any surface, instrument, or material contaminated with feces may serve as a reservoir for the *C. difficile* spores, which can survive for months on surfaces. In a healthcare facility, the endospores could be transferred to patients via the hands of healthcare personnel who have touched a contaminated surface or object.

In the colon, the pathogen produces two toxins, an enterotoxin that causes fluid loss and diarrhea, and a **cytotoxin** that causes further mucosal injury to **pseudomembranous colitis**, a severe inflammation of the colon that can lead to a grossly dilated bowel (**toxic megacolon**) and rupture (FIGURE 11.15). Symptoms of pseudomembranous colitis can begin within 1 to 2 days after beginning an antibiotic, or may not occur until several weeks after discontinuing the antibiotic. Because most patients develop only a mild diarrhea, stopping antibiotic therapy, if clinically possible, together with fluid replacement, usually results in rapid improvement. In 2012, the U.S. Food and Drug Administration (FDA) approved fidaxomicin for *C. difficile* diarrhea. For more severe cases like pseudomembranous colitis, anticlostridial antibiotics, such as metronidazole or vancomycin, may be required. Earlier in this chapter, fecal transplants (bacteriotherapy) were mentioned as a curative measure (Figure 11.4).

Prior to 2000, CDIs was almost always found in older adults in the hospital or healthcare setting. However, in 2008, a new hypervirulent strain of *C. difficile* emerged and has been the cause of

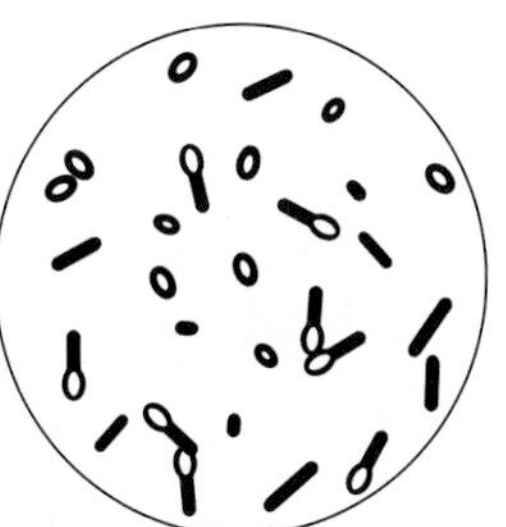

Clostridium difficile

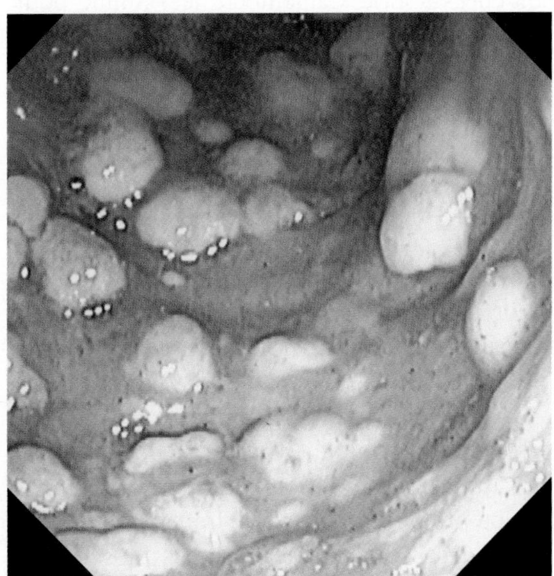

FIGURE 11.15 **Pseudomembranous Colitis.** An endoscope view of an inflamed colon caused by a *Clostridium difficile* infection. The white plaques represent mucus and dead cells that have built up on the colon walls. © David M. Martin, M.D./Photo Researchers, Inc. »» What factors are produced by *C. difficile* that trigger pseudomembranous colitis?

community-acquired infections among a wider age range of patients, including children, with no recent antibiotic therapy. This new strain causes a more severe illness and the hypervirulence appears to be due to its ability to produce greater quantities of the two toxins and perhaps produces more spores. Historically, mortality from CDIs was less than 2% but in some cases this has increased to over 10% with the hypervirulent strain.

Listeriosis. An important public health concern in the United States today is **listeriosis**, a serious GI infection and foodborne illness.

Causative Agent and Epidemiology. Listeriosis is caused by *Listeria monocytogenes*, a small, facultatively anaerobic, gram-positive rod (FIGURE 11.16). The CDC estimates that there are about 800 laboratory-confirmed cases of listeriosis in the United States annually (821 cases reported in 2010). The disease primarily affects older adults, pregnant women, newborns, and adults with weakened immune systems.

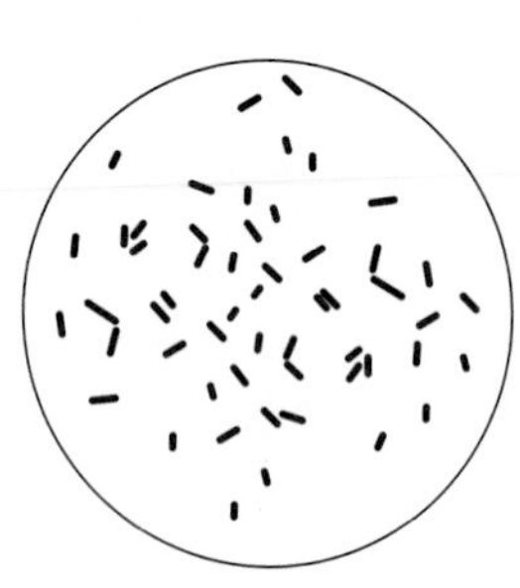

Listeria monocytogenes

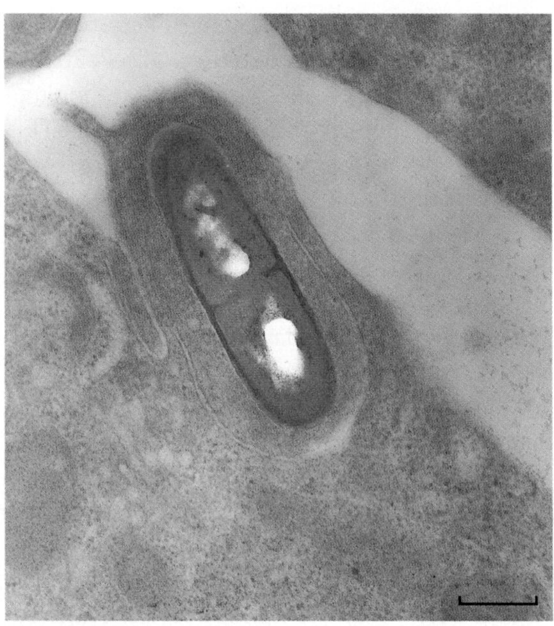

FIGURE 11.16 ***Listeria monocytogenes.*** A false-color transmission electron micrograph of *L. monocytogenes* entering a macrophage. (Bar = 1 μm.) Courtesy of Dr. Balasubr Swaminathan and Peggy Hayes/CDC. »» What advantage does *L. monocytogenes* gain by invading macrophages?

Listeriosis is transmitted to humans through food contaminated with *L. monocytogenes*. However, healthy persons may consume such contaminated foods without becoming ill. Because the pathogen is a psychrotroph, refrigerator temperatures will not suppress growth. Thus, it can contaminate delicatessen cold cuts, as well as soft cheeses (Brie, Camembert, feta), milk, poultry, and seafood. In 2011, *Listeria*-contaminated whole cantaloupes sickened some 140 people in 28 states and caused the death of at least 30 individuals, making it the worst deadly foodborne outbreak in the United States since 1924.

Most *Listeria* infections remain asymptomatic in healthy people. However, in elderly or immunocompromised individuals, the infection can lead to life-threatening complications, including septicemia and meningitis. The bacterial cells live inside specific white blood cells called macrophages and, as such, are hidden from immune responses and can multiply within these white blood cells.

Clinical Presentation. The incubation period can be prolonged, anywhere from 2 to 6 weeks. Infection can lead to three outcomes. Most healthy individuals experience no symptoms and recover completely. In pregnant women, a potentially fatal perinatal infection in the fetus or newborn may lead to miscarriage or stillbirth. If infected macrophages cross the blood-brain barrier, inflammation of the CNS occurs and **listerial meningitis** can develop. This form of meningitis has no clear symptoms or signs to distinguish it from other forms of community-acquired meningitis. Thus, listerial meningitis presents as high fever, headache, and stiff neck and symptoms may also include confusion, loss of balance, or convulsions. Listerial meningitis can be fatal in some 20% of patients or lead to a permanent disability, so early diagnosis is critical.

Treatment and Prevention. Treatment must be started as soon as possible in the course of the disease. Appropriate antibiotic treatment (for 2 weeks) reduces the risk of serious complications or death. The general guidelines for the prevention of listeriosis include thoroughly cooking raw food from animal sources and washing hands, knives, and cutting boards used to prepare raw foods.

Vibriosis. Infections by species other than *Vibrio cholerae* can be acquired through the ingestion of contaminated seafood. In 2010, the CDC reported almost 846 cases of **vibriosis**.

The leading cause of seafood-associated gastroenteritis in the United States is *Vibrio parahaemolyticus*, a curved, gram-negative, halophilic rod that is naturally found in warm marine waters. The CDC estimates there are more than 4,500 cases each year in the United States, with many more cases not reported or detected. After a 24-hour incubation period, the secreted enterotoxin, causes acute abdominal pain, vomiting, watery diarrhea, and nausea.

In 1996, the CDC reported an outbreak of intestinal illness due to another *Vibrio* species, *V. vulnificus*. This species, the most virulent of the vibrios, occurs naturally in brackish and seawaters, where oysters and clams live. People who consume these mollusks raw are at risk, especially immunocompromised individuals and those who suffer from liver disease or low stomach acid. (Indeed, taking an antacid after a meal of any contaminated food may neutralize stomach acid and facilitate the passage of the pathogen to the bloodstream.) The gastrointestinal infection involves fever, nausea, and severe abdominal cramps. The infection can become systemic and cause necrotic skin lesions and cellulitis. The mortality rate for the disease can be as high as 50%. There are some 100 reported cases in the United States each year, usually limited to the Gulf Coast.

Intestinal Anthrax. Besides cutaneous and inhalation anthrax, another but much rarer form of anthrax is **intestinal anthrax**, which is also caused by *Bacillus anthracis*. Consumption of contaminated or undercooked meat can trigger an acute inflammatory gastroenteritis. Initial signs include nausea, loss of appetite, fever, and vomiting. This is followed by abdominal pain, vomiting of blood, and severe diarrhea. Intestinal anthrax results in death in 5% to 60% of untreated cases. However, like the other forms of anthrax, if caught early, it is almost always successfully treated with antibiotics.

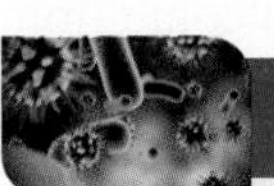

Chapter Challenge E

Bacterial gastroenteritis, inflammatory and invasive, can be experienced through a number of bacterial infections transmitted through food or water.

QUESTION E:

For inflammatory gastroenteritis, what types of precautions should be taken to prevent the establishment of an infection?

Answers can be found on the Student Companion Website in **Appendix D**.

Several Bacterial Species Can Cause an Invasive Gastroenteritis

Besides those bacterial species that cause inflammatory gastroenteritis, additional species possess virulence factors that allow them to invade layers of the lower GI tract beyond the epithelium. A few become systemic and affect other body systems.

Typhoid Fever. Among the classical diseases (the "slate wipers") that have ravaged human populations for generations is **typhoid fever**. The disease was first studied by Karl Eberth and Georg Gaffky, both coworkers of Robert Koch.

There are some 2,500 unique serotypes of *Salmonella*. Serotypes are used for *Salmonella* instead of species because of the uncertain relationships existing among the organisms. Typhoid fever is caused by *S. enterica* serotype Typhi, a motile, nonspore-forming, facultatively anaerobic, gram-negative rod. For convention sake, we will refer to it as S. Typhi.

The cells of S. Typhi display high resistance to environmental conditions outside the body, which enhances their ability to remain alive for long periods in water, sewage, and certain foods exposed to contaminated water. The pathogen is transmitted by the five Fs: flies, food, fingers, feces, and fomites (FIGURE 11.17). Humans are the only host for S. Typhi.

In 2010, more than 460 cases of typhoid fever were reported to the CDC. However, about 85% of infections are acquired during international travel to endemic areas, especially south Asia.

The organism is acid resistant, and with the buffering effect of food and beverages, it survives passage through the stomach. During the 5- to 21-day incubation period, it invades the small intestine, causing deep ulcers, bloody stools, and abdominal pain. Blood invasion leads to a systemic illness and the patient experiences mounting fever, lethargy, and delirium. In about 30% of cases, the abdomen becomes covered with a faint rash (**rose spots**), indicating blood hemorrhage in the skin. Symptoms last for 3 to 4 weeks. If untreated, about 15% of individuals die.

Treatment is generally successful with the antibiotic ceftriaxone. About 5% of recoverers are carriers and continue to harbor and shed the organisms for a year or more. Because food handlers can be carriers of disease, public health

■ Cellulitis: A bacterial infection of the skin and tissues just below the skin.

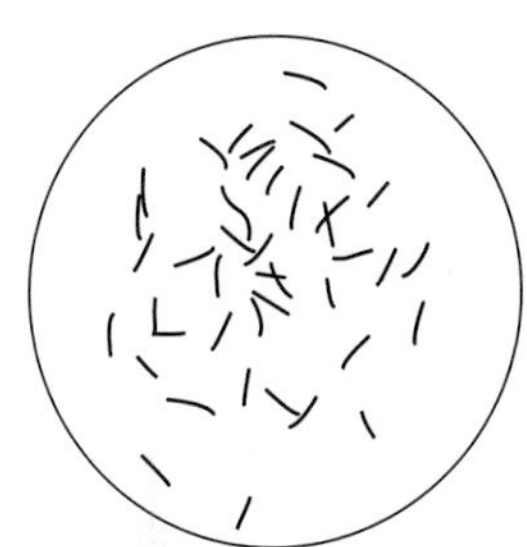

Salmonella Typhi

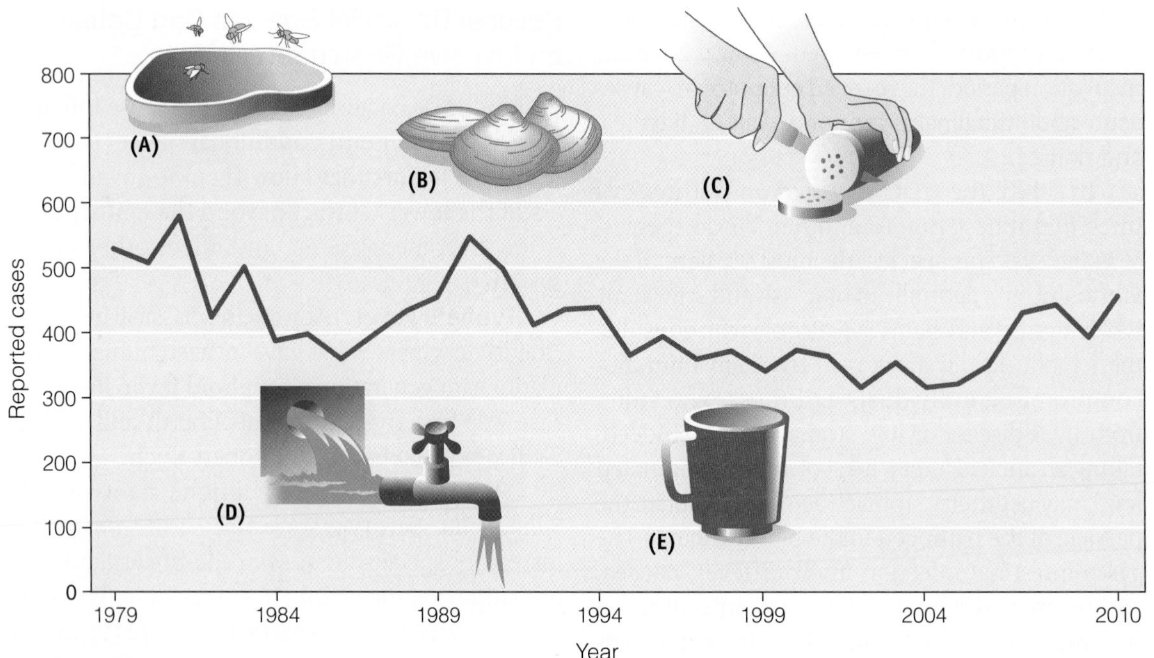

FIGURE 11.17 **The Incidence of Typhoid Fever.** Reported cases of typhoid fever in the United States by year, 1979–2010. Data from CDC, Summary of Notifiable Diseases, 2010. »» What do the five images (A–E) portray? Name them.

departments usually monitor the activities of carriers. The experiences of one of history's most famous carriers, Typhoid Mary, are recounted in MicroFocus 11.5.

The CDC recommends that travelers going to areas with increased S. Typhi incidence be vaccinated with one of the two available vaccines. Neither is 100% effective and they lose effectiveness after a few years, so booster shots may be needed for future travel.

Salmonellosis. The CDC estimates that each year there are more than 1.4 million cases of **salmonellosis** in the United States of which most go unreported. For example, in 2010 there were some 54,000 reported cases. Although the most common serotypes involved with illness are *S. enterica* serotype Enteritidis and *S. enterica* serotype Typhimurium (FIGURE 11.18A), there is an increasing incidence of Newport, Mississippi, and Javiana serotypes.

With increased awareness and modern methods of detection, salmonellosis has been linked to a broad variety of foods, including unpasteurized milk and poultry products. *Salmonella* serotypes commonly infect chickens and turkeys when the normal bacterial species of the gut are absent (FIGURE 11.18B). In 2005, more than 1,000 people in Spain came down with salmonellosis after consuming contaminated precooked chicken. This was the largest outbreak of the disease in Spain's history.

Contaminated produce eaten raw has increasingly become a vehicle for transmission of *Salmonella* and other pathogens. In April 2008, an outbreak of salmonellosis in the United States was traced to fresh, raw jalapeno or serrano peppers. More than 1,400 individuals were infected. In 2010, 500 million eggs were recalled due to S. Enteritidis contamination that caused more than 1,500 illnesses. However, exposure does not have to come from food. In 2010–2011, the CDC reported 132 cases of human S. Parathyphi in 18 states arising from pet turtle exposures. Sale of pet turtles has been prohibited in the United States since 1975 because of the potential for *Salmonella* transmission. No deaths were reported.

A relatively large infectious dose (100,000 cells) is required to initiate an illness. After an incubation period of 6 to 48 hours, cells invade the mucosa and submucosa, where the inflammation causes fever, nausea, vomiting, diarrhea, and abdominal cramps. Intestinal ulceration is usually less severe than in typhoid fever, and blood invasion is uncommon. The symptoms typically

MICROFOCUS 11.5: History

Typhoid Mary

By 1906, typhoid fever was claiming about 25,000 lives annually in the United States. During the summer of that year, a puzzling outbreak occurred in the town of Oyster Bay on Long Island, New York. One girl died and five others contracted typhoid fever, but local officials ruled out contaminated food or water as sources. Eager to find the cause, they hired George Soper, a well-known sanitary engineer from the New York City Health Department.

Soper's suspicions centered on Mary Mallon, a seemingly healthy family cook. But she had disappeared 3 weeks after the disease surfaced. Soper was familiar with Robert Koch's theory that infections like typhoid fever could be spread by people who harbor the organisms. Quietly, he began to search for the woman who would become known as Typhoid Mary.

Soper's investigations led him back over the 10 years' time during which Mary Mallon cooked for several households. Twenty-eight cases of typhoid fever occurred in those households, and each time, the cook left soon after the outbreak.

Soper tracked Mary Mallon through a series of leads from domestic agencies and finally came face-to-face with her in March 1907. She had assumed a false name and was now working for a family in which typhoid had broken out. Soper told her he believed she was a carrier of typhoid and pleaded with her to be tested for typhoid bacilli. When she refused to cooperate, the police forcibly brought her to a city hospital on an island in the East River off the Bronx shore. Tests showed her stools teemed with typhoid bacilli, but fearing her life was in danger, she adamantly refused the gall bladder operation (where the bacilli reside) that would eliminate them. As news of her imprisonment spread, Mary became a celebrity. Soon public sentiment led to a health department policy deploring the isolation of carriers. She was released in 1910.

But Mary's saga had not ended. In 1915, she turned up again at New York City's Sloane Hospital working as a cook under a new name. Eight people had recently died of typhoid fever, most of them doctors and nurses. Mary was taken back to the island, this time in handcuffs. Still, she refused the operation and vowed never to change her profession. Doctors placed her in isolation in a hospital room while trying to decide what to do. The weeks wore on.

Eventually, Mary became less incorrigible and assumed a permanent residence in a cottage on the island. She gradually accepted her fate and began to help out with routine hospital work. However, she was forced to eat in solitude and was allowed few visitors. Mary Mallon died in 1938 at the age of 70 from the effects of a stroke. She was buried without fanfare in a local cemetery.

last 4 to 7 days. Dehydration may occur in some patients, necessitating fluid replacement.

Shigellosis. Members of the genus *Shigella* are gram-negative rods closely related to *E. coli.*

Shigellosis in the United States is caused primarily by *Shigella sonnei,* which continues to cause outbreaks in daycare centers and accounts for the majority of shigellosis cases (less than 15,000 in 2010). Another species, *S. dysenteriae,* causes deadly epidemic dysentery in the developing world where there are 65 million illnesses and 1 million deaths annually.

Humans are the primary reservoir of *Shigella,* which is consumed in contaminated water as well as ingested in many contaminated foods, especially eggs, vegetables, shellfish, and dairy products contaminated through handling.

An infectious dose requires as few as 100 cells, which penetrate the mucosa of the colon and, after 2 to 3 days, produce sufficient exotoxins (**Shiga toxin**) to trigger gastroenteritis. Infection of the large intestine results in fever, abdominal pain, and bloody mucoid stools that sometimes produce a fatal "bacterial dysentery."

Most cases of shigellosis subside within a week and usually produce few complications. However, patients who lose excessive fluids must be given salt tablets, oral solutions, or intravenous injections of salt solutions for rehydration. Careful hygiene is most important although antibiotics are sometimes effective in reducing the duration of illness and the number of bacilli shed, but many strains of *Shigella* are becoming more resistant to antibiotics. Recoverers generally

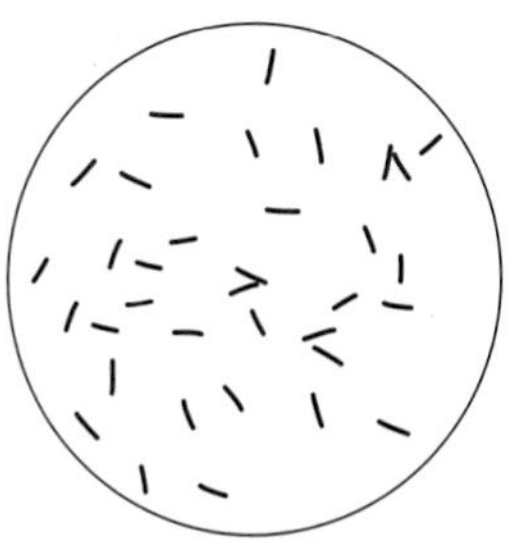

Shigella sonnei

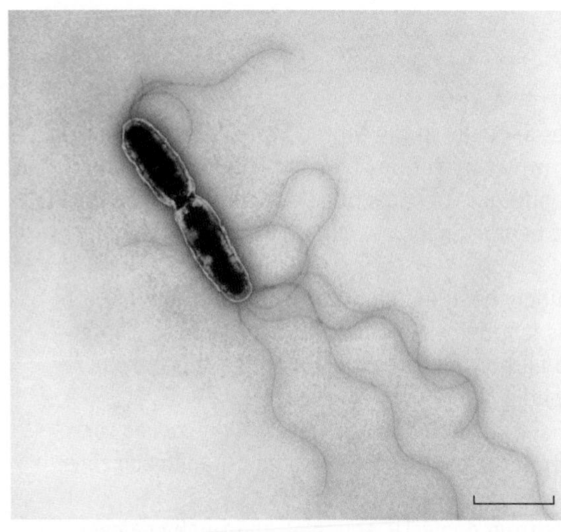

(A)

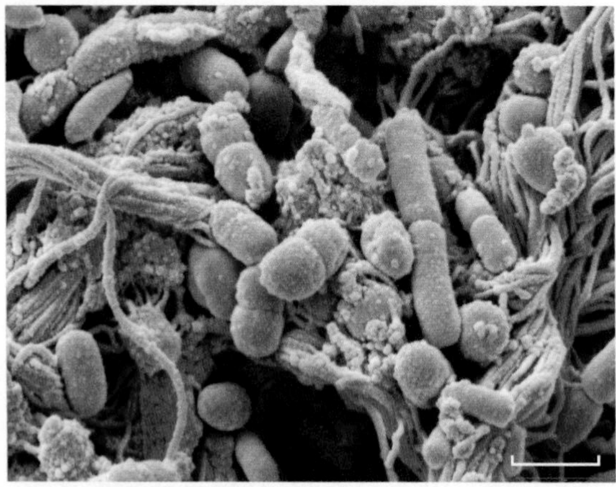

(B)

FIGURE 11.18 *Salmonella enterica.* Two views of *S. enterica.* (**A**) A light micrograph photo of *S. enterica* serotype Typhi cells. Note the long length of the flagella relative to the cell. (Bar = 3 µm.) © Kwangshin Kim/Photo Researchers, Inc. (**B**) A false-color scanning electron micrograph of *S. enterica* serotype Typhimurium cells (blue) observed on the collagen fibers of muscle tissue from an infected chicken. (Bar = 3 µm.) © Scimat/Photo Researchers, Inc. »» How might *Salmonella* cells attach to the collagen fibers?

are carriers for a month or more and continue to shed the bacilli in their feces. Vaccines are not available.

Hemorrhagic Colitis. Besides the enterotoxigenic (ETEC) and enteropathogenic (EPEC) strains of *E. coli* described earlier that were responsible for an inflammatory gastroenteritis, there also is **enterohemorrhagic *E. coli*** (**EHEC**). These strains cause disease by producing a toxin called Shiga toxin and the strains are called "**Shiga toxin-producing" *E. coli***, or **STEC**. The most commonly identified and the most dangerous STEC in North America is *E. coli* O157:H7. Epidemiologists at the CDC have reported that in 2010 there were more than 5,400 cases of STEC, of which most were due to O157:H7. A particularly noteworthy outbreak occurred in 2006 when *E. coli* O157:H7 contaminated fresh, bagged spinach and sickened almost 200 Americans across the nation (FIGURE 11.19). One death was reported.

The reservoir for STEC is cattle and the organism is often transmitted from contaminated and undercooked ground beef. However, many other sources are appearing, including unpasteurized milk and juice, sprouts, lettuce, and salami. In these instances, the contaminated product had come from contact with cattle. The prevailing wisdom is that *E. coli* O157:H7 normally exists in the intestines of healthy cattle but causes no disease in these animals. Slaughtering brings *E. coli* to beef products, and excretion to the soil accounts for transfer to plants and fruits. The source of the *E. coli* contamination in the bagged spinach outbreak was traced to cattle feces from a cattle ranch within 2 kilometers of the produce fields. TEXTBOOK CASE 11 presents another example. Waterborne transmission can also occur by swimming in contaminated water parks and swimming pools, or drinking water inadequately chlorinated.

The STEC strain is particularly pathogenic because it is acid-tolerant and some 100 bacilli can establish an infection if they make it to the colon. There, the pathogen produces large amounts of the Shiga toxin that blocks protein synthesis. After

FIGURE 11.19 **Fresh Bagged Spinach.** The 2006 outbreak of *E. coli* O157:H7 contaminated bagged spinach and forced the recall of all bagged products. © Jeff Chiu/AP Photos. »» Propose a mechanism by which the spinach became contaminated with the *E. coli* strain.

Textbook CASE 11

Outbreak of *E. coli* 0157:H7 Infection

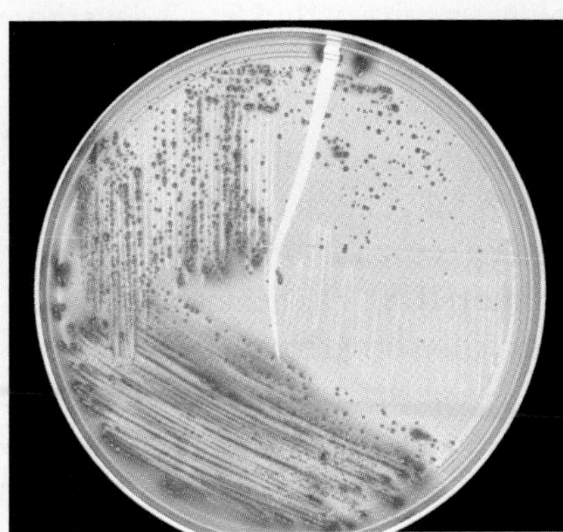

E. coli 0157:H7 growing on a culture plate. Courtesy of CDC.

1. On October 6, 1996, family A from a large city in Connecticut decided to take a drive in the country. Along the way they stopped at a general store for a bite of lunch. The father and two daughters had apple cider; the mother had a soda.
2. Three days later, the father and children began to experience serious abdominal pains and vomiting. Moreover, there was blood in their stools. The mother had no symptoms.
3. One of the daughters of family A became worse and had to be admitted to the hospital. The presence of bloody diarrhea was noted by the doctor.
4. On October 11, the Connecticut Department of Public Health (DPH) was notified of the three illnesses plus five more with disease onset during the same period. A case definition was defined and a stool sample from the daughter was sent to the clinical lab for identification.
5. Meanwhile, the kidneys of family A's daughter were suffering and the doctor advised dialysis to assist the kidney function.
6. The laboratory results identified and confirmed *E. coli* 0157:H7 as the cause of infection (see figure).
7. Health officials were notified, and they began a telephone survey to find out if anyone else was similarly infected. Over two dozen cases were found. All were asked about food consumption during the 7 days preceding the illness.
8. Based on the interviews, increased risk of illness was associated with drinking fresh apple cider from cider plant A.
9. When inspectors visited the cider plant, they were told the apples were taken from a pasture where cattle, sheep, and wild deer grazed. The apples were picked directly from the trees. What most interested investigators was hearing many apples also were picked up from the ground.
10. Appropriate control measures were instituted immediately to prevent further cases.

Questions

(Answers can be found on the Student Companion Website in **Appendix D.**)

A. What would be the case definition defined by the DPH?

B. Hearing that the cause of infection was *E. coli* 0157:H7, what types of food might the DPH investigators be most interested in from the phone survey?

C. What is the infection complication exhibited by family A's daughter?

D. Why were the DPH investigators most interested in the "drop" apples collected from the soil surface of the pasture?

E. What control measures were instituted to prevent further outbreak cases?

For additional information, see: www.cdc.gov/mmwr/preview/mmwrhtml/00045558.htm.

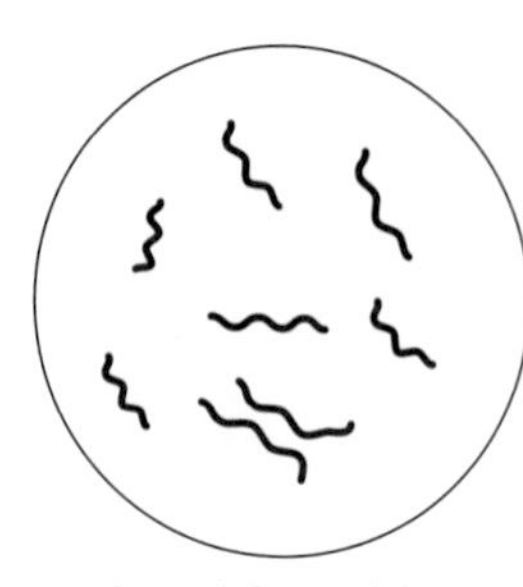

Campylobacter jejuni

■ Glomerulus: A ball-like cluster of interconnected capillaries in the kidney that filter body waste to be excreted as urine.

an incubation period of 1 to 8 days, O157:H7 produces **hemorrhagic colitis**, which is characterized by diarrhea, abdominal cramps, mild fever, and bloody diarrhea. In uncomplicated cases, the symptoms resolve within 5 to 7 days. Young children and the elderly are more likely to develop this type of gastroenteritis.

About 5% to 10% of patients with STEC infections, especially children, develop a potentially life threatening complication called **hemolytic uremic syndrome** (**HUS**). The hemorrhagic colitis leads to the destruction of red blood cells. Once this process begins, the damaged red blood cells clog the kidney glomerulus and can cause kidney failure. Without treatment, life-threatening kidney failure can occur.

In addition to *E. coli* O157:H7, other STEC strains account for another 113,000 illnesses each year in the United States. These other forms are sometimes called "non-O157 STEC." One is *E. coli* O104:H4 that had not been associated with any major disease outbreak until 2011. On May 22, the German government reported a significant number of patients with HUS. In fact, it became one of the world's most severe outbreaks of *E. coli*. By the end of the outbreak in August 2011, more than 3,800 people were sickened with some 850 exhibiting HUS that ultimately killed 54 people in 15 countries. Another 100 are expected to need either a kidney transplant or lifelong dialysis. The source of the outbreak was traced back to eating fenugreek sprouts contaminated with O104:H4.

The O104:H4 is a newly evolving enteroaggregative *E. coli* (EAEC) strain. It seems to be especially sticky to the intestinal wall, may produce an excessive amount of toxin, and is resistant to several classes of antibiotics. Scientists believe these properties were acquired through horizontal gene transfer.

Treatment regimens for HUS require prompt treatment that may include fluid replacement, red blood cell transfusions, and kidney dialysis. Prompt treatment usually leads to a full recovery for most patients. Antibiotics are not usually used as they can actually increase the risk of developing HUS. Research continues on a vaccine for all *E. coli* diarrheas.

Campylobacteriosis. Since the early 1970s, **campylobacteriosis** has emerged from an obscure disease in animals to being one of the most commonly reported bacterial causes of invasive gastroenteritis in the United States. The illness affects over 2.5 million persons and causes about 100 deaths each year.

The pathogen, *Campylobacter jejuni*, is a microaerophilic, curved (*campylo* = "bending"), gram-negative rod (FIGURE 11.20). Reservoirs for the organism are the intestinal tracts of many warm-blooded animals, including dairy cattle, chickens, and turkeys. In fact, chickens raised commercially are colonized with *C. jejuni* by the fourth week of life. Unpasteurized dairy products, including raw milk, also can be a source of infection.

C. jejuni primarily is transmitted via the fecal-oral route through contact or exposure to contaminated foods or water. During an incubation period of 2 to 7 days, the bacterial cells colonize the small or large intestine. Invasion of the mucosa leads to inflammation and occasional mild ulceration. However, the signs and symptoms of campylobacteriosis are not unique as they range from mild diarrhea due to production of an enterotoxin to severe gastroenteritis with bloody diarrhea due to production of a cytotoxin. Most patients recover in less than a week without treatment, but some have high fevers and bloody stools for prolonged periods. Erythromycin therapy hastens recovery.

Some people may develop a rare immunoreactive sequela several weeks after the diarrheal illness. This condition, called **Guillain-Barré syndrome (GBS)**, results from an autoimmune reaction where antibodies attack the body's own nerves. The resulting peripheral nerve damage can

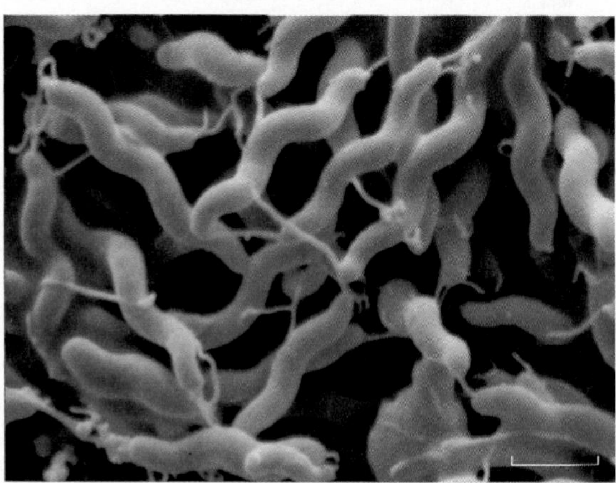

FIGURE 11.20 ***Campylobacter jejuni.*** A false-color scanning electron micrograph of *C. jejuni* cells. (Bar = 0.5 μm.) © Medical-on-Line/Alamy. »» What shape are these cells?

cause paralysis lasting several weeks and usually requires intensive care. Approximately 1 in every 1,000 reported cases of campylobacteriosis leads to GBS and up to 40% of GBS cases in the United States might be caused by campylobacteriosis.

Raw *Campylobacter*-contaminated poultry often is the source of infection. A common route of infection is to cut chicken on a cutting board and then use the unwashed cutting board to prepare other raw foods or vegetables. Therefore, hand washing and washing of cutting boards and utensils after contact with raw poultry are important prevention measures.

Yersiniosis. Another emerging foodborne illness is **yersiniosis**, caused by *Yersinia enterocolitica*. This bacterial species is motile at room temperature and is a member of a family of gram-negative rods that includes *Escherichia*, *Salmonella*, and *Shigella*. *Y. enterocolitica* is widely distributed in animals, which can become asymptomatic carriers of the disease. Only a few strains of *Y. enterocolitica* cause illness in humans. Infections (primarily in children) occur by consuming food that came in contact with infected domestic animals, raw or undercooked pork products, or by ingesting contaminated water or milk.

The *Y. enterocolitica* cells cause an invasive gastroenteritis through tissue destruction in the ileum, followed by multiplication in Peyer patches (FIGURE 11.21). After a 1- to 2-day incubation, affected individuals experience fever, diarrhea, and abdominal pain. The symptoms last 1 to 3 weeks and, unless the illness becomes systemic, antibiotic therapy is unnecessary.

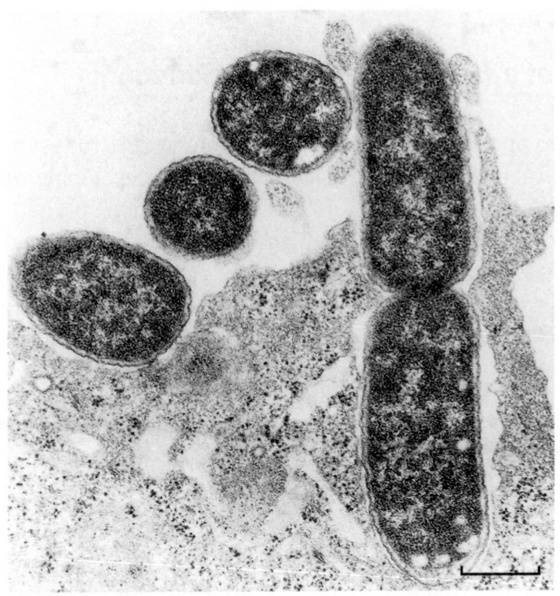

FIGURE 11.21 ***Yersinia enterocolitica*.** Transmission electron micrographs of invasive *Y. enterocolitica*. A number of bacterial cells attached to the plasma membrane of a host cell. One bacterial cell appears to be undergoing cell division and entering the cell at the same time. (Bar = 1 μm.) Reprinted with permission from the American Society for Microbiology (ASM News, January 2002, p. 20–24.) Photo courtesy of Doctor Virginia L. Miller, University of North Carolina. »» What types of foods or beverages might be contaminated by *Y. enterocolitica*?

Gastric Ulcer Disease Can Be Spread Person to Person

Approximately 25 million Americans suffer from peptic ulcers during their lifetime. For decades, scientists believed such ulcers resulted from "excess stomach acid" due to factors such as nervous stress, smoking, alcohol consumption, diet, and physiological stress. However, the work of two Australian gastroenterologists, Barry Marshall and J. Robin Warren, made it clear that the bacterial species *Helicobacter pylori* is primarily involved (MicroInquiry 11).

This microaerophilic, gram-negative curved rod infects half the world's population, yet only 2% are afflicted with **gastric ulcer disease**. One percent also develop gastric cancer, which makes *H. pylori* the only known bacterial carcinogen.

■ Carcinogen: A substance or chemical that can lead to tumor formation.

It is uncertain how *H. pylori* is transmitted. Most likely, it is spread person to person through the fecal-oral route. How *H. pylori* manages to survive in the intense acidity of the stomach is interesting. When *H. pylori* penetrates the stomach mucous layer, it attaches to the stomach wall. There it secretes the enzyme urease that digests urea in the area and produces ammonia as an end product (FIGURE 11.22). The ammonia neutralizes acid in the vicinity of the infection. The ammonia and an *H. pylori* cytotoxin cause destruction of the mucous-secreting cells, exposing the underlying connective tissue to the stomach acid. Over several years, a sore up to 12 cm in diameter appears. The pain is severe and is not relieved by food or an antacid.

Infection with *H. pylori* is also the major cause of gastric cancer, which represents the fourth most common cancer and the second leading cause of cancer deaths in the world (about 1 million deaths each year). In fact, the presence of *H. pylori* confers

MICROINQUIRY 11

The Bad and Good of a *Helicobacter pylori* Infection

Several idioms and phrases apply to the stomach. You probably have heard someone say "from the pit of my stomach"; or "I can't stomach this"; or, as Napoleon said, "An army marches on its stomach." In fact, we are beginning to appreciate the fact that the stomach is quite a center of activity, not only as part of our digestive process, but also behaviorally through one particular microbe—*Helicobacter pylori*.

From this chapter, you know that *H. pylori* can be a pathogen. About 40% of humans acquire the species during childhood and it then proceeds to establish a persistent presence in the stomach. Unfortunately, for some of us the infection can lead to gastric ulcers and, for a few, to an increased risk of gastric cancer. So, it would seem we would all do better without the microbe. In fact, today antibiotics are used with great success to cure people with gastric ulcers or as a precaution to prevent the development of the ulcers. Therefore, it would appear that *H. pylori* is something we can do without.

Within the last two generations, the incidence of *H. pylori* in developed nations has dropped from 80% to just a few percent, while in the developing world about 50% of the population still harbors the pathogen. This sounds like good news, especially if you live in the developed nations. However, evidence suggests that people without the pathogen are at greater risk of acid (gastroesophageal) reflux disease and esophageal cancer. So, is colonization by *H. pylori* a good or bad situation? Don't decide yet!

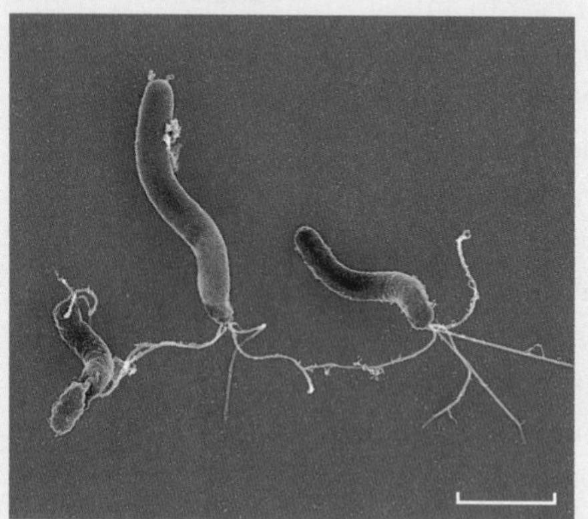

Helicobacter pylori. (Bar = 1 μm.) © SPL/Photo Researchers, Inc.

Several recent research papers report that *H. pylori* has a role beyond the stomach. Studies have suggested that people with Parkinson's disease (a brain disorder that leads to tremors and difficulty with walking and coordination) are more likely to have ulcers and to be infected with *H. pylori* than healthy people. Then in 2011, researchers at Louisiana State University Health Sciences Center reported that middle-aged mice, but not young mice, infected with *H. pylori* developed over several months abnormal, Parkinson-like movements. Further, the middle-aged mice made less dopamine (a neurotransmitter produced in the brain), suggesting the dopamine-making cells were dying just as they do in humans with Parkinson's disease. And more intriguing, the living bacterial cells are not needed to cause the effect in mice. Feeding mice dead *H. pylori* cells produced the same result, suggesting some chemical or bacterial component is the trigger. So, is colonization by *H. pylori* a good or bad situation? Don't decide yet!

Also in 2011, immunologists and allergy experts in Switzerland and Germany reported that infection with *H. pylori* provides protection from allergy-induced asthma. They showed that neonate and adult mice exposed to an asthma-inducing allergen and challenged later with the same allergen would develop asthma. However, if they were first infected with *H. pylori*, the adult mice and especially the neonate mice had much milder symptoms. So, is colonization by *H. pylori* a good or bad situation? Don't decide yet!

In 2012, researchers at New York University School of Medicine announced that the presence of *H. pylori* may be linked to adult type 2 diabetes (the most common form of diabetes where the body does not respond properly to insulin). The scientists looked at the levels of glycosylated hemoglobin, which is a known marker for blood glucose levels and diabetes, in participants of two national health surveys. Those infected with *H. pylori*, especially obese individuals with a high body mass index (BMI), consistently had higher levels of the glycosylated hemoglobin than those with a lower BMI. The researchers concluded that *H. pylori* eradication may be beneficial for older, obese individuals as a precaution to developing type 2 diabetes.

All these studies need to be confirmed and studied in more detail. Perhaps the more eye-opening take-home message is that it often is not a simple matter of labeling a microbe as friend or foe.

Discussion Point

So, is colonization by H. pylori *a good or bad situation? If you had an ulcer, would you submit to antibiotic treatment to eliminate the infection knowing that* H. pylori *eradication may have other consequences?*

Answers (and comments) can be found on the Student Companion Website in **Appendix D**.

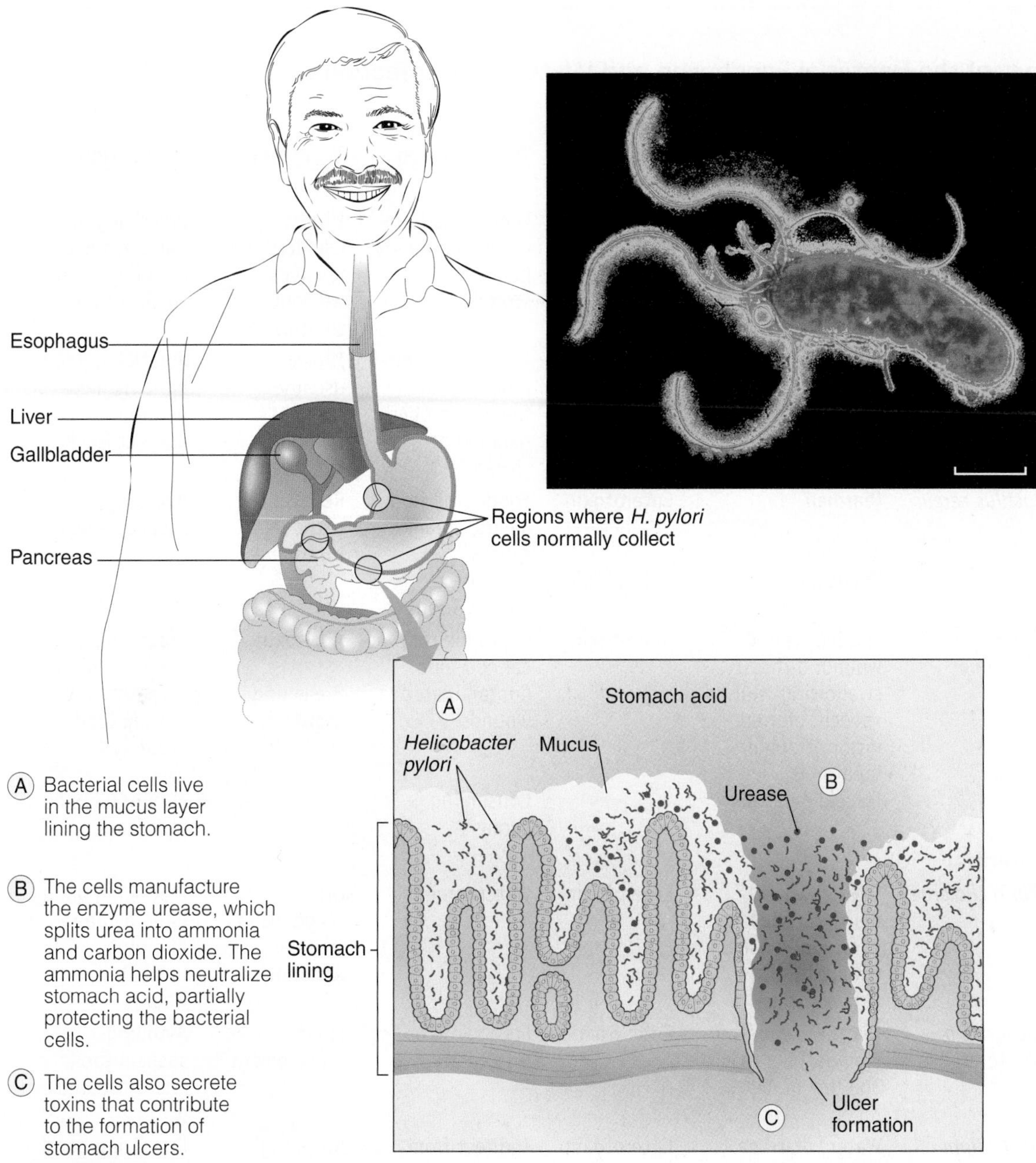

FIGURE 11.22 The Progression of Gastric Ulcers. The majority of peptic ulcers are caused by *Helicobacter pylori* [inset]. (Bar = 1 µm.) This figure illustrates how they cause an ulcer. © P. Hawtin/Photo Researchers, Inc.
»» How might peptic ulcer disease lead to stomach cancer?

an approximately six-fold increased risk of developing gastric cancer. Exactly how the pathogen contributes to cancer development remains to be identified. Some scientists suggest that infections of *H. pylori* may cause DNA damage in the stomach lining that cannot always be repaired correctly. This would then lead to mutations or cell death that damages the stomach epithelium—and possibly to cancer.

Thanks to Marshall and Warren, doctors have revolutionized the treatment of ulcers by prescribing antibiotics such as amoxicillin, tetracycline, or clarithromycin (Biaxin®), along with omeprazole (Prilosec®) for acid suppression. They have achieved cure rates of up to 94%; relapses are uncommon.

The bacterial GI tract diseases discussed in this chapter are summarized in TABLE 11.3.

TABLE

11.3 A Summary of the Bacterial Foodborne and Waterborne Infections

Disease	Causative Agent	Signs and Symptoms	Toxin Involved	Transmission	Treatment	Prevention
Noninflammatory Gastroenteritis						
Staphylococcal food poisoning	*Staphylococcus aureus*	Abdominal cramps, nausea, vomiting, and diarrhea	Enterotoxin	Foodborne from foods improperly handled or stored	Illness usually resolves without treatment	Practicing good hand hygiene Avoiding suspect foods
Clostridial food poisoning	*Clostridium perfringens*	Abdominal cramping, watery diarrhea	Enterotoxin	Foodborne from protein-rich foods improperly handled or stored	Illness usually resolves without treatment	Practicing good hand hygiene Avoiding suspect foods
Bacillus cereus food poisoning	*Bacillus cereus*	Diarrhea Vomiting	Enterotoxin Enterotoxin	Foodborne from meats, cream sauces Foodborne starchy foods	Recovery without treatment	Avoiding suspect foods
Botulism	*Clostridium botulinum*	Foodborne and wound: difficulty swallowing, slurred speech, blurred vision, trouble breathing Flaccid paralysis	Neurotoxin	Contaminated canned food Contaminated wounds Consuming spores	Antitoxin Breathing assistance	Practicing good home canning Preparing and storing food properly Avoiding honey in infants
Inflammatory Gastroenteritis						
Cholera	*Vibrio cholerae*	Severe, watery diarrhea, nausea, vomiting, muscle cramps, and dehydration	Enterotoxin	Waterborne	Oral rehydration therapy Antibiotics	Practicing good hand hygiene Avoiding untreated water
ETEC, EPEC	*Escherichia coli* serotypes	Diarrhea, vomiting, cramps, nausea, and low-grade fever	Enterotoxins	Foodborne or waterborne	Fluid replacement	Avoiding suspect foods and untreated water
Clostridium difficile infection	*Clostridium difficile*	Watery diarrhea, Pseudomembranous colitis	Enterotoxin Cytotoxin	Indirect from contaminated hands or materials	Stopping antibiotic therapy Anticlostridial antibiotic therapy	Practicing good hand hygiene Keeping bathrooms and kitchens disinfected
Listeriosis	*Listeria monocytogenes*	Headache, stiff neck, confusion, loss of balance, convulsions	Not established	Food contaminated with fecal matter Contaminated animal foods	Ampicillin	Practicing good hand washing Washing and preserving food properly
Vibriosis	*Vibrio parahaemolyticus*	Acute abdominal pain, vomiting, watery diarrhea	Enterotoxin	Foodborne in contaminated seafood (oysters and raw shellfish)	None Antibiotic therapy for severe or prolonged illnesses	Cooking seafood thoroughly, especially oysters

TABLE

11.3 A Summary of the Bacterial Foodborne and Waterborne Infections *(Continued)*

Disease	Causative Agent	Signs and Symptoms	Toxin Involved	Transmission	Treatment	Prevention
	Vibrio vulnificus	Fever, nausea, severe abdominal cramps	Cytotoxin	See above	Immediate antibiotic therapy	Avoiding raw oysters and clams
Intestinal anthrax	*Bacillus anthracis*	Nausea, loss of appetite, fever, vomiting, abdominal pain, bloody vomiting, and severe diarrhea	Enterotoxins	Foodborne	Early treatment with antibiotics	Avoiding meat that is not properly slaughtered and cooked
Invasive Gastroenteritis						
Typhoid fever	*Salmonella* Typhi	Bloody stools, abdominal pain, fever, lethargy, delirium	Endotoxin	Foodborne from person shedding *S.* Typhi Foodborne and waterborne from contaminated sewage	Antibiotics	Avoiding risky foods and drinks Getting vaccinated
Salmonellosis	*Salmonella* serotypes	Fever, diarrhea, vomiting, and abdominal cramps	Not established	Foodborne in a broad variety of foods	Fluid replacement Antibiotic therapy	Practicing good hand hygiene and food preparation
Shigellosis	*Shigella sonnei*	Diarrhea, dysentery	Shiga enterotoxin	Foodborne and waterborne	Antibiotics Fluid and salt replacement	Practicing good hand hygiene
EHEC, STEC	*Escherichia coli* O157:H7	Severe, bloody diarrhea	Enterotoxin Shiga-like enterotoxin	Foodborne through ingestion of undercooked meat or contaminated fruits and vegetables	For HUS: red blood cell and platelet transfusions Kidney dialysis	Practicing good hand hygiene Cooking foods thoroughly Washing fruits and vegetables; avoiding unpasteurized milk
Campylobacteriosis	*Campylobacter jejuni*	Diarrhea, fever	Enterotoxin Cytotoxin	Foodborne from contaminated foods or water	None Antibiotic therapy for severe or prolonged illnesses	Practicing good hand hygiene and food preparation
Yersiniosis	*Yersinia enterocolitica*	Fever, diarrhea, and abdominal pain	Enterotoxin	Foodborne from contaminated foods (pork and pork products)	None Antibiotic therapy for severe or prolonged illnesses	Avoiding raw or undercooked pork or pork products Practicing good hand hygiene
Gastric ulcer disease	*Helicobacter pylori*	Aching or burning pain in abdomen, nausea, vomiting, bloating, bloody vomit or stools	Cytotoxin	May be transmitted person to person through direct or indirect saliva contact	Antibiotics and acid suppression medications	Practicing good hand hygiene Not sharing utensils or glasses

CONCEPT AND REASONING CHECKS 5

a. How is the mechanism of action of cholera toxin similar to that of other enterotoxins mentioned in this chapter?
b. How is the cause of diarrhea different between ETEC and EPEC serotypes?
c. Why would *C. difficile* be considered an opportunistic pathogen?
d. How does *L. monocytogenes* make its way to the brain and meninges?
e. How might seafood become contaminated with either *Vibrio parahaemolyticus* or *V. vulnificus*?
f. How might one contract intestinal anthrax?
g. Why is typhoid fever often called enteric fever?
h. What is the best way to protect oneself from salmonellosis?
i. What control measure (hygiene practice) would most likely decrease transmission rates of shigellosis?
j. Why are children susceptible to EHEC?
k. Explain how *C. jejuni* could be responsible for traveler's diarrhea.
l. Why is *Y. enterocolitica* considered an invasive pathogen?
m. Summarize how an *H. pylori* infection causes stomach ulceration.

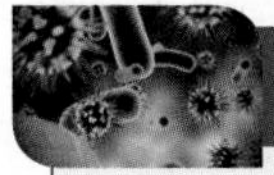

Chapter Challenge F

Several bacterial species can cause some form of invasive gastroenteritis that can infect the submucosa of the intestine. One of the most common is caused by serotypes of *Salmonella enterica*. Contaminated products such as eggs, peanut butter, raw turkey burgers, ice cream, and chicken salad are among the food products that have caused outbreaks in the recent past in the United States.

QUESTION F:

Propose ways to prevent: (1) egg contamination on the farm; (2) peanut butter contamination at the manufacturing plant; (3) raw turkey contamination at the manufacturing plant; (4) ice cream contamination in delivery trucks previously used for hauling other raw products; and (5) chicken salad contamination at a local restaurant?

Answers can be found on the Student Companion Website in **Appendix D.**

■ **In conclusion**, we learned in this chapter that a variety of bacterial species can invade the GI tract and produce some form of gastroenteritis, the most common symptom being diarrhea. How about another symptom being obesity? Researchers are discovering that our gut microbiota may be an important player in human obesity. There is increasing evidence that the types of bacterial species and the relative abundance between these microbes can influence one's body weight, either toward obesity or thinness. Exactly how this occurs is being studied, but resident bacterial species can influence how well energy is absorbed from foods and thus influence our appetite. Just as fascinating is the suggestion that foodborne illnesses could produce chronic after effects. Minor cases of fever, vomiting, or diarrhea may lead to arthritis, urinary tract complications, or vision problems. In 2000, the drinking water in Walkerton, Ontario, Canada became contaminated with *E. coli* O157:H7 due to manure being washed into the town's aquifer. About 2,300 people in the town developed diarrhea and fever. A study published in 2010 found that those individuals who became ill during the outbreak had greater than a 30% likelihood of developing high blood pressure and had statistically significant increases in heart problems and kidney impairment compared to those who had no symptoms. The full long-term consequences of foodborne infections remain to be clearly understood.

SUMMARY OF KEY CONCEPTS

11.1 The Digestive System Has an Extensive Indigenous Microbiota

- The **digestive system** includes the **gastrointestinal (GI) tract** and the **accessory digestive organs** (teeth, tongue, salivary glands, liver, gallbladder, and pancreas). (Fig. 11.2)
- Digestive system defenses include: salivary mucins, lysozyme, lactoferrin, and defensins; antibodies; stomach acid; mucus on intestinal surfaces; peristalsis; exfoliation of epithelial cells; bile and proteolytic enzymes; **Peyer patches.**
- The human intestinal **microbiome** contains an enormous population of normal microbiota that provides protection through microbial antagonism. (Fig. 11.3)

11.2 Bacterial Diseases of the Oral Cavity Can Affect One's Overall Health

- **Dental caries**
  1. *Streptococcus mutans. S. sobrinus*
- **Gingivitis** and **periodontitis**
  2. *Bacteroides* and several other genera

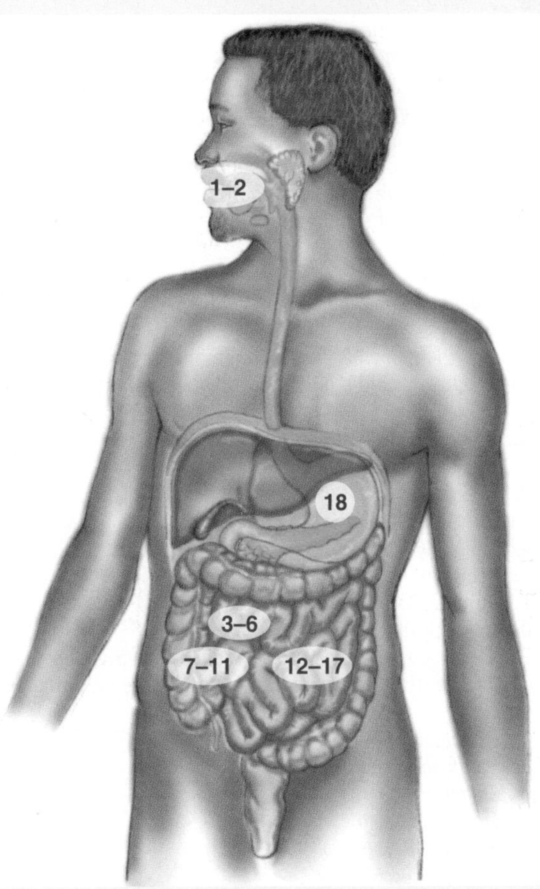

11.3 Bacterial Diseases of the GI Tract Are Usually Spread Through Food and Water

- **Intoxications** represent a form of **noninflammatory gastroenteritis** caused by bacterial toxins, while **inflammatory** and **invasive gastroenteritis** are infections and diseases arising from bacterial growth in the GI tract.
- Contaminated food and water arising from unsanitary procedures often are the source of the intoxication or infection. (Fig. 11.9)

11.4 Some Bacterial Diseases Are the Result of Foodborne Intoxications

- **Staphylococcal food poisoning**
  3. *Staphylococcus aureus*
- **Clostridial food poisoning**
  4. *Clostridium perfringens*
- ***Bacillus cereus* food poisoning**
  5. *Bacillus cereus*
- **Botulism**
  6. *Clostridium botulinum*

11.5 GI Infections Can Be Caused By Several Bacterial Pathogens

- **Cholera**
  7. *Vibrio cholerae*
- **ETEC, EPEC**
  8. *Escherichia coli*
- **Pseudomembranous colitis**
  9. *Clostridium difficile*
- **Listeriosis**
  10. *Listeria monocytogenes*
- **Other foodborne infections**
  11. *Vibrio parachaemolyticus, V. vulnificus, Bacillus anthracis*

Invasive

- **Typhoid fever**
  12. *Salmonella* Typhi
- **Salmonellosis**
  13. *Salmonella* serotypes
- **Shigellosis**
  14. *Shigella sonnei*
- **EHEC**
  15. *Escherichia coli* 0157:H7
- **Campylobacteriosis**
  16. *Campylobacter jejuni*
- **Yersiniosis**
  17. *Yersinia enterocolitica*
- **Gastric ulcer disease**
  18. *Helicobacter pylori*

CHAPTER SELF-TEST

For **STEPS A–D**, answers to even-numbered questions and problems can be found in **Appendix C** on the Student Companion Website at **http://microbiology.jbpub.com/10e**. In addition, the site features eLearning, an online review area that provides quizzes and other tools to help you study for your class. You can also follow useful links for in-depth information, read more MicroFocus stories, or just find out the latest microbiology news.

STEP A: REVIEW OF FACTS AND TERMS

Read each question carefully, then select the ***one*** answer that best fits the question or statement.

Multiple Choice

1. Which one of the following is NOT a digestive organ of the gastrointestinal (GI) tract?
A. Large intestine
B. Oral cavity
C. Liver
D. Small intestine

2. What part of the GI tract contains the largest population of microorganisms (microbiota)?
A. Colon
B. Jejunum
C. Duodenum
D. Stomach

3. Which one of the following does NOT apply to dental plaque?
A. It is an example of a biofilm.
B. It is most noticeable on the molars.
C. It can lead to gingivitis and periodontal disease.
D. It is dominated by aerobic bacterial species.

4. What type of periodontal disease occurs when plaque bacteria build up between teeth and gums?
A. Canker sore
B. Dental caries
C. Gingivitis
D. Periodontitis

5. Gastroenteritis can result in _______.
A. an intestinal inflammation
B. an infection
C. an intoxication
D. All the above (**A–C**) are correct.

6. Foodborne microbes can be found in _______.
A. cattle carcasses
B. fresh fruits and vegetables
C. healthy animals
D. All the above (**A–C**) are correct.

7. Which one of the following bacterial species is NOT a cause of food poisoning (noninflammatory gastroenteritis)?
A. *Escherichia coli*
B. *Bacillus cereus*
C. *Staphylococcus aureus*
D. *Clostridium perfringens*

8. One of the most excessive diarrheas of the GI tract is associated with which of the following poisonings or diseases?
A. Staphylococcal food poisoning
B. Typhoid fever
C. Cholera
D. Campylobacteriosis

9. *Escherichia coli* is a common gram _____ that can be a cause of _____.
A. positive rod; hemorrhagic colitis
B. negative rod; traveler's diarrhea
C. positive coccus; typhoid fever
D. negative rod; cholera

10. *Clostridium difficile* is _______.
A. the cause of pseudomembranous colitis
B. traveler's diarrhea
C. meningoencephalitis
D. undulant fever

11. This gram-positive rod is a psychrotroph that can multiply within macrophages.
A. *Bacillus cereus*
B. *Listeria monocytogenes*
C. *Clostridium perfringens*
D. *Escherichia coli*

12. This species is the most virulent of the vibrios.
A. *V. vulnificus*
B. *V. cholerae*
C. *V. enterocolitica*
D. *V. parahaemolyticus*

13. Typhoid fever is characterized by _______.
A. the production of an exotoxin
B. hemolytic uremic syndrome
C. rose spots on chest and abdomen
D. All the above (**A–C**) are correct.

14. The symptoms of salmonellosis usually last about _______.
A. 24 hours
B. 48 hours
C. 5 days
D. 14 days

15. What is the name of the syndrome of fever, abdominal cramps, and bloody mucoid stools caused by *Shigella* species?
A. Bacterial dysentery
B. Typhoid fever
C. Pseudomembranous colitis
D. HUS

16. Enterohemorrhagic *E. coli* (EHEC) can cause _______.
A. undulant fever
B. hemolytic uremic syndrome
C. Guillain-Barré syndrome
D. stomach ulcers

17. The most commonly reported cause of invasive bacterial gastroenteritis is associated with which of the following bacterial genera?
A. *Campylobacter*
B. *Staphylococcus*
C. *Shigella*
D. *Clostridium*

18. Yersiniosis is caused by _______.
A. *Yersinia pestis*
B. *Yersinia pseudotuberculosis*
C. *Yersinia entercolitica*
D. All of the above (**A–C**) cause the illness.

19. Gastric ulcer disease is caused by _______.
A. *Helicobacter pylori*
B. *Yersinia entercolitica*
C. *Escherichia coli*
D. *Salmonella* Typhi

Term Selection

For each choice, circle the word or term that best completes each of the following statements.

20. To treat patients who have botulism, large doses of (antitoxin, antibiotic) must be administered.

21. Disease associated with *Shigella* species can produce (diarrhea, dysentery), which is identified by the presence of cramps and bloody stools.

22. (Neurotoxins, Cytotoxins), such as those found with botulism, can cause flaccid paralysis.

23. Many foodborne and waterborne bacterial diseases have ill-defined (symptoms, syndromes), making pathogen identification difficult.

24. Only a small percentage of those who recover from typhoid fever remain (carriers, free) of the bacterial cells.

25. Dental caries is caused by (*Streptococcus mutans, Staphylococcus aureus*), a gram-positive sphere.

26. The diarrheal disease (traveler's diarrhea, cholera) is caused by a motile, gram-negative, curved rod.

27. The genus *Salmonella* is made up of many (genotypes, serotypes); that is, there are many closely related groups that are identified by a specific antibody reaction.

28. An infection of the blood, which can occur with listeriosis, is referred to as (septicemia, gastroenteritis.)

29. If one has ingested botulism exotoxins, the individual is considered to be (infected, intoxicated).

STEP B: CONCEPT REVIEW

30. Identify the organs composing the (a) **gastrointestinal (GI) tract** and (b) the **accessory digestive organs**, and summarize the digestive system defenses against pathogen colonization and infection. (**Key Concept 1**)

31. Estimate the population size and diversity of the human intestinal **microbiome.** (**Key Concept 1**)

32. Assess the seriousness of **dental caries** occurring in the **oral cavity.** (**Key Concept 2**)

33. Describe the role played by oral bacterial species in causing **periodontal disease.** (**Key Concept 2**)

34. Differentiate between a bacterial **intoxication** and a bacterial **infection.** (**Key Concept 3**)

35. Identify ways that foods and water become contaminated with bacterial pathogens. (**Key Concept 3**)

36. Distinguish between the four bacterial species causing intoxications. (**Key Concept 4**)

37. Summarize the clinical significance of *Vibrio cholerae* and assess the impact of **cholera** pandemics. (**Key Concept 5**)

38. Compare and contrast the **ETEC** and **EPEC** forms of *E. coli* infections. (**Key Concept 5**)

39. Explain why *Clostridium difficile* is of concern today. (**Key Concept 5**)

40. Explain why a *Listeria* infection is most dangerous to newborns, the elderly, and pregnant women. (**Key Concept 5**)

41. Summarize the clinical significance of *Salmonella* Typhi infections. (**Key Concept 5**)

42. Identify what foods are most likely to become contaminated with *Salmonella enterica* serotypes. (**Key Concept 5**)

43. Describe how *Shigella sonnei* is transmitted and how it infects humans. (**Key Concept 5**)

44. Describe the consequences of **EHEC** infection. (**Key Concept 5**)

45. Explain why **campylobacteriosis** has become the most commonly reported form of bacterial gastroenteritis. (**Key Concept 5**)

46. Discuss how *Yersinia enterocolitica* infects the digestive system. (**Key Concept 5**)

47. Diagram the steps involved in **gastric ulcer disease.** (**Key Concept 5**)

STEP C: APPLICATIONS AND PROBLEMS

48. You are doing the supermarket shopping for the upcoming class barbecue. What are some precautions you can take to ensure that the event is remembered for all the right reasons?

49. You read in the newspaper that botulism was diagnosed in 11 patrons of a local restaurant. The disease was subsequently traced to mushrooms bottled and preserved in the restaurant. What special cultivation practice enhances the possibility that mushrooms will be infected with the spores *Clostridium botulinum?*

50. In preparation for a summer barbecue, your roommate cuts up chickens on a wooden carving board. After running the board under water for a few seconds, he uses it to cut up tomatoes, lettuce, peppers, and other salad ingredients. What sort of trouble may occur?

51. The state department of health received reports of illness in 18 workers at a local pork processing plant. All the affected employees worked on the plant's "kill floor." All had gram-negative rods in their blood. Their symptoms included fever, chills, fatigue, sweats, and weight loss. Which disease was pinpointed in the workers?

52. A classmate plans to travel to a tropical country for spring break. To prevent traveler's diarrhea, she was told to take 2 ounces or 2 tablets of Pepto-Bismol four times a day for 3 weeks before travel begins. Short of turning pink, what better measures can you suggest she use to prevent traveler's diarrhea?

STEP D: QUESTIONS FOR THOUGHT AND DISCUSSION

53. In 1997, researchers in Boston reported that *Helicobacter pylori* accumulates in the gut of houseflies after the flies feed on food containing the pathogen. What are the implications of this research?

54. Some years ago, the CDC noticed a puzzling trend: Reported cases of salmonellosis seemed to soar in the summer months and then drop radically in September. Can you venture a guess as to why this is so?

55. Most physicians agree that the illness called "stomach flu" is not influenza at all. They say the cramps, diarrhea, and vomiting can be due to a variety of bacterial species. Which organisms in this chapter might be good candidates?

56. A frozen-food manufacturer recalls thousands of packages of jumbo stuffed shells and cheese lasagna after a local outbreak of salmonellosis. Which parts of the pasta products would attract the attention of inspectors as possible sources of salmonellosis? Why?

12

CHAPTER PREVIEW

12.1 Several Soilborne Bacterial Diseases Develop from Endospores

12.2 Some Bacterial Diseases Can Be Transmitted by Arthropods

Investigating the Microbial World 12: The Lungs as Microbial Playgrounds

12.3 Rickettsial and Ehrlichial Diseases Are Arthropodborne

Textbook Case 12: An Outbreak of Tickborne Relapsing Fever

MicroInquiry 12: Dreams of Conquest Stopped by a Microbe

Image courtesy of Dr. Fred Murphy/CDC.

Soilborne and Arthropodborne Bacterial Diseases

Father abandoned child, wife husband, one brother another . . . And I, Agnolo di Tura . . . buried my five children with my own hands. . . . So many died that all believed that it was the end of the world.

—Agnolo di Tura, describing bubonic plague in his chronicle (*The Plague of Siena*) in 1348

As the above quote attests, bubonic plague, commonly known as "the Black Death," was probably the greatest catastrophe ever to strike Europe. It swept back and forth across the continent for almost a decade, each year increasing in ferocity. By 1348, two-thirds of the European population was stricken and half of the sick had died. Houses were empty, towns were abandoned, and a dreadful solitude hung over the land. The sick died too quickly for the living to bury them, so victims often were buried in "plague pits" (FIGURE 12.1). At one point, the Rhône River was consecrated as a graveyard for plague victims. Contemporary historians wrote that posterity would not believe such things could happen, because those who saw them were themselves appalled. The horror was almost impossible to imagine; to many people, including Agnolo di Tura, "it was the end of the world."

Before the century concluded, the Black Death visited Europe at least five more times in periodic reigns of terror. During one epidemic in Paris, an estimated 800 people died each day; in Siena, Italy, the population dropped from 42,000 to 15,000; and in Florence, almost 75% of the citizenry perished. Flight was the chief recourse for people who could afford it, but ironically, the escaping travelers spread the disease. Those who remained in the cities were locked in their homes until they succumbed or recovered.

The immediate effect of the plague was a general paralysis in Europe. Trade ceased and wars stopped. Bewildered peasants who survived encountered unexpected prosperity because landowners had to pay higher wages to obtain help. Land values declined and class relationships were upset, as the system of feudalism gradually crumbled. However, medical practices became increasingly sophisticated, with new standards of sanitation and a 40-day period of detention (quarantine) imposed on vessels docking at ports.

The graveyard of plague left fertile ground for the renewal of Europe during the Renaissance. To many historians, the Black Death remains a major turning point in Western civilization.

FIGURE 12.1 **The Plague Pit.** This painting shows the unloading of dead bodies during the plague of 1665. These pits were no more than mass graves used to bury the plague victims that had been gathered up from the streets on "dead carts." © Photos.com. »» Why would these plague victims be buried in mass graves rather than by traditional funerals—and at night?

Plague is a disease caused by *Yersinia pestis,* an organism found in rodents and their fleas in many areas around the world. Although there are sporadic outbreaks of plague today around the world, there is another reason to be concerned about the potential horrors of the disease. Many microbiologists and government officials see *Y. pestis* as a possible bioterror agent. Used in an aerosol attack, the pathogen could cause cases of pneumonic plague, an infection of the lungs. One to six days after becoming infected with the bacilli, people would develop fever, weakness, and rapidly developing pneumonia with shortness of breath, chest pain, cough, and sometimes bloody or watery sputum. Nausea, vomiting, and abdominal pain may also occur. Without early treatment, pneumonic plague usually leads to respiratory failure, shock, and rapid death.

Because of the delay between being exposed to *Y. pestis* and becoming sick, people could travel over a large area before becoming contagious and possibly infecting others. Once people have the disease, the bacterial cells can spread to others who have close contact with them. Several types of antibiotics are effective for curing the disease provided they are given within 24 hours of the first symptoms.

A bioweapon carrying *Y. pestis* is possible because the pathogen occurs in nature and could be isolated and grown in an appropriately equipped laboratory, although weaponizing *Y. pestis*—that is, making it easily transmissible through the air—would require advanced knowledge and technology.

Besides plague, other diseases like typhus and relapsing fever are transmitted by arthropods, and both can be interrupted by arthropod control. Neither is a major problem in our society, but a substantial number of cases of other arthropodborne diseases, such as tularemia, Rocky Mountain spotted fever, and Lyme disease are reported each year. We shall study each of these diseases in this chapter.

■ Arthropods: Animals having jointed appendages and segmented bodies (e.g., ticks, lice, fleas, mosquitoes).

First we will examine a number of soilborne diseases where organisms enter the body through a cut, wound, or abrasion, or by inhalation. Among these are anthrax, another feared disease in bioterrorism, and tetanus, a concern to anyone who has stepped on a nail or piece of glass. We also will study other diseases receiving wider recognition as detection methods improve. The soilborne diseases, as well as the arthropodborne diseases, are primarily problems of the blood.

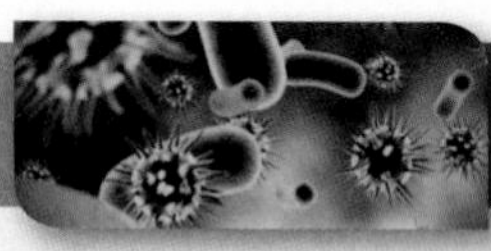

Chapter Challenge

© qcontrol/ShutterStock, Inc.

The soil is home to a diverse group of organisms. In fact, microbial ecologists estimate that up to 25% of all Earth's species inhabit the soil. Among the microorganisms, the vast majority of species is not a threat to human health, but rather is beneficial to the soil and the recycling of elements. However, soils do contain some pathogens capable of causing diseases in humans. The pathogens responsible for causing such diseases can be divided into two groups: those pathogens that are true soil-dwellers and those that are soil-transmitted. As we progress through this chapter, let's see if we can identify to which group each of the soilborne, and even arthropodborne, bacterial organisms belong.

KEY CONCEPT 12.1 Several Soilborne Bacterial Diseases Develop from Endospores

Soilborne bacterial diseases are those whose agents are transferred from the soil to the unsuspecting individual. To remain alive in the soil, the bacterial cells must resist environmental extremes, and often the cells form **endospores**, as the first three diseases illustrate.

Anthrax Is an Enzootic Disease

Anthrax was the first bacterial disease shown by Koch to be due to a germ. Anthrax is primarily an enzootic disease of large, domestic herbivores, such as cattle, sheep, and goats. Animals ingest the spores from the soil during grazing, and soon they are overwhelmed with vegetative bacterial cells as their organs fill with bloody black fluid (*anthrac* = "coal"; the disease name is thus a reference to the blackening of the blood). About 80% of untreated animals die.

■ Enzootic: Refers to a disease endemic to a population of animals.

Anthrax is caused by *Bacillus anthracis*, a spore-forming, aerobic, gram-positive rod (FIGURE 12.2). Endospores germinate rapidly on contact with human tissues to produce vegetative cells. The thick capsule of the cells impedes phagocytosis and the organisms produce three exotoxins that work together to cause disease. Capsule and toxins are coded by genes carried on two plasmids.

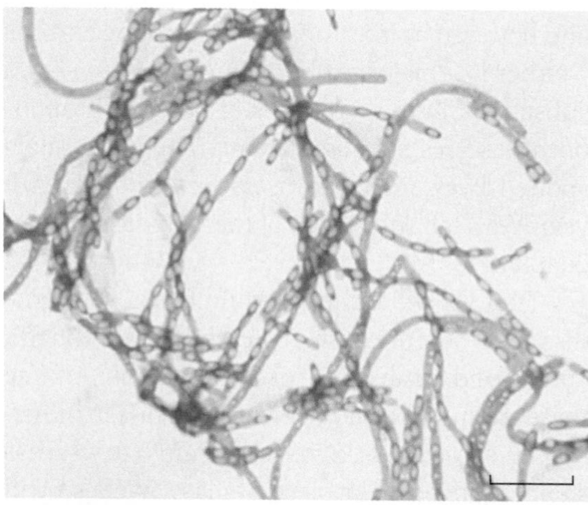

(A)

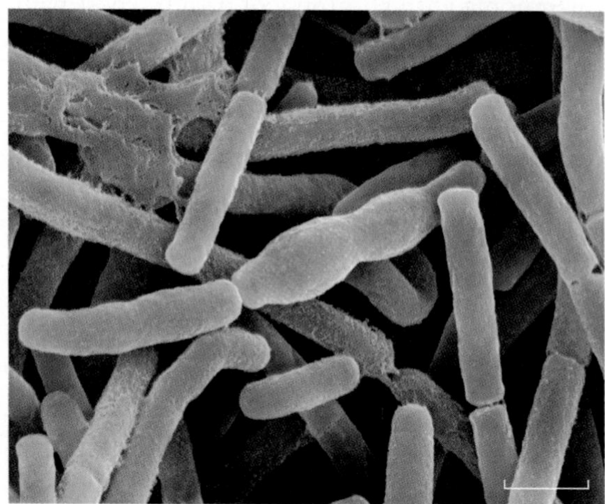

(B)

FIGURE 12.2 *Bacillus anthracis*. *B. anthracis* is the cause of anthrax. (**A**) Spores (white ovals) in vegetative cells can be seen in this photomicrograph. (Bar = 10 µm.) © Michael Abbey/Photo Researchers, Inc. (**B**) A false-color scanning electron micrograph of vegetative cells. It is from such cells that the exotoxins are produced. Pink cells represent cells undergoing endospore formation. (Bar = 2 µm.) © Dennis Kunkel Microscopy, Inc./Phototake/Alamy Images. »» What advantage is provided to the organism by producing endospores?

Humans acquire anthrax from infected animal products, contaminated dust, or directly from the soil. This can happen in one of three ways.

Inhalation Anthrax. Workers who tan hides, shear sheep, or process wool may inhale the spores and contract **inhalational (pulmonary) anthrax** as a form of pneumonia called **woolsorter's disease**. It initially resembles a common cold (fever, chills, cough, chest pain, headache, and malaise). After several days, the symptoms may progress to severe breathing problems and shock. Inhalation anthrax is usually fatal without early treatment.

Intestinal Anthrax. Consumption of contaminated and undercooked meat may lead to **gastrointestinal anthrax**. It is characterized by an acute inflammation of the intestinal tract. Initial signs include nausea, loss of appetite, vomiting, and fever. This is followed by abdominal pain, vomiting of blood, and severe diarrhea. Intestinal anthrax results in death in 25% to 60% of untreated cases.

Cutaneous Anthrax. Skin abrasions with spore-contaminated animal products, including violin bows, shaving bristles, goatskin drumheads, and leather jackets, can lead to **cutaneous anthrax**. Skin infection begins as a papule, but within 1 to 2 days it develops into a pustule of black, necrotic (dying) tissue that eventually crusts over (FIGURE 12.3). Lymph glands in the adjacent area may be invaded and swell. Cutaneous anthrax accounts for more than 95% of all anthrax infections. About 20% of untreated cases will result in death.

Overall, in the United States there may be one or two human cases reported each year. However, the World Health Organization (WHO) estimates there are some 2,000 to 20,000 human cases of anthrax globally each year.

B. anthracis infections can be treated with penicillin or ciprofloxacin. At present, there is no vaccine for civilian use, although there is a cell-free vaccine for veterinarians and others who work with livestock.

Anthrax also is considered a threat in bioterrorism and in biological warfare (MicroFocus 12.1). The seriousness of using biological agents as a means for bioterrorism was underscored in October 2001 when *B. anthracis* spores were distributed intentionally through the United States mail. In all, 22 cases of anthrax (11 inhalation and 11 cutaneous) were identified, making the case-fatality rate among patients with inhalation anthrax 45% (5/11). The six other individuals with inhalation anthrax and all the individuals with cutaneous anthrax recovered. Had it not been for antibiotic therapy, many more might have been stricken.

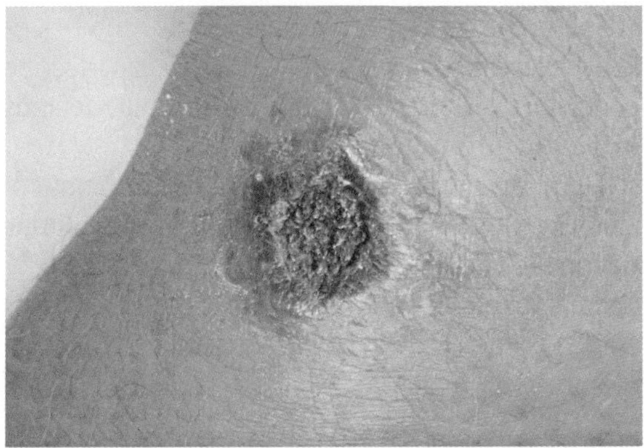

FIGURE 12.3 **An Anthrax Lesion.** This cutaneous lesion is a result of infection with anthrax bacilli. Lesions like this one develop when anthrax spores contact the skin, germinate to vegetative cells, and multiply. Courtesy of James H. Steele/CDC. »» Why was a skin ulceration like this given the name "anthrax"?

Tetanus Causes Hyperactive Muscle Contractions

Causative Agent and Epidemiology. Tetanus is one of the most dangerous human diseases. *Clostridium tetani*, the bacterial species causing the disease, is an anaerobic, gram-positive bacillus that forms endospores. The organism is typically found in barnyard and garden soils containing animal manure.

- Papule: A raised itchy bump that resembles an insect bite.
- Pustule: A papule containing pus.

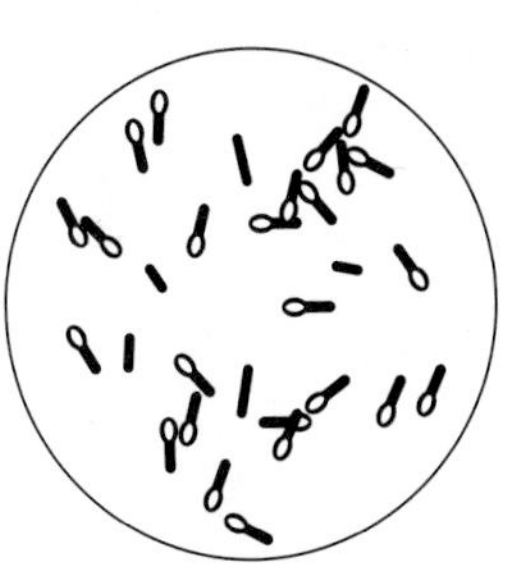

Clostridium tetani

The United States has had a steady decline in the incidence of tetanus, with 26 cases confirmed in 2010. Older Americans are primarily affected because they either have not been immunized or kept up their booster immunizations.

In other parts of the world, tetanus remains a major health problem. Globally, neonatal tetanus accounts for the majority of cases and deaths, often the result of the umbilical stump becoming infected from non-sterile instruments or dressings.

Clinical Presentation. Spores enter the body through a deep puncture wound resulting from a fracture, gunshot, animal bite, or puncture by a piece of glass, a thorn, or a rusty nail. Even illicit drugs may contain spores (MicroFocus 12.2). In dead, oxygen-free tissue of the wound, spores germinate into vegetative bacilli that produce several toxins. The most important of these toxins is the neurotoxin **tetanospasmin**, the second most powerful toxin known to science (after the botulism toxin).

MICROFOCUS 12.1: History

The Legacy of Gruinard Island

In 1941, the specter of airborne biological warfare hung over Europe. Fearing that the Germans might launch an attack against civilian populations, British authorities performed a series of experiments to test their own biological weapons. Anthrax spores were seen as an agent that could be aerosolized and released unobtrusively. Drifting over a large city, they would be undetectable and could infect thousands of individuals.

To test this possibility, investigators placed 80 tethered sheep on Gruinard Island, a 2 kilometer-long patch of land off the coast of Scotland. Anthrax spore-containing bombs were exploded near the tethered sheep. Within days, all the sheep were dead.

Gruinard Island. © Scotsman Publ Ltd.

Warfare with biological weapons never came to reality in World War II, but the contamination of Gruinard Island remained. A series of tests in 1971 discovered anthrax spores still viable at and below the upper crust of the soil. Fearing they could be spread by earthworms, British officials posted signs warning people not to set foot on the island (see figure) but did little else.

Then a strange protest occurred in 1981. Activists demanded that the British government decontaminate the island. They backed their demands with packages of soil taken from the island. Notes led government officials to two sealed packages of island soil; one package contained spore-laden soil.

Partly because of the protests, the British government instituted a decontamination of the island in 1986. Technicians used a powerful brushwood killer, combined with burning and treatment with formalin in seawater. Finally, they managed to rid the soil of anthrax spores. By April 1987, sheep were once again grazing on the island. In 1990, the British government announced that the island was safe and that it was sold back to the heirs of the original owner for £500.

At the neuromuscular junction, tetanospasmin prevents the release of neurotransmitters needed to inhibit muscle contraction. Without any inhibiting influence, volleys of spontaneous impulses arise in the motor neurons, causing muscle spasms and stiffness.

Symptoms of tetanus intoxication develop rapidly, often within hours of exposure. A patient first experiences generalized muscle stiffness, especially in the facial and swallowing muscles. Spasms of the jaw muscles cause the teeth to clench and bring on a condition called **trismus**, or **lockjaw**. Severe cases are characterized by a "fixed smile" (risus sardonicus) and muscle spasms cause an arching of the back (**opisthotonus**; FIGURE 12.4). Spasmodic inhalation and seizures in the diaphragm and rib muscles leads to reduced ventilation, and patients often experience violent deaths.

Treatment and Prevention. Patients are treated with sedatives and muscle relaxants and are placed in quiet, dark rooms as noise and bright light can trigger muscle spasms. Physicians prescribe penicillin to destroy the organisms and inject tetanus antitoxin into a vein to neutralize the toxin.

Immunization for children involves injections of tetanus toxoid in the **diphtheria-tetanus-acellular pertussis (DTaP) vaccine**. Booster injections of tetanus toxoid in the **Td vaccine** (a "tetanus shot") are recommended every 10 years to keep the level of immunity high.

Gas Gangrene Causes Massive Tissue Damage

Gangrene (*gangren* = "a sore") develops when the blood flow ceases to a part of the body, usually as a result of blockage by dead tissue. The body

MICROFOCUS 12.2: Public Health

Tetanus Outbreak Among Injecting Drug Users

When one thinks about tetanus, the typical image coming into mind is stepping on a rusty nail. Such nails do pose a threat because spores cling to the rough edges of the nail and the nail may cause extensive tissue damage as it penetrates. Actually, most cases in the developed nations of the world are found in older women who become infected by contaminated soil while gardening. However, there are other ways spores can be introduced into the body.

Between July 2003 and January 2004, 25 cases of tetanus in injecting drug users (IDUs) were reported to public health officials in the United Kingdom. The method of injection was subcutaneous injection of heroin. Thirteen women and 12 men between the ages of 20 and 53 years were identified as having clinical tetanus. Sixteen were hospitalized with severe generalized tetanus. In the end, two patients died. Among the 23 survivors, 2 had mild disease and 21 required intensive treatment for a median of 40 days.

Twenty-two of the 23 IDUs had either not been immunized against tetanus or kept up immunization boosters.

The source of the tetanus infection has remained unclear. Because all cases were clustered in a short period, the most likely source was the drug or the adulterant (a substance added to the drug to make it less pure).

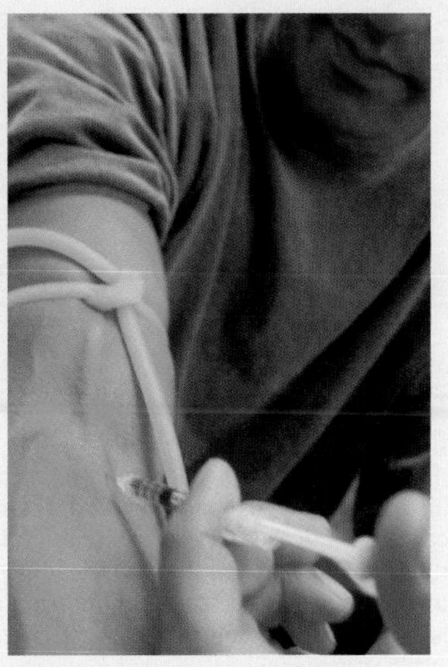

Subcutaneous injection. © Oscar Knott/FogStock/Alamy Images.

part, generally an extremity, becomes dry and shrunken, and the skin color changes to purplish or black. The gangrene may spread as enzymes from broken cells destroy other cells, and the tissue may have to be debrided or the body part amputated.

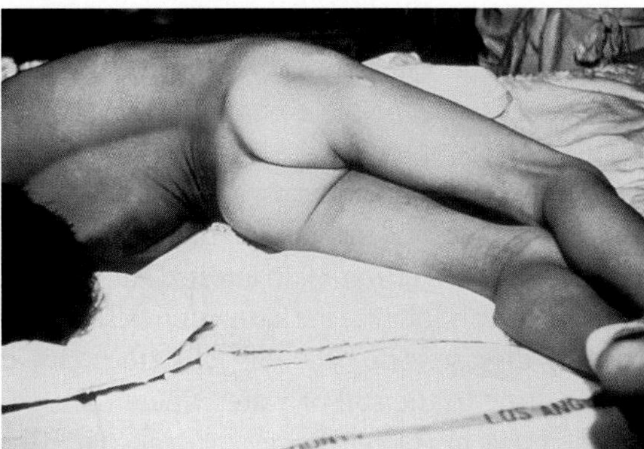

FIGURE 12.4 **Opisthotonos.** This patient exhibits a condition in which there is a curvature of the body posture (called opisthotonos) caused by a *Clostridium tetani* exotoxin. Courtesy of CDC. »» What part of the CNS appears to be targeted in this condition?

Causative Agent. **Gas gangrene**, or **myonecrosis** (*myo* = "muscle"; *necros* = "death") is caused primarily by *Clostridium perfringens,* an anaerobic, spore-forming, gram-positive rod. *C. perfringens* also is a common cause of food poisoning due to the production of an enterotoxin.

Clostridium perfringens

■ Debrided: Referring to the removal of dead, damaged, or infected tissue.

Clinical Presentation. After endospores in contaminated dirt are introduced through a severe, open wound, the spores germinate and the vegetative cells multiply rapidly in the anaerobic environment. As they grow, they ferment muscle carbohydrates and decompose the muscle proteins (thus the term "myonecrosis"). Large amounts of gas may result from this metabolism, causing a crackling sound as the gas accumulates under the skin. The gas also presses against blood vessels, thereby blocking the flow and forcing cells away from their blood supply. In the infection process, the organisms secrete at least 12 exotoxins. The most important is α-toxin, which damages plasma membranes and disrupts cell tissues, facilitating the passage of bacterial cells and spread of infection.

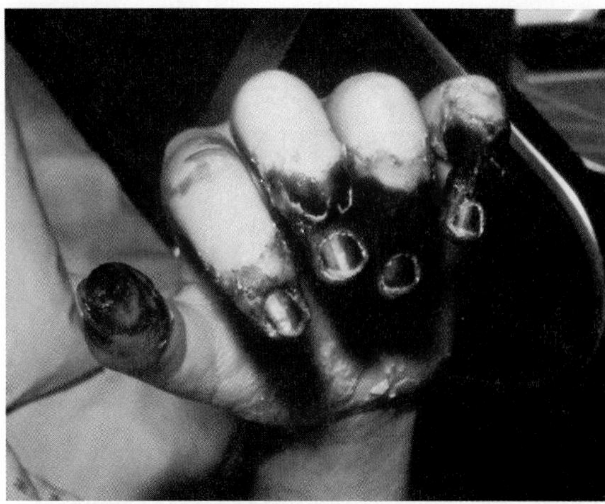

FIGURE 12.5 **Gas Gangrene of the Hand.** A severely infected hand showing gangrene (blackened tissue necrosis). This gangrene developed from an infection from an accident while the patient was scaling fish. Antibiotic drugs may prevent the infection leading to gangrene, but in this advanced stage, amputation of the hand may be necessary. Courtesy of Dr. Jack Poland/CDC.
»» What properties of *Clostridium perfringens* would cause the tissue necrosis?

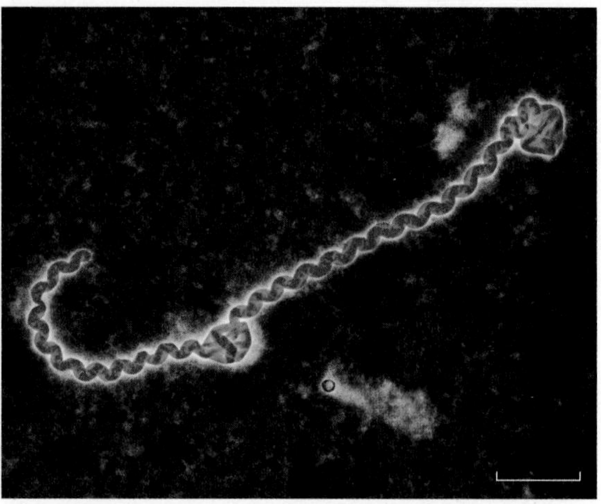

FIGURE 12.6 ***Leptospira interrogans.*** A false-color transmission electron micrograph of *L. interrogans,* the agent of leptospirosis. Note the tightly coiling spirals. (Bar = 2 μm.) © Eye of Science/Photo Researchers, Inc.
»» Besides the coil shape, what are the other notable features of this bacterial species?

The symptoms of gas gangrene include a foul odor, and intense pain and swelling at the wound site. Initially the site turns dull red, then green, and finally blue black (FIGURE 12.5). Anemia is common, and bacterial toxins may damage the heart and nervous system. Treatment consists of antibiotic therapy as well as debridement, amputation, or exposure in a hyperbaric oxygen chamber. However, without treatment, disease spreads rapidly, and death frequently occurs within days of gangrene initiation.

■ Hyperbaric oxygen: Refers to oxygen pressures (concentrations) higher than normally found in the body.

Leptospirosis Is an Emerging Zoonotic Disease

Causative Agent and Epidemiology. **Leptospirosis** (*lepto* = "thin;" *spir* = "spiral") is the most widespread zoonotic disease in the world. The agent of leptospirosis is *Leptospira interrogans,* a thin, aerobic, gram-negative, motile spirochete with a hook at one end resembling a question mark; hence the name interrogans (*roga* = "ask") (FIGURE 12.6). The undulating movements, caused by contractions of endoflagella, allows the cell to burrow into tissue.

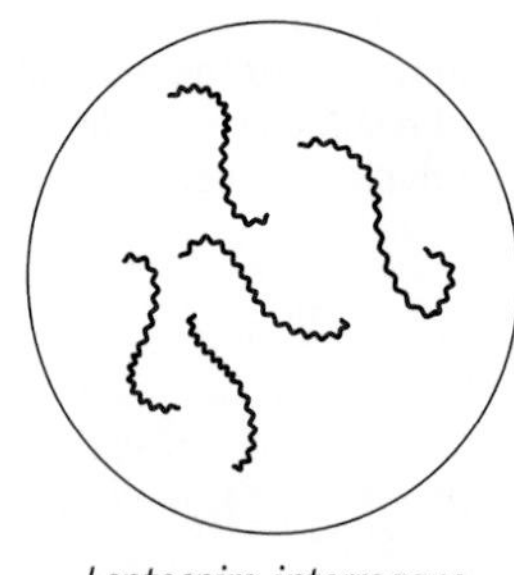

Leptospira interrogans

Globally, leptospirosis is considered to be an emerging infectious disease. There are more than 500,000 cases per year with a 10% mortality rate. Infections are highest in warm, subtropical regions such as Southeast Asia and South America, which is of particular concern to adventure travelers to these exotic locales (MicroFocus 12.3). There are usually between 40 and 100 cases of leptospirosis reported each year in the United States (50% of cases occur in Hawaii). In 1998, the largest ever outbreak of leptospirosis in the continental United States occurred among athletes after a triathlon competition in Springfield, Illinois. About 100 of the more than 800 athletes reported the illness, probably from contact with *Leptospira*-contaminated water from Lake Springfield during the swim portion of the competition. The water contamination most likely came from water runoff during a rainstorm prior to the competition. An animal reservoir for the contamination was not identified.

Household dogs and cats, as well as barnyard and wild animals and marine mammals, can become infected with *L. interrogans.* In infected animals, the spirochetes can colonize the kidney tubules without causing disease. The bacterial cells are excreted in the urine to the soil or water where they can remain viable in a biofilm for several weeks. Humans acquire it by direct contact with these animals or, as indicated above, indirectly from soil, food, or water contaminated with the urine from the infected animals.

MICROFOCUS 12.3: Public Health/Environmental Microbiology

A Real Eco-Challenge!

It started as a headache on the plane back from Borneo to the States. Within 3 days, Steve went to a hospital emergency room in Los Angeles complaining of fever and chills, muscle aches, vomiting, and nausea.

The Eco-Challenge 2000 in Sabah, Borneo was the site for the annual adventure race. Some 304 participants composing 76 teams from 26 countries competed in the 10-day endurance event, which was designed to push participants and their teams beyond their athletic limits. During Eco-Challenge 2000, teams would kayak on the open ocean, mountain bike into the rainforest, spelunk in hot caves, and swim in local rivers (see figure). Of the 76 teams starting the challenge, 44 teams finished.

A river in Borneo. © Andrea Seemann/ShutterStock, Inc.

About the time Steve arrived at the emergency room, the Centers for Disease Control and Prevention (CDC) in Atlanta received calls from the Idaho Department of Health, the Los Angeles County Department of Health Services, and the GeoSentinel Network (a network of international travel clinics) reporting cases of a febrile illness similar to Steve's.

The CDC quickly carried out a phone questionnaire to 158 participants in the Eco-Challenge. Many reported symptoms similar to Steve's, including chills, fever, headache, diarrhea, and conjunctivitis. Twenty-five respondents had been hospitalized. Within a few days, antibiotic therapy had Steve recovering. In fact, all 135 affected participants completely recovered.

The similar symptoms suggested leptospirosis and laboratory tests either confirmed the presence of *Leptospira* antibodies or positive culture of the organism from serum samples collected from ill participants.

To identify the source and the exposure risk, information was gathered from participants about various portions of the race course. Analysis identified swimming in and kayaking on the Sagama River as the probable source.

Several participants who did not become ill had taken doxycycline as a preventative for malaria and leptospirosis, as race organizers had advised. Unfortunately, Steve had not heeded those words and suffered a real "eco-challenge." Asked if he would participate in the 2001 Eco-Challenge in New Zealand, he said, "Heck yes! It just adds to pushing the limits."

Clinical Presentation. As a result of a person swimming or wading in contaminated water, *L. interrogans* enters the human body through the mucus membranes of the eyes, nose, and mouth, or through skin abrasions. The bacterial cells multiply rapidly and 90% of infected individuals experience vague flu-like symptoms with muscle aches and conjunctivitis. Episodes of fever and chills occur for 4 to 9 days, but then disappear.

Depending on the strain of infecting *L. interrogans* and the level of exposure, a more severe form of leptospirosis, called **Weil syndrome**, may develop. This late phase of the infection occurs as the spirochetes directly invade and infect various organs and the immune system reacts to the infection. A fever returns and there will be a persistent yellowing of the skin and whites of the eyes (jaundice). Aseptic meningitis may develop and last a few days or up to 2 weeks. Most importantly, inflammation may occur in the liver and lungs, leading to liver and kidney dysfunction. The illness can last from a few days to more than 3 weeks. Without treatment, recovery may take several months and complicated cases can be life-threatening; in fact, Weil syndrome carries a mortality rate of 5% to 10%.

■ Conjunctivitis: An inflammation of the conjunctiva of the eye.

Treatment and Prevention. Treatment of severe leptospirosis should be started as soon as possible. Despite the numerous tissues involved,

TABLE

12.1 A Summary of Soilborne Bacterial Diseases

Disease	Causative Agent	Signs and Symptoms	Transmission	Treatment	Prevention
Anthrax	*Bacillus anthracis*	Fever, chills, cough, chest pain, headache, and malaise Severe breathing and shock can develop	Airborne endospores	Penicillin Ciprofloxacin	Avoiding contact with infected livestock and animal products
Tetanus	*Clostridium tetani*	Muscle stiffness in jaw and neck, trismus	Wounds contaminated with soil, dust, and animal feces	Tetanus antitoxin Penicillin	Immunizing with toxoid vaccine
Gas gangrene	*Clostridium perfringens*	Foul odor and intense pain and swelling at the wound site	Soil Endogenous transfer	Debridement Surgery Cephalosporin Hyperbaric oxygen	Cleaning wounds Debridement
Leptospirosis	*Leptospira interrogans*	Acute headache Muscle aches Vomiting and nausea Fever and chills	From contaminated soil and water	Antibiotics	Avoiding contaminated water

doxycycline is used with success. Prevention involves prophylactic doxycycline for high-risk individuals and public health awareness to the occupational and recreational risks. Vaccines to some strains are available for cattle and dogs.

The soilborne bacterial diseases are summarized in TABLE 12.1.

CONCEPT AND REASONING CHECKS 1

a. What cellular factors make *B. anthracis* a dangerous pathogen?
b. What are the definitive symptoms of tetanus?
c. What characteristic of *C. perfringens* is being "attacked" by placing a gas gangrene patient in a hyperbaric chamber?
d. What bacterial and host factors make *L. interrogans* a potent pathogen?

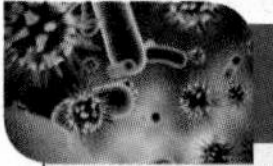

Chapter Challenge A

True soil-dwelling organisms are called euedaphic (*eu* = "true"; *edaph* = "soil"). In this first section of the chapter, we have discussed the bacterial species that are the causative agents for anthrax, tetanus, gas gangrene, and leptospirosis.

QUESTION A:

Which of these species would represent euedaphic pathogens and which are soil-transmitted pathogens?

Answers can be found on the Student Companion Website in **Appendix D**.

KEY CONCEPT 12.2 Some Bacterial Diseases Can Be Transmitted by Arthropods

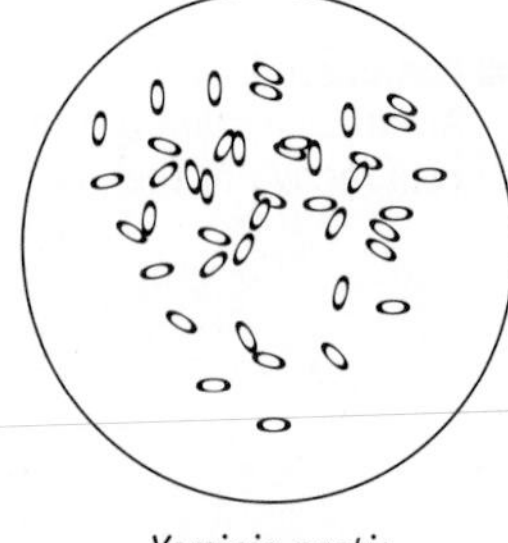

Yersinia pestis

A living organism such as an arthropod that transmits disease agents is called a **vector** (*vect* = "carried"). Arthropods transmit diseases to humans usually by taking a blood meal from another animal and themselves becoming infected. Then, they pass the organisms to another individual during the next blood meal. Arthropodborne diseases occur primarily in the bloodstream, and they often are characterized by a high fever and a body rash.

Plague Can Be a Highly Fatal Disease

Causative Agent and Epidemiology. Plague is caused by *Yersinia pestis* (*pestis* = "plague") (FIGURE 12.7A). This facultative, gram-negative

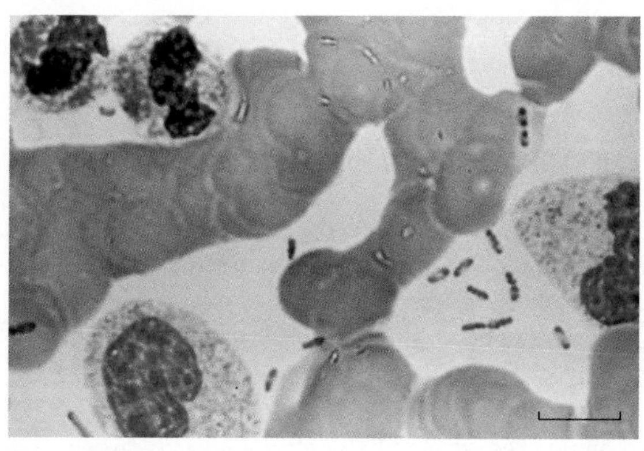

(A)

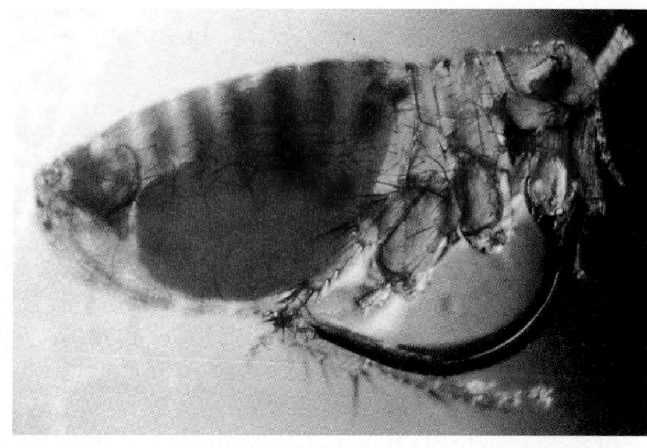

(B)

FIGURE 12.7 *Yersinia Pestis* and the Flea Vector. (**A**) This light micrograph shows the bipolar staining of the bacterial cells. (Bar = 1 μm.) (**B**) *Xenopsylla cheopis* (oriental rat flea) with clotted *Y. pestis* mass (red foregut). Courtesy of CDC.
»» How does clotting in the foregut lead to human infections?

rod stains heavily at the poles of the cell, giving it a safety-pin appearance and a characteristic called **bipolar staining** when direct smears from infected specimens are observed.

Few diseases have had the rich and terrifying history as bubonic plague, nor can any match the array of social, economic, and religious changes wrought by this disease (see this chapter introduction).

The first documented pandemic of plague probably occurred in China in 224 BC. In Europe, the first pandemic occurred during the reign of the Roman emperor Justinian in 542 AD. It lasted 60 years, killed millions, and contributed to the downfall of Rome. The second pandemic was known as the **Black Death** because of the purplish-black splotches on victims and the terror it evoked in the 1300s (MicroFocus 12.4). The pandemic killed an estimated 25 million people in Europe, almost one-third of the population on the continent (FIGURE 12.8). A deadly epidemic also occurred in London in 1665, where 70,000 people succumbed to the disease.

The third pandemic occurred in the late 1800s, when Asian warfare facilitated the spread of a Burmese origin of plague, and migrations brought infected individuals to China and Hong Kong. During an epidemic in 1894, the causative organism was isolated by Alexandre Yersin and, independently, by Shibasaburo Kitasato. Plague first appeared in the United States in San Francisco in 1900, carried by rats on ships from Asia. The disease spread to ground squirrels, prairie dogs, and other wild animals, and it is now endemic in the southwestern states, where it is commonly called "sylvatic plague" (*sylva* = "forest"). MicroFocus 12.5 recounts a case associated with wild animals.

Plague remains a threat in endemic areas of the world (Africa, America, Asia) where wild rodents are infected. Between 2004 and 2008, globally there were more than 12,500 cases of human plague, including some 850 deaths. The Centers for Disease Control and Prevention (CDC) reported two cases in the United States in 2010.

Urban plague also is a zoonosis but primarily resides in rats. If infected rodents are bitten by fleas, the plague bacilli in the blood meal clot the fleas' digestive system, starving the insects (FIGURE 12.7B). This causes the fleas to become even more voracious in finding a blood meal. Then, as septicemic rats die, the fleas may jump to another animal, such as humans, in an attempt to feed in the skin. In this process, bacterial cells are regurgitated into the human bloodstream.

Clinical Presentation. There are three major forms of plague. **Bubonic plague** starts when the bacterial cells multiply in the submucosa and eventually localize in the lymph nodes, especially those of the armpits, neck, and groin. Hemorrhaging in the lymph nodes causes painful and substantial swellings called **buboes** (*bubon* = "the groin"). Dark, purplish splotches from hemorrhages also can be seen through the skin. Blood pressure drops and, without treatment, mortality reaches 60%.

From the buboes, the bacilli may spread to the bloodstream, where they cause **septicemic plague** (**Black Death**). Symptoms include high fever, diarrhea, and abdominal pain. The septicemia can lead to **plague meningitis**. Nearly 100% of untreated cases are fatal.

Human-to-human transmission of plague during epidemics is spread by respiratory droplets when

MICROFOCUS 12.4: History

Catapulting Terror

Bodies came flying across the walls. People in panic either fled in horror or tried to remove the dead, decaying bodies. Caffa was in chaos.

One of the most horrendous emerging infectious diseases was starting to spread to Europe, North Africa, and the Near East in the mid-fourteenth century. It was the Black Death—bubonic plague—which historians believe moved out of the lands north of the Caspian and Black Seas.

Caffa (today Feodosija, Ukraine) was a port city situated on the north shore of the Black Sea. Through an agreement with the local Tartars (Mongols) who controlled the area, Caffa was placed under control of Genoa, Italy, and Christian merchants were allowed to trade goods with the Far East. In 1343, a group of Italian merchants from Genoa found themselves trapped behind the walls of Caffa after a brawl between the Italians and Tartars. The dreaded Tartars laid siege to the city and over the next 5 years, Genoa lost and regained control of the city several times.

During the siege of 1346, the Tartars were unable to drive the Italians and other Christians from the city. Then, plague broke out. Large numbers of Tartars started dying. Losing interest in the siege, the Tartars had the bodies of their dead plague victims placed in catapults and lobbed over the walls of Caffa and into the city. The Tartars hoped the stench would kill everyone in Caffa; and, in fact, plague soon was sweeping through Caffa. The townspeople were terrified: Either the plague would kill them inside the walls, or the Tartars would kill them outside the walls. But the Tartars were equally terrified of the plague and they were withdrawing.

Sensing an opportunity to escape, the merchants ran for their ships and sailed off to Genoa, Venice, and other homeports in the Mediterranean. Unfortunately, their voyage home would be a voyage of death. Many died of the plague onboard, and the survivors spread the disease wherever they stopped to replenish their supply of food and water.

Could such a tale be true? Could the dead diseased bodies catapulted into Caffa transmit plague? Almost certainly they could. City defenders would have carried away the dead, mangled bodies, which would spread the disease by contact. Poor sanitation and health of its citizens in Caffa would make transmission even easier and more widespread, especially if pneumonic plague broke out.

The attempted siege of Caffa in 1346 represents the most spectacular early episode of biological warfare, with the Black Death as its consequence. It demonstrates the very essence of terrorism as defined today—the intentional or threatened use of biological agents to cause fear in or actually inflict death or disease upon a large population for political, religious, or ideological reasons. The siege of Caffa shows us the horrifying consequences that can come from the use of infectious disease as a weapon.

septicemic infections progress to the lungs. In this form, the disease is called **pneumonic plague** and is highly contagious. Lung symptoms are similar to pneumonia, with headache, malaise, and extensive coughing. Hemorrhaging and fluid accumulation are common. Many suffer cardiovascular collapse with death common within 2 to 3 days of the onset of symptoms. Mortality rates for pneumonic plague approach 100%. How the plague bacilli overcome the lung immune defenses is examined in Investigating the Microbial World 12.

Treatment and Prevention. When detected early, plague can be treated with streptomycin or doxycycline, reducing mortality to less than 10%. A vaccine consisting of dead *Y. pestis* cells is available for high-risk groups.

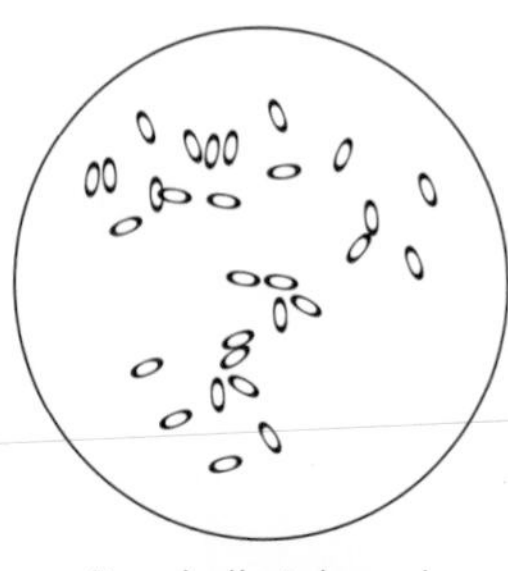

Francisella tularensis

Tularemia Has More Than One Disease Presentation

Causative Agent and Epidemiology. Tularemia is one of several microbial diseases first identified in the United States (others include St. Louis encephalitis, Rocky Mountain spotted fever, Lyme disease, and Legionnaires' disease). The CDC reported 124 cases in 2010.

The causative organism, *Francisella tularensis*, is a very small, aerobic, gram-negative rod that often exhibits bipolar staining. It is extremely virulent because as few as 50 cells can cause disease. It is a zoonotic disease that can be transmitted by a broad variety of wild animals, especially rodents, and it is particularly prevalent in rabbits

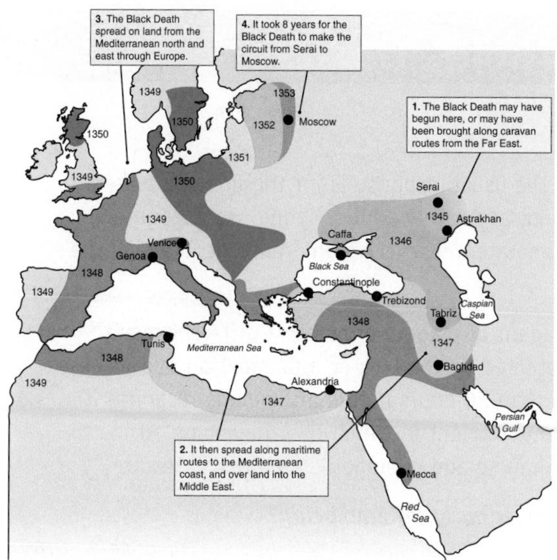

FIGURE 12.8 The Spread of Bubonic Plague (Black Death) in 14th Century Europe. Over a period of 8 years, the plague bacillus spread clockwise from the Black Sea through Europe to Moscow. »» If such an outbreak occurred today, would it spread faster or slower than in the 14th century? Explain.

(in which case it is known as **rabbit fever**). Dogs may acquire the organisms during romps in the woods, and humans are infected by arthropods from the fur of animals. Ticks are important vectors in this regard, as evidenced by an outbreak of 20 tickborne cases in South Dakota in the 1980s. Other methods of transmission include consumption of infected rabbit meat, drinking contaminated water, or inhaling infectious aerosols.

Clinical Presentation. Various forms of tularemia exist, depending on where the bacilli enter the body. An arthropod bite, for example, may lead to swollen lymph nodes (**glandular tularemia**) and a skin ulcer at the bite site (FIGURE 12.9). Individuals typically experience flu-like symptoms. Inhalation of *F. tularensis* cells may lead to respiratory disease (**inhalational tularemia**) and produce swollen lymph nodes, a dry cough, and pain under the breast bone.

Treatment and Prevention. Tularemia usually resolves on treatment with streptomycin, and few

MICROFOCUS 12.5: Public Health

"What a Tragedy!"

"Just an incredible day," he must have been thinking, as he trudged along. It felt wonderful to be in the great outdoors. Great to get away from the books for a while. Final exams were only a few weeks away, then summer, then college, then who knows? Maybe a career in physical education? Maybe a coaching job? Three miles completed, 5 miles to go. He had to keep moving to finish the hike and make it back home for dinner.

He didn't mind the solitude of the Arizona landscape. It was cool in Flagstaff, just right for hiking. And there were plenty of animals to see and interesting plants to examine. He would stop now and then for some water or to watch the horizon—or feed a colony of prairie dogs as they bustled about the terrain. Then he continued on home.

He enjoyed wrestling as well, and he was the team captain. Two days after the hike, he sustained a groin injury while wrestling. When the ache remained, he went to the doctor. The doctor noted the groin swelling and wondered whether it had anything to do with the fever the young man was experiencing. Was it just a coincidence—the ache, the swelling, the fever? This was no ordinary groin injury. "We'll watch it for a couple of days," the doctor said as he gave the young man a pain reliever.

Tragedy struck the next morning. All his mother remembered was a loud thud from the bathroom. She didn't remember her terrified shriek or dialing the emergency number. The emergency medical technicians were there in a flash, but it was too late. He was dead.

The investigation that followed took public health officials down many dead ends. "It wasn't the wrestling injury," they concluded. Still, the groin swelling made them suspicious. "That hiking trail," they asked his mother, "where is it?"

They set out in search of an elusive answer. About 6 kilometers out, they came upon a colony of prairie dogs. The animals didn't look well. In fact, some were dead nearby. Carefully, they trapped a sick animal and carried it back to the lab. Two days later, the lab report was ready—the animal was sick with plague. Then the young man's tissues were tested—again, plague. The investigators shook their heads. "What a tragedy!"

Investigating the Microbial World 12

The Lungs as Microbial Playgrounds

When the human lungs become infected with a bacterial pathogen, the usual response is for the immune system to mount an immediate attack in an attempt to rid the infection. And, as healthy adults, should we encounter one of those pathogens, the immune system usually works fine to generate a restrictive environment in which the pathogens cannot survive—and are eliminated.

- **OBSERVATIONS:** Infection of the lungs with *Yersinia pestis*, the agent of plague, produces a different result. The infection does not trigger an immediate immune response as detected by the lack of inflammation and disease symptoms. Rather, the plague bacilli can grow unmolested for more than 36 hours before an immune response is mounted. Somehow the bacterial cells have produced a permissive environment in which they have ample time to grow and multiply to levels that the immune system usually cannot handle and pneumonic plague will result.

- **QUESTION:** ***Why does* Y. pestis, *unlike other pathogens, not trigger early inflammation?***

- **HYPOTHESIS:** The lack of an inflammatory response is because the plague bacilli suppress the early immune response making the otherwise restrictive environment into a permissive one. If so, then coinfecting the lungs with *Y. pestis* and other microbes will also produce a permissive environment in which those coinfected microbes, which are usually destroyed, can grow.

- **EXPERIMENTAL DESIGN:** Rats were used as experimental organisms and were infected or coinfected intranasally with *Y. pestis, Y. pestis* mutants, or other bacterial species. Bacterial growth was then determined by counting the numbers of bacterial cells (as colony-forming units; CFUs) in lung tissue.

- **EXPERIMENT 1:** Separate groups of rats were infected or coinfected (inoculated) with 10^4 CFU of wild type (wt) *Y. pestis* and a nonpathogenic *Y. pestis* mutant (mut) that is normally destroyed in the lungs. The CFUs per lung was determined at 24, 48, and 60 hours postinoculation.

- **EXPERIMENT 2:** Separate groups of rats were infected or coinfected with 10^4 CFU of wild-type *Y. pseudotuberculosis* or *Klebsiella pneumoniae,* both of which generate lung infections similar to *Y. pestis,* or nonpathogenic mutants of these species that again normally are destroyed in the lungs. The CFUs per lung was determined at 48 hours post-inoculation.

- **RESULTS:**

  EXPERIMENT 1: see figure 1.

  EXPERIMENT 2: see figure 2.

- **CONCLUSIONS:**

  QUESTION 1: ***Was the hypothesis validated? Explain.***

  QUESTION 2: ***What information can you deduce from the results of experiment 1 (figure 1)?***

  QUESTION 3: ***What additional information is provided from the data obtained from experiment 2 (figure 2)?***

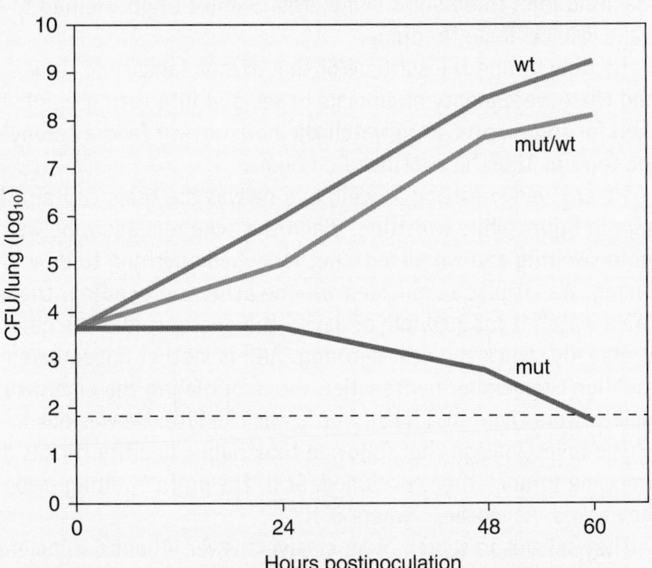

FIGURE 1 CFUs formed by wild-type (wt) and a nonpathogenic mutant (mut) of *Y. pestis,* and mut CFUs formed from a coinfection with wt (mut/wt). Modified from: Price, P. A., Jin, J., and Goldman, W. E. (2012). *Proc. Natl. Acad. Sci.* USA 109 (8): 3083–3088.

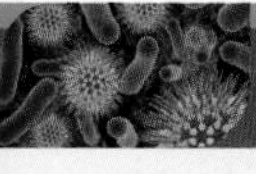

Investigating the Microbial World 12 (continued)

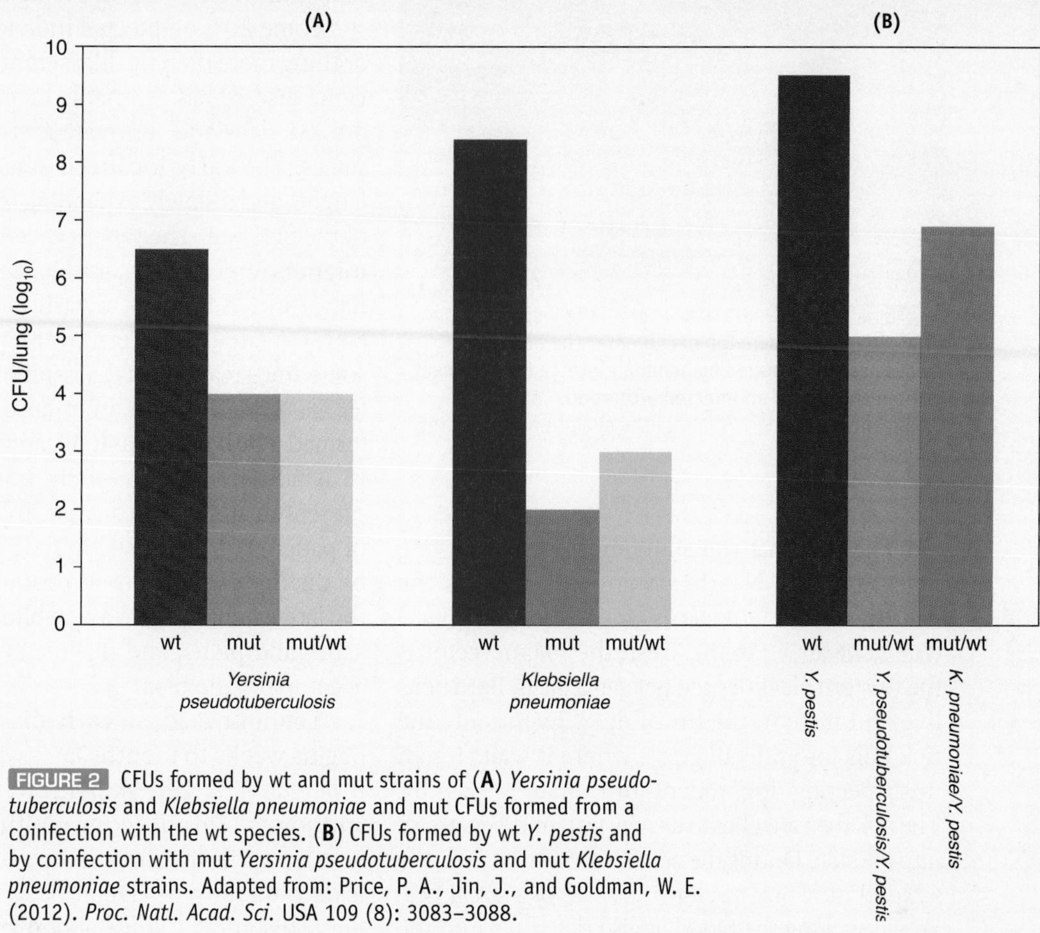

FIGURE 2 CFUs formed by wt and mut strains of (**A**) *Yersinia pseudotuberculosis* and *Klebsiella pneumoniae* and mut CFUs formed from a coinfection with the wt species. (**B**) CFUs formed by wt *Y. pestis* and by coinfection with mut *Yersinia pseudotuberculosis* and mut *Klebsiella pneumoniae* strains. Adapted from: Price, P. A., Jin, J., and Goldman, W. E. (2012). *Proc. Natl. Acad. Sci.* USA 109 (8): 3083–3088.

Note: *Y. pestis* and *Y. pseudotuberculosis* are evolutionarily very closely related. Studying the genetic and molecular differences between these two species may identify the mechanism that *Y. pestis* uses to produce the permissive environment.

Answers can be found on the Student Companion Website in **Appendix D.**

Adapted from: Price, P. A., Jin, J., and Goldman, W. E. (2012). *Proc. Natl. Acad. Sci. USA* **109** (8):3083–3088. Icon image © Tischenko Irina/ShutterStock, Inc.

people die of the disease. Epidemics are unknown, and evidence suggests that tularemia may not be communicable among humans despite the many modes of entry into the body. Avoiding infected animals, aerosols, and ticks are the best prevention methods.

Lyme Disease and Relapsing Fever Are Transmitted by Spirochetes

Lyme Disease. One of the major emerging infectious diseases is **Lyme disease**, currently the most commonly reported arthropodborne illness in the United States. Lyme disease is named for Old Lyme, Connecticut, the suburban community where the first cluster of cases was observed in 1975. The disease is caused by the spirochete *Borrelia burgdorferi* (FIGURE 12.10A), named after Willy Burgdorfer, the microbiologist who studied the spirochete in the gut of infected ticks.

Although 95% of cases occur in the northeastern, Mid-Atlantic, and north-central states, cases also have been reported in the Pacific coast states and the Southeast. In 2010, there were more than 30,000 Lyme disease cases reported to the CDC.

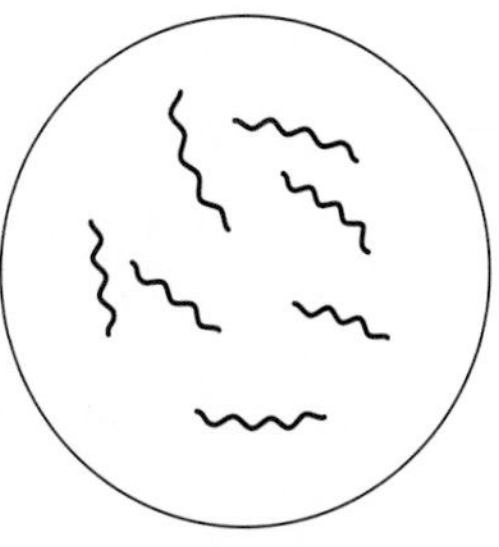

Borrelia burgdorferi

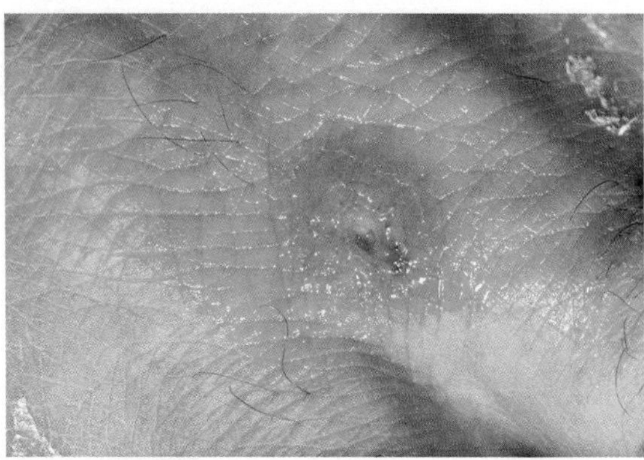

FIGURE 12.9 A Tularemia Lesion. The lesions of tularemia occur where the bacilli enter the body. The disease can be acquired by handling infected rabbit meat or from the bite of an infected arthropod. Courtesy of Dr. Brachman/CDC. »» What are the symptoms associated with lesion formation?

The tick that transmits most cases of Lyme disease in the Northeast and Midwest is the deer tick *Ixodes scapularis* (formerly *I. dammini*) (FIGURE 12.10B); in the West, the major vector is the western black-legged tick *I. pacificus*. Both ticks live and mate in the fur of deer. Eventually the tick falls into the tall grass, where it waits for an unsuspecting dog, rodent, or human to pass by. The tick then attaches to its new host and penetrates into the skin. During the next 24 to 36 hours, it takes a blood meal and swells to the size of a small pea. While sucking the blood, it also defecates into the wound, and, if the tick is infected, spirochetes are transmitted (FIGURE 12.11). If the tick is observed on the skin, it should be removed with forceps or tweezers, and the area thoroughly cleansed with soap and water before applying an antiseptic.

Some 20% of infected individuals experience nothing more than flu-like symptoms. For others, Lyme disease has a variable incubation period of 3 to 31 days and, untreated, typically has three stages. The **early localized stage** involves fever, fatigue, and a slowly expanding red rash at the site of the tick bite. The rash is called **erythema** (red) **migrans** (expanding), or **EM**. Beginning as a small flat or raised lesion, the rash increases in diameter in a circular pattern over a period of weeks, sometimes reaching a diameter of 10 to 15 inches (FIGURE 12.12). With an intense red border, it is termed a **bull's-eye rash**. It can vary in shape and is usually hot to the touch, but it need not be present in all cases of disease. Indeed, about 20% of patients do not develop EM. The tick bite can be distinguished from a mosquito bite because the latter itches, while a tick bite does not. Fever, aches and pains, and flu-like symptoms usually accompany the rash.

Left untreated, an **early disseminated stage** begins weeks to months later with the spread of *B. burgdorferi* to the skin, heart, nervous system, and joints. On the skin, multiple smaller EMs develop while invasion of the nervous system can lead to meningitis, facial palsy, and peripheral nerve disorders. Cardiac abnormalities are the most common, such as brief, irregular heartbeats. Joint and muscle pain also occur.

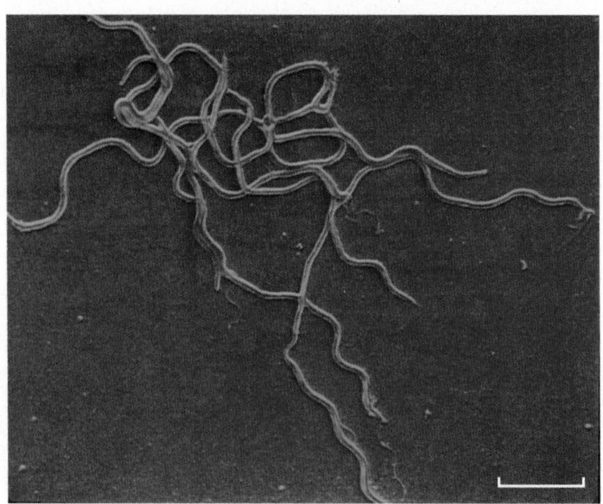

(A)

(B)

FIGURE 12.10 *Borrelia burgdorferi* and Its Tick Vector. (**A**) A false-color scanning electron micrograph showing the *B. burgdorferi* spirochetes, the agents of Lyme disease. (Bar = 1 µm.) © Phototake/Alamy Images. (**B**) A photograph of *Ixodes scapularis*, a tick species that transmits Lyme disease. Courtesy of Scott Bauer/USDA. »» Although not evident in the electron micrograph, what structure typical of spirochetes contributes to their motility?

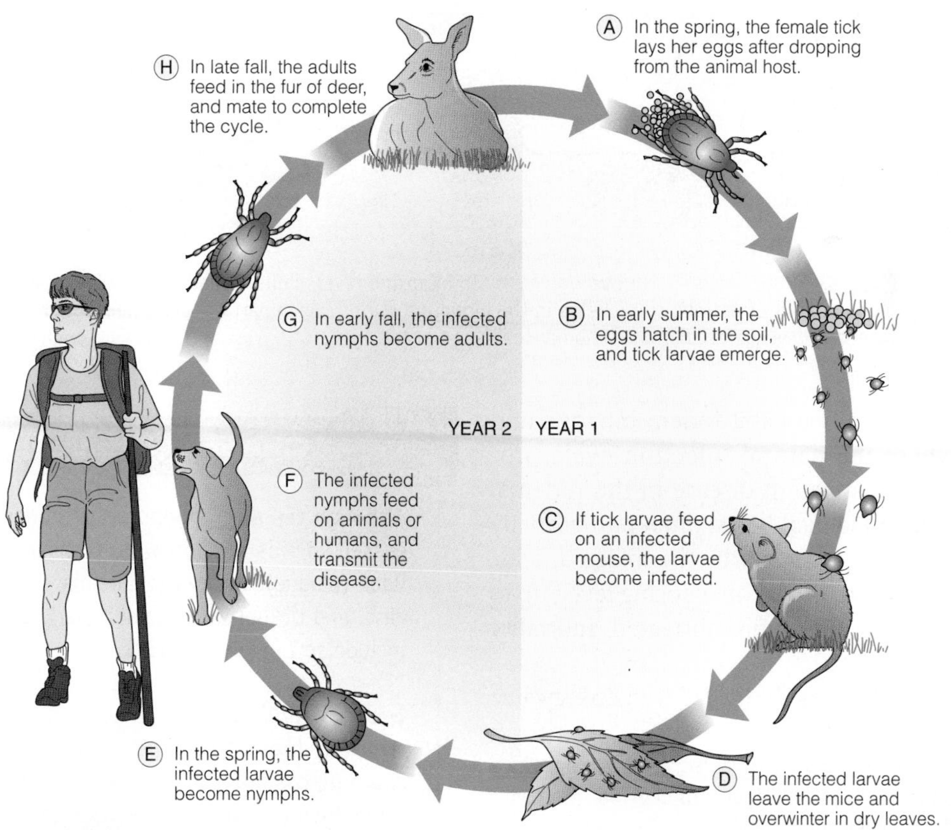

FIGURE 12.11 Life Cycle of Ixodid Ticks. The life cycle of *Ixodes scapularis* and *I. pacificus*, the ticks that transmit Lyme disease. »» Why is Lyme disease called a zoonotic disease?

If still left untreated, a **late stage** occurs months to years later. About 10% of patients develop chronic arthritis with swelling in the large joints, such as the knee. Although Lyme disease is not known to have a high mortality rate, the overall damage to the body can be substantial.

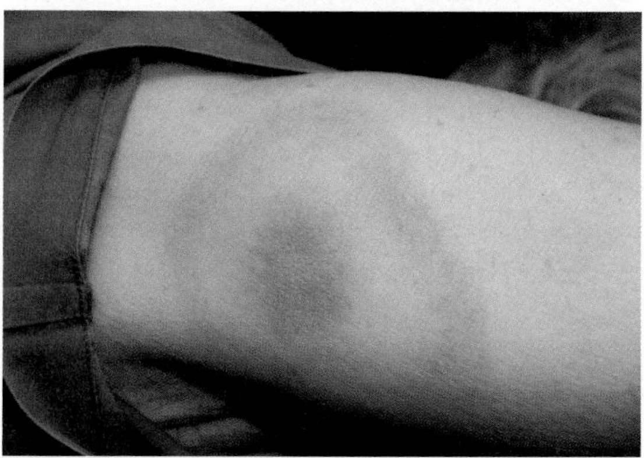

FIGURE 12.12 Erythema Migrans. Erythema migrans (EM) is the rash accompanying 80% of cases of Lyme disease. The rash consists of a large patch with an intense red border. It is usually hot to the touch, and it expands with time. Courtesy of James Gathany/CDC. »» What is the common name for the rash?

Effective treatment depends on the stage of infection but it can usually be treated with amoxicillin or doxycycline. Avoiding ticks and wearing protective clothing are preventative measures. A vaccine for dogs (LymeVax®) has been licensed and is in routine use. A vaccine for humans (LYMErix®) was removed from the market in February 2002 apparently because of poor sales, a complicated immunization schedule, and side effects.

Relapsing Fever. Another spirochete is responsible for a form of **relapsing fever** in which the infected individual goes through periods of fever and chills when many spirochetes are present in the blood. As the spirochetes decline in number, the individual recovers for several days before a recurrence of the symptoms (FIGURE 12.13).

Tickborne (endemic) relapsing fever in the United States is caused by *Borrelia hermsii*, which is transmitted to humans by the bite of soft ticks from a rodent host. Cabins in wilderness areas of the northwest and southwest United States are favorable nesting sites for infected rodents and their ticks, especially rustic cabins where rodents have access (TEXTBOOK CASE 12).

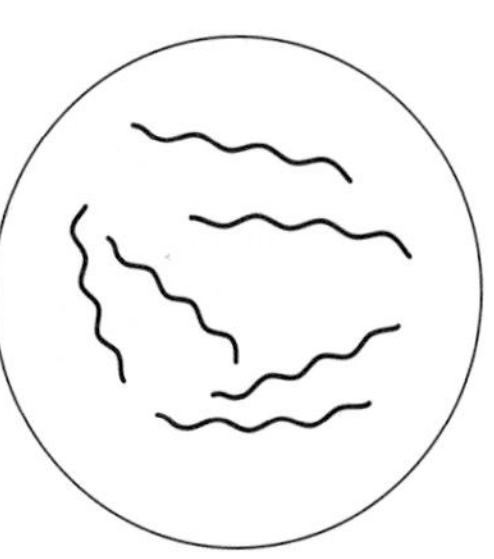

Borrelia hermsii

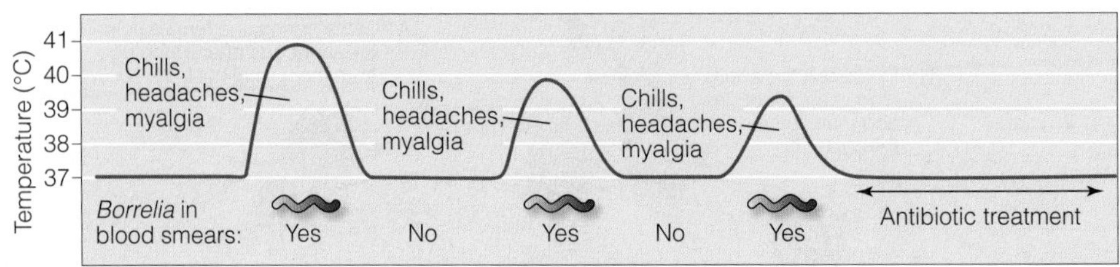

FIGURE 12.13 **The Cycles of Relapsing Fever.** In relapsing fever, chills, headache, and fever peak with high numbers of *Borrelia* spirochetes in the blood. »» During which part of each cycle would specific antibodies against *Borrelia* be produced?

Ticks normally inhabit rodent burrows and nests, where the natural infection cycle proceeds without apparent disease in the rodents. Humans are incidental hosts, often bitten briefly (5 to 20 minutes) at night by the infected ticks. With this tickborne-form of relapsing fever, up to 13 relapses can occur and, untreated, mortality can be 2% to 5%.

Relapsing fever can be treated with doxycycline to hasten recovery.

CONCEPT AND REASONING CHECKS 2

a. Explain how bubonic plague can develop into communicable pneumonic plague.

b. Propose a hypothesis to explain why the *Francisella tularensis* infectious dose is so small.

c. Describe the three clinical stages of untreated Lyme disease.

d. Explain the reason that infected individuals experience relapsing fevers.

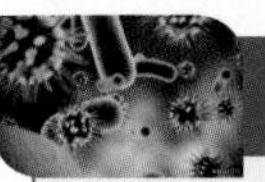

Chapter Challenge B

Although the arthropodborne pathogens discussed in this section were transmitted by fleas (plague) or ticks (tularemia, Lyme disease, and tickborne relapsing fever), they still include soil involvement.

QUESTION B:

Which of these arthropodborne pathogens could be considered euedaphic pathogens and which are soil-transmitted pathogens based on the vector used to transmit the pathogen?

Answers can be found on the Student Companion Website in **Appendix D.**

KEY CONCEPT 12.3 Rickettsial and Ehrlichial Diseases Are Arthropodborne

In 1909, Howard Taylor Ricketts, a University of Chicago pathologist, described a new organism in the blood of patients with Rocky Mountain spotted fever and showed that ticks transmit the disease. A year later, he located a similar organism in the blood of animals infected with Mexican typhus, and discovered that fleas were the important vectors in this disease. Unfortunately, in the course of his work, Ricketts fell victim to the disease and died. When later research indicated that Ricketts had described a unique group of microorganisms, the name "rickettsiae" was coined to honor him.

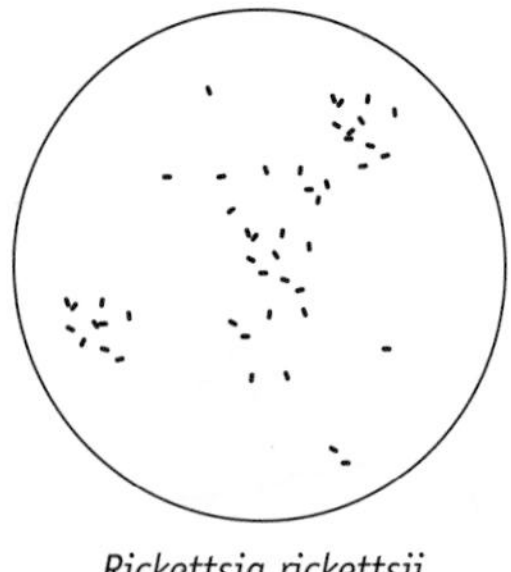

Rickettsia rickettsii

Rickettsial Infections Are Transmitted by Arthropods

The rickettsiae are very small, gram-negative, obligate, intracellular parasites. Most infections are transmitted by ticks, lice, or fleas. All illnesses can be treated effectively with doxycycline or chloramphenicol.

Rocky Mountain Spotted Fever (RMSF). RMSF is caused by *Rickettsia rickettsii* (FIGURE 12.14A) and is transmitted by hard ticks. About 1,000 cases were reported annually in the United States in the early 1980s, but public education about the disease, along with improved methods of diagnosis and treatment, caused a drop until 1998. Reported cases have since been increasing with almost 2,000 cases reported in 2010. Contrary to its name, RMSF is not commonly reported in western states any longer. However, in 2011, there were 54 cases and 11 deaths reported in northern Arizona. It remains a problem in Oklahoma, Arkansas, Texas, and many southeastern and

Textbook CASE 12

An Outbreak of Tickborne Relapsing Fever

1 In late July 2002, a family reunion was planned at a remote, previously uninhabited cabin in the mountains of northern New Mexico. Three days before the reunion, three family members arrived to clean the cabin. After the reunion, about half of the 39 family members slept in the cabin overnight.

2 Four days after the reunion, one of the family members who cleaned the cabin arrived at a local hospital complaining of fever, chills, muscle aches, and a rash on the forearms. The symptoms had started 2 days previously.

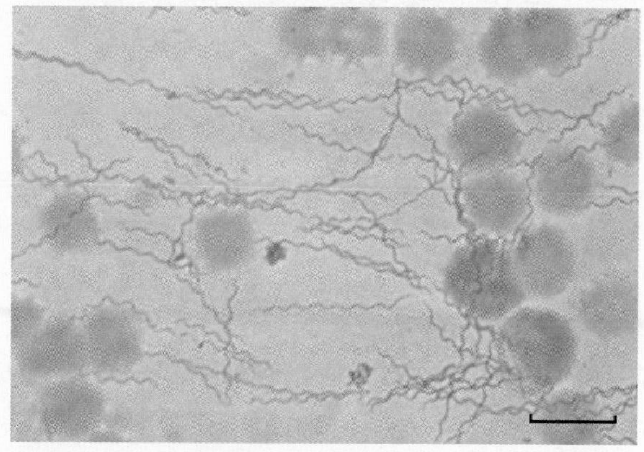

Tickborne relapsing fever spirochetes. (Bar = 8 um.) © Eric V. Grave/Photo Researchers, Inc.

3 A hospital lab technician took a blood sample from the patient and identified spirochetes on a blood smear (see figure). The immediate diagnosis was tickborne relapsing fever (TBRF). On August 2, the New Mexico Department of Health and the Indian Health Services were notified.

4 By August 7, 39 family members had sought medical care or were visited by a public health nurse. A study was begun to determine risk factors for infection, better understand the outbreak, and provide prevention measures.

5 A case of TBRF was defined. Based on the definition, 14 family members suffering the illness were identified.

6 Blood samples were taken from all 14 symptomatic patients. Spirochetes were identified in the blood smears of nine patients.

7 Samples from 13 patients were sent to the Centers for Disease Control and Prevention (CDC) in Atlanta. Eleven samples, including two samples negative for spirochetes by blood smear, were cultured and all grew *Borrelia hermsii*.

8 All 14 patients received antibiotic therapy and recovered.

9 Risk analysis revealed that all three family members who cleaned the cabin and eight of the other 36 were more likely to be TBRF patients.

10 Inspection of the cabin by Indian Health Service workers identified abundant rodent nesting materials and droppings within the walls of the cabin.

Questions:

(Answers can be found on the Student Companion Website in **Appendix D.**)

A. Why was such a prompt diagnosis and notification of health services necessary?

B. How would you define a case of TBRF?

C. Why were spirochetes only found in 9 of the 14 patients?

D. Why did all three who cleaned the cabin become ill?

E. Of the other eight patients, what is most likely the reason for them contracting the disease?

For additional information see www.cdc.gov/mmwr/preview/mmwrhtml/mm5234a1.htm.

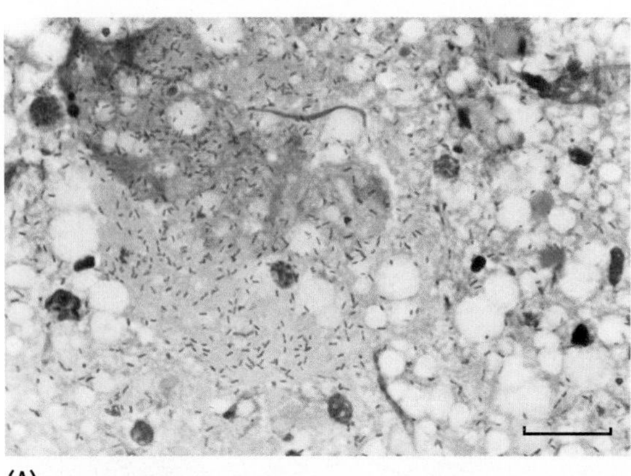

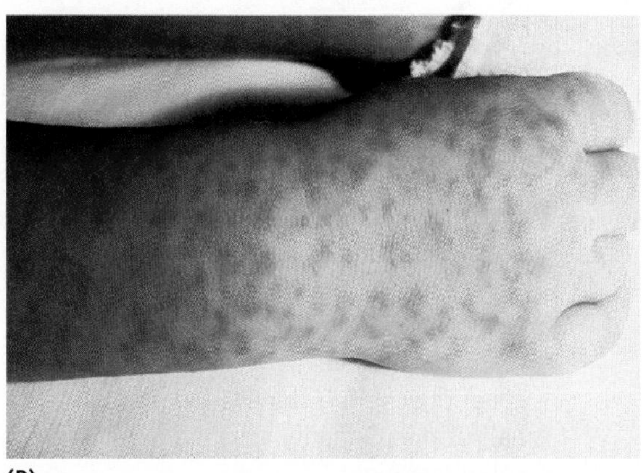

(A) (B)

FIGURE 12.14 ***Rickettsia rickettsii* and Symptoms of Rocky Mountain Spotted Fever.** (**A**) *R. rickettsii* in the cytoplasm of a kidney cell. (Bar = 0.5 μm.) Courtesy of Billie Ruth Bird/CDC. (**B**) Child's right hand and wrist displaying the characteristic spotted rash of Rocky Mountain spotted fever. Courtesy of CDC. »» What is happening in the host to produce a spotted rash?

Atlantic Coast states. Children are its primary victims because of their contact with ticks.

Following a tick bite, the hallmarks of RMSF are a high fever lasting for many days, severe headaches, and a skin rash reflecting damage to the small blood vessels. The rash begins as pink spots (macules), and progresses to pink-red papules (FIGURE 12.14B). Where the spots fuse, they form a maculopapular rash, which becomes dark red and then fades without evidence of scarring. As the disease progresses, the rash appears on the palms of the hands and soles of the feet and progressively spreads to the body trunk. Mortality rates of untreated cases are about 30%. Antibiotic treatment reduces this rate significantly and prevention involves prompt tick removal.

■ Maculopapular: Refers to a lesion with a broad base that slopes from a raised center.

Epidemic typhus (also called **typhus fever**) is one of the most notorious of all bacterial diseases and has been chronicled in Hans Zinsser's classic book *Rats, Lice and History* (MicroFocus 12.6). MicroInquiry 12 recounts how epidemic typhus has influenced the course of history.

Epidemic typhus is caused by *Rickettsia prowazekii*. The rickettsiae are transmitted to humans by body lice where hygiene is poor. The rickettsiae are excreted in feces from the lice, so scratching the bite will facilitate infection into the wound (rather than through the bite itself). Following an incubation period of 1 to 3 weeks, the characteristic fever and rash are particularly evident. The maculopapular rash, unlike the rash of RMSF, appears first on the body trunk and progresses to the extremities. Intense fever, sometimes reaching 40°C, remains for many days as the patient hallucinates and becomes delirious. Some patients suffer permanent damage to the blood vessels, heart, kidney, and lungs, and over 75% of sufferers die in epidemics. Again, antibiotic therapy reduces this percentage substantially, and the disease is rare in the United States.

Today microbiologists believe **Brill-Zinsser disease** is a relapse of an earlier case of epidemic typhus in which *R. prowazekii* remains dormant in the patient for many years. However, symptoms are usually milder.

Murine Typhus. The agent of **murine (endemic) typhus** (*murin* = "of the mouse") is *Rickettsia typhi* and the disease is prevalent in rodent populations where fleas abound (such as rats and squirrels). Cats and their fleas are involved also, and lice may harbor the bacilli. Most recent cases have occurred in south Texas where more than 80 cases have been reported between 2004 and 2009.

When an infected flea feeds in the human skin, it deposits the rickettsiae into the wound. Murine typhus is usually characterized by a high fever, persistent headache, chills, nausea, and a maculopapular rash that spreads from trunk to extremities. Often the recovery is spontaneous, without the need of drug therapy. However, lice may transport the rickettsiae to other individuals and initiate an epidemic.

Scrub typhus occurs in Asia and the Southwest Pacific. It is so named because the chigger mite transmitting the disease lives in scrubland where the soil is sandy or marshy and vegetation is poor.

MICROFOCUS 12.6: History

Students and Typhus Fever

Hans Zinsser once remarked, "*. . . [the students] force a teacher continually to renew the fundamental principles of the sciences from which his specialty takes off. So while we are, technically speaking, professors, we are actually older colleagues of our students, from whom we often learn as much as we teach them.*"

Hans Zinsser was born in New York City in 1878. He achieved fame for isolating the bacterial agent of epidemic typhus and was a pioneer in the study of autoimmunity and how microbes could induce disease, as in rheumatic fever. Zinsser was equally well known for his writing and wrote some of the most uniquely personal, wise, and witty prose of the twentieth century.

Zinsser also was a dedicated teacher. Writing at a time when most medical students were male, he remarked, "*. . . as we grow wiser we learn that the relatively small fractions of our time which we spend with well-trained, intelligent young men are more of a privilege than an obligation.*"

Perhaps the culmination of his writing was his most famous book entitled (with tongue in cheek) *Rats, Lice, and History: Being a Study in Biography, Which, After Twelve Preliminary Chapters Indispensable for the Preparation of the Lay Reader, Deals with the Life History of Typhus Fever.* It was an international best seller.

Dr. Zinsser. © National Library of Medicine.

Here is a sample of Zinsser's writing from his 1935 book: "*Soldiers have rarely won wars. They often mop up after the barrage of epidemics. And typhus, with its brothers and sisters—plague, cholera, typhoid, dysentery—has decided more campaigns than Caesar, Hannibal, Napoleon, and all the other generals of history. The epidemics get the blame for defeat, the generals the credit for victory. It ought to be the other way 'round. . . .*" Zinsser died in 1940 from lymphatic cancer at the age of 61.

If you think you'd enjoy learning more about disease and its effect on civilization, Zinsser's book is still in print. It would be a worthwhile investment of your time.

Scrub typhus also is called **tsutsugamushi fever**, from the Japanese words tsutsuga for "disease" and mushi for "mite." The causative agent, *Orientia* (formerly *Rickettsia*) *tsutsugamushi*, enters the skin during mite infestations and soon causes fever and a rash resembling epidemic typhus. Outbreaks may be significant, as evidenced by the 7,000 U.S. servicemen affected in the Pacific during World War II.

Rickettsialpox was first recognized in 1946 in an apartment complex in New York City. Investigators traced the disease to mites in the fur of local house mice, and named the disease rickettsialpox because the skin rash was similar to a chickenpox rash. Rickettsialpox is now considered a benign disease. It is caused by *Rickettsia akari* (*acari* = "mite") and transmitted by mite bites. Fever and rash are typical symptoms; fatalities are rare.

Other Tickborne Zoonoses Are Emerging Diseases in the United States

Ehrlichiosis and **anaplasmosis** are now recognized as two similar rickettsial tickborne diseases. **Human monocytic ehrlichiosis** (**HME**) is caused by *Ehrlichia chaffeensis* (because the first case was observed at Fort Chaffee, Arkansas) and **human granulocytic anaplasmosis** (**HGA**) is caused by *Anaplasma phagocytophilum* (FIGURE 12.15). The CDC reported over 700 cases of HME and over 1,700 cases of HGA in 2010. However, this may be an underestimate because identification and reporting of human cases are incomplete at the state level. Still, HGA cases predominate in the upper Midwest and Northeast while HME cases predominate in New York, North Carolina, and the central Midwest.

MICROINQUIRY 12

Dreams of Conquest Stopped by a Microbe

In the spring of 1812, Napoleon Bonaparte was at the height of his power. He was emperor to 45 million French and through conquering much of Europe had made France the wealthiest and most influential empire on the European continent. France's power was based on its Grande Armée against which no opposing army could match. With such power, Napoleon set out to expand his European conquests by invading Russia. On June 22, 1812, Napoleon's troops crossed the Niemen River in Poland and marched into Russia with 600,000 soldiers. Napoleon was confident that Tsar Alexander's outnumbered Russian troops would collapse when faced with the Grande Armée. However, a third, but microscopic enemy was lurking on the Russian frontier. This virulent and swift microscopic foe, this "war fever" bug, along with battle wounds and weather, would take the lives of all but a few thousand soldiers.

Having crossed the Niemen River and moving into Poland, progress became difficult for Napoleon's army as the roads were poor and getting needed supplies of food and water to the front lines was increasingly difficult. Because the Russians had destroyed much of Poland in their retreat, Napoleons' troops were forced to search for food in the countryside, pillaging from the peasants who lived in filthy conditions and were infested with bed bugs, fleas, and lice. Soon, many soldiers began to develop high fevers and pink rashes that developed on the body trunk and spread to the extremities. Some developed hallucinations and became delirious only to then feel like their heart had briefly stopped. The virulent microbe had begun its attack on the Grande Armée.

Within the first month of the campaign, Napoleon had lost more than 80,000 soldiers to the infection and, along with dysentery, the mortality and morbidity numbers soon reduced the Grande Armée to 130,000 soldiers. Napoleon's doctors could not cope with the magnitude of the disease and were powerless to explain what had struck down so many soldiers. Perhaps it was a miasma, as was a common belief for disease spread at the time.

Through World War I, infections killed more soldiers than did the enemy's swords and bullets. "War fever," one of the most common infections, is the informal name for the infectious disease that struck Napoleon's troops. Hans Zinsser, the American bacteriologist and author of *Rats, Lice and History* (1935), said that historically this fever *"had come to be the inevitable and expected companion of war and revolution; no encampment, no campaigning army, and no besieged city escaped it."* In fact, the conditions of combat are perfect for "war fever" to explode into a raging epidemic because filthy living conditions, crowding, and malnutrition encouraged its spread.

Napoleon's retreat from Russia left many dead from "war fever." Drawing by Raffet and engraved by J Smith. © Hulton Archive/Stringer/Getty Images.

Napoleon's forces entered Moscow in September with only 100,000 men. In Moscow, most Russian citizens had fled the city with their belongings and deliberately set fire to the city on their retreat. There was little habitable shelter in Moscow and no food—and Tsar Alexander refused to surrender. So, on October 19, Napoleon's remaining 95,000 soldiers started the 1,000 mile retreat as the storms of the Russian winter approached. The filthy, dirty clothing of the soldiers would be a warm home for the lice carrying the disease.

Once infected with the "war fever" microbe, a soldier experienced a high fever that continued for some 2 weeks during which time symptoms might also include severe headaches, bronchial disturbances, and mental confusion. After about 6 days, a red rash appeared on the torso, then on the hands, feet, and face. The soldiers also slept in large groups in confined spaces and here the lice could quickly move to uninfected soldiers, spreading the disease. Soldiers would wake from sleeping with a horrendous tingling as lice covered and crawled over their bodies.

Thousands more died from "war fever" as the first snows began to fall on November 3 and the temperature became bitterly cold. Many soldiers froze to death during the return march while other disease victims huddled together on rotten straw mixed with their own excrement to keep warm. They were delirious from hunger and "war fever" spread throughout the countryside that winter. By the end of December, the remnants of the Grande Armée finally crossed the Niemen. It is estimated that of the original 600,000 soldiers, more than 50% of the deaths may have been solely from "war fever." It should be mentioned that the Russian army also suffered from malnutrition, exposure, dysentery, and "war fever." Estimates suggest that more than 100,000 Russian soldiers died from wounds and disease. The actual number probably was much higher.

In the end, the clash between two imposing and remarkable forces meeting on the road to Moscow resulted in a microbial victory. French military strength would never fully recover and Napoleon's illustrious career was crushed, partly the result of microbes. And then the microbe disappeared—and by 1814 it was no longer a major epidemic killer in armed conflict.

12. From the description of the battle and the symptoms of infection, what was the disease that caused "war fever"?

In 2001, construction work in Vilnius, the capital of Lithuania, unearthed a mass grave containing between 2,000 and 3,000 corpses from Napoleon's Russian campaign. Dr. Didier Raoult and colleagues of the University of the Mediterranean in Marseille, France, examined the grave site and were able to amplify DNA sequences from the remains of lice and from dental pulp from the soldiers' teeth. The analysis identified the presence of *Rickettsia prowazekii* in some of the teeth (positive signals were found in three out of 35 different soldiers) as well as *Bartonella quintana*, the causative agent of trench foot, in some teeth as well as in lice.

Discussion Point

Why did "war fever" become an explosive epidemic in the Russian campaign, killing tens of thousands, if not hundreds of thousands of troops and citizens? With filthy clothing and uniforms, how did the lice transmit the "war fever" microbe?

Answers (and comments) can be found on the Student Companion Website in **Appendix D**.

If you want to read the whole story, check out a copy of Stephan Talty's *The Illustrious Dead: The Terrifying Story of How Typhus Killed Napoleon's Army*. Crown Publishers: New York. 2009.

HME is transmitted by the lone star tick (prevalent in the South), while HGA is transmitted by the blacklegged tick, which also transmits Lyme disease (prevalent in the Northeast). Patients with HME or HGA suffer from headache, malaise, and high fever, with some liver disease and, infrequently, a maculopapular rash. Indeed, both HGA and HME are quite similar to Lyme disease, except the symptoms come on faster in HGA and in HME, they clear more quickly, and the rash is infrequent. Because *E. chaffeensis* reproduces in the body's monocytes (hence "monocytic") and *A. phagocytophilum* reproduces in neutrophils (a type of granulocyte, hence "granulocytic"), both bring about a lowering of the white blood cell count (**leukopenia**).

Doxycycline is the drug of choice for both diseases.

The systemic bacterial diseases are summarized in TABLE 12.2.

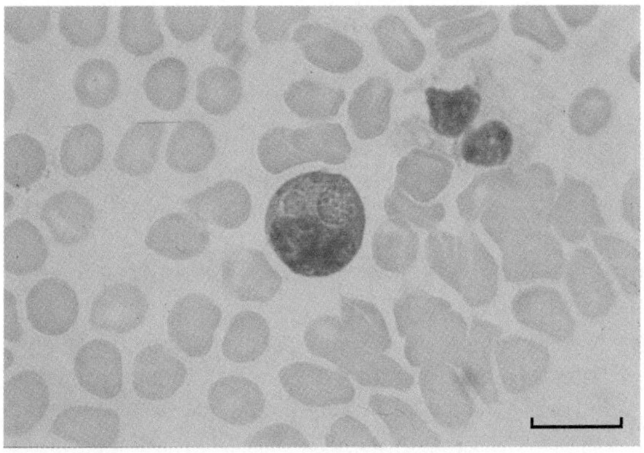

FIGURE 12.15 ***Anaplasma phagocytophilum* in a Blood Smear.** *A. phagocytophilum* causes human granulocytic ehrlichiosis (HGA). This tickborne organism multiplies in human white blood cells (stained purple) called granulocytes. The granulocytes are neutrophils. (Bar = 10 µm.) Courtesy of Armed Forces Institute of Pathology. »» What are the predominant light pink cells in this blood smear?

CONCEPT AND REASONING CHECKS 3

a. What distinguishes the different rickettsial diseases from one another?

b. Draw a concept map for the rickettsial diseases indicating how the diseases differ from one another.

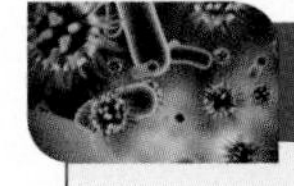

Chapter Challenge C

The rickettsial and ehrlichial arthropodborne pathogens discussed in this final section were transmitted by ticks (RMSF, ehrlichiosis, anaplasmosis), lice (epidemic typhus), and fleas (murine typhus).

QUESTION C:

Which of these arthropodborne pathogens could be considered euedaphic pathogens and which are soil-transmitted pathogens based on the vector used to transmit the pathogen?

Answers can be found on the Student Companion Website in **Appendix D**.

TABLE

12.2 A Summary of Arthropodborne Bacterial Diseases

Disease	Causative Agent	Signs and Symptoms	Transmission	Treatment	Prevention
Plague	*Yersinia pestis*	Bubonic: Sudden onset of fever and chills, headache, fatigue, muscle aches, and buboes Septicemic: High fever, abdominal pain, diarrhea Pneumonic: Headache, malaise, extensive coughing	Infected flea bite Inhaled infectious droplets from person or animal	Streptomycin and gentamicin given intravenously or intramuscularly	Avoiding contact with sick or dead animals Flea control
Tularemia	*Francisella tularensis*	Flu-like symptoms Own set of symptoms depending on body site	Several modes of transmission: ticks, sick or dead animals, airborne, contaminated food or water	Streptomycin	Avoiding and preventing tick bites Avoiding sick or dead animals
Lyme disease	*Borrelia burgdorferi*	Rash, flu-like symptoms, joint pain	Bite of infected deer tick	Amoxicillin	Avoiding and preventing tick bites
Tickborne/ Relapsing fever	*Borrelia hermsii*	Periods of fever and chills	Bite of infected tick	Doxycycline	Avoiding and preventing tick bites
Rocky Mountain spotted fever	*Rickettsia rickettsii*	High fever, severe headache, skin rash of red spots	Bite of infected hard tick	Doxycycline or tetracycline	Avoiding and preventing tick bites
Epidemic typhus	*Rickettsia prowazekii*	Fever and rash	Scratching bites from body lice	Doxycycline	Avoiding and preventing lice infestation
Murine typhus	*Rickettsia typhi*	Mild fever, persistent headache, rash	Bite of infected flea	Doxycycline	Avoiding and preventing flea bites
Ehrlichiosis and Anaplasmosis	(HME) *Ehrlichia chaffeensis* (HGA) *Anaplasma phagocytophilum*	Headache, malaise, fever	Bite of infected lone star tick Bite of blacklegged tick	Tetracycline or doxycycline	Avoiding and preventing tick bites

SUMMARY OF KEY CONCEPTS

12.1 Several Soilborne Bacterial Diseases Develop from Endospores

1. **Anthrax** is an acute infectious disease caused by *Bacillus anthracis*. Human contact can be by inhalation, consumption, or skin contact with spores. Inhalation produces symptoms of respiratory distress and causes a blood infection. Consumption and skin contacts lead to boil-like lesions. (Figs. 12.2, 12.3)
2. *Clostridium tetani* is the causative agent of **tetanus**. Symptoms of generalized muscle stiffness and **trismus** lead to convulsive contractions with an unnatural fixed smile. Antitoxin and antibiotics can be used to neutralize the toxin and kill the bacterial cells. (Fig 12.4)
3. **Gas gangrene** is caused by the anaerobic endospore-forming *Clostridium perfringens*. Symptoms include intense pain, swelling, and a foul odor at the wound site. (Fig. 12.5)
4. **Leptospirosis** is a disease spread from animals to humans (**zoonosis**) by *Leptospira interrogans*. Infected individuals have flu-like symptoms. Up to 10% of patients experience a systemic form of the disease. (Fig. 12.6)

12.2 Some Bacterial Diseases Can Be Transmitted by Arthropods

5. *Yersinia pestis* is the causative agent of **plague**. This highly fatal infectious disease is transmitted to humans by the bites of infected fleas. **Bubonic plague** is characterized by the formation of **buboes**. Spreading of the bacilli to the blood leads to **septicemic plague**. Localization in the lungs is characteristic of **pneumonic plague**, which can be spread person to person. Without treatment, septicemic and pneumonic plague are nearly 100% fatal. (Fig. 12.7)
6. **Tularemia** is caused by *Francisella tularensis*, which is highly infectious at low doses. Various forms of the disease occur depending on where the bacilli enter the body. Skin ulcers, eye lesions, and pulmonary symptoms can result. (Fig. 12.9)

7. **Lyme disease** results from an infection by the spirochete *Borrelia burgdorferi*. It involves three stages: the **erythema migrans (EM)** rash; neurological and cardiac disorders of the central nervous system; and migrating arthritis. (Fig. 12.11)
8. **Relapsing fever** results in recurring attacks of high fever caused by ticks carrying *Borrelia hermsii* (**endemic relapsing fever**). Symptoms include substantial fever, shaking chills, headache, and drenching sweats. (Fig 12.13)

12.3 Rickettsial and Ehrlichial Diseases Are Arthropodborne

9. Rickettsial and ehrlichial infections usually involve a skin rash, fever, or both.
   - **Rocky Mountain spotted fever** is caused by *Rickettsia rickettsii*, an intracellular parasite. Carried by ticks, the symptoms include a high fever for several days, severe headache, and a maculopapular skin rash. (Fig. 12.14)
   - **Epidemic typhus** is a potentially fatal disease occurring in unsanitary conditions and in overcrowded living conditions. It is caused by *Rickettsia prowazekii* that is carried by body lice. A maculopapular rash progresses to the extremities and an intense fever, hallucinations, and delirium are characteristic symptoms of the disease. In epidemics, mortality can be as high as 75%.
   - **Murine typhus** is transmitted by fleas. The causative agent, *Rickettsia typhi,* causes a mild fever, headache, and maculopapular rash. Recovery often is spontaneous.
   - Other diseases caused by rickettsiae include **scrub typhus** (*Orientia tsutsugamushi*) and **rickettsialpox** (*Rickettsia akari*).
10. Ehrlichiosis has been recognized in two forms: **human monocytic ehrlichiosis** (*Ehrlichia chaffeensis*) and **human granulocytic anaplasmosis** (*Anaplasma phagocytophilum*).

CHAPTER SELF-TEST

For **STEPS A–D**, answers to even-numbered questions and problems can be found in **Appendix C** on the Student Companion Website at **http://microbiology.jbpub.com/10e**. In addition, the site features eLearning, an online review area that provides quizzes and other tools to help you study for your class. You can also follow useful links for in-depth information read more MicroFocus stories, or just find out the latest microbiology news.

STEP A: REVIEW OF FACTS AND TERMS

Multiple Choice

Read each question carefully, then select the ***one*** answer that best fits the question or statement.

1. Woolsorter disease applies to the _____ form of _____.
   - A. inhalation; tularemia
   - B. toxic; myonecrosis
   - C. intestinal; anthrax
   - D. inhalation; anthrax
2. Which one of the following describes the mode of action of tetanospasmin?
   - A. It inhibits muscle contraction.
   - B. It damages and lyses red blood cells.
   - C. It disrupts cell tissues.
   - D. It inhibits muscle relaxation.
3. A crackling sound associated with myonecrosis is due to _____.
   - A. respiratory distress due to plague
   - B. nerve contractions due to tetanus
   - C. lymph node swelling due to plague
   - D. gas produced by *C. perfringens*
4. *Leptospira interrogans* has all the following characteristics *except* _____.
   - A. endoflagella
   - B. aerobic metabolism
   - C. exotoxin production
   - D. a hook at one end of the cell
5. A characteristic of *Y. pestis* is a/an _____ staining.
   - A. gram-positive
   - B. bipolar
   - C. acid-fast
   - D. gram-variable
6. Skin ulcers are a common lesion resulting from being bitten by _____.
   - A. a tick infected with *B. burgdorferi*
   - B. fleas infected with *Y. pestis*
   - C. a tick infected with *F. tularensis*
   - D. lice infected with *C. tetani*
7. Erythema migrans is typical of the _____ stage of Lyme disease.
   - A. early localized
   - B. early disseminated
   - C. late
   - D. recurrent
8. A brief tick bite and a small number of recurring periods of fever and chills is typical of _____.
   - A. relapsing fever
   - B. ehrlichiosis
   - C. Rocky Mountain spotted fever
   - D. epidemic typhus
9. Rocky Mountain spotted fever is most common in the _____.
   - A. southeastern United States
   - B. Rocky Mountains
   - C. Pacific northwest
   - D. New England
10. A lowering of the white blood cell count is characteristic of _____.
    - A. plague
    - B. anthrax
    - C. ehrlichiosis
    - D. RMSF

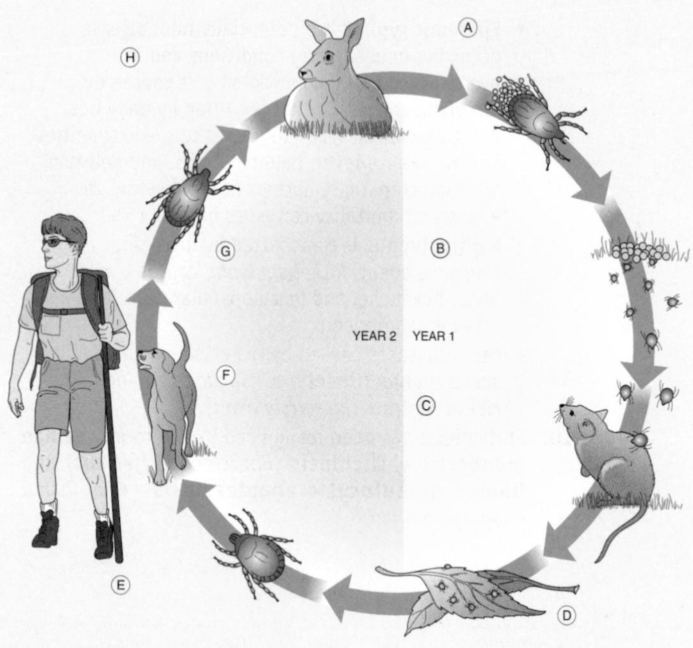

11. In the figure to the left, outline the life cycle (A-H) of the Ixodid tick responsible for the transmission of Lyme disease.

Matching

Match the statement on the left with the disease on the right by placing the correct letter in the available space. A letter may be used once, more than once, or not at all.

Statement

___ **12.** Accompanied by erythema migrans.

___ **13.** Transmitted by lice; caused by *Rickettsii prowazekii.*

___ **14.** Affects monocytes in the body; transmitted by ticks.

___ **15.** Caused by a spore-forming rod that produces hyaluronidase and α-toxin.

___ **16.** Primarily endemic in large herbivores, such as cattle, sheep, and goats.

___ **17.** Treated with antitoxins; caused by an anaerobic spore former.

___ **18.** Bubonic, septicemic, and pneumonic stages.

___ **19.** Maculopapular rash beginning on extremities and progressing to body trunk.

___ **20.** Caused by a spirochete that infects kidney tissues in pets and humans.

___ **21.** Occurs in small game animals, especially rabbits.

___ **22.** Caused by a gram-negative rod with bipolar staining; transmitted by the rat flea.

___ **23.** Up to 13 attacks of substantial fever, joint pains, and skin spots; *Borrelia* involved.

___ **24.** Pulmonary, intestinal, and skin forms possible; due to a *Bacillus* species.

___ **25.** Most commonly reported tickborne disease in the United States.

Disease

A. Anthrax
B. Ehrlichiosis
C. Epidemic typhus
D. Gas gangrene
E. Leptospirosis
F. Lyme disease
G. Murine typhus
H. Plague
I. Relapsing fever
J. Rickettsialpox
K. Rocky Mountain spotted fever
L. Tetanus
M. Tularemia

STEP B: CONCEPT REVIEW

26. Summarize the clinical significance of the three forms of **anthrax.** (**Key Concept 1**)
27. Identify how *Clostridium tetani* causes **tetanus** and list the symptoms of the disease. (**Key Concept 1**)
28. Explain why **gas gangrene** is referred to as **myonecrosis** and identify the toxins and enzymes involved with the disease. (**Key Concept 1**)
29. Distinguish between the more mild and systemic forms of **leptospirosis.** (**Key Concept 1**)
30. Contrast between **bubonic, septicemic,** and **pneumonic plague.** (**Key Concept 2**)
31. Summarize the clinical significance of glandular and inhalation **tularemia.** (**Key Concept 2**)
32. Distinguish between the three stages of untreated **Lyme disease.** (**Key Concept 2**)
33. Compare and contrast the two forms of **relapsing fever.** (**Key Concept 2**)
34. Identify the hallmarks of **Rocky Mountain spotted fever** and how it differs from **epidemic** and **murine typhus.** (**Key Concept 3**)
35. Discuss the characteristics of the two forms of **ehrlichiosis.** (**Key Concept 3**)

STEP C: APPLICATIONS AND PROBLEMS

36. In February 1980, a patient was admitted to a Texas hospital complaining of fever, headache, and chills. He also had greatly enlarged lymph nodes in the left armpit. A sample of blood was taken and Gram stained, whereupon gram-positive diplococci were observed. The patient was treated with cefoxitin, a drug for gram-positive organisms, but soon thereafter he died. On autopsy, *Yersinia pestis* was found in his blood and tissues. Why was this organism mistakenly thought to be diplococci, and what error was made in the laboratory? Why were the symptoms of plague missed?
37. Some estimates place epidemic typhus among the all-time killers of humans; one listing even has it in third place behind malaria and plague. In 1997, during a civil war, an outbreak of epidemic typhus occurred in the African country of Burundi. The outbreak was estimated to be the worst since World War II. What conditions may have led to this epidemic?
38. Leptospirosis has been contracted by individuals working in such diverse locales as subway tunnels, gold mines, rice paddies, and sewage-treatment plants. As an epidemiologist, what precautions would you suggest these workers take to protect themselves against the disease?
39. A young woman was hospitalized with excruciating headache, fever, chills, nausea, muscle pains in her back and legs, and a sore throat. Laboratory tests ruled out meningitis, pneumonia, mononucleosis, toxic shock syndrome, and other diseases. On the third day of her hospital stay, a faint pink rash appeared on her arms and ankles. By the next day, the rash had become darker red and began moving from her hands and feet to her arms and legs. Can you guess the eventual diagnosis?
40. In Chapter 9 of the Bible, in the Book of Exodus, the sixth plague of Egypt is described in this way: "Then the Lord said to Moses and Aaron, 'Take a double handful of soot from a furnace, and in the presence of Pharaoh, let Moses scatter it toward the sky. It will then turn into a fine dust over the whole land of Egypt and cause festering boils on man and cattle throughout the land.' " Which disease in this chapter is probably being described?

STEP D: QUESTIONS FOR THOUGHT AND DISCUSSION

41. Although the tetanus toxin is second in potency to the toxin of botulism, many physicians consider tetanus to be a more serious threat than botulism. Would you agree? Why?
42. Murine typhus was observed in five members of a Texas household. On investigation, epidemiologists learned that family members had heard rodents in the attic, and 2 weeks previously they had used rat poison on the premises. Investigators concluded that both the rodents and the rat poison were related to the outbreak. Why?
43. Centuries ago, the habit of shaving one's head and wearing a wig probably originated in part as an attempt to reduce lice infestations in the hair. Why would this practice also reduce the possibilities of certain diseases? Which diseases?
44. Even before October 2001, people from various government and civilian agencies were concerned about a terrorist attack using anthrax spores. In various scenarios, try to paint a picture of how such an attack might happen. Then, using your knowledge of microbiology, present your vision of how agencies might deal with such an attack.
45. At various times, local governments are inclined to curtail deer hunting. How might this lead to an increase in the incidence of Lyme disease?
46. In autumn, it is customary for homeowners in certain communities to pile leaves at the curbside for pickup. How might this practice increase the incidence of tularemia, Lyme disease, and Rocky Mountain spotted fever in the community?

13

CHAPTER PREVIEW

13.1 Portions of the Female and Male Reproductive Systems Contain an Indigenous Microbiota

13.2 Many Sexually Transmitted Diseases Are Caused by Bacteria

TEXTBOOK CASE 13: A Case of Chlamydia with Tragic Consequences

13.3 Urinary Tract Infections Are the Second Most Common Body Infection

Investigating the Microbial World 13: Does Cranberry Juice Cure Urinary Tract Infections?

13.4 Contact Diseases Can Be Caused by Indigenous Bacterial Species

13.5 Contact Diseases Can Also Be Caused by Exogenous Bacterial Species

MICROINQUIRY 13: Sexually Transmitted and Contact Disease Identification

13.6 Several Contact Diseases Affect the Eye

Image courtesy of Dr. Fred Murphy/CDC.

Sexually Transmitted and Contact Transmitted Bacterial Diseases

I just froze. Then I closed the door, and went in my room and cried.
—A soft-spoken, 38-year-old woman recalling her reaction when she was visited by a health official and told she had syphilis

What do Abraham Lincoln, Adolf Hitler, Friedrich Nietzsche, Oscar Wilde, Ludwig van Beethoven, and Vincent van Gogh have in common? Very likely all suffered from syphilis if Deborah Hayden's research is correct—and it more than likely is. In 2003, she wrote a book entitled *Pox: Genius, Madness, and the Mysteries of Syphilis* (New York: Basic Books) in which she looks at 14 eminent figures from the 15th to 20th centuries whose behavior, careers, or personalities were more than likely shaped by this sexually transmitted disease.

Syphilis originally was called the Great Pox to separate it from smallpox and until the introduction of penicillin in 1943, was untreatable. It caused a chronic and relapsing disease that could disseminate itself throughout the body, only to reappear later as so-called tertiary syphilis. In this most dangerous and terminal form, the disease produces excruciating headaches, gastrointestinal pains, and eventually deafness, blindness, paralysis, and insanity.

Yet sometimes ecstasy and fierce creativity were part of the "symptoms." As Deborah Hayden says, "one of the 'warning signs' of tertiary syphilis is the sensation of being serenaded by angels." In fact, writer Karen Blixen (Isak Dinesen) once said that "Syphilis sold her soul to the devil for the ability to tell stories." Deborah Hayden believes it is just such emotions that provided much of the creative spark for many of the notable historical figures she describes in her book. Perhaps the most intriguing is the debated proposal that syphilis may have driven Hitler mad and that he was dying of syphilis when he committed suicide in his Berlin bunker in the final days of World War II.

If Deborah Hayden's arguments are true, it is amazing how a bacterial organism has affected the body and mind in different ways, shaping the thoughts of writers and philosophers, the creative genius of artists, composers, and scientists—and yes, the madness of dictators.

Sexually transmitted diseases (STDs) remain a health problem in the United States and around the world, the seriousness of which is underscored by the number of infections. The Centers for Disease Control and Prevention (CDC) estimate that in 2007, more than 19 million Americans contracted an STD, almost half among the 15 to 24 year age group. Importantly, more than one STD can be acquired at the same time. In addition, individuals infected with an STD are two to five times more likely to acquire HIV than uninfected individuals if they are exposed to the virus through sexual contact.

The STDs discussed in this chapter make up four of the top 10 infectious diseases in the United States (FIGURE 13.1). However, the increase in STDs is but one example of how changing social patterns can affect the incidence of other bacterial diseases discussed in this chapter. For example, the incidence of leprosy in the United States has risen because in the last decades immigrant groups have brought the disease with them. Toxic shock syndrome was first recognized widely in 1980 when a new brand of high-absorbency tampon appeared on the commercial market. These and other contact and miscellaneous diseases will be addressed in this chapter. We will start with diseases associated with the reproductive system.

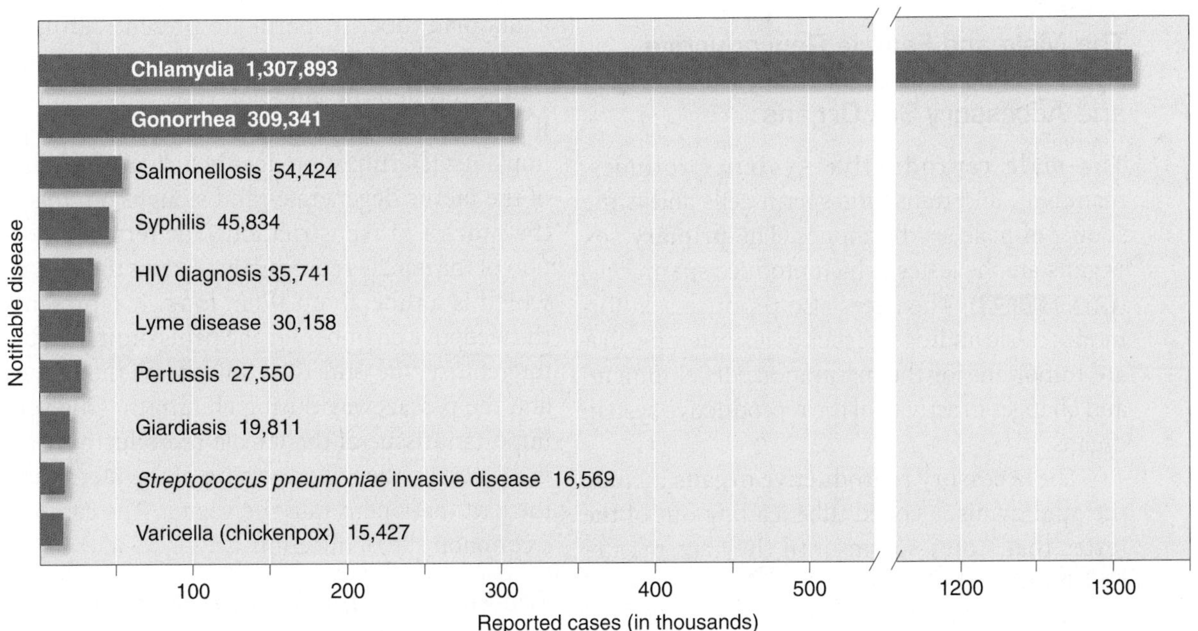

FIGURE 13.1 **Reported Cases of Notifiable Diseases in the United States, 2010.** Three of the top five most reported microbial diseases in the United States are sexually transmitted. »» How many of the top 10 diseases are of bacterial origin?

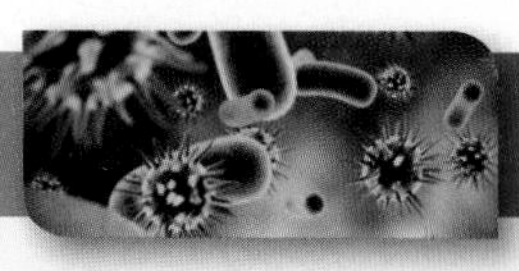

Chapter Challenge

This chapter on the infections and diseases caused by bacterial species brings together several different parts of human anatomy, including the female and male reproductive systems, the urinary system, the skin, and the eye. So, in this chapter, separate challenges will be offered for the different systems and organs discussed.

KEY CONCEPT 13.1 Portions of the Female and Male Reproductive Systems Contain an Indigenous Microbiota

Sexually transmitted diseases (STDs) are a group of infectious diseases usually transmitted through sexual contact and affect the male and female reproductive systems. They often are considered to be "hidden epidemics" because some individuals may not realize they are infected or are reluctant to disclose their condition because of the social stigma of these diseases. Look back again at Figure 13.1 to remind yourself of the significant numbers of reported infections causing chlamydia, gonorrhea, and syphilis.

To understand the diseases, we need to first briefly review the male and female reproductive systems.

The Male and Female Reproductive Systems Consist of Primary and Accessory Sex Organs

The **male reproductive system** produces, maintains, and transports sperm cells and is the source of male sex hormones. The **primary sex organs** are the testes, which produce sperm cells (FIGURE 13.2A). The testes also produce sex hormones called androgens, such as testosterone, that are important for the maturation, development, and changes in activity of the reproductive system organs.

The **accessory reproductive organs** include the epididymis, a coiled tube leading out of the testes that stores sperm until they are mature and motile. The epididymis terminates in the vas deferens, which combines with the seminal vesicle and terminates in the ejaculatory duct. The paired seminal vesicles secrete a viscous fluid that nourishes the sperm. The prostate gland is a walnut-shaped structure at the base of the urethra. It also contributes a slightly milky fluid that contains nutrients for sperm health and an antibacterial compound that combats urinary tract infections.

The external organs are the scrotum, containing the testes, and the penis, the cylindrical organ that houses the urethra and through which urine passes and semen is ejaculated.

The **female reproductive system** consists of the primary sex organs, the ovaries, and several accessory sex organs, specifically the fallopian tubes (also called uterine tubes), uterus, vagina, and vulva (FIGURE 13.2B). During ovulation, an egg is released from one of the ovaries and enters a fallopian tube. If sperm are present, fertilization may take place after which the fertilized egg moves through the fallopian tube to the uterus. It is in the uterine lining where the fertilized egg implants. If fertilization does not occur, the lining of the uterus degenerates and sloughs off; this is the process of menstruation. The terminal portion of the female reproductive tract is the vagina, which is a tube about 9 cm long. It represents the receptive chamber for the penis during sexual intercourse, the outlet for fluids from the uterus, and the passageway during childbirth. One very important tissue of the female reproductive tract is the cervix, the lower portion of the uterus and the part that opens into the vagina. The cervix is a common site of infection.

The Female Reproductive System Is Prone to More Infections than the Male Reproductive System

In the male reproductive system, only the urethra is colonized by indigenous microbes. In addition, the reproductive system can mount local immune responses, providing antibodies along the entire length of the urethra and in the seminal fluid.

As mentioned above, the female reproductive system is structured to bear a fetus. This anatomical organization therefore makes the female reproductive system more prone to infection. For example, while the vagina, vulva, and cervix harbor an indigenous microbiota, the rest of the system is sterile. To defend against infection, the vagina produces an acidic pH in which *Lactobacillus* species thrive and contribute to the acid condition by converting sugars to lactic acid. The acidic environment discourages the growth of many potential pathogens.

The mucosa of the vagina and cervix also contain immune defense mechanisms. Because the cervix represents a potential portal of entry for microbes from the lower genital tract, the cervix contains several antimicrobial defense mechanisms, including: a mucociliary escalator; mucus that contains a variety of antimicrobial chemicals, including lysozyme and lactoferrin; and antibacterial peptides that can kill or inhibit the growth of many bacterial species. One interesting note with the female reproductive system is that these chemicals and peptides fluctuate in concentration in the cervix and vagina during the menstruation cycle. As the concentrations of lactoferrin and IgA antibodies drop, the concentrations of human defensins and lysozyme rise (FIGURE 13.3). Thus, there is a constant chemical defense against potential microbial invasion.

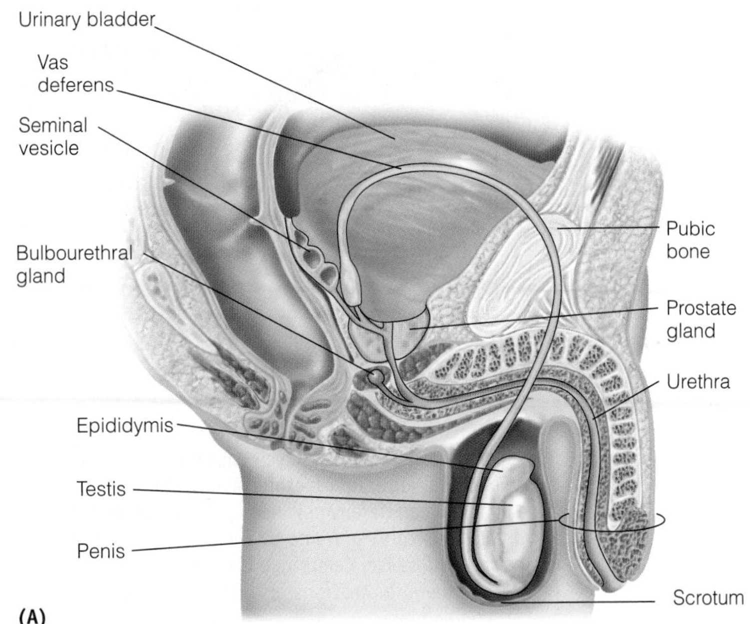

(A)

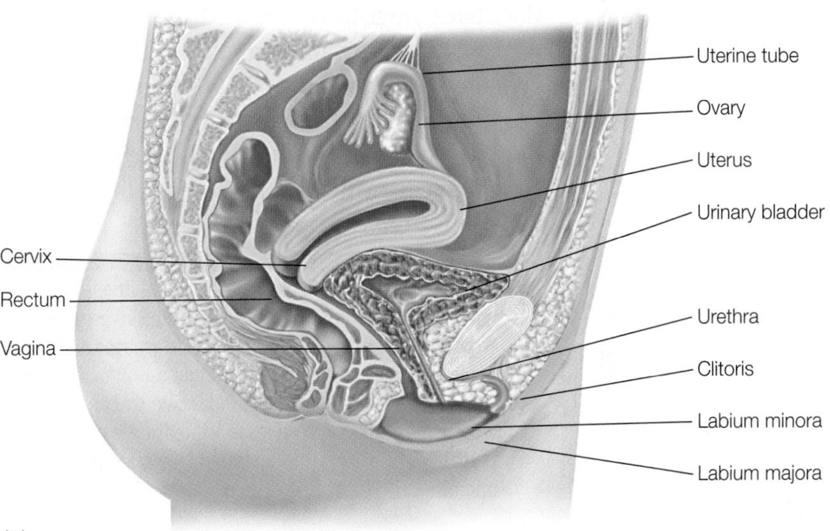

(B)

FIGURE 13.2 **Reproductive Systems of Males and Females.** The **(A)** male reproductive system and **(B)** female reproductive system showing the primary and secondary sex organs. »» Identify the primary and secondary sex organs in males and females.

Common Vaginal Infections Come from Indigenous Microbiota

Not all bacterial infections of the reproductive system originate from pathogens that are sexually transmitted. There is one noninflammatory condition of the female reproductive system that is normally caused by the indigenous microbiota.

Bacterial Vaginosis. The most common vaginal infection in women of child-bearing age is **bacterial vaginosis (BV)**, which occurs when there is a disruption in the normal balance of vaginal microbiota. Although most women are unaware of the condition, the lactobacilli that normally produce lactic acid and various bacteriostatic and bactericidal compounds in the vagina are replaced by a mixture of other indigenous bacteria, including *Gardnerella vaginalis, Prevotella,* and *Mobiluncus* (FIGURE 13.4). The reason for the switch in dominance from lactobacilli to the other polymicrobial species is not clearly understood. Menstruation normally causes a rise in vaginal pH to about 7 and this perhaps is one factor why the population changes.

For 50% of women with BV, there are no outward symptoms of the condition. If a woman does have symptoms, they include a foul-smelling

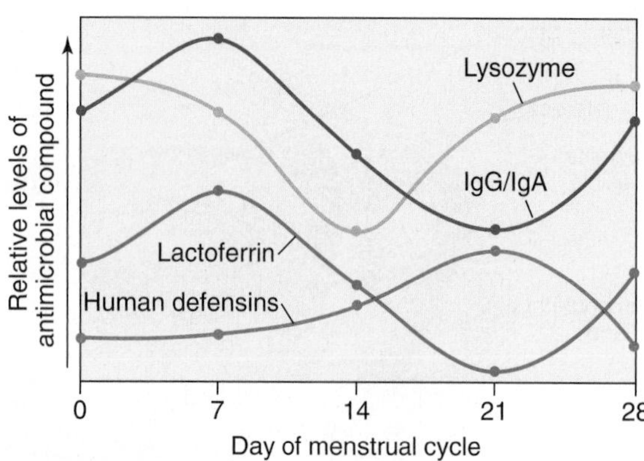

FIGURE 13.3 Fluctuations in Antimicrobial Chemicals and Proteins in the Vagina During the Menstrual Cycle. During the female menstrual cycle, as some antimicrobial compounds fall (IgG/IgA, lactoferrin), others rise (lysozyme, human defensins). Adapted from Wilson, M. *Microbial Inhabitants of Humans: Their Ecology and Role in Health and Disease*. Cambridge University Press. 2005. »» Why are there fluctuations in the levels of these antimicrobial compounds?

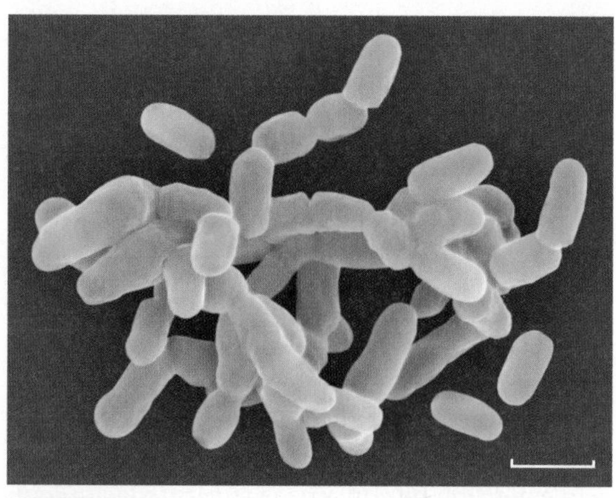

FIGURE 13.4 False-Color Scanning Electron Microscope Image of *Gardnerella vaginalis*. These *G. vaginalis* rods are part of the polymicrobial overgrowth during bacterial vaginosis. (Bar = 2 μm.) © Dennis Kunkel Microscopy, Inc./Phototake/Alamy Images. »» Could *G. vaginalis* be isolated from a healthy woman not suffering from bacterial vaginosis? Explain.

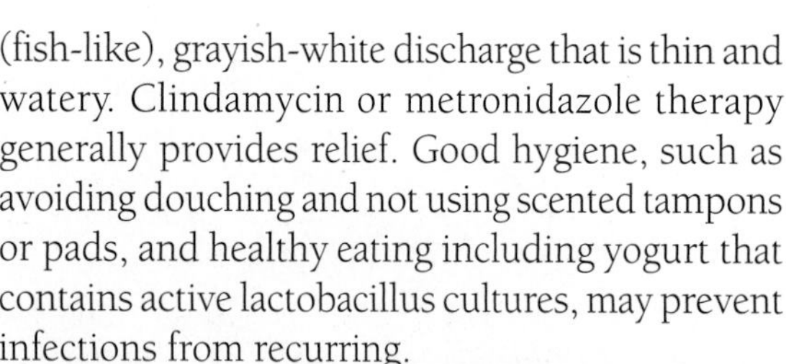

(fish-like), grayish-white discharge that is thin and watery. Clindamycin or metronidazole therapy generally provides relief. Good hygiene, such as avoiding douching and not using scented tampons or pads, and healthy eating including yogurt that contains active lactobacillus cultures, may prevent infections from recurring.

CONCEPT AND REASONING CHECKS 1

a. Trace the path of (a) a sperm cell and (b) an egg as they move through their respective reproductive systems.
b. Why does there appear to be a much greater and more diverse defense system in the female reproductive tract as compared to the male reproductive tract?
c. What factors appear to be responsible for the overgrowth of the vagina by other members of the indigenous microbiota?

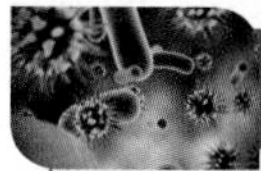

Chapter Challenge A

In this section, it was stated that the female reproductive system is prone to more infections than is the male reproductive system.

QUESTION A:

Do you agree that the female reproductive system is more prone to infections? Provide evidence to support your answer.

Answers can be found on the Student Companion Website in **Appendix D.**

KEY CONCEPT 13.2 Many Sexually Transmitted Diseases Are Caused by Bacteria

Until the 1990s, STDs were commonly known as **venereal diseases** (*venerea* = referring to Venus, the Roman goddess of love) because they are a group of infectious diseases usually transmitted through sexual contact. STDs also are known as **sexually transmitted infections (STIs)**, denoting the fact that a person may be infected—and can potentially infect others—without showing signs of disease.

STDs continue to be a major public health challenge in the United States. Although the incidence of syphilis and gonorrhea has remained relatively stable since 1995, the incidence of chlamydia over this period has continued to increase (FIGURE 13.5). The CDC estimates that almost 20 million new infections occur each year—and 50% of these infections occur in young people between the ages 15 to 24. This is in spite of the recent advances in preventing, diagnosing, and treating certain STDs. Despite the fact that STDs are extremely widespread, most people in the United States remain unaware of the risk and consequences of all but the most prominent STD—acquired immunodeficiency disease (AIDS).

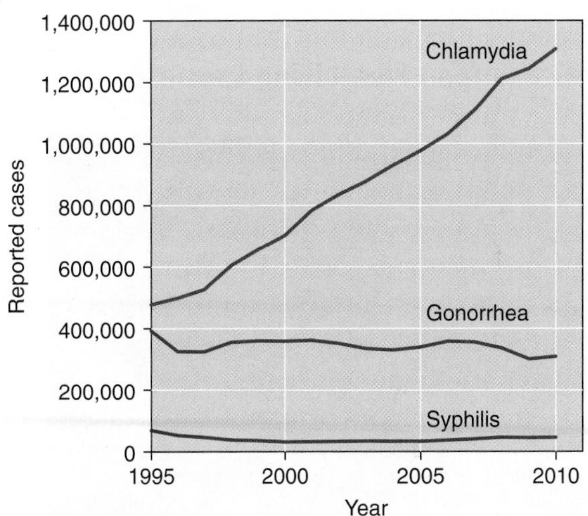

FIGURE 13.5 **Reported Cases of Chlamydia, Gonorrhea, and Syphilis—United States and U.S. Territories, 1995–2010.** The syphilis cases represent all stages of the disease. »» Why are STDs considered a health challenge?

Before discussing the bacterial species causing STDs, it is useful to mention a few distinctive characteristics that these pathogens possess. Because these organisms are dependent on sexual intercourse (which may be sporadic) for transmission, the pathogens must:

- Persist in an infective form until that person has sexual intercourse.
- Hide themselves from recognition and elimination by the immune system.
- Be ready to be transmitted when the infected individual has sexual intercourse.

In this section, we will consider the common STDs and the bacterial species that cause them. We will start with the big three: chlamydia, gonorrhea, and syphilis.

Chlamydial Urethritis Is the Most Frequently Reported STD

Causative Agent and Epidemiology. The disease "**chlamydia**" or **chlamydial urethritis** is caused by *Chlamydia trachomatis,* an exceptionally small (0.35 μm), round to ovoid-shaped organism with a cell membrane and outer membrane, but without any peptidoglycan. This obligate, intracellular parasite also has one of the smallest genomes (MicroFocus 13.1).

Chlamydial urethritis is the most common STD globally and the most frequently reported STD in the United States (Figure 13.1). In fact, the 2010 case count is the highest since reports began in the mid-1980s. About 75% of these cases occur in women between 15 and 29 years of age. Part of the reason for the rise is due to the increased number of screening programs available and the development of better diagnostic tests.

C. trachomatis has a biphasic and unique reproductive cycle (FIGURE 13.6). There is a nonreplicating, extracellular, infectious **elementary body** (**EB**) and a replicating, intracellular, noninfectious **reticulate body** (FIGURE 13.7). Humans appear to be the only host for the organism.

Chlamydial urethritis represents one of several diseases collectively known as **nongonococcal urethritis**, or **NGU**. NGU is a general term for a condition in which people without gonorrhea have a demonstrable infection of the urethra usually characterized by inflammation, and often accompanied by a discharge. Key components to *C. trachomatis'* ability to evade the host immune system and persist in the body include the ability of the bacterial cells to vary the proteins found on the outer membrane, preventing an attack by antibodies. The pathogen also produces several enzymes that inhibit the host inflammatory response.

Clinical Presentation. *C. trachomatis* causes a gonorrhea-like disease transmitted by vaginal, anal, or oral sex and any sexually active individual can be infected. The disease has an incubation period of about 1 to 3 weeks. Chlamydia often is referred to as the "silent disease" because the organism does not cause extensive tissue injury directly. Thus, some 85% to 90% of infected individuals have no symptoms, often do not seek treatment, and can unknowingly pass the disease on to others. If symptoms do occur, they are due to the inflammation caused by the immune system's response to limit the spread of the infection.

Females often note a slight cervical discharge, as well as inflammation of the cervix. Burning pain also is experienced on urination, reflecting infection in the urethra. In complicated cases, the disease may spread to, block, and inflame the fallopian tubes, causing **salpingitis**. About 40% of untreated infections progress to **pelvic inflammatory disease** (**PID**), which is an inflammation of the uterus, fallopian tubes, and/or ovaries. PID from chlamydia and gonorrhea is believed to affect about 50,000 women in the United States annually. Often, however, there are few symptoms of disease before the salpingitis manifests

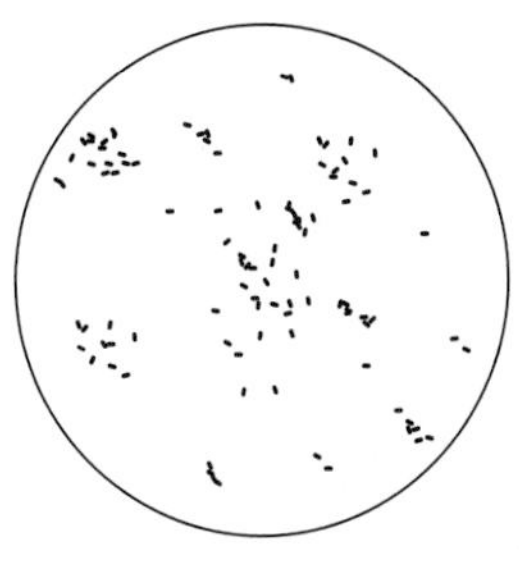

Chlamydia trachomatis

MICROFOCUS 13.1: Evolution

Why Do Intracellular Parasites Have Fewer Genes Than Free-Living Species?

The members of the phylum Chlamydiae are among the oldest described human pathogens, and have long fascinated researchers because of their obligate intracellular lifestyle, their small genome size, and the number of diseases they cause. *Chlamydia trachomatis* (see figure) is the causative agent of a number of diseases and has been divided into three groups: serotypes A–C can lead to trachoma, the leading cause of infectious blindness in the world; serotypes D–K cause sexually transmitted diseases of the urogenital tract, such as chlamydial urethritis; and serotype L causes another STD, lymphogranuloma venereum (LGV), which is a more invasive serotype than A–K.

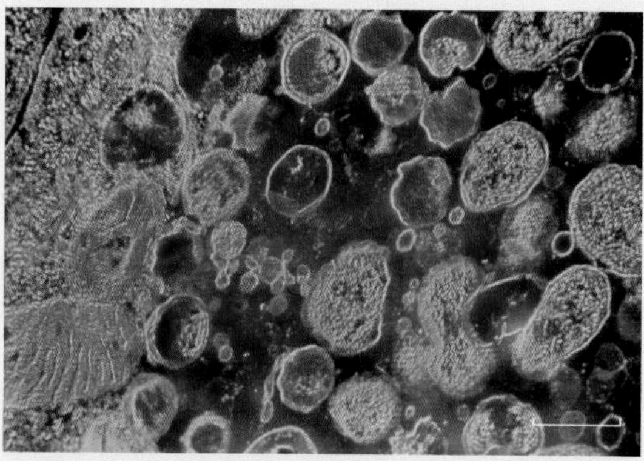

False-color transmission electron microscope image of *Chlamydia trachomatis* cells. (Bar = 0.5 µm.) © CNRI/Photo Researchers, Inc.

Genomics has provided important insights into the evolution of chlamydial types. Complete genome sequences have been produced for *C. trachomatis* biotypes A, D, and L. All three genomes consist of only about 1 million base pairs, while its free-living primitive cousin, *Protochlamydia amoebophila,* contains about 2.4 million base pairs. All three *C. trachomatis* genomes have some 846 protein-coding genes—its primitive cousin has about 2,000 protein-coding genes. All three chlamydial serotypes also share similar gene locations and nucleotide sequence similarities, suggesting that horizontal gene transfer (HGT) has not had significant impact on the genomes or disease pathology within *C. trachomatis*—HGT has had a significant role in the evolution and pathogenesis of other free-living pathogens. So why the small genome?

Once an organism "settles in" as an obligate intracellular parasite, there are gene functions that can be eliminated if the organism can get by without those gene products or by accessing them ready made from the host. Indeed, analysis of the *C. trachomatis* genomes shows that gene loss has occurred relative to its ancestors. As one example, several important metabolic pathways, including the citric acid cycle, are incomplete. Yet the organism can still make ATP. These examples of metabolic streamlining are consistent with those in many other pathogens that depend on an intracellular lifestyle. So, all-in-all, sequence analysis supports the large-scale deletion of genes from the organism. But blanket statements can be dangerous to make. A puzzling discovery is that all *C. trachomatis* types have retained the set of genes to synthesize and assemble peptidoglycan—even though the organisms apparently have no peptidoglycan in their cell walls. If genome reduction is correlated with no longer needing a product, why keep these genes, especially because the organism remains susceptible to penicillin? One would have to conclude that peptidoglycan synthesis of some sort is still involved with chlamydial binary cell division.

So, sequence analysis suggests the evolutionary history of this obligate intracellular pathogen has undergone significant genome reduction accompanied by loss of metabolic functions. In the steady-state environment of a host cell, obligate pathogens may no longer have a need for the diverse genes that allow free-living organisms to rapidly adapt to environmental conditions or, if needed, can be obtained directly from the host.

To really stir up the pot—one of the most surprising observations from genome analyses has been the identification of genes with a sequence similarity to plant sequences! These "plant genes" appear to be derived not from plants directly but rather from a cyanobacterial endosymbiont, reflecting an ancient evolutionary relationship between the Chlamydiae, Cyanobacteria, and the chloroplast. Perhaps further sequence analyses will uncover additional information regarding the evolution of the Chlamydiae from free-living bacteria to obligate parasites.

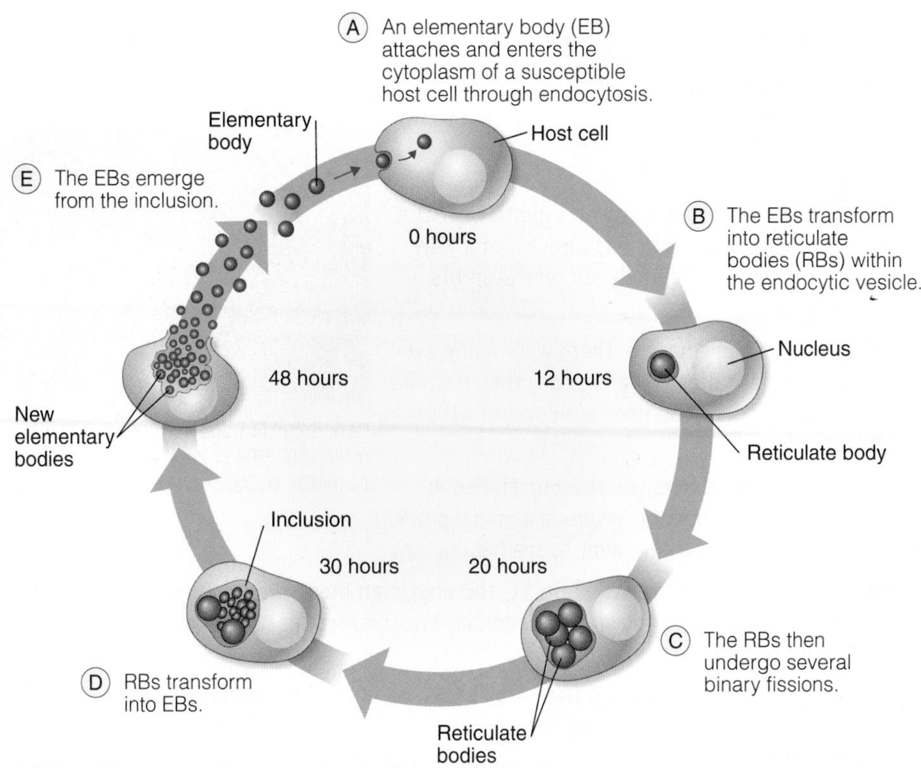

FIGURE 13.6 **The Chlamydial Life Cycle.** The reproductive cycle of the Chlamydiae involves two types of cells. After infection, nonreplicating elementary bodies (EBs) reorganize into reticulate bodies (RBs), which divide to form additional RBs. Within 30 hours, the RBs begin to reorganize into EBs within an inclusion. The EBs are released directly from the inclusion body or by the spontaneous lysis of the inclusion and host cell. »» Because *C. trachomatis* only reproduces inside cells, what cellular products might the bacterium require to survive and grow?

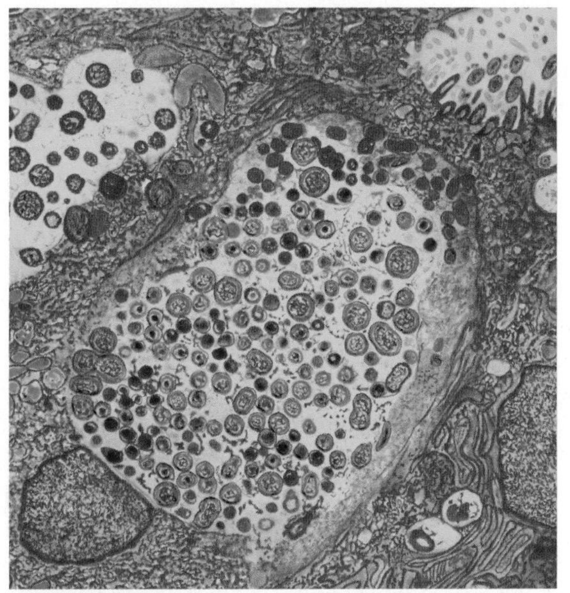

FIGURE 13.7 **A Chlamydial Inclusion.** False-color transmission electron micrograph of cells from a woman's fallopian tube infected with *Chlamydia trachomatis*. Spherical cells (green/brown) are elementary bodies and reticulate bodies seen inside an inclusion body (yellow). Cell cytoplasm is reddish-brown. © Dr. R. Dourmashkin/Photo Researchers, Inc. »» What complications can arise from an infection of the fallopian tubes?

itself, thus adding to the danger of infertility and ectopic pregnancy (TEXTBOOK CASE 13). Chlamydial urethritis is the number one cause of first trimester pregnancy-related deaths in the United States.

■ Ectopic pregnancy: Development of a fertilized egg outside the womb, often in a fallopian tube.

In males, chlamydia is characterized by painful urination and a discharge that is more watery and less copious than in gonorrhea. The discharge often is observed after urinating for the first time in the morning. Tingling sensations in the penis are generally evident. Inflammation of the epididymis may result in sterility, but this complication is uncommon. Chlamydial pharyngitis or proctitis is possible as a result of oral or anal intercourse.

■ Proctitis: An inflammation of the rectum.

Newborns may contract *C. trachomatis* during delivery from an infected mother and develop a disease of the eyes known as **neonatal conjunctivitis**.

Treatment and Prevention. Chlamydial infections may be treated successfully (95% cure rate) with one dose of azithromycin or 7 days with doxycycline. If a woman is pregnant, erythromycin is substituted because doxycycline affects bone formation in newborns.

Textbook CASE 13

A Case of Chlamydia with Tragic Consequences

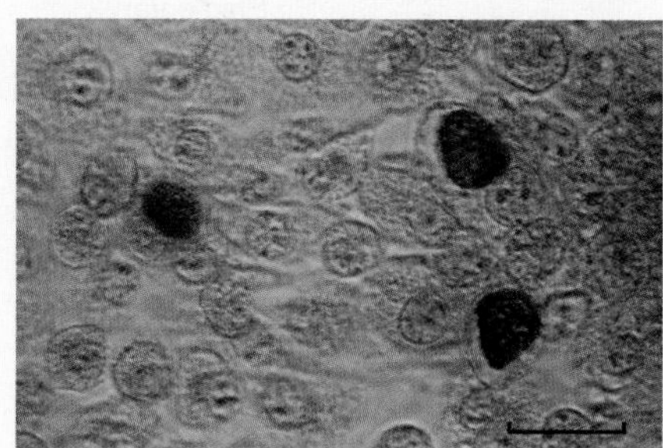

Dark inclusion bodies typical of a chlamydial infection. (Bar = 2 μm.) Courtesy of Dr. E. Arum/Dr. N. Jacobs/CDC.

1 An educated, professional woman met a gentleman at a friend's house one evening. She was director of a New York law firm. The man was equally successful in his professional career.

2 The couple hit it off immediately. There were many evenings of quiet candlelit dinners, and soon, they became sexually intimate. Neither one used condoms or other means of protection.

3 Three days after having intercourse, the woman began experiencing fever, vomiting, and severe abdominal pains. She immediately made an appointment to see her doctor.

4 Assuming the illness was an intestinal upset, the physician prescribed appropriate medication. The woman was actually suffering from chlamydia, but the fact that she was sexually active did not come up during the examination.

5 Feeling better, the woman continued her normal routine. But 6 months later, with no apparent warning, she collapsed on a New York City sidewalk.

6 The woman was rushed to a local hospital, where doctors diagnosed a chlamydial infection by observing the dark inclusion bodies typical of chlamydia (see figure). They performed emergency surgery: her uterus and fallopian tubes were badly scarred. She recovered; however, the scarring left her unable to have children.

Questions:

(Answers can be found on the Student Companion Website in **Appendix D.**)

A. Would condom use block the transfer of *C. trachomatis* cells?

B. Does fever, vomiting, and abdominal pains indicate a chlamydial infection?

C. What "complications" were developing during the 6 months when the woman was in improved health?

D. What are inclusion bodies?

E. How would scarring lead to infertility?

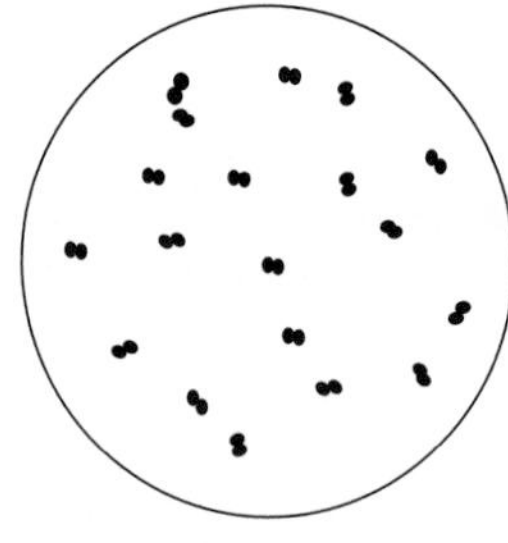

Neisseria gonorrhoeae

Timely treatment of a patient's sexual partners is critical to prevent reinfection. Sexual partners are evaluated, tested, and treated if they have had contact with the patient in the previous 60 days.

Gonorrhea Can Be an Infection in Any Sexually Active Person

Causative Agent and Epidemiology. Gonorrhea is the second most frequently reported notifiable STD in the United States (see Figure 13.1) with the highest percentage of infections being in persons 15 to 29 years of age (slightly more cases in females). Epidemiologists suggest that 3 to 4 million cases go undetected or unreported each year.

The agent of gonorrhea is *Neisseria gonorrhoeae*, a small, unencapsulated, nonmotile, gram-negative diplococcus named for Albert Neisser, who isolated it in 1879 (FIGURE 13.8). The organism, commonly known as the **gonococcus**, has a characteristic double-bean shape. *N. gonorrhoeae* is a very fragile organism susceptible to most antiseptics and disinfectants. Being sensitive to dehydration, it survives only a brief period outside the body and is rarely contracted from a dry surface such as a toilet seat. The great majority of cases of gonorrhea therefore are transmitted during sexual intercourse. (Gonorrhea is sometimes called "the clap," from the French "clappoir" for "brothel.")

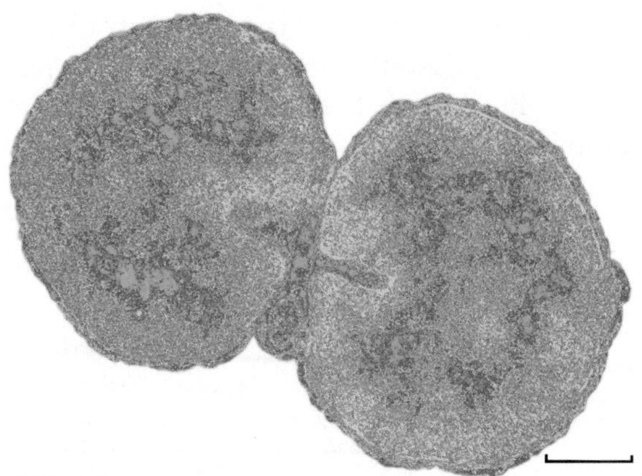

FIGURE 13.8 *Neisseria gonorrhoeae.* A false-color transmission electron micrograph of *N. gonorrhoeae* cells. (Bar = 5 µm.) © Dr. David Phillips/Visuals Unlimited/Corbis. »» What is the unique feature of these cells?

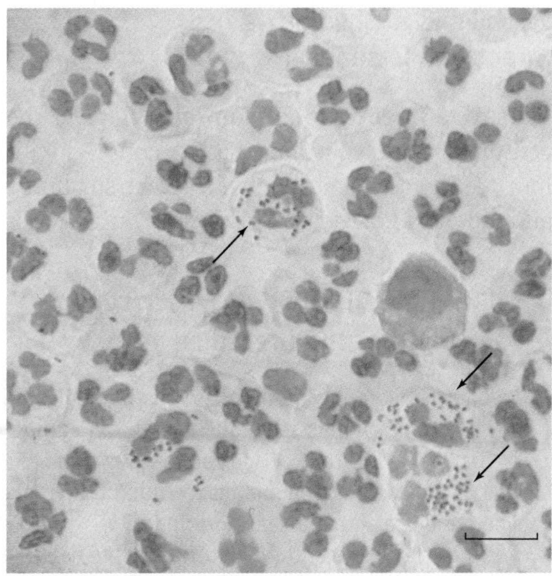

FIGURE 13.9 Gonococcal Urethral Smear. A Gram-stained smear of discharge from the male urethra showing the gonococci (arrows) in the cytoplasm of white blood cells. (Bar = 10 µm.) Courtesy of Joe Miller/CDC. »» What is the gram reaction for *N. gonorrhoeae*?

To attach to the genital tract, *N. gonorrhoeae* uses its pili to adhere to host epithelial cells. However, the pili will provoke a strong antibody response. Therefore, to escape detection and persist in the body the pathogen has a virtually limitless number of pilus protein variants that it can generate so pilin-specific antibodies never have a chance to bind to the cells.

Clinical Presentation. The incubation period for gonorrhea ranges from 2 to 6 days. In females, the gonococci invade and attach by pili to the epithelial surfaces of the cervix and the urethra. The cervix may be reddened, and a discharge may be expressed by pressure against the pubic area. Patients often report abdominal pain and a burning sensation on urination, and the normal menstrual cycle may be interrupted.

In some females, gonorrhea also spreads to the fallopian tubes. As these thin passageways become riddled with pouches and adhesions, salpingitis and PID may occur. Sterility may result from scar tissue remaining after the disease has been treated, or a woman may experience an ectopic pregnancy. It should be noted that symptoms are not universally observed in females, and an estimated 50% of affected women exhibit no symptoms. Such asymptomatic women may spread the disease unknowingly.

Symptoms tend to be more acute in males than in females, and males thus tend to seek diagnosis and treatment more readily. When gonococci infect the mucus membranes of the urethra, symptoms include a tingling sensation in the penis, followed in a few days by pain when urinating. There is also a thin, watery discharge at first, and later a more obvious yellow, thick fluid resembling semen (FIGURE 13.9). Frequent urination and an urge to urinate develop as the disease spreads further into the urethra. The lymph nodes of the groin may swell, and sharp pain may be felt in the testicles. Unchecked infection of the epididymis may lead to sterility.

Gonorrhea does not restrict itself to the urogenital organs. **Gonococcal pharyngitis**, for example, may develop in the pharynx if bacterial cells are transmitted by oral-genital contact; patients complain of sore throat or difficulty in swallowing. Infection of the rectum, or **gonococcal proctitis**, also is observed, especially in individuals performing anal intercourse. Transmission to the eyes may occur by fingertips or towels, developing into keratitis.

■ Keratitis: An inflammation and swelling of the cornea.

Gonorrhea is particularly dangerous to infants born to infected women. The infant may contract gonococci during passage through the birth canal and develop neonatal conjunctivitis.

Gonorrhea can be treated effectively with ceftriaxone. However, antibiotic resistance has become an important challenge to controlling gonorrhea such that the cephalosporin antibiotics have been the only class used. In 2011, the first strains resistant to the cephalosporins were identified. Their rise could transform the infection into a potential global public health threat.

Syphilis Is a Chronic, Infectious Disease

Causative Agent and Epidemiology. Over the centuries, Europeans have had to contend with four pox diseases: chickenpox, cowpox, smallpox, and the Great Pox, a disease now known as syphilis. The first European epidemic was recorded in the late 1400s, shortly after the conquest of Naples by the French army. For decades the disease had various names, but by the 1700s, it had come to be called **syphilis** (MicroFocus 13.2).

Syphilis is currently ranked among the top five most reported microbial diseases in the United

MICROFOCUS 13.2: History
The Origin of the Great Pox

Among the more intriguing questions in medical history are how and why syphilis suddenly emerged in Europe in the late 1400s. Writers of that period tell of an awesome new disease that swept over Europe and on to India, China, and Japan. But where did the actual disease originate?

Since the 16th century, historians have debated the geographic origins of syphilis. For most of the past 500 years, Christopher Columbus has been alternately blamed or exonerated for bringing syphilis to Europe. In the 20th century, critics of the "Columbian theory" proposed that syphilis had always been present in the Old World, perhaps coming from Spain and Portugal through the slave trade with Africa.

Now, a genetic analysis of the disease-causing *Treponema pallidum* carried out by Kristin Harper and colleagues at Emory University in Atlanta suggests that today's syphilis is a close cousin to another South American tropical disease called yaws, a tropical infection of the skin, bones, and joints.

To see if Columbus and his men introduced syphilis to Europe after "catching yaws" in the Americas, Harper and her colleagues in 2007 collected species of *T. pallidum* from across the globe (Africa, Europe, Asia, the Middle East, the Americas, and the Pacific Islands) to determine a phylogenetic family tree—which microbes gave rise to which—and perhaps determine where syphilis originated. They also sequenced and compared genes from *Treponema pertenue,* which causes yaws, from Africa and South America. By comparing genetic sequences between the *Treponema* species, the researchers found the species that cause syphilis originated relatively recently, with their closest relatives collected in South America and responsible for yaws.

However, yaws is spread through simple skin contact, not sexual intercourse. So how did a simple skin disease in South America become an STD in Europe? One suggestion is that syphilis became sexually transmitted only when it reached Europe, where the climate was not as hot and humid as it was in the tropics, and where people wore more clothing, limiting skin contact. Because skin contact would not be a viable transmission mechanism, the bacterium evolved to use sexual intercourse as the transmission mechanism.

This woodcut depicts a man infected with syphilis as evidenced by the lesions on the face and legs. The woodcut refers to syphilis as the result of an unfavorable planetary alignment in 1484. © Mary Evans Picture Library/Alamy Images.

To completely confirm the "New World" theory, more genetic analyses need to be done, including the sequencing of the entire DNA of the related *Treponema* species. Professor George Armelagos, one of co-workers on the project said, "Understanding [*T. pallidum's*] evolution is important not just for biology, but for understanding social and political history. It could be argued that syphilis is one of the important early examples of globalization and disease, and globalization remains an important factor in emerging diseases."

States (see Figure 13.1). Statistics indicate more than 45,000 people are afflicted with the disease annually; 12 million cases are reported each year worldwide. Taken alone, these figures suggest the magnitude of the syphilis epidemic, but some public health microbiologists believe for every case reported, as many as nine cases go unreported.

Syphilis is caused by *Treponema pallidum,* (*trepo* = "turn"; *nema* = "thread"; *pallid* = "pale"; thus literally "pale turning thread"). This spirochete moves by means of endoflagella. Humans are the only host for *T. pallidum,* so the organism must spread by direct human-to-human contact, usually during sexual intercourse.

T. pallidum is able to evade immune defenses because it has very few integral outer membrane proteins that the immune system is capable of recognizing. The lack of these proteins makes the cells virtually invisible to the immune system and allows the pathogen to persist in the body.

The pathogen penetrates the skin surface through the mucous membranes of the genitalia or via a wound, abrasion, or hair follicle. The variety of clinical symptoms accompanying the stages, and their similarity to other diseases, have led some physicians to call syphilis the "great imitator." Untreated, the disease can progress through a number of stages.

Clinical Presentation. The incubation period for syphilis varies greatly, but it averages about 3 weeks. **Primary syphilis** is the first stage of the disease. It is characterized by a lesion, called a **chancre**, which is a painless circular, purplish ulcer with a small, raised margin with hard edges (FIGURE 13.10A). The chancre develops at the site of entry of the spirochetes, often the genital organs. However, any area of the skin may be affected, including the pharynx, rectum, or lips. The chancre teems with spirochetes and represents the stage that is most infectious. It persists for 3 to 6 weeks, and then heals spontaneously. However, the infection has not been eliminated, as the spirochetes have spread through the blood and lymph to other body organs.

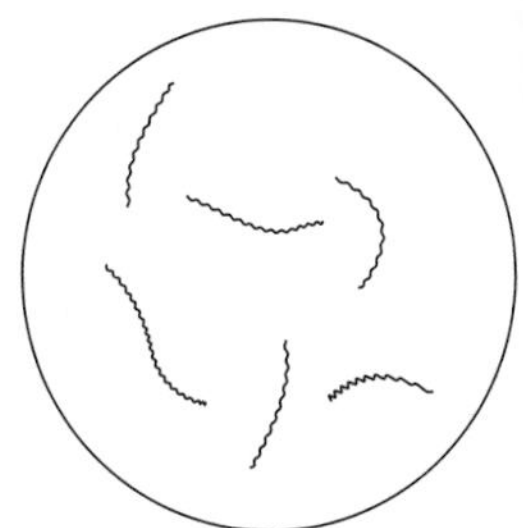
Treponema pollidum

Several weeks after the chancre has healed, the untreated patient experiences **secondary syphilis**. Symptoms include fever and a flu-like illness as well as swollen lymph nodes. The skin rash that develops, which may be mistaken for measles, rubella, or chickenpox, appears as reddish-brown spots on the palms, face, and trunk (FIGURE 13.10B). Transmission can occur if there are moist lesions.

In untreated patients, the symptoms resolve after several weeks. Most patients recover, but they bear pitted scars from the healed lesions and remain "pockmarked." In 2010, there were more than 13,700 cases of primary and secondary syphilis reported to the CDC—the first decline in reported cases since 2000.

These individuals now enter a latent period that can last several years. Many patients will have relapses of secondary syphilis during which time they remain infectious. Within 4 years, the relapses cease and the disease is no longer infectious (except in pregnant females). Patients either remain asymptomatic or slowly progress to the third stage.

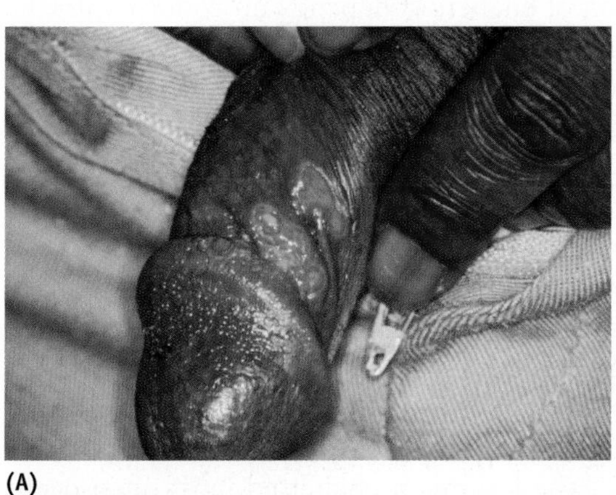
(A)

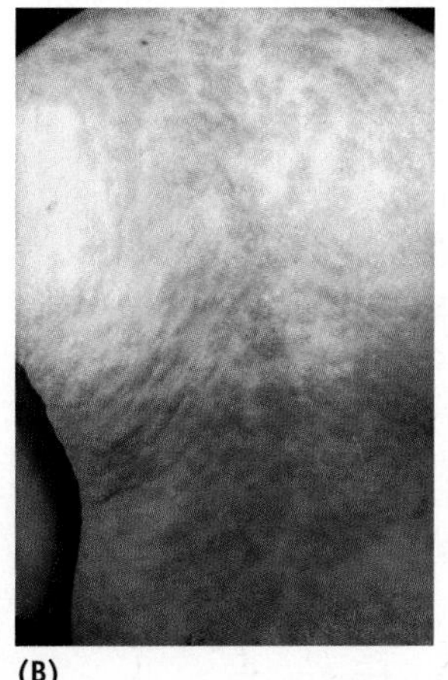
(B)

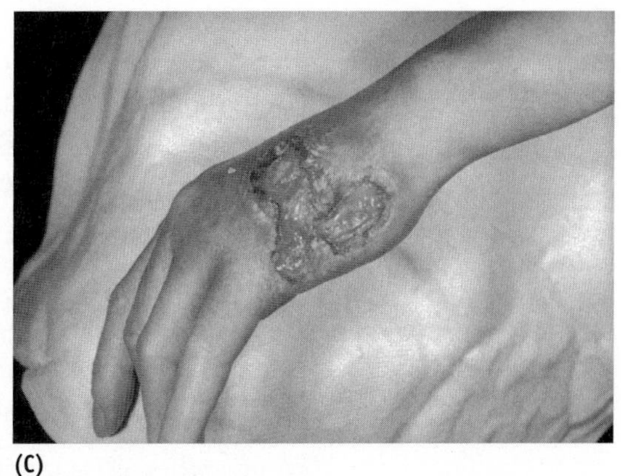
(C)

FIGURE 13.10 **The Stages of Syphilis.** (**A**) The chancre of primary syphilis as it occurs on the penis. The chancre has raised margins and is usually painless. Courtesy of M. Rein, VD/CDC. (**B**) A skin rash is characteristic of secondary syphilis. Courtesy of Dr. Gavin Hart/CDC. (**C**) The gumma that forms in tertiary syphilis is a granular, diffuse lesion compared with the primary chancre. Courtesy of Susan Lindsley/CDC. »» From which of these stages could the spirochete be most contagious?

About 40% of untreated patients eventually develop **tertiary syphilis**. This stage occurs in many forms, but most commonly it involves the skin, skeletal, or cardiovascular and nervous systems. The hallmark of tertiary syphilis is the **gumma**, a soft, painless, gummy noninfectious granular lesion (FIGURE 13.10C). In the cardiovascular system, gummas weaken the major blood vessels, causing them to bulge (aneurysm) and burst; in the spinal cord and meninges, gummas lead to degeneration of the tissues and paralysis; and in the brain, they alter the patient's personality and judgment, and cause insanity so intense that for many generations people with tertiary syphilis were confined to mental institutions (but read chapter introduction). Damage can be so serious as to cause death.

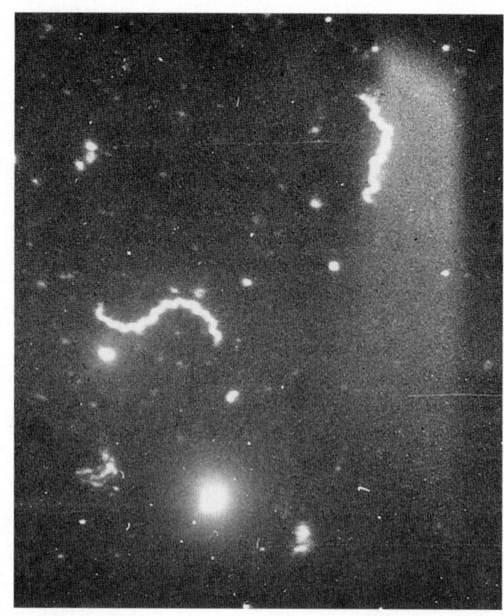

FIGURE 13.11 **The *Treponema pallidum* Spirochete. A dark-field microscope view of the syphilis spirochete seen in a sample taken from the chancre of a patient. Courtesy of the CDC.** »» Why are all photos of *T. pallidum* taken from patient samples and not from samples grown in culture?

Syphilis is a serious problem in pregnant women because the spirochetes penetrate the placental barrier after the third or fourth month of pregnancy, causing **congenital syphilis** in the fetus. Infection can lead to death (stillbirth). Surviving infants develop skin lesions and open sores. Affected children often suffer poor bone formation, meningitis, or **Hutchinson's triad**, a combination of deafness, impaired vision, and notched, peg-shaped teeth. In 2010, there were 377 congenital syphilis cases reported to the CDC, a number that has been declining since 1991.

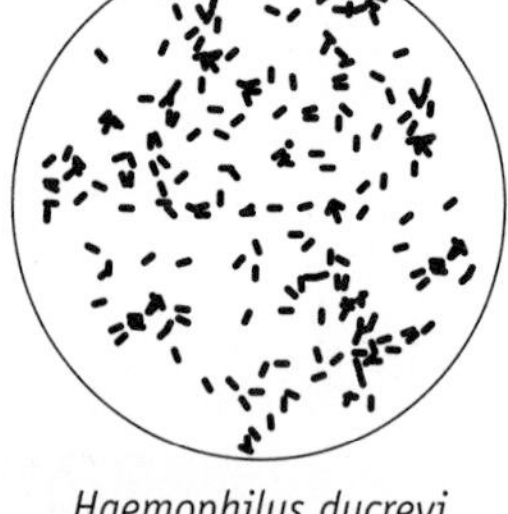

Haemophilus ducreyi

■ Erythema: A redness of the skin resulting from inflammation.

Treatment and Prevention. There is no vaccine for syphilis, so the cornerstone of syphilis control is safe sex practices, and the identification and treatment of the sexual contacts of patients. Penicillin is the drug of choice and a single dose usually is sufficient to cure primary and secondary syphilis. Because *T. pallidum* cannot be cultivated on laboratory media, diagnosis in the primary stage depends on the observation of spirochetes from the chancre using fluorescence or dark-field microscopy (FIGURE 13.11). As the disease progresses, a number of tests to detect syphilis antibodies becomes useful, including the rapid plasma reagin (RPR) test and the Venereal Disease Research Laboratory (VDRL) test.

Other Sexually Transmitted Diseases Also Exist

A few other STDs merit brief mention.

Chancroid. An STD believed to be more prevalent worldwide (especially in tropical and semitropical climates) than gonorrhea or syphilis is **chancroid**. The disease is endemic in many developing nations, and it is common in tropical climates and where public health standards are low. The disease has been decreasing in the United States since 1987 and there were 24 cases reported in 2010. However, the organism is difficult to culture and therefore the disease could be substantially underdiagnosed.

The causative agent of chancroid is *Haemophilus ducreyi*, a small gram-negative rod. The transmission of chancroid depends on contact with the lesion, although sexual contact is not required. Contact with open ulcers or their fluid can spread the disease. After a 3 to 5 day incubation period, a tender papule surrounded by erythema forms at the entry site. The papule quickly becomes pus filled and then breaks down, leaving a shallow, saucer-shaped ulcer that bleeds easily and is painful in men. The ulcer has ragged edges and soft borders, a characteristic distinguishing it from the primary lesion of syphilis. For this reason, the disease is often called **soft chancre**.

The lesions in chancroid most often occur on the penis in males and the labia or clitoris in females (see Figure 13.2). Substantial swelling in

TABLE
13.1 A Summary of Sexually Transmitted Diseases Caused by Bacteria

Disease	Causative Agent	Signs and Symptoms	Transmission	Treatment	Prevention
Chlamydia	*Chlamydia trachomatis*	Pain on urination Watery discharge Salpingitis	Sexual intercourse	Doxycycline Erythromycin	Practicing safe sex Limiting sex partners Screening for STDs
Gonorrhea	*Neisseria gonorrhoeae*	Pain on urination Discharge Salpingitis	Sexual intercourse	Ceftriaxone	Practicing safe sex Avoiding oral sex
Syphilis	*Treponema pallidum*	Chancre Skin lesions Gumma	Sexual intercourse	Penicillin	Practicing safe sex
Chancroid	*Haemophilus ducreyi*	Soft chancre Erythema Swollen inguinal lymph nodes	Sexual intercourse	Azithromycin Erythromycin Ceftriaxone	Practicing safe sex
Lymphogranuloma venereum (LGV)	*Chlamydia trachomatis*	Swollen lymph nodes Proctitis	Sexual intercourse	Doxycycline	Practicing safe sex

the lymph nodes of the groin may be observed. However, the disease generally goes no further; in fact, in women, it often goes unnoticed. The clinical picture in chancroid makes the disease recognizable, but definitive diagnosis depends on isolating *H. ducreyi* from the lesions.

Azithromycin, erythromycin, or ceftriaxone drugs are useful for therapy, but the disease often disappears in 10 to 14 days without treatment.

Lymphogranuloma venereum. A systemic STD caused by a different serotype of *C. trachomatis* is **lymphogranuloma venereum (LGV)**. LGV is more common in males than females, and is accompanied by fever, malaise, and swelling and tenderness in the lymph nodes of the groin. Females may experience infection of the rectum (proctitis), if the chlamydial cells pass from the genital opening to the nearby rectum. LGV is prevalent in Southeast Asia and Central and South America. Sexually active individuals returning from these areas may show symptoms of the disease, but treatment with doxycycline leads to rapid resolution.

Recently, outbreaks of LGV among men who had sex with men in Canada, the Netherlands, and other European countries were reported.

TABLE 13.1 summarizes the STDs caused by bacterial agents.

CONCEPT AND REASONING CHECKS 2

- **a.** Why is chlamydia referred to as the "silent disease?"
- **b.** Contrast the symptoms of gonorrhea in females and males.
- **c.** Describe the three stages of syphilis.
- **d.** Propose an explanation as to why all the STD-causing walled bacteria are gram-negative.

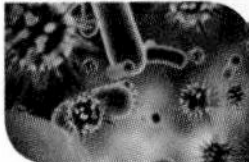

Chapter Challenge B

Most STDs have an acute phase that causes pain, produces a visible lesion, and generates an inflammatory reaction at the site of entry. They also may have a chronic phase where the infection may spread to more distant sites and generate worse damage to host tissues.

QUESTION B:

For chlamydia, gonorrhea, and syphilis, identify the acute phase of the infection and the chronic phase for each STD.

Answers can be found on the Student Companion Website in **Appendix D.**

KEY CONCEPT 13.3 Urinary Tract Infections Are the Second Most Common Body Infection

If not for the urinary system, waste products from all the other body systems would accumulate in the blood producing such toxic levels that the waste products would kill us. However, because all parts of the urinary tract are joined by a fluid medium, infection at any site may spread to involve other areas of the system.

The Urinary System Removes Waste Products from the Blood and Helps Maintain Homeostasis

The organs of the urinary system are two kidneys, two ureters, a urinary bladder, and a urethra (FIGURE 13.12). The kidneys filter waste products from the bloodstream and convert the filtrate into urine. The ureters, urinary bladder, and urethra are collectively known as the **urinary tract** because they transport the urine from the kidneys through the ureters to the urinary bladder. **Urination** involves expulsion of urine from the bladder through the urethra to the exterior of the body.

When considering infections and diseases of the urinary system, it is important to understand that the urethral opening in females is closer to the anus than in males and is also close to the vaginal opening, making these heavily colonized sites important sources of potential microbial colonizers. These factors combine to generate significant differences not only in the types of microbes colonizing the urinary system of males and females, but also their relative susceptibility to infection.

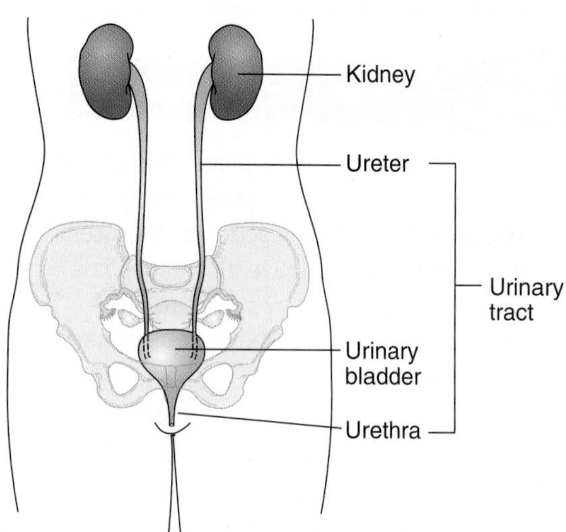

FIGURE 13.12 **The Female Urinary Tract.** After the kidneys remove excess liquid and waste from the blood as urine, the ureters carry the urine to the bladder in the lower abdomen. Stored urine in the bladder is emptied through the urethra. »» Explain how an infection of the kidney would arise.

Part of the Urinary Tract Harbors an Indigenous Microbiota

The kidneys and ureters of healthy individuals are normally sterile due to normal urine flow. The urethra and possibly the bladder are colonized by microbes. As already mentioned, the urethra of females is much shorter than that of males, making ascending infections into the bladder and kidneys more frequent in females.

One of the main antimicrobial defense mechanisms in the urethra is the shedding of the outermost cells of the mucosa with their adherent microbial populations. In addition, the secretion of mucus forms a layer on the epithelial surface that entraps microbes as well as prevents their adhesion to the urethral epithelium. Microbial colonization of the urethra also is hindered by the flushing action of urine, and its normally acidic pH and high urea content can be microbicidal and microbiostatic. The urine also contains a number of antimicrobial proteins, including defensins and secretory antibodies.

Many of the defense mechanisms operating in the male urethra are similar to those in the female urethra, including shedding of epithelial cells, urinary flow, and the presence in urine of antimicrobial and anti-adhesive factors. In addition, the prostate—the genital organ that surrounds and communicates with the first segment of the male urethra—secretes zinc, which in prostatic secretions can inhibit the growth of potential pathogens such as *Escherichia coli*.

The main indigenous bacterial genera colonizing both the female and male urethra are *Corynebacterium*, *Streptococcus*, *Staphylococcus*, *Bacteroides*, and a variety of lesser known genera including *Peptostreptococcus*, *Prevotella*, and *Porphyromonas*.

Elimination of bacterial cells from the urethra occurs mainly during urination. In females and males who do not have a urinary tract infection (UTI), the number of viable bacterial cells in the urine can be anywhere from 0 to approximately 10^5 cells per milliliter, with most individuals having between 10 and 1,000 cells per milliliter of urine. The diagnosis of a UTI depends on counting only the infecting bacterial cells in urine, not the indigenous population. This is accomplished by allowing the initial urinary flow to flush out the urethral microbiota. Then a "midstream" urine sample is taken for analysis.

Although, like urine, semen is sterile when first produced, it will accumulate microbes during its passage through the urethra. In fact, the semen of approximately 80% of males contains between one and nine different bacterial species. Therefore, sexual intercourse is likely to provide an avenue by which urethral microbes can be disseminated.

Urinary Tract Infections Occur Primarily in the Urethra and Bladder

It has been estimated that up to 50% of humans will suffer a **urinary tract infection (UTI)** at some time during their lives. In the United States, the CDC reports that UTIs account for about 4 million ambulatory-care visits each year, representing about 1% of all outpatient visits. UTIs account for about 10 million doctor visits each year. The major bacterial agents causing UTIs are shown in FIGURE 13.13.

Women are most at risk of developing a UTI. In fact, half of all women will develop a UTI during their lifetimes, and many will experience more than one (MICROFOCUS 13.3).

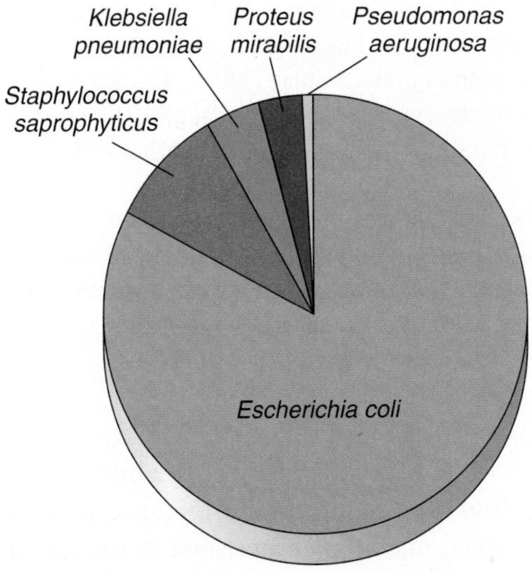

UTI infections in males remain rare through the fifth decade of life, when enlargement of the male prostate begins to interfere with emptying of the bladder. Infection rates can then rise 20%.

Most UTIs occur in the lower tract and involve the urethra, bladder, or prostate. Infections of the upper urinary tract involve the kidneys.

Urethritis. An inflammation and infection of the urethra is called **urethritis**. In women, *E. coli* can travel from the lower intestine and anus to the vagina. From there, due to the close proximity, the infecting organisms can travel to the urethra. Infections in men are far less frequent due to the greater distance between the urethra and anus. Sexually transmitted organisms, such as *N. gonorrhoeae* and *C. trachomatis*, can also spread to the urethra during sexual intercourse with an infected partner.

Infections confined to the urethra are characterized by painful urination in women and men, and discharge of mucoid or purulent material from the urethral orifice in men.

■ Purulent: Relating to, containing, or consisting of pus.

Treatment of urethritis depends on the cause of the infection. Antibiotics are given for bacterial infections. Sexually transmitted diseases that cause urethritis can be prevented by using a condom.

Cystitis. **Cystitis** is an infection of the bladder most often caused by *E. coli*. Community-acquired cystitis is particularly common and recurring in women between 30 and 50 years of age. Pregnancy can increase the chances of cystitis because the pregnancy can interfere with emptying of the bladder. Use of a diaphragm with a spermicide also can increase the chances of cystitis because the spermicide inhibits the vaginal microbiota and allows cystitis-causing organisms to grow.

Cystitis in men is less common. Community-acquired infections often begin in the urethra, move to the prostate, and then to the bladder. In fact, the most common cause of cystitis in men is a prolonged bacterial infection in the prostate.

Hospital-acquired (nosocomial) bladder infections can occur in patients in a medical care facility who have had a urinary catheter placed through the urethra and into the bladder to collect urine.

The symptoms of cystitis are similar to those of urethritis and result from an irritation of the mucosal surface of the urethra as well as the bladder. Unlike urethritis, cystitis is produced by the multiplication of *E. coli* in the bladder urine. It is

FIGURE 13.13 **Common Causes of UTIs.** Several bacterial species can cause UTIs. *E. coli* that causes the vast majority of uncomplicated UTIs is the strain that is normally found as part of the microbiota of the gastrointestinal tract. »» How would *E. coli* get from the gastrointestinal tract to the urinary tract?

MICROFOCUS 13.3: Clinical Microbiology

UTIs—Like Salmon Migrating Upstream

© Dan Guravich/Photo Researchers, Inc.

The upstream migration every year of salmon to reach their spawning grounds is one of nature's most exciting dramas. It is a long and strenuous journey, fraught with predators (bears and eagles) and man-made obstacles (dams and pollution) along with a river current that tries to push the salmon back downstream (see figure). The starting population at the ocean is not the same as the one that completes the journey after multiple rounds of ascension of the river.

So, what's this have to do with UTIs? Well, in fact, this yearly salmon spectacle is not all that much different from the way *Escherichia coli* cells make their way up the urinary system. Up to 95% of UTIs require pathogens, like *E. coli,* to progress in an ascending manner, beginning with bacterial colonization of the urethral area, infection of the bladder (cystitis), and if not treated, often an ascension of the ureters leading to kidney infection (pyelonephritis). How do *E. coli* cells make the long journey from urethra to bladder to kidney?

In a 2012 study, mice were infected with 10 slightly different clones of *E. coli* that could be identified by special "tags" incorporated into the DNA sequences. The study showed that different clones dominate different sites along the journey. After passing through the urethra, the *E. coli* cells invade the bladder and infect the cytoplasm of the surface bladder cells. There, they create a biofilm called an intracellular bacterial community (IBC) that allows *E. coli* to evade the immune system and serves as a reservoir for subsequent UTIs. Single bacterial cells of different clones (mixed population) not in IBCs then spread, like salmon swimming upstream, from the bladder toward the ureter and kidneys. For *E. coli,* the mixed population of clones swims with their flagella against a flow of urine that is trying to push them back downstream. To ascend the urine stream, multiple rounds of ascension and dissemination are required. During the process, cells die by being attacked by their "predators," the immune cells, changing the composition of the population. In addition, bacterial cells are washed back down the ureter by urine flow. But, like salmon, more ascend (perhaps a different mix of *E. coli* clones from the bladder IBC), only to be washed back down. Over many cycles, like salmon reaching their spawning ground, some cells of one or more clones finally make it upstream to a kidney and establish an infection. In the end, lineages (clones) constantly increased and decreased in abundance as they migrated from one organ to another and battled the downstream urine flow.

So, the infection process from ureter to kidney is not a simple process any more than salmon migration upstream is a simple process. But unlike the way that understanding salmon migration can provide for a sustainable salmon industry, understanding the dynamics of *E. coli* populations may provide novel ways to treat and prevent UTIs in the future.

■ Hematuria: The presence of blood in the urine.

clinically distinguished from urethritis by a more acute onset, more severe symptoms, the presence of bacteria in the urine, and, in approximately half of cases, hematuria.

Symptoms of cystitis often disappear without treatment. People with frequent bladder infections caused by bacterial species often take antibiotics regularly at low doses. Symptoms usually disappear within a few days of treatment; however, antibiotics may need to be taken for up to a week, depending on the severity of the infection. The bacterial species causing hospital-acquired bladder infections often are resistant to the common antibiotics used to treat community-acquired bladder infections. Therefore, different types of antibiotics and different treatment approaches may be needed. Investigating the Microbial World 13 looks into cranberry juice for preventing UTIs.

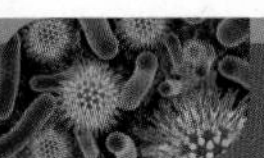

Investigating the Microbial World 13

Does Cranberry Juice Cure Urinary Tract Infections?

Urinary tract infections (UTIs), such as cystitis, are an unpleasant illness. Besides the increased urge to urinate, there often is a burning sensation when one does urinate. Although the infection and symptoms can resolve without medical treatment, there is a 24% chance of a recurrence within 6 months.

■ **OBSERVATIONS:** The cranberry and especially cranberry products, such as cranberry juice or cranberry capsules, have been considered or even touted by many as an effective home treatment for UTIs or to prevent recurring UTIs. Proponents say that cranberry products work by acidifying the urine, which would make the urinary tract less hospitable to pathogens like *Escherichia coli,* the most common cause of UTIs. Also the sugar (fructose) and proanthocyanidins in cranberries may interfere with the ability of the pili on *E. coli* cells to adhere to the cells lining the urinary tract. Opponents say the evidence is less than compelling and too anecdotal. In addition, good quality, randomized, double-blind, and placebo-controlled studies on the effects of cranberries have not been undertaken.

■ **QUESTION:** ***Does cranberry juice prevent recurrent UTIs?***

■ **HYPOTHESIS:** Regular drinking of cranberry juice cocktail (CJC) will reduce the likelihood of recurrent UTIs. If so, then a randomized, double-blind comparison of the efficacy of CJC and placebo juice on women with an acute UTI should reduce the rate and duration of UTI symptoms.

■ **EXPERIMENTAL DESIGN:** Out of 419 college women enrolled, 319 had a positive urine culture for a UTI. The experimental juice consisted of a formulated low-calorie CJC (27% juice) with a standardized proanthocyanidins component. The placebo juice was formulated to imitate the flavor (sugar and acidity) and color of cranberry juice but without any cranberry or proanthocyanidin content.

■ **EXPERIMENT:** The 319 women were randomly split into two groups. One group (155 women) drank two 8 oz. glasses of CJC twice daily for 6 months. The other group (164 women) drank two 8 oz. glasses of placebo juice twice daily for 6 months. Neither the participants nor investigators knew which group was drinking which juice. Compliance was based on self-reporting.

The clinical assessment consisted of analyzing clean-catch, mid-stream urine specimens from the participants at the beginning of the study, and at 3 and 6 months. Self-collected vaginal and rectal specimens were cultured for *E. coli* pathogens. Participants also completed questionnaires at the beginning of the study, and at 3 and 6 months, regarding any UTI symptoms as well as other pertinent medical information.

■ **RESULTS:** See figure. Of the 319 participants that started the study, 230 completed the entire study (116 in the CJC group and 114 in the placebo group). The presence of urinary and vaginal symptoms over the course of the study was similar between the two groups. A positive UTI was based on a combination of symptoms and a urine culture positive for a known uropathogen. Gastrointestinal symptoms were twice as frequent in the placebo group as in the CJC group.

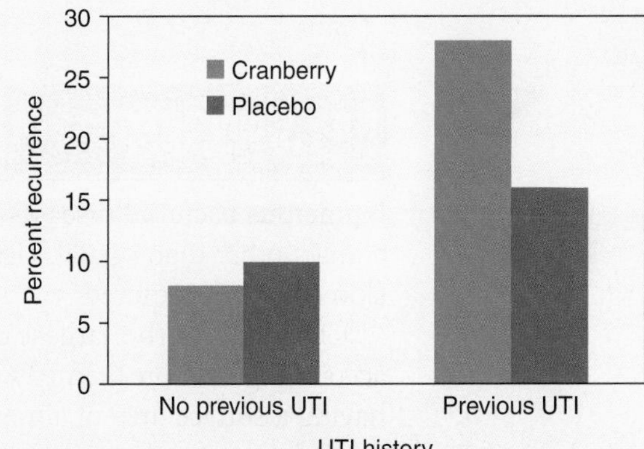

FIGURE Risk of a recurring UTI by history and juice assignment.

■ **CONCLUSIONS:**

QUESTION ***1: Was the hypothesis validated? Explain using the figure.***

QUESTION ***2: Explain why this was a (a) randomized, (b) double-blind, and (c) placebo-controlled study.***

QUESTION ***3: Can you think of any problems or caveats in the set up and performance of this study that could make the results questionable?***

Answers can be found on the Student Companion Website in **Appendix D.**

Adapted from: Barbosa-Cesnik, C. *et al.* (2011). *Clin Infect Dis* **52** (1):23–30. Icon image © Tischenko Irina/ShutterStock, Inc.

Prostatitis. An inflammation and infection of the prostate, called **prostatitis**, typically develops in young men and in older men following placement of an indwelling catheter. Prostatitis has been classified by the National Institutes of Health (NIH) into one of two categories if it involves a bacterial infection. In **acute bacterial prostatitis**, symptoms come on suddenly and may include fever and chills, and the signs and symptoms of cystitis, including increased urinary urgency and frequency. In **chronic bacterial prostatitis**, signs and symptoms develop more slowly and usually are not as severe as those of acute prostatitis. Antibiotics are usually the choice of treatment for both categories of prostatitis.

Pyelonephritis. A UTI involving one or both kidneys is called **pyelonephritis**. If not treated properly, kidney infection can permanently damage the kidneys or spread to the bloodstream and cause a life-threatening infection. Thus, prompt medical attention is essential.

Again, pyelonephritis is more common in women than in men. About 90% of cases among individuals living in the community are caused by *E. coli*. Infections usually ascend from the genital area through the urethra to the bladder, then up through the ureters and into the kidneys. Importantly, it has been estimated that 20% to 50% of pregnant women with acute pyelonephritis give birth to premature infants, making pyelonephritis one of the most serious consequences of a UTI.

Symptoms of pyelonephritis often begin suddenly with chills, fever, pain in the lower part of the back, nausea, and vomiting. About one-third of people with pyelonephritis also have symptoms of cystitis. Antibiotics are the first line of treatment and are started as soon as the doctor suspects pyelonephritis. Usually, the signs and symptoms of kidney infection begin to clear up within a few days of treatment. However, patients need to continue antibiotics for a week or more.

The bacterial diseases of the urinary system are summarized in TABLE 13.2.

CONCEPT AND REASONING CHECKS 3

a. How does the urinary system differ from the urinary tract and why is the urinary tract more likely to become infected?
b. Why is the urethra the only part of the urinary system normally colonized by indigenous microbiota?
c. How can you diagnose the different UTIs based on signs and symptoms?

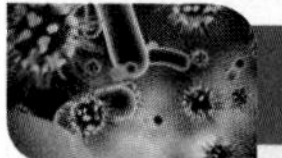

Chapter Challenge C

An earlier challenge asked you to decide whether or not you believe the female reproductive system is prone to more infections than is the male reproductive system. Now, what about the female urinary system?

QUESTION C:

Do you believe the female urinary system would be more prone to infections than the male urinary system? Again back up your decision with evidence.

Answers can be found on the Student Companion Website in **Appendix D.**

KEY CONCEPT 13.4 Contact Diseases Can Be Caused by Indigenous Bacterial Species

Numerous bacterial diseases are transmitted by contact other than sexual. Usually, some form of skin contact is required.

The skin is the largest organ of the body, accounting for more than 10% of body weight and having a surface area of almost 2 square meters. Being the interface between environment and internal body tissues and organs, the skin plays a critical role in protecting the body against trauma as well as physical and biological injury. Should anything go wrong with skin function, or should there be damage to—or wounding or puncturing of—the skin, inflammation, infection, and disease may occur.

The Skin Protects Underlying Tissues from Microbial Colonization

The skin, along with its accessory structures (nails, hair, sweat glands, and sebaceous glands) constitutes the **integumentary system** (FIGURE 13.14).

The upper, bloodless portion of skin that you see is called the **epidermis** (*epi* = "over"). It consists of layers of **keratinocytes**, the most com-

TABLE
13.2 Urinary Tract Infections Caused by Bacteria

Disease	Causative Agent	Signs and Symptoms	Transmission	Treatment	Prevention
Urethritis	*Escherichia coli* *Chlamydia trachomatis* *Mycoplasma* Other bacterial species	Persistent urge to urinate Burning sensation when urinating Blood in the urine Cloudy, strong smelling urine	Bacterial entry from the large intestine to the urethra Sexual intercourse	Antibiotics	Drinking plenty of liquids (water) Urinating frequently Keeping the genital area clean
Cystitis	*Escherichia coli*	Strong, persistent urge to urinate Burning sensation when urinating Blood in the urine Passing cloudy or strong-smelling urine	Bacterial entry from the large intestine through the urethra to the bladder Sexual intercourse	Antibiotics	Drinking plenty of liquids (water) Urinating frequently Keeping the genital area clean
Prostatitis	*Escherichia coli* *Staphylococcus* *Proteus* *Pseudomonas*	Fever and chills Flu-like symptoms Pain in the prostate gland, lower back, or groin Increased urinary urgency and frequency Pain when urinating Blood-tinged urine	Bacteria from the large intestine From bladder Sexual intercourse	Antibiotics	Practicing good hygiene Keeping the penis clean Drinking enough fluids to cause regular urination
Pyelonephritis	*Escherichia coli* *Enterococcus* *Klebsiella* *Pseudomonas*	Frequent urination Persistent urge to urinate Burning sensation or pain when urinating Abdominal pain or pressure Cloudy urine with a strong odor Pus or blood in urine	Bacteria from the large intestine enter urethra From bloodstream	Antibiotics	Practicing good hygiene Drinking enough fluids to cause regular urination

mon cell type in the epidermis. Keratinocytes are produced in the stratum basale and, as they push upward toward the surface, they produce a tough, fibrous protein called **keratin**. The keratinocytes eventually die in the stratum corneum, and are worn away and replaced by newer cells that are continually pushing up from below. This thickened, superficial layer is dry and waterproof, and, along with epidermal antimicrobial peptides and dendritic (Langerhans) cells, prevent most bacterial cells, viruses, and other foreign substances from entering the underlying tissues.

■ Dendritic cells: Immune cells that recognize and process foreign material and present it to other immune cells.

Beneath the epidermis lies the **dermis**, a complex, thick layer of fibrous and elastic connective tissue. The dermis contains sweat glands, the sweat being composed of water, salt, and a few other chemical substances. Specialized sweat glands, called apocrine sweat glands, in the armpit

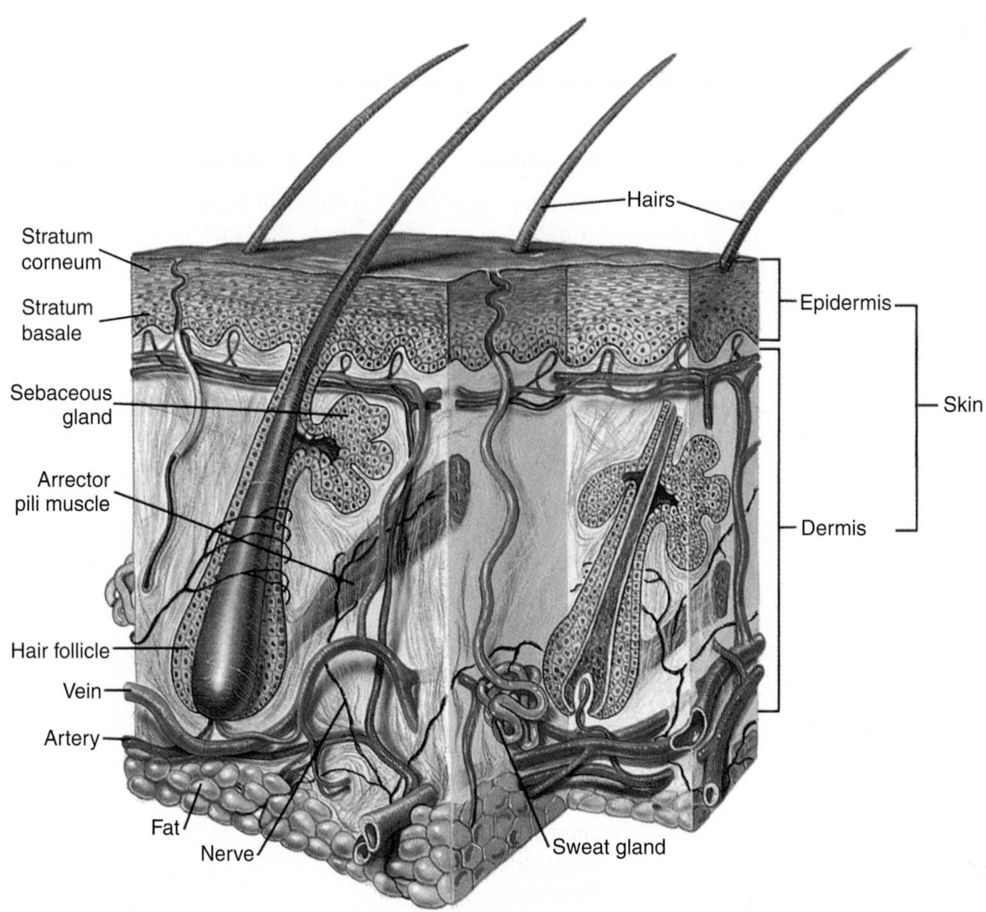

FIGURE 13.14 **Layers of the Skin.** A diagrammatic sectional view through the epidermis and dermis layers. »» Identify the accessory structures most susceptible to infection.

and genital region secrete a thick, oily sweat that produces a characteristic body odor when the sweat is digested by the resident skin microbiota in those areas (MicroFocus 13.4).

Hair covers much of the skin surface and consists of dead keratinized cells emerging from hair follicles that originate deep in the dermis. Associated with these follicles are **sebaceous glands** that are primarily found on the face, scalp, and upper torso. These glands produce an oily substance, called **sebum**, which keeps the skin and hair soft and moist. By being open to the environment as infoldings of the epidermis, the most susceptible target in the dermis for microbial colonization is the hair follicle and its associated sebaceous gland.

■ Microbiome: The complete set of microbes found in a specific environment.

The Skin Harbors an Indigenous Microbiota

The skin can be envisioned as an ecosystem composed of various habitats or specialized niches colonized by a resident **microbiota**, a unique community of indigenous and diverse bacterial and fungal species and viruses. Most of these residents are harmless or actually provide some vital functions, such as building a barrier to keep out more pathogenic organisms. Much of what we know about the skin microbiome today is coming from research carried out by the U.S. National Institutes of Health Common Fund Human Microbiome Project (HMP) and the International Human Microbiome Consortium (IHMC), which aim to characterize the human microbiota and its role in health and disease. Because most of the bacterial species cannot yet be grown in culture, the HMP and IHMC have used genomic methods to identify the presence of these species.

The most recent studies suggest that the skin microbiome is dominated by four bacterial phyla: the gram-positive Actinobacteria and Firmicutes, and the gram-negative Bacteroidetes and Proteobacteria. Furthermore, when analyzing the skin microbiota by skin site, the diversity of

MICROFOCUS 13.4: Public Health

"Smelly Sweat"

In February 2010, a man was removed from an Air Canada flight due to his body odor. The man was escorted off the plane after multiple passengers complained about his smell. Now, sweating by any animal, including humans, has an important function—it carries excess heat from the body and when the sweat evaporates (heat of vaporization), the excess heat is carried away, cooling off the body. However, sweating can make a person smell awful as in the case of the airline passenger. You certainly have experienced this by being unlucky enough to walk past a "smelly" pedestrian or get a whiff of the "stinky" person next to you on a bus. Such individuals need to take a shower to get rid of the "smelly sweat;" that is, the stinky body odor (BO).

Foot odor is due to the bacterial microbiota (false-color scanning electron micrograph) and the acid products they produce (molecular models = isovaleric acid). © Scimat/Photo Researchers, Inc.

Body odor, called bromhidrosis (brom = "stench;" hidro = "sweat;" sis = "the act of"), is due to bacterial species on the skin giving off smelly odors. When bacterial cells on the skin multiply in the sweat, they break down some of the proteins in the sweat. For example, the propionibacteria in the sebaceous glands break down amino acids into propionic acid, which has a pungent vinegar-like smell. Likewise, *Staphylococcus epidermidis* on the skin degrades the amino acid leucine into isovaleric acid, which has the typical "locker room" smell (see figure). The typical human has 3 to 4 million sweat glands and the sweat or perspiration from these glands is most likely to occur on your face, in your armpits, on your palms, and on the soles of your feet. The apocrine glands, found in the genital area, breasts, armpits, and eyelids, become active with puberty and contribute to apocrine bromhidrosis through the breakdown of fatty sweat. Most often, the bacterial breakdown of apocrine sweat is the major cause of body odor.

So, for most cases of sweat, just make sure you maintain good hygiene; take regular showers or baths and use a deodorant (designed to eliminate odor by turning the skin pH acidic, making for a more hostile environment for bacterial growth) or antiperspirant (designed to prevent odor and reduce sweat production by blocking the sweat pores).

these phyla and specific species varies. The skin microbiota can be broken down into three skin sites: sebaceous, moist, and dry (FIGURE 13.15).

- **Sebaceous sites.** Bacterial diversity appears lowest at sebaceous sites, such as the forehead and back. The predominant organisms are species of *Propionibacterium* (phylum Actinobacteria) that are capable of metabolizing the sebum that is toxic to many other bacterial species. *P. acnes* chiefly inhabits hair follicles where the microaerophilic environment is optimal for growth. In higher numbers, as we will discuss below, *P. acnes* can be associated with skin conditions like acne.
- **Moist sites.** Analysis of the microbiota in moist sites, such as the navel, groin, sole of the foot, back of the knee, and the inner elbow, reveals that *Staphylococcus* and *Corynebacterium* species are the most dominant. These organisms prefer areas with high humidity. The "**coryneforms**," also called "diphtheroids," do not form spores, are non-acid-fast, and are gram-positive rods. This includes several species of *Corynebacterium* (*coryn* = "club") that usually do not cause disease. However, immunocompromised patients can have a higher rate of *Corynebacterium* colonization and infection than healthy individuals. Another coryneform, *Brevibacterium*, is an obligate aerobe. Several species are the contributors to foot odor and, in fact, may produce odors that attract mosquitoes, accounting for the insects' tendency to bite around the feet and ankles.

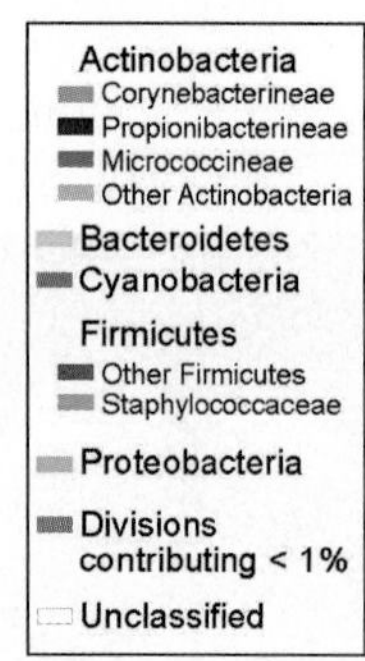

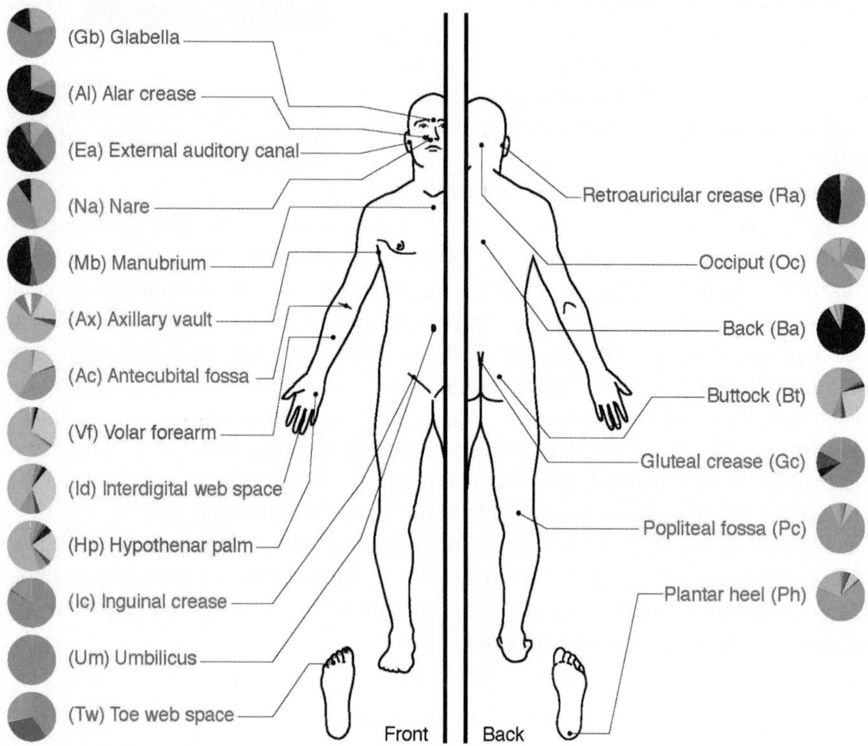

FIGURE 13.15 Topographical Distribution of the Skin Microbiota. The presence of specific bacterial taxa is governed by the microenvironment of each of the three skin sites. Courtesy of Darryl Leja, National Human Genome Research Institute (NHGRI). »» What are the most common organisms on each of the three skin sites?

- **Dry sites.** The areas with the highest diversity of microorganisms are the dry sites, which include the forearm, buttocks, and hand. These sites have a diverse population of all four bacterial phyla. From culture studies, it was previously thought that these dry sites were dominated by gram-positive bacterial species, which would be better at enduring the dry skin environment. In fact, the genomic studies indicate that there is an abundance of gram-negative species as well.

One of the cautions in doing these molecular studies is defining which species are statistically significant; that is, are the identified species unique to just certain individuals and their body sites? Are any of the species actually transient microbiota that just happened to be "caught" in the skin study done?

The antagonistic activities of the indigenous microbiota toward exogenous or transient microbes also are the consequence of their community interactions. Take, for example, the microaerophilic propionibacteria. Their survival depends on other aerobic and facultative species, such as *Micrococcus*, *Acinetobacter*, and *Staphylococcus*, consuming oxygen and creating the optimal microaerophilic conditions for growth of the propionibacteria. Thus, most pathogens entering the cutaneous microbial environment can't compete effectively for space or nutrients; they find themselves in a very hostile environment and usually are out-competed by the resident microbiota.

Acne Is the Most Common Skin Condition in the Developed World

Whether you call it zits, blackheads, or pimples, **acne** can be an annoying and persistent malady that is especially stressful for many teenagers—and sometimes adults (FIGURE 13.16A). Happily, today there are some treatments available that can reduce the severity of and lessen the damage caused by acne.

Causative Agent and Etiology. The development of acne, referred to medically as "acne vulgaris," is not the result of poor hygiene or eating greasy foods. Rather, it involves several interacting factors. The most common microbial factor is *Propionibacterium acnes*, a small, slow-growing, aerotolerant, gram-positive rod (FIGURE 13.16B). However, it is not considered to be the cause of acne because *P. acnes* cells also are part of the stable skin microbiome of acne-free individuals. Thus, other factors also contribute to the development of acne. These include excess sebum (skin oil) production, abnormal follicular development, and the individual's immune response to the buildup of *P. acnes* cells. In fact, a major underlying cause is one's family history. If both a mother and father have suffered with acne, there is a better than 75% chance their children will develop acne as well. Therefore, many medical experts refer to acne as a "physiological condition" rather than a disease.

Epidemiology. About 85% of adolescents and young adults develop acne. In these individuals, it is more common in males than females due to testosterone increase during puberty. Yet the condition can continue into adulthood as some 5% of males and females still have acne at age 45. In addition, flare-ups may occur in young women during menstruation or pregnancy. Overall, acne affects 40 to 50 million Americans each year and, on a global scale, it is one of the most common skin conditions with high prevalence in the developed nations.

Clinical Presentation. The predominant microbial species colonizing the sebaceous sites (e.g., face, upper chest, and back) of the skin is *P. acnes* (FIGURE 13.17). During puberty, sex hormone changes (especially testosterone in males) cause the sebaceous glands to enlarge and produce excessive amounts of sebum. The *P. acnes* cells degrade these lipids and adhere to the fatty acid products, aiding colonization and population growth.

The mildest form and initial step of acne occurs when hair follicles become swollen with sebum and keratinocytes (squames). The *P. acnes*

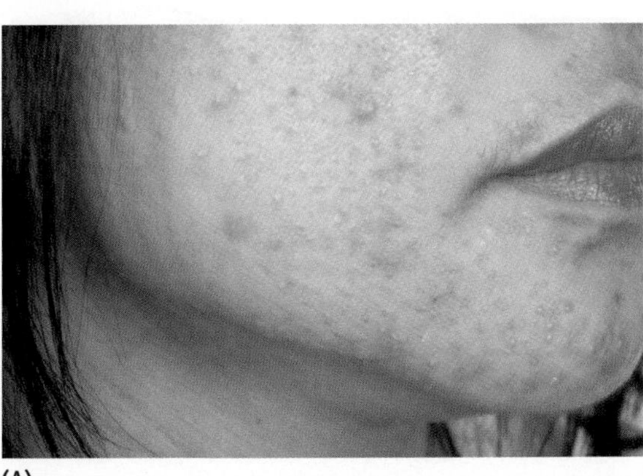

(A)

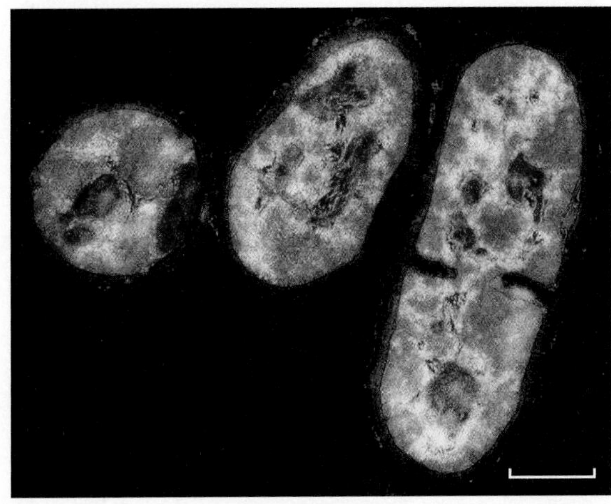

(B)

FIGURE 13.16 **Acne Vulgaris.** **(A)** Acne, a typical affliction of teenagers and young adults, primarily affects the face, upper torso, and back. © Medical-on-Line/Alamy Images. **(B)** *Propionibacterum acnes* bacteria. This false-color transmission electron micrograph shows three of the rod-shaped bacterial cells. (Bar = 1 µm.) © Kwangshin Kim/Science Photo Library. »» Why are the face, upper torso, and back the sites most likely to develop acne?

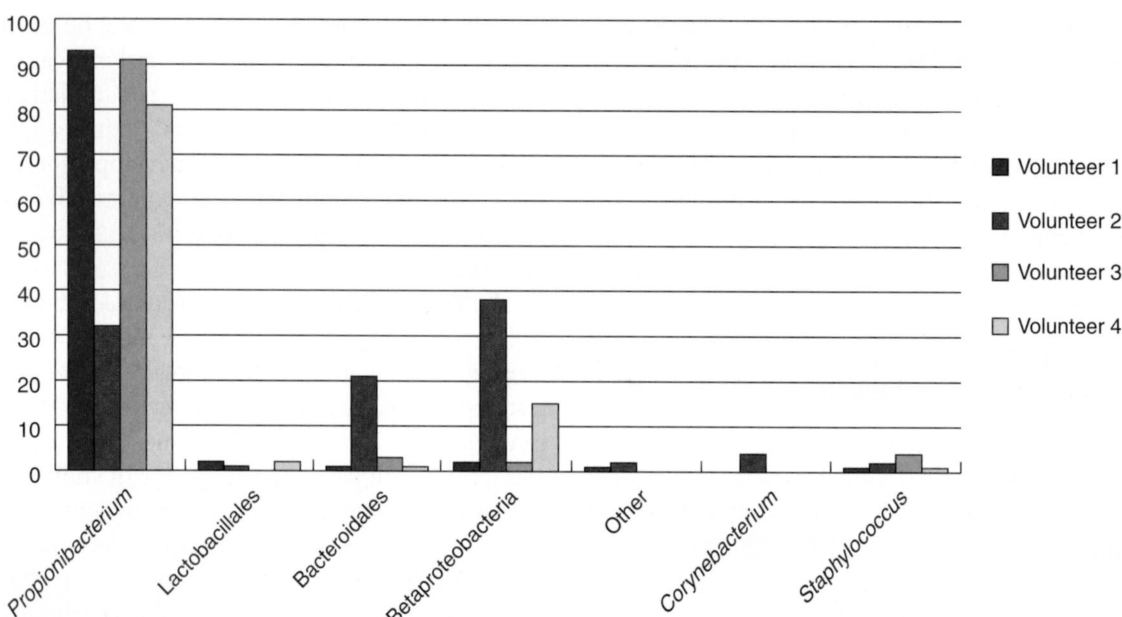

FIGURE 13.17 **Variation in Skin Microbiome of the Back.** The distribution of skin microbiota (by bacterial taxa) differs between these four volunteers. However, *Propionibacterium* is the predominant organism. Data from: Grice, E. A. and J. A. Segre (2010). *Nature Rev. Microbiol.* 9:244–252. »» What is unusual about volunteer 2?

cells exacerbate the condition by causing the keratinocytes to become sticky, plugging the follicles. Such plugged sebaceous glands, called **comedones** (sing., comedo), can exist in two noninflammatory forms (FIGURE 13.18):

- Open comedones ("blackheads") represent follicles where the sebum plug does not completely block the pore. A chemical reaction oxidizes melanin, giving the material in the follicle a black color.
- Closed comedones ("whiteheads") represent follicles in which pore blockage is complete. Because air cannot reach the melanin, it is not oxidized and the follicle remains white.

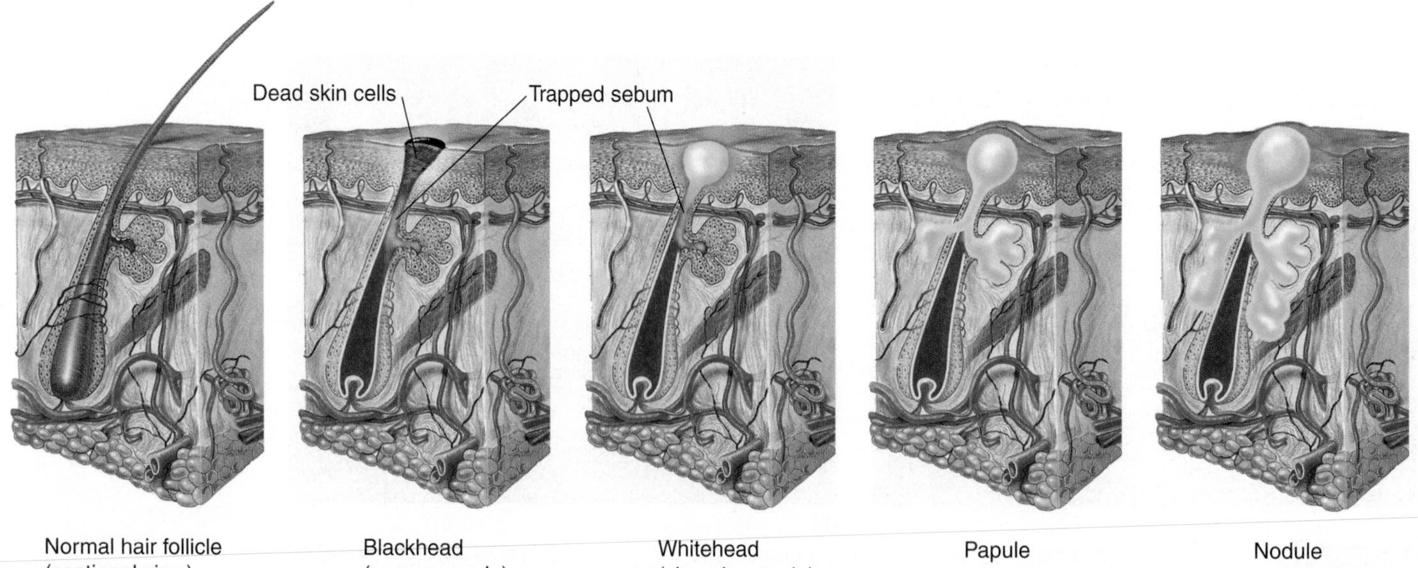

FIGURE 13.18 **Skin Lesions of Acne.** Plugged sebaceous glands can develop into open or closed comedones, papules, or nodules. Adapted from McKinley, M. and V. O'Loughlin. *Human Anatomy, Second edition.* McGraw-Hill, 2007. »» Why does *P. acnes* prefer to grow in a hair follicle?

Within the closed comedones, the increased population of *P. acnes* cells triggers the attraction of neutrophils and lymphocytes. These cells accumulate around and damage the swollen follicles, allowing the fatty acids and microbial antigens to erupt into the dermis. This now presents as local inflammation, forming papules and pustules.

Some 20% of individuals with acne will develop severe conditions where numerous nodules or cysts develop deeper in the dermis. Such **nodulocystic acne** can be quite painful and cysts can join together to form large abscesses that can result in physical scarring.

Treatment and Prevention. Acne is not communicable and therefore it is not a preventable disease. Most mild cases can be treated effectively at home, with good daily skin care using mild soap and, if necessary, over-the-counter treatments, such as benzoyl peroxide. Benzoyl peroxide has antibacterial properties and may decrease the inflammatory reaction at the hair follicle.

More severe cases may require antibiotics, such as clindamycin and erythromycin that are effective against *P. acnes*. By reducing the numbers of bacterial cells, the production of inflammatory substances will also be reduced. However, the widespread use of topical antibiotics for treatment of acne has resulted in an increased rate of antibiotic resistance by *P. acnes* and, by killing other skin microbiota, may leave the skin vulnerable to opportunistic infections. Topical retinoids, such as tretinoin (Retin-A®), are used to prevent blockage of follicles and to promote extrusion of the plugged material in the follicles of comedones. Systemic retinoids, such as isotretinoin (Accutane®) is only used for severe nodulocystic cases. It is not prescribed for women who are or may become pregnant as the compound can cause birth defects.

Although acne is not preventable, washing acne-prone areas of the skin twice a day with a gentle cleanser, avoiding irritants, like corticosteroids or anabolic steroids, and shying away from heavy foundation makeup and cosmetics can limit or control the extent of the condition.

Indigenous Microbiota Can Form Biofilms

Should a wound occur, not only could exogenous microbes enter and cause an infection but also indigenous microbiota. **Chronic wounds** are the result of predisposing conditions, such as peripheral vascular or metabolic diseases where there is a lack of blood flow. Here, dead skin breaks and produces an open sore, resulting in a leg ulcer or pressure sores (bed sores) that become infected. Although the bacterial microbiota does not cause the initial wounding, it can contribute to the lack of healing and chronic inflammation. Skin microbiome studies have not identified a unique organism to a particular type of chronic wound. In fact, most such indigenous infections probably are **polymicrobial**; that is, they are the result of more than one infectious agent (MICROFOCUS 13.5).

■ Papule: A solid, raised lesion lacking fluid.

■ Pustule: A raised, inflamed lesion containing pus.

An **acute wound** arises from cuts, lacerations, or surgical procedures on or to the skin. In such cases, indigenous microbes can enter the deeper layers of tissue and cause an infection. For example, each year in the United States intravenous catheters, used in about 50% of hospitalized patients, cause an estimated 80,000 bloodstream infections, resulting in almost 30,000 deaths among patients in intensive care units. Because the device must be inserted through the skin, there is a potential that skin bacteria, such as *Staphylococcus epidermidis*, can be mechanically "injected" into the tissues when the catheter is inserted. *S. epidermidis* then can colonize the catheter and form a **biofilm**.

■ Intravenous catheters: Tubes inserted into the patient's body cavity or a blood vessel where they deliver medications or fluids to the body or drain fluids from the body.

Implanted medical devices, such as hip and knee prostheses, pacemakers, and heart valves, can also be subject to the same biofilm scenario as a result of the surgical invasion.

To try to limit such infections, rigorous cleaning and disinfection of the skin surface must be done to minimize the resident microbiota. In addition, the device or implant needs to be checked for sterility. Still, bacterial cells can remain "hidden" in hair follicles and can infect the incision and form a biofilm on the device.

CONCEPT AND REASONING CHECKS 4

a. How does the epidermis and dermis help protect the body from microbial colonization and infection?

b. What are the three skin sites colonized by the indigenous microbiota and what are some examples of the resident microbes?

c. How does *Propionibacterium acnes* contribute to acne?

d. How do indigenous bacterial species colonize and survive on an invasive medical device?

MICROFOCUS 13.5: Genomics
Keeping Wounds Only Skin Deep

Skin infections are common and may be caused by bacteria, fungi, or viruses. Breaks in the skin integrity, particularly those that inoculate pathogens into the dermis, frequently cause or exacerbate skin infections. Upon injury to the skin, a set of complex biochemical events takes place in a closely orchestrated cascade to repair the damage. This process, however, is susceptible to interruption or failure leading to the formation of chronic, nonhealing wounds. In some patients, such as diabetics, even simple skin wounds can result in open, infected sores that are slow to heal (see figure). Part of the reason is because in diabetics clogged arteries lead to poor blood flow and therefore immune defenses are slow to respond.

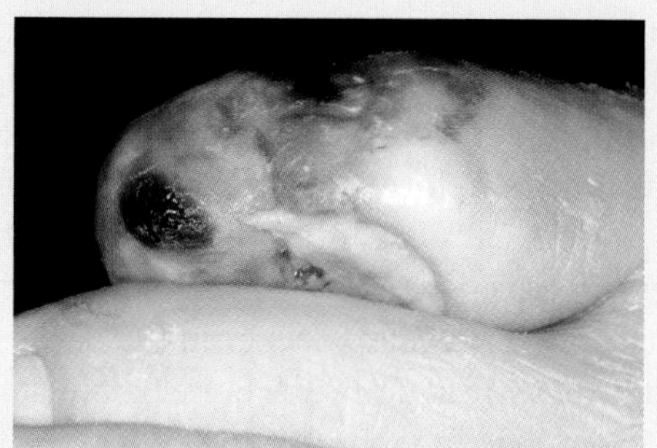

An infected skin wound on the big toe of a diabetes patient does not heal rapidly. © Dr. P. Marazzi/Photo Researchers, Inc.

In humans, such skin wounds are polymicrobial, consisting of *Corynebacterium, Enterococcus, Pseudomonas, Staphylococcus,* and *Streptococcus* species. Yet these same bacterial species are found on normal skin. So, could the skin microbiome, the diversity of bacterial species and other microbes normally found on the skin, play a role in wound healing?

Using normal and diabetic mice as test models, Elizabeth Grice and colleagues at the National Human Genome Research Institute in Bethesda, Maryland, discovered that diabetic mice have 40-times more bacterial cells on their skin surface than normal mice but less species diversity. The researchers also made a small wound on the shaved backs of the two groups of mice and then collected bacterial and skin samples during wound healing. It took about 2 weeks for the wounds to heal on the backs of the normal mice but more than a month for wound healing on the diabetic mice, which remained inflamed around the wound area.

The bacterial populations in the wounds were also different. In the diabetic mice, *Staphylococcus* and rod-shaped bacterial species were dominant. In normal mice, *Streptococcus* and *Clostridium* species dominated. The researchers suggest that *Streptococcus* and *Clostridium* species may simply outcompete the other species or they may produce compounds that fight off the competitors.

The picture is surely more complex. Other researchers have shown that staphylococci on the epidermis can limit the extent of inflammation in response to skin injury but induce inflammation (accidental pathogen?) when present in a wound infection below the dermis. In addition, how diabetic and normal mice establish differences in skin microbiota remains to be determined and if such differences are relevant to normal and diabetic humans.

KEY CONCEPT 13.5 Contact Diseases Can Also Be Caused by Exogenous Bacterial Species

When exogenous skin pathogens infect the body, they either colonize the skin directly or reach the skin through the bloodstream. Such bloodborne involvement often is observed on the skin surface as a rash, often the result of a bacterial toxin. Therefore, a physician must consider many possible diseases when evaluating a skin infection and may need to identify the presence or absence of specific signs and symptoms before coming to a diagnosis.

Staphylococcal Contact Diseases Have Several Manifestations

Staphylococci, gram-positive spheres that form clustered cell arrangements, are normal inhabitants of the human skin, mouth, nose, and throat. Although they generally live in these areas without causing harm, they can initiate disease when they penetrate the skin barrier or the mucous membranes. *Staphylococcus aureus* is the species often involved in these contact diseases.

Localized Skin Infections. The hallmark of a *S. aureus* skin infection is **folliculitis**, which produces pus-filled pockets at the base of a hair follicle (FIGURE 13.19A). Because infection is often associated with sweat gland activity and areas of the body subjected to rubbing or abrasion, it typically occurs on the neck, face, armpits, and buttocks. *Pseudomonas aeruginosa*, an aerobic gram-negative rod, also can cause folliculitis. In fact, infection by *P. aeruginosa* has become more common with people who frequent hot tubs or whirlpool baths that are not kept clean and chlorinated.

A more serious infection by *S. aureus* involves the formation of an **abscess**, a circumscribed pus-filled lesion. A **furuncle** (boil) is a warm, painful abscess that develops in the region of a hair follicle. Rupturing of the abscess can lead to infection of the surrounding tissue. Treatment requires a physician to open and drain the pus, and then rinse the abscess with sterile saline solution to remove any remaining bacterial cells. The disease needs to be treated with caution because staphylococci in a trivial skin boil can invade the blood and be transported to other organs, including the lungs, heart, brain, and kidneys.

Carbuncles are a group of connected, deeper abscesses (FIGURE 13.19B). Skin contact with other people spreads the disease. Thus, food handlers should be aware that staphylococci from furuncles and carbuncles can be transmitted to food, where they can cause food poisoning if transferred to others. Infected individuals often are tired and may have a fever. Because bacterial cells can enter the blood, antibiotic treatment and debridement may be needed.

Toxin-Generated *S. aureus* Contact Diseases. A more widespread and highly contagious staphylococcal skin disease is **impetigo**. Following a scrape or cut, the infection develops as thin-walled blisters oozing a yellowish fluid that forms a honey-colored, flaky crust (FIGURE 13.20A). Usually the blisters occur on the exposed parts of the body, especially on the face and limbs. *S. aureus* produces two toxins that cause the blistering skin.

Impetigo is commonly treated with penicillin, but resistant strains of *S. aureus* are well known, and physicians may need to test a series of alternatives before an effective antibiotic is selected. In fact, today multidrug-resistant *Staphylococcus aureus* has appeared in many hospitals. These resis-

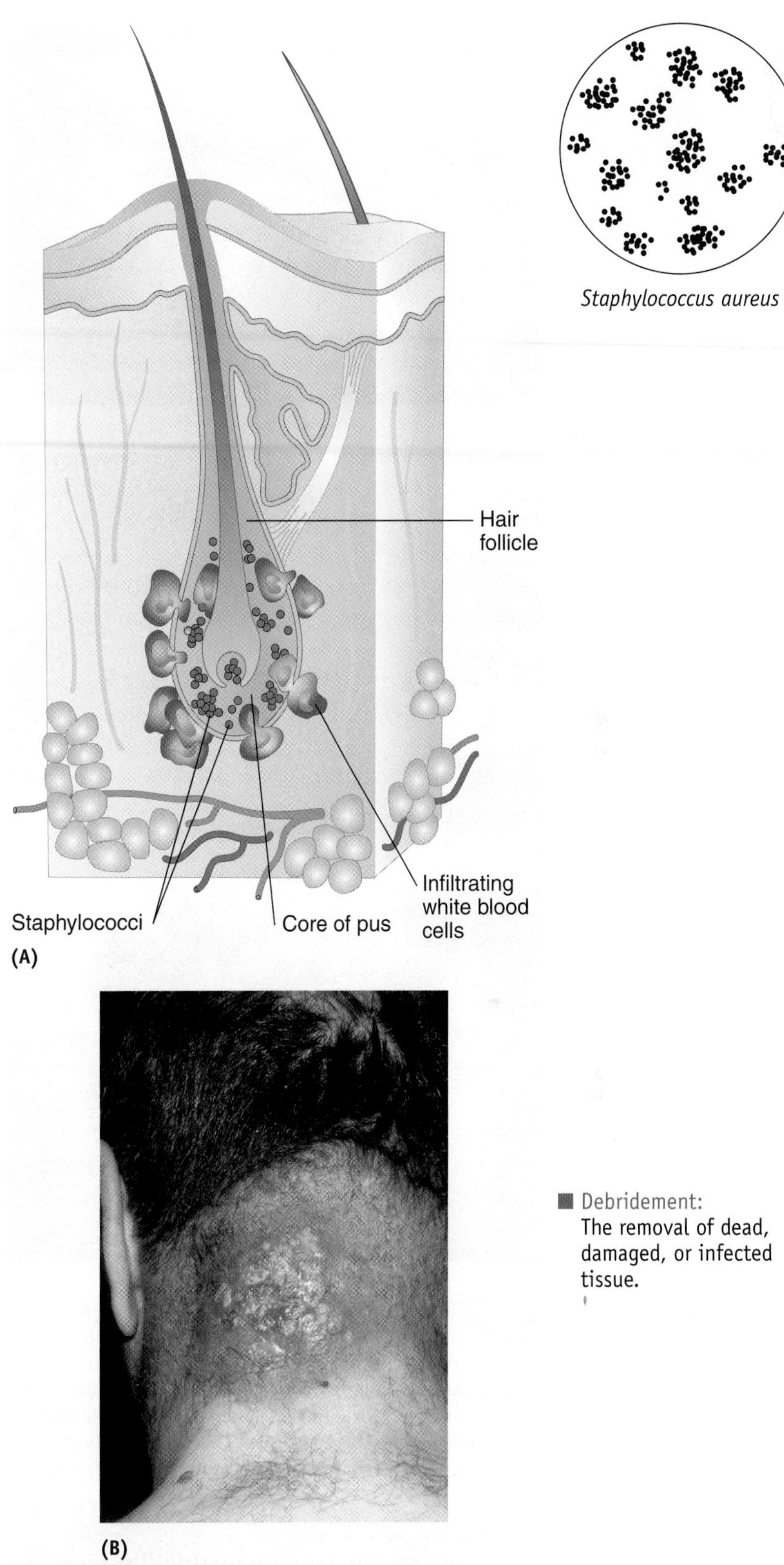

FIGURE 13.19 Staphylococci and Skin Abscesses. (**A**) Staphylococci infect the base of a hair follicle during the development of a skin lesion. White blood cells that engulf bacterial cells have begun to collect at the site as pus accumulates, and the skin has started to swell. (**B**) A severe carbuncle on the back of the head/neck. © Medical-on-Line/Alamy. »» How would a severe carbuncle be treated?

■ Debridement: The removal of dead, damaged, or infected tissue.

■ Superantigen: A pathogen substance capable of directly activating T cells.

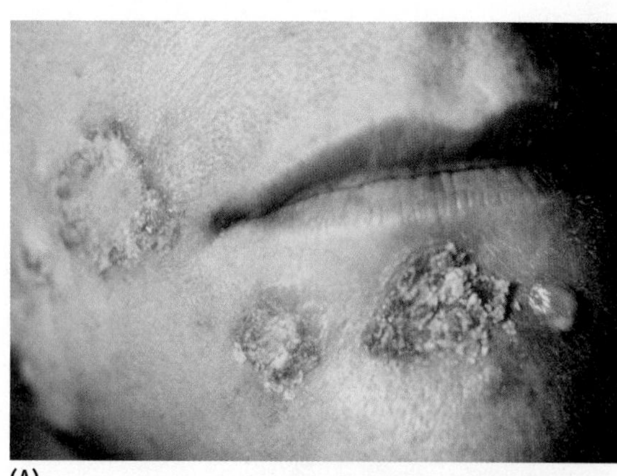

(A)

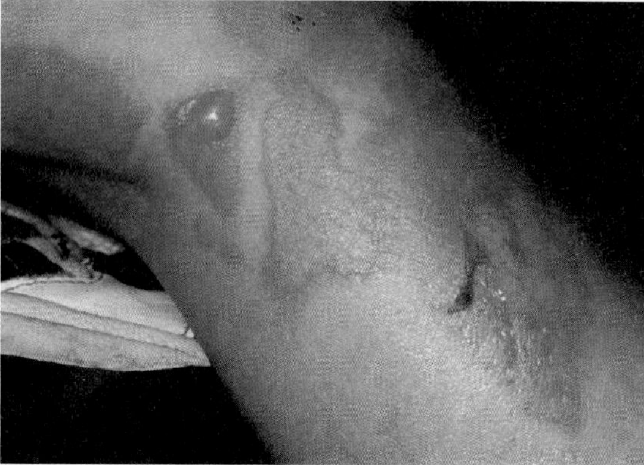

(B)

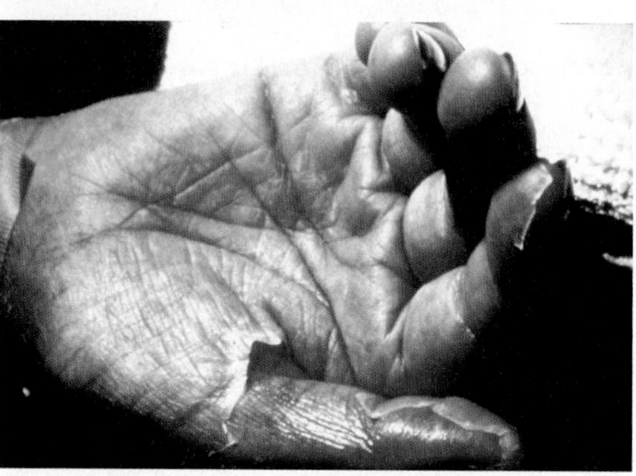

(C)

FIGURE 13.20 **Staphylococcal Skin Diseases.** **(A)** A patient with impetigo on the cheeks and chin. © Dr. Ken Greer/Visuals Unlimited, Inc. **(B)** An individual with scalded skin syndrome. © SPL/Photo Researchers, Inc. **(C)** A patient with toxic shock syndrome. In all cases, note the peeling of the skin. Courtesy of CDC. »» What chemical factors are responsible for the pathology observed?

tant strains are referred to as **methicillin-resistant *S. aureus* (MRSA)**. To help reduce the chances of antibiotic resistance, doctors may drain an abscess rather than treat the infection with drugs.

Scalded skin syndrome is usually seen in infants and young children under the age of 5. Exotoxins produced at the infection site by some strains of *S. aureus* travel through the bloodstream to the skin. The skin becomes red, wrinkled, and tender to the touch, with a sandpaper appearance. The epidermis may then peel off just above the stratum basale (FIGURE 13.20B). Mortality rates in children are very low unless there is a bacterial invasion of the lungs or blood. Cefazolin is the drug of choice for treatment.

Toxic shock syndrome (TSS) is the term for a disease characterized by sudden fever and circulatory collapse. TSS remained in relative obscurity until the fall of 1980, when a major outbreak occurred in menstruating women who used a particular brand of highly absorbent tampons.

TSS is a rare illness caused by another toxin-producing strain of *S. aureus*. The toxins act as superantigens and trigger T-cell activation and cytokine release. The earliest symptoms of disease include a rapidly rising fever, accompanied by vomiting and watery diarrhea. Patients then experience a sore throat, severe muscle aches, and a sunburn-like rash. Between the third and seventh days peeling of the skin, especially on the palms of the hands and soles of the feet, occurs (FIGURE 13.20C). A sudden drop in blood pressure can possibly lead to shock and heart failure. Antibiotics, such as semisynthetic penicillins or clindamycin, may be used to inhibit bacterial growth, but measures such as blood transfusions must be taken to control the shock.

The staphylococci involved in TSS can be the result of a skin abscess, surgery, or even tattooing (MicroFocus 13.6). In 2010, 82 cases of TSS were reported to the CDC.

Streptococcal Diseases Can Be Mild to Severe

Streptococci are a large and diverse group of encapsulated, nonmotile, facultatively anaerobic, gram-positive cocci. The bacterial cells divide in one plane and cling together to form pairs or chains of various lengths (FIGURE 13.21A).

Microbiologists classify the streptococci by two widely accepted systems. The first system divides streptococci into hemolytic (*hemo* = "blood") groups, depending on how they affect sheep red blood cells when the streptococci are plated on blood agar. The **α-hemolytic** streptococci turn blood agar an olive-green color as a secreted toxin partially destroys red blood cells in

MICROFOCUS 13.6: Public Health

"It Seemed Like a Good Idea!"

The idea of a tattoo seemed okay. All her friends had them, and a tattoo would add a sense of uniqueness to her personality. After all, she was already 22. It took some pushing from her friends, but she finally made it into the tattoo parlor that day in Fort Worth, Texas.

Two weeks later the pains started—first in her stomach, then all over. Her fever was high, and now a rash was breaking out; it looked like her skin was burned and was peeling away. There was one visit to the doctor, then immediately to the emergency room of the local hospital. The gynecologist guessed it was an inflammation of the pelvic organs (pelvic inflammatory disease, they called it), so he gave her an antibiotic and sent her home.

But it got worse—the fever, the rash, the peeling, the pains. Back she went to the emergency room. This time they would admit her to the hospital, give her intravenous blood transfusions and antibiotics, keep her for 11 days, and discover a severe blood infection due to *Staphylococcus aureus*. And there was an unusual diagnosis: toxic shock syndrome. Don't women get that from tampons? Most do, she was told, but a few get it from staphylococci entering a skin wound—a wound that can be made by a contaminated tattooing needle.

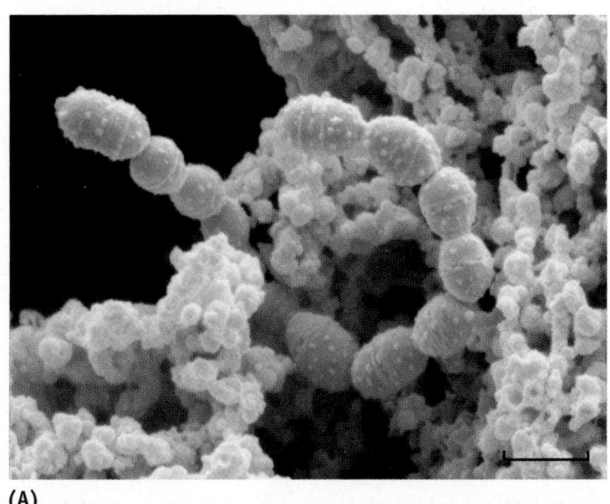

(A)

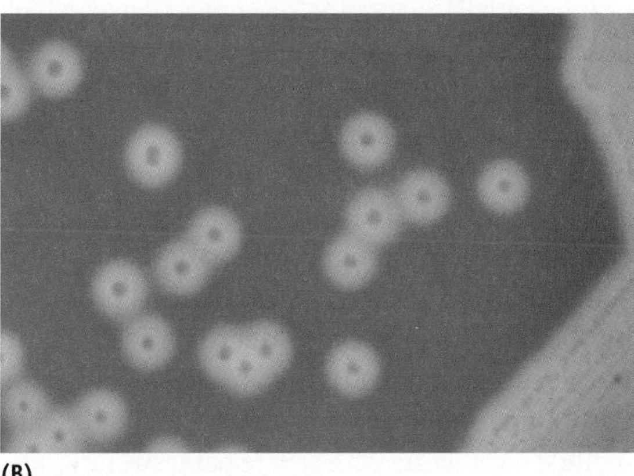

(B)

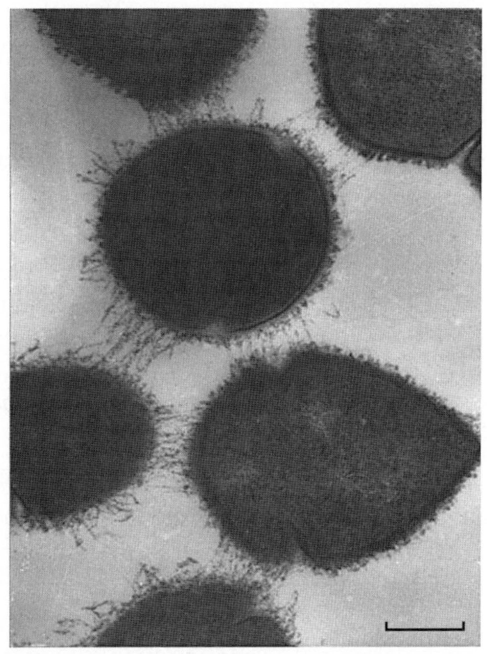

(C)

FIGURE 13.21 **Streptococci.** (**A**) A false-color scanning electron micrograph of streptococci growing as long chains. (Bar = 2 μm.) © Medical-on-Line/Alamy Images. (**B**) An example of β-hemolysis caused by a toxin released from *Streptococcus pyogenes* cells. © Dr. Gladden Willis/Visuals Unlimited/Corbis. (**C**) A close-up view of *S. pyogenes* cells showing strands of M protein protruding through the capsule. (Bar = 0.5 μm.) Courtesy of Vincent A. Fischetti, Ph.D., Head of the Laboratory of Bacterial Pathogenesis at Rockefeller University. »» What function does M protein play in pathogenesis?

the medium; colonies of **β-hemolytic** streptococci produce clear, colorless zones around the colonies due to the complete destruction of red blood cells (FIGURE 13.21B); and **nonhemolytic** streptococci have no effect on red blood cells and thus cause no change in blood agar.

The second classification system is based on variants of a carbohydrate located in the cell walls of some β-hemolytic streptococci. Groups A and B are the most important to human disease and *Streptococcus pyogenes* is the most common species. This β-hemolytic organism is generally implied when physicians refer to **group A streptococci** (GAS).

Streptococcus pyogenes

The pathogenicity of *S. pyogenes* is enhanced by the presence of virulence factors, such as a capsule and the **M protein**. This protein is anchored in the cell wall and appears as fibrils protruding from the cell surface (FIGURE 13.21C). The protein helps the cells adhere to tissue and inhibits phagocytosis. Over 100 serotypes of M protein have been identified.

■ Serotype:
A closely related group of microorganisms or structures distinguished by a specific set of immune-stimulating structures.

The GAS are among the most common human bacterial pathogens. They can be carried asymptomatically but also can be responsible for a variety of diseases; most are relatively mild, such as **streptococcal pharyngitis**, popularly known as **strep throat**. Although *S. aureus* is the most common cause of impetigo, *S. pyogenes* also can cause the infection. The clinical signs are similar.

S. pyogenes is a major cause of **cellulitis**, a potentially serious skin infection often found in infants, children, and the elderly. A superficial form of cellulitis involving the dermis is **erysipelas** (*erysi* = "red"; *pela* = "skin"). The streptococci enter the skin through minor trauma, insect bites, eczema, or surgical incisions; or from the individual's own nasal passages. Within 48 hours, symptoms appear and include a high fever, shaking and chills, and headache. The skin lesion enlarges rapidly, invading and spreading through the lymphatic vessels. It appears as a fiery red, swollen, warm, and painful rash usually on the lower limbs or face (FIGURE 13.22). Treatment usually involves oral or intravenous antibiotics, such as penicillin or clindamycin. The symptoms may resolve in a few days, although the skin may take weeks to return to normal.

■ Eczema:
A skin inflammation involving reddening and itching along with formation of scaly or crusty patches.

■ Fascia:
The connective tissue covering or binding together parts of the body such as muscles or organs.

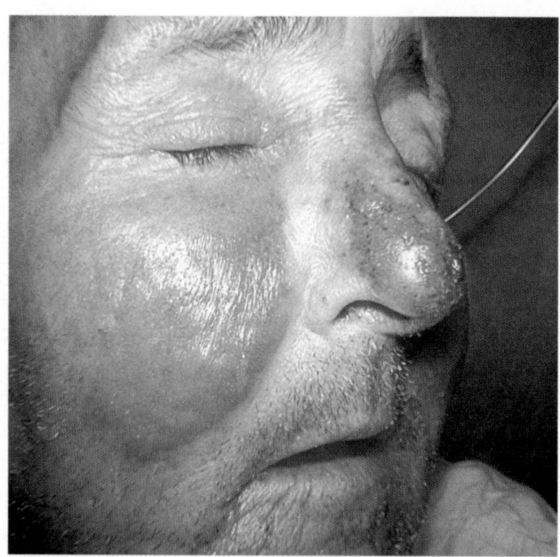

FIGURE 13.22 **Facial Erysipelas.** Infection of the skin and subcutaneous tissue with *Streptococcus pyogenes* produces a bright red rash of the affected areas, in this case the cheek and nose. Courtesy of Dr. Thomas F. Sellers, Emory University/CDC. »» What is most likely the source of this patient's infection?

GAS also causes more life-threatening diseases. **Streptococcal toxic shock syndrome (STSS)** is caused by toxins structurally similar to those in staphylococcal TSS that also function as superantigens triggering T-cell activation and cytokine release. In 2010, 142 cases of STSS were reported to the CDC. In STSS, early signs and symptoms include fever, dizziness, confusion, and a large flat rash over several areas of the body. Infection can cause a sudden drop in blood pressure, causing organs such as the kidneys, liver, and lungs to fail. Because more than half of STSS patients die without treatment, dealing with STSS requires prompt action with high doses of penicillin and clindamycin.

Necrotizing fasciitis (occasionally described in the media as "the flesh-eating disease") is a rare but dangerous infection that destroys muscles, fat, and skin tissue (FIGURE 13.23). GAS reach the subcutaneous tissue through a wound or trauma to the skin surface. The enzymes and toxins produced by the streptococci cause **necrosis** (cell death) of the subcutaneous tissue and adjacent fascia. Early signs and symptoms of necrotizing fasciitis include fever along with severe pain, swelling, and redness at the wound site. Widespread damage to the surrounding tissue along with blockage of small subcutaneous vessels produces additional dermal cell death. For persons with necrotizing fasciitis, debridement and surgery often are needed to remove damaged tissue; in severe cases, amputation may be the only recourse. Therefore, early diagnosis and treatment are critical to preventing devastating tissue destruction. Failing to recognize necrotizing fasciitis and its severity can lead to death; in fact, about 20% of patients with necrotizing fasciitis die.

Other Wounds Also Can Cause Skin Infections

The risk of infection due to a **traumatic wound**, such as a battlefield injury, compound fracture, or thermal burn, depends on several factors, including the extent of potential contamination, the contaminating dose of bacterial cells, and their virulence. The physical and physiologic nature of the wound—that is, are there areas of necrosis, or poor blood and oxygen supply—also are important factors affecting bacterial colonization and infection.

Gas gangrene is often the result of a traumatic wound. Because it is a soilborne disease, it was discussed in another chapter.

Burns are one of the most common and devastating forms of trauma. In the United States, the National Center for Injury Prevention and Control reports approximately 2 million fires each year, resulting in 1.2 million burn injuries. In a burn injury, the skin has been mechanically damaged and the underlying tissues are open to potential infection (FIGURE 13.24). Thus, individuals with moderate to severe burn injuries require hospitalization, which accounts for 100,000 infectious cases each year. About 5,000 hospitalized patients die each year from burn-related complications.

Burn wounds can be classified as **wound cellulitis**, which involves the unburned skin at the margin of the burn, or as an **invasive wound infection**, which is characterized by microbial invasion of viable tissue beneath the burn.

Although there is a variety of bacterial species that can cause infection (TABLE 13.3), one of the most likely infective agents in many burn centers is *Pseudomonas aeruginosa*. This aerobic, gram-negative rod is widely distributed in soil, water, plants, and animals (including humans).

P. aeruginosa can form mature biofilms within about 10 hours after colonization. Therefore, limiting burn wound infections and patient mortality requires rapid burn debridement and wound

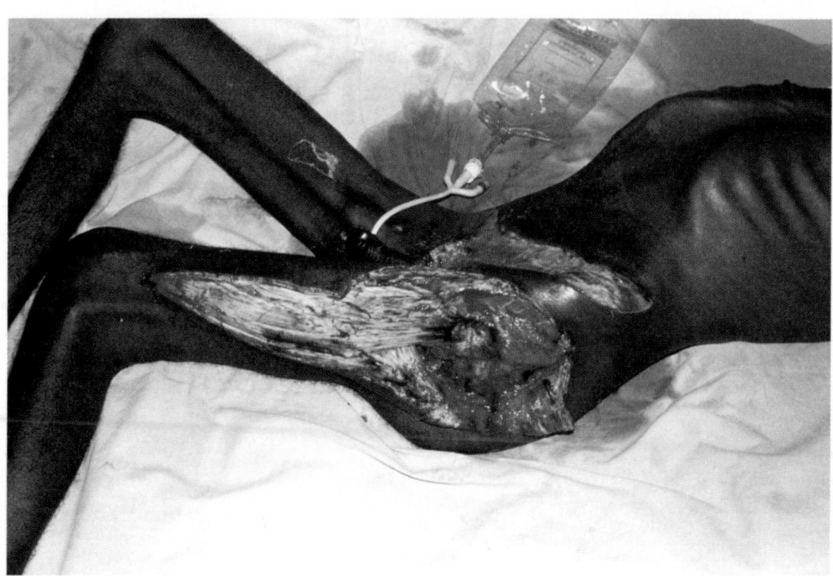

FIGURE 13.23 **Necrotizing Fasciitis.** The extensive loss of connective tissue can be seen in the leg of a 15-year-old AIDS patient. Extensive wound debridement was required. © Dr. M.A. Ansary/Photo Researchers, Inc. »» What tissues are most affected by the GAS?

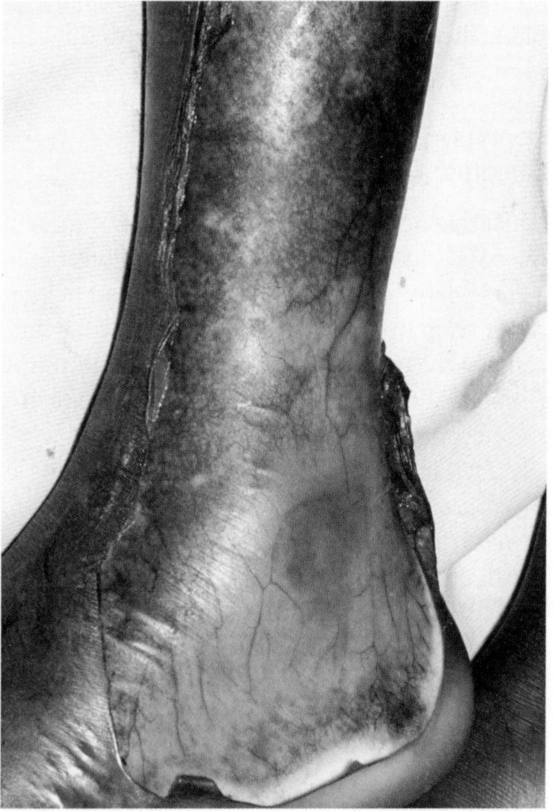

FIGURE 13.24 **A Burn Trauma to the Leg.** Burns to this patient's lower leg and ankle have reached underlying fat and muscle tissue, making the tissue extremely susceptible to infection. © Dr. M.A. Ansary/Photo Researchers, Inc. »» What types of bacteria are likely to infect this patient?

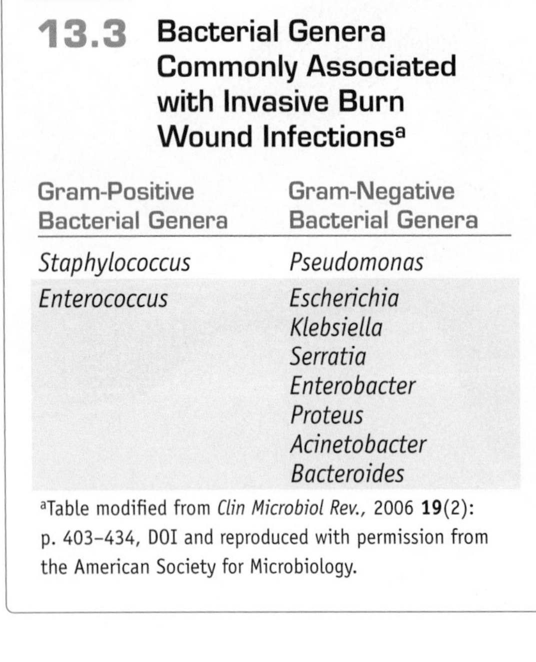

TABLE

13.3 Bacterial Genera Commonly Associated with Invasive Burn Wound Infections[a]

Gram-Positive Bacterial Genera	Gram-Negative Bacterial Genera
Staphylococcus	*Pseudomonas*
Enterococcus	*Escherichia* *Klebsiella* *Serratia* *Enterobacter* *Proteus* *Acinetobacter* *Bacteroides*

[a]Table modified from *Clin Microbiol Rev.*, 2006 **19**(2): p. 403–434, DOI and reproduced with permission from the American Society for Microbiology.

closure. In addition, patients with serious invasive burns require immediate topical and systemic antibiotic therapy to further minimize morbidity and mortality.

Leprosy (Hansen Disease) Is a Chronic, Systemic Infection

Causative Agent and Epidemiology. For most of the past 2,600 years, **leprosy** has been considered a curse of the damned. It did not kill, but neither did it seem to end. Instead, it lingered for years, causing the tissues to degenerate and deforming the body. As a result, in biblical times, the afflicted were ostracized from the community.

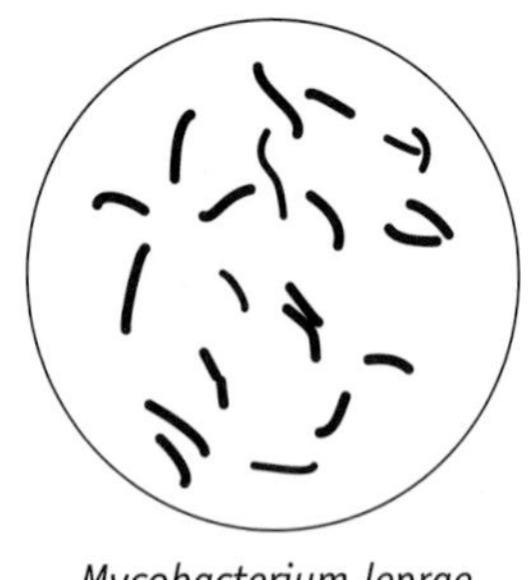

Mycobacterium leprae

The agent of leprosy or Hansen disease is *Mycobacterium leprae,* an acid-fast rod related to *M. tuberculosis.* Being an obligate intracellular parasite, it cannot be cultivated in artificial laboratory media. In 1960, researchers at the CDC succeeded in cultivating the bacillus in the footpads of mice, and in 1969, scientists discovered it would grow in the skin of nine-banded armadillos.

You might believe that leprosy is a disease of the past. In fact, it still is a disease that strikes many individuals. In the United States, there were 98 new cases reported in 2010 and there are some 3,600 patients being treated for the chronic condition. About 70% of the new cases occur in susceptible individuals who traveled to a leprosy-endemic area abroad. Of concern to health experts is a new strain of *M. leprae* that was first identified in 2011. This strain appears to be spread by zoonotic transfer, specifically from infected armadillos to native-born Americans living in the southern United States.

Clinical Presentation. Leprosy is hard to transmit because about 95% of the world's population has a natural immunity to the disease. It is spread by contact with nasal secretions, which are taken up through the upper respiratory tract. The disease has an unusually long incubation period of 3 to 6 years, a factor making diagnosis very difficult. Because the organisms are heat sensitive, the symptoms occur in the skin and peripheral nervous system in the cooler parts of the body, such as the hands, feet, face, and earlobes.

The World Health Organization (WHO) categorizes leprosy by the type and number of skin areas affected. **Paucibacillary (tuberculoid) leprosy** is a more limited disease with few bacterial cells in the skin and nerves. No bacilli would be detectable in lesion samples from patients with this form of leprosy, making it the milder and less contagious form of the disease. Thus, paucibacillary leprosy is characterized by a few (less than five) flat or slightly raised skin lesions of various sizes that are typically pale or slightly red, and numb to touch because the bacterial cells damage the underlying nerves. **Multibacillary (lepromatous) leprosy** is a more widespread, contagious disease, involving larger numbers of infecting bacilli producing six or more skin lesions. This form of the disease is more severe, as there are more areas of numbness and certain muscle groups may become weak. Multibacillary leprosy involves the skin and thickening of many peripheral nerves but may also involve the eyes, nose, and kidneys.

The most severe symptoms result from infection of the peripheral nerves, resulting in nerve deterioration such that the person cannot feel pain or temperature. Inattentive individuals may pick up a pot of boiling water without flinching. Repeated damage may eventually lead to loss of fingers and toes. The damage to peripheral nerves may cause muscle weakness that can result in deformities. For example, the fingers may be weakened, causing them to curve inward (like a claw). Damage to the nasal passages can result in a chronically stuffy nose and, if untreated, complete

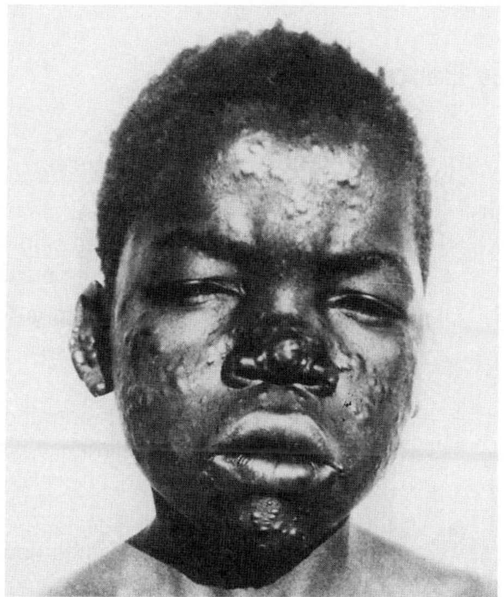

(A)

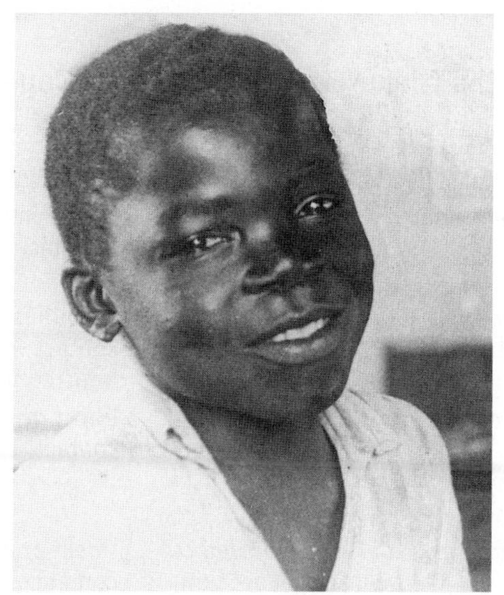

(B)

FIGURE 13.25 **Treating Leprosy.** The young boy with multibacillary leprosy is pictured (**A**) before treatment with dapsone and (**B**) some months later, after treatment. Note that the lesions of the ear and face and the swellings of the lips and nose have largely disappeared. Courtesy of American Leprosy Mission, www.leprosy.org. »» Why is the disfiguring caused by *Mycobacterium leprae* limited to the body extremities?

erosion of the nose (FIGURE 13.25A). Damage to the eyes may lead to glaucoma or blindness.

Treatment and Prevention. Until recently, one of the principal drugs for the treatment and cure of leprosy was a sulfur compound called Dapsone. In many cases, such as the one shown in FIGURE 13.25B, the results were dramatic. Today, treatment involves multidrug therapy with dapsone, rifampacin, and clofazimine.

In 1991, the WHO began a campaign to "eliminate" leprosy by 2000. However, elimination meant "controlling" the number of registered cases to one registered case in every 10,000 people in a population. In 2000, this goal was achieved. In fact, over the past 20 years, more than 14 million leprosy patients have been cured.

TABLE 13.4 summarizes the bacterial contact diseases. MicroInquiry 13 presents four cases for study involving sexually transmitted and contact diseases.

CONCEPT AND REASONING CHECKS 5

a. Summarize the characteristics of toxigenic *S. aureus* infections.

b. Assess the clinical significance of group A streptococci (GAS) in causing streptococcal toxic shock syndrome and necrotizing fasciitis.

c. Why is debridement necessary for most burn wounds?

d. Distinguish between paucibacillary (tuberculoidal) and multibacillary (lepromatous) leprosy.

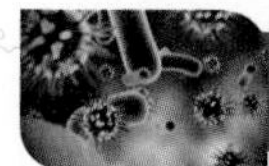

Chapter Challenge D

The human skin is the largest organ of the integumentary system. It weighs about 4.5 kilograms and has a surface area of almost two square meters. Being that it covers such a large area, it is not surprising that there are numerous potential pathogens (bacterial, viral, fungal) that can impact the human skin. Thus, one of the most important roles of the skin is to act as a barrier to these outside pathogens.

QUESTION D:

Explain the difference between indigenous skin diseases and those caused by exogenous pathogens and why the exogenous ones are more dangerous to human health.

Answers can be found on the Student Companion Website in **Appendix D**.

TABLE

13.4 A Summary of the Major Contact Diseases Caused by Bacterial Species

Inflammation or Disease	Causative Agent	Signs and Symptoms	Transmission	Treatment	Prevention
Acne	*Propionibacterium acnes*	Comedones (whiteheads and blackheads)	Part of indigenous skin microbiota	Benzoyl peroxide Antibiotics Isotretinoin	Gentle washing of affected skin Benzoyl peroxide
Furuncle and Carbuncle	*Staphylococcus aureus*	Painful single or cluster of boils	Autoinfection Contact with infected person	Drainage Debridement	Practicing good hygiene
Impetigo	*Staphylococcus aureus* *Streptococcus pyogenes*	Thin-walled blisters forming a crust	Direct or indirect contact	Skin cleansing Topical antibiotic	Practicing good hygiene
Scalded skin syndrome	*Staphylococcus aureus*	Red, wrinkled and tender skin Epidermis may peel	Direct or indirect contact	Cefazolin	Practicing good hygiene
Toxic shock syndrome (TSS)	*Staphylococcus aureus*	Fever, vomiting, watery diarrhea, sore throat, muscle aches, and sunburn-like rash	Vaginal tampons Skin wounds Surgery	Supportive care Antibiotics	Avoiding highly absorbent vaginal tampons
Erysipelas	*Streptococcus pyogenes*	High fever, shaking and chills, headache leading to fiery rash on lower limbs and face	Minor skin trauma Eczema Surgical incisions	Oral or intravenous antibiotics	Avoiding dry skin Preventing cuts and scrapes
Streptococcal TSS (STSS)	*Streptococcus pyogenes*	Fever, dizziness, confusion, and flat body rash	Direct contact with patients or carriers	Clean any wounds Debridement	Cleaning wounds
Necrotizing fasciitis	*Streptococcus pyogenes*	Fever with pain and swelling at the wound site	Trauma to skin surface	Broad spectrum antibiotics Debridement Surgery	Cleaning skin after a cut, scrape, or other deep wound
Burn infections	*Pseudomonas aeruginosa* Other bacterial species	Difficult to diagnose May be absent, minimal or late developing	Nosocomial	Antibiotic therapy Debridement	Protecting burn patients Practicing high levels of disinfection and sterilization
Leprosy (Hansen disease)	*Mycobacterium leprae*	Disfiguring of skin and bones, loss of pain sensation, loss of facial features	Nasal secretions	Multidrug therapy with dapsone, rifampin, and clofazimine	Avoiding contact where endemic

MICROINQUIRY 13

Sexually Transmitted and Contact Disease Identification

Below are several descriptions of sexually transmitted, contact, and miscellaneous bacterial diseases based on material presented in this chapter. Read the case history and then answer the questions posed. Answers can be found on the Student Companion Website in **Appendix D.**

Case 1

The patient is a 17-year-old woman who comes to the clinic indicating that several days ago she started feeling nauseous but had not experienced any vomiting. She tells the physician that the day before coming to the clinic she had a fever and chills; she also has been urinating more frequently and the urine has a foul smell. She is diagnosed as having a urinary tract infection.

13.1a. What types of bacterial species could be responsible for her illness?

13.1b. Why are these types of diseases more prevalent in women than they are in men?

13.1c. What types of urinary infections can occur?

13.1d. How could this patient attempt to avoid another UTI?

13.1e. What role do biofilms play in UTIs?

Case 2

A 19-year-old unwed mother arrives at the emergency room of the county hospital complaining of having cramps and abdominal pain for several days. She says she had never had a urinary tract infection and could not have gonorrhea, as she was treated and cured of that 2 years ago. She has not experienced nausea or vomiting. When questioned, she tells the emergency room nurse that she has a single male sexual partner and condoms always are used. Based on further examination, the patient is diagnosed with pelvic inflammatory disease (PID). An endocervical swab is used for preparing a tissue culture. Staining results indicate the presence of cell inclusions.

13.2a. What bacterial species can be associated with PID? What disease does she most likely have?

13.2b. Why was a tissue culture inoculum ordered? Describe the reproductive cycle of this organism.

13.2c. What other tests could be ordered for the patient's infection?

13.2d. Why was the emergency room concerned about her sexual activity?

13.2e. What misconception does the patient have about her past gonorrhea infection?

Case 3

A 17-year-old man comes to a free neighborhood clinic. He says that he noticed some white pus-like discharge and a tingling sensation in his penis. Since yesterday, he has had pain on urinating. He tells the physician that he has been sexually active with several female partners over the past eight months, but no one has had any sexually transmitted disease. Examination determines that there is no swelling of the lymph nodes in the groin or pain in the testicles. A Gram stain indicates the presence of gram-negative diplococci. The patient is given antibiotics, instructed to tell his female partners they both should be medically examined, and then he is released.

13.3a. Based on the clinic findings, what disease does the patient have and what bacterial species is responsible for the infection?

13.3b. Why is it important for his sexual partners to be medically examined, even if they experience no symptoms? What complications could arise if they are infected?

13.3c. For which other organisms is this patient at increased risk? Why?

13.3d. What significance can be drawn from the fact that the patient does not have any swelling of the lymph nodes in the groin or pain in the testicles?

13.3e. What antibiotics most likely would be given to the patient?

KEY CONCEPT 13.6 Several Contact Diseases Affect the Eye

Although the eye has a multilayered defense against infection, eye diseases do occur. As we will discover, trachoma is the most prevalent infectious eye disease in the world. However, in the United States and in other developed nations, eye infections typically involve inflammation of the eyelid, conjunctiva, or cornea (FIGURE 13.26). The most common diseases of the eye surface are described below.

Some Bacterial Eye Infections Can Cause Blindness

Several eye infections are caused by a few bacterial species. *S. aureus* can infect the eyelid and the cornea. When it infects the eyelid margin, a painful red inflammation called **blepharitis** develops. The inflammation sometimes leads to the formation of a **stye**. Treatment usually involves warm

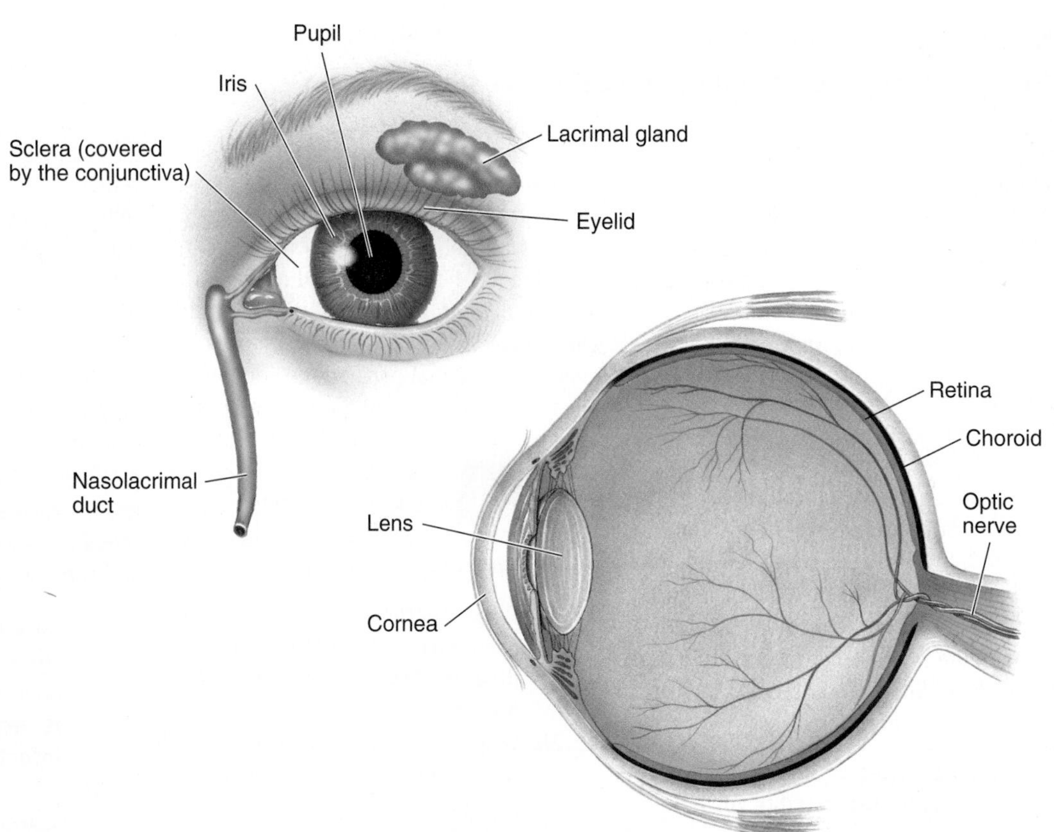

FIGURE 13.26 **Eye Anatomy.** External infections are most often associated with the eyelid, conjunctiva, or cornea. »» Why would the conjunctiva and cornea be most susceptible to infection?

compresses and a topical antibiotic, such as one containing bacitracin.

One form of **bacterial conjunctivitis**, commonly called **pink eye** or **red eye**, is an eye redness due to dilation of the conjunctival blood vessels (FIGURE 13.27). It is common in childhood, but the inflammation can occur in people of any age. The organisms that most commonly cause bacterial conjunctivitis are staphylococci, pneumococci, or streptococci that are picked up by contact with the eye.

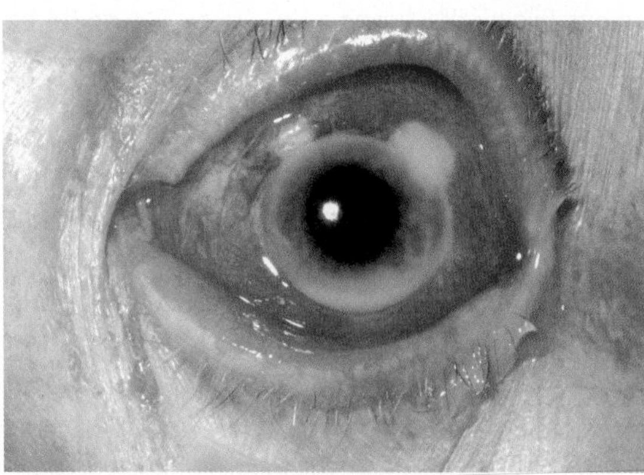

FIGURE 13.27 **Conjunctivitis.** The inflammation on the conjunctiva, also called pink eye, causes swelling on the surface. © Medical-on-Line/Alamy Images. »» What causes the redness in the sclera?

Bacterial conjunctivitis usually causes no long-term eye or vision damage. Still, it is important to see a doctor because some types require treatment with antibiotic eye drops or ointment. Prevention includes frequent hand washing with soap and warm water. Infected individuals should not share eye drops, tissues, eye makeup, washcloths, towels, or other objects that come in contact with the eye.

A form of **chronic bacterial conjunctivitis**, leading to keratitis, also is caused by *S. aureus*. The bacterium invades the cornea after some form of eye trauma that causes a break in the corneal epithelium. The resulting ulcers are painful and are treated with antibiotic drops.

The most common and severe form of **hyperacute bacterial conjunctivitis** is caused by *N. gonorrhoeae*. Left untreated, it can progress to keratitis and corneal perforation, the latter providing an entry point for a blood infection. Ceftriaxone can be used to treat the infection. However, any delay in treatment can lead to corneal damage or eye loss.

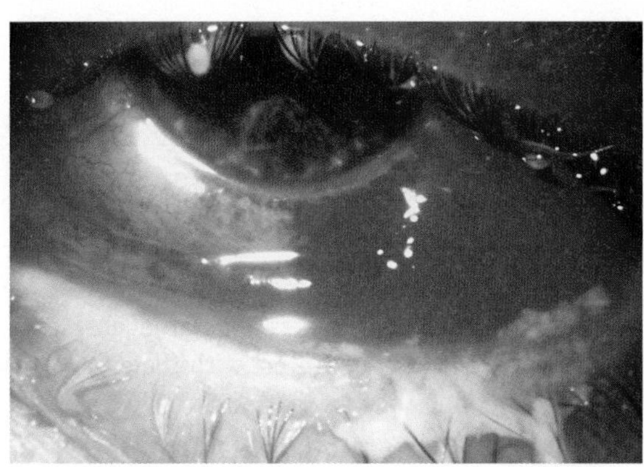

FIGURE 13.28 **Gonorrheal Conjunctivitis.** This case of gonorrheal conjunctivitis resulted in partial blindness due to the spread of *N. gonorrhoeae* bacteria. Courtesy of CDC. »» How does a newborn become infected with *N. gonorrhoeae*?

Neonatal conjunctivitis is an inflammation of the conjunctiva of the newborn. The inflammation, also called **ophthalmia of the newborn**, results from contact with the bacterium during passage through the birth canal of a mother infected with *N. gonorrhoeae* or *C. trachomatis*. Infection with *N. gonorrhoeae* is the most serious and, if untreated, can lead to blindness (FIGURE 13.28). It usually heals without permanent eye damage. Treatment for both forms of conjunctivitis involves the antibiotics doxycycline or erythromycin. In the United States, prevention involves the use of antimicrobial drugs put into the eyes of all newborns after delivery. Silver nitrate drops are still used in some parts of the world.

Trachoma is the world's leading cause of preventable blindness. It occurs in hot, dry regions of the world and it is prevalent in Mediterranean countries, parts of Africa and Asia, and in the southwestern United States in Native American populations (FIGURE 13.29). There are 500 million infections, mostly children, worldwide and 7 to 9 million individuals have been blinded by trachoma.

Trachoma is caused by serotypes A, B, and C of *C. trachomatis* that are not sexually transmitted but rather by personal contact with contaminated fingers, towels, and optical instruments. Face-to-face contact and flies also are important modes of transmission.

The chlamydiae multiply in the conjunctiva of the eye. A series of tiny, pale nodules form on this membrane, giving it a rough appearance (*trach* = "rough"; *oma* = "tumor"). An initial infection typically heals without permanent damage. However, the initial infection sets up a hypersensitive state, such that repeated infections result in a chronic inflammation. Over 10 to 15 years, scarring of the conjunctiva occurs as the eyelashes turn inward and abrade the cornea, eventually leading to blindness.

Azithromycin helps reduce the symptoms of trachoma, but in many patients, the relief is only temporary because chlamydiae reinfect the tissues (FIGURE 13.30). In 1997, the WHO established the Alliance for Global Elimination of Trachoma by 2020 (GET 2020). Since then, 10 national programs, making up 50% of the global trachoma

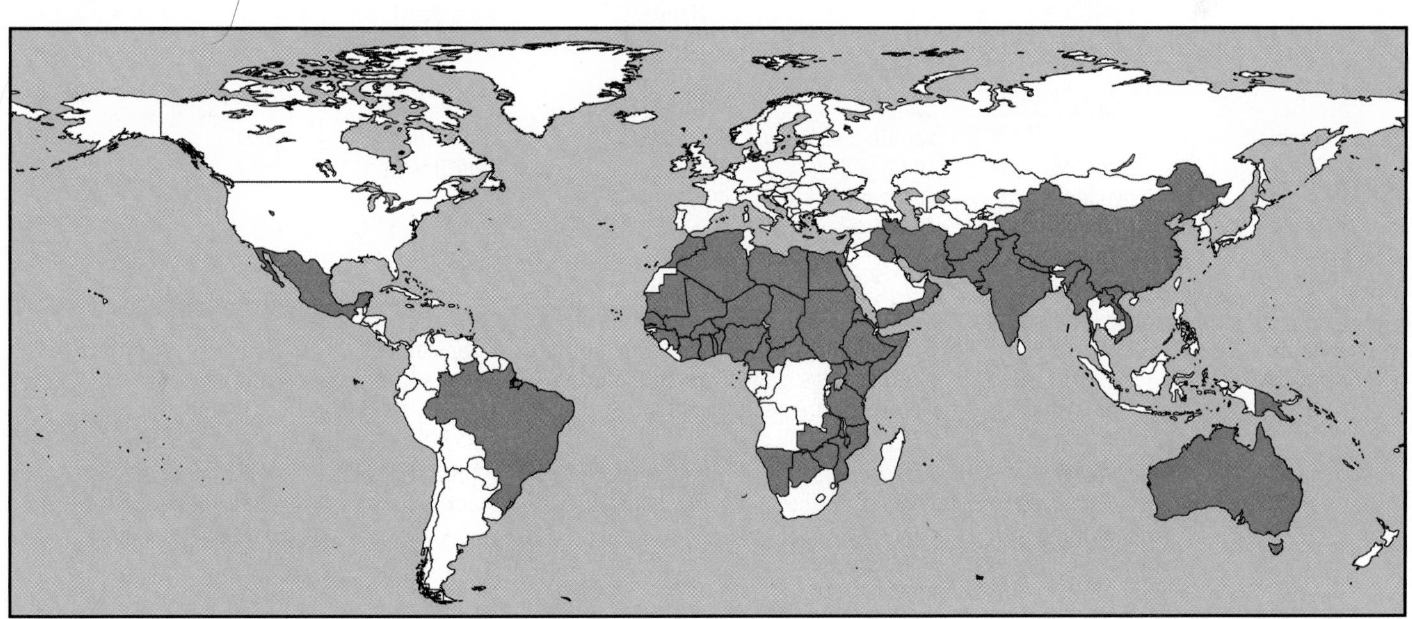

FIGURE 13.29 **Worldwide Trachoma.** Blinding or suspected blinding trachoma affects individuals worldwide, but especially in Africa, Southeast Asia, Mexico, and parts of South America. (Map by Silvio Mariotti/WHO.) »» How can trachoma be treated?

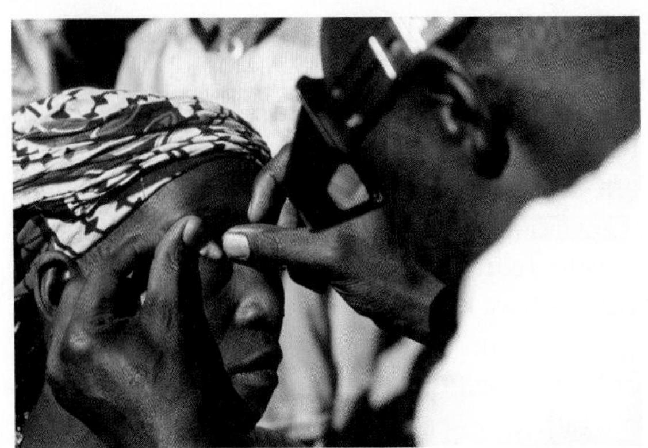

FIGURE 13.30 **Trachoma.** An opnthalmic surgeon in rural Malawi, Africa, attends to a trachoma patient. © Suzanne Porter/Alamy. »» How is trachoma spread?

burden, have reduced acute infections in children by 50%. This has involved using a "**SAFE strategy**"; that is, **S**urgery of the eyelids; **A**ntibiotics for acute infections; **F**acial hygiene improvements; and **E**nvironmental access to safe water.

TABLE 13.5 summarizes the bacterial eye diseases.

CONCEPT AND REASONING CHECKS 6

a. What are the challenges facing health officials trying to eliminate trachoma?

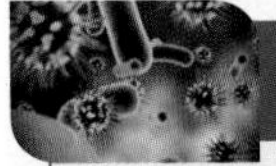

Chapter Challenge E

The human eye is exposed to the environment and all the microbes and pathogens that might come in contact with the eye. The infectious agents that potentially can cause disease can be introduced to the eye either directly or indirectly.

QUESTION E:

For blepharitis, bacterial conjunctivitis (pink eye), and trachoma, identify if the infection is the result of direct or indirect contact. As such, what is the best way to protect your eyes from potential infections?

Answers can be found on the Student Companion Website in **Appendix D.**

TABLE

13.5 A Summary of Infectious Bacterial Eye Diseases

Inflammation or Disease	Causative Agent	Signs and Symptoms	Transmission	Treatment	Prevention
Blepharitis (stye)	*Staphylococcus aureus*	Burning sensation in eye	Contaminated instruments Hands Shared towels Droplets	Warm compress Antibiotic medication	Practicing good hygiene
Bacterial conjunctivitis	*Staphylococcus aureus* *Streptococcus pyogenes* *Haemophilus influenzae* *Neisseria gonorrhoeae*	Eye pain, swelling, redness, and a yellow or greenish discharge	Direct or indirect contact	Broad-spectrum antibiotic	Practicing good hygiene
Neonatal conjunctivitis	*Neisseria gonorrhoeae* *Chlamydia trachomatis*	Eye swelling and pus discharge Watery discharge	Infected mother to child during childbirth	Topical and oral antibiotics	Using silver nitrate or antibiotics Screening mother
Trachoma	*Chlamydia trachomatis* (serotype A-C)	Tiny, pale nodules on the conjunctiva Upper eyelid abrasion can cause blindness	Indirect contact Mechanical vector	Topical or oral antibiotics	Washing face Controlling flies Freshwater source

■ **In conclusion**, we have now finished our survey of many bacterial infections and diseases that affect the human body. If you go to the CDC website and visit the diseases and conditions page (www.cdc.gov/diseasesconditions/), you will find an index of many more diseases than the ones described in this and previous chapters. In addition, you can visit the weekly *Morbidity and Mortality Report* (*MMWR*) at (www.cdc.gov/mmwr/). This is the agency's "primary vehicle for scientific publication of timely, reliable, authoritative, accurate, objective, and useful public health information and recommendations." The information provided in the *MMWR* is based on weekly reports compiled by the CDC from information supplied by state health departments. It is written at a level that you, as a "microbiology student," can comprehend. When you read about an infectious disease outbreak in the news media, more than likely there will be one or more articles that will soon appear in the *MMWR*. Also of note is a similar publication put out by the WHO called the *Weekly Epidemiological Record* (*WER*; www.who.int/wer/en/) that on a global perspective "serves as an essential instrument for the rapid and accurate dissemination of epidemiological information on cases and outbreaks of diseases under the International Health Regulations and on other communicable diseases of public health importance, including emerging or re-emerging infections." So, if you want to keep up with the latest in disease outbreaks, epidemics, and sometimes pandemics, these are the places to go.

■ SUMMARY OF KEY CONCEPTS

13.1 Portions of the Female and Male Reproductive Systems Contain an Indigenous Microbiota

- The **primary sex organs** in the male are the testes, while the epididymis, vas deferens, seminal vesicles, prostate, and penis are **accessory reproductive organs.** In females, the primary sex organs are the ovaries, while the accessory organs consist of the fallopian tubes, uterus, vagina, and vulva. (Fig. 13.2)
- Antimicrobial defenses in the reproductive tracts include the urethral mucosa, the vagina, vulva, and cervix in females. *Lactobacillus* species in the vagina as well as antibodies and other antimicrobial products of the systems produce an environment not favorable for pathogen colonization. (Fig. 13.3)
- Nonsexually transmitted illnesses include:
  - **Bacterial vaginosis**
    - *Gardnerella vaginalis*
    - *Prevotella*
    - *Mobiluncus*
    - Other indigenous bacterial species

13.2 Many Sexually Transmitted Diseases Are Caused by Bacteria

- **Chlamydial urethritis**
  1. *Chlamydia trachomatis*
- **Gonorrhea**
  2. *Neisseria gonorrhoeae*
- **Syphilis**
  3. *Treponema pallidum*
- **Chancroid**
  4. *Haemophilus ducreyi*
- **Lymphogranuloma venereum**
  5. *Chlamydia trachomatis*

13.3 Urinary Tract Infections Are the Second Most Common Body Infection

- The organs of the **urinary system** that are susceptible to infection are the kidneys and the urinary tract (ureters, urinary bladder, and urethra). The kidneys, ureters, and bladder are normally sterile due to normal urine flow. However, the urethra is colonized by microbes either along its whole length or near the terminus. (Fig. 13.12)
- Bacterial diseases include:
  - **Urethritis**
    6. *Escherichia coli*
    7. *Chlamydia trachomatis*
    8. *Neisseria gonorrhoeae*
  - **Cystitis**
    9. *Escherichia coli*
  - **Prostatitis** (in males—not shown)
    - *Escherichia coli*
    - *Staphylococcus aureus*
    - *Proteus*
    - *Pseudomonas aeruginosa*
  - **Pyelonephritis**
    10. *Escherichia coli*
    11. *Staphylococcus aureus*
    12. *Klebsiella pneumoniae*
    13. *Pseudomonas aeruginosa*

13.4 Contact Diseases Can Be Caused by Indigenous Bacterial Species

- Many bacterial diseases are caused by contact with the skin, which normally protects the underlying tissues from bacterial colonization and infection.
- **Acne**
  14. *Propionibacterium acnes*
- Burn infections
  15. *Pseudomonas aeruginosa*

13.5 Contact Diseases Can Also Be Caused by Exogenous Bacterial Species

- **Furuncles** (boils) and carbuncles
  16. *Staphylococcus aureus*
- **Impetigo**
  17. *Staphylococcus aureus*
  18. *Streptococcus pyogenes*
- **Scalded skin** and **toxic shock syndromes**
  19. *Staphylococcus aureus*
- **Erysipelas** and **necrotizing fasciitis**
  20. *Streptococcus pyogenes*
- Leprosy (Hansen disease)
  21. *Mycobacterium leprae*

13.6 Several Contact Diseases Affect the Eye

- **Conjunctivitis**
  22. *Neisseria gonorrhoeae*
  23. *Chlamydia trachomatis*
- **Trachoma**
  24. *Chlamydia trachomatis*

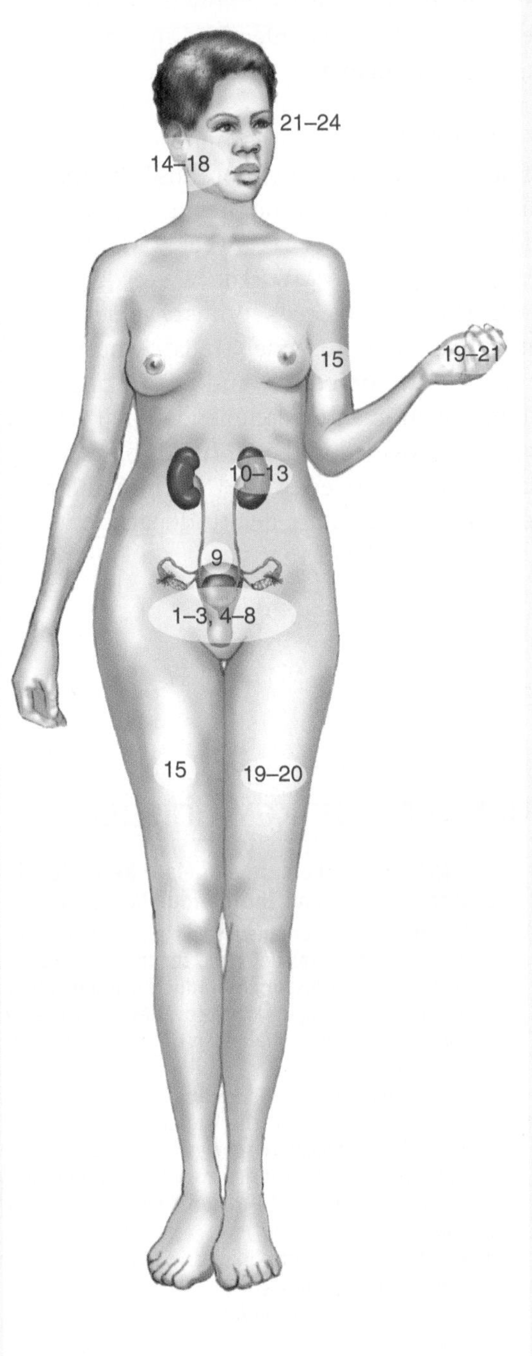

CHAPTER SELF-TEST

For **STEPS A–D**, answers to even-numbered questions and problems can be found in **Appendix C** on the Student Companion Website at **http://microbiology.jbpub.com/10e**. In addition, the site features eLearning, an online review area that provides quizzes and other tools to help you study for your class. You can also follow useful links for in-depth information read more MicroFocus stories, or just find out the latest microbiology news.

STEP A: REVIEW OF FACTS AND TERMS

Read each question carefully, then select the ***one*** answer that best fits the question or statement.

Multiple Choice

1. The primary sex organs of the female reproductive system is/are the ______.
A. uterus
B. vagina
C. fallopian tubes
D. ovaries

2. What part or parts of the male and female reproductive systems are typically colonized by indigenous microbiota?
A. Male: ureters; female: vagina and ovaries
B. Male: testes and epididymis
C. Male: urethra; female: vagina, vulva, and cervix
D. Male: bladder and ureters; female: fallopian tubes and cervix

3. Which one of the following microbes is NOT associated with vaginosis?
A. *Mobiluncus*
B. *Gardnerella*
C. *Prevotella*
D. *Staphylococcus*

4. Which one of the following statements is NOT correct concerning the reproductive cycle of *Chlamydia*?
A. Reticulate bodies are infectious.
B. Reticulate bodies reorganize into elementary bodies.
C. Elementary bodies infect host cells.
D. Elementary bodies transform into reticulate bodies.

5. Salpingitis is associated with ______ and can lead to ______.
A. syphilis; gumma formation
B. gonorrhea; sterility
C. chlamydia; ophthalmia
D. chancroid; soft chancre

6. A chancre is typical of which stage of syphilis?
A. Primary
B. Secondary
C. Tertiary
D. Chronic, latent

7. Besides chlamydia urethritis, what other STD is associated with another serotype of *Chlamydia trachomatis*?
A. Lymphogranuloma venereum (LGV)
B. Genital warts
C. Gonorrhea
D. Chancroid

8. Which one of the following is NOT part of the urinary tract?
A. Urethra
B. Bladder
C. Kidneys
D. Ureters

9. What bacterial species is most often associated with cystitis?
A. *Treponema pallidum*
B. *Escherichia coli*
C. *Chlamydia trachomatis*
D. *Pseudomonas aeruginosa*

10. What type of immune defensive cell is found in the sublayers of the epidermis?
A. Keratinocyte
B. Dendritic (Langerhans) cell
C. Neutrophil
D. Basophil

11. The skin is ______.
A. dominated by gram-negative bacterial cells
B. free of bacterial cells
C. without a microbiota
D. dominated by gram-positive bacterial cells

12. Which one of the following statements is NOT true of acne?
A. Acne is caused by *Propionibacterium acnes*.
B. Plugged sebaceous glands are called erythemas.
C. Whiteheads are completely blocked follicles.
D. Acne is not a preventable disease.

13. An acute wound could be due to ______.
A. surgical procedures
B. cuts
C. lacerations
D. All the above (A–C) are correct.

14. In children, this skin disease is characterized by the production of thin-walled blisters oozing a yellowish fluid and forming yellowish-brown flakes.
A. Toxic shock syndrome
B. Scalded skin syndrome
C. Erysipelas
D. Impetigo

15. Which one of the following skin diseases is NOT caused by *Streptococcus pyogenes?*
A. Necrotizing fasciitis
B. Toxic shock syndrome
C. Gas gangrene
D. Erysipelas

16. The most common cause of an invasive wound infection, such as a burn, is ______.
A. gram-positive bacterial species
B. *Treponema pallidum*
C. *Pseudomonas aeruginosa*
D. *Escherichia coli*

17. Leprosy can be contracted by contact with ______.
A. contaminated water
B. insects
C. nasal secretions
D. contaminated food

18. The SAFE strategy has greatly reduced the global burden of what disease?
A. Trachoma
B. Neonatal conjunctivitis
C. Leprosy
D. Blepharitis

Fill-in

Answer each of the following by filling in the blank with the correct word or phrase.

19. __________ is the disease caused by *Haemophilus ducreyi.*

20. Lymphogranuloma venereum (LGV) is prevalent in __________-Asia.

21. *Gardnerella* is the cause of __________.

22. __________ are a group of connected, deep abscesses caused by *S. aureus*.

23. __________ is the most common agent of urinary tract infections.

24. Sufferers of UTIs have a __________ on urination.

25. Dapsone, rifampin, and clofazimine are drugs used to treat __________.

STEP B: CONCEPT REVIEW

26. Trace the pathway of (a) a sperm cell through the **male reproductive system** and (b) an egg through the **female reproductive system**. (**Key Concept 1**)
27. Identify the defenses in the male and female reproductive systems that normally prevent colonization and infection. (**Key Concept 1**)
28. Describe the causes of **bacterial vaginosis.** (**Key Concept 1**)
29. Distinguish between the signs and symptoms of **chlamydial urethritis** in males and females. (**Key Concept 2**)
30. Describe (a) the possible complications resulting from **gonorrhea** in females and (b) explain the danger of gonorrhea in pregnant females. (**Key Concept 2**)
31. Distinguish among the three possible stages of **syphilis.** (**Key Concept 2**)
32. Describe the signs and symptoms of **chancroid.** (**Key Concept 2**)
33. Assess the role of **urination** and indigenous microbiota to maintaining a healthy urinary system free from infection. (**Key Concept 3**)
34. Differentiate among the various forms of UTIs: **urethritis, cystitis, prostatitis**, and **pyelonephritis**. (**Key Concept 3**)
35. Identify the skin layers and other chemical skin defenses protecting underlying tissues from microbial colonization. (**Key Concept 4**)
36. Name the significant members of the indigenous skin microbiota that normally out-compete most transient microbiota and pathogens. (**Key Concept 4**)
37. Assess the role of *Propionibacterium acnes* in triggering **acne** and describe the follicle-associated lesions. (**Key Concept 4**)
38. Estimate the significance of infections by indigenous microbiota resulting from surgical procedures. (**Key Concept 4**)
39. Assess the role of *Staphylococcus aureus* as an agent for contact diseases. (**Key Concept 5**)
40. Describe the contrast skin infections caused by *Streptococcus pyogenes.* (**Key Concept 5**)
41. Summarize the types of diseases caused by **traumatic wounds**, including burns. (**Key Concept 5**)
42. Summarize the clinical significance of **leprosy** (**Hansen disease**). (**Key Concept 5**)
43. Explain why **trachoma** is a major cause of blindness in many developing nations. (**Key Concept 6**)

STEP C: APPLICATIONS AND PROBLEMS

44. Suppose a high incidence of leprosy existed in a particular part of the world. Why is it conceivable that there might be a correspondingly low level of tuberculosis?
45. An African patient reports to a local hospital with an upper lip swollen to about three times its normal size. Probing with a safety pin at facial points where major nerve endings terminate showed that the area to the left of the nose and above the lip was without feeling. When a biopsy of the tissue was examined, it revealed round reservoirs of immune system cells called granulomas within the nerves. On bacteriological analysis, acid-fast rods were observed in the tissue. What disease do all these data suggest?
46. Certain microscopes have the added feature of a small hollow tube that fits over the eyepieces or oculars. Viewers are encouraged to rest their eyes against the tube and thereby block out light from the room. Why is this feature hazardous to health?
47. A woman suffers two miscarriages, each after the fourth month of pregnancy. She then gives birth to a child, but impaired hearing and vision become apparent as it develops. Also, the baby's teeth are shaped like pegs and have notches. What medical problem existed in the mother?

STEP D: QUESTIONS FOR THOUGHT AND DISCUSSION

48. One of the major problems of the current worldwide epidemic of AIDS is the possibility of transferring the human immunodeficiency virus (HIV) among those who have a sexually transmitted disease. Which diseases in this chapter would make a person particularly susceptible to penetration of HIV into the bloodstream?
49. Studies indicate that most cases of *Staphylococcus*-related impetigo occur during the summer months. Why do you think this is the case?
50. In some African villages, blindness from trachoma is so common that ropes are strung to help people locate the village well, and bamboo poles are laid to guide farmers planting in the fields. What measures can be taken to relieve such widespread epidemics as this?
51. At a specified hospital in New York City, hundreds of patients pay a regular visit to the "neurology ward." Some sign in with numbers; others invent fictitious names. All receive treatment for leprosy. Why do you think this disease still carries such a stigma?
52. Several years ago the Rockefeller Foundation offered a $1 million prize to anyone who could successfully develop a simple and rapid test to detect chlamydia and/or gonorrhea. The test had to use urine as a test sample, and be performed and interpreted by someone with a high school education. No one ever claimed the prize. Can you guess why?

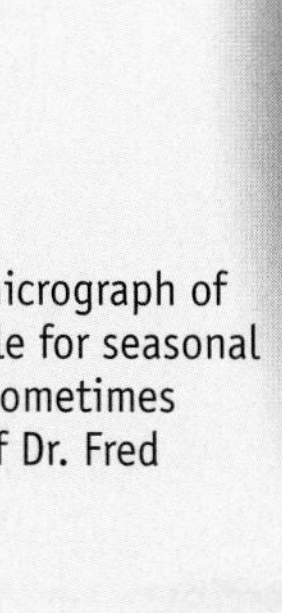

A false-color transmission electron micrograph of several flu viruses that are responsible for seasonal outbreaks, as well as epidemic, and sometimes pandemic human disease. Courtesy of Dr. Fred Murphy/Cynthia Goldsmith/CDC.

PART 4 Viruses and Eukaryotic Microorganisms

CHAPTER 14 The Viruses and Virus-Like Agents

CHAPTER 15 Viral Infections of the Respiratory Tract and Skin

CHAPTER 16 Viral Infections of the Blood, Lymphatic, Gastrointestinal, and Nervous Systems

CHAPTER 17 Eukaryotic Microorganisms: The Fungi

CHAPTER 18 Eukaryotic Microorganisms: The Parasites

The bacterial species we have examined in the previous chapters are but one of several groups of microbial agents interwoven with the lives of humans. Other prominent groups are the viruses, fungi, and parasites. Knowledge of these groups developed slowly during the early 1900s, partly because they were generally more difficult to isolate and cultivate than bacterial organisms. Also, the established methods for research into bacterial growth were more advanced than for other microorganisms, and investigators often chose to build on established knowledge rather than pursue uncharted courses of study. Moreover, the urgency to learn about the other groups was not as great because they did not appear to cause such great epidemics and pandemics.

This perception changed in the second half of the 1900s. Many bacterial diseases came under control with the advent of vaccines and antibiotics, and the increased funding for biological research allowed attention to shift to other infectious agents. The viruses finally were identified and cultivated, and microbiologists laid the foundations for their study. Fungi gained prominence as tools in biological research, and scientists soon recognized their significance in ecology and industrial product manufacturing. As remote parts of the world opened to trade and travel, public health microbiologists realized the global impact of parasitic diseases.

In Part 4, we examine the viruses and eukaryotic microorganisms. Chapter 14 is devoted to a study of the viruses and viral-like agents, while Chapters 15 and 16 outline the multiple diseases caused by these infectious particles. In Chapter 17, the discussion moves to fungi, while in Chapter 18, the area of interest is the protists and the multicellular parasites. Throughout these chapters, the emphasis is on human disease. You will note some familiar diseases, such as hepatitis, chickenpox, and malaria, as well as some less familiar ones, such as dengue fever, toxoplasmosis, and schistosomiasis. Today, the spectrum of diseases continues to unfold as scientists develop new methods for the detection, isolation, and cultivation of viruses and eukaryotic microorganisms.

MICROBIOLOGY PATHWAYS

Virology

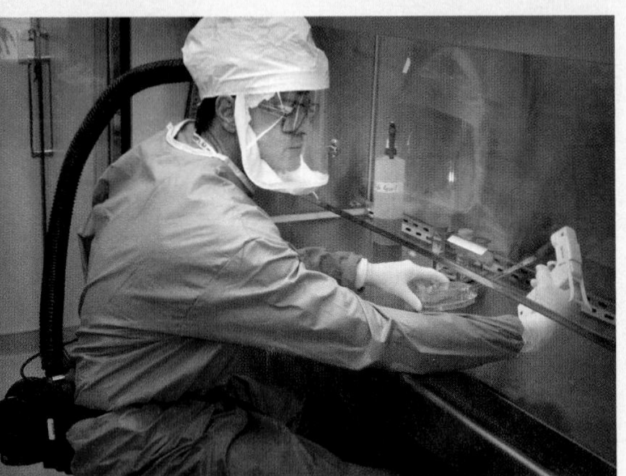

Courtesy of James Gathany/CDC.

When Ed Alcamo was in college, he was part of a group of twelve biology majors. Each of them had a particular area of "expertise." John was going to be a surgeon, Jim was interested in marine biology, Walt was a budding dentist, and Ed was the local virologist. He was fascinated with viruses, the ultramicroscopic bits of matter, and at one time wrote a term paper summarizing arguments for the living or nonliving nature of viruses. (At the time, neither side was persuasive, and even his professor gracefully declined to take a stand.)

Ed never quite made it to being a virologist, but if your fascination with these infectious particles is as keen as his, you might like to consider a career in virology.

Virologists investigate dreaded diseases such as AIDS, polio, and rabies, while others (epidemiologists) investigate disease outbreaks. Virologists also concern themselves with many types of cancer, and others study the chemical interactions of viruses with various tissue culture systems and animal models.

Virologists also are working to replace agricultural pesticides with viruses able to destroy mosquitoes and other pests. Some virologists are inserting viral genes into plants and are hoping the plants will produce viral proteins to lend resistance to disease. One particularly innovative group is trying to insert genes from hepatitis B viruses into bananas. They hope that one day we can vaccinate ourselves against hepatitis B by having a banana for lunch.

If you wish to consider the study of viruses, an undergraduate major in biology would be a good choice. Because the biology of viruses is related to the biology of cells, courses in biochemistry and cell biology will be required.

Following completion of college, most virologists study for an M.D. or Ph.D. degree. M.D.s pursue virology research in the context of patients or disease and become investigators with an interest in infectious disease or epidemiology. Most Ph.D.s pursue more basic questions with academic institutions, or in industrial or governmental organizations.

A great way to find out if you have the "research bug" is to work in a college or university laboratory. Many colleges and universities with research programs employ undergraduate students (near minimum wage!) in the research laboratory. So, here is a good way to "get your feet wet." If you find it interesting, it also will enhance your chances to be accepted by a top-flight graduate school.

Not interested in lab bench research? Virologists also can find careers in full-time teaching. In addition, your knowledge can be used to pursue a career in communications, serving as a science writer or reporter. They also pursue careers in business administration or law, especially involving the pharmaceutical industry or patent law.

14

CHAPTER PREVIEW

14.1 **Filterable Infectious Agents Cause Disease**

14.2 **Viruses Have a Simple Structural Organization**
Investigating the Microbial World 14: Finding the Correct Address

14.3 **Viruses Can Be Classified by Their Genome**

14.4 **Viral Replication Follows a Set of Common Steps**

14.5 **Viruses and Their Infections Can Be Detected in Various Ways**
MicroInquiry 14: The One-Step Growth Curve

14.6 **Some Viruses Are Associated with Human Tumors and Cancers**

14.7 **Emerging Viruses Arise from Genetic Recombination and Mutation**
Textbook Case 14: Multistate Outbreak of Monkeypox

14.8 **Virus-Like Agents Include Viroids and Prions**

The Viruses and Virus-Like Agents

It's just a piece of bad news wrapped up in protein.
—Nobel Laureate Peter Medawar (1915–1987) describing a virus

Under the Naica hills in northern Mexico lies the Cave of Crystals. Buried some 300 meters below the desert surface, in temperatures of 45°C to 50°C and saturating humidity, these caves appeared devoid of life, especially microbial life. Still, scientists collected water samples from the cave pools for analysis. The scientists, led by Dr. Curtis Suttle, wanted to discover if any organisms could survive in the harsh conditions of the caves. When the water samples were taken back to the lab in British Columbia, the samples were prepared for microscopy. What the samples contained stunned the researchers.

In these gigantic caverns, some the length of a football field and two stories high and containing 50 ton crystals of gypsum that were formed 500,000 years ago (FIGURE 14.1A), were bacterial cells. And more surprisingly, in those samples were tiny, geometric forms; forms that had been cut off from the outside world for perhaps millions of years. The geometric forms were viruses! And there were lots of viruses, perhaps up to 200 million per drop of water.

Almost halfway around the world in Pozzuoli, Italy in acidic hot springs (85°C to 95°C; pH 1.5) microbiologists discovered that enrichment cultures of several archaeal organisms contained virus particles. Much further south in the Antarctic's Lake Limnopolar, a freshwater lake that is frozen nine months of the year, scientists have found bacterial cells, protists, and—yes—viruses. Again huge numbers are seen, representing perhaps 10,000 different types, making the lake one of the most diverse virus communities in the world. Even the world's oceans are filled with viruses. For example, Rachel Parsons and her collaborators have detected enormous virus populations in the western Atlantic (FIGURE 14.1B).

The number of bacterial viruses alone on Earth exceeds 10^{30}, and scientists and microbiologists are finding them faster than they can

Image courtesy of Dr. Fred Murphy/CDC.

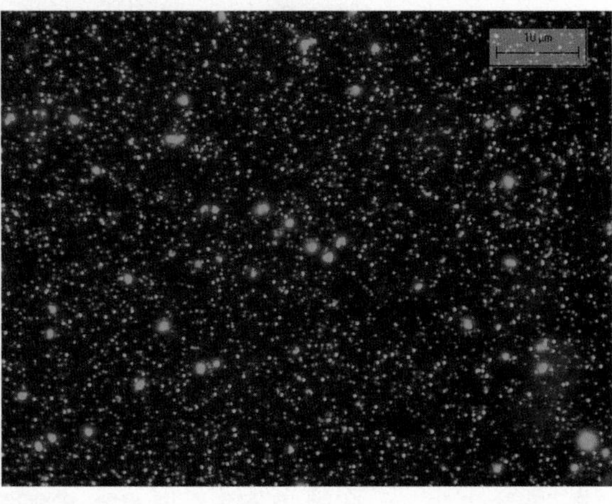

(A) (B)

FIGURE 14.1 **The Virosphere.** Viruses can be found wherever life exists. This includes isolated environments deep in the earth such as (**A**) the Cave of Crystals and (**B**) huge, open expanses such as the world's oceans. The larger, fluorescent dots are bacterial cells and the smaller spots marine bacterial viruses. (A) © Javier Trueba/MSF/Photo Researchers, Inc. and (B) Courtesy of Rachel Parsons, Bermuda Institute of Ocean Science. »» Does it seem strange to you that huge numbers of viruses are found in places like these? Explain.

make sense of them. In fact, the planet's most abundant "coterrestrials" are the viruses. This so-called **virosphere** exists, as Suttle's lab explains, "*wherever life is found; it is a major cause of* [morbidity and] *mortality, a driver of global geochemical cycles, and a reservoir of the greatest unexplored genetic diversity on Earth.*" And they are a major driving force for evolution because each virus infection has the potential to introduce new genetic information into just about any organism in the domains Bacteria, Archaea, and Eukarya, or into their own progeny viruses. The virosphere is huge, it is incredibly diverse, and it produces tremendous impact beyond infection and disease.

In this chapter, we study the properties of viruses, especially bacterial and animal (human) viruses, focusing on their unique structure and mechanism for replication. We will see how they are classified, how they are cultured and identified, and how some are associated with tumors and cancer. The chapter then discusses even more bizarre virus-like agents that can cause disease in plants and animals.

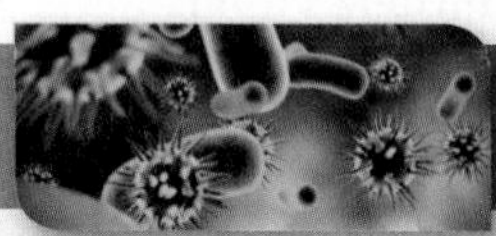

Chapter Challenge

Viruses are often described as being on the edge of life, in some limbo land between "life" and "nonlife." But on which side of the edge are they? Most biology textbooks use certain emergent properties of life to define something as a living organism. These properties include an ability to:

- Grow and develop
- Reproduce
- Establish a complex organization
- Regulate its internal environment (maintain homeostasis)
- Transform energy (light to chemical; chemical to cellular)
- Respond to the environment
- Evolve by adapting to a changing environment

So, are they alive? Let's consider this question as we study the viruses, their structure, and behavior in this chapter.

© qcontrol/ShutterStock, Inc.

KEY CONCEPT 14.1 Filterable Infectious Agents Cause Disease

The development of the germ theory recognized disease patterns associated with a specific bacterial species. However, some diseases resisted identification and many of these would turn out to be of viral origin.

Many Scientists Contributed to the Early Understanding of Viruses

In 1882, a German agricultural chemist, Adolf Mayer, was asked by local tobacco growers to study why their tobacco plants were becoming diseased with what Mayer called tobacco mosaic disease (TMD; FIGURE 14.2A). Carrying out what was probably the first experiments with viruses, Mayer took the filtered juice from diseased plants and transferred it healthy ones. The inoculated plants soon developed TMD. Even though Mayer could not find any bacterial cells by microscopy, he believed the infectious agent was a very small bacterial cell that could pass through the filter or perhaps was an enzyme. Apparently unaware of that observation, in 1892 a Russian pathologist Dimitri Ivanowsky filtered the crushed leaves of a TMD plant, hoping to trap what he believed was a bacterial infectious agent on the filter (FIGURE 14.2B). Rather, to his surprise he discovered that the clear liquid passing through the filter, called the "filtrate," contained the infectious agent. Then, in 1898 a Dutch microbiologist, Martinus Beijerinck, once again repeated the experiments first carried out by Mayer by using a porcelain filter that should trap even the tiniest bacterial cells. However, his findings again indicated that the extracts from diseased plants would pass through the filter. In addition, he reported that (1) no bacterial growth occurred during three months of storage; (2) the extracts remained infective during this time; and (3) the agent multiplies and is part of the active metabolism in the plant tissues. Beijerinck's concept of TMD was that it was caused by a filterable *contagium vivum fluidum* (contagious living fluid), which did not fit the prevailing germ theory. Throughout his studies, Beijerinck often referred to the infectious agent as the "virus" (*virus* = "poison").

■ Tobacco mosaic disease: A viral disease causing tobacco leaves to shrivel and assume a mosaic appearance.

In 1898, foot-and-mouth disease also was suspected as being caused by a virus, implying that a virus could be transmitted among animals as well as plants. Three years later, the American Walter Reed and his group in Cuba provided evidence linking yellow fever with a virus, so viruses could be associated with human disease as well.

■ Foot-and-mouth disease: A highly contagious viral disease of cloven-hoofed animals (i.e., cattle, sheep, deer).

■ Yellow fever: A mosquito-borne viral disease of the human liver and blood.

In 1915, English bacteriologist Frederick Twort discovered viruses that infected bacterial

(A)

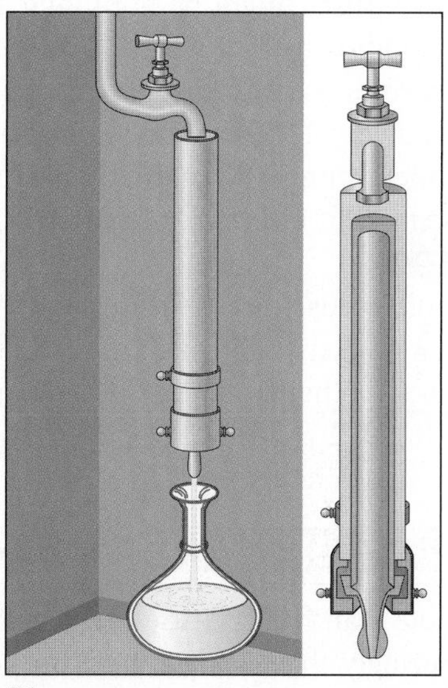

(B)

FIGURE 14.2 **Tobacco Mosaic Disease (TMD).** **(A)** An infected tobacco leaf exhibiting the mottled or mosaic appearance caused by the disease. Courtesy of Clemson University - USDA Cooperative Extension Slide Series, Bugwood.org. **(B)** A filter similar to the one used to produce the filterable liquid containing the TMD virus. »» Why was it impossible for Ivanowsky and his contemporaries to see viruses with the light microscope?

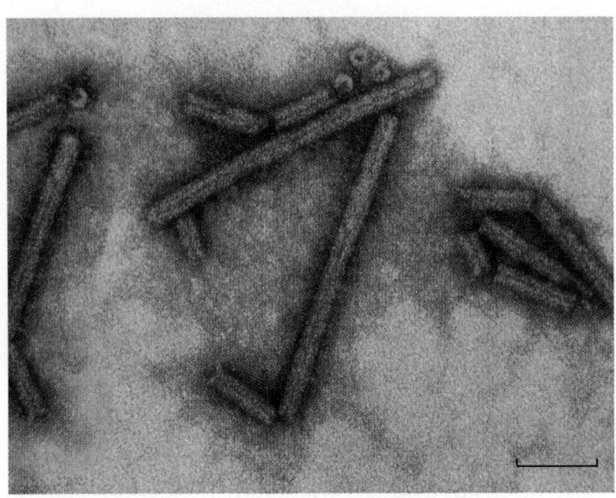

FIGURE 14.3 **The Tobacco Mosaic Virus (TMV).** This false-color transmission electron micrograph of TMV shows the rod-shaped structure of the virus particles. (Bar = 80 nm). © Dennis Kunkel Microscopy, Inc./Phototake/Alamy Images. »» From this micrograph, why would virologists call viruses "crystallizable" particles?

cells. Two years later, such viruses were identified by French-Canadian scientist Felix d'Herrelle. He called them **bacteriophages** (*phage* = "eat"), or simply **phages**, for their ability to destroy the bacterial cells they infected. When a drop of phages was placed in a broth culture of bacterial cells, the cells soon disintegrated.

By the early 1930s, viruses were thought to be living microorganisms below the resolving power of light microscopes. However, in 1935 the tobacco mosaic virus (TMV) was crystallized, suggesting viruses might be nonliving agents of disease. Additional work with TMV revealed the virus was composed exclusively of nucleic acid and protein. By 1960, advances in staining of viruses, including TMV, made their study with the transmission electron microscope possible (FIGURE 14.3).

Because viruses are parasites, they will not grow on a nutrient agar plate the way bacterial cells do. Some other form for virus cultivation was needed. In 1931, Alice M. Woodruff and Ernest W. Goodpasture described how fertilized chicken eggs could be used to cultivate some viruses. The shell of the egg was a natural culture dish containing nutrient medium, and viruses multiplied within the chick embryo tissues.

Another key development occurred in the 1940s as a result of the national polio epidemic. Attempts at vaccine production were stymied by the inability to cultivate polioviruses outside the body, but John Enders, Thomas Weller, and Frederick Robbins of Children's Hospital in Boston solved that problem. Meticulously, they developed a test tube medium of nutrients, salts, and pH buffers in which living animal cells would remain alive. In these living cells, polioviruses replicated to huge numbers, and by the late 1950s, Jonas Salk and Albert Sabin had adapted the technique to produce massive quantities of virus for their independently developed polio vaccines.

CONCEPT AND REASONING CHECKS 1

a. Describe the major events leading to the recognition of viruses as pathogens.

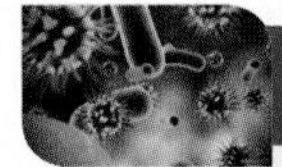

Chapter Challenge A

We have learned a little about the discovery of viruses and that viruses, like TMV, can be crystallized. In addition, viruses cannot be grown in culture by themselves the way many bacterial and other microbial organisms can be grown. Rather, they need to be cultivated in living cells.

QUESTION A:

Based on these observations and the characteristics of life, where would you place the viruses? Are they living or nonliving, and why?

Answers can be found on the Student Companion Website in **Appendix D**.

KEY CONCEPT 14.2 Viruses Have a Simple Structural Organization

Today more than 5,000 viruses have been identified. Amazingly, this is only a small proportion of the estimated 400,000 different viruses virologists believe may exist—their total numbers making viruses, as we saw in the chapter introduction, the most abundant biological entities on Earth.

Viruses Are Tiny Infectious Agents

Viruses are small, obligate, intracellular particles (FIGURE 14.4); that is, most can be seen only with the electron microscope and they must infect and take over a host cell in order to replicate

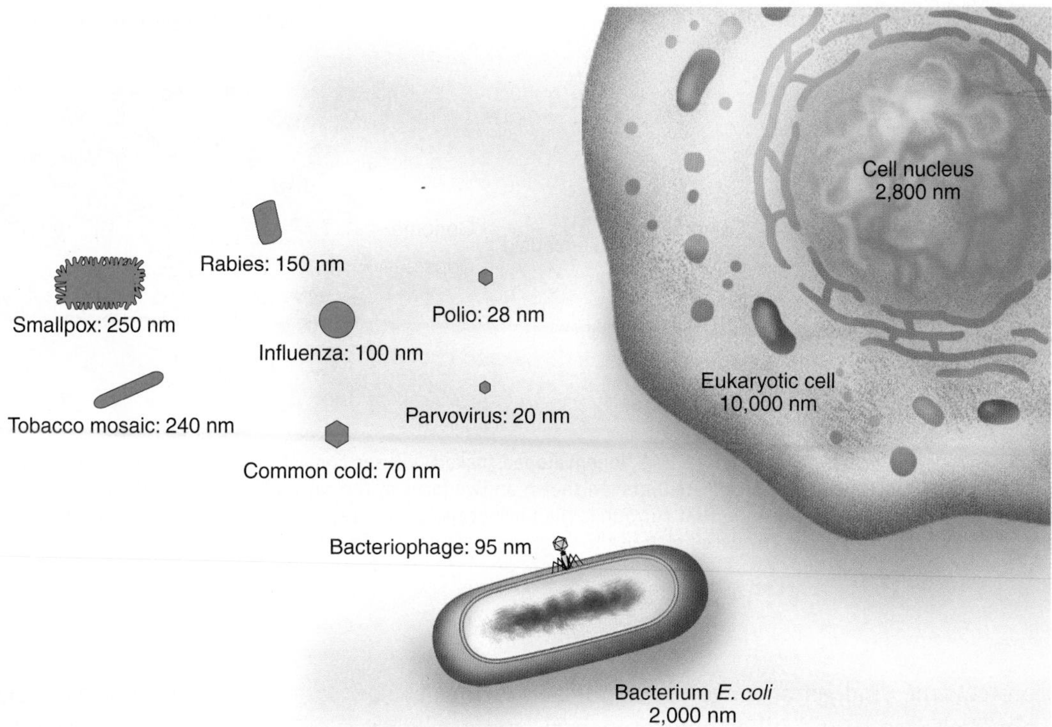

FIGURE 14.4 **Size Relationships Among Cells and Viruses.** The sizes (not drawn to scale) of various viruses relative to a eukaryotic cell, a cell nucleus, and the bacterium *E. coli*. Viruses range from the very small poliovirus to the much larger smallpox virus. »» Propose a hypothesis to explain why viruses are so small.

(FIGURE 14.5). This is because they lack the chemical machinery for generating energy and synthesizing large molecules. Viruses, therefore, must find an appropriate host cell in which they can replicate—and, as a result, often cause disease.

FIGURE 14.5 **The Viral Replication Cycle.** To produce more viruses, an inert virus must infect a host cell in which the viral genome controls the production of more virus particles. The virus particles are released into the environment, where each initiates a new infectious cycle with another host cell. »» What do you predict happens to the infected host cell when the viruses are released?

Viruses have some unique features not seen with the living microorganisms. They have no organelles, no cell wall, no cytoplasm, and no cell nucleus or nucleoid. Instead, they are comprised of two basic components: an indispensable nucleic acid core and a surrounding coat of protein; thus, as Peter Medawar remarked (chapter opening quote), a virus is "*just a piece of bad news* [meaning they cause disease] *wrapped up in protein.*"

The **viral genome** of almost all viruses contains either DNA or RNA, and the nucleic acid occurs in either a double-stranded or a single-stranded form. Usually the nucleic acid is a linear or circular molecule, although in some instances (as with influenza viruses) it exists as separate, nonidentical segments. The viral genome consists of relatively few genes, ranging from less than ten to a few hundred.

The protein coat of a virus particle, called the **capsid**, gives shape or symmetry to the virus (FIGURE 14.6). Generally, the capsid is subdivided into individual protein subunits called **capsomeres** (the organization of capsomeres yields the viral symmetry) and the capsid with its enclosed genome is referred to as a **nucleocapsid**. The capsid

Website Animation:
Virus structure

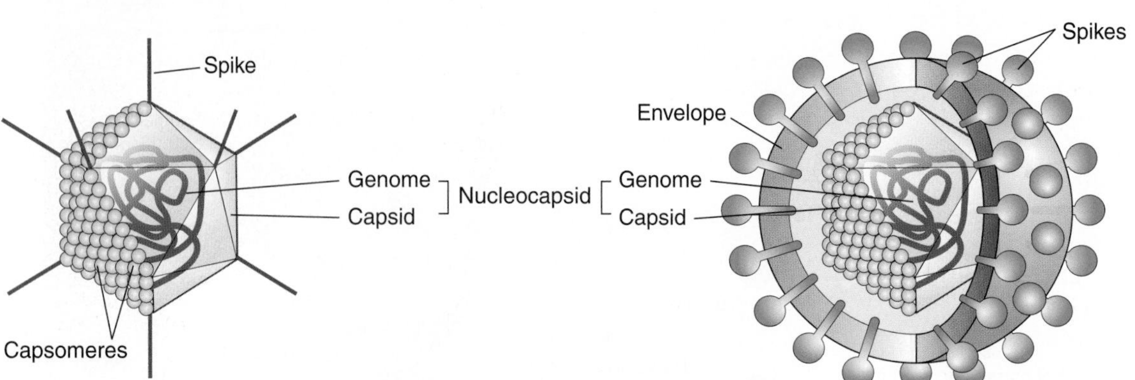

FIGURE 14.6 **The Components of Viruses.** (**A**) Nonenveloped (naked) viruses consist of a nucleic acid genome (either DNA or RNA) and a protein capsid. Capsomere units are shown on two faces of the capsid from which spikes may protrude. (**B**) Enveloped viruses have an envelope that surrounds the nucleocapsid. Again spikes usually are present on the envelope. »» What important role do spikes play in the infective behavior of viruses?

also protects the viral genome against chemical and physical agents, and other environmental fluctuations.

Viruses composed solely of a nucleocapsid are sometimes referred to as nonenveloped ("naked") viruses. On some viruses, either special capsid proteins called **spikes** or receptor-binding proteins protrude from the surface, helping attach the virus to the host cell and facilitating entry into the cell.

The nucleocapsids of many other viruses are surrounded by a flexible membrane known as an **envelope**; such viruses are referred to as enveloped viruses (Figure 14.5B). The envelope is composed of lipids and protein, similar to the host cell membrane; in fact, it is acquired from the host cell during replication and is unique to each type of enveloped virus. These viruses may lose their infectivity if the envelope is destroyed. Also, when the envelope is present, the symmetry of the capsid may not be apparent because the envelope is generally a loose-fitting structure over the nucleocapsid. Many enveloped viruses also contain protein spikes projecting from the envelope. These proteins also function for attachment and host cell entry. Enveloped viruses also contain a layer or two of protein between the capsid and envelope called the **matrix** that holds the nucleocapsid to the envelope.

A completely assembled and inert infectious virus particle outside its host cell is known as a **virion**. MicroFocus 14.1 reveals some particularly unusual viruses.

Viruses Are Grouped by Their Shape

Viruses can be separated into groups based on their nucleocapsid symmetry; that is, their three-dimensional shapes. Certain virus capsids, such as rabies and tobacco mosaic viruses, exist in the form of a rod or filament and are said to have helical symmetry (FIGURE 14.7A). The helix is a tightly wound coil resembling a corkscrew or spring.

Other viruses, such as the herpes viruses and the polio viruses, have capsids in the shape of a polyhedron with 20 triangular sides and hence, **icosahedral** (*icos* = "twenty," *edros* = "side") symmetry (FIGURE 14.7B).

Some viruses have capsids with a **complex** pattern, meaning they have several parts with different shapes (FIGURE 14.7C). Some bacteriophages, for example, have an icosahedral head with a collar and tail assembly in the shape of a helical sheath. Poxviruses, by contrast, are brick shaped, with submicroscopic filaments occurring in a swirling pattern at the surface of the virus.

Before we leave viral shapes, it is worth mentioning that many archaeal viruses, all of which to date have a DNA genome, are often morphologically unique, often having a spindle shape. Others have the typical icosahedral or helical shape.

MICROFOCUS 14.1: Evolution

Hypothesis: Giant Viruses Arose from Cellular Life

In 2002, researchers at the Centre National de la Recherche Scientifique (CNRS) in France were looking for the bacterium *Legionella* in freshwater samples when they stumbled upon a "microbe" in a waterborne amoeba that they thought was bacterial. The "microbe" had double-stranded DNA and protein fibrils that extended from the surface, making the "microbe" some 750 nm (0.75 µm) in diameter. However, it was discovered that this infectious microbe was actually a nonenveloped virus. It was the first virus discovered that was large enough to be seen with an ordinary light microscope and was named the Mimivirus ("microbe-mimicking" virus).

The Mimivirus genome is a linear DNA molecule that contains more than 1,000 **protein-coding genes**; the influenza viruses and HIV each have around ten genes. It is even larger than the genome in some intracellular bacterial parasites, which everyone agrees are living organisms. Like other viruses though, it cannot convert energy or replicate on its own. But that's where the similarities end.

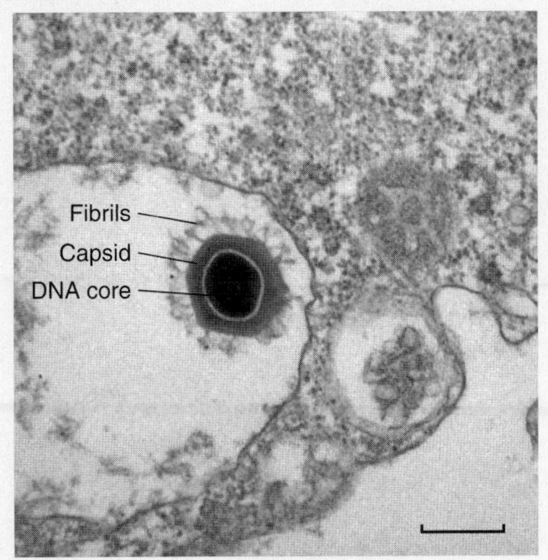

Transmission electron micrograph of the mimivirus. (Bar = 400 nm.) Courtesy of Didier Raoult, Rickettsia Laboratory, La Timone, Marseille, France.

The Mimivirus contains both DNA *and* RNA, something viruses by definition are not supposed to contain. It also has several genes shared between all three domains of life—Bacteria, Archaea, and Eukarya. When the virus replicates in an amoeba, it constructs a gigantic "replication factory" and uses all its own genes and proteins for the assembly of new virions. In addition, scientists have discovered that the Mimivirus can be "infected" by its own virus, called Sputnik. This so-called **virophage** (a virus that attacks other viruses) uses the Mimivirus replication factory for its own replication.

Since then other giant viruses or "giruses" have been discovered, one called Mamavirus and another, a phycodnavirus (a DNA virus that infects algae) in an Antarctic lake. However, in 2011 the biggest virus yet discovered was found in the sea off Chile. Called Megavirus, it has more than 1,100 genes. Its host organism has not yet been identified.

These discoveries certainly blur the lines between viruses and single-celled organisms. Because these viruses are believed to have a very old lineage, some scientists have hypothesized that such giruses evolved from ancient eukaryotic cells that gradually lost genes, resulting in an agent that replicates like a virus. Other scientists propose an even more startling idea that a girus gave rise to the eukaryotic cell nucleus (read MicroFocus 14.4).

Studies of the giruses may shed light on the origin of all life forms and the possibility that they represent a fourth domain of life.

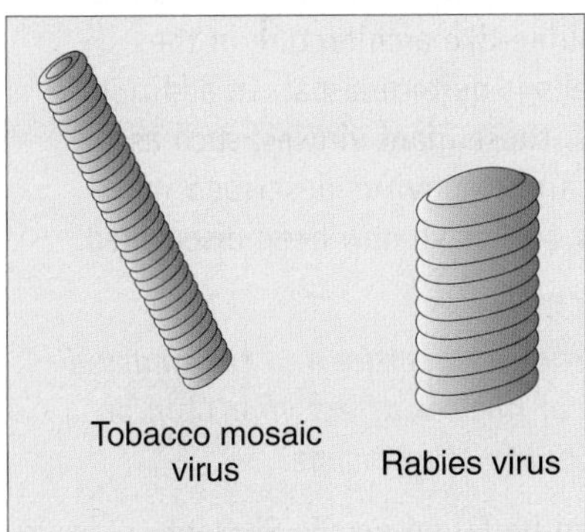

(a) Helical viruses

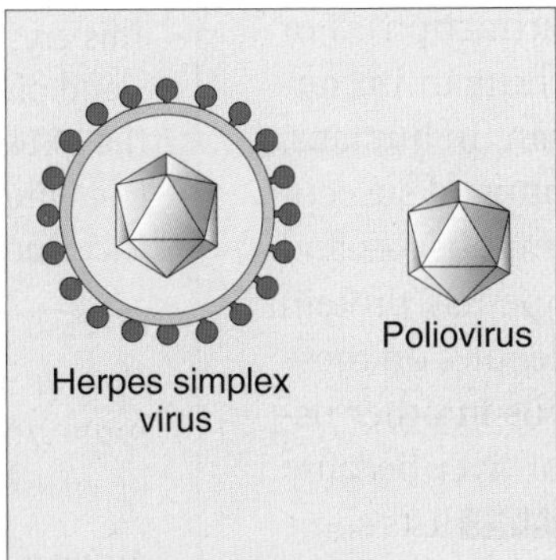

(b) Icosahedral viruses

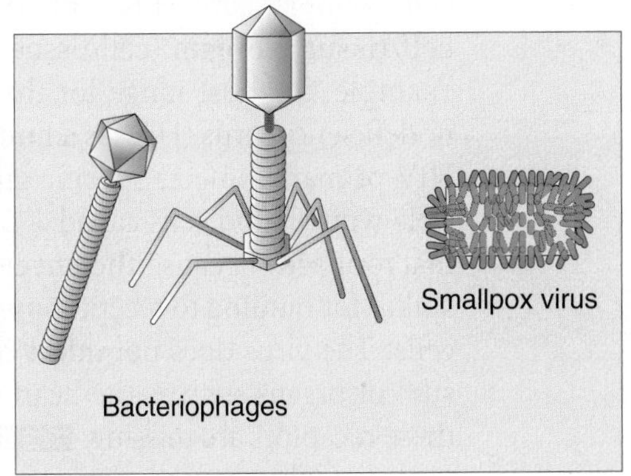

(c) Complex viruses

FIGURE 14.7 **Viral Shapes.** Viruses exhibit variations in shape. The nucleocapsid may have helical symmetry (**A**) in the tobacco mosaic and rabies viruses or (**B**) icosahedral symmetry typical of the herpesviruses and polioviruses. In other viruses (**C**), a complex pattern exists. »» Although the bacteriophages are classified as complex, what two symmetries do these viruses exhibit?

TABLE
14.1 Viral Cell Tropism

Virus(es)	Cell Type
HIV	CD4+ T lymphocytes, macrophages
Rabies	Muscle, neurons
Human papilloma	Differentiating keratinocytes
Hepatitis A, B, C	Liver (hepatocytes)
Human herpes simplex 1 and 2	Mucoepithelium
Influenza A	Respiratory epithelium
Rotavirus	Intestinal epithelium
Norovirus	Intestinal epithelium
Cytomegalovirus	Epithelium, monocytes, lymphocytes
Rhinovirus	Nasal epithelium
Poliovirus	Intestinal epithelium
Epstein-Barr	B cells

Viruses Have a Host Range and Tissue Specificity

As a group, viruses can infect almost any cellular organism. There are viruses that infect bacterial cells, while others infect protists, fungi, plants, or animals. A virus' **host range** refers to what organisms (hosts) the virus can infect and it is based on a virus' capsid or envelope structure. Most viruses have a very narrow host range. A specific bacteriophage, for example, only infects specific bacterial species and the smallpox virus only infects humans. A few viruses may have a broader host range, as the rabies viruses infect humans and most warm-blooded animals (e.g., bats, skunks, dogs).

Even within its host range, many viruses only infect certain cell types or tissues within a multicellular organism. This limitation is called the **cell/tissue tropism** (cell/tissue attraction). For example, the host range for the human immunodeficiency virus (HIV) is a human. In humans, HIV primarily infects specific groups of susceptible white blood cells called T lymphocytes and macrophages because the envelope has protein spikes for binding to receptor molecules on these cells. The virus does not infect cells in other tissues or organs such as the heart or liver because these receptors are missing. TABLE 14.1 lists some viruses and their cell tropisms. Therefore, if a potential host cell lacks the appropriate receptor or the virus lacks the complementary protein, the virus usually cannot infect that cell. How we know that susceptible host cells have receptors that bind specific viruses is explored in the Investigating the Microbial World 14.

CONCEPT AND REASONING CHECKS 2

a. Identify the role of each structure found on (a) a nonenveloped and (b) an enveloped virus.

b. What shapes can viruses have and what structure determines that shape?

c. How does viral structure determine host range and tissue tropism?

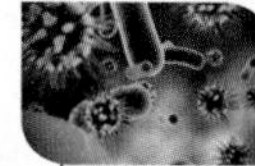

Chapter Challenge B

We have seen that the viruses do have a tendency to have a particular geometrical shape, especially at the level of the capsid. This crystalline-like architecture in the world of cells is quite unusual. In addition, the giruses, those giant viruses, such as Mimivirus and Megavirus described in MicroFocus 14.1 have now been discovered.

QUESTION B:
How do viral architecture and the unusual properties of giruses affect your opinion of viruses being alive or not?

Answers can be found on the Student Companion Website in **Appendix D.**

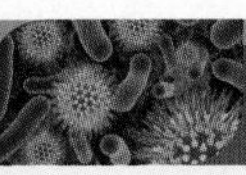

Investigating the Microbial World 14

Finding the Correct Address

■ **OBSERVATION:** The host range and cell tissue tropism allow viruses to find the appropriate host and then the correct susceptible cell or tissue to infect. Several such receptors are listed in the table to the right.

■ **QUESTION:** ***Being that one cannot see cell surface receptors with microscopy, how do we know these viruses actually bind to some type of receptor on the susceptible host cells?***

■ **HYPOTHESIS:** Viruses cannot infect cells if the correct host membrane receptor molecule is absent. If so, then removing and blocking the receptor molecules, or altering the cell surface, on susceptible host cells will prevent virus infection.

TABLE

Cell Surface Receptors Used by Viruses to Attach and Enter Cells

Virus	Cell Surface Receptor
Influenza A	Sialic acid
HIV-1	CD4 and co-receptors (CXCR5, CCR4)
Hepatitis C	Low-density lipoprotein receptor
Rabies	Acetylcholine receptor, neural cell adhesion molecule, nerve growth factor, gangliosides, phospholipids
Rhinovirus	Intracellular adhesion molecule 1 (ICAM-1)
Hepatitis B	IgA receptor
Adenovirus	Integrins
Poliovirus	Immunoglobulin superfamily protein (CD155)

■ **EXPERIMENTAL DESIGN:** Appropriate susceptible host cells are treated in one of three ways to remove, block, or alter their receptors:

■ **EXPERIMENT 1:** Susceptible respiratory host cells are treated with neuraminidase (removes sialic acid).

■ **EXPERIMENT 2:** Susceptible respiratory cells are treated with an antibody that binds to ICAM-1.

■ **EXPERIMENT 3:** The gene (PVR) coding for the poliovirus receptor is genetically engineered into mouse cells that normally do not bind poliovirus.

■ **RESULTS:**

EXPERIMENT 1

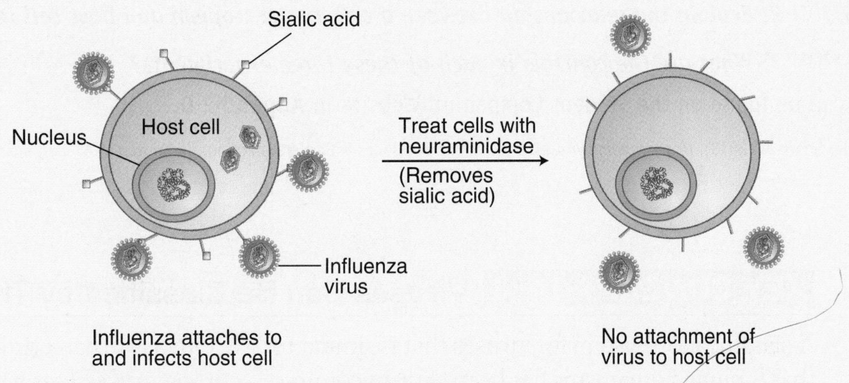

EXPERIMENT 2

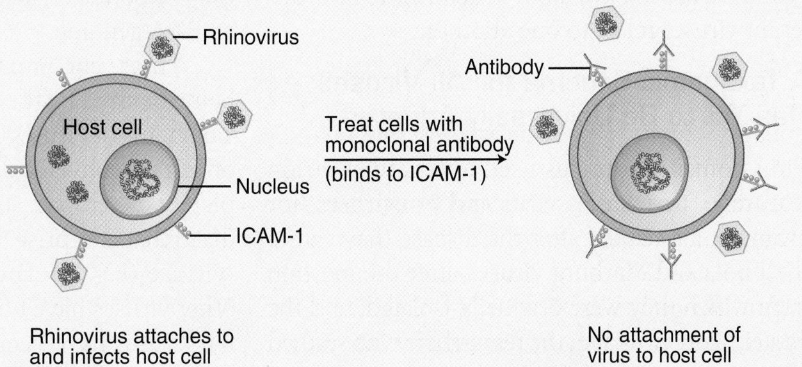

(continued on next page)

Investigating the Microbial World 14 (continued)

EXPERIMENT 3

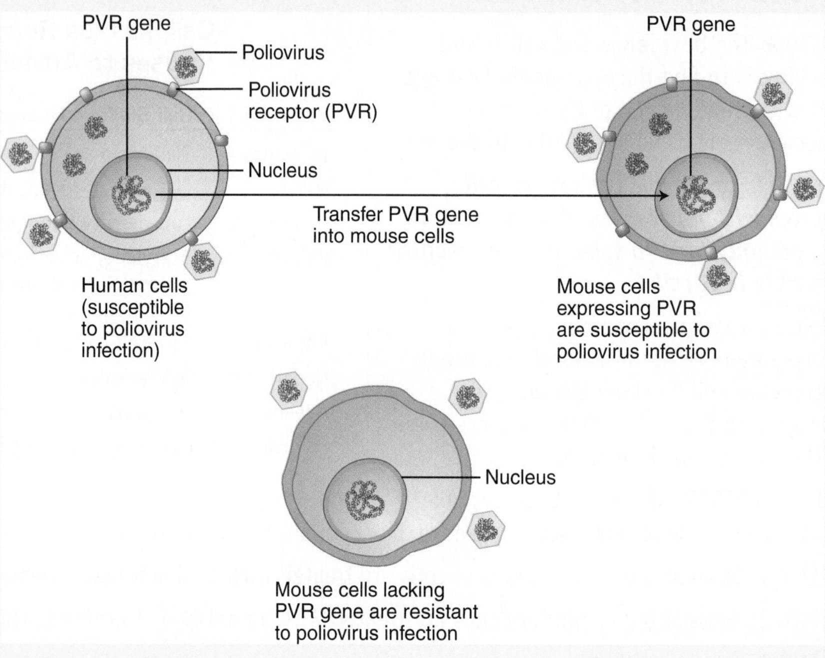

■ **CONCLUSION:**

QUESTION ***1: Is the hypothesis supported? Answer by analyzing the results from each of the three experiments.***

QUESTION ***2: Explain the relationship between a cell/tissue tropism and host cell receptors.***

QUESTION ***3: What are the controls in each of these three experiments?***

Answers can be found on the Student Companion Website in **Appendix D.**

Adapted from: Shors, T. (2013). *Understanding Viruses,* Second Edition. Jones & Bartlett Learning, Burlington, MA. Icon image © Tischenko Irina/ShutterStock, Inc.

KEY CONCEPT 14.3 Viruses Can Be Classified by Their Genome

A classification system for viruses that is similar to that for living organisms has been slow in coming, in part because sufficient data on virus origins and evolution are not available to determine how different viruses relate to one another.

A Taxonomic Scheme for All Viruses Has Yet to Be Universally Adopted

Viral nomenclature has used a variety of virion features. The measles virus and poxviruses, for example, are named after the disease they cause; the Ebola and Marburg viruses after the location from which they were originally isolated; and the Epstein-Barr virus after the researchers who studied it. Others are named after morphologic factors—the coronaviruses (*corona* = "crown") have a crown-like capsid and the picornaviruses (*pico* = "small"; *rna* = "ribonucleic acid") are very small viruses with an RNA genome.

A more encompassing classification system is being devised by the International Committee on Taxonomy of Viruses (ICTV). At this writing, higher order taxa (phyla and classes) have not been, completely developed. To date, six orders are recognized that comprise 87 families, each ending with -viridae (e.g., **Herpesviridae**). However, many other viruses have not yet been assigned to a family. Viruses have been categorized into hundreds of genera; each genus name ends with the suffix -virus (e.g., human herpes virus). In this text, we mostly use the common names (e.g., herpes simplex virus) for specific viruses.

Animal viruses are usually classified into two groups based on their genome (DNA or RNA) and then split into separate families based on characteristics such as strand type (double stranded and single stranded) and presence or absence of an envelope. A selection of viral families affecting humans, together with some of their characteristics and diseases, is presented in FIGURE 14.8.

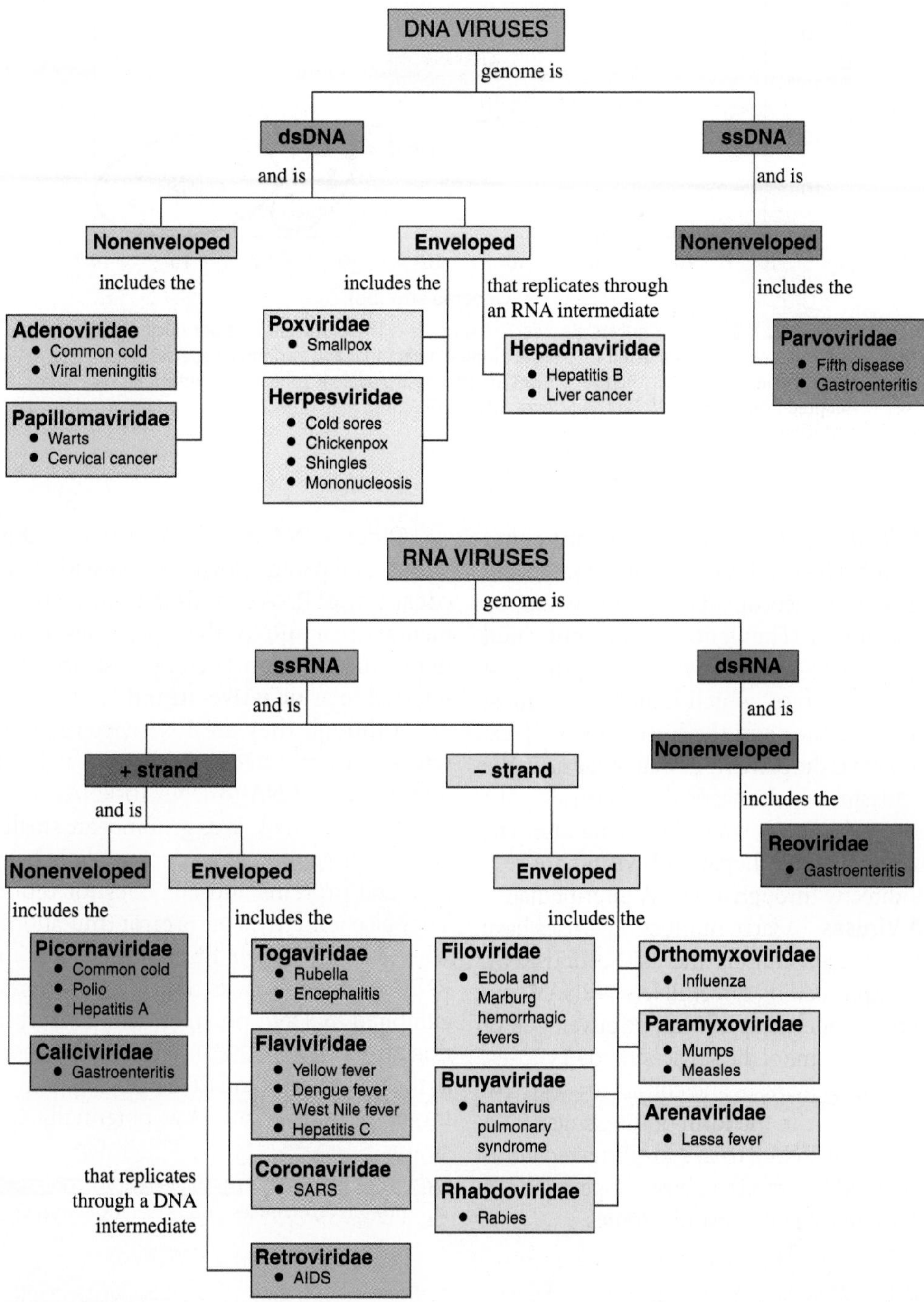

FIGURE 14.8 **A Classification for the Human DNA and RNA Viruses.** The human (animal) virus families have been classified based on genome type [double stranded (ds) or single stranded (ss)], enveloped or nonenveloped, and, for the ssRNA viruses, (+) strand or (–) strand. In this concept map, some of the distinctive human diseases associated with each family are indicated. »» According to this concept map, does it mean, for example, that the Herpesviridae are closely related to the Poxviridae? Explain.

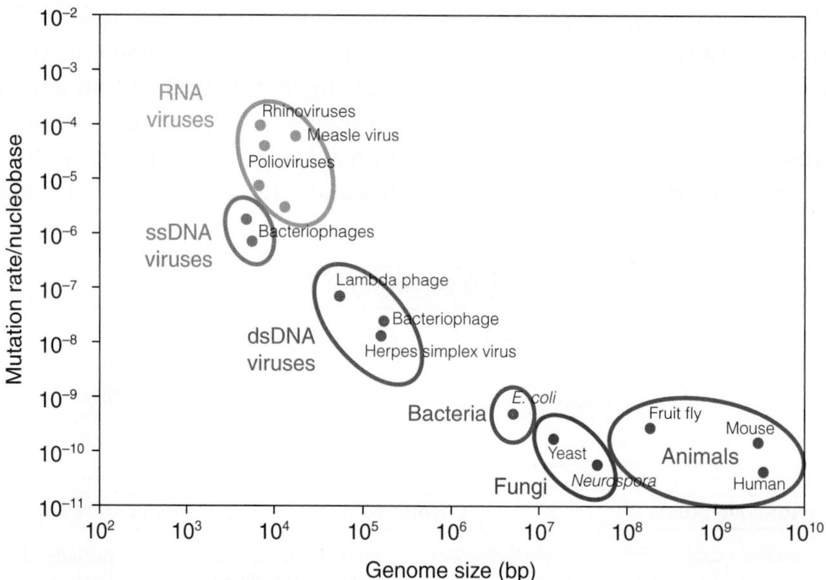

FIGURE 14.9 **Mutation Rate Versus Genome Size.** The mutation rate per nucleobase versus genome size is plotted for several viruses, a bacterial and two fungal species, and three animals. Note that the RNA viruses are the most prevalent to undergo mutation. Adapted from Gago *et al.* (2009) *Science* **323**, 1308. »» Why are RNA viruses so prone to mutations?

DNA Viruses. Many viruses contain either single-stranded (ss) or double-stranded (ds) DNA genomes with nucleocapsids that are enveloped or nonenveloped. The genomes are replicated by direct DNA-to-DNA copying using the host cell DNA polymerase, which requires that most DNA viruses replicate in the host cell's nucleus. One exception is the poxviruses that replicate in the host cytoplasm, which means these viruses must carry the gene for their own DNA polymerase. The Hepadnaviridae (e.g., hepatitis B virus) are replicated indirectly through a ssRNA intermediate.

RNA Viruses. A large number of viruses have RNA genomes consisting of either ssRNA or dsRNA, which are replicated by direct RNA-to-RNA copying. Again, the nucleocapsids can be enveloped or nonenveloped. Some of the single-stranded viruses, such as the picornaviruses and coronaviruses, have their RNA genome in the form of messenger RNA (mRNA). These RNA viruses are referred to as **positive-strand** (+ strand) because the genome can be directly translated by host ribosomes.

Other ssRNA viruses, such as the orthomyxoviruses and paramyxoviruses, have RNA genomes consisting of RNA strands that would be complementary to a mRNA; these genomes, which cannot be directly translated by host ribosomes, are referred to as **negative-strand** (– strand).

Although they are RNA viruses, the retroviruses are replicated indirectly through a dsDNA intermediate (RNA-to-DNA-to-RNA).

As a rule, RNA virus genomes are smaller than DNA virus genomes and depend more heavily on host cell proteins and enzymes for replication. They also tend to be more error (mutation) prone when copying their RNA genome because the RNA polymerase lacks the efficient proofreading exhibited by DNA polymerase to correct replication errors (FIGURE 14.9). Thus, RNA viruses, such as the flu viruses, tend to "genetically drift," evolving more rapidly into new, potentially epidemic strains.

CONCEPT AND REASONING CHECKS 3

a. How do DNA viruses differ from RNA viruses?

KEY CONCEPT 14.4 Viral Replication Follows a Set of Common Steps

The process of viral replication is one of the most remarkable events in nature (FIGURE 14.10). A virus (1) encounters and attaches to the appropriate host cell; (2) invades the host cell a thousand or more times its size; (3) hijacks the metabolism of the cell to produce viral parts that are then (4) assembled

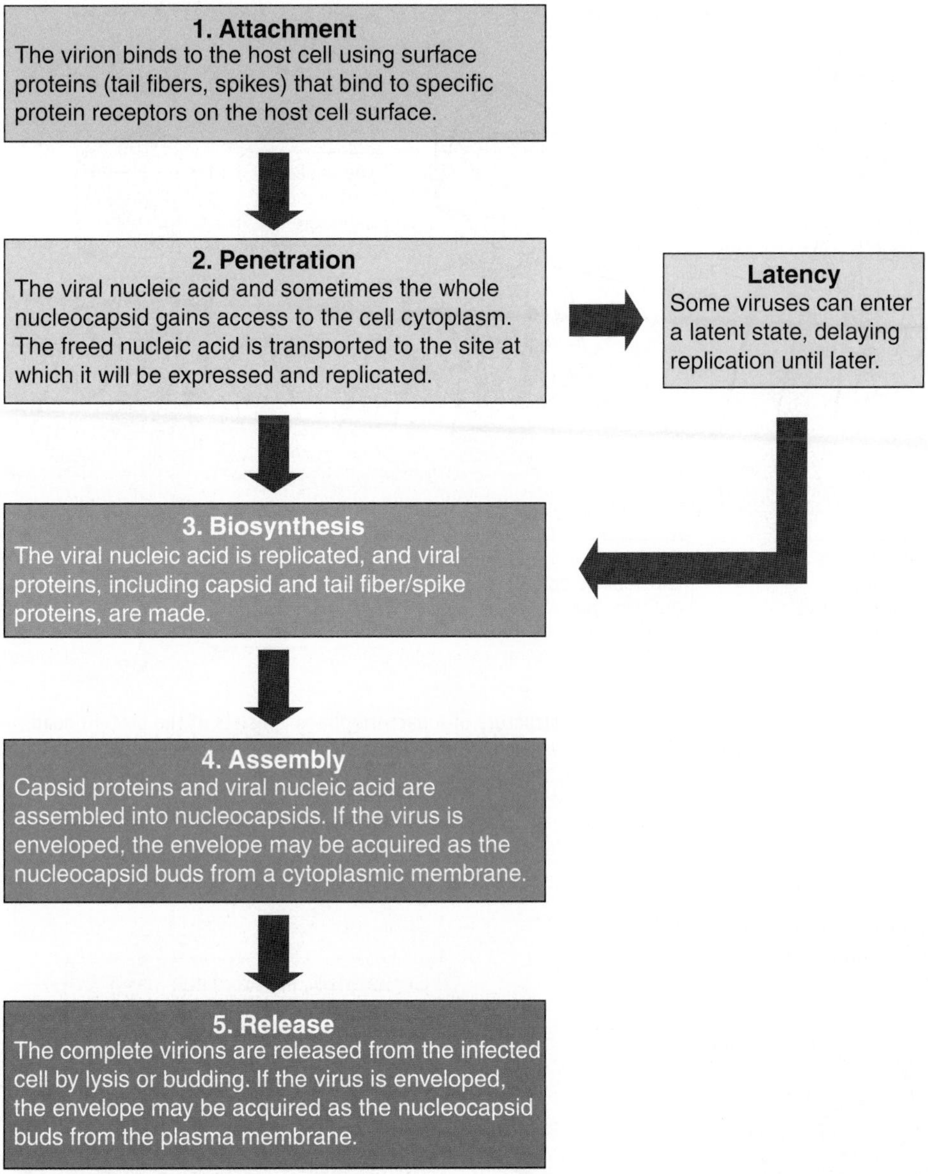

1. Attachment
The virion binds to the host cell using surface proteins (tail fibers, spikes) that bind to specific protein receptors on the host cell surface.

2. Penetration
The viral nucleic acid and sometimes the whole nucleocapsid gains access to the cell cytoplasm. The freed nucleic acid is transported to the site at which it will be expressed and replicated.

Latency
Some viruses can enter a latent state, delaying replication until later.

3. Biosynthesis
The viral nucleic acid is replicated, and viral proteins, including capsid and tail fiber/spike proteins, are made.

4. Assembly
Capsid proteins and viral nucleic acid are assembled into nucleocapsids. If the virus is enveloped, the envelope may be acquired as the nucleocapsid buds from a cytoplasmic membrane.

5. Release
The complete virions are released from the infected cell by lysis or budding. If the virus is enveloped, the envelope may be acquired as the nucleocapsid buds from the plasma membrane.

FIGURE 14.10 **The Steps of Virus Replication.** These five steps are common to all viruses.
»» Why can't viruses carry out these steps outside a host cell?

into new virions; and (5) new virions then escape the cell, often destroying the host cell in the process.

Replication has been studied in a wide range of viruses and their host cells. Before we examine the animal (human) viruses, we will look at the ubiquitous bacteriophages because much of what we know of animal viruses comes from phage studies. Also, phages are the most prevalent and numerous biological agents in the world.

The Replication of Bacteriophages Can Follow One of Two Cycles

Bacteriophages exhibit two strategies of infection, a lytic (virulent) infection and a lysogenic (temperate) infection.

The Lytic Cycle. One of the best studied processes of replication is that carried out by bacteriophages of the T-even group (T for "type"). Bacteriophages T2, T4, and T6 are in this group. They are large, complex, DNA virions with a characteristic head and tail structure (FIGURE 14.11). They contain tail fibers, which function similar to spikes on animal viruses and identify what bacterial species the phage can infect. The T-even phages are **virulent** viruses meaning they lyse the host cell while carrying out a **lytic cycle** of infection.

We shall use phage T4 replication in *Escherichia coli* as a model for the lytic cycle. An overview of the five-step process is presented in FIGURE 14.12.

Website Animation: Bacteriophage Replication (Lytic Cycle)

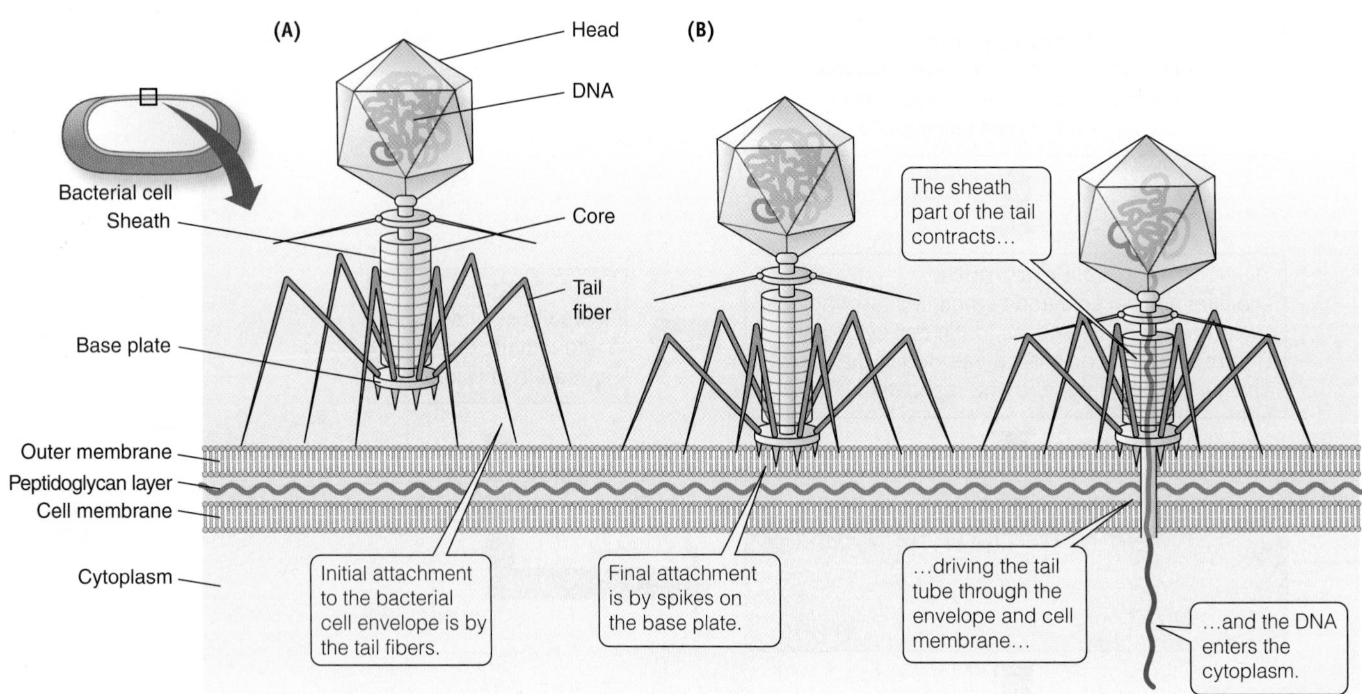

FIGURE 14.11 **Bacteriophage Structure and Genome Penetration.** (**A**) The structure of a bacteriophage consists of the protein head, inside of which is the nucleic acid, and the protein tail. The hollow core of the tail allows transfer of the nucleic acid during infection while the tail fibers attach the phage to receptors on the cell envelope. (**B**) Following attachment, the tail core contracts and is driven through the cell wall.
»» Is the phage in this figure infecting a gram-positive or gram-negative cell? Explain.

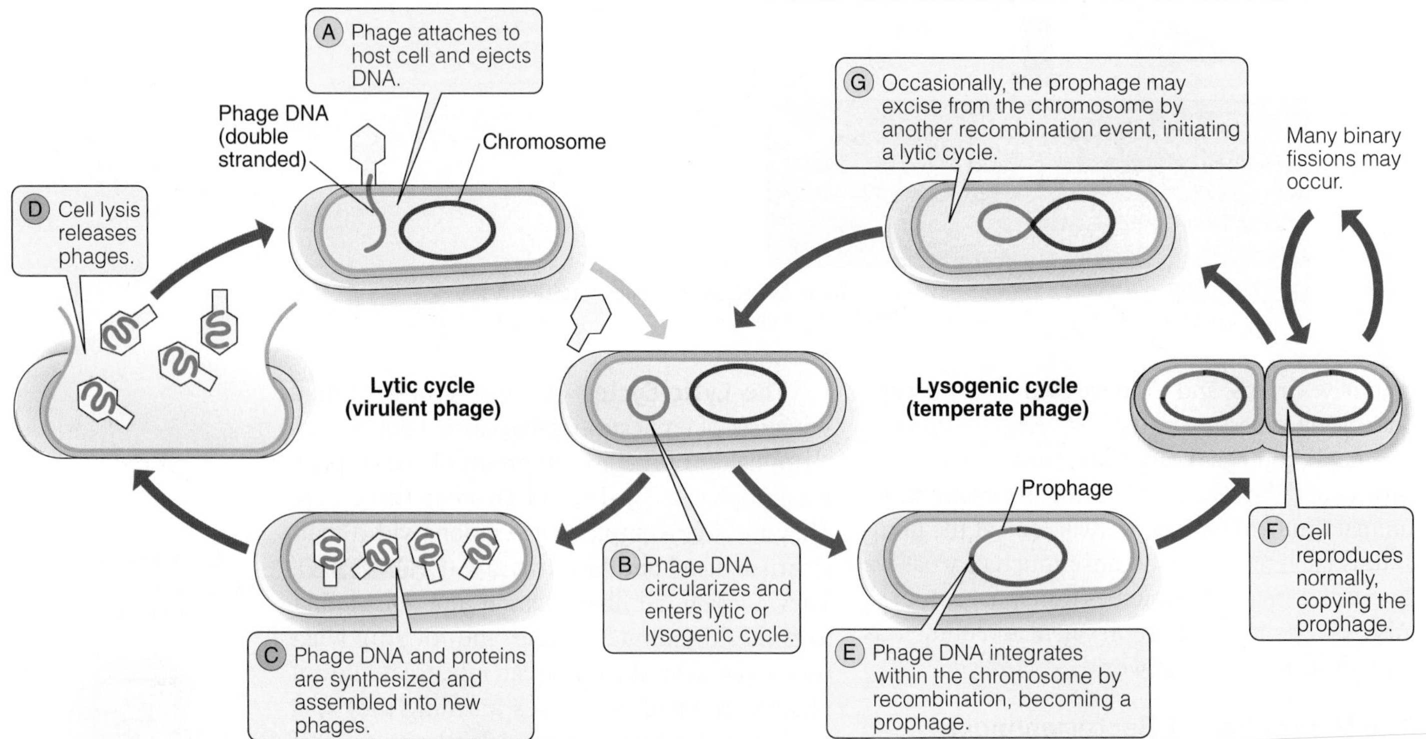

FIGURE 14.12 **Bacteriophage Replication.** The pattern of replication in bacteriophages (tail fibers not shown) can involve a lytic or lysogenic cycle.
»» What might determine whether a temperate phage follows a lytic or lysogenic cycle?

1. **Attachment.** The first step in the replication cycle of a virulent phage occurs when phage and bacterial cells randomly collide. If sites on the phage's tail fibers match with a complementary receptor site on the cell envelope of the bacterial cell, attachment will occur. The actual attachment (adsorption) consists of a weak chemical union between phage and receptor site.
2. **Penetration.** Following attachment, the phage DNA must get across the cell envelope (Figure 14.11). The tail of the phage releases lysozyme, an enzyme that dissolves a portion of the bacterial cell wall. The tail sheath then contracts and the tail core drives through the cell wall. As the tip of the hollow core reaches the cell membrane below, a protein plug blocking the core is removed and the DNA is ejected through the core and into the bacterial cytoplasm. The ejection process takes less than two seconds and the unneeded capsid remains outside the wall.
3. **Biosynthesis.** Having entered the cytoplasm, production of new phage genomes and capsid parts begins. It is important to note that the DNA in the T4 phage contains only the genes needed for viral replication. Therefore, as phage genes code for the disruption of the host chromosome, the phage DNA uses bacterial nucleotides and enzymes to synthesize multiple copies of its genome. Messenger RNA molecules transcribed from phage genes appear in the cytoplasm, and the biosynthesis of phage enzymes and capsid proteins begins. The host's ribosomes, amino acids, and enzymes are all enlisted for biosynthesis.
4. **Assembly.** Once the phage parts are made, they are assembled into complete virus particles. The enzymes encoded by viral genes guide the assembly in step-by-step fashion. In one area of the host cytoplasm, phage heads and tails are assembled from protein subunits; in another area, the heads are packaged with DNA; and in a third area, the tails are attached to the heads.
5. **Release.** For some phages, lysozyme, encoded by the bacteriophage genes late in the replicative cycle, degrades the bacterial cell wall. Mature phage particles now burst out from the ruptured bacterial shell and are set free to infect more bacterial cells.

The lytic cycle of infection represents a **productive infection** because many virions are produced.

MicroFocus 14.2 describes the use of phages as a way to combat bacterial diseases.

The Lysogenic Cycle. Other phages interact with bacterial cells in a slightly different way, called a **lysogenic cycle** (Figure 14.12). For example, lambda (λ) phage also infects *E. coli* but may not immediately enter a lytic cycle of infection. Instead, the phage DNA integrates into the bacterial chromosome as a **prophage**. Bacteriophages participating in this cycle are known as **temperate** phages. The bacterial cell survives the infection and continues to grow and divide normally. As the bacterial cell goes through its cell cycle, the prophage is copied and vertically transferred to daughter cells as part of the replicated bacterial chromosome. Because the prophage remains "inactive" by not lysing the cell, the infection is referred to as a **latent** (inactive) **infection**.

Website Animation: Bacteriophage Replication (Lysogenic Cycle)

Such binary fissions can continue for an undefined period of time. Usually at some point, the bacterial cells become "stressed" (e.g., lack of nutrients, presence of noxious chemicals). This triggers the prophage to excise itself from the bacterial chromosomes (Figure 14.12) and initiate a lytic cycle, lysing the bacterial cells as new λ phage are released.

Animal Virus Replication Often Results in a Productive Infection

Website Animation: Virus Entry

Like bacteriophages, animal viruses also lead often brief but eventful "lives" as they produce more viruses as a result of infection. Such a productive infection retains the five replication steps described for the bacteriophages. An overview of the dsDNA and ssRNA viruses is presented here.

1. **Attachment.** Animal viruses infect host cells by binding to receptors on the host cell's plasma membrane. This binding is facilitated by capsid proteins or spikes distributed over the surface of the capsid (e.g., adenovirus) or envelope (e.g., HIV).
2. **Penetration.** Some viruses, such as HIV and the adenoviruses, require a second receptor, called a "co-receptor," for viral entry into the cytoplasm. Viral entry also differs from that in phages, in that animal viruses often are taken into the cytoplasm as intact nucleocapsids. For viruses like HIV, the viral envelope fuses with the plasma membrane and releases the nucleocapsid into the cytoplasm (FIGURE 14.13A).

MICROFOCUS 14.2: History/Public Health/Biotechnology

Phage Therapy

When bacteriophages were identified in the early 1900s, some scientists, including one of their discoverer's, Felix d'Herrelle, started to promote phages as therapy for curing dreaded bacterial diseases such as cholera and bubonic plague.

D'Herrelle and others believed that if phages could destroy bacterial cells (see figure) in test tubes, why not try using them to treat bacterial diseases in the human body? So, in 1919, they treated four children suffering from bacterial dysentery. All four began to recover within 24 hours. Phage therapy, however, never was deeply embraced in the United States, partly due to the rise of antibiotics in the 1940s and 1950s.

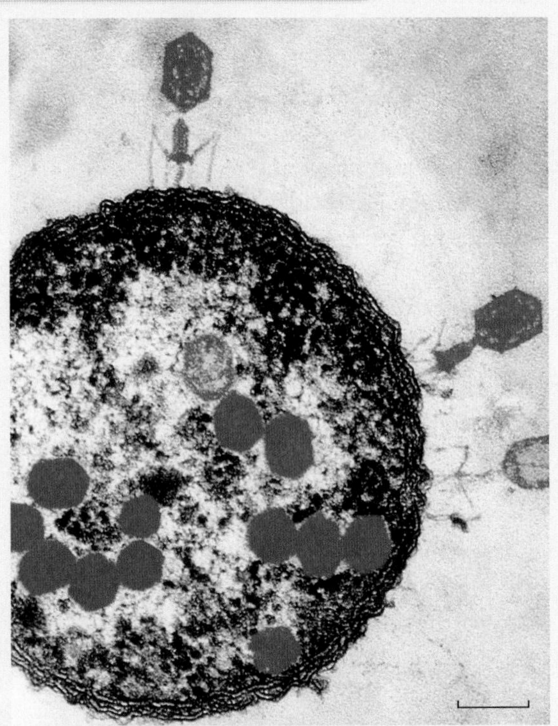

Phage infection of an *E. coli* cell. (Bar = 100 nm.)

Fast-forward to the modern era.

- 1980s: A Polish microbiologist, Stefan Slopek, injected a solution of phages into 550 patients with a blood infection. Each patient benefited from the treatment, and several patients seemed completely cured.
- 2004: Researchers in Vienna, Austria engineered bacteriophages that would kill bacteria but not lyse the cells. More surprising was the discovery that these phages were ten times more lethal than the natural phages.
- 2006: The U.S. Food and Drug Administration (FDA) approved a "cocktail" of six purified phages (ListShield™) that can be safely sprayed on luncheon meats (cold cuts, hot dogs, sausages) to kill any of 170 strains of *Listeria monocytogenes,* which cause a serious foodborne disease in newborns, pregnant women, and immunocompromised individuals.
- 2007: James Collins and his team at Boston University engineered a phage to attack and destroy bacterial biofilms that are often the cause of human infections and disease.
- 2007: A group of Italian researchers identified a phage that cured 97% of mice infected with a deadly form of methicillin-resistant *Staphylococcus aureus* (MRSA).
- 2008: The FDA approved the first clinical trial using phages with people having venous leg ulcers. No harm to the volunteers was reported.

So, at a time when we are seeing increased bacterial resistance to antibiotics, phage therapy envisioned by d'Herrelle is on the rebound and holds great promise as a complementary approach for the treatment of acute and chronic infections, as well as for use as a preservative in the food industry.

Website Animation: Herpesvirus Replication

For other animal viruses, such as the nonenveloped adenoviruses and enveloped influenzavirus, the virion is taken into the cell by endocytosis. At the attachment site, the cell engulfs the virion within a vacuole and brings it into the cytoplasm (FIGURE 14.13B). Once in the cell, regardless of penetration mechanism, the capsid disassembles from the genome in a process called **uncoating** and the genome is transported to the site where transcription or replication will occur.

3. and 4. Biosynthesis and Assembly. One way we split the virus families was based on whether they have DNA or RNA as their genetic infor-

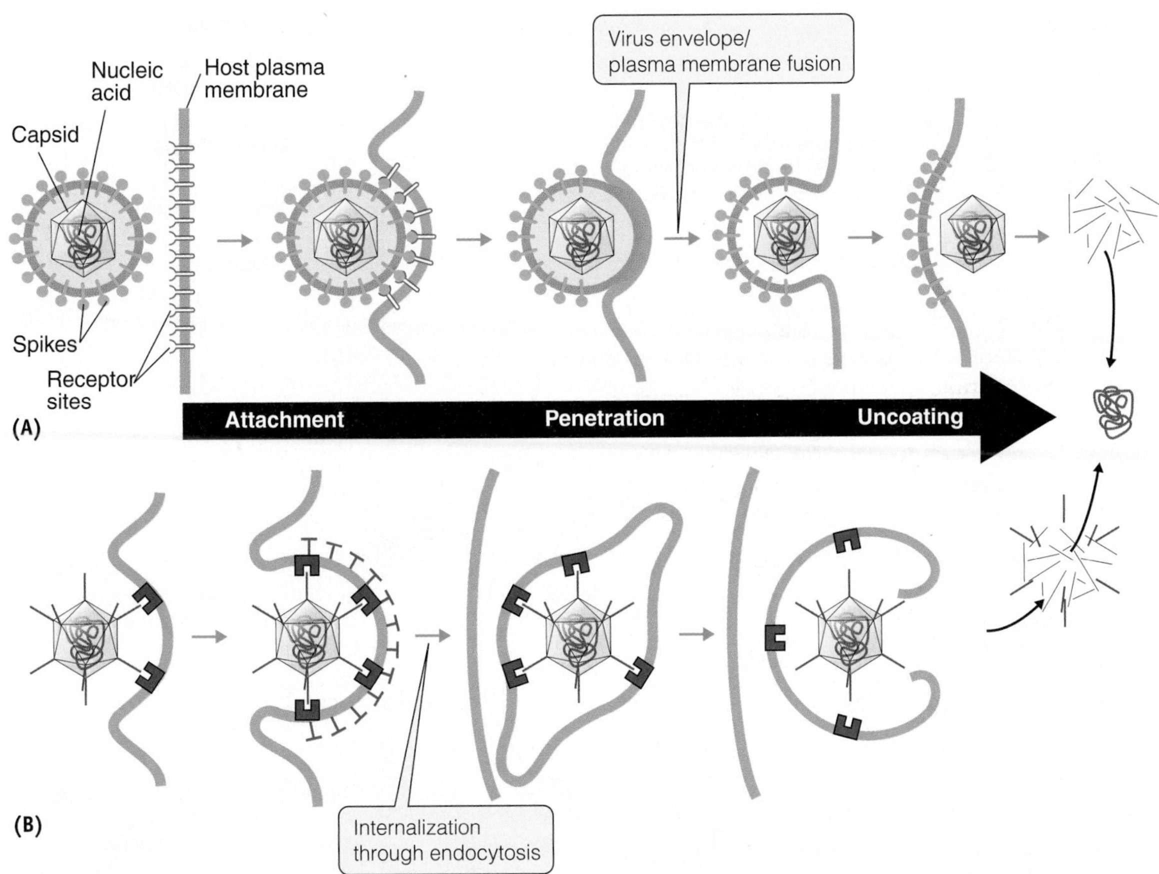

FIGURE 14.13 **The Entry of Animal Viruses into Their Host Cells.** Animal viruses enter their host cells by one of two major routes. (**A**) An enveloped virus, such as HIV, contacts the plasma membrane and the spikes interact with receptor sites on the membrane surface. This action is highly specific for the viral spikes and receptor sites. The envelope fuses with the plasma membrane, and the nucleocapsid passes into the host cell's cytoplasm. Then, the nucleic acid is released (uncoated). (**B**) A second entry method also involves a specific interaction between spikes and receptor sites on the plasma membrane. However, for this nonenveloped adenovirus, it undergoes endocytosis into a vacuole forming around the virus. The vacuole pinches off into the cytoplasm. Loss of the vacuole membrane liberates the nucleocapsid into the cytoplasm and uncoating occurs. »» Propose a hypothesis to explain why two different entry mechanisms exist for virion entry into a host cell.

mation. The DNA of a DNA virus supplies the genetic codes for enzymes that synthesize viral parts from available building blocks. Although the poxviruses replicate entirely in the host cell cytoplasm, most of the DNA viruses employ a division of labor: DNA genomes are synthesized in the host cell nucleus, and capsid proteins are produced in the cytoplasm (FIGURE 14.14). The proteins are then transported to the nucleus and join with the nucleic acid molecules for maturation. Adenoviruses and herpesviruses follow this pattern.

RNA viruses follow a slightly different pattern. Because the RNA in +ssRNA viruses acts as a messenger RNA, following uncoating, the RNA is directly translated into viral proteins as genome replication occurs. Other −ssRNA viruses, such as the influenzavirus, use their RNA as a template to synthesize a complementary (+) strand of RNA. An RNA replicase is present in the virus to synthesize the (+) strand. The synthesized +ssRNA then is used as a messenger RNA molecule for protein synthesis as well as the template to form the −ssRNA genome.

The final step of assembly for some enveloped viruses is the acquisition of an envelope. In this step, envelope proteins (spikes) are synthesized and, depending on the virus, incorporated into a nuclear, cytoplasmic, or plasma membrane.

5. **Release.** In the final stage, the nucleocapsids of some viruses push through the plasma

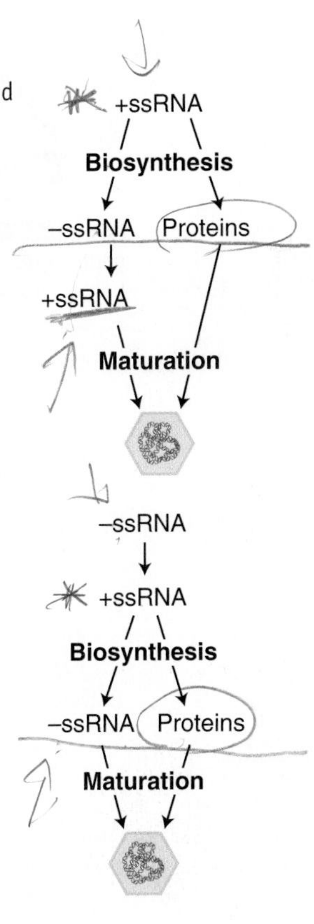

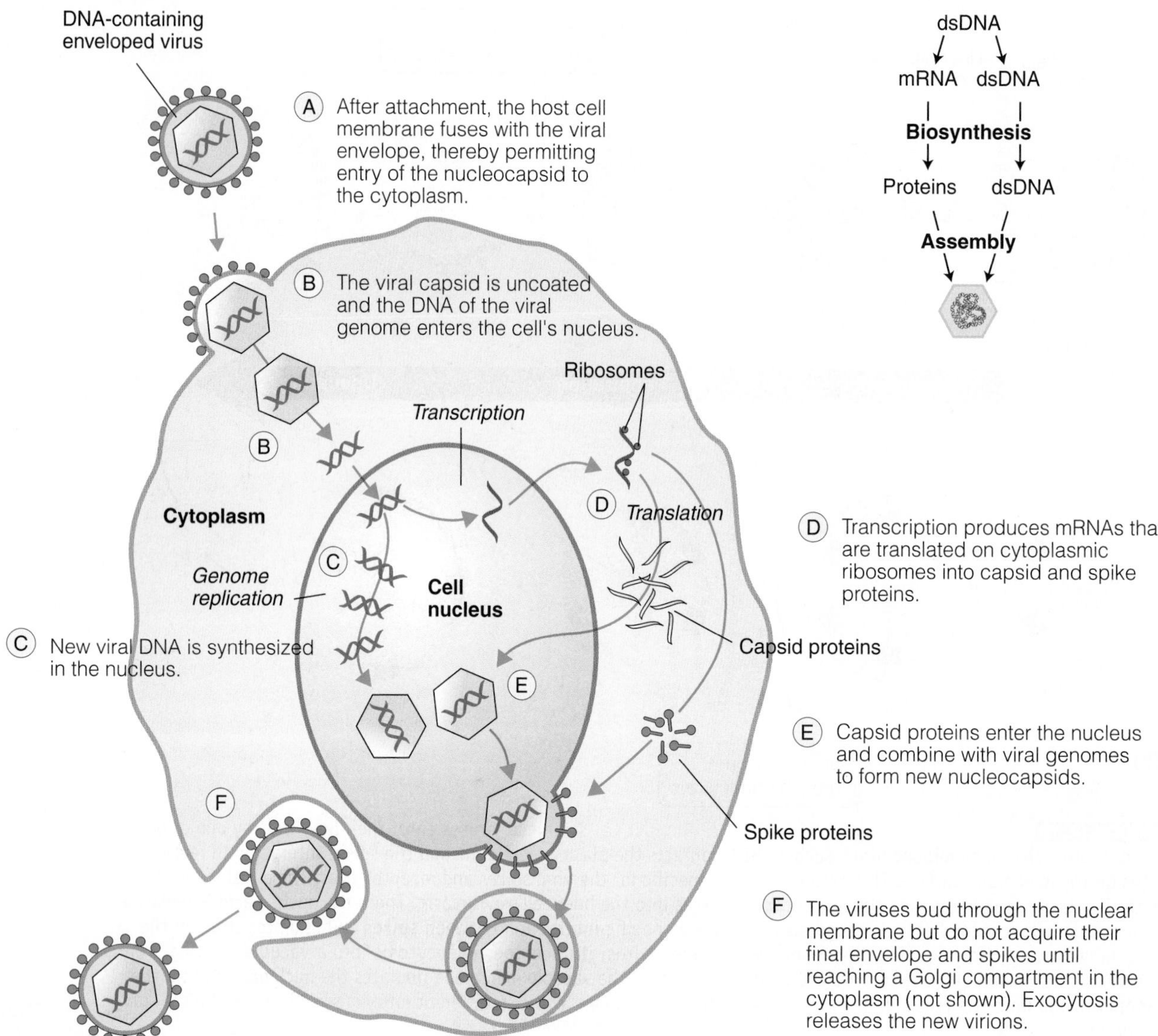

FIGURE 14.14 Replication of a Double-Stranded DNA Virus. The replication cycle for a DNA virus, such as a herpesvirus, involves genome replication and assembly in the cell nucleus and biosynthesis in the cytoplasm (inset). »» Why must the viral DNA of a herpesvirus enter the cell nucleus to replicate?

membrane, forcing a portion of the membrane ahead of and around the virion, resulting in an envelope. Other enveloped viruses, like the herpesvirus, fuse with the plasma membrane, releasing the virion. This process, called **budding**, need not necessarily kill or damage the cell during release. The same cannot be said for nonenveloped viruses. They leave the cell when the cell membrane ruptures, a process that generally results in cell lysis and death.

Once released, new progeny virions are capable of attaching to additional host cells and starting another replication cycle. Virus transmission from one host to another can be by direct contact (e.g., hand shaking, coughing, kissing, sexual contact) or through indirect contact involving vectors (e.g., infected insects) or fomites (e.g., contaminated blankets, utensils, syringe needles).

But what about virion transmission within the body, specifically from one cell to another? For something like the influenza virus, is it simply a matter of flu virions being released and then attaching to neighboring, uninfected epithelial cells of the respiratory tract? The operative word here is "uninfected." How does a virion "know"

MICROFOCUS 14.3: Medical Microbiology

Follow the "Bouncing Virus" to an Uninfected Cell

Bacteriophages are somewhat "smart" viruses. Take for example a lysogenic infection in which the lambda (λ) phage has integrated its DNA as a prophage in the bacterial chromosome of an infected *Escherichia coli* cell. While in a dormant or latent state, what keeps another λ phage from infecting the same *E. coli* cell—and perhaps triggering a lytic infection? The simple answer is that the prophage is not quite dormant. It produces a repressor protein that prevents all other λ phage viral genes from being expressed. Therefore, if another λ phage comes along and penetrates the cell, none of its genes can be expressed and it cannot cause a lytic or lysogenic infection.

Immunofluorescence micrograph of a cell infected with vaccinia virus particles. Actin filaments, which the virus uses to propel the newly formed particles out of the cell, are the green protrusions. (Bar = 5 um.) © Dr. Dan Guravich/Photo Researchers, Inc.

In the animal virus world there also is a "virus intelligence" at work to find only uninfected cells. The prevailing belief had been that viruses infect, replicate, and release new virions that randomly come in contact with other, hopefully uninfected, cells. Contact then triggers another cycle of infection. But what if the contacted cell is already infected? It would be inefficient for the virus to "spend time" infecting an infected cell. The vaccinia virus, a poxvirus used to produce the smallpox vaccine, has found a way around this predicament.

When a cell is infected, the vaccinia virus triggers the formation of actin filaments that help propel the new virions toward other cells (see figure). Geoffrey Smith's lab at the Imperial College of London has discovered that actin projections "protect" infected cells from being reinfected. Smith's group used a culture of monkey liver cells. When infected, the first infected cell releases virions as described above. When the surrounding uninfected cells are infected, they produce two proteins on their cell surface that mark the cell as "infected." Should a vaccinia virion land on such a cell, the surface protein complex triggers the formation of actin protrusions in the host cell, which repel the virions away from the cell and toward potentially uninfected cells. So, a vaccinia virion could "bounce off" multiple infected cells before it contacts an uninfected cell. From the view of the vaccinia virus, this hopping of virions is extremely efficient, as it accelerates virus spread. The virions do not spend time associated with an already infected cell and more quickly "find" uninfected cells.

It remains to be seen if other virus infections can behave this way—and if the virus would act in a similar way in the human body. But at least in culture, you can follow the "bouncing vaccinia virions" to an uninfected cell.

---

if the cell it comes in contact with has already been infected? MICROFOCUS 14.3 looks at one clever way.

Some Animal Viruses Produce a Latent Infection

Unlike most RNA viruses that go through a productive infection, many of the DNA viruses and the retroviruses can establish a latent infection, characterized by repression of most viral genes. Thus, the virus lies "dormant." For example, some herpesviruses, such as herpes simplex virus-1 (HSV-1), can generate a productive or latent infection. In an infected sensory neuron, HSV-1 undergoes latency as the viral dsDNA enters the neuron's cell nucleus and circularizes. No viral particles are produced for months or

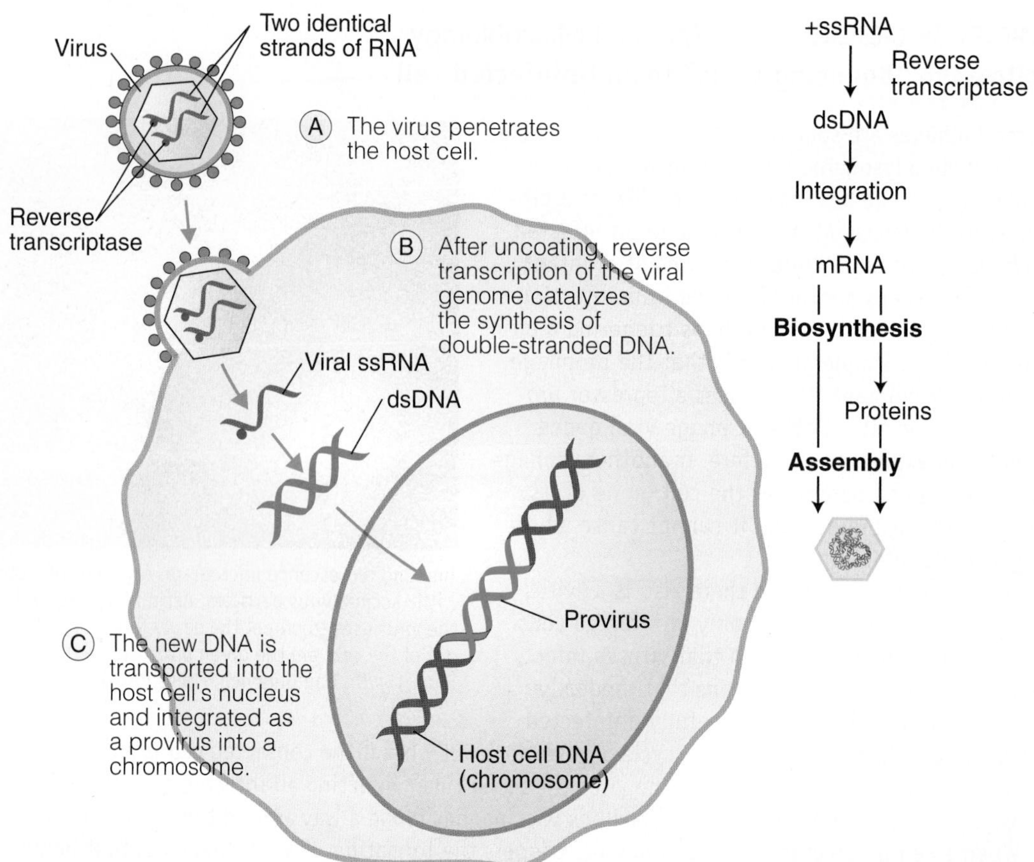

FIGURE 14.15 **The Formation of a Provirus by HIV.** Retroviruses can incorporate their genome into a host cell chromosome. Shown here, is the human immunodeficiency virus (HIV), a retrovirus causing HIV disease and AIDS. The single-stranded RNA genome can be reverse transcribed into double-stranded DNA by a reverse transcriptase enzyme. The DNA then enters the cell nucleus and integrates into a chromosome. »» Propose a reason why genome integration is an advantage for the virus.

years until some stress event reactivates the viral dsDNA and a new productive infection will be initiated.

Retroviruses, such as HIV, also carry out a latent infection. However, in this group of viruses, the virus carries an enzyme, called **reverse transcriptase**, which is used to reverse transcribe its +ssRNA into dsDNA (FIGURE 14.15). The dsDNA then enters the host cell nucleus and, like the temperate phage DNA inserted into a bacterial chromosome, becomes integrated randomly into the DNA of one chromosome. This integrated viral genome is referred to as a **provirus** and represents a unique and stable association between the viral DNA and the host genome. The advantage for the virus is that every time the host cell divides, the provirus will be replicated along with the host genome and be present in all progeny cells. In addition, as a provirus, it is protected from attack by antiretroviral drugs. However, at any time, the provirus can be reactivated and a productive infection involving biosynthesis and assembly of new virions will ensue.

MicroFocus 14.4 describes a hypothesis for the origin of cellular DNA arising from viruses. FIGURE 14.16 summarizes the outcomes for animal virus infections.

CONCEPT AND REASONING CHECKS 4

a. Why would it be advantageous for a phage to carry out a lysogenic cycle rather than a lytic cycle?

b. Distinguish between the replication events in DNA and RNA viruses.

c. How do viruses cause disease?

d. Why don't RNA viruses, like the poliovirus or influenzavirus, undergo provirus formation?

MICROFOCUS 14.4: Evolution

A Hypothesis: DNA Arose from Viruses

As we have seen on many occasions in this text, there are three domains of life—the Bacteria, Archaea, and Eukarya. One characteristic they all have in common is the presence of a DNA genome. Because many researchers believe life started in an "RNA world," where did DNA come from?

Some evolutionary biologists propose that in the "RNA-world organisms" some fragments of genetic material (genes) and macromolecular assemblies in these organisms "escaped" and evolved into RNA viruses (see figure **A**). However, when these viruses infected an RNA-based cell, the viral RNA could be attacked and perhaps destroyed by defensive host cell enzymes. On the other hand, if a virus contained DNA (like the giruses Mimivirus and Megavirus; MicroFocus 14.1), it could avoid host defenses because the host cell would not have yet evolved any defense against this novel nucleic acid molecule. The DNA viruses would have a selective advantage.

The hypothesis. So, a hypothesis for the "DNA" world says that different DNA (founder) viruses became integrated into a host cell (see figure **B**), where they could coexist (like a plasmid or transposon today). Then, at some point if the DNA viruses lost the ability to "escape" the host cell but replicate much like plasmids do today, they could become the depository for new genes from RNA viruses that attempted to infect the host cell. Slowly the "DNA chromosome" enlarged until the cell became totally dependent on DNA for survival and being more stable and less prone to mutation than their RNA predecessors, the "DNA world" was born.

In many ways, members of the Archaea and Eukarya are more similar to each other than to the Bacteria. Therefore, two somewhat similar founder DNA viruses might have taken up residence in the ancestors of these domains, one within the Archaea/Eukarya branch and another less related DNA founder virus within an ancestor of the Bacteria (see figure **C**).

Since we cannot go back to the RNA world, making sense of an RNA to DNA transition requires careful thinking based on what we know today about the domains of life and the virosphere. It is very controversial and subject to great interpretation, discussion, and debate by members of the science community.

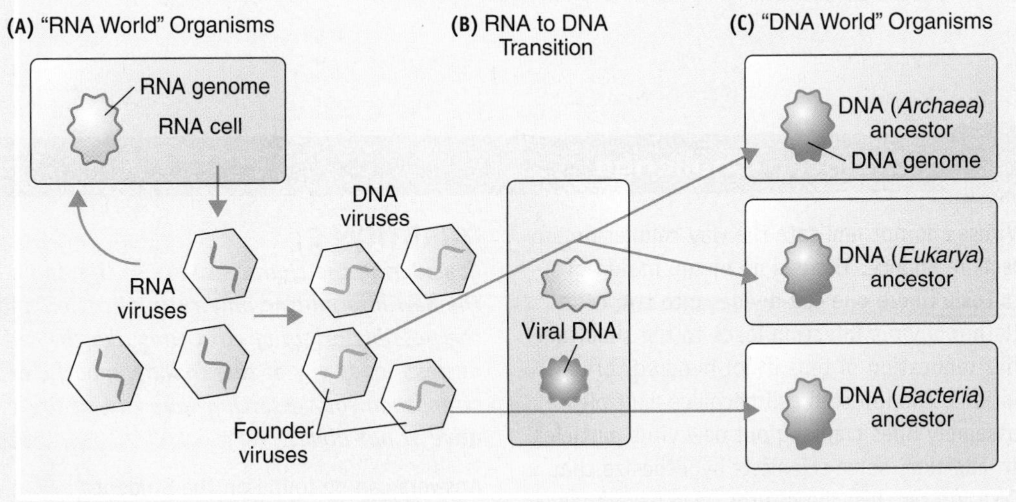

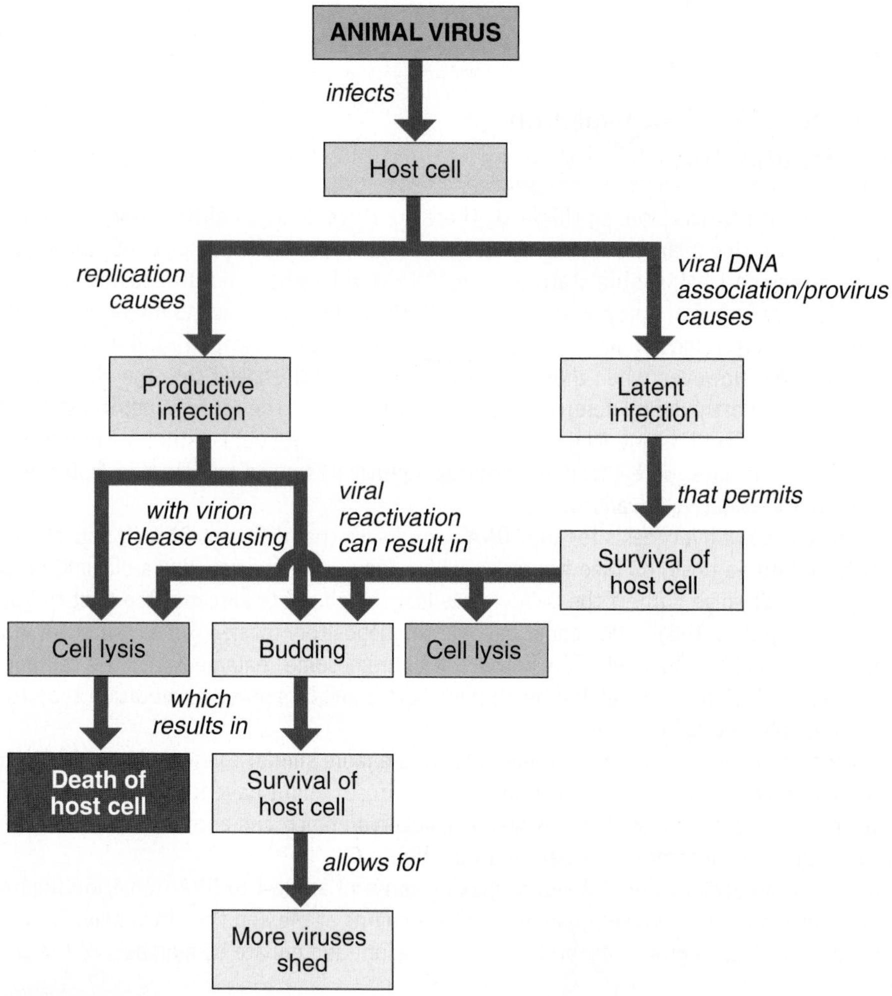

FIGURE 14.16 **Potential Outcomes for Animal Virus Infection of a Host Cell.** As depicted in this concept map, depending on the virus and host cell interaction, the host cell may be killed or survive. »» Identify a specific virus that follows each of the three outcomes (productive infection; cell death or survival; latent infection).

Chapter Challenge

C

Viruses do not replicate the way cellular organisms reproduce. There is no binary fission or mitosis where one cell divides into two cells. Rather, a virus infection leads to the simultaneous replication of tens if not hundreds of new virions. The infected cell becomes multiple assembly lines cranking out new virus particles. In addition, some scientists hypothesize that DNA viruses, like the giruses, may have been the source for the evolution of the eukaryotic cell nucleus (MicroFocus 14.4).

QUESTION C:

Considering the characteristics of life and the new information on virus replication and the possible origin of structures like the cell nucleus, does any of this change, modify, or strengthen your assertion that viruses are alive or not alive?

Answers can be found on the Student Companion Website in **Appendix D.**

KEY CONCEPT 14.5 Viruses and Their Infections Can Be Detected in Various Ways

If an individual contracts a viral disease, there are various ways the viral agent can be detected for identification and eventual diagnosis. A prompt identification often is necessary for selecting possible antiviral therapy.

Detection of Viruses Often Is Critical to Disease Identification

The diagnosis of viral diseases like the flu and a cold are usually straightforward and do not require further laboratory confirmation. In some cases, viral infections leave their mark on the infected individuals. Measles is accompanied by **Koplik spots**, which are a series of bright red patches with white pimple-like centers on the lateral mouth surfaces. Swollen salivary glands and teardrop-like skin lesions are associated with mumps and chickenpox, respectively.

However, virus detection is not always so straightforward and the proper identification has a bearing on relating a particular virus to a particular disease. The clinical microbiology laboratory or reference laboratory diagnosis of a viral disease can be carried out using light or electron microscopy to examine cells obtained from body tissues or fluids (FIGURE 14.17). Other common methods used to detect viral infections use immunological diagnostic procedures to search for viral antibodies in a patient's serum. However, animal viruses (like all viruses) cannot be grown in culture but rather must be grown in cells in culture.

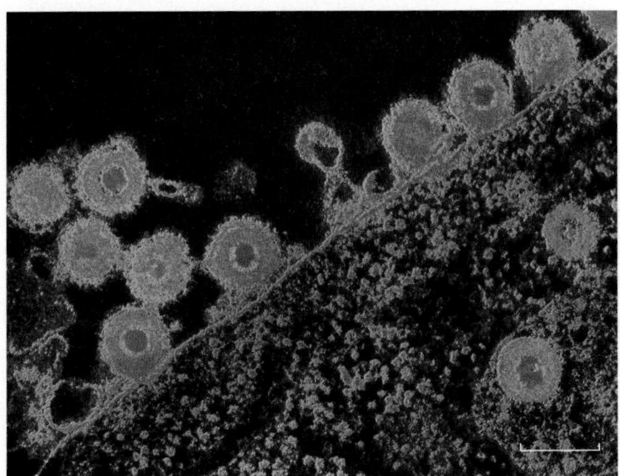

FIGURE 14.17 **Cells and Viruses.** A false-color transmission electron micrograph of tissue cells and associated herpesviruses (orange). The envelope of the virus is seen as a ring at the viral surface, and the nucleocapsid of the virus is the darker center. (Bar = 200 nm.) © Dennis Kunkel Microscopy, Inc./Phototake/Alamy Images »» How might these herpesviruses enter the host cell?

One common method once used to cultivate and detect viruses was to use fertilized (embryonated) chicken eggs. A hole is drilled in the egg shell and a suspension of the material is introduced (FIGURE 14.18A). Today, only the influenza viruses are routinely cultivated by this method, usually to produce high concentrations of viruses for vaccine production.

The most common method for detecting viruses is to infect animal cell cultures. In such a **primary cell culture**, the cells are suspended in a solution of nutrients, growth factors, pH buffers, and salts. The different cell types in a primary cell culture can be separated enzymatically and isolated as a single cell type, called a **cell line**. The type of cell culture used will depend on the virus species to be cultivated as a single layer of cells called a **monolayer**. Viruses are then introduced into the culture (FIGURE 14.18B).

When viruses replicate in a cell line, they often cause a noticeable change to the cells, called the **cytopathic effect (CPE)**. Viruses often cause a change in cell structure (TABLE 14.2). For example, infectious mononucleosis is characterized by large numbers of lymphocytes with a "foamy-looking," highly vacuolated cytoplasm. Paramyxoviruses cause host cells to fuse together into multinucleate giant cells called **syncytia**. Viruses sometimes produce cell inclusions. The brain tissue of a rabid animal can contain cytoplasmic nucleoprotein inclusions called **Negri bodies**, and cells from patients with herpes infections contain nuclear granules. Such CPEs, often detected by electron microscopy, can be key to virus identification.

Viruses also can be detected by the formation of plaques in a cell culture. A **plaque** is a clear zone within the cloudy "lawn" of bacterial cells or monolayer of animal cells (FIGURE 14.18C). The viruses infect and replicate in the cells, thereby destroying them and forming plaques. Epidemiologic investigations sometimes use

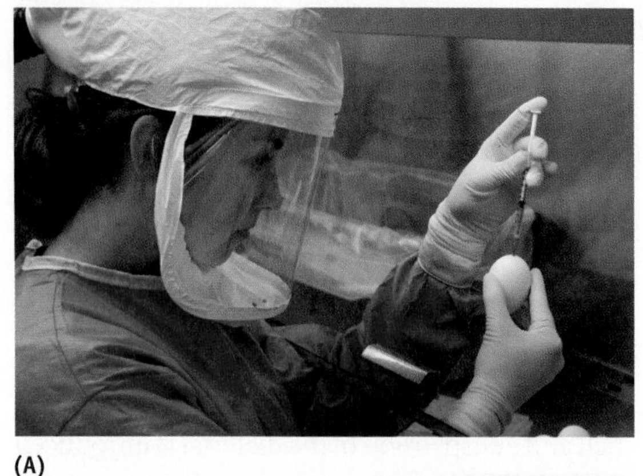

(A)

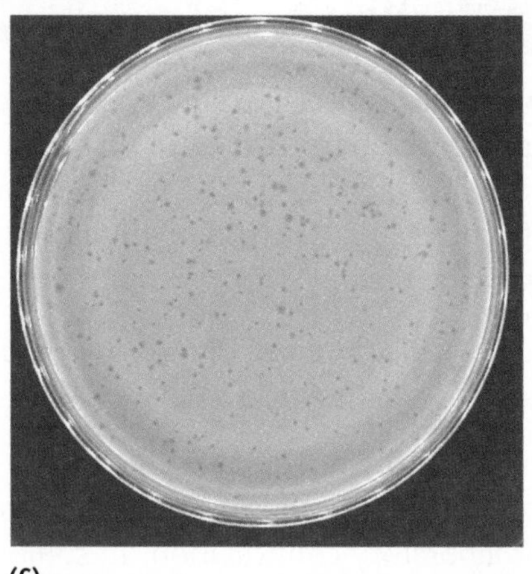

(C)

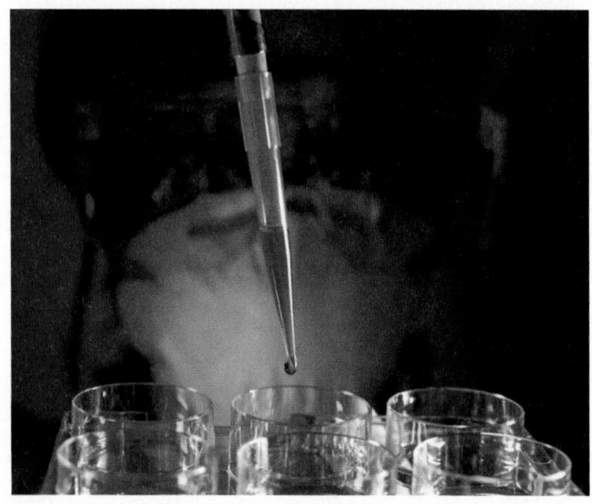

(B)

FIGURE 14.18 Infection of Cells in Embryonated Eggs and Cell Culture. (**A**) Inoculation of fertilized eggs. Techniques such as these are standard practice for growing the flu virus. Courtesy of Greg Knobloch/CDC. (**B**) A masked researcher uses a pipette to infect a culture of human cells with a virus. The dishes contain a culture medium, which allows the cells to survive outside the body. The effect of the virus on the infected cells will then be studied. © James King-Holmes/Photo Researchers, Inc. (**C**) Plaque formation in a cell culture. Plaques are evident as the clear areas in the dish. Courtesy of Giles Scientific Inc, CA, www.biomic.com.
»» Why must a face guard or mask be worn when infecting embryonated eggs and cells in culture?

TABLE

14.2 Examples of Virus Cytopathic Effects

Virus Family	Cytopathic Effect
	Changes in cell structure
Picornaviridae	Shrinking of cell nucleus
Papovaviridae	Cytoplasmic vacuoles
Paramyxoviridae, Coronaviridae	Cell fusion (syncytia)
Herpesviridae	Chromosome breakage
Herpesviridae, Adenoviridae, Picornaviridae, Rhabdoviridae	Rounding and detachment of cells from culture
	Cell inclusions
Adenoviridae	Virions in nucleus
Rhabdoviridae	Virions in cytoplasm (Negri bodies)
Poxviridae	"Viral factories" in cytoplasm
Herpesviridae	Nuclear granules

MICROINQUIRY 14

The One-Step Growth Cycle

In the research laboratory, we can follow the replication of viruses (productive infection) by generating a one-step growth curve. Realize that we are not really looking at growth, but the replication and increase in number of virus particles. There are several periods associated with a virus growth cycle that you should remember because you will need to identify the periods in the growth curve. The "eclipse period" is the time when no virions can be detected inside cells (intracellular) and the "latent phase" is the time during which no extracellular virions can be detected. Also, the "burst size" is the number of virions released per infected cell.

To generate our growth curve, we start by inoculating our viruses onto a susceptible cell culture. There are 100,000 (10^5) cells in each of 10 cultures. We then add ten times the number of viruses (10^6) in a small volume of liquid to each culture to make sure all the cells will be infected rapidly. After 60 minutes of incubation, we wash each culture to remove any viruses that did not attach to the cells and add fresh cell growth medium. Then, at 0 hour and every 4 hours after infection, we remove one culture, pour off the growth medium, and lyse the cells. The number of viruses in the growth medium and in the lysed cells are determined. Because we cannot see viruses, we measure any viruses in the growth medium or in the lysed cells using a plaque assay. We assume that one plaque (plaque-forming unit or PFU) resulted from one infected cell.

The curves drawn below plot the results from our growth experiment.

Answer the following questions based on the one-step growth curve. Answers can be found on the Student Companion website in **Appendix D**.

14.1a. How long is the eclipse period for this viral infection? Explain what is happening during this period.

14.1b. How long is the latent period for this infection? Explain what is happening during this period.

14.1c. What is the burst size?

14.1d. Explain why the growth curve shows a decline in phages between 0 and 4 hours.

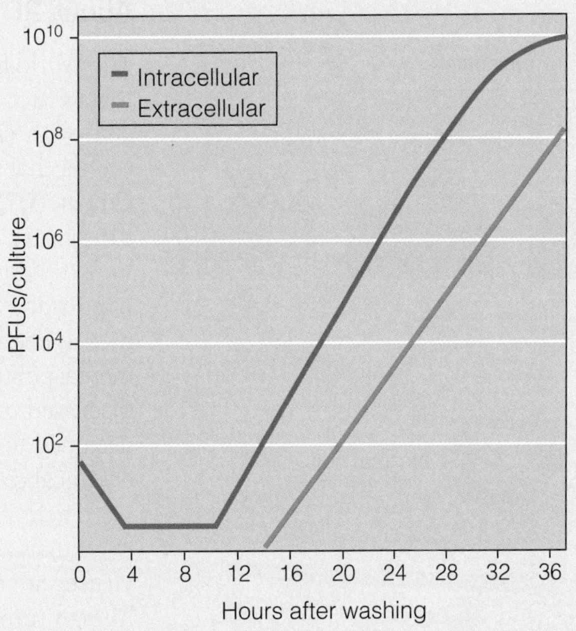

A One-Step Growth Cycle

phage typing to identify plaques characteristic of a specific strain of a bacterial disease, such as staphylococcal food poisoning. MICROINQUIRY 14 uses the plaque assay to monitor intracellular virus production and extracellular virus release during a productive infection of animal cells in culture.

CONCEPT AND REASONING CHECKS 5

a. How can some viruses be identified simply by observing an infected cell culture?

KEY CONCEPT 14.6 Some Viruses Are Associated with Human Tumors and Cancers

Cancer is indiscriminate. It affects humans and animals, young and old, male and female, rich and poor. In the United States, over 557,000 people die of cancer annually, making the disease the leading cause of death after heart disease and stroke. Worldwide, over 7 million people die of cancer each year. Cancer is a very complex topic, so we will only summarize the important points related to viruses.

Cancer Is an Uncontrolled Growth and Spread of Cells

Cancer results, in part, from the uncontrolled reproduction (mitosis) of a single cell. The cell escapes the cell cycle's controlling factors and, as it continues to multiply, a cluster of cells soon forms. Eventually, the cluster yields a clone of abnormal cells referred to as a **tumor**. Normally, the body will respond to a tumor by surrounding it with a capsule of connective tissue. Such a local tumor is designated **benign** because it usually is not life threatening.

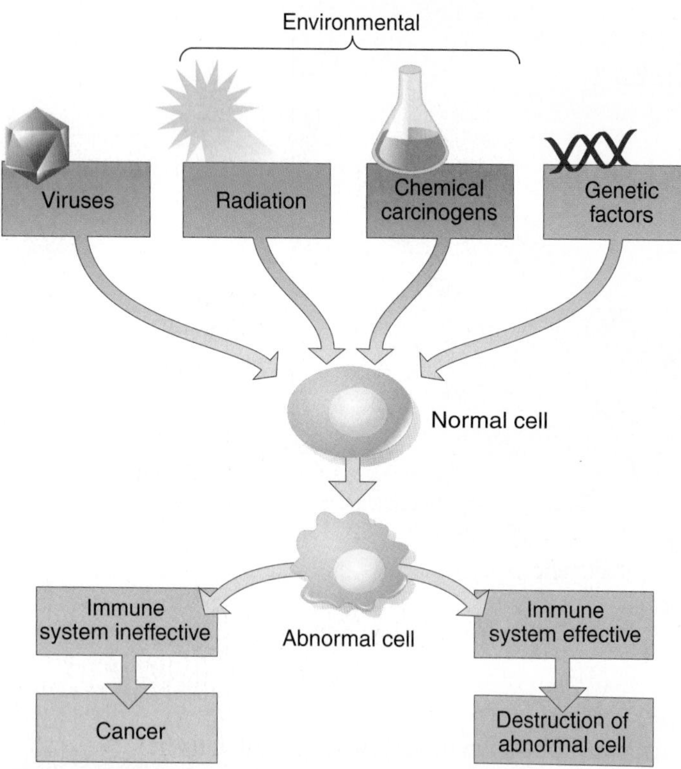

FIGURE 14.19 **The Onset of Cancer.** Viruses and other factors can induce a normal cell to become abnormal. When the immune system is effective, it destroys abnormal cells and no cancer develops. However, when the abnormal cells evade the immune system, a tumor may develop and become malignant, spreading to other tissues in the body. »» What factors are required for a tumor to become malignant?

Additional changes can occur to tumor cells that release them from their specific confines. Often they stick together less firmly than normal cells and fail to stop dividing when cells come in contact with one another. They may break out of the capsule and **metastasize**, a spreading of the cells to other tissues of the body. Such a tumor now is described as **malignant** and the individual now has **cancer** (*cancer* = "crab"; a reference to the radiating spread of cells, which resembles a crab). So, a series of changes must occur to a healthy cell before it can become a tumor or cancer cell.

How can such a mass of cells bring illness and misery to the body? By their sheer numbers, cancer cells invade and erode local tissues, interrupt normal functions, and choke organs to death by robbing them of vital nutrients. Thus, the cancer patient will commonly experience weight loss even while maintaining a normal diet.

Viruses Are Associated with About 20% of Human Tumors

The World Health Organization (WHO) estimates that 60% to 90% of all human cancers are associated with **carcinogens**, chemicals and physical agents that produce cellular changes leading to cancer (FIGURE 14.19). Physical agents include ultraviolet (UV) light and X rays. Among the known chemical carcinogens are the hydrocarbons found in cigarette smoke as well as asbestos, nickel, certain pesticides and dyes, and environmental pollutants in high amounts. The **Ames test**, discussed in the chapter on microbial genetics, is designed to use bacterial cells to detect potential chemical carcinogens in humans.

There also are several viruses that cause tumors in animals and humans. At present, at least seven viruses are associated directly or indirectly with human tumors and cancers (TABLE 14.3). When these **oncogenic** (tumor-causing) viruses are transferred to test animals or cell cultures, an observable cellular **transformation** occurs; that is, a change in the morphological, biochemical, or growth patterns of normal cells takes place. Such a change to a normal cell involves a complex, multistep sequence of events, of which viruses play one part. Examples of such oncogenic viruses include the Epstein-Barr virus, which is associated with infectious mononucleosis and Burkitt lymphoma, and the hepatitis

TABLE

14.3 Human Oncoviruses and Their Effects on Cell Growth

Oncogenic Virus	Benign Disease	Effect	Associated Cancer
DNA Tumor Viruses			
Papillomaviridae			
• Human papilloma virus (HPV)	Benign warts Genital warts	Encodes genes that inactivate cell growth regulatory proteins	Cervical, penile, and oropharyngeal cancer
Polyomaviridae			
• Merkel cell polyoma virus (MCPV)	Unknown	Being investigated	Merkel cell carcinoma
Herpesviridae			
• Epstein-Barr virus (EBV)	Infectious mononucleosis	Stimulates cell growth and activates a host cell gene that prevents cell death	Burkitt lymphoma Hodgkin lymphoma Nasopharyngeal carcinoma
• Herpesvirus 8 (HV8)	Growths in lymph nodes	Forms lesions in connective tissue	Kaposi sarcoma
• Cytomegalovirus (CMV)	Mononucleosis syndrome	Stimulates genes involved with tumor signaling	Salivary gland cancers
Hepadnaviridae			
• Hepatitis B virus (HBV)	Hepatitis B	Stimulates overproduction of a transcriptional regulator	Liver cancer
RNA Tumor Viruses			
Flaviviridae			
• Hepatitis C virus (HCV)	Hepatitis C	Chromosomal aberrations, including enhanced chromosomal breaks and sister chromatid exchanges	Liver cancer
Retroviridae			
• Human T-cell leukemia virus (HTLV-1)	Weakness of the legs	Encodes a protein that activates growth-stimulating gene expression	Adult T-cell leukemia/lymphoma

B virus, which is linked to chronic liver disease and liver cancer. Of special note is cervical cancer, which along with liver cancer, are responsible for 80% of virus-associated cancers. Cervical cancer is the second most common cancer in women under age 35 and is caused by several subtypes of the human papilloma virus (HPV; Papillomaviridae). In one study, 71% of the women developing cervical cancer had HPV present in their Pap smear. There is now a vaccine, called Gardasil®, that provides almost 100% protection against infection by the two most common HPV strains responsible for 70% of cervical cancers and two other strains responsible for 90% of genital warts.

Oncogenic Viruses Transform Infected Cells

As Table 14.3 shows, oncogenic viruses are found only in some families composing the DNA viruses (DNA tumor viruses), which have dsDNA genomes, and the retroviruses of the RNA viruses (RNA tumor viruses), which have dsDNA when integrated as a provirus in a host cell.

One way that the cellular transformation process is mediated is by proteins encoded by **oncogenes**, which are genes capable of converting a normal cell into a tumorous or cancerous cell. These genes usually affect growth control, cell division, or chemical signaling involved in cell growth. **Cellular oncogenes** (*c-onc*) are mutated forms of normal cellular genes called **proto-oncogenes**. These proto-oncogenes normally control cell growth by triggering cell division when necessary and remaining silent when cell division is not required (FIGURE 14.20). If they become mutated (*c-onc*), they are continually "on," growth control is lost, and the cells continue to divide, leading to a tumor.

■ Pap smear: A test to detect cancerous or precancerous cells of the cervix, allowing for early diagnosis of cancer.

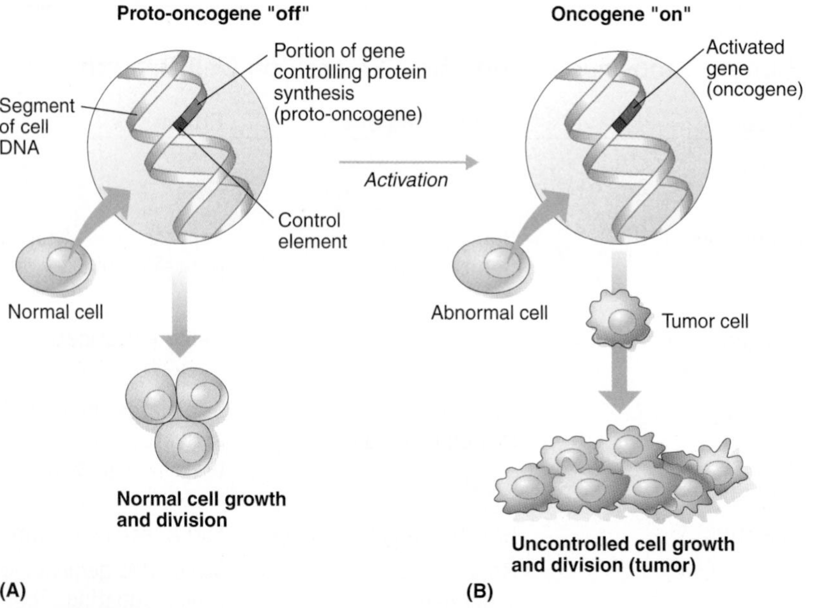

FIGURE 14.20 **The Oncogene Theory.** The oncogene theory helps explain the process of tumor development. (**A**) The normal cell grows and divides without complications. Within its DNA it contains proto-oncogenes that are "turned off" when cells are not dividing. (**B**) When the genes are activated by viruses, mutations, or other factors, they revert to oncogenes, which "turn on." An abnormal tumor cell results. The oncogenes encode proteins that regulate the development of a tumor and may contribute to transformation from a normal cell to a cancer cell. »» What properties are common to tumor cells?

Another way cellular transformation can occur is through the loss of gene activity. Cells contain several **tumor suppressor genes** whose normal function is to inhibit transformation; that is, if a cell converts to an abnormal cell, the tumor suppressor proteins trigger a programmed cell death called **apoptosis** (FIGURE 14.21A). Loss of tumor suppressor protein function can result in uncontrolled cell divisions. Here is how each group of tumor viruses operates.

RNA Tumor Viruses. As described earlier, all retrovirus infections involve reverse transcribing the ssRNA into dsDNA, which is then integrated randomly into a chromosome as a provirus. If provirus insertion lies near a tumor suppressor gene, that gene may become inactivated (**insertional inactivation**); if near a proto-oncogene or *c-onc*, it may become activated (**insertional activation**). In both cases, provirus integration leads to a disruption of normal growth control and a stimulation of cell division, leading to tumor formation. A third possibility is that the retrovirus carries a viral oncogene (*v-onc*) as part of its genome. This is usually a slightly altered *c-onc* that was incorporated into the viral genome during transduction many generations previous (FIGURE 14.22). On infection of another cell, integration of the *v-onc* can destabilize the host cell's growth control, leading again to increased cell divisions.

DNA Tumor Viruses. Although there is a variety of DNA tumor viruses, they are all similar in that the oncogenes they carry are essential viral genes needed for replication; there is no corresponding *c-onc*. The viral oncogene (*v-onc*) products mainly are nuclear proteins that disrupt host DNA replication control by binding to and inactivating the proteins coded by tumor suppressor genes (FIGURE 14.21B). The loss of tumor suppressor protein function will again result in excessive cell divisions and tumor development. Generally, cancers developing from DNA tumor viruses require a persistent infection for years or decades. Examples would include the hepatitis B virus and HPV, both of which may result in cancer many years after the initial infection.

This ability of viruses to get inside cells and deliver their viral information has been manipulated through genetic engineering to deliver genes to cure genetic illnesses. MicroFocus 14.5 looks at the pros and cons of using viruses for this purpose.

CONCEPT AND REASONING CHECKS 6

a. Explain how the oncogene theory is related to virus infection by RNA and DNA tumor viruses.

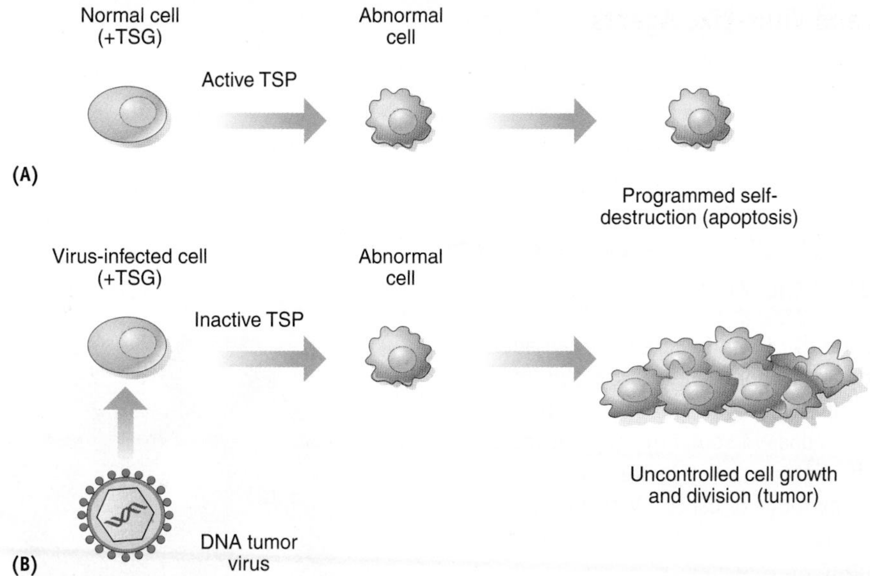

FIGURE 14.21 **Viral Oncogenes and Tumor Suppressor Gene Action.** (**A**) All human cells contain tumor suppressor genes (TSG) that function to produce tumor suppressor proteins (TSP) that will trigger programmed cell death (apoptosis) if the cell converts to a tumor cell. (**B**) Infection by a DNA tumor virus can carry an oncogene whose protein product will inactivate TSPs, leading to uncontrolled cell division and tumor formation. »» What advantage is it to the DNA tumor virus to have infected cells divide endlessly?

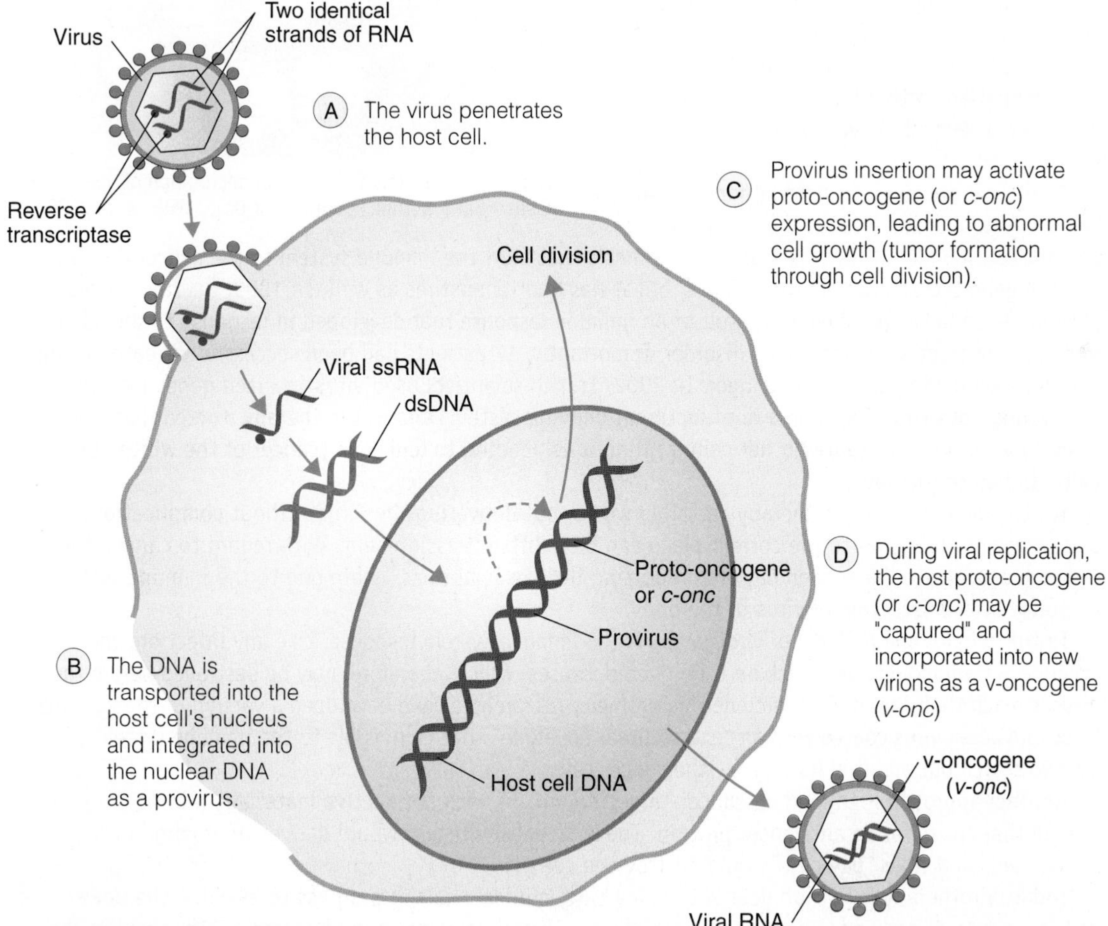

FIGURE 14.22 **Retrovirus Stimulation of a Tumor and v-Oncogene Formation.** An oncogenic retrovirus may integrate its genome into a host chromosome adjacent to a proto-oncogene or cellular oncogene (*c-onc*). This may activate the expression of the proto-oncogene or *c-onc*. During replication, the provirus along with the proto-oncogene (*c-onc*) will be replicated and incorporated into new virions as a viral oncogene (*v-onc*). »» How would the oncogenic virus with the v-oncogene lead to tumor formation?

MICROFOCUS 14.5: Biotechnology

The Power of the Virus

How would you like to be injected with an adenovirus (see figure) or herpesvirus? How about the virus used to make the smallpox vaccine or a retrovirus? This doesn't sound like a great idea, but in fact it might be a way to treat genetic diseases and many forms of cancer. **Virotherapy**, as it is called, may be useful in curing disease rather than causing it.

Ever since the power of genetic engineering made it possible to transfer genes between organisms, scientists and physicians have wondered how they could use viruses to cure disease.

One potential way is to use viruses against cancer. In 1997, a mutant herpesvirus was produced that could only replicate in tumor cells. Because viruses make ideal cellular killers, perhaps they would kill the infected cancer cells. Used on a terminal patient with a form of brain cancer, the herpesvirus seems to have worked, as the patient has survived.

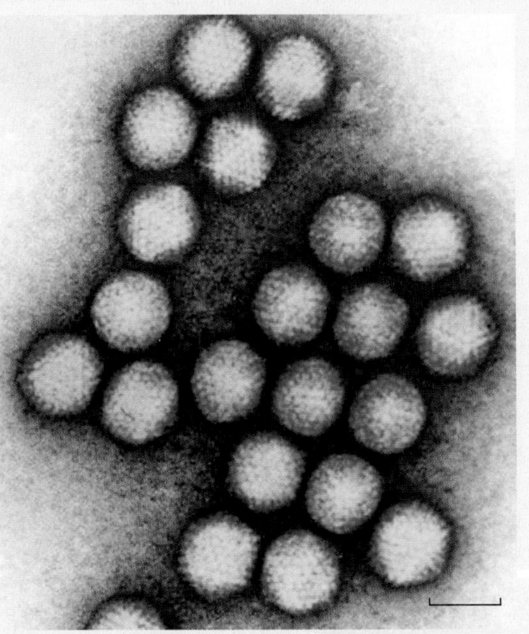

False-color transmission electron micrograph of adenoviruses. (Bar = 50 nm.) Courtesy of Dr. G. William, Jr./CDC.

Additional oncolytic (cancer-killing) viruses are being developed; some cause the cancer cells to commit cell suicide, while others deliver chemicals that alert the immune system to the cancer danger.

Viral gene therapy has shown promise, but it has had its misfires as well. In 1999, an 18-year-old patient, Jesse Gelsinger, died as a result of an immune response that developed in response to the adenovirus used to treat a rare metabolic disorder. Importantly, 17 patients had been successfully treated using the same adenovirus prior to Gelsinger. In 2002, French scientists used virus-inserted genes to cure four young boys suffering from a nonfunctional immune system. Again, the therapy worked, but the virus-inserted gene disrupted other cellular functions, leading to leukemia (cancer of the white blood cells) in two of the boys.

So, the idea of using virotherapy certainly works, but delivering the virus without complications and targeting the genes to the correct place can be a difficult assignment. With regard to cancer, the only adverse effects have been patients reporting flu-like symptoms, again due to the immune system's response to the detection of virus in the body.

In this regard, one problem of virotherapy is the immune system response. Like any infection, the immune system mounts an attack on the injected viruses, which, therefore, may be destroyed before the viruses reach their target. To overcome this problem, researchers have developed a version of the vaccinia virus and adenovirus coated with an "extracellular envelope" that is invisible to the immune system yet able to search out and find its target—the cancer cells.

Another approach is to load up cancer-targeting viruses with radioactive materials, a toxic drug, or a gene that codes for an anticancer protein. These "stealth viruses" would deliver their cargo only to cancer cells and their "payload" would destroy the cells.

Today, virotherapy is in high gear with many early clinical trials in progress to examine the power of the virus to directly or indirectly kill cancer cells. Virotherapy may soon become another tool in the fight against cancer.

KEY CONCEPT 14.7 Emerging Viruses Arise from Genetic Recombination and Mutation

Almost every year a newly emerging influenza virus descends upon the human population, 2009–2010 not withstanding with the "swine flu" pandemic. Other viruses not even heard of a few decades ago, such as HIV, Ebolavirus, Hantavirus, West Nile virus, and SARS virus, are often in the news. Where are these viruses coming from?

Emerging Viruses Usually Arise Through Natural Phenomena

The United States is at greater risk than ever from **zoonotic diseases**, which are diseases transmitted from other animals to humans (TABLE 14.4). Many of these **emerging infectious diseases** are the result of viruses appearing for the first time in a population or rapidly expanding their host range with a corresponding increase in detectable disease. Many are transmitted by insect vectors; that is, they are transmitted to humans primarily by ticks, fleas, or mosquitoes. One example is the West Nile virus that marched across the United States between 1999 and 2009. It is now endemic across the continental United States. Dengue fever, another mosquito-borne disease, causes thousands of cases of illness in U.S. territories and American travelers, and 100 million infections worldwide every year. Looking ahead, the Centers for Disease Control and Prevention (CDC) is preparing for the potential emergence of chikungunya disease that is sustained by human-mosquito-human transmission of the chikungunya virus. It infected more than 265,000 people in an outbreak on the French island of Réunion in the Indian Ocean, as well as 1,400,000 people in India in 2006. In 2007, transmission was reported for the first time in Europe, in a localized outbreak in northeastern Italy. With so many Americans traveling around the globe, it is just a matter of time before the chikungunya virus reaches North America.

But no matter how these viruses are transmitted, what caused their emergence?

One way "new" viruses arise is through **genetic recombination**. Take for example influenza. Genetic recombination allows different influenza viruses to reassort genome segments. The "swine flu" that broke out in Mexico in 2009 was the result of the reassortment of genome segments from a strain of avian flu virus, a human flu virus, and a swine flu virus.

Viruses also arise as a result of the second force driving evolution—**mutation**. For example, when a single nucleotide is altered (point mutation) in an RNA virus genome in the host cytoplasm, there is no way to "proofread" or correct the mistake during replication. Occasionally, one of these mutations

TABLE

14.4 Examples of Emerging Viruses

Virus	Family	Emergence Factor
Influenza	Orthomyxoviridae	Mixed pig and duck agriculture, mobile population
Dengue fever	Flaviviridae	Increased population density, environments that favor breeding mosquitoes
Sin Nombre (Hantavirus)	Bunyaviridae	Large deer mice population and contact with humans
Ebola/Marburg	Filoviridae	Human contact with fruit bats
HIV	Retroviridae	Increased host range, blood and needle contamination, sexual transmission, social factors
West Nile	Flaviviridae	Mosquito transported unknowingly to New York City
Nipah/Hendra	Paramyxoviridae	Human contact with flying foxes (bats)
SARS-associated	Coronaviridae	Contact with horseshoe bat
Chikungunya	Togaviridae	Spread through new mosquito vectors and global travel

may be advantageous. In the case of HIV, a beneficial mutation could generate a new virus strain resistant to an antiviral drug. With a rapid replication rate and burst size, it does not take long for a beneficial mutation to establish itself within a population.

Even if a new virus has emerged, it must encounter an appropriate host to replicate and spread. It is believed that smallpox and measles both evolved from cattle viruses, while flu probably originated in ducks and pigs. HIV almost certainly has evolved from a monkey (simian) immunodeficiency virus. Consequently, at some time such viruses had to make a species jump. What could facilitate such a jump?

Our proximity to animals and pathogens makes such a jump possible. Today, population pressure is pushing the human population into new areas where potentially virulent viruses may be lurking. Evidence shows the Machupo and Junin viruses (Arenaviridae viruses that cause hemorrhagic fevers in South America) jumped from rodents to humans as a result of increased agricultural practices that, for the first time, brought infected rodents into contact with humans.

An increase in the size of the animal host population carrying a viral disease also can "explode" as an emerging viral disease. The spring of 1993 in the American Southwest was a wet season, providing ample food for deer mice. The expanding deer mice population brought them into closer contact with humans. Leaving behind mouse feces and dried urine containing the hantavirus made infection in humans likely. The deaths of 14 people with a mysterious respiratory illness in the Four Corners area that spring eventually were attributed to this newly recognized virus.

So, emerging viruses are not really new. They are simply evolving from existing viruses and, through human changes to the environment, are given the "opportunity" to spread or to increase their host range.

TEXTBOOK CASE 14 describes the transmission of viruses via the pet trade and the emerging disease that can result.

CONCEPT AND REASONING CHECKS 7

a. Distinguish the ways that emerging viral diseases arise.

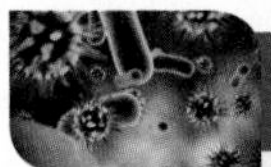

Chapter Challenge D

The emergence of "new" viruses or the reemergence of others was described as the product of genetic recombination and mutation. As such, viruses can adapt to changing environmental conditions by inheriting characteristics that enhance their survival in a particular environment. Adaptation was one of the characteristics of life.

QUESTION D:

Because viruses evolve, is that enough to classify them as living? Can nonliving objects, like a car, or better yet a computer virus, evolve? Explain.

Answers can be found on the Student Companion Website in **Appendix D**.

KEY CONCEPT 14.8 Virus-Like Agents Include Viroids and Prions

When viruses were discovered, scientists believed they were the ultimate infectious particles. It was difficult to conceive of anything smaller than viruses as agents of disease in plants and animals. However, the perception was revised as scientists discovered new disease agents—the subviral particles referred to as "virus-like agents;" that is, the viroids and prions.

Viroids Are Infectious RNA Particles

Viroids are tiny fragments of nucleic acid known to cause diseases in crop plants. In the 1960s, Theodore O. Diener and colleagues at the U.S. Department of Agriculture in Beltsville, Maryland were investigating a suspected viral disease, potato spindle tuber (PST), which results in long, pointed potatoes shaped like spindles. Nothing would destroy the disease agent except an RNA-degrading enzyme, and in 1971, the group postulated that a fragment of single-stranded, "naked" RNA was involved. Diener called the agent a viroid, meaning "virus-like."

Today, more than two dozen crop diseases have been related to viroids. The largest of these particles is about one-fifteenth of the size of the smallest virus (FIGURE 14.23). The RNA chain of the PST viroid has a known molecular sequence (359 nucleotides), but it contains so few genetic sequences that it is not understood how the viroids replicate. Diener speculated the viroids originated as introns, the sections of RNA spliced

Textbook CASE 14

Multistate Outbreak of Monkeypox

1 On April 9, 2003, a Texas animal importer received a shipment of 800 small mammals from Accra, Ghana. This included rope squirrels, Gambian giant rats, and dormice. On April 21, 2003, an Illinois animal distributor received several Gambian rats and dormice from the Texas distributor. These animals were caged with prairie dogs the distributor had previously caught. All the animals were then distributed to six states (Illinois, Indiana, Kansas, Missouri, Ohio, and Wisconsin).

2 In June 2003, a Wausau, Wisconsin couple went to a local swap meet. At a stall selling exotic pets, they purchased two prairie dogs. Back home, the animals were introduced to their 3-year-old daughter. Two days later, the young girl was bitten on the finger while playing by one of the prairie dogs.

3 Within several days the young girl was taken to the local hospital with a fever of 40°C that would not abate. She had developed skin pustules, small, round, raised areas of inflamed skin filled with pus (see figure). She was hospitalized, put on an IV, and started on antibiotics. A biopsy from the pustules failed to test positive for any known bacterial or viral pathogen.

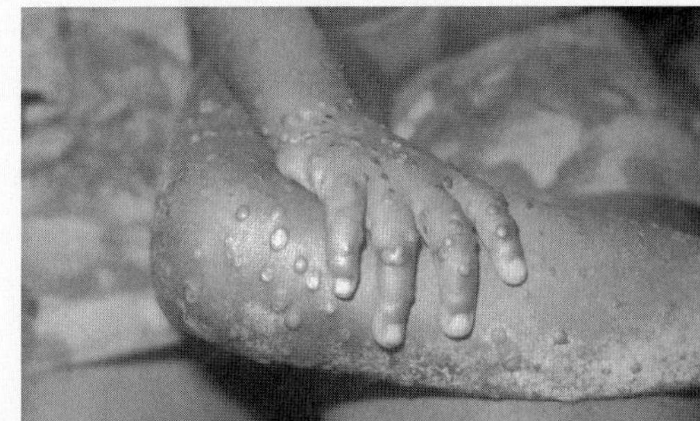

A photograph of an African child showing the pustules typical of a monkeypox infection. Courtesy of CDC.

4 Then her mother started to get itchy, red pustules on her body. The illness progressed from bumps to cold sweats at night. A few days later, her husband also developed symptoms. Both recovered and a biopsy from the mother's pustules was sent to the Centers for Disease Control and Prevention (CDC) in Atlanta for identification. Meanwhile, one of the purchased prairie dogs died.

5 After 7 days in the hospital, the daughter started improving and was sent home where she recovered completely. The day after the daughter went home, the CDC report was received. A pox virus was identified, specifically monkeypox, a disease formerly found only in Central and West Africa.

6 Alarmed by this potentially emerging virus, the CDC carried out an investigation to see how many other people exhibited the symptoms of monkeypox. The July 8, 2003, report identified a total of 71 cases of monkeypox from Wisconsin (39), Indiana (16), Illinois (12), Missouri (2), Kansas (1), and Ohio (1). The number of cases increased from May 15 through the week ending June 8 and then declined with no additional cases reported after June 20, 2003. All patients recovered completely.

7 A traceback investigation was undertaken to determine the source of the confirmed human cases of monkeypox.

Questions:

(Answers can be found on the Student Companion Website in **Appendix D.**)

A. Why was the hospital unable to identify the pathogen causing the pustules in the daughter?

B. How did the daughter become exposed to the monkeypox virus?

C. How did the mother and father become exposed to the monkeypox virus?

D. Based on the information provided, what did the traceback identify as the source of monkeypox cases?

For additional information see www.cdc.gov/mmwr/preview/mmwrhtml/mm5227a5.htm.

■ Dementia:
A loss of memory, judgment, and intellect.

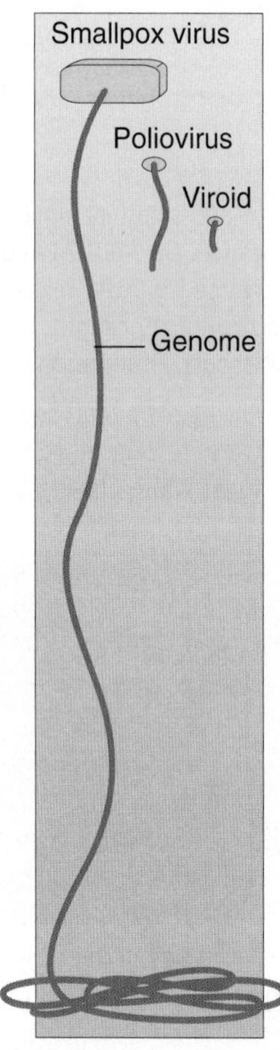

FIGURE 14.23 **Genome Relationships.** The size relationships of a smallpox virus, poliovirus, and viroid genome. The genome of the potato spindle tuber viroid has 359 nucleotides, while that of the smallpox virus has almost 200,000. »» Estimate whether viroid RNA is large enough to code for a variety of proteins.

out of messenger RNA molecules before the messengers are able to function. Because the viroid RNA encodes no proteins, another hypothesis suggests the viroid RNA interacts with host cell RNA, inactivating proteins that bring about disease through loss of cell function. A similar proposal suggests the viroid RNA "silences" host cell "target RNA," again bringing about disease through loss of cell regulation.

Prions Are Infectious Proteins

In 1986, cattle in Great Britain began dying from a mysterious illness. The cattle experienced weight loss, became aggressive, lacked coordination, and were unsteady in their gait. These detrimental effects became known as "**mad cow disease**" and were responsible for the eventual death of these animals. A connection between mad cow disease and a similar human disease surfaced in Great Britain in 1995 when several young people died of a human brain disorder resembling mad cow disease. Symptoms included dementia, weakened muscles, and loss of coordination. Health officials suggested the human disease was caused by eating beef that had been processed from cattle with mad cow disease. It appeared the disease agent was transmitted from cattle to humans. This resulted in the identification of more than 180,000 infected cattle and the slaughter of 4.4 million animals.

Besides mad cow disease, similar neurologic degenerative diseases have been discovered and studied in other animals and humans. These include scrapie in sheep and goats, chronic wasting disease in elk and deer, and Creutzfeldt-Jakob disease in humans. All are examples of a group of rare diseases called **transmissible spongiform encephalopathies (TSEs)** because, like mad cow disease (bovine spongiform encephalopathy; BSE), they can be transmitted to other animals of the same species and possibly to other animal species, including humans, and the disease causes the formation of "sponge-like" holes in brain tissue (FIGURE 14.24A).

Many scientists originally believed these agents were a new type of virus. However, in the early 1980s, Stanley Prusiner and colleagues isolated an unusual protein from scrapie-infected tissue, which they thought represented the infectious agent. Prusiner called the proteinaceous infectious particle a **prion** (pronounced pree-on). The sequencing of the protein led to the identification of the coding gene, called *PrP*. The *PrP* gene is primarily expressed in the brain.

This led Prusiner and colleagues to propose the **protein-only hypothesis**, which predicts that prions are composed solely of protein and contain no nucleic acid. The protein-only hypothesis further proposes there are two types of prion proteins (FIGURE 14.24B, C). Normal cellular prions (PrP^{C}) are found in normal brain and nerve tissue. These proteins have important roles to play in everything from neuron protection to the renewal of blood tissue in the bone marrow. Only when they undergo mutation do they become abnormal, misfolded, disease-causing prions (PrP^{SC}). More definitive proof came in 2004 when Prusiner's group demonstrated that purified prions can cause disease when injected into brains of genetically engineered mice.

Many researchers believe prion diseases—that is, TSEs—are spread by the infectious PrP^{SC} being taken up intact from the digestive tract and transported to the nervous tissue. They then bind to normal PrP^{C}, causing the latter to change shape (FIGURE 14.25). In a domino-like scenario, the

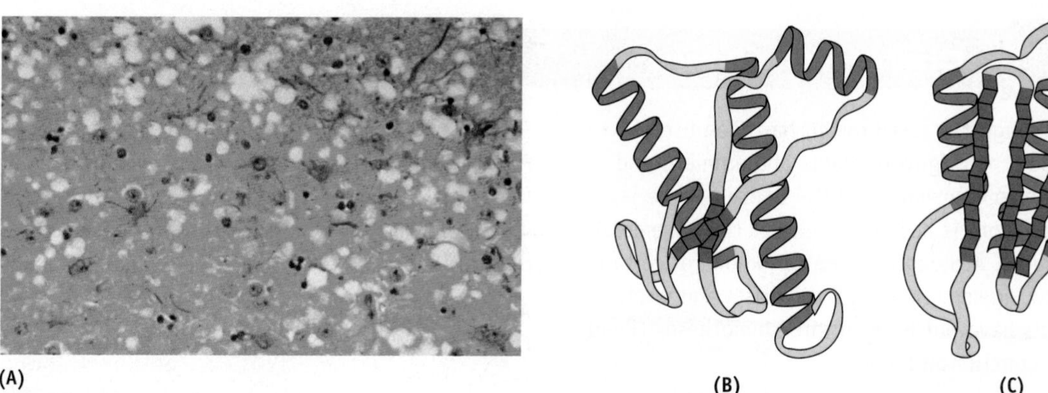

FIGURE 14.24 **Prion "Infection" and Structure.** (**A**) A photomicrograph showing the vacuolar ("spongy") degeneration of gray matter characteristic of human and animal prion diseases. Courtesy of APHIS photo by Dr. Al Jenny/CDC. (**B**) This drawing shows the tertiary structure of the normal prion protein (PrP^C). The helical regions of secondary structure are green. (**C**) A misfolded prion (PrP^{SC}). This infectious form of the prion protein results from the helical regions in the normal protein unfolding and an extensive pleated sheet secondary structure (blue ribbons) forming within the protein. The misfolding allows the proteins to clump together and react in ways that contribute to disease. »» How does a change in tertiary structure affect the functioning of a protein?

newly converted PrP^{SC} proteins, in turn, would cause more PrP^C proteins to become abnormal. The PrP^{SC} proteins then form insoluble protein fibers that aggregate, forming sponge-like holes left where groups of nerve cells have been killed. Importantly, PrP^{SC} does not trigger an immune response. Death of the animal occurs from the numerous nerve cell deaths that lead to loss of speech and brain function.

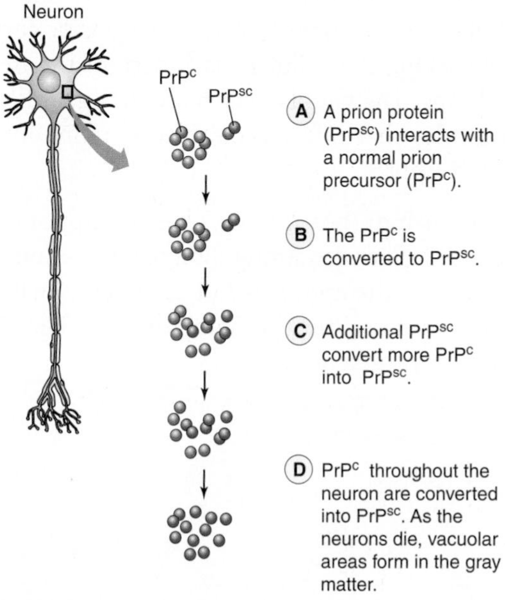

Neuron
PrP^c
PrP^{sc}
(A) A prion protein (PrP^{sc}) interacts with a normal prion precursor (PrP^c).
(B) The PrP^c is converted to PrP^{sc}.
(C) Additional PrP^{sc} convert more PrP^c into PrP^{sc}.
(D) PrP^c throughout the neuron are converted into PrP^{sc}. As the neurons die, vacuolar areas form in the gray matter.

FIGURE 14.25 **Prion Formation and Propagation.** Prion proteins (PrP^{SC}) bind to normal prion precursors (PrP^C) in neurons, causing these proteins to the misfolded PrP^{SC} conformation. This process is repeated over and over, leading to the accumulation of PrP^{SC} protein fibers that cause progressive neuron cell death. »» How could these areas of PrPSC protein fiber accumulation be observed in a brain autopsy from a person who has died from variant CJD?

The human form of TSE appears similar to the classical and spontaneous form of Creutzfeldt-Jakob disease (CJD). The new form of CJD, called **variant CJD (vCJD)**, is characterized clinically by neurologic abnormalities such as dementia. Neuropathology shows a marked spongiform appearance throughout the brain and death occurs within 3 to 12 months after symptoms appear.

The number of confirmed and probable deaths from vCJD in the United Kingdom (UK) peaked in 2000 with 28 reported victims. As of September 2012, there have been a total of 176 vCJD deaths in the UK since 1995. As such, BSE does not appear to be highly contagious considering the thousands of individuals who must have ingested contaminated beef. Experts believe BSE is almost extinct because there was only one reported case in cattle in the UK in 2012.

In North America, the first mad cow was discovered in Canada in 2003, and in December the first such cow was reported in the United States. In all, there have been 19 BSE cattle in Canada and four in the United States (through August 2012). No human cases of vCJD have been reported. Protection measures include condemning those animals having signs of neurologic illness and holding any cows suspected of having BSE until test results are known. In addition, "downer cattle"—those unable to walk on their own—cannot be used for human food or animal feed.

CONCEPT AND REASONING CHECKS 8

a. How do viroids differ from the RNA viruses?

b. How are prions (a) similar to and (b) different from viruses?

Chapter Challenge E

Although the fossil record has been useful to study the origins of plants and animals, and even some members of the domain Bacteria, the record is not much help for the ephemeral viruses. Although no one can know for certain how viruses originated, scientists and virologists have put forward three hypotheses (Read **In Conclusion** below).

QUESTION E:

From your experiences with viruses and virus-like agents in this chapter, would any of these hypotheses fit with the idea that viruses are living infectious agents?

Answers can be found on the Student Companion Website in **Appendix D.**

In conclusion, although the fossil record has been useful to study the origins of plants and animals, and even some members of the domain Bacteria, the record is not much help for the ephemeral viruses. Although no one can know for certain how viruses originated, scientists and virologists have put forward three hypotheses based on modern-day viruses and their genomes. The best guess is that they appeared some 2 to 3 billion years ago, making them about as old as the oldest life forms on Earth. Because they depend on host cells for replication, one could argue they appeared soon after the first cells. Here are the three main hypotheses.

Regressive Evolution Hypothesis. Viruses are degenerate life forms; that is, they are derived from small cells that functioned as intracellular parasites. Over time they lost many functions that could be supplied by their hosts and have retained only those genes essential for directing their parasitic way of life.

Cellular Origins Hypothesis. Viruses are derived from subcellular components (DNA and RNA) and functional assemblies of macromolecules that have escaped their cellular origins by being able to replicate autonomously in host cells. Modern day bacterial plasmids that replicate independently from the bacterial chromosome, and transposons, the jumping genes that can bound around a genome or be transferred to another genome, lend support to this hypothesis.

Coevolution Hypothesis. Viruses coevolved with cellular organisms from the self-replicating molecules believed to have existed in the primitive prebiotic earth. This hypothesis would suggest viruses arose at the same time as cellular life and transitioned to a parasitic form by hijacking the cellular machinery of the new life forms. The viroids, which are the small RNA molecules that infect some plants, represent a similar agent.

While each of these theories has its supporters, the topic generates strong disagreements among experts. In the end, it is not so much a matter of how viruses arose, but rather how we become infected with them.

SUMMARY OF KEY CONCEPTS

14.1 Filterable Infectious Agents Cause Disease

1. The concept of a **virus** as a filterable agent of disease was first suggested by Ivanowsky, while Twort and d'Herrelle identified viruses as infective agents in bacterial organisms. In the 1940s, the electron microscope enabled scientists to see viruses and innovative methods allowed researchers to cultivate them.

14.2 Viruses Have a Simple Structural Organization

2. Viruses are small, obligate, intracellular particles composed of nucleic acid as either DNA or RNA and in either a **single-stranded** or **double-stranded** form. The genome is surrounded by a protein **capsid**, and many viruses have an **envelope** surrounding the **nucleocapsid. Spikes** protruding from the capsid or

envelope are used for attachment to host cells. (Fig. 14.6)

3. Viruses have **helical, icosahedral**, or **complex symmetry**. (Fig. 14.7)
4. The **host range** of a virus refers to what organisms (hosts) the virus can infect. Many viruses only infect certain cell types or tissues within the host referred to as a **tissue tropism** (tissue attraction).

14.3 Viruses Can Be Classified by Their Genome

5. Two broad classes of viruses can be organized based on their genome and strand type: the **single-stranded** or **double-stranded DNA viruses** and the **single-stranded** or **double-stranded RNA viruses.** (Fig. 14.8)

14.4 Viral Replication Follows a Set of Common Steps

6. **Bacteriophages** undergo either a **lytic cycle**, involving attachment, penetration, biosynthesis, maturation, and release stages, or a **lysogenic cycle** where the phage genome integrates into the bacterial chromosome as a **prophage.** (Figs. 14.11 and 14.12)
7. Animal viruses also progress through the same stages of replication. However, penetration can occur by membrane fusion or endocytosis, and the biochemistry of nucleic acid synthesis varies among DNA and RNA viruses. (Fig. 14.13)
8. Many of the DNA viruses and the retroviruses can incorporate their viral DNA independently within the host's cell nucleus, or as a **provirus.** (Fig.14.14)

14.5 Viruses and Their Infections Can Be Detected in Various Ways

9. Various detection methods for viruses are based on the cultivation of animal viruses using cell cultures. The detection of viruses can be determined by characteristic cytologic changes referred to as the **cytopathic effect** and by **plaque** formation.

14.6 Some Viruses Are Associated with Human Tumors and Cancers

10. **Tumor** formation is a complex condition in which cells multiply without control and possibly develop into **cancers.** (Fig. 14.19)
11. There are at least seven **carcinogenic viruses** known to cause human tumors. Many of these tumors may develop into **cancer.** (Table 14.3)
12. Viruses may bring about tumors by converting **proto-oncogenes** into tumor **oncogenes.** If such genes are carried in a virus, they are called **viral oncogenes.** (Figs. 14.20, 14.21, 14.22)

14.7 Emerging Viruses Arise from Genetic Recombination and Mutation

13. Viruses use **genetic recombination** and **mutations** as mechanisms to evolve. Human population expansion into new areas and increased agricultural practices in previously forested areas have exposed humans to existing animal viruses. (Table 14.4)

14.8 Virus-Like Agents Include Viroids and Prions

14. **Viroids** are infectious particles made of RNA. They infect a few plant species and cause disease. Viroids lack protein and a capsid, but can replicate themselves inside the host.
15. **Prions** are infectious particles made of protein. They are capable of causing a number of animal diseases, including **mad cow disease** in cattle and **variant Creutzfeldt-Jakob disease (vCJD)** in humans. Prions cause disease by folding improperly and in the misfolded shape, cause other prions to misfold. These and other similar diseases in animals are examples of a group of rare diseases called **transmissible spongiform encephalopathies (TSEs).** (Figs. 14.24, 14.25)

CHAPTER SELF-TEST

For **STEPS A–D**, answers to even-numbered questions and problems can be found in **Appendix C** on the Student Companion Website at **http://microbiology.jbpub.com/10e**. In addition, the site features eLearning, an online review area that provides quizzes and other tools to help you study for your class. You can also follow useful links for in-depth information, read more MicroFocus stories, or just find out the latest microbiology news.

STEP A: REVIEW OF FACTS AND TERMS

Multiple Choice

Read each question carefully, then select the ***one*** answer that best fits the question or statement.

1. Which one of the following scientists was NOT involved with discovering viruses?
   A. Felix d'Herrelle
   B. Dimitri Ivanowsky
   C. Robert Fleming
   D. Martinus Beijerinck
2. Viral genomes consist of _____.
   A. DNA only
   B. RNA only
   C. DNA or RNA
   D. DNA and RNA
3. A nucleocapsid can have _____ symmetry.
   A. radial
   B. icosahedral
   C. vertical
   D. bilateral
4. Tissue tropism refers to _____.
   A. what tissues grow due to a viral infection
   B. what tissues are resistant to viral infection
   C. what organisms a virus infects
   D. what cells or tissues a virus infects
5. An RNA virus genome in the form of messenger RNA is referred to as a _____ RNA.
   A. (+) strand
   B. double-stranded
   C. (–) strand
   D. reverse strand

6. A virulent bacteriophage will _____.
   A. carry out a lytic cycle
   B. integrate its genome in the host cell
   C. remain dormant in a bacterial cell
   D. exist as a prophage
7. The release of the viral genome from the capsid is called _____.
   A. uncoating
   B. endocytosis
   C. penetration
   D. maturation
8. Provirus formation is possible in members of the _____.
   A. single-stranded DNA viruses
   B. retroviruses
   C. double-stranded RNA viruses
   D. single-stranded (– strand) RNA viruses
9. Cytopathic effects would include all the following *except* _____.
   A. changes in cell structure
   B. plaque formation
   C. vacuolated cytoplasm
   D. syncytia formation
10. A benign tumor _____.
    A. will metastasize
    B. represents cancer
    C. is malignant
    D. is a clone of dividing cells
11. Which of the following is NOT a carcinogen?
    A. Genetic factors
    B. UV light
    C. Certain chemicals
    D. X rays
12. The oncogene theory states that transforming genes _____.
    A. normally occur in the host genome
    B. can exist in viruses
    C. are not of viral origin
    D. All of the above (A–C) are correct.
13. Newly emerging viruses causing human disease can arise from _____.
    A. species jumping
    B. mutations
    C. genetic recombination
    D. All of the above (A–C) are correct.
14. Viroids contain _____.
    A. RNA and DNA
    B. only RNA
    C. DNA and a capsid
    D. RNA and an envelope
15. Which one of the following statements about prions is FALSE?
    A. Prions are infectious proteins.
    B. Prions have caused disease in Americans.
    C. Human disease is called variant CJD.
    D. Prions can be transmitted to humans from infected beef.

Identification

Use the following syllables to compose the term that answers each clue from virology. The number of letters in the term is indicated by the dashes, and the number of syllables in the term is shown by the number in parentheses. Each syllable is used only once.

A BAC CAP CO COS DRON EN GENE HE I I I LENT MOR O ON ON ONS OPE PHAGE PRI SID TER TU U VEL VIR VIR

16. Viral protein coat (2) __ __ __ __ __ __
17. Bacterial virus (5) __ __ __ __ __ __ __ __ __ __ __ __ __ __
18. Clone of abnormal cells (2) __ __ __ __ __
19. Cancer gene (3) __ __ __ __ __ __ __ __
20. Phages that lyse bacterial cells (3) __ __ __ __ __ __ __ __
21. Shape of poliovirus (5) __ __ __ __ __ __ __ __ __ __ __ __
22. Completely assembled virus (3) __ __ __ __ __ __
23. May surround the capsid (3) __ __ __ __ __ __ __ __
24. Infectious protein particles (2) __ __ __ __ __ __

STEP B: CONCEPT REVIEW

25. Identify the important historical developments in the identification and role of **viruses.** (**Key Concept 1**)
26. Distinguish between the structures common to all viruses and those that separate **nonenveloped** from **enveloped** viruses. (**Key Concept 2**)
27. Contrast viral **host range** with **tissue tropism.** (**Key Concept 2**)
28. Explain the difference between **DNA viruses** and **RNA viruses.** (**Key Concept 3**)
29. Identify the five stages of and explain the events occurring in a phage replication cycle. State how a **lysogenic cycle** differs from a **lytic cycle** in **bacteriophages.** (**Key Concept 4**)
30. Describe how animal viruses cause a **productive infection.** (**Key Concept 4**)
31. Explain how some DNA viruses and retroviruses establish a **latent infection.** (**Key Concept 4**)
32. Assess the role of the **cytopathic effect** for virus detection. (**Key Concept 5**)
33. Define how **benign** and **malignant tumors** arise. (**Key Concept 6**)
34. Identify three viruses that are involved in causing human tumors.
35. Explain how viruses can cause tumors. (**Key Concept 6**)
36. Identify the two mechanisms by which emerging viruses arise and how an animal virus could jump to humans. (**Key Concept 6**)
37. Summarize the properties of **viroids.** (**Key Concept 7**)
38. Summarize the properties of and diseases resulting from **prions.** (**Key Concept 8**)

STEP C: APPLICATIONS AND PROBLEMS

39. As part of a lab exercise, you and your lab partner are to estimate the number of bacteriophages in a sample of sewer water using a plaque assay with *Escherichia coli*. You add the phages to a lawn of bacteria and let the plates incubate for 24 hours before counting the number of plaques. This then got your lab partner thinking. She says, "If all these bacteriophages are in the sewer, they must have come from human waste flushed into the sewer. If they indeed did come from humans, why haven't the phage wiped out all the microbiota in the human digestive tract?" How would you respond to her question?
40. A person appears to have died of rabies. As a coroner, you need to verify that the cause of death was rabies. What procedures could you use to confirm the presence of rabies virus in the brain tissue of the deceased?
41. As a young genetic engineer with a biotech company, you have been given a virotherapy project. The hepatitis B virus is a dsDNA virus that during its replicative cycle often produces different forms of virus particles. The ones that are filamentous or spherical completely lack a viral genome and are thus defective; they are just assembled capsids but with the same spikes as the infective virions. The biotech company wants you to come up with a plan to use these defective particles for biotechnology applications. What might you include on your list of applications?

STEP D: QUESTIONS FOR THOUGHT AND DISCUSSION

42. If you were to stop 1,000 people on the street and ask if they recognize the term "virus," all would probably respond in the affirmative. If you were then to ask the people to describe a virus, you might hear answers like, "It's very small" or "It's a germ," or a host of other colorful but not very descriptive terms. As a student of microbiology, how would you describe a virus?
43. Oncogenes have been described in the recent literature as "Jekyll and Hyde genes." What factors may have led to this label, and what does it imply? In your view, is the name justified?
44. Researchers studying bacterial species that live in the oceans have long been troubled by the question of why these microorganisms have not saturated the oceanic environments. What might be a reason?
45. When discussing the multiplication of viruses, virologists prefer to call the process replication, rather than reproduction. Why do you think this is so? Would you agree with virologists that replication is the better term?
46. Bacterial species can cause disease by using their toxins to interfere with important body processes; by overcoming body defenses, such as phagocytosis; by using their enzymes to digest tissue cells; or other similar mechanisms. Viruses, by contrast, have no toxins and produce no digestive enzymes. How, then, do viruses cause disease?
47. How have revelations from studies on viruses, viroids, and prions complicated some of the traditional views about the principles of biology?

15

CHAPTER PREVIEW

15.1 Viruses Account for Most Upper Respiratory Tract Infections
Investigating the Microbial World 15: Is Obesity Infectious?

15.2 Viral Infections of the Lower Respiratory Tract Can Be Severe
MicroInquiry 15: Drifting and Shifting—How Influenza Viruses Evolve
Textbook Case 15: The 2002–2003 Outbreak of SARS

15.3 Herpesviruses Cause Several Human Skin Diseases

15.4 Several Other Viral Diseases Affect the Skin

Viral Infections of the Respiratory Tract and Skin

Sometime soon we will face a biological attack that has nothing to do with terrorists and everything to do with viruses mutating in the chicken farms of southeast Asia.
—*NewScientist magazine,* January 10, 2004

Every year there are seasonal flu outbreaks that may become **epidemic**; that is, they spread quickly through a large population. However, once in awhile there is a **pandemic**, a worldwide epidemic. The biggest influenza pandemic, called the "Spanish flu," occurred in 1918 to 1919. A fifth of the world's population was infected by a virus thought to have originated in birds. Estimates suggest between 20 and 100 million people died, young adults being the worst hit. As we will see in this chapter, such pandemics can occur when a new infectious virus appears for which the human population has no immunity.

Unfortunately, influenza continues to be an ongoing problem in the twenty-first century. Health experts and microbiologists have been saying that we are overdue for another serious and perhaps very deadly flu pandemic. In fact, one may be "brewing." An avian flu virus has been circulating throughout Asia and into Europe and Africa. By December 2011, this influenza virus had killed at least 332 people of the 566 infected in Asia and Southeast Asia since 2003. At this time, it is still difficult for people to contract the disease because it is spread mainly from infected fowl to humans. Human-to-human transmission has not yet been documented. However, it might only take a few mutations to transform the avian flu virus into a form that can become highly contagious, as the chapter opening quote suggests.

■ Contagious: Capable of being easily transmitted from one person to the next.

Another potential flu strain caught the world's attention in 2009. In April 2009, the Centers for Disease Control and Prevention (CDC) identified two children in California that had been infected with a new strain

Image courtesy of Dr. Fred Murphy/CDC.

of influenza virus. It was soon discovered that this strain was causing the major outbreak of flu in Mexico. By the end of April, cases of this new flu strain, originally called "swine flu" but eventually known as "novel H1N1," were being reported in cities around the world. The quick spread caused the World Health Organization (WHO) to raise the global flu pandemic alert to 5 (6 is the highest) and it appeared the first flu pandemic in over 40 years was emerging.

As the cases continued to mount up in the spring and early summer of 2009, the actual symptoms for most infected individuals remained relatively mild even though the WHO raised the pandemic alert to 6 in June. The fear was that there might be another second "flu wave" that would be much more devastating, which was the case for the Spanish flu. Luckily, that second wave never materialized. In fact, the death toll in the United States was estimated to be about 13,000 over the period, which was less than the 23,000 that usually die every year from the seasonal flu (FIGURE 15.1).

There were at least two major take-home lessons from this "mini-flu pandemic." One was

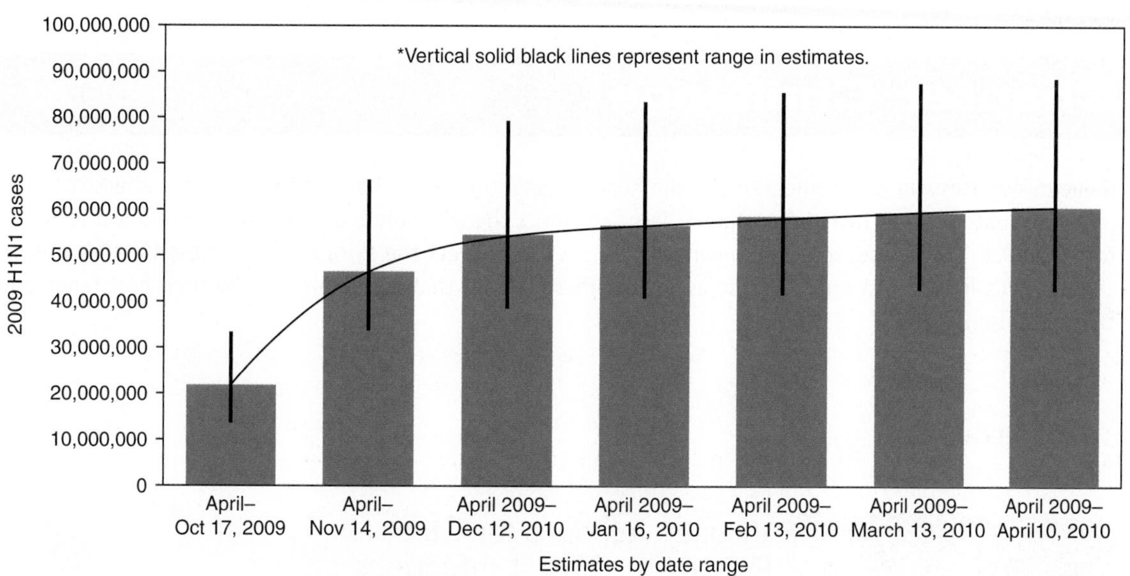

(A)

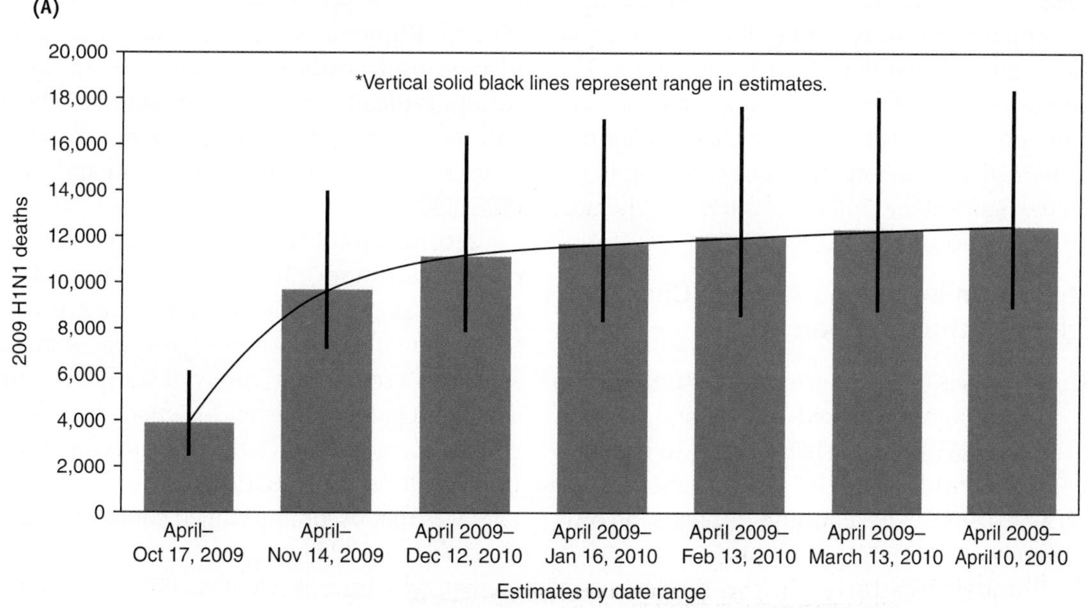

(B)

FIGURE 15.1 **Cumulative 2009 H1N1 Cases (A) and Deaths (B) in the United States by Date, April 2009–April 10.** The vertical black lines indicate the range in 2009 H1N1 estimates for each time period. Reproduced from CDC Estimates of 2009 H1N1 Influenza Cases, Hospitalizations and Deaths in the United States. »» From these graphs, when did (a) the greatest increase in 2009 H1N1 cases and (b) greatest increase in 2009 H1N1 deaths occur?

that in this age of globalization, new viral threats can come from anywhere and spread with the speed of jet travel. The second lesson was that we need to be on "virological alert" worldwide for new virus strains. Most flu outbreaks and epidemics start in Southeast Asia—this one surprised epidemiologists by coming from Mexico. So, this time we dodged the bullet—next time we may not be so lucky.

In this chapter we focus on several viral diseases, besides influenza, which affect the respiratory tract—the so-called "pneumotropic diseases." We also examine viral diseases affecting the skin—the "dermotropic diseases." Note: The division of the pneumotropic and dermotropic viral diseases is an artificial classification simply for grouping convenience. Therefore, the symptoms described may go beyond the respiratory tract or skin, respectively.

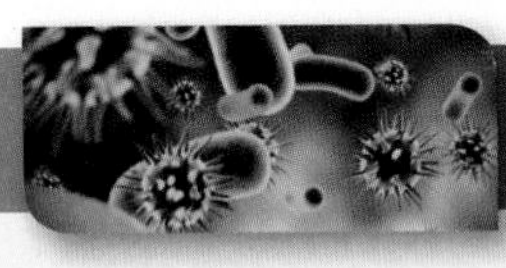

Chapter Challenge

Numerous viruses can cause infections in the human respiratory tract. Many of these viruses affect the upper respiratory tract (URT) only while others first infect the URT before progressing to the lower respiratory tract (LRT). Likewise, there are a variety of viruses that affect the skin. As we progress through this chapter, let's investigate some specific aspects of these viruses that are determined by their host range and tissue tropism.

KEY CONCEPT 15.1 Viruses Account for Most Upper Respiratory Tract Infections

The **upper respiratory tract (URT)** consists of the nose, sinuses, pharynx (throat), and larynx. The various versions of the common cold are the most common viral infections of the URT, accounting for more than 1 billion infections each year in the United States alone. In fact, it is the world's most common infectious disease.

Rhinovirus

Rhinovirus Infections Are the Chief Cause of the Common Cold

Rhinoviruses (*rhino* = "nose") are a broad group of over 100 different naked, icosahedral, +ssRNA viruses (FIGURE 15.2). They belong to the family Picornaviridae (*pico* = "small"; hence, small-RNA-viruses). The genome consists of only 10 genes.

■ Croup: An inflammation of the larynx and trachea, causing a cough, hoarseness, and breathing difficulties.

Rhinoviruses thrive in the human nose, where the temperature is a few degrees cooler (33°C) than in the rest of the body. They are transmitted through airborne droplets or by contact with an infected person or contaminated objects. Rhinoviruses account for 30% to 50% of **common colds**, also called "head colds." Adults typically suffer two or three colds and children six to eight colds per year. The rhinoviruses are most common in the fall and spring (FIGURE 15.3).

A common cold is a viral infection of the lining of the nose, sinuses, throat, and upper airways. One to 3 days after infection, a sequence of symptoms (common-cold syndrome) begins. These include sneezing, a sore throat, runny or stuffy nose, mild aches and pains, and a mild-to-moderate hacking cough. Some children suffer from croup. The illness usually lasts 7 to 10 days. One of the old wives tales was that becoming chilled could cause colds (MicroFocus 15.1).

So, why cannot scientists find a cure for the common cold? The answer is quite simple. There are more than 200 viruses and strains—some 100 rhinoviruses alone—that cause common colds. Therefore, it would seem impractical to develop

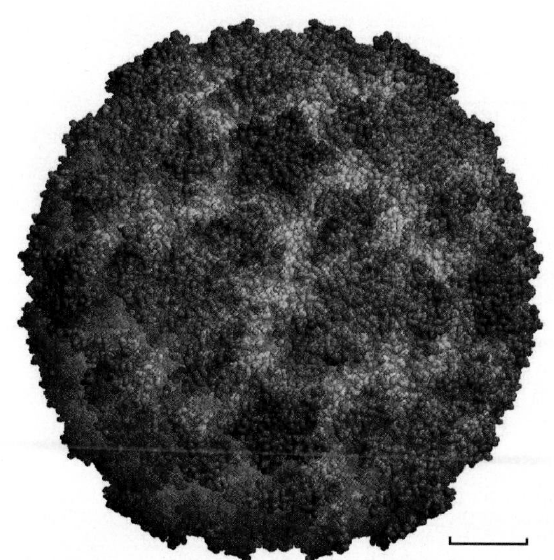

FIGURE 15.2 **A Rhinovirus.** A computer-generated image of the surface (capsid) of human rhinovirus 16. (Bar = 3 nm.) Image by J.Y. Sgro, UW-Madison. »» What is unique about the rhinoviruses that makes vaccine development impractical?

a vaccine to immunize people against all different types of rhinoviruses.

However, researchers have identified a set of core genes that are common to all rhinoviruses. If a vaccine can be made against a protein product from this core set, perhaps disease prevention will be possible.

One group of human **enteroviruses** can also cause a mild URT illness. These enteroviruses are also in the Picornaviridae family and closely related to the human rhinoviruses. However, one isolate, called EV68, is being increasingly recognized as sometimes causing severe illness in children and requiring hospitalization. Since 2008 to 2010, clusters of cases were reported in the Philippines, the Netherlands, and, in the United States, in Georgia, Pennsylvania, and Arizona.

Antibiotics will not prevent or cure a cold. Antihistamines can sometimes be used to treat the symptoms of a cold; however, they do not shorten the length of the illness. Because rhinoviruses spread by respiratory droplets, washing hands remains an important hygiene measure to decrease transmission of the viruses.

Adenovirus Infections Also Produce Symptoms Typical of a Common Cold

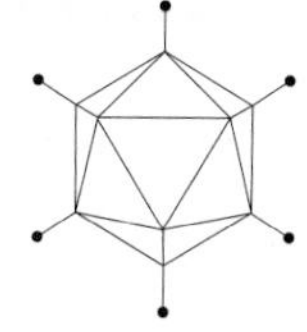

Adenoviruses (family Adenoviridae) are a group of over 50 types of nonenveloped, icosahedral virions having double-stranded DNA (FIGURE 15.4), which has the potential to encode 30 to 40 genes. The viruses take their name from the adenoid tissue from which they were first isolated.

■ Adenoid tissue: Refers to the pharyngeal tonsil located in the upper rear of the pharynx.

Some adenoviruses are frequent causes of acute URT infections often symptomatic of a common cold in infants and young children. Transmitted through respiratory droplets, the viruses most often cause distinctive symptoms because the fever is substantial, the throat is very sore (**acute** febrile **pharyngitis**), and the cough is usually severe. In addition, the lymph nodes of the neck swell and a whitish-gray material appears over the throat surface.

■ Febrile: Relates to fever.

Some adenoviruses may produce a form of conjunctivitis called **pharyngoconjunctival**

■ Conjunctivitis: An inflammation of the conjunctiva of the eye.

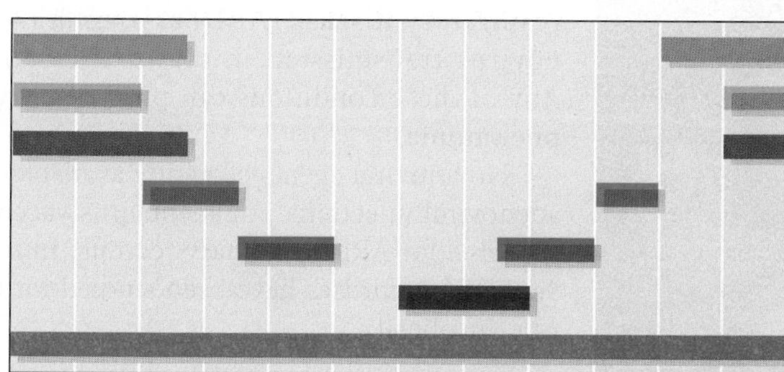

FIGURE 15.3 **The Seasonal Variation of Viral Respiratory Diseases.** This chart shows the seasons associated with various viral diseases of the respiratory tract (and their annual percentage). Enteroviruses cause diseases of the gastrointestinal tract as well as respiratory disorders. »» Hypothesize as to why different cold viruses cause diseases at different times of the year.

MICROFOCUS 15.1: Being Skeptical

Catching a Chill: Can It Cause a Cold?

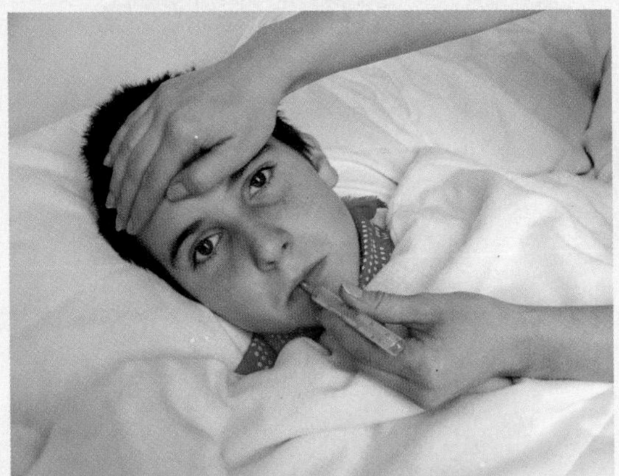

© Tihis/ShutterStock, Inc.

How many times can you remember your mom or a family member saying to you, "Bundle up or you will catch a cold!" Can you actually "catch" a cold from a body chill?

During the last 50 years of research, scientists have claimed that one cannot get a cold from a chill. Rather, colds result from more people being cooped up indoors during the winter months, making virus transmission from person-to-person very likely. At least, that was the scientific dogma until 2005.

In the November 2005 issue of *Family Practice,* British researchers Claire Johnson and Professor Ron Eccles at Cardiff University's Common Cold Center announced that a drop in body temperature can allow a cold to develop.

The researchers signed up 180 volunteers between October and March to participate in the study. Split into two groups, one group put their bare feet into bowls of ice cold water for 20 minutes. The other group put their bare feet in similar but empty bowls.

Over the next 5 days, 29% of individuals who had their feet chilled developed cold symptoms, while just 9% of the control group developed symptoms.

Professor Eccles suggests that colds may develop not because the volunteers actually "caught" a cold virus, but rather that they harbored the virus all along. Chilling lowers the person's immunity, providing the opportunity for the virus to produce more severe symptoms.

Chilling causes a constriction of blood vessels in the nose, limiting the warm blood flow supplying white blood cells to eliminate or control the cold viruses present. Professor Eccles states that, "A cold nose may be one of the major factors that causes common colds to be seasonal."

The verdict: Well, the results of one study do not make for a general consensus. More definitive studies may help verify or refute the "cold nose" claim.

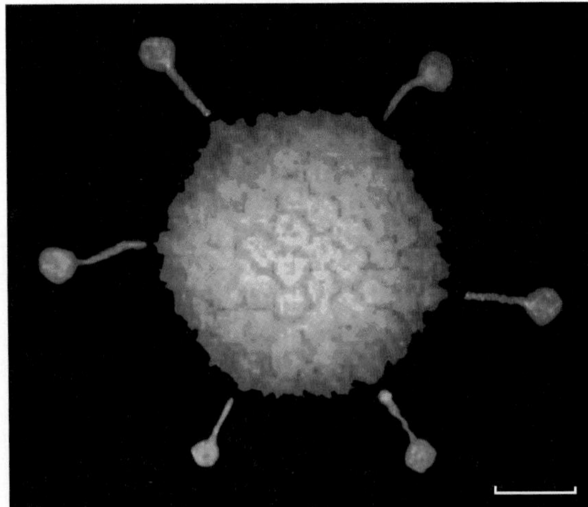

FIGURE 15.4 **An Adenovirus.** This false-color transmission electron micrograph of an adenovirus shows the icosahedral symmetry. The rough surface of the virus is due to the presence of capsomeres. (Bar = 25 nm.) © Dr. Linda Stannard, UCT/Photo Researchers, Inc. »» What are the projections from the capsid surface?

fever, which is most commonly contracted by swimming in virus-contaminated water. There were more than 42,000 cases reported in Japan in 2006. New military recruits may suffer **acute respiratory disease** (**ARD**) as a result of adenovirus transmission in crowded locations. Any of these conditions can progress to **viral pneumonia**.

No antiviral agents currently available treat adenoviral infections. An adenovirus vaccine is available for ARD in military recruits. In recent years, adenoviruses have been studied for their role in obesity (Investigating the Microbial World 15).

Although bacterial infections of the larynx are extremely rare, the most common cause of **laryngitis** is a viral infection of the URT caused by a common cold or flu virus. Symptoms are an unnatural change of voice, such as hoarseness,

Investigating the Microbial World 15

Is Obesity Infectious?

According to the Centers for Disease Control and Prevention, in the past 20 years there has been a dramatic increase in obesity in the United States. Today, more than one-third of U.S. adults (35.7%) and approximately 17% (or 12.5 million) of children and adolescents aged 2–19 years are obese. The reasons for this are many but most involve a poor diet and lack of exercise. Yet, if those were the only reasons, why are some individuals in our "obesigenic" environment less vulnerable to becoming obese?

- **OBSERVATIONS:** Besides diet and exercise, there is emerging evidence that viruses may be at least partly to blame for causing human obesity. In particular, adenovirus 36 (AD36) has been found to increase body fat in several different types of test animals. In addition, antibodies to AD36 (indicating a previous adenovirus infection) were found in many of these animals.
- **QUESTION:** ***Is AD36 antibody present in obese children and adolescents?***
- **HYPOTHESIS:** Obese children and adolescents (referred to as "children") are more likely to have AD36 antibodies than nonobese children. If so, then AD36 antibodies (serum antibody) should be more prevalent in obese children (AD36-positive) than nonobese children (AD36-negative).
- **EXPERIMENTAL DESIGN:** Children between the ages of 8 and 18 were recruited from primary care clinics. Those enrolled were classified according to their body mass index (BMI), which is calculated using the individuals, weight and height and is a reasonable indicator of body fatness for most children and teens. An obese individual was defined as having a BMI at or above the 95th percentile for children of the same age and sex and a normal (but possibly overweight) individual as having a BMI below the 95th percentile. Of 124 children enrolled, 67 were obese and 57 were normal weight.
- **EXPERIMENT 1:** A blood test to detect AD36 antibody was performed on each of the 124 participants. From those:
  - How many of the 67 obese children and 57 normal weight children were AD36-positive?
  - What was the median BMI for the AD36-positive and AD36-negative groups? A healthy BMI is 19-25.
- **EXPERIMENT 2:** Within the obese group of 67, a secondary analysis was performed to determine the degree of obesity (mean BMI) and weight difference within the AD36 subgroups.
- **RESULTS:** Differences between AD36-negative and AD36-positive children were not reflected in gender, race, or ethnicity.

  EXPERIMENT 1: See table A.

  EXPERIMENT 2: See table B.
- **CONCLUSIONS:**

  QUESTION ***1: Was the hypothesis validated? Explain using the tables.***

  QUESTION ***2: As a public citizen, would you be inclined to encourage researchers to look more at viruses as a factor contributing to obesity? Explain.***

  QUESTION ***3: Can you think of any problems or caveats in the design of these experiments? Hint: The BMI for healthy weight children in this age group has a range of about 16 to 23 and for overweight children a range of about 19 to 29.***

Note: If obesity does have an infectious component, it is not communicable!

Answers can be found on the Student Companion Website in **Appendix D.**

TABLE

A AD36 Results

Participants	AD36-Positive	AD36-Negative
Total = 124	19	105
• Of the 67 obese	15	52
• Of the 57 nonobese	4	53
• Median BMI	33.7	25.4

TABLE

B AD36-Specific Antibodies in Obese Subgroup

Participants	AD36-Positive	AD36-Negative
Total = 67	15	52
• Mean BMI	36.4	31.8
• Weight	AD36-positive averaged 16.1 kg heavier than AD36-negative	

Adapted from: Gabbert, C. et al. (2010). *Pediatrics* 126 (4):721–726.

TABLE

15.1 A Summary of the Major Viral URT Diseases

Disease	Causative Agent	Signs and Symptoms	Transmission	Treatment	Prevention
Common colds (rhinitis)	Rhinoviruses Adenoviruses Other viruses	Sneezing, sore throat, runny and stuffy nose, hacking cough	Respiratory droplets	Pain relievers Decongestants	Practicing good hygiene
Laryngitis	Cold viruses	Hoarseness Loss of voice	Respiratory droplets	Rest voice	Drinking plenty of water

or even loss of voice that develops within hours to a day or so after infection. The throat may tickle or feel raw, and a person may have a constant urge to clear the throat. Fever, malaise, difficulty swallowing, and a sore throat may accompany severe infections. Treatment of viral laryngitis depends on the symptoms. Resting the voice (by not speaking), drinking extra fluids, and inhaling steam relieve symptoms and help healing.

The viral diseases of the URT are summarized in TABLE 15.1.

CONCEPT AND REASONING CHECKS 1

a. Why do we get colds over and over again?

b. Why isn't an adenovirus vaccine available to the general public?

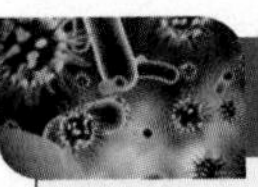

Chapter Challenge A

Most upper respiratory tract infections are caused by viruses. The result is a usually mild condition involving a cough and often fever, headache, and aches and pains.

QUESTION A:

What areas of the upper respiratory tract are most subject to infection and what is the best way to prevent such infections?

Answers can be found on the Student Companion Website in **Appendix D**.

KEY CONCEPT 15.2 Viral Infections of the Lower Respiratory Tract Can Be Severe

The **lower respiratory tract** (**LRT**) in humans consists of the trachea (windpipe), bronchial tubes, and the alveoli.

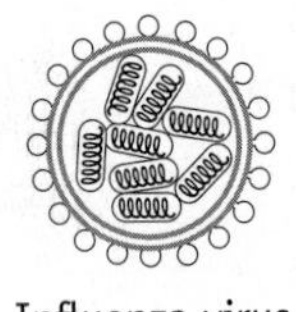

Influenza virus

Influenza Is a Highly Communicable Acute Respiratory Infection

Epidemiology. **Influenza** (the "flu") is a highly contagious and acute URT disease that is transmitted by airborne respiratory droplets. However, severe (complicated) cases involve the LRT as well. The disease is believed to take its name from the Italian word for "influence," a reference either to the influence of heavenly bodies, or to the *influenza di freddo,* "influence of the cold." Since the first recorded epidemic in 1510, scientists have described 31 pandemics. The most notable pandemic of the twentieth century was the "Spanish" flu in 1918–1919; others took place in 1957 (the "Asian" flu) and in 1968 (the "Hong Kong" flu). Today, there are about 23,000 deaths in the United States and 250,000 to 500,000 deaths worldwide annually related to seasonal influenza infections.

Causative Agent. There are three types of influenza viruses, fittingly named **influenza A**, **influenza B**, and **influenza C**. They all share similar structural features in that they are spherical shaped in culture, enveloped, and have a segmented genome consisting of several single-stranded (–strand) RNA molecules. Because influenza A is responsible for most cases of human flu (TABLE 15.2), the remaining discussion will center on this type.

The genome of the influenza A virus consists of eight (–ssRNA) segments, each wound helically and surrounded by a protein capsid (FIGURE 15.5). The entire nucleocapsid core is surrounded by a

TABLE
15.2 Comparison of the Influenza Viruses

Characteristic	Influenza A	Influenza B	Influenza C
Reservoir	Humans, pigs, birds	Only humans	Only humans
Epidemiology			
• Epidemics	Yes	Yes	No
• Pandemics	Yes	No	No
Occurrence	Seasonal	Seasonal	Nonseasonal
Antigenic changes	Drift/shift	Drift	Drift
Potential severity of illness	Severe	Moderate	Mild
Seasonal vaccine preparation	Two subtypes	One subtype	No vaccine
Genome	Eight segments	Eight segments	Seven segments

matrix protein, which in turn is surrounded by an envelope. Projecting through the envelope are some 500 protein spikes. About 80% of the spikes are **hemagglutinin** (**H**) proteins (**H spikes**) that facilitate the attachment and entry of influenza A into the epithelial host cells lining the nasopharynx and tracheobronchial tree. The rest of the spikes are **neuraminidase** (**N**) proteins (**N spikes**) that assist in the release of the newly replicated virions from the host cells.

Influenza A is divided into subtypes based on the 17 antigenically different types of H spike proteins (H1–H17) and 10 different N spike proteins (N1–N10). For example, the current seasonal flu subtypes in humans are A(2009 H1N1) and A(H3N2).

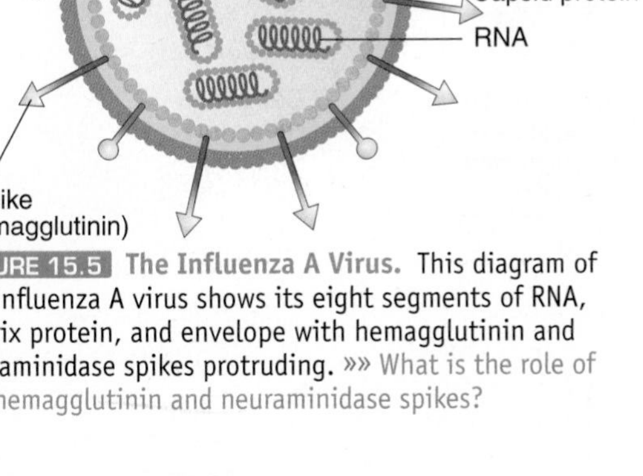

FIGURE 15.5 **The Influenza A Virus.** This diagram of the influenza A virus shows its eight segments of RNA, matrix protein, and envelope with hemagglutinin and neuraminidase spikes protruding. »» What is the role of the hemagglutinin and neuraminidase spikes?

Each year a slightly different seasonal flu strain evolves based, in part, on the changes to H and/or N spikes. Therefore, say you are immune to H1N1. If the next flu season has a H5N1 subtype, you are not protected from being infected because this subtype is antigenically different and not immediately recognized by your immune system. Sometimes new subtypes are quite mild and may not even cause an epidemic, while in other years the subtype may be more dangerous—and potentially pandemic. How do influenza A subtypes arise? MicroInquiry 15 examines their evolution.

■ Tracheobronchial tree: The trachea, bronchi, and bronchioles that form the airways leading to the lungs.

■ Antigenically: Referring to a substance that stimulates an antibody response by the immune system.

Clinical Presentation. Because some other viral agents and bacterial pathogens can produce an influenza-like respiratory illness, a diagnosis of influenza, if necessary, is based on several factors. This includes the pattern of spread in the community, observation of disease signs and symptoms, laboratory isolation of viruses, or hemagglutination of red blood cells. MicroFocus 15.2 examines the similarities and differences between the common cold and flu.

The onset of influenza A is abrupt after an incubation period of 1 to 4 days. The individual with an uncomplicated illness develops sudden chills, fatigue, headache, and pain most pronounced in the chest, back, and legs. Over a 24-hour period, body temperature can rise to 40°C, and a severe cough develops. Individuals may experience sore throat, nasal congestion, sneezing, and tight chest, the latter a probable reflection of viral invasion of tissues of the trachea and bronchi. Despite these severe symptoms, influenza is normally short-lived and has a favorable prognosis. The disease is self-limiting and usually resolves in 7 to 10 days.

MICROINQUIRY 15

Drifting and Shifting—How Influenza Viruses Evolve

Influenza viruses evolve in two different ways. Both involve **antigenic variation**, a process in which chemical and structural changes occur periodically in hemagglutinin (H) and neuraminidase (N) spike proteins (antigens), thereby yielding new virus strains.

Antigenic drift involves small changes to the virus (**Panel A** below). These changes involve minor point mutations resulting from RNA replication errors. The mutations will be expressed in the new virions produced. Although such mutations may be detrimental, on occasion one might confer an advantage to the virus, such as being resistant to a host's immune system. For example, a spike protein might have a subtle change in shape (that is, the structural shape has "drifted") so immunity to the flu virus is incomplete. This is what happens prior to most flu seasons. The virus spikes are different enough from the previous season that the host's antibodies fail to properly recognize the new strain. Both influenza A and B viruses can undergo antigenic drift.

Antigenic shift is an abrupt, major change in structure to influenza A viruses. Antigenic shift may give rise to new strains that can now jump to another species, including humans (**Panel B** below). (that is, the spike structure has "shifted") to which everyone is totally defenseless, and from which a pandemic may ensue (see figure, next page).

Two mechanisms account for antigenic shift.

The "Spanish" flu was the introduction of a completely new flu strain (H1N1) into the human population from birds. The H1N1 strain jumped directly to humans and adapted quickly to replicate efficiently in humans (**Panels B, C**).

The second mechanism involves "gene swapping" or reassortment between different flu viruses (**Panels B, D, E**). The 1957 influenza virus ("Asian" influenza; H2N2), for example, was a reassortment, where the human H1N1 virus acquired new spike genetic segments (H2 and N2) from an avian species. The 1968 influenza virus ("Hong Kong" influenza; H3N2) was the result of another reassortment; the human H2N2 strain acquired a new hemagglutinin genetic segment (H3) from another avian species.

In this second transmission mechanism, pigs usually are the reassortment "vessels" (or intermediate host) because they can be infected by both avian and human flu viruses. The 2009 swine flu is another example (**Panel F**).

Discussion Point

The current avian (or bird) flu (H5N1strain) is lethal to domestic fowl and is very inefficiently transmitted from fowl to human. Since 2003, of the approximately 600 reported human infections, 60% have died. In late 2011 two research groups, one at the University of Wisconsin, Madison and the other from the Erasmus Medical Center in the Netherlands, each identified the few mutations needed to make H5N1 readily transmissible to humans. Both reports were initially blocked from publication because critics feared that making the methodology and resulting mutation sequences available might give terrorists the "recipe" to make a "doomsday weapon" or the virus might escape from a research lab. Proponents for publication say knowing what mutations must occur would allow epidemiologists to "be on the lookout" for such naturally occurring mutations in H1N1 and to test the efficacy of existing vaccines and antiviral drugs. So what do you think? Should the papers be published? Note: Both papers have now been published.

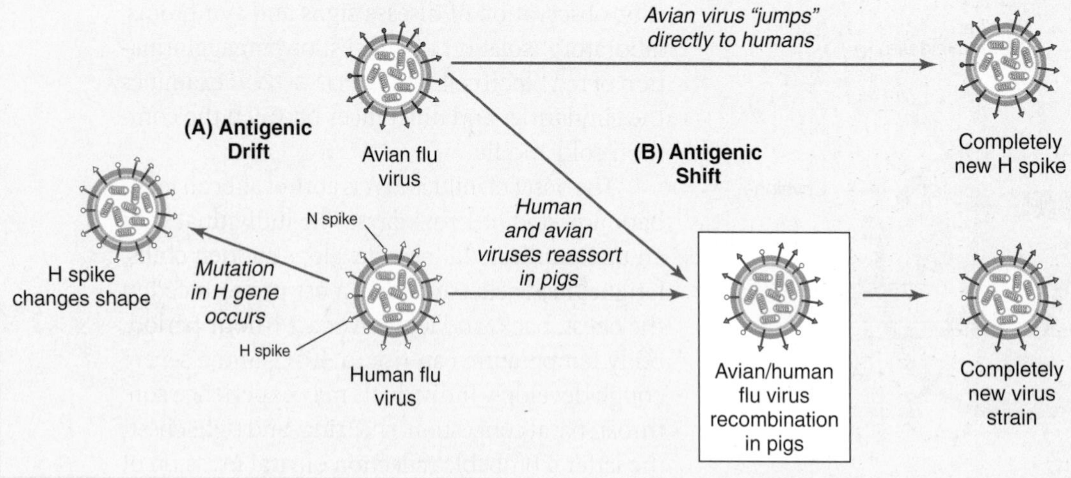

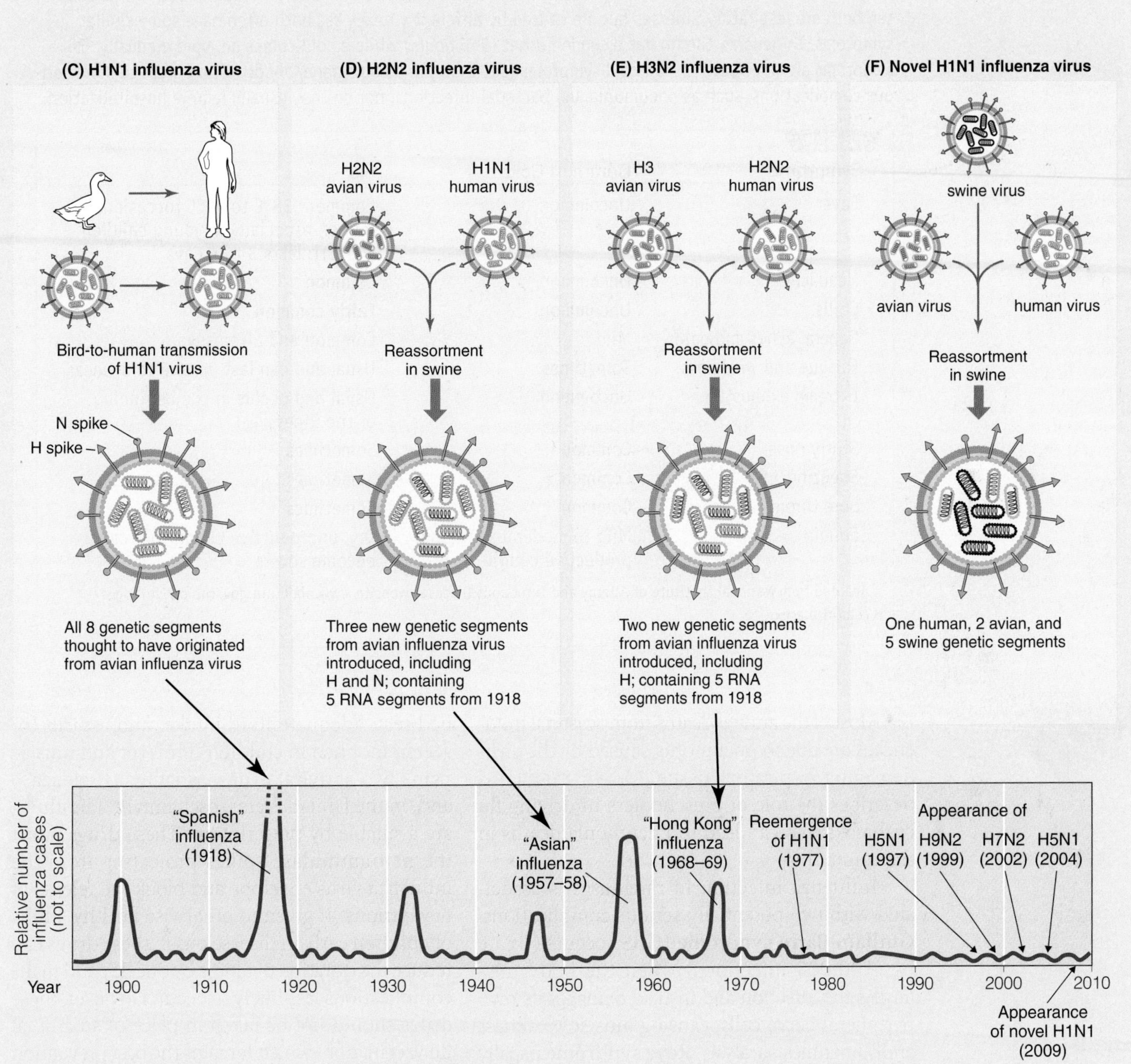

The Major Influenza Pandemics and Novel H1N1 Appearance in Humans. The four pandemics were the result of antigenic shifts.

MICROFOCUS 15.2: Public Health

Is It a Cold Or the Flu?

Do you know the differences in symptoms between a common cold and the flu? As described in this chapter, both are respiratory illnesses but are caused by different viruses. Yet both often have some similar symptoms. In general, the flu has a sudden onset (3–6 hours) while a cold comes on more gradually. Flu symptoms are more severe than cold symptoms (see table) and colds generally do not progress to more serious complications, such as pneumonia and bacterial infections, nor do they usually require hospitalization.

TABLE

Symptoms	Common Cold	Flu
Fever	Uncommon	Common; 38°C to 39°C [occasionally higher, especially in young children (40°C)]; lasts 3 to 4 days
Headache	Uncommon	Common
Chills	Uncommon	Fairly common
General aches and pains	Mild	Common and often severe
Fatigue and weakness	Sometimes	Usual and can last up to 2 to 3 weeks
Extreme exhaustion	Uncommon	Usual and occurs at the beginning of the illness
Stuffy nose	Common	Sometimes
Sneezing	Common	Sometimes
Sore throat	Common	Sometimes
Cough	Mild to moderate hacking, productive cough	Dry, unproductive cough that can become severe

Adapted from National Institute of Allergy and Infectious Diseases website www.niaid.nih.gov/topics/Flu/Pages/coldOrFlu.aspx.

Most of the annual deaths from seasonal influenza A are due to pneumonia caused by the virus damaging respiratory tissue. MICROFOCUS 15.3 describes the role of preschoolers in driving flu outbreaks that can lead to pneumonia deaths in the community.

Influenza infection in rare cases is associated with two potentially serious complications. **Guillain-Barré syndrome (GBS)** occurs (1 case per 100,000 infections) when the body mistargets the infection and instead damages its own peripheral nerve cells, causing muscle weakness and sometimes paralysis. **Reye syndrome** usually makes its appearance in young people after they are given aspirin to treat fever or pain associated with influenza. It begins with nausea and vomiting, but the progressive mental changes (such as confusion or delirium) may occur. Thankfully, very few children develop Reye syndrome (less than 0.03–1 case per 100,000 persons younger than 18 years).

Treatment and Prevention. Treatment of uncomplicated cases of influenza involves bed rest, adequate fluid intake, and aspirin (or acetaminophen in children) for fever and muscle pain. Two antiviral drugs, zanamivir (Relenza®) and, in the United States, oseltamivir (Tamiflu®), are available by prescription. These drugs target the neuraminidase spikes projecting from the influenza virus envelope and block the release of new virions. If given to otherwise healthy adults or children early in disease onset, these drugs can reduce the duration of illness by one day and make complications less likely to occur. However, these drugs should not be taken in place of an annual flu vaccination, which remains the best prevention strategy. Each year's batch of vaccine is based on the previous year's predominant influenza A and B viruses and is about 75% effective.

Some Paramyxovirus Infections Can Cause Serious Respiratory Disease

Several viruses of the Paramyxoviridae represent important human pathogens (FIGURE 15.6). All these viruses are enveloped, nonsegmented

MICROFOCUS 15.3: Public Health

Preschoolers Drive Flu Outbreaks

Every October to November in the Northern Hemisphere, we brace for another outbreak or epidemic spread of the seasonal flu. Many people go to their physician, clinic, or even grocery store for a flu shot. Others of us will "catch" the flu and suffer through several days of agony. For some 35,000, especially the elderly and immunocompromised, contracting the flu will lead to pneumonia and death. Wouldn't it be great if there was some way to predict pneumonia and flu deaths in a population? That now may be possible.

Researchers at the Children's Hospital of Boston and Harvard Medical School reported biosurveillance data suggesting that otherwise healthy preschoolers (3- to 4-year-olds) drive flu epidemics. The researchers found that by late September, kids in this age group were the first to develop flu symptoms.

Current immunization policies suggest that infants 6 to 23 months old be vaccinated against the flu because their immune systems are not yet fully developed. Policy also suggests that older children, including preschoolers, only be vaccinated if they have high medical-risk conditions. That means most will be vulnerable to the flu. And with many being in daycare and preschool, close contact makes spreading the flu effortless as those exposed bring the infection home.

The data also demonstrate that when preschoolers start sneezing, flu viruses are transmitted to the unvaccinated elderly who then become ill. Thus, the research suggests vaccinating those individuals who are driving and transmitting flu to others—the preschoolers. They are the sentinels by which an ensuing flu outbreak or epidemic can be identified. Immunizing preschoolers will decrease flu transmission and limit adult mortality in the unvaccinated.

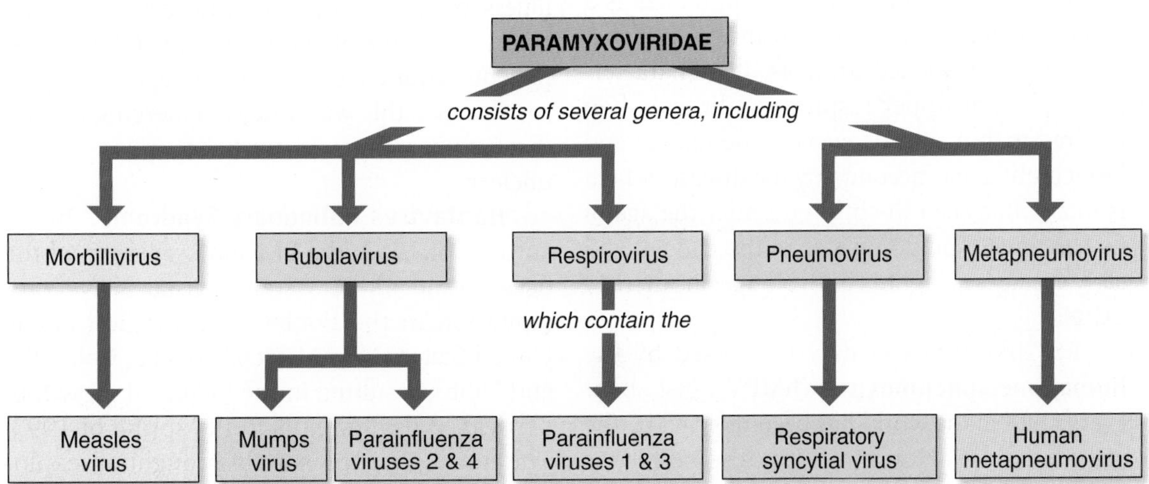

FIGURE 15.6 **The Relationships between the Human Paramyxoviruses.** This concept map illustrates the relationships between the paramyxoviruses that cause human respiratory and skin diseases. »» Which virus is most closely related to the respiratory syncytial virus?

–ssRNA viruses. The measles and mumps viruses will be discussed in the next section.

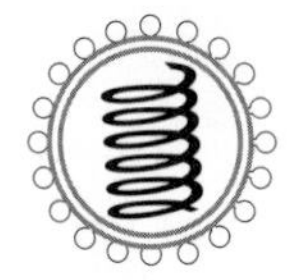

Respiratory syncytial virus

Respiratory syncytial (RS) disease is caused by the respiratory syncytial virus (RSV). Since 1985, RS disease has been the most common lower respiratory tract disease affecting infants and children. On a global scale, there are 33.8 million RSV infections and 200,000 deaths each year.

Some virologists believe that up to 95% of all children have been exposed to the disease by the age of 5, and CDC epidemiologists estimate there are 75,000 to 125,000 hospitalizations and 4,500 deaths in infants and children each year in the United States as a result of RS disease.

Community outbreaks of RSV are spread by respiratory droplets or virus-contaminated hands. Infection takes place in the bronchioles and air sacs of the lungs, and the disease is often described as **viral pneumonia**. When the virus infects tissue cells, the latter tend to fuse together, forming giant multinucleate cells called **syncytia**.

■ Bronchiole: A narrow tube inside of the lungs that branches off of the main air passages (bronchi).

Maternal antibodies passed from mother to child probably provide protection during the first few months of life, but the risk of infection increases as these antibodies disappear. Indeed, researchers have successfully used preparations of antibodies to lessen the severity of established cases of RS disease. Aerosolized ribavirin, an antiviral drug, also has been used with success.

Parainfluenza (*para* = "near") infections are caused primarily by human parainfluenza viruses 1 and 3 (see Figure 15.6). They account for 40% of acute respiratory infections in children. Although as widespread as influenza, parainfluenza is a much milder disease and is transmitted by direct contact or aerosolized droplets. It is characterized by minor upper respiratory illness, often referred to as a cold. However, pneumonia and bronchiolitis may accompany the disease, which is most often seen in children under the age of 6. The disease predominates in the late fall and early spring (see Figure 15.3). No specific therapy exists.

■ Bronchiolitis: An inflammation of the bronchioles.

RSV-like illnesses may be caused by the **human metapneumovirus (hMPV)**. Just about every child in the world has been infected by the virus by age 10. Human MPV is responsible for 12% of LRT infections and 15% of common colds in children. It appears to be milder than RS disease and, like RS disease, can be treated with ribavirin.

Other Respiratory Viruses Also Produce Pneumonia

A few other viruses can cause LRT infections in adults.

SARS. Severe acute respiratory syndrome (SARS) was first reported in southeastern China in spring 2003, and quickly spread through Southeast Asia and to 29 countries. It is an example of how fast an emerging disease can spread (Textbook Case 15).

Scientists at the CDC and other laboratories identified in SARS patients a previously unrecognized coronavirus, which they named the SARS coronavirus (SARS-CoV). Being a member of the Coronaviridae, it is a nonsegmented +ssRNA virus with helical symmetry and a spiked envelope (FIGURE 15.7).

SARS-CoV can be spread through close person-to-person contact by touching one's eyes, nose, or mouth after contact with the skin of someone with SARS. Spreading also comes from contact with objects contaminated through coughing or sneezing with infectious droplets by a SARS-infected individual. Whether SARS can spread through the air or in other ways remains to be discovered. Bats are the reservoir of SARS-CoV.

■ Reservoir: The location or organism where pathogens exist and maintain their ability for infection.

Many people remain asymptomatic after contacting SARS-CoV. However, in affected individuals, moderate URT illness may occur and include fever, headache, an overall feeling of discomfort, and body aches. After 2 to 7 days, SARS patients may develop a dry cough and have trouble breathing. In those patients progressing to a severe LRT illness, pneumonia develops with insufficient oxygen reaching the blood. In 10% to 20% of cases, patients require mechanical ventilation.

Because this was a newly emerging disease that has not reappeared, treatment options remain unclear.

Hantavirus Pulmonary Syndrome. In the autumn of 1992, the El Niño oscillation of the ocean-atmosphere system caused heavy precipitation in the Four Corners region of the United States (New Mexico, Arizona, Colorado, and Utah), resulting in the increased growth of berries, seeds, and nuts in the spring of 1993. The increased food supply brought an explosion in the rodent population in this area. Then, a cluster of sudden and unexplained deaths in previously healthy young adults occurred in

Textbook CASE 15

The 2002–2003 Outbreak of SARS

1 On February 11, 2003, the Chinese Ministry of Health notified the World Health Organization (WHO) of a mystery respiratory illness that had been occurring since November 2002 in Guangdong province in southern China. However, officials of China refused to allow WHO officials to investigate.

2 February 21, a 64-year-old doctor from Guangdong came to Hong Kong to attend a wedding. He stayed at a Hong Kong hotel, infecting 16 people who spread the disease to Hanoi, Vietnam; Singapore, and Toronto, Canada (see figure).

3 February 23, a Canadian woman tourist checked out of the same Hong Kong hotel and returned to Toronto, where her family greeted her. She died 10 days later as five family members were hospitalized.

4 February 28, a WHO doctor in Hanoi, Carlo Urbani, treated one of the people infected in Hong Kong and realized this was a new disease, which he named "severe acute respiratory syndrome" (SARS). He died of the disease 29 days later.

5 March 15, the WHO declared SARS a worldwide health threat. To block the chain of transmission, isolation and quarantine were instituted. Over half of those infected were healthcare workers.

6 April 16, the identification was made by 13 laboratories around the world that a new, previously unknown, coronavirus caused SARS.

7 June 30, the WHO announced there had been no new cases of SARS for 2 weeks. During the 114-day epidemic, more than 8,000 people from 29 countries were infected and 774 died.

Questions:

(Answers can be found on the Student Companion Website in **Appendix D.**)

A. Why might Chinese officials be reluctant to allow a WHO investigation into the mystery illness?

B. Justify the use of quarantine and isolation to break the chain of SARS transmission.

C. Propose an explanation as to why there was such a disproportionately high number of infections in healthcare workers.

D. How could such a large number of infections in healthcare workers have been prevented?

E. Why is SARS a textbook case for an emerging infectious disease?

For additional information see www.cdc.gov/mmwr/preview/mmwrhtml/mm5226a4.htm.

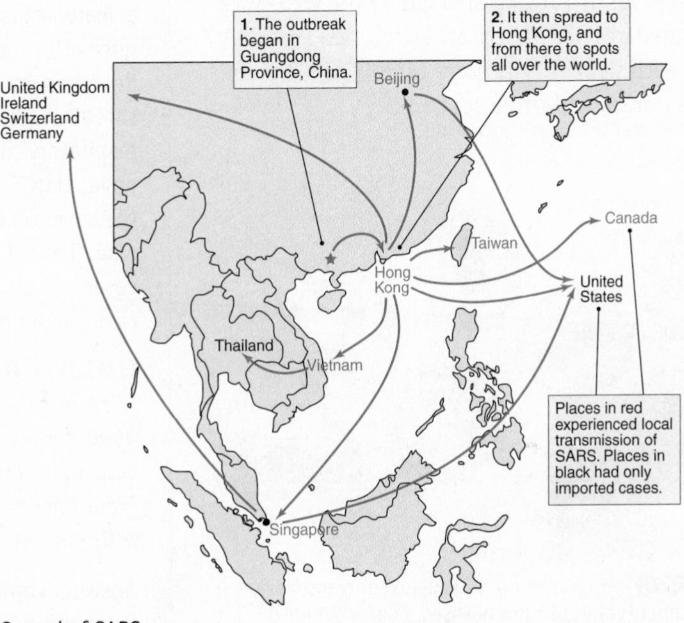

Spread of SARS.

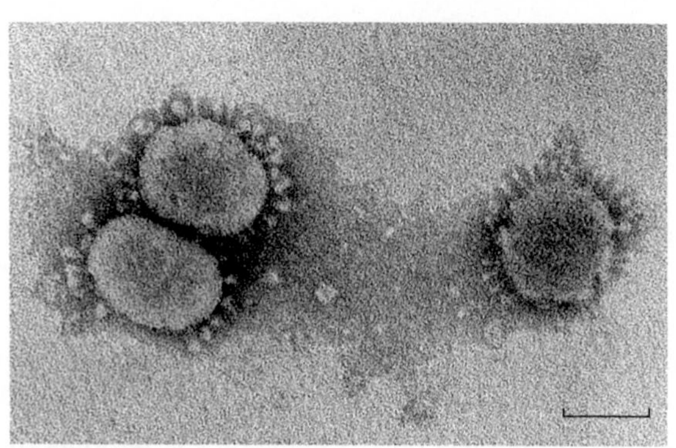

FIGURE 15.7 **Coronaviruses.** False-color transmission electron micrograph of three human coronaviruses. The spikes can be seen clearly extending from the viral envelope. Viruses similar to these are responsible for severe acute respiratory syndrome (SARS). (Bar = 100 nm.) © Dr. Gary D. Gaugler/Phototake/Alamy Images. »» By looking at the micrograph, explain why the virus is referred to as a "corona" virus.

rural New Mexico and the Four Corners region. The CDC identified a hantavirus, now called Sin Nombre virus (SNV), as the infectious agent and termed the pulmonary disease **hantavirus pulmonary syndrome (HPS)**. HPS has been identified in 31 states with more than 50% of reported cases outside the Four Corners region. Between 1993 and 2011, there have been more than 550 HPS cases reported in the United States. In September 2012, nine confirmed cases of HPS were reported in people who recently visited Yosemite National Park.

The hantaviruses are members of the Bunyaviridae. Their name is derived from the Hantaan River in South Korea where the virus was first isolated in 1978. These are enveloped, –ssRNA viruses with helical symmetry (FIGURE 15.8). The genome consists of three segments.

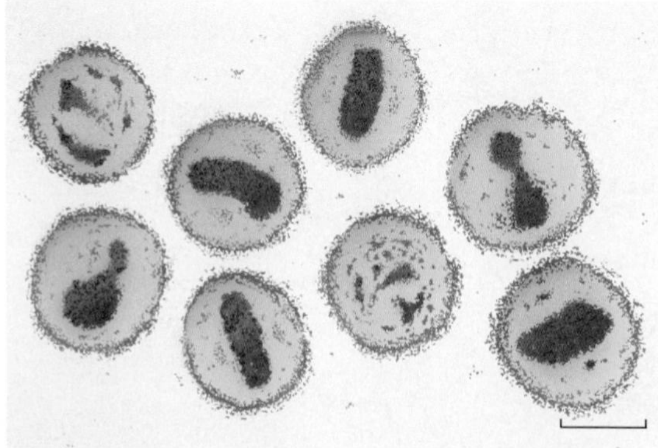

FIGURE 15.8 **The Hantavirus.** False-color transmission electron micrograph of hantaviruses. (Bar = 90 nm.) © Chris Bjornberg/Photo Researchers, Inc. »» What do the red geometric shapes represent in the hantaviruses?

The deer mouse is the reservoir for SNV, and it sheds the virus in saliva, urine, and feces. Humans usually are infected by breathing the infectious aerosolized dried urine or feces.

In 1 to 5 weeks after exposure, early symptoms of infection include fatigue, fever, and muscle aches. About half of all HPS patients experience headaches, dizziness, difficulty breathing, and low blood pressure that can lead to respiratory failure as the lungs fill with fluid.

Prevention consists of eliminating rodent nests and minimizing contact with them. There is no vaccine for hantavirus infection.

HPS has now become an established disease in the lexicon of medicine not only in the United States but throughout much of the Americas.

TABLE 15.3 summarizes the viral diseases affecting the LRT.

CONCEPT AND REASONING CHECKS 2

a. From MicroInquiry 15, why is the influenza A virus the cause of most flu epidemics and pandemics?
b. Identify the common relationships between the paramyxoviruses.
c. How do the SARS coronavirus and the hantavirus differ in their spread between individuals?

Chapter Challenge B

During the 2009–2010 novel H1N1 flu pandemic, French scientists reported that the novel H1N1 influenza A virus did not become a major cause of disease among the French until after the fall rhinovirus season declined. Researchers found that the percentage of throat swabs from French individuals with respiratory illness that tested positive for novel H1N1 declined in September, while at the same time rhinovirus, which causes colds, rose. Then, in late October as the rhinovirus cases declined, the number of novel H1N1 cases again rose.

QUESTION B:
With viral tissue tropism in mind, hypothesize why having a common cold may have protected these children from becoming infected with the influenza A virus.

Answers can be found on the Student Companion Website in **Appendix D**.

TABLE

15.3 A Summary of the Major Viral LRT Diseases

Disease	Causative Agent	Signs and Symptoms	Transmission	Treatment	Prevention
Influenza	Influenza A, B, and C viruses	Chills, fatigue, headache Chest, back, and leg pain	Respiratory droplets	Bed rest and fluids Antiviral medications (oseltamivir or zanamivir)	Getting annual flu vaccination
Respiratory syncytial (RS) disease	Respiratory syncytial virus (RSV)	Influenza-like	Respiratory droplets Hand contact	Fever-reducing medications Ribavirin for severe cases	Practicing good hygiene
Parainfluenza	Human parainfluenza viruses 1 and 3	Cold-like	Respiratory droplets Direct contact	No specific therapy	Practicing good hygiene
RSV-like illness	Human metapneumovirus	Cold-like	Respiratory droplets Direct contact	Fever-reducing medications Ribavirin for severe cases	Practicing good hygiene
SARS	SARS coronavirus (SARS CoV)	Fever, headache, body aches, dry cough, and breathing difficulty	Respiratory droplets and airborne particles Direct contact	No effective treatment	Practicing good hygiene
Hantavirus pulmonary syndrome	Hantavirus (Sin Nombre virus)	Fatigue, fever, muscle aches, headache, dizziness, breathing difficulty	Aerosolized droplets of rodent saliva, urine, feces	Supportive care	Eliminating rodent nests Minimizing contact

KEY CONCEPT 15.3 Herpesviruses Cause Several Human Skin Diseases

Several viral diseases of the skin are caused by members of the Herpesviridae. Some, such as herpes simplex, remain epidemic in contemporary times; others such as chickenpox are being brought under control through effective vaccination programs. However, there are relatively few antiviral drugs for, and prevention programs remain a major course of action in, dealing with these diseases.

Presently, there are eight known viral species in the family Herpesviridae that infect humans (FIGURE 15.9). However, the small number of viruses should not be an indication of their significance. For example, some virologists believe over 90% of Americans have been exposed to the herpes simplex virus (HSV) by age 18 and more than 90% of adults worldwide have been infected with the Epstein-Barr virus (EBV). In addition, genital herpes is one of the most common sexually transmitted diseases (STDs) in the world. The word *herpes* is Greek for "creeping," referring to the spreading of herpes infections through the body after contact has been made.

All human herpesviruses are large virions with a nonsegmented, double-stranded DNA genome. They have icosahedral symmetry and an envelope with spikes. Another shared characteristic is their ability to establish a latent infection and then reactivate at some later period. The herpesviruses can infect several types of cells in epithelial and neural tissue, causing a variety of skin and related diseases.

Human Herpes Simplex Infections Are Widespread and Often Recurrent

Two of the most prevalent herpesviruses are herpes simplex viruses 1 and 2 (HSV-1, HSV-2; FIGURE 15.10A).

Herpes simplex virus

Cold sores are caused primarily by HSV-1 and they are contagious. Such sores and blisters of herpes infections have been known for centuries. In ancient Rome, an epidemic was so bad

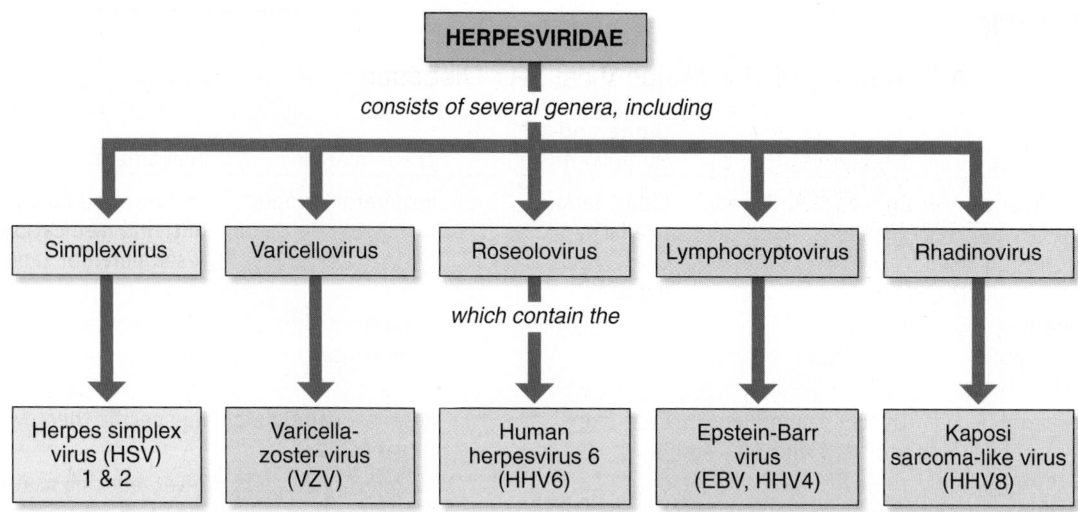

FIGURE 15.9 **The Relationships Between the Human Herpesviruses.** This concept map illustrates the relationships between six of the eight viral species in the family Herpesviridae that cause human disease. Only EBV and HHV8 are known to be oncogenic. »» What does it mean for a virus to be oncogenic?

that the Emperor Tiberius banned kissing and Shakespeare, in *Romeo and Juliet,* writes of "blisters o'er ladies lips."

Cold sores, also called fever blisters or **herpes simplex** labialis, start as a tingling sensation and the presence of a small, hard spot on the lip. Within a couple of days red, fluid-filled blisters appear (FIGURE 15.10B). The blisters eventually break, releasing the fluid containing infectious virions. A yellow crust forms and then peels off without leaving a scar. Cold sores generally clear up without treatment in 7 to 10 days. A person is most likely to transmit the infection from the time the blisters appear until they have completely dried and crusted over.

■ Ganglia: Structures containing a dense cluster of nerve cells.

After the acute (primary) infection, the viruses enter the sensory neurons and become latent in the nearby sensory ganglia. Viral reactivation and movement to the epithelia can trigger another round of cold sores. Reactivation of the dormant viruses often occurs after some form of trauma, often in response to stressful

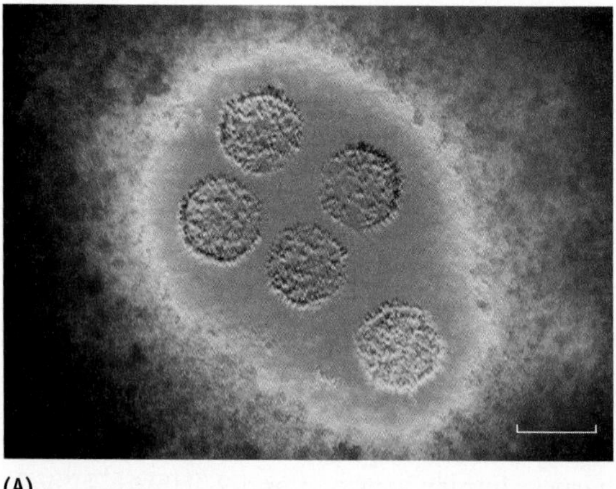

(A)

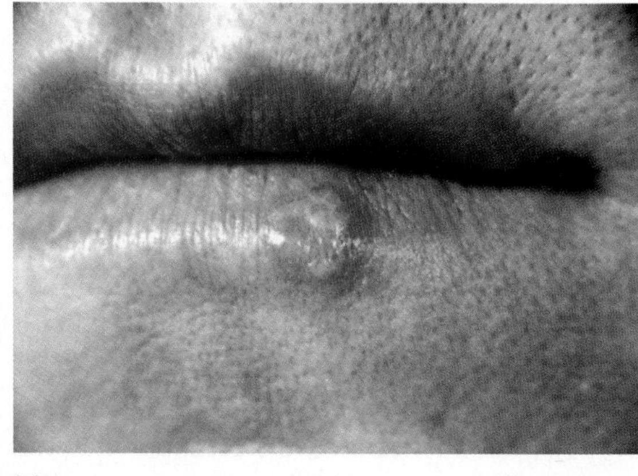

(B)

FIGURE 15.10 **Herpes Simplex Virus and Cold Sores.** (**A**) A false-color transmission electron micrograph of a cell infected with herpes simplex viruses. (Bar = 150 nm.) © Phototake/Alamy Images. (**B**) The cold sores (fever blisters) of herpes simplex erupting as tender, itchy papules and progressing to vesicles that burst, drain, and form scabs. Contact with the sores accounts for spread of the virus. Courtesy of Dr. Hermann/CDC. »» How would the release phase of the HSV replication cycle lead to blisters and vesicles on the lips?

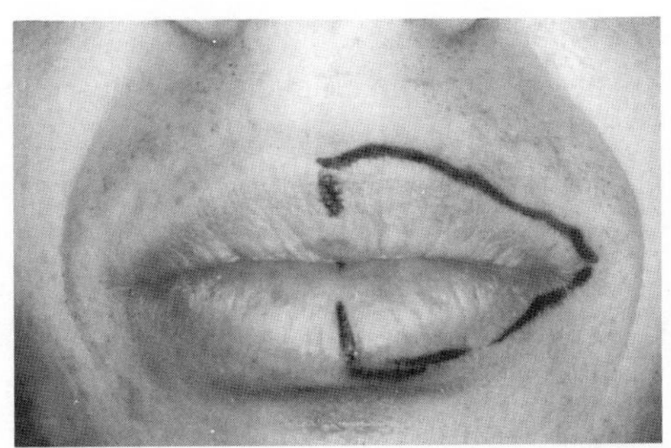

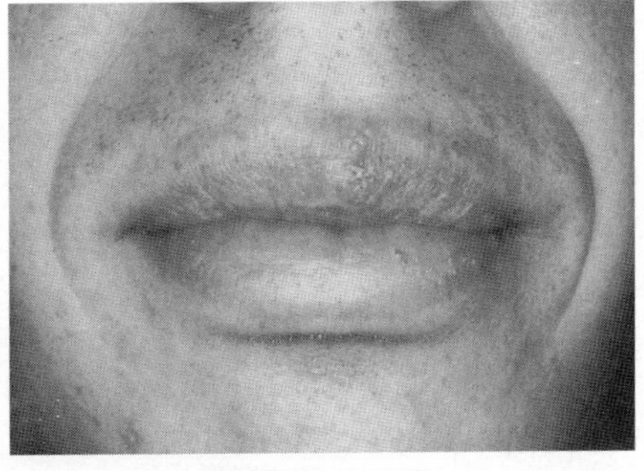

(A) (B)

FIGURE 15.11 **Ultraviolet Light and Herpes.** An experiment showing the effects of ultraviolet (UV) light on the formation of herpes simplex sores of the lips. This patient usually experienced sores on the left upper lip. (**A**) The patient was exposed to UV light from a retail cosmetic sunlamp on the left upper and lower lips in the area designated by the line. The remainder of the face was protected by a sunscreen. (**B**) Sores formed on the left upper lip. Reproduced from *J. Clin. Microbiol.*, 1985, vol. 22, pp. 366–368, DOI and reproduced with permission from the American Society for Microbiology. Photo courtesy of Woody Spruance, M.D, Professor of Internal Medicine, at University of Utah. »» What would you conclude from these experiments regarding the reactivation of herpesviruses?

triggers, such as fever, menstruation, or emotional disturbance. Even environmental factors like sunburn (exposure to ultraviolet light) can trigger reactivation (FIGURE 15.11). In rare cases, reactivation can lead to an infection of the brain (**herpes encephalitis**).

Preventing the transmission of cold sores means not kissing others while the blisters are present. Washing one's hands often and not touching other areas of the body also help limit virus spread. For example, the eyes can become infected through touching with a contaminated finger. Infection of the eye, called **herpes keratitis**, causes scarring of the cornea and is a leading cause of blindness in the United States. Some 400,000 Americans suffer from a form of ocular herpes.

Genital Herpes. Although it is not a notifiable disease, the CDC estimates that over 45 million Americans (14 to 49 years of age) are infected with **genital herpes**, with a slightly higher rate among women. The infection rate is continuing to increase with a 30% jump over the last three decades, accounting for more than 600,000 new cases diagnosed each year. Alarmingly, about 80% of individuals do not know they are infected.

Signs of a primary (acute) infection generally appear within a few days of sexual contact, often as itching or throbbing in the genital area. This is followed by reddening and swelling of a small area where painful, thin blisters erupt on the vulvar or penile skin (FIGURE 15.12). The blisters crust over and the sores disappear, usually within about 3 weeks. The signs often are very mild and may go unnoticed. Fifty percent of infected individuals experience only one outbreak in their lifetime.

However, for many infected individuals, a latent period occurs during which time the virus remains in nerve cells near the blisters. In the majority of cases, the virus becomes reactivated again by stressful situations and symptoms reappear. The cycle of latency and recurrent

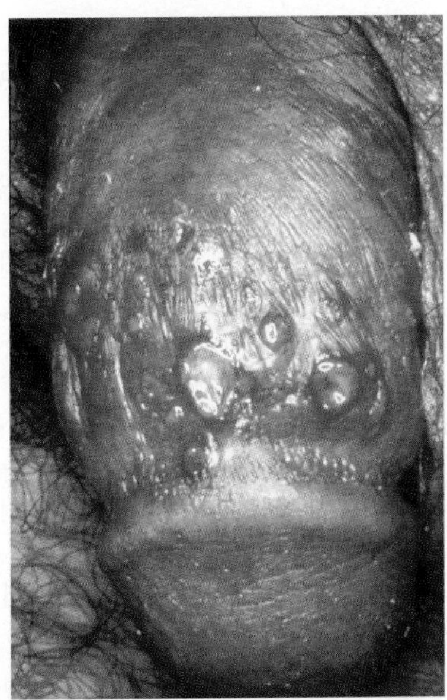

FIGURE 15.12 **Genital Herpes Rash on the Penile Shaft.** Signs of male genital herpes usually appear as blisters on or around the genitals or rectum. Courtesy of Dr. N.J. Flumara and Dr. Gavin Hart/CDC. »» How might this male patient have gotten genital herpes?

infections can occur three to eight times a year. People with active herpes lesions are highly infectious and can pass the viruses to others during sexual contact.

In healthy individuals, genital herpes usually causes no serious complications. Importantly though, a person with the infection has an increased risk of transmitting or contracting other sexually transmitted diseases, including AIDS.

Prevention is similar to that for any STD—abstain from sexual activity or limit sexual contact to only one person who is infection-free. Although there is no cure for genital herpes, antiviral drugs such as acyclovir can help heal the sores sooner and reduce the frequency of recurrent infections.

Neonatal herpes is a devastating and life-threatening disease transmitted by infected mothers to newborns during childbirth. Although most pregnant women with genital herpes have healthy babies, a small number spread the infection to their newborns during labor and delivery. Even with adequate treatment for the infected newborn, mental development can be delayed, blindness can occur, and persistent seizures may result. If symptoms and a diagnosis of an active genital herpes infection are made prior to delivery, the obstetrician may recommend birth by cesarean section.

Chickenpox Is No Longer as Prevalent a Disease in the United States

In the centuries when pox diseases regularly swept across Europe and other parts of the world, people had to contend with the Great Pox (syphilis), the smallpox, the cowpox, and the chickenpox. Today, chickenpox, and its associated disease shingles, still causes significant infectious disease.

Causative Agent and Epidemiology. Chickenpox and shingles are two different diseases caused by the same virus called the varicella-zoster virus (VZV), another virus in the Herpesviridae. **Chickenpox (varicella)** is a highly communicable disease that usually attacks just once in a lifetime and most often in children. **Shingles (herpes zoster)** is usually an adult disease. Anyone who has had chickenpox as a child is at risk of developing shingles later in life from the latent VZV infection. A person with an active case of shingles can pass VZV to anyone who has not had chickenpox; however, that person will develop chickenpox, not shingles.

■ Dorsal root ganglion: A group of nerve cell bodies of sensory neurons that lie along the vertebral column of the spine.

Prior to the availability of a vaccine in 1995, about 4 million children contracted chickenpox each year in the United States. In 2010, the CDC reported some 15,000 cases. Shingles develops in only about 10% of adults having a latent VZV infection. Still, the CDC reports that there are more than 1 million cases of shingles each year with about 50% of those cases occurring in men and women 60 years of age or older.

Clinical Presentations. VZV is transmitted by respiratory droplets and skin contact, and it has an incubation period of about 2 weeks. The disease begins in the respiratory tract, with fever, headache, and malaise. Viruses then pass into the bloodstream and localize in the peripheral nerves and skin. As it multiplies in the cutaneous tissues, VZV produces a red, itchy rash on the face, scalp, chest, and back, although it can spread across the entire body. The rash quickly turns into small, teardrop-shaped, fluid-filled vesicles (FIGURE 15.13A). The vesicles in chickenpox develop over 3 or 4 days in a succession of "crops." They itch intensely and eventually break open to yield highly infectious virus-laden fluid. In chickenpox, the vesicles form crusts that fall off. A person who has chickenpox can transmit the virus for up to 48 hours before the telltale rash appears and remains contagious until all spots crust over. Chickenpox usually lasts 2 weeks or less and rarely causes complications.

The most common complication of chickenpox in adults is a bacterial infection of the skin, pneumonia, or an inflammation of the brain (encephalitis). Reye syndrome may occur during the recovery period, so aspirin should not be used to reduce fever in both children and adults.

After having chickenpox, some VZV remains dormant (latent) in the dorsal root ganglia. Many years later, the virus can reactivate, travel down the nerves to the skin of the body trunk, and resurface as shingles (FIGURE 15.13B). Here, blisters with blotchy patches of red appear that encircle the trunk (*zoster* = "girdle"). Many sufferers also experience a series of headaches as well as facial paralysis and sharp "ice-pick" pains described as among the most debilitating known. The condition can occur repeatedly and is linked to emotional and physical stress (such as radia-

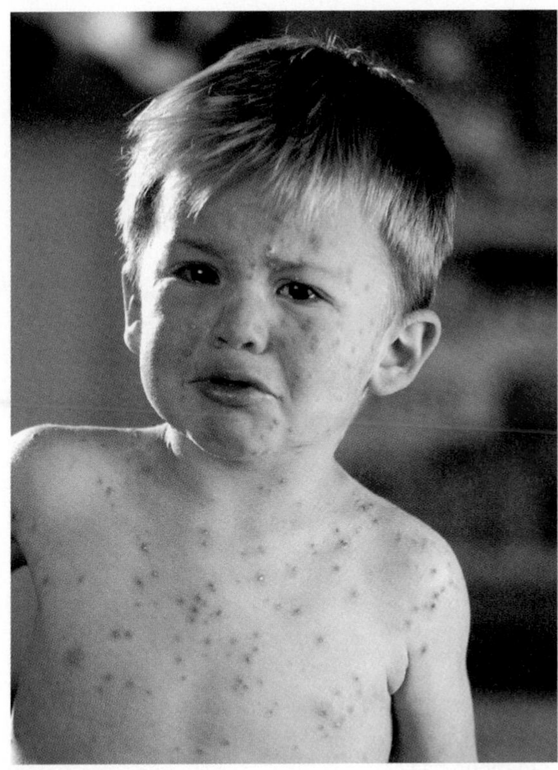

(A)

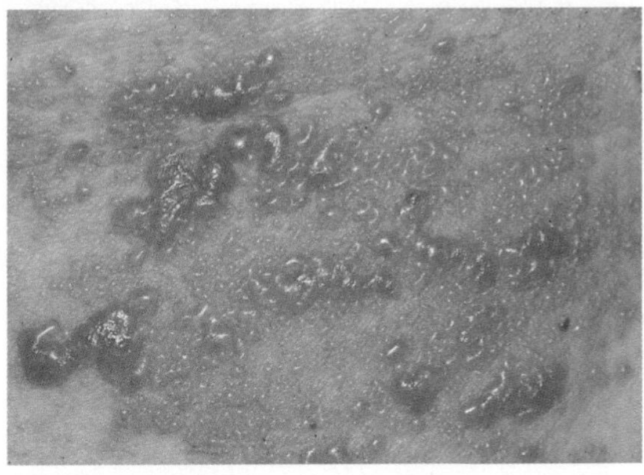

(B)

FIGURE 15.13 **The Lesions of Chickenpox and Shingles.** **(A)** A typical case of chickenpox. The lesions may be seen in various stages, with some in the early stage of development and others in the crust stage. © SW Productions/age fotostock. **(B)** Dermal distribution of shingles lesions on the skin of the body trunk. The lesions contain less fluid than in chickenpox and occur in patches as red, raised blotches. © Science Photo Library/Photo Researchers, Inc. »» Summarize how varicella latency could lead to shingles.

tion therapy) as well as to a suppressed immune system or aging.

Shingles can lead to its own complication—a condition in which the pain of shingles persists for years after the blisters disappear. This complication, called **postherpetic** neuralgia, can be severe. Most cases occur in people over age 60.

Treatment and Prevention. Acyclovir has been used in high-risk groups to lessen the symptoms of chickenpox and hasten recovery. A varicella-zoster immune globulin (VZIG), which contains antibodies to the chickenpox virus, may also be used. The chickenpox vaccine (Varivax®) is the safest, most effective way to prevent chickenpox and its possible complications. The varicella vaccine is 85% effective in disease prevention and appears to lessen the risk of shingles later in life. In 2006, a live, attenuated vaccine, called Zostavax®, was approved that can reduce the incidence of shingles by more than 50% and the number of postherpetic neuralgia by more than 60% in people over 60 years of age. Acyclovir therapy also lessens the symptoms. Some unconventional methods might also help (MicroFocus 15.4).

Other Herpesviruses Also Cause Human Disease

Human herpesvirus 6 (HHV-6) belongs to a different genus of herpesviruses from HSV-1 and HSV-2 (see Figure 15.9). HHV-6 primarily affects infants and young children 6 months to 3 years old. The virus causes **roseola**, an acute, self-limiting condition marked by high fever. This often is followed by a red body rash (FIGURE 15.14). Roseola can spread from person to person through contact with an infected person's respiratory secretions or saliva. The infection usually lasts about 1 week and treatment requires bed rest, fluids, and fever-reducing (nonaspirin) medications.

Some 30% to 60% of bone marrow transplant recipients suffer a HHV-6 viremia during the first few weeks after transplantation. The recurrence can potentially lead to pneumonia or encephalitis. No drugs have been approved for HHV-6 infections.

■ Viremia: The presence of viruses in the blood.

Kaposi sarcoma (KS) is a highly angiogenic tumor of the blood vessel walls. It is most commonly seen in individuals with a weakened immune system, such as those with HIV/AIDS. The malignancy is caused by human herpesvirus 8 (HHV-8) or Kaposi sarcoma–associated herpesvirus. KS, which is marked by dark or purple skin lesions, has become one of the most common tumors in AIDS patients (FIGURE 15.15).

■ Angiogenic: Referring to the generation of many blood vessels.

■ Neuralgia: Severe pain in a part of the body through which a particular nerve runs.

MICROFOCUS 15.4: Being Skeptical

Can Chinese Tai Chi Prevent Shingles?

© Leah-Anne Thompson/ ShutterStock, Inc.

We all know that exercise is a good thing. It builds strength, endurance, and cardiovascular health. One Chinese form of physical and aerobic exercise for attaining good health is through Tai Chi. Tai Chi involves a series of 20 slow and deliberate body movements that purport to produce a calm and tranquil mind. In 2003, researchers announced that Tai Chi Chih, a low-impact form of Tai Chi, boosts shingles immunity in the elderly. Can Tai Chi actually do this?

The theory goes that when a person's immunity is compromised or wanes as one ages, the individual is susceptible to more infections. For example, it is true that individuals over 50 are more prone to recurrent shingles attacks as the varicella-zoster virus (VZV) reactivates because it may no longer be kept in check by the immune system. Stress and other trauma also can intensify the reactivation. Therefore, Michael R. Irwin and colleagues at UCLA's Neuropsychiatric Institute decided to examine if Tai Chi could reduce the levels of stress and thereby boost a person's immunity—and reduce the chances for shingles.

Thirty-six men and women over 60 were enrolled in the study. Half were started on a 15-week program of Tai Chi Chih. The other half were asked to postpone the program for 15 weeks.

After the 15-week period, analysis indicated the Tai Chi group had a 50% increase over the control group in immune memory function for the shingles virus. The researchers state the increase was sufficient to actually help prevent shingles from reoccurring. Because Tai Chi did not improve physical movement within the group, Irwin speculates that Tai Chi's calming influence reduces stress levels, which could boost immune system function—including keeping VZV in check.

Such studies are part of a growing field called "psychoneuroimmunology," which looks at the relationships between the nervous system and the immune system. Professor Irwin believes if memory to the VZV can be maintained, memory to other infections might also be maintained as one ages.

The verdict: This was a small study involving only 36 volunteers. A more thorough study would be to follow groups of Tai Chi Chih volunteers and monitor how many actually report a flare-up of shingles compared to the general population. Still, this study does support the mind-body connection as a way to maintain good health.

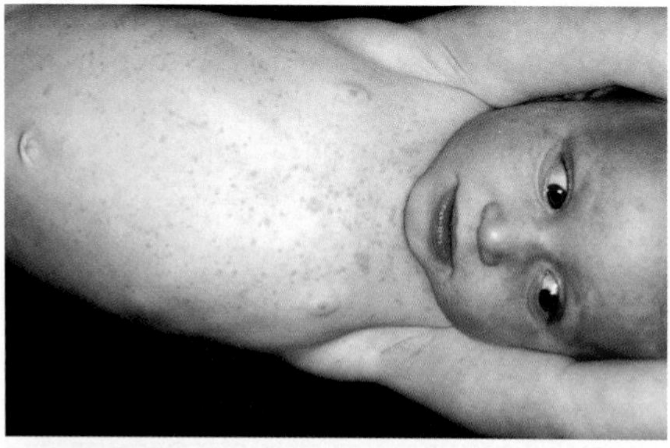

FIGURE 15.14 **Roseola.** Following a high fever, a red body rash may appear on the neck and trunk of infants and young children. © Scott Camazine & Sue Trainor/Photo Researchers, Inc. »» Why should aspirin not be given to infants or young children suffering a viral infection such as roseola?

Treatment of smaller skin lesions involves the use of liquid nitrogen, low-dose radiation, or chemotherapy applied directly to the lesion.

TABLE 15.4 summarizes the characteristics of the herpesviruses that affect the skin.

CONCEPT AND REASONING CHECKS 3

a. Explain how a primary infection and a latent infection differ for genital herpes.
b. Assess the consequences of someone having shingles being able to transmit the virus to a susceptible person.
c. Determine why a bone-marrow transplant recipient is at risk of contracting an HHV-6 infection.
d. Although all herpesviruses discussed in this section appear to have a latent state, why is only HHV-8 potentially carcinogenic?

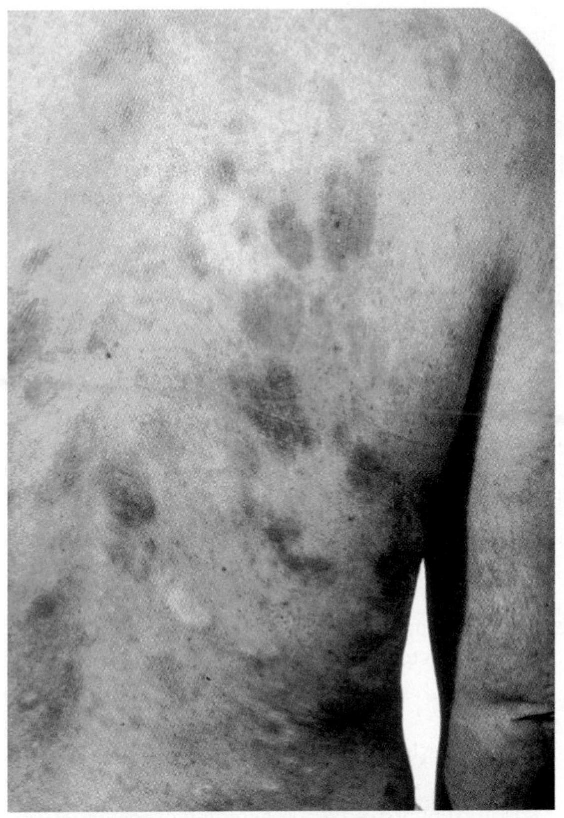

FIGURE 15.15 **Kaposi Sarcoma.** This man is suffering from AIDS. Kaposi sarcoma is an opportunistic disease, appearing as dark lesions on the skin. Courtesy of National Cancer Institute. »» What is meant by an opportunistic disease?

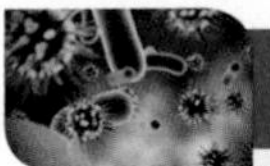

Chapter Challenge C

One of the most common oral herpes virus infections is herpes labialis, or more commonly called cold sores or fever blisters. The oral infection is caused by herpes simplex virus type 1 (HSV-1) and most people in the United States are infected with this virus by age 20. Yet, one does not need to have a cold or a fever to develop cold sores or fever blisters.

QUESTION C:

Based on the mechanism for reactivation of latent HSV-1, provide an explanation as to how these oral lesions got the name of "cold sores" or "fever blisters."

Answers can be found on the Student Companion Website in **Appendix D.**

KEY CONCEPT 15.4 Several Other Viral Diseases Affect the Skin

There are a number of other viral infections and diseases associated with the skin. Some are common childhood diseases, such as measles, while another, smallpox, has been eradicated but was once a worldwide scourge.

A Few Viruses Cause Typical Childhood Illnesses

Measles (**rubeola**) is caused by the measles virus, a single-stranded (–strand) RNA virus in the Paramyxoviridae family (see Figure 15.6). The disease usually occurs in children where it is highly contagious with an infection rate approaching 95%. In the United States, prior to the development of a measles vaccine, some 4 million children came down with measles annually, resulting in about 500 deaths and another 1,000 children experiencing chronic, life-long disabilities. In 2000, measles transmission was eliminated in the United States and has been maintained through childhood immunizations with the measles-mumps-rubella (MMR) vaccine. In 2010, there were 63 reported cases. Most of these outbreaks were import associated and transmitted by respiratory droplets from infected visitors (or returning residents) to unvaccinated children. Globally, the WHO reports that measles deaths worldwide have fallen by almost 75% since 2000 due to vaccination efforts.

Symptoms commonly include a hacking cough, runny nose, eye redness, and a high fever. Diagnostic red patches with white grain-like centers (**Koplik spots**) appear along the gum line in the mouth 2 to 4 days after the onset of symptoms (FIGURE 15.16A). The characteristic red facial rash appears about 2 days after the first evidence of Koplik spots. Beginning as pink-red pimple-like spots (maculopapules), the rash breaks out at the hairline, then covers the face and spreads to the trunk and extremities (FIGURE 15.16B). Within 3 to 5 days, the rash turns brown and fades.

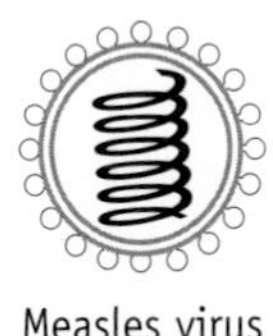

Measles virus

TABLE

15.4 A Summary of Viral Diseases Caused by the Herpesviruses

Disease	Causative Agent	Signs and Symptoms	Transmission	Treatment	Prevention
Cold sores Herpes simplex labialis	Herpes simplex virus 1 (HSV-1)	Small, red hard spot on lip	Direct or indirect contact	Lidocaine Acyclovir	Avoiding skin contact Not sharing personal items
Genital herpes	Herpes simplex virus 2 (HSV-2)	Itching or throbbing in the genital area Reddening and swelling of a small area where painful, thin blisters erupt	Sexual intercourse	Antiviral drugs (acyclovir)	Abstaining from sexual activity Limiting sexual contact to only one person who is infection free
Neonatal herpes	Herpes simplex virus 2	Mental development can be delayed Blindness can occur Persistent seizures	From infected mother to newborn during childbirth	Acyclovir Supportive therapy	Cesarean section if the maternal infection is recognized Screening women considered to be at high risk
Chickenpox (varicella)	Varicella-zoster virus (VZV)	Fever, headache, malaise with red, itchy rash on face, scalp, chest, and back	Droplet contact	Supportive care Acyclovir	Chickenpox vaccine
Shingles (herpes zoster)	Varicella-zoster virus (VZV)	Blisters on body trunk with intense pain		Acyclovir	Shingles vaccine
Roseola	Human herpesvirus 6	Red rash on neck and trunk	Contact	Supportive care	Avoiding exposure to infected child
Kaposi sarcoma	Human herpesvirus 8	Dark lesion on skin	Contact (sexual and nonsexual)	Antiretroviral therapy	Using anti-HIV medications

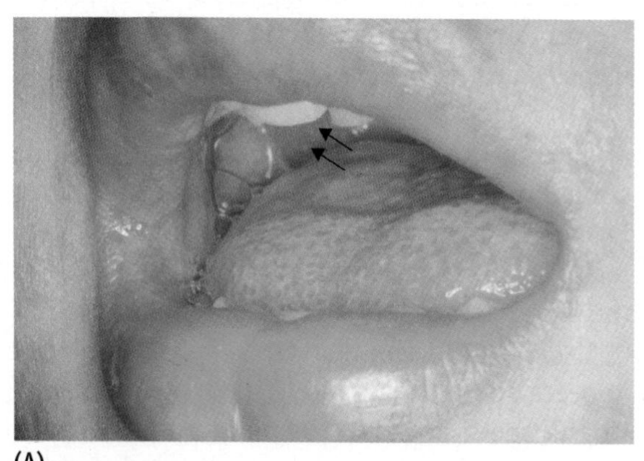
(A)

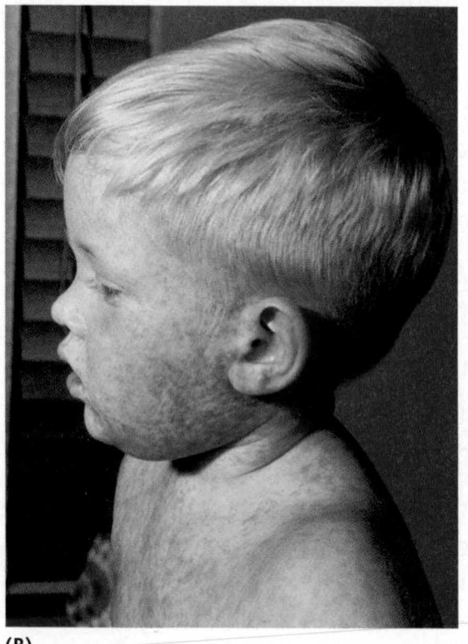
(B)

FIGURE 15.16 **Koplik Spots and the Measles Rash.** (**A**) Koplik spots on the inside cheek (arrows) of a child suffering from measles. The red spots with white centers are a frequent symptom. © Medical-on-Line/Alamy Images. (**B**) A child with measles, showing the typical rash on face and torso. Courtesy of CDC.
»» Based on the rash, how would you distinguish measles from chickenpox?

Measles usually is characterized by complete recovery. In developing nations, a small percentage of cases develop an opportunistic bacterial ear infection or pneumonia. There are two other possible complications from measles. One is **subacute sclerosing panencephalitis** (SSPE), a rare brain disease characterized by a decrease in cognitive skills and loss of nervous function. It can occur anytime from 1 month to 25 years after clinical measles (usually within 7 to 10 years). Another complication is **Paget's disease of the bone**, a disease characterized by enlarged, fragile bones that cause pain and are subject to fracture. A measles virus nucleocapsid protein appears to trigger the condition that affects 1 million Americans.

There is no prescribed treatment for a measles virus infection. Prevention is accomplished with the measles vaccine, which usually is given as part of the MMR vaccine.

Mumps is caused by another member of the Paramyxoviridae (see Figure 15.6). Its name comes from the English "to mump," meaning to be sullen or to sulk. The characteristic sign of the disease is enlarged jaw tissues arising from swollen salivary glands, especially the parotid glands (FIGURE 15.17). **Infectious parotitis** is an alternate name for the disease.

Mumps is spread by respiratory droplets or contact with contaminated objects; it is considered less contagious than measles or chickenpox. About 20% of people infected with the mumps virus have no signs or symptoms. When signs and symptoms do appear, swollen and painful salivary glands are typical, which causes the puffy cheek appearance.

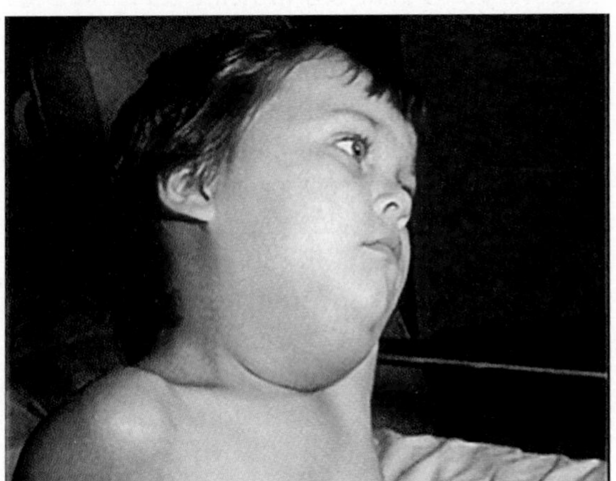

FIGURE 15.17 **Mumps.** Close-up of a young child with mumps (infectious parotitis). Courtesy of NIP/Barbara Rice/CDC. »» What causes the swelling of the jaw region?

Obstruction of the ducts leading from the parotid glands retards the flow of saliva, which causes the characteristic swelling. The skin overlying the glands is usually taut and shiny, and patients experience pain when the glands are touched as well as when chewing or swallowing.

In male patients, the mumps virus may pose a threat to the reproductive organs causing swelling and damage to the testes, a condition called **orchitis** (*orchi* = "the testicles"). The sperm count may be reduced, but sterility is not common. An estimated 25% of mumps cases in post-adolescent males develop into orchitis.

Prevention of mumps is achieved in developed nations through vaccination with the MMR vaccine.

Rubella (*rube* = "reddish"; *ella* = "small") or **German measles** (*germanus* = "similar") is caused by the rubella virus, a single-stranded (+strand) RNA virus of the Togaviridae (*toga* = "cloak") family. In 1969, physicians reported 58,000 cases of rubella, but by 2010, the number was down to 5 reported cases and rubella is no longer considered an endemic disease in the United States. Viral transmission generally occurs by respiratory droplets.

Rubella virus

■ Parotid glands: One of three pairs of salivary glands, located below and in front of the ears.

■ Endemic: Referring to a disease occurring within a specific area, region, or locale.

■ Transplacental: Refers to movement from mother to fetus.

Rubella is accompanied by fever with a pale-pink maculopapular rash beginning on the face and spreading to the body trunk and extremities. The rash develops rapidly, often within a day, and fades after another 2 days.

Rubella is dangerous to the developing fetus in a pregnant woman. Called **congenital rubella syndrome**, transplacental virus transmission to the fetus can lead to destruction of the fetal capillaries, and blood insufficiency follows. The organs most often affected are the eyes, ears, and cardiovascular organs, and children may be born with cataracts, glaucoma, deafness, or heart defects.

Rubella symptoms are so mild that no treatment is required. Children in developed nations are immunized with the MMR vaccine.

Fifth disease or **erythema infectiosum** is caused by human parvovirus B19, which is a small, single-stranded DNA virion of the Parvoviridae (*parv* = "small") family.

Human parvovirus B19

In the late 1800s, numbers were assigned to diseases accompanied by skin rashes. Disease I was measles, II was scarlet fever, III was rubella, IV was Duke's disease (also known as roseola and now recognized as any rose-colored rash), and V was erythema infectiosum.

Community outbreaks occur worldwide primarily among children and transmission appears to

FIGURE 15.18 **Fifth Disease.** The fiery red rash of a child with fifth disease (erythema infectiosum). The confluent red rash is seen. © Dr. P. Marazzi/Photo Researchers, Inc. »» Why is fifth disease sometimes called "slapped cheek" disease?

be by respiratory droplets. Although most infections are asymptomatic, if symptoms occur, the outstanding characteristic is a red facial rash (FIGURE 15.18). The rash fades within 7 to 10 days.

Parvovirus B19 only infects humans (dog and cat parvoviruses do not infect humans). However, fifth disease is not limited to children. Nonimmune adults can suffer from painful joints especially in the fingers, wrists, knees, and ankles, similar to the symptoms of rheumatoid arthritis. Although antibody preparations (immune globulin) are available for treatment, the symptoms usually resolve spontaneously.

Some Human Papillomavirus Infections Cause Warts

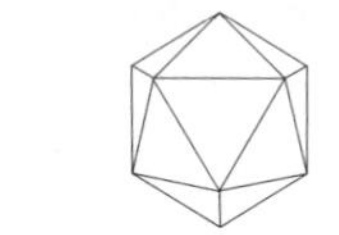

Human papilloma virus

Common Warts. Common (dermal) warts are small, usually benign skin growths resulting from infection by a specific strain of the human papillomavirus (HPV). These viruses represent a collection of over 100 different types of icosahedral, naked, double-stranded DNA virions of the Papovaviridae family.

HPV-1 to HPV-4 commonly cause warts on the hands or fingers, and **plantar warts** occur on the soles of the feet. In most cases, these skin warts are white or pink and cause no pain. Common warts can be acquired through direct contact with HPV from another person or by direct contact with a towel or object used by someone who has the virus. Therefore, prevention requires maintaining proper cleanliness and not picking at them, which can lead to their spread.

Common warts can be difficult to treat and eradicate. A physician may try freezing or minor surgery, or prescribe a chemical treatment. Recently, a new and innovative treatment has shown great success in eliminating dermal warts (MicroFocus 15.5).

Genital Warts. According to the CDC, there are some 1 million new cases of **genital warts** each year in the United States. Although a third of the known HPV types is sexually transmitted, about 90% of genital warts are caused by only two types, HPV-6 and -11.

Being an STD, the virus is most commonly transmitted through oral, vaginal, or anal sex with someone who has an HPV infection. Most people who acquire these types never develop warts or any other symptoms. When the warts do occur, they appear as small, flat, flesh-colored bumps or tiny, cauliflower-like bumps anywhere in the genital region or areas around the anus (FIGURE 15.19). The warts are sometimes called **condylomata** (*condylo* = "knob"), a reference to the bumpy appearance of the warts.

Although genital warts are not life threatening, there is no cure for the HPV infection because it can remain dormant in a latent form. Visible warts can be removed by freezing or electrosurgical excision, or by using specific chemicals available only to doctors. No one method is better than the others. Abstinence is the only sure way to prevent infection (although see below), as all other methods carry some risk. But, even after treatment, genital warts can come back. In fact, 25% of cases recur within 3 months.

HPV-Associated Cancers. Oncogenic HPVs are associated with most all cervical cancers and many vulvar, vaginal, penile, and oropharyngeal cancers. CDC analyses indicate that there are some 26,000 new HPV-associated cancer cases each year, with 18,000 in women and 8,000 in men. Although HPV-6 and -11 are considered a "low risk" potential as cancer-causing agents, other HPV types are strongly associated with precancerous changes and cervical cancers (FIGURE 15.20). HPV-16 is responsible for about 50% of cervical cancers, and together with types 18, 31, and 45, account for 80% of the cancers. Once cervical cells begin to change or become precancerous, it typically takes 10 to 15 years before invasive cervical cancer develops. All HPV viruses can result in abnormal Pap smears, so Pap tests are a critical screening procedure for all women, especially women who may be infected with HPV.

MICROFOCUS 15.5: Being Skeptical
Skin Test Antigens Eliminate Warts

Many of us suffer from allergies and have had skin tests done to discover what substances (allergens) we are sensitive to. Researchers now suggest that injections of these same allergens can cure common or genital warts. So, what's the evidence?

Researchers at the University of Arkansas School for Medical Sciences reported in 2005 that injecting substances representing skin test antigens could eliminate warts—all of them! All that was required was to inject a single wart.

The researchers stated that 50% of volunteers injected with the antigens had complete eradication of all body warts, including genital warts. The volunteers were wart free!

How could this be? The research team suggests that the antigen injection into a single wart (see figure) stimulates an attack by the immune system. This attack is not only on the antigens injected, but also the human papillomaviruses in the mix of antigens.

The verdict: It is hard to argue against results that completely abolish the infection. And besides being effective, it is safe and relatively painless.

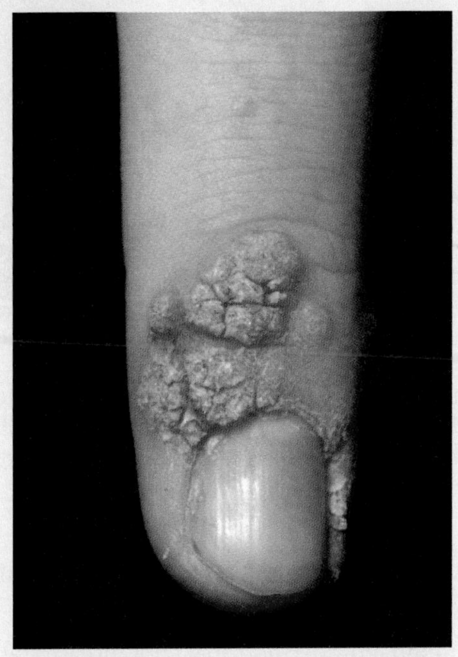

Skin warts on finger. © Medical-on-Line/Alamy Images.

A vaccine, called Gardasil®, has now been approved for use in females 9 to 26 years of age. The vaccine is aimed at the most prominent types of HPV causing genital warts and cancer; in fact, the U.S. Food and Drug Administration (FDA) reports the vaccine to be essentially 100% effective against HPV types 16 and 18, which cause approximately 70% of cervical cancers and against HPV types 6 and 11, which cause approximately 90% of genital warts. In 2009, the vaccine was approved for boys and men, as HPV infections can lead to penile cancer. Importantly, the vaccine is not a treatment as it does not eliminate the virus in already infected individuals.

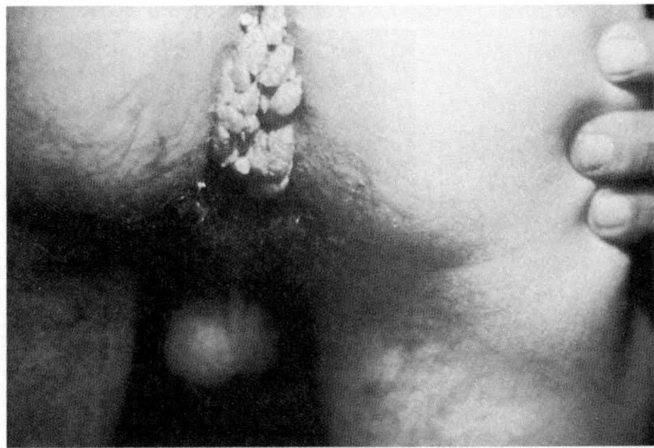

FIGURE 15.19 Genital Warts. Genital warts are caused by the human herpes simplex virus types 6 or 11. The warts typically appear as bumps on the genitalia, or in this case on the anal region of this male patient. Courtesy Dr. Wiesner/CDC. »» What must the herpesvirus infection do to the infected skin and mucosal surfaces to produce the wart appearance?

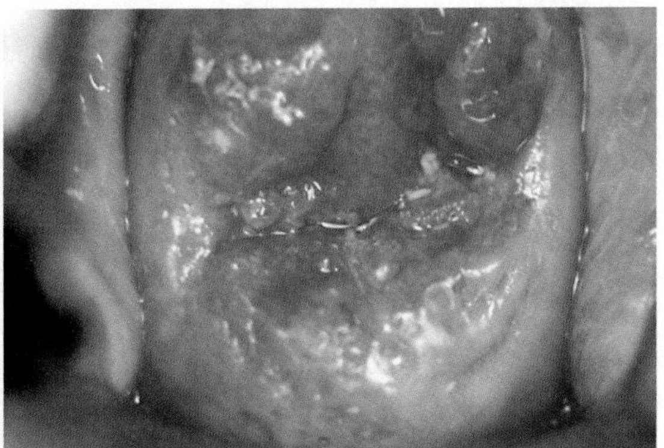

FIGURE 15.20 An Early Stage of Cervical Cancer. This photograph shows the precancerous cervix of a patient typified by erosion to her cervix. Such abnormal changes can lead to invasive cervical cancer. Courtesy of CDC. »» Why is it most likely that this patient is in her 20s rather than her 50s?

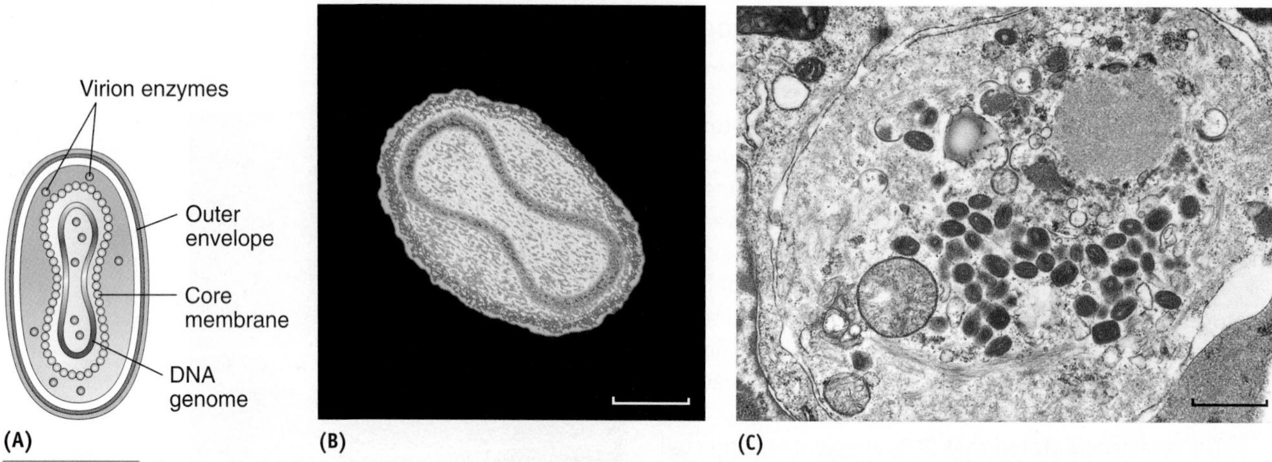

FIGURE 15.21 **The Smallpox Virus.** (**A**) A drawing of the smallpox virus, showing its complex features. (**B**) A false-color transmission electron micrograph of the smallpox virus. (Bar = 50 nm.) © Science VU/Visuals Unlimited/Getty. (**C**) A false-color transmission electron micrograph of a cell infected with smallpox viruses. Rectangular mature virions (red) can be observed. (Bar = 800 nm.) © Science Source/Photo Researchers, Inc. »» Although variola virus has a DNA genome, it replicates totally in the cell cytoplasm. What enzyme must it carry to ensure proper DNA replication in the cytoplasm?

Poxvirus Infections Have Had Great Medical Impacts on Populations

Though most dermotropic viral diseases tend not to be life-threatening, a few, such as smallpox, have exacted heavy tolls of human misery.

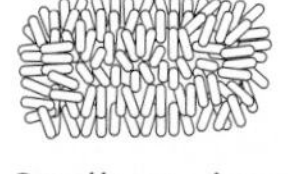
Smallpox virus

Causative Agent. **Smallpox** is caused by a brick-shaped double-stranded DNA virus of the Poxviridae family (FIGURE 15.21). It is one of the largest virions, approximately the size of chlamydiae. The nucleocapsid is surrounded by a series of fiber-like rods with an envelope.

Epidemiology. Smallpox is a contagious and sometimes fatal disease that until recently had ravaged people around the world since prebiblical times. Few people escaped the pitted scars accompanying the disease, and children were not considered part of the family until they had survived smallpox. Thanks to Edward Jenner, the first attempts at a vaccine were begun in the late 1700s.

Clinical Presentation. Transmission is by respiratory droplets or contact with viruses from skin lesions. Following an incubation period of 7 to 17 days, the initial symptoms are high fever, headache, vomiting, and general body weakness. Pink-red spots, called **macules**, soon follow, first on the face and then on the body trunk. The spots become pink pimples, called **papules** followed by fluid-filled vesicles so large and obvious that the disease is also called variola (*varus* = "vessel"; FIGURE 15.22). The vesicles become deep **pustules**,

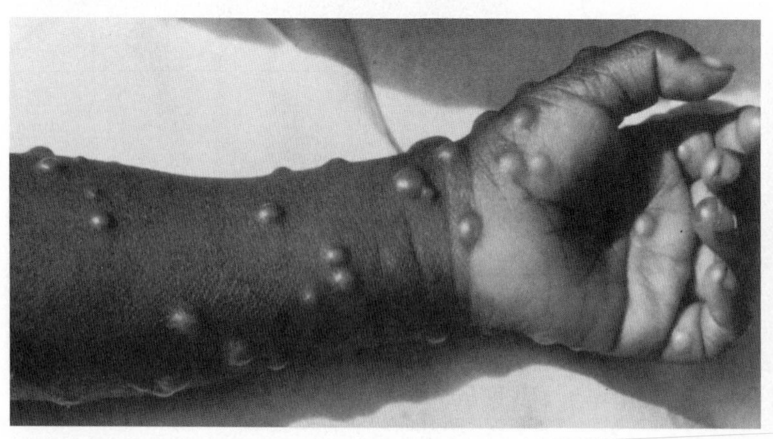

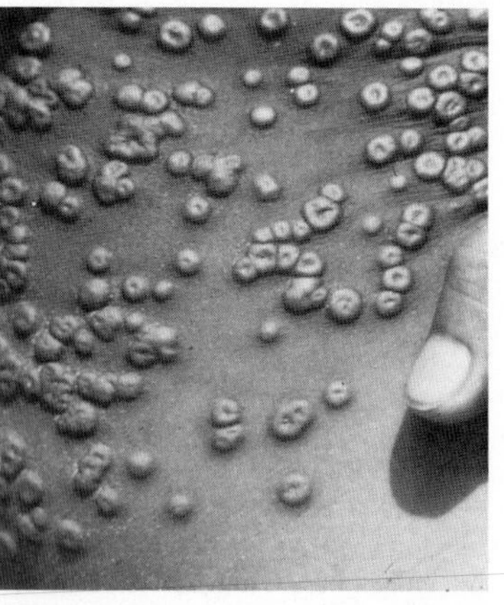

(A) (B)

FIGURE 15.22 **The Lesions of Smallpox.** (**A**) The smallpox lesions are raised, fluid-filled vesicles similar to those in chickenpox. For this reason, cases of chickenpox sometimes are misdiagnosed as smallpox. Courtesy of World Health Organization; Diagnosis of Smallpox Slide Series/CDC. Later, the lesions will become pustules (**B**) and then form pitted scars, the pocks. Courtesy of James Hicks/CDC. »» What is the difference between a vesicle and a pustule?

which break open and emit pus. If the person survives, the pustules crust and scab over, leaving pitted scars, or **pocks**.

Prevention. Smallpox vaccination has been hailed as one of the greatest medical and social advances because it was the first attempt to control disease on a national scale. It was also the first effort to protect the community rather than the individual. In 1966, the WHO received funding to attempt the global eradication of smallpox. With help from international service organizations, such as Rotary International, the Smallpox Eradication Programme attempted to isolate every known smallpox victim, and all contacts were vaccinated. The eradication was aided by the fact that smallpox viruses do not exist anywhere in nature except in humans. On October 26, 1977, healthcare workers reported isolation of the last case. In 1979, the WHO announced worldwide smallpox eradication, the first such claim made for any disease.

There are two known stocks of smallpox virus, one at the CDC in Atlanta and the other at a similar facility in Russia. The destruction of the remaining smallpox stocks at the CDC and in Russia has been planned by the WHO. However, it has been postponed several times because of the controversy over the value of keeping smallpox stocks (MicroFocus 15.6).

Because vaccinations against smallpox stopped in the United States in 1972 and elsewhere soon

MICROFOCUS 15.6: Public Health

"Should We or Shouldn't We?"

One of the liveliest global debates in microbiology is whether the last remaining stocks of smallpox viruses in Russia and the United States should be destroyed. Here are some of the arguments.

For Destruction

- People are no longer vaccinated, so if the virus should escape the laboratory, a deadly epidemic could ensue.
- The DNA of the virus has been sequenced and many cloned fragments are available for performing research experiments; therefore, the whole virus is no longer necessary.
- The elimination of the remaining stocks of laboratory virus will eradicate the disease and complete the project.
- No epidemic resulting from the theft or accidental release of the virus can occur if the remaining stocks are destroyed.
- If the United States and Russia destroy their smallpox stocks, it will send a message that biological warfare will not be tolerated.

Against Destruction

- Future studies of the virus are impossible without the whole virus. Indeed, certain sequences of the viral genome defy deciphering by current laboratory means.
- Insights into how the virus causes disease and affects the human immune system cannot be studied without having the genome and whole virus. The virus research may identify better therapeutic options that can be applied to other infectious diseases.
- Mutated viruses could cause smallpox-like diseases, so continued research on smallpox is necessary in order to be prepared.
- No one actually knows where all the smallpox stocks are located. Smallpox virus stocks may be secretly retained in other labs around the world for bioterrorism purposes, so destroying the stocks may create a vulnerability in protecting the public. Smallpox viruses also may remain active in buried corpses.
- Destroying the virus impairs the scientists' right to perform research, and the motivation for destruction is political, not scientific.
- Today, it is possible to create the smallpox virus from scratch. So why bother to destroy it?
- Because the smallpox virus (vaccinia) may have evolved from camelpox, who is to say that such evolution could not happen again from camelpox?

Discussion Point

Now it's your turn. Can you add any insights to either list? Which argument do you prefer? Note: In 2011, the World Health Assembly of the World Health Organization (WHO) met to again consider the evidence for retention or destruction of the smallpox stocks. WHO decided to postpone a decision until 2014.

after, a majority of people lack immunity to the disease. This makes smallpox one of the most dangerous weapons of bioterrorism, even though those who were vaccinated prior to 1972 still have some level of immunity. In addition, the United States has stated that it now has adequate stockpiles of smallpox vaccine to vaccinate every American, if necessary.

Molluscum contagiosum is another viral disease that forms mildly contagious, wart-like skin lesions. The virus of molluscum contagiosum is an enveloped, double-stranded DNA virion of the Poxviridae family. Transmission is generally by sexual contact.

The lesions are firm, waxy, and elevated with a depressed center. When pressed, they yield a milky, curd-like substance. Although usually flesh toned, the lesions may appear white or pink. Possible areas involved include the facial skin and eyelids in children, and the external genitals in adults. The lesions may be removed by excising them (cutting them out) and pose no public health threat. A characteristic feature of the disease is the presence of large cytoplasmic bodies called **molluscum bodies** in infected cells from the base of the lesion.

TABLE 15.5 presents a summary of these other dermotropic viral diseases.

TABLE

15.5 A Summary of Other Viral Diseases of the Skin

Disease	Causative Agent	Signs and Symptoms	Transmission	Treatment	Prevention
Measles (rubeola)	Measles virus	Cough, nasal discharge, eye redness, and high fever; Koplik spots	Droplet contact	Supportive care	MMR vaccine
Mumps	Mumps virus	Swollen and painful parotid glands Pain on chewing and swallowing	Person-to-person in infected saliva	Supportive care	MMR vaccine
Rubella (German measles)	Rubella virus	Fever with pale-pink maculopapular rash spreading to body trunk	Droplet contact	Supportive care	MMR vaccine
Fifth disease (erythema infectiosum)	Parvovirus 19	Maculopapular rash on cheeks and ears	Direct and indirect contact	Supportive hygiene	Practicing good care
Warts	Human papillomaviruses types 1–4 (HPV-1 to HPV-4)	White or pink skin growth on hands and feet	Direct and indirect contact	Freezing Minor surgery	Practicing proper care of affected areas
Genital warts	Human papillomavirus types 6 and 11 (HPV-6, HPV-11)	Small, flat, flesh-colored bumps or tiny, cauliflower-like bumps anywhere in the genital region or areas around the anus	Oral, vaginal, or anal sex with someone who has HPV	Removal by freezing, electro-surgical excision, or by using specific chemicals	Abstaining from sexual activity HPV vaccine
HPV-associated cancers	Human papillomavirus types 16 and 18 (HPV-16, HPV-18)	Vaginal bleeding and pelvic pain Precancerous growth on the cervix or penis	Oral, vaginal, or anal sex with someone who has HPV	Surgery Radiation therapy Chemotherapy	HPV vaccine Pap test
Smallpox (variola)	Variola virus	Fever, macules that became papules, then vesicles and pustules	Droplet contact Fomites	Immediate vaccination	Vaccine if available
Molluscum contagiosum	Molluscum contagiosum virus (MCV)	Flesh toned, wart-like lesions	Direct and sexual contact	Removal of papules	Avoiding touching papules Avoiding sexual contact

CONCEPT AND REASONING CHECKS 4

a. How do measles and mumps differ as diseases?
b. Why is rubella called three-day measles?
c. What is the unique feature of fifth disease?
d. Why would a woman be considered to be at risk of developing cervical cancer if she has genital warts?
e. Describe how one could identify smallpox from chickenpox based on the rash formed.

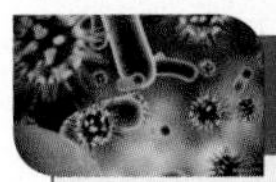

Chapter Challenge D

Many of the viral infections discussed in this section (including measles, mumps, and rubella) primarily affect children. In fact, the first of the two doses of the MMR vaccine should be given to infants 12 to 15 months of age. Although adults can contract these diseases, it is rare.

QUESTION D:

Assuming no one has been immunized, what is it about adults that make them more resistant to infection while young children are most prone to contracting these diseases? Hint: Think about what vaccines (in general) are meant to do.

Answers can be found on the Student Companion Website in **Appendix D**.

In conclusion, you might wonder where all the new viral diseases come from? Certainly for AIDS, we know that it originated in chimpanzees and "jumped" to humans perhaps as early as the 1950s. This ability to jump species is not unique to HIV. Other viral diseases, including influenza and SARS that were covered in this chapter, have also emerged from an animal reservoir (e.g., influenza from wild birds and SARS from horseshoe bats). With this ability to jump species, a new viral disease could become pandemic quickly because of air travel or simply by jumping from wild animals to domestic ones and then on to humans. Therefore, today it is important to establish a global surveillance network to monitor viruses (and other microbial pathogens) in wild animals. In fact, the Global Viral Forecasting Initiative (GVFI) has been established among 100 scientists and public health officials around the world (primarily in tropical regions which is the reservoir for many animal pathogens) to track dangerous pathogens and their potential movement from animals to humans. And importantly, this is not a one-way street. Human diseases can be transmitted back to animals. For example, measles has been transmitted to mountain gorillas, yellow fever to South American monkeys, and polio to chimpanzees—the latter two diseases being part of the viral community of pathogens we will discuss elsewhere in the text.

SUMMARY OF KEY CONCEPTS

15.1 Viruses Account for Most Upper Respiratory Tract Infections

- **Common colds**
  1. Rhinoviruses, adenoviruses, and others
- **Laryngitis**
  2. Rhinoviruses

15.2 Viral Infections of the Lower Respiratory Tract Can Be Severe

- **Influenza**
  3. Influenza A and B viruses
- **Respiratory syncytial (RS) disease**
  4. Respiratory syncytial virus
- **Parainfluenza**
  5. Human parainfluenza viruses 1 and 3
- RSV-like illness
  6. Human metapneumovirus

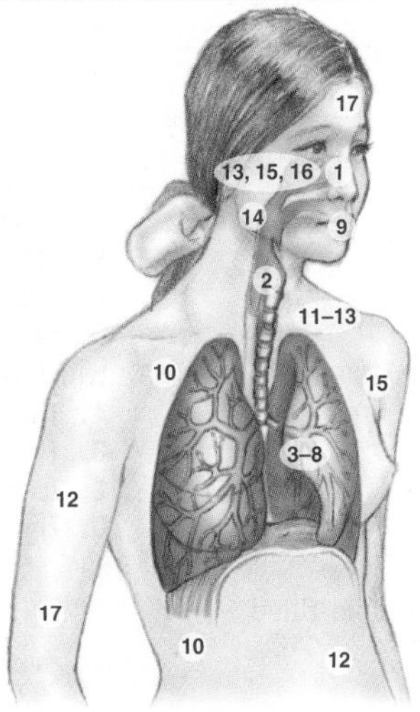

- SARS
  7. SARS coronavirus
- **Hantavirus pulmonary syndrome (HPS)**
  8. Hantavirus

15.3 Herpesviruses Cause Several Human Skin Diseases

- **Cold sores**
  9. Herpes simplex 1
- **Genital herpes** (not shown)
  - Herpes simplex 2
- **Chickenpox** and **shingles**
  10. Varicella zoster
- **Roseola** (children)
  11. Human herpesvirus 6
- **Kaposi sarcoma**
  12. Human herpesvirus 8

15.4 Several Other Viral Diseases Affect the Skin

- **Measles**
  13. Measles virus
- **Mumps**
  14. Paramyxoviruses
- **Rubella**
  15. Rubella virus
- **Fifth disease**
  16. Parvovirus B19
- **Warts** (extremities; not shown)
  Papillomaviruses
- **Smallpox**
  17. Variola

CHAPTER SELF-TEST

For **STEPS A–D**, answers to even-numbered questions and problems can be found in **Appendix C** on the Student Companion Website at **http://microbiology.jbpub.com/10e.** In addition, the site features eLearning, an online review area that provides quizzes and other tools to help you study for your class. You can also follow useful links for in-depth information read more MicroFocus stories, or just find out the latest microbiology news.

STEP A: REVIEW OF FACTS AND TERMS

Multiple Choice

Read each question carefully, then select the ***one*** answer that best fits the question or statement.

1. There are more than _____ different rhinoviruses, which belong to the _____ family.
   A. 50; Orthomyxoviridae
   B. 100; Adenoviridae
   C. 30; Paramyxoviridae
   D. 100; Picornaviridae
2. All of the following are diseases caused by the adenoviruses *except* _____.
   A. viral pneumonia
   B. acute respiratory disease
   C. parainfluenza
   D. common cold
3. Which one of the following statements is NOT true of the influenza viruses?
   A. They have a segmented genome.
   B. The genome is double-stranded DNA.
   C. The viruses have an envelope.
   D. There are three types of flu viruses.
4. Which one of the following is NOT a member of the Paramyxoviridae.
   A. RSV
   B. Human metapneumovirus
   C. SARS-CoV
   D. Parainfluenza virus
5. SARS is _____.
   A. a skin infection
   B. spread through close person-to-person contact
   C. a mild, respiratory infection
   D. most often seen in young children
6. Cold sores and genital herpes can be caused by _____.
   A. HSV-1
   B. HHV-6
   C. VZV
   D. HHV-8
7. A red, itchy rash that forms small, teardrop-shaped, fluid-filled vesicles is typical of _____.
   A. measles
   B. rubella
   C. chickenpox
   D. mumps
8. HHV-6 that causes roseola primarily affects _____.
   A. the elderly
   B. pregnant mothers
   C. infants
   D. teenagers
9. Kaposi sarcoma is a tumor of the _____.
   A. blood vessels
   B. liver
   C. lymph nodes
   D. kidneys
10. _____ are diagnostic for measles.
    A. Koplik spots
    B. Wart-like lesions
    C. Swollen lymph nodes
    D. Blisters on the body trunk
11. The characteristic sign of rubella is _____.
    A. orchitis
    B. pale-pink maculopapular rash
    C. fiery red rash on cheeks and ears
    D. high fever and sensitivity to light
12. Fifth disease _____.
    A. is hospital acquired
    B. causes white skin warts
    C. causes benign skin growths
    D. is transmitted by respiratory droplets
13. Some papillomaviruses are capable of causing _____.
    A. a "lacy" rash on the skin
    B. lung cancer
    C. pneumonia
    D. cervical cancer
14. Which one of the following statements applies to smallpox?
    A. The disease is associated with animal contact.
    B. The disease has been eradicated worldwide.
    C. It can be sexually transmitted.
    D. The virus can lie dormant in host cells.

Term Selection

For each choice, circle the word or term that best completes each of the following statements.

15. Rhinoviruses are a collection of (RNA, DNA) viruses having (helical, icosahedral) symmetry and the ability to infect the (air sacs, nose), causing (mild, serious) respiratory symptoms.
16. Herpes simplex is a viral disease transmitted by (breathing contaminated air, contact) and is characterized by thin-walled (blisters, ulcers) that often appear during periods of (emotional stress, exercising).
17. In children, the skin lesions of chickenpox occur (all at once, in crops) and resemble (teardrops, pitted scars), but in adults the lesions are known as (shingles, erythemas).
18. The complications of influenza include (Reye, Koplik) syndrome; for mumps, the complication is a disease of the (testes, pancreas) called (colitis, orchitis).
19. One of the early signs of (smallpox, measles) is a series of (Koplik spots, inclusion bodies) occurring in the (lungs, mouth) and signaling a (red, purple) rash is forthcoming.
20. Antigenic variation among (mumps, influenza) viruses seriously hampers the development of a highly effective (vaccine, treatment), and a life-threatening situation can occur if secondary infection due to (fungi, bacteria) complicates the primary infection.
21. Respiratory syncytial disease is caused by a (DNA, RNA) virus infecting the (lungs, intestines) of (adults, children) and inducing cells to (fuse together, cluster) and form giant cells called (syncytia, tumors).
22. SARS is caused by a (coronavirus, orthomyxovirus), a/an (naked, enveloped) virus spread by (sexual, person-to-person) contact.
23. Adenoviruses include a collection of (DNA, RNA) viruses and are responsible for (yellow fever, common colds), as well as infections of the (eye, ear).
24. Genital herpes is caused by a (helical, icosahedral) virus most often (HSV-1, HSV-2) and causes blisters with (thick, thin) walls that disappear in about 3 (days, weeks), only to reappear when (stress, physical injury) occurs.

STEP B: CONCEPT REVIEW

25. Explain why a vaccine against **rhinoviruses** is not feasible. (**Key Concept 1**)
26. List the types of diseases associated with **adenovirus** infections. (**Key Concept 1**)
27. Identify the major **influenza viruses** and the structures involved in generating subtypes. (**Key Concept 2**)
28. Organize the **paramyxoviruses** into related species. (**Key Concept 2**)
29. Distinguish how **severe acute respiratory syndrome** (SARS) differs from other respiratory tract infections including **hantavirus pulmonary syndrome** (**HPS**). (**Key Concept 2**)
30. Describe the infections caused by herpes simplex virus-1 and herpes simplex virus-2. (**Key Concept 3**)
31. Explain why the incidence of **chickenpox** has declined in the United States and how the varicella-zoster virus causes **shingles**. (**Key Concept 3**)
32. Summarize the diseases caused by human herpesvirus-6. (**Key Concept 3**)
33. Describe the relationship between human herpesvirus-8 and HIV/AIDS. (**Key Concept 3**)
34. Summarize the characteristics of the paramyxovirus infections causing (1) **measles** and (2) **mumps**. (**Key Concept 4**)
35. State the potential complication to a pregnant mother who has contracted **rubella**. (**Key Concept 4**)
36. Identify the characteristics associated with **fifth disease**. (**Key Concept 4**)
37. Distinguish between **common** and **genital warts**, including possible complications. (**Key Concept 4**)
38. Summarize the clinical and social significance of **smallpox**. (**Key Concept 4**)

STEP C: APPLICATIONS AND PROBLEMS

39. The CDC reports an outbreak of measles at an international gymnastics competition. A total of 700 athletes and numerous coaches and managers from 51 countries are involved. What steps would you take to avert a disastrous international epidemic?
40. A child experiences "red bumps" on her face, scalp, and back. Within 24 hours, they have turned to tiny blisters and become cloudy, some developing into sores. Finally, all become brown scabs. New "bumps" keep appearing for several days, and her fever reaches 39°C by the fourth day. Then the blisters stop coming and the fever drops. What disease has she had?
41. A man experiences an attack of shingles and you warn him to stay away from children as much as possible. Why did you give him this advice?

STEP D: QUESTIONS FOR THOUGHT AND DISCUSSION

42. Thomas Sydenham was an English physician who, in 1661, differentiated measles from scarlet fever, smallpox, and other fevers, and set down the foundations for studying these diseases. How would you go about distinguishing the variety of look-alike skin diseases discussed in this chapter?
43. A Boeing 737 bound for Kodiak, Alaska, developed engine trouble and was forced to land. While the airline rounded up another aircraft, the passengers sat for 4 hours in the unventilated cabin. One passenger, it seemed, was in the early stages of influenza and was coughing heavily. By the week's end, 38 of the 54 passengers on the plane had developed influenza. What lessons about infectious disease does this incident teach?
44. In the United Kingdom, the approach to rubella control is to concentrate vaccination programs on young girls just before they enter the childbearing years. In the United States, the approach is to immunize all children at the age of 15 months. Which approach do you believe is preferable? Why?

16

CHAPTER PREVIEW

16.1 **Viral Infections Can Affect the Blood and the Lymphatic System**

16.2 **Some Viral Diseases Cause Hemorrhagic Fevers**

Investigating the Microbial World 16: How Is Yellow Fever Transmitted?

16.3 **Viral Infections of the Gastrointestinal Tract Are Major Global Health Problems**

Textbook Case 16: Hepatitis A Outbreak

16.4 **Viral Diseases of the Nervous System Can Be Deadly**

MicroInquiry 16: The Last Push to Polio Eradication?

Image courtesy of Dr. Fred Murphy/CDC.

Viral Infections of the Blood, Lymphatic, Gastrointestinal, and Nervous Systems

Every day at least 150 people die of rabies—$8 vaccines could save them.

—The Global Alliance for Rabies Control

The following is a digest of a story posted on the Centers for Disease Control and Prevention's (CDC) Rabies Web Page "That's Just for Kids!"

Sean was 11 years old when he went on a class trip to the Okeefenokee National Wildlife Refuge in Georgia with his teachers and fifth grade class. On the first day, they set up tents in which they would sleep and helped the teachers unpack the vans with all the food and other camping gear. The class went on many expeditions that included canoeing up the river and just "hanging out around the camp."

On the fourth night, Sean woke up in the middle of the night with a sharp pain in his arm. When he looked at his arm, there were two bite marks that drew blood. Hearing a rustling sound, Sean looked up and saw a raccoon running out of his tent. He immediately called for his teacher and they cleaned out the bite with soap, hot water, and a disinfectant.

First thing in the morning Sean was taken to the hospital in Homerville, Georgia. The attending physician took Sean into a room where he would get the immune globulin shots for rabies. While waiting, various hospital

FIGURE 16.1 **Rabies Risk Worldwide.** Rabies causes more than 55,000 deaths worldwide, half of which are in India. *Source*: World Health Organization. Reproduced from WHO, Essential Rabies Maps. »» How do some countries stay at low risk?

staff stopped in to see Sean and tell him, "Well, you know that you are a lucky boy because 10 years ago you had to get 26 shots in the stomach and boy did they hurt."

Sean was given two shots of immune globulin, one shot for tetanus, and the rabies vaccine. Sean said he had to wait 1 hour "to make sure that I did not have an allergic reaction to any of it." With no adverse effects, he got to go home. After the initial shots, Sean received four more shots over a period of 1 month and he fully recovered.

Fortunately for Sean, today there is a vaccine that, given in time, is 100% effective in preventing rabies. Unfortunately, these vaccines are not available worldwide and rabies takes the lives of more than 55,000 people each year (FIGURE 16.1). Notably, 30% to 50% of the deaths occur in children under the age of 15 years.

The most common global source of rabies in humans is from uncontrolled rabies in dogs, which is why children are often at greatest risk. To get the message out concerning rabies, in 2006 a group of researchers and medical professionals, along with the CDC, formed the Global Alliance for Rabies Control (GARC). The GARC was created to alleviate the burden of rabies across the world by promoting and implementing rabies control and human rabies prevention. This culminated in the creation of World Rabies Day, which now occurs every September 28. (This is the date in 1895 when Louis Pasteur, who produced the first rabies vaccine, died.) Since the 2007 inaugural World Rabies Day, more than 182 million people have been educated about rabies and 7.7 million dogs have been vaccinated.

Rabies is by no means the only human viral infection, so we also examine several other diseases of concern. Overall, the diseases addressed in this chapter fall into three general categories. Some illnesses, such as hepatitis B, yellow fever, and mononucleosis are diseases of the blood, while others, including the noroviruses, affect the digestive system. Still others, such as rabies, polio, and West Nile encephalitis affect the nervous system.

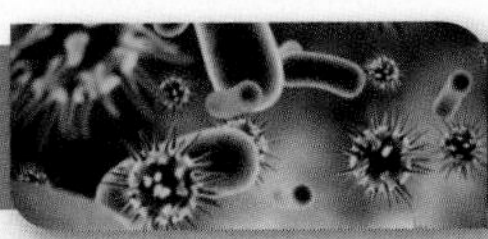

Chapter Challenge

There are a large number of viruses that can infect various tissues and organs of the blood, gastrointestinal, and nervous system. The diseases they cause can be very distinct. As you read through this chapter, see if you can identify the disease and answer a few other questions based on the case histories provided.

KEY CONCEPT 16.1 Viral Infections Can Affect the Blood and the Lymphatic System

Several viral diseases are of the blood and the lymphatic systems. To reach these areas, the viruses generally are introduced into the body tissues by mechanical means or sexual contact.

Two Herpesviruses Cause Blood Diseases

■ Lymphoma: A malignant tumor originating in the lymphatic system.

Epstein-Barr virus

Although most members of the human herpesviruses primarily affect the skin, two species are associated with blood infections and disease.

Epstein-Barr Virus Disease. The Epstein-Barr virus (EBV), being a member of the Herpesviridae, is a relatively large, enveloped double-stranded, DNA virus. It most commonly causes **infectious mononucleosis** (or "mono") in young adults who often refer to it as the "kissing disease" because EBV is spread person to person via saliva or by saliva-contaminated objects, such as table utensils and drinking glasses. The disease strikes an estimated 100,000 people annually in the United States and up to 90% of the world population between 35 and 40 years of age has been infected with EBV.

Infectious mononucleosis is a blood disease and the virus primarily targets epithelial cells of the oropharynx and then spreads to nearby B lymphocytes (B cells). Such an infection can have different outcomes for the infected individual. In children less than 5 years old, EBV may trigger no symptoms or the symptoms may be indistinguishable from other typical childhood illnesses. If the primary infection occurs during adolescence or young adulthood, the individual runs a 35% to 50% chance of suffering infectious mononucleosis. Major symptoms are sore throat, enlargement of the lymph nodes ("swollen glands"), and fever, giving the disease another name—**glandular fever**. Mononucleosis usually runs its course in 3 to 4 weeks but the virus persists as a latent infection for the life of the individual.

The diagnostic procedures for mononucleosis include detection of an elevated lymphocyte count and the observation of **Downey cells**, the damaged B cells with vacuolated and granulated cytoplasm. The **Monospot test** can be used to detect heterophile antibodies to EBV (FIGURE 16.2). No drugs have proven effective in treating mononucleosis and no vaccine is available to prevent infection.

■ Heterophile antibody: Immune protein nonspecifically reacting with other proteins or cells from unrelated animal species.

In immunocompromised individuals, a primary infection may be more serious and lead to the development of a tumor or cancer. In fact, EBV was the first human virus associated with a malignancy. In China and Southeast Asia, EBV causes a **nasopharyngeal carcinoma** of the nasopharynx and in areas of equatorial Africa EBV can cause **Burkitt** lymphoma, a tumor of the connective tissues of the jaw that is especially prevalent in children. Recent research suggests that in healthy people infected with EBV, the immune response and what is called the "oncogenic stress" response destroy most of the EBV-infected B cells. However, in immunocompromised individuals, EBV is not destroyed and the virus overcomes the stress response. The result is a continued proliferation of infected B cells leading to the tumor.

Virologists continue to search for reasons why EBV is associated with tumors on one continent and infectious mononucleosis on another. Some cancer specialists theorize the malaria parasite, common in central Africa, acts as an irritant of the lymph gland tissue, thereby stimulating activation of the virus genome and Burkitt lymphoma. In addition, some 90% of Burkitt lymphoma cases contain a chromosomal translocation that activates an oncogene which, through uncontrolled mitosis, can result in a cancerous condition.

In addition, there is growing evidence suggesting a possible link between EBV and **Hodgkin lymphoma**, a cancer of the lymph nodes and spleen. Also, evidence is mounting for an association between EBV and an increased risk of developing **multiple sclerosis**, a chronic, muscle-weakening disease of the central nervous system.

Cytomegalovirus Disease. The cytomegalovirus (CMV) is the largest member of the Herpesviridae (human herpesvirus-5; HHV-5). The virus takes its name from the enlarged cells (*cyto* = "cell"; *megalo* = "large") found in infected tissues. Usually, these are cells of the salivary glands, epithelium, or liver. After the primary infection, CMV undergoes lifelong latency.

Cytomegalovirus disease is among the most common diseases in U.S. communities. Infection may be asymptomatic or produce a mononucleosis-like syndrome, involving fever and malaise. Most patients recover uneventfully.

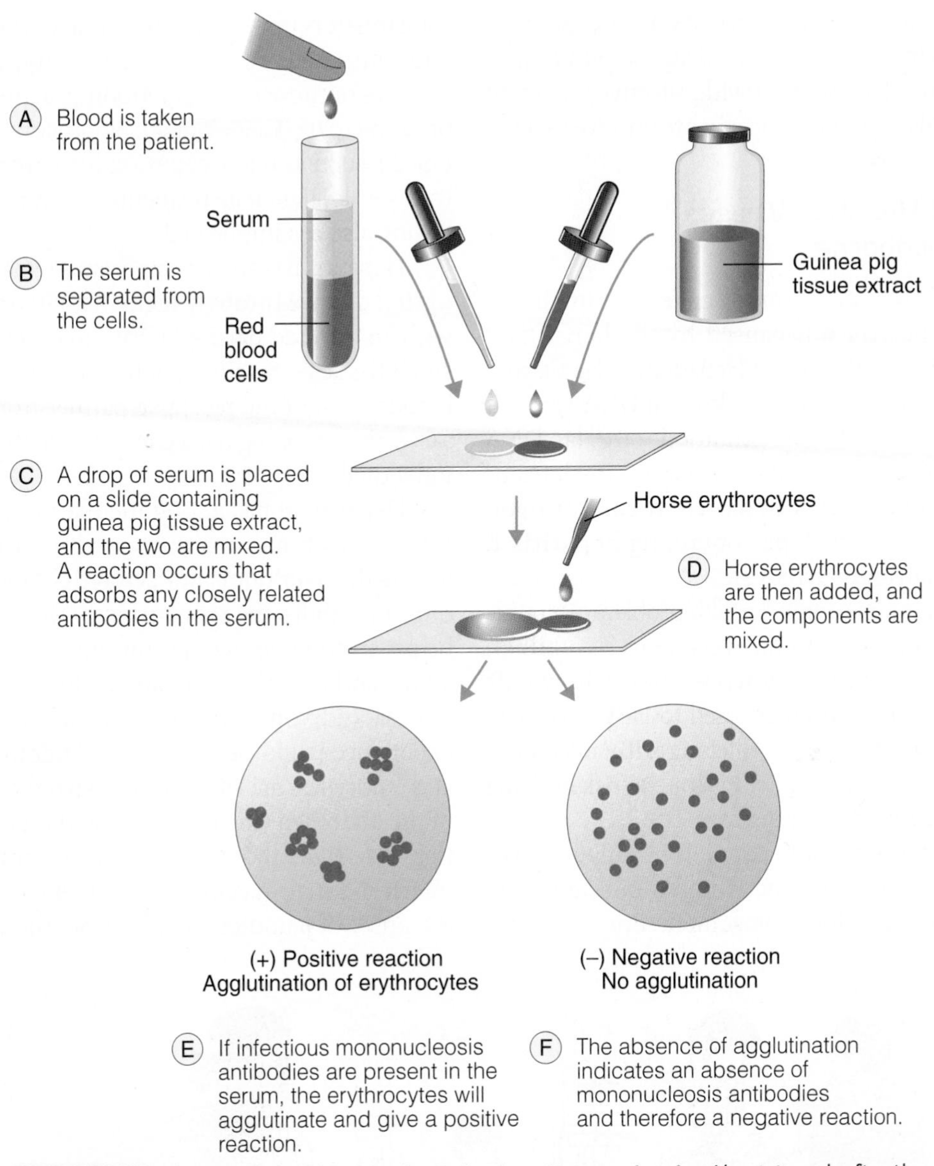

FIGURE 16.2 The Monospot Slide Test for Infectious Mononucleosis. About 1 week after the onset of infection by EBV, many patients develop heterophile antibodies. The antibodies peak at weeks two to five and may persist for several months to 1 year. The Monospot test is performed by first mixing samples of the patient's serum with a guinea pig tissue extract, which binds to and removes any closely related antibodies. Horse erythrocytes then are added and the red blood cells are observed for agglutination. Suitable controls (not shown) must also be included.
»» What types of controls should be included?

However, if a CMV-infected woman is pregnant, a serious congenital disease may ensue in 5% to 10% of infants if the virus passes into the fetal bloodstream and damages the fetal tissues. Mental impairment is sometimes observed among the 40,000 congenital CMV-associated infections in the United States and Europe every year.

CMV infection and latency generate an immune response that keeps the virus in check. However, in immunocompromised individuals, CMV can reactivate. Prior to the development of highly active antiretroviral therapy (HAART), up to 25% of AIDS patients experienced CMV-induced retinitis. CMV also can accelerate the progression to AIDS and infect the lungs, liver, brain, and kidneys and cause death. Patients undergoing cancer therapy or receiving organ transplants may be susceptible to CMV disease because immunosuppressive drugs often are administered to "knock out" the immune system's ability to reject the transplant.

■ **Retinitis:** A serious infection of the retina that can lead to blindness.

A vaccine for congenital CMV-associated infections is being studied. Several drugs, including acyclovir and valacyclovir provide effective treatment for transplant patients and as a preemptive therapy to CMV disease.

■ Endoscopes: Instruments consisting of a fiber-like strand that are inserted through an incision for diagnostic or surgical procedures.

Several Hepatitis Viruses Are Bloodborne

Hepatitis B. Formerly called "serum hepatitis," **hepatitis B** is caused by the hepatitis B virus (HBV) in the family Hepadnaviridae (*hepa* = "liver") and is the smallest known DNA virus. It has a partially double-stranded, circular DNA genome. Normal virions consist of a nucleocapsid surrounded by a **hepatitis B core antigen (HBcAg)** and envelope containing **hepatitis B surface antigen (HBsAg)**.

Hepatitis B virus

■ Cytolytic: Refers to the destruction (lysis) of cells.

Hepatitis B is a global health problem, accounting for 1 million deaths every year. Two billion people, representing almost one-third of the world's population, have been exposed to HBV and some 350 million have chronic (lifelong) HBV infections. About 20% of these individuals are at risk of dying from HBV-related liver disease.

■ Jaundice: A condition in which bile pigments seep into the circulatory system, causing the skin and whites of the eyes to have a dull yellow color.

Transmission of hepatitis B usually involves direct or indirect contact with an infected body fluid such as blood or semen. For example, transmission may occur by contact with blood-contaminated needles, such as hypodermic syringes or those used for tattooing, acupuncture, or ear piercing (FIGURE 16.3). Blood-contaminated objects such as endoscopes, saliva-contaminated (nonsterile) dental instruments, and renal dialysis tubing also are implicated.

Hepatitis B is also an important sexually transmitted disease through vaginal, anal, or oral sex with an infected partner. HBV can enter through small tissue tears, allowing the virus to enter the bloodstream of the receptive partner from blood, saliva, semen, or vaginal secretions from the infected individual.

Hepatitis B has a long incubation period of 4 weeks to 6 months during which time HBV infects the liver but is not cytolytic. Among children and adults, the primary (acute) infection may be asymptomatic. Symptoms are more likely in adults and include fever, fatigue, loss of appetite, nausea and vomiting, and dark urine. Patients experience jaundice weeks later. Abdominal pain and tenderness are felt in the upper-right quadrant of the abdomen (the liver is located on the upper right side of the abdomen). Recovery from an acute infection usually occurs about 3 to 4 months after the onset of jaundice. By this time, the virus has

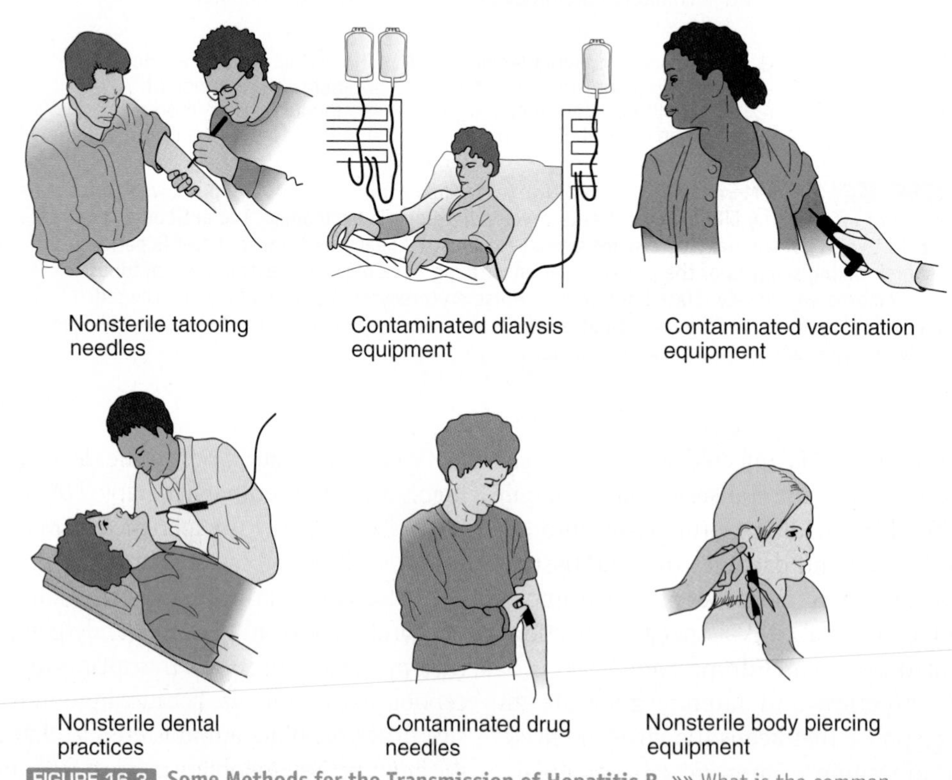

FIGURE 16.3 **Some Methods for the Transmission of Hepatitis B.** »» What is the common denominator in all these methods of transmission?

been cleared from the blood and liver, and the individuals develop immunity to reinfection.

About 10% of patients develop chronic infections and may or may not have symptoms. In rare cases, **cirrhosis**, an extensive hepatocellular injury, can occur due to immune system reactions to the infection. In addition, chronically-infected carriers run a 100-times higher chance of developing liver cancer or **hepatocellular carcinoma (HCC)** than noncarriers. The cellular and molecular reasons for carcinoma development are not completely understood.

Hepatitis B can be prevented by immunization with any of several hepatitis B vaccines. These vaccines consist of HBsAg produced by genetically engineered yeast cells. Recommended for all age groups (including infants), they are particularly valuable for healthcare workers who might be exposed to blood from patients. For infant use, it is combined with the Hib vaccine as Comvax®. As a result of child and adolescent vaccinations, there has been a steady decline in reported cases of hepatitis B in the United States (3,374 cases in 2010).

Injections of interferon alfa (Intron A®) can influence the course of hepatitis B. Moreover, injections of HBV immune globulin can be used for needle-stick exposures and for persons without known immunity who have come in contact with HBV.

Hepatitis C. Another form of hepatitis, **hepatitis C**, is caused by the hepatitis C virus (HCV), an enveloped, +ssRNA virus of the Flaviviridae family. The liver disease affects 170 million people worldwide, with the highest proportion in Asia and Africa. Although some 3 million Americans are infected, new infections have declined almost 90% since 1992 to 849 reported cases in 2010.

HCV is primarily transmitted by blood, injection drug use, or blood transfusions (in countries without a blood screening program; FIGURE 16.4). However, it also has been spread inadvertently through medical intervention, as MicroFocus 16.1 clearly demonstrates.

There are few symptoms associated with the acute (primary) infection and 15% to 25% of patients fully recover, although no permanent immunity is generated. Most cases (75% to 85%) develop a symptomless, insidious chronic infection, which takes the lives of almost 15,000 Americans who die from cirrhosis, HCC, or other complications every year. Indeed, cirrhosis from hepatitis C is the primary reason for liver transplants in the United States. In 2012, the CDC recommended anyone born between 1945 and 1965 be tested for hepatitis C because of the risk of having become infected with the virus as a result of a blood transfusion or organ transplant prior to the start of blood supply screening for HCV in the early 1990s.

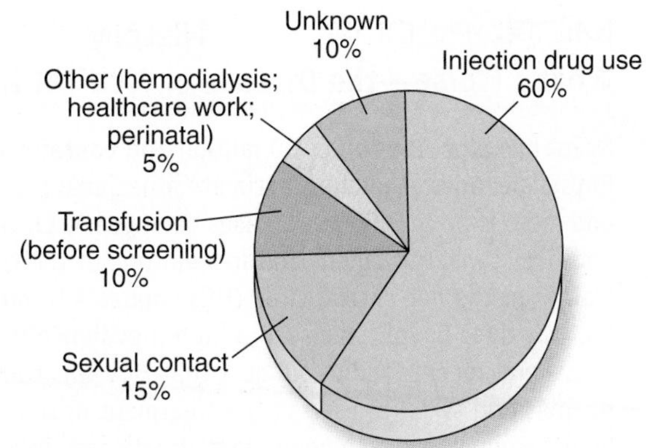

FIGURE 16.4 Hepatitis C Infections—United States, by Source. There has not been a reported case of transfusion-related hepatitis C in the United States since 1993. Data from: CDC. »» In developed nations, what would you suspect are the two most likely sources of HCV infection? In developing nations?

Interferon alfa and ribavirin, standard therapies for the disease, cure less than 50% of those infected but hopes are high for a newly approved drug called boceprevir that, when used with the standard therapies, increases the cure rate to 75%. No vaccine is available.

Hepatitis D. **Hepatitis D** is caused by two viruses: HBV and the hepatitis D virus (HDV). The latter virus consists of a protein fragment called the delta antigen and a segment of RNA, and can only cause liver damage when HBV is present. HDV requires the outside coat of HBV to infect cells. Therefore, one cannot become infected with hepatitis D unless he or she already is infected with HBV. Chronic liver disease with cirrhosis is 2 to 6 times more likely in a co-infection.

Hepatitis G. Another chronic liver illness is **hepatitis G**. The hepatitis G virus (HGV) is an enveloped, +ssRNA virus of the Flaviviridae. Like HBV and HCV, HGV is transmitted by blood and blood products. It appears to cause persistent infections in 15% to 30% of infected adults.

TABLE 16.1 summarizes the viral diseases associated with the blood and the lymphatic system.

CONCEPT AND REASONING CHECKS 1

a. Identify the serious complications that may result from (i) EBV and (ii) CMV infection.

b. Summarize the similarities in symptoms between hepatitis B and C.

MICROFOCUS 16.1: History

What's Worse—the Disease or Medical Intervention?

Egypt has a population of 80 million and contains the highest prevalence of hepatitis C in the world. Physicians and researchers estimate a national prevalence rate of at least 12% (or 9.6 million people and more than 500,000 new cases each year). Chronic hepatitis C is the main cause of liver cirrhosis and liver cancer in Egypt, and indeed, one of the top five leading causes of death. The highest concentration of the hepatitis C virus (HCV) appears in farming people living in the Nile delta and rural areas. So, why does Egypt have such a high prevalence rate of HCV?

Several recent studies suggest that HCV was transmitted through the contamination of reusable needles and syringes used in the treatment of schistosomiasis, a disease caused by a blood parasite. Schistosomiasis is a common parasitic disease in Egypt and can cause urinary or liver damage over many years. Farmers and rural populations are at greatest risk of acquiring the disease through swimming or wading in contaminated irrigation channels or standing water.

■ Emetic: A chemical substance that induces vomiting.

Prior to 1984, the treatment for schistosomiasis was intravenous tartar emetic. Between the 1950s and the 1980s, hundreds of thousands of Egyptians received this standard treatment, called parenteral antischistosomal therapy (PAT). Today, drugs for schistosomiasis are administered in pill form.

Evidence suggests inadequately sterilized needles used in the PAT campaign contributed to the transmission of HCV and set the stage for the world's largest transmission of bloodborne pathogens resulting from medical intervention. The PAT treatment campaign was conducted with the best of intentions, using accepted sterilization techniques of the time. However, much of the PAT campaign was carried out before disposable syringes and needles were available. In addition, no one was aware of HCV prior to the 1980s or the dangers associated with blood exposure.

Further evidence for the correlation between the PAT campaign and hepatitis C was the drop in the hepatitis C rate when PAT injections were replaced with oral medications. Sadly, in part because of the high number of people who were infected and inadequate health care, the transmission of HCV in the Egyptian population continues to spiral upward.

TABLE

16.1 Viral Diseases Associated with the Blood and Lymphatic Systems

Disease	Causative Agent	Signs and Symptoms	Transmission	Treatment	Prevention
Infectious mononucleosis	Epstein-Barr virus	Sore throat, fever, swollen lymph nodes in neck and armpits	Person-to-person via saliva or saliva-contaminated objects	Bed rest and adequate fluid intake	Avoiding kissing Not sharing food or personal objects
Cytomegalovirus disease	Cytomegalovirus	Asymptomatic Fever, malaise, swollen glands	Contact with the body fluids of an infected person	None	Practicing good hand hygiene Avoiding kissing Not sharing food or personal objects
Hepatitis B	Hepatitis B virus	Fatigue, loss of appetite, nausea, vomiting Jaundice and abdominal tenderness	Direct or indirect contact with body fluids	Interferon and other drugs	Receiving the hepatitis B vaccine
Hepatitis C	Hepatitis C virus	Early: no signs or symptoms May experience fatigue, loss of appetite, nausea, vomiting, low-grade fever, jaundice	Direct or indirect contact with body fluids Blood transfusions	Interferon and other drug therapies	Avoiding illegal drugs, tattooing

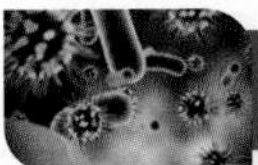

Chapter Challenge **A**

A 27-year-old male with a history of intravenous and over-the-counter drug abuse comes to the emergency room complaining of nausea, vomiting, headache, and abdominal pain. He indicates he has had these symptoms for 2 days. His vital signs are normal and he presents with slight jaundice. Physical examination indicates discomfort in the upper right quadrant. Serologic tests for HBsAg and serum antibodies to HAV are negative.

QUESTION A:

(i) ***Based on these symptoms and clinical findings, what disease does the patient have and what virus is the cause?***

(ii) ***What is the probable source of the infection?***

(iii) ***To what other diseases is the patient at risk?***

Answers can be found on the Student Companion Website in **Appendix D.**

KEY CONCEPT 16.2 Some Viral Diseases Cause Hemorrhagic Fevers

In March 2005, the Angola Ministry of Health and the World Health Organization (WHO) reported 63 hemorrhagic deaths (mostly children along with three healthcare workers) at the Uige Provincial Hospital. By late July, there were a total of 374 cases of which 329 died from an outbreak of Marburg hemorrhagic fever. This is just one of several illnesses called **viral hemorrhagic fevers (VHFs)** caused by four families of RNA viruses. The illnesses, characterized by vascular system damage (rash, bleeding gums and mucous membranes, internal bleeding), occur sporadically and are rare in the United States.

Flaviviruses Can Cause a Terrifying and Severe Illness

The Flaviviridae are enveloped, icosahedral, +ssRNA virions. They also are referred to as **arboviruses** because they are *ar*thropod-*bo*rne viruses. The two most virulent are the yellow fever and dengue fever viruses.

Yellow Fever. The first human disease associated with a virus was **yellow fever**. As a result of the slave trade from Africa, the disease spread rapidly in large regions of the Caribbean and tropical Americas, and was common in the southern and eastern United States for many generations in the 18th and 19th centuries. In 1901, a group led by the American Walter Reed identified mosquitoes as the agents of transmission as Investigating the Microbial World 16 recounts. With widespread mosquito control, the incidence of the disease gradually declined. Still, yellow fever is endemic in 33 countries in Africa and South America, and causes over 200,000 cases and 30,000 deaths annually.

■ Endemic: Referring to a disease occurring within a specific area, region, or locale.

Sylvatic (jungle) yellow fever occurs in monkeys and other jungle animals, where the virus is taken up by blood-feeding mosquitoes. Epidemics occur when a person is bitten by an infected mosquito in the forest. If that person then travels to urban areas, the virus can be passed via a blood meal to another mosquito species, *Stegomyia* (formerly *Aedes*) *aegypti*. This species then transmits the virus, now causing an "urban yellow fever" among humans.

The *S. aegypti* mosquito injects the virus into the human bloodstream. After 3 to 6 days, acute phase symptoms of headache, fever, and muscle pain appear and last 3 to 5 days. Most patients recover at this stage.

In 15% of patients, the illness reappears in a much more severe form, causing high fever, severe nausea, uncontrollable hiccups, and a violent, black vomit (blood hemorrhage). Liver damage produces jaundice (the disease often is called "yellow jack") and major hemorrhaging from the mouth, eyes, and nose occur as the patient becomes delirious. Up to 50% of patients go into a coma and die from internal bleeding. The rest recover without signs of major organ damage.

Yellow fever virus

As a zoonosis, yellow fever cannot be eradicated. It can be prevented by immunization with either of two vaccines. To this end, the WHO and

■ Zoonosis: A disease transmitted from animals to humans.

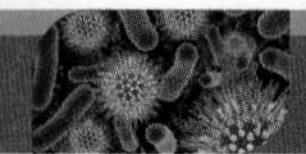

Investigating the Microbial World 16

How Is Yellow Fever Transmitted?

In 1900, neither the causative agent for nor the mode of transmission of yellow fever was known. In 1881, Carlos Finlay, a Cuban physician, had suggested from his investigations that mosquitoes transmitted the yellow fever agent. Unfortunately, he did not have good experimental technique and his mosquito idea was not accepted. Rather, the medical profession believed yellow fever was transmitted by infected bedding used by yellow fever patients and thus could be transmitted through the air (like a miasma). In 1900, the American Surgeon General sent Major Walter Reed and a team of young doctors (the Reed Commission) to Cuba to investigate the disease that in the past decades had produced severe outbreaks in many port cities in the United States.

- **OBSERVATIONS:** The Reed Commission was able to identify infected mosquitoes as the transmission agent for yellow fever through a series of experiments using American soldiers, Spanish immigrants, and even team members as volunteers. They also found that the yellow fever agent would pass through a filter, so it was probably a virus. But to really prove the case, Reed needed to show that the other modes of transmission (contaminated bedding and the air) were not valid.

- **QUESTION:** ***Is yellow fever transmitted by other methods than mosquitoes?***

- **HYPOTHESIS:** Yellow fever is only transmitted by infected mosquitoes. If so, then volunteers exposed only to yellow fever-contaminated bedding or "infected" air should not contract yellow fever.

- **EXPERIMENTAL DESIGN:** Two small frame houses (1 and 2) were constructed. House 1 was used for testing yellow fever-contaminated bedding and house 2 for testing "infected" air and mosquitoes. House 2 was further divided by a wire screen into two rooms (A and B) through which air could pass but mosquitoes could not. Room A contained volunteers and infected mosquitoes while room B only had volunteers.

- **EXPERIMENT 1:** House 1 was filled with boxes containing sheets, pillowcases, blankets, and other materials that had been in contact with individuals who had yellow fever and contained "discharges" (black vomit, urine, fecal matter) from such cases. Three nonimmune volunteers (A–C) thoroughly handled and made contact with all the bedding materials daily for 20 days. The experiment was repeated twice more, each time with two more nonimmune volunteers (D and E, F and G) for 20 days each.

- **EXPERIMENT 2:** House 2 contained only articles that had been steam disinfected. In room A, 15 infected mosquitoes were released and then one volunteer (H) entered the room for 15 minutes to be bitten by the mosquitoes. This was repeated with the same volunteer two more times during the day. During each visit by volunteer H, two nonimmune volunteers (J and K) were in room B where they stayed and slept for 18 nights.

- **RESULTS:** If any volunteers exhibited typical signs of yellow fever, the board of experts would confirm or refute the presence of yellow fever in the volunteer.

  EXPERIMENTS 1 AND 2: See table.

- **CONCLUSIONS:**

  QUESTION 1: ***Was the hypothesis validated? Explain using the table.***

  QUESTION 2: ***Is the size of the volunteer population large enough to be statistically significant? Remember, one is dealing with a disease that the volunteers know could be deadly. Would there be concerns trying to do similar experiments with human volunteers today?***

  QUESTION 3: ***Provide three conclusions you can make from the Reed Commission experiments on yellow fever transmission.***

Answers can be found on the Student Companion Website in **Appendix D**.

Investigating the Microbial World 16 (continued)

TABLE

Results of Experiments 1 and 2 Concerning Non-Mosquito Modes of Transmission

Experiment (volunteers)	Number of Volunteers: Diagnosed with yellow fever	Number of Volunteers: Remaining healthy
1 (A–C)	0	3
1 (D and E)	0	2
1 (F and G)	0	2
2 (H)	1	0
2 (J and K)	0	2

Note: Although the mode of transmission of the yellow fever virus was decided, it wouldn't be until 1927 that the virus was finally isolated.

Adapted from: Reed, W.R., Carroll, J., and Agramonte, A. (1901). *J. Amer. Med. Assoc.* 36 (7):431–440. Icon image © Tischenko Irina/ShutterStock, Inc.

health charities plan to immunize almost 50 million people in West Africa in 2012 to reestablish immunity to yellow fever. Except for supportive therapy, no treatment exists for yellow fever.

Dengue Fever. Another hemorrhagic fever, **dengue fever**, takes its name from the Swahili word *dinga*, meaning "cramp-like attack," a reference to the symptoms. There are four serotypes (DENV-1, DENV-2, DENV-3, and DENV-4) of the dengue fever virus (FIGURE 16.5), all transmitted by mosquitoes. Dengue fever is the most common arboviral disease of humans, as globally there are some 50 million new infections and over 20,000 deaths annually. Infections traditionally were confined to Southeast Asia and West Africa but have now spread globally to over 100 countries, perhaps triggered in part by climate change. In fact, the WHO now regards dengue fever as a major international health concern. In the Americas, outbreaks have occurred in Puerto Rico and Venezuela in 2007 and Brazil in 2008. Dengue fever is now becoming endemic in the United States. A 2005 outbreak in Brownsville, Texas resulted in 25 hospital cases (associated with large outbreaks in neighboring Mexican cities) and in 2009 the CDC reported the first three locally acquired cases in Key West, Florida. In 2010, a study reported that nearly 5% of the population living in Key West, Florida, exhibited post-exposure to DENV infection.

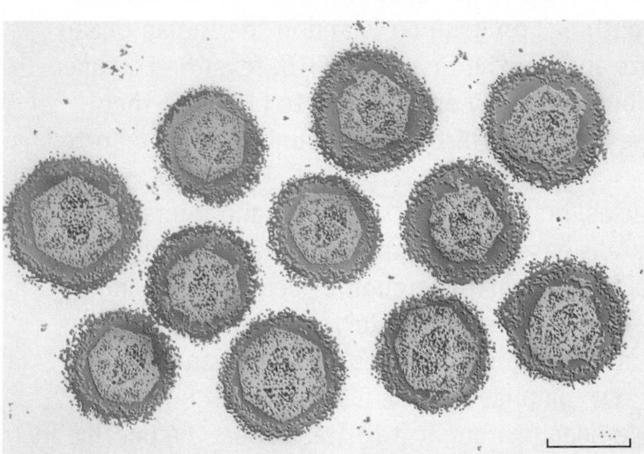

FIGURE 16.5 **The Dengue Fever Virus.** False-color transmission electron micrograph of dengue fever viruses. (Bar = 50 nm.) © Chris Bjornberg/Photo Researchers, Inc.
»» What do the yellow geometric shapes represent in the dengue fever viruses?

While taking a blood meal, infected mosquitoes inject the viruses into the blood where they multiply on lipid droplets in monocytes and macrophages. Most infections are asymptomatic or involve a mild fever with or without a rash. Cases of dengue fever present as a sudden high fever, headache, joint and muscle pain, and a maculopapular rash. The joint and muscle pain gives patients the sensation that their bones are breaking; thus, the disease has been called **breakbone fever**. After about a week, the symptoms fade.

Serious complications or death are uncommon unless one of the other three virus serotypes later enters the body of a recovered patient. Then, a condition called **dengue hemorrhagic fever** (DHF) may occur. In this condition, the immune system reacts to the memory of the first dengue infection, allowing the new one to replicate unchallenged. A rash from skin hemorrhages appears on the face and extremities, and severe vomiting and shock ensue (**dengue shock syndrome**) as blood pressure decreases dramatically.

Several vaccines are in clinical trials but no successful antiviral therapy has been identified. Vector control, using insecticides and draining standing water, are the main methods of prevention. A newly emerging and dangerous dengue-like hemorrhagic fever is described in MicroFocus 16.2.

MICROFOCUS 16.2: Public Health

A Newly Emerging Hemorrhagic Fever

In 2006, the Centers for Disease Control and Prevention (CDC) reported 37 cases of a unique hemorrhagic fever in U.S. travelers returning from destinations in the Indian Ocean and India—that is 34 more cases than had occurred in the previous 15 years. These travelers experienced fever, headache, fatigue, nausea, vomiting, muscle pain, and a skin rash—typical symptoms of dengue fever. However, unlike dengue, these patients also had incapacitating joint pain. The symptoms typically lasted a few days to a few weeks, although the joint pain sometimes lasted for many months. In the CDC cases, all recovered.

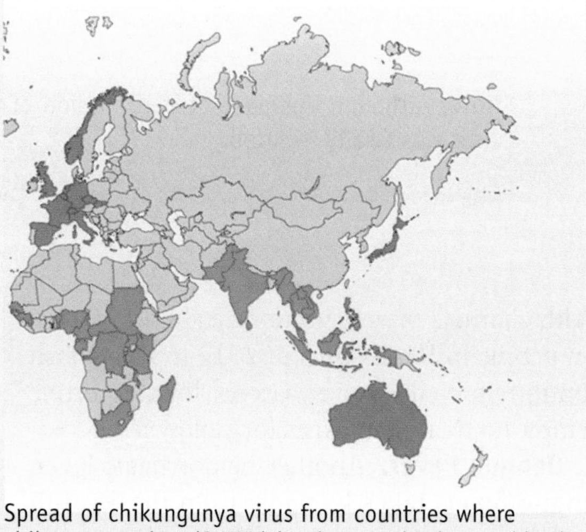

Spread of chikungunya virus from countries where chikungunya virus (CHIKV) has been endemic or epidemic (orange) to more temperate climates (purple). Modified from: Powers, A., et al., *J Gen Virol* 8 (2007): 2363–2377.

The disease experienced by these travelers was chikungunya (CHIK) fever (chikungunya means "to walk bent over," referring to the severe joint pain). CHIK fever is caused by the chikungunya virus (CHIKV), the only arbovirus in the family Togaviridae and is endemic to tropical East Africa and regions rimming the Indian Ocean. This single-stranded (+strand), enveloped RNA virus is transmitted by mosquitoes and, since its first isolation in 1953 in Tanzania, has caused numerous CHIK fever epidemics in humans. The 2006 outbreak on Réunion Island in the Indian Ocean affected more than 300,000 of the 780,000 inhabitants and, for the first time, CHIK fever had claimed a substantial number of lives; 240 fatalities were attributed directly or indirectly to CHIKV. It then spread to India where more than 1.5 million cases were reported. Thirty-seven countries have reported CHIK fever cases (see map).

CHICKV spreads through the blood to the liver, muscles, brain, lymphatic tissues, and joints. There is no specific antiviral treatment for CHIK fever and care is based on symptoms. Prevention consists of protecting individuals from mosquito bites and controlling the vector through insecticide spraying.

What makes this emerging disease especially worrisome is the report in July 2007 of a CHIK fever outbreak in northern Italy, a temperate country far from the tropical confines of the Indian Ocean. More than 200 cases were reported; one person died from complications. It is believed an infected individual from a state in India where the disease was epidemic brought it to Italy. There, he became ill and was bitten by a *Stegomyia* (tiger) mosquito, which, once infected, could transmit CHIKV to other humans. Many believe CHIK fever may now be endemic in southern Europe.

Such movements only underscore the ease with which an emerging disease can be transported in the era of jet travel. The possibility of CHIKV reaching North America, where competent tiger mosquitoes also exist, must be seriously considered.

Members of the Filoviridae Produce Severe Hemorrhagic Lesions

Ebola and Marburg hemorrhagic fevers are severe illnesses caused by viruses in the Filoviridae (*filo* = "thread") family. They consist of long thread-like, –ssRNA viruses (FIGURE 16.6).

Ebola Hemorrhagic Fever (EHF). The first reported cases of **Ebola hemorrhagic fever (EHF)** captured headlines in 1976 and 1979 when outbreaks occurred in Sudan and Zaire (now the Democratic Republic of Congo; DRC). When it was over, 88% (280) of infected people died. Through 2012, there have been 21 confirmed outbreaks of EHF in Africa. Over 2,200 cases have been reported and 67% have died.

EHF is caused by infection with one of the four infectious strains of ebolavirus (Ebola is the river in the DRC where the illness was first recognized). Recent evidence suggests the virus is zoonotic and is normally maintained in an animal host native to Africa. Fruit bats have been identified as a possible reservoir.

Uninfected individuals, including healthcare workers, can be exposed to the virus by direct contact with the blood and/or secretions of an infected person or animal, or through contact with contaminated objects, such as needles. Following an incubation period of a few days to 2 weeks, symptoms include fever, headache, joint and muscle aches, sore throat, and weakness. This is followed in 2 to 3 days by diarrhea, vomiting of blood, and stomach pain. A rash, red eyes, hiccups and internal and external bleeding may be seen in some patients.

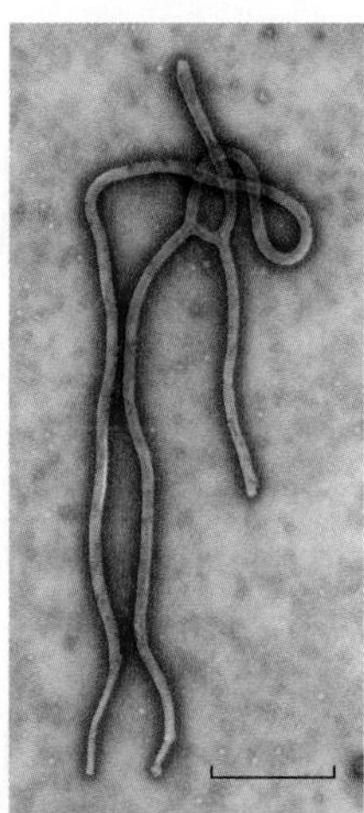

FIGURE 16.6 **The Ebolavirus.** False-color transmission electron micrograph of an ebolavirus. Here the viral filament is seen looping back on itself. (Bar = 140 nm.) Courtesy of Cynthia Goldsmith/CDC. »» Why are these viruses placed in the Filoviridae family?

The ebolavirus damages endothelial cells, causing massive internal bleeding and hemorrhaging. Viral replication masks the viral RNA so there is a slow immune response. Thus, the patient bleeds to death internally before a reasonable immunologic defense can be mounted.

■ Endothelial: Refers to the layer of cells lining body cavities, such as the veins and arteries.

Marburg Hemorrhagic Fever (MHF). Named for Marburg, West Germany, where an outbreak occurred in 1967, **Marburg hemorrhagic fever (MHF)** was first identified in tissues of green monkeys imported from Africa. How transmission of the marburgvirus occurs is unclear, as the reservoir for MHF has not been identified.

After an incubation period of 5 to 10 days, MHF develops symptoms similar to EHF but usually has a lower fatality rate.

For both EHF and MHF, prevention measures have not been established. A subunit vaccine is being tested that completely protects monkeys from infection by both viruses.

■ Reservoir: The natural host or source of a pathogen.

Members of the Arenaviridae Are Associated with Chronic Infections in Rodents

Lassa Fever. First reported in the town of Lassa, Nigeria, in 1969 **Lassa fever** is caused by a zoonotic, –ssRNA virus of the Arenaviridae family that has sandy-looking granules in the virion (*arena* = "sand"). Lassa fever is responsible for about 5,000 deaths per year in West Africa.

Transmission of the Lassa fever virus is shown in FIGURE 16.7. Infection leads to severe fever, exhaustion, and patchy blood-filled hemorrhagic lesions of the throat. The fever persists for weeks and profuse internal hemorrhaging is common. The case fatality rate is 15% to 20%.

Ribavirin is effective in treating the disease. No vaccine is available, so prevention involves avoiding contact with rodents and keeping homes clean.

Other viral hemorrhagic fevers, all caused by Bunyaviridae or Arenaviridae are Crimean-Congo hemorrhagic fever, which occurs worldwide; Oropouche fever, which affects regions of Brazil; and Junin and Machupo, the hemorrhagic fevers of Argentina and Bolivia, respectively. An arenavirus

FIGURE 16.7 **Transmission of Lassa Fever Virus.** Transmission of Lassa fever is primarily through aerosol or direct contact with excretions from infected rodents, or from contaminated food. »» Why might Lassa fever-infected rodents be a health hazard around homes or villages?

called the Sabia virus has caused hemorrhagic illnesses in Brazil, while the Guanarito virus is associated with Venezuelan hemorrhagic fever.

TABLE 16.2 summarizes the diseases caused by the hemorrhagic fever viruses.

CONCEPT AND REASONING CHECKS 2

- **a.** Compare the similarities and differences between yellow fever and dengue fever.
- **b.** Why is EHF such a deadly disease?
- **c.** How are humans most likely infected by the Lassa fever virus?

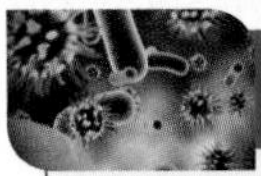

Chapter Challenge B

A 31-year-old woman comes to a neighborhood clinic with a fever, backache, headache, and bone and joint pain. She indicates these symptoms appeared about 2 days ago. She also complains of eye pain. Questioning the woman reveals that she has just returned from a trip to Bangladesh where she was doing ecological research in the tropical forests. Skin examination shows remnants of several mosquito bites, which the woman corroborates. She indicates this was her first trip to Southeast Asia. She also reports that she had been taking some antibiotics that she had been given.

QUESTION B:

(i) *Based on the woman's symptoms, what viral disease does she most likely have? What clues lead you to this conclusion?*

(ii) *Explain why the antibiotics did not help the woman's condition.*

(iii) *Why might the woman want to seriously consider not returning to Southeast Asia to continue her ecological studies?*

Answers can be found on the Student Companion Website in **Appendix D.**

TABLE

16.2 Viral Diseases Causing Hemorrhagic Fevers

Disease	Causative Agent	Signs and Symptoms	Transmission	Treatment	Prevention
Yellow fever	Yellow fever virus	Acute phase: Headache, fever, muscle pain Toxic phase: Severe nausea, black vomit, jaundice, hemorrhaging	Bite from a *Stegomyia aegypti* mosquito	No antiviral medications Supportive care	Vaccination Avoiding mosquito bites in endemic areas
Dengue fever	Dengue fever virus	Sudden high fever, headache, nausea, vomiting	Bite from an infected *Stegomyia aegypti* mosquito	No specific treatment available	Avoiding mosquito bites in endemic areas
Dengue hemorrhagic fever	A different serotype of dengue fever virus	Decrease in platelets, skin hemorrhage	Bite from an infected *Stegomyia* mosquito with another dengue virus	No specific treatment available	Avoiding mosquito bites in endemic areas
Ebola/Marburg hemorrhagic fevers	Ebola and Marburg viruses	Fever, headache, joint and muscle aches, sore throat, weakness Internal bleeding and hemorrhaging	Bite of infected fruit bat Blood transfer through cut, abrasion, or infected animal bite	No specific treatment available	Avoiding dead animals and bats in outbreak areas
Lassa fever	Lassa fever virus	Severe fever, exhaustion, hemorrhagic lesions on throat	Aerosol and direct contact with excreta from infected rodents	Ribavirin	Avoiding dead or infected rodents Maintaining good home sanitary conditions

KEY CONCEPT 16.3 Viral Infections of the Gastrointestinal Tract Are Major Global Health Problems

Several human viruses are responsible for illnesses and diseases of the digestive (gastrointestinal) system. These illnesses include hepatitis and viral gastroenteritis.

Hepatitis Viruses A and E Are Transmitted by the Gastrointestinal Tract

Hepatitis A. As an acute inflammatory disease of the liver, **hepatitis A** is caused by the hepatitis A virus (HAV), which is a small, nonenveloped, +ssRNA virion belonging to the Picornaviridae family (FIGURE 16.8). HAV is commonly transmitted by food or water contaminated by the feces (fecal–oral route) of an infected individual. Approximately 1.5 million cases of hepatitis A occur each year worldwide and it remains one of the most frequently reported vaccine-preventable diseases. In 2010, there were less than 1,700 cases reported to the CDC.

Transmission of hepatitis A (formerly referred to as "infectious hepatitis") often involves an infected food handler and outbreaks have been traced also to daycare centers where workers contacted contaminated feces. Interestingly, children often serve as the principal reservoir because their infections usually are asymptomatic. In addition, the disease may be transmitted by raw shellfish such as clams and oysters because these animals filter and concentrate the viruses from contaminated seawater.

Hepatitis A virus

The incubation period is usually between 2 to 4 weeks. Because the primary site of replication is the gastrointestinal tract, initial symptoms in older children and adults include anorexia, nausea, vomiting, and low-grade fever. The virus then is transported to the liver, its major site of replication. Discomfort in the upper-right quadrant of the abdomen follows as the liver enlarges. Considerable jaundice usually follows the onset

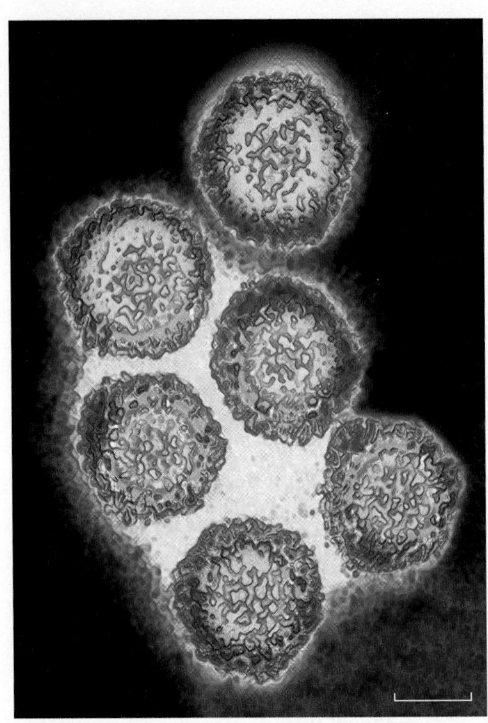

FIGURE 16.8 **Hepatitis A Viruses.** A false-color transmission electron micrograph of hepatitis A viruses. (Bar = 15 nm.) © James Cavallini/Photo Researchers, Inc.
»» In what organ would one most likely find these viruses replicating?

of symptoms by 1 or 2 weeks (the urine darkens, as well), but many cases are without jaundice. Viruses from infected liver cells are shed through the intestines and into the feces.

The symptoms may last for several weeks, and relapses commonly occur in up to 20% of cases. A long period of convalescence generally is required, during which alcohol and other liver irritants are excluded from the diet. Recovery brings life-long immunity. There are no chronic infections with HAV. The largest ever outbreak of hepatitis A in the United States occurred in late 2003 (Textbook Case 16).

There is no treatment for hepatitis A except for prolonged rest and relieving symptoms. In those exposed to HAV, it is possible to prevent development of the disease by administering hepatitis A immune globulin within 2 weeks of infection. This preparation consists of antiviral antibodies obtained from blood donors. The blood supply is routinely screened for hepatitis antibodies.

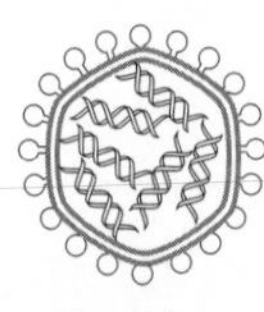

Rotavirus

Maintaining high standards of personal hygiene (hand washing), and removing the source of contamination are essential to preventing the spread of hepatitis A. Two safe and highly effective vaccines, known commercially as Havrix® and Vaqta®, are available in the United States for people over 1 year of age. Also available for individuals over 18 years of age is Twinex®, a combination vaccine for hepatitis A and B.

Hepatitis E. An opportunistic, emerging disease caused by a naked, +ssRNA virus of the Caliciviridae family is **hepatitis E**. It shares many clinical characteristics and symptoms with hepatitis A and is primarily transmitted via the fecal-oral route.

The disease affects immunocompromised individuals, with pregnant women in developing nations being particularly susceptible to illness. Mortality can be as high as 30%. No evidence of chronic infection has been noted.

The impact of hepatitis E can be appreciated from two 2004 outbreaks coming from contaminated water. One outbreak was in the Greater Darfur region of Sudan, where almost 7,000 cases and 87 deaths were reported; the other was in neighboring Chad, where there were some 1,500 cases and 46 deaths in refugee camps.

Several Unrelated Viruses Can Cause Viral Gastroenteritis

Viral gastroenteritis is a general name for a common illness occurring in both epidemic and endemic forms. It affects all age groups worldwide and may include some of the frequently encountered traveler's diarrheas. Public health officials believe gastroenteritis is second only to the common cold in frequency among infectious illnesses affecting people in the United States. In developing nations, gastroenteritis is estimated to be the second leading killer of children under the age of 5, accounting for 23% of all deaths in this age group.

Clinically, viral gastroenteritis diseases vary, but usually they have an explosive onset with some combination of watery diarrhea, nausea, vomiting, low-grade fever, cramping, headache, and malaise. The condition can be severe in infants, the elderly, and patients whose immune systems are compromised by other illnesses. Some people mistakenly call it "stomach flu."

Rotavirus Gastroenteritis. One of the world's deadliest forms of gastroenteritis in children is **rotavirus gastroenteritis**. The diarrhea-related illness is associated with 25 million clinic visits, 2 million hospitalizations, and more than

Textbook CASE 16

Hepatitis A Outbreak

1 On November 5, 2003, an outbreak of hepatitis A was confirmed through lab analysis reported by the Pennsylvania Health Department. At that time, there were 34 cases confirmed at a mall restaurant (restaurant A) with 10 customers and 12 restaurant employees reporting symptoms of hepatitis A infection.

2 By Friday, November 7, 130 people had contracted hepatitis A (see figure) and the health department provided injections of immunoglobulin as a precaution for anyone who had eaten at the restaurant between September 22 and November 2.

3 At this time, state officials and arriving CDC investigators suspected the virus was spread by an infected worker who failed to wash his or her hands before handling food.

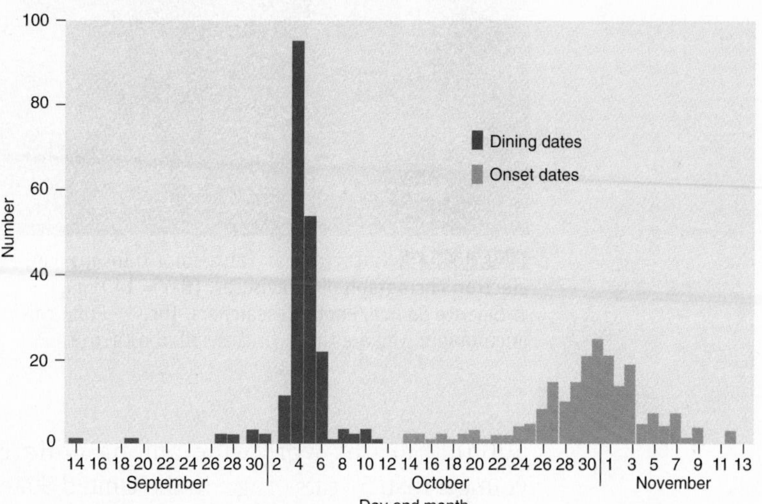

Number of hepatitis A cases by date of eating at restaurant A and illness onset.

4 On November 8, the first fatality from the hepatitis A outbreak was reported. The person died from liver failure. The outbreak had risen to 240 confirmed cases. Officials still believed the problem centered on an infected food worker.

5 On November 12, health officials announced the confirmed case count had risen to 340. Mention also was made of a recent multi-state (Tennessee, Georgia, and North Carolina) outbreak of hepatitis A in late September and early October. These 250 cases resulted from eating contaminated green onions at a few local restaurants.

6 By November 13, 410 illnesses were reported. Transmission from an infected food worker was ruled out as all employees became ill after the outbreak started. Interviews of restaurant A patrons began in order to identify what and how much they ate at the mall restaurant.

7 By November 15, two more people had died from the hepatitis A outbreak and more than 500 illnesses had been reported.

8 Patron interviews and further menu item investigations pointed to Mexican salsa containing green and white onions as the prime source of illness.

9 Ninety-eight percent of restaurant A patrons who became ill reported eating salsa containing raw green onions. Illness was not associated with eating salsa containing raw white onions.

10 In all, more than 550 people had been stricken during the hepatitis A outbreak. Genetic analysis of the virus implicated raw green onions imported from three firms in Mexico.

Questions:

(Answers can be found on the Student Companion Website in **Appendix D.**)

A. Why would the infections of restaurant employees be significant?

B. Explain why immunoglobulin injections were recommended.

C. How common are deaths from hepatitis A?

D. Why was it important to discover what and how much food the affected restaurant customers ate?

E. Provide some ways that the green onions may have initially become contaminated with the hepatitis A virus.

For additional information see www.cdc.gov/mmwr/preview/mmwrhtml/mm5247a5.htm.

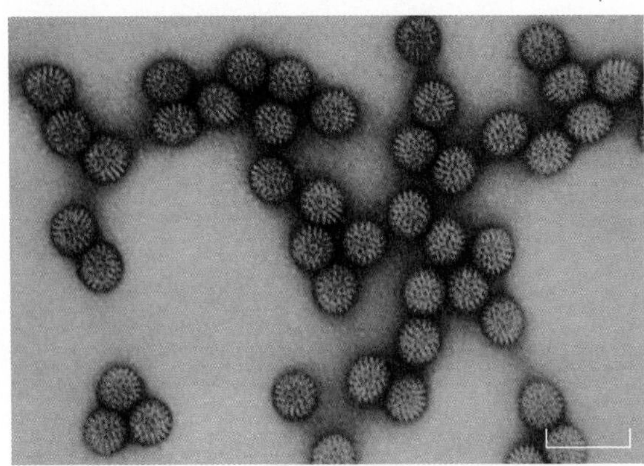

FIGURE 16.9 **Rotaviruses.** False-color transmission electron micrograph of rotaviruses. (Bar = 140 nm.) © Science Source/Photo Researchers, Inc. »» From this micrograph, why are these viruses called rotaviruses?

500,000 deaths worldwide among children younger than 5 years of age. In the United States, there are an estimated 75,000 hospitalizations and 20 to 40 childhood deaths each year.

The rotavirus (*rota* = "wheel") is a naked, circular-shaped virion whose genome contains 11 segments of dsRNA (FIGURE 16.9). It is a member of the Reoviridae family. Rotavirus infections tend to occur in the cooler months (October–April) in the United States and are often referred to as "winter diarrhea."

Coxsackie virus

Transmission occurs by ingestion of contaminated food or water (fecal-oral route), or from contaminated surfaces. The rotaviruses make their way to the small intestine where they infect enterocytes, inducing diarrhea, vomiting, and chills. The disease normally runs its course in 3 to 8 days. Treatment requires oral rehydration salt solutions.

■ Enterocytes: Cells of the intestinal epithelium.

Recovery from the infection does not guarantee immunity, as many children have multiple rounds of reinfection.

The 500,000 childhood deaths worldwide (90% in developing nations) make development of a safe and effective rotavirus vaccine a high priority objective. In early 2006, a new oral vaccine, RotaTeq®, was licensed in the United States while another, Rotarix®, was licensed in 2008. With the use of these vaccines, hospitalizations for severe diarrhea have been reduced 70% to 80%.

Norovirus Infections. The most likely cause of nonbacterial gastroenteritis in adults is a **norovirus infection**. Noroviruses (formerly called the Norwalk-like viruses) belong to the family Caliciviridae and are naked, icosahedral virions with a single-stranded (+strand) RNA genome. Currently, there are four known norovirus groups. All are highly contagious, and as few as 10 virions can lead to illness in an individual. Humans are the only known reservoir.

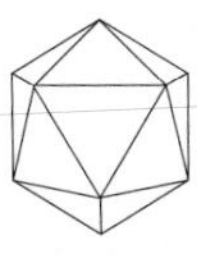

Norovirus

Noroviruses are transmitted primarily through the fecal-oral route, either by consumption of contaminated food or water, or by direct person-to-person spread. Contamination of surfaces also may act as a source of infection as the viruses are extremely stable in the environment. The CDC estimates that almost 60% of all foodborne outbreaks are caused by norovirus infections. The infections occur most frequently during the summer months, giving the disease the name "summer diarrhea."

The incubation period for norovirus gastroenteritis is 15 to 48 hours. Typical gastrointestinal symptoms of fever, diarrhea, abdominal pain, and extensive vomiting last about 24 hours and recovery is complete. Dehydration is the most common complication. The only treatment for norovirus gastroenteritis is fluid and electrolytes. Washing hands and having safe food and water are important prevention measures.

Enterovirus Infections. Viral gastroenteritis is also caused by enteroviruses, small, +ssRNA virions of the Picornaviridae family.

One enterovirus is the Coxsackie virus, first isolated from a patient residing in Coxsackie, New York. The virus occurs in many strains, with B4 and B5 associated most commonly with a form of gastroenteritis called **hand, foot, and mouth disease**. This is a moderately contagious disease typically affecting infants and young children in the spring to fall. Symptoms include fever, poor appetite, malaise, and a sore throat. A rash may develop on the palms of the hands and the soles of the feet. Most patients recover in 7 to 10 days. There is no specific treatment.

The second enterovirus is the echovirus, which gets its name from *e*nteric (intestinal), *c*ytopathogenic (pathogenic to cells), *h*uman (human host), and *o*rphan (a virus without a famous disease). Echoviruses occur in many strains and cause infantile diarrhea.

TABLE 16.3 summarizes the characteristics of the virally caused gastrointestinal illnesses.

CONCEPT AND REASONING CHECKS 3

a. Compare and contrast hepatitis A and E.

b. Explain why oral rehydration salt solutions are the treatment for most forms of viral gastroenteritis.

TABLE

16.3 Viral Infections of the Gastrointestinal Tract

Disease	Causative Agent	Signs and Symptoms	Transmission	Treatment	Prevention
Hepatitis A	Hepatitis A virus	Nausea, vomiting, low-grade fever	Indirect through food and raw shellfish	No specific treatment	Receiving the hepatitis A vaccine Practicing good hygiene
Hepatitis E	Hepatitis E virus	Nausea, muscle pain, low-grade fever	Indirectly through water Zoonosis	No effective treatment	Avoiding untreated water
Viral gastroenteritis	Rotavirus	Diarrhea, nausea, vomiting, low-grade fever, stomach cramping, headache, malaise	Indirect through food or water Contaminated surfaces	Self care	Practicing good hand hygiene Avoiding shared items
	Norovirus	Nausea, vomiting, diarrhea, stomach cramping	Indirect through food or water Person-to-person	Self care Oral rehydration fluids	Practicing good hand hygiene Washing fruits and vegetables
	Enterovirus	Fever, mild rash, mild upper respiratory tract illness	Indirect through food or water Person-to-person	Self care	Practicing good hand hygiene Cleaning contaminated surfaces

Chapter Challenge C

A 59-year-old man visits the local hospital emergency room complaining of nausea, abdominal discomfort with vomiting, and fever. On physical examination, he has a temperature of 38°C and appears jaundiced. He indicates the symptoms appeared abruptly 2 weeks ago after he had returned from a tour to Egypt. While there, the patient indicated he lived like the locals, ate in local restaurants, and often drank the local water. He did not have any vaccinations before leaving on his trip.

QUESTION C:

(i) *Name two viral diseases discussed in this chapter that the patient might have.*

(ii) *What is this patient's most likely disease? Why would you make that conclusion?*

(iii) *Identify a possible transmission route for this virus.*

(iv) *If the patient had sought medical intervention earlier in the illness, what treatment could have been prescribed?*

Answers can be found on the Student Companion Website in **Appendix D.**

KEY CONCEPT 16.4 Viral Diseases of the Nervous System Can Be Deadly

Several viral diseases affect the human nervous system, which can suffer substantial damage when viruses replicate in the tissue. Rabies, polio, and West Nile virus disease are perhaps the most recognized diseases.

The Rabies Virus Is of Great Medical Importance Worldwide

Causative Agent and Epidemiology. Rabies (*rabies* = "madness") is notable for having the

MICROFOCUS 16.3: Public Health

The "Milwaukee Protocol"

On September 12, 2004, 15-year-old Jeanna Giese of Fond du Lac, Wisconsin, was bitten on the finger as she tried to rescue a bat outside her church. Not realizing the bat was rabid, Jeanna did not seek immediate treatment and became gravely ill a month later.

Normally, once the symptoms of rabies set in, recovery is not an option and death occurs within a week. Jeanna was rushed to Children's Hospital in Milwaukee where she presented symptoms of fever, slurred speech, and profuse salivation. The next day, doctors administered an unproven treatment. She was placed in a chemically induced coma and given a combination of antiviral drugs (ketamine, midazolam, ribavirin, and amantadine), sedatives, and anesthetics in their effort to save her life.

Nine days later, Jeanna was brought out of the coma. She was paralyzed, unable to speak or walk, and without sensation, as she suffered from the effects of rabies on her nervous system. Physicians detected brain-wave activity, but were unsure what was ahead after the drugs would wear off. However, when they did wear off, Jeanna demonstrated some eye movements and reflexes—a real positive sign. Progress continued and on January 1, 2005, Jeanna was released from the hospital in a wheelchair. She had a long road to recovery, if indeed recovery would be complete. She would need to regain her faculties, including her ability to speak and walk.

What followed was two years of rehabilitation during which time she regained her ability to walk and talk, and nerves damaged by rabies recommunicated with muscles. Jeanna returned to finish high school, and, in 2011, graduated from Lakeland College with a bachelor's degree.

Jeanna's lead physician, Dr. Rodney E. Willoughby, Jr. at Children's Hospital and the Medical College of Wisconsin, has had his so-called "Milwaukee protocol" used over 30 times with some success. Jeanna is now one of six who have survived rabies through the Milwaukee protocol. The others were in Colombia, Brazil, Peru, Qatar, and most recently, in 2011, an 8-year-old girl in California. Unfortunately, the survivors in Columbia and Peru died during rehabilitation.

Importantly, although the Milwaukee protocol may be a potential cure for human rabies, medical experts agree that the focus needs to remain on rabies prevention. To that end, in the summer of 2011, Jeanna and her family went to an island in the Philippines where rabies had recently been eradicated. There she visited schools and learned about the efforts that were made to prevent rabies.

highest mortality rate of any human infectious disease, once the symptoms have fully materialized; there are an estimated 55,000 deaths annually, mostly in rural areas of Africa and Asia (see chapter opener). Even though such deaths are preventable through postexposure immunization, few people in history have survived rabies, and in those who did, it is uncertain whether the symptoms were due to the disease or the therapy (MicroFocus 16.3).

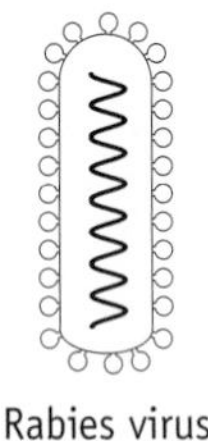
Rabies virus

The rabies virus is a –ssRNA virion of the Rhabdoviridae family with a meager five genes in its genome. It is rounded on one end, flattened on the other, and looks like a bullet (FIGURE 16.10).

Animal rabies can occur in most warm-blooded animals, including dogs, cats, prairie dogs, and bats. In 2010, more than 4,300 wildlife cases were reported throughout the United States, with raccoons, foxes, skunks, and bats accounting for the majority of the cases (FIGURE 16.11). An "outbreak" in raccoons in New York City's Central

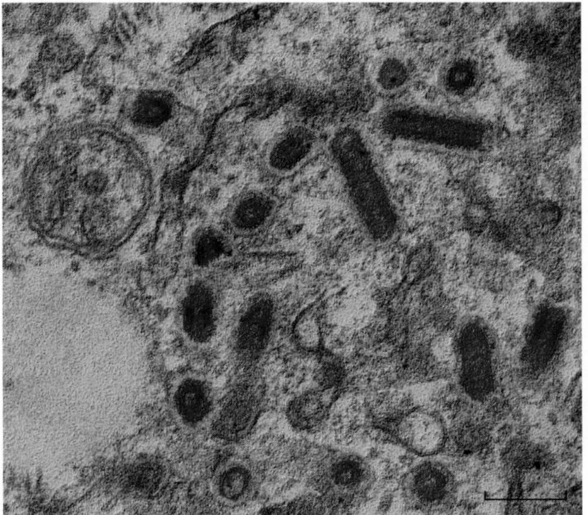
FIGURE 16.10 **The Rabies Virus.** An accumulation of rabies viruses in the salivary gland tissue of a canine. In this false-color transmission electron micrograph, the viruses are red. (Bar = 100 nm.) © Eye of Science/Photo Researchers, Inc. »» In this micrograph, identify the viruses that most closely have the shape characteristic of rhabdoviruses.

FIGURE 16.11 **Reported Cases of Animal Rabies.** This geographic map identifies the location of animal rabies reported in 49 states, the District of Columbia, and Puerto Rico. *Source*: CDC. Available at www.cdc.gov/ncidod/dvrd/rabies/Epidemiology/Epidemiology.htm.
»» Why are so many of the animal rabies cases found in the Northeast?

Park in 2010 led to a 400% increase in the number of reported rabid animals.

Clinical Presentation. The virus enters the tissue and peripheral nervous system through a skin wound contaminated with the saliva from a rabid animal. The incubation period for rabies varies according to the amount of virus entering the tissue and the wound's proximity to the central nervous system. As few as 6 days or as long as a year may elapse before symptoms appear. A bite from a rabid animal does not ensure transmission, however, because experience shows that only 5% to 15% of bitten individuals develop the disease.

Early signs of rabies encephalitis are abnormal sensations such as tingling, burning, or coldness at the site of the bite. Fever, nausea, and vomiting are additional symptoms. The acute (neurological) phase is characterized by increased muscle tension as the patient becomes hyperactive and aggressive. Soon there is paralysis, especially in the swallowing muscles of the pharynx, and saliva drips from the mouth. Brain inflammation, together with an inability to swallow, increases the violent reaction to the sight, sound, or thought of water. The disease therefore has been called **hydrophobia**—literally, the fear of water. Death usually comes within days from respiratory paralysis.

Treatment and Prevention. An estimated 10 million people worldwide receive **postexposure immunization** each year after being bitten by rabies-suspect animals. As described in the chapter introduction, this entails five injections on days 0, 3, 7, 14, and 28 in the shoulder muscle. These injections are preceded by thorough cleansing and one dose of

■ Encephalitis: An inflammation of the brain.

rabies immune globulin to provide immediate antibodies at the site of the bite. This usually is accompanied by a tetanus booster, but the latter is omitted if exposure is not certain. For high-risk individuals such as veterinarians, trappers, and zoo workers, a preventive immunization of three injections may be given.

Rabies historically has been a major threat to animals. One form, called **furious rabies**, is accompanied by violent symptoms as the animal becomes wide eyed, drools, and attacks anything in sight. In the second form, **dumb rabies**, the animal is docile and lethargic, with few other symptoms. Health departments now are conducting a novel campaign to immunize wild animals using vaccine air-dropped within biscuit-sized baits of dog food and fish meal.

The Polio Virus May Be the Next Infectious Disease Eradicated

Poliovirus

The name **polio** is a shortened form of **poliomyelitis** (*polio* = "gray"; *myelo* = for "spinal cord"), referring to the "gray matter," which is the nerve tissue of the spinal cord and brain in which the virus infects.

Causative Agent. The polioviruses, being in the Picornaviridae family, are among the smallest virions, measuring 27 nm in diameter. They are composed of a nonenveloped capsid containing a single-stranded (+ strand) RNA genome (FIGURE 16.12).

Clinical Presentation. Polioviruses usually enter the body by the fecal-oral route through contaminated water and food. About 90% to 95% of cases are asymptomatic. The most common clinical presentation is an **abortive poliomyelitis** where the patient experiences fever, headache, nausea, and sore throat. There is no entry into the CNS and complete recovery occurs.

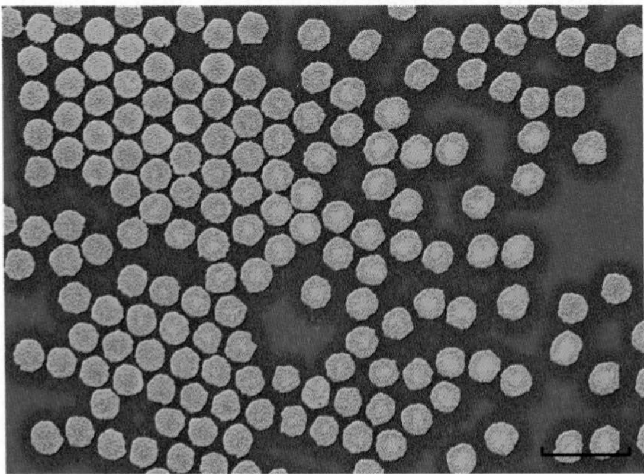

FIGURE 16.12 **Polioviruses.** A false-color transmission electron micrograph of polioviruses. With a diameter of about 27 nm, these are among the smallest viruses that cause human disease. (Bar = 100 nm.) © CDC/Visuals Unlimited/Corbis. »» How can such a small virus cause such a paralytic disease?

In perhaps 1% of cases the poliovirus infection passes across the blood-brain barrier and localizes in the meninges. One result may be aseptic meningitis, which is a **nonparalytic poliomyelitis** characterized by fever, headache, and stiff neck. Complete recovery usually occurs. The other outcome in the most severe infections is **paralytic poliomyelitis**, which occurs as the virus infects the brain. Paralysis of the arms, legs, and body trunk are affected. Swallowing is difficult as paralysis develops in the tongue, facial muscles, and neck. Paralysis of the diaphragm muscle causes labored breathing and may end in death. Those that do recover often have permanent paralysis.

Treatment. Virologists have identified three types of poliovirus: type I causes a major number of epidemics and sometimes paralysis; type II occurs sporadically but invariably causes paralysis; and type III usually remains in the intestinal tract. In the 1950s, a team led by Jonas Salk grew large quantities of the viruses and inactivated them with formaldehyde to produce the first polio vaccine in 1955. Albert Sabin's group subsequently developed a vaccine containing attenuated (weakened) polioviruses. This vaccine was in widespread use by 1961 and could be taken orally (OPV) as compared with Salk's vaccine, which had to be injected (IPV). Both are referred to as **trivalent** vaccines because they contain all three strains of virus. One drawback of the Sabin vaccine is that being attenuated, a few cases of vaccine-caused polio have occurred (1 reported case in every 2.4 million vaccinations).

The vaccines have contributed substantially to the reduction of polio. In 1988, the forty-first World Health Assembly, with funds raised by Rotary International, launched a global initiative to finally eradicate polio (MicroInquiry 16).

A discouraging legacy of the polio epidemics is **postpolio syndrome** (**PPS**). Apparently, many people who had polio decades ago now are experiencing the initial ailments they had with polio, including muscle weakness and atrophy, general fatigue and exhaustion, muscle and joint pain, and breathing or swallowing problems. The National Center for Health Statistics estimates there are

MICROINQUIRY 16

The Last Push to Polio Eradication?

A global effort to eradicate polio began in the 1988 as the Global Polio Eradication Initiative. Stimulated by the recent success in eradicating smallpox from the globe and polio from the Americas in 1979, the World Health Organization, Rotary International, the Centers for Disease Control and Prevention, and UNICEF spearheaded the effort to rid the world by 2000 of a virus responsible for the majority of paralysis and disability in children. Although this goal was not met, significant progress was made. By 2000, annual reports of polio cases had fallen by more than 99%, to fewer than 1,000 and the disease was only endemic in four countries—Afghanistan, Pakistan, India, and Nigeria. Thus, the Initiative succeeded in reducing cases through routine immunization of infants with the trivalent oral polio vaccine.

Eradication has not been completed for several reasons. Remaining reservoirs of naturally occurring (wild-type) polio virus have been difficult to eliminate in the endemic countries. Conflicts and wars have made it hard to move freely through the countries to ensure immunization has been accomplished. In India, the problem has been trying to vaccinate an enormously overcrowded and growing population, where poor sanitation and diarrheal diseases abound.

Basic biology has also hampered efforts to eradicate the virus. A new, circulating vaccine-derived poliovirus has emerged that reverted back to a virulent form similar to the parent strain from which the vaccine was originally produced. Although the outbreaks have been controlled through additional immunization campaigns, the emergence of these new virus strains is worrisome.

For these reasons, and that an increasing number of polio-free areas were becoming reinfected, the stakeholders in the global eradication of polio launched the Strategic Plan 2010–2012 for eradicating wild polio virus. It calls for a 3-year campaign to target and eliminate remaining wild polio virus reservoirs in endemic countries, pushes for mass immunizations using the new bivalent (types I and III) OPV, and addresses the need to strengthen health system weaknesses in the remaining polio-prone countries.

Hopefully, the Strategic Plan 2010–2012 will soon bring the end to another devastating human disease.

Discussion Point

According to the Global Polio Eradication Initiative, $2.6 billion is needed to fight polio for the Strategic Plan 2010–2012. What might happen if polio is not eradicated and immunizations were to stop? What does it say about potential future immunization campaigns for eradicating other diseases if the polio initiative fails?

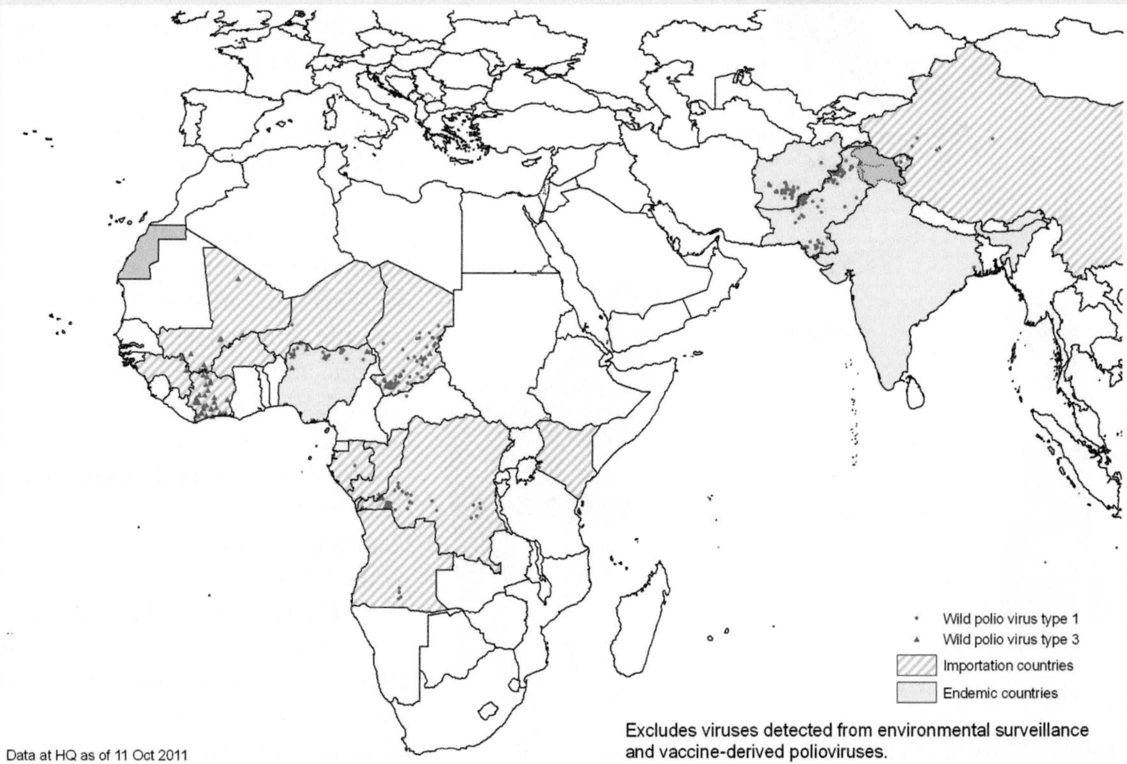

Global Polio Cases, 2010. Cases are limited to the four endemic countries and outbreaks in neighboring countries due to importation. *Source:* WHO; www.polioeradication.org/content/general/casemap.shtml.

440,000 polio survivors in the United States at risk of developing PPS, of whom 25% to 50% may be affected.

Several theories have been put forward to explain the cause of PPS, including a reactivated virus or an autoimmune reaction, which over the years has caused the body's immune system to attack motor neurons as if they were foreign substances.

Arboviruses Can Cause a Type of Primary Encephalitis

Viruses that are transmitted by mosquitoes and ticks (arboviruses; arbo = *ar*thropod-*bo*rne) have been the cause of a fairly rare primary encephalitis called **arboviral encephalitis**. Arboviral encephalitis is an example of a **zoonosis**, a disease transmitted by a vector from another vertebrate host to humans. Mosquitoes are the most common vector and most cases occur during warmer weather, when the insects are more active.

The **primary transmission cycle** involves a mosquito taking a blood meal from an infected bird, such as a crow (FIGURE 16.13). The infected mosquito can then infect other birds when taking another blood meal. An infected mosquito also can take a blood meal from another animal, such as a horse or human. These represent **incidental infections** that may cause illness in what is called a "dead-end" host; that is, a host that does not produce significant viremia and does not contribute to the primary transmission cycle. Most arboviral encephalitis cases occur during warmer weather of late spring to fall, when the insects are most active. Because all these arboviral infections involve bites from infected mosquitoes, prevention includes using insect repellent (N, N-diethyl-meta-toluamide; DEET) and wearing protective clothing (MicroFocus 16.4).

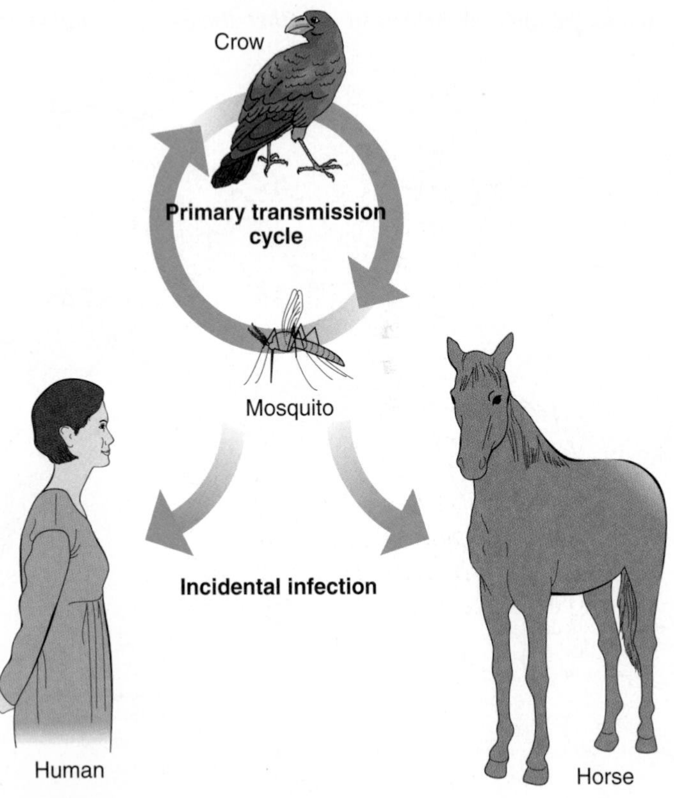

FIGURE 16.13 **The Transmission of Arboviral Encephalitis.** Shown here is the generalized pattern of viral encephalitis transmission among various animals, including humans. »» Although it may be difficult to interrupt viral encephalitis transmission, why would mosquito elimination be the best bet for disease interruption?

Arboviral encephalitis may be sporadic or epidemic, and the diseases have a global distribution (FIGURE 16.14). In the United States, the most reported types are St. Louis encephalitis (SLE), Eastern equine encephalitis (EEE), Powassan virus disease, La Crosse (LAC) encephalitis, and most notably, West Nile virus disease.

In humans, arboviral encephalitis is characterized by sudden, very high fever and a severe headache. Normally, the patient experiences pain and stiffness in the neck, with general disorientation. Patients become drowsy and stuporous, and may experience a number of convulsions before lapsing into a coma. Paralysis and mental disorders may afflict those who recover. Mortality rates are generally high.

Arboviral encephalitis is also a serious problem in horses, causing erratic behavior, loss of coordination, and fever. The disease can be transmissible from horses to humans by ticks, mosquitoes, and other arthropods. Important forms are EEE, Western equine encephalitis (WEE), and Venezuelan equine encephalitis (VEE).

West Nile Virus Disease. The most prominent emerging arboviral disease in the Western Hemisphere is West Nile virus disease. It is caused by the West Nile virus (WNV), another member of the Flaviviridae (FIGURE 16.15).

Before the outbreak in the United States in 1999, WNV was established in Africa, western Asia, and the Middle East. It is closely related to St. Louis encephalitis virus. The virus has a somewhat broad host range and can infect humans, birds, mosquitoes, horses, and some other mammals. After the first outbreak in 1999 in New York City, each

MICROFOCUS 16.4: Public Health
Avoiding Mosquitoes

Have you ever felt that you were "picked on" by mosquitoes while others were essentially left alone? Well, your feelings are correct and avoiding the diseases carried by them can be difficult because mosquitoes "love" some of us more than others (see figure).

There are more than 170 species of mosquitoes in the United States. Luckily, not all of them like humans and, of those that do, only the females take a blood meal from their victim. So, what makes for a good victim?

Mosquito feeding on human. Courtesy of Robert S. Craig/CDC.

Compounds, such as lactic acid and carbon dioxide gas, as well as some perfumes attract mosquitoes. If you are hot and sweaty, you are a more likely target than someone whose body temperature is cooler and dry. Mosquitoes tend to be drawn to the face, ankles, and hands. Men are more attractive than women and adults more often are bitten than children.

One's daily physiology may influence a mosquito bite. Women, for example, at certain points in their menstrual period are more likely to be attacked. Walking outside in the early morning or late evening makes one more likely to be bitten as that is when many mosquito species are most active.

So, it is best to wear insect repellent, such as one containing at least 20% to 35% DEET, and cover the extremities as best as possible. Also, eliminate any standing water around your home where mosquitoes may breed. As the CDC says, "Tell mosquitoes to buzz off!"

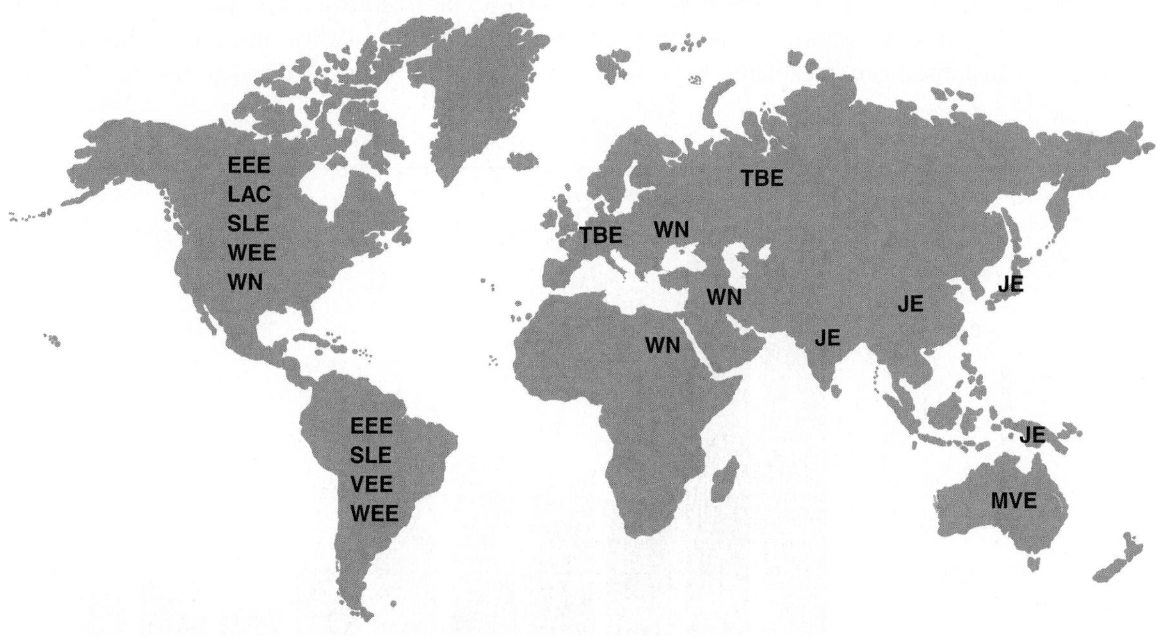

EEE Eastern equine encephalitis
JE Japanese encephalitis
LAC La Crosse encephalitis
MVE Murray Valley encephalitis
SLE St. Louis encephalitis
TBE Tickborne (Powassan) virus disease
WEE Western equine encephalitis
WN West Nile virus disease
VEE Venezuelan equine encephalitis

FIGURE 16.14 **Global Distribution of Major Arboviral Encephalitides.** The various diseases caused by arboviruses are shown. Data from: CDC. »» Propose a reason why there is such a large variety of arboviral encephalitides in the Western Hemisphere.

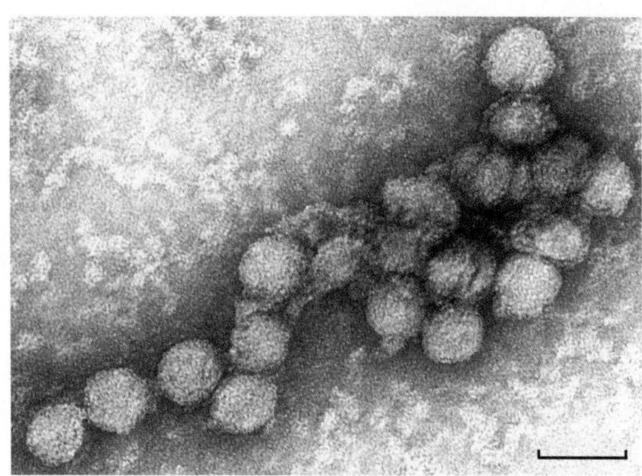

FIGURE 16.15 **West Nile Virus.** False-color transmission electron micrograph of West Nile viruses (WNV). WNV is transmitted by mosquitoes and is known to infect both humans and animals (such as birds). (Bar = 100 nm.) Courtesy of Cynthia Goldsmith/CDC. »» How would you describe the shape of the West Nile virus?

year WNV moved farther west across the United States. By 2006, cases had been reported in every state except Alaska and Hawaii. Now, experts believe the virus is here to stay and will cause seasonal epidemics that flare up in the summer and continue into the fall (FIGURE 16.16). In 2011, there were 712 cases reported and 43 deaths.

Health officials reported a very small number of cases where WNV was transmitted through blood transfusions, organ transplants, breastfeeding, or even during pregnancy from mother to baby. The virus is not spread through casual contact such as touching or kissing a person who is infected.

Although 80% of infected individuals remain asymptomatic, WNV represents a potentially serious illness. The other 20% of infected individuals typically develop symptoms between 3 and 14 days after they are bitten by the infected mosquito. Most of these infected individuals display **West Nile virus disease**, consisting of a mild fever, headache, body aches, and extreme fatigue. They also may develop a skin rash on the chest, stomach, and back. Symptoms typically last a few days.

Less than 1% of people infected with WNV will develop **West Nile encephalitis** or **meningitis** that affects the central nervous system. Symptoms can include high fever, headache, stiff neck, stupor, disorientation, coma, convulsions, muscle weakness, vision loss, numbness, paralysis, and coma. These symptoms may last several weeks to over a year, and the neurologic effects may be permanent. Death can result.

There is no vaccine yet available nor is there any specific treatment for human WNV infections. In cases of encephalitis and meningitis, people may need to be hospitalized so they can receive

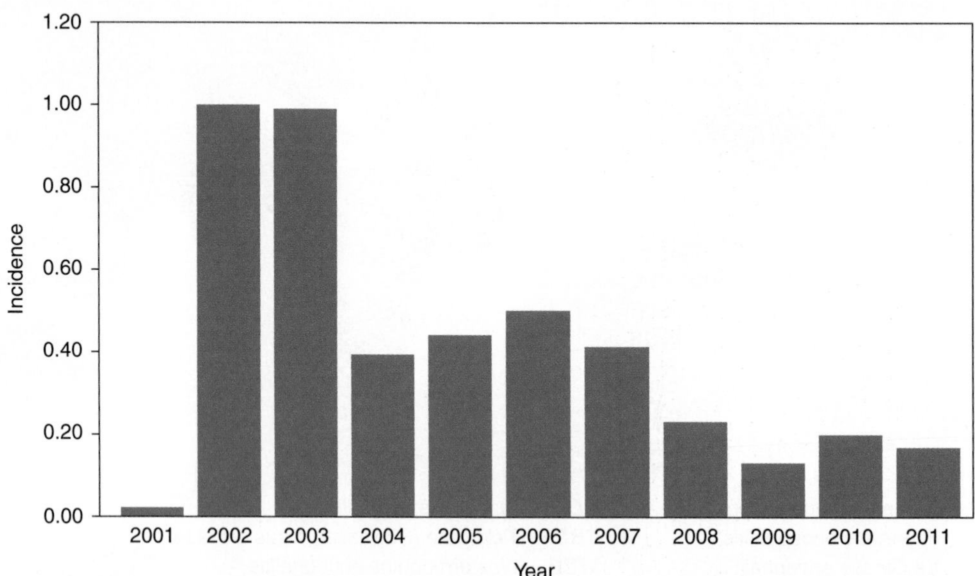

FIGURE 16.16 **Incidence of West Nile Virus—United States, 2001–2011.** This figure shows the incidence (reported cases per 100,000 population) of neuroinvasive disease since 2001. WNV incidence was stable between 2004 and 2007 and then decreased through 2009. Reproduced from: *Morbidity and Mortality Weekly Report*, 59(53);1–116. »» What factors may have brought about the decline in incidence between 2006 and 2009?

TABLE

16.4 A Summary of the Viral Diseases of the Nervous System

Disease	Causative Agent	Signs and Symptoms	Transmission	Treatment	Prevention
Rabies	Rabies virus	Tingling, burning, coldness at bite site Fever, headache, increased muscle tension Paralysis and hydrophobia	Bite from rabid animal	Rabies immune globulin Rabies vaccine	Avoiding rabid animals Thoroughly washing the bitten area Pre-exposure vaccination when needed
Polio	Poliovirus	Asymptomatic Abortive Nonparalytic Paralytic	Fecal-oral route	Supportive treatments	Polio vaccine Practicing good personal hygiene
Arboviral encephalitis	Arboviruses (St. Louis encephalitis, Eastern equine, Western equine, La Crosse, West Nile)	Sudden high fever, severe headache, stiff neck, disorientation, paralysis, mental disorders Convulsions and coma	Mosquitoes and ticks	Supportive treatment	Avoiding mosquitoes in endemic areas Insect repellant Covering exposed skin

supportive treatment including intravenous fluids, help with breathing, and nursing care. People who spend a lot of time outdoors are more likely to be bitten by an infected mosquito. These individuals should take special care to avoid mosquito bites. Also, people over the age of 50 are more likely to develop serious symptoms from a WNV infection if they do get sick, so they too should take special care to avoid mosquito bites.

TABLE 16.4 summarizes the neurotropic viral diseases.

CONCEPT AND REASONING CHECKS 4

a. Why is the incubation period so variable for the rabies virus?

b. Explain why the American polio vaccination schedule has dropped the OPV as part of the vaccination.

c. Assess the consequences of being infected with the West Nile virus.

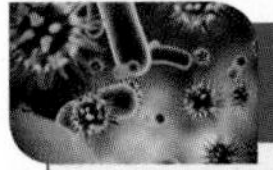

Chapter Challenge D

A 29-year-old man visited a neurologist with 2 days of increasing right arm pain and abnormal sensations. The neurologist diagnosed atypical neuropathy. The next day the man returned complaining of hand spasms and sweating on the right side of the face and trunk. The patient was discharged twice from an emergency department but symptoms worsened. After developing difficulties swallowing, having excessive saliva production, exhibiting agitation, and presenting generalized muscle twitching, the patient was admitted to a local hospital. Vital signs and blood tests were normal, but within hours he became confused. The patient was placed on broad-spectrum antibiotics and mechanical ventilation. Renal failure developed and the patient died 3 days later.

QUESTION D:

(i) *Based on the man's symptoms, what important signs did the doctor miss in trying to produce a diagnosis? What is the one important question that was not asked of the patient?*

(ii) *What disease did the man die from?*

(iii) *What medical treatment, if started immediately, might have saved the patient's life?*

Answers can be found on the Student Companion Website in **Appendix D**.

In conclusion, we have examined several viral diseases that affect various body systems. As pointed out elsewhere in the text, students often wonder where these virus originated? Again, there is good evidence that many have "jumped" species from an animal to humans at some time in the past. For example, hepatitis B jumped from apes, yellow fever came from African primates, and dengue fever originated in Old World primates. Virologists are interested in understanding how these transformations from animal to human occur and have conceived a "five-step program" for the development of such diseases.

- Stage 1: The virus has been found in animals but not in humans.
- Stage 2: The virus can be transmitted to humans but not between humans (e.g., rabies).
- Stage 3: The virus can be transmitted to humans and briefly between humans (brief outbreak) before "dying out" (e.g., Ebola hemorrhagic fever).
- Stage 4: Transmission to humans and then between humans leads to extended outbreaks (e.g., dengue fever).
- Stage 5: Since being transmitted to humans, the virus now has evolved into a pathogen exclusively in humans and outbreaks are solely the result of human-to-human transmission (e.g., AIDS).

SUMMARY OF KEY CONCEPTS

16.1 Viral Infections Can Affect the Blood and the Lymphatic System

- **Infectious mononucleosis**
  1. Epstein-Barr virus
- **Cytomegalovirus disease**
  2. Cytomegalovirus
- **Hepatitis B**
  3. Hepatitis B Virus
- **Hepatitis C**
  4. Hepatitis C Virus

16.2 Some Viral Diseases Cause Hemorrhagic Fevers

- **Hemorrhagic fevers**
  5. Yellow fever virus
  6. Dengue fever virus
  7. Ebola virus/Marburg virus
  8. Lassa fever virus

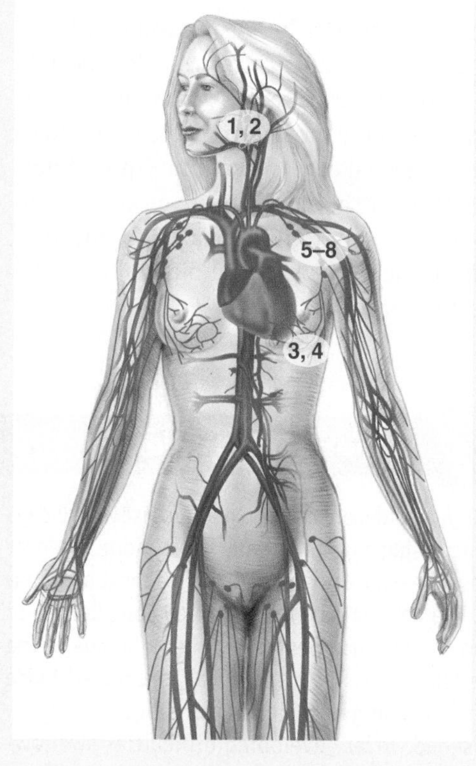

16.3 Viral Infections of the Gastrointestinal Tract Are Major Global Health Problems (not shown)

- **Hepatitis A**
  Hepatitis A virus
- **Hepatitis E**
  Hepatitis E virus
- **Viral gastroenteritis**
  Rotavirus, norovirus, enterovirus

16.4 Viral Diseases of the Nervous System Can Be Deadly

- **Rabies**
  9. Rabies virus
- **Polio**
  10. Poliovirus
- **Arboviral encephalitis**
  11. St. Louis encephalitis virus
  12. Eastern equine encephalitis virus
  13. Western equine encephalitis virus
  14. La Crosse virus
  15. West Nile virus

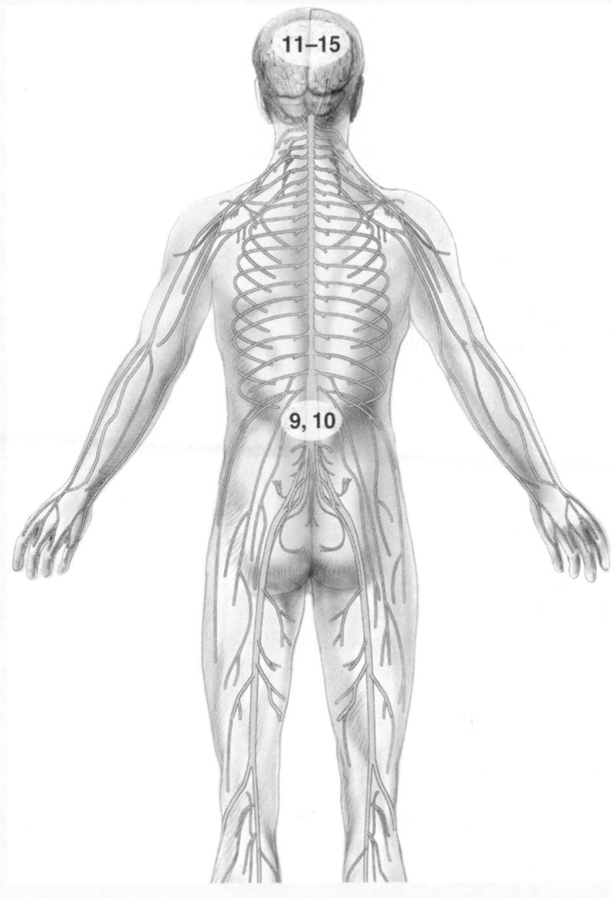

CHAPTER SELF-TEST

For **STEPS A–D**, answers to even-numbered questions and problems can be found in **Appendix C** on the Student Companion Website at **http://microbiology.jbpub.com/10e**. In addition, the site features eLearning, an online review area that provides quizzes and other tools to help you study for your class. You can also follow useful links for in-depth information, read more MicroFocus stories, or just find out the latest microbiology news.

STEP A: REVIEW OF FACTS AND TERMS

Multiple Choice

Each of the following questions is designed to assess your ability to remember or recall factual or conceptual knowledge related to this chapter. Read each question carefully, then select the ***one*** answer that best fits the question or statement.

1. Mononucleosis is an infection of _____ cells by the _____.
A. T; cytomegalovirus
B. B; Epstein-Barr virus
C. lung; cytomegalovirus
D. red blood; Epstein-Barr virus

2. Which of the following is NOT a transmission mechanism for hepatitis B?
A. Sexual contact
B. Nonsterile body piercing equipment
C. Fecal-oral route
D. Blood-contaminated needles

3. Symptoms of headache, fever, and muscle pain lasting 3 to 5 days, followed by a 2 to 24 hour abating of symptoms is characteristic of _____.
A. yellow fever
B. hepatitis C
C. dengue fever
D. Ebola hemorrhagic fever

4. A long thread-like RNA virus is typical of the _____ viruses.
A. hepatitis C
B. Ebola
C. polio
D. West Nile

5. The reservoir for Lassa fever is _____.
A. rats
B. mosquitoes
C. ticks
D. sandflies

6. Which one of the following characteristics pertains to hepatitis A?
   A. Transmission is by the fecal-oral route.
   B. The incubation period is 2 to 4 weeks.
   C. It is an acute, inflammatory liver disease.
   D. All of the above (**A–C**) are correct.
7. _____ are the single most important cause of diarrhea in infants and young children admitted to American hospitals.
   A. Noroviruses
   B. Echoviruses
   C. Hepatitis A viruses
   D. Rotaviruses
8. Hydrophobia is a term applied to _____.
   A. rotavirus infections
   B. West Nile fever
   C. arboviral encephalitis
   D. rabies
9. These viruses multiply first in the tonsils and then the lymphoid tissues of the gastrointestinal tract.
   A. Rabies virus
   B. Rotavirus
   C. Polio virus
   D. Hepatitis A virus
10. Arboviral encephalitis is an example of a _____.
    A. disease causing gastroenteritis
    B. disease spread by the fecal-oral route
    C. zoonosis
    D. type of hepatitis

True-False

Each of the following statements is true (T) or false (F). If the statement is false, substitute a word or phrase for the underlined word or phrase to make the statement true.

11. _____ Both yellow fever and dengue fever are caused by a DNA virus transmitted by the mosquito.
12. _____ Eighty percent of people infected by West Nile virus experience flu-like symptoms.
13. _____ Downey cells are a characteristic sign of the viral disease infectious mononucleosis.
14. _____ The term "hydrophobia" means "fear of water," and it is commonly associated with patients who have encephalitis.
15. _____ Norovirus and rotavirus are both considered to be agents of viral encephalitis.
16. _____ The Epstein-Barr virus is the cause of infectious mononucleosis.
17. _____ Hepatitis is primarily a disease of the liver.
18. _____ A vaccine is available to prevent yellow fever.
19. _____ The Salk and Sabin vaccines are used for immunizations against hepatitis.
20. _____ Hepatitis B is most commonly transmitted by contact with infected semen or infected blood.
21. _____ One of the most important causes of diarrhea in infants and young children admitted to hospitals is the Epstein-Barr virus.
22. _____ Filoviruses are long, thread-like viruses that cause hemorrhagic fevers and include the marburgvirus.
23. _____ Enlargement of the lymph nodes, sore throat, mild fever, and a high count of B-lymphocytes are characteristic symptoms in people who have polio.

STEP B: CONCEPT REVIEW

After understanding the textbook reading, you should be capable of writing a paragraph that includes the appropriate terms and pertinent information to answer the objective.

24. Describe the symptoms of **mononucleosis** and the potential complications arising from the illness. (**Key Concept 1**)
25. Compare the similarities and differences between the nature of the hepatitis B and C viruses and the illnesses they cause. (**Key Concept 1**)
26. Discuss the potential complications of a **yellow fever** illness and explain why **dengue fever** is considered the most important **arboviral disease** of humans. (**Key Concept 2**)
27. Summarize the symptoms of **Ebola** and **Marburg hemorrhagic fevers**. (**Key Concept 2**)
28. Describe how **Lassa fever** is transmitted. (**Key Concept 2**)
29. Explain how the hepatitis A virus is spread and prevented. (**Key Concept 3**)
30. Explain why rotavirus infections are so deadly in children and describe how noroviruses are transmitted. (**Key Concept 3**)
31. Assess the outcome to someone bit by a rabid animal and recommend treatment if **rabies** symptoms have not yet appeared. (**Key Concept 4**)
32. Explain how the polio virus causes disease and identify the two types of **polio** vaccines. (**Key Concept 4**)
33. List the possible outcomes from an infection with the West Nile virus. (**Key Concept 4**)

STEP C: APPLICATIONS AND PROBLEMS

34. Written on some blood donor cards is the notation "CMV." What do you think the letters mean, and why are they placed there?
35. Sicilian barbers are renowned for their skill and dexterity with razors (and sometimes their singing voices). French researchers studied a group of 37 Sicilian barbers and found that 14 had antibodies against hepatitis C, despite never having been sick with the disease. By comparison, when a random group of 50 blood donors was studied, none had the antibodies. As an epidemiologist, what might account for the high incidence of exposure to hepatitis C among the barbers?
36. As a state health inspector, you are suggesting all restaurant workers should be immunized with the hepatitis A vaccine. Why would restaurant owners agree or disagree with your idea?
37 An epidemiologist notes that India has a high rate of dengue fever but a very low rate of yellow fever. What might be the cause of this anomaly?

STEP D: QUESTIONS FOR THOUGHT AND DISCUSSION

38. Health authorities panicked when an outbreak of Ebola hemorrhagic disease occurred among imported macaques in a quarantine facility in Reston, Virginia, in 1989. What sparks such a dramatic response when a disease like Ebola fever breaks out?

39. A diagnostic test has been developed to detect hepatitis C in blood intended for transfusion purposes. Obviously, if the test is positive, the blood is not used. However, there is a lively controversy as to whether the blood donor should be informed of the positive result. What is your opinion? Why?

40. In the southwestern United States, abundant rain and a mild winter often bring conditions that encourage a burgeoning rodent population. Under these circumstances, what viral disease would health officials anticipate and what precautions should they give residents?

41. Disney World and 20 swampy counties in Florida use "sentinel chickens" strategically placed on the grounds to detect any signs of viral encephalitis. Why do you suppose they use chickens? Why are Disney World and many Florida counties particularly susceptible to outbreaks of viral encephalitis? What recommendations might be offered to tourists if the disease broke out?

17

CHAPTER PREVIEW

17.1 The Kingdom Fungi Includes the Molds and Yeasts

17.2 Fungi Have Evolved into a Variety of Forms

Investigating the Microbial World 17: Bats, White Noses, and Koch's Postulates

MicroInquiry 17: Evolution of the Fungi

17.3 Some Fungi Cause Intoxications

17.4 Some Fungi Can Invade the Skin

17.5 Many Fungal Pathogens Cause Lower Respiratory Tract Diseases

Textbook Case 17: An Outbreak of Coccidioidomycosis

Image courtesy of Dr. Fred Murphy/CDC.

Eukaryotic Microorganisms: The Fungi

...to see mold growing everywhere. Green, black, and blue mold growing on every surface in the houses. If the water line had been 8 feet up the wall, the mold was growing 10 feet up. There was mold on TVs, bookcases, countertops, dining tables, chairs, stove tops. Literally every surface had mold growing on it.

—Richard, a photojournalist describing the aftermath of Hurricane Katrina to Cynde Lee, Content Manager for MedicineNet.com

In August 2005, Hurricane Katrina left unimaginable devastation everywhere along the south-central coast of the United States. In New Orleans, thousands of homes were flooded and left sitting in feet of water for weeks. Many health experts were concerned about outbreaks of infectious diseases like cholera, West Nile fever, and gastrointestinal illnesses. Homes sitting in stagnant water could become breeding grounds for microorganisms.

Thankfully, most of these infectious disease scenarios did not occur. However, what did break out in many of the parishes in New Orleans was mold. There was mold on walls, ceilings, cabinets, clothes, and just about anything that provided a source of moisture and nutrients (FIGURE 17.1). It formed carpets of spore-forming colonies everywhere.

If you see small spots of mold in your home, a bleach solution will do a great job to kill and eliminate the problem. But what if an entire home and its contents are one giant "culture dish"? More than likely, most of these homes have been demolished (or stripped to the framing) and furniture and other home contents destroyed. Many home items, like beds, couches, or cabinets that were above the water level contained mold due to the prolonged humidity, and had to be replaced. Most health officials told residents to follow the same slogan used for potentially spoiled food: "When in doubt, throw it out."

(A)

(B)

FIGURE 17.1 **A Wall of Mold.** **(A)** Every circular spot seen on this wall and **(B)** on the ceiling and other objects represents a mold colony. Each colony started from a single spore; now each mature colony contains millions of spores. **(A)** © Scott Threlkeld, The Times-Picayune/Landov. **(B)** © Reuters/Charles W. Luzier/Landov. »» How do you think molds reproduce?

At Tulane Hospital, the first floor was covered with mold, which made cleanup and reopening of many wards a very difficult chore. In homes as well as offices and hospitals, molds were discovered growing in ventilation systems and the ventilation ducts. If the ventilation fans were turned on, literally hundreds of billions of spores would be blown and spread to new areas to germinate and grow. In fact, 6 weeks after Katrina, the mold spore count in the air at various sites in New Orleans was as high as 102,000 spores per cubic meter—twice the number considered normal for New Orleans.

What about illnesses or disease from breathing the mold spores? By November 2005, many New Orleans residents who had returned were suffering from upper respiratory problems—the residents "affectionately" called it the "Katrina cough." In fact, residents with asthma, bronchitis, and allergies who had left New Orleans were asked not to return just yet. Although no respiratory outbreaks have occurred, in 2008 a 5-year study began to investigate if workers in New Orleans are at risk from inhaling mold spores or bacteria stemming from the Hurricane Katrina floods.

This discussion indicates that molds grow in many natural and constructed environments. They often break down dead or decaying matter, such as the cases described in New Orleans. Many molds though are of great importance to the natural recycling in the biosphere. Still others can act as pathogens and cause some dangerous and debilitating diseases.

■ Biosphere: The whole area of Earth's surface, atmosphere, and sea that is inhabited by living things.

From the description of the situation in New Orleans, you can appreciate fungi as producers of massive numbers of **spores**, representing microscopic cells for disseminating the organisms to new territories and environments where they germinate, grow, and again reproduce.

Molds and yeasts are fungi, and they contain many species some of which we survey and study in this chapter. We will encounter many beneficial fungi such as those used to make antibiotics or in commercial and industrial processes. We also will identify and discuss several widespread human diseases caused by fungi. Many are of concern to immunocompromised individuals.

■ Immunocompromised: Refers to the lack of an adequate immune response resulting from disease, exposure to radiation, or treatment with immunosuppressive drugs.

Our study begins with a focus on the structures, growth patterns, and life cycles of fungi—something quite unique from the other groups of microorganisms.

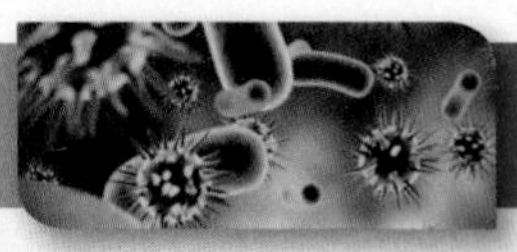

Chapter Challenge

This chapter on the fungi covers their basic characteristics and classification in the domain Eukarya, and the major human diseases they cause. Because these are diverse topics, this chapter's challenge will be a set of mini-challenges for each section.

KEY CONCEPT 17.1 The Kingdom Fungi Includes the Molds and Yeasts

The **fungi** (sing., fungus) are a diverse group of eukaryotic microorganisms. Some 75,000 species have been described, although as many as 1.5 million may exist. For many decades, fungi were classified as plants, but laboratory studies have revealed at least four properties that distinguish fungi from plants:

- Fungi lack chlorophyll, while plants have this pigment.
- Fungal cell walls contain a carbohydrate called chitin; plant cell walls have cellulose.
- Most fungi are not truly multicellular like plants.
- Fungi are heterotrophic, while plants are autotrophic.

Mainly for these reasons, fungi are placed in their own kingdom Fungi, within the domain Eukarya of the "tree of life." The study of fungi is called **mycology** (*myco* = "fungus") and a person who studies fungi is a mycologist.

Fungi Share a Combination of Characteristics

Fungi generally have life cycles involving two phases: a growth (vegetative) phase and a reproductive phase. A major group of fungi, the **molds**, which you have come to appreciate already from the chapter opener, grow as long, tangled filaments of cells that give rise to visible colonies (FIGURE 17.2A). Another group, the **yeasts**, are unicellular organisms whose colonies on agar visually resemble bacterial colonies (FIGURE 17.2B). Yet other forms are **dimorphic**; usually at ambient temperature (25°C) they grow as filamentous molds, but at body temperature (37°C) they convert to unicellular, pathogenic, yeast-like forms.

The molds consist of masses of intertwined, tubular filaments called **hyphae** (sing., hypha). The hyphae are the morphological unit of a filamentous

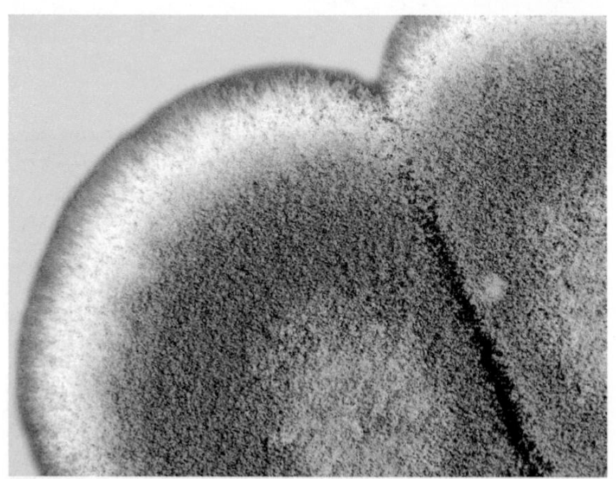

(A)

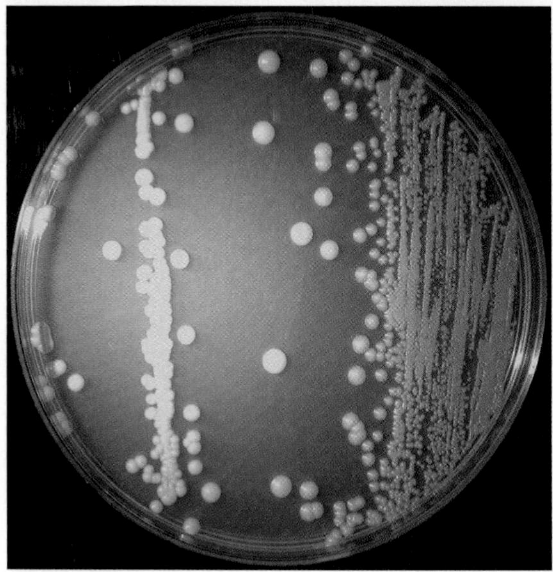

(B)

FIGURE 17.2 **Fungal Colonies.** (**A**) On growth media, molds, such as this *Penicillium* species, grow as colonies visible to the naked eye. The spores are the gray-green regions of the colonies. Courtesy of Jeffrey Pommerville. (**B**) Petri dish culture showing colonies of the yeastlike fungus *Torulopsis glabrata*. Courtesy of John Pitt of CSIRO Food Science, Australia. »» What does the white fuzzy growth in (A) represent?

fungus and individual hyphae usually are visible only with the aid of a microscope (FIGURE 17.3A). Hyphae have a broad diversity of forms and can be highly branched. A thick mass of hyphae is called a **mycelium** (pl., mycelia). This mass is usually large enough to be seen with the unaided eye, and generally it has a rough, cottony texture (Figure 17.2A). The mycelium along with any reproductive structures would represent the fungal organism.

Being eukaryotic organisms, fungi have one or more nuclei as well as a range of organelles including mitochondria, an endomembrane system, ribosomes, and a cytoskeleton. The cell wall is composed of large amounts of chitin. **Chitin** is a carbohydrate polymer of acetylglucosamine units; that is, glucose molecules containing amino and acetyl groups. The cell wall provides rigidity and strength, which, like the cell wall of all microorganisms, allows the cells to resist bursting due to high internal water pressure.

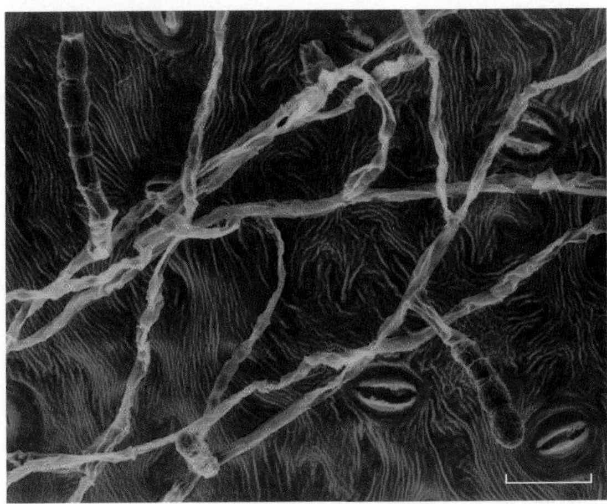

(A)

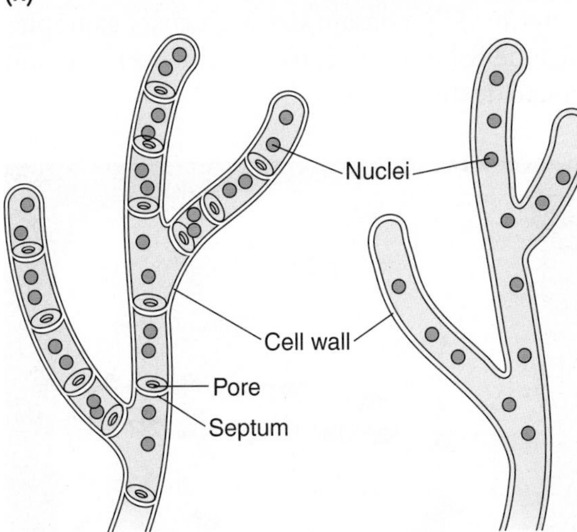

(B)

FIGURE 17.3 **Hypha Structure.** (**A**) A false-color scanning electron micrograph of fungal hyphae growing on a leaf surface. (Bar = 40 μm.) © Dennis Kunkel Microscopy, Inc./Visuals Unlimited/Corbis. (**B**) Molds have hyphae that are either septate or nonseptate. Septa compartmentalize hyphae into separate cells, although the septa have a pore through which cytoplasm and nuclei can move. »» What unique structural features are presented in this figure?

In many species of fungi, hyphal cross walls, called **septa** (sing., septum), divide the cytoplasm into separate "cells" (FIGURE 17.3B). Such fungi are described as **septate**. In the blue-green mold *Penicillium chrysogenum,* the septa are incomplete, however, and pores in the cross walls allow adjacent cytoplasms to mix. In other fungal species, such as the common bread mold *Rhizopus,* the filaments are without septa. Such hyphae without septa are considered **coenocytic**, meaning they contain many nuclei in a common cytoplasm.

Because fungi absorb preformed organic matter, they are described as **chemoheterotrophs**. Most are **saprobes**, feeding on dead or decaying organic matter. Together with many bacterial species, these fungi make up the **decomposers**, recycling vast quantities of organic matter (MicroFocus 17.1). By living off the remains of dead organisms, they break down complex biological compounds into simpler raw materials (e.g., carbon, minerals) that can then be used by the photoautotrophs.

Other fungi are pathogens, living on plants or animals, and often causing disease. Most such fungi are **opportunistic** and only cause serious disease in immunocompromised individuals.

Fungal Growth Is Influenced by Several Factors

Fungi acquire their nutrients through absorption, either as saprobes or pathogens. Being mostly terrestrial organisms, the molds and yeasts secrete enzymes into the surrounding environment that break down (hydrolyze) complex organic compounds into simpler ones. As a result of this extracellular digestion, simpler compounds, like glucose and amino acids, can be absorbed by the hyphae.

The mycelium formed by a mold represents the "feeding network" for these fungi. In some cases the mycelium can form a tremendously large surface area for nutrient absorption. For the yeasts, nutrients are absorbed across the cell surface, similar to the way bacterial cells obtain their nutrients.

Mycelial growth occurs by the elongation of the hyphae. The nuclei divide by mitosis and growth of these mycelial fungi involves lengthening and branching of the hyphae at the hyphal tips. Yeast

MICROFOCUS 17.1: Evolution

When Fungi Ruled the Earth

About 250 million years ago, at the close of the Permian period, a catastrophe of epic proportions visited the Earth. Scientists believe that over 90% of animal species in the seas vanished. The great Permian extinction, as it is called, also wreaked havoc on land animals and cleared the way for dinosaurs to inherit the planet.

But land plants managed to survive, and before the dinosaurs came, they spread and enveloped the world. At least, that is what paleobiologists traditionally believed.

A mushroom species growing on a rotting log. © Izatul Lail bin Mohd Yasar/ShutterStock, Inc.

Now, however, they are revising their theory and finding a significant place for the fungi. In 1996, Dutch scientists from Utrecht University presented evidence that land plants were decimated by the extinction and that for a brief geologic span, dead wood covered the planet. During this period, they suggest the fungi emerged and wood-rotting species experienced a powerful spike in their populations (see figure). Support for their theory is offered by numerous findings of fossil fungi from the post-Permian period. The fossils are bountiful and they come from all corners of the globe. Significantly, they contain fungal hyphae, the active feeding forms rather than the dormant spores.

And so it was that fungi proliferated wildly and entered a period of feeding frenzy where they were the dominant form of life on Earth. It's something worth considering next time you kick over a mushroom growing on a rotting log.

growth in number occurs by cell division as the cells undergo many rounds of mitosis and cytokinesis to form a large population of individual cells.

Fungal growth is influenced by many factors in the environment. Besides the availability of chemical nutrients, oxygen, temperature, and pH influence growth.

Oxygen. The majority of fungi are aerobic organisms, with the notable exception of the facultative yeasts, which can grow in either the presence of oxygen or under fermentation conditions.

Temperature. Most fungi grow best at about 23°C, a temperature close to normal room temperature. Notable exceptions are the pathogenic fungi, which grow optimally at 37°C, which is body temperature. As mentioned, dimorphic fungi grow as yeastlike cells at 37°C and a mycelium at 23°C. Psychrophilic fungi grow at still lower temperatures, such as the 5°C found in a normal refrigerator.

■ Psychrophilic: **Refers to organisms that prefer to grow at cold temperatures.**

pH. Many fungi thrive under mildly acidic conditions at a pH between 5 and 6. Acidic soil therefore may favor fungal turf diseases, in which case lime (calcium carbonate) is added to neutralize the soil. Mold contamination also is common in acidic foods such as sour cream, citrus fruits, yogurt, and most vegetables. Moreover, the acidity in breads and cheese encourages fungal growth. Blue (Roquefort) cheese, for example, consists of milk curds in which the mold *Penicillium roqueforti* has been added for flavor and texture (FIGURE 17.4).

Normally, high concentrations of sugar are conducive to growth, and laboratory media for fungi usually contain extra glucose; examples include Sabouraud dextrose (glucose) agar and potato dextrose agar.

FIGURE 17.4 **Roquefort Cheese. Roquefort cheese is made from cow's milk and contains the mold *Penicillium roquefortii*. © Jones & Bartlett Learning. Photographed by Kimberly Potvin.** »» Why would a mold be added to the ripening process of cheeses such as this?

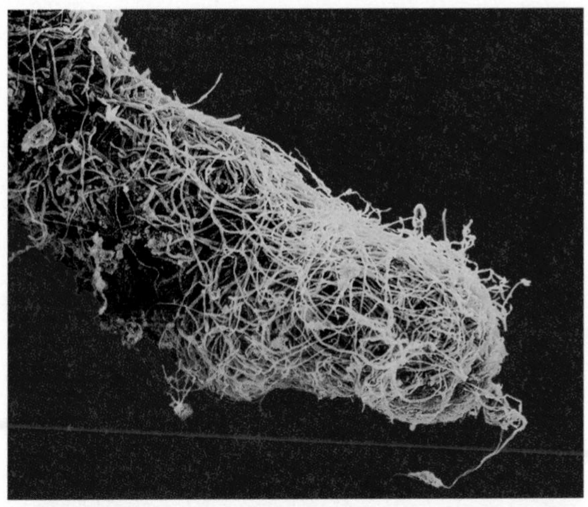

(A)

(B)

FIGURE 17.5 **Mycorrhizae and Their Effect on Plant Growth.** (**A**) Mycorrhizae surround the root of a *Eucalyptus* tree in this false-color scanning electron micrograph. These fungi are involved symbiotically with their plant host, such as aiding in mineral metabolism. © Dr. Gerald Van Dyke/Visuals Unlimited/Getty. (**B**) An experiment analyzing the mycorrhizal effects on plant growth. © Science VU/R.Roncadori/Visuals Unlimited. »» Which plant or plants (CK, GM, GE) represent(s) the control and experimental set ups?

Fungal growth in nature forms important links in ecological cycles because fungi, along with bacterial species, rapidly decompose animal and plant matter. Working in immense numbers, fungi release carbon and minerals back to the environment, making them available for recycling.

Many fungi also live in a mutually beneficial relationship with other species in nature through a symbiotic association called mutualism. Fungi called **mycorrhizae** (*rhiza* = "root") live harmoniously with plants where the hyphae of these fungi invade or envelop the roots of plants (FIGURE 17.5A). These mycorrhizae consume some of the carbohydrates produced by the plants, but in return act as a second root system, contributing essential minerals and water to promote plant growth (FIGURE 17.5B). Mycorrhizae have been found in plants from salt marshes, deserts, and pine forests. In fact, these beneficial fungi live in and around the roots of 95% of examined plant species.

Besides the mycorrhizae, most plants examined also contain fungal **endophytes**, which are fungi living and growing entirely within plants, especially leaf tissue. They do not cause disease but, rather like mycorrhizae, they provide better or new growth opportunities for the plant. In the southwestern Rocky Mountains, for instance, a fungus thrives on the blades of a species of grass, producing a powerful poison that can put horses and other animals to sleep for about a week (the grass is called "sleepy grass" by the locals). Thus, the grass survives while other species are nibbled to the ground. MicroFocus 17.2 describes a few more remarkable examples.

Reproduction in Fungi Involves Spore Formation

Sporulation is the process of spore formation. It usually occurs in structures called **fruiting bodies**, which represent the part of a fungus in which spores are formed and from which they are released. These structures may be asexual and invisible to the naked eye, or sexual structures, such as the macroscopic mushrooms.

■ Mutualism: A symbiotic relationship between two organisms of different species that benefits both.

Asexual Reproduction. Asexual reproductive structures develop at the ends of specialized hyphae. As a result of mitotic divisions, thousands of spores are produced, all genetically identical.

Many asexual spores develop within sacs or vessels called **sporangia** (sing., sporangium; *angio* = "vessel") (FIGURE 17.6A). Appropriately, the spores are called **sporangiospores**. Other fungi produce spores on supportive structures called **conidiophores** (FIGURE 17.6B). These unprotected, dust-like spores are known as **conidia** (sing., conidium; *conidio* = "dust"). Fungal spores are extremely light and are blown about in huge numbers by wind currents. In yet other fungi, spores may form simply by fragmentation of the hyphae yielding **arthrospores** (*arthro* = "joint"). The fungi that cause athlete's foot multiply in this manner.

Many yeasts reproduce asexually by **budding**. In this process, the cell becomes swollen at one edge, and a new cell called a **blastospore**, (*blasto* = "bud") develops (buds) from the parent cell (FIGURE 17.7). Eventually, the spore breaks free

MICROFOCUS 17.2: Environmental Microbiology

Fungi as Protectors

What would you think if you saw wheat growing in the deserts of California or Arizona? You probably would think it was a mirage or you were hallucinating from the lack of water. Well, in the near future, such an apparition might indeed be real.

It turns out that mycorrhizae are not the only fungi providing protection and growth benefits for hundreds of species of plants. The hyphae of endophytes grow within plant tissue and between cells of many healthy leaves where nutrients are exchanged between host and fungus. No damage is done to the leaves and only the reproductive structures make it to the surface to release spores into the wind.

In cacao leaves, the more mature (green) leaves gain more benefit from endophyte associations than young (red) leaves. Courtesy of Dr. A. Elizabeth Arnold, Assistant Professor & Curator, Gilbertson Mycological Herbarium, Department of Plant Sciences, University of Arizona.

One example of endophyte benefit was discovered by Regina Redman and Russell Rodriguez of the U.S. Geological Survey in Seattle, Washington. They took perennial grasses normally growing in hot soils around geysers and grew them in the lab. Those plants exposed to hot soil but lacking endophytes died while those living with endophytes survived. Most interesting, the fungi by themselves also died in the hot soils. Apparently, there needs to be a give and take between host and fungus—representing a true mutualistic relationship.

Redman and Rodriguez then teamed up with Joan Henson at Montana State University to show that these same endophytes protect other plants. At least in the lab, when endophyte spores were placed on tomato, watermelon, or wheat seedlings, the tomato and watermelon seedlings with endophytes survived the stresses of high temperatures (50°C) or drought conditions. Although the wheat seedlings died, the plants survived about a week longer than those without endophytes.

Meanwhile in Panama, another group of researchers, led by A. Elizabeth Arnold, now at the University of Arizona, discovered that chocolate-tree (cacao) leaves harboring endophytes are more resistant to pathogen attack, while fungus-free leaves become diseased. In their studies, when leaves were purposely inoculated with one of the major pathogens of cacao, leaves associated with several different endophytes were less likely to be invaded and would most likely survive. Although young leaves benefited from the symbiosis, the older leaves appeared to benefit the most from the endophyte association (see figure).

So, at least in the lab, endophytic fungi play various but important roles in plant survival. As more is discovered, perhaps the day will come when wheat will grow in the desert.

to live independently. The parent cell can continue to produce additional blastospores.

Once free of the fruiting body, spores landing in an appropriate environment have the capability of germinating to reproduce new unicellular yeast cells or a new hypha (FIGURE 17.8). Continued mold growth will eventually form a mycelium.

■ Mating types: Separate mycelia of the same fungus or separate hyphae of the same mycelium.

Sexual Reproduction. Many fungi also produce spores by sexual reproduction. In this process, opposite mating types come together and fuse (FIGURE 17.9). Because the nuclei are genetically different in each mating type, the fusion cell represents a **heterokaryon** (*hetero* = "different"; *karyo* = "nucleus"); that is, a cell with genetically dissimilar nuclei existing for some length of time in a **dikaryotic** cytoplasm. Eventually the nuclei fuse and a diploid cell is formed. The chromosome number soon is halved by meiosis, returning the cell or organism to a haploid condition.

A visible fruiting body often results during sexual reproduction and it is the location of the haploid spores. Perhaps the most recognized fruiting body from which spores are produced is the mushroom.

CONCEPT AND REASONING CHECKS 1

a. Assess the role of hyphae to fungal growth.
b. Estimate the value of mycorrhizae and fungal endophytes to plant survival.
c. Assess the role of asexual and sexual reproduction in fungi.

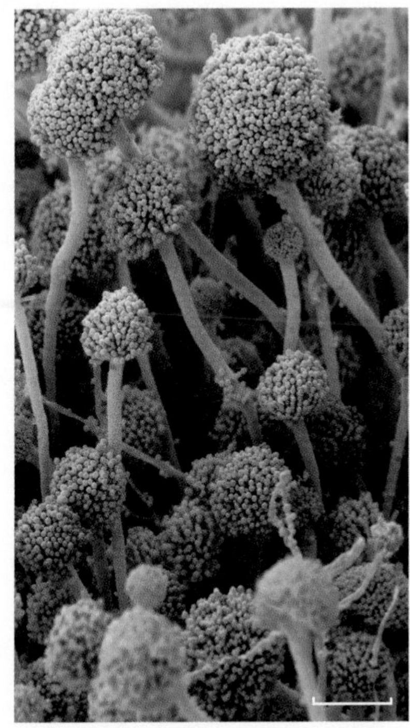

(A)

(B)

FIGURE 17.6 **Fungal Fruiting Bodies.** False-color scanning electron micrographs of sporangia and conidia. (**A**) Sporangia of the common bread mold *Rhizopus*. Each round sporangium contains thousands of sporangiospores. (Bar = 20 µm.) © Andrew Syred/Photo Researchers, Inc. (**B**) The conidiophores and conidia in the mold-like phase of *Penicillium roquefortii*. Many conidiophores are present within the mycelium. Conidiophores (orange) containing conidia (blue) are formed at the end of specialized hyphae (green). (Bar = 20 µm.) © Dennis Kunkel Microscopy, Inc./Phototake/Alamy Images. »» Why must fungal spores be elevated on the tips of hyphae?

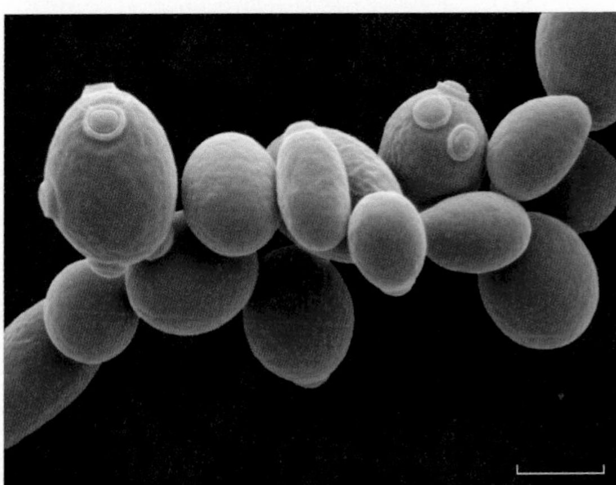

FIGURE 17.7 **Yeast Budding.** False-color scanning electron micrograph of the unicellular yeast *Saccharomyces cerevisiae*. (Bar = 3 µm.) © Medical-on-Line/Alamy Images. »» Propose what the circular areas represent on the parent cell at the left and right center of the micrograph.

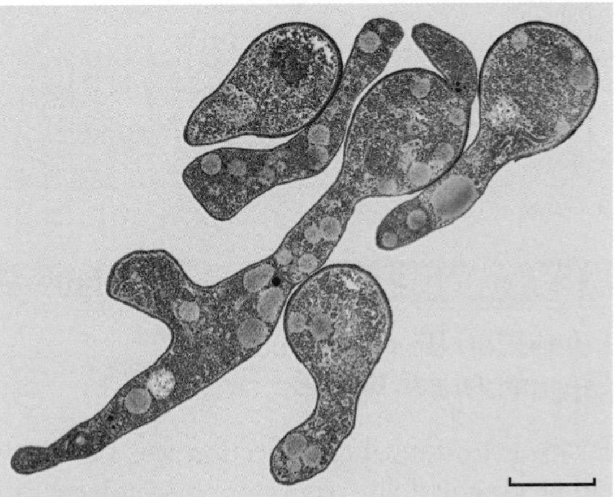

FIGURE 17.8 **Germinating Fungal Spores.** A false-color transmission electron micrograph of germinating fungal spores. (Bar = 3 µm.) © Science Photo Library/Photo Researchers, Inc. »» What do the elongated structures protruding from the round spore represent?

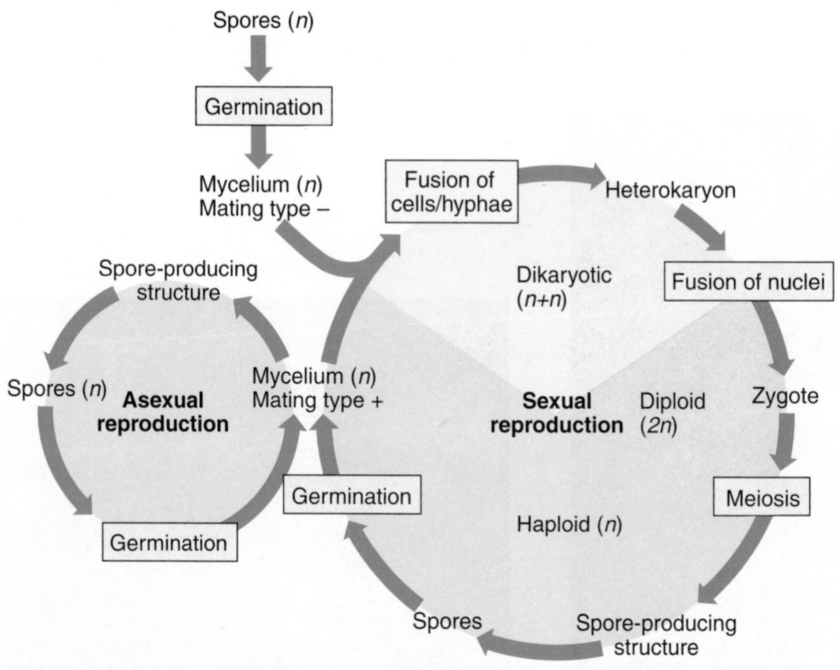

FIGURE 17.9 **A Typical Life Cycle of a Mold.** Many fungi have both an asexual and sexual reproduction characterized by spore formation. The unique phase in the life cycle is the presence of a heterokaryon where nuclei from two different mating types remain separate (dikaryotic) in a common cytoplasm. »» How could an organism like a mold survive without a sexual cycle?

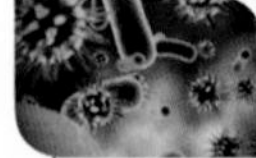

Chapter Challenge A

This section examined the basic characteristics of the fungi. One of the most unique characteristics is the filamentous (mycelial) growth of the molds.

QUESTION A:

What is the advantage of growing as a large network of hyphae? But when doing so, how can the organism exist as a coenocytic (multinucleate) structure? Most organisms, humans included, would die if cells contained more than one cell nucleus.

Answers can be found on the Student Companion Website in **Appendix D.**

KEY CONCEPT 17.2 Fungi Have Evolved into a Variety of Forms

Fungi Can Be Classified into Several Major Groups

Historically, fungal classification was based on either structural differences, or physiological and biochemical patterns. However, DNA analyses and genome sequencing are becoming an important tool for drawing new evolutionary relationships among fungal groups (FIGURE 17.10).

A fungus can be cataloged into one of several groups, including the Microsporidia, Chytridiomycota, Zygomycota, Glomeromycota, Ascomycota, and the Basidiomycota. If the fungus lacks a recognized sexual cycle, it is placed into an informal group called the mitosporic fungi.

The Cryptomycota and Microsporidia. In 2011, scientists in England, Spain, and the United States announced that they had discovered a diverse, prevalent, and previously undescribed group of eukaryotic organisms that on the phylogenetic tree appears to branch with the fungi. Tentatively called the **Cryptomycota**, these small, flagellated cells are unique in lacking a chitin cell wall. As such, the researchers believe the

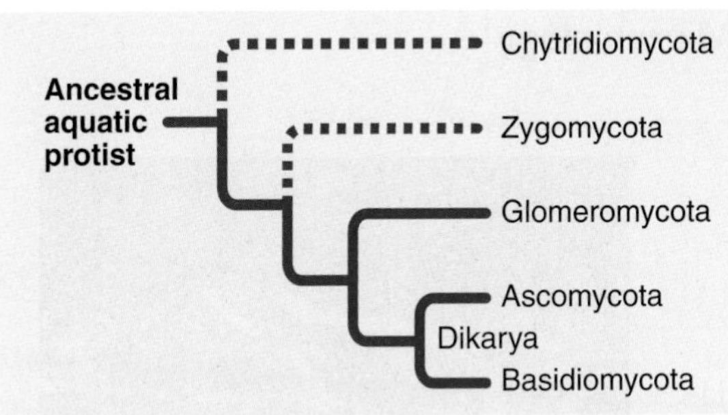

FIGURE 17.10 **A Phylogeny for the Kingdom Fungi.** This generalized scheme for the relationships between fungal groups is based on molecular findings. The dashed lines represent an as yet unclear lineage for the chytrids and zygomycetes. »» Because the chytrids are the only fungal group with flagella, indicate on the phylogenetic tree where you would place the phrase *"Loss of flagella."*

Cryptomycota must be a very ancient group of microorganisms. How they are related to the other fungal groups will require more study.

Another group of eukaryotic microbes, the **Microsporidia**, is now thought to be a sister group of the fungi. These organisms are unicellular parasites that live within the cells or tissues of protists and animals and are among the smallest known eukaryotes, some being only 1µm in diameter. The cells do contain chitin cell walls but lack true mitochondria. As with the Cryptomycota, more studies are underway to better clarify the evolutionary relationships within the group and with other fungal groups.

The Chytridiomycota and Glomeromycota. The oldest known fungi are related to certain members of the **Chytridiomycota**, commonly called **chytrids**. Members of the phylum have given us clues about the possible origin of fungi. First, chytrids are predominantly aquatic, and not terrestrial, organisms. This means the fungi originated in water probably from a flagellated, protistan ancestor. Second, being aquatic, chytrids have flagellated reproductive cells called **zoospores** that no other fungal phylum has. Finally, like other fungi, chytrids have chitin strengthening their cell walls. Until recently, few chytrids had any noticeable impact—for good or bad (MicroFocus 17.3).

The **Glomeromycota** form what some consider the most extensive symbiosis on Earth. These fungi represent a group of **endomycorrhizae** that exist within the roots of more than 80% of the world's land plants. They do not kill the plants but rather interact mutualistically by providing essential phosphate and other nutrients to the plant. In return, the fungi receive needed organic compounds from the plant. In fact, some mycologists believe that plant evolution onto land more than 400 million years ago depended on the symbiosis with the ancestral Glomeromycota, which provided plants with needed nutrients from the soil.

The Zygomycota. Like the Chytridiomycota, the **Zygomycota** is of uncertain evolutionary origin and may represent members from other fungal groups. However, for now, the Zygomycota is a useful category to describe these organisms.

The phylum Zygomycota ("zygomycetes") consists of a group of fungi inhabiting terrestrial environments. Familiar representatives include fast-growing bread molds and other molds typically growing on spoiled fruits with high sugar content or on acidic vegetables (FIGURE 17.11). On these and similar materials, the saprobic fungi typically grow inside their food, dissolving the substrate with extracellular enzymes, and taking up nutrients by absorption.

Members of the phylum make up about 1% of the described species of fungi. The zygomycetes have chitinous cell walls and grow as a coenocytic mycelium. During sexual reproduction, sexually opposite mating types fuse, forming a unique, heterokaryotic, diploid **zygospore** (FIGURE 17.12). After a period of dormancy, the zygospore releases haploid sporangiospores from a sporangium. Elsewhere in the mycelium, thousands of asexually produced sporangiospores are produced within sporangia. Both sexually produced and asexually produced spores are dispersed on wind currents. Several members can cause infections and disease in humans.

The last two groups of fungi are evolutionarily related in that both have a dikaryotic stage. Together they form the clade **Dikarya**.

The Ascomycota. Members of phylum **Ascomycota** (*asco* = "sac") or sac fungi, commonly are called the "ascomycetes." They are very diverse and account for about 75% of all known fungi. The phylum contains many common and useful fungi, including *Saccharomyces cerevisiae* (Baker's yeast), *Morchella esculentum* (the edible morel), and *Penicillium chrysogenum* (the mold that produces penicillin) (FIGURE 17.13). The phylum also has several members associated with illness and disease. *Aspergillus flavus* produces **aflatoxin**, a fungal contaminant of nuts and stored grain that

MICROFOCUS 17.3: Environmental Microbiology

The Day the Frogs Died

Amphibians are the oldest class of land-dwelling vertebrates, having survived the effects that brought the extinction of the dinosaurs. Now the amphibians are facing their own potential extinction from a different, infectious effect. In 2010, scientists estimated that more than 100 frog species have gone extinct worldwide and many others are facing the same outcome. Why are the frogs dying?

A species of harlequin frog, one of many species being wiped out by a chytrid infection. © Bruce Coleman Inc./Alamy images.

In the early 1990s, researchers in Australia and Panama started reporting massive declines in the number of amphibians in ecologically pristine areas. As the decade progressed, massive die-offs occurred in dozens of frog species and a few species even became extinct. Once filled with frog song, the forests were quiet. "They're just gone," said one researcher.

By 1998, infectious disease was identified as one of the reasons for the decline. More than 100 amphibian species on four continents, including Central and South America and Australia, were infected with a chytrid called *Batrachochytrium dendrobatidis.* This fungus, the only one known to infect vertebrates, uses the frog's keratinized skin as a nutrient source. The skin infection alters the frog's skin permeability, which leads to a fatal osmotic imbalance and heart failure. Roughly one third of the world's 6,000 amphibian species are now considered under threat of extinction due to the disease chytridiomycosis (see figure). But, where did the chytrid come from?

Researchers believe the chytrid may have originated in African frogs exported around the world. The hypothesis is that some infected African frogs escaped (they do not die from the infection) and passed the fungus to hardier ones, like bullfrogs, which in turn infected more susceptible frog species. The global amphibian trade was the mode of transmission.

Are frog deaths from chytridiomycosis a sign of a yet unseen shift in the ecosystem, much like a canary in a coal mine? Some believe this type of "pathogen pollution" may be as serious as chemical pollution. It represents an alarming example of how an emerging pathogen could potentially wipe out a whole species and eventually a whole ecosystem.

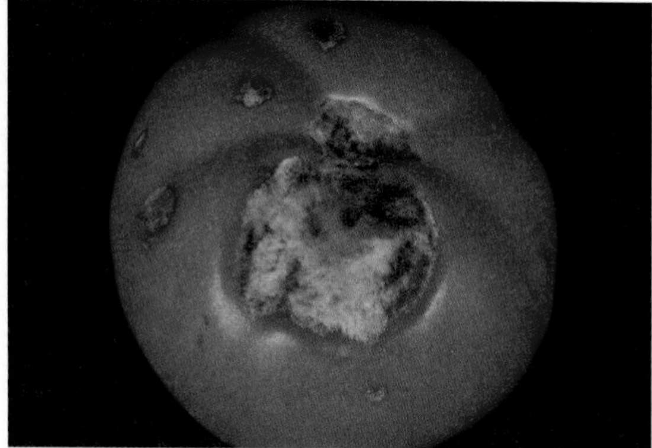

FIGURE 17.11 **Mold Growing on a Tomato.** Zygomycete molds typically grow and reproduce on overripe fruits or vegetables, such as this tomato. © Jones & Bartlett Learning. Photographed by Kimberly Potvin. »» Identify the black structures and the white fuzzy growth on the tomato.

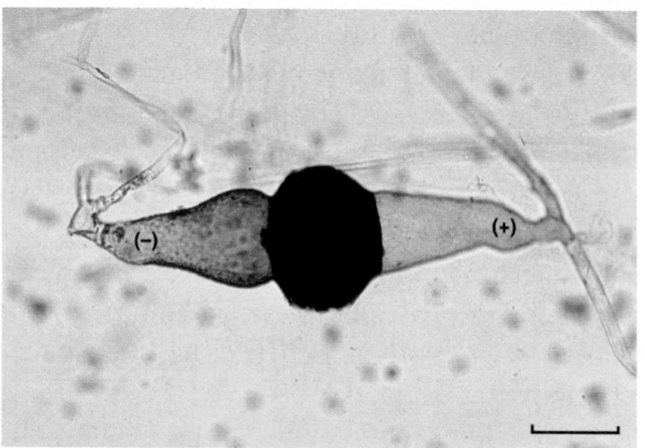

FIGURE 17.12 **Sexual Reproduction in the Zygomycota.** Sexual reproduction between hyphae of different mating types (+ and –) produces a darkly pigmented zygospore. (Bar = 30 μm.) © Phototake/Alamy Images. »» When the zygospore germinates, what will be produced from the structure?

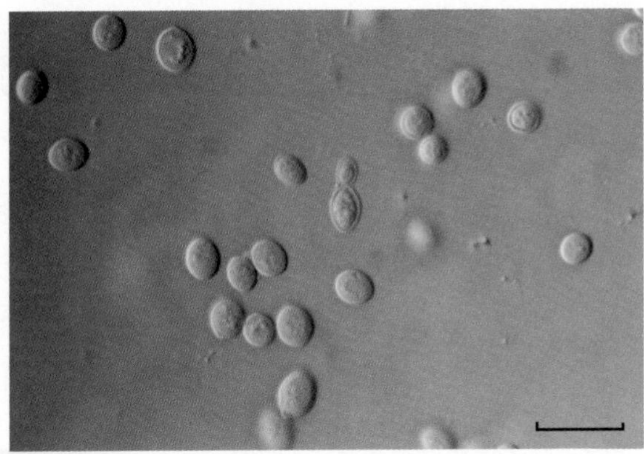

(A)

(B)

(C)

FIGURE 17.13 **Representative Ascomycetes.** (**A**) Light micrograph of yeast cells. The cell in the center is undergoing budding. (Bar = 10 μm.) © Michael Abbey/Photo Researchers, Inc. (**B**) The edible morel, an ascomycete prized for its delicate taste. © Liane Matrisch/Dreamstime.com. (**C**) A *Penicillium* species growing on an orange. Courtesy of Jeffrey Pommerville. »» Where is the growing edge of the mycelium on the orange?

is both a toxin and the most potent known natural carcinogen and *Candida albicans* cause of thrush, diaper rash, and vaginitis ("yeast infection"). In addition, ascomycetes have caused serious non-human diseases as well. Most notably is *Cryphonectria parasitica,* the main cause of chestnut blight, a devastating disease that has wiped out more than 4 billion chestnut trees since the early 1900s. Another ascomycete has a more recent association with infectious disease in bats as Investigating the Microbial World 17 explores.

The Ascomycota is a sister group to the Basidiomycota because the hyphae of both phyla are septate; both have cross walls dividing the hyphae into segments, but with large pores allowing a continuous flow of cytoplasm. The hyphae, like other filamentous fungi, form a mycelium to obtain nutrients from dead or living organisms. In fact, their biggest ecological role is in decomposing and recycling plant material.

The mycelial ascomycetes have the ability to form conidia through asexual reproduction or **ascospores** through sexual reproduction. Ascospores are formed within a reproductive structure called an **ascus** (pl., asci), within which eight haploid ascospores form (FIGURE 17.14). Many asci and other hyphae form the fruiting body.

Some ascomycetes form symbioses with plant roots (mycorrhizae) or the leaves and stems of plants (endophytes). Ascomycetes also are the most frequent fungal partner in lichens. A **lichen** is a mutualistic association between a fungus and a photosynthetic organism (**photobiont**) such as a cyanobacterium or green alga (FIGURE 17.15A, B). Most of the visible body of a lichen is the fungus with the hyphae penetrating the cells of the photosynthetic partner to receive carbohydrate nutrients. The photosynthetic organism receives fluid from the water-husbanding fungus.

Lichen asexual reproduction occurs through the formation of **soredia** that consist of a group of photobionts surrounded by hyphae. Soredia are carried by wind currents and deposited on a new surface, where a new lichen then forms.

Lichens often are grouped by appearance into leafy lichens (foliose), shrubby lichens (fruticose), and crusty lichens (crustose) (FIGURE 17.15C). Together, the organisms form an association that readily grows in environments where neither organism could survive by itself

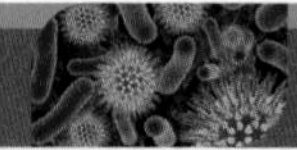

Investigating the Microbial World 17

Bats, White Noses, and Koch's Postulates

White-nose syndrome (WNS) is a disease associated with the deaths of up to 6 million North American bats. The disease causes a distinctive white lesion visible on the nose (and muzzle), wings, tail, and ears of hibernating bats. According to the United States Fish and Wildlife Service, since it was first detected in 2006 in New York, WNS has spread rapidly and has now been found among bats in more than 115 caves and mines from Canada to Indiana and Missouri, and south to Alabama. As far as is known, there is no evidence of a health threat to humans.

- **OBSERVATIONS:** An ascomycete fungus, called *Geomyces destructans,* has been isolated from infected bats. However, its role in causing WNS has been uncertain because the characteristic white lesions are the only consistent pathological finding associated with the disease. In addition, the same fungus has been found to colonize the skin of European bats without causing any reported deaths.

- **QUESTION:** ***Is* G. destructans *the cause of WNS?***

- **HYPOTHESIS:** *G. destructans* is the cause of WNS. If so, then using Koch's postulates, it should be shown that susceptible bats will contract WNS from *G. destructans.*

- **EXPERIMENTAL DESIGN:** Twenty-five little brown bats that were naturally infected with WNS were collected in New York. Additional healthy little brown bats collected from an area in Wisconsin outside the known range for WNS were housed in the lab under hibernation conditions. Pure cultures of the fungus were grown and the conidia collected from the plates. Test samples in a phosphate-buffered solution (PBS) contained 5×10^5 conidia.

- **EXPERIMENT 1:** A group of 29 healthy bats had a sample of conidia applied to one wing and another sample applied to the fur between the eye and the ear. The group consisted of 34 healthy bats that had PBS only applied to the same regions. The experiment was run for up to 102 days.

- **EXPERIMENT 2:** The group of 25 infected bats (positive control) were placed in physical contact with 18 healthy bats for up to 102 days.

- **EXPERIMENT 3:** A group of 36 healthy bats were placed in mesh cages next to the 25 infected bats for up to 102 days.

- **RESULTS:** Any bats that died and all surviving bats after 102 days were examined histologically by examination of the muzzle and skin from each wing. The *G. destructans* fungus from infected bats was re-isolated in culture from wing skin and compared genomically to the applied or natural strain. All strains were identical.

EXPERIMENTS 1–3: See table.

TABLE

Development of WNS in Experimental Bat Groups

Experiment	Group	Number with WNS lesions	Number without WNS lesions	Total
1	Conidia treated	29	0	29
	PBS treated	0	34	34
2	Contact exposure	16	2	18
	Positive control	25	0	25
3	Airborne exposure	0	36	36

- **CONCLUSIONS:**

QUESTION 1: ***Was the hypothesis validated? Explain using the table.***

QUESTION 2: ***What is the important epidemiological and disease management implication that can be drawn from experiments 2 and 3?***

QUESTION 3: ***Using experiment 1, reconstruct the four steps of Koch's postulates for WNS.***

Answers can be found on the Student Companion Website in **Appendix D.**

Adapted from: Lorch, J. M. *et al.* (2011). *Nature* 480(7377):376–378. Icon image © Tischenko Irina/ShutterStock, Inc.

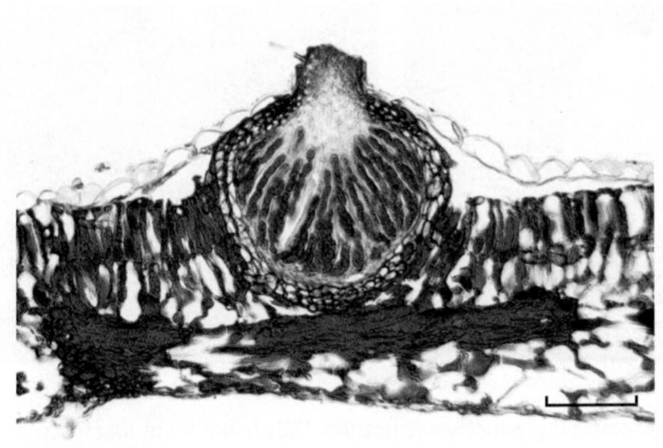

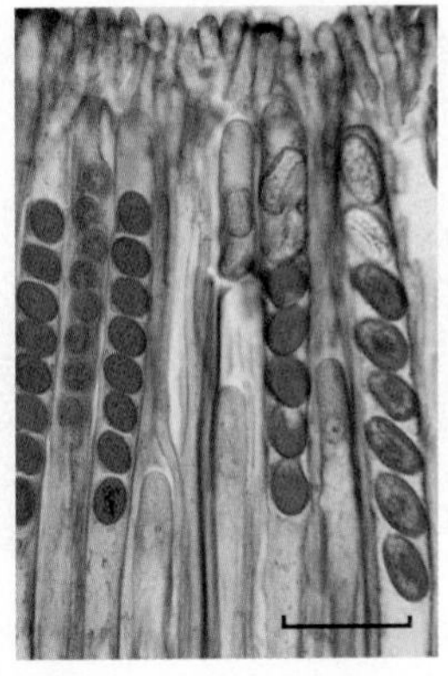

(A) (B)

FIGURE 17.14 **Ascomycetes and Spores.** (**A**) Cross section of a stained fruiting body on an apple leaf. (Bar = 20 μm.) © Biodisc/Visuals Unlimited/Corbis. (**B**) A higher magnification of several asci, each containing eight ascospores. (Bar = 5 μm.) © Dr. John D. Cunningham/Visuals Unlimited/Corbis. »» Ascospores are representative of what stage of the ascomycete life cycle?

(e.g., rock surfaces). Indeed, in some harsh environments, lichens support entire food chains. In the Arctic tundra, for example, reindeer graze on carpets of reindeer moss, which actually is a type of lichen.

The Basidiomycota. Members of the phylum **Basidiomycota**, commonly known as "basidiomycetes," are club fungi. The Basidiomycota contains about 30,000 identified species, representing 37% of the known species of true fungi. Members of the Basidiomycota can be unicellular or multicellular and have sexual and asexual cycles. The most recognized members are the mushrooms and puffballs (FIGURE 17.16A, B). Some basidiomycetes are important saprobes, decomposing wood and other plant products. Other members form mycorrhizae while still others are important plant pathogens, such as the so-called "rusts" and "smuts" that infect cereals and other grasses. Rust fungi are so named because of the rusty, orange-red color of the infected plant, while smut fungi are characterized by sooty black masses of spores forming on leaves and other plant parts. Some Basidiomycota cause serious diseases in animals and humans.

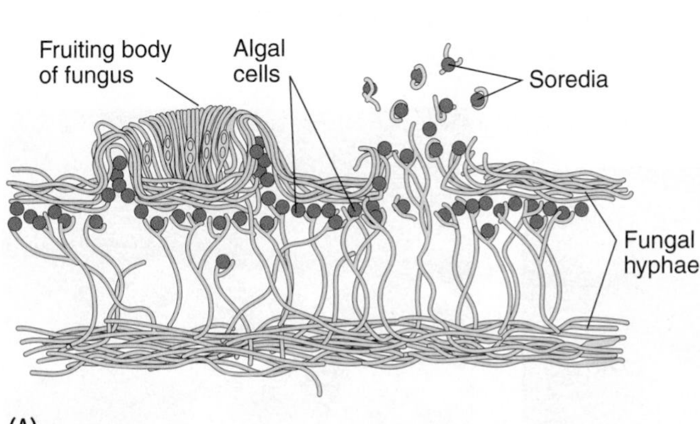

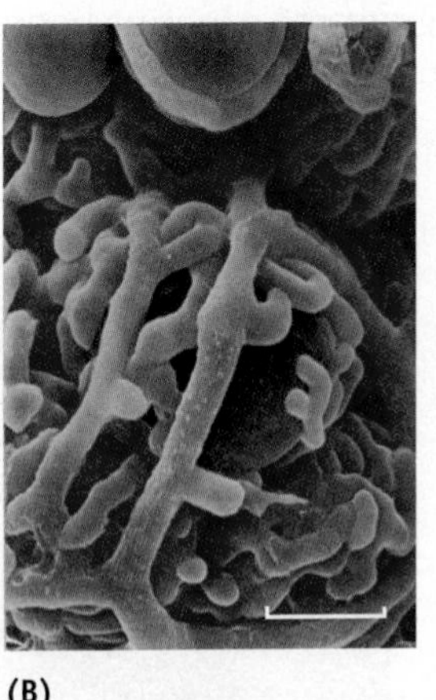

(A) (B) (C)

FIGURE 17.15 **Lichens.** (**A**) A cross section of a lichen, showing the upper and lower surfaces where tightly coiled fungal hyphae enclose photosynthetic algal cells. On the upper surface, a fruiting body has formed. Airborne clumps of algae and fungus called soredia are dispersed from the ascocarp to propagate the lichen. Loosely woven fungi at the center of the lichen permit the passage of nutrients, fluids, and gases. (**B**) A false-color scanning electron micrograph of the close, intimate contact between fungal hyphae (orange) and an alga cell (green). (Bar = 2 μm.) © V. Ahmadjian/Visuals Unlimited. (**C**) A typical lichen growing on the surface of a rock. Lichens are rugged organisms that can tolerate environments where there are few nutrients and extreme conditions. The brown fruiting bodies can be seen. Courtesy of Jeffrey Pommerville. »» What attributes of the fungus and alga permit the lichen to withstand extreme environmental conditions?

(A)

(B)

(C)

FIGURE 17.16 **The Basidiomycota.** The basidiomycetes are characterized by sexual reproductive structures that usually are macroscopic. These include (**A**) the mushrooms, such as this species of *Amanita muscaria,* and (**B**) the puffballs. (**A–B**) Courtesy of Jeffrey Pommerville. (**C**) The outward spreading of a fungal mycelium can be visually detected by the formation of a ring of mushrooms, often called a "fairy ring." © Anthony Collins/Alamy Images. »» Knowing how a mycelium grows, where on the mycelium (fairy ring) in (C) do the mushrooms form?

The name basidiomycete refers to the reproductive structure on which sexual spores are produced. In many mushrooms, the underside of the cap is lined with "gills" on which club-shaped **basidia** (sing., basidium; *basidium* = "small pedestal") are formed (FIGURE 17.17). Within these basidia, the haploid, sexual spores, called **basidiospores**, are produced.

In the soil, the basidiospores germinate and grow as a haploid mycelium. When mycelia of different mating types come in contact, they fuse into a heterokaryon containing genetically different haploid nuclei (FIGURE 17.18). Under appropriate environmental conditions, some of the hyphae become tightly compacted and force their way to the surface and grow into a fruiting body (basidiocarp) typically called a **mushroom**. Often a ring of mushrooms forms, which has been called a "fairy ring" because centuries ago people thought the mushrooms appeared where "fairies" had danced the night before (FIGURE 17.16C).

MicroInquiry 17 looks at the relationship of the fungi to the other eukaryotic kingdoms.

The Mitosporic Fungi. Certain fungi lack a known sexual cycle of reproduction; consequently, they are labeled with the term **mitosporic fungi** because the asexual spores are the product of mitosis. Many mitosporic fungi are reclassified when a sexual cycle is observed or comparative genomics identifies a close relationship to a known phylum.

Among the mitosporic fungi, a few are pathogenic for humans. These fungi usually reproduce by

(A)

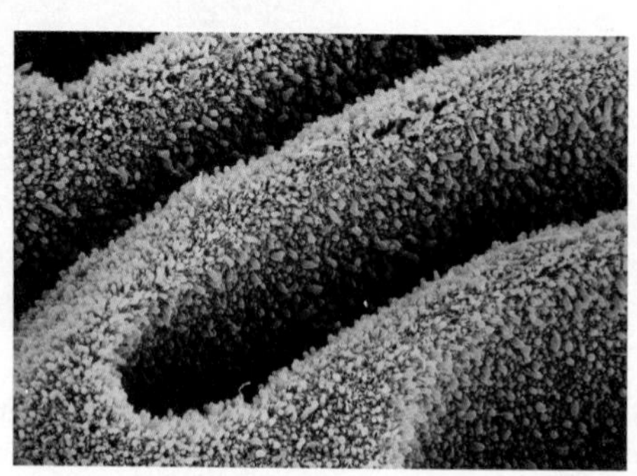
(B)

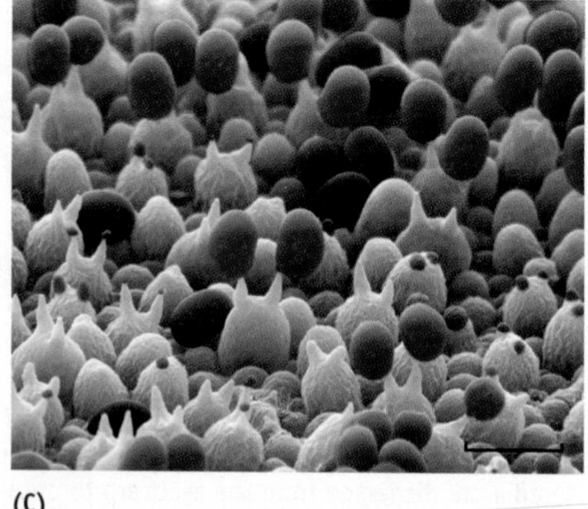
(C)

FIGURE 17.17 **Mushroom Gills and Basidiospores.** (**A**) A group of *Agaricus bisporis* mushroom (Portobello) caps, showing the gills on their underside. Courtesy of Jeffrey Pommerville. (**B**) A false-color scanning electron micrograph of the mushroom gills. The rough appearance represents forming basidia on which basidiospores will form. © Dr. Jeremy Burgess/Photo Researchers, Inc. (**C**) Another scanning electron micrograph showing the basidiospores (dark brown) attached to basidia. (Bar = 20 μm.) © Eye of Science/Photo Researchers, Inc. »» Why is the specific epithet for this species of *Agaricus* called *bisporis*?

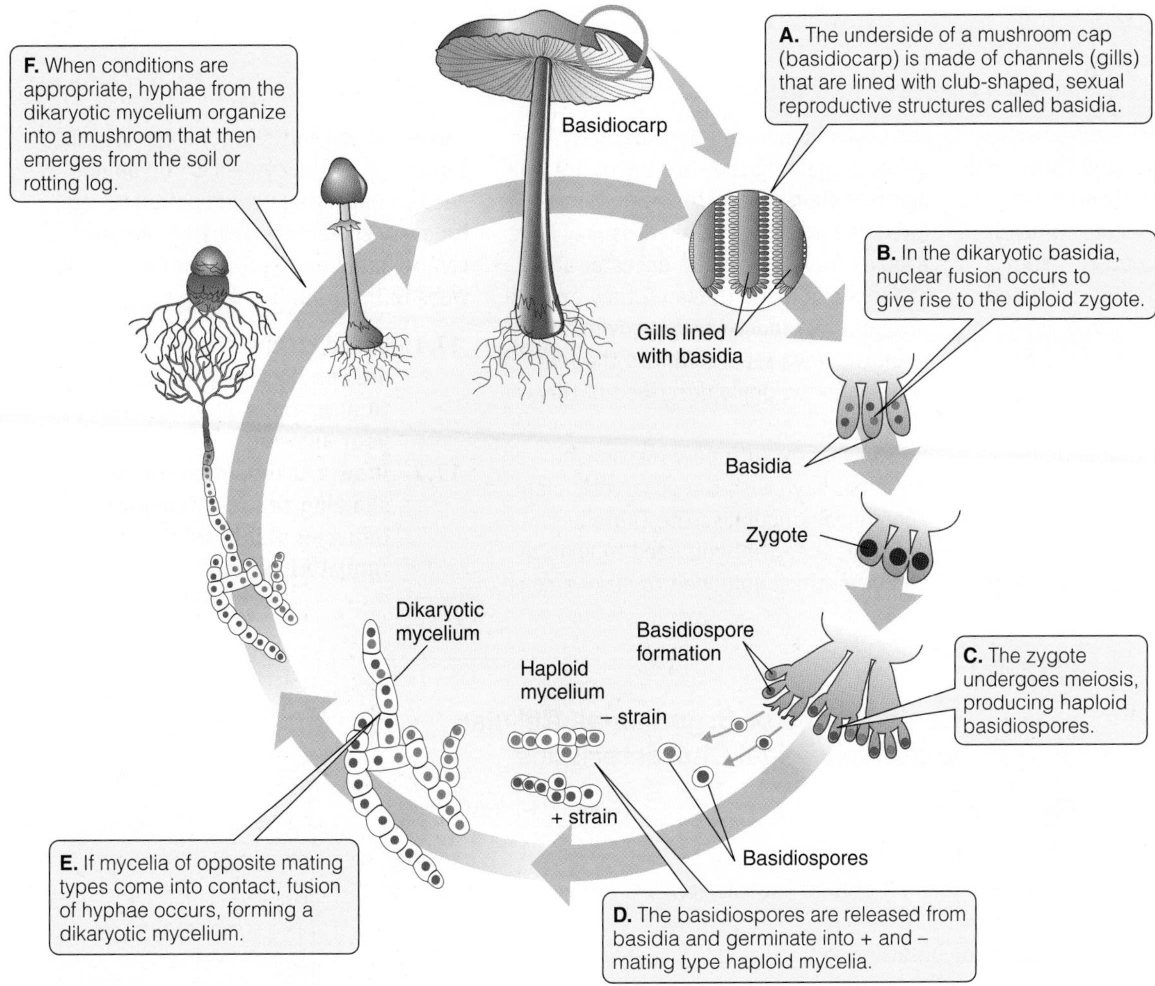

FIGURE 17.18 The Sexual Reproduction Cycle of a Typical Basidiomycete. The basidiomycetes include the mushrooms from which basidiospores are produced. »» Identify the portion of the cycle that exists in a dikaryotic state.

asexual spores, budding, or fragmentation, where the hyphal segments are commonly blown about with dust or deposited on environmental surfaces. For example, fragments of the athlete's foot fungus are sometimes left on towels and shower room floors.

The three phyla of fungi important to human health are compared in TABLE 17.1.

Yeasts Represent a Term for Any Single-Celled Stage of a Fungus

The word "yeast" refers to a large variety of unicellular fungi (as well as the single-cell stage of any fungus). Included in the group are nonspore–forming yeasts of the mitosporic fungi, as well as certain yeasts belonging to the Basidiomycota and Ascomycota. The yeasts we consider here are the species of *Saccharomyces* used extensively in industry and research.

Saccharomyces (*saccharo* = "sugar") is a fungus with the ability to ferment sugars. The most commonly used species of *Saccharomyces* are *S. cerevisiae* and *S. ellipsoideus,* the former used for bread baking (Baker's yeast) and alcohol production, the latter for alcohol production. Yeast cells reproduce chiefly by budding, as described previously (see Figure 17.13A), but a sexual cycle also exists in which cells fuse and form an enlarged structure (an ascus) containing spores (ascospores). *S. cerevisiae* therefore is an ascomycete.

The cytoplasm of *Saccharomyces* is rich in B vitamins, a factor making yeast tablets valuable nutritional supplements (ironized yeast) for people with iron-poor blood. The baking industry relies heavily upon *S. cerevisiae* to supply the texture in breads. During the dough's rising period, yeast cells break down glucose and other carbohydrates, producing

MICROINQUIRY 17

Evolution of the Fungi

The fossil record provides evidence that fungi and plants "hit the land" about the same time. There are fossilized fungi that are over 450 million years old, which is about the time that plants started to colonize the land. In fact, as described in this chapter, some fossilized plants, perhaps representing some of the first to colonize land, appear to have mycorrhizal associations.

Taxonomists believe that a single ancestral species gave rise to the five fungal phyla (Chytridiomycota, Glomeromycota, Zygomycota, Ascomycota, and Basidiomycota). Only the Chytridiomycota (chytrids) have flagella, suggesting they are the oldest line and that fungal ancestors were flagellated and aquatic. The chytrids may have evolved from a protistan ancestor and, as fungi colonized the land, they evolved different reproductive styles, which taxonomists have separated into the four nonflagellated phyla described in this chapter.

So, what evolutionary relationships do the fungi have with the eukaryotic plant and animal kingdoms? The **Table** below has some data that you need to analyze and from which you need to make a conclusion as to what relationships are best supported by the evidence. You do not need to understand the function or role for every characteristic listed. Answers can be found on the Student Companion Website in **Appendix D**.

17.1. Based on the data presented, are fungi more closely related to animals or plants? Justify your answer.

17.2. Draw a taxonomic scheme showing the relationships between plant, animal, and fungal kingdoms.

TABLE

Comparisons of Organismal, Cellular, and Biochemical Characteristics

Characteristic	Fungi	Plants	Animals
Protein sequence similarities	✓		✓
Elongation factor 3 protein (translation) similarities	✓		✓
Types of polyunsaturated fatty acids			
Alpha linolenic		✓	
Gamma linolenic	✓		✓
Cytochrome system similarities	✓		✓
Mitochondrial UGA codes for tryptophan	✓		✓
Plate-like mitochondrial cristae	✓		✓
Presence of chitin	✓		some
Type of glycoprotein bonding			
O-linked	✓		✓
N-linked	✓		✓
Absorptive nutrition	✓		some
Glycogen reserves	✓		✓
Pathway for lysine biosynthesis			
Aminoadipic acid pathway	✓		
Diaminopimelic pathway		✓	
Type of sterol intermediate			
Lanosterol	✓		✓
Cycloartenol		✓	
Mitochondrial ribosome 5S RNA present		✓	
Presence of lysosomes	✓		✓

TABLE

17.1 Comparisons of the Fungal Phyla Important to Human Disease

Phylum	Common Name	Cross Walls (Septa)	Sexual Structure	Sexual Spore	Asexual Spore	Representative Genera
Zygomycota	Zygomycetes	No	Zygospore	Zygospore	Sporangiospore	*Rhizopus*
Ascomycota	Ascomycetes, sac fungi	Yes	Ascus	Ascospore	Conidia	*Saccharomyces* *Aspergillus* *Penicillium*
Basidiomycota	Basidiomycetes, club fungi	Yes	Basidium	Basidiospore	Conidia Hyphal fragments	*Agaricus* *Amanita*

carbon dioxide gas. The carbon dioxide expands the dough, causing it to rise. Protein-digesting enzymes in yeast partially digest the gluten protein of the flour to give bread its spongy texture.

Yeasts are plentiful where there are orchards or fruits (the haze on an apple is a layer of yeasts). In natural alcohol fermentations, wild yeasts of various *Saccharomyces* species are crushed with the fruit; in controlled fermentations, *S. ellipsoideus* is added to the prepared fruit juice. The fruit juice bubbles profusely as carbon dioxide evolves. When the oxygen is depleted, the yeast metabolism shifts to fermentation and the pyruvate from glycolysis changes to consumable ethyl alcohol. The huge share of the American economy taken up by the wine and spirits industries is testament to the significance of the fermentation yeasts. A fuller discussion of fermentation processes is presented in another chapter.

S. cerevisiae probably is the most understood eukaryotic organism at the molecular and cellular levels. Its complete genome was sequenced in 1997, the first eukaryotic model organism to be completely sequenced. *S. cerevisiae* contains about 6,000 genes. It might appear initially that *S. cerevisiae* would have little in common with human beings. However, both are eukaryotic organisms with a cell nucleus, chromosomes, and a similar mechanism for cell division (mitosis).

S. cerevisiae, therefore, has been used to better understand not only cell function in animals, but also as a model for human disease. In fact, about 20% of human disease genes have counterparts in yeast. For example, the chemical and signaling process by which a cell prepares itself for mitosis was either first discovered using yeast cells or major contributions to the understanding came from research with yeast cells. Research with yeast has identified the presence of prions, which have helped in the understanding of human prion diseases.

Potential drugs useful in disease treatment also can be screened using yeast cells. A yeast mutant, for example, with the equivalent of a human disease-causing gene, can be treated with potential therapeutic drugs to identify a compound able to restore normal function to the yeast cell gene. Such drugs, or modifications of them, might also be useful in humans.

CONCEPT AND REASONING CHECKS 2

a. What are the unique features of the Chytridiomycota and Glomeromycota?
b. Describe the unique properties of the Zygomycota.
c. Summarize the properties of the Ascomycota.
d. Identify the properties of the Basidiomycota.
e. Distinguish the characteristics of the mitosporic fungi.
f. Summarize the importance of yeasts to commercial interests and research.

Chapter Challenge B

The fungi, together with the plants and animals, make up the domain Eukarya and, with the domains Bacteria and Archaea, form the phylogenetic "tree of life."

QUESTION B:

Draw a simple "tree of life" showing the three domains. Then, in the Eukarya, show (using separate lines coming off the main branch) how the fungi are related to the plants and animals. Consider MicroInquiry 17 when constructing your tree.

Answers can be found on the Student Companion Website in **Appendix D.**

■ Model organism: A relatively simple organism used to study general principles of biology.

KEY CONCEPT 17.3 Some Fungi Cause Intoxications

Although there are no major infectious diseases of the digestive system caused by fungi, some mushrooms, molds, and yeasts produce toxins called **mycotoxins**. If these toxins are ingested, they may have poisonous effects distant from the digestive system. We will describe several mycotoxin-associated intoxications here.

Some Fungi Can Be Poisonous or Even Deadly When Consumed

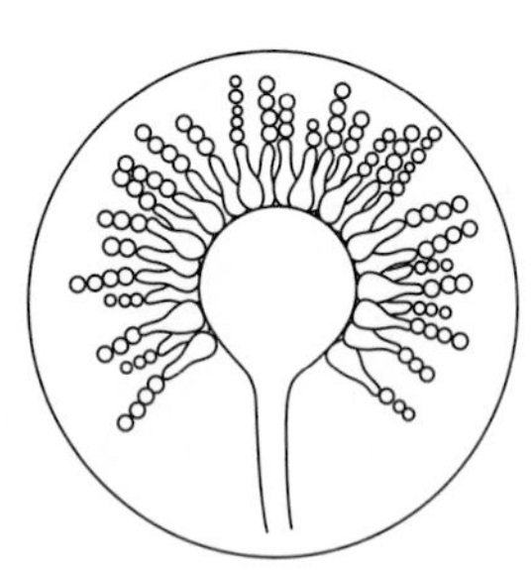
Aspergillus flavus

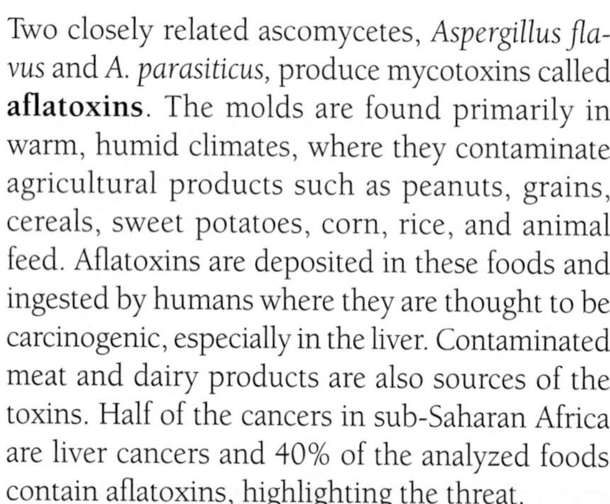

Two closely related ascomycetes, *Aspergillus flavus* and *A. parasiticus*, produce mycotoxins called **aflatoxins**. The molds are found primarily in warm, humid climates, where they contaminate agricultural products such as peanuts, grains, cereals, sweet potatoes, corn, rice, and animal feed. Aflatoxins are deposited in these foods and ingested by humans where they are thought to be carcinogenic, especially in the liver. Contaminated meat and dairy products are also sources of the toxins. Half of the cancers in sub-Saharan Africa are liver cancers and 40% of the analyzed foods contain aflatoxins, highlighting the threat.

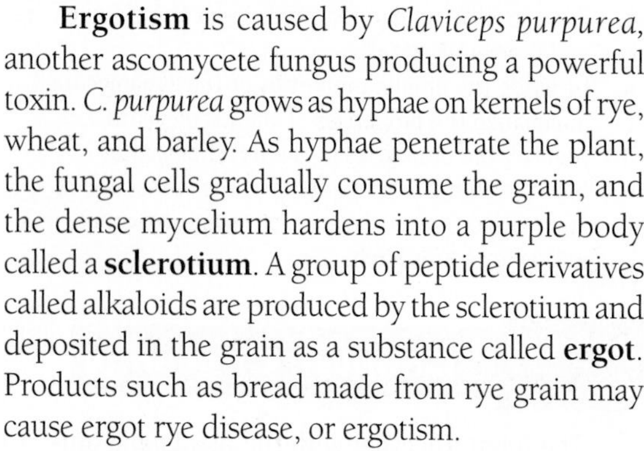

Ergotism is caused by *Claviceps purpurea*, another ascomycete fungus producing a powerful toxin. *C. purpurea* grows as hyphae on kernels of rye, wheat, and barley. As hyphae penetrate the plant, the fungal cells gradually consume the grain, and the dense mycelium hardens into a purple body called a **sclerotium**. A group of peptide derivatives called alkaloids are produced by the sclerotium and deposited in the grain as a substance called **ergot**. Products such as bread made from rye grain may cause ergot rye disease, or ergotism.

Symptoms may include numbness, hot and cold sensations, convulsions with epileptic-type seizures, and paralysis of the nerve endings. Lysergic acid diethylamide (LSD) is a derivative of an alkaloid in ergot. Commercial derivatives of these alkaloids are used to cause contractions of the smooth muscles, such as to induce labor or relieve migraine headaches. MicroFocus 17.4 relates another possible effect of ergotism.

Some Mushrooms Produce Mycotoxins

Amateur mushroom hunting is a popular pastime in many parts of the world. However, every year in the United States such hunting results in some 9,000 cases of mushroom poisoning being reported to the American Association of Poison Control Centers; children under 6 years of age account for the majority of cases. A poisonous mushroom refers to any mushroom that produces various toxins that can cause adverse or harmful reactions when eaten. By this definition, of the more than 5,000 species of identified mushrooms in the United States, about 100 are responsible for mushroom poisoning but less than a dozen species are considered deadly.

Mushroom poisoning, or **mycetism**, can occur from mushrooms that produce mycotoxins that affect the human body. Some are neurotoxins that: (1) mimic the action of acetylcholine, affecting the peripheral parasympathetic nervous system; (2) generate psychotropic effects (hallucinations); or (3) produce a condition resembling alcohol intoxication. Deaths are rare, but in severe cases the intoxication may cause cardiac or respiratory failure. Further, in children, accidental consumption of large quantities of some mushrooms may cause convulsions, coma, and other neurologic problems for up to 12 hours.

Amanita phalloides is referred to as the "death cap" because it accounts for about 90% of deaths from eating poisonous mushrooms (FIGURE 17.19A). About 10 to 14 hours after ingesting just one to three mushrooms, enough of the protein toxins, called **amatoxins** and **phallotoxins**, are present in the body to produce gastrointestinal symptoms of severe abdominal cramping, vomiting, and watery diarrhea. The symptoms last for about 24 hours and can cause dehydration and low blood pressure. A period of remission of symptoms lasts for 1 to 2 days during which the patient feels better, but blood tests begin to show evidence of liver and kidney deterioration. Liver and kidney failure then develops and either results in death within about a week or recovery within 2 to 3 weeks.

Numerous species of mushrooms, including *Agaricus, Amanita, Boletus, Lactarius, Lepiota, Lycoperdon, Polyporus,* and *Russula,* contain mycotoxins that can cause a noninflammatory gastroenteritis (FIGURE 17.19B). Within 30 to 90 minutes of ingestion, sudden severe vomiting and mild to severe diarrhea with abdominal cramps occur. The symptoms typically last about 6 hours. The main diagnostic identification of these intoxications is a rapid onset of symptoms. Fatalities are rare but if they do occur they are associated with dehydra-

MICROFOCUS 17.4: History

"The Work of the Devil"

A witch trial in Massachusetts, 1692. © North Wind Picture Archives/Alamy Images.

As an undergraduate, Linda Caporael was missing a critical history course for graduation. Little did she know that through this class she was about to provide a possible answer for one of the biggest mysteries of early American history: the cause of the Salem Witch Trials. These trials in 1692 led to the execution of 20 people who had been accused of being witches in Salem, Massachusetts (see figure).

Linda Caporael, now a behavioral psychologist at New York's Rensselaer Polytechnic Institute, in preparation of a paper for her history course had read a book where the author could not explain the hallucinations among the people in Salem during the witchcraft trials. Caporael made a connection between the "Salem witches" and a French story of ergot poisoning in 1951. In Pont-Saint-Esprit, a small village in the south of France, more than 50 villagers went insane for a month after eating ergotized rye flour. Some people had fits of insomnia, others had hallucinogenic visions, and still others were reported to have fits of hysterical laughing or crying. In the end, three people died.

Caporael noticed a link between these bizarre symptoms, those of Salem witches, and the hallucinogenic effects of drugs like LSD, which is a derivative of ergot. Could ergot possibly have been the perpetrator in Salem too?

During the Dark Ages, Europe's poor lived almost entirely on rye bread. Between 1250 and 1750, ergotism, then called "St. Anthony's fire," led to miscarriages, chronic illnesses in people who survived, and mental illness. Importantly, hallucinations were considered "the work of the devil."

Toxicologists know that eating ergotized food can cause violent muscle spasms, delusions, hallucinations, crawling sensations on the skin, and a host of other symptoms—all of which, Linda Caporael found in the records of the Salem witchcraft trials. Ergot thrives in warm, damp, rainy springs and summers, which were the exact conditions Caporael says existed in Salem in 1691. Add to this that parts of Salem village consisted of swampy meadows that would be the perfect environment for fungal growth and that rye was the staple grain of Salem—and it is not a stretch to suggest that the rye crop consumed in the winter of 1691–1692 could have been contaminated by large quantities of ergot.

Caporael concedes that ergot poisoning can't explain all of the events at Salem. Some of the behaviors exhibited by the villagers probably represent instances of mass hysteria. Still, as people reexamine events of history, it seems that just maybe ergot poisoning did play some role—and, hey, not bad for an undergraduate history paper!

tion and electrolyte imbalances in debilitated, very young, or very old patients. In children, dehydration may be severe enough to require hospital treatment.

If people suspect they have eaten a poisonous mushroom, it is not wise to wait for symptoms to appear, as symptoms may not develop until several days later. It should be treated as a medical emergency.

As far as prevention goes, all mushrooms not bought at the grocery store should be considered potentially dangerous—and no antidote exists. Early replacement of lost body fluids (water and electrolytes) is important in improving recovery rates. Still, according to some medical reports death rates are 20% to 30%, with a higher mortality rate of 50% in children less than 10 years old. According to the Minnesota Mycological Society, "*There are old mushroom hunters and bold mushroom hunters. But there are no old, bold mushroom hunters.*"

CONCEPT AND REASONING CHECKS 3

a. Summarize the unique features of aflatoxins and ergot.

b. In the case of *A. phalloides* mushroom poisoning, what is probably happening during the 2 days when symptoms are in remission?

(A) (B)

FIGURE 17.19 Poisonous Mushrooms. (**A**) ***Amanita phalloides.*** Referred to as the "death cap" mushroom, these fungi can produce a deadly mycotoxin. © Niels-DK/Alamy Images. (**B**) ***Boletus santanas.*** Referred to as the "Satan's mushroom," if eaten raw, this mushroom can cause gastrointestinal symptoms of nausea and vomiting. Courtesy of Misterzin, www.flickr.com/photos/misterzin. »» To what fungal phylum do these mushrooms belong?

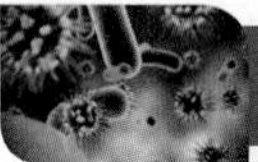

Chapter Challenge C

Here is a more relaxing challenge. Wild mushroom hunting or gathering usually for eating is an activity (hobby?) that many Europeans, Asians, and Americans have participated in for centuries.

QUESTION C:

Knowing what you now know about these macroscopic fungi, would you eat mushrooms that were picked by your friend's grandmother who says she has done this since she was growing up in Europe in the 1940s? Explain your decision.

Answers can be found on the Student Companion Website in **Appendix D.**

KEY CONCEPT 17.4 Some Fungi Can Invade the Skin

Trichophyton

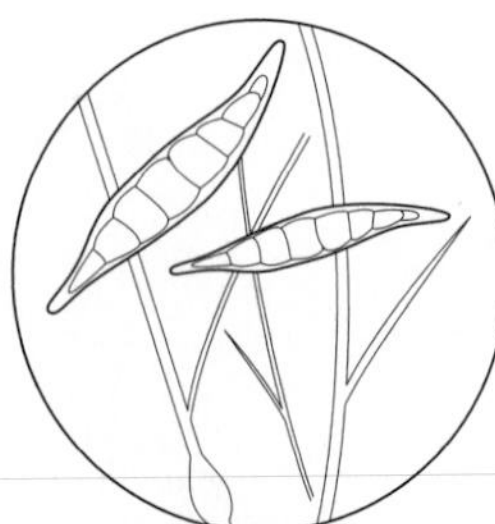

Microsporum

In humans, fungal diseases, called **mycoses**, often affect many body regions. Some of these affect the skin or body surfaces. For example, several diseases, including ringworm and athlete's foot, involve the skin areas. One disease, candidiasis, may take place in the oral cavity, intestinal tract, skin, vaginal tract, or other body locations depending on the conditions stimulating its development.

Dermatophytosis Is an Infection of the Skin, Hair, and Nails

Dermatophytosis (*dermato* = "skin"; *phyto* = "plant," referring to the days when fungi were grouped with plants) is a general name for a fungal disease of the hair, skin, or nails. The diseases are commonly known as **tinea infections** (*tinea* = "worm") because in ancient times, worms were thought to be the cause. Tinea infections compose several forms of **ringworm**, including: athlete's foot (tinea pedis); head ringworm (tinea capitis); body ringworm (tinea corporis); groin ringworm or "jock itch" (tinea cruris); and nail ringworm (tinea unguium).

The causes of dermatophytosis are a group of fungi called **dermatophytes**. *Epidermophyton* currently is considered a mitosporic fungus, while species of *Trichophyton* and *Microsporum* are ascomycetes (FIGURE 17.20A).

If protected from dryness, the dermatophytes live for weeks on wooden floors of shower rooms or on mats. People transmit the fungi by contact and on towels, combs, hats, and numerous other types of **fomites** (inanimate objects). Because tinea

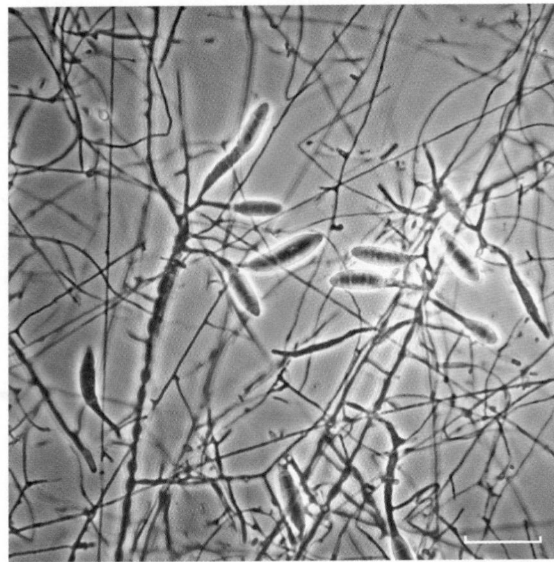

(A)

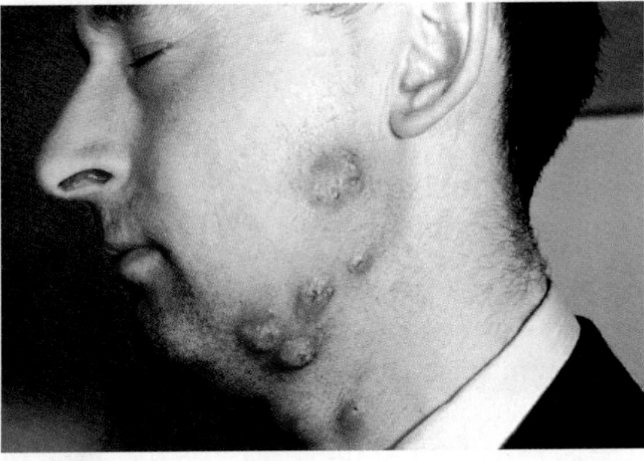

(B)

FIGURE 17.20 **Ringworm.** Ringworm of the skin or body (tinea corporis). (**A**) Light microscope photograph of *Microsporum,* one of the fungi causing ringworm on the scalp and body. (Bar = 15 μm.) Courtesy of Dr. Libero Ajello/CDC. (**B**) A case of body ringworm on the face and neck that was "caught" from a cat. Courtesy of CDC. »» Why is the shape of each ringworm skin lesion in (B) circular?

diseases affect cats and dogs, they can transmit the fungi to humans (FIGURE 17.20B).

Dermatophytosis is commonly accompanied by blister-like lesions appearing along the webs of the toes (*Epidermophyton*), on the scalp or skin (*Microsporum*), or in the nail plate (*Trichophyton*). Often a thin, fluid discharge exudes when the blisters are scratched or irritated. As the blisters dry, they leave a scaly ring. Infection is limited due to the immune system's inflammatory response.

Treatment of dermatophytosis often is directed at changing the conditions of the skin environment. Commercial powders dry the diseased area, while ointments change the pH to make the area inhospitable for the fungus. Certain acids such as undecylenic acid (Desenex®) and mixtures of acetic acid and benzoic acid (Whitfield's ointment) are active against the fungi. Also, tolnaftate (Tinactin®) and miconazole (Micatin®) are useful as topical agents for infections not involving the nails and hair. Griseofulvin, administered orally, is a highly effective chemotherapeutic agent for severe dermatophytosis.

Candidiasis Often Is a Mild, Superficial Infection

Candida albicans often is present in the skin, mouth, vagina, and intestinal tract of healthy humans, where it lives without causing disease. The organism is a small mitosporic yeast that forms filaments called pseudohyphae when cultivated in laboratory media. When immune system defenses are compromised, or when changes occur in the normal microbial population in the body, *C. albicans* flourishes and causes numerous forms of **candidiasis**.

■ Pseudohyphae: Cells formed by budding that are more elongated than typical oval cells.

One form of candidiasis occurs in the vagina and is often referred to as **vulvovaginitis**, or a "yeast infection." There are some 20 million cases reported every year in the United States. Symptoms include itching sensations (pruritis), burning internal pain, and a white "cheesy" discharge. Reddening (erythema) and swelling of the vaginal tissues also occur. Diagnosis is performed by observing *C. albicans* in a sample of vaginal discharge or vaginal smear (FIGURE 17.21A), and by cultivating the organisms on laboratory media. Treatment is usually successful with nystatin (Mycostatin®) applied as a topical ointment or suppository. Miconazole, clotrimazole, and ketoconazole are useful alternatives.

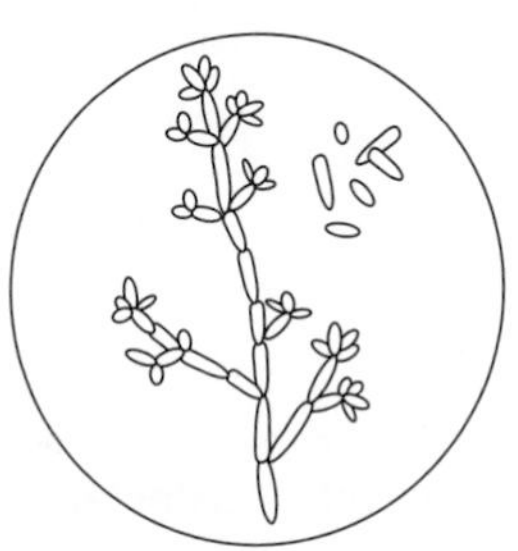

Candida albicans

Vulvovaginitis is considered a sexually transmitted disease. In addition, studies have shown that excessive antibiotic use may encourage loss of the rod-shaped lactobacilli normally present in the vaginal environment. Without lactobacilli as competitors, *C. albicans* flourishes. Other predisposing factors are the contraceptive intrauterine device (IUD), corticosteroid treatment, pregnancy, diabetes, and tight-fitting garments, which increase the local temperature and humidity.

Oral candidiasis is known as **thrush**. This disease is accompanied by small, white flecks that appear on the mucous membranes of the oral cavity and then grow together to form soft, crumbly, milk-like curds (FIGURE 17.21B). When scraped off, a red, inflamed base is revealed. Oral

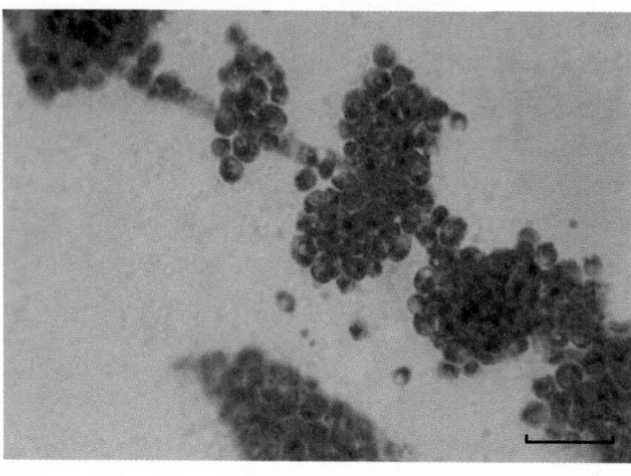

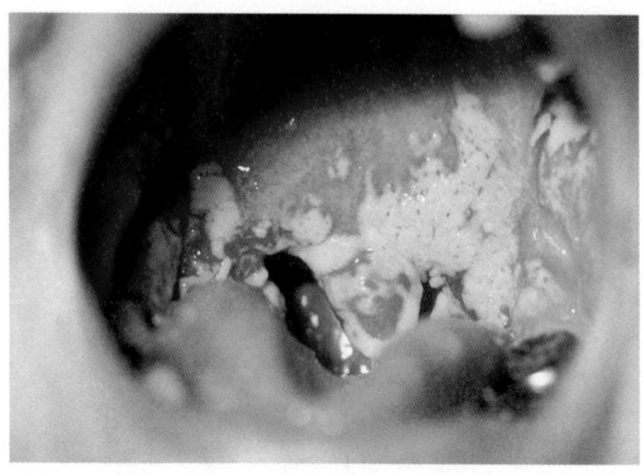

(A) (B)

FIGURE 17.21 *Candida albicans.* (**A**) A light micrograph of stained *C. albicans* cells from a vaginal swab. (Bar = 60 μm.) Courtesy of Dr. Godon Roberstad/CDC. (**B**) A severe case of oral candidiasis, showing a thick, creamy coating over the tongue. Courtesy of CDC. »» From photo (A), how do you know this specimen of *C. albicans* was from a human infection?

suspensions of gentian violet and nystatin ("swish and spit") are effective for therapy. The disease is common in newborns, who acquire it during passage through the vagina (birth canal) of infected mothers. Children also may contract thrush from nursery utensils, toys, or the handles of shopping carts. Candidiasis may be related to a suppressed immune system. Indeed, thrush may be an early sign of AIDS in an adult patient.

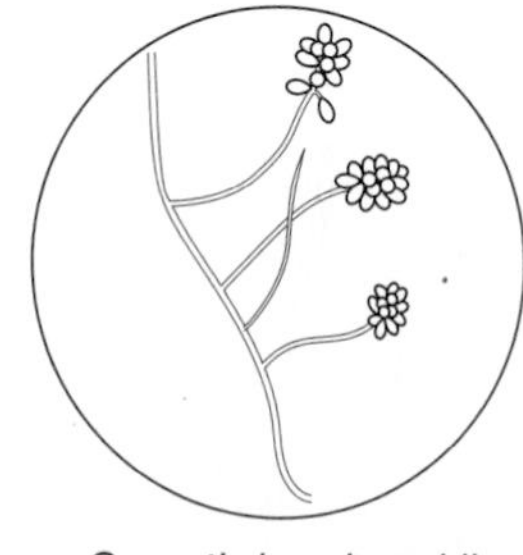

Sporothrix schenckii

Candidiasis in the intestinal tract is closely tied to the use of antibiotics. Certain drugs destroy the bacterial cells normally found there and allow *C. albicans* to flourish. In the 1950s, yogurt became popular as a way of replacing the bacterial cells. Today when intestinal surgery is anticipated, the physician often uses antifungal agents to curb *Candida* overgrowth. Moreover, people whose hands are in constant contact with water may develop a hardening, browning, and distortion of the fingernails called **onychia**, also caused by *C. albicans*.

Sporotrichosis Is an Occupational Hazard

People who work with wood, wood products, or the soil may contract **sporotrichosis**. Handling sphagnum (peat) moss used to pack tree seedlings or skin punctures by rose thorns (rose thorn disease) can lead to the disease as the result of infection by conidia from *Sporothrix schenckii* (FIGURE 17.22A). Pus-filled purplish lesions form

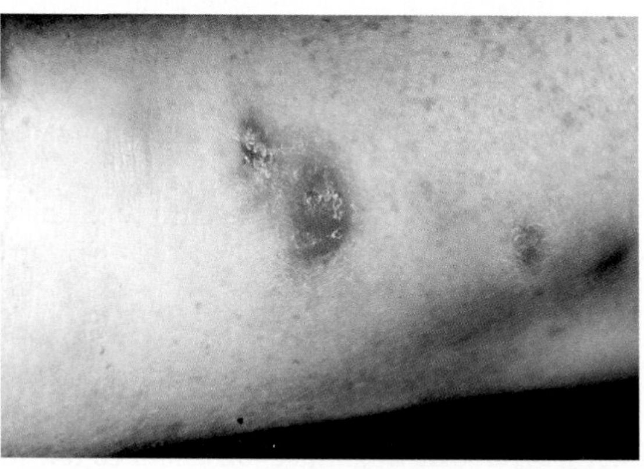

(A) (B)

FIGURE 17.22 *Sporothrix schenckii.* (**A**) A false-color scanning electron micrograph of hyphae (orange) and conidia (purple) formed on conidiophores. (Bar = 8 μm.) © E. Gueho - CNRI/Photo Researchers, Inc. (**B**) A patient showing the lesions of sporotrichosis on an infected arm. Characteristic "knots" can be felt under the skin. © Everett Beneke/Visuals Unlimited. »» Explain how a rose thorn harboring *S. schenckii* conidia can cause a skin disease.

TABLE

17.2 A Summary of the Fungal Skin Diseases

Inflammation or Disease	Causative Agent	Signs and Symptoms	Transmission	Treatment	Prevention
Dermatophytosis (Tineas)	*Epidermophyton* *Microsporum* *Trichophyton*	Blister-like lesions	Direct or indirect contact Soil	Antifungal medications	Keeping skin dry Not sharing personal items
Vaginal candidiasis	*Candida albicans*	Vaginal itching, irritation, white, thick discharge	Disruption of normal vaginal microbiota	Antifungal medications	Avoiding baths and hot tubs Avoiding douches
Thrush	*Candida albicans*	White flecks on mucous membranes	Passage through the birth canal	No treatment in children	Practicing good oral hygiene Limiting sugar intake
Sporotrichosis	*Sporothrix schenckii*	Pus-filled, purplish lesions	From plant material harboring the fungus	Itraconazole Amphotericin B	Wearing long sleeves and gloves while working with suspect materials or vegetation

at the site of entry, and "knots" may be felt under the skin (FIGURE 17.22B). Dissemination, though rare, may occur to the bloodstream, where blockages may cause swelling of the tissues (edema). In one outbreak, 84 cases of cutaneous sporotrichosis occurred in people who handled conifer seedlings packed with sphagnum moss from Pennsylvania. Cutaneous infections are controlled with itraconazole, but systemic infections require amphotericin B therapy.

TABLE 17.2 summarizes the fungal diseases of the skin.

CONCEPT AND REASONING CHECKS 4

a. Describe the characteristics of dermatophytosis.
b. Describe the different forms of candidiasis based on body location.
c. Describe the characteristics of sporotrichosis.

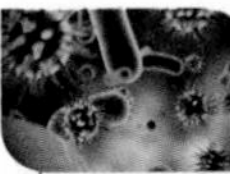

Chapter Challenge D

Dermatophytes comprise a group of fungi that can cause skin, hair, and nail infections. Infections caused by these fungi are also sometimes known as "ringworm" or "tinea."

QUESTION D:

Suppose there is a ringworm outbreak in your child's school/daycare center. What should you do?

Answers can be found on the Student Companion Website in **Appendix D**.

KEY CONCEPT 17.5 Many Fungal Pathogens Cause Lower Respiratory Tract Diseases

Additional mycoses affect other body parts in humans, with a primary infection in the lungs that can spread to other body areas. In many of these fungal diseases, a weakened immune system contributes substantially to the occurrence of the infection.

Cryptococcosis Usually Occurs in Immunocompromised Individuals

Cryptococcosis is caused by an oval-shaped yeast known as *Cryptococcus neoformans*, and is a member of the Basidiomycota. The organism

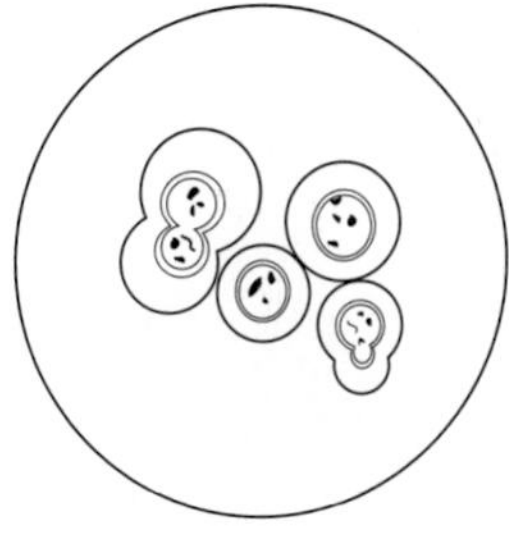

Cryptococcus neoformans

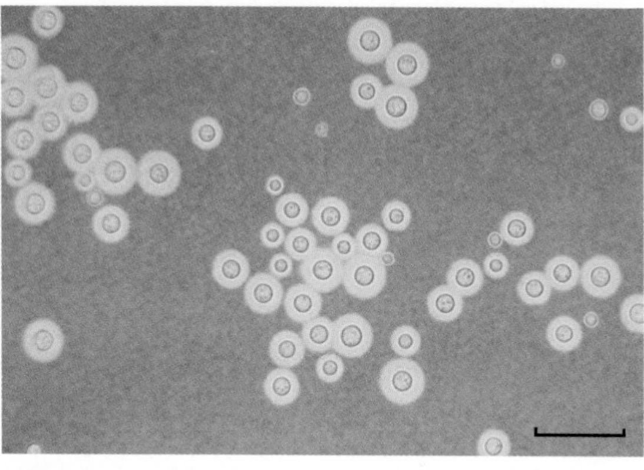

(A)

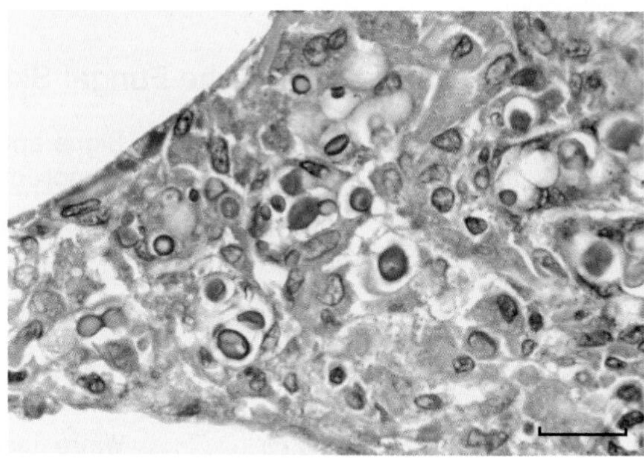

(B)

FIGURE 17.23 *Cryptococcus neoformans.* (**A**) A light microscope photomicrograph of *C. neoformans* cells. The white halo surrounding the cells is the capsule. (Bar = 20 μm.) Courtesy of Dr. Leanor Haley/CDC. (**B**) A stained photomicrograph of *C. neoformans* cells (red) from lung tissue of an AIDS patient. The capsule surrounding the cells provides resistance to phagocytosis and enhances the pathogenic tendency of the fungus. (Bar = 20 μm.) © Scott Camazine/Alamy. »» Why would *C. neoformans* be a serious health threat to AIDS patients?

is found in the soil of urban environments and grows actively in the droppings of pigeons, but not within the pigeon tissues.

Cryptococcosis is among the most dangerous fungal diseases in humans. As an opportunistic mycosis, it affects the lungs and the meninges and is estimated to account for over 25% of all deaths from fungal disease. Cryptococci may become airborne when dried bird droppings are stirred up by gusts of wind, and the organisms subsequently enter the respiratory passageways of humans.

■ Meninges: The membranes surrounding and protecting the brain and spinal cord.

The *C. neoformans* cells, which have a diameter of about 5 to 6 μm, are embedded in a thick, gelatinous capsule that provides resistance to phagocytosis (FIGURE 17.23). Infection usually produces a mild or asymptomatic pneumonia. However, in immunocompromised patients the cryptococci may pass into the bloodstream and localize in the meninges and brain, causing the patient to experience piercing headaches, stiffness in the neck, and paralysis. Untreated cryptococcal meningitis may be fatal. However, intravenous treatment with the antifungal drug amphotericin B is usually successful, even in severe cases.

Another species of *Cryptococcus*, *C. gattii*, emerged in British Columbia in 1999 and soon spread to the Pacific Northwest (Washington, Idaho, Oregon, and California) and western Canada. This subtropical fungus has thus adapted to a more temperate climate. The fungus is not associated with bird droppings but rather is found in the soil around eucalyptus and Douglas fir trees. The fungal spores are thought to be inhaled, where they can cause a lung infection and pneumonia in 70% to 80% of otherwise healthy individuals. Symptoms include chest pain, a persistent cough, fever, and weight loss. Blindness can be an outcome in *C. gattii*-infected individuals. This form of cryptococcosis also infects domestic and wild animals, including dogs, cats, pet birds, ferrets, horses, and porpoises.

Although antifungal drugs (amphotericin B) are available, up to 25% of individuals who develop an infection die, so public health officials must exert increased vigilance for this potentially serious, perhaps life-threatening, infection. Presently there is no vaccine or preventive medications.

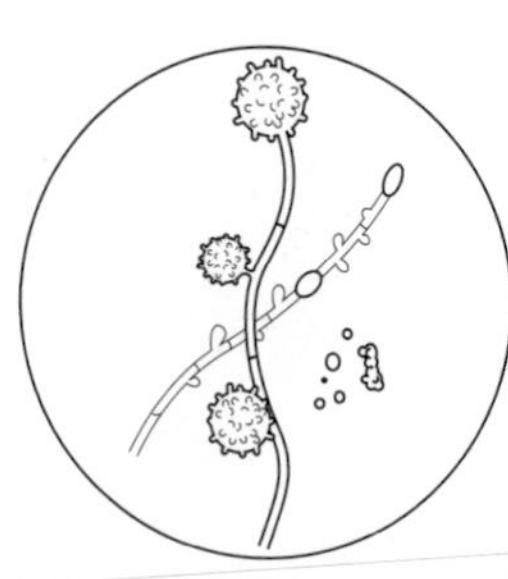

Histoplasma capsulatum

Histoplasmosis Can Produce a Systemic Disease

Histoplasmosis is a lung disease that occurs worldwide. It is endemic in the Ohio and the Mississippi River valleys where it is often called "summer flu." The causative agent is *Histoplasma capsulatum*. The lungs are the primary portal of entry; thus infection usually occurs from the inhalation of spores present in dry, dusty soil or found in the air of chicken coops and bat caves (FIGURE 17.24). Being a dimorphic fungus, it grows as a yeast form at 37°C.

Most people remain asymptomatic or experience mild influenza-like illness and recover with-

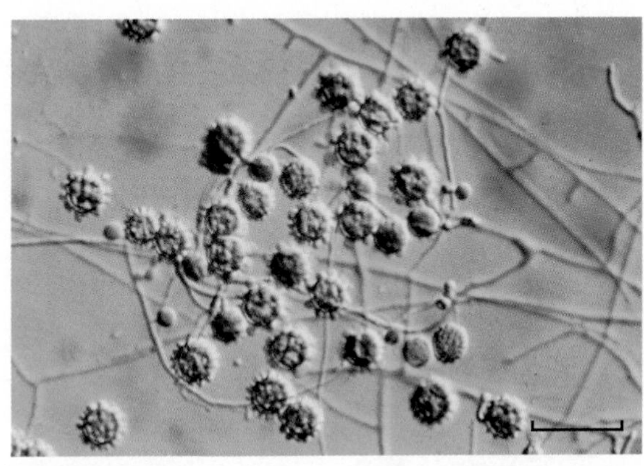

FIGURE 17.24 *Histoplasma* Mycelium and Spores. *H. capsulatum* grows as a mycelium in soil enriched by animal excrement. The spores are produced from the hyphal tips. (Bar = 20 µm.) © Science Source/Photo Researchers, Inc. »» Which dimorphic form of the fungus is associated with human infections?

out treatment. However, in immunocompromised people a disseminated form with tuberculosis-like lesions of the lungs and other internal organs may occur, making AIDS patients especially vulnerable. Amphotericin B or ketoconazole may be used in treatment. Avoiding known endemic areas is the best prevention.

Blastomycosis Usually Is Acquired Via the Respiratory Route

Blastomycosis occurs principally in Canada, the Great Lakes region, and areas of the United States from the Mississippi River to the Carolinas. The pathogen is *Blastomyces dermatitidis*, a dimorphic fungus producing conidia that are inhaled. Within the lungs, the conidia germinate as the yeast form.

Acute blastomycosis is associated with enriched, dusty soil and bird droppings, particularly in moist soils near barns and sheds. Inhalation leads to lung lesions with persistent cough and chest pains in immunocompromised patients. Chronic pneumonia is the most common clinical manifestation. The disseminated form of blastomycosis may involve many internal organs (bones, liver, spleen, or central nervous system) and may prove fatal. Amphotericin B is used in therapy.

Coccidioidomycosis Can Become a Potentially Lethal Infection

Coccidioidomycosis, known more commonly as "valley fever" or "desert fever," is caused by *Coccidioides immitis* and *C. posadasii*. Travelers to the San Joaquin Valley of California and dry regions of the southwestern United States may contract this fungal disease. During most of the 1980s, about 450 annual cases of coccidioidomycosis were reported to the CDC. In 1991, that number jumped to over 1,200 cases, and in 2010, the number of reports was almost 16,000, the majority occurring in California and Arizona.

In the soil, the ascomycete fungus produces highly infectious arthrospores by a unique process of spherule development and endospore formation (FIGURE 17.25). When inhaled into the human lungs, the arthrospores develop into large spherules. Infection induces an influenza-like disease, with a dry, hacking cough, chest pains, and high fever.

Although some 40% of cases develop a self-limiting pulmonary disease, a small number of cases become disseminated and involve skin, bone, and the central nervous system, including the meninges of the spinal cord. Recovery brings lifelong immunity. Amphotericin B is prescribed for severe cases, and another drug, nikkomycin, and a vaccine are in the early stages of development. TEXTBOOK CASE 17 describes an outbreak in northeastern Utah.

Pneumocystis Pneumonia Can Cause a Lethal Pneumonia

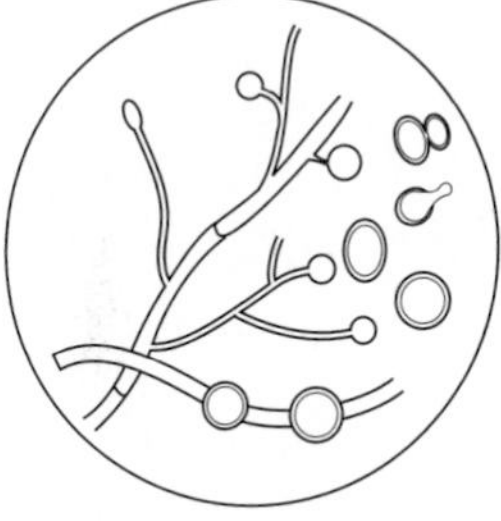

Blastomyces dermatitidis

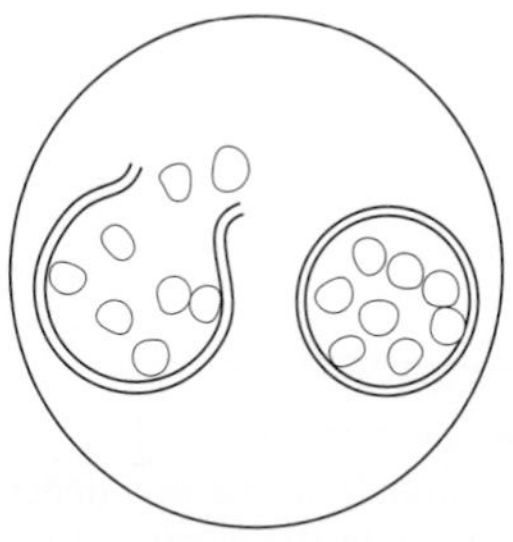

Pneumocystis jiroveci

***Pneumocystis* pneumonia** (**PCP**) currently is the most common cause of nonbacterial pneumonia in Americans with suppressed immune systems. The causative organism, *Pneumocystis jiroveci* (previously called *Pneumocystis carinii*) produces an atypical pneumonia that remained in relative obscurity until the 1980s, when it was recognized as the cause of death in over 50% of patients dying from the effects of AIDS.

P. jiroveci is transmitted person to person by droplets from the respiratory tract, although transmission from the environment also can occur. A wide cross section of individuals harbors the organism without symptoms, mainly because of the control imposed by the immune system.

P. jiroveci has a complex life cycle taking place entirely in the alveoli of the lung. A feeding stage, called the **trophozoite**, swells to become a precyst stage, in which up to eight sporozoites develop in forming a mature cyst. The cyst then ruptures

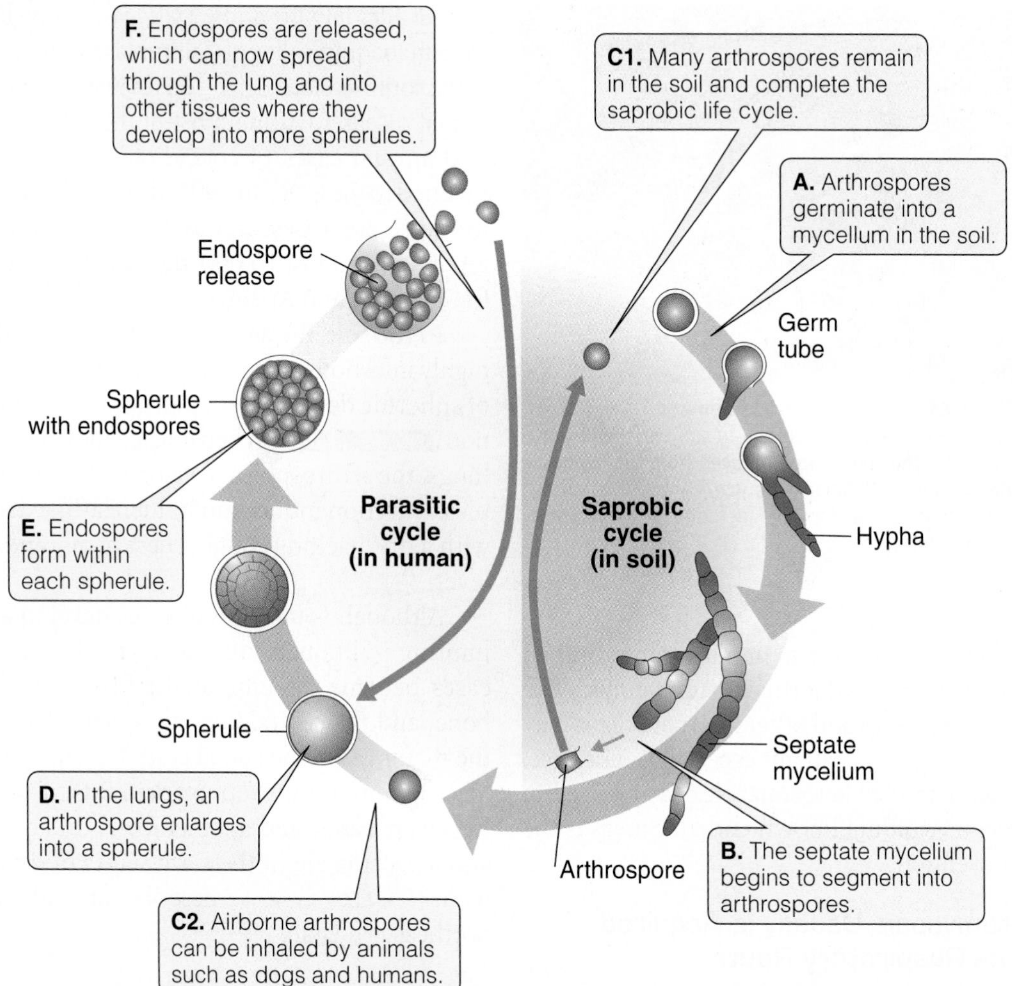

FIGURE 17.25 **The Life Cycle of *Coccidioides*.** Outside the body, the fungus goes through a saprobic cycle. However, the arthrospores in the respiratory tract go through a parasitic cycle, producing endospores capable of forming more spherules and infecting other tissues such as the skin, bone, and central nervous system.
»» Explain why *Coccidioides* is considered a dimorphic fungus.

and liberates the sporozoites, which enlarge and undergo further reproduction and maturation to trophozoites.

When the immune system is suppressed, as in AIDS patients, *Pneumocystis* trophozoites and cysts fill the alveoli and occupy the air spaces. A nonproductive cough develops, with fever and difficult breathing. Progressive deterioration leads to consolidation of the lungs and, respiratory failure. The current treatment for severe PCP is trimethoprim-sulfamethoxazole (co-trimoxazole) and corticosteroid therapy.

■ Consolidation: Formation of a firm dense mass in the alveoli.

Other Fungi Also Cause Mycoses

A few other ascomycete fungal diseases deserve brief mention because they are important in certain parts of the United States or they affect individuals in certain professions. Generally the diseases are mild, although complications may lead to serious tissue damage in severely immunocompromised patients.

Invasive **aspergillosis** is a unique disease because the fungus enters the body as conidia and then grows as a mycelium. Disease usually occurs in an immunosuppressed host or where an overwhelming number of conidia has entered the tissue. The most common cause is *Aspergillus fumigatus,* which is found in decaying leaves, rotting vegetables, and stored grain. An opportunistic infection of the lung may yield a round ball of mycelium called a **pulmonary aspergilloma** that produces a bloody cough,

Textbook CASE 17

An Outbreak of Coccidioidomycosis

1 On June 18, 2001, two archeologists with the National Park Service (NPS) directed a team of six student volunteers and two leaders at an archeological site in Dinosaur National Monument (DNM), a 515 square mile area in northeastern Utah and northwestern Colorado.

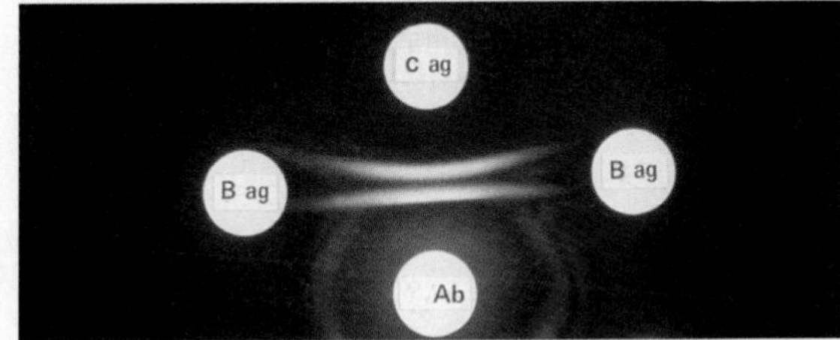

Courtesy of Dr. Errol Reiss/CDC, modified.

2 The volunteers and leaders laid stone steps, built a retaining wall, and sifted dirt for artifacts. Peak dust exposure occurred on June 19, the day most sifting occurred. Workers did not wear protective facemasks.

3 Between June 29 and July 3, all eight team members and the two NPS archeologists sought medical care at a local hospital emergency department for a flu-like illness with fever, cough, headache, rash, and muscle aches. Chest X rays showed that all 10 individuals had fluid filling the pulmonary airspaces; eight individuals were hospitalized with pneumonia.

4 NPS closed the work site to all visitors and staff, and the TriCounty Health Department alerted the public. On July 2, the TriCounty Health Department, the Utah Department of Health, and the Centers for Disease Control and Prevention (CDC) initiated an investigation to identify the risk factors, cause, and extent of the outbreak.

5 During July 2–4, a total of 18 persons (the 8 team members and 10 archeologists) with potential exposure to dust at the work site in June were interviewed to determine symptoms and previous activities. Hospital records were reviewed to ascertain clinical information. A case definition was defined for a person working at DNM.

6 Results of blood cultures from the hospitalized persons were negative for bacterial pathogens. Initial serologic tests were negative for antibodies to *Francisella tularensis, Yersinia pestis, Mycoplasma* species, *Histoplasma capsulatum,* and *Blastomyces dermatitidis.* Further serological tests to detect IgM antibodies (typically found in an early immunological response) to *Coccidioides immitis* showed positive precipitin lines indicative of IgM antibodies in 9 of the 10 acute serum specimens from patients, confirming the diagnosis of acute coccidioidomycosis (see figure). All hospitalized patients were treated with fluconazole.

7 A serosurvey of 40 other park employees was conducted to identify other infected persons and to guide prevention and control measures. None of the 40 had detectable antibodies to *C. immitis.*

8 The DNM site, which reopened on September 28, is located approximately 200 miles north of areas where *C. immitis* is normally found.

Questions:

(Answers can be found on the Student Companion Website in **Appendix D.**)

A. What would be present in the dust from the sifted dirt that would lead to the described illness?

B. A case definition is the method public health professionals use to define who is included as a "case" in an outbreak investigation. Provide a case definition for the illness described in this textbook case.

C. Why was a serosurvey seen as an important part of the outbreak analysis? What would health officials be looking for in a serosurvey; that is, in serum collected from the 40 other individuals?

D. What conclusions can you draw from the fact that this coccidioidomycosis outbreak was much farther north than where the disease is usually found?

E. How could these cases of coccidioidomycosis have been prevented?

For additional information see www.cdc.gov/mmwr/preview/mmwrhtml/mm5045a1.htm.

TABLE
17.3 A Summary of the Major Fungal LRT Diseases

Disease	Causative Agent	Signs and Symptoms	Transmission	Treatment	Prevention
Cryptococcosis	*Cryptococcus neoformans* *Cryptococcus gattii*	Asymptomatic Opportunistic infection leads to severe headache, stiff neck, paralysis	Airborne yeast cells	Amphotericin B	Maintaining strong immune system
Histoplasmosis	*Histoplasma capsulatum*	Mild influenza-like illness Can disseminate to other organs	Airborne spores	Amphotericin B or ketoconazole for systemic disease	Wearing face mask in contaminated areas
Blastomycosis	*Blastomyces dermatitidis*	Persistent cough Chest pains Chronic pneumonia	Airborne spores	Amphotericin B	Wearing face mask in contaminated areas
Coccidioidomycosis	*Coccidioides immitis* *Coccidioides posadasii*	Dry, hacking cough Chest pains High fever	Airborne arthrospores	Amphotericin B	Limiting exposure where infection is highest
Pneumocystis pneumonia	*Pneumocystis jiroveci*	Nonproductive cough Fever Breathing difficulty	Airborne droplets	Trimethoprim-sulfa-methoxazole	Maintaining strong immune system
Aspergillosis	*Aspergillus fumigatus*	Bloody cough Chest pain Wheezing Shortness of breath	Airborne spores	Voriconazole	Staying away from sources of mold

chest pain, and wheezing with shortness of breath.

The most deadly form of aspergillosis—**invasive aspergillosis**—occurs when the fungal infection spreads beyond the lungs to the other organs, such as the skin, heart, kidneys, or brain. Signs and symptoms depend on which organs are affected, but, in general, include headache, fever with chills, bloody cough, shortness of breath, and chest or joint pain.

In 2012, some steroidal medications contaminated with A. *fumigatus* and/or a similar fungus called *Exserohilum rostratum* infected over 500 healthy individuals being treated for back, neck, and joint pain.

More than 35 of these individuals treated for back and neck pain succumbed to this rare form of **fungal meningitis**.

Treatment usually involves antifungal drugs such as voriconazole. It can be difficult to avoid *Aspergillus* spores in the environment. Staying away from obvious sources of mold, such as compost piles and damp places, can help prevent infection in susceptible individuals.

TABLE 17.3 summarizes the fungal diseases of the lower respiratory tract.

CONCEPT AND REASONING CHECKS 5

a. Explain why cryptococcosis is such a dangerous fungal disease.
b. How does histoplasmosis differ in "healthy" individuals versus in an immunocompromised individual?
c. Why is blastomycosis a dangerous disease in immunocompromised individuals?
d. What is the unique feature of a coccidioidomycosis infection?
e. What are the unique features of *P. jiroveci?*
f. Summarize the unique features of aspergillosis.

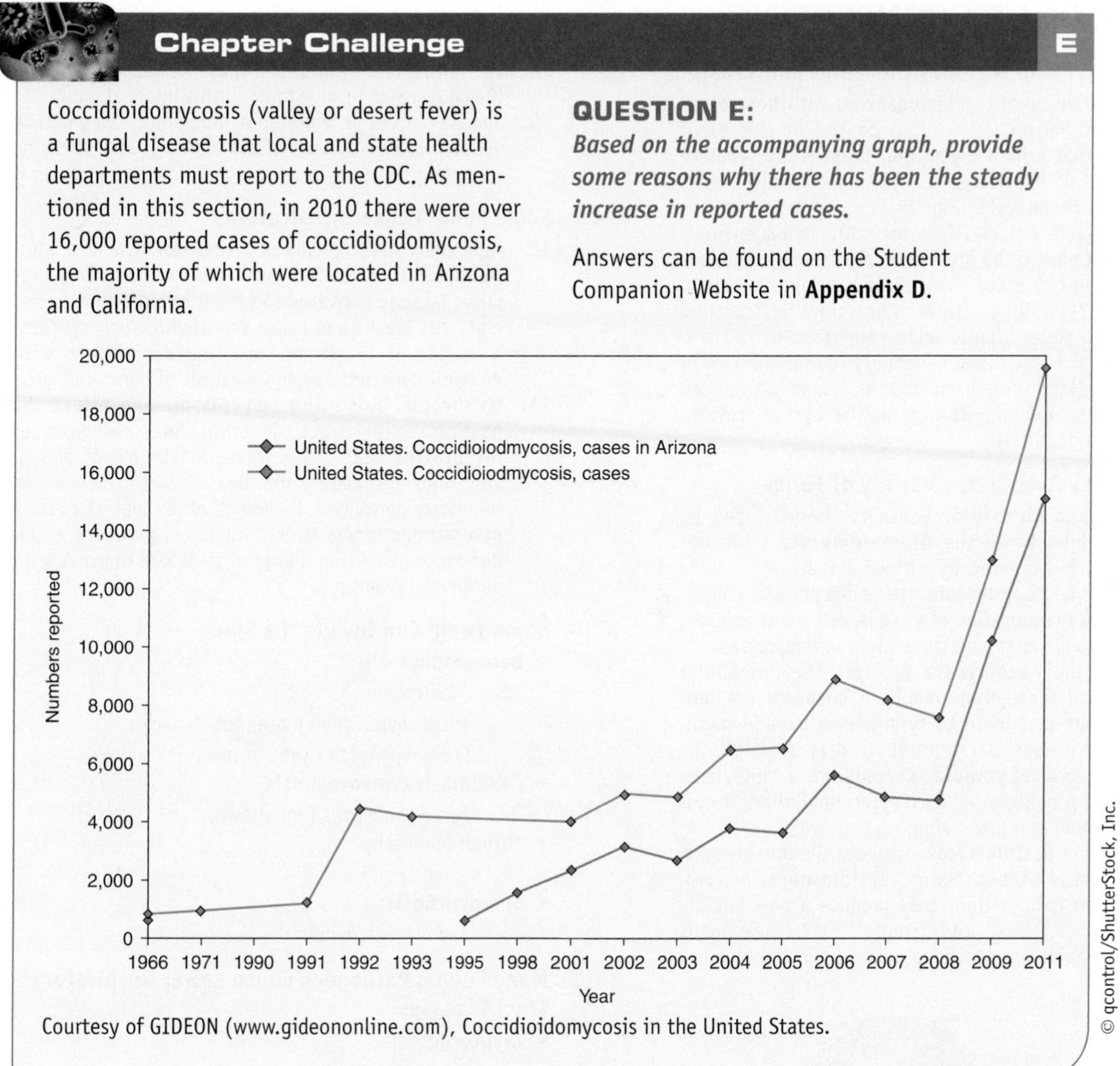

Chapter Challenge E

Coccidioidomycosis (valley or desert fever) is a fungal disease that local and state health departments must report to the CDC. As mentioned in this section, in 2010 there were over 16,000 reported cases of coccidioidomycosis, the majority of which were located in Arizona and California.

QUESTION E:

Based on the accompanying graph, provide some reasons why there has been the steady increase in reported cases.

Answers can be found on the Student Companion Website in **Appendix D.**

Courtesy of GIDEON (www.gideononline.com), Coccidioidomycosis in the United States.

■ **In conclusion**, we have learned about several of the fungal diseases that affect humans. In the global picture, fungi infect billions of people every year, yet their significance to worldwide infectious disease remains unrecognized. Although there is no accurate record keeping for the diseases caused by the fungal pathogens of humans, experts say that taken as a whole these pathogens kill as many people every year as tuberculosis or malaria. Part of the increasing incidence of fungal diseases is due to immunosuppressive diseases, like AIDS, that weaken the immune system's response to opportunistic infections such as cryptococcosis and other respiratory diseases we discussed in this chapter. And this increased incidence of fungal diseases is not something that is limited to humans. In the last 20 years, there has been an extraordinary number of fungal diseases that have caused severe die-offs and extinctions. As described in this chapter, frog populations, especially in Central America, are being decimated by a fungal disease and the bat population in North America could be wiped out by a deadly fungus causing white-nose syndrome. In addition, collapse of honeybee colonies in 2010 was linked to a fungus and insect virus coinfection that may have been responsible for 40% to 60% of disappearing beehives since 2005. So, other animals, as well as humans, are at risk for life-threatening fungal infections.

SUMMARY OF KEY CONCEPTS

17.1 The Kingdom Fungi Includes the Molds and Yeasts

1. Fungi are eukaryotic microorganisms with **heterotrophic** metabolism. Most fungi consist of masses of **hyphae** that form a **mycelium. Cross-walls** separate the cells of hyphae in many fungal species. Most fungi are **coenocytic.** (Fig. 17.3)
2. Fungi secrete enzymes into the surrounding environment and absorb the breakdown products. Tremendous absorption can occur when there is a large mycelium surface. Most fungi are aerobic, grow best around 25°C, and prefer slightly acidic conditions.
3. Reproductive structures generally occur at the tips of hyphae. Masses of asexually- or sexually-produced **spores** are formed within or at the tips of **fruiting bodies.** (Fig. 17.9)

17.2 Fungi Have Evolved into a Variety of Forms

4. The phylum **Chytridiomycota** is characterized by motile cells, while the **Glomeromycota** represent fungi living symbiotically with land plants.
5. In the phylum **Zygomycota**, the sexual phase is characterized by the formation of a **zygospore**, which releases haploid spores that germinate into a new mycelium.
6. The phylum **Ascomycota** includes the unicellular yeasts and filamentous molds. **Ascospores** are produced that germinate to form a new haploid mycelium, while asexual reproduction is through the dissemination of **conidia. Lichens** are a mutualistic association between an ascomycete and either a cyanobacterium or a green alga.
7. The phylum **Basidiomycota** includes the **mushrooms.** Within these fruiting bodies, **basidiospores** are produced. On germination, they produce a new haploid mycelium. "Rusts" and "smuts" that cause many plant diseases are additional members of the phylum.
8. The **mitosporic fungi** lack a sexual phase. Many human fungal diseases involve fungi in this informal group.
9. *Saccharomyces* is a notable unicellular ascomycete **yeast** involved in baking and brewing, and scientific research.

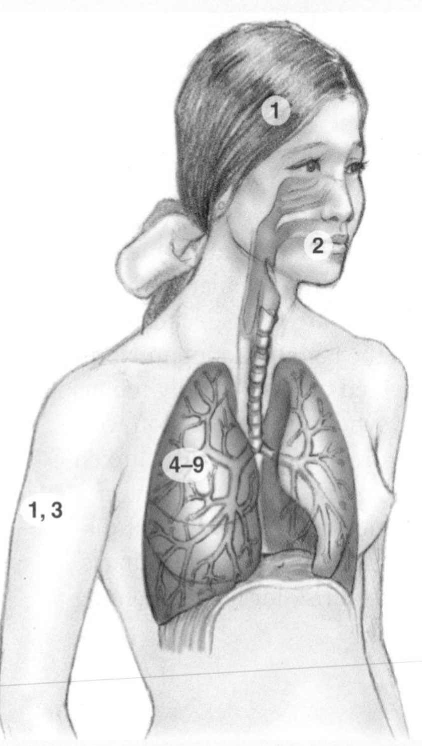

17.3 Some Fungi Cause Intoxications

10. *Aspergillus flavus* produces a **mycotoxin**, called **aflatoxin**, which is carcinogenic. *Claviceps purpurea* produces alkaloid derivatives in grain called **ergot.** Eating ergotized breads can cause **ergotism**, which produces a variety of symptoms including convulsions with epileptic-type seizures and paralysis of nerve endings.
11. **Mushroom poisoning** (**mycetism**) can affect the gastrointestinal tract. *Amanita phalloides* produces **mycotoxins** that cause severe abdominal cramping, vomiting, and watery diarrhea when the raw mushrooms are consumed. Numerous other species cause a noninflammatory gastroenteritis that produces a sudden severe vomiting and mild to severe diarrhea with abdominal cramps.

17.4 Some Fungi Can Invade the Skin

- **Dermatophytosis**
  1. *Microsporum*
     Trichophyton (nail plate; not shown)
     Epidermophyton (webs of toes; not shown)
- **Candidiasis (vulvovaginitis)**
  Candida albicans (not shown)
- **Thrush** (children)
  2. *Candida albicans*
- **Sporotrichosis**
  3. *Sporothrix schenckii*

17.5 Many Fungal Pathogens Cause Lower Respiratory Tract Diseases

- **Cryptococcosis**
  4. *Cryptococcus neoformans*
- **Histoplasmosis**
  5. *Histoplasma capsulatum*
- **Blastomycosis**
  6. *Blastomyces dermatitidis*
- **Coccidioidomycosis**
  7. *Coccidioides immitis, Coccidioides posadasii*
- *Pneumocystis* **pneumonia**
  8. *Pneumocystis jirovecii*
- Aspergillosis
  9. *Aspergillus fumigatus*

CHAPTER SELF-TEST

For **STEPS A–D**, answers to even-numbered questions and problems can be found in **Appendix C** on the Student Companion Website at **http://microbiology.jbpub.com/10e**. In addition, the site features eLearning, an online review area that provides quizzes and other tools to help you study for your class. You can also follow useful links for in-depth information, read more MicroFocus stories, or just find out the latest microbiology news.

STEP A: REVIEW OF FACTS AND TERMS

Multiple Choice

Read each question carefully, then select the ***one*** answer that best fits the question or statement.

1. Which one of the following statements about fungi is NOT true?
A. Some fungi are dimorphic.
B. Fungi have cell walls made of chitin.
C. Fungi are photosynthetic organisms.
D. Fungi consist of the yeasts and molds.

2. Which one of the following best describes the growth conditions for a typical fungus?
A. pH 3; 23°C; no oxygen gas present
B. pH 8; 30°C; no oxygen gas present
C. pH 6; 30°C; oxygen gas present
D. pH 3; 23°C; oxygen gas present

3. All the following are examples of asexual spore formation *except* _____.
A. arthrospores
B. conidia
C. sporangiospores
D. basidiospores

4. An organism without a known sexual stage would be classified in the _____.
A. Mitosporic fungi
B. Zygomycota
C. Basidiomycota
D. Ascomycota

5. Yeasts of the species *Saccharomyces* _____.
A. are used in bread making
B. reproduce by budding
C. are members of the ascomycetes
D. All of the above (**A–C**) are correct.

6. Aflatoxin is produced by _____ and is _____.
A. *Sporothrix schenckii;* a hallucinogen
B. *Aspergillus flavus;* carcinogenic
C. *Claviceps purpurea;* a hallucinogen
D. *Aspergillus niger;* a mycotoxin

7. Mushroom poisoning is _____.
A. always fatal
B. characterized by immediate symptoms
C. most common in children under 6 years of age
D. treatable with antifungal drugs

8. This fungal disease can cause a blister-like lesion on the scalp.
A. Candidiasis
B. Dermatophytosis
C. Cryptococcosis
D. Histoplasmosis

9. ______ causes more than 20 million cases each year and symptoms include an itching sensation and burning internal pain.
A. Thrush
B. "Jock itch"
C. Vulvovaginitis
D. Sporotrichosis

10. From the following, sporotrichosis would most likely be transmitted from _____.
A. peat moss
B. bat caves
C. mushrooms
D. dusty soil

11. Which one of the following fungi would most likely be found in pigeon droppings?
A. *Pneumocystis*
B. *Cryptococcus*
C. *Coccidioides*
D. *Sporothrix*

12. Moving to the Ohio or Mississippi River valleys might make one susceptible to _____.
A. PCP
B. valley fever
C. dermatophytosis
D. histoplasmosis

13. This dimorphic fungus produces conidia that are inhaled from dusty soil or bird droppings.
A. *Aspergillus fumigatus*
B. *Pneumocystis jiroveci*
C. *Amanita phalloides*
D. *Blastomyces dermatitidis*

14. The formation of arthrospores and spherules is characteristic of _____.
A. coccidioidomycosis
B. histoplasmosis
C. candidiasis
D. aspergillosis

15. This fungal disease is the most common cause of nonbacterial pneumonia in immunocompromised individuals.
A. Blastomycosis
B. *Pneumocystis* pneumonia
C. Aspergillosis
D. Coccidioidomycosis

16. The most deadly form of aspergillosis is _____.
A. a pulmonary form
B. a toxigenic form
C. a blood form
D. an invasive form

Matching

Match the statement on the left to the organism on the right by placing the correct letter in the available space. A letter may be used once, more than once, or not at all.

Statement

___ **17.** Produces a widely used antibiotic.

___ **18.** Used for bread baking.

___ **19.** Causes "valley fever" in the southwestern U.S.

___ **20.** Common white or gray bread mold.

___ **21.** Agent of rose thorn disease.

___ **22.** Associated with the droppings of pigeons.

___ **23.** Agent of ergot disease in rye plants.

___ **24.** Cause of vaginal yeast infections.

___ **25.** One of the causes of dermatophytosis.

___ **26.** Often found in chicken coops and bat caves.

___ **27.** Produces a toxic aflatoxin.

___ **28.** Reproduction includes a spherule.

___ **29.** Involves a trophozoite stage.

Organism

A. *Agaricus* species
B. *Aspergillus flavus*
C. *Aspergillus* species
D. *Blastomyces dermatitidis*
E. *Candida albicans*
F. *Claviceps purpurea*
G. *Coccidioides immitis*
H. *Cryptococcus neoformans*
I. *Epidermophyton* species
J. *Histoplasma capsulatum*
K. *Penicillium notatum*
L. *Pneumocystis jiroveci*
M. *Rhizopus stolonifer*
N. *Saccharomyces cerevisiae*
O. *Saccharomyces ellipsoideus*
P. *Sporothrix schenckii*

STEP B: CONCEPT REVIEW

30. Differentiate between **molds**, **yeasts**, and **dimorphic** fungi and summarize the structure and function of fungal **hyphae**. (**Key Concept 1**)

31. Identify and describe the physical factors governing fungal growth and discuss examples of fungal symbioses, including the **mycorrhizae** and the fungal **endophytes**. (**Key Concept 1**)

32. Distinguish between the forms of asexual and sexual spores produced by fungi and describe the generalized sexual life cycle of a mold. (**Key Concept 1**)

33. Summarize the characteristics of the **Chytridiomycota** and their relevance to fungal taxonomy. Identify the key characteristics of the **Zygomycota**, **Ascomycota**, and **Basidiomycota**. (**Key Concept 2**)

34. Assess the usefulness of yeasts, such as *Saccharomyces*, to industry and scientific research. (**Key Concept 2**)

35. Summarize the effects of (a) *Aspergillus flavus* intoxication and (b) *Claviceps purpurea* ingestion. (**Key Concept 3**)

36. Justify why it can be dangerous to eat non-store-bought mushrooms. (**Key Concept 3**)

37. Summarize the symptoms and treatment of **dermatophytosis**. (**Key Concept 4**)

38. Describe the major types of *Candida* infections. (**Key Concept 4**)

39. Identify common mechanisms for transmission of *Sporothrix schenckii*. (**Key Concept 4**)

40. Summarize the symptoms and complications of **cryptococcosis**. (**Key Concept 5**)

41. Discuss the consequences of a **histoplasmosis** infection in an immunocompromised individual. (**Key Concept 5**)

42. Explain the dimorphic nature of *Blastomyces dermatitidis*. (**Key Concept 5**)

43. Review the symptoms of and the complications arising from **coccidioidomycosis**. (**Key Concept 5**)

44. Evaluate the potential seriousness of a ***Pneumocystis* pneumonia** infection. (**Key Concept 5**)

45. Summarize the possible affects of an *Aspergillus* species infection. (**Key Concept 5**)

STEP C: APPLICATIONS AND PROBLEMS

46. You decide to make bread. You let the dough rise overnight in a warm corner of the room. The next morning you notice a distinct beer-like aroma in the air. What did you smell, and where did the aroma come from?

47. In a Kentucky community, a crew of five workers demolished an abandoned building. Three weeks later, all five required treatment for acute respiratory illness, and three were hospitalized. Cells obtained from the patients by lung biopsy revealed oval bodies and epidemiologists found an accumulation of bat droppings at the demolition site. As the head epidemiologist, what disease did the workers contract?

48. A woman comes to you complaining of a continuing problem of ringworm, especially of the lower legs in the area around the shins. Questioning her reveals she has five very affectionate cats at home. What disease does she have and what would be your suggestion to her?

49. Using the diagram below, explain why these fungal diseases are considered to be systemic mycoses.

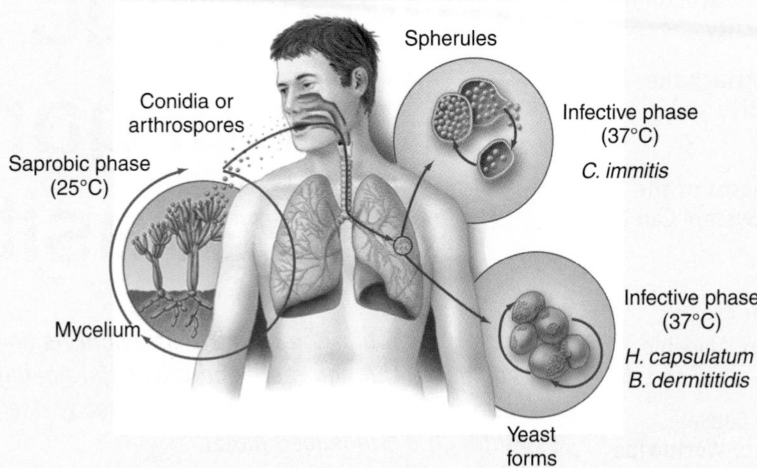

STEP D: QUESTIONS FOR THOUGHT AND DISCUSSION

50. In a suburban community, a group of residents obtained a court order preventing another resident from feeding the flocks of pigeons that regularly visited the area. Microbiologically, was this action justified? Why?

51. Mr. A and Mr. B live in an area of town where the soil is acidic. Oak trees are common, and azaleas and rhododendrons thrive in the soil. In the spring, Mr. A spreads lime on his lawn, but Mr. B prefers to save the money. Both use fertilizer, and both have magnificent lawns. Come June, however, Mr. B notices that mushrooms are popping up in his lawn and that brown spots are beginning to appear. By July, his lawn has virtually disappeared. What is happening in Mr. B's lawn and what can Mr. B learn from Mr. A?

52. Residents of a New York community, unhappy about the smells from a nearby composting facility and concerned about the health hazard posed by such a facility, had the air at a local school tested for the presence of fungal spores. Investigators from the testing laboratory found abnormally high levels of *Aspergillus* spores on many inside building surfaces. Is there any connection between the high spore count and the composting facility? Is there any health hazard involved?

53. On January 17, 1994, a serious earthquake struck the Northridge section of Los Angeles County in California. From that date through March 15, 170 cases of coccidioidomycosis were identified in adjacent Ventura County. This number was almost four times the previous year's number of cases. What is the connection between the two events?

18

CHAPTER PREVIEW

18.1 Protists Exhibit Great Structural and Functional Diversity

18.2 Protistan Parasites Attack the Skin, and the Digestive and Urinary Tracts

18.3 Many Protistan Diseases of the Blood and Nervous System Can Be Life Threatening

Investigating the Microbial World 18: Love Your Enemy

Textbook Case: Primary Amoebic Meningoencephalitis

18.4 Parasitic Helminths Cause Substantial Morbidity Worldwide

MicroInquiry 18: Parasites as Manipulators

Image courtesy of Dr. Fred Murphy/CDC.

Eukaryotic Microorganisms: The Parasites

It races through the bloodstream, hunkers down in the liver, then rampages through red blood cells before being sucked up by its flying, buzzing host to mate, mature, and ready itself for another wild ride through a two-legged motel.

—The editor of *Discover* magazine describing, in flowery terms, the life cycle of the malarial parasite

Approximately 1 million people die every year from malaria, an infection caused by a protozoan parasite and carried from person to person by mosquitoes. The disease is one of the most severe public health problems worldwide and is a leading cause of death and disease in many developing countries. Yet in 1957, a global program to eradicate the parasite commenced only to end in failure 21 years later. What happened?

Believe it or not, malaria once was an infectious killer in the United States. In the late 1880s, malaria was quite common in the American plains and southeast with epidemics reaching as far north as Montreal, Canada. Malaria was a major source of casualties in the American Civil War and until the 1930s, malaria remained endemic in the southern states.

To eliminate malaria, American officials established the National Malaria Eradication Program on July 1, 1947. This was a cooperative undertaking between the newly established Communicable Disease Center (the original CDC and a new component of the U.S. Public Health Service) and state and local health agencies of the 13 malaria-affected southeastern states. In 1947, 15,000 malaria cases were reported. However, by 1950, after more than 4,650,000 homes had been sprayed with the DDT (dichlorodiphenyltrichloroethane) pesticide to kill the mosquitoes, only 2,000 malaria cases were reported. By 1952, the United States was malaria free and the program ended.

Encouraged by the success of the American eradication effort, in 1957 the World Health Organization (WHO) began a similar effort to eradicate malaria worldwide. These efforts involved house spraying with

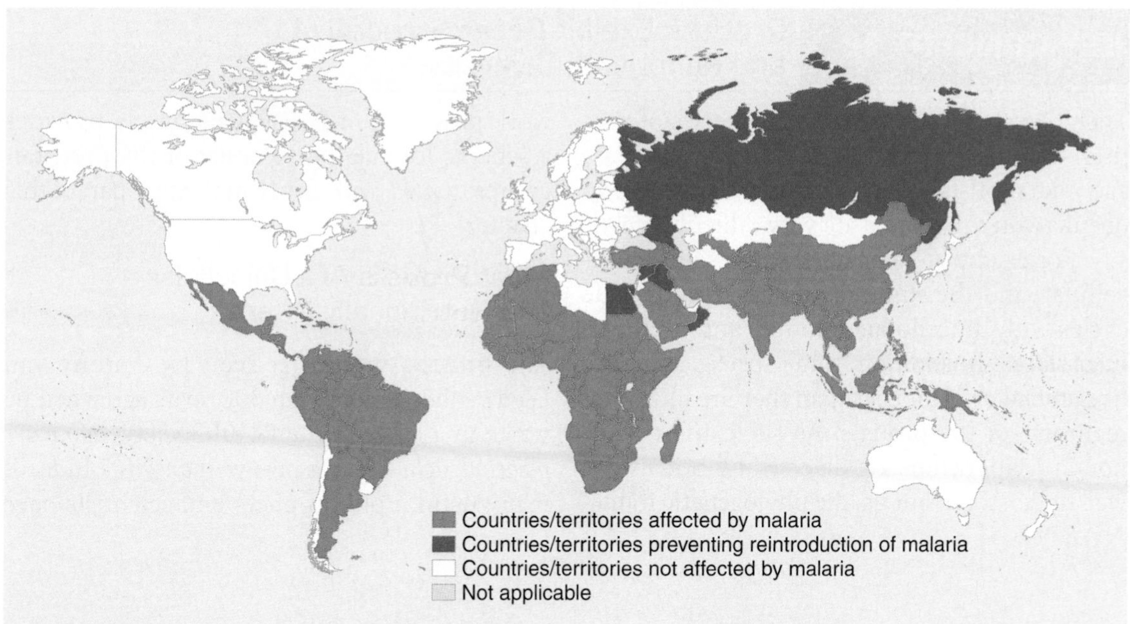

FIGURE 18.1 **Countries and Territories Affected by Malaria—2010.** Malaria occurs in 99 countries and puts an estimated 3.3 billion persons at risk. Reproduced from WHO Global Health Observatory. »» Where do you think the majority of the 216 million estimated malaria cases occur?

insecticides, antimalarial drug treatment, and surveillance. Successes were made in some countries, but the emergence of drug resistance, widespread mosquito resistance to DDT, wars, and massive population movements made the eradication efforts unsustainable. The eradication campaign was eventually abandoned in 1978.

Now, more than 30 years later, a substantial portion of the world's population is still at risk of malaria (FIGURE 18.1). The good news is another global campaign, initiated by the WHO, several United Nations agencies, and the World Bank, is ending more positively. Named "Roll Back Malaria," the program called for a 50% reduction in the burden of malaria by 2010. As we will see, that program has been a great success.

Malaria is just one of a number of human diseases caused by **parasites**, organisms that must live in or on a different species to get their nourishment. There are two different groups of eukaryotic parasites of concern to microbiology because of their ability to cause infectious disease.

One group contains single-celled **protists**. Some of the diseases they cause are familiar to us, such as malaria. Others affect the intestine, blood, or other organs of the body.

The second group are multicellular parasites, referred to as **helminths** (*helminth* = "worm"). These include the flatworms and roundworms, which together probably infect more people worldwide than any other group of organisms. In the strict sense, flatworms and roundworms are animals, but they are studied in microbiology because of their ability to cause disease. Together with the parasitic protists, they are the subject of study of the biological discipline known as **parasitology**.

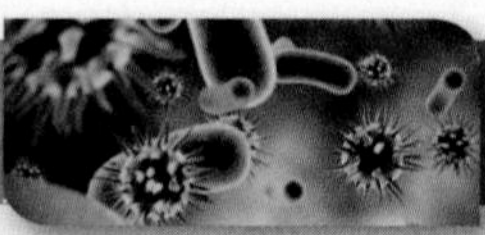

Chapter Challenge

The protists are by far the most diverse of any eukaryotic (or prokaryotic) group of organisms, which illustrates the broad evolutionary relationships among these microorganisms. So, the term protist does not describe a close-knit taxonomic group but rather it is an "umbrella" term for any eukaryotic organism that is not plant, animal, or fungus. In this chapter's challenges, you will examine a few events about the protists and then finish with a challenge that includes the parasitic helminths.

KEY CONCEPT 18.1 Protists Exhibit Great Structural and Functional Diversity

There are about 200,000 named species of protists, some seen by Leeuwenhoek over 300 years ago. The protists are an extremely diverse group of eukaryotes that often are very difficult to classify. For example, some algal species are multicellular, and the slime molds have unique life cycles with unicellular, colonial, and multicellular stages. In addition, some species are more like animal or plant cells than they are like other members of the protists. As such, their taxonomic relationships are diverse and not always well understood. For us, the phylogenetic framework presented in FIGURE 18.2 simply provides a scheme for cataloging some of the protistan groups that we will survey in the first part of this chapter.

Most Protists Are Unicellular and Nutritionally Diverse

The protists were first seen by Antony van Leeuwenhoek more than 300 years ago when he wrote in a letter to the Royal Society, "*No more pleasant sight has met my eye than this.*" Indeed, many natural philosophers and scientists have

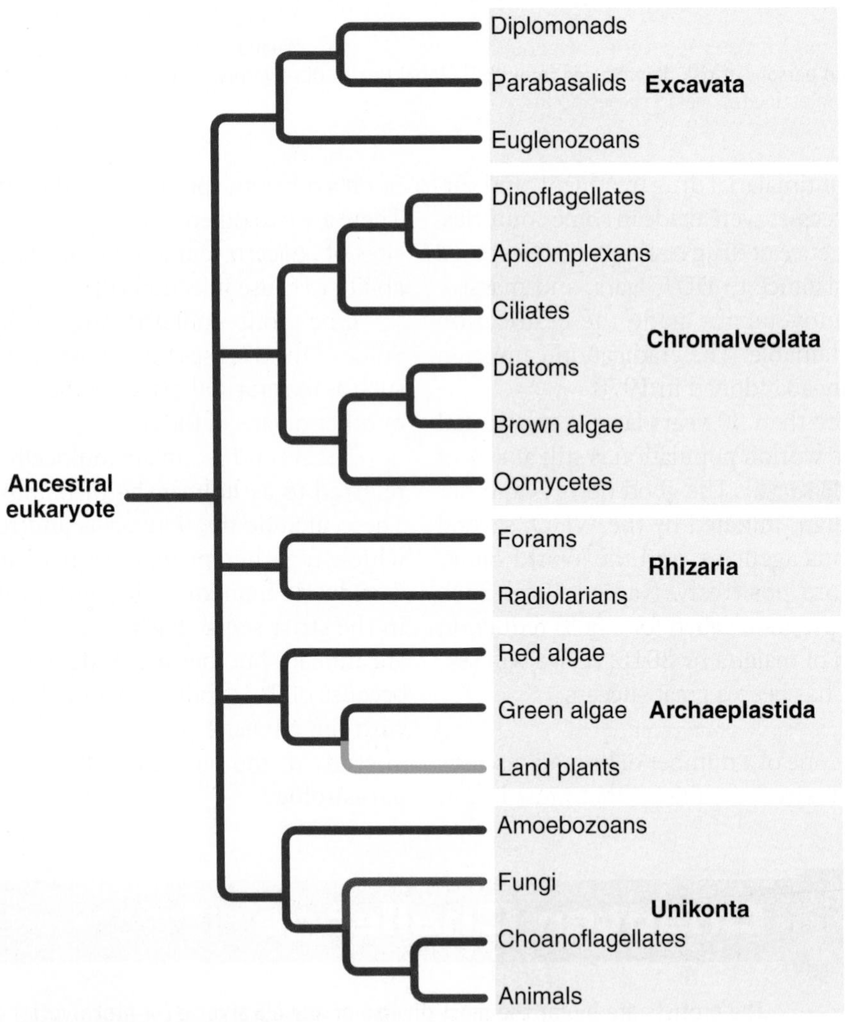

FIGURE 18.2 **A Tentative Phylogeny for the Domain Eukarya.** The phylogeny includes the five eukaryotic "supergroups" (in boldface). Everything besides the land plants and animals can be considered microbial and everything besides the land plants, animals, and fungi would be considered in the kingdom Protista. »» What kingdom of organisms is most closely related to the animals?

continued to study this structurally and functionally diverse assortment of eukaryotes.

The protists are primarily unicellular; however, the functions of the single cell bear a resemblance to the functions of a multicellular organism rather than to those of an isolated cell from the organism. Most protists are free-living and thrive where there is water. They may be located in damp soil, mud, drainage ditches, and puddles. Some species remain attached to aquatic plants or rocks, while other species swim about.

Protists also are very diverse nutritionally. Many are heterotrophic and obtain their energy and organic molecules from other organisms, often obtaining these materials by forming a parasitic relationship with a susceptible host. Others, including the red and green **algae**, contain chloroplasts and carry out photosynthesis similar to green plants (FIGURE 18.3A). Some protists, such as the **dinoflagellates**, are part of the freshwater and marine phytoplankton and include both heterotrophic and photosynthetic species, making them important primary producers in the world's oceans. Some dinoflagellates cause the infamous red tides, which are discussed in more detail in another chapter. One dinoflagellate, *Pfiesteria piscicida*, may be linked to extensive fish kills in waters from Alabama to Delaware.

Other marine protists also are part of the marine phytoplankton. The **radiolarians** have highly sculptured and glassy silica plates with radiating cytoplasmic arms to capture prey (FIGURE 18.3B). The skeletons of dead radiolarians litter the ocean floor, forming deposits, sometimes hundreds of meters thick, called "radiolarian ooze." The **foraminiferans** (**forams**) are marine protists

- Phytoplankton: Microscopic free-floating communities of cyanobacteria and unicellular algae.
- Primary producer: Organism that produces organic compounds from carbon dioxide gas.
- Red tide: Brownish-red discoloration in seawater that is caused by an increased presence of dinoflagellates.

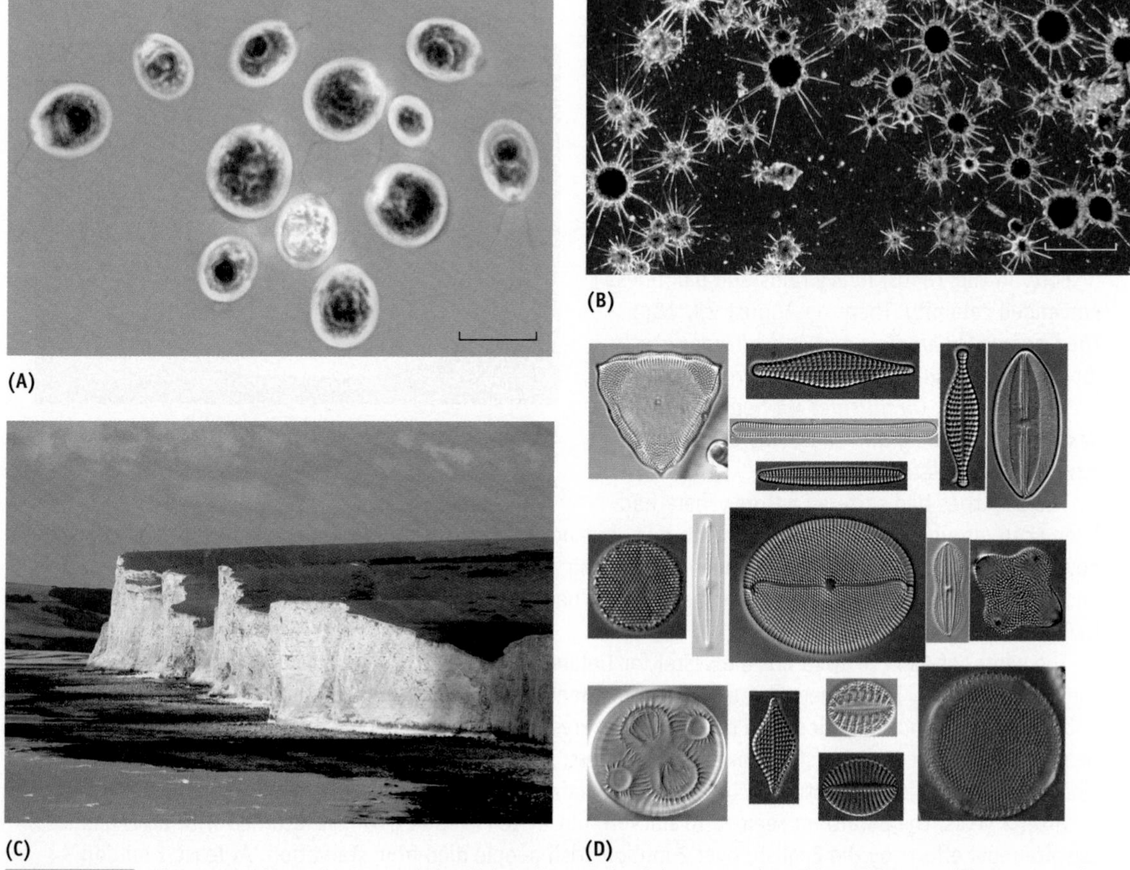

FIGURE 18.3 **Algae and Phytoplankton.** (**A**) A light micrograph of the green alga *Chlamydomonas*. (Bar = 10 μm.) © M I (Spike) Walker/Alamy Images. (**B**) Light micrograph of an assortment of radiolarians. Radiolarians build hard skeletons made of silica around themselves as they float in seas with other plankton. (Bar = 120 μm.) © M. I. Walker/Photo Researchers, Inc. (**C**) The white chalk cliffs at Beachy Head in Sussex, England, consist of the ancient shells of foraminiferans. © Mark Bond/ShutterStock, Inc. (**D**) The diversity in shape of the diatoms. Courtesy of Eduardo A. Morales, Ph.D., The Academy of Natural Sciences of Philadelphia. »» What characteristics are similar in the algae and phytoplankton?

that have chalky skeletons, often in the shape of snail shells with openings between sections (*foram* = "little hole"). The shells of dead forams form sediments hundreds of meters thick. When brought to the surface by geologic upthrusting, massive white cliffs have formed (FIGURE 18.3C).

The **diatoms** are another group of single-celled protists. By being encased in a two-part, hard-shelled, silica wall (FIGURE 18.3D), the cells can withstand great pressures and are not easily crushed or destroyed by predators. Diatoms carry out photosynthesis and compose an important part of the phytoplankton found in marine and freshwater environments. The massive accumulations of fossilized diatom walls are mined and ground up into **diatomaceous earth**, which has many useful applications, such as a filtering agent in swimming pools and a mild abrasive in household products, including toothpastes and metal polishes. Diatomaceous earth also can be used as a pesticide, because it grinds holes in the exoskeleton of crawling insects, causing the animals to desiccate.

Yet other protists are fungal-like. The **oomycetes** are completely heterotrophic and absorb extracellularly digested food materials. However, these protists resemble fungi because they produce a filamentous growth characteristic of the molds. Some are plant pathogens. *Phytophthora ramorum,* for example, infects the California coastal live oak and causes sudden oak death. *P. infestans* is the agent of late blight in potatoes. It was responsible for the Irish famine of the 19th century and its infection had great ecological impact on humans and society (MicroFocus 18.1).

MICROFOCUS 18.1: History
The Great Irish Potato Famine

Ireland of the 1840s was an impoverished country of 8 million people. Most were tenant farmers paying rent to landlords who were responsible, in turn, to the English property owners. The sole crop of Irish farmers was potatoes, grown season after season on small tracts of land. What little corn was available was usually fed to the cows and pigs.

Early in the 1840s, heavy rains and dampness portended calamity. Then, on August 23, 1845, *The Gardener's Chronicle and Agricultural Gazette* reported: *"A fatal malady has broken out amongst the potato crop. On all sides we hear of the destruction. In Belgium, the fields are said to have been completely desolated."*

Potatoes infected with *Phytophthora infestans* (dark brown regions). © Holt Studios International Ltd/ Alamy Images.

The potatoes had suffered before. There had been scab, drought, "curl," and too much rain, but nothing was quite like this new disease. It struck down the plants like frost in the summer. Beginning as black spots, it decayed the leaves and stems, and left the potatoes a rotten, mushy mass with a peculiar and offensive odor (see figure). Even the harvested potatoes rotted.

The winter of 1845 to 1846 was a disaster for Ireland. Farmers left the rotten potatoes in the fields, and the disease spread. The farmers first ate the animal feed and then the animals. They also devoured the seed potatoes, leaving nothing for spring planting. As starvation spread, the English government attempted to help by importing corn and establishing relief centers. In England, the potato disease had few repercussions because English agriculture included various grains. In Ireland, however, famine spread quickly.

After 2 years, the potato rot seemed to slacken, but in 1847 ("Black '47") it returned with a vengeance. Despite relief efforts by the English, over 2 million Irish people died from starvation. At least 1 million Irish people left the land and moved to cities or foreign countries. During the 1860s, great waves of Irish immigrants came to the United States—and in the next century, their Irish American descendants would influence American culture and politics. And to think—it all resulted from a filamentous protist, *Phytophthora infestans,* which remains a difficult organism to control today.

The protists also include many motile, predatory, or parasitic species that absorb or ingest food. These protists traditionally have been called the **protozoa** (*proto* = "first"; *zoo* = "animal"), referring to their animal-like properties that incorrectly suggested to biologists that protozoa were close evolutionary ancestors of the first animals. Though often studied by zoologists, protozoa also interest microbiologists because they are unicellular, most have a microscopic size, and several are responsible for infectious disease. Therefore, these are the microbial parasites we will emphasize next.

The Protists Encompass a Variety of Parasitic Lifestyles

A tentative taxonomy, based on comparative studies involving genetic analysis and genomics, places those protists (protozoa) that are human parasites in one of three informal "supergroups." Here, we briefly consider the biological features of the three groups.

Supergroup Excavata. This group contains species that are single-celled and possess flagella for motility. Some members in the group may represent organisms whose ancestors were the earliest forms of eukaryotes.

Members of the **parabasalids** lack mitochondria but have another membrane-enclosed structure called a **hydrogenosome**. This organelle can make ATP through a fermentation-like process that produces acetate and hydrogen gas as end products. As such, these protists live in low oxygen or anaerobic environments. Several species, including *Trichonympha,* are found in the guts of termites where the symbionts participate in a mutualistic relationship (FIGURE 18.4A). The cells contain hundreds of flagella with the characteristic arrangement of microtubules typically found in all

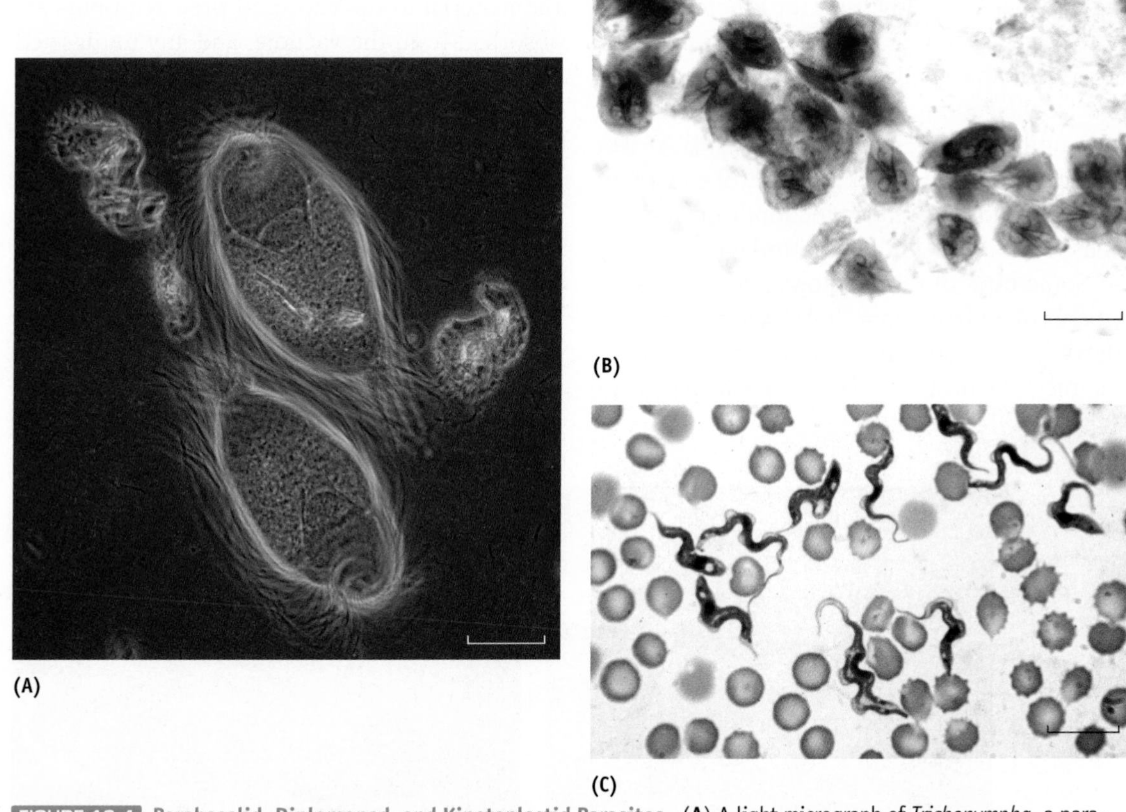

FIGURE 18.4 **Parabasalid, Diplomonad, and Kinetoplastid Parasites.** (**A**) A light micrograph of *Trichonympha,* a parabasalid parasite found in the gut of termites. Each thin, wispy line represents a flagellum used for motility. (Bar = 25 μm.) © Wim van Egmond/Visuals Unlimited/Getty. (**B**) Another light micrograph of stained *Giardia intestinalis* cells. The pear-shaped cell body of this diplomonad and flagella are typical features. (Bar = 5 μm.) © Visuals Unlimited/Corbis. (**C**) Light micrograph of *Trypanosoma* in a blood smear. The wavy cell appearance is due to the wavy membrane and flagellum. (Bar = 10 μm.) Courtesy of the Laboratory of Parasitology, University of Pennsylvania School of Veterinary Medicine. »» If the parabasalids and diplomonads lack (or have modified) mitochondria, what metabolic pathway remains for synthesizing ATP?

eukaryotic flagella. Undulations sweep down the flagella to the tip, and the lashing motion forces water outward to provide locomotion. Another species, *Trichomonas vaginalis,* is a human parasite and is transmitted through sexual intercourse.

The **diplomonads** have two haploid nuclei and three pairs of flagella at the anterior end and one pair at the posterior end, giving the cell bilateral symmetry. Reproduction is only asexual by binary fission. These protists have modified mitochondria and extract energy from anaerobic metabolic pathways.

■ Bilateral symmetry: A form of symmetry where an imaginary plane divides an object into right and left halves, each side being a mirror of the other.

The most notable species is *Giardia intestinalis* (FIGURE 18.4B). It is spread through water and, thus, affects the gastrointestinal tract. The diplomonad can survive outside the anaerobic environment of the intestine by forming a **cyst**, which is a dormant, highly resistant stage. So many lakes and rivers in the United States are contaminated with the cysts that hikers and campers must first boil or filter the water before drinking.

Another set of protists in the Excavata is the **Euglenozoa**. Among these are the **kinetoplastids**, another ancient lineage of heterotrophic species. A unique characteristic of these species is a single, posterior flagellum that is attached to the cell's wavy, undulating membrane (FIGURE 18.4C). The kinetoplastids have the typical array of eukaryotic organelles and the single mitochondrion contains a mass of DNA called the **kinetoplast**.

Some 60% of the kinetoplastid species are trypanosomes (*trypano* = "hole"; *soma* = "body"), referring to the hole the organism bores to enter and infect the host. Two *Trypanosoma* species are transmitted by insects and cause forms of human sleeping sickness in Africa and South America, affecting millions of people. Species of *Leishmania,* which is also transmitted by insects, can produce a skin disease or an often fatal visceral infection.

Supergroup Unikonta. One group of unikonts, the **amoebozoans**, are mostly free-living, single-celled organisms that can be as large as 1 mm in diameter. They usually live in freshwater or marine environments and reproduce by binary fission. The amoebozoans are soft bodied organisms that have the ability to change shape (*amoeba* = "change") as their cytoplasm flows into temporary formless cytoplasmic projections called **pseudopods** (*pseudo* = "false"; *pod* = "a foot"); thus, the motion is called **amoeboid motion**. Pseudopods also capture bacterial cells and other protistan cells through the ingestive process of **phagocytosis** (FIGURE 18.5A). The pseudopods enclose the particles to form an organelle called a **food vacuole**, which then joins with lysosomes. The lysosomes contain digestive enzymes to digest the material in the captured prey. Nutrients are absorbed from the vacuole, and any undigested residue is eliminated from the cell.

The genus *Entamoeba* can be far more serious, as all species are parasitic. In humans, amoebic dysentery or encephalitis may result from drinking water or consuming food contaminated with amoebal cysts. Amoebic dysentery is the third leading cause of death due to a parasitic infection.

Supergroup Chromalveolata. This group is very diverse, and includes the dinoflagellates and diatoms described earlier. The ciliated protists,

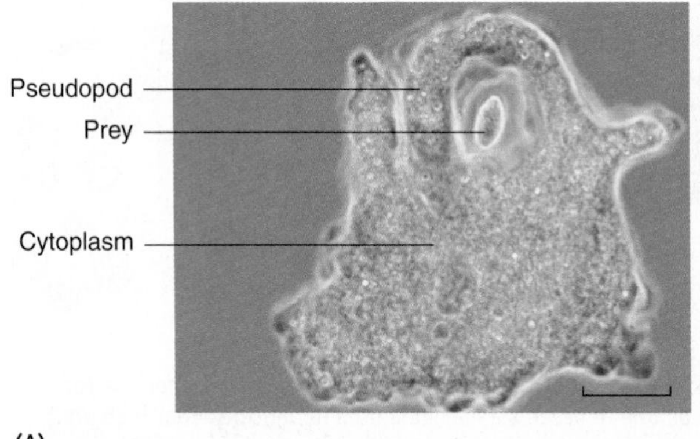

(A)

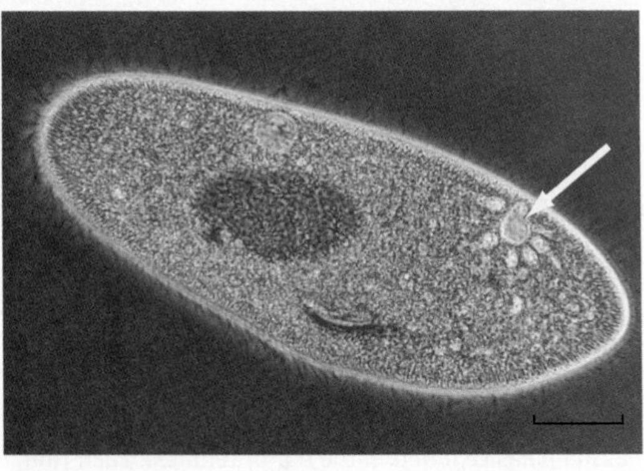

(B)

FIGURE 18.5 **An Amoeba and a Ciliate.** (**A**) A light micrograph of *Amoeba proteus.* The pseudopodia are extending around the prey, beginning the process of phagocytosis. (Bar = 100 μm.) © M I (Spike) Walker/Alamy Images. (**B**) This light micrograph of the ciliate *Paramecium* shows the contractile vacuole (arrow). (Bar = 50 μm.) © Roland Birke/Phototake/Alamy Images. »» What is the function of the contractile vacuole?

or **ciliates**, are among the most complex cells on Earth and have been the subject of biological investigations for many decades. They are found in almost any pond water sample. They have a variety of shapes and can exhibit elaborate and controlled behavior patterns. The cytoplasm contains the typical eukaryotic organelles.

Ciliates range in size from a microscopic 10 μm to a huge 3 mm. All ciliates are covered with **cilia** (sing., cilium) in longitudinal or spiral rows. Cilia beat in a synchronized and coordinated pattern, the organized "rowing" action moving the ciliate along in one direction.

Ciliates, such as *Paramecium*, are heterotrophic by ingestion through a primitive gullet, which sweeps in food particles for digestion. In addition, freshwater protozoa continually take in water by the process of osmosis and eliminate the excess water via organelles called **contractile vacuoles** (FIGURE 18.5B). These vacuoles expand with water drawn from the cytoplasm and then appear to "contract" as they release water through a temporary opening in the cell membrane.

Asexual reproduction in the ciliates occurs by binary fission. The complexity of ciliates however is illustrated by the nature of sexual reproduction. Ciliates have two types of nuclei; there is a single large **macronucleus** that only has the genes for cell metabolism, and one **micronucleus** that contains a complete set of genes. During sexual recombination, called **conjugation**, two cells make contact, and a cytoplasmic bridge forms between them (FIGURE 18.6). A micronucleus from each cell undergoes two divisions to form four micronuclei, of which three disintegrate and only one remains to undergo mitosis. Now a "swapping" of micronuclei takes place, followed by a union to reform the normal micronucleus.

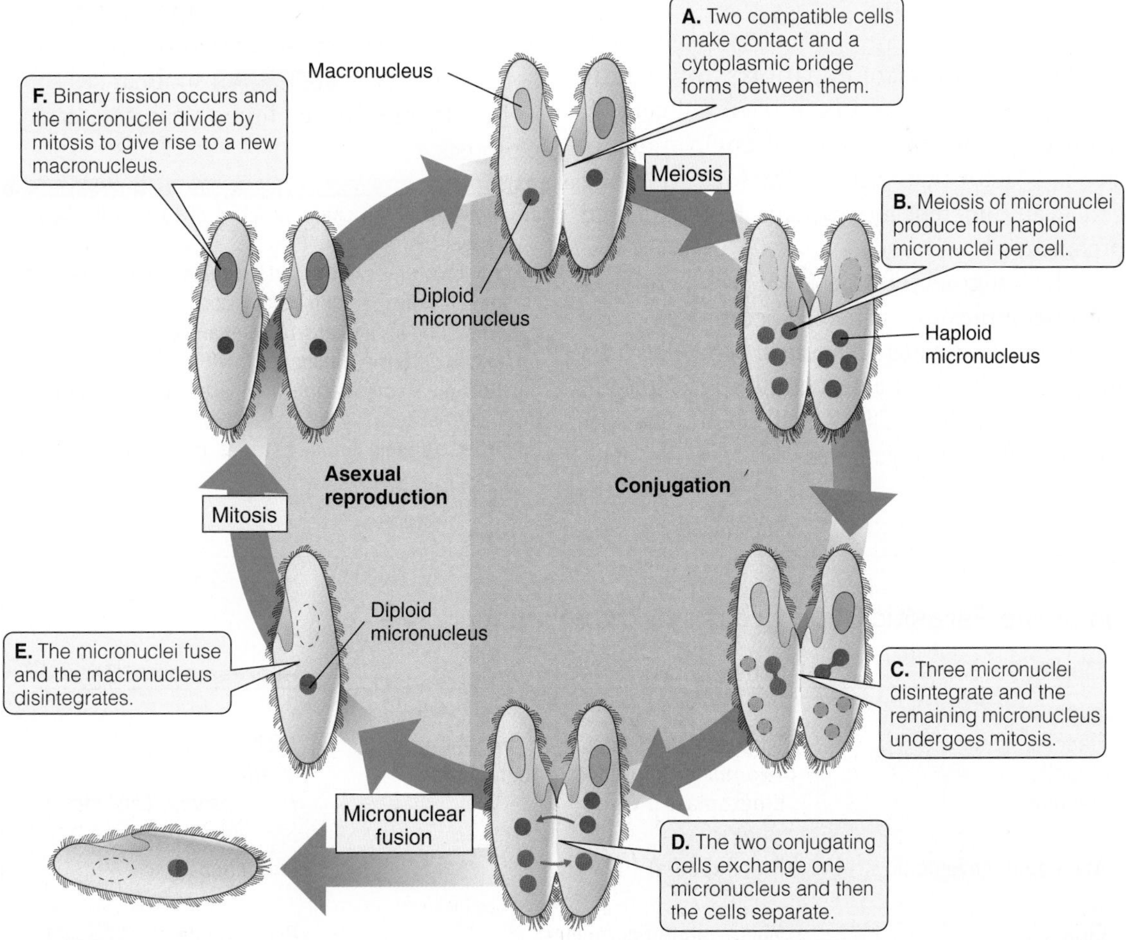

FIGURE 18.6 **Conjugation and Reproduction in *Paramecium*.** Ciliates, such as *Paramecium*, reproduce sexually by the process of conjugation. In this process, an exchange of micronuclei gives rise to new macronuclei. »» Provide a hypothesis to explain why ciliates, unlike most other organisms, must have two types of nuclei.

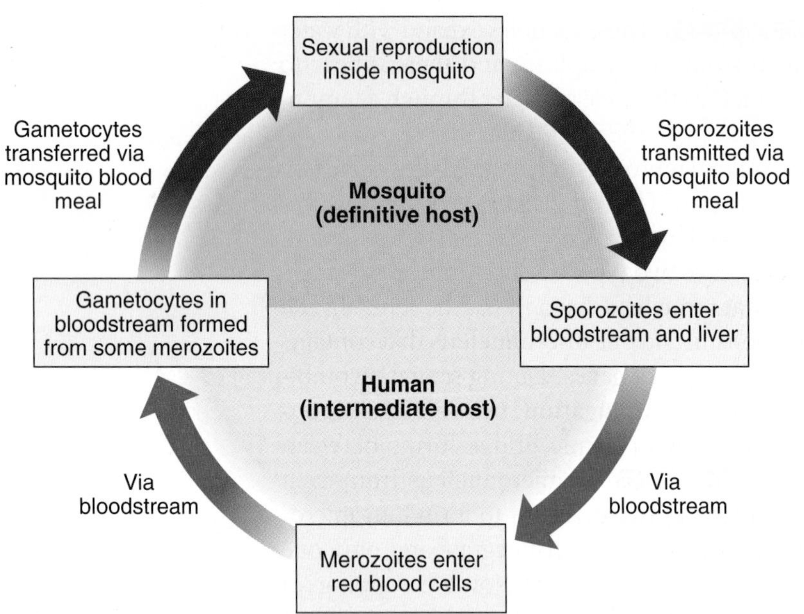

FIGURE 18.7 **The Life Cycle of *Plasmodium*.** The malaria parasite requires two different hosts (definitive and intermediate) in which specific stages of the life cycle occur. »» In trying to prevent malaria, which stage of the *Plasmodium* life cycle would appear to be the most amenable to blocking transmission of the parasite?

This genetic recombination is outwardly analogous to what occurs in bacterial cells. It is observed during periods of environmental stress, a factor that suggests the formation of a genetically different and, perhaps, better-adapted organism.

The last group of protists we will mention is the **apicomplexans**, so named because the *api*cal tip of the cell contains a *complex* of organelles used for penetrating host cells. Adult apicomplexan cells have no cilia or flagella, although a few species have flagellated gametes.

These animal parasites of the Chromalveolata have a complex life cycle including alternating sexual and asexual reproductive phases. These phases often occur in different hosts. Two parasitic species, *Toxoplasma* and *Plasmodium*, are of special significance because the first is associated with AIDS and the second causes one of the most prolific killers of humans—malaria.

Although most parasitic protists require only one host for the completion of their life cycle, the apicomplexans typically require two or more different hosts to complete their life cycle. The host organism in which the sexual cycle occurs is called the **definitive host** while the host in which the asexual cycle occurs is the **intermediate host** (FIGURE 18.7).

For example, the parasite *Plasmodium* produces infective **sporozoites** in mosquitoes (definitive host). These cells enter the human body. Later, within the human body (intermediate host), plasmodial gametes, called **gametocytes**, are produced. These are taken up by mosquitoes in a blood meal where sexual reproduction forms more sporozoites. TABLE 18.1 summarizes some of the attributes of the human parasitic groups of protists.

CONCEPT AND REASONING CHECKS 1

a. Summarize the characteristics of the plant-like and fungal-like protists.
b. What features can be used to separate the parabasalids, diplomonads, and kinetoplastids?
c. What unique cellular structures and behaviors are associated with the Unikonta?
d. How does conjugation in ciliates differ from that in bacterial cells?
e. What are some unique features of the apicomplexans?

TABLE

18.1 Comparison of the Parasitic Supergroups of the Protists

Supergroup	Motility	Other Characteristics	Representative Genera
Excavata			
Parabasalids	Flagella	Thousands of flagella; no mitochondria	*Trichomonas*
Diplomonads	Flagella	Two nuclei; modified mitochondria	*Giardia*
Kinetoplastids	Flagella	Kinetoplast DNA	*Trypanosoma, Leishmania*
Unikonta			
Amoebozoans	Amoeboid movement	Pseudopodia	*Entamoeba*
Chromalveolata			
Ciliates	Cilia	Macro- and micronuclei	*Paramecium*
Apicomplexans	Flagella (gametes only)	Apical complex; multiple hosts	*Plasmodium, Toxoplasma*

Chapter Challenge A

As you have seen, the protists are a diverse group of eukaryotic microorganisms. Until recently, they had been placed in the kingdom Protista, which was a catch-all for any eukaryotic organism that was not a plant, animal, or fungus. Some taxonomists today want to split the protists into a loose grouping of 30 or more dissimilar phyla based on structure and function.

QUESTION A:

Provide some examples of how structure and function could be used to assign these microorganisms to separate and unique groups. What structures and functions do they have in common?

Answers can be found on the Student Companion Website in **Appendix D.**

KEY CONCEPT 18.2 Protistan Parasites Attack the Skin, and the Digestive and Urinary Tracts

Protistan diseases occur in a variety of systems of the human body. For example, some diseases, such as amoebiasis and giardiasis, take place in the digestive system, while others, such as trichomoniasis, develop in the urogenital tract. We start with a cutaneous protistan disease.

Leishmania Can Cause a Cutaneous or Visceral Infection

Leishmaniasis is a rare disease in the United States, but it occurs in 88 countries on four continents with an at-risk population of 350 million people. About 12 million people are affected and it carries the ninth highest infectious disease burden globally. The responsible protists are in the kinetoplastid group and include several species of *Leishmania*, including *L. major* and *L. donovani* (FIGURE 18.8A). Transmission is by the bite of an infected female sand fly of the genus *Phlebotomus* (FIGURE 18.8B). The sand fly vector, which is only about one third the size of a mosquito, becomes infected by biting an infected animal, such as a rodent, dog, another human, or another mammalian reservoir.

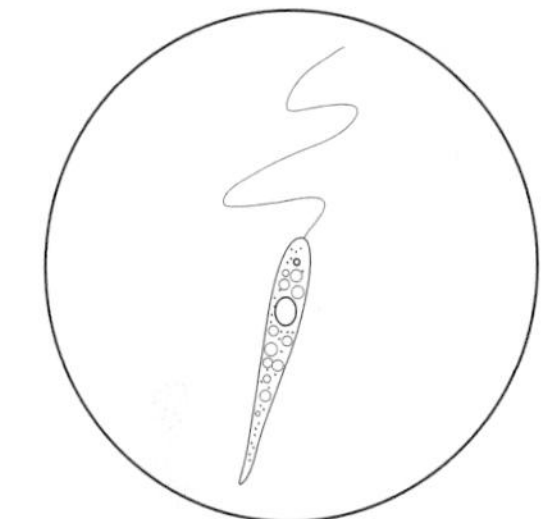

Leishmania donovani

■ Vector:
An insect that transmits pathogens or parasites from an infected animal to humans.

There are two main forms of leishmaniasis. *L. major* causes a disfiguring **cutaneous** (skin) **disease**. Within a few weeks after being bitten, a

(A)

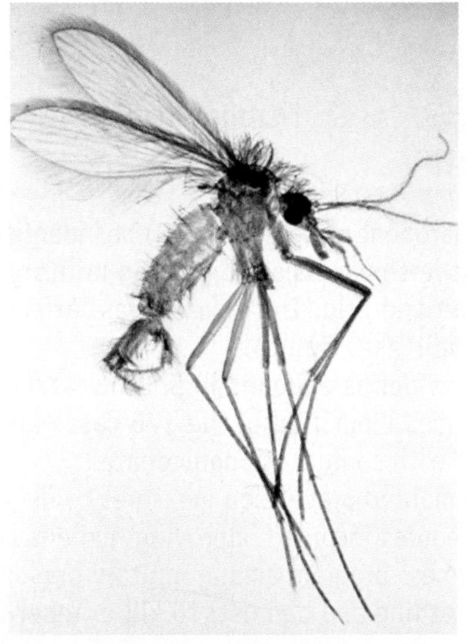

(B)

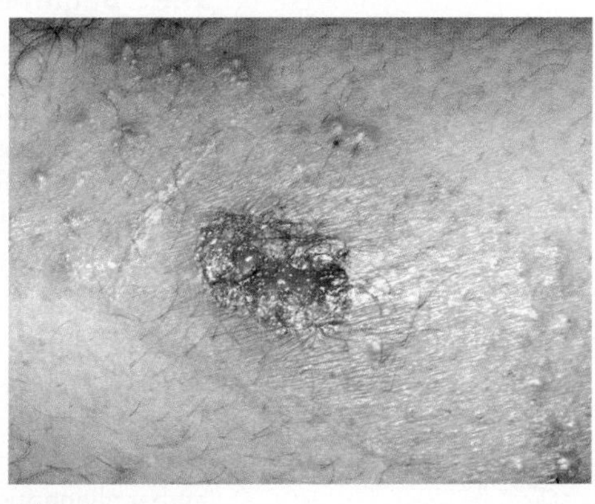

(C)

FIGURE 18.8 **Leishmaniasis.** (**A**) A false-color scanning electron micrograph showing a cluster of *Leishmania* cells, which are long, thin, and flagellated. (Bar = 20 μm.) © Dennis Kunkel Microscopy, Inc./Phototake/Alamy Images. (**B**) The sand fly, *Phlebotomus,* which is the vector for transmission of leishmaniasis. Courtesy of WHO/CDC. (**C**) Skin lesion due to cutaneous leishmaniasis. © Medical-on-Line/Alamy Images. »» Explain how the parasite can bring about the physical skin lesion.

sore appears on the skin. The sore then expands and ulcerates to resemble a volcano with a raised edge and central crater (FIGURE 18.8C). Some sores may be painless and become covered by a scab. There are over 1 million new cases of cutaneous leishmaniasis each year worldwide.

The other form of leishmaniasis is a **visceral** (body organ) **disease** called **kala azar**, meaning "black fever." It is caused by *L. donovani*. Symptoms do not appear until several months after being bitten by a sand fly. Infection of the white blood cells leads to irregular bouts of fever, swollen spleen and liver, progressive anemia, and emaciation. About 90% of cases are fatal, if not treated. There are about 1.5 million new cases globally each year, 90% occurring in India, Bangladesh, Nepal, Sudan, and Brazil. American soldiers have been infected during the Iraq conflicts with the cutaneous or visceral form (MICROFOCUS 18.2).

Entamoeba histolytica

Control of the sand fly remains the most important method for preventing outbreaks of leishmaniasis. The antimony compound, stibogluconate, is used to treat established cases.

Several Protistan Parasites Cause Diseases of the Digestive System

There are three groups of parasite diseases caused by different supergroups of protists, including the amoebozoans, the apicomplexans, and the kinetoplastids.

Amoebiasis. The second leading cause of death from parasite diseases, only surpassed by malaria, is **amoebiasis**. This parasitic form of **gastroenteritis** occurs worldwide and primarily affects children and adults who are undernourished and living in unsanitary conditions. Although an intestinal illness at first, it can spread to various organ systems. Some 40,000 to 100,000 people die each year from amoebiasis.

The causative agent of amoebiasis is *Entamoeba histolytica*. In nature, the protist exists in the cyst form, which enters the body by food or water contaminated with human or animal feces, or by direct contact with feces. The cysts pass through the stomach and emerge as amoebae, called trophozoites, in the distant portion of the small intestine and in the large intestine (FIGURE 18.9).

Most infected individuals remain asymptomatic but are carriers, shedding cysts in the stool. About 10% of people who are infected with *E. histolytica* become sick from the infection and develop **amoebic dysentery**. The *E. histolytica* amoebae have the ability to destroy tissue (*histo* = "tissue"; *lyt* = "loosened"). Using their protein-digesting enzymes, they can penetrate the wall of the large intestine, causing lesions and deep ulcers. Patients experience stomach pain, bloody stools, and fever. In rare cases, the parasites invade the blood and cause an invasive infection in the liver, lung, or brain, where fatal abscesses may develop.

Metronidazole and paromomycin commonly are used to treat amoebiasis, but the drugs do not affect the cysts, and repeated attacks of amoebiasis may occur for months or years as the patient

MICROFOCUS 18.2: Public Health
The "Baghdad Boil"

The United States Department of Defense (DoD) has identified hundreds of cases of cutaneous leishmaniasis (CL) among military personnel serving in Afghanistan and Iraq. The disease was "affectionately" called the "Baghdad boil" (see figure).

Leishmania major, which is endemic in Southwest/Central Asia, was the parasitic species identified in the 176 cases analyzed. Patients were treated with sodium stibogluconate.

The DoD has implemented prevention measures to decrease the risk of CL. These procedures included improving hygiene conditions, instituting a CL awareness program among military personnel, using permethrin-treated clothing and bed nets to kill or repel sand flies, and applying insect repellent containing 30% DEET to exposed skin. These measures, according to the Department of Defense's Medical Surveillance Monthly Report, have reduced the number of reported cases. Since 2002, there have been more than 2,000 reported cases, mostly of the cutaneous form.

Cutaneous leishmaniasis. © Leslie E. Kossoff/AP Photos.

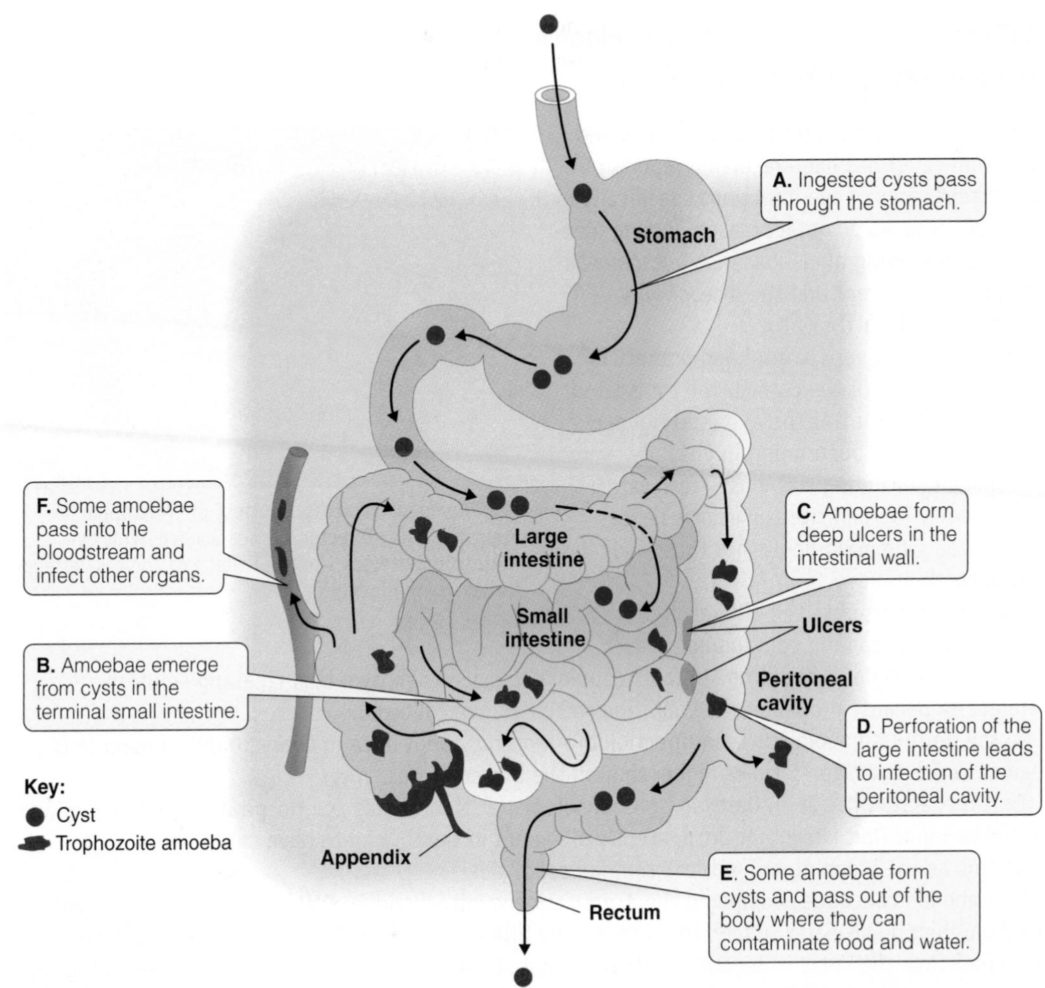

FIGURE 18.9 **The Course of Amoebiasis Due to *Entamoeba histolytica*.** The parasite enters the body as a cyst, which develops into an amoeboid form that causes deep ulcers. »» What is the advantage for the parasite to enter the body as a cyst?

continues to shed cysts in the feces. The best prevention is not eating potentially contaminated food or drinking unpasteurized milk or other dairy products. Bottled water or boiled water should be consumed in countries where amoebiasis occurs.

MicroFocus 18.3 describes a parasitic amoebal infection of the eye.

Giardiasis. In 2010, the Centers for Disease Control and Prevention (CDC) received reports of almost 20,000 cases of **giardiasis**, making it the most common intestinal parasitic infection reported in the United States. Because not all cases are reported to the CDC, the disease is estimated to cause 1 million infections annually. The disease is sometimes mistaken for viral gastroenteritis and is considered a type of traveler's diarrhea.

The causative agent is the diplomonad *Giardia intestinalis* (also called *G. lamblia*). This organism is distinguished by four pairs of anterior flagella and two nuclei that stain darkly to give the appearance of eyes on a face. The cell can be divided equally along its longitudinal axis and is therefore said to display bilateral symmetry.

Giardiasis is commonly transmitted by food or water containing *Giardia* cysts stemming from cross-contamination of drinking water with sewage as well as by the fecal-oral route. The cysts pass through the stomach and the trophozoites emerge as flagellated cells in the duodenum. They multiply rapidly by binary fission and adhere to the intestinal lining using a sucking disk located on the lower cell surface (FIGURE 18.10).

Acute giardiasis develops after an incubation period of about 7 days. The patient feels nauseous, experiences gastric cramps and flatulence, and emits a foul-smelling watery diarrhea

■ Duodenum: The first short section of the small intestine immediately beyond the stomach.

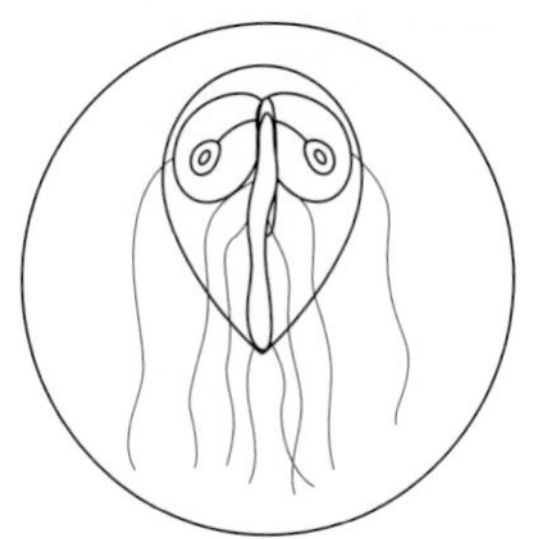

Giardia intestinalis

MICROFOCUS 18.3: Public Health
What's Growing in Your Plumbing?

Since 1986, the Centers for Disease Control and Prevention (CDC) has been receiving reports of intermittent cases of eye infections specifically in individuals who wore contact lenses. Testing identified the free-living amoebozoan *Acanthamoeba* as the infecting agent and the disease was *Acanthamoeba* keratitis (AK).

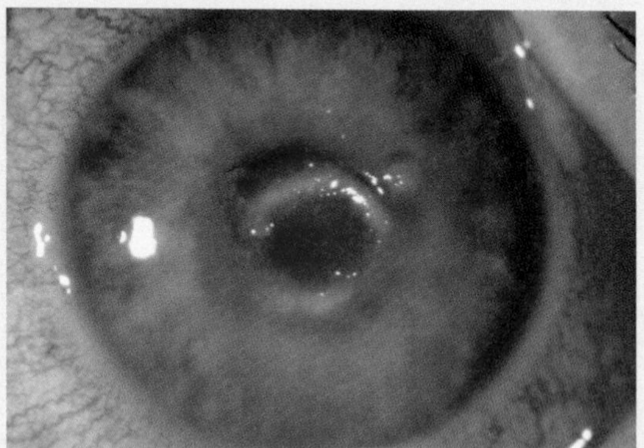

Acanthamoeba keratitis. Courtesy of Christopher J. Rapuano, M.D.; Director, Cornea Service, Wills Eye. Professor, Jefferson Medical College of Thomas Jefferson University, Philadelphia, PA.

AK is a rare but very painful and potentially blinding infection of the cornea, the transparent covering at the front of the eye (see figure). *Acanthamoeba* has been found in virtually every environment, including soil, dust, freshwater, and seawater. The parasite sometimes resides in untreated swimming pools, hot tubs, and even in bottled water.

The cases reported to the CDC resulted from contact lenses becoming contaminated after improper cleaning and handling. In fact, people who make their own lens cleaning solution are more at risk because proper sterile conditions often are not followed.

In the United Kingdom (UK), an infection of AK occurs about once in every 30,000 contact lens wearers, which is a rate 15 times higher than in the United States. Why?

That is exactly what John Dart, an ophthalmologist at Moorefields Eye Hospital in London, England, wanted to know. Dart knew that until recently all homes in the UK had to have cold-water storage tanks. Could this be the breeding ground for *Acanthamoeba?*

Dart and his colleagues compared the *Acanthamoeba* mitochondrial DNA from eight patients with that from their home water supply. In six cases, the DNAs were identical, indicating the organisms must have come from the home water supply. However, not all patients stored their contact lenses using the home tap water, so Dart believes the source of the eye infections comes either from the patients having washed their faces with water while wearing their contact lenses or handling the lenses with wet hands. A program for better hygiene practices has been introduced.

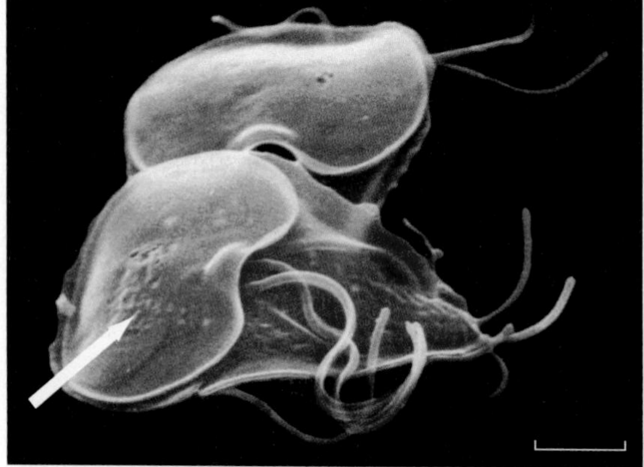

FIGURE 18.10 ***Giardia intestinalis.*** False-color scanning electron micrograph of *G. intestinalis* in the human small intestine. The pear-shaped cell body and sucker device (arrow) are evident, as are the flagella of this diplomonad. (Bar = 1 μm.) © Jerome Paulin/Visuals Unlimited. »» What is the function of the sucker device?

sometimes lasting for 1 or 3 weeks. Infectious cysts are excreted in the feces.

Treatment of giardiasis may be administered with drugs such as metronidazole or tinidazole. However, these drugs have side effects and the physician may wish to let the disease run its course without treatment. Those who recover often become carriers and excrete the cysts for years.

Cryptosporidiosis. Since 1976, outbreaks of **cryptosporidiosis** have been reported in several countries. The most remarkable outbreak occurred in 1993 in Milwaukee, Wisconsin, where more than 400,000 people were affected (and 54 died), making it the largest waterborne infection ever recorded in the United States. In 2010, there were 9,000 reported cases, with the

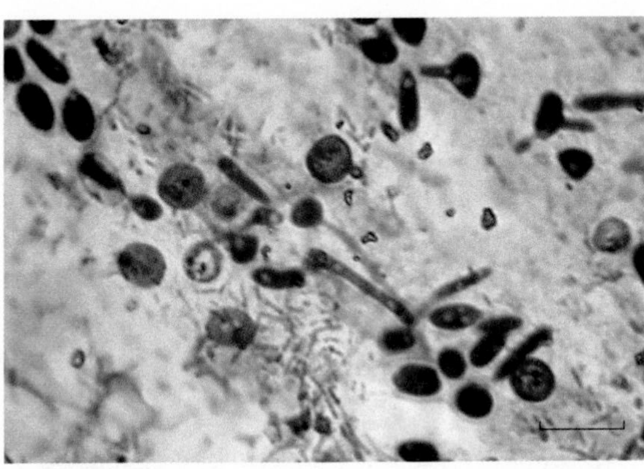

FIGURE 18.11 *Cryptosporidium* **Oocysts.** A light micrograph of a stained fecal smear. *Cryptosporidium* oocysts are red. (Bar = 2 µm.) Reproduced from Ma, P. and R. Soave, J Infect Dis. 147:5 (1983): 824-828. With permission from University of Chicago Press. »» What is an oocyst?

highest number of reported cases in children 1 to 9 years of age.

Human cryptosporidiosis is caused by the apicomplexans *Cryptosporidium parvum* and *C. hominis*. The parasites have a complex life cycle involving trophozoite, sexual, and oocyst stages (FIGURE 18.11). Transmission occurs mainly through contact with contaminated water (such as drinking or recreational water containing oocysts). Physical contact also can transmit *Cryptosporidium* oocysts, making children in day-care centers at risk.

Cryptosporidiosis has an incubation period of about 1 week. Patients with competent immune systems appear to suffer limited diarrhea lasting 1 or 2 weeks during which time newly-formed infectious oocysts are excreted in the feces.

In immunocompromised individuals, such as AIDS patients, *Cryptosporidium* is an opportunistic infection. Patients experience cholera-like profuse diarrhea that can be severe and irreversible. These patients undergo dehydration and emaciation, and often die of the disease.

Cyclosporiasis. In the late 1990s, public health officials in the United States identified a series of clusters of intestinal disease related to the consumption of raspberries imported from Guatemala. In 2004 and 2005, outbreaks in Texas, Illinois, and Florida sickened over 400 people after eating raw basil. In all cases, the outbreaks were related to the apicomplexan *Cyclospora cayetanensis*.

Fresh produce and water can serve as vehicles for transmission of oocysts, which are ingested in contaminated food or water. The oocysts are similar to, but larger than, those of *Cryptosporidium*. Differential diagnosis is important because *C. cayetanensis* responds to the drug combination of trimethoprim-sulfamethoxazole, whereas *Cryptosporidium* does not.

Cyclosporiasis has an incubation period of 1 week. Symptoms of the disease include watery diarrhea, nausea, abdominal cramping, bloating, and vomiting. Treatment is successful with the drugs noted above, but the symptoms often return. Moreover, the symptoms often remain for over 1 month during the first illness.

How fresh produce becomes contaminated during outbreaks is not clear. Tainted water used for washing the produce may be the source, the produce may be handled by someone whose hands are contaminated, or the produce may be contaminated during shipping. Regardless of the source, better control measures focusing on improved water quality and better sanitation methods on local farms have resulted in only 179 reported cases by the CDC in 2010.

■ Oocyst: The thick-walled fertilized gamete of apicomplexans.

A Protistan Parasite Also Infects the Urinary Tract

Trichomoniasis is among the most common parasitic diseases in men and women in industrialized countries, including the United States, where an estimated 7.4 million new cases occur annually. The disease is transmitted primarily by sexual contact and is considered a sexually transmitted disease (STD).

Trichomonas vaginalis, the causative agent, is a pear-shaped, flagellated protist of the parabasalids (FIGURE 18.12). The organism has only a trophozoite stage and its only host is humans, where it thrives and replicates by binary fission in the slightly acidic environment of the female vagina and the male urethra. Establishment may be encouraged by physical or chemical trauma, including poor hygiene, drug therapy, diabetes, or mechanical contraceptive devices such as the intrauterine device (IUD).

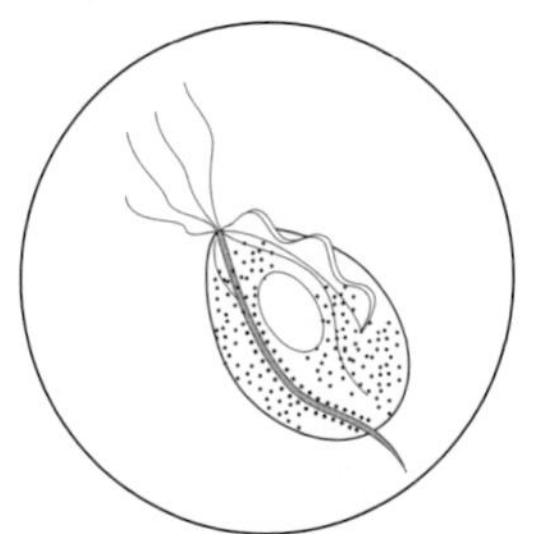

Trichomonas vaginalis

The incubation period is 5 to 28 days. In females, trichomoniasis is accompanied by intense itching (pruritis) and discomfort during urination and sexual intercourse. Usually, a yellow-green, frothy discharge also is present. The symptoms are frequently worse during menstruation, and erosion of the cervix may occur. In males, the disease

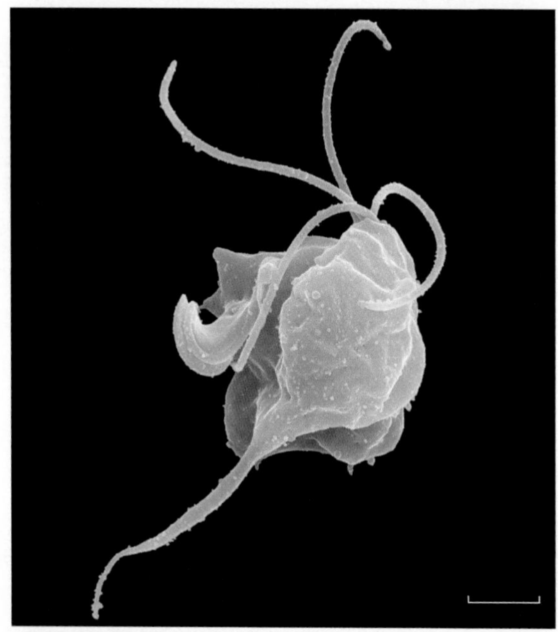

FIGURE 18.12 ***Trichomonas vaginalis.*** False-color scanning electron micrograph of a *T. vaginalis* trophozoite. (Bar = 5 μm.) © Dr. Dennis Kunkel/Visuals Unlimited/ Corbis. »» What are the whip-like structures evident in the micrograph?

may be asymptomatic. Symptoms occur primarily in the urethra, with slight pain on urination and a thin, mucoid discharge. The disease can occur concurrently with gonorrhea.

The drug of choice for treatment is orally administered metronidazole or tinidazole, and both patient and sexual partners should be treated concurrently to prevent transmission or reinfection. Drug resistance has become of increasing concern.

TABLE 18.2 summarizes the protistan diseases of the skin, and the digestive and urinary tracts.

CONCEPT AND REASONING CHECKS 2

a. Summarize the two types of leishmaniasis and their health consequences.
b. Why is amoebiasis considered an example of "invasive" gastroenteritis?
c. Identify why most cases of giardiasis are reported in the summer to early fall.
d. Compare and contrast cryptosporidiosis and cyclosporiasis.
e. Why is trichomoniasis such a common protozoan disease?

Chapter Challenge B

Here is a map showing the incidence of giardiasis in 2010. A *Giardia* infection can result from swallowing cysts picked up from surfaces (such as bathroom handles, changing tables, diaper pails, or toys) that contain stool from an infected person or animal; drinking *Giardia*-contaminated water (for example, untreated or improperly treated water from lakes, streams, or wells); swallowing water while swimming or playing in water where *Giardia* may live, especially in lakes, rivers, springs, ponds, and streams.

QUESTION B:

Can you draw any conclusions from this map concerning the distribution of the disease state-by-state?

Answers can be found on the Student Companion Website in **Appendix D.**

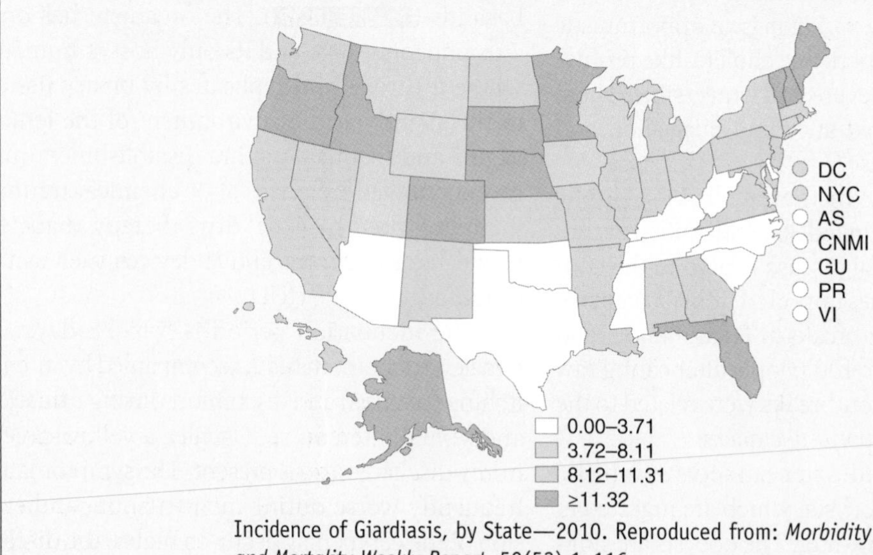

Incidence of Giardiasis, by State—2010. Reproduced from: *Morbidity and Mortality Weekly Report*, 59(53);1–116.

TABLE

18.2 A Summary of Protistan Diseases of the Skin, and the Digestive and Urinary Tracts

Disease	Causative Agent	Signs and Symptoms	Transmission	Treatment	Prevention
Cutaneous leishmaniasis	*Leishmania major*	Skin sore that ulcerates	Sand fly bites	Stibogluconate	Protecting skin from sand fly bites
Visceral leishmaniasis (kala azar)	*Leishmania donovani*	Fever, swollen spleen and liver, anemia, emaciation	Sand fly bites	Stibogluconate	Protecting skin from sand fly bites
Amoebiasis	*Entamoeba histolytica*	Bloody stools, stomach pain and cramping	Indirect through food or water Direct contact with feces	Metronidazole Paromomycin	Avoiding contaminated food, unpasteurized milk
Giardiasis	*Giardia intestinalis*	Nausea, gastric cramps, flatulence, foul-smelling watery diarrhea	Fecal-oral route	Metronidazole	Practicing good hand hygiene Avoiding untreated water
Cryptosporidiosis	*Cryptosporidium parvum* *Cryptosporidium hominis*	Watery diarrhea, dehydration, vomiting, nausea, stomach cramps, fever, malaise	Contact with contaminated water	Nitazoxanide, paromomycin Fluid replacement	Practicing good hand hygiene Avoiding untreated water
Cyclosporiasis	*Cyclospora cayetanensis*	Watery diarrhea, nausea, abdominal cramping, bloating, vomiting	Contaminated fresh produce or water	Trimethoprim-sulfamethoxazole Fluid replacement	Avoiding untreated water and contaminated fresh produce
Trichomoniasis	*Trichomonas vaginalis*	Intense itching Discomfort on urination and sexual intercourse Yellow-green frothy discharge with strong odor	Oral, vaginal, or anal sex with someone who is infected	Metronidazole or tinidazole	Abstaining from sexual activity Limiting sexual partners Patients and partner treatment

KEY CONCEPT 18.3 Many Protistan Diseases of the Blood and Nervous System Can Be Life Threatening

We finish our discussion of the protist parasites by examining those that cause infections in the blood or in the nervous system. This includes two of the most prevalent diseases, malaria and sleeping sickness.

Plasmodium Can Be a Deadly Blood Parasite

Causative Agent and Epidemiology. **Malaria** is caused by four species of the apicomplexan genus *Plasmodium: P. vivax, P. ovale, P. malariae,* and *P. falciparum.* The most serious infections that can be life threatening are caused by *P. falciparum.*

The disease has been infecting humans for more than 5,000 years. During the 1700s, Europeans suffered wave after wave of malaria and few regions were left untouched. Even American pioneers settling in the Mississippi and Ohio valleys suffered great losses from the disease.

Today, between 300 and 500 million of the world's population suffer from malaria, which exacts its greatest toll in Africa. The WHO estimates that more than 800,000 people die from malaria every year. Children are the most susceptible because their immune system has not had time to develop partial immunity from mild infections. So, one child dies every 30 seconds from malaria! No infectious disease of contemporary times can claim such a dubious distinction. Even the United States is involved in the ongoing

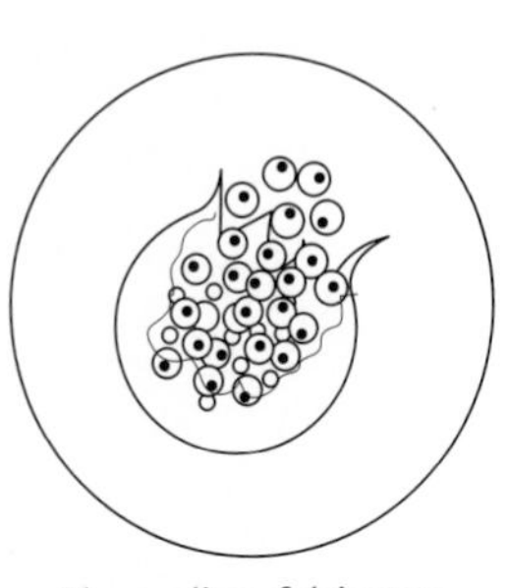
Plasmodium falciparum

malaria pandemic—over 1,700 imported cases were reported in 2010.

All four species are transmitted by the female *Anopheles* mosquito, which consumes human blood to provide chemical components for her eggs. The life cycle of the parasites has three important stages: the sporozoite, the merozoite, and the gametocyte. Each is a factor in malaria.

The mosquito (definitive host) sucks blood from a person with malaria and acquires gametocytes, the form of the protist found in the blood (FIGURE 18.13). Within the insect, sexual reproduction occurs and a transition to sporozoites takes place. The sporozoites then migrate to the salivary gland.

Clinical Presentation. Sporozoite infection causes the female *Anopheles* mosquito to increase its biting frequency. When the mosquito bites another human (intermediate host), several hundred sporozoites enter the person's bloodstream and quickly migrate to the liver. After several hours, the transformation of sporozoites to **merozoites** is completed, and the merozoites emerge from the liver to invade the red blood cells (RBCs). This triggers the blood infection stage and the pathology characteristic of malaria.

While in the RBCs, the merozoites can synthesize any of about 150 proteins that attach to RBC membranes and cause the RBCs to cluster in the blood vessels. By constantly switching among

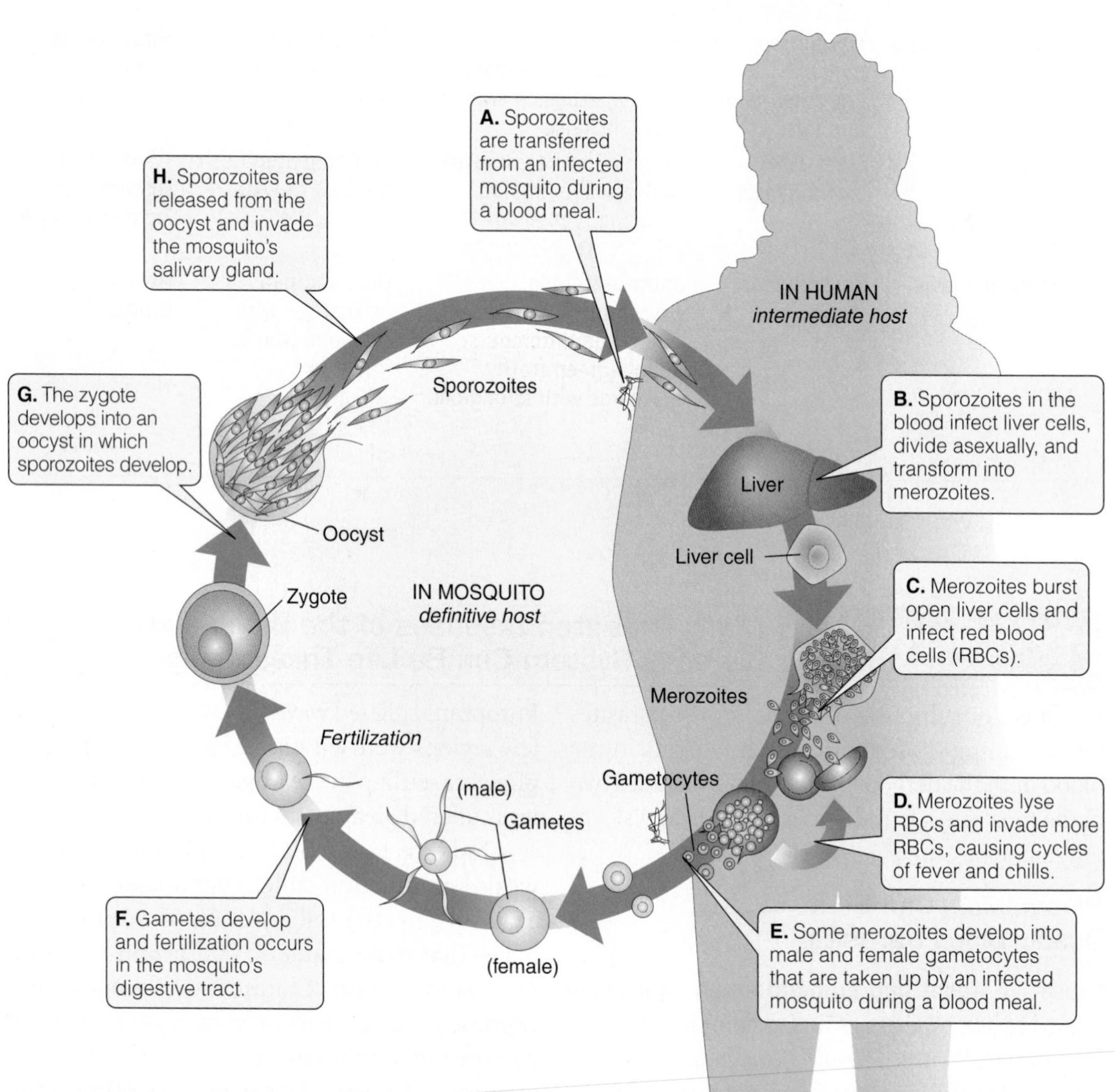

FIGURE 18.13 **Malaria and the *Plasmodium* Life Cycle.** The propagation of the *Plasmodium* parasite requires two hosts, mosquitoes and humans. »» Why are two hosts required to complete the *Plasmodium* life cycle?

150 genes (for 150 proteins), the malarial parasite avoids detection by the body's immune system.

Within RBCs, the merozoites undergo another series of transformations resulting in several gametocytes and thousands of new merozoites being formed. In response to a biochemical signal, thousands of RBCs rupture simultaneously releasing the parasites and their toxins in a 48 to 72 hour cycle.

Now the excruciating malaria attack begins. First, there is intense cold, with shivers and chattering teeth. The temperature then rises rapidly to 40°C, and the sufferer develops intense fever, headache, and delirium. After 2 to 3 hours, massive perspiration ends the hot stage, and the patient often falls asleep, exhausted. During this quiet period, the merozoites enter a new set of RBCs and repeat the cycle of transformations.

Death from malaria is due to a number of factors related to the loss of red blood cells. Substantial anemia develops, and the hemoglobin from ruptured blood cells darkens the urine; malaria is, therefore, sometimes called **blackwater fever**. Cell fragments and RBC clustering accumulate in the small vessels of the brain, kidneys, heart, liver, and other vital organs and cause clots to form. Heart attacks, cerebral hemorrhages, and kidney failure are common.

Treatment and Prevention. Since its discovery around 1640, quinine has been the mainstay for treating malaria. During World War II, American researchers developed the drug chloroquine, which remained an important mode of therapy until recent years, when drug resistance began emerging in *Plasmodium* species. Another drug, artemisinin, is effective in curing malaria especially when combined with other drugs to limit the development of drug resistance. Many other drugs are in clinical trials.

Since 1989, mefloquine has been recommended as a preventative drug for individuals entering malaria regions of the world. However, serious medical side effects, including cognitive disorders, have been associated with some people taking the drug. Malavone is a newer antimalarial drug now recommended for falciparum malaria.

Experimental vaccines directed against the sporozoite or merozoite stage are being tested.

Though progress has been made in the control of malaria, the mortality and morbidity figures remain appallingly high. MicroFocus 18.4 looks at a program to reduce this "malaria burden."

The *Trypanosoma* Parasites Can Cause Life-Threatening Systemic Diseases

Trypanosomiasis is a general name for two diseases caused by parasitic species of the kinetoplastid *Trypanosoma* (FIGURE 18.14A). The two diseases caused by trypanosomes are traditionally known as human African sleeping sickness and Chagas disease.

African Trypanosomiasis. African trypanosomiasis or **human African sleeping sickness** is endemic in 36 African countries and exerts a level of mortality greater than that of HIV disease/AIDS. The WHO estimates there are more than 500,000 new cases every year with some 50,000 deaths.

Trypanosoma circulates between humans and the tsetse fly (FIGURE 18.14B). If the fly bites an infected person or animal, the trypanosomes localize in the insect's salivary gland. After a 2-week development, transmission occurs during a bite. The point of entry becomes painful and swollen in several days and a chancre similar to that in syphilis is observed. Invasion of, and multiplication in, the bloodstream follows (stage 1). It then spreads to the central nervous system (stage 2).

Two types of African sleeping sickness exist. A chronic form, common in central and western Africa, is caused by *Trypanosoma brucei* variety *gambiense*. It is accompanied by chronic bouts of fever, as well as severe headaches, changes in sleep patterns and behavior, and a general wasting away. As the trypanosomes invade the brain, the patient slips into a coma (hence the name, "sleeping sickness"). The second form, common in eastern and southern Africa, is due to *Trypanosoma brucei* variety *rhodesiense*. The disease is more acute with high fever and rapid coma preceding death.

Trypanosoma brucei

Prevention involves clearing brushlands and treating areas where the tsetse flies breed. Patients are treated with either the drug pentamidine, melarsoprol, or eflornithine, depending on the form and stage of the disease.

American Trypanosomiasis. American trypanosomiasis or **Chagas disease** is found in Mexico and 17 countries in Central and South America. A recent estimate put the number of cases in South and Central America at 12 million with approximately 200,000 new cases every year and some 15,000 deaths.

Chagas disease is caused by *Trypanosoma cruzi* and is transmitted by the reduviid bug. The insect feeds at night and bites where the skin is thin, such

MICROFOCUS 18.4: Public Health

Roll Back Malaria

The World Health Organization (WHO) estimates there are over 300 million cases of malaria each year, resulting in more than 800,000 deaths.

The United Nations General Assembly designated 2001–2010 the decade to "Roll Back Malaria" in developing nations by reducing the world's malaria burden by 50%. The program was designed to help countries take effective and sustained action against malaria by providing their citizens with rapid and effective treatment, and attempting to prevent and control malaria in pregnant women.

Children in Guayoquil, Ecuador receiving malaria tablets. © Mark Pink/Alamy Images.

Because over 70% of all malarial deaths occur in children under 5 years of age, a child's most vulnerable period for contracting malaria starts at 6 months, when the mother's protective immunity wears off and before the infant has established its own fully functional immune system. During this window of vulnerability, a child's condition can deteriorate quickly and the child can die within 48 hours after the first symptoms appear. Therefore, a key part of Roll Back Malaria was aimed at children (see figure). Now in 2012, how well did the program work?

- At least 11 African countries where malaria was endemic have reported a 50% drop in the number of confirmed malaria cases.
- At least 32 malaria-endemic countries outside Africa have reported a 50% drop in the number of confirmed malaria cases. Eight other malaria-endemic countries outside Africa have reported a 25% to 50% drop in the number of confirmed malaria cases.
- More than 75 million people have been protected from malaria by indoor residual spraying for mosquitoes and some 42% of African homes had at least one insecticide-treated mosquito net.
- Malaria was eliminated in Morocco and Turkmenistan.
- Cheap and accurate diagnostic tests are available so a patient can be confirmed as having malaria before antimalarial drugs are prescribed.

Many nations, celebrities, activists, and nongovernmental organizations, such as the Gates Foundation, have in total increased money and funding for malaria five-fold since 2006. This has encouraged the World Health Organization to suggest in its 2010 World Malaria Report that "*By maintaining these essential gains, we can end malaria deaths by 2015.*" A 5,000 year war would finally be won.

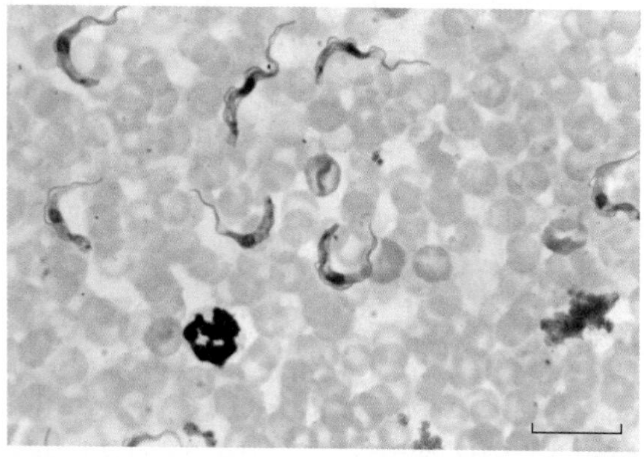

(A)

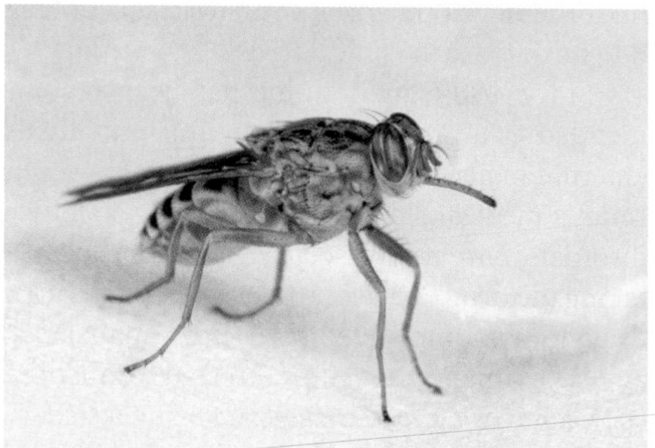

(B)

FIGURE 18.14 ***Trypanosoma.*** **(A)** A light micrograph of stained *Trypanosoma* among red blood cells. (Bar = 15 µm.) Courtesy of Dr. Mae Melvin/CDC. **(B)** The tsetse fly, a vector for *Trypanosoma*. Courtesy of Peggy Greb/USDA ARS. »» What is unique about the *Trypanosoma* cells?

as on the forearms, face, or lips. For this reason, it is called the "kissing bug."

Parasites in the feces of the reduviid bug enter the bite wound. Once in the blood, the trypanosomes invade many cell types and undergo multiple rounds of binary fission. During this acute phase, the individual experiences high parasite numbers even though most infections are asymptomatic. Following the acute phase, the parasite number declines. In 20% to 30% of infected individuals, a chronic irreversible disease occurs that in 10 to 30 years can develop clinical symptoms, which vary by geographical region. Individuals may experience widespread tissue damage including intestinal tract abnormalities and extensive cardiac nerve destruction that is so thorough the victim experiences sudden heart failure. Benznidazole and nifurtimox have proved useful for acute disease; no effective drug is available for chronic infections.

In 2006, two American heart transplant patients developed acute Chagas disease after receiving donor hearts. So in January 2007, the American Red Cross and Blood Systems started screening the U.S. blood supply for Chagas disease and *T. cruzi*. About one in 30,000 donors test positive. Some 300,000 people in the United States are infected.

Babesia Is an Apicomplexan Parasite

Babesiosis is a malaria-like disease caused by *Babesia microti*. The protists live in ticks of the genus *Ixodes*, and are transmitted when these arthropods feed in human skin. Tickborne transmission has occurred primarily in the Northeast and upper Midwest of the United States during the spring and summer. However, the geographical distribution is spreading such that it may soon rival Lyme disease as the most common tickborne illness in the United States.

B. microti penetrates human red blood cells. As the cells disintegrate, a mild anemia develops. Piercing headaches accompany the disease and, occasionally, meningitis occurs. A suppressed immune system appears to favor establishment of the disease. However, babesiosis is rarely fatal and drug therapy is not recommended. Carrier conditions may develop in recoverers, and spread by blood transfusion is possible. Travelers returning from areas of high incidence therefore are advised to wait several weeks before donating blood to blood banks. Tick control is considered the best method of prevention.

Babesia has a significant place in the history of American microbiology. In the late 1800s, Theobald Smith located *B. bigemina* in the blood of cattle suffering from Texas fever. His report was one of the first linking protists to disease, and, in part, it necessitated that the then-prevalent "bacterial" theory of disease be modified to include eukaryotic microorganisms.

Toxoplasma Causes a Relatively Common Blood Infection

Toxoplasmosis affects up to 50% of the world's population, including 50 million Americans. Thus, the causative agent, *Toxoplasma gondii*, is regarded as a universal parasite. Some researchers believe it is the most common parasite of humans and other vertebrates.

T. gondii, an apicomplexan, exists in three forms: the trophozoite, the cyst, and the oocyst. Trophozoites are crescent-shaped or oval organisms without flagella (FIGURE 18.15). Located

Toxoplasma gondii

FIGURE 18.15 ***Toxoplasma gondii.*** A false-color scanning electron micrograph of numerous crescent-shaped *T. gondii* trophozoites (orange). (Bar = 5 µm.) © Dennis Kunkel Microscopy, Inc./Phototake/Alamy Images. »» What infective stage is represented by these cells?

Babesia microti

in tissue during the acute stage of disease, they force their way into all mammalian cells, with the notable exception of erythrocytes. To enter cells, the parasites form a ring-shaped structure on the host cell membrane and then pull the membrane over themselves, much like pulling a sock over the foot. Cysts develop from the trophozoites within host cells and may be the source of repeated infections. Muscle and nerve tissue are common sites of cysts. Oocysts are oval bodies that develop from the cysts by a complex series of asexual and sexual reproductive processes.

This apicomplexan can exist in nature in the cyst and oocyst forms. Grazing animals acquire these forms from the soil and pass them to humans via contaminated beef, pork, or lamb (FIGURE 18.16). Rare hamburger meat is a possible source. Domestic cats acquire the cysts from the soil or from infected birds or rodents. Oocysts then form in the cat. Humans are exposed to the oocysts when they forget to wash their hands after contacting cat feces and changing the cat litter. Touching the cat also can bring oocysts to the hands, and contaminated utensils, towels, or clothing can contact the mouth and transfer oocysts. Investigating the Microbial World 18 looks at how *Toxoplasma* manipulates one of its natural intermediate hosts, the rat.

Toxoplasmosis develops after trophozoites are released from the cysts or oocysts in the host's gastrointestinal tract. *T. gondii* rapidly invades the intestinal lining and spreads throughout the body via the blood. However, for most healthy individuals, the parasite causes no serious illness even

FIGURE 18.16 **The Cycle of Toxoplasmosis in Nature.** Humans become infected through contact with the feces of an infected cat or consumption of undercooked, contaminated beef. »» What form of the *Toxoplasma* parasite is passed to humans?

Investigating the Microbial World 18

Love Your Enemy

What happens when a healthy rat smells its nemesis, a cat (actually it's smelling the cat's urine)? It scrams or it might be caught and be the feline's next meal. Now, what happens if a rat infected with *Toxoplasma* smells a cat? It sticks around and continues to sniff the odors—a dangerous decision on the part of the rat! But it seems it cannot help itself. *Toxoplasma* is controlling the rat's behavior.

- **OBSERVATIONS:** Often the parasite *Toxoplasma* uses a cat's (domestic or otherwise) digestive system for sexual reproduction. As a result, oocysts are shed in the cat's feces in large numbers and end up in soil or water. Intermediate hosts, like rats, ingest the oocyst-contaminated soil or water and become infected with the oocysts. The oocysts transform into tachyzoites that localize in the rat's neural and muscle tissue and develop into tissue cysts. Now, in order to continue the life cycle, the *Toxoplasma* cysts have to get back into a cat, so cats have to catch and eat the infected rats.
- **QUESTION:** ***How does* Toxoplasma *"manipulate" rat behavior to ensure the parasite is ingested by a cat?***
- **HYPOTHESIS:** *Toxoplasma* affects the limbic system (a part of the brain that controls emotions) of the rat brain. If so, then cat urine should show increased activity in this brain region.
- **EXPERIMENTAL DESIGN:** Male rats were put into one of four groups and exposed to cat urine or estrous female odor. Immediately after a 20-minute exposure to the urine/odor, rats were sacrificed and brain activity analyzed immunochemically for c-Fos activation (an indirect indicator of neural activity).
- **EXPERIMENT 1:** Uninfected male rats were exposed to a towel containing 1 ml of cat urine or exposed to estrous female rat odor (estrous female separated from males by a plastic barrier, with holes).
- **EXPERIMENT 2:** Infected male rats were exposed to a towel containing 1 ml of cat urine or exposed to estrous female rat odor (estrous female separated from males by a plastic barrier, with holes).
- **RESULTS:**

  EXPERIMENTS 1 AND 2: See figures and table.

- **CONCLUSIONS:**

  QUESTION 1: ***Was the hypothesis validated? Explain using the figures and table.***

  QUESTION 2: ***How does brain activity in uninfected male rats exposed to estrous female odor compare to brain activity in infected male rats exposed to cat urine?***

  QUESTION 3: ***What can you conclude about the effect of* Toxoplasma *on the innate fear response?***

Note: About 33% of the human population has been exposed to *Toxoplasma* and several studies find that the infection increases the risk for schizophrenia and obsessive compulsive disorder.

Answers can be found on the Student Companion Website in **Appendix D**.

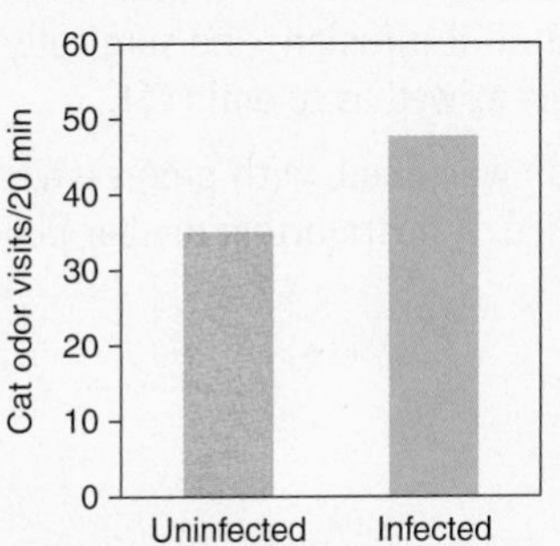

FIGURE A Time spent exploring cat urine. Modifed from: House, P.K., Vyas, A., and Sapolsky, R. (2011). *PLoS ONE* 6(8): e23277.

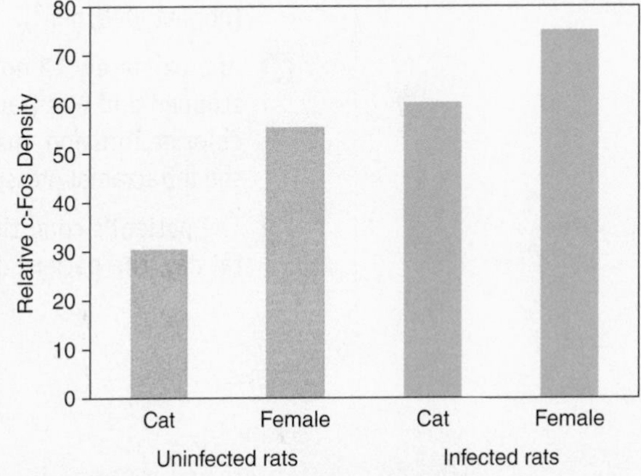

FIGURE B Neural activity (c-Fos activity) in rats exposed to cat urine (bars 1 and 3) or to female estrous odor (bars 2 and 4). Modifed from: House, P.K., Vyas, A., and Sapolsky, R. (2011). *PLoS ONE* 6(8): e23277.

TABLE

Rat Behavioral Brain Responses

Experimental treatment	Relative limbic activity involved with: Defensive behavior	Reproductive behavior
Uninfected exposure to		
• Urine	Increased	No increase
• Estrous odor	Diminished	Increased
Infected exposure to		
• Urine	Unchanged	Increased
• Estrous odor	Diminished	Increased

Adapted from: House, P. K., Vyas, A., and Sapolsky, R. (2011). *PLoS ONE* 6(8):e23277.

Textbook CASE 18

Primary Amoebic Meningoencephalitis

1 In late August, a previously healthy boy was evaluated in a local emergency department for a 2-day history of headache and vomiting; he had a fever and was lethargic. There were no signs of meningitis, and a lumbar puncture identified no infectious organism. Four days before onset of illness, the patient had attended a social event and had swum in a freshwater river with a group of friends in southern Georgia.

2 The patient was started on intravenous antibiotics for suspected bacterial meningitis. Within several hours of admission, he had spontaneous nonpurposeful movements, was unable to follow verbal commands, and was transferred to a children's hospital intensive care unit (ICU). En route to the ICU, he had a 30-minute right-sided seizure. A computed tomography (CT) scan of the head on admission to the ICU showed edema of the midbrain, and cranial magnetic resonance imaging (MRI) demonstrated areas of meningeal enhancement in the brainstem suggestive of meningitis.

3 No organisms were observed on a Gram-stained smear of cerebrospinal fluid (CSF); CSF antigen-detection tests were negative for bacterial pathogens. Fresh preparation of CSF revealed: no amoebae, a red blood cell count of 1,550/mm^3 (normal: 0/mm^3); and a white blood cell count of 13,650/mm^3 (normal: 0–5/mm^3).

4 Approximately 12 hours after admission to the ICU, the patient had episodes when his breathing stopped and was tracheally intubated. Treatment included hyperventilation, hypertonic sodium chloride infusion, mannitol infusion, and surgically placing a tube through the skull to monitor the intracranial pressure as well as to drain CSF.

5 The patient's condition worsened, with progressive neurologic deterioration. On the fourth hospital day, the patient died. A postmortem lumbar puncture demonstrated a few motile amoebae.

though one remains infected for life. The highly contagious nature of the parasite stems from its need to get back into its definitive host, the cat, where oocysts are produced.

Pregnant women, however, are at risk of developing a dangerous toxoplasmosis infection because the parasite may cross the placenta and infect the fetal tissues (Figure 18.16). Neurologic damage, lesions of the fetal visceral organs, or spontaneous abortion may result. Congenital infection is least likely during the first trimester, but damage may be substantial when it occurs. By contrast, congenital infection is more common if the woman is infected in the third trimester, but fetal damage is less severe. Lesions of the retina are the most widely documented complication in congenital infections.

T. gondii also is known to cause severe disease in immunosuppressed individuals. For example, in patients with AIDS the parasite attacks the brain tissue, causing an inflammation and swelling that often results in cerebral lesions, seizures, and death.

Naegleria fowleri

■ Meningoencephalitis: An inflammation of the brain and meninges.

Naegleria Can Infect the Central Nervous System

Primary amoebic meningoencephalitis (PAM) is a rare disease with less than 200 cases reported worldwide since 1965. However, it is the most deadly disease of the central nervous system after rabies. Ninety-five percent of patients die within 4 to 5 days of infection (Textbook Case 18).

PAM is caused by several species of thermophilic parasites in the genus *Naegleria,* especially *N. fowleri* (FIGURE 18.17). It also can be caused by *Acanthamoeba* and *Hartmannella* species.

Naegleria is an opportunistic pathogen of humans, causing meningoencephalitis when inhaled, often by swimming in contaminated warm surface water. The free-living trophozoites

6 Autopsy findings revealed acute primary amoebic meningoencephalis (PAM) caused by *N. fowleri* (see figure).

7 An epidemiologic investigation discovered that the boy was one of five children who had participated in water activities that included swimming, swimming under water, wrestling in the water, and diving into the water.

8 The environmental investigation revealed a high ambient temperature (>90°F [>32°C]) and water temperature (91°F [33°C]) in the river at the time of the exposure. In addition, the river level was low and the river was flowing slowly. Bacteriologic testing of the river water demonstrated that fecal coliform levels were within acceptable limits.

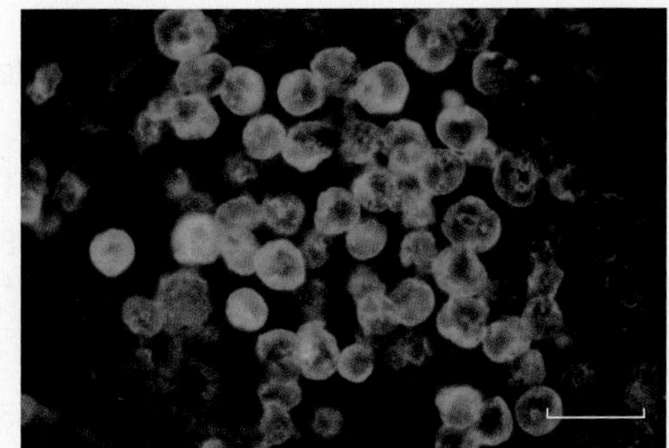

Fluorescent antibody stain of *Naegleria fowleri* amoebae. (Bar = 40 μm.) Courtesy of CDC.

Questions:

(Answers can be found on the Student Companion Website in **Appendix D.**)

A. Would the absence of microbes in the initial lumbar puncture mean the young boy did not have meningitis? Explain.

B. Why was a follow-up lumbar puncture ordered?

C. How might the patient have become infected with the parasite?

D. Why were no other children infected?

E. What precautions should be taken to prevent another incident?

For additional information see www.cdc.gov/mmwr/preview/mmwrhtml/mm5240a4.htm.

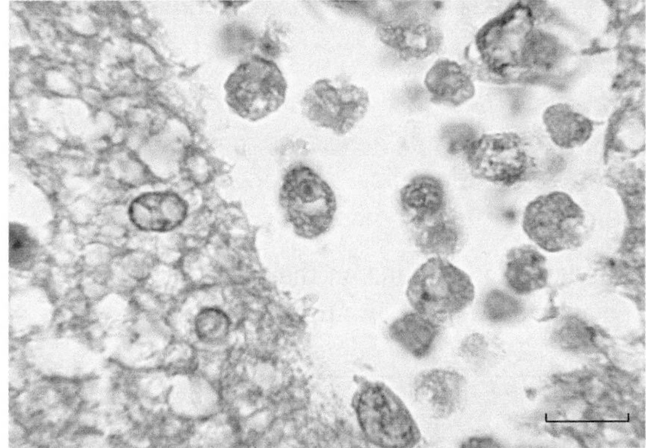

FIGURE 18.17 ***Naegleria fowleri.*** A light micrograph of brain tissue showing the infection with *N. fowleri* amoebae (round cells). (Bar = 25 μm.) Courtesy of Dr. Govinda S. Visvesvara/CDC. »» How might these aquatic amoebae infect brain tissue?

appear to enter the body through the mucous membranes of the nose and then follow the olfactory tracts to the brain.

The symptoms resemble those in other forms of encephalitis and meningitis. Nasal congestion precedes piercing headaches, fever, delirium, neck rigidity, and occasional seizures. Treatment with amphotericin B, miconazole, and rifampin has had limited success.

TABLE 18.3 summarizes the protistan diseases of the blood and nervous system.

CONCEPT AND REASONING CHECKS 3

a. Summarize the life cycle of the *Plasmodium* parasite.

b. What is the vector difference between African and American trypanosomiasis?

c. Why is babesiosis considered to be a malaria-like disease?

d. Assess the impact of *Toxoplasma* to pregnant women and immunocompromised people.

e. What similarities exist between *Naegleria* and *Entamoeba?*

TABLE

18.3 A Summary of Protistan Diseases of the Blood and Nervous System

Disease	Causative Agent	Signs and Symptoms	Transmission	Treatment	Prevention
Malaria	4 *Plasmodium* species	Moderate to severe shaking chills, high fever, profuse sweating with chills, malaise	Bite of infected *Anopheles* mosquito	Chloroquine, quinine, mefloquine, malarone	Mefloquine, malarone
African trypanosomiasis (African sleeping sickness)	*Trypanosoma brucei*	Chronic bouts of fever, severe headache, change in sleep patterns and behavior	Bite of infected tsetse fly	Pentamidine, melarsoprol, eflornithine	Clearing areas where tsetse flies breed
Chagas disease (American trypanosomiasis)	*Trypanosoma cruzi*	Acute phase: Redness and swelling at the site of infection, asymptomatic Chronic phase: widespread tissue damage	Bite of an infected reduviid bug	Acute phase: benznidazole and nifurtimox Chronic phase: depends on signs and symptoms	Avoiding endemic residences Insecticides
Babesiosis	*Babesia microti*	Mild anemia, piercing headache	Bite of infected tick	Azithromycin and atovaquone	Avoiding ticks in endemic areas
Toxoplasmosis	*Toxoplasma gondii*	Usually asymptomatic	Contact with contaminated cat feces, ingestion of contaminated food, water, or fomites	Pyrimethamine and sulfadiazine for acute disease	Practicing good hygiene Avoiding raw or undercooked meat and unpasteurized goat's milk
Primary amoebic meningoencephalitis (PAM)	*Naegleria fowleri*	Piercing headache, fever, delirium, neck rigidity	Inhalation of contaminated water	Early with amphotericin B, miconazole, and rifampin	Avoiding swimming in contaminated water

Chapter Challenge C

A number of protistan diseases are transmitted by "flying insects." This includes malaria (mosquitoes), African sleeping sickness (tsetse flies), and leishmaniasis (sand flies). In the next section, you will see that lymphatic filariasis is also transmitted by mosquitoes.

QUESTION C:

Why have these diseases evolved to use a flying insect as the means of transmission?

Hint: Think about how the parasite is carried through the human host.

Answers can be found on the Student Companion Website in **Appendix D**.

KEY CONCEPT 18.4 Parasitic Helminths Cause Substantial Morbidity Worldwide

The **helminths** are among the world's most common animal parasites. For example, 2 billion people—approximately 33% of the human population—are infected with soil-transmitted helminths! Therefore, in concluding this chapter on the parasites, we are concerned with medical helminthology and the diseases caused by the parasitic worms.

As mentioned in the introduction to this chapter, such parasites are of interest to microbiologists because of the parasites' ability to cause an enormous level of morbidity worldwide. However, different from many of the bacterial and fungal pathogens we have discussed, most parasitic helminths are dependent on the host or hosts for sustenance, so it is to the helminth's benefit that the host stays alive. Therefore, the helminths tend to cause diseases of debilitation and chronic morbidity resulting from physical factors related to the **helminthic load** (number of worms present) or location in the body.

There Are Two Groups of Parasitic Helminths

The helminths of medical significance are the flatworms and the roundworms.

Flatworms. Animals in the phylum Platyhelminthes (*platy* = "flat"; "*helmin*" = "worm") are the **flatworms**. As multicellular animals, they have tissues functioning as organs in organ systems. However, they have no specialized respiratory or circulatory structures, and they lack a digestive tract. The gut (gastrovascular cavity) simply consists of a sac with a single opening, thus placing the worm in close contact with its surroundings. Complex reproductive systems are found in many species within the phylum, and a large number of species are hermaphroditic. Two groups of flatworms are of concern regarding human disease.

The **trematodes**, includes the **flukes**, which have flattened, broad bodies (FIGURE 18.18A). The animals exhibit bilateral symmetry. Trematodes have a complex life cycle that may include encysted egg stages and temporary larval forms. Sucker devices are commonly present to enable the parasite to attach to its host. In many cases, two hosts exist: an intermediate host, which harbors the larval form, and a definitive host, which harbors the mature adult form. In this chapter, we are concerned with parasites whose definitive host is a human.

The life cycle of a fluke often contains several phases. In the human host, the parasite produces fertilized eggs generally released in the feces. When the eggs reach water, they hatch and develop into tiny ciliated larvae called **miracidia** (sing., miracidium). The miracidia penetrate snails (the intermediate host) and go through a series of asexual reproductive stages. The cyst makes its way back to humans.

The trematode lifestyle requires the parasite to evade the host's immune system. It accomplishes this by having its surface resemble the surface of the host cells, so the immune system "sees" the worm as a "normal" cell, not an invader. This mimicry is quite effective as some flukes can remain in a human host for 40 years or more.

The other group of flatworms is the **cestodes**, which includes the **tapeworms**. These parasitic

■ Hermaphroditic: Refers to an organism having both male and female reproductive organs.

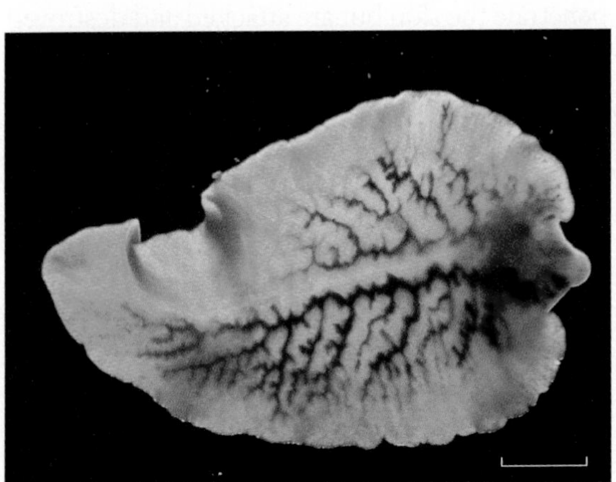

(A)

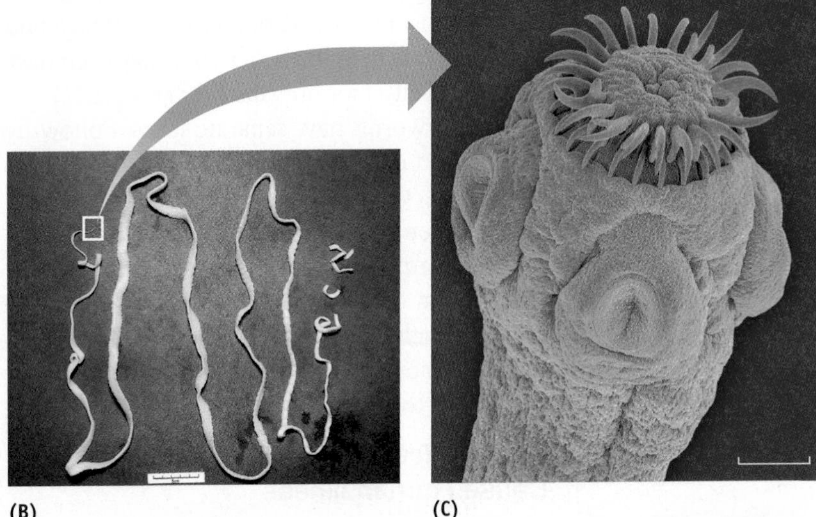

(B) (C)

FIGURE 18.18 **Flukes and Tapeworms.** **(A)** Light micrograph of a liver fluke, *Fasciola hepatica*, a parasite infecting sheep, cattle, and humans. (Bar = 50 µm.) © Sinclair Stammers/Photo Researchers, Inc. **(B)** Photograph of a coiled beef tapeworm, *Taenia saginata*, which has grown to several meters in length. © Medical-on-Line/Alamy Images. **(C)** A false-color scanning electron micrograph of the head, scolex, of *T. saginata* showing the suckers and hooks. (Bar = 1 mm.) © Dennis Kunkel Microscopy, Inc./Phototake/Alamy Images. »» What does the dark branched structures form in the liver fluke (A) and what is the purpose of the suckers and hooks on the scolex head of the tapeworm (C)?

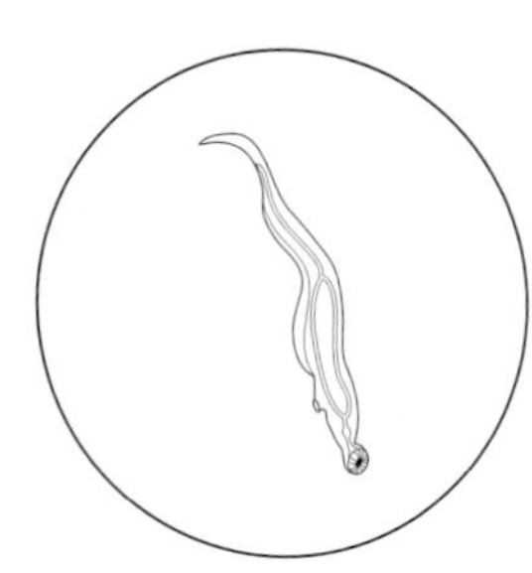
Schistosoma mansoni

worms have a head region, called the **scolex**, and a ribbon-like body consisting of segments called **proglottids** (FIGURE 18.18B, C). The proglottids most distant from the scolex are filled with fertilized eggs. As the proglottids break free, they spread the eggs.

Tapeworms generally live in the intestines of a host organism. In this environment, they are constantly bathed by nutrient-rich fluid, from which they absorb food already digested by the host. Tapeworms have adapted to a parasitic existence and have lost their intestines, but they still retain well-developed muscular, excretory, and nervous systems.

Tapeworms are widespread parasites infecting practically all mammals, as well as many other vertebrates. Because they are more dependent on their hosts than flukes, tapeworms have precarious life cycles. Tapeworms have a limited range of hosts, and the chances for completing the cycle are often slim. With rare exceptions, tapeworms require at least two hosts. Humans often become infected by eating undercooked meat containing tapeworm cysts, which then develop into mature adult worms.

Roundworms. Among the most prevalent animals are the **roundworms** in the phylum Nematoda (*nema* = "thread"). These parasites have a thread-like body and occupy every imaginable habitat on Earth. They live in the sea, in freshwater, and in soil from polar regions to the tropics. Good topsoil, for example, may contain billions of nematodes per acre. They parasitize every conceivable type of plant and animal, causing both economic crop damage and serious disease in animals. Yet, they may even have a beneficial effect for humans (MicroFocus 18.5).

Roundworms have separate sexes. Following fertilization of the female by the male, the eggs hatch to larvae that resemble miniature adults. Growth then occurs by cellular enlargement and mitosis. Damage in hosts is generally caused by large worm burdens in the blood vessels, lymphatic vessels, or intestines (FIGURE 18.19). Also, the infestation may result in nutritional deficiency or damage to the muscles.

Several Trematodes Can Cause Human Illness

Schistosomiasis. The WHO estimates that 200 million people in 74 countries suffer from schistosomiasis, which kills approximately 280,000 every year. There are even about 400,000 individuals in the United States who suffer from a mild form of the disease.

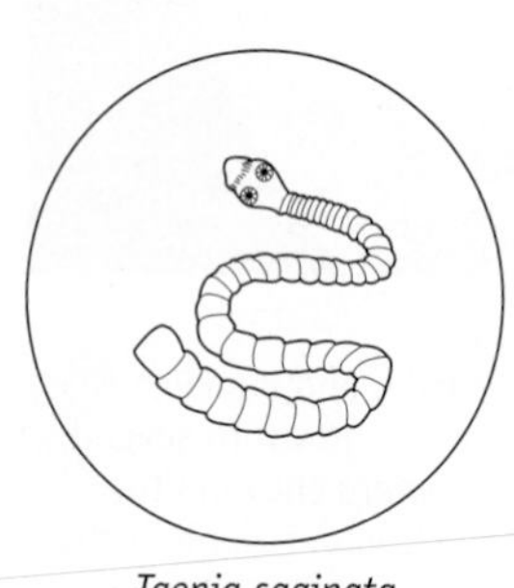
Taenia saginata

Schistosomiasis is caused by several species of blood flukes, including *Schistosoma mansoni* (Africa and South America), *S. japonicum* (Asia), and *S. haematobium* (Africa and India). In some regions, the term **bilharziasis** is still used for the disease; it comes from the older name for the genus, *Bilharzia*.

Species of *Schistosoma* measure about 10 mm in length. The eggs hatch in freshwater to produce miracidia, which then make their way to snails. In the snails, miracidia convert to a second larval form called **cercariae** (FIGURE 18.20). The cercariae escape from the snails and attach themselves to the bare skin of humans wading in contaminated water. Cercariae infect the blood and mature into adult flukes, which cause fever and chills. However, the major effects of disease are due to eggs: carried by the bloodstream to the liver, they cause substantial liver damage; in the intestinal wall, ulceration, diarrhea, and abdominal pain occur; and in the bladder, egg infection causes bloody urine and pain on urination. Male and female species mate in the human liver and produce eggs that are released in the feces or urine.

The antihelminthic drug praziquantel is used for treatment.

Certain species of *Schistosoma* penetrate no farther than the skin because the definitive hosts are birds instead of humans. Often after swimming in schistosome-contaminated water, the cercariae penetrate the skin but are attacked and destroyed by the body's immune system. However, they release allergenic substances that cause an itching and body rash, commonly known as **swimmer's itch**. The condition, not a serious threat to health, is common in northern lakes in the United States.

MicroInquiry 18 examines some amazing phenomena of how parasitic flukes can manipulate their host for their own benefit—and survival.

Tapeworms Survive in the Human Intestines

Beef and Pork Tapeworm Diseases. Approximately 50,000 people die each year from beef and pork tapeworm diseases. Humans are the definitive hosts for both the beef tapeworm *Taenia saginata* and the pork tapeworm *Taenia solium*. Humans acquire the tapeworm cysts by eating poorly

MICROFOCUS 18.5: Being Skeptical

Eat Worms and Cure Your Ills—and Allergies!

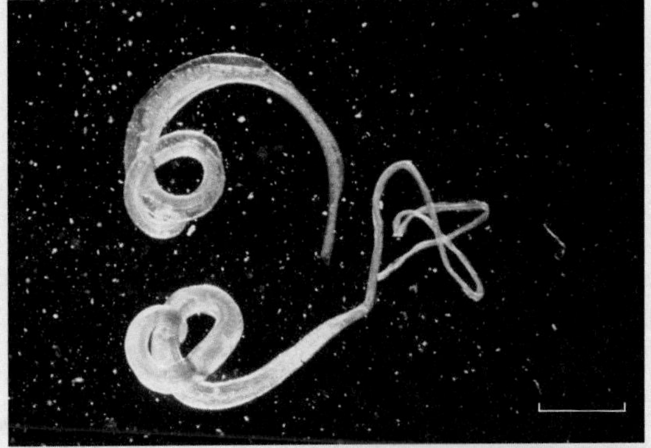

Adult whipworms (male top; female bottom). (Bar = 1 cm.)

Doctors and researchers have shown that eating worms (actually drinking worm eggs) can fight disease. What! Are you nuts? Drink worm eggs!

Joel Weinstock, a gastroenterologist at the University of Iowa, discovered that as allergies and other diseases have increased in Western countries, infections by roundworm parasites have declined. This is not the case in other countries where allergies are rare and worm infections are quite common. Weinstock wondered if there was a correlation between allergy increase and parasite decline.

To test his hypothesis, Weinstock "brewed" a liquid concoction consisting of thousands of pig whipworm eggs (ova). The whipworms are called *Trichuris suis,* so his product with ova is called TSO. This roundworm (see figure) was chosen because once the ova hatch, they will not survive long in the human digestive system and will be passed out in the feces.

One woman in Iowa was suffering from incurable ulcerative colitis, which is caused by the immune system overreacting. Immune cells start attacking the person's own gut lining, making it bleed. The symptoms are severe cramps and acute, intense diarrhea. In a trial run, Weinstock gave the woman a small glass full of his TSO. Every 3 weeks she downed another glass. Guess what? Her ulcerative colitis is in remission and she no longer suffers any disease symptoms.

Further trials in 2004 involved 100 people suffering the same disease and a further 100 suffering Crohn's disease, which is another type of inflammatory bowel disease related to immune function. In this study, 50% of the volunteers suffering ulcerative colitis and 70% of those suffering Crohn's disease went into remission, as identified by no symptoms of abdominal pain, ulcerative bleeding, and diarrhea.

Weinstock believes some parasites are so intimately adapted with the human gut that if they are eradicated, bad things may happen, such as the bowel disorders mentioned. He says the immune system has become so involved with defending against parasites that if you take them away, the immune system overreacts to other events.

Is there something to this? Alan Brown, an academic researcher in the United Kingdom (UK) picked up a hookworm infection while on a field trip outside the UK. Being that he was a well-nourished Westerner, the 300 hookworms in his gut caused no major problem. However—since being infected, his hayfever allergy has disappeared!

The result: I need more evidence before I would drink worm eggs for a gastrointestinal disease.

cooked beef or pork. The beef tapeworm may reach 10 meters in length, while the pork tapeworm length is 2 to 8 meters.

Attachment via the scolex occurs in the small intestine and obstruction of this organ may result. In most cases, however, there are few symptoms other than mild diarrhea, and a mutual tolerance may develop between parasite and host. Each tapeworm may have up to 2,000 proglottids and infected individuals will expel numerous gravid proglottids daily. The proglottids accumulate in the soil and are consumed by cattle or pigs. Embryos from the eggs travel to the animal's muscle, where they form cysts.

In rare instances, infection can lead to very bizarre behaviors (MicroFocus 18.6).

Echinococcosis. Dogs and other canines such as wolves, foxes, and coyotes are the definitive hosts for dog tapeworms belonging to the genus *Echinococcus* (FIGURE 18.21A). Eggs reach the soil in feces and spread to numerous intermediary hosts, one of which is humans. Contact with a dog also may account for transmission. In humans, the parasites travel by the blood to

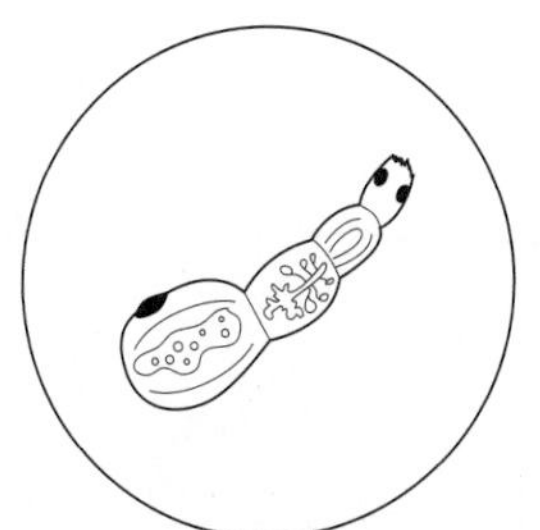

Echinococcus granulosus

■ Gravid:
Refers to carrying eggs.

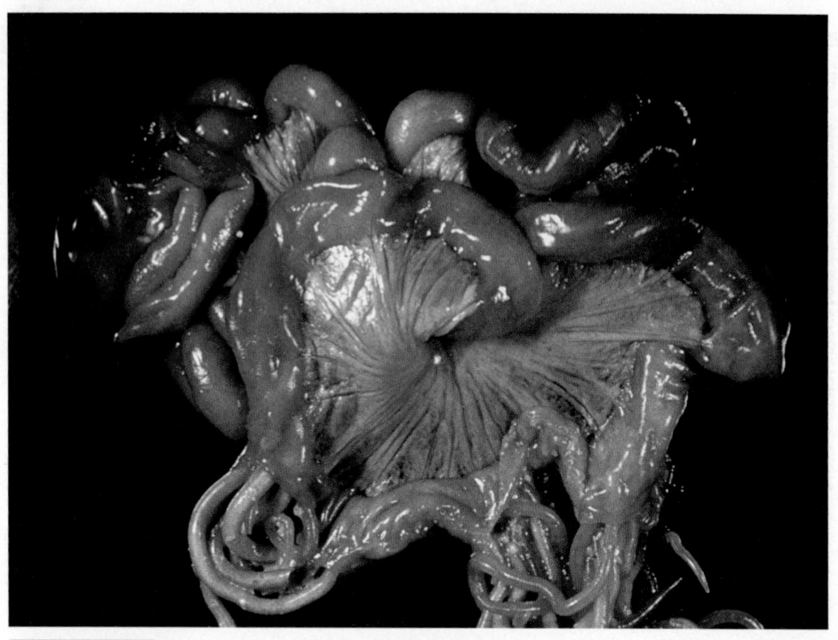

FIGURE 18.19 **Roundworms.** Photograph of threadworms seen on the surface of, and burrowing into, the gut wall of a pig. © R.F. Ashley/Visuals Unlimited. »» How do the nematodes differ from the flatworms?

the liver, where they form thick-walled **hydatid cysts** (FIGURE 18.21B). Common symptoms include abdominal and chest pain, and coughing up blood. The rupture of cysts can trigger an anaphylactic reaction: Treatment may require an antiparasite drug or surgical removal of the cysts. Prevention includes deworming dogs and practicing good hygiene.

Humans Are Hosts to at Least 50 Roundworm Species

Here, we discuss a few significant parasitic infections caused by roundworms.

Pinworm Disease. The most prevalent helminthic infection in the United States is **pinworm disease**, where an estimated 30% of children and 16% of adults serve as hosts.

Pinworm disease is caused by *Enterobius vermicularis*. The male and female worms live in the distant part of the small intestine and in the large intestine, where the symptoms of infection

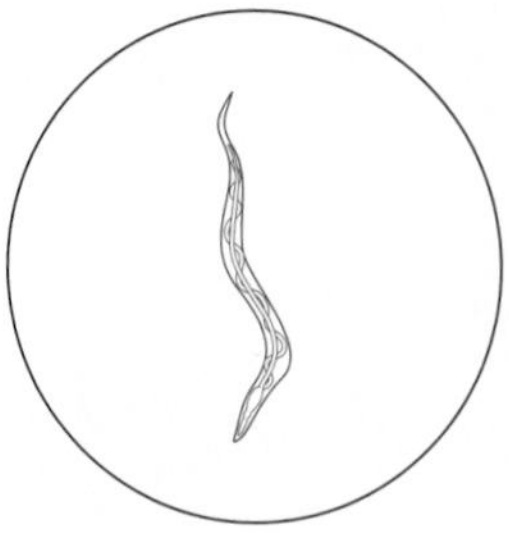

Enterobius vermicularis

E. Cercariae pass into the bloodstream. Females deposit eggs that are eliminated in feces or urine.

A. Eggs of the parasite are deposited in water via human feces or urine.

Eggs

B. Eggs of the parasite hatch and release miracidia that swim to aquatic snails and penetrate host tissue.

Miracidium

Human (definitive host)

Snail (intermediate host)

D. The cercariae penetrate the skin of a human host.

C. In snails, miracidia develop into cercariae that are released into the water.

Cercaria

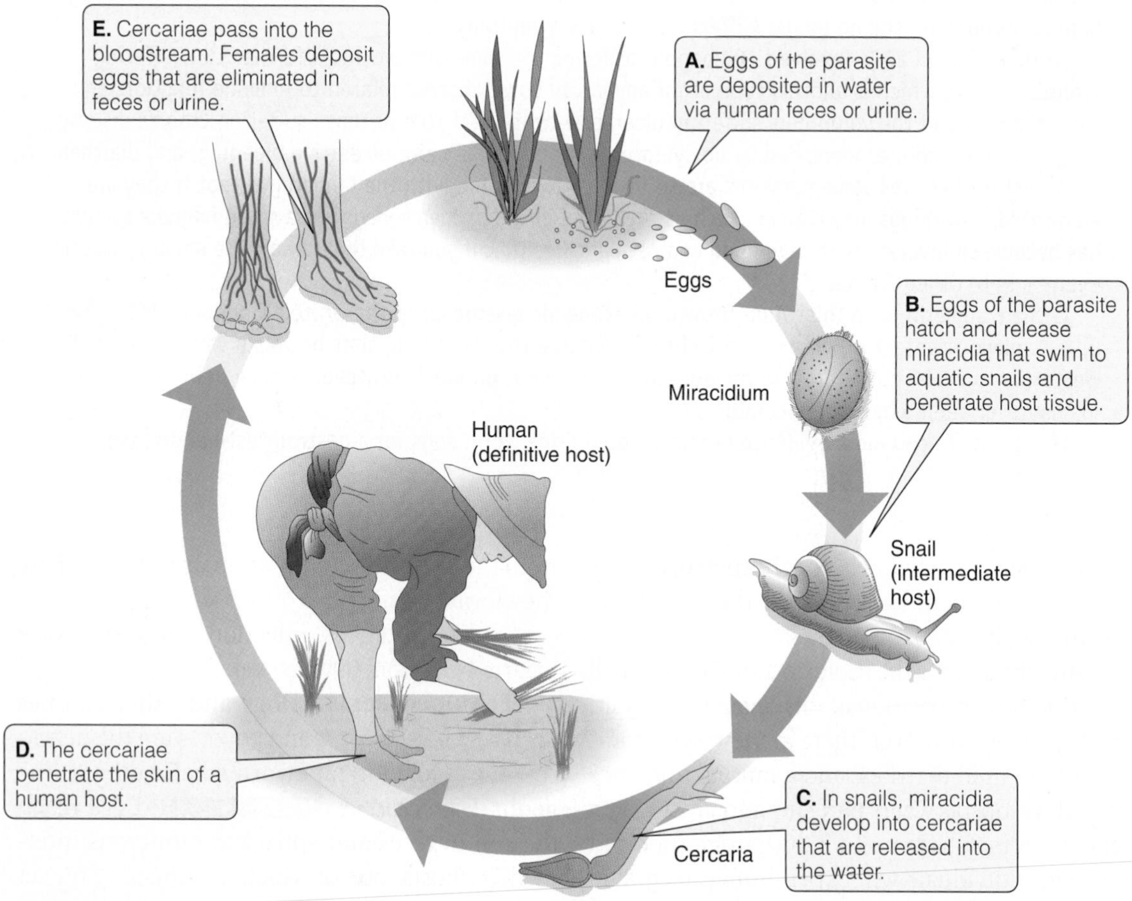

FIGURE 18.20 **The Life Cycle of the Blood Fluke *Schistosoma mansoni*.** The life cycle of *S. mansoni* involves an intermediate host, the snail, and a definitive host, a human. »» What defines a definitive and intermediate host?

MICROINQUIRY 18

Parasites as Manipulators

Over the last several chapters, we have been discussing viral, bacterial, and fungal pathogens, and protist and multicellular parasites. For the latter parasites, you might conclude they make their living "sponging off" their hosts with the ultimate aim to reproduce. In fact, for a long time scientists thought these organisms simply evolved so they could take advantage of what their hosts had to offer. In reality, the parasites may be quite a bit more sophisticated than previously thought. Two cases involving multicellular parasites are submitted for your inquiry.

Case 1

The lancet fluke, *Dicrocoelium dendriticum,* infects cows. Because the infected cow is the definitive host, the fluke needs to get out of the cow to form cysts in another animal. If the cysts can then be located on grass blades, another cow may eat those cyst-containing grass blades and another round of infection by *D. dendriticum* can occur.

18.1a. As a parasite, how do you ensure cysts will be on grass blades that cows eat?

Infected cows excrete dung containing fluke eggs. Snails (intermediate host #1) forage on the dung and in the process ingest the fluke eggs. The eggs hatch and bore their way from the snail gut to the digestive gland where fluke larvae are produced. In an attempt to fight off the infection, the snail smothers the parasites in balls of slime coughed up into the grass.

Ants (intermediate host #2) come along and swallow these nutrient-rich slime balls. Now, here is the really interesting part, because the lancet fluke has yet to get cysts onto grass blades so they will be eaten by more cows. In infected ants, the larvae travel to the ant's head and specifically the nerves that control the ant's mandibles. While most of the larvae then return to the abdomen where they form cysts, a few flukes remain in the head and control the ant's behavior. As evening comes and the temperature cools, the head-infected ants with a belly full of cysts leave their colony, climb to the tip of grass blades, and hold on tight with their mandibles (see figure). The lancet fluke has taken control of the ants and placed them in a position most favorable to be eaten by a grazing cow.

18.1b. What happens if no cow comes grazing that evening?

If a grazing cow does not come by, the next morning the flukes release their influence on the ants, which return to their "normal duties." But, the next evening, the flukes again take control and the ants march back up to the tips of grass blades. Should a cow eat the grass with the ants, the fluke cysts in the ant abdomen quickly hatch and another cycle of reproduction begins in the cow. Now, that is quite a behavioral driving force to control your destiny; in this case, to ensure fluke cysts are positioned so they will be eaten by the cow.

Case 2

(This case is based on research carried out in a coastal salt marsh by Kevin Lafferty's group at the University of California at Santa Barbara.) Another fluke, *Euhaplorchis californiensis,* uses shorebirds as its definitive host. Infected birds drop feces loaded with fluke eggs into the marsh. Horn snails (intermediate host) eat the droppings containing the eggs. The eggs hatch and castrate the snails. Larvae are produced in the water and latch onto the gills of their second intermediate host, the California killifish. In the killifish, the larvae move from the blood vessels to a nerve that carries the larvae to the brain where the larvae form a thin layer on top of the brain. There they stay, waiting for the fish to be eaten by a shorebird. Once in the bird's gut, the adults develop and produce another round of fertilized eggs. Here, we have two questions.

18.2a. What is the purpose for the larvae castrating the snails?

An experiment was set up to answer this question. Throughout the salt marsh, Lafferty set up cages with uninfected snails and other cages with infected ones. The results were as expected—the uninfected snails produced many more offspring than did the infected snails. Lafferty believes a marsh full of uninfected snails would reproduce so many offspring as to deplete the algae and increase the population of crabs that feed on the snails. By castrating the snails, the parasite actually is controlling the snail population and keeping the salt marsh ecosystem balanced—again for its benefit.

An ant clutching a grass blade with its feet and mandibles. © Eye of Science/Photo Researchers, Inc.

18.2b. Why do the larvae in the killifish take up residence on top of the fish's brain?

Through a series of experiments, Lafferty's group discovered that infected fish underwent a swimming behavior not observed in the uninfected fish. The infected killifish darted near the water surface, a very risky behavior that makes it more likely the fish would be caught by shorebirds. In fact, experiments showed infected fish were 30 times more likely to be caught by shorebirds than uninfected fish. Now, from the parasite's perspective, they want to be caught so another round of reproduction can start in the shorebird. So again, the fluke has manipulated the situation to its benefit.

Parasites may cause disease, but they also maximize their chances for ensuring infection. Pretty amazing!

MICROFOCUS 18.6: Public Health

Man Leaps Off Speeding Car, Report Says

It was truly a bizarre death. The chief financial officer (CFO) for the city of Phoenix, Arizona, was a very professional, quiet man who had a way with numbers. He cared about others and worked with the underprivileged. In fact, he and his wife had formed a foundation to help working women and children in need find health insurance. Life was good.

So, why in 2004, wearing tattered jeans and no shoes, would he climb out the window of his car going 55 mph on a crowded street, stand on top of the car with his arms outstretched—and then jump to his death?

A few days before his death, the finance officer had taken a sick day to see his doctor complaining of feeling "tired and worn out." But other than that, there was no indication of any abnormal behavior.

However, it happens that 2 years earlier, he had fallen ill after returning from a trip to Mexico. At that time, nobody outside of the family knew of his illness.

According to his wife, he had been suffering from cysticercosis, an infection caused by the larval stage of the pork tapeworm *Taenia solium.* He had been taking medication for the parasitic infection and had been considering changing the dosage. In fact, after his death it was discovered he had been having flare-ups of the parasite.

Apparently, the CFO had been infected in Mexico probably from eating vegetables or fruits contaminated with pig feces containing the parasite. On very rare occasions, the larvae can move to the brain and cause frontal lobe disinhibition, which means that they can make an individual do bizarre things, including impairing one's decision-making abilities.

So, it appears, like MicroInquiry 18 describes, parasites can cause strange behaviors. In this case, the result was a very sad and tragic consequence.

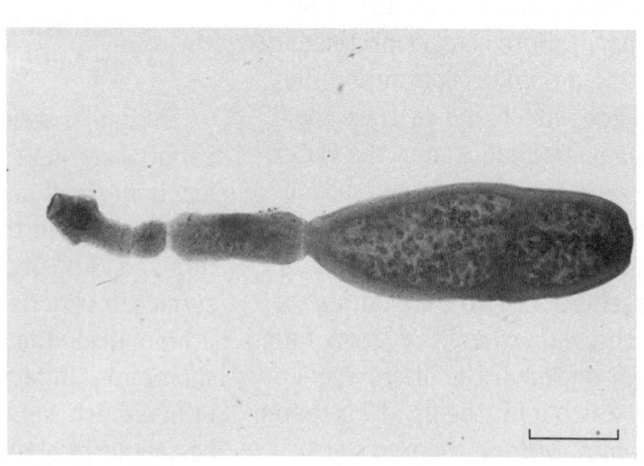

(A)

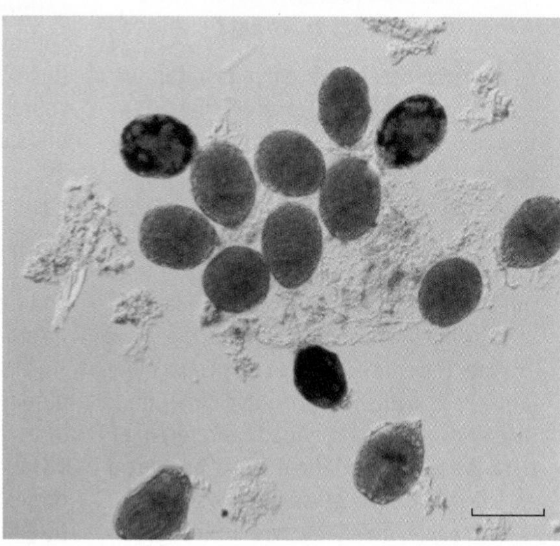

(B)

FIGURE 18.21 The Dog Tapeworm. (**A**) A light micrograph of the dog tapeworm, *Echinococcus.* A large section of the worm (right) contains numerous eggs. (Bar = 2 mm.) © Biodisc/Visuals Unlimited/Corbis. (**B**) In humans, the tapeworm migrates to the liver or lungs, where they form slow-growing hydatid cysts (red). (Bar = 200 μm.) © Alfred Pasieka/Photo Researchers, Inc. »» Identify the scolex in the light micrograph of *Echinococcus.* What was the identifying feature?

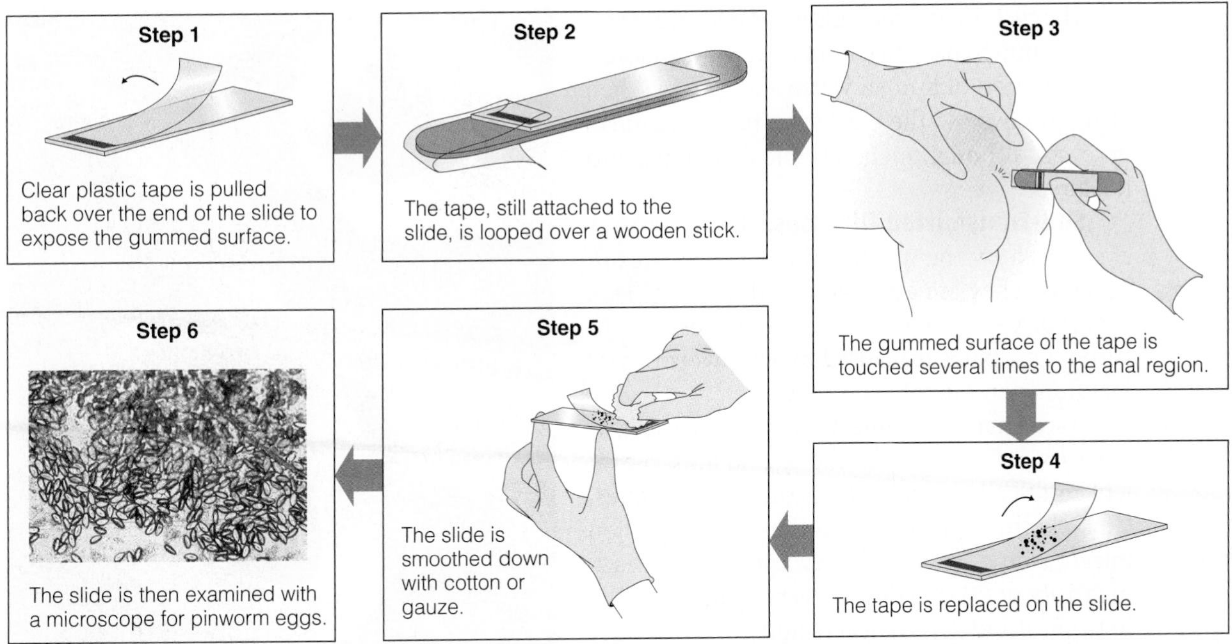

FIGURE 18.22 **Diagnosing Pinworm Disease.** The transparent tape technique used in the diagnosis of pinworm disease. Inset © Pr. Bouree/Photo Researchers, Inc. »» What time of day might be best to use the tape technique?

include diarrhea and itching in the anal region. The female worm is about 10 mm long, and the male is about half that size.

The life cycle of the pinworm is relatively simple. Females migrate to the anal region at night and lay a considerable number of eggs. The area itches intensely and scratching contaminates the hands and bed linens with eggs. Reinfection can take place if the hands are brought to the mouth or if eggs are deposited in foods by the hands. The eggs are swallowed, whereupon they hatch in the duodenum and mature in the regions beyond.

Diagnosis of pinworm disease may be made accurately by applying the sticky side of cellophane tape to the area about the anus and examining the tape microscopically for pinworm eggs (FIGURE 18.22). Mebendazole is effective for controlling the disease, and all members of an infected person's family should be treated because transfer of the parasite probably has taken place. Even without medication, however, the worms will die in a few weeks, and the infection will disappear as long as reinfection is prevented.

Trichinellosis. Most of us are familiar with the term **trichinellosis** because packages of pork often contain warnings to cook the meat thoroughly to avoid this disease. The disease is rare in the United States.

Trichinellosis often is caused by the small roundworm *Trichinella spiralis*. The worm lives in the intestines of pigs and several other mammals. Larvae of the worm migrate through the blood and penetrate the pig's skeletal muscles, where they remain in cysts. When raw or poorly cooked pork is consumed, the cysts pass into the human intestines and the worms emerge. Intestinal pain, vomiting, nausea, and constipation are common symptoms.

Complications of trichinellosis occur when *T. spiralis* adults mate and the female produces larvae in the intestinal wall. The tiny larvae migrate to the muscles primarily in the tongue, eyes, and ribs where they form cysts (FIGURE 18.23). The patient commonly experiences pain in the breathing muscles of the ribs and loss of eye movement.

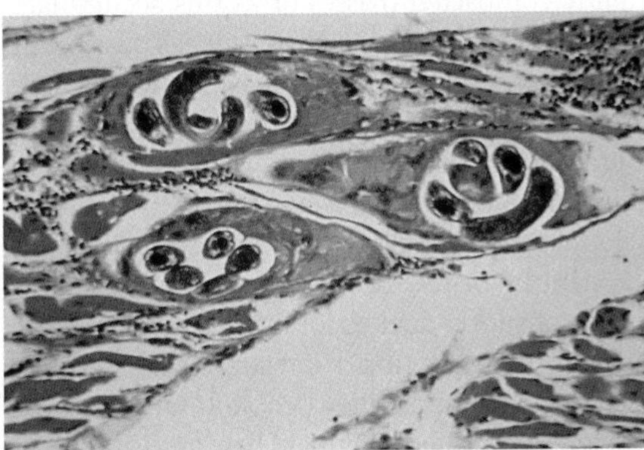

FIGURE 18.23 ***Trichinella spiralis.*** A stained light micrograph of a *T. spiralis* larva (dark red) coiled inside a cyst in muscle tissue (red). Courtesy of CDC. »» What is the purpose of forming a cyst in muscle, which represents a good nutrient source for a parasite?

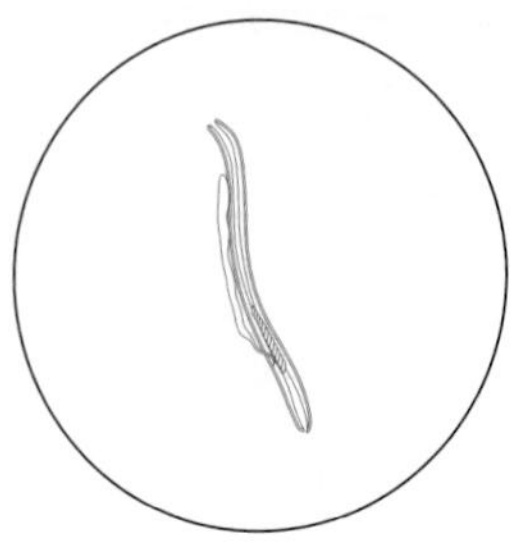

Trichinella spiralis

The cycle of trichinellosis is completed as cysts are transmitted back to nature in the human feces. Consumption of human waste and garbage then brings the cysts to the pig. Drugs have little effect on cysts, although mebendazole can be used to kill larvae.

Soil-Transmitted Diseases. These diseases are caused by the most significant parasites in humans. They are associated with poverty, lack of adequate sanitation and hygiene, and overpopulation. One parasite, caused by whipworms, was mentioned in MicroFocus 18.5.

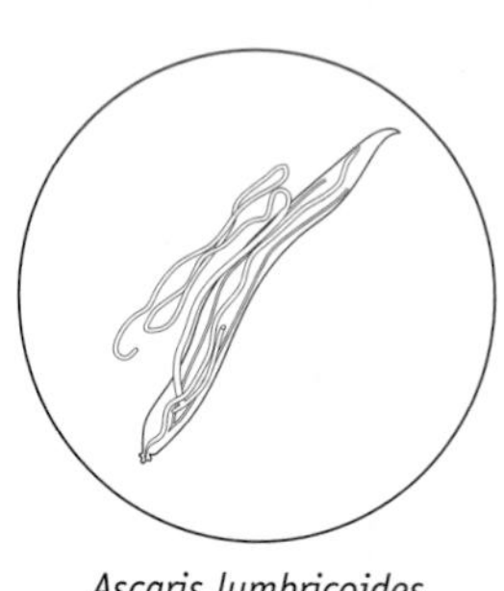

Ascaris lumbricoides

Ascariasis is an infection with *Ascaris lumbricoides,* a parasite that is the second most prevalent multicellular parasite in the United States. Globally, the WHO estimates there are 1.4 billion infections and 380 million cases worldwide, leading to about 60,000 deaths every year, especially in tropical and subtropical regions.

The parasite resembles an earthworm and is one of the largest intestinal nematodes; females may be up to 30 cm long, and males 20 cm long. A female *Ascaris* is a prolific producer of eggs, sometimes generating over 200,000 per day. The eggs are fertilized and passed to the soil in the feces, where they can remain viable for several months. Unfortunately, in many parts of the world, human feces, called "night-soil," are used as fertilizer for crops, which adds to the spread of the parasite. Contact with contaminated fingers and consumption of water containing soil runoff are other possible modes of transmission.

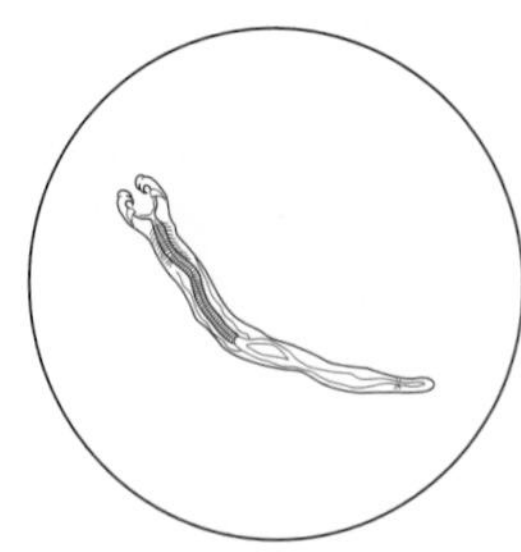

Necator americanus

When ingested in contaminated food, each egg releases a larva that grows but does not multiply in the small intestine. Abdominal symptoms develop as the worms reach maturity in about 2 months. Intestinal blockage may be a consequence when tightly compacted masses of worms accumulate, and perforation of the small intestine is possible. In addition, roundworm larvae may pass to the blood and infect the lungs, causing pneumonia. If the larvae are coughed up and then swallowed, intestinal reinfection occurs. Mebendazole is the drug for treatment and good hygiene can prevent infection.

Hookworms are roundworms with a set of hooks or sucker devices for firm attachment to tissues of the host's upper intestine (FIGURE 18.24). Approximately 1.3 billion people around the globe are infected by hookworms. There are 150 million cases and 65,000 deaths each year from this disease.

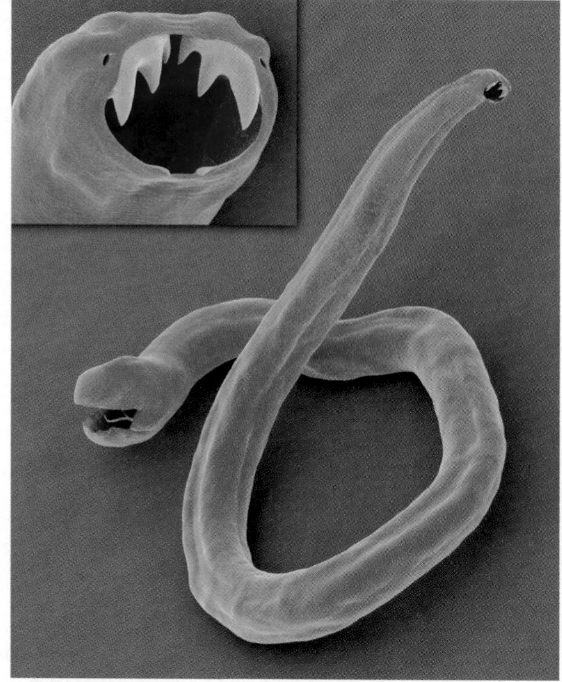

FIGURE 18.24 **A Hookworm.** False-color scanning electron micrograph of the parasitic hookworm *A. duodenale.* Inset: the head of the hookworm. © Dennis Kunkel Microscopy, Inc./Phototake/Alamy Images. »» Why are these parasites called hookworms?

Two hookworms, both about 10 mm in length, may be involved in human disease. The first is the Old World hookworm, *Ancylostoma duodenale,* which is found in Europe, Asia, and the United States; the second is the New World hookworm, *Necator americanus,* which is prevalent in the Caribbean islands.

These parasites live in the human intestine, where they suck blood from the ruptured capillaries. **Hookworm disease** therefore is accompanied by blood loss and is generally manifested by anemia. Cysts also may become lodged in the intestinal wall, and ulcer-like symptoms may develop.

The life cycle of a hookworm involves only a single host, the human (FIGURE 18.25). Female hookworms can release 5,000 to 20,000 eggs, which are excreted to the soil and remain viable for months. Eventually, larvae emerge as long, rod-like **rhabditiform** (*rhabdo* = "rod"; *form* = "shape") larvae. These later become thread-like **filariform** (*filum* = "thread") larvae that attach themselves to vegetation in the soil. When contact with bare feet is made, the filariform larvae penetrate the skin layers and enter the bloodstream. Soon, they localize in the lungs and are carried up to the pharynx in secretions, and then swallowed into the intestines.

Hookworms are common where the soil is warm, wet, and contaminated with human feces and the disease is prevalent where people go barefoot. Mebendazole may be used to reduce the worm burden and the diet may be supplemented

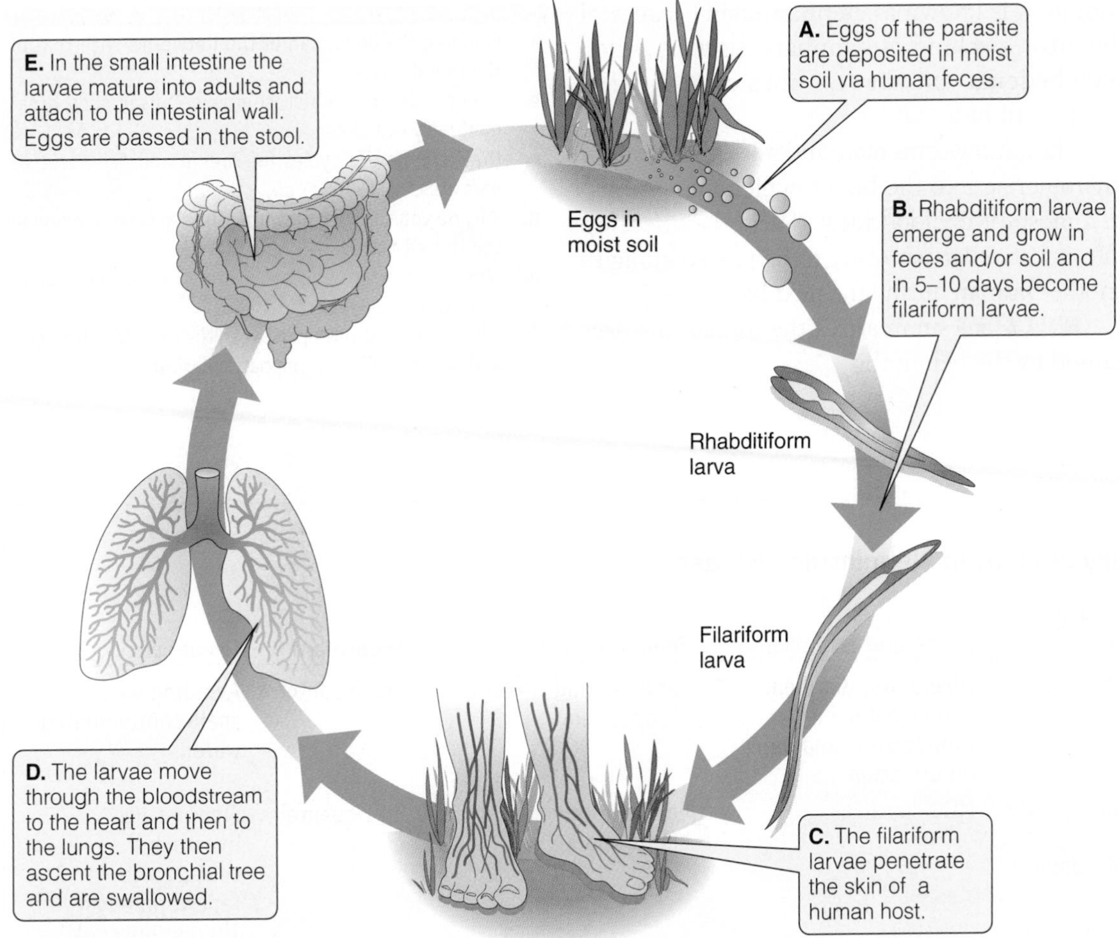

FIGURE 18.25 **The Life Cycle of the Hookworms *Ancylostoma duodenale* and *Necator americanus*.** The filariform larvae are the infective form. »» What other helminth can be coughed up and swallowed?

with iron to replace that in the blood loss. It should be noted that dogs and cats also harbor hookworm eggs and pass them in the feces.

Roundworms Also Infect the Lymphatic System

Lymphatic filariasis is a parasitic disease affecting over 130 million people in 80 countries throughout the tropics and subtropics. The disease in Africa and South America is caused by *Wuchereria bancrofti,* and in Asia by *Brugia malayi.* Both are transmitted by mosquitoes.

The female roundworm is about 100 mm long and carries larvae called **microfilariae**. Injected in a blood meal, the larvae grow to adults and infect the lymphatic system where they can survive for up to 7 years, causing extensive inflammation and damage to the lymphatic vessels and lymph glands. After years of infestation, the arms, legs, and scrotum swell enormously and become distorted with fluid (FIGURE 18.26). This

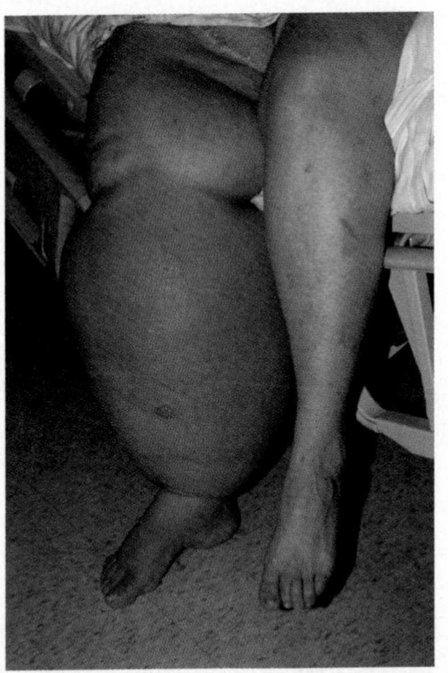

FIGURE 18.26 **Elephantiasis.** The severely swollen leg in a patient with lymphatic filariasis (elephantiasis) caused by *Wuchereria bancrofti*. © John Greim/Photo Researchers, Inc. »» Why is the disease called elephantiasis?

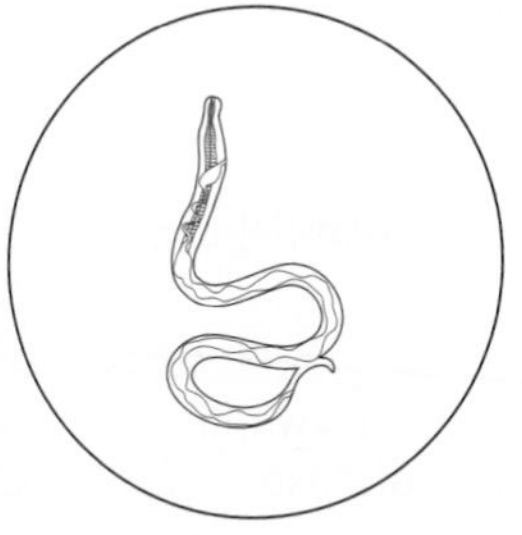

Wuchereria bancrofti

condition is known as **elephantiasis** because of the gross swelling of lymphatic tissues, called **lymphedema**, and the resemblance of the skin to elephant hide.

The adult worms mate and release millions of microfilariae into the blood, which are ingested in a mosquito's blood meal. There they develop into infective larvae, ready to be passed along to another human during the next blood meal.

TABLE 18.4 summarizes the human diseases caused by the helminths.

CONCEPT AND REASONING CHECKS 4

a. Compare the body plan of the flatworms with that of the roundworms.
b. Summarize how schistosomiasis causes such great morbidity worldwide.
c. Draw simple life cycles for *Taenia* and *Echinococcus* parasites.
d. Why do you suppose pinworm disease is so common in the United States?
e. Why are ascariasis and hookworm disease so prevalent worldwide?
f. What is the relationship between the nematode infection and limb swelling in lymphatic filariasis?

TABLE

18.4 A Summary of Human Helminthic Diseases

Disease	Causative Agent	Signs and Symptoms	Transmission	Treatment	Prevention
Schistosomiasis	*Schistosoma* species	Ulceration, diarrhea, abdominal pain Bloody urine and pain on urination	Larval spread through bloodstream	Praziquantel	Avoiding wading in snail-contaminated water
Beef tapeworm disease Pork tapeworm disease	*Taenia saginata* *Taenia solium*	Mild diarrhea	Eating poorly cooked beef or pork	Praziquantel	Eliminating livestock contact with tapeworm eggs Thoroughly cook or freeze all meat and pork
Dog tapeworm disease	*Echinococcus granulosus*	Abdominal chest pain, coughing up blood	Transmitted from domestic dogs and livestock	Praziquantel	Avoiding infected animals
Pinworm disease	*Enterobius vermicularis*	Anal or vaginal itching, irritability and restlessness, intermittent abdominal pain and nausea	Ingestion of pinworm eggs from contaminated food, drink, or hands	Albendazole	Practicing good hygiene and household cleaning
Ascariasis	*Ascaris lumbricoides*	Vague abdominal pain, nausea, vomiting, diarrhea or bloody stools	Ingestion of contaminated food	Mebendazole Albendazole	Practicing good hygiene
Trichinellosis	*Trichinella spiralis*	Intestinal pain, vomiting nausea, constipation	Consumption of raw or undercooked pork	Mebendazole	Avoiding under-cooked pork or wild animal meat
Hookworm disease	*Ancylostoma duodenale* *Necator americanus*	Raised rash at skin entry site Abdominal pain, loss of appetite, diarrhea, weight loss	Larvae in soil	Mebendazole	Improving sanitation Avoiding contact with contaminated soil
Lymphatic filariasis	*Wuchereria bancrofti*	Swelling of arms, legs, scrotum	Lymphatic vessels from bite of infected mosquito	Diethyl-carbamazine Albendazole	Avoiding mosquitoes in endemic areas

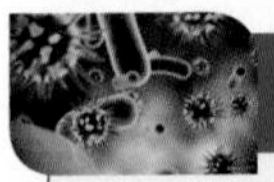

Chapter Challenge D

Helminths are anything but microorganisms. Some tapeworms are meters long and many are visible with the naked eye, making them macroscopic.

QUESTION D:

Why are the human helminth diseases discussed with other true "microscopic diseases" (e.g., viral, bacterial, and protistan diseases)?

Answers can be found on the Student Companion Website in **Appendix D.**

In conclusion, we have now finished up the chapters describing the major pathogens (bacterial, viral, fungal, protistan, and helminthic) that affect humans. One of medicine's goals has been to eradicate as many of the highly dangerous and deadly diseases as humanly possible. So far, that goal has only been reached for smallpox—although in 2011 a viral disease called rinderpest (cattle plague) became the first animal disease to be eradicated. Other human diseases are being targeted by global health workers. Polio is near its end and hopefully it will be in the next few years. All these have been targeted with vaccines and regional or global vaccination programs. Others have to be eradicated in a different fashion. Take for example Guinea worm, which has been widespread across Africa, the Middle East, and parts of Asia. Infection causes paralysis and a gruesome affliction that eventually produces a searing acid causing skin blisters that feel like one has been stabbed with a red-hot needle; all this as the worms try to emerge from the skin. The disease now exists in only four African countries and there were only about 1,000 cases in 2011. The key to eradication here is clean water and the hopes are that by 2013 or 2014 Guinea worm will be eradicated. Hopefully other parasitic diseases including lymphatic filariasis and river blindness will not be far behind.

SUMMARY OF KEY CONCEPTS

18.1 Protists Exhibit Great Structural and Functional Diversity

1. Protists are a diverse group. The majority are unicellular and free-living organisms inhabiting moist areas or water. Some protists, such as the **green algae** are photosynthetic and other protists, like the **dinoflagellates, radiolarians**, and **foraminiferans**, are part of the marine **phytoplankton**. Other protists are heterotrophic and have a fungus-like structure.
2. The human parasitic protists have many lifestyles but can be cataloged into one of three super groups that contain human parasites. The **parabasalids** and **diplomonads** are motile but lack mitochondria; the **kinetoplastids** are motile, have an undulating membrane, and a mass of DNA called the **kinetoplast**; the amoebas move by means of pseudopods and reproduce by binary fission; the **ciliates** have their cell surfaces covered by cilia and contain two different types of nuclei—**macronuclei**, which control metabolic events and **micronuclei**, which play a critical role in genetic recombination through **conjugation**; and **apicomplexans**, which are nonmotile as adults and are obligate parasites. Some parasitic protists (and most helminths) require at least two hosts to complete their life cycle, the **definitive host**, where sexual reproduction occurs, and the **intermediate host**, where asexual reproduction takes place.

18.2 Protistan Parasites Attack the Skin, and the Digestive and Urinary Tracts

- **Cutaneous leishmaniasis**
  1. *Leishmania major*
- **Visceral leishmaniasis**
  2. *Leishmania donovani*
- **Amoebiasis**
  3. *Entamoeba histolytica*
- **Giardiasis**
  4. *Giardia intestinalis*
- **Cryptosporidiosis**
  5. *Cryptosporidium parvum, C. hominis*
- **Cyclosporiasis**
  6. *Cyclospora cayetanensis*
- **Trichomoniasis** (not shown)
  Trichomonas vaginalis

18.3 Many Protistan Diseases of the Blood and Nervous System Can Be Life Threatening

- **Malaria**
  7. *Plasmodium* species
- **Trypanosomiasis (Human African sleeping sickness)**
  8. *Trypanosoma brucei*
- **Trypanosomiasis (American trypanosomiasis; Chagas disease)** (not shown)
  Trypanosomiasis cruzi
- **Babesiosis** (not shown)
  Babesia microti
- **Toxoplasmosis**
  9. *Toxoplasma gondii*
- **Primary amoebic meningoencephalitis**
  10. *Naegleria fowleri*

18.4 Parasitic Helminths Cause Substantial Morbidity Worldwide

Among the **helminths**, there are two groups responsible for parasitic diseases in humans—**flatworms** and **roundworms.** Flatworms are multicellular animals lacking respiratory and circulatory structures, and having a gastrovascular cavity. **Tapeworms** can infect all mammals and generally are transmitted to humans in foods. Roundworms (nematodes) are among the most prevalent animals worldwide. They are multicellular with separate sexes. Disease usually is the result of large worm burdens within the individual.

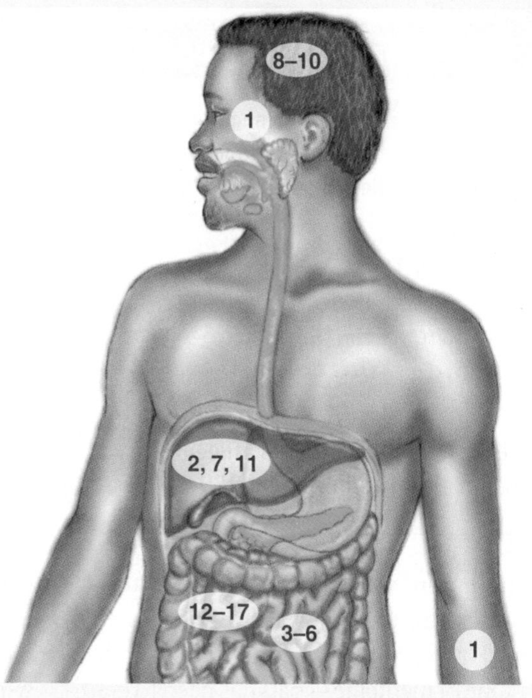

- **Schistosomiasis**
  11. *Schistosoma mansoni*
- **Beef/Pork Tapeworm disease**
  12. *Taenia saginata, T. solium*
- **Echinococcosis**
  13. *Echinococcus granulosus*
- **Pinworm disease**
  14. *Enterobius vermicularis*
- **Trichinellosis**
  15. *Trichinella spiralis*
- **Ascariasis**
  16. *Ascaris lumbricoides*
- **Hookworm disease**
  17. *Ancylostoma duodenale, Necator americanus*
- **Filariasis** (not shown)
  Wuchereria bancrofti, Brugia malayi

CHAPTER SELF-TEST

For **STEPS A–D**, answers to even-numbered questions and problems can be found in **Appendix C** on the Student Companion Website at **http://microbiology.jbpub.com/10e**. In addition, the site features eLearning, an online review area that provides quizzes and other tools to help you study for your class. You can also follow useful links for in-depth information, read more MicroFocus stories, or just find out the latest microbiology news.

STEP A: REVIEW OF FACTS AND TERMS

Multiple Choice

Read each question carefully, then select the ***one*** answer that best fits the question or statement.

1. The _____ are members of the _____.
   A. green algae; dinoflagellates
   B. dinoflagellates; phytoplankton
   C. radiolarians; fungus-like protists
   D. ciliates; photosynthetic protists
2. This group of protists has a single mitochondrion with a mass of DNA.
   A. Kinetoplastids
   B. Ciliates
   C. Apicomplexans
   D. Diplomonads
3. The _____ are a group of protists that have food vacuoles and pseudopods.
   A. amoebas
   B. parabasalids
   C. kinetoplastids
   D. ciliates
4. Which one of the following is NOT found in the ciliates?
   A. A contractile vacuole
   B. Macronuclei and micronuclei
   C. A complex of organelles in the tip
   D. Mitochondria
5. An intermediate host is _____.
   A. where parasite asexual cycle occurs
   B. always a nonhuman host
   C. where parasite sexual cycle occurs
   D. the human host between two other animal hosts
6. The vector transmitting leishmaniasis is the _____.
   A. mosquito
   B. sand fly
   C. tsetse fly
   D. sand flea
7. _____ enters the human body as a cyst and develops into a trophozoite in the small intestine; a severe form of dysentery may occur.
   A. *Cryptosporidium parvum*
   B. *Entamoeba histolytica*
   C. *Giardia intestinalis*
   D. *Cyclospora cayetanensis*
8. Sucker-like devices allow this protist to adhere to the intestinal lining.
   A. *Cyclospora cayetanensis*
   B. *Entamoeba histolytica*
   C. *Giardia intestinalis*
   D. *Cryptosporidium parvum*
9. This disease sickened more than 400,000 residents of Milwaukee in 1993.
   A. Giardiasis
   B. Cryptosporidiosis
   C. Trypanosomiasis
   D. Cyclosporiasis
10. This genus of parabasalid affects over 7.4 million Americans annually and is considered a sexually transmitted disease.
    A. *Toxoplasma*
    B. *Cryptosporidium*
    C. *Cyclospora*
    D. *Trichomonas*
11. The _____ form of the malarial parasite enters the human blood while the _____ enters the mosquito.
    A. sporozoites; merozoites
    B. merozoites; gametocytes
    C. merozoites; sporozoites
    D. sporozoites; gametocytes
12. Chagas disease is caused by _____.
    A. *Trypanosome cruzi*
    B. *Toxoplasma gondii*
    C. *Babesia microti*
    D. *Trypanosoma brucei*
13. Babesiosis is carried by _____ and infects _____.
    A. mosquitoes; RBCs
    B. fleas; the kidneys
    C. ticks; RBCs
    D. mosquitoes; the liver
14. This opportunistic protist causes primary meningoencephalitis (PAM).
    A. *Plasmodium vivax*
    B. *Toxoplasma gondii*
    C. *Naegleria fowleri*
    D. *Paragonimus westermani*

Matching

Read the statement concerning the helminths, and then select the answer or answers that best apply to the statement. Place the letter(s) next to the statement.

___ **15.** Transmitted by an arthropod.
   A. Filariasis
   B. Trichinosis
   C. Hookworm disease

___ **16.** Beef tapeworm species.
   A. *Echinococcus granulosus*
   B. *Schistosoma mansoni*
   C. *Taenia saginata*

___ **17.** Type of fluke.
   A. *Schistosoma*
   B. *Necator*
   C. *Echinococcus*

___ **18.** Cause of pinworm disease.
   A. *Trichinella*
   B. *Enterobius*
   C. *Ascaris*

___ **19.** Infects the human intestines.
   A. *Trichinella spiralis*
   B. *Ascaris lumbricoides*
   C. *Taenia saginata*

___ **20.** Type of tapeworm.
   A. *Taenia*
   B. *Echinococcus*
   C. *Necator*

___ **21.** Snail is the intermediate host.
   A. Blood fluke
   B. *Echinococcus*
   C. Intestinal fluke

___ **22.** Attaches to host tissue by hooks.
   A. *Necator*
   B. *Trichinella*
   C. *Enterobius*

___ **23.** Affects pigs as well as humans.
A. *Wuchereria*
B. *Trichinella*
C. *Schistosoma*

___ **24.** Life cycle includes miracidia and cercaria.
A. *Schistosoma*
B. *Echinococcus*
C. *Ascaris*

___ **25.** Acquired by consuming contaminated pork.
A. *Taenia solium*
B. *Echinococcus*
C. *Necator americanus*

___ **26.** Classified in the phylum Nematoda.
A. *Ascaris*
B. *Wuchereria*
C. *Schistosoma*

___ **27.** Causes inflammation and damage to the lymphatic vessels.
A. *Echinococcus*
B. *Ascaris*
C. *Wuchereria*

___ **28.** Male and female forms exist.
A. *Taenia*
B. *Ascaris*
C. *Schistosoma*

___ **29.** Forms hydatid cysts.
A. Blood fluke
B. *Enterobius*
C. *Echinococcus*

Label Identification

30. Label the two parasite life cycles for (1) malaria and (2) schistosomiasis, using the term lists provided.

(1) *Plasmodium* Life Cycle Term List

Definitive host
Gametes
Gametocyte
Intermediate host
Merozoites
Oocyst
Sporozoites
Zygote

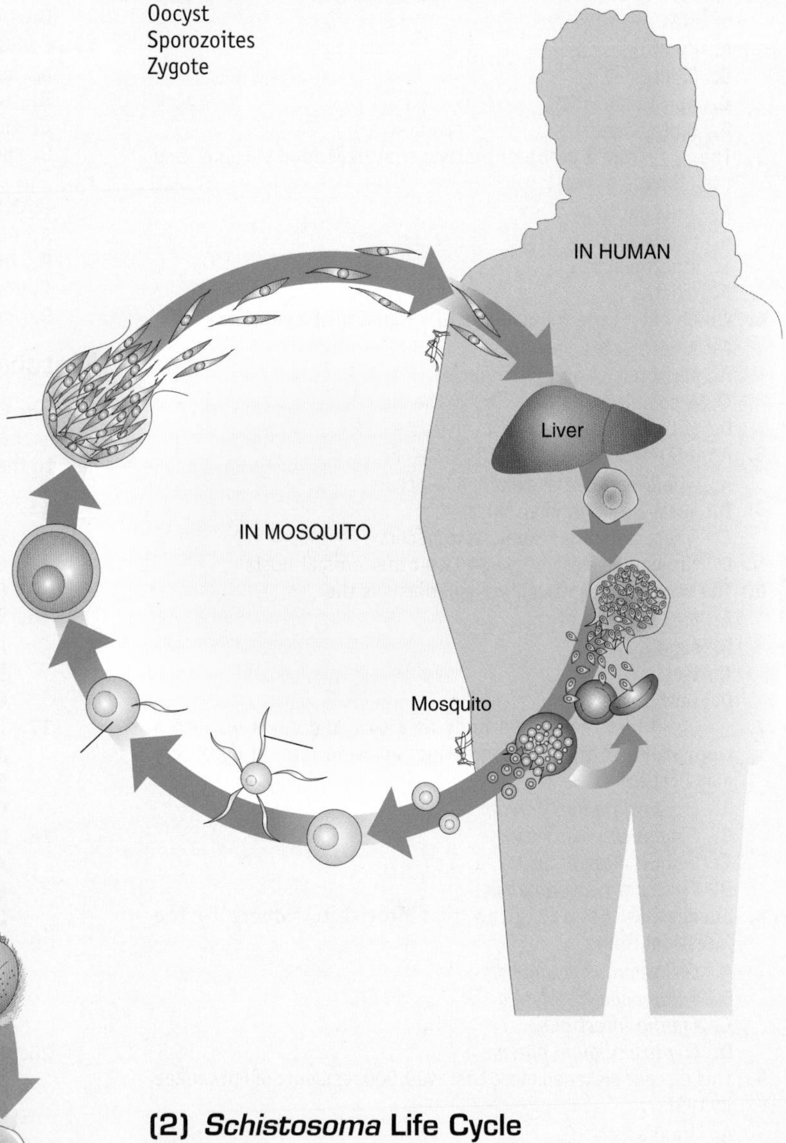

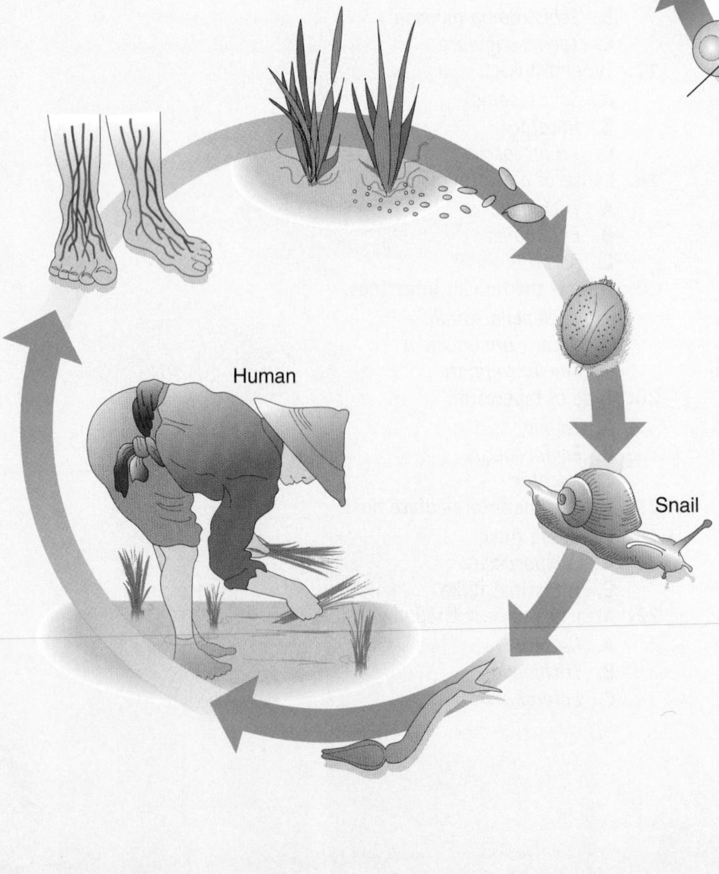

(2) *Schistosoma* Life Cycle Term List

Cercaria
Definitive host
Eggs
Intermediate host
Miracidium

STEP B: CONCEPT REVIEW

31. List the characteristics of the **protista**, including those of the **green algae** and members of the **phytoplankton**. (**Key Concept 1**)
32. Identify the characteristics of the **parabasalids, diplomonads,** and **kinetoplastids.** (**Key Concept 1**)
33. Describe the mechanism of locomotion and food capture by the **Unikonta.** (**Key Concept 1**)
34. Explain how **ciliates** move in a watery environment. (**Key Concept 1**)
35. Differentiate between a **definitive host** and an **intermediate host.** (**Key Concept 1**)
36. Describe the mode of transmission and forms of **leishmaniasis.** (**Key Concept 2**)
37. Summarize the characteristics of **amoebiasis, giardiasis, cryptosporidiosis,** and **cyclosporiasis** as human diseases. (**Key Concept 2**)
38. Explain how **trichomoniasis** is sexually transmitted. (**Key Concept 2**)
39. Construct a simple life cycle for *Plasmodium*. (**Key Concept 3**)
40. Compare and contrast **human African sleeping sickness** and **Chagas disease.** (**Key Concept 3**)
41. Infer why **babesiosis** is a malaria-like disease. (**Key Concept 3**)
42. Identify the role of **trophozoites, cysts,** and **oocysts** to the infection cycle of *Toxoplasma*. (**Key Concept 3**)
43. Name the agent and summarize the characteristics of **primary amoebic meningoencephalitis** (**PAM**). (**Key Concept 3**)
44. Identify the two groups of **helminths** of medical importance and summarize their characteristics. (**Key Concept 4**)
45. Explain the life cycle and infection process of *Schistosoma*. (**Key Concept 4**)
46. Compare and contrast **beef** and **pork tapeworm diseases.** (**Key Concept 4**)
47. Assess the significance of **pinworm disease** and **trichinellosis** in the United States, and describe their infectious cycles. (**Key Concept 4**)
48. Describe the soil-transmitted helminthic diseases, **ascariasis** and **hookworm disease.** (**Key Concept 4**)
49. Discuss how *Wuchereria bancrofti* causes **filariasis** (**elephantiasis**). (**Key Concept 4**)

STEP C: APPLICATIONS AND PROBLEMS

50. You and a friend who is 3 months pregnant stop at a hamburger stand for lunch. Based on your knowledge of toxoplasmosis, what helpful advice can you give your friend? On returning to her home, you notice she has two cats. What additional information might you share with her?
51. Cardiologists at a local hospital hypothesized that a few patients with a certain protozoal disease easily could be lost among the far larger population of heart disease sufferers patronizing county clinics. They proved their theory by finding 25 patients with this protozoal disease among patients previously diagnosed as having coronary heart disease. Which protozoal disease was involved?
52. Federal law stipulates that food scraps fed to pigs must be cooked to kill any parasites present. It also is known that feedlots for swine are generally more sanitary than they have been in the past. As a result of these and other measures, the incidence of trichinellosis in the United States has declined, and the acceptance of "pink pork" has increased. Do you think this is a dangerous situation? Why?

STEP D: QUESTIONS FOR THOUGHT AND DISCUSSION

53. A newspaper article written in the 1980s asserted that parasitology is a "subject of low priority in medical schools because the diseases are exotic infections only occurring in remote parts of the world." Would you agree with that statement today? Explain.

54. Many historians believe malaria contributed to the downfall of ancient Rome. Over the decades, malaria incidence increased with expansion of the Roman Empire, which stretched from the Sahara desert to the borders of Scotland, and from the Persian Gulf to the western shores of Portugal. How do you suspect the disease and the expansion are connected?

55. It has been said that until recent times, many victims of a particular protozoal disease were buried alive because their life processes had slowed to the point where they could not be detected with the primitive technology available. Which disease was probably present?

56. The World Health Organization has reported that, after malaria, schistosomiasis and filariasis (200 million and 90 million annual cases, respectively) are the most prevalent tropical diseases. How do you believe the incidence of these diseases can be reduced on a global scale?

57. Some restaurants offer a menu item called steak tartare, which is a dish served with raw ground beef. What hazard might this meal present to the restaurant patron?

58. Justify this statement: "Perhaps the most important reason for discussing parasitical diseases is that they highlight just how enmeshed we are in the web of life."

59. Many of the parasitic diseases described in this chapter, including schistosomiasis, ascariasis, filariasis, trypanosomiasis, leishmaniasis, and some bacterial diseases mentioned in previous chapters, such as trachoma and leprosy, are often considered "neglected diseases." Why do you think these diseases have been somewhat "neglected" by the medical community, especially in the developed world?

False-color scanning electron micrograph of human immunodeficiency viruses (green) associated with human lymphocytes (red). Courtesy of C. Goldsmith, P. Feorino, E. L. Palmer, W. R. McManus/CDC.

Interactions and Impact of Microorganisms with Humans

PART 5

CHAPTER 19 Infection and Disease
CHAPTER 20 Resistance and the Immune System: Innate Immunity
CHAPTER 21 Resistance and the Immune System: Adaptive Immunity
CHAPTER 22 Immunity and Serology
CHAPTER 23 Immune Disorders and AIDS
CHAPTER 24 Antimicrobial Drugs

In past centuries, the spread of disease appeared to be willfully erratic. Illnesses would attack some members of a population while leaving others untouched. A disease that for many generations had taken small, steady tolls would suddenly flare up in epidemic proportions. And strange, horrifying plagues descended unexpectedly on whole nations.

Scientists now know humans live in a sometimes precarious equilibrium with the microorganisms surrounding them. Generally, the relationship is harmonious as most are nonpathogens and humans can develop resistance to most pathogens they encounter. However, when the natural resistance is unable to overcome the aggressiveness of some pathogens, disease sets in. In other instances, the resistance is diminished by a pattern of human life that gives microorganisms the edge. For example, during the Industrial Revolution of the 1800s, many thousands of Europeans moved from rural areas to the cities. They sought new jobs, adventure, and prosperity. Instead, they found endless labor, unventilated factories, and wretched living conditions—and they found disease.

In Part 5 of this text, we shall explore the infectious disease process and the mechanisms by which the body responds to disease. Chapter 19 opens with an overview of the host–microbe relationship and the factors contributing to the establishment of disease. Epidemiology and diseases within populations also will be explored. In Chapters 20 and 21, the discussion turns to nonspecific and specific methods by which body resistance develops, with emphasis on the immune system. Various types of immunity are explored in Chapter 22, together with a survey of laboratory methods using the immune reaction in the diagnosis of disease. In Chapter 23 the discussion centers on immune disorders leading to serious problems in humans. This includes an extensive discussion of AIDS. Finally, in Chapter 24, we move to treating the patient by discussing antimicrobial drugs, including antibiotics. In these chapters, we uncover the roots of infectious disease and resistance, and come to understand the interactions and impact that microorganisms and viruses have on humans at the fundamental level.

MICROBIOLOGY PATHWAYS

Epidemiology

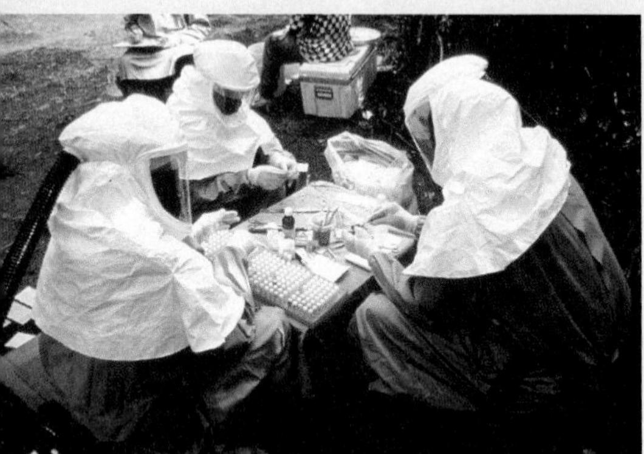

Courtesy of Ethleen Lloyd/CDC.

Flying to an impoverished African country on your second day of work to battle Ebola, one of the most deadly viruses, isn't most people's idea of a dream assignment. But it was for Marta Guerra. In fact, the trip to Uganda in 2000 was the assignment she had been coveting. "I wasn't that worried," Guerra says. "This particular strain has only a 65% death rate instead of the Congo strain which is 85%."

Guerra is a disease epidemiologist, popularly known as a "disease detective," with the Epidemic Intelligence Service (EIS) of the Centers for Disease Control and Prevention (CDC) in Atlanta.

Growing up in the multicultural environments of Havana, Cuba, and Washington, D.C., Marta Guerra developed a keen interest in other cultures and teaching people about health risks. She was fascinated by stories her professors told about working overseas. After seeing the movie *Outbreak,* "I thought, that's what I want to do—help contain a deadly epidemic," recalls Guerra. So she obtained a Master's in public health and a Ph.D. in tropical medicine.

Like all EIS officers, Guerra's job is to isolate the cause of an outbreak, prevent its spread, and get out public health messages to people who could have been exposed. When Guerra flew to Uganda in November 2000, the Ebola outbreak had already been identified, so her task was to go from village to village, trying to locate family members and friends, and educate them about symptoms and treatment. "In every corner of Africa people know the word Ebola, and they are terrified of it," Guerra explains. "Sometimes they hide sick family members; sometimes they're frightened of survivors."

The job of a disease detective can be difficult—and dangerous. Although Guerra seldom considered she might acquire a disease she was investigating, she was concerned about political violence. In Uganda, Guerra's team needed military escorts on their travels through villages. In Ethiopia, while on a polio eradication mission in the summer of 2001, she recalls, "There was rebel activity in all the areas we traveled through—plus land mines. It was pretty scary."

Perhaps you might be interested in a similar, and sometimes adventurous, career. "You have to be highly motivated, with the ability to think fast on your feet and make quick decisions. You have to be able to walk into a chaotic situation and deal with whatever is thrown at you," Guerra says. "Sometimes I barely drop my bags at home before I'm called out again. Being adaptable is really essential."

To get started, you need a bachelor's degree in a biological science. In addition to required courses in chemistry and biology, undergraduates should study microbiology, mathematics, and computer science. A master of science in epidemiology or public health also is required; many have a Ph.D. or a medical or veterinary degree. Most American disease epidemiologists then apply to EIS's two-year, post-graduate program of service and on-the-job training, where they work with mentors like Marta. She says, "I like the fact that I am contributing to science in the sense that what I do will affect people far into the future."

(Essay modified from *Disease Detective* by Carol Sonenklar in www.MedHunters.com)

Infection and Disease

"Health care matters to all of us some of the time, public health matters to all of us all of the time"
—C. Everett Koop, former Surgeon General of the United States

By late July 1999, crows were literally dropping out of the sky in New York City and dead crows were found in surrounding areas as well. By early September, officials at the Bronx Zoo discovered other birds, including a cormorant, two red Chilean flamingoes, and an Asian pheasant, at the zoo had died of the same brain and heart inflammations as found in the crows.

On August 23, 1999, an infectious disease physician at a hospital in northern Queens reported to the New York City Department of Health that two patients had been admitted with encephalitis. In fact, on further investigation, the health department identified a cluster of six patients with encephalitis. Testing by the Centers for Disease Control and Prevention (CDC) of these initial cases for antibodies to common North American **arboviruses**—viruses transmitted by arthropods, such as insects—was positive for St. Louis encephalitis (SLE) virus, which is carried by mosquitoes. These findings prompted the health department to begin aerial and ground application of insecticides.

News of SLE and spraying caught the attention of the Bronx Zoo officials. If the birds were dying from the same encephalitis disease as in humans, it could not be caused by the SLE virus because birds do not contract SLE.

So, a reinvestigation by the CDC of virus samples taken from humans, birds, and mosquitoes was undertaken. Results indicated all viral isolates were closely related to West Nile virus (WNV), which had never been isolated in the western hemisphere. It soon became evident that, indeed, the disease in birds and humans was caused by WNV.

19

CHAPTER PREVIEW

19.1 The Host and Microbe Form an Intimate Relationship in Health and Disease

19.2 Establishment of Infection and Disease Involves Host and Pathogen

Investigating the Microbial World 19: Can a Yogurt a Day Keep the Doctor Away?

19.3 Infectious Disease Epidemiology Includes Frequency and Spread of Disease

MicroInquiry 19: Epidemiological Investigations

Textbook Case 19: Locally Acquired Dengue Fever

Image courtesy of Dr. Fred Murphy/CDC.

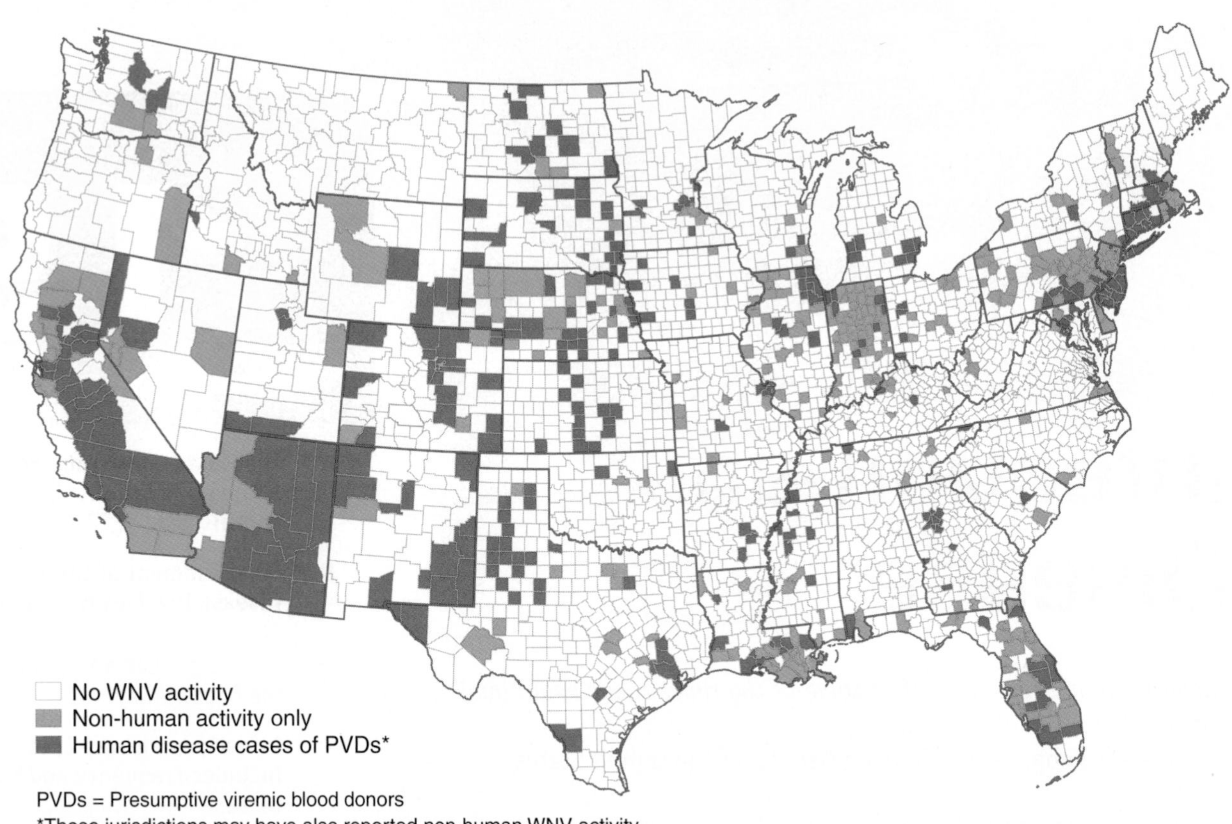

FIGURE 19.1 **West Nile Virus Activity, United States—2010.** The map indicates the geographical distribution of nonhuman activity and human infections by county in 2010 (a county is shaded no matter in what area of the county the disease case(s) occurred). Source: CDC. »» On the map, what is meant by nonhuman activity?

By early fall, mosquito activity waned and the number of human cases declined. In all, 61 people would be infected and 7 would die. Although New York City and the surroundings could breathe a sigh of relief, the WNV outbreak was only the beginning of a virus march across the continental United States. Since 2006, human cases of West Nile encephalitis have been reported across America (FIGURE 19.1).

The outbreak of West Nile encephalitis is one example describing the epidemiology of infection and disease; that is, the scientific (and medical) study of the causes, transmission, and prevention of disease within a population. Modern epidemiology can trace its origins back to John Snow, the English surgeon who studied a cholera outbreak in London in 1854.

Today, the CDC, the World Health Organization (WHO), and other agencies throughout the world have built on the historical work of John Snow, Ignaz Semmelweis, Joseph Lister, Louis Pasteur, Robert Koch, and many others. Yet, the epidemiologists employed by these organizations seek the same goals as John Snow: to identify and investigate disease outbreaks; conduct research to enhance prevention; and then devise prevention strategies.

In this chapter, we discuss the mechanisms underlying the spread and development of infectious disease. Our purpose is to bring together many concepts of disease and synthesize an overview of the host–microbe relationship. We summarize much of the important terminology used in medical microbiology and outline some of the factors used by microorganisms to establish themselves in tissues. An understanding of the topics concerning the host–microbe relationship will be essential preparation for the detailed discussion of host resistance (immune) mechanisms.

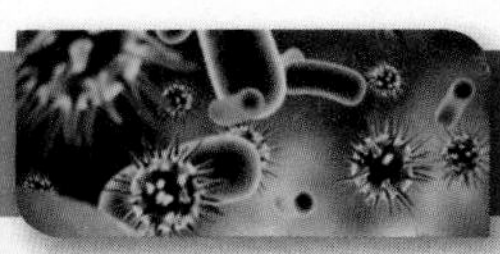

Chapter Challenge

Patients, especially older adults, receiving medical care can contract serious, sometimes life-threatening infections called healthcare-associated infections (HAIs). One very serious HAI is caused by *Clostridium difficile* (commonly called *C. diff*) that produces a spectrum of lower gastrointestinal disease. In mild cases, this gram-positive, anaerobic, spore-forming rod causes watery diarrhea three or more times a day for two or more days and mild abdominal cramping and tenderness. In severe cases, *C. diff* causes colitis (an inflamed colon) or pseudomembranous colitis (patches of raw tissue that can bleed or produce pus). Because the bacterial cells are found in the feces, poor hygiene can result in items or surfaces being contaminated with *C. diff*. Touching such items and then touching the mouth or mucous membranes can lead to *C. diff*-associated infection (CDAI). Healthcare workers can spread the bacterial cells to patients or contaminate surfaces through hand contact. In fact, severe cases of *C. diff* infections account for about 14,000 American deaths each year. So, what is it about *C. diff* that makes it a "poster child" for infection and disease, the subject of this chapter? Let's find out!

KEY CONCEPT 19.1 The Host and Microbe Form an Intimate Relationship in Health and Disease

By the early 1970s, the Surgeon General of the United States claimed we could "close the books on infectious diseases." The development and use of antibiotics, and vaccines, would make the threat of infectious disease of little consequence. However, antibiotic resistance and new emerging diseases, including Legionnaires' disease, AIDS, Lyme disease, hantavirus pulmonary syndrome, and SARS, have thwarted such optimism. In 2010, of the approximately 57 million humans who died worldwide, more than 25% (15 million) died from infectious diseases, making them the second leading cause of death (behind cardiovascular disease) (FIGURE 19.2). In fact, infectious diseases are the leading cause of death in children under 5 years of age.

The Human Body Maintains a Symbiosis with Its Microbiota

Infection refers to the multiplication of a microbe in a host and, in the process, produces inflammation in the infected tissue. (Note: in this chapter, for simplicity of discussion, "microbe" includes the viruses.) A host whose resistance is strong remains healthy, and the microbe is either driven from or assumes a transient relationship with the host. By contrast, if the infection leads to tissue or organ damage or dysfunction, disease develops. The term **disease** therefore refers to any change from the general state of good health. It is important to note that disease and infection are not synonymous; a person may be infected without suffering a disease.

Whether host or microbe gets the upper hand often is due in part to the estimated 100 trillion microbes found on and in the human body. This

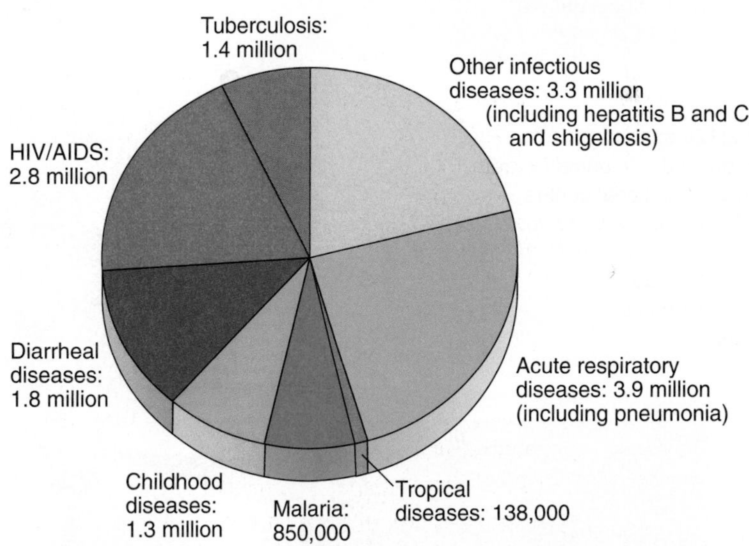

FIGURE 19.2 **Infectious Disease Deaths Worldwide.** This pie chart depicts the leading causes of infectious diseases and the number of worldwide deaths as reported by the World Health Organization. Tropical diseases: African trypanosomiasis, Chagas disease, schistosomiasis, leishmaniasis, filariasis, and onchocerciasis. Childhood diseases: diphtheria, measles, pertussis, polio, and tetanus. »» Only one of the "pie slices" has grown explosively in mortality numbers since 1993. Identify the disease and explain why that is so.

remarkable number amounts to almost 1.5 kg of human weight and represents the so-called human **microbiome**, the population of microorganisms (99% bacterial) normally residing in the body without directly causing disease. Some, referred to a the **indigenous microbiota**, establish a permanent relationship with various parts of the body, while others, the **transient microbiota**, are more temporary and found only for limited periods of time. In the large intestine of humans, for example, *Escherichia coli* and *Candida albicans* are almost always found, but species of *Streptococcus* are transient.

Our microbiota resides on several body tissues (FIGURE 19.3). These include the skin, the external ears and eyes, and upper respiratory tract. Most of the digestive tract, from oral cavity to rectum, is heavily populated with different communities of indigenous microbes as are the urogenital orifices in both males and females. Most other tissues of the body remain sterile; the blood, cerebrospinal fluid, joint fluid, and internal organs, such as the kidneys, liver, muscles, bone, and brain, are sterile unless disease is in progress.

■ Sterile: Devoid of living microorganisms, viruses, and spores.

One of the important roles that our microbiota plays is to protect us from pathogen invasion. By forming a physical barrier over the epithelium of the respiratory and digestive tracts, pathogens cannot establish a foothold for colonization. In addition, the microbiota successfully competes for the available nutrients and secretes antimicrobial compounds that make the environment less hospitable for pathogens. Thus, the relationship between the body and its microbiota is an example of a **symbiosis**, or living together. If the symbiosis is beneficial to both the host and the microbe, the relationship is called **mutualism**. For example, species of *Lactobacillus* live in the female vagina and derive nutrients from the environment while producing acid to prevent the overgrowth of other organisms.

A symbiosis also can be beneficial only to the microbe and the host is unaffected, in which case the symbiosis is called **commensalism**. *E. coli*

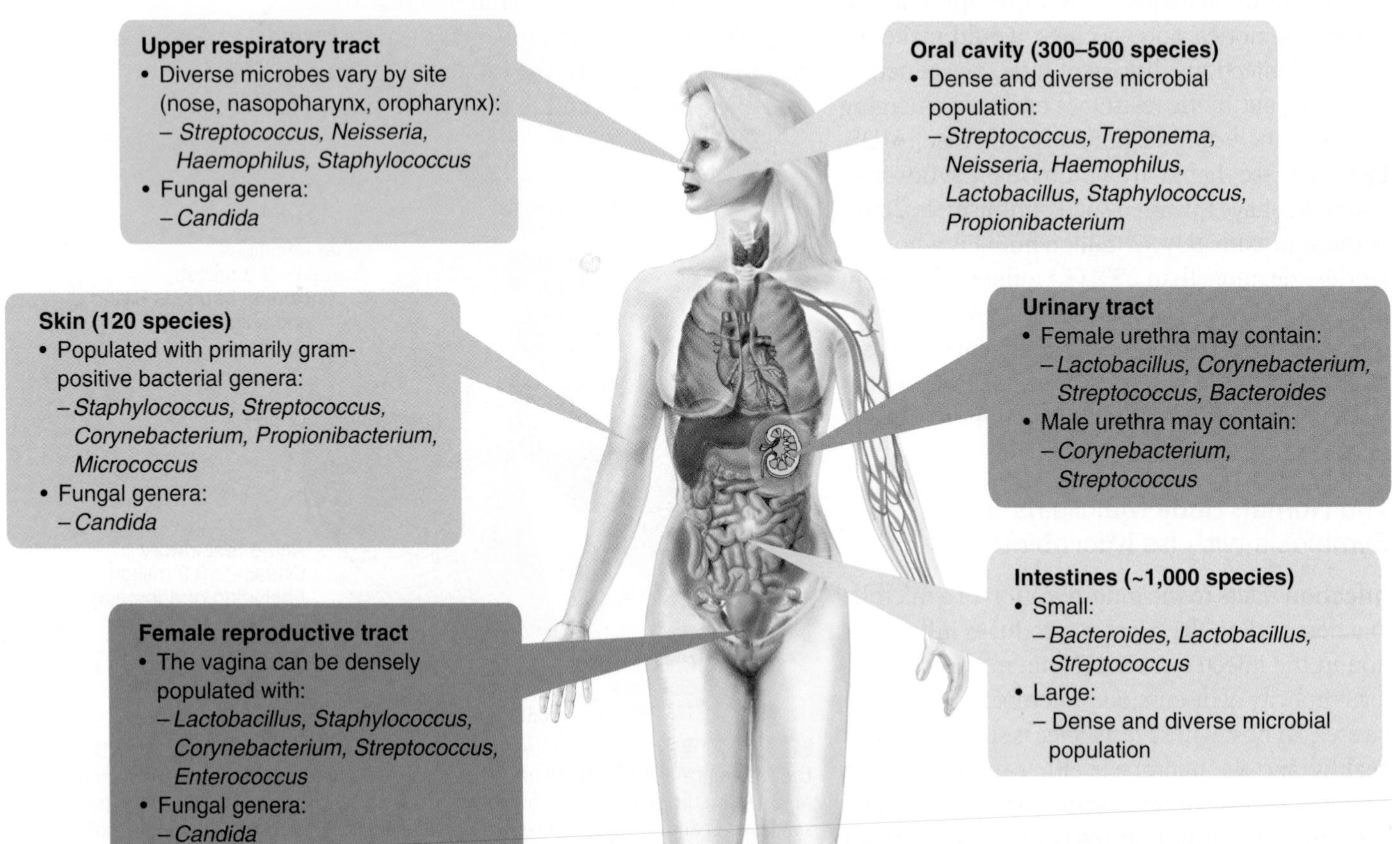

FIGURE 19.3 **A Sampling of the Indigenous Microbiota of the Human Body.** In reality there are thousands of microbial species on and within some human body systems. »» Explain why some body systems (e.g., circulatory system, nervous system) normally remain sterile?

is generally presumed to be a commensal in the human intestine, although some evidence exists for mutualism because the bacterial cells produce nonessential amounts of vitamins B and K.

Our appreciation for and understanding of the human microbiome rapidly expanded with the launching in 2007 of the "Human Microbiome Project" (HMP). The HMP used metagenomics to identify and characterize the microbial communities in healthy and diseased humans; this includes the microbiome of the oral cavity, skin, and gastrointestinal tract. Researchers also are interested in determining if all humans share a core set of microbes and whether fluctuations in the human microbiome reflect changes in human health.

Although our microbiota is capable of maintaining itself, physical, chemical, or biological injury to the resident microbes can sometimes overextend its repair abilities. Taking antibiotics for an infection is one example that can destroy and injure the gut microbiota. In some cases, this allows opportunistic pathogens, such as *Clostridium difficile,* to overrun the remaining bowel microbiota.

Recent research has suggested that ingesting living microbes can help reestablish the normal microbiota. Called **probiotics**, these products are composed of nonpathogenic, avirulent microbes similar to those normally found in the gut. Their presence in adequate amounts is thought to compete with any pathogens and help restore the normal microbiota. Although they are of questionable benefit to a healthy individual, physicians may suggest taking probiotics during or after antibiotic treatment. One of the most familiar food products that may help as a probiotic is yogurt. Those containing *Lactobacillus bulgaricus* and *Streptococcus thermophilus* are said to contain "live, active cultures." Other probiotic yogurts may have additional species, especially *Bifidobacterium lactis.* Investigating the Microbial World 19 looks at whether probiotic yogurt species affect the normal microbiota in the gut.

The Human Microbiome Begins at Birth

As we just saw, humans (and other animals too) have coevolved with a diverse set of microorganisms. However, in the womb, a healthy human fetus is sterile; there are no microbes present. Then, by adulthood, the body contains 10 times more microbial cells than human body cells, some 100 trillion microbial cells. So where did all these microbes making up the microbiota in the body come from?

Indigenous microbiota are initially introduced when the newborn passes through the birth canal or from the environment in a cesarean birth (FIGURE 19.4). Research reports state that babies born by C-section can have a markedly different microbiota on their skin, nose, mouth, and rectum than those born vaginally. Additional organisms enter upon first feeding where nursing or formula can influence what microbes colonize the newborn's gut.

■ Metagenomics: The study of the collective genomes from a group of microbial species in a specific environment (e.g., the intestine or upper respiratory tract).

During the following weeks, additional contact with the mother and other individuals will expose the infant to additional intestinal microbes. Besides the gut, the skin will be colonized by many different bacterial and fungal species. The oral cavity and upper respiratory tract will be covered with a diverse group of bacterial species while the lower urethra will be populated by bacterial and fungal organisms, as well as a few potential pathogens.

By one year, the infant's indigenous microbiota is adult-like and remains throughout life, undergoing small changes in response to the internal and external environment of the individual.

Pathogens Differ in Their Ability to Cause Disease

There also are symbiotic relationships, called **parasitism**, where the pathogen causes damage to the host and disease can result. Microbiologists once believed microbes were either pathogenic or nonpathogenic; they either caused disease or they did not. We now know that is not true.

Pathogenicity refers to the ability of a microorganism to gain entry to the host's tissues and bring about a physiological or anatomical change, resulting in altered health and leading to disease. Certain pathogens, such as the cholera, plague, and typhoid bacilli, are well known for their ability to cause serious human disease. Others, such as common cold viruses, induce milder illnesses.

Whether a disease is mild or severe depends on the pathogen's ability to do harm to a susceptible host. Thus, the degree of pathogenicity, called **virulence** (*virul* = "poisonous"), depends on the host–microbe interaction. For example, an organism invariably causing disease, such as the typhoid bacillus, is said to be "highly virulent." By comparison, an organism sometimes causing disease,

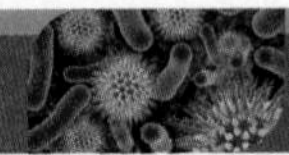

Investigating the Microbial World 19

Can a Yogurt a Day Keep the Doctor Away?

Anecdotal evidence represents nonscientific observations, usually based solely on people's personal experiences. Such stories do not provide proof yet they often are used in place of clinical or scientific evidence to support a claim. Take for example the probiotic properties of yogurt.

- **OBSERVATION:** For a long time, people have touted the health benefits of soured milk and then cultured dairy products for human digestion. In reality, the validity of such claims to human health remains unknown. But what such claims often do is stimulate research to verify or refute the claim.

- **QUESTION:** ***Can the bacterial species in probiotic yogurt affect the diverse population of normal microbes in the human gut?***

- **HYPOTHESIS:** Daily consumption of yogurt significantly alters the composition of the gut's microbiome. If so, then test subjects will show a changed bacterial composition after consuming the yogurt.

- **EXPERIMENTAL DESIGN:**

- **EXPERIMENT 1:** The fecal bacterial composition in seven healthy, adult female identical twins was analyzed before, during, and after consuming a specific brand of yogurt (with five known bacterial strains) referred to as a fermented milk product (FMP) over a 4-month period. All participants consumed two servings of FMP per day for 7 weeks.

- **EXPERIMENT 2:** The fecal bacterial composition in gnotobiotic mice was analyzed before, during, and after being inoculated with the same five bacterial strains found in the FMP. Note: gnotobiotic mice are born with germ-free guts and then "humanized" by rearing them such that their guts only contain 15 members of a typical human gut's microbiota (representing the three major bacterial phyla found in the human gut).

- **RESULTS:** In the twins experiment, consumption of the FMP did not change the bacterial species of the gut microbial community. Using the "humanized" mice, again there was no change in the composition of the 15 bacterial species and the five FMP bacterial species did not take up residence in the gut of the mice. However, in both the twins experiment and the "humanized" rat experiment, study analysis showed that consumption of the FMP or exposure to the five bacterial species in the FMP did result in a significant change in the way the gut microbiota metabolically processed carbohydrates.

- **CONCLUSION:**

  QUESTION ***1: Do these two experiments support the hypothesis? Explain.***

  QUESTION ***2: Why were the mice in experiment 2 "humanized?"***

  QUESTION ***3: Propose what might have happened if these experiments were done using individuals that had a gut microbiota that was out of balance, say with a significant* C. difficile *population.***

Answers can be found on the Student Companion Website in **Appendix D**.

Adapted from: McNulty, N. P. *et al.* (2011). *Science Translational Medicine* **3**: 1–14. Icon image © Tischenko Irina/ShutterStock, Inc.

such as *Candida albicans*, is labeled "moderately virulent." The level of virulence then depends partly on the structural and/or chemical factors, called **virulence factors**, that the microbe possesses. In general, the more factors present, the more dangerous (virulent) the microbe. Certain organisms, described as **avirulent**, are not regarded as disease agents because they usually lack the virulence factors in that host. In other words, a microbe could be virulent in one host and avirulent in another.

The level of virulence also depends on the host because host immunity through vaccination can negate virulence. Someone who has been vaccinated against smallpox renders the microbe as completely avirulent in that individual.

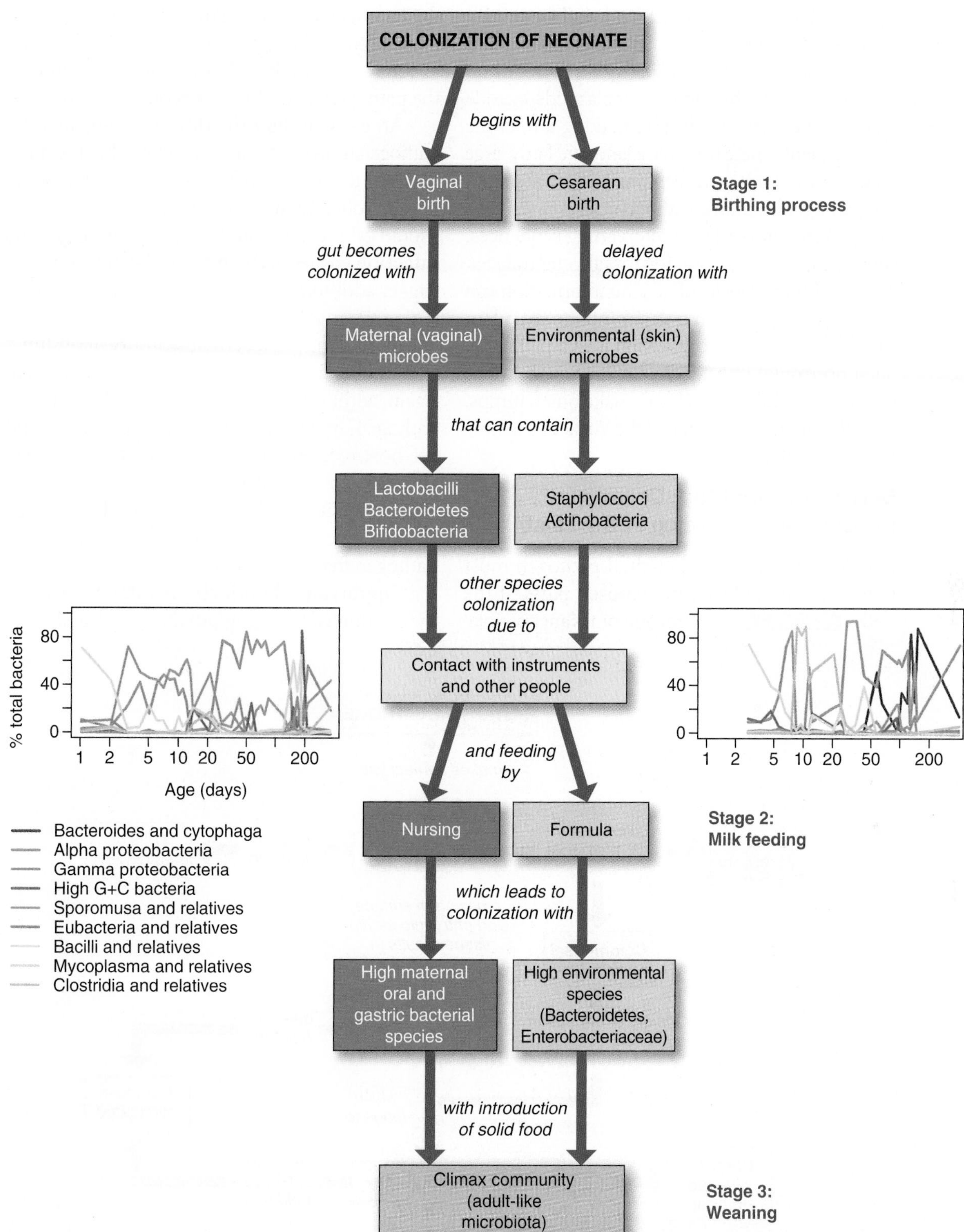

FIGURE 19.4 **Bacterial Colonization of the Neonate.** This concept map diagrams possible gut colonization of two newborns based on the three major stages of microbiota acquisition. The graphs (inserts) plot fluctuations in bacterial microbiota (left: vaginal birth; right: cesarean birth) over the first 200 days. Graphs modified from Palmer, C., et al. (2007), *PLoS Biology* Vol. 5, No. 7, el77 doi:10.1371/journal.pbio.0050177. »» How would each form of birth and feeding introduce microbiota into the newborn?

In recent years, a new term, **pathogenicity islands**, has been used to refer to clusters of genes responsible for virulence. The genes, present on the bacterial cell's chromosome or plasmids, encode many of the virulence factors making a microbe more virulent. These unstable islands are fairly large segments of a pathogen's genome and are absent in nonpathogenic strains. Pathogenicity islands have many of the properties of intervening sequences, suggesting they move by horizontal gene transfer. A copy of these blocks of genetic information can move from a pathogenic strain into an avirulent (harmless) organism, converting it to a pathogen. Such horizontal transfer processes show how the evolution of pathogenicity can make quick jumps.

We will describe some of the virulence factors later in the chapter.

Several Events Must Occur for Disease to Develop in the Host

For disease to occur, a potential pathogen must first come in contact with exposed parts of the body (FIGURE 19.5). Several outcomes are possible: the pathogen may be lost to the environment or it may colonize the normal microbiota and remain as a transient member. Depending on the nature of the pathogen, it could also become a commensal.

An **exogenous infection** is established if a pathogen from the environment breaches the host's external defenses and enters the host. Likewise, if a microbial member of the normal microbiota should gain access to sterile tissue, an **endogenous infection** ensues. In both cases, the infection may trigger additional host defenses capable of eliminating the invader.

Should the pathogen cause injury or dysfunction to host tissues, then a disease is established. Again, additional host defenses may eliminate the pathogen, in which case the disease declines and the host recovers. In other cases, the pathogen and host reach a stalemate where neither has the advantage. Tuberculosis is an example of such a chronic state. Lastly, the inability to eliminate the pathogen may lead to death.

Opportunistic infections often are caused by commensals taking advantage of a shift in

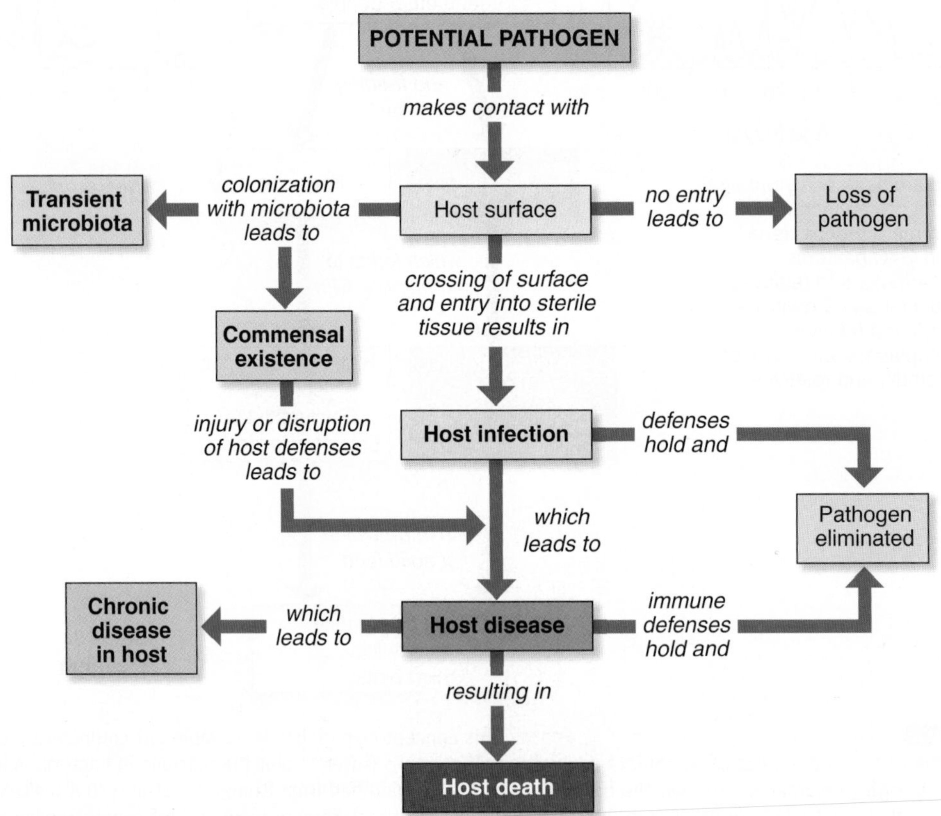

FIGURE 19.5 **The Progression and Outcomes of Infection and Disease.** A concept map illustrating possible outcomes resulting from the contact between host and pathogen. »» Propose some ways that a pathogen would gain entry into sterile tissue.

the body's delicate balance to one favoring the microbe. If the indigenous microbiota is reduced or the host's immune system is weakened, some commensals seize the "opportunity" to invade the tissues and cause disease. AIDS is an example where crippling of the immune system makes the patient highly susceptible to opportunistic organisms.

Infections may develop in one of two ways. A **primary infection** occurs in an otherwise healthy body, while a **secondary infection** develops in an individual weakened by the primary infection. In the influenza pandemic of 1918 and 1919, hundreds of millions of individuals contracted influenza as a primary infection and many developed pneumonia as a secondary infection. Numerous deaths in the pandemic were due to pneumonia's complications. Often it is the case that a secondary infection will be more serious due to the immune system being worn down by the primary infection or the microbiota being damaged.

As the term implies, **local diseases** are restricted to a single area of the body, representing the **focus of infection. Systemic diseases** are those where the focus of infection has spread via the blood to deeper organs and systems. Thus, a staphylococcal skin boil beginning as a localized skin lesion may become more serious if staphylococci spread and cause systemic disease of the bones, meninges, or heart tissue.

The transient appearance of living bacterial cells in the blood is referred to as **bacteremia. Septicemia** refers to the proliferation and spread of bacterial cells in the blood, which can be a life-threatening condition (MicroFocus 19.1). Other microbes also are disseminated. **Fungemia** refers to the spread of fungi, **viremia** to the spread of viruses, and **parasitemia** to the spread of protists and multicellular worms through the blood.

CONCEPT AND REASONING CHECKS 1

a. Identify the importance of the human microbiota in fighting infection and disease.
b. How might the route of birth affect a newborn's health?
c. Distinguish between pathogenicity and virulence.
d. Contrast exogenous, endogenous, and opportunistic infections.

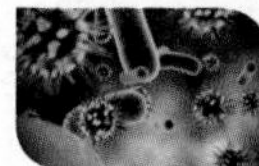

Chapter Challenge A

Most at risk of a CDAI are those who take antibiotics as part of their medical care. In fact, *C. diff* has become more resistant to many of the antibiotics normally used to treat other HAIs. Therefore, standard antibiotic treatment could be inducing the CDAI and contributing to the outbreaks that have been widely reported in hospitals and other healthcare settings.

QUESTION A:

Based on the material presented in Key Concept 19.1, suppose a healthy person carrying* C. diff *transmits the* C. diff *to a patient on antibiotic therapy. The healthy person does not get ill but the patient soon contracts a CDAI. How can that be if the patient is on antibiotics? In addition, what have the antibiotics done to the patient's gut that gave the pathogen an even easier passage into the intestinal cells for both colonizing the tissue and establishing an infection?

Answers can be found on the Student Companion Website in **Appendix D**.

KEY CONCEPT 19.2 Establishment of Infection and Disease Involves Host and Pathogen

Disease is the result of a dynamic series of events expressing the competition between host and pathogen. To overcome host defenses and bring about the anatomic or physiologic changes leading to disease, the pathogen must possess unusual abilities.

In this section, we outline the stages of disease from which we can explore the processes and factors determining whether disease can occur.

Diseases Progress Through a Series of Stages

In most instances, there is a recognizable pattern in the progress of the disease following the entry of the pathogen into the host. Often these periods are identified by **signs**, which represent evidence of disease detected by an observer (e.g., physician). Fever or bacterial cells in the blood would

MICROFOCUS 19.1: Evolution

Sepsis and Septic Shock

The presence of living, transient bacteria in the bloodstream is called **bacteremia**. Because the blood is sterile, any bacterial cells detected in the blood are cause for alarm. In most situations, however, only a small number of bacterial cells gain entry and no symptoms develop because the transients are rapidly removed by white blood cells. Such temporary bacteremia may occur in healthy individuals during dental procedures or tooth brushing, when bacterial species living on the gums around the teeth enter the bloodstream through trauma to the gums. However, in a vulnerable host, such as a person with heart valve disease, prevention of bacteremia may include taking antibiotics prior to surgery or dental procedures to prevent any bacterial cells from colonizing the heart.

If more cells enter the bloodstream than can be effectively removed, an infection will develop. A more serious, but rare condition is **septicemia** (or **sepsis**) that occurs when the bacterial cells multiply and spread throughout the bloodstream (see figure A). Often sepsis results from another infection in the body or from surgery on an infected area. In the United States, sepsis is the leading cause of death in noncoronary intensive care unit (ICU) patients, and the tenth most common cause of death overall according to data from the CDC.

Sepsis is also called **systemic inflammatory response syndrome (SIRS)** because it involves an overreaction of the host immune system to released bacterial toxins (see figure B). It is not a direct result of pathogen infection. Such a systemic response is very serious and can be life threatening. The manifestations of SIRS include two or more of the following conditions: high or low body temperature, increased heart rate, rapid breathing, and an elevated leukocyte count. In addition, some bacterial proteins cause blood clotting. In attempting to dissolve the clot, overactive neutrophils release a chemical that over dilates the blood vessels, causing them to become leaky so that fluid is lost from the blood into the surrounding tissues. Leakage and swelling also can develop in the lungs, causing difficulty breathing (respiratory distress).

Another dangerous effect of bacterial toxins is widespread clotting of the blood within the small blood vessels. This is called **disseminated intravascular coagulation (DIC)**, and can be fatal if not treated quickly. As clotting factors are used up, blood vessel leakage can lead to hemorrhaging. Often the person now has a condition called **severe sepsis**, which is associated with at least one acute organ dysfunction, decreased blood flow, or low blood pressure.

The risk of death now becomes high and requires immediate, aggressive treatment with antibiotics. A delay starting antibiotic treatment greatly decreases the person's chances of survival. Often two or three antibiotics will be given together to increase the chances of killing the bacterial cells.

Without treatment, septicemia often develops into **septic shock**, which is characterized by a dangerously low drop in blood pressure and multiple organ system failure (see figure B). In the United States, septic shock is the number one cause of death in ICUs and the 13th most common cause of death. It occurs most often in newborns and people with a weakened immune system.

The loss of fluid from the blood may be so great that the normal circulation (the rate the heart pumps at) cannot be maintained and blood pressure drops. Persistent hypotension reduces the blood flow and supply of oxygen to major organs such as the heart, kidneys, and brain. Signs of septic shock include a rapid and very weak pulse, reduced urine flow, confusion, and collapse—that is, multiple organ failure.

Septic shock is a medical emergency and is normally treated in an ICU. Large doses of antibiotics, along with infusions of fluids, are given to fight off the infection and maintain blood pressure. Drugs are given to increase blood flow to the brain, heart, and other organs. If the lungs fail, the person may need a mechanical ventilator to help breathing. Despite all efforts, some 30% of people with septic shock die.

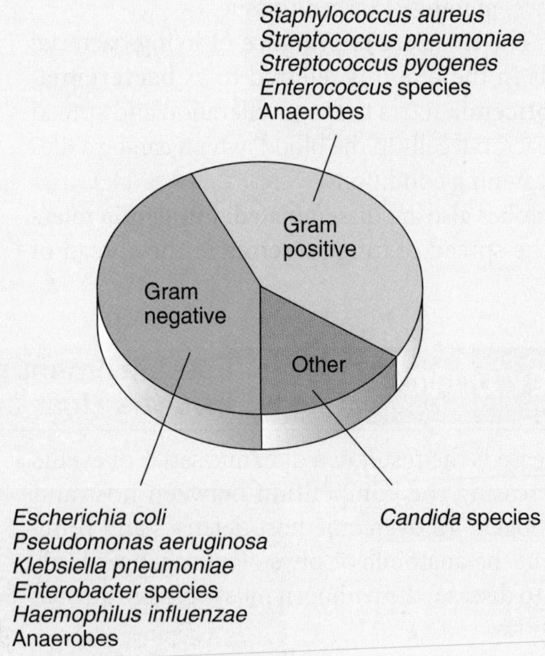

Staphylococcus aureus
Streptococcus pneumoniae
Streptococcus pyogenes
Enterococcus species
Anaerobes
Gram positive
Gram negative
Other
Escherichia coli
Pseudomonas aeruginosa
Klebsiella pneumoniae
Enterobacter species
Haemophilus influenzae
Anaerobes
Candida species

FIGURE A Microbes causing septicemia.

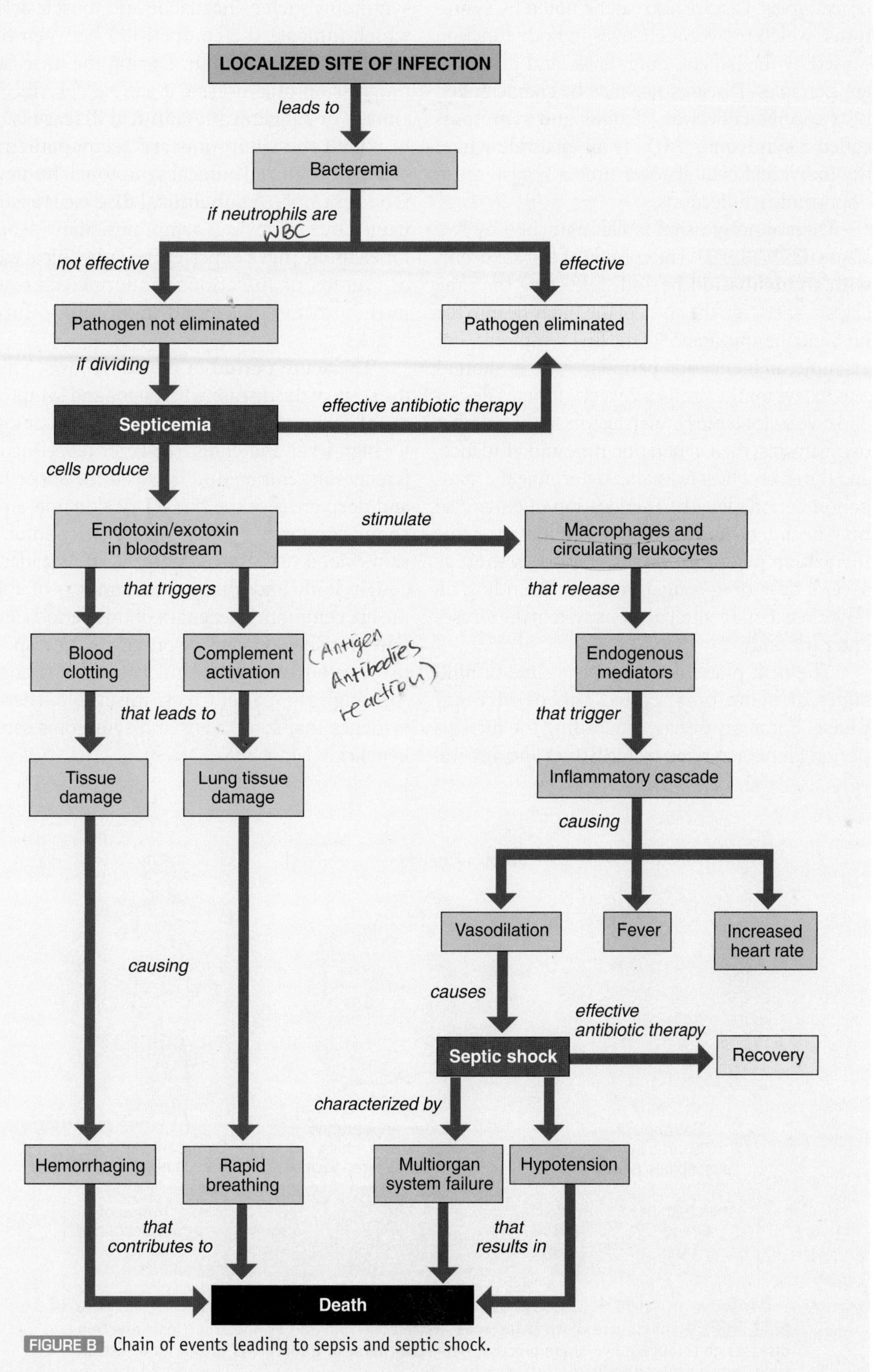

FIGURE B Chain of events leading to sepsis and septic shock.

be examples. Disease also can be noted by **symptoms**, which represent changes in body function sensed by the patient. Sore throat and headache are examples. Diseases also may be characterized by a specific collection of signs and symptoms called a **syndrome**. AIDS is an example where the individual exhibits over time a typical set of opportunistic infections.

Disease progression is distinguished by five stages (FIGURE 19.6). The episode of disease begins with an **incubation period**, reflecting the time elapsing between the entry of the microbe into the host and the appearance of the first symptoms. For example, an incubation period may be as short as 2 to 4 days for the flu; 1 to 2 weeks for measles; or 3 to 6 years for leprosy. Such factors as the number of organisms, their generation time and virulence, and the level of host resistance determine the incubation period's length. The location of entry also may be a determining factor. For instance, the incubation period for rabies may be as short as several days or as long as a year, depending on how close to the central nervous system the viruses enter the body.

The next phase in disease is a time of mild signs or symptoms, called the **prodromal phase**. For many diseases including the flu, this period is characterized by indistinct and general symptoms such as headache and muscle aches, which indicate the competition between host and microbe has begun. During the onset and progression of a disease, it can be described as clinical or subclinical. A **clinical disease** is one in which the symptoms are accompanied by clearly recognized clinical symptoms. Influenza is one example. A **subclinical disease** is accompanied by few obvious symptoms. Many people, for example, have experienced subclinical cases of mumps or infectious mononucleosis, and in the process developed immunity to future attacks.

The **acute period** or **climax** follows. This is the stage of the disease when signs and symptoms are of greatest intensity. For the flu, patients suffer high fever and chills, the latter reflecting differences in temperature between the superficial and deep areas of the body. Dry skin and a pale complexion may result from constriction of the skin's blood vessels to conserve heat. A headache, cough, body and joint aches, and loss of appetite are common. The length of this period can be quite variable, depending on the body's response to the pathogen and the virulence of the pathogen. Although the patient feels miserable, there is evidence that some signs and symptoms can be beneficial (MicroFocus 19.2).

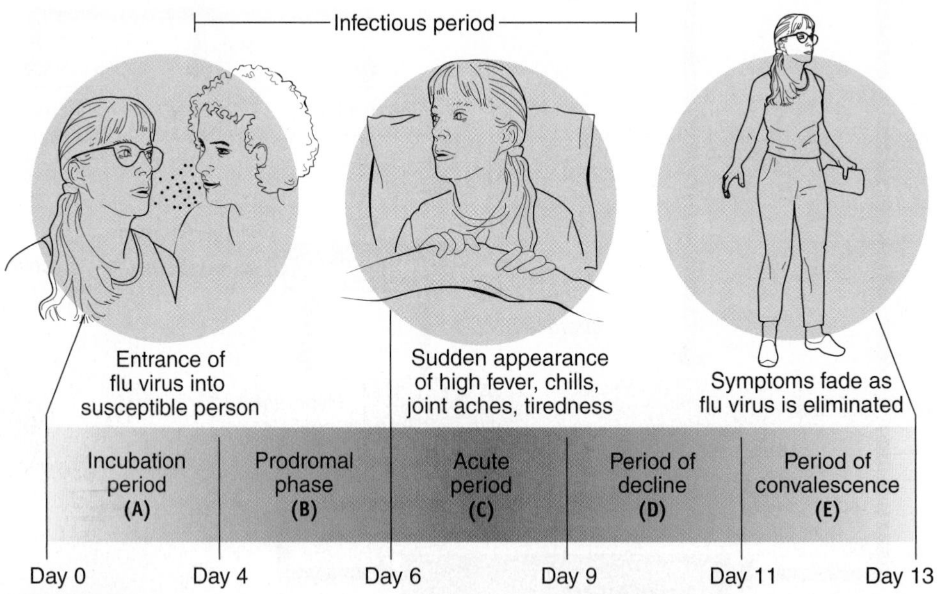

FIGURE 19.6 **The Course of an Infectious Disease, as Typified by the Flu.** Most infectious diseases go through a five-stage process. Here, the course of infection is shown for a susceptible person exposed to flu viruses. »» How would the course of the flu differ from that for a common cold?

MICROFOCUS 19.2: Public Health
Illness May Be Good for You

For most of the twentieth century, medicine's approach to infectious disease was relatively straightforward: Note the symptoms and eliminate them. However, that approach may change in the future, as Darwinian medicine gains a stronger foothold. Proponents of Darwinian medicine ask why the body has evolved its symptoms, and question whether relieving the symptoms may leave the body at greater risk.

Consider coughing, for example. In the rush to stop a cough, we may be neutralizing the body's mechanism for clearing pathogens from the respiratory tract. Nor may it be in our best interest to stifle a fever (at least a low grade fever), because fever enhances the immune response to disease. Many physicians view iron insufficiency in the blood as a symptom of disease, yet many bacterial species (e.g., tubercle bacilli) require this element, and as long as iron is sequestered out of the blood in the liver, the bacterial cells cannot grow well. Even diarrhea can be useful—it helps propel pathogens from the intestine and assists the elimination of the toxins responsible for the illness.

Darwinian biologists point out that disease symptoms have evolved over the vast expanse of time and probably have other benefits waiting to be understood. They are not suggesting a major change in how doctors treat their patients, but they are pushing for more studies on whether symptoms are part of the body's natural defenses. So, don't throw out the Nyquil®, Tylenol®, or Imodium® quite yet.

As the signs and symptoms begin to subside, the host enters a **period of decline**. Sweating may be common as the body releases excessive amounts of heat and the normal skin color soon returns as the blood vessels dilate. The sequence comes to a conclusion after the body passes through a **period of convalescence**. During this time, the body's systems return to normal.

When studying the course of a disease, it often can be defined by its severity or duration. An **acute disease**, like the flu, develops rapidly, is usually accompanied by severe symptoms, comes to a climax, and then fades rather quickly. A **chronic disease**, by contrast, often lingers for long periods of time. The symptoms are slower to develop, an acute period is rarely reached, and convalescence may continue for several months. Hepatitis A and infectious mononucleosis are examples of chronic diseases. Sometimes an acute disease may become chronic when the body is unable to rid itself completely of the microbe. For example, one who has contracted a parasitic disease, such as giardiasis or amoebiasis, may experience sporadic symptoms for many years.

Pathogen Entry into the Host Depends on Cell Adhesion and the Infectious Dose

With this understanding of the disease stages, we now can examine several factors required for the establishment of disease, as outlined in FIGURE 19.7.

A portal of entry refers to the characteristic route by which an exogenous pathogen enters the host. It varies considerably for different organisms and is a key factor leading to the establishment of disease. These entry routes include: (1) a respiratory portal (inhalation of pathogens in air); (2) a gastrointestinal portal—the **fecal–oral route** (ingestion of contaminated food or water); (3) a sexually-transmitted portal; and (4) parenteral portals (piercing the skin or mucous membranes through cuts, animal/insect bites, wounds, injections).

■ Parenteral: Referring to direct entry into an organism by a nonoral route.

Having reached the appropriate portal of entry, the ability of a pathogen to establish an infection and possible disease usually depends on the **infectious**

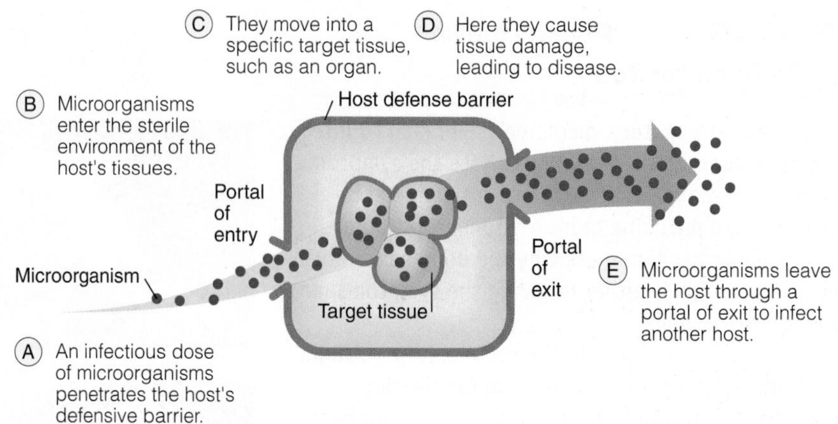

FIGURE 19.7 **The Generalized Events for a Local Disease.** The infectious dose and adhesion to cells or tissues are required to initiate infection and disease. »» Identify which events would not occur if an infection but not a disease occurred.

dose, the number of pathogens that need to be taken into the body to cause infection—and possibly disease. Using food poisoning as an example, the consumption of a few hundred *Campylobacter* cells will lead to disease while 100,000 *Salmonella* bacilli must be ingested if salmonellosis is to be established. One explanation for the difference is the higher resistance of the *Campylobacter* cells to the acidic conditions in the stomach, in contrast to the low resistance of *Salmonella* bacilli. Also, it may be safe to eat fish when the seawater contains hepatitis A viruses, but eating raw clams from the same seawater can be dangerous because clams are filter-feeders, concentrating hepatitis A viruses in their bodies to a concentration capable of providing an infectious dose.

Often the host is exposed to low doses of a pathogen and, as a result, develops immunity. For instance, many people can tolerate low numbers of mumps viruses without exhibiting disease. They may be surprised to find they are immune to mumps when it breaks out in their family at some later date.

Many pathogens enter at a particular portal of entry because they contain on their surface "sticky" factors, called **adhesins**, that only adhere to specific tissues. A variety of adhesins often are associated with bacterial capsules, flagella, or pili (FIGURE 19.8). For example, many gram-negative

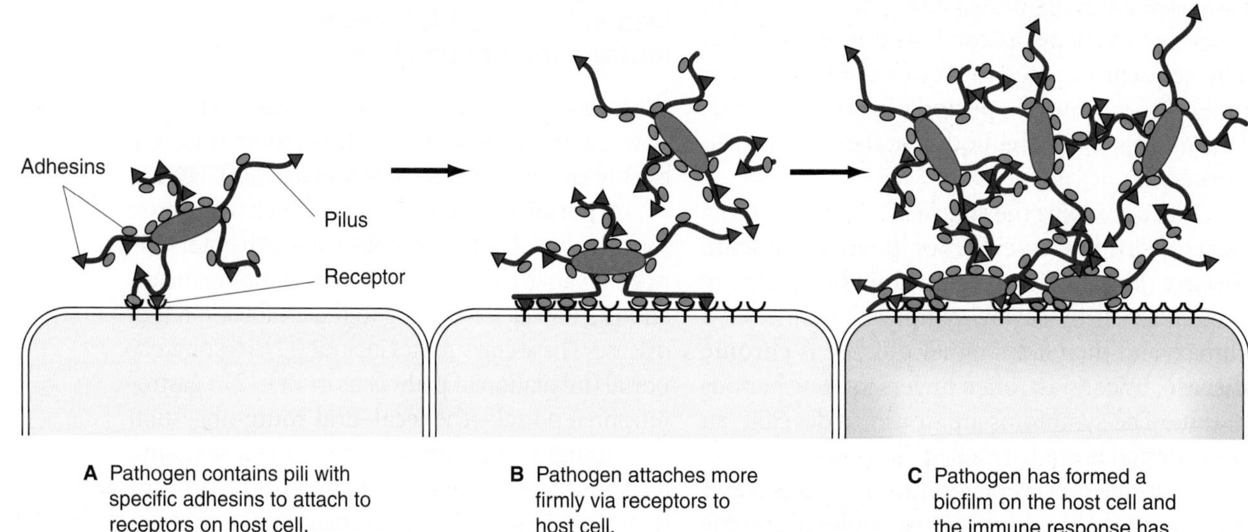

FIGURE 19.8 **Pathogen Adhesion to a Host Cell.** Many gram-negative bacterial species and some gram-positive species contain pili that can adhere to a specific host cell or tissue based on the adhesins located on the pili. Adapted from: *Telford, J. L. et al. 2006. Nature Reviews Microbiology* **4**(7): 509–519. »» What is the advantage of the pathogen forming a biofilm on the host cell surface?

species and some gram-positive species, such as *C. difficile* attach by means of pili to specific receptor sites found only on the appropriate target cells. The host cell is often an active partner in the adhesion because the pathogen triggers the host cell to express on its surface target receptor sites for adhesin binding. Also, many viruses have spikes on the capsid or envelope, allowing for attachment.

Some pathogens have multiple portals of entry. The tubercle bacillus, for instance, may enter the body in respiratory droplets, contaminated food or milk, or skin wounds. The bacterial species causing Q fever can enter by any of these portals, as well as by an arthropod bite. The tularemia bacillus may enter the eye by contact, the skin by an abrasion, the respiratory tract by droplets, the intestines by contaminated meat, or the blood by an arthropod bite.

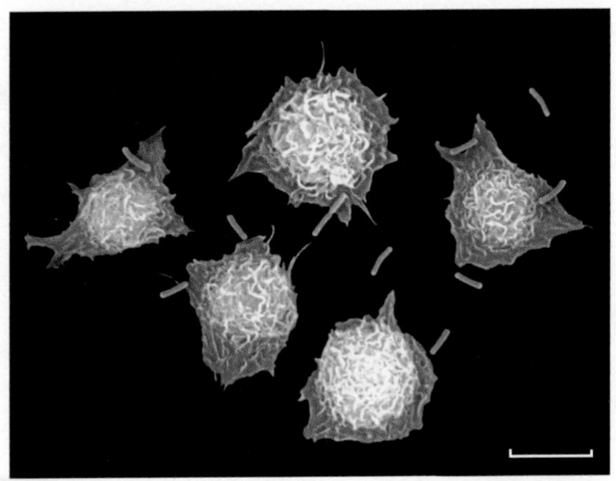

FIGURE 19.9 **Macrophages Undergoing Phagocytosis.** In this false-color scanning electron micrograph, macrophages, a type of immune cell, are capturing bacterial cells (blue) for phagocytosis. (Bar = 10 µm.) © Dr. K. G. Murti/ Visuals Unlimited/Corbis. »» Why might phagocytosis be an important activity of the human immune system when infection occurs?

Breaching the Host Barriers Can Establish Infection and Disease

Some pathogens do not need to penetrate cells or tissues to cause disease. The pertussis bacillus, for example, adheres to the surface layers of the respiratory tract while producing the toxins causing disease. Likewise, the cholera bacillus attaches to the surface of the intestine, where it produces toxins.

However, the virulence of many pathogens requires the cells to penetrate tissues and cause structural damage or physiological change to the host cells. The ability to penetrate and spread is called **invasiveness** and the bacilli of typhoid fever and the protist causing amoebiasis are examples of pathogens that depend on their invasiveness. By penetrating the tissue of the gastrointestinal tract, these microorganisms cause ulcers and sharp, appendicitis-like pain characteristic of the respective diseases.

Invasiveness often is facilitated through the pathogen's purposeful internalization by the host cell (FIGURE 19.9). Some immune cells, including macrophages, undergo **phagocytosis** by engulfing pathogens, taking them into the cell cytoplasm in vacuoles, and then attempting to destroy them with lysosomes. In addition, other bacterial pathogens induce nonphagocytic cells to undergo phagocytosis. In both cases, if the pathogen can evade destruction by lysosomes, the cell interior provides a protective niche or a vehicle to spread undetected by the immune system. Here is one example.

Some cytoplasm-invading pathogens, such as *Shigella flexneri* (shigellosis), *Rickettsia prowazekii* (epidemic typhus), and *Listeria monocytogenes* (listeriosis), have cell membrane adhesive proteins that form a zipper-like binding of pathogen to the host cell (FIGURE 19.10). As a result of this molecular adhesion and cross-talk, the pathogen triggers the host cell to undergo phagocytosis. Once in the cell, the pathogens either escape from the vacuole or block lysosome fusion, eliminating any chance of their destruction. In the cytoplasm, the bacterial cells trigger the host cell to synthesize an actin tail on the bacterial cells, which forcefully propels the cells through the cytoplasm. When a bacterial cell "bumps" against the host's plasma membrane, it distorts and indents the adjacent cell, bridges the junction between the two cells, and enters the next cell (somewhat like moving from train car to train car through connecting doors). The system allows bacterial invasion to occur without the bacterial cells leaving the cellular environment and, importantly, not alerting the immune system.

■ Actin:
A cytoskeletal protein essential for cell movement and the maintenance of cell shape in most eukaryotic cells.

■ Macrophage:
A large white blood cell that removes waste products, microorganisms, and foreign material from the bloodstream.

Successful Invasiveness Requires Pathogens to Have Virulence Factors

When a pathogen enters a host cell, the pathogen is confronted with a profoundly different environment in which to survive. The pathogen must be able to adapt and to spread while withstanding the resistance put forward by the host. Often several virulence factors may be needed and can be a mixture of enzymes and toxins.

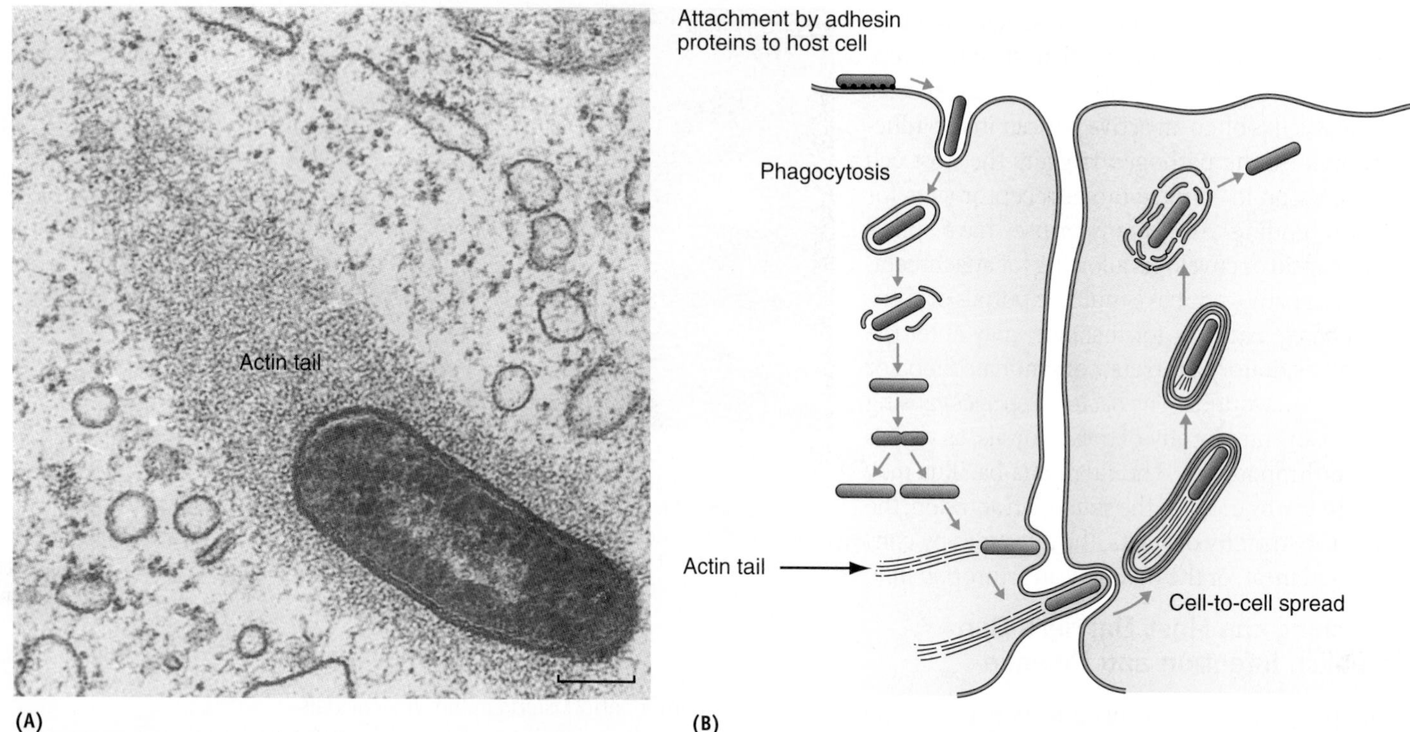

FIGURE 19.10 **Invasion by *Listeria monocytogenes*.** (**A**) A transmission electron micrograph of a *Listeria* cell with its actin tail. (Bar = 0.3 μm.) Reproduced from *Trends in Microbiology*, Vol. 1, Tilney, L. G. and Tilney, M. S., © 1993, with permission from Elsevier. [http://www.sciencedirect.com/science/journal/0966842X. (**B**) After entering a host cell by phagocytosis, the bacterial cell loses the surrounding membrane and divides to form a larger population. Invasion of neighboring cells uses the actin tail to "drive" the cells into the adjacent cell. In that cell, the cell again loses the surrounding membranes and initiates another infection. »» What invasion process is negated by the *Listeria* cell's ability to generate an actin tail?

Enzymes. Some virulence factors are secreted enzymes that defensively or offensively respond to structural and immune defenses. A few examples illustrate how bacterial enzymes act on host cells and interfere with certain functions or barriers meant to retard invasion.

Some species of *Staphylococcus* secrete a defensive enzyme called **coagulase** that coats the surface of the cells and the cluster of cells (FIGURE 19.11A). Coagulase catalyzes the conversion of plasma fibrinogen into fibrin, which then sticks to and forms a clot around the staphylococci cells.

■ Pulmonary embolism: A blockage of the main artery of the lung or one of its branches by a blood clot formed at one place in the circulation and then lodging in another.

$$\text{Fibrinogen} \xrightarrow{\textit{Coagulase}} \text{Fibrin}$$

Part of the walling-off process observed in a staphylococcal skin boil is due to clot formation that then protects the pathogen from phagocytosis by white blood cells. In the diagnostic identification of staphylococci, the **coagulase test** is a test tube assay using rabbit plasma to distinguish between pathogenic *S. aureus* (coagulase-positive = plasma clots) and other species like *S. epidermidis* (coagulase-negative = no clot formed).

In an attempt to block pathogen spread, the body often defends against the pathogen by producing a clot as a barrier to isolate an infected area and prevent further spread of the infection. Many streptococci and staphylococci have the ability to produce an offensive fibrinolytic enzyme, **streptokinase** or staphylokinase, which dissolves the fibrin clot and thus allows further tissue invasion by the pathogen. Streptokinase is used medically as an effective clot-dissolving medication for some cases of heart attacks and pulmonary embolisms.

Hyaluronidase is sometimes called the "spreading factor" because it enhances penetration of a pathogen through the tissues. The offensive enzyme digests hyaluronic acid, a polysaccharide that binds cells together in a tissue, into fragments (FIGURE 19.11B). The term "tissue cement" is occasionally applied to the polysaccharide.

$$\text{Hyaluronic acid} \xrightarrow{\textit{Hyaluronidase}} \text{Hyaluronic acid fragments}$$

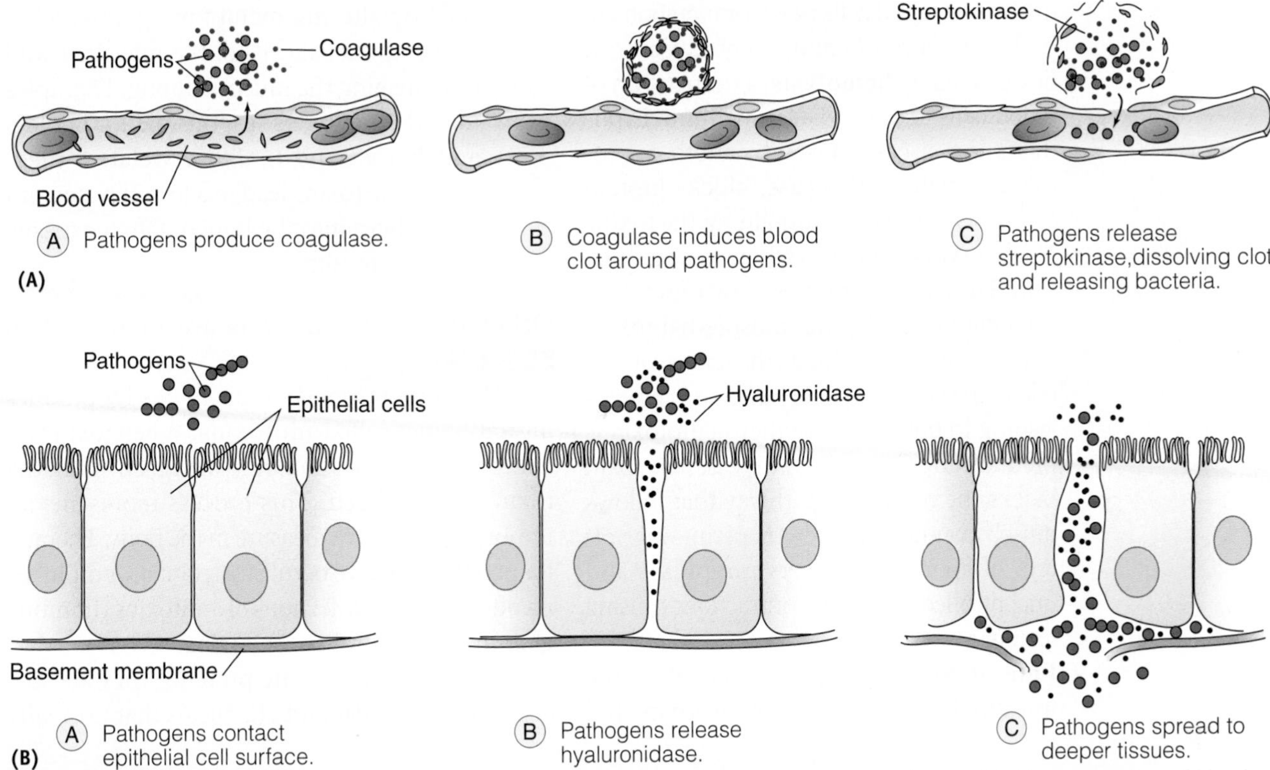

FIGURE 19.11 **Enzyme Virulence Factors.** **(A)** Some bacterial cells produce the enzyme coagulase, which triggers clotting of blood plasma. Bacterial cells within the clot can break free by producing streptokinase. **(B)** Some invasive bacterial species produce the enzyme hyaluronidase, which degrades the cementing polymer holding cells of the intestinal lining together. »» From these examples, how do virulence factors protect bacterial cells and increase their virulence?

Hyaluronidase is an important virulence factor in pneumococci and certain species of streptococci and staphylococci. In addition, a *C. difficile* hyaluronidase releases nutrients that promote pathogen colonization in the gut.

Although not truly a virulence factor, if a pathogen exists in a **biofilm**, its virulence can be enhanced because here it can resist body defenses and drugs. A biofilm is a sticky layer of extracellular polysaccharides and proteins enclosing a colony of bacterial cells at the tissue surface. Phagocytes and antibodies have difficulty reaching the microorganisms in this slimy conglomeration of armor-like material. Moreover, microorganisms often survive without dividing in a biofilm. This makes them impervious to the antibiotics that attack dividing cells. (Indeed, the antibiotics do not penetrate the biofilm easily.) CDC officials have estimated that 65% of human infections involve biofilms.

Toxins. Microbial poisons, called **toxins**, can profoundly affect the establishment and course of disease. The ability of pathogens to produce toxins is referred to as **toxigenicity**, while toxins present in the blood is called **toxemia** and such a person is considered "intoxicated." Two types of toxins are recognized: exotoxins and endotoxins.

Exotoxins are heat-sensitive protein molecules, manufactured during bacterial metabolism. They are produced by gram-positive and gram-negative bacterial cells, many of the toxins being coded by plasmids or prophages. The exotoxins are released into the host environment or, with some gram-negative bacteria, injected directly into the host cell. The toxins act locally or diffuse to their site of activity and are categorized into one of three groups based on their primary target.

- **Cytotoxins** kill cells. Some staphylococci, streptococci, and pneumococci produce **leukocidins** that form pores in and lyse those white blood cells (leukocytes) whose job it is to phagocytize and destroy the pathogens. **Hemolysin**, produced by staphylococci and streptococci, lyse red blood cells (erythrocytes), which gives the pathogens access to the iron in hemoglobin that the bacterial cells need for metabolism. In the laboratory, hemolysin producers can be detected by **hemolysis**, a destruction of blood cells in a blood agar medium. Diphtheria is a disease caused by another cytotoxin, which, in this case, blocks protein synthesis in the cells containing the toxin.

- **Neurotoxins** interfere with nerve transmission. The exotoxin produced by *Clostridium botulinum* is among the most lethal toxins known. The toxin inhibits the release of acetylcholine at the synaptic junction, a process leading to botulism. Another neurotoxin is produced by *C. tetani*. In this case, the toxin blocks the relaxation pathway that follows muscle contraction, thereby permitting volleys of spontaneous nerve impulses and uncontrolled muscular contractions causing tetanus.
- **Enterotoxins** are exotoxins that affect the intestinal tract. These protein toxins kill cells by altering membrane permeability of the epithelial cells of the intestinal wall or damaging the mucosal lining. Examples include the toxins responsible for cholera and *E. coli*- and *C. difficile*-related infections. The toxins lead to a loss of water and electrolytes into the lumen. Often extreme diarrhea results.

Other types of exotoxins are identified in TABLE 19.1.

The body responds to exotoxins by producing antibodies called **antitoxins**. When toxin and antitoxin molecules combine with each other, the toxin is neutralized. This process represents an important defensive measure in the body. Therapy for people who have botulism, tetanus, or diphtheria often includes injections of antitoxins (immune globulin) to neutralize the toxins.

Because exotoxins are proteins, they are susceptible to the heat and chemicals that normally

TABLE

19.1 Characteristics and Effects of Some Bacterial Exotoxins

Exotoxin (Type)	Organism	Gene Location	Disease	Effect
Alpha toxin (Cytotoxin)	*Clostridium perfringens*	Chromosomal	Gas gangrene	Hemolysis; membrane lysis
Anthrax toxin (Cytotoxin)	*Bacillus anthracis*	Plasmid	Anthrax	Hemorrhaging and pulmonary swelling
Botulism toxin (Neurotoxin)	*Clostridium botulinum*	Prophage	Botulism	Respiratory paralysis
Cholera toxin (Enterotoxin)	*Vibrio cholerae*	Prophage	Cholera	Severe diarrhea
Diphtheria toxin (Cytotoxin)	*Corynebacterium diphtheriae*	Prophage	Diphtheria	Inhibits protein synthesis; cell death
Enterotoxin Enterotoxin A	*Escherichia coli* *Staphylococcus aureus*	Plasmid Prophage	Diarrhea Food poisoning	Diarrhea Diarrhea and nausea
Erythrogenic toxin (Cytotoxin)	*Streptococcus pyogenes*	Prophage	Scarlet fever	Capillary destruction; rash
Pertussis toxin	*Bordetella pertussis*	Chromosomal	Whooping cough (pertussis)	Interferes with host cell communication
Pyrogenic toxin (Enterotoxin)	*Staphylococcus aureus*	Prophage	Toxic shock syndrome	Fever, shock
Tetanus toxin (Neurotoxin)	*Clostridium tetani*	Plasmid	Tetanus	Rigid paralysis; respiratory failure
Clostridium difficile toxins A and B (Enterotoxins)	*Clostridium difficile*	Chromosomal	Colitis Pseudomembranous colitis	Mucosal lining destruction; diarrhea

TABLE

19.2 A Comparison of Exotoxins and Endotoxins

Characteristic	Exotoxins	Endotoxins
Source	Living gram-positive and gram-negative bacterial cells	Lysed gram-negative bacterial cells
Location	Released from living cells	Part of cell wall
Chemical composition	Protein	Lipopolysaccharide
Heat sensitivity	Unstable (60°C)	Stable (60°C)
Immune reaction	Strong	Weak
Conversion to toxoid	Possible	No
Fever	No	Yes
Toxigenicity	High	Low
Representative diseases	Botulism Diphtheria Cholera	Salmonellosis Typhoid fever Meningococcal meningitis

denature proteins. A chemical such as formaldehyde may be used in the laboratory to alter the toxin and destroy its toxicity without hindering its ability to elicit an immune response. The result is a **toxoid**. When the toxoid is injected into the body, the immune system responds with antitoxins, which circulate and provide a measure of defense against disease. Toxoids are used for diphtheria and tetanus immunizations in the diphtheria-tetanus-acellular pertussis (DTaP) vaccine.

Endotoxins all have similar effects and usually are released upon disintegration of gram-negative cells. They are a structural component in the outer membrane of many gram-negative bacilli and are part of the **lipopolysaccharide** (**LPS**) complex. The lipid portion of the LPS is the toxic agent. Endotoxins do not stimulate a strong immune response in the body, nor can they be altered to prepare toxoids. They function by activating a blood-clotting factor to initiate blood coagulation and by influencing the complement system. The toxins of plague bacilli are especially powerful.

Endotoxins all have similar toxic effects. At high concentrations, they manifest their presence by certain signs and symptoms. Usually an individual experiences an increase in body temperature, substantial body weakness and aches, and diarrhea. Damage to the circulatory system and shock may occur. In this case, the permeability of the blood vessels changes and blood leaks into the intercellular spaces. The tissues swell, the blood pressure drops, and the patient may lapse into a coma. This condition, commonly called **endotoxin shock**, may accompany antibiotic treatment because endotoxins are released as the bacterial cells are killed by the antibiotic.

Like exotoxins, endotoxins add to the virulence of a microbe and enhance its ability to establish disease. TABLE 19.2 summarizes the characteristics of the bacterial toxins.

Pathogens Must Be Able to Leave the Host to Spread Disease

At the conclusion of its pathogenicity cycle, pathogens exit the host through some suitable **portal of exit** (FIGURE 19.12). Perhaps most common is coughing and sneezing, which easily spread nasal secretions, saliva, and sputum as respiratory droplets or aerosols. This is of more than passing importance because easy transmission permits the pathogen to continue its pathogenic existence in the world.

CONCEPT AND REASONING CHECKS 2

a. Assign the following signs and symptoms of the flu to the appropriate disease stage (fever, headache, chills, cough, sore throat, fatigue, and muscle aches).

b. What is the relationship between the portals of entry and adhesins in establishing an infection?

c. Assess phagocytosis as an invasiveness mechanism used by pathogens.

d. Evaluate the role of enzymes and toxins as important virulence factors in the establishment of disease.

e. Are portal of entry and exit always the same? Explain.

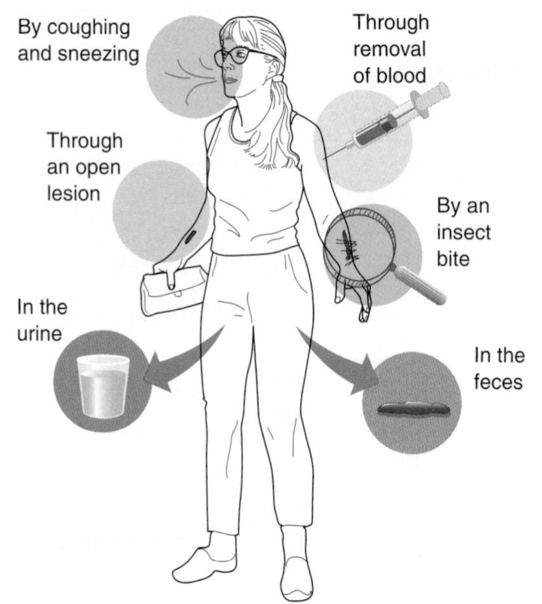

FIGURE 19.12 **Six Different Portals of Exit from the Body.** »» Why do pathogens need specific portals of exit?

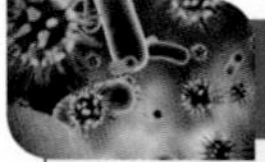

Chapter Challenge B

So, our patient has a CDAI. Although the organism is highly resistant to many front-line antibiotics, being a highly infectious pathogen (high virulence), it must have an array of virulence factors that make it a dangerous pathogen.

QUESTION B:

From your reading in Key Concept 19.2, identify what enzymatic and toxin virulence factors that C. diff possesses. How does each of these factors contribute to the organism's overall virulence?

Answers can be found on the Student Companion Website in **Appendix D.**

KEY CONCEPT 19.3 Infectious Disease Epidemiology Includes Frequency and Spread of Disease

Infectious disease epidemiology is concerned with how infectious diseases are distributed in a population and the factors influencing or determining that distribution. In this final section, we examine the factors putting population groups at risk of contracting infectious disease.

We also look at special environments, such as healthcare settings, and the public agencies saddled with the job of emerging disease identification, control, and prevention.

Epidemiologists Often Have to Identify the Reservoir of an Infectious Disease

To cause an infection, pathogens have to be transferred from a source to a susceptible host (FIGURE 19.13). For many diseases to perpetuate themselves, the disease-causing microbes must exist somewhere in the environment. These ecological niches or sources where microbes live, multiply, and spread are called **reservoirs** of infection. There are three types of potential reservoirs.

■ Niche: Environmental area that ensures an organism's survival.

Animals are one type of reservoir. A domestic dog that has rabies, for example, can transmit the virus to humans through a bite. How did the dog get infected? Perhaps the animal was bit by a rabid bat, which would represent another example of a reservoir. In this example, the human would not be a reservoir because that individual cannot pass on the virus; the human is called a **dead end host**. A disease that is transmitted from animals to humans is called a **zoonosis** (pl. zoonoses). Besides rabies,

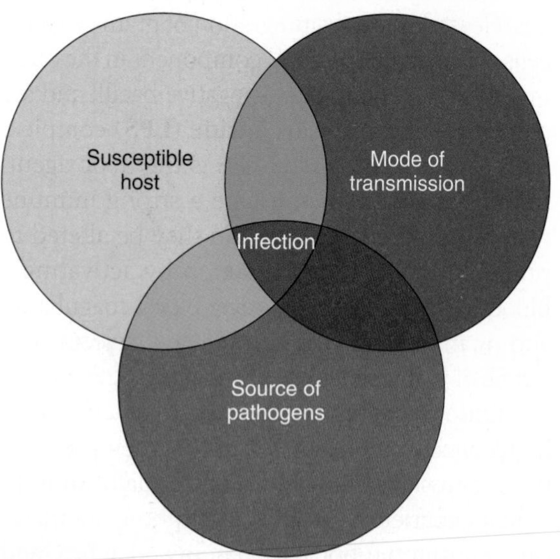

FIGURE 19.13 **Infectious Disease Elements.** Assuming there is a susceptible host, to cause an infection the pathogenic organism must have a way to be transmitted to that susceptible host. »» What is meant by a susceptible host?

the plague, West Nile fever, and malaria would be examples of zoonoses (the reservoir for plague could be rats and the reservoir for West Nile fever and malaria is mosquitoes).

Another special type of reservoir is a **carrier**, which is a person who has recovered from the disease but continues to shed the disease agents. For instance, a person who has recovered from typhoid fever or amoebiasis becomes a carrier for many weeks after the symptoms of disease have subsided. The feces of this individual may spread the bacterial cells or cysts to others via contaminated food or water. Further, not all diseases have nonhuman reservoirs. Until its elimination from the world, the smallpox virus only existed in humans, so humans would be the reservoir. This is why the WHO by locating all human reservoirs was able to curtail the spread of the virus through vaccination and eradicate smallpox from the world.

A third type of reservoir is the nonliving type, specifically food, soil, and water. These can be reservoirs because they can become contaminated with disease agents through poor sanitation and hygiene. Water is perhaps the most dangerous, acting as a reservoir for the agents of many diseases including polio, hepatitis A, cholera, traveler's diarrhea, amoebiasis, and cryptosporidiosis, to name just a few.

Epidemiologists Have Several Terms that Apply to the Infectious Disease Process

Most diseases studied in this text are **communicable diseases**; that is, infectious diseases, like tuberculosis and AIDS, that are transmissible among susceptible individuals in a population. As expected, human reservoirs are the source for most communicable diseases. Certain communicable diseases are described as being **contagious** because they pass with particular ease among hosts and are highly infectious. Chickenpox and measles fall into this category.

Noncommunicable diseases are singular events in which the agent is acquired directly from the environment and is not transmitted to other susceptible individuals. In tetanus, for example, penetration of soil containing *Clostridium tetani* spores into the anaerobic tissue of a wound must occur before this disease develops. It cannot be spread person-to-person.

Infectious Diseases Can Be Transmitted in Several Ways

Diseases can be transmitted from reservoir to host in a variety of ways (FIGURE 19.14).

Direct Contact. Person-to-person or **horizontal transmission** implies close or personal contact with someone who is infected or who has the disease. Handshaking or kissing an infected person can spread bacterial cells, viruses, or other pathogens to an uninfected person. For zoonotic diseases, such as rabies and Lyme disease, direct contact with an animal is necessary. An animal bite or scratch by an infected animal can spread the disease to an uninfected person. The exchange of body fluids, such as through sexual contact, is another example of direct contact transmission for diseases like gonorrhea and AIDS.

Direct transmission also can involve the violent expulsion of **respiratory droplets** through sneezing, coughing, or simply talking (FIGURE 19.15). This form of transmission through the air requires the "recipient" be close to an infected individual. In a sneeze, the droplets are fairly large (50–100 μm) and fall out of the air within about 1 meter of their source. If an uninfected person is within that distance, the eyes, mouth, or nose may be portals of entry for the airborne pathogens.

Direct contact called **vertical transmission** includes the spread of pathogens, such as HIV, from a pregnant mother to her unborn child. Also, transmission of gonorrhea from mother to newborn can occur during labor or delivery.

Indirect Contact. Indirect contact transmission can be the result of contact with a nonliving object or a vector (Figure 19.14). **Fomites** are inanimate objects on or in which disease organisms linger for some period of time. For instance, doorknobs or drinking glasses may be contaminated with flu or cold viruses and transmitted to another person when touching the object. This shows the importance of hand washing.

Vehicle transmission involves the indirect spread of disease through contaminated food and water, or air. Foods can be contaminated during processing or handling, or they may be dangerous when made from diseased animals. Poultry products, for example, are often a reservoir for

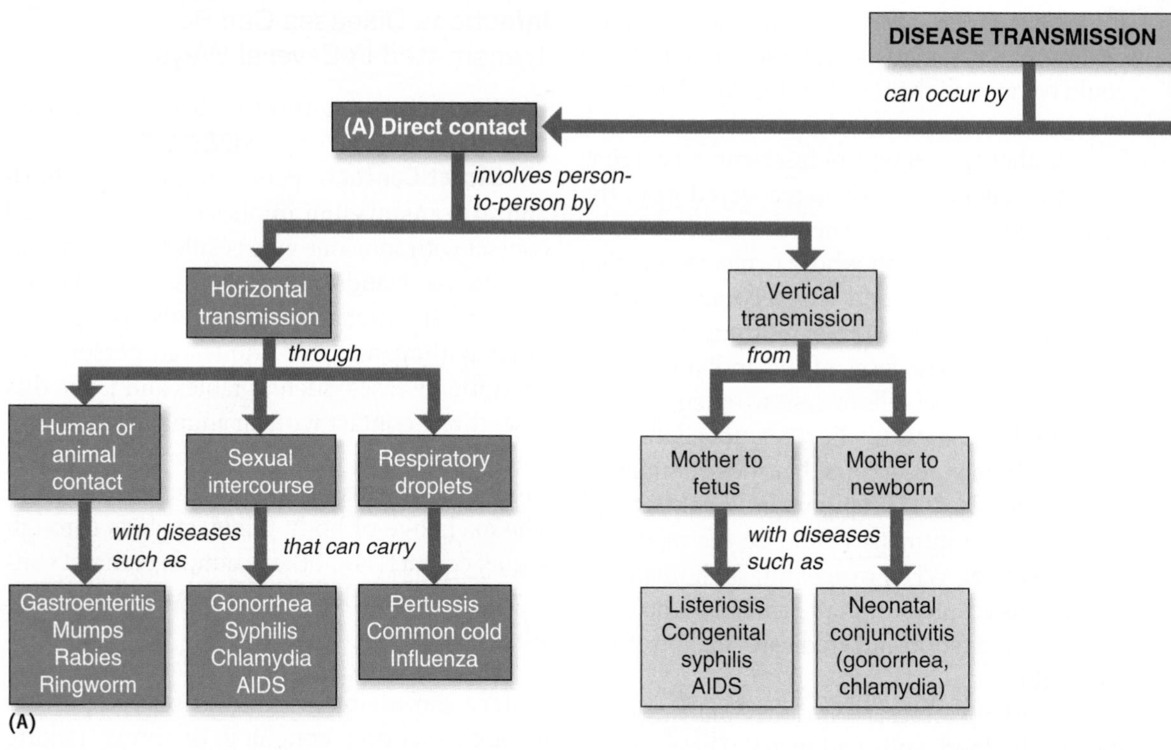

FIGURE 19.14 **Transmission of Microorganisms and Viruses.** (**A**) Pathogens can be transmitted by direct contact involving horizontal or vertical transmission. (**B**) Pathogens also can be transmitted by fomites, contaminated food and water, and vectors. »» Would indirect contact transmission represent horizontal or vertical transmissions? Explain.

salmonellosis because *Salmonella* species frequently infect chickens and can lead to cross contamination during food preparation.

Pathogens also are transmitted through the air on smaller particles (<3 μm) called **aerosols**. With aerosols, the particles can remain suspended in the air for indefinite periods of time and can be moved some distance by air currents. The virus of SARS and the bacterial cells of tuberculosis are two pathogens that can be carried in the air by droplets or aerosols. MicroFocus 19.3 describes long-range effects of airborne transmission.

FIGURE 19.15 **Droplet Transmission.** Sneezing or coughing represents a method for airborne transmission of pathogens. © John Lund/age fotostock. »» Identify some portals of entry for pathogen-containing droplets.

Arthropods represent another indirect, zoonotic method of transmission. Many pathogens hitch a ride on arthropods, such as mosquitoes, ticks, fleas, and lice, which act as **vectors**, living organisms acting as reservoirs and carrying disease agents from one host to another. **Mechanical vectors** represent arthropods passively transporting microbes on their legs and other body parts. For example, house flies can carry diseases picked up on their feet. In other cases, arthropods represent **biological vectors**, where the pathogen must multiply in the insect before it can infect another host. The malarial protist and the West Nile virus infect and reproduce in mosquitoes and accumulate in their salivary glands, from which the pathogens are injected during the next blood meal.

MicroFocus 19.4 describes two transmission modes: airborne and formites.

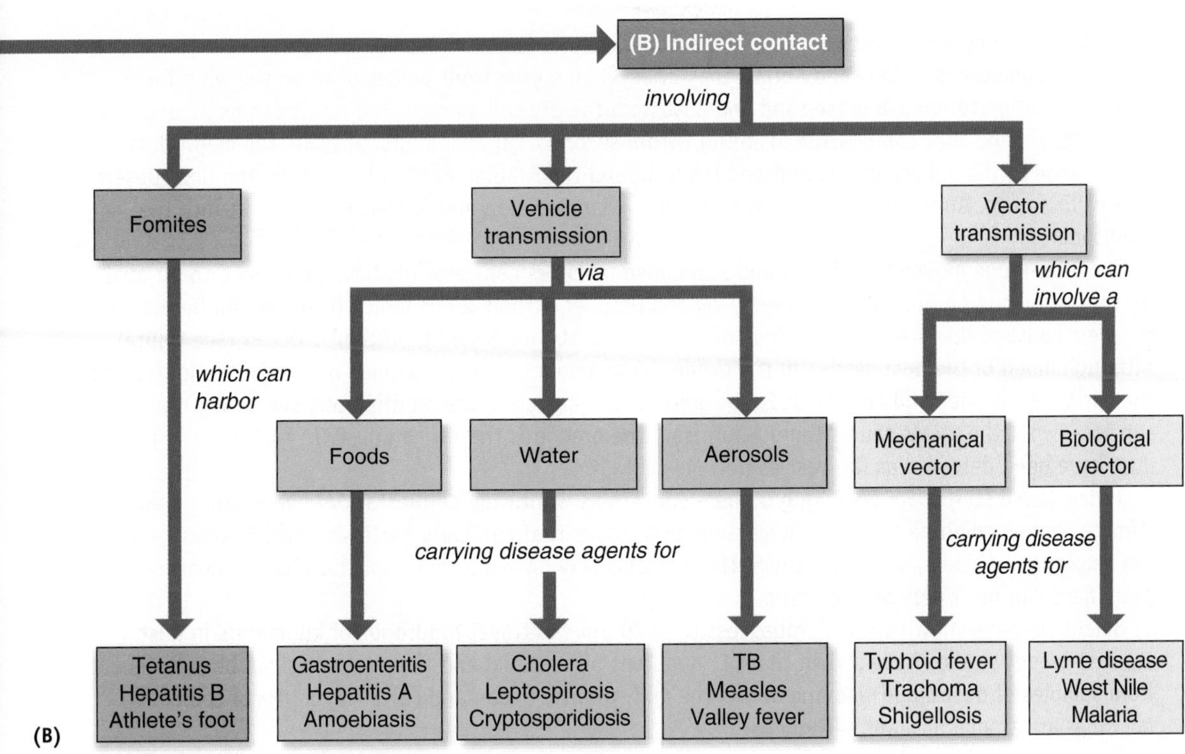

(B)

Diseases Also Are Described by How They Occur Within a Population

When epidemiologists investigate an infectious disease, they need to determine if it is localized or spread through a community or region. **Endemic** refers to a disease habitually present at a low level in a certain geographic area. Plague in the American Southwest is an example.

By comparison, an **epidemic** refers to a disease that occurs in excess of what is normally found within that population. Influenza often causes widespread epidemics. This should be contrasted with an **outbreak**, which is a more contained epidemic. An abnormally high number of measles cases in one American city would be classified as an outbreak. Not only do epidemiologic investigations look at current outbreaks, they also consider future outbreaks, including bioterrorism (MicroFocus 19.5).

A **pandemic** is a worldwide epidemic, affecting populations around the globe. The most obvious example here would be AIDS and the 2009 H1N1 flu.

As in the opening quote, "*Health care matters to all of us some of the time, public health matters to all of us all of the time,*" maintaining vigilance against infectious disease is extremely important. For this reason, national and international public health organizations, such as the CDC and the WHO, learn a lot about diseases by analyzing disease data reported to them. The CDC, for example, has a list of infectious diseases that physicians must report to state health departments, which then report them to the CDC (TABLE 19.3). These are published in the *Morbidity and Mortality Weekly Report.*

MicroInquiry 19 explores the use of epidemiological data as a tool for understanding disease occurrence.

Nosocomial Infections Are Serious Health Threats Within the Healthcare System

Nosocomial infections represent that portion of **healthcare-associated infections** (**HAIs**) associated with hospitals and account for an estimated 1.7 million infections (1 in 20 patients receiving

(text continues on page 658)

MICROFOCUS 19.3: Environmental Microbiology

Riders on the Storm

Dust storms can be a relatively common meteorological phenomenon in arid and semi-arid regions of the world. Sometimes called sandstorms, they arise when a gust front passes or when the wind force is strong enough to remove loose sand and dust from the dry soil surface. The result can be an awesome wall of sand that can obscure visibility within seconds. On a more global scale, the major dust storms arise in the Sahara Desert and arid lands around the Arabian Peninsula. In Asia, the Gobi Desert is a major source. Annually, such storms carry more than 3 billion metric tons of dust aloft into the atmosphere.

The long-range movement of dust and suspended particles can certainly have an impact on air quality that often can be observed. Between May and October, strong winds blow off the Sahara Desert and the west coast of North Africa, carrying soil and dust westward across the Atlantic Ocean (see figure). Although much of the dust settles in the ocean, large amounts stay suspended and make the journey to the Caribbean islands and Florida. It is not unusual to wake up in the morning and see a fiery orange sunrise—what the locals call a "tequila sunrise." The orange is the African dust. In fact, traces of such dust have been detected as far west as New Mexico!

Today, these storms are becoming of more concern as worldwide deforestation, overgrazing, and climate change combine to generate massive dust clouds that can carry particles aloft. Besides causing respiratory distress, such as asthma, these storms may have another impact on human health—pathogens can be "riders on the storm."

Scientists have shown that microbes less than 20 μm can travel thousands of kilometers in dust storms. Recent studies have shown that (1) hundreds of bacterial and fungal species can be cultured from samples of dust clouds moving across the mid-Atlantic Ocean and (2) 20% to 30% of these microbes are animal or plant pathogens.

More research is needed to define the role of dust storms as an indirect transmission mechanism spreading pathogens across the globe. This is important work because these "riders on the storm" might become more prevalent and more frequent as climate change expands arid regions around the globe.

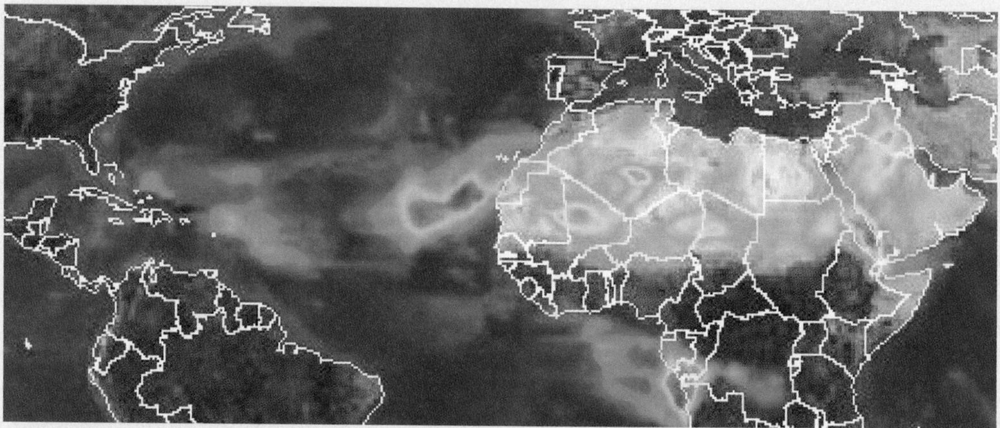

Satellite image of African dust storm spreading westward across the Atlantic Ocean toward the Caribbean Islands and North America. Courtesy of Jay Herman/NASA.

MICROFOCUS 19.4: Public Health

Planes, Trains, and —Ambulances

Vehicles that can take us to great destinations at supersonic speeds, or save our lives, can also harbor and transmit infectious disease.

Planes and Trains. In August 2004, a New Jersey man returned home from a trip to West Africa. Within hours after taking the train home from the airport, the man was stricken with fever, chills, a severe sore throat, diarrhea, and back pain. The family rushed him to a local hospital. However, despite intensive care, the gentleman continued to decline and died a few days later. Clinical and postmortem specimens were sent to the Centers for Disease Control and Prevention (CDC) for a specific cause of death. The finding: Lassa fever.

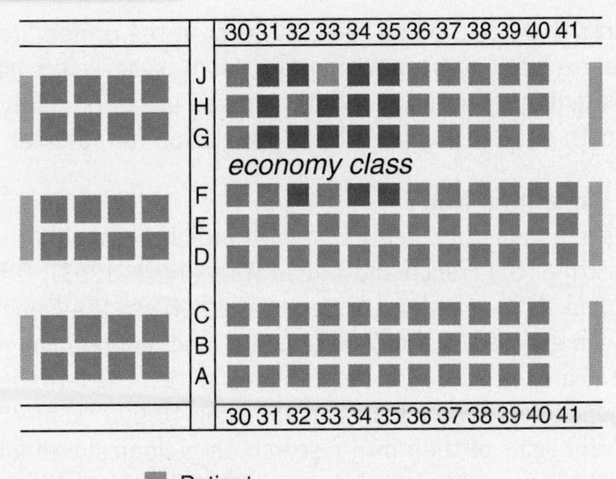

Passenger seating and contact status.

The alarms went off! Lassa fever is an acute viral disease, rarely seen outside West Africa where the disease is endemic. In West Africa, the virus, which has its reservoir in rodents, infects 100,000 to 300,000 people every year, and kills 5,000. The case of the New Jersey man was a lethal health risk as the virus can be spread person to person if a susceptible person comes into contact with the virus in the blood, tissue, or excretions from infected individuals. Although the virus cannot be spread through casual contact (including skin-to-skin contact without exchange of body fluids), CDC epidemiologists contacted all passengers seated nearby the man (see figure) and others who could have been exposed. These individuals were asked to monitor their body temperatures for fever. Luckily, there were no reported cases.

The scary part is that the man was traveling while ill and potentially exposed some 200 people to the virus: 19 family members, 139 healthcare workers, 16 lab workers, 18 airplane passengers, and numerous commuters on the New Jersey train. This time we dodged the bullet.

Ambulances. From November 2004 to April 2005, a decontamination firm in the United Kingdom examined the ambulances from 12 firms for microbial contamination. They swabbed several fomites, including stretcher rails, the stretcher tracks below the stretcher, the paramedic's utility bag, and five other sites within the vehicle. The swabs were streaked on nutrient agar plates to see what bacterial species would grow.

Examination of the plates indicated the ambulances were heavily contaminated with a diverse group of bacterial species. In fact, in many cases, there were so many bacterial colonies present, they could not be counted. The bacterial species included antibiotic-resistant *Staphylococcus aureus* and a variety of species typically found in the human colon. More surprising, after the ambulances were cleaned by standard procedures, there was little reduction in the numbers of bacterial cells present. In fact, another study showed that cleaning actually spread the bacterial cells onto previously "clean" surfaces. Such contaminated fomites could be dangerous to a person with open wounds in the ambulance.

Since that initial 2005 report, emergency medical services have evaluated and improved their best practices for cleaning and disinfecting surfaces on patient care equipment.

MICROFOCUS 19.5: History/Public Health

Bioterrorism: The Weaponization and Purposeful Transmission of Human Pathogens

The anthrax attacks that occurred on the East Coast of the United States in October 2001 confirmed what many health and governmental experts had been saying for over 10 years—it is not if bioterrorism would occur but when and where. Bioterrorism represents the intentional or threatened use of primarily microorganisms or their toxins to cause fear in or actually inflict death or disease upon a large population for political, religious, or ideological reasons.

Is Bioterrorism Something New?

Bioterrorism has a long history, beginning with infectious agents that were used for biowarfare. In the United States, during the aftermath of the French and Indian Wars (1754–1763), British forces, under the guise of goodwill, gave smallpox-laden blankets to rebellious tribes sympathetic to the French. The disease decimated the Native Americans, who had never been exposed to the disease before and had no immunity. Between 1937 and 1945, the Japanese established Unit 731 to carry out experiments designed to test the lethality of several microbiological weapons as biowarfare agents on Chinese soldiers and civilians. In all, some 10,000 "subjects" died of bubonic plague, cholera, anthrax, and other diseases. After years of their own research on biological weapons, the United States, the Soviet Union, and more than 100 other nations in 1973 signed the Biological and Toxin Weapons Convention, which prohibited nations from developing, deploying, or stockpiling biological weapons. Unfortunately, the treaty provided no way to monitor compliance. As a result, in the 1980s the Soviet Union developed and stockpiled many microbiological agents, including the smallpox virus, and anthrax and plague bacteria. After the 1991 Gulf War, the United Nations Special Commission (UNSCOM) analysts reported that Iraq had produced 8,000 liters of concentrated anthrax solution and more than 20,000 liters of botulinum toxin solution. In addition, anthrax and botulinum toxin had been loaded into SCUD missiles.

In the United States, several biocrimes have been committed. **Biocrimes** are the intentional introduction of biological agents into food or water, or by injection, to harm or kill groups of individuals. The most well known biocrime occurred in Oregon in 1984 when the Rajneeshee religious cult, in an effort to influence local elections, intentionally contaminated salad bars of several restaurants with the bacterium *Salmonella*. The unsuccessful plan sickened over 750 citizens and hospitalized 40. Whether biocrime or bioterrorism, the 2001 events concerning the anthrax spores mailed to news offices and to two U.S. congressmen only increases our concern over the use of microorganisms or their toxins as bioterror agents.

What Microorganisms Are Considered Bioterror Agents?

A considerable number of human pathogens and toxins have potential as microbiological weapons. These "Tier 1 agents" include bacterial organisms, bacterial toxins, and viruses. The seriousness of the agent depends on the severity of the disease it causes (virulence) and the ease with which it can be disseminated. Tier 1 agents can be spread by aerosol contact, such as anthrax and smallpox, or added to food or water supplies, such as the botulinum toxin (see table below).

Why Use Microorganisms?

At least 15 nations are believed to have the capability of producing bioweapons from microorganisms. Such microbiological weapons offer clear advantages to these nations and terrorist organizations in general. Perhaps most important, biological weapons represent "The Poor Nation's Equalizer." Microbiological weapons are cheap to produce compared to chemical and nuclear weapons, and provide those nations with a deterrent every bit as dangerous and deadly as the nuclear weapons possessed by other nations. With biological weapons, you get high impact and the most "bang for the

TABLE

Some Tier 1 Agents and Perceived Risk of Use

Type of Agent	Disease (Microbe Species or Virus Name)	Perceived Risk
Bacterial	Anthrax (*Bacillus anthracis*)	High
	Plague (*Yersinia pestis*)	Moderate
	Tularemia (*Francisella tularensis*)	Moderate
Viral	Smallpox (Variola virus)	Moderate
	Hemorrhagic fevers (Ebola, Marburg virus)	Low
Toxin	Botulinum toxin (*Clostridium botulinum*)	Moderate

buck." In addition, microorganisms can be deadly in minute amounts to a defenseless (nonimmune) population. They are odorless, colorless, and tasteless, and unlike conventional and nuclear weapons, microbiological weapons do not damage infrastructure, yet they can contaminate such areas for extended periods. Without rapid medical treatment, most of the select agents can produce high numbers of casualties that would overwhelm medical facilities. Lastly, the threatened use of microbiological agents creates panic and anxiety, which often are at the heart of terrorism.

How Would Microbiological Weapons Be Used?

All known microbiological agents (except smallpox) represent organisms naturally found in the environment. For example, the bacterium causing anthrax is found in soils around the world (see left figure). Assuming one has the agent, the microorganisms can be grown (cultured) easily in large amounts. However, most of the select agents must be "weaponized"; that is, they must be modified into a form that is deliverable, stable, and has increased infectivity and/or lethality. Nearly all of the microbiological agents in category A are infective as an inhaled aerosol. Weaponization, therefore, requires the agents be small enough in size so inhalation would bring the organism deep into the respiratory system and prepared so that the particles do not stick together or form clumps. Several of the anthrax letters of October 2001 involved such weaponized spores.

Dissemination of biological agents by conventional means would be a difficult task. Aerosol transmission, the most likely form for dissemination, exposes microbiological weapons to environmental conditions to which they are usually very sensitive. Excessive heat, ultraviolet light, and oxidation would limit the potency and persistence of the agent in the environment. Although anthrax spores are relatively resistant to typical environmental conditions, the bacterial cells causing tularemia become ineffective after just a few minutes in sunlight. The possibility also exists that some nations have developed or are developing more lethal bioweapons through genetic engineering and biotechnology. The former Soviet Union may have done so. Commonly used techniques in biotechnology could create new, never before seen bioweapons, making the resulting "designer diseases" true doomsday weapons.

Conclusions

Ken Alibek, a scientist and defector from the Soviet bioweapons program, has suggested the best biodefense is to concentrate on developing appropriate medical defenses that will minimize the impact of bioterrorism agents. If these agents are ineffective, they will cease to be a threat; therefore, the threat of using human pathogens or toxins for bioterrorism, like that for emerging diseases such as 2009 H1N1 flu and West Nile fever, is being addressed by careful monitoring of sudden and unusual disease outbreaks. Extensive research studies are being carried out to determine the effectiveness of various antibiotic treatments (see right figure) and how best to develop effective vaccines or administer antitoxins. To that end, vaccination perhaps offers the best defense. The United States has stated it has stockpiled sufficient smallpox vaccine to vaccinate the entire population if a smallpox outbreak occurred. Other vaccines for other agents are in development.

This primer is not intended to scare or frighten; rather, it is intended to provide an understanding of why microbiological agents have been developed as weapons for bioterrorism. We cannot control the events that occur in the world, but by understanding bioterrorism, we can control how we should react to those events—should they occur in the future.

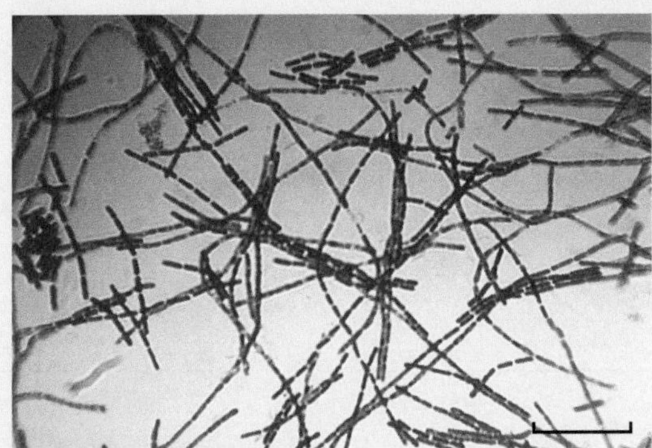

Light micrograph of gram-stained *Bacillus anthracis*, the causative agent of anthrax. There is concern that terrorists could release large quantities of anthrax spores in a populated area, which potentially could cause many deaths. (Bar = 20 μm.) Courtesy of CDC.

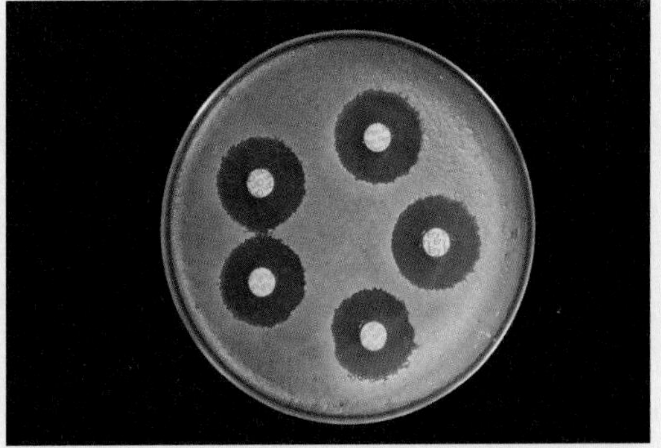

Antibiotic drugs in paper discs are used to test the sensitivity of anthrax bacteria (*Bacillus anthracis*) cultured on an agar growth medium. The clear zone surrounding each disc indicates the bacterial cells are sensitive to the antibiotic. © CNRI/Photo Researchers, Inc.

TABLE

19.3 CDC's Summary of Notifiable Diseases in the United States in 2012

Anthrax
Arboviral diseases, neuroinvasive and nonneuroinvasive
Babesiosis
Botulism
Brucellosis
Chancroid
Chlamydia trachomatis infection
Cholera
Cryptosporidiosis
Cyclosporiasis
Dengue virus infection
Diphtheria
Ehrlichiosis/Anaplasmosis
Giardiasis
Gonorrhea
Haemophilus influenzae, invasive disease
Hansen disease (leprosy)
Hantavirus pulmonary syndrome
Hemolytic uremic syndrome, post-diarrheal
Hepatitis A, B, C
HIV infection
Influenza-associated pediatric mortality
Legionellosis
Listeriosis
Lyme disease
Malaria
Measles
Meningococcal disease
Mumps
Novel influenza A virus infections
Pertussis
Plague
Polio infection, nonparalytic
Poliomyelitis, paralytic
Psittacosis
Q fever, acute and chronic
Rabies, animal and human
Rubella
Rubella, congenital syndrome
Salmonellosis
Severe acute respiratory syndrome-associated coronavirus (SARS-CoV) disease
Shiga toxin-producing *Escherichia coli* (STEC)
Shigellosis
Smallpox
Spotted fever rickettsiosis
Streptococcal toxic-shock syndrome
Streptococcus pneumoniae, invasive disease
Syphilis
Tetanus
Toxic-shock syndrome (other than streptococcal)
Trichinellosis
Tuberculosis
Tularemia
Typhoid fever
Vancomycin-intermediate *Staphylococcus aureus* (VISA) infection
Vancomycin-resistant *Staphylococcus aureus* (VRSA) infection
Varicella (morbidity)
Varicella (mortality)
Vibriosis
Viral hemorrhagic fevers
Yellow fever

Source: CDC *Morbidity and Mortality Weekly Report.*

MICROINQUIRY 19

Epidemiological Investigations

Infectious disease epidemiology is a scientific study from which health problems are identified. In this inquiry, we are going to look at just a few of the applications and investigative strategies for analyzing the patterns of illness. Answers can be found on the Student Companion Website in **Appendix D**.

One of the important measures is to assess disease occurrence. The **incidence** of a disease is the number of new cases within a set population in a given time frame. **Figure A** is a line graph showing the incidence of primary and secondary syphilis per year in the United States since 1993.

The **prevalence** of a disease refers to the total number of infected individuals (new and old) within a population at a given time.

19.1a. What was the incidence of syphilis in 1993 and 2008?

19.1b. How has the prevalence of syphilis changed between 2000 and 2008, assuming the U.S. population has remained at 300 million?

Descriptive epidemiology describes activities (time, place, people) regarding the distribution of diseases within a population. Once some data have been collected on a disease, epidemiologists can analyze these data to characterize disease occurrence. Often a comprehensive description can be provided by showing the disease trend over time, its geographic extent (place), and the populations (people) affected by the disease.

Characterizing by Time

Traditionally, drawing a graph of the number of cases by the date of onset shows the time course of an epidemic. An epidemic curve, or "epi curve," is a histogram providing a visual display of the magnitude and time trend of a disease.

Look at the epi curve in **Figure B** for an Ebola outbreak in Africa. One important aspect of a bar graph is to consider its overall shape. An epi curve with a single peak indicates a **common source** (or "point source") **epidemic** in which people are exposed to the same source over a relatively short time. If the duration of exposure is prolonged, the epidemic is called a **propagated epidemic**, and the epi curve will have a plateau instead of a peak. Person-to-person transmission is likely and its spread may have a series of plateaus one incubation period apart.

19.2a. Identify the type of epi curve drawn in Figure B and explain what the onset says about the nature of disease spread.

19.2b. Is there more than one plateau? Explain the significance that multiple plateaus might have in interpreting the spread of the Ebola hemorrhagic fever outbreak.

Characterizing by Place

Analysis of a disease or outbreak by place provides information on the geographic extent of a problem and may show clusters or patterns that provide clues to the identity and origins of the infectious agent. It is a simple and useful technique to look for geographic patterns where the affected people live, work, or may have been exposed. A geographic distribution

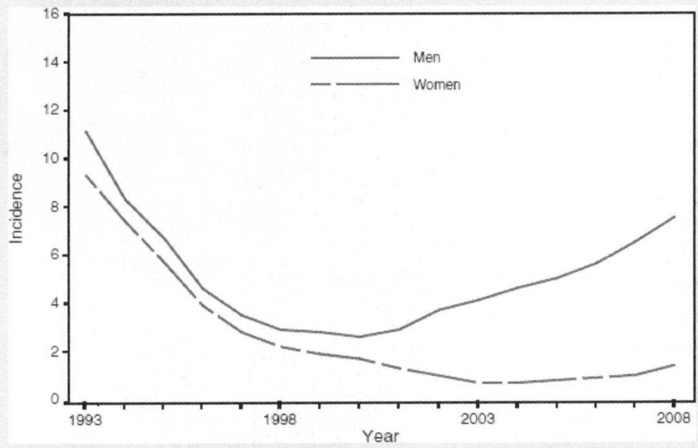

FIGURE A **Incidence of Primary and Secondary Syphilis—United States 1993–2008.** Incidence numbers are per 100,000 population. *Source:* CDC.

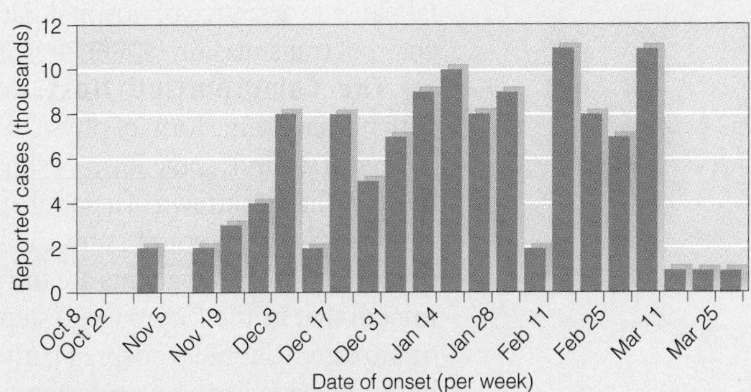

FIGURE B **Ebola Hemorrhagic Fever (Congo and Gabon).** Number of cases of Ebola hemorrhagic fever by week from October 2001 to March 2002. *Data from: Weekly Epidemiological Record,* No. 26, June 27, 2003.

for Lyme disease is shown in **Figure C.** This is a spot map, where each reported case of a disease in a state is shown to reflect clusters or patterns of disease.

19.3a. From this spot map, what inferences can you draw with regard to the reported cases of Lyme disease?

Characterizing by Person

Populations at risk for a disease can be determined by characterizing a disease or outbreak by person. Persons also refer to populations identified by personal characteristics (e.g., age, race, gender) or by exposures (e.g., occupation, leisure activities, drug intake). These factors are important because they may be related to disease susceptibility and to opportunities for exposure.

Age and gender often are the characteristics most strongly related to exposure and to the risk of disease. For example, **Figure D** is a histogram showing the number of pertussis (whooping cough) cases in the United States in 2008.

19.4a. Look at the histogram and describe what important information is conveyed in terms of the majority of cases in 2008.

19.4b. As a healthcare provider, what role do you see for vaccinations and booster shots with regard to this disease?

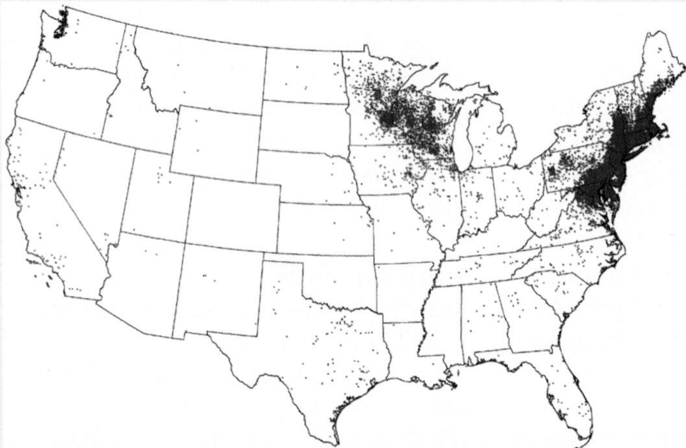

FIGURE C **Lyme Disease.** Each dot represents one reported case of Lyme disease in 2007. *Source:* CDC.

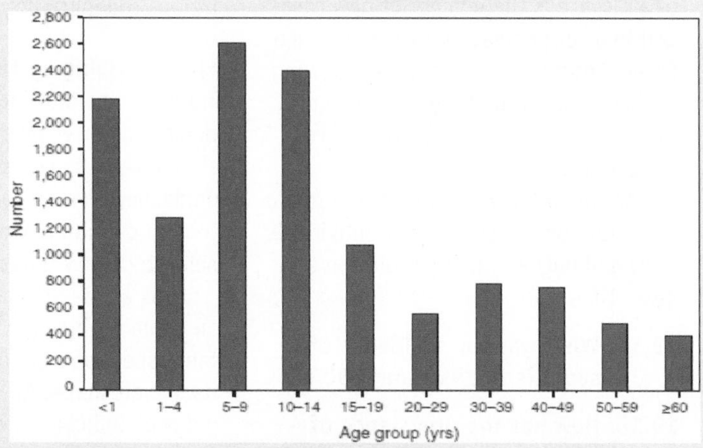

FIGURE D **Pertussis.** The reported number of cases of pertussis by age group in the United States in 2008. *Source:* CDC.

medical care) and, according to the CDC, some 99,000 deaths each year in the United States. Like all infections, nosocomial infections involve three elements: a compromised host (the hospital patient), a source of hospital pathogens, and a chain of transmission (FIGURE 19.16).

The Compromised Host. Most hospital patients have some form of physical injury, such as a surgical wound, some form of skin trauma like a burn, or the breakdown of a portal of entry through which a pathogen could enter the body. In addition, many hospital patients are **immunocompromised**; that is, their immune system is weakened or impaired. Should a hospital pathogen enter the body, the patient's immune system may be unable to mount an attack and eliminate it. The most common sites of infection are listed in FIGURE 19.17.

The Hospital Pathogens. Hospital personnel attempt to maintain a sanitary and clean hospital environment. Still, the facility can be a reservoir for human pathogens. Some pathogens come from

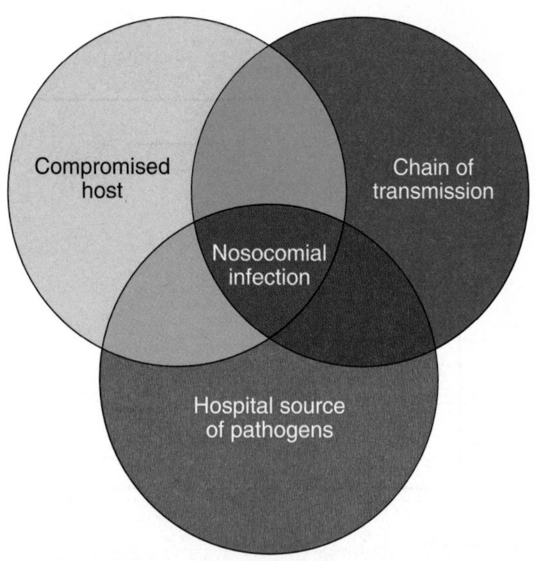

FIGURE 19.16 **Nosocomial Infection Elements.** For a nosocomial infection to occur, there needs to be a susceptible (compromised) host, pathogenic organisms within the hospital setting, and a chain of transmission. »» How does this nosocomial infection figure compare to that for an infectious disease (see Figure 19.13)?

other patients being treated for an infectious disease, from patient visitors, or by transmission from healthcare staff. Many nosocomial infections are caused by **opportunistic** agents, microbes that do not normally cause illness in healthy individuals, but given the "opportunity," can infect an immunocompromised patient. More than 50% of surgical infections come from bacterial cells that patients normally carry as opportunists in their nose or on their skin. Importantly, proper methods of patient screening, scrubbing with a skin disinfectant, and pretreating with an antibiotic ointment can reduce infections by 60%.

Patients also can become infected from contaminated indwelling medical devices such as catheters. The common skin bacterial species *Staphylococcus epidermidis* can form biofilms as a result of being carried along with the insertion of an indwelling catheter or other medical device. Again proper skin degerming and extensive cleaning of medical devices before and after use can greatly reduce these sources of nosocomial pathogens.

The most common microorganisms responsible for nosocomial infections are listed in TABLE 19.4. While examining this table, note that

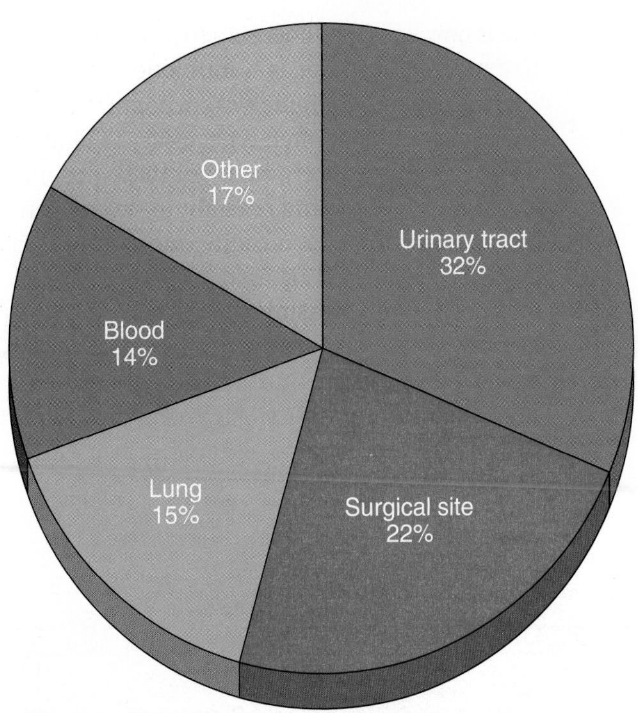

FIGURE 19.17 **Sites of Nosocomial Infections.** This pie chart shows the sites for the estimated 1.7 million healthcare-associated infections in American hospitals. *Source:* CDC. Division of Healthcare Quality Promotion (2010). »» What might be the primary source for the infections among the top four sites?

TABLE

19.4 Top 10 Infectious Agents Involved in Nosocomial Infections

Microorganism	% of Total Infections	% Antibiotic Resistant[1]	Nosocomial Infections
Coagulase-negative staphylococci	15.3%	Not reported (historically greater than 80%)	Blood and surgical site
Staphylococcus aureus	14.4%	56%	Urinary, blood, lung, and surgical site
Enterococcus species	12.1%	3%–90%	Urinary tract, blood, and surgical site
Candida species	10.7%	Not reported	Urinary tract
Escherichia coli	9.6%	1%–30%	Urinary and surgical site
Pseudomonas aeruginosa	7.9%	6%–33%	Urinary, blood, and surgical site
Klebsiella pneumoniae	5.8%	3%–27%	Lung, blood, and surgical site
Enterobacter species	4.8%	Not reported	Urinary and surgical site
Acinetobacter baumannii	2.7%	26%–37%	Urinary, blood, lung, and surgical site
Klebsiella oxytoca	1.1%	3%–17%	Urinary, blood, lung, and surgical site

[1]Means for different species and different nosocomial infections.

Source: CDC. *National Healthcare Safety Network Annual Update* (2008).

there is now another dimension to the virulence of these potential pathogens—antibiotic resistance. With the patient's immune system compromised, the use of antibiotics often is necessary to fight an infection. Unfortunately, many of these hospital pathogens are becoming resistant to several generations of antibiotics, meaning more toxic and expensive drugs must be used.

■ Central line: A tube placed in a large vein of a patient's neck or chest to provide medical treatment.

The Chain of Transmission. The key to nosocomial disease and its prevention stems from the way the agents are transmitted to the patient. These chains of transmission may involve direct contact between patients or between healthcare staff and patient. Indirect contact can also be part of the chain of transmission (FIGURE 19.18). Perhaps one of the most common chains of transmission is through the use of indwelling instruments that are not sterile or have not been cleaned thoroughly. For example, some 41,000 bloodstream infections strike hospital patients with central lines each year. About 25% of these septicemic patients die.

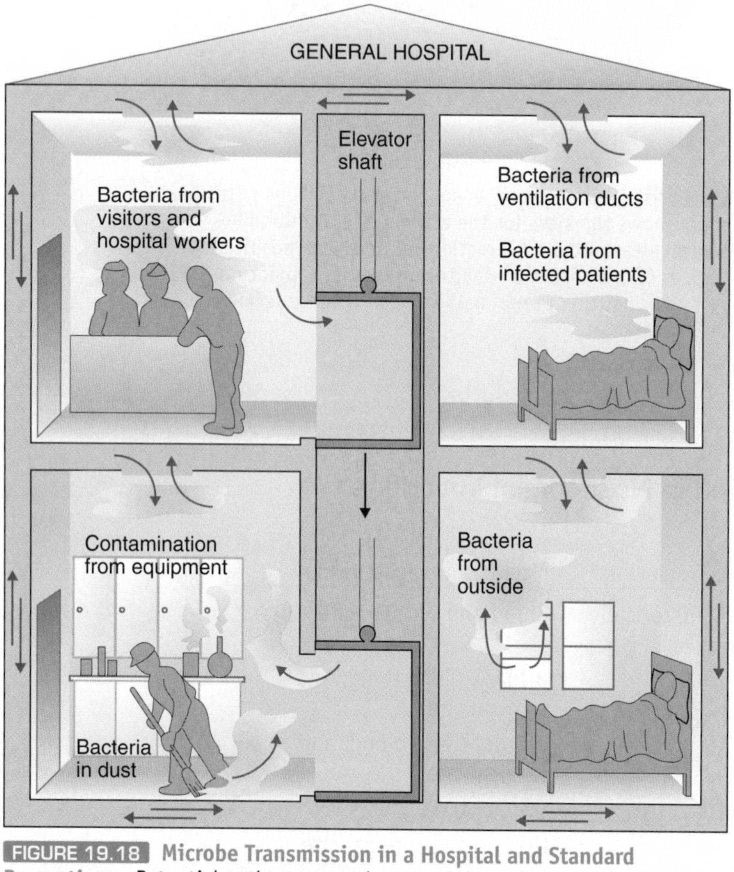

FIGURE 19.18 **Microbe Transmission in a Hospital and Standard Precautions.** Potential pathogens can be spread through several means within the hospital or healthcare environment. Protecting patients and other healthcare workers also means using standard precautions if working with a patient who may be infected. The precautions when handling blood or other body fluids that may harbor pathogens (HIV, hepatitis B or C) include:

- Washing hands
- Wearing personal protective equipment (gloves, mask, and eye protection).
- Handling and disposing of sharps (hypodermic needles) properly.
- Disposing of all hazardous and contaminated materials in approved and labeled biohazard containers.
- Cleaning up all spills with disinfectant or diluted bleach solution to kill any pathogens present.

»» Why is hand washing always at the top of the list for preventing disease transmission?

Therefore, the key to reducing nosocomial infections is to break the chain of transmission. Besides the use of **standard precautions** when working with blood or other body fluids (Figure 19.18), the CDC has published preferred methods for cleaning, disinfecting, and sterilizing patient-care medical devices and general methods for cleaning and disinfecting the healthcare environment. The proper use of chemical disinfectants, and sterilization when possible, is essential. The CDC stresses that these chemical and physical methods must be used properly to reduce the risk for infection associated with both invasive and noninvasive medical and surgical devices. And, importantly, it all starts with good hand hygiene on the part of healthcare providers and visitors while in the hospital.

Infectious Diseases Continue to Challenge Public Health Organizations

In an era of supersonic jet travel and international commerce, it is not possible to adequately protect the health of any nation without focusing on diseases and epidemics elsewhere in the world. In 2012, we only have to look at the 2009 H1N1 (swine) flu to realize the seriousness of infectious diseases and their threats to global health FIGURE 19.19.

Emerging infectious diseases are caused by newly identified pathogens or by pathogens appearing in new areas of the world for the first time. TEXTBOOK CASE 19 describes a recent outbreak in Key West, Florida. **Reemerging (resurgent) infectious diseases** are those that previously were controlled or declining, but now are showing increased incidence. Globally, there have been more than 40 new diseases and at least 20 resurgent diseases identified since 1980 FIGURE 19.20.

There are more than 1,400 known human pathogens. However, less than 100 are specialized within humans; over half of the 1,400 represent zoonoses. Some 335 (24%) represent emerging or resurgent diseases, with the largest single

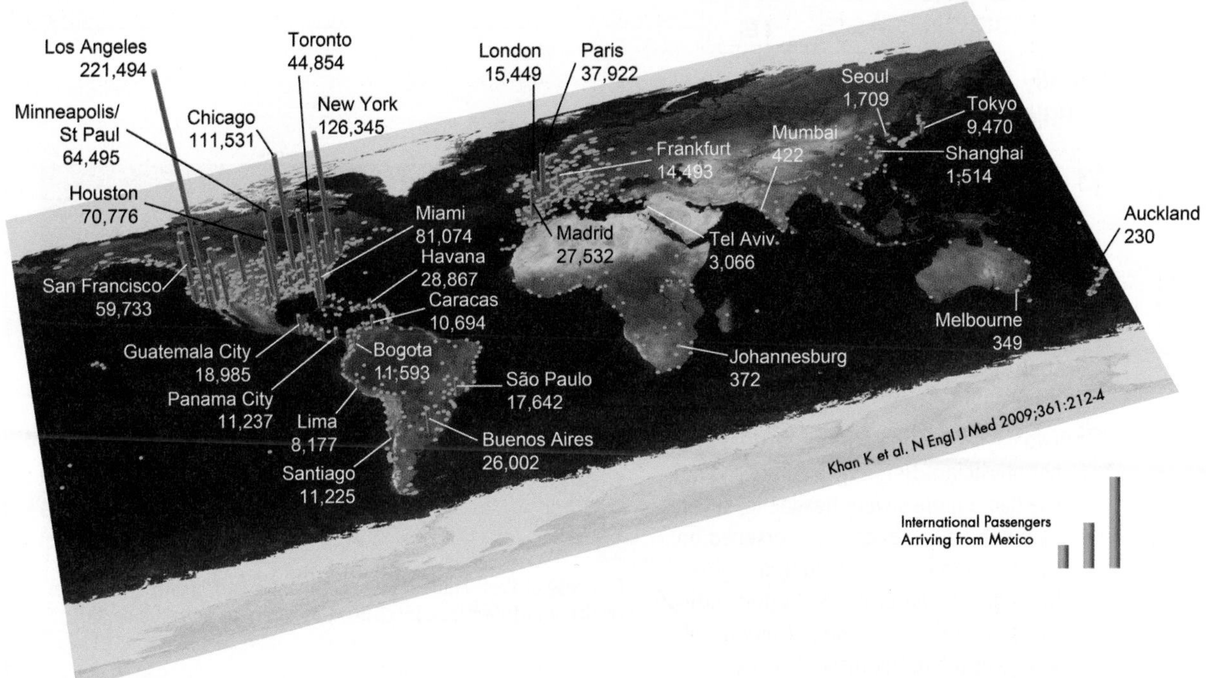

FIGURE 19.19 **Spreading the Flu.** This map shows the destination cities around the world to which airline passengers flew (and the volume of passengers) on departure from Mexico during March and April, 2008. Reproduced from *N Engl J Med.* 2009 Jul 9;361(2):212–4. Printed with permission from the Massachusetts Medical Society. »» Since this map is for the year before the 2009 H1N1 flu broke out in Mexico, why are these destinations and passenger volumes important?

number (54%) being bacterial species, followed by viruses (25%), most of which are RNA viruses (FIGURE 19.21).

Numerous reasons help explain how disease emergence and resurgence are driven (in decreasing rank of disease involvement).

1. **Changes in land use or agriculture practices**. Urbanization, deforestation, and water projects can bring new or reemergent diseases (dengue fever and schistosomiasis).
2. **Changes in human demographics**. The migration of many peoples or whole societies from agrarian to dense, urban settings has brought new diseases that can be easily transmitted to a susceptible population (malaria).
3. **Poor population health**. In many developing nations, large numbers of people suffer from malnutrition or poor public health infrastructure, making disease eruption much more likely (cholera, diphtheria).
4. **Pathogen evolution**. Pathogens have developed resistance to antibiotics and antimicrobial drugs, typical of many resurgent diseases (tuberculosis, yellow fever).
5. **Contamination of food sources and water supplies**. Substandard application of, or lapses in, sanitation practices can bring a resurgence of disease (listeriosis).
6. **International travel**. The number of people traveling internationally can spread diseases to other parts of the globe quickly (SARS, West Nile fever, 2009 H1N1 flu).
7. **Failure of public health systems**. Failure of immunization programs can bring a resurgence of disease (diphtheria).
8. **International trade**. The global transport of produce can introduce new or resurgent foodborne diseases (hepatitis A, cyclosporiasis). Likewise, wildlife trade (legal and illegal) provides new mechanisms for disease transmission (monkeypox).
9. **Climate change**. Global changes in weather patterns bring new diseases to new latitudes and elevations (hantavirus pulmonary syndrome).

By understanding these drivers, health organizations such as the CDC and WHO can develop plans to limit or stop emerging disease threats.

Textbook CASE 19

Locally Acquired Dengue Fever

Courtesy of Prof. Frank Hadley Collins, Director, Center for Global Health and Infectious Diseases, University of Notre Dame/CDC.

1 On August 11, 2009, a previously healthy, 34-year-old woman in Rochester, New York, went to her primary care physician complaining of fever, headache, malaise, and chills. She told the physician that the symptoms had appeared 24 hours earlier. A urine sample was taken for analysis.

2 Two days later, the patient returned to her physician. Her fever had abated but she had a more severe headache, severe pain behind the eyes that worsened on eye movement, and a feeling of light-headedness. Her urinalysis report indicated bacterial cells and red blood cells were present in the urine. She was referred to a local hospital emergency department.

3 The emergency room evaluation showed all vital signs were normal. A complete blood cell workup revealed a low white blood cell and platelet count and a normal hematocrit. A CT scan and cerebral spinal fluid (CSF) from a lumbar puncture were normal. Because her light-headedness disappeared, she was discharged from the emergency department.

4 On August 17, the patient returned to her primary care physician expressing the feeling that she just didn't feel good. Although all vital signs were normal, petechiae (tiny purplish-red spots due to blood hemorrhages) were noted on her lower extremities.

5 A consultation with an infectious disease specialist suggested the patient could have dengue fever. Questioning the patient, it was determined that she had not traveled to any dengue-endemic area in the world. She did state that prior to the onset of symptoms she had just returned from a trip to Key West, Florida and, while there, had been bitten several times by mosquitoes. A serum sample from the patient was tested for antibodies to dengue fever virus. The results were positive. Confirmatory testing of serum and CFS samples was done by the Centers for Disease Control and Prevention (CDC) in Atlanta. Both samples were positive for antibodies against dengue fever virus.

6 On August 19, the patient reported to her physician that she was feeling much better. She had completely recovered when interviewed by the Monroe County (Florida) Health Department on September 1.

7 Further investigation identified another 24 cases of dengue fever, all locally acquired in the Key West area.

Questions:

(Answers can be found on the Student Companion Website in **Appendix D.**)

A. Why was a urine sample taken for analysis?

B. Why didn't the original CSF sample taken on August 13 indicate a dengue fever infection?

C. What sign indicated to the infectious disease specialist that the patient might have dengue fever?

D. Considering the number of dengue fever cases in Key West, what measures should be taken to lessen and control the outbreak?

For additional information see www.cdc.gov/mmwr/preview/mmwrhtml/mm5919a1.htm.

Emerging and re-emerging infectious diseases, and pathogens, in North, Central, and South America

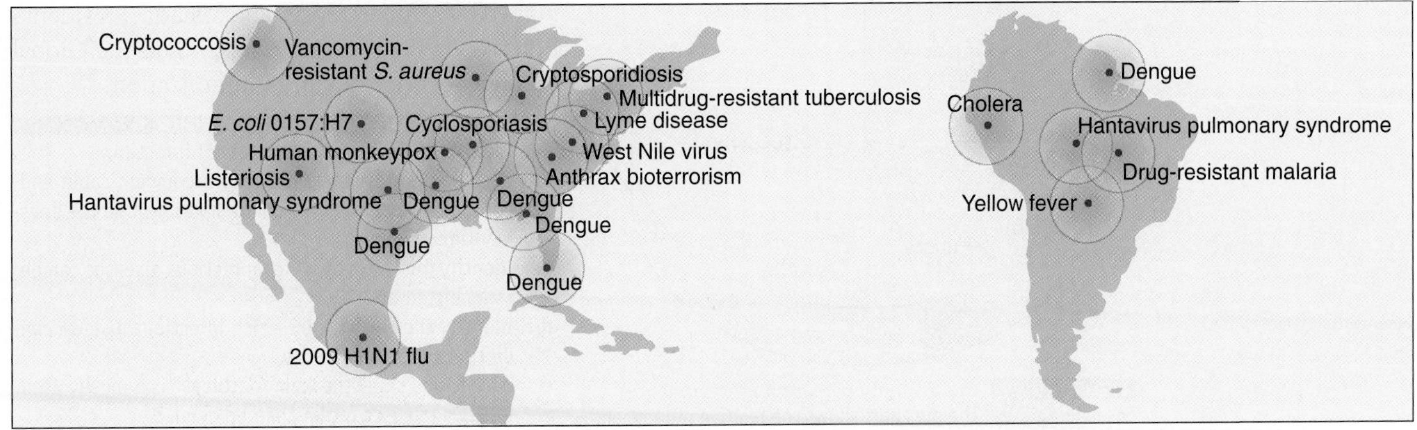

Emerging and re-emerging infectious diseases, and pathogens, in Europe, Africa, Asia, and Australia

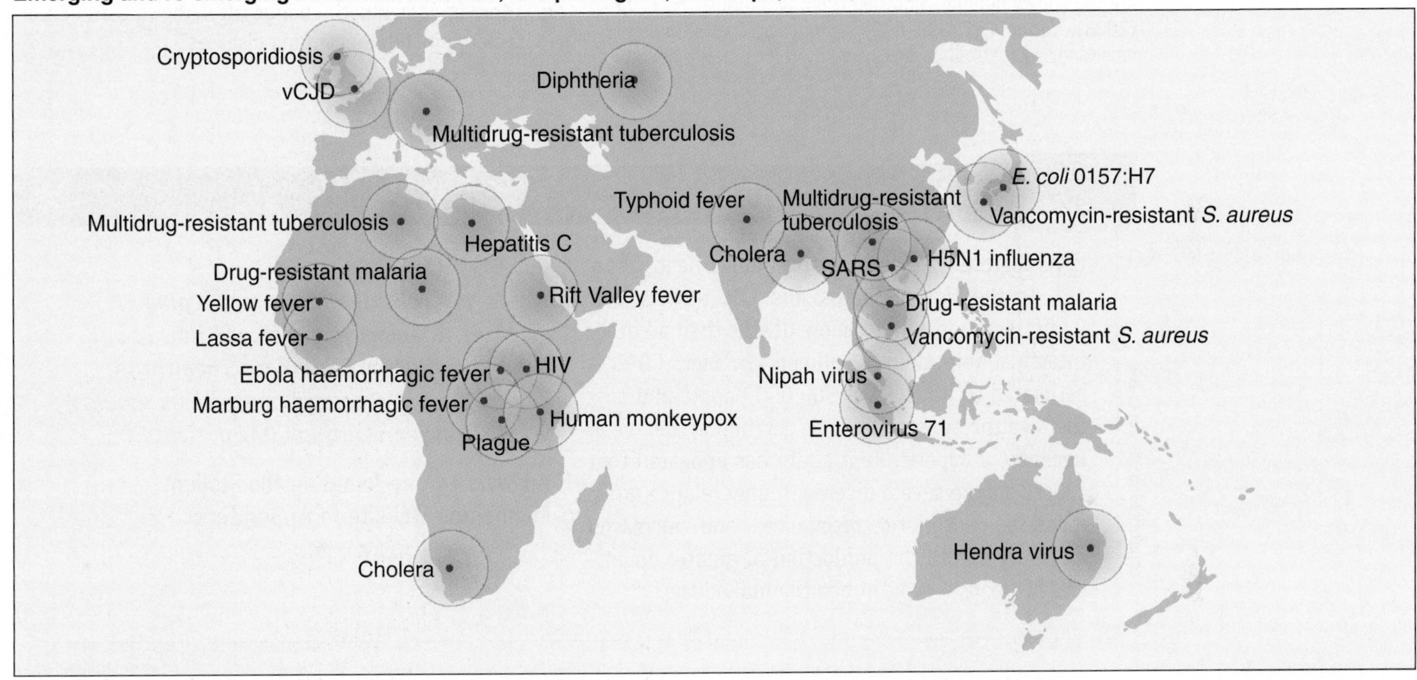

FIGURE 19.20 **The Global Occurrence of Emerging and Re-Emerging Disease.** Modified from American College of Microbiology. *Clinical Microbiology in the 21st Century: Keeping the Pace.* ASM Press, 2008, Washington, D.C. »» Can you suggest a reason why many of the diseases have appeared in North America?

The CDC has had a history of serving American public health. Regarding emerging and resurgent diseases, the CDC's priority areas include:

- International outbreak assistance to host countries to maintain control of new pathogens when an outbreak is over.
- A global approach to disease surveillance by establishing a global "network of networks" for early warning of emerging health threats.
- Applied research on diseases of global importance.
- Global disease control through initiatives to reduce HIV disease/AIDS, malaria, and tuberculosis.
- Public health training that supports the establishment of International Emerging Infections Programs in developing nations.

However, a new global health movement requires the involvement of more governmental and nongovernmental organizations. Happily, today there is an array of new organizations committed to this cause, including: the Bill and

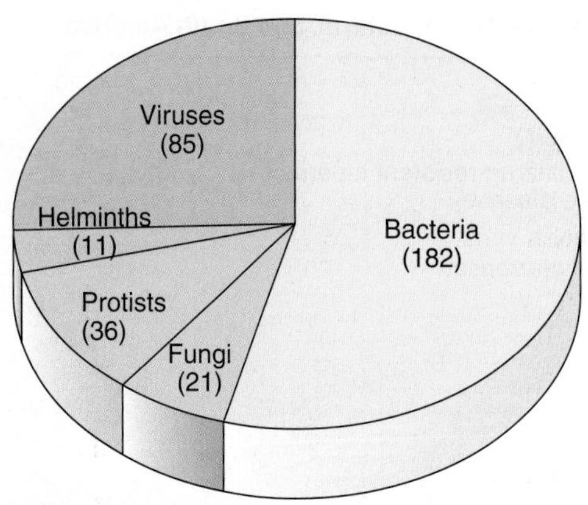

FIGURE 19.21 **The Diversity of Emerging/Resurgent Disease Agents.** The pie chart shows the relative proportions of emerging/resurgent diseases caused by bacteria, fungi, protists, helminths, and viruses. The numbers in parentheses are the known number. Modified from: Woolhouse, M. E. J. (2008). *Nature* **451**: 898–899. »» Name several prominent emerging/resurgent diseases caused by each group of microbes.

Melinda Gates Foundation; The Global Fund to Fight AIDS, Tuberculosis, and Malaria; President's Emergency Plan for AIDS Relief; and the Global Alliance for Vaccines and Immunization.

CONCEPT AND REASONING CHECKS 3

a. Identify the different reservoirs of infection.

b. Explain the difference between a communicable and contagious disease. Provide examples beyond the ones mentioned above.

c. Identify four ways by which infectious diseases can be transmitted directly.

d. Identify the methods by which infectious disease can be transmitted indirectly.

e. Why do you think the term "outbreak" is typically used in news releases rather than epidemic?

f. How do standard precautions limit the chains of transmission?

g. How might climate change alter the occurrence of emerging and re-emerging diseases?

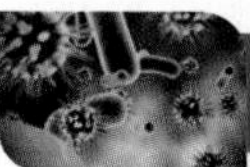

Chapter Challenge C

CDAI represents a considerable public health hazard. In the United States, the CDC reports that *C. diff* is responsible for more deaths than all other intestinal infections combined. In fact, CDAI is estimated to affect over 500,000 people and cost the healthcare system over $3 billion per year. Recently, a hypervirulent strain has appeared that causes a more severe disease, higher relapse rates, increased antibiotic resistance, and increased mortality due to the production of greater quantities of toxins—even in healthy individuals.

QUESTION C:

Would you consider CDAI an emerging disease? A reemerging disease? How about the hypervirulent strain? As a healthcare provider, what standard precautions should you take when treating CDAI patients?

Answers can be found on the Student Companion Website in **Appendix D**.

In conclusion, with the appearance of emerging and resurgent infectious diseases and the declining power of antibiotics, our ability to live a healthy, germ-free life seems unattainable. One of the luminaries in microbiology, the French-American microbiologist René Dubos (1901–1982), devoted much of his professional life to the study of microbial diseases and the mechanisms of resistance to infection. Dubos also considered the environmental and social factors that affect human health. In 1959, he wrote *Mirage of Health* in which he argues that "*Complete freedom from disease . . . is almost incompatible with the process of living.*" He argues that the possibility of a germ-free existence is a dream because science and medicine have ignored the dynamic process of adaptation that every microorganism (and every living organism) must face (e.g., antibiotic resistance) in a constantly changing environment (medicine's use of antibiotics). Dubos argues that infectious disease is not simply a battle between the "good guys" (humans) fighting the "bad guys" (pathogens). And, over 50 years after the publication of Dubos' book, we are losing the battle against many pathogens. We may not have lost the war, but we need to be aware of other factors, besides the causative organism, to effect good health. One factor is our immune system, which, along with sound sanitary practices, perhaps gives us our best chance to shield ourselves from infectious diseases.

SUMMARY OF KEY CONCEPTS

19.1 The Host and Microbe Form an Intimate Relationship in Health and Disease

1. **Infection** refers to competition between host and microbe for supremacy, while **disease** results when the microbe wins the competition. However, the human body contains large resident populations of **indigenous microbiota**, which usually outcompete invading pathogens. **Transient microbiota** are temporary residents of the body. (Fig. 19.3)
2. The human microbiota is established at birth and is influenced by the method of birth and feeding. (Fig. 19.4)
3. **Parasitism** is the symbiotic relationship occurring when the microbe does harm to the host. **Pathogenicity**, the ability of the pathogen to cause disease, and resistance go hand-in-hand. **Virulence** refers to the degree of pathogenicity a microbe displays.
4. Infections may come from an exogenous or endogenous source. If the immune system is compromised, microbes normally acting as commensals may cause **opportunistic infections.** A **primary infection** is an illness caused by a pathogen in an otherwise healthy host, while a **secondary infection** involves the development of other diseases as a result of the primary infection lowering host resistance. **Local diseases** are restricted to a specific part of the body while **systemic diseases** spread to several parts of the body and deeper tissues. (Fig. 19.5)

19.2 Establishment of Infection and Disease Involves Host and Pathogen

5. Most diseases have certain **signs** and **symptoms**, making it possible to follow the course of a disease. (Fig. 19.6)
6. To cause disease, most pathogens must enter the body through an appropriate **portal of entry**. The **infectious dose** represents the number of pathogens taken into the body that can cause a disease. Bacterial **adhesins** or viral spikes allow bacterial cells or viruses to adhere to specific cells. (Figs. 19.7, 19.8)
7. The possibility of disease is enhanced if a microbe can penetrate host tissues. The pathogen's ability to penetrate tissues and cause damage is called **invasiveness.** Virulence and invasiveness are strongly dependent on the spectrum of **virulence factors** a pathogen possesses. Adhesion or attachment usually leads to **phagocytosis** of the pathogens. (Figs. 19.9, 19.10)
8. Many bacterial species produce enzymes to overcome the body's defenses. These include **coagulase**, **streptokinase**, and **hyaluronidase**. In addition, some species produce **exotoxins** that are proteins released by gram-positive and gram-negative cells. Their effects depend on the host cells or tissues affected. **Endotoxins** are the lipopolysaccharides released from dead gram-negative cells. Their effects on the host are more universal. (Fig. 19.11 and Table 19.2)
9. To efficiently spread the disease to other hosts, the pathogen also must leave the body through an appropriate **portal of exit.** (Fig. 19.12)

19.3 Infectious Disease Epidemiology Includes Frequency and Spread of Disease

10. **Reservoirs** include humans, who represent carriers of a disease, arthropods, and any food and water in which some parasites survive.
11. A disease may be **communicable**, such as measles, or **noncommunicable**, such as tetanus.
12. Diseases may be transmitted by **direct** or **indirect** methods. Indirect methods include consumption of contaminated food or water, contaminated inanimate objects (**fomites**), and arthropods. Arthropods can be **mechanical** or **biological vectors** for the transmission of disease. (Fig. 19.14)
13. The occurrence of diseases falls into three categories: **endemic**, **epidemic**, and **pandemic.** An "**outbreak**" is essentially the same as an epidemic, although usually an outbreak is more confined in terms of disease spread.
14. **Nosocomial infections**, or **healthcare-associated infections**, are infections acquired as a result of being treated for some other injury or medical problem. **Standard precautions** limit the **chain of transmission.** (Figs. 19.16, 19.17)
15. Public health organizations, such as the CDC, are charged with the duty of limiting or stopping disease threats, including **emerging** and **re-emerging diseases.** (Fig. 19.20)

CHAPTER SELF-TEST

For **STEPS A–D**, answers to even-numbered questions and problems can be found in **Appendix C** on the Student Companion Website at **http://microbiology.jbpub.com/10e**. In addition, the site features eLearning, an online review area that provides quizzes and other tools to help you study for your class. You can also follow useful links for in-depth information, read more MicroFocus stories, or just find out the latest microbiology news.

STEP A: REVIEW OF FACTS AND TERMS

Multiple Choice

Read each question carefully, then select the ***one*** answer that best fits the question or statement.

1. Which one of the following is NOT part of the indigenous human microbiota?
   A. *Lactobacillus*
   B. *Candida*
   C. *Clostridium*
   D. *Escherichia*
2. A newborn _____.
   A. contains indigenous microbiota before birth
   B. remains sterile for many weeks after birth
   C. becomes colonized soon after conception
   D. is colonized with many common microbiota within minutes after birth
3. Factors affecting virulence may include _____.
   A. the presence of pathogenicity islands
   B. their ability to penetrate the host
   C. the infectious dose
   D. All the above (**A–C**) are correct.
4. A healthy person can be diagnosed as having a _____ infection with _____, the multiplication of bacterial cells in the blood.
   A. primary; bacteremia
   B. primary; viremia
   C. primary; septicemia
   D. secondary; parasitemia
5. Changes in body function sensed by the patient are called _____.
   A. symptoms
   B. syndromes
   C. prodromes
   D. signs
6. Adhesins can be found on _____.
   A. host cells
   B. viruses
   C. bacterial pili and capsules
   D. cells at the portal of entry
7. In the body, bacterial invasiveness can be limited by _____.
   A. fever
   B. phagocytosis
   C. enzyme production
   D. toxin production
8. Which one of the following is NOT true of exotoxins?
   A. They are proteins.
   B. They are part of cell wall structure.
   C. They are released from live bacterial cells.
   D. They trigger antibody production.
9. A portal of exit would be _____.
   A. the feces
   B. an insect bite
   C. blood removal
   D. All of the above (**A–C**) are correct.
10. If a person has recovered from a disease but continues to shed disease agents, that person is a _____.
    A. vector
    B. fomite
    C. vehicle
    D. carrier

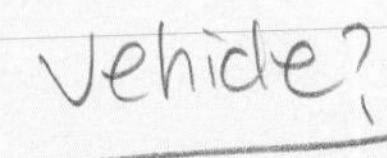

11. All of the following are examples of communicable diseases *except* _____.
    A. chickenpox
    B. measles
    C. the common cold
    D. tetanus
12. Which one of the following is an example of transmission by indirect contact?
    A. Coughing
    B. Aerosols
    C. Sexual intercourse
    D. Mother to fetus
13. Fifty cases of hepatitis A during one week in a community would most likely be described as a/an _____.
    A. outbreak
    B. pandemic
    C. endemic disease
    D. epidemic
14. The most common nosocomial infection involves _____.
    A. the blood
    B. the lungs
    C. the urinary tract
    D. a surgical site
15. Emerging infectious diseases can be the result of _____.
    A. pathogen evolution
    B. international trade
    C. agricultural practices
    D. All of the above (**A–C**) are correct.

True-False

Each of the following statements is true (T) or false (F). If the statement is false, substitute a word or phrase for the underlined word or phrase to make the statement true.

_____ **16.** An epidemic disease occurs at a low level in a certain geographic area.

_____ **17.** Among the microbial toxins able to destroy blood cells are hemolysins and leukocidins.

_____ **18.** The term disease refers to a symbiotic relationship between two organisms and the competition taking place between them for supremacy.

_____ **19.** Organs of the human body lacking a normal microbiota include the blood and the small intestine.

_____ **20.** Commensalism is a form of symbiosis where the microbe benefits and causes no damage to the host.

_____ **21.** A biological vector is an arthropod that carries pathogenic microorganisms on its feet and body parts.

_____ **22.** Organisms causing disease when the immune system is depressed are known as opportunistic organisms.

_____ **23.** The human body responds to the presence of exotoxins by producing endotoxins.

_____ **24.** The term bacteremia refers to the spread of bacteria through the bloodstream.

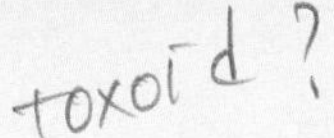

25. A toxoid is an immunizing agent prepared from an exotoxin.

26. Few symptoms are exhibited by a person who has a subclinical disease.

27. Indirect methods of disease transmission include kissing and handshaking.

28. A chronic disease develops rapidly, is usually accompanied by severe symptoms, and comes to a climax.

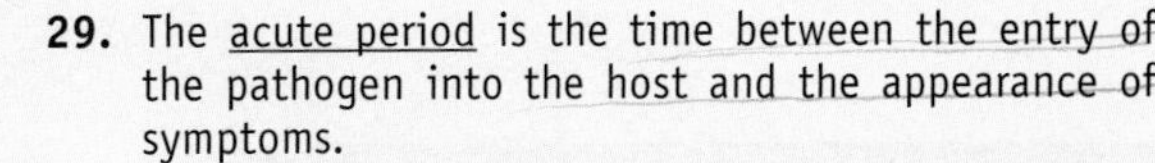

29. The acute period is the time between the entry of the pathogen into the host and the appearance of symptoms.

30. Symptoms are changes in body function detected by a physician.

STEP B: CONCEPT REVIEW

31. Distinguish between **infection** and **disease**, and between **indigenous** and **transient microbiota**. (**Key Concept 1**)
32. Explain how the human microbiota is first established. (**Key Concept 1**)
33. Contrast **pathogenicity** and **virulence**, explaining how each affects the establishment of disease. (**Key Concept 1**)
34. Discuss the consequences of **exogenous** and **endogenous** (including opportunistic) infections on the progression and outcomes of infection and disease. Distinguish between (a) **primary** and **secondary infections**, (b) **local** and **systemic diseases**, and (c) **bacteremia** and **septicemia**. (**Key Concept 1**)
35. Explain the differences between **signs**, **symptoms**, and **syndromes**. Identify the characteristics that compose the five stages in the course of disease development. (**Key Concept 2**)
36. Assess the role of the **infectious dose** and pathogen adhesion to establishing an infection and disease. (**Key Concept 2**)
37. Discuss the importance of **invasiveness** to the establishment of an infection. (**Key Concept 2**)
38. Name three enzymes and describe their roles as **virulence factors**. Summarize the differences between **exotoxins** and **endotoxins** as virulence factors associated with disease. (**Key Concept 2**)
39. Summarize the characteristics of **reservoirs** as applied to infectious disease. (**Key Concept 3**)
40. Distinguish between a **communicable**, **contagious**, and **noncommunicable disease** and give an example of each. (**Key Concept 3**)
41. Distinguish between the **direct** and **indirect contact** methods of disease transmission. (**Key Concept 3**)
42. Discuss the three types of disease occurrence within populations. (**Key Concept 3**)
43. Explain how **nosocomial infections** can be controlled or eliminated through using the **standard precautions** to break the **chain of transmission**. (**Key Concept 3**)
44. Identify the drivers responsible for **emerging** and **resurgent infectious diseases**. (**Key Concept 3**)

STEP C: APPLICATIONS AND PROBLEMS

45. The transparent covering over salad bars is commonly called a "sneeze guard" because it helps prevent nasal droplets from reaching the salad items. As a community health inspector, what other suggestions might you make to prevent disease transmission at the salad bar?
46. While slicing a piece of garden hose, your friend cut himself with a sharp knife. The wound was deep, but it closed quickly. Shortly thereafter, he reported to the emergency room of the community hospital, where he received a tetanus shot. What did the tetanus shot contain, and why was it necessary?
47. After reading this chapter, you decide to make a list of the ten worst "hot zones" in your home. The title of your top ten list will be "Germs, Germs Everywhere." What places will make your list, and why?
48. As a state epidemiologist responsible for identifying any disease occurrences, would an epidemic disease or an endemic disease pose a greater threat to public health in the community? Explain.

STEP D: QUESTIONS FOR THOUGHT AND DISCUSSION

49. In 1840, Great Britain introduced penny postage and issued the first adhesive stamps. However, politicians did not like the idea because it deprived them of the free postage they were used to. Soon, a rumor campaign was started, saying that these gummed labels could spread disease among the population. Can you see any wisdom in their contention? Would their concern "apply" today?
50. In 1892, a critic of the germ theory of disease named Max von Pettenkofer sought to discredit Robert Koch's work by drinking a culture of cholera bacilli diluted in water. Von Pettenkofer suffered nothing more than mild diarrhea. What factors may have contributed to the failure of the bacilli to cause cholera in von Pettenkofer's body?
51. A man takes a roll of dollar bills out of his pocket and "peels" off a few to pay the restaurant tab. Each time he peels, he wets his thumb with saliva. What is the hazard involved?
52. When Ebola fever broke out in Africa in 1995, disease epidemiologists noted how quickly the responsible virus killed its victims and suggested the epidemic would end shortly. Sure enough, within 3 weeks it was over. What was the basis for their prediction? What other conditions had to apply for them to be accurate in their guesswork?
53. A woman takes an antibiotic to relieve a urinary tract infection caused by *Escherichia coli*. The infection resolves, but in 2 weeks, she develops a *Candida albicans* ("yeast") infection of the vaginal tract. What conditions may have caused this to happen?

20

CHAPTER PREVIEW

20.1 The Immune System Is a Network of Cells and Molecules to Defend Against Foreign Substances

20.2 Surface Barriers Are Part of Innate Immunity

Investigating the Microbial World 20: Who Goes There?

20.3 Coordinated Cellular Defenses Respond to Pathogen Invasion

20.4 Effector Molecules Damage Invading Pathogens

Textbook Case 20: A Clinical History of Bacterial Infections

MicroInquiry 20: Visualizing the Bridge Between Innate and Adaptive Immunity

Image courtesy of Dr. Fred Murphy/CDC.

Resistance and the Immune System: Innate Immunity

". . . medical science today has set itself the task of attempting to prevent disease. In order to achieve this aim one must attempt, on the one hand to find the disease germ and destroy it, and on the other hand to give the body the strength to resist attack."
—K. A. H. Mörner, Rector of the Royal Caroline Institute on presenting the 1908 Nobel Prize in Physiology or Medicine to Elie Metchnikoff and Paul Ehrlich for their work on the theory of immunity

Edward Jenner's epoch-making moment in history came in 1798 when he demonstrated the protective action of vaccination against smallpox. As great as Jenner's discovery was, it did not advance the development or understanding of **immunity**, which is how the human body can generate resistance to a particular disease, whether by recovering from the disease or as a result of vaccination, such as devised by Jenner.

It would be another 90 years before Russian zoologist and immunologist Elie Metchnikoff devised the first experiments to study immunity by investigating how an organism could destroy a disease-causing microbe in the body. Through landmark studies with invertebrates, Metchnikoff proposed that certain types of cells could attack, engulf, and destroy foreign material, including infectious microbes. These were among many experiments from a chain of investigations that studied immunity and then applied that knowledge to mammals and humans. Metchnikoff's work culminated in the theory of phagocytosis, which suggested certain types of human white blood cells could capture and destroy disease-causing microbes that had penetrated the body.

Around the same time, Louis Pasteur developed vaccines for chicken cholera and rabies, further stimulating an interest in the mechanism for protective immunity. In Pasteur's lab, Émile Roux and Alexandre Yersin discovered that the bacterial species causing diphtheria actually produced a toxin causing the disease. Then, Koch's coworker Emil von Behring modified the diphtheria toxin to produce a substance with "antitoxic activity" that successfully immunized animals for a short period against diphtheria. The substance gave the body the power to resist infection.

With these early discoveries, the science of immunology was off and running. Today, Metchnikoff's work forms part of the basis for "innate immunity," which consists of several nonspecific defenses present in most animals. The discoveries of Roux, Yersin, and von Behring are the basis for part of what is referred to today as "adaptive immunity." This form of protective immunity or resistance is found only in vertebrates. It is a response to a specific pathogen and is directed only against that pathogen. Together, innate and adaptive immunity are equally important arms that form a "microbiological umbrella" protecting us against the torrent of potential microbial pathogens to which we are continually exposed (FIGURE 20.1).

This chapter examines innate immunity, the body's immediate response to pathogens, toxins, and other foreign substances. It will serve as the foundation for adaptive immunity, which is described in another chapter.

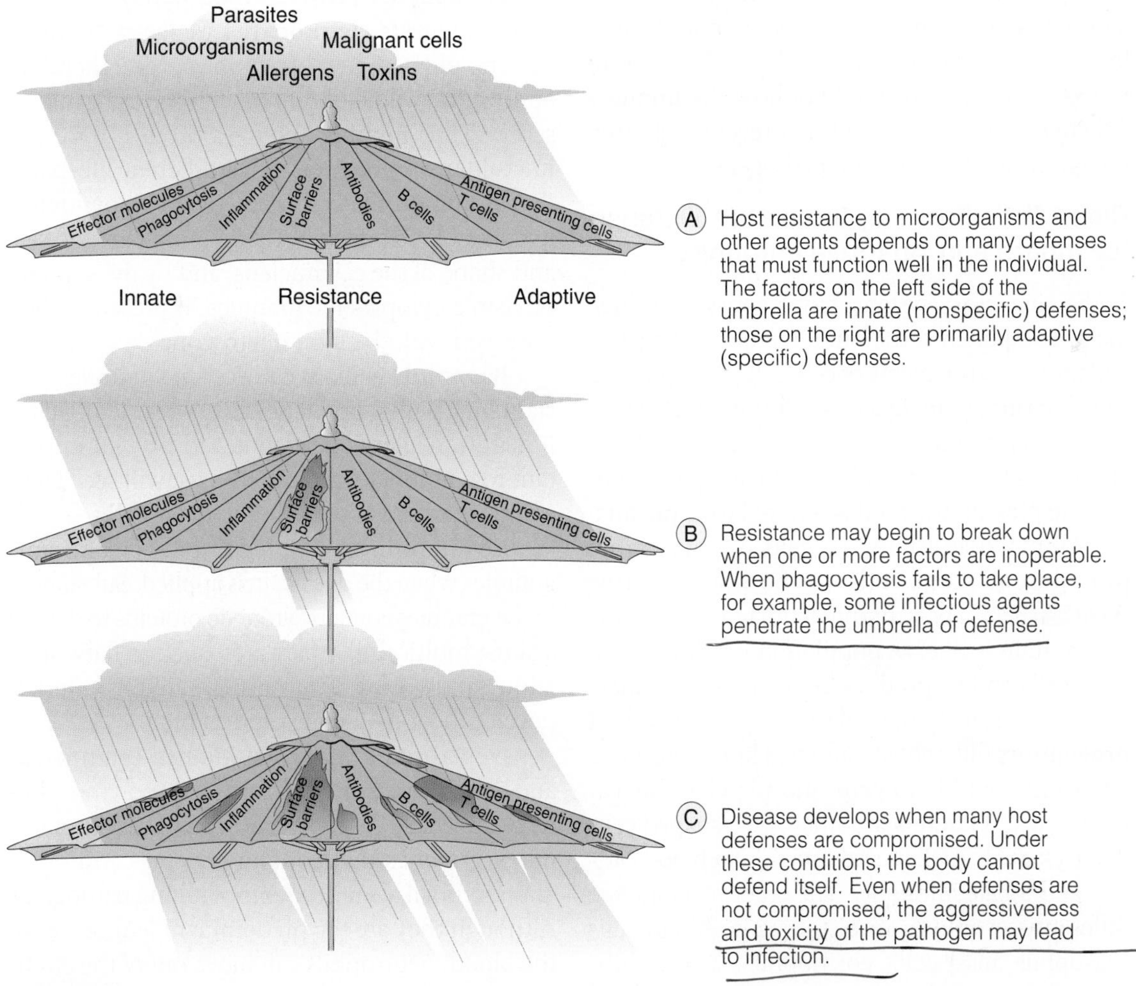

FIGURE 20.1 The Relationship Between Host Resistance and Disease. Host resistance can be likened to a "microbiological umbrella" that forms a barrier or defense against infectious disease. »» Why are the defenses on the left side of the umbrella considered innate and those on the right adaptive?

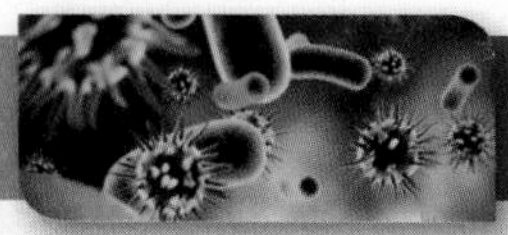

Chapter Challenge

Influenza, commonly called the flu, is a contagious respiratory illness caused by influenza viruses that infect the nose, throat, and eventually the lungs. The type A flu virus is responsible for about 70% of human seasonal flu cases. The virus has two types of surface protein spikes, called hemagglutinin (H) and neuraminidase (N). The H spikes are needed for the virus to attach to and penetrate host cells of the respiratory tract while the N spikes are needed for the release of the virus from the infected cells. The disease can be mild to severe and at times can lead to death. So, what happens immunologically when we are exposed to the type A flu virus? Let's examine the innate immune response as we proceed through this chapter. Warning: we won't have recovered from the flu by the end of this chapter! We will have such a severe case that we will need to continue the investigation into the chapter on adaptive immunity before we get better. So, let's catch the flu!

KEY CONCEPT 20.1 The Immune System Is a Network of Cells and Molecules to Defend Against Foreign Substances

The work of Metchnikoff, Roux, Yersin, and von Behring as well as that of their contemporaries and followers, opened up the whole field of **immunology**, the scientific study of how the immune system functions in the body to prevent or destroy foreign material, including pathogens.

Blood Cells Form an Important Defense for Innate and Adaptive Immunity

In the circulatory system, blood consists of three major components: the fluid, the clotting agents, and the formed elements or cells. The fluid portion, called **serum**, is an aqueous solution of minerals, salts, proteins, and other organic substances. When clotting agents, such as fibrinogen and prothrombin, are present, the fluid is referred to as **plasma**.

All blood and immune cells originate from **pluripotent stem cells** in the bone marrow (FIGURE 20.2). These undifferentiated cells divide to maintain a reserve population of pluripotent stem cells and to produce cells that differentiate into myeloid and lymphoid cell lines. The **myeloid progenitors** differentiate into two lines. One gives rise to the red blood cells and platelets (megakaryocytes). **Platelets** are small, disc-shaped cells that lack a cell nucleus and function in blood clotting and innate immunity. The other myeloid line differentiates into most of the white blood cells (basophils, mast cells, eosinophils, neutrophils, monocytes, and dendritic cells). The **lymphoid progenitors** differentiate into B and T lymphocytes, natural killer (NK) cells, and some dendritic cells.

Leukocytes (White Blood Cells). As their name suggests, **leukocytes** (*leuko* = "white") have no pigment in their cytoplasm and therefore appear gray when observed with the light microscope. They number about 4,000 to 12,000 per microliter of blood and have different lifespans, depending on the cell type. Most can be identified in a blood smear or tissue sample by the size and shape of the cell nucleus, and by the staining of visible cytoplasmic granules, if present, when observed with the light microscope.

Basophils contain cytoplasmic granules that stain blue with methylene blue. These leukocytes function in allergic reactions and play an important role in innate immunity as we will see. Their counterparts in tissues are the **mast cells**.

■ Methylene blue: A blue cationic (basic) dye.

Eosinophils exhibit red or pink cytoplasmic granules when the dye eosin is applied. Substances in the granules contain cytotoxic proteins to defend against multicellular parasites, such as flatworms and roundworms. They also are involved in the development of allergies and asthma.

■ Eosin: A red anionic (acidic) dye.

Neutrophils have a multilobed cell nucleus and therefore are also referred to as "polymorphonuclear" cells (PMNs). Their cytoplasm contains many granules that contain enzymes and other antimicrobial agents capable of killing pathogens. Although they are the predominant leukocytes in the blood, neutrophils can move out of the circulation to engulf by phagocytosis foreign particles or pathogens (bacterial cells, viruses) at the site of an infection. Their lifespan is short, only 1 to

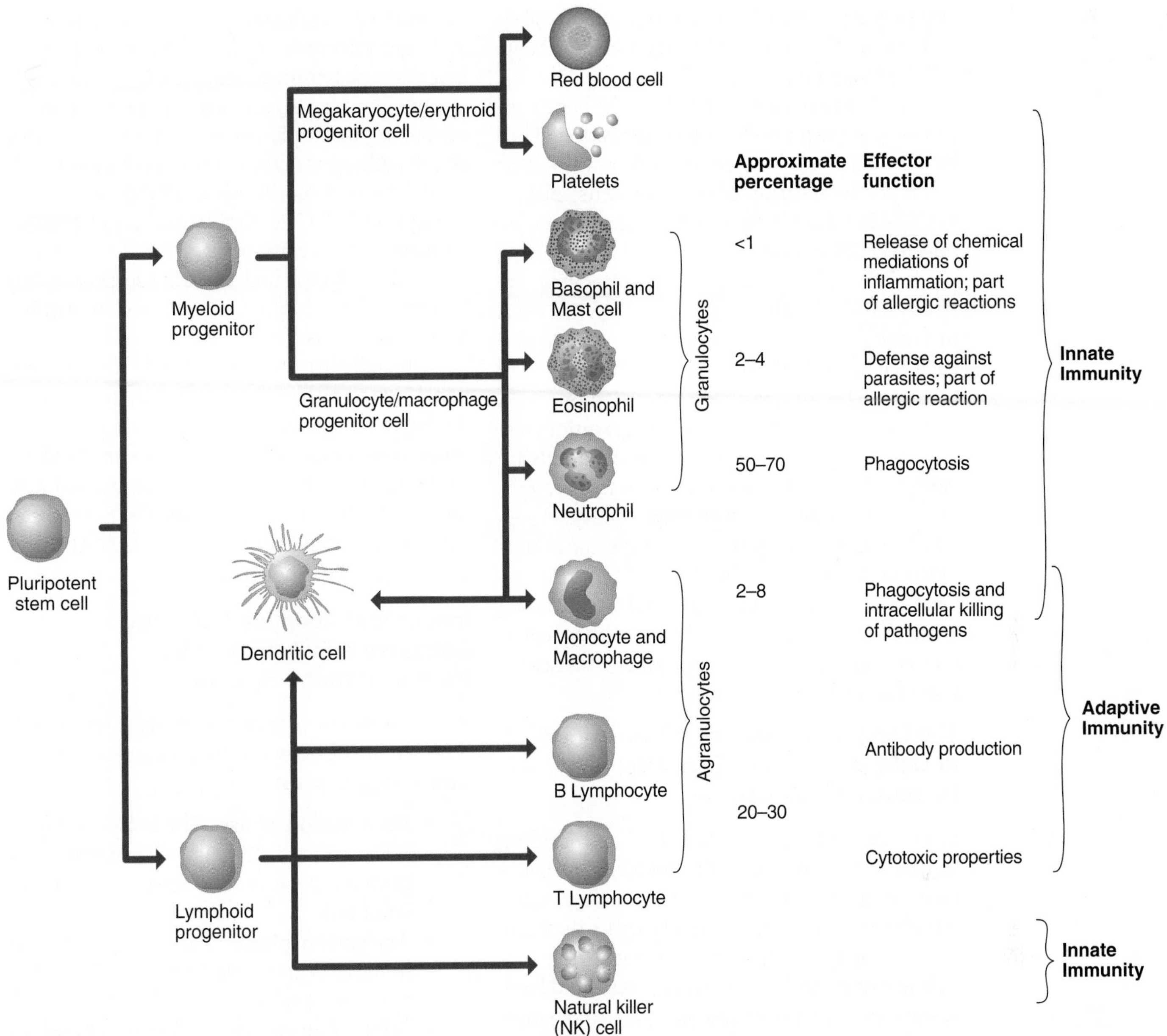

FIGURE 20.2 **Major Types of Leukocytes and Their Origin.** All blood cells, including the white blood cells, arise from pluripotent stem cells in the bone marrow. The approximate percentage in the blood and their effector functions are given. The dendritic cell numbers are tissue specific. »» Explain what may have occurred if an individual lacked neutrophils, basophils, eosinophils, and monocytes/macrophages.

2 days, so they are continually replenished from the bone marrow.

Basophils, eosinophils, and neutrophils are concentrated in the skin, lungs, and the gastrointestinal tract, locations where breaks in the skin, inhalation, or ingestion might be portals of entry for pathogens. Because these leukocytes contain cytoplasmic granules, they often are called **granulocytes**.

Another major group of leukocytes is the **monocytes**. These cells circulate in the blood, have a single, bean-shaped cell nucleus, and lack visible cytoplasmic granules. In infected tissues, monocytes mature into **macrophages**. In contrast to the 2-week lifespan of monocytes, macrophages may live for several months. Macrophages (*macro* = "big"; *phage* = "eat") are more effective at phagocytosis of bacterial cells and viruses than neutrophils and are one of the key cells in both innate and acquired immunity. As such, they are tissue residents in the spleen, lymph nodes, and thymus, and are common immune cells in the lungs, skin, and nervous system. Because both neutrophils and macrophages carry out phagocytosis, they are called **phagocytes**.

The **lymphocytes** migrate from the bone marrow to the lymph nodes after maturation. They have a single, large nucleus and no granules. One cell type, the **natural killer** (**NK**) **cells**, plays a key role in innate immunity by destroying virus-infected cells and tumor cells. The two other cell types, **B lymphocytes** and **T lymphocytes**, are key cells of adaptive immunity. They increase in number dramatically during the course of a bacterial or viral infection but have very different functions. Lacking visible granules, monocytes and lymphocytes often are called **agranulocytes**.

The last group of leukocytes is the **dendritic cells** (DCs). The name comes from their resemblance to the long, thin extensions (dendrites) seen on neurons. DCs are positioned at specific portals in the body, including the skin and mucosal surfaces, where they can rapidly respond to invading bacterial or viral pathogens. As active phagocytes, they are crucial to innate immunity and the activation of adaptive immunity.

The Lymphatic System Is Composed of Cells and Tissues Essential to Immune Function

In the human body, the clear fluid surrounding the tissue cells and filling the intercellular spaces is called "interstitial fluid." Part of that fluid is the **lymph** (*lymph* = "water"), which bathes the body cells, supplying oxygen and nutrients, while collecting wastes, and transporting excess fluid back to the blood. To move these materials, the lymph must be pumped through tiny vessels by the contractions of skeletal muscle cells. Eventually the lymph vessels unite to form larger lymphatic vessels. Along the way, lymph nodes filter the lymph before the lymph returns via the thoracic duct to the bloodstream.

It is the lymphatic tissues in which lymphocytes mature, differentiate, and divide (FIGURE 20.3A). The **primary lymphoid tissues** consist of the **thymus**, which lies behind the breast bone, and the **bone marrow**. Both are sites where T and B lymphocytes form or mature.

The **secondary lymphoid tissues** are sites where mature immune cells interact with pathogens and carry out the adaptive immune response. The **spleen**, a flattened organ at the upper left of the abdomen, contains immune cells to monitor and fight infectious microbes entering the body. Likewise, the **lymph nodes**, prevalent in the neck, armpits, and groin, are bean-shaped structures containing macrophages and dendritic cells that engulf pathogens in the lymph, and lymphocytes, which respond specifically to foreign substances brought in from the circulation (FIGURE 20.3B). Because resistance mechanisms are closely associated with the lymph nodes, it is not surprising they become enlarged (often called "swollen glands") during infections.

Other organ systems in the body contain additional secondary lymphoid tissues. In the intestine, the Peyer patches and appendix form part of the **mucosa-associated lymphoid tissue** (**MALT**), while secondary lymphoid tissues associated with the respiratory tract are the **tonsils**. Specialized lymphocytes and dendritic cells of the MALT help defend against pathogen infection.

Innate and Adaptive Immunity Compose a Fully Functional Human Immune System

When the invader (pathogen, allergen, tumor cell) is present in the body, the immune system initiates a three-step response:

- **Recognition phase**. The body's immune cells need to distinguish normal body cells from the invader (self versus nonself recognition);
- **Activation phase**. The appropriate immune cells and molecules are mobilized to fight the invader;
- **Effector phase**. The mobilized cells (effector cells) and molecules (effector molecules) attempt to eliminate the invader.

The immune system consists of two overlapping defenses, innate and adaptive immunity, when responding to an invader, such as a pathogen.

Innate Immunity. The first response to a pathogen stimulates a nonspecific defense called **innate immunity** because genetically encoded molecules, present in the body from birth, and immune cells, are capable of recognizing microbial features common to diverse pathogens and foreign substances. Therefore, in the case of an infection, innate immunity represents the host's early-warning system that responds within minutes or a few hours after infection.

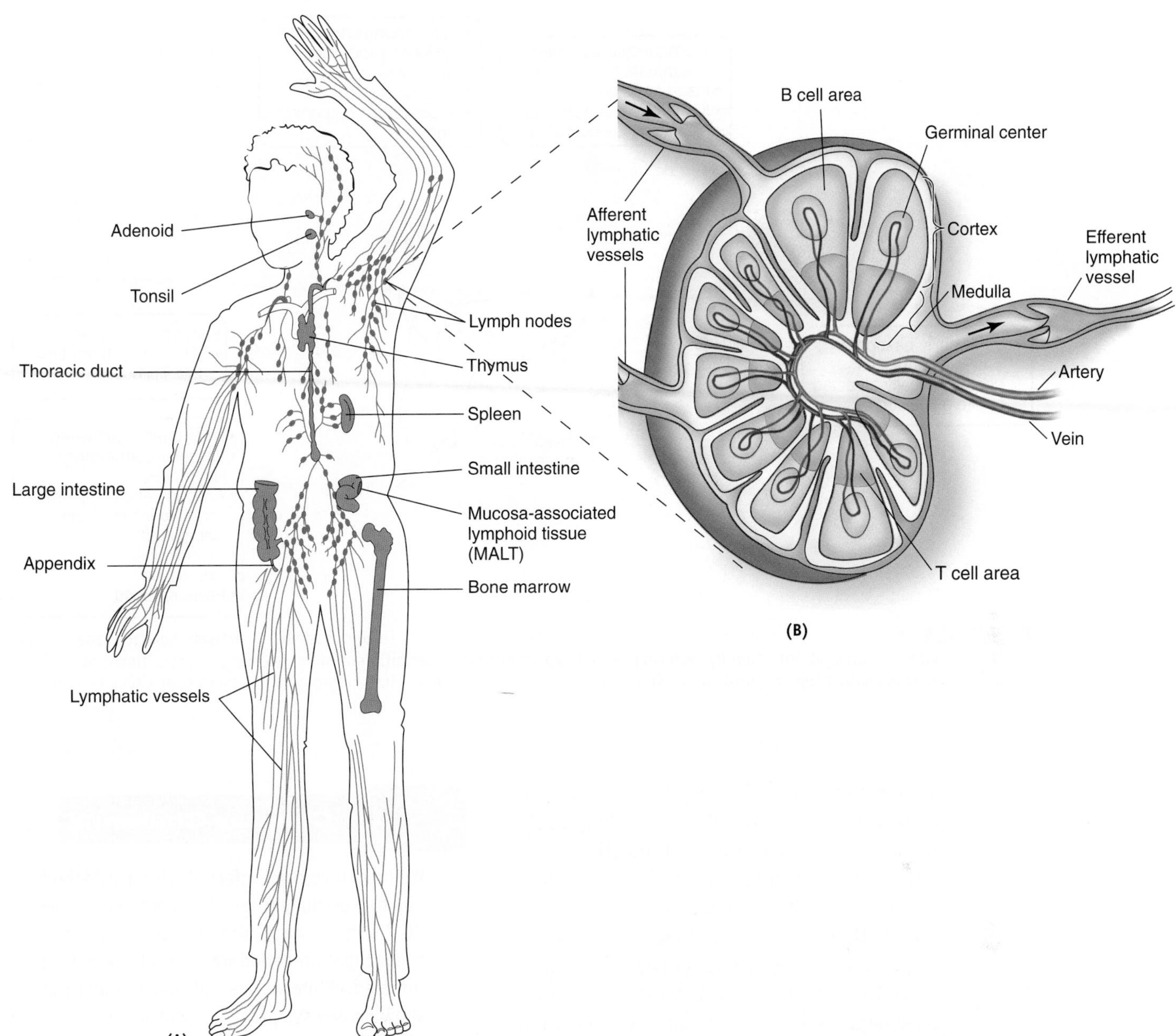

FIGURE 20.3 **The Human Lymphatic System.** (**A**) The human lymphatic system consists of lymphocytes, lymphatic organs, lymph vessels, and lymph nodes located along the vessels. The lymphatic organs are illustrated, and the preponderance of lymph nodes in the neck, armpits, and groin is apparent. (**B**) A lymph node is a filtering site for circulation of lymph. Any microbes present are removed by macrophages and dendritic cells before the "cleaned" lymph exists at an efferent lymphatic vessel. The outer cortex and germinal centers are sites rich in B cells while the deep cortex and medulla contain large numbers of T cells. »» Why are there so many lymph nodes distributed throughout the body?

Upon mobilization, not only do the effector cells and molecules try to eliminate or hold an infection in check, they also send additional chemical signals to tissues involved with initiating adaptive immunity.

Adaptive Immunity. This form of defense is called **adaptive immunity** because the system must first adapt to (acquire) information about the invading pathogen. Therefore, the response is relatively slow compared to the innate response; in fact, the effector phase (antibodies and lymphocytes), takes several days to more than a week to generate an effective response and a protective defense.

Much of this chemical communication that is vital to innate and adaptive immunity involves signaling molecules produced at the onset and throughout the duration of infection. These molecules, called **cytokines** (*kine* = "movement")

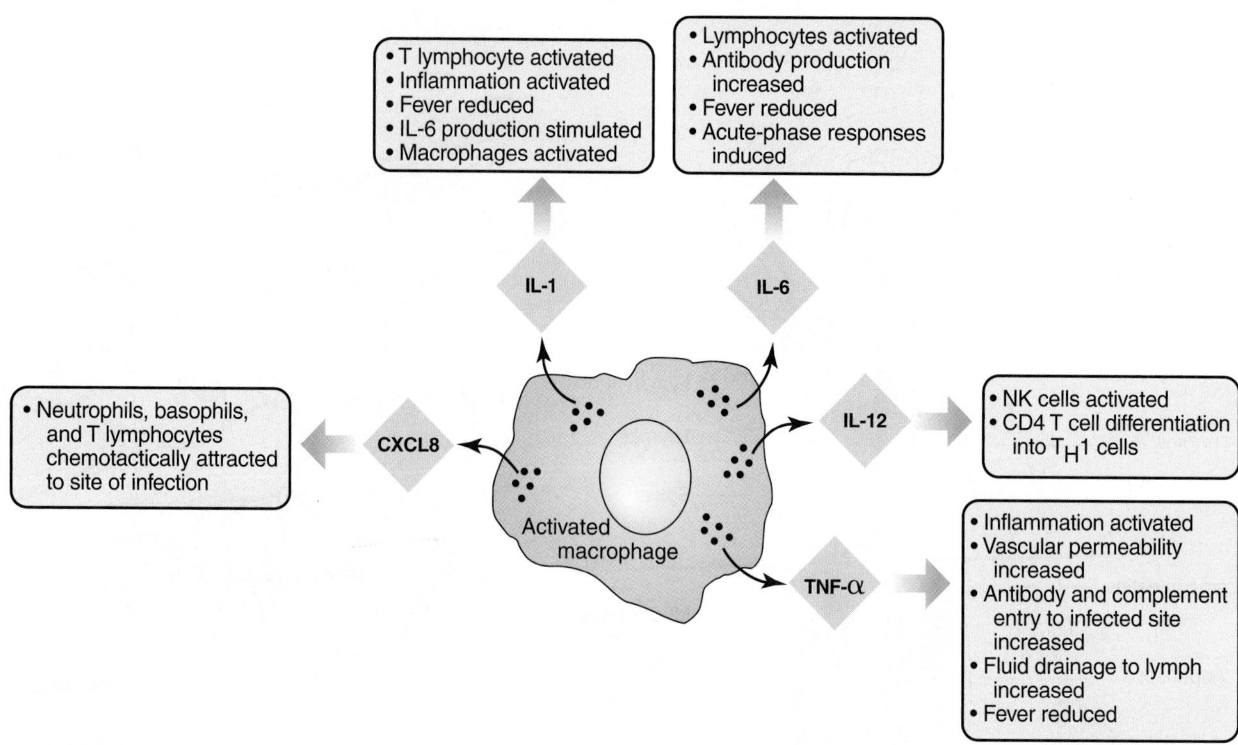

FIGURE 20.4 **An Example of the Diverse Cytokine Secretion by Immune Cells.** In this example, an activated macrophage is capable of responding to pathogen infection by secreting several cytokines that target different immune defensive cells. Tumor necrosis factor-α (TNF-α) is another type of cytokine. »» From the description of function, what subgroup of cytokines does CXCL8 belong?

are regulatory and effector proteins that operate at extremely low concentrations. Many of these mediators are called **interleukins (ILs)** because they mediate reactions between leukocytes locally or systemically. They are identified by number (such as IL-1, IL-2, IL-12), and these effector proteins bind to cytokine receptors on target cells, activating genes that alter the behavior and function of those target cells. Another subgroup of cytokines is the **chemokines** that are involved in triggering the attraction (chemotaxis) of leukocytes to sites of infection. Although many immune and nonimmune cells can produce cytokines, the principal cytokine producers are T lymphocytes, dendritic cells, and macrophages (FIGURE 20.4). We will become more familiar with cytokine action as we examine the innate immune defenses in this chapter.

CONCEPT AND REASONING CHECKS 1

a. In a blood smear, how could you tell if a patient was deficient in neutrophils?
b. Assess the role of the lymphatic system to defense against pathogens.
c. Summarize the need for chemical communication between immune defensive cells.

Chapter Challenge A

We have now been infected with a sufficient number of flu viruses that contracting the disease is looming, although at this point in the incubation period we do not know we are infected. While we are still healthy and not showing any symptoms, consider this:

QUESTION A:

Most viral infections, including the flu, are controlled by the innate and adaptive arms of the immune system. Make a checklist of the types of white blood cells discussed in this section of the chapter that would most likely be involved in identifying flu viruses and activated to fight the infection. What role will the lymphatic system play in optimizing the response?

Answers can be found on the Student Companion Website in **Appendix D.**

KEY CONCEPT 20.2 Surface Barriers Are Part of Innate Immunity

The innate defenses to infection first involve a broad group of nonspecific surface barriers through which a pathogen would have to pass.

Host Defensive Barriers Limit Pathogen Entry

There are several portals of entry by which pathogens can enter the body. One is the intact skin and the mucous membranes, which extend over the body surface or into the body cavities, providing barriers to colonization and infection of deeper tissues.

Physical Barriers. The **skin** is a multilayered tissue and, as such, forms a tough and impenetrable barrier to infection FIGURE 20.5. In addition, the thin outer epidermis is constantly being sloughed off, and with it goes any attached microorganisms. The skin also is a poor source of nutrition for invading pathogens and the low water content of the skin presents a veritable desert. Unless pathogens can penetrate the skin barrier, disease is rare, although some fungal dermatophytes can establish localized diseases like athlete's foot and ringworm on the skin surface.

The **mucous membranes** (**mucosa**) are the moist epithelial lining of the digestive, urogenital, and respiratory tracts. Because these exposed sites represent potential portals of entry, these surfaces provide protection from pathogen invasion. For example, cells of the mucous membranes that line the respiratory passageways secrete **mucus**, a sticky secretion of glycoproteins that traps heavy particles, microorganisms, and viruses (MICROFOCUS 20.1). If the pathogens are taken deeper into the upper respiratory tract, they may still be trapped in the respiratory mucus. However, the epithelial cilia then move the microbe-laden mucus up to the throat, where the material is swallowed, coughed or sneezed out.

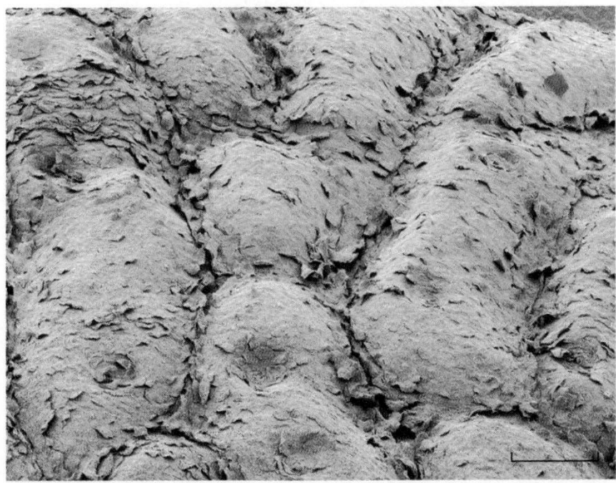

FIGURE 20.5 **The Outer Surface of the Human Skin.** This false-colored scanning electron micrograph shows the top layer (epidermis) of the skin. It is composed of dead cells that are continuously sloughed off and replaced with new cells from below. (Bar = 10 um.) © Steve Gschmeissner/Photo Researchers, Inc. »» From a potential colonization standpoint, what is the advantage of dead skin cells being sloughed off?

Chemical Barriers. Several innate effector molecules also provide an antimicrobial defense to infection by inhibiting microbial growth or viral replication. Resistance in the vaginal tract is enhanced by the low pH. This develops when resident *Lactobacillus* species break down glycogen to various acids. Many researchers believe the disappearance of lactobacilli during antibiotic treatment allows opportunistic diseases such as candidiasis (vaginal yeast infection) to develop.

A natural barrier to the gastrointestinal tract is provided by stomach acid, which, with a pH of approximately 2.0, destroys most pathogens. However, there can be notable survivors, including polio and hepatitis A viruses, the typhoid and tubercle bacilli, and *Helicobacter pylori,* a major cause of peptic ulcers. Bile from the gallbladder enters the system at the duodenum and serves as an inhibitory substance. In addition, duodenal enzymes hydrolyze the proteins, carbohydrates, lipids, and other large molecules of microorganisms.

■ Bile: A yellowish-green fluid, which in the small intestine plays an essential role in emulsifying fats.

The chemical barriers also consist of numerous small, cationic, antimicrobial peptides, called **defensins** that damage membranes, killing the pathogens through cell lysis. The defensins are produced by immune cells, such as neutrophils and NK cells, as well as epithelial cells of the respiratory, gastrointestinal, and urogenital tracts. The sebum on the skin surface contains several defensins. In the body, some defensins are continually produced, while others are induced by microbial products. They are active against many gram-negative and gram-positive bacterial cells, fungi, and

■ Sebum: An oily substance secreted by the sebaceous glands to lubricate the hair and skin.

MICROFOCUS 20.1: Miscellaneous
Who Turned On the Spigot?

© nazira_g/ShutterStock, Inc.

Every time we get a cold, the flu, or a seasonal allergy, we often end up with the sniffles or a truly raging runny nose. When this happens, it is simply your body's response to what it believes is an infection—or a hypersensitivity to cold temperatures or spicy food. As a defensive barrier against infection, innate immunity uses mucus flow as the best way to wash respiratory pathogens out of the airways.

You always are producing—and swallowing—mucus. Most people are not aware of their mucus production until their body revs up mucus secretion in response to a cold or flu virus, or allergens. So how much mucus is produced?

In a healthy individual, glands in the nose and sinuses are continually producing clear and thin mucus—often more than 200 milliliters each day! An individual is not aware of this production because the mucus flows down the throat and is swallowed. Now if that individual comes down with a cold or the flu, the nasal passages often become congested, forcing the mucus to flow out through the nostrils of the nose. This requires clearing by blowing the nose or (to put it nicely) expectorating from the throat. With a serious cold or flu, the mucus may become thicker and gooier, and have a yellow or green color. The revved up mucus flow in such cases can amount to about 200 milliliters (a little less than a cup) every hour; if you blow your nose 20 times an hour, each blow could amount to anywhere from 2 to 10 milliliters of mucus. If you have watery eyes as well, then that tear liquid can enter the nasal passages and combine with the mucus, producing an even larger "flow per blow."

So, although a runny nose is usually just an annoyance, make sure you drink plenty of water to make up for the lost mucus from the runny spigot.

viruses. Investigating the Microbial World 20 examines how specific these peptides are using fruit flies as test subjects.

In a somewhat similar manner, **lysozyme** is a chemical inhibitor found in human tears, mucus, and saliva. The enzyme disrupts the cell walls of gram-positive bacterial cells by weakening the peptidoglycan. Osmotic lysis results in bacterial cell death.

Other soluble proteins, such as the complement components and interferon, are important physiological barriers in innate immunity and will be described later in this chapter.

Cellular Barriers. Innate defensive barriers also include the normal microbiota of the body surfaces. These indigenous, nonpathogenic microbes form a cellular barrier by outcompeting pathogens for nutrients and attachment sites on the skin and mucosa. The human microbiota was discussed in another chapter.

CONCEPT AND REASONING CHECKS 2

a. Explain how the skin and mucous membranes provide a physical and chemical defense against infection.

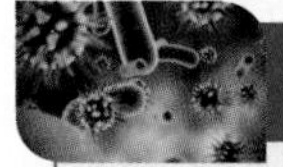

Chapter Challenge B

So, we have been exposed to an infectious dose (100 to 1,000 particles) of flu virus transmitted via air droplets. As such, the droplet-containing flu viruses may infect the mucosa of the respiratory tract.

QUESTION B:

What surface barriers of innate immunity attempt to keep the flu viruses out of the respiratory tract? What mechanisms are involved in the attempted elimination?

Answers can be found on the Student Companion Website in **Appendix D**.

© qcontrol/ShutterStock, Inc.

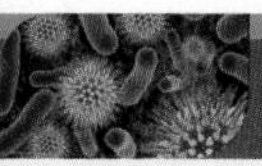

Investigating the Microbial World 20

Who Goes There?

Pathogens that manage to breach the body's surface barriers comprising the skin and mucous membranes are now in a position to initiate colonization and infection. To prevent this from happening, innate defenses must recognize the intruder as being an invader and activate a response to eliminate the pathogen. In other words, the host defenses must know "who goes there"—friend or foe.

- **OBSERVATION:** The identification of a pathogen by innate immunity is often accomplished by recognizing unique molecular markers on the surface of the pathogen that are not found on the surface of host cells. This recognition results in the production and secretion of several types of antimicrobial peptides.
- **QUESTION:** ***Will the same antimicrobial peptide defend against different pathogens?***
- **HYPOTHESIS:** The antimicrobial peptides function against a diverse set of pathogen molecular markers. If so, then a single antimicrobial peptide should be effective against two different types of pathogens.
- **EXPERIMENTAL DESIGN:** Fruit flies are excellent experimental organisms because mutant fruit flies can be produced with just about any specific phenotype. In this case, mutant flies were generated that can recognize pathogens but cannot respond by producing any antimicrobial peptides (AMP⁻).

Four groups of fruit flies were used. Group 1 consisted of wild type flies that can respond by producing antimicrobial peptides (AMP⁺) and group 2 was the mutant strain (AMP⁻) mentioned above. The other two groups were AMP⁻ flies that had been genetically engineered to produce significant amounts of the antimicrobial peptide defensin (Group 3; Def⁺) or drosomycin (Group 4; Dro⁺).

- **EXPERIMENT 1:** One set of 20 flies from all four groups were infected with the fungus *Neurospora crassa* and monitored over 5 days.
- **EXPERIMENT 2:** Another set of 20 flies from all four groups were infected with the bacterial species *Micrococcus luteus* and monitored over 5 days.
- **RESULTS:**

  EXPERIMENT 1: See figure A.

  EXPERIMENT 2: See figure B.

- **CONCLUSION:**

  QUESTION 1: ***In experiments 1 and 2, what do the results with AMP⁺ (group 1) and AMP⁻ (group 2) flies tell you about the flies' ability to mount an antimicrobial peptide defense?***

  QUESTION 2: ***In experiments 1 and 2, what do the results with Def⁺ (group 3) and Dro⁺ (group 4) flies tell you about the flies' ability to use the same antimicrobial peptide against two different pathogens?***

  QUESTION 3: ***Was the hypothesis supported? Explain.***

Answers can be found on the Student Companion Website in **Appendix D.**

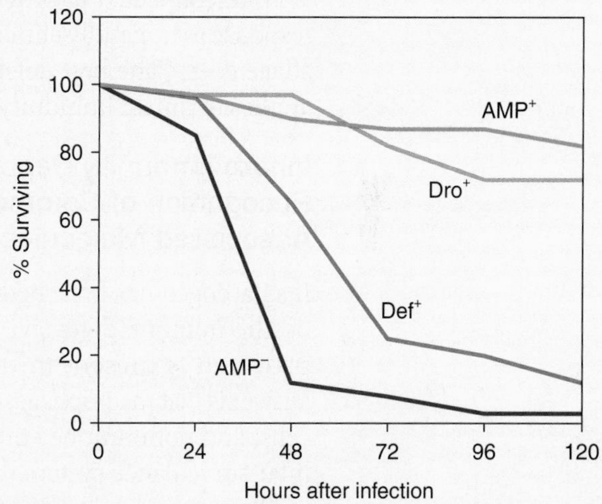

FIGURE A Percent fly survival after infection with *Neurospora crassa*. Modified from: Tzou, P. *et al.* (2002). *Proc. Natl. Acad. Sci. USA* 99:2152–2157

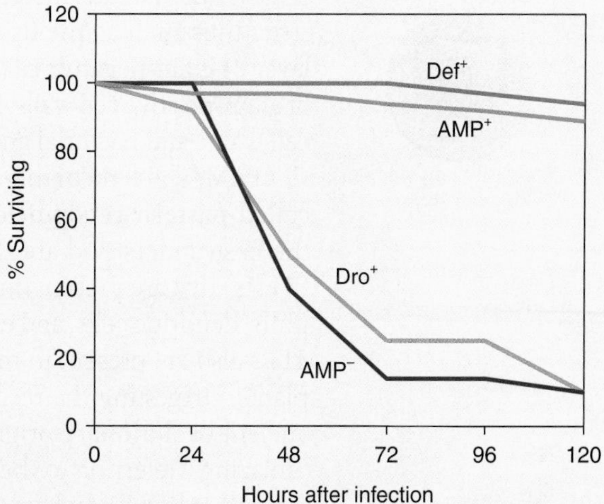

FIGURE B Percent fly survival after infection with *Micrococcus luteus*. Modified from: Tzou, P. *et al.* (2002). *Proc. Natl. Acad. Sci. USA* 99:2152–2157

Adapted from: Tzou, P. *et al.* (2002). *Proc. Natl. Acad. Sci. USA* **99**: 2152–2157. Icon image © Tischenko Irina/ShutterStock, Inc.

KEY CONCEPT 20.3 Coordinated Cellular Defenses Respond to Pathogen Invasion

As we just described, the surface barriers of innate immunity guard against pathogens trying to colonize or penetrate the body surfaces at the skin and mucous membranes. Should one of these barriers be breached by a pathogen, say by a cut finger, a skin puncture, a burn, or a bite from an infected insect, then a group of coordinated innate defenses become activated. These defenses involve immune defensive cells and the cytokine communication that occurs between them to generate an effective and strong response to infection. Although we will examine each defense separately, realize that they are interconnected to provide a potent and rapid response to hopefully eliminate a pathogen; as the adage goes, "The best defense is a good offense," applies to innate immunity too.

■ Zymosan:
A carbohydrate-protein complex in fungal cell walls.

Innate Immunity Depends on Receptor Recognition of Common Pathogen-Associated Molecules

If a pathogen has breached a surface barrier, how do the immune defensive cells "know" that the pathogen is present in the body tissues? The answer is that macrophages, neutrophils, dendritic cells, and nonimmune epithelial cells have a cellular surveillance system that recognizes unique, highly conserved (invariant) molecular features in or on the pathogen. Such **pathogen-associated molecular patterns (PAMPs)** are absent from host cells. PAMPs include molecular features present in structures such as the lipopolysaccharide (LPS) layer of gram-negative cell walls, peptidoglycans of gram-positive cell walls, flagella proteins, fungal cell walls, and bacterial and viral nucleic acids.

Website Animation: Phagocytosis

PAMPs are recognized by cellular sensors called **pattern recognition receptors**. Among the best understood are the **toll-like receptors (TLRs)** that are present on macrophages, neutrophils, dendritic cells, and endothelial cells. In fact, TLRs also are present in many other animals and plants, suggesting the recognition system has an ancient evolutionary origin. (*Toll* is German for "amazing," referring to the toll receptors first discovered in fruit flies where they are a fly's "amazing" defense against fungal infections.)

The TLR family in humans consists of at least ten protein receptors, each working to mediate a specific response to distinct PAMPs. Specific TLRs are located on the cell's plasma membrane and the membrane compartments called "endosomes" (FIGURE 20.6). For example, several plasma membrane TLRs can bind to unique bacterial lipopeptides, zymosan, LPS, or bacterial flagellar proteins. TLR-3 found on the endosome membrane binds to double-stranded RNA that is unique to some viruses. TABLE 20.1 lists the known human TLRs and their PAMP targets.

Once bound, pattern recognition activates several defensive processes (phagocytosis and inflammation) and releases several antimicrobial peptides and proteins (defensins, complement, interferon) that attack the invader.

Phagocytosis Is a Nonspecific Defense Mechanism to Clear Microbes from Infected Tissues

One of Metchnikoff's key observations involved the larvae of starfish. When jabbed with a splinter of wood, motile cells in the starfish gathered around the splinter. From observations like these, Metchnikoff suggested cells could actively seek out and engulf (ingest) foreign particles in the body. Metchnikoff's theory of phagocytosis laid the foundation for one of the most important functions of innate and adaptive immunity.

Phagocytosis (*phago* = "eat") is the capturing and digesting of foreign particles by phagocytes (neutrophils, macrophages, and dendritic cells). This may occur at the site of the infection as well as in the lymphoid tissues.

If a bacterial or viral pathogen breaches the surface barriers, the intruder usually is recognized by its PAMPS, ingested, and killed by neutrophils and macrophages residing in the tissue. On binding the pathogen, the phagocytes release cytokines that will trigger an inflammatory response (to be described next) as well as activate an adaptive immune response. Secreted chemokines stimulate the migration of neutrophils and monocytes to the infection site.

The process begins with the attachment of the microbe to the cell surface of the phagocyte (FIGURE 20.7). Macrophages do not simply "bump" into their victim. Rather, they extend thin cytoplasmic protrusions, called **filopodia**, that wave like a fishing lure. If a filopodium "catches" its

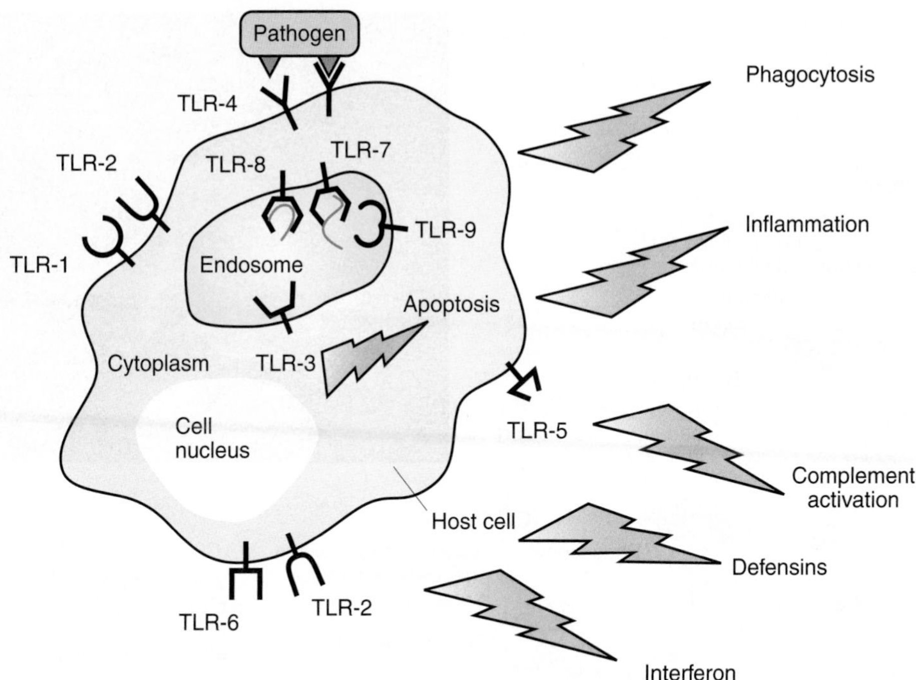

FIGURE 20.6 **Pattern Recognition Receptors and the Innate Response.** Proteins called toll-like receptors (TLRs) are located on the plasma membrane and endosome membrane. Binding to a pathogen-associated molecular pattern (PAMP) [e.g., bacterial wall PAMP on plasma membrane with TLR-4 or viral single-stranded RNA (purple lines) in endosome with TLR-7 and -8] initiates several processes associated with the innate immune response. »» Why do some TLRs have to exist in the endosome?

victim by latching onto a PAMP, the filopodium rapidly contracts. The plasma membrane then surrounds the microbe and an invagination, or folding in, of the membrane produces an internalized vacuole, called a **phagosome**. The phagosome now becomes acidified and the lowered pH aids in killing or inactivating the pathogen. Lysosomes then fuse with the phagosome, forming a **phagolysosome**, and lysosome enzymes, such as lysozyme and acid hydrolases, digest the pathogen. In addition, phagocytosis triggers a so-called **respiratory burst** that produces toxic metabolites that destroy the pathogen membrane. These metabolites include hydrogen peroxide (H_2O_2), nitric oxide (NO), superoxide anions (O_2^-), and hypochlorous acid (OCl^-; bleach).

TABLE

20.1 The Recognized Human Toll-Like Receptors

Toll-Like Receptor	Location	PAMP (Molecular Sequences)
1 and 2	Cell surface	Bacterial cells (lipoprotein)
2 and 6	Cell surface	Gram-positive bacterial cell walls (peptidoglycan) Gram-positive bacterial cell walls (lipoteichoic acid) Fungal cell walls (zymosan)
3	Endosome membrane	Viruses (double-stranded RNA)
4	Cell surface	Gram-negative bacterial cell walls (LPS)
5	Cell surface	Bacterial flagellar protein
7 and 8	Endosome membrane	Viruses (single-stranded RNA)
9	Endosome membrane	Bacterial and viral double-stranded DNA
10	Cell surface	Bacterial pili

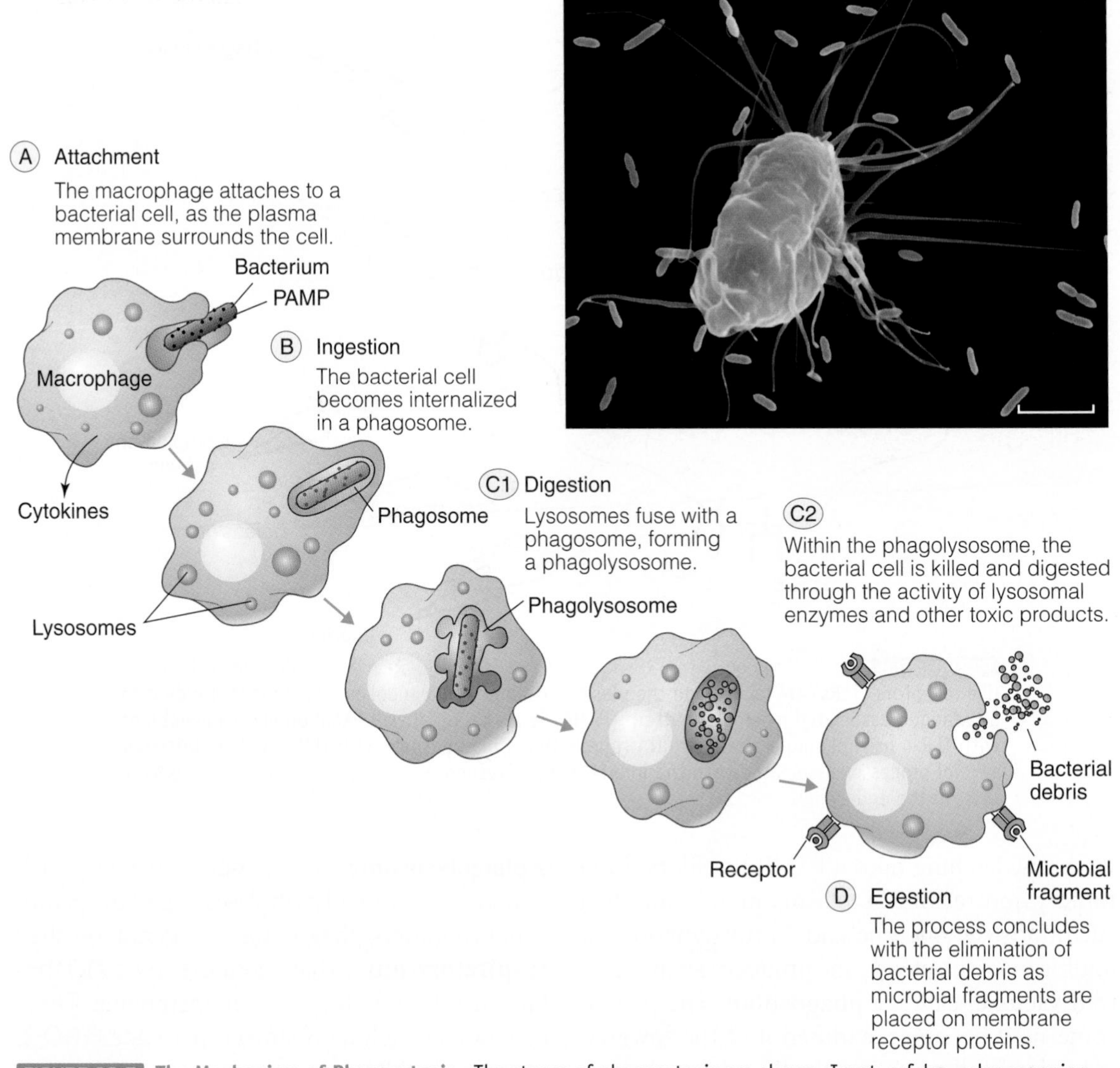

FIGURE 20.7 **The Mechanism of Phagocytosis.** The stages of phagocytosis are shown. Inset: a false-color scanning electron micrograph of a macrophage (blue) attaching to *E. coli* cells (pink) by means of filopodia. (Bar = 5 µm.) © Dennis Kunkel Microscopy, Inc./Visuals Unlimited/Corbis. »» What might the bacterial debris or waste products be in (D)?

Neutrophils are usually the first cell type arriving at the infection site. Because neutrophils are short-lived cells, monocytes attracted to the site quickly mature into the longer-lived macrophages.

Besides phagocytosis, many neutrophils can combat infections by forming extracellular fiber-like networks called **neutrophil extracellular traps (NETs)**. When such a neutrophil binds a PAMP from any one of a variety of bacterial or fungal pathogens, the cell undergoes a loss of nuclear continuity, disintegration of the nuclear envelope, and mixing of the nuclear proteins and DNA with cytoplasmic antimicrobial proteins and enzymes (FIGURE 20.8). This mixing forms the NETs that are released as the neutrophil plasma membrane breaks and the cell dies. The released NETs have antimicrobial activity, and by binding to and killing the pathogens, prevent further spreading of the pathogen in the body.

Platelets also have the ability to bind to pathogens, recognize PAMPs by TLRs, and bind to and activate neutrophils, which then quite rapidly (in minutes) form NETs. The accumulation of platelets and neutrophils at a wound form a clot, which along with NETs, can again serve as a physical barrier preventing further spread of the pathogen. So, besides cytokine signaling and phagocytosis, neutrophils have a third, independent weapon to attack pathogens.

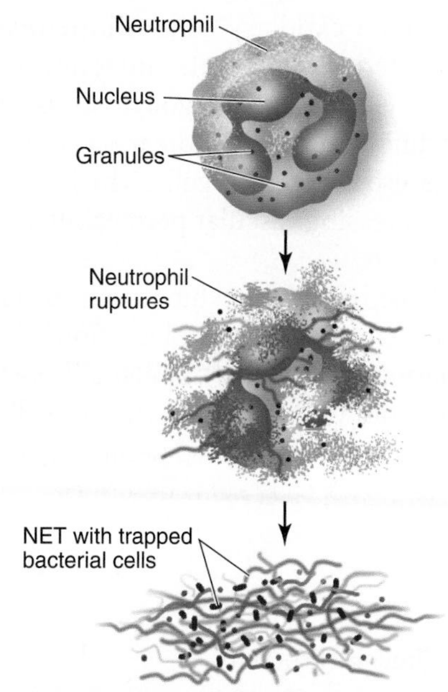

FIGURE 20.8 **Formation of NETs.** A neutrophil becomes activated, leading to the release of DNA and granule proteins that, on rupture, form into NETs. Adapted from Lee, Warren, L. and Grinstein, S. (2004). *Science* 303: 1477–1478. »» How is innate immunity improved by generating NETs?

In the end, the host clears the pathogen and the infection ends, as waste materials are eliminated from the phagocytes. During phagocytosis, macrophages and dendritic cells place some of the digested fragments from the pathogen on proteins located on the phagocyte's cell surface (Figure 20.7). This "presenting of antigen" will be critical to the activation of adaptive immunity.

The host-microbe relationship is always evolving to give either the host or the microbe the advantage. Sometimes the pathogen gets the upper hand by "using" phagocytosis to its advantage (MicroFocus 20.2).

Inflammation Plays an Important Role in Fighting Infection

Acute inflammation is an immediate, but nonspecific defensive response by the body to trauma. It develops after a mechanical injury, such as an injury or blow to the skin, or from exposure to a chemical agent, such as acid or bee venom. Inflammation also may be a response to an infection.

MICROFOCUS 20.2: Evolution
Avoiding the "Black Hole" of the Phagocyte

Lysosomes contain quite a battery of enzymes, proteins, and other substances capable of efficiently digesting, or being toxic to, almost anything with which they come in contact. However, some bacterial species can avoid a digestive death.

Some species, such as *Streptococcus pneumoniae,* have a thick capsule, which is not easily internalized by phagocytes. Other pathogens simply invade in such large numbers they overwhelm the host defenses and the ability of phagocytes to "sweep" them all up. More interesting are pathogens that have evolved strategies to evade the lysosome—the "black hole" of the phagocyte (see figure).

Listeria monocytogenes is internalized by phagocytes. However, before lysosomes can fuse with the phagosome membrane, the bacterial cells release a pore-forming toxin that lyses the phagosome membrane. This allows the *L. monocytogenes* cells to enter and reproduce in the phagocyte cytoplasm. The bacterial cells then spread to adjacent host cells by the formation of actin tails.

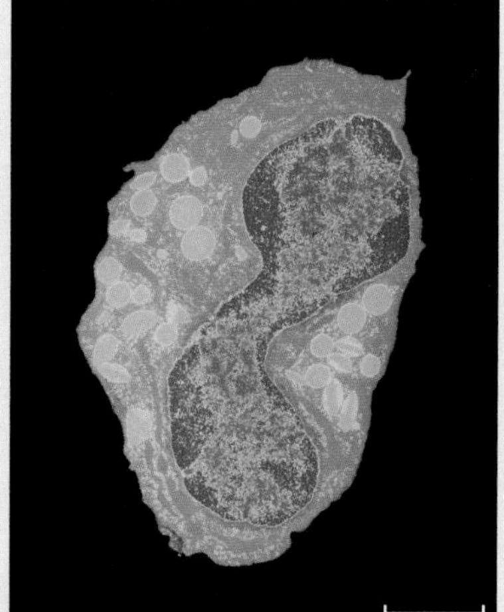

False-color transmission electron micrograph of lysosomes (yellow) in a macrophage. (Bar = 4 μm.) © Dr. Gopal Murti/Photo Researchers, Inc.

Other pathogens prevent lysosome fusion with the phagosome in which they are contained. The causative agent of tuberculosis, *Mycobacterium tuberculosis,* is internalized by macrophages. However, once in a phagosome, the bacterial cells prevent fusion of lysosomes with the phagosome. In fact, the tubercle bacteria actually reproduce within the phagosome. A similar strategy is used by *Legionella pneumophila.*

Toxoplasma gondii, the protist responsible for toxoplasmosis, evades the lysosomal "black hole" by enclosing itself in its own membrane vesicle that does not fuse with lysosomes. Avoiding digestion by lysosomes is key to survival for these pathogens.

■ Histamine: A chemical mediator causing contraction of smooth muscle and dilation of blood vessels.

In the case of an infection, the pathogen's presence and tissue injury sets into motion an innate process to limit the spread of the pathogen and localize the infection (FIGURE 20.9). At the infection site, resident tissue macrophages secrete several cytokines triggering a local dilation of the blood vessels (**vasodilation**) and increasing capillary permeability. This allows the flow of plasma into and fluid accumulation (**edema**) at the site of infection. Chemokines attract phagocytes (neutrophils and monocytes) toward the infected tissue. These cells adhere to the blood vessels near the infection and then migrate between capillary cells (**diapedesis**) into the infected tissue. At the infection site, the neutrophils augment phagocytosis as new monocytes differentiate into more macrophages. Chemokines also attract mast cells, which secrete histamine, increasing vascular permeability and vasodilation.

The inflamed area thus exhibits four cardinal signs: redness (from increased blood flow); heat (from the warmth of the blood); swelling (from the accumulation of fluid); and pain (from swelling putting pressure on nerves). Sometimes itching may occur, as MicroFocus 20.3 describes.

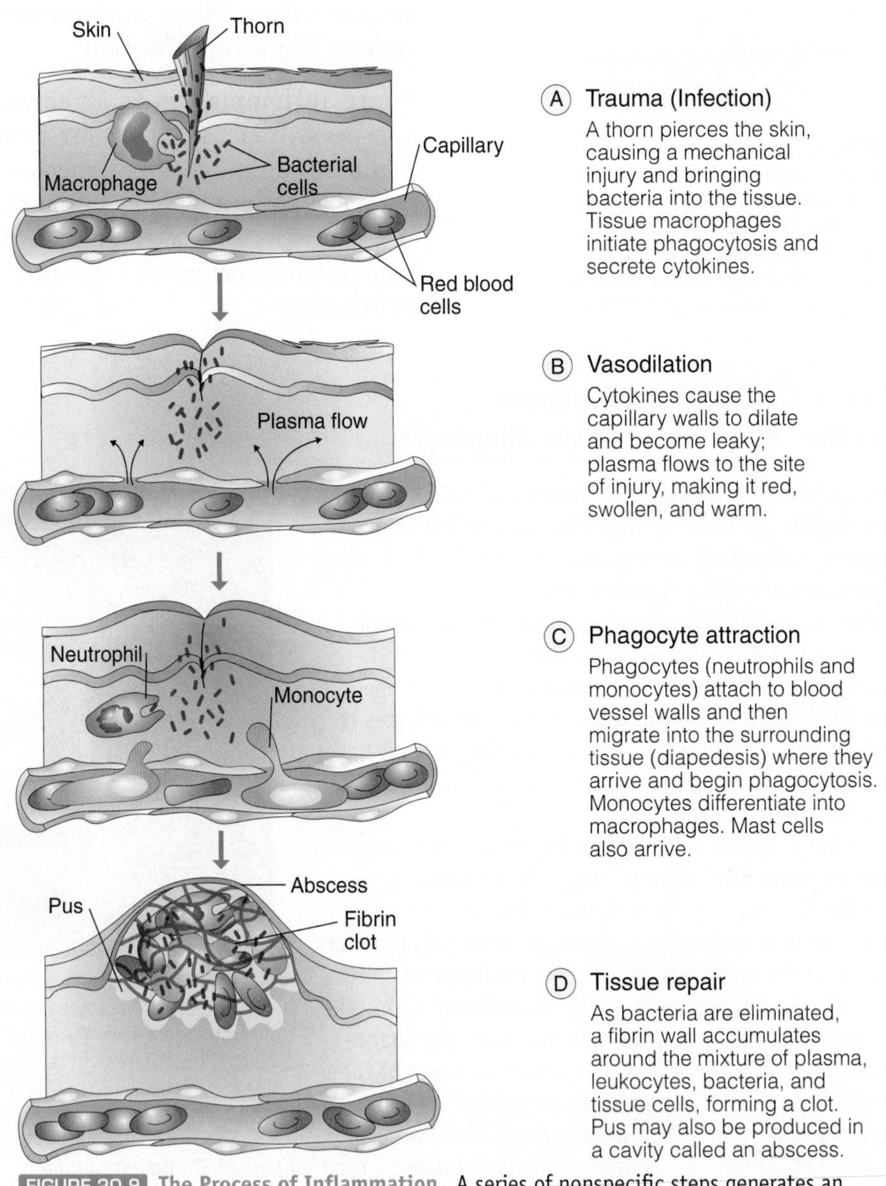

FIGURE 20.9 **The Process of Inflammation.** A series of nonspecific steps generates an inflammatory response to some form of trauma or infection, such as being pierced by a plant thorn harboring bacterial cells. »» Why does the inflammatory response generate the same set of events no matter what the nature of the foreign substance?

MICROFOCUS 20.3: Evolution

But It Doesn't Itch!

How many of us at some time have had a small red spot or welt on our skin that looks like an insect bite—but it doesn't itch? When most of us are bitten by an insect, such as an ant, bee, or mosquito, the site of the bite starts itching almost immediately. Yet, there are cases where bites do not itch, such as a tick bite. So, what's the difference?

Most insect bites trigger a typical inflammatory reaction. Take a mosquito as an example. When it bites and takes a blood meal, saliva injected into the skin contains substances so the blood does not coagulate. The bite causes tissue damage and activation of chemical mediators that trigger the typical characteristics of an inflammatory response: redness, warmth, swelling, and pain at the injured site. One of the major chemical mediators is histamine, released by the damaged tissue and immune cells within the tissue. An itch actually is a form of pain and arises when histamine affects nearby nerves. Itching is the result.

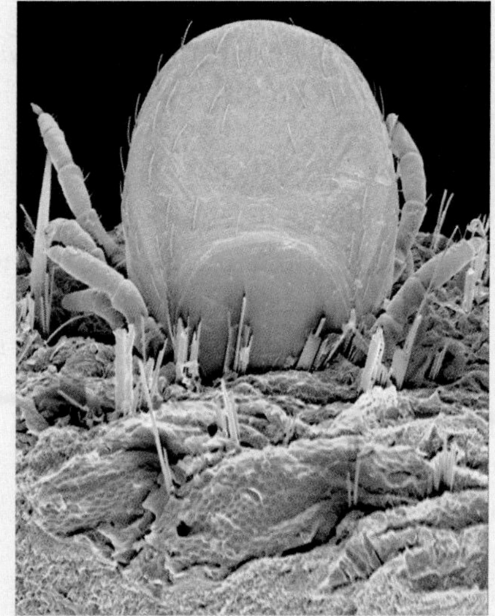

False-color scanning electron micrograph of a tick (orange) feeding head-down in human skin. © Volker Steger/Photo Researchers, Inc.

In a tick bite, the scenario is slightly different. The bite triggers the same set of inflammatory reactions and histamine is produced. However, in the tick saliva is another molecule called histamine-binding protein (HBP). HBP binds to histamine, preventing the mediator from affecting the nearby nerves—and no itching occurs.

So why is that an advantage for the tick? Probably because ticks take long blood meals that can last for 3 to 6 days. By the bite being painless and not itching, there is less chance of the ticks being noticed and removed.

During inflammation a fibrin clot forms, which prevents the spread of pathogens to the blood. Pus may be a product during inflammation and when it becomes enclosed in a wall of fibrin through the clotting mechanism, an **abscess**, or boil, may form.

To limit tissue damage, the body tries to regulate the number of phagocytes and the level of cytokine production in the defense against infection. Therefore, the inflammatory response must be managed so it is only short-lived. Should the inflammatory stimulus not be cleared, **chronic inflammation** may occur. The persisting cytokines can then produce excessive damage, leading to a number of chronic degenerative diseases, including rheumatoid arthritis, coronary heart disease, atherosclerosis, and cancer.

■ Atherosclerosis: An arterial disease in which degeneration and cholesterol deposits (plaques) form on the inner surfaces of the arteries.

Another consequence of an overwhelming "cytokine storm" is **septic shock**, characterized by a collapse of the circulatory and respiratory systems. In some gram-negative bacterial infections, large amounts of cell wall lipopolysaccharide (LPS) are released, which is referred to as an endotoxin. The systemic spread of LPS through the body triggers macrophages to secrete the "overdoses" of cytokines. Shock can lead to death.

Moderate Fever Benefits Host Defenses

Fever is an abnormally high body temperature that remains elevated above the normal 37°C. Fever supports the immune system response in trying to gain an advantage over the pathogens by making the body a less favorable environment. On the other hand, fever is metabolically costly; keeping the body temperature elevated is an energy drain and makes the heart work harder.

Fever-producing substances, called **pyrogens**, include endogenous cytokines produced by activated macrophages and other leukocytes, and exogenous microbial fragments from bacterial cells, viruses, and other microorganisms. These pyrogens move through the blood and

■ Hypothalamus: The part of the brain controlling involuntary functions.

■ MHC proteins: A class of cell-surface proteins found on mammalian cells.

affect the anterior hypothalamus such that the body temperature becomes elevated. As this takes place, cell metabolism increases and blood vessels constrict, thus denying blood to the skin and keeping its heat within the body. Patients thus experience cold skin and chills along with the fever.

A low to moderate fever in adults may be beneficial to immune defense because the elevated temperature inhibits the rapid growth of many microbes, inactivates toxins, encourages rapid tissue repair, and heightens phagocytosis. However, if the temperature rises above 40.6°C (105°F), convulsions and death may result from host metabolic inhibition. Also, infants with a fever above 38°C (100°F), or older children with a fever of 39°C (102°F), may need medical attention. MicroFocus 20.4 examines the saying that it is good to "feed a cold and starve a fever."

Natural Killer Cells Recognize and Kill Abnormal Cells

Besides the phagocytes, natural killer (NK) cells are another type of defensive leukocyte. Derived from lymphoid progenitors, NK cells are formed in the bone marrow and then migrate to the tonsils, lymph nodes, and spleen, where they await activation. On stimulation, they secrete several cytokines, triggering adaptive immune responses by macrophages and other immune cells. While those events are mobilized, the NK cells move into the blood and peripheral tissues where they act as potent effector cells that can destroy tumor cells and virus-infected cells.

NK cells are not phagocytic; rather, they contain on their surfaces a set of special receptor sites capable of forming cell-to-cell interactions with a target cell (FIGURE 20.10). If the receptor sites match up with a group of class I MHC proteins on the target cell, the NK cell recognizes the target cell as one of the body's own (self) and leaves it alone.

However, when these MHC proteins are absent or in reduced amounts (as on a cancer cell or virus-infected cell), then the matchup is incomplete and the NK cell secretes cytotoxic mediators that damage and eliminate the tumor or virus-infected cell. These cytolylic media-

MICROFOCUS 20.4: Being Skeptical
Feed a Cold, Starve a Fever?

One of the typical symptoms of the flu is a high fever (39°C to 40°C) while a cold seldom produces a fever. So, is there any truth to the old adage, "You should feed a cold and starve a fever?" The definite answer is—yes!

There are three reasons you should reduce food intake with a fever. First, food absorbed by the intestines could be misidentified by the body as an allergen, worsening the illness and fever. Also, with the body already under stress, heavy eating can contribute on rare occasions to body seizures, collapse, and delirium.

Perhaps most common is that eating stimulates digestion of the foods. During times of increased physiologic stress, digestion can overstimulate the parasympathetic nervous system, when the sympathetic nervous system (involved in constricting blood vessels and conserving heat = fever) already is active. In other words, eating tends to drop body temperature, possibly eliminating fever as a protective mechanism.

So, with the flu or other illness-induced fever, stick to bed rest and just drink plenty of liquids.

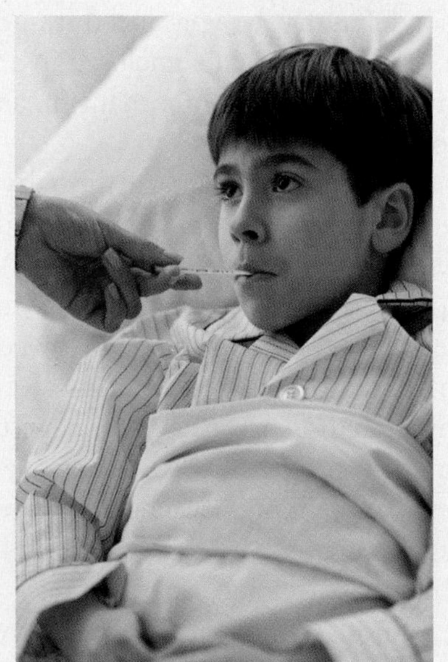

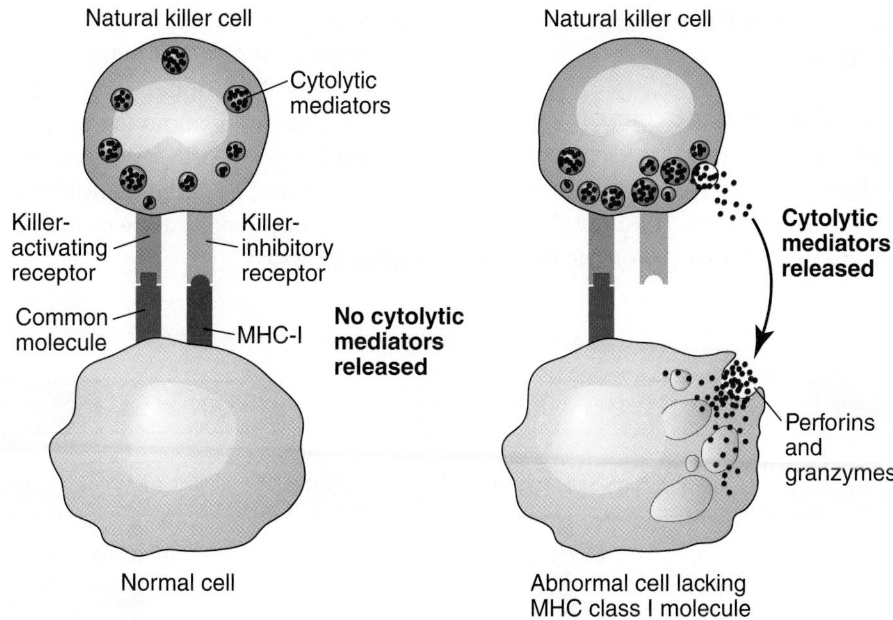

FIGURE 20.10 **Natural Killer (NK) Cell Recognition.** NK cells have the ability to destroy any cell they "see" as abnormal (for example, tumor cells or virus-infected cells). (Adapted from Delves, P. J., and Roitt, I. M. (2000). *N Engl J Med* 343: 37–49.) »» How does the presence or absence of class I MHC receptors affect NK cell activity?

tors include proteins called **perforins**, which drill holes in the target cell, and **granzymes**, which are cytotoxic enzymes that pass through the holes, and induce the target cell to undergo apoptosis. The DNA condenses, the cell shrinks, and breaks into tiny pieces that can be mopped up by wandering phagocytes.

■ Apoptosis:
A type of cell death activated by an internal death program.

CONCEPT AND REASONING CHECKS 3

a. Justify the need for the spectrum of toll-like receptors found on and in human immune cells.
b. Discuss the importance of phagocytosis to innate immunity.
c. Explain how inflammation serves as an innate immune defense.
d. Assess the advantage of a low grade fever during an infection.
e. How do natural killer cells assist the other innate defense mechanisms during a viral infection?

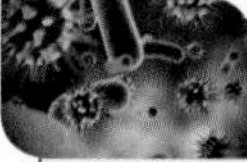

Chapter Challenge C

Unfortunately the surface barriers did not hold and the flu viruses have breached the mucous membranes of the nose and upper respiratory tract. In fact, we are now starting to feel the signs and symptoms: sudden onset of fever, body aches, headache, and tiredness.

QUESTION C:
What cellular defenses of innate immunity are now activated in trying to limit the spread of the infection? How is this related to headaches and muscle aches?

Answers can be found on the Student Companion Website in **Appendix D.**

KEY CONCEPT 20.4 Effector Molecules Damage Invading Pathogens

Besides the cellular defenses of phagocytosis and inflammation, innate defenses also produce and release a variety of effector peptides and proteins to attack pathogens. We have already mentioned the defensin peptides. In this last section, we examine the complement proteins and interferons, which, unlike the defensins, are unique molecules specific to vertebrate immune systems.

Complement Marks Pathogens for Destruction

Complement (also called the complement system) is a group of nearly 30 inactive proteins that are produced in the liver and circulate in the bloodstream and tissues. If a microbe penetrates the body, complement proteins become sequentially activated through a cascade of steps that enhance the inflammatory response and phagocytosis. Additional complement proteins in the cascade bring about the destruction of the invading bacterial cells or viruses through cell or virion lysis. The three complementary pathways for complement activation are illustrated in FIGURE 20.11.

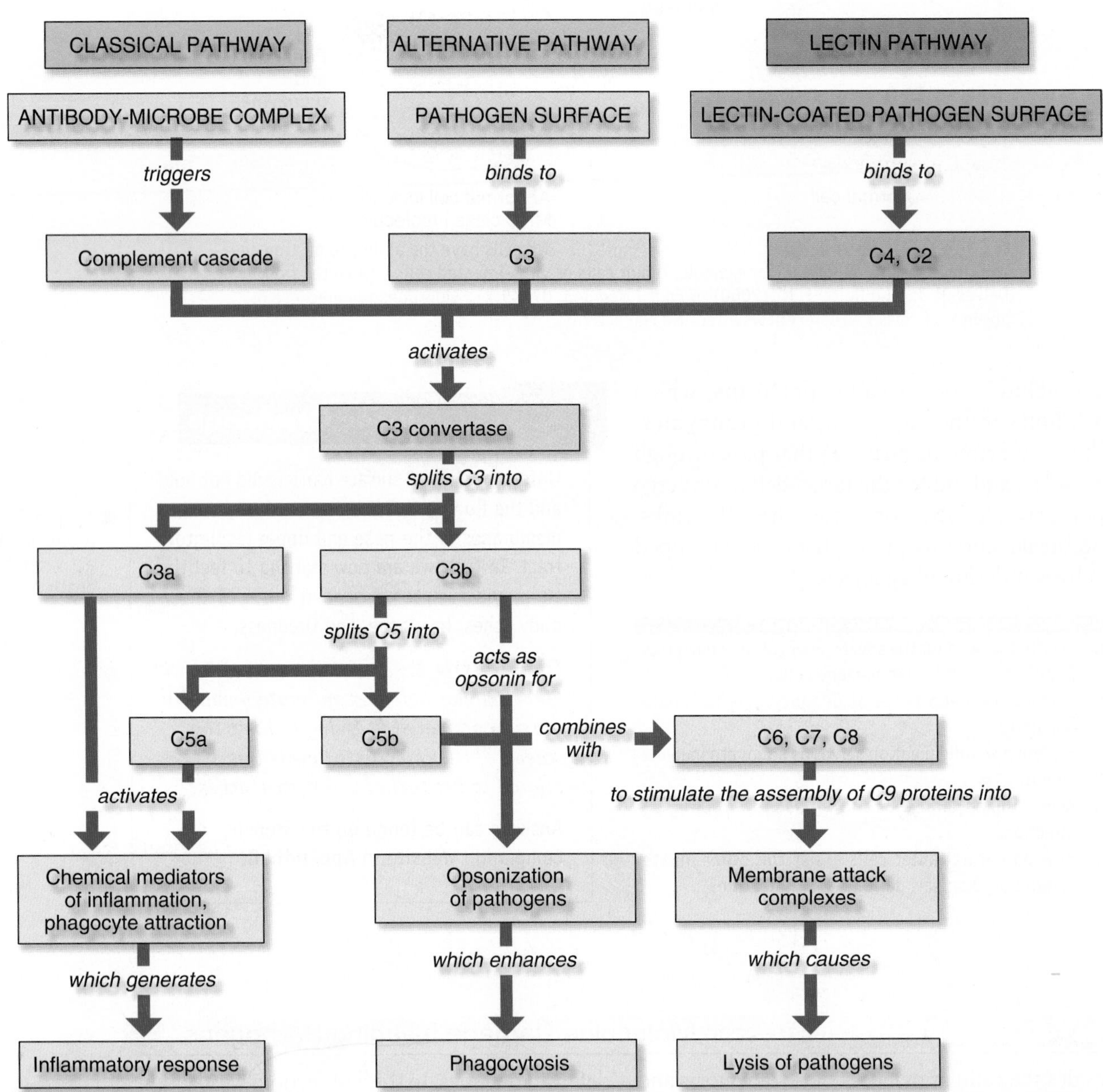

FIGURE 20.11 **A Concept Map of the Complement Pathways.** The early events of all three pathways of complement activation culminate in the activation of C3 convertase. Complement fragment C3b binds to the membrane and opsonizes bacterial cells, allowing phagocytes to internalize them (phagocytosis). The small fragments, C5a and C3a, are mediators of local inflammation. C5b also activates another complement cascade, whereby the terminal C9 proteins assemble into a membrane-attack complex (MAC) that can damage the membrane of gram-negative cells and enveloped viruses. »» What is the advantage of having all three pathways leading to activation of C3 convertase?

All these pathways—the classical, alternative, and lectin pathway—culminate in the activation of an enzyme called C3 convertase, which splits complement protein C3 into a C3a and C3b fragment. C3b activates another complement protein C5, which also is hydrolyzed into C5a and C5b fragments. Complement fragments C3a and C5a trigger an inflammatory response through vasodilation and increased capillary permeability. C3b also acts as an opsonin by binding to the pathogen surface (opsonization) and enhancing phagocytosis.

C5b plays one more important role. The complement fragment triggers another complement cascade that triggers the assembly of C9 proteins into what are called **membrane attack complexes** (**MACs**). These C9 complexes form large holes in the membranes of many microbes (FIGURE 20.12). The MACs are so prevalent that lysis occurs, killing the pathogen.

An inflammation-like response often seen in very ill patients results when neutrophils and macrophages secrete a set of cytokines that cause the liver to synthesize a set of so-called **acute phase proteins**. Among the proteins stimulated are cytokines as well as clotting and vasodilation proteins. One protein, C-reactive protein (CRP), can activate macrophages and bind to bacterial cell surfaces, triggering the classical complement pathway and inflammation. This has made CRP a useful marker of inflammation in medical pathology.

This teamwork within the complement system and inflammation provides a formidable obstacle to pathogen invasion. However, sometimes genetic abnormalities can affect innate immunity and predispose one to infectious disease (TEXTBOOK CASE 20).

Interferon Puts Cells in an Antiviral State

Infection of cells with viruses triggers yet another nonspecific, innate immune response. Virus-infected cells produce a set of regulating cytokines called **interferons** (**IFNs**) that "alert" surrounding cells to the viral threat. Two IFNs, IFN-alpha and IFN-beta, trigger a nonspecific reaction via TLR-3 designed to protect against infecting viruses.

Website Animation: Interferon action

IFN-alpha and IFN-beta do not interact directly with viruses, but rather with the cells they alert. The IFNs are produced when a virion releases its genome into the cell. High concentrations

Activated complement proteins C5b-8 attach to the pathogen's membrane in step-by-step sequence, triggering formation of a membrane attack complex composed of C9 proteins.

The attack complex results in pores in the membrane, causing cell lysis.

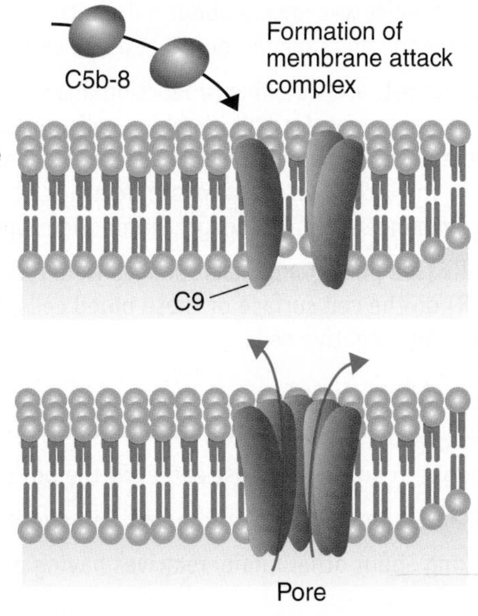

(A)

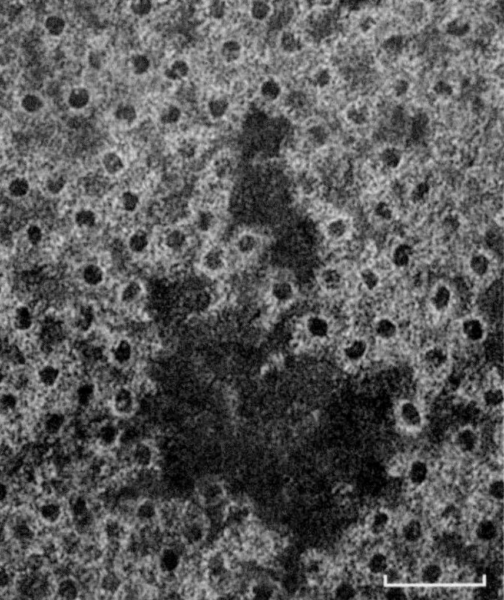

(B)

FIGURE 20.12 **A Membrane Attack Complex (MAC).** (**A**) Activated complement proteins C5b, 6, 7, and 8 guide the formation of a tubular pore (MAC) by C9 through the pathogen cell membrane. (**B**) Large numbers of MACs, as seen by transmission electron microscopy, riddle the membrane, causing cell lysis. (Bar = 50 nm.) Reproduced from Carlos Rosado et al, *Science* 317 (2007): 1548–1551. Reprinted with permission from AAAS. »» Would bacterial species that are gram positive or gram negative, and viruses that are enveloped or nonenveloped, be most sensitive to MAC formation? Explain.

Textbook CASE 20

A Clinical History of Bacterial Infections

1 On March 17, 2008, a 4-year-old boy in Galveston, Texas was taken by his parents to his pediatrician. The boy was suffering another bacterial infection, a recurrent problem over the last few years, including two hospitalizations for bacterial pneumonia. The pediatrician took several specimens for testing, placed the young boy on antibiotics, and asked the parents to bring back their son in 3 days for the test results.

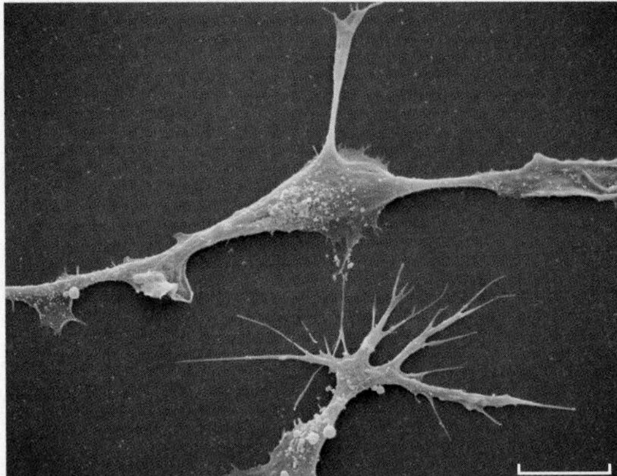

A false-colored scanning electron micrograph of two dendritic cells showing the long cell surface projections. (Bar = 20 μm.) © David Scharf/Photo Researchers, Inc.

2 On his return, the pediatrician informed the parents that their son had a *Haemophilus influenzae* type b infection but that the antibiotics once again should take care of the infection. However, the recurrence of infections, all due to gram-negative bacterial pathogens, was unusual so the pediatrician asked an infectious disease (ID) specialist to consult.

3 A clinical history report for the boy was given to the ID specialist and she was told the boy's bacterial infections had always been cleared with long-term antibiotic therapy and that he was up-to-date on his vaccinations.

4 The ID specialist asked if the boy had suffered any increased susceptibility to viral or fungal infections. The answer was no; they had all been gram-negative bacterial infections. She then asked the parents if any other members in or relatives of the family had ever had a similar history of recurrent infections. The parents remarked that two relatives, one female and one male, had had a similar condition. The ID specialist was unsure about a definitive cause for the boy's illnesses but suggested a battery of blood and immunological tests be carried out.

5 A blood cell count, serum antibody test, and cytokine production analysis were done at varying time intervals. All results were normal, suggesting all his blood cells, including neutrophils, monocytes, and lymphocytes were in the normal range and appeared to be functioning properly. The ID specialist was intrigued by the lack of viral or fungal infections, so she ordered tests to examine the ability of the boy's immune system to recognize infectious agents.

6 Functional tests for dendritic cells (see photo) and macrophages were run. The results confirmed a lack of a toll-like receptor (TLR) on the cell surface of these blood cells, thus "blinding" the white blood cells in recognizing gram-negative cells.

Questions:

(Answers can be found on the Student Companion Website in **Appendix D.**)

A. Why hadn't the pediatrician brought in an ID specialist earlier in the boy's medical history?

B. Why would the ID specialist ask about previous viral and fungal infections?

C. What was the significance in asking about other family relatives having a similar recurrent condition?

D. In the analysis of dendritic cells and macrophages, which TLR (using Table 20.1) is most likely to be missing?

For additional information see www.nature.com/nrmicro/journal/v3/n1/full/nrmicro1068.html.

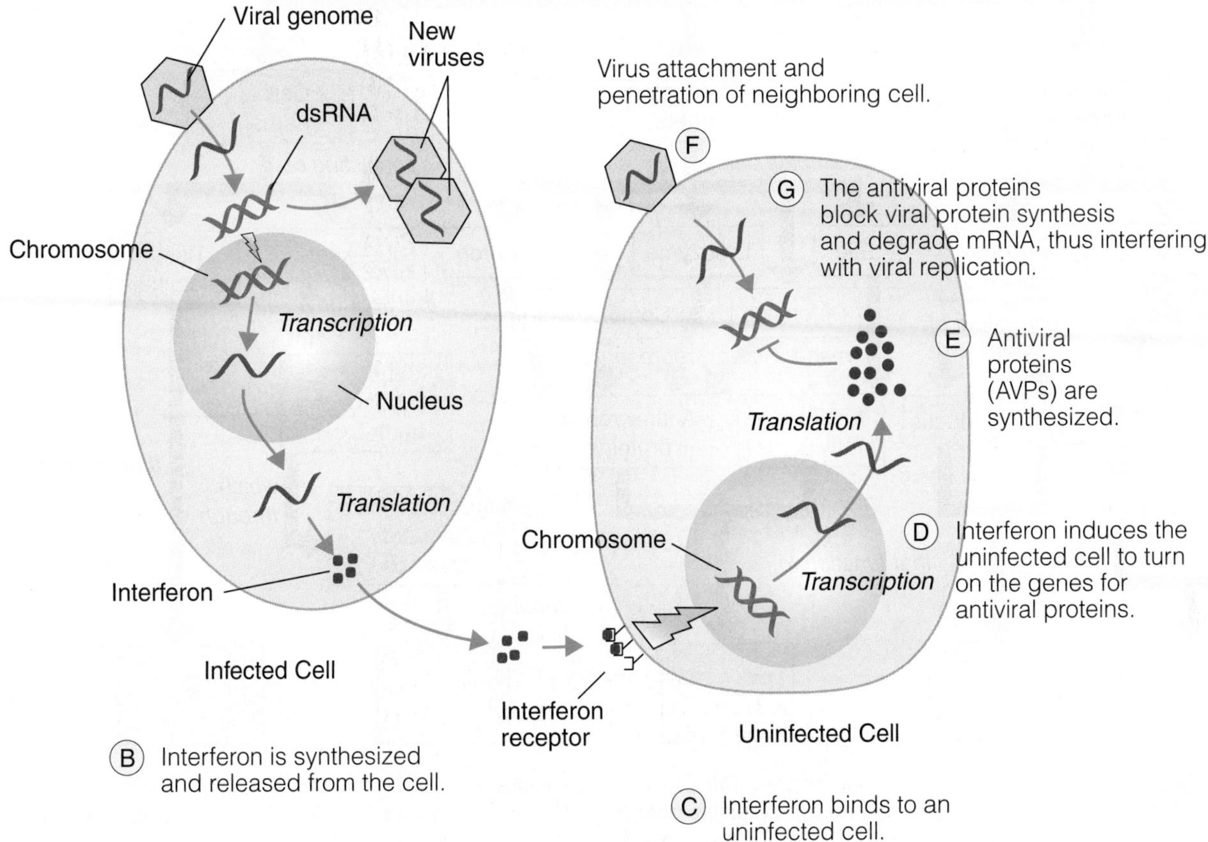

FIGURE 20.13 The Production and Activity of Interferon. Left: A host cell produces interferon following viral infection. The virus replicates in the same cell. Right: The interferon reacts with receptors at the surface of a neighboring uninfected cell, inducing the cell to produce antiviral proteins. The uninfected cell enters an activated, antiviral state capable of inhibiting viral replication. »» Why would interferon itself not be considered an antiviral compound?

of double-strand RNA formed during the replicative process in the infected cell induces the cell to synthesize and secrete IFNs (FIGURE 20.13). These bind to specific IFN receptor sites on the surfaces of adjacent cells. Binding triggers a signaling pathway that turns on new gene transcription involved in the synthesis of **antiviral proteins (AVPs)** within those cells. Such cells are now in an **antiviral state** and capable of inhibiting viral replication. The AVPs inhibit translation and degrade mRNA—both viral and host if the cells are infected.

Obviously, the interferons are not a magic bullet for preventing viral infections; we all still get colds, the flu, and other virally caused illnesses. Today, the use of recombinant DNA technology has made it possible for pharmaceutical companies to mass produce synthetic IFN-alpha for disease treatments. The FDA has approved INF-alpha (alfa) 2a for use, along with other drugs, in treating hepatitis B and C, and genital warts—all caused by viruses.

FIGURE 20.14 summarizes the innate barrier mechanisms.

CONCEPT AND REASONING CHECKS 4

a. Explain the outcome to someone who has an infectious disease but has a genetic defect making the person unable to produce C5 complement protein.

b. Distinguish between the role of interferon and antiviral proteins.

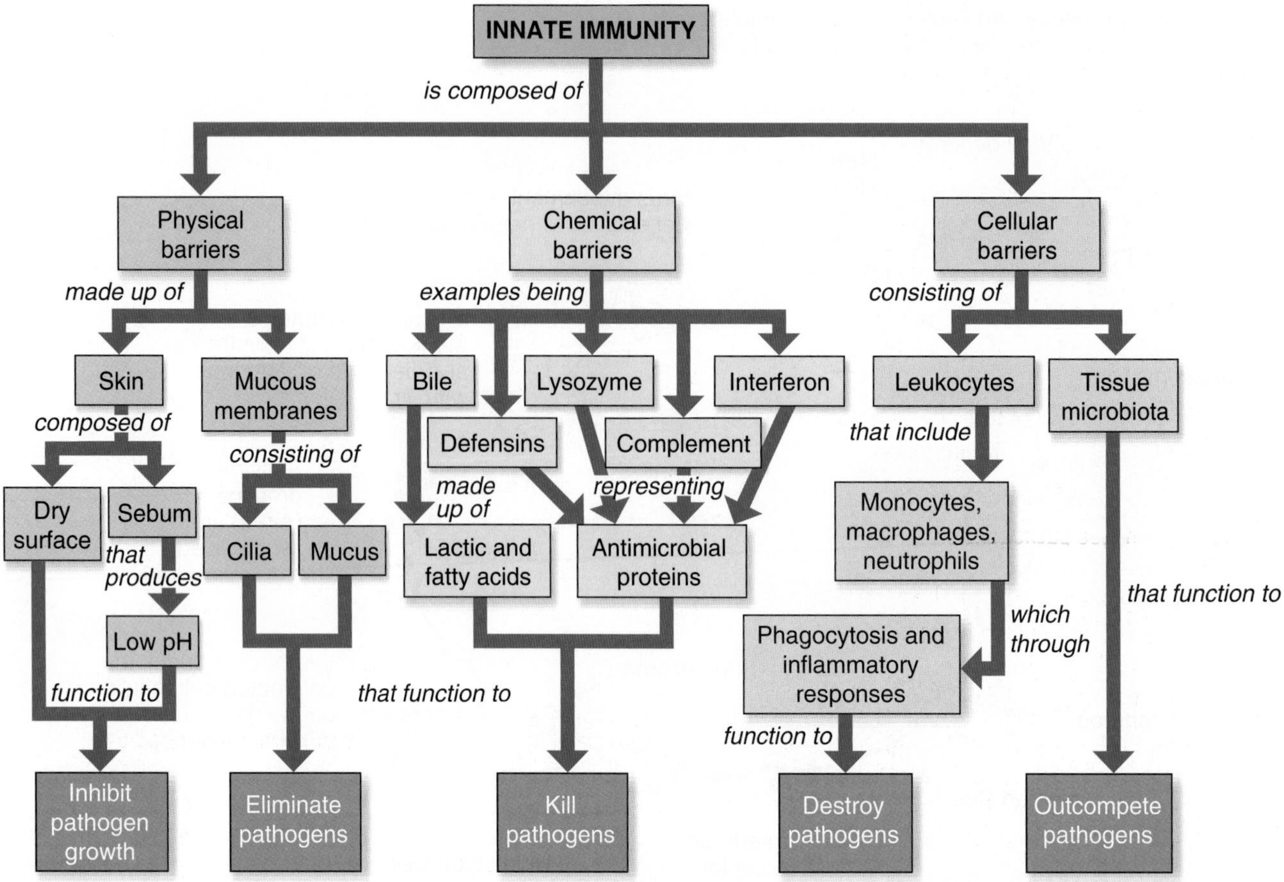

FIGURE 20.14 **Innate Defense Barriers to Pathogens.** This concept map summarizes the ways innate immunity is poised to inhibit, eliminate, destroy, and compete against an invading pathogen. »» Why does there need to be such a diversity of defense mechanisms?

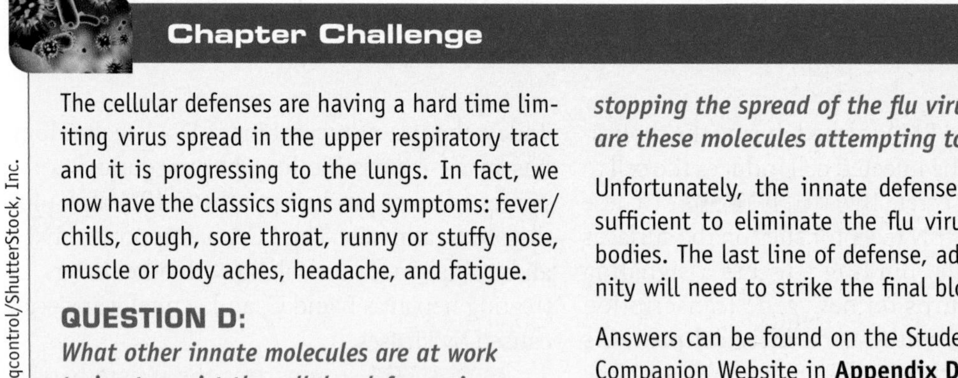

Chapter Challenge D

The cellular defenses are having a hard time limiting virus spread in the upper respiratory tract and it is progressing to the lungs. In fact, we now have the classics signs and symptoms: fever/chills, cough, sore throat, runny or stuffy nose, muscle or body aches, headache, and fatigue.

QUESTION D:

What other innate molecules are at work trying to assist the cellular defenses in stopping the spread of the flu viruses? How are these molecules attempting to work?

Unfortunately, the innate defenses will not be sufficient to eliminate the flu viruses from our bodies. The last line of defense, adaptive immunity will need to strike the final blow.

Answers can be found on the Student Companion Website in **Appendix D**.

■ **In conclusion**, physical barriers, immune cells, antimicrobial chemicals, and signaling chemicals are needed to protect the human host, generate an effective innate immune response, and induce the acquired immune response. MICROINQUIRY 20 presents an overview linking innate and adaptive immunity.

MICROINQUIRY 20

Visualizing the Bridge Between Innate and Adaptive Immunity

If the physical barriers of skin or mucous membranes are breached, innate immunity involves several cells and chemical mediators:

- Macrophages, neutrophils, and dendritic cells are involved with phagocytosis and cytokine production and secretion;
- Natural killer cells can search out and destroy virus-infected cells using perforins and granzymes;
- Defensive proteins (defensins, complement, acute phase proteins, interferon) can lyse pathogens or interfere with virus replication; and
- Signaling receptors (the toll-like receptors) and proteins (the cytokines) are instrumental in stimulating and coordinating a sustained immune defense.

Still, many pathogens in just sheer numbers, or by having appropriate virulence factors, can win the battle against the innate immune response. So, it is critical for innate immunity to muster the apparatus to activate the last and ultimate defense against pathogens—the adaptive immune response. But how does the innate response do this? How does it bridge these two overlapping components of the immune system into a seamless and directed immune response?

The major link between innate and adaptive immunity actually is quite simple—and therefore rather easy to subvert if you are the human immunodeficiency virus (HIV). The link involves the toll-like receptors and phagocytosis of the pathogen during the innate immune response (see figure, part **A**). During phagocytosis, the process not only destroys the pathogen, it also "presents" to the adaptive immune system cells (T lymphocytes) fragments of the pathogen on the phagocyte's cell surface (see figure, part **B**). This fragment presentation is primarily carried out by dendritic cells, macrophages, and monocytes, so these cells are often called "antigen presenting cells" (APCs). The environment in which the T cell recognizes a pathogen fragment on an APC, the amount of pathogen fragments presented, and for how long the fragments are presented will affect the nature and strength of the T-cell response. The activated dendritic cells are the most effective triggers of T-cell and B-cell activity because the dendritic cell population has a high expression of pathogen fragment presentation and generates stimulatory molecules necessary for maximal T-cell stimulation—and adaptive immunity in general (see figure, part **C**).

Discussion Point

Innate immunity is integral to adaptive immunity. The HIV virus can "disconnect" the bridge between the two components of the immune system by infecting and destroying T cells. How does this affect the overall function of the immune system and what does it tell us about the ability of the innate immune system to control and eliminate viral infections?

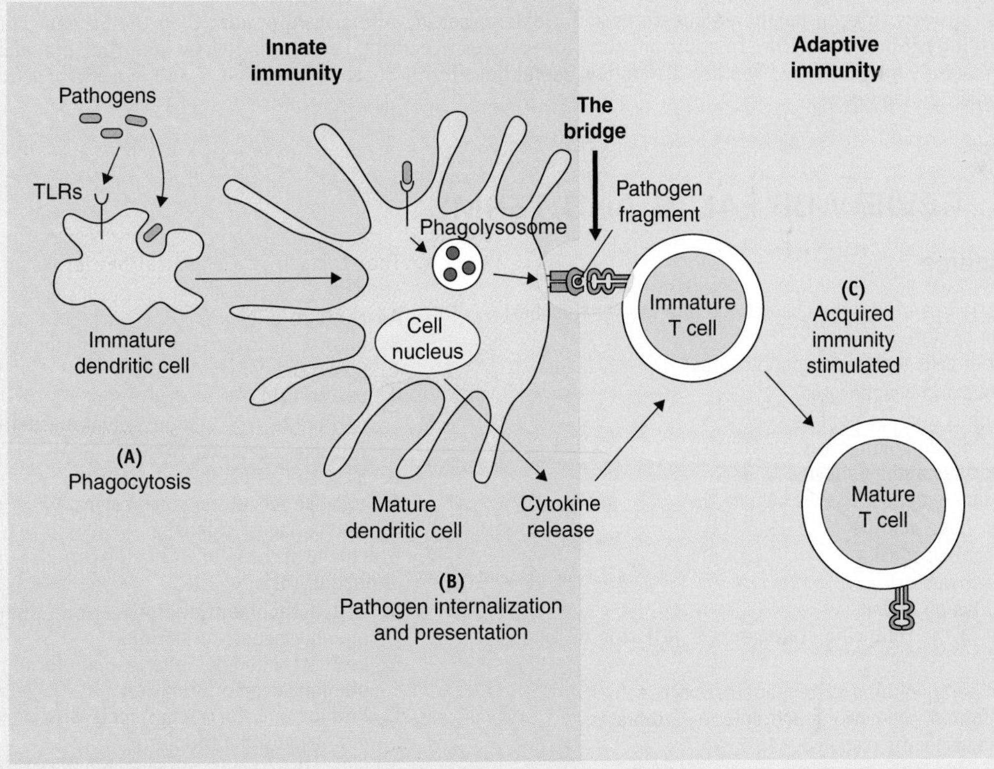

The Bridge Between Innate and Acquired Immunity. Pathogen recognition receptors and phagocytosis are important in establishing the bridge to activation of adaptive immunity.

SUMMARY OF KEY CONCEPTS

20.1 The Immune System Is a Network of Cells and Molecules to Defend Against Foreign Substances

1. In terms of infection, the most important cells are the **leukocytes**, which consist of the **neutrophils, eosinophils, basophils, macrophages, dendritic cells**, and **lymphocytes** (**NK cells, B lymphocytes**, and **T lymphocytes**). (Fig. 20.2)
2. The **lymphatic system** functions in disease to filter and trap pathogens in **lymph nodes** or other secondary lymphoid tissues populated with B cells, T cells, macrophages, and dendritic cells. (Fig. 20.3)
3. A fully functional immune defense system consists of the **innate immune response** that is genetically encoded from birth and the **acquired immune responses** that one develops during one's lifetime through exposure to pathogens. Communication within and between responses is dependent on **cytokines**. (Fig. 20.4)

20.2 Surface Barriers Are Part of Innate Immunity

4. The **skin** and **mucous membranes** are mechanical barriers, while chemical defenses include the acidic environment in the urinary tract and stomach. **Defensins** damage cell membranes, and **lysozyme** and interferon destroy bacterial cells and block virus replication, respectively. The normal **microbiota** limits pathogen spread and infection.

20.3 Coordinated Cellular Defenses Respond to Pathogen Invasion

5. **Toll-like receptors** are responsible for the ability of innate immunity to recognize **pathogen-associated molecular patterns (PAMPs)**. (Fig. 20.6)
6. **Phagocytosis** by **phagocytes** (neutrophils, macrophages, dendritic cells) internalizes and kills pathogens. (Fig. 20.7)
7. **Inflammation** can be a defensive mechanism to tissue injury by pathogens. Chemical mediators, including cytokines, trigger **vasodilation** and increase capillary permeability at the site of infection. The inflamed area exhibits redness, warmth, swelling, and pain. (Fig. 20.9)
8. Moderate **fever** inhibits the growth of pathogens while increasing metabolism for tissue repair and enhancing phagocytosis.
9. **Natural killer (NK) cells** are lymphocytes acting nonspecifically on virus-infected cells and abnormal host cells. (Fig. 20.10)

20.4 Effector Molecules Damage Invading Pathogens

10. **Complement** proteins enhance innate defense mechanisms and lyse cells with **membrane attack complexes**. (Figs. 20.11, 20.12)
11. **Interferon** represents a group of antiviral substances naturally produced by cells when infected. Secretion of interferon stimulates uninfected neighboring cells to enter an **antiviral state** by producing antiviral proteins. (Fig. 20.13)

CHAPTER SELF-TEST

For **STEPS A–D**, answers to even-numbered questions and problems can be found in **Appendix C** on the Student Companion Website at **http://microbiology.jbpub.com/10e**. In addition, the site features eLearning, an online review area that provides quizzes and other tools to help you study for your class. You can also follow useful links for in-depth information, read more MicroFocus stories, or just find out the latest microbiology news.

STEP A: REVIEW OF FACTS AND TERMS

Multiple Choice

Read each question carefully, then select the *one* answer that best fits the question or statement.

1. Which pair of cells represents granulocytes?
   - **A.** Basophils and lymphocytes
   - **B.** Neutrophils and eosinophils
   - **C.** Eosinophils and monocytes
   - **D.** Lymphocytes and monocytes
2. The secondary lymphoid tissues include the _____ and _____.
   - **A.** thymus; bone marrow
   - **B.** bone marrow; tonsils
   - **C.** spleen; thymus
   - **D.** spleen; lymph nodes
3. Which one of the following statements is NOT true of innate immunity?
   - **A.** It is an early-warning system against pathogens.
   - **B.** It is a form of immunity found only in vertebrates.
   - **C.** It is a nonspecific response.
   - **D.** It responds within minutes to many infections.
4. The stomach is a chemical barrier to infection because the stomach _____.
   - **A.** contains bile
   - **B.** harbors *Helicobacter pylori,* a member of the host microbiota
   - **C.** possesses defensive cells
   - **D.** has an acid pH
5. _____ bind to _____ on microbial invaders.
   - **A.** Toll-like receptors; PAMPs
   - **B.** Mast cells; histamine
   - **C.** Toll-like receptors; complement
   - **D.** Macrophages; defensins
6. Which one of the following is the correct sequence for the events of phagocytosis?
   - **A.** Cell attachment, acidification, phagosome formation, phagolysosome formation
   - **B.** Cell attachment, phagosome formation, acidification, phagolysosome formation
   - **C.** Phagosome formation, cell attachment, acidification, phagolysosome formation
   - **D.** Cell attachment, phagosome formation, phagolysosome formation, acidification

7. Which characteristic sign of inflammation is NOT correctly associated with its cause?
   A. Edema—nerve damage
   B. Heat—blood warmth
   C. Swelling—fluid accumulation
   D. Redness—blood accumulation
8. Pyrogens are _____.
   A. proteins affecting the hypothalamus
   B. bacterial fragments
   C. fever-producing substances
   D. All the above (**A–C**) are correct.
9. Natural killer (NK) cells kill by secreting _____.
   A. lysozymes
   B. granzymes
   C. defensins
   D. interferons
10. Which one of the following is NOT a function of complement?
   A. Stimulation of inflammation
   B. Stimulation of antibody formation
   C. Formation of membrane attack complexes
   D. Heightened level of phagocytosis
11. Which one of the following statements about interferon is FALSE?
   A. Interferon is produced in response to a viral infection.
   B. Interferon is a naturally produced protein.
   C. Interferon puts uninfected cells in an antiviral state.
   D. Interferon is a protein that binds to RNA virus genomes.

Matching

Match the statement on the left to the term on the right by placing the correct letter in the available space. A term may be used once, more than once, or not at all.

Statement

_____ 12. Peyer patches are part of this tissue.
_____ 13. Recognizes tumor and virus-infected cells.
_____ 14. Produced in response to viral infection.
_____ 15. A flattened organ found in the upper left of the abdomen.
_____ 16. Small protein released by various defensive cells in response to an activating substance.
_____ 17. A cytotoxic enzyme that induces apoptosis.
_____ 18. A T cell or B cell is one.
_____ 19. A monocyte matures into one of these.
_____ 20. Found in human tears and sweat.
_____ 21. Also called a PMN.
_____ 22. Forms holes in a target cell membrane.
_____ 23. A phagocyte that circulates in the blood and survives for several months.
_____ 24. Refers to a substance causing fever.
_____ 25. A short-lived phagocyte.

Term

A. Agranulocyte
B. Cytokine
C. Dendritic cell
D. Granulocyte
E. Granzyme
F. Interferon
G. Lymphocyte
H. Lysozyme
I. Macrophage
J. MALT
K. Mast cell
L. Monocyte
M. Natural killer (NK) cell
N. Neutrophil
O. Perforin
P. Pyrogen
Q. Spleen
R. Thymus

STEP B: CONCEPT REVIEW

26. Summarize the functions for the six groups of **leukocytes**. (**Key Concept 1**)
27. Compare and contrast the **primary** and **secondary lymphoid tissues** in terms of function. (**Key Concept 1**)
28. Distinguish between **innate** and **adaptive immunity**. (**Key Concept 1**)
29. Describe how physical, chemical, and cellular barriers protect against pathogen invasion. (**Key Concept 2**)
30. Explain the role of **toll-like receptors** to innate immunity. (**Key Concept 2**)
31. Assess the importance of **phagocytosis** as a nonspecific defense mechanism. (**Key Concept 2**)
32. Explain why **inflammation** is a key factor in the identification of and defense against an invading microbe. (**Key Concept 2**)
33. Assess the value of a low grade **fever** in the nonspecific responses to infection. (**Key Concept 2**)
34. Discuss how **natural killer (NK) cells** would identify and kill virus-infected cells in a host. (**Key Concept 2**)
35. Propose what microbes would most likely be susceptible to **membrane attack complexes**. (**Key Concept 2**)
36. Diagram the pathway by which **interferon** stimulates an **antiviral state**. (**Key Concept 2**)

STEP C: APPLICATIONS AND PROBLEMS

37. A roommate cuts his finger and develops an inflammation at the cut site. Having taken microbiology, he asks you to explain exactly what is causing the throbbing pain and the warmth at the cut site. How would you reply to your roommate's question?
38. On a windy day, some dust blows in your eye and your eye waters. Why does your eye water when something gets into your eye and how does this relate to immune defenses?
39. A friend is ill with an infection and asks you why she has "swollen glands" behind the jaw. As a microbiology student, what would you tell her?

STEP D: QUESTIONS FOR THOUGHT AND DISCUSSION

40. The opening of this chapter suggests that for many diseases, a penetration of the mechanical barriers surrounding the human body must take place. Can you think of any diseases where penetration is not a prerequisite to illness?
41. It has been said that no other system in the human body depends and relies on signals as greatly as the immune system. From the discussion of innate immunity, what evidence can you offer to support or reject this concept?
42. Phagocytes have been described as "bloodhounds searching for a scent" as they browse through the tissues of the body. The scent they usually seek is a chemotactic factor, a peptide released by bacterial cells. Does it strike you as unusual that bacterial cells would release a substance to attract the "bloodhound" that will eventually lead to the bacterial cell's demise? Explain.

21

CHAPTER PREVIEW

21.1 **The Adaptive Immune Response Targets the Specific Invading Pathogen**

21.2 **Humoral Immunity Is an Antibody Response to Pathogens in Body Fluids**

Investigating the Microbial World 21: It's in the Blood

MicroInquiry 21: Mucosal Immunity

Textbook Case 21: An Immunodeficiency That Has Mostly Been Asymptomatic

21.3 **Cell-Mediated Immunity Detects and Eliminates Intracellular Pathogens**

Resistance and the Immune System: Adaptive Immunity

Identifying HIV was the critical first step in defining the cause of AIDS, but, as Robert Koch so elegantly pointed out more than a century ago, showing that a particular infectious agent causes a specific disease can be an arduous process.
—Stanley Prusiner in *Historical Essay: Discovering the Cause of AIDS*

The great breakthroughs in any field of science are infrequent. Although most experiments advance our understanding of a phenomenon, more often than not they are the result of a "yes" or "no" answer. But sometimes an experiment or observation not only gives a yes or no answer, it also is unexpected and makes a giant leap forward in scientific understanding. The observations and work carried out by Pasteur and Koch on the germ theory are just a few historical breakthroughs that come to mind.

In 1981, an immunodeficiency disease among gay men was reported. Once there was an alarming rise in the number of new cases of what became known as acquired immunodeficiency syndrome (AIDS), the race was on to discover its cause and how the disease propagated itself.

The eventual appearance of AIDS in distinctly different populations including homosexual and heterosexual individuals, children, intravenous drug users, hemophiliacs, and blood transfusion recipients suggested it must be an infectious agent. And as the number of AIDS cases continued to rise, so did the hypotheses about its possible causation.

Luc Montagnier, Françoise Barré-Sinoussi, and their colleagues at the Pasteur Institute in Paris were the first to report the discovery of the virus now called the **human immunodeficiency virus** (**HIV**) associated with AIDS. Coupled with this breakthrough was the work of Robert Gallo and

Image courtesy of Dr. Fred Murphy/CDC.

his research team at the National Cancer Institute in Maryland. Gallo's group showed that the virus identified by Montagnier was the cause of AIDS.

So, what was the virus doing to cause the disease? As the clinical descriptions of AIDS solidified, it became apparent that the virus infected T lymphocytes and caused a dramatic decrease in their number. (FIGURE 21.1).

Finding the cause of AIDS led to the development of a blood test to identify HIV, preventing millions of people from being infected through the transfusion of tainted blood. Knowing how the virus interacts and replicates in T cells stimulated the identification and design of many antiretroviral drugs to combat the disease, allowing infected individuals to live longer and more productive lives.

Most scientists today would agree that the discoveries made by Montagnier and Gallo rank as one of the major scientific breakthroughs of the 20th century. In 1987, both Montagnier and Gallo were given equal credit for the discovery. However, in 2008, the Nobel Prize in Physiology or Medicine went to Barré-Sinoussi and Montagnier.

If one set out to design a virus to cripple the immune system and the very cells designed to defend against the virus, one could not do better than the naturally evolved HIV. By destroying T cells, the entire immune system is crippled and unable to properly respond to any infectious disease exposure.

How is this possible? How can a simple virus inactivate a complex and powerful immune defense?

This chapter provides the answer as we examine how the actions of innate immunity, supply the adaptive immune response with the needed boost to fight pathogens.

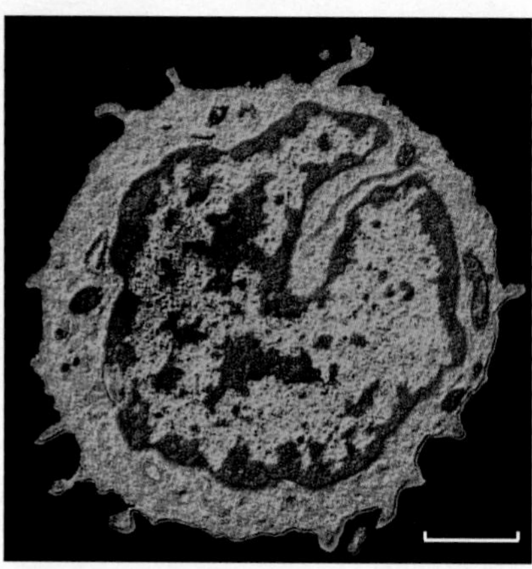

(A)

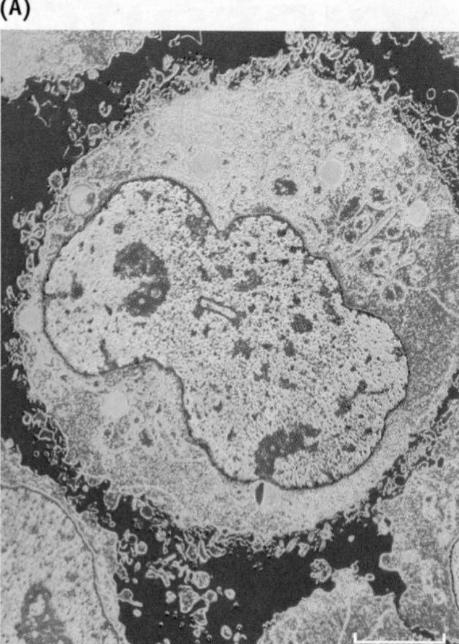

(B)

FIGURE 21.1 **Uninfected and HIV-Infected T Lymphocytes.** False-color transmission electron micrographs of T lymphocytes. (**A**) An uninfected T cell showing the typical large cell nucleus. © Dr. Klaus Boller/Photo Researchers, Inc. (**B**) An infected T cell showing HIV virions (red) budding off the cell surface. (A & B: Bar = 2 µm.) © NIBSC/Photo Researchers, Inc. »» Does HIV replication show a cytopathic effect in infected cells? Explain.

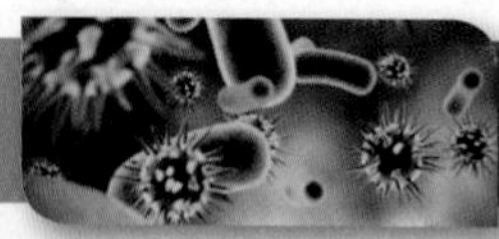

Chapter Challenge

As you remember, we have the flu and the cellular defenses described in the text on innate immunity were unable to limit virus spread in the upper respiratory tract, which has progressed to the lungs. We are now exhibiting the classic signs and symptoms: fever/chills, cough, sore throat, runny or stuffy nose, muscle or body aches, headache, and fatigue. Because innate immunity was unable to stem the spread of the flu viruses, the last line of defense, adaptive immunity, needs to provide the force to eradicate the disease. So, let's recover from the flu!

KEY CONCEPT 21.1 The Adaptive Immune Response Targets the Specific Invading Pathogen

With regard to infectious disease, innate immunity is designed to nonspecifically eliminate any of a diverse group of pathogens—or at least limit their spread until the other half of the immune response, **adaptive** (specific) **immunity**, can mobilize the body's cells and molecules to eliminate the identified pathogen. Before we examine the components and events of adaptive immunity, let's summarize its key features.

The Ability to Eliminate Pathogens Requires a Multifaceted Approach

Adaptive immunity defends against a tremendous variety of potential pathogens. This defense is based on four important attributes.

1. Specificity. Innate immunity has the ability to recognize a diversity of foreign nonself substances, including pathogens. Such microbes or their molecular parts that are capable of mobilizing the immune system and provoking an immune response are called **antigens**.

A typical antigen, like an *Escherichia coli* cell or a flu virus, contains numerous sites on its surface to which immune cells or antibodies can bind (FIGURE 21.2A). These discrete regions of the antigen are called **epitopes** (or **antigenic determinants**). For example, a structure such as a bacterial flagellum may have several epitopes, each having a characteristic and distinct three-dimensional shape (FIGURE 21.2B). It is the job of adaptive immunity to recognize these specific microbial "fingerprints" and generate the specificity needed.

Proteins are the most potent antigens because their amino acids have the greatest array of building blocks. Carbohydrates are less potent antigens because they lack chemical diversity and rapidly break down in the body. Nucleic acids and lipids also can be antigenic. It is estimated that the human immune system can respond to about 10^{14} diverse epitopes. If the body loses its ability to respond to antigens and antigenic determinants, an **immune deficiency** will occur such as AIDS.

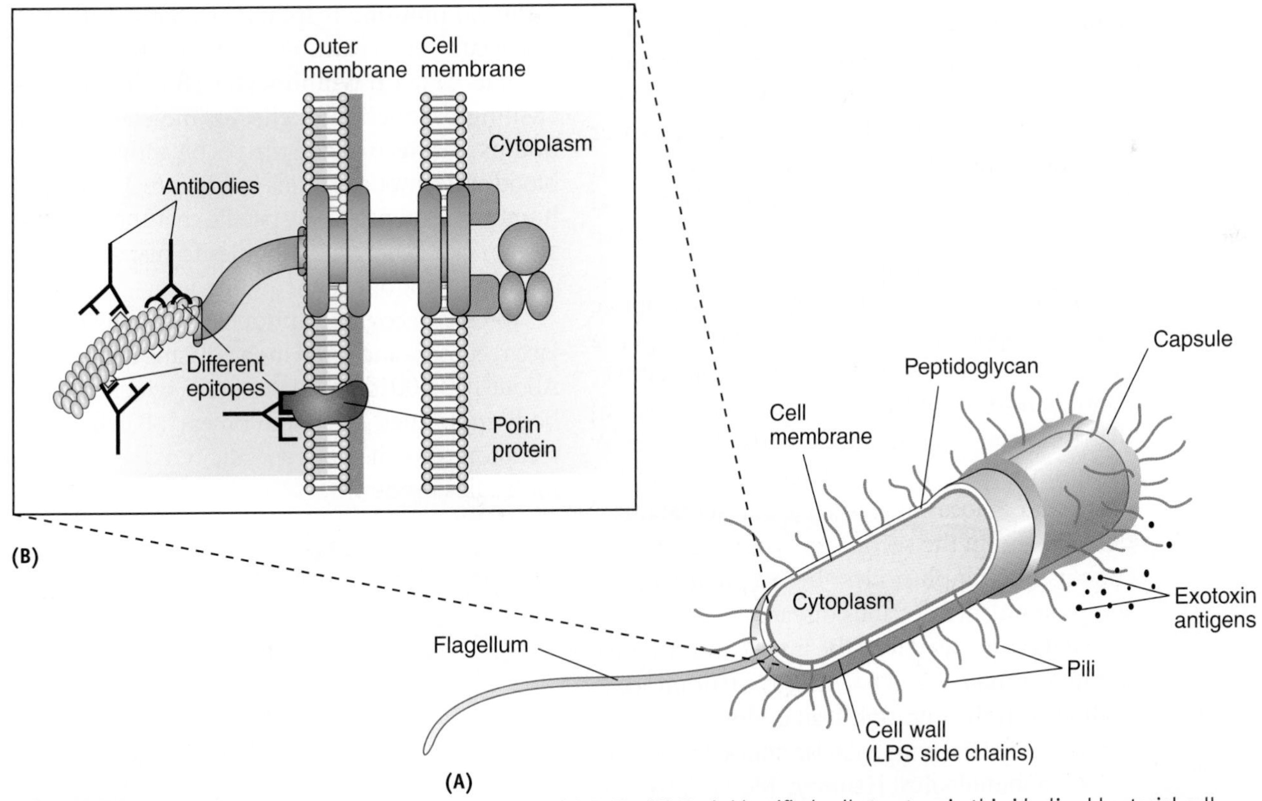

FIGURE 21.2 **The Various Epitopes Possible on a Bacterial Cell.** (**A**) Each identified cell structure in this idealized bacterial cell (antigen) is capable of stimulating an immune response. (**B**) Examples of epitopes include sites on the flagellum and cell membrane to which specific antibodies can bind. »» What organic molecule is common to most of the bacterial structures identified in (**A**)?

2. Tolerance of "Self." Under normal circumstances, one's own cells and molecules ("self") with their own molecular determinants do not stimulate an adaptive immune response. This **self tolerance** can be established in several ways. Among them is **immune suppression** of a "self" immune response by **regulatory T cells**, which prevent autoreactive T cells from attacking self. A second mechanism of self tolerance is through **clonal deletion**. Should any autoreactive cells (the self-recognizing clone of immune cells) bind to a self antigen, the cells are triggered to undergo cell death through apoptosis; that is, the self recognizing clone is eliminated. So, the immune system develops a tolerance of "self" and remains extremely sensitive to "nonself" epitopes present on pathogens, tumor cells, and cells of other individuals of the same species. However, should some cells escape suppression or elimination, self-tolerance may break down, causing an **autoimmune disorder**, such as lupus erythematosus or rheumatoid arthritis.

■ Apoptosis:
A process of programmed cell death.

Many nucleotides, hormones, peptides, and other molecules are recognized by immune cells but do not stimulate an immune system response; that is, they are not **immunogenic**. However, when these nonimmunogenic molecules, called **haptens** (*hapt* = "fastened") are linked (fastened) to carrier proteins, the larger combination may be recognized as "nonself" and trigger an immune system response to both hapten and the carrier protein. Examples of haptens include penicillin molecules, molecules in poison ivy plants, and molecules in certain cosmetics and dyes. These often are the cause of allergies.

3. Minimal "Self" Damage. In an adaptive immune response, it is important that the response be strong enough to eliminate the pathogen, yet controlled so as not to cause extensive damage to the tissues and organs of the body. That is why most of the encounters between pathogen and host occur at local sites, such as the actual infection site and in the secondary lymphatic tissues, such as the lymph nodes. In fact, many of the symptoms of an infectious disease are due to collateral damage caused by the innate and adaptive immune responses and not directly to the invading pathogen. If the immune system does overreact, a potentially fatal anaphylactic shock can occur.

■ Anaphylactic shock:
A sudden drop in blood pressure, itching, swelling, and difficulty in breathing typical of some allergic reactions.

4. Immunological Memory. Most of us during our lives have become immune to certain diseases, such as chickenpox and measles from which we have recovered, or to which we have been immunized. This precise long-term ability to "remember" past pathogen exposures is called **immunological memory** and is a hallmark of adaptive immunity. Thanks to immunological memory, a second or ensuing exposure to the same antigenic variation of a pathogen produces such a rapid and vigorous immune response that the person does not even know they were exposed. Loss of immunological memory is one reason why some people suffer frequent and repeated infections.

Adaptive Immunity Generates Two Complementary Responses to Most Pathogens

The cornerstones of adaptive immunity are the **lymphocytes**, small 10 to 20 μm in diameter cells, each with a large cell nucleus taking up almost the entire space of the cytoplasm. Viewed with the microscope, all lymphocytes look similar. However, two types of lymphocytes can be distinguished on the basis of their complementary adaptive immune responses, cellular function, and unique surface markers.

The Humoral Immune Response. One of the immune behaviors toward antigens is called the **humoral immune response** (*humor* = "a fluid"). The result of this reaction is the activation of effector cells called **B lymphocytes** (**B cells**) and the ensuing production of effector molecules, **antibodies**, that recognize epitopes on antigens in the blood and lymphatic fluids (see Figure 21.2B). The humoral response is so specific that the immune system can generate antibodies to just about any antigen or epitope it encounters.

B cells recognize epitopes by means of surface receptors anchored in the plasma membrane. About 100,000 identical receptors, which are antibody molecules, are found on each B cell. These receptors have the same specificity as the antibody molecules produced.

The Cell-Mediated Immune Response. Should pathogens leave the body fluids and enter the cytoplasm of host cells, antibodies are useless as they cannot get inside cells. Therefore, the complementary adaptive immune response, called the **cell-mediated immune response**, becomes activated. **T lymphocytes (T cells)** control and regulate this response and provide resistance through a variety of interactions with infected or otherwise abnormal cells (for example, virus-infected cells or tumor cells).

T cells, like B cells, have receptor proteins in the plasma membrane. Again, there can be some 100,000 identical receptors on a single T cell. MicroFocus 21.1 suggests how people might help boost their T cell population and stay healthier.

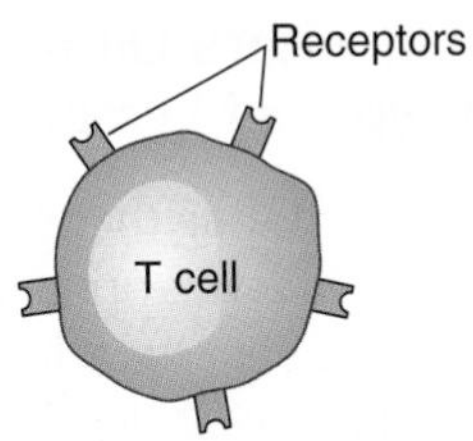

The presence of highly specific receptor proteins on the lymphocyte plasma membrane implies that even before an antigen enters the body, B and T cells are present and waiting "in the wings" to recognize epitopes. For example, an individual may never experience malaria, yet that person has B and T cells with surface receptor proteins for recognizing and binding to the epitopes of malaria parasites should they enter the body.

MICROFOCUS 21.1: Being Skeptical

Tea Protects Us from Infectious Disease

The healthy aspect of tea has been appreciated for centuries, but within the last 15 years scientific evidence has accumulated to back up the age-old claim. It is known that tea contains flavonoids and antioxidants that researchers believe can inhibit the formation of cancer cells and have a positive protective effect on the cardiovascular system by preventing atherosclerosis.

In 2003, Jack Bukowski at Harvard Medical School's Brigham and Women's Hospital in Boston published a paper suggesting tea also can help defend against infectious disease. It seems just three to five cups of ordinary tea made from tea leaves each day will do the trick. Really? What's the evidence?

Bukowski's team was studying the actions of a special subset of T cells called gamma delta T cells. When exposed to ethylamine, the T cells secrete large amounts of interferon, which is involved in inhibition of viral replication. Ethylamine also is found in some bacterial cells, fungi, and parasites. Looking for other sources of ethylamine, Bukowski discovered that tea contains this chemical.

Further experiments mimicking an infection showed that exposing blood to ethylamine in the test tube could increase the numbers of gamma delta T cells by up to five times. In contrast, human blood cells not exposed to the chemical showed a much less significant cell response.

Using human volunteers, the investigators had 21 non-tea drinkers drink either five to six small cups of black tea or five to six small cups of instant black coffee daily for four weeks. Coffee was used as a control because it doesn't contain ethylamine. The results indicated the blood of the tea-drinking volunteers had a fivefold increase in interferon over the course of the study; drinking coffee had no such effect.

The verdict? It appears drinking tea can bring gamma delta T cells to a state of readiness. Should infection then occur, these T cells are available and waiting to defend the body. Whether such stimulation would actually be useful in fighting a real infection remains to be discovered. But these early findings do present intriguing evidence that tea can affect the immune system.

■ Undifferentiated: Referring to cells that have yet to acquire specific characteristics and functions.

Lymphoid Progenitors Differentiate into Several Types of Lymphocytes

All immune cells, including the phagocytes, dendritic cells, and lymphocytes originate from pluripotent stem cells in the bone marrow (FIGURE 21.3). These undifferentiated cells develop into two cell types: **myeloid progenitors**, which become red blood cells and most of the white blood cells, and **lymphoid progenitors**, some of which become lymphocytes.

The B cells (B for bone) mature in the bone marrow. They then move through the circulation to colonize organs of the lymphoid system. B cells are a key effector cell for the humoral arm of adaptive immunity.

Other lymphoid progenitors take a different course. From the bone marrow, some of the cells proceed to the thymus where they develop into T cells (T for thymus) and further differentiate into effector cell subsets, such as **T helper cells** and **cytotoxic T cells**. These cells are key to the cell-mediated arm of adaptive immunity.

Clonal Selection Activates the Appropriate B and T Cells

Adaptive immunity is called a "specific response" because only the specific B and T cells having receptors that recognize epitopes or antigenic fragments are committed. The vast majority of the B and T cells remain unreactive. The process is called **clonal selection** and is diagrammed for B cells in FIGURE 21.4.

Recognition by the appropriate B cells occurs when antigens enter a lymphoid organ, bringing the epitopes close to the appropriate B cells where they bind with B cell receptor proteins. This, along with cytokine stimulation from T helper cells, activates B cells to divide and form a **clone**; that is, a population of genetically identical B cells. A similar set of events occurs to activate T cells, which will be described in Section 21.3.

Both B- and T-cell clones contain activated lymphocytes that will develop into effector cells and memory cells. For example, the B cell clones develop into **plasma cells**, which then synthesize

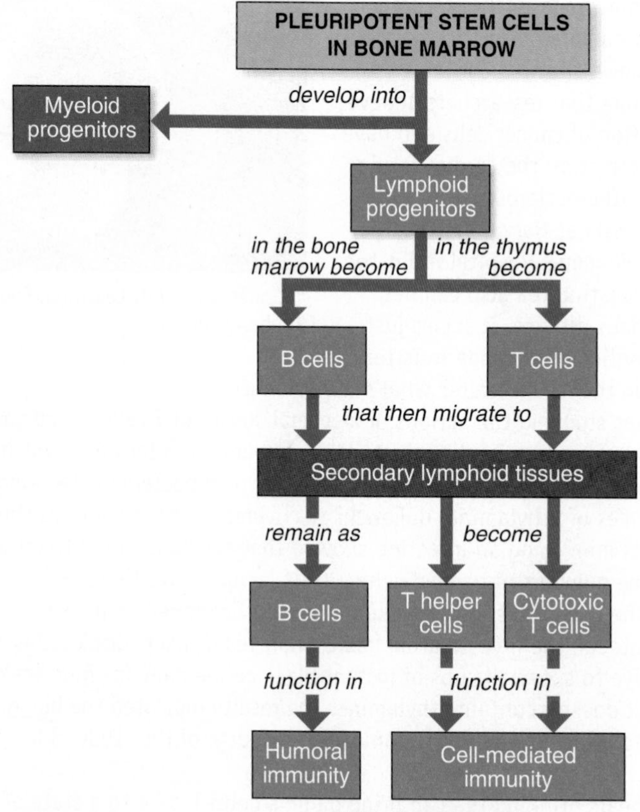

FIGURE 21.3 **The Fate of Lymphoid Progenitors.** Lymphocytes arise in the bone marrow and mature in the bone marrow (B cells) or thymus (T cells). »» Why are two different lineages of lymphoid progenitors needed?

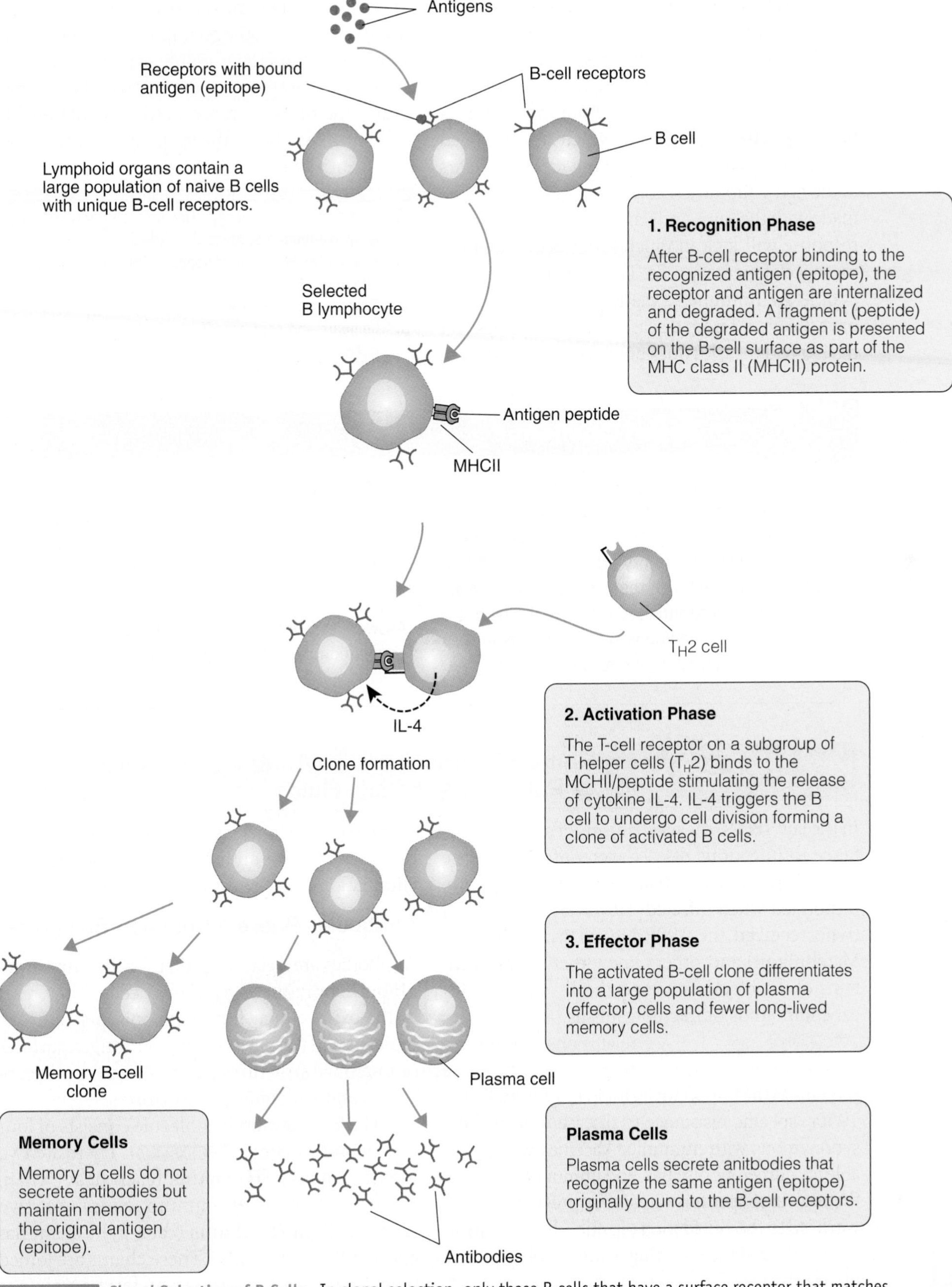

FIGURE 21.4 **Clonal Selection of B Cells.** In clonal selection, only those B cells that have a surface receptor that matches the shape of an epitope will be stimulated to divide, producing a clone of identical B cells. »» Why are the plasma cells called effector cells?

and secrete antibodies matching the invading pathogen epitopes.

The clonal selection process also is important because the events are critical to the memory characteristic of immunity. In the immune response, **memory cells** are produced, which are long-lived B and T cells capable of surviving for many years in the body. Should a second or ensuing contact be made with the same antigen, the adaptive immune response will kick in much faster because of the presence of memory cells. This explains why most people are immune to many diseases once they have recovered from diseases like the measles. Yet having had measles does not prevent a child from contracting chickenpox. Likewise, immunization against a specific pathogen like the diphtheria bacterium or hepatitis B virus only stimulates an adaptive immune response to the agent in the vaccine. Again, the protection acquired by experiencing (adapting to) one of these agents is specific for that agent alone.

CONCEPT AND REASONING CHECKS 1

a. Justify the statement, "The four characteristics of adaptive immunity are really variations of specificity."
b. How do the two arms of adaptive immunity differ from one another?
c. Describe the fates for the lymphoid progenitors.
d. Compare and contrast effector cells and memory cells in the clonal selection of B cells.

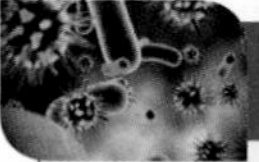

Chapter Challenge **A**

Adaptive immunity consists of two balancing responses to infection. Humoral immunity attacks antigens (pathogens) in body fluids, while adaptive immunity attacks cells that have become infected with the pathogen with the aim to destroy these infected cells. At the height of the flu, we have millions of flu viruses being replicated and released from respiratory tissues.

QUESTION A:

What is the response that (a) humoral immunity and (b) adaptive immunity will mount to try to eliminate the infection of flu viruses?

Answers can be found on the Student Companion Website in **Appendix D**.

KEY CONCEPT 21.2 Humoral Immunity Is an Antibody Response to Pathogens in Body Fluids

Website Animation: Humoral Immunity

In the late 1800s, the mechanisms of specific resistance to infectious disease were largely obscure because no one was really sure how the body responded when infected. However, Paul Ehrlich (who received the 1908 Nobel Prize with Elie Metchnikoff) and others knew that certain proteins of the blood unite specifically with chemical compounds of microorganisms. Investigating the Microbial World 21 relates one of the most significant experiments toward identifying these proteins as the important factor in immunity. By the 1950s, specific resistance to disease was virtually synonymous with immunity. Vaccines were available for numerous diseases, and immunologists saw themselves as specialists in disease prevention. This work led to the elucidation of antibody structure in the 1960s, and the maturing of immunology to one of the key scientific disciplines of our modern times.

In this section, we study adaptive immunity as it relates to the humoral immune response and to the structure and function of antibodies. How the interaction of antibody and antigen leads to elimination of the antigen is also discussed.

Antibodies Share a Common Structure

Antibodies are a class of proteins called **immunoglobulins** produced by plasma cells. They react in the extracellular fluid with epitopes on toxin molecules in the bloodstream, on antigen surfaces or microbial structures (such as flagella, pili, capsules), and with epitopes on viruses.

The basic antibody molecule consists of four polypeptide chains (FIGURE 21.5). There are two identical **heavy (H) chains** (each heavy chain consists of about 400 amino acids) and two identical **light (L) chains** (each light chain has about 200 amino acids). These chains are joined together by disulfide bridges to form a Y-shaped structure, which represents the flexible, monomeric unit.

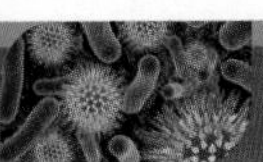

Investigating the Microbial World 21

It's in the Blood

From the experiments of Jenner and Pasteur, where immunity to a disease (smallpox and rabies) could be generated through vaccination, scientists in the late 1800s thought that blood could in some way render toxic substances harmless; that is, the blood provided immunity to diseases like tetanus and diphtheria. In 1890, Emil von Behring and Shibasaburo Kitasato (both in Koch's research group) published a set of experiments that today is considered one of the most important contributions that microbiology has made to medicine. These experiments provided the foundation for the concept of antibodies as the neutralizing agent in blood.

- **OBSERVATION:** Experiments indicated that the blood of immune animals could render the tetanus and diphtheria toxins harmless, making the animals immune to the diseases.
- **QUESTION:** ***What is it about the blood that confers the immunity? Specifically, is it the blood cells or something in the serum (blood fluid without white or red blood cells)?***
- **HYPOTHESIS:** The blood cells of immunized animals neutralize the tetanus toxin. If so, then the blood serum from immunized animals will not protect against a later exposure to the same toxin because the blood cells are absent.
- **EXPERIMENTAL DESIGN:** Immune and nonimmune rabbits and nonimmune mice were used as the test animals and the tetanus toxin was used as the test substance. Immune rabbits were resistant to the tetanus toxin.
- **EXPERIMENT 1:** Nonimmune rabbits and nonimmune mice were injected with a potentially lethal dose of the tetanus toxin. The rabbits and mice were then observed for the effect of the toxin.
- **EXPERIMENT 2:** Immune rabbits were injected with a potentially lethal dose of tetanus toxin. Blood from the surviving rabbits was injected into nonimmune mice, which were then injected with the same dose of tetanus toxin as used in experiment 1. The rabbits and mice were then observed for the effect of the toxin.
- **EXPERIMENT 3:** Serum from the surviving mice in experiment 2 was injected into a group of nonimmune mice, which were then injected with either the same dose of tetanus toxin as used in experiment 1 or an equally lethal dose of diphtheria toxin. The mice were then observed for the effect of the two toxins.
- **RESULTS:** See figure.
- **CONCLUSIONS:**

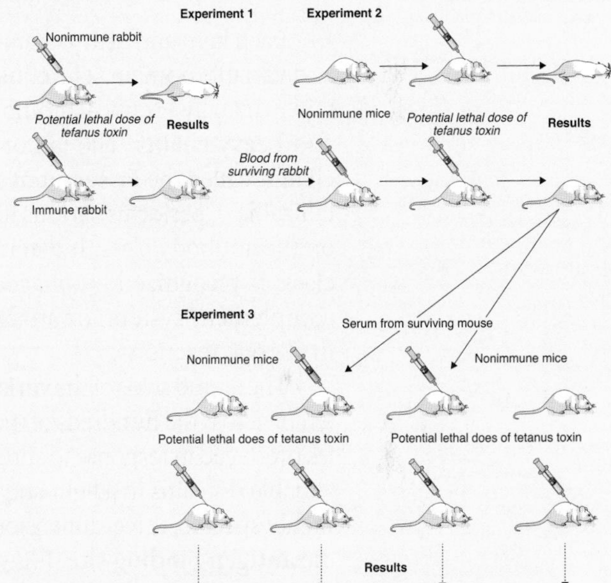

FIGURE The three experiments to determine how the blood confers immunity. Adapted from: Brock, T. D. (1999). in *Milestones in Microbiology: 1546 to 1940*. ASM Press: Washington, D.C.

QUESTION 1: ***What affect did the tetanus toxin have on the nonimmune rabbits and mice (experiment 1)?***

QUESTION 2: ***Explain why the rabbits and mice in experiment 2 survived the lethal dose of tetanus toxin.***

QUESTION 3: ***Explain why the mice in experiment 3 survived or did not survive the lethal dose of tetanus and diphtheria toxins.***

QUESTION 4: ***Was the hypothesis supported? Explain.***

Answers can be found on the Student Companion Website in **Appendix D.**

Note: Emil von Behring won the Nobel Prize in Physiology or Medicine in 1901 for his sole work with the diphtheria toxin. Although the tetanus work with Kitasato was the key investigation, diphtheria was a more serious threat to children of the time—thus the award to von Behring.

Adapted from: Brock, T. D. (1999). in *Milestones in Microbiology: 1546 to 1940*. ASM Press: Washington, D.C.

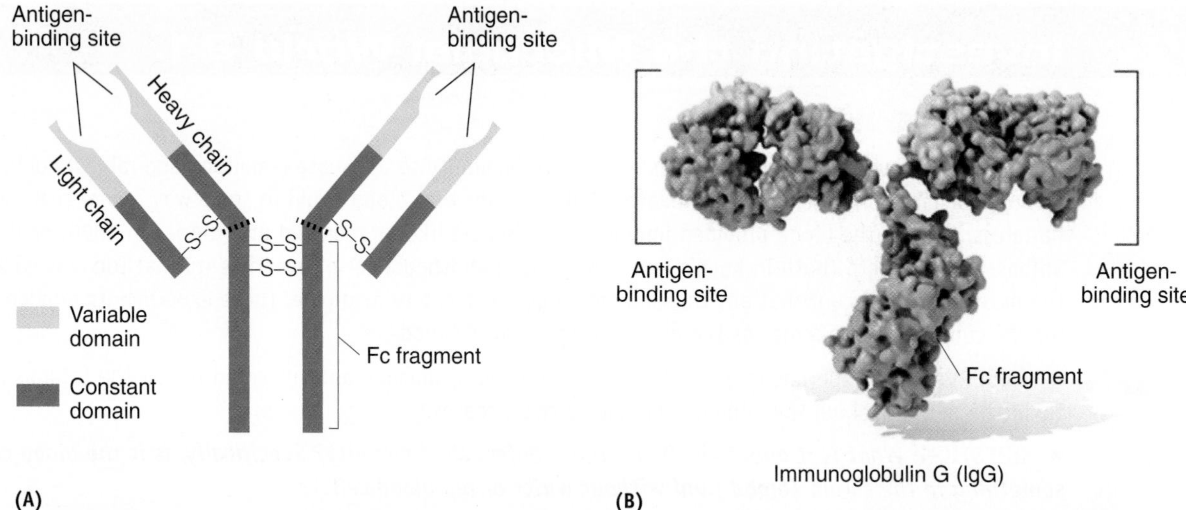

FIGURE 21.5 **Structure of an Antibody.** **(A)** An immunoglobulin monomer consists of two light and two heavy polypeptide chains. The variable domains in each light and heavy chain form a pocket called the antigen-binding site. **(B)** Molecular model of the antibody immunoglobulin G (IgG). *Source*: http://ghr.nlm.nih.gov/handbook/illustrations/igg »» Using the illustration in (A), identify the light and heavy chains in (B).

Each light and heavy chain has both a constant and variable domain. The **constant domain**, contains virtually identical amino acids in both light and heavy chains. The tail or stem of the heavy chains, called the **Fc segment** (for "fragment crystallizable"), performs various functions depending on the antibody class. It determines the antibody class, is recognized by phagocytes, activates the complement system, or attaches to certain cells in allergic reactions.

The amino acids of the **variable domain** are different across the hundreds of thousands of antibodies produced in response to antigens (epitopes). The variable domains in a light and heavy chain form a highly specific, three-dimensional structure, called the **antigen-binding site**. It is to this region on the antibody that the antigen with its epitope binds.

The antigen-binding site is uniquely shaped to "fit" a specific epitope. Moreover, the two "arms" of the antibody are identical, so a single monomeric antibody can combine with two identical epitopes on the same or separate pathogens. As we will see, these combinations may lead to a complex of antibody and antigen molecules, which become targets for elimination by phagocytes.

There Are Five Immunoglobulin Classes

Classes of immunoglobulins are based on whether the antibody will be secreted into the bloodstream, attached onto a cell as a receptor, or deposited in body secretions. Using the abbreviation Ig (immunoglobulin), the five classes are designated IgG, IgM, IgA, IgE, and IgD. The five classes are outlined in TABLE 21.1.

IgG. The antibody commonly referred to as gamma globulin represents the **IgG** class. This antibody is the major circulating antibody, comprising about 80% of the antibody in serum. IgG is primarily produced in response to protein antigens and appears about 24 to 48 hours after antigenic stimulation. Booster injections of a vaccine raise the level of this antibody considerably in the serum. IgG also is the maternal antibody that crosses the placenta and confers additional immunity to the fetus.

IgM. The **IgM** (M stands for macroglobulin) class is the largest antibody molecule and, in serum, consists of a pentamer (five monomers) whose tail segments are connected inwards by a glycoprotein. It is primarily produced in response to polysaccharides and is the first, albeit short-lived, immunoglobulin to appear in the circulation after antigenic stimulation. Thus, the presence of IgM in the serum of a patient indicates a very recent infection. Because of its large size, most IgM remains in circulation. As a monomer, it is a receptor on the B-cell surface.

IgA. About 10% of the total antibody in normal serum is **IgA**. However, the dimeric form, **secretory IgA**, is the predominant immunoglobulin secreted through specialized epithelial cells into the external environment at mucosal surfaces such as the gastrointestinal tract. Thus, IgA

TABLE

21.1 The Five Immunoglobulin Classes and Functional Properties

Property	Immunoglobulin Class: IgG	IgM	IgA	IgE	IgD
Number of monomers	1	5	2	1	1
Antibody in serum (%)	80	5–8	10–15	0.004	0.2
Half-life[1] in serum (days)	7–24	5–10	4–7	1–5	2–8
Activates complement	Yes	Yes	No	No	No
Crosses placenta	Yes	No	No	No	No
Neutralizes bacterial toxins	Yes	Yes	Yes	No	No
Function	Principal antibody of secondary antibody response	First antibody formed in a primary antibody response	Monomer in serum; dimer secreted onto mucosal surfaces	Role in allergic reactions; effective against parasitic worm infections	Receptor on B cell surface to initiate humoral response

[1]The time for 50% of the antibodies to be eliminated.

is an important part of **mucosal immunity** where it provides a first-line defense against potential pathogens (MicroInquiry 21). IgA also is located in tears and saliva, and in the colostrum, the first milk secreted by a nursing mother. When consumed by an infant, the antibodies provide added resistance to potential gastrointestinal pathogens.

IgE. The **IgE** class is another monomeric immunoglobulin. It plays a major role in allergic reactions by sensitizing mast cells and basophils to certain antigens.

IgD. The last class of immunoglobulins is the **IgD** antibody, which also exists as a cell surface receptor on B cells. It has no known effector role in serum.

Textbook Case 21 describes a common antibody deficiency.

Antibody Responses to Pathogens Are of Two Types

The first time the immune system encounters a pathogen or antigen, a **primary antibody response** occurs (FIGURE 21.6). B cells are activated and effector cells, the plasma cells, start producing and secreting antibodies. There is a lag of several days before there is a measurable antibody **titer**; that is, the concentration of antibodies circulating in the bloodstream. The IgM antibodies are the first on the scene, but they are soon replaced by a longer-lasting IgG response. For an otherwise healthy adult, it can take 10–14 days to mount a "full-blown" antibody attack.

During the primary antibody response, memory cells also are produced, which provide the immunological memory needed for subsequent encounters with the same antigen. Therefore, a second or subsequent infection by the same pathogen or antigen produces a more powerful and longer lasting **secondary antibody response**. Due to the presence of memory cells, a rapid response to antigen leads to the production of IgG, the principal antibody. The secondary antibody response occurs much faster (4–7 days) and is more vigorous than the primary response. It also provides a much longer period of resistance against disease.

Antibody Diversity Is a Result of Gene Rearrangements

For decades, immunologists were puzzled as to how antibodies could recognize millions of different epitopes from the limited number of genes associated with the immune system—the human genome has only about 25,000 genes for all organism functions.

The antibody diversity problem has been solved by showing that developing B cells contain

MICROINQUIRY 21

Mucosal Immunity

Most of this chapter discusses the systemic immune system; that is, the body's immune cells and chemical signals located in the blood and lymphatic systems. Until recently, however, not much was known about immunity at the body surfaces, where specialized cells and antibodies protect the body at its vulnerable portals of entry, such as the mucous membranes lining the respiratory, urinary, and gastrointestinal tracts. This defense is called mucosal immunity and is constructed to guard *against infection* rather than the systemic immune system's response to *resolve an infection* that is already occurring. Let's briefly look at the structure and function of mucosal immunity, using the gastrointestinal (GI) tract as our example.

The Mucosal Surface

Mucosal immunity represents an integrated network of tissues, cells, and effector molecules to protect the host from infection of the mucous membrane surfaces (see figure and numbers as we proceed). If the epithelium is damaged by gastrointestinal pathogens such as *Vibrio cholerae*, enterotoxigenic *Escherichia coli* (ETEC), *Shigella* sp., or noroviruses, the pathogen may gain access to the mucosal tissues ①. Therefore, the mucosal surface must protect itself against such possibilities.

In the GI tract, mucosal immunity involves the **mucosa-associated lymphoid tissue** (**MALT**), the largest immune organ of the body. The intestinal epithelium contains specialized **M cells** that are important to uptake, transport, processing, and possibly presentation of microbial antigens within the intestinal lumen. The epithelium also contains **intraepithelial lymphocytes** (**IELs**). The lamina propria includes secondary lymphoid structures, such as the Peyer's patches, and a large concentration of macrophages, dendritic cells (DCs), plasma cells, and B and T lymphocytes. In a healthy human adult, MALT houses almost 80% of all immune cells.

Although not discussed here, the mucosa must exhibit "oral tolerance." In an environment where immune cells are continuously exposed to large numbers of foreign but nonpathogenic antigens from foods and endogenous microbiota, inflammatory responses would damage and destroy the intestinal lining. The phenomenon of oral tolerance is a unique feature of the mucosal immune system.

MALT Function: Epithelial Layer

One function of the mucosal immune system is to protect the mucous membranes and intestinal epithelium against invasion and colonization by potentially dangerous infectious agents that may be encountered in the intestinal lumen.

If a pathogen makes it through the acidic environment of the stomach, most often it will end up in the intestinal lumen or trapped in the mucus layer ②. The MALT, therefore, forms a defensive line against the pathogenic microbes, which tend to invade intestinal spaces devoid of microbiota, such as epithelial crypts and epithelial cells. The epithelial cells make up most of the cells of the intestinal lining, so they have the most contact with pathogens in the lumen. The epithelial cells can ingest and present antigen fragments to IELs scattered through the epithelium and secrete cytokines to stimulate development of the IELs. The IELs contribute to the removal of injured or infected epithelial cells.

The M cells are located over Peyer's patches and contain channels or passageways through which lymphocytes and DCs get closer to the luminal surface ③. Antigens taken up by M cells can be shuttled to, or directly captured by, DCs, B cells, and macrophages. These antigen fragments can be presented to typical helper T cells and cytotoxic T cells located nearby in the Peyer's patches.

MALT Function: Lamina Propria

The lamina contains large numbers of plasma cells, phagocytes, DCs, and B and T cells. The DCs have the ability to sample the contents of the intestinal lumen by extending finger-like projections between epithelial cells ④.

Peyer's patches are somewhat like lymph nodes in that they contain several follicles populated with B and T cells ⑤. The B cells are primarily committed to producing secretory IgA (S-IgA). The IgA-expressing plasma cells move to the crypts where they secrete the dimeric IgA that is then transported across the epithelial cell and deposited into the intestinal lumen ⑥. More S-IgA is produced at the mucosal surface than all other antibody classes (i.e., IgG, IgM, IgD, and IgE) in the body combined.

S-IgA protects the mucosa by binding to potential antigens. For example, S-IgA reduces influenza virus attachment and prevents virus penetration at epithelial surfaces. S-IgA also neutralizes bacterial toxins.

⑦ B cells, T cells, and DCs can leave the site of initial encounter with antigen, transit through the lymph, enter local lymph nodes, and then re-establish themselves in the mucosa of origin, or at other mucosal sites, where they differentiate into memory or effector cells. Thus, there appears to be a common mucosal immune system where immune cells activated at one site can disseminate that immunity to other mucosal tissues.

Discussion Point

Now that scientists are beginning to better understand mucosal immunity, they are considering ways to produce newer and better vaccines to boost mucosal immunity against enteric infections caused by bacteria and viruses. Traditional vaccines targeting the systemic immune system do not always confer strong mucosal immunity; that is, the IgA antibody response is not as strong as one would hope. Discuss what would be the best way to administer a mucosal vaccine and identify the "human physiology" hurdles in getting a vaccine to the mucosal surface.

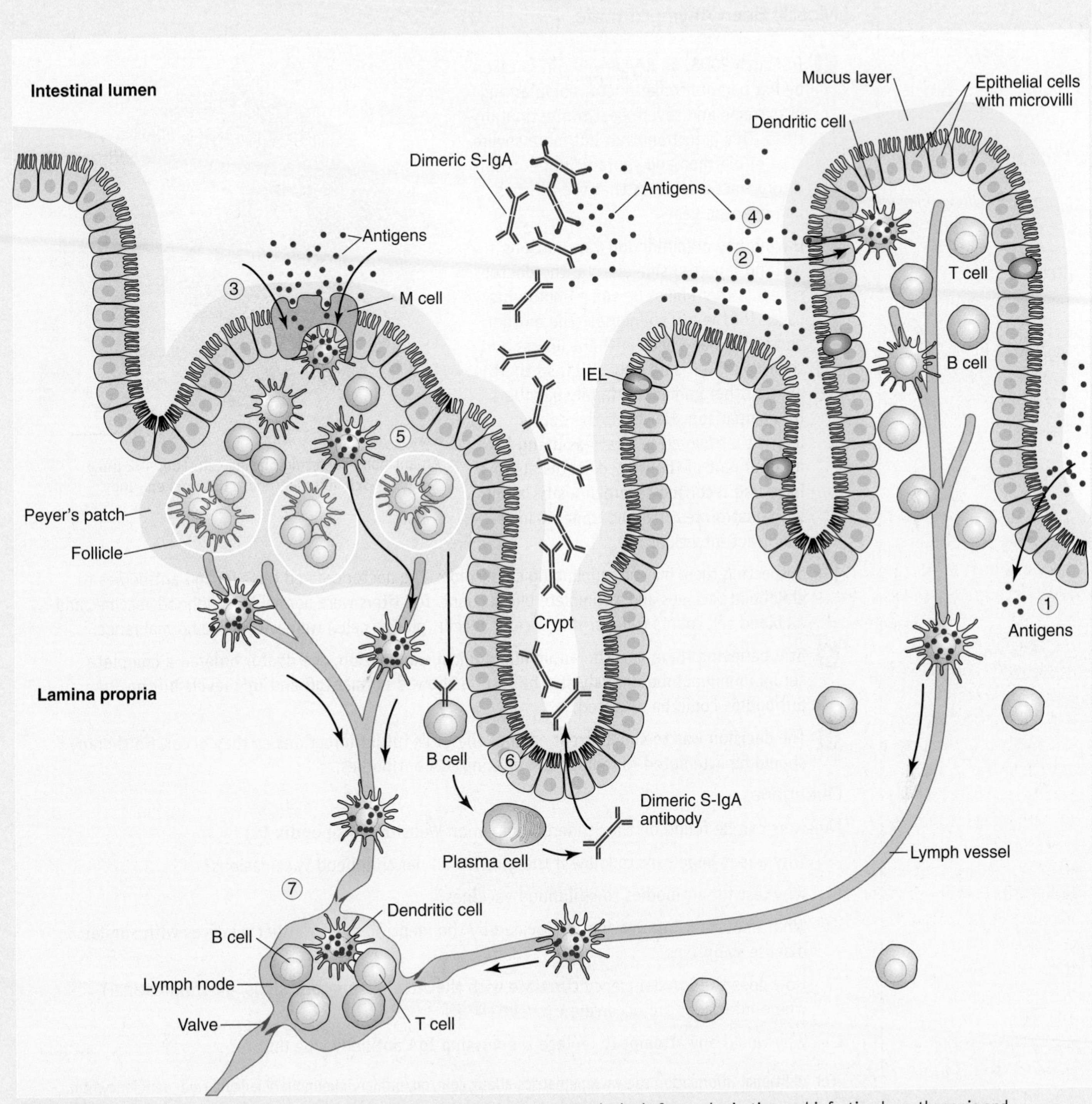

FIGURE The Mucosa of the Digestive Tract. Mucosal immunity helps protect the body from colonization and infection by pathogenic and commensal microorganisms. (Adapted from Wells, Jerry M. and Mercenier, Annick, *Nat Rev Microbiol* 6 (2008): 349–362.)

Textbook CASE 21

An Immunodeficiency That Has Mostly Been Asymptomatic

1 In March 2008, an 8-year-old girl is taken by her parents to her doctor complaining of extreme and severe gastrointestinal distress. This is just another but more severe case of diarrhea and sporadic upper respiratory tract infections that the girl has had over the years.

2 Having had all childhood vaccinations with no reported side effects, the doctor suspects there must be some underlying cause for the girl's illnesses. The mother and father have been without illness and their daughter is an only child, so there are no other immediate family members for comparison. However, the doctor learns that two close relatives living in another part of the United States also have had recurring severe sinusitis along with gastrointestinal and upper respiratory tract infections.

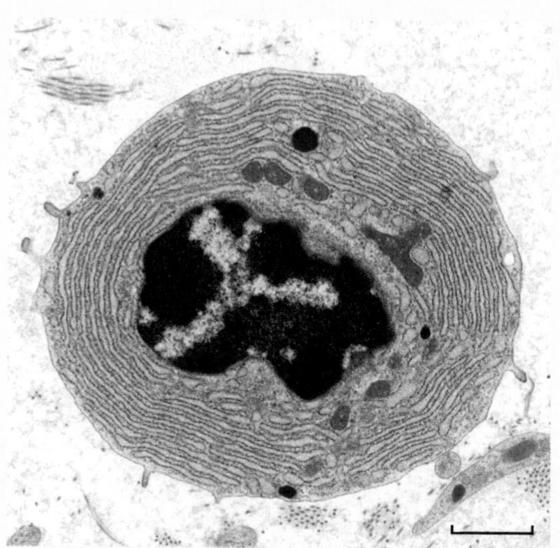

An antibody-secreting plasma cell. (Bar = 2 μm.) © Steve Gschmeissner/Photo Researchers, Inc.

3 Suspecting there may an immunological disorder, the doctor ordered tests for IgG antibodies to childhood vaccines and a complete blood count. IgG titers were normal to childhood vaccines and all blood cell counts (phagocytes, T cells, B cells, plasma cells) were within the normal range.

4 Still believing there must be an immunological explanation, the doctor ordered a complete serum immunoglobulin analysis. The results showed normal IgG and IgM levels but no IgA antibodies could be detected.

5 The decision was to simply treat each of the girl's future infections as they occur. No therapy should be attempted to replace the missing IgA antibodies.

Questions:

(Answers can be found on the Student Companion Website in **Appendix D.**)

A. Why was it important to know if the girl had all her childhood vaccinations?

B. Why test for antibodies to childhood vaccines?

C. What important information is provided by the identification of other relatives with similar disease symptoms?

D. How does the IgA deficiency correlate with the sites (respiratory, sinus, gastrointestinal) where infections are occurring?

E. Why would any attempt to replace the missing IgA antibodies be unsafe?

For additional information see www.pediatrics.about.com/od/primaryimmunodeficiency/a/iga_deficiency.htm.

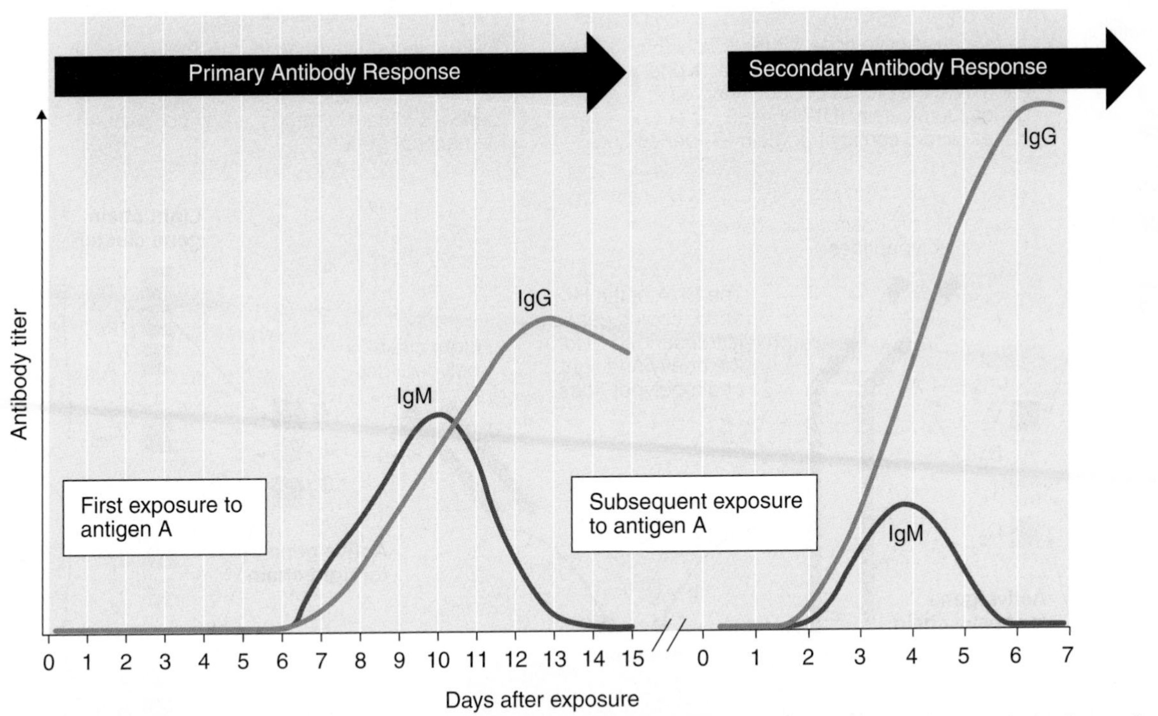

FIGURE 21.6 **The Primary and Secondary Antibody Responses.** After the initial antigenic stimulation, IgM is the first antibody to appear in the circulation. Later, IgM is supplemented by IgG. On a subsequent exposure to the same antigen (antigen A), the production of IgG is more rapid, and the concentration in the serum reaches a higher level than in the primary response. »» Hypothesize why the IgM titer in the secondary antibody response is weaker than in the primary antibody response.

about 300 genetic segments, which are shuffled like transposons and combined in each B cell as unique arrangements. This process, known as **somatic recombination**, is a random mixing and matching of gene segments to form unique antibody genes. Information encoded by these genes then is expressed in the surface receptor proteins of B cells, and in the antibodies later expressed by the effector plasma cells.

The process of somatic recombination uses gene segments coding for the light and heavy chains of an antibody that are located on different chromosomes. The light and heavy chains are synthesized separately, then joined to form the antibody. One of 8 constant gene, one of 6 joiner gene, one of 30 diversity gene, and one of up to 300 variable gene clusters can be used to form a heavy-chain gene. A 300 variable gene cluster is selected and combined with one of 5 joiner gene clusters and a constant gene to form the active light-chain gene. After deletion of intervening sequences (introns), the new gene can function in protein synthesis, as shown in FIGURE 21.7.

Current evidence indicates more than 600 different antibody gene segments exist per cell. Additional diversity is generated through imprecise recombination and somatic mutation. Therefore, the total antibody diversity produced by the B cells is in the range of 10^{14} immunoglobulin possibilities for the variable region of an antibody.

Antibody Interactions Mediate the Disposal of Antigens (Pathogens)

Website Animation: Antibody Function and Antigen Clearance

In order for specific resistance to develop, antibodies must interact with antigens so that the antigen is changed in some way. Thus, the formation of **antigen-antibody complexes** may result in death to the microorganism possessing the antigen, inactivation of the antigen, or increased susceptibility of the antigen to other body defenses. FIGURE 21.8 summarizes these mechanisms.

Certain antibodies can inhibit viral attachment to host cells through **viral inhibition**. By reacting with and covering capsid proteins or spikes, antibodies prevent viruses from attaching to host receptors and entering the host cells.

Neutralization represents a mechanism of defense against toxins and microbes. Because binding to a host cell surface is a necessary step toward infection, so-called "neutralizing antibodies" (IgG and IgA) can prevent microbes or toxins from binding to cells. Viral inhibition and neutralization also

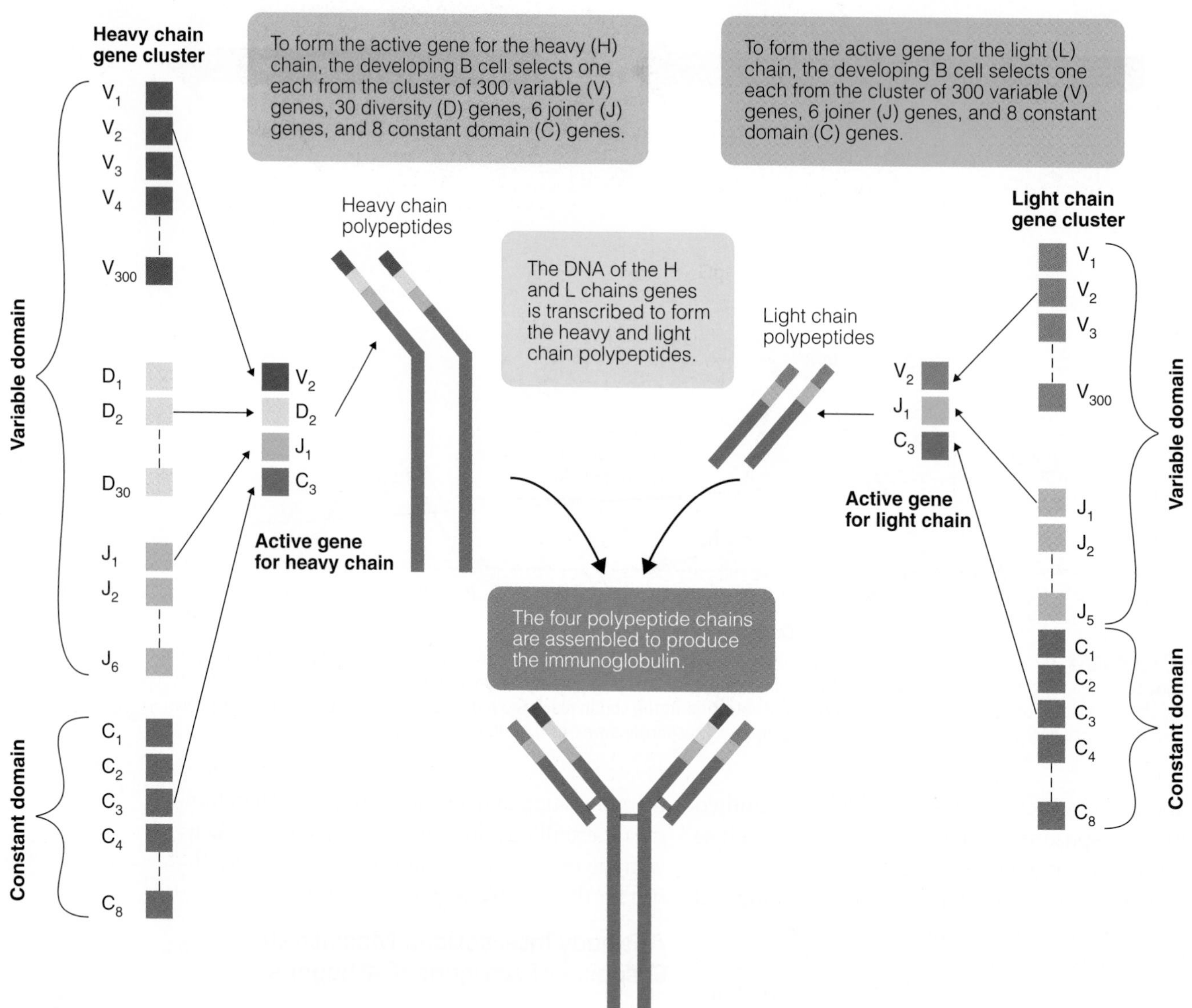

FIGURE 21.7 **The Source of Antibody Diversity.** Almost any substance can stimulate the production of a specific antibody. Immunoglobulin gene rearrangements provide the mechanism for this diversity. Note: Not all potential heavy and light chain segments are shown in the diagram. »» How can gene arrangements generate such antibody diversity?

increase the size of the antigen-antibody complex, thus encouraging phagocytosis while lessening the infective agent's ability to diffuse through the tissues.

As discussed earlier in the chapter, antibodies can bind to bacterial cells, preventing bacterial attachment to host cells. This process, called **opsonization** (the antibody is the **opsonin**) marks the pathogen for elimination by phagocytes.

Because antibodies are bivalent, they can cross-link two separate antigens having the same epitope. This action causes clumping, or **agglutination**, of the bacterial cells. Movement of motile cells is inhibited if antibodies react with antigens on the flagella of microorganisms. The reaction of antibodies with pilus antigens prohibits attachment of an organism to the tissues while agglutinating them and increasing their susceptibility to phagocytosis.

Antibodies also can react with dissolved antigens and convert them to solid precipitates (**precipitation**). In all these disposal interactions, the antigen-antibody complexes formed enhance phagocytosis and thus the removal of the infecting pathogen or antigen (along with attached antibodies) from the body.

A final example of antigen-antibody interaction involves the three pathways of the comple-

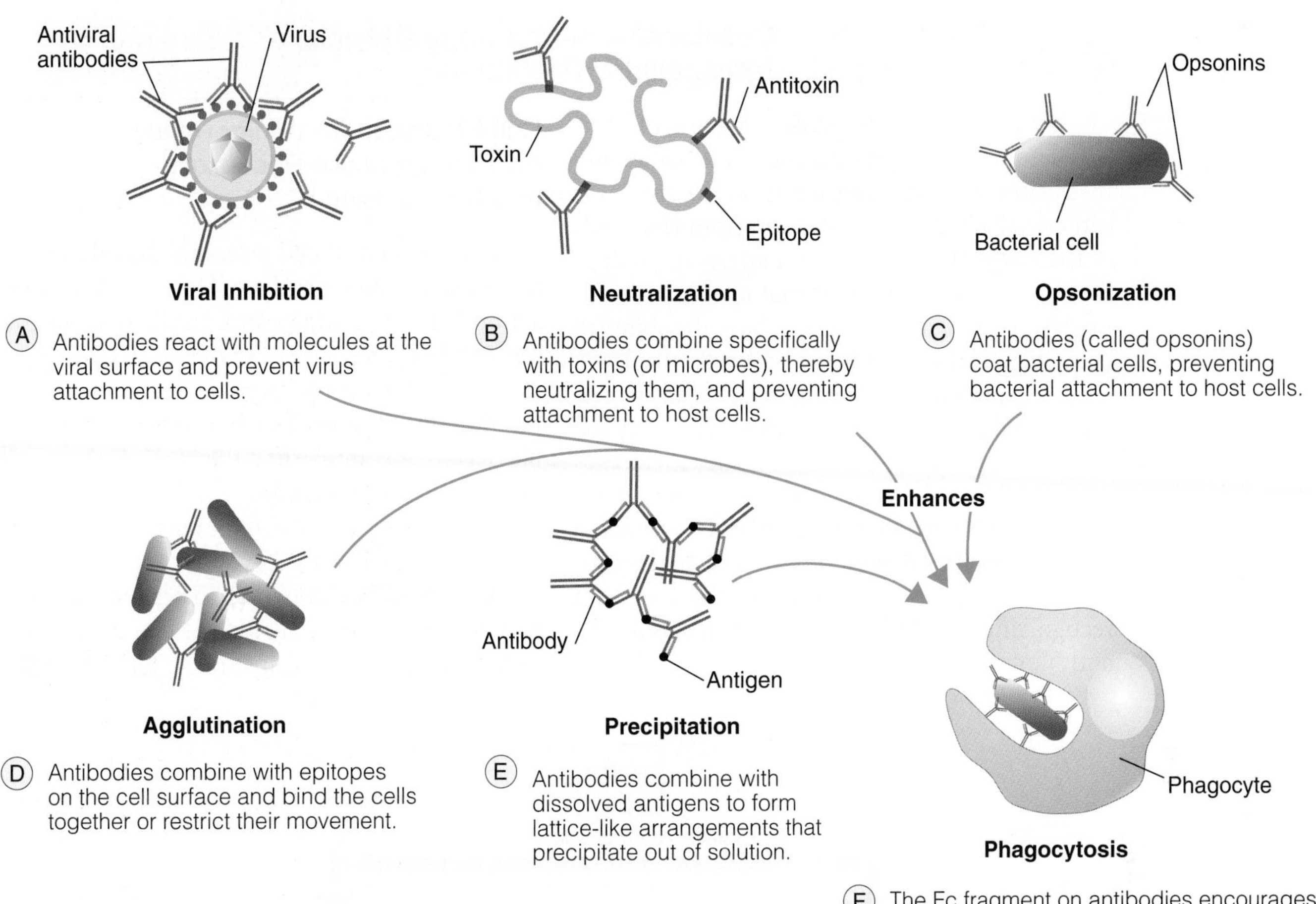

FIGURE 21.8 Mechanisms of Antigen Clearance. In all cases, the antibody-antigen complex facilitates phagocytosis by phagocytes.
»» Reorganize these mechanisms for antigen clearance into two basic strategies.

ment system, described in the text on innate immunity. One set of reactions leads to lysis of bacterial cells through a complement cascade at the cell surface, resulting in the formation of **membrane attack complexes**. These complexes form pores in the membrane, increasing cell membrane permeability and inducing the cell to undergo lysis through the unregulated flow of salts and water.

CONCEPT AND REASONING CHECKS 2

a. How is an antibody's antigen binding site similar to the active site on an enzyme?
b. Why are there five different classes of antibodies in the body?
c. Justify the need for the two antibody responses to pathogens.
d. Why don't the constant genes for an antibody show the same diversity as the variable genes?
e. To this point, summarize the roles that phagocytes have had in the acquired immune response.

Chapter Challenge B

By the time we are experiencing the most severe flu signs and symptoms, the humoral response is well underway. In fact, the response should be near 100% activated with effector cells fully operational by the time the acute phase of the flu infection has occurred.

QUESTION B:

Diagram the humoral response to flu viruses and explain how antibodies can be useful even though there is little if any virus actually in the blood.

Answers can be found on the Student Companion Website in **Appendix D**.

KEY CONCEPT 21.3 Cell-Mediated Immunity Detects and Eliminates Intracellular Pathogens

The humoral immune response is one arm of adaptive immunity. The production of antibodies and their association in antigen-antibody complexes can effectively clear an infection from the fluids of the body. However, many pathogens such as flu viruses and certain bacterial species (including *Mycobacterium tuberculosis*), have the ability to enter inside cells. Once in the host cell cytoplasm, they are hidden from the onslaught that would otherwise be leveled by antibodies, which are too large to enter cells.

The body's defense against microorganisms infecting cells and other "nonself" cells is centered in the cell-mediated immune response. In this final section, we describe the roles for the T cells in recognizing and eliminating virus-infected cells and "nonself" cells.

Cell-Mediated Immunity Relies on T Lymphocyte Receptors and Recognition

As we have seen, B-cell receptors directly recognize epitopes on an antigen. However, the interaction between antigen and T cells depends on sets of surface receptors capable of recognizing a multitude of antigenic (peptide) fragments. T-cell receptors are capable of such recognition because, like B cells and antibodies, they too are the result of somatic recombination.

As mentioned earlier, there are two major subpopulations of T cells.

Cytotoxic T cells (CTLs). These effector cells have T-cell receptors (TCRs) and coreceptor proteins, called **CD8**, on their cell surface (FIGURE 21.9A).

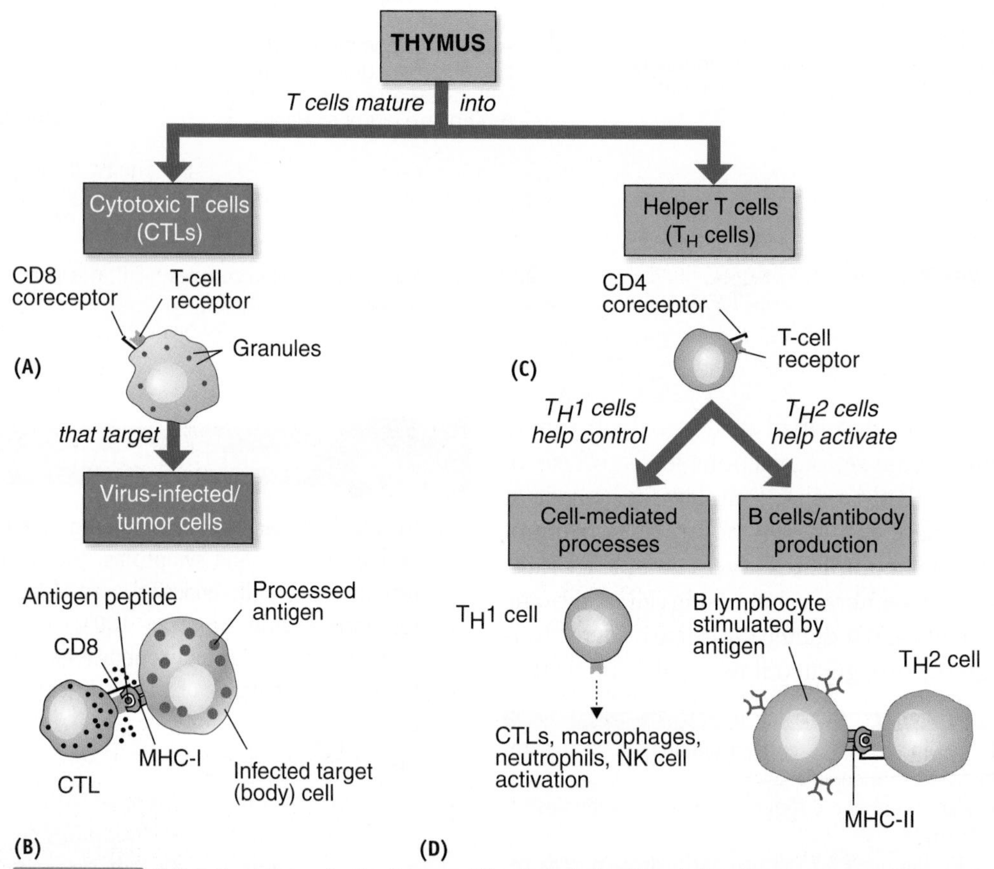

FIGURE 21.9 **T-Cell Receptors.** In the thymus, (**A**) cytotoxic T cells gain T-cell receptors and CD8 coreceptors that recognize antigen peptides on virus-infected cells and tumor cells (**B**). (**C**) Helper T cells gain T-cell receptors and CD4 coreceptors, one class helping eliminate pathogens from infected antigen-presenting cells and the other class helping activate B cells and antibody production (**D**). »» What is common to the receptors of all T cells?

This combination allows the CTLs to recognize and eliminate nonself antigens, such as virus-infected cells and tumor cells.

Helper T (T_H) cells. The other group of T cells, the so-called helper T cells, have TCRs and coreceptors called **CD4** (FIGURE 21.9B). These cells are divided into two effector classes, one (T_H1) that "helps" control many cell-mediated processes, such as CTL, macrophage, neutrophil, and NK cell activation, and the other (T_H2) that "helps" activate B cells in the humoral immune response that we have already discussed.

How do these T cells recognize nonself? The combination of TCRs and coreceptors allows T cells to recognize and bind to another set of nonself protein molecules called the **major histocompatibility complex** (**MHC**). MHC proteins are embedded in the membranes of most cells of the body. At least 20 different genes encode MHC proteins, and at least 50 different variants of those genes exist. Thus, the variety of MHC proteins existing in the human population is enormous (MHC is also called **human leukocyte antigen; HLA**), and the chance of two people having the same MHC proteins is incredibly small. (The notable exception is in identical twins.) The MHC proteins define the uniqueness of the individual and play a role in the immune response. Recent evidence also suggests MHC proteins may play more "amorous" roles (MICROFOCUS 21.2).

There are two important classes of MHC proteins (TABLE 21.2).

MHC Class I (**MHC-I**) molecules are found on all nucleated cells of the body. The MHC-I molecules fold such that they can bind a small antigen peptide that has been internalized. If these proteins contain an antigen fragment from an infecting virus, CTLs will bind to the "presented" MHC-I/antigen fragment using the TCR/CD8 receptor complex (FIGURE 21.9C).

MICROFOCUS 21.2: Being Skeptical

Of Mice and Men—Smelling a Mate's MHC

Would you pick a mate by sight—or smell? At least in the nonhuman world, it might be by smell.

The major histocompatibility complex (MHC) is a group of immune system molecules essentially unique to each individual. Recent research suggests mammals, including humans, might consciously or unconsciously select a mate who has a different MHC than themselves. What's the evidence?

In mice, there is a region of the nose that detects chemicals in the air that are important to mouse reproduction. Frank Zufall and colleagues at the University of Maryland at College Park have discovered that mice also use the nasal region to recognize MHC molecules found in mouse urine. The researchers believe mice might use such "markers" to identify an individual through smell—and perhaps a suitable mate as well. In fact, mice prefer to smell other mice that have a different MHC than themselves and prefer mating with mice of another MHC.

Now to the human "experience." In a human study, people were given sweaty T-shirts to smell. In recorded responses, the participants indicated they preferred smells that the researchers then linked back to different MHCs.

Zufall's theory is that in both mice and humans, individuals preferring a different MHC are preventing inbreeding. Therefore, mating between two individuals with different MHCs will produce offspring having a unique set of MHCs that might make that individual more resistant to factors like infectious disease.

However, most humans don't pick their mates by smelling sweaty T-shirts. So, Craig Roberts at the University of Liverpool wondered if there was a relationship between human faces and MHC. After determining the MHC for a large group, he selected 92 women and showed them photographs of six men, three with somewhat similar MHCs and three with very different MHCs as compared to the women. The women were asked if they would prefer a long- or short-term relationship with each.

Roberts was surprised to find that the women preferred the faces of men who had similar MHCs for a long-term relationship.

The verdict? There certainly is much, much more research to be done. However, Roberts proposes two mechanisms may apply in the selection of a mate. First, select a mate facially with a similar MHC. Then, smell will make sure the mate is not too similar, so more "successful" offspring will be produced.

TABLE

21.2 Antigen-Presenting Cells and T-Cell Recognition

Cell Type	Antigen Presented	MHC Class	T-Cell Recognizing	T-Cell Surface Coreceptor
Nucleated cells	Protein fragments	I	Cytotoxic T cell (CTL)	CD8
Macrophages, dendritic cells, B cells	Protein fragments	II	Helper T cell (TH cell)	CD4

MHC Class II (**MHC-II**) proteins are found primarily on the surface of B cells, macrophages, and dendritic cells. Again, the MHC-II molecules fold to form a pocket in which small antigen (peptide) can bind. Thus, these leukocytes also "present" the MHC-I antigen peptide to CD4 T cells, which will bind to them with their TCR/CD4 receptor complex.

Confused? The next section explains how these responses occur and what they accomplish. As a break, read MicroFocus 21.3, which talks about the possibility of "thinking healthy."

Naive T Cells Mature into Effector T Cells

The actual adaptive immune response originates with the innate immune response and the entry of antigens into the body. At their site of entry or within the lymphatic system, the antigens are phagocytized by the APCs (dendritic cells and macrophages). As the cells migrate to the lymph nodes or other secondary lymphoid tissues, the antigens are broken down in the cytoplasm and antigen peptides are displayed on the surface as MHC-II/peptide complexes (FIGURE 21.10A).

When the APCs enter the lymphoid tissue, they mingle among the myriad groups of naive T cells (T_H0), searching for the cluster having the surface receptors that recognize the MHC-II/peptide. This process requires considerable time and energy because only one cluster of T lymphocytes may have matching T cell/CD4 receptors.

Website Animation: Cell-Mediated Immunity

It is important to remember that recognition is between the MHC-II proteins on the APC surface and receptors on the T-cell surface. It is as if the T lymphocyte must first ensure it and the APC are from the same body (same MHC proteins) before it will respond. The CD4 coreceptor enhances binding of the naive T cell to the APC.

Binding to the APC stimulates T-cell activation (FIGURE 21.10B). The naive T cells secrete other cytokines, including **interleukin-2** (**IL-2**), which stimulate cell division of that T cell and others activated by the APCs. The result is the production of a clone of antigen-specific activated T helper cells. Some of these mature into memory T helper cells and await a future encounter with the same MHC-II/peptide.

Activated T cells mature into one of several types of effector T cells, depending on the influence of specific cytokines (FIGURE 21.10C). Some mature into helper T2 (T_H2) cells that "help" in the activation of humoral immunity, which we discussed earlier.

Other activated T cells mature into helper T1 (T_H1) cells. These effector cells control other adaptive and innate immune responses; that is, they trigger macrophage and neutrophil cell division as well as boost CTL division. So, T_H cells are the master controllers, ensuring that an appropriate immune response is generated to the invading pathogen.

Activated Cytotoxic T Cells Destroy Virus-Infected Cells

Because body cells also can become infected with viruses, the immune system has evolved a way to attack virus-infected cells. During the infection, the host cells manage to degrade some of the viral antigens into small peptides. These are attached onto MHC-I proteins and transported to the cell surface, where they are displayed like "red flags" to denote an infected cell.

Activation of naive CD8 T cells occurs through interaction with dendritic cells or macrophages displaying a MHC-I/peptide (FIGURE 21.11A). CD8

MICROFOCUS 21.3: Being Skeptical

Can Thinking "Well" Keep You Healthy?

In 2010, psychological scientists at the University of Kentucky and the University of Louisville wanted to see how optimism affects the immune system. Studying law students, they were interested in knowing if student expectations about their future (optimistic or not) was reflected in their immune responses. The 6-month study showed an optimistic disposition made no difference in their immune responses. However, as each student had highs and lows in law school, their immune response showed a similar response. In optimistic times, their immune system was quicker to respond to an immunological challenge and in pessimistic times, their immune system was slower to respond to a similar challenge. The scientists suggest that an optimistic outlook may promote a stronger immune system.

The idea that mental states can influence the body's susceptibility to, and recovery from, disease has a long history. The Greek physician Galen thought cancer struck more frequently in melancholy women than in cheerful women. During the twentieth century, the concept of mental state and disease was researched more thoroughly, and a firm foundation was established linking the nervous system and the immune system. As a result of these studies, a new field called "psychoneuroimmunology" has emerged.

The outcome of these discoveries is the emergence of a strong correlation between a patient's mental attitude and the progress of disease. Rigorously controlled studies conducted in recent years have suggested that the aggressive determination to conquer a disease can increase the lifespan of those afflicted. Therapies consist of relaxation techniques, as well as using mental imagery suggesting that disease organisms are being crushed by the body's immune defenses. Behavioral therapies of this nature can amplify the body's response to disease and accelerate the mobilization of its defenses.

In a 2009 study, scientists at Ohio State University concluded that women who practiced yoga regularly had lower levels of interleukin-6 (IL-6) in their blood than those women who did not practice yoga. Besides the benefits of IL-6 signaling as a proinflammatory cytokine produced by macrophages to stimulate an immune response to trauma and the acute phase response, when unregulated (under stress conditions), IL-6 signaling has been shown to contribute to the onset of rheumatoid arthritis, inflammatory bowel disease, osteoporosis, multiple sclerosis, and some types of cancer. In the yoga study, after a stressful experience, the women doing yoga showed smaller increases in IL-6 than their nonpractitioners.

Few reputable practitioners of behavioral therapies believe such therapies should replace drug therapy. However, the psychological devastation associated with many diseases, such as AIDS, cannot be denied, and it is this intense stress that the "thinking well" movement attempts to address. Very often, for instance, a person learning of a positive HIV test goes into severe depression, and because depression can adversely affect the immune system, a double dose of immune suppression ensues. Perhaps by relieving the psychological trauma, the remaining body defenses can adequately handle the virus.

As with any emerging treatment method, there are numerous opponents of behavioral therapies. Some opponents argue that naive patients might abandon conventional therapy; another argument suggests therapists might cause enormous guilt to develop in patients whose will to live cannot overcome failing health. Proponents counter with the growing body of evidence showing that patients with strong commitments and a willingness to face challenges—signs of psychological hardiness—have relatively greater numbers of T cells than passive, nonexpressive patients. To date, no study has proven that mood or personality has a life-prolonging effect on immunity. Still, doctors and patients are generally inspired by the possibility of using one's mind to help stave off the effects of infectious disease.

T cells then divide into a clone of activated cytotoxic T cells (CTLs) (FIGURE 21.11B). CTL memory cells also are produced.

The CTLs leave the lymphoid tissue and enter the lymph and blood vessels. They circulate until they come upon their target cells—the infected cells displaying the tell-tale MHC-I/peptide on their surface (FIGURE 21.11C). There they proliferate through additional stimulation by IL-2 from T_H1 cells.

The effector CTLs bind to the MHC-I/peptide on the virus-infected cell surface and release a number of active substances. Toxic proteins, called **perforins**, insert into the membrane of

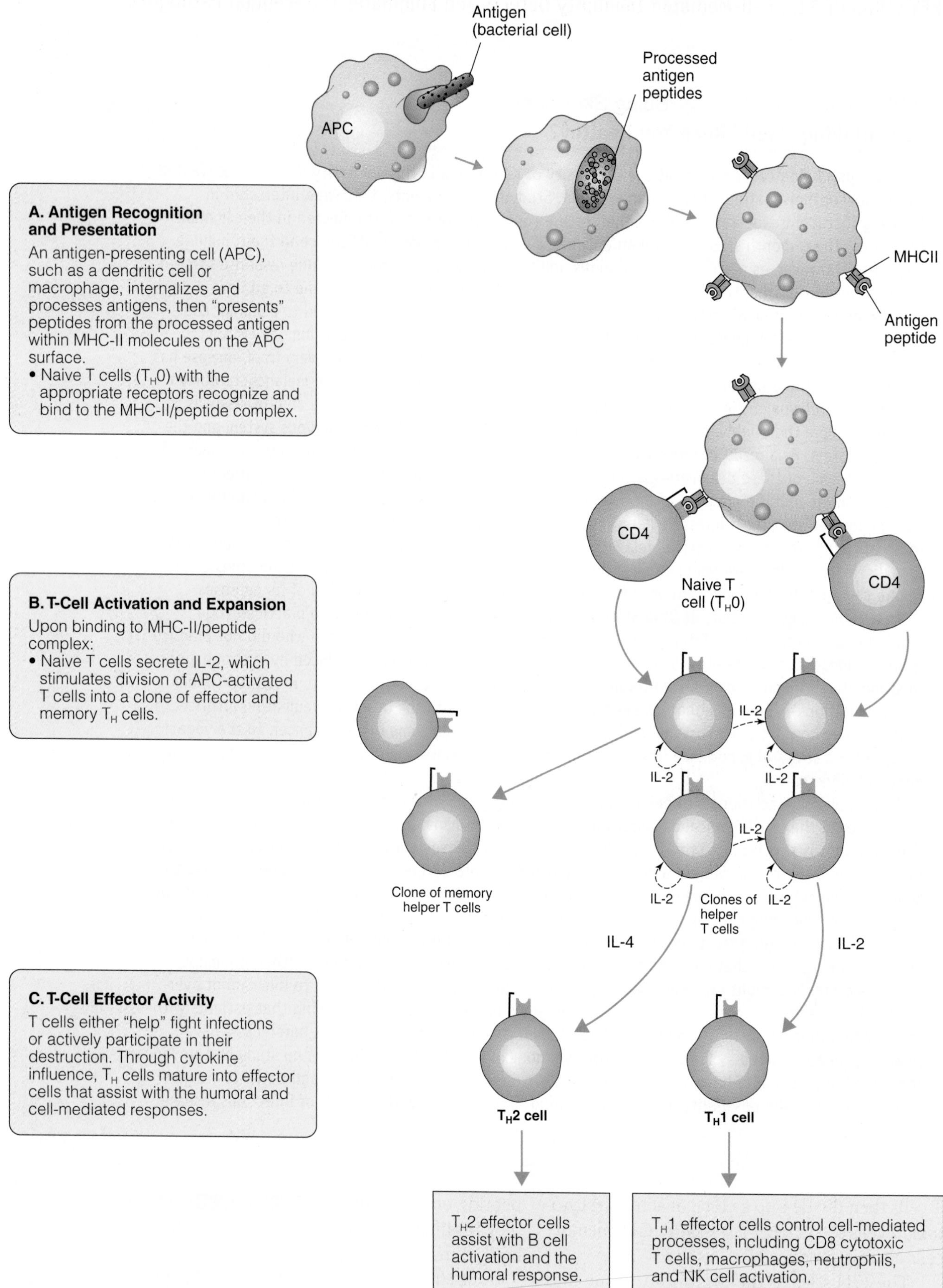

FIGURE 21.10 **Cell-Mediated Immunity: Helper T Cell Activation and Effector Activity.** Triggered by antigen peptide presentation, naive T cells become activated and develop into specialized subtypes of Th cells important to adaptive (humoral and cell-mediated) immunity and innate immunity. »» How do T_H1 cells and T_H2 cells differ in their control of immune responses?

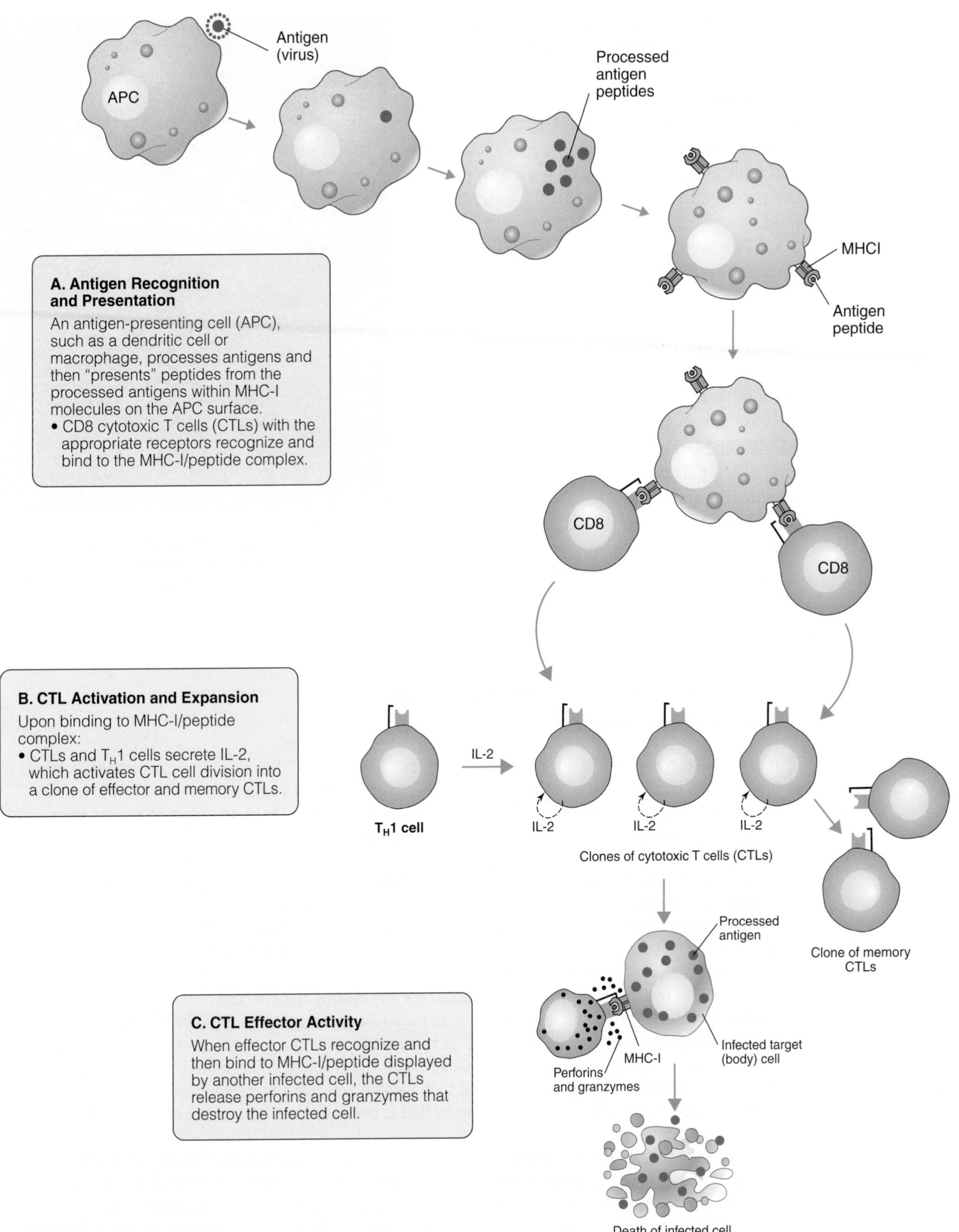

FIGURE 21.11 **Cell-Mediated Immunity: Cytotoxic T Cell Activation and Effector Activity.** Triggered by antigen peptide presentation, naive cytotoxic T cells (CTLs) become activated and primed to destroy virus-infected or tumor cells. »» Why are CTLs also called killer T cells?

the infected cell, forming cylindrical pores in the membrane. This "lethal hit" releases ions, fluids, and cell structures. In addition, the CTLs release **granzymes** that enter the target cell and trigger cell suicide through apoptosis. Cell death not only deprives the viral pathogen of a place to survive and replicate, but through cell lysis, also exposes the pathogen to antibodies in the extracellular fluid.

CTLs also are active against tumor cells because these cells often display distinctive molecules on their surfaces. The molecules are not present in other body cells, so they are viewed as foreign antigen peptides. Harbored within MHC-I proteins at the cell surface, the antigen peptides react with receptors on CTLs and the tumor cells are subsequently killed through apoptosis (FIGURE 21.12). However, some tumors reduce the level of MHC-I proteins at the cell surface, which impedes the ability of CTLs to "find" the abnormal cells. Thus, tumor cells can escape immunologic surveillance and survive.

MicroFocus 21.4 describes how the "brakes" can be applied to the cell-mediated response.

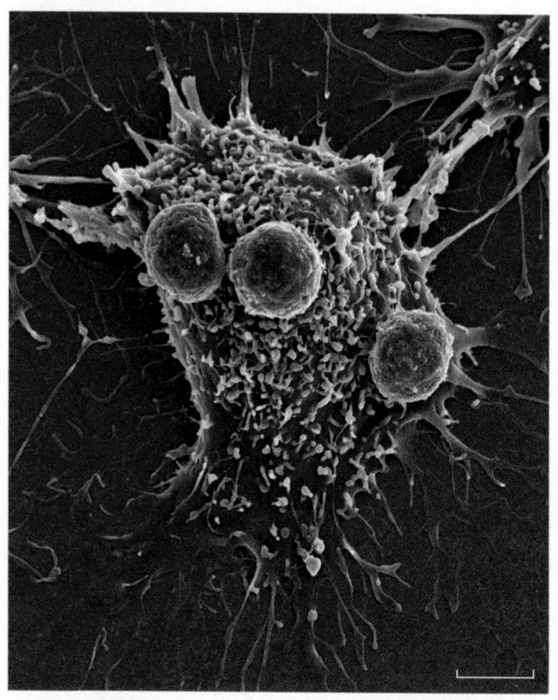

FIGURE 21.12 **A "Lethal Hit."** A false-color scanning electron micrograph of three cytotoxic T cells (pink) attacking a tumor cell. (Bar = 10 μm.) © Steve Gschmeissner/Photo Researchers, Inc. »» How do cytotoxic T cells "find" a tumor cell?

MICROFOCUS 21.4: Miscellaneous

Putting on the "Immune System" Brakes

Why does the misery of a common cold or the flu only last so long? The obvious answer would be that the body has eliminated the cold or flu virus. Yet the symptoms of a cold or flu are on the wane long before the virus has been eliminated. As mentioned in this chapter, immune system responses must be strong enough to hopefully eliminate the pathogen yet not be so extreme as to harm the body. So there must be more to an immune response than simply an "on and off switch."

In 2007, scientists at Johns Hopkins University School of Medicine reported that they had identified a chemical necessary to control the immune machinery used to fight infections and foreign invaders. The researchers discovered a protein molecule called Carabin that is produced by T cells during a viral infection. When someone gets a cold or the flu, the virus infects the cells of the respiratory tract and uses the cells as viral factories to make more viruses. The immune system mounts a response to the infection by activating cytotoxic T cells that attack and destroy the infected cells. Viral infection also triggers Carabin synthesis in the same T cells to moderate the immune response to infection. In fact, in lab experiments, the researchers found that the more Carabin in a cell, the more severely was its action to "damp down" T-cell activity. So, Carabin acts like a "built-in timer" for the immune system. The immune response only lasts so long and then Carabin acts like an internal brake to slow down the speed and intensity of the immune response so that it does not go to an extreme or get out of control and attack healthy cells.

Carabin appears to act on the same protein receptor target as immune suppressing drugs used in transplant patients to block transplant rejection. If further research proves promising, Carabin may be useful in limiting the immune system's attack following transplant surgery. In addition, it could help control "autoimmune" disorders, such as multiple sclerosis and rheumatoid arthritis, by again dampening down the immune system's attack on self.

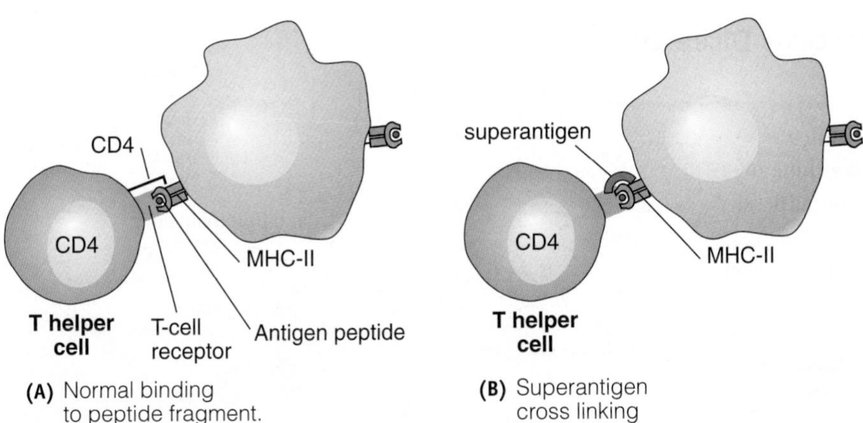

FIGURE 21.13 **Superantigen Binding.** (**A**) Normal binding between APC and T cell requires an antigen peptide that was processed within the APC. (**B**) The presence of a superantigen cross links the T-cell receptor and the MHC-II receptor on the APC, triggering T-cell activation.
»» What type of molecule (carbohydrate, lipid, protein, or nucleic acid) is a superantigen? Explain.

Some Antigens Are T-Cell Independent

All the examples of antigens we have discussed so far are called **T-dependent antigens** because they require presentation by T helper cells. However, a few are **T-independent antigens**; these substances (such as in bacterial capsules and flagella) do not require the help of T cells. For example, some such antigens bind directly with the receptor proteins on the B-cell surface and stimulate the cells to produce plasma cells and secrete antibodies. However, the immune response is generally weaker, primarily producing IgM antibodies and no memory cells.

Another group of antigens worth mentioning are the **superantigens**. "Regular" antigens must be broken down and processed to antigen peptides before they are presented on the APC's cell surface (FIGURE 21.13). Superantigens, such as certain viral proteins and bacterial exotoxins, crosslink a MHC protein and T-cell receptor that has already bound a peptide fragment. Thus, massive numbers of activated T cells form and proliferate, with an unusually high secretion of cytokines. The result is an extremely vigorous and excessive immune response called a "cytokine storm," that can lead to shock and death (MicroFocus 21.5). Superantigens and massive cytokine release are associated with the staphylococcal toxins causing food poisoning and toxic shock syndrome.

FIGURE 21.14 summarizes the humoral and cell-mediated immune responses of adaptive immunity. Note that the T helper cells are at the heart of almost all responses. Should these cells be infected and killed by HIV, humoral immunity is adversely affected, stimulation of innate immunity does not occur, and the ability to fully activate cytotoxic T cells is limited. That is why the chapter opener referred to HIV as the perfectly designed virus to cripple the immune response. As the T-cell population drops, cell-mediated immunity becomes less able to respond to pathogens. Eventually, there are so few T cells that the entire immune response collapses.

CONCEPT AND REASONING CHECKS 3

a. How can T cells be separated based on receptors and function?
b. Justify the need for two populations of "helper T cells."
c. Summarize the events leading to apoptosis of targeted cells.
d. Why are some antigens said to be T-independent?

Chapter Challenge C

Besides the activation of the humoral response to the flu virus, the cell-mediated arm of adaptive immunity also is well underway.

QUESTION C:
What two types of effector T cells will be critical in the attempt to rid the body of the flu viruses? Explain how each cell type works.

And happily to report, we have made it through the acute phase of infection and disease thanks to our immune system. We are well on our way through the period of convalescence and we'll be back in class on Monday!

Answers can be found on the Student Companion Website in **Appendix D**.

MICROFOCUS 21.5: Infectious Disease

The "Over-Perfect" Storm

You get a cold or the flu, or a bacterial infection, and you suffer for several days. We think of pathogens as our worst enemy during an outbreak of influenza or pneumonia and we rely on our immune system to get us over the illness. However, sometimes the immune system is worse than the pathogen when it overreacts to an infection. It can result in a potentially fatal immune response—an "over-perfect storm" in terms of disease elimination.

For reasons scientists do not completely understand, too many immune cells can be sent to the infection site and the cells produce more cytokines, which brings even more immune cells to the site of infection. This uncontrolled immune response propagates what is referred to as a cytokine storm where excessive numbers of immune cells are caught in an endless positive feedback loop that brings more and more immune cells to fight the infection. This "over-perfect" storm causes inflammation in the tissue surrounding the infection and this is what makes the immune reaction potentially so deadly.

For example, a cytokine storm in response to a lung infection can potentially cause permanent lung damage, resulting in multiple organ system failure due, in part, to an inadequacy of oxygen reaching the body's tissues. It explains why the 1918 Spanish flu pandemic, which killed an estimated 50 to 100 million people worldwide, took the lives of a disproportionate number of young adults—their healthy immune systems triggered a cytokine storm. A large number of the 774 individuals killed during the 2003 SARS epidemic also were the result of a cytokine storm.

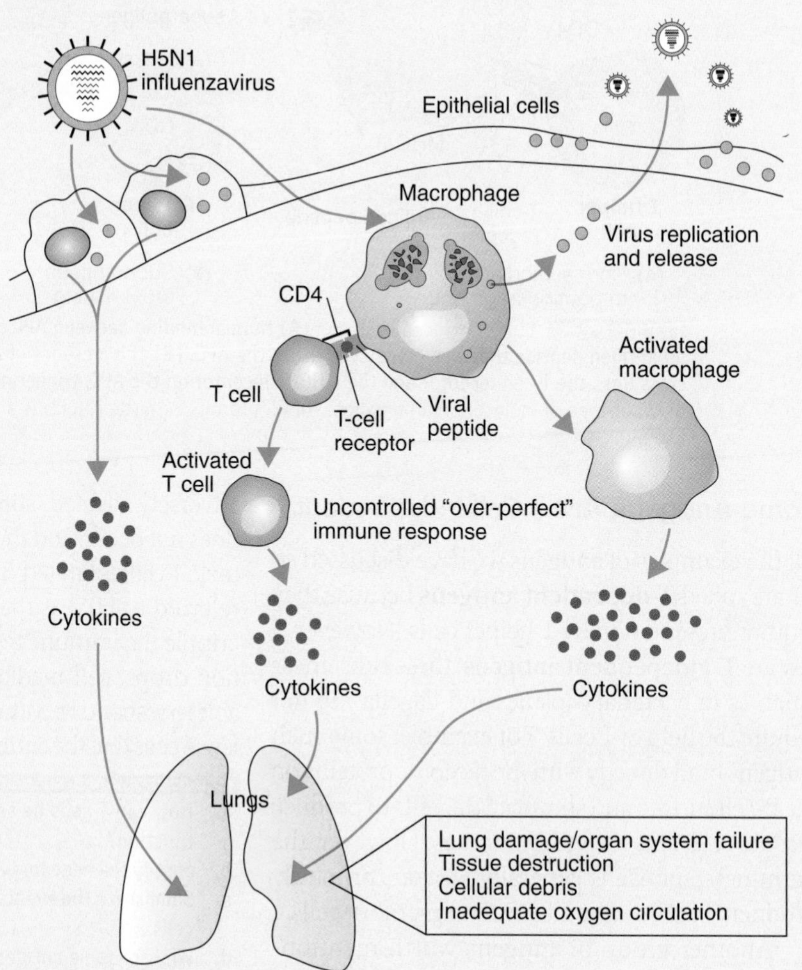

How the H5N1 influenzavirus triggers a cytokine storm. Adapted from Osterholm, M. T., *N Engl J Med.* 352 (2005): 1839–1842.

Today, we are seeing two more flu illnesses: the 2009–2010 pandemic of H1N1 (swine) flu, which did not trigger a cytokine storm; and the H5N1 avian flu that does trigger a flood of cytokines and has been responsible for many of the 334 deaths among the more than 569 people (60%) that have been infected with H5N1 as of October 2011 (see figure). For some reason, H5N1 viruses are "more potent inducers" of cytokines in infected epithelial cells, macrophages, and T cells than are the H1N1 viruses responsible for seasonal pandemic flu. In fact, autopsies of H5N1 flu victims in Southeast Asia have revealed that these victims had lungs filled with debris from the excessive inflammation triggered by the virus and cytokines. In lab studies, the H5N1 virus produced 10-times higher levels of cytokines from human cells in culture than did seasonal flu viruses.

Research is continuing to understand the nature of this "over-perfect" storm and how best to treat affected individuals.

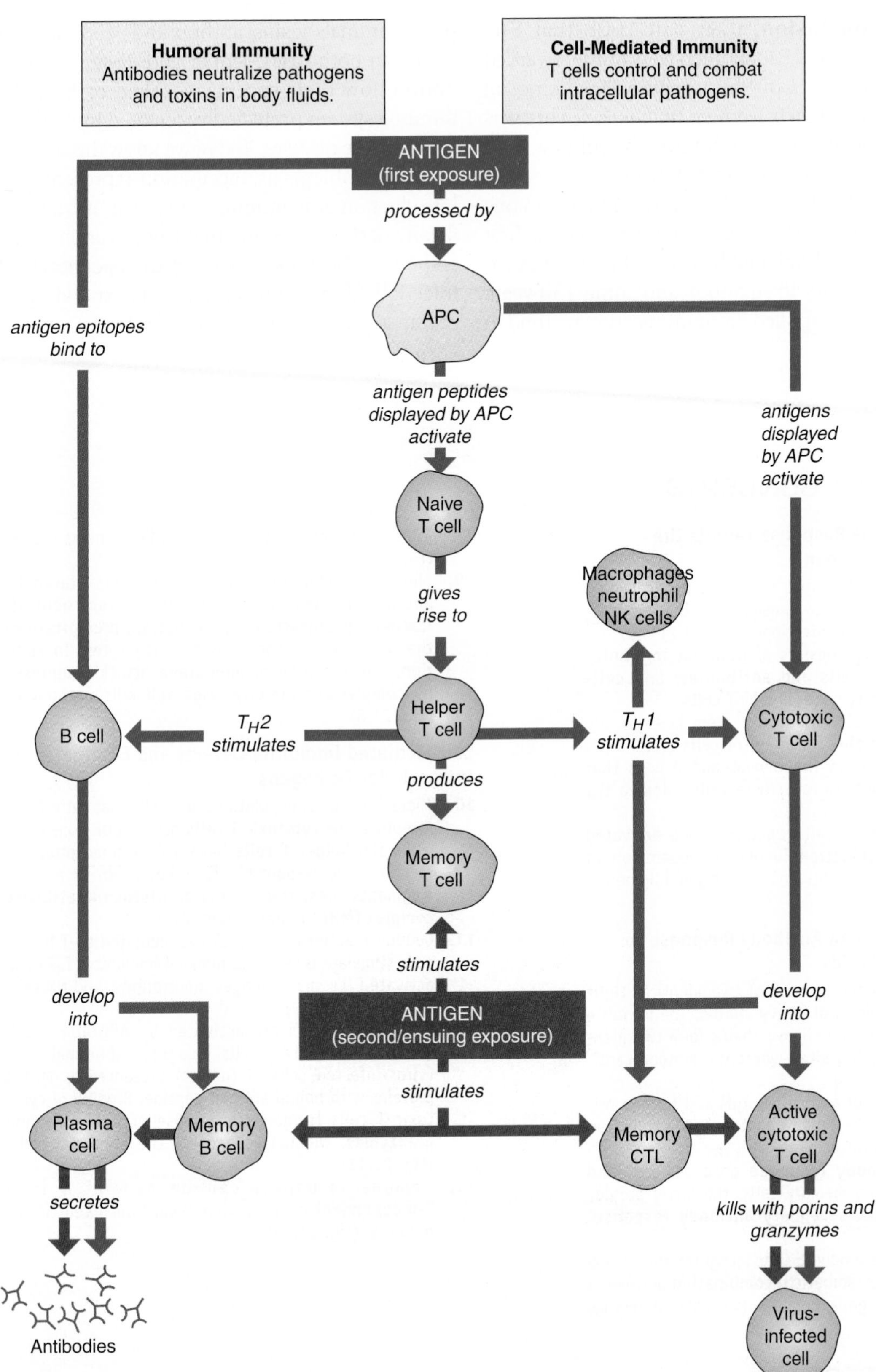

FIGURE 21.14 **A Concept Map Summarizing Adaptive Immunity.** Humoral immunity produces antibodies that respond and bind to antigens, while cell-mediated immunity stimulates helper T cells that activate B cells and cytotoxic T cells that bind to and kill infected cells or other abnormal cells (blue lines). Memory B and T cells are important in a second or ensuing exposure to the same antigen (red lines). »» How would an HIV infection affect the immune response illustrated in this figure?

In conclusion, it was in 1908 that Elie Metchnikov and Paul Ehrlich were jointly awarded the Nobel Prize in Physiology or Medicine "*in recognition of their work on immunity.*" This was groundbreaking research into a brand new field that in just over 100 years has made great progress in understanding how the body fights infectious disease. It has been over 200 years since the first vaccine was developed by Edward Jenner to provide immunity to smallpox and some 130 years since Pasteur applied his immunization method to protect animals against anthrax and people against rabies. In both cases, Jenner and Pasteur had no notion how their vaccines worked or how the immune system protected vaccinated individuals against these diseases. Today, we know the answers to these problems as vaccination represents the introduction of an immunogen; that is, it is an artificially derived infection that triggers an immune response. Vaccines are one of the backbones of infectious disease prevention today and are part of our study of immunity and serology.

SUMMARY OF KEY CONCEPTS

21.1 The Adaptive Immune Response Targets the Specific Invading Pathogen

1. **Adaptive immunity** requires specificity for **epitopes**, tolerance of and minimal damage to "self," and generates memory of past infections. (Fig. 21.2)
2. Adaptive immunity consists of **humoral immunity** maintained by **B cells** and **antibodies**, and **cell-mediated immunity** controlled by **T cells.**
3. Immune system cells arise from stem cells in the bone marrow. **Lymphoid progenitor cells** give rise to T cells that mature in the thymus and B cells that mature in the bone marrow. These cells colonize the lymphoid organs. (Fig. 21.3)
4. Appropriate B and T cell populations are activated through **clonal selection** involving recognition of specific epitopes on antigens or antigen fragments. (Fig. 21.4)

21.2 Humoral Immunity Is an Antibody Response to Pathogens in Body Fluids

5. Monomeric antibodies consist of two identical **light chains** and two identical **heavy chains.** The **variable regions** of the light and heavy chains form two identical **antigen-binding sites** where the epitope (antigen) binds. (Fig. 21.5)
6. Of the five classes of antibodies, **IgM** and **IgG** are primary disease fighters, while secretory **IgA** is found on body (mucosal) surfaces. (Table 21.1)
7. A **primary antibody response** produces IgM and IgG antibodies and memory cells, the latter providing immunity in a **secondary antibody response.** (Fig. 21.6)
8. Antibodies can be produced that recognize almost any antigen. To do this, **somatic recombination** between a large number of gene clusters (variable, diversity, joiner, constant) for the heavy and light chains occurs. (Fig. 21.7)
9. The actual clearance of antigen is facilitated by antibody-antigen interactions (**inhibition, neutralization, opsonization, agglutination, precipitation**) that result in phagocytosis by phagocytes. In addition, the formation of **membrane attack complexes** by complement directly lyses and kills pathogens. (Fig. 21.8)

21.3 Cell-Mediated Immunity Detects and Eliminates Intracellular Pathogens

10. There are two subpopulations of T cells that carry T-cell receptors. The **cytotoxic T cells** have a **CD8** coreceptor while the **helper T cells** have a **CD4** coreceptor. The receptors are responsible for recognition of antigen fragments presented in **major histocompatibility complex** (**MHC**) proteins. (Fig. 21.9)
11. Cellular reactions and cytokines mediated by T helper cells stimulate B cells and humoral immunity (**T_H2**) and activate CTLs, macrophages, neutrophils, and NK cells (**T_H1**). (Fig. 21.10)
12. **Cytotoxic T cells** are activated by APC/CD4 T cell combination. Active cells recognize abnormal cells (virus-infected cells or tumors) presented as MHC-I proteins with bound antigen peptide. Binding of cytotoxic T cells triggers the release of **perforins** and **granzymes**, which lyse and kill the abnormal cells. (Fig. 21.11)
13. Immune responses can be generated in a **T-independent** manner, where help from T cells is not required. (Fig. 21.13)

CHAPTER SELF-TEST

For **STEPS A–D**, answers to even-numbered questions and problems can be found in **Appendix C** on the Student Companion Website at **http://microbiology.jbpub.com/10e**. In addition, the site features eLearning, an online review area that provides quizzes and other tools to help you study for your class. You can also follow useful links for in-depth information, read more MicroFocus stories, or just find out the latest microbiology news.

STEP A: REVIEW OF FACTS AND TERMS

Multiple Choice

Read each question carefully, then select the ***one*** answer that best fits the question or statement.

1. All the following are immunogenic *except* _____.
   A. bacterial flagella
   B. haptens
   C. bacterial pili
   D. viral spikes
2. _____ cells are associated with _____ immunity while _____ cells are part of _____ immunity.
   A. B; cell-mediated; T; innate
   B. T; humoral; B; cellular
   C. T; cell-mediated; B; humoral
   D. T; humoral; B; nonspecific
3. Which one of the following cell types is NOT derived from lymphoid progenitors?
   A. Macrophages
   B. B cells
   C. Helper T cells
   D. Cytotoxic T cells
4. Clonal selection includes _____.
   A. antigen-receptor binding on B cells
   B. antibody secretion recognizing same epitope as on B cell receptors
   C. differentiation of B cells into plasma cells and memory cells
   D. All the above (**A–C**) are correct.
5. An antigen binding site on the IgG antibody is a combination of _____.
   A. one variable region from a light chain and one from a heavy chain
   B. two variable regions from two light chains
   C. two variable regions from two heavy chains
   D. one variable region from a constant region and one from a variable region
6. This dimeric antibody class often occurs in secretions of the respiratory and gastrointestinal tracts.
   A. IgE
   B. IgM
   C. IgA
   D. IgG
7. The presence of IgM antibodies in the blood indicates _____.
   A. an early stage of an infection
   B. a chronic infection
   C. an allergic reaction is occurring
   D. humoral immunity has yet to start
8. Antibody diversity results from _____.
   A. apoptosis
   B. antigenic shift
   C. somatic recombination
   D. complement binding
9. A/an _____ mechanism facilitates the clearance of toxins from the body.
   A. opsonization
   B. precipitation
   C. agglutination
   D. neutralization
10. MHC class I proteins would be found on _____ whereas MHC class II proteins would be found on _____.
    A. nucleated cells; plasma cells
    B. nucleated cells; macrophages
    C. dendritic cells; neutrophils
    D. only white blood cells; red blood cells
11. T_H1 cells activate _____.
    A. B cells
    B. CTLs, macrophages, and NK cells
    C. antibodies
    D. humoral immunity
12. Perforins and granzymes are found in _____.
    A. helper T cells
    B. antigen-presenting cells
    C. cytotoxic T cells
    D. B cells
13. T-independent antigens include _____.
    A. bacterial capsules
    B. superantigens
    C. bacterial flagella
    D. All the above (**A–C**) are correct.

True-False

Each of the following statements is true (T) or false (F). If the statement is false, substitute a word or phrase for the underlined word or phrase to make the statement true.

_____ **14.** Cell-mediated immunity involves T lymphocyte activity.
_____ **15.** Small molecules called antigens are not immunogenic.
_____ **16.** Cytokine is an alternate name for an antibody.
_____ **17.** The end of an antibody molecule where an antigen binds is called the Fc fragment.
_____ **18.** Antigenic materials are classified as "nonself."
_____ **19.** Cells that secrete antibody molecules are B cells.
_____ **20.** IgM consists of five monomers.
_____ **21.** A secondary antibody response primarily involves IgG.
_____ **22.** IgD has four heavy chains in the antibody molecule.
_____ **23.** Epitope is an alternate name for an antigenic determinant.
_____ **24.** Basophils phagocytize microorganisms and begin an immune response.
_____ **25.** Dendritic cells are involved in an immune response.
_____ **26.** Antibodies are transported in the blood.
_____ **27.** There are four polypeptide chains in a monomeric antibody molecule.
_____ **28.** Secretory IgA has two antigen binding sites.
_____ **29.** The humoral immune response depends on antibody activity.
_____ **30.** The Fc fragment of an antibody consists of light chains.
_____ **31.** Lysozyme is secreted by cytotoxic T cells.
_____ **32.** IgA crosses the placenta.
_____ **33.** Monocytes secrete antibody.
_____ **34.** The IgE antibody is found on the surface of B lymphocytes.
_____ **35.** T_H2 cells activate B cell activation.

STEP B: CONCEPT REVIEW

36. Identify and describe the four characteristics of an **adaptive immune response.** (**Key Concept 1**)
37. Distinguish between the cells responsible for a **humoral immune response** and a **cell-mediated immune response.** (**Key Concept 1**)
38. Trace the fates of the **lymphoid progenitors.** (**Key Concept 1**)
39. List the steps in **clonal selection** of B and T cells. (**Key Concept 1**)
40. Draw the structure of a monomeric **antibody** and label the parts. (**Key Concept 2**)
41. Summarize the characteristics for each of the five **immunoglobulin** classes. (**Key Concept 2**)
42. Differentiate between a **primary** and **secondary antibody response** in terms of antibodies and responding cells. (**Key Concept 2**)
43. Explain how antibody diversity is generated through gene arrangements. (**Key Concept 2**)
44. Distinguish between the antibody mechanisms used to clear antigens (pathogens) from the body. (**Key Concept 2**)
45. Compare receptors, receptor binding, and function for **cytotoxic T cells** and **helper T cells.** (**Key Concept 3**)
46. Summarize how naive T cells are activated and identify how these effects contribute to cell-mediated immunity. (**Key Concept 3**)
47. Discuss how cytotoxic T cells are activated and how they eliminate virus-infected cells. (**Key Concept 3**)
48. Explain how **superantigens** stimulate an immune response. (**Key Concept 3**)

STEP C: APPLICATIONS AND PROBLEMS

49. A microbiology professor suggests that an antigenic determinant arriving in the lymphoid tissue is like a parent searching for the face of a lost child in a crowd of a million children. Do you agree with this analogy? Why or why not?
50. In the book and classic movie, *Fantastic Voyage*, a group of scientists is miniaturized in a submarine (the Proteus) and sent into the human body to dissolve a blood clot. The odyssey begins when the miniature submarine carrying the scientists is injected into the bloodstream. Today, microscopic robots called nanorobots are being designed that would be injected into the body to fight diseases, including infectious ones. What do you think about this future microscopic robot technology?
51. As a consultant for the company Acme Nanobots, what immunological hurdles need to be considered before such microscopic robots could be fully developed?

STEP D: QUESTIONS FOR THOUGHT AND DISCUSSION

52. The ancestors of modern humans lived in a sparsely settled world where communicable diseases were probably very rare. Suppose that by using some magical scientific invention, one of those individuals was thrust into the contemporary world. How do you suppose he or she would fare in relation to infectious disease? What is the immunological basis for your answer?
53. Your brother's high school biology text contains the following statement: "T cells do not produce circulating antibodies. Rather, they carry cellular antibodies on their surface." What is fundamentally incorrect about this statement?
54. Some time ago, an immunologist reported that cockroaches injected with small doses of honeybee venom develop resistance to future injections of venom that would ordinarily be lethal. Does this finding imply that cockroaches have an immune system? What might be the next steps for the research to take? What does this research tell you about the cockroach's ability to survive for 3 or 4 years, far longer than most other insects?

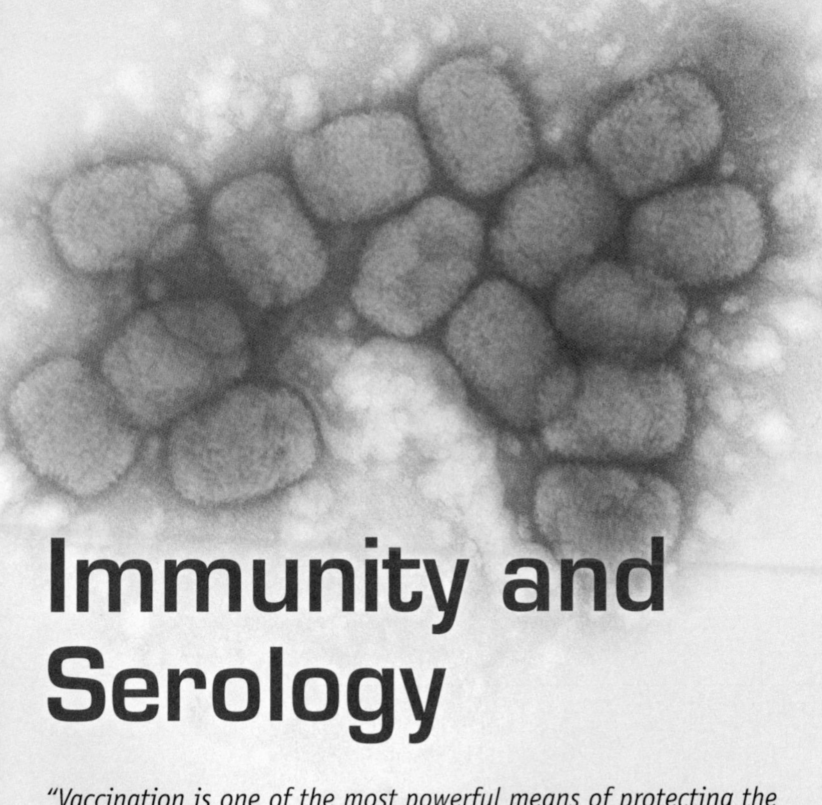

Immunity and Serology

"Vaccination is one of the most powerful means of protecting the public health."
—Marta A. Balinska, Institut National de Prévention et d'Education pour de Santé, Vanves, France

CHAPTER PREVIEW

22.1 Immunity to Disease Can Be Generated Naturally or Artificially
Investigating the Microbial World 22: The Salk Polio Vaccine Field Trials

22.2 Serological Reactions Can Be Used to Diagnose Disease
MicroInquiry 22: Applications of Immunology: Disease Diagnosis
Textbook Case 22: A Case of Tickborne Encephalitis

22.3 Monoclonal Antibodies Are Used for Immunotherapy

Image courtesy of Dr. Fred Murphy/CDC.

Prior to the age of modern vaccines, the only way one could become immune to a disease was to catch the disease and hope for recovery. Unfortunately, the symptoms often were severe and perhaps disfiguring. There also was the risk of complications, where if the disease did not kill, a complication might. In addition, ill individuals could be contagious and spread the disease to other unsuspecting and susceptible individuals. Epidemics, and even pandemics, could result.

As early as the eleventh century, Chinese doctors ground up smallpox scabs and blew the powder into the noses of healthy people to protect them against the ravages of smallpox. Jenner improved on this technique by using a preparation from cowpox. Other diseases of past or recent history, including diphtheria and polio, were equally deadly to large numbers of people.

However, with the development of modern vaccines, these and many other infectious diseases have been either eradicated (smallpox), almost eradicated (polio), or brought under control (diphtheria) in much of the world. Equally important, vaccination is a technique to prevent disease from occurring—and preventing a disease from occurring is much safer and cheaper than trying to cure a disease after it has already struck.

Vaccination is the most cost-effective medical intervention. Cases of diseases like polio, measles, and rubella have been controlled or nearly eliminated globally by vaccinating infants and children and maintaining high vaccine coverage from infancy through childhood (FIGURE 22.1A). In fact, every year, vaccines prevent more than 3 million deaths and save 750,000 children from disability due to infectious disease.

Unfortunately, too many people around the world have not been vaccinated against vaccine-preventable diseases. Many live in regions of the world where vaccines are not available, while others have elected not to use the

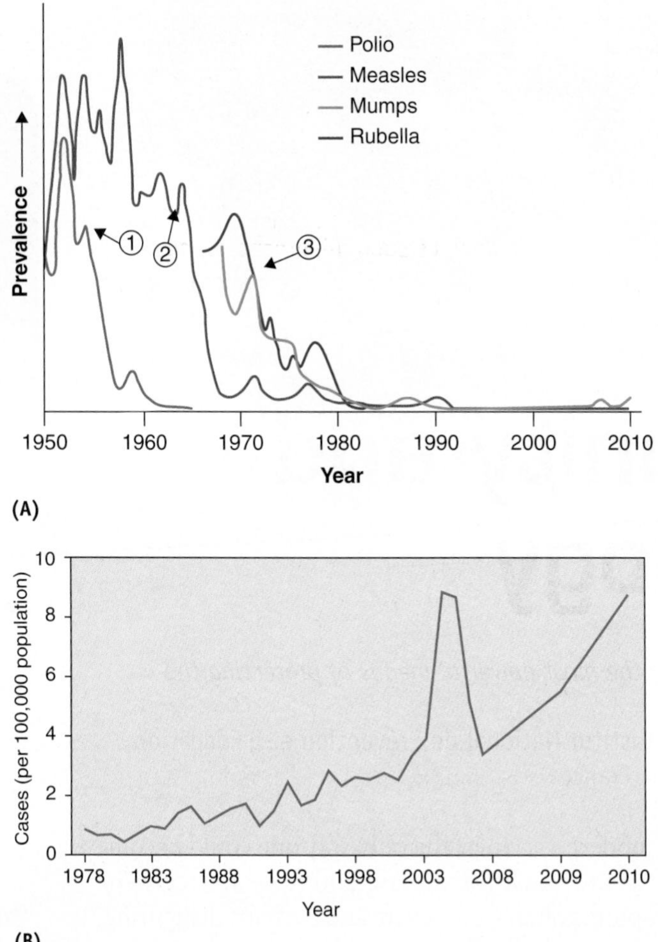

FIGURE 22.1 **Effect of Vaccination on Disease Prevalence.** (**A**) In the United States and most developed nations, there has been a swift decline in infectious such as polio, measles, mumps, and rubella with the introduction of vaccines to protect against those diseases. Vaccine introduction: ① Salk polio vaccine (1955); ② measles vaccine (1963); ③ measles, mumps, rubella (MMR) vaccine (1971). Reproduced from The American Academy of Microbiology, *Adult Vaccines: A Grown Up Thing to Do*. 2012. (**B**) However, not being vaccinated or keeping up on vaccinations can result in outbreaks of disease, such as the current outbreak of pertussis (whooping cough) in many states. Data from: CDC, *Summary of Notifiable Diseases*—United States, 2010. »» Provide a rebuttal to counter the argument that rubella was on the decline before the introduction of the MMR vaccine in 1971.

opportunity of vaccination to protect themselves and their family. Yet vaccines are a key to public health and to the goal of preventing infectious disease. In 2010, the Gates Foundation pledged $10 billion over the next 10 years to help research, develop, and deliver vaccines to the poorest developing nations.

However, some vaccine-preventable diseases once under control in the United States are resurfacing. Thanks to vaccination against pertussis (whooping cough), cases in the United States reached an all-time low in 1981 (FIGURE 22.1B). Now, the Centers for Disease Control and Prevention (CDC) predicts that the United States is on track in 2012 to have the most reported cases of whooping cough since 1959.

In this chapter, we will study the role of vaccines in the immune response and examine the four major mechanisms by which immunity comes about. We also will examine how antibodies may be detected in a patient by a variety of laboratory tests. These diagnostic procedures help the physician understand the disease and prescribe a course of treatment. Immune mechanisms generally protect against disease, but when they fail, these laboratory tests provide a clue about what is taking place in the patient.

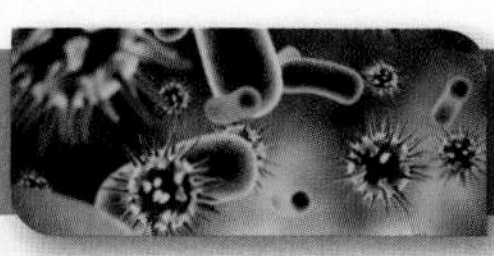

Chapter Challenge

Children have always been more susceptible to infectious diseases than their parents. The major reason for this is because kids have had limited exposure to the pathogens causing the diseases and therefore haven't yet built the immunologic memory required to protect themselves. However, many of the historical childhood infections, such as polio, diphtheria, pertussis (whooping cough), rubella (German measles), mumps, chickenpox, and measles are no longer major threats, at least in developed nations, because childhood vaccines are available to generate the needed immunity. But suppose you can't remember if you ever had one of these diseases (let's take measles), you have no immunization record of having been vaccinated, and your parents do not remember if you ever had measles during your childhood. Knowing that 90% of susceptible individuals will get the disease if exposed, what should you do now that you are an adult and 25 years old? Let's determine a course of action as we proceed through this chapter on vaccines and the blood tests available to detect infectious diseases.

KEY CONCEPT 22.1 Immunity to Disease Can Be Generated Naturally or Artificially

In biology, **immunity** refers to a condition under which an individual is protected from disease. However, it does not mean one is immune to all diseases, but only those the individual has recovered from or been vaccinated against. It is the result of **adaptive immunity** and depends on the presence of antibodies, T lymphocytes, and other factors originating in the immune system in response to a specific "nonself" antigen. Although the types of adaptive immunity discussed here focus on humoral immunity, remember that cell-mediated immunity also is an important and essential arm of resistance to infectious disease. Four types of adaptive immunity are recognized.

Adaptive Immunity Can Result by Actively Producing Antibodies to an Antigen

Active immunity occurs when antigens enter the body and the individual's immune system actively responds by producing antibodies and specific lymphocytes. This exposure to antigens may be unintentional, as when one becomes infected with a pathogen, or intentional, when one is purposely exposed to an antigen via a vaccine.

Before vaccines were developed, the only way to become immune to a disease was by suffering the disease and recovering. Thus, such **naturally acquired active immunity** follows a bout of illness and occurs in the "natural" scheme of events (FIGURE 22.2A). However, this is not always the case, because subclinical diseases also may result in immunity. For example, many people have adaptive immunity from subclinical cases of mumps or from subclinical fungal diseases such as cryptococcosis. So, this active production of antibodies represents the primary antibody response.

Memory cells residing in the lymphoid tissues are responsible for the production of antibodies in a secondary antibody response. The cells remain active for many years and produce IgG almost immediately upon a subsequent exposure to the same antigen or pathogen that triggered the primary antibody response.

Artificially acquired active immunity is less risky and represents an easier way to become immune to an infectious disease. This form of active immunity develops after the immune system produces antibodies and memory cells following an intentional exposure to antigens; that is, through **vaccination** (FIGURE 22.2B). Because the antigens are contained in a vaccine, the exposure is called "artificial."

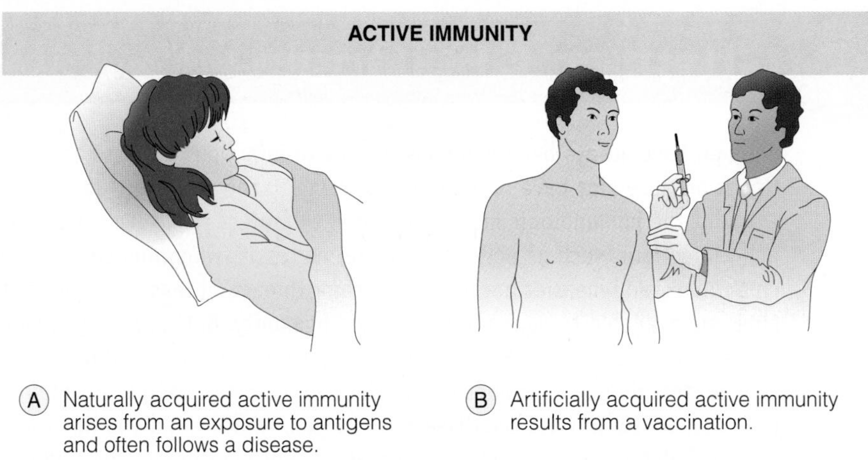

(A) Naturally acquired active immunity arises from an exposure to antigens and often follows a disease.

(B) Artificially acquired active immunity results from a vaccination.

PASSIVE IMMUNITY

(C) Naturally acquired passive immunity stems from the passage of IgG across the placenta from the maternal to the fetal circulation.

(D) Artificially acquired passive immunity is induced by a transfer of antibodies (antitoxins) taken from the circulation of an animal or another person.

FIGURE 22.2 Four Ways to Acquire Immunity. Immunity can be generated by natural or artificial means and acquired in an active or passive form. »» Why do these forms of immunity represent an adaptive response rather than an innate response to an antigen?

Vaccines are substances that prepare the immune system to recognize and respond to a pathogen (antigen), resulting in protection (immunity) to that pathogen. Vaccines are composed of treated microorganisms or viruses, chemically altered toxins, or molecular parts of microorganisms. However, because these pathogen look-alikes have been altered in some way (see below), the vaccine does not trigger the disease and the person vaccinated does not become ill. Importantly, memory cells are usually formed. If the vaccinated person is exposed to the same pathogen at some later time, the memory cells act swiftly and produce a secondary antibody response, stopping the infection before it can make the individual sick. Vaccines then are simply "educating" the body's natural infection fighting system.

Vaccines may be administered by injection, oral consumption, or, as is used for some influenza vaccines, nasal spray.

Let's now examine the different types of vaccines and find out how they work.

Whole Agent Vaccines Contain Weakened or Inactivated Antigens

The viral and bacterial vaccines currently in use in the United States are summarized in TABLE 22.1. Most of these vaccines were developed by using the whole bacterium, virus, or toxin as the antigen.

Attenuated Vaccines. Some pathogens can be weakened in the lab such that they should not cause disease. These **attenuated** pathogens multiply only at low rates in the body, increasing the dose of antigen to which the immune system

TABLE

22.1 The Principal Bacterial and Viral Vaccines Currently in Use

Disease	Route of Administration	Recommended Vaccine Usage/Comments
Contain Killed Whole Bacteria		
Cholera	Subcutaneous (SQ) injection	For travelers; short-term protection
Typhoid	SQ and intramuscular (IM)	For travelers only; variable protection
Plague	SQ	For exposed individuals and animal workers; variable protection
Contain Live, Attenuated Bacteria		
Tuberculosis (BCG)	Intradermal (ID) injection	For high-risk occupations only; protection variable
Subunit Bacterial Vaccines (Capsular Polysaccharides)		
*Meningitis (meningococcal)	SQ	Part of childhood immunization schedule
Meningitis (*H. influenzae*)	IM	Part of childhood immunization schedule; may be administered with DTaP
*Pneumococcal pneumonia	IM or SQ	Part of childhood immunization schedule; moderate protection
Pertussis	IM	Part of childhood immunization schedule
Subunit Bacterial Vaccine (Protective Antigen)		
Anthrax	SQ	For lab workers and military personnel
Toxoids (Formaldehyde-Inactivated Bacterial Exotoxins)		
Diphtheria	IM	Part of childhood immunization schedule
Tetanus	IM	Part of childhood immunization schedule; highly effective
Botulism	IM	For high-risk individuals, such as laboratory workers
Contain Inactivated Whole Viruses		
Polio (Salk)	IM	Part of childhood immunization schedule; highly effective
Rabies	IM	For individuals sustaining animal bites or otherwise exposed
Influenza	IM	Part of childhood immunization schedule
*Hepatitis A	IM	Part of childhood immunization schedule; protection for travelers and anyone at risk
Contain Attenuated Viruses		
Adenovirus infection	Oral	For immunizing military recruits
Measles (rubeola)	SQ	Part of childhood immunization schedule; highly effective
Mumps (parotitis)	SQ	Part of childhood immunization schedule; highly effective
Polio (Sabin)	Oral	Possible vaccine-induced polio
Smallpox (vaccinia)	Pierce outer layers of skin	For lab workers, military personnel, healthcare workers
Rubella	SQ	Part of childhood immunization schedule; highly effective
Chickenpox (varicella)	SQ	Part of childhood immunization schedule; immunity can diminish over time
*Shingles (zoster)	SQ	Prevention in individuals 60 years and older
Yellow fever	SQ	For travelers, military personnel in endemic areas
Influenza	IM	In a nasal spray form, may be more effective than inactivated form; only for healthy, non-pregnant individuals 2–50 years old
*Rotavirus infection	Oral	Part of childhood immunization schedule
Recombinant Viral Vaccine		
Hepatitis B	IM	Part of childhood immunization schedule; medical, dental, laboratory personnel; highly effective
*Genital warts/cervical cancer	IM	Preventative vaccine for individuals 9–26 years old

*Vaccines licensed since 2000.

will respond. Such vaccines are the closest to the natural pathogens and, therefore, they generate the strongest immune response. Often, the person vaccinated will have lifelong immunity. Also, attenuated organisms can be transmitted from the vaccinated individual to other people who become reimmunized, or, in fact, immunized for the first time.

The downside of attenuated vaccines results from their continued multiplication. Because the vaccines contain dividing bacterial cells or replicating viruses, there is a remote chance one of them could cause a mild form of the disease. Usually a healthy person with a fully functioning immune system (**immunocompetent**) clears the infection without serious consequence. However, individuals with a compromised immune system, such as patients with AIDS or pregnant women, should not be given attenuated vaccines, if possible.

Today, there are many viral vaccines that consist of attenuated viruses. The Sabin oral polio vaccine, as well as the measles, mumps, and chickenpox vaccines, contain attenuated viruses. To avoid multiple injections of immunizing agents, it is sometimes advantageous to combine vaccines into a **single-dose vaccine**. The measles-mumps-rubella (MMR) vaccine is one example. In 2005, the U.S. Food and Drug Administration (FDA) approved a combination vaccine (ProQuad®) for children 12 months to 12 years old. This single-dose vaccine protects against chickenpox, measles, mumps, and rubella.

Making a vaccine with attenuated bacterial cells is more difficult. In fact, there are no such vaccines routinely used in the United States. The BCG tuberculosis vaccine, which is used in some countries with a high burden of tuberculosis, is composed of attenuated *Mycobacterium bovis* bacterial cells.

On a global scale, attenuated vaccines may not be the vaccine strategy of choice. These vaccines require refrigeration to retain their effectiveness, which could present a problem in many developing nations lacking widespread refrigeration facilities.

Inactivated Vaccines. Another strategy for preparing vaccines is to "kill" the pathogen. These vaccines are relatively easy to produce because the pathogen is killed by simply using certain chemicals, heat, or radiation. However, the inactivation process alters the antigen's structure and shape so it produces a weaker immune response.

The Salk polio vaccine and the hepatitis A vaccine typify such preparations of inactivated whole viruses. In the case of influenza, the virus changes genetically from year to year, so a different vaccine must be provided annually. MicroFocus 22.1 describes how the components of the flu vaccine are decided each year.

Some whole organism (bacterial) vaccines are used for short-term protection. For instance, bubonic plague and cholera vaccines are available to limit an epidemic. In these cases, the immunity lasts only for several months because the material in the vaccine is weakly antigenic.

Compared to attenuated vaccines, inactivated vaccines are safer as they cannot mutate and therefore cannot cause the disease in a vaccinated individual. The vaccines can be stored in a freeze-dried form at room temperature, making them a vaccine of choice in developing nations.

Toxoid Vaccines. For some bacterial diseases, such as diphtheria and tetanus, a bacterial exotoxin is the main cause of illness. So, a third immunization strategy is to inactivate these toxins and use them as a vaccine. Such toxins can be inactivated with formalin, and the resulting inactivated toxin is called a **toxoid**. Immunity induced by a toxoid vaccine allows the body to generate antibodies and memory cells to recognize the natural toxin, should the individual again come in contact with it. Because toxoid vaccines are inactivated products, booster shots are necessary (see below).

■ Formalin: A solution of formaldehyde in water.

Single-dose vaccines include diphtheria-pertussis-tetanus vaccine (DPT) and the newer diphtheria-tetanus-acellular pertussis (DTaP) vaccine. For other vaccines, however, a combination single-dose vaccine may not be useful because the antibody response may be lower for the combination than for each vaccine taken separately.

Newer Vaccines Contain Only Subunits or Fragments of Antigens

Although attenuated vaccines often provide lifetime immunity, the drawback is that sometimes there is a slight chance of a vaccinated individual actually contracting the disease from the vaccination. Therefore, second-generation vaccines have been developed that contain only a subunit or fragment of the bacterial cell or virus. With only

MICROFOCUS 22.1: Public Health/Tools
Preparing for Battle

Chicken eggs being "inoculated" with a flu virus.

Each flu season approximately 10% to 20% of Americans get the flu. Of this number, some 200,000 are hospitalized and anywhere from 6,000 to 15,000 die from the complications of flu. Many of these hospitalizations and deaths could be prevented with a yearly flu shot, especially for those people at increased risk (people over 50, immunocompromised individuals, and health-care workers in close contact with flu patients). Because influenza viruses change often, the influenza vaccine is updated each year to make sure it is as effective as possible. How is each year's vaccine designed?

Each flu season, information on circulating influenza strains and epidemiological trends is gathered by the World Health Organization (WHO) Global Influenza Surveillance Network. The network consists of 128 national influenza centers in 99 countries and four WHO Collaborating Centres for Reference and Research on Influenza located in Atlanta, United States; London, United Kingdom; Melbourne, Australia; and Tokyo, Japan. The national influenza centers sample patients with influenza-like illness and submit representative isolates to WHO Collaborating Centres for immediate strain identification.

Twice a year (February: northern hemisphere; September: southern hemisphere), WHO meets with the directors of the Collaborating Centres and representatives of key national laboratories to review the results of their strain identifications and to recommend the composition of the influenza vaccine for the next flu season.

In the United States, the Food and Drug Administration (FDA) or the Centers for Disease Control and Prevention (CDC) provide the identified viral strains to vaccine manufacturers in February. Each virus strain is grown separately in chicken eggs (see figure). After the virus has replicated many times, the fluid containing the viruses is removed, and the viruses purified and attenuated or inactivated. Then, the appropriate strains are mixed together with a carrier fluid. Production usually is completed in August and ready for shipment by October.

a small piece of the antigen, a vaccinated individual cannot possibly contract the disease, but these vaccines usually fail to generate a cytotoxic T cell response.

Subunit Vaccines. Unlike the whole agent attenuated or inactivated vaccines, the strategy for a subunit vaccine is to have the vaccine contain only those parts or subunits of the antigen that stimulate a strong immune response. For example, the subunit vaccine for pneumococcal pneumonia contains 23 different polysaccharides from the capsules of 23 strains of *Streptococcus pneumoniae*.

One way of producing a subunit vaccine is to use recombinant DNA technology, where the resulting vaccine is called a **recombinant subunit vaccine**. The hepatitis B vaccine (Recombivax HB® or Engerix-B®) is an example. Several hepatitis B virus genes are isolated and inserted into yeast cells, which then synthesize the antigens. These antigens are collected and purified to make the vaccine.

Adverse reactions to such subunit vaccines are very rare and the subunits cannot produce disease in the person vaccinated. Also, the vaccine is not made from blood fragments (as was a previous hepatitis B vaccine), so it dispels the fear of contracting human immunodeficiency virus (HIV) from contaminated blood.

Conjugate Vaccines. *Haemophilus influenzae* b (Hib), which is responsible for a form of childhood meningitis, produces a capsule. Because the capsular polysaccharides by themselves are not strongly immunogenic, the strategy is to conjugate (attach) these polysaccharides (haptens) to the tetanus or diphtheria toxoid (carrier protein), which will stimulate a strong immune response. The result is the Hib **conjugate vaccine**, which has been instrumental in reducing the incidence of *Haemophilus* meningitis in American children under 5 years of age from 18,000 cases annually in 1986 to 38 cases in 2009. In 2005, a conjugate vaccine (MCV4) was licensed for a meningococcal vaccine against meningitis caused by *Neisseria meningitidis.*

By 6 years of age, children should be vaccinated against more than a dozen infectious diseases. FIGURE 22.3 outlines the licensed vaccines recommended by the CDC.

DNA Vaccines. Part of the renaissance in vaccine development is a strategy to use DNA as a vaccine. These investigational **DNA vaccines** consist of plasmids engineered to contain one or more protein-encoding genes from a viral or bacterial pathogen (FIGURE 22.4). Injected intramuscularly in a saline solution, cells in the body will take up injected foreign DNA, commence to make proteins encoded by the DNA, and display these antigens on its surface, much like an infected cell presents antigen peptides to T cells. Such a display should stimulate a strong antibody (humoral) and cell-mediated immune response if the individual comes in contact with the actual pathogen. Someone vaccinated with a DNA vaccine could not contract the disease.

Thus far, few experimental trials have produced an immune response equivalent to first- or second-generation vaccines. However, in July 2005, a veterinary DNA vaccine to protect horses from West Nile virus became the world's first licensed DNA vaccine. Plasmid-based DNA vaccines are being studied for a variety of infectious diseases in humans, including AIDS, influenza, and hepatitis C, and for several cancers.

Some Vaccines Are Specifically Recommended for Adults

Many adults think vaccinations are only of concern for children and, as adults, no further vaccinations are necessary. In fact, more than 40,000 adults in the United States die each year from diseases that

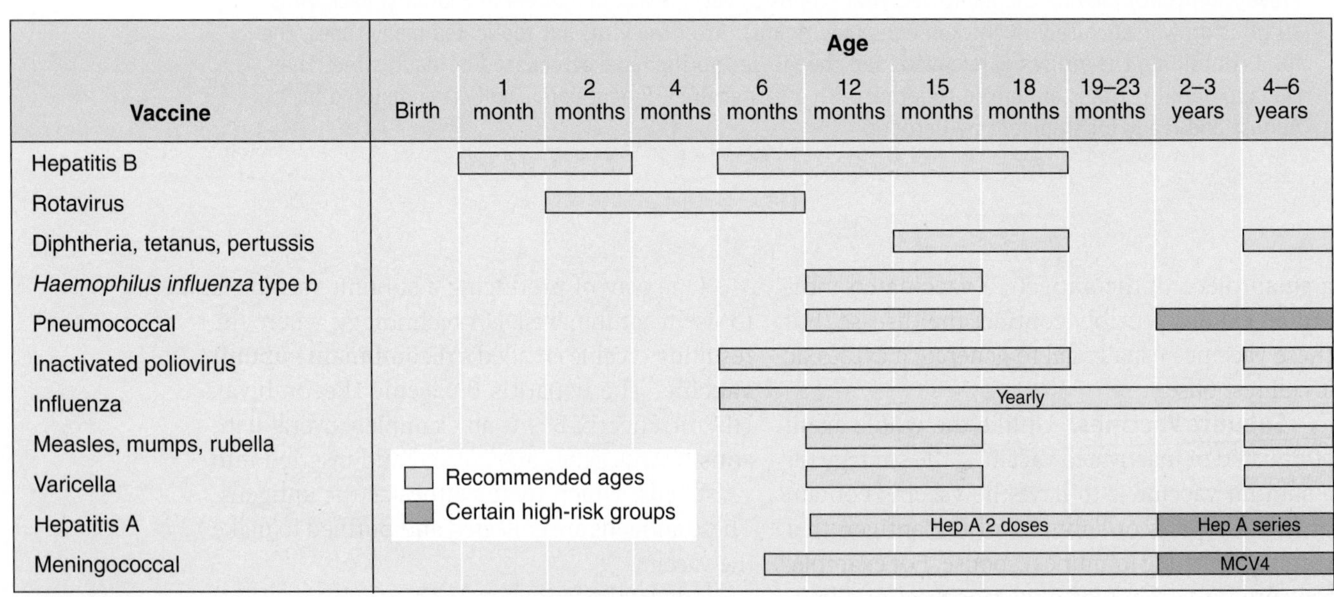

FIGURE 22.3 **Childhood (0–6 years) Immunization Schedule—2012.** Each year the Advisory Committee on Immunization Practices of the CDC reviews the childhood immunization schedule. This schedule indicates the recommended ages for routine administration of licensed vaccines. Additional recommendations concerning the schedule are available from the CDC at www.cdc.gov/vaccines/schedules/downloads/child/0-6yrs-schedule-pr.pdf. Source: CDC. »» If you have children, or younger brothers or sisters, are they current on these vaccinations?

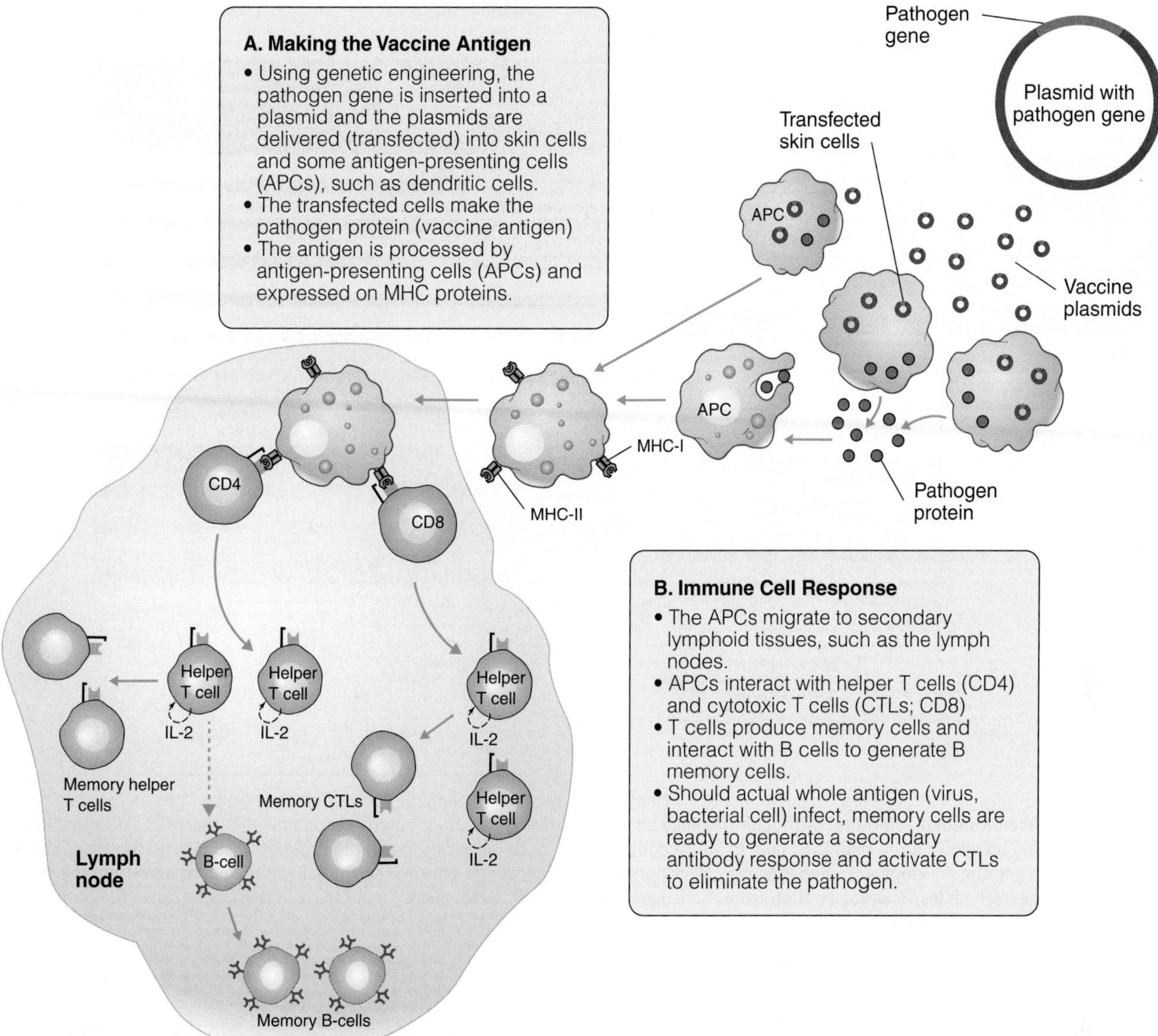

FIGURE 22.4 **Making a DNA Vaccine.** A DNA plasmid with the gene of interest can be inserted into immune cells that can trigger an immune response. The memory cells produced within the vaccinated individual are capable of quickly responding to an actual pathogen infection. »» Describe the memory cells' response if such a vaccinated individual were infected with the actual pathogen.

are vaccine preventable. FIGURE 22.5 identifies the adult vaccines recommended by the CDC.

There are several reasons why adult vaccinations are needed. First, some adults may have not experienced a childhood disease, such as chickenpox or measles, or were never vaccinated as children. Adult chickenpox can lead to serious pneumonia complications while adult measles can produce serious complications, including arthritis and encephalitis (brain inflammation). Second, newer vaccines, such as hepatitis B and shingles, may not have been available when some adults were children. Shingles is a rare illness in children, so the shingles vaccine is recommended only for adults 60 years and older. Third, as adults age and their immune systems start to decline in strength, they become more susceptible to serious diseases caused by common infections, such as flu and tetanus. Fourth, some vaccines simply do not afford protection beyond one "infectious season." The flu virus changes every season so a vaccine taken one season may not protect for the next flu season.

One other reason for adult vaccinations stems from the fact that many vaccines do not provide

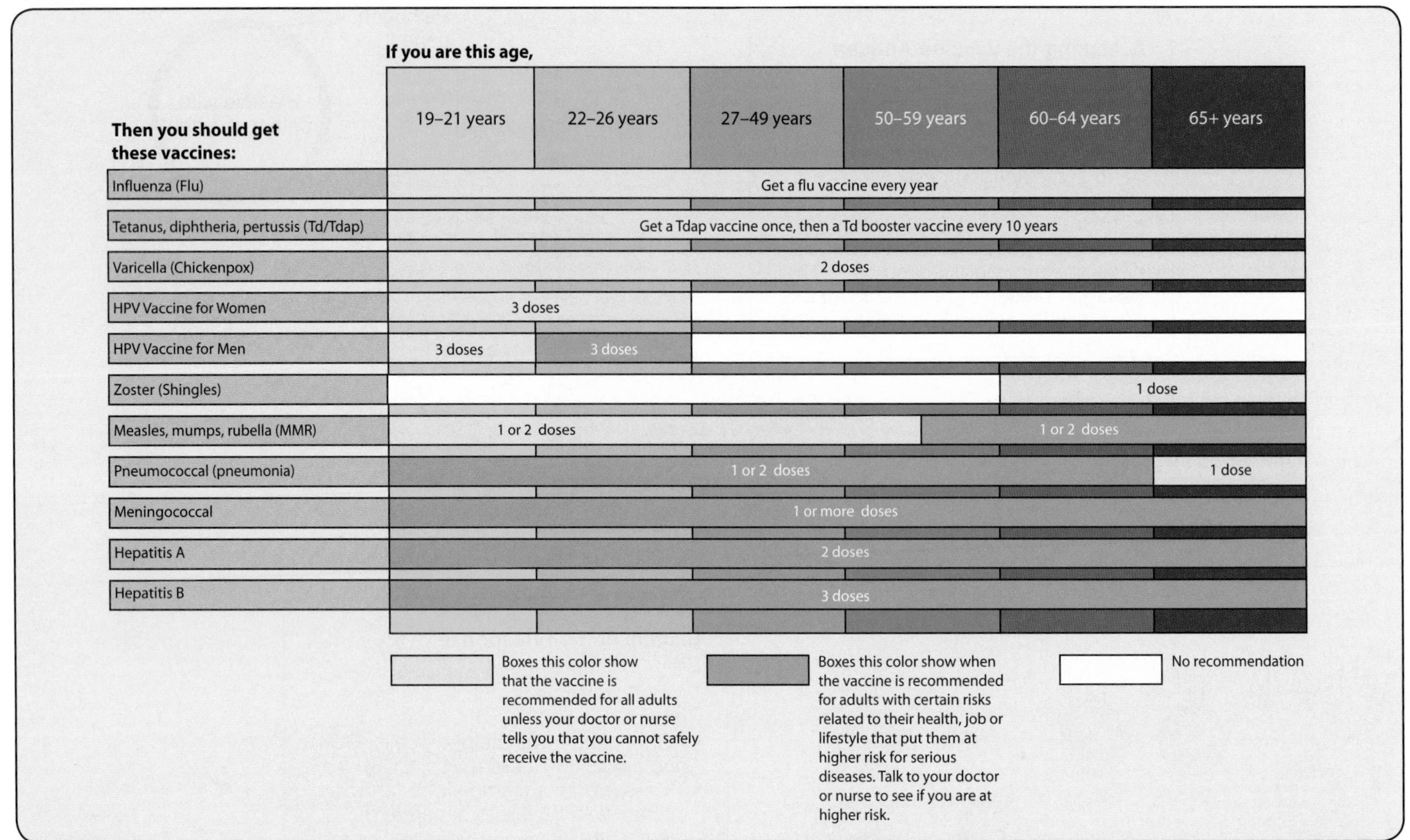

If you are this age,

Then you should get these vaccines:	19–21 years	22–26 years	27–49 years	50–59 years	60–64 years	65+ years
Influenza (Flu)	Get a flu vaccine every year					
Tetanus, diphtheria, pertussis (Td/Tdap)	Get a Tdap vaccine once, then a Td booster vaccine every 10 years					
Varicella (Chickenpox)	2 doses					
HPV Vaccine for Women	3 doses					
HPV Vaccine for Men	3 doses	3 doses				
Zoster (Shingles)					1 dose	
Measles, mumps, rubella (MMR)	1 or 2 doses			1 or 2 doses		
Pneumococcal (pneumonia)	1 or 2 doses					1 dose
Meningococcal	1 or more doses					
Hepatitis A	2 doses					
Hepatitis B	3 doses					

FIGURE 22.5 Recommended Adult Immunization Schedule—2012. Many adults are under the impression that the vaccines they received as children will continue to protect them through adulthood. Although this may be true in some cases, it is important to remember that: (1) Not all adults were vaccinated as children; (2) immunity often begins to fade in adulthood; and (3) adults become more susceptible to serious disease caused by secondary infections (e.g., flu and the development of pneumonia). Additional recommendations concerning the schedule are available from the CDC at http://www.cdc.gov/vaccines/schedules/downloads/adult/adult-schedule-easy-read.pdf. Reproduced from CDC. »» Look at the table and determine if you are current on all your adult vaccinations.

lifelong immunity and the immune system's memory starts to fade after several years. Therefore, we may need **booster shots**, which are additional doses of a vaccine given periodically to "refresh" the immune system's antibodies and memory cells. For example, some form of the tetanus, diphtheria (Td), and perhaps pertussis (Tdap) vaccine is recommended for adults every 10 years.

Lastly, although many vaccines provide effective immunity to protect the individual from infection, some may just lessen the severity of the illness. For example, shingles can be a very painful and debilitating disease. Getting the shingles vaccine will not protect one from getting shingles (the viral DNA lays latent in peripheral nerves) but it will lessen the severity of the symptoms.

In some cases, killed pathogens do not stimulate a strong immune response and substances need to be added to increase their effectiveness (MicroFocus 22.2). And then there are some cases where vaccine development has just been a tough "nut to crack." In this latter case, we still have no vaccine against diseases like AIDS, even though it has been more than 30 years since the AIDS virus, the human immunodeficiency virus (HIV), was discovered (MicroFocus 22.3).

Adaptive Immunity Also Can Result by Passively Receiving Antibodies to an Antigen

Recall that in this section we have been discussing how one develops adaptive immunity. The previous discussion looked at the active forms of adaptive immunity. We finish this section by examining the ways one acquires antibodies in a passive and temporary manner.

Passive immunity develops when antibodies enter the body from an outside source (in contrast

MICROFOCUS 22.2: Tools
Enhancing Vaccine Effectiveness

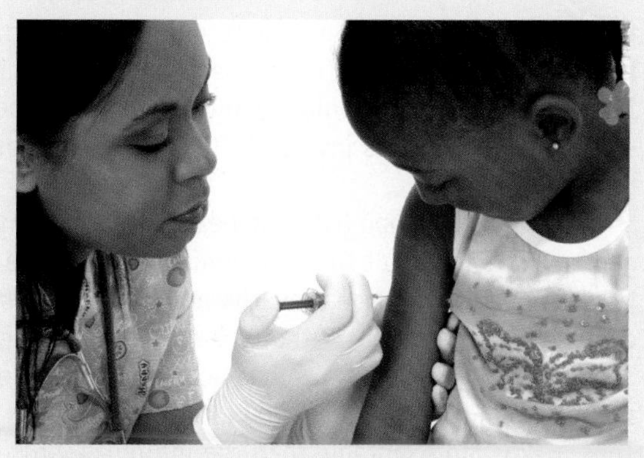

© Jaimie Duplass/ShutterStock, Inc.

Vaccines are like any other medical product. They contain active and inactive ingredients. All the ingredients are usually suspended in water and salt (a saline solution).

Active ingredients: The main component in a vaccine is the antigen that is prepared in such a way as to stimulate the immune response. Sometimes another component, called an **adjuvant** (*adjuvare* = "help"), is added to the antigen preparation to increase the immune response and, in so doing, make the vaccine more effective. By provoking a more sustained immune response, adjuvants also may allow the vaccine manufacturer to reduce the actual amount of antigen needed in each dose. The adjuvants licensed for general use in human vaccines in the United States are aluminum salts, specifically aluminum hydroxide or phosphate formulations simply called alum.

Inactive ingredients: Vaccines may contain stabilizers that act like preservatives. These are added to ensure the vaccine does not break down and remains effective until ready to use. Some vaccines may contain antibiotics that are added during the manufacturing process to prevent potential bacterial contamination. Low levels of formaldehyde that are safe for humans may also be used. Other stabilizers, such as sugars or amino acids, may be added to protect the vaccine from damage that can be caused by light, heat, or humidity.

One preservative that had been used since the 1930s in some vaccines (not MMR) is **thimerosal**, a mercury-containing compound. Although thimerosal was safe at the low levels used in vaccines, the American Academy of Pediatrics, the U.S. Public Health Service, and vaccine manufacturers stated in 1999 that, as a precautionary measure, thimerosal should be eliminated in all vaccines except tetanus-diphtheria (Td) vaccine, which is given to children 7 years and older and some flu vaccines. The seasonal flu vaccine often is packaged as multidose vials and contains thimerosal to protect the vaccine from possible contamination once it is opened. The single-dose vaccine units and the nasal spray vaccine do not contain thimerosal. The preservative is not found in any of the pediatric vaccines given to children younger than 6 years of age.

Vaccines may also contain traces of leftover ingredients from the manufacturing process. The flu viruses used in the manufacture of the seasonal flu vaccine, for example, are replicated in chicken eggs (see MicroFocus 22.1). Although the vaccine may contain traces of egg protein, an individual may still be able to receive the flu shot. Speaking with one's physician before getting a flu vaccination can determine if the flu shot is a viable option for flu protection.

to active immunity, in which individuals synthesize their own antibodies). Again, the source of antibodies may be unintentional, such as a fetus receiving antibodies from the mother, or intentional, such as the transfer of antibodies from one individual to another (see Figure 22.2C and D).

Naturally acquired passive immunity, also called "congenital immunity," develops when antibodies pass into the fetal circulation from the mother's bloodstream via the placenta and umbilical cord. The process occurs in the "natural" scheme of events.

The maternally passed IgG antibodies remain with the child for approximately 3 to 6 months after birth and play an important role during these months of life in providing additional resistance to diseases such as pertussis, staphylococcal infections, and viral respiratory diseases. Certain antibodies, such as measles antibodies, remain for 12 to 15 months.

Maternal antibodies also pass to the newborn through the first milk, or **colostrum**, of a nursing mother as well as during future breastfeedings. In this instance, IgA is the predominant antibody,

MICROFOCUS 22.3: Public Health

An AIDS Vaccine—Why Isn't There One After All These Years?

Producing an AIDS vaccine might appear rather straightforward: Cultivate a huge batch of human immunodeficiency virus (HIV), inactivate it with chemicals, purify it, and prepare it for marketing. In fact, this was the mentality and approach in the mid-1980s. Unfortunately, things are not quite so simple when HIV is involved. Here are the major reasons why a vaccine remains elusive after all these years.

The effects of a bad batch of weakened or inactivated vaccine would be catastrophic, and people generally are reluctant to be immunized with whole HIV particles, no matter how reassuring the scientists may be. In addition, how the body actually protects itself from pathogens such as HIV is not yet completely understood. Without that understanding, an effective vaccine cannot be developed.

Vaccine development also has been slow because of the high mutation rate of the virus. Some HIV strains around the world vary by as much as 35% in the capsid and envelope proteins they possess. Thus, HIV is more mutable than the influenza viruses, and no vaccine has been developed yet to make one completely immune to influenza. It is hard to make a vaccine targeted at one strain when the virus keeps changing its coat.

Another reason why a vaccine has not been successfully developed is because a short-lived vaccine would protect an individual only for a short time, necessitating an endless series of scheduled booster shots to which few would adhere. In addition, a vaccine also might act like dengue fever, where vaccination could actually make someone more at risk if they actually were to contract the disease. In 2002, an HIV patient whose body was holding the virus in check became infected with another strain of HIV through unprotected sex. This patient subsequently became "superinfected," meaning his immune system could keep the original strain in check, but was powerless to control the new strain. So, could a vaccine produce the same result if the individual was infected with another strain of HIV? Almost every AIDS vaccine researcher around the world is concerned about the unknown factors concerning an AIDS vaccine.

As of 2010, no vaccine trials have been shown to stimulate cell-mediated immunity to a level necessary to destroy HIV. The number of cytotoxic T cells and memory T cells produced is simply not up to the job. In March 2002, Anthony Fauci, Director of the National Institute for Allergy and Infectious Diseases, reported to the Presidential Advisory Council on HIV and AIDS that a "broadly effective AIDS vaccine could be a decade or more away." By 2012, minimal progress has been reported.

although IgG and IgM also have been found in the milk. The antibodies accumulate in the respiratory and gastrointestinal tracts of the child and support increased disease resistance.

Artificially acquired passive immunity arises from the intentional transfer of antibody-rich serum into the patient's circulation. The exposure to antibodies is thus "artificial." In the decades before the development of antibiotics, such an infusion was an important therapeutic tool for the treatment of disease. The practice still is used for viral diseases such as Lassa fever and arthropodborne encephalitis, and for bacterial diseases in which a toxin is involved. For example, established cases of botulism, diphtheria, and tetanus are treated with serum containing their respective antitoxins.

Various terms are used for the serum that renders artificially acquired passive immunity. **Antiserum** is one such term. Another common term, **gamma globulin**, takes its name from the fraction of blood serum in which most antibodies are found. Gamma globulin usually consists of a pool of sera from different human donors, and thus contains a mixture of antibodies (usually IgG), including those for the disease to be treated.

Passive immunity must be used with caution because in many individuals, the immune system recognizes foreign serum proteins as nonself antigens and synthesizes antibodies against them in an allergic reaction. When antibodies interact with the proteins, a series of chemical molecules called **immune complexes** may form and, with the activation of complement, the person develops a condition called **serum sickness**. This often is characterized by a hive-like rash at the infusion site, accompanied by labored breathing and swollen joints.

Although artificially acquired passive immunity provides substantial and immediate protection against disease, it is only a temporary measure. The immunity developing from antibody-rich serum usually wears off within weeks or months. For example, a person traveling to a country where hepatitis A is prevalent can obtain an antiserum preparation of hepatitis A antibodies several weeks prior to departing. The half-life for the antibodies is about 13 weeks.

Herd Immunity Results from Effective Vaccination Programs

In 2000, measles was declared eliminated from the United States. However, in 2011 there were 17 outbreaks of measles affecting 222 individuals. This is four times higher than the usual number of outbreaks (four outbreaks) and cases (60 cases) in the decade. According to the CDC, most of these cases were in children and teens whose parents had exempted them from being vaccinated as part of the mandatory school vaccinations with the MMR vaccine. Two hundred of the cases were the result of viral infections linked to importation; that is, to an American citizen who was infected in a foreign country or to foreign visitors in the Unites States (FIGURE 22.6). Luckily, there were no deaths although a third of the patients were hospitalized.

This example shows what can happen if small numbers of individuals or pockets of unvaccinated individuals come in contact with a pathogen. A disease outbreak is likely. Vaccinations are never meant to reach 100% of the population. However, if the vast majority of the population is vaccinated, then it is unlikely that an infectious individual will come in contact with a susceptible person and spread the disease. This indirect protection, called **herd (community) immunity**, implies that if enough people in a population are immunized against certain diseases, then it is very difficult for those diseases to spread (FIGURE 22.7).

Microbiologists and epidemiologists believe that when greater than 85% of the population is vaccinated (**herd immunity threshold**), the spread of the disease is effectively stopped. The rest of the "herd" or population remains susceptible, which is allowable because it then becomes very hard for pathogens "to find someone" who isn't vaccinated. Susceptible individuals are protected from catching the disease, and if one of these individuals should catch the disease, there are so many vaccinated people in the "herd" that it is unlikely the person could easily spread it.

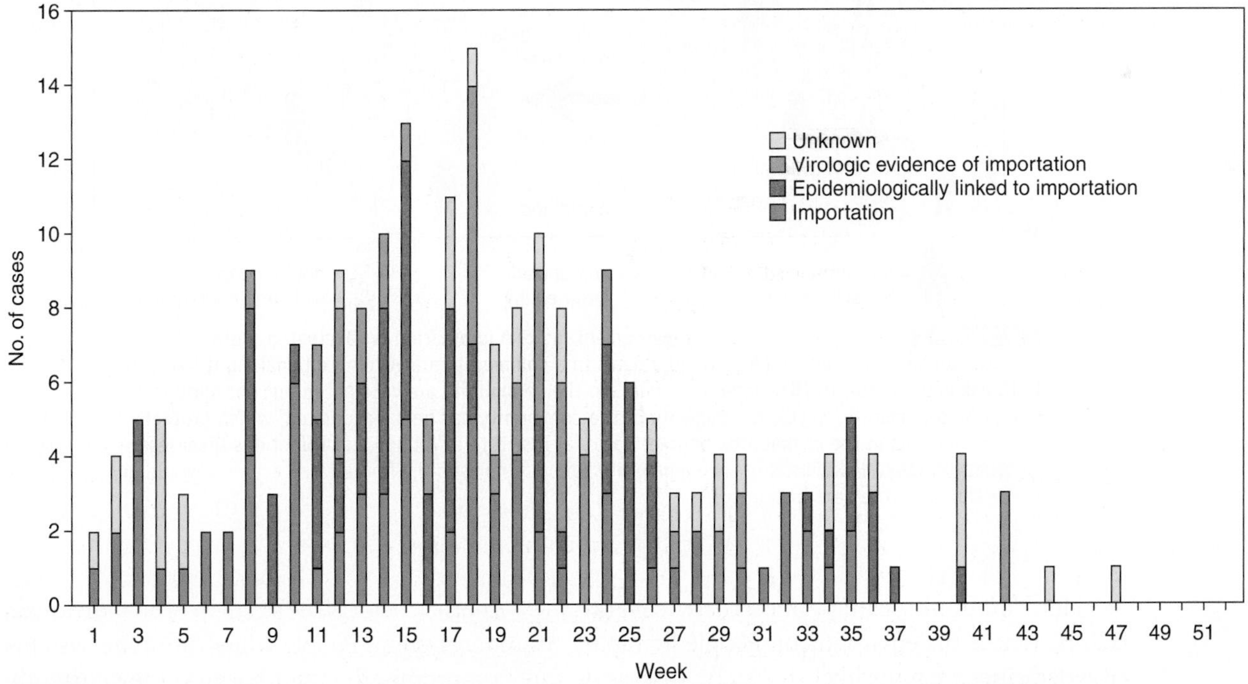

FIGURE 22.6 **Measles Cases, by Import Status and Week of Rash Onset—United States, 2011.** The majority of measles cases (200 of the 222) were associated with importations, most commonly from France, Italy, Romania, Spain, and Germany. Reproduced from: *Morbidity and Mortality Weekly Report*, 61(15);253–257. »» Only 72 cases developed in people who had traveled abroad. So, why were the other 128 cases considered the result of importation?

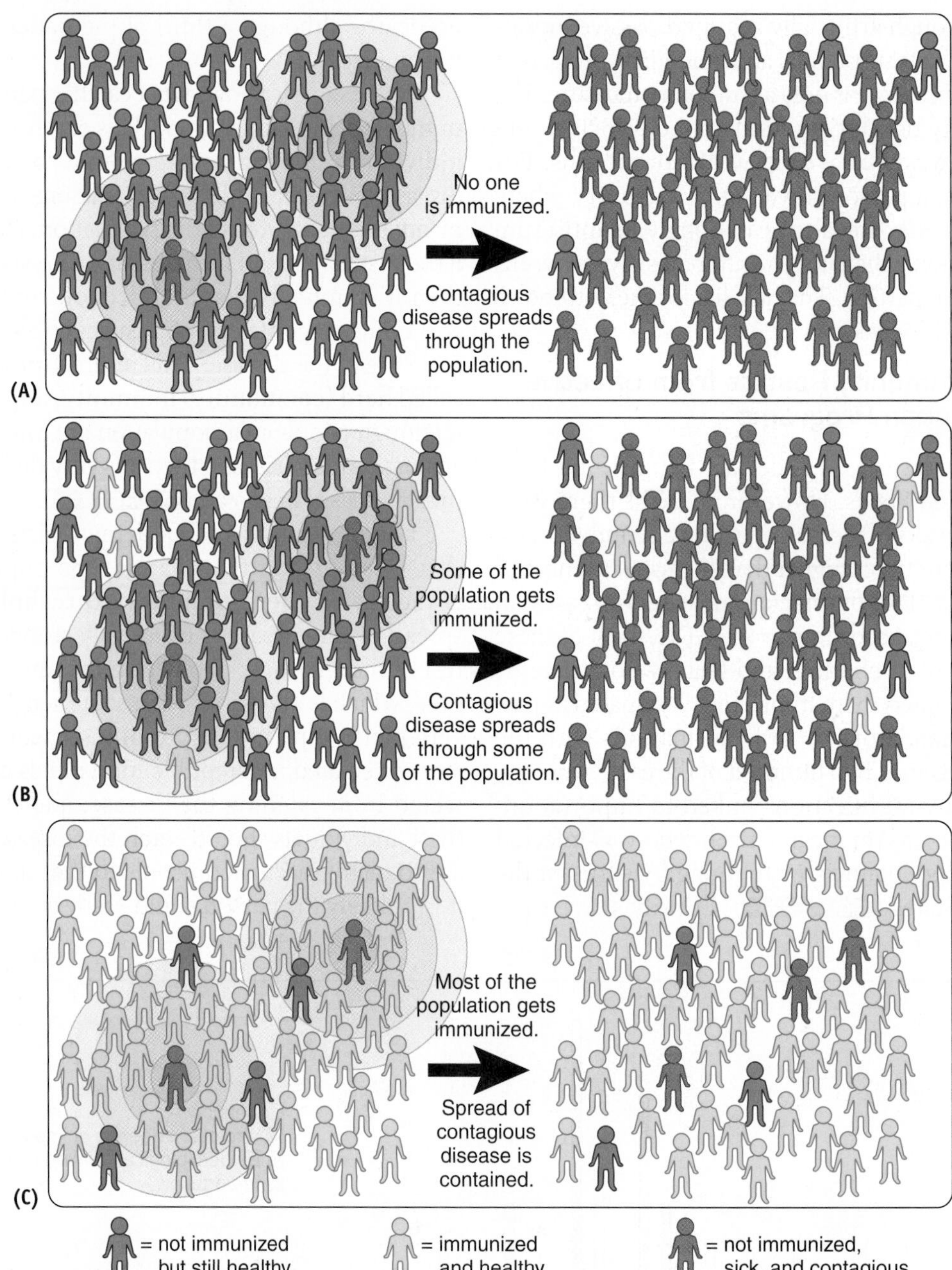

FIGURE 22.7 **Herd Immunity.** If a large percentage of a population is vaccinated, outbreaks of disease can be prevented. In (**A**), no individuals in a community are vaccinated, making it likely an outbreak could occur. In (**B**), some individuals in the community are vaccinated but not enough to confer herd immunity. In (**C**), the majority of the community has been vaccinated, which protects most individuals in the community. Source: National Institute of Allergy and Infectious Diseases.
»» What groups of individuals in the community, who are not eligible for certain vaccines, would get protection from the herd immunity?

Herd immunity can be affected by several factors. One is the environment. People living in crowded cities are more likely to catch a disease if they are not vaccinated than nonvaccinated people living in rural areas because of the constant close contact with other people in the city.

Another factor is the strength of an individual's immune system. People whose immune systems are compromised, either because they currently have a disease or because of medical treatment (anticancer or antirejection drugs), may not be able to be immunized. They are at a greater

MICROFOCUS 22.4: Public Health
The Risk of Not Vaccinating: Community and Healthcare Settings

Data from epidemiological studies clearly show that community vaccinations essentially eliminate outbreaks of infectious disease by maintaining high herd immunity. Yet, about half the states in the United States allow parents to apply for vaccination waivers prior to enrolling their children in school. Although there are some valid reasons why a few children should not be vaccinated, the Centers for Disease Control and Prevention (CDC) reported in 2010 that more than 75% of American children have been vaccinated on schedule against the nine diseases mandated by federal law. The CDC said that although coverage varied widely among states and major cities, there were areas in the country where too many children are not up-to-date on their shots. In addition, only a small percentage of adults in the community receive booster shots for diseases like tetanus and diphtheria.

Likewise, vaccination of healthcare workers (HCWs) against diseases like influenza is just as important to maintain patient safety and quality-of-care. HCWs are frequently exposed to influenza and even if they are asymptomatic, they can still transmit the virus to patients, especially the vulnerable elderly and immunocompromised. Influenza vaccination of HCWs not only reduces the disease burden by maintaining high herd immunity but also has been shown to reduce the rate of influenza disease and overall mortality in the patients cared for by HCWs. Therefore, since 1981, the CDC has recommended priority vaccinations for all HCWs, unless they have had a previous adverse reaction to the inactivated influenza vaccine.

Yet, despite local and national efforts to encourage influenza vaccination, the overall vaccination rate among HCWs in the United States remains unacceptably low at approximately 40%. Unvaccinated HCWs surveyed said they did not get the influenza vaccine because they feared the side effects and needles, they were skeptical of vaccine efficacy, they believed in their own innate ability to resist infection, or they reported an inability to access the vaccine. As with community vaccinations, unless there are extenuating circumstances, HCWs, as part of their "duty of care," should be vaccinated against influenza to prevent illness to themselves and transmission to others.

risk of catching the disease to which others are immunized.

MicroFocus 22.4 looks at the consequences of declining vaccinations and herd immunity.

Do Vaccines Have Dangerous Side Effects?

Many people today are too young to remember the prevaccine era, the large number of infectious cases, and the terrible damage some infectious diseases caused—birth defects from rubella (German measles); central nervous system paralysis from polio; and the uncontrollable, violent coughing and breathing struggles from whooping cough (TABLE 22.2).

If you look back at Figure 22.3, notice the current immunization schedule in the United States for young children consists of up to 12 shots of five vaccines by 6 months of age and an additional four shots of six more vaccines by 18 months. Records show that such vaccinations have been very successful at eliminating or greatly reducing the incidence of childhood diseases in the United States. Still, many parents wonder if this number of vaccines can have adverse effects on a developing young child, and if so, should they avoid taking their child to the doctor for routine vaccinations?

TABLE 22.2 Decline in Disease Cases, United States

	Mean U.S. Cases per Year (20th century)		
Disease	Pre-vaccine	2010	% Decrease
Diphtheria	21,053	0	100
Measles	530,217	61	>99
Mumps	162,344	2,528	98
Pertussis (whooping cough)	200,752	21,291	89
Polio	16,316	0	100
Rubella	47,745	6	>99
Smallpox	29,005	0	100

Source: CDC.

The chance of experiencing a serious side effect, such as an allergic reaction, from a vaccine is less than one in a million. What is more likely is to experience short-lived symptoms, such as a slight inflammation (soreness, redness, and some swelling) at the injection site. But realize these minor side effects are telling you that the vaccine worked, as the symptoms are due to the immune system recognizing and responding to the microbial agent in the vaccine. The Food and Drug Administration (FDA) requires vaccine manufacturers to follow extensive safety procedures when producing vaccines used in the United States to ensure they are safe and effective. After lab and animal tests have been completed, a promising vaccine must undergo thorough human clinical trials before being licensed by the FDA for use.

To "catch" any rare occurrences of a serious side effect, the FDA and the CDC have established the **Vaccine Adverse Events Reporting System (VAERS)** to which anyone, including doctors, patients, and parents, can report adverse vaccine reactions. The FDA and CDC researchers regularly monitor the system for any suspicious trends.

One vaccine that was permanently removed from the U.S. market was RotaShield®, which was administered to infants to provide immunity against rotavirus infections. In 1999, the Advisory Committee on Immunization Practices (ACIP) suggested that the vaccine licensed in the United States should no longer be recommended for infants based on a review of scientific studies indicating the virus might be associated with intussusception among some infants during the first 1 to 2 weeks following vaccination. Although these scientific studies may have been in error, a new and better vaccine, RotaTeq®, is now available and very effective against rotavirus disease.

■ Intussusception: A rare, potentially life-threatening condition where bowel sliding creates swelling and intestinal obstruction.

In 2002–2003, there was much discussion weighing the risks of the smallpox vaccine's potential side effects (mostly mild, but potentially deadly for a very few) versus the risk of a smallpox bioterrorist attack. Although the smallpox vaccine is not available to the general public, in the case of a bioterrorist attack, the U.S. government purportedly has sufficient vaccine to vaccinate every American.

Perhaps the vaccine of most concern regarding standard childhood vaccines is MMR. In 1998, a paper was published suggesting there might be a link between the MMR vaccine and autism in children. The idea was put forward that perhaps the sheer number of vaccinations could overwhelm a child's immune system and cause neurological damage. To make a long story short, as of 2012, numerous studies carried out by other scientists and researchers around the world have found no evidence to support the autism claim.

■ Autism: A neurodevelopmental disturbance in which the use of communication, as well as normal behavior and social relationships, are not fully established and follow unusual patterns.

Through vaccination and booster shots, vaccines have been essential to maintaining public health and controlling infectious diseases (MicroFocus 22.5). Although a few of the millions of people vaccinated each year suffer a serious consequence from vaccination, the risks of contracting a disease (especially in infants and children) from *not* being vaccinated are thousands of times greater than the risks associated with any vaccine. In addition, the licensed vaccines are always being reexamined for ways to improve their safety and effectiveness.

Investigating the Microbial World 22 recounts the field studies carried out to test the effectiveness of the Salk polio vaccine.

CONCEPT AND REASONING CHECKS 1

a. How do the naturally acquired and artificially acquired forms of active immunity differ?
b. List the benefits and drawbacks of attenuated, inactivated, and toxoid vaccines.
c. How do subunit and conjugate vaccines differ from one another?
d. Explain why there are several vaccines specifically for adults.
e. How do the naturally acquired and artificially acquired forms of passive immunity differ?
f. What are the major factors required for a population to maintain high herd immunity?
g. Assess the need for and effectiveness of maintaining vaccinations.

Chapter Challenge A

You cannot remember if you ever had measles or the vaccination as a child. So, you do some background research.

QUESTION A:

What is the schedule for childhood measles vaccination, what is the combination formulation used for measles vaccination, what are the antigens normally found in the vaccine, and how is the vaccine administered?

Answers can be found on the Student Companion Website in **Appendix D**.

MICROFOCUS 22.5: Public Health
Vaccine Development: An Upward Climb

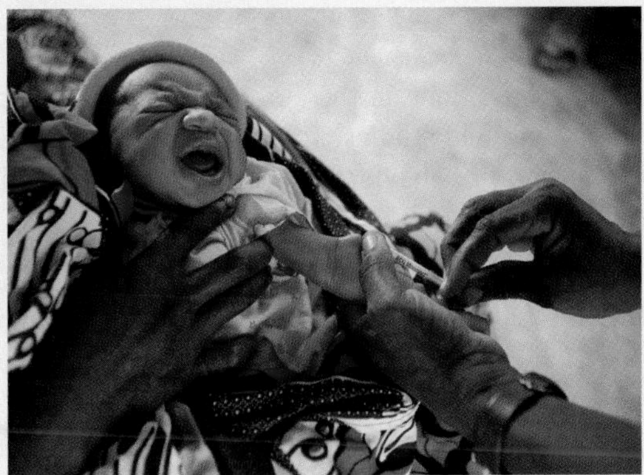
A child receives vaccination against tuberculosis, Benin, Africa. © Olivier Asselin/Alamy Images.

Even though most agree that vaccination is the most cost-effective way to prevent infectious disease, there has recently been a lack of enthusiasm for vaccine development among the pharmaceutical companies. In fact, the number of companies in the United States supplying vaccines has dropped from 30 to five since 1980. Such a drop has been partially responsible in recent years for shortages of influenza, tetanus-diphtheria, MMR, and other vaccines.

Like any drug, vaccine development is a multimillion-dollar (often billion-dollar) investment. Therefore, a pharmaceutical company that invests its time and money in development expects a large return on its investment. Unfortunately, vaccines (and antibiotics too) often do not provide that payback as compared to other pharmaceutical drugs. A vaccine may be administered only once or twice in a lifetime to an individual, whereas many pharmaceutical drugs, such as Lipitor® (lowers "bad" cholesterol), Singulair® (treats or helps prevent asthma), and Fosamax® (prevents and treats osteoporosis), are long-term, often lifetime, medications for individuals.

In addition, an adverse reaction to a vaccination may bring legal action against the company. The potential exposure to legal liability is not an attractive business opportunity, so pharmaceutical companies have been reluctant to invest the time and money in vaccine development.

Many developing nations suffer from the burden of malaria, AIDS, cholera, meningitis, tuberculosis, and many other diseases less well known in the developed world (see figure). These developing nations cannot afford the costs of vaccines, and because the developed world already has vaccines available for most of their vaccine-preventable diseases, the incentive is not there for drug companies to spend the huge sums of money in vaccine development that are needed for treating and preventing infectious disease in the developing world.

Today, much of the vaccine research and development occurs through public and private partnerships and organizations. For example, the Gates Foundation supports the Medicines for Malaria Venture with its work to develop more affordable and effective malaria treatments and vaccines; the Global Alliance for Vaccines and Immunizations and their efforts to provide new vaccines to the millions of children who die every year from potentially vaccine-preventable diseases; the Aeras Global TB Vaccine Foundation to aid in the development of new vaccines to prevent tuberculosis; and the Advance Market Commitment (AMC) for Pneumococcal Vaccines, which is a new initiative launched by the Gates Foundation and five nations to promote the development of life-saving vaccines.

Recently, increased funding and higher profits are bringing the pharmaceutical companies back into the game. Recently, three new vaccines were approved: for the human papilloma virus, rotavirus, and herpes zoster; so "big pharma" is a player once again. The elimination of disease can be accomplished with new vaccines, so getting vaccine design and development back on track at the top of the global public health agenda is key.

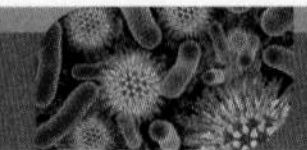

Investigating the Microbial World 22

The Salk Polio Vaccine Field Trials

Experiments are not always done in the lab. Many science investigations take place outside the lab and compose what are called field studies. These research projects still use science inquiry methods to collect data about some phenomenon or behavior. They also may involve clinical trials that involve observational studies concerning the effects of some intervention on the health of a population. The Salk polio vaccine field studies in the 1950s were one of the most famous and largest statistical studies ever conducted.

■ **OBSERVATION:** The following time series provides the approximate number of polio cases in the United States between 1930 and 1955, the year the Salk polio vaccine was released.

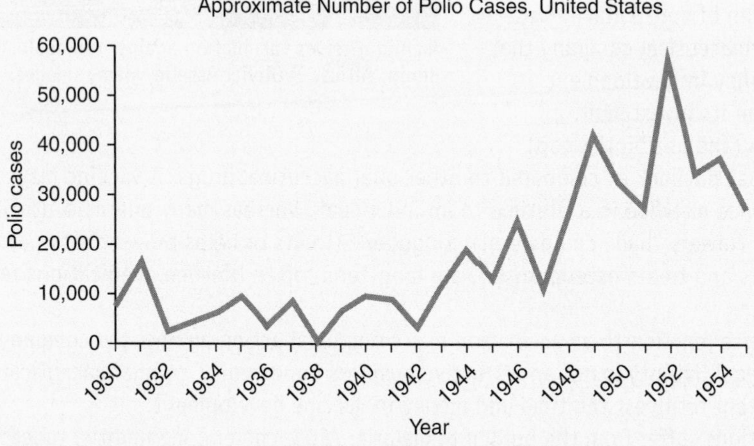

Adapted from: Francis, T. *et al.* (1955). *Amer. J. Pub. Health* 45 (suppl.): 1–50.

The need for a vaccine was urgent and two labs, those of Jonas Salk and Albert Sabin, were involved in the search. Looking at the Salk investigation, a vaccine was eventually prepared from a heat- and formaldehyde-killed sample of the polio virus. The key in the making of the vaccine was to ensure the viruses were "killed" (cannot replicate in host cells) but the treatment was not so severe that it destroyed the immunogenic properties of the antigens. Once the vaccine was thought to be safe according to lab experiments, it needed to be field tested for its effectiveness.

■ **QUESTION:** ***Was the vaccine effective? Did it prevent polio?***

■ **HYPOTHESIS:** The vaccine is effective. If so, then administering the vaccine to a treatment group should show a statistically significant reduction in polio cases compared to a nontreatment (control) group.

■ **EXPERIMENTAL DESIGN:** Designing the epidemiological field studies is often fraught with problems. The treatment and control groups had to be as identical as possible. The individuals in each group needed to come from the same region and the studies needed to be done at the same time of year. Large numbers of individuals were needed because the incidence of paralytic polio in the United States was about 50 cases per 100,000 individuals. And diagnosis of mild polio cases (nonparalytic) was tricky because the symptoms of fever and weakness could be typical of other diseases as well.

The field trials, carried out in 1954, involved two distinct designs:

■ **FIELD TRIAL 1:** Almost 950,000 first- through third-graders in 127 selected schools in 33 states were enlisted (with parents' consent). The second-graders at selected schools across the country were vaccinated (treatment group) while the first- and third-graders in the same schools were not vaccinated (control group).

■ **FIELD TRIAL 2:** More than 400,000 children in 84 selected schools in 11 states were enlisted (with parent's consent). One group was given the vaccine while a second group was given a placebo injection (a saline solution).

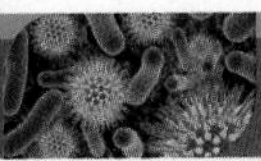

Investigating the Microbial World 22 (continued)

- **RESULTS:**

TABLE

Field trial	Group	Number of students	Paralytic Polio cases	Incidence per 100,000
1	Treatment	221,998	38	17.1
	Control	725,173	330	45.5
2	Treatment	200,745	33	16.4
	Placebo	201,229	115	57.1

Adapted from: Francis, T. *et al.* (1955). *Amer. J. Pub. Health* 45 (suppl.): 1–50.

- **CONCLUSION:**

QUESTION 1: ***Were the number of paralytic polio cases in the control groups about equal to the actual incidence in the United States? Explain.***

QUESTION 2: ***Did treatment with the Salk polio vaccine reduce the incidence of paralytic polio? Explain.***

QUESTION 3: ***Was the hypothesis supported? Explain.***

Answers can be found on the Student Companion Website in **Appendix D.**

Adapted from: Francis, T. *et al.* (1955). *Amer. J. Pub. Health* **45** (suppl.): 1–50. Icon image © Tischenko Irina/ShutterStock, Inc.

KEY CONCEPT 22.2 Serological Reactions Can Be Used to Diagnose Disease

Antigen-antibody interactions studied in the immunology section of the clinical microbiology lab are known as **serological reactions** because they commonly involve serum from a patient. Such reactions first were adapted to laboratory tests used in the diagnosis of disease. The principle was simple and straightforward: If the patient had an abnormal level of a specific antibody in the serum, a suspected disease agent probably was present. Today, **serology**, the study of blood serum and its constituents, and especially its role in protecting the human body against disease, have diagnostic significance as well as more broad-ranging applications. For example, they are used to confirm identifications made by other procedures and to detect pathogens in body tissues. In addition, they help the physician to follow the course of disease and determine the immune status of the patient, and they aid in determining taxonomic groupings (serotypes) of microorganisms below the species level.

Serological Reactions Have Certain Characteristics

Diagnostic microbiology makes great use of the serological reactions between antigen and antibody. Thus, immunodiagnostic procedures can be used to identify infectious diseases by detecting antigens, or the antigens can be used to detect antibodies present in a blood (serum) sample.

Successful serological reactions require the antigen or antibody solution to be greatly diluted in order to reach a concentration at which a reaction will be most favorable. The process, called **titration**, may be used to the physician's advantage because the dilution series is a valuable way of determining the titer of antibodies. The **titer** is the most dilute concentration of serum antibody yielding a detectable reaction with its specific antigen (FIGURE 22.8). This number is expressed as the denominator of the dilution (for example, a 1:50 dilution, titer = 50). For an influenza infection, the titer of influenza antibodies may rise from 20 to 320 as an episode

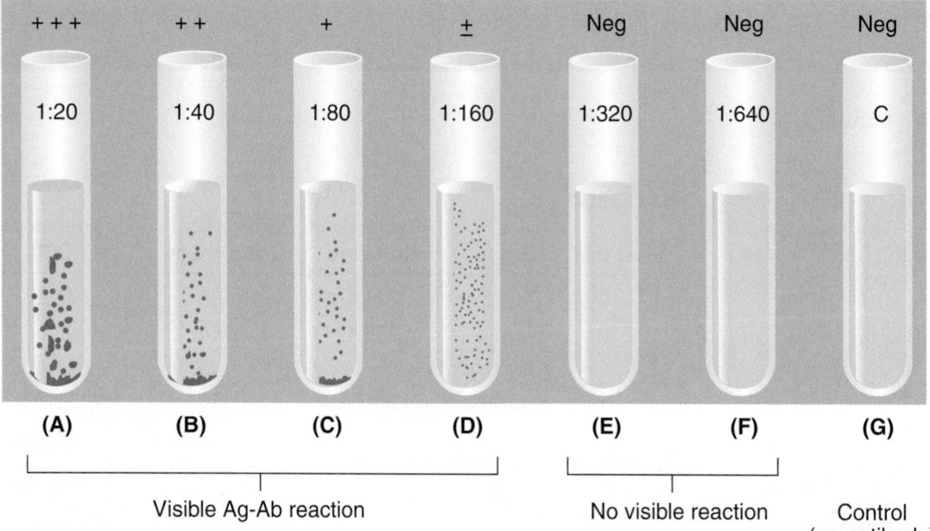

FIGURE 22.8 **The Determination of Titer.** A sample of antibody (Ab)-containing serum was diluted in saline solution to yield the dilutions shown. An equal amount of antigen (Ag) was then added to each tube, and the tubes were incubated. An antigen-antibody interaction may be seen in tubes (**A**) through (**D**), but not in tubes (**E**) or (**F**), or the control tube (**G**). The titer of antibody is the highest dilution of serum antibody in which a reaction is visible. »» What is the titer of antibody in this example?

of influenza progresses, then continue upward and stabilize at 1,280 as the disease reaches its peak. A rise in the titer is evidence that an individual has a disease, an important factor in diagnosis. Serology has become a highly sophisticated, and often automated, branch of immunology, as we will see in the following tests.

Neutralization Involves Antigen-Antibody Reactions

Neutralization is a serological reaction used to identify toxins and antitoxins, as well as viruses and viral antibodies. Normally, little or no visible evidence of a neutralization reaction is present, so the test mixture must therefore be injected into a laboratory animal to determine whether neutralization has taken place.

The detection of botulism toxin is an example of a neutralization test (FIGURE 22.9). Normally, the toxin is lethal to a laboratory animal, and if a sample of the food contains the toxin, the animal will succumb after an injection. However, if the food is first mixed with botulism antitoxins, the antitoxin neutralizes the toxin, and the mixture has no effect on the animal.

Conversely, if another lethal toxin was produced by some other bacterial food pathogen, no neutralization will occur, and the mixture will still be lethal to the animal. A similar test for diphtheria is the **Schick test**, in which a person's immunity to diphtheria can be determined by injecting diphtheria toxin intradermally. No skin reaction will occur if the person has neutralizing antibodies. A local red, swollen area occurs if no antidiphtheria antibodies are present, indicating the person is susceptible to diphtheria.

Precipitation Requires the Formation of a Lattice Between Soluble Antigen and Antibody

Precipitation reactions are serological reactions involving thousands of antigen and antibody molecules cross-linked at multiple determinant sites, forming a lattice. The lattices are so large that they precipitate in a form that can be observed visually.

Precipitation tests are performed in either fluid media or gels. In fluids, the antibody and antigen solutions are layered over each other in a thin tube. The molecules then diffuse through the fluid until they reach a **zone of equivalence**, the ideal concentration for precipitation to occur. A visible mass of particles now forms either at the interface or at the bottom of the tube.

In **immunodiffusion**, the diffusion of antigens and antibodies takes place through a semisolid gel,

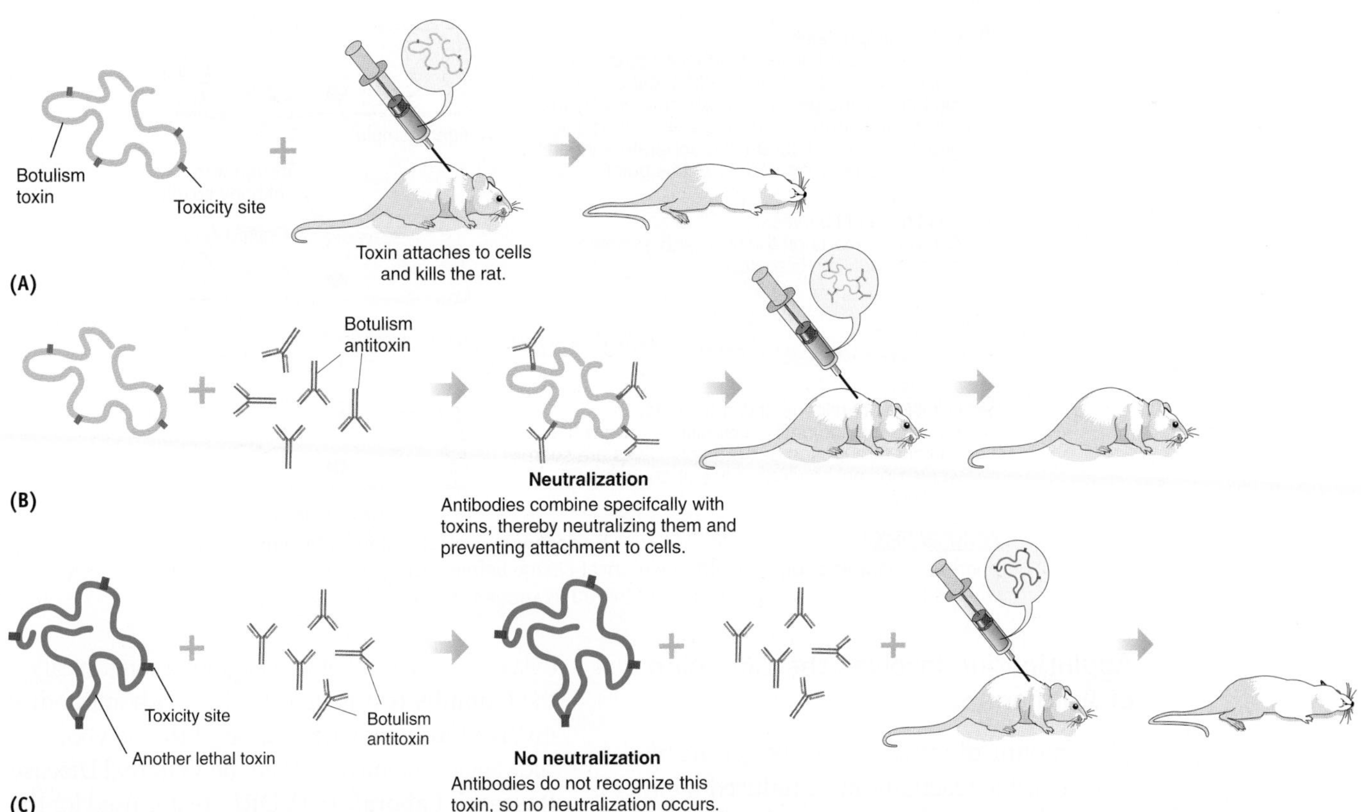

FIGURE 22.9 **Neutralization Test.** (**A**) The botulism toxin injected into a rat will attach to nerve cells and kill the rat. (**B**) Mixing of toxin and antibotulism antibodies (antitoxin) neutralizes the toxin and on injection of the mix, the rat lives. (**C**) Mixing another lethal toxin with botulism antitoxin does not lead to a serological reaction and the rat dies after injection of the mixture, demonstrating that the toxin is not the botulism toxin. »» In (B), how is the toxin-antitoxin complex removed from the circulation?

such as agar. As the molecules diffuse through the gel, they eventually reach the zone of equivalence, where they interact and form a visible precipitate. Variations of this technique are called the Ouchterlony plate technique, named for Orjan Ouchterlony, who devised it in 1953, and the **double diffusion assay**, so named because both reactants diffuse.

In immunodiffusion, antigen and antibody solutions are placed in wells cut into agar in Petri dishes. The plates are incubated and precipitation lines form at the zone of equivalence (FIGURE 22.10). The test has been used to detect fungal antigens of *Histoplasma*, *Blastomyces*, and *Coccidioides*.

In the antigen-detecting procedure known as **immunoelectrophoresis**, gel electrophoresis, and diffusion are combined. A mixture of antigens is placed in a reservoir on an agarose slide, and an electrical field is applied to the ends of the slide. The different antigens then move through the agarose at different rates of speed, depending on their electrical charges (FIGURE 22.11). A trough is then cut into the agarose along the same axis, and a known antibody solution is added. During incubation, antigens and antibodies diffuse toward each other and precipitation lines form, as in the immunodiffusion technique.

■ Agarose:
A polysaccharide derived from seaweed and part of the structure of agar.

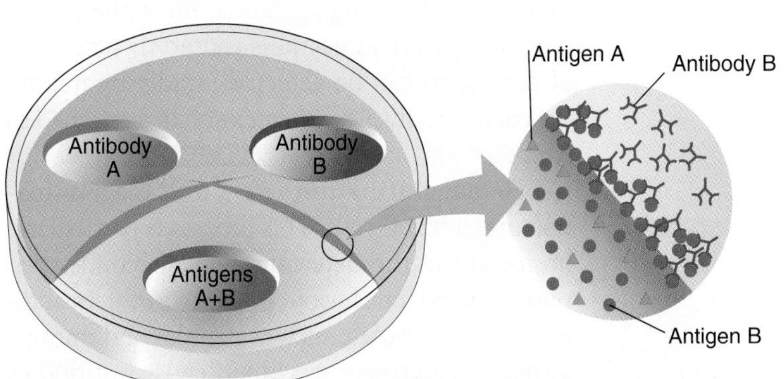

FIGURE 22.10 **A Precipitation Test.** Wells are cut into a plate of purified agar. Different known antibodies then are placed into the two upper wells, and a mixture of unknown antigens is placed into the lower well. During incubation, the reactants diffuse outward from the wells, and cloudy lines of precipitate form. The lines cross each other because each antigen has reacted only with its complementary antibody. »» Why does the precipitin line form where it does?

(A) Gel electrophoresis:
On a gel-coated slide, an antigen sample is placed in a central well. An electrical current is run through the gel to separate antigens by their electrical charge (electrophoresis). Unlike the depiction in the diagram, the separate antigens cannot be detected visually at this point.

(B) Addition of antibodies:
A trough is made on the slide and a known antibody solution is added.

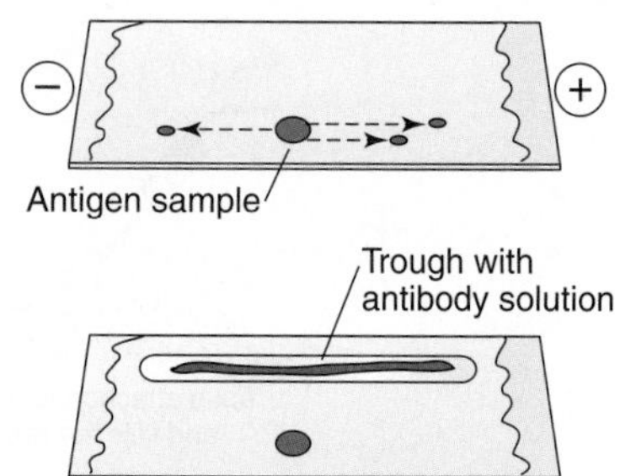

(C) Diffusion of antigens and antibodies:
As antigens and antibodies diffuse toward one another through the gel, precipitin lines are seen where optimal concentrations of antigen and antibodies meet.

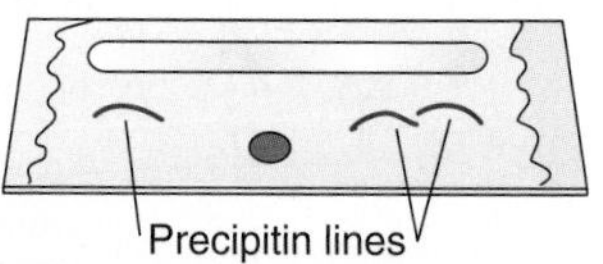

FIGURE 22.11 **Immunoelectrophoresis.** By applying an electrical field, immunoelectrophoresis separates antigens by their electrical charge before antibodies are added. »» In (A), what is the electrical charge (+ or -) on the three antigens shown?

Agglutination Involves the Clumping of Antigens

The amount of antibody or antigen needed to form a visible reaction can be reduced if either is attached to the surface of an object. The result is a clumping together, or **agglutination**, of the linked product. Agglutination procedures are performed on slides or in tubes. For bacterial agglutination, emulsions of unknown bacterial species are added to drops of known antibodies on a slide, and the mixture is observed for clumping.

Passive agglutination is a modern approach to traditional agglutination methods. Most often, antigens are adsorbed onto the surface of latex spheres or polystyrene particles (FIGURE 22.12A). Serum antibodies can be detected rapidly by observing agglutination of the carrier particle. Bacterial infections caused by a number of pathogens can be detected rapidly by mixing the bacterial sample with latex spheres attached to streptococcus antibody (FIGURE 22.12B, C).

Hemagglutination refers to the agglutination of red blood cells. This process is particularly important in the determination of blood types prior to blood transfusion (FIGURE 22.12D). In addition, certain viruses, such as measles and mumps viruses, agglutinate red blood cells. Antibodies for these viruses may be detected by a procedure in which the serum is first combined with laboratory-cultivated viruses and then added to the red blood cells. If serum antibodies neutralize the viruses, agglutination fails to occur. This test is called the **hemagglutination inhibition (HAI) test** (MicroFocus 22.6). A hemagglutination test called the **Coombs test** is used to detect Rh antibodies involved in hemolytic disease of the newborn. A slide agglutination test called the **Venereal Disease Research Laboratory (VDRL) test** is used for the rapid screening of patients to detect syphilis.

Complement Fixation Can Detect Antibodies to a Variety of Pathogens

The **complement fixation test** is performed in two parts. The first part—the test system—heats the patient's serum to destroy any complement present in the serum. Next, carefully measured amounts of antigen and guinea pig complement are added to the serum (FIGURE 22.13A). This test system then is incubated at 37°C for 90 minutes. If antibodies specific for the antigen are present in the patient's serum, an antibody-antigen interaction takes place, and the complement is used up, or "fixed." However, there is no visible sign of whether or not a reaction has occurred.

Now, in the second part, the indicator system (sheep red blood cells and antisheep antibodies) is added to the tube, and the tube is reincubated (FIGURE 22.13B). If the complement was previously fixed, lysis of the sheep red blood cells cannot take place. The blood cells therefore would remain intact, and when the tube is centrifuged, the technician observes clear fluid with a "button" of blood cells at the bottom. Conclusion: The patient's serum contained antibodies that reacted with the antigen and fixed the complement.

If the complement was not fixed in the test system, it will still be available to react with the

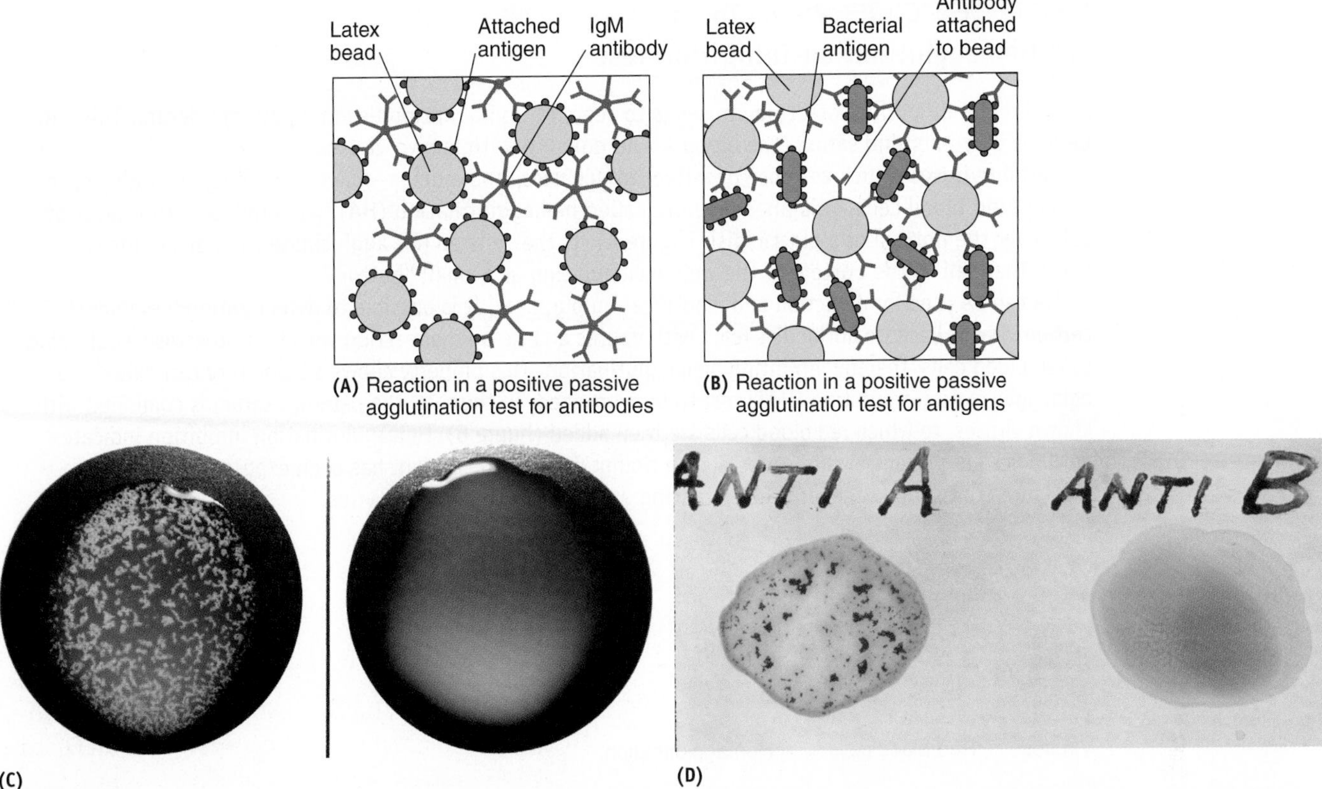

FIGURE 22.12 **Passive Agglutination Tests.** (**A**) When particles are bound with antigens, agglutination indicates the presence of antibodies, such as the IgM shown here. (**B**) When particles are bound with antibodies, agglutination indicates the presence of antigens. (**C**) Latex agglutination test for *Staphylococcus*. If a specific staph antigen is present, it will bind to antibody-coated latex beads and agglutinate them for a positive test (left). A negative test is shown for comparison on the right. Courtesy of Gunnar Flåten, Bionor Laboratories AS. (**D**) Blood type can be determined using the agglutination test. Red blood cells are diluted with saline and mixed with anti-A agglutinin (at left), and anti-B agglutinin (at right). For some minutes they are allowed to react. The clumping (agglutination) is due to antibody-antigen reactions. These tests use antigens or antibodies adsorbed onto the surface of latex spheres. © Ed Reschke/Getty. »» What is the blood type in the example shown in (D)?

antibodies bound to sheep red blood cells and, as a result, the sheep red blood cells will lyse. When the tube is centrifuged, the technician sees red fluid, colored by the hemoglobin of the broken blood cells, and no evidence of blood cells at the bottom of the tube. Conclusion: The serum lacked antibodies for the antigen tested.

The complement fixation test is valuable because it may be adapted by varying the antigen. In this way, tests may be conducted for such diverse diseases as encephalitis, Rocky Mountain spotted fever, meningococcal meningitis, and histoplasmosis. The versatility of the test, together with its sensitivity and relative accuracy, has secured its continuing role in diagnostic medicine.

Labeling Methods Are Used to Detect Antigen-Antibody Binding

The detection of antigen-antibody binding can be enhanced (amplified) visually by attaching a label to the antigen or antibody. This tag may be a fluorescent dye, a radioisotope, or an enzyme.

Immunofluorescence Antibody Test. The detection of antigen-antibody binding can be done on a slide using an **immunofluorescence antibody (IFA) test**. Two commonly used dyes are fluorescein, which emits an apple-green glow, and rhodamine, which gives off orange-red light.

The IFA test may be direct or indirect. In the direct method, the fluorescent dye is linked to a known antibody. After combining with particles having complementary antigens, the three components react, causing the complex to glow on illumination with ultraviolet (UV) light and viewed with a fluorescence microscope.

For example, suppose you want to know if spirochetes were present in a serum sample. The sample would be combined with anti-spirochete-labeled antibodies. If spirochetes are present, the tagged antibodies accumulate on the particle surface and the particle glows when viewed with fluorescence microscopy (FIGURE 22.14).

With the indirect IFA test, the fluorescent dye is linked to an antibody recognizing human

MICROFOCUS 22.6: Tools

The Hemagglutination-Inhibition Test

An indirect method for detecting viruses is to search for viral antibodies in a patient's serum. This can be done by combining serum (the blood's fluid portion) with known viruses.

Certain viruses—such as those of influenza, measles, and mumps—have the ability to agglutinate (clump) red blood cells. This phenomenon, called hemagglutination (HA), will produce a thin layer of cells over the bottom of a plastic dish (figure A). If the cells do not agglutinate (no hemagglutination), they fall to the bottom of the well accumulating as a small "button."

HA can be used for detection and identification purposes. It is possible to detect antibodies against certain viruses because antibodies react with viruses and tie up the reaction sites that otherwise would bind to red blood cells, thereby inhibiting hemagglutination. This property allows a laboratory test called the hemagglutination-inhibition (HAI) test to be performed. In the test, the patient's serum is combined with known viruses, to which red blood cells are then added (figure B). Hemagglutination inhibition indicates antibodies are present in the serum. Such a finding implies the patient has been exposed to the virus.

Figure C presents a scenario for detecting exposure to the measles virus.

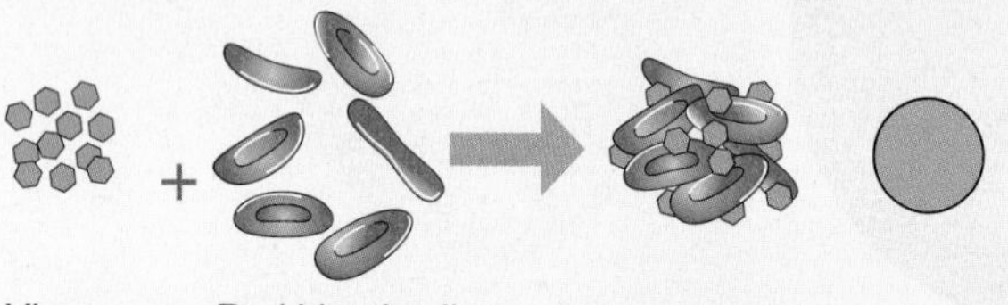

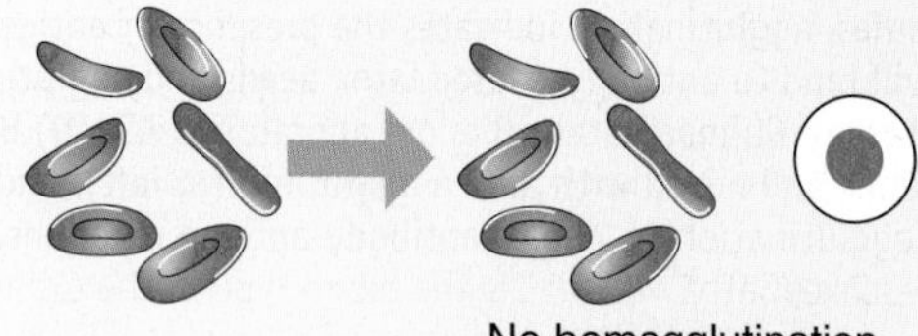

(A)

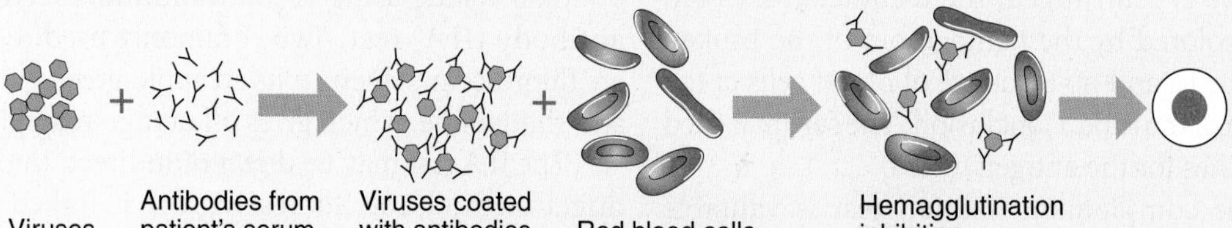

(B)

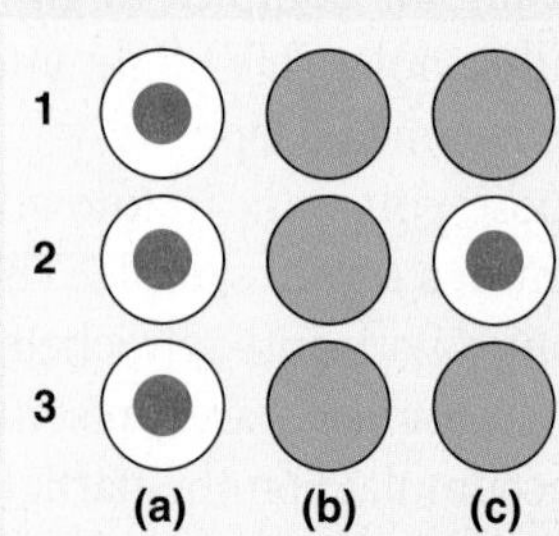

(C)

Tests for Detecting Viruses (A) In hemagglutination, viruses of certain diseases, such as measles and mumps, can agglutinate (clump) red blood cells, forming a film of cells over the bottom of a plastic well. No agglutination produced a small pellet or "button" at the bottom of the cell. **(B)** In the hemagglutination-inhibition (HAI) test, known viruses are combined with serum from a patient. If the serum contains antibodies for that virus, the antibodies coat the virus, and when the viruses are then combined with red blood cells, no agglutination will take place.
(C) In December, three young children (1, 2, 3) who attended the same daycare center were admitted to the hospital with nonbacterial pneumonia. All three were tested for the presence of influenza A antibodies using the HAI test (see left diagram). Which child or children likely had the flu? (a) = positive control; (b) = negative control; (c) = patient's serum.

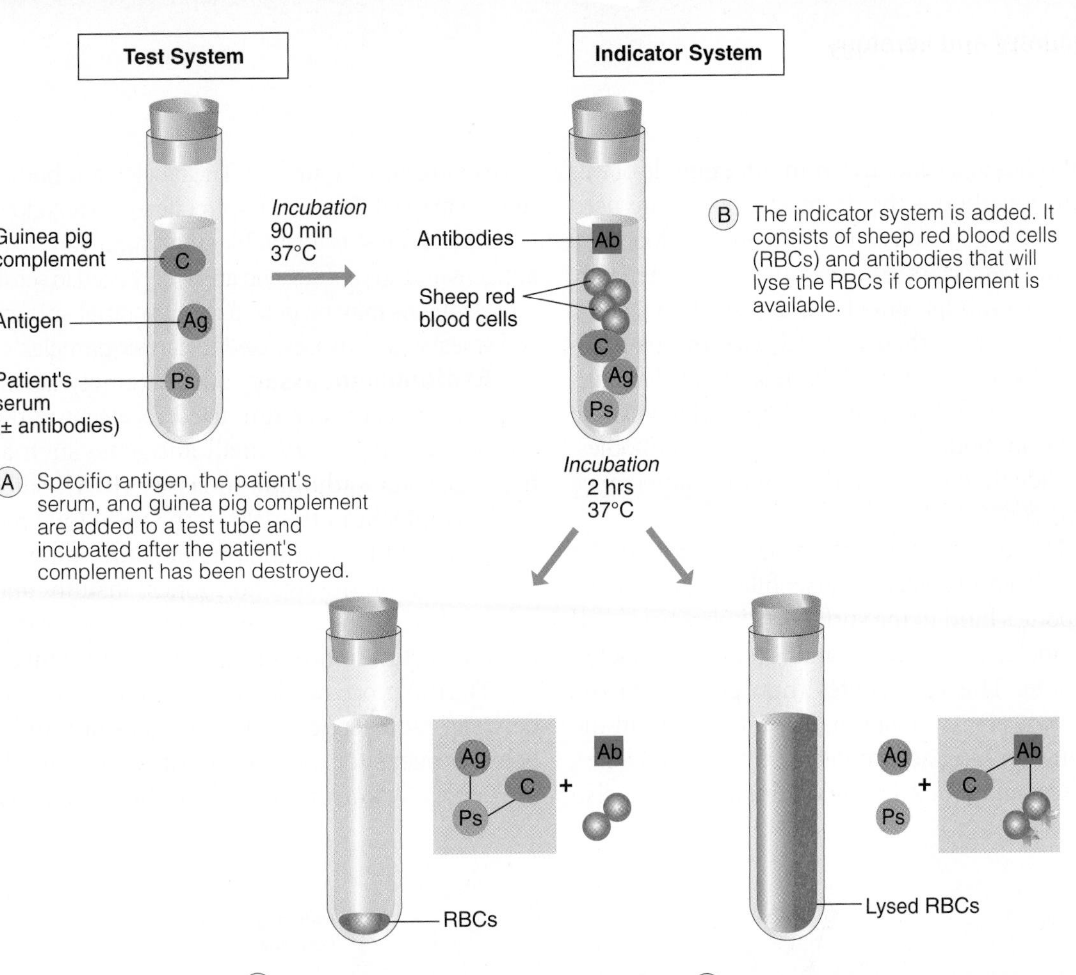

FIGURE 22.13 **The Complement Fixation Test.** Complement fixation tests are carried out in two stages involving (A) a test system and (B) an indicator system. »» What would happen if the patient's complement proteins were not inactivated?

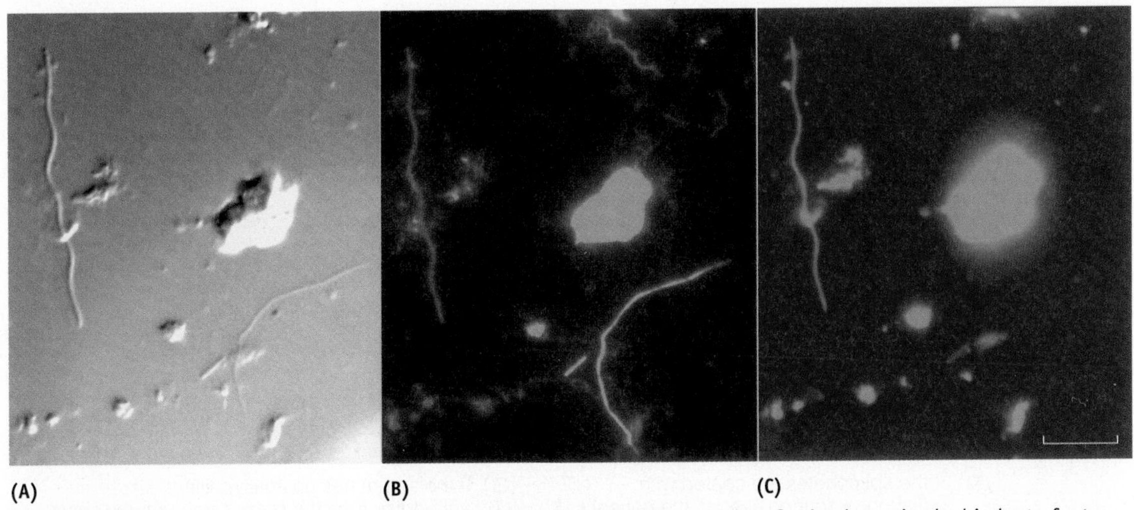

FIGURE 22.14 **Direct Immunofluorescence Antibody (IFA) Test.** Three views of spirochetes in the hindgut of a termite identified by a direct IFA test. (**A**) An unstained area displayed by differential interference contrast microscopy. (**B**) The same viewing area seen by fluorescence microscopy using rhodamine B as a stain. (**C**) The same area viewed after staining with fluorescein. (Bar = 10 μm.) Reproduced from *Appli. Environ. Microbiol,* 1996, vol. 62(2), p. 347–352, DOI and reproduced with permission from the American Society for Microbiology. Photo courtesy of Doctor Bruce J. Paster, Senior Member at Department of Molecular Genetics, The Forsyth Institute.»» Why is this method referred to as direct fluorescent antibody technique?

antibodies in a patient's serum. An example of this indirect method is the diagnostic procedure used for detecting syphilis antibodies in the blood of a patient (FIGURE 22.15). A sample of commercially available syphilis spirochetes is placed on a slide, and the slide is then flooded with the patient's serum. Next, a sample of fluorescein-labeled antiglobulin (antihuman) antibodies is added. These are the antibodies recognizing human antibodies. The slide then is observed using the fluorescence microscope.

The test is interpreted as follows. If the patient's serum contains antisyphilis antibodies, the antibodies bind to the surfaces of spirochetes and the labeled antiglobulin antibodies are attracted to them. The spirochetes then glow from the dye. However, if no antibodies are present in the serum, nothing accumulates on the spirochete's surface, and labeled antiglobulin antibodies also fail to gather on the surface. The labeled antibodies remain in the fluid and the spirochetes do not glow.

The IFA test is adaptable to a broad variety of antigens and antibodies, and are widely used in serology. Antigens may be detected in bacterial smears, cell smears, and viruses fixed to carrier particles.

Radioimmunoassay. An extremely sensitive serological procedure used to measure the concentration of very small antigens, such as haptens, is the **radioimmunoassay (RIA)**. Since its development in the 1960s, the technique has been adapted for quantitating hepatitis antigens as well as reproductive hormones, insulin, and certain drugs. One of its major advantages is that it can detect trillionths of a gram of a substance.

The RIA procedure is based on the competition between radioactive-labeled antigens and unlabeled antigens for the reactive sites on antibody molecules. A known amount of the radioactive

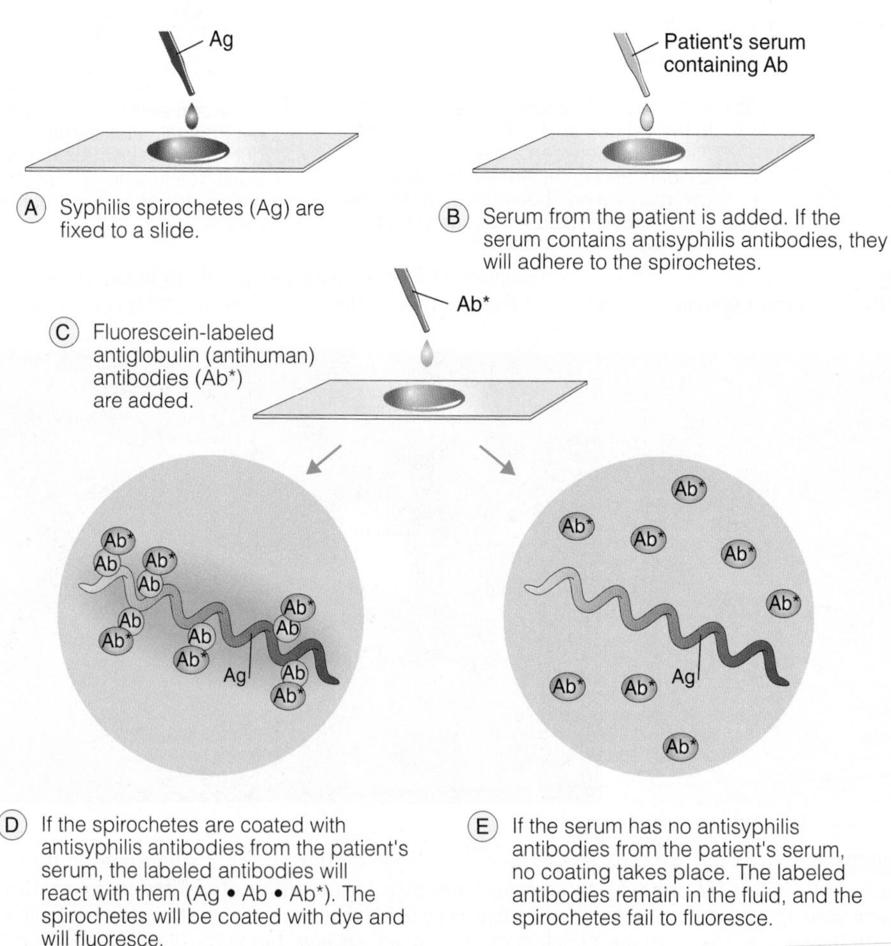

FIGURE 22.15 Indirect Immunofluorescence Antibody Test (IFA) for Diagnosing Syphilis. With the indirect IFA test, a larger visible signal (fluorescence) is observed. »» Why is this method referred to as the indirect fluorescent antibody technique?

(labeled) antigens is mixed with a known amount of specific antibodies and an unknown amount of unlabeled antigens. The antigen-antibody complexes that form during incubation then are separated out, and their radioactivity is determined. By measuring the radioactivity of free antigens remaining in the leftover fluid, one can calculate the percentage of labeled antigen bound to the antibody. This percentage is equivalent to the percentage of unlabeled antigen bound to the antibody because the same proportion of both antigens will find spots on antibody molecules. The concentration of unknown unlabeled antigen then can be determined by reference to a standard curve.

Radioimmunoassay procedures require substantial investment in sophisticated equipment and carry a certain amount of risk because radioactive isotopes are used. For these reasons, the procedure is not routinely used in clinical microbiology laboratories.

The **radioallergosorbent test** (**RAST**) is an extension of the radioimmunoassay. The test may be used, for example, to detect IgE antibodies in the serum of a person possibly allergic to compounds like penicillin.

To detect IgE against penicillin, penicillin antigens are attached to a suitable plastic device, such as a plastic well (FIGURE 22.16). Serum is then added. If the serum contains antipenicillin IgE, it will combine with the penicillin antigens on the surface of the plastic well. Now another antibody—one that will react with human antibodies—is added. This antiglobulin antibody carries a radioactive label. The entire complex, therefore, will become radioactive if the antiglobulin antibody combines with the IgE. By contrast, if no IgE was present in the serum, no reaction with the antigen on the well will take place, and the radioactive antibody will not be attracted to the antigen. When tested, the well will not show radioactivity.

The RAST is commonly known as a "sandwich" technique. There is no competition for an active site as in RIA, and the type of unknown antibody, as well as its amount, may be learned by determining the amount of radioactivity deposited.

Enzyme-Linked Immunosorbent Assay. Another test to detect antigens or antibodies is the **enzyme-linked immunosorbent assay** (**ELISA**), also called the **enzyme immunoassay** (**EIA**). It has virtually the same sensitivity as radioimmunoassay and the RAST, but does not require expensive equipment or involve radioactivity. The procedure involves attaching antibodies or antigens to a solid surface and combining (immunosorbing) the coated

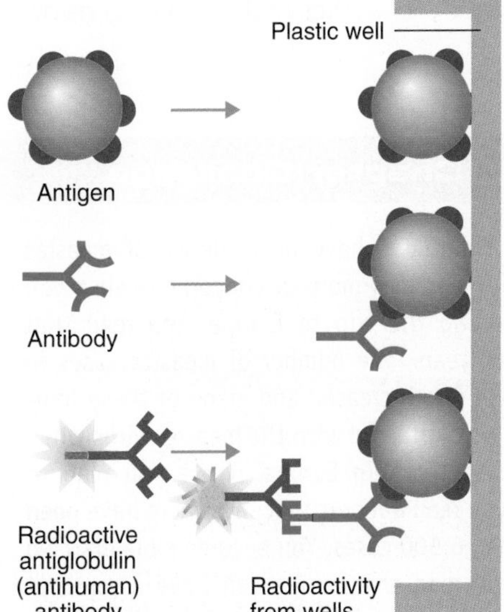

FIGURE 22.16 **The Radioallergosorbent Test (RAST).** This technique provides an effective way of detecting very tiny amounts of antibody in a preparation. A known antigen is used. The objective is to determine whether the complementary antibody is present. »» Would this test represent direct or indirect radioactive antibody labeling? Explain.

surfaces with the test material. An enzyme system then is linked to the complex, the remaining enzyme is washed away, and the extent of enzyme activity is measured. This gives an indication that antigens or antibodies are present in the test material.

An application of the ELISA is found in the highly efficient clinical laboratory test used to detect antibodies against the human immunodeficiency virus (HIV) (FIGURE 22.17). A serum sample is obtained from the patient and mixed with a solution of plastic or polystyrene beads coated with antigens from an HIV antigen. If antibodies to HIV are present in the serum, these "primary antibodies" will adhere to the antigens on the surface of the beads. The beads then are washed and incubated with antiglobulin (antihuman) antibodies chemically tagged with molecules of horseradish peroxidase or a similar enzyme. These antibodies are referred to as the "secondary antibody." The preparation is washed, and a solution of substrate molecules for the peroxidase enzyme is added. Initially the solution is clear, but if enzyme molecules react with the substrate, the solution will become yellow-orange in color. The enzyme molecules will be present only if HIV antibodies are present in the serum. If no HIV antibodies are in the serum, no enzyme molecules could concentrate on the bead surface, no change in the substrate molecules could occur, and no color change would be observed.

ELISA procedures may be varied depending on whether one wishes to detect antigens or antibodies. The solid phase may consist of beads, paper disks, or other suitable supporting mechanisms. Additionally, alternate enzyme systems, such as the alkaline phosphatase system may be used, which produce a different color solution in a positive reaction. In addition, the results of the test may be quantified by noting the degree of enzyme-substrate reactions as a measure of the amount of antigen or antibody in the test sample. The availability of inexpensive ELISA kits has brought the procedure into the doctor's office and serological laboratory. Importantly, a positive ELISA for HIV or for any pathogen being tested only indicates that antibodies to the pathogen are present. A positive ELISA does not necessarily mean the person has the disease. The presence of antibodies only says the person has been exposed to the pathogen. However, in the case of HIV, a positive ELISA probably does indicate an infection, as no one has ever been known to be cured of an HIV infection. MicroInquiry 22 looks at the ELISA test and Textbook Case 22 uses serological tests to detect a case of tickborne encephalitis.

CONCEPT AND REASONING CHECKS 2

a. What does titration accomplish?
b. Explain why the Schick test is an example of neutralization.
c. What do lines of precipitation indicate?
d. Summarize the uses for neutralization, precipitation, and agglutination tests.
e. Describe the uses for the complement fixation test.
f. Assess the need to label antigen or antibody in the RAST and ELISA procedures.

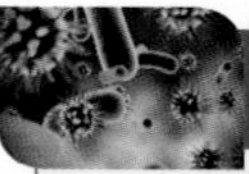

Chapter Challenge B

The fact that you have no evidence of measles immunity becomes more of concern because you are planning the trip to Europe. You read that in recent years, the number of measles cases in Americans has increased and many of these individuals were infected with the measles virus while traveling abroad. In Europe, where you plan to travel, you read a report that says there have been more than 6,500 cases. You become more alarmed when you read that adults can have a rare but serious measles brain inflammation called acute measles encephalitis. The report says that people of all ages should be up to date on MMR vaccination before traveling abroad. Therefore, you wonder: Are there any blood (serological) lab tests that can be used to determine if you have ever had measles or been vaccinated?

QUESTION B:

Identify three serological tests that could be used to determine if you have antibodies to the measles virus? How does each test work?

Answers can be found on the Student Companion Website in **Appendix D.**

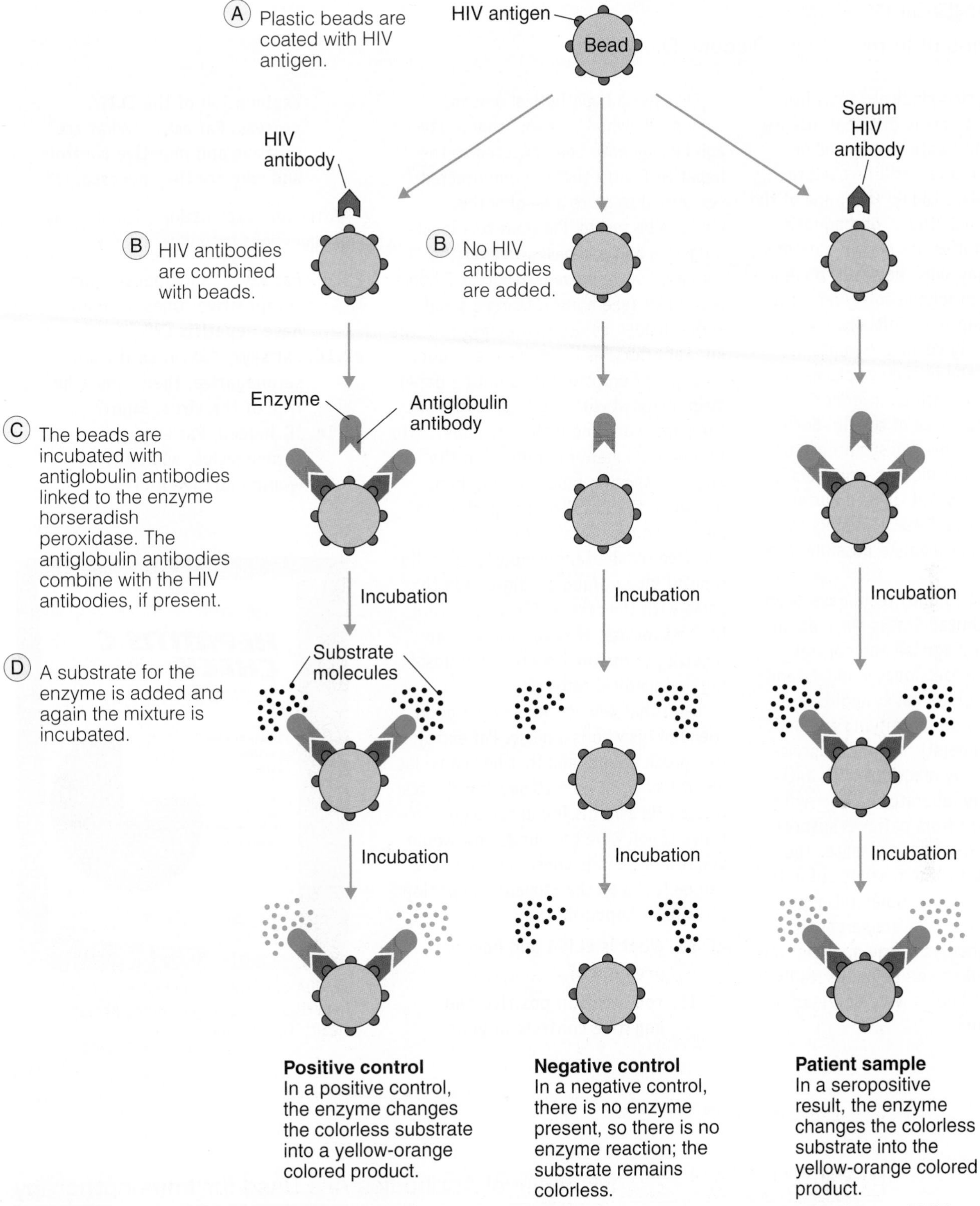

FIGURE 22.17 **The ELISA.** The enzyme-linked immunosorbent assay (ELISA) as it is used to detect antibodies (in this case HIV) in a patient's serum. »» What are positive and negative controls, and why are they needed in the test?

MICROINQUIRY 22

Applications of Immunology: Disease Diagnosis

Ancient Egyptian medical papyri from 1500 BC refer to many different disease symptoms and treatments. Some of these symptoms can still be used to identify diseases today. Thus, one of the most traditional "tools" of diagnosis over the centuries has been a patient's signs and symptoms. However, there are problems when relying solely on these signs and symptoms. Initially, many diseases display common symptoms. For example, the initial symptoms of a hantavirus or anthrax infection are very similar to those of the flu. Some diseases do not display symptoms for perhaps weeks or months—or years in the case of AIDS. Yet, it is important to identify these diseases rapidly so appropriate treatment, if possible, can be started.

Serological (blood) tests have been used in the United States since about 1910 to both diagnose and control infectious disease. Today's understanding of immunology has brought newer tests that rely on identifying antibody-mediated (humoral) immune responses; that is, antibody reactions with antigens. Serology laboratories work with serum or blood from patients suspected of having an infectious disease. The lab tests look for the presence of antibodies to known microbial antigens. Serological tests that are seropositive indicate antibodies to the microbe were detected, while a seronegative result means no antibodies were detected in a patient's serum.

Let's use a hypothetical person, named Pat, who "believes" that 7 days ago he may have been exposed to the hepatitis C virus through unprotected sex. Afraid to go to a neighborhood clinic to be tested, Pat goes to a local drugstore and purchases an over-the-counter, FDA-approved, hepatitis C home testing kit (see figure). Using a small spring-loaded device that comes with the kit, Pat pricks his little finger and puts a couple of drops of blood onto a paper strip included with the kit. He fills out the paperwork and mails the paper strip in a prepaid mailer (supplied in the kit) to a specific blood testing facility where the ELISA test is performed. In 10 business days, Pat can call a toll-free number anonymously, identify himself by a unique 14-digit code that came with the testing kit, and ask for his test results. If necessary, he can receive professional post-test counseling and medical referrals.

Later that day, Pat runs into a good friend of his who is a nurse. Pat explains his "predicament" and that he is anxious about having to wait 10 days for the test results. He asks his friend some questions. If you were the nurse, how would you respond to his questions? Answers can be found on the Student Companion Website in **Appendix D**.

22.1a. What is ELISA and how is it performed?

22.1b. You mention positive and negative controls in your explanation of the ELISA process. Pat asks, "What are positive and negative controls and why are they necessary?"

After your explanations, Pat still has more questions for you to answer.

22.1c. Pat says, "So suppose I am seropositive. Does it mean I have hepatitis C?"

22.1d. Pat says, "Okay, so if I am seronegative, then I must be free of the virus. Right?"

22.1e. If, indeed, Pat was seronegative, what advice would you give him?

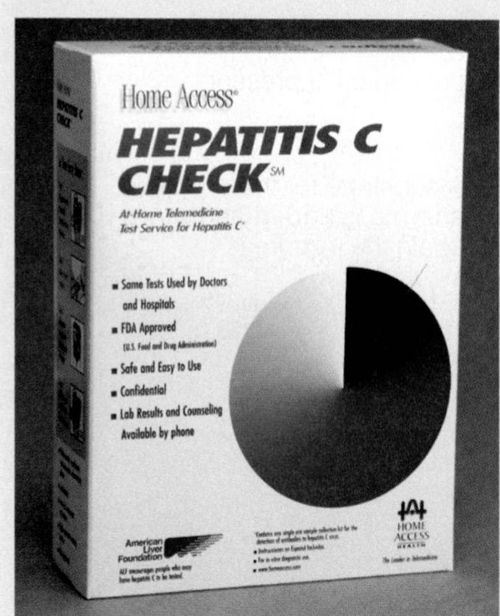

An FDA-approved home testing kit for hepatitis C. Courtesy of Home Access Health Corporation.

KEY CONCEPT 22.3 Monoclonal Antibodies Are Used for Immunotherapy

The variety of laboratory tests used to diagnose infectious disease continues to expand as new biochemical and molecular techniques are adapted to special clinical situations. Monoclonal antibodies represent one of the promising newer developments.

Monoclonal Antibodies Are Becoming a "Magic Bullet" in Biomedicine

Usually pathogens contain several different epitopes. Therefore, when an infection occurs, a different B-cell population is activated for each epitope

Textbook CASE 22

A Case of Tickborne Encephalitis

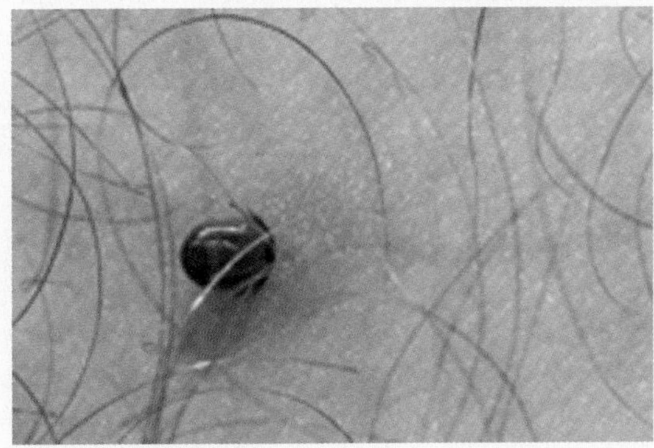

Ixodes ricinus taking a blood meal on the human skin. © Graham Ella/Alamy.

1 On July 11, 2001, a 57-year-old man in previous good health came to a Utah hospital emergency room complaining of fever and right arm tremors. Further examination indicated a fever of 39.3°C (102.7°F), an excessively rapid heartbeat, and right-side rigidity. He appeared disoriented and showed a lack of mental alertness. He was immediately admitted into the hospital.

2 On questioning the patient, he indicated he had traveled to eastern Russia in June. Further questioning discovered that the patient had been bit multiple times by ticks. On June 26, the patient had developed fever, muscle pain, and nausea, but the symptoms improved with a broad-spectrum antibiotic.

3 On July 7, the patient developed headache, a stiff neck, and confusion, which became worse over the next few days. The worsening symptoms are what triggered the patient's return to the hospital on July 11.

4 A serum sample was taken. Examination of a cerebrospinal fluid sample showed abnormal numbers of lymphocytes ($112/mm^3$; normal = $0–5/mm^3$) and elevated protein (84 mg/dl; normal = 15–45 mg/dl). These findings were indicative of an infectious condition, so further microbiological testing was done. Bacterial cultures were negative, as were herpes simplex virus and enterovirus detection by PCR.

5 An MRI was ordered, which revealed left-sided cerebral edema with a restriction in blood supply in the thalamus and striatum. A diagnosis of encephalitis of unknown cause was made and treated with broad-spectrum antibiotics and corticosteroids. The patient exhibited no fever and demonstrated normal mental status by the next day.

6 The patient was released after 14 days with residual right-sided weakness and impaired cognition. Although the patient's motor skills returned within 6 months, he continued to have mild cognitive impairment.

7 Serum collected on admission tested positive for tickborne encephalitis virus (TBEV) IgM and TBEV-specific neutralizing antibodies. Serological tests for other arboviruses [St. Louis encephalitis virus and West Nile virus] and pathogens [*Borrelia burgdorferi* (Lyme disease), *Leptospira* (leptospirosis), and *Rickettsia* (typhus, Rocky Mountain spotted fever; RMSF)] were negative.

Questions:

(Answers can be found on the Student Companion Website in **Appendix D.**)

A. What might the right arm tremors, right-sided rigidity, disorientation, and lack of mental alertness indicate?

B. For the patient, a diagnosis of encephalitis of unknown cause was made. What was missed such that the encephalitis was "of unknown cause"?

C. Tickborne encephalitis is called a biphasic disease, meaning the symptoms occur in two waves. Describe the biphasic nature of tickborne encephalitis.

D. Serum collected on admission tested positive for tickborne encephalitis virus (TBEV) IgM and TBEV-specific neutralizing antibodies. What does the presence of IgM antibodies indicate and how would the presence of neutralizing antibodies identify TBEV?

E. What types of serological tests might have been done to rule out the other arboviruses and pathogens?

For additional information see www.cdc.gov/mmwr/preview/mmwrhtml/mm5911a3.htm.

and different antibodies specific to each of the epitopes are produced. Serum from such a patient would contain **polyclonal antibodies** because the antibodies are derived from several different clones of B cells. However, in the laboratory it is possible to produce populations of identical antibodies that bind to only one epitope.

In 1975, César Milstein of Argentina and Georges J. F. Köhler of then West Germany developed a method for the laboratory production of **monoclonal antibodies (MAbs)**; that is, antibodies recognizing only one epitope. The key was using **myelomas**, cancerous tumors that arise from the uncontrolled division of plasma cells. Although these myeloma cells have lost the ability to produce antibodies, they can be grown indefinitely in culture.

There are two basic steps to generating MAbs. In the first step, a mouse is injected with the antigen of interest (FIGURE 22.18), triggering humoral immunity. Antibody-secreting plasma cells are then removed from the mouse's spleen and mixed with myeloma cells from a cell culture. By forcing the fusion of plasma cells with myeloma cells, hybrid cells are formed. These so-called **hybridomas** can multiply indefinitely (a myeloma-cell characteristic), producing antibodies recognizing but one epitope (a B-cell characteristic).

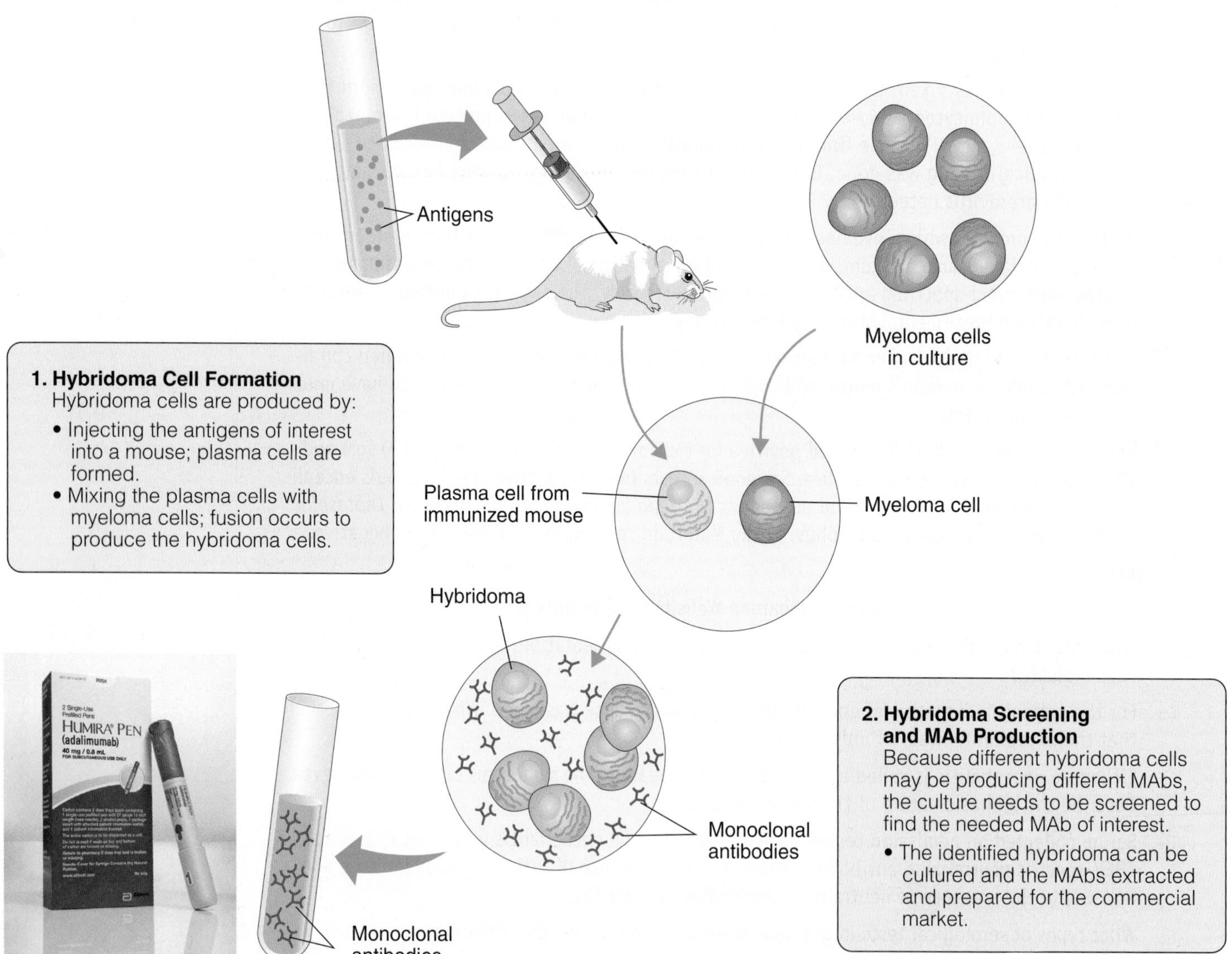

FIGURE 22.18 **The Production of Monoclonal Antibodies.** The production of a monoclonal antibody (MAb) requires fusing a myeloma cell with a plasma cell. After screening for the MAb needed, those clones can be cultivated to produce more of the same MAb. (inset) Humira® was the first fully human MAb approved for use in the United States. Image courtesy of Abbott Laboratories. »» What are the unique features of the plasma cell and the myeloma cell?

Because the antigen originally injected may have several epitopes, different hybridomas may be producing different antibodies. Therefore, the second step is to screen each hybridoma clone for the synthesis of the desired antibody. The hybridoma producing the desired MAb can then be propagated (cloned) in tissue culture.

■ Screen: Referring to testing or examining something for a particular characteristic or disease.

Currently, more than 100 MAbs are in drug clinical trials and some 20 have been approved for use in the United States. So far, the most prominent MAb targeted at infectious disease is palivizumab (Synagis®), which is designed to prevent respiratory syncytial virus infections in high-risk groups. Monoclonal antibodies also are in the pipeline for treating hepatitis B and C infections.

Monoclonal antibodies also are being used for **immunotherapy**, which is controlling overactive inflammatory responses. In fact, the first truly human MAb product, called adalimumab (Humira®), is directed against inflammatory responses resulting in rheumatoid arthritis.

Perhaps most prominent are those MAb drugs targeted at some forms of cancer. Monoclonal antibody products have been approved for treating malignant lymphoid tumors, leukemia, and both colorectal and pancreatic cancers. Many additional products are in clinical trials.

More recent developments have devised ways to add a "payload" to a MAb. For example, scientists can link a toxin onto the Fc fragment of a MAb. When the antibodies are injected back into the patient, they act like "stealth missiles." They react specifically with the tumor cells, and the toxin kills the cells without destroying other tissue cells.

The use of MAbs represents one of the most elegant expressions of modern biotechnology—and their potential uses in the clinic and medicine remain to be fully appreciated.

CONCEPT AND REASONING CHECKS 3

1. Explain how monoclonal antibodies are produced.

Chapter Challenge C

All the serological tests to detect measles antibodies were negative. Therefore, it appears you did not ever have measles. You still plan to go to Europe though.

QUESTION C:

To be on the safe side, before traveling to Europe, what should you do to ensure you will be protected should you be exposed to someone infected with the measles virus?

Answers can be found on the Student Companion Website in **Appendix D**.

■ **In conclusion**, we have now seen how vaccines work to protect an individual from infection by many human pathogens. It would seem that we should be less susceptible to disease as more vaccines are developed and made available. In fact, it was through a worldwide vaccine campaign in the 1960s and 1970s that led to the eradication of smallpox, the first (and only) human pathogen that medicine has been able to eradicate. But here is an interesting twist. Can removal of one pathogen through global vaccination produce the opportunity for other pathogens to flourish? In 2010, researchers at Virginia's George Mason University suggested that the smallpox vaccine gave some protection against HIV and the development of AIDS. They propose that with the eradication of smallpox, no one is being vaccinated against the pathogen and this has, inadvertently, helped boost the AIDS pandemic. In tests with white blood cells, the investigators report that the smallpox vaccine somehow interfered with the ability of HIV to replicate, cutting its replication five-fold. A link? Both the smallpox virus and HIV use the same surface receptor on white blood cells for attachment and host cell entry.

■ SUMMARY OF KEY CONCEPTS

22.1 Immunity to Disease Can Be Generated Naturally or Artificially

1. In **naturally acquired active immunity** antibodies are produced, and lymphocytes activated, in response to an infectious agent. Immunity arises from either contracting the disease and recovering, or from a subclinical infection. **Artificially acquired active immunity** is when antibodies and lymphocytes are produced as a result of a **vaccination** and immunity is established for some length of time. (Fig. 22.2)
2. There are several strategies for producing vaccines, including **attenuated**, **inactivated**, **toxoid**, **subunit**, and **conjugate vaccines**.

3. Adult vaccines can return immunological memory and immune cells to protective levels. (Fig. 22.5)
4. **Naturally acquired passive immunity** comes from the passage of antibodies from the mother to fetus, or mother to newborn through **colostrum**, which confers immunity for a short period of time. **Artificially acquired passive immunity** is when one receives an **antiserum** (antibodies) produced in another human or animal. This form confers immunity for a short period of time. Serum sickness can develop if the recipient produces antibodies against the antiserum.
5. **Herd immunity** results from a vaccination program that lowers the number of susceptible members within a population capable of contracting a disease. If there are few susceptible individuals, the probability of disease spread is minimal. (Fig. 22.7)
6. Established vaccines are quite safe, although some people do experience side effects that may include mild fever, soreness at the injection site, or malaise after the vaccination. Few people suffer serious consequences.

22.2 Serological Reactions Can Be Used to Diagnose Disease

7. Serological reactions consist of antigens and antibodies (serum). Successful reactions require the correct dilution (**titer**) of antigens and antibodies. (Fig. 22.8)
8. **Neutralization** is a serological reaction in which antigens and antibodies neutralize each other. Often there is no visible reaction, so injection into an animal is required to see if the reaction has occurred. (Fig. 22.9)
9. In a **precipitation** reaction, antigens and antibodies react to form a matrix that is visible to the naked eye. Different microbial antigens can be detected using this technique. (Fig. 22.10)
10. **Agglutination** involves the clumping of antigens. The cross-linking of antigens and antibodies causes the complex to clump. Some infectious agents can be detected readily by this method. (Fig. 22.12)
11. **Complement fixation** can detect antibodies to a variety of pathogens. This serological method involves antigen-antibody complexes that are detected by the fixation (binding) of complement. (Fig. 22.13)
12. The **fluorescent antibody technique** involves the addition of fluorescently tagged antibodies to a known pathogen to a slide containing unknown antigen (pathogen). If the antibody binds to the antigen, the pathogen will glow (fluoresce) when observed with a fluorescence microscope, and the pathogen is identified. The **radioimmunoassay** (**RIA**) uses radioactivity to quantitate radioactive antigens through competition with nonradioactive antigens for reactive sites on antibody molecules. The **enzyme-linked immunosorbent assay** (**ELISA**) can be used to detect if a patient's serum contains antibodies to a specific pathogen, or to detect or measure antigens in serum. A positive ELISA is seen as a colored reaction product. (Figs. 22.14–22.17)

22.3 Monoclonal Antibodies Are Used for Immunotherapy

13. **Monoclonal antibodies** result from the fusion of a specific B cell (plasma cell) with a **myeloma cell** to produce a **hybridoma** that secretes an antibody recognizing a single epitope. (Fig. 22.18)

CHAPTER SELF-TEST

For **STEPS A–D**, answers to even-numbered questions and problems can be found in **Appendix C** on the Student Companion Website at **http://microbiology.jbpub.com/10e**. In addition, the site features eLearning, an online review area that provides quizzes and other tools to help you study for your class. You can also follow useful links for in-depth information, read more MicroFocus stories, or just find out the latest microbiology news.

STEP A: REVIEW OF FACTS AND TERMS

Multiple Choice

Read each question carefully, then select the ***one*** answer that best fits the question or statement.

1. Exposure to the flu virus, contracting the flu, and recovering from the disease would be an example of ______.
   A. artificially acquired passive immunity
   B. naturally acquired active immunity
   C. artificially acquired active immunity
   D. naturally acquired passive immunity
2. An attenuated vaccine contains ______.
   A. inactive toxins
   B. living, but slow growing (replicating) antigens
   C. killed bacteria
   D. noninfective antigen subunits
3. Which one of the following vaccinations would most likely require booster shots periodically throughout life?
   A. Tetanus
   B. Measles
   C. Hepatitis B
   D. Mumps
4. Immune complex formation and serum sickness are dangers of ______.
   A. artificially acquired passive immunity
   B. naturally acquired active immunity
   C. artificially acquired active immunity
   D. naturally acquired passive immunity
5. Herd immunity is affected by ______.
   A. the percentage of a population that is vaccinated
   B. the strength of an individual's immune system
   C. the number of susceptible individuals
   D. All the above (**A–C**) are correct.

6. Approximately _____ of 100,000 vaccinated individuals are likely to suffer a serious reaction to the vaccination.
   A. 1
   B. 50
   C. 100
   D. 500
7. Titer refers to ______.
   A. the most concentrated antigen-antibody concentration showing a reaction
   B. the first diluted antigen-antibody concentration showing a reaction
   C. the precipitation line formed between an antigen-antibody reaction
   D. the most dilute antigen-antibody concentration showing a reaction
8. ______ is a serological reaction that produces little or no visible evidence of a reaction.
   A. Precipitation
   B. ELISA
   C. Neutralization
   D. Agglutination
9. The serological reaction where antigens and antibodies form an extensive lattice of large particles is called ______.
   A. fixation
   B. precipitation
   C. neutralization
   D. agglutination
10. When antigens are attached to the surface of latex beads and then reacted with an appropriate antibody, a/an ______ reaction occurs.
    A. inhibition
    B. agglutination
    C. neutralization
    D. precipitation
11. What serological test requires sheep red blood cells and a preparation of antibodies that recognizes the sheep red blood cells?
    A. ELISA
    B. Radioimmunoassay
    C. Immunodiffusion
    D. Complement fixation test
12. In an ELISA, the primary antibody represents ______.
    A. the patient's serum
    B. the antibody recognizing the secondary antibody
    C. the enzyme-linked (labeled) antibody
    D. the antibodies having been washed away
13. A hybridoma cell ______.
    A. secretes monoclonal antibodies
    B. presents antigens on its surface
    C. secretes polyclonal antibodies
    D. is an antigen-presenting cell

Completion

On completing Section 22.1 (Immunity to Disease Can Be Generated Naturally or Artificially), test your comprehension of the section's contents by filling in the following blanks with two terms that answer the description best.

14. Two general forms of immunity: ________________ and ________________.
15. Two types of natural immunity: ________________ and ________________.
16. Two diseases that MMR is used against: ________________ and ________________.
17. Two diseases that DPT is used against: ________________ and ________________.
18. Two types of passive immunity: ________________ and ________________.
19. Two names for antibody-containing serum: ________________ and ________________.
20. Two ways newborns have acquired maternal antibodies: ________________ and ________________.
21. Two types of viruses in viral vaccines: ________________ and ________________.
22. Two bacterial diseases for which toxoids are used: ________________ and ________________.
23. Two methods for administering vaccines: ________________ and ________________.
24. Two functions of antibodies in antiserum: ________________ and ________________.
25. Two bacterial diseases where passive immunity is used: ________________ and ________________.

STEP B: CONCEPT REVIEW

After understanding the textbook reading, you should be capable of writing a paragraph that includes the appropriate terms and pertinent information to answer the objective.

26. Identify how **artificially acquired active immunity** produces immunity. (**Key Concept 1**)
27. Differentiate between vaccines consisting of live, **attenuated** antigens and those containing **inactivated** antigens. Provide examples for each. (**Key Concept 1**)
28. Explain why **booster shots** are needed for some vaccines. (**Key Concept 1**)
29. Distinguish between **naturally acquired passive immunity** and **artificially acquired passive immunity.** (**Key Concept 1**)
30. Assess the importance of **herd immunity** to community and national health. (**Key Concept 1**)
31. Judge the safety of vaccines and the reporting system to identify possible adverse affects. (**Key Concept 1**)
32. Identify the role of **titration** in identifying disease. (**Key Concept 2**)
33. Compare and contrast (a) **neutralization**, (b) **precipitation**, and (c) **agglutination** as examples of serological reactions. (**Key Concept 2**)
34. Explain how the **complement fixation test** works. (**Key Concept 2**)
35. Differentiate between **fluorescent antibody**, **RAST**, and **ELISA** as labeling methods to detect antigen-antibody binding. (**Key Concept 2**)
36. Summarize the characteristics of **monoclonal antibodies** and assess their role in identifying infectious disease. (**Key Concept 3**)

STEP C: APPLICATIONS AND PROBLEMS

37. A friend is heading on a trip to a foreign country where hepatitis A outbreaks are quite common. For passive immunity, he gets a shot of an antiserum containing IgG antibodies. Why do you suppose IgM is not used, especially since the immunoglobulins are the important components of the primary antibody response?

38. A complement fixation test is performed with serum from a patient with an active case of syphilis. In the process, however, the technician neglects to add the syphilis antigen to the tube. Would lysis of the sheep red blood cells occur at the test's conclusion? Why?

39. Suppose the titer of mumps antibodies from your blood was higher than that for your fellow student. What are some of the possible reasons that could have contributed to this? Try to be imaginative on this one.

40. Given a choice, which of the four general types of adaptive immunity would it be safest to obtain? Why?

41. An immunocompromised child is taken by his mother to the local clinic to receive a belated measles-mumps-rubella (MMR) vaccination. As the nurse practitioner, you decline to give the child the vaccine. Explain to the mother why you refused to vaccinate the child.

STEP D: QUESTIONS FOR THOUGHT AND DISCUSSION

42. It is estimated that when at least 90% of the individuals in a given population have been immunized against a disease, the chances of an epidemic occurring are very slight. The population is said to exhibit "herd immunity," because members of the population (or herd) unknowingly transfer the immunizing agent to other members and eventually immunize the entire population. What are some ways by which the immunizing agent can be transferred?

43. When children are born in Great Britain, they are assigned a doctor. Two weeks later, a social services worker visits the home, enrolls the child on a national computer registry for immunization, and explains immunization to the parents. When a child is due for an immunization, a notice is automatically sent to the home, and if the child is not brought to the doctor, the nurse goes to the home to learn why. Do you believe a method similar to this can work (or should be used) in the United States to achieve uniform national immunization?

44. Since 1981, the incidence of pertussis has been increasing in the United States, with the greatest increase found in adolescents and adults. Why are adolescents and adults targets of the bacterial pathogen, considering these individuals were usually considered immune to the disease?

45. Children between the ages of 5 and 15 are said to pass through the "golden age of resistance" because their resistance to disease is much higher than that of infants and adults. What factors may contribute to this resistance?

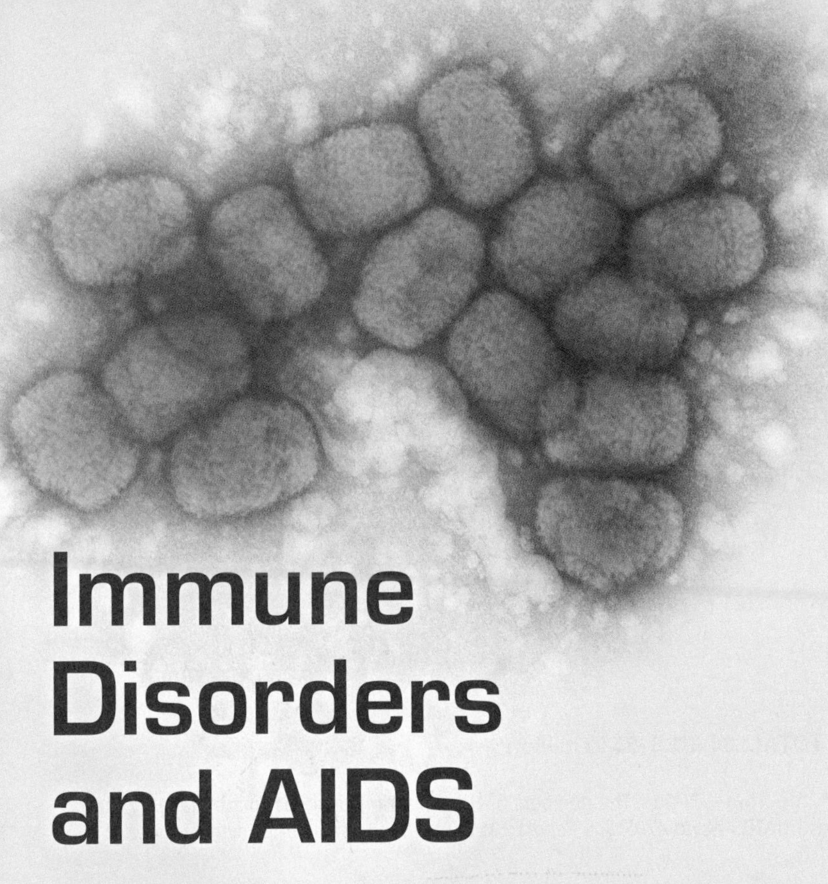

Immune Disorders and AIDS

23

CHAPTER PREVIEW

23.1 Type I Hypersensitivity Represents a Familiar Allergic Response

Investigating the Microbial World 23.1: Microbes and the Allergic Response

MicroInquiry 23: Allergies, Microbes, and the Hygiene Hypothesis

23.2 Other Types of Hypersensitivities Represent Immediate or Delayed Reactions

23.3 Autoimmune Disorders and Transplantation Are Immune Responses to "Self"

23.4 Immunodeficiency Disorders Can Be Inherited or Acquired

Textbook Case 23: *Pneumocystis* Pneumonia

Investigating the Microbial World 23.2: How Does HIV Kill T Cells?

I don't think we are losing the war, but we're certainly not finished with the war.
—Ronald Valdiserri, Deputy Director of the National Center for HIV, STD, and TB Prevention

2011 marked 30 years into the pandemic of acquired immunodeficiency syndrome (AIDS) and health officials are still reporting little progress toward finding a cure for the disease. The Joint United Nations Programme on HIV/AIDS (UNAIDS) reported there are some 33 million people living with AIDS (FIGURE 23.1)—only about half know they are infected—and between 25 million and 35 million have died of AIDS since 1981. UNAIDS also reported that worldwide about 7,000 people still become infected with HIV each day—that is one million new infections every 142 days.

Although these numbers are still staggering and a cure seems distant, in 2012 more positive news was reported on the AIDS pandemic. Globally, between 2001 and 2009 there was a 25% reduction in cases per year with some countries reporting drops of new infections by 35% to 50%. These are the lowest levels of infection since 1997. The largest reductions have been in sub-Saharan Africa and Southeast Asia. Unfortunately, Eastern Europe, the Middle East, and North Africa have reported increases in infection rates. Almost 6.6 million individuals in developing nations are now on antiretroviral therapy (out of some 16 million who are eligible for treatment), which is up almost 22 times from 2001. Further, in 2011 more than 420,000 children were on antiretroviral therapy, which is up 50% from 2009. In all, some 5 million lives have been saved with the therapy. Still, for every one person put on antiretrovirals, two new people become infected and UNAIDS

Image courtesy of Dr. Fred Murphy/CDC.

FIGURE 23.1 **Adults and Children Living with HIV Infection—2011.** The numbers of HIV infections is an estimate made by the Joint United Nations Programme on HIV/AIDS. Source: UNAIDS *World AIDS Day Report* 2011. »» Provide some reasons why these numbers represent only estimates.

reports that 20% of patients on drug therapy stop taking the drugs within 1 year.

Although there is no safe and effective vaccine yet developed for AIDS, research studies show that people who adhere to their antiretroviral regimen are 96% less likely to retransmit the disease to sexual partners. This also means that HIV-positive pregnant mothers on antiretroviral therapy are less likely to transmit the virus to their newborns; in fact, compared to 2001, there were 26% fewer newborns infected with HIV in 2009 from infected mothers on antiretroviral therapy.

Such reductions in AIDS infections, the spread of antiretroviral drug therapy, and the ability of such therapy to severely stop the transmission of the disease has led UNAIDS in November 2011 to proclaim that for AIDS "The end is in sight." However, that end depends on rapidly scaling up preventative programs globally because most of the 40% on antiretroviral therapy are in developing nations. In addition, another 16 million individuals who are infected have no symptoms and have not been put on the eligible list for receiving treatment. UN Secretary-General Ban Ki-moon wants to have 13 million people on antiretrovirals by 2015.

In the end, even if a highly successful vaccine is not discovered, it is quite possible that a combination of vaccine and antiviral drugs may just do the trick—and eradicate HIV.

Previous chapters examined the host defenses (innate and adaptive immune responses) that continually try to protect the body from, among other things, infectious diseases. These responses work together to provide a powerful and intricate system that usually works without fail. However, sometimes due to genetic errors or an infection, the immune system defenses do not function properly, leading to some type of immune disorder.

Although AIDS is a major topic in this chapter, it is by no means the only immune disorder. Hypersensitivities and allergies represent immune disorders that many believe can be at least partly the result of children not fighting the types of infectious diseases experienced by past generations. Also included in the broad category of immune disorders are the autoimmune diseases that often have a microbial component associated with them. We survey each type of immune disorder in the following sections.

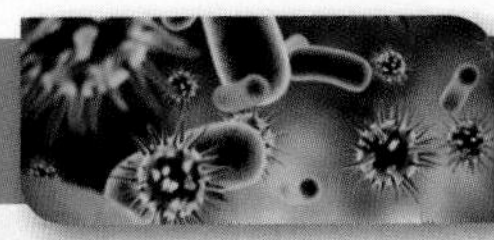

Chapter Challenge

Carla is a 20-year-old long distance runner for her college track team. Recently she has started to develop what appears to be an allergic reaction to many unrelated events. In high school when she was 17 years old, she began to develop hives all over her skin surface every time she ran. Hives, also called urticaria, is a sudden outbreak of swollen, pale red bumps (wheals and flares) on the skin that most often are an allergic reaction to a drug or food item. But exercise? For Carla, the itchy bumps would disappear in a couple of hours, but would reappear every time she ran. Now in college, the same reaction still occurs. What has made her more concerned is that while taking an anatomy and physiology exam, hives started appearing all over her body, just like when she ran. But this time they lasted for 5 months and itched intensely. What is going on with Carla's body? Is this some type of rare allergy caused by her immune system? Let's find out as we progress through this chapter on immune disorders.

KEY CONCEPT 23.1 Type I Hypersensitivity Represents a Familiar Allergic Response

Hypersensitivity is a multistep phenomenon triggered by exposure to substances like pollen, peanuts, or jewelry. It consists of a dormant (latent) stage, during which an individual becomes sensitized, and an active (hypersensitive) stage following a subsequent exposure to the same substance. The process may involve elements of humoral or cell-mediated immunity, or sometimes both.

Hypersensitivities can be classified into one of four types. **Immediate hypersensitivity** refers to an immune reaction that is experienced within minutes to 24 hours after exposure to the antigen. Immediate hypersensitivities include (FIGURE 23.2):

- Type I IgE-mediated hypersensitivity, which is a process involving IgE, mast cells, basophils, and cell mediators inducing smooth muscle contraction;
- Type II cytotoxic hypersensitivity, which involves IgG, IgM, complement, and the destruction of host cells; and
- Type III immune complex hypersensitivity, which involves IgG, IgM, complement, and the formation of antigen-antibody immune complexes in the tissues.

The fourth type is a **delayed hypersensitivity**, where a cell-mediated immune response develops over 2 to 3 days. This type IV cellular hypersensitivity involves cytokines and T lymphocytes.

Realize that a hypersensitivity is an exaggerated or inappropriate immune defense that is causing the problems in an affected individual.

Type I Hypersensitivity Is Induced by Allergens

A **type I hypersensitivity** is an exaggerated (hypersensitive) or inappropriate immune defense. For pathogens (e.g., viruses, bacterial, fungal, and protist cells), this involves the direct killing of the infectious agent and the destruction of infected cells. Besides IgG and IgM antibodies, several types of immune cells, especially helper T cells and cytotoxic T cells are involved. The immune system is also primed to defend against macroparasites, such as the helminths and ticks. Here, because of their size, these invaders are harder to kill immunologically, so the immune system uses expulsion as the best defense. Although most individuals in the developed world seldom experience these animal parasites, IgE and IgG antibodies and other groups of white blood cells, including eosinophils, mast cells, and basophils, stand ready to respond.

One idea proposed for type I hypersensitivities is that noninfectious, but often noxious, chemical substances we are exposed to in the environment also must be expelled. These antigens, called **allergens**, include insect venoms, mosquito and tick salivary excretions, and irritants like dust. Therefore, an allergic reaction to an allergen

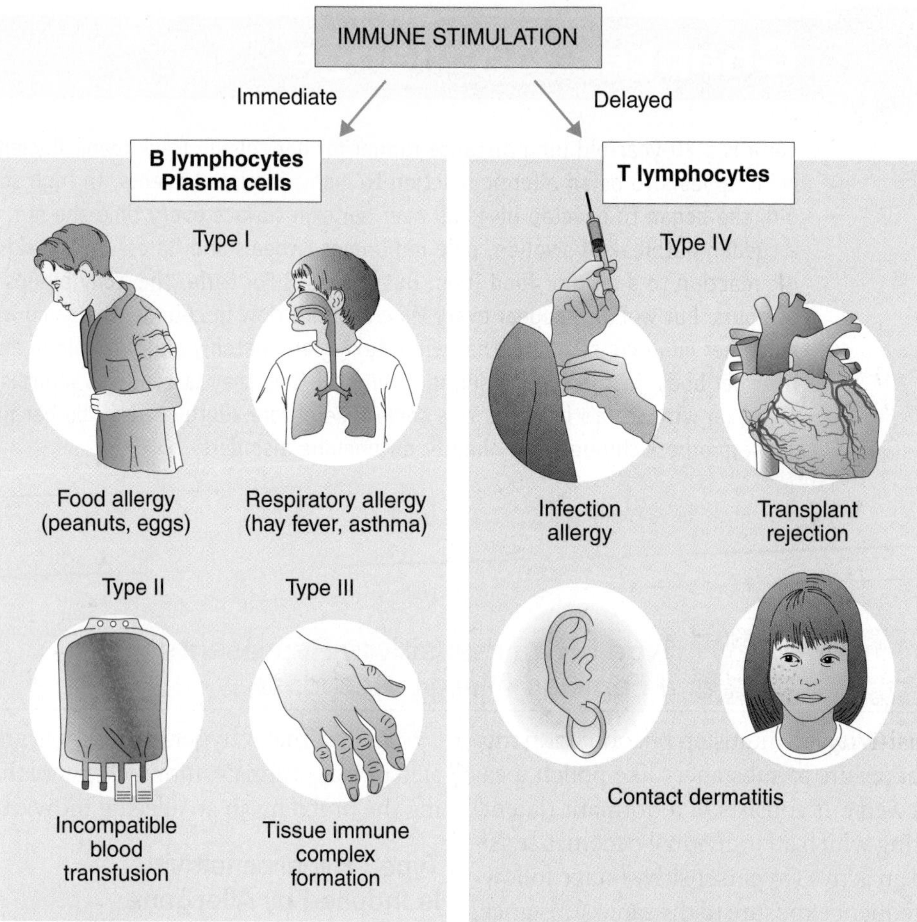

FIGURE 23.2 **The Various Forms of Hypersensitivities.** Immune system hypersensitivities may be related to B lymphocytes or T lymphocytes. Stimulation of the immune system is the starting point for all these disorders. »» Why are all these examples of immune stimulation referred to as hypersensitivities?

provokes a similar response; one experiences the "expulsion responses" of coughing, sneezing, vomiting, and itching that go along with an allergy attack. However, if appropriately targeted, the allergic reaction is beneficial and provides protection from the allergen. For example, an allergic response to bee venom may have evolved as a protection against the sometimes severe tissue damaging and life-threatening effects caused by enzymes in the venom. Unfortunately, an overreaction, if triggered systemically, can also be life-threatening.

But what about allergens like plant pollens? Certainly they are not noxious or dangerous substances. As mentioned above, a complete understanding of the hypersensitivity response remains elusive.

A type I hypersensitivity has all the characteristics of a humoral immune response. It begins with the recognition of an **allergen**, which triggers an allergic reaction in the body called an **allergy**. The allergen may be any of a wide variety of materials such as plant pollen, certain foods, bee venom, serum proteins, or a drug, such as penicillin. Doses of allergen as low as 0.001 mg have been known to sensitize a person. Allergists refer to this first exposure as the **sensitizing dose**.

The immune system responds to the allergen as if it was a dangerous antigen, such as a pathogen. The allergen is taken up by antigen-presenting cells (APCs) and fragments presented to **helper T2 (T_H2) cells** (FIGURE 23.3). As in humoral immunity, T_H2 cells stimulate B cells to mature into plasma cells, which produce IgE antibodies. The antibodies enter the circulation and attach by the Fc segment to the surface of mast cells and basophils. **Mast cells** are connective tissue cells found in the respiratory and

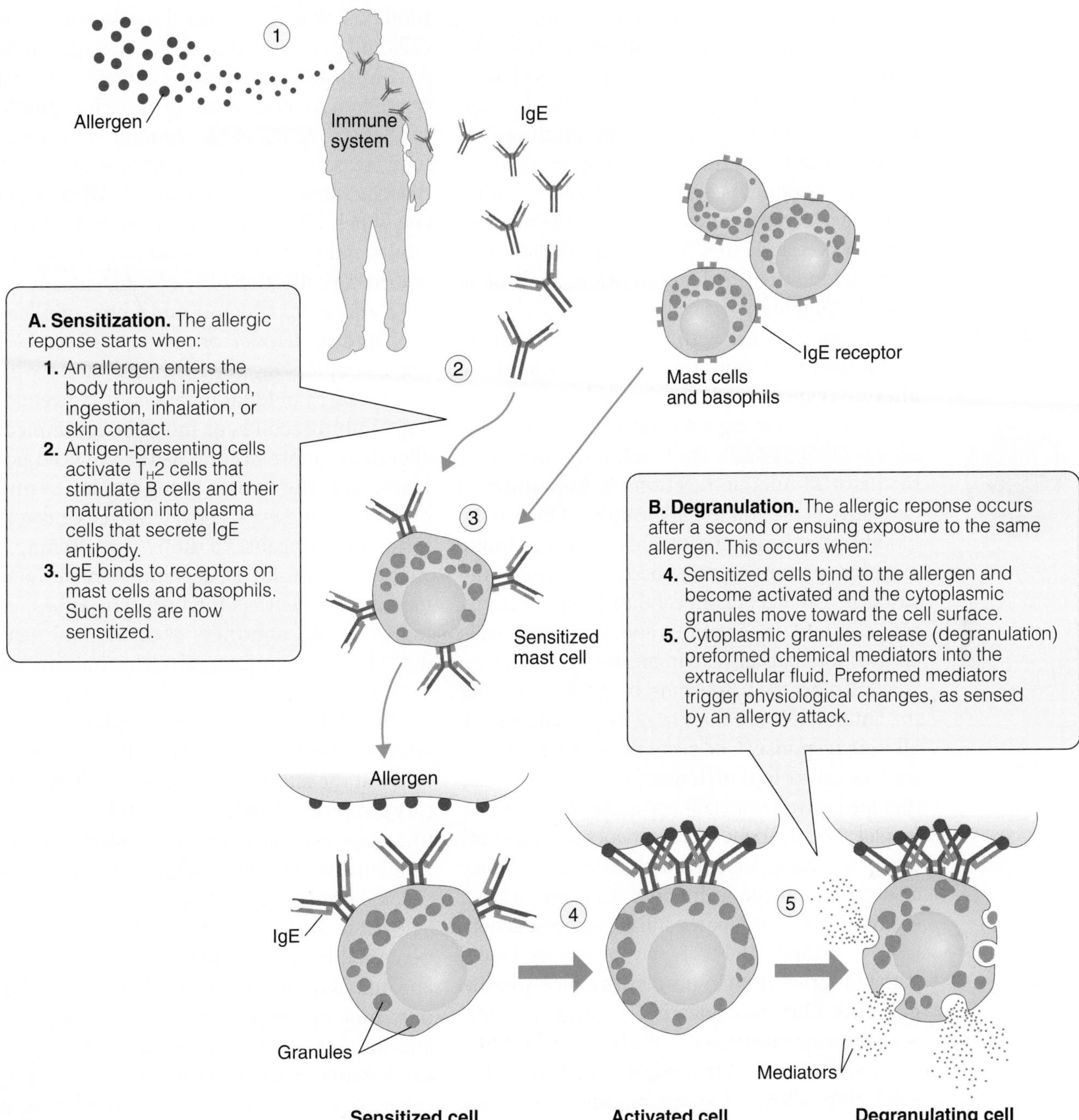

FIGURE 23.3 **Type I IgE-Mediated Hypersensitivity.** Basophils, mast cells, and IgE antibodies are involved in typical allergies like hay fever. »» Does a person experience the symptoms of an allergy on the first or ensuing exposure to an allergen? Explain.

gastrointestinal tracts, and near the blood vessels. They measure about 10 μm to 15 μm in diameter and are filled with 500 to 1,500 granules containing histamine and other physiologically active substances.

Basophils are circulating leukocytes in the blood that also are rich in granules. They represent about 1% of the total leukocyte count in the circulation and measure about 15 μm in diameter. Mast cells and basophils each have over 100,000 receptor sites where IgE antibodies can attach by the Fc segment. As IgE antibodies attach to mast cells and basophils, the individual becomes "sensitized." Multiple stimuli by allergen molecules may be required to sensitize a person fully. This is why penicillin often must be taken several times before a penicillin allergy manifests itself.

Sensitization usually requires a minimum of 1 week, during which time millions of molecules of IgE attach to thousands of mast cells and basophils. Because attachment occurs at the Fc fragment of the antibody, the allergen binding sites point outward from the cell (see Figure 23.3).

On subsequent exposure to the same allergen, the allergen molecules bind to the allergen binding sites and cross-link IgE antibodies. This cross-linking triggers **degranulation**, a release of granule contents at the cell surface.

Website Animation: Type 1 Hypersensitivity

As granules fuse with the plasma membrane of the basophils and mast cells, an **immediate allergic response** releases a number of chemical mediators having substantial pharmacologic activity (FIGURE 23.4). The best known preformed mediator of allergic reactions is **histamine**, a derivative of the amino acid histidine. Once in the bloodstream, histamine attaches to the histamine receptors present on most body cells, triggering vascular (capillary) permeability, tissue swelling, redness, mucus secretion, and smooth muscle contraction that can lead to breathing difficulties.

Still other mediators must be synthesized after the antigen-IgE reaction, triggering a **late-phase allergic response**. One example is a series of substances called **leukotrienes** (so named because they are derived from leukocytes and have a triene [triple] chemical bond). Leukotrienes are extremely potent at causing smooth muscle constriction. They also cause leakage in blood vessels and attract eosinophils to continue the inflammatory reaction.

The second family of newly formed mediators is the group of human hormones called **prostaglandins**. One prostaglandin, prostaglandin D2, is a powerful constrictor of the bronchial tubes.

Cytokines also are thought to be involved in a late-phase allergic response and have actions that both stimulate and inhibit inflammation. One cytokine, interleukin-4 (IL-4), promotes IgE production by B cells; another, interleukin-5 (IL-5), encourages the maturation and activity of eosinophils.

Type I Hypersensitivities Can Be Localized or Systemic

■ Allergenic: A substance causing an allergic reaction.

There are two forms of hypersensitivities that are also referred to as **anaphylactic** (*ana* = "without"; *phylaxi* = "protection") **reactions**.

Localized Anaphylaxis. The vast majority of hypersensitivities represent a **localized anaphylaxis**; that is, a confined sensitization of mast cells and a controlled production of IgE antibody. According to public health estimates, more than 60 million Americans test positive for some form of allergen (FIGURE 23.5). Among these are common (seasonal) allergies or what is called **atopic disease** (*atopo* = "out of place"). **Allergic rhinitis**, which affects between 10% and 30% of all adults and up to 40% of children in the United States, is a collection of symptoms, mostly in the nose and eyes, which occur when an allergen, such as dust, dander, or pollen, are breathed in. For example, "ragweed rhinitis" starts when a sensitized individual breathes in ragweed pollen. Within seconds or minutes, an immediate allergic response occurs. As the ragweed pollen reacts with the tissues lining the inner surface of the nose and eyes, activated mast cells release histamine that causes a runny nose, itching, and sneezing. In the lungs, degranulation of chemical mediators constricts the breathing tubes, causing wheezing, shortness of breath, and coughing. These symptoms tend to subside after about an hour.

Several hours later a late-phase response occurs as eosinophils, neutrophils, and mast cells arrive at the affected tissue and intensify respiratory symptoms through released chemical mediators. Repeated inhalations of the allergen can lead to recurrent responses, which can develop into chronic or persistent allergic asthma.

There also are year-round allergies. These perennial hypersensitivities usually result from chronic exposure to such substances as house dust, mold spores, dust mites, detergent enzymes, and the particles of animal skin and hair ("dander"). Actually, dander itself is not the allergen; the actual allergens are proteins deposited in the dander from the animal's saliva when it grooms itself.

A variety of foods can cause **food allergies**, which usually are accompanied by symptoms in the mouth and the gastrointestinal tract, including swollen lips, abdominal cramps, nausea, and diarrhea. The skin may break out in a rash consisting of a central puffiness, called a **wheal**, surrounded by a zone of redness known as a **flare**. Such a hive-like rash is called **urticaria** (*urtica* = "stinging needle"). Allergenic foods include fish, shellfish, eggs, wheat, cow's milk, soy, tree nuts, and even peanuts (MicroFocus 23.1).

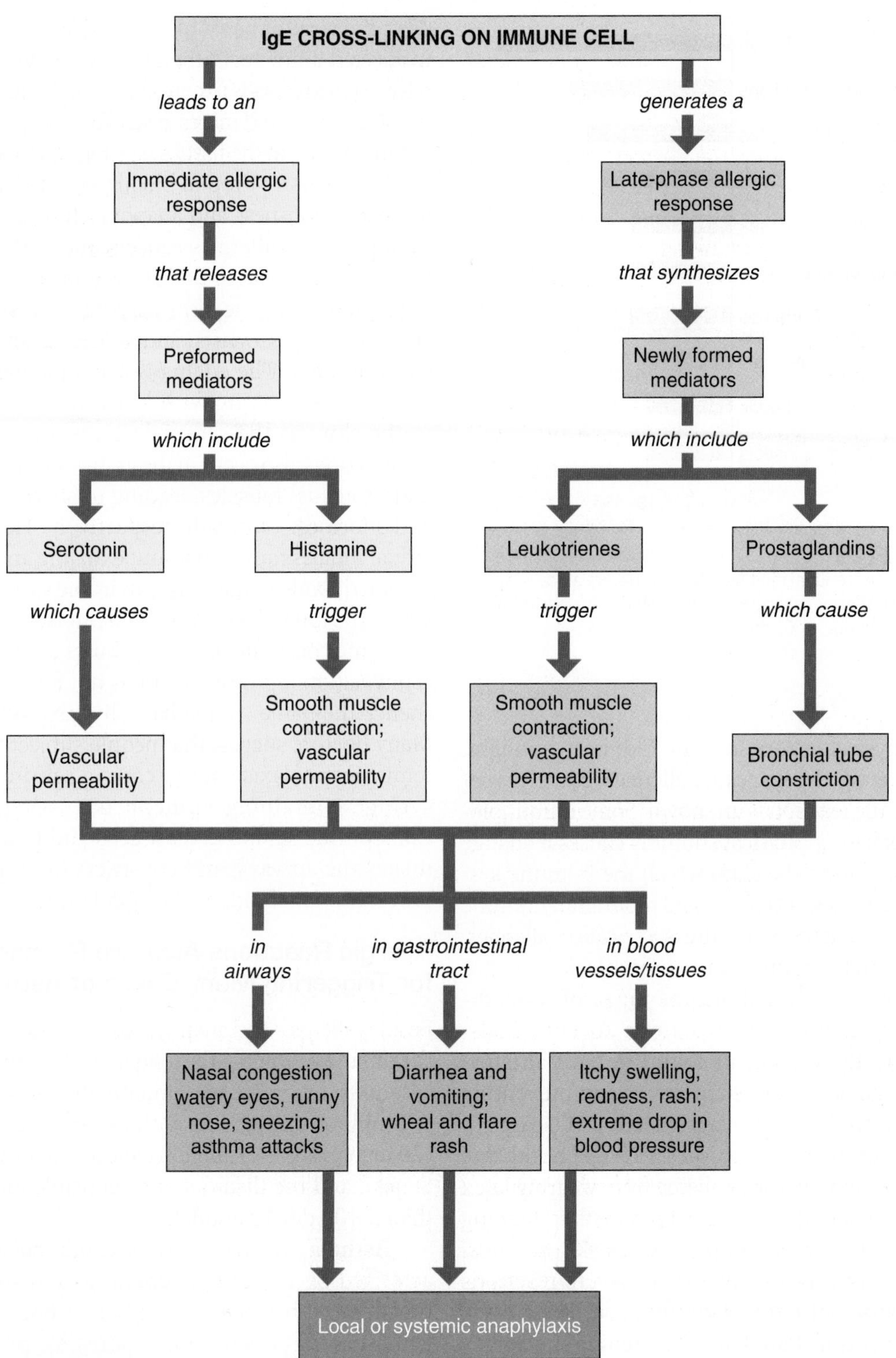

FIGURE 23.4 **Concept Map for Type I Hypersensitivity and Mediator Substance Production.** Several chemical mediators are stimulated upon IgE cross-linking on mast cells and basophils that have various effects in the body. »» Because IgE is an antibody normally targeting parasitic helminths, what might these mediator substances have done if it was a parasite infection rather than an allergen?

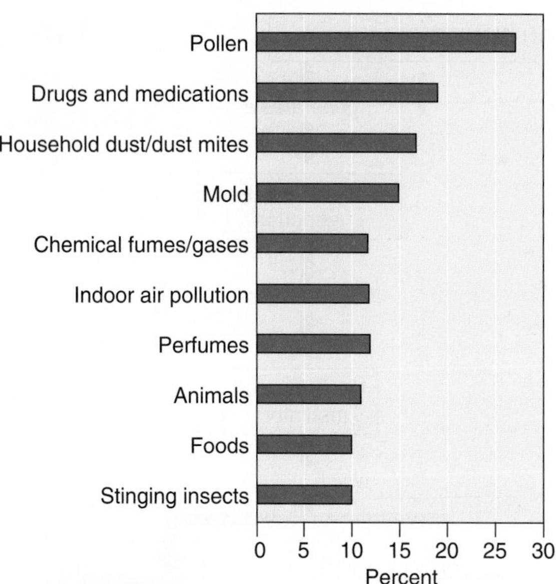

FIGURE 23.5 **The Top 10 Causes of Allergies.** These allergens were identified by allergy sufferers as causing the most allergic reactions. Source: Marist Institute for Public Opinion. »» Of these allergens, which ones, if any, cause your allergies?

Physical factors such as cold, heat, sunlight, or sweating also can cause allergies. Exactly what causes the reaction is unknown. Some immunologists believe a physical stimulus causes a change in a protein in the skin, which the immune system then "sees" as foreign and mounts an immune response. Exercise sometimes also causes allergies, usually in the form of an asthma attack.

Mention should also be made of immediate hypersensitivity reactions to chemicals, specifically to proteins found in natural latex. Although latex gloves are effective in preventing the transmission of many infectious diseases to healthcare workers, more than 10% of healthcare workers exhibit a latex allergy from wearing latex gloves or contacting medical products containing latex. A mild reaction presents as skin redness, itching, or a rash, while more severe reactions can cause sneezing, a runny nose, itchy eyes, and wheezing. Rarely, life-threatening shock may occur. Today, most latex products have suitable alternatives.

Systemic Anaphylaxis. One of the potentially most dangerous reactions of type I hypersensitivity is **systemic anaphylaxis**. This reaction involves allergens, such as bee venom, penicillin, nuts, and seafood, that are directly deposited into the bloodstream or via the digestive tract, triggering degranulation of basophils throughout the body. In affected individuals, the principal activity of the released mediators is to contract smooth muscles in the small veins and increase vascular permeability, forcing fluid into the tissues (see Figure 23.4). The resulting massive drop in blood pressure can lead to unconsciousness and anaphylactic shock. For example, food allergies cause some 150 to 200 fatalities every year in the United States.

In addition, the skin may become swollen around the eyes, wrists, and ankles, a condition called **edema**. The edema is accompanied by a hive-like rash, along with burning and itching in the skin, as the sensory nerves are excited. Contractions also occur in the gastrointestinal tract and bronchial muscles, leading to sharp cramps and shortness of breath, respectively. The individual inhales rapidly without exhaling and traps carbon dioxide in the lungs, an ironic situation in which the lungs are fully inflated but lack oxygen. Death may occur in 10 to 15 minutes as a result of asphyxiation if prompt action is not forthcoming (hence the name "immediate" hypersensitivity). Many doctors suggest that people subject to systemic anaphylaxis should carry a self-injecting syringe containing epinephrine (adrenalin). Epinephrine stabilizes mast cells and basophils, dilates the airways, and constricts the capillary pores, keeping fluids in the circulation.

Allergic Reactions Also Are Responsible for Triggering Many Cases of Asthma

Besides allergies and anaphylaxis, asthma is part of the same syndrome—the tendency of the immune system to overreact to common allergens. Some 300 million people live with asthma worldwide (20 million adults, 7 million children in the United States), and the disorder is responsible for more than 3,400 deaths annually.

Asthma, which is a chronic condition, is characterized by wheezing, coughing, and stressed breathing. It has been blamed on airborne allergens such as pollen, mold spores, or products from insects like dust mites. In fact, about 50% of asthma cases have an environmental origin. Attacks also can be induced by physical exercise or cold temperatures, situations that irritate and inflame the airways (FIGURE 23.6).

Importantly, microbes can influence the onset of asthma as Investigating the Microbial World 23.1 explores.

MICROFOCUS 23.1: **Miscellaneous**

Warning: This Product Contains Peanuts

Americans love peanuts—salted, unsalted, oil roasted, dry roasted, Spanish, honey crusted, in shells, out of shells, and on and on.

Yet of all food allergies the one that is most worrisome is the peanut allergy, which is the most lethal of all food allergies. More than 1.5 million Americans have peanut allergies and hospital emergency departments treat about 30,000 cases of food-related allergic anaphylaxis each year. Peanuts are responsible for about 50% to 60% of all food allergy deaths each year in the United States.

Courtesy of USDA-ARS.

People allergic to peanuts also worry because they often do not need to come into direct contact with peanuts to develop an allergic reaction. Peanut particle allergens can travel by air or be "trace contaminants" in other foods processed on the same manufacturing line used for peanuts. Although medical experts believe it is unlikely that sensitized individuals would go into anaphylactic shock merely from inhaling or ingesting these minute quantities of peanut particles, the sensitized individuals often do develop allergic symptoms, like a runny nose or itchy, watery eyes, in the presence of these particles.

The good news for peanut allergy suffers is that new immunotherapeutic strategies are being developed. This includes a peanut vaccine that so far is effective in protecting sensitized mice against peanut proteins. Other strategies are trying to alter the immune system's response to the peanut allergens, including ways to prevent the activation of T_H2 cells, which regulate the IgE response. In 2010, two immunotherapy trials were reported. Researchers from the Duke University and Arkansas Children's Hospital in Little Rock reported the development of a treatment regimen for children that suffer from peanut allergy. The treatment started with minute doses of peanuts that eventually increased to about 15 peanuts a day. In the study, 33 children with a peanut allergy received the daily desensitization therapy, which was given as a peanut powder sprinkled on food. Most of the children tolerated the therapy without developing an allergic reaction. In fact, the treatment was stopped for five children after 2 and a half years because they could now tolerate peanuts in their regular diet. However, four children dropped out because they could not tolerate the treatment. In another report from Cambridge, England also used desensitization therapy. In this study, children also were exposed to peanut allergens in small increments over time. In one trial, 21 out of 23 children on desensitization therapy tolerated the peanut product (peanut flour) without experiencing allergic reactions.

Obviously, much more research must be done. However, the research is looking promising and some researchers hope that by 2015 there will be some form of immunotherapy generally available for individuals suffering from peanut allergies.

Asthma primarily is an inflammatory disorder that occurs in two parts. First, there is an immediate allergic response to allergen exposure where T_H2 cells control the IgE response. However, unlike a common allergy, the mediators are not released in the nose or eyes, but rather from mast cells in the lower respiratory tract. The resulting bronchial constriction, vasodilation, and mucus buildup can make breathing difficult (FIGURE 23.7).

The synthesis and release of other cytokines result in the recruitment of eosinophils and neutrophils into the lower respiratory tract. These events represent a late-phase allergic response because they occur hours after the initial exposure to an allergen. The presence of the eosinophils and neutrophils can cause tissue injury and potentially cause blockage of the airways (bronchioles). Bronchodilators can be used to open the airways by widening the bronchioles. More recently, the use of antiinflammatory agents such as inhaled steroids or nonsteroidal cromolyn sodium have been prescribed. Cromolyn sodium blocks degranulation by mast cells.

FIGURE 23.6 **Using an Asthma Inhaler.** For some asthma sufferers, simple exercise might bring on an asthma attack. Using an inhaler that delivers medication directly to the lungs allows sufferers to live active lives without fear of an attack. © Michael Donne/Photo Researchers, Inc.
»» How could physical exercise trigger an asthma attack?

Why Do Only Some People Have Allergies?

Not everyone suffers from allergies. An interesting avenue of research was opened when it was discovered that the B cells responsible for IgE and IgA production lie close to one another in the lymphoid tissue, and that the IgA level and its corresponding lymphocytes are greatly reduced in atopic individuals. Immunologists have suggested that in nonallergic individuals, the B lymphocytes and plasma cells that produce IgA shield lymphocytes producing IgE from antigenic stimulation, and therefore atopic people may lack sufficient IgA-secreting lymphocytes to block the allergens.

Another theory of atopic disease maintains that an allergy results from a breakdown of feedback mechanisms in the immune system. Research findings indicate that IgE synthesis may be controlled by **suppressor T lymphocytes**. In nonallergic individuals, a rising IgE concentration causes suppressor T-cells to shut off IgE synthesis. However, in atopic individuals the mechanism malfunctions, possibly because the T cells are defective, and IgE is produced in massive quantities. Allergic people are known to possess almost 100 times the IgE level of people who do not have allergies.

Most interestingly, hygiene appears to affect the development of allergies and asthma, especially in young children. But not the way you

Investigating the Microbial World 23.1

Microbes and the Allergic Response

Allergies and asthma have long been suffered by many individuals in various populations. The causes of the disorders have also been studied for decades and yet we still have an incomplete picture of how the process is triggered even as the numbers of allergy and asthma cases continue to increase. Recently though, researchers have been getting some very interesting clues to the causes.

■ **OBSERVATION:** Children on farms appear to suffer many fewer allergic attacks and less asthmatic lung disease than children who live in more urban areas and large cities (see MicroInquiry 23). In addition, it has been proposed that the overuse of antibiotics could be a predisposing factors leading to allergies and asthma. Therefore, some researchers are investigating whether the bacterial populations (microbiota) in our airways and gut influence the onset of an allergic airway disease (AAD) such as asthma. It is known that germ-free animals have defects in their immune responses that produce an allergic airway response that mimics the clinical features seen in humans.

■ **QUESTION:** ***Does the removal of natural bacterial populations (microbiota) from the gut lead to the development of AAD?***

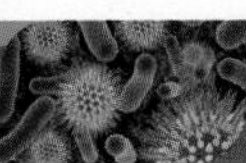

Investigating the Microbial World 23.1 (continued)

■ **HYPOTHESIS:** Established bacterial populations in the gut lessen the chances of developing AAD. If so, then in test animals the removal of the gut microbial population and exposure to a set of airway allergens will cause AAD response.

■ **EXPERIMENTAL DESIGN:** Groups of pathogen-free mice were split into three groups. An AAD response was characterized by increased numbers of immune cells (eosinophils, mast cells, neutrophils) and immune molecules (IgE antibody and cytokines).

Experimental group (group A): One group of mice was treated for 5 days with a broad-spectrum antibiotic to decrease the total bacterial microbiota and then given a single oral dose of the yeast *Candida albicans* to establish a low level of yeast in the microbiota, which typically occurs as a result of human antibiotic therapy. On days 2 and 9, the mice were exposed intranasally to mold spores from *Aspergillus fumigatus,* a common indoor allergen affecting humans. Mouse tissues were examined on day 12.

Control groups (groups B and C): One group of mice (group B) received the antibiotic and yeast exposures, but no exposure to mold spores. The other group (group C) were not exposed to antibiotic or yeast, but were exposed to mold spores on days 2 and 9.

■ **RESULTS:** See figure. Note: the intranasal exposure to fungal spores did not cause any type of infection or fungal disease in any of the three groups of mice.

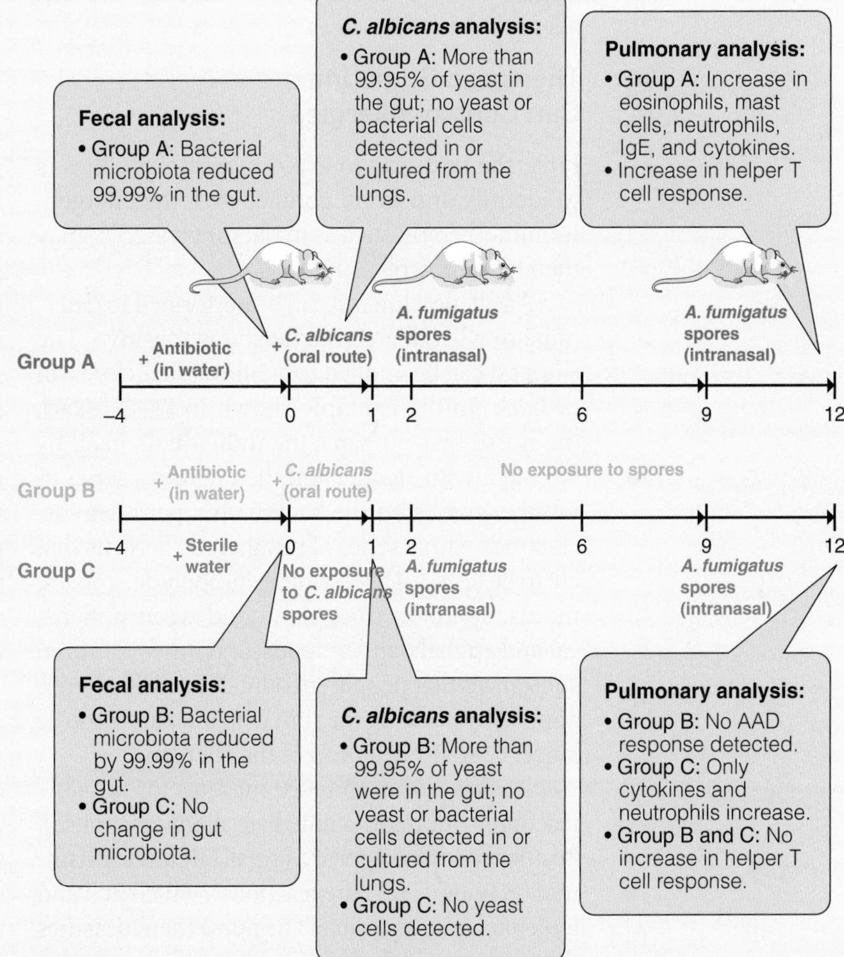

■ **CONCLUSIONS:** Based on the results, clinical features mimicking an AAD were noted.

QUESTION 1: ***Did the experimental study support the hypothesis being tested? Explain.***

QUESTION 2: ***Which group(s) of mice exhibited the characteristics of AAD? Explain your selection.***

QUESTION 3: ***Only animals with active helper T cells show the characteristics of AAD. Why were active T cells only found in Group A?***

Answers can be found on the Student Companion Website in **Appendix D.**

Adapted from: Noverr, M. C. *et al.* (2004). *Infect. Immun.* **72** (9): 4996–5003.

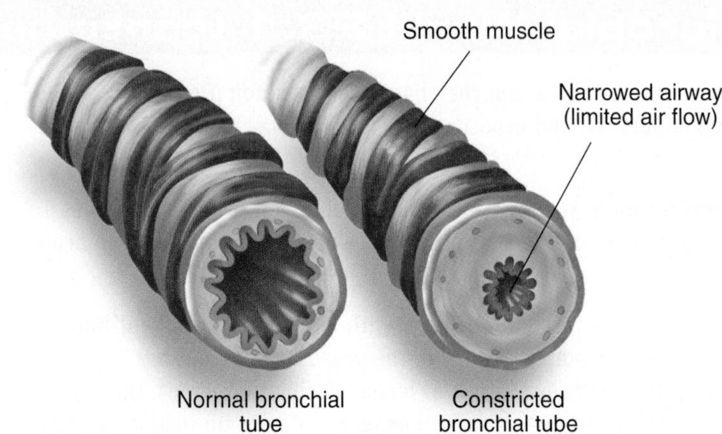

FIGURE 23.7 **Airway Narrowing During an Asthma Attack.** During an asthma attack, chemical mediators (cytokines) cause the smooth muscle layers to spasm, constrict, and form a thick mucus that can block the airway. »» What types of cells are responsible for releasing the asthma-causing chemical mediators?

might think as microbes may actually prevent allergies (MicroInquiry 23).

Therapies Sometimes Can Control Allergies

First, the best way to avoid hypersensitivities is to identify and avoid contact with the allergen. Immunotherapy, such as in MicroFocus 23.1, may then help control allergies.

An allergy skin test can be performed to determine to what allergens a person is sensitive. The skin test can be applied to an individual's forearm or back. In the example shown in FIGURE 23.8, the nurse first prepares the individual's back by wiping it with alcohol and then makes a series of number marks on the back with a pen that corresponds with a series of numbered allergens that are to be tested. The test allergens include: grasses; molds; common tree, plant, and weed pollens; cat and animal dander; foods; and other common substances that people encounter. Each allergen is applied by pricking the skin with a separate sterile lancet containing a drop of the allergen.

It takes about 15 to 20 minutes for possible reactions to develop (immediate allergic response). Positive reactions appear as small circles of inflammation (wheal and flare reactions), which itch and look like mosquito bites. The nurse then measures the diameter of any wheals with a millimeter ruler. To be sensitive, a wheal must be at least 3 mm larger than the reaction to the negative control. Based on wheal size, allergen sensitivity can be assessed as a mild, moderate, or severe reaction. Once all the test allergens are tabulated, the specific allergens to which the person is sensitive can be deduced and possible treatments discussed with the allergist.

Allergy blood tests are often used for people who cannot have skin tests or for those who have had severe allergic reactions (anaphylaxis) in the past. The most common type of blood test used is the **enzyme-linked immunosorbent assay (ELISA)**. It measures the level of IgE in a person's blood in response to certain allergens. IgE levels four times the normal level usually indicate sensitivity to the allergen or set of allergens.

Obviously, avoiding the allergen or allergens is the most effective way of preventing an allergic reaction. However, often that might not be possible, so various medications are available that "cover up" allergic symptoms or prevent them temporarily. There are numerous over-the-counter (OTC) and prescription drugs used to relieve the symptoms of allergies or prevent the release of the chemical mediators that cause the allergies (Figure 23.4).

Antihistamines are OTC and prescription drugs, such as Allegra® and Claritin®, that are used systemically to relieve or prevent the symptoms of hay fever and other types of common allergies by blocking the effects of histamine. Steroids, known medically as **corticosteroids**, help reduce inflammation associated with allergies. These medicines, such as Flonase® and Advair® (for asthma) are sprayed or inhaled into the nose or mouth. Mast cell stabilizers, such as cromolyn, help control allergic symptoms by inhibiting the release of chemical mediators from mast cells. These products, therefore, are only effective where they are applied (nasal passages or eyes). Leukotriene modifiers are used to treat asthma and nasal allergy symptoms. Drugs, such as Singulair®, block the effects of leukotrienes that cause smooth muscle contractions.

When medications fail to adequately control allergy symptoms, an allergist may recommend **allergen immunotherapy** (commonly called **allergy shots**) to reduce sensitivity to allergens. It is especially helpful for individuals sensitive to seasonal allergens (allergic rhinitis), indoor allergens, and insect stings. As described in MicroFocus 23.1, therapy involves a series of shots containing the allergen extract given regularly for several years. The shots over the first 3 to 7 months contain very tiny, but increasing amounts of the allergen(s) that are just sufficient to stimulate the immune system but not to cause an allergic reaction. Once the effective level

MICROINQUIRY 23

Allergies, Microbes, and the Hygiene Hypothesis

For reasons that are not completely understood, the incidence and severity of asthma—and allergies in general—are increasing in developed nations. Today, about 20 million Americans suffer from asthma, including more than 7 million children and adolescents. Many scientists have suggested the increase is in large part due to our overly clean urban lifestyle. We use disinfectants for almost everything in the home, and antibacterial products have flooded the commercial markets. In other words, maintaining overly good hygiene is making us sick. We need to "get down and dirty"!

The "hygiene hypothesis," first proposed in 1989, suggests that a lack of early childhood exposure to diverse microbes and other infectious agents can lead to immune system weakness and an increased risk of developing asthma and allergies. In past centuries, infants and their immune systems had to battle all sorts of infectious diseases—from typhoid fever and polio to diphtheria and tuberculosis—as well as ones that were more mundane. Such interactions and recovery "pumped up" the immune system and prepared it to act in a controlled manner. Today, most children in developed nations are exposed to far fewer pathogens, and their immune systems remain "wimpy," often unable to respond properly to nonpathogenic substances like pollen and cat dander. Their immune systems have not had the proper "basic training."

The hygiene hypothesis has been the subject of debate since it was proposed in 1989. But new research studies are providing evidence that may make the hygiene hypothesis a theory. In 2003, researchers at the National Jewish Medical and Research Center in Denver reported that mice infected with the bacterium *Mycoplasma pneumoniae* had less severe immunological responses when challenged with an allergen. However, if mice were exposed to allergens first, they developed more severe allergic responses. Also, the allergy-producing mediators were at lower levels in mice first exposed to *M. pneumoniae*. So, early "basic training" of the immune system seems to temper allergic responses.

Two other studies also bolster the hygiene hypothesis. One study used data from the Third National Health and Nutrition Examination Survey conducted by the Centers for Disease Control and Prevention (CDC). The survey included 33,994 American residents ranging from 1 year to older than 90. The CDC analysis concluded that humans who were seropositive for hepatitis A virus, *Toxoplasma gondii,* and herpes simplex virus type 1—that is, markers for previous microbial exposures—were at a decreased risk of developing hay fever, asthma, and other atopic diseases.

A second study used data collected from 812 European children ages 6 to 13 who either lived on farms or did not live on farms. Using another marker—an endotoxin found in dust samples from bedding—the investigators reported that children who did not live on farms were more than twice as likely to have asthma or allergies than did children growing up on farms. Presumably, on farms, children are exposed to more and a wider range of "immune-strengthening" microbes.

So, all in all, microbial challenges to the immune system as it develops in young children can drive the system to a balanced response to allergens. If the hypothesis proves correct, eating dirt or moving to a farm is not a practical solution, nor is a return to pre-hygiene days. However, a number of environmental factors can help lower incidence of allergic disease early in life. These include the presence of a dog or other pet in the home before birth, attending day care during the first year of life, and simply allowing children to do what comes naturally—play together and get dirty.

Dirt may be good for you. © Jean Schweitzer/Dreamstime.com.

Discussion Point

Do you suffer from common allergies? Discuss whether or not your allergies fit the description in this MicroInquiry.

is reached, for the next 3 to 5 years or longer, the person receives a maintenance dose every month. Allergy shots hopefully allow the individual's body to stop producing as much IgE antibody, and, therefore, the individual will not have as severe an allergic response when exposed to the natural allergen in the environment. This process is called **desensitization**. After a course of allergy shots, 80% to 90% of patients have less severe allergic reactions and may even have their allergies completely resolved.

CONCEPT AND REASONING CHECKS 1

- **a.** Summarize the events occurring during a first and second exposure to an allergen.
- **b.** Distinguish between the different forms of atopic disorders.
- **c.** Explain why systemic anaphylaxis is life-threatening.
- **d.** How can asthma be a life-threatening allergic reaction?
- **e.** Summarize the factors that can contribute to the development of allergies.
- **f.** Evaluate the medications and treatments available for allergies.

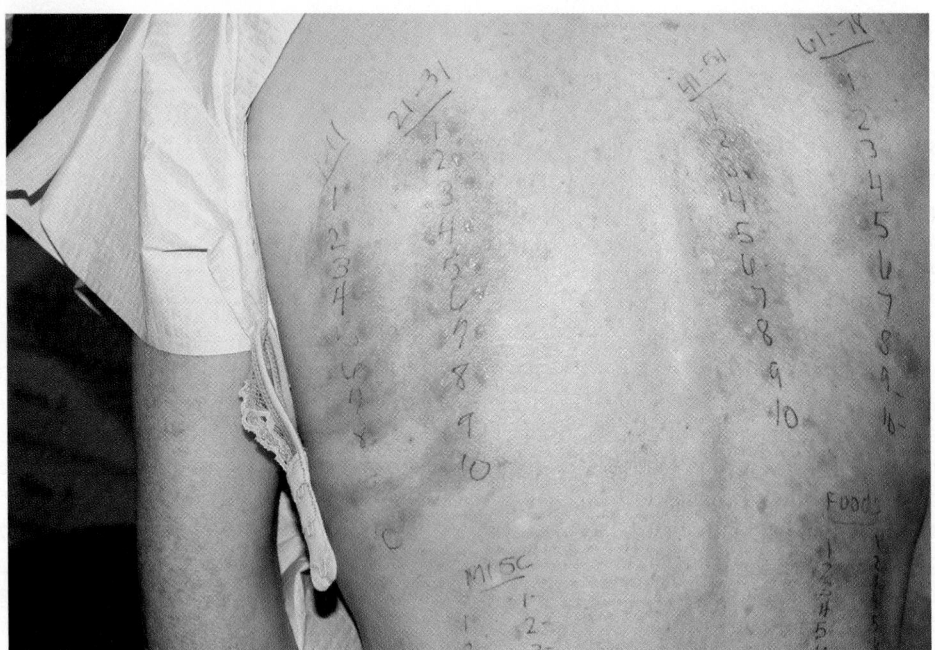

FIGURE 23.8 **An Allergy Skin Test.** The back of the individual is marked and then pricked with a large variety of test allergens. In about 20 minutes, the nurse notes if a skin wheal has developed in response to each allergen. The diameter of the wheal is then measured. Courtesy of Jeffrey Pommerville. »» Would you consider this patient to have an overall skin reaction that is mild, moderate, or severe?

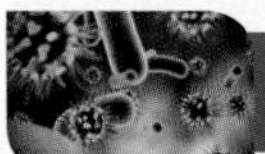

Chapter Challenge A

Carla's outbreak of hives continues. She decides to visit an allergist. The allergist asks her if she is on any kind of medication or taking antibiotics. He also asks Carla if she can remember eating any specific type of food before the onset of hives. Carla says no and this condition has been reoccurring for 3 years anyway.

QUESTION A:

What type of localized anaphylaxis does the allergist suspect Carla might have? Does this hypersensitivity seem to match her symptoms? What might you suspect could be the cause of Carla's skin reaction?

Answers can be found on the Student Companion Website in **Appendix D.**

KEY CONCEPT 23.2 Other Types of Hypersensitivities Represent Immediate or Delayed Reactions

Immunological responses involving IgG antibodies or T cells also can lead to adverse hypersensitivity reactions.

Type II Hypersensitivity Involves Cytotoxic Reactions

A **type II hypersensitivity** is a cytotoxic, cell-damaging response that occurs when IgG reacts with antigens on the surfaces of cells (FIGURE 23.9). Complement often is activated and IgM may be involved, but IgE does not participate, nor is there any degranulation of mast cells. Rather, antibody-bound cells are subject to the formation of complement-stimulated membrane attack complexes (MACs). Alternatively, natural killer (NK) cells recognize antibody bound to cells, causing perforin and granzyme release that causes cell lysis.

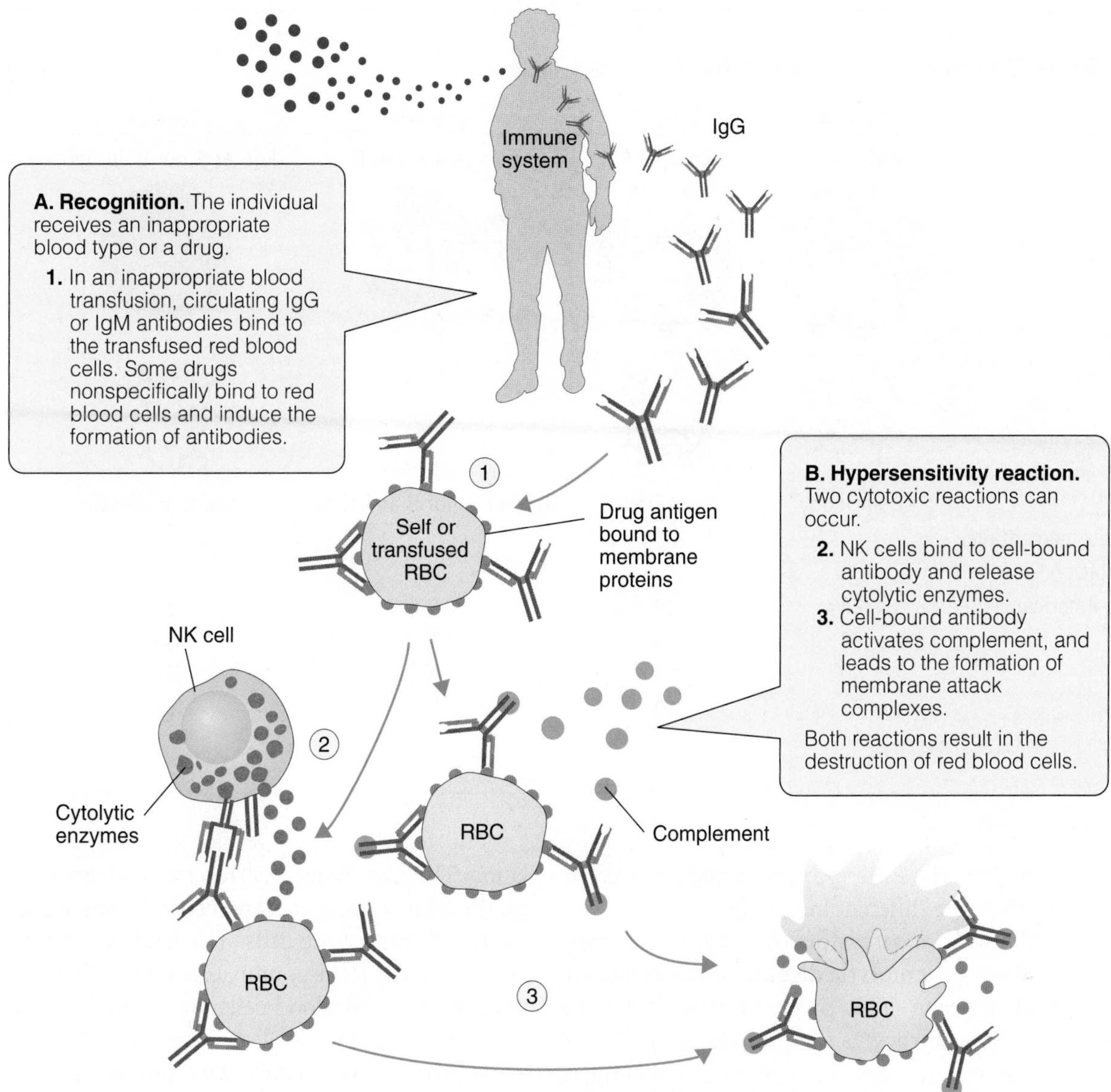

FIGURE 23.9 **Type II Cytotoxic Hypersensitivity.** Type II hypersensitivities involve the lysis of red blood cells due to incorrect blood cell typing or to antibody formation to an antibiotic. »» In an antibiotic-sensitive patient, what will happen when the drug is removed?

A well-known example of cytotoxic hypersensitivity is the transfusion reaction arising from the mixing of incompatible blood types (TABLE 23.1).

A person receiving blood of the incorrect type can have a series of reactions that destroy the red blood cells (RBCs; erythrocytes) and produce an inflammatory response that can lead to kidney failure or even death. For example, if a person with type A blood donates to a recipient with type O blood, the A antigens on the donor's erythrocytes will react with anti-A antibodies in the recipient's plasma, and the cytotoxic effect will be expressed as agglutination of donor erythrocytes and activation of complement in the recipient's circulatory system. Most blood banks cross-match the donor's RBCs with the recipient's serum to ensure compatibility.

When some antibiotic drugs, such as penicillin, are taken by patients, the drug may adsorb nonspecifically to membrane proteins on erythrocytes. The complex in some patients leads to a rare **drug-induced hemolytic anemia** if the complex triggers antibody formation. Binding to the attached antibodies induces complement-mediated lysis and thus progressive anemia.

TABLE

23.1 Some Characteristics of the Major Blood Types

	Type A	Type B	Type AB	Type O
	Antigen A	Antigen B	Antigens A and B	Neither A nor B antigens
Red blood cells				
Serum				
	B antibody	A antibody	Neither A nor B antibody	A and B antibodies
Approximate percentage, U.S. population*				
African-American	26	19	4	51
Asian	28	25	7	40
Caucasian	40	11	4	45
Hispanic	31	10	2	57

*Percentages from the American Red Cross.

MicroFocus 23.2 examines a microbial reason why there are different human blood types.

Another expression of cytotoxic hypersensitivity is **hemolytic disease of the newborn**, or **Rh disease**. This problem arises from the fact that erythrocytes of approximately 85% of Caucasian Americans contain a surface antigen, first described in rhesus monkeys and therefore known as the Rh antigen. Such individuals are said to be Rh-positive. The 15% who lack the antigen are considered Rh-negative.

The ability to produce the Rh antigen is a genetically inherited trait. When an Rh-negative woman marries an Rh-positive man, there is a 3 to 1 chance (or 75% probability) that the trait will be passed to the child, resulting in an Rh positive child. During the birth process, a woman's circulatory system is exposed to her child's blood, and if the child is Rh positive, the Rh antigens enter the woman's blood and stimulate her immune system to produce Rh antibodies (FIGURE 23.10). If a succeeding pregnancy results in another Rh-positive child, IgG antibodies (from memory cell activation) will cross the placenta (along with other resident IgG antibodies) and enter the fetal circulation. These antibodies will react with Rh antigens on the fetal erythrocytes and cause complement-mediated lysis of the cells. The fetal circulatory system rapidly releases immature erythroblasts to replace the lysed blood cells, but these cells are also destroyed. The result may be stillbirth or, in a less extreme form, a baby with jaundice.

Modern treatment for hemolytic disease of the newborn consists of the mother receiving an injection of Rh antibodies (**RhoGAM**). The injection is given within 72 hours of delivery of an Rh-positive child (no injection is necessary if the child is Rh negative). Antibodies in the preparation interact with Rh antigens on any fetal red blood cells in the mother and remove them from the circulation, thereby preventing the cells from stimulating the woman's immune system.

Type III Hypersensitivity Involves an Immune Complex Reaction

A **type III hypersensitivity** occurs when IgG antibodies combine with antigens, forming soluble immune complexes that accumulate in blood vessels or on organ surfaces (FIGURE 23.11). These immune complexes activate those complement

MICROFOCUS 23.2: Evolution

Why Are There Different Blood Types in the Human Population?

The human population contains four different blood types: A, B, AB, and O. Is this just a matter of divergent evolution producing different populations with different blood types? Or is there another reason for the difference?

Robert Seymour and his group at University College London believe that blood types are an evolutionary response to balancing defenses against viruses and bacteria. Remember: As shown in Table 23.1, type A people have anti-B antibodies and type B people have anti-A antibodies; Type AB have neither while type O have both antibodies in the circulation.

A photograph of blood types being stored. © Stockbyte Silver/Getty Images.

Seymour's group associates blood types with virus transmission. For example, when measles viruses break out of infected cells, they carry on their envelope the chemical group (antigen) identifying the blood type of that individual. Therefore, measles viruses from people with blood type A or B would be neutralized by type O blood because O blood has antibodies to antigens A and B. In reverse, measles viruses emerging from someone with blood type O carries neither the A nor B antigen, so the viruses would not be neutralized by the blood of people with blood types A, B, or AB. Therefore, people with blood type O are better prepared to defend against viruses coming from people with other blood types—and they are better at transmitting viruses to those blood types.

To make matters more complex, Alexandra Rowe and her colleagues at the University of Edinburgh have reported that, at least in Africa, children suffering from mild forms of malaria are three times more likely to be blood type O than those with severe malaria. The malaria parasite needs to attach to specific sugars on the surface of red blood cells, and type O blood, compared to A and B, has far fewer of these sugars.

If this is all true, then why isn't type O blood universal? Apparently, because of other pathogens. Seymour's group suggests that because there would be more type O individuals, probability says they would more likely be attacked by other pathogens, especially bacterial. That is supported by Rowe's group who says that type O individuals are more susceptible to other diseases, such as stomach ulcers and cholera. In fact, in Latin America where the majority of people are type O, people suffering from severe cholera are eight times more likely to be type O than those suffering a milder form of the disease.

If all this is validated with much more work, here is yet another way that microbes and viruses have affected the human species. Once again, they do rule the world!

proteins that increase vascular permeability and exert a chemotactic effect on phagocytic neutrophils. At the target site, neutrophils release lysosomal enzymes, which cause tissue damage. Local inflammation is common, and fibrin clots may complicate the problem.

Serum sickness is a common manifestation of type III hypersensitivity. It develops when the immune system produces IgG against residual proteins in a serum preparation. The IgG then reacts with the proteins, and immune complexes gather in the kidney over a period of days. The problem is compounded when IgE, also from the immune system, attaches to mast cells and basophils, thereby inducing a type I hypersensitivity. The sum total of these events is kidney damage, along with hives and swelling in the face, neck, and joints.

Another form of type III hypersensitivity is the **Arthus phenomenon**, named for Nicolas Maurice Arthus, the French physiologist who described it in 1903. In this immune complex reaction, excessively large amounts of IgG form complexes with antigens, either in the blood vessels or near the site of antigen entry into the body. Antigens in dust from moldy hay and in dried pigeon feces are known to cause

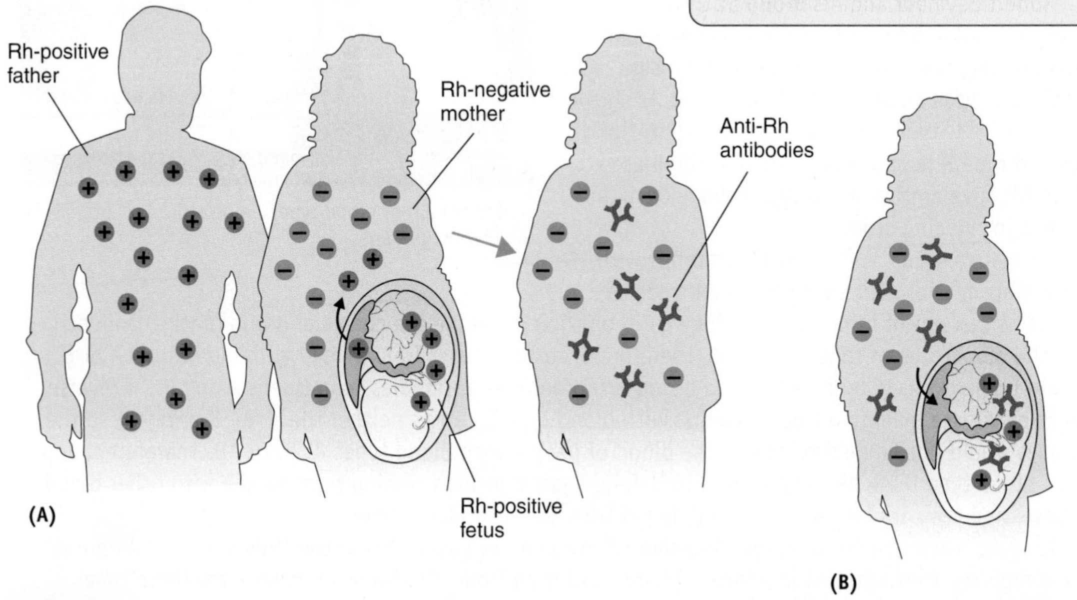

FIGURE 23.10 **Hemolytic Disease of the Newborn.** If a mother and a second child are Rh incompatible, maternal antibodies cause the lysis and removal of the fetal red blood cells. »» Why is this type II hypersensitivity also called erythroblastosis fetalis?

■ Thrombosis: The formation or presence of a blood clot that partially or completely blocks one or more arteries.

this phenomenon. The names "farmer's lung" and "pigeon fancier's disease" are applied to the conditions, respectively. Thromboses in the blood vessels may lead to oxygen starvation and cell death.

Several microbial diseases also are complicated by immune complex formation. For example, rheumatic fever and glomerulonephritis that follow streptococcal diseases appear to be consequences of immune complex formation in the heart and kidneys, respectively. In these cases, the deposit of complexes relates to common antigens in streptococci and the tissues. Other immune complex complications and inflammatory injury are associated with autoimmune disorders such as rheumatoid arthritis and systemic lupus erythematosus (see below).

Type IV Hypersensitivity Is Mediated by Antigen-Specific T Cells

A **type IV hypersensitivity** or "delayed-type hypersensitivity" is an exaggerated cell-mediated immune response. The adjective "cell-mediated" is used because T_H1 cells and their activated macrophages interact with antigen leading to cytokine secretion. The hypersensitivity is a delayed reaction because maximal effect is not seen until 24 to 72 hours after exposure to soluble antigen. It is characterized by a thickening and drying of the skin tissue, a process called **induration**, and a surrounding zone of **erythema** (redness). TABLE 23.2 compares this delayed hypersensitivity with type I immediate hypersensitivity.

Two major forms of type IV hypersensitivity are infection allergy and contact dermatitis.

Infection Allergy. When the immune system responds to certain microbial agents, sensitized T_H1 cells migrate to the antigen site and release cytokines. The cytokines attract phagocytes and encourage phagocytosis. Sensitized lymphocytes then remain in the tissue and provide immunity to successive episodes of infection. Among the microbial agents stimulating this type of immunity are

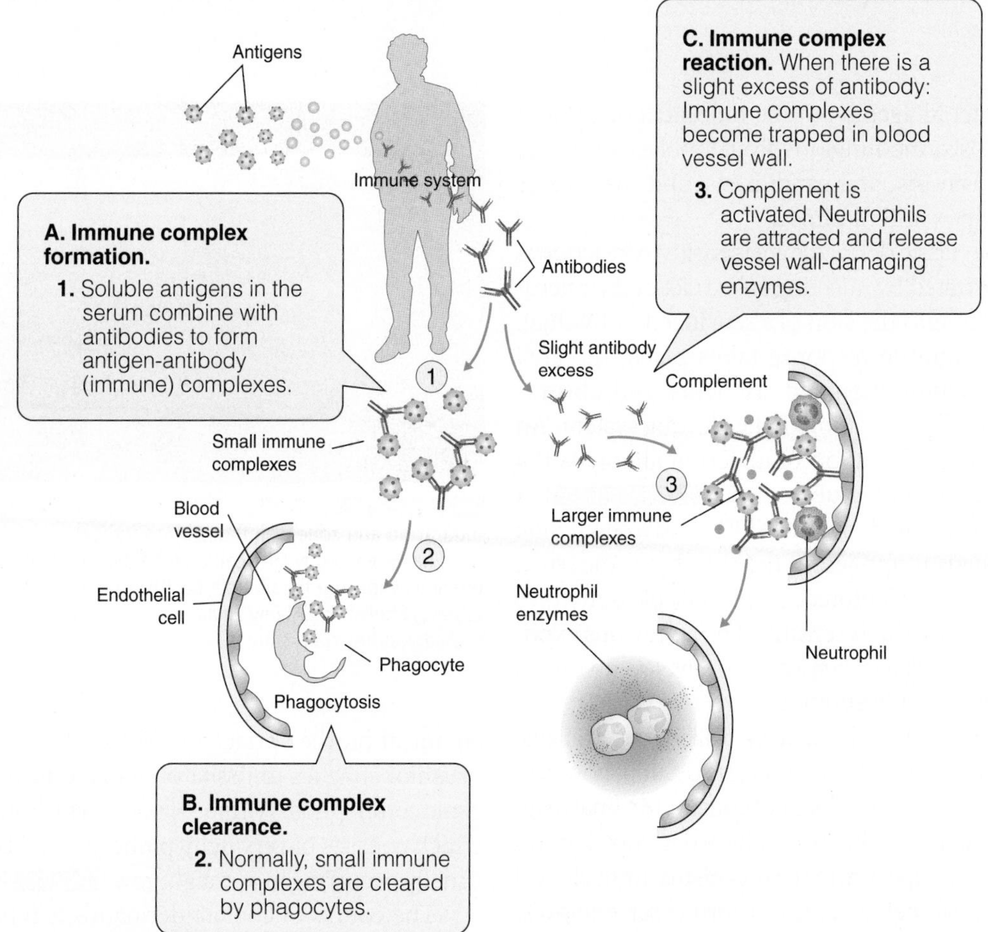

FIGURE 23.11 **Type III Immune Complex Hypersensitivity.** Type III hypersensitivities occur when excess antibodies combine with antigens to form aggregates (immune complexes) that accumulate in blood vessels or on tissue surfaces. »» What types of physical problems can develop from immune complex formation?

TABLE

23.2 Immediate and Delayed Hypersensitivities Compared

	Type I Immediate Hypersensitivity	Type IV Delayed Hypersensitivity
Clinical state:	Hay fever Asthma Urticaria Allergic skin conditions Anaphylactic shock	Drug allergies Infectious allergies Tuberculosis Rheumatic fever Histoplasmosis Trichinosis Contact dermatitis
Onset:	Immediate	Delayed
Duration:	Short: hours	Prolonged: days or longer
Allergens:	Pollen Molds House dust Danders Drugs Antibiotics Soluble proteins and carbohydrates Foods	Drugs Antibiotics Microorganisms: bacteria, viruses, fungi, animal parasites Poison ivy/oak and plant oils Plastics and other chemicals Fabrics, furs Cosmetics
Passive transfer of sensitivity:	With serum	With cells or cell fractions of lymphoid series

the bacterial agents of tuberculosis, leprosy, and brucellosis; the fungi involved in blastomycosis, histoplasmosis, and candidiasis; and the viruses of smallpox and mumps.

The classic delayed hypersensitivity reaction is demonstrated by injecting an extract of the microbial agent into the skin of a sensitized individual. As the immune response takes place, the area develops induration and erythema, and fibrin is deposited by activation of the clotting system. An important application of infection allergy is the **tuberculin test** for tuberculosis (FIGURE 23.12). A purified protein derivative (PPD) of *Mycobacterium tuberculosis* is applied to the skin by intradermal injection (the Mantoux test) or multiple punctures (tines). Individuals sensitized by a previous exposure to *Mycobacterium* species develop a vesicle, erythema, and induration.

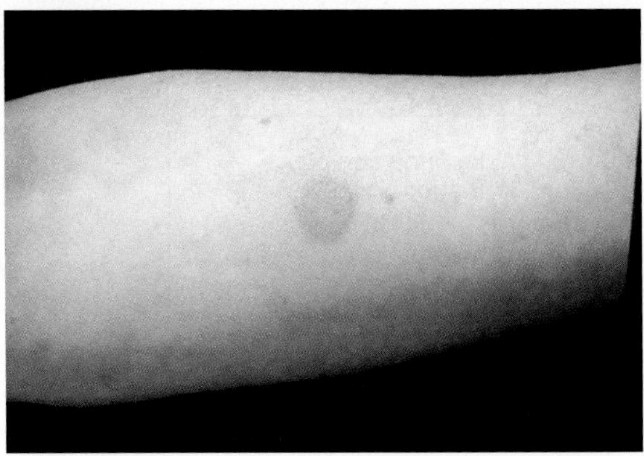

FIGURE 23.12 **A Positive Tuberculin Test.** The raised induration and zone of inflammation indicate that antigens have reacted with T cells, probably sensitized by a previous exposure to tubercle bacilli. © Bart's Medical Library/Phototake/Alamy Images. »» Does this test mean the person has tuberculosis? Explain.

Skin tests based on infection allergy are available for many diseases. Immunologists caution, however, that a positive result does not constitute a final diagnosis. Rather, sensitivity may have developed from a subclinical exposure to the organisms, from clinical disease years before, or from a former screening test in which the test antigens elicited a T-cell response.

Contact Dermatitis. After exposure to a variety of antigens, a delayed-type hypersensitivity reaction occurs. In poison ivy, the allergen has been identified as urushiol, a low-molecular-weight chemical on the surface of the leaf. In the body, urushiol attaches to tissue proteins to form allergenic compounds. Within 48 hours, a rash appears, which consists of very itchy pinhead-sized blisters usually occurring in a straight row (FIGURE 23.13).

The course of contact dermatitis is typical of the type IV reaction. Repeated exposures cause a drying of the skin, with erythema and scaling. Examples of other type IV responses are seen on the scalp when allergenic shampoo is used, on the hands when contact is made with detergent

(A) When a sensitized person touches poison ivy, a substance called urushiol stimulates T lymphocytes in the skin; within 48 hours, a type IV reaction takes place.

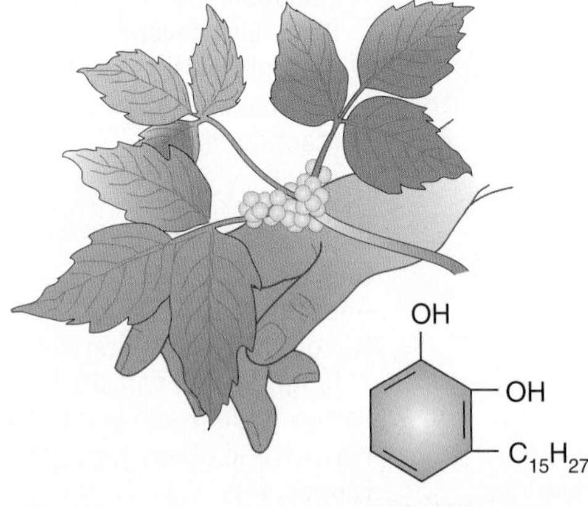

(B) The reaction is characterized by pinhead-sized blisters that usually occur in a straight row.

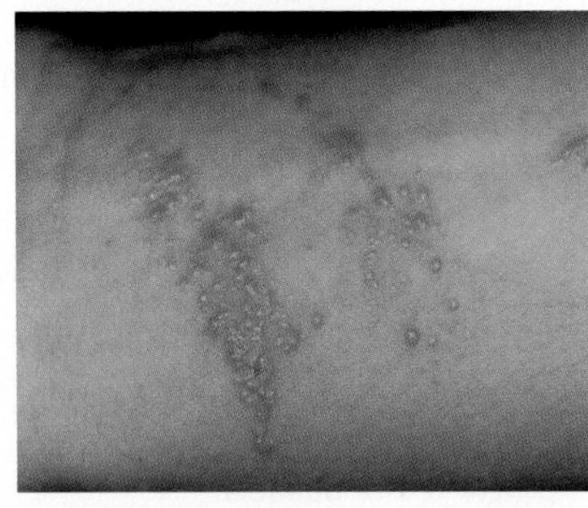

FIGURE 23.13 **The Poison Ivy Rash.** When sensitized skin makes contact with substances like urushiol in poison ivy, the chemical combines with tissue proteins to form allergenic compounds. The result is a rash on the skin surface. © Bill Beatty/Getty. »» Why is this form of hypersensitivity called delayed hypersensitivity?

TABLE

23.3 Overview of the Hypersensitivity Reactions

Hypersensitivity Type	Origin of Hypersensitivity	Antibody Involved	Cells Involved	Mediators Involved	Evidence of Hypersensitivity	Examples
Type I IgE-mediated	B lymphocytes	IgE	Mast cells Basophils	Histamine Serotonin Leukotrienes Prostaglandins	30 minutes or less	Hay fever Systemic anaphylaxis Asthma
Type II Cytotoxic	B lymphocytes	IgG IgM	RBC WBC	Complement	5–8 hours	Transfusion reactions Hemolytic disease of newborns
Type III Immune complex	B lymphocytes	IgG	Host tissue cells	Complement	2–8 hours	Serum sickness Arthus phenomenon SLE Rheumatic fever Rheumatoid arthritis
Type IV Cellular	T lymphocytes	None	Host tissue cells	Cytokines	1–3 days	Contact dermatitis Infection allergy

enzymes, and on the wrists when an allergy to costume jewelry exists. Contact dermatitis also may occur on the face where contact is made with cosmetics, on areas of the skin where chemicals in permanent-press fabrics have accumulated, and on the feet when there is sensitivity to dyes in leather shoes. Factory workers exposed to photographic materials, hair dyes, or sewing materials may experience allergies. The list of possibilities is endless.

Applying a sample of the suspected substance to the skin (a skin patch test) and leaving it in place for 24 to 48 hours will help pinpoint the source of the allergy. Relief generally consists of avoiding the inciting agents. TABLE 23.3 summarizes the four types of hypersensitivities.

CONCEPT AND REASONING CHECKS 2

a. Summarize the factors responsible for type II hypersensitivities.
b. Summarize the factors responsible for type III hypersensitivities.
c. Summarize the factors responsible for type IV hypersensitivities.

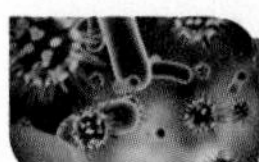

Chapter Challenge B

Carla's outbreak of hives has become a chronic condition that baffles the allergist, who has even called in an infectious disease specialist to consult. Allergy skin tests were unable to pinpoint a specific cause. In the meantime, the itching has gotten so bad at times that Carla cannot sleep at night. Over-the-counter anti-itch (antihistamine) creams do not help much nor are combinations of prescription antihistamines much better.

QUESTION B:

Do any of the other forms of hypersensitivity, especially type IV, appear to correlate with Carla's symptoms? Explain.

Answers can be found on the Student Companion Website in **Appendix D**.

KEY CONCEPT 23.3 Autoimmune Disorders and Transplantation Are Immune Responses to "Self"

One of the properties of the immune system is tolerance of "self"; that is, one's own cells and molecules with their antigenic determinants do not stimulate an immune response. Should self-tolerance break down, an **autoimmune disorder** may occur.

An Autoimmune Disorder Is a Failure to Distinguish Self from Nonself

Autoimmunity is critical to the normal development and functioning of the human immune system—and essential to the development of immunological tolerance. Each individual's immune system must have a mechanism to differentiate "self" from "nonself," so it will only react to foreign antigens, such as pathogens. Such tolerance is thought to develop in the following ways (FIGURE 23.14):

- **Clonal Deletion Theory**. This theory says that self-reactive lymphoid cells are destroyed through apoptosis during the development of the immune system in an individual.
- **Clonal Anergy Theory**. This theory proposes that self-reactive T cells become inactivated in the normal individual and cannot differentiate into effector cells when presented with antigen.
- **Regulatory T Cell Theory**. According to this theory, specific regulatory T cells function to suppress exaggerated immune responses.

In fact, all of these theories may be correct and, as such, several mechanisms could actively contribute to the development of immunological tolerance.

Up to 8% of the American population (23.5 million individuals) suffers from an autoimmune disorder, and far more women than men are affected. So, what causes the loss of tolerance? Part of the answer lies in human heredity. Various gene mutations have been identified that affect cell division and apoptosis—and therefore clonal deletion, anergy, and regulatory T cell activity. There also are several other ways autoimmune disorders can be triggered.

- **Access to privileged sites**. Certain parts of the body (e.g., brain, eye) are "hidden" from the immune system. If such sites become "accessible" to the immune system through injury, an immune response will be mounted.
- **Antigenic mimicry**. A foreign substance or microbe may enter the body that closely resembles (mimics) a similar body substance. The immune system then sees the "self" substance as foreign and attacks it. Rheumatoid arthritis and type I diabetes may be examples.

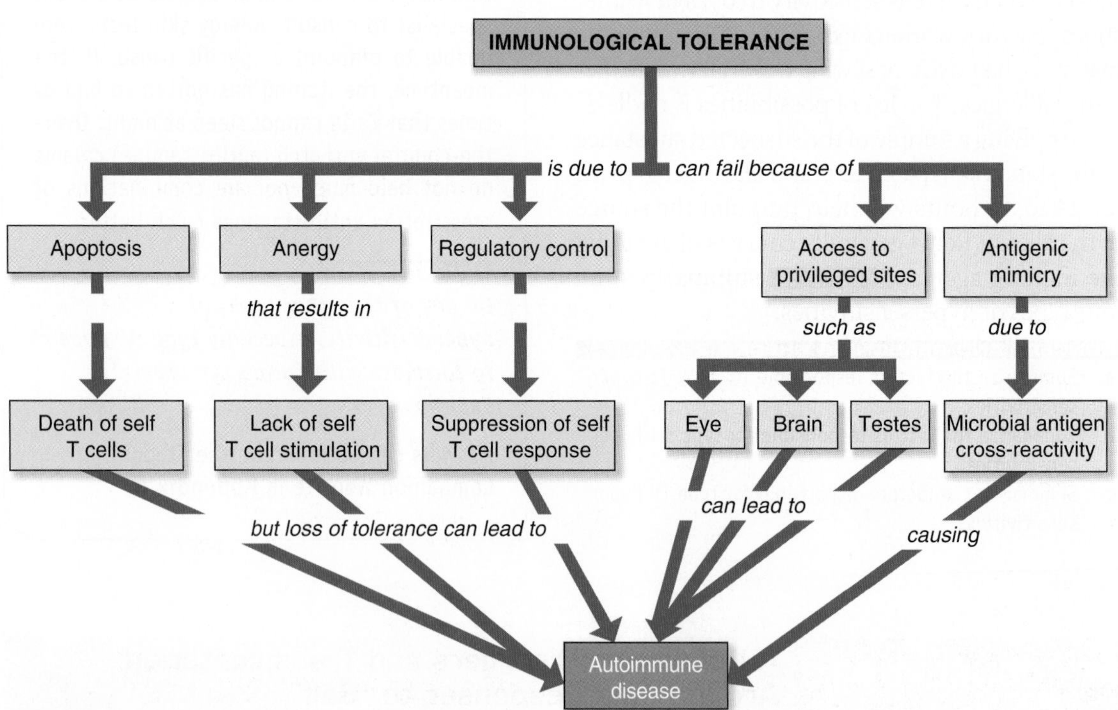

FIGURE 23.14 **Mechanisms of Tolerance and Failure.** Immunological tolerance involves killing self T cells, blocking stimulation, or suppressing a self T cell response. Autoimmune diseases result from a loss of tolerance, T-cell access to privileged sites, or antigenic mimicry through cross reaction between and exogenous and self-antigen. »» Why might T cells have gained access to privileged sites?

As a result of this loss of tolerance, the immune system attempts to mount an immune response against its own cells and tissues. Although many such disorders cause fever, specific symptoms depend on the disorder and the part of the body affected. The resulting inflammation and tissue damage can cause pain, deformed joints, jaundice, chronic itching and urticaria, breathing difficulties, and even death. To date, some 100 clinically distinct autoimmune disorders have been identified (TABLE 23.4).

Among the more notable organ-specific disorders is **myasthenia gravis**. In this disorder, **autoantibodies** (antibodies against self) react with acetylcholine receptors on membranes covering the muscle fibers. This interaction reduces nerve impulse transfer to the fibers and results in a loss of muscle activity, manifested as weakness and fatigue.

In **type I diabetes**, which usually arises in children and teenagers, the islet cells of the pancreas are destroyed, resulting in a lack of insulin production. Without insulin, the blood glucose level skyrockets and cells are starved for energy; blindness, kidney failure, or other complications can occur. Because type I diabetes typically appears after an infection, the immune response to insulin-producing cells may result from the immune system's response to an earlier viral infection. More than 300 million adults and 430,000 children worldwide have type 1 diabetes.

A systemic autoimmune disorder primarily in women is **systemic lupus erythematosus** (**SLE**), also known as "lupus." With SLE, plasma cells produce IgG upon stimulation by nuclear components of disintegrating white blood cells. Immune complexes accumulate in the skin and body organs, and complement is activated. The patient experiences a butterfly rash, a facial skin condition across the nose and cheeks, and body rashes (FIGURE 23.15). Lesions also form in the heart, kidneys, and blood vessels. Another systemic disorder is **rheumatoid arthritis** (**RA**). Unlike osteoarthritis, which results from wear and tear on joints, RA is an inflammatory condition resulting in the accumulation of immune complexes in the joints. Some researchers suspect rheumatoid arthritis is triggered through infection

TABLE

23.4 A Summary of Some Autoimmune Disorders

Disorder	Target	Effect	Female:Male Ratio
Autoantibody Mechanism			
Thrombocytopenia	Blood platelet extracellular matrix protein	Red blood cell lysis and anemia	3:1
Goodpasture syndrome	Kidney extracellular matrix collagen	Glomerulonephritis	1:6
Myasthenia gravis	Neuromuscular junction acetylcholine receptor	Muscle weakness	3:1
Graves disease	Thyroid stimulating hormone	Hyperthyroidism	5–10:1
Systemic lupus erythematosus	Nuclear antigens in skin, kidneys, joints	Skin rash, glomerulonephritis, arthritis	9:1
Self T-Cell Mechanism			
Multiple sclerosis	Nerve myelin basic protein	Partial or complete paralysis	Children—3:1 Adults over 50—1:1
Self T-Cell/Autoantibody Mechanism			
Hashimoto disease	Thyroglobulin	Destruction of thyroid gland	10–20:1
Type 1 (insulin dependent) diabetes	Pancreatic islet cells	Failure to produce insulin	1:1
Rheumatoid arthritis	Synovium of joints	Joint inflammation Loss of movement	3–5:1

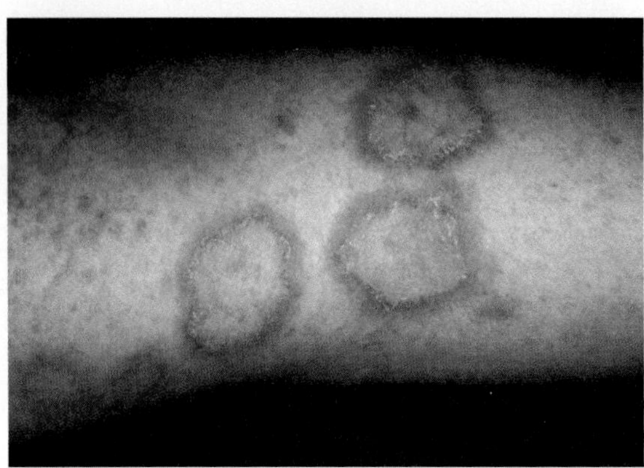

FIGURE 23.15 **Skin Lesions of Systemic Lupus Erythematosus.** This autoimmune disorder can affect the skin and other body organs. © Scott Camazine/Alamy Images. »» What causes the inflammation seen with lupus?

by a virus or bacterial pathogen and may involve antigenic mimicry.

Treatment of most autoimmune disorders requires suppressing the immune system, which means interrupting the system's ability to fight infectious disease. Immunosuppressants include corticosteroids, such as prednisone. Some disorders resolve as spontaneously as they appear while others become chronic, life-long disorders. Thus, the prognosis depends on the particular disorder.

Transplantation of Tissues or Organs Is an Important Medical Therapy

Modern techniques for the transplantation of tissues and organs trace their origins to Jacques Reverdin, who in 1870 successfully grafted bits of skin to wounded tissues. Enthusiasm for the technique rose after his reports were published, but it waned when doctors found that most transplants were rapidly rejected by the body. Then, in 1954, a kidney was transplanted between identical twins, and again, interest grew. The graft survived for several years, until ultimately it was destroyed by a recurrence of the recipient's original kidney disease. During that time, attempts to transplant kidneys between unrelated individuals were less successful.

Transplantation technology improved considerably during the next few decades, and today, four types of transplantations, or grafts, are recognized, depending upon the genetic relationship between donor and recipient. A graft taken from one part of the body and transplanted to another part of the same body is called an **autograft**. This graft is never rejected because it is the person's own tissue. A tissue taken from an identical twin and grafted to the other twin is an **isograft**. This, too, is not rejected because the genetic constitutions of identical twins are very similar.

Rejection mechanisms become more vigorous as the genetic constitutions of donor and recipient become more varied (MicroFocus 23.3). For instance, grafts between brothers and sisters, or between fraternal twins, may lead to only mild rejection because of their genetic similarity. Grafts between cousins may be rejected more rapidly, and as the relationship becomes more distant, the vigor of rejection increases proportionally. **Allografts**, or grafts between genetically different members of the same species, such as two humans, have variable degrees of success. Most transplants are allografts. **Xenografts**, or grafts between members of different species, such as a pig and a human, are rarely successful without a high degree of immunosuppression.

Mechanism of Allograft Rejection. With an allograft, the transplanted donor tissue or organ is rejected by the recipient if the immune system of the recipient interprets the transplant as "nonself." This recognition of nonself is stimulated by the recipient's immune system recognizing foreign **major histocompatibility complex** (**MHC**) proteins on the surface of the transplant. These proteins, typically referred to as **human leukocyte antigens** (**HLAs**) in the transplantation field, confer uniqueness to each individual. In the case of an allograft, the HLAs of donor and recipient can be quite distinct.

The mechanism of rejection is defined by the speed of the rejection mechanism. **Hyperacute rejection** can occur within minutes after transplantation as preformed antibodies attack the foreign HLA transplant. The antibodies activate complement and immediately block and destroy blood vessels in the donor transplant. In **acute rejection**, over a period of 10 to 30 days, T_H1 cells respond to the foreign HLA transplant by activating cytotoxic T cells (CTLs) that search out and destroy transplant cell tissue. T_H2 cells activate B cells and antibody production, which, as in hyperacute rejection, activate complement and lead to transplant destruction.

The closer the match between donor and recipient HLA proteins, the greater the chance of a successful transplant. The matching of donor and recipient is performed by tissue typing (FIGURE 23.16). In this procedure, the laboratory uses standardized anti-HLA antibodies for

MICROFOCUS 23.3: Miscellaneous Acceptance

Ordinarily, a woman's body will reject a foreign organ, such as a kidney or heart, but it will accept the fetus growing within her womb. This acceptance exists even though half of the fetus' genetic information has come from a "foreigner"—namely, the father. Has her immune system failed?

Apparently not. It seems that the sperm carries an antigenic signal that induces the woman's immune system to produce a series of so-called blocking antibodies. The blocking antibodies form a type of protective screen that protects the fetus and prevents its antigens from stimulating the production of rejection antibodies by the mother. In addition, regulatory T cells (Treg), which congregate in the womb, dampen the activity of the lymphocytes, helping protect the fetus from rejection.

© Photodisc.

Sometimes, however, a rejection in the form of a miscarriage occurs. Research indicates the level of blocking antibodies and Treg cells in some pregnant women are too low to protect the fetus. Ironically, the low immune resistance may be because the father's tissue is very similar to the mother's. In such a case, the sperm's antigens elicit a weak antibody and Treg response, too low to protect the fetus.

Physicians are now attempting to boost the level of blocking antibodies as a way of preventing miscarriage. They inject white blood cells from the father into the mother, thereby stimulating her immune system to produce antibodies to the cells. These antibodies exhibit the blocking effect.

In other experiments, injections of blocking antibodies are administered to augment the woman's normal supply. Both approaches have been successful in trial experiments, and continuing research has given cause for optimism that cases of fetal rejection can give way to acceptance and a full-term delivery.

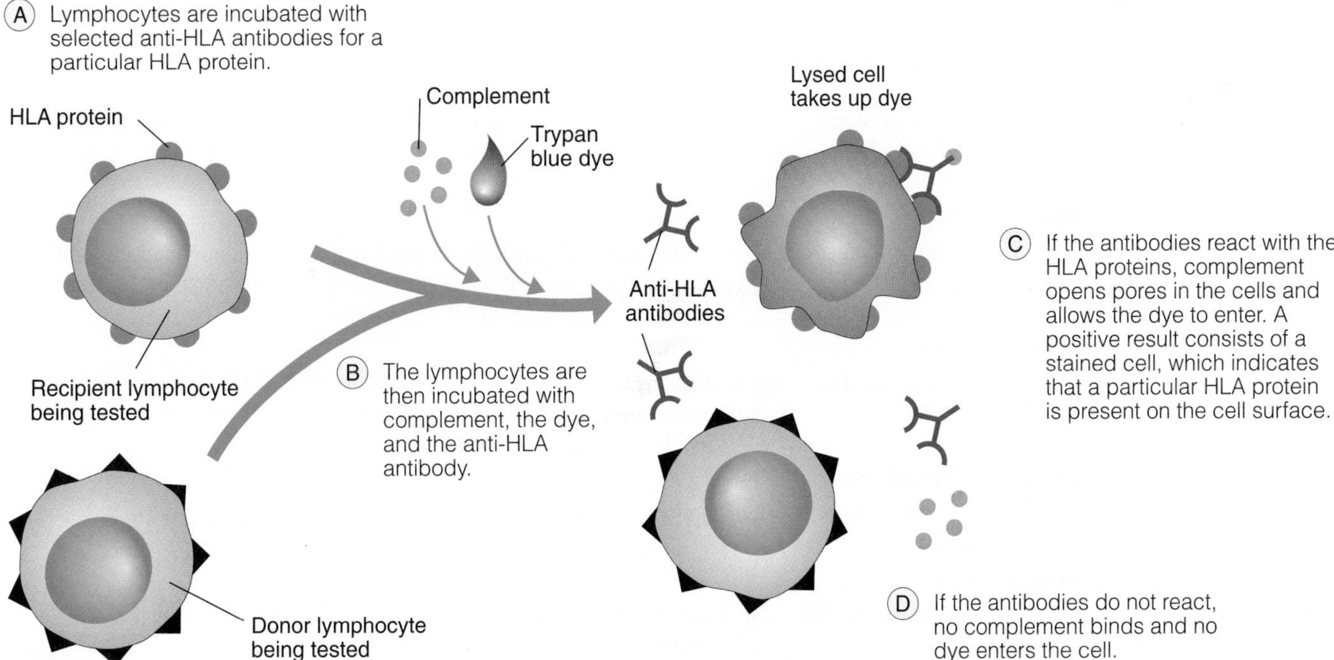

FIGURE 23.16 Tissue Typing for MHC Proteins. Dye staining and antibodies can be used to identify potentially successful tissue transplants.
»» If the Ⓓ lymphocyte is from a donor and Ⓒ is from a potential recipient, is this a good tissue match? Explain.

particular HLA proteins. Lymphocytes from the potential donor are incubated with a selected type of HLA antibodies. Complement and a dye, such as trypan blue, are then added. If the selected HLA antibodies react with the HLA proteins of the lymphocytes, the cell becomes permeable and dye enters the cells (living cells do not normally take up the dye). Similar tests then are performed with lymphocytes from the person needing the transplant to determine which HLA proteins are present and how closely the tissues match one another.

Even with tissue matching, **chronic rejection** can occur over a period of months to years as T cells and antibodies respond to very minor HLA differences. The result can be tissue lysis and eventual transplant rejection.

A rejection mechanism of a completely different sort is sometimes observed in bone marrow transplants. In this case, the transplanted marrow may contain immune system cells that form immune products against the host after the host's immune system has been suppressed during transplant therapy. Essentially, the graft is rejecting the host; it is a "reverse rejection." This phenomenon is called a **graft-versus-host reaction** (**GVHR**). It sometimes can lead to fatal consequences in the host body.

Immunosuppressive Agents Prevent Allograft Rejection

To inhibit rejection, it is necessary to suppress activity of the immune system. Because there is no clinical procedure that can be used to eliminate the rejection of allografts, all transplant patients require daily treatment with **immunosuppressive agents**, which are drugs that in some way suppress or inhibit T cells and antibody production. Because these drugs are injected and spread systemically through the body, they all have drawbacks that are related either to their toxicity and side effects in the body or to a failure to produce sufficient immune system suppression. In addition, immunosuppression can leave the body open to tumor (cancer) formation and opportunistic infections, so often transplant patients also need to be on antibiotics to minimize the latter risk. TABLE 23.5 lists some immunosuppressive agents used to prevent transplant rejection.

CONCEPT AND REASONING CHECKS 3

a. Identify the common attributes to all autoimmune disorders.

b. Assess the need for tissue typing before a transplant is performed.

c. Explain why immunosuppression leaves a transplant patient susceptible to opportunistic infections.

TABLE

23.5 The Types of Drugs Used to Limit Transplant Rejection

Type	Drug Example	Comments
Corticosteroids		
Anti-inflammatory drugs suppress T-cell activity	Prednisone	Given by intravenous (IV) at time of transplantation at high dosage; then monthly at reduced dosage indefinitely
Immunoglobulins		
Antibodies lower circulating lymphocyte populations	Antilymphocyte Antithymocyte	Given by IV or used with other immunosuppressants so that these latter drugs can be started later or at reduced dosage
Antimetabolites		
Drugs suppress humoral or cell-mediated immunity	Azathioprine Methotrexate	Given by IV or orally at time of transplantation at high dosage; then monthly at reduced dosage indefinitely
Immune cell inhibitors		
Drugs inhibit T-cell activity	Cyclosporin A	Given first by IV and later orally; used with prednisone or azathioprine
Monoclonal antibodies		
Target and suppress specific immune cells	Basiliximab Infliximab Muromonab	Given by IV at time of transplantation or if a rejection event is initiated

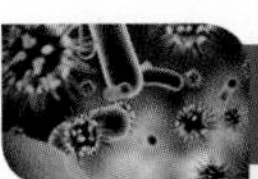

Chapter Challenge **C**

On further conversations with Carla, the allergist discovers that Carla's mother has recently experienced similar episodes of hives when she swims in her pool that has quite cold water. Carla tells her allergist that she leaned from her mother that even her grandmother and one of her aunts also experienced hives to sun exposure earlier in their lives, but such reactions have since disappeared. The allergist says "Oh, now I know what your disorder is! The good news is that you have chronic idiopathic urticaria. The bad news," he said, "is that we do not know its precise cause, which is what idiopathic means." The allergist prescribes a drug that seems to relieve the itch in others with Carla's condition and says he hopes the condition will spontaneously resolve as it did in her grandmother and aunt.

QUESTION C:

What type of disorder do Carla and her mother have that eventually resolved in her grandmother and aunt? Provide a hypothesis as to how the disorder developed.

Answers can be found on the Student Companion Website in **Appendix D**.

KEY CONCEPT 23.4 Immunodeficiency Disorders Can Be Inherited or Acquired

The spectrum of immunodeficiency disorders ranges from relatively minor deficiency states to major abnormalities that are life-threatening. The latter may be serious in populations where malnutrition and frequent contact with pathogenic organisms are common. Diagnostic techniques for determining immune deficiency diseases include measurements of antibody types, detection tests for B-cell function, and enumeration of T cells. Assays of complement activity and phagocytosis are also useful in diagnosis.

Immunodeficiencies Can Involve Any Aspect of the Immune System

If there is an immune system malfunction or developmental abnormality in immune system function, an **immunodeficiency disorder** may result. Such disorders hamper the immune system's ability to provide a strong response to any viral or microbial infection. Immunodeficiency disorders are identified by the part of the innate or adaptive immune system affected. Thus, the disorders may involve B cells and antibodies, T cells, both B and T cells, phagocytes, or complement proteins. The affected immune component may be absent, present in reduced numbers or amounts, or functioning abnormally.

The immunodeficiency may be congenital (**primary immunodeficiency**) as a result of a genetic abnormality. Such disorders present from birth are rare, although more than 70 different congenital disorders have been documented.

An immunodeficiency may be acquired later in life (**secondary immunodeficiency**). The most common acquired immunodeficiency is AIDS.

Primary Immunodeficiencies. People with primary immunodeficiencies have immune systems that do not work properly. Such individuals are subject to multiple infections, especially recurring respiratory infections. Bacterial infections are common, often severe, and lead to complications.

An example of a humoral (B cell) defect is **X-linked (Bruton) agammaglobulinemia**, first described by Ogden C. Bruton in 1952. In this disease, B cells fail to develop in the bone marrow. The patient's lymphoid tissues lack mature B cells and plasma cells, all five classes of antibodies are either low in level or absent, and antibody responses to infectious disease are undetectable. Infections from staphylococci, pneumococci, and streptococci are common in immunodeficient individuals between the ages of 6 months and 2 years. As the name suggests, the disorder is a sex-linked inherited trait, much more frequently observed in males than in females. Artificially acquired passive immunity is used to treat infections in these patients.

DiGeorge syndrome is a cell-mediated defect in which T cells fail to develop. The deficiency is linked to failure of the thymus gland to mature in the embryo. Cell-mediated immunity is defective in such individuals, and susceptibility is high to many fungal and protistan diseases and certain viral diseases. Grafts of thymus tissue attempt to correct the disorder.

Perhaps the most severe and life-threatening primary immunodeficiency is one combining both humoral and cell-mediated immunity disorders. **Severe combined immunodeficiency disease (SCID)** involves lymph nodes depleted in both B and T cells. Without a population of B and T cells, all acquired immune functions are suppressed. One form of the disorder is caused by an enzyme deficiency, and individuals with the disorder have to be kept in strict isolation to prevent infections and diseases from occurring. If such measures are not taken, most children die before the age of 2. Gene therapy is now being used to correct the gene defect (FIGURE 23.17).

■ Gene therapy: A genetic treatment to insert a normal or genetically altered gene into cells in order to replace or make up for the nonfunctional or missing gene.

Other immune deficiency diseases are linked to leukocyte malfunction. In the rare disease known as **Chédiak-Higashi syndrome**, the phagocytes fail to kill microorganisms because of the inability of lysosomes to release their contents into a phagosome. In **chronic granulomatous disease**, the phagocytes fail to produce hydrogen peroxide and superoxide radicals needed to kill ingested bacterial cells and fungi. Antibiotics and interferon are given to reduce the number and severity of infections.

Deficiencies in the complement system may be life-threatening. Complement is a series of proteins, any of which the body may fail to produce. For reasons not currently understood, many patients with complement deficiencies suffer SLE or an SLE-like syndrome. Meningococcal and pneumococcal diseases are often observed in patients who lack C3, probably as a result of poor opsonization.

Even though primary immunodeficiencies can present as a diverse set of disorders, the lack of a fully functional immune system means immunodeficiency patients all share a high prevalence of chronic or recurrent infections. Without antibodies, encapsulated bacterial species, such as *Streptococcus pneumoniae* and *Haemophilus influenzae*, are among the most common pathogens in patients with primary antibody deficiencies (PADs). These patients commonly develop both upper and lower respiratory tract infections, and complications causing chronic lung disease are a major cause of morbidity and mortality. PAD patients also commonly suffer from gastrointestinal complications involving chronic or recurrent diarrhea. Agammaglobulinemic patients may also

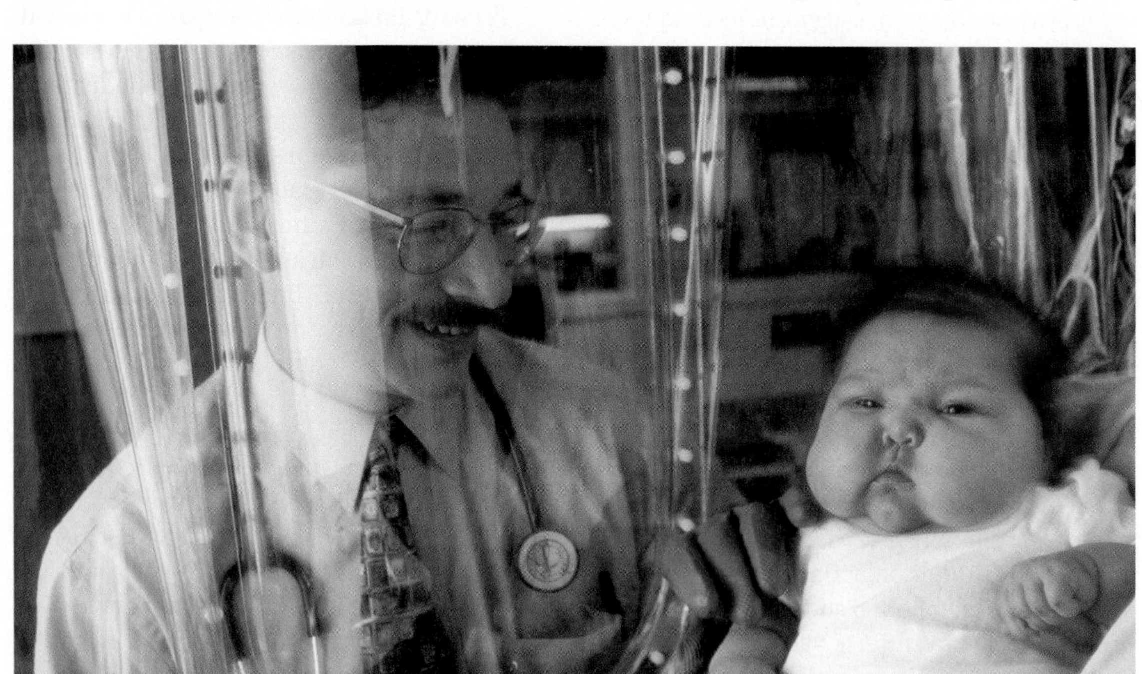

FIGURE 23.17 **SCID and Gene Therapy.** A 5-month-old baby suffering from severe combined immunodeficiency disease (SCID) is protected in a sterile tent to prevent infection. He is receiving gene therapy to insert a gene for this disease into stem cells from his bone marrow. The stem cells are then transplanted back into the baby. With this enzyme restored, these stem cells may produce normal immune system blood cells.
»» Why are stem cells in the bone marrow being used?

develop sepsis caused by *Pseudomonas*, *S. pneumoniae*, and *H. influenzae*, arthritic conditions due to *Mycoplasma* and *Ureaplasma* infections, and often fatal infections caused by the enteroviruses (echoviruses, polioviruses, and coxsackieviruses). It is imperative that early disease recognition and treatment be available for PAD patients.

Secondary Immunodeficiencies. Autoimmune disorders acquired later in life are more common than primary immunodeficiency disorders. Cases often result from immunosuppressive treatments such as chemotherapy and radiation therapy that reduce or eliminate populations of white blood cells. Often such disorders are the result of a viral infection. Certainly acquired immunodeficiency syndrome (AIDS), caused by the human immunodeficiency virus (HIV), is the most recognizable and is the subject of the last section of this chapter.

The Human Immunodeficiency Virus (HIV) Is Responsible for HIV Infection and AIDS

In 1981, physicians in the United States first reported a syndrome involving the development of certain opportunistic infections, including fungal pneumonia and cytomegalovirus infections (TEXTBOOK CASE 23). These illnesses, along with immune system deficiencies, represented the signs and symptoms for a disease that became known as **acquired immunodeficiency syndrome** (**AIDS**).

By 1983, the most plausible factor responsible for AIDS was a virus. In that year, Luc Montagnier and his French group isolated the infectious agent from a patient and Robert Gallo's group in the United States discovered how to grow the virus in culture and published convincing evidence that HIV causes AIDS. By 1986, the virus was given its current name of **human immunodeficiency virus type 1** (**HIV-1**).

Textbook CASE 23

Pneumocystis Pneumonia

Note: On June 5, 1981, the CDC published a report in *Morbidity and Mortality Weekly Report* about homosexual men in Los Angeles who had contracted *Pneumocystis carinii* pneumonia. This was the first published report of what, a year later, would become known as acquired immunodeficiency syndrome (AIDS). These descriptions are taken almost verbatim from the report (see reference at end of case).

1 Between October 1980–May 1981, five young men, all active homosexuals, arrived at three different hospitals in Los Angeles, California and were treated for biopsy-confirmed *Pneumocystis carinii* (now called *P. jiroveci*) pneumonia. All five patients also had laboratory-confirmed cases of a previous or current infection with cytomegalovirus (CMV) and with the fungus *Candida albicans* (mucosal candidiasis infection).

- The patients did not know each other, had no known common contacts or knowledge of sexual partners who had had similar illnesses, and had no comparable histories of sexually transmitted disease.
- Two of the five patients reported having frequent homosexual contacts with various partners.
- All five patients reported using inhalant drugs, and one reported injection drug abuse.

2 Patient 1: A previously healthy 33-year-old man developed *P. carinii* pneumonia and oral mucosal candidiasis in March 1981 after a 2-month history of fever associated with elevated liver enzymes, low leukocyte count, and CMV in the urine. The serum complement-fixation CMV titer in October 1980 was 256; in May 1981 it was 32. The patient's condition deteriorated despite courses of treatment with the drugs trimethoprim-sulfamethoxazole (TMP/SMX), pentamidine, and acyclovir. He died May 3, 1981 and postmortem examination showed residual *P. carinii* and CMV pneumonia.

3 Patient 2: A previously healthy 30-year-old man developed *P. carinii* pneumonia in April 1981 after a 5-month history of fever each day and of elevated liver-function tests, CMV in the urine, and documented antibodies to CMV in immunofluorescence tests. He also had a low white blood cell count and mucosal candidiasis. His pneumonia responded to a course of intravenous TMP/SMX, but, as of the latest reports, he continues to have a fever each day.

(continued on next page)

4 Patient 3: A 30-year-old man was well until January 1981 when he developed esophageal and oral candidiasis that responded to amphotericin B treatment. He was hospitalized in February 1981 for *P. carinii* pneumonia that responded to oral TMP/SMX. His esophageal candidiasis recurred after the pneumonia was diagnosed, and he was again given amphotericin B. The CMV complement-fixation titer in March 1981 was 8. Material from an esophageal biopsy was positive for CMV.

5 Patient 4: A 29-year-old man developed *P. carinii* pneumonia in February 1981. He had had Hodgkin's disease 3 years earlier, but had been successfully treated with radiation therapy alone. He did not improve after being given intravenous TMP/SMX and corticosteroids and died in March. Postmortem examination showed no evidence of Hodgkin's disease, but *P. carinii* and CMV were found in lung tissue.

6 Patient 5: A previously healthy 36-year-old man with a clinically diagnosed CMV infection in September 1980 was seen in April 1981 because of a 4-month history of fever, breathing difficulties, and a cough. On admission he was found to have *P. carinii* pneumonia, oral candidiasis, and CMV retinitis. A complement-fixation CMV titer in April 1981 was 128. The patient has been treated with two short courses of TMP/SMX that have been limited because of a sulfa-induced neutropenia. He is being treated for candidiasis with topical nystatin.

Questions:

(Answers can be found on the Student Companion Website in **Appendix D.**)

A. If the patients did not know each other, how did they all end up with essentially the same set of infectious diseases?

B. What does the huge reduction in the serum complement-fixation CMV titer indicate in patient 1?

C. Do you believe patient 2 had been cured of his illness? Explain.

D. Knowing today that all the infectious diseases described were due to HIV weakening their immune system, what is the most likely way that the patients became infected with HIV?

For additional information see www.cdc.gov/mmwr/PDF/wk/mm4534.pdf.

Structure of HIV. HIV is a member of the Retroviridae. The virion contains two copies of a single-stranded (+ strand) RNA (FIGURE 23.18). Unique to these RNA viruses, the genome is packaged with several enzymes, one of which is called **reverse transcriptase** that is needed to copy single-stranded RNA into double-stranded DNA. This reversal of the usual mode of genetic information transfer (transcription) gives the virus its name, retrovirus (*retro* = "backward") and the enzyme its name, reverse transcriptase.

The HIV genome is surrounded by a cone-shaped icosahedral capsid. Between the capsid and envelope is a protein **matrix** that facilitates viral penetration. Like other enveloped viruses, the HIV envelope contains protein **spikes** for attachment and entry into the host cell. One spike protein, gp120, is used to attach to an appropriate host cell (see below) while the other, gp41, promotes fusion of the viral envelope with the host plasma membrane.

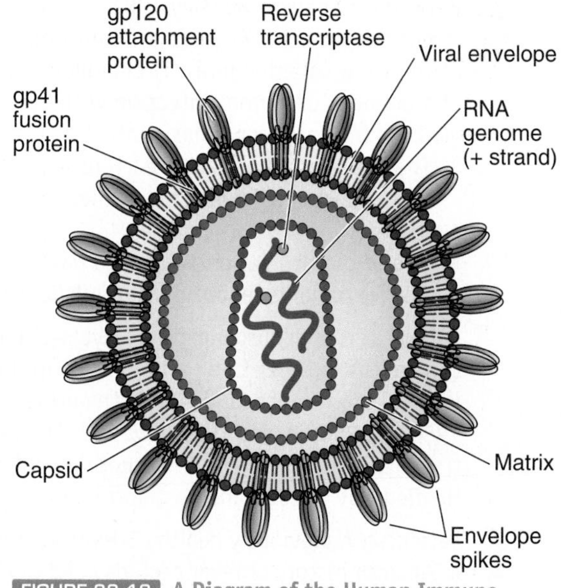

FIGURE 23.18 **A Diagram of the Human Immunodeficiency Virus (HIV).** The virus consists of two molecules of RNA and reverse transcriptase. A protein capsid surrounds the genome and an envelope with spikes of protein lies outside the capsid and matrix. »» What is the most unique feature about the structure of HIV?

In 1986, a second type of HIV, called HIV-2, was isolated from AIDS patients in West Africa. Comparative studies indicate both types have the same mode of transmission and are associated with similar infections. However, HIV-2 appears to develop more slowly so people infected with HIV-2 are less infectious early in the course of infection. There are few reported cases of HIV-2 in the United States. It is important to note that for both types, AIDS is the end result of an HIV infection. MicroFocus 23.4 discusses the origins of HIV.

Replication of HIV. The viral replication cycle begins when gp120 proteins on the HIV spikes contact CD4 receptors present on the plasma membrane of a host cell (FIGURE 23.19). If the infection is transmitted through sexual intercourse, the initial cells encountered are the dendritic cells (DCs) and CD4 T cells associated with the mucosa of the genital tract. At least 50% of the body's CD4 T cells are located in the mucosal lining.

Following entry of the capsid into the host cell cytoplasm, which is facilitated by the gp41 spike proteins, uncoating occurs and the reverse transcriptase synthesizes a molecule of DNA from the viral single-stranded RNA. The reverse transcriptase

Website Animation: HIV Structure and Replication

MICROFOCUS 23.4: Evolution

HIV's Family Tree

For years, scientists and researchers have debated the origin of the human immunodeficiency virus (HIV). Did it come from contaminated polio vaccine? Was it a government secret project that went awry? Where did the virus arise?

HIV is a member of the genus *Lentivirus,* all of which produce slow (*lent* = "slow"), incessant infections of the immune system. These viruses have been found in several animals, including cats, sheep, horses, and cattle, and most importantly, monkeys, where the simian immunodeficiency virus (SIV) has been isolated. It is now generally accepted that HIV is a descendant of SIV because certain strains of SIV bear very close gene sequences to HIV-1 and HIV-2, the two types of HIV. HIV-2 corresponds to the SIV strain found in the sooty mangabey (or green monkey), which is indigenous to western Africa.

A young chimpanzee. © Dana Ward/ShutterStock, Inc.

Until recently, the origin for the more virulent HIV-1 was more difficult to place. The closest counterpart was to SIV found in chimpanzees. However, this virus still had certain significant differences from HIV-1. In 2002, a group of researchers from the University of Alabama announced that they found a type of SIV almost identical to HIV-1 in a frozen sample taken from a subgroup of chimpanzees once common in west central Africa. They concluded that wild chimps had been infected simultaneously with two different SIVs, which had recombined to form a third virus that could be passed on to other chimps and, more significantly, was capable of infecting humans and causing AIDS.

These two different viruses were traced back to an SIV that infected red-capped mangabeys and one found in greater spot-nosed monkeys. The Alabama researchers believe that the hybridization took place inside chimps that had become infected with both strains of SIV after they hunted and killed the two smaller species of monkeys. (Yes, some chimps do hunt and kill animals!)

The most likely scenario is that sometime between 1902 and 1921 HIV-1 jumped to humans who were eating "bush meat" (the term used for monkey meat). In fact, transfer of retroviruses from primates to hunters can still be documented. HIV-2 jumped from the sooty mangabey in the 1960s.

Gene sequencing analysis suggests that HIV arrived in the United States around 1969. Considering that it takes about 10 years for an infection to produce symptoms, it is not surprising that the first cases did not show up until 1981. In fact, some researchers believe that by then there were already some 100,000 infections.

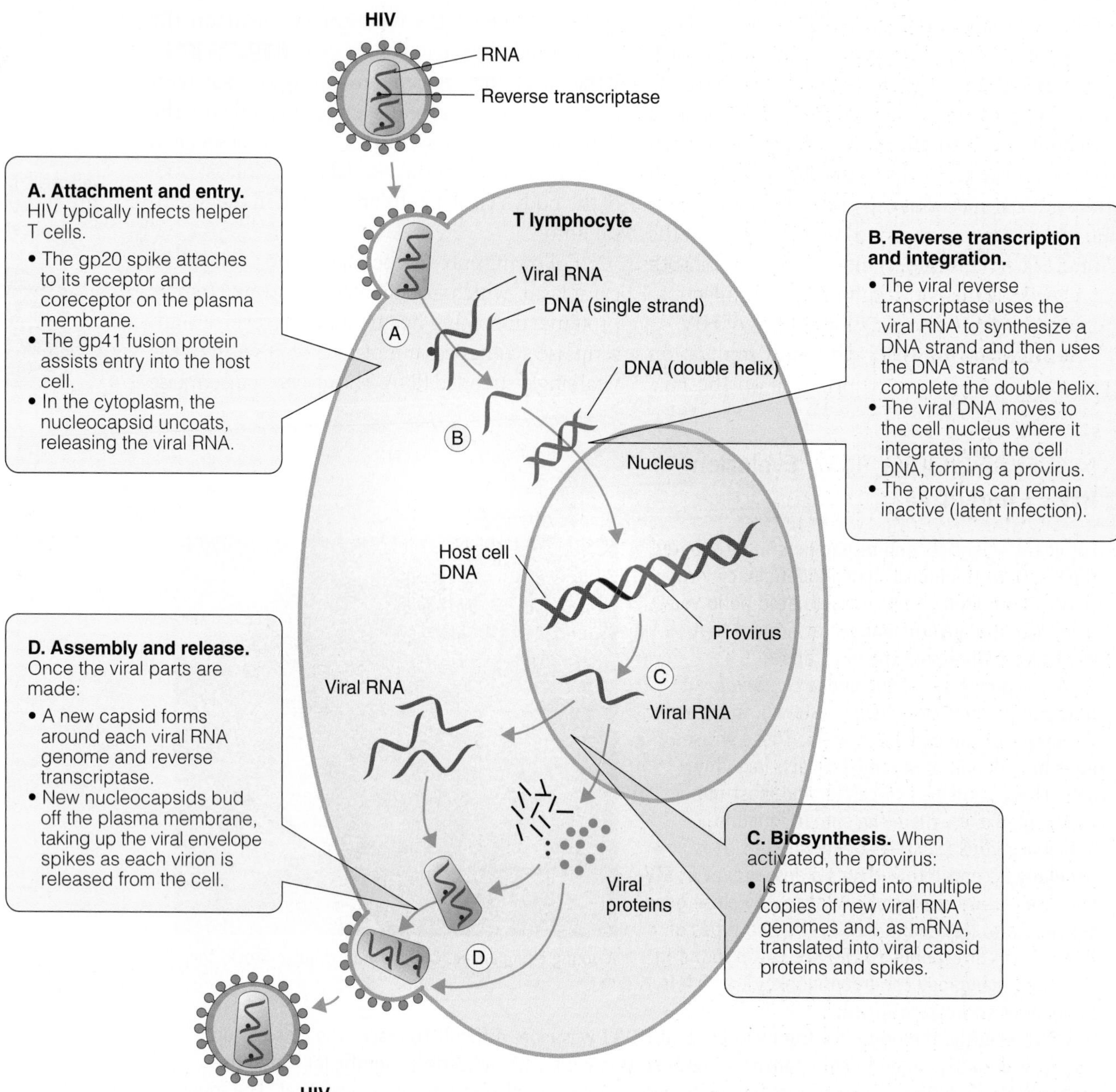

FIGURE 23.19 **The Replication Cycle of the Human Immunodeficiency Virus (HIV).** Replication is dependent on the presence and activity of the reverse transcriptase enzyme. »» What is the role of the reverse transcriptase enzyme?

makes many replication errors, which occurs to such a degree that years into an infection, the virus is genetically different from the one that initiated the infection years earlier (see treatment below).

Within 72 hours of infection of a host cell, the DNA molecule has integrated into one of the host chromosomes as a **provirus**. If the infected cell is actively dividing, the provirus initiates DNA transcription and translation, resulting in the biosynthesis and maturation of new virions. The new virions then "bud" from the host cell to infect other cells. If the infected cell is not dividing, the provirus remains dormant. In either situation, the infected individual now has a "primary HIV infection."

Typically, within 2 to 3 weeks after infection, HIV has destroyed half of the immune cell population, including CD4 T cells, in the mucosal lining. If transmission occurs via the blood, CD T cells, macrophages, and DCs are among the first cells infected, primarily in the lymphoid organs. How the CD4 T cells are killed is the subject in Investigating the Microbial World 23.2.

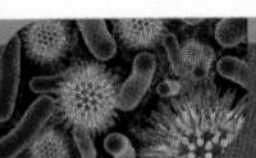

Investigating the Microbial World 23.2

How Does HIV Kill T Cells?

Once investigators determined that AIDS was caused by a virus, HIV, that infected CD4 T cells, the scientists wondered how the virus killed and brought about a progressive depletion in T-cell number as the years of infection passed. Certainly, many of the infected cells were killed directly as a result of a virus productive infection within the infected cells. But is that the only way?

■ **OBSERVATIONS:** More than 98% of CD4 T cells are found in lymphoid organs, such as the tonsils and spleen, and it is within these organs that HIV replicates in an infected individual. However, using human lymphoid cultures (HLCs) prepared from tonsillar tissue, which closely mimic the cellular environment normally experienced by HIV in humans:

- Less than 5% of the CD4 T cells show a productive infection (produce HIV virions), yet 99% of all the CD4 T cells in those cultures die. Therefore, direct killing cannot explain all the T-cell deaths. For some reason, the 94% of surrounding cells, called "bystander cells," are also dying even though they produce few, if any virus particles.
- If HLCs are treated with the antiretroviral (ARV) drug enfuvirtide (see figure) to block virus fusion with the CD4 T cell plasma membrane, no bystander cell deaths are observed after HIV challenge, so the bystander cells must be "infected" for death to occur.
- If HLCs are treated with another ARV drug, raltegravir (see figure) to prevent HIV DNA from inserting into the host cell DNA, bystander deaths are observed after HIV challenge.

■ **QUESTION:** ***How are infected bystander CD4 T cells being killed in these lymphoid organs if there has to be viral entry but no resulting provirus formation or productive infection?***

■ **HYPOTHESIS:** Bystander CD4 T cells die from indirect killing involving an incomplete (abortive) infection that triggers cell suicide. If so, then appropriate ARV drugs that block retrovirus replication will pinpoint where HIV infection is aborted and the products triggering cell suicide.

■ **EXPERIMENTAL DESIGN:** HCLs were treated with specific ARV drugs and challenged with HIV (see figure). After 9 days, the viability of the cells was measured.

EXPERIMENT 1: HLCs were not treated with any ARV drug but were challenged with HIV.

EXPERIMENT 2: HLCs were treated with AZT and then challenged with HIV.

EXPERIMENT 3: HLCs are treated with efavirenz and nevirapine (inhibit the ability of reverse transcriptase to copy the viral ssRNA into dsDNA) and then challenged with HIV.

■ **RESULTS:**

TABLE

Experiment	ARV Drug: Mode of Action	Result
1	None used	HLCs depleted of CD4 T cells after 9 days
2	AZT: blocks the elongation stage of DNA replication but allows up to 250 base pairs in the early initiation stage of DNA replication to occur	HLCs depleted of CD4 T cells after 9 days
3	Efavirenz and nevirapine: inhibit the ability of reverse transcriptase to copy the viral ssRNA into dsDNA	HLCs not depleted of CD4 T cells after 9 days

■ **CONCLUSIONS:**

The bystander cells, which are the major cause of CD4 T-cell depletion in lymphoid organs, become infected but due to an abortive infection, the accumulated products trigger cell death.

(continued on next page)

Investigating the Microbial World 23.2 (continued)

QUESTION ***1: By adding different ARV drugs that target precise steps of the HIV replicative cycle, where does the abortive step of viral replication occur?***

QUESTION ***2: What are the accumulated products that trigger CD4 T cell death?***

Answers can be found on the Student Companion Website in **Appendix D**.

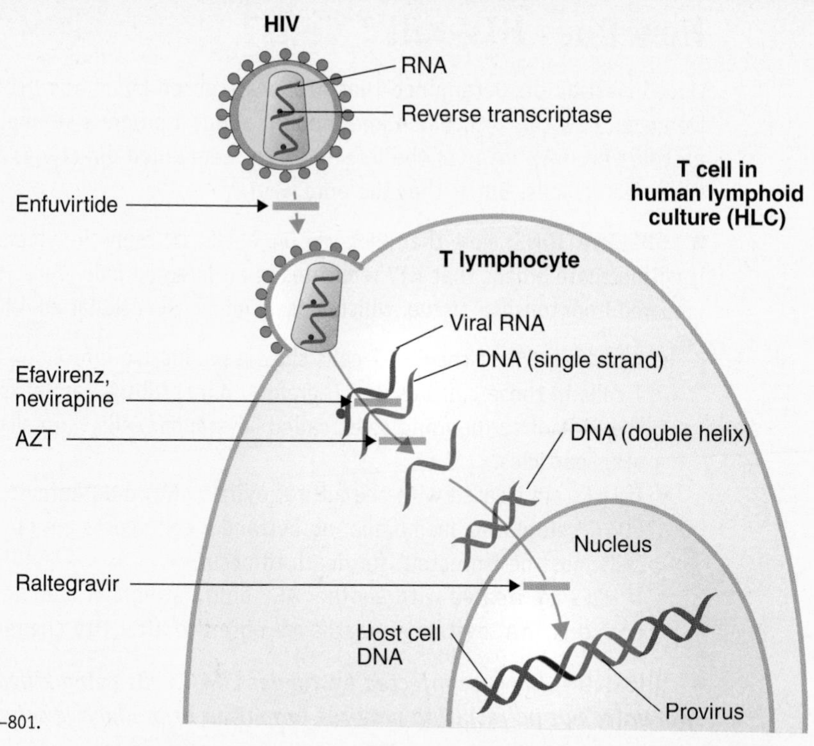

Adapted from: Doitsh, G. *et al.* (2010). *Cell* **143** (5): 789–801.

Clinical Disease Progression. Although the rate of clinical disease progression varies among individuals, realize the following stages are a continuum.

Stage I: Primary HIV Infection. Many people are asymptomatic when they first become infected with HIV. However, about 70% experience an **acute-phase** HIV infection, which produces a flu-like illness within a month or two of HIV exposure. The symptoms of fever, chills, rashes, and night sweats usually last no longer than a few days.

During this period, the immune system is at war with HIV as the virus infects lymph nodes ①, replicates, and releases virions Ⓐ into the bloodstream (FIGURE 23.20).

The immune system starts responding to the infection by producing antibodies and T cytotoxic cells directed against HIV. Such a process in the body is referred to as **seroconversion**, meaning HIV antibodies can now be detected in the blood. This usually occurs within 1 to 3 months postinfection Ⓑ.

Stage II: Asymptomatic Stage. Without treatment, the infected individual often remains free of major diseases and is quite healthy for many years, although early signs of immune system decline may occur as indicated by periods of swollen lymph nodes. The individual can also infect others. During this time that is often called "clinical latency," the level of HIV in the blood slowly rises Ⓒ as the number of CD4 T cells steadily declines ②.

As the immune system continues to decline, the affected individual will start to experience mild HIV disease symptoms, such as skin rashes, night sweats, fatigue, some weight loss, and often fungal skin infections. As the disease progresses, more serious conditions may occur, including recurrent herpes blisters on the mouth, diarrhea, fever, and significant weight loss.

Stage III: AIDS. Over time, the immune system loses the fight against HIV. At full bore, HIV can produce more than 10 billion new virions per day Ⓓ. Thus, lymph nodes and tissues become damaged from the battles and the immune system

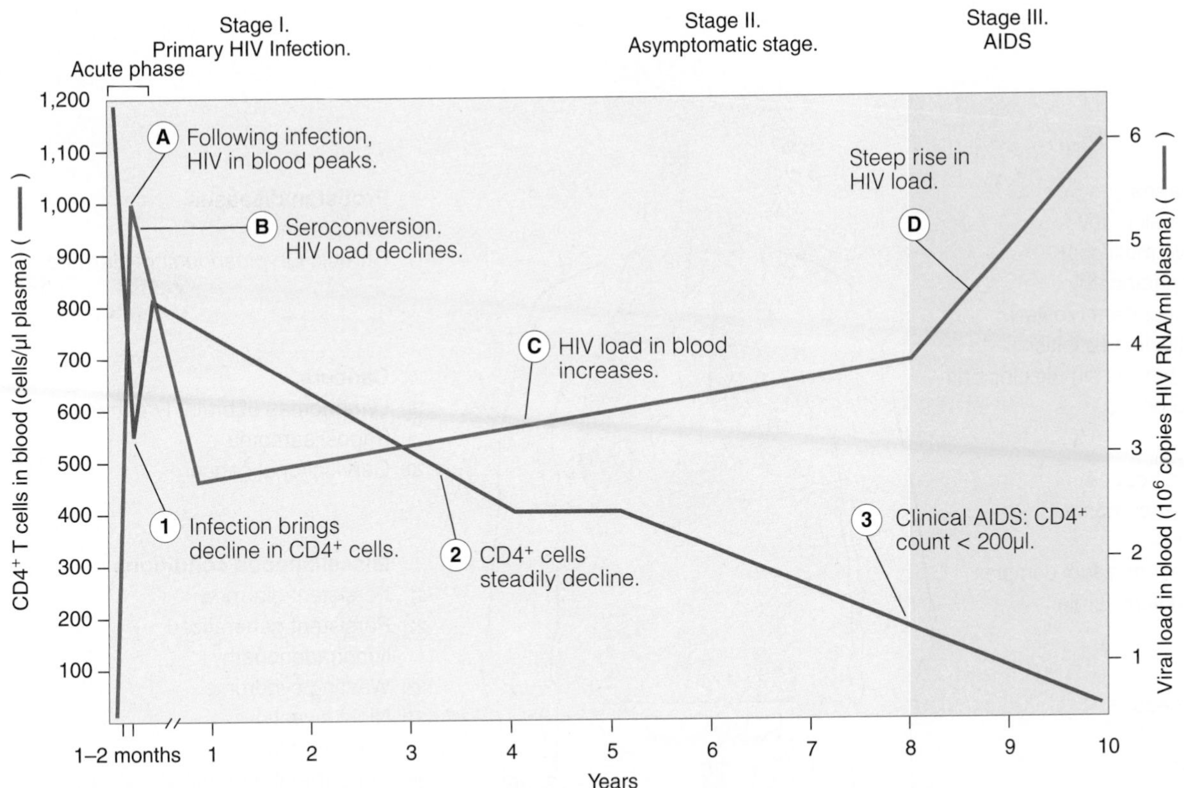

FIGURE 23.20 **HIV Infection and AIDS.** Once infected with HIV, an immune response brings about an abrupt drop in the blood HIV population, which rises as the immune system fails (**A–D**). Without antiretroviral therapy, the T cell count slowly drops (1–3). Once the T cell count is below 200/ul, the person is said to have AIDS. »» Explain why it takes so long for the T-cell number to drop.

can no longer replace T cells at the rate they are destroyed. In addition, HIV may have mutated into a form that is more pathogenic and more aggressive in destroying T cells.

As the patient's immune system deteriorates, the symptoms worsen and **opportunistic infections** develop. These infections can occur in many parts of the body, but the defining conditions are listed in FIGURE 23.21.

Infected individuals do not have AIDS until they reach this late stage of HIV disease. The CDC defines an HIV-infected individual as having AIDS if their CD4 T cell count is below 200 cells per microliter of blood (healthy adults usually have CD4 T cell counts of 1,000 or higher) ③ or have one or more of the defining conditions listed in Figure 23.21.

Opportunistic infections now can be severe and eventually fatal because the immune system is so damaged by HIV that the body cannot fight off multiple pathogen infections. Some of the most common are *Pneumocystis* pneumonia, cytomegalovirus infections, toxoplasmosis, and candidiasis. People with AIDS also are likely to develop various cancers, especially those caused by viruses, such as Kaposi sarcoma and cervical cancer, or cancers of the lymphatic system known as lymphomas.

Scientists have been studying those few individuals who have been infected with HIV for 10 or more years and yet HIV is virtually undetectable in their bodies. Scientists are trying to discover what factors allow these "long-term nonprogressors" to keep the viral load down. One intriguing finding concerns the way the virus enters cells. The HIV gp120 spike needs to bind to two receptors on host helper T cells: the CD4 receptor and a coreceptor called CCR5 (FIGURE 23.22). HIV-infected individuals having helper T cells with both receptors tend to develop AIDS. However, about 1% of the Caucasian population lacks the CCR5 receptor so the virus cannot attach and enter the cells. These individuals would thus have natural resistance to HIV infection.

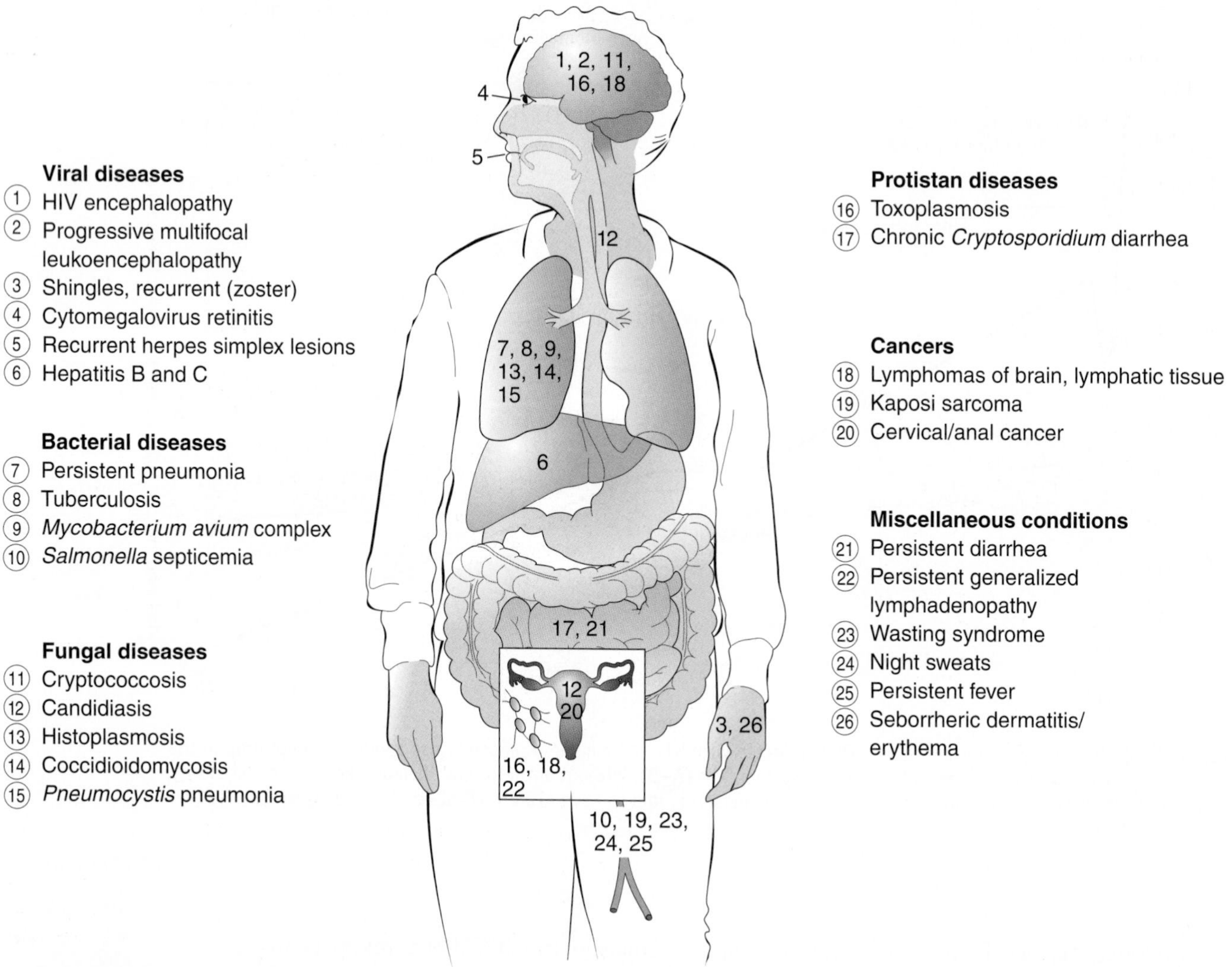

Viral diseases
1 HIV encephalopathy
2 Progressive multifocal leukoencephalopathy
3 Shingles, recurrent (zoster)
4 Cytomegalovirus retinitis
5 Recurrent herpes simplex lesions
6 Hepatitis B and C
Bacterial diseases
7 Persistent pneumonia
8 Tuberculosis
9 Mycobacterium avium complex
10 Salmonella septicemia
Fungal diseases
11 Cryptococcosis
12 Candidiasis
13 Histoplasmosis
14 Coccidioidomycosis
15 Pneumocystis pneumonia
Protistan diseases
16 Toxoplasmosis
17 Chronic Cryptosporidium diarrhea
Cancers
18 Lymphomas of brain, lymphatic tissue
19 Kaposi sarcoma
20 Cervical/anal cancer
Miscellaneous conditions
21 Persistent diarrhea
22 Persistent generalized lymphadenopathy
23 Wasting syndrome
24 Night sweats
25 Persistent fever
26 Seborrheic dermatitis/ erythema
1, 2, 11, 16, 18
4
5
12
7, 8, 9, 13, 14, 15
6
17, 21
12 20
16, 18, 22
3, 26
10, 19, 23, 24, 25

FIGURE 23.21 **Defining Conditions in AIDS Patients.** The variety of opportunistic illnesses that affect the body as a result of infection with HIV. Note the various systems that are affected and the numerous organisms involved. »» Why is there such a diverse group of potential pathogens that can cause opportunistic infections?

HIV
CCR5
Helper T cell
CD4
(A)
(B)

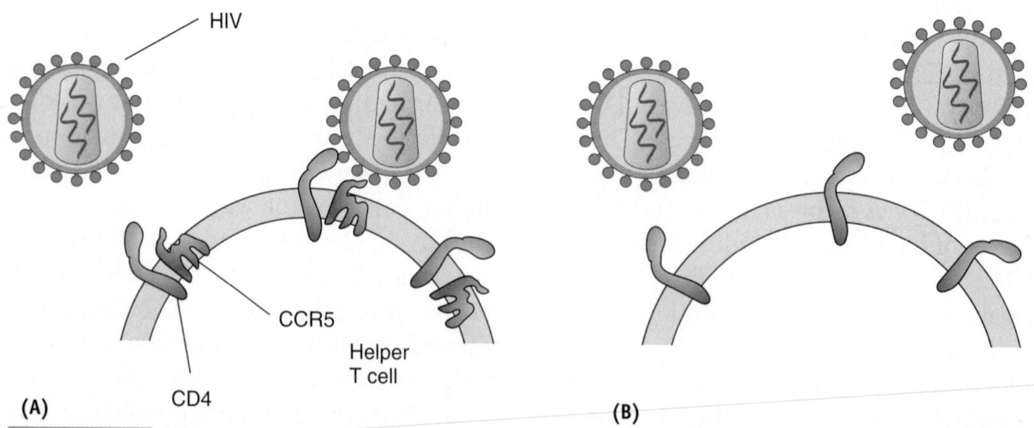

FIGURE 23.22 **HIV Attachment.** (**A**) In order to infect a host cell, like a helper T cell, the HIV spikes must recognize and bind to two receptors, CD4 and CCR5, on the host cell plasma membrane. (**B**) A few individuals naturally lack the CCR5 coreceptor and, as such, the virus cannot attach to the cells. »» What would happen if the helper T cells lacked the CD4 receptor?

Transmission. HIV can be transmitted by "risky behaviors," such as sharing of blood-contaminated needles with an HIV-infected person; having unprotected sexual contact, including vaginal, anal, or oral with an infected individual; or having unprotected intercourse with a person of unknown HIV status (FIGURE 23.23). During intercourse, rectal tissues bleed and lesions, cuts, or abrasions of the vaginal tract occur, which give HIV access to cells and tissues. Importantly, an individual who has a sexually transmitted disease (STD), including gonorrhea, syphilis, chlamydia, or genital herpes, is more susceptible to getting HIV infection during sex with infected partners. The use of condoms has been shown to decrease the transmission of HIV significantly.

Donated red blood cells and blood factor concentrates can contain the virus, but extensive tests now are performed in the United States and many other countries to preclude HIV being transmitted through a blood transfusion or blood products.

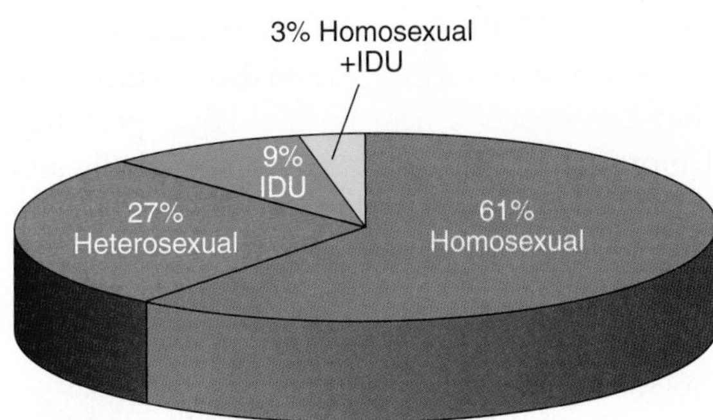

(A) United States

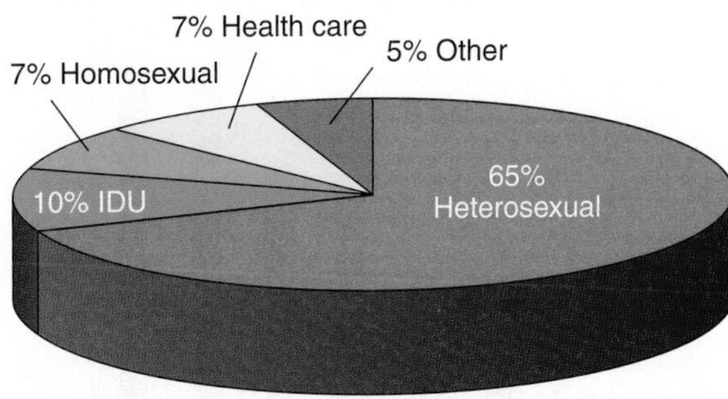

(B) World

FIGURE 23.23 **Number of New HIV Infections, by Mode of Transmission.** (**A**) Estimated number of new HIV infections in the United States has been roughly stable since the 1990s. The number of cases due to injected drug use (IDU) has dropped by 80% since 1990. (**B**) Globally, most cases are transmitted heterosexually. Data from: (A) www.cdc.gov/nchhstp/newsroom/docs/HIV-infections_2006-2009.pdf (B) www.avert.org/worlstatinfo.htm. »» In (B), what might be a transmission mechanism in healthcare settings?

Healthcare workers are at risk of acquiring HIV during their professional activities, such as through an accidental needle stick. Individuals at high risk always should practice established infection-control procedures (**standard precautions**) even though chances are rare that an accidental stick with a contaminated needle or other medical instrument would transmit HIV.

HIV also can be transferred from an infected mother to her fetus (transplacental transfer) or baby at birth. An infected mother also can transmit HIV to her baby through breast milk.

Diagnosis. Because early HIV infection may be asymptomatic and some 20% of HIV-positive Americans are unaware of their infection, increased opportunities for HIV testing are vital. As of 2009, six FDA-approved, rapid HIV antibody screening tests were available. These ELISA-type tests can help reduce unrecognized infections by improving access to testing in both clinical and community settings and by increasing the proportion of those tested who "immediately" (20 to 40 minutes) then learn the results (MicroFocus 23.5).

For HIV antibody tests, if a person is tested and the result is negative, it may mean the person was tested prior to seroconversion, which is too early to detect HIV antibodies. A repeat test at a later time is recommended. The Western blot analysis is used to confirm a positive test result.

■ Western blot: A technique to identify protein constituents, such as from a virus.

Other tests include the **APTIMA assay**, which amplifies and detects HIV RNA, as a diagnosis of primary HIV-1 infection and the **viral load test**, which detects the RNA of HIV and is available to monitor HIV-1 virus circulating in the blood of patients with established infections.

Treatment. The first drug used for treatment was azidothymidine, commonly known as AZT. AZT interferes with reverse transcriptase activity and acts as a chain terminator as it inhibits DNA synthesis. Other antiretroviral drugs are discussed in the chapter on antimicrobial drugs. They include reverse transcriptase inhibitors and protease inhibitors that interfere with viral genome replication and the processing step of capsid production, respectively. Other antiretroviral drugs include the fusion inhibitors, which work by blocking viral entry into the CD4 cells and integrase that blocks provirus formation.

HIV can become resistant to any of these drugs when the drug is used singly; that is, as **monotherapy** (FIGURE 23.24). Therefore, a more effective treatment requires a combination (or "cocktail") of drugs.

MICROFOCUS 23.5: Tools

Checking for HIV Infection—At Home

If you thought you might be infected with HIV, would you be more likely to: (a) be too afraid to be tested; (b) go to a clinic or family physician to be tested; (c) go to the drugstore and purchase an HIV testing kit that is sent to a lab for processing; or (d) go to your medicine cabinet and do the test right at home? The latter is now possible.

A quick HIV test, called the OraQuick® In-Home HIV Test, is 99.9% accurate in people not infected with HIV and 91.7% accurate in people infected with HIV.

The test works as follows: Oral fluid is collected by gently swabbing completely around the outer gums, both upper and lower, one time around, using the flat pad of the detection device. The device is then inserted into the vial of developer solution. The result is visually readable in about 20 to 40 minutes.

If one reddish band appears on the device at the control (C) location, the test result is negative for HIV antibodies (see figure). If two reddish bands appear on the device at both the control (C) location and the test (T) location, the test is "positive," meaning the result is a preliminary positive for HIV antibodies. Importantly, this is a "preliminary" result and does not definitely mean the person is HIV-positive. Rather, the individual would need to go to a clinic to have a second test done to confirm the result. *The person is considered HIV-positive only if the confirmatory test result is positive.*

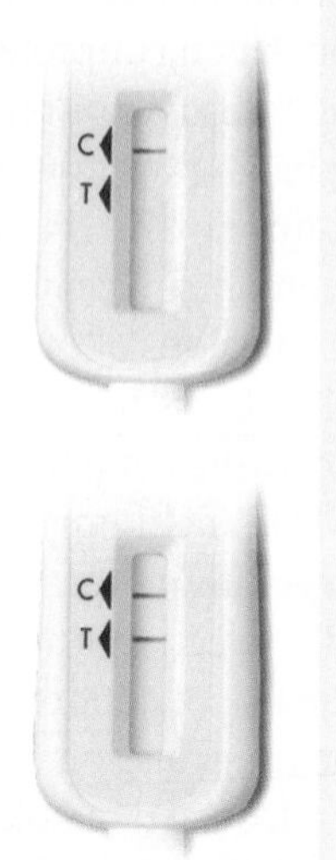

The OraQuick® test stick. Courtesy of OraSure Technologies, Inc.

When three or more drugs are used together, the combination is referred to as **highly active antiretroviral therapy** (**HAART**). Although HAART is not a cure, it has been significant in reducing the risk of HIV transmission as well as the number of deaths. Overall, antiretroviral therapy has extended the life of HIV/AIDS patients. In fact, many patients are being told that if they stay on their medications, they can expect a near normal life expectancy. Unfortunately, the drugs are scarce globally and available to too few individuals. In addition, HIV can still hide out from drug attacks (MICROFOCUS 23.6).

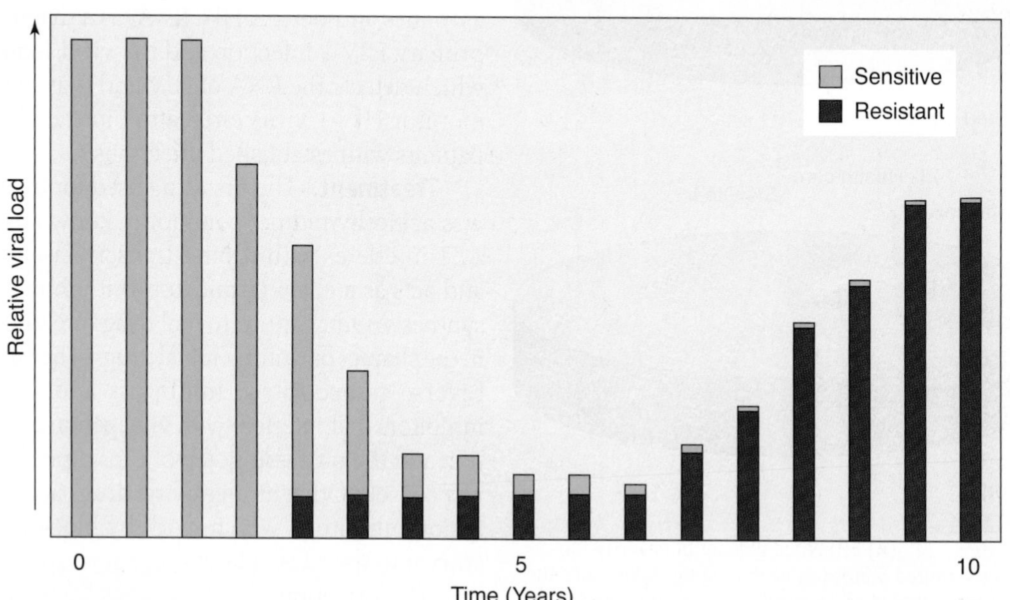

FIGURE 23.24 **Viral Load and Drug Resistance.** Because HIV replicates at a high rate and the reverse transcriptase enzyme is error prone, mutations will occur. After several years of drug monotherapy, the antiretroviral drug may have eliminated most of the original strain. However, some mutations will have conferred drug resistance and these viruses can now replicate free of any inhibition to that drug. »» What is the most logical way to combat drug resistance to monotherapy treatment?

MICROFOCUS 23.6: Miscellaneous

The HIV Hideouts

Even though the levels of HIV in an individual on HAART are extremely low, the person is not cured because antiretroviral therapy cannot completely eradicate the virus from the body. Therefore, patients who can afford the drugs must keep taking drugs for their entire life. If they stop, the virus comes back with a vengeance.

One place HIV hides out is in memory T cells, a subgroup of T cells that gives us immunity to diseases, such as measles and chickenpox, we have previously experienced in our lives. Many of the infected cells are destroyed by the infection itself or by attack from cytotoxic T cells. However, some memory T cells remain inactive and the HIV provirus in such cells remains latent. These resting T cells will not be destroyed by antiretroviral drugs because the drugs only are effective in actively dividing cells. Should the memory T cells reactivate, the provirus also becomes active and starts making more viruses.

Other hideouts also exist (see figure). One is the central nervous system (CNS) where HIV-associated dementia arises from the production of neurotoxins released by infected macrophages in the brain. Such macrophages usually become infected in some other tissue but then cross the blood-brain barrier and take up residence in the CNS, where new viruses can now infect specialized macrophages in the CNS called microglia. Thus, even individuals on HAART can develop this form of dementia because the blood-brain barrier prevents the entry of potentially effective drugs.

HIV also hides out in other areas where antiretroviral drugs seem to be ineffective. This includes macrophages and dendritic cells in the gastrointestinal tract and genital tract. Interestingly, the semen in male HIV patients often contains HIV genomes even though the blood appears to be clear of the virus. If these and other hideouts are to be destroyed, new drugs must be developed that can seek out and destroy the HIV hideouts.

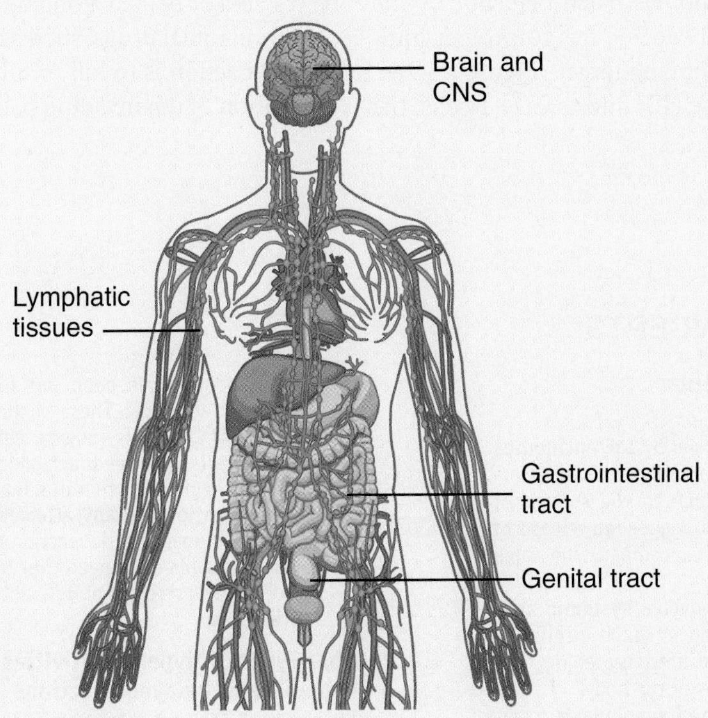

Prevention. Until recently, the only way to prevent HIV infection was to avoid behaviors putting a person at risk, such as sharing needles or having unprotected sex. However, in 2012, the FDA approved the first medication to "prevent" HIV infection. The drug consists of two reverse transcriptase inhibitors, tenofovir and emtricitabine (Truvada®), that can be taken once daily by adults who are HIV-negative but are at risk of becoming infected. Importantly, the drug combination, along with other safe sex practices, reduces substantially (up to 75%) the risk of acquiring an HIV-1 infection; it is not a 100% effective prevention strategy.

A successful vaccine for HIV disease and AIDS has yet to be developed. Two types of vaccines are being considered. **Preventive vaccines** (such as Truvada®) for HIV-negative individuals (health-care providers) could be given to prevent infection. **Therapeutic vaccines**, for HIV-positive individuals, would help the individual's immune system control HIV either by blocking virus entry, provirus formation, or replication such that the disease could not progress or be transmitted to others.

There are two major reasons why a successful vaccine has been lacking. First, HIV continually mutates and recombines. Such behavior by the virus means that a vaccine has to protect individuals against a moving target (Figure 23.24). In addition, because HIV infects CD4 T cells, the vaccine needs to activate the very cells infected by the virus. Although at times it appears that vaccine research is an uphill battle, scientists are optimistic that a safe, effective, affordable, and stable HIV vaccine can be produced. Importantly, progress in basic and clinical research is moving forward and scientists are inching closer to identifying products suitable for a successful HIV vaccine.

CONCEPT AND REASONING CHECKS 4

a. Compare and contrast the symptoms of the primary immunodeficiency disorders.
b. Distinguish between the three stages of HIV disease and AIDS.

■ **In conclusion**, we have now examined in full circle how the immune system functions to prevent and eliminate infections in the body. We have learned how vaccines can be used to prevent many infectious diseases, AIDS being one of the major exceptions. And we have seen what happens when the immune system overreacts or is deficient in immune reactions to a foreign agent. It is amazing though to realize that with the number of daily "pathogen attacks," our immune system wins the battle almost every time. Once in a while it might falter and often the weakened immune system can be rescued or helped along by the administration of antimicrobial drugs, such as antibiotics, whose mode of action is to kill or inhibit the growth or replication of the invading pathogen.

■ SUMMARY OF KEY CONCEPTS

23.1 Type I Hypersensitivity Represents a Familiar Allergic Response

1. **Type I hypersensitivity** is caused by **IgE antibodies** produced in response to certain antigens called **allergens**. The antibodies can attach to the surface of **mast cells** and **basophils**, and trigger the release of the mediators on subsequent exposure to the same allergen. (Fig. 23.3)
2. Type I hypersensitivities can involve **systemic anaphylaxis**, a whole-body reaction in which a series of mediators induces vigorous and life-threatening contractions of the smooth muscles of the body.
3. **Atopic diseases** involve **localized anaphylaxis** (common allergies), such as hay fever or a food allergy. (Fig. 23.5)
4. **Asthma** is an inflammatory disease involving an early response to an allergen and a late response by eosinophils and neutrophils that cause tissue injury and airway blockage.
5. Several ideas have been put forth concerning the nature of allergies. These include the shielding of IgE-secreting B cells (plasma cells) by IgA-secreting plasma cells and feedback mechanisms where IgE limits its own production in nonatopic individuals.
6. **Desensitization therapy** attempts to limit the possibility of an anaphylactic reaction through the injection of tiny amounts of allergen over time. Therapy also can involve the presence of IgG antibodies as **blocking antibodies.**

23.2 Other Types of Hypersensitivities Represent Immediate or Delayed Reactions

7. In **type II cytotoxic hypersensitivity**, the immune system produces IgG and IgM antibodies, both of which react with the body's cells and often destroy the latter. (Fig. 23.9)
8. No cells are involved in **type III hypersensitivity.** Rather, the body's IgG and IgM antibodies interact with dissolved antigen molecules to form visible

immune complexes. The accumulation of immune complexes in various organs leads to local tissue destruction in such illnesses as **serum sickness, Arthus phenomenon**, and **systemic lupus erythematosus.** (Fig. 23.11)

9. **Type IV cellular hypersensitivity** involves no antibodies, but is an exaggeration of cell-mediated immunity based in T lymphocytes. **Contact dermatitis** and **infection allergies** are manifestations of this hypersensitivity.

23.3 Autoimmune Disorders and Transplantation Are Immune Responses to "Self"

10. **Autoimmune disorders** can occur through defects in **clonal deletion, clonal anergy**, or **regulatory T-cell activity**. Human heredity, as well as access to **privileged sites** and **antigenic mimicry**, also can trigger an autoimmune response. (Fig. 23.14)
11. Four types of grafts or transplants can be performed: **autografts, isografts, allografts**, and **xenografts.** Allografts are the most common. Rejection of grafts or transplants involves cytotoxic T cells and antibodies. The graft also can be rejected by immune cells in the graft that reject the recipient (**graft-versus-host reaction**).
12. Prevention of rejection is strengthened by using immunosuppressive agents, including antimetabolites, immune cell inhibitors, anti-inflammatory drugs, and monoclonal antibodies.

23.4 Immunodeficiency Disorders Can Be Inherited or Acquired

13. **Immunodeficiency disorders** may be congenital (**primary immunodeficiencies**) or acquired later in life (**secondary immunodeficiencies**).
14. **AIDS** (**acquired immunodeficiency syndrome**) is the final stage of the HIV (human immunodeficiency virus) infection. HIV infects and destroys CD4 T cells, eventually leading to an inability of the immune system to fend off opportunistic diseases. Transmission is through blood, blood products, contaminated needles, or unprotected sexual intercourse. Many **antiretroviral drugs** are available to slow the progression of disease. (Figs. 23.19, 23.20)

CHAPTER SELF-TEST

For **STEPS A–D**, answers to even-numbered questions and problems can be found in **Appendix C** on the Student Companion Website at **http://microbiology.jbpub.com/10e**. In addition, the site features eLearning, an online review area that provides quizzes and other tools to help you study for your class. You can also follow useful links for in-depth information, read more MicroFocus stories, or just find out the latest microbiology news.

STEP A: REVIEW OF FACTS AND TERMS

Multiple Choice

Read each question carefully, then select the ***one*** answer that best fits the question or statement.

1. All the following are types of immediate hypersensitivities *except* _____.
   A. asthma
   B. contact dermatitis
   C. food allergies
   D. hay fever
2. Systemic anaphylaxis is characterized by _____.
   A. contraction of smooth muscles
   B. a red rash
   C. blood poisoning
   D. hives
3. Which of the following is NOT a type I hypersensitivity?
   A. Food allergies
   B. Contact dermatitis
   C. Allergic rhinitis
   D. Exercise-induced allergies
4. The early response in asthma is due to _____ activity.
   A. cytotoxic T cell
   B. basophil
   C. T_H2 cell and NK cell
   D. dendritic cells
5. What type of immune cell may control IgE-mediated hypersensitivities?
   A. Suppressor T cells
   B. Plasma cells
   C. Cytotoxic T cells
   D. Neutrophils
6. Desensitization therapy can involve _____.
   A. the use of blocking antibodies
   B. injections of small amounts of allergen
   C. allergen injections of several months
   D. All the above (**A–C**) are correct.
7. A cytotoxic hypersensitivity would occur if blood type _____ is transfused into a person with blood type _____.
   A. A; AB
   B. O; AB
   C. A; O
   D. O; B
8. Serum sickness is a common symptom of _____.
   A. contact dermatitis
   B. hemolytic disease of the newborn
   C. immune complex hypersensitivity
   D. food allergies

9. Which one of the following allergens is NOT associated with contact dermatitis?
A. Foods
B. Cosmetics
C. Poison ivy
D. Jewelry

10. Immunological tolerance to "self" is established by _____.
A. destruction of self-reactive lymphoid cells
B. clonal anergy
C. clonal deletion
D. All the above (**A–C**) are correct.

11. A _____ is a graft between genetically different members of the same species.
A. xenograft
B. autograft
C. allograft
D. isograft

12. Immunosuppressive agents used in preventing transplant rejection primarily affect _____.
A. macrophages
B. neutrophils
C. dendritic cells
D. T cells

13. What immunodeficiency disorder is associated with a lack of T and B cells and complete immune dysfunction?
A. DiGeorge syndrome
B. Severe combined immunodeficiency disease
C. Chronic granulomatous disease
D. Chédiak-Higashi syndrome

14. A patient with a CD4 T cell count of 1,000/ul would be in the _____ of HIV disease/AIDS.
A. Secondary infection stage
B. Acute phase
C. AIDS stage
D. Asymptomatic phase

Rearrangement

This chapter has summarized some of the disorders associated with the immune system. To gauge your understanding, rearrange the scrambled letters to form the correct word for each of the spaces in the statements.

15. The simple compound _______________ is one of the major mediators released during allergy reactions.

I T M E H I A S N

16. An immune deficiency called _____________ syndrome is characterized by the failure of T lymphocytes to develop.

I D O G G E R E

17. Cases of rheumatoid arthritis are accompanied by immune complex formation in the body's _______________.

N I S T J O

18. In a _______________ hypersensitivity, antibodies unite with cells and trigger a reaction that results in cell destruction.

X Y O T C C I T O

19. Hay fever is an example of an _____________ disease, one in which a local allergy takes place.

O C T I A P

20. Immune complex hypersensitivities develop when antibody molecules interact with _______________ molecules and form aggregates in the tissues.

E N I G N A T

21. The skin test for _______________ relies on a response by T lymphocytes to PPD placed in the skin tissues.

U U I C T R L S B E O S

22. Mast cells and _______________ are the two principal cells that function in anaphylactic responses.

S S B I O H P A L

23. A key element in transplant acceptance or rejection is a set of molecules abbreviated as ____________________ proteins.

L A H

24. Urticaria is a form of skin ____________________ occurring in a person having an allergic reaction.

A H S R

25. The HIV DNA integrated into a chromosome is called a ____________________.

S R V R U P O I

STEP B: CONCEPT REVIEW

26. Summarize the events occurring in and the role of humoral immunity in **type I hypersensitivities.** (**Key Concept 1**)

27. Explain the initiation and outcomes of **systemic anaphylaxis.** (**Key Concept 1**)

28. Distinguish between the different forms of **atopic disease.** (**Key Concept 1**)

29. Compare **asthma** to other forms of type I hypersensitivities. (**Key Concept 1**)

30. Discuss the reasons why allergies develop. (**Key Concept 1**)

31. Identify the therapies and treatments available for type I hypersensitivities. (**Key Concept 1**)

32. Summarize the characteristics of **type II hypersensitivity** and its relationship to blood transfusions and **hemolytic disease of the newborn.** (**Key Concept 2**)

33. Summarize the characteristics of **type III hypersensitivity** and its relationship to **serum sickness** and the **Arthus reaction.** (**Key Concept 2**)

34. Summarize the characteristics of **type IV hypersensitivity** and its relationship to **infection allergies** and **contact dermatitis.** (**Key Concept 2**)

35. Identify the ways in which an **autoimmune disease** can arise. (**Key Concept 3**)

36. Describe the immunological reasons why **organ transplants** are rejected and list the four types of grafts (transplants). (**Key Concept 3**)

37. Assess the usefulness of **immunosuppressive drugs** by transplant patients. (**Key Concept 3**)

38. Contrast **primary** and **secondary immunodeficiencies** and list several primary disorders and their accompanying immunological deficiencies. (**Key Concept 3**)

39. Diagram how the **human immunodeficiency virus** (**HIV**) infects a cell and identify the prevention and treatment methods used for **HIV disease** and **AIDS.** (**Key Concept 4**)

STEP C: APPLICATIONS AND PROBLEMS

40. During war and under emergency conditions, a soldier whose blood type is O donates blood to save the life of a fellow soldier with type B blood. The soldier lives, and after the war becomes a police officer. One day he is called to donate blood to a brother officer who has been wounded and finds that it is his old friend from the war. He gladly rolls up his sleeve and prepares for the transfusion. Should it be allowed to proceed? Why?

41. Coming from the anatomy lab, you notice that your hands are red and raw and have begun peeling in several spots. This was your third period of dissection. What is happening to your hands, and what could be causing the condition? How will you solve the problem?

42. "He had a history of nasal congestion, swelling of his eyes, and difficulty breathing through his nose. He gave a history of blowing his nose frequently, and the congestion was so severe during the spring he had difficulty running." The person in this description is former President Bill Clinton, and the writer is an allergist from Little Rock, Arkansas. What condition (technically known as allergic rhinitis) is probably being described?

STEP D: QUESTIONS FOR THOUGHT AND DISCUSSION

43. As part of an experiment, one animal is fed a raw egg while a second animal is injected intravenously with a raw egg. Which animal is in greater danger? Why?

44. A woman is having the fifth injection in a weekly series of hay fever shots. Shortly after leaving the allergist's office, she develops a flush on her face, itching sensations of the skin, and shortness of breath. She becomes dizzy, then faints. What is taking place in her body, and why has it not happened after the first four injections?

45. You may have noted that brothers and sisters are allowed to be organ donors for one another, but that a person cannot always donate to his or her spouse. Many people feel bad about being unable to help a loved one in time of need. How might you explain to someone in such a situation the basis for becoming an organ donor and why it may be impossible to serve as one?

46. The immune system is commonly regarded as one that provides protection against disease. This chapter, however, seems to indicate that the immune system is responsible for numerous afflictions. Even the title is "Immune Disorders." Does this mean that the immune system should be given a new name? On the other hand, is it possible that all these afflictions are actually the result of the body's attempts to protect itself? Finally, why can the phrase "immune disorder" be considered an oxymoron?

47. In many diseases, the immune system overcomes the infectious agent, and the person recovers. In other diseases, the infectious agent overcomes the immune system, and death follows. Compare this broad overview of disease and resistance to what is taking place with AIDS, and explain why AIDS is probably unlike any other disease encountered in medicine.

24

CHAPTER PREVIEW

24.1 Antimicrobial Agents Are Chemical Substances Used to Treat Infectious Disease

24.2 Synthetic Antibacterial Agents Primarily Inhibit DNA Synthesis and Cell Wall Formation

24.3 Beta-Lactam Antibiotics Inhibit Bacterial Cell Wall Synthesis

24.4 Other Bacterially Produced Antibiotics Inhibit Some Aspect of Metabolism

24.5 Other Antimicrobial Drugs Target Viruses, Fungi, and Parasites

24.6 Antimicrobial Drug Resistance Is a Growing Challenge

Textbook Case 24: Severe Community-Acquired MRSA Pneumonia

Investigating the Microbial World 24: The Source of Antibiotic Resistance

MicroInquiry 24: Testing Drugs—Clinical Trials

Image courtesy of Dr. Fred Murphy/CDC.

Antimicrobial Drugs

We are facing a crisis because doctors are pressured to prescribe antibiotics for the common cold and inner ear infection, yet we know that it is not prudent to do so. We must collectively inform our patients about the reasons why overprescribing antibiotics will not help patients return to work sooner, and that in the long run, could make them more susceptible to drug-resistant diseases.

—Richard E. Besser, M.D., former Acting Director, Centers for Disease Control and Prevention

For centuries, physicians believed heroic measures were necessary to save patients from the ravages of infectious disease. They prescribed frightening courses of purges (bowel emptying), enormous doses of strange chemical concoctions, blood-curdling ice water baths, deadly starvations, and bloodlettings. These treatments probably complicated an already bad situation by reducing the natural body defenses to the point of exhaustion. In fact, the death of George Washington in 1799 is believed to have been due to a streptococcal infection of the throat, perhaps exacerbated by the bloodletting treatment that removed almost two liters of his blood within a 24-hour period.

When the germ theory of disease emerged in the late 1800s, insights about microorganisms added considerably to the understanding of disease and increased the storehouse of knowledge available to the doctor. However, it did not change the fact that little, if anything, could be done for the infected patient. Then, in the 1940s, antimicrobial agents, including antibiotics, burst on the scene, and another revolution in medicine began.

Doctors were astonished to learn they could kill microorganisms in the body without doing substantial harm to the body itself. Medicine had a period of powerful, decisive growth, as doctors found they could successfully alter the course of infectious disease. Antibiotics effected a radical change in medicine and charted a new course for treating infectious disease. Since the 1940s, millions of lives have been saved. In 1969, then U.S. Surgeon General William Stewart and many pharmaceutical companies believed it was time to "close the books on infectious diseases."

Unfortunately, it soon became clear that bacterial species and other microorganisms could quickly develop resistance to antimicrobials. This is no clearer today than the rise to public attention of multidrug resistance in *Mycobacterium tuberculosis* and *Neisseria gonorrhoeae,* and in *Clostridium difficile* in the healthcare environment along with the spread of methicillin-resistant *Staphylococcus aureus* (MRSA) from the hospital to community. (FIGURE 24.1). In fact, the World Health Organization (WHO) sees antibiotic resistance as one of the greatest threats to human health.

There have been several reasons for the increase in antibiotic resistance. The general public often misuses or abuses antibiotics. Individuals, for example, might stop taking an antibiotic as soon as they start feeling better; they do not finish the course of antibiotic treatment; they take an antibiotic for a viral infection like a cold or the flu; they take an antibiotic that was prescribed for some other illness; or they take an antibiotic that was prescribed for someone else.

In addition, many doctors still prescribe antibiotics for diseases that are untreatable with such drugs. Again, this includes viral infections such as colds and the flu. Today in the United States, doctors prescribe an estimated 10 million ineffective and unnecessary prescriptions for childhood infections. Often this is at the insistence of the patient, as the opening quote for this chapter states.

We cannot live without antimicrobials, but we need to use these drugs wisely.

In this chapter, we discuss the antimicrobial drugs as the mainstays of our healthcare delivery system to treat bacterial, viral, fungal, and parasitic infections and diseases. Many drugs have been known for decades, while some are of recent development. We explore their discovery and examine their uses—and misuses.

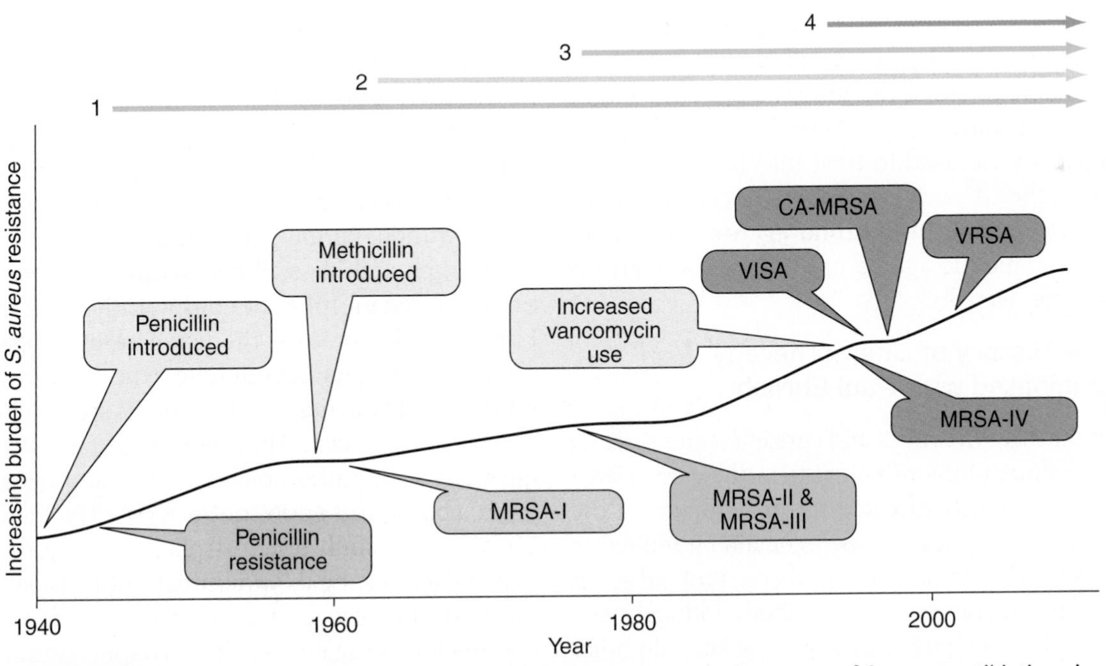

FIGURE 24.1 **Evolution of *Straphylococcus aureus* Antibiotic Resistance.** Four waves of *S. aureus* antibiotic resistance have added to the burden of antibiotic resistance. Wave 1 began soon after penicillin was introduced and such strains remain today. Wave 2 followed the introduction of methicillin with MRSA-I, which no longer exists today. Wave 3 began in the late 1970s as other strains evolved from MRSA-I and other hospital strains (MRSA-II and -III), which have caused outbreaks in hospitals and healthcare facilities. Wave 4 has included additional strains (MRSA-IV) and, with the use of vancomycin to treat MRSA infections, vancomycin-intermediate strains (VISA). Notable is the emergence of new MRSA strains in the community (CA-MRSA) and several vancomycin-resistant (VRSA) strains. Modified from: Chambers and DeLeo (2009). *Nature Reviews Microbiology* 7: 629–641. »» What types of "genetic events" could be responsible for the appearance of these new antibiotic-resistant strains of *S. aureus*?

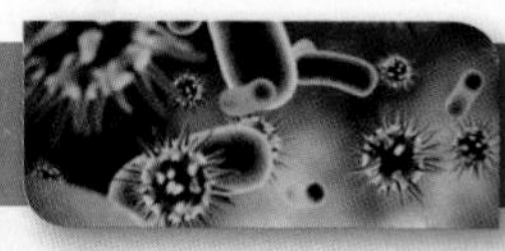

Chapter Challenge

Over the last few years, there has been an increasing number of reports of superbugs, microbial species that have become resistant to multiple antimicrobial agents. Many of these are bacterial species, such as methicillin-resistant *Staphylococcus aureus* (MRSA), multidrug-resistant *Mycobacterium tuberculosis,* and a prominent nosocomial pathogen *Pseudomonas aeruginosa.* But now a bigger threat may loom. A strain of *Klebsiella pneumoniae,* a bacterial species that can cause pneumonia and bloodstream infections in hospitalized patients, was isolated from the urine of a hospitalized patient who had visited India. This strain contained a transmissible genetic element encoding multiple drug resistance. In fact, a bacterial cell carrying this genetic element, called NDM-1, along with its already established multidrug resistance, makes the cell essentially resistant to all available antibiotics. In August 2010, a Belgian man, who had become infected with *K. pneumoniae* after a car accident in Pakistan, died in Belgium despite being administered an antibiotic, colistin, which should have cured the infection. It was discovered that the pathogen carried a plasmid with the NDM-1 genetic element. To date, strains of *Klebsiella, Escherichia,* and *Acinetobacter* are known to possess the NDM-1 genetic element. Let's find out what all the fuss is about concerning NDM-1 as we explore antimicrobial drugs.

KEY CONCEPT 24.1 Antimicrobial Agents Are Chemical Substances Used to Treat Infectious Disease

As described in this chapter opener, beginning in the late 1940s, a group of chemicals became available to assist the immune system to fight infection and disease. These **chemotherapeutic agents** were used to treat infections, diseases, and other disorders, such as cancer. In microbiology, the **antimicrobial agents**, those chemotherapeutic agents used to treat infectious disease, were key.

The History of Chemotherapy Originated with Paul Ehrlich

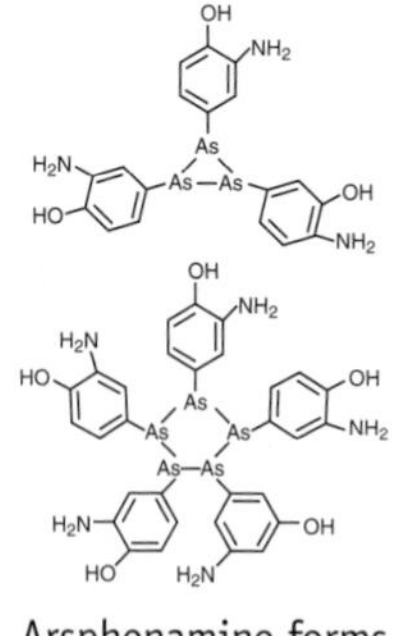

Arsphenamine forms

In the drive to control and cure infectious disease, the efforts of microbiologists in the early 1900s were primarily directed toward enhancing the body's natural defenses. Sera containing antibodies lessened the impact of diphtheria, typhoid fever, and tetanus; and effective antibody-inducing vaccines for smallpox and rabies (and later, diphtheria and tetanus) reduced the incidence of these diseases.

Among the leaders in the effort to control disease was an imaginative German investigator named Paul Ehrlich. Ehrlich knew that specific dyes would stain specific bacterial species. Therefore, he believed there must be specific chemicals that would be toxic to these species. This "selective toxicity" concept was developed in the early 1900s when Ehrlich thought he could discover molecules that would be "magic bullets"—specific chemicals that would seek out and destroy specific disease organisms in infected tissues without harming those tissues.

Ehrlich and his staff had synthesized hundreds of arsenic-phenol compounds. One of Ehrlich's collaborators, the Japanese investigator Sahachiro Hata, set out to test the chemicals for their ability to destroy the syphilis spirochete *Treponema pallidum.* After months of painstaking study, Hata's attention focused on arsphenamine, compound #606 in the series. Hata and Ehrlich successfully tested arsphenamine against *T. pallidum* in animals and human subjects, and in 1910, they gave a derivative of the drug to doctors for use against syphilis. Arsphenamine, the first modern synthetic antimicrobial agent, was given the brand name Salvarsan because it offered salvation from syphilis and contained arsenic.

Salvarsan met with mixed success during the ensuing years. Its value against syphilis was without question, but local reactions at the injection site, and indiscriminate use by some physicians, brought adverse publicity. Ehrlich's death in 1915, together with the general ignorance of organic chemistry and the impending first World War, further eroded enthusiasm for chemotherapy. However, the team approach to drug discovery used by Ehrlich would become the model for modern pharmaceutical research.

Over the next 20 years, German chemists continued to synthesize and manufacture dyes for fabrics and other industries, and they routinely tested their new products for antimicrobial qualities. In 1932, one of these products was a red dye, trademarked as Prontosil.

Prontosil had no apparent effect on bacterial cells in culture. However, things were different in animals, where the drug is converted into an active antimicrobial form. When Gerhard Domagk, a German pathologist and bacteriologist, tested Prontosil in animals, he found a pronounced inhibitory effect on staphylococci, streptococci, and other gram-positive bacterial species. In February 1935, Domagk injected the dye into his daughter Hildegard, who had become gravely ill with septicemia after pricking her finger with a needle. Hildegard's condition gradually improved and her arm did not have to be amputated. Many historians see her recovery as setting into motion the age of modern chemotherapy. For his discovery, Domagk was awarded the 1939 Nobel Prize in Physiology or Medicine.

■ Chemotherapy: The use of chemical agents to treat diseases or disorders.

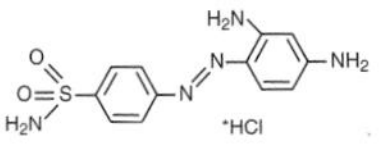

Prontosil

Fleming's Observation of the Penicillin Effect Ushered in the Era of Antibiotics

One of the first to postulate the existence and value of antibiotics was the British microbiologist Alexander Fleming (FIGURE 24.2A). During his early years, Fleming experienced the excitement of the classical Golden Age of microbiology and spoke up for the therapeutic value of Salvarsan. In 1928, Fleming was performing research on staphylococci at St. Mary's Hospital in London. On one of these nutrient agar plates, he noted a blue-green mold contaminating the plate. Surrounding the mold no bacterial colonies were growing (FIGURE 24.2B). Intrigued by the failure of staphylococci to grow near the mold, Fleming isolated the mold, identified it as a species of *Penicillium*, and found it produced a substance that kills gram-positive bacterial species. Though he failed to isolate the elusive substance, he named it "penicillin."

(A)

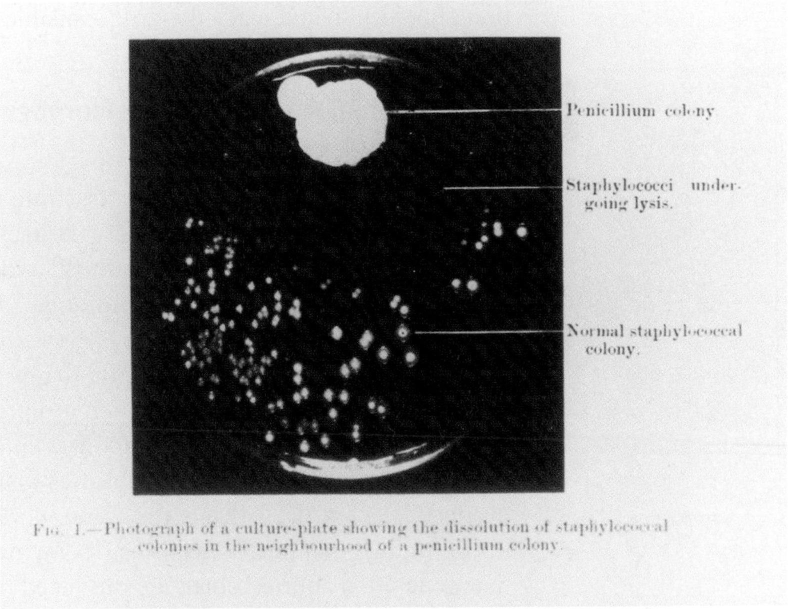

Penicillium colony

Staphylococci undergoing lysis.

Normal staphylococcal colony.

Fig. 1.—Photograph of a culture-plate showing the dissolution of staphylococcal colonies in the neighbourhood of a penicillium colony.

(B)

FIGURE 24.2 Alexander Fleming and His Culture of *Penicillium*. (**A**) Alexander Fleming reported the existence of penicillin in 1928 but was unable to purify it for use as an antimicrobial agent. © Science Source, photo by Dean Pausett/Photo Researchers, Inc. (**B**) The actual photograph of Fleming's culture plate shows how staphylococci in the region of the *Penicillium* colony have been killed (they are "undergoing lysis") by some unknown substance produced by the mold. Fleming called the substance penicillin. © National Library of Medicine. »» Why haven't the staphylococcal colonies farther from the mold also undergone lysis?

Fleming was not the first to note the antibacterial qualities of *Penicillium* species. Joseph Lister had observed a similar phenomenon in 1871; John Tyndall did likewise in 1876; and a French medical student, Ernest Duchesne, wrote a research paper on the subject in 1896. However, it was Fleming who first proposed that penicillin could be used to eliminate gram-positive bacteria from mixed cultures. Further, he unsuccessfully tried to use a filtered broth from the fungus on infected wound tissue. Unfortunately, biochemistry was not sufficiently advanced to make complex separations possible, and Fleming's discovery soon was forgotten.

In 1935, Gerhard Domagk's dramatic announcement of the antimicrobial effects of Prontosil again fueled speculation that chemicals could be used to fight disease in the body. Then, in 1939, a group at England's Oxford University, led by pathologist Howard Florey and biochemist Ernst Chain, reisolated Fleming's penicillin and conducted trials with highly purified samples. They discovered that penicillin was effective against a large variety of diseases, including gonorrhea, meningitis, tetanus, and diphtheria. However, England was already involved in World War II, so a group of American companies developed the techniques for the large-scale production of penicillin and made the drug available for commercial use.

■ Efficacy: The ability to produce the necessary or desired results.

Antimicrobial Agents Have a Number of Important Properties

The preceding historical accounts illustrate the origins for the two groups of antimicrobial agents. Those drugs, like Salvarsan and Prontosil and all antiviral and many antibacterial, antifungal, and antiparasite agents, which are made (synthesized) in the laboratory, are called **synthetic drugs**. By contrast, drugs like penicillin, which are products of, or derived from, the metabolism of living microorganisms are called **antibiotics** (*anti* = "against"; *biosis* = "life"); that is, the agents work to kill or inhibit other microorganisms. Today, many new generation antimicrobials are produced by a process that chemically modifies an antibiotic of microbial origin. These are called **semisynthetic drugs**. For discussion, all these antimicrobials will be referred to as antibiotics.

The antibiotic drugs have important properties that need to be considered when prescribing a drug for an infection or disease.

Selective Toxicity. Ehrlich's idea of a magic bullet was based on **selective toxicity**, which says that an antimicrobial drug should harm the pathogen but not the host. Today, two terms are used when considering the toxicity of a drug. The **toxic dose** refers to the concentration of the drug causing harm to the host. The **therapeutic dose** refers to the concentration of the drug that effectively destroys (**microbicidal**) or inhibits (**microbiostatic**) the pathogen. Together these are used to formulate the **chemotherapeutic index**, which is the highest concentration (per kilogram of body weight) of the drug tolerated by the host divided by the lowest concentration (per kilogram body weight) of the drug that will eliminate the infection or disease agent.

The chemotherapeutic index of each antimicrobial drug must be considered, and the efficacy in eliminating disease and providing symptom relief must outweigh the associated toxicity and adverse events (FIGURE 24.3).

As we will see below, the best way to accomplish this is to develop drugs that target a specific component of microbial cells, such as the cell wall or a certain metabolic pathway, which is absent in human cells.

Antimicrobial Spectrum. Another important property in prescribing an appropriate drug is identifying the range of pathogens to which a particular drug will work. This range of antimicrobial action is the **antimicrobial spectrum**. Using the domain Bacteria as an example, all the antibacte-

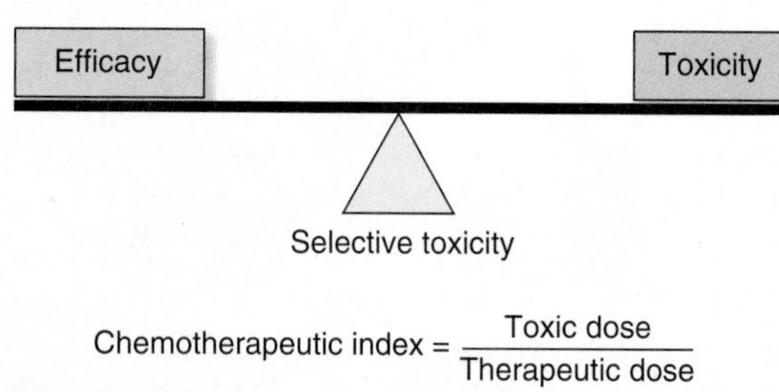

FIGURE 24.3 **A Representation of the Chemotherapeutic Index.** Finding the correct drug means considering the efficacy of the drug and its toxicity to the pathogen being considered. »» Determine whether a drug with a high or low chemotherapeutic index would be the more desirable.

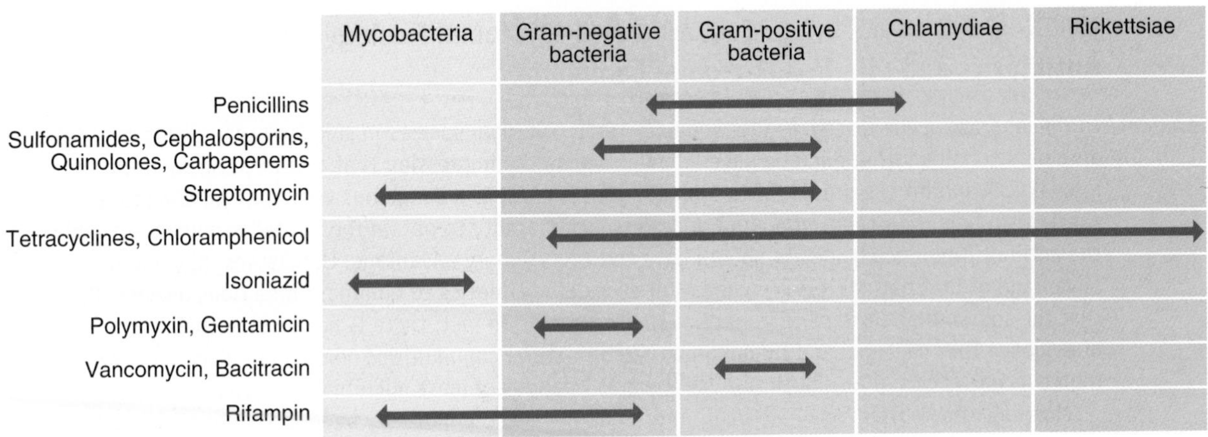

FIGURE 24.4 **The Antimicrobial Spectrum of Activity.** Antimicrobial drugs have a limited range of activity in the domain Bacteria. »» Using the figure, which drugs would have a broad spectrum and which ones would have a narrow spectrum?

rial drugs affecting many taxonomic groups are considered as having a **broad spectrum** of action, while those affecting few pathogens have a **narrow spectrum** of action (FIGURE 24.4). We will identify the drug spectrum with each antimicrobial drug discussed.

Antibiotics Are More Than Agents of Natural Biological Warfare

Before we proceed to examine specific antibiotics, it is worth noting the reasons why some bacterial and fungal species produce antibiotics. Antibiotics represent **secondary metabolites**—that is, compounds not directly used for growth and reproduction. They often are produced during the stationary phase of growth, when an antibiotic would be less likely to have an adverse effect on the antibiotic producer's growth, yet when competition for limited nutrients is intense. It has been assumed that in the natural environment (soil, water), there can be fierce competition between the immense number of microbes for limited nutrients. Therefore, if a bacterial or fungal species has the ability to secrete an antibiotic, it may kill or inhibit the growth of those competitors sensitive to the chemical. Antibiotic production thus gives the producer a selective advantage for nutrients and space.

However, if antibiotics are so destructive, how can 1,000 different bacterial species still remain in the soil and not be killed? Maybe it is not always necessary to kill your neighbors. MICROFOCUS 24.1 describes some alternatives.

CONCEPT AND REASONING CHECKS 1

a. Evaluate the early discoveries of and uses for chemotherapeutic agents for selective toxicity.

b. Identify how penicillin was discovered and later developed into a useful drug.

c. Assess the importance of the chemotherapeutic index and the antimicrobial drug spectrum in prescribing treatment for an infectious disease.

d. What advantage is afforded by bacterial cells that secrete antibiotics in the soil?

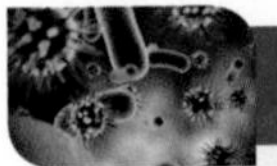

Chapter Challenge 

The multidrug-resistant *K. pneumoniae* strains, combined with the genetic element NDM-1, are simply adding to the battery of resistance genes that can be used to provide resistance against antibiotics. In addition, these superbug bacterial species are gram-negative (members of the Enterobacteriaceae), which, even without the presence of NDM-1, are historically harder to kill than their gram-positive brethren.

QUESTION A:

(a) What is it about the cell wall structure of the gram-negative species that makes them harder to treat and (b) what does this tell you about the chemotherapeutic index in trying to balance efficacy and toxicity in the affect patient?

Answers can be found on the Student Companion Website in **Appendix D.**

MICROFOCUS 24.1: Evolution and Microbial Ecology

Antibiotics and Life in a Microbial Community

In the laboratory, the discovery that some bacterial or fungal species in soil can use antibiotics as a weapon to kill another microbe gave microbiologists the impression that antibiotics have but one role—kill or inhibit the growth of competitors. After all, it is a dangerous world out there and if one species can eliminate or severely limit another species' ability to obtain the limited resources available, the antibiotic-producing species should have the evolutionary advantage. Understandably, doctors took advantage of that natural role by using high doses of antibiotics to eliminate infectious disease. But is that the sole role of antibiotics in natural environments? In fact, there is not an overwhelming amount of evidence that bacterial and fungal organisms in natural communities normally produce relatively high concentrations of antibiotics or use them as weapons of mass microbial destruction.

More microbiologists today are beginning to appreciate the mutually beneficial roles that low (sub-lethal) concentrations of antibiotics can play in the everyday life of a microbial community. Here are a few examples of what microbiologists are discovering (see figure).

Antibiotics cue gene activity. If *Salmonella typhimurium* is exposed to low concentrations of antibiotics, like streptomycin and rifampicin, the antibiotics influence the control of hundreds of genes involved in metabolism and cell behavior. The chemical communication molecules cue the recipient to activate some genes and repress others.

Antibiotics signal biofilm formation. Nystatin, an antifungal agent used medically to treat yeast infections, is produced by *Streptomyces noursei*. Naturally, nystatin is an intracellular trigger for *S. noursei* and an intercellular trigger for *Bacillus subtilis* to form biofilms. Other antibiotics that are microbicidal at high concentrations, such as streptomycin and tetracycline, can, at low concentrations, also turn on genes for biofilm formation. Such signaling benefits both the sender and recipient.

Antibiotics can manipulate microbial interactions. When *Pseudomonas aeruginosa* is exposed to tobramycin, an antibiotic produced by *Streptomyces tenebrarius*, the *Pseudomonas* cells become more motile and flee the area. The result is a lowering of the competition without the killing of the competition. The sender manipulated the recipient's behavior.

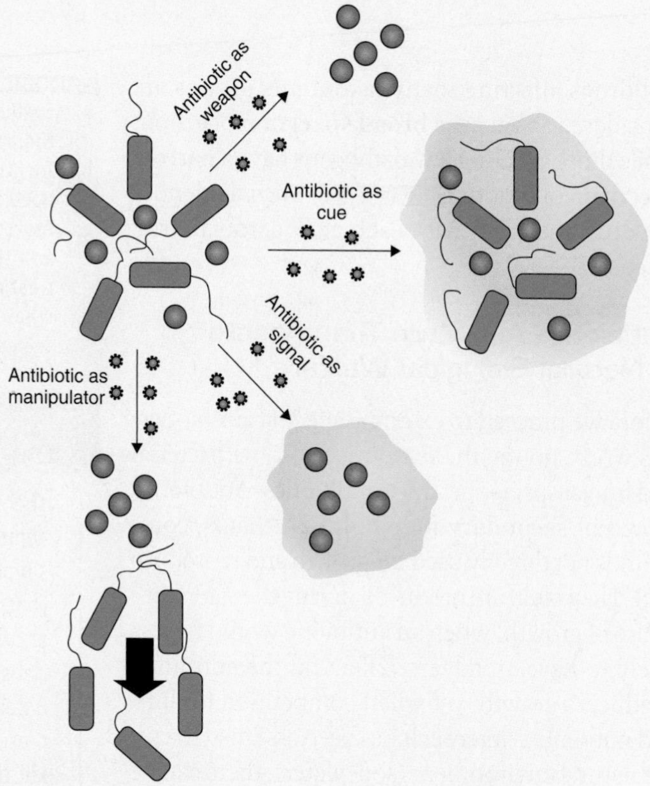

Possible roles of antibiotics. Adapted from: Ratcliff, W. C. and Denison, R. F. (2011). *Science* 332 (6029): 547–548.

So, why then do other bacterial species in the soil have antibiotic resistance genes? One hypothesis suggests that the products of these resistance genes fine tune, or modulate, the antibiotic levels to produce a more balanced signal; or, more radically, to mute the signal completely. Perhaps a biofilm is one way the organism reduces future exposure to the antibiotic.

As more is learned about the diverse role of antibiotics in nature, we will certainly come to better understand and appreciate antibiotics in the daily life of a microbial community, which may not be as dangerous a place for microbes as we once thought.

KEY CONCEPT 24.2 Synthetic Antibacterial Agents Primarily Inhibit DNA Synthesis and Cell Wall Formation

The work of Gerhard Domagk in identifying the usefulness for Prontosil set the stage for the identification of many other synthetic antimicrobials. In 1935, a group at the Pasteur Institute isolated the active form of Prontosil. They found it to be a chemical called sulfanilamide, which was highly active against gram-positive bacterial species; it quickly became a mainstay for treating wound-related infections during World War II.

Sulfonamides Target Specific Metabolic Reactions

Sulfanilamide was the first of a group of broad-spectrum synthetic agents known as **sulfonamides**. These so-called "sulfa drugs" interfere with the metabolism (bacteriostatic) of gram-positive and gram-negative bacterial cells without damaging body tissues. Here's how they work.

Bacterial cells synthesize an important growth factor called **folic acid** for use in nucleic acid synthesis. To produce folic acid, a bacterial enzyme joins together three important components, one of which is **para-aminobenzoic acid (PABA)**. This molecule is similar in chemical structure to the sulfonamide sulfamethoxazole (SMZ) (FIGURE 24.5). Therefore, treating patients with SMZ allows the drug to compete with PABA for the active site in the bacterial enzyme. Such **competitive inhibition** blocks folic acid synthesis, which prevents nucleic acid synthesis and DNA replication. However, drug resistance is now quite common as mutations have arisen that allow the resistant species to absorb folic acid from outside sources.

Newer formulations of sulfonamides often combine two drugs. For example, co-trimoxazole (Bactrim®) is a combination trimethoprim and

Trimethoprim

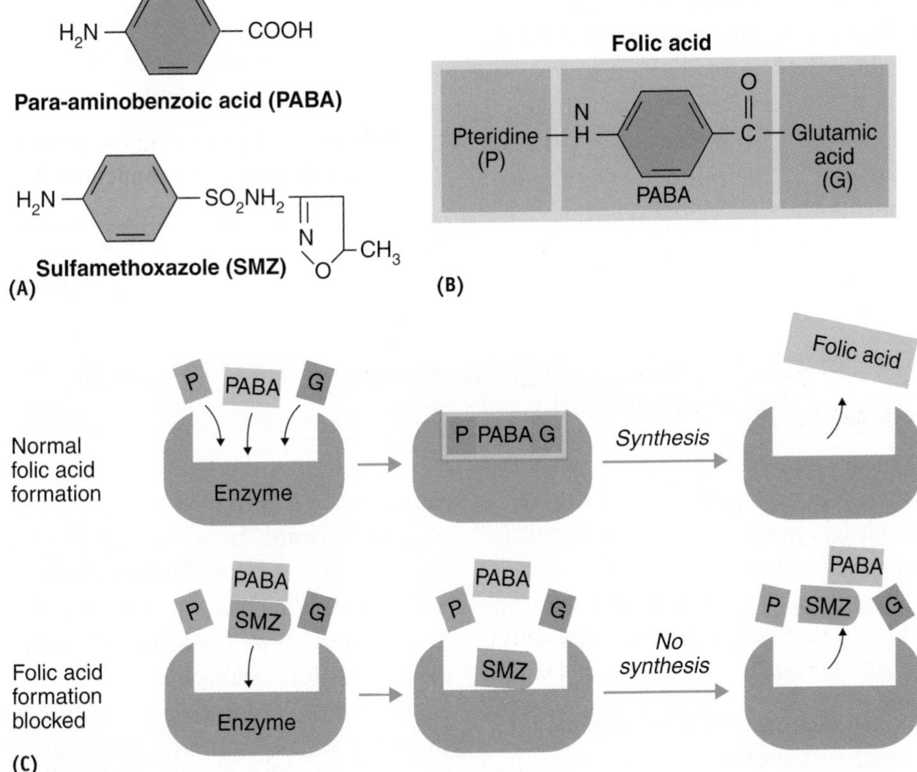

FIGURE 24.5 **The Disruption of Folic Acid Synthesis by Competitive Inhibition.** **(A)** The chemical structures of para-aminobenzoic acid (PABA) and sulfamethoxazole (SMZ) are very similar. **(B)** Folic acid is made up of three components: pteridine (P), PABA, and glutamic acid (G). **(C)** In the normal synthesis of folic acid, a bacterial enzyme joins the three components to form folic acid. However, in competitive inhibition **(D)**, SMZ competes for the active site because of its great abundance. SMZ assumes the position normally reserved for PABA, and folic acid cannot form.
»» Why does treatment with SMZ require a high therapeutic dose?

Levofloxacin

sulfamethoxazole (TMP-SMZ) and is used to treat infections of the urinary tract, lungs (pneumonia), and ears. This combination of drugs is an example of **drug synergism**, meaning the two drugs together are of more benefit than either drug alone. Because each drug is in competition with the bacterial cell's natural substrate, the therapeutic dose for each is relatively high. However, in combination, the therapeutic dose is much lower and makes the generation of resistance less likely.

Two other important synthetic drugs blocking PABA metabolism in *Mycobacterium* species are *p*-aminosalicylic acid (PAS), which is used for treating tuberculosis, and dapsone (diaminodiphenylsulfone), which is effective against leprosy.

Isoniazid

Other Synthetic Antimicrobials Are Commonly Prescribed

There are also other synthetic agents currently in wide use.

Isoniazid (isonicotinic acid hydrazide, or INH) has a very narrow, bacteriostatic drug spectrum as the active form of the drug specifically interferes with cell wall synthesis in *Mycobacterium* species by inhibiting the production of mycolic acid, a component of the wall. Isoniazid often is combined in therapy with such drugs as rifampin and ethambutol and remains one of the first-line drugs for patients with latent tuberculosis.

Another group of synthetic, bactericidal drugs is the **quinolones**, which block DNA synthesis in gram-positive and gram-negative bacterial cells. The **fluoroquinolones**, one of the most prescribed antibiotics in the United States, are used to treat urinary tract infections, gonorrhea and chlamydia, and intestinal tract infections. Examples include levofloxacin and ciprofloxacin (Cipro). Alarmingly, the extensive use of fluoroquinolones has lead to the rapid emergence of resistance worldwide.

CONCEPT AND REASONING CHECKS 2

a. Explain how sulfanilamide interferes with DNA replication.

b. Describe the mode of action for (i) isoniazid and (ii) the fluoroquinolones.

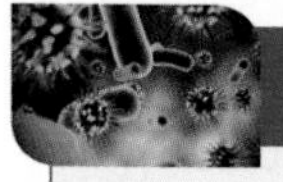

Chapter Challenge B

Among the antibiotic resistance in these superbugs that is conferred by non-NDM-1 genes is inactivation of ciprofloxacin.

QUESTION B:

Is this antibiotic a synthetically or naturally produced drug? What is its mode of action on the superbugs?

Answers can be found on the Student Companion Website in **Appendix D.**

KEY CONCEPT 24.3 Beta-Lactam Antibiotics Inhibit Bacterial Cell Wall Synthesis

One of the most common mechanisms of antibiotic action is blocking the synthesis of the bacterial cell wall. Besides the synthetic drugs isoniazid and ethambutol, which target mycobacterial species, there is a large number of natural and semisynthetic antibiotics that target the assembly of the peptidoglycan component of bacterial cell walls.

Penicillin Has Remained the Most Widely Used Antibiotic

Thanks to the purification and production work of Florey, Chain, and coworkers in the 1940s, **penicillin** has saved the lives of millions of individuals. Its high chemotherapeutic index has made it the drug of choice in eradicating many infections.

Penicillin's mode of action targets the bacterial cell wall. It inhibits the peptide cross-linking of carbohydrates between peptidoglycan layers during wall formation. This results in such a weak wall that internal osmotic pressure allows the cell to swell and burst. Penicillin is therefore bactericidal in rapidly multiplying bacterial cells (as in an infection). Where bacterial cells are multiplying slowly or are dormant, the drug may have only a bacteriostatic effect or no effect at all.

Natural Penicillins. Penicillin G has been the most popular penicillin antibiotic and is usually the one intended when doctors prescribe "penicillin." It is sensitive to acid, so it is primarily given intravenously. Penicillin V is more acid resistant

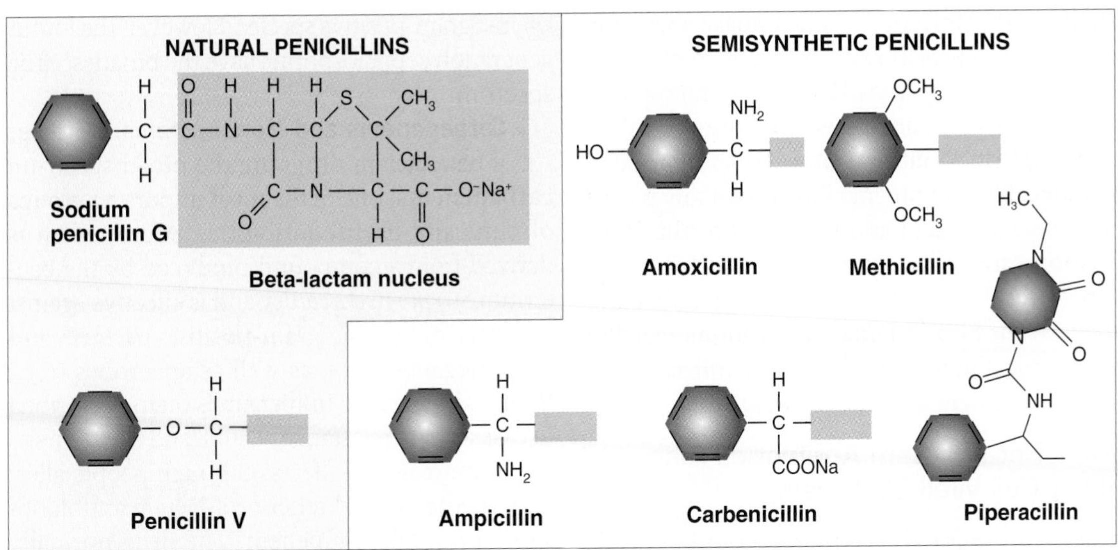

FIGURE 24.6 Some Members of the Penicillin Group of Antibiotics. The beta-lactam nucleus is common to all the penicillins. Different penicillins are formed by varying the side group on the molecule. »» Why is there a need for so many semisynthetic penicillins?

and can be given orally. Both forms of penicillin have a beta-lactam nucleus but differ in the side groups attached (FIGURE 24.6).

The **natural penicillins** have a narrow drug spectrum and are most useful against a variety of gram-positive bacterial species, including staphylococci, streptococci, clostridia, and pneumococci.

Over the years, two major drawbacks to the use of natural penicillin have surfaced. The first is the anaphylactic reaction occurring in allergic individuals. This allergy applies to all compounds related to penicillin. Swelling around the eyes or wrists, flushed or itchy skin, shortness of breath, and a series of hives are signals of a hypersensitivity; penicillin therapy should cease immediately if these symptoms occur.

The second disadvantage is the evolution of penicillin-resistant bacterial species. Many of these organisms produce **beta-lactamase**, which inactivates beta-lactam antibiotics converting drugs like penicillin G into harmless penicilloic acid (FIGURE 24.7). Thus, a process of natural selection takes place, and the rapid multiplication of beta-lactamase-producing bacterial cells yields organisms over which penicillin has no effect.

Semisynthetic Penicillins. In the late 1950s, the beta-lactam nucleus of the penicillin molecule was identified and synthesized, and scientists found they could attach various groups to this nucleus and create new penicillins that are more resistant to beta-lactamases and have a wider drug spectrum. In the following years, numerous **semisynthetic penicillins** emerged (Figure 24.6). Ampicillin and amoxacillin are broader spectrum drugs of value against some gram-negative rods (*Escherichia, Proteus, Haemophilus*) as well as gonococci and meningococci. The drugs resist stomach acid and are absorbed from the intestine after oral consumption. Other semisynthetic penicillins, such as carbenicillin and piperacillin, are effective against an even broader range of gram-negative bacteria and

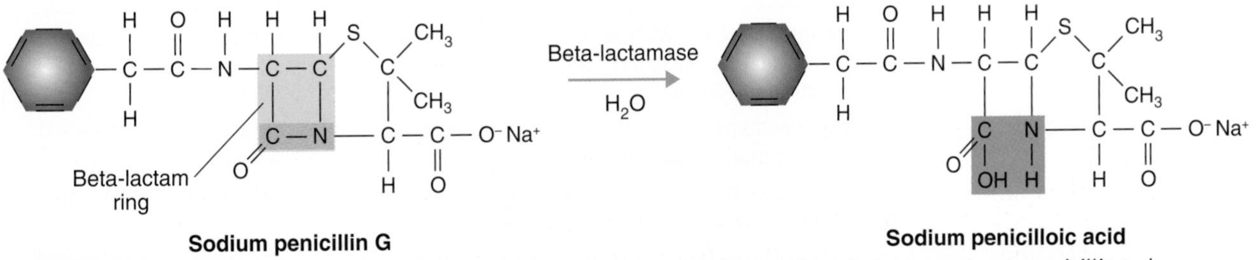

Beta-lactamase H_2O

Beta-lactam ring

Sodium penicillin G

Sodium penicilloic acid

FIGURE 24.7 The Action of Beta-Lactamase on Sodium Penicillin G. The enzyme beta-lactamase converts penicillin to harmless penicilloic acid by opening the beta-lactam ring and inserting a hydroxyl group to the carbon and a hydrogen to the nitrogen. »» What does the action of beta-lactamase tell you about the role of the beta-lactam ring in the structure of penicillin?

Meropenem

can be used for infections of the urinary tract. Still, resistance develops as exemplified by methicillin-resistant *S. aureus* (MRSA) that is resistant to a wide range of penicillins and other antibiotics.

Penicillin Combinations. Penicillins, such as amoxicillin and piperacillin, are usually paired with clavulanic acid and taxobactam (the combinations are called Augmentin® and Zosyn®, respectively). Clavulanic acid and taxobactam inactivate the beta-lactamase, allowing amoxicillin and piperacillin to affect cell wall synthesis. Here we have another example of synergism.

Other Beta-Lactam Antibiotics Also Inhibit Cell Wall Synthesis

Cephalosporin

Besides the penicillins, there are other wall-inhibiting antibiotics with a beta-lactam ring.

Cephalosporins. Another group of antibiotics were also first isolated from a fungus, *Cephalosporium acremonium*. As such, the **cephalosporins** also inhibit cell wall synthesis and resemble penicillins in chemical structure, except the beta-lactam nucleus has a slightly different composition. They are used as alternatives to penicillin where resistance is encountered, or in cases where penicillin allergy exists. Cephalosporins also have a broader bactericidal spectrum against gram-negative bacterial pathogens, are longer lasting in the body, and in some cases may be resistant to beta-lactamases.

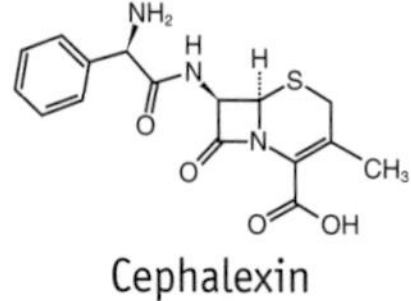

Cephalexin

The cephalosporin nucleus has been modified to produce chemically-related antibiotics with different properties. As such, the cephalosporins can be separated into "generations" by their drug spectrum (TABLE 24.1). Each succeeding generation has a greater activity against gram-negative species than the preceding generation, often with less activity against gram-positive species. However, the fourth generation cephalosporins have the broadest drug spectrum.

Carbapenems and Monobactams. Another set of beta-lactam drugs are the broad spectrum **carbapenems**, one of the most important groups of clinically useful antibiotics. Meropenem is derived from a compound produced by the bacterium *Streptomyces cattley* and is effective against a variety of aerobic gram-positive bacteria and gram-negative rods, as well as anaerobes (e.g., *Bacteroides fragilis*). Imipenem is often prescribed in cases where resistance occurs, and it appears to have minimal side effects, although people allergic to penicillin and other beta-lactam antibiotics should not take imipenem. The drug normally is degraded by the kidneys before it can effect its action in the body. Therefore, imipenem usually is prescribed in combination with cilastatin. Cilastatin prevents the kidneys from degrading the drug. The imipenem/cilastatin combination, known as Primaxin®, is active against some 98% of gram-positive and gram-negative clinical isolates.

A group of narrow-spectrum antibiotics first produced in the early 1980s is the **monobactams**, the first being isolated from the bacterium *Chromobacter violaceum*. The synthetic monobactam, like aztreonam, has a ring structure not recognized by beta-lactamase. The drug is useful against *Pseudomonas* infection in cystic fibrosis patients.

CONCEPT AND REASONING CHECKS 3

a. Explain how penicillin works and how beta-lactamase affects penicillin.

b. List the advantages of the cephalosporins over the penicillins.

c. List the advantages of the carbapenems and monobactams.

TABLE

24.1 The "Generations" of Cephalosporins

Generation	Description	Example
First	Narrow spectrum with activity against many gram-positive bacterial species and some gram-negative species	Cefazolin, cephalexin
Second	Expanded spectrum with increase activity against gram-negative rods	Cefaclor, cefuroxime
Third	Broad spectrum with more activity against gram-negative species and *Pseudomonas*	Cefotaxime, ceftriaxone
Fourth	Extended spectrum with increased activity against gram-negative species	Cefepime

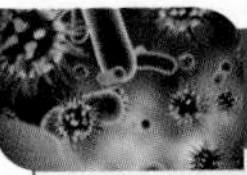

Chapter Challenge C

The original *K. pneumoniae* strain expressing the NDM-1 enzyme had the ability to destroy most known beta-lactam antibiotics, including the penicillins and cephalosporins. What was most alarming was that the strain could also hydrolyze the carbapenems, among the last resort antibiotics for treating infected patients.

QUESTION C:

One of the monobactams that could still be used was aztreonam, which was introduced to the market in 1986. Hypothesize why it was still effective compared to the more established and older drugs.

Answers can be found on the Student Companion Website in **Appendix D**.

KEY CONCEPT 24.4 Other Bacterially Produced Antibiotics Inhibit Some Aspect of Metabolism

Besides antibiotics that target folic acid synthesis and the bacterial cell wall, there are many others that affect the cell membrane, protein synthesis, or nucleic acid synthesis (FIGURE 24.8). However, we begin our discussion with a few more drugs affecting cell walls.

Vancomycin Also Inhibits Cell Wall Synthesis

Vancomycin, a bactericidal cell wall inhibitor, is a product of *Amycolatopsis* (formerly *Streptomyces*) *orientalis*. It is administered by intravenous injection against diseases caused by gram-positive bacterial species, especially severe staphylococcal diseases where penicillin allergy or bacterial resistance is found. It also is used against *Clostridium* and *Enterococcus* species (enterococci).

As drug resistance has developed and spread among staphylococci (MRSA), the choice of antibiotics has gradually diminished, and vancomycin has emerged as a key treatment in therapy; it often is referred to as the "drug of last resort." However, there are now *S. aureus* strains that are resistant to vancomycin (VRSA; see Figure 24.1). In addition, resistance to vancomycin among other bacterial groups such as the enterococci has also been observed. In fact, enterococci account for over 10% of all hospital acquired infections in the United States.

Polypeptide Antibiotics Affect the Cell Envelope

Both bacitracin and polymyxin B are polypeptide antibiotics produced by *Bacillus* species. These antibiotics are quite toxic internally and can cause kidney damage. Therefore, they generally are restricted to topical use, such as on the skin.

Bacitracin is a cyclic polypeptide that is bactericidal and interferes with the transport of cell wall precursors through the cell membrane. Because it is

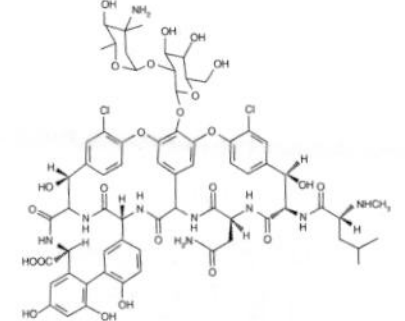

Vancomycin

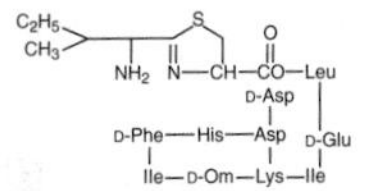

Bacitracin

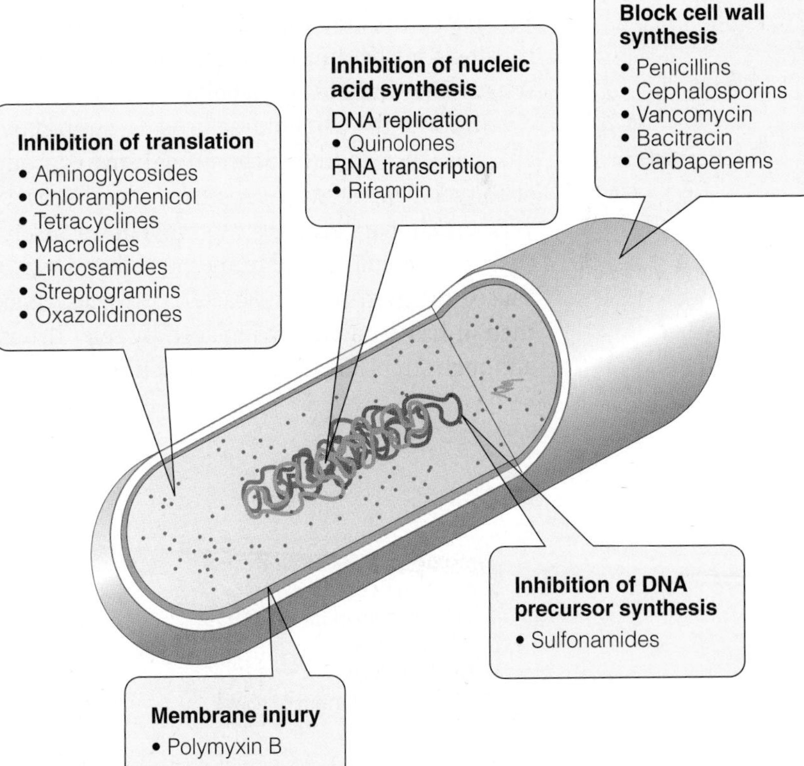

FIGURE 24.8 **The Targets for Antibacterial Agents.** There are six major targets for antibacterial agents: the cell wall, cell membrane, ribosomes (protein synthesis), nucleic acid synthesis (RNA and DNA synthesis), and metabolic reactions. »» Hypothesize why there are so many antibiotics that affect the cell wall.

Gentamicin

very toxic when taken internally, bacitracin is only available in ointments for topical treatment of skin infections caused by gram-positive bacteria or for the prevention of wound infections. When combined with neomycin (see below) and polymyxin B, it is sold under the brand Neosporin®.

Polymyxins are cyclic polypeptides that insert into the cell membrane. Acting like detergents, these bactericidal drugs increase permeability and lead to cell death. Polymyxin B is valuable against *Pseudomonas aeruginosa* and other gram-negative bacilli, particularly those causing superficial infections in wounds, abrasions, and burns. The two antibiotics, bacitracin and polymyxin B, are combined with gramicidin (which increases permeability of the cell membrane) in Polysporin®.

■ Cystic fibrosis: A hereditary disease starting in infancy and affecting various glands, resulting in the buildup of thick mucus in the lungs that encourages bacterial infections.

Many Antibiotics Affect Protein Synthesis

There are several groups of bacterially produced antibiotics that target the protein synthesizing machinery in other bacterial species by binding to the 30S (small) or 50S (large) subunit (FIGURE 24.9).

Aminoglycosides. The **aminoglycosides** are bactericidal antibiotic compounds that attach irreversibly to the 30S subunit of bacterial ribosomes, causing a misreading of the genetic code on messenger RNA (mRNA) molecules.

Streptomycin

In 1943, the first aminoglycoside, streptomycin, was discovered by Selman Waksman's group at Rutgers University. At the time, the discovery was sensational because streptomycin was useful against tuberculosis and numerous other diseases caused by gram-negative bacterial species. Since then, it has been largely replaced by safer drugs, although streptomycin still is prescribed on occasion for tuberculosis. The major side effect is damage to the auditory branch of the nerve extending from the inner ear—deafness may result.

Gentamicin, a still-useful aminoglycoside, is administered for serious infections of the urinary tract caused by gram-negative bacterial species. The antibiotic is produced by a species of *Micromonospora*, a bacterial species related to *Streptomyces*. As already mentioned, neomycin, which also was discovered in Waksman's lab, is available in combination with polymyxin B and bacitracin as Neosporin®. Physicians use an aerosolized version of another aminoglycoside, tobramycin (Tobi®), to treat *Pseudomonas*-caused respiratory infections in patients with cystic fibrosis.

Tetracyclines. The **tetracyclines** are broad-spectrum, bacteriostatic antibiotics that inhibit translation by blocking attachment of the tRNA to the 30S subunit. There are naturally occurring chlortetracyclines, isolated from species of *Streptomyces*, and semisynthetic tetracyclines, such as minocycline and doxycycline (FIGURE 24.10A). All have the four benzene ring chemical structure.

Tetracycline antibiotics may be taken orally, a factor that led to their indiscriminate use in the 1950s and 1960s. The antibiotics were consumed in huge quantities by tens of millions of people, and in some people, the normal microbiota of the intestine was destroyed. Tetracyclines also cause a yellow-gray-brown discoloration of teeth and stunted bones in children (FIGURE 24.10B). These problems are minimized by restricting use of the antibiotic in pregnant women and children through the teen years.

Despite these side effects, tetracyclines, like doxycycline, remain the drugs of choice for most rickettsial and chlamydial diseases. They are used

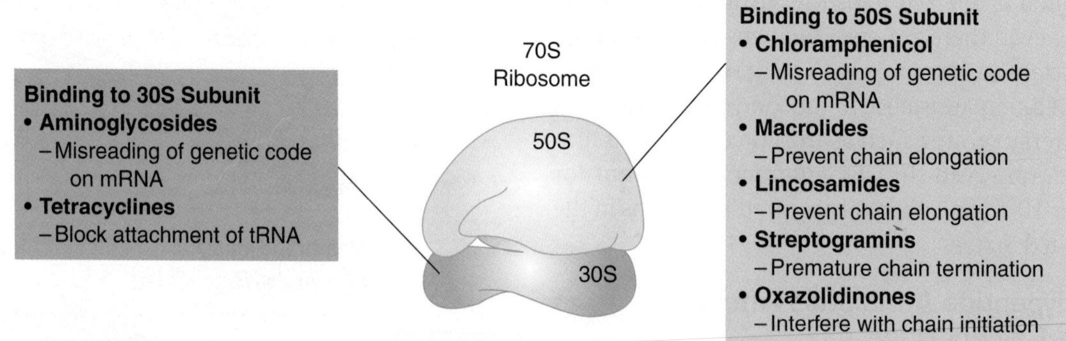

FIGURE 24.9 **Antibiotics and Their Affect on Protein Synthesis.** Many antibiotics interfere with a translation step on the 30S (small) or 50S (large) ribosome subunit. »» Why are there so many antibiotics that affect translation?

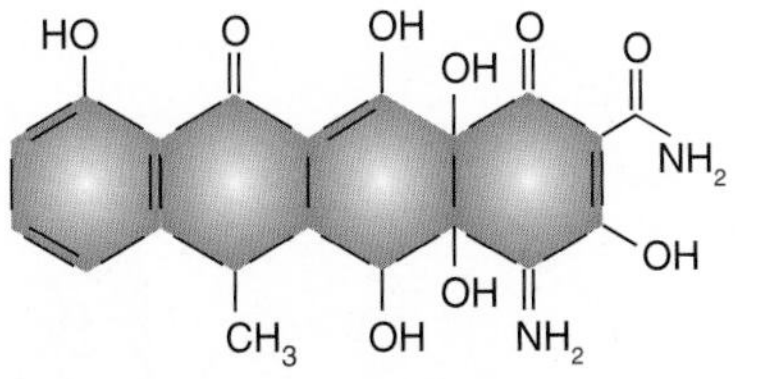

HO O OH OH O O NH2 OH CH3 OH OH NH2

Doxycycline

(A)

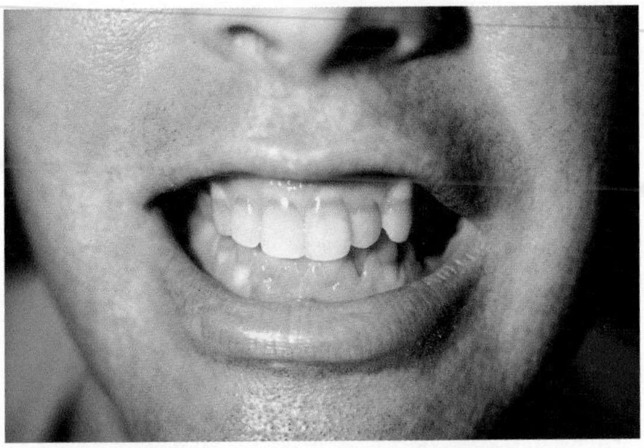

(B)

FIGURE 24.10 **The Tetracyclines.** The staining of teeth associated with tetracycline use. © Kenneth E. Greer/Visuals Unlimited. »» What chemical feature characterizes all tetracycline antibiotics?

against a wide range of gram-negative bacterial species, and are valuable for treating primary atypical pneumonia, syphilis, gonorrhea, and pneumococcal pneumonia. Evidence indicates that tetracycline may have been present in beverages of ancient people (MicroFocus 24.2).

In 2005, the U.S. Food and Drug Administration (FDA) approved a new class of antibiotics related to the tetracyclines, called the **glycylcyclines**. The new drug, tigecycline (Tygacil®) is effective against MRSA infections.

Chloramphenicol. An antibiotic with a broad spectrum is **chloramphenicol**. Its isolation from *Streptomyces venezuelae* was hailed as a milestone in microbiology because the drug is capable of inhibiting a wide variety of gram-positive and gram-negative bacterial species, as well as several species of *Rickettsia* and *Chlamydia.*

Chloramphenicol causes a bacteriostatic effect by interfering with peptide bond formation in the 50S ribosome subunit. It remains the drug of choice in the treatment of typhoid fever and is an alternative to tetracycline for epidemic typhus and Rocky Mountain spotted fever.

These drugs usually are reserved for treating serious and life-threatening infections because of their side effects. Chloramphenicol also accumulates in the blood of newborns, causing a toxic reaction and sudden breakdown of the cardiovascular system known as the **gray syndrome**.

Macrolides. The **macrolides** are a group of bacteriostatic antibiotics consisting of large carbon rings attached to unusual carbohydrate molecules. They block protein synthesis by binding to the 50S subunit and inhibiting chain elongation.

In 1949, Abelardo Aguilar, a Filipino scientist, sent some soil samples to his employer Eli Lilly. In the soil was the species *Streptomyces erythreus* (now called *Saccaropolyspora erythraea*), from which the Lilly scientists isolated a drug called erythromycin. In the 1970s, researchers discovered that erythromycin was effective for treating primary atypical pneumonia and Legionnaires' disease. It is recommended for use against gram-positive bacterial species in patients with penicillin allergy and against both *Neisseria* and *Chlamydia* species affecting the eyes of newborns. Although it has few side effects, at higher doses it sometimes causes nausea, vomiting, and diarrhea.

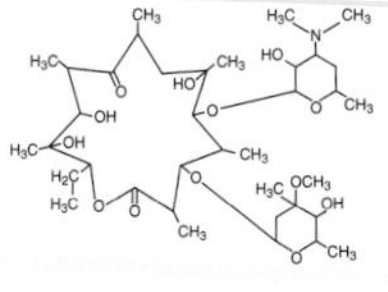

Erythromycin

Other macrolides with a broader spectrum include clarithromycin (Biaxin®) and azithromycin (Zithromax®), one of the world's best-selling antibiotics. Both macrolides are semisynthetic drugs associated with causing serious allergic and dermatologic reactions.

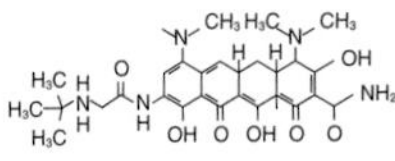

Tigecycline

Lincosamides. The **lincosamides** also bind to the 50S subunit and inhibit chain elongation. *Streptomyces lincolnensis* produces an antibiotic called lincomycin from which the semisynthetic drug clindamycin is derived. This bacteriostatic drug is an alternative in cases where penicillin resistance is encountered. Clindamycin is active against aerobic, gram-positive cocci and anaerobic, gram-negative bacilli (e.g., *Bacteroides* species). Use of the antibiotic is limited to serious infections because the drugs eliminate competing organisms from the intestine and permit *Clostridium difficile* to overgrow the area. The clostridial toxins then may induce a potentially lethal condition called **pseudomembranous colitis**, in which membranous lesions cover the intestinal wall.

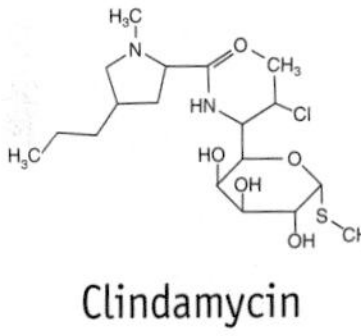

Clindamycin

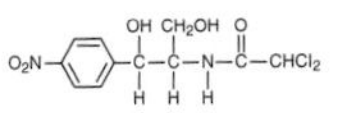

Chloramphenicol

Streptogramins. Another group of cyclic peptides are the **streptogramins**, discovered in 1962 in yet another species of *Streptomyces.* Synthetic versions of these antibiotics are prescribed as a combination of two cyclic peptides, called quinupristin-dalfopristin (Synercid®). Both

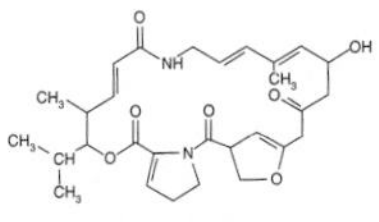

Streptogramin A

MICROFOCUS 24.2: History
Fortified Beer

In September 1980, a chance observation led to the discovery that antibiotics were protecting humans from disease long before anyone had suspected.

The remarkable find was made by Debra L. Martin, a graduate student at Detroit's Henry Ford Hospital. After preparing thin bone sections for microscopic observation, Martin placed her slides under a fluorescence microscope because no other microscope was available for her use at the time. When illuminated with ultraviolet light, the sections glowed with a peculiar yellow-green color. Her colleagues attributed the glow to the antibiotic tetracycline.

These were no ordinary bone sections. Rather, they were from the mummified remains of Nubian people (350–550 CE) excavated along the floodplain of the Nile River. Anthropologists from the University of Massachusetts, led by George Armelagos, had previously established that the Nubian population was remarkably free of infectious disease, and now Martin's discovery gave a possible reason why.

Streptomyces species are very common in desert soil, and the anthropologists postulated that the bacterial cells may have contaminated the grain bins and deposited tetracycline. Bread or beer made from the antibiotic-rich grain might then confer freedom from disease.

A modern-day Nubian mother and child.

In 2010, Armelagos, now at Emory University in Atlanta, reported that the most likely source of tetracycline was in the beer the ancient people drank. Presumably, the grain used to ferment the beer contained one of the *Streptomyces* species that produces tetracycline. A colleague, Mark Nelson of Paratek Pharmaceuticals, Inc., found bone samples from the ancient Nubians to be "saturated with tetracycline." Based on the levels of the antibiotic in the bones, he believes this was no accidental contamination but rather that the Nubians were consciously producing the drug in the beer fermentation process, which would make this one of the earliest examples of biotechnology!

Armelegos says he is convinced the Nubians used "empirical evidence to develop therapeutic agents"—in this case, some 1,600 years before the "modern discovery" of tetracycline.

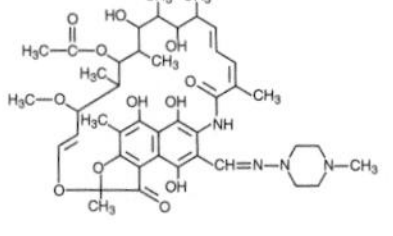

Rifampin

Linezolid

components interfere with protein synthesis on the 50S subunit, the interference being synergistic and bactericidal. The drugs are effective against a broad range of gram-positive bacterial species, including *S. aureus*, and respiratory pathogens.

Oxazolidinones. After the identification of the streptogramins in 1962, no new structural classes of antibiotics were discovered until 2000. Then in 2001, the **oxazolidinones**, were approved for clinical use. These drugs interfere with chain initiation by the 50S subunit. One member of the group, linezolid (Zyvox®), is effective in treating gram-positive bacterial species, including MRSA. However, it can produce allergic reactions and is toxic to mitochondria. Therefore, the oxazolidinones are drugs of last resort, being used only where every other antibiotic has failed.

Some Antibiotics Inhibit Nucleic Acid Synthesis

The synthetic quinolones that inhibit bacterial DNA synthesis have already been described.

Other drugs affect transcription. **Rifampin**, a semisynthetic bacteriostatic drug derived from *Streptomyces mediterranei*, interferes with RNA synthesis. Because mycobacterial resistance to rifampin can develop quickly, it is prescribed in combination with isoniazid and ethambutol for

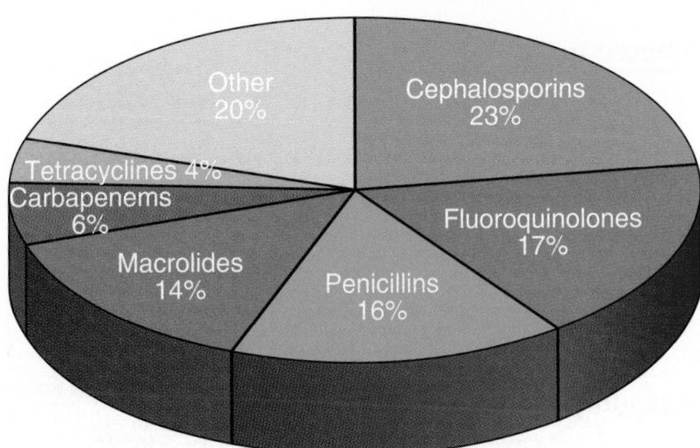

FIGURE 24.11 Global Antibiotics Market, by class—2009 The global antibiotics market has been dominated by four classes. *Source:* BCC Research.

»» What percent of the antibiotic market targeted (a) bacterial cell wall synthesis, (b) protein synthesis, and (c) DNA synthesis?

tuberculosis and leprosy patients. It also is administered to carriers of *Neisseria* and *Haemophilus* species that cause meningitis and as a prophylactic when exposure has occurred.

FIGURE 24.11 identifies the top classes of antibiotics on the global market. TABLE 24.2 reviews those antibacterial drugs currently in use.

CONCEPT AND REASONING CHECKS 4

a. What problems are associated with vancomycin use?
b. Why are the polypeptide antibiotics only used topically?
c. Identify the bacterial source for the aminoglycosides, chloramphenicol, tetracyclines, macrolides, clindamycin, and the streptogramins.
d. To what enzyme must rifampin bind to inhibit transcription?

Chapter Challenge D

Additional antibiotic resistance in these superbugs that is conferred by non-NDM-1 genes includes inactivation of erythromycin, rifampin, chloramphenicol, and polymyxins.

QUESTION D:

From these first four challenges (A-D), use the diagram in Figure 24.8 to identify the resistance groups (major targets) to which* K. pneumoniae *is resistant. Would you consider this organism a superbug? Explain.

Answers can be found on the Student Companion Website in **Appendix D.**

KEY CONCEPT 24.5 Other Antimicrobial Drugs Target Viruses, Fungi, and Parasites

If you look at the Centers for Disease Control and Prevention (CDC) website that lists diseases and conditions, about 36% of the diseases have a bacterial origin, 33% have a viral origin, 6% are fungal, and 25% are due to parasites (FIGURE 24.12). If we look at the developing nations, the percentage of viral and especially parasitic diseases would be much higher. Although these data are important for monitoring trends and for targeting research, prevention, and control efforts, they do not include non-notifiable diseases, such as colds, forms of gastroenteritis, yeast infections, etc. Still, we can make the following generalization: We have dozens more antibiotics in our antimicrobial arsenal to fight bacterial infections than we have drugs to fight viral, fungal, and parasitic diseases. In addition, the antiviral drugs we do have are useful against only a very limited number of viral diseases and most have been developed to fight HIV disease/AIDS. Importantly, unlike antibiotics, there are no antiviral drugs that will definitely cure a person of any viral disease. Therefore, whether it is an antiviral drug directed against HIV, herpes, or influenza, antiviral chemotherapy only lessens the effects of the infection.

The ability to use antimicrobial agents effectively against viruses and fungi presents another problem not encountered with bacterial pathogens. The major problem is finding drugs affecting the viral or fungal pathogen but not the host tissue. Because fungi are members of the Eukarya, they possess much of the cellular machinery common to animals and humans and viruses make use of that machinery. Thus, many drugs targeting viruses and fungi will also target host tissues and potentially be quite toxic, making it difficult to maintain selective toxicity.

Antiviral Drugs Interfere with Viral Replication

A limited number of antiviral drugs are of value for treating diseases caused by picornaviruses,

TABLE

24.2 A Summary of Major Antibacterial Drugs (by Mode of Action)

Antibacterial Agent	Source	Antibacterial Spectrum	Side Effects
Competitive inhibitors of DNA precursor synthesis			
Sulfonamides Sulfanilamide Sulfamethoxazole/ Trimethoprim	Synthetic	Broad	Kidney and liver damage Allergic reactions
Inhibitors of cell wall synthesis			
Isoniazid	Synthetic	Tubercle bacilli	Liver damage
Penicillins Penicillin G Ampicillin Amoxicillin Piperacillin	*Penicillium notatum* and *Penicillium chrysogenum* Some semisynthetic	Narrow/broad (depending on drug)	Allergic reactions Anemia
Cephalosporins Cephalothin Cephalexin	*Cephalosporium* species Some semisynthetic	Broad	Allergic reactions Kidney damage
Carbapenems Meropenem	*Streptomyces cattley*	Broad	Few reported
Vancomycin	*Streptomyces orientalis*	Gram-positive bacteria, especially staphylococci	Ear and kidney damage
Bacitracin	*Bacillus subtilis*	Gram-positive bacteria, especially staphylococci	Kidney damage, if injected
Inhibitors of cell membrane function			
Polymyxin B	*Bacillus polymyxa*	Gram-negative rods, especially in wounds	Kidney damage
Inhibitors of protein synthesis			
Aminoglycosides Streptomycin Gentamicin Neomycin	*Streptomyces* species *Micromonospora* species	Primarily gram-negative bacteria and staphylococci	Hearing defects Kidney damage
Tetracyclines Chlortetracycline Minocycline Doxycycline	*Streptomyces* species Some semisynthetic	Broad, rickettsiae, chlamydiae	Discoloration of teeth Kidney and liver damage
Chloramphenicol	*Streptomyces venezuelae*	Broad	Allergic reactions Gray syndrome
Macrolides Erythromycin Clarithromycin Azithromycin	*Streptomyces erythreus* Some semisynthetic	Broad	Gastrointestinal distress Liver damage
Lincosamides Clindamycin	*Streptomyces lincolnesis*	Gram-positive bacteria	Pseudomembranous colitis
Streptogramins Quinupristin/ Dalfopristin	*Streptomyces* species	Resistant strains of *Staphylococcus* and *Enterococcus*	Muscle aches, rash, headache
Oxazolidinones Linezolid	Synthetic	Last resort antibiotics, gram-positive bacteria	Toxic to mitochondria
Inhibitors of DNA synthesis			
Fluoroquinolones Levofloxacin	Synthetic	Broad	Gastrointestinal distress Allergic reactions
Inhibitors of RNA synthesis			
Rifampin	*Streptomyces mediterranei* Semisynthetic	Tubercle bacilli Some gram-negative bacteria	Liver damage Allergic reactions

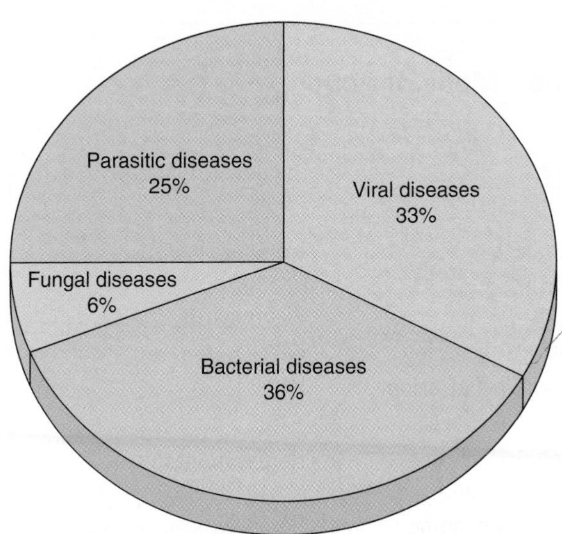

FIGURE 24.12 **Types of Human Infectious Disease.** The percentage of bacterial and viral disease is about equal. »» Provide a reason as to why there are fewer fungal diseases than there are diseases of the other groups.

herpesviruses, hepatitis B and C viruses, HIV, and influenza viruses. These drugs affect viral penetration/uncoating, genome replication, or viral maturation/release (TABLE 24.3).

Penetration/Uncoating Inhibitors. Some antivirals block the ability of viruses to penetrate or uncoat in the host cell. Amantadine and rimantadine prevent the influenza A viruses from escaping the endocytotic vacuole in which they are contained. Antiviral drugs such as pleconaril integrate into the picornavirus capsid and prevent capsid uncoating. HIV penetration is blocked by enfuvirtide, which blocks the fusion of the HIV envelope with the host cell plasma membrane.

Genome Replication Inhibitors. Most of the antiviral drugs designed target the DNA polymerase of the herpesviruses or the reverse transcriptase of HIV and hepatitis B virus. Antiviral drugs like valacyclovir represent **base analogs** and insert themselves into the replicating DNA strand, blocking the ability of the virus to continue replicating its genome. **Reverse transcriptase inhibitors**, such as nevirapine and efavirenz, bind directly to reverse transcriptase and inhibit reverse transcription, thereby preventing the synthesis of retroviral DNA.

Maturation/Release Inhibitors. Other antiviral drugs work by blocking the maturation process or release of the virions from the host cells. **Protease inhibitors**, such as lopinavir, react with the HIV protease, the enzyme that trims viral proteins down to working size for the construction of the capsid. In the type A influenza viruses, neuraminidase is a spike enzyme that helps the viruses spread to other cells. The **neuraminidase inhibitors** zanamivir and oseltamivir block the action of neuraminidase, preventing release of new virions—and thereby limiting disease spread in the body. The development of all these drugs is based on biochemical knowledge of virus function coupled with human ingenuity.

Several Classes of Antifungal Drugs Cause Membrane Damage

As mentioned above, drugs used for fungal infections lack a high level of selective toxicity and can generate severe side effects in the host.

Plasma Membrane Inhibitors. The **polyenes** are large, ring-shaped organic compounds that bind to ergosterol, a sterol found in the fungal plasma membrane. This results in increased permeability, causing the cell's contents to leak out. Human and animal cells lack this sterol.

For infections of the intestine, vagina, or oral cavity due to *Candida albicans*, physicians often prescribe nystatin. This product of *Streptomyces noursei* is commercially sold in ointment, cream, or suppository form. Often it is combined with antibacterial antibiotics to retard *Candida* overgrowth of the intestines during the treatment of bacterial diseases.

Amphotericin B

For serious systemic fungal infections, the drug of choice is amphotericin B. This broad-spectrum drug, extracted from *Streptomyces nodosus*, is used intravenously with immunocompromised patients, and for treating aspergillosis, cryptococcosis, and candidiasis. However, it causes a wide variety of side effects, including kidney damage, and therefore is used only in progressive and potentially fatal cases.

Synthetic antifungal drugs, called **azoles**, inhibit an enzyme needed to form ergosterol in the fungal plasma membrane. Again, the lack of the sterol causes the cell's contents to leak out and, depending on the fungus and the drug, either brings about an inhibition of fungal growth (fungistatic) or cell death (fungicidal).

Miconazole

The **imidazoles** include clotrimazole, miconazole, and ketoconazole. Clotrimazole (Gyne-Lotrimin®) is used topically for *Candida* skin infections, while the other drugs are used

TABLE

24.3 Examples of Antiviral Drugs, Their Mode of Action, and Targeted Viruses

Antiviral Drug (Trade Name)	Mode of Action	Targeted Viruses
Blocking of Penetration/Uncoating		
Amantadine (Symmetrel®) Rimantadine (Flumadine®)	Block viral uncoating	Influenza A viruses
Pleconaril (Picovir®)	Blocks uncoating	Picornavirus
Enfuvirtide (Fuzeon®)	Blocks penetration/uncoating	HIV
Maraviroc (Selzentry®)	Blocks attachment/penetration	HIV
Inhibition of Genome Replication		
Acyclovir (Zovirax®)	Base analog (mimics guanine) inhibiting DNA replication	Herpesviruses
Valacyclovir (Valtrex®)	Base analog (mimics guanine) inhibiting DNA polymerase	Herpesviruses
Ribavirin (Virazole®)	Base analog (mimics guanine) inhibiting viral replication	Hepatitis B and C viruses
Azidothymidine (AZT; Retrovir®)	Base analog (mimics adenine) terminating DNA chain elongation	HIV
Vidarabine (Vira-A®)	Terminates DNA chain elongation	Herpesviruses
Idoxuridine (Stoxil®) Trifluridine (Viroptic®)	Base analogs (mimic thymidine) causing replication errors (mutations)	Herpesviruses
Dideoxycytidine (Hivid®) Dideoxyinosine (Videx®)	Base analogs (mimic cytosine and adenine, respectively) terminating DNA chain elongation	HIV
Nevirapine (Viramune®) Efavirenz (Sustiva®)	Reverse transcriptase inhibitors	HIV
Raltegravir (Isentress®)	Block provirus integration	HIV
Inhibition of Virion Maturation/Release		
Lopinavir (Kaletra®) Atazanavir (Reyataz®) Saquinavir (Invirase®)	Inhibit HIV protease	HIV
Oseltamivir (Tamiflu®) Zanamivir (Relenza®)	Inhibit viral release	Influenza A viruses

topically as well as internally for systemic diseases. Side effects are uncommon. Miconazole is commercially available as Micatin® for athlete's foot and Monistat® for yeast infections. Ketoconazole has been used to treat fungal infections in immunocompromised patients, including those with AIDS.

The **triazoles**, such as fluconazole and itraconazole, are less toxic than the imidazoles.

Cell Wall Inhibitors. Because fungi have cell walls, this structure represents a unique site for selective toxicity. The semisynthetic **echinocandins** inhibit the synthesis of the fungal cell wall and are used to treat invasive candidiasis and aspergillosis.

Other Antifungal Drugs. The antimetabolite drug **flucytosine** is converted in fungal cells to an inhibitor that interrupts nucleic acid synthesis. The drug is active against some strains of *Candida* and *Cryptococcus.* The drug usually is used in combination with amphotericin B.

Griseofulvin is a product of *Penicillium griseofulvum* and is taken orally. Griseofulvin interferes with mitosis by binding to microtubules. It is used

H N O N F NH_2

Flucytosine

OCH_3 OCH_3 O O CH_3O O Cl CH_3

Griseofulvin

for fungal infections of the skin, hair, and nails, such as ringworm and athlete's foot.

The Goal of Antiprotistan Agents Is to Eradicate the Parasite

Because protists also are members of the domain Eukarya, antiprotistan drugs attempt to target unique aspects of nucleic acid synthesis, protein synthesis, or metabolic pathways.

Aminoquinolines. The **aminoquinolines** are antimalarial drugs that accumulate in parasitized red blood cells and are toxic to the malaria parasite, interfering with the parasite's ability to break down and digest hemoglobin. The parasite thus starves or the drug causes the accumulation of toxic products resulting from the degradation of hemoglobin in the parasite.

Quinine was one of the first natural antimicrobials and is derived from the bark of the South American cinchona tree (MicroFocus 24.3). It was the primary agent used to treat malaria until substantial resistance to quinine developed. More effective synthetic drugs, such as chloroquine, mefloquine, and primaquine, are now used. Chloroquine remains the drug of choice to treat all species of *Plasmodium* while mefloquine (Lariam®) is primarily used for malaria caused by *P. falciparum*. Reports have been made that mefloquine can have rare but serious side effects, including severe depression, anxiety, paranoia, nightmares, and insomnia.

Sulfonamides. Similar to bacterial cells, protistan cells require folic acid for the synthesis of nucleic acids, but are unable to absorb it from the environment. Therefore, the same sulfonamides that are used against bacterial pathogens will produce a similar result with the protists. Another drug, diaminopyrimidine, can be used with trimethoprim to achieve a synergistic effect. These two drugs in combination with sulfamethoxazole are effective in treating toxoplasmosis.

Nitroimidazoles. One of the most commonly used antiprotistan drugs is metronidazole (Flagyl®) which interferes with anaerobic metabolism. These

Quinine

Metronidazole

MICROFOCUS 24.3: History
The Fever Tree

Rarely had a tree caused such a stir in Europe. In the 1500s, Spaniards returning from the New World told of its magical powers on malaria patients, and before long, the tree was dubbed "the fever tree." The tall evergreen grew only on the eastern slopes of the Andes Mountains (see figure). According to legend, the Countess of Chinchón, wife of the Spanish ambassador to Peru, developed malaria in 1638 and agreed to be treated with its bark. When she recovered, she spread news of the tree throughout Europe, and a century later, Linnaeus named it *Cinchona* after her.

Leaves of a cinchona tree. Photo by Forest & Kim Starr.

For the next two centuries, cinchona bark remained a staple for malaria treatment. Peruvian Indians called the bark quina-quina (bark of bark), and the term *quinine* gradually evolved. In 1820, two French chemists, Pierre Pelletier and Joseph Caventou, extracted pure quinine from the bark and increased its availability still further. The ensuing rush to stockpile the chemical led to a rapid decline in the supply of cinchona trees from Peru, but Dutch farmers made new plantings in Indonesia, where the climate was similar. The island of Java eventually became the primary source of quinine for the world.

During World War II, Southeast Asia came under Japanese domination, and the supply of quinine to the West was drastically reduced. Scientists synthesized quinine shortly thereafter, but production costs were prohibitive. Finally, two useful substitutes were synthesized in chloroquine and primaquine. Today, as resistance to these drugs is increasingly observed in the malarial parasites, scientists once again are looking for another "fever tree" to help control malaria.

drugs are effective in the treatment of amebiasis, giardiasis, and trichomoniasis.

Heavy Metals. Arsenic and antimony derivatives have been used since ancient times. The arsenic derivative melarsoprol is used to treat African trypanosomiasis, while antimoniate is effective against leishmaniasis. Although the drugs are toxic in the nervous system and kidneys, they primarily affect the intense metabolism of protistan cells.

Other Antiprotistan Drugs. Several of the antibiotics that affect bacterial protein synthesis also inhibit protein synthesis in protists. Clindamycin and the tetracyclines, such as doxycycline, are used to treat malaria. Another protein synthesis inhibitor, paromomycin, can be used against cryptosporidiosis.

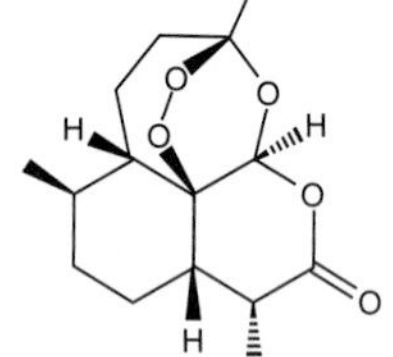

Artemisinin

Pentamidine, a drug of unknown action, has been used against the parasites causing leishmaniasis. **Artemisinin** is used to treat multidrug resistant strains of *P. falciparum*. The drug, isolated from the shrub *Artemisia annua* (sweet wormwood) has long been used by Chinese herbalists (FIGURE 24.13). In red blood cells, the drug releases free radicals that destroy the malarial parasites. There are now synthetic forms that are more potent than the natural drug.

■ Free radical: A highly reactive atom or group of atoms with unpaired electrons.

FIGURE 24.13 **The Source of Artemisinin.** *Artemisia* shrubs are grown in Africa, as indicated in this harvest by a Kenyan farmer. © Jack Barker/Alamy Images. »» What is the mode of action of artemisinin?

Antihelminthic Agents Target Nondividing Helminths

The helminths are eukaryotic, multicellular parasites. Although most antimicrobial drugs are targeted at actively dividing cells of the pathogen, the mechanism of action of most antihelminthic drugs targets the nondividing organisms.

Praziquantel is thought to change the permeability of the parasite plasma membranes of cestodes and trematodes. The permeability change causes the inflow of calcium ions, which leads to muscle contraction and paralysis in the parasite. Parasites then cannot feed and eventually die. Praziquantel has been the drug of choice for mass therapy campaigns, including the treatment of schistosomiasis.

Praziquantel

Mebendazole inhibits uptake of glucose and other nutrients by adult and larval worms from the host intestine where helminths are located. Without the ability to carry out ATP synthesis, the parasites will die. The drug has a wide spectrum, affecting many nematodes and cestodes. The drug also disrupts microtubules and cell division.

The **avermectins** are antihelminthic drugs derived from *Streptomyces avermitilis*. The drugs are effective in extremely low doses and against a wide variety of roundworms. The avermectins affect the nematode nervous system such that muscle paralysis results. Ivermectin (Stromectol®) is one drug currently used to treat onchocerciasis (river blindness) and is being investigated for treatment of a number of nematode infections. TABLE 24.4 summarizes the characteristics of the antifungal and antiparasitic agents.

CONCEPT AND REASONING CHECKS 5

a. Why have most of the antiviral agents synthesized been targeted against viral genome replication?
b. Assess the emphasis for developing antifungal drugs targeting the fungal plasma membrane or cell wall.
c. Identify the cellular structures targeted by the antiprotistan agents.
d. Provide several reasons why there are fewer effective antifungal and antiparasitic drugs to treat infections.

TABLE

24.4 A Summary of Antifungal and Antiparasitic Drugs

Antifungal Agent	Source	Antifungal Spectrum	Mode of Action	Side Effects
Polyenes Nystatin	*Streptomyces noursei*	*Candida albicans*	Cell membrane function	Few reported
Amphotericin B	*Streptomyces nodosus*	Systemic infections	Cell membrane function	Fever, gastrointestinal distress
Azoles Clotrimazole Ketoconazole Miconazole Fluconazole Itraconazole	Synthetic	Superficial infections Superficial and systemic infections	Inhibit sterol synthesis	Few reported
Echinocandins	Semisynthetic	*Aspergillus, Candida*	Cell wall synthesis	Low incidence
Flucytosine	Synthetic	*Candida, Cryptococcus*	Nucleic acid synthesis	Adverse renal and liver effects
Griseofulvin	*Penicillium griseofulvum*	Superficial infection	Microtubules	Occasional allergic reactions

Antiparasitic Agent	Source	Antiparasitic Spectrum	Mode of Action	Side Effects
Aminoquinolines Quinine Chloroquine Mefloquine Primaquine	Synthetic	*Plasmodium* species	Parasite digestion of hemoglobin	Severe neurological and behavioral side effect possible with mefloquine
Sulfonamides	Synthetic	*Toxoplasma, Plasmodium*	Folic acid metabolism	Kidney and liver damage; allergic reactions
Nitroimidazoles Metronidazole	Synthetic	*Trichomonas*	DNA synthesis	Tumors in mice
Artemisinin	*Artemisia annua*	*Plasmodium* species	Free radicals destroy parasites	Few reports of adverse effects
Praziquantel	Synthetic	*Schistosoma*	Membrane permeability	Result from killing of parasites
Mebendazole	Synthetic	Broad	Glucose uptake	Diarrhea, stomach pain
Ivermectin	*Streptomyces avermitilis*	*Ascaris, Enterobius*	Nervous system	Low incidence

KEY CONCEPT 24.6 Antimicrobial Drug Resistance Is a Growing Challenge

The substantial variety of antibiotics and chemotherapeutic agents developed since the 1930s necessitates that medical professionals know which one is best for the patient under the circumstances of the infection. Accordingly, antibiotic sensitivity assays can be performed. These assays are used (1) to measure the susceptibility of a bacterial strain to one or more antibiotics and (2) to monitor the evolution of bacterial resistance.

There Are Several Antibiotic Susceptibility Assays

Two general methods are in common use to test the susceptibility of a bacterial pathogen to specific antibiotics. These are the tube dilution method and the agar strip/disk diffusion method.

The **tube (broth) dilution method** determines the lowest concentration of antibiotic

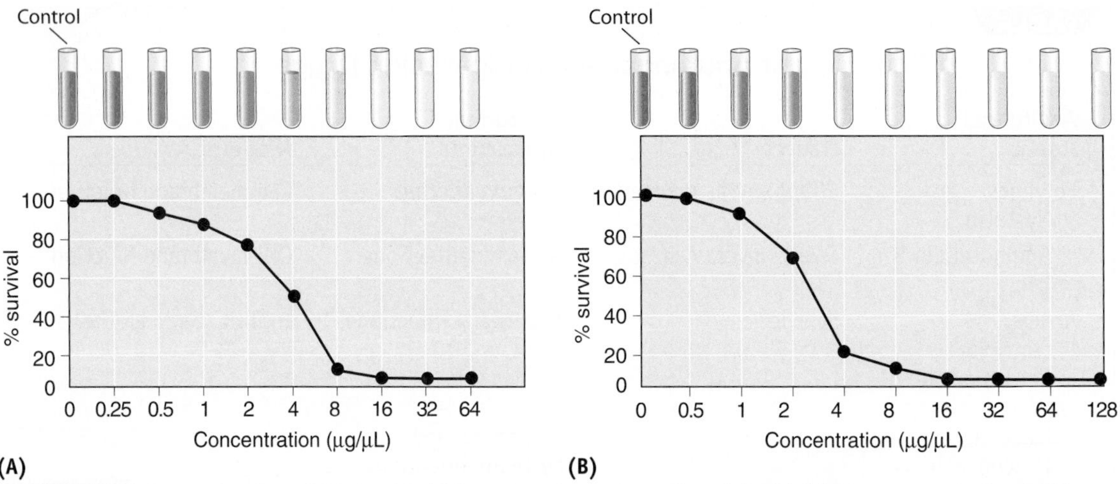

FIGURE 24.14 **Determination of Minimal Inhibitory Concentration.** The minimal inhibitory concentration (MIC) for ampicillin against (**A**) *Staphylococcus aureus* and (**B**) *Escherichia coli*. From the tube dilutions (top), the MIC can be plotted and determined. *Source:* Data modified from: Atwal, R. (2003). *Int. JARVM* 1(3). »» What are the MICs for *Staphylococcus aureus* and *Streptococcus pyogenes*?

that will prevent growth of the pathogen. This amount is known as the **minimum inhibitory concentration** (**MIC**). To determine the MIC, the microbiologist prepares a set of tubes with different concentrations (dilutions) of a particular antibiotic (FIGURE 24.14). Each tube is inoculated with an identical number of cells, incubated, and examined for the growth of bacterial cells. The extent of growth diminishes as the concentration of antibiotic increases, and eventually an antibiotic concentration is observed at which growth no longer occurs. This is the MIC.

The second method is an antibiotic susceptibility test, which operates on the principle that an antibiotic will diffuse from a paper or plastic-coated strip or disk into an agar medium containing a test organism. In one version of the test, called the **Etest®**, a strip impregnated with a marked gradient of antibiotic is placed on the plate. As the drug diffuses into the agar culture, the higher drug concentrations will inhibit bacterial growth (FIGURE 24.15A). The inhibition of growth is observed as a clear, oval halo on the plate. By reading the number on the strip (antibiotic concentration) where growth intersects the strip, the MIC can be determined.

In another version of the test, called the **disk diffusion method**, inhibition of bacterial growth is again observed as a failure of a susceptible bacterial species to grow. A common application of the agar disk diffusion method is the **Kirby-Bauer test**, named after W. M. Kirby and A. W. Bauer, who developed it in the 1960s. This procedure determines the susceptibility of a bacterial species to a group of antibiotics each added to a paper disk and placed on an agar plate containing the test organism. After a 24h to 48h incubation, each antibiotic has diffused into the agar and may have affected bacterial growth. If the bacterial cells are susceptible to a particular antibiotic, a

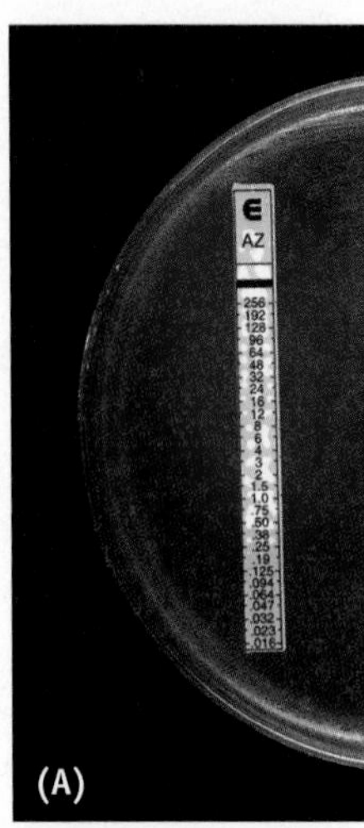

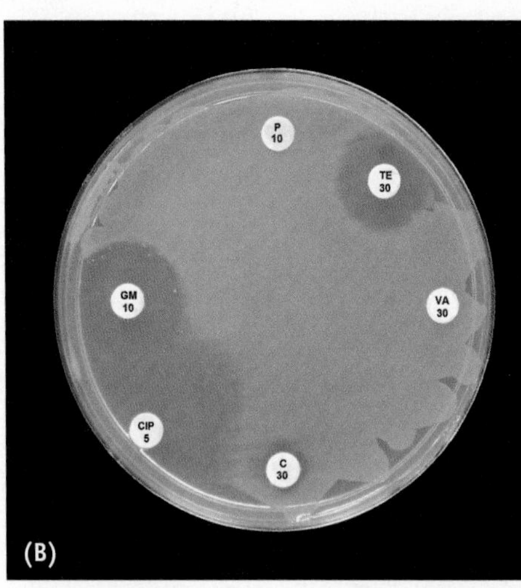

FIGURE 24.15 **Antimicrobial Susceptibility Testing.** (**A**) With the Etest®, the antibiotic-containing strip can determine categorical results (susceptible, intermediate, resistant), as well as the minimum inhibitory concentration (MIC). Courtesy of bioMerieux, Inc. (**B**) The results of a disk diffusion assay (Kirby-Bauer test) using *Micrococcus luteus*. The antibiotics tested were: chloramphenicol (C); ciprofloxacin (CIP); gentamicin (GM); penicillin G (P); tetracycline (TE); and vancomycin (VA). The numbers refer to the drug concentration in micrograms or units (penicillin). Courtesy of bioMerieux, Inc. »» In (A), what is the MIC and in (B) to which antibiotics was *M. luteus* susceptible? Note: Intermediate means that there was some inhibition but would not be sufficient to inhibit growth of the pathogen in the body.

clear halo (no growth) called a **zone of inhibition**, surrounds the paper disk (FIGURE 24.15B). No halo or a zone less than a certain diameter (using a standard table for comparison) indicates the strain was resistant to the antibiotic. Basically, a "susceptible" category indicates that the strain could be treated with the dosage used in the disk, while a "resistant" category indicates the strain would not be effective at that dosage.

Antibiotic Resistance Can Develop and Spread in Several Ways

Between the late 1930s and the 1960s, most all major groups of antibiotics were introduced. Since then and until 2000, the intervening decades represent an "innovation gap" where no new antibiotic classes were introduced (FIGURE 24.16). Importantly, during this gap, an increasing number of bacterial species became resistant to many of these antibiotics; that is, the bacterial cells are not effectively inhibited or killed, allowing the resistant cells to continue multiplying in the presence of therapeutic levels of an antibiotic. Resistance develops in the following two ways.

Mutations. Sometimes resistance is of its own making, such as through **mutations**, those rare, spontaneous changes in a cell's genetic material (FIGURE 24.17A). For example, in 2007, scientists reported that one strain of *Staphylococcus aureus* in an infected patient evolved resistance to vancomycin over a period of 30 months. During that time the *S. aureus* strain accumulated 18 sequential mutations, making it resistant not only to vancomycin but also to several other antibiotics. The scary part is that the patient was never treated with any of these antibiotics! The mutations produced "spontaneous" resistance. Such resistant organisms are increasingly responsible for human diseases of the intestinal tract, lungs, skin, and urinary tract. Widespread diseases like bacterial pneumonia, tuberculosis, streptococcal sore throat, and gonorrhea, which until recently were susceptible to a single dose of antibiotics, are now among the most difficult to treat. Patients in intensive care units and burn wards are particularly vulnerable, as are infants, the elderly, and the infirm.

Horizontal Gene Transfer. Antibiotic resistance often comes from bacterial cells picking up resistance genes from other bacterial cells by means of **horizontal gene transfer**. These transfer mechanisms can allow bacterial cells to acquire

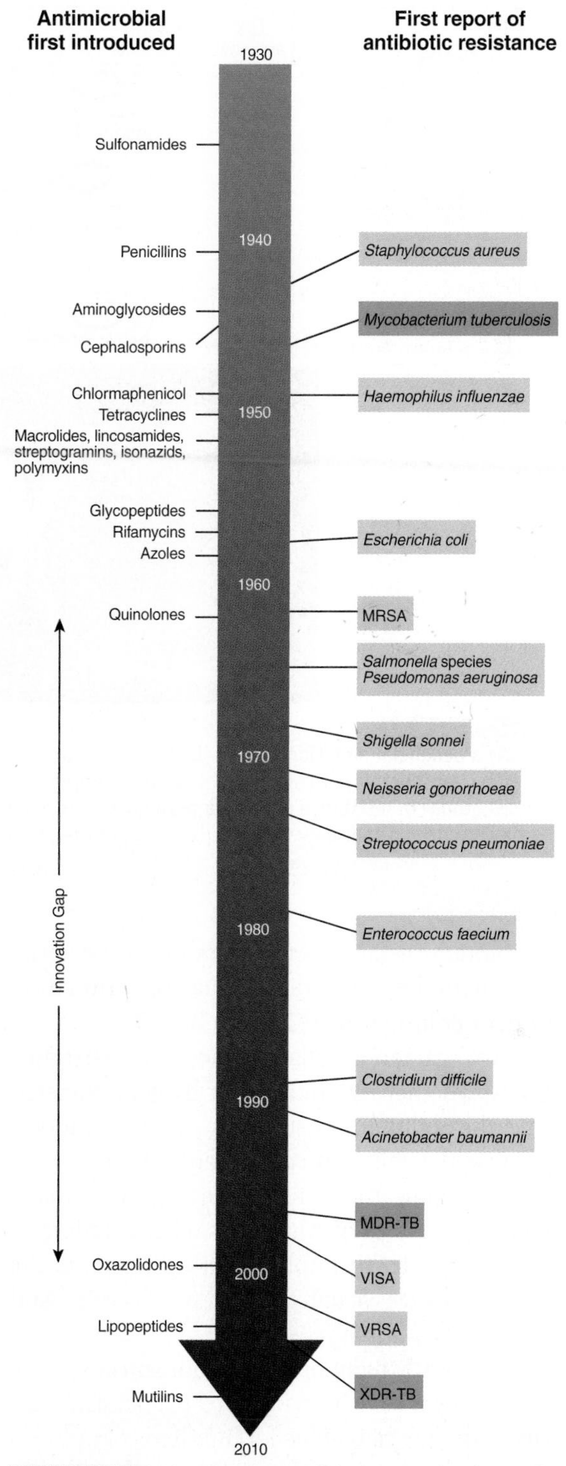

FIGURE 24.16 **Timeline for Antibiotic Introduction and Appearance of Antibiotic Resistance.** This timeline shows the time when major antibiotics were introduced and the approximate dates when some notable bacterial pathogens (pink = gram-negative species; purple = gram-positive species; red = acid-fast species) were identified as being antibiotic resistant. MRSA (VISA, VRSA) = methicillin-resistant (vancomycin intermediary resistant; vancomycin-resistant) *Staphylococcus aureus;* MDR (XDR)-TB = multidrug-resistant (extensively drug-resistant) tuberculosis. »» What are the two ways the bacterial species gained resistance?

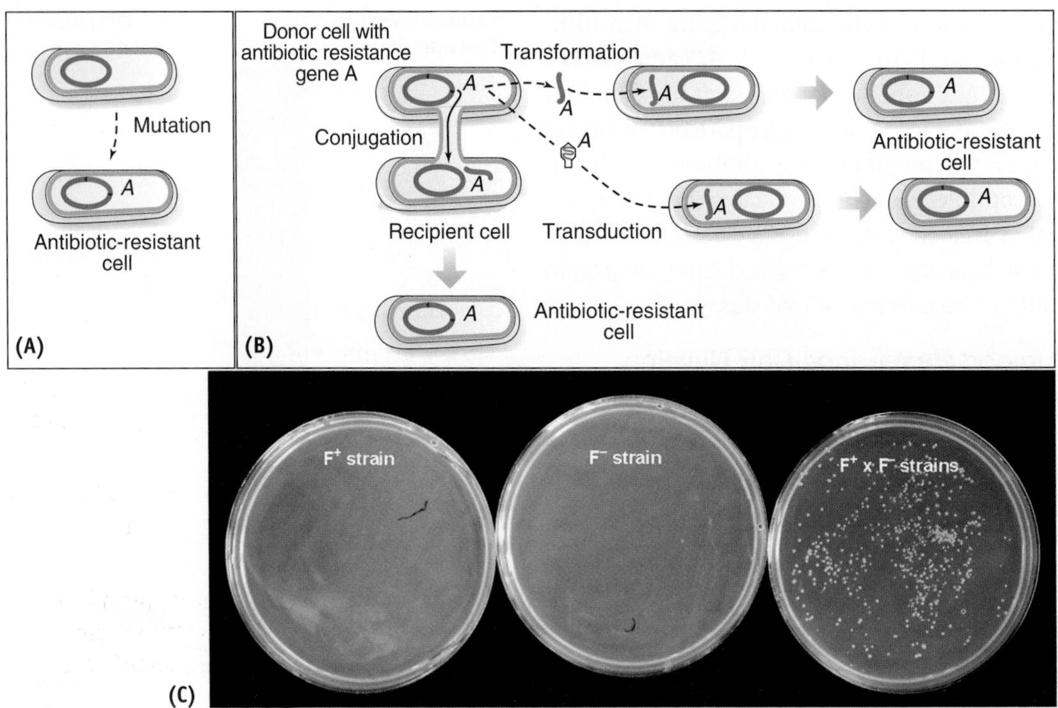

FIGURE 24.17 **Mechanism of Antibiotic Resistance Acquisition.** Antibiotic resistance can be the result of (**A**) a beneficial mutation and/or (**B**) transfer of an antibiotic resistance gene to a recipient cell by transformation, transduction, or conjugation. (**C**) In a conjugation experiment, an F^+ strain of *Escherichia coli* carrying a streptomycin resistance gene on a plasmid was mixed with a recipient F^- strain lacking the plasmid. The medium is designed so that only the F^- cells receiving the plasmid from the F^+ can grow. Courtesy of Jeffrey Pommerville.
»» What does each colony on the mixed ($F^+ \times F^-$ strains) plate represent?

antibiotic resistance genes from donor bacterial cells by undergoing **transformation**, **transduction**, or **conjugation** (FIGURE 24.17B,C).

Any bacterial cells that acquire resistance genes, whether by mutation or by gene transfer, have the ability to resist one or more antibiotics. Because the cells can collect multiple resistance traits over time, they can become resistant to many different families of antibiotics. One example of major concern to public health officials is the bacterial species *S. aureus* (TEXTBOOK CASE 24 and MICROFOCUS 24.4).

■ Phosphorylation, acetylation, or adenylation: **Referring to the addition, respectively, of a phosphate group, an acetyl group, or an adenyl group to a molecule.**

As already mentioned, microorganisms in the environment have been involved in chemical warfare with their neighbors for hundreds of millions of years. To counteract the production of antibiotics by some microbes, others developed self-protective (resistance) mechanisms to the natural antibiotics. Thus, it is not surprising that microbes quickly develop resistance to the natural antibiotics used to treat infectious diseases. Antibiotic resistance arising from mutations or horizontal gene transfer involves four major mechanisms—two offensive and two defensive (FIGURE 24.18).

Antibiotic Hydrolysis. Resistance can arise from the microorganism's ability to enzymatically split apart (hydrolyze) the antibiotic. The production of a beta-lactamase enzyme by penicillin-resistant species is an example. By breaking the beta-lactam ring, penicillin- or cephalosporin-resistant bacterial cells prevent inhibition of cell wall synthesis.

Antibiotic Modification. Another offensive mechanism involves modifying the antibiotic. For example, streptomycin-resistant bacterial cells can enzymatically modify (inactivate) streptomycin by phosphorylation, acetylation, or adenylation of the antibiotic so the drug cannot bind to ribosomes.

Membrane Modification. A defensive resistance mechanism involves the ability of microbes to prevent drug entry into the cytoplasm. For example, changes in membrane permeability in penicillin-resistant *Pseudomonas* prevent the antibiotic from entering the cytoplasm. Bacterial species such as *E. coli* and *S. aureus*, which are resistant to tetracyclines, actively export (pump out) the drug. Cytoplasmic and membrane proteins in

Textbook CASE 24

Severe Community-Acquired MRSA Pneumonia

1 On December 17, 2006, a previously healthy eight-year-old girl was taken to her primary care provider. Her parents told the physician she had developed a fever 3 days ago and it continues. The physician noted a fever of 39.4°C (103°F). The girl also had spells of forceful coughing followed immediately by vomiting. She was treated in the provider's office with azithromycin, dexamethasone to decrease any swelling of the larynx, and aerosolized albuterol to dilate the bronchi.

2 Even with these interventions, her condition worsened and the girl was rushed to a local emergency department, where she received intravenous ceftriaxone and nebulized albuterol. A chest radiograph revealed pneumonia in the right lower lobe. She was immediately transported to a referral hospital.

3 At the hospital, the physicians noted that the patient had low blood pressure and low blood oxygenation. She was intubated and provided with both cardiac and respiratory support oxygen. During intubation, she had cardiac arrest but was resuscitated.

4 Viral and sputum cultures had been collected during the visit to the primary care provider on December 17. The viral culture tested positive for influenza and the sputum culture tested positive for MRSA; blood cultures were negative for MRSA. The patient had not had an influenza vaccination.

5 Over the next 3 weeks in the hospital, her situation worsened and antibiotic therapy appeared hopeless. The patient suffered renal and hepatic failure, and an abscess formed in the space between the pleura and diaphragm (subpulmonary).

6 The patient died on January 7, 2007, a total of 24 days after onset of symptoms. Cause of death was listed as pneumonia, respiratory distress, and MRSA sepsis.

Questions:

(Answers can be found on the Student Companion Website in **Appendix D.**)

A. What was the point of providing azithromycin, dexamethasone, and aerosolized albuterol?

B. Why is this considered a community-acquired MRSA?

C. Why didn't antibiotics like azithromycin and ceftriaxone work on the patient?

D. What do the viral, sputum, and blood cultures tell you about the spread of the infections?

E. What does the short duration between respiratory symptom onset and death (3 weeks) indicate?

For additional information see www.cdc.gov/mmwr/preview/mmwrhtml/mm5614a1.htm.

these bacterial cells act as pumps to remove the antibiotic before it can affect the ribosomes in the cytoplasm.

Target Modification. Another defensive mechanism involves altering the cell structure targeted by the antibiotic. Some streptomycin-resistant bacterial species have modified the structure of their ribosomes so that the antibiotic cannot bind to the ribosome but protein synthesis is not inhibited. Other targets that can be modified include RNA polymerase and enzymes involved in DNA replication. TABLE 24.5 summarizes the resistance mechanisms in relation to the antibiotics mode of action and class of drugs.

Antibiotic Resistance Is of Vital Concern in the Medical Community

The rise in antibiotic resistance is partly the result of improper use of antibiotics. For example, drug companies promote antibiotics heavily, patients pressure doctors for quick cures, and physicians sometimes misdiagnose infections or write prescriptions to avoid ordering costly tests to pinpoint the patient's illness. In addition, people may

MICROFOCUS 24.4: Public Health
Methicillin-Resistant *Staphylococcus aureus* (MRSA)

Staphylococcus aureus, often referred to simply as "staph," is a gram-positive bacterial species (see figure) that commonly colonizes the skin or the nose of some 25% to 30% of healthy people—without causing an infection. However, if it invades the body, the organism can cause life-threatening infections, such as septicemia, pneumonia, endocarditis, and meningitis. In fact, *S. aureus* is involved in more than 250,000 infections in Americans each year, primarily involving patients in hospitals and nursing homes.

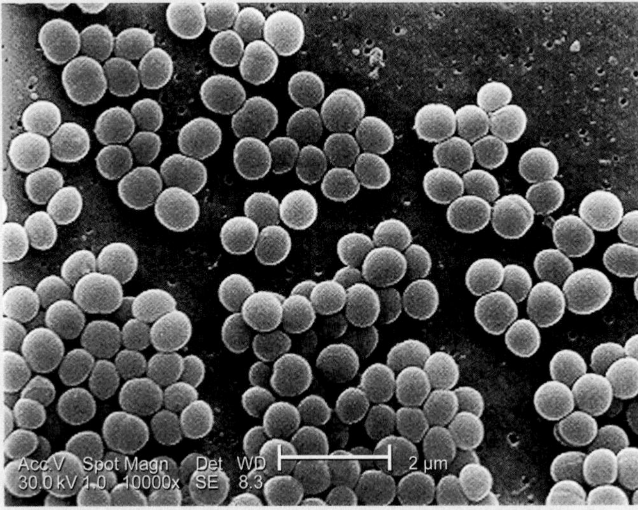

A false-color scanning electron microscope image of clusters of *Staphylococcus aureus* cells. (Bar = 2 µm.) Courtesy of Janice Haney Carr/CDC.

Of major concern to health officials is that over the years strains of *S. aureus* have developed resistance to penicillin and other broad-spectrum drugs. Then, several decades ago, *S. aureus* strains were identified that were resistant to the action of methicillin and related beta-lactam antibiotics (e.g., penicillin and cephalosporin). Termed MRSA, for methicillin-resistant *S. aureus,* the strains could only be treated with more potent, expensive, and often toxic drugs, like vancomycin.

In 1997, a MRSA strain was identified with intermediate (partial) vancomycin resistance; it was named VISA, for vancomycin-intermediate resistant *S. aureus.* Although researchers found useful alternative drug combinations, they were worried that soon *S. aureus* would be completely resistant to vancomycin, often the drug of "last resort" for MRSA patients. Indeed, in 2001, reports of vancomycin-resistant *S. aureus* (VRSA) were reported in many countries, indicating global spread. Vancomycin resistance is still a rare occurrence. Unfortunately, VRSA may also be resistant to meropenem and imipenem, two other antibiotics used to treat MRSA patients.

Prevalence. In 2007, a report in the *Journal of the American Medical Association* estimated that in 2005 the number of Americans developing a serious MRSA infection surpassed 94,000 and more than 18,000 died. While most were invasive MRSA infections that were traced to a hospital stay or some other healthcare exposure, about 15% of invasive infections occurred in people with no known health-care risk. Two-thirds of the 85% of MRSA infections that could be traced to hospital stays or other healthcare exposures occurred among people who were no longer hospitalized. People over age 65 were four times more likely than the general population to get a MRSA infection.

Today, MRSA is often subcategorized as hospital-associated MRSA (HA-MRSA) or community-associated MRSA (CA-MRSA), depending upon how the strain was acquired. Based on current data, these are distinct strains of *S. aureus.*

HA-MRSA. The hospital-associated strain occurs most frequently among hospital patients who undergo invasive medical procedures, or who have weakened immune systems and are being treated in hospitals and healthcare facilities, such as nursing homes and dialysis centers. HA-MRSA in healthcare settings commonly causes serious and potentially life-threatening infections, such as septicemia, surgical site infections, or pneumonia.

Transmission commonly comes from HA-MRSA patients who already have an infection or who carry the bacterial cells on their bodies (colonized) but are asymptomatic. The main mode of transmission to other patients is through human hands, especially those of healthcare workers after contact with infected or colonized patients. If appropriate hand hygiene, such as washing with soap and water or using an alcohol-based hand sanitizer, is not performed, the bacterial cells can be spread when the healthcare worker touches other patients. The good news is that cases of HA-MRSA appear to be declining since 2005. This might be due to increasing education about MRSA and better infection control, such as hand hygiene.

CA-MRSA. A community-associated infection is defined as a MRSA infection occurring in otherwise healthy people who have not been recently (within the past year) hospitalized, or had a medical procedure (such as dialysis, surgery, catheters). Studies have shown that rates of CA-MRSA infection are growing fast. One study of children in south Texas found that cases of CA-MRSA had a 14-fold increase between 1999 and 2001. In 2007, at least three school-age children died from MRSA infections, causing the schools to close while extensive disinfection measures were carried out and control measures instituted.

The CA-MRSA strain, called USA 300, is very contagious. However, about 75% of CA-MRSA infections are usually skin infections, causing abscesses, boils, and other pus-filled lesions that can be treated effectively. Health experts worry though that USA 300 will soon display enhanced virulence and cause more severe illness to vital organs, leading to widespread infection (sepsis), toxic shock syndrome, and pneumonia. It is not known why some healthy people develop CA-MRSA skin infections that are treatable whereas others infected with the same strain develop severe, fatal infections. CA-MRSA skin infections have been identified among certain populations that share close quarters or experience more skin-to-skin contact. Examples are team athletes, military recruits, and prisoners. However, more and more CA-MRSA infections are being seen in the general community as well.

CA-MRSA also appears to infect younger people. In a Minnesota study published in the *Journal of the American Medical Association,* the authors reported that the average age of people with MRSA in a hospital or healthcare facility was 68; the average age of a person with CA-MRSA was only 23.

Protection. With the spread of antibiotic resistance, there are procedures that you can take to lessen the chances of a MRSA infection whether in the community or as a patient in a healthcare facility (see table)—of course, hand washing is at the top of the list. Recently, scientists have been working on a *S. aureus* vaccine that is showing good success in mice. Hopefully, such results will also be seen when human studies are started.

TABLE

Measures to Protect Yourself in Community and Healthcare Settings

Community
Wash hands frequently.
Cover cuts and scrapes with a bandage.
Do not touch wounds (or bandages) on other people.
Do not share personal items such as towels and razors.
Healthcare
Request all hospital staff to wash their hands before touching you.
Ask what the hospital/healthcare facility is doing to prevent MRSA infections.
Make sure IV tubes and catheters are inserted and removed under sterile conditions.

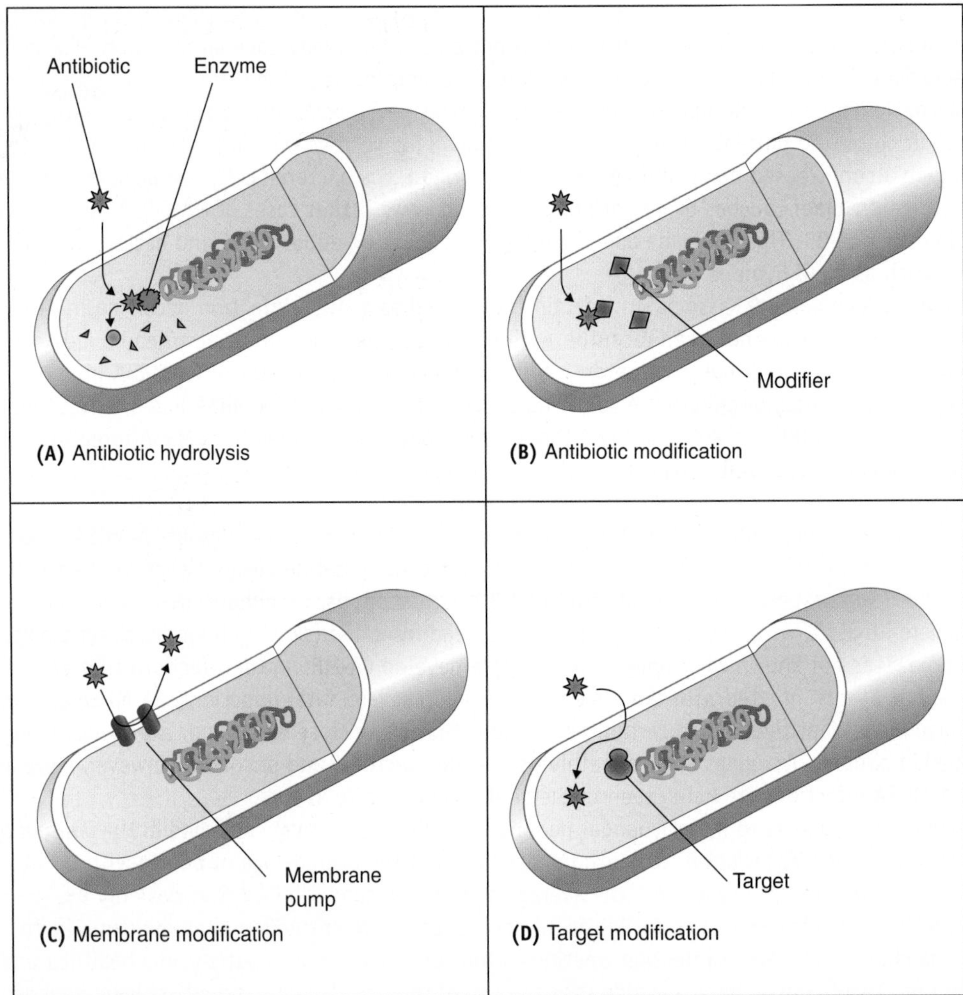

FIGURE 24.18 **Antibiotic Resistance Tactics.** Antibiotic resistance can develop by: (**A**) beta-lactamases breaking the structure of antibiotics like penicillin; (**B**) chemical groups inactivating an antibiotic like streptomycin; (**C**) membrane pumps pumping out an antibiotic like tetracycline; and (**D**) cell targets such as ribosomes changing their structure so antibiotics like streptomycin will not bind. »» Which resistance mechanisms appear to be more "defensive" and which more "offensive" in action? Explain.

TABLE

24.5 A Summary of Bacterial Resistance Mechanisms

Antibiotic Target	Antibiotic Class	Resistance Mechanisms
Inhibition of cell wall synthesis	Penicillins, cephalosporins, carbapenems, vancomycin	Altered wall composition; drug destruction by beta-lactamases
Membrane permeability and transport	Polypeptide antibiotics	Altered membrane structure
Inhibition of DNA synthesis	Fluoroquinolones	Inactivation of drug; altered drug target
Inhibition of RNA synthesis	Rifampin	Altered drug target
Inhibition of protein synthesis	Aminoglycosides, chloramphenicol, tetracyclines, macrolides, clindamycin, streptogramins, oxazolidinones	Altered membrane permeability; drug pumping; antibiotic inactivation; altered drug target
Inhibition of DNA precursor synthesis	Sulfonamides	Alternate metabolic pathway; altered drug target

diagnose their own illness and take leftover antibiotics from their medicine cabinet for ailments where antibiotics are useless. Moreover, many people fail to complete the course of their prescription, and some organisms remain alive and evolve into resistant forms. The survivors proliferate well because they face reduced competition from susceptible organisms FIGURE 24.19.

Hospitals are another forcing ground for the emergence of resistant bacterial species. In many cases, physicians use unnecessarily large doses of antibiotics to prevent infection during and following surgery. This increases the possibility that resistant strains will replace the susceptible normal microbiota destroyed by the antibiotic, causing a **superinfection**. As bacterial species have evolved and acquired more resistance to antibiotics, more and more species are becoming **multidrug resistant** (**MDR**). For example: many hospital and community strains of *S. aureus* (MRSA) are multidrug resistant; there are multidrug resistant and extensively drug-resistant

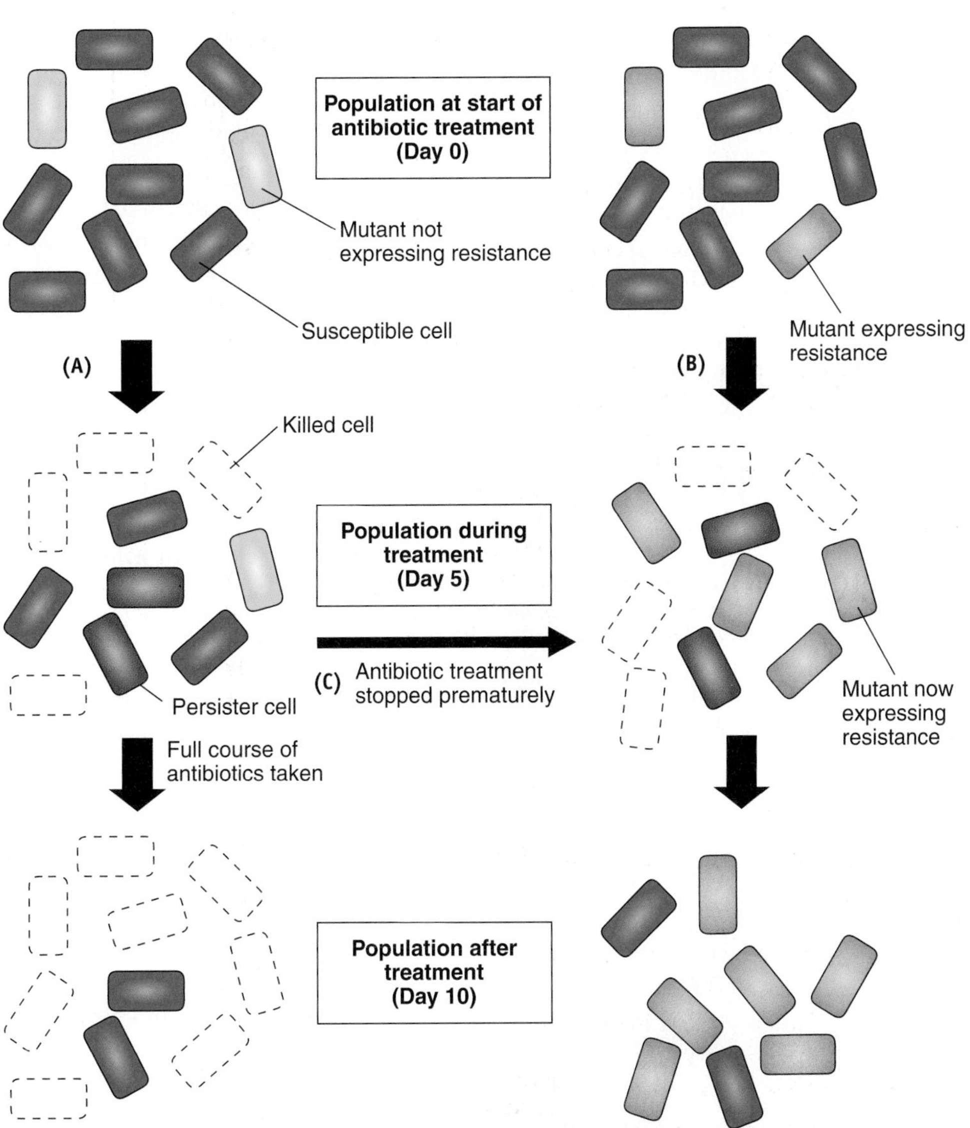

FIGURE 24.19 **The Possible Outcomes of Antibiotic Treatment.** (**A**) Ideally, with a complete course of antibiotics, all pathogens will be destroyed, although a few persister cells (purple) may survive. (**B**) If there are some resistant cells in the infecting population, they will survive and grow without any competition. If the person "feels better" and prematurely stops taking the antibiotic (**C**), any mutant cells may have had the opportunity to express antibiotic resistance, survive, and again grow along with persister cells without any competition. »» Would the persister cells be antibiotic resistant? Explain.

strains of *Mycobacterium tuberculosis* (MDR-TB and XDR-TB); and multidrug resistant strains of *Clostridium difficile* and *Acinetobacter baumannii* have increased in hospitals.

Antibiotics also are abused in developing countries where they often are available without prescription (FIGURE 24.20). Some countries permit the over-the-counter sale of potent antibiotics, and large doses encourage resistance to develop. Between 1968 and 1971, some 12,000 people died in Guatemala from shigellosis attributed to antibiotic-resistant *Shigella dysenteriae*.

Perhaps most disturbing is the nontherapeutic use of antibiotics in animal feeds (MicroFocus 24.5).

Antibiotics have been known as miracle drugs but the development of resistance questions their long-term use. In fact, did antibiotic resistance exist before the modern era of antibiotic therapy or is it a result of contemporary antibiotic use? Investigating the Microbial World 24 looks at this important question.

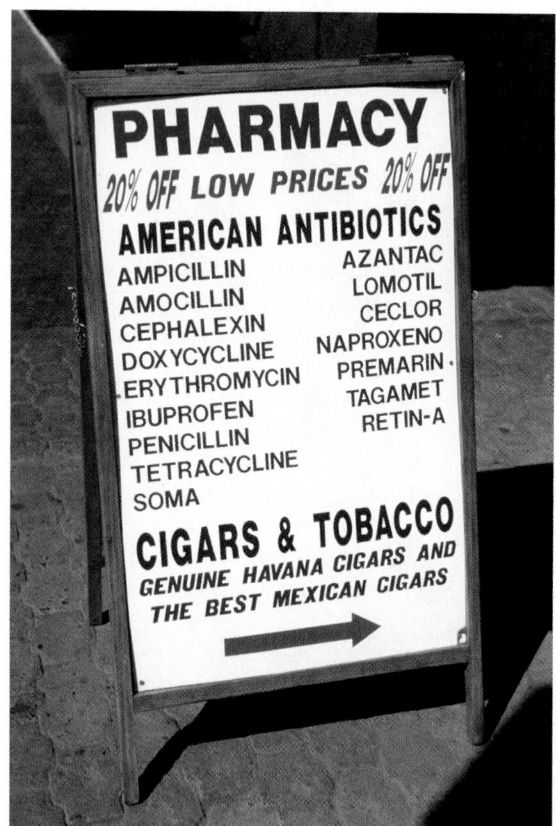

FIGURE 24.20 Antibiotics On Sale. A sign in front of a foreign pharmacy selling antibiotics over the counter. © Stephen Saks Photography/Alamy Images. »» What are the risks of buying such over-the-counter drugs?

New Approaches to Antibiotic Therapy Are Needed

Combating antibiotic resistance requires a two-pronged approach. First, as already described, we need to preserve the antibiotics we still have by using them prudently. That is the relatively easy part toward preventing abuse and misuse of the antimicrobial drugs, and the public can help limit the spread of resistance by taking appropriate steps (MicroFocus 24.6).

The second part is a bit more involved and challenging and that is to stimulate the development and supply of new antimicrobial drugs. Due to the perception in the late 1960s that antibiotics would cure all infectious diseases, pharmaceutical companies and research organizations stopped or severely slowed the development of new antibiotics. Unfortunately, many microbes and pathogens continued to become resistant to them. Yet despite this increase in antimicrobial resistance, the development of new antimicrobial agents is declining in a time when there is a pressing need for new agents (FIGURE 24.21).

There are several reasons why the pharmaceutical industry has lagged behind in developing and bringing new antimicrobial drugs to market. As mentioned, there was the perception that new drugs were no longer needed even in the face of emerging and resurgent infectious diseases. Second, the cost of development is staggering, requiring close to $1 billion over 10 years or more (MicroInquiry 24). Pharmaceutical companies say that the relatively large research cost and time for development is lost if the medication does not make it to market. Third, there is a much higher financial reward to pharmaceutical companies in developing and marketing medications for the treatment of chronic illnesses (e.g., depression, hypertension, diabetes, cancer, cholesterol, arthritis) because, unlike antibiotics, which are usually given for a short 5 to 14 day period and then discontinued, medications for chronic illnesses may be taken for a lifetime.

The Infectious Disease Society of America (IDSA) and most health experts agree that it is time to focus on the development of new antibacterial drugs. As the IDSA notes, today there are "few antimicrobial drugs in the pipeline that offer benefits over existing drugs and few drugs moving forward that will treat infections due to

MICROFOCUS 24.5: Evolution

Antibiotics in the Feedlots

© Dario Sabljak/ShutterStock, Inc.

For decades, many farmers have been using animal feeds mixed with low amounts of antibiotics. In fact, of the more than 500 metric tons of antibiotics that are produced each year globally, more than 40% of these antibiotics are used in the United States, and some 50% to 80% of those are not for clinical purposes but rather dispensed on large commercial feedlots to promote the faster growth of healthy cattle, swine, and poultry. The bottom line is that subtherapeutic doses of antibiotics fatten up livestock faster and get the animals to market sooner.

Such indiscriminate use of antibiotics might not be as dangerous if these drugs were different from the ones used in the clinic and hospital. But they are not. If fact, many of these growth-promoting antibiotics are close relatives to the very ones used in human medicine. This includes the fluoroquinolones, penicillins, and tetracyclines.

By using these antibiotics unselectively, many feedlots are becoming breeding grounds for the evolution of antibiotic-resistant bacterial species. Numerous scientific studies have shown that using subtherapeutic levels of antibiotics in animals kills many bacterial species but not all. Those that survive are the resistant ones, which, as they reproduce, will spread their resistance genes through horizontal gene transfer to many similar and dissimilar species, including any pathogens that happen to pass by. And eventually these antibiotic-resistant strains will and actually have found their way into people. Indeed, some resistant strains in humans match the resistant strains found in the farm animals.

Some farmers have voiced concern that eliminating the antibiotics will harm animal health (animals are usually closely penned) and result in economic loss. Fair enough if that was true. But look what Denmark has done. In 1995, Denmark enacted tough rules for the use of subtherapeutic antibiotics in livestock, including pigs. For example, the Danish government banned the veterinary use of avoparcin as a growth promoter because resistance to avoparcin also generates resistance to vancomycin. The farmers worried that their livelihood would be harmed without the use of antibiotics. However, they soon discovered that by leaving their piglets with their mothers for a few extra weeks bolstered the piglets' immune system. The result: animal health on the large and intensively run farms (similar to those in the United States) was not harmed, pigs gained as much weight as with antibiotics, and the litter size actually increased. The only necessity was to maintain high hygiene practices and reduce overcrowding in the feedlots. Today, productivity of the Danish pork market is at an all-time high, making Denmark the world's leading exporter of pork. Does the elimination or reduction in animal feed antibiotics reflect lowered antibiotic resistance? In 2001, it was reported that vancomycin-resistant enterococci positive samples in pigs decreased from 20% to 6%. This all culminated in 2006 with the European Union banning antibiotics as growth promoters in farm animal feed.

In 1977, the U.S. Food and Drug Administration (FDA) stated that the overuse of antibiotics in livestock could generate resistance and make these drugs less effective in humans. So, the agency issued an order to ban nonmedical use of penicillin and tetracycline in farm animals. However, due to governmental inaction, the rule has never been applied. Then in 2005, the FDA prohibited the use of fluoroquinolones in poultry in the hopes of reducing fluoroquinolone resistance and, in January 2012, the agency prohibited the use of certain cephalosporins for use in livestock. In March 2012, a federal court judge ordered the FDA to take the necessary steps to ban the two antibiotics for routine use in animal feed.

Investigating the Microbial World 24

The Source of Antibiotic Resistance

Following the discovery and development of the first antibiotics for clinical use in the 1940s, it did not take many bacterial species long to express antibiotic resistance. But did the development and use of antibiotics in medicine produce the "driving force" for bacterial species to evolve and spread resistance genes that then transformed much of the natural microbiota? Or is resistance an ancient phenomenon that spread from the natural microbiota to clinically important pathogenic species?

■ **OBSERVATIONS:** Some surveys comparing soil samples from the pre-antibiotic era (~1940) with contemporary soil samples (2008) report that the relative abundance of antibiotic resistance genes has increased in contemporary soils. Such surveys suggest resistance genes are modern evolutionary consequences of antibiotic development and use. Other investigations analyzing microbial DNA sequences from Pleistocene permafrost sediments (dated to be 30,000 years old) report the identification of resistance genes to many different classes of antibiotics. These investigations suggest that antibiotic resistance is quite ancient, reflecting a rich and diverse reservoir of resistance genes.

■ **QUESTION:** ***Is the presence of resistance genes in microbes the result of contemporary development and use of antibiotics?***

■ **HYPOTHESIS:** Antibiotic resistance is prevalent in microbial populations isolated from contact with human sources of antibiotics. If so, then investigating an environment that has never been exposed to contemporary antibiotics should result in finding resistance genes among the environment's microbiome.

■ **EXPERIMENTAL SITE AND DESIGN:** A region of Lechuguilla Cave, located in Carlsbad Caverns National Park, New Mexico, was selected for study. The cave is 300 to 400 meters below the surface and is believed to have been isolated from surface water and human exposure for more than 4 million years. Of the 500 unique bacterial isolates collected from three deep, remote sample sites in the cave (see figure A), 93 grew readily in tryptic soy broth. Ribosomal RNA gene sequencing classified 33% as gram-positive and 63% as gram-negative genera.

■ **EXPERIMENT:** To determine if any of the 93 strains contained antibiotic resistance, each was tested for growth in the presence and absence of up to 26 different antibiotics (20 µg/ml), representing natural, semisynthetic, and synthetic drugs.

■ **RESULTS:**

EXPERIMENT: See figure B. Note: Apramycin = an aminoglycoside; minocycline = a tetracycline derivative; telithromycin = macrolide; novobiocin = blocks DNA synthesis; rifampicin = rifampin; fosfomycin = inhibits DNA synthesis; daptomycin = disrupts cell membrane function.

FIGURE A Lechuguilla Cave. Courtesy of Max Wisshak www.speleo-foto.de.

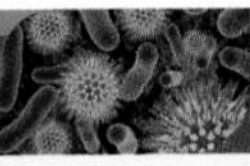

Investigating the Microbial World 24 (continued)

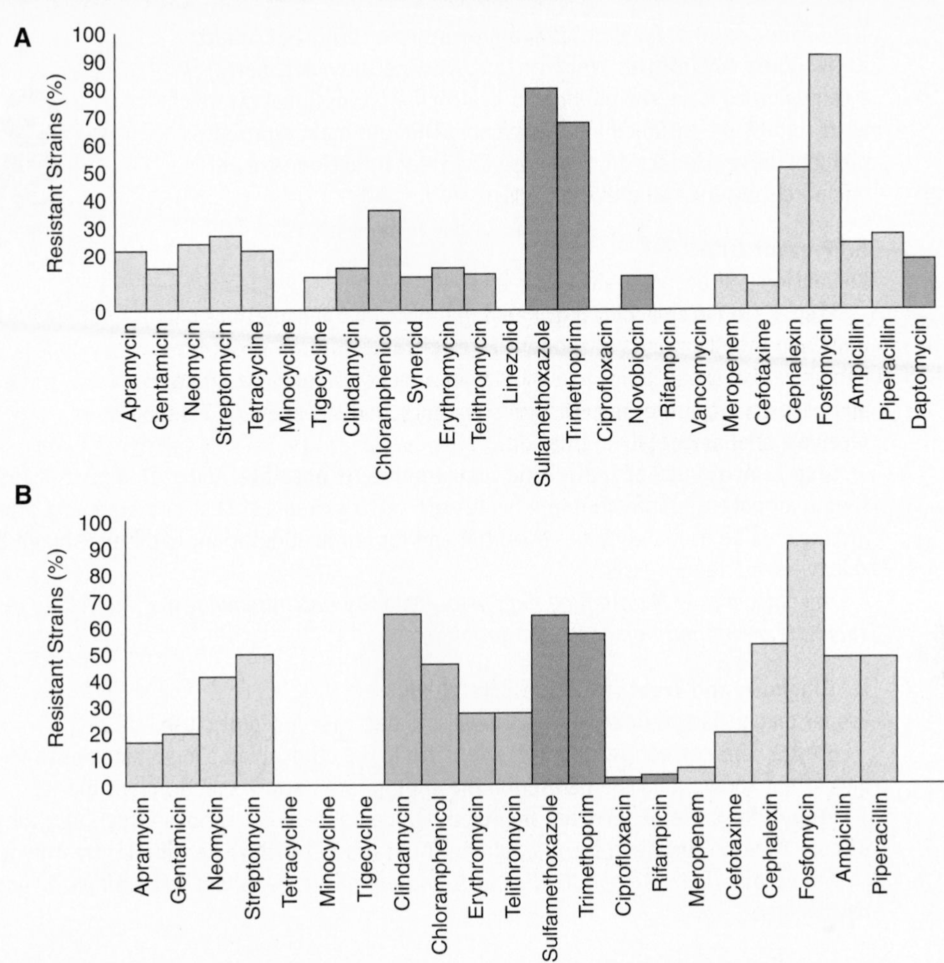

FIGURE B Resistance levels against various antibiotics. (**A**) = gram-positive isolates; (**B**) = gram-negative isolates. Modified from: Bhullar, K. *et al.* (2012). *PLoS ONE*, 7(4): 1–11.

QUESTION ***1: From figure B, do the antibiotics used in this study target the diverse bacterial cell structures and processes that are potentially susceptible to antibiotics? Explain. (Also refer to Figure 24.8.)***

QUESTION ***2: Daptomycin is a new class of antibiotics that was approved for clinical use in the late 1980s. Is it surprising that almost 20% of the gram-positive isolates (see figure) are resistant to this drug? Does this finding favor the idea that antibiotic resistance is ancient? Explain.***

QUESTION ***3: Is the hypothesis supported? Explain.***

Answers can be found on the Student Companion Website in **Appendix D**.

Adapted from: Bhullar, K. *et al.* (2012). *PLoS ONE*, **7**(4): 1–11. Icon image © Tischenko Irina/ShutterStock, Inc.

MICROFOCUS 24.6: Public Health
Preventing Antibiotic Resistance: Steps You Can Take

The Centers for Disease Control and Prevention's (CDC) **Get Smart: Know When Antibiotics Work** (http://www.cdc.gov/getsmart/) is a campaign to make the public and healthcare professional aware of ways to prevent antimicrobial resistance. The four main strategies include: prevent infection, diagnose and treat infection, use antimicrobials wisely, and prevent transmission.

Courtesy of CDC.

A. Prevent Infection
Prevention will decrease antibiotic use—if:

Step 1. You are vaccinated. Keeping up with all vaccinations will limit any infection by a pathogen. In addition, children should be immunized on schedule, people should get annual influenza vaccinations and keep up with other required booster shots, and those over 60 should receive a pneumococcal vaccination.

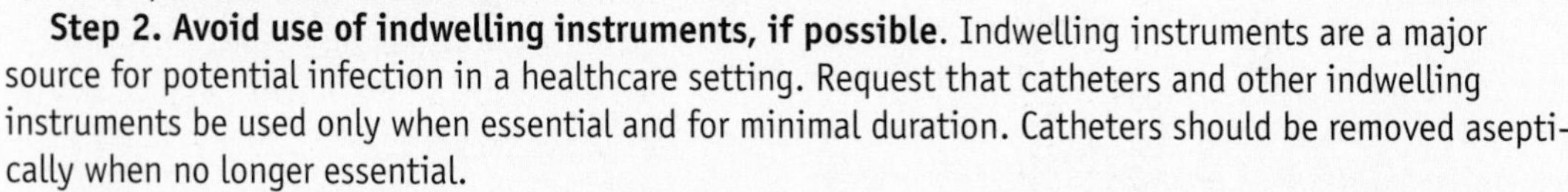
Step 2. Avoid use of indwelling instruments, if possible. Indwelling instruments are a major source for potential infection in a healthcare setting. Request that catheters and other indwelling instruments be used only when essential and for minimal duration. Catheters should be removed aseptically when no longer essential.

The bottom line: Maintaining high herd immunity and minimal use of indwelling instruments will eliminate or minimize any need for antibiotics.

B. Diagnose and Treat Infection Effectively
Proper diagnosis and infection treatment will decrease antibiotic use—if:

Step 3. The correct pathogen is identified. The clinical lab should attempt to identify the pathogen, if not known, and then prescribe the appropriate narrow-spectrum antibiotic.

Step 4. Reliable experts are involved. For complicated or serious infections, consult an infectious disease expert. Consider a second opinion if unsatisfied with the antibiotic treatment or recovery.

The bottom line: Appropriate diagnosis and antibiotic treatment will limit the chances of resistance development.

C. Use Antimicrobials Wisely
Proper use of antibiotics will decrease antibiotic use—if:

Step 5. You are an informed "drug user." Minimize use of broad-spectrum antibiotics as these can lead to more rapid antibiotic resistance. Also, avoid chronic or long-term antimicrobial use.

Step 6. You take and stop antimicrobial treatment when indicated. When cultures are negative and infection is unlikely stop antibiotic treatment. If positive, do not stop taking the antibiotic treatment prematurely, but once the infection has resolved with the prescribed course of antibiotic treatment, antibiotic use should stop.

The bottom line: Appropriate use of and completion of antibiotic treatment will limit potential resistance.

Prevent Transmission
Preventing transmission will decrease antibiotic use—if:

Step 7. The chain of contagion is broken. Stay home when sick and cover your mouth when you cough or sneeze. Educate family members and coworkers as to proper hygiene.

Step 8. You and others perform hand hygiene. Use alcohol-based handrubs or wash your hands often, especially when sick. Encourage the same for all family members and coworkers.

The bottom line: Proper hygiene will limit pathogen transmission and minimize the need for antibiotics.

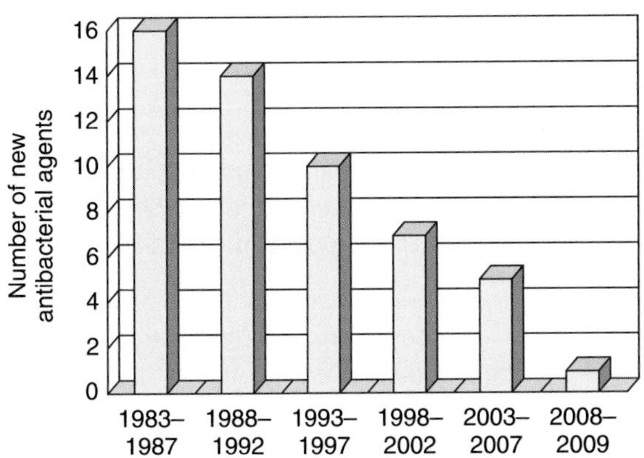

FIGURE 24.21 **Antibiotic Approvals, 1983–2009.** The number of antibiotic approvals has been declining over the past 25 years. Modified from: Spellberg et al, 2009. *Clinical Infectious Diseases* 9(38): 1279–1286. »» What are some reasons for the lack of new antibiotics coming through the development pipeline?

the so-called "ESKAPE" pathogens (***E****nterococcus faecium*, ***S****taphylococcus aureus*, ***K****lebsiella pneumoniae*, ***A****cinetobacter baumannii*, ***P****seudomonas aeruginosa*, and ***E****nterobacter* species), which currently cause the majority of U.S. hospital infections and effectively "escape" the effects of approved antibacterial drugs." To this end, the IDSA has launched a new collaboration titled the "10 × 20" initiative, which aims to reverse the downward trend in antibiotic development by stimulating the development of 10 new, safe, and effective antibiotics by 2020.

Scientists and pharmaceutical companies need to hasten their search for new approaches to antibiotic therapy as well as identifying new targets to which pathogens are susceptible. A few ideas are briefly considered below.

- One targeted approach is to focus on the lipopolysaccharide (LPS) in the outer membrane of gram-negative bacteria. The LPS contains several unusual carbohydrates representing possible targets for new drugs.
- Another approach is to interfere with an enzyme called DNA adenine methylase, which bacterial cells use to coat DNA with methyl groups and regulate DNA replication and repair. Researchers have significantly reduced the virulence of a *Salmonella* species by disabling the gene encoding the methylating enzyme. The researchers now hope to develop a chemical compound (synthetic antibiotic?) to neutralize the enzyme, thereby pinpointing a protein found only in the bacterial cell.
- The current explosion in microbial genome sequencing affords an opportunity to discover essential bacterial genes that may be targets for antimicrobial drugs and to identify other forms of antibiotic resistance.
- Because many of the newer antibiotics are simply minor modifications of existing drugs, there is a need to find new and unique antibiotics. The marine microbial community has barely been tapped as a source of new antibiotics. In addition, because only a very small portion of bacterial species can be cultured, 90% to 99% of the marine metagenome has not been sampled—and there may be many beneficial antibiotics awaiting discovery.
- Many microbiology labs around the world are sequencing the genomes of dozens of different pathogens thought to contain genes for undiscovered antibiotics. For example, the genus *Streptomyces* is the source of more than 65% of the antibiotics in current use. Almost certainly there are more antibiotics produced by this genus. Through genome sequencing, researchers have identified clusters of genes in *Streptomyces coelicolor* that may be sources for new antibiotic agents. What other, perhaps novel, natural products exist is an area of active research.

Although antimicrobial drugs probably are not the magic bullets once perceived by Ehrlich, new drug discovery may provide us with many new antimicrobial agents to fight pathogens and infectious disease.

CONCEPT AND REASONING CHECKS 6

a. Explain the significance of knowing a drug's minimal inhibitory concentration.
b. How do the resistance mechanisms confer resistance to an antibiotic?
c. Summarize the misuses and abuses of antibiotics.
d. Assess the need for new and novel antibiotics and targets for those antibiotics.

MICROINQUIRY 24
Testing Drugs—Clinical Trials

Before any antibiotic or chemotherapeutic agent can be used in therapy, it must be tested to ensure its safety and efficacy. The process runs from basic biomedical laboratory studies to approval of the product to improve health care. On average it takes some 8–12 years to study the drug in the laboratory, test it in animals, and finally run human clinical trials.

In this MicroInquiry, we examine a condensed version of the steps and time estimates through which a typical antibiotic would pass. Some questions asked are followed by the answer so you can progress further in your analysis. Please try to answer these questions before moving forward through the scenario. Also, methods are used that you have studied in this and previous chapters. A real pharmaceutical company might have other, more sophisticated methods available. Answers can be found on the Student Companion Website in **Appendix D**.

1. Discovery
You are head of a drug research group with a pharmaceutical company that has isolated a chemical compound (let's call it FM04) that is believed to have antibacterial properties and thus potential chemotherapeutic benefit. As head researcher, you must move this drug through the development pipeline from testing to clinical trials, evaluating, at each step, whether to proceed with further testing.

2. Preclinical Testing (4 years)
Many experiments need to be done before the chemical compound can be tested in humans. Because FM04 is "believed" to have antibacterial properties,

24.1a. What would be the first experimental tests to be carried out?

You would experimentally test the drug on bacterial cells in culture to ascertain its relative strength and potency as an antibacterial agent. This testing might include the agar disk diffusion method described in this chapter. The figure to the right shows three diffusion disk plates, each plated with a different bacterial species (*Escherichia coli*, *Staphylococcus aureus*, and *Pseudomonas aeruginosa*) and disks with different concentrations of FM04 (0, 1, 10, 100 μg).

24.1b. What can you conclude from these disk diffusion studies? Should drug testing continue? Explain.

Let's assume the bacterial studies look promising. FM04 now would be tested on human cells in culture to see if there are any toxic effects on metabolism and growth. Table A presents the results using the same drug concentrations from the bacterial studies. (The drug was prepared as a liquid stock solution to prepare the final drug concentration in culture.)

24.1c. What would you conclude from the studies of human cells in culture? Should drug testing continue? Explain.

24.1d. What was the point of including a solution with 0 μg of the drug?

When drugs are injected or ingested into the whole body, concentrations required for a desired chemotherapeutic effect may be very different from the experimental results with cells. Higher concentrations often are required and these may have toxic side effects. In animal testing, efforts should be made to use as few animals as possible and they should be subjected to humane and proper care. Let's assume the research studies of FM04 on mice used oral drug concentrations of 1, 10, 100, and 1,000 mg. Table B presents the simplified results.

24.2a. What would you conclude from the mouse studies? Should drug testing continue? Explain.

Because the results look positive and higher doses can be administered without serious toxic side effects (except at 1,000 mg), additional studies would be carried out to determine how much of the drug is actually absorbed into the blood, how it is degraded and excreted in the animal, and if there are any toxic breakdown products produced. These studies could take a few years to complete. Again, for the sake of the exercise, let's say the studies with FM04 remain promising.

Although mice are quite similar to humans physiologically, they are not identical, so human clinical trials need to be done. This is the reason the U.S. Food and Drug Administration (FDA) requires all new drugs to be tested through a *three-phase clinical trial period.*

First, as head of the research team, you must provide all the preclinical and animal testing results as part of an FDA *Investigational New Drug* (IND) *Application* before human tests can begin. The application must spell out how the clinical investigations will be conducted. As head of the investigational team, you must design the protocol for these human trials.

24.2b. How would you conduct the human tests to see if FM04 has positive therapeutic effects as an antibiotic in humans?

Briefly, there are many factors to take into consideration in designing the protocol.

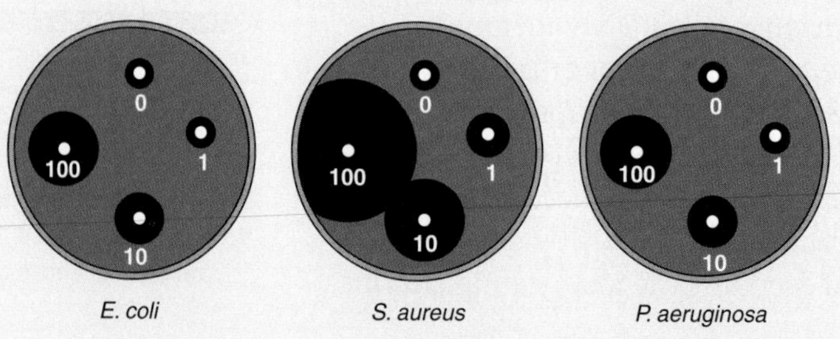

(i) How will the drug be administered? Let's assume it will be oral in the form of pills.

(ii) You need to recruit groups of patients (volunteers) of similar age, weight, and health status (healthy patients as well as patients with the infectious condition to which the drug may be therapeutic). Note that the bacterial studies suggested FM04 had its most potent effect on *Staphylococcus aureus,* a gram-positive species. Perhaps patients with a staph infection would be recruited.

(iii) Eventually, volunteers (patients) need to be split into two groups, one that receives the drug and one that receives a placebo (an inactive compound that looks like the FM04 pill).

(iv) The clinical studies should be carried out as "double blind" studies. In these studies, neither the patients nor the investigators know which patients are getting the drug or placebo. This prevents patients and investigators from "wishful thinking" as to the outcome of the studies.

In actual clinical studies there are even more factors to consider, but we will assume that FM04 wins approval for clinical testing.

3. Clinical Trials (6 years)
In *Phase 1 studies,* a small number (20 to 100) of healthy volunteers are treated with the test drug at different concentrations (doses) to see if there are any adverse side effects. These studies normally take at least several months. If unfavorable side effects are minimal, Phase 2 trials begin.

Phase 2 studies use a larger population of patient volunteers (several hundred) who have the disease the drug is designed to treat. Here, it is especially important to split the patients into the test and control groups because the major purpose of these trials is to study drug effectiveness. There needs to be a control group to accurately contrast effectiveness. These trials can take anywhere from several months to 2 years to complete. Provided there are no serious side effects or toxic reactions, or a lack of effectiveness, Phase 3 trials begin. There is an ethical question that can arise at this stage in the clinical trials regarding making the drug available, especially if the drug shows signs of reversing an illness that might be life-threatening. Even medical experts disagree as to the answer.

24.3 Is it ethical to give ill patients placebos when effective treatment appears available?

Phase 3 studies include a few thousand patient volunteers. The trials can take several years to complete because the purpose is to evaluate safety, dosage, and effectiveness of the drug.

4. FDA Review and Approval (2 years)
If the clinical trials are positive, as head researcher you can recommend that the pharmaceutical manufacturer apply to the FDA for a New Drug Application (NDA). The application is reviewed by an internal FDA committee that examines all clinical data, proposed labeling, and manufacturing procedures. The most important questions that need to be addressed are:

(i) Were the clinical studies well controlled to provide "substantial evidence of effectiveness"?

(ii) Did the clinical trials demonstrate that the drug was safe; that is, do the benefits outweigh the risks?

Based on its findings, the committee then makes its recommendation to the FDA. If approved, an approval letter is given to the company to market the drug.

Note that very few drugs (perhaps five in 5,000 tested compounds) actually make it to human clinical trials. Then, only about one of those five are found to be safe and effective enough to reach the pharmacy.

TABLE A **Adverse Cellular Effects Based on FM04 Drug Concentration**

Final Drug Concentration (μg/culture)	Adverse Cellular Effects
0	None
1	None
10	None
100	Abnormal cells and cell death

TABLE B **Toxicity Results on 50 Mice Given Different Concentrations of FM04**

Drug Concentration (mg)	Number of Mice	Adverse Reactions
0	10	None
1	10	None
10	10	None
100	10	None
1,000	10	Tremors and seizures in 8 mice

Chapter Challenge E

So the *K. pneumoniae* strain containing NDM-1 that was first identified in 2008 is a real dangerous superbug. By 2009, a study in India showed that there were already 22 Enterobacteriaceae strains harboring NDM-1. Ten were *Klebsiella* species, 9 were *Escherichia coli* strains, and 2 were *Enterobacter* species.

QUESTION E:

The fact that the NDM-1 genetic element spread so fast argues against mutation as the source of resistance. Explain how the resistance must have been spread so rapidly among the strains of Enterobacteriaceae?

Note: Not only are the superbugs containing the NDM-1 genetic element turning up in the clinic, they are also being found in common environments in South Asia, including in tap water, water puddles, and pools where children play. As of this writing, such environmental sources have not yet caused any known outbreaks [fingers crossed]. In early 2012, NDM-1 had been reported in patients in more than 35 countries, many of whom were "medical tourists" who traveled to India or Pakistan for inexpensive medical care or surgery.

Answers can be found on the Student Companion Website in **Appendix D**.

In conclusion, at the beginning of the 20th century, we saw that infectious diseases were among the most common causes of death in the United States and worldwide. However, by the 1950s, the "golden era" of antibiotic development and use was in full swing, with many new classes of antibiotics being introduced. These new antibiotics were viewed as being so powerful in destroying infectious disease that by the late 1960s some experts believed that infectious diseases would soon be historical events of the past. Unfortunately, since the early 1980s, we have seen and heard about an unprecedented number of emerging and resurging infectious diseases on a global scale. Whether antibiotic resistant or not, infectious disease remains a major cause of mortality and morbidity worldwide.

SUMMARY OF KEY CONCEPTS

24.1 Antimicrobial Agents Are Chemical Substances Used to Treat Infectious Disease

1. Ehrlich saw **antibiotics** as a chemotherapeutic approach to alleviating disease. One of his students isolated the first chemotherapeutic agent, **Salvarsan**. Domagk used **Prontosil** to treat bacterial infections.
2. Penicillin was discovered by Fleming, and purified and prepared for chemotherapy by Florey and Chain.
3. All antimicrobial agents represent **synthetic drugs** or **antibiotics,** demonstrate **selective toxicity**, and have a **broad** or **narrow drug spectrum**. (Fig. 24.3)
4. Antibiotics are natural compounds toxic to some competitive bacterial species.

24.2 Synthetic Antibacterial Agents Primarily Inhibit DNA Synthesis and Cell Wall Formation

5. The **sulfonamides** interfere with the production of **folic acid** through **competitive inhibition**. Modern sulfonamides are trimethoprim and sulfamethoxazole. (Fig. 24.5)
6. **Isoniazid** is an antituberculosis drug affecting wall synthesis. **Ciprofloxacin** is a fluoroquinolone that interacts with DNA to inhibit replication.

24.3 Beta-Lactam Antibiotics Inhibit Bacterial Cell Wall Synthesis

7. **Penicillin** interferes with cell wall synthesis in gram-positive bacterial cells. It can cause an anaphylactic reaction in sensitive individuals. Numerous synthetic and semisynthetic forms of penicillin are more resistant to **beta-lactamase** activity. (Fig. 24.7)
8. Certain **cephalosporin** drugs are first-choice antibiotics for penicillin-resistant bacterial species and a wide variety of these drugs is currently in use. The **carbapenems** (meropenem) inhibit cell wall synthesis and are broad-spectrum drugs.

24.4 Other Bacterially Produced Antibiotics Inhibit Some Aspect of Metabolism

9. **Vancomycin** inhibits cell wall synthesis.
10. **Bacitracin** and **polymyxin B** affect the permeability of the cell membrane.
11. **Aminoglycosides** (gentamicin, neomycin) inhibit protein synthesis in gram-negative bacterial cells. **Chloramphenicol** is a broad-spectrum antibiotic used against gram-positive and gram-negative bacterial species. Less severe side effects accompany tetracycline use, and this antibiotic is recommended against gram-

negative bacterial species as well as rickettsiae and chlamydiae. The **macrolides, lincosamides,** and **streptogramins** inhibit protein synthesis.

12. Besides the quinolones that inhibit DNA replication, **rifampin** interferes with transcription.

24.5 Other Antimicrobial Drugs Target Viruses, Fungi, and Parasites

13. Antiviral drugs interfere with viral entry, replication, or maturation.
14. The **polyenes** (nystatin, amphotericin B) and **imidazoles** are valuable against fungal infections. The **echinocandins** inhibit fungal cell wall synthesis, **flucytosine** interrupts nucleic acid synthesis, and **griseofulvin** interferes with mitosis.
15. Antiprotistan agents include the **aminoquinolines** (quinine and chloroquine) that are used to treat malaria; **sulfonamides** interfere with folic acid synthesis; and **nitroimidazoles** inhibit nucleic acid synthesis.
16. Antihelminthic drugs include **praziquantel, mebendazole,** and **avermectins.**

24.6 Antimicrobial Drug Resistance Is a Growing Challenge

17. Antibiotic assays include the **tube dilution assay,** which measures the **minimal inhibitory concentration,** and the **agar disk diffusion method,** which determines the susceptibility of a bacterial species to a series of antibiotics. (Figs. 24.14, 24.15)
18. Bacterial species have developed resistance to antimicrobial agents by altering metabolic pathways, inactivating antibiotics, reducing membrane permeability, or modifying the drug target. (Figs. 24.17, 24.18)
19. Arising from any of several sources such as changes in microbial biochemistry, **antibiotic resistance** threatens to put an end to the cures of infectious disease that have come to be expected in contemporary medicine. (Fig. 24.19)
20. New approaches include the discovery of new drug targets and developing new antibiotics to fight infectious disease and antibiotic resistance.

CHAPTER SELF-TEST

For **STEPS A–D**, answers to even-numbered questions and problems can be found in **Appendix C** on the Student Companion Website at **http://microbiology.jbpub.com/10e**. In addition, the site features eLearning, an online review area that provides quizzes and other tools to help you study for your class. You can also follow useful links for in-depth information, read more MicroFocus stories, or just find out the latest microbiology news.

STEP A: REVIEW OF FACTS AND TERMS

Multiple Choice

Read each question carefully, then select the ***one*** answer that best fits the question or statement.

1. Ehrlich and Hata discovered _____ that was used to treat _____.
   A. Salvarsan; syphilis
   B. penicillin; surgical wounds
   C. Salvarsan; malaria
   D. Prontosil; malaria
2. The reisolation and purification of penicillin was carried out by _____.
   A. Waksman
   B. Florey and Chain
   C. Domagk
   D. Fleming
3. The concentration of an antibiotic causing harm to the host is called the _____.
   A. toxic dosage level
   B. therapeutic dosage level
   C. minimal inhibitory concentration
   D. chemotherapeutic index
4. In the soil, antibiotics are produced by some _____.
   A. fungal species and viruses
   B. bacterial species and viruses
   C. viruses and helminthic species
   D. fungal and bacterial species
5. Trimethoprim and sulfamethoxazole are examples of _____ that block _____ synthesis.
   A. sulfonamides; PABA
   B. penicillins; cell wall
   C. sulfonamides; folic acid
   D. macrolides; protein
6. Isoniazid is used to treat _____.
   A. cholera
   B. influenza
   C. MRSA
   D. tuberculosis
7. Penicillins are useful in treating _____.
   A. gram-positive infections
   B. leprosy
   C. gram-negative infections
   D. tuberculosis
8. All the following are drugs or drug classes blocking cell wall synthesis *except* _____.
   A. cephalosporins
   B. carbapenems
   C. monobactams
   D. tetracyclines
9. Vancomycin inhibits _____________ synthesis.
   A. protein
   B. DNA
   C. bacterial cell wall
   D. RNA
10. Two cyclic polypeptide antibiotics are _____.
    A. vancomycin and streptomycin
    B. penicillin and cephalosporin
    C. bacitracin and polymyxins
    D. gentamicin and chloramphenicol
11. Which one of the following is NOT an inhibitor of protein synthesis?
    A. Clindamycin
    B. Macrolides
    C. Rifampin
    D. Chloramphenicol

12. Antibiotics inhibiting nucleic acid synthesis include _____.
   A. rifampin and quinolones
   B. aminoglycosides and tetracyclines
   C. lincosamides and streptogramins
   D. macrolides and aminoglycosides
13. Antiviral drugs that are base analogs inhibit _____.
   A. viral entry
   B. genome replication
   C. uncoating
   D. naturation
14. Antifungal drugs, such as _____, inhibit proper formation of _____.
   A. miconazole; a plasma membrane
   B. griseofulvin; DNA
   C. ketoconazole; a cell wall
   D. flucytosine; a microtubule
15. All of the following antiprotozoal drugs have been used to treat malaria *except* _____.
   A. melarsoprol
   B. quinine
   C. mefloquine
   D. chloroquine
16. This antihelminthic agent makes the membrane permeable to calcium ions.
   A. Mebendazole
   B. Pentamidine
   C. Ivermectin
   D. Praziquantel
17. The _____ is used to determine an antibiotic's minimal inhibitory concentration (MIC).
   A. Ames test
   B. tube (broth) dilution method
   C. agar disk diffusion method
   D. Kirby-Bauer test
18. The phosphorylation of an antibiotic is an example of which mechanism of resistance?
   A. Target modification
   B. Reduced permeability
   C. Antibiotic inactivation
   D. Altered metabolic pathway
19. A superinfection could arise from _____.
   A. using antibiotics when not needed
   B. using unnecessarily large doses of antibiotics
   C. stopping antibiotic treatment prematurely
   D. All of the above (**A–C**) are correct.
20. New approaches to antibiotic therapy include _____.
   A. carbohydrate targets in the wall LPS
   B. unregulating DNA replication
   C. discovering new and unique antibiotics
   D. All of the above (**A–C**) are correct.

True-False

On completing your study of these pages, test your understanding of antimicrobial drugs deciding whether the following statements are true or false. If the statement is true, write "True" in the space. If false, substitute a word for the underlined word to make the statement true.

_____ 21. Tetracycline has <u>four</u> benzene rings in the molecule.
_____ 22. Sulfonamides block folic acid formation in <u>fungi</u>.
_____ 23. Gentamicin is an antibiotic that interferes with <u>transcription</u>.
_____ 24. A side effect of <u>chloramphenicol</u> is a discoloration of the teeth.
_____ 25. An enzyme that breaks the structure of penicillin is known as <u>alpha-lactamase</u>.
_____ 26. <u>Polymyxin B</u> affects membrane permeability.
_____ 27. <u>Folic acid</u> synthesis is inhibited by rifampin.
_____ 28. <u>Penicillin</u> has no effect on viruses.
_____ 29. <u>Artemisinin</u> is a drug that releases free radicals.
_____ 30. Neosporin contains bacitracin, polymyxin B, and <u>tetracycline</u>.

Identification

Identify the bacterial structure inhibited or affected by each of the antibiotics in the description below.

Descriptions

31. ___________ Sulfonamides act here.
32. ___________ Polymyxin B affects this structure.
33. ___________ Rifampin blocks this process.
34. ___________ Quinolones act here.
35. ___________ Tetracyclines inhibit the function of this structure.
36. ___________ Cephalosporins target this structure.
37. ___________ The macrolides act here.
38. ___________ Vancomycin inhibits the assembly of this structure.
39. ___________ Aminoglycosides inhibit this structure.
40. ___________ Penicillin acts here.

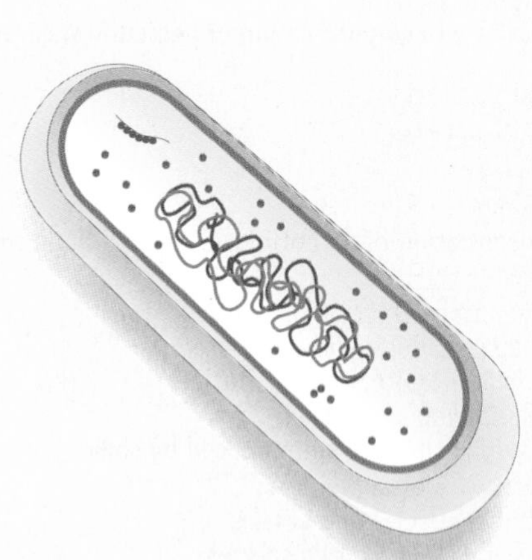

STEP B: CONCEPT REVIEW

41. Summarize the accomplishments of Paul Ehrlich and Gerhard Domagk to early advances in antimicrobial **chemotherapy**. (**Key Concept 1**)
42. Describe the important advances made by Fleming and Florey and Chain toward chemotherapy. (**Key Concept 1**)
43. Identify the important properties of antimicrobial agents. (**Key Concept 1**)
44. Explain why some microbes produce **antibiotics** in the environment. (**Key Concept 1**)
45. Explain how **sulfonamide** drugs block the **folic acid** metabolic pathway in bacterial cells. (**Key Concept 2**)
46. Compare and contrast the mechanism of action for **isoniazid** and the **quinolones**. (**Key Concept 2**)
47. Summarize the mechanism of action of **penicillin** and its semisynthetic derivatives. (**Key Concept 3**)
48. Discuss the other beta-lactam antibiotics that affect cell wall synthesis. (**Key Concept 3**)
49. Indicate how **vancomycin** differs from the beta-lactam antibiotics. (**Key Concept 4**)
50. Distinguish between the mechanisms of action for the polypeptide antibiotics. (**Key Concept 4**)
51. Hypothesize why there is such a large number of natural antibiotics that target the ribosome and protein synthesis. (**Key Concept 4**)
52. Identify the antibiotics affecting RNA synthesis. (**Key Concept 4**)
53. Describe the targets for antiviral drugs. (**Key Concept 5**)
54. Indicate the targets for antifungal drugs. (**Key Concept 5**)
55. Identify the drug targets in the protistan parasites. (**Key Concept 5**)
56. Identify several antihelminthic drugs and their targets. (**Key Concept 5**)
57. Contrast between the two antibiotic susceptibility assays. (**Key Concept 6**)
58. Describe the four mechanisms used by bacterial species to generate resistance to antibiotics. (**Key Concept 6**)
59. Justify the statement that **antibiotic resistance** has resulted from their misuse and abuse. (**Key Concept 6**)
60. Identify ways researchers are attempting to discover new drug targets and antibiotics. (**Key Concept 6**)

STEP C: APPLICATIONS AND PROBLEMS

61. In 1877, Pasteur and his assistant Joubert observed that anthrax bacilli grew vigorously in sterile urine but failed to grow when the urine was contaminated with other bacilli. What was happening?
62. In May 1953, Edmund Hillary and Tenzing Norgay were the first to reach the summit of Mount Everest, the world's highest mountain. Since that time over 150 other mountaineers have reached the summit, and groups have gone to Nepal from all over the world on expeditions. The arrival of "civilization" has brought a drastic change to the lifestyle of Nepal's Sherpa mountain people. For example, half of all Sherpas used to die before the age of 20, but partly due to antibiotics available to fight disease, the population has grown from 9 million to more than 23 million. Medical enthusiasts are proud of this increase in the life expectancy, but population ecologists see a bleaker side. What do you suspect they foresee, and what does this tell you about the impact antibiotics have on a culture?
63. Why would a synthetically produced antibiotic be more advantageous than a naturally occurring antibiotic? Why would it be less advantageous?

STEP D: QUESTIONS FOR THOUGHT AND DISCUSSION

64. According to historians, 2,500 years ago the Chinese learned to treat superficial infections such as boils by applying moldy soybean curds to the skin. Can you suggest what this implies?
65. Most naturally occurring antibiotics appear to be products of the soil bacteria belonging to the genus *Streptomyces*. Can you draw any connection between the habitat of these organisms and their ability to produce antibiotics?
66. Of the thousands and thousands of types of organisms screened for antibiotics since 1940, only five bacterial genera appear capable of producing these chemicals. Does this strike you as unusual? What factors might eliminate potentially useful antibiotics?
67. Is an antibiotic that cannot be absorbed from the gastrointestinal tract necessarily useless? How about one that is rapidly expelled from the blood into the urine? Explain your answers?
68. The antibiotic resistance issue can be argued from two perspectives. Some people contend that because of side effects and microbial resistance, antibiotics will eventually be abandoned in medicine. Others see the future development of a superantibiotic, a type of "miracle drug." What arguments can you offer for either view? Which direction in medicine do you support?
69. History shows that over and over, creativity is a communal act, not an individual one. How does the 1945 Nobel Prize to Fleming, Florey, and Chain reflect this notion?

Appendix A: Metric Measurement

TABLE

	Fundamental Unit	Quantity	Symbol	Numerical Unit	Scientific Notation
Length	Meter (m)				
		kilometer	km	1,000 m	10^3
		centimeter	cm	0.01 m	10^{-2}
		millimeter	mm	0.001 m	10^{-3}
		micrometer	μm	0.000001 m	10^{-6}
		nanometer	nm	0.000000001 m	10^{-9}
Volume (liquids)	Liter (l)				
		milliliter	ml	0.001 l	10^{-3}
		microliter	μl	0.000001 l	10^{-6}
Mass	Gram (g)				
		kilogram	kg	1,000 g	10^3
		milligram	mg	0.001 g	10^{-3}
		microgram	μg	0.000001 g	10^{-6}

Appendix B: Temperature Conversion Chart

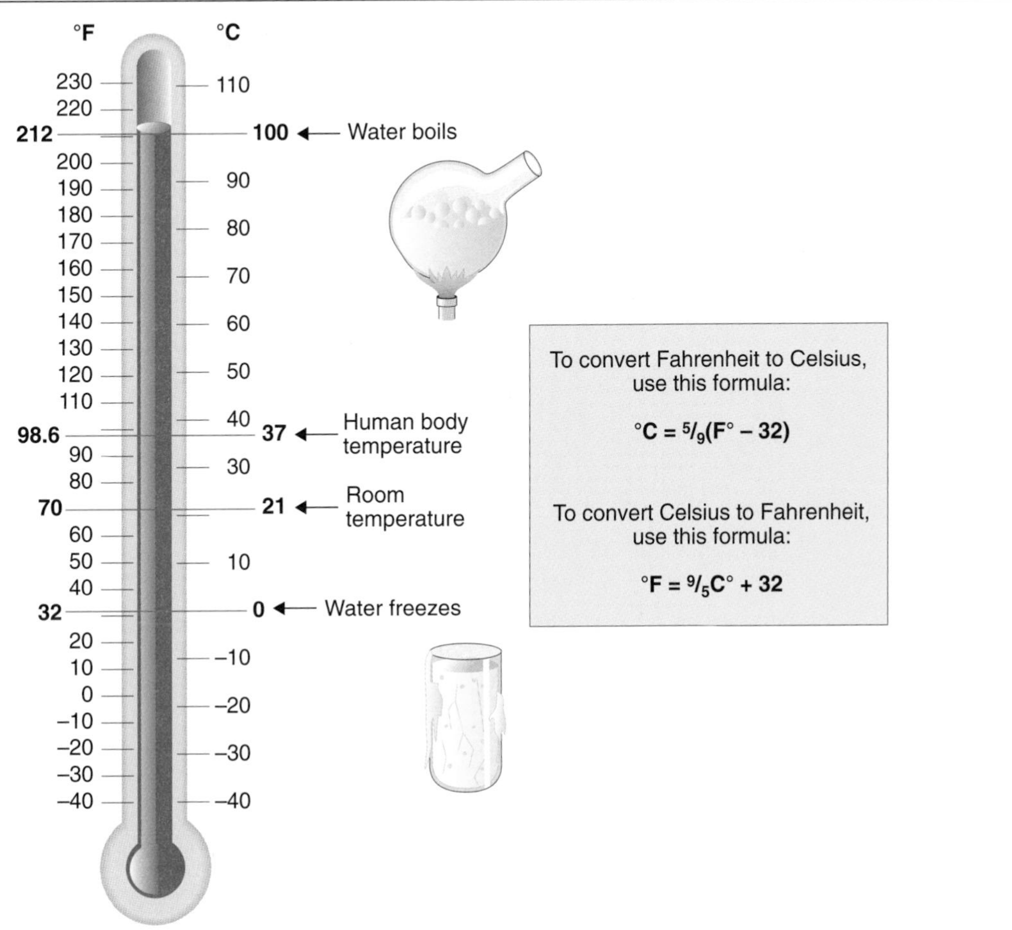

To convert Fahrenheit to Celsius, use this formula:

$$°C = \frac{5}{9}(F° - 32)$$

To convert Celsius to Fahrenheit, use this formula:

$$°F = \frac{9}{5}C° + 32$$

Glossary

This glossary contains concise definitions for microbiological terms and concepts only. **Please refer to the index for specific infectious agents, infectious diseases, anatomical terms, specific antimicrobial drugs, and immune disorders.**

A

abscess A circumscribed pus-filled lesion characteristic of staphylococcal skin disease; also called a boil.

abyssal zone The environment at the bottom of oceanic trenches.

acetyl CoA One of the starting compounds for the citric acid cycle.

acid A substance that releases hydrogen ions (H^+) in solution; *see also* **base**.

acid-fast technique A staining process in which mycobacteria resist decolorization with acid alcohol.

acidic dye A negatively charged colored substance in solution that is used to stain an area around cells.

acidophile A microorganism that grows best at acidic pHs below 4.

Actinobacteria A phylum in the domain Bacteria that exhibits fungus-like properties when cultivated in the laboratory.

actinomycete A soil bacterium that exhibits fungus-like properties when cultivated in the laboratory.

activated sludge Aerated sewage containing microorganisms added to untreated sewage to purify it by accelerating its bacterial decomposition.

activation energy The energy required for a chemical reaction to occur.

active immunity The immune system responds to antigen by producing antibodies and specific lymphocytes.

active site The region of an enzyme where the substrate binds.

active transport An energy-requiring movement of substances from an area of lower concentration across a biological membrane to a region of higher concentration by means of a membrane-spanning carrier protein.

acute disease A disease that develops rapidly, exhibits substantial symptoms, and lasts only a short time.

acute inflammation An immediate, but nonspecific innate immune defense to trauma.

acute period The phase of a disease during which specific symptoms occur and the disease is at its height.

acute phase protein A defensive blood protein secreted by liver cells that elevates the inflammatory and complement responses to infection.

acute rejection A transplant rejection response that takes 10 to 30 days.

acute wound One that arises from cuts, lacerations, bites, or surgical procedures.

adaptive immunity A response to a specific immune stimulus that involves immune defensive cells and frequently leads to the establishment of host immunity.

adenosine diphosphate (ADP) A molecule in cells that is the product of ATP hydrolysis.

adenosine triphosphate (ATP) A molecule in cells that provides most of the energy for metabolism.

adhesin A protein in bacterial pili that assists in attachment to the surface molecules of cells.

adjuvant An agent added to a vaccine to increase the vaccine's effectiveness.

ADP *See* **adenosine diphosphate**.

aerobe An organism that uses oxygen gas (O_2) for metabolism.

aerobic respiration The process for transforming chemical energy to ATP in which the final electron acceptor in the electron transport chain is oxygen gas (O_2).

aerosol A small particle transmitted through the air.

aerotolerant A bacterium not inhibited by oxygen gas (O_2).

aflatoxin A toxin produced by *Aspergillus flavus* that is cancer causing in vertebrates.

agar A polysaccharide derived from marine seaweed that is used as a solidifying agent in many microbiological culture media.

agglutination A type of antigen-antibody reaction that results in visible clumps of organisms or other material.

agranulocyte A white blood cell lacking visible granules; includes the lymphocytes and monocytes; *see also* **granulocyte**.

alcoholic fermentation A catabolic process that forms ethyl alcohol during the reoxidation of NADH to NAD^+ for reuse in glycolysis to generate ATP.

alga (pl. **algae**) An organism in the kingdom Protista that performs photosynthesis.

algal bloom An excessive growth of algae on or near the surface of water, often the result of an oversupply of nutrients from organic pollution.

alginate A sticky substance used as a thickener in foods and beverages.

alkaliphile An organism that grows best at pH 9 or higher.

allergen An antigenic substance that stimulates an allergic reaction in the body.

allergen immunotherapy A procedure involving a series of allergen shots over several years to reduce the sensitivity to allergens.

allergy shot *See* **allergen immunotherapy**.

allograft A tissue graft between two members of the same species, such as between two humans (not identical twins).

alpha helix A spiral portion of a polypeptide consisting of amino acids stabilized by hydrogen bonds.

alpha (α) hemolytic Referring to those bacterial species that when plated on blood agar cause a partial destruction of red blood cells as seen by an olive green color in the agar around colonies.

amatoxin A group of toxic compounds found in several genera of poisonous mushrooms.

Ames test A diagnostic procedure used to detect potential cancer-causing agents in humans by the ability of the agent to cause mutations in bacterial cells.

amino acid An organic acid containing one or more amino groups; the monomers that build proteins in all living cells.

aminoglycoside An antibiotic that contains amino groups bonded to carbohydrate groups that inhibit protein synthesis; examples are gentamicin, streptomycin, and neomycin.

amoeboid motion A crawling type of movement caused by the flow of cytoplasm into plasma membrane projections; typical of the amoebas.

amoebozoan A member of the super group Unikonta characterized by amoeboid movement and pseudopodia.

anabolism An energy-requiring process involving the synthesis of larger organic compounds from smaller ones; *see also* **catabolism**.

anaerobe An organism that does not require or cannot use oxygen gas (O_2) for metabolism.

anaerobic respiration The production of ATP where the final electron acceptor is an inorganic molecule other than oxygen gas (O_2); examples include nitrate, sulfate, and carbonate.

animalcule A tiny, microscopic organism observed by Leeuwenhoek.

anion An ion with a negative charge; *see also* **cation**.

anoxygenic photosynthesis A form of photo-synthesis in which molecular oxyten (O_2) is not produced.

antibiotic A substance naturally produced by a few bacterial or fungal species that inhibits or kills other microorganisms.

antibody A highly specific protein produced by the body in response to a foreign substance, such as a bacterium or virus, and capable of binding to the substance.

anticodon A three-base sequence on the tRNA molecule that binds to the codon on the mRNA molecule during translation.

antigen A chemical substance that stimulates the production of antibodies by the body's immune system.

antigen-antibody complex The interaction of an antigen with antibodies.

antigen binding site The region on an antibody that binds to an antigen.

antigen presenting cell (**APC**) A macrophage or dendritic cell that exposes antigen peptide fragments on its surface to T cells.

antigenic determinant A section of an antigen molecule that stimulates antibody formation and to which the antibody binds; also called epitope.

antigenic drift A minor variation over time in the antigenic composition of influenza viruses.

antigenic shift A major change over time in the antigenic composition of influenza viruses.

antigenic variation Referring to changes in the spike proteins on the flu viruses.

antihistamine A drug that blocks cell receptors for histamine, preventing allergic effects such as sneezing and itching.

antimicrobial agent (**drug**) A chemical that inhibits or kills the growth of microorganisms.

antimicrobial spectrum The range of antimicrobial drug action.

antisepsis The use of chemical methods for eliminating or reducing the growth or replication of infectious agents.

antiseptic A chemical used to reduce or kill pathogenic microorganisms on a living object, such as the surface of the human body.

antiserum (pl. **antisera**) A blood-derived fluid containing antibodies and used to provide temporary immunity.

antitoxin An antibody produced by the body that circulates in the bloodstream to provide protection against toxins by neutralizing them.

antiviral protein A protein made in response to interferon and that blocks viral replication.

antiviral state A cell capable of inhibiting viral protein synthesis due to interferon activation.

apicomplexan A protist containing a number of organelles at one end of the cell that are used for host penetration; no motion is observed in adult forms.

apoptosis A programmed cell death.

aqueous solution One or more substances dissolved in water.

arbovirus A virus transmitted by arthropods (e.g., insects).

Archaea The domain of living organisms that excludes the Bacteria and Eukarya.

arthrospore An asexual fungal spore formed by fragmentation of a septate hypha.

Arthus phenomenon An immune complex hypersensitivity when large amounts of IgG antibody for complexes with antigens in blood vessels or near the site of antigen entry.

artificially acquired active immunity The production of antibodies by the body in response to antigens in a vaccination; or the passive transfer of antibodies formed in one individual or animal to another susceptible person.

artificially acquired passive immunity The temporary immunity resulting from the transfer of antibodies from one individual or animal to another.

Ascomycota A phylum of fungi whose members have septate hyphae and form ascospores within saclike asci, among other notable characteristics.

ascospore A sexually produced fungal spore formed by members of the ascomycetes.

ascus (pl. **asci**) A saclike structure containing ascospores; formed by the ascomycetes.

asepsis The process or method of bringing about a condition in which no unwanted microbes are present.

asexual reproduction The form of reproduction that maintains genetic constancy while increasing cell numbers.

atom The smallest portion into which an element can be divided and still enter into a chemical reaction.

atomic nucleus The positively charged core of an atom, consisting of protons and neutrons that make up most of the mass.

atomic number The number of protons in the nucleus of an atom.

ATP *See* **adenosine triphosphate**.

ATP/ADP cycle The cellular processes of synthesis and hydrolysis of ATP.

ATP synthase The enzyme involved in forming ATP by using the energy in a proton gradient.

attenuated Referring to the reduced ability of bacterial cells or viruses in a vaccine to do damage to the exposed individual.

attractant A substance that attracts cells through motility.

autoantibody An antibody made against a self antigen.

autoclave An instrument used to sterilize microbiological materials by means of high temperature using steam under pressure.

autograft Tissue taken from one part of the body and grafted to another part of the same body.

autoimmune disorder A reaction in which antibodies react with an individual's own chemical substances and cells.

autotroph An organism that uses carbon dioxide (CO_2) as a carbon source; *see also* **chemoautotroph** *and* **photoautotroph**.

autotrophy The process by which an organism makes its own food.

auxotroph A mutant strain of an organism lacking the ability to synthesize a nutritional need; *see also* **prototroph**.

avirulent Referring to an organism that is not likely to cause disease.

B

bacillé Calmette Guérin (BCG) A strain of attenuated *Mycobacterium bovis* used for immunization against tuberculosis and, on occasion, leprosy.

bacillus (pl. **bacilli**) (1) Any rod-shaped bacterial or archaeal cell. (2) When referring to the genus *Bacillus*, it refers to an aerobic or facultatively anaerobic, rod-shaped, endospore-producing, gram-positive bacterial genus.

bacteremia The presence of live bacterial cells in the blood.

Bacteria The domain that includes all organisms not classified as Archaea or Eukarya.

bacterial growth curve The events occurring over time within a population of growing and dividing bacterial cells.

bactericidal Referring to any agent that kills bacterial cells.

bacteriochlorophyll A pigment located in the membrane systems of purple sulfur bacteria that upon excitement by light, loses electrons and initiates photosynthetic reactions.

bacteriocin One of a group of bacterial proteins toxic to other bacterial cells.

bacteriology The scientific study of prokaryotes in the domain Bacteria and Archaea.

bacteriophage (phage) A virus that infects and replicates within bacterial cells.

bacteriostatic Referring to any substance that prevents the growth of bacteria.

bacterium (pl. **bacteria**) A single-celled microorganism lacking a cell nucleus and membrane-enclosed compartments, and often having peptidoglycan in the cell wall.

barophile A microorganism that lives under conditions of high atmospheric pressure.

base A chemical compound that accepts hydrogen ions (H^+) in solution; *see also* **acid**.

base analog A nitrogenous base with a similar structure to a natural base but differing slightly in composition.

basic dye A positively charged colored substance in solution that is used to stain cells.

Basidiomycota The phylum of fungi whose members have septate hyphae and form basidiospores on supportive basidia, among other notable characteristics.

basidiospore A sexually produced fungal spore formed by members of the basidiomycetes.

basidium (pl. **basidia**) A club-like structure containing basidiospores; formed by the basidiomycetes.

basophil A type of white blood cell with granules that functions in allergic reactions.

B cell *See* **B lymphocyte**.

benign Referring to a tumor that usually is not life threatening or likely to spread to another part of the body.

benthic zone The environment at the bottom of a deep river, lake, or sea.

benzoic acid A chemical preservative used to protect beverages, catsup, and margarine.

beta (β) hemolytic Referring to those bacterial species that when plated on blood agar completely destroy the red blood cells as seen by a clearing in the agar around the colonies.

beta-lactamase The enzyme that converts the beta-lactam antibiotics (penicillins, cephalosporins, and carbapenems) into inactive forms.

beta oxidation The breakdown of fatty acids during cellular metabolism through the successive removal from one end of two carbon units.

binary fission An asexual process in bacterial and archaeal cells by which a cell divides to form two new cells while maintaining genetic constancy.

bioburden The estimated number of viable microbes in or on a medical device or surface before sterilization.

biocatalysis The metabolic reactions of microorganisms carried out at an industrial scale.

biochemical oxygen demand (BOD) A number referring to the amount of oxygen used by the microorganisms in a sample of water during a 5-day period of incubation.

biocrime The intentional introduction of biological agents into food or water, or by injection, to harm or kill individuals.

biofilm A complex community of microorganisms that form a protective and adhesive matrix that attaches to a surface, such as a catheter or industrial pipeline.

biological pollution The presence of microorganisms from human waste in water.

biological safety cabinet A cabinet or hood used to prevent contamination of biological materials.

biological vector An infected arthropod, such as a mosquito or tick, that transmits disease-causing organisms between hosts; *see also* **mechanical vector**.

biomass The total weight of all organisms in a defined area or environment.

bioreactor A large fermentation tank for growing microorganisms used in industrial production; also called a fermentor.

bioremediation The use of microorganisms to degrade toxic wastes and other synthetic products of industrial pollution.

biotechnology The commercial application of genetic engineering using living organisms.

bioterrorism The intentional or threatened use of biological agents to cause fear in or actually inflict death or disease upon a large population.

bipolar staining A characteristic of *Yersinia* and *Francisella* species in which stain gathers at the poles of the cells, yielding the appearance of safety pins.

bisphenol A combination of two phenol molecules used in disinfection.

blanching A process of putting food in boiling water for a few seconds to destroy enzymes.

blastospore A fungal spore formed by budding.

B lymphocyte (B cell) A white blood cell that matures into memory cells and plasma cells that secrete antibody.

booster shot A repeat dose of a vaccine given some years after the initial course to maintain a high level of immunity.

B period The growth phase in a bacterial cell cycle when the cells are increasing in size and mass.

bright-field microscopy An optical configuration of the light microscope that magnifies an object by passing visible light directly through the lenses and object.

broad spectrum Referring to an antimicrobial drug useful for treating many groups of microorganisms, including gram-positive and gram-negative bacteria; *see also* **narrow-spectrum**.

broth A liquid growth medium.

bubo A swelling of the lymph nodes due to inflammation.

budding (1) An asexual process of reproduction in fungi, in which a new cell forms as a swelling at the border of the parent cell and then breaks free to live independently. (2) The controlled release of virus particles from an infected animal cell.

buffer (1) A compound that minimizes pH changes in a solution by neutralizing added acids and bases. (2) Refers to a solution containing such a substance.

bull's eye rash A circular lesion on the skin with a red border and central clearing; characteristic of Lyme disease.

burst size The number of virus particles released from an infected bacterial cell.

C

cancer A disease characterized by the radiating spread of malignant cells that reproduce at an uncontrolled rate.

capnophilic Referring to a prokaryotic cell requiring low oxygen gas (O_2) and a high concentration of carbon dioxide gas (CO_2) for metabolism.

capsid The protein coat that encloses the genome of a virus.

capsomere Any of the protein subunits of a capsid.

capsule A layer of polysaccharides and small proteins covalently bound some prokaryotic cells; *see also* **slime layer** *and* **glycocalyx**.

carbohydrate An organic compound consisting of carbon, hydrogen, and oxygen that is an important source of carbon and energy for all organisms; examples include simple sugars, starch, and cellulose.

carbon cycle A series of interlinked processes involving carbon compound exchange between living organisms and the nonliving environment.

carbon-fixing reaction A chemical reaction in the second part of photosynthesis in which carbohydrates are formed.

carbuncle An enlarged abscess formed from the union of several smaller abscesses or boils.

carcinogen Any physical or chemical substance that causes tumor formation.

carrier An individual who has recovered from a disease but retains the infectious agents in the body and continues to shed them.

casein The major protein in milk.

casing soil A non-nutritious soil used to provide moisture for mushroom formation.

catabolism An energy-liberating process in which larger organic compounds are broken down into smaller ones; *see also* **anabolism**.

cation A positively charged ion; *see also* **anion**.

CD4 A coreceptor on the cell surface of a T helper cell.

CD8 A coreceptor on the cell surface of a cytotoxic T cell.

cell cycle The periods through which a cell passes between one division and the next.

cell envelope The cell wall and cell membrane of a bacterial or archaeal cell.

cell line A group of identical cells in culture and derived from a single cell.

cell-mediated immune response The body's ability to resist infection through the activity of T-lymphocyte recognition of antigen peptides presented on macrophages and dendritic cells and on infected cells.

cell membrane A thin bilayer of phospholipids and proteins that surrounds the prokaryotic cell cytoplasm. *See also* **plasma membrane**.

cell theory The tenet that all organisms are made of cells and arise from preexisting cells.

cell (tissue) tropism Refers to the specific cells or tissues that a virus infects.

cellular chemistry The chemical study of cells and cellular molecules.

cellular oncogene (***c-onc***) A mutated form of a normal gene controlling cell growth.

cellular respiration The process of converting chemical energy into cellular energy in the form of ATP.

cell wall A carbohydrate-containing structure surrounding fungal, algal, and most bacterial and archaeal cells.

central dogma The doctrine that DNA codes for RNA through transcription and RNA is converted to protein through translation.

centrosome The microtubule-organizing center of a eukaryotic cell.

cercaria (pl. **cercariae**) A tadpole-like larva form in the life cycle of a trematode.

cesspool Concrete cylindrical rings with pores in the walls that is used to collect human waste.

cestode A flatworm, commonly known as a tapeworm, that lives as a parasite in the gut of vertebrates.

CFU *See* **colony forming unit**.

chain of transmission How infectious diseases can be spread from human to human (or animal to human).

chancre A painless, circular, purplish hard ulcer with a raised margin that occurs during primary syphilis.

chaperone A protein that ensures a polypeptide folds into the proper shape.

chemical bond A force between two or more atoms that tends to bind those atoms together.

chemical element Any substance that cannot be broken down into a simpler one by a chemical reaction.

chemically defined medium A substance in which the form and quantity of each nutrient is known; *see also* **complex medium**.

chemical pollution The presence of inorganic and organic waste in water.

chemical reaction A process that changes the molecular composition of a substance by redistributing atoms or groups of atoms without altering the number of atoms.

chemiosmosis The use of a proton gradient across a membrane to generate cellular energy in the form of ATP.

chemoautotroph An organism that derives energy from inorganic chemicals and uses the energy to synthesize nutrients from carbon dioxide gas (CO_2).

chemoheterotroph An organism that derives energy from organic chemicals and uses the energy to synthesize nutrients from carbon compounds other than carbon dioxide gas (CO_2).

chemokine A protein that prompts specific white blood cells to migrate to an infection site and carry out their immune system functions.

chemotaxis A movement of a cell or organism toward a chemical or nutrient.

chemotherapeutic agent A chemical compound used to treat diseases and infections in the body.

chemotherapeutic index A number that represents the highest level of an antimicrobial drug tolerated by the host divided by the lowest level of the drug that eliminates the infectious agent.

chemotherapy The process of using chemical agents to treat diseases and infections, or other disorders, such as cancer.

chitin A polymer of acetylglucosamine units that provides rigidity in the cell walls of fungi.

Chlamydae A phylum of extremely small bacterial cells that can be cultured only in living host cells.

chlorination The process of treating water with chlorine to kill harmful organisms.

chlorophyll A green or purple pigment in algae and some bacterial cells that functions in capturing light for photosynthesis.

chloroplast A double membrane-enclosed compartment in algae that contains chlorophyll and other pigments for photosynthesis.

cholera toxin An enterotoxin that triggers an unrelenting loss of fluid.

Chromalveolata One of the super groups of the protists that includes the dinoflagellates, diatoms, ciliates, and the apicomplexans.

chromosome A structure in the nucleoid or cell nucleus that carries hereditary information in the form of genes.

chronic disease A disease that develops slowly, tends to linger for a long time, and requires a long convalescence.

chronic inflammation A long-lasting, but nonspecific innate immune response to trauma.

chronic rejection A form of transplant rejection that takes months to years to develop.

chronic wound An open sore, such as a leg ulcer or bed sore.

chytrid A fungus in the phylum Chytridiomycota.

Chytridiomycota A phylum of predominantly aquatic fungi.

ciliate A protistan organism that moves with the aid of cilia.

cilium (pl. **cilia**) A hair-like projection on some eukaryotic cells that along with many others assist in the motion of some protozoa and beat rhythmically to aid the movement of a fluid past the respiratory epithelial cells in humans.

cirrhosis Extensive injury of cells of the liver.

citric acid cycle A metabolic pathway in which acetyl groups are completely oxidized to carbon dioxide gas and some ATP molecules are formed. Also called Krebs cycle.

class A category of related organisms consisting of one or more orders.

classification The arrangement of organisms into hierarchal groups based on relatedness.

climax *See* **acute period**.

clinical disease A disease in which the symptoms are apparent.

clonal deletion A process by which the immune system eliminates autoreactive cells.

clonal selection The theory that certain lymphocytes are activated from the mixed population of B or T lymphocytes when stimulated by antigen or antigen peptide fragments.

clone A population of genetically identical cells (or plasmids).

cloning vector A plasmid used to introduce genes into a bacterial cell.

coagulase An enzyme produced by some staphylococci that catalyzes the formation of a fibrin clot.

coagulase test A diagnostic procedure to distinguish between *Staphylococcus aureus* and *S. epidermidis*.

coccus (pl. **cocci**) A spherical-shaped bacterial or archaeal cell.

codon A three-base sequence on the mRNA molecule that specifies a particular amino acid insertion in a polypeptide.

coenocytic Referring to a fungus containing no septa (cross-walls) and multinucleate hyphae.

coenzyme A small, organic molecule that forms the nonprotein part of an enzyme molecule.

coenzyme A (**CoA**) A small, organic molecule of cellular respiration that functions in release of carbon dioxide gas (CO_2) and the transfer of electrons and protons to another coenzyme.

cofactor A inorganic substance that acts with and is essential to the activity of an enzyme; examples include metal ions and some vitamins.

coliform bacterium A gram-negative, nonsporeforming, rodshaped cell that ferments lactose to acid and gas and usually is found in the human and animal intestine; high numbers in water is an indicator of contamination.

colony A visible mass of microorganisms of one type growing on or in a solid growth medium.

colony forming unit (**CFU**) A measure of the viable cells by counting the number of colonies on a plate; each colony presumably started from one viable cell.

colostrum The yellowish fluid rich in antibodies secreted from the mammary glands of animals or humans prior to the production of true milk.

comedo (pl. **comedones**) A plugged sebaceous gland.

commensalism A close and permanent association between two species of organisms in which one species benefits and the other remains unharmed and unaffected.

commercial sterilization A canning process to eliminate the most resistant bacterial spores.

communicable disease A disease that is readily transmissible between hosts.

comparative genomics The comparison of DNA sequences between organisms.

competence Referring to the ability of a cell to take up naked DNA from the environment.

competitive inhibition The prevention of a chemical reaction by a chemical that competes with the normal substrate for an enzyme's active site; *see also* **noncompetitive inhibition**.

complement A group of blood proteins that functions in a cascading series of reactions with antibodies to recognize and help eliminate certain antigens or infectious agents.

complement fixation test A serological procedure to detect antibodies to any of a variety of pathogens by identifying antibody-antigen-complement complexes.

complex Referring to one form of symmetry found in some viral capsids.

complex medium A chemically undefined medium in which the nature and quantity of each component has not been identified; *see also* **chemically defined medium**.

compound A substance made by the combination of two or more different chemical elements.

compound microscope *See* **light microscope**.

condensation reaction *See* **dehydration synthesis reaction**.

conidiophore The supportive structure on which conidia form.

conidium (pl. **conidia**) An asexually produced fungal spore formed on a supportive structure without an enclosing sac.

conjugate vaccine An antigen preparation consisting of the antigen bound to a carrier protein.

conjugation (1) In Bacteria and Archaea, a unidirectional transfer of genetic material from a live donor cell into a live recipient cell during a period of cell contact. (2) In the ciliates, a sexual process involving the reciprocal transfer of micronuclei between cells in contact.

conjugation pilus (pl. **pili**) A protein filament essential for conjugation between donor and recipient bacterial cells.

constant domain The invariable amino acids in the light and heavy chains of an antibody.

contact dermatitis A delayed hypersensitivity reaction occurring after contact with an allergen.

contagious Referring to a disease whose agent passes with particular ease among susceptible individuals.

contractile vacuole A membrane-enclosed structure within a cell's cytoplasm that regulates the water content by absorbing water and then contracting to expel it.

contrast Refers to the ability to see an object against the background in the microscope.

control That part of an experiment not exposed to or treated with the factor being tested.

Coombs test An antibody test used to detect Rh antibodies involved in hemolytic disease of the newborn.

C period The DNA replication phase of a bacterial cell cycle.

corticosteroid A synthetic drug used to control allergic disorders by blocking the release of chemical mediators.

covalent bond A chemical linkage formed by the sharing of electrons between atoms or molecules.

Crenarchaeota A group within the domain Archaea that tend to grow in hot or cold environments.

critical control point (**CCP**) In the food processing industry, a place where contamination of the food product could occur.

Cryptomycota A group of fungi with small, flagellated cells without a chitin cell wall.

Cyanobacteria A phylum of oxygen-producing, photosynthetic bacterial cells.

cyst A dormant and very resistant form of a protozoan and multicellular parasite.

cytochrome A compound containing protein and iron that plays a role as an electron carrier in cellular respiration and photosynthesis; *see also* **electron transport chain**.

cytokine Small proteins released by immune defensive cells that affects other cells and the immune response to an infectious agent.

cytopathic effect (**CPE**) Visible effect that can be seen in a virus-infected host cell.

cytoplasm The complex of chemicals and structures within a cell; in plant and animal cells excluding the nucleus.

cytoskeleton (1) The structural proteins in a prokaryotic cell that help control cell shape and cell division. (2) In a eukaryotic cell, the internal network of protein filaments and microtubules that control the cell's shape and movement.

cytosol The fluid, ions, and compounds of a cell's cytoplasm excluding organelles and other structures.

cytotoxic hypersensitivity A cell-damaging or cell-destroying hypersensitivity that develops when IgG reacts with antigens on the surfaces of cells.

cytotoxic T cell (**CTL**) The type of T lymphocyte that searches out and destroys infected cells.

cytotoxin A chemical that is poisonous to cells.

D

dark-field microscopy An optical system on the light microscope that scatters light such that the specimen appears white on a black background.

dead end host A host that cannot transmit a pathogen.

deamination A biochemical process in which amino groups are enzymatically removed from amino acids or other organic compound.

decimal reduction time (**D valve**) The time required to kill 90% of the viable organisms at a specified temperature.

decline phase The final portion of a bacterial growth curve in which environmental factors overwhelm the population and induce death; also called death phase.

decomposer A bacterial or fungal organism that breaks down and recycles dead or decaying matter.

defensin An antimicrobial peptide present in white blood cells that plays a role in the prevention or elimination of infection.

definitive host An organism that harbors the adult, sexually mature form of a parasite.

degerm To mechanically remove organisms from a surface.

degranulation The release of cell mediators from mast cells and basophils.

dehydration synthesis reaction A process of bonding two molecules together by removing a water molecule and joining the open bonds.

delayed hypersensitivity An immune response to an allergen that takes 2 to 3 days to develop.

denaturation A process caused by heat or pH in which proteins lose their function due to changes in their 3-D structure.

dendritic cell A white blood cell having long finger-like extensions and found within all tissues; it engulfs and digests foreign material, such as bacterial cells and viruses, and presents antigen peptides on its surface.

dental plaque A biofilm consisting of salivary proteins, food debris, and bacteria attached to the tooth surface.

deoxyribonucleic acid (**DNA**) The genetic material of all cells and many viruses.

dermatophyte A pathogenic fungus that affects the skin, hair, or nails.

desensitization A process in which minute doses of antigens are used to remove antibodies from the body tissues to prevent a later allergic reaction.

detergent A synthetic cleansing substance that dissolves dirt and oil.

dextran A water-soluble glucose polymer.

diapedesis A process by which phagocytes move out of the blood vessels by migrating between capillary cells.

diatom One of a group of microscopic marine algae that performs photosynthesis.

dichotomous key A method of deducing the correct species assignment of a living organism by offering two alternatives at each juncture, with the choice of one of those alternatives determining the next step.

differential medium A growth medium in which different species of microorganisms can be distinguished visually.

differential staining procedure A technique using two dyes to differentiate cells or cellular objects based on their staining; *see also* **simple stain technique**.

Dikarya The group of fungi containing the Ascomycota and Basidiomycota.

dikaryotic Referring to fungal cytoplasm in which two genetically different haploid nuclei closely pair.

dimorphic Referring to pathogenic fungi that take a yeast form in the human body and a filamentous form when cultivated in the laboratory.

dinoflagellate A microscopic photosynthetic marine alga that forms one of the foundations of the food chain in the ocean.

dipicolinic acid An organic substance that helps stabilize the proteins and DNA in a bacterial spore, thereby increasing spore resistance.

diplobacillus (pl. **diplobacilli**) A pair of rod-shaped bacterial or archaeal cells.

diplococcus (pl. **diplococci**) A pair of spherical-shaped bacterial or archaeal cells.

diplomonad A protist that contains four pair of flagella, two haploid nuclei, and live in low oxygen or anaerobic environments; most members are symbiotic in animals.

direct contact The form of disease transmission involving close association between hosts; *see also* **indirect contact**.

direct microscopic count Estimation of the number of cells by observation with the light microscope.

disaccharide A sugar formed from two single sugar molecules; examples include sucrose and lactose.

disease Any change from the general state of good health.

disinfectant A chemical used to kill or inhibit pathogenic microorganisms on a lifeless object such as a tabletop.

disinfection The process of killing or inhibiting the growth of pathogens.

disk diffusion method A procedure for determining bacterial susceptibility to an antibiotic by determining if bacterial growth occurs around an antibiotic disk; also called the Kirby-Bauer test.

disseminated intravascular coagulation (DIC) A systemic clotting of blood in small blood vessels.

disulfide bridge A covalent bond between sulfur-containing R groups in amino acids.

DNA *See* **deoxyribonucleic acid**.

DNA helicase An enzyme that unwinds and separates the two polynucleotide strands during DNA replication.

DNA ligase An enzyme that binds together DNA fragments.

DNA microarray A series of DNA segments used to study changes in gene expression.

DNA polymerase An enzyme that catalyzes DNA replication by combining complementary nucleotides to an existing strand.

DNA probe A short segment of single stranded DNA used to locate a complementary strand among many other DNA strands.

DNA replication The process of copying the genetic material in a cell.

DNA vaccine A preparation that consists of a DNA plasmid containing the gene for a pathogen protein.

domain The most inclusive taxonomic level of classification; consists of the Archaea, Bacteria, and Eukarya.

double diffusion assay Another name for the diffusion of antibodies and antigens.

double helix The structure of DNA, in which the two complementary strands are connected by hydrogen bonds between complementary nitrogenous bases and wound in opposing spirals.

Downey cell A swollen lymphocyte with foamy cytoplasm and many vacuoles that develops as a result of infection with infectious mononucleosis viruses.

D period The binary fission phase of a bacterial cell cycle.

droplet transmission The movement of an airborne particle of mucus and sputum from the respiratory tract.

drug synergism Referring to two antimicrobial drugs that work better together than seperately.

dysbiosis A change to the normal (healthy) composition of the gut microbiota.

E

eclipse period The period of a viral infection when no viruses can be found inside the infected cell.

edema A swelling of the tissues brought about by an accumulation of fluid.

effector cell An activated immune cell targeting a pathogen.

electron A negatively charged particle with a small mass that moves around the nucleus of an atom.

electron microscope An instrument that uses electrons and a system of electromagnetic lenses to produce a greatly magnified image of an object; *see also* **transmission electron microscope** *and* **scanning electron microscope**.

electron shell An energy level surrounding the atomic nucleus that contains one or more electrons.

electron transport chain A series of proteins that transfer electrons in cellular respiration to generate ATP.

elementary body An infectious form of *Chlamydia* in the early stage of reproduction.

ELISA *See* **enzyme-linked immunosorbent assay**.

elongation (1) The addition of complementary nucleotides to a parental DNA strand. (2) The addition of addition of amino acids onto the forming polypeptide during translation.

emerging infectious disease A new disease or changing disease that is seen within a population for the first time; *see also* **reemerging infectious disease**.

endemic Referring to a disease that is constantly present in a specific area or region.

endergonic reaction A chemical process that requires energy; *see also* **exergonic reaction**.

endocytosis The process by which many eukaryotic cells take up substances, cells, or viruses from the environment.

endoflagellum (pl. **endoflagella**) A microscopic fiber located along cell walls in certain species of spirochetes; contractions of the filaments yield undulating motion in the cell.

endogenous infection A disorder that starts with a microbe or virus that already was in or on the body as part of the microbiota; *see also* **exogenous infection**.

endomembrane system A cytoplasmic set of membranes that function in the transport, modification, and sorting of proteins and lipids in eukaryotic cells.

endophyte A fungus that lives within plants and does not cause any known disease.

endoplasmic reticulum (ER) A network of membranous plates and tubes in the eukaryotic cell cytoplasm responsible for the synthesis and transport of materials from the cell.

endospore An extremely resistant dormant cell produced by some gram-positive bacterial species.

endosymbiosis The idea that mitochondria and chloroplasts originated from free-living bacterial cells and cyanobacteria that took up residence in a primitive eukaryotic cell.

endotoxin A metabolic poison, produced chiefly by gram-negative bacteria, that are part of the bacterial cell wall and consequently are released on cell disintegration; composed of lipid-polysaccharide-peptide complexes.

endotoxin shock A drop in blood pressure due to an endotoxin.

energy-fixing reaction A reaction in the first part of photosynthesis where light energy is converted into chemical energy in the form of ATP.

enriched medium A growth medium in which special nutrients must be added to get an species to grow.

enterotoxin A toxin that is active in the gastrointestinal tract of the host.

enterotype Refers to a specific composition of bacterial species in the gut.

envelope The flexible membrane of protein and lipid that surrounds many types of viruses.

enzyme A reusable protein molecule that brings about a chemical change while itself remaining unchanged.

enzyme immunoassay (EIA) *See* **enzyme-linked immunosorbent assay**.

enzyme-linked immunosorbent assay (ELISA) A serological test in which an enzyme system is used to detect an individual's exposure to a pathogen.

enzyme-substrate complex The association of an enzyme with its substrate at the active site.

eosinophil A type of white blood cell with granules that stains with the dye eosin and plays a role in allergic reactions and the body's response to parasitic infections.

epidemic Referring to a disease that spreads more quickly and more extensively within a population than normally expected.

epidemiology The scientific study of the source, cause, and transmission of disease within a population.

epitope *See* **antigenic determinant**.

ergotism A condition inducing convulsions and hallucinations from fungal toxins in rye.

erythema A zone of redness in the skin due to a widening of blood vessels near the skin surface.

erythema migrans (EM) An expanding circular red rash that occurs on the skin of patients with Lyme disease.

erythrogenic Referring to a streptococcal poison that leads to the rash in scarlet fever.

E-test An antibiotic sensitivity test using a paper strip containing a marked gradient of antibiotic.

ethylene oxide A chemical gas that is used to sterilize many objects and instruments.

Euglenozoa A group of protists in the Excavata.

Eukarya The taxonomic domain encompassing all organisms having a cell nucleus and membrane-bound compartments.

eukaryote An organism whose cells contain a cell nucleus with multiple chromosomes, a nuclear envelope, and membrane-bound compartments; *see also* **prokaryote**.

eukaryotic Referring to a cell or organism containing a cell nucleus with multiple chromosomes, a nuclear envelope, and membrane-bound compartments.

Euryarchaeota A group within the domain Archaea that contains the methanogens and extreme halophiles.

exanthema A maculopapular rash occurring on the skin surface.

Excavata One of the super groups of the protists, many lacking true mitochondria.

exergonic reaction A chemical process releasing energy; *see also* **endergonic reaction**.

exogenous infection A disorder that starts with a microbe or virus that entered the body from the environment; *see also* **endogenous infection**.

exon The coding sequence in a split gene; *see also* **intron**.

exotoxin A bacterial metabolic poison composed of protein that is released to the environment; in the human body, it can affect various organs and systems.

experiment A test or trial to verify or refute a hypothesis.

extreme acidophile An archaeal organism living at an extremely acidic pH.

extreme halophile A microbial species living at high salt (NaCl) concentration.

extremophile A microorganism that lives in extreme environments, such as high/low temperature, high acidity, or high salt.

extrinsic factor An environmental characteristic that influences the growth of food microbes. *See also* **intrinsic factor**.

F

facilitated diffusion The movement of substances from an area of higher concentration across a biological membrane to a region of lower concentration by means of a membrane-spanning channel or carrier protein.

facultative anaerobe Referring to an organism that grows in the presence or absence of oxygen gas (O_2).

family A category of related organisms consisting of one or more genera.

Fc segment The stem portion of an antibody molecule that combines with phagocytes, mast cells, or complement.

fecal transplant The process of transferring fecal microbiota from a healthy individual into a recipient suffering a stubborn intestinal disease.

feedback inhibition The slowing down or prevention of a metabolic pathway when excess end product binds noncompetitively to an enzyme in the pathway.

fermentation A metabolic pathway in which carbohydrates serve as electron donors, the final electron acceptor is not oxygen gas (O_2), and NADH is reoxidized to NAD^+ for reuse in glycolysis for generation of ATP.

fermentor *See* **bioreactor**.

fever An abnormally high body temperature that is usually caused by a bacterial or viral infection.

F factor A plasmid containing genes for plasmid replication and conjugation pilus formation.

field The circular area seen when looking in the light microscope.

filariform Referring to the thread-like shape of the larvae of the hookworm.

filopodium (pl. **filopodia**) A thin protrusion from a cell, such as a phagocyte.

filtration A mechanical method to remove microorganisms by passing a liquid or air through a filter.

Firmicutes A phylum in the domain Bacteria that contains many of the gram-positive species.

flaccid paralysis A loss of voluntary movement in which the limbs have little tone and become flabby.

flagellum (pl. **flagella**) A long, hair-like appendage composed of protein and responsible for motion in microorganisms; found in some bacterial, archaeal protistan, algal, and fungal cells.

flare A spreading zone of redness around a wheal that occurs during an allergic reaction; *also see* **wheal**.

flash pasteurization method A treatment in which milk is heated at 71.6°C for 15 seconds and then cooled rapidly to eliminate harmful bacteria; also called HTST ("high temperature, short time") method.

flatworm A multicellular parasite with a flattened body; examples include the tapeworms.

flavin adenine dinucleotide (FAD) A coenzyme that acts as an electron carrier for ATP production.

floc A jelly-like mass that forms in a liquid and made up of coagulated particles.

flocculation (1) A serological reaction in which particulate antigens react with antibodies to form visible aggregates of material. (2) The formation of jelly-like masses of coagulated material in the water-purification process.

fluid mosaic model The representation for the cell (plasma) membrane where proteins "float" within or on a bilayer of phospholipid.

fluke *See* **trematode**.

fluorescence microscopy An optical system on the light microscope that uses ultraviolet light to excite dye-containing objects to fluoresce.

fluorescent antibody technique A diagnostic tool that uses fluorescent antibodies with the fluorescence microscope to identify an unknown organism.

flush A burst of mushroom growth.

focus of infection The area in the body where an infection is located.

folic acid The organic compound in bacteria whose synthesis is blocked by sulfonamide drugs.

fomite An inanimate object, such as clothing or a utensil, that carries disease organisms.

food vacuole A membrane-enclosed compartment in some eukaryotic that results from the intake of large molecules, particles, or cells, for digestion.

foraminiferan A shell-containing amoeboid protozoan having a chalky skeleton with window-like openings between sections of the shell.

formalin A solution of formaldehyde used as embalming fluid, in the inactivation of viruses, and as a disinfectant.

F plasmid A DNA plasmid in the cytoplasm of an F^+ bacterial cell that may be transferred to a recipient bacterial cell during conjugation.

fruiting body The general name for a reproductive structure of a fungus from which spores are produced.

functional genomics The identification of gene function from a gene sequence.

functional group A group of atoms on hydrocarbons that participates in a chemical reaction.

fungemia The dissemination of fungi through the circulatory system.

Fungi One of the four kingdoms in the domain Eukarya; composed of the molds and yeasts.

fungicidal Referring to any agent that kills fungi.

fungistatic Referring to any substance that inhibits the growth of fungi.

furuncle An infection of a hair follicle.

G

gametocyte The stage in the life cycle of the malaria parasite during which it reproduces sexually in the blood of a mosquito.

gamma globulin A general term for antibody-rich serum.

gamma ray An ionizing radiation that can be used to sterilize objects.

gangrene A physiological process in which the enzymes from wounded tissue digest the surrounding layer of cells, inducing a spreading death to the tissue cells.

gastroenteritis Infection of the stomach and intestinal tract often due to a virus.

gas vacuole A cytoplasmic compartment in some bacterial and archaeal cells used to regulate buoyancy.

gene A segment of a DNA molecule that provides the biochemical information for a polypeptide or for a functioning RNA molecule.

gene probe A small, single-stranded DNA fragment labeled for identification of a specific DNA segment.

generalized transduction A process by which a bacteriophage carries a bacterial chromosome fragment from one cell to another; *see also* **specialized transduction**.

generation time The time interval for a cell population to double in number.

genetically modified organism (GMO) An organism produced by genetic engineering.

genetic code The specific order of nucleotide sequences in DNA or RNA that encode specific amino acids for protein synthesis.

genetic engineering The use of bacterial and microbial genetics to isolate, manipulate, recombine, and express genes.

genetic recombination The process of bring together different segments of DNA.

genome The complete set of genes in a virus or cell.

genome annotation The identification of gene locations and function in a gene sequence.

genomic island A series of up to 25 genes absent in other strains of the same prokaryotic species.

genomics The study of an organism's gene structure and gene function in viruses and organisms.

genus (pl. **genera**) A rank in the classification system of organisms composed of one or more species; a collection of genera constitute a family.

germicide Any agent that kills microorganisms.

germ theory of disease The principle formulated by Pasteur and proved by Koch that microorganisms are responsible for infectious diseases.

Glomeromycota A group of mycorrhizal fungi that exist within the roots of most land plants.

glucose A six-carbon sugar used as a major energy source for metabolism.

glutaraldehyde A liquid chemical used for sterilization.

glycocalyx A viscous polysaccharide material covering many prokaryotic cells to assist in attachment to a surface and impart resistance to desiccation; *see also* **capsule** *and* **slime layer**.

glycolysis A metabolic pathway in which glucose is broken down into two molecules of pyruvate with a net gain of two ATP molecules.

Golgi apparatus A stack of flattened, membrane-enclosed compartments in eukaryotic cells involved in the modification and sorting of lipids and proteins.

graft-versus-host (GVH) reaction A phenomenon in which a tissue graft or transplant produces immune substances against the recipient.

gram-negative Referring to a bacterial cell that stains red after Gram staining.

gram-positive Referring to a bacterial cell that stains purple after Gram staining.

Gram stain technique A staining procedure used to identify bacterial cells as gram-positive or gram-negative.

granulocyte A white blood cell with visible granules in the cytoplasm; includes neutrophils, eosinophils, and basophils; *see also* **agranulocyte**.

granuloma A small lesion caused by an infection.

granzyme A cytotoxic enzyme that causes infected cells to undergo a programmed cell death.

groundwater Water originating from deep wells and subterranean springs; *see also* **surface water**.

group A streptococci (GAS) Contains those streptococcal species that are β-hemolytic organisms.

Guillain-Barré syndrome (GBS) A complication of influenza and chickenpox, characterized by nerve damage and polio-like paralysis.

gumma A soft, granular lesion that forms in the cardiovascular and/or nervous systems during tertiary syphilis.

H

HAART *See* **highly active antiretroviral therapy**.

HACCP *See* **hazard analysis critical control point**.

halogen A chemical element whose atoms have seven electrons in their outer shell; examples include iodine and chlorine.

halophile An organism that lives in environments with high concentrations of salt.

halotolerant A microbe that grows best without salt but can tolerate low concentrations.

hapten A small molecule that combines with tissue proteins or polysaccharides to form an antigen.

Hazard Analysis Critical Control Point (HACCP) A set of federally enforced regulations to ensure the dietary safety of seafood, meat, and poultry.

healthcare-associated infection (HAI) An infection resulting from treatment for another condition while in a hospital or healthcare facility.

heat fixation The use of warm temperatures to prepare microorganisms for staining and viewing with the light microscope.

heavy (H) chain The larger polypeptide in an antibody.

heavy metal A chemical element often toxic to microorganisms; examples include mercury, copper, and silver.

helical One form of symmetry found in some viral capsids.

helminth A term referring to a multicellular parasite; includes roundworms and flatworms.

helminthic load The number of worms in the human body.

helper T (T_H) cell A type of lymphocyte involved in regulation of cell-mediated immunity and in helping B cells to produce antibody.

helper T1 (T_H1) cell A T lymphocyte that enhances the activity of B lymphocytes.

helper T2 (T_H2) cell A T lymphocyte that stimulates destruction of macrophages infected with bacterial cells.

hemagglutination The formation of clumps of red blood cells.

hemagglutination-inhibition (HAI) test A test using a patient's serum to detect the presence of antibodies against a specific infectious agent.

hemagglutinin (1) An enzyme composing one type of surface spike on influenza viruses that enables the viruses to bind to the host cell. (2) An agent such as a virus or an antibody that causes red blood cells to clump together.

hemolysin A bacterial toxin that destroys red blood cells.

hemolysis The destruction of red blood cells.

hepatitis B core antigen (HBcAG) An antigen located in the inner lipoprotein coat enclosing the DNA of a hepatitis B virus.

hepatitis B surface antigen (HBsAg) An antigen located in the outer surface coat of a hepatitis B virus.

herd (community) immunity The proportion of a population that is immune to a disease; also called community immunity.

herd immunity threshold The percent of the population that must be vaccinated to effectively stop disease spread.

heterokaryon A fungal cell that has two or more genetically different nuclei.

heterotroph An organism that requires preformed organic matter for its energy and carbon needs; *see also* **photoheterotroph** *and* **chemoheterotroph**.

heterotrophy The process by which an organism uses preformed organic compounds for metabolism.

high-efficiency particulate air (HEPA) filter A fibrous filter that traps particles, microbes, and spores in the air.

high frequency of recombination (Hfr) Referring to a bacterial cell containing an F factor incorporated into the bacterial chromosome.

highly active antiretroviral therapy (HAART) The combination of several (typically three or four) antiretroviral drugs for the treatment of infections caused by retroviruses, especially the human immunodeficiency virus.

highly perishable Referring to foods that spoil easily.

histamine A mediator in type I hypersensitivity reactions that is released from the granules in mast cells and basophils and causes the contraction of smooth muscles.

holding (batch) method A pasteurization process that exposes a liquid to 63° for 30 min.

homeostasis The tendency of an organism to maintain a steady state or equilibrium with respect to specific functions and processes.

hops The dried petals of the vine *Humulus lupulus* that impart flavor, color, and stability to beer.

horizontal gene transfer (HGT) The movement of genes from one organism to another within the same generation; also called lateral gene transfer.

horizontal transmission The spread of disease from one person to another.

host An organism on or in which a microorganism lives and grows, or a virus replicates.

host range The variety of species that a disease-causing microorganism can infect.

human genome The complete set a genetic information in a human cell.

human leukocyte antigen (HLA) Cell surface antigens involved with tissue transplantation.

humoral immune response The immune reaction of producing antibodies directed against antigens in the body fluids.

hyaluronidase An enzyme that digests hyaluronic acid and thereby permits the penetration of pathogens through connective tissue.

hybridoma The cell resulting from the fusion of a B cell and a cancer cell.

hydatid cyst A thick-walled body formed in the human liver by *Echinococcus granulosus* and containing larvae.

hydrocarbon An organic molecule containing only hydrogen and carbon atoms that are connected by a sharing of electrons.

hydrogen bond A weak attraction between a positively charged hydrogen atom (covalently bonded to oxygen or nitrogen) and a covalently bonded, negatively charged oxygen or nitrogen atom in the same molecule or separate molecules.

hydrogen peroxide An unstable liquid that readily decomposes in water and oxygen gas (O_2).

hydrolysis reaction A process in which a molecule is split into two parts through the interaction of H^+ and $(OH)^-$ of a water molecule.

hydrophilic Referring to a substance that dissolves in or mixes easily with water; *see also* **hydrophobic**.

hydrophobia An emotional condition ("fear of water") arising from the inability to swallow as a consequence of rabies.

hydrophobic Referring to a substance that does not dissolve in or mix easily with water; *see also* **hydrophilic**.

hyperacute rejection An immediate rejection response to a transplant.

hypersensitivity An immunological response to exposure to an allergen or antigen.

hyperthermophile A prokaryote that has an optimal growth temperature above 80°C.

hypha (pl. **hyphae**) A microscopic filament of cells representing the vegetative portion of a fungus.

hypothesis An educated guess or answer to a properly framed question.

I

icosahedral Referring to a symmetrical figure composed of 20 triangular faces and 12 points; one of the forms of symmetry found in some viral capsids.

IgA The class of antibodies found in respiratory and gastrointestinal secretions that help neutralize pathogens.

IgD The class of antibodies found on the surface of B cells that act as receptors for binding antigen.

IgE The class of antibodies responsible for type I hypersensitivities.

IgG The class of antibodies abundant in serum that are major diseases fighters.

IgM The first class of antibodies to appear in serum in helping fight pathogens.

immediate hypersensitivity Immune reactions to an allergen that occur within minutes to a few hours.

immediate response An allergic reaction that occurs within seconds or minutes.

immune complex A combination of antibody and antigen capable of complement activation and characteristic of type III hypersensitivity reactions.

immune complex hypersensitivity An immune response involving antibodies combining with antigens, the complex accumulating in blood vessels or on tissue surfaces.

immune deficiency The lack of an adequate immune system response.

immune suppression A process whereby the immune system does not respond to self antigens.

immunity The body's ability to resist infectious disease through innate and acquired mechanisms.

immunization The process of making an individual resistant to a particular disease by administering a vaccine; *see also* **vaccination**.

immunocompetent The ability of the body to develop an immune response in the presence of a disease-causing agent.

immunocompromised Referring to an inadequate immune response as a result of disease, exposure to radiation, or treatment with immunosuppressive drugs.

immunodeficiency disorder The inability, either inborn or acquired, of the body to produce an adequate immune response to fight disease.

immunodiffusion The movement of antigen and antibody toward one another through a gel or agar to produce a visible precipitate.

immunoelectrophoresis A laboratory diagnostic procedure in which antigen molecules move through an electric field and then diffuse to meet antibody molecules to form a precipitation line; *see also* **electrophoresis**.

immunofluorescence antibody (IFA) test The use of a fluorescently labeled antibody to detect an antigen.

immunogenic Capable of generating an immune response.

immunoglobulin (Ig) The class of immunological proteins that react with an antigen; an alternate term for antibody.

immunological memory The long-term ability of the immune system to remember past pathogen exposures.

immunology The scientific study of the structure and function of the immune system.

immunosuppressive agent A drug given to block transplant rejection by the immune system.

immunotherapy The use of monoclonal antibodies to control an overactive inflammatory response.

incidence The number of reported disease cases in a given time frame.

incineration To burn to ashes, or cause something to burn to ashes.

inclusion (1) A granule-like storage structure found in the prokaryotic cell cytoplasm. (2) A virus in the cytoplasm or nucleus of an infected cell.

incubation period The time that elapses between the entry of a pathogen into the host and the appearance of signs and symptoms.

index of refraction A measure of the light bending ability of a medium through which light passes.

indicator organism A microorganism whose presence signals fecal contamination of water.

indigenous microbiota The microbial agents that are associated with an animal for long periods of time without causing disease; *see also* **transient microbiota**.

indirect contact The mode of disease transmission involving nonliving objects; *see also* **direct contact**.

induced mutation A change in the sequence of nucleotide bases in a DNA molecule arising from a mutagenic agent used under controlled laboratory conditions.

induration A thickening and drying of the skin tissue that occurs in type IV hypersensitivity reactions.

industrial fermentation Any large scale industrial process, with or without oxygen gas (O_2), for growing microorganisms; *see also* **fermentation**.

infection The relationship between host and pathogen and the competition for supremacy that takes place between them.

infection allergy A type IV hypersensitivity reaction in which the immune system responds to the presence of certain microbial agents.

infectious disease A disorder arising from a pathogen invading a susceptible host and inducing medically significant symptoms.

infectious dose The number of microorganisms needed to bring about infection.

inflammation A nonspecific defensive response to injury; usually characterized by redness, warmth, swelling, and pain.

initiation (1) The unwinding and separating of DNA strands during replication. (2) The beginning of translation.

innate immunity An inborn set of the preexisting defenses against infectious agents; includes the skin, mucous membranes, and secretions.

insertional activation Referring to the "turning on" of a gene by a provirus.

insertional inactivation Referring to the "turning off" of a gene by a provirus.

insertion sequence (IS) A segment of DNA that forms a copy of itself, after which the copy moves into areas of gene activity to interrupt the genetic coding sequence.

interferon An antiviral protein produced by body cells on exposure to viruses and which trigger the synthesis of antiviral proteins.

interleukin A chemical cytokine produced by white blood cells that causes other white blood cells to divide.

intermediate host The host in which the larval or asexual stage of a parasite is found.

intermittent sterilization A method to kill all living microbes and spores over three successive days using free-flowing steam. Also called tyndallization.

intoxication The presence of microbial toxins in the body.

intraepithelial lymphatocyte (IEL) An intestinal epithelial cell that removes injured or infected cells.

intrinsic factor A characteristic of a food product that influences microbial growth. *See also* **extrinsic factor**.

intron A non-coding sequence in a split gene; *see also* **exon**.

in use test A procedure used to determine the value of a disinfectant or antiseptic.

invasiveness The ability of a pathogen to spread from one point to adjacent areas in the host and cause structural damage to those tissues.

invasive wound infection Microbial invasion of living tissue beneath a burn.

iodophor A complex of iodine and detergents that is used as an antiseptic and disinfectant.

ion An electrically charged atom.

ionic bond The electrical attraction between oppositely charged ions.

ionizing radiation A type of radiation such as gamma rays and X rays that causes the separation of atoms or a molecule into ions.

isograft Tissue taken from one identical twin and grafted to the other twin.

isomer A molecule with the same molecular formula but different structural formula.

isotope An atom of the same element in which the number of neutrons differs; *see also* **radioisotope**.

K

keratin The tough, fibrous protein produced by keratinocytes.

keratinocyte The most common cell type of the epidermis.

kinetoplast The mass of DNA found in the mitochondria of the kinetoplastids.

kinetoplastid A protozoan with one flagellum; most are parasitic in aerobic or anaerobic environments.

kingdom The most inclusive taxonomic rank below the domain level.

Kirby-Bauer test *See* **disk diffusion method**.

Koch's postulates A set of procedures by which a specific pathogen can be related to a specific infectious disease.

Koplik spots Red patches with white central lesions that form on the gums and walls of the pharynx during the early stages of measles.

Krebs cycle *See* **citric acid cycle**.

L

lactic acid fermentation A catabolic process that produces lactic acid during the reoxidation of NADH to NAD^+ for reuse in glycolysis to generate ATP.

lactone A chemical compound used as a flavoring ingredient in foods and beverages.

lagering The secondary aging of beer.

lagging strand During DNA replication, the new strand that is synthesized discontinuously; *see also* **leading strand**.

lag phase A portion of a bacterial growth curve encompassing the first few hours of the population's history when no growth occurs.

latency A condition in which a virus integrates into a host chromosome without immediately causing a disease.

latent infection A viral infection where the viral DNA remains "dormant" within a host chromosome.

latent phase The period during a viral infection when virus particles cannot be found outside the infected cells.

late-phase anaphylaxis After an acute IgE mediated reaction, a second response occurs many hours after the initial response, and appears to be based on the activity of eosinophils.

leading strand During DNA replication, the new strand that is synthesized continuously; *see also* **lagging strand**.

leproma A tumor-like growth on the skin associated with leprosy.

leukocidin A bacterial enzyme that destroys phagocytes, thereby preventing phagocytosis.

leukocyte Any of a number of types of white blood cells.

leukopenia A condition characterized by an abnormal drop in the normal number of white blood cells.

leukotriene A substance that acts as a mediator during type I hypersensitivity reactions; formed from arachidonic acid after the antigen-antibody reaction has taken place.

lichen An association between a fungal mycelium and a cyanobacterium or alga.

light (L) chain A smaller polypeptide in an antibody.

light microscope An instrument that uses visible light and a system of glass lenses to produce a magnified image of an object; also called a compound microscope.

lipid A nonpolar organic compound composed of carbon, hydrogen, and oxygen; examples include triglycerides, phospholipids, and sterols.

lipid A A component in the outer membrane of the gram-negative cell wall.

lipopolysaccharide (LPS) A molecule composed of lipid and polysaccharide that is found in the outer membrane of the gram-negative cell wall of bacterial cells.

littoral zone The environment along the shoreline of an ocean.

local disease A disease restricted to a single area of the body.

localized anaphylaxis A type I hypersensitivity that remains in one part of the body; *see also* **systemic anaphylaxis**.

lockjaw *See* **trismus**.

logarithmic (log) phase The portion of a bacterial growth curve during which active growth leads to a rapid rise in cell numbers.

looped domain structure The term used to describe organization and packing of the prokaryotic chromosome.

lymph The tissue fluid that contains white blood cells and drains tissue spaces through the lymphatic system.

lymphedema An abnormal swelling due to the loss of normal lymph vessel drainage of the affected part; typical of elephantiasis.

lymph node A bean-shaped organ located along lymph vessels that is involved in the immune response and contains phagocytes and lymphocytes.

lymphocyte A type of white blood cell that functions in the immune system.

lymphoid progenitor A bone marrow cell that gives rise to lymphocytes; *see also* **myeloid progenitor**.

lyophilization A process in which food or other material is deep frozen, after which its liquid is drawn off by a vacuum; also called freeze-drying.

lysis The rupture of a cell and the loss of cell contents.

lysogenic cycle The events of a bacterial virus infection that result in the integration of its DNA into the bacterial chromosome.

lysosome A membrane-enclosed compartment in many eukaryotic cells that contains enzymes to degrade or digest substances.

lysozyme An enzyme found in tears and saliva that digests the peptidoglycan of gram-positive bacterial cell walls.

lytic cycle A process by which a bacterial virus replicates within a host cell and ultimately destroys the host cell.

M

macrolide An antibiotic that blocks protein synthesis.

macromolecule A large, chemically-bonded substance built from smaller building blocks.

macronucleus The larger of two nuclei in most ciliates that is involved in controlling metabolism; *see also* **micronucleus**.

macrophage A large cell derived from monocytes that is found within various tissues and actively engulfs foreign material, including infecting bacterial cells and viruses.

macule A pink-red skin spot associated with infectious disease.

magnetosome A cytoplasmic inclusion body in some bacterial cells that assists orientation to the environment by aligning with the magnetic field.

magnification The increase in the apparent size of an object observed with a microscope.

major histocompatibility complex (MHC) A set of genes that controls the expression of MHC proteins; involved in transplant rejection; *see also* **MHC class I** *and* **MHC class II**.

malignant Referring to a tumor that invades the tissue around it and may spread to other parts of the body.

malting The process whereby barley grains are treated to digest starch to maltose.

Mantoux test The infection of a tuberculosis purified protein derivative into the forearm.

mashing A fermentable mixture of hot water and barley grain from which alcohol is distilled.

mass The amount of matter in a sample.

mass number The total number of protons and neutrons in an atom.

mast cell A type of cell in connective tissue which releases histamine during allergic attacks.

matrix protein A protein shell found in some viruses between the genome and capsid.

matter Any substance that has mass and occupies space.

M cell An intestinal epithelial cell that processes foreign antigens.

mechanical vector A living organism, usually an insect, that transmits disease agents on its surface; *see also* **biological vector**.

membrane attack complex A set of complement proteins that forms holes in a bacterial cell membrane and leads to cell destruction.

membrane filter A pad of cellulose acetate or polycarbonate.

memory cell A cell derived from B lymphocytes or T lymphocytes that reacts rapidly upon re-exposure to antigen.

meningitis A general term for inflammation of the covering layers of the brain and spinal cord due to any of several bacteria, fungi, viruses, or protozoa.

merozoite A stage in the life cycle of the malaria parasite that invades red blood cells in the human host.

mesophile An organism that grows in temperature ranges of 20°C to 40°C.

messenger RNA (mRNA) An RNA transcript containing the information for synthesizing a specific polypeptide.

metabolic engineering An interdisciplinary field that designs and optimizes ways to alter a set of enzymes (metabolic pathway) to produce a desired product.

metabolic pathway A sequence of linked enzyme-catalyzed reactions in a cell.

metabolism The sum of all biochemical processes taking place in a living cell; *see also* **anabolism** *and* **catabolism**.

metachromatic granule A polyphosphate-storing granule commonly found in *Corynebacterium diphtheriae* that stains deeply with methylene blue; also called volutin.

metagenome The collective genomes from a population of organisms.

metagenomics The study of genes isolated directly from environmental samples.

metastasize Referring to a tumor that spreads from the site of origin to other tissues in the body.

methanogen An archaeal organism that lives on simple compounds in anaerobic environments and produces methane during its metabolism.

MHC class I A group of glycoproteins found on all nucleated cells to which cytotoxic T cells can bind.

MHC class II A group of glycoproteins found on B cells, Macrophages, and dendritic cells that are recognized by helper T cells.

miasma An ill-defined idea of the 1700s and 1800s that suggested diseases were caused by an altered chemical quality of the atmosphere.

microaerophile An organism that grows best in an oxygen-reduced environment.

microbe *See* **microorganism**.

microbial forensics The discipline involved with the recognition, identification, and control of a pathogen.

microbial genomics The discipline of sequencing, analyzing, and comparing microbial genomes.

microbicidal Referring to any agent that kills microbes.

microbiology The scientific study of microscopic organisms and viruses, and their interactions with other organisms and the environment.

microbiome The community of microorganisms in an environment or body location.

microbiostatic Referring to any agent that inhibits microbial growth.

microbiota The population of microorganisms that colonize various parts of the human body and do not cause disease in a healthy individual.

microcompartment A region in some bacterial cells surrounded by a protein shell.

microenvironment A cell's or organism's physical and chemical surroundings.

microfilaria (pl. **microfilariae**) The larva stage of the parasitic nematode *Wuchereria bancrofti*.

micrometer (μm) A unit of measurement equivalent to one millionth of a meter; commonly used in measuring the size of microorganisms.

micronucleus The smaller of the two nuclei in most ciliates that contains genetic material and is involved in sexual reproduction; *see also* **macronucleus**.

microorganism (microbe) A microscopic form of life including bacterial, archaeal, fungal, and protistan cells.

Microsporidia A sister group of fungi that are parasitic in protists and animals.

mineralization The conversion of organic compounds to inorganic compounds and ammonia.

minimum inhibitory concentration (MIC) The lowest concentration of an antimicrobial agent that will inhibit its growth.

miracidium (pl. **miracidia**) A ciliated larva representing an intermediary stage in the life cycle of a trematode.

mismatch repair A mechanism to correct mismatched bases in the DNA; *see also* **excision repair**.

missense mutation A base substitution that codes for an incorrect amino acid.

mitochondrion (pl. **mitochondria**) A double membrane-enclosed compartment in eukaryotic cells that carries out aerobic respiration.

mitosporic fungus A fungus without a known sexual stage of reproduction.

mold A type of fungus that consists of chains of cells and appears as a fuzzy mass of thin filaments in culture.

molecular formula The representation of the kinds and numbers of each atom in a molecule.

molecular taxonomy The systematized arrangement of related organisms based on molecular characteristics, such as ribosomal DNA nucleotide sequences.

molecular weight The sum of the atomic masses of all atoms in a molecule.

molecule Two or more atoms held together by a sharing of electron pairs.

monoclonal antibody A type of antibody produced by a clone of hybridoma cells, consisting of antigen-stimulated B cells fused to myeloma cells.

monocyte A circulating white blood cell with a large bean-shaped nucleus that is the precursor to a macrophage.

monolayer (1) A single layer of cultured cells; (2) A single layer of phospholipids.

monomer A simple organic molecule that can join in long chains with other molecules to form a more complex molecule; *see also* **polymer**.

monosaccharide A simple sugar that cannot be broken down into simpler sugars; examples include glucose and fructose.

monospot test A method used to detect the presence of heterophile antibodies, which is indicative of infectious mononucleosis.

monotherapy The use of a single drug to treat a disease.

morphology Refers to the form (shape) and structure of cells and organisms.

most probable number (**MPN**) A laboratory test in which a statistical evaluation is used to estimate the number of bacterial cells in a sample of fluid; often employed in determinations of coliform bacteria in water.

M protein A protein that enhances the pathogenicity of streptococci by allowing organisms to resist phagocytosis and adhere firmly to tissue.

mucin The major glycoprotein in mucus.

mucociliary clearance A defensive mechanism to clear microbes or foreign material from the airways via the ciliated epithelium.

mucosal immunity The immune defenses on the mucosal surfaces, such as the gastrointestinal tract.

mucous membrane (**mucosa**) A moist lining in the body passages of all mammals that contains mucus-secreting cells and is open directly or indirectly to the external environment.

mucus A sticky secretion of glycoproteins.

multidrug-resistant (**MDR**) Referring to microbes that are resistant to the effects of multiple antibiotics.

mushroom A spore-bearing fruiting body typical of many members of the basdiomycetes.

must The juice resulting from crushing grapes.

mutagen A chemical or physical agent that causes a mutation.

mutant An organism carrying a mutation.

mutation A permanent alteration of a DNA sequence.

mutualism A close and permanent association between two populations of organisms in which both benefit from the association.

mycelium (pl. **mycelia**) A mass of fungal filaments from which most fungi are built.

mycetism Mushroom poisoning.

mycolic acid A waxy lipid composing the cell wall of mycobacterial species.

mycology The scientific study of fungi.

mycorrhiza (pl. **mycorrhizae**) A close association between a fungus and the roots of many plants.

mycosis (pl. **mycoses**) A disease caused by a fungus.

mycotoxin A poison produce by a fungus that adversely affects other organisms.

myeloid progenitor A bone marrow cell that gives rise to red blood cells and all white blood cells except lymphocytes; *see also* **lymphoid progenitor**.

myeloma A malignant tumor that develops in the blood-cell–producing cells of the bone marrow; the cells are used in the procedure to produce monoclonal antibodies.

myonecrosis *See* **trismus**.

myxobacteria A group of soil-dwelling bacterial species that exhibit multicellular behaviors.

N

nanometer (**nm**) A unit of measurement equivalent to one billionth of a meter; the unit used in measuring viruses and the wavelength of energy forms.

narrow spectrum Referring to an antimicrobial drug that is useful for a restricted group of microorganisms; *see also* **broad spectrum**.

natural killer (**NK**) **cell** A type of defensive body cell that attacks and destroys cancer cells and infected cells without the involvement of the immune system.

naturally acquired active immunity A host response resulting in antibody production as a result of experiencing the disease agent, or a passive response resulting from the passage of antibodies to the fetus via the placenta or the milk of a nursing mother.

naturally acquired passive immunity The form of immunity that results from the passage of IgG antibodies from mother to fetus.

necrosis Cell or tissue death.

negative control A form of gene regulation where a repressor protein binds to an operator and blocks transcription.

negative selection A method for identifying mutations by selecting cells or colonies that do not grow when replica plated.

negative stain technique A staining process that results in colorless bacterial cells on a stained background when viewed with the light microscope.

negative strand Referring to those RNA viruses whose genome cannot be directly transcribed into protein.

Negri body A cytoplasmic inclusion that occurs in brain cells infected with rabies viruses.

neuraminidase An enzyme composing one type of surface spike of influenza viruses that facilitates viral release from the host cell.

neuraminidase inhibitor A compound that inhibits neuraminidase, preventing the release of flu viruses from an infected cell.

neurotoxin A poison that interferes with nerve transmission.

neutralization A type of antigen-antibody reaction in which the activity of a toxin is inactivated.

neutron An uncharged particle in the atomic nucleus.

neutrophil The most common type of white blood cell; functions chiefly to engulf and destroy foreign material, including bacterial cells and viruses that have entered the body.

neutrophil extracellular traps (**NETS**) The nuclear DNA and antimicrobial proteins of dead neutrophils that form a fibrous mesh trapping pathogens.

neutrophile An organism that grows best at pH 7.

nicotinamide adenine dinucleotide (**NAD**) A coenzyme that acts as an electron carrier for ATP production.

nitrogen cycle The processes that convert nitrogen gas (N_2) to nitrogen-containing substances in soil and living organisms, then reconverted to the gas.

nitrogen fixation The chemical process by which microorganisms convert nitrogen gas (N_2) into ammonia.

noncommunicable disease A disease whose causative agent is acquired from the environment and is not transmitted to another individual.

noncompetitive inhibition The prevention of a chemical reaction by a chemical that binds elsewhere than to active site of an enzyme; *see also* **competitive inhibition**.

nonhalophile A microorganism that cannot grow in the presence of added sodium chloride.

nonhemolytic Referring to those bacterial species that when plated on blood agar cause no destruction of the red blood cells.

nonperishable Referring to foods that are least likely to spoil.

nonpolar molecule A covalently bonded substance in which there is no electrical charge; *see also* **polar molecule**.

nonsense mutation A base substitution that codes for a stop codon.

nosocomial infection A disorder acquired during an individual's stay at a hospital or chronic care facility.

nucleic acid A high-molecular-weight molecule consisting of nucleotide chains that convey genetic information and is found in all living cells and viruses; *see* **deoxyribonucleic acid** *and* **ribonucleic acid**.

nucleobase Any of five nitrogen-containing compounds found in nucleic acids, including adenine, guanine, cytosine, thymine, and uracil.

nucleocapsid The combination of genome and capsid of a virus.

nucleoid The chromosomal region of a bacterial and archaeal cell.

nucleoid-associated protein (NAP) A protein that causes the DNA double helix to twist (supercoil).

nucleotide A component of a nucleic acid consisting of a carbohydrate molecule, a phosphate group, and a nitrogenous base.

nucleotide excision repair A process that removes a thymine dimer, along with adjacent nucleotides, and replaces them with the correct sequence.

O

obligate aerobe An organism that requires oxygen gas (O_2) for metabolism.

obligate anaerobe An organism that cannot use oxygen gas (O_2) for metabolism.

observation The use of the senses or instruments to gather information on which science inquiry is based.

Okazaki fragment A segment of DNA resulting from discontinuous DNA replication.

oncogene A segment of DNA that can induce uncontrolled growth of a cell if permitted to function.

oncogenic Referring to any agent such as viruses that can cause tumors.

operator A sequences of bases in the DNA to which a repressor protein can bind.

operon The unit of bacterial DNA consisting of a promoter, operator, and a set of structural genes.

opisthotonus An arching of the back that is characteristic of tetanus.

opportunist A microorganism that invades the tissues when body defenses are suppressed.

opportunistic infection A disorder caused by a microorganism that does not cause disease but that can become pathogenic or life-threatening if the host has a low level of immunity.

opsonin An antibody or complement component that encourages phagocytosis.

opsonization Enhanced phagocytosis due to the activity of antibodies or complement.

oral rehydration therapy A procedure that uses a mixture of blood salts and glucose in water to restore their normal levels in the body.

order A category of related organisms consisting of one or more families.

organelle A specialized compartment in cells that has a particular function.

organic acid A short-chain chemical used as a preservative.

organic compound A substance characterized by chains or rings of carbon atoms that are linked to atoms of hydrogen and sometimes oxygen, nitrogen, and other elements.

origin of transfer The fixed point on an F plasmid (factor) where one strand is nicked and transferred to a recipient cell.

osmosis The net movement of water molecules from where they are in a high concentration through a semipermeable membrane to a region where they are in a lower concentration.

osmotic lysis The bursting of a cell due to the inflow of water.

osmotic pressure The force that must be applied to a solution to inhibit the inward movement of water across a membrane.

outbreak A small, localized epidemic.

outer membrane A bilayer membrane forming part of the cell wall of gram-negative bacteria.

oxidation A chemical change in which electrons are lost by an atom; *see also* **reduction**.

oxidation lagoon A large pond in which sewage is allowed to remain undisturbed so that digestion of organic matter can occur.

oxidative phosphorylation A series of sequential steps in which energy is released from electrons as they pass from coenzymes to cytochromes, and ultimately, to oxygen gas (O_2); the energy is used to combine phosphate ions with ADP molecules to form ATP molecules.

oxygenic photosynthesis A form of photosynthesis in which molecular oxygen (O_2) is produced.

P

pandemic A worldwide epidemic.

papule A pink pimple on the skin.

parabasalid A protist that contains numerous of flagella and lives in low oxygen or anaerobic environments; most members are symbiotic in animals.

parasite A type of heterotrophic organism that feeds on live organic matter such as another organism.

parasitemia The spread of parasitic protists and multicellular worms through the circulatory system.

parasitism A close association between two organisms in which one (the parasite) feeds on the other (the host) and may cause injury to the host.

parasitology The scientific study of parasites.

paroxysm A sudden intensification of symptoms, such as a severe bout of coughing.

particle transmission The movement of particles suspended in the air.

passive agglutination An immunological procedure in which antigen molecules are adsorbed to the surface of latex spheres or other carriers that agglutinate when combined with antibodies.

passive immunity The temporary immunity that comes from receiving antibodies from another source.

pasteurization A heating process that destroys human pathogens in a fluid such as milk and lowers the overall number of bacterial cells in the fluid.

pasteurizing dose The amount of irradiation used to eliminate pathogens.

pathogen A microorganism or virus that causes illness or disease.

pathogen-associated molecular pattern (PAMP) A unique microbial molecular sequence recognized by innate immune system receptors.

pathogenicity The ability of a disease-causing agent to gain entry to a host and bring about a physiological or anatomical change interpreted as disease.

pathogenicity island A set of adjacent genes that encode virulence factors.

pattern recognition receptor A protein sensor that recognizes pathogen-associated molecular patterns.

penicillin Any of a group of antibiotics derived from *Penicillium* species or produced synthetically; effective against gram-positive bacteria and several gram-negative bacteria by interfering with cell wall synthesis.

peptide bond A covalent linkage between the amino group on one amino acid and the carboxyl group on another amino acid.

peptidoglycan A complex molecule of the bacterial cell wall composed of alternating units of *N*-acetylglucosamine and *N*-acetylmuramic acid cross linked by short peptides.

perforin A protein secreted by cytotoxic T lymphocytes and natural killer cells that forms holes in the plasma membrane of a targeted infected cell.

period of convalescence The phase of a disease during which the body's systems return to normal.

period of decline The phase of a disease during which symptoms subside.

periplasm A metabolic region between the cell membrane and outer membrane of a gram-negative bacterial cell.

peroxide A compound containing an oxygen–oxygen single bond.

persister cell A bacterial cell that has temporarily ceased cell division but maintains a very low metabolic rate.

pH An abbreviation for the power of the hydrogen ion concentration $[H^+]$ of a solution.

phage *See* **bacteriophage**.

phage typing A procedure of using specific bacterial viruses to identify a particular strain of a bacterial species.

phagocyte A white blood cell capable of engulfing and destroying foreign materials, including bacterial cells and viruses.

phagocytosis A process by which foreign material or cells are taken into a white blood cell and destroyed.

phagolyososme A membrane-enclosed compartment resulting from the fusion of a phagosome with lysosomes and in which the foreign material is digested.

phagosome A membrane-enclosed compartment containing foreign material or infectious agents that the cell has engulfed.

phallotoxin A group of chemical compounds present in the mushroom *Amanita phalloides*.

phase-contrast microscopy An optical system on the light microscope that uses a special condenser and objective lenses to examine cell structure.

phenol A chemical compound that has one or more hydroxyl groups attached to a benzene ring and derivatives are used as an antiseptic or disinfectant; also called carbolic acid.

phenol coefficient (PC) A number that indicates the effectiveness of an antiseptic or disinfectant compared to phenol.

phenotype The visible (physical) appearance of an organism resulting from the interaction between its genetic makeup and the environment.

phospholipid A water-insoluble compound containing glycerol, two fatty acids, and a phosphate head group; forms part of the membrane in all cells.

phosphorylation The addition of a phosphate group to a molecule.

photoautotroph An organism that uses light energy to synthesize nutrients from carbon dioxide gas (CO_2).

photobiont The photosynthetic partner in a lichen.

photoheterotroph An organism that uses light energy to synthesize nutrients from organic carbon compounds.

photophobia Sensitivity to bright light.

photophosphorylatyion The generation of ATP through the trapping of light.

photosynthesis A biochemical process in which light energy is converted to chemical energy, which is then used for carbohydrate synthesis.

photosystem A group of pigments that act as a light trapping system for photosynthesis.

pH scale A range of values that extends from 0 to 14 and indicates the degree of acidity or alkalinity of a solution.

phycology The scientific study of algae.

phylogenetic tree A diagram illustrating the inferred relationships among organisms.

phylogeny The evolutionary history of a species or group of species.

phylum (pl. **phyla**) A category of organisms consisting of one or more classes.

physical pollution The presence of particulate matter in water.

phytoplankton Microscopic free-floating communities of cyanobacteria and unicellular algae.

pilus (pl. **pili**) A short hair-like structure used by bacterial cells for attachment.

planktonic bacteria Referring to bacterial cells that live as individual cells.

plaque A clear area on a lawn of bacterial cells where viruses have destroyed the bacterial cells.

plasma The fluid portion of blood remaining after the cells have been removed; *see also* **serum**.

plasma cell An antibody-producing cell derived from B lymphocytes.

plasma membrane The phospholipid bilayer with proteins that surrounds the eukaryotic cell cytoplasm; *see also* **cell membrane**.

plasmid A small, closed-loop molecule of DNA apart from the chromosome that replicates independently and carries nonessential genetic information.

platelet A small disc-shaped cell without a cell nucleus that functions in blood clotting and innate immunity.

pleated sheet The zig-zag secondary structure of a polypeptide in a flat plane.

pluripotent stem cell An undifferentiated cell from which specialized cells arise.

pneumonia An inflammation of the bronchial tubes and one or both lungs.

point mutation The replacement of one base in a DNA strand with another base.

polar molecule A substance with electrically-charged poles; *see also* **nonpolar molecule**.

polyclonal antibodies Antibodies produced from different clones of B cell.

polymer A substance formed by combining smaller molecules into larger ones; *see also* **monomer**.

polymerase chain reaction (PCR) A technique used to replicate a fragment of DNA many times.

polymicrobial disease A clinical or pathological condition caused by more than one infectious agent.

polynucleotide A chain of linked nucleotides.

polypeptide A chain of linked amino acids.

polysaccharide A complex carbohydrate made up of simple sugars linked into a branched or chain structure; examples include starch and cellulose.

polysome A cluster of ribosomes linked by a strand of mRNA and all translating the mRNA.

porin A protein in the outer membrane of gram-negative bacterial cells that acts as a channel for the passage of small molecules.

portal of entry The site at which a pathogen enters the host.

portal of exit The site at which a pathogen leaves the host.

positive selection A method for selecting mutant cells by their growth as colonies on agar.

positive strand Referring to the RNA viruses whose genome consists of a mRNA molecule.

post-exposure immunization The receiving of a vaccine after contracting the pathogen.

post-polio syndrome A condition that affects polio survivors years after recovery from an initial acute attack by the polio virus.

potability Referring to water that is safe to drink because it contains no harmful material or microbes.

pour plate method A process by which a mixed culture can be separated into pure colonies and the colonies isolated; *see also* **streak plate method**.

precipitation A type of antigen-antibody reaction in which thousands of molecules of antigen and antibody cross-link to form visible aggregates.

prevalence The percentage of the population affected by a disease.

prevacuum autoclave An instrument that uses saturated steam at high temperatures and pressure for short time periods to sterilize materials.

preventative vaccine A vaccine that could be given to HIV-negative individuals to prevent infection to HIV.

primary antibody response The first contact between an antigen and the immune system, characterized by the synthesis of IgM and then IgG antibodies; *see also* **secondary antibody response**.

primary cell culture Animal cells separated from tissue and grown in cell culture.

primary immunodeficiency An autoimmune disorder present from birth. *See also* **secondary immunodeficiency**.

primary infection A disease that develops in an otherwise healthy individual.

primary metabolite A small molecule essential to the survival and growth of an organism; *see also* **secondary metabolite**.

primary structure The sequence of amino acids in a polypeptide.

prion An infectious, self-replicating protein involved in human and animal diseases of the brain.

probiotic Living microbes that help reestablish or maintain the human microbiota of the gut.

prodromal phase The phase of a disease during which general symptoms occur in the body.

product A substance or substances resulting from a chemical reaction.

productive infection The active assembly and maturation of viruses in an animal cell.

progenote A hypothetical first life form.

proglottid One of a series of segments that make up the body of a tapeworm.

prokaryote A microorganism in the domain *Bacteria* or *Archaea* composed of single cells having a single chromosome but no cell nucleus or other membrane-bound compartments.

prokaryotic Referring to cells or organisms usually having a single chromosome but no cell nucleus and few subcompartments.

promoter The region of a template DNA strand or operon to which RNA polymerase binds.

prophage The viral DNA of a bacterial virus that is inserted into the bacterial DNA and is passed on from one generation to the next during binary fission.

propionic acid A chemical preservative used in cheese, breads, and other bakery products.

prostaglandin A hormone-like substance that acts as a mediator in type I hypersensitivity reactions.

protease inhibitor A compound that breaks down the enzyme protease, inhibiting the replication of some viruses, such as HIV.

protein A chain or chains of linked amino acids used as a structural material or enzyme in living cells.

protein-only hypothesis The idea that prions are composed solely of protein and contain no nucleic acid.

protein synthesis The process of forming a polypeptide or protein through a series of chemical reactions involving amino acids.

Proteobacteria A phylum of gram-negative, chemoheterotrophic species in the domain Bacteria that are defined primarily in terms of their ribosomal RNA (rRNA) sequences; examples include *Escherichia coli*, *Salmonella*, and the rickettsiae.

protist An informal term used to describe a eukaryote that is not animal, plant, or fungus.

proton A positively charged particle in the atomic nucleus.

proto-oncogene A region of DNA in the chromosome of human cells; they are altered by carcinogens into oncogenes that transform cells.

prototroph An organism that contains all its nutritional needs; *see also* **auxotroph**.

protozoan (pl. **protozoa**) A term formally used to describe a single-celled eukaryotic organism that lacks a cell wall and usually exhibits chemoheterotrophic metabolism.

protozoology The scientific study of protozoa.

provirus The viral DNA that has integrated into a eukaryotic host chromosome and is then passed on from one generation to the next through cell division.

pseudomembrane An accumulation of mucus, leukocytes, bacteria, and dead tissue in the respiratory passages of diphtheria patients.

pseudopeptidoglycan A complex molecule of some archaeal cell walls composed of alternating units of *N*-acetylglucosamine and *N*-acetyltalosamine uronic acid.

pseudopod A projection of the plasma membrane that allows movement in members of the amoebozoans.

psychrophile An organism that lives at cold temperature ranges of 0°C to 20°C.

psychrotolerant Referring to microorganisms that grow at 0°C but have a temperature optima of 20° to 40°C.

psychrotroph *See* **psychrotolerant**.

pure culture An accumulation or colony of microorganisms of one species.

pus A mixture of dead tissue cells, leukocytes, and bacteria that accumulates at the site of infection.

pyrogen A fever-producing substance.

pyruvate The end product of the glycolysis metabolic pathway.

Q

quat *See* **quaternary ammonium compound**.

quaternary ammonium compound (quat) A positively charged detergent with four organic groups attached to a central nitrogen atom; used as a disinfectant.

quaternary structure The association of two or more polypeptides in a protein.

question A statement written from an observation and used to formulate a hypothesis.

quorum sensing The ability of microbial cells to chemically communicate and coordinate behavior via signaling molecules.

R

radioallergosorbent test (RAST) A type of radioimmunoassay in which antigens for the unknown antibody are attached to matrix particles.

radioimmunoassay (RIA) An immunological procedure that uses radioactive-tagged antigens to determine the identity and amount of antibodies in a sample.

radioisotope A unstable form of a chemical element that is radioactive; *see also* **isotope**.

radiolarian A single-celled marine organism with a round silica-containing shell that has radiating arms to catch prey.

reactant A substance participating in a chemical reaction.

recombinant DNA molecule A DNA molecule containing DNA from two different sources.

recombinant F$^-$ cell A recipient cell that received a few chromosomal genes and partial F factor genes from a donor cell during conjugation.

recombinant subunit vaccine The synthesis of antigens in a microorganism using recombined genes for the purpose of producing a vaccine.

reduction The gain of electrons by a molecule; *see also* **oxydation**.

redundancy Referring to multiple codons coding for the same amino acid.

reemerging infectious disease A disease showing a resurgence in incidence or a spread in its geographical area; *see also* **emerging infectious disease**.

regulatory gene A DNA segment that codes for a repressor protein.

regulatory T cell A population of lymphocytes that prevent other T lymphocytes from attacking self.

replication factory The location in a cell where DNA synthesis occurs.

replication fork The point where complementary strands of DNA separate and new complementary strands are synthesized.

replication origin The fixed point on a DNA molecule where copying of the molecule starts.

repressor protein A protein that when bound to the operator blocks transcription.

reservoir The location or organism where disease-causing agents exist and maintain their ability for infection.

resolution The numerical value of a lens system indicating the size of the smallest object that can be seen clearly when using that system.

respiratory burst A phagocytic response producing toxic metabolites.

respiratory droplet Small liquid droplets expelled by sneezing or coughing.

restriction endonuclease A type of enzyme that splits open a DNA molecule at a specific restricted point; important in genetic engineering techniques.

resurgent infectious disease *see* **reemerging infectious disease**.

reticulate body The replicating, intracellular, noninfectious stage of *Chlamydia trachomatis*.

reverse transcriptase An enzyme that synthesizes a DNA molecule from the code supplied by an RNA molecule.

reverse transcriptase inhibitor A compound that inhibits the action of reverse transcriptase, preventing the viral genome from being replicated.

revertant Referring to a mutant organism or cell that has reacquired its original phenotype or metabolic ability.

R group The side chain on an amino acid.

rhabditiform Referring to the elongated, rod-like shape of the larvae of the hookworm.

RhoGAM Rh-positive antibodies.

ribonucleic acid (RNA) The nucleic acid involved in protein synthesis and gene control; also the genetic information in some viruses.

ribosomal RNA (rRNA) An RNA transcript that forms part of the ribosome's structure.

ribosome A cellular structure made of RNA and protein that participates in protein synthesis.

ribozyme An RNA molecule that can catalyze a chemical reaction in a cell.

rice-water stool A colorless, watery diarrhea containing particles of intestinal tissue in cholera patients.

Rickettsiae Short gram-negative rods that are obligate intracellular human pathogens.

RNA *See* **ribonucleic acid**.

RNA polymerase The enzyme that synthesizes an RNA polynucleotide from a DNA template.

rolling circle mechanism A type of DNA replication in which a strand of DNA "rolls off" the loop and serves as a template for the synthesis of a complementary strand of DNA.

rose spots A bright red skin rash associated with diseases such as typhoid fever and relapsing fever.

roundworm A multicellular parasite with a round body; examples include the nematodes.

R plasmid A small, circular DNA molecule that occurs frequently in bacterial cells and carries genes for drug resistance.

S

salt An ionic compound resulting from the electrostatic attraction between positively and negatively charged atoms.

sanitization To remove microbes or reduce their populations to a safe level as determined by public health standards.

sanitize Referring to the reduction of a microbial population to a safe level.

saprobe A type of heterotrophic organism that feeds on dead organic matter, such as rotting wood or compost.

sarcina (pl. **sarcinae**) (1) A packet of eight spherical-shaped prokaryotic cells. (2) A genus of gram-positive, anaerobic spheres.

saturated Referring to a water-insoluble compound that cannot incorporate any additional hydrogen atoms; *see also* **unsaturated**.

scanning electron microscope (SEM) The type of electron microscope that allows electrons to scan across an object, generating a three-dimensional image of the object.

Schick test A skin test used to determine the effectiveness of diphtheria immunization.

schmutzdecke In water purification, a slimy layer of microorganisms that develops in a slow sand filter.

science The organized body of knowledge that is derived from observations and can be verified or tested by further investigation.

scientific inquiry The way a science problem is investigated by formulating a question, developing a hypothesis, collecting data about it through observation, and experiment, and interpreting the results; also called scientific method.

sclerotium (pl. **sclerotia**) A hard purple body that forms in grains contaminated with *Claviceps purpurea*.

scolex The head region of a tapeworm.

sebum An oily substance produced by the sebaceous glands that keep the skin and hair soft and moist.

secondary antibody response A second or ensuing response triggered by memory cells to an antigen and characterized by substantial production of IgG antibodies; *see also* **primary antibody response**.

secondary immunodeficiency An autoimmune disorder acquired later in life. *See also* **primary immunodeficiency**.

secondary infection A disorder caused by an opportunistic microbe as a result of a primary infection weakening the host.

secondary metabolite A small molecule not essential to the survival and growth of an organism; *see also* **primary metabolite**.

secondary structure The region of a polypepetide folded into an alpha helix or pleated sheet.

secretory IgA (S-IgA) The form of IgA antibodies secreted into the intestinal lumen.

sedimentation The removal of soil particulates from water.

selective medium A growth medium that contains ingredients to inhibit certain microorganisms while encouraging the growth of others.

selective toxicity A property of many antimicrobial drugs that harm the infectious agent but not the host.

self tolerance The ability of the immune system to not respond to one's own molecular determinants.

semiconservative replication The DNA copying process where each parent (old) strand serves as a template for a new complementary strand.

semiperishable Referring to foods that spoil less quickly.

semisynthetic drug A chemical substance synthesized from natural and lab components used to treat disease.

sense codon A nucleotide sequence that specifies an amino acid.

sensitizing dose The first exposure to an allergy-causing antigen.

sepsis The growth and spreading of bacteria or their toxins in the blood and tissues.

septate Referring to the cross-walls formed in the filaments of many fungi.

septic shock A collapse of the circulatory and respiratory systems caused by an overwhelming immune response.

septicemia A growth and spreading of bacterial cells in the bloodstream.

septic tank An enclosed concrete box that collects waste from the home.

septum (pl. **septa**) A cross-wall in the hypha of a fungus.

seroconversion The time when antibodies to a disease agent can be detected in the blood.

serological reaction An antigen-antibody reaction studied under laboratory conditions and involving serum.

serology A branch of immunology that studies serological reactions.

serum (pl. **sera**) The fluid portion of the blood consisting of water, minerals, salts, proteins, and other organic substances, including antibodies; contains no clotting agents; *see also* **plasma**.

serum sickness A type of hypersensitivity reaction in which the body responds to proteins contained in foreign serum.

sexually transmitted disease (STD) A disease such as gonorrhea or chlamydia that is normally passed from one person to another through sexual activity.

sexually transmitted infection (STI) *See* **sexually transmitted disease**.

Shiga toxin A bacterial poison that inhibits protein synthesis in target cells.

sign An indication of the presence of a disease, especially one observed by a doctor but not apparent to the patient; *see also* **symptom**.

silent mutation A change in a base sequence that produces no change to the protein made.

simple stain technique The use of a single cationic dye to contrast cells; *see also* **differential stain technique**.

single-dose vaccine The combination of several vaccines into one measured quantity.

S-layer The cell wall of most archaeal species consisting of protein or glycoprotein assembled in a crystalline lattice.

slide agglutination test *See* **VDRL** test.

slime layer A thin, loosely bound layer of polysaccharide covering some prokaryotic cells; *see also* **capsule** *and* **glycocalyx**.

sludge The solids in sewage that separate out during sewage treatment.

sludge tank The area in which secondary water treatment occurs.

somatic recombination The reshuffling of antibody genetic segments in a B lymphocyte as it matures.

soredium (pl. **soredia**) The disseminated group of fungal and photosynthetic cells formed by a lichen.

sour curd The acidification of milk, causing a change in the structure of milk proteins.

spawn A mushroom mycelium used to start a new culture of the fungus.

specialized transduction The transfer of a few bacterial genes by a bacterial virus that carries the genes to another bacterial cell; *see also* **generalized transduction**.

species The fundamental rank in the classification system of organisms.

specific epithet The second of the two scientific names for a species.

spike A protein projecting from the viral envelope or capsid that aids in attachment and penetration of a host cell.

spiral A shape of many bacterial and archaeal cells.

spirillum (pl. **spirilla**) A bacterial cell shape characterized by a twisted or curved rod.

Spirochaetes A phylum in the domain Bacteria whose members possess a helical cell shape.

spirochete A twisted bacterial rod with a flexible cell wall containing endoflagella for motility.

spontaneous generation The doctrine that nonliving, decaying matter could spontaneously give rise to living organisms.

spontaneous mutation A mutation that arises from natural phenomena in the environment.

sporangiospore Asexual spore produced by many fungi.

sporangium (pl. **sporangia**) The structures in fungi in which asexual spores are formed.

spore (1) A reproductive structure formed by a fungus. (2) A highly resistant dormant structure formed from vegetative cells in several genera of bacteria, including *Bacillus* and *Clostridium*; *see also* **endospore**.

sporozoite A stage in the life cycle of the malaria parasite that enters the human body.

sporulation The process of spore formation.

sputum Thick, expectorated matter from the lower respiratory tract.

stabilizing protein A protein that keeps the DNA template strands separated during DNA replication.

standard plate count procedure A direct method to estimate the number of cells in a sample dilution spread on an agar plate.

standard precautions Using those measures to avoid contact with a patient's bodily fluids; examples include wearing gloves, goggles, and proper disposal of used hypodermic needles.

staphylococcus (pl. **staphylococci**) An arrangement of bacterial cells characterized by spheres in a grapelike cluster.

start codon The starting nucleotide sequence (AUG) in translation.

stationary phase The portion of a bacterial growth curve in which the reproductive and death rates of cells are equal.

sterile Free from living microorganisms, spores, and viruses.

sterilization The removal of all life forms, including bacterial spores.

sterol A type of lipid containing several carbon rings with side chains; examples include cholesterol.

stop codon The nucleotide sequence that terminates translation.

streak plate method A process by which a mixed culture can be streaked onto an agar plate and pure colonies isolated; *see also* **pour plate method**.

streptobacillus (pl. **streptobacilli**) A chain of bacterial rods.

streptococcus (pl. **streptococci**) A chain of bacterial spheres.

streptokinase An enzyme that dissolves blood clots; produced by virulent streptococci.

stridor A high-pitched wheezing sound during breathing.

structural formula A chemical diagram representing the arrangement of atoms and bonds within a molecule.

structural gene A segment of a DNA molecule that provides the biochemical information for a polypeptide.

subclinical disease A disease in which there are few or inapparent symptoms.

substrate The substance or substances upon which an enzyme acts.

substrate-level phosphorylation The formation of ATP resulting from the transfer of phosphate from a substrate to ADP.

subunit vaccine A vaccine that contains parts of microorganisms, such as capsular polysaccharides or purified fimbriae.

sulfur cycle The processes by which sulfur moves through and is recycled in the environment.

sulfur dioxide A chemical preservative used in dried fruits.

superantigen An antigen that stimulates an immune response without any prior processing.

supercoiled domain A loop of wound DNA consisting of 10,000 bases.

supercoiling The process by which a chromosome is twisted and packed.

superinfection The overgrowth of susceptible strains by antibiotic resistant ones.

suppressor T cell A group of lymphocytes that regulate IgE antibody production.

surface water The water in lakes, streams, and shallow wells.

surfactant A synthetic chemical, such as a detergent, that emulsifies and solubilizes particles attached to surfaces by reducing the surface tension.

symbiosis An interrelationship between two populations of organisms where there is a close and permanent association.

symptom An indication of some disease or other disorder that is experienced by the patient; *see also* **sign**.

syncytium (pl. **syncytia**) A giant tissue cell formed by the fusion of cells infected with respiratory syncytial viruses.

syndrome A collection of signs or symptoms that together are characteristic of a disease.

synthetic biology A field of study that attempts to "build" new living organisms by combining parts of other species.

synthetic drug (agent) A substance made in the lab to prevent illness or treat disease.

systematics The study of the diversity of life and its evolutionary relationships.

systemic anaphylaxis The release of cell mediators throughout the body. *See also* **localized anaphylaxis**.

systemic disease A disorder that disseminates to the deeper organs and systems of the body.

systemic inflammatory response (SIRS) *See* **sepsis**.

T

tapeworm *See* **cestode**.

taxonomy The science dealing with the systematized arrangements of related living things in categories.

T cell *See* **T lymphocyte**.

T-dependent antigen An antigen that requires the assistance of T_H2 lymphocytes to stimulate antibody-mediated immunity.

T-independent antigen A foreign substance not requiring T helper cells for an immune response.

teichoic acid A negatively charged sugar-alcohol polymer in the cell wall of gram-positive bacterial cells.

temperate Referring to a bacterial virus that enters a bacterial cell and then the viral DNA integrates into the bacterial cell's chromosome.

termination (1) The completion of DNA or RNA synthesis. (2) The release of a polypeptide from a ribosome during translation.

termination factor A protein that triggers the release of a polypeptide from a ribosome.

terminator A set of nucleobases that stops RNA synthesis.

terminus The point where RNA synthesis stops.

tertiary structure The folding of a polypeptide back on itself.

tetanospasmin An exotoxin produced by *Clostridium tetani* that acts at synapses, thereby stimulating muscle contractions.

tetrad An arrangement of four spherical bacterial cells in a cube shape.

theory A scientific explanation supported by many experiments done by separate individuals.

therapeutic dose The concentration of an antimicrobial drug that effectively destroys an infectious agent.

therapeutic vaccine A vaccine that could be given to HIV-positive individuals to control HIV disease.

thermal death point (TDP) The temperature required to kill a bacterial population in a given length of time.

thermal death time (TDT) The length of time required to kill a bacterial population at a given temperature.

thermoduric Referring to an organism that tolerates the heat of the pasteurization process.

thermophile An organism that lives at high temperature ranges of 40°C to 90°C.

thimerosal A stabilizer put in some vaccines as a preservative.

thioglycollate broth A microbiological medium containing a chemical that binds oxygen from the atmosphere and creates an environment suitable for anaerobic growth.

three domain system The classification scheme placing all living organisms into one of three groups based, in part, on ribosomal RNA sequences.

thylakoid membrane The phospholipid bilayer where photosynthesis occurs in cyanobacteria.

T-independent antigen A foreign substance not requiring T helper cells for an immune response.

titer A measurement of the amount of antibody in a sample of serum that is determined by the most dilute concentration of antibody that will yield a positive reaction with a specific antigen.

titration A method of calculating the concentration of a dissolved substance, such as an antibody, by adding quantities of a reagent of known concentration to a known volume of test solution until a reaction occurs.

T lymphocyte (T cell) A type of white blood cell that matures in the thymus gland and is associated with cell-mediated immunity.

toll-like receptor (TLR) A signaling molecule on immune cells that recognizes a unique molecular pattern on an infectious agent.

toxemia The presence of toxins in the blood.

toxic dose (1) The amount of toxin need to cause a disease. (2) The amount of an antimicrobial drug that causes harm to the host.

toxigenicity The ability of an organism to produce a toxin.

toxin A poisonous chemical substance produced by an organism.

toxoid A preparation of a microbial toxin that has been rendered harmless by chemical treatment but that is capable of stimulating antibodies; used as vaccines.

transcription The biochemical process in which RNA is synthesized according to a code supplied by the template strand of a gene in the DNA molecule.

transduction The transfer of a few bacterial genes from a donor cell to a recipient cell via a bacterial virus.

transfer RNA (tRNA) A molecule of RNA that unites with amino acids and transports them to the ribosome in protein synthesis.

transformation (1) The transfer and integration of DNA fragments from a dead and lysed donor cells to a recipient cell's chromosome. (2) The conversion of a normal cell into a malignant cell due to the action of a carcinogen or virus.

transient microbiota The microbial agents that are associated with an animal for short periods of time without causing disease; *see also* **indigenous microbiota**.

translation The biochemical process in which the code on the mRNA molecule is converted into a sequence of amino acids in a polypeptide.

transmissible spongiform encephalopathy (TSE) A group of progressive conditions that affect the brain and nervous system.

transmission electron microscope (TEM) The type of electron microscope that allows electrons to pass through a thin section of the object, resulting in a detailed view of the object's structure.

transposable element A fragment of DNA called an insertion sequence or transposon that can cause mutations.

transposase An enzyme that moves insertion sequences to a new DNA location.

transposon A segment of DNA that moves from one site on a DNA molecule to another site, carrying information for protein synthesis.

traumatic wound A deep cut, compound fracture, or thermal burn.

trematode A flatworm, commonly known as a fluke, that lives as a parasite in the liver, gut, lungs, or blood vessels of vertebrates.

trismus A sustained spasm of the jaw muscles, characteristic of the early stages of tetanus; also called lockjaw.

trivalent vaccine A vaccine consisting of three components, each of which stimulates immunity.

trophozoite The feeding form of a microorganism, such as a protozoan.

tube (broth) dilution method A procedure for determining bacterial susceptibility to an antibiotic by determining the minimal amount of the drug needed to inhibit growth of the pathogen; *see* **minimal inhibitory concentration**.

tubercle A hard nodule that develops in tissue infected with *Mycobacterium tuberculosis*.

tuberculin test A procedure performed by applying purified protein derivative from *Mycobacterium tuberculosis* to the skin and noting if a thickening of the skin with a raised vesicle appears within a few days; used to establish if someone has been exposed to the bacterium.

tumor An abnormal uncontrolled growth of cells that has no physiological function.

tumor suppressor gene A normal gene that inhibits tumor formation.

turbid (turbidity) The cloudiness of a broth culture due to bacterial growth.

tyndallization *See* **intermittent sterilization**.

U

ultra high temperature (URT) method A treatment in which milk is heated at 140°C for 3 seconds to destroy pathogens.

ultrastructure The detailed structure of an cell, virus, or other object when viewed with the electron microscope.

ultraviolet (UV) light A type of electromagnetic radiation of short wavelengths that damages DNA.

uncoating Referring to the loss of the viral capsid inside an infected eukaryotic cell.

Unikonta A group of protists that includes the amoebozoans.

unsaturated Referring to a water-insoluble compound that can incorporate additional hydrogen atoms; *see also* **saturated**.

urticaria A hive-like rash of the skin.

use dilution test A method to evaluate the effectiveness of a chemical agent on microbes dried on steel cylinders.

V

vaccination Inoculation with weakened or dead microbes, or viruses, in order to generate immunity; *see also* **immunization**.

vaccine A preparation containing weakened or dead microorganisms or viruses, treated toxins, or parts of microorganisms or viruses to stimulate immune resistance.

vaccine adverse events reporting system (VAERS) A reporting system designed to identify any serious adverse reactions to a vaccination.

variable That part of an experiment exposed to or treated with the factor being tested.

variable domain The different amino acids in different antibody light and heavy chains.

variolation An obsolete method to protect a susceptible person from smallpox by infecting the person with dried material from a smallpox vesicle.

vasodilation A widening of the blood vessels, especially the arteries, leading to increased blood flow.

VBNC *See* **viable but noncultured**.

VDRL test A screening procedure used in the detection of syphilis antibodies.

vector (1) An arthropod that transmits the agents of disease from an infected host to a susceptible host. (2) A plasmid used in genetic engineering to carry a DNA segment into a bacterium or other cell.

vehicle transmission The spread of disease through contaminated food and water.

venereal disease *See* **sexually transmitted disease**.

vertical gene transfer The passing of genes from one cell generation to the next; *see also* **horizontal gene transfer**.

vertical transmission The spread of disease from mother to fetus or newborn.

viable but noncultured (VBNC) Referring to microbes that are alive but not dividing.

viable count The living cells identified from a standard plate count.

vibrio (1) A prokaryotic cell shape occurring as a curved rod.

viral inhibition The prevention of a virus infection by antibodies binding to molecules on the viral surface.

viral load test A method used to detect the RNA genome of HIV.

viremia The presence and spread of viruses through the blood.

virion A completely assembled virus outside its host cell.

viroid An infectious RNA segment associated with certain plant diseases.

virology The scientific study of viruses.

virosphere Refers to all places where viruses are found or interact with their hosts.

virulence The degree to which a pathogen is capable of causing a disease.

virulence factor A structure or molecule possessed by a pathogen that increases its ability to invade or cause disease to a host.

virulent Referring to a virus or microorganism that can be extremely damaging when in the host.

virus An infectious agent consisting of DNA or RNA and surrounded by a protein coat; in some cases, a membranous envelope surrounds the coat.

volutin *See* **metachromatic granule**.

W

wheal An enlarged, hive-like zone of puffiness on the skin, often due to an allergic reaction; *see also* **flare**.

white blood cell *See* **leukocyte**.

wild type The form of an organism or gene isolated from nature.

wort A sugary liquid produced from crushed malted grain and water to which is added yeast and hops for the brewing of beer.

X

xenograft A tissue graft between members of different species, such as between a pig and a human.

X ray An ionizing radiation that can be used to sterilize objects.

Y

yeast (1) A type of unicellular, nonfilamentous fungus that resembles bacterial colonies when grown in culture. (2) A term sometimes used to denote the unicellular form of pathogenic fungi.

Z

zone of equivalence The region in a precipitation reaction where ideal concentrations of antigen and antibody occur.

zone of inhibition The area around a chemically soaked paper disk where growth is inhibited.

zoonosis (pl. **zoonoses**) An animal disease that may be transmitted to humans.

zoonitic disease A disease spread from another animal to humans.

zoospore A flagellated asexual cell in the Chytridiomycota.

Zygomycota A phylum of fungi whose members have coenocytic hyphae and form zygospores, among other notable characteristics.

zygospore A sexually produced spore formed by members of the Zygomycota.

Index

NOTE: page numbers followed by *b* indicate Investigating the Microbial World, MicroInquiry, MicroFocus boxes; *f*, figures, and *t*, tables. Page numbers in **bold** indicate chapters online only.

A

Abc (*Acinetobacter calcoaceticus-A. baumannii*) complex, 329
Abdalati, Waleed, 149*b*
Abnormal behavior, parasitic disease and, 616*b*
Abortive poliomyelitis, 542
Abscesses, 433, 433*f*, 682*f*, 683
Abyssal zones, **881**
Acanthamoeba, 598*b*, 608
Accessory digestive organs, 345, 345*f*
Accessory reproductive organs, 408
Accutane, 431
Acetic acid, 573, **864**, **869**, **871**
Acetobacter aceti, **870**
Acetone, 183, 184*f*, 279*b*, **901**
Acetyl CoA, citric acid cycle and, 171*f*, 173
Acetylation, 828
Acetylcholine, 359
N-Acetylglucosamine (NAG), 49, 50, 116, 119*f*
N-Acetylmuramic acid (NAM), 50, 116, 119*f*
N-Acetyltalosamine uronic acid (NAT), 118
Acid-fast staining technique, 89*f*, 90–91, 118, 120*t*
Acidic (anionic) dyes, 87–88, 88*f*
Acidity (acids). *See also* pH
  characteristics of, 45
  dental caries and, 49*b*
  fungal growth and, 556
  microbial life and, 134*t*
  organic, **864**
  sterilization and, 199*b*
  vaginal, 409, 410*f*
Acidophiles, 148
Acidophilus milk, **871**
Acinetobacter spp., 115*f*, 429, 438*t*, **918*b***
Acinetobacter baumannii, 90*b*, 659*t*, 834, 839
Acinetobacter calcoaceticus-A. baumannii (Abc) complex, 329
Ackerman, Jennifer, 64
Acne, 429–431, 429*f*, 440*t*
Acquired immunity. *See* Adaptive immunity
Acquired immunodeficiency syndrome. *See* AIDS
Actin, 469*b*
Actin tail, 643, 644*f*
Actinobacillus, 354*t*
Actinobacteria, 101, 311, 312*f*, 314, 426–427
Actinomycetes, 101, **879**
  cell walls, 116, 118
Actinomycin, 26
Activated sludge, **887**, **887*f***
Activation energy, 166, 166*f*
Activation phase, of immune system, 672
Active immunity, 727, 728*f*
Active sites, 165
Active transport, 121
Active vaccine ingredients, 735*b*
Acute bacterial meningitis (ABM), 319*t*
  causative agents and epidemiology, 316–318, 316*f*, 317*f*
  clinical presentation, 318, 318*f*
  treatment and prevention, 318–319
Acute bacterial prostatitis, 424
Acute bronchitis, 328
Acute diseases, 641
Acute febrile pharyngitis, 493
Acute inflammation, 681–683
Acute otitis externa, 315–316, 319*t*
Acute otitis media (AOM), 316, 319*t*
Acute period of disease, 640–641, 640*f*
Acute phase proteins, 687
Acute rejection of allografts, 784
Acute respiratory disease (ARD), 494, 631*f*
Acute sinusitis, 315
Acute wounds, 431
Acyclovir, 253–254, 508, 509, 526, 822*t*
Adalimumab, 757
Adaptive immunity. *See also* Vaccines
  bridge between innate immunity and, 691*b*
  cell-mediated immune response
    activated CTLs destroy virus-infected cells, 714–715, 716–717*f*, 718
    Carabin and, 718*b*
    characteristics, 699
    naive T cell maturation into effector T cells, 714
    T-cell receptors and recognition, 712–714, 712*f*, 714*t*
    T-independent antigens and, 719
  clonal selection and, 700
  concept map of, 721*f*
  as defense against disease, 673–674
  defined, 669
  humoral immune response
    antibody common structure and, 702, 704, 704*f*
    antibody diversity and, 705, 709, 710*f*
    antigen-antibody complexes and, 709–711
    characteristics, 698
    immunoglobulin classes, 704–705, 705*t*
    primary and secondary antibody responses, 705, 709*f*
  immunological memory and, 698
  minimal self damage with, 698
  passive, 734–737
  self tolerance with, 698
  as specific resistance, 697
  uncontrolled response, 720*b*
Adenine (A), 52, 52*f*
Adenoid tissue, 493
Adenosine diphosphate (ADP)
  ATP/ADP cycle and, 169–170, 169*f*
  in chemiosmosis, 177–178*b*
Adenosine triphosphate (ATP)
  aerobic respiration and, 171, 171*f*
  anaerobic respiration and, 171, 171*f*, 182–183
  chemiosmosis and synthesis of, 176*f*, 177, 177–178*b*, 179*f*, 180
  energy for metabolism and, 168–170, 169*f*
  glucose and production of, 170–171, 171*f*
  in glycolysis, 172*f*, 173
  nucleotides and formation of, 53
  oxidative phosphorylation and, 175–177, 176*f*
  photosynthesis and, 187–188, 187*f*
Adenoviridae (adenoviruses)
  cell/tissue tropism, 459–460*b*
  classification of, 461*f*
  cytopathic effects of, 474*t*
  icosahedral symmetry of, 494*f*
  infections, 493–494, 493*f*, 496, 496*t*
  obesity and, 495*b*
  virotherapy and, 480*b*
Adenylation, 828
Adhesins, 110, 642–643, 642*f*, 644*f*
Adjuvant vaccine components, 735*b*
Adult T-cell leukemia/lymphoma, 477*t*
Adults, vaccinations for, 732–734, 734*f*
Advair, 772
Advance Market Commitment (AMC) for Pneumococcal Vaccines, 741*b*
Adverse reactions
  to recombinant subunit vaccines, 731
  to vaccines, 739–740, 739*b*, 741*b*
Advisory Committee on Immunization Practices (ACIP), 740
Aeras Global TB Vaccine Foundation, 741*b*
Aerobic respiration, 171, 171*f*
Aerosols, 650, 651*f*, 652*b*. *See also* Airborne bacterial diseases
Aerotaxis, 113, 186*b*
Aerotolerant organisms, 146
Aflatoxins, 563, 570, **855**
Africa
  cholera outbreaks in, 344*f*, 361*b*
  drinking water in, **884**
  dust storm spreading westward from, 652*b*
  emerging and reemerging diseases in, 663*f*
  HIV infections in, 761, 762*f*
  plague in, 389
  sub-Saharan
    childhood mortality from pneumonia in, 333*b*
    meningitis belt in, 317, 317*f*, 319
African trypanosomiasis (sleeping sickness), 603, 610*t*
Agammaglobulinemia, 788–789
Agar disk diffusion method, 825, 826
Agar or agar plates. *See also* Culture media
  MacConkey, 152*f*
  mutant identification using, 260–261, 261*f*
  precipitation tests using, 744–745, 745*f*
  for solid culture media, 18, 19*b*, 148*f*
Agaricus spp., 570–571
Agaricus bisporis, 566*f*, **915*f***
Agarose slides, 745
Agglutination, 710, 711*f*, 746, 748*b*
Aging
  of bacterial cells, 137*b*
  for beer production, **907–908**, **909*f***
  for wine production, **910**, **910*f***
Agranulocytes, 672
Agriculture
  antibiotics used in, 835*b*
  carbon cycle and, **893–894**, **893*f***
  changes in, disease emergence or resurgence and, 661
  crop rotation and, **896–897**
  fallow fields and, **895*b***

genetic engineering and, 286–287, 286*f*
metagenomics and, 299
Agriculture, U.S. Department of
Food Safety and Inspection Service, **848**
FoodNet and, **850**
Agrobacterium tumefaciens, 286–287, 286*f*, **913**, **915**
Aguilar, Abelardo, 817
AIDS (acquired immunodeficiency syndrome). *See also* HIV
discovery of, 695–696
first reports of, 789–790*b*
as late stage HIV disease, 794–795
pandemic, 761–762, 762*f*
Pneumocystis pneumonia and, 578
toxoplasmosis and, 608
tuberculosis and, 326
vaccine for, 286, 736*b*, 800
worldwide deaths due to, 631*f*
Air pollution, indoor, allergies to, 768*f*
Airborne bacterial diseases. *See also* Aerosols; Respiratory diseases; Respiratory droplets
spread of, 307–308, 652*b*, 653*b*
Albendazole, 620*t*
Alcaligenes, **857**
Alcamo, I. Edward, 450
Alcanivorax, **918*b***
Alcoholic beverages
beer, **907–908**, **908*b***, **909*f***
discovery of, **901**
distilled spirits, **911**
wine, **908**, **910**, **910*f***, **911*b***
yeasts for production of, 567
Alcoholic fermentation, 183, 184*f*
natural, 569
Alcohols, as skin antiseptics, 219, 223*t*
Aldehydes, 221, **871**, **911**
Alexander the Great, **854*b***
Algae
brown, 588*f*
copper as inhibitor of, 218–219
cyanobacteria as, 103–104
green, 23, 23*f*, 588*f*, 589, 589*f*
lichens and, 563, 565*f*
as photoautotrophs, 189
photosynthesis by, 187*f*
protist characteristics of, 589
red, 36*f*, 588*f*, 589
Algal blooms, **879**. *See also* Blooms; Red tides
Alginate, **906**
Alibek, Ken, 655*b*
Alien life forms, 35
Aliivibrio fischeri, 69*b*
Alimentary canal, 345
Alkaliphiles, 148
Alkalis (bases), 134*t*, 199*b*. *See also* pH
Allergen immunotherapy, 772–773
Allergenic foods, 766, 768
Allergens, 763–764
Allergic rhinitis, 766
Allergies and allergic reactions. *See also* Hypersensitivity(ies)
haptens and, 698
hygiene and, 770, 770–771*b*, 772, 773*b*
immune system response to, 764–766
microbes and, 770–771*b*
roundworms for treatment of, 613*b*
skin test for, 772
therapies for, 772–773
top 10 causes of, 768*f*
Allergy shots, 772–773
Allografts, 784
Alpha helix, 54, 55*f*
Alpha toxin, 646*t*
α-toxin, 385
Alphaproteobacteria, **918*b***
Alternative pathway of complement activation, 686–687, 686*f*
Alteromonas, 235–236*b*
Altman, Sidney, 168*b*
Amanita muscaria, 566*f*
Amanita phalloides, 570, 572*f*
Amantadine, 822*t*
Amatoxins, 570
Ambulances, disease transmission on, 653*b*
American trypanosomiasis, 603, 605, 610*t*
Ames, Bruce, 261
Ames test, 261, 262*f*, 263, 476
Amino acids
exons and, 243
nitrogen cycle and, **894**
protein synthesis and, 54, 55*f*
Amino group, 47*t*
Aminoglycosides, 815*f*, 816, 816*f*, 820*t*
resistance to, 827*f*, 832*t*, 836–837*b*
Aminoquinolines, 823, 825*t*
p-Aminosalicylic acid (PAS), 812
Ammonia, 57–58*b*, **894**
Ammonium chloride, 219
Ammonium ions, 54*b*, 65, 220*f*. *See also* Quaternary ammonium compounds
Amoeba, **879**
Amoeba proteus, 592*f*
Amoebiasis, 596–597, 597*f*, 598*b*, 601*t*, **865**
Amoebic dysentery, 592, 596
Amoeboid motion, 592
Amoebozoans, 588*f*, 592, 594*t*, 598*b*
Amoxicillin, 315, 373, 395, 813, 814, 820*t*
Amphibians, fungal disease and, 562*b*
Amphotericin B, 822, 825*t*
for blastomycosis, 577
broad spectrum uses of, 821
for coccidioidomycosis, 577
for cryptococcosis, 576
for histoplasmosis, 577
Ampicillin, 813, 813*f*, 820*t*
resistance to, 282*b*, 837*b*
amp^r, 282–283*b*
Amycolatopsis orientalis, 815
Amylases, **905**
Anabaena, 23*f*, 103*f*
Anabolism, 163, 164*t*, 167
Anaerobic organisms, 146, 146*f*, 148*f*, 638*b*, **888**
Anaerobic respiration, 171, 171*f*, 182–183
Anaerobic–aerobic degradation, **916**
Anaphylactic shock, 698, 779*t*
Anaphylaxis (anaphylactic reaction), 766, 767*f*, 768, 813
Anaplasma phagocytophilum, 399, 401*f*, 402*t*
Anaplasmosis, 399, 401, 402*t*
Ancylostoma duodenale, 618, 619*f*, 620*t*
Anemia, drug-induced hemolytic, 775
Angiogenic, use of term, 509
Animal hides, curing of, **906*b***
Animal rabies, 540–542, 541*f*
Animal viruses
latent infections from, 469–470
replication of, 465–469, 467*f*, 472*f*
Animalcules, 6–7, 7*f*, 104
Animalia (kingdom), 73
Animals. *See also* Zoonoses; *specific animals*
allergies to, 768*f*
antibiotics in feed for, 834, 835*b*
as disease reservoirs, 648–649
as eukaryotic organisms, 588*f*
Anionic (acidic) dyes, 87–88, 88*f*
Anions, 38
Annan, Kofi, **877**
Anne, Queen of England, 305
Anopheles mosquito, 602, 602*f*
Anoxic environments, 183
Anoxygenic photosynthesis, 188
Antarctica, freshwater lakes beneath, 149*b*
Anthrax. *See also Bacillus anthracis*
as biological weapon, 654*b*, 655*b*
bioterrorism and, 295, 384*b*
characteristics, 382–383, 383*f*, 388*t*
chlorine dioxide decontamination of, 222*b*
inhalational, 337, 338*t*
intestinal, 365
Koch's research on, 17
Pasteur's research on, 19*f*
toxin, 646*t*
treatment for, 812
Antibacterial soaps, 220*b*
Antibiotic resistance. *See also* Drug resistance; Methicillin; Vancomycin
bacteria and, 102*b*
biofilms and, **889**
case studies, 829*b*
cholera and, 361*b*
emerging and reemerging diseases and, 664
gonorrhea and, 415
as homeostasis, 66
horizontal gene transfer and, 274
increases in, 267, 267*t*
mechanisms of, 827–829, 828*f*
medical community concerns, 829, 833–834
to natural penicillins, 813
nosocomial infections and, 660
P. acnes and, 431
prevention of, 838*b*
Shigella and, 367
source of, 836–837*b*
staphylococcal pneumonia and, 329
susceptibility assays for, 825–827, 826*f*
in third Golden Age of microbiology, 29
timeline for appearance of, 827*f*
Antibiotics
approvals 1983–2009, 839*f*
for bacterial bronchitis, 328
beta-lactam family, 812–814, 813*f*, 814*t*
candidiasis and, 574
cell biology studies and development of, 109
for chronic granulomatous disease, 788
colds and, 493
discovery of, 25–26

DNA or RNA function and, 53
in feedlots, 834, 835*b*
fungal diseases and, 819
gene expression interference by, 246
genetic engineering and, 266–267
global market for, 819*f*
gram-negative vs. gram-positive bacteria and, 90
hydrolysis and resistance to, 828
improper use of, 829, 833
for intestinal anthrax, 365
irradiation of, 206
for listeriosis, 364
modifications in, drug resistance and, 828, 832*f*
as natural biological warfare agents, 809, 810*b*
new approaches to, 834, 839
overuse, allergies and asthma and, 770–771*b*
production of, **912–913**
as revolutionary, 804–805
synthetic drugs vs., 808
for urinary tract infections, 422, 424
viruses and, 819
Antibodies. *See also* Hypersensitivity(ies)
blocking, 785*b*
gene rearrangements for diversity of, 705, 709, 710*f*
microorganism identification using, 83
neutralizing toxins in blood, 703*b*
structure of, 702, 704, 704*f*
Antifungal agents (drugs), 821–823, 825*t*
Antigen-antibody complexes, 709–711. *See also* Monoclonal antibodies
Antigen-binding sites, 704, 704*f*
Antigenic determinants (epitopes)
in adaptive immunity, 698, 701*f*, 702
antigen-binding sites and, 704
antigens and, 697, 697*f*
Antigenic drift, 498–499*b*
Antigenic mimicry, 782, 782*f*
Antigenic shift, 498–499*b*
Antigenic variation, 498–499*b*
Antigen-presenting cells (APCs), 691*b*, 712–714, 712*f*, 714*t*, 716*f*, 717*f*
Antigens, 697. *See also* Antigen-antibody complexes; Antigenic determinants
Antihelminthic agents, 824
Antihistamines, 772
Antilymphocyte, 786*t*
Antimetabolites, 786*t*
Antimicrobial agents (drugs). *See also* Antibiotics; Drug resistance
antifungal, 819, 821–823, 825*t*
antihelminthic, 824
antiparasitic, 819, 823–824, 823*b*, 824*f*, 825*t*
antiviral, 819, 821, 822*t*
bacterial targets for, 815*f*
cell metabolism and, 815–819
cell wall synthesis inhibitors, 812–815, 813*f*, 814*t*
clinical trials for development of, 840–841*b*
control of bacterial diseases with, 305, 804–805
Ehrlich and history of, 806–807
Fleming's penicillin discovery and development of, 807–808, 807*f*
properties of, 808–809, 809*f*
synthetic, 811–812
uses of, 806
Antimicrobial spectrum, 808–809, 809*f*
Antimoniate, 824
Antiretroviral drugs, 525, 761–762, 793–794*b*, 797, 798
Antisense molecules, 248*b*
Antisepsis, 17, 17*f*
Antiseptics
biofilms and, **889**
disinfectants vs., 211*f*
evaluation for effectiveness, 213–214, 213t, 214*f*
in kitchen pantry, 215*b*, 216*b*
phenol coefficients of, 213*t*
properties of, 209–210
Antisera (antiserum), 83, 736
Antithymocyte, 786*t*
Antitoxins, 384, 646
Antiviral drugs, 819, 821, 822*t*
Antiviral proteins (AVPs), 689, 689*f*
Antiviral state, 689
Apicomplexans, 588*f*, 594, 594*t*. *See also* *Plasmodium* spp.
Babesia spp., 605
Toxoplasma gondii, 606
Apoptosis, 478, 479*f*, 685, 698, 782*f*
Appendix, 345*f*, 346, 346*b*
Appert, Nicholas, **858–859**, ***858b***
Apples, GI tract and, 348–349*b*
Apramycin, 836–837*b*
APTIMA assay, for HIV RNA, 797
Aqueous solutions, 44
Aquifex spp., 104
Arboviruses. *See also* Arthropodborne diseases
arboviral encephalitis, 544, 544*f*, 546–547, 547*t*
dengue fever, 531–532
in New York City, 629–630
yellow fever, 529, 530–531*b*, 531
Archaea (archaeal organisms). *See also* Prokaryotes
Bacteria and Eukarya differences from, 105*b*
brief survey of, 23*f*, 24
cell membrane, 120–121, 121*f*
cell shapes and arrangements, 106, 107*f*
cell structure
cell envelope
cell membrane, 121, 122*f*
cell walls, 118, 120
components of, 116
cell function and, 108–109, 108*f*
concept map for studying, 108*f*
cytoplasm, 122–126, 122*f*
external, 110–116
flagella, 110, 112*f*
pili, 110, 110*f*
summary of, 127–128*t*
chromosome characteristics, 237*t*
diversity among, 100, 104, 105*f*
DNA replication in, 238
as domain, 77–78
extremophiles among, 99–100
gamma-radiation resistant, 257*b*
genome size among, 233, 235*f*
nutritional pattern of, 188
on phylogenetic tree of life, 78*f*, 101*f*
prokaryotic cellular organization of, 25
reproduction, 135. *See also* Binary fission
restriction enzymes from, 280
RNA transcription in, 242–243
temperature and growth of, 144*f*, 145
in three-domain system, 129*b*
transcription and translation in, 128–129
viable but noncultured, 155
viruses, shapes of, 456
Archaebacteria, 77, 100
Archaeplastida (supergroup), 588*f*
Arenaviridae (arenaviruses), 461*f*, 533–534
Argentina hemorrhagic fever, 533
Aristander, ***854b***
ARMAN (Archaeal Richmond Mine Acidophilic Nanoorganism), 85*b*
Armelagos, George, 416*b*, 818*b*
Arnold, A. Elizabeth, 558*b*
Aroma additives, **906**. *See also* Odors
Arrhythmias, 314
Arsenic, in groundwater, **885**
Arsenic derivatives, 824
Arsphenamine, 806–807
Artemisia annua, 824, 824*f*, 825*t*
Artemisinin, 603, 824, 825*t*
Arthritis. *See also* Rheumatoid arthritis
agammaglobulinemia and, 789
M. chelonae and, 326
Arthrobacter, **905**, **919**
Arthroderma, 572
Arthropodborne diseases
bacterial
Lyme disease, 393–395, 394*f*, 395*f*
plague, 388–390, 390*b*, 391*b*
relapsing fever, 395–396, 396*f*
rickettsial, 396, 398–401, 400–401*b*
summary of, 402*t*
tularemia, 390–391, 393, 394*f*
as blood diseases, 381
multiple portals of entry for, 643
vectors and, 388
viral. *See* Arboviruses
Arthropods, defined, 101, 381
Arthrospores, 557
Arthus, Nicolas Maurice, 777
Arthus phenomenon, 777–778
Artificially acquired active immunity, 728, 728*f*
Artificially acquired passive immunity, 728*f*, 736–737, 787
Ascariasis, 618, 620*t*
Ascaris lumbricoides, 618, 620*t*
Asci (ascus), 563
Ascomycota, 560
Ascomycota (ascomycetes), 561, 563, 563*f*, 564*b*, 565*f*
dermatophytes as, 572–573
disease and drug research using, 569*t*
Ascospores, 563, 565*f*
Asepsis, 211
Aseptic technique, 155
Ashbya gossypii, **905**
Asia
childhood mortality from pneumonia in, 333*b*
cholera outbreaks in, 344*f*
emerging and reemerging diseases in, 663*f*
HIV infections in, 761, 762*f*
plague in, 389
SARS in, 502
typhoid fever as endemic to, 365
Asian flu (1957), 496, 498–499*b*

Asimov, Isaac, 35
Aspergillosis, 578, 580, 580*t*
Aspergillus flavus, 563, 570, **855**
Aspergillus fumigatus, 580, 580*t*, 771*b*
Aspergillus niger, **904**, **904*f***, **905**
Aspergillus oryzae, **870**, **905**, **906**
Aspergillus parasiticus, 570
Aspirin, Reye syndrome and, 500, 508
Assembly
  HIV, 792*f*
  in viral replication, 463*f*, 464*f*, 465, 466–467, 468*f*
Associative learning, by prokaryotes, 66
Asthma, allergy-induced, 764*f*, 768–769, 770*f*, 779*t*
  H. pylori infection and, 372*b*
  microbes and, 770–771*b*
Asticcacaulis excentris, **914**
Atazanavir, 822*t*
Atherosclerosis, 683
Athlete's foot, 567, 572–573
Atmosphere, microbial life in, 4*f*, 5
Atomic nucleus, 37–38, 37*f*
Atomic number, 38
Atoms, 37–39, 37*f*, 39*f*, 56*f*, 84*f*
Atoms for Peace, Eisenhower's, **865**
Atopic disease, 766, 770. *See also* Allergies and allergic reactions
ATP synthase, 177, 177–178*b*
ATP/ADP cycle, 169*f*, 170, 170*f*. *See also* Adenosine triphosphate
Attachment, in viral replication, 463*f*, 464*f*, 465, 467*f*
Attachment pili. *See* Pili
Attenuated vaccines, 729*t*, 730
Attractants, for bacterial and archaeal cells, 113, 113*f*
Atypical pneumonia, 332, 338*t*
Augmentin, 814
Australia, emerging and reemerging diseases in, 663*f*
Autism, 740
Autoantibodies, 783
Autoclaves, 200–201, 201*f*, 202*b*, 203, 203*f*, 209*t*, 221
Autografts, 784
Autoimmune diseases or disorders, 698, 718*b*, 781–784, 782*f*, 783*t*
Autolytic enzymes, 116
Autotrophs, 188–189, 189*f*
Autotrophy, 188–189, 189*f*
Auxotrophs, 260
Avermectins, 824
Avian influenza (flu virus), 481, 490, 498–499*b*
Avirulent organisms, 634, 636
Avoparcin, 835*b*
Axial Seamount (Pacific deep-sea volcano), 277*b*
Azathioprine, 786*t*
Azidothymidine (AZT), 793–794*b*, 797, 822*t*
Azithromycin, 413, 419, 443, 817, 820*t*
Azoles, 821, 825*t*, 827*f*
Azotobacter, **896**

B

B lymphocytes (B cells)
  adaptive immunity and, 698
  clonal selection of, 700, 701*f*, 702
  immune system and, 670, 671*f*, 672
  immunodeficiency disorders and, 787, 788
  infectious mononucleosis and, 524
  mucosal immunity and, 706–707*b*
  origins of, 700, 700*f*
  type I hypersensitivity and, 764, 764*f*
B period, 135, 136*f*, 139
Babesia bigemina, 605
Babesia microti, 605, 610*t*
Babesiosis, 605, 610*t*
Bacille Calmette-Guérin (BCG) vaccine, 326, 730
Bacilli (bacillus) shape, 106
Bacillus spp., 101, **855**, **857**, **859**, **896**, **912**
Bacillus amyloliquefaciens, 280*t*
Bacillus anthracis. *See also* Anthrax
  anthrax caused by, 382–383, 383*f*, 388*t*
  anthrax toxin, 646*t*
  as biological weapon, 655*b*
  as encapsulated pathogen, 116
  endospores of, 141, 142, 142*f*, 382*f*
  plasmids of, 238
  rectangular-shaped cells of, 106
Bacillus cereus, 23*f*, 356, 374*t*
Bacillus licheniformis, 136*f*
Bacillus megaterium, 138*t*
Bacillus polymyxa, 820*t*
Bacillus sphaericus, 144*b*, **914**
Bacillus subtilis
  bacitracin and, 820*t*
  cell shape, 125, 127*f*
  chlorine dioxide decontamination of, 222*b*
  endospore formation in, 143*f*
  genetic engineering using, **919**
  nystatin and, 810*b*
  protease production and, **906**
  in space, 134*t*
  transcription and translation in, 128*f*, 251, 251*f*
Bacillus thuringiensis, **913**, **914*f***
Bacillus thuringiensis toxin (Bt-toxin), **913–914**, **914*f***
Bacitracin, 809*f*, 815–816, 815*f*, 820*t*
Bacteremia, 637, 638–639*b*
Bacteria (*Bacteria*). *See also* Antibiotics; Bacterial diseases; Gram-negative bacteria; Gram-positive bacteria; Prokaryotes
  antimicrobial agents' targets in, 815*f*
  Archaea and Eukarya differences from, 105*b*
  batteries run by, 179*b*
  brief survey of, 23–24, 23*f*
  cell structure
    cell envelope
      cell membrane, 120–121, 120*f*, 122*f*
      cell walls, 49, 116, 118, 118*f*
    cell function and, 108–109, 108*f*
    cell membrane, 120–121, 120*f*, 121*f*
    cell shapes and arrangements, 106, 107*f*
    concept map for studying, 108*f*
    cytoplasm, 122–124*f*, 122–126, 125*b*, 126*b*
    external, 110–116
      flagella, 110, 112*f*, 113–115
      glycocalyx, 115–116, 115*f*
      pili, 110
    of idealized cell, 109*f*
    summary of, 127–128*t*
  chromosome characteristics, 237*t*
  demystifying facts about, 102*b*
  diversity among, 100–101, 103–104, 103*f*
  as domain, 77
  in dust storms, 652*b*
  Ehrenberg's naming of, 14
  enteric, rapid identification of, 82*b*
  in extreme environments, 134*t*
  extremophiles among, 99
  gamma-radiation resistant, 257*b*
  genome size among, 233, 235*f*
  growth. *See* Microbial growth
  as kingdom, 77
  magnetotactic, 125*b*
  major chemical elements of, 37, 37*t*
  mitochondria, chloroplasts, microbial eukaryotes and, 75*t*
  nutritional pattern of, 188
  on phylogenetic tree of life, 78*f*, 101*f*
  prokaryotic cellular organization of, 25
  protein synthesis in, 246–247*f*
  reproduction, 135, 136*f*, 137*b*, 138. *See also* Microbial growth
  restriction enzymes from, 280
  taxonomic classification of, 80*t*
  temperature and growth of, 144*f*, 145
  in three-domain system, 129*b*
  transcription in, 128–129, 128*f*, 242–243, 244*f*
  translation in, 128–129, 128*f*
  vaccines currently in use for, 728, 729*t*
  viable but noncultured, 155, 156*b*
  viral genomes in lungs, 311
  viruses and, 451–452, 452*f*
Bacterial diseases
  airborne
    lower respiratory, 320–338, 326
    upper respiratory, 311–319, 319*t*
  arthropodborne, 388–396
  blood types and, 777*b*
  as challenges to humans, 305
  contact
    exogenous bacteria and, 432–434, 433–434*f*, 435*b*
    of the eye, 441–444, 442–444*f*, 444*t*
    indigenous microbiota and, 429–431, 429*f*
    summary of, 430*t*
  foodborne, 355–373
  GI tract, 355–373, **882–883**
  HIV/AIDS and, 796*f*
  indigenous microbiota and, 431
  oral cavity, 350–354, 351*f*, 352*b*, 353*f*, 354*b*
  publications on, 445
  sexually transmitted, 406–407, 410–419
  soilborne, 382–388
  spread of, 307–308
  urinary tract, 421–422, 421*f*, 423*b*, 424, 425*t*
  waterborne, 355–373, **882–883**
Bacterial growth curve, 139–140, 139*f*
Bactericidal agents, 197
Bacteriochlorophylls, 188
Bacteriocins, 238
Bacteriology, 19, 22*f*
Bacteriophages
  as bacterial diseases therapy, 466*b*
  naming of, 454
  in oceanic microbes, 277*b*
  replication of, 463, 464*f*, 465
  transduction and, 274, 276*f*, 277

typing of, for disease identification, 475
Venter and assembly of, 12*b*
Bacteriostatic agents, 197
Bacteriostatic food preservation methods, 208
Bacteroidales, 430*f*
Bacteroides, 311, 312*f*, 347, 354*t*, 420, 438*t*
Bacteroides fragilis, 814
Bacteroidetes, 347, 426–427
Bactrim, 275*b*, 811–812. *See also* Trimethoprim-sulfamethoxazole
Baculovirus, **914**
Baghdad boils, 596*b*
Baker's yeast, 561, 567
Bakery products, contamination of, **855**
Baking soda, sterilizing with, 200
Balfour Declaration (1917), 279*b*
Balinska, Marta A., 725
*Bam*HI, 280*t*
Barophiles (barophilic organisms), 148, **881**
Barré-Sinoussi, Françoise, 695, 696
Bartonella quintana, 401*b*
Basal body, 112*f*, 113
Base analogs, 253–254, 254*f*, 821
Base pairs
deletion or insertion in, 254–255, 255*f*
mismatch repairs to, 255–256, 256*f*
nucleotide excision repairs to, 256, 256*f*, 257*b*
substitutions in, 254, 255*f*
Bases (alkalis). *See also* pH
characteristics of, 45
sterilization and, 199*b*
Basic (cationic) dyes, 87, 88*f*
Basidia (basidium), 566, 566*f*, 567*f*
Basidiocarps, 566, 567*f*
Basidiomycota (basidiomycetes), 560, 565–566, 566*f*, 569*t*, 575–576
Basidiospores, 566, 566*f*, 567*f*
Basiliximab, 786*t*
Basophils, 670, 671, 763, 765, 765*f*, 781*t*
Batch method of pasteurization, 203–204, **860**
Batrachochytrium dendrobatidis, 562*b*
Bats, white-nose disease among, 564*b*, 580
Batteries, bacterial cells-powered, 179*b*
Bauer, A. W., 826
BCG vaccine, 326, 730
Beadle, George, 25
Bee stings, 768
Beef. *See also* Cattle; Cows
E. coli O157:H7 contamination of, 368
irradiation, 206
toxoplasmosis and, 606, 606*f*
Beef tapeworm disease, 612–613
Beef tapeworms, 611*f*
Beer production, 818*b*, **905*t***, **907–908**, **908*b***, **909*f***
Beethoven, Ludwig van, 406
Beggiatoa, **894**
Behavior, abnormal, parasitic disease and, 616*b*
Behavioral therapy, immune system and, 715*b*
Behring, Emil von, 21*t*, 669, 703*b*
Beijerinck, Martinus W., 20, 22, 453, **894**, **895*b***
Bell, Robin, 149*b*
Benign tumors, 476
Benthic zones, **881**
Benzalkonium chloride, 219, 220*f*
Benzoic acid, 573, **864**
Benzoyl peroxide, 219, 223*t*, 431
Berg, Paul, 279–280
Bergey, David H., 79–80
Bergey's Manual, 90
Bergey's Manual of Determinative Bacteriology, 80
Bergey's Manual of Systematic Bacteriology, 80, 83
Berliner, G. S., **913**
Besser, Richard E., 804
Beta oxidation, 182
Betadine, 217–218
Beta-lactam antibiotics, 812–814, 813*f*
Beta-lactamases, 813, 813*f*, 828, 832*f*
Betaproteobacteria, 430*f*
Biaxin, 373, 817
Bifidobacterium lactis, 633
Bilateral symmetry, 592
Bile, 346, 675
Bilharzia, 612
Bilharziasis, 612
Bill and Melinda Gates Foundation, 604*b*, 663–664, 726, 741*b*
Binary fission
for bacterial reproduction, 137*b*
bacteriophage replication and, 464*f*, 465
cell cycle and, 135
by ciliates, 593
for diplomonad reproduction, 592
in vertical gene transfer, 268, 269*f*
Binomial nomenclature, 78–79, 79*b*
Bioburden, 212
Biocatalysis, **901**, **902*f***
Biochemical oxygen demand (BOD), **882**
Biochemical tests for microorganism identification, 80, 83*f*
Biochemistry, emergence as scientific discipline, 164*b*
Biocrimes, 654*b*
Biodefense, metagenomics and, 299
Biofilms
cell communication and, 66, 67*f*, 68, 68*f*, 70
chronic otitis media and, 316
dental plaque as, 49*b*, 351
in disease, 70*f*
E. cloacae infection and, 117*b*
intravenous catheters and, 431
as microbial communities, 24, 30, 30*f*, **888–889**, **888*f***
as mixed culture, 155
nosocomial infections and, 659
phage therapy against, 466*b*
pili and, 110
resistance of, 102*b*
in sewage treatment, **887**
virulence and, 645
Biofilters, **887–888**
Biogenesis, 9
Biolog MicroPlate, 83*f*
Biological indicators of sterilization, 203, 203*f*
Biological pollution of water, **882**
Biological safety cabinet, 205–206, 205*f*
Biological vectors, 650, 651*f*
Biological warfare. *See* Bioterrorism and bioweapons
Biomass, microorganisms as component of, 5
Bioreactors, **902**
Bioremediation
biofilms and, 30, 30*f*, 70, **889**
D. radiodurans used for, 257*b*
Deepwater Horizon oil spill and, 41*b*
genetic engineering and, 284
industrial, **916**, **917–918*b***, **918–919**, **918*f***, **920*b***
metagenomics and, 299
Biosensors, ribozymes as, 168*b*
Biospheres, 35, 65, 553
Biosynthesis
HIV, 792*f*
in viral replication, 463*f*, 464*f*, 465, 466–467, 468*f*
Biotechnology. *See also* Genetic engineering; Industrial microbiology
commercial and practical applications of, 281, 282–283*b*, 284, 285*f*, 286–287
industrial processes and, **902**, **902*f***
laboratory production of monoclonal antibodies, 756–757, 756*f*
metagenomics and, 299
Bioterrorism and bioweapons
anthrax and, 295, 337, 382, 384*b*
genome transplantation and, 300*b*
microbial forensics and, 295
pathogens as weapons in, 654–655*b*
plague and, 381, 390*b*
smallpox and, 518
in third Golden Age of microbiology, 29
Biotypes, 79
Bipolar staining, 389
Birds. *See also* Avian influenza
droppings, disease transmission and, 576, 615*b*
psittacosis and, 337
Birdseye, Clarence, **862**
Bisphenols, 218
Bivalent polio vaccines, 543*b*
Bizio, Bartholemeo, **854*b***
Black Death, 253*b*, 380–381, 389, 390*b*, 391*f*. *See also* Plague
Black fever, 596
Black rot, **855**
Blackheads, 430–431, 430*f*
Blacklegged tick, 401
Black-water fever, 603
Blakemore, Richard P., 125*b*
Blanching foods, **859**
Blastomyces spp., 745
Blastomyces dermatitidis, 577, 580*t*
Blastomycosis, 577, 580*t*, 780
Blastospores, 557
Blepharitis, 441–442, 444*t*
Blepharospasm, 359
Blindness
infectious, 412*b*
river, 621
Blixen, Karen, 407
Blocking antibodies, 785*b*
Blood, human. *See also* Antibodies; B lymphocytes; Bloodborne diseases; Serological reactions; T lymphocytes
adaptive immunity experiments with, 703*b*
allergy tests, 772
clots, group A streptococci and, 313*b*
HIV transmission and, 797
infection, phage therapy against, 466*b*
innate and adaptive immunity and, 670–672, 671*f*

nosocomial infections and, 659*f*
transfusion reactions, 764*f*, 775
types, 747*f*, 776*f*, 777*b*
Blood agar, 140*f*. *See also* Agar or agar plates
Blood clotting factors, genetically engineered, 285*f*
Bloodborne diseases
parasitic
babesiosis, 605
malaria, 601–603, 602*f*
toxoplasmosis, 605–606
trypanosomiasis, 603, 604*f*, 605
viral. *See also* HIV
CMV disease, 524–526
Ebola hemorrhagic fever, 533
Epstein-Barr virus disease, 524, 525*f*
hepatitis viruses, 526–527, 526*f*, 527*f*, 528*b*
summary of, 528*t*
Blood-brain barrier, meningitis and, 318
Blooms, **881**, **881*f***. *See also* Red tides
algal, **879**
defined, 103
Boceprevir, 527
BOD (biochemical oxygen demand), **882**
Body odor, offensive, 427*b*
Body ringworm, 572
Body temperature. *See also* Fever
colds and, 494*b*
Boiling water, microbial control using, 198, 200, 209*t*
Boils (skin), 433. *See also* Furuncles
Boletus spp., 570–571
Boletus santanas, 572*f*
Bolivian hemorrhagic fever, 533
Bone marrow, 672, 673*f*
Bone marrow transplants
graft-versus-host reaction and, 786
HHV-6 viremia and, 509
Booster shots, 730, 734
Bordet, Jules, 21*t*
Bordetella bronchiseptica, 296, 297*f*
Bordetella parapertussis, 296, 297*f*
Bordetella pertussis, 296, 297*f*, 338*t*. *See also* Pertussis
cells, 308*f*, 321*f*
outbreak of, 307–308
pertussis toxin, 646*t*
pili adhesins on, 110
resistant strain of, 321
Borrelia spp., 104
Borrelia burgdorferi, 393–395, 394*f*, 402*t*
Borrelia hermsii, 395–396, 396*f*, 397*b*, 402*t*
Botrytis cinerea, **910**
Botulinum toxin (Botox), 359–360, 646*t*, 654*b*
Botulism
clostridial endospores and, 142
exposure to, 736
as foodborne intoxication, 357, 359, 374*t*, **865**
home-canned foods and, **866–867**
naming of, **854**
neutralization test for, 744
specialized transduction and, 277
Bovine growth hormone (BGH), 204*b*
Bovine somatotropin (rBST), 284
Bovine spongiform encephalopathy (BSE), 484, 485
Boyer, Herbert, 280, **919**
Boyer, Paul D., 177*b*
Brandy, **911**
Brazilian hemorrhagic virus, 534
Bread
decaying, *S. marcescens* and, **854*b***
molds, 561
sourdough, **856*b***
spoilage of, **855**
yeasts for production of, 567, 569
Breakbone fever, 532
Breast milk
colostrum, 705, 735–736, **857**
HIV transmission and, 797
Brevibacterium, 427, **871**, **905**
Brewer's yeast, 80*t*
Brewing. *See* Beer production
Bright-field microscopy, 85, 89*f*, 95*t*
Brill-Zinsser disease, 398
Broad-spectrum antimicrobials, 809. *See also* Antimicrobial agents
Brokaw, Tom, 222*b*
Bromhidrosis, 427*b*
5-Bromouracil (5-BU), 253, 254*f*
Bronchiolitis, 502
Bronchitis, 328, 328*f*, 338*t*
Bronchodilators, 769
Bronchopneumonia, 332
Broth dilution method, for antibiotic susceptibility assay, 825–826
Broth(s)
for liquid culture media, 7, 155
thioglycollate, 146, 148*f*
Brown, Alan, 613*b*
Brown algae, 588*f*
Bruce, David, 20, 21*t*
Brucellosis, 780, **867**
Brugia malayi, 619–620
Bruton, Ogden C., 787
Bruton agammaglobulinemia, 787
BSE (bovine spongiform encephalopathy), 484, 485
Bt toxin, **913–914**, **914*f***
Buboes, 389
Bubonic plague, 253*b*, 380–381, 389, 391*f*, 402*t*, 730. *See also* Plague
Buchner, Eduard and Hans, 164*b*
Budding, 468, 557, 559*f*, 563*f*, 567
Buffers, 45–46, 46*f*
Bukowski, Jack, 699*b*
Bull's-eye rash, 394, 395*f*
Bunyaviridae (bunyaviruses), 461*f*, 481*t*, 504, 533–534
Burgdorfer, Willy, 393
Burkitt lymphoma, 476, 477*t*, 524
Burn infections, 437, 437*f*, 440*t*
Burns, Harry, 195
Butanol, 183, 184*f*
Butterfat, **856**
Buttermilk, **871**
Butyl alcohol, **901**

C

C. diff-associated infections (CDAIs), 631, 646. *See also Clostridium difficile*
C period, 135, 136*f*, 139
C3 convertase, 686*f*, 687
C5a and C5b complement proteins, 686*f*, 687
Cacao leaves, fungal endophytes and, 558*b*
Caesar salad, egg spoilage and, **860*b***
Caffa, siege of (1346), 390*b*
Calcium (Ca), 37*t*
Calcium gluconate, **905**
Caliciviridae (caliciviruses), 461*f*, 536
Calmette, Albert, 21*t*, 326
Calories, 171
Camembert cheese, **871–872**
Campylobacter spp., 207*b*, 642, **882**
Campylobacter jejuni
cells, 370*f*
food poisoning caused by, 145*b*
garlic organosulfur compounds and, 216*b*
invasive gastroenteritis due to, 370–371
microbial genomics studies of, 294–295
in raw milk, **867–868**
Campylobacteriosis, 370–371
Cancer, human. *See also specific types of cancer*
genetic engineering and, **921**
HIV/AIDS and, 795, 796*f*
monoclonal antibodies and, 757
onset of, 476*f*
viruses and, 476–478, 477*t*, 478*f*, 479*f*, 480*b*
Candida spp., 347*f*, 638*b*, 659*t*
Candida albicans
allergies and asthma and, 771*b*
daughter cells budding from mother cells, 163*f*
fermenting form of, 162–163, 183
as indigenous microbiota, 632
infections, 573–574, 575*t*
micrograph of, 574*f*
as sac fungi, 563
Candida milleri, **856*b***
Candidiasis, 573–574, 575*t*, 780, 789–790*b*, 796*f*
Canning foods, **859**, **859*f***
botulism precautions, **866–867**
Cano, Raul, 144*b*
Capnophiles, 147
Caporael, Linda, 571*b*
Capsid, 465
in HIV, 790, 790*f*
host range and, 458
viral replication and, 463*f*, 465, 468*f*
viral shape and, 455–456, 456*f*
Capsomeres, 455–456, 456*f*
Capsules, 115–116, 642
Carabin, 718*b*
Carbapenems
antimicrobial spectrum of, 809*f*
global market for, 819*f*
K. pneumoniae resistant to, 259*b*
properties of, 814, 815*f*, 820*t*
resistance to, 832*t*
Carbenicillin, 813–814, 813*f*
Carbohydrates
antigenic properties of, 697
in bacterial cells, 47*f*
characteristics and groups of, 47–50
as chemical energy sources, 180, 181*f*
Carbolic acid. *See* Phenol and phenolic compounds
Carbon (C)
chemical reactions and, 39
as essential element, 37, 37*f*, 37*t*

Carbon cycle, **892–894**, **893*f***
Carbon dioxide
  in citric acid cycle, 173, 174*f*
  molecular formula of, 40
Carbon fixation, **892–894**
Carbonate, anaerobic respiration and, 183
Carbon-fixing reactions, 188
Carboxyl group, 47*t*
Carboxysomes, 124, 185*f*, 188, **892*b***
Carbuncles, 433, 433*f*, 440*t*
Carcinogens, 261, 262*f*, 371, 476, 476*f*, **864**. *See also* Cancer, human
Cardini, Caesar, **860*b***
Career preparation
  for clinical microbiology, 306
  for epidemiology, 628
  for food inspectors, **848**
  for science, 2
  for virology, 450
Carriers, disease, 649
Casein, **856**, **857**
Casing soil, for mushroom farming, **915**
Catabolism
  anabolism vs., 164*t*
  characteristics of, 163–164
  glucose
    ATP synthesis and, 177–178*b*
    citric acid cycle, 173, 174*f*, 175, 175*f*
    glycolysis, 171, 172*f*, 173, 175*f*
    other nutrient sources for, 180–182
    oxidative phosphorylation, 175–177, 180
  in metabolic pathways, 167
Catarrhal stage, of whooping cough, 321
Catheters
  antibiotic resistance prevention and, 831*t*, 838*b*
  biofilms and, 431, **888**, **888*f***
  intravenous, infections with, 275*b*, 431, 660
Cationic (basic) dyes, 87, 88*f*
Cationic detergents, 219, 220*f*, 223*t*
Cations, 38
Cats
  leptospirosis and, 386
  tinea infections and, 572–573
  toxoplasmosis and, 606, 606*f*, 608
Cattle. *See also* Cows
  anthrax and, 337, 382
  antibiotics in feed for, 834, 835*b*
  babesiosis in, 605
  C. jejuni in, **868**
  plague in, 621
  transmissible spongiform encephalopathies in, 484–485
Caulobacter crescentus, 126, 127*f*
Caventou, Joseph, 823*b*
Cavities. *See* Dental caries
CCR5 coreceptor, 795, 796*f*
CD4 coreceptor, 712*f*, 713
CD4 receptor, 796*f*
CD4 T cells, 791, 792, 793–794*b*, 794, 795, 795*f*
CD8 coreceptor, 712, 712*f*, 714t
CDIs (*Clostridium difficile* infections), 363–364
Cech, Thomas R., 168*b*
Ceepryn, 220*f*
Cefaclor, 814*t*
Cefazolin, 434, 814*t*
Cefepime, 814*t*
Cefotaxime, 318, 837*b*
Ceftriaxone
  for chancroid, 419
  for gonorrhea, 415
  for hyperacute bacterial conjunctivitis, 442
  for meningitis, 318
  for neonatal meningitis, 319
  as third-generation cephalosporin, 814*t*
  for typhoid fever, 365
Cefuroxime, 814*t*
Cell(s). *See also* Eukaryotes; Prokaryotes
  as fundamental unit of life, 70
  Hooke's description of, 6
  organizational patterns, 25
    common, 71, 71*f*
    of prokaryotes vs. eukaryotes, 72–73
  ultrastructure visualization in, 93
Cell cycle, binary fission and, 135, 136*f*
Cell death. *See also* Programmed cell death
  HIV and, 793–794*b*
Cell envelope, 119*f*, 127*t*
  cell membrane
    archaeal, 121, 122*f*
    bacterial, 120–121, 120*f*, 122*f*
  cell walls
    archaeal, 118, 120
    bacterial, 116, 118, 119*f*, 120*t*
Cell lines, cultivation of, 473
Cell lysis, 118*f*. *See also* Lytic cycle; Osmotic lysis
  fluid mosaic model of cell membrane and, 120
  of pathogens, complement activation and, 686
  water balance and, 72
Cell membrane, 71, 71*f*
  archaeal, 121, 122*f*, 127*t*
  bacterial, 120–121, 120*f*, 122*f*, 127*t*
  energy metabolism by prokaryotic cells and, 72
  permeability, drug resistance and, 120–121, 121*f*, 828–829, 832*f*, 832*t*
Cell morphology, 87, 106, 107*f*
Cell surface receptors
  viral binding to hosts and, 459–460*b*
Cell theory, 70
Cell tropism, viral, 458, 458t, 459–460*b*
Cell walls
  of archaeal species, 127*t*
  of bacteria, 116, 119*f*, 120*t*, 127*t*
  of fungi, 49
    inhibitors, 822
  of prokaryotes and some eukaryotes, 72–73
  synthesis inhibitors
    antibacterial agents, 815*f*
    beta-lactam antibiotics, 812–814, 832*t*
Cell-mediated immune response, 699. *See also* Adaptive immunity
Cellular barriers to disease, 676, 690*f*. *See also* Indigenous microbiota
Cellular chemistry
  defined, 36
  pH and, 45–46, 46*f*
Cellular oncogenes (*c-onc*), 477–478, 478*f*, 479*f*
Cellular Origins Hypothesis, for viruses, 486
Cellular respiration
  aerobic, 171, 171*f*
  anaerobic, 171, 171*f*, 182–183
  batteries run by, 179*b*
  carbon cycle and, **893**
  energy metabolism and, 72
Cellulases, **905*t***
Cellulitis, 365, 436, 437
Cellulomonas, **879**
Cellulose, 49
Centers for Disease Control and Prevention (CDC)
  annual flu vaccine development and, 731*b*
  childhood immunization schedule, 732*f*
  on emerging viruses, 481
  Epidemic Intelligence Service, 628
  epidemiological tracking by, 15*b*, 630
  on foodborne illnesses, **850**
  Get Smart: Know When Antibiotics Work, 838*b*
  on irradiation of food, 207*b*
  Morbidity and Mortality Weekly Report, 445, 651
  notifiable diseases in 2012 to, 656*t*
  priorities on emerging and resurging diseases, 661, 663
  on sexually transmitted diseases, 410
Central America
  emerging and reemerging diseases in, 663*f*
  HIV infections in, 762*f*
Central dogma, 242, 242*f*
Central lines, 660. *See also* Catheters
Central nervous system, 799*b*
Centrifuges, 123
Centrosome, 72
Cépacol, 219
Cephalexin, 814*t*, 820*t*, 837*b*
Cephalosporins
  antibiotic properties of, 814, 814*t*, 815*f*, 820*t*
  antimicrobial spectrum of, 809*f*
  in farm animals, FDA on, 835*b*
  global market for, 819*f*
  for gonorrhea, 415
  resistance to, 827*f*, 832*t*
Cephalosporium acremonium, 814
Cephalothin, 820*t*
Cercariae, 612, 614*f*
Cerumen, 315–316
Cervical cancer, 477, 477*t*, 514–515, 515*f*, 795, 796*f*
Cervical lymphadenitis, 327
Cesium-137, for food irradiation, 206
Cesspools, **885–886**
Cestodes, 611–612
Cetylpyridinium chloride, 213t, 219, 220*f*
Chadwick, Edwin, **877–878**, **886**
Chagas disease, 603, 605, 610*t*
Chain, Ernst, 26, 27*b*, 808
Chain elongation, in protein synthesis, 245, 247*f*
Chain initiation, in protein synthesis, 245, 246*f*
Chain termination/release, in protein synthesis, 245, 247*f*
Chamberland, Charles, 18, 204
Chancre
  soft (chancroid), 418–419
  of syphilis, 417, 417*f*
  in trypanosomiasis, 603
Chancroid, 418–419, 419*t*
Chaperones, 245
Chapin, Charles, 21*t*
Chase, Martha, 25, 59*b*
Chatton, Edouard, 77, 129*b*
Chaudhuri, Swades, 179*b*

Cheddar cheese, **871**
Chédiak-Higashi syndrome, 788
Cheese. *See also* Milk
  L. monocytogenes contamination of, 364
  microbial ecology of, **872–873*b***
  production, 183, 184*f*, **857**, **871–872**, **871*f***. *See also* Whey
  Q fever and, 337
Chemical(s)
  DNA or RNA function and, 53
  food preservation with, **864**
  fumes or gases, allergies to, 768*f*
  immediate hypersensitivity to, 768
  water pollution and, **882**
Chemical barriers to disease, 675–676, 690*f*
Chemical bonds, 39–43, 43*t*
Chemical composition, food spoilage and, **851**, **852*f***, **853**
Chemical elements, 37, 37*f*, 56*f*
Chemical evolution, 57–58*b*
Chemical mutagens or carcinogens, 253–254, 254*f*, 476, 476*f*
Chemical reactions, 42–43, 164–165. *See also* Metabolism
Chemically defined culture media, 151
Chemiosmosis, 176–177, 176*f*, 177–178*b*, 187–188
Chemoautotrophs, 189, 189*f*
Chemoheterotrophs, 189*f*, 190*b*, 191, 555
Chemokines, 674, 678, 682
Chemotaxis, 113, 113*f*
Chemotherapeutic agents, 806–807. *See also* Drugs, therapeutic
Chemotherapeutic index, 808, 808*f*
Chesapeake Bay, fish handler's disease and pollution in, **883*b***
Chicken. *See also* Eggs; Poultry
  contamination of, 366, **855**
Chickenpox (varicella)
  infections, 508–509, 509*f*, 512*t*
  reported U.S. cases in 2010, 407*f*
  vaccine for, 733
Chigger mites, 398–399
Chikungunya (CHIK) fever, 532*b*
Chikungunya virus (CHIKV), 481*t*, 532*b*
Childbed fever, 13–14, 13*f*
Children
  acute otitis media in, 316
  bones of, tetracyclines and, 816
  CDC immunization schedule for, 732*f*
  cervical lymphadenitis in, 327
  fifth disease in, 513–514, 514*f*
  mortality due to infectious diseases in, 306
  N. meningitis group A and, 317
  paramyxovirus infections in, 502
  parental concerns on vaccines for, 739*b*
  pneumonia in, 332, 333*b*
  roseola in, 509
  scalded skin syndrome in, 434, 434*f*
  scarlet fever in, 312
Chitin, 49, 554–555, 561, 568*b*
Chlamydia and chlamydial diseases
  characteristics, 411, 413–414, 413*f*, 419*t*
  HIV transmission and, 797
  pharyngitis, 413
  reported U.S. cases, 407*f*, 411*f*
  sterility due to, 414*b*
  treatment for, 812
  urethritis, 411, 412*b*, 413
Chlamydia psittaci, **855**
Chlamydia trachomatis
  characteristics, 411, 413–414, 413*f*, 419*t*
  eye diseases due to, 443–444
  genome size of, 235*f*
  lymphogranuloma venereum due to, 419, 419*t*
  serotypes, 412*b*
  urethritis and, 421, 425*t*
Chlamydiae (phyla), 101*f*, 104
Chlamydial pneumonia, 338*t*
Chlamydomonas, 589*f*
Chlamydophila pneumoniae, 328, 337, 338*t*
Chlamydophila psittaci, 337, 338*t*
Chloramines, 213t, 217, 223*t*
Chloramphenicol
  antimicrobial spectrum of, 809*f*
  gene expression interference by, 246, 816*f*
  properties of, 815*f*, 817, 820*t*
  resistance testing, 826*f*
  resistance to, 827*f*, 832*t*, 837*b*
  for rickettsial infections, 396
Chlorhexidine, 218, 223*t*
Chlorination, **884–885**, **885*f***, **887**, **887*f***
Chlorine (Cl), 37*t*, 217, 217*f*, 223*t*
Chlorine dioxide, 221, 223*t*
Chlorophyll, 66
Chlorophyll *a*, 186–187
Chloroplasts
  characteristics, 72, 75*t*
  Chlamydiae and, 412*b*
  evolution of, 74*b*, 75*f*, 104, 147*b*
Chloroquine, 823, 823*b*, 825*t*
Chlortetracyclines, 816, 820*t*
CHNOPS, 37, 40
Choanoflagellates, 588*f*
Chocolates, soft-centered, **906**
Choke points, food contamination, **853**
Cholera. *See also Vibrio cholerae*
  blood type O and, 777*b*
  clinical characteristics of, 362*t*
  epidemics, 15*b*, 17, 333–334, 361*b*, **878**
  inflammatory gastroenteritis and, 360, 361*b*, 362
  Pasteur's research on, 18
  portal of entry for, 643
  sanitation and outbreaks of, 196, 196*f*
  shellfish contamination by, **854**
  Snow's work on, 14, 14*f*
  spread of, 361*b*
  vaccine for, 730
  as waterborne disease, 343–344
Cholera toxin, 360, 646*t*
Cholesterol, **911*b***
Chromalveolata (supergroup), 588*f*, 592–594, 594*t*
Chromobacter violaceum, 814
Chromosomes, 53, 71, 233, 237*f*, 237*t*. *See also* DNA
Chronic bacterial conjunctivitis, 442
Chronic bacterial prostatitis, 424
Chronic bronchitis, 328
Chronic diseases, 641
Chronic granulomatous disease, 788
Chronic inflammation, 683
Chronic obstructive pulmonary disorder (COPD), 311
Chronic otitis media (COM), 316
Chronic rejection, of allografts, 786
Chronic sinusitis, 315
Chronic wasting disease, 484
Chronic wounds, 431
Chroococcidiopsis, 257*b*
Churchill, Winston, 279*b*
Chytridiomycosis, 562*b*
Chytridiomycota (chytrids), 560, 561, 562*b*, 568*b*
Ciguatera fish poisoning, **883**
Cilastatin, 814
Cilia (cilium), 72, 309, 310*f*, 593
Ciliates, 588*f*, 593, 594*t*
Cinchona (tree), 823, 823*b*
Cinnamon, as antiseptic, 215*b*
Ciprofloxacin (Cipro), 382, 812, 826*f*, 837*b*
Cirrhosis, 527
Citric acid (citrate)
  in citric acid cycle, 174*f*
  as secondary metabolite, **904**, **904*f***
Citric acid cycle, 171*f*, 173, 174*f*, 175, 183, 412*b*
Cladophora, 186*b*
Clams, raw, vibriosis and, 365
Clarithromycin, 373, 817, 820*t*
Class, in taxonomic classification, 79
Classical Golden Age of Microbiology, 16–20, 21*t*, 22*f*
Classical pathway of complement activation, 686–687, 686*f*
Claviceps purpurea, 570, **855**, **906**
Clavulanic acid, 814
Climate change, 29–30, 661
Climax, disease, 640–641
Clindamycin
  for acne, 431
  for bacterial vaginosis, 410
  properties of, 817, 820*t*
  protein synthesis in protists and, 824
  resistance to, 832*t*, 837*b*
  for streptococcal skin diseases, 436
  for toxic shock syndrome, 434
Clinical diseases, 640
Clinical microbiology lab (CML), 80, 151, 306
Clinical trials, for drug development, 840–841*b*
Cloaca Maxima ("Great Sewer"), **886**
Clofazimine, 439
Clonal Anergy Theory, 782, 782*f*
Clonal deletion, 698
Clonal Deletion Theory, 782, 782*f*
Clonal selection, 700, 701*f*, 702
Clones and cloning vectors. *See also* DNA probes; Polymerase chain reactions
  for clonal selection, 700, 701*f*
  genetic engineering and, 280, 281, 282–283*b*, 284, 285*f*, 286
Clostridial food poisoning, 356, 374*t*, **865**
Clostridium spp.
  as anaerobic species, 146
  bacteriostatic control of, 208
  endospores of, 142, 142*f*
  fermentation by, 184*f*
  as Firmicutes, 101
  food spoilage and, **857**, **859**
  pectinase production and, **905**, **906*b***

in polluted water, **881**
wound healing in diabetics and, 432*b*
Clostridium acetobutylicum, 279*b*, **901**
Clostridium botulinum. *See also* Botulism
as anaerobe, **853**
endospore formation by, 141
food poisoning caused by, **849–850**
foodborne intoxications and, 357, 359
meat contamination by, **853–854**
neurotoxins, 646
nitrites and, **864**
types of, 359
Clostridium difficile
healthcare-associated infections and, 631, 646
infections, 363–364
infections, fecal transplants and, 349, 350*f*, 363
intestinal, clindamycin and, 817
multidrug-resistant, 805, 834
pili, entry into host and, 643
Clostridium difficile toxins A and B, 646*t*
Clostridium perfringens, 356, 384–386, 388*t*, 646*t*
Clostridium tetani, 383–384, 388*t*, 646, 646*t*
Clotrimazole, 573, 821, 825*t*
Club fungi, 565–566
Coagulase, 644
Coagulase test, 644
Cobalt-60, for food irradiation, 206
Cocci (coccus) shape, 106
Coccidioides spp., 745
Coccidioides immitis, 577, 578*f*, 579*b*, 580*t*
Coccidioides posadasii, 577, 578*f*, 580*t*
Coccidioidomycosis, 577, 578*f*, 579*b*, 580*t*, 796*f*
Cocoa, microbial fermentation and, **870**
Codons, 242, 243–245
Coenocytic hyphae, 555
Coenzyme A (CoA), in citric acid cycle, 173
Coenzymes, 166
Coevolution Hypothesis, for viruses, 486
Cofactors, 166
Coffee, microbial fermentation and, **870**
Coghill, Robert H., **912**
Cohen, Stanley, **919**
Cohn, Ferdinand, 21*t*
Cold environments. *See also* Refrigeration
allergic reactions and, 768
for food preservation, 208, **861–862**
microbial growth and, 144–145, 144*f*
microbial life and, 133–134, 134*t*
Cold sores, 505–507, 506*f*, 512*t*
Colds, common, 492–493, 494*b*, 496*t*, 500*b*, 676*b*
Coliform bacteria
canned foods and, **859**
as indicator organism, **889**, **890–891*b***
in milk, tests for, **861*t***
in polluted water, **879**, **880*f***
water quality testing, gene probe and PCR for, 287
Colitis
hemorrhagic, 368, 370
pseudomembranous, 363, 363*f*, 817
ulcerative, 613*b*
Collins, James, 466*b*
Colonies, bacterial, 17–18, 205, 205*f*
Colony-forming units (CFUs), 155, 392–393*b*
Colostrum, 705, 735–736, **857**
Comedones, 430, 430*f*
Commensalism, 632–633, 636–637, 636*f*
Commercial sterilization, **859**
Common colds, 492–493, 494*b*, 496*t*, 500*b*, 676*b*
Common source epidemics, 657*b*
Common warts, 477*t*, 514, 515*b*
Communicable diseases, 649
Community immunity, 737
Community-acquired pneumonia (CAP), 329–330, 332, 336, 338*t*
Community-associated MRSA (methicillin-resistant *S. aureus*, CA-MRSA), 829*b*, 830–831*b*, 833
Comparative genomics, 296–297, 297*f*
Competence, of a cell, 271
Competitive inhibition, 167*f*, 168, 811
Complement (complement system)
antigen-antibody complexes and, 710–711
cytotoxic hypersensitivity and, 774, 775, 775*f*
deficiencies, 788
immune complex hypersensitivity and, 776–777, 779*f*
innate immunity and, 686–687, 686*f*, 687*f*
transplantation rejection and, 784
Complement fixation test, 746–747, 749*f*
Complex culture media, 151, 151*t*
Complex viruses, 456, 457*f*
Compounds, 40
Comvax, 527
c-onc (cellular oncogenes), 477–478, 478*f*, 479*f*
Condensation, 43. *See also* Dehydration synthesis reactions
Condylomata, 514. *See also* Genital warts
Congenital immunity, 735. *See also* Naturally acquired passive immunity
Congenital rubella syndrome, 513
Congenital syphilis, 418
Congo-Crimea hemorrhagic fever, 533
Conidia (conidium), 557, 559*f*
Conidiophores, 557, 559*f*
Conjugate vaccines, 732
Conjugation
antibiotic resistance and, 828, 828*f*
in ciliates, 593–594, 593*f*
horizontal gene transfer and, 271–272, 272*f*, 273*f*, 274, 275*b*
Conjugation pili (pilus), 110, 272, 272*f*
Conjunctivitis
adenoviruses and, 493–494
bacterial, 442–444, 442*f*, 444*t*
leptospirosis and, 387, 387*b*
neonatal, 413, 415, 443
Consolidation, in lungs, 578
Consumption, 322
Contact bacterial diseases. *See also* Sexually transmitted diseases
exogenous bacteria and, 432
staphylococcal, 432–434, 433*f*, 434*f*, 435*b*
streptococcal, 434, 435*f*, 436–437, 436*f*
of the eye, 441–444, 442–444*f*, 444*t*
identifying cases of, 441*b*
indigenous microbiota and, 424
acne, 429–431, 429*f*
biofilms, 431
summary of, 440*t*
Contact dermatitis, 764*f*, 779*t*, 780–781, 780*f*
Contact lenses, 219, 598*b*, **888**
Contagious diseases, 490, 649
Contagium vivum fluidum, 453
Contamination, sterilization vs., 197, 197*f*
Contractile vacuoles, 592*f*, 593
Contrast, for bright-field microscopy, 87
Controls, in experiments, 10*b*
Convalescence, 640*f*, 641
Convalescent stage, of whooping cough, 321
Convex, use of term, 6
Cook, Captain, 305
Coombs test, 746
Copland, Herbert, 77
Copper (Cu), 37, 218, 223*t*
Copper sulfate, 218–219, 223*t*
Cordite, 279*b*. *See also* Gunpowder manufacturing
Co-receptors, for animal virus replication, 465
Corn, genetic engineering of, **913–914**
Corn steep liquor, **905**, **912**
Coronary artery disease (CAD), **911*b***
Coronary heart disease, 683
Coronaviridae (coronaviruses)
classification of, 461*f*
cytopathic effects of, 474*t*
as emerging viruses, 481*t*
infections, 493*f*
micrograph of, 504*f*
naming of, 460
Corticosteroids, 772, 786*t*. *See also* Steroids
Corynebacteria, 311
Corynebacterium spp., 420, 427, 430*f*, 432*b*
Corynebacterium diphtheriae. *See also* Diphtheria
as airborne bacterial disease, 312, 319*t*
club-shaped cells of, 106
diphtheria toxin, 646*t*
pathogenicity islands of, 296, 297*f*
specialized transduction and, 277
Coryneforms, 427
Co-trimoxazole, 578, 811–812
Cotton bollworms, **920**
Coughing. *See also* Pertussis
beneficial effects of, 309, 641*b*
Covalent bonds, 40, 41*f*, 42, 43*t*, 49, 52–53
Cowpox, 9
Cows. *See also* Cattle; Dairy products; Milk
contamination of carcasses, 356
as intermediate hosts, 615*b*
methane gas formation in, 40
methanogens in, 104
Q fever and, 337
Coxiella burnetii, 203–204, 336–337, 336*f*, 338*t*
Coxsackie virus, 538, **883**
Cranberry juice, urinary tract infections and, 423*b*
C-reactive protein (CRP), 687
Crenarchaeota, 104
Creosote remediation, **920*b***
Crescentin, 126
Creutzfeldt-Jakob disease, 484, 485
Crick, Francis H. C., 25, 53, 231, 233, 234*b*
Critical control points (CCPs), **868**
Crohn's disease, 349, 613*b*
Cromolyn sodium, 769, 772
Crop diseases
GMOs and, 285*f*
viroids and, 482
Croup, 492

Crusty lichens (crustose), 563, 565
Cryphonectria parasitica, 563
Cryptococcosis, 575–576, 580*t*, 796*f*
Cryptococcus gattii, 576, 580*t*
Cryptococcus neoformans, 575–576, 580*t*
Cryptoendoliths, 134*t*
Cryptomycota, 560–561
Cryptosporidiosis, 598–599, 601*t*, 796*f*
Cryptosporidium, **883**
Cryptosporidium hominis, 599, 601*t*
Cryptosporidium oocysts, 599*f*
Cryptosporidium parvum, 599, 601*t*
Crystal violet stain, 88, 88*f*
Culture media. *See also* Agar or agar plates
  differential, 153*b*
  great plate count anomaly, 156*b*
  Koch and introduction of agar for, 18, 19*b*
  measuring growth in, 155, 157–158*f*, 158
  mutant identification using, 260–261, 261*f*, 262*f*
  pure, 18, 155
  selective or differential, 151–152, 152*f*
  types of, 151, 151*t*, 152*t*
Curiosity (Mars rover), 134
Cutaneous anthrax, 382
Cutaneous diseases. *See* Skin diseases
Cyanobacteria
  carbon cycle and, **892*b***
  Chlamydiae and, 412*b*
  lichens and, 563
  nitrogen fixation and, **896**
  as photoautotrophs, 189
  photosynthesis by, 147*b*, 185*f*, 186, 187*f*
  on phylogenetic tree of life, 101*f*
  types of, 103–104
Cyanocobalamin, **905**
Cyclospora cayetanensis, 599, 601*t*
Cyclosporiasis, 599, 601*t*
Cyclosporin A, 786*t*
Cysteine, **855**, **894**
Cystic fibrosis, 68, 311, 816
Cysticercosis, 616*b*
Cystitis, 421–422, 425*t*
Cysts
  amoebal, 596–597, 597*f*, **854**
  fluke, 611
  Giardia, 592, 596–597
  hydatid, 614, 616*f*
  protistan, microbial control of, 200
  tapeworm, 612
  Toxoplasma gondii, 605, 606
  Trichinella spiralis, 617
Cytochromes, 175, 177–178*b*, 183
Cytokine storms, 683, 719, 720*b*
Cytokines, 309, 310*f*
  asthma and, 769
  immune response and, 673–674, 674*f*
  inflammation and, 682, 682*f*, 683, 687
  late allergic response and, 766
  natural killer cells and, 684
  phagocytosis and, 678
Cytolytic infections, 526
Cytomegalovirus (CMV), 458*t*, 477*t*, 789–790*b*, 795
Cytomegalovirus disease, 524–526, 528*t*
Cytopathic effects (CPEs), 473, 474*f*
Cytoplasm
  of bacterial and archaeal cells
    components of, 122
    inclusions, 124–125
    microcompartments, 123–124, 124*f*
    nucleoids, 122–123, 123*f*
    plasmids, 123
    ribosomes, 123, 124*f*
  of halophiles, 148
  sporulation and, 141
Cytosine (C), 52, 52*f*
Cytoskeleton
  in eukaryotes and prokaryotes, 72
  homologous, in bacterial and archaeal cells, 125–126, 127*f*, 128*t*
  without a cell wall, 126*b*
Cytosol, 71, 122, 175, 176, 176*f*, 177–178*b*
Cytotoxic (type II) hypersensitivity, 763, 764*f*, 774–776, 775*f*
Cytotoxic T cells (CTLs)
  antigen-presenting cells and, 714t
  Carabin as timer for, 718*b*
  cell-mediated immune response and, 700, 712–713, 714–715, 717*f*, 718
  transplantation rejection and, 784
  tumor cells and, 718, 718*f*
  type I hypersensitivity and, 763
Cytotoxins, 363, 645–646, 646*t*

D

D period, 135, 136*f*, 139
D value, 199*b*
Dairy products. *See also* Cheese; Milk
  cheese flavors for, **906**
  contamination of, 358, **855–857**
    aflatoxins, 570
    Shigella, 367
  fermentation and, 183, 184*f*
  genetic engineering and, 284
Daltons, 44*b*
Danovaro, Roberto, 272*b*
Dapsone, 439, 439*f*, 812
Daptomycin, 836–837*b*
Dark-field microscopy, 91*f*, 92, 95*t*
Dart, John, 598*b*
Darwin, Charles, 30–31
Daschle, Tom, 222*b*
Davies, Paul, 35
DDT (dichlorodiphenyltrichloroethane), 586, 587
De Bary, Anton, 20, 22
de Kruif, Paul, 306
Dead end hosts, 648
Deamination, 180
Debridement, tissue, 385
Decimal reduction time (DRT), 199*b*
Decline period (disease), 640*f*, 641
Decline phase (microbial growth), 139*f*, 140
Decomposers, 4*f*, 5, 24, 182*b*, 555, **893**
Deepwater Horizon oil spill, 41*b*, **916**
Deer mice, hantavirus and, 482, 504
Defensins, 309, 345, 409, 410*f*, 675–676
Definitive hosts, 594*f*, 602, 602*f*, 611
Degerm, definition of, 211
Degranulation, 765*f*, 766
Dehydration, 208, 570–571. *See also* Drying foods
Dehydration synthesis reactions
  described, 43
  DNA synthesis and, 233
  enzyme-substrate complexes and, 166
  of fatty acids, 50
  of nucleotides, 52–53, 52*f*
  of proteins, 54, 55*f*
Deinococcus radiodurans, 134*t*, 256, 257*b*, **918–919**
Delayed hypersensitivity, 763, 764*f*, 778, 779*t*
Delbrück, Max, 25
Dell'Anno, Antonio, 272*b*
Dementia, 484, 799*b*
Demographics, diseases and, 355, 658*b*, 661
Denaturation, 57, 198, 200, 218
Dendritic cells (DCs), 425, 670, 671*f*, 672, 706–707*b*
Dengue fever, 481, 481*t*, 531–532, 535*t*, 662*b*, 736*b*
Dengue fever virus, 531*f*
Dengue hemorrhagic fever (DHF), 532
Dengue shock syndrome, 532
Denmark, on antibiotics in feedlots, 835*b*
Dental caries (cavities), 49*b*, 350–352, 351*f*, 354*t*, **889**
Dental pellicle, 350, 353*f*
Dental plaque, 49*b*, 351, 353*f*
Deoxyribonucleic acid, 52–53, 233. *See also* DNA
Deoxyribose, 52, 233
Dermatophytes, 572–573
Dermatophytosis, 575*t*
Dermis, 425–426, 426*f*
Descriptive epidemiology, 657*b*
Desenex, 573
Desensitization therapy, 769*b*, 773
Desert fever, 577, 578*f*
Desulfovibrio, 183, **881**, **894**
Detergents, 219, 220*f*, 223*t*, **905*t***
Developing countries
  antibiotic resistance and, 834, 834*f*
  HIV infections in, 761
  vaccine development and, 741*b*
  viral diseases in, 819
  water purification in, **886*b***
Dextran, 49, 49*b*, 116
D'Herelle, Felix, 454, 466*b*
Diabetes. *See also* Insulin
  H. pylori infection and, 372*b*
  oral health and, 354*b*
  type I, 280, 782, 783, 783*t*
  wound healing and, 432*b*
Diacetyl, **871**
Diaminodiphenylsulfone, 812. *See also* Dapsone
Diaminopimelic acid (DAP), **905**
Diaminopyrimidine, 823
Diapedesis, 682
Diarrhea, 111–112*b*, 362–363, 538, 641*b*. *See also* Cholera; Dysentery; Gastroenteritis
Diatomaceous earth, 590
Diatoms, 588*f*, 590, **881**, **881*f***
Dichotomous keys, 80, 81*b*
Dicrocoelium dendriticum, 615*b*
Dideoxycytidine, 822*t*
Dideoxyinosine, 822*t*
Diener, Theodore O., 482
Diethylcarbamazine, 620*t*
Differential culture media, 151–152, 152*t*, 153*b*

Differential staining procedure, 88, 89*f*, 90–91, 90*b*
Diffusion
  carbon cycle and, **893*f***
  facilitated, 121, 121*f*
  for water balance in cels, 120
DiGeorge syndrome, 787–788
Digestive system, 345–346, 345*f*, 347*f*. *See also* Gastrointestinal (GI) tract
Dihydroxyacetone phosphate (DHAP), 172*f*
Dikarya (clade)
  Ascomycota, 561, 563, 563*f*, 564*b*, 565*f*
  Basidiomycota, 565–566, 566*f*
Dikaryotic cytoplasm, 558, 560*f*
Dimorphic fungi, 554
Dinesen, Isak, 407
Dinoflagellates, 588*f*, 589, **881**, **881*f***
Diphtheria. *See also Corynebacterium diphtheriae*
  as airborne bacterial disease, 312, 314, 314*f*, 319*t*
  exposure to, 736
  Schick test for immunity to, 744
  spread of, 305
  vaccine for, 739*t*
Diphtheria toxin, 646*t*
Diphtheria-pertussis-tetanus (DPT) vaccine, 730
Diphtheria-tetanus-acellular pertussis (DTaP) vaccine, 314, 321, 384, 647, 730, 734. *See also* Tdap
Diphtheria-tetanus-pertussis (DTP) vaccine, 321
Diphtheroids, 427
Dipicolinic acid, 141
Diplobacilli cell arrangement, 106, 107*f*
Diplococci cell arrangement, 106, 107*f*
Diplomonads, 588*f*, 591*f*, 592, 594*t*
Direct contact, for disease transmission, 649, 650*f*
Direct immunofluorescence antibody test, 747, 749*f*
Direct microscopic count, of bacterial growth, 155, 157*f*
Direct Observation Treatment System (DOTS), 327*b*
Directed evolution, 168*b*
Disaccharides, 48–49, 48*f*, 180
Disease(s). *See also* Host-microbe relationships; Infections or infectious diseases
  beneficial effects of, 641*b*
  genetic engineering and, **921**
  host resistance to, 669, 669*f*
  human, fungal phyla and, 569*t*
  microbes and, 4*f*, 5
  notifiable, to CDC in 2012, 656*t*
  pathogen contact and development of, 636–637, 636*f*
  stages of, 637, 640–641, 640*f*
  transmission of
    direct contact for, 649, 650*f*
    early theories of, 13–15, 13*f*, 14*f*
    host barriers and, 643, 643*f*, 644*f*
    indirect contact for, 649–650, 651*f*, 652*b*
    occurrence descriptions, 651
    pathogen entry and, 641–643, 642*f*
    portals of exit and, 647, 648*f*
    vaccination preventing, 9, 12–13
    virulence factors and, 643–647, 645*f*, 646*t*, 647*t*
  use of term, 631
  yeasts in research on, 569
Disinfectants, 209–210, 211*f*, 213–214, 213*t*, 214*f*, **889**
Disinfection, 209, 223*t*, 224*f*, **884–885**, **885*f***. *See also* Microbial control, chemical agents for
Disk diffusion method, 214, 214*f*, 826–827, 826*f*
Disseminated intravascular coagulation (DIC), 638–639*b*
Disseminated tuberculosis, 325
Distilled spirits, **911**
Disulfide bridges, 54, 55*f*
DNA. *See also* Genetics; Microbial genomics; Mutations
  ancient, viability of, 144*b*
  central dogma of gene expression and, 241–242, 242*f*
  characteristics, 52–53
  discovery of, 233, 234*b*
  evolution from viruses, hypothesis of, 471*b*
  Hershey-Chase experiment on, 59*b*
  replication
    lagging strand synthesis, 240–241, 240*f*
    leading strand synthesis, 240, 240*f*
    and segregation in bacteria, 135, 136*f*
    stages of, 238, 239*f*, 240
  RNA compared to, 241*t*
  sporulation and, 141
  synthesis inhibition, drug resistance and, 832*t*
DNA adenine methylase, 839
DNA chips, 290–291*b*
DNA double helix, 52*f*, 53, 233, 237*f*
DNA helicase, 240
DNA ligase, 240*f*, 241, 280, 281*f*
DNA microarrays, 287, 290–291*b*, 294–295
DNA polymerase, 240–241, 240*f*, 288*b*. *See also* Base pairs
DNA probes, 287, 289*f*, 290*b*
DNA tumor viruses, 477–478, 477*t*
DNA vaccines, 732, 733*f*
DNA viruses, 461, 461*f*, 462, 465–469
DNA world hypothesis, RNA world hypothesis vs., 471*b*
Dog tapeworms, 616*f*
Dogs
  leptospirosis and, 386
  tapeworms of, 613–614
  tinea infections and, 572–573
  tularemia and, 391
Domagk, Gerhard, 26, 807, 808
Domagk, Hildegard, 807
Donkin, Bryan, **859**
Dorsal root ganglia, 508
Double diffusion assay, 745
Double helix, DNA, 52*f*, 53, 233, 237*f*
Double pneumonia, 332
Double-stranded (ds) DNA, 461*f*, 462
  DNA tumor viruses and, 478
  replication of, 465–469, 468*f*
Doubling time, 135, 138. *See also* Generation time
Downey cells, 524
Doxycycline
  for chlamydial pneumonia, 337
  for ehrlichiosis and anaplasmosis, 401
  for leptospirosis, 388
  for leptospirosis and malaria prevention, 387*b*
  for Lyme disease, 395
  for neonatal conjunctivitis, 443
  for plague, 390
  properties of, 816–817, 820*t*
  protein synthesis in protists and, 824
  for relapsing fever, 396
  for rickettsial infections, 396
Drug companies, 741*b*, 829, 834
Drug resistance. *See also* Antibiotic resistance
  DNA transformation and, 271
  fluoroquinolones and, 812
  HIV monotherapy and, 797–798, 798*f*
  malaria and, 603
  Moringa oleifera and, **886*b***
  sulfonamides and, 811
  triclosan and, 218
Drug synergism, 812
Drug users, injecting
  AIDS and, 789–790*b*
  hepatitis B and, 526
  hepatitis C and, 527
  HIV transmission and, 797, 797*f*
  tetanus and, 385*b*
Drug-induced hemolytic anemia, 775
Drugs, therapeutic. *See also* Pharmaceuticals
  allergies to, 768*f*
  new, ribozymes and identification of, 168*b*
  yeasts in research on, 569
Dry skin sites, microbiota of, 428
Drying foods, 208, 209*t*, **862–863**, **863*b***
ds-DNA (double-stranded DNA), 461*f*, 462
  DNA tumor viruses and, 478
  replication of, 465–469, 468*f*
ds-RNA (double-stranded RNA), 461*f*, 462
Dubos, René, 664
Duchesne, Ernest, 808
Dumb rabies, 542
Dunaliella salina, 148
Duodenum, 345*f*, 597
Dust mite allergies, 766, 768*f*
Dust storms, disease transmission from, 652*b*
Dysbiosis, 346–347
Dysentery
  amoebic, 592, 596
  bacterial, 367
  spread of, 305
Dysport, 359–360
Dystonias, 359

E

E. coli. *See Escherichia coli*
EAEC (enteroaggregative *E. coli*), 370
Early disseminated stage, of Lyme disease, 394
Early localized stage, of Lyme disease, 394
Ears
  anatomy, 315*f*
  infections of, 315–316
Eastern equine encephalitis (EEE), 544, 545*f*, 547*t*
Eberth, Karl, 365
Ebola hemorrhagic fever (EHF), 533, 535*t*
Ebola virus, 460, 481*t*, 628
Eccles, Ron, 494*b*
Echinocandins, 822, 825*t*
Echinococcosis, 613–614
Echinococcus granulosus, 620*t*
Echinococcus spp., 613–614, 616*f*

Echoviruses, 538
Eclipse period, in virus replication, 475*b*
Eco-Challenge 2000, Borneo, 387*b*
Ecology
  fungi as tool for, 449, 556
  metagenomics and, 299
  microbial, 22, 30, 30*f*, 65, **872–873*b***
*Eco*RI, 280, 280*t*, 281*f*
Ecotypes, 65
Ectopic pregnancy, 413
Eczema, 436
Edam cheese, **871**
Edema, 682, 768
Efavirenz, 793–794*b*, 822*t*
Effector cells
  B lymphocyte, 698, 701*f*
  cytotoxic T cells, 700, 712–713, 714–715, 717*f*, 718
  helper T cells, 700, 712*f*, 713, 716*f*, 717*f*
Effector phase, of immune system, 672, 673
Efficacy, of antimicrobial agents, 808
Eflornithine, 603
Eggs
  allergies to, 766
  calcium gluconate and, **905**
  contamination of, 367, **854–855**
  pasteurization of, **860*b***
  for virus cultures, 454, 473, 474*f*, 731*b*
Eggs (parasitic)
  Ascaris lumbricoides, 618
  Echinococcus spp., 613–614
  fluke, 611
  hookworm, 618, 619*f*
  pinworm, 617
  Schistosoma spp., 612, 614*f*
  tapeworm, 612
Egypt
  ancient
    alcoholic beverage in, 818*b*, **907–908**, **908*b***
    medical practices in, 754*b*
    microbial control in, 209
  hepatitis C infections in, 528*b*
  living endospores in mummies from, 144*b*
EHEC (enterohemorrhagic *E. coli*), 368
Ehrenberg, Christian, 14
Ehrlich, Paul, 21*t*, 25–26, 702, 722, 806
Ehrlichia chaffeensis, 399, 401, 402*t*
Ehrlichiosis, 399, 401, 402*t*
Eisenhower, Dwight, **865**
El Tor serogroup, *V. cholerae* serogroup 01, 361*b*
Electrolytes, 360
Electron microscopes
  cellular organization studies using, 25
  development and types of, 93–95, 94*f*, 95*t*
  schematic of, 93*f*
Electron shells, 37*f*, 38–39
Electron transport chains, 175, 176*f*
Electrons, 37*f*, 38–39
  photosynthesis and, 185–188, 187*f*
Elementary bodies (EBs), 411, 413*f*
Elements
  chemical, 37, 37*f*, 56*f*
  transposable, 256, 258, 258*f*
Elephantiasis, 619*f*, 620
ELISA (enzyme-linked immunosorbent assay), 751–752, 753*f*, 754*b*, 772
Elongation, in DNA replication, 238, 239*f*, 240, 243*f*
EM (erythema migrans), 394, 395*f*
Emergency medical equipment, disease transmission on, 653*b*
Emerging diseases
  infectious, 28, 29*f*, 660–661, 663–664, 663*f*
  viral, 481–482, 481*t*, 483*b*
Emetic substances, 528*b*
Emtricitabine, 800
Enamel, tooth, 351
Encapsulated pathogens, 116
Encephalitis. *See also* Meningoencephalitis
  arboviral, 544, 544*f*, 546–547, 547*t*, 736
  chickenpox and, 508
  complement fixation test for, 747
  herpes, 507
  HHV-6 viremia and, 509
  rabies, 541
  tickborne, 755*b*
Encyclopedia of DNA Elements (ENCODE), 293
End products of fermentation, 183–184, 184*f*
Endemic diseases
  chikungunya fever, 532*b*
  defined, 651
  polio, 543*b*
  relapsing fever, 395–396, 397*b*
  typhus, 398–399
  yellow fever, 529
Endergonic reactions, 163, 164*t*
Enders, John, 454
Endocytosis, 466, 467*f*
Endoflagella, 115, 115*f*
Endogenous infections, 636
Endogenous viral elements (EVEs), 294
Endomembrane system, 72, 74*b*
Endomycorrhizae, 561
Endophthalmitis, 202*b*
Endophytes, fungal, 557, 558*b*, 563
Endoplasmic reticulum (ER), 72
Endoscopes, 526
Endosomes, 678
Endospores
  anthrax and, 382–383
  bacterial, 141–142, 142*f*, 143*f*
  chemical agent selection for disinfection and, 213
  coccidioidomycosis and, 577, 578*f*
  food poisoning and, 356
  irradiation and, 207*b*
  tetanus and, 385
Endosymbiont model, 74*b*, 75*f*
Endosymbiosis, 78, 101. *See also* Symbiosis
Endothelial cells, 533
Endotoxic shock, 647
Endotoxins, 118, 647, 647*t*, 683
Energy
  and enzymes in metabolism, 163–170
  metagenomics and, 299
Energy-fixing reactions, 185–188, 187*f*
Enfuvirtide, 793–794*b*, 821, 822*t*
Engelmann, Theodor W., 186*b*
Engerix-B vaccine, 286, 731
Enriched culture media, 152, 152*t*
Entamoeba, 592
Entamoeba histolytica, 596, 597*f*, 601*t*, **883**
Enteric bacterial species, 82*b*, 183–184, 184*f*
Enteroaggregative *E. coli* (EAEC), 370
Enterobacter spp.
  as ESKAPE pathogen, 839
  fermentation by, 184*f*, **869**
  food spoilage and, **857**
  NDM-1 strain, 842
  nosocomial infections and, 659*t*
  in polluted water, **879**
  septicemia due to, 638*b*
  traumatic wound infections and, 438*t*
Enterobacter aerogenes, **857**, **870**, **905**
Enterobacter cloacae, 117*b*
Enterobius vermicularis, 614, 620*t*
Enterococcus spp., 425*t*, 432*b*, 438*t*, 638*b*
Enterococcus faecalis, **873*b***
  vancomycin-resistant, 275*b*, 815
Enterococcus faecium, 839
  vancomycin-resistant, 815
Enterocytes, 538
Enterohemorrhagic *E. coli* (EHEC), 368
Enteropathogenic *E. coli* (EPEC), 363
Enterotoxigenic *E. coli* (ETEC), 362–363
Enterotoxin A, 646*t*
Enterotoxins, 358, 646–647, 646*t*, **851**, **862**
Enterotube II, 82*b*
Enterotypes, 347, 349
Enterovirus diseases or infections, 493, 493*f*, 538, 539*t*, 789
Envelope, viral, 456, 456*f*, 461*f*, 463*f*, 467–468
Enveloped viruses, 456, 456*f*, 461*f*, 466, 467*f*
Environment, metagenomics studies of, 299
Environmental health officers
  preparation for careers as, **848**
Environmental microbiology
  bacteriological water analysis, **889–890**, **889*f***, **890–891*b***
  in cycles of nature, **891–892**
    carbon cycle, **892–894**, **893*f***
    nitrogen cycle, **894**, **895*f***, **896–897**
    sulfur cycle, **894**, **894*f***
  dust storms transmitting disease, 652*b*
  genetic engineering and, 284
  Great Sanitary Movement and, **877–878**
  microbial populations in water, **879**, **880*f***, **881**
  sewage treatment, **885–888**, **886*b***, **887*f***
  water pollution
    types of, **881–882**
  waterborne diseases, **882–883**, **882*f***, **883*b***
Enzootic diseases, 382. *See also* Zoonoses
Enzyme immunoassay (EIA), 751–752
Enzyme-linked immunosorbent assay (ELISA), 751–752, 753*f*, 754*b*, 772
Enzymes
  Buchner brothers' discovery of, 164*b*
  catalytic effects of RNA on, 168, 168*b*
  chemical reactions in cells and, 164–165
  and energy in metabolism, 163–170
  environmental inhibition of, 167
  enzyme-substrate complexes, 165–166
  enzyme-substrate complexes and, 165*f*
  functional groups and, 47
  inactivating antibiotics, 828, 832*f*
  industrial production of, **905–906**, **905*t***
  metabolic pathways and, 166–167
  metabolic pathways and inhibition of, 167–168, 167*f*

proteins as, 53
virulence and, 644–645, 645*f*
Enzyme-substrate complexes, 165–166, 165*f*
Eosin (dye), 670
Eosinophils, 670, 671, 671*f*, 763, 766, 769
EPEC (enteropathogenic *E. coli*), 363
Epidemic curve (epi curve), 657*b*
Epidemic Intelligence Service (EIS), 628
Epidemic typhus, 398, 399*b*, 402*t*, 817
Epidemics. *See also* Pandemics; *specific diseases*
characterizing, 657–658*b*
cholera, 15*b*, 17, 343–344, 361*b*, **878**
defined, 490, 651
international travel and, 492
SARS, 503*b*, 720*b*
Epidemiology
of cholera epidemic in Haiti, 15*b*
disease occurrence descriptions for, 651
infectious disease, 13–14, 13*f*, 14*f*
elements in, 648–649, 648*f*
terminology for, 649
transmission methods, 649–650, 650–651*f*
investigations, 657–658*b*
preparation for careers in, 628
of West Nile virus outbreak, 629–630
Epidermis, 424–425, 426*f*
Epidermophyton spp., 572, 573, 575*t*
Epiglottitis, 314, 319*t*
Epinephrine, 768
Epithelial cells, 309, 310*f*
Epitopes (antigenic determinants)
in adaptive immunity, 698, 701*f*, 702
antigen-binding sites and, 704
antigens and, 697, 697*f*
Epstein-Barr virus (EBV)
cell/tissue tropism, 458*t*
infections, 524, 528*t*
naming of, 460
as oncogenic agent, 476, 477*t*
other Herpesviridae and, 506*f*
Ergosterol, 50*f*, 51, 821
Ergot, 570
Ergotism, 570, 571*b*, **855**
Ermengem, Emile van, 21*t*
Ertapenem, 259*b*
Erysipelas, 436, 436*f*, 440*t*
Erysipeloid infections, **882**
Erysipelothrix rhusiopathiae, **882**
Erythema, 418, 778
Erythema infectiosum (fifth disease), 513–514, 514*f*, 518*t*
Erythema migrans (EM), 394, 395*f*
Erythrogenic toxin, 646*t*
Erythromycin
for acne, 431
for campylobacteriosis, 370
for chancroid, 419
for chlamydial infections, 413
for chlamydial pneumonia, 337
for diphtheria, 314
gene expression interference by, 246
isolation of, 817
for neonatal conjunctivitis, 443
for pertussis, 321
for primary atypical pneumonia, 332
properties of, 820*t*
resistance to, 837*b*
Escherich, Theodore, 21*t*, 79*b*
Escherichia spp., 184*f*, 438*t*
Escherichia coli
antibacterial soaps and, 220*b*
antibiotic-resistant, 828*f*
associative learning by, 66
ATP synthesis and, 177–178*b*
capsule of, 115*f*
cell shape in, 125
cells, 362*f*
conjugation between other genera and, 274
conjugation in, 271–272, 272*f*
coupled transcription and translation in, 248*f*
diarrhea and, 111–112*b*, 362–363
dividing, 26*f*
DNA probes and PCR in detection of, 287
DNA replication in, 239*f*
DNA studies using, 233
DNA synthesis in, 240
enterotoxin, 646, 646*t*
filtration of, 205*f*
food spoilage and, 356, **857**
generation time in, 137*b*, 138, 138*t*
genetic engineering using, 282–283*b*, **919**
genome of, 123, 235*f*, 291
as indigenous microbiota, 632
K1 strain, 319
light micrograph of, 103*f*
localization sequences in mRNA of, 251
lysine production and, **905**
macrophage attaching to, 680*f*
as molecular biology model system, 25
mutation repair mechanisms, 255–256
naming of, 79*b*
NDM-1 strain, 842
nosocomial infections and, 659*t*
nucleoids in, 123*f*
O104:H4, 370, **850**
O157:H7
foodborne disease and, 207*b*
infections in U.S., **850**, **850*f***
outbreak of, 369*b*
pathogenicity islands of, 297, 297*f*
in raw milk, **867**
as Shiga toxin-producing, 368, 370
operon model of protein synthesis in, 250*b*
pasteurization and, 204
phage T4 replication in, 463
with pili, 110*f*
in polluted water, **879**
restriction enzymes, 280, 280*t*
septicemia due to, 638*b*
supercoiled DNA of, 234, 237, 237*f*
taxonomic classification of, 80*t*
tetracycline-resistant, 828–829
urinary tract infections and, 421, 421*f*, 422*b*, 424, 425*t*
waterborne diseases and, **882**
ESKAPE pathogens, 839
Ester linkages, of archaeal cell membrane lipid tails, 121, 122*f*
ETEC (enterotoxigenic *E. coli*), 362–363
Etest, 826, 826*f*
Ethambutol, 812, 818–819
Ether linkages, of archaeal cell membrane lipid tails, 121, 122*f*
Ethics
clinical trials, for drug development and, 841*b*
Ethyl alcohol (ethanol)
distilled spirits production and, **911**
fermentation and, 184*f*, **869**, **901**
microbial control using, 219, 223*t*
phenol coefficient, 213t
wine production and, **908**
Ethylene oxide, 221, 223*t*
Etiology, use of term, 355
Eucalyptus tree, mycorrhizae and, 558*b*
Euglena, **879**
Euglenozoa, 588*f*, 592
Euhaplorchis californiensis, 615*b*
Eukarya (domain), 77, 78. *See also* Animals; Fungi; Plants
Bacteria and Archaea differences from, 105*b*
chromosome characteristics, 237*t*
evolution of, 101
on phylogenetic tree of life, 78*f*, 101*f*
phylogeny for, 588*f*
in three-domain system, 129*b*
Eukaryotes (eukaryotic cells). *See also* Cell(s); Parasites
cellular organization of, 25, 26*f*
DNA replication in, 238
evolution, 74*b*, 75*f*
genome size among, 233
mitochondria, chloroplasts, bacteria and, 75*t*
organizational patterns
of prokaryotes and, 71, 71*f*
of prokaryotes vs., 72–73, 73*t*
RNA transcription in, 242–243, 244*f*
S. cerevisiae as model organism of, 569
temperature and growth of, 144*f*, 145
viral genomes in lungs, 311
Europe
emerging and reemerging diseases in, 663*f*
HIV infections in, 761, 762*f*
Euryarchaeota, 104
Eustachian tubes, 315, 315*f*
EV68 (enterovirus), 493
Evolution
of antibiotic resistant *S. aureus*, 805*f*
appearance of life on Earth and, 231–232, 232*f*
bacteria and, 102*b*
blood types and, 777*b*
chemical, 57–58*b*
of Chlamydiae, 412*b*
cyanobacteria and, 147*b*
directed, 168*b*
emerging viruses and, 482
of fungi, 568*b*
of HIV, 791*b*
of influenza viruses, 498–499*b*
microbial, 30–31
of pathogens or pathogenicity, 296, 661
RNA world hypothesis vs. DNA world hypothesis of, 471*b*
of viruses, hypotheses for, 486
Yersinia spp., 253*b*
Excavata (supergroup), 588*f*, 591–592, 594*t*
Excision repairs. *See* Nucleotides, excision repair by
Exercise-induced allergies, 768, 770*f*
Exergonic reactions, 164, 164*t*

Exogenous infections, 636
Exons, 243
Exotoxins, 385, 645–647, 646*t*, 647*t*
Experiments, scientific, 7, 9, 10*b*, 11*f*
Extensively drug-resistant tuberculosis (XDR-TB), 326
Extreme acidophiles, 148
Extreme halophiles, 104, 105*f*, 148
Extremely drug-resistant TB (XXDR-TB), 326
Extremophiles
  Archaea as, 104
  cold environments and, 144, 144*f*
  environmental conditions for, 99–100, 133–134
  examples of, 134*t*
  in freshwater lakes beneath Antarctica, 149*b*
  study of, 36, 36*f*
Extrinsic factors, in spoilage and, **852*f***, **853**
Exxon Valdez oil spill, **916**, **918*f***
Eyes
  anatomy, 442*f*
  bacterial infections of, 441–444

F

F factors (plasmids), 237, 272, 274
Facilitated diffusion, 121, 121*f*
Facultative microbes, 146
FAD (flavin adenine dinucleotide), 166
Fairy rings, 566, 566*f*
Family/families, in taxonomic classification, 79
Farmer's lung, 778
Fascia, 436
Fasciola hepatica, 611*f*
Fastidious organisms, 152
Fats, metabolism of, 180–182, 181*f*
Fauci, Anthony, 736*b*
Fc fragment, 704, 704*f*, 764, 765
Febrile, use of term, 493
Fecal transplants, 349, 350*f*, 363
Fecal–oral route, 641
Feedback inhibition, in metabolic pathways, 167–168, 167*f*
Female reproductive system, 408–409, 409*f*, 410*f*, 632*f*
Female urinary tract, 420, 420*f*
Fermentation. *See also* Alcoholic beverages
  as anaerobic metabolism, 171
  antibiotics production and, **912–913**
  of *Candida albicans*, 162–163
  as chemical process, 164*b*
  discovery of, **901**
  energy production by, 183–184, 184*f*
  foods produced by, **869–870**, **869*f***
  glycolysis and, 171*f*
  industrial, **901**, **902**, **902*f***, **903*f***
  maltose as byproduct of, 49
  Pasteur on yeast cells and, 16, 16*f*
Fermentors, **902**, **902*f***
Ferroplasma acidarmanus, 134*t*
Fever, 683–684, 684*b*. *See also specific types of fever*
Fever blisters, 506–507, 506*f*
Fever tree, 823*b*
Fibrin clots, 644, 682*f*, 683
Fidaxomicin, 363
Field, of microscope, 87
Fifth disease (erythema infectiosum), 513–514, 514*f*, 518*t*
50S ribosomal subunit, 123, 124*f*, 246–247*f*
  antibiotics and, 816, 816*f*
Filariform larvae, of hookworms, 618, 619*f*
Filobasidiella neoformans, 575–576
Filopodia, 678–679, 680*f*
Filoviridae (filoviruses), 461*f*, 481*t*, 533
Filters and filtration
  in antibiotics production, **912**
  microbial control using, 204–206, 205*f*, 209*t*
  virus research using, 453–454, 453*f*
  in water purification, **884**, **885*f***
Fimbriae (fimbria), 110. *See also* Pili
Finlay, Carlos, 530*b*
Fire, Andrew Z., 248*b*
Firmicutes, 101, 311, 312*f*, 314, 426–427. *See also specific genera*
Fish. *See also* Seafood; Shellfish
  allergies to, 766
  disease and, **883**
Fish handler's disease, **883*b***
Flaccid paralysis, 359
Flagella (flagellum)
  cell motility using, 72, 110, 112*f*, 113, 115, 127*t*
  of chytrids, 561, 568*b*
  of diplomonads, 591*f*, 592
  origin of, 114*b*
  sticky, pathogen entry into host and, 642
Flame, direct, sterilization using, 198, 200*f*
Flares, 766, 772
Flash pasteurization method, 204, **860–861**
Flatworms, 611–612. *See also* Helminths
Flavin adenine dinucleotide (FAD), 166, 174*f*, 175
Flaviviridae (flaviviruses), 461*f*, 477*t*, 481*t*, 527
Flavorings, for food, **906**
Flax retting, for linen production, **905**, **905*t***, **906*b***
Fleas, disease transmitted by
  murine typhus, 398
  plague, 253*b*, 389
Fleming, Alexander, 26, 27*f*, 807–808, 807*f*
Flies
  sand fly, 595, 595*f*, 596
  tsetse, 603, 604*f*
Flocculation, **884**, **885*f***
Flocs, **884**
Flonase, 772
Floppy baby syndrome, 359
Florey, Howard, 26, 27*b*, 808
Flu. *See* Influenza
Fluconazole, 822, 825*t*
Flucytosine, 822, 825*t*
Fluid mosaic model, 120, 120*f*
Flukes
  characteristics, 611, 611*f*
  intermediate hosts for, 615*b*
  lancet, 615*b*
  life cycle of, 614*f*
  schistosomiasis due to, 612
Flumadine, 822*t*
Fluorescence microscopy, 91*f*, 92–93, 95*t*
Fluorescent antibody technique, 92–93, 92*f*
Fluoridated water, **885**
Fluoroquinolones, 326, 812, 819*f*, 820*t*, 832*t*, 835*b*
Flush, in mushroom farming, **915**
Focus of infection, 637
Folic acid, 811
Foliose lichens, 563, 565
Folliculitis, 433
Fomites, 572–573, 649, 651*f*
Food. *See also* Foodborne diseases
  allergies to, 764*f*, 766, 768, 768*f*, 769*b*
  commercial, sterilizing, 199*b*
  contamination of, 355–356, 661
    bacteria and, 23*f*, 24
    breads and bakery products, **855**
    conditions for, **851**, **852*f***, **853**
    fermentation and, **869**
    grains, **855**
    meat and seafood, **853–854**
    microbial products and, **853**
    microbial sources for, **851**
    milk and dairy products, **855–857**
    poultry and eggs, **854–855**
    spices and, **851**
    stone crabs' avoidance of, 182*b*
  as disease reservoir, 649, 651*f*
  fermented, **869–872**, **869–872*f***, **872–873*b***
  fevers and, 684*b*
  flavoring ingredients for, **906**
  genetically modified, 284, 285*f*
  microbial genomics and production of, 294–295
  poisoning. *See* Food poisoning
  preservation
    chemical, **864**
    drying, **862–863**, **863*b***
    eggs for Caesar salad, **860*b***
    heat effects on, **859–861**, **859*f***
    history of, **858–859**
    irradiation, 206, 207*b*, 208, **864–865**, **865*t***
    low temperatures and, **861–862**
    for Napoleon's army, **858*b***
    osmotic pressure, **863–864**
    physical methods for, 208
  safety, quiz on, **867*b***
  safety monitoring, HACCP and, **868**
  transgenic plants for, 286
Food and Drug Administration (FDA)
  annual flu vaccine development and, 731*b*
  on antibiotics in feedlots, 835*b*
  clinical trials, for drug development and, 840–841*b*
  FoodNet and, **850**
  on genetic engineering, 284, 284*f*
  on honey as antiseptic, 215*b*
  on irradiation of food, 206, 207*b*
Food inspectors, preparation for careers as, **848**
Food poisoning
  bacterial intoxications and, 357–360, 357*f*, 358*f*
  C. botulinum causing, **849–850**
  C. jejuni causing, 145*b*
  fungal intoxications, 570–571, 571*b*, 572*f*
  specialized transduction and, 277
Food vacuoles, of pseudopods, 592
Food web, 4*f*
Foodborne diseases
  bacterial cause for, 355–356
  bacterial inflammatory gastroenteritis
    C. difficile infections, 363–364, 363*f*
    cholera, 360, 361*b*, 362, 362*f*, 363*t*
    E. coli diarrhea, 362
    enteropathogenic *E. coli*, 363

enterotoxigenic *E. coli*, 362–363
intestinal anthrax, 365
L. monocytogenes, 364, 364*f*
vibriosis, 364–365
bacterial intoxications, 51*b*, 357–360, 357*f*, 358*f*
bacterial invasive gastroenteritis
campylobacteriosis, 370–371
hemorrhagic colitis, 368, 369*b*, 370
salmonellosis, 366–367
shigellosis, 367–368
typhoid fever, 365–366, 366*f*, 367*b*
yersiniosis, 371
FoodNet and, **850**
fungal intoxications, 570–571, 571*b*, 572*f*
imported foods and, ***866b***
incidence in U.S., in 2010, 356*f*
as infections or intoxications, **865–868**
microbial forensics and cases involving, 295
parasitic
amoebiasis, 596–597, 597*f*, 598*b*
cyclosporiasis, 599
giardiasis, 597–598, 598*f*
viral
hepatitis A virus, 535–536, 536*f*, 537*b*
norovirus infections, 538
rotavirus infections, 536, 538
Foodborne Diseases Active Surveillance Network (FoodNet), **850**
Foot odor, 427*b*
Foot-and-mouth disease, 198, 453
Foraminiferans (forams), 588*f*, 589–590, 589*f*, **881**
Forensics, microbial, 295
Formaldehyde, 223*t*, 646–647, 735*b*
Formalin, 213t, 221, 223*t*, 730
Formylmethionine (fMet), 245, 245*t*
Fortified wines, **910**
Fosfomycin, 836–837*b*
Fossil microbes, 232*f*
Fox, George, 77
Fractional sterilization. *See* Intermittent sterilization
Francisella tularensis, 390–391, 402*t*
Franklin, Rosalind, 53, 233, 234*b*
Fraser, Claire, 292*b*
Free radicals, 824
Freeze-drying food, **863**, **863*b***, **863*f***
Freezers
food preservation using, 208, 209*t*, **861–862**
microbial growth and, **853**
Fried-egg colonies, of *M. pneumoniae*, 332, 334*f*
Frogs, fungal disease and, 562*b*, 580
Fructose, 166, 166*f*
Fructose-1,6-bisphosphate, 172*f*, 173
Fructose-6-phosphate, 172*f*
Fruit, contamination of, 356
Fruiting bodies
fungal, 557, 559*f*, 565*f*
nutrient starvation and, 69*b*
Fruticose lichens, 563, 565
Fuels, metabolic engineering of, 296
Functional genomics, 296
Functional groups, 47, 47*t*
Fung, Daniel Y. C., 215*b*
Fungemia, 637
Fungi
brief survey of, 23*f*, 24
cell walls of, 49
characteristics, 554–555
classification of
Ascomycota, 561, 563, 563*f*
Basidiomycota, 565–566, 566*f*
Chytridiomycota, 561
Cryptomycota, 560–561
Glomeromycota, 561
as kingdom, 77, 554
Microsporidia, 561
Zygomycota, 561, 562*f*
disease and drug research using, 569*t*
in dust storms, 652*b*
early research on, 20, 22
as eukaryotic organisms, 25, 588*f*
evolution of, 568*b*
growth of, 555–557, 559*f*, 560*f*
HIV/AIDS and, 796*f*
intoxications by, 570–571, 571*b*, 572*f*
lower respiratory diseases caused by, 575–580
mitosporic, 566–567
as pesticides, **915**
plant protection by, 558*b*
plants compared to, 554
psychrotolerant, 145
reproduction of, 557–558
skin diseases caused by, 572–575
study of, 449
white-nose syndrome and, 564*b*
yeasts, 567, 569
Fungicidal agents, 197
Fungistatic agents, 197
Furious rabies, 542
Furuncles, 433, 440*t*
Fusion inhibitors, 797
Fusobacteria, 312*f*
Fusobacterium, 353, 354*t*
Fuzeon, 822*t*

G

Gaffky, Georg, 21*t*, 365
β-Galactosidase, 250*b*, 282*b*
X-Galactosidase, 282–283*b*
Galen, 715*b*
Gallo, Robert, 695–696, 789
Gambierdiscus toxicus, **883**
Gametocytes, *Plasmodium*, 594, 594*f*
Gamma globulin, 736
Gamma rays
bacteria resistant to, 257*b*
destroying microorganisms with, 206
for irradiation of food, 207*b*, 209*t*, **864–865**
wavelength of, 206*f*
Ganglia, 506
Gangrene, 384–386. *See also* Gas gangrene
Gardasil, 286, 477, 515
Gardnerella vaginalis, 409, 410*f*
Garlic, antimicrobial chemicals in, 216*b*
Gas gangrene, 142, 384–386, 386*f*, 388*t*, 437
Gas vacuoles, 124, 128*t*
Gases
laboratory creation of life and, 57–58*b*
sterilizing, 221
GasPak system, 146, 146*f*
Gastric cancer, 371, 373
Gastric ulcer disease, 371, 372*b*, 373, 373*f*
Gastroenteritis
inflammatory, 360–365, 361*b*, 362–364*f*, 363*t*, 374*t*
invasive, 360
noninflammatory, 357–360, 374*t*
parasitic, 596–597
rotavirus, 536
types of, 355
viral, 536, 538, 539*t*
waterborne, **883**
Gastrointestinal (GI) tract. *See also* Digestive system
apple-eating and, 348–349*b*
bacterial diseases
anthrax, 382
foodborne intoxications, 357–360, 357*f*, 358*f*
gastric ulcer disease, 371, 372*b*, 373, 373*f*
inflammatory gastroenteritis, 360–365, 361*b*, 362–364*f*, 363*t*
intoxications or infections and, 355
invasive gastroenteritis, 365–371, 367*b*
summary of, 374*t*
chemical barriers to disease in, 675
digestive system and, 345–346, 345*f*
fecal transplants and, 349, 350*f*
fungal intoxications, 570–571, 571*b*, 572*f*
indigenous microbiota of, 102*b*, 346–347, 346*b*, 347*f*, 632*f*, 634*b*
mucosal immunity in, 706–707*b*
as portal of disease entry, 641
protist parasites
amoebiasis, 596–597, 597*f*, 598*b*, **865**
cryptosporidiosis, 598–599, 796*f*
cyclosporiasis, 599
giardiasis, 407*f*, 597–598, **865**
summary of, 601*t*
viral diseases
hepatitis A virus, 535–536, 536*f*, 537*b*
hepatitis E, 536
summary of, 539*t*
viral gastroenteritis, 536, 538
Gates (Bill and Melinda) Foundation, 604*b*, 663–664, 726, 741*b*
GAVI Alliance, 332
Geison, Gerald, 20*b*
Gel electrophoresis, 745, 746*f*
Gelsinger, Jesse, 480*b*
Gene cloning. *See* Genetic engineering
Gene expression
antibiotic interference with, 246
central dogma of, 241–242, 242*f*
concept map for, 252*f*
control mechanisms, 246, 249, 249*f*
operon model of, 250*b*
DNA microarrays for study of, 290*b*
localized in nucleoid, 251, 251*f*
transcription during, 242–245, 243*f*, 244*f*
translation during, 245–246, 246–247*f*
Gene swapping. *See* Genetic recombination
Gene therapy, 788, 788*f*
Genera (genus), 78–79
Generalized transduction, 274, 276*f*, 277, 277*b*
Generation time, 135, 137*b*, 138, **903**
Genes
DNA in, 53, 233
regulatory, 249, 249*f*
structural, 249, 249*f*

Genetic abnormalities. *See also* Mutations
  genetic engineering and, **921**
  immune system, 688*b*
Genetic code. *See also* Genetic recombination; Genetics
  central dogma on, 241–242, 242*f*
  three-letter words in, 243–245, 245*t*
Genetic engineering
  antibiotics production and, 266–267
  applications of, 281, 282–283*b*, 284, 285*f*, 286–287
  defined, 267
  DNA probes, 287, 289*f*, 290*b*
  genetic recombination and, 279–281
  industrial, **919–921**
  of insecticides, **913**
  microbial forensics and, 295
  polymerase chain reactions, 288*b*
  virotherapy and, 480*b*
Genetic recombination
  emerging viruses and, 481
  genetic engineering and, 279–281
  horizontal gene transfer
    concept map of, 278*f*
    conjugation, 271–272, 272*f*, 273*f*, 274, 275*b*
    marine snow, 272*b*
    transduction, 274, 276*f*, 277
    transformation, 268, 269–270*b*, 271, 271*f*
    in world's oceans, 277*b*
  influenza virus evolution and, 498–499*b*
  vertical gene transfer, 268, 269*f*
Genetically modified organisms (GMOs), 267, 285*f*
Genetics
  DNA replication, 238, 239*f*, 240–241
  gene expression. *See* Gene expression
  mutations. *See* Mutations
  RNA compared to DNA, 241*t*
  study of, 233
Gengou, Octave, 21*t*
Genital herpes, 507, 507*f*, 512*t*
Genital warts, 477, 477*t*, 514, 515*b*, 515*f*, 518*t*
Genome annotation, 296
Genome reduction, 233, 235–236*b*
Genome(s)
  defined, 123, 233
  human, potential microbial ancestors for, 293–294
  size of, mutation rate vs., 462, 462*f*
  transplantation of, 299–300*b*
  of viruses, 455, 460–462, 461*f*
Genomic islands, 296–297
Genomics. *See also* Microbial genomics
  of *Chlamydia trachomatis* serotypes, 412*b*
  defined, 267
Gentamicin, 809*f*, 816, 820*t*, 826*f*, 837*b*
Gentian violet, 574
Geobacillus stearothermophilus, 203, 203*f*
Geographic distribution of disease, 657–658*b*
Geomyces destructans, 564*b*
George, Lloyd, 279*b*
Germ theory of disease. *See also* Koch, Robert; Pasteur, Louis
  disease control and, 17, **847**
  early research on, 16
  Gordon's experiments on, 334*b*
German measles. *See* Rubella
Germicides, 197
Germs, Pasteur's naming of, 16
GI tract. *See* Gastrointestinal tract
Giardia intestinalis, 591*f*, 592, 597–598, 598*f*, 600*b*, 601*t*
Giardia lamblia, 597, **883**
Giardiasis, 407*f*, 597–598, 601*t*, **865**
Giese, Jeanne, 540*b*
Gingivitis, 352–353, 353*f*, 354*t*
Giovannoni, Stephen, 5
Giruses (giant viruses), 457*b*
Glandular fever, 524
Glandular tularemia, 391
Gliding motility, 110
Global Alliance for Rabies Control (GARC), 523
Global Alliance for Vaccines and Immunization, 332, 664, 741*b*
Global Fund to Fight AIDS, Tuberculosis, and Malaria, 664
Global Influenza Surveillance Network, 731*b*
Global Polio Eradication Initiative, 543*b*
Global Viral Forecasting Initiative (GVFI), 519
Global warming. *See* Climate change
Glomeromycota, 560, 561
Glomerulonephritis, 778
Glomerulus, 370
Gluconic acid (gluconate), **904–905**
Gluconobacter, **905**
Glucose
  catabolism, 170–171, 171–172*f*, 173, 174–176*f*, 175–177, 178–179*f*, 180
  fungal growth and, 556
  hydrolysis of sucrose and, 166, 166*f*
  molecular formula of, 40
  as monosaccharide, 48, 48*f*
Glucose-6-phosphate, 172*f*, 173
Glutamic acid (glutamate), **905**
Glutaraldehyde, 221, 223*t*
Glyceraldehyde-3-phosphate (G3P), 172*f*, 187*f*, 188
Glycerol, 50, 50*f*, **905**
Glycocalyx, 115–116, 115*f*, 127*t*
Glycogen, 49, 180
Glycolysis, 171, 171*f*, 172*f*, 175*f*
Glycopeptides, resistance to, 827*f*
Glycylcyclines, 817
Goat cheese, **871**, **872*f***
Goats
  anthrax and, 337, 382
  contamination of carcasses, 356
  Q fever and, 337
Goblet cells, 309
Golden Ages of microbiology
  classical (first), 16–20, 21*t*, 22, 22*f*, 807
  Second, 25–26
  Third, 1, 28–31, 30*f*, 294–296
Golgi apparatus, 72, 74*b*
Gonococcal pharyngitis, 415
Gonococcal proctitis, 415
Gonorrhea. *See also Neisseria gonorrhoeae*
  characteristics, 414–415, 415*f*, 419*t*
  HIV transmission and, 797
  reported U.S. cases in 2010, 407*f*
  treatment for, 812
Gonorrheal conjunctivitis, 442, 443*f*
Gonyaulax, **881**
Gonyaulax catanella, **883**
Goodpasture, Ernest W., 454
Goodpasture syndrome, 783*t*
Gordon, Mervyn Henry, 334*b*
Gouda cheese, **871**
Gould, Stephen J., 99
gp41, 790, 790*f*, 791, 792*f*
gp120, 790, 790*f*, 791, 792*f*, 795
Graft-versus-host reaction (GVHR), 786
Graham, Sarah, 179*b*
Grain, spoilage of, **855**
Gram, Hans Christian, 21*t*, 88
Gram stain technique, 88, 88*f*, 89*f*, 90*b*
Gramicidin, 816
Gram-negative bacteria
  cell walls, 118, 119*f*, 120*t*
  conjugation between genera of, 274
  on differential culture media, 152*f*
  staining to identify, 88, 88*f*, 89*f*, 90
  traumatic wound infections and, 438*t*
Gram-positive bacteria
  attachment pili of, 110
  cell walls, 116, 119*f*, 120*t*
  classification of, 101, 101*f*
  on differential culture media, 152*f*
  mycoplasmas as, 126*b*
  staining to identify, 88, 88*f*, 89*f*, 90
  transformation in, 271*f*
  traumatic wound infections and, 438*t*
Granulocytes, 671, 671*f*
Granulomas, **883**
Granzymes, 685, 718, 774
Grape fanleaf virus (GFLV), **914–915**
Graves disease, 783*t*
Gravid proglottids, 613
Gray rot, **857**
Gray syndrome, 817
Great plate count anomaly, 156*b*
Great Pox, 406–407, 416*b*. *See also* Syphilis
Great Sanitary Movement, 14, **877–878**
Green algae, 23, 23*f*, 588*f*, 589, 589*f*
Green Revolution, **919–920**
Green rot, **855**, **857**
Green sulfur bacteria, 188
Grice, Elizabeth, 432*b*
Griffith, Frederick, 269–270*b*, 271
Griseofulvin, 573, 822–823, 825*t*
Groin ringworm, 572
Groundwater, **879**, **885**
Group A streptococci (GAS), 311, 313*b*, 436
Growth. *See also* Microbial growth
  of bacterial and archaeal cells, 108–109
  of fungi, 555–557, 559*f*, 560*f*
Growth hormone, human, 284, 285*f*
Gruby, David, 20
Gruinard Island, anthrax contamination on, 384*b*
Guanarito virus, 534
Guanine (G), 52, 52*f*
Guérin, Camille, 21*t*, 326
Guerra, Marta, 628
Guillain-Barré syndrome (GBS), 370–371, 500
Guinea worm, 621
Gulf of Mexico, Deepwater Horizon oil spill, 41*b*, **916**
Gulf War of 1991, bioremediation after, **919**
Gummas, syphilitic, 417*f*, 418

Gunpowder manufacturing, **901**. *See also* Cordite
Gymnodinium, **881**
Gyne-Lotrimin, 821

H

H spikes, 497
H1N1 influenza virus, 491, 491*f*, 497, 498–499*b*, 655*b*, 720*b*
H2N2 influenza virus, 498–499*b*
H3N2 influenza virus, 497, 498–499*b*
H5N1 influenza virus, 497, 498–499*b*, 720*b*
H7N2 influenza virus, 499*b*
H9N2 influenza virus, 499*b*
HAART (highly reactive antiretroviral therapy), 525, 798
HACCP (Hazard Analysis and Critical Control Points), **868**
Haden, Arthur, 164*b*
Haeckel, Ernst H., 73, 76
Haemophilus spp., 311
Haemophilus ducreyi, 418–419
Haemophilus influenzae
  airborne bacterial diseases and, 319*t*, 338*t*
  bronchitis and, 328
  ear infections and, 316
  epiglottitis and, 314
  genome sequencing, 291, 292*b*
  immunodeficiency disorders and, 788, 789
  meningitis and, 316
    vaccine for, 319
  restriction enzymes, 280*t*
  septicemia due to, 638*b*
  type B (Hib)
    genetic abnormalities of immune system and, 688*b*
    meningitis due to, 317–318, 319*t*
    pneumonia due to, 333*b*
    vaccine for, 732
HAIs (healthcare-associated infections), 651. *See also* Hospital-acquired infections
  C. diff-associated, 631, 646
Haiti, cholera epidemic in, 15*b*, 343–344, 361*b*
Haldane, J. B. S., 57*b*
Halobacterium salinarium, 148
Halogens, 217–218
Halophiles, 148, **881**
Halotolerant species, 148
Hamburg, Germany, cholera epidemic in, **878**
Hamon, Mélanie, 102*b*
Hand, foot, and mouth disease, 538
Hand washing
  with antibacterial soaps, 220*b*
  antibiotic resistance prevention and, 831*t*, 838*b*
  common colds and, 493
  disease transmission and, 15*b*
  effectiveness of, 223*t*
  food handling and, **866**
  HA-MRSA and, 831*b*
  hepatitis A and, 536
  reducing pneumonia incidence with, 333*b*
  Semmelweis on, 13–14, 13*f*
Hansen, Gerhard, 21*t*
Hansen disease, 438–439, 440*t*. *See also* Leprosy
Hantavirus, 481*t*, 482
Hantavirus pulmonary syndrome (HPS), 502, 504, 505*t*
Haploid cells, 123
Haptens, 698
Harlequin frogs, chytrid infections and, 562*b*
Harper, Kristin, 416*b*
Hartmannella, 608
Hashimoto disease, 783*t*
Havrix, 536
Hay, Mark, 182*b*
Hay fever, 764*f*, 772, 773*b*, 779*t*, 781*t*
Hayden, Deborah, 406, 407
Hazard Analysis and Critical Control Points (HACCP), **868**
Head colds, 492–493
Head ringworm, 572
Health, oral health and, 354*b*
Healthcare workers
  HA-MRSA transmission and, 831*b*
  HIV transmission and, 797, 797*f*
  risk of not vaccinating, 739*b*
Healthcare-acquired pneumonia (HCAP), 329, 338*t*
Healthcare-associated infections (HAIs), 651
  C. diff-associated, 631, 646
Heart disease, 354*b*, 683, ***911b***
Heart transplantation, Chagas disease and, 605
Heat
  allergic reactions and, 768
  endospore viability and, 141
  food preservation and, **859–861**, ***859f***
  microbial control using
    hot-air ovens, 198
    incineration, 198, 200*f*
    killing effect of, 198
    moist
      boiling water, 198, 200
      pasteurization, 203–204, 203*f*, 204*b*
  microbial life and, 134*t*
Heat fixation, 87
Heatley, Norman, 27*b*
Heavy (H) chains, 702, 704, 704*f*, 709, 710*f*
Heavy metals, 218–219, 824, ***920b***
Helical symmetry, of viruses, 456, 457*f*
Helicobacter pylori
  as friend or foe, 372*b*
  gastric ulcer disease and, 371, 373, 373*f*
  red wine and resveratrol effect on, ***911b***
Helminthic load, 611
Helminths
  allergies and, 613*b*
  illness caused by
    agents for treatment of, 824
    ascariasis, 618
    echinococcosis, 613–614
    hookworm disease, 618–619
    lymphatic filariasis, 619–620
    pinworm disease, 614, 617, 617*f*
    schistosomiasis, 612, 614*f*
    summary of, 620*t*
    tapeworms, 612–613, 616*f*
    trichinellosis, 617–618, 617*f*
  type I hypersensitivity and, 763
  as type of parasite, 587, 610–611
Helper T (T_h) cells, 713, 714*t*, 719, 763
Helper T1 (T_h1) cells, 712*f*, 713, 714*t*, 716*f*
Helper T2 (T_h2) cells
  allergic reactions and, 764, 765*f*
  cell-mediated immune response and, 712*f*, 713, 714*t*, 716*f*
  transplantation rejection and, 784
Hemagglutination inhibition (HAI) test, 746, 748*b*
Hemagglutinin, 746
Hemagglutinin (H) spikes, 497, 497*f*, 498–499*b*
Hematuria, 422
Hemicellulases, ***905t***
Hemodialysis, *E. cloacae* infection and, 117*b*
Hemolysins, 645–646
Hemolysis, 646
Hemolytic disease of the newborn, 776, 778*f*
Hemolytic uremic syndrome (HUS), 370
α-Hemolytic streptococci, 311, 434, 436
β-Hemolytic streptococci, 311, 435*f*, 436
Hemorrhagic colitis, 368, 370
Hemorrhagic fevers, viral, 529, 531–534, 535*t*
Hendra virus, 481*t*
Henle, Jacob, 14–15
Henson, Joan, 558*b*
Hepadnaviridae (hepadnaviruses), 461*f*, 477*t*, 526
Hepatitis A virus
  allergic reactions and, 773*b*
  cell/tissue tropism, 458*t*
  as chronic disease, 641
  infections, 535–536, 539*t*
  infectious dose of, 642
  micrograph of, 536*f*
  outbreak of, 537*b*
  shellfish contamination by, **854**
  vaccine for, 730
  waterborne, **883**
Hepatitis B core antigen (HBcAg), 526
Hepatitis B surface antigen (HBsAg), 526, 527
Hepatitis B virus (HBV)
  as bloodborne infection, 526–527, 526*f*, 528*t*
  cell/tissue tropism, 458*t*, 459–460*b*
  HIV/AIDS and, 796*f*
  as oncogenic agent, 476–477, 477*t*, 478
  vaccine for, 286, 731, 733
Hepatitis C virus (HCV)
  cell/tissue tropism, 458*t*, 459–460*b*
  HIV/AIDS and, 796*f*
  home testing kit, 754*b*
  infections, 527, 528*b*, 528*t*
  as oncogenic agent, 477*t*
Hepatitis D virus infections, 527
Hepatitis E virus infections, 536, 539*t*
Hepatitis G virus infections, 527
Hepatocellular carcinoma (HCC), 527
Herbicides, 285*f*, **893**, **913–915**
Herd immunity, 737–739, 737*f*, 738*f*, 739*b*, 838*b*
Herd immunity threshold, 737
Heredity. *See* DNA
Hermaphroditic organisms, 611
Herpes encephalitis, 507
Herpes keratitis, 507
Herpes simplex labialis, 506, 512*t*
Herpes simplex virus (HSV), 506*f*, 507*f*
Herpes simplex virus-1 (HSV-1), 469–470, 505–507, 512*t*, 773*b*, 796*f*
Herpes simplex virus-2 (HSV-2), 512*t*
Herpes zoster (shingles), 508–509, 509*f*, 510*b*, 512*t*
  vaccine for, 733, 734, 741*b*

Herpesviridae (herpesviruses)
blood diseases caused by, 524–526, 525*f*
classification of, 461*f*
cytopathic effects of, 474*t*
envelope of, 473*f*
as oncoviruses, 477*t*
relationships between species of, 506*f*
skin diseases caused by, 509–510, 510*f*, 512*t*
treatment for, 253–254
Herpesvirus 8, 477*t*
Hershey, Alfred, 25, 59*b*
Hesse, Fanny, 19*b*
Hesse, Walther, 19*b*
Heterokaryons, 558, 560*f*
Heterophile antibodies, 524
Heterotrophs, 189, 189*f*, 191
Heterotrophy, 188, 189*f*
Hexachlorophene, 218, 218*f*, 223*t*
Hibiclens, 218
High frequency of recombination (Hfr) cells, 274
High-efficiency particulate air (HEPA) filters, 205–206, 205*f*
Highly perishable foods, **851**, **852*f***
Highly reactive antiretroviral therapy (HAART), 525, 798
High-throughput DNA sequencing, 289–290*b*
*Hind*III, 280*t*
Hinnebusch, Joseph, 253*b*
His⁻ strains. *See* Mutants
Histamine, 682, 683*b*, 766, 767*f*
Histamine-binding protein (HBP), 683*b*
Histoplasma spp., 745
Histoplasma capsulatum, 576–577, 577*f*, 580*t*
Histoplasmosis, 576–577, 580*t*, 747, 779*t*, 780, 796*f*
Hitler, Adolf, 406
HIV (human immunodeficiency virus). *See also* AIDS
as AIDS cause, 789
antisense molecules and inhibition of, 248*b*
bridge between innate immunity and adaptive immunity and, 691*b*
CD4 T cell depletion by, 793–794*b*
cell/tissue tropism, 458, 458*t*
diagnosis, 797, 798*b*
DNA probes and PCR in detection of, 287
ELISA test for, 752, 753*f*
as emerging virus, 481*t*
evolution of, 482
hideouts in human body for, 799*b*
home testing kit, 798*b*
host range, 458
origins of, 791*b*
pandemic, 761–762, 762*f*
prevention of, 800
progression of, 794–795
provirus formation by, 470, 470*f*
psychological distress and, 715*b*
replication of, 791–792, 792*f*
reported U.S. cases in 2010, 407*f*
smallpox vaccine and, 757
structure of, 790–791, 790*f*
T helper cells and, 719
transmission of, 797, 797*f*
treatment for, 797–798, 798*f*, 799*b*
tuberculosis and, 326
type 1 (HIV-1), 789, 791*b*
cell/tissue tropism, 459–460*b*
type 2 (HIV-2), 791, 791*b*
vaccine for, 800
Hivid, 822*t*
Hodgkin lymphoma, 477*t*, 524
Hoffman, Erich, 21*t*
Holding method of pasteurization, 203–204, **860**
Home testing kits
for hepatitis C virus, 754*b*
for HIV, 798*b*
Homeostasis, 66, 67*f*
Homologs, 125–126
Honey
as antiseptic, 215*b*
infant botulism and, 359
Honeybee colony collapse, 580
Hong Kong flu (1968), 496, 498–499*b*
Hoof-and-mouth disease, 20
HOOH (toxic hydrogen peroxide), 235–236*b*
Hooke, Robert, 6, 7
Hookworm disease, 618–619, 619*f*, 620*t*
Hookworms, 618, 618*f*
Hops, **907**, **909*f***
Horizontal gene transfer (HGT)
antibiotic resistance and, 827–829
comparative genomic studies of, 296
concept map of, 278*f*
conjugation, 271–272, 272*f*, 273*f*, 274, 275*b*
marine snow and, 272*b*
process of, 268, 269*f*
transduction, 274, 276*f*, 277
transformation, 268, 269–270*b*, 271, 271*f*
Horizontal transmission of disease, 649, 650*f*
Hormones, irradiation of, 206
Horn snails, 615*b*
Horses
arboviral encephalitis and, 544, 544*f*
DNA vaccine for West Nile virus in, 732
Hospital-acquired infections (HAIs), 196, 259*b*, 295
Hospital-associated MRSA (methicillin-resistant *Staphylococcus aureus*) (HA-MRSA), 830–831*b*, 833
Hospitals
antibiotic resistance and, 833–834
copper fixtures in, 219
pathogen-contaminated operating room, 211*f*
Host range, viral, 458, 459–460*b*
Host-microbe relationships
disease outcomes after pathogen contact, 636–637, 636*f*
human microbiota and, 631–633, 632*f*
pathogen entry and, 641–643, 642*f*
pathogenicity and, 634, 636
phagocytosis and, 678–681, 680*f*
Hosts. *See also* Immunity; Immunocompromised individuals
for apicomplexans, 594, 594*f*
resistance to disease by, 669, 669*f*
for rickettsial reproduction, 101
Hot environments. *See* Heat
Hot-air ovens, microbial control using, 198, 209*t*
HTST (high temperature, short time) pasteurization, **860**
Hudson River (New York) cleanup, **916**
Human(s). *See also* Indigenous microbiota
major chemical elements of, 37, 37*t*
taxonomic classification of, 80*t*
Human African sleeping sickness, 603
Human endogenous retroviruses (HERVs), 293–294
Human genome, 291, 293–294
Human Genome Project, 291
Human granulocytic anaplasmosis (HGA), 399
Human growth hormone, genetically engineered, 284, **919**
Human Gut Microbiome Project (HGMI), 346–347
Human herpesvirus 6 (HHV-6), 509, 512*t*
Human herpesvirus 8 (HHV-8), 509–510, 512*t*
Human herpesvirus (HHV), 506*f*
Human immunodeficiency virus. *See* HIV
Human leukocyte antigens (HLAs), 713, 784. *See also* Major histocompatibility complex proteins
Human lymphoid cultures (HCLs), 793–794*b*
Human metapneumovirus (hMPV), 501*f*, 502, 505*t*
Human microbiome, 346, 347*f*
Human Microbiome Project (HMP), 633
Human monocytic ehrlichiosis (HME), 399, 401
Human papilloma viruses (HPV)
cell/tissue tropism, 458*t*
DNA probe for detection of, 287
oncogenic, 477, 477*t*, 478, 514–515, 518*t*
vaccine for, 741*b*
warts due to, 477*t*, 514, 518*t*
Human T-cell leukemia virus-1 (HTLV-1), 477*t*
Humira, 757
Humoral immune response, 698. *See also* Adaptive immunity
Humoral immunodeficiencies, 787
Humulin, 281
Humulus lupulus, **907**
Hurricane Katrina, mold formed after, 552–553
Hutchinson's triad, 418
Hyaluronidase, 644–645, **906**
Hybridomas, 756–757, 756*f*
Hydatid cysts, 614, 616*f*
Hydrocarbons, 40, 41*b*, 41*f*
Hydrogen (H), 37, 37*f*, 37*t*, 57–58*b*, 188
Hydrogen bonds, 42, 42*f*, 43*t*, 54, 57
Hydrogen peroxide (H_2O_2), 219, 679
toxic (HOOH), 235–236*b*
Hydrogen sulfide, 188, **855**, **857**, **894**
Hydrogenosome, 591
Hydrolysis reactions, 43, 47, 165–166, 165*f*, 166*f*
Hydrophilic properties, 44, 118, 120, 120*f*, 121*f*
Hydrophobia, 541
Hydrophobic properties, 44, 50, 51, 118, 120, 120*f*
Hydrostatic pressure, 148
Hydroxybenzoic acid, **864**
Hydroxyl group, 47*t*
Hygiene
allergies and, 770, 770–771*b*, 772, 773*b*
disease transmission and, 13–14, 15*b*, 17*f*
microbial forensics and cases involving, 295
public concerns regarding, 195–196, 196*f*
Hyperacute bacterial conjunctivitis, 442
Hyperacute rejection of allografts, 784
Hyperhidrosis, 360

Hypersensitivity(ies)
  characteristics and types of, 763, 764*f*
  overview of, 781*t*
  predisposition for, 770, 772
  therapies for, 772–773
  type I IgE-mediated
    asthma, 768–769, 770–771*b*, 770*f*
    characteristics, 763–766, 764*f*, 765*f*
    concept map for, 767*f*
    hygiene and, 773*b*
    localized or systemic, 766, 768
    therapies for, 772–773, 774*f*
  type II cytotoxic, 774–776, 775*f*
  type III immune complex, 776–778, 779*f*
  type IV "delayed," 778
Hyperthermophiles, 101*f*, 104, 105*b*, 144*f*, 145
Hypertonic condition, 116
Hyphae (hypha), 554, 555*f*
Hypochlorous acid, 679
Hypothalamus, 684
Hypothesis, in scientific inquiries, 10*b*

I

Icosahedral symmetry, of viruses, 456, 457*f*, 494*f*
Idoxuridine (IDU), 822*t*
IFNs. *See* Interferons
IgA
  characteristics, 704–705, 705*t*
  deficiency of, 708*b*
  in female reproductive system, 409, 410*f*
  as maternal antibody, 735–736
  mucosal immune system and, 345–346
  secretory, 706–707*b*
IgD, 705, 705*t*
IgE
  characteristics, 705, 705*t*
  immune complex hypersensitivity and, 777
  in type I hypersensitivity, 763–770, 765*f*, 767*f*, 772–774, 781*t*
IgG
  characteristics, 704, 705*t*
  cytotoxic hypersensitivity and, 774–776
  in female reproductive system, 410*f*
  immune complex hypersensitivity and, 776–778
  as maternal antibody, 735, 736
  type I hypersensitivity and, 763
IgM, 704, 705*t*, 736, 774
Ignicoccus hospitalis, 235*f*
Imidazoles, 821–822
Imipenem, 814, 830*b*
Immediate hypersensitivity, 763, 764*f*, 766, 767*f*
  delayed hypersensitivity vs., 779*t*
Immune complex (type III) hypersensitivity, 763, 764*f*, 776–778, 779*f*
Immune complexes, 736
Immune deficiency, 697. *See also* Immunodeficiency disorders
Immune globulins
  as antitoxins, 646
  for hepatitis B, 527
  Parvovirus B19, 514
  for rabies, 522–523, 542
  varicella-zoster, 509
Immune suppression, 698
Immune system. *See also* Immunity; Immunocompromised individuals
  abnormal cells, cancer and, 476*f*
  bacterial invasion bypassing, 643, 643*f*
  Carabin as timer for, 718*b*
  genetic abnormalities of, 688*b*
  nervous system and, 715*b*
  overview of, 670–674, 671*f*, 673*f*, 674*f*
  privileged sites for, autoimmune disorders and, 782, 782*f*
  trematode evasion of, 611
  ulcerative colitis and, 613*b*
  virotherapy and, 480*b*
  weakened, tuberculosis and, 326
Immunity. *See also* Adaptive immunity; Innate immunity
  defined, 727
  development of understanding about, 668–669
  herd, 737–739, 737*f*, 738*f*, 739*b*
  infectious dose and development of, 642
Immunization. *See also* Vaccinations
  control of bacterial diseases with, 305
  public health and, **847**
Immunocompetent individuals, 730
Immunocompromised individuals. *See also* AIDS; HIV
  attenuated vaccines and, 730
  bacterial GI diseases and, 355
  blastomycosis and, 577
  cryptococcosis in, 576
  cryptosporidiosis and, 599
  fungi and, 553
  hepatitis E and, 536
  herd immunity and, 738–739
  nosocomial infections and, 658
  opportunistic infections and, 580
  oral candidiasis in, 574
  Pneumocystis pneumonia and, 577–578
  staphylococcal pneumonia and, 329
  toxoplasmosis and, 608
Immunodeficiency disorders. *See also* AIDS; HIV
  asymptomatic, 708*b*
  types of, 787–789
Immunodiffusion tests, 744–745
Immunoelectrophoresis, 745, 746*f*
Immunofluorescence antibody test (IFA), 747, 749*f*, 750
Immunogenic responses, 698
Immunoglobulins. *See also* Antibodies; *under Ig*
  classes of, 704–705, 705*t*
  deficiency of, 708*b*
  structure of, 702, 704, 704*f*
  transplantation acceptance and, 786*t*
Immunological memory, 698
Immunology, 22, 669, 670. *See also* Psychoneuroimmunology
Immunosuppression. *See also* AIDS; HIV
  aspergillosis and, 581
  CMV disease and, 525
  secondary immunodeficiencies and, 789
  toxoplasmosis and, 608
Immunosuppressive agents, 786, 786*t*
Immunotherapy, 757, 769*b*, 772
Impetigo, 433–434, 434*f*, 436, 440*t*
Import inspectors, preparation for careers as, **848**
Inactivated vaccines, 730
Inactive vaccine ingredients, 735*b*
Incidence of disease, 307, 657*b*
Incidental infections, of arboviral encephalitis, 544, 544*f*
Incineration, microbial control using, 198, 209*t*
Inclusions (inclusion bodies), 124–125, 128*t*, 413*f*, 414*b*, 474*t*
Incubation period
  for campylobacteriosis, 370
  defined, 138, 355, 640, 640*f*
  for listeriosis, 364
  for tuberculosis, 322–323, 324*f*
  for typhoid fever, 365
  for yersiniosis, 371
Index of refraction, 87
India ink (dye), 87–88
Indicator organisms, **889**
Indigenous microbiota
  airborne bacterial diseases and, 319*t*
  colonization of, 633
  of female reproductive system, 409–410, 410*f*
  of GI tract, 346–347, 346*b*, 347*f*, 349, 350*f*
    apple-eating and, 348–349*b*
    probiotic yogurt and, 634*b*
  human body and, 631–633, 632*f*
  of skin, 426–429, 428*f*
    acne and, 429–431, 429*f*
  of upper respiratory tract, 309
  of urinary tract, 420–421
Indirect contact, for disease transmission, 649–650, 651*f*, 652*b*
Indirect immunofluorescence antibody test, 747, 750*f*
Induced mutations, 252
Inducers, 250*b*
Induration, 325, 325*f*, 778
Industrial fermentation, **901**, **902**, ***902f***, ***903f***
Industrial microbiology. *See also* Biotechnology
  alcoholic beverages, **907–908**, ***911b***
  antibiotics production, **912–913**
  bacteria as tool in, 102*b*
  bioremediation, **916**, ***917–918b***, **918–919**, ***918f***
  enzyme production, **905–906**, ***905t***
  fungi as tool for, 449
  genetic engineering, **919–921**
  mushrooms, **915**, ***915f***
  organic compounds produced by, **903–905**
  pesticides, **913–915**
  sterilization methods, 199*b*
  uses of, 5
Industrial Revolution, **847**, **877**
Infants
  botulism and, 359
  congenital rubella in, 513
  congenital syphilis in, 418
  enteropathogenic *E. coli* and diarrhea in, 363
  mortality, global, by infectious diseases, 305
  newborn
    conjunctivitis in, 413, 415, 443, 444*t*
    gray syndrome in, 817
    hemolytic disease of, 776, 778*f*
    herpes simplex in, 508, 512*t*
    meningitis in, 319
    microbiota colonization in, 633, 635*f*
    oral candidiasis in, 574

tetanus in, 383
toxoplasmosis and, 608
roseola in, 509
RS disease in, 502
scalded skin syndrome in, 434, 434*f*
transplacental HIV transmission to, 797
transplacental viral transmission to, 513
Infection allergy, 778, 779*t*, 780, 780*f*
Infections or infectious diseases. *See also* Bacterial diseases; Disease(s); Host-microbe relationships
antibiotics discovery and, 25–26
emerging and reemerging, 28–29, 29*f*
epidemiology. *See* Epidemiology, infectious disease
global mortality from (2011), 28*f*
healthcare-associated (2008), 267*t*
immunodeficiency disorders and, 788–789
inflammation and, 681–683
intoxications compared to, 355
nosocomial. *See* Nosocomial infections
obesity and, 495*b*
pathogen contact and development of, 636–637, 636*f*
pathogen identification, 289–290*b*
public concerns regarding, 195–196
as public health challenge, 660–661, 663–664
stages of, 637, 640–641, 640*f*
types of, in percentages, 821*f*
use of term, 631
worldwide deaths due to, 631*f*
Infectious bronchitis, 328, 328*f*, 338*t*
Infectious Disease Society of America (IDSA), 834, 839
Infectious dose, 355, 366, 641–642
Infectious hepatitis, 535
Infectious mononucleosis, 476, 524, 525*f*, 528*t*, 641
Infectious parotitis, 513
Inflammation
complement activation and, 686, 686*f*, 687
histamine-binding protein and, 683*b*
innate immunity and, 681–683
uncontrolled immune response and, 720*b*
Inflammatory gastroenteritis, 360–365, 361*b*, 362–364*f*, 363*t*, 374*t*
Infliximab, 786*t*
Influenza (virus)
characteristics, 496–497, 497*t*, 505*t*
colds vs., 500*b*
as emerging virus, 481, 481*t*
evolution of, 482, 498–499*b*
immunizing preschoolers and, 501*b*
pandemic of 1918 and 1919, 490, 498–499*b*, 637, 720*b*
seasonal variations in, 493*f*
as twenty-first century problem, 490–491
vaccines for, 730, 731*b*, 733, 735*b*, 739*b*
Influenza A, 458*t*, 459–460*b*, 496, 497*t*, 500
Influenza B, 496, 497*t*
Influenza C, 496, 497*t*
Inhalational anthrax, 337, 338*t*, 382
Inhalational tularemia, 391
Inhaled drugs, AIDS and, 789–790*b*
Initiation, in DNA replication, 238, 239*f*, 240, 242
Injecting drug users, tetanus and, 385*b*
Innate immunity
bridge between adaptive immunity and, 691*b*
cellular barriers and, 676
chemical barriers and, 675–676
complement and, 686–687, 686*f*, 687*f*
concept map of, 690*f*
as defense against disease, 672–673
defined, 669
fever and, 683–684
inflammation and, 681–683
mucus production and, 676*b*
natural killer cells and, 684–685, 685*f*
pathogen identification and, 677*b*
pathogen-associated molecular patterns and, 678, 679*f*, 679*t*
phagocytosis and, 678–681
physical barriers and, 675, 675*f*
Insecticides, 893, **913–915**
Insects. *See also* Arboviruses; *specific species*
inflammatory response from bite of, 683*b*
resistance in, genetic engineering and, 285*f*
stinging, allergies to, 766
Insertion sequences (IS), 256, 258*f*
Insertional activation, 478
Insertional inactivation, 478
Insidious diseases, 332
Insulin, 280, 281, **919**
Integral membrane proteins, 121
Integumentary system, 424–426. *See also* Skin
Interferons (IFNs)
for chronic granulomatous disease, 788
genetically engineered, 284, 285*f*, **919**
for hepatitis, 527
tea drinking and production of, 699*b*
as URT defense, 309
viral infections and, 687, 689, 689*f*
Interleukin-2 (IL-2), 714, 716*f*
Interleukin-4 (IL-4), 766
Interleukin-5 (IL-5), 766
Interleukin-6 (IL-6), 715*b*
Interleukins (ILs), 674
Intermediate filaments (IF), 126
Intermediate hosts, 594*f*, 602, 602*f*, 607*b*, 611, 615*b*
Intermittent sterilization, 201*f*, 203, 209*t*
International Committee on Taxonomy of Viruses (ICTV), 460
International Human Microbiome Consortium (IHMC), 426
Intestinal anthrax, 365, 382
Intestines. *See* Gastrointestinal (GI) tract
Intoxications
defined, 355–356
foodborne, 357–360
fungal, 570–571, 571*b*, 572*f*
Intraepithelial lymphocytes (IELs), 706–707*b*
Intravenous catheters, 431, 831*t*
Intrinsic factors, for food spoilage, **851**, **852*f***
Intron A (IFN-alpha drug), 527
Introns, 243
Intussusception, 740
Invasive aspergillosis, 580
Invasive gastroenteritis, 360
Invasive wound infections, 437, 438*t*
Invasiveness
defined, 643
virulence factors and, 643–647, 645*f*
Invertase, **905*t***, **906**
Investigational New Drug (IND) *Application*, FDA's, 840–841*b*
Invirase, 822*t*
Iodine
disinfection with, 217–218
in Gram stain technique, 88, 88*f*
tincture of, 209, 213t, 223*t*
Iodophors, 217–218, 223*t*
Ionic bonds, 39–40, 40*f*, 43*t*, 54, 55*f*, 57
Ionizing radiations, 206, 206*f*, 253
Ions, 38, 38*f*, 40*f*, 47*f*
Irish potato famine, 590, 590*b*
Iron (Fe), 37
Irradiation of food, 206, 207*b*, 208, **864–865**, **865*t***
Irritable bowel syndrome, 349
Irwin, Michael R., 510*b*
Isentress, 822*t*
Isografts, 784
Isoniazid (INH)
antimicrobial spectrum of, 809*f*
resistance to, 827*f*
rifampin and, 818–819
for tuberculosis, 326, 812, 820*t*
Isopropyl alcohol, 223*t*
Isotretinoin, 431
Israel, founding of, 279*b*
Itching, inflammation and, 682, 683*b*
Itraconazole, 822, 825*t*
Ivanowsky, Dimitri, 453
Ivermectin, 825*t*
Ixodes spp., 605
Ixodes pacificus, 394, 395*f*
Ixodes ricinus, 755*f*
Ixodes scapularis, 394, 394*f*, 395*f*

J

Jacob, François, 249
Janssen, Zacharias, 6
Japanese encephalitis, 545*f*
Jaundice, 387, 526, 529, 535–536
Jefferson, Thomas, 12
Jenner, Edward, 9, 12–13, 516, 668, 722, 725
Jewish homeland in Palestine, 279*b*
Jock itch, 572
Johnson, Claire, 494*b*
Joint United Nations Programme on HIV/AIDS (UNAIDS), 761–762
Juices
E. coli O157:H7 contamination of, 368, 369*b*
production of, **905*t***
Jumping genes. *See* Genetic recombination
Jungle yellow fever, 529
Junin virus, 482

K

Kala azar, 596, 601*t*
Kaletra, 822*t*
Kanamycin, 326, **919**

Kaposi sarcoma (KS), 477*t*, 509–510, 511*f*, 512*t*, 795, 796*f*
Kaposi sarcoma-like virus, 506*f*
Kepler, Johannes, 133
Keratin, 425
Keratinocytes, 424–425
Keratitis, 415, 442, 507, 598*b*
Ketoconazole, 573, 577, 821–822, 825*t*
Ketones, **871**
Kilbourne, F., 21*t*
Kinetoplastids, 591*f*, 592, 594*t*
King Tut Ale, **908*b***
Kingdoms, in taxonomic classification, 76–77
Kirby, W. M., 826
Kirby-Bauer test, 826–827, 826*f*
Kirin Brewery Company Ltd., **908*b***
Kissing bug, 605
Kitasato, Shibasaburo, 21*t*, 389, 703*b*
Klebsiella spp., 425*t*, 438*t*, 842, **857**, **869**
Klebsiella oxytoca, 275*b*, 659*t*
Klebsiella pneumonia, 329
Klebsiella pneumoniae
  carbapenem-resistant, 259*b*
  characteristics, 329
  as ESKAPE pathogen, 839
  immune system and, 392–393*b*
  NDM-1 strain, 842
  nosocomial infections and, 659*t*
  septicemia due to, 638*b*
  urinary tract infections and, 421*f*
Koa tree, Hawaii, **896*b***
Koch, Robert, 17, 18–19, 322, 360, 361*b*
Koch's postulates, 18, 18*f*, 154*b*
Koji, **870**
Koop, C. Everett, 629
Koplik spots, 473, 511, 512*f*
Krebs cycle, 173. *See also* Citric acid cycle

L

La Crosse (LAC) encephalitis, 544, 545*f*, 547*t*
Lactalbumin, **856**
Lactarius, 570–571
Lactase, **905*t***
Lactic acid
  cheese production and, **903**
  dental plaque and, 49*b*
  fermentation and, 183, 184*f*, **869**, **871**
  food preservation and, **864**
  as secondary metabolite, **904**
Lactobacillales, 430*f*
Lactobacillus spp.
  acid-tolerance of, 148
  fermentation and, 184*f*, **870**
  food spoilage and, **855**, **856–857**
  meat contamination by, **854**
  symbiosis with human body, 632
  vaginal acidity and, 409
Lactobacillus acidophilus, 138*t*, **871**
Lactobacillus bulgaricus, 633, **871**, **904**
Lactobacillus curvatus, **873*b***
Lactobacillus plantarum, **873*b***
Lactobacillus sanfranciscensis, **856*b***
Lactococcus lactis, **872–873*b***
Lactoferrin, 309, 345, 409, 410*f*
Lactones, **906**
Lactose, 49, 180, **856**, **912**
Lactose (Lac) operon, 250*b*
lacZ gene, 282–283*b*
Lafferty, Kevin, 615*b*
Lag phase, in microbial growth, 139, 139*f*
Lagenidium giganteum, **915**
Lagering, in beer production, **908**
Lagging strands, in DNA replication, 240–241, 240*f*
Laginex, **915**
Lake Calumet, Illinois, pH level in, 46*b*
Lake Vostok, below Antarctic, 149*b*
Lakes. *See also* Oceans
  microorganisms in, 592
Lamb. *See also* Sheep
  irradiation, 206
  toxoplasmosis and, 606
Lambda (λ) phage, 465, 469*b*
Lamina propria, 706–707*b*
Lancet fluke, 615*b*
Land use changes, disease emergence or resurgence and, 661
Langerhans cells, 425
Large (50S) ribosomal subunit, 123, 124*f*, 246–247*f*, 816, 816*f*
Lariam, 823
Laryngitis, 494, 496, 496*t*
Lassa fever, 533, 534*f*, 535*t*, 653*b*, 736
Last universal common ancestor (LUCA), 78, 78*f*
Latency (latent phase)
  in animal virus replication, 469–470
  of cytomegalovirus, 524–525
  of herpesviruses, 506–508
  of HIV, 794
  of tuberculosis, 323, 323*f*
  of varicella-zoster virus, 508
  in virus replication, 463*f*, 465, 469*b*, 475*b*
Late-phase allergic response, 766, 767*f*
Latex allergy, 768
Laveran, Charles, 20, 21*t*
Lax, Eric, 27*b*
Leading strands, in DNA replication, 240, 240*f*
Leafy lichens (foliose), 563, 565
Leahy, Patrick, 222*b*
Leary, James, **896*b***
Leather manufacturing, **906*b***
Lechuguilla Cave, Carlsbad Caverns National Park, New Mexico, 836–837*b*
Lectin pathway of complement activation, 686–687, 686*f*
Lederberg, Joshua, 271
Leeuwenhoek, Anton van, 1, 6–7, 104, 588
Leftover food, spoilage of, **866**
Legionella pneumophila
  characteristics, 332, 335*b*, 336*f*
  identification of, 154*b*
  phagocytosis and, 681*b*
  as waterborne infection, **882**, **882*f***
Legionellosis, 336
Legionnaires' disease, 154*b*, 332, 335*b*, 336
Legumes, **896–897**, **897*f***
Leishmania spp., 592, 595*f*
Leishmania donovani, 595–596, 601*t*
Leishmania major, 595–596, 596*b*, 601*t*
Leishmaniasis, 595–596, 595*f*, 596*b*
Lentivirus spp., 791*b*
Lepiota spp., 570–571
Lepromatous leprosy, 438
Leprosy, 407, 438–439, 439*f*, 440*t*, 780
Leptospira interrogans, 386–388, 386*f*, 388*t*
Leptospirosis, 386–388, 387*b*, 388*t*
Leuconostoc, **854**, **869**, **870**
Leuconostoc citrovorum, **871**
Leuconostoc mesenteroides, **873*b***
Leukemia/lymphoma, adult T-cell, 477*t*
Leukocidins, 645
Leukocytes (white blood cells), 670–672, 671*f*
Leukopenia, 401
Leukotrienes, 766, 767*f*
Levofloxacin, 812, 820*t*
Liang, Arthur P., **849**
Lice, disease transmitted by, 398, 400–401*b*, 402*t*
Lichens, 134*t*, 563, 565, 565*f*
Licorice root, as antiseptic, 215*b*
Life. *See also* Evolution; Microbial life
  on Earth, appearance of, 231–232, 232*f*
  generation of, 12*b*, 12*f*
  genome transplantation and, 299–300*b*
  laboratory creation of, 57–58*b*
  major chemical elements of, 37, 37*t*
  origins of, 35
  phylogenetic tree of, 78*f*
  water and, 43–46
Light (L) chains, 702, 704, 704*f*, 709, 710*f*
Light microscopes. *See also* Microscopes and microscopy
  alternate optical configurations for, 91–93, 91*f*, 92*f*
  features, 85, 86*f*, 87
  types compared, 95*t*
Light-dependent reactions, 185–188, 187*f*. *See also* Photosynthesis
Limburger cheese, **871**
Lincomycin, 817
Lincosamides, 815*f*, 816*f*, 817, 820*t*, 827*f*
Linen production, **905**, **906*b***
Linezolid, 818, 820*t*
Linnaeus, Carolus, 73, 78, 79
Lipases, **905*t***
Lipid A, 118, 119*f*
Lipids, 50–51, 50*f*, 72, 697
Lipopeptides, resistance to, 827*f*
Lipopolysaccharide (LPS), 118, 119*f*, 120*t*, 679*t*, 683, 839
Lipopolysaccharide (LPS) complex, 647
Lipoproteins, low-density, red wine and, **911*b***
Lister, Joseph, 17, 17*f*, 218, 808
Listeria spp., 207*b*, **853**
Listeria monocytogenes
  ABM and, 316
  characteristics, 364, 364*f*
  as cytoplasm-invading pathogen, 643, 644*f*
  low temperatures and, **861**
  neonatal meningitis and, 319
  nitrites and, **864**
  phage therapy against, 466*b*
  phagocytosis and, 681*b*
  red wine and resveratrol effect on, **911*b***

Listerial meningitis, 364
Listeriosis, **867**
Littoral zones, **881**
Liver cancer, 477, 477*t*
Liver disease, chronic, 477
Livestock. *See* Cattle; Pigs
Lobar pneumonia, 332
Local diseases, 637
Localized anaphylaxis, 766, 767*f*, 768
Lockjaw, 384
Löeffler, Friedrich, 21*t*
Logarithmic (log) phase (exponential), 139*f*, 140
Lone star tick, 401
Looped domain structures, 237, 237*f*
Lopinavir, 821, 822*t*
Lovely, Derek, 179*b*
Lower respiratory tract (LRT), 309. *See also* Respiratory diseases, lower
LSD (lysergic acid diethylamide), 570, 571*b*
LTLT (low temperature, long time) pasteurization, **860**
LUCA (last universal common ancestor), 78, 78*f*
Lung diseases. *See* Respiratory diseases, lower
Lupus erythematosus, systemic, 778, 783, 783*t*, 784*f*, 788
Luria, Salvador, 25
Lycoperdon spp., 570–571
Lyme disease, 393–395, 394*f*, 395*f*, 402*t*, 407*f*
LYMErix, 395
LymeVax, 395
Lymph, 672
Lymph nodes, 672, 673*f*
Lymphatic filariasis, 619–620, 620*t*
Lymphatic system, 672, 673*f*
Lymphedema, 620
Lymphocryptovirus, 506*f*
Lymphocytes, 672, 698–699. *See also* B lymphocytes; T lymphocytes
Lymphocytic choriomeningitis virus (LCMV), 290*b*
Lymphogranuloma venereum (LGV), 419, 419*t*
Lymphoid progenitors, 670, 671*f*, 700, 700*f*
Lymphomas, 796*f*
Lyophilization, **863**, ***863b***, ***863f***
Lysergic acid diethylamide (LSD), 570, 571*b*
Lysine production, **905**
Lysogenic cycle, 463, 464*f*
Lysol, 213*t*
Lysosomes, 72, 679, 680*f*, 681*b*, 788. *See also* Phagolysosomes
Lysozyme
  bacteriophage replication and, 465
  as chemical barrier to disease, 676, 679
  in eggs, **855**
  in female reproductive system, 409
  protein content of, 166
  in saliva, 345
  viral replication and, 465
Lytic cycle, 463, 464*f*

M

M cells, 706–707*b*
M proteins, 312, 435*f*, 436
MacConkey agar plates, 152*f*
Machupo virus, 482
Macrolides
  antibiotic characteristics, 815*f*, 816*f*, 817, 820*t*
  global market for, 819*f*
  resistance to, 827*f*, 832*t*, 836–837*b*
Macronucleus, in ciliates, 593
Macrophages
  immune system and, 671*f*, 672–673
  inflammation and, 682
  phagocytosis and, 643, 643*f*, 678–679, 680, 680*f*, 681
Macules, 516
Maculopapular rash, 398, 398*f*
Mad cow disease, 484, 485
Magnesium (Mg), 37*t*
Magnesium ion (Mg^{+2}), 38*f*
Magnetosomes, 124–125, 125*b*, 128*t*
Magnetotaxis, 125*b*
Magnification, in microscopy, 85
Major histocompatibility complex (MHC) proteins, 713
  Class I, 684, 685*f*, 712*f*, 713, 714, 714*t*
  Class II, 712*f*, 713–714, 714t
  mate selection and, 713*b*
  superantigens and, 719, 719*f*
  transplantation and, 784
Malaria. *See also Plasmodium* spp.
  blood type O and, 777*b*
  early research on, 20
  fever tree and, 823*b*
  as global concern, 196, 586–587, 587*f*, 604*b*
  as parasitic disease, 601–603, 602*f*, 610*t*
  worldwide deaths due to, 631*f*
Malavone, 603
Males
  mumps and, 513
  reproductive system, 408, 409*f*
  urinary tract, 420–421
Malignant tumors, 476
Mallon, Mary, 367*b*
Malting, in beer production, **907**, ***909f***
Maltose, 49
Mamavirus, 457*b*
Manganese (Mn), 37
Mantoux test, 325–326, 780, 780*f*
Manuka honey, 215*b*
Maraviroc, 822*t*
Marburg hemorrhagic fever (MHF), 533, 535*t*
Marburg virus, 460, 481*t*
Margulis, Lynn, 74*b*
Marine microorganisms, 3, 5, 839
Marine snow, 272*b*. *See also* Oceans
Marinobacter, ***918b***
Marison, Ian, ***886b***
Mars, life on, 133–134
Marshall, Barry, 371, 373
Martin, Debra L., 818*b*
Mashing, in beer production, **907**, ***909f***
Mass, 37
Mass number, 38, 44*b*
Mast cells
  asthma and, 769
  cromolyn sodium and, 772
  epinephrine and, 768
  immune system and, 670, 671*f*
  inflammation and, 682
  serum sickness and, 777
  type I hypersensitivity and, 763, 764–765, 765*f*, 767, 767*f*, 781*t*
Masuda, Hedeki, 215*b*
Mate selection, MHC proteins and, 713*b*
Maternal antibodies, 502, 704, 735
Mating types, fungal, 558
Matrix, viral, 456
Matrix protein, 790, 790*f*
Matter, 37
Mayer, Adolf, 453
McCoy, George, 21*t*
MCV4 (conjugate vaccine), 732
Measles (rubeola)
  evolution of, 482
  herd immunity and, 737, 737*f*
  infections, 511, 512*f*, 513, 518*t*
  maternal antibodies for, 735
  Morbillivirus and, 501*f*
  vaccinations and incidence, mortality from, 725, 726*f*
  vaccine for, 733, 739*t*
Measles-mumps-rubella (MMR) vaccine, 511, 513, 730, 740
Meat
  contamination of, **853–854**
    aflatoxins, 570
    bacterial causes, 356, 358
    intestinal anthrax and, 365
  cooking, disease transmission control and, **866**
  HACCP inspections of, **868**, ***868f***
  irradiation of, 206
  production, enzymes and, ***905t***
Mebendazole, 617, 618–619, 620*t*, 824, 825*t*
Mechanical vectors, 650, 651*f*
Medawar, Peter, 2, 451, 455
Medical negligence, microbial forensics and, 295
Medicine
  genetic engineering and, 284, 286
  metagenomics and, 298
Medicines for Malaria Venture, 741*b*
Mefloquine, 603, 823, 825*t*
Megavirus, 457*b*
Melarsoprol (Mel B), 603, 824
Mello, Craig C., 248*b*
Membrane attack complexes (MACs)
  antigen-antibody complexes and, 711
  complement activation and, 686*f*, 687, 687*f*
  cytotoxic hypersensitivity and, 774
Membrane filters, 204–205, **884**, **889**, ***889f***
Memory cells, 700, 701*f*, 702, 727
Memory T cells, 799*b*
Menactra, 319
MenAfriVac, 319
Meninges, 318, 576
Meningitis
  aseptic, leptospirosis and, 387
  bacterial
    acute, 316–319
    diagnosis of, 90*b*
    plague meningitis, 389
  complement fixation test for, 747
  cryptococcal, 576
  listerial, 364
  vaccines for, 318–319
  West Nile virus and, 546–547
Meningitis belt, Africa, 317, 317*f*, 319
Meningococcal diseases. *See* Meningitis
Meningoencephalitis, 608–609

Menomune, 318
Menstrual cycle, 409, 410*f*
Merbromin, 218, 223*t*
Mercuric chloride, 223*t*
Mercurochrome, 213t, 218
Mercury (Hg), 218, 223*t*
Merkel cell carcinoma, 477*t*
Merkel cell polyoma virus (MCPV), 477*t*
Meropenem, 814, 820*t*, 830*b*, 837*b*
Merozoites, 594*f*, 602–603, 602*f*
Merthiolate, 218, 223*t*, 321
Mesophiles, 142, 144*f*, 145–146
Messenger RNAs (mRNAs)
  antisense molecules and inhibition of, 248*b*
  transcription of, 242, 243*f*, 244–245, 245*f*
Metabolic pathways
  ATP/ADP cycle and, 170*f*
  enzyme inhibition and, 167–168, 167*f*
  enzymes and, 166–167
  metabolic engineering and, 295–296
Metabolism. *See also* Photosynthesis
  ATP for energy in, 168–170, 169*f*
  cellular, 71, 72
  compartmentalism of bacterial and archaeal cells for, 108
  defined, 163
  diversity in, 188–189, 189*f*, 190*b*, 191
  enzymes and chemical reactions for, 164–165
  enzymes and energy in, 163–170
  glucose catabolism and, 170
    citric acid cycle, 173, 174*f*, 175, 175*f*
    glycolysis, 171, 172*f*, 173, 175*f*
    oxidative phosphorylation, 175–177, 176*f*, 177–178*b*, 179*f*, 180
  other potential energy sources for
    anaerobic respiration, 182–183
    fermentation, 183–184, 184*f*
    nutrients, 180–182, 181*f*
Metabolites, from industrial production, **903–904**
Metachromatic granules, 124, 128*t*
Metagenome, 298
Metagenomics, 298–299, 298*f*, 299–300*b*, 633
Metals
  allergies to, 764, 781
  extraction vs. smelting of, **920*b***
  heavy, for microbial control, 218–219, 824
Metapneumovirus, 501*f*
Metastasis, 476
Metazoans, 126
Metchnikoff, Elie, 21*t*, 668, 678, 722
Methane gas
  anaerobic organisms and, 146, 147*b*, 183
  carbon cycle and, **893**
  covalent bonds in, 40, 40*f*
  Deepwater Horizon oil spill and, 41*b*
  laboratory creation of life and, 57–58*b*
  sewage treatment and, **887*f***, **888**
Methanogens, 104, 105*b*, 133, 183
Methanospirillum hungatei, 105*f*
Methicillin, 813*f*, 814. *See also Staphylococcus aureus*, antibiotic-resistant, methicillin
Methionine (Met), 245, **894**
Methyl red test, 183–184
Methylene blue, 670
Methylketones, **906**
Metronidazole
  for amoebiasis, 596
  as antiparasitic drug, 823, 825*t*
  for bacterial vaginosis, 410
  for giardiasis, 598
  for pseudomembranous colitis, 363
  for trichomoniasis, 600
MHC Class I (MHC-I) proteins, 684, 685*f*, 712*f*, 713, 714, 714t
MHC Class II (MHC-II) proteins, 712*f*, 713–714, 714t
Miasma, 9, 13*f*
Micatin, 573, 822
Miconazole, 573, 821–822, 825*t*
Microaerophiles, 146, 146*f*
Microbe Hunters (de Kruif), 306
Microbes. *See also* Microorganisms
  diversity of, 3
  fossil, 232*f*
Microbial control. *See also* Microbial growth
  chemical agents for
    alcohols, 219
    concept map of, 224*f*
    garlic organosulfur compounds, 216*b*
    general principles of, 209–211
    halogens, 217–218, 217*f*
    heavy metals, 218–219
    history of, 209
    peroxides, 219
    phenol and phenolic compounds, 218, 218*f*
    range of, 214, 215*b*
    selection considerations, 212–213, 212*f*
    soaps and detergents, 219, 220*b*, 220*f*
    as sterilants, 221, 222*b*
  general principles of, 197
  physical methods
    concept map of, 210*f*
    filtration, 204–206, 205*f*
    heat, 198–204
    irradiation, 206, 207*b*, 208
    for retarding food spoilage, 208
    summary of, 209*t*
    ultraviolet light, 206
  public concerns regarding, 195–196, 196*f*
Microbial ecology
  of cheese, **872–873*b***
  pioneers of, 22
  studying phytoplankton communities, 65
  in third Golden Age of microbiology, 30, 30*f*
Microbial evolution, 30–31, 65
Microbial genomics. *See also* Genomics
  ancestors in human genome segments, 293–294
  comparative genomics, 296–297, 297*f*
  DNA microarrays as tool for, 290–291*b*
  metagenomics, 298–299, 298*f*
  microbial forensics and, 295
  sequencing, 291, 292*b*, 293, 293*f*
    of pathogens, 289–290*b*
    targets for antimicrobial drugs and, 839
  Third Golden Age of microbiology and, 294–296
Microbial growth. *See also* Microbial control
  bacterial growth curve, 139–140, 139*f*
  bioremediation and, **916**, **917–918*b***, **918–919**
  in culture media, 151–152, 153*b*, 154*b*, 155
  physical factors in
    hydrostatic and osmotic pressure, 148
    oxygen, 146–147, 146*f*, 148*f*
    pH, 147–148
    temperature, 142, 144–146, 144*f*
    types of microbes based on, 150*f*
  unfavorable environments for
    endospores and, 141–142, 142*f*, 143*f*
    persister cells and, 140–141, 141*f*
Microbial life. *See also* Extremophiles
  in freshwater lakes beneath Antarctica, 149*b*
  study of, 36
Microbial reproduction. *See* Reproduction
Microbicidal agents, 197, 210, 808
Microbiology
  classical Golden Age of, 16–19*f*, 16–20, 21*t*, 22*f*, 807
  disciplines of, 22*f*
  discovery of microbes and, 6–9, 7*f*
  disease transmission theories and, 9, 13–15, 13*f*, 14*f*
  global pioneers in, 20, 22
  microbial forensics and, 295
  other global pioneers in, 20, 21*t*, 22
  preparation for careers in, 306, 450
  scientific inquiry and, 10*b*, 11*f*
  Second Golden Age of, 25–26
  Third Golden Age of, 1, 28–31, 30*f*, 294–296
Microbiome
  human, 632, 633
    intestinal, 5, 346–347, 347*f*
  skin, 426–427, 428*f*, 430*f*
Microbiostatic agents, 197, 210, 808
Microbiota
  colonization of neonate, 633, 635*f*
  indigenous
    airborne bacterial diseases and, 319*t*
    of GI tract, 346–347, 346*b*, 347*f*, 349, 350*f*
      apple-eating and, 348–349*b*
    of skin, 426–429, 428*f*
      acne and, 429–431, 429*f*
    of upper respiratory tract, 309
    of urinary tract, 420–421
  tetracyclines and destruction of, 816
Micrococcus, 429, **857**, **905**
Micrococcus luteus, 106, 677*b*, 826*f*
Microcompartments, 72, 123–124, 128*t*
Microfilariae, 619–620
Micrographia (Hooke), 6
Micromonospora, 816
Micronucleus, in ciliates, 593
Microorganisms. *See also* Viable but noncultured organisms
  as biological weapons, 654–655*b*
  classification of
    cataloging attempts, 73, 76–77
    concept map of development of, 76*f*
    hierarchical system for, 79–80
    identification methods, 80, 83
    kingdoms and domains, 77–78, 78*f*
    nomenclature, 78–79, 79*b*
  day in life of, 3, 4*f*, 5
  disease transmission and, 9, 12–13
  diversity of, 3, 36
  in oceans, 64–65, 65*f*
  size relationships among, 84, 84*f*, 85*b*
  in unpolluted and polluted water, **879**, **880*f***, **881**

Microscopes and microscopy
  alternate optical configurations for, 91–93, 91*f*, 92*f*
  electron
    cellular organization studies using, 25
    development and types of, 93–95, 93*f*, 94*f*, 95*t*
  improved optics of, 14
  Leeuwenhoek's use of, 1, 6–7, 7*f*
  staining techniques for contrast in, 87–88, 89*f*, 90–91, 90*b*
  units of measurement for, 83–84
Microsporidia, 560, 561
Microsporum spp., 572, 573, 573*f*, 575*t*
Microwaves
  cooking frozen foods in, **862**
  microbial control using, 208
  pasteurization using, **860*b***
  in spectrum, 208
Middle East, HIV infections in, 761, 762*f*
Migraine headaches, chronic, 360
Miliary tuberculosis, 325
Milk. *See also* Dairy products; Pasteurization
  allergies to, 768
  assays for quality of, **861*t***
  contamination of, **855–857**, **857*f***
    E. coli O157:H7, 368
    L. monocytogenes, 364
    Salmonella, 366
  disease transmitted by, 337, **867–868**
  fermentation of, **871–872**, **871*f***
  pasteurization of, 203–204, 203*f*, 204*b*
Milk sugar, 49
Miller, Stanley, 57–58*b*
Milstein, César, 756
Milwaukee protocol, for rabies, 540*b*
Mimiviridae (mimivirus, mimicking virus), 457*b*
Mineralization, **894**
Minimum inhibitory concentration (MIC), 826, 826*f*
Minocycline, 816, 820*t*, 836–837*b*
Miracidia (miracidium), 611, 614*f*
Mirage of Health (Dubos), 664
Miscarriages, immune system and, 785*b*
Mismatch repairs, 255–256
Missense mutations, 254, 255*f*
Mitchell, Margaret, 352*b*
Mitchell, Peter, 176
Mitochondria (mitochondrion)
  cellular respiration and, 72
  chloroplasts, bacteria, microbial eukaryotes and, 75*t*
  evolution of, 74*b*, 75*f*
  linezolid toxicity and, 818
Mitosporic fungi, 566–567
MMR (measles-mumps-rubella) vaccine, 511, 513, 730, 740
Mobiluncus, 409
Model organism, eukaryotic, 569
Moist skin sites, microbiota of, 427
Molan, Peter, 215*b*
Mold in Dr. Florey's Coat, The (Lax), 27*b*
Mold-ripened cheeses, **871–872**
Molds. *See also* Fungi; Spores
  aflatoxins and, 570
  allergies to, 766, 768*f*, 771*b*
  Hurricane Katrina and, 552–553
  life cycle of, 554, 560*f*
  wine production and, **910**
  zygomycetes as, 561, 562*f*
Mole, 171
Molecular formulas, 40, 41*f*
Molecular taxonomy, 83
Molecular weights, 44*b*
Molecules, 40, 84*f*
Molluscum bodies, 518
Molluscum contagiosum, 518, 518*t*
Monera (kingdom), 77
Moneres, 76
Monistat, 822
Monkey immunodeficiency virus, 482
Mono Lake, California, 36*f*
Monobactams, 814
Monoclonal antibodies (MAbs)
  biomedical uses of, 754, 756–757
  production of, 756*f*
  transplantation acceptance and, 786*t*
Monocytes, 671, 671*f*, 678, 680
  inflammation and, 682, 682*f*
Monod, Jacques, 249
Monolayer
  for cell cultures, 473, 474*f*
  lipid, in archaeal cell membrane, 121, 122*f*
Monomers, 47, 48*f*, 163
Mononucleosis, infectious, 476, 524, 525*f*, 528*t*, 641
Monosaccharides, 48, 48*f*, 49, 180
Monosodium glutamate, **905**
Monospot test, 524, 525*f*
Monotherapy, for HIV, 797–798
Montagnier, Luc, 695, 696, 789
Mood, immunity and, 715*b*
Morbidity and Mortality Weekly Report (CDC), 445, 651
Morbillivirus, 501*f*
Morchella esculentum (morels), 561, 563*f*
Moringa oleifera, **886*b***
Moritella, 134*t*
Mosquitoes
  arboviral encephalitis and, 544, 544*f*, 545*f*, 546–547, 547*t*
  avoiding, 545*b*
  chikungunya fever and, 532*b*
  dengue fever and, 531–532, 662*b*
  emerging viruses and, 481, 481*t*
  inflammatory response from bite of, 683*b*
  lymphatic filariasis and, 619–620
  malaria and, 586–587, 602, 602*f*
  microbial control of, **914**, **915**
  transgenic, **920**
  viruses able to destroy, 450
  West Nile virus and, 629–630
  yellow fever and, 453, 529, 530–531*b*
Most probable number (MPN) test, 155, **890**, **890*b***, **917*b***
MRSA. *See Staphylococcus aureus*, antibiotic-resistant, methicillin
Mucins, 309, 345
Mucociliary clearance, 309
Mucor, **906**
Mucosa-associated lymphoid tissue (MALT), 346, 672, 673*f*, 706–707*b*
Mucosal immunity, 705, 706–707*b*
Mucosal surface (mucous membranes), 110, 346, 675
Mucus, 310*f*, 675, 676*b*
Muenster cheese, **871**
Mullis, Kary, 288*b*
Multibacillary leprosy, 438
Multicellular microbial communities. *See* Biofilms
Multidrug-resistant (MDR) species
  healthcare-associated infections and, 267*t*
  increases in, 267, 805, 833–834
  S. aureus, 433–434, 828
  tuberculosis (MDR-TB), 326, 805, 834
Multiple sclerosis, 524, 783*t*
Mummies
  Egyptian, living endospores in, 144*b*
  Nubian, tetracycline in, 818*b*
Mumps. *See also* MMR vaccine
  infection allergy and, 780
  infections, 513, 518*t*
  Rubulavirus and, 501*f*
  vaccinations and incidence, mortality from, 726*f*
  vaccine for, 739*t*
Murine typhus, 398, 402*t*
Muromonab, 786*t*
Murray Valley encephalitis, 545*f*
Mushrooms. *See also* Fungi
  as basidiomycetes, 566, 566*f*
  commercial farming of, **915**, **915*f***
  poisoning by, 570–571
Must, in wine production, **908**, **910*f***
Mutagens, 252
Mutants, identifying
  Ames test for, 261, 262*f*, 263
  plating techniques for, 260, 261*f*, 262*f*
Mutations
  antibiotic resistance and, 827, 828–829, 828*f*
  defined, 252
  DNA replication errors and, 240
  emerging viruses and, 481–482
  genome size vs. rate of, 462, 462*f*
  as heritable changes in a genome, 252–254
  plague's pathogenicity and, 253*b*
  point mutations, 254–255, 255*f*
  repair mechanisms, 255–256
  study of, 25
  transposable elements, 256, 258, 258*f*
  vertical gene transfer and, 268
Mutilins, resistance to, 827*f*
Mutualism
  defined, 632
  fungi and, 556–557, 558*b*
  lichens and, 563
  nitrogen fixation and, **896–897**, **896*b***, **897*f***
  protists and, 591–592
Myasthenia gravis, 783, 783*t*
Mycelia (mycelium), 554
Mycetism, 570–571
Mycobacterium spp., 101, 116, 118
Mycobacterium avian complex (MAC), 327, 796*f*
Mycobacterium bovis, 326, 730
Mycobacterium chelonae, 326
Mycobacterium haemophilum, 326
Mycobacterium kansasii, 326–327
Mycobacterium leprae, 438, 440*t*
Mycobacterium marinum, **882–883**, **883*b***

Mycobacterium tuberculosis. See also Tuberculin reaction; Tuberculosis
generation time in, 138*t*
genome size of, 235*f*
multidrug-resistant, 326, 805, 834
pasteurization and, 203–204
persister cells and, 140–141
phagocytosis and, 681*b*
thermal death time for, 198
waxes in, 51
Mycolic acid, 118, 120*t*
Mycology, 22, 22*f*
Mycoplasma spp. (mycoplasmas)
agammaglobulinemia and, 789
as Firmicutes, 101
scanning electron micrograph of, 103*f*
sterols in, 51
urethritis and, 425*t*
without a cell wall, 118, 126*b*
Mycoplasma capricolum, 299–300*b*
Mycoplasma genitalium, 126*b*, 291, 292*b*
Mycoplasma laboratorium, 300*b*
Mycoplasma mycoides genome, 299–300*b*
Mycoplasma pneumoniae
allergic reactions and, 773*b*
bronchitis and, 328
lower respiratory diseases caused by, 338*t*
pleomorphic shape and colonies of, 334*f*
wall-less cytoskeleton of, 126*b*
Mycorrhizae, 557, 557*f*, 563, 565
Mycoses
of lower respiratory tract, 575–580
of the skin, hair, and nails, 572–575
Mycostatin, 573
Mycotoxins, 570–571
Myeloid progenitors, 670, 671*f*, 700, 700*f*
Myelomas, 756
Myonecrosis, 385
Myxobacteria, 69*b*
Myxococcus xanthus, 69*b*, 79*b*

N

N spikes, 497
NAD+. *See* Nicotinamide adenine dinucleotide
NADH. *See* Nicotinamide adenine dinucleotide
NADPH, 185, 187*f*, 188
Naegleria fowleri, 608–609, 608–609*b*, 609*f*, 610*t*
NAG (*N*-Acetylglucosamine), 49, 50, 116, 119*f*
Nails
candidiasis of, 574
ringworm of, 572
Naive T cells, 714
Naked viruses, 456, 456*f*
NAM (*N*-Acetylmuramic acid), 50, 116, 119*f*
Names, scientific, 78–79, 79*b*
viral, 460–461, 461*f*
Nannizzia, 572
Nanoarchaeum equitans, 235*f*
Nanometer (nm), 84
Napoleon Bonaparte, 400–401*b*, **858*b***
Narrow-spectrum antimicrobials, 809
Nasal infections, 314–315, 315*f*
Nasopharyngeal carcinoma, 477*t*, 524
Nasopharynx, 497
National Malaria Eradication Program, 586
National Nosocomial Infections Surveillance System, 267
Natural killer (NK) cells
cytotoxic hypersensitivity and, 774, 775*f*
immune system and, 670, 671*f*, 672
innate immunity and, 684–685, 685*f*
Natural penicillins, 812–813, 813*f*
Naturally acquired active immunity, 727, 728*f*
Naturally acquired passive immunity, 728*f*, 735
NDM-1 genetic element, 842
Necator americanus, 618, 619*f*, 620*t*
Necrosis, 436
Necrotizing fasciitis, 436–437, 437*f*, 440*t*
Needham, John, 7, 9
Needle contamination, 526, 526*f*, 528*b*, 533. *See also* Drug users, injecting
Negative control, 249, 249*f*
Negative stain technique, 87–88, 89*f*
Negative-strand (– strand) RNA viruses, 462, 467
Negri bodies, 473
Neisser, Albert L. S., 21*t*, 79*b*
Neisseria spp., 101, 147, 311
Neisseria gonorrhoeae
cells, 415*f*
on culture media, 152
diplococcal shape of, 106
gonorrhea caused by, 414–415
hyperacute bacterial conjunctivitis and, 442
multidrug resistance by, 805
naming of, 79*b*
neonatal conjunctivitis and, 443
nucleoid and ribosomes of, 124*f*
penicillin-resistant, 252
pili adhesins on, 110
silver nitrate and, 219
urethritis and, 421
Neisseria meningitidis, 106, 316, 316*f*, 318, 319*t*, 732
Nelson, Mark, 818*b*
Nematoda, 612
Neomycin, 816, 820*t*, 837*b*
Neonatal conjunctivitis, 413, 415, 443, 444*t*
Neonatal herpes, 508, 512*t*
Neonatal meningitis, 319
Neosporin, 816
Nervous system
immune system and, 715*b*
parasitic diseases of, 610*t*
babesiosis, 605
malaria, 601–603
primary amoebic meningoencephalitis, 608–609, 608–609*b*
trypanosomiasis, 603, 604*f*, 605
viral diseases of, 547*t*
arboviral encephalitis, 544, 544*f*, 546–547
HIV, 799*b*
polio, 542, 542*f*, 543*b*, 544
rabies, 539–542
Nestlé Company, **863*b***
Neuralgia, postherpetic, 509
Neuraminidase (N) spikes, 497, 497*f*, 498–499*b*
Neuraminidase inhibitors, 821
Neurospora, 25
Neurospora crassa, 677*b*
Neurotoxins, 359, 570–571, 646, 646*t*
Neutral spirits, **911**
Neutralization, 709–711, 711*f*, 744, 745*f*
Neutrons, 38
Neutrophil extracellular traps (NETs), 680, 681*f*
Neutrophiles, 147–148
Neutrophils, 670–671, 671*f*, 678
asthma and, 769
inflammation and, 682, 682*f*
phagocytosis and, 680
Nevirapine, 793–794*b*, 822*t*
Niches, ecological, 252, 648
Nicolle, Charles, 21*t*
Nicotinamide adenine dinucleotide (NAD+)
in citric acid cycle, 174*f*, 175
as coenzyme, 166
in glycolysis, 172*f*
in oxidative phosphorylation, 175, 176*f*, 177
Nietzsche, Friedrich, 406
Night soil, 618, **916**
Nigrosin (dye), 87–88
Nikkomycin, 577
Nipah virus, 481*t*
Nitrate, **894**
Nitric oxide (NO), 679
Nitrites, **864**
Nitrobacter spp., 189, **894**
Nitrogen (N), 37
Nitrogen cycle, 189, **894**, **895*f***, **896–897**
Nitrogen fixation
bacterial chemistry and, 54*b*
genetically engineered, **920**
in koa trees, **896*b***
nitrogen cycle and, **894**, **895*b***
on roots, **897*f***
Winogradsky on, 22
Nitrogenase, 54*b*
Nitroimidazoles, 823–824, 825*t*
Nitrosamines, **864**
Nitrosomonas, 189, **894**
Nitrous acid, 253, 254*f*
Nodulocystic acne, 431
Nomenclature
binomial system of, 78–79, 79*b*
viral, 460–461, 461*f*
Noncommunicable diseases, 649
Noncompetitive inhibition, in metabolic pathways, 167–168, 167*f*
Nonenveloped viruses, 456, 456*f*, 461*f*, 466, 467*f*
Nongonococcal urethritis (NGU), 411
Nonhalophiles, 148
Noninflammatory gastroenteritis, 357–360, 374*t*
Non-O157 STEC (Shiga toxin-producing *E. coli*), 370
Nonparalytic poliomyelitis, 542
Nonperishable foods, **852*f***, **853**
Nonpolar covalent bonds, 40, 41*f*
Nonpolar molecules, 40
Nonself. *See* Self tolerance
Nonsense mutations, 254, 255*f*
Norovirus
cell/tissue tropism, 458*t*
infections, 538, 539*t*, **883**
North America
emerging and reemerging diseases in, 663*f*
HIV infections in, 762*f*

Nosocomial infections
  chain of transmission, 660, 660*f*
  elements in, 651, 658–660, 658*f*
  levels of cleanliness and, 196
  sites of, 659*f*
  10 most common agents in, 659*t*
Novobiocin, 836–837*b*
Nubian mummies, tetracycline in, 818*b*
Nucleic acids
  antigenic properties of, 697
  in bacterial cells, 47*f*
  characteristics of, 51–53
  hydrogen bonds in, 42
  microorganism identification using, 83
  nitrogen fixation and, 54*b*
  synthesis, as antibacterial agent target, 815*f*, 818–819, 820*t*
Nucleobases, 52
Nucleocapsids, 455–456, 456*f*, 467–468
Nucleoid-associated proteins NAPs), 234, 237, 237*f*
Nucleoids
  of bacterial and archaeal cells, 122–123, 123*f*, 128*t*, 233
  DNA organization within, 233–234, 237
  transcription and translation localized in, 251, 251*f*
Nucleotides, 52, 52*f*, 233
  excision repair by, 256, 256*f*, 257*b*
Nutrient starvation, fruiting bodies and, 69*b*
Nutrition, balanced, reducing pneumonia incidence with, 333*b*
Nutritional mutants, 260
Nuts, allergic reactions to, 768, 769*b*
Nuttall, George, 21*t*
Nystatin, 573, 574, 810*b*, 821, 825*t*

O

O polysaccharide, 118, 119*f*
Obermeier, Otto, 21*t*
Obesity, adenoviruses and, 494, 495*b*
Obligate aerobes, 146, 171
Obligate anaerobes, 146
Obligate and intracellular parasites, 101
Obligate symbionts, genome size among, 233, 235*f*
Observation, as scientific technique, 7, 10*b*, 13
Obsessive compulsive disorder, 607*b*
Oceans
  gene transfers in, 277*b*
  marine snow in, 272*b*
  microbes in, 3, 4*f*, 5, 190*b*
  photosynthesis in, 64–65, 65*f*
Odors
  aroma additives and, **906**
  body, 427*b*
  of food spoilage, **853**, **855**
    avoidance of, 182*b*
  mate selection and, 713*b*
Oil industry
  bioremediation for spills by, 41*b*, **916**, **918*b***
  exploration, **920*b***
Okazaki fragments, 240*f*, 241
Old World hookworm, 618
Omeprazole, 373
Oncogenes, 477–478, 478*f*
Oncogenic viruses (oncoviruses), 476–478, 477*t*, 514–515
Onychia, 574
Oocysts, 599, 599*f*, 602*f*, 605, 606
Oomycetes, 588*f*, 590
Oparin, Aleksandr, 57*b*
Operating room, pathogen-contaminated, 211*f*
Operators, 249
Operons, 249, 249*f*, 250*b*
Ophthalmia of the newborn, 443
Opisthotonus, 384, 385*f*
Opportunistic pathogens
  fungal, 555, 576, 580
  infections due to
    commensals and, 636–637
    HIV/AIDS and, 795, 796*f*
    immunocompromised individuals and, 580
    nosocomial, 659
Opportunity (Mars rover), 134
Opsonins, 710
Opsonization, 710, 711*f*
Optimism, immune system and, 715*b*
Oral cavity
  bacterial diseases of, 354*t*
    dental caries, 350–352, 351*f*
    periodontal disease, 352–354
  digestive system and, 345, 345*f*
  microbiota of, 347, 347*f*, 632*f*
  yeast (candidiasis) infection, 573–574, 574*f*, 575*t*
Oral rehydration therapy, 362
Oral tolerance, 706*b*
OraQuick Advance Rapid HIV-1/2 Antibody Test, 798*b*
Orchitis, 513
Orders, in taxonomic classification, 79
Organelles, 72
Organic acids, **864**
Organic compounds
  in bacterial cells, 47*f*
  carbohydrates, 47–50, 48*f*, 49*b*
  concept map of, 56*f*
  functional groups on, 47*t*
  industrial production of, **903–905**
  lipids, 50–51, 50*f*
  nucleic acids, 51–53, 52*f*
  proteins, 53–54, 56*f*, 57
Organosulfur compounds (OSCs), 216*b*
Orientia (formerly *Rickettsia*) *tsutsugamushi*, 399
Origin of replication (*ori*C), 239*f*
Origin of transfer (*ori*T), 272
Ornithosis, 337
Oropharyngeal cancer, 477*t*, 514–515
Oropouche fever, 533
Orthomyxoviridae (orthomyxoviruses), 461*f*, 481*t*
Orthophenylphenol, 218, 218*f*
Oseltamivir, 821, 822*t*
Osmosis, 72, 208, **864**
Osmotic lysis, 116, 118*f*. *See also* Cell lysis
Osmotic pressure, 208, **863–864**
Otitis externa, 319*t*
Ouchterlony, Orjan, 745
Outbreaks of disease, 651
Outer membrane, 118, 119*f*, 120*t*
Oxacillin, 813*f*
Oxaloacetate, 174*f*, **904*f***
Oxazolidinones, 815*f*, 816*f*, 818, 820*t*
  resistance to, 827*f*, 832*t*
Oxidation, 173, 175
Oxidation lagoons, **886**
Oxidative phosphorylation, 175–177, 176*f*
Oxygen (O). *See also* Aerobic respiration
  as essential element, 37, 37*f*, 37*t*
  fermentation and, **902**
  food spoilage and, **853**
  fungal growth and, 555
  microbial growth and, 146–147, 146*f*, 148*f*, 150*f*
  microbial production of, 4*f*
  oxidative phosphorylation and, 175
  as photosynthesis byproduct, 185, 186
  as toxin, 147*b*
Oxygen revolution, cyanobacteria and, 104, 147*b*
Oxygenic photosynthesis, 185, 189
Oysters, raw, 365, **854**

P

Pace, Norm, 129*b*
Pacini, Filippo, 14
Paget's disease of the bone, 513
PAH (polynuclear aromatic hydrocarbons), **920*b***
Palindromes, 280
Palisade cell arrangement, 312
Pandemics. *See also* Epidemics
  cholera, 360, 361*b*
  defined, 651
  influenza, 496
    of 1918 and 1919, 490, 498–499*b*, 637, 720*b*
    antigenic variations and, 498–499*b*
  plague, 380–381, 389
Pap smear, 477, 514
Paper manufacturing, **905*t***
Papillomaviridae (papilloma viruses), 286, 461*f*. *See also* Human papilloma viruses
Papovaviridae (papovaviruses), 474*t*, 477*t*
Papules, 382, 430*f*, 431, 516
Para-aminobenzoic acid (PABA), 811
Parabans, **864**
Parabasalids, 588*f*, 591, 591*f*, 594*t*
Parainfluenza, 493*f*, 501*f*, 502, 505*t*
Paralytic poliomyelitis, 542
Paramecium, 592*f*, 593, 593*f*, **879**
Paramyxoviridae (paramyxoviruses)
  classification of, 461*f*
  cytopathic effects of, 474*t*
  as emerging viruses, 481*t*
  infections, 500, 502, 511, 513, 518*t*
Parasitemia, 637
Parasites. *See also* Helminths; Protists
  antiprotistan agents and, 823–824
  of blood and nervous system
    babesiosis, 605
    malaria, 601–603
    primary amoebic meningoencephalitis, 608–609, 608–609*b*
    summary of, 610*t*
    toxoplasmosis, 605–606
    trypanosomiasis, 603, 604*f*, 605
  chemoheterotrophs as, 191
  eradication of, 621
  genome size among, 233, 235*f*
  as global concern, 586–587, 587*f*

international travel and diseases caused by, 449
as manipulators, 615*b*
of skin and digestive and urinary tracts
amoebiasis, 595–596, 597*f*, 598*b*
cryptosporidiosis, 598–599, 599*f*
cyclosporiasis, 599
giardiasis, 597–598, 598*f*
leishmaniasis, 595–596
summary of, 601*t*
trichomoniasis, 599–600, 600*f*
types of, 587
Parasitism, 633–634, 636
Parasitology, 22*f*, 587
Parenteral antischistosomal therapy (PAT), 528*b*
Parenteral portal of disease entry, 641
Parkinson's disease, 372*b*
Paromomycin, 596, 824
Parotid glands, mumps and, 513
Parotitis, infectious, 513
Paroxysmal stage, of whooping cough, 321
Paroxysms, 321
Particle transmission of disease. *See* Aerosols
Parvoviridae (parvoviruses), 461*f*, 513–514, 514*f*
PAS (*p*-aminosalicylic acid), 812
Passive agglutination, 746, 747*f*
Passive immunity, 728*f*, 734–737
Pasteur, Louis
competition with Koch, 18–19
on garlic as antiseptic, 216*b*
germ theory of disease and, 16–17, 19*f*
pasteurization method and, **859–860**
private side of, 20*b*
scientific method of, 2
spontaneous generation experiments by, 9, 10*b*, 11*f*
vaccines developed by, 669, 722
on wine, **911*b***
World Rabies Day and, 523
on yeast cells and fermentation, 16, 16*f*, 20, **901**
Pasteurization
of eggs, **860*b***
microbial control using, 209*t*
of milk, 203–204, 203*f*, 204*b*
of wine and milk, 16, **859–860**
Pasteurizing dose, 207*b*, 208
Patch tests, of skin, 781. *See also* Skin, allergy tests
Pathogen-associated molecular patterns (PAMPs), 678, 679, 679*f*, 679*t*, 680*f*
Pathogenicity
evolution and, 296
use of term, 633–634
Pathogenicity islands, 296–297, 297*f*, 634, 636
Pathogens. *See also* Immunity; Viable but noncultured organisms
bacteria as, 182*b*
chemoheterotrophs as, 191
Chlamydiae as, 104
contaminating an operating room, 211*f*
disease development and, 636–637, 636*f*
DNA microarrays for identification of, 291*b*
emerging, surveillance and diagnosing of, 289–290*b*
emerging and reemerging, 660–661
hospital source of, 658–659, 658*f*
innate immunity and identification of, 677*b*
microbes as, 4*f*, 5
in milk, tests for, **861*t***
similar-looking, differentiating, 153*b*
Pattern recognition receptors, 678, 679*f*
Paucibacillary leprosy, 438
PCV13 (pneumococcal conjugate vaccine), 319, 332
Peanut allergy, 768, 769*b*
Pébrine (silkworm disease), 17
Pectinases, **905**, **905*t***, **906*b***
Pelagibacter ubique, 3–4, 5, 235*f*
Pelletier, Pierre, 823*b*
Pelvic inflammatory disease (PID), 411, 413, 415, 441*b*
Penetration
inhibitors, 821, 822*t*
in viral replication, 463*f*, 464*f*, 465, 465–466, 467*f*
Penicillin(s). *See also Penicillium* spp.
allergic reactions to, 765
for anthrax, 382
antimicrobial spectrum of, 809*f*
cell wall synthesis and, 812–814, 815*f*, 820*t*
for diphtheria, 314
discovery of, 807–808, 807*f*
in farm animals, FDA on, 835*b*
global market for, 819*f*
industrial production of, **912**
for meningitis, 318
N. gonorrhoeae resistance to, 252
for pneumococcal pneumonia, 332
preservation during World War II of, 27*b*
resistance to, 827*f*, 832*t*
for streptococcal skin diseases, 436
for syphilis, 418
systemic anaphylaxis and, 768
testing allergies to, 751
for tetanus, 384
for toxic shock syndrome, 434
Penicillin G, 812, 813, 813*f*, 820*t*, 826*f*
Penicillin V, 812–813, 813*f*
Penicillium spp.
antibiotics production and, **912**
on bread, 23*f*
Fleming's discovery of, 26, 27*f*, 807–808, 807*f*
on growth medium, 554*f*
on orange, 563*f*
Penicillium camemberti, **871–872**
Penicillium chrysogenum, 555, 561, 820*t*, **912**
Penicillium griseofulvum, 822–823, 825*t*
Penicillium notatum, 820*t*, **912**
Penicillium roquefortii, 556, 556*f*, 559*f*, **873*b***, **906**
Penicilloic acid, 813, 813*f*
Penile cancer, 477*t*, 514–515
Pentamidine, 603, 824
Peptic ulcers, 373*f*
Peptide bonds, 54
Peptidoglycan(s)
C. trachomatis and, 412*b*
cell wall and, 116, 118, 118*f*, 119*f*, 120*t*
PAMPs and, 679*t*
sporulation and, 141
Peptostreptococcus spp., 353, 354*t*, 420
Perchlorate (ClO_4^-), 70
Perennial hypersensitivities, 766
Perforins, 685, 715, 718, 774
Perfumes. *See also* Odors
allergies to, 768*f*
production of, **906**
Period of convalescence, 640*f*, 641
Period of (disease) decline, 640*f*, 641
Periodontal disease (PD), 352–354
Periodontitis, 353–354, 353*f*, 354*t*
Peripheral membrane proteins, 121
Periplasm, 118, 119*f*, 120*t*
Perkin, William, 279*b*
Peroxides, 219
Persister cells, 140–141, 141*f*
Personality, immunity and, 715*b*
Pertussis (whooping cough)
as airborne bacterial disease, 320–321, 338*t*
cases reported in CA, 1945–2011, 308*f*
incidence in U.S., 1990–2010, 320*f*
maternal antibodies for, 735
outbreak of, 307–308
portal of entry for, 643
reported U.S. cases in 2010, 407*f*
vaccinations and incidence, mortality from, 726–727
vaccine for, 739*t*
Pertussis toxin, 646*t*
Pessimism, immune system and, 715*b*
Pesticides, **893**, **913–915**
Petri, Julius, 19*b*, 204
Petri dish, 19*b*, 151. *See also* Agar or agar plates; Culture media
Petrochemical wastes, **920*b***
Petroleum spills, 41*b*, **916**, **918*b***
Pets, exotic, jungle viruses and, 483*b*
Peyer's patches, 346, 706–707*b*
Pfeiffer, Richard, 21*t*
Pfiesteria piscicida, 589
pH
as acidity measure, 45, 46*b*
biological indicators of, 203, 203*f*
cellular chemistry and, 45–46, 46*f*
chemical agent selection for disinfection and, 212
dental plaque and, 49*b*
fermentation and, **902**
food spoilage and, **851**, **852*f***
fungal growth and, 556–557, 573
microbial growth and, 147–148, 150*f*
phagocytosis and, 679
scale, 45, 45*f*
of soaps, 219
of stomach, 346
vaginal, 409
Phage lambda (λ), 465, 469*b*
Phage therapy, 466*b*
Phage typing, 475
Phages, 277, 454, 464*f*, 465. *See also* Bacteriophages
Phagocytes, 672
Phagocytosis
antigen-antibody complexes and, 710, 711*f*
biofilms and, **889**
complement activation and, 686, 686*f*
encapsulated pathogens and, 116
infection allergy and, 778
inflammation and, 682, 682*f*
link between innate immunity and adaptive immunity and, 691*b*
Metchnikoff and theory of, 668
as nonspecific defense, 678–681, 680*f*
pathogen evasion strategies and, 681*b*

pathogen invasiveness and, 643, 643*f*, 644*f*
by pseudopods, 592, 592*f*
Phagolysosomes, 679, 680*f*, 691*b*
Phagosomes, 679, 680*f*
Phallotoxins, 570
Pharmaceutical industry
antibiotic development by, 839
antimicrobial drug development by, 834
vaccine development and, 741*b*
Pharmaceuticals. *See also* Drugs, therapeutic
as ergot derivatives, **906**
metabolic engineering of, 296
Pharyngitis
acute febrile, 493
chlamydial, 413
gonococcal, 415
streptococcal, 311–312, 319*t*, 436
Pharyngoconjunctival fever, 493–494
Phase-contrast microscopy, 91, 91*f*, 95*t*
Phases, clinical trial, 841*b*
Phenol and phenolic compounds, 218, 218*f*, 223*t*, **864**
Phenol coefficients (PC), 213, 213*t*
Phenolics, in wine vs. beer, **911*b***
Phenotype, defined, 260
pHisoHex, 218
Phlebotomus spp., 595–596, 595*f*
Phoenix Mars Lander, 134
Phosphatase test, of milk quality, **861*t***
Phosphate groups, 47*t*
3-Phosphoglycerate (3PG), 187*f*, 188
Phospholipids, 47*f*, 50, 50*f*, 119*f*, 120–121, 120*f*
Phosphorus (P), 37, 37*f*, 37*t*
Phosphorylation, 169*f*, 170, 173, 174*f*, 175–177, 828
Photoautotrophs, 188–189, 189*f*
Photobionts, 563
Photoheterotrophs, 189, 189*f*
Photophosphorylation, 188
Photorhabdus luminescens toxin (Pht), **914**
Photosynthesis
by algae, 589, 881
carbon cycle and, **892–894**, **893*f***
carbon-fixing reactions of, 188
chemical energy and, 185–186
by cyanobacteria, 64, 65*f*, 103–104, 147*b*
by diatoms, 590, 881
energy metabolism and, 72
energy-fixing reactions of, 185–188, 187*f*
glucose synthesis and, 48
microbes and, 4*f*
microbes that perform, 185*f*
oxygenic, 189
Photosystems, 186–187, 187*f*, 188
Phototaxis, 113
Pht (*Photorhabdus luminescens* toxin), **914**
Phycoanavirus, 457*b*
Phycology, 22, 22*f*
Phyla (phylum), in taxonomic classification, 79
Phylochip, 291*b*
Phylogenetic relationships, DNA microarray assessment of, 291*b*
Phylogenetic trees, 77, 78*f*, 101*f*
Phylogenetics, microbial forensics and, 295
Phylogeny, 77
Physical barriers to disease, 675, 675*f*, 690*f*
Physical characteristics, for microorganism identification and classification, 80
Physical mutagens or carcinogens, 252–253, 476, 476*f*
Physical pollution of water, **881–882**
Physical structure, food spoilage and, **851**, **852*f***
Physicians, antibiotic resistance and, 829, 833–834
Phytophthora infestans, 590, 590*b*
Phytophthora ramorum, 590
Phytoplankton, 64, 589
Picnics, food spoilage at, **866**
Picornaviridae (picornaviruses)
classification of, 461*f*
cytopathic effects of, 474*t*
enterovirus infections, 538, 539*t*
hepatitis A virus, 535–536, 536*f*, 537*b*
polioviruses, 542, 543*b*, 544
rhinoviruses, 492–493
Picovir, 822*t*
Pig whipworms, 613*b*
Pigeon fancier's disease, 778
Pigs. *See also* Pork; Swine flu
antibiotics in feed for, 835*b*
contamination of carcasses, 356
Pili (pilus)
adhesins and, 642–643, 642*f*
of bacterial and archaeal cells, 110, 110*f*, 127*t*
role of, 111–112*b*, 642–643, 642*f*
Pink eye, 442, 442*f*
Pinworm disease, 614, 617, 617*f*, 620*t*
Piperacillin, 813–814, 813*f*, 814, 820*t*, 837*b*
Plague. *See also Yersinia pestis*
in 14th century Europe, 380–381, 381*f*, 390*b*, 391*f*
in Arizona, 391*b*
as biological weapon, 654*b*
characteristics, 388–390, 402*t*
genetic changes in, 253*b*
vaccine for, 730
Plague meningitis, 389
Plague Pit, The (painting), 381*f*
Planktonic bacteria, **888–889**
Plantar warts, 514
Plants. *See also* Carbon cycle; Nitrogen cycle
fungi compared to, 554
genetically engineered, 286–287, 286*f*
land, 588*f*
mycorrhizae and, 557, 557*f*
Plaque (dental). *See* Dental plaque
Plaque (viral), 473, 474*f*, 475
Plasma, blood, 670, **863*b***
Plasma cells, 700, 701*f*, 702, 764, 764*f*
Plasma membrane, 71*f*, 120–121. *See also* Cell membrane
inhibitors, 821–822
Plasmids
of bacterial and archaeal cells, 123, 128*t*
genetic engineering using, 280–281, 281*f*, 282–283*b*, **919**
of microbial cells, 237–238
micrograph of, **921*f***
Plasminogen, 313*b*
Plasmodium spp., 594, 594*f*, 602*f*, 610*t*, 823. *See also* Malaria
Plasmodium falciparum, 291*b*, 601, 823, 824
Plasmodium malariae, 601
Plasmodium ovale, 601
Plasmodium vivax, 601
Plastics, carbon cycle and, **893**
Platelets, 670, 671*f*, 680
Plating techniques. *See* Culture media
Platyhelminthes, 611–612
Pleated sheets, of proteins, 54
Pleconaril, 821, 822*t*
Pluripotent stem cells, 670, 671*f*, 700, 700*f*
Pneumococcal meningitis, 317, 319
Pneumococcal pneumonia, 269–270*b*, 731
Pneumocystis carinii pneumonia, 789*b*
Pneumocystis jiroveci, 577–578, 580*t*
Pneumocystis pneumonia (PCP), 577–578, 580*t*, 795, 796*f*
Pneumonia
blastomycosis and, 577
community-acquired, 329–330, 332, 336, 338*t*
HA-MRSA and, 830*b*
hantavirus pulmonary syndrome and, 504
healthcare-acquired, 329, 338*t*
HHV-6 viremia and, 509
HIV/AIDS and, 796*f*
intracellular parasites and, 336–337
Klebsiella, 329
oral health and, 354*b*
pneumococcal, 269–270*b*, 731
Pneumocystis, 577–578, 580*t*, 789*b*
SARS and, 502
types of, 328–329
viral, 494, 502
Pneumonic plague, 253*b*, 390, 392–393*b*, 402*t*. *See also* Plague
Pneumovax 23, 332
Pneumovirus, 501*f*
Pocks, 517
Point mutations, 254–255, 255*f*. *See also* Mutations
Point source epidemics, 657*b*
Poison ivy rash, 780, 780*f*
Polar covalent bonds, 42, 42*f*
Polar molecules, 42
Polio (poliomyelitis), 542, 543*b*, 544, 547*t*, 621
vaccinations and incidence, mortality from, 725, 726*f*
Polio vaccines, 454, 542, 543*b*, 739*t*
Polioviruses
cell/tissue tropism, 458*t*, 459–460*b*
icosahedral symmetry of, 457*f*
micrograph of, 542*f*
viroid genome vs. genome of, 484*f*
waterborne, **883**
Pollen allergies, 768*f*
Polychlorinated biphenyls (PCBs), **916**
Polyclonal antibodies, 756
Polyenes, 821, 825*t*
Polymerase chain reactions (PCR), 287, 288*b*
Polymers, 47, 48*f*, 163
Polymicrobial diseases, 28, 431, 432*b*
Polymorphonuclear leukocytes (PMNs), 670–671
Polymyxin B, 815, 815*f*, 820*t*
Polymyxins, 120, 809*f*, 815, 827*f*
Polynuclear aromatic hydrocarbons (PAH), **920*b***
Polynucleotides, 52*f*, 233
Polyomaviridae (polyoma viruses), 477*t*
Polypeptide antibiotics, 815–816, 832*t*

Polypeptides in protein synthesis, 54
  translation and, 245, 246–247*f*
Polyporus, 570–571
Polysaccharides, 48*f*, 49, 180. *See also* Lipopolysaccharide
Polysomes, 246
Polysporin, 816
Pommerville Jeffrey C., 2
Pontiac fever, 336
Populations. *See also* Demographics, diseases and
  characterizing disease by, 658*b*
Porcelain filters, 204
Porins, 118, 119*f*, 120*t*
Pork. *See also* Pigs
  irradiation, 206
  methicillin-resistant *S. aureus* in, 358
  tapeworm disease, 612–613, 616*f*, 620*t*
  toxoplasmosis and, 606, 606*f*
  trichinellosis and, 617–618
  Y. enterocolitica contamination of, 371
Porphyromonas, 311, 353, 354*t*, 420
Port du Salut cheese, **871**
Portals of (disease) entry, 641–643, 642*f*
Portals of (disease) exit, 647, 648*f*
Positive selection plating technique, 260–261, 262*f*
Positive-strand (+ strand) RNA viruses, 462, 467, 468*f*
Postexposure immunization, 541
Postherpetic neuralgia, 509
Postpolio syndrome (PPS), 542, 544
Potability, water, **879**
Potassium (K), 37*t*
Potato blight, Ireland, 590, 590*b*
Potato spindle tuber (PST), 482
Potatoes, Peruvian preservation of, **863*b***
Poultry. *See also* Eggs; Turkey
  contamination of, 356, **854–855**
    C. jejuni, 371
    L. monocytogenes, 364
    Salmonella, 366
  fluoroquinolones banned in, 835*b*
  HACCP inspections of, **868**
  irradiation, 206
Pour-plate isolation method, 155, 157*f*
Powassan virus disease, 544, 545*f*
Pox: Genius, Madness, and the Mysteries of Syphilis (Hayden), 406
Poxviridae (poxviruses)
  classification of, 461*f*
  complex structure of, 456
  cytopathic effects of, 474*t*
  infections
    molluscum contagiosum, 518, 518*t*
    smallpox, 516–518, 516*f*, 517*b*
PPSV (pneumococcal polysaccharide vaccine), 319, 332
Praziquantel, 612, 620*t*, 824, 825*t*
Precipitation, 710, 711*f*, 744–745, 745*f*, 746*f*
Prednisone, 786*t*. *See also* Corticosteroids
Pregnancy
  acne and, 429
  chlamydia and, 413
  cystitis and, 421
  cytomegalovirus and, 525
  direct transmission of disease and, 649
  ectopic, 413
  gonorrhea and, 415
  hepatitis E and, 536
  HIV transmission and, 797
  immune system, miscarriages and, 785*b*
  periodontitis and, 354
  pyelonephritis and, 424
  rubella and, 513
  syphilis and, 417, 418
  systemic retinoids and, 431
  toxoplasmosis and, 608
Preschoolers, influenza vaccines for, 501*b*
President's Emergency Plan for AIDS Relief, 664
Prespores, 141
Pressure. *See also* Osmotic pressure
  atmospheric, microbial life and, 134*t*
  hydrostatic and osmotic, microbial growth and, 148, 150*f*
  killing endospores with, 142
Pressurized steam, for microbial control, 200–201, 209*t*
Prevacuum autoclaves, 203
Prevalence of disease, 657*b*
Preventive vaccines, 800
Prevnar 13, 332
Prevotella, 311, 347, 353, 354*t*, 409, 420
Prilosec, 373
Primaquine, 823, 823*b*, 825*t*
Primary active TB disease, 323, 325
Primary amoebic meningoencephalitis (PAM), 608–609, 608–609*b*, 610*t*
Primary antibody deficiencies (PADs), 788–789
Primary antibody response, 705, 709*f*, 752
Primary atypical pneumonia, 332
Primary cell cultures, 473
Primary HIV infection, 794
Primary immunodeficiencies, 787
Primary infections, 637
Primary lymphoid tissues, 672
Primary metabolites, **903**
Primary producers, 589
Primary sewage treatment, **886–887**, **887*f***
Primary sex organs, male and female, 408
Primary structure, of proteins, 54, 55*f*
Primary syphilis, 417, 417*f*
Primary TB infection, 323
Primary transmission cycle, of arboviral encephalitis, 544, 544*f*
Primaxin, 814
Prions, 484–485, 485*f*
Private Science of Louis Pasteur, The (Geison), 20*b*
Probiotics, 633, 634*b*
Prochlorococcus, 64, 65*f*, 66, 235–236*b*
Proctitis
  chlamydial, 413
  gonococcal, 415
Prodromal phase of disease, 640, 640*f*
Produce contamination
  E. coli O157:H7, 368
  Salmonella, 366
Productive infections, 465–469
Products. *See also* Industrial microbiology
  in chemical reactions, 42–43, 165, 165*f*
  of fermentation, 183–184, 184*f*
Progenote, 78
Proglottids, 612
  gravid, 613
Programmed cell death, 478, 479*f*
Prokaryotes (prokaryotic cells). *See also* Archaea; Bacteria; Cell(s)
  behaviors, 66, 67*f*, 68, 70
  cell shapes and arrangements, 106
  cellular organization, 26*f*
  characteristics, 65
  organizational patterns
    of eukaryotes and, 71, 71*f*
    of eukaryotes vs., 72–73, 73*t*
  as term for bacterial and archaeal cells, 129*b*
Promoters, 242, 249
Prontosil, 26, 807
Proof numbers, **911**
Propagated epidemics, 657*b*
Prophages, 464*f*, 465
Propionibacteria, 428–429
Propionibacterium spp., 184*f*, 427, 430*f*, **871**, **905**
Propionibacterium acnes, 429–431, 429*f*, 440*t*
Propionic acid, 183, 184*f*, **864**
ProQuad vaccine, 730
Prostaglandins, 766, 767*f*
Prostatitis, 424, 425*t*
Protease inhibitors, 797, 821
Proteases, **905*t***, **906**
Protein-coding genus, 457*b*
Protein-only hypothesis, 484–485
Proteins
  antigenic properties of, 697
  in bacterial cells, 47*f*
  in enzymes, 166
  as genetic material, Hershey-Chase experiment on, 59*b*
  hydrogen bonds in, 42
  metabolism, 180–181
  as organic compounds, 53–54, 56*f*, 57
  stabilizing, in DNA replication, 240, 240*f*
  synthesis
    as antibacterial agent target, 816–818, 816*f*
    in bacteria and eukaryotes, 71
    central dogma of gene expression and, 241–242, 242*f*
    inhibition, drug resistance and, 829, 832*t*
    translation in, 245–246, 246–247*f*
  transport in prokaryotic and eukaryotic cells, 72
Proteobacteria
  chemoautotrophic, 190*b*
  as extremophiles, 134*t*
  as indigenous microbiota, 311, 312*f*
  as phylum, 101, 101*f*
  skin microbiome and, 426–427
Proteus spp., 425*t*, **855**, **857**, **879**
Proteus mirabilis, 421*f*
Protists. *See also* Cysts; Parasites
  antiprotistan agents and, 823–824
  brief survey of, 23*f*, 24
  classification of, 588, 588*f*
  diversity among, 588–591
  eukaryotic cellular organization of, 25
  microbial control of, 200
  of Protista kingdom, 76–77
  supergroups of, 588*f*, 591–592, 594*t*
  as type of parasite, 587
  use of term, 587
Protochlamydia amoebophila, 412*b*

Protons, 38
in chemiosmosis, 175–177, 176*f*, 177–178*b*
in photosynthesis, 185–188
Proto-oncogenes, 477–478, 478*f*, 479*f*
Prototrophs, 260
Protozoa. *See also* Protists
cellular organization of, 26*f*
early research on, 20
as parasites, 591
Protozoology, 22, 22*f*
Providencia stuartii restriction enzyme, 280*t*
Provirus, 470, 470*f*, 792, 792*f*
Provolone cheese, **871**
Prp gene, 484
Prusiner, Stanley, 484
Pseudohyphae, 573
Pseudomembrane, 314
Pseudomembranous colitis, 363, 363*f*, 817
Pseudomonas spp.
agammaglobulinemia and, 788–789
cyanocobalamin production and, **905**
Deep Water oil spill and, **918*b***
ear infections and, 316, 319*t*
food spoilage and, **853**, **855**, **857**
nitrogen fixation and, **896**
penicillin-resistant, 828
in polluted water, **879**
as resistant indigenous microbiota, 311
traumatic wound infections and, 438*t*
urinary tract infections and, 425*t*
wound healing in diabetics and, 432*b*
Pseudomonas aeruginosa
burn infections and, 437–438, 440*t*
cystic fibrosis pneumonia and, 68
electron microscopy of, 94*f*
endophthalmitis due to, 202*b*
as ESKAPE pathogen, 839
folliculitis and, 433
nosocomial infections and, 659*t*
pneumonia due to, 329
septicemia due to, 638*b*
tobramycin and, 810*b*
urinary tract infections and, 421*f*
Pseudopeptidoglycan, 118
Pseudopods, 592, 592*f*
Psittacosis, 337, 338*t*, **855**
*Pst*I, 280*t*
Psychoneuroimmunology, 715*b*
Psychrophiles (psychrophilic organisms), 142, 144, 556, **881**
Psychrotrophic microorganisms, **861**
Psychrotrophic or psychrotolerant organisms, 144–145, 144*f*
Public health
deficiencies, disease emergence or resurgence and, 661
public transportation and, 653*b*
sanitation and, 195–196, 196*f*
sterilization and sanitization and, 197, 197*f*
vaccinations and, 725–726, 726*f*
Public health inspectors, **848**
Public transportation. *See also* Travel, international
disease transmission on, 653*b*
Pulmonary anthrax, 382
Pulmonary aspergilloma, 580
Pulmonary embolisms, 644
Pulp and paper manufacturing, **905*t***
Pure cultures, 18, 155
Purified protein derivative (PPD) test, 325*f*, 780, 780*f*
Purines, 52, 52*f*
Purple nonsulfur bacteria, 188
Purple sulfur bacteria, 188
Purulent discharges, 421
Pus, 682*f*, 683
Pustules, 382, 431, 516–517
Pyelonephritis, 424, 425*t*
Pyrimidines, 52, 52*f*
Pyrogenic toxin, 646*t*
Pyrogens, 683–684
Pyruvate
in citric acid cycle, 173
fermentation and, 183, 184*f*
glycolysis and, 171, 171*f*, 172*f*
as primary metabolite, **903**, **903*f***, **904*f***

Q

Q fever, 336–337, 338*t*, 643, **867**
Quarantines, 9
Quaternary ammonium compounds, 219, 220*f*, 223*t*
Quaternary structure, of proteins, 55*f*, 57
Quats, 219, 220*f*
Questions, in scientific inquiries, 10*b*
Quinine, 603, 823, 823*b*, 825*t*
Quinolones, 809*f*, 812, 815*f*, 827*f*
Quinupristin-dalfopristin, 817–818, 820*t*
Quorum sensing (QS), 68, 68*f*, 69*b*, 141, **889**

R

R groups, 54, 55*f*
R plasmids, 123
R strain pneumococci, 269–270*b*, 271
Rabbit fever, 391
Rabies
cases reported in U.S., 541*f*
global risk for, 523*f*
immune globulin shots for, 522–523
infections, 539–542, 540*b*, 547*t*
Rabies virus
in canine salivary tissue, 540*f*
cell/tissue tropism, 458*t*, 459–460*b*
helical symmetry of, 457*f*
vaccine for, 18–19, 523
Radiation
as carcinogen, 476, 476*f*
endospores and, 141–142
microbial control using
for food preservation, 206, 207*b*, 208, **864–865**, **865*t***
ultraviolet light, 206
microbial life and, 134*t*
spectrum for, 206*f*
virotherapy using, 480*b*
Radioactive dump sites, bioremediation of, **918–919**
Radioallergosorbent test (RAST), 751, 751*f*
Radioimmunoassay (RIA), 750–751
Radioisotopes, 38
Radiolarian ooze, 589
Radiolarians, 588*f*, 589, 589*f*, **881**
Radura symbol, **865**, **865*f***
Ragweed rhinitis, 766
Raltegravir, 793–794*b*, 822*t*
Raoult, Didier, 401*b*
Rapid plasma reagin (RPR) test, 418
Rats, Lice, and History (Zinsser), 398, 399*b*, 400*b*
Rats, toxoplasmosis and, 606, 606*f*, 607*b*
Raw foods, bacterial species on, 356, 365
Reactants, chemical, 42–43
RecA, 257*b*
Recognition phase, of immune system, 672
Recombinant DNA molecules, 279–280, 281*f*, 284, 285*f*
Recombinant F^- cells, 274
Recombinant subunit vaccines, 731
Recombivax HB vaccine, 286, 731
Red algae, 36*f*, 588*f*, 589
Red eye, 442
Red rot, **855**, **857**
Red tides, 589, **881**. *See also* Algal blooms; Blooms
Red wine, **911*b***
Redi, Francesco, 7, 8*b*
Redman, Regina, 558*b*
Reductase test, of milk quality, **861*t***
Reduction, 173, 175
Redundancy, in genetic code, 244
Reduviid bugs, 603, 605
Reed, Walter, 20, 21*t*, 453, 529, 530*b*
Reemerging (resurgent) infectious diseases, 28–29, 29*f*, 660–661, 663–664, 663*f*
Refractive index, 87
Refrigeration. *See also* Cold environments
for food preservation, 208, 209*t*, **861–862**, **866**
microbial growth and, **853**
Regressive Evolution Hypothesis, for viruses, 486
Regulatory genes, 249, 249*f*
Regulatory T cell theory, 782, 782*f*
Regulatory T cells, 698, 785*b*
Reindeer moss, 565
Rejection, transplantation, 764*f*, 784, 786
Relapsing fever, 395–396, 396*f*, 397*b*, 402*t*
Relaxation techniques, 715*b*
Release
HIV, 792*f*
in viral replication, 463*f*, 464*f*, 465, 467–468
Relenza, 500, 822*t*
Rems (Roentgen Equivalent in Man), 141–142
Rennet, **871**
Rennin, **871**, **871*f***
Reoviridae (reoviruses), 538
Replication factories, for DNA, 239*f*, 240, 240*f*
Replication forks, 239*f*, 240
Replication origin (*oriC*), 240
Repressor proteins, 249, 249*f*
Reproduction
asexual. *See also* Binary fission
bacterial sporulation, 141–142, 143*f*
fungal sporulation, 557–558, 560*f*
by bacterial and archaeal cells, 108–109
cell cycle and, 135, 136*f*, 137*b*, 138
by fungi, 557–558
in men and women, 409*f*

sexual, 599
conjugation in ciliates, 593–594
Cryptosporidium, 599
in fungi
basidiomycetes, 566, 566*f*, 567*f*
sporulation, 558, 560*f*
zygomycetes, 561, 562*f*
in men and women, 408–409
Reservoirs of disease (infection), 358, 502, 533, 648–649
Resistance. *See* Antibiotic resistance; Drug resistance
Resistance factors, 238. *See also* R plasmids
Resolution (resolving power), in microscopy, 85, 87, 87*b*
Respiration. *See* Cellular respiration
Respiratory allergies, 764*f*. *See also* Anaphylaxis; Asthma; Hay fever
Respiratory burst, phagocytosis and, 679
Respiratory diseases
identifying, 339*b*
lower
bacterial
anthrax, 337, 338*t*, 382
bronchitis, 328, 328*f*
M. chelonae, 326
parasitic, 336–337
pertussis, 320–321
pneumonia, 328–330, 332, 336
summary of, 338*t*
tuberculosis, 321–323, 324*f*, 325–326, 327*b*
fungal
blastomycosis, 577
coccidioidomycosis, 577, 578*f*, 579*b*
cryptococcosis, 575–576
histoplasmosis, 576–577
summary of, 580*t*
viral
hantaviruses, 502
influenza, 490, 496–497, 498–499*b*, 500, 500*b*, 501*b*
SARS coronavirus, 502, 503*b*, 504*f*
summary of, 505*t*
nosocomial, 659*f*
testing for transmission of, 334*b*
upper
bacterial, 314
diphtheria, 312, 314, 314*f*
of the ear, 315–316
epiglottitis, 314
meningitis, 316–319
of the nose, 314–315, 315*f*
pharyngitis, 311–312, 413–415, 436
summary of, 319*t*
tularemia, 391
viral
adenoviruses, 493–494, 496
paramyxoviruses, 500, 501*f*, 502
rhinoviruses, 492–493, 494*b*
seasonal variations in, 493*f*
summary of, 496*t*
viral, maternal antibodies and, 735
worldwide deaths due to, 631*f*
Respiratory droplets
direct disease transmission and, 649, 650*f*
fifth disease and, 514
meningococcal meningitis and, 317
parainfluenza and, 502
pertussis and, 321
plague and, 389–390
strep throat and, 311–312
varicella-zoster virus and, 508
viral URT diseases and, 493, 496*t*
Respiratory syncytial (RS) disease, 493*f*, 502, 505*t*
Respiratory syncytial virus (RSV), 501*f*, 502, 505*t*
Respiratory system
anatomy, 309, 310*f*, 311
as portal of disease entry, 641
Respirovirus, 501*f*
Restriction endonuclease enzymes, 280, 280*t*, 281*f*
Resveratrol, **911*b***
Reticulate bodies (RBs), 411, 413*f*
Retin-A, 431
Retinitis, CMV-induced, 525
Retrovir, 822*t*
Retroviridae (retroviruses)
classification of, 461*f*
as emerging viruses, 481*t*
human endogenous, 293–294
as oncoviruses, 477*t*, 479*f*
provirus formation by, 470, 470*f*
replication of, 462
Retting flax, for linen, **905**, **906*b***
Reverdin, Jacques, 784
Reverse transcriptase, 462, 470, 479*f*, 790, 790*f*, 791–792, 792*f*
Reverse transcriptase inhibitors, 797, 821
Revertants, 261
Reyataz, 822*t*
Reye syndrome, 500, 508
Rh disease, 776, 778*f*
Rhabditiform larvae, of hookworms, 618, 619*f*
Rhabdoviridae (rhabdoviruses), 461*f*, 474*t*, 539–542, 540*b*
Rhadinovirus, 506*f*
Rheumatic fever, 312, 778, 779*t*
Rheumatic heart disease, 312
Rheumatoid arthritis (RA), 683, 757, 778, 782, 783–784, 783*t*
Rhinitis, 314, 766
Rhinoviruses, 492–493, 493*f*, 494*b*, 496*t*
cell/tissue tropism, 458*t*, 459–460*b*
Rhizaria (supergroup), 588*f*
Rhizobium, **896**, **896*b***, **897*f***, **920**
Rhizopus, 555, 559*f*
Rhodobacteraceae, **918*b***
Rhodoferax ferrireducens, 179*b*
RhoGAM, 776
Ribavirin, 502, 527, 533, 822*t*
Riboflavin, **905**
Ribonucleic acid (RNA), 52–53. *See also* RNA
Ribose, 52
Ribosomal RNAs (rRNAs)
gene abundance testing, **917–918*b***
transcription of, 242, 244–245, 245*f*
translation and, 245–246, 246–247*f*
Ribosomes, 71, 123, 124*f*, 128*t*. *See also* Large (50S) ribosomal subunit; Small (30S) ribosomal subunit
modifications in, drug resistance and, 829, 832*f*
Ribozymes, 168*b*, 245
Ribulose 1,5-bisphosphate carboxylase/oxidase (RuBisCO), 188
Ribulose 1,5-bisphosphate (RuBP), 188
Rice-water stools, 362
Ricketts, Howard Taylor, 21*t*, 396
Rickettsia akari, 399
Rickettsia prowazekii, 398, 401*b*, 402*t*, 643
Rickettsia rickettsii, 396, 398–401, 398*f*, 402*t*
Rickettsia typhi, 398, 402*t*
Rickettsiae (rickettsia), 101, 396, 398–401, 398*f*, 400–401*b*
Rickettsialpox, 399
Rifampacin, 439
Rifampicin, 836–837*b*
Rifampin
antimicrobial spectrum of, 809*f*
nucleic acid synthesis and, 815*f*, 818–819, 820*t*
resistance to, 832*t*, 836–837*b*
for tuberculosis, 326, 812
Rifamycins, 827*f*
Riftia pachyptila, 190*b*
Rimantadine, 822*t*
Rinderpest, 621
Ringworm, 572, 573*f*
River blindness, 621
Rivers. *See also* Oceans; Water, polluted
death of, **880*f***
microorganisms in, 592
RNA. *See also* Messenger RNAs; Ribosomal RNAs; RNA viruses; Transfer RNAs
catalytic effects on enzymes of, 168, 168*b*
central dogma of gene expression and, 241–242, 242*f*
characteristics, 52–53
DNA compared to, 241*t*
replication errors, antigenic variation and, 498–499*b*
synthesis inhibition, drug resistance and, 832*t*
RNA interference (RNAi), 248*b*
RNA polymerase, 242, 243*f*, 251, 251*f*, 829
RNA tumor viruses, 477–478, 477*t*
RNA viruses. *See also* HIV
classification of, 461, 461*f*, 462
DNA evolution hypothesis and, 471*b*
replication of, 465–469
RNA world hypothesis, DNA world hypothesis vs., 471*b*
Robbins, Frederick, 454
Roberts, Craig, 713*b*
Rocky Mountain spotted fever (RMSF), 101, 396, 398, 398*f*, 402*t*
chloramphenicol for, 817
complement fixation test for, 747
Rodriguez, Russell, 558*b*
Roll Back Malaria, 587, 604*b*
Rolling-circle mechanism, 272, 274
Root-Bernstein, Robert S., 2
Rosalind Franklin: The Dark Lady of DNA (Maddox), 234*b*
Rose spots, 365
Roseola, 509, 510*f*, 512*t*
Roseolovirus, 506*f*
Ross, Ronald, 21*t*
Rotarix vaccine, 538
Rotary International, 542, 543*b*

RotaShield vaccine, 740
RotaTeq vaccine, 538, 740
Rotavirus
  cell/tissue tropism, 458*t*
  gastroenteritis, 536
  infections, 539*t*
  vaccine for, 538, 740, 741*b*
  waterborne disease and, **883**
Roundworms, 612, 613*b*, 614*f*. *See also* Helminths
  ascariasis and, 618
  hookworm disease and, 618–619, 619*f*
  lymphatic filariasis and, 619–620
  pinworm disease and, 614, 617, 617*f*
  trichinellosis and, 617–618
Roux, Émile, 18–19, 21*t*, 669
Rowe, Alexandra, 777*b*
RS (respiratory syncytial) disease, 493*f*
RSV-like illness, 502, 505*t*
Rubella (German measles). *See also* MMR vaccine
  infections, 513, 518*t*
  vaccinations and incidence, mortality from, 725, 726*f*
  vaccine for, 739*t*
RuBisCO (ribulose 1,5-bisphosphate carboxylase/oxidase), 188
RuBP (ribulose 1,5-bisphosphate), 188
Rubrobacter, 257*b*
Rubulavirus, 501*f*
Rum, **911**
Ruminococcus, 347
Ruska, Ernst, 93
Russula spp., 570–571
Rust fungi (rusts), 565–566

S

S strain pneumococci, 269–270*b*, 271
Sabia virus, 534
Sabin, Albert, 454, 542
Sac fungi, 561, 563, 563*f*, 565, 565*f*
Saccharopolyspora erythraea, 817
Saccharomyces spp., 183, 184*f*, **908**, **911**, **919**
Saccharomyces carlsbergensis, **907**
Saccharomyces cerevisiae, 79*b*, 561, 567, 569, **907**
Saccharomyces ellipsoideus, 567, 569, **910**
Sack, David, 343
SAFE strategy, for trachoma, 444
Safranin (red cationic dye), 88, 88*f*
St. Louis encephalitis (SLE), 544, 545*f*, 547*t*, 629
Salem witches, ergotism and, 571*b*
Sales, microbiology and careers in, 306
*Sal*I, 280*t*, 282*b*
Saliva, 352*b*, 705
Salivary gland cancers, 477*t*
Salk, Jonas, 454, 542
Salk polio vaccine, 730, 742–743*b*
Salmon, Daniel E., 21*t*
Salmonella spp.
  antibacterial soaps and, 220*b*
  antisera of, 83
  bacteriostatic control of, 208
  as biological weapon, 654*b*
  Caesar salad, egg spoilage and, **860*b***
  conjugation between other genera and, 274
  DNA adenine methylase and, 839
  foodborne disease and, 207*b*
  in frozen foods, **862**
  HIV/AIDS and, 796*f*
  infections in U.S., **850**, **850*f***
  infectious dose of, 642
  micrograph of, **862*f***
  pasteurization and, 204
  as Proteobacteria, 101
Salmonella enterica
  microcompartments of, 124*f*
  red wine and resveratrol effect on, **911*b***
  serotype Enteritidis, 51*b*, 366, **855**, **855*f***
  serotype Javiana, 366
  serotype Mississippi, 366
  serotype Newport, 366
  serotype Parathyphi, 366
  serotype Typhi
    cells, 368*f*
    on growth medium, 140*f*
    invasive gastroenteritis due to, 365–366
    phenol coefficients and, 213, 213t
    rod-shaped cells of, 106
  serotype Typhimurium
    Ames test and, 261, 262*f*
    antibiotics and, 810*b*
    cells, 368*f*
    salmonellosis due to, 366
    in unpasteurized milk, **867**
Salmonellosis, 366–367, 407*f*
Salpingitis, 411, 413, 415
Salt(s)
  extreme halophiles and, 104, 105*b*
  fermentation and, **869**
  for food preservation, 208, **864**
  formation of, 39
  microbial life and, 134*t*
Salvarsan, 26, 806–807
Samuel, Delwen, **908*b***
Sand filters, 204
Sand fly, 595, 595*f*, 596
Sandwich technique, 751
Sanitation
  Great Sanitary Movement, 14, **877–878**
  HACCP inspections for, **868**
  public health and, 195–196, 196*f*, **847**
Sanitization, 197, 211, 295
Saprobes, 191, 555, 565
Saquinavir, 822*t*
Sarcina arrangement, of cocci, 106
Sargasso Sea, 5
SARS (severe acute respiratory syndrome), 502, 503*b*, 505*t*
SARS (severe acute respiratory syndrome) coronavirus (SARS-CoV), 502, 504*f*, 505*t*
SARS-associated virus, 481*t*
Saturated fatty acids, 50, 50*f*
Sauerkraut, **864**, **869**
Sausage poisoning, **854**
Sausages, fermented, **870**
Scalded skin syndrome, 434, 434*f*, 440*t*, 719
Scanning electron microscope (SEM), 94, 95*t*
Scarlet fever, 277, 312, 313*f*, 319*t*
Schaudinn, Fritz, 21*t*
Schiaparelli, Giovanni, 133
Schick test, 744
Schistosoma spp., 620*t*, **920**
Schistosoma haemotobium, 612
Schistosoma japonicum, 612
Schistosoma mansoni, 612, 614*f*
Schistosomiasis, 528*b*, 612, 614*f*, 620*t*
Schizophrenia, 607*b*
Schleiden, Matthias, 70
Schmutzdecke, **884**
Schwann, Theodor, 70, 79*b*
Science. *See also* Microbiology
  preparation for careers in, 2
  study of, 36
Scientific inquiry or scientific method, 10–11*b*
Sclerotium, 570
Scolex, 612, 613
Scorpions, insecticides from toxin of, **914**
Scrapie, 484
Screening tests, 261, 262*f*, 757
Scrub typhus, 398–399
Scurvy, **869**
Seafood. *See also* Fish; Shellfish
  allergic reactions to, 768
  HACCP inspections of, **868**
  L. monocytogenes contamination of, 364
  spoilage of, **853–854**
Seasonal flu. *See* Influenza
Sebaceous glands, 426, 426*f*
Sebaceous sites, skin microbiota and, 427
Sebum, 426, 675–676
Secondary active TB disease, 323, 325
Secondary antibody response, 705, 709*f*, 752
Secondary immunodeficiencies, 787, 789
Secondary infections, 637
Secondary lymphoid tissues, 672
Secondary metabolites, 809, **903–905**
Secondary sewage treatment, **887–888**, **887*f***
Secondary structure, of proteins, 54, 55*f*
Secondary syphilis, 417, 417*f*
Secretory IgA (S-IgA), 704–705, 706–707*b*
Sedimentation, in water purification, **884**, **885*f***
Selective culture media, 151, 152*t*
Selective toxicity, 806, 808
Self tolerance, 698. *See also* Autoimmune diseases or disorders; Transplantation
Selzentry, 822*t*
Semiconservative DNA replication, 240
Semiperishable foods, **851**, **852*f***, **853**
Semisynthetic drugs, 808, **912–913**
Semisynthetic penicillins, 813–814, 813*f*
Semmelweis, Ignaz, 13–14, 13*f*, 223*t*
Senescence, 137*b*
Sense codons, 244
Sensitization, by allergens, 764–766, 765*f*
Sensitizing dose, 764
Sepsis. *See* Septicemia
Septa (septum), 135, 136*f*, 555
Septate fungi, 555
Septic shock, 638–639*b*, 683
Septic tanks, **886**
Septicemia, 210–211, 637, 638–639*b*, 789, 830*b*
Septicemic plague, 253*b*, 389, 402*t*. *See also* Plague
Sequela, of disease, 312
Sequence-based metagenomics, 298
Seroconversion, 794, 795*f*
Serogroups, 317
Serological reactions
  agglutination, 746, 747*f*, 748*b*
  characteristics, 743–744

complement fixation test, 746–747, 749*f*
as disease diagnosis tool, 754*b*, 755*b*
labeling methods
enzyme-linked immunosorbent assay, 751–752, 753*f*
immunofluorescence antibody test, 747, 749*f*, 750, 750*f*
radioallergosorbent test, 751, 751*f*
radioimmunoassay, 750–751
for microorganism identification and classification, 80, 83
neutralization, 744, 745*f*
precipitation, 744–745, 745*f*, 746*f*
Serology, 743
Serotonin, 767*f*
Serotypes, 436
Serrati, Serafino, **854*b***
Serratia spp., 274, 438*t*
Serratia marcescens, 334*b*, **854*b***, **855**, **857**
Serum, blood, 670. *See also* Blood, human
Serum sickness, 736, 777
70S ribosomes, 123, 124*f*
Severe acute respiratory syndrome (SARS), 502, 503*b*, 505*t*
Severe acute respiratory syndrome (SARS) coronavirus (SARS-CoV), 502, 504*f*, 505*t*
Severe combined immunodeficiency (SCID) disease, 788, 788*f*
Severe sepsis, 638–639*b*
Sewage treatment, 14, **877–878**, **885–888**, **887*f***
Sexually transmitted diseases (STDs). *See also* Contact bacterial diseases
bacterial
chancroid, 418–419
chlamydia, 411, 412*b*, 413–414, 413*f*, 414*b*
gonorrhea, 414–415, 415*f*
lymphogranuloma venereum, 419
as public health challenge, 410–411
summary of, 419*t*
syphilis, 416–418, 416*b*, 417*f*, 418*f*
fungal (vulvovaginitis), 573
identifying cases of, 441*b*
portal of entry for, 641
protozoal parasitic (*Trichomonas vaginalis*), 592, 599–600
as U.S. and world health problem, 407
viral
genital herpes, 507–508, 507*f*, 512*t*
genital warts, 477, 477*f*, 514, 515*b*, 515*f*
hepatitis B, 526
HIV, 789–800
molluscum contagiosum, 518
Sexually transmitted infections (STIs), 410
Seymour, Robert, 777*b*
Shadow biosphere, 35
Sheep. *See also* Lamb
anthrax and, 337, 382
Q fever and, 337
Shellfish. *See also* Seafood
allergies to, 768
contamination of, **854**
Gonyaulax catanella and, **883**
Shewanella, 134*t*
Shiga, Kiyoshi, 21*t*
Shiga toxin, 367
Shiga toxin-producing *E. coli* (STEC), 368, 370, **850**
Shigella spp., 101, 274, 356
Shigella dysenteriae, 367–368, 834
Shigella sonnei, 367–368
Shigella flexneri, 643
Shigellosis, 367–368
Shingles (herpes zoster), 508–509, 509*f*, 510*b*, 512*t*
HIV/AIDS and, 796*f*
vaccine for, 733, 734, 741*b*
Shock. *See also* Toxic shock syndrome
endotoxic, 647
septic, 638–639*b*
Shrubbery lichens (fruticose), 563, 565
Signs, 637, 640. *See also* Symptoms
Silent mutations, 254, 255*f*
Silkworms, pébrine disease and, 17
Silver nitrate ($AgNO_3$), 219, 223*t*, 443
Simian immunodeficiency virus (SIV), 482, 791*b*
Simian virus-40 (SV40), 279–280
Simple stain technique, 87, 89*f*
Simplexvirus, 506*f*
Sin Nombre virus (SNV), 481*t*, 504, 505*t*
Single-dose vaccines, 730
Single-stranded (ss) DNA, 461*f*, 462
Single-stranded RNA (ss-RNA), 461*f*, 462
replication of, 465–469
RNA tumor viruses and, 478
viroids and, 482, 484
Singulair, 772
Sinus infections, 314
Sinuses, 314, 315*f*
Sinusitis, 314, 319*t*
Skin
allergic reactions and, 767*f*, 776, 779*t*
allergy tests, 772, 774*f*, 781
anatomy, 424–426, 426*f*
indigenous microbiota of, 426–429, 428*f*, 430*f*, 632*f*
limiting pathogen entry, 675, 675*f*
Skin diseases
bacterial
acne, 429–431, 429*f*, 430*f*
leprosy, 438–439, 439*f*
M. chelonae and, 326
staphylococcal, 432–434, 433*f*
CA-MRSA, 831*b*
toxin-generated, 433–434, 434*f*, 435*b*
streptococcal, 434, 435*f*, 436–437, 436*f*
summary of, 440*t*
traumatic wounds and, 437–438, 437*f*
fungal
candidiasis, 573–574
dermatophytosis, 572–573
sporotrichosis, 574–575, 574*f*
summary of, 575*t*
protist parasites (leishmaniasis), 595–596, 595*f*, 596*b*, 601*t*
viral
chickenpox, 508–509, 509*f*
fifth disease (erythema infectiosum), 513–514, 514*f*, 518*t*
herpes simplex virus 1 or 2, 505–508, 507*f*
Kaposi sarcoma, 509–510, 511*f*
measles (rubeola), 511, 512*f*, 513
molluscum contagiosum, 518
mumps, 513
oncogenic, 514–515, 515*f*
roseola, 509, 510*f*
rubella (German measles), 513
shingles (herpes zoster), 508–509, 509*f*, 510*b*
smallpox, 516–518, 516*f*, 517*b*
summary of, 512*t*, 518*t*
warts, 514, 515*b*, 515*f*
S-layer, of archaeal species, 118
Sleeping sickness, 20, 603, 610*t*. *See also* *Trypanosoma* spp.
Slime, on foods, **854**
Slime layer, 115–116
Slopek, Stefan, 466*b*
Sludge, **887**, **887*f***, **888**
Small (30S) ribosomal subunit, 123, 124*f*, 246–247*f*, 816, 816*f*
Smallpox (variola)
as biological weapon, 654*b*, 655*b*
eradication debate, 517*b*
evolution of, 482
HIV and vaccine for, 757
infection allergy and, 780
infections, 516–518, 516*f*, 518*t*
microbes and, 4*f*
vaccine for, 9, 739*t*, 740
viroid genome vs. genome of, 484*f*
virus, 23*f*, 457*f*, 516*f*
Smells. *See also* Odors
of food spoilage, **853**, **855**
avoidance of, 182*b*
locker room, 427*b*
mate selection and, 713*b*
Smith, Geoffrey, 469*b*
Smith, Hamilton, 292*b*
Smith, Theobald, 21*t*, 605
Smoking, for food preservation, **864**
Smuts (smut fungi), 565–566
Snow, John, 13, 13*f*, 343, 361*b*, **878**
Soaps, 219, 220*b*
Social networking, bacterial, 69*b*
Sodium (Na), 37*t*. *See also* Salt(s)
Sodium chloride (NaCl), 39, 40, 40*f*
Sodium hypochlorite, 217, 223*t*
Soft chancre, 418–419
Softened water, **885**
Soil
bacteria in, 23*f*
as disease reservoir, 649
microbes in, 4*f*, 5, 22
Soilborne diseases
bacterial
anthrax, 382–383, 383*f*, 384*b*
gas gangrene, 384–386
leptospirosis, 386–388, 387*b*, 387*f*
summary of, 388*t*
tetanus, 384–386, 385*b*, 385*f*
as blood diseases, 381
parasitic, 620*t*
ascariasis, 618
hookworm disease, 618–619, 619*f*
Solfatara Crater, Pozzuoli, Italy, sulfurous steam vents of, 100*f*
Solutes, 44
Solvents, 44, 183, 184*f*

Somatic recombination, 709, 710*f*, 712
Soper, George, 367*b*
Soredia, 563, 565*f*
Sørenson, Søren P. L., 45
Sour curd, **857**
Sourdough bread starter, **856*b***
South America
  emerging and reemerging diseases in, 663*f*
  HIV infections in, 762*f*
Soy allergies, 768
Soy sauce production, **870**, **870*f***
Space, microbial life in, 134*t*, 149*b*
Spallanzani, Lazzaro, 9, **858**
Spanish flu (1918-1919), 490, 496, 498–499*b*, 720*b*
Sparkling wines, **910**
Spawn, for mushroom farming, **915**
Specialized transduction, 276*f*, 277
Species, 78–79
Specific epithet, 78–79
Specific immunity, 697. *See also* Adaptive immunity
Spherules, 577, 578*f*
Spikes
  of enveloped viruses, 456, 456*f*, 463*f*
  in HIV, 790, 790*f*
  viral penetration of cells and, 465–466, 467, 467*f*, 468*f*
Spiral shape, of bacterial cells, 106
Spirilla (spirillum) shape, of bacterial cells, 106, 107*f*
Spirillum, **896**
Spirit (Mars rover), 134
Spirochaetes (phyla), 101*f*, 104
  flagella, 113, 115, 115*f*
  as indigenous microbiota, 312*f*
Spirochete shape, of bacterial cells, 106, 107*f*
Spirochetes, disease transmitted by
  Lyme disease, 393–395, 394*f*, 395*f*
  relapsing fever, 395–396, 396*f*, 397*b*
  syphilis, 416–418, 418*f*
Spirogyra, 186*b*
Spiroplasma citri, 126*b*
Spleen, 672, 673*f*
Spontaneous generation, 7, 8*b*, 10*b*, 11*f*. *See also* Life
Spontaneous mutations, 252. *See also* Mutations
Sporangia (sporangium), 557
Sporangiospores, 561
Spore stain technique, 89*f*, 91
Spores. *See also* Endospores; Prespores; Sporulation
  bacterial, microbial control of, 200
  drying foods and, **862**
  fungal, germinating, 559*f*
  fungal, microbial control of, 200
  post-Hurricane Katrina distribution of, 552–553
Sporothrix schenckii, 574–575, 574*f*, 575*t*
Sporotrichosis, 574–575, 574*f*, 575*t*
Sporozoites, 578, 594, 594*f*, 602, 602*f*
Sporulation
  bacterial, 141–142, 143*f*
  fungal, 557–558, 559*f*
Spreading factor, 644
Sprouts, contamination of
  E. coli O104:H4, 370
  E. coli O157:H7, 368
Sputnik (virus), 457*b*
ss-DNA (single-stranded DNA), 461*f*, 462
ss-RNA (single-stranded RNA), 461*f*, 462
  replication of, 465–469
  RNA tumor viruses and, 478
  viroids and, 482, 484
Stabilizing proteins, in DNA replication, 240, 240*f*
Staining techniques, for contrast in microscopy, 87–88, 88*f*, 90–91, 90*b*
Standard plate count (SPC) procedure, 155, 158*f*, **861*t***, **889–890**
Standard precautions, 660, 660*f*
Staphylococcal contact diseases, 432–434, 433*f*
  toxin-generated, 433–434, 434*f*, 435*b*
Staphylococcal food poisoning, 357–358, 374*t*, **862**, **864**, **865**
Staphylococci (staphylococcus) shape, 106, 107*f*
Staphylococcus
  coagulase and invasiveness of, 644
  differentiating species of, 153*b*
  ear infections and, 316, 319*t*
  on epithelial cells of trachea, 310*f*
  as Firmicutes, 101
  as halotolerant, 148
  latex agglutination test for, 747*f*
  prostatitis and, 425*t*
  as skin microbiota, 427, 430*f*
  skin microbiota and, 429
  traumatic wound infections and, 438*t*
  in urethra, 420
  wound healing in diabetics and, 432*b*
Staphylococcus aureus
  antibiotic-resistant, 805, 805*f*
    membrane permeability and, 828–829
    methicillin (MRSA)
      appearance of, 805*f*
      community-acquired, 829*b*
      contact diseases and, 434
      hemodialysis infection and, 275*b*
      phage therapy against, 466*b*
      public health and, 814, 830–831*b*
      in raw pork, 358
    multidrug, 828, 833
    mutations and, 827
    vancomycin (VRSA), 275*b*, 805*f*, 815, 835*b*
    vancomycin-intermediate resistant, 830*b*
  blepharitis due to, 441–442
  cells, 357*f*
  contact diseases caused by, 440*t*
  enterotoxin A, 646*t*
  as ESKAPE pathogen, 839
  food poisoning due to, 357–358
  generation time in, 138
  genome size of, 235*f*
  as grape-like cluster, 106
  as indigenous microbiota, 311
  lower respiratory diseases caused by, 338*t*
  lysis of, 118*f*
  micrograph of, 103*f*
  nosocomial infections and, 338*t*, 659*t*
  phenol coefficients and, 213, 213t
  pneumonia due to, 329
  pyrogenic toxin, 646*t*
  septicemia due to, 638*b*
  simplicity of, 67*f*
  skin infections and, 432–434, 433*f*, 434*f*, 435*b*
  typing of, 357*f*, 358
Staphylococcus epidermidis, 79*b*, 427*b*, 431, 659
Staphylococcus equorum, **873*b***
Staphylococcus saprophyticus, 421*f*
Staphylokinase, 644
Starch, 49, 180
Start codons, 244
Stationary phase, in microbial growth, 139*f*, 140
Steam. *See also* Canning foods; Pasteurization
  pressurized, in microbial control, 200–201
STEC (Shiga toxin-producing *E. coli*), 368, 370, **850**
Steel slag in water, pH level and, 46*b*
Stegomyia (tiger) mosquito, 532*b*
Stegomyia aegypti, 529
Stem cells, pluripotent, 670, 671*f*, 700, 700*f*
Sterile tissues
  in human body, 632
  of living animals, **853**
Sterility
  chlamydia and, 413, 414*b*
  gonorrhea and, 415
  mumps and, 513
Sterilization. *See also* Microbial control
  autoclaves for, 200–201, 201*f*, 202*b*, 203, 203*f*, 221
  chemical agents for, 221, 222*b*, 224*f*
  commercial, **859**
  food preservation and, 199*b*
  sanitization vs., 197
  without pressurized steam, 203
Steroids, 769. *See also* Corticosteroids
Sterols, 50*f*, 51
Stewart, Eric, 137*b*
Stewart, William, 805
Stibogluconate, 596
Sticky ends, 280, 281*f*, 282–283*b*
Stilton blue cheese, **872**, **872–873*b***
Stomach flu, 536
Stone crabs, avoidance of rotten food by, 182*b*
Stop codons, 244
Stoxil, 822*t*
Strabismus, 359
Strawberry tongue, 313*f*
Streak-plate isolation method, 155, 157*f*
Strep throat, 311–312, 436
Streptobacilli (streptobacillus) cell arrangement, 106, 107*f*
Streptococcal contact diseases, 434, 435*f*, 436–437, 436*f*
Streptococcal pharyngitis, 311–312, 319*t*, 436
Streptococcal toxic shock syndrome (STSS), 436, 440*t*
Streptococci (streptococcus)
  as indigenous microbiota, 311
  shape, 106, 107*f*
Streptococcus spp.
  acid-tolerance of, 148
  as capnophiles, 147
  ear infections and, 316, 319*t*
  as Firmicutes, 101
  food spoilage and, **855**, **856–857**, **862**
  in polluted water, **879**
  prontosil for, 26
  as resistant indigenous microbiota, 311

slime on meats and, **854**
as transient microbiota, 632
in urethra, 420
wound healing in diabetics and, 432*b*
Streptococcus agalactiae, 319
Streptococcus cremoris, **871**
Streptococcus lactis, 138*t*, 183, **861**
Streptococcus mutans
as cocci in a chain, 106
conjugation and, 274
dental cavities and, 49*b*, 351, 352*b*, 354*t*
slime layer of, 116
wasabi and, 215*b*
Streptococcus pneumoniae
acute bacterial meningitis and, 316, 317, 318, 319*t*
agammaglobulinemia and, 788–789
bronchitis and, 328
cells, 330*f*
ear infections and, 316, 319*t*
as encapsulated pathogen, 116
immunodeficiency disorders and, 788
invasive, reported U.S. cases in 2010, 407*f*
lower respiratory diseases caused by, 338*t*
as molecular biology model system, 25
phagocytosis and, 681*b*
pneumococcal pneumonia and, 269–270*b*, 329–330, 332, 333*b*
septicemia due to, 638*b*
subunit vaccine from, 731
transmission studies, 330–331*b*
Streptococcus pyogenes
airborne bacterial diseases and, 311–312, 319*t*
blood clotting to control, 313*b*
cells, 435*f*
as cocci in a chain, 106
erythrogenic toxin, 646*t*
pathogen secretion from, 108–109
septicemia due to, 638*b*
skin infections and, 436, 440*t*
Streptococcus sobrinus, 351, 354*t*
Streptococcus thermophilus, 633, **871**
Streptogramins, 815*f*, 816*f*, 817–818, 820*t*
resistance to, 827*f*, 832*t*
Streptokinase, 313*b*, 644, **906**
Streptomyces spp.
as Actinobacteria, 101
for antibiotics production, 839
annual production, **912**
plasmids and, 238
cyanocobalamin production and, **905**
genetic engineering using, **919**
micrograph of, 103*f*, ***913f***
Streptomyces albus restriction enzyme, 280*t*
Streptomyces avermitilis, 825*t*
Streptomyces cattley, 814, 820*t*
Streptomyces coelicolor, 235*f*, 266, 267, 267*f*, 839
Streptomyces erythreus, 817
Streptomyces griseus, ***913f***
Streptomyces lincolnensis, 817, 820*t*
Streptomyces mediterranei, 818–819, 820*t*
Streptomyces nodosus, 825*t*
Streptomyces noursei, 810*b*, 821, 825*t*
Streptomyces orientalis, 815, 820*t*
Streptomyces tenebrarius, 810*b*
Streptomyces venezuelae, 817, 820*t*
Streptomycin
antimicrobial spectrum of, 809*f*
discovery of, 26
as genetic cue or weapon, 810*b*
for plague, 390
protein synthesis and, 123, 816, 820*t*
resistance to, 829, 832*f*, 837*b*
for tularemia, 391, 393
Stridor, 314
Stromatolites, 147*b*
Structural formulas, 40
Structural genes, 249, 249*f*
Styes, 441–442, 444*t*
Subacute sclerosing panencephalitis (SSPE), 513
Subclinical diseases, 640, 727
Subculturing, 155
Substrate-level phosphorylation, 173, 174*f*
Substrates, 80, 83*f*, 165, 165*f*
Subunit vaccines, 730–732
Succinate, 174*f*
Sucrase, 166, 166*f*
Sucrose, 48–49, 166, 166*f*, 180
Sudden oak death, 590
Sugar. *See also* Carbohydrates
beer production and, **907**, ***907f***
for food preservation, 208
wine production and, **908**, **910**
Sulfamethoxazole (SMZ), 811, 820*t*, 823
Sulfanilamide (SFA), 820*t*
Sulfhydryl group, 47*t*
Sulfites, **864**, **908**
Sulfolobus, 105*f*
Sulfonamides (sulfa drugs), 820*t*
antimicrobial spectrum of, 809*f*
competitive inhibition and, 811–812
DNA synthesis and, 815*f*
protistan cells and, 823
resistance to, 827*f*, 832*t*
Sulfur (S), 37, 37*t*
Sulfur cycle, ***894f***
Sulfur dioxide, **864**, **908**
Summer flu, 576–577
Sunlight, allergic reactions and, 768
Superantigens, 434, 719, 719*f*
Superbugs, 29
Supercoiled DNA, 234, 237
Superfund, ***920b***
Superinfections, 833
Superoxide anions (O_2^-), 679
Suppressor T lymphocytes, 770
Surface water, **879**
Surfactants, 219
Surgery
HA-MRSA and, 830*b*
nosocomial infections and, 659*f*
Sustiva, 822*t*
Suttle, Curtis, 451, 452
Svedberg units (S), 123
Swaart, Charlie, 162–163
Sweating, allergic reactions and, 768
Sweet acidophilus milk, **871**
Sweet curdling, of milk, **857**
Sweet wines, **910**
Swimmer's ear, 315–316
Swimmer's itch, 612
Swine flu, 295, 481, 491, 498–499*b*. *See also* H1N1 influenza virus
Swiss cheese, 183, **871**
Sylvatic plague, 389. *See also* Plague
Sylvatic yellow fever, 529
Symbiosis (symbiotic relationships). *See also* Endosymbiosis
ascomycetes and plants, 563
Glomeromycota and plants, 561
microbes and deep-sea hydrothermal vent giant tube worms, 190*b*
microbiota of the human body and, 631–633, 632*f*
nitrogen fixation and, ***896b***
parasitism as, 633–634, 636
Symbiotic species, **896**
Symmetrel, 822*t*
Symptoms, 640
Syncytia (syncytium), 473, 502
Syndromes, 359, 640
Synechococcus elongatus, ***892b***
Synercid, 817–818
Synthetic biology, 12*b*
Synthetic broth media, 151*t*
Synthetic drugs, 808, 811–812
Syphilis. *See also Treponema pallidum*
ancient methods for treating, 209
characteristics, 416–418, 417*f*, 418*f*, 419*t*
famous people with, 406–407
measuring incidence and prevalence of, 657*b*
salvarsan for, 26, 806–807
spread of, 305
testing for, 750, 750*f*
theories on origin of, 416*b*
Systema Naturae (Linnaeus), 78
Systematics, 77
Systemic anaphylaxis, 767*f*, 768
Systemic diseases, 637
Systemic inflammatory response syndrome (SIRS), 638–639*b*
Systemic lupus erythematosus (SLE), 778, 783, 783*t*, 784*f*, 788

T

T helper (T_h) cells, 713, 714*t*, 719, 763. *See also* T_h1 cells; T_h2 cells
T lymphocytes (T cells). *See also* Cytotoxic T cells
adaptive immunity and, 699
HIV, dendritic cells and, 691*b*
HIV and, 696, 696*f*
immune system and, 670, 671*f*, 672
immunodeficiency disorders and, 787–788
mucosal immunity and, 706–707*b*
origins of, 700, 700*f*
regulatory
autoimmune disorders and, 782, 782*f*
miscarriages and, 785*b*
self tolerance and, 782
Taenia saginata, 611*f*, 612–613, 620*t*
Taenia solium, 612–613, 616*b*, 620*t*
Tai Chi, shingles immunity and, 510*b*
Tail fibers, viral, 456, 463, 463*f*, 464*f*, 465
Tamiflu, 822*t*

Tapeworms, 611–613, 611*f*
Tattoos, toxic shock syndrome caused by, 435*b*
Tatum, Edward, 25, 271
Taxa (taxon), 77
Taxobactam, 814
Taxonomy, 77–78, 79–80, 80*t*, 81*b*. *See also* Nomenclature
  molecular, 83
  of viruses, 460, 461*f*
T-cell receptors (TCRs), 712–713, 712*f*
Td vaccine, 384
Tdap (vaccine), 321, 734. *See also* Diphtheria-tetanus-acellular pertussis vaccine
T-dependent antigens, 719
Tea drinking, interferon production and, 699*b*
Teeth, tetracycline staining on, 816, 817*f*
Teichoic acids, 116, 119*f*, 120*t*
Telithromycin, 836–837*b*
Temperate phages, 463, 464*f*, 465
Temperature. *See also* Body temperature; Cold environments; Heat
  disinfection and, 212
  fermentation and, **902**
  food microbiology and, **862*f***
  food spoilage and, 356, 359, **852*f*, 853**
  fungal growth and, 556
  microbial growth and, 142, 144–146, 144*f*, 150*f*
  physical control of microorganisms and, 201*f*
Temporal sensing, 113
10 × 20 initiative, on new antibiotics, 839
Tenofovir, 800
Termination, in DNA replication, 238, 239*f*, 240
Termination factors, 245, 247*f*
Terminators, 242
Tertiary sewage treatment, **887*f*, 888**
Tertiary structure, of proteins, 54, 55*f*
Tertiary syphilis, 417*f*, 418
Tetanospasmin, 383–384
Tetanus
  characteristics, 383–384, 385*f*, 388*t*
  clostridial endospores and, 142
  exposure to, 736
  injecting drug users and, 385*b*
  vaccines for, 314, 321, 384, 730, 734
Tetanus toxin, 646*t*
Tetanus-diphtheria (Td) vaccine, 734, 735*b*
Tetracyclines
  antimicrobial spectrum of, 809*f*
  discovery of, 818*b*
  in farm animals, FDA on, 835*b*
  for gastric ulcers, 373
  gene expression interference by, 246
  genetic engineering using, **919**
  global market for, 819*f*
  properties of, 816–817, 816*f*
  protein synthesis and, 123, 820*t*
  resistance testing, 826*f*
  resistance to, 827*f*, 832*f*, 832*t*, 837*b*
  translation and, 815*f*
Tetrad arrangement, of cocci, 106, 107*f*
Tetrahymena, 168*b*
Tetrahymena pyriformis, 336*f*
Tetramune, 318
T-even bacteriophages, 463
Texas fever, 605
Textile industry, microorganisms in, **905**, **905*t***, **906*b***
T_h1 (helper T1) cells, 712*f*, 713, 714, 716*f*
T_h2 (helper T2) cells
  allergic reactions and, 764, 765*f*, 769
  cell-mediated immune response and, 712*f*, 713, 714, 716*f*
  transplantation rejection and, 784
The Institute for Genomic Research (TIGR), 291
Theory, in science, 10*b*
Therapeutic dose, 808
Therapeutic vaccines, 800
Thermal death point, 198
Thermal death time (TDT), 198, 199*b*
Thermoactinomyces spores, 144*b*
Thermococcus gammetolerans, 257*b*
Thermoduric organisms, **861**
Thermophiles, 142, 144*f*
Thermotoga spp., 104
Thermotoga maritima, 296
Thimerosal, 218, 735*b*
Thiobacillus, **894**
Thioglycollate broth, 146, 148*f*
Thiomargarita namibiensis, 85*b*
Thiothrix, **894**
30S ribosomal subunit, 123, 124*f*, 246–247*f*, 816, 816*f*
Three-domain system, 77–78, 78*f*, 129*b*
Thrombocytopenia, 783*t*
Thrush, 573–574, 575*t*
Thylakoid membranes, 185
Thymine (T), 52, 52*f*, 253, 254*f*
Thymus, 672, 673*f*, 788
Ti (tumor-inducing) plasmid, 286–287, 286*f*
Ticks
  disease transmitted by
    anaplasmosis, 399, 401
    babesiosis, 605
    ehrlichiosis, 399, 401
    encephalitis, 755*b*
    Lyme disease, 393–395, 394*f*, 395*f*
    Powassan virus disease, 545*f*
    relapsing fever, 395–396, 397*b*, 402*t*
    Rocky Mountain spotted fever, 396, 398, 398*f*
    tularemia, 391, 393, 394*f*
  inflammatory response from bite of, 683*b*
  type I hypersensitivity and, 763
Tier 1 bioterror agents, 654*b*
Tigecycline, 817, 837*b*
Tiger (*Stegomyia*) mosquito, 532*b*
Tinactin, 573
Tinctures, 209. *See also* Iodine, tincture of
T-independent antigens, 719
Tinea capitis, 572
Tinea corporis, 572
Tinea cruris, 572
Tinea infections, 572–573, 575*t*
Tinea pedis, 572
Tinea unguium, 572
Tinidazole, 598, 600
Tissue cement, 644
Tissue immune complex formation, 764*f*. *See also* Immune complex hypersensitivity
Tissue plasminogen activator (TPA), 285*f*
Tissue tropism, of viruses, 458, 458t
Tissue typing, 784, 785*f*
Titer, 705, 743–744, 744*f*
Titration, 743
TM7 (indigenous microbiota), 312*f*
TMP-SMZ. *See* Trimethoprim-sulfamethoxazole
TNT (trinitrotoluene), **919**
Tobacco mosaic disease (TMD), 453, 453*f*
Tobacco mosaic virus (TMV), 20, 454, 454*f*, 457*f*
Tobramycin (Tobi), 810*b*, 816
Togaviridae (togaviruses), 461*f*, 481*t*, 532*b*
Tolerance. *See* Self tolerance
Toll-like receptors (TLRs), 678, 679*f*, 679*t*, 688*b*, 691*b*
Tolnaftate, 573
Tonsils, 672, 673*f*
Tooth anatomy, 351*f*
Tooth decay. *See* Dental caries
Torulopsis glabrata, 554*f*
Total magnification, 85
Totally drug-resistant TB (TDR-TB), 326
Toxemia, 645
Toxic dose, 355, 808
Toxic hydrogen peroxide (HOOH), 235–236*b*
Toxic shock syndrome (TSS)
  changing social patterns and, 407
  cytokine storms and, 719
  staphylococcal, 434, 434*f*, 435*b*, 440*t*
  streptococcal, 436, 440*t*
  superantigens and, 719
Toxic-waste sites. *See* Bioremediation
Toxigenicity, 645
Toxins. *See also* Endotoxins; Enterotoxins; Exotoxins; Intoxications
  of gram-negative vs. gram-positive bacterial species, 90
  irradiation and, 207*b*
  oxygen as, 147*b*
  plasmids and, 238
  virotherapy using, 480*b*
  virulence and, 645–647, 646*t*
Toxoid vaccines, 730
Toxoids, 314, 647
Toxoplasma spp., 594
Toxoplasma gondii
  allergic reactions and, 773*b*
  characteristics, 605–606, 605*f*, 610*t*
  intermediate hosts for, 607*b*
  phagocytosis and, 681*b*
Toxoplasmosis, 605–606, 606*f*, 610*t*, 795, 796*f*
Trace elements, 37
Tracers, radioactive, 38
Tracheobronchial tree, 497
Trachoma, 412*b*, 443–444, 443*f*, 444*f*, 444*t*
Trade, international, 449, 562*b*, 660–661, **866*b***. *See also* Travel, international
Transcription
  in bacterial and archaeal cells, 128–129, 128*f*
  central dogma of gene expression and, 241–242, 242*f*
  of genetic information into complementary RNA, 242–243, 243*f*, 244*f*, 245*f*
  localized in nucleoid, 251, 251*f*
  negative control and, 246, 249, 249*f*
  rifampin and, 815*f*, 818–819
  translation in *E. coli* and, 248*f*
  in viral replication, 468*f*

Transduction
antibiotic resistance and, 828, 828*f*
generalized, 274, 276*f*, 277, 277*b*
specialized, 276*f*, 277
Transfer RNAs (tRNAs)
structure of, 243*f*
transcription of, 242–243, 244–245, 245*f*
translation and, 245, 246–247*f*
Transformation
antibiotic resistance and, 828, 828*f*
of cells into cancer cells, 476–478
horizontal gene transfer and, 268, 269–270*b*, 271, 271*f*
in marine snow, 272*b*
Transfusion reactions, 764*f*, 775
Transgenic plants, 286, **920**, **921*f***
Transient microbiota, 632, 636, 636*f*
Translation
as antibacterial agent target, 815*f*
in bacterial and archaeal cells, 128–129, 128*f*
central dogma of gene expression and, 242, 242*f*
localized in nucleoid, 251, 251*f*
transcription in *E. coli* and, 248*f*
in viral replication, 468*f*
Transmissible spongiform encephalopathies (TSEs), 484–485
Transmission electron microscope (TEM), 93–94, 95*t*
Transplacental HIV transmission, 797
Transplacental viral transmission, 513
Transplantation
Carabin and acceptance of, 718*b*
drugs to limit rejection of, 786, 786*t*
rejection of, 764*f*, 784, 786
technology for, 784, 785*f*
Transport systems. *See also* Flagella
magnetotaxis, 125*b*
Transposable elements (TEs), 256, 258, 258*f*
Transposase, 256, 258*f*
Transposons, 256, 258*f*, 274
Traumatic wounds, 437–438, 437*f*, 438*t*
Travel, international. *See also* Trade, international
disease transmission and, 653*b*, 661*f*
emerging and reemerging diseases and, 660–661
flu outbreaks and epidemics and, 492
parasitic diseases and, 449, 605
SARS outbreak and, 503*b*
tickborne encephalitis and, 755*b*
Traveler's diarrhea, 363
Trees
Cinchona, malaria and, 823, 823*b*
Eucalyptus, mycorrhizae and, 558*b*
koa tree, Hawaii, **896*b***
Trematodes, 611
Tremblaya princeps, 235*f*
Trench foot, 401*b*
Treponema, 354*t*
Treponema pallidum
arsphenamine for, 806
endoflagella of, 115*f*
generation time in, 138*t*
as microaerophile, 146
micrograph of, 103*f*
on phylogenetic tree of life, 416*b*
as spirochete, 104
syphilis caused by, 416–418, 417*f*
Treponema pertenue, 416*b*
Tretinoin, 431
Triazoles, 822
Trichinella spiralis, 206, 617, 617*f*, 620*t*
Trichinellosis (trichinosis), 617–618, 620*t*, 779*t*
Trichloroethylene (TCE), **916**, **918**
Trichomonas vaginalis, 592, 600*f*, 601*t*
Trichomoniasis, 601*t*
Trichonympha, 591–592, 591*f*
Trichophyton spp., 572, 573, 575*t*
Trichuris suis, 613*b*
Trickling filters, in sewage treatment, **887–888**, **887*f***
Triclosan, 218, 218*f*, 220*b*, 223*t*
Trifluridine, 822*t*
Triglycerides, 50, 50*f*
Trimethoprim-diaminopyrimidine, 823
Trimethoprim-sulfamethoxazole, 275*b*, 315, 578, 599, 811–812, 820*t*. *See also* Bactrim
Trinitrotoluene (TNT), **919**
Tripedia, 321
Trismus, 384
Trivalent vaccines, 542
Trophosome, 190*b*
Trophozoites
amoebal, 596, 596*f*
Cryptosporidium, 599
Giardia, 597
Naegleria fowleri, 608–609
Pneumocystis jiroveci, 578
Toxoplasma gondii, 605–606, 605*f*
Trichomonas vaginalis, 599–600, 600*f*
Truvada, 800
Trypanosoma spp., 23*f*, 591*f*, 592, 603, 604*f*
Trypanosoma brucei, 610*t*
variety *gambiense*, 603
variety *rhodesiense*, 603
Trypanosoma cruzi, 603, 605, 610*t*
Trypanosomes, 592
Trypanosomiasis, 603, 605
Tryptophan (*trp*) operon, 250*b*
TSEs (transmissible spongiform encephalopathies), 484–485
Tsetse flies, 603, 604*f*
Tsutsugamushi fever, 399
Tube dilution method, for antibiotic susceptibility assay, 825–826
Tubercle, 323, 325
Tuberculin reaction (skin test)
AIDS and, 326
infection allergy and, 780, 780*f*
for tuberculosis, 323, 325–326, 325*f*, 339*b*
Tuberculoid leprosy, 438
Tuberculosis (TB). *See also Mycobacterium tuberculosis*
as airborne bacterial disease, 338*t*
causative agents and epidemiology, 322, 322*f*
clinical presentation, 322–323, 324*f*, 325
concept map for, 323*f*
detection of, 325–326, 325*f*
HIV/AIDS and, 796*f*
infection allergy and, 779*t*, 780
inflammation and, 683
multiple portals of entry for, 643
prevention, 326
pulmonary, 325*f*
spread of, 305
treatment for, 326, 812, 820*t*
worldwide deaths due to, 631*f*
Tularemia, 390–391, 393, 394*f*, 402*t*, 643
Tumor cells, cytotoxic T cells and, 718, 718*f*
Tumor necrosis factor α (TNF-α), 674*f*
Tumor suppressor genes, 478, 479*f*
Tumors. *See also* Cancer, human
use of term, 476
Turbidity, bacterial growth and, 151, 155, 158, 158*f*
Turkey. *See also* Poultry
contamination of, 366, **855**
Turtles, pet
Salmonella transmission and, 366
Twinex, 536
Twitching motility, type IV pili and, 110
Twort, Frederick, 453–454
Tygacil, 817
Tyndall, John, 203, 808
Tyndallization, 203
Type I diabetes, 782, 783, 783*t*
Type I IgE-mediated hypersensitivity, 781*t*
asthma, 768–769, 770–771*b*, 770*f*
characteristics, 763–766, 764*f*, 765*f*
concept map for, 767*f*
hygiene and, 773*b*
localized or systemic, 766, 768
therapies for, 772–773, 774*f*
Type II cytotoxic hypersensitivity, 763, 764*f*, 774–776, 775*f*
Type III immune complex hypersensitivity, 763, 764*f*, 776–778, 779*f*
Typhoid bacilli, shellfish contamination by, **854**
Typhoid fever, 365–366, 366*f*, 817, **865**
Typhoid Mary, 367*b*
Typhus fever, 101, 398

U

Ulcerative colitis, 613*b*
Ulcers
gastric ulcer disease, 371, 372*b*, 373, 373*f*
venous leg, phage therapy against, 466*b*
Ultra high temperature (UHT) pasteurization method, 204, 204*b*, **861**
Ultrastructure, visualization of, 93
Ultraviolet (UV) light
as carcinogen, 476
DNA damage by, 53
food preservation and, **864**
herpes simplex and, 507*f*
for mail decontamination, 222*b*
microbial control using, 206, 209*t*
as physical mutagen, 252–253, 253*f*
wavelength of, 206*f*
Uncoating, 466, 467, 467*f*
inhibitors, 821, 822*t*
Undecylenic acid, 573
Undifferentiated cells, 700
Unikonta (supergroup), 588*f*, 592, 594*t*
United Nations. *See also* World Health Organization
Millennium Declaration, **878**
Unsaturated fatty acids, 50, 50*f*
Upper respiratory tract (URT), 309, 310*f*, 311, 49?
632*f*. *See also* Respiratory diseases, up

Uracil (U), 52*f*
Urban plague, 389. *See also* Plague
Urban yellow fever, 529
Ureaplasma, 789
Urethritis
  characteristics, 421, 425*t*
  gonococcal. *See* Gonorrhea
  nongonococcal (chlamydial), 411, 412*b*
Urey, Harold, 57–58*b*
Urinary tract
  anatomy, 420, 420*f*
  indigenous microbiota of, 420–421, 632*f*
  infections (UTIs)
    biofilms and, **888**, **889**
    cranberry juice and, 423*b*
    cystitis, 421–422
    identifying cases of, 441*b*
    nosocomial, 659*f*
    process of, 422*b*
    prostatitis, 424
    pyelonephritis, 424
    statistics and risk for, 421, 421*f*
    summary of, 425*t*
    urethritis, 421
  parasites (*T. vaginalis*), 599–600
Urination, 420–421
Urticaria, 766, 779*t*
Urushiol, 780, 780*f*
USA 300 strain, CA-MRSA, 831*b*
Use dilution test, 213–214
Ustilago zeae, **906**

V

Vaccinations. *See also* Immunization; *specific diseases*
  for adults, 732–734, 734*f*
  antibiotic resistance prevention and, 838*b*
  for artificially acquired active immunity, 728
  for children, 732*f*
  cost-benefit from, 725
  disease prevention by, 9, 12–13
  herd immunity and, 737–739, 738*f*, 739*t*
  incidence, mortality and, 725–726, 726*f*
  waivers for, 739*b*
Vaccine Adverse Events Reporting System (VAERS), 740
Vaccines. *See also specific diseases*
  for adults, 732–734, 734*f*
  for children, 732*f*
  conjugate, 732
  currently in use, 729*t*
  development of, 741*b*
  disease decline and, 739*t*
  DNA, 732, 733*f*
  field testing, 742–743*b*
  genetically engineered, 284, 285*f*
  ingredients in, 735*b*
  irradiation of, 206
  mechanism of action, 728
  side effects from, 739–740
  subunit vaccines, 730–732
  whole agent, 728, 730
Vaccinia virus, 469*b*, 517*b*
Vacuoles
  contractile, 592*f*, 593
  food, 592
  gas, 124, 128*t*
Vagina
  candidiasis of, 573, 575*t*
  chemical barriers to disease in, 675
  T. vaginalis of, 599–600
Vaginosis, bacterial, 409–410
Valacyclovir, 526, 821, 822*t*
Valley fever, 577, 578*f*
Valtrex, 822*t*
Van Gogh, Vincent, 406
Vancomycin
  antimicrobial spectrum of, 809*f*
  properties of, 815, 815*f*, 820*t*
  for pseudomembranous colitis, 363
  resistance testing, 826*f*
  resistance to, 832*t*
Vancomycin-intermediate resistant *S. aureus* (VISA), 830*b*
Vancomycin-resistant *E. faecalis*, 275*b*, 815
Vancomycin-resistant *E. faecium*, 815
Vancomycin-resistant *S. aureus* (VRSA), 275*b*, 805*f*, 815, 835*b*
Vaqta, 536
Variable domains, 704, 704*f*
Variables, in experiments, 10*b*
Variant Creutzfeldt-Jakob disease (vCJD), 485
Varicella-zoster immune globulin (VZIG), 509
Varicella-zoster virus (VZV), 506*f*, 508, 510*b*, 512*t*
Varicellovirus, 506*f*
Variola. *See* Smallpox
Variolation, 9
Varivax, 509
Vasodilation, 682, 682*f*
VBNC (viable but noncultured) organisms, 155, 156*b*, 159, 295, 298
Vectors. *See also* Flies; Mosquitoes; Ticks
  of arthropodborne diseases, 388, 650, 651*f*
  control of, 532
  plasmids as, 123
Vegetables
  contamination of, 356, 367
  fermented, **869–870**
Vegetalia (kingdom), 73
Vehicle transmission of disease, 649–650, 651*f*
Veillonella, 311
Venereal Disease Research Laboratory (VDRL) test, 418, 746
Venereal diseases, 410. *See also* Sexually transmitted diseases
Venezuelan equine encephalitis (VEE), 544, 545*f*
Venezuelan hemorrhagic fever, 534
Venipuncture, 219
Venter, J. Craig, 12*b*, 291, 292*b*, 295, 298, 299–300*b*
Vertical gene transfer, 268, 269*f*
Vertical transmission of disease, 649, 650*f*
Vesicle bioreactors, 12*f*
Vesicles
  chickenpox, 508, 509*f*
  cold sore, 506*f*
  of Golgi apparatus, 72
  magnetosomes, 125*b*
  smallpox, 516–517, 516*f*
Viable but noncultured (VBNC) organisms, 155, 156*b*, 159, 295, 298, 839
Vibrio spp., 101
Vibrio cholerae
  cells, 362*f*
  cholera toxin, 646*t*
  classic, 79
  curved rod shape of, 106
  El Tor strain, 79, 361*b*
  glycocalyx of, 115–116
  isolation of, 343, 360
  naming of, 79*b*
  pH and growth of, 147
  salt and growth of, 148
Vibrio fischeri, 69*b*
Vibrio parahaemolyticus, 364–365
Vibrio shape, of bacterial cells, 106, 107*f*
Vibrio vulnificus, 365, **882**, **883**
Vibriosis, 364–365
Vidarabine, 822*t*
Videx, 822*t*
Vinegar, **864**, **870**
Vira-A, 822*t*
Viral gastroenteritis, 536, 538, 539*t*
Viral hemorrhagic fevers (VHFs), 529, 531–534, 535*t*
Viral infections. *See* Viruses, infections due to
Viral inhibition, 709–711, 711*f*
Viral load test, 797
Viral oncogenes (*v-onc*), 478, 479*f*
Viral pneumonia, 494, 502
Viramune, 822*t*
ViraPap test, 287
Virazole, 822*t*
Virchow, Rudolph, 9, 70
Viremia, 509, 637
Virion(s)
  complement activation and, 686
  defined, 456
  endocytosis and, 466, 467*f*
  in productive infection, 465
  viral nomenclature and features of, 460
  viral replication and, 462–463, 463*f*, 468–469
Viroids, 482, 484, 484*f*
Virology, 20, 22*f*, 450, 453–454
Virophages, 457*b*
Viroptic, 822*t*
Virosphere, 451–452, 452*f*
Virotherapy, 480*b*
Virulence, defined, 633–634
Virulence factors, 110, 116, 634, 643–647, 645*f*, **906**
Virulent phages, 463, 464*f*
Viruses. *See also* HIV; Virology
  antiviral drugs, 819, 821, 822*t*
  bacterial genomes in lungs, 311
  brief survey of, 23*f*, 24
  cancer and, 476–478, 477*t*, 478*f*, 479*f*
  cell-mediated immune response to, 714–715, 716–717*f*, 718
  cell/tissue tropism, 458, 458*t*, 459–460*b*
  cellular organization in, 25
  classification of, 460–462, 461*f*
  components of, 455–456, 456*f*
  detection of, 473, 474*f*, 474*t*, 475
  DNA evolution from, hypothesis on, 471*b*

early research on, 20, 22
emerging, 481–482, 481*t*
endogenous, 294
environments for, 451–452, 452*f*
as filterable infectious agents, 22*f*, 453–454, 453*f*
genetic information in, 53
giant forms of, 457*b*
Global Viral Forecasting Initiative and, 519
hemorrhagic fevers due to, 529, 531–534, 535*t*
host range, 458, 459–460*b*
hypotheses on origin of, 486
infections due to
antisense molecules and inhibition of, 248*b*
blood and lymphatic system, 524–528
blood types and, 777*b*
Carabin and, 718*b*
five-step program for, 548
gastrointestinal, 535–539
HIV/AIDS and, 796*f*
interferon activation and, 687, 689, 689*f*
nervous system, 539–547
respiratory, 493–505
skin, 505–518
waterborne, **883**
irradiation and, 207*b*
microorganism sizes and, 84*f*
pest-control using, **914–915**
radioactive labeling of, 59*b*
replication cycle, 454–455, 455*f*
replication of
animal viruses, 465–469, 467*f*, 472*f*
antiviral drugs and, 819, 821
bacteriophages, 463, 464*f*, 465
common steps in, 462–470, 463*f*
latent infections from, 469–470
one-step growth cycle in, 475*b*
shapes of, 456, 457*f*
size relationships among, 455*f*
study of, 449
transduction and, 274, 276*f*, 277, 277*b*
ultrastructure visualization in, 93
vaccines for, 728, 729*t*
Virus-like agents, 482, 484–486, 484*f*, 485*f*
Visceral diseases, 596, 601*t*
Vital force, belief in, 7
Vitamin B_2 production, **905**
Vitamin B_{12} production, **905**
Vitamins, irradiation of, 206
Vitis vinifera, **908**
Voges-Proskauer test, 184
Volutin granules, 124
Volvox, 23*f*
von Behring, Emil, 21*t*, 669, 703*b*
v-onc (viral oncogenes), 478, 479*f*
Voriconazole, 580
Vostok Lake, below Antarctic, 149*b*
VRSA (vancomycin-resistant *S. aureus*), 275*b*, 805*f*, 815, 835*b*
Vulvar cancer, 514–515
Vulvovaginitis, 573

W

Waksman, Selman, 26, 816
Walking pneumonia, 332
War fever, 400–401*b*
Warren, J. Robin, 371, 373
Warts, 477*t*, 514, 515*b*, 515*f*, 518*t*
Wasabi, as antiseptic, 215*b*
Washington, George, 804
Wasting disease, in elk and deer, 484
Water. *See also* Oceans; Waterborne diseases
in bacterial cells, 47*f*
bacteriological analysis of, **889–890**, **889*f***, **890–891*b***
balance in prokaryotic and eukaryotic cells, 72–73
cell membrane and diffusion of, 120
contamination of, 355–356, 661
dependence of organisms on, 43–46
as disease reservoir, 649, 651*f*, **882–883**, **882*f***
food spoilage and, **851**, **852*f***
laboratory creation of life and, 57–58*b*
molecular formula of, 40
as polar molecule, 42, 42*f*
polluted
fish handler's disease and, **883*b***
microorganisms in, **879**, **880*f***, **881**
types of, **881–882**
wine as alternative to, **908**
properties of, 44
as protist environment, 589
purification
diseases and, **877–878**
Moringa oleifera and, **886*b***
steps in, **884–885**, **885*f***
quality testing, gene probe and PCR for, 287
safe, populations lacking access to, **878**, **878*f***
types of, **879**
unpolluted, microorganisms in, **879**, **881**
Water vapor, microbes and, 4*f*
Waterborne diseases
bacterial
cholera, 343–344, 344*f*. *See also* Cholera
fish handler's disease, **883*b***
leptospirosis, 386–388, 387*b*
tularemia, 391
parasitic
amoebiasis, 596–597, 597*f*, 598*b*
cryptosporidiosis, 598–599, 599*f*
cyclosporiasis, 599
giardiasis, 597–598, 598*f*, 600*b*
primary amoebic meningoencephalitis, 608–609, 608–609*b*
pharyngoconjunctival fever, 493–494
types of, **882–883**, **882*f***
Watson, James D., 53, 231, 233, 234*b*
Waxes, 51
Weekly Epidemiological Report (WHO), 445
Weil syndrome, 387
Weinstock, Joel, 613*b*
Weizmann, Chaim, 279*b*, **901**
Welch, William, 21*t*
Weller, Thomas, 454
Wescodyne, 217
West Nile encephalitis, 546–547, 629–630
West Nile virus disease, 544, 545*f*, 546–547, 547*t*
West Nile virus (WNV)
as biological weapon, 655*b*
as emerging virus, 481, 481*t*
in horses, DNA vaccine for, 732
incidence in U.S., 546*f*
micrograph of, 546*f*
in New York City, 629–630
in U.S. in 2010, 630*f*
Western blot analysis, 797
Western equine encephalitis (WEE), 544, 545*f*, 547*t*
Wheals, 766, 772, 774*f*
Wheat allergies, 768
Whey, **856**, **857**, **871**, **903**, **904**
Whipworms, 613*b*, 618
Whiskey, 209, **911**
White blood cells (leukocytes), 670–672, 671*f*
White plague, 322. *See also* Tuberculosis
White wine, **911*b***
Whiteheads, 430, 430*f*
White-nose syndrome (WNS), 564*b*, 580
Whitfield's ointment, 573
Whittaker, Robert H., 77
Whooping cough, 320. *See also* Pertussis
Wild type cells, 252
Wild types, 260
Wilde, Oscar, 406
Wilkins, Maurice, 53, 234*b*
Willoughby, Rodney E., Jr., 540*b*
Wilson, E. O., 30
Wine
enzymes and, **905*t***
Pasteur's research on, 16
production of, **908**, **910**, **910*f***
red, resveratrol and, **911*b***
Winogradsky, Sergius, 22
Winter diarrhea, 538
Woese, Carl, 77, 99–100
Woodruff, Alice M., 454
Woolsorters' disease, 337, 382
World Health Organization (WHO)
on anthrax, 382
on cholera, 360
Direct Observation Treatment System of, 327*b*
epidemiological tracking by, 15*b*, 28, 630
Global Influenza Surveillance Network, 731*b*
malaria eradication and, 604*b*
on measles, 511
SARS outbreak and, 503*b*
on smallpox eradication, 12
smallpox eradication and, 517, 517*b*
on trachoma cases, 443–444, 443*f*
Weekly Epidemiological Report, 445
World Rabies Day, 523
Wort, **907**, **909*f***
Wounds
acute or chronic, indigenous microbiota and, 431
botulism and, 359
chronic, nonhealing, 432*b*
infected, *M. chelonae* and, 326
traumatic, 437–438, 437*f*
Wuchereria bancrofti, 619–620, 620*t*

X

X rays, 206, 206*f*, 209*t*, 233, 234*b*, 476
Xenografts, 784
X-linked agammaglobulinemia, 787
Xylanases, **905*t***

Y

Yaws, 416*b*
Yeast(s)
  ascomycetes, 563*f*
  characteristics, 567, 569
  commercial uses of, **915**
  genetic engineering using, **919**
  life cycle of, 554
  nitrogen fixation and, **896**
  Pasteur's research on, 16, 16*f*, 20
  vinegar production and, **870**
  in wine production, **908**, **910**
Yeast infection, 573
Yellow fever, 453, 529, 530–531*b*, 531, 535*t*
Yellow jack, 529
Yellowstone Park, Grand Prismatic Spring, 36*f*, 100*f*
Yersin, Alexandre, 21*t*, 389, 669
Yersinia spp., 101, **882**
Yersinia enterocolitica, 371, 371*f*, **861**
Yersinia pestis, 109, 114*b*, 253*b*, 381, 388–390, 402*t*. *See also* Plague
Yersinia pseudotuberculosis, 253*b*, 392–393*b*
Yersiniosis, 371
Yogurt, 346*b*, 574, 633, 634*b*, **864**

Z

Zanamivir, 500, 821, 822*t*
Zephiran, 219, 220*f*
Zinc (Zn), 37
Zinsser, Hans, 398, 399*b*, 400*b*
Zithromax, 817
Zone of equivalence, 744
Zone of inhibition, 214, 214*f*, 827
Zoogloea ramigera, **887**
Zoonoses (zoonosis, zoonotic diseases)
  arboviral encephalitis as, 544, 544*f*, 546–547
  Ebola hemorrhagic fever as, 533
  emerging infectious diseases due to, 481
  exotic pets and, 483*b*
  Lassa fever, 533, 534*f*
  leprosy, 438
  leptospirosis, 386–388, 387*b*, 388*t*
  pneumonia-causing, 336–337
  rabies, 539–542, 540*b*
  reservoirs for, 648–649
  yellow fever, 529
Zoospores, 561
Zostavax, 509
Zoster. *See* Herpes zoster
Zosyn, 814
Zovirax, 822*t*
Zufall, Frank, 713*b*
Zygomycota (zygomycetes), 560, 561, 562*f*, 569*t*
Zygospores, 561
Zymosan, 678, 679*t*
Zyvox, 818

Pronouncing Organism Names (*continued*)

Leuconostoc citrovorum lü-kū-nos'tok sit-rō-vôr'um
L. mesenteroides mes-en-ter-oi'dēz
Listeria monocytogenes lis-te'rē-ä mo-nō-sī-tô'je-nēz

Methanobacterium meth-a-nō-bak-tėr'ē-um
Methanococcus jannaschii meth-a-nō-kok'kus jan-nä'shē-ē
Micrococcus luteus mī-krō-kok'kus lū'tē-us
Micromonospora mī-krō-mō-nos'pōr-ä
Microsporum racemosum mī-krō-spô'rum ras-ē-mōs'um
Morchella esculentum môr-che'lä es-kyū-len'tum
Mucor mū'kôr
Mycobacterium avium mī-kō-bak-ti'rē-um ā'vē-um
M. bovis bō'vis
M. cheloni kē-lō'ē
M. haemophilum hē-mo'fil-um
M. kansasii kan-sä-sē'ī
M. leprae lep'rī
M. marinum mãr'in-um
M. tuberculosis tü-bėr-kū-lō'sis
Mycoplasma genitalium mī-kō-plaz'mä jen'i-tä-lē-um
M. hominis ho'mi-nis
M. pneumoniae nu-mō'nē-ī
Myxococcus xanthus micks-ō-kok'kus zan'thus

Naegleria fowleri nī-gle'rē-ä fou'lėr-ē
Nannizzia nan'nė-zė-ä
Necator americanus ne-kā'tôr ä-me-ri-ka'nus
Neisseria gonorrhoeae nī-se'rē-ä go-nôr-rē'ī
N. meningitidis me-nin ji'ti-dis
Neurospora nū-ros'pōr-ä
Nitrobacter nī-trō-bak'tėr
Nitrosomonas nī-trō-sō-mō'näs
Nocardia asteroids nō-kär'dē-a as-tėr-oi'dēz
Nostoc nos'tok

Paragonimus westermani pãr-ä-gōn'e-mus we-stėr-ma'nē
Paramecium pãr-ä-mē'sē-um
Pasteurella multocida pas-tyėr-el'lä mul-tō'si-dä
Pedomicrobium ped-ō-mī-krō'bē-um
Penicillium camemberti pen-i-sil'lē-um kam-am-bėr'tē
P. chrysogenum krī-so'gen-um
P. griseofulvin gris-ē-ō-fül'vin
P. notatum nō-tä'tum
P. roqueforti rō-kō-fôr'tē

Pfiesteria piscicida fes-ter'ē-ä pis-si-sē'dä
Photorhabdus luminescens fō-tō-rab'dus lü-mi-nes'senz
Phytophthora infestans fī-tof'thô-rä in-fes'tans
P. ramorum rä-môr'-um
Picrophilus pik-rä'fil-us
Plasmodium falciparum plaz-mō'dē-um fal-sip'är-um
P. malariae mä-lā'rē-ī
P. ovale ō'va'lē
P. vivax vī'vaks
Plesiomonas shigelloides ple-sē-ō-mō'näs shi-gel-loi'dēs
Pneumocystis jiroveci nü-mō-sis'tis jėr-ō-vek'ē
Prochlorococcus prō-klôr-ō-kok'kus
Porphyromonas gingivalis pôr'fī-rō-mō-näs jin-ji-val'is
Propionibacterium acnes prō-pē-on'ē-bak-ti-rē-um ak'nēz
Proteus mirabilis prō'tē-us mi-ra'bi-lis
Providencia stuartii prō-vi-den'sē-ä stū-är'tē-ē
Pseudomonas aeruginosa sū-dō-mō'näs ã-rü ji-nō'sä
P. cepacia se-pā'sē-ä
P. marginalis mär-gin-al'is
P. pseudomallei sū-dō-mal'lē-ī
P. putida pyū-tē'dä
Pyrolobus fumarii pī-rol'ō-bus fū-mär'ē-ē

Rhizobium rī-zō'bē-um
Rhizopus stolonifer rī'zo-pus stō-lon-i-fėr
Rhodobacter capsulatus rō-dō-bac'tėr kap-sü-la'tus
Rhodoferax ferrireducens rō-dō-fer'aks fer-i-re-dü'sens
Rhodospirillum rubrum rō-dō-spī-ril'um rūb'rum
Rickettsia akari ri-ket'sē-ä ä-kãr'ī
R. prowazekii prou-wa-ze'kē-ē
R. rickettsii ri-ket'sē-ē
R. tsutsugamushi tsü-tsü-gäm-ü'shē
R. typhi tī'fē
Rochalimaea quintana rōk-ä-li-mē'ä kwin-tä'nä

Saccharomyces carlsbergensis sak-ä-rō-mī'sēs kä-rls-bėr-gen'sis
S. cerevisiae se-ri-vis'ē-ī
S. ellipsoideus ē-lip-soi'dē-us
Saccharopolyspora erythraea sak-kãr-ō-pol'ē-spo-rä ē-rith'rä-ē
Salmonella enterica säl-mōn-el'lä en-tėr-i'kä
S. enterica serotype Enteritidis en-tėr-i-tī'dis
S. enterica serotype Typhi tī'fē
S. enterica serotype Typhimurium tī-fi-mur'ē-um